THE BUILDING ESTIMATOR'S REFERENCE BOOK

A Reference Book Setting Forth Detailed Procedures And Cost Guidelines For Those Engaged in Estimating Building Trades

Editor
Scott Siddens

Technical Editors:
Jerrold Ratner, C.P.E., C.C.A.
William H. Spradlin, Jr., C.P.E.

Published By
FRANK R. WALKER COMPANY

Eugene R. Callahan, President

P.O. Box 3180
Lisle, Illinois 60532
(800) 458-3737
www.frankrwalker.com

We will be grateful to readers of this volume who will kindly call our attention to any errors, typographical or otherwise, discovered herein. We also invite constructive criticism and suggestions that will assist in making future editions more complete and useful.

FRANK R. WALKER COMPANY
Publishers of
Walker's Practical Accounting and Cost Keeping for Contractors
Walker's Insulation Techniques and Estimating Handbook
Walker's Quantity Survey & Basic Construction Estimating
F.R. Walker's Remodeling Reference Book
Walker's Manual for Construction Cost Estimating
Walker's Pocket Estimator
Walker's Practical® Forms for Contractors

ISBN: 0-911592-26-1
Library of Congress Catalog Card Number 15-23586

Printed in the U.S.A.

IMPORTANT

How To Use This Book

In a book of estimating cost data that is used in all parts of the U.S. and in many foreign countries, it is impossible to quote material prices and labor costs that apply in every locality. To make the best use of a reference like this one, it takes some computation on your part.

All labor cost tables give the quantity of work that a worker or crew should perform per hour or per day, together with the number of hours required to complete a certain unit of work. We have taken this data and multiplied by wage rates that are the national average to determine labor unit prices for a given item.

The wage rates used in the labor cost tables found throughout this book are U.S. averages. Each rate is an average of the total wage and benefits that resulted from collective bargaining for that trade throughout the U.S. The rate shown may or may not reflect wage and benefit rates in your area.

You will want to insert your local wage scales and material prices into the cost tables to ensure accurate estimates. A standard method of tabulation has been used throughout this book, an example of which is given below:

When framing and placing wood floor joists up to 2" x 8" (50 x 200 mm) in buildings of regular construction, a carpenter should frame and place 550-600 b.f. (1.30-1.42 cu.m) per 8-hr. day at the following labor cost per 1,000 b.f. (2.36 cu.m):

	Hours	Rate	Total	Rate	Total
Carpenter	13.9	$....	$....	$26.44	$367.52
Labor	4.5	....	....	21.04	94.68
			$....		$462.20

The blank spaces are for you to insert your local wage rates and extend the prices as shown below:

	Hours	Rate	Total	Rate	Total
Carpenter	13.9	$16.75	$ 232.83	$26.44	$367.52
Labor	4.5	13.32	59.94	21.04	94.68
			$ 292.77		$462.20

Note that there is no factor included for overhead and profit on any of the costs in this book, because this is a variable item that each contractor must establish for each individual project.

Because of wide variance in state sales tax laws, this cost has not been included, and the user must apply the prevailing rate to obtain the total material cost.

To accommodate the maximum number of contractors, the insurance and taxes applicable to labor costs have not been included in any of the labor unit costs contained in this book, because some contractors prefer to incorporate these costs as part of overhead, while others maintain them as direct costs.

Using the Metric System

In many ways metric is more coherent and efficient than the standard English (Imperial) inch-pound system. The metric system avoids the need for much tedious conversion. For example, lineal measurements are expressed in the meter and its decimal multiples. Compared with all the various units of measure—inches, feet, yards—in the English system, conversions in metric are accomplished by simply moving a decimal point to the left or right.

Rather than create two separate editions of the same book, one metric and the other inch-pound, we feel that our users are better served by this dual English-Metric guide. All contractors, even ones who do mostly government work, will continue to do a substantial amount of work in the traditional system.

It is difficult to make a comprehensive and general statement about the conversion of data in this book to metric, but for the most part, it has been what is called "soft" converted. Nothing changes about a material item itself. An inch-pound measurement is simply expressed in metric. For example, 2" x 4" lumber is 50 x 100 mm. For lineal dimensions the key conversions are 1 inch = 25 mm and 1 foot = 300 mm.

The metric conversion of data is this book was only possible with the able assistance of Mr. Suresh Ceyyur.

Neither the Frank R. Walker Company nor the editors guarantees or warrants the correctness or sufficiency of the information contained in this book; nor do either assume any responsibility or liability in connection with the use of this book or of any products, process, or pricing described in this book.

TABLE OF CONTENTS

1. GENERAL REQUIREMENTS

5. METALS
CSI DIVISION 5

6. CARPENTRY
CSI DIVISION 6

7. THERMAL AND MOISTURE PROTECTION
CSI DIVISION 7

8. WINDOWS AND DOORS
CSI DIVISION 8

9. FINISHES
CSI DIVISION 9

10. SPECIALTIES
CSI DIVISION 10

11. EQUIPMENT
CSI DIVISION 11

12. FURNISHINGS

13. SPECIAL CONSTRUCTION
CSI DIVISION 13

CHAPTER 1

GENERAL REQUIREMENTS

THE ROLE OF THE CONTRACTOR
CONSTRUCTION FINANCE
DRAWINGS AND SPECIFICATIONS
SETTING UP THE ESTIMATE
CONSTRUCTION SCHEDULING
TOOLS AND EQUIPMENT

Buildings are built by the efforts of several parties. Traditionally, these have been the owner, the architect, and the contractor. The owner might have as consultants a realtor, for land purchase and building management, and financiers, who provide short-term cash for building construction and long-term mortgage when the project is complete.

The architect consults with structural and mechanical engineers, site planners, and interior space planners, any of whom might be part of his immediate office. He also consults with the governmental agencies that control zoning, building codes, and city planning, though usually he is not the one who takes out the building permit. He also relies on the cooperation of the producers who manufacture the products that he designs into a structure.

The contractor has on his team the various subcontractors and the producers and suppliers of the equipment, materials, services, appliances, tools, and machinery called for by the architect and required to complete the job. The contractor must also cooperate with the governing officials and must take out all building permits and conduct his operation without violating local laws that govern everything from blocking traffic to waking up the neighbors. And then there are the trade unions, the local utilities, and the ever-necessary bankers.

When land becomes scarce, money rates increase, and ecological concerns grow, the exclusive club of owner-architect-contractor is forced to let in others, and the triangle becomes more of a circle. On large projects today, the owner may be the large landowner who wishes to develop his holdings, a government that is working to rebuild the inner city, or a large insurance company that needs to keep its investments active. The architect's concern may no longer be the individual building but an entire complex, a neighborhood, or a whole new town. And the general contractor may find himself in demand less for craftsmanship abilities than for managerial capabilities. But no matter how small or how complex the project, the owner has a need, the architect develops that need into satisfactory plans, and the contractor interprets the plans into an actual building.

THE ROLE OF THE CONTRACTOR

Erecting a building is a complex undertaking, and seldom is one firm capable of doing all phases of the work. Yet the owner or developer usually prefers to let one contract and make one firm responsible for the completion of the project. That firm is then known as the *General Contractor* (or *general*). Usually, he assumes this role when the owner asks for firm price bids. He is the one who will compile all material, labor and service, and other costs called for on the architect's drawings and specifications. Where his own firm is not able to do work called for, he will turn to other contractors who have expertise in those fields. These are the *Subcontractors* (or *subs*), and if the project is awarded to the general, they will enter into contracts with him and be responsible to him rather than directly to the owner. The general will ask for bids from many subcontractors in order to put together not only the best price but also the best team.

The amount of work performed by specialty subcontractors varies to suit each project's needs, as determined by the general contractor. This major decision is not predicated solely upon economic factors. Each time some portion of the work is assigned to a subcontractor, there is introduced into the construction team another member, whose strengths and weaknesses translate directly into the total performance plan. The general who attempts to "broker" a project by maximum use of subcontractors in order to limit financial responsibility will find that other hazards have been created. There will be a considerable loss in coordination and production, because the general has relinquished the right and ability of direct control in the assignment of personnel, materials, and equipment. He might be subjected to a subcontractor who does not have his same dedication to the project.

Initially, it may appear that the large dollar value assigned for subcontractor performance insulates the general from financial loss on that portion of the project, but this is usually found to be false economy. Upon signing the subcontract document, the general contractor has inherited any faults and deficiencies that exist in the subcontractor's company. The extent of these faults becomes apparent during the life of the contract, and they adversely affect the project schedule, subjecting the general to delay claims from other subcontractors as well as from the owner. In addition, his own direct costs will increase in direct proportion to delayed completion.

In the most severe instances, the general may be forced to terminate the subcontract, and here again, excess costs will be incurred due to reprocurement costs and associated delays. It is an accepted practice to employ certain subcontractors who possess expertise in a particular discipline. However, they must be carefully selected on the basis of financial capacity, technical ability and the proven ability to perform consistent with the requirements of the contract documents. This policy can ensure a successful project, a credit to the owner, the A/E, and the contracting team.

The average percent of work performed by subcontractors cannot be precisely determined, but surveys conducted by *The Associated General Contractors of America* indicate from 40% to 70%.

The general contractor must retain the financial ability to perform the work volume he has under contract, including allowances for unanticipated delays in receiving payments for work performed. Such delays seem to be increasing at an alarming rate, placing an additional burden on the financial resources of the general. There must be in-house personnel who are experienced in the particular type of projects in which the firm is engaged. Of equal importance, there must exist managerial skills to coordinate the field construction activities and the administrative functions into a total effort. The general is expected to provide many services to the owner and the architect, before the contract award as well as during construction. His knowledge and contribution may be the deciding factor in his receiving a contract award or in his expanding a project under construction.

How much of the contract will be sublet will vary with each general contractor. The *American Subcontractor's Association* claims that 90% of the work force in the building construction industry is employed by subcontractors. It is certainly quite possible for a subcontractor to erect a greater portion of the project than the general, and in some cases, the general is forced to sublet work. Some manufacturers insist that their products be installed by contractors licensed to them. For example, in curtain wall construction the specifications will call for the supplier to install the product in order to put the responsibility for any corrective work on one source. Obviously, if anything like 90% of the contract is to be sublet to others, the general contractor must have the managerial skills to schedule and coordinate all the firms involved so that the work proceeds without delay.

In addition to building and managerial skills, he must be financially responsible. Not only must he meet his own payroll and overhead costs, but he also must pay his subcontractors and suppliers in advance of being reimbursed, hopefully with some profit, by the owner. Usually the general will put in a *request for payment* each month for that work completed the month before. These requests may have to be accompanied by *waivers of lien* for the amount requested from each sub and supplier. In submitting such a waiver, the sub gives up his right to file a claim against the property. To obtain a waiver, the general will have had to pay the sub or supplier at least the major portion of his request. Usually there is a *retainage* of at least 5% to 10% by the general contractor to his subs and in turn by the owner to the general contractor. Retainage is not completely released until the project is substantially complete, but it has become common practice to reduce the retainage by 50% when the project is 50% complete.

A contractor whose financial affairs are in good order and who has a contract with a responsible client, who in turn has secured adequate construction and mortgage loans, should have little trouble in securing a loan from his bank to finance the month-to-month payments. The contract will serve as the loan security. The amount needed for this interim financing is often stated as one dollar for every ten dollars in the contract. But before quoting a final price for a project, the contractor should determine the length of the period he must finance, estimate the amount involved, and add the cost of the money to the job cost. It should not have to come out of the profit.

The successful general contractor is one who has a broad knowledge of building construction. His initial contact with a prospective client may be about site selection, sources of construction finance, or the advantages of one type of construction over another. He must develop contacts with architects, bankers, realtors, mortgage brokers, soil engineers, and the local enforcers of building and zoning codes. His knowledge in these allied fields may be the deciding factor in his being considered for the job.

Most municipalities and many states require subcontractors and general contractors to be licensed. There is no general rule governing licensing in all jurisdictions. A state license may not fully qualify a contractor to practice in a municipality within that state. Many suburban areas will require additional fees from out-of-town contractors. One must always inquire about the laws governing the site of construction.

Often a license or permit bond is required guaranteeing a public body that the contractor will comply with all applicable statutes and ordinances. If land development is part of the contract, subdivision bonds may be required guaranteeing proper installation of roads, sewers and other utilities, today even extending to provision of schools and park systems.

And to practice, one must also enter into proper relationship with the unions who may also require a wage bond guaranteeing union scale and welfare fund payments.

Finally, one must also consider the type of business relationship that defines the firm; that is, whether to practice as an individual, a partnership, or a corporation. Consult your lawyer and banker.

Construction Management

In *Construction Management (CM)*, a general contractor or engineering company enters into a contract with the owner prior to the bidding period and acts in an advisory role. Bids are taken under the construction manager's supervision, but separate contracts are let for each phase directly between the contractor for that phase and the owner.

The construction manager is paid a flat fee by the owner, and this fee is completely separate from other job expenses. CM fees are determined in one of several ways:

1. A stated number of dollars plus expenses. The expense items would include the hourly wages paid the principals billed at an agreed upon fixed rate, plus employees' time billed at an agreed upon multiplier to cover overhead and profit, plus a multiple of expense for any outside consultants.

2. Same as above but without a fixed fee, all billing being on a multiple of direct expense.

3. An agreed upon fixed fee.

4. A percentage of the total construction cost.

Services rendered vary, but generally the CM is hired early in the planning and will work with the architect in setting up job budgets, bidding and construction schedules, site development, and selecting building systems. He might make detailed value engineering studies. Not only will he determine what material or procedure is least costly, but which ones will give optimum quality within the budget. He will cooperate with the architect in arranging specifications to include the proper extent of work for each phase to be bid. He will prepare prequalification criteria for the bidders and will conduct prebid conferences to acquaint the bidders with the extent of work covered under other contracts what facilities are available to them at the job site and what they must furnish themselves, the time scheduling, and special procedures to be followed.

When the bids are received, the CM will evaluate them and recommend that one be accepted or that the phase should be modified and rebid.

When the construction begins, the CM will coordinate all work on the site to coincide with the agreed to schedules, budgets, and quality standards. He will recommend corrective procedures and job changes that he feels are necessary, but his decisions are subject to the approval of the architect and owner. He will maintain an office at the job site with sufficient personnel to inspect the work, record job progress, keep up-to-date records of drawing and specification changes, shop drawings, samples, operating manuals and other records which will be delivered to the owner upon completion of the project. He will set up and maintain the project accounting system, which on larger jobs might be computer based and which will give instant reports on the project cost status, payment status, and an analysis of each contract and the project cash flow. The construction management approach has gained wide acceptance, especially in federally sponsored construction. It gives the owner access to complete data on all phases of the contract.

It frees the contractor from many of the risks of a lump sum contract. But if it limits his losses, it also limits his profits, because any job savings that the CM's expertise may bring about accrue to the owner.

The CM format is not attractive to all owners, especially not to the ones who desire to minimize their direct activity with the project during the construction period, in which case it is more desirable to employ the services of a general contractor on a lump sum contract basis.

Bidding for a Contract

Construction contracts are usually awarded in one of two ways—*competitive bidding* or *negotiation*. Competitive bidding is the method used for most contracts, but it is only truly effective when complete working drawings and specifications are available and when contractors are screened so all who bid are of the same caliber. Bids based on sketchy plans and vague specifications are a waste of time. You will almost always be underbid by those who sacrifice quality for price. One of the most valuable contacts an aspiring contractor can make is to get on the bidding list of an architect who produces detailed drawings and tight specifications and insists on their being carried out. The architect has already spent many hours working with the owner to make most of the critical decisions, and by the time bids are taken, the contractor's responsibilities are clearly set forth. He can proceed, if awarded the contract, knowing that the finished project will most probably satisfy the owner, and satisfied owners are probably the most valuable asset a contractor can have.

In private work the architect, often with the owner's help, will limit the list of bidders. You can judge your competition and know that if your price is low and reasonable you will get the contract. Public work is usually open to any "qualified" bidder, but qualifications are generally concerned more with financial status than reputation for quality. In order to propose a low bid against unknown competition, a good contractor must be certain his organization can bring to the project unique experience and efficiencies. Otherwise, he most certainly will be underbid by those desperate enough for work to bid without profit just to get a job.

Private work is sometimes announced in the newspaper or listed in trade publications. A real estate agent or a banker can often steer a contractor to jobs. Public work is almost always publicly advertised. Often, to get on a bidding list, a contractor must be prequalified. There are various ways this is done, by organizations, governmental bodies, and banks, but in general, the information that must be submitted will follow that contained in *AIA Document A-305 Contractor's Qualifications Statement*. This requires that a firm—whether corporation, partnership, or individual—provide details on:

length of time in business; names of principals; percent of work performed by own staff; record of and reasons for any past failures to complete a job; major projects now under construction; major projects completed in past five years; a record of experience of firm employees; bank references; trade references; name of bonding company; and a statement of financial conditions including current assets, liabilities, and the auditor's name. However, often just being a member in good standing of a local contractor's association and a bank reference will satisfy an inquiry.

Once on a bidding list, a contractor will receive an *Invitation to Bidders* for each prospective job. This form will include such information as job location, type of proposal, the number of sets of drawings and specifications to be furnished to bidders, what to do should discrepancies or errors be found in drawings and specifications, bulletins to be issued later, when and how bids will be opened, and when and where bids are due. A proposal sheet is usually attached so only the blanks need be filled out, but sometimes the contractor must furnish his own.

Often the invitation to bid is accompanied by an *Instruction to Bidders* further defining the job restrictions such as completion dates, visiting job site, special conditions, etc.

Bid deposits may be required, the amount being either a lump sum or a percentage of the proposal. The *bid bond* guarantees that the bidding contractor, if awarded the contract, will enter into the contract and furnish a *performance bond*, if required. If he does not honor his bid, he forfeits the amount of the bond, or the difference between his bid and the amount the owner can contract the work for with another contractor, whichever sum is less. Bid bonds—and the later performance, material, labor, maintenance, completion, supply, and subcontractor bonds—are often encountered in public work, but often are waived in private work, where the contractor's reputation is deemed sufficient and the cost of bonds unwarranted. The cost of the bid bond is a cost borne by the contractor. Performance bonds, unless made a part of the contract, are usually billed to the owner. Bonds are further discussed under "Definitions" in this chapter.

Negotiating a Contract

The other type of contract, one that is *negotiated*, allows the owner control over the selection of his general contractor. It is often used on complex work, for example, in remodeling jobs where the space is occupied and must be done piecemeal; interior work where strict control of subcontractors is desired; contracts in which mechanical equipment is a major factor and the owner may prefer one manufacturer over another; work which must proceed in great haste where the architect may not be able to produce complete plans in the time allotted and work will be let in sections (i.e., foundation work, then steel framing, then masonry). Often the contractor will proceed as with bid jobs in submitting a lump sum price, usually on the safe side and known as an *upset* or *guaranteed figure*. This will sometimes include the contractor's profit, but often he will be reimbursed for that which he has spent in completing the work. His profit and overhead will be a stated fixed dollar amount or a percentage of the job total.

A variation of the negotiated contract is *phased construction*. In this arrangement the project may be started before all the plans are fully developed. Each phase of the job, such as foundation, masonry, carpentry, etc., is bid separately, just before the phase is required to be installed. The advantages are:

1. The project can be started earlier, rather than wait until all the design details are settled.

2. The bidders need not add contingencies to cover unknowns such as wage and material cost rises or adverse job conditions. The work will proceed immediately, and these factors will already be known.

3. The bidding time is increased from three or four weeks to several months, and more "shopping around" is possible.

4. There is better control of the budget. If any phase comes in over the budget, it can be redesigned and rebid. If an earlier phase is contracted over the budget, later stages can be modified so that the total job can still conform to the budget.

As the phases are bid, the successful contractors may be assigned to a general contractor in the same manner as a lump-sum contract; or the general may act in the role of a project manager, in which case each contractor for each phase will have a direct contract with the owner. This variation is referred to as *multiple bidding*.

Contract Documents

Once it is determined which contracting firm is to do the job, a formal contract will be drawn up. The Contract Documents usually should include the Owner-Contractor Agreement; the General Conditions of the Contract; Supplementary Conditions of the Contract (if any); the Working Drawings, giving all sheet numbers with revisions; Specifications, giving page numbers; and Addenda or Bulletins issued prior to contract. After the contract is signed, any changes are added as Modifications to Contract and would take the form of written amendments, Change Orders or architectural interpretations.

The *Owner-Contractor Agreement* is a legal matter, but for many jobs standard forms are used. If the architect furnishes the Owner-Contractor Agreement, he will most probably use one printed by the American Institute of Architects. These are:

1. The Stipulated Sum Agreement A-101, four pages long or A-107, a shortened form agreement for small construction contracts. A special form A-10-111/CM is a stipulated sum agreement for construction management.

2. The Cost of the Work Plus a Fee Agreement, Form A-111, eight pages long.

3. Agreement Between Contractor and Subcontractor, Form A-401, seven pages long.

Agreements A-101 and A-111 assume that the *AIA Document A-201 General Conditions of the Contract for Construction*, 18 pages long, is made a part of the contract. These 14 articles are discussed later in this chapter; a shortened version of the General Conditions is included in Document A-107.

The Stipulated Sum Agreement will include the date of the agreement, the owner's name, contractor's name, the description of work, the architect's name, the contract sum, the time of commencement and completion, final payment, and a list of the contract documents.

The Cost Plus a Fee Agreement will include date of agreement, owner's name, contractor's name, description of work, architect's name, the contractor's duties and status, time of commencement and completion, cost of work and guaranteed maximum cost (this may be deleted and if included may be supplemented by a provision for distribution of savings), the contractor's fee, provisions for changes in the work, reimbursable costs, non reimbursable costs, discounts, rebates and refunds, subcontractors, accounting records, applications for payment, payments to contractor, termination of contract, and a list of the contract documents.

For less complicated *stipulated sum* contracts, where the contractor furnishes the form, one of the forms published by the Frank R. Walker Company may be useful. These include a single-page proposal Form #147, which includes an acceptance paragraph for the owner to sign to make a short contract, and a one-page contract agreement Form #158. If the contract is for a construction manager, AIA Form B-801 may be used.

Large corporations and governmental bodies will usually have contracts drawn up by their legal staffs. The contractor may want his own lawyer to review these contracts.

A third type of contract, but one only for both experienced owners and contractors, is the contract for a *cost plus a variable sum*. Here, a fixed cost for the building is agreed upon. If the contractor is able to complete a project at less than the agreed upon figure, the savings is split between owner and contractor; if the cost runs higher, the overage is also split. Often, not only cost is included with penalties and rewards, but also completion dates.

A bonus is given the contractor for each day of delivery earlier than the specified date, but a penalty is deducted for each day of failure to deliver on the specified date. Not only must a contractor be knowledgeable about his ability to perform under these conditions, he should also be well acquainted with the owner's reputation. What constitutes "completion" by a certain date can be a matter for the courts.

The General Conditions of the Contract usually refers to the American Institute of Architects Document A-201. These are often accompanied by Supplementary Conditions which modify and extend the General Conditions. Sometimes these two are combined and rewritten in a form more exactly fitting the special job conditions. But in any case, they set forth the rights, responsibilities, and relations of all parties to the contract and every contractor should be thoroughly acquainted with them.

The latest edition of the AIA form contains fourteen articles. Briefly they cover the following:

Article 1. Defines the contract documents, the work, the project, the execution, correlation, intent and interpretations of the contract documents, and sets forth the copies and ownership of these documents.

Article 2. Defines the architect and his administration of the contract. (A further discussion of administration is included later in this chapter.)

Article 3. Defines the owner, the information and services required of him, and his right to stop work and to carry out work if contractor defaults.

Article 4. Defines the contractor; his responsibility to review the contract documents; his supervision and construction procedures; his furnishing and control of labor and materials; his responsibility for a warranty, for taxes, for permits, fees, and notices; cash allowances, if any; the furnishing of a superintendent; his responsibility for those performing the work; the preparation of a progress schedule; the maintenance on the job site of drawings and specifications; the preparation of shop drawings and furnishing of samples; the use of the site; cutting and patching of work; cleaning up; proper channel for communications, and the indemnification of both owner and architect against all claims, damages, losses, and expenses within the limits set forth.

Article 5. Defines the subcontractors, the award of subcontracts, subcontractual relations, and payments to subcontractors.

Article 6. Sets forth the owner's right to award separate contracts, the mutual responsibilities of contractors, the problems of cutting and patching under separate contracts, and the owner's right to clean up.

Article 7. Includes miscellaneous provisions such as the law of the place, successors and assigns, what constitutes written notice, how claims for damages are to be made, right of owner to require performance and labor and material bonds and how they will be paid for rights and remedies, payments of royalties and patents, tests that may be required, interest on money not paid when due, and arbitration of claims and disputes.

Article 8. Defines contract time, progress and completion, and delays and extensions.

Article 9. Discusses payments and completion sum, the filing of a schedule of values to be used as a basis for contractor's applications, progress payments, certificates for payment, payment withheld, failure of payment, and substantial completion and final payment.

Article 10. Covers protection of persons and property including safety precautions and programs, safety of persons and property, and emergency precautions.

Article 11. Lists the insurance requirements including contractor's liability insurance, owner's liability insurance, property damage insurance, and loss of use insurance.

Article 12. Covers changes in the work, including change orders, how to file claims for additional cost, and minor changes in the work and field orders.

Article 13. Discusses uncovering and correction of work and acceptance of defective or nonconforming work by the owner.

Article 14. Discusses termination of the contract by the contractor or by the owner.

The Supplementary Conditions may include two lists: those that amend the General Conditions and those that make additions to it. A special edition AIA Document A-201/CM is available for jobs using Construction Management. For federally assisted construction projects, the standard A-201 form is supplemented by form A-201/SC, which adds three articles that modify the 14 conditions and include conditions such as equal opportunity employment and special insurance requirements.

Setting Up the Job

Having been awarded the contract, the general contractor must then proceed to sign up his subcontractors. The standard AIA Owner-Contractor Agreement calls for the successful bidder to submit a list of subcontractors for approval prior to award of contract, and the owner and architect must object at that time to any firm on the list. If there is an objection, the contractor has two alternatives: submit an acceptable substitute with the change in cost, if any, in which case the owner may at his discretion accept or reject the bid; or the contractor may withdraw his bid without penalties.

Additional expenses occasioned by owner's or architect's rejection of subcontractors after award of contract will be added by change order to the contract sum and paid for by the owner. There is conflict on this issue, and extreme caution is to be observed.

The contractor must also submit a progress schedule and a schedule of values based on his subcontracts.

The AIA agreement further sets forth the contents of agreements between subcontractors and the general contractor and the method of paying the subcontractors. AIA Document A-401 may be used, or the shorter Frank R. Walker Forms.

The general contractor, with the cooperation of his subs, the architect, and owner, will file for and usually pay for all permits. Often it is necessary to have selected certain subcontractors before filing for permit, because the permit may require the license number of certain trades.

The Supplementary General Conditions will usually set forth the initial site requirements—office, telephone, toilets, water, light, heat, barricades, temporary partitions—that the general contractor must provide. He must also make arrangements with local utilities.

Carrying Out the Work

The administration of the contract is clearly set forth in the AIA standard contractor agreements. The general contractor is responsible for executing the work in strict accordance with the contract documents and for turning the completed project over to the owner in full conformance with them. He must provide the continual supervision necessary to check all labor, materials, and procedures to accomplish this.

The architect serves in an advisory capacity to both contractor and owner, interpreting drawings, and specifications, approving shop drawings and samples, checking conformity of the job with contract documents, making periodic inspections to determine progress and approving payments to the contractors, and approving final acceptance of the contract. The architect will also have his consultants visit the site at certain intervals.

FORM 591
MFD IN U.S.A

SWORN STATEMENT FOR CONTRACTOR AND SUBCONTRACTOR TO OWNER

State of_____ } —

County of_____

The affiant,_____ being first duly sworn, on oath deposes
and says that he is (1)_____

_____ owner____ for

contract with (2)_____

(3)_____
on the following described premises in said County, to-wit:_____

That, for the purpose of said contract, the following persons have been contracted with, and have furnished, or are furnishing and preparing
materials for, and have done or are doing labor on said improvement. That there is due and to become due them, respectively, the amounts set
opposite their names for materials or labor as stated. That this statement is made to said owner____ for the purpose of procuring from said
owner____ (4) Partial—Final Payment on said contract, and is a full, true and complete statement of all such persons, and of the amounts paid,
due and to become due them.

(1) A member of the firm of, or officer of the corporation of, naming name. If a subcontractor so state and name the contractor. (2) Name of the owner or
owners. (3) What the contract or subcontract is for. (4) Partial or Final Payment.

NAME AND ADDRESS	CONTRACT FOR	AMOUNT OF CONTRACT	TOTAL PREVIOUS REQUESTS	AMOUNT OF THIS REQUEST	BALANCE TO COMPLETE	

AMOUNT OF ORIGINAL CONTRACT	$	
EXTRAS TO CONTRACT	$	
TOTAL CONTRACT AND EXTRAS	$	
CREDITS TO CONTRACT	$	
NET AMOUNT OF CONTRACT	$	

TOTAL AMOUNT REQUESTED		$
LESS % RETAINED		$
NET AMOUNT EARNED		
AMOUNT OF PREVIOUS PAYMENTS		$
AMOUNT DUE THIS PAYMENT		$
BALANCE TO COMPLETE		$

It is understood that the total amount paid to date plus the amount requested in this application shall not exceed _____% of the cost of
work completed to date.
I agree to furnish Waivers of Lien for all materials under my contract when demanded.

Signed_____

_____

Subscribed and sworn to before me this_____ day of_____ 19____

PRACTICAL
© FRANK R. WALKER CO. PUBLISHERS.
CHICAGO   The above sworn statement should be obtained by the owner before each and every payment.   Notary Public

Walker Form 591 – Sworn Statement for Contractors and Subcontractor to Owner

The owner often wants more supervision than the architect allows for in the standard contract and a "clerk of the works" is hired. He is usually paid by the owner but works under the architect. He keeps complete job records and reports daily progress, but he does not supplant the architect, whose decisions are the final ones, and does not relieve the contractor of responsibilities of supervision.

In addition, public building inspectors will probably visit the site periodically. Any objections they raise to apparent contradictions between codes and the contract documents should be reported in writing to the architect. At certain periods the contractor will be required to make tests, and he must notify the proper authorities and the architect when these are to occur.

Getting Paid

Applications for payment are usually made monthly. The Frank R. Walker Form #591 is an excellent one for this. It provides a column for the list of subcontractor names, what work the contract is for, the total amount of each contract, the amount previously requested, and finally the balance to complete. The bottom of the form provides a summary of total amounts, less percentages retained, and a space for notarizing. Waivers of lien showing that the contractor has paid for all materials, labor, and subcontractor billings are usually required to accompany the request for payment.

The architect then checks the application, and if it agrees with what he feels to be proportionate to the amount of the contract, he will issue a certificate of payment. His decision in this may be reviewed by either the company writing the performance bond or the company supplying the interim financing. All of this takes time for the contractor to prepare, the architect to check and approve, and the owner to process, so the contractor must keep all his billings and payments current. With the high cost of money today, short term loans to tide a contractor over for a month because he didn't get around to filing a proper or complete request for payment can eat away at profits as surely as having to correct mistakes on the job.

When standard AIA contract forms are used, then the AIA *application for payment* will also be used to retain the same document format. Upon final completion and acceptance, the contractor should be prepared to file final waivers along with any written guarantees, certificates of inspections, operating instructions, etc., that may have been called for.

Definitions

Standard contract forms use certain terms that have special meanings within the profession. Some of these are explained below:

Addenda: Modifications to the contract documents issued by the architect/engineer's office during the bidding period. Sometimes referred to as "bulletins". The addendum can be written, graphic, or both.

Alternates: Additions or subtractions to a contract sum for substitutions asked for by the architect, which the contractor must submit with his proposal.

Approved Equal: A contract term that allows the contractor to substitute another product for one specified by the architect, providing it is "equal" in all respects. Sometimes, the "or equal" must be accompanied by a proposal with a list of substitutions. The owner or the architect has the final say as to what is equal.

Arbitration: A method of settling a dispute whereby parties in disagreement submit their differences to a neutral third party. It is an alternative to formal litigation.

Bid: A proposal stating the sum for which the contractor will complete a project.

Bid Bond: A bond furnished by the contractor that guarantees he will sign a contract on the basis of his bid.

Cash Allowance: Sums set forth by the architect for specific items that the contractor must include in his proposal. Upon completion of a job, the contractor must make an accounting of these sums, returning any unspent amounts to the owner. Often used for buying hardware, face brick, light fixtures, and other special items.

Certificate of Occupancy: Issued by a government authority, authorizing occupation of a building.

Change Order: A written authorization by the architect to change the basic contract, plans, or specifications because of additions or deductions from it.

Contract Time: The period of time established in the contract to complete the contract or specific tasks. It can be expressed in calendar days, weeks, months, or final completion date.

Cost Breakdown (or Schedule of Values): A schedule prepared by the contractor showing how the contract sum is divided among the various divisions of the work, which is filed before work is started. The architect will use this as a basis for approving applications for payment and waivers. It should be accurate and kept up to date, reflecting any changes in the contract.

Extras: Construction costs outside the original contract amount. A change order usually is for additional work, or reduction of work scope.

Final Acceptance: When the owner accepts a project as complete.

Payment and Performance Bond: A bond assuring that the contractor will complete a project in accordance with contract documents including discharging all financial obligations.

Letter of Intent: Written prior to drawing up complete contract documents to tell the contractor a formal agreement is forthcoming. Often this authorizes certain preliminary work such as filing for building permits and site clearance.

Liens: Legal claims against an owner for labor and materials for construction of a project.

Liquidated Damages: A sum established by the owner, and stipulated in the contract as a charge to the contractor for damages suffered by the owner because of the contractor's failure to fulfill the contract obligations.

Maintenance Bond: Guarantees owner that contractor will rectify defects in workmanship or materials reported to him within a specified period following final acceptance of the work under contract. Often this is included as part of the Performance Bond and usually runs for one year. It may also be issued for longer periods covering specific phases of the work, such as roofing, curtain wall, paving, etc.

Punch List: A list, established by the owner or the owner's representative, of work yet to be done or to be corrected by the contractor.

Retainage: A percentage set forth in the Owner-Contractor Agreement to be withheld from each payment to the contractor. Usually 10%, reduced to a total of 5% upon substantial completion.

Separate Contract: A contract let by the owner directly to a contractor other than the general contractor. Usually for special equipment, landscaping, interior furnishings, etc.

Shop Drawings: Drawings prepared at the contractor's expense showing how special items shall be fabricated or installed. These usually are prepared by the subcontractors and are standard for such items as structural and ornamental steel, reinforcement steel, cut stone, millwork, partitions, door frames, fire protection, electrical fixtures, HVAC, etc. Shop drawings must be submitted to the architect or engineer for approval.

Subcontractor Bonds: A performance bond given by a subcontractor to the general contractor guaranteeing performance of contract and payment of labor and material bills.

Substantial Completion: When the project reaches a state where sufficiently complete to allow the owner to occupy it but not necessarily finally accept it.

Superintendent: Contractor's representative at the job site. This individual should not to be confused with the architect's superintendent (administration) or owner's clerk of the works.

Supplier: Sometimes referred to as a vendor. An organization that supplies materials, fabrication, or equipment for the project.

Supply Affidavit (Bond): Written notarized statement to the owner by the manufacturer that the materials delivered to the project comply with contract documents.

Unit Prices: Amounts asked for on the proposal for furnishing materials per unit of measurement. That is, cost of concrete per cu. yd. (cu. m, m^3), floor tile per s.f. (sq. m, m^2), or steel per ton (kilogram, kg).

Upset Price: An amount agreed to by the contractor as a maximum cost to perform a specific project. Used on Cost Plus a Fee jobs.

Warranty: A guarantee furnished by the contractor or a manufacturer covering quality, workmanship, and performance for a specified period.

Value Engineering

A primary objective of management is to achieve the maximum *value* for every expenditure. Many cost control programs have been devised to assist in this quest. None has been as easy to apply, and with as sure an expectation of substantial cost reduction, as *value engineering (VE)*.

The value methodology is based on identification of function. In essence, it asks, What does the client really need in order to accomplish an objective? This is more difficult to quantify than it might appear. In order to be sure that the basic function is uncovered and clearly defined, a multidisciplinary team is assembled. The team should be composed of persons who can address all sides of an issue and usually includes the following:

1. certified value specialist
2. architect
3. structural engineer
4. mechanical engineer
5. electrical engineer
6. cost estimator

In the VE process, a specific time is allotted for study. A typical project might require one week for the *pre-workshop* phase, one week for the *workshop* phase, and another week for the *post-workshop* phase.

During the pre-workshop phase, the design documents are studied. When practical, a site visit is made. If a cost estimate exists, and if in the opinion of the team it does not represent the true cost of the design as it then exists, it is modified. If there is no cost estimate, then one is produced by the team cost estimator.

Cost/Worth Model. A *cost model* is produced by the cost estimator and the certified value specialist. This gives high cost visibility to the functions that compose the total project. Sometimes these models follow the CSI format—site, foundations, superstructure, etc. At other times, they might highlight design features, such as penthouse, annex, or parking. The purpose of the cost model is to graphically show how the costs are distributed by function.

The team next introduces a subjective factor called *worth*. This factor may modify the value of each *cost* by providing a comparative cost based on the experience of the team. For example, the team may perceive that the cost shown for HVAC in the cost model is much higher than for comparable projects. They determine what they consider the lowest cost possible to accomplish the basic function.

The cost model is then modified by giving each entry a worth, and it is then a *cost/worth model,* which will enable the team to see where the cost exceeds the worth by a significant amount. These become *targets* for the VE workshop.

VE Workshop. After a thorough review of the design documents, a site visit, and the preparation of a cost/worth model, the VE team convenes a workshop. Because this usually lasts a full week, it is called a *40-hour VE workshop.*

The workshop is the key intervention of value engineering. A multidiscipline team of experienced design and construction professionals work to seek cost effective alternatives to the present design. Each discipline sees the project from a different perspective, one of the important advantages of VE. However, it also means a potential problem in communication. This is solved by having a strong team leader, the certified value specialist as the *VE team coordinator (VETC)*. To lead the team, the VETC uses the *VE job plan.*

VE Job Plan. The VE job plan is a logical series of phases that guides the team:

1. information phase
2. functional phase
3. creative phase
4. analytical phase
5. development phase
6. presentation phase

Information Phase. The information phase asks two questions of a component or system: What is it? What does it cost? During this phase the cost estimator works with each team member to produce a cost estimate for every target of investigation. Only the components involved in the system to be investigated are included. Simple forms are completed to clearly establish the present design for each target to be challenged.

Functional Phase. Now that the team has picked its targets and determined the costs, the functions are carefully analyzed. The questions asked in this phase are: What does the system do? What must the system do? The differences between answers to these two questions lead the team to fresh solutions that eliminate the functions that may be desirable but are not essential to the design.

Creative Phase. Each function is addressed to seek cost effective alternatives. Often this can be done by discussions among the members, and an effective technique is "brainstorming", where ideas are thrown out by any member as they occur, by association, contiguous thoughts, opposite concepts or the like.

What is sought in this phase is a great number of rough ideas, not engineered solutions, the reason being that an idea advanced by one member automatically triggers other ideas from other members of the team. Creativity seems to work best when negative expressions are not allowed to surface. The subconscious mind is at its best with positive direction, and it is this subconscious resource that we are trying to tap.

Analytical Phase. The long list of ideas generated during the creative phase is examined. Similar ideas are grouped so that a comparative analysis can be made to determine those ideas that must be discarded and those that show merit and should be developed. Several techniques are used to assist in this comparison. The experience of the team enables them to grade ideas with reasonable accuracy. The forms keep the team on track and expedite the comparative analysis.

Development Phase. One or two alternatives are selected by the team for development into the preferred solution. Closer cost estimates are made to compare with the present design.

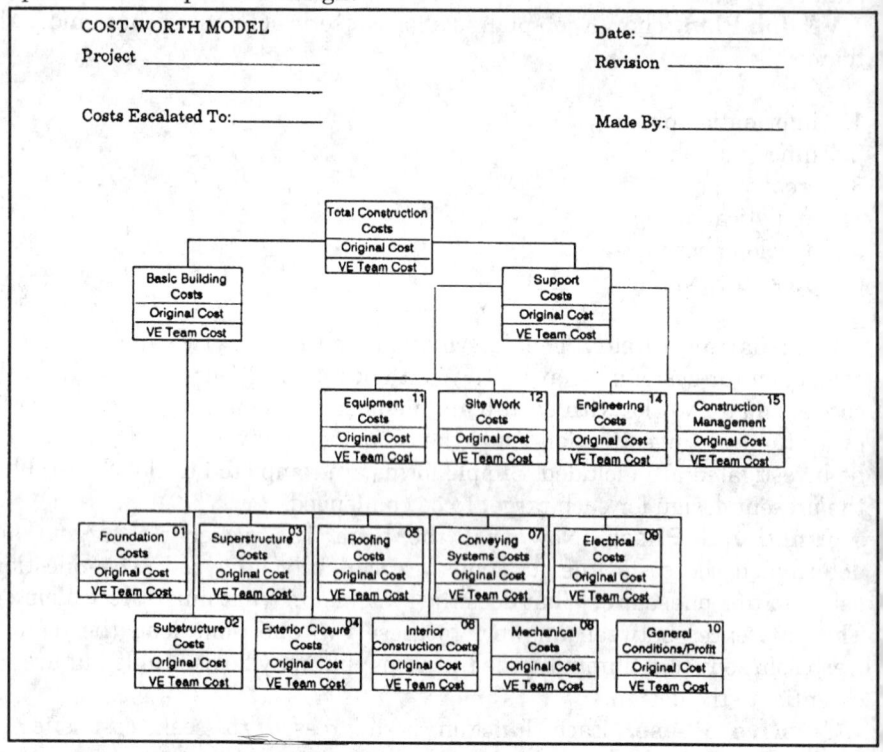

Courtesy of James Hudson Associates
Spotsylvania, VA

The team must be aware of its responsibility. As professionals they are expected to propose concepts, systems, and components that fulfill the basic functions required to properly satisfy the project. This is *not* a cheapening process! Anyone can use lower cost materials to reduce the cost. The goal of value engineering is to propose alternatives that completely fulfill the *needed* functions.

Often the savings are in *life cycle cost (LCC)*. The alternative may have the same or higher initial cost but may have significant cost savings over the life of the facility, product, etc. These LCC are highly desirable for most clients, though in some projects the "first costs" are the primary consideration. The VE team seeks this type of preliminary guidance during the pre-workshop phase to ascertain the real interest of the decision makers in LCC.

It is the duty of the VE team to make very specific proposals. The designers do not need a "laundry list" of good ideas. Only a clear, well-defined alternative can be of use in modifying the design. The proposal must be clearly written, with sketches, to convey the concept.

Presentation Phase. *Alternatives are useless unless they are implemented.* It is imperative that the VE team prepare a written report, complete with all of the back-up worksheets prepared during the workshop, so that the decision makers fully understand the reasoning. In addition, whenever possible a verbal presentation should be given.

While the project is fresh in their minds, members of the VE team can give a complete oral recitation of their findings, proposals, and emotional reaction to the present design and to the proposed design. Often, on the last day of a five-day workshop, the owner, designer, user, and other decision makers are asked to attend a presentation. During the presentation, each team member is given the opportunity to express without reservations the feelings that underlie the proposals. This might include minority opinions on some of the proposals. A properly led workshop seeks truth, not spectacular cost savings. This usually results in many cost effective alternatives, but it might also reveal serious deficiencies in the design. Great care is taken to divulge such findings with diplomacy and compassion.

Effectiveness of Value Engineering. VE has been used in construction for many years. When properly conducted, it has produced average savings of more than 10% of the estimated cost without sacrificing quality. VE has proven effective in controlling the cost of design and construction and is being used widely by several federal government agencies.

CONSTRUCTION FINANCE

Before getting into the details of estimating the materials and labor that go into a project, let us take a quick look at the "material" that makes it all possible—money.

The Cost of Money

Let's say a house that cost $45,000 ten years ago now costs $125,000. But add finance charges, based on a 20% down payment and a mortgage for twenty-five years, with interest charges of 9% ten years ago and 10% today. The house really cost $99,640 ten years ago and costs $272,700 today. The increase in the initial cost of the house in ten years was 74%, but the increase in the cost of financing was 120%. And in tight money markets, mortgages are not at 10%. Obviously, the point is that money must be used as efficiently as any other commodity. And by efficiently we mean not only spent wisely but carefully planned for, shopped for, drawn upon only at the exact time it is needed, and paid for on time.

Our concern here is not with the money that the contractor needs to run his own business affairs, but the money to pay for the project itself. This breaks down into two types of financing—immediate cash to pay for the labor and materials as they are placed on the job, and once the project is completed, a loan against the project with repayment scheduled over many years. These are known as the *short term* or the construction loan and the *long term* or mortgage loan.

Project financing is not usually the direct concern of the contractor. It is the owner's or project developer's responsibility to secure both the short and long term loans. But it is essential that a contractor understand all phases of the project financing and his relationship to it.

First, few projects today are paid for from funds the owner already has accrued. He must find someone willing to loan the money to make the project possible, and that someone must have enough faith in the project to be willing to wait for repayment of his loan from money generated by the success of the project. In order to "sell" his project to a loan source, the owner must prepare a complete analysis of all aspects, including reliable cost projections, and will often turn to a contractor for this information. It is important that the contractor understands what such an estimate could commit him to, for what the lender "sees" is what the contractor later "gets" if the project loan is approved.

Second, the construction loan is the source from which the contractor will be paid, and he will have to work closely with the lender and his job inspectors to get his money on time. If the contractor is inefficient in compiling his requests for payment, his men and suppliers will go unpaid unless the contractor borrows the money until the next period the lender makes his payouts. The interest the contractor will have to pay will cut into his profit as surely as if he had underestimated the materials for the job.

Third, the owner may be very naive in construction finance. He may build only one project in his entire life. The local bank may be his only financial contact. The local banks are an excellent starting place to seek construction funds; they know the owner and the community and the owner's best interests are presumably the bank's best interests. But the bank may have other business loans with the owner, and they may not be willing, or even legally able, to make an additional loan of the size, at the rate, and for the length of time the owner requires. Thus the project may die in the bank's board room. The owner may lack the time and the sophistication to search further, and so the contractor loses the job as surely as if he had been underbid by a competitor. But there are other money sources besides banks, and an aggressive contractor will develop these other contacts so that he can steer the owner to them and salvage a good job that otherwise might never get off the drawing board.

Fourth, as we mentioned at the very beginning of this chapter, the triangle of owner-architect-contractor is nowadays often more of a circle, with each overlapping the other's territory. The contractor might also be the owner, either as the project developer or as the official landlord who then leases the project with perhaps an option to buy. Such arrangements can be very profitable and can be used in slack times to keep a contractor's force busy. But a good contractor does not always make a good landlord or even a good client for his company.

Sources of Money

The main sources of money are:

1. Commercial Banks—short and intermediate term financing
2. Savings and Loans—small residential loans
3. Insurance Companies—restricted long term loans
4. Pension and Trust Funds—long term loans. These funds are often administered by commercial banks.
5. Real Estate Investment Trusts (REITS) —long and short term loans, land loans, second mortgages, sell and leaseback deals.
6. Governmental Agencies—FHA and VA loans for residential work (guarantee loans)
7. Government Sponsored Bonds—industrial revenue bonds for projects that increase employment. Local governments issue tax-exempt bonds secured by the project income.
8. Mortgage Banking Firms—long and short term loans, land loans, second mortgages, sell and leaseback deals.

Each source has an area it is most interested in, but they compete to a degree. Many projects will draw on two or more sources. Short term loans are for three years or less. Long term loans are seldom for less than ten years and may run up to thirty years.

One thing all loan sources have in common is that they are "selling" their money, not "buying" your project. When the project is presented to them, they not only want to know the physical details but how the project will generate the money to pay back the loan. The essence of a good presentation is the proof that the loan can be repaid.

Mortgage Loans

While it is the construction, or short term, loan that is drawn on first, and land and "front money" loans may precede it, the mortgage, or long term, loan is shopped for first. Once the owner has the mortgage loan lined up, the construction loan is more easily secured, because it will eventually be paid by the mortgage loan when the project is completed. Further, the interest charged for the mortgage loan may decide whether the project can generate the income to make the necessary payments. If one fails to get a commitment

Constant Annual Percent

Monthly Payment In Arrears

Description: This table shows the percent of the principal amount of a loan needed each year to pay off the loan when the actual payments are monthly and are paid in arrears. Divide the percent by 12 to get the level monthly payment per $100 that includes both interest and principal.

Example: The Constant Annual Percent needed to pay off a 15%, 30 year loan if payments are made monthly and in arrears is 15.18%. Divide by 12 to get the actual monthly payment. The constant annual payment for a $50,000 loan is $7,590. The monthly payment is $632.50.

Interest Rate	5 yr	6 yr	7 yr	8 yr	9 yr	10 yr	11 yr	12 yr	13 yr	14 yr	15 yr	16 yr	17 yr	18 yr	19 yr	20 yr
10.00	25.50	22.24	19.93	18.21	16.90	15.86	15.03	14.35	13.78	13.30	12.90	12.56	12.26	12.00	11.78	11.59
10.25	25.65	22.39	20.08	18.37	17.06	16.03	15.20	14.52	13.96	13.48	13.08	12.74	12.45	12.20	11.98	11.78
10.50	25.80	22.54	20.24	18.53	17.23	16.20	15.37	14.69	14.14	13.67	13.27	12.93	12.64	12.39	12.18	11.99
10.75	25.95	22.69	20.39	18.69	17.39	16.37	15.54	14.87	14.31	13.85	13.46	13.12	12.84	12.59	12.37	12.19
11.00	26.10	22.85	20.55	18.86	17.56	16.54	15.72	15.05	14.50	14.03	13.64	13.31	13.03	12.79	12.57	12.39
11.25	26.25	23.00	20.71	19.02	17.72	16.71	15.89	15.23	14.68	14.22	13.83	13.51	13.23	12.98	12.78	12.60
11.50	26.40	23.15	20.87	19.18	17.89	16.88	16.07	15.40	14.86	14.41	14.02	13.70	13.42	13.18	12.98	12.80
11.75	26.55	23.31	21.03	19.34	18.06	17.05	16.24	15.58	15.04	14.59	14.21	13.89	13.62	13.39	13.18	13.01
12.00	26.70	23.47	21.19	19.51	18.23	17.22	16.42	15.77	15.23	14.78	14.41	14.09	13.82	13.59	13.39	13.22
12.25	26.85	23.62	21.35	19.67	18.40	17.40	16.60	15.95	15.42	14.97	14.60	14.29	14.02	13.79	13.60	13.43
12.50	27.00	23.78	21.51	19.84	18.57	17.57	16.78	16.13	15.60	15.16	14.80	14.49	14.22	14.00	13.80	13.64
12.75	27.16	23.94	21.67	20.01	18.74	17.75	16.96	16.32	15.79	15.36	14.99	14.68	14.42	14.20	14.01	13.85
13.00	27.31	24.09	21.84	20.17	18.91	17.92	17.14	16.50	15.98	15.55	15.19	14.88	14.63	14.41	14.22	14.06
13.25	27.46	24.25	22.00	20.34	19.08	18.10	17.32	16.69	16.17	15.74	15.39	15.09	14.83	14.62	14.44	14.28
13.50	27.62	24.41	22.16	20.51	19.28	18.28	17.50	16.87	16.36	15.94	15.58	15.29	15.04	14.83	14.65	14.49
13.75	27.77	24.57	22.33	20.68	19.43	18.46	17.68	17.06	16.55	16.13	15.78	15.49	15.25	15.04	14.86	14.71
14.00	27.93	24.73	22.49	20.85	19.61	18.64	17.87	17.25	16.75	16.33	15.99	15.70	15.45	15.25	15.08	14.93
14.25	28.08	24.89	22.66	21.02	19.78	18.82	18.05	17.44	16.94	16.53	16.19	15.90	15.66	15.46	15.29	15.15
14.50	28.24	25.05	22.83	21.19	19.96	19.00	18.24	17.63	17.14	16.73	16.40	16.11	15.87	15.68	15.51	15.36
14.75	28.40	25.22	22.99	21.37	20.14	19.18	18.43	17.82	17.33	16.93	16.60	16.32	16.09	15.89	15.72	15.59
15.00	28.55	25.38	23.16	21.54	20.31	19.37	18.62	18.02	17.53	17.13	16.80	16.53	16.30	16.11	15.94	15.81
15.25	28.71	25.54	23.33	21.71	20.49	19.55	18.80	18.21	17.73	17.33	17.01	16.74	16.51	16.32	16.16	16.03
15.50	28.87	25.71	23.50	21.89	20.67	19.73	18.99	18.40	17.93	17.53	17.21	16.95	16.72	16.54	16.38	16.25
15.75	29.03	25.87	23.67	22.06	20.85	19.92	19.19	18.60	18.12	17.74	17.42	17.16	16.94	16.76	16.60	16.48
16.00	29.19	26.04	23.84	22.24	21.04	20.11	19.38	18.79	18.33	17.94	17.63	17.37	17.16	16.98	16.83	16.70
16.25	29.35	26.20	24.01	22.42	21.22	20.29	19.57	18.99	18.53	18.15	17.84	17.58	17.37	17.20	17.05	16.93
16.50	29.51	26.37	24.18	22.59	21.40	20.48	19.76	19.19	18.73	18.36	18.05	17.80	17.59	17.42	17.27	17.15
16.75	29.67	26.53	24.35	22.77	21.58	20.67	19.96	19.39	18.93	18.56	18.26	18.01	17.81	17.64	17.50	17.38
17.00	29.83	26.70	24.53	22.95	21.77	20.86	20.15	19.59	19.14	18.77	18.47	18.23	18.03	17.86	17.72	17.61
17.25	29.99	26.87	24.70	23.13	21.95	21.05	20.35	19.79	19.34	18.98	18.69	18.45	18.25	18.08	17.95	17.84
17.50	30.15	27.04	24.88	23.31	22.14	21.24	20.54	19.99	19.55	19.19	18.90	18.66	18.47	18.31	18.17	18.06
17.75	30.31	27.21	25.05	23.49	22.33	21.43	20.74	20.19	19.75	19.40	19.11	18.88	18.69	18.53	18.40	18.29
18.00	30.48	27.37	25.23	23.67	22.51	21.63	20.94	20.39	19.96	19.61	19.33	19.10	18.91	18.76	18.63	18.52
18.25	30.64	27.54	25.40	23.86	22.70	21.82	21.14	20.60	20.17	19.82	19.55	19.32	19.13	18.98	18.86	18.76
18.50	30.80	27.71	25.58	24.04	22.89	22.01	21.34	20.80	20.38	20.04	19.76	19.54	19.36	19.21	19.09	18.99
18.75	30.97	27.89	25.76	24.22	23.08	22.21	21.54	21.01	20.59	20.25	19.98	19.76	19.58	19.44	19.32	19.22
19.00	31.13	28.06	25.93	24.41	23.27	22.41	21.74	21.21	20.80	20.47	20.20	19.98	19.81	19.67	19.55	19.45
19.25	31.30	28.23	26.11	24.59	23.46	22.60	21.94	21.42	21.01	20.68	20.42	20.21	20.03	19.89	19.78	19.69
19.50	31.46	28.40	26.29	24.78	23.65	22.80	22.14	21.63	21.22	20.90	20.64	20.43	20.26	20.12	20.01	19.92
19.75	31.63	28.57	26.47	24.96	23.84	23.00	22.34	21.84	21.43	21.11	20.86	20.65	20.49	20.35	20.24	20.16
20.00	31.80	28.75	26.65	25.15	24.04	23.20	22.55	22.04	21.65	21.33	21.08	20.88	20.72	20.58	20.48	20.39
20.25	31.96	28.92	26.83	25.34	24.23	23.39	22.75	22.25	21.86	21.55	21.30	21.10	20.94	20.82	20.71	20.63
20.50	32.13	29.10	27.01	25.52	24.42	23.59	22.96	22.46	22.08	21.77	21.53	21.33	21.17	21.05	20.95	20.88
20.75	32.30	29.27	27.20	25.71	24.62	23.80	23.16	22.68	22.29	21.99	21.75	21.56	21.40	21.28	21.18	21.10

Interest Rate	21 yr	22 yr	23 yr	24 yr	25 yr	26 yr	27 yr	28 yr	29 yr	30 yr	31 yr	32 yr	33 yr	34 yr	35 yr	40 yr
10.00	11.41	11.26	11.13	11.01	10.91	10.82	10.73	10.66	10.59	10.54	10.48	10.44	10.39	10.36	10.32	10.19
10.25	11.62	11.47	11.34	11.22	11.12	11.03	10.95	10.88	10.82	10.76	10.71	10.66	10.62	10.58	10.55	10.43
10.50	11.82	11.68	11.55	11.43	11.34	11.25	11.17	11.10	11.04	10.98	10.93	10.89	10.85	10.81	10.78	10.67
10.75	12.03	11.88	11.76	11.65	11.55	11.46	11.39	11.32	11.26	11.21	11.16	11.12	11.08	11.05	11.02	10.91
11.00	12.23	12.09	11.97	11.86	11.77	11.68	11.61	11.54	11.48	11.43	11.39	11.35	11.31	11.28	11.25	11.14
11.25	12.44	12.30	12.18	12.08	11.98	11.90	11.83	11.77	11.71	11.66	11.62	11.58	11.54	11.51	11.48	11.38
11.50	12.65	12.51	12.40	12.29	12.20	12.12	12.05	11.99	11.94	11.89	11.85	11.81	11.77	11.74	11.72	11.62
11.75	12.86	12.73	12.61	12.51	12.42	12.35	12.28	12.22	12.16	12.12	12.08	12.04	12.01	11.98	11.95	11.87
12.00	13.07	12.94	12.83	12.73	12.64	12.57	12.50	12.44	12.39	12.35	12.31	12.27	12.24	12.22	12.19	12.11
12.25	13.28	13.16	13.05	12.95	12.87	12.79	12.73	12.67	12.62	12.58	12.54	12.51	12.48	12.45	12.43	12.35
12.50	13.50	13.37	13.26	13.17	13.09	13.02	12.96	12.90	12.85	12.81	12.78	12.74	12.71	12.69	12.67	12.59
12.75	13.71	13.59	13.48	13.39	13.31	13.24	13.18	13.13	13.09	13.05	13.01	12.98	12.95	12.93	12.91	12.84
13.00	13.93	13.81	13.71	13.62	13.54	13.47	13.41	13.36	13.32	13.28	13.25	13.22	13.19	13.17	13.15	13.08
13.25	14.14	14.03	13.93	13.84	13.77	13.70	13.64	13.59	13.55	13.51	13.48	13.45	13.43	13.41	13.39	13.32
13.50	14.36	14.25	14.15	14.07	13.99	13.93	13.87	13.83	13.79	13.75	13.72	13.69	13.67	13.65	13.63	13.57
13.75	14.58	14.47	14.37	14.29	14.22	14.16	14.11	14.06	14.02	13.99	13.96	13.93	13.91	13.89	13.87	13.81
14.00	14.80	14.69	14.60	14.52	14.45	14.39	14.34	14.30	14.26	14.22	14.19	14.17	14.15	14.13	14.11	14.06
14.25	15.02	14.92	14.82	14.75	14.68	14.62	14.57	14.53	14.49	14.46	14.43	14.41	14.39	14.37	14.36	14.30
14.50	15.24	15.14	15.05	14.98	14.91	14.86	14.81	14.77	14.73	14.70	14.67	14.65	14.63	14.61	14.60	14.55
14.75	15.47	15.37	15.28	15.21	15.14	15.09	15.04	15.00	14.97	14.94	14.91	14.89	14.87	14.86	14.84	14.80
15.00	15.69	15.59	15.51	15.44	15.37	15.32	15.28	15.24	15.21	15.18	15.15	15.13	15.12	15.10	15.09	15.04
15.25	15.92	15.82	15.74	15.67	15.61	15.56	15.51	15.48	15.45	15.42	15.40	15.38	15.36	15.34	15.33	15.29
15.50	16.14	16.05	15.97	15.90	15.84	15.79	15.75	15.72	15.69	15.66	15.64	15.62	15.60	15.59	15.58	15.54
15.75	16.37	16.28	16.20	16.13	16.08	16.03	15.99	15.95	15.93	15.90	15.88	15.86	15.85	15.83	15.82	15.79
16.00	16.59	16.50	16.43	16.37	16.31	16.27	16.23	16.19	16.17	16.14	16.12	16.10	16.09	16.08	16.07	16.03
16.25	16.82	16.74	16.66	16.60	16.55	16.50	16.47	16.43	16.41	16.38	16.36	16.35	16.33	16.32	16.31	16.28
16.50	17.05	16.97	16.89	16.83	16.78	16.74	16.71	16.67	16.65	16.63	16.61	16.59	16.58	16.57	16.56	16.53
16.75	17.28	17.20	17.13	17.07	17.02	16.98	16.94	16.92	16.89	16.87	16.85	16.84	16.82	16.81	16.80	16.78
17.00	17.51	17.43	17.36	17.31	17.26	17.22	17.19	17.16	17.13	17.11	17.10	17.08	17.07	17.06	17.05	17.02
17.25	17.74	17.66	17.60	17.54	17.50	17.46	17.43	17.40	17.38	17.36	17.34	17.33	17.32	17.31	17.30	17.27
17.50	17.97	17.90	17.83	17.78	17.74	17.70	17.67	17.64	17.62	17.60	17.59	17.57	17.56	17.55	17.55	17.52
17.75	18.20	18.13	18.07	18.02	17.97	17.94	17.91	17.88	17.86	17.85	17.83	17.82	17.81	17.80	17.79	17.77
18.00	18.44	18.37	18.31	18.26	18.21	18.18	18.15	18.13	18.11	18.09	18.08	18.06	18.05	18.05	18.04	18.02
18.25	18.67	18.60	18.54	18.49	18.45	18.42	18.39	18.37	18.35	18.34	18.32	18.31	18.30	18.29	18.29	18.27
18.50	18.91	18.84	18.78	18.73	18.69	18.66	18.64	18.61	18.60	18.58	18.57	18.56	18.55	18.54	18.54	18.52
18.75	19.14	19.07	19.02	18.97	18.94	18.90	18.88	18.86	18.84	18.83	18.81	18.80	18.80	18.79	18.78	18.77
19.00	19.37	19.31	19.26	19.21	19.18	19.15	19.12	19.10	19.09	19.07	19.06	19.05	19.04	19.04	19.03	19.02
19.25	19.61	19.55	19.50	19.45	19.42	19.39	19.37	19.35	19.33	19.32	19.31	19.30	19.29	19.28	19.28	19.26
19.50	19.85	19.79	19.74	19.69	19.66	19.63	19.61	19.59	19.58	19.56	19.55	19.55	19.54	19.53	19.53	19.51
19.75	20.08	20.02	19.98	19.94	19.90	19.88	19.86	19.84	19.82	19.81	19.80	19.79	19.79	19.78	19.78	19.76
20.00	20.32	20.26	20.22	20.18	20.15	20.12	20.10	20.08	20.07	20.06	20.05	20.04	20.03	20.03	20.02	20.01
20.25	20.56	20.50	20.46	20.42	20.39	20.36	20.34	20.33	20.32	20.30	20.30	20.29	20.28	20.28	20.27	20.26
20.50	20.80	20.74	20.70	20.66	20.63	20.61	20.59	20.57	20.56	20.55	20.54	20.54	20.53	20.53	20.52	20.51
20.75	21.03	20.98	20.94	20.90	20.88	20.85	20.84	20.82	20.81	20.80	20.79	20.78	20.78	20.77	20.77	20.76

for the mortgage interest and interest rates soar, the debt service may take all the projected profits.

Small projects can usually be financed both long and short term at local banks and savings and loan associations. These institutions make no direct charge for reviewing proposals and can usually give a yes or no answer in a few days, so one can shop several sources at the same time with no obligation. Once a loan is approved and accepted, there are certain charges called closing costs, and a part of these costs, known as *points, origination fees,* or *the discount,* cover the cost of setting up the loan. These points are a one-time charge made as a percentage of the loan and are deducted from the amount of the loan. If the points are 2%, then the amount of two dollars is deducted from every hundred dollars of the loan. The usual range of points is from 1% to 3%, but in states where there are legal limits set on the interest that may be charged, points have been quoted as high as 7% as a way to get around the usury laws. Such points are common to all loan sources and are one of the important items to shop for.

Mortgage Banker

For larger projects and projects that generate income, one should check out all the money sources. It isn't likely that even the most gregarious contractor knows all the institutional sources, or even if he did, could keep abreast of their current interests and capabilities. One must consult a specialist, and that specialist is the mortgage banker, as we shall call him here to distinguish him from a commercial banker.

The mortgage broker is an intermediator between borrowers and lenders, although his firm may have some funds of its own to invest. He is a "matchmaker" who brings parties together. To do this he must carefully evaluate both sides, shopping for the best terms for the owner and for projects most compatible with the portfolio of the lender. For consummating the transaction, he charges a flat fee of around 2% of all the loans placed by him. This fee is in addition to the usual closing costs that are charged by the lender. Both expenses are paid by the borrower. Many lenders require a 1% fee at the time the loan application is filled. This fee is refundable if the lender rejects the loan but becomes earned once the lender approves the application and is credited to the fee due the lender at the time of settlement.

There are other benefits in working with a mortgage broker. He can advise on the techniques of financing as well as the sources. There are numerous options available, and the owner will want to review them all with his accountant and lawyer and select what may suit his particular situation best. Decisions to be made would include the desired length of the mortgage; whether to select a loan at lower rates but "locked in" for many years or one at higher rates but with prepayment provisions; what personality—that is, items not part of the building itself such as equipment, carpet, drapes and the like—should be included in the mortgage; what owner assets should be pledged; what existing property should be made part of the new mortgage (wrap-around mortgage); what "kickers" might be offered, such as sharing income or equity with the lender if certain income or worth is exceeded; what advantages second and third mortgages might have; what method of repayment of the mortgage loan should be adopted (usually a constant covering interest and principal); payments of interest only, with the principal paid when the note is due; a constant payment of principal with interest figured on the declining balance; no payments on either interest or principal until the note is due but with the interest compounding; or a combination of the preceding with refinancing at certain periods to fit the owner's needs. There are no fixed answers that cover all situations. These decisions can contribute as much to the success of the project as decisions about room arrangements or types of building materials.

Having determined the best package for the owner, the mortgage banker will then shop his various money sources for the best deal. Each source will be looking for certain features, and the final deal probably will be a compromise.

The mortgage banker's opinions are useful in other ways. Having reviewed numerous similar projects at the preliminary stage, he can often suggest refinements in the physical arrangements, as they relate to potential income, that he has seen work out well in other situations. And finally, in accepting or refusing a project, he gives his opinion as to its probable success. If two or three mortgage bankers turn a project down when they are placing others, the owner should review his figures. The risk elements must be greater than those in the normal market. The mortgage broker need not be just a necessary evil. He can be a creative part of the building team.

Selling the Lender

Whether one works directly with a loan organization or through a broker, one must be prepared to make an effective presentation of his project. While each type of project will emphasize different elements, all should include the following:

1. physical analysis of the project including a surveyor's plot plan, an architect's sketch plans and specifications, an artist's rendering of the developed site, and a contractor's estimate of the cost;

2. a market feasibility study;

3. a financial analysis of the ownership including personal statements;

4. a financial analysis of the project.

A great deal of thought must go into the presentation of the physical plant. Once the project is approved, it is difficult to make major changes. A too glamorous presentation may backfire. It is utility not beauty that sells the lender. An impressive entrance framed by a tree lined drive circling a large fountain may be just an artist's window dressing and will be the first thing abandoned when the final budgets are reviewed. But that may be the feature a lender remembers and insists be built. Another lender, with his eye on the cash flow, will see such embellishments as so much maintenance expense and as a detriment rather than asset.

In addition to the contractor's estimate, the lender will make one of his own and may ask the borrower to have an independent source make up a third. This is usually done by a member of the American Institute of Real Estate Appraisers, whose members carry the title MAI (Member Appraisal Institute), or senior member of an appraisal society such as the American Society of Appraisers.

The owner's financial statement may be prepared by his certified public accountant, but often the lender will have his own standard form to fill out. The proposed contractor may also have to file a financial statement. Some loans will only be made if performance bonds are posted. Before presenting his proposal an owner will want to check with his lawyer and accountant as to the best form of ownership for the project—individual, corporate, or partnership. Some buildings are owned "at arms length". A third party actually owns the building, and the promoter of the project becomes the tenant.

The market feasibility study shows the compatibility of the project with the surroundings, means of public access, the effect on the environment, and compliance with zoning regulations. Similar successful projects in the area should be noted, and the continuing need for additional facilities proven. And finally, the adaptability to another occupancy and resale value should be evaluated.

The most carefully scrutinized section of the presentation will probably be the projected value of the project based on income, or whether the project will generate enough income to repay the loan on time with interest. What is included here will vary with the type of project. Where rents are involved, rent rolls should be projected. The lender in such cases may stipulate that a certain rent level be reached, usually around 80% of the projection, before the mortgage becomes final. If not, the mortgage will be for a lower amount. This is called *setting a floor* for the mortgage. The spread may be as much as 20%, and if the owner does not reach the agreed upon rent level, he must find another source to cover this amount. This will be discussed later.

The Commitment

Once a project is accepted by a lending organization, a commitment paper is issued stating that the lender will lend a certain amount of money for a certain length of time at a certain rate of interest, providing the borrower fulfills certain stated conditions and closes the loan by a certain date. This commitment, if the owner accepts, will become the basis for the first mortgage on the project. There may be other mortgages either on the project as a whole, or on some phase of it, such as furnishings, equipment, or the land itself. But the holder of the first mortgage takes precedence over all other claims except taxes, and he has the authority to take possession of the project upon proven default and to sell it and deduct the amount due him from the selling price, which is usually determined by a public auction. If any excess cash remains, it then goes to the holder of the second mortgage, then to the third, if any, and so on, with the original developer the last to receive any distribution. Holders of mortgages other than the first cannot instigate the sale of the project. They are subordinate to the first mortgage holder and therefore carry a higher risk and charge a higher rate of interest.

The commitment will be quite specific and should not be thought of as a preliminary proposal, but rather, as the basis of a contract that one must live with for many years. This is why the presentation of the project must be carefully thought out and based on realistic goals.

Short Term Loans

Having received a commitment, the owner must then work fast to finalize his plans. The commitment will carry a deadline date for acceptance that may be only a couple of weeks away. It will also carry an expiration date by which time all closing documents shall be submitted to the loan company so that the final mortgage loan may go into effect. If this date is not met, the owner forfeits an agreed upon deposit, which is usually around 1% of the mortgage amount and is deposited with the lender's agent upon accepting the commitment. One must then proceed immediately to secure a construction loan, gap financing (if it is needed), and "front money".

The construction loan is a short term loan to cover the building costs during the erection of the project. With the mortgage loan commitment in hand, the construction loan repayment is guaranteed providing that the project is completed in a satisfactory manner, so it is not necessary to sell the feasibility of the project all over again. But the construction loan lender does take the risk that the owner or the contractor will not get in financial difficulties during the construction period, or that the projected budget will not prove inadequate. The lender may then have to complete the project himself. Because the mortgage amount may be less than the stated amount if certain income levels are not reached, the construction loan will be based on the mortgage "floor" rather than the "ceiling" figure, which may deduct as much as 20% from the amount needed to complete the construction. Further, the construction loan is seldom for more than 80% of the floor figure, and like all loans, is subject to points and other closing costs, so perhaps another 2% will be deducted from the loan.

While shopping around might improve one's position, the construction loan will still fall short of the amount needed to complete the project. The missing funds must be supplied by the owner and represent his *equity* in the project. If the borrower can pledge sufficient liquid assets, that is cash, stocks or bonds, or other unencumbered property, he might be able to raise the additional funds from the same source as the construction loan. If not, he will have to take in a partner or borrow from those who take high risks. One such source is those who loan the difference between the floor and the ceiling of the mortgage. This is known as *gap financing*. They will issue a commitment to pay this difference if the income generated by the project is less than that specified to qualify for the full or ceiling amount. To obtain such a commitment one must pay in advance a flat fee, usually around 5% of the amount to be loaned. If the project reaches the income level to qualify for the full mortgage, and the gap loan is not needed, the fee is not refundable. If it is needed, the loan becomes a second mortgage against the project and is repaid with interest.

In setting up the project budget, there are other costs to be met besides construction costs. First, the land must be obtained. While this is sometimes leased or paid for under a mortgage arrangement, the first mortgage commitment may ask that the land title be *unencumbered*. In addition to land costs, there may be land development expenses such as surveys, soil investigation borings, utility extensions and connections, and access roads or sidings. There also may be purchasing expenses, including attorney and real estate agent fees, taxes and permits, and zoning fees. If the zoning for the land is to be changed, it can be both time-consuming and expensive.

And the architect must be paid. His fee will run from around 4% for a large industrial plant up to 15% for a fine residence or interior remodeling job. Some of this fee will not become due until the job is completed.

When the loan commitments are accepted, they may require a good faith deposit of around 1% of the loan. Consulting fees must be paid to accountants and lawyers. At loan closings legal and recording fees and transaction taxes may be charged, and if asked for, a performance bond obtained.

When the work is started, there is always a lapse of time between paying for permits, materials, and salaries and being reimbursed for these items from the construction loan. Usually, each request for payment from the construction loan must be accompanied by waivers of lien. In actual practice the contractor may finance this period rather than the owner, but the lender may still demand that the owner show he has the money to cover these start-up costs.

How much front money, or equity, an owner will need to launch his project will vary with the type of project, the money market, and the owner's reputation. It is often said that an owner with a proven need, a piece of property free of debt, and an architect's set of plans can obtain all the financing he will need. We have used the term "owner" in our preceding discussions as if the one who builds the project were the one who will occupy it. Today even large corporations may choose to be tenants rather than owners and will rely on developers to put the project together. The contractor may be asked to be the developer or at least to become a partner in it. Often the owner of the real estate and the architect will be included as *limited partners*, their limits of both liability and equity being proportional to the cost of the services or land contributed by them. When the construction loan is paid off by the mortgage loan, they are also often bought out and the ownership reverts to the general partner. However, some mortgage commitments guard against such arrangements by adding a penalty fee if there is any change of ownership within a stipulated time.

Rates of interest on construction loans vary from week to week and community to community but are usually tied to the prime rate, which is the interest charged by commercial banks to their preferred customers for short term loans. Currently construction loans are quoted at 2-1/2% to 4% over the prime. High risk loans will have considerably higher rates.

It is not our intention to encourage the contractor to become the project entrepreneur. But with the cost of money tied so closely to the success of the project, from an income-producing standpoint a contractor must know all the procedures in order to follow the time and budget schedules, to process his monthly draws on time, and to turn the completed project over to the owner in such condition as to be completely acceptable to the mortgage lender.

Interim Financing

This particular type of financing has nothing to do with paying for the construction costs of a project. Nor does it have to do with long-term mortgage financing, which is repaid over an extended period from the proceeds of the revenue generated from the project.

The definition of the word interim is *provisional, intervening, interlude or relating to a time period*. This is the type of financial assistance required by virtually all contractors, regardless of size, for short-term working capital to compensate for unanticipated delays in receiving payments from owners for work performed and properly invoiced.

The source is usually a commercial bank that is responsive to the needs of the contractor. Not all banks are receptive to this type of business, and the contractor must establish banking relations with a bank that has personnel who possess intimate knowledge of the financial structure found in the construction industry. Once an acceptable financial institution has been selected, then the contractor should identify a particular person within the bank with whom a personal relationship can be established.

Next, it is the obligation of the contractor to prove, to the complete satisfaction of the bank, the credibility of the technical and financial assets of the company. This can be accomplished by furnishing *certified (CPA) financial statements* of the company along with a written detailed resume of the qualifications of each person directing the destiny of the company. Progress, at this point, should place the contractor in a position to obtain from the bank a general commitment as to the limit and terms under which they would participate in granting short term loans (usually 30 to 90 days). The ideal situation is to obtain from the bank a letter of credit outlining its position so that at any time during the next 12 months, a loan could be arranged without further review from a committee. This avoids days, or even weeks, of delay from the time the loan application is presented until the proceeds are available.

At the time a loan request is made, the position of the contractor can be significantly enhanced by presentation of a cash flow analysis of all current projects, projecting anticipated expenditures and income on a monthly basis so that the amount of money required and the availability of sufficient income to repay the loan can be clearly defined.

Expect the interest on this type of loan to be from 1% to 6% above the current prime rate. Banks consider this high risk business. If the current prime rate is 16% and the loan is assigned 3% over prime, this means the contractor is paying 19% for the use of the money. Obviously, a contractor is forced to impose restrictions on the use of such costly money. It cannot be used as continuous working capital, which at best has an earning capacity of about 10%. It has to be justified by large discounts. For example, prompt payment of equipment invoices would affect an early delivery date that would equate with a time and cost saving justifying the interest rate.

The availability of interim financing provides a measure of security for the contractor, who is assured that external financial assistance is available to augment working capital during periods of depressed cash flow.

DRAWINGS AND SPECIFICATIONS

The owner's dreams and the architect's visions must eventually be put down in a form the contractor can use as a basis for bidding the project and for constructing in enduring materials. This form has developed into the dual one of *working drawings* and *specifications*. Basically, the drawings present the design, the location, and the dimensions of a project. The specifications give the quality required.

The drawings are done in pencil or ink on tracing paper or cloth, which is easily reproducible, and prints are delivered to the contractor. The architect retains the original drawings in his office. The specifications, except on very small jobs for which they may be incorporated on the drawings, are printed and bound separately. Both are made available to each contractor bidding the job. As bid periods are usually quite limited, it is the contractor's ability to interpret them that enables him to present a realistic and complete bid. This chapter presents an outline of what information a contractor should expect to find in a bid set and where it can be found.

The Working Drawings

Arrangement. A typical complete project will find several drawing sheets, usually all of uniform size, bound into one job set. The first sheet should include a site plan, sometimes incorporating the roof plan, plus the complete job title, and sometimes, an index of the drawings. This will be followed by the sheets showing architectural details, usually in the following order: basement plan, first floor plan, upper floor plans, exterior elevations, sections, interior elevations and details.

The architectural sheets are followed by the structural set including footing and foundation plans, basement framing plans, floor framing plans, roof framing plans, and then the structural details. These are followed by the heating and ventilating plans; then the plumbing plans, and finally, the electrical plans, each starting with the lowest floor and working upward.

Each sheet should have a title block in the lower right-hand corner with the sheet number (usually "A-" for architectural, "S-" for structural, "M-" for heating and ventilating, "P-" for plumbing and "E-" for electrical); the number of sheets in each set (i.e., A-1 of 7); the date made plus each date it has been revised, and the initials of the person or persons who drew and approved the sheet.

Revisions are sometimes listed outside the title block. Some architects, rather than revising a drawing once it is printed, will issue supplementary drawings which must be attached to the appropriate set as they are issued.

Type of Drawings. Most working drawings for building construction are based on *orthographic projection*, which is a parallel projection to a plane by lines perpendicular to the plane. In this way, all dimensions will be true. If the plane is horizontal, the projection is a plan; if vertical, it is an elevation for outside the building, or a sectional elevation if through the building.

The only descriptive drawing that presents a building as the eye sees it is the perspective. But perspectives are used mainly to study the building and present it to the client in a form he can easily understand. A perspective is seldom useful for presenting information on working drawings.

However, other pictorial presentations are helpful to the builder. Two of these are the *isometric* and the *cabinet projection*. Isometrics are drawings in which all horizontal and vertical lines have a true length, and those lines parallel on the object are also parallel on the drawing. Vertical lines are vertical, but horizontal lines are set at 30 or 60 degrees.

A GARDEN SHED
6' x 6' x 8' HIGH
WITH PEAKED ROOF
SCALE: 1/16"=1'0"

SHOWN AS AN ISOMETRIC,

LEFT SIDE   FRONT   RIGHT SIDE   REAR

FLOOR PLAN   ROOF PLAN

IN ORTHOGRAPHIC PROJECTION,

45°   1/2 SCALE

AND AS A CABINET DRAWING

COPPER FINIAL
SLATE ROOF
FACE BRICK
METAL LINTLE
WOOD CASEM'T
BRICK SILL
GRADE
CONC. FND'N

ASPHALT SHINGLES
D.H. WOOD SASH
WOOD SHUTTERS & TRIM
WOOD SIDING
GRADE
CONC. FND'N

WEST ELEVATION
SCALE: 1/8" = 1'-0"

W.I. VANE
WOOD SHAKES
STONE LINTLE & SILL
METAL WIND.
RAND. ASHLAR
GRADE
CONC. FND'N

THE GARDEN SHED SHOWING THE SAME WEST (LEFT SIDE) ELEVATION BUT WITH THREE DIFFERENT MATERIAL INDICATIONS AS MIGHT BE FOUND ON A WORKING DRAWING.

Isometrics are often used for piping diagrams and to show complicated intersections, such as on sloped roofs.

Cabinet drawings are those with the front face shown in true shape and size, as if it were an orthographic projection, but they simulate a perspective. The sides are shown receding at 45 degrees and at 1/2 scale (1:20). Variations of this are *oblique drawings*, where the angle and side scale may be anything that best shows the object, and *cavalier drawings,* where the side scale is at the same scale as the front. Cabinet drawings get their name from the fact that they are often used for cabinet work.

Scale. The architect's scale, with the inch divided into 1/8S, 1/16S, 1/32S, is standard for building construction in the United States. The engineer's scale, with the inch divided into tenths, is sometimes used in structural work or on site plans. However, it is advantageous to have architectural, structural, and mechanical drawings all at the same scale for a job.

The metric scale is divided into centimeters and millimeters, 2.54 centimeters equalling one inch. It is the common scale throughout the world. The SI metric scale is based on a decimal with the meter representing a unit of 1.0. The Metric Conversion Act of 1975, as amended by the Omnibus Trade and Competitiveness Act of 1988, establishes the modern metric system (System International, S.I.) as the preferred system of measurement in the United States.

Unit	Abbreviation	Length, meters	Approximate U.S Equivalent
kilometer	km	1,000	0.62 miles
hectometer	h	100	109.36 yards
decameter	dkm	10	32.81 feet
meter	m	1	39.37 inches
decimeter	dm	0.1	3.94 inches
centimeter	cm	0.01	0.39 inches
millimeter	mm	0.001	0.04 inches

Metric drawing scales are expressed in nondimensional ratios. Preferred scales are: 1:1 (one to one, which is full size), 1:5, 1:10, 1:20, 1:50, 1:100, 1:200, and 1:500.

Graphic scales are those devised by the draftsman and shown on the drawings with subdivisions comparable to the other scale. This is convenient if the drawing is to be reduced in size in reproduction and is meant to be scaled rather than dimensioned. It is seldom used on working drawings.

Plans and elevations are usually drawn at no less than 1/8" = 1'-0" (1:100); plot plans at 1/16" = 1'-0" (1:200). Complicated plan areas, such as toilet rooms, are usually repeated at 1/4" (1:50) to 1/2" (1:20) scale. Wall sections and cabinet work must show more detail and are from 3/4" (1:200) to 1-1/2" (1:10) scale. Certain areas of sections will be further detailed at 3" (1:5) scale, while moldings and decorative items might even be drawn at half to full scale.

Dimensioning. While scales should be clearly identified for each drawing, either in the title block or beneath each individual layout, contractors should never scale drawings but should insist that the architect furnish them with complete stated dimensions. This is the duty of the architect, and the contractor is foolish to take responsibility for it. Blueprints are subject to stretching and shrinking with humidity and some reproductive methods reduce the drawings in size.

Notes on Drawings. The drawings should not be cluttered with notes, which should be in the specifications, but architects will vary on this. The draftsman will often put notes on drawings to alert his own specification writer rather than the contractor. It is the specification that should govern, and all notes on drawings should be checked against the specifications and cross-referenced. If a contractor finds discrepancies during bidding, he should report this to the architect and ask for a written statement, known as an *Addendum,* to clarify it.

Finish Schedules. Schedules may appear either as part of the drawings or the specifications. Room finish schedules are often on the drawings next to the floor plan to which they refer. Schedules for lintels, columns, footings, doors, and frames are usually on the drawings, while hardware schedules are usually in the specifications. Fixture schedules for mechanical and electrical work may be found either place.

Reproduction. *Blueprints*, with lines showing in white on a blue background, may be required by many building departments for permits, because they are difficult to alter. Notes can only be added in colored pencil, either red or yellow. The contractor should limit the notes he makes on his construction set and authorize only one person to add them so that somebody's doodles don't end up in brick and mortar.

Today, many architects have their own printing equipment. These are usually of the ammonia process which produce black, blue, or maroon lines on a white background. They are much more easily marked on, although the added note will not be as obvious as on a blueprint, unless made in color. Small prints may be run off on one of the many types of office duplicating machines.

If it is desirable to reduce or enlarge a drawing, a *photostat*, which is a photographic process that produces negatives as positives, can be used. There are processes that transfer blueprints to cloth drawings or to *sepia prints*, which can be altered, added to, and printed just like an original tracing.

Interpreting Lines. The architect will use various lines in making his drawings. Thick solid lines are used to outline the edges of the plans and elevations. Lines consisting of long and short dashes in sequence denote center lines. Short dashes usually indicate hidden lines but are also used on windows to show where they are hinged if they are operable. Thin lines are used for dimensions. These terminate in arrows to show the limit of the given dimensions. Repeat marks or the notation "do" indicates the repetition of a given dimension. Dimension lines are often broken by a zigzag which shows the object in the drawing has been shortened.

Symbols

Material indications and items related to plumbing, heating and ventilation, and electrical work have been standardized. Some architects may have a symbol schedule on their drawings, but most no longer do. The following are many of the usual symbols a contractor may encounter, but always check with the specifications for building materials, because symbols may vary from draftsman to draftsman.

EARTH

ROCK

CINDER FILL

STONE CONCRETE

CINDER CONCRETE

CEMENT

CONCRETE BLOCK

TILE

GYPSUM BLOCK

COMMON BRICK

FACE BRICK

FIRE BRICK

CUT STONE

RUBBLE STONE

MARBLE

ROUGH WOOD

FINISH WOOD

METAL

REINFORCING BARS

LOOSE INSULATION

SOLID INSULATION

GENERAL OUTLETS:

CEILING WALL

Ceiling	Wall	
◯	─◯	Fixture Outlet
Ⓑ	─Ⓑ	Blanked Outlet
Ⓒ	─Ⓒ	Clock Outlet
Ⓓ		Drop Cord Outlet
Ⓕ	─Ⓕ	Fan Outlet
Ⓙ	─Ⓙ	Junction Box
Ⓡ	─Ⓡ	Recessed Outlet
Ⓢ	─Ⓢ	Pull Switch
Ⓧ	─Ⓧ	Exit

Surface Fluorescent

Recessed Fluorescent

Bare Lamp Fluorescent

CONVENIENCE OUTLETS:

Duplex

Waterproof

Radio

Range

Floor

Special Purpose

SWITCH OUTLETS:

S	Single Pole Switch
S_2	Double Pole Switch
S_3	Three Way Switch
S_D	Automatic Door Switch
S_E	Electrolier Switch
S_K	Key Operated Switch
S_{CB}	Circuit Breaker
S_{MC}	Momentary Contact
S_{RC}	Remote Control
S_{WP}	Weatherproof Switch
S_F	Fused Switch

PANELS & CIRCUITS:

	Lighting Panel
	Power Panel
	Two Wire Branch Circuit
	Three Wire Branch Circuit
	Under Floor Duct & Box
Ⓖ	Generator
Ⓜ	Motor
	Controller

SIGNALLING SYSTEMS:

- Push Button
- Buzzer
- Bell
- Chime
- Annunciator
- Electric Door Opener
- Telephone Switchboard
- Interconnecting Telephone
- Outside Telephone
- Fire Alarm Bell
- Fire Alarm Station
- City Fire Alarm
- Automatic Fire Alarm
- Watchman's Station
- Horn
- Nurse's Signal Plug
- Maid's Signal Plug
- Radio Outlet
- Television Outlet
- Interconnection Box

PIPE FITTING SYMBOLS:

———————	Low Pressure Steam
—/——/——/—	Medium Pressure Steam
—//——//——//—	High Pressure Steam
— — — — —	Return Line
>—o—o—o—	Condensate Line
·—·——·——·—	Make Up Water
—— — —— ——	Air Relief Line
———FOF———	Fuel Oil Flow
———FOR ——	Fuel Oil Return
———FOV ——	Fuel Oil Vent
———A ———	Compressed Air
———RL ——	Refrigerant Liquid
———RD ——	Refrigerant Discharge
——— C ———	Condenser Water Flow
— — —CR— — —	Condenser Water Return
———CH ———	Chilled Water
— — —CHR— — —	Chilled Water Return
—·— H —·—	Humidification Line
———D ———	Drain Line
——— B ———	Brine Supply

HEATING SYMBOLS:

Exposed Radiator

Flush Enclosed Radiator

Projecting Enclosed Radiator

Unit Heater (Prop. Fan)

Unit Heater (Cent. Fan)

Unit Ventilator

Thermostatic Trap

Float Trap

Boiler Return

Pressure Reducing Valve

Air Line Valve

Lock & Shield Valve

Diaphragm

Air Eliminator

Strainer

Thermometer

Thermostat

Gage

Pressure Switch

Expansion Valve

DUCTWORK SYMBOLS:

Duct and Direction of Flow

Inclined Drop

Inclined Rise

Access Door

Adjustable Blank Off

Adjustable Plaque

Automatic Damper

Canvas Connections

Deflecting Damper

Volume Damper

Supply Duct in Section

Exhaust Duct in Section

Supply & Exhaust Wall
    Grilles

Vanes

PIPING SYMBOLS:

—‖—	*Flanged Joint*
—┼—	Screwed Joint
—⊂—	Bell & Spigot Joint
—✕—	Welded Joint
—⊖—	Soldered Joint
⊙‖	Flanged Elbow-Up
⊖‖	Flanged Elbow-Down
‖⊙‖	Flanged Tee-Up
‖⊖‖	Flanged Tee-Down
‖▷	Flanged Reducer-Concentric
‖◺	Flanged Reducer-Eccentric
‖◁▷‖	Flanged Gate Valve
‖◀▶‖	Flanged Globe Valve
‖◥‖	Flanged Check Valve
‖⬡‖	Flanged Cock Valve
‖◁⊠▷‖	Flanged Safety Valve
‖⊏⊐‖	Flanged Expansion Joint
‖◻—	Reducing Flange
—‖‖—	Flanged Union
‖···‖	Flanged Sleeve

Abbreviations. The following is a list standard abbreviations used within the construction industry. The list is alphabetized by the abbreviations.

A	ampere; air	BC	bottom of curb
AB	anchor bolt	BD	board
ABAND	abandoned	BDL	bundle
ABT	about	BEL	below
ABUT	abutment	BET	between
ABV	above	BF	board foot, board feet
AC	alternating current; acoustical	BH	baseboard heater
		BIT	bituminous
A/C	air conditioning	BJT	bed joint
ACC	access	BL	building line
ACFL	access floor	BLDG	building
ACPL	acoustical plaster	BLK	block
ACR	acrylic plastic	BLKG	blocking
ACS	access	BLW	below
ACT	acoustical tile	BLW FL	below floor
AD	area drain; access door	BM	bench mark; beam
ADD	addendum; additional	BOT	bottom
ADH	adhesive	BP	back plaster; base plate; bearing pile
ADJ	adjacent; adjustable		
ADJT	adjustable	BPL	bearing plate
ADMIN	administration	BRG	bearing
AFF	above finished floor	BRK	brick
AGG	aggregate	BRKR	breaker
AL	aluminum	BRZ	bronze
ALT	alternate	BS	both sides
AMP	amperage; ampere	BSMT	basement
ANC	anchor	BT	bent
ANOD	anodized	BTM	bottom
AP	access panel	BTU	British Thermal Unit
APPR	approved	BTUH	British Thermal Unit per hour
APPROX	approximate		
APX	approximate	BU	built up
ARCH	architect; architectural	BUR	built up roofing
ASB	asbestos	BVL	beveled
ASC	above suspended ceiling	BW	both ways
ASPH	asphalt		
ASSY	assembly	C	course
AT	asphalt tile	C & G	curb and gutter
AUTO	automatic	CA	compressed air
AUX	auxiliary	CAB	cabinet
AVG	average	CAD	cadmium
AWG	American Wire Gauge	CANTIL	cantilever
		CAP	capacity
BAL	balance	CB	catch basin
BBD	bulletin board	CBLSTN	cobblestone
BBL	barrel	CCF	hundred cubic feet

GENERAL REQUIREMENTS 1.49

CEM	cement	CSMT	casement
CENT	centrifugal	CST	cast stone
CER	ceramic	CT	ceramic tile
CF	cubic foot; cellular floor	C to C	center to center
CFL	counterflashing	CTR	counter
CFM	cubic foot per minute	CTSK	countersunk screw
CFS	cubic foot per second	CTWT	counter weight
CG	corner guard	CU	cubic
CHAM	chamfer	CUFT	cubic foot
CHBD	chalkboard	CULV	culvert
CHEM	chemical	CU M	cubic meter
CHT	ceiling height	CUYD	cubic yard
CI	cast iron	CWT	hundred pounds
CIP	cast in place; cast-iron pipe	CY	cubic yard
		CX	connection
CIPC	cast-in-place concrete	D	drain; depth; discharge
CIR	circle	d	penny (nail size, i.e., 10d = 10 penny nail)
CIRC	circ; circumference		
CJT	control joint	DA	doubleacting
CK	caulk	DB	direct burial
CLG	ceiling	DBL	double
CLJ	controlled joint	DC	direct current
CLL	contract limit line	DEFL	deflect; deflection
CLOS	closure	DEG	degree
CLR	clear	DEM	demolish
CLS	closure	DEMOB	demobilization
CM	centimeter	DEP	depressed
CMP	corrugated metal pipe	DEPT	department
CMT	ceramic mosaic tile	DET	detail
CMU	concrete masonry unit	DF	drinking fountain
COL	column	DH	double hung
COMB	combination	DHW	domestic hot water
COMP	compression	DI	drop inlet; ductile iron
COMPO	composition	DIA	diameter
COMPT	compartment	DIAG	diagonal
CONC	concrete	DIAM	diameter
CONST	construction	DIFF	diffuser
CONT	continuous	DIM	dimension
CONTR	contract	DIP	ductile iron pipe
COORD	coordinate	DIR	direction
CORR	corrugated	DISC	disconnect
CPM	critical path method	DISCH	discharge
CPR	copper	DIV	division
CPT	carpet	DK	deck
CR	chromium	DL	dead load
CRG	cross grain	DMH	drop manhole
CRS	course	DML	demolition; demolish
CS	countersink; combined sewer	DMT	demountable
		DN	down

DP	dampproofing	EXMP	expanded metal plate
DPR	damper or dispenser	EXP	exposed
DR	door; drive	EXS	extra strong
DRB	drainboard	EXT	exterior
DS	downspout	EXTN	extension
DT	drain tile	F	female; fill
DTA	dovetail anchor	f	fahrenheit
DTL	detail	FA	fire alarm
DTS	dovetail anchor slot	FAB	fabricate
DUP	duplicate	FAS	fasten
DW	dumbwaiter	FB	face brick
DWG	drawing	FBD	fiberboard
DWL	dowel	FBO	furnished by others
DWR	drawer	FBRK	fire brick
		FC	foot-candle
E	east	FD	floor drain; fire damper
EB	expansion bolt	FDN	foundation
EE	each end	FDR	feeder
EF	each face	FE	fire extinguisher
EJ	expansion joint	FEC	fire extinguisher cabinet
EL	elevation	FF	factory finish;
ELEC	electric		factory fresh;
ELEV	elevator; elevation		far face
EM	expanded metal	FFE	finished floor elevation
EMER	emergency	FFL	finished floor line
EMH	electrical manhole	FGL	fiberglass
EMT	electrical metallic	FH	fire hydrant
	conduit;	FHC	fire hose cabinet
	thin wall conduit	FHMS	flathead machine screw
ENC	enclose	FHS	fire hose station
ENG	engine	FHWS	flathead wood screw
ENGR	engineer	FIG	figure
ENT	entering	FIN	finish
ENTR	entrance	FJT	flush joint
EP	electrical panelboard;	FL	floor
	explosion proof; edge	FLCO	floor cleanout
	of pavement	FLG	flashing; flange
EQ	equal	FLR	floor
EQPT	equipment	FLSH	flashing
EQUIP	equipment	FLUR	fluorescent
ESC	escalator	FLX	flexible
EST	estimate	FN	fence
E to E	end to end	FND	foundation
EW	each way	FO	fiber optics
EWC	electric water cooler	FOC	face of concrete
EXC	excavation; excavate	FOF	face of finish
EXCA	excavate	FOM	face of masonry
EXG	existing	FOS	face of studs
EXH	exhaust	FP	fireproof

FPL	fireplace or floor plate
FPM	foot (feet) per minute
FR	frame
FRA	fresh air
FRC	fire-resistant coating
FRG	forged
FRT	fire-retardant
FS	full size; far side
FT	foot; feet
FTG	footing
FT LB	foot-pound
F to F	face to face
FUR	furred
FURN	furnish; furniture
FUT	future
FW	fire water
G	gauge; gas; gas main or service
g	gram
GA	gauge
GAL	gallon
GALV	galvanized
GB	grab bar
GC	general contractor
GCMU	glazed concrete masonry units
GD	grade
GEN	generator; general
GF	ground face
GI	galvanized iron
GKT	gasket
GL	glass or glazing
GLB	glass block
GLF	glass fiber
GOVT	government
GP	galvanized pipe
GPD	gallons per day
GPDW	gypsum dry wall
GPH	gallons per hour
GPL	gypsum lath
GPM	gallons per minute
GPPL	gypsum plaster
GPT	gypsum tile
GR	grade
GRN	granite
GSS	galvanized steel sheet
GST	glazed structural tile
GT	grout
GV	galvanized
GVL	gravel
GYP	gypsum
H	height
HB	hose bibb
HBD	hardboard
HC	handicapped; hollow core; high capacity
HD	heavy duty
HDR	header
HDW	hardware
HES	high early-strength cement
HH	handhole
HJT	head joint
HK	hook
HM	hollow metal
HOR	horizontal
HORIZ	horizontal
HP	horse power; high pressure
HPR	high pressure return
HPS	high pressure steam
HR	hour
HT	height
HTG	heating
HV	high voltage; heating/ventilating unit
HVAC	heating/ventilating/air conditioning
HW	hot water
HWD	hardwood
HWH	hot water heater
HWR	hot water return
HX	hexagonal
HYD	hydrant
HZ	hertz
ID	inside diameter; identification
ILK	interlock
IN	inch
INCIN	incinerator
INCL	inclusive
INS	insulation
INSC	insulating concrete
INSF	insulating fill
INSTL	install

INSUL	insulation; insulated; insulator; insulating	LOC	locate; local
		LONG	longitudinal
INT	interior	LP	lightproof; light pole
INTM	intermediate	LPT	low point
INV	invert	LS	lump sum
IP	iron pipe	LT	light
IPS	iron pipe size	LTL	lintel
IR	irrigation	LV	low voltage
		LVR	louver
J	joist	LW	lightweight
JB	junction box	LWC	lightweight concrete
JC	janitor's closet		
JCT	junction	M	meter; thousand
JF	joint filler	MACH	machine
JT	joint	MAINT	maintenance
		MAS	masonry
KCPL	Keene's cement plaster	MATL	material
KG	kilogram	MAX	maximum
KIT	kitchen	MB	machine bolt
KLF	kips per lineal foot	MBF	thousand board feet
KM	kilometer	MBH	thousand BTU per hour
KO	knockout	MBR	member
KPL	kickplate	MC	medicine cabinet
KSF	kips per square foot	MCC	motor control center
KSI	kips per square inch	MECH	mechanical
KW	kilowatt	MED	medium
		MET	metal
L	length; left; liter; long	MFD	metal floor deck
		MFG	manufacture
LAB	laboratory	MFR	manufacturer
LAD	ladder	MGD	million gallons per day
LAM	laminated	MH	manhole
LAV	lavatory	MHW	mean high water
LB	lag bolt	MI	mile; malleable iron
LBL	label	MID	middle
LBS	pounds	MIN	minimum
LBS/HR	pounds per hour	MIR	mirror
LBS/LF	pounds per lineal foot	MISC	miscellaneous
LBS/ SQIN	pounds per square inch	MJ	mechanical joint
		MK	mark
LC	light control	MLD	moulding
LF	lineal foot; lineal feet	MLW	mean low water
LG	long; length	MM	millimeter
LH	left hand	MMB	membrane
LIQ	liquid	MO	month; masonry opening
LL	live load	MOB	mobilization
LLH	long leg horizontal	MOD	modular
LLV	long leg vertical	MON	monument
LMS	limestone	MOV	movable

MPA	megapascals
MPH	miles per hour
MR	mop receptor
MRB	marble
MRD	metal roof deck
MSL	mean sea level
MTD	mounted
MTFR	metal furring
M³	cubic meter
MTHR	metal threshold
MTL	material
M²	square meter
MULL	mullion
MWK	millwork
MWP	membrane waterproofing
N	north
NA	not applicable
NAT	natural
NEC	national electrical code
NF	near face
NI	nickel
NIC	not in contract
NL	nailable
NMT	nonmetallic
NO	number
NOM	nominal
NR	noise reduction
NRC	noise reduction coefficient
NTS	not to scale
OA	overall
O & P	overhead and profit
OBS	obscure
OC	on center
OD	outside diameter
OF	outside face
OH	overhead
OHMS	ovalhead machine screw
OHWS	ovalhead wood screw
OJ	open-web joist
OP	opaque
OPNG	opening
OPH	opposite hand
OPP	opposite
OPS	opposite surface
O to O	out to out

OZ	ounce
P	pole; power; pump; page
PA	pascal
PAR	parallel
PB	panic bar; push button
PBD	particle board
PCC	precast concrete
PCF	pounds per cubic foot
PE	porcelain enamel
PED	pedestal
PEJ	premolded expansion joint
PERF	perforated
PERI	perimeter
PFB	prefabricated
PFL	pounds per lineal foot
PFN	prefinished
PG	plate glass
PK	parking
PL	plate or property line
PLAM	plastic laminate
PLAS	plaster
PLY	plywood
PLYWD	plywood
PNL	panel
PNT	painted
PR	pair
PRC	precast reinforced concrete
PREFAB	prefabricated
PRESS	pressure
PRF	preformed
PROV	provide
PSC	prestressed concrete
PSF	pounds per square foot
PSI	pounds per square inch
PSIG	pounds per square inch gauge
PT	point; point of tangent
PTC	post-tensioned concrete
PTD	paper towel dispenser
PTN	partition
PTR	paper towel receptor
PV	paving
PVC	polyvinyl chloride
PVMT	pavement
PWR	power

QT	quarry tile	SFGL	safety glass
		SG	sheet glass
R	riser; radius	SH	shelf
RA	return air	SHO	shoring
RAD	radius	SHT	sheet
RB	rubber base	SHTH	sheathing
RBL	rubble stone	SIM	similar
RBT	rabbet; rebate;	SK	sketch
	rubber tile	SKL	skylight
RBR	rubber	SL	sleeve; slope
RCP	reinforced concrete pipe	SMH	sanitary manhole
RCPT	receptacle	SNT	sealant
RD	roof drain	SP	soundproof
RE	reinforced	SPC	spacer
REF	reference	SPEC	specification
REG	register; regular	SPK	speaker
REINF	reinforcement	SPL	special
REM	remove	SPR	sprinkler
REQD	required	SQ	square; 100 square feet
RES	resilient	SQ FT	square foot; square feet
RET	return	SQ M	square meter
REV	revision or revised	SQ YD	square yard
RFG	roofing	SS	sanitary sewer
RFH	roof hatch	SSK	service sink
RFL	reflected	SST	stainless steel
RH	right hand	ST	steel
RL	railing; roof ladder	STA	station
RM	room	STD	standard
RND	round	STG	seating
RO	rough opening	STL	steel
ROW	right of way	STO	storage
RR	railroad	STR	structural
RVS	reverse	STRL	structural
RVT	rivet	STY	story
R/W	right of way	SUBS	subcontractors
RWC	rainwater conductor	SURF	surface
		SUS	suspended
S	south; sanitary sewer;	SVCE	service
	supply; sign	S/W	sidewalk
SAN	sanitary	SWGR	switchgear
SBSTA	substation	SY	square yard
SC	solid core	SYM	symmetrical
SCH	schedule	SYMM	symmetrical
SCHED	schedule	SYN	synthetic
SCN	screen	SYS	system
SCT	structural clay tile		
SD	storm drain	T	tread; top; tangent;
SEC	section		telephone
SF	square foot	T & B	top and bottom

T & G	tongue and groove	VAC	vacuum
TB	towel bar	VAR	varnish
TC	terra cotta; top of curb	VAT	vinyl asbestos tile
TCP	terra cotta pipe	VB	vapor barrier or
TEL	telephone		vinyl base
TEMP	temperature	VC	vertical curve
TERM	terminal	VEL	velocity
THHN	nylon jacked wire	VENT	ventilate
THK	thickness	VERT	vertical
THR	threshold	VF	vinyl fabric
THRU	through	VG	vertical grain
THW	insulated strand wire	VIB	vibration
THWN	nylon jacketed wire	VIN	vinyl
TKBD	tackboard	VJ	v-joint
TKS	tackstrip	VLF	vertical lineal feet
TOC	top of concrete	VNR	veneer
TOL	tolerance	VOL	volume
TOS	top of steel	VRM	vermiculite
TOT	total	VT	vinyl tile
TPC	top of pile cap		
TPD	toilet paper dispenser		
TPTN	toilet partition	W	west; wide; watt;
TR	transom		water main; wire
T/R	top of rail	W/	with
TRANS	transformer	WB	wood base
TRK	track	WC	water closet
TS	top of steel; top of stone	WD	wood
TSL	top of slab	WDO	window
TV	television	WE	water elevation
TW	top of wall	WF	wide flange
TYP	typical	WG	wired glass; water gage
TZ	terrazzo	WH	wall hung
		WHB	wheel bumper
UC	undercut	WI	wrought iron
U/C	under construction	WIN	window
UCI	uniform construction	WM	wire mesh; water meter
	index	WO	without
UG	underground	W/O	without
UL	underwriters laboratory	WP	waterproofing
UNF	unfinished	WPT	working point
UNIF	uniform	WR	water repellent
UNO	unless otherwise noted	WS	waterstop
UR	urinal	WSCT	wainscot
USGS	United States Geological	WST	waste
	Survey	WT	watertight; weight
UTIL	utility	WTR	water
		WTW	wall to wall
V	voltage; volt; vent;	WV	water valve
	valve	WWF	welded wire fabric

		Y	wye
XARM	cross arm	YD	yard
XFR	transfer	YHYD	yard hydrant
XFMR	transformer	YI	yard inlet
XHD	extra heavy duty	YR	year

Specifications

Arrangement. The *Uniform System* developed by the Construction Specifications Institute (CSI) has become the accepted industry standard for setting up specifications. This system not only sets forth an arrangement for specifications, but it is equally applicable to data filing (most of the manufacturers' catalogs carry the CSI symbol, a rectangle enclosing three ellipses, and the Uniform System number) and to cost accounting systems.

This format is ideal for developing estimates and estimating check lists, and later on in this chapter, when we discuss setting up the estimate, we will refer the reader back to the list below for this purpose.

The CSI format is divided into four parts:

1. Bidding Requirements
2. Contract Forms
3. General Conditions
4. Specifications

The first three parts—bidding requirements, contract forms and general conditions—are not specifications sections. The CSI Masterformat refers to these as documents rather than sections. Divisions 1 through 16 contain the specifications for work items.

BIDDING REQUIREMENTS, CONTRACT FORMS AND CONDITIONS OF THE CONTRACT

00010	**Pre-Bid Information**
00020	Invitation to Bid
00030	Advertisement for Bids
00040	Prequalification Forms
00100	**Instruction to Bidders**
00120	Supplementary Instruction to Bidders
00130	Pre-Bid Conference
00200	**Information Available to Bidders**
00210	Preliminary Schedules
00220	Geotechnical Data
00230	Existing Conditions
00240	Project Financial Information
00300	**Bid Forms**

00400	**Supplements to Bid Forms**
00410	Bid Security Forms
00420	Bidder Qualification Forms
00430	Subcontractors List
00440	Substitution List
00450	Equipment Suppliers List
00460	List of Alternates/Alternatives
00470	List of Estimated Quantities
00480	Noncollusion Affidavit
00500	**Agreement Forms**
00600	**Bonds and Certificates**
00610	Performance Bonds
00620	Payment Bonds
00630	Warranty Bonds
00640	Maintenance Bond
00650	Certificate of Insurance
00660	Certificates of Compliance
00700	**General Conditions**
00800	**Supplemental Conditions**
00810	Modification to General Conditions
00820	Additional Articles
00830	Wage Determination Schedule
00900	**Addenda**

This system is the generally accepted form for specifications today. This book is organized according to the system. What follows is a complete listing of the 16 CSI divisions and their subdivisions:

01	**GENERAL REQUIREMENTS**
01010	**Summary Of Work**
01020	**Allowances**
01021	Cash Allowances
01024	Quantity Allowances
01025	**Measurement of Payment**
01030	**Alternates/Alternatives**
01035	**Modifications Procedures**
01040	**Coordination**
01041	Project Coordination
01042	Mechanical and Electrical Coordination
01043	Job Site Adminstration
01045	Cutting and Patching
01050	**Field Engineering**
01060	**Regulatory Requirements**
01070	**Identification Systems**
01090	**References**

01091	Reference Standards
01092	Abbreviations
01093	Symbols
01094	Definitions
01100	**Special Project Procedures**
01200	**Project Meetings**
01210	Preconstruction Conferences
01220	Progress Meetings
01245	Installation Meetings
01300	**Submittals**
01310	Progress Schedules
01320	Progress Reports
01330	Survey and Layout Data
01340	Shop Drawings, Product Data and Samples
01360	Quality Control Submittals
01380	Construction Photographs
01400	**Quality Control**
01410	Testing Laboratory Services
01420	Inspection Services
01425	Field Samples
01430	Mock-ups
01440	Contractor's Quality Control
01445	Manufacturer's Field Services
01500	**Construction Facilities and Temporary Controls**
01505	Mobilization
01510	Temporary Utilities
01520	Temporary Construction
01525	Construction Aids
01530	Barriers and Enclosures
01540	Security
01550	Access Roads and Parking Areas
01560	Temporary Controls
01570	Traffic Regulation
01580	Project Identification and Signs
01590	Field Offices and Sheds
01600	**Material and Equipment**
01610	Delivery, Storage and Handling
01620	Installation Standards
01630	Product Options and Substitutions
01650	**Facility Startup/Commissioning**
01655	Starting of Systems
01660	Testing, Adjusting and Balancing of Systems
01670	Systems Demonstrations
01700	**Contract Closeout**
01710	Final Cleaning
01720	Project Record Documents
01730	Operation and Maintenance Data
01740	Warranties and Bonds
01750	Spare Parts and Maintenance Materials

03410	Structural Precast Concrete–Plant Cast
03420	Structural Precast Post-Tensioned Concrete–Plant Cast
03430	Structural Precast Concrete–Site Cast
03450	Architectural Precast Concrete–Plant Cast
03460	Architectural Precast Concrete–Site Cast
03470	Tilt-up Precast Concrete
03480	Precast Concrete Specialties
03500	**Cementitious Decks and Toppings**
03510	Gypsum Concrete
03520	Insulating Concrete Deck
03530	Cementitious Wood Fiber Systems
03540	Composite Concrete and Insulation Decks
03550	Concrete Toppings
03600	**Grout**
03700	**Concrete Restoration and Cleaning**
03710	Concrete Cleaning
03720	Concrete Resurfacing
03730	Concrete Rehabilitation
03800	**Mass Concrete**
04	**MASONRY**
04100	**Mortar and Masonry Grout**
04150	**Masonry Accessories**
04200	**Unit Masonry**
04210	Clay Unit Masonry
04220	Concrete Unit Masonry
04230	Reinforced Unit Masonry
04235	Pre-assembled Masonry Panel Systems
04240	Non-reinforced Masonry Systems
04270	Glass Unit Masonry
04280	Gypsum Unit Masonry
04290	Adobe Unit Masonry
04400	**Stone**
04410	Rough Stone
04420	Cut Stone
04440	Flagstone
04450	Stone Veneer
04455	Marble
04460	Limestone
04465	Granite
04470	Sandstone
04475	Slate
04500	**Masonry Restoration and Cleaning**
04510	Masonry Cleaning
04520	Masonry Restoration
04550	**Refractories**
04555	Flue Lines
04560	Combustion Chambers
04565	Firebrick

05820	Slide Bearings
05830	Bridge Expansion Joint Assemblies
05900	**Hydraulic Structures**
05910	Penstocks
05915	Bulkheads
05920	Trashracks
05925	Manifolds
05930	Bifurcations
06	**WOOD and PLASTIC**
06050	**Fasteners and Adhesives**
06100	**Rough Carpentry**
06105	Treated Wood Foundations
06110	Wood Framing
06115	Sheathing
06120	Structural Panels
06125	Wood Decking
06128	Mineral Fiber Reinforced-Cement Panels
06130	**Heavy Timber Construction**
06132	Mill-framed Structures
06133	Pole Construction
06135	Timber Trusses
06140	Timber Decking
06145	Timber Bridges and Trestles
06150	**Wood and Metal Systems**
06170	**Prefabricated Structural Wood**
06180	Glued-laminated Construction
06190	Wood Trusses
06195	Prefabricated Wood Beams and Joists
06200	**Finish Carpentry**
06220	Millwork
06240	Laminates
06250	Prefinished Wood Paneling
06255	Prefinished Hardwood Paneling
05260	Board Paneling
06300	**Wood Treatment**
06310	Preservative Treatment
06320	Fire Retardant Treatment
06330	Insect Treatment
06400	**Architectural Woodwork**
06410	Custom Casework
06420	Panelwork
06430	Stairwork and Handrails
06440	Wood Ornaments
06450	Standing and Running Trim
06460	Exterior Frames
06470	Screens, Blinds and Shutters
06480	Custom Wood Turning
06500	**Structural Plastics**

07550	Protected Membrane Roofing
07560	Roof Maintenance and Repairs
07570	**Traffic Coatings**
07572	Pedestrian Traffic Coatings
07576	Vehicular Traffic Coatings
07600	**Flashing and Sheet Metal**
07610	Sheet Metal Roofing
07620	Sheet Metal Flashing and Trim
07630	Sheet Metal Roofing Specialties
07650	Flexible Flashing
07700	**Roof Specialties and Accessories**
07710	Manufactured Roof Specialties
07720	Roof Accessories
07800	**Skylights**
07810	Plastic Unit Skylights
07820	Metal-framed Skylights
07900	**Joint Sealers**
07910	Joint Fillers and Gaskets
07920	Sealants and Caulks
08	**DOORS and WINDOWS**
08100	**Metal Doors and Frames**
08110	Steel Doors and Frames
08120	Aluminum Doors and Frames
08130	Stainless Steel Doors and Frames
08140	Bronze Doors and Frames
08200	**Wood and Plastic Doors**
08210	Wood Doors
08220	Plastic Doors
08250	**Door Opening Assemblies**
08255	Packaged Steel Door Assemblies
08260	Packaged Wood Door Assemblies
08265	Packaged Plastic Door Assemblies
08300	**Special Doors**
08305	Access Doors
08310	Sliding Doors and Grilles
08315	Pressure Resistant Doors
08320	Security Doors
08325	Cold Storage Doors
08330	Coiling Doors and Grilles
08350	Folding Doors and Grilles
08355	Chain Closures
08360	Section Overhead Doors
08365	Vertical Lift Doors
08370	Industrial Doors
08375	Hanger Doors
08380	Traffic Doors
08385	Sound Control Doors
08390	Storm Doors

09205	Furring and Lathing
09210	Gypsum Plaster
09215	Veneer Plaster
09220	Portland Cement Plaster
09225	Adobe Finish
09230	Plaster Fabrications
09250	**Gypsum Board**
09260	Gypsum Board Systems
09270	Gypsum Board Accessories
09300	**Tile**
09310	Ceramic Tile
09320	Thin Brick Tile
09330	Quarry Tile
09340	Paver Tile
09350	Glass Mosaics
09360	Plastic Tile
09370	Metal Tile
09380	Cut Natural Stone Tile
09400	**Terrazzo**
09410	Portland Cement Terrazzo
09420	Precast Terrazzo
09430	Conductive Terrazzo
09440	Plastic Matrix Terrazzo
09450	**Stone Facing**
09500	**Acoustical Treatment**
09510	Acoustical Ceilings
09520	Acoustical Wall Treatment
09525	Acoustical Space Units
09530	Acoustical Insulation and Barriers
09540	**Special Wall Surfaces**
09545	**Special Ceiling Surfaces**
09550	**Wood Flooring**
09560	Wood Strip Flooring
09565	Wood Block Flooring
09570	Wood Parquet Flooring
09580	Wood Composition Flooring
09590	Resilient Wood Floor System
09600	**Stone Flooring**
09610	Flagstone Flooring
09615	Marble Flooring
09620	Granite Flooring
09625	Slate Flooring
09630	**Unit Masonry Flooring**
09635	Brick Flooring
09640	Pressed Concrete Unit Flooring
09650	**Resilient Flooring**
09660	Resilient Tile Flooring
09665	Resilient Sheet Flooring
09670	Fluid Applied Resilient Flooring

10120	Tackboards
10130	Operable Board Units
10140	Display Tracking System
10145	Visual Aid Board Units
10150	**Compartments and Cubicles**
10160	Metal Toilet Compartments
10165	Plastic Laminated Toilet Compartments
10170	Plastic Toilet Compartments
10175	Particleboard Toilet Compartments
10180	Stone Toilet Compartments
10185	Shower and Dressing Compartments
10190	Cubicles
10200	**Louvers and Vents**
10210	Metal Wall Louvers
10220	Louvered Equipment Enclosures
10225	Metal Door Louvers
10230	Metal Vents
10240	**Grilles and Screens**
10250	**Service Wall Systems**
10260	**Wall and Corner Guards**
10270	**Access Flooring**
10272	Rigid Grid Access Floor Systems
10274	Snap-on Stringer Access Floor Systems
10276	Stringerless Access Floor Systems
10290	**Pest Control**
10292	Rodent Control
10294	Insect Control
10296	Bird Control
10300	**Fireplaces and Stoves**
10305	Manufactured Fireplaces
10310	Fireplace Specialties and Accessories
10320	Stoves
10340	**Manufactured Exterior Specialties**
10342	Steeples
10344	Spires
10346	Cupolas
10348	Weathervanes
10350	**Flagpoles**
10352	Ground Set Flagpoles
10354	Wall Mounted Flagpoles
10356	Automatic Flagpoles
10358	Nautical Flagpoles
10400	**Identifying Devices**
10410	Directories
10415	Bulletin Boards
10420	Plaques
10430	Exterior Signs
10440	Interior Signs
10450	**Pedestrian Control Devices**

10452	Portable Posts and Railings
10454	Rotary Gates
10456	Turnstiles
10458	Detection Specialties
10500	**Lockers**
10505	Metal Lockers
10510	Wood Lockers
10515	Coin-operated Lockers
10518	Glass Lockers
10520	**Fire Protection Specialties**
10522	Fire Extinguishers, Cabinets and Accessories
10526	Fire Blankets and Cabinets
10528	Wheeled Fire Extinguisher Units
10530	**Protective Covers**
10532	Walkway Covers
10534	Car Shelters
10536	Awnings
10538	Canopies
10550	**Postal Specialties**
10551	Mail Chutes
10552	Mail Boxes
10554	Collection Boxes
10556	Central Mail Delivery Boxes
10600	**Partitions**
10605	Wire Mesh Partitions
10610	Folding Gates
10615	Demountable Partitions
10630	Portable Partitions, Screens and Panels
10650	**Operable Partitions**
10652	Folding Panel Partitions
10655	Accordion Folding Partitions
10660	Sliding Partitions
10665	Coiling Partitions
10670	**Storage Shelving**
10675	Metal Storage Shelving
10680	Storage and Shelving Systems
10683	Mobile Storage System
10685	Wire Shelving
10688	Prefabricated Wood Storage Shelving
10700	**Exterior Protection Devices for Openings**
10705	Exterior Sun Control Devices
10710	Exterior Shutters
10715	Storm Panels
10750	**Telephone Specialties**
10755	Telephone Enclosures
10760	Telephone Directory Units
10765	Telephone Shelves
10800	**Toilet and Bath Accessories**
10810	Toilet Accessories

10820	Bath Accessories
10880	**Scales**
10900	**Wardrobe and Closet Specialties**
11	**EQUIPMENT**
11010	**Maintenance Equipment**
11012	Vacuum Cleaning Systems
11014	Window Washing Systems
11016	Floor and Wall Cleaning Equipment
11018	Housekeeping Carts
11020	**Security and Vault Equipment**
11022	Vault Doors and Day Gates
11024	Security and Emergency Systems
11026	Safes
11028	Safe Deposit Boxes
11030	**Teller and Service Equipment**
11032	Service and Teller Window Units
11034	Package Transfer Units
11036	Automatic Banking Systems
11038	Teller Equipment Systems
11040	**Ecclesiastical Equipment**
11042	Baptisteries
11044	Chancel Fittings
11050	**Library Equipment**
11052	Book Theft Protection Equipment
11054	Library Stack Systems
11056	Study Carrels
11058	Book Depositories
11060	**Theater and Stage Equipment**
11062	Stage Curtains
11064	Rigging Systems and Controls
11066	Acoustical Shell Systems
11068	Folding and Portable Stages
11170	**Instrumental Equipment**
11072	Organs
11074	Carillons
11076	Bells
11080	**Registration Equipment**
11090	**Checkroom Equipment**
11100	**Mercantile Equipment**
11102	Barber and Beauty Shop Equipment
11104	Cash Registers and Checking Equipment
11106	Display Cases
11108	Food Processing Equipment
11110	**Commercial Laundry and Dry Cleaning Equipment**
11112	Washers and Extractors
11114	Dry Cleaning Equipment
11116	Drying and Conditioning Equipment
11118	Finishing Equipment

11335	Sedimentation Tank Equipment
11340	Scum Removal Equipment
11345	Chemical Equipment
11350	Sludge Handling and Treatment Equipment
11360	Filter Press Equipment
11365	Trickling Filter Equipment
11370	Compressors
11375	Aeration Equipment
11380	Sludge Digestion Equipment
11385	Digester Mixing Equipment
11390	Package Sewage Treatment Plants
11400	**Food Service Equipment**
11405	Service Storage Equipment
11410	Food Preparaton Equipment
11415	Food Delivery Carts and Conveyors
11420	Food Cooking Equipment
11425	Hood and Ventilation Systems
11430	Food Dispensing Equipment
11435	Ice Machines
11440	Cleaning and Disposal Equipment
11445	Bar and Soda Fountain Equipment
11450	**Residential Equipment**
11452	Residential Appliances
11454	Built-in Ironing Boards
11458	Disappearing Stairs
11460	**Unit Kitchens**
11470	**Darkroom Equipment**
11472	Transfer Cabinets
11474	Darkroom Processing Equipment
11476	Revolving Darkroom Doors
11480	**Athletic, Recreational and Therapeutic Equipment**
11482	Scoreboards
11484	Backstops
11486	Gym Dividers
11488	Bowling Alleys
11490	Gymnasium Equipment
11492	Exercise Equipment
11494	Therapy Equipment
11496	Shooting Ranges
11500	**Industrial and Process Equipment**
11600	**Laboratory Equipment**
11650	**Planetarium Equipment**
11660	**Observatory Equipment**
11680	**Office Equipment**
11700	**Medical Equipment**
11710	Medical Sterilization Equipment
11720	Examination and Treatment Equipment
11730	Patient Care Equipment
11740	Dental Equipment

11750	Optical Equipment
11760	Operating Room Equipment
11770	Radiology Equipment
11780	**Mortuary Equipment**
11850	**Navigation Equipment**
11870	**Agricultural Equipment**
12	**FURNISHINGS**
12050	**Fabrics**
12100	**Artwork**
12110	Murals
12120	Wall Decorations
12140	Sculpture
12160	Ecclesiastical Artwork
12170	Stained Glass Work
12300	**Manufactured Casework**
12301	Metal Casework
12302	Wood Casework
12304	Plastic Laminated Faced Casework
12345	Laboratory Casework
12350	Medical Casework
12360	Educational Casework
12370	Residential Casework
12380	Specialty Casework
12500	**Window Treatment**
12510	Blinds
12515	Interior Shutters
12520	Shades
12525	Solar Control Film
12530	Curtain Hardware
12540	Curtains
12600	**Furniture and Accessories**
12605	Portable Screens
12610	Open Office Furniture
12620	Furniture
12640	Furniture Systems
12650	Furniture Accessories
12670	**Rugs and Mats**
12675	Rugs
12680	Foot Grilles
12690	Floor Mats and Frames
12700	**Multiple Seating**
12705	Portable Audience Seating
12710	Fixed Audience Seating
12730	Stadium and Arena Seating
12740	Booths and Tables
12750	Multiple Use Fixed Seating
12760	Telescoping Stands

12770	Pews and Benches
12775	Seat and Table Systems
12800	**Interior Plants and Planters**
12810	Interior Plants
12815	Artificial Plants
12820	Interior Planters
12825	Interior Landscape Accessories
12830	Interior Plant Maintenance
13	**SPECIAL CONSTRUCTION**
13010	**Air-supported Structures**
13020	**Integrated Assemblies**
13025	Integrated Ceilings
13030	**Special Purpose Rooms**
13032	Athletic Rooms
13034	Sound Condition Rooms
13036	Clean Rooms
13038	Cold Storage Rooms
13040	Hyperbaric Rooms
13042	Insulated Rooms
13046	Shelters and Booths
13048	Planetariums
13050	Prefabricated Rooms
13052	Saunas
13054	Steam Baths
13056	Vaults
13080	**Sound, Vibration and Seismic Control**
13090	**Radiation Protection**
13100	**Nuclear Reaction**
13120	**Pre-engineered Structures**
13121	Pre-engineered Buildings
13122	Metal Building Systems
13123	Glazed Structures
13124	Portable and Mobile Buildings
13125	Grandstand and Bleachers
13130	Observatories
13132	Prefabricated Dome Structures
13135	Cable-supported Structures
13140	Fabric Structures
13142	Log Structures
13145	Modular Mezzanines
13150	**Aquatic Facilities**
13152	Swimming Pools
13160	Aquariums
13165	Aquatic Park Facilities
13170	Tubs and Pools
13175	**Ice Rinks**
13180	**Site Construction Incinerators**
13182	Sludge Incinerators

13830	Clock Control Systems
13835	Elevator Monitoring and Control Systems
13840	Escalators and Moving Walks Monitoring and Control Systems
13845	Alarm and Detection Systems
13850	Door Control Systems
13900	**Fire Suppression and Supervisory Systems**
13950	**Special Security Construction**
14	**CONVEYING SYSTEMS**
14100	**Dumbwaiters**
14110	Manual Dumbwaiters
14120	Electric Dumbwaiters
14140	Hydraulic Dumbwaiters
14200	**Elevators**
14210	Electric Traction Elevators
14240	Hydraulic Elevators
14300	Escalators and Moving Walks
14310	Escalators
14320	Moving Walks
14400	**Lifts**
14410	People Lifts
14420	Wheelchair Lifts
14430	Platform Lifts
14440	Sidewalk Lifts
14450	Vehicle Lifts
14500	**Material Handling Systems**
14510	Automatic Transport Systems
14530	Postal Conveying Systems
14540	Baggage Conveying and Dispensing Systems
14550	Conveyors
14560	Chutes
14570	Feeder Equipment
14580	Pneumatic Tube Systems
14600	**Hoists and Cranes**
14605	Crane Rails
14610	Fixed Hoists
14620	Trolley Hoists
14630	Trolley Hoists
14640	Gantry Cranes
14650	Jib Cranes
14670	Tower Cranes
14680	Mobile Cranes
14690	Derricks
14700	**Turntables**
14800	**Scaffolding**
14810	Suspended Scaffolding
14820	Rope Climbers
14830	Telescoping Platforms

15530	Refrigerant Piping
15535	Refrigerant Specialties
15540	Hvac Pumps
15545	Chemical Water Treatment
15550	**Heat Generation**
15555	Boilers
15570	Boiler Accessories
15575	Breechings, Chimneys and Stacks
15580	Feedwater Equipment
15590	Fuel Handling Systems
15610	Furnaces
15620	Fuel-fired Heaters
15650	**Refrigeration**
15655	Refrigeration Compressors
15670	Condensing Units
15680	Water Chillers
15710	Cooling Towers
15730	Liquid Coolers
15740	Condensers
15750	**Heat Transfer**
15755	Heat Exchangers
15760	Energy Storage Tanks
15770	Heat Pumps
15780	Packaged Air Conditioning Units
15790	Air Coils
15810	Humidifiers
15820	Dehumidifiers
15830	Terminal Heat Transfer Units
15845	Energy Recovery Units
15850	**Air Handling**
15855	Air Handling Units with Coils
15860	Centrifugal Fans
15865	Axial Fans
15870	Power Ventilators
15875	Air Curtain Units
15880	**Air Distribution**
15885	Air Cleaning Devices
15890	Ductwork
15910	Ductwork Accessories
15920	Sound Attenuators
15930	Air Terminal Units
15940	Air Inlets and Outlets
15950	**Controls**
15955	Building Systems Control
15960	Energy Management and Conservation Systems
15970	Control Systems
15980	Instrumentation
15985	Sequence of Operation
15990	**Testing, Adjusting and Balancing**

16485	Contactors
16490	Switches
16500	**Lighting**
16501	Lamps
16502	Luminaire Accessories
16510	Interior Luminaries
16520	Exterior Luminaries
16535	Emergency Lighting
16545	Underwater Lighting
16580	Theatrical Lighting
16600	**Special Systems**
16610	Uninterruptible Power Supply Systems
16620	Package Engine Generator Systems
16630	Battery Power Systems
16640	Cathodic Protection
16650	Electromagnetic Shielding Systems
16670	Lightning Protection Systems
16680	Unit Power Conditioners
16700	**Communications**
16720	Alarm and Detection Systems
16730	Clock and Program Systems
16740	Voice and Data Systems
16770	Public Address and Music Systems
16780	Television Systems
16785	Satellite Earth Station Systems
16790	Microwave Systems
16850	**Electrical Resistance Heating**
16855	Electric Heating Cables and Mats
16880	Electric Radiant Heaters
16900	**Controls**
16910	Electrical Systems Control
16915	Lighting Control Systems
16920	Environmental Systems Control
16930	Building Systems Control
16940	Instrumentation
16950	**Testing**
16960	Electrical Equipment Testing
16970	Electrical System Startup/Commissioning
16980	Demonstration of Electrical Equipment

Subdivision. Each section of each division of the specifications is divided into the *Scope* or *Work Included, Materials* listing the quality of all materials to be furnished, and finally, *Fabrication and Erection*, defining the quality of workmanship desired.

Scope includes items to be furnished and installed, items to be furnished but installed by others, items furnished by others and installed under the section, and finally, related work in other divisions of the specifications. Scope will also list some general requirements such as those for shop drawings, samples, tests, methods of delivery, and storage that may be required. No matter how detailed the scope section, it will not include each and every item. The drawings must always be checked against scope, which often carries the phrase "including, but not limited to, the following".

Materials will list the materials to be used in one of several ways, often found in combination. The *closed specification* will list a single trade name, and the specified product must be furnished. The *contractor's option specification* (or bidder's choice) lists more than one trade name, and the contractor may choose from those listed. The *substitute bid specification* might list a choice of trade names and the bid must be based on one of the products included, but the contractor is allowed to suggest a substitute at the time of submitting a bid, naming the amount he would subtract for using the alternate product.

A variation is the *product approval specification* which asks the contractor to submit any substitutions prior to submitting a bid. If the architect approves the substitute, it will be put in an addenda sent to all contractors. This "or approved equal" type specification is the most common. It provides the widest competition and rewards the contractor who knows his products, prices, and sources.

However, it is the architect who decides whether the product is equal. If the contractor has any doubts about a substitute's being approved, then he should bring it to the architect's attention upon being selected as low bidder. In this way, any adjustments can be made before award of contract should the architect rule that the substitution is not equal.

The *product description specification* is used to describe bulk materials and some manufacturers' articles, the latter particularly in government work. The *performance specification* describes not the material but what work is required to produce strength, mechanical ability, or similar measurable results.

Specifications based on *reference standards* refer to those which are published by industry associations and organizations. They are encountered often, are general in nature, and usually apply only in part to the specific job. A contractor should build a library of these published specifications.

Below is a list of construction associations, many of which offer commonly referred standards. In addition, many architects will refer to the published specifications of a named manufacturer. These are readily available and generally are included in installation procedures.

Fabrication and Erection will set performance standards for both shop and field work and will include, if required, sections on protection, cleaning, guarantees, warranties, maintenance, and operating instructions. The specifications of trade organizations and manufacturers' installation instructions are often made a part of this section.

Construction Organizations and Associations

AA
The Aluminum Association
Washington, DC 20006
Web Site: Aluminum.org

AAA
American Arbitration Association
New York, NY 10020-1203
Web Site: adr.org

AACE
American Association of Cost Engineers International
Morgantown WV 26507-1557
Web Site: ACEI.org

AAMA
Architectural Aluminum Manufactures Association
Palatine, IL 60067

AASHTO
American Association of State Highway and
Transportation Officials
Washington, DC 20001
Web Site: aashto.org

ABC
Associated Air Balance Council
Washington, DC 20005
Web Site: www.aabchq.com

ACCA
Air Conditioning Contractors of America
Washington, DC 20036

ACI
American Concrete Institute
Detroit, MI 48219
Web Site: aci-int.org

AFA
American Fence Association
Stone Mountain, GA 30083-3154
Web Site: americanfenceassoc.org

ACPA
American Concrete Pavement Association
Skokie, IL 60077-1059
Web Site: pavement.com

ACPA
American Concrete Pumping Association
Benicia, CA 94510

ACEC   American Consulting Engineers Council
Washington, DC 20005

ADC   Air Diffusion Council
Chicago, IL 60601-4298

AFPA   American Forest and Paper Association
Washington, DC 20036
Web Site: afandpa.org

AGC   Associated General Contractors of America
Washington, DC 20006
Web Site: agc.org

AHA   American Hardboard Association
Palatine, IL 60067

AHAM   Association of Home Appliance Manufacturing
Chicago, IL 60606

AI   Asphalt Institute
Lexington, KY 40512-4052
Web Site: asphaltinstitute.org

AIA   American Institute of Architects
Washington, DC 20006
Web Site: aiaonline.com

AIA   American Insurance Association
Washington, DC 20036
Web Site: aiadc.org

AIA/NA   Asbestos Information Association of North America
Arlington, VA 22202
Web Site: social.com

AISC   American Institute of Steel Construction
Chicago, IL 60601-2001
Web Site: aisc.org

AISI   American Iron and Steel Institute
Washington, DC 20036-4700
Web Site: steel.org

AITC   American Institute of Timber Construction
              Englewood CO 80112
              Web Site: aitc-glulam.org

ALCA   Associated Landscape Contractors of America
              Falls Church, VA 22046

AMCA   Air Movement & Control Association
              Arlington Heights, IL, 60004-1893
              Web Site: amca.org

ANMC   American National Metric Council
              Bethesda, MD 20814-4411

ANSI   American National Standards Institute
              New York, NY 10036
              Web Site: web.ansi.org

APA/TEWA   American Plywood Association
              The Engineered Wood Association
              Tacoma, WA 98411
              Web Site: www.apawood.org

APA   Architectural Precast Association
              Winter Park, FL 32789-2106
              Web Site: archprecast.org

ARI   Air Conditioning and Refrigeration Institute
              Arlington, VA 22203
              Web Site: ari.org

ARTBA   American Road and Transportation Builders
              Washington, DC 20024

ASA   Acoustical Society of America
              Woodbury, NY 11797
              Web Site: asa.aip.org

ASA   American Subcontractors of America
              Alexandria, VA 22314

ASCE   American Society of Civil Engineers
Washington, DC 20005
Web Site: asce.org

ASHRAE   American Society of Heating, Refrigeration
and Air Conditioning Engineers
Atlanta, GA 30329
Web Site: ashrae.org

ASLA   American Society of Landscape Architects
Washington, DC 20001-3736
Web Site: asla.org

ASME   American Society of Mechanical Engineers
New York, NY 10036
Web Site: asme.org

ASPA   American Sod Producers Association
Rolling Meadows, IL 60008

ASPE   American Society of Professional Engineers
Wheaton, MD 20902

ASTM   American Society for Testing and Materials
Philadelphia, PA 14828-2959
Web Site: astm.org

AWCI   Association of the Wall and Ceiling Industries
Alexandria, VA 22314
Web Site: awci.org

AWPA   American Wood Preservers' Association
Granbury, TX 76049-7979
Web Site: awpa.com

AWPI   American Wood Preservers Institute
Fairfax, VA 22031-4312
Web Site: awpi.org

AWI   Architectural Woodwork Institute
Reston, VA 20190
Web Site: awinet.org

AWS   American Welding Society
               Miami, FL 33126
               Web Site: amweld.org
               E - mail: webmasterAwsparc.org

AWWA   American Water Works Association
               Washington, DC 20005
               Web Site: awwa.org

BGA   Barre Granite Association
               Barre, VT 05641

BHMA   Builders Hardware Manufactures Association
               New York, NY 10017
               Web Site: ikiiki.htm

BIA   Brick Institute of America
               Reston, VA 22091
               Web Site: bia.org

BOCA   Building Officials and Code Administrators International
               Country Club, IL 60478-5795
               Web Site: bocai.org

BOMA   Building Owners and Managers Association
               Washington, DC 20005
               Web Site: boma.org

BSI   Building Stone Institute
               New York, NY 10006

CIPI   Cast Iron Soil Pipe Institute
               Chattanooga, TN 37421
               Web Site: cispi.org

CISCA   Ceilings and Interior Systems Construction Association
               St. Charles, IL 60174
               Web Site: cisca.org

CMAA   Construction Management Association of America
               McLean, VA 22102
               Web Site: access.digex.net

CRA	California Redwood Association Novato, CA 94949 Web Site: calredwood.org
CRI	Carpet and Rug Institute Dalton, GA 30722 Web Site: carpet-rug.org
CRSI	Concrete Reinforcement Steel Institute Schaumburg, IL 60173-4758 Web Site: crsi.org
CSI	Construction Specifications Institute Alexandria, VA 22314-1797 Web Site: csinet.org
DHI	Door and Hardware Institute Chantilly, VA 22021-2223 Web Site: dhi.org
ESCSI	Expanded Shale, Clay, and Slate Institute Salt Lake City, UT 84117 Web Site: concreteworld.com/escsi
FGMA	Flat Glass Marketing Association Topeka, KS 66611-2279
FHVA/AWMA	Fine Hardwoods Veneer Association American Walnut Manufactures Association Evansville, IN 46077
GA	Gypsum Association Washington, DC 20002 Web Site: gypsum.org
HPVA	Hardwood, Plywood and Veneer Association Reston, VA 22090 Web Site: erols.com/hpva
ICBO	International Conference of Building Officials Whittier, CA 90601 Web Site: constructionmaterials.com

IEEE	Institute of Electrical and Electronics Engineers Piscataway, NJ 08855-0459 Web Site: ieee.org
IESNA	Illuminating Engineering Society of North America New York, NY 10017 Web Site: iesna.org
ILIA	Indiana Limestone Institute of America Bedford, IN 47421 Web Site: tri-statecutstone.com
IMI	International Masonry Institute Washington, DC 20005 Web Site: imiweb.org
KCMA	Kitchen Cabinet Manufactures Association Reston, VA 22091-4327 Web Site: kcma.org
LIA	Lead Industries Association New York, NY 10017 Web Site: arcat.org
MBMA	Metal Building Manufactures Association Cleveland, OH 44115-2851 Web Site: taol.org
MCAA	Mechanics Contractor Association of America Rockville, MD 20850 Web Site: mcaa.org
MFMA	Maple Flooring Manufactures Association Northbrook, IL 60062
MHI	Material Handling Institute Charlotte, NC 28217 Web Site: mhia.org
NAMF	National Association of Metal Finishes Chicago, IL 60611

NADC	National Association of Demolition Contractors Doylestown, PA 18901 Web Site: recycle.net
NAPA	National Asphalt Pavement Association Lauham, MD 20706-4413
NAWIC	National Association of Woman in Construction Fort Worth, TX 76104 Web Site: nawic.org
NBGQA	National Building Granite Quarries Association, Inc Washington, DC 20005 Web Site: nbgqa.com
NCAC	National Council of Acoustical Consultants Springfield, NJ 07081-1409 Web Site: ncac.com
NCMA	National Concrete Masonry Association Herndon, VA 22071-3406 Web Site: ncma.org
NEA	National Erectors Association Arlington, VA 22209
NECA	National Electrical Contractors Association Bethesda, MD Web Site: necanet.org
NAEC	National Association of Elevator constructors Conyers, GA 30207-3872 Web Site: naec.org
NEMA	National Electrical Manufactures Association Rosslyn, VA 22209 Web Site: nema.org
NFPA	National Fire Protection Association Quincy, MA 02269-9101 Web Site: nfpa.org
NLBA	National Lumber Builders Association Washington, DC 20003

NHLA   National Hardwood Lumber Association
                Memphis, TN 38184-0518
                Web Site: natlhardwood.org

NIBS   National Institute of Building Sciences
                Washington, DC 20005-4905
                Web Site: nibs.org

NLA   National Lime Association
                Arlington, VA 22203
                Web Site: lime.org

NOFMA   National Oak Flooring Manufactures Association
                Memphis, TN 38173-0007
                Web Site: nofma.org

NPA   National Particleboard Association
                Gaithersburg, MD 20879
                Web Site: pbmdf.com

NPCA   National Paint and Coating Association
                Washington, DC 20005
                Web Site: paint.org

NPCA   National Precast Concrete Association
                Indianapolis, IN 46290-1081
                Web Site: precast.org

NRCA   National Roofing Contractors Association
                Rosemont, IL 60018-5607
                Web Site: roofonline.org

NRDCA   National Roof Deck Contractors Association
                Chicago, IL 60613
                Web Site: nrdca.org

NSA   National Stone Association
                Washington, DC 20007

NSPA   National Swimming Pool Foundation
                Merrick, NY 11566
                Web Site: nspf.com

NSSEA	National School Supply and Equipment Association Silver Spring, MD 20910 Web Site: nssea.org
NTMA	National Terrazzo and Mosaic Association Des Plains, IL 60018 Web Site: ntma.org
PCA	Portland Cement Association Skokie, IL 60077 Web Site: portcement.org
PCI	Precast/prestressed Concrete Institute Chicago, IL 60604 Web Site: pci.org
PDCA	Painting and Decorating Contractors of America Fairfax, VA 22030 Web Site: pdca.org
PDI	Plumbing and Drainage Institute South Easton, MA 02375 Web Site: pdionline.org
PEI	Porcelain Enamel Institute Washington, DC 20036-2294 Web Site: porcelainenamal.com
PI	Perlite Institute Staten Island, NY 10306
PPI	Plastic Pipe Institute Washington, DC 20006-1301 Web Site: plasticpipe.org
PWC	Professional Woman in Construction New York, NY 10173
RIS	Redwood Inspection Service Novato, CA 94949
RTI	Resilient Tile Institute Rockville, MD 20850

SAVE   Society of American Value Engineers
           Northbrook, IL 60062
           Web Site: www.value-eng.com

SBCCI   Southern Building Code Congress International
           Birmingham, AL 35213-1206
           Web Site: www.sbcci.org

SCEA   Society of Cost Estimating and Analysis
           Alexandria, VA 22304
           Web Site: erols.com/scea

SDI   Steel Deck Institute
           Fox River Grove, IL 60021-0025
           Web Site: www.sdi.org

SIGMA   Sealed Insulating Glass Manufactures Association
           Chicago, IL 60611
           Web Site: sigma.org

SFPA   The Southern Forest Products Association's
           Kenner, LA 70064-1700
           Web Site: sfpa.org

SGAA   Stained Glass Association of America
           Kansas City, MO 64113
           Web Site: www.stainedglass.org

SJI   Steel Joist Institute
           Myrtle Beach, SC 29577-6760
           Web Site: www.steeljoist.org

SMACNA   Sheet Metal and Air Conditioning Contractors
           National Association
           Chantilly, VA 22021
           Web Site: www.smacna.org

SPI   Society of the Plastics Industry
           Washington, DC 20005
           Web Site: socplas.org

SPIB	Southern Pine Inspection Bureau Pensacola, FL 32504 Web Site: spib.org
SSPC	Steel Structures Painting Council Pittsburgh, PA 15213-2683 Web Site: sspc.org
SWI	Steel Window Institute Cleveland, OH 44115-2851
TCA	Tile Council of America Clemson, SC 29633 Web Site: www.tileusa.com
TUCA	Tilt-Up Concrete Association Mount Vernon IA Web Site: tilt-up.org
TPI	Truss Plate Institute Madison, WI 53719
UBPVLS	Uniform Boiler and Pressure Vessel Laws Society Louisville, KY 40243-1076 Web Site: www.uboiler.com
UL	Underwriters Laboratories Northbrook, IL 60062 Web Site: ul.com
TVA	The Vermiculite Association Columbus, OH 43235 Web Site: www.vermiculite.org
WCLIB	West Coast Lumber Inspection Bureau Tigard., OR 97223
WSC	Water Systems Council Glen Ellyn, IL 60137 Web Site: wschome.org

WRCLA	Western Red Cedar Lumber Association
	Vancouver, BC Canada V&X 1S7
	Web Site: cofi.org

WRI	Wire Reinforcement Institute
	Washington, DC 20036-4303
	Web Site: bright.net

WTCA	Wood Truss Council of America
	Madison, WI 53711-4125
	Web Site: trusspro.com

WMMPA	Wood Molding and Millwork Producers Association
	Woodland, CA 95695
	Web Site: wmmpa.com

WWPA	Western Wood Products Association
	Portland, OR 97204-2122
	Web Site: www.wwpa.org

Alternates. At the time of bidding, the architect often wants to investigate the cost, either upward or downward, of a substitute product, material, or procedure, or the cost of adding or subtracting work from the contract. Such requests for alternate prices are usually noted at the end of the section of the specifications to which they refer, and are repeated at the end of Division 1 of the specifications and on the proposal form. On the typical lump sum proposal form, the Alternate follows the Statement of the Lump Sum price in a form such as the following:

If the following Alternates are accepted,	Add:	Deduct:
Alternate No. 1	$___	$
Alternate No. 2	$___	$

Obviously, a request for lots of alternates complicates bidding. Sometimes, the extent of the alternate is unclear. Since the architect will accept the price change for an alternate with the same responsibility as attached to the lump sum figure, the contractor must be certain of all the possible contingencies involved should the change be accepted.

Often an alternate is put only under the one section that makes the basic substitution, but other subcontractors' work is also affected by the change. It is the contractor's responsibility to foresee this. As an example, under Carpentry, an architect might include:

"Alternate No. 1—For substituting one 3'-0"x6'-8"x1-3/4" (900 x 2 000 x 44 mm) door with frame for the two 3'-0"x6'-8"x1-3/4" (900 x 2 000 x 44 mm) doors and frame shown between the living and dining rooms, deduct $___."

Obviously, the door cost will be less, but the wall construction will be more. Yet there might be no mention of this under the division for the latter.

Cash Allowances. Sometimes, the architect does not have a final decision from the owner on certain items, and rather than leave them out of the lump sum proposal, he gives a definite budget amount in the specification that is to be included in the bid. This is often encountered with the selection of brick, hardware, and other finish items. As an example:

"Facing brick shall be included under an allowance of $75 per M, fob, job site."

Every bidder will include the same material cost, rather than just guessing what type of face brick the owner may choose and using that cost. Cash allowances are almost always only for the cost of material and applicable taxes delivered to the site. Labor to set and overhead and profit on the material should be included in the lump sum price unless specifically noted otherwise.

Upon completion of the job, the actual cost of the items furnished under Allowance shall be tabulated with any savings credited to the owner, any overruns added to the contract cost.

Unit Prices. Where quantity of materials is in doubt, but quality is known, the specification may ask for unit prices. As opposed to allowances, unit prices are set by the bidder, not the architect; they are firm and binding on the bidder, the lump sum price being adjusted up or down, only according to variation in the quantity actually used on the job.

Unit Prices are included on the proposal form for each material per unit of measure. For example, unit prices are often asked for concrete per square yard (square meter), piling per lineal foot (meter), partition block per square foot (millimeters squared), etc. These prices should be complete with all costs, profit, and overhead included.

Unit Prices must be quoted with care, even if the amount of work covered by them is small, because they will be compared directly with those submitted by other bidders. The owner may expect the same unit prices to hold on additional work.

Addenda

Few plans and specifications are perfect when they leave the architect's office, and contractors bidding the job should call to the architect's attention all discrepancies they note. These, plus changes the architect and owner may wish to make after the plans and specifications have been issued but before bids are turned in, are incorporated in the *Addenda*. This usually takes the form of a printed sheet or sheets the same size as the specification to which it should be attached. Addenda eventually becomes a part of the Contract Documents.

Even changes to the drawings often can be described and included this way. But sometimes additional drawings or sketches must be issued, or a drawing is reissued with a revision date. Reissued drawings, however, usually should be avoided during bidding.

Each addendum is given a number, and the changes it covers are listed in the same order as the divisions in the specification, with drawing changes last.

Sometimes, addenda must be acknowledged by signing a receipt; often, they are sent by registered mail. Usually, they are acknowledged on the proposal form in a space provided for it. The single addendum or several addenda are also included in the Owner-Contractor Agreement.

Upon receipt of an addendum, the contractor should mark his specifications and drawings with appropriate notes, or with a stamp on each affected page saying "See Addendum #_____." The Addendum will usually identify the change by division and page number and a reference to location on the page. As an example:

3:01—**Scope:** Add item "h) Remove existing wire fence along north edge of property."

6:02—**Materials:** Delete paragraph referring to concrete masonry lintels and add "...lintels in block work shall be units of sizes and shapes shown on the lintel schedule."

Dwg. Sheet A-2: Move rolling shutter door from center of Bay C-D to just north of Column B.

Dwg. Sheet A-4; delete ladder to roof.

Change Orders

Change orders are modifications issued after the contract is signed. Errors on drawings and specifications are usually uncovered during bidding. A contractor who sees where a drawing is not complete but does not include in his price what he knows must be furnished may very well end up having to pay for the item anyway to complete the job or face a lengthy lawsuit.

Some change orders are always necessary due to job conditions, changes required when plans are submitted for permits, substitutions made to satisfy the owner or his mortgage holder, products not being available, or because as the job progresses, certain changes for the benefit of the job become obvious.

The contractor must cooperate in making these changes. This is set forth in Article 12 of the General Conditions. Sometimes no cost is involved. Most often, an extra is called for. This may be handled by submitting a firm price change, or by order to proceed on a time and material basis. The change order is issued by the architect but signed by the owner and contractor. The format of the change order will follow closely that of the addendum, except that costs of changes and tabulation of the new contract amount are added. The Frank R. Walker Form L-101 is a typical contract change order format.

Summary

As can be imagined, what appears on drawings will sometimes differ with what is stated in specifications. There is no legal precedent that one is more binding than the other, although some specifications will state which is to be followed in case of inconsistency.

In general, sizes, quantities, design, and location will be best taken from the drawings. Materials, quality, procedures, and general requirements are most reliably defined in the specifications. Never rely solely on one or the other. A draftsman's delineation of a stone wall in elevation may not show the coursing called for in the specifications at all, while the specifier might not make special mention of a stone sill condition carefully detailed on the drawings. One must always complement the other.

SETTING UP THE ESTIMATE

Estimating is one of the most important aspects of a building contractor's business. In nearly every instance it is necessary for a contractor either to estimate the cost of the work, or to bid in competition with others before a contract for the work is awarded. If a contractor's bid is too low and a contract is awarded on the basis of it, the contractor often must complete the work without profit and sometimes at a loss in time and money.

A contractor must know the size, complexity, and quality of work that his firm can efficiently handle. Exceeding these limits often leads to a poor contract, which can be costly not only monetarily but also in terms of dissatisfied customers and loss of future opportunities for bidding.

Most contractors fall into categories as to the type of work which their organizations can handle efficiently, expeditiously, and profitably. A heavy construction contractor specializing in structures such as airport runways, highways, bridge abutments, etc., usually should forego bidding on such work as churches, banks, hospitals, or modern office buildings. A contractor whose organization and personnel are geared to a particular category of construction can find himself in difficulty if he strays from his specialty.

Many contractors in business today are former employees of contracting firms who have gone into business for themselves. Many of these people doubtless were exceptional mechanics, foremen or superintendents, and had a thorough understanding of how the work should be carried out in the field, but sometimes they lack business ability and training to estimate costs accurately. Some of these contractors make a reasonable profit on their work and remain in business, but many more of them fail and eventually go back to work for someone else.

There are various reasons for these failures, but probably the most common one is the inability of the person estimating costs to come up with realistic and profitable estimates. The procedures involved in preparing an accurate, detailed cost estimate necessary for a realistic bid are relatively simple. Yet almost all instances of failure of contracting firms are the result of poor estimating practices, resulting in the submission of unsound and unprofitable bids. There is no substitute for complete and detailed material take-off and pricing and for cross-checking estimated costs with actual costs so that adjustments in pricing can be made for future bidding.

Another common reason for failure in the contracting business is overexpansion. Again, the contractor must know the capabilities of his office and supervisory staff and how much they can adequately handle at any one time. The successful contractor does not take on more work than his staff can take care of.

Bonding and insurance companies have certain minimum requirements that contractors applying for surety bonds or contractual liability insurance must meet. Some of the most important considerations that such companies make before bonding an applicant are:

1. the technical ability of the contractor applying for a bond, whether or not he understands his business thoroughly, and his capabilities in the preparation of detailed cost estimates;

2. the reputation of the contractor for honesty and his standing among those with whom he does business;

3. the financial ability of the contractor, and whether or not he has sufficient capital at his disposal to carry on his business, purchase materials and supplies, meet his payrolls and current expenses, etc.

Bonding companies seldom get the opportunity of bonding the best contracting firms. On public work projects the bidding is usually open to all bidders and in some instances as many as fifty or more bids may be received on one job. In most cases, the lowest bidder is awarded the contract, and he often may be the poorest risk because of substandard estimating procedures, lack of technical ability to perform the work as specified, and lack of financial responsibility. On privately owned projects, the risk to the bonding company may not be as great, because the bidding list can be controlled. Only contractors with the proper credentials are asked to submit proposals. Architectural and engineering firms that handle work for private owners are often asked to select contractors to bid on a project. They will attempt to choose bidders whom they know can perform the work equally well and will carry out the intent of the drawings and specifications in the best interest of the owner.

Others besides the contractor attach importance to a correct and detailed estimate. The success of a contractor depends as much on a well prepared estimate as it does on the manner in which he purchases materials, lets his subcontracts, or carries out his work on the job.

The Detailed Estimate

If twenty estimators were furnished the same set of plans and specifications and told to prepare a cost estimate, it is safe to say that not more than two of the estimates would be prepared on the same basis or using the same units. And an estimator doing the same estimate a year later would most likely arrive at different final costs the second time.

There are so many different method of estimating and a lack of detailed information regarding costs. Each contractor has developed a system of his own, making it difficult for one contractor to check another's figures. Some contractors use a unit manhour estimating procedure. A given unit of work takes a certain amount of time. For example, it might take 8 manhours to set 100 lineal feet (30 meters) of 1/2" (15 mm) rigid conduit. Other contractors use the *crew* method, i.e., a concrete pouring crew of 8 workers will place 100 cubic yards (75 cu. meters) of concrete per day.

Each contractor develops his own unit prices from cost records that reflect his particular experience and expertise, which varies considerably among contractors. This is not to say that one estimate is correct and another is incorrect. In the final analysis of estimates and bids, they should be accepted as representing the projected final costs for performing certain prescribed work within a stipulated time schedule, taking into consideration all of the variables that a prudent contractor can anticipate during a construction contract based on information available at the time of bidding.

To prove the validity of the bid, one must consider the extreme risks that the contractor is willing to assume by furnishing payment and performance bonds and by entering into a legal contract dedicating the technical and financial resources of the firm to a successful completion of the contract in hopes of realizing a modest profit.

Estimating is a laborious and costly operation and under no circumstances should an estimator or contractor produce a less than totally professional estimate that will stand the test of proving each element contained in the estimate.

There are many types of cost estimates that an owner may require before the bid estimate is needed. These estimates can be performed by or in conjunction with the owner's staff, contractors, professional estimators, or the estimating section of an engineering or architectural firm. Each type of cost estimate should serve the purpose for which it was intended.

Walker Form 514 – General Estimate

Types of Cost Estimates

Budget Estimate. The budget estimate, once it has been developed, is effectively cast in stone. No matter what happens throughout the life of the project, the costs stated in the budget estimate are what everyone will quote until the project is completed. If the budget estimate does not accurately represent the work to be performed, then it might mean one of two things.

On the one hand, if costs have been overstated, the project might be eliminated before it gets started. On the other hand, if the project starts with insufficient funding, the work might have to stop in midstream, or it will be necessary to borrow funds from other sources, delaying the start of other work.

While the budget estimate does not reduce the importance of the other types of cost estimates, it is usually the governing cost estimate of the project. At this stage the project either gets the green light or dies in committee. (Contingency percentage is 30% to 50%)

Pre-design Estimate. The purpose of this type of estimate is to provide approximate values for design work packages to be awarded A/E firms selected by the owner. At this stage of the process, there still has not been much increase in the design information available. (Contingency percentage is 30% to 50%)

Conceptual Estimate. For this level of estimating, the estimator has been supplied with sketches. These "drawings", in conjunction with discussions and meetings with the owner's representatives and the A/E, are used to form the estimate outline for costing.

The estimator formulates a list of pay items. These pay items have total unit costs applied to them. The conceptual estimate normally becomes the estimate that all other "working" estimates are compared with for a) design changes, b) scope changes, and c) cost changes. (Contingency percentage is 30%)

Preliminary Estimate 10%. At this stage the estimator has preliminary scope and drawing information. On the basis of information from the previous types of cost estimates, the estimator uses this new information, along with discussions with the owner's staff and the A/E, to expand to this level of estimating.

The item list generated in and expanded upon in the previous estimates is used. Individual items of work are broken down, where possible, into units of cost for labor, equipment, material, and subcontractor.

Prior to the presentation of the preliminary cost estimate, the estimator makes a site tour. The basic assumptions about how the work will be performed are reviewed and refined or revised as required. A written report on the site conditions, and on any effect that these conditions have on the estimate, is prepared for review by the owner.

At this stage of design development, cost items generally fall into the subcontract or "plug" cost column. As information becomes available during this or any later stage of the estimate, this "plug" cost can be expanded into a more detailed breakdown. All changes in quality, scope, etc. are fully identified and justified as the estimate progresses. (Contingency percentage is 25%)

Design Review Estimate 40%. This cost estimate is an expansion of the previous 10% cost estimate. Items of work are identified, to the extent possible for this level of estimating, in units of cost for labor, equipment, materials, and subcontract. In accordance with the details, as requested by the owner or his A/E, the 40% estimate provides backup to support labor rates, crew mixes, etc. Where possible, labor rates should include the contractor's indirect costs. Otherwise, indirect costs must be identified and enumerated separately. (Contingency percentage is 20%)

Pre-construction Estimate 70%. At the 70% estimate, the estimator completes the quantity take-offs on all defined items of work. Each identified item is quantified in terms of unit cost and total cost for material, labor hours, and labor costs. The estimator adjusts the estimate to present either lump sum or unit price information, to conform to the type of contract that is being bid. This estimate uses the same format as the previous type of cost estimate. (Contingency percentage is 10%)

Construction Cost Estimate 100%. The estimator now starts an entirely new cost estimate based on the items of work identified in the contract specifications and drawings. All items of work are broken down into units of labor, equipment, material, and subcontract when possible. This is in the same format that has been used for all other cost estimates to date. Using a standard format allows for identification of cost and design changes when the 100% estimate is completed and compared with cost estimates from the previous stages.

The important difference between this estimate and those from all previous stages is that the estimator develops and completes the 100% estimate entirely separate from all the others. Only after the 100% estimate is complete is it then compared with estimates from the previous stages.

In addition, at this stage all vendor quotes, subcontractor costs and equipment prices are identified. All line items that the general contractor will perform are spread out into units of labor, equipment and material. (Contingency percentage is 5%)

The use of this detailed estimate extends far beyond being the basis of a bid. It serves as the guide for the purchase of materials and the award of subcontracts. It also provides the information for the establishment of a cost control system so that project progress can be monitored to determine if the estimated material and the labor costs are consistent with the actual costs incurred during performance of the contract. As an example, assume that the detailed estimate reflects the following labor cost allowances for each of the listed items of work:

	Labor Allowance	Quantity
Footing Excavation	$3,000.00	300 c.y. (230 cu.m)
Form Footings	$1,200.00	1,200 s.f. (110 sq.m)
Pour Footing Concrete	$1,500.00	300 c.y. (230 cu.m)

As each of these phases of work starts, the labor chargeable to each is allocated on a project daily time sheet (Walker Form 505) so that as work progresses, there is an accurate cost record that can be reviewed daily, weekly, or monthly, depending on the magnitude of the operation. In this way, periodic projections can be made. For example, the 50% completion review of each of the items reflects the following conditions:

Item	Labor Allowance	Expended To Date	Remaining Cost	+ - (loss)
Footing Excavation	$3,000.00	$1,800.00	$1,800.00	($600.00)
Form Footings	1,200.00	400.00	400.00	400.00
Pour Footings	1,500.00	600.00	600.00	300.00

The contractor should not view the combination of these three work items as reflecting a net $100.00 savings above the estimated costs but rather should be alerted to a serious overrun of 20% in the labor budget for footing excavation. This item should be reviewed in detail to determine if a remedy is available to improve labor production for the remaining 50% of work. Should it be found that there is no solution to the problem, then it is to be assumed that the project anticipated profits will be reduced by the $600.00 additional cost. However, valuable cost information has been developed, and these actual cost records become the basis of unit costs for estimating future projects. The above example indicates that excavation should be increased from $10.00 to $12.00 per c.y. ($13.00-16.00 per cu.m) and unit prices for each of the other items could reduced, producing a more accurate and competitive bid.

An effective estimating format can be established using several pre-printed forms from the Frank R. Walker Company.

Post-estimating Requirements. An estimating system must be responsive to the construction industry. Every estimate should be compared with the ultimate cost of the work—the contract price—as determined by the successful bidder. It is normal practice for an estimator to review bids on a project and to compare them with the estimate of the work. He or she attempts, where possible, to identify variations from the estimate and variations among the bidders. Significant differences are investigated to check the estimating system and to identify possible alternative methods of performing the work. The usual output of such a review would be a recommendation to the owner to either award a contract or not award one. (Contingency percentage is 0%)

Review of Unit Price Estimates. In examining unit price contracts, the estimator examines the total price as well as the individual unit quantities and prices. Significant variations from the estimate, or from other bidders, should be further investigated by checking on estimated quantities, computations, and basic assumptions about construction methods. Where the bid of the apparent responsible low bidder is within an acceptable range of the estimate, the estimator should focus on the unit prices in order to identify imbalances in unit vs. time, "front loading", or in creating a "windfall" where there might be significant variations in the actual quantities of a particular item of work.

Review of Lump Sum Bids. In reviewing bids on lump sum contracts, the estimator should initially determine whether the apparent low bid varies significantly from his estimate, or from the other bidders. The estimator then prepares for the contractor's submission of his detailed breakdown, which is compared with the detailed estimate to verify the estimate and to identify possible imbalances. Where there is a significant variation between bids and the estimate, or where there are wide variations among the bidders, the estimator attempts to determine the cause.

The Estimator

The estimator is the most important person in a contractor's organization, because it is up to him to prepare estimates that will either make or break the contractor. The most efficient job organization or the best purchasing department in existence cannot make money on contracts taken below cost.

On the other hand, it does not take a good salesman or contract man to obtain all the work you can handle if your estimates are way below your competitor's. Remember, the low man is wrong more often than he is right. The efficient organization is one that can get the work at the right price. It is not an easy task, but a business built upon the principles of service, dependability, good materials and workmanship, and fair price will be in business many years after the get-the-work-at-any-price contractor has made his last assignment in favor of his creditors.

The estimator should be practical and must possess a thorough understanding of job conditions, of how the work will be carried on in the field, and of the operations necessary to assemble the materials and put them in place in the structure. He must be able to visualize and must be able to take a set of blueprints and draw a mental picture of the building. He must know every branch of the work to be handled, the materials required, and the labor operations necessary to convert a pile of gravel, sand, brick, cement, lumber, steel, and glass into a completed building. The job superintendent or foreman may be able to take a set of blueprints and lay the entire job out in his mind and picture the building progressing from the foundation to the finished structure, but unless he possesses the knowledge necessary to compute the quantity of materials and the labor cost of putting them in place, he will never be able to prepare an accurate estimate of cost. The following requisites are essential for the making of a good estimator:

1. Able to read plans and measure them accurately.
2. Possess a good knowledge of arithmetic, including addition, subtraction, multiplication, and division, together with a knowledge of the decimal system. Nearly all measurements and computations in the average building estimate are made in lineal feet (millimeters), square feet or yards (square millimeters or meters), and cubic feet or yards (cubic meters). These material quantities are then usually multiplied at a certain unit price, so a knowledge of decimals is necessary to obtain the various costs. More and more drawings will carry dimensions based on the metric system, and one must be able to convert from one to the other.
3. Able to visualize. He should be able to take a set of blueprints and draw a mental picture of the building—how it will look during construction and at completion. He must be able to place his equipment so that the work can be handled economically. In short, he must first construct the building on paper.
4. Possess an intimate knowledge of job conditions, the most practical methods of handling materials and labor on the job and the most economical methods of construction. While the estimator is generally an office man, he must possess much the same knowledge of construction methods as the superintendent or foreman.
5. Know how and be able to assemble materials into workable units. A thousand brick in the wall means more than just the brick. It means a certain amount of sand, lime and cement, and possibly mortar color to make the mortar. A cubic yard of concrete means that exact quantities of cement, sand, and gravel must be computed. The same applies to nearly every complete unit in the building. The competent estimator must have this knowledge at his finger tips.
6. Possess an intimate knowledge of labor performances and operations and be able to convert them into costs in dollars and cents. This requires a working knowledge of what a worker will do under given conditions and is obtained only by a careful study of different jobs completed under varying conditions.
7. Have at least an average amount of common sense.

The American Society of Professional Estimators (ASPE), with local chapters all over the country, has several categories of membership, including student and affiliate. Education programs are promoted, conventions are held on local and national levels, and newsletters are printed. ASPE has a certification program, and examinations for certified construction estimator are given periodically.

Office Overhead Expense

The item of overhead expense is neglected by many contractors, but it is of such importance that every contractor should give it careful consideration. Every contractor has a certain fixed expense that must be paid regardless of the amount of work done or contracts received, and these items should be charged to *office overhead*.

This includes such items as office rent, fuel, lights, telephone and telegraph, stationery, office supplies, advertising, trade journals and magazines, donations, legal and accounting expenses not directly chargeable to any one job, fire and liability insurance for the office, club and association dues, office employees, such as bookkeeper, stenographer, clerks, estimator, janitor services and salaries of executives, along with travel and entertainment expenses of a good will nature and depreciation on office furnishings and equipment. These items should be estimated for a year and then reduced to a certain percentage of the total annual business, as follows:

Non-reimbursable Salaries* .... $	$70,000
Employee Benefits*	21,000
Legal & Accounting Fees	5,000
Insurance*	10,000
Office Rent	5,000
Office Utilities	3,000
Office Supplies	2,000
Job Procurement & Promotion	7,000
Depreciation*	2,000
Maintenance*	2,000
Job Management* (incl. transportation)	5,000
Education/Regulations	2,000
Miscellaneous	6,000
	$140,000

*Indirect expenses only, that is, those not reimbursed by fees from individual jobs.

The list at the left includes many more expenses than a small contractor will have and considerably less than those carried by larger contractors, but it serves as an example of the items that should be included as overhead or general expense. For example, suppose you have an overhead expense of $140,000 per year, and do an annual business approximating $1,500,000. On this basis your overhead or general expense would average 9.3% of the total annual business.

Overhead may run 6% for some firms. For others, it might be 10% to 15% of the total annual business, but every contractor can approximate his overhead based on his average yearly business. Even an approximation is far better than completely ignoring office overhead.

Many smaller contractors do not maintain an office and probably attend to the greater part of their office duties themselves, so only such items of expense as they actually incur should be included.

Proposals for jobs taken on a time and material basis and provisions for extra work on jobs taken on a stipulated sum basis may require the contractor to state the percentage to be added for overhead, plus that for profit. In a job bid from five medium-sized contractors, these varied as follows: 8+4, 8+7, two at 10+5, and one at 10+10. While overhead percentage varied only from 8% to 10%, profit varied from 4% to 10%.

Whether these same percentages are used in figuring competitive bids is, of course, each contractor's secret. But certainly, some addition to cover office overhead should be made, unless the organization is so successful that the year's overhead has already been absorbed by other jobs, or the work is of such nature that it is primarily subcontracted and office expenses are minor.

All manufacturers, jobbers, and retailers add their overhead expense to the cost of their products when computing the selling price, and there is no reason why contractors should not do the same, always keeping their eye on the competition.

Office Furniture and Equipment. All expenditures for office furniture, typewriters, calculating machines, filing cabinets, etc., should be kept as a separate account and a certain percentage of this amount charged off each year as depreciation. The depreciation, however, should be charged as a part of the overhead expense of conducting the business, as the cash value of the equipment decreases each year.

In a completely efficient office, construction equipment would be constantly utilized on the job and would be charged as a job cost, not office overhead. However, there are always slack periods when equipment is idle, or it might have been bought for a specific job but still have a useful life. It can be depreciated and charged to overhead until it can be put to use again. The Internal Revenue Department publishes fixed schedules of allowed depreciation.

Some contractors set up a "rental" cost for equipment owned by the firm but chargeable to individual jobs. Today the trend is to rent tools and equipment from outside firms specializing in this, which simplifies bookkeeping—all costs are directly charged to the job and no depreciation need be carried—and frees the contractor's capital for more lucrative investment. A discussion of plant and equipment is included later in this chapter.

Insurance. Insurance costs directly chargeable to office overhead would include fire and theft for office property, liability for office premises, workmen's compensation for office employees, and automobile insurance for trucks and cars owned by the firm.

Liability insurance is a must in today's construction. This is often included as an office overhead, a necessary cost of doing business. However, if the policy has rather low limits, the contractor will find most specifications will require that he take out additional coverage, and this would be a job expense rather than overhead expense.

Surety bonds, performance bonds, property insurance for work under construction, and worker's compensation for on-job employees can be charged directly to the job costs and are discussed under job expenses. The cost of bid bonds for work awarded to another contractor would obviously have to be absorbed by office overhead.

Job Overhead

Job overhead includes three divisions. First, those items the general contractor must furnish for the conduct of the job as a whole are usually known as *General Conditions*, often supplemented with modifications and additions under *Special Conditions*.

Secondly, there are items the general contractor may furnish to expedite the job which may benefit all or just a single trade. These are carried under *Plant and Equipment*.

Finally, the contractor may be asked to include in his work certain auxiliary functions which require outside advisors. This may range from a simple plot survey to complete design services. These expenses are carried under *Professional Services*.

Job Overhead: General Conditions

Each job will entail certain expenses that are chargeable to the job as a whole. These are set forth in the General Conditions (the 14 articles of the AIA usually made a part of the contract) and the Supplementary Conditions and/or General Requirement Section of the Specifications. These will include such items as superintendence, temporary buildings, enclosures, barricades, offices, toilets, utilities, protection, clean-up, permits, surveys, photographs, tools and equipment, insurance and benefits, sales taxes, surety bonds, and warranties.

Sometimes, these are covered only in a general way, and it is up to the contractor to visit the site to determine what is needed. Often the unions and local laws may be more demanding than the architect's specifications, and the contractor must know what items he is liable for.

Job overhead and office overhead are best kept separately, although occasionally employees' insurance, tools and equipment, and other items may be divided between office expense and job expense.

Many contractors figure a certain percentage of the cost of the job to take care of all overhead expense and plant charges, but this is not the correct method to use and does not give the contractor or estimator a reliable basis for computing the cost of these expenses. Each item entering into the cost of job overhead expense should be estimated the same as any other branch of work.

Superintendence. This should include the salary and traveling expenses of the job superintendent, assistant superintendent, engineers, timekeepers, material clerks, watchmen, water boys, labor foremen, and all other foremen who are carried on the general contractor's payroll.

The proper method of computing these costs is to estimate the salaries of the various personnel required at a certain price per week or per month for the entire estimated time they will be required on the job.

Temporary Buildings and Enclosures. This should include temporary fences, bridges or platforms, temporary sheds and storerooms, temporary enclosures over hoisting engines, concrete mixers, temporary runways, ladders and stairs in the building before the permanent stairs are installed, temporary dustproof partitions to isolate work areas from building space still occupied, and temporary closures for doors and windows before the permanent doors and windows are in place.

Temporary Office. This should include the temporary job office for use by the contractor and architect during construction, together with all necessary expense in maintaining it, such as telephone, heat, light, stationery, and other items of expense. These are usually trailers, which may be rented. An 8' x 30' (2 400 x 9 000 mm) trailer, furnished, will cost around $5,000 but can be rented for around $200.00 per month.

Barricades and Signal Lights. This should cover the cost of building and maintaining barricades during construction, cost of maintaining signal lights, etc.

Temporary Toilet. Practically every job must have some provision for sanitary facilities. Local codes usually govern what must be supplied. Chemical portable toilets usually suffice. On large jobs, however, a more elaborate setup is usually specified requiring a complete water supply, drain, vent, and fixture installation. This work is sometimes included in the plumbing specifications. Chemical toilets can be rented for about $65.00 to $80.00 per month.

In the case of a remodeling job where permission is granted to use the existing facilities, an item should be included in the estimate for periodic and final cleaning of the toilet room and its appurtenances.

Temporary Water. On most jobs water for construction purposes must be provided by the general contractor for all trades. This item can become a high-cost operation, especially in barren, undeveloped localities where either a well must be drilled or water must be hauled to the job. When a water main is available, a plumber must be hired to tap it, install water meter, and run a water line to strategic locations on the job. This work is sometimes included in the plumbing specifications. Also included under this heading are the charges made by the local water department for the volume of water used. Often hydrants can be used at a set fee from the municipality.

Temporary Light and Power. This item can vary, from a small charge for tapping into existing adjacent power lines and running entrance cable into a meter setup from which extension cords may be run, to complex installations requiring long run pole lines, transformers, and switch gear, such as for electric mixers, hoists, and other heavy duty electric power equipment. This work is sometimes included in the electrical specifications. Also included in this item are the power company charges for energy consumed and rental of transformers, etc.

Temporary Heat. If a job is to run into or through the winter, it is very likely that some temporary heat will be required to maintain proper working conditions. On small jobs a few salamanders may be enough to fulfill the requirements. Large jobs, however, usually require temporary hookup of radiators to the new and as yet uncompleted heating system.

Repairing Streets and Pavements. This item should cover the repair and replacement of all sidewalks and pavements damaged or destroyed during construction.

Damage to Adjoining Buildings. There is always a chance of damage to adjacent structures, such as to windows, to foundations and walls improperly shored, or to skylights, and these items should be considered when preparing the estimate.

Protecting New Work from Damage During Construction. During construction there are certain classes of work that must be protected from damage, such as cut stone, marble, terra cotta, and granite. Cover newly-finished cement or wood floors to prevent damage.

Repairing New Work Damaged During Construction. This item should cover patching of damaged plaster, replacing broken lights of glass, etc. Quite often architects specify the cost of such repair work to be prorated among the various contractors working on the job unless specific responsibility for the damages can be charged to an individual contractor.

Cutting and Patching for Other Trades. Many specifications require the general or mason contractor to do all cutting and patching of walls and floors for the plumbing, heating, and electrical contractors, where it is necessary for the mechanical trades to place sleeves or pipes through the floors or walls after the permanent work is in place.

Removing and Replacing Public Utilities. In the larger cities it may be necessary to remove or relocate water hydrants, telephone and utility poles and wires, sewers and drains, or other underground pipes. In such instances it will be necessary for the contractor to pay for any damage incurred.

Removing Rubbish, Cleaning Floors, Windows, Etc. During the construction of every building there is always a certain amount of rubbish, such as old concrete, broken bricks, and tile, that must be removed at intervals. The most satisfactory method is to estimate so many loads of rubbish at a certain price per load.

If finish floors are to be swept or scrubbed before acceptance of the building by the owner, it is advisable to make a certain allowance to provide for this work. This should cost about $3.50 per 100 s.f. ($0.40 per sq.m).

Some architects require the contractor to clean all windows before the building will be accepted. To arrive at this cost, count the windows and allow a certain price for cleaning each window, generally around $.09 per s.f., 2 sides (0.90 per sq.m).

Building Permits. The cost of building permits varies from city to city. Some cities base the cost of the permit on the total cost of the work, or the cubic content of the building, while others charge for water consumed, street frontage occupied, and the number of plumbing fixtures, for which an inspection charge is made.

As a general rule a building permit will cost from 1/10 to 1/4 of 1% of the total cost of the building. The following are examples of permit fees. Check with local governing agencies for costs in the area where you are building.

1. Single car garage, two car garage and shelter sheds: For each 1,000 cubic feet (30 cu.m) or fractional part, $4.00; minimum charge, $10.00.
2. New buildings or other structures, except those included in a): For each 1,000 cubic feet (30 cu.m) of volume or fractional part, $1.25; minimum charge, $20.00. Volume computations under a) and b) shall include every part of the building from the basement to the highest point of the roof and include bay windows and other projections.
3. Alterations and repairs to any structures: For the first $1,000.00 of estimated cost, $7.75; for each additional $1,000.00 of estimated cost, $3.00; minimum charge, $20.00.
4. Shoring, raising, underpinning, or moving any building: For each 1,000 cubic feet (30 cu.m) of volume or fractional part, $0.40; minimum charge, $20.00.
5. Wrecking any building or other structure more than one story in height: For each 25 lineal feet (7 500 mm) of frontage or fractional part, $13.00; minimum charge, $20.00.
6. Fire Escape, erection or alteration: Four stories or less in height, $20.00; each story above four, $2.75.
7. Canopy or marquee, erection or alteration, $20.00.
8. Chimneys, isolated or over 50 feet (15 000 mm) above any roof, $20.00.
9. Tanks, above roof or tower: 400 gallon (1 500 liter) capacity or less, $13.00; over 400 gallon (1 500 liter) capacity, $25.00; structural supports for tank over 400 gallon (1 500 liter) capacity, $25.00.
10. Elevators and escalators, power operated, five floors or less in height, installed or altered, $20.00; each additional floor above five, $4.00; hand operated, $13.50; levelators, $13.50; theater curtains, $50.00; stage or orchestra platforms, $20.00. Dumbwaiters, power operated, five floors or less in height, installed or altered, $20.00; each additional

floor above five, $4.00. Hand operated, five floors or less in height, $13.00; each additional floor above five, $2.75. Temporary construction towers over 50 feet (15 000 mm) in height $20.00.

11. Amusement devices, mechanical riding, sliding, sailing or swinging: portable, for each assembly or installation, $12.50. Permanent, installed or altered, $105.00. Temporary seating stands, $70.00.

12. Ventilating systems, mechanical supply or exhaust: Capacity, 3,000 c.f. (85 cu.m) of air per minute, $10.50; for each additional 1,000 c.f. (30 cu.m) per minute, $3.00; increase in capacity, each 1,000 c.f. (30 cu.m) per minute, $3.00. Capacity shall be the sum of supply and exhaust. Warm air furnaces, $12.50 each.

13. Fences over 5 feet high, 100 l.f. (30 m), $10.00; each additional 100 l.f. (30 m) or part thereof, $3.00.

14. Tanks for flammable liquids: Classes I and II, capacity 121 to 500 gals. (460 to 1 900 liters), $21.00; each additional 1,000 gals. (3 800 liters) or fraction thereof, $0.65; Classes III and IV, capacity 551 to 1,000 gals. (2 100 to 3 800 liters), $21.00; each additional 1,000 gals. (3 800 liters) or fraction thereof, $0.40.

15. Temporary platforms for public assembly units, $21.00.

16. Roof of any building, recoating or recovering, $6.50.

17. Billboards and signboards—erection, construction or alteration: up to 150 s.f. (14 sq.m) $5.00; 151 to 225 s.f. (14 to 20 sq.m) $7.75; 226 to 375 s.f. (20 to 35 sq.m) $13.00; for each additional 375 s.f. (35 sq.m) or part thereof $13.00.

A fee of $20.00 is paid to a city collector for the approval of plans and for inspection and test of any plumbing within any building containing not more than five plumbing fixtures. An additional fee of $3.00 shall be paid for every plumbing fixture in excess of five within such building.

Surveys. Where the contractor must furnish a survey of the property on which the building is to be erected, a charge of $500 to $1,000 is usually made for a lot up to 100' x 100'. In the downtown sections of large cities this charge will be substantially more. A topographical survey with 2' (0.6 m) contours will cost around $300 for an acre (4 050 sq.m); $1700 for 10 acres (40 470 sq.m); and about $100 per acre from 10 to 100 (4 050 to 40 470 sq.m).

Photographs. Where it is necessary to have photographs taken at intervals to show job progress, the cost varies with the number of views made at one time. It will cost from $250 to $300 per visit, including shots of all building corners and two black and white 8" x 10" (200 x 250 mm) prints of each shot. Additional black and white 8" x 10" (200 x 250 mm) prints cost about $10.00 each. In smaller towns and cities, this cost may be reduced.

Social Security Tax. The Federal Insurance Contributions Act, F.I.C.A., requires contractors to pay a tax on all payrolls, both job and office. At the present time the employer's share of this tax is 6.20% of the first $60,600 of taxable wages paid each employee for the year.

The employer is also obliged to deduct 6.20% from each employee's salary or wages, but except for the cost of additional clerical help in maintaining the records, this portion of the tax does not increase the contractor's cost. In many organizations extra employees have been required to keep these records.

In addition to the social security tax, the employer must pay a medicare tax of 1.45% on wages or salary and deduct 1.45% from the employee's wages or salary:

Regular FICA	6.20% of wages up to $65,400 annually
Medicare Tax	1.45% of wages (no limit)
Total	7.65%

Federal Unemployment Tax. This tax applies to the first $7,000 of wages paid each employee during a calendar year. The rate is 6.2 but you may take a credit up to a maximum 5.4% of total wages for contributions you paid into State Unemployment Funds. Federal Unemployment Tax is imposed on employers and must not be deducted from wages of employees.

State Unemployment Tax. This will vary from state to state. Note above that a credit is allowed against the federal rate.

Sales Taxes. Many states have sales or use taxes applicable to building construction as well as to other commodities. This tax varies with the different states, and there are different methods of levying it. In some states, the tax is levied on all building materials and is paid by the contractor when he pays for his materials. In other states the tax is applicable to the entire cost of materials entering into the construction of the building and is paid by the contractor directly to the state. In either case, the contractor should add this cost to his estimate. Otherwise, it will make a big dent in estimated profits. Note that some jobs will be tax exempt, which could save as much as 6% or more on materials. However, it is up to the client to establish his tax exemption, and the contractor should qualify his bid accordingly.

Job Overhead: Plant and Equipment

Plant overhead includes equipment costs, the cost of transporting equipment to and from the job site, and the costs of operating it. These costs may be in the form of depreciation allowances if owned, or as actual monthly or daily charges if rented. Generally, equipment used by only one trade will be furnished as part of that subcontract and will not appear as general contractor's overhead.

At the other extreme, items such as outside hoists and elevators on a high-rise building will be used by all trades and would be charged similarly to a general conditions expense. But there are many items that will be used by several, though not all, trades.

As an example, a job may require a rolling scaffold that will be used by the mechanical trades and electricians installing above-ceiling work; by metal lathers and plasterers installing a fancy coved lobby ceiling; by acoustical tilers installing office ceilings; by drywall men installing storage room ceilings; and by painters decorating ceilings. If each trade brought its own scaffold, there would be 7 transportation charges, and individual trades would probably be renting on a daily rate. If the general contractor furnishes the scaffold, one transport charge is involved with a much cheaper monthly rate. This savings could lower the bid and win the job.

If the contractor owns the scaffold, he can rent to his subs at the going rate and make a profit for himself, as well as lowering the job cost. Another advantage in the general contractor's furnishing plant equipment is that he can control quality and safety on the job. With Occupational Safety and Health Administration (OSHA) rules in effect, it is crucial that the general contractor control the quality of equipment on his job or face stiff fines.

If the contractor owns the equipment, it may be depreciated for tax purposes. However, if the inflation rate is high, the depreciation allowed by the Internal Revenue Service might not be adequate to cover replacement costs, and the tax savings might not be so attractive. On the other hand, a high inflation rate might make it possible to buy equipment, use it, and resell it for what one paid for it.

The advantages in renting (a short term agreement) or leasing (a long term agreement) include:

1. eliminates tying up capital or borrowing against bank credit for equipment that most probably will not return as much money as if those assets were applied toward financing another job or taking advantage of cash discounts;
2. offers flexibility in choosing equipment; one can try one product on one job and another on a different job, until one finds just the right piece of equipment for each situation, or if the item does not perform to expectations, it can be returned;
3. eliminates personal property taxes on the equipment;
4. eliminates maintenance costs and the expense of training personnel to handle those tasks and minimizes possible losses due to obsolescence and vandalism;
5. eliminates costs of shipping, handling and storage, and tying up trucks and drivers to move this equipment, as this will be done by the lessor.

The question then becomes whether to rent or to buy and depreciate plant equipment. In a large community where several rental sources are available, renting is probably preferable, leaving the contractor's money for other forms of investment, such as building up capital in order to take on larger jobs or investing in land on which a contractor cannot only hope to make a profit on resale but also can insist his firm be used as contractor for any development. If the contractor does elect to purchase equipment, he should set up separate records for each item. Be sure to consider the Internal Revenue Service's depreciation schedule allowed, the probable life, probable maintenance and storage cost, actual cost of transporting the item to and from the job, and the costs of operating the equipment on the job. Enter an hourly rate chargeable for the use of the item on each job estimate.

Charges should be sufficient to replace the item when it is worn out, as well as to cover interest charges if a loan is taken out to purchase it and a separate fund, often known as a *sinking fund*, set aside to cover future replacement. These charges should not be carried as profit. If the equipment is idle for long periods, the costs could be transferred to general overhead, but this is rather dodging the issue. If the equipment costs are accurately charged against each job, a contractor can easily tell whether his costs of ownership exceed current rental costs, always keeping in mind that the funds tied up in plant ownership could be making money elsewhere. Plant and equipment costs are discussed in greater detail later on in this chapter, under "Tool and Equipment".

Job Overhead: Prepurchase and Prepayment of Materials

With the cost of building materials being pushed ever upward, and with spot shortages of some raw materials a constant threat, a contractor may more and more be confronted with owners who want to protect themselves and order out materials months in advance of their being installed on the job. If this is a stated part of the contract, the estimator must make allowances for it.

In most cases the owner will reimburse the contractor upon his properly presenting his request for payment; but if the contractor is not to be paid until the material is installed, the estimator must add the costs of interim financing. Then the material must be suitably stored. Even if a material is weatherproof, few job sites today are vandalproof. Sometimes the material may arrive before the construction site is available. If the contractor must provide a shed, a watchman, or an off-site warehouse, or if the manufacturer is to store the material until it is needed at the site, these costs must also be figured and added to the bid or specifically excluded and made part of an extra work order. There are insurance and handling costs that must also be assigned either to the bid or to the owner. And it must be kept in mind that these costs will grow if the job schedule is not adhered to.

Job Overhead: Professional Services

Sometimes an owner may approach a contractor to act as his advisor on matters other than field construction. This can include finding a suitable site; lining up financing; handling rezoning or other legal and governmental problems such as sewer and road extensions, utilities and tax assessments; hiring of architects and engineers to design the project, landscapers to beautify the grounds, and decorators to furnish the interiors.

Some of the initial time spent on such consultation will be charged as "good will" under general office expense; then, if the job progresses, the contractor should be reimbursed not only for fees for outside services, but also for his and his staff's time spent in administering these services.

A contractor should move slowly in assuming these auxiliary functions. All too often they require much more time, responsibility and cash investment than anticipated, and his major responsibility, the contracting business, is left without adequate supervision and in a cash bind because reimbursement for preliminary phases has been held up until ground is broken and the interim loan is in effect.

Further, a contractor relies on the good will of the community for his future business. Choosing one real estate office, architect, lawyer, banker, or one decorator over another may limit his future prospects more than the immediate job can justify.

However, a good contractor does tend to be a good citizen, and as his experience grows he is a valuable source of information on building in the community. This knowledge is an asset, and it is entirely proper to charge for it on an hourly basis as for any other skill.

Bonds

Bonds are usually issued to cover a single job and are a direct job cost. Some cover only a single trade and may be included in a subcontract sum, such as warranties or roofing bonds. Bonds taken out by the general contractor are usually carried under *Overhead: General Conditions*. Some of the types of bonds encountered are discussed below.

Bonds are usually issued by insurance companies or surety companies specializing in such bonds. A division of the U.S. Department of Labor, Small Business Administration, has been set up to provide surety bonding for small and minority contractors. Surety bonds are credit guarantees of contractors and are issued after careful scrutiny by underwriters of a contractor's financial statements, skills, work in progress, past work, and general reputation. Bonds are not purchased by contractors, but rather, issued for the payment of a fee based on contract cost.

Bid Bonds. Bids are invited by advertisements, and the bidder may have to submit with his bid a certified check, usually for 5% of the bid, or a bid bond, usually for 10% of the bid. This requirement is made to insure that bids will be submitted only by qualified contractors who will not only enter into the contract but qualify for performance and payment bonds.

The bid bond guarantees that as the contract is awarded, the principal will, within a specified time, sign the contract and furnish the required performance and payment bonds. If the principal fails to furnish the required performance bond, there will be a default under the bid bond, and the damage will be the difference between the principal's bid and that of the next highest bidder, not exceeding the penalty of the bid bond.

Surety companies generally follow the practice of authorizing a bid bond only after the performance bond on the contract has been underwritten and approved, so there are few defaults under bid bonds.

Performance and Payment Bonds. Some owners and most public work require that each contractor to whom a contract is awarded shall furnish bonds which contain two obligations: a performance bond to indemnify the owner against loss resulting from the failure of the contractor to complete the work in accordance with the plans and specifications; and a payment bond to guarantee payment of all bills incurred by the contractor for labor or materials for the work.

The federal government, under the Miller act, requires that a contractor furnish two separate bonds, one for the performance and one for the payment of labor and materials. Some states follow the same procedure.

The risk of the surety under the construction contract bond is if the contractor fails in the performance of his obligation to complete the job, or to pay the bills for labor or materials, then it assumes that the responsibility for completing the project and paying the bills.

The underwriting skill is designed to weigh the hazards involved against the contractor's qualifications. If the contractor's ability is deemed sufficient to warrant the assumption that he will be able to complete the work and pay the bills, the bond is issued. If not, it is declined.

The general practice is to underwrite each bond on an individual basis. The contractor cannot assume that every request will be favorably received by underwriters. The contractor's volume of work that is still outstanding at the time of the bond request is a key factor.

Some of the most common causes of failure on the part of construction contractors are abnormal labor costs, fire or other loss not adequately covered by insurance, unforeseen restriction or withdrawal of credit, weather or subsurface conditions that increase the cost but not the contract price, over extension beyond the contractor's financial capacity, unavailability of materials, failure of subcontractors, death or disability of key men in the contractor's organization, a poor accident record, diversion of funds from one job to pay off overdue accounts on other jobs, inexperience in the type of work involved, or dishonesty on the part of the contractor's employees.

Costs of performance bonds depend on amount of contract and length of coverage, as well as the contractor's reputation and whether the coverage is for a portion or complete cost of work.

Bonds are usually issued for a 24 month period and might run on an average job at $12 per $1,000 of contract cost for the first $500,000; $7.25 per $1,000 for the next $2,000,000; $5.75 per $1,000 for the next $2,500,000; $5.25 per $1,000 for the next $2,500,000 and $4.85 per $1,000 over $7,500,000. Should the bond need to be extended beyond two years, figure 1% of the regular premium per month.

Maintenance Bonds. These may be required to guarantee an owner that the general contractor agrees to correct all defects of workmanship and materials for a specified time following completion. Often the above performance bond can be arranged to cover this period which usually lasts one to three years.

License or Permit Bonds. These are bonds provided by a contractor to a public body, guaranteeing compliance with local codes and ordinances. If the contractor regularly operates within an area requiring such bonds, this cost should be carried under office overhead, because it is a normal cost of doing business. But if it is required for a specific job, it can be charged directly as a job expense.

Subdivision Bonds. With so many restrictions being applied to uses of sewers and water supplies, with shortages of natural gas, and with restrictions on fuel emissions, more and more public bodies require bonds be filed guaranteeing that all site improvements, utilities, and buildings themselves will be constructed in accordance with governing codes and ordinances. Unless the contractor is the developer, this is usually filed directly by the owner, who in turn is already protected by the contractor's performance bond to the extent that the drawings and specifications comply with the ordinances.

Specifications may include catchall phrases such as "and in accordance with all local codes and ordinances..." following detailed instructions of procedures that may be contrary to such codes. In addition, there is a growing legal history on developments that have been refused licenses after completion because they do not comply. Consequently, the contractor should inform himself of all local requirements before entering into any contract.

Surety underwriters very carefully review requests for subdivision bonds, and these bonds are issued only when the underwriter is satisfied that the contractor is able to meet all obligations required by authorities.

Supply and Subcontractor Bonds. These are guarantees from manufacturers and installers for specific items and workmanship furnished, the cost of which is usually borne by them and included in the subcontract price for that trade rather than in overhead.

Insurance

A reliable insurance agent is one of the most valuable "subs" a contractor can add to his team. Almost everything can be insured and has been—at a price. Some types of insurance are required by law. Some are made part of the contract. Others may be indicated because of some job peculiarity. The following list discusses the more usual types encountered. Rates will obviously vary all over the country and from one part of town to the other, from one suburb to the next. There will be some overlap between insurance costs chargeable to office overhead and to job overhead, but it is best to keep them as separate as possible.

Insurance is a significant cost component in construction and is currently estimated to range between 3% and 5% of the "hard costs" of a contract. Construction contracts ordinarily contain a section titled *insurance* that specifies the insurance coverages that must be purchased by the contractor and those that are to be provided by the owner, usually coverage for damage to the project itself. This section of the contract usually sets insurance coverage conditions, terms, limits, and special wordings, such as for underground damage and contractual liability. Often, this section requires that the contractor name the owner as an additional insured and waives subrogation (the right to sue on one's behalf) in favor of the owner, its agents, and its employees. Sometimes, those who finance the project seek the same insurance protection. Insurance companies are reluctant to make such amendments, and if they do, the contractor can expect an additional premium charge.

Sometimes, construction contracts, particularly those for larger jobs, state that the owner will provide all insurance, except for specific types. Contractors will be required to bid *ex insurance* and the normal contractor overhead allowance for certain insurance will not be granted. This type of insurance is referred to as the *owner controlled program* and also as *wrap up insurance.*

Since its inception, owner controlled insurance has been the subject of much controversy and in many states is prohibited for public work. There are some notable exceptions. For instance, some states allow this type of insurance for certain *public authorities* but not for all. It is normally for private projects that exceed certain sums.

When faced with owner controlled insurance, a contractor should carefully review the insurance section and other pertinent parts of the contract in order to properly identify the risks and the extent to which the owner's program provides protection. It is not to recommend that the contractor exclude the project from an existing program, but rather to have a contractor's program, the excess over valid and collectible insurance provided by the owner. This is not automatic and must be specifically negotiated with insurance companies.

Worker's Compensation. This insurance provides the contractor's employees with the benefits specified by the various state and federal compensation laws in the event of injury or death arising out of the employment. Most compensation acts contain at least some provisions specifying benefits for the most common occupational diseases.

Rates vary widely among the several states, the average being from less than 3% in Indiana to over 22% in the District of Columbia. Sixteen states, primarily in the Southeast, average from 4% to 6%; eight states, primarily in the West, average over 10% and the remaining twenty-five average from 6% to 10%. In those states where medical benefits are limited, it is advisable to carry full or extra-legal medical coverage.

Certain compensation laws are restricted and provide benefits for only specified occupations. The workmen's compensation policy may be broadened to voluntarily pay compensation to employees injured in other occupations.

In addition to providing the statutory benefits, the standard workmen's compensation policy affords employer's liability insurance against claims for bodily injury or disease arising out of employment in the states named in the policy. This coverage against liability not within the scope of the compensation law is subject to $25,000 standard limit which can be increased for a moderate premium charge.

In the event of the death of an injured employee, his dependents generally have the right to choose between the benefits of the state where the accident occurred or the state where the employee was hired. In addition, a number of states impose penalties upon an employer who though subject to the compensation act has failed to provide insurance coverage. Accordingly, it is important that you discuss these points with your insurance representative before entering a state not covered by your policy and before hiring anyone in another state.

The general insurance rates in the preceding table are representative rates to be used as a guide only. Rates and experience modifications are subject to frequent and sometimes radical changes and should be verified with an insurance representative for each specific project. Major job classifications are listed in the table, but you must consult your insurance representative for specific classifications.

Reference has been made herein to base or manual rates and to experience modification of the manual rates or premium. A program of "retrospective rating" is available on an optional basis. This rating plan, superimposed on experience rated premiums, is usually attractive and appropriate to the larger contractors. Subject to certain state limitations as to the combinations permissible, it is available on an intrastate or interstate basis for worker's compensation and the liability lines, either separately or in combination.

Your insurance representative will be glad to explain the details of this rating plan, as well as any other features of coverage or rating on these various lines of insurance.

Workers Compensation Manual Rates per $100 of Payroll
For Various States

Code No. & Type of Work	CA	DC	FL	IL	NJ	NY	TX
5022 Masonry	14.55	25.81	28.18	23.22	10.77	18.61	23.12
5040 Struc Steel Erection	23.93	55.08	40.27	78.98	22.52	27.11	50.47
5057 Steel Erection – NOC	21.91	56.98	48.25	70.79	12.96	24.09	31.01
5102 Ornamental Metal	13.55	44.32	17.82	25.56	8.03	14.12	14.29
5183 Plumbing	11.59	16.13	13.50	14.48	5.84	9.03	12.88
5190 Electrical Wiring	9.93	10.28	11.51	9.82	4.23	6.89	11.60
5213 Concrete Work	13.18	24.27	31.32	36.76	10.83	24.92	25.59
5348 Ceramic Tile	7.74	17.73	11.58	15.71	5.74	10.56	9.94
5403 Carpentry	28.28	21.60	29.77	23.02	10.40	14.50	28.29
5474 Painting	19.45	13.97	28.47	14.93	11.49	12.55	18.29
5480 Plastering	18.12	14.44	26.52	15.82	7.88	13.03	20.99
5551 Roofing	39.79	28.69	52.41	37.32	29.41	32.78	47.24
6217 Excavation	7.88	18.06	14.88	10.43	7.95	9.69	18.16

Contractor's Public Liability Insurance. This policy protects against liability for bodily injury to a member of the public arising out of the contractor's operations and against property damage liability. It can be had under either a schedule general liability form policy or a comprehensive general liability form.

Under the schedule form, the several hazards to be insured are specifically named, and the coverage may be applied to specified locations and operations of the insured contractor. The comprehensive form is broader in nature and covers all the usual hazards of a general liability nature other than those specifically excluded.

Insurance companies are almost always willing to expand coverage conditions to cover excluded risks or to amend with broader terms the definition of an exclusion. The *broad form comprehensive general liability endorsement* is an example of a large package of additional coverage.

Under either form, consideration should be given to at least the lines of coverage briefly outlined below:

Owners, Landlords and Tenants, Premises Exposure: This hazard includes such exposures as offices, warehouses and fixed locations in contrast to specific contracting operations as such. The usual rating basis is a charge per 100 s.f. of area, and normally, the premium for this hazard is relatively minor compared with that for the operations hazard.

Products Liability: Completed operations coverage applying to accidents which occur after the contractor's operations have been completed. Rates per

$1,000 of receipts or contract price vary over a wide range, according to the operations being conducted. These rates are frequently the subject of individual account rate determination rather than the application of any base manual rates.

Sometimes, an owner specifically requires that the contractor carry product/completed operations coverage for a period of time after completion of a project. The completion of the work is the trigger that begins coverage, which applies to acts or errors, failures or negligence committed during the construction period. The need to provide such coverage after job completion is redundant. The key factor in the granting of coverage is when the negligent act or omission occurred, not when the accident happened. Of course, if the contractor has, or is required to provide, a new form of coverage called *claims made*, then the above can be ignored, because claims made coverage applies only for claims made against the insured during the term of the policy. With claims made insurance, the contractor is obligated to carry insurance protection for as many years subsequent to completion of the project or for as long as the contractor deems an exposure exists. It should be noted that claims made insurance has a like effect on all other coverage sections of the general liability policy, and contractors should carefully weigh all potential risks and exposures before accepting this type of coverage.

Owner's Protective: Covering the liability of the owner for injuries to the public arising out of operations of independent contractors. Characteristic rates for general construction operations in most states are .018 bodily injury and .011 property damage per $100 of contract cost. These rates apply to the first $500,000 of cost on each specific project. Corresponding rates for the next $500,000 of contract price are .009 and .006; for contract cost over $1,000,000, it is .002 and .002. Rates in Illinois and New York are somewhat higher.

Contractor's Protective: Covering liability of the contractor for injuries to the public arising out of operations of subcontractors. Rating basis and rate levels stated above for owners protective are applicable under this line to sublet contract costs.

Additional features can be incorporated into the general liability policy, either for application on an individual exposure basis or to be applied on an automatic or blanket basis.

Contractual Coverage: Applying to the liability of others assumed by the contractor. Rates depend on the nature of exposure and degree of liability assumed.

Accident Defined Coverage: Broadening the coverage to apply to continuous or repeated exposure giving rise to injury or damage, in contrast to the "caused by accident" basis on which most General Liability Policies are written. Cost is usually expressed as a percentage of policy premium.

Personal Injury Coverage: Broadening the usual "bodily injury" nature to include such personal injury as libel, slander, false arrest, malicious prosecution, etc. Pricing varies with the degree of exposure and extent of coverage afforded.

Malpractice Coverage: Covering contractor for liability arising out of error or mistake in professional treatment in on-the-job hospital or medical facilities under his control. This line is individually rated.

Increased Limits of Liability: The general liability rates referred to herein and in subsequent pages are at standard limits of $5,000 per person, $10,000 per accident for bodily injury and at $5,000 per accident, $25,000 aggregate for property damage. The percentage increase factors for various higher limits can be secured from your insurance representative.

Automobile Liability Coverage: Bodily injury and property damage coverage for owned vehicles, non-owned vehicles, hired vehicles, or independent contractor's can be secured under a schedule form or a comprehensive form of automobile liability policy. Coverage can be selected either on the basis of a listing of owned automobiles or on an automatic basis whereby coverage applies to all owned vehicles. The rating basis for owned vehicles is based on size and weight, usage of the vehicle, and geographical territory involved.

Physical damage coverage (fire, theft, and collision) on owned vehicles may be either under the policy for automobile liability coverage or as a separate policy.

Property Damage Insurance: The contractor should make a careful study of what his responsibilities are in regard to damage to other property and to the work completed so that his policies will mesh with those carried by the owner.

If provision is not made in the specifications that fire insurance will be carried by the owner during construction, it is important that the contractor protect his interests by carrying some type of builders' risk insurance. This is normally written on a standard fire policy together with an extended coverage endorsement that extends the fire and lightning provisions of the standard fire policy to include windstorm, hail, explosion, riot, smoke damage, and damage from aircraft and vehicles. An additional endorsement, vandalism and malicious mischief, may be obtained. Some companies afford a broader coverage of an all risk nature although there are a number of exclusions.

There are two common builders' risk forms that may be written:

Builders' Risk Reporting Form: This form affords coverage for the contractor from the time material is placed on the job site until the building is completed. Under this form the contractor reports to the insurance company the actual value of the building at the end of each month as construction progresses and pays a premium each month on the additional values that go into the building.

Completed Value Builders' Risk Form: This form of coverage is normally preferred by a contractor, because it gives 100% coverage without the necessity of keeping records as to the value of the building as it progresses or reporting at the end of each month.

This form is written for the completed value of the building. The normal rate charged is about 55% of the rate on the basis that at the beginning of the job there is little value and at the end you have the full value. Normally the premium under the two types of policies is remarkably close.

The interests of a general contractor can be covered and he may exclude or include the interests of subcontractors who also may be employed in the construction of the building.

Temporary structures, tools, and equipment that belong to the contractor and are on the building site may also be included in the coverage for the hazards insured.

Today, many contracts that require the contractor to provide the coverage call for coverage on all risk of loss basis, naming the owner, lender, and numerous other parties, as well as the contractor, as insureds in the policy.

In addition, many contractors require the inclusion of specified exposures related to the project. Delay in opening and bank financing changes are two of the more frequently seen "soft cost" coverage requirements.

On occasion, contracts with clauses that are not contained in the insurance section of the contract but that are found in other sections, such as *force majeure,* risk of loss, indemnification, and work place safety, leave the contractor with liabilities that may or may not be covered by insurance. For example, the insurance section might specify limits of liability to be purchased for policies provided by the contractor, but another section of the contract or specification, contractual indemnification or risk loss, may specifically state that the contractor's liability is full and not limited to or by any insurance requirements otherwise stated in the contract.

Costs of such insurance vary by location, type of construction, and available public fire protection and may be obtained from your local insurance representative.

Contractor's Equipment Floater: Many contractors possess many thousands of dollars in contracting machinery and equipment. To properly protect these values, such equipment and machinery is usually insured under a contractor's equipment floater policy.

Insurance under this form is quite flexible so that many situations may be insured to fit the contractor's particular needs. The insurance is normally written on a named peril policy which includes loss or damage by fire or lightning, explosion, windstorm, flood, earthquake, collapse of bridges and culverts, theft, landslide, collision, and upset. Or if desired, a policy may be issued protecting from practically all risks of physical loss or damage. Such contracts usually write a deductible clause.

Installation Floater: The contractor who specializes in plumbing, heating, air conditioning, or electrical equipment may find an installation floater more suited to his needs. This type of policy may be written for an individual job or may be written to cover all the jobs he may secure during a year's time.

Coverage may be on a so-called "named peril" policy, usually including the hazards of transportation of his material to the job site and covering fire and lightning, windstorm and hail, explosion, riot and civil commotion, smoke damage, and aircraft and vehicle damage at the job site. He may also include coverage for vandalism and theft if desired.

Profit

The amount of profit to be added to the estimated cost of the work is a question that a contractor must answer individually for each bid. There is no set amount that can be added. It all depends on local conditions, competition, and how badly the job is wanted.

On small jobs, alterations, remodeling, and similar work, a contractor is justified in adding 20% to 25% profit to the actual cost, but he must ask himself whether he can actually obtain this amount.

On new work, where it is possible to estimate cost with a fair degree of accuracy, a contractor is entitled to 10% on the actual cost of the work (job overhead included in the actual cost of the job), but it is safe to say that competitive figures submitted for many jobs show a 5% instead of a 10% basis. A contractor is entitled to 10% profit, but getting it is another matter.

Another topic of considerable discussion is whether or not a contractor should add a certain percentage to his estimate to take care of contingencies. The answer is both yes and no. Yes, because it is not necessary to be quite so careful in the preparation of the estimate, and if any items have been overlooked, there is a fund to take care of omissions. But a contractor who is not careful in the preparation of his estimates never knows how much should be added to provide for forgotten items or contingencies. No, because if you add a percentage for contingencies, it is quite possible that one or more of your competitors has not, and another job is lost.

The only safe way is to be as careful as possible in listing all items from the plans and specifications. Include them all, and it will not be necessary to add for contingencies. This includes everything to be furnished and all labor costs, including job overhead expense and a percentage for profit.

An item often overlooked, but which can become a major expense, is the finishing up of a job and making corrections after the job is completed. Often, there is confusion on the job at the end and it is difficult to say who chipped the paint, scratched the door, broke the window, stole the light bulb, broke the door handle, or backed into the fencing. Most of the trades are on the job doing touch-up work, the owner is moving in, his decorator is hanging drapes, and his children are running wildly through the corridors. But the general contractor gets the complaint first, and in the interest of having a satisfied client, will probably assume the responsibility for correcting this. In a warehouse, such problems may be minimal; in a fine residence, you might be called back daily for weeks.

Other factors an estimator must evaluate in setting up his bid are the weather conditions that are apt to prevail at the time of construction; possible rises in labor rates and the possibility of strikes if labor contracts expire during the construction period; material shortages and strike actions at the manufacturers level; the money supply and its cost; and public relations with the community in which the project is located and possible job stoppage or slowdowns because of environmental or safety concerns which the drawings and specifications fail to cover.

Labor Inefficiencies on Construction Projects Due to Overtime

Number of Days Per Week	Hours Per Day	Hours Per Week	Lost Efficiency Factors %
5.00	8.00	40.00	0
5.00	9.00	45.00	0-5
5.00	10.00	50.00	2-8
5.00	11.00	55.00	5-11
5.00	12.00	60.00	8-14
6.00	8.00	48.00	2-3
6.00	9.00	54.00	10-13
6.00	10.00	60.00	16-17
6.00	11.00	66.00	20-21
6.00	12.00	72.00	22-25
7.00	8.00	56.00	8-15
7.00	9.00	63.00	16-16
7.00	10.00	70.00	21-22
7.00	11.00	77.00	24-25
7.00	12.00	84.00	28-29

Percentage

Hours worked per week

Lost Efficiency Factors

Estimate Check List

Too many estimates are done under the pressure of a deadline. If at all possible, a second party should check all arithmetic on the original estimate sheets and the original figures against those copied on the final bid sheet. It is not unusual to receive a bid from a subcontractor with an obvious error in the figures. This is an indication that the bid was not checked prior to tendering, and it could severely affect the company offering the bid and the general contractor using the bid as part of the general estimate.

Storage, transportation, and handling costs are often overlooked. It is often a good idea to qualify a bid as to the time of year a job is to be started. A bid given in March anticipating an April start and October finish should say so. If the contract is delayed for any reason, so that the job starts in October and runs through the winter, allowances must be added for temporary enclosures and heat, idle days or weeks, and less efficient work conditions. Furthermore, inflation is a major factor today, and labor and material costs can escalate 4% to 6% in six months.

The estimate should be carefully checked for plant equipment costs, with each subcontractor being carefully informed as to what will be furnished to him. Architect specifications are usually very general on this matter, often stating in Work Included "...all labor, materials, and related items..." to cover plant equipment. It is left up to the general contractor to have a clear understanding with his subs on this.

Check your profit. How much of this is certain, and how much contingent on optimism? Set forth a definite percentage for contingencies and try to hold to it. Keep the profit apart. If there is a questionable item, alert the owner and architect, and then, qualify your bid accordingly. Most proposals provide a space for this. Give a firm price on what is clear. Set up an allowance to cover what is in doubt with a proposal, stating what is to be charged against the allowance and why.

Finally, make certain all items are covered. Check not only the drawings and specifications, but also any bulletins addenda. An addendum changing wood paneling from full height to wainscot in an office means a savings in millwork. But now, of course, the drywall above needs to be taped and finished for painting. So both the drywall and painting contractors must be alerted, even though the addendum only refers to a change in the millwork.

For a comprehensive check list of estimate items, based on the Construction Specifications Institute (CSI) Uniform System, refer to the *Specifications* section earlier in this chapter.

The list is intended to cover every type of structure from the smallest residence to the largest skyscraper, commercial or industrial plant. Compare it with your estimate before submitting your bid, because it might be the means of including some items that would otherwise be omitted from your estimate. This may require a few minutes extra time, but always remember, it is much better to be safe than sorry. Forgetting has cost many a contractor his profit.

Determining Costs

The detailed estimate is a projection into the future of the anticipated costs of a project. It reflects all of the known conditions at the time of preparation. It is important to maintain a cost system that uses as a base each of the components contained in the estimated cost. This is necessary in order to compare estimated costs with actual costs incurred during construction. If the estimated costs are realistic and accurate, they will produce the anticipated profit.

Equally important, the periodic review of actual cost versus estimated cost will serve as a data base that can be adjusted, either by increasing or decreasing the unit costs, so that future estimates will be produced more competitively.

Cost control systems can take many forms. They can be relatively simple and still produce accurate information, or they can be designed in an elaborate and detailed format to provide information relating to cost for every item of material and labor productivity for the entire project.

The selection of the system is one of personal choice, but it must be remembered that any cost system requires considerable expenditure of time in accumulating the material and labor records, in distributing to the proper categories of work, and then, in calculating the costs of each and recording this information on progressive cost record forms so that the actual costs can be monitored while the work is being performed. When this information is available, it is then possible, should an overrun appear, to review the situation to determine if a remedy is possible, such as adjusting crew balance, pursuing another method of construction, or allocating additional supervision to a specific operation. The system selected should be one that serves the needs of the company. Anything in excess of this is a burden and wasteful in both time and cost.

CONSTRUCTION SCHEDULING

In order to complete a construction project within the estimated cost, the contractor must insure that individual work activities are completed within the time allowed by the cost estimate. A schedule must be prepared and followed. The overall process can be separated into three distinct steps:

1. Planning
2. Scheduling
3. Monitoring

There are three traditional methods of construction time scheduling:

1. Critical Path Method (CPM)
2. Bar (or Gantt) Charts
3. Program Evaluation and Review Technique (PERT)

In *CPM scheduling*, there are two methods of drawing logic diagrams: arrow-diagram and precedence network. Both CPM methods involve identifying individual work activities within a project, determining the relationships among the activities, and presenting the information in a graphical network diagram.

Bar Charts are similar. Bars reflect start and end dates and duration of individual activities, but they do not show the relationship among activities. *PERT*, like CPM, is a network diagramming technique, but it is more suitable for research and development than as an actual scheduling method. The differences among and varying uses of these scheduling techniques are discussed in detail later. For now, suffice it to say that CPM, PERT, and Bar Charts all involve the same principal considerations.

Cost Estimating and Scheduling. The cost estimator is, in fact, the initial project scheduler. The quantities of work determined and the labor and equipment productivity estimated in the cost estimate will budget the time that can be expended on each individual work activity. The work plan anticipated in the cost estimate is actually used as the estimating unit (months, weeks, days) for job site overhead and general condition costs. The estimator must understand the concepts and principles of construction planning and scheduling in order to prepare an accurate cost estimate. The interdependency of cost control and production control reflects the relationship between estimating and scheduling. The field effort must accomplish the productivity estimated in order to complete the activity within the cost allocated.

Objectives of Construction Scheduling. The principal objective of construction scheduling is to efficiently manage the resources used in the construction process. These resources include labor and supervision, material and supplies, equipment, general condition or jobsite overhead support items, and subcontractors (including material suppliers). The objective of the management effort is to effectively use the resources so as to accomplish the project and its individual activities within the budgeted cost.

The construction schedule provides the principal measuring tool for evaluating progress. Circumstances and situations encountered in the construction process will affect the work that remains. An accurate, updated schedule allows the contractor to identify and evaluate alternative plans in responding to the changes in project conditions that will result in the best project decisions.

How to Plan. Many builders plan out their work mentally while preparing the quantity take-off and cost estimate. The work items described in the estimating process are defined by the quantity of work (or materials) to be installed and the cost allocated to complete the installation. The labor cost estimate anticipates a worker or crew productivity, which is a principal consideration in planning construction work. Because the critical path method (CPM) is one of the most widely used planning and scheduling tools, we will use it as the basis for our discussion of planning.

The initial planning step is to identify the work items or activities. Often, these activities differ from the work items as listed in the estimator's checklist. In the planning process the contractor determines the relationship between one activity and other activities, or series of activities. *Sequential* activities require that one activity be substantially complete before the next one begins. These are usually identified first. Then, *simultaneous* activities, or chains of activities, that are not directly dependent can be planned to take place at the same time. Planning also identifies the labor, equipment, and material resource requirements for each activity or chain of activities.

How to Schedule. An accurate construction schedule estimates a *duration* for each activity. When individual activity durations are totaled for a path of activities that occur in sequence, the duration for that chain of activities has been calculated. The highest sum of activity durations that form a continuous chain of sequential activities through the planned project, allowing contingency time for weather and other delaying factors, is the scheduled duration of the entire project. Then, when the start date is identified, the schedule can be converted from work days to calendar days, identifying the actual dates that each activity will begin and end, and indicating when each resource will be required.

The cost estimate for many equipment items, job site overhead items, and general condition items is often expressed as cost per unit of time (for example, dollars per month). The cost estimate assumes a time duration for the overall project and for certain chains of activity.

Monitoring and Reporting. Planning and scheduling are the only valid approach to understanding process and productivity of a project. Monitoring the progress of individual activities according to a valid construction schedule and promptly adjusting to changing conditions means maximum profit.

Most builders are capable of making good decisions when good, timely information is available. A system of reviewing and recording the cost and production of individual activities and chains of activities usually results in good control of construction time and costs. An active system of monitoring and reporting is even better.

Planning

The level of planning detail varies, depending on the size and type of project, on the amount of work that will be subcontracted, and on the method of scheduling used. The planning process described here is appropriate for Bar Chart, CPM, or PERT methods.

Planning is the most time consuming and critical element of the construction schedule process. The construction plan for a project requires an understanding of all the field requirements to complete the project. The sequence of the identified work activities must be planned in sufficient detail to account for all of the particular requirements of the contract documents, project site, anticipated availability of labor-equipment-material, anticipated weather conditions, and the anticipated performance of subcontractors and suppliers. Although some builders develop a checklist of activities for particular types of construction projects, it is generally recommended that the sequence of work for each portion of each project be considered unique.

When bar chart schedules are used, the description of work usually involves broad scope activities that have long durations. An individual activity, represented by one bar on the chart, usually includes a category of work, often categorized by craft or material, and does not always have a direct finish-start relationship with the other bars. It is not possible to develop accurate resource control based on typical bar charts. Their principal use is to plot progress for review.

CPM usually involves a more detailed breakdown of work items. There are no specific rules for the selection of activities, but the principal considerations are:

1. Activity Responsibility: Subcontract work is segregated into separate activities.
2. Activity Craft/Crew Requirements: The manpower requirements for individual activities must be identified.
3. Activity Material Requirements: The material requirements for individual activities must be identified.
4. Activity Equipment Requirements: The equipment requirements for an individual activity or series of activities must be identified.
5. Location of Work: The site conditions will often require similar work to be performed with different crews or at different times.
6. Subdivisions of Work: A particular project component often requires a breakdown of sequential activities to adequately plan and schedule the work. For example, structural slab is broken down into shoring, formwork, reinforcing, and other activities required to place structural slabs.
7. Cost Control Breakdown: The tracking of actual cost incurred is often linked to the planning and scheduling activities.

The level of detail of activities for a construction schedule depends on the anticipated use of the schedule. The *overall control* schedule defines the project by broad scope activities. *Short interval* schedules for production control are based on specific, short duration work activities. A successful schedule depends on an appropriate selection of activities.

Subcontract Activities. Numerous project activities will be completed under subcontract to trade and specialty contractors. The prime contractor spreads the financial risk and overhead cost by subcontracting work but often encounters difficulty managing the time aspect of the portions that have been subcontracted.

CPM has become the standard tool for resource control. One of the prime benefits of the initial CPM schedule is to identify which subcontractors are required on the project at specific times and to reveal the impact of subcontractor delays on the project schedule. The scheduling of subcontracted work requires both the subcontractors' input and their commitment to the work schedule. When using activity duration information furnished by a trade or specialty contractor, it is important for the general (prime) contractor to know the size and anticipated productivity of the trade crew in order to accurately monitor the progress of work. As an example, consider the CPM network in Figure 1.

The *Electrical Rough-in* Activity (33-34) on the sample project has an activity duration estimate of 3 days.

If the electrical subcontractor assumed using a crew of 4 electricians to complete the rough-in when the activity duration was established, then the prime contractor must anticipate a delay in the project if a smaller crew is available when it is time to perform the work.

The contractor cannot monitor the subcontract work and react promptly and effectively to prevent delay if the crew size and productivity assumed in the estimate of the activity duration is unknown.

The prime contractor should be certain that critical items and low-float activities that are subcontracted are identified in the performance terms of a subcontract. (*Float* measures the amount of time an activity can be delayed and still allow the project to finish on schedule; the concept of float is discussed in greater detail later.) The construction schedule is a useful tool in providing for notification in advance of the start of subcontract activities.

Procurement Activities. The planning process must also recognize and identify procurement activities. The commencement of field work for a particular activity or series of activities depends on the delivery of the material. The delivery of material often depends on other activities that involve the vendor, contractor, designer, or subcontractors.

Long-lead items and material procurement that require approval of submittals or shop drawings are often critical on the construction schedule. It is especially important that these activities be identified in the planning process. It must be noted that the monitoring and updating of procurement activities is often the most critical early project management task.

Network Sequence. Networking is an accurate, graphical method of recording the planned sequence of activities. A physical diagram of the work plan is a useful communication tool. Figure 1 is a typical network, with activity durations noted, for a construction project.

The network provides a visual reference for the flow of the work activities. CPM and PERT diagrams graphically display the finish to start relationship of activities that are sequential. The network is a communication tool that allows the construction planners to see the construction plan and to evaluate, critique, and revise it prior to preparing the schedule.

Construction Logic. One of the primary questions to be resolved in planning is to decide the logical progression of work. Although there may be numerous approaches to completing a given project, some are usually more efficient, and therefore more profitable, than others. The planning process involves evaluating alternate approaches for completing the work and determining reasonable alternatives that can be considered if the project schedule is disrupted during construction.

The relationship of one activity to another is often directly sequential. For example, concrete cannot be placed until the formwork and reinforcing activities have been completed. The logic of construction prohibits altering this sequence of activities. Likewise, the construction logic will often determine that there need not be a direct sequential relationship and that several activities can take place simultaneously. Separate chains of activities may have a direct, sequential relationship to the same starting or finishing activity.

Resource Constraints. Often, construction projects require multiple repetition of the same sequence of construction activities in different locations, as is usually the case in high-rise construction. Although sequences are independent according to construction logic, they are related if the same crews, equipment, or supplies (e.g. formwork reuse) will be required on similar activities within repetitive work sequences.

The procurement and delivery of material on time to commence the work activity for which it is required is as important to meeting estimated cost as is purchasing the material. Costly project delays or revision of work sequences result when materials are not available in time. Planning for storage and on-site material handling must be based on understanding not only the work method and installation duration anticipated, but also the lead time required and the method and volume of delivery.

Material procurement is often established as a separate activity for each chain of work activities and appropriate durations are assigned. Within each of these individual procurement activities is a sequence of activities commencing with a purchase order or subcontract, often including the preparation and approval processing of shop drawings or submittals, through delivery to the site. These activities are best managed on a separate, expediting schedule, which then becomes a sub network of the master network. The lead time constraint for items needed early in the project (e.g. foundation reinforcing, anchor bolts, and layout templates) and long-lead items involving shop drawing approval and/or off-site fabrication (e.g. structural steel) can result in a procurement activity being critical.

Management of the equipment resources is similar to materials management in that timely availability is critical to the project schedule. Management of the equipment resources is similar to management of labor, because like labor, it must be productive while on site for the budget allocated in the cost estimate to be sufficient to cover the actual cost.

It is prudent to evaluate the equipment, supplies such as scaffolding, and possibly tools on a *sub network* schedule. The identification of equipment-dependent activities, their anticipated start dates, and the anticipated duration of each activity or chain of activities becomes the equipment monitoring and management tool.

Equipment for construction is either owned by the contractor, furnished by the subcontractor, or must be rented or leased. In building construction, contractors usually maintain a limited inventory of particular pieces of versatile, multi-use equipment (e.g. construction tractor with loader, box scraper, backhoe) and specific-use equipment (e.g. erection) that will be used enough to justify the investment. To a large extent, there can be more control of the availability of owned equipment than of rental equipment. The contractor must order rental equipment with sufficient lead time to assure availability and delivery so as not to delay the project.

As is the case with material and labor, the contractor reduces a cost risk when required equipment, tools, and supplies are subcontracted. But some control, including the availability of equipment, is also entrusted to the subcontractor.

The labor resource constraint is the principal culprit in inaccurate planning and insufficient project progress. The duration of most activities depends primarily on the productivity of the crew, or crews, assigned to complete the work. The planning phase should always include a resource logic evaluation of the crews that might be required simultaneously according to the construction logic and the actual number of crews that are available to the project. Activities to be accomplished by the same crews show a *resource logic constraint*. The activities must have a finish-to-start relationship dependency; that is, one activity must finish before the next can start. This constraint must be indicated on the network.

Master Network/Sub Networks/Short Interval Plans. A bar chart is a *master network*. The direct logic and duration of individual activities are not shown. The diagram in Figure 2 identifies the anticipated start and finish dates of the principal work sequences required to complete the project. Milestone events identifying significant accomplishments in the progress— completion of the foundation, topping out of the structure, weatherproof enclosure—are easily identified and the principal benefits of the bar chart.

Master networks are not developed to a sufficient level of detail to accomplish control of resources. The proper objective of a master network is to serve as a reference for those participants (owner, A/E, home office) not directly responsible for productivity and progress on the site.

Sub networks provide information for a particular task (e.g. expediting and material management, equipment management), a particular craft or subcontractor, or a specific sequence of work. Sub networks are developed to a level of detail to accomplish the start of work activities as scheduled.

Short interval plans are developed into short interval schedules during the actual construction phase. From one to several weeks in scope, the short interval plan and schedule provide an in-depth evaluation of the upcoming work and resource requirements. These networks have the greatest level of detail.

Network diagrams (CPM and PERT) are best at describing the interrelationship of individual project activities. However, these diagrams are not the most effective job-site communication tool. Supervisors, subcontractors, and craft personnel commonly use the Bar Chart as the schedule communication tool, because its format is easy and quick to use, but it is important to recognize that for a Bar Chart to be accurate, it should be based on CPM computations. Our discussion of scheduling begins with the complete development of CPM for a project.

Figure 1 – CPM Network

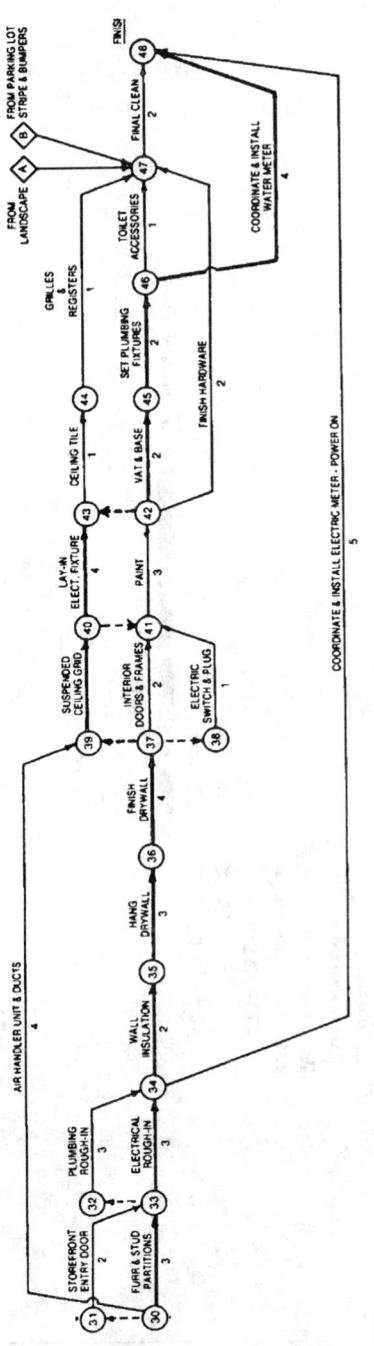

Figure 1 Continued – CPM Network

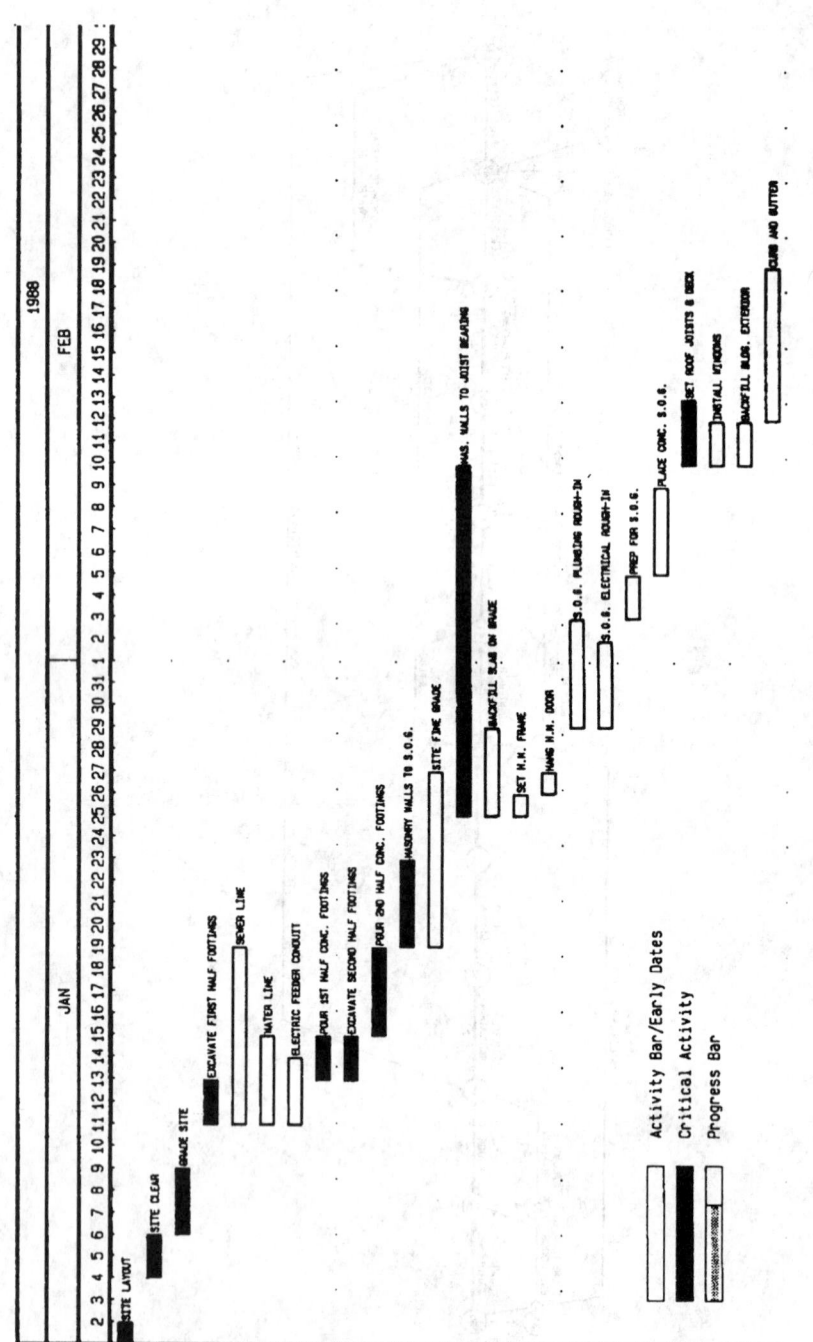

Figure 2

Scheduling

Scheduling involves the development of duration estimates for individual activities and the mathematical application of those durations to the network. Often, the evolution of a valid and accurate construction schedule involves revisions in duration estimates and network logic.

The actual project scheduling is accomplished by mathematical computation and the application of the calendar in the CPM process. Computer programs that are based on CPM scheduling can do the computations, can facilitate development of sub networks and short interval schedules, can evaluate the impact of alternate methods or plans, and will ease schedule updating. Computer applications have made the scheduling process simpler and more accurate.

The scheduler must work to an appropriate level of detail to achieve the level of project control desired. The time (duration) required to complete an individual construction activity is based on the amount of work required and the productivity of the labor and equipment to be used. The following is taken from the sample project in Figure 1.

The *Masonry Walls to S.O.G. (Slab on Grade)* Activity (15-16), *Masonry Walls to Joist Bearing Activity* (16-21) and *Top Masonry Wall* Activity (26-27), in Figure 1 have durations of 4, 12, and 2 days respectively, or 18 days total duration. The activities are *critical;* that is, they are on the critical path of activities that must be completed in sequence and on time in order for the project as a whole to finish on schedule. The project schedule depends on accurate duration estimates for these activities.

The estimator can reference unit costs under "Section 04220 Concrete Unit Masonry" in this book. Using lightweight 12" (300 mm) concrete masonry units (CMU), the productivity of one mason with one laborer or hod carrier per 8 hour day is:

Below grade	120-160 units per day
Above grade	110-130 units per day

If the quantity take-off lists,

Below grade 12" (300 mm) CMU	4,480 units
CMU grade to joist bearing	11,520 units
CMU above joist	1,920 units

then the duration estimates require an 8 mason crew:

CMU		Crew x Blocks per day		Duration
		(8 x 140)		4 days
4,480	÷		=	

11,520	÷	(8 x 120)	=	12 days
1,920	÷	(8 x 120)	=	2 days
Sequential Activity Duration			=	18 days

If only 6 masons are available, the durations will be extended:

CMU		Crew x Blocks per day		Duration
4,480	÷	(6 x 140)	=	5.3 days
11,520	÷	(6 x 120)	=	16 days
1,920	÷	(6 x 120)	=	2.7 days
Sequential Activity Duration			=	24 days
Project Delay (24 minus 18)			=	6 days

Alternately, if 12 masons are available, the durations will be shortened:

CMU		Crew x Blocks per day		Duration
4,480	÷	(12 x 140)	=	2.7 days
11,520	÷	(12 x 120)	=	8 days
1,920	÷	(12 x 120)	=	1.3 days
Sequential Activity Duration			=	12 days
Project Acceleration (18 minus 12)			=	6 days

The construction schedule includes activities that will be completed by the contractor's own forces and activities that will be accomplished by subcontractors. The method of determining the duration of the individual activities is the same regardless of the craft, equipment, or material required. As in the cost estimate, an accurate and complete determination of work quantities is essential in order to prepare a valid construction schedule.

The productivity of the labor and equipment is the essential element in the activity duration estimate. Productivity is a contractor's prime project risk; when the field productivity is less than estimated, cost overruns occur. Many contractors use the schedule time duration as their primary check of the labor estimate. As an example, assume that the contractor's labor estimate for the sample project in Figure 1 is as follows:

Item	Quantity	Unit Labor Cost	Total Labor Cost
5/8" (16 mm) Gypsum Board	19,360 SF	$0.11 per SF	$2,129
Fireproof Columns	1,430 SF	0.22 per SF	315
			$2,444

The contractor's schedule provides 3 days to complete the Hang Drywall Activity (35-36):

$2,444 ÷ 3 days = $815 per Activity Day
$16.97 per manhour x 8 hours per day = $135.76 per Man Day
$815 per Activity Day ÷ $135.76 per Man Day = 6 Worker Crew

The total labor cost estimate budgets a crew of 6 carpenters given the 3-day duration. The contractor can evaluate the labor cost estimate by considering the amount of work normally accomplished by a 6-worker crew in 3 days as compared to the amount of work required for the project.

The CPM process of transforming a plan into a schedule is based on computations summing the individual activity durations. An accurate estimate of each individual activity duration is critical to development of an accurate construction schedule. Since the activity duration is calculated based on the amount of work required to complete the activity and the anticipated productivity of the crew and/or equipment, the scheduler must have and use good information. The amount of work for each activity can be quantified by methods described elsewhere in this book. Likewise, anticipated productivity levels for a number of crews and tasks are also included.

CPM (and PERT) Computations. The computations used to develop construction schedules from the planning network are simply summations of the durations of interrelated activities. The early start time in work days or calendar dates is the earliest time that the activity can commence as soon as all preceding activities can be completed; that is, the early start time for each activity is dependent on the finish time of the last preceding activity to be completed.

As an example of how a CPM logic network is developed, a short interval job for reinforced concrete slab has been subdivided into nine activities with their respective durations and activity numbers:

Activity	Duration	Number
Start	0	10
Excavate	5	20
Build Forms	4	30
Procure Reinf. Steel	15	40
Fine Grade	2	50
Set Forms	3	60
Place Reinf. Steel	1	70
Place & Finish Concrete	2	80
Cure Concrete	7	90

The sequence of events is as follows. The procurement of reinforcing steel, the excavation, and the building of forms are all opening activities that can proceed at the same time and independently of one another. Fine grading follows excavation, but forms cannot be set until both the excavation and form building are completed. The placing of reinforcing steel cannot start until fine grading, form setting, and steel procurement have been completed. Concrete is placed and finished after the steel has been placed, and curing is the terminal operation. Figure 3 shows a precedence diagram for the job.

The scheduler must determine four limiting times for each network activity. The *early start* (ES) of an activity is the earliest time the activity can possibly start, allowing for the time required to complete preceding activities. The *early finish* (EF) of an activity is the earliest possible time by which it can be completed; it is determined by adding the activity's duration to its early start time. The *late finish* (LF) of an activity is the very latest it can finish and still allow the project to be completed by a designated time or date. The *late start* (LS) of an activity is the latest possible time that it can be started and still allow the targeted completion date of the project to be met; the LS is obtained by subtracting the activity's duration from its LF time.

The network computations to sum individual activity durations are known as *forward pass* and *backward pass*.

Forward Pass. The early start and early finish computations proceed from project start to project finish, left to right. Each activity starts just as soon as the last of its predecessor activities is finished. The ES of each activity is determined first. The project must start on Day Zero and that value is entered in the upper left hand corner of its activity box. (Notice in Figure 3 that each activity has its own box.) The EF is then obtained by adding the activity duration to its ES value; the EF is recorded in the upper right corner of the activity box. Because Activity 10 has a duration of zero, the EF value is zero. Activities 20, 30, and 40 can all start after Activity 10 is completed. Because Activity 10 has an EF equal to zero, each of the following three activities can start as early as Day Zero. The EF of each of these three activities is then obtained by adding their respective durations to the ES value of zero. All remaining forward pass calculations for the job continue in the same manner, moving from left to right on the network diagram.

Activity 70 is an example of a *merge activity*. Its start depends on the completion of two or more antecedent activities; in this case, it is Activities 50 and 60. The computation rule for a merge activity on the forward pass is that its earliest possible start time is equal to the latest (or largest) EF value of the activities immediately preceding it. Figure 4 shows the completed forward pass.

Backward Pass. With the forward pass complete, project calculations are now "turned around" and a second set of calculations is performed to compute the late start (LS) and late finish (LF) times for each activity. These calculations start at the projected finish of the project and proceed backward through the network in a right to left manner. The supposition during backward pass is that each activity finishes as late as possible without delaying project completion. The LF value of each activity is obtained first and is entered into the lower right portion of the activity box. The LS is obtained by subtracting the activity duration from the LF value and is entered in the lower left corner of the activity box.

Activity 20 is known as a *burst activity*, which has more than one activity following it. The rule for burst activities is that the LF value for such an activity is equal to the earliest (smallest) LS for the activities that follow it. The backward pass computation proceeds until the start of the project is reached. A completed backward pass is shown in Figure 5.

Figure 3

Float. Time leeway exists in the scheduling of some activities but not in others. This leeway is a measure of the time available for a given activity above and beyond its estimated duration. This extra time is called *float*. There are two types of float.

Figure 4

The *total float* of an activity is obtained by subtracting its ES from its LS time. Subtracting the EF time from its LF time gives the same result. The total float of an activity is entered into the activity box in the space between ES and EF values. The importance of total float is with respect to a path of activities. The activities in a chain share total float. The use of total float for one activity will reduce it by a like amount in others.

The *free float* of an activity is found by subtracting its early finish time from the earliest start time of the activity or activities that directly follow it. Free float is the amount of time by which an activity can be delayed without affecting the early start of the following activity, and also affecting any other activities in the network.

An activity with zero float is called a *critical activity*. An activity with no float time cannot be delayed, or otherwise, the project is delayed. These activities are on the *critical path* of the schedule network. The significance of float is that it indicates the degree to which an activity is critical; that is, it measures the amount of time that an activity can be delayed without hindering scheduled completion of the project.

Figure 5

PERT Computations. The Program Evaluation and Review Technique is use on some complex projects of long duration. Pert is similar to CPM in that it is based on a network diagram and the computations are developed from activity duration estimates. PERT, however, uses three separate estimations for the duration of each activity:

1. Optimistic Activity Duration
2. Probable Activity Duration
3. Pessimistic Activity Duration

The project and activity chain duration forecasts can then be computed using a variety of anticipated conditions given a single network diagram, and a range of "what if" situations can be easily evaluated. However, the method is complex and is not necessary on less complex projects of shorter duration.

Project (Subproject) Durations. The principal objective of the planning and scheduling process is to forecast the time required to complete the project and portions of the project. As shown in the previous example, the duration of a particular chain of activities is the sum of the durations along the longest sequential path of activities.

Those activities are the critical activities in that if any delay occurs in any of the activities, the projected completion will be delayed. In the sample project, it was forecast that 18 days will be required from the beginning of excavation/form building/procurement through the placement and finishing of concrete.

Contingency Time. Planning and scheduling of a project must include consideration of the impact of weather and other potential delaying factors. The General and/or Supplemental Conditions of the Contract outline requirements for the performance and time extensions; these requirements are a principal consideration in preparing a realistic project schedule.

The development of individual activity durations is based on the concept that the actual productivity for each activity can be monitored and controlled to that duration. The actual time required for certain activities is affected by the time of year (and hence, weather conditions) when the work is scheduled for; this is not always known at the time of planning.

The best and most consistent method of planning for potential and anticipated extra days, for which contract time extensions will not be permitted, is to include time contingencies as a separate and specific consideration. The contractor should review a particular chain of activities that is likely to be delayed, review the overall duration of that chain of activities, anticipate the probable time impact, and include a time contingency as a separate activity at the end of the chain.

This process places float on the critical path to accommodate non-productive project time and assures that the scheduled start date for activities occurring later in the project are more accurate than if contingency time were not included.

Schedule Revisions. The initial CPM schedule is usually developed as the early start schedule, which anticipates that each activity will start on the computed early start date. The resource requirements for accomplishing an early start schedule are most often uneconomical and inefficient.

Individual activities and chains of activities with float can start later than the computed early start date without delaying the project. The actual project schedule is usually a revised early start schedule in which start dates of non-critical activities are adjusted by using the float time. When the project is in progress, adjustments in productivity and project conditions may require revisions in activities that have been completed or have not started. The contractor can adjust the resources and revise the duration estimate for an activity or chain of activities, or the network logic can be revised to shorten or lengthen the chain of activities.

The purpose of the initial schedule is to develop an effective resource utilization plan. The schedule updates provide the information to revise the schedule so that it accomplishes the specific needs and objectives of the contractor.

Milestone Schedules. A milestone schedule lists the dates anticipated for the start or completion of key and critical project activities and work sequences as a measurement of project progress. Although not valuable as a resource management tool, a milestone schedule is excellent for progress communication with concerned parties, such as architect or owner, who do not have frequent project interaction.

The milestones are events rather than activities, and therefore, the schedule date for the milestone most often indicates the beginning or end of some segment of the construction project.

Materials Management. Proper management of materials is essential to project progress. Whenever labor is employed on an hourly basis and short-term layoffs are undesirable, the lost productivity from waiting for or double handling of materials is expensive. Unless that loss is anticipated in the construction cost estimate, profits are reduced.

One category of materials that requires special attention is items that require the preparation, submission, and approval of shop drawings and submittals. Many contractors use a spreadsheet approach to management of procurement activities that involve an approval process. The process calls for all items requiring shop drawings or submittals to be identified as early in the construction phase as possible and for the procurement process to commence as soon as feasible. The expediting effort includes time management of the activities of both the vendor (preparation) and the project designer (approval).

The procurement activity for materials required in the initial segment of construction (e.g. reinforcing steel, anchor bolts and templates, approval of borrow/fill material) are often on the critical path. This portion of the work is also the most sensitive to adverse weather and the associated costs of the resulting delay. Attention to prompt ordering, fabrication and shipping time agreements in purchase orders, and expediting is essential to successful material management.

The site utilization plan, storage and security provisions, weather protection, and the cost and impact of material re-handling must be considered in sequencing delivery of material with the installation activities. Timely delivery is the arrival of materials on the project site so as to be installed undamaged at the appropriate time.

Monitoring

A common excuse to justify informal scheduling in lieu of a formal process is that construction schedules too often become invalid due to continuing changes in conditions. Until the recent availability of economical and efficient computer hardware and software packages, the updating process required for construction schedules prohibited prompt and frequent information for resource management needs. There continues to be some resistance in the field to formal scheduling.

The key to successful scheduling is prompt, accurate, and appropriate reporting. The essential data for scheduling falls under the following categories:

1. Delaying Factors
2. Activities Started
3. Activities Completed
4. Ongoing Activities
5. Crew Sizes

The data input and processing is most often accomplished weekly, after reviewing work progress over the past week. When combined with cost records, and more specifically with labor cost, the resulting information, reports, and forecasts can be integrated into a project management system for project control.

Material procurement and delivery is often the culprit in project delay. The contractor must anticipate and prepare for handling and storage in the event of early material arrival and additional labor costs in case of late arrival. Material management, including control of the shop drawing and submittal approval process, is a key element of project time management. It should be noted that off-the-shelf software for scheduling often requires the support of specific task electronic or manual spreadsheets for applications such as material management.

Collecting Information. The level of detail in the schedule indicates the level of detail of information to be collected for updating and monitoring the schedule. The activity durations are indicative of the frequency of information collection and reporting. Computers are beneficial in processing reported project progress and forecasting remaining durations. The principal time items to review at any point in the project are:

1. Anticipated Delays
2. Project Productivity
3. Time Remaining on Current Activities

The key to a successful schedule updating effort is to collect information on as simple a basis as possible and to collect only the information needed to process time and cost reports. Given the variety of information reporting efforts available from different project management teams, the collection system must be well-defined and consistent.

Reports. When CPM is a requirement of the general or supplemental conditions of a project, an updated report is required with each periodic, usually monthly, pay request. But a useful time management and resource control system as well as cost control reports must be more frequent, most often weekly. A good reporting system will accurately forecast the remaining duration of the ongoing activities.

A percent complete computed by time rather than cost for each activity is an essential element in the collection of information. A standard approach is to update the project schedule using the same scheduling system (Bar Chart, CPM, PERT) originated at the start of the project.

Schedule Updates. The concept of project time management is based on resource management to complete project activities in the sequence and within duration planned. A good system for resource control is one that provides for "management by exception" and for advance notice of delays or other time factors.

The principal complaint of many builders and owners is that the schedule is outdated before it is even seen. Infrequent updates are not useful to protect management. A bar chart allows for write-over updating on a current basis. CPM requires reprocessing. The update report provides vision of the impact of current progress on upcoming resource needs. Looking forward two weeks or even two months in advance will result in the opportunity to anticipate the delay impact of individual problems and to reduce the cost. Changes and occurrences in the field that depart from the assumptions in project planning result in the need for continual updating of the project schedule. Ideally, the impact of variance from planned productivity would be dated on the short interval schedule and automatically integrated into a project milestone schedule.

Schedule Recovery Plan. The update anticipates the impact of recent project events on the scheduled completion of the project and on individual milestones. When delays are forecast, it is often desired, or even required, to accelerate the project so as to get back on schedule; the first step in the process is to identify, review, and make decisions based on alternate schedule recovery plans. The concept of schedule recovery is to complete work sequences within the time frame that was originally planned for certain chains of activities. Given that the duration is based on the amount of work required or remaining and the amount of available resources, the alternative duration reduction methods include:

1. Increase manpower or crews
2. Add crews
3. Add equipment
4. Work overtime—Extra hours or days
5. Multiple shifts

Obviously, the cost of completing the work at an accelerated rate often exceeds the cost of normal rate work. The schedule recovery plan is the method of comparing the increased cost of acceleration for individual activities and chains of activities. The ability to have control of scheduled completion at the most efficient cost is founded in the ability to promptly and accurately implement schedule recovery when actual project conditions differ from those anticipated.

Resource Planning. The objective of project management is the effective use of resources through the duration of the project. The resources include capital, labor, material, equipment, and subcontractors. The planning of resources for efficient use is required to successfully and profitably complete the project. The material management plan must control not only the quality and quantity of materials, but must identify when the materials are needed and how materials will be handled and stored on site. One of the principal causes of delay and reduced productivity is that materials are unavailable when needed. The project schedule is the control tool for material scheduling. Likewise, equipment must be available when required for the scheduled duration of the activity or chain of activities. The equipment plan is an essential cost control tool.

Labor productivity is critical to the financial success of the project. The contractor, in establishing the estimate and activity duration, assumes productivity of the crew or crews. The resource control is based on accomplishing the work within the scheduled time with planned crew.

Subcontractors must often perform on a number of sites for several different contractors simultaneously. The contractor needs a commitment to the schedule from all subcontractors. The subcontractor needs advance notice of schedule requirements and any changes in the schedule. The management of the subcontract resources is critical to the successful accomplishment of schedule requirements.

TOOLS AND EQUIPMENT

There may be no other part of contracting that puts the estimator's ability to visualize future job conditions to test more than planning for equipment and plant. There was a time when the estimator would only include that equipment specified in the general and special conditions, such as barricades, scaffolds, dust chutes, and the like. It was assumed that each subcontractor would furnish the additional equipment he might need. This is still true in certain trades. The excavator bases his bid on furnishing and using specific equipment that allows him to complete the work expeditiously; the concrete firm furnishes its own mixers and vibrators; the mason provides his own saws and grinders.

But all trades will have certain needs in common. The estimator should foresee these needs and lump all shared equipment under the responsibility of the general. This eliminates duplication of machinery and operators, unnecessary transport to and from the site, and the cost of setup and dismantling. Further savings will be realized by taking advantage of reduced rental weekly and monthly rental rates that apply to extended periods. Further, the more equipment the general sets up, the more control he has over the safety conditions at the job site. Many a contractor has been fined, or worse, has had his job closed down, because a government inspector ruled job conditions unsafe.

Caution

Owning, renting, or using tools and equipment subjects the contractor to compliance with specific safety requirements, which may be imposed by cities, states, and the federal government. It is not unusual to detect considerable conflict or variance in the degree of protection required under the several sources of regulations. In such instances, it has been held that the contractor is required to comply with the maximum requirement.

A national safety guide is the Occupational Safety and Health Act (OSHA), established by the Department of Labor in 1970. All tools and equipment used on a construction project should meet OSHA standards. Purchase orders or rental agreements should contain a specific clause relative to this compliance. Failure to meet the provisions of the safety laws can subject the contractor to fines and costly litigation.

Determining Hourly Rate

The general contractor should consult with his subs to determine what their needs will be and then tell them what he will be furnishing. He may then have the subs deduct these costs from their bids or make an agreed upon back charge for rental rates.

Whether the equipment is rented or owned, the estimator should charge for it at a fixed hourly rate. This rate would include the basic rental plus delivery, setup, and dismantling charges. It would then be divided by the working hours that the equipment stayed at the site. In some cases one might elect to charge only for the hours that the equipment was in use at the job site. The remainder of the time would then be charged to office overhead. Operating costs would be added as a separate item only for the actual in-use hours.

Equipment rental rates vary greatly throughout the country, depending on local practices and conditions. They might also vary according to how the equipment is used, such as for hard service in rocky soil or for use in sand.

It is standard practice to base rates upon one shift of 8 hours per day, 40 hours per week, or 176 hours per month for a 30 consecutive day period. In some urban areas, union building trades work 7 hours per day, 35 hours per week.

The lessor usually bears the cost of repairs due to normal wear and tear on non-tractor equipment. On rubber-tired hauling and tractor equipment, the lessee may be required to bear all costs for maintenance and repair.

The equipment rented should be delivered to the lessee in good operating condition and should be returned to the lessor in the same condition as delivered, less normal wear.

National average rates quoted herein are f.o.b. the lessor's warehouse or shipping point. The lessee pays all freight or delivery charges from shipping point to destination and return. Lessee also pays all charges for unloading, assembling, dismantling, and loading where required.

Rentals are usually paid in advance and subject to the terms and conditions of the lessor's rental contract. Rates do not include contractor's insurance, license fees, sales taxes, or use taxes.

Where the contractor owns the equipment, the estimator must arrive at an hourly cost to substitute for the basic rental. To do so, he must consider the following:

a) original capital investment
b) anticipated life
c) depreciation
d) estimate of maintenance
e) estimate of taxes, insurance, interest, and storage

Items c), d), and e) are expressed in percentages of the original capital investment, item a). The estimator will need the cooperation of the firm's accountant to determine the hourly cost figure, as well as some knowledge of the possible future use of the equipment. The capital investment cost should include all original delivery, test, and start-up charges. The expected life may be based on Internal Revenue tables. It is usually entered as 2000 hours per year (40 hour week, 50 week year).

Depreciation cost is the original capital investment less any resale or scrap value. This is often so small that depreciation on all but highway trucks may be figured at 100%. However, in times of high inflation, for equipment held only a short time, the inflation rate may approximate the depreciation rate. The equipment would appear as having zero depreciation.

Maintenance allowances will vary from 40% to 100% of the investment cost, but most equipment will fall in the 65% to 75% bracket. Those items falling below 50% would include air tools, gas engines, small motors, mortar mixers, vibrators, drills, screw jacks, small power tools, towers, small tractors, trailers, and trucks. Those items for which over 75% of investment cost should be used would include bituminous equipment, hoppers, crushers, derricks, electric generators, hydraulic jacks, pumps, and rollers. Taxes, insurance, and interest percentages should be worked out with the accountant. Ten percent used to be an acceptable figure for these items, but with today's rates 15% would probably be more realistic, and 3% is generally added for storage. A formula to arrive at this entry would be:

$$\frac{p^* (n + 1)}{2} + p^{**} \text{ (useful life)} = \text{total percentage}$$

p^* is percentage for taxes, insurance, and interest.
p^{**} is percentage for storage.

If the useful life is five years, p^* is 15% and p^{**} is 3%, then the total percentage to be entered under item e) would be:

$$\frac{15(5+1)}{2} + 3x5 = 45 + 15 = 60\%$$

Once items a) through e) are determined, the hourly rate can be figured as follows:

$$\frac{\text{capital investment}}{\text{anticipated life in hours}} \times \text{sum of \% for items c), d), and e)}$$

As an example, assume the contractor has purchased a $100,000 piece of equipment. The following is a reasonable cost projection:

a) capital investment..$100,000.00
b) anticipated life 5 yrs at 2000 hrs... 10,000 hrs.
c) depreciation $5,000 scrap value...95%
d) maintenance...70%
e) taxes, insur., int. & stor..60%

Hourly Use Charge =
$100,000 ÷ 10,000 hrs. = $10.00 per hr.
$10.00 x 95% + 70% + 60% = $22.50 per hr.

Material Handling

Manpower. Labor is too expensive today to use where machines can be used instead, but sometimes it is necessary. A laborer is generally rated as being able to handle a maximum of 100 lbs. (45 kg) For loads over that, or for oversized material, additional help must be scheduled.

One worker should be able to unload and pile on a site 300 to 500 standard bricks per hour using his hands, or up to 1000 using special tongs. A forklift truck handling a palletted load can handle well over twenty times as many units in the same time and can provide a wider coverage than a man. One can allot considerable expense for temporary access for mechanical equipment, if much material is involved, and still show substantial savings over using manual labor.

Labor for unloading and piling other materials can be estimated as 80-120 per hour for sacks of cement; 100-200 per hour for block or tile; and 1000-2500 b.f. (190 to 500 50-mm pcs.) per hour for lumber.

Hand shovelling of bulk materials will vary as to whether the load is transferred at the same or lower elevation or raised to a higher one. The following table is based on the amount one worker can handle in one hour under dry conditions.

	Even Transfer (tons/hr)	Even Transfer (metric t/hr)	Upward Transfer (tons/hr)	Upward Transfer (metric t/hr)
Sand	3.00	2.70	2.60	2.35
Stone	2.12	1.90	1.80	1.65
Gravel	2.30	2.10	2.00	1.80
Top soil	2.60	2.35	2.35	2.15

Wheelbarrows. Distribution over the job site will still often require hand powered wheelbarrows, trucks, and buggies. Rental costs can be figured at $12.00 per day or $50.00 per month.

Power buggies are widely used on larger construction projects and they have proved to be a more economical method of transporting materials. The walk-behind models rent for $180.00 per day or $550.00 per month, not including gas.

Sizes, Weights and Capacities of Contractor's Wheelbarrows

Struck Capacity cu. ft.	Struck Capacity cu. m	Heaped Capacity cu. ft.	Heaped Capacity cu. m	Approx. Weight lbs.	Approx Weight kg	Type of Materials Used for
2-1/2	0.070	3-1/2	0.100	74	35.40	Dry Materials
3	0.080	4	0.113	78	35.40	Dry Materials
3-1/2	0.100	4-1/4	0.120	75	34.00	Concrete-Mortar
4	0.113	5	0.142	79	35.80	Concrete-Mortar
4-1/2	0.127	6	0.170	83	37.60	Concrete-Mortar

It requires about 150 shovels (No. 2 shovel) for one cu. yd.

Chutes and Conveyors. The cheapest energy the estimator can plan for is gravity. He should see that the job is laid out so that, as much as possible, material distribution will be by gravity. Straight line chutes can be rented in

10' (1 m) sections for $25.00 to $35.00 per week or may be built up out of plywood at the site. Prefabricated metal vertical chutes will cost $40.00 to $60.00 per ft. (300 mm) erected. Gravity steel roller conveyors can be purchased in 10' (3 000 mm) lengths for around $550.00 for 18" (450 mm) widths. Powered conveyor belts will run two and a half times as much, plus one must add for operating costs.

Portable Belt Conveyors. Some applications require the use of portable belt conveyors of various sizes and capacities. The use of this equipment is flexible and does not limit the user to a fixed receiving or discharging point. When used in series, the flexibility to carry materials over long horizontal distances, over obstructions, and around corners is even greater.

Portable conveyors are available in boom lengths to 120 ft. (36 000 mm) and belt widths of 18", 24", 30", and 36" (450 mm, 600 mm, 750 mm, and 900 mm). Swivel wheels permit building of high capacity radial stock piles or discharge into any one of several bins or hoppers. Both single- and multi-deck vibrating screens may be used at the discharge end to separate the material into two or three sizes. Feeder-trap and feeder-hopper units mounted at the tail end permit charging by front end loaders, dozers, cranes, draglines, shovels, or other conveyors.

Portable conveyors are used for conveying earth, clay, sand, gravel, stone, bulk cement, and concrete from stock piles or to convey and elevate materials to trucks. They can be used for conveying wet concrete from mixer to elevated structures where the incline is not too steep.

Conveyors are furnished with either plain or cleated belts, but the cleated belts should not be used with wet materials. Listed below are capacities of belt conveyors, wheelbarrows, mixers, and concrete carts.

Capacities of Standard Conveyors in Tons Per hour (Based on
the Assumption of Uniform and Continuous Loading)

Belt Width	Belt Speed ft. per min.	Earth Dry	Clay Dry	Sand	Stone Gravel	Bulk Cement
18"	300	112	92	150	160	150
24"	300	215	185	290	315	290
30"	300	350	300	470	510	470
36"	300	520	450	690	750	690

Capacities of Standard Conveyors in Metric Tons Per hour
(Based on the Assumption of Uniform and Continuous Loading)

Belt Width	Belt Speed meter/min.	Earth Dry	Clay Dry	Sand	Stone Gravel	Bulk Cement
450 mm	90	100	85	140	145	140
600 mm	90	195	165	265	286	265
750 mm	90	320	270	425	465	425
900 mm	90	470	410	625	680	625

Hoisting Towers

Tubular steel hoisting towers are used extensively by contractors in all parts of the country for hoisting all types of building materials. They are also used for concreting because of the ease with which they are erected and dismantled, their great strength and stability, and their economy. Helicopters are playing an increasingly important role in modern construction, particularly in materials handling.

Tubular hoisting towers are designed for handling material and concrete, with or without chuting. They are built in any height and with one or more compartments. Their durability, low maintenance, and remarkable adaptability make them ideal as low cost equipment for contractors.

A tubular hoisting tower consists of a bottom section, the necessary number of intermediate tubular sections, each 6'-6" (1950 mm) in height, and a top section containing the cathead with sheaves. All intermediate sections of a given type of tower are identical, and the top and bottom sections can be attached to any one of them.

The single and double (two compartment) towers are classified as follows:

	Heavy Type	Light Type
Size of legs	3-in.	77mm
	2-in.	50 mm
Maximum height	350'-0"*	105 m
	201'-0"	60 m
Live load capacity	5,000 lbs.	2,265 kg
	3,200 lbs.	1,450 kg
Maximum size concrete bucket	35 c.f.	1.00 cu.m

*Higher towers require special material.

The light type towers are known as class 3200 and the heavy type towers are known as class 5000.

The size of the cage platform in the single tower is 5'-10-1/2" x 7'-6-1/2"(1760 x 2235 mm). For double towers the size of one cage platform will be 5'-10-1/2" x 7'-6-1/2" (1760 x 2235 mm) and the other platform will be 5'-10" x 7'-2" (1750 x 2150 mm).

Erecting and Dismantling Steel Tubular Towers. When estimating the cost of erecting and dismantling a steel tubular tower, the number of manhours involved depends upon the height of the tower and the general conditions. For instance, it will take considerably longer to erect a tower 208'-0" (60 m) high than one 65'-0" (20 m) high, not only in time consumed but also in manhours per section, taking into consideration the amount of material involved as well as the height to which the equipment must be raised.

Time given includes erection, dismantling, and handling in manhours per lin. ft. of height.

Light Type Single Tubular Towers ....1-1.5 hr.
Light Type Double Tubular Towers ....1.5-2 hr.
Heavy Type Single Tubular Towers..1.5-2 hr.
Heavy Type Double Tubular Towers...2-2.5 hr.

Prices for light or heavy duty hoisting towers should be obtained from the manufacturer. Quotations for monthly rental charges can also be obtained for a specific project. A 50' (15 m) tower will rent for about $1,500 per month, with each additional 10' (3 m) section about $40 per month.

Scaffolding

Tubular Steel Scaffolding. The most versatile type scaffolding is galvanized steel tubing with a male fitting on one end and a female fitting on the other to form interlocking units. These units are then coupled together with a patented coupling device. It can be adapted to a variety of uses and its simplicity has found general acceptance among the building trades. Basically, it consists of these parts: interlocking steel tubes of various sizes, a base plate or caster, and an adjustable or right angle coupler. All parts are galvanized as a permanent protection against rust.

This galvanized carbon steel tubing or pipe is approximately 2" (50 mm) o.d.—1-1/2" (37 mm) standard pipe—and weighs 2-3/4 lbs. per l.f. (4 kg per m). Other sizes are 2-1/2" (62 mm) o.d.—2" (50 mm) standard pipe size—which weighs about 3-3/4 lbs. per l.f. (1.70 kg/300 mm).

Standard lengths for tube are 6'-0", 8'-0", 10'-0", 13'-0", 16'-0", and 20'-0" (1.8, 2.5, 3.0, 4.0, and 5.0 m) and for pipe are 6'-0", 8'-0", 10'-0", and 13'-0" (1.8, 2.5, 3.0 and 4.0 m). Other sizes can be obtained, but those mentioned are the sizes generally used by the building trades. Each section is fitted with a bayonet locking device.

This locking device between members is permanently affixed to the tubes, which insures a perfectly rigid attachment that has been tested to a load of 20,000 lbs. (9070 kg) in tension. The assembled tube and fittings make up a complete unit so designed as to have flush joints.

Standard couplers are made of drop-forged parts, except for the steel hinge pins on which the catch bolts hinge. The most widely used couplers are 2" x 2" (50 x 50 mm), both adjustable and standard, and 2" x 2-1/2" (50 x 62 mm), both adjustable and standard, and 2-1/2" x 2-1/2" (62 x 62 mm), adjustable and standard. This type of scaffold is the most versatile of all built-up scaffolds.

Estimating costs of erecting scaffolding depend on many conditions: the type of job to be done, whether interior or exterior; ground conditions; height and width, as well as load to be carried; and length of time it will be in use. Costs of purchasing or renting "exterior tubular steel" scaffolding are quoted on the square footage of wall area actually covered and may be obtained from the equipment manufacturer. Low buildings will cost about $40.00 per hundred s.f. (10 m² installed, rental). For buildings over 5 stories, it will cost about $60.00 per hundred s.f. (10 m²) of surface installed, rental).

Costs for purchasing or renting "interior tubular steel" scaffolding are quoted on the number of cubic feet (cubic meters) actually occupied by the scaffolding and may be obtained from the manufacturer or supplier of the equipment.

Sectional Steel Scaffolding. Sectional steel scaffolding or prefabricated scaffolding is readily assembled and installed and is well adapted to various kinds of building and maintenance work. Various sizes of 5'-0" (1500 mm) wide frame units, trusses, and ladder scaffold components make it simple to erect a rigid and safe scaffold under any conditions.

The scaffold components are manufactured to resist corrosion and rusting. Rugged construction prevents damage during assembly and dismantling. This is a lighter type scaffold than the tube-and-coupler type and has an approved working height limit of approximately 125'-0" (40 m).

The tubing used in the frames and trusses is carbon steel, which reduces the weight without sacrificing strength.

Sectional steel scaffolding has found wide acceptance among bricklayers, plasterers, and other craft workers, because it adapts to rolling towers and to exterior and interior construction and maintenance.

Tubular Steel Scaffolding for Exterior Work

The spacing between frames may be either 3'-0", 5'-0", 6'-0", 7'-0", or 10'-0" (900, 1500, 1800, 2100, and 3000 mm), and the frames vary from 3'-0" to 10'-0" (900 to 3000 mm) in height, allowing for a wide variety of uses. Braces are attached to the frames by means of wing nuts or automatic locks. Costs of the component parts of sectional steel scaffolding may be obtained from the manufacturer or supplier.

Suspended Scaffolding

The most economical method of scaffolding for masonry work on buildings with a structural steel or reinforced concrete skeleton framework is suspended scaffolding. The working platform is supported by regularly spaced scaffolding machines, each consisting of a pair of cable drum hoist mechanisms with connecting steel putlog, suspended by steel wire rope from steel or aluminum outrigger beams anchored to roof or floor framing above.

Standard width machines accommodate a 5'-0" (1500 mm) width of solid planking. Machines are available to accommodate 8'-0" (2400 mm) wide decking to permit wheeling of material on scaffold platform. Machines are also available for mechanized material handling with the inboard cable drums set back from the wall to provide a 5'-0" (1500 mm) wide platform for material handling traffic and stockpiles on one side and a separate 20" (500 mm) wide working platform for masons on the other. All machines are available with or without steel frames to support overhead protective planking. For maximum safety, overhead planking should be used, together with toe boards and guard rails along outer edge of platform.

Scaffolding machines are usually spaced 7'-0" (2100 mm) apart, permitting the use of 16'-0" (4800 mm) scaffold plank. A ratchet device on the drum hoist mechanisms provides positive locking while still permitting easy raising or lowering of scaffold to maintain bench-high location of work for maximum efficiency and best workmanship.

Outrigger beams are usually anchored to the roof construction 7'-0" (2100 mm) on centers, if building is not over 150'-0" (45 m) high, the maximum lift capacity of scaffolding machines. For structures exceeding 150'-0" (45 m) high, one or more additional installations will be required on lower floors. There are various methods of anchoring beams to building framework, such as securing with steel straps to anchor bolts set in concrete slabs or beams, bolting to inserts set in concrete, or anchoring to beams or joists with U-bolts and bars through holes in concrete slabs.

Conditions affecting the installation of suspended scaffolding vary so widely it is impossible to give unit costs that will be of any value in estimating. The manufacturer and erector will gladly quote a firm price on any specific job.

Ladder Type Scaffolding. For interior scaffolding, particularly for low to medium height work, ladder type scaffolds are considered very efficient. The type with horizontal bracing is most desirable for interior work, because it spans over furniture, machinery, or equipment. Scaffolding that can be assembled without bolts or nuts can be erected or dismantled more quickly and easily. Platforms that can be adjusted for height permit workers to stand at the most efficient distance from their work, a feature that is important for electricians, umbers, heating and ventilating workers, acoustical erectors, painters, maintnance personnel, and others doing off-the-floor jobs.

Ladder type scaffolds are sold as complete scaffold units, which include two ladders, two platform support trusses, and one plywood platform (exterior type plywood, metal bound). The platforms are adjustable for height every three inches. Each unit is 6'-0" high by 6'-0" long (or longer if desired) by 2'-5" wide. They can be used separately, or in combinations to scaffold larger areas. Casters can be furnished either with or without locks.

One of the outstanding advantages of scaffolds of this type is their maneuverability over permanent or semi-permanent objects, such as church pews, machinery, or furniture. It isn't necessary to scaffold an entire room or auditorium because the scaffolds can be readily moved from place to place.

Scaffolds of this type sell at approximately $160.00 to $215.00 per unit, including the platform, plus the casters.

Sidewalk Bridges

Where construction is adjacent to heavy traffic areas and it is impracticable to reroute sidewalk traffic into the street, a sidewalk bridge must be erected to protect the public. Rigid standards must be met in the design and erection of these structures.

They are built of 4-1/2" (113 mm) o.d. steel columns, set on heavy timber mudsills and supporting 8" (200 mm) steel I-beams bridging the sidewalk. Beams are securely attached to columns by means of special steel column head castings. Columns are tied together longitudinally with two rows of pipe girts and with an occasional diagonal pipe brace as required. Wood timbers, cut to a taper and set to slope towards the street, are fastened to the cross beams with U-bolts to support 6" (150 mm) steel beams, spaced approximately 16" (400 mm) on center, running the length of the sidewalk bridge. Two layers of 2" (50 mm) or 3" (75 mm) planking are then laid over the beams, the bottom layer running laterally and the top layer longitudinally. A parapet for the street side may be added if desired.

Most contractors lease sidewalk bridges on an erected and dismantled basis, and the supplier usually quotes a lump sum figure for a specific job. Before submitting a bid, a firm quote covering this work should be obtained.

For preliminary estimating purposes only, under average conditions sidewalk bridges can be figured as follows:

Erection and dismantling, per l.f. (0.3 m)	$20.00-$36.00
Sidewalk bridge rental, per l.f. (0.3 m), per month	4.75
Erect and dismantle parapet, per l.f. (0.3 m)	9.00
Parapet rental, per l.f. (0.3 m), per month	2.00

Air Compressors

One sales and rental source for air compressors advertises over 1800 models. Add to it all the various attachments that are made to work with the compressor, and one can see one must give some thought to matching the unit to the tools for the most efficient operation.

Two and more tools can be operated from the same compressor. The rating of the compressor should be the delivery, but leaking hoses, poor tools, and adverse job conditions could lower the cfm output. In some gang operations not all units will be in operation at the same time so that a unit sized for two may be adequate for three, providing the compressor is set up for more than two hose connections.

The following list will give some idea of air requirements for the more commonly used tools:

Breakers
>35# - 30 cfm* (15.9 kg - 0.014 m³/s**)
>60# - 40 cfm (27.2 kg - 0.019 m³/s)
>80# - 50 cfm (36.3 kg - 0.024 m³/s)

Diggers
>20# - 20 cfm (9.1 kg - 0.009 m³/s)
>25# - 25 cfm (11.3 kg - 0.012 m³/s)
>35# - 35 cfm (15.9 kg - 0.014 m³/s)

Air Drills
>1" - 35 cfm (25 mm - 0.017 m³/s)
>2" - 50 cfm (50 mm - 0.024 m³/s)
>4" - 60 cfm (100 mm - 0.028 m³/s)

Wagon Drills
>3" - 160 cfm (75 mm - 0.076 m³/s)
>3-1/2" - 200 cfm (88 mm - 0.094 m³/s)
>4" - 250 cfm (100 mm - 0.118 m³/s)

Jack Hammers
>25# - 50 cfm (11.3 kg - 0.024 m³/s)
>45# - 100 cfm (18.1 kg - 0.047 m³/s)
>75# - 110 cfm (34.0 kg - 0.052 m³/s)

Chipping
>light - 20 cfm (0.009 m³/s)
>heavy - 30 cfm (0.014 m³/s)

Rivet
>5/8" - 25 cfm (16 mm - 0.012 m³/s)
>7/8" - 35 cfm (22 mm - 0.017 m³/s)

Circular Saws
>12" - 35 cfm (300 mm - 0.017 m³/s)

Chain Saws
>30" - 35 cfm (750 mm - 0.017 m³/s)
>36" - 35 cfm (900 mm - 0.017 m³/s)
>48" - 35 cfm (1200 mm - 0.017 m³/s)

Tampers
>35# - 30 cfm (15.9 kg - 0.014 m3/s)
>60# - 40 cfm (27.2 kg - 0.019 m³/s)
>80# - 50 cfm (36.3 kg - 0.024 m³/s)

Vibrators
>3/4" - 35 cfm (19 mm - 0.017 m³/s)
>1-3/4" - 50 cfm (44 mm - 0.024 m³/s)
>2-1/2" - 25 cfm (63 mm - 0.012 m³/s)
>3" - 45 cfm (75 mm - 0.021 m³/s)

5" - 80 cfm (125 mm - 0.038 m^3/s)

Wrenches (Impact)
5/8" - 15 cfm (16 mm - 0.007 m^3/s)
3/4" - 30 cfm (19 mm - 0.014 m^3/s)
1-1/2" - 75 cfm (38 mm - 0.035 m^3/s)

Drills
less than 1" - 50 cfm (less than 25 mm - 0.024 m^3/s)
more than 1" - 80 cfm (more than 25 mm - 0.038 m^3/s)

Air Nailers - 10 cfm (0.005 m^3/s)
Grinders small, light - 50 cfm (0.024 m^3/s) heavy duty - 80 cfm (0.038 m^3/s)
Hoist light duty - 150 cfm (0.071 m^3/s) heavy duty - 175 cfm (0.083 m^3/s)
*cfm = cubic foot per minute
*m^3/s = cubic meters per second

Compressor rentals will vary as to the condition and rated size of the equipment, but should be available for following approximations:

fuel	cfm*	m^3/s**	per day	per week	per month
gas	85	0.040	$35	$130	$380
"	100	0.047	40	130	400
"	125	0.059	40	145	440
"	150	0.071	40	150	470
"	175	0.083	40	170	500
diesel	250	0.118	70	340	1,000
"	375	0.177	90	430	1,300
"	600	0.283	130	670	2,000
"	750	0.354	150	750	2,300
"	1,000	0.472	200	1,000	3,000
"	1,200	0.566	270	1,300	4,000
"	1,600	0.755	280	1,400	4,200

*cfm = cubic feet per minute
**m^3/s = cubic meters per second

Tools, hoses, oilers, noise abatement controls, and other attachments must be added separately.

CHAPTER 2

SITE WORK

CSI DIVISION 2
02010 SUBSURFACE EXPLORATION
02100 CLEARING
02060 DEMOLITION
02200 EARTHWORK
02250 SOIL TREATMENT
02300 PILE FOUNDATIONS
02350 CAISSONS
02400 SHORING
02500 SITE DRAINAGE
02550 SITE UTILITIES
02600 PAVING AND SURFACING
02700 SITE IMPROVEMENTS
02800 LANDSCAPING

02010 SUBSURFACE EXPLORATION

When preparing estimates on any structures involving foundations or earthwork, there is a need to obtain all information available concerning the conditions and characteristics of the soils to be encountered in the subsurface. Frequently some of this information is made available on the working drawings as an aid to the contractors, but complete reliance on such limited information places the contractor at risk, especially when the drawings provide for a disclaimer of responsibility by the architect/engineer for the accuracy or completeness of the subsurface information. A site investigation is mandatory, because many conditions can be determined through adequate visual observations of the particular site involved and adjacent property. There can be discovery of previous stream beds, evidence of prior dumping of debris, existence of abandoned mining operations, existence of flood plains, appearance of rock outcroppings, and many other situations that would strongly influence the construction methods and the costing involved.

In urban areas records are usually available from utility companies and highway authorities concerning subsurface conditions affecting their installations.

It must be recognized that any penetration of the ground surface has attendant risk, and the accumulation of knowledge of the conditions to be encountered during excavation will tend to control the risk to a reasonable level thus avoiding exposure to a catastrophic situation.

Excavating test pits or caissons produces accurate information in cohesive soils, insofar as this method permits visual inspection of the soil in place, and samples for laboratory tests may be carved out of the pit walls in a relatively undisturbed condition. Costs of this work are high, however, and will be about the same as excavating costs for comparable pits and caissons.

However, most subsoil investigations require studies of the soil at greater depths than can be economically reached by test pits. For these projects, various methods of soil boring and sampling are used. A relatively small diameter hole is sunk into the ground from which soil samples are obtained.

Hand Auger Boring Method. The hand auger boring method, satisfactory for highway explorations at shallow depths, can be used only for preliminary investigations for foundations. Auger borings are used in cohesive soils and cohesionless soils above ground water. Depths up to 20 feet (6 meters) may be reached. Soil samples obtained through this method are disturbed to such an extent that little or no information is furnished about the character of the soil in its natural state. Where the auger hole is filled with water, sample disturbance is particularly acute. Auger holes are most useful for ground water determination.

Hand auger borings generally cost from $10.00 to $15.00 per lin. ft. ($32.00 to 50.00 per m) depending on the soil, the amount of boring necessary, and the diameter of the hole, which will vary from 4" to 8" (100 to 200 mm).

Split Spoon Sampling. A split barrel sampler known as a split spoon is frequently used for obtaining representative samples. This method is described in detail in ASTM Method D-1586.

In securing split spoon samples, the drill hole is opened and cleaned to the desired sampling level by drilling bits and wash water, or alternatively, the hole may be drilled with a regular power auger. Drill casing or drilling mud is used when it is necessary to prevent the soil above the sampling level from closing the drill hole. A drive pipe is attached to the upper end of the drill rod and a slip jar weight placed on the drive pipe.

The sampling spoon is then driven into the soil by repeated blows of a drop hammer. The number of blows required to drive the spoon are recorded in the logs. The spoon is then brought to the surface, and the sample is removed, classified, and placed in a glass jar.

The ASTM method specifies a 2-inch (50 mm) OD x 1-3/8 inch (33 mm) ID split spoon driven 18" to 24" (450 to 600 mm) by a 140 lb. (64 kg) weight falling 30 inches (750 mm). Various other sizes of split spoons and drop weights and lengths of drops are employed. The split spoon method is most useful in granular or dense soils.

Thin-Wall Tube Sampling. The method that yields the least disturbed samples is thin-wall tube sampling. This method is described in detail in ASTM Method D-1587.

In this method the hole is opened and cleaned as in the split spoon sampling method, but a thin-wall tube, usually with an outside diameter of

2", 3", or 5" (50, 75 or 125 mm) is then attached to the end of the drill rod and lowered to the bottom of the boring. The tube is then pushed into the soil by a hydraulic piston arrangement with a rapid and continuous motion.

The thin-wall tube is then raised to the surface, cleaned, labeled, and sealed to prevent loss of moisture. It is then taken to the laboratory for testing. Here it is usually cut into short lengths, and the sample in each length is ejected and tested.

Samples are normally taken at 5-foot (1500 mm or 1.5 m) intervals or every time a soil change is expected, determined by watching the overflowing wash water and feeling the resistance to penetration of the drill bit.

Cost of drilling and sampling is usually between $6.00 and $16.00 per lin. ft. ($20-53 per m) depending on the depth, diameter, and difficulty of drilling, plus on and off charges for the equipment and workers. Laboratory testing is performed at an additional cost, depending on what tests are needed--for density, moisture content, compression, or mechanical analysis.

Core Borings. In addition to soil samples, information is frequently wanted on the nature or thickness of underlying rock. Core borings are made for this purpose with rotating coring tools such as diamond drills, shot drills, or carbide bits. Continuous samples are generally recovered for examination and further testing. ASTM Standard D-2113, Diamond Core Drilling for Site Investigation, should be called for in performing core borings.

The cost of making core borings will depend upon the location of the work, quantity, and the difficulty encountered in getting to the holes.

The drilling costs per lin. ft. (300 mm) should be estimated at from $18.00 to $22.00 ($60-73 per m), with the lower costs on jobs of larger size. To the above prices, add the on and off charges for the equipment and workers and the testing charges.

Other Methods. There are many methods used to investigate soil conditions for foundation purposes. Some are reliable and sufficient to use for design purposes while others produce inadequate or dangerously deceptive data. In problem soils the gathering of the basic information under controlled conditions is important, but to obtain definitive and factual information there is a need for professional engineering interpretation. In such instances the best approach is to obtain the services of a soils engineer with the laboratory facilities to perform a complete subsurface investigation and to provide a written report of the findings and recommendations for remedy of the problems to be encountered.

Engineering and laboratory services are cheap. However, when balanced against the magnitude of potential loss that could be encountered without this advance knowledge, the cost is usually justified.

Engineering costs vary widely by geographic location and services required. There is enough competition to allow the contractor to solicit proposals from several engineering firms, who provide services that are tailored to the specific project.

Seismic resistivity, vane shear tests, cone penetration test, and nuclear methods are more frequently being used in subsurface investigations, performed by specially qualified engineers having the necessary equipment and regularly engaged in this type of work. Cost estimates will be tendered to the contractor predicated upon the services required.

02100 CLEARING

Clearing of a construction site by the removal of trees and vegetation has become more costly because of federal and state environmental protection regulations. Most areas prohibit the burning of trees. If the wood does not have a market value, the use of chipping machines is employed. This method requires additional personnel to cut the trees into manageable lengths, remove the branches, and transport them to the chipping machine. Some of the cost can be recovered by selling the by-product to nurseries and landscape companies to use as mulch. Stump disposal is very expensive, particularly where stumps cannot be buried at some location on the site and have to be transported to a distant disposal area.

Clearing, Cut & Chip	per acre	per cu.m
Light, 6" (150mm) Diam.	$1,650-2,090	$0.41-0.52
Medium, 12" (300mm) Diam.	2,750-3,300	0.68-0.81
Heavy, 16" (400mm) Diam.	3,850-4,950	0.95-1.22
Grub Stump Plus Haul Charge, per ea.	$27.50-55	

If chipping is not required, average prices can be reduced by 25%.

Selective Cutting Down and Removing Trees

The cost of cutting down and removing trees will vary with their size and the method used in cutting and disposing of them.

The following production times are based on a crew of 3 or 4 workers, cutting down trees by hand, removing branches, and cutting the trunk into 4'-0" (1.2 m) lengths. Labor costs of digging down around roots and removing a stump are given extra. It is assumed the limited quantity would not require chipping.

Approx. Dia. of Tree	Approx. Hgt of Tree	Labor Hours	Extra Labor Hrs. Removing Stump
8"-12" (200-300 mm)	20'-25' (6.0-7.5 mm)	9 to 12	8
14"-18" (350-450 mm)	30'-40' (9-12 m)	12 to 16	10
20"-24" (500-600 mm)	45'-50' (13.5-15 m)	20 to 24	12

If necessary to cut trunks and stumps into small pieces and load into trucks, add extra for this work. Add also for removal from the site.

Where many trees are to be removed, a power or chain saw will speed the work and reduce costs considerably. A chain saw will cut within 2 to 4 inches (50 to 100 mm) of the ground, but if the stump must be removed, it is usually done by hand.

Chain saws are gas-, electric-, or air-powered; 24", 36", 48" (600, 900, 1200 mm) and larger; and rent for about $40.00 per day.

Labor required using a chain saw, with 3 workers in the crew cutting trees, removing branches, and cutting the trees into 2'-0" (600 mm) lengths, should average as follows:

Approx. Dia. of Tree	Approx. Hgt of Tree	Labor Hours	Extra Hrs. Removing Stump
8"-12" (200-300 mm)	20'-25' (6-7.5 mm)	2-1/4 to 3	8
14"-18" (350-450 mm)	30'-40' (9-12 m)	3 to 4	10
20"-24" (500-600 mm)	45'-50' (13.5-15 m)	5 to 6	12

A faster and more economical method of removing trees is by using a bulldozer or loader, provided it does not cause damage to adjacent trees.

Trees up to 15 inches (375 mm) diameter or trees such as pine, which have small root systems, can be rapidly uprooted by raising the dozer blade or shovel bucket to maximum height and pushing against the tree. It may be necessary to strike the tree a few times to loosen it.

Trees with diameters of 24 inches (600 mm) or more, or trees like elms, maples, and oaks, which have heavy, wide-flaring root systems, usually require root cutting before they can be pushed over. First make a trench-cut with the dozer blade or shovel bucket across the roots on the side opposite the direction in which the tree must be felled. Backfill the trench and form a ramp up to the trunk so the machine can strike as high on the tree as possible. Unless the tree is exceptionally large or well-rooted, it can now be pushed over. If not, try cutting the roots on each side of the tree perpendicular to the first cut, and if necessary, the roots on the side toward which it will be felled.

Once the tree is felled, it can be pushed or dragged to a disposal area where it can be sawed into lengths for hauling or burning (where permitted).

A Word of Caution—Always watch for falling limbs. In the case of pines, if the whipping action is too great, the top may snap off and kick back on the machine or operator. Also, as the tree begins to fall, the machine must be backed up quickly so that the upturning root system will not come up under the machine, possibly damaging the unit or hanging it up.

An experienced operator, using a medium to large tractor, 80-HP (60 kg) and up if crawler type, 160-HP (120 kg) and up if rubber-tired, with ground conditions affording good traction, and no problems regarding steep slopes, should fell trees at the following rates:

Approx. Diam. of Tree	Approx. Height of Tree	Minutes Req'd. for Felling
8"–12" (200–300 mm)	20'–25' (6–7.5 m)	5 to 10
15"–24" (375–450 mm)	30'–50' (9–15 m)	15 to 20
27"–36" (675–900 mm)	55'–75' (16.5–22.5 m)	25 to 30

Add for pushing felled trees to disposal area and for disposing of them.

Stripping & Storing Topsoil

Following the site clearing operation, the stripping and storing of topsoil, if available, will be the next phase. In the interest of economy, the recovery of the maximum quantity of existing topsoil is important. However, any area cleared of trees and stumps will have a loss factor of approximately 15% to 50% depending on the number and size of stumps removed. The remaining topsoil should be stripped, cleaned of foreign materials, stored in a convenient location, and protected from erosion or contamination.

The depth of the topsoil influences the production rate. Less quantity would be available in a given area if it were 4" (100 mm) in depth as opposed to 6" (150 mm) and yet the same area would have to be stripped by the dozer. The selection of the storage area is important. It should not only provide protection for the topsoil material, but it also should minimize the haul time involved in the storing and redistribution of the material to the final location.

Strip & Store Topsoil

Description	per cu.yd.	per cu.m
With dozer, 4"–6" (100–150 mm) depth 100' (30 m) haul–average conditions	$0.99–1.21	$1.32–1.54
(erosion protection additional)		
With dozer, 4"–6" (100–150 mm) depth 300' (90 m) haul–average conditions	3.85–4.68	$16.50–20.10
(erosion protection additional)		

If adverse conditions are present increase unit price by 20% to 33%. Price range depends on terrain and capacity of equipment.

The redistribution of topsoil usually requires some preparation of the ground prior to placing the topsoil.

Construction activities cause rutting, depressions, and general unevenness of the site area, and to avoid excessive quantities of topsoil being used, the site should be graded to the designed contours less the depth of the topsoil. Any preparatory work such as scarifying, raking, or rolling should be accomplished before the introduction of the topsoil. Requirements for soil conditioning, fertilizer, etc., become cost items in addition to the basic spreading cost. When specifications have rigid conditions concerning the quality of topsoil, it may be necessary to submit samples to a laboratory for analysis to establish compliance.

Off-site purchase of topsoil is very expensive and can vary from $10.00 to $30.00 per cu. yd. ($13.00–39.25 per cu. m) depending on the quality and availability.

Spread and Prepare Topsoil

Description	Unit	Avg. Cost	Unit	Avg. Cost
Scarify subsoil–average conditions	Sq. Ft.	$0.02	Sq. M	$0.22
Spread, 300' (90 m) with dozer, average cond.	Cu. Yd.	1.50–2.50	Cu. M	2.00–3.30
Hand dress topsoil, ($20.00 hourly rate)	Cu. Yd.	3.00–4.00	Cu. M	3.90–5.23
Rolling, push roller	Sq. Ft.	0.05–0.07	Sq. M	0.54–0.75

Caution is to be exercised in determining if the specified depth of topsoil is loose or compacted depth. Depending on the characteristics of the material involved, topsoil will compact from 33% to 50%. This could mean to obtain 4" (100 mm) compacted depth would require 6" (150 mm) of loose material. An estimator not taking this into consideration would discover a 50% shortage from the estimated quantities.

If topsoil is to be placed in a series of small restricted areas or at the perimeter of buildings, walks, drives, or other obstructions, additional hand labor will be required to place, grade, and hand compact the material.

It is absolutely essential that the estimator become familiar with local laws regarding disposal. Silt fence for containment during stripping and spreading operations costs about $0.75 to $3.00 per lin. ft. (300 mm). If salt hay is used, costs vary greatly depending on the availability of material, usually between $0.50 and $5.00 per lin. ft. (300 mm).

02060 DEMOLITION

Demolition can be divided into several main types—demolition of entire buildings or structures, controlled removals for historic preservation work, and general building demolition of items such as doors and frames, windows, floors, ceilings, and partitions.

In most areas of the country, there are specialty contractors whose sole business is the demolition of entire buildings and structures. These contractors will use a wide range of heavy equipment, even explosives, for the controlled demolition work. It is always best to solicit a quotation from one of these firms for this type of demolition. However, the following per cubic foot (cubic meter) unit prices will provide some budget guidelines in the absence of quotes:

Building Type	Unit Price per Cu. Ft.	Unit Price per Cu. M
Low Rise–Steel Frame	$0.28	$9.74
Low Rise–Concrete Frame	0.39	13.60
Low Rise–Masonry Wall Bearing	0.33	11.66

Demolition associated with historical preservation work is highly specialized, and since each project has its own particular set of intricacies, it is extremely difficult to produce accurate historical cost data that can then be saved for use in future estimates.

Of less intricate nature than the controlled demolition of historical preservation work, but closely related, is general building demolition. It is this type of work that we will address for the various types and items of demolition and their appropriate labor production tables and costs.

The estimator must always take into consideration factors such as working space, is the building occupied or not, method and route in removing debris to the trash receptacle or truck, special equipment requirements, temporary bracing and shoring, dust and security control partitions, length of the haul to the dump site, and special use permits, EPA permits, and dump fees.

Demolition Using Compressed Air Tools

Compressed air is used extensively in all classes of construction work including such operations as concrete breaking, masonry chipping, excavating, compacting earth, and driving sheet pilings. To reduce the overall operating cost (ownership or rental rates), it is always wise to use an air compressor that is capable of handling at least two pneumatic tools. The use of two tools lowers the rental rate per hour for the machine down to a level half of what it would be if only one operator was using the machine to do an operation. For example, if you had 1,000 sq. ft. (90 sq.m) of concrete sidewalk to demolish, using one compressor with two hammers and two workers would require the same total number of labor hours as one worker. However, your equipment rental hours would be half of that required if you used only one worker on the air compressor.

The following chart shows the classification of work to be demolished by an air hammer, the daily production of work per crew of two laborers at 8 hours each, and the total number of manhours required per 100 sq. ft. (10 sq.m) or per lin. ft. (mm), whichever is applicable. Keep in mind that the production rates shown in the following table only include the material break-up, not transport/removal to the trash receptacle.

Item Description	Crew Daily Production	Manhours per 100 s.f. (10 sq.m)	Manhours per 100 l.f. (30 m)
Wall Footing–			
2' wide x 1' thick (0.6 x 0.3 m)	125 l.f (38 m)	--	12.80
3' wide x 1'–6" thick (0.9 x 0.45 m)	65 l.f. (20 m)	--	24.62
Slab on Grade – 4" (100 mm) with mesh	450 s.f. (42 sq.m)	3.56	--
Slab on Grade – 8" (200 mm) with mesh	250 s.f. (23 sq.m)	6.40	--
Concrete Foundations Walls			
Lightly Reinforced:			
6" (150 mm) Thick	200 s.f. (19 sq.m)	8.00	--
8" (200 mm) Thick	160 s.f. (15 sq.m)	10.00	--
10" (250 mm) Thick	135 s.f. (13 sq.m)	11.85	--
12" (300 mm) Thick	110 s.f. (10 sq.m)	14.55	--
Header Curb	225 l.f. (69 m)	--	7.11
Curb & Gutter	175 l.f. (53 m)	--	9.14

Partition Removal

The following labor production table gives the daily crew (2 laborers, 8 hours each) output and the manhours required per 100 s.f. (9.3 cu.m) for the demolition of various types of partitions. The labor production hours do not include the transport/removal of demolished materials to the trash receptacle.

Partitions 10' High (3000 mm)

Type of Partition	Daily Prod. sq. ft.	Manhours per 100 s.f.	Daily Prod. sq.m	Manhours per 10 sq.m
Metal Stud & Drywall	850	1.88	79	2.00
Plaster & Lath on				
Wood Stud	650	2.46	60	2.60
4" (100 mm) Concrete				
Masonry Units	650	2.46	60	2.65
8" (200 mm) Concrete				
Masonry Units	500	3.20	46	3.45
12" (300 mm) Concrete				
Masonry Units	400	4.00	37	4.30
4" (100 mm) Brick	500	3.20	46	3.45

Partitions 15' High (4500 mm)

Type of Partition	Daily Prod. sq. ft.	Manhours per 100 s.f.	Daily Prod. sq.m	Manhours per 10 sq.m
Metal Stud & Drywall	800	2.00	79	2.20
Plaster & Lath on				
Wood Stud	600	2.67	60	2.90
4" (100 mm) Concrete				
Masonry Units	600	2.67	60	2.90

Type of Partition	Daily Prod. sq. ft.	Manhours per 100 s.f.	Daily Prod. sq.m	Manhours per 10 sq.m
8" (200 mm) Concrete Masonry Units	450	3.56	46	3.80
12" (300 mm) Concrete Masonry Units	375	4.27	37	4.60
4" (100 mm) Brick	450	3.56	46	3.80

Door & Frame Removal

The following labor production rates are for the removal of doors and frames from walls to be demolished. Keep in mind that if a frame is to be removed from a wall that is to remain in place, more care must be used to prevent needless damage and costly repairs to the wall. As a guide the hours shown in the following table could be doubled for the removal of metal door frames from masonry walls that will remain in place. The production times shown in the table does not include any transport of the demolished materials to the trash receptacle.

Item Description	Manhours per unit
Single wood door & frame in stud partition	1.00
Pair – wood doors & frame in stud partition	1.25
Single door and frame in masonry partition	1.25
Pair doors and frame in masonry partition	1.50
Single hollow metal door and frame in stud partition	1.50
Pair hollow metal doors & frame in masonry partition	2.50
Single hollow metal door & frame in masonry partition	2.25

Window Removal

The following table shows various types of window materials and the required man hours per unit for their removal, including removal of trim. The production times shown in the table does not include transport of the demolished materials to the trash receptacle.

Type	Size	Manhours per unit
Wood	up to 3' x 5' (900 x 1500 mm)	0.75
Wood	up to 5' x 6' (1 500 x 1800 mm)	1.25
Aluminum	up to 3' x 5' (900 x 1500 mm)	1.25
Aluminum	up to 5' x 6' (1500 x 1800 mm)	1.75
Steel	up to 3' x 5' (900 x 1500 mm)	2.00
Steel	up to 5' x 6' (1500 x 1800 mm)	2.50

Floor Finishes Removal

The following table shows various types of finish flooring materials and the required manhours per 100 sq.ft. (10 sq. m) for their removal. The production times shown in the table does not include transport of the demolished materials to the trash receptacle.

Type of Flooring	Manhours per 100 s.f.	Manhours per 10 sq.m
Vinyl composition tile	2.00	2.20
Wood parquet	2.50	2.70
Wood strip	2.75	3.00
Carpet & pad – tack strip	0.40	0.43
Carpet – glued down	1.00	1.10
Ceramic tile	2.50	2.70

Ceiling Finishes Removal

The following table shows various types of ceiling finish materials and the required manhours per 100 sq. ft. (10 sq. m) for their removal. The production times shown in the table does not include transport of the demolished materials to the trash receptacle.

Type of Material	Manhours per 100 s.f.	Manhours per 10 sq.m
Drywall	1.00	1.10
Suspended plaster	3.00	3.20
Suspended acoustical tile and grid	1.25	1.35
Plywood	1.00	1.10

If the ceiling finishes are more than 10' (3 m) above the finished floor, add a 10% labor factor multiplier for every 2' (0.60 m) increment over 10' (3 m).

Wall Finish Removal

The following table shows various types of wall finish materials and the required manhours per 100 sq. ft. (10 sq. m) for their removal. The production times shown in the table does not include transport of the demolished materials to the trash receptacle.

Type of Material	Manhours per 100 s.f.	Manhours per 10 sq.m
Drywall	0.60	0.65
Plaster	1.00	1.10
Wood paneling	0.85	0.90
Ceramic tile	2.25	2.40

Roofing Material Removal

The following table shows various types of roofing materials and the required manhours per 100 sq. ft. (10 sq. m) of their removal. The production times shown in the table do not include transport of the demolished materials to the trash receptacle.

Item Description	Manhours per 100 s.f.	Manhours per 10 sq.m
Built up roofing–4 ply w/gravel	3.50	3.75
Rigid insulation board	0.75	0.80
Shingles – composition strip	0.90	1.00
Shingles – wood	1.25	1.35
Shingles – slate	1.30	1.40
Sheathing – plywood	0.95	1.00
Sheathing – 1" (25 mm) plank	1.10	1.20
Sheathing – 2" (50 mm) plank	1.50	1.60

Siding Removal

The following table shows various types of siding materials and the required manhours per 100 sq. ft. (10 sq. m) for their removal. The production times shown in the table do not include transport of the demolished materials to the trash receptacle.

Item Description	Manhours per 100 Sq. Ft.	Manhours per 10 Sq. M
Metal – horizontal	2.75	2.95
Metal – vertical	3.00	3.20
Wood – horizontal	3.00	3.20
Wood – vertical	3.25	3.50
Shingle – wood or asbestos	3.50	3.75
Plywood	1.50	1.60

Core Drilling

One of the quickest and neatest ways of cutting a hole in a concrete slab or wall is by core drilling. This particular rig uses a circular-shaped diamond drill bit that is operated by a large drilling machine mounted on a stand. The drill requires continuous water circulated into the core drilling bit to provide lubrication and cooling. The water should not be a problem. The amount of water used is controllable, either by damming or suctioning, to prevent damage to surrounding finishes.

Core drill bits come in diameters from 1" to 18" (25 to 450 mm) in size and the depth of cut allowed by the average coring bit is about 12" (300 mm).

However, special bits are made to allow boring through columns or walls up to 30" (750 mm) in thickness or more.

Core drilling is usually performed by specialty subcontractors whose only business is to provide this service to general contractors. It is best to obtain a firm price quote for any core drilling that is required for a project, but if a quick budget estimate is needed, core drilling is usually priced by the unit inch, which is the diameter in inches x the depth in inches. For example a hole 4" (100 mm) in diameter and 10" (250 mm) deep would be equal to 40 unit inches. Currently the going rate per unit inch is approximately $2.00 to $2.50 depending on the quantity of holes to be drilled, location, and set-up time required. Horizontal holes will cost about 10% to 15% more than a vertical hole. If reinforcement steel is to be encountered, add 50% to 100% more.

02200 EARTHWORK

It is not difficult to compute the number of cubic yards of earth to be excavated or removed from the premises on a specific project, but it is quite another matter to estimate the actual cost of performing this work, because of the many factors and possible unknown items entering into the cost. This includes such items as type of soil, whether or not water will be encountered, pumping, whether the banks of the excavated portion will be self-supporting or whether it will be necessary to brace and sheet them, disposal of excavated earth, length of haul to dump, type of material, etc.

The prudent estimator will examine the equipment that is available when preparing the cost estimate. The accurate choice of equipment at the bidding stage will determine the final costs. Type of equipment selected, as well as the size of equipment, should be based on several factors—subsurface information, site conditions, volume of material to be moved, and availability of specific equipment.

Estimating Quantities of Excavation. Excavating is measured by the cubic yard, cu.yd. (cubic meter, cu.m, m³). One cubic yard contains 27 cu. ft.

General or Mass Excavation. When computing the volume or cubic contents of any basement or portion to be excavated, take all measurements from the outside face of wall footings, add 6" to 1'-0" (150 to 300 mm) on each side to allow for placing and removing forms, and then allow for sloping the sides of the excavation so they will stand reasonably stable without being supported by bracing or sheet piling. The amount of slope necessary to provide a "safe" hole will vary with the kind of soil to be excavated, depth of excavation, etc.

On construction projects the Federal Occupational Safety and Health Administration (OSHA), as well as state and local agencies, require for excavations of five feet deep and over soil support of sheeting or some other type. If none is used, then the side banks of the excavation shall be sloped back to the soils natural angle of repose so that the excavation sides will remain stable.

Digging in previously undisturbed material and assuming that no water or unstable conditions exist, most estimators, when taking off quantities for excavation, use a 1:1 slope; that is, one horizontal to one vertical, for sand and gravel; a 1:2 slope for ordinary clay and a 1:3 or 1:4 slope for stiff clay.

In some cases job conditions may not permit the sloping of banks and then it is usually necessary to sheet and brace the banks to prevent accidents that result in damage to adjoining property, or worse, worker injury or death.

On jobs with column footings that project beyond the wall footings, additional cuts must be made into the banks at the column locations to accommodate the additional width of the footings. The volume of this cut is computed by taking the column footing length plus 1'-0" or 2'-0" (300 or 600 mm) for work space plus "layback" of banks on both ends and multiplying by the amount of projection beyond the wall footing to obtain the area of the additional cut, which is then multiplied by the depth of the cut to obtain the volume of additional excavation required.

The depth of the general cut is usually taken from the underside of the topsoil, previously removed, to the underside of the floor slab. Where sand, gravel, or other fill material is to be placed under the concrete floor, the depth should be taken to the underside of the fill material.

Example: Assume you are figuring the general or mass excavation for a building 100'-0" x 100'-0" (30 x 30 m) at grade. The wall footings extend 6" (150 mm) beyond the face of the foundation walls, and on each side there are 6 column footings that are 6'-0" (1.8 m) long and project 2'-0" (0.6 m) beyond the face of the wall footings.

Also assume the depth of the excavation to be 8'-0" (2.4 m) and that the soil will stand at a slope of 1:2, one horizontal to two vertical. The volume of the excavation should be figured as follows:

Building length is 100'-0" (30 m) plus 2 x 0'-6" (2 x 0.15 m) for wall footings plus 2 x 0'-6" (2 x 0.15 m) for work space plus an average of 4'-0" (1.2 m) for slope equals 106' (31.8 m) length. Since the building is square, the main volume would be 106'-0" x 106'-0" x 8'-0" or 89,888 cu. ft. (31.8 x 31.8 x 2.4 = 2,427 cu.m).

Alternate example: 100'-0" x 100'-0" (30 x 30 m) at grade. The wall footings extend 6" (0.15 m) beyond the face of the foundation walls on each side. Building length is 100'-0" + (2 x 6") for wall footings + (2 x 6") for work space equals 102'x102' (30.6 x 30.6). Next, add the slope area, remembering to add additional slope area at the corners.

Base Excavation: Length x Width x Height
Side Slope: Area of triangle = (base x height) divided by 2; multiply by 4 sides, plus 4 corners

Elevation

	(metric)
102' x 102' x 8' = 83,232	30.6 x 30.6 x 2.4 = 2,247
(4' x 8') ÷ 2 x 102' x 4 = 6,528	(1.2 x 2.4) ÷ 2 x 30.6 x 4 = 176
(4' x 8') ÷ 2 x 2 x 4 = 128	(1.2 x 2.4) ÷ 2 x 2 x 4 = 11
83,232 + 6,528 + 128 = 89,888 c.f.	2,247 + 176 + 11 = 2,434 cu.m

For the 24 column footings which project beyond the wall footings, the length of each additional cut would be the 6'–0" (1.8 m) footing length plus 2 x 0'–6" (2 x 0.15 m) for work space plus an average of 4'–0" (1.2 m) for slope or 11'–0" (3.3 m) and the width of each cut would be the 2'–0" (0.6 m) projection. The additional excavation required to accommodate the column footing projections would be:

24 x 11'–0" x 2'–0" x 8'–0" = 4,224 c.f.
(24 x 3.3 x 0.6 x 2.4 = 114 cu.m)

The total volume required to be excavated would be:

89,888 + 4,224 = 94,112 c.f. or 3,486 cu. yds.
(2,434 + 114 = 2,548 cu.m)

In some cases a portion of or the entire building site may be high and require a grading cut to produce the required contours. In these instances the grading cut should be figured first and the general or mass excavation computed afterwards taking the depth from the underside of the grading cut to the bottom of the floor slab or fill under the floor.

Trench and Pit Excavation. To complete the building excavation quantity take-off, there are usually additional items of excavation to figure for footings, foundations, pits, trenches, etc., which extend below the level of the general or mass excavation, or which occur in unexcavated areas. This kind of excavation is always estimated separately from the general work as it is usually a more expensive operation.

In many instances a good portion of this work may be done with power equipment, with some handwork required to clean out corners and grade the bottoms. On the other hand, there might be portions of the work that must be done entirely by hand, which would be much costlier than machine work. With such a big difference in unit cost between machine work and handwork, the estimator should decide, when taking off the quantities, into which category each portion of the work will fall and list the machine work separately from the handwork so that each may be properly priced.

Typical Examples of Swell Factors

Material	Percentage Swell
Gravel	
Dry	15
Wet	15
Clay	
Dry	20
Wet	20
Rock, Blasted	
Blasted, well	20
Blasted, average	25
Blasted, poorly	50
Earth	
Dry	20
Wet	10
Gravel	
Dry	10
Dry, 1/4" to 2" (6 to 50 mm)	15
Wet, 1/4" to 2" (6 to 50 mm)	15
Topsoil	30

Quantities for this work are measured and listed in the same manner as for general excavation, except the slope factor at times may be ignored, unless the additional depth exceeds 5'–0" (1.5 m) or if the ground will not stand safely for this depth with vertical banks. Under these conditions, allowances for sloping the banks should be made in the same manner as for general excavation, or allowances for soil support, such as sheeting and bracing, need to be included.

The estimator must remember that the plan dimensions shown and used to compute the amount of material to be excavated are sometimes referred to as pit measured or bank cubic yards (cubic meters); that is, the material as it rests in its natural state. After the material is removed, and either stored or loaded into trucks, it swells. The same material is now referred to as "loose cubic yards (cubic meters)". The swell factors will vary with the type of materials being excavated.

Backfill. After all excavating items have been figured, the amount of backfill should be computed. An easy method is to compute the displacement volume of the construction that is to be built within the limits of the excavation; that is, the volume of footings, piers, the basement volume figured from the underside of the fill under the floor to the elevation of the top of the general cut, etc., and deducting this volume from the sum of the general or mass excavation and the trench and pit excavation. The remainder is the volume of backfill required.

Another method of figuring backfill, usually more difficult, is to compute the volume from the actual dimensions of the spaces requiring backfill.

Some specifications require interior backfills to be made with selected materials, such as sand, gravel, bankrun gravel, etc., while the exterior backfill may be, if suitable and approved, the excavated material. In this case it is necessary to keep the two items separate as the interior backfill material will probably have to be purchased and brought in from outside sources.

Backfill very often is specified to be compacted by mechanical tampers and a degree of compaction is also specified. In these instances the backfill operations must be considered very carefully, because this is expensive work.

When calculations for the backfill are complete, the estimator should consider the compaction factor. This is the measured difference between the truck—loose cubic yardage—delivered quantity and the final in-place measured quantity. Backfill material purchased in loose cubic yards (cubic meters) after compaction will not yield the same in-place quantity. Therefore, additional yardage (meters) will be required. For example, assume a hole to be backfilled that is 9'0" x 9'0" x 3'0" (2.7 x 2.7 x 0.9 m) deep; measured in place, 9 cu. yds. (6.5 cu.m) of material will be required. After compaction, assuming a compaction loss of 15%. Therefore, 9 ÷ 0.85 = 10.59 cu. yds. (8.1 cu.m).

Excavation and Fill for New Site Grades. Excavation and filling for new site grades, including cuts and fills for drives, walks, and parking areas, vary from simple to complex according to the ground conditions. Estimating quantities for this work may be rather difficult and some surveying or engineering knowledge will greatly help.

There are a number of methods to determine volumes of cut and fill required. The method that is most adaptable to the particular problem will be chosen by the estimator after considering the problem, with the information given and the accuracy desired.

Cross Sectional Method

One of the easiest and most frequently used methods of computing grading cuts and fills when the plot plan shows both original and proposed contours is that of gridding or dividing up the area to be graded into squares, rectangles, triangles, or combinations of these figures, of regular and convenient dimensions, thus forming a series of truncated prisms. The depths of cut or fill at the corners of these prisms are computed and the volume of each prism may be determined by averaging these depths and multiplying the average by the cross sectional area of the prism. The results of these computations are totaled, keeping the cuts separate from the fills, and the total excavation and/or fill required is obtained.

Example: Assume the contour sketch shown on the next page to be a 400' x 200' (120 x 60 m) portion of a site grading plan showing both present and proposed contour lines. Since both the present and proposed surfaces have fairly regular slopes, as indicated by the even spacing of contour lines, a grid spacing of 50 feet in both directions has been chosen and grid lines drawn as shown, labeled alphabetically for the vertical lines and numerically for the horizontal lines. The next operation is to figure the present ground elevation for each grid line intersection by interpolating between the present contour lines. The results are marked on the drawing in the upper right quadrant of each respective intersection.

Following this, the proposed ground elevations are similarly figured for each intersection, interpolating between the proposed contour lines and entering the results in the upper left quadrant of each respective intersection.

When this is completed, the depth of cut or fill at each intersection can be readily determined by subtracting one elevation from the other and noting whether a cut or fill is required. This determination should be made for each intersection and the results entered in the proper quadrant, using the lower left for cuts and the lower right for fills.

When all intersections have a depth value entered for either cut or fill, the plan should be examined closely to see where the grading operation will change from cut to fill or vice versa. In the sketch it will be seen that advancing from left to right on horizontal grid line "1" a change from fill to cut occurs between vertical grid lines "D" and "E" and by interpolating between the fill and cut depths the zero point "a" can be established approximately 36 feet to the right of intersection "D-1."

Similarly, zero points "b, d, e, and g" are established on the horizontal grid lines and zero points "c and f" on the vertical grid lines. These points are then connected by a line "a, b, c, d, e, f, g" and for all practical purposes it may be assumed that all points on this line are at the proposed elevation and require neither cut nor fill. In this case the line forms the boundary between cuts and fills. The change from cut to fill and vice versa could, of course, occur many times in actual problems depending on the irregularity of the present and proposed contour lines.

After the previously described operations have been completed, the actual volume of cut and/or fill required can be computed.

The volume of cut or fill in each prism in the grid system is obtained by averaging the corner depth values of each prism and multiplying by its cross sectional area. Thus, for the square prism "A-1, B-1, B-2, A-2" the volume of fill would be 1/4 x (2.4' + 1.6' + 1.6' + 2.5') x 50.0' x 50.0' or 5,063 cu. ft. (143 cu.m). All grid prisms are similarly treated, keeping cut and fill results separated, except those prisms through which the zero line passes. So, for square prism "D-1, E-1, E-2, D-2" it will be seen that this volume has been divided into two trapezoidal prisms "D-1, a, b, D-2" and "a, E-1, E-2, b". To compute the respective fill and cut volumes of these prisms they must be divided into triangular prisms first and then the corner depths may be averaged and multiplied by their respective cross sectional areas. Thus for the triangular prism "D-1, a, b" the volume of fill would be 1/3 x (0.5' + 0.0' + 0.0') x 1/2 x 50.0' x 36.0' or 150 cu. ft. (4.25 cu.m). Similarly, the volume of fill for the triangular prism "D-1, b, D-2" would be 1/3 x (0.5' + 0.0' + 0.3') x 1/2 x 50.0' x 21.0' or 140 cu. ft. (4 cu.m). The volume of cut required for the trapezoidal prism "a, E-1, E-2, b" is figured in the same manner. In the case of square cross section "C-2, D-2, D-3, C-3" it will be seen that the zero line crosses one corner of the square and divides it into a triangle "c, D-3, d" and a five-sided figure "C-2, D-2, c, d, C-3". The five-sided figure should be further divided into three triangles "C-2, D-2, c", "C-2, c, C-3" and "C-3, c, d" and the volumes of fill for each triangular prism would be as follows:

1/3 x (0.7' + 0.3' + 0.0') x 1/2 x 50.0' x 30.0' or 250 c.f. (7 cu.m)
1/3 x (0.7' + 0.0' + 0.5') x 1/2 x 50.0' x 50.0' or 500 c.f. (14 cu.m)
1/3 x (0.5' + 0.0' + 0.0') x 1/2 x 36.0' x 20.0' or 60 c.f. (1.7 cu.m)

When there are several prisms with the same cross section to be either cut or filled the volume of such can be computed as one solid by assembling them as follows. Multiply each corner depth by the number of typical prisms in which it occurs and then add these results and divide by 4. This is then multiplied by the cross sectional area of one prism. For example the volume bounded by A-1, D-1, D-2, C-2, C-4, B-4, B-5, A-5, A-1 can be obtained by one computation because it is composed of a series of prisms having the same cross section. In the summation of depths, those at A-1, D-1, D-2, C-4, B-5 and A-5 are taken but once, those at B-1, C-1, C-3, A-4, A-3 and A-2 are multiplied by 2, at C-2 and B-4 the depths are multiplied by 3 and at B-2 and B-3 they are multiplied by 4. All of these values are then totaled and divided by 4 and then multiplied by the cross sectional area of one prism giving the volume of fill required.

When a volume has been computed for each grid division of the site, either individually or collectively, it is only a matter of summing up the cuts and fills to obtain the total excavation or fill required to bring the site to the proposed grades.

End Area Method

It often happens that a cut and fill operation is to be performed over a long and narrow area such as for roads, levees, ditches, etc. In this case the usual method of computing the earthwork quantities is by means of end areas.

In jobs of this sort a profile plan and a topographic plan is generally furnished to the contractor as part of the bidding documents. Using this information, sections are taken perpendicular to the longitudinal centerline of the work, usually at each 100-foot (30-meter) station, at each radical change in either present or proposed grades and where the cut changes to fill or vice versa. These sections are plotted to scale showing the original grades as obtained from the topographic plan and the proposed grades as obtained from the profile plan and design details of the surface required. After this has been done, it will be apparent that areas are enclosed by the plotted lines representing the present and proposed grades at each section and these areas should be computed.

To obtain the volume of cut and fill between each section, the usual and easiest method is to average the areas of two adjacent sections and multiply by the distance between them, keeping the cuts and fills separate. This method is not quite accurate, but the results are on the safe side and in practically every case are close enough for estimating purposes.

Example: Assume the following sketch to be a 300-foot (90-meter) portion of the topographic plan and profile plan of a proposed road location. The topographic plan shows the undulations of the present ground surface by the contours and the location of the road by the center and shoulder lines. The profile plan shows the relation between the present ground surface and the proposed road bed grade at the centerline of the road. Also assume the road bed is to be 20.0 feet wide and the shoulder slopes are to be 1-1/2 : 1, that is 1-1/2 horizontal to 1 vertical.

To compute the volumes of earthwork involved in this proposed grading operation, by means of the end area method, sections must be taken perpendicular to the center-line of the proposed road and the present and proposed grade lines plotted. In this case sections have been taken and plotted, as illustrated in the sketch, at even stations 15+00, 16+00, 17+00, and 18+00. Sections have also been taken at stations 15+07, 15+37, 16+50 and 16+65 because at these points the cut or fill at a shoulder of the proposed road is zero. A section is taken at station 17+65 because at this point the present grade makes a radical change in slope.

In plotting the sections any convenient scale may be used and if cross-section paper is used the job will be much easier.

If a greater degree of accuracy is required the prismoidal formula must be employed which necessitates figuring the middle area between two adjacent sections. This middle area is obtained by averaging the corresponding dimensions of both end areas and computing an area from the average dimensions; it should not be taken as the average of the two end areas. To obtain the volume between adjacent sections using the prismoidal formula, multiply the length between sections by one-sixth of the sum of the two end areas and 4 times the middle area combined.

Most specifications call for the fills to be placed in layers and then specify some degree of compaction to be attained. This should be considered by the estimator, because additional equipment such as sheepsfoot rollers, pneumatic rollers, or tandem rollers will be required plus labor for the operation.

As each section is plotted, the area enclosed by the present and proposed grade lines is computed and the result tabulated. Final computations are made as follows:

Station	Distance Between Stations	Actual Cut	Average Cut	Actual Fill	Average Fill	Cu. Ft. Cut	Cu. Ft. Fill
15+00		24.38		--			
	7.0 ft.		20.41		--	142.87	--
15+07		16.44		--			
	30.0 ft.		8.22		7.46	246.60	223.80
15+37		--		14.92			
	63.0 ft.		--		51.23	--	3,227.49
16+00		--		87.54			
	50.0 ft.		--		53.45	--	2,672.50
16+50		--		19.36			
	15.0 ft.		7.56		9.68	113.40	145.20
16+65		15.11		--			
	35.0 ft.		26.72		--	935.20	--
17+00		38.33		--			
	65.0 ft.		59.77		--	3,885.05	--
17+65		81.20		--			
	35.0 ft.		62.34		--	2,181.90	--
18+00		43.48		--			
						7,505.02	6,268.99

Area of Sections in Sq. Ft.

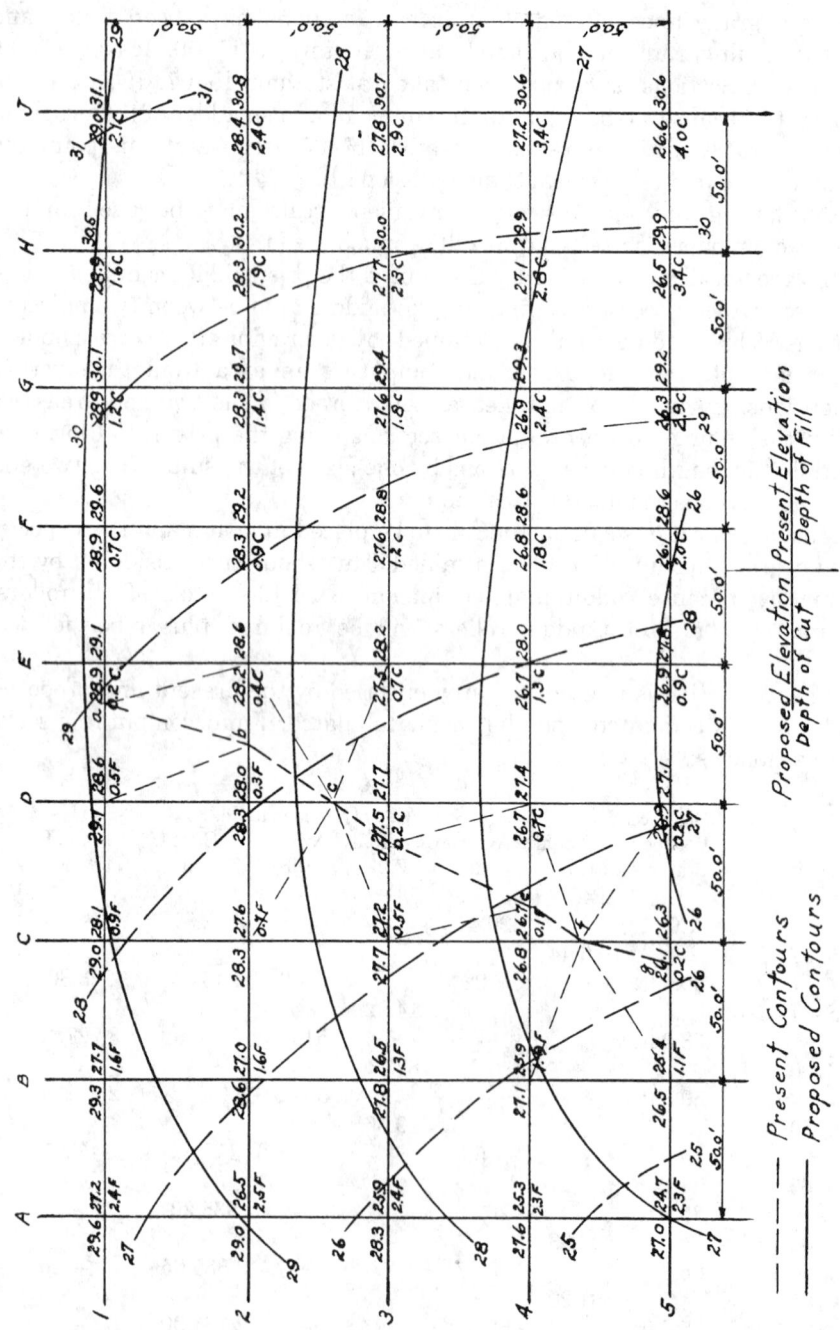

Method of Cross Sectioning a Contour Plan

Topographic and Profile Plans

Method of Plotting End Areas

Clear and Grub. Prior to stripping of top soil or to performance of rough grading, it may be necessary to clean and grub the area. This is the removal of all trees, bushes, and other growth, with roots to be grubbed out of the ground, in order for top soil to be removed in an efficient manner.

The smallest trees and their roots can be chipped into a chipping machine and then efficiently trucked off the site or sold to a garden supply company, or used in erosion control.

Rough Grading of Site. An item for rough grading, consisting of the total area cut and/or filled, less buildings, roads, walks, etc., should be included in the estimate to cover the expense of dressing up the surface after the required grades have been approximately attained and preparatory to the spreading of top soil, if required.

Spreading Top Soil. In most instances the specifications call for spreading top soil over site areas not covered by other construction, usually to a depth of 6 inches.

Finish Grading of Site. Fine grading of top soil, usually including a hand raking operation, is generally called for and should be listed in the estimate as an item, the quantity being the area used in computing the volume of top soil to be spread.

Other Grading. Another grading item to be considered is grading the subgrade for floor slabs, walks, paving, etc. Some contractors also include an item for grading bottoms of footings, when they are to be dug by machine, and include an amount for labor squaring and cleaning the holes.

Disposal of Excess or Borrow. After all excavation, backfill, grading cuts and fills, etc., have been computed and listed in the estimate, the total cuts and the total fills should be compared to determine whether there will be an excess of materials to dispose of or a deficiency of materials to be made up by borrowing or purchasing from outside sources. The results of this comparison should be listed in the estimate. The top soil comparison should be kept separate from the other materials, because the cost of any top soil that must be purchased usually is much more expensive than ordinary fill material. Also, any excess of top soil may find a ready market and be sold, reducing the overall cost of the earthwork. The difference between the cuts and fills will give the bank or compacted volume required. In case of a deficiency of materials, this compacted volume quantity must be adjusted for shrinkage due to compaction, because the material will probably be bought by loose measure. In the same manner all excess bank measurement volumes must be adjusted for swell due to bulking to obtain the true volume that must be handled in the disposal operation. The amount of increase for gravels and sand is from 5% to 12% and for clays and loams from 10% to 30%.

Pumping or De-watering. Finally, careful consideration should be given to the probability of pumping or de-watering operations. In some parts of the country this is no problem whatsoever, quite the reverse being the case.

However, in most localities ground or climatic conditions make it necessary to include pumping or de-watering as an item in the estimate.

This item is quite variable, being affected by the season of the year as well as the locale. If only rainwater run-off is expected, an allowance for a few small pumps may be sufficient. If, however, the job is in close proximity to a body of water or springs are present, it might be necessary to de-water part or all of the operational area by means of a wellpoint or deep well system, in which case the contractor should consult a company that specializes in this work. (See the section on wellpoint systems in this chapter.)

Estimating the Labor Cost of Excavation. Before an excavating estimate is made up or a bid submitted, the estimator should visit the site and ascertain the exact conditions surrounding the work to be performed. The natural grade of the lot should be noted to compute the depth of the excavation from grade to underside of floor slab; the nature of the soil should be determined, whether sand, loam, clay, or rock, and whether or not water is likely to be encountered. Water usually means an additional cost—pumping out the hole day and night until the foundations are completed. Rock usually means blasting, with a greatly increased cost for drilling, dynamiting, excavating, and loading.

The safest method of determining the kind of soil is to dig a test hole the approximate depth of the basement, or take a set of borings. From this one can obtain a fairly accurate idea of the conditions to be encountered.

In most areas the soil strata, as well as the ground-water table, is fairly consistent, and its character may be ascertained by talking to contractors or engineers who have worked in the area for some time. In most larger jobs, the plans will show the results of test borings ordered by the owner or architect/engineer.

If the lot is low and the basement does not cover the entire lot, part of the excavated soil may be used for filling. On the other hand, if the natural grade of the lot is above the sidewalk or established grade, it may be necessary to remove a certain amount of soil from the entire lot.

In large structures having basements and sub-basements, the excavation may extend 40 to 50 feet (12 to 15 m) below grade. This requires the construction of steep runways, haul roads, or ramps so that trucks may get in and out of the hole, and an extra truck or bulldozer may be required to help the loaded trucks out of the hole. These items must all be taken into consideration when pricing any excavating job.

If it is necessary to remove the excavated soil from the premises, locate the nearest available landfill, and ascertain just how the soil will be handled at the landfill (for example, whether it is necessary to spread it).

National and local environmental protection agency rules may require the contractor to test material to be excavated and disposed of, or even to be placed back into the same area, after excavation. Local EPA rules at times require local disposal site operators not to accept material without a certification from the contractor stating that no contaminants are in the material. Materials excavated that have contaminants must be disposed of at designated disposal sites. The types and amounts of contaminants will determine the amount of additional costs.

These are all items that affect the cost and should be given careful consideration by the estimator.

In most localities there are contractors who specialize in excavation and are familiar with the kind of soil encountered in different parts of the area. These contractors usually quote a price per cu. yd. (cu. m) for all of the general excavation, or they may quote a lump sum for the entire job.

Trench, pier, and all hand excavating is usually performed by the general contractor.

METHODS OF EXCAVATING

Excavating costs will vary with the kind of soil and the method used in removing and hauling to a final destination.

Except on very small jobs, practically all excavation today is performed by power operated equipment—power shovels, cranes with dragline or clamshell buckets, bulldozers, tractor excavators, front end loaders, scrapers, and trenching machines. Grading, spreading, and backfilling are performed by bulldozers, tractor excavators, and front end loaders.

In shallow excavations, where the excavated earth may be spread over the site not more than 50 feet (15 m) from the excavation, bulldozers or loaders (a tractor equipped with a shovel that may be elevated to load directly into trucks or over the sides of the excavation) are frequently used. This type of excavator is also used for excavating basements, loading directly into trucks.

What was at one time referred to as a "back-digger" power shovel is now on track or rubber tire and commonly called a "backhoe", which is extensively used for basement and trenching excavation. It is equipped with a bucket that "pulls" or "crowds" toward the machine as it excavates, which is always located on the bank above the excavation, working from the sides and ends of the excavation.

In large excavations, the use of a regular power shovel or hydraulic front shovel is the most satisfactory and economical method to use. The power shovel will excavate and load the trucks as fast as they can be handled in and out of the hole and will enable the job to proceed much faster.

If it is desirable to keep the backhoe on the bank above the excavation, but the depth is greater than the hoe can reach efficiently, a crane equipped with a dragline bucket or clam bucket is used.

For individual pier footings, small pits, and hard-to-reach spots, a crane equipped with a clamshell bucket or a small combination backhoe/loader can be used.

On medium to large grading jobs, self-propelled or tractor-hauled scrapers are used quite often and perform cutting and filling operations very creditably.

Quantities and production times given in the tables below are average. Experienced excavators will perform more work than given, but an inexperienced worker will perform less, so average conditions have been used.

Hydraulic power excavating equipment has become very extensive. For the most part, hydraulic-operated excavating equipment has replaced the old cable-operated machines and is used almost 100 percent on the smaller excavation machines.

HAND EXCAVATION

Most construction projects require extensive use of hand labor to accomplish the selective types of excavation which cannot be performed by machine excavators.

In the excavation of footings, it is frequently found to be cost effective to combine the use of machines for the bulk excavation and then to "hand dress" the bottoms, sides, and corners to produce the precise horizontal and vertical planes required. Similarly, this is true in finalizing the subgrade elevation under slabs on grade, walks, and patio slabs.

The quantities and production rates given in the following table are for average conditions.

Material & Operation	Avg. No. Cu. Yd. per 8–Hr Day	Manhours per Cu. Yd.	Avg. No. Cu. M 8–Hr Day	Manhours per Cu. M
Sandy loam				
Small footings, hand	6	1.33	4.60	1.33
Trenches 3'–4' (.9–1.2 m) deep, hand	6	1.33	4.60	1.33
Piers 6' (1.8 m) deep, hand 1 lift	5	1.60	3.80	1.60
Piers 12' (3.7 m) deep, hand 2 lifts	3	2.66	2.30	2.66
Heavy soil and clay				
Small footings, hand	5	1.60	3.80	1.60
Trenches 3'–4' (.9–1.2 m) deep, hand	5	1.60	3.80	1.60
Piers 6' (1.8 m) deep, hand, 1 lift	4	2.00	3.00	2.00
Piers 12' (3.7 m) deep, hand, 2 lifts	2	4.00	1.50	4.00
Excavation with ½-cu.yd. (.40 cu. m) backhoe, casting to side				
Trench footing	300	--	230	--
Pier footing 6' (1.8 m) deep	200	--	153	--
Pier footing 12' (3.7 m) deep	150	--	115	--

Hand Trimming Trench, Piers And Slabs After Machine Excavation

Area	SF per 8-Hr Day	SF per Hour	Sq. M per 8-Hr Day	Sq. M per Hour
Bottom and side areas	240	30	22.40	2.80
Fine grade under slabs	600	75	56.00	7.00

Backfilling By Hand With Compaction Material From Stockpile

Ordinary Light Soils	Cu. Yd. per 8-Hr Day	Manhours per Cu. Yd.	Cu. M per 8-Hr Day	Manhours per Cu. M
Soil distribution laborer	12	0.67	9.20	0.90
Compactor operator	12	0.67	9.20	0.90

No allowance for compactor equipment cost.

MACHINE EXPLORATION

Excavating Using a 1 Cu. Yd. (.76 Cu. M) Tractor Shovel

Class of Work	Avg. No. Cu. Yds. per 8-Hr Day	Tractor & Operator Hours per 100 Cu. Yds.	Avg. No Cu. M per 8-Hr Day	Tractor & Operator Hours per 76 Cu. M
Excavating ordinary soil and placing in piles on premises	325-375	2.30	250–290	2.30
Excavating heavy soil and clay and placing in piles on premises	250–300	2.90	190–230	2.90
Loading loose sand or gravel into trucks	400–450	1.90	305–345	1.90
Excavating ordinary soil and loading into trucks	400–450	1.90	305–345	1.90
Excavating heavy soil and clay and loading into trucks	300–350	2.50	230–270	2.50
Backfilling loose earth—Bulldozing	500–600	1.45	380–460	1.45

The above table is based upon a 50-foot haul from excavation to dump. For each additional 50 feet in length of haul, add 1-1/2 to 2 hrs. tractor time per 100 cu. yds. or consider the use of trucks.

Excavating Using a 2-1/4 Cu. Yd. (1.70 Cu. M) Tractor Shovel

Class of Work	Avg. No. Cu. Yds. per 8-Hr Day	Tractor & Operator Hours per 100 Cu. Yds.	Avg. No. Cu. M per 8-Hr Day	Tractor & Operator Hours per 76.0 Cu. M
Excavating ordinary soil and placing in piles on premises	600–650	1.30	460–500	1.30

Excavating heavy soil and clay and
placing in piles on premises .... 450–500 .... 1.70 .... 345–380 .... 1.70
Loading loose sand or gravel
into trucks .... 775–825 .... 1.00 .... 595–630 .... 1.00
Excavating ordinary soil and loading
into trucks .... 775–825 .... 1.00 .... 595–630 .... 1.00
Excavating heavy soil and clay and
loading into trucks .... 575–625 .... 1.35 .... 440–480 .... 1.35
Backfilling loose earth—bulldozing... 925-975 .... 0.85 .... 710-745 .... 0.85

The above table is based upon a 50-foot haul from excavation to dump. For each additional 50 feet in length of haul, add 3/4 to 1 hr. tractor time per 100 cu. yds. or consider the use of trucks.

Excavating Using Hydraulic Tractor (Crawler) Backhoe
1 Cu. Yd. (.76 Cu. M) Bucket

Type of Material	Avg. No. Cu. Yds. per 8-Hr Day	Backhoe & Operator Hours per 100 Cu. Yds.	Avg. No. Cu. M per 8-Hr Day	Backhoe & Operator Hours per 76.0 Cu. M
Sandy Clay/Moist Loam	720	1.10	550	1.10
Gravel/Sand	690	1.20	530	1.20
Clay, Hard	570	1.40	435	1.40
Rock				
Well Blasted	450	1.80	345	1.80
Poorly Blasted	300	2.70	230	2.70

Above production is based on the following: project conditions, average to above average; operator, average; obstructions, none; swing, to 60 degrees or 1/4 turn.

Excavating Using Hydraulic Tractor (Crawler) Backhoe
1/4-Cu. Yd. (.20 Cu. M) Bucket

Type of Material	Avg. No. Cu. Yds. per 8-Hr Day	Backhoe & Operator Hours per 100 Cu. Yds.	Avg. No. Cu. M per 8-Hr Day	Backhoe & Operator Hours per 76.0 Cu. M
Sandy Clay/Moist Loam	900	0.90	690	0.90
Gravel/Sand	840	0.95	640	0.95
Clay, Hard	570	1.40	435	1.40
Rock				
Well Blasted	450	1.80	344	1.80
Poorly Blasted	300	2.70	230	2.70

Above production is based on the following: project conditions, average to above average; operator, average; obstructions, none; swing, to 60 degrees or 1/4 turn.

Excavating Using Hydraulic Tractor (Crawler) Backhoe
1-1/2 Cu. Yd. (1.15 Cu. M) Bucket

Type of Material	Avg. No. Cu. Yds. per 8-Hr Day	Backhoe & Operator Hours per 100 Cu. Yds.	Avg. No. Cu. M per 8-Hr Day	Backhoe & Operator Hours per 76.0 Cu. M
Sandy Clay/Moist Loam	1080	0.75	825	0.75
Gravel/Sand	1020	0.80	780	0.80
Clay, Hard	870	0.90	665	0.90
Rock				
Well Blasted	660	1.20	505	1.20
Poorly Blasted	420	1.90	320	1.90

Above production is based on the following: project conditions, average to above average; operator, average; obstructions, none; swing, to 60 degrees or 1/4 turn.

Excavating Using Hydraulic Tractor (Crawler) Backhoe
1-3/4 Cu. Yd. (1.34 Cu. M) Bucket

Type of Material	Avg. No. Cu. Yds. per 8-Hr Day	Backhoe & Operator Hours per 100 Cu. Yds.	Avg. No. Cu. M per 8-Hr Day	Backhoe & Operator Hours per 76.0 Cu. M
Sandy Clay/Moist Loam	1260	0.60	965	0.60
Gravel/Sand	1200	0.70	915	0.70
Clay, Hard	1020	0.80	780	0.80
Rock				
Well Blasted	750	1.05	575	1.05
Poorly Blasted	510	1.60	390	1.60

Above production is based on the following: project conditions, average to above average; operator, average; obstructions, none; swing, to 60 degrees or 1/4 turn.

Excavating Using Hydraulic Tractor (Crawler) Backhoe
2 Cu. Yd. (1.52 Cu. M) Bucket

Type of Material	Avg. No. Cu. Yds. per 8-Hr Day	Backhoe & Operator Hours per 100 cu. yds.	Avg. No. Cu. M per 8-Hr Day	Backhoe & Operator Hours per 76.0 cu.m
Sandy Clay/Moist Loam	1440	0.55	1100	0.55
Gravel/Sand	1380	0.60	1055	0.60
Clay, Hard	1140	0.70	871	0.70
Rock				
Well Blasted	870	0.90	665	0.90
Poorly Blasted	720	1.10	550	1.10

Above production is based on the following: project conditions, average to above average; operator, average; obstructions, none; swing, to 60 degrees or 1/4 turn.

Excavating Using Hydraulic Tractor (Crawler) Backhoe
3 Cu. Yd. (2.30 Cu. M) Bucket

Type of Material	Avg. No. Cu. Yds. per 8-Hr Day	Backhoe & Operator Hours per 100 cu. yds.	Avg. No. Cu. M per 8-Hr Day	Backhoe & Operator Hours per 76.0 cu.m
Sandy Clay/Moist Loam	1800	0.45	1376	0.45
Gravel/Sand	1710	0.45	1307	0.45
Clay, Hard	1440	0.55	1100	0.55
Rock				
Well Blasted	1080	0.75	825	0.75
Poorly Blasted	720	1.10	550	1.10

Above production is based on the following: project conditions, average to above average; operator, average; obstructions, none; swing, to 60 degrees or 1/4 turn.

ESTIMATING THE COST OF OWNING AND OPERATING CONSTRUCTION EQUIPMENT

The following information on the costs of owning and operating power shovels, hoes, cranes, truck cranes, tractors, and scrapers will give a general idea of the costs involved in making up an estimate for this type of work.

The equipment purchase prices given are only a relative guideline to be used in the tables. What is most importance is the format for determining these costs, not the price of the equipment listed, which will vary greatly depending on locality, method of purchase, number of pieces bought at one time, etc.

The price f.o.b. factory should cover the complete machine with all variable equipment and accessories, such as various attachments, light plant, magnet generators, and clamshell bucket.

Total Investment or Cost of Equipment. Depreciation, interest, taxes, and insurance are directly related to the initial investment in construction equipment. In addition, certain other costs may be estimated by their normal relationship to this figure. Therefore, the proper determination of the total cost is a basic requirement.

Approximate Prices of Power Shovels, Draglines, Clamshells, and Lift Cranes

The following are approximate prices, f.o.b. factory, and will give the estimator some idea of the investment required in shovels and cranes of various sizes:

Power Shovel
Crawler

Size of Shovel
3/4 cu. yd (0.57 cu. m)...$100,100.00
1 cu. yd (0.75 cu. m)..105,600.00
1-1/2 cu. yd (1.15 cu. m)..160,600.00
2 cu. yd (1.50 cu. m)...341,000.00
3-1/4 cu. yd (2.50 cu. m)..631,400.00
4-1/4 cu. yd (3.25 cu. m)..740,000.00
5-1/2 cu. yd (4.20 cu. m)..880,000.00
6 cu. yd (4.60 cu. m)..1,050,000.00

Backhoe Crawler

Bucket Size of Hoe
3/4 cu. yd. (0.57 cu. m).. $125,400.00
1 cu yd. (0.75 cu. m)..165,000.00
1-1/4 cu. yd. (0.95 cu. m)..170,500.00
1-1/2 cu. yd. (1.15 cu. m)..258,500.00
2 cu. yd. (1.50 cu. m)...298,100.00
2-1/2 cu. yd. (1.90 cu. m)..440,000.00
4-1/2 cu. yd. (3.45 cu. m)..1,020,000.00

Dragline Excavators Including a Medium Duty Bucket

Size of Dragline
1/2 cu. yd (0.38 cu. m)...$89,100.00
3/4 cu. yd (0.57 cu. m)...99,000.00
1 cu. yd (0.75 cu. m)..145,200.00
1-1/4 cu. yd (0.95 cu. m)..292,500.00
2-1/2 cu. yd (1.90 cu. m)..517,000.00
3 cu. yd (2.30 cu. m)...640,200.00
3-1/2 cu. yd (2.67 cu. m)..770,000.00

Clamshell Excavators Including a Medium Duty Bucket

Size of Clamshell
1/2 cu. yd (0.38 cu. m)...$88,000.00
3/4 cu. yd (0.57 cu. m)...99,000.00
1 cu. yd (0.75 cu. m)..145,200.00
1-1/4 cu. yd (0.95 cu. m)..288,200.00
2-1/2 cu. yd (1.90 cu. m)..517,000.00
3 cu. yd (2.30 cu. m)...640,200.00
3-1/2 cu. yd (2.67 cu. m)..770,000.00

Lift Cranes - Crawler Mounted

Capacity Tons	Kips	Kg (kilograms)	Cost
15	30	13,600	$165,000.00
20	40	18,145	203,500.00
30	60	27,215	216,700.00
40	80	36,290	231,000.00
45	90		271,700.00
50	100	45,360	302,500.00
70	140	63,500	418,000.00
100	200	90,720	597,300.00
150	300	136,080	715,000.00
200	400	181,400	880,000.00
400	800	362,880	1,320,000.00

Approximate Rental Costs For Excavating Equipment With Operator

Type	Size Cu. Yds.	Cu. M	Day	Week	Month
Crawler Mounted Backhoe	1/2	0.38	$453	$1,020	$6,770
	1	0.75	600	2,240	8,000
	2	1.50	1,020	3,700	12,500
	2-1/2	1.90	1,300	4,600	15,300
Compactor (with operator)	1000# Ram	454 kg Ram	57	275	980
	1000# Vib	454 kg Vib	47	195	915
	5000# Vib	2,268 kg Vib	220	915	3,400
	2000# Drum	907 kg Drum	100	430	1,950
Crane-Cable Crawler (with operator plus oiler)					
	30 T 60 Kips	27,215 kg	600	2,660	9,310
	40 T 80 Kips	36,290 kg	660	2,920	10,100
	60 T 120 Kips	54,430 kg	880	3,800	12,780
	100 T 200 Kips	90,720 kg	1,060	4,570	15,100
	150 T 300 Kips	136,080 kg	1,170	6,000	19,450
Crane-Cable Truck (with operator plus oiler)					
	40 T 80 Kips	36,290 kg	790	3,260	10,950
	60 T 120 Kips	54,430 kg	840	3,440	11,500
	80 T 160 Kips	72,575 kg	1,080	4,350	14,150
	100 T 200 Kips	90,720 kg	1,260	5,100	16,300
Crane-Hydraulic Truck (with operator plus oiler)					
	15 T 30 Kips	13,610 kg	500	2,170	7,440
	30 T 60 Kips	27,215 kg	740	3,130	10,740
	60 T 120 Kips	54,430 kg	1,140	4,790	15,600
Drill (with laborer)	(light)35 lb rock	18 kg	25	75	200
	(medium) 65 lb rock	30 kg	30	85	240
	(heavy) 90 lb rock	40 kg	40	95	300
Scraper (with operator)	11 cy	8.40 cu. m	500	2,250	9,000
	22 cy	16.8 cu. m	850	3,900	15,800
Tractor - Dozer (with operator)	105 hp	78 kw	400	1,900	7,600
	180 hp	134 kw	450	2,100	8,600
Loader (with operator)	1-1/2 cy	1.15 cu. m	400	1,900	7,600
	2-1/4 cy	1.70 cu. m	480	2,300	9,100

Type	Cu. Yds.	Cu. M	Day	Week	Month
Trencher (with operator)..........................	16 cy	12.25 cu. m	330	1,400	4,500
	24 cy	18.34 cu. m	400	1,600	5,800

The above costs do not include move charges or permits, and they do not include fuel or other daily operating expenses.

Basic Power Shovel Unit and Front End Attachments. Naturally, the ideal situation would be to have the most suitable and economical power shovel, backhoe, crane, or dragline for each job, but this is seldom practical or profitable. More frequently, the contractor must take into consideration the handling of a wide range of work with varying conditions, requiring more or less frequent and easy conversion (for example, shovel to dragline or crane to clamshell) by simply changing attachments on the basic unit.

Power cranes and shovels are designed to accommodate a wide range of front-end attachments and working tools to meet the various job requirements. The attachments are classified into three basic groups: the shovel boom, the crane boom, and the hoe boom. These attachments generally are interchangeable in the field.

Since the basic purpose of the shovel boom is for shovel work and the hoe boom for hoe work, the crane-type boom serves for all other materials handling tools, such as dragline buckets, clamshell buckets, orange peel buckets, hooks, hook blocks, tongs, grabs, clamps, grapples, concrete buckets, skull crackers, and pile driving leads.

Crane type booms can be extended for extremely high lifts by inserting standard sections in the center of the boom and by boom tip extensions called jibs. These jibs are used primarily to extend horizontal reach with boom raised vertically close to maximum elevation. They also provide an increase in vertical lifting range. Since the jib adds weight at the outer lifting radius and extends the working range, it has a limiting effect on lifting capacity.

Average Useful Life of Power Shovels, Hoes, and Cranes. A reasonable allowance for the exhaustion, wear, and tear of property used in business, including a reasonable allowance for obsolescence, is permitted in figuring depreciation on your equipment.

Average Useful Life of Power Shovels and Hoes

3/8–3/4 cu. yds. (.28-.57 cu. m)..5 years or 10,000 hours
1–1-1/2 cu. yds. (.76-1.15 cu. m)..6 years or 12,000 hours
2 cu. yds. and over (1.53 cu. m and over) ....8 years or 16,000 hours

Average Useful Life of Draglines, Clamshells and Cranes

	Tons	Years	Hours
3/8–3/4 c.y (0.28-0.57 cu.m)	2-1/2–5	5	10,000
1–1-1/2 c.y. (0.76-1.15 cu.m)...................	10-15	9	18,000
2 c.y (1.52 cu.m) and over.................	20 and over	12	24,000

The grouping shown for lifting cranes is a separate grouping and does not necessarily relate to the machines by size in cu. yds. For example, many 3/4 cu. yd. excavators (clamshell or dragline) would have a greater rating than 5 tons and might be rated in tons to fall in a different group than they do as excavators.

Depreciation. The straight line method of figuring depreciation is used. This method is in general adequate, but if it is desired to take changing prices for equipment into account in estimating, an annual appraisal to determine a current value should be considered.

Therefore, for the years and hours used, the following percentage of depreciation for the various groups are used:

Years	Hours	% per Year	% per Hour
5	10,000	20.00%	0.01%
6	12,000	16.67%	0.00833%
8	16,000	12.50%	0.00625%
9	18,000	11.11%	0.00555%
10	20,000	10.00%	0.005%
12	24,000	8.33%	0.00416%

Interest, Taxes and Insurance. Interest, taxes, insurance, and storage are usually charged at 20% of the average investment, of which 15% is interest (to be adjusted for current rates) and 5% covers taxes, insurance, storage, and incidentals.

Average investment must be established, based on the number of years used for depreciation. Since the first year is considered as 100% and since the investment is considered at the beginning of each year, the method of calculating average investment for a 5-year depreciation period is as follows:

1st year	100% of total investment
2nd year	80% of total investment
3rd year	60% of total investment
4th year	40% of total investment
5th year	20% of total investment
	300% of total investment

Average investment equals 300% divided by 5 years equals 60% of total investment.

The percent of Total Investment to use for Average Investment for 5, 6, 8, 9, 10, and 12 year periods are as follows:

No. Years Depreciation	Percent Average Investment per Year
5	60%
6	58.33%
8	56.25%

9	55.56%
10	55%
12	54.17%

Therefore, the percentage of Total Investment per year and per hour to use for interest, insurance, taxes, and storage, based on 20% of average investment for the groups considered here, is as follows:

Interest, Insurance, Taxes, Storage
Based on 20 percent of Average Investment per Year

Avg. Investment	Years	Hours	Percent of Total Investment Per Year	Per Hour
60%	5	10,000	12.000%	0.006000%
58.33%	6	12,000	11.666%	0.005833%
56.25%	8	16,000	11.250%	0.005625%
55.56%	9	18,000	11.112%	0.005556%
55%	10	20,000	11.000%	0.005500%
54.17%	12	24,000	10.834%	0.005417%

Fixed Costs, Repairs, Maintenance, and Supplies. These are difficult to estimate. The figures used are not based on actual records but are believed to be representative. In the case of shovels and hoes, this cost is based on 100% of the total investment spread over the life of the machine and includes repairs, replacement parts and repair labor for normal operation, and such items as ropes, dipper and bucket teeth, oil and grease for lubricating machinery, as well as seasonal overhaul. It does not include engine fuel or lubricating oil.

In the case of draglines and clamshells, 80% of the total investment spread over the economic life of the machine is a fair average cost of these items.

In the case of lifting cranes, 60% of the total investment spread over the economic life of the machinery may be used.

The owner of an excavator should anticipate three things in regard to repairs and maintenance. First, the average figures used presume proper maintenance. Without this, repair bills will greatly exceed the estimates given here. Second, it is assumed the equipment will be used within its specified rating. Third, these expenses are not uniform each year.

Maintenance is then further defined in two major categories. The first is normal maintenance as a daily operating expense, which includes regular servicing on the job, such as fuel, changing oil, filters, greasing, brakes and hydraulic fluid levels, minor tuneups and the like.

The second category is defined as major maintenance, which requires that the equipment be idle for an extended period while repairs are made in the shop. An example of major maintenance would be the undercarriage of a dozer.

Provision should be made for a periodic general overhaul, and it should be expected that both the amount of repairs required and the losses caused by shutdowns will increase as the equipment becomes older.

Repairs, Maintenance, and Supplies, Including Labor Associated With Them

Shovels and Hoes

Size	Years	Hours	Percent Of Total Investment Per Year	Per Hour
3/8–3/4 c.y. (.28-.57 m3)	5	10,000	20.00%	0.01000%
1–1-1/2 c.y. (.76-1.15 m3)	6	12,000	16.67%	0.00833%
2–3-1/2 c.y. (1.53-2.68 m3)	8	16,000	12.50%	0.00625%
4-1/4 c.y. (3.25 m3) or more	10	20,000	10.00%	0.00500%

Based on 100% of total investment spread over life of machine.

For example, assume a 3/4-c.y. (0.60 m³) shovel cost $125,000.00. Repairs, maintenance, and supplies equal $125,000.00 x .0001 = $12.50 per hr.

Draglines and Clamshells

Cu. Yds.	Years	Hours	Percent Of Total Investment Per Year	Per Hour
3/8–3/4 c.y. (.28-.57 m3)	5	10,000	16.00%	0.00800%
1–1-1/2 (.76-1.15 m3)	9	18,000	8.89%	0.00445%
2–3-1/2 (1.53-2.68 m3)	12	24,000	6.66%	0.00333%

Based on 80% of Total Investment Spread over Life of Machine.

Lifting Cranes

Tons	Years	Hours	Percent Of Total Investment Per Year	Per Hour
2-1/2–5	5	10,000	12%	.006%
10–15	9	18,000	6.67%	.00333%
20 and over	12	24,000	5%	.0025%

Based on 60% of Total Investment Spread over Life of Machine.

Operating Costs, Engine Fuel, and Lubricating Oil. A formula for estimating the approximate amount of diesel fuel consumption for equipment of this type is as follows:

BHP x Factor x lbs. fuel per HP hr.
Weight of fuel per gallon (liter)

= Gallons (Liters) per Hour, where

BHP = Brake HP of engine—or rated HP
Factor = Factor for this use 50 to 60%
Diesel = 0.5 lbs. per brake horsepower hour.
Diesel = 7.3 lbs. per gallon (U.S.)

From this formula we obtain an approximate diesel fuel consumption of 0.034 to 0.041 gals. (0.13-0.15 liters) per horsepower hour (based on 50% to 60% factor respectively), and suggest using an average of about 0.040 gal. per horsepower hour.

Shovels normally will consume a greater amount of fuel than the other types of machines considered here. Therefore the larger consumption rate indicated should be used for estimating fuel for shovels and the smaller rate for draglines and clamshells. Machines used for lifting crane service only usually operate intermittently and fuel consumption is difficult to estimate.

Example: A 100-horsepower engine in shovel service is estimated to consume per hour: 100 x 0.040 = 4.0 gals. (15 liters) of diesel fuel per hr.

Lubricating oil is considered with fuel, because it varies with the size and type of engine. It usually includes a complete change every 100 hours plus make-up oil between changes. Allow 15% of fuel costs.

Labor Operating Costs. The labor rates, as well as the number in the operating crew, vary in different parts of the country and on different jobs. The contractor must estimate this cost. Costs related to crew costs, besides rates of pay, are employer contributions for Federal Insurance Contribution Act, Workmen's Compensation Insurance, Unemployment Compensation, Overtime, Paid Holidays, and Contractor's Contributions to Union Welfare and Pension Funds.

Power Shovel Yardages. To estimate yardage production on a job is a real problem. If all conditions were the same on all jobs, it would be a simple matter, but jobs are as different as night and day. There are many factors which must be considered, many of which can be learned only through experience. The type of material, the depth of cut for maximum effect, delays in operation, a 90° swing for the shovel. All affect the output.

The following table gives an approximate idea of the difference in maximum output possible, subject to conditions such as those listed above.

Hourly Shovel Output in Cubic Yards

Class of Material	3/8	1/2	3/4	1	1-1/4	1-1/2	1-3/4	2	2-1/2
					Shovel Capacity				
Moist Loam, Light Sandy Soil	85	115	165	205	250	285	320	355	405
Sand and Gravel	80	110	155	200	230	270	300	330	390
Common Earth	70	95	135	175	210	240	270	300	350
Clay, Hard and Tough	50	75	110	145	180	210	235	265	310
Rock, Well Blasted	40	60	95	125	155	180	205	230	275
Common Earth, with Rocks and Roots	30	50	80	105	130	155	180	200	245
Clay, Wet and Sticky	25	40	70	95	120	145	165	185	230
Rock, Poorly Blasted	15	25	50	75	95	115	140	160	195

Hourly Shovel Output in Cubic Meters

Class of Material	0.3	0.4	0.6	0.8	1.0	1.1	1.3	1.5	1.9
					Shovel Capacity				
Moist Loam, Light Sandy Soil	65	88	126	157	191	218	245	271	310
Sand and Gravel	61	84	119	153	176	206	230	252	298
Common Earth	54	73	103	134	160	183	206	230	268
Clay, Hard and Tough	38	57	84	111	138	161	180	203	237
Rock, Well Blasted	31	46	73	96	119	138	157	176	210
Common Earth, with Rocks and Roots	23	38	61	80	99	119	138	153	187
Clay, Wet and Sticky	19	31	54	73	92	111	126	141	176
Rock, Poorly Blasted	11	19	38	57	73	88	107	122	149

The quantities in the above table are based on bank measure, which means cubic yards removed from the bank rather than cubic yards (cubic meters) in the hauling unit. There is a big difference between a yard of dirt in the bank and a yard of dirt in the truck. This is because of the swell of the material or its increase in volume due to voids when it is dug or loosened. For instance, common excavation will swell from 10% to 30%. Another condition is that the machine is working in a depth of cut suitable for maximum digging efficiency. The optimum depth of cut for various sizes of shovel may be defined as that depth which produces the greatest output and at which the dipper comes up with a full load.

The output figures are based on continuous operations, 60 minutes per working hours, without any delays for adjustments, lubrication, operator stopping for any reason, etc.

The figures are based on each dipper-full of material being swung through an arc of 90° before dumping. This is important, because a swing of either a lesser number or greater number of degrees than 90° will either save or consume time and affect output capacity.

Tables are based on all materials being loaded into hauling units, with the further understanding that these hauling units are properly sized to shovel capacity and are provided in sufficient quantities to take away all the material that the shovel can dig.

While these figures represent a very comprehensive job and as realistic a picture as can be presented on this involved subject, a word of caution is in order. These figures are general only and should not be considered as guaranteed outputs, or used by anyone as a basis for figuring and bidding jobs.

How Degrees of Swing and Depth of Cut Affect Shovel Yardage Output. The quantities in the above table are based on the optimum depth of cut and a swing of 90° was specified.

Because variations in either of these conditions will have considerable effect on the yardage output, the following chart shows what happens when swing and depth of cut vary from the conditions on which the previous basic hourly yardages were established.

Table Giving Effect of Depth of Cut and Angle of Swing
on Power Shovel Output

Depth of Cut in % of Optimum, l.f. (m)	45°	60°	75°	90°	120°	150°	180°
40' (12.2m)	0.93	0.89	0.85	0.80	0.72	0.65	0.59
60' (18.3m)	1.10	1.03	0.96	0.91	0.81	0.73	0.66
80' (24.4m)	1.22	1.12	1.04	0.98	0.86	0.77	0.69
100' (30.5m)	1.26	1.16	1.07	1.00	0.88	0.79	0.71
120' (36.6m)	1.20	1.11	1.03	0.97	0.86	0.77	0.70
140' (42.7m)	1.12	1.04	0.97	0.91	0.81	0.73	0.66
160' (48.8m)	1.03	0.96	0.90	0.85	0.75	0.67	0.62

Dragline Yardages. As in the case of power shovels and hoes, dragline yardage production is affected by the type of material to be excavated, the depth of cut, the swing before unloading, the type of unloading (unloaded into hauling units, cast onto spoil banks, etc.), the degree of continuity in the operation, etc.

The following table gives an approximate idea of the difference in maximum production possible subject to conditions listed:

Hourly Short Boom Dragline Output in Cubic Yards

Bucket Capacity	3/8	1/2	3/4	1	1-1/4	1-1/2	1-3/4	2	2-1/2
Class of Material									
Moist Loam, Light Sandy Clay	70	95	130	160	195	220	245	265	305
Sand and Gravel	65	90	125	155	185	210	235	255	295
Common Earth	55	75	105	135	165	190	210	230	265
Clay, Hard and Tough	35	55	90	110	135	160	180	195	230
Clay, Wet and Sticky	20	30	55	75	95	110	130	145	175

Hourly Short Boom Dragline Output in Cubic Meters

Bucket Capacity	.28	.38	.57	.76	.96	1.15	1.34	1.53	1.91
Class of Material									
Moist Loam, Light Sandy Clay	54	73	99	122	149	168	187	203	233
Sand and Gravel	50	69	96	119	141	161	180	195	225
Common Earth	42	57	80	103	126	145	160	175	203
Clay, Hard and Tough	27	42	69	84	103	122	138	150	175
Clay, Wet and Sticky	15	23	42	57	73	84	99	110	134

The dragline is working in the optimum depth of cut for maximum efficiency.

The dragline is working a full 60 minutes each hour—no delays.

The dragline is making a 90° swing before unloading.

The bucket loads are being dumped into "properly sized" hauling units.

The proper type bucket is being used for the job.

The dragline is being used within the working radius recommended by the manufacturer for machine stability.

How Degrees of Swing and Depth of Cut Affect Dragline Yardage Output. The following table gives the effect of depth of cut and angle of swing on dragline yardage production. Variations in degrees of swing, in particular, have a marked effect on production and this is important to keep in mind when laying out a job as the shorter the swing the more the yardage.

Table Giving Effect of Depth of Cut (in feet and meters)
and Angle of Swing on Dragline Output

Depth of Cut in % of Optimum	Angle of Swing in Degrees							
	30	45	60	75	90	120	150	180
20' (6.1m)	1.06	0.99	0.94	0.90	0.87	0.81	0.75	0.70
40' (12.2m)	1.17	1.08	1.02	0.97	0.93	0.85	0.78	0.72
60' (18.3m)	1.24	1.13	1.06	1.01	0.97	0.88	0.80	0.74
80' (24.4m)	1.29	1.17	1.09	1.04	0.99	0.90	0.82	0.76
100' (30.5 m)	1.32	1.19	1.11	1.05	1.00	0.91	0.83	0.77
120' (36.6m)	1.29	1.17	1.09	1.03	0.98	0.90	0.82	0.76
140' (42.7m)	1.25	1.14	1.06	1.00	0.96	0.88	0.81	0.75
160' (48.8m)	1.20	1.10	1.02	0.97	0.93	0.85	0.79	0.73
180' (54.9m)	1.15	1.05	0.98	0.94	0.90	0.82	0.76	0.71
200' (61.0m)	1.10	1.00	0.94	0.90	0.87	0.79	0.73	0.69

Clamshell Production. The clamshell excavator is not to be considered a high production machine but rather a machine to be used where the work is beyond the scope of other types of equipment. For example, a condition which usually requires a clamshell is where digging is vertical or practically straight down as in digging pier holes or shafts. Digging in trenches that are sheathed and cross-braced generally calls for a clamshell, because the vertical action of the bucket enables it to be worked through the cross-bracing. Jobs requiring accurate dumping or disposal of materials are usually clamshell jobs. Also, for high dumping jobs, whether it be charging a bin, building a stockpile, or wherever the material must be dumped well above the machine level, the clamshell is well adapted. In general, the clamshell can operate vertically and dig or spot dump below, at, or above the level of the machine.

The clamshell is only effective where the materials to be handled are relatively soft or loose.

Conditions are so variable on clamshell operation that a table showing typical production has little value.

However, in an effort to establish some point to work from, the following table has been prepared showing maximum production to be expected from clamshell excavators operating under the following conditions:

The clamshell is engaged in open digging such as a basement or large footing, with the permissible cut at least a full bucket depth.

1. The depth of the excavation is not more than 10'-0" (3 m).

2. The quantities are in terms of bank measure.

3. There is no wasted time, 60 minutes of digging each hour.

4. The clamshell is making a 90° swing before unloading.

5. The bucket loads are being dumped into "properly sized" hauling units.

6. The clamshell is being used within the working radius recommended by the manufacturer for machine stability.

Hourly Short Boom Clamshell Handling Capacity, Cubic Yards

Bucket Capacity	3/8	1/2	3/4	1	1-1/4	1-1/2	1-3/4	2	2-1/2
Class of Material									
Moist Loam or Sandy Clay	50	65	95	120	140	155	170	190	225
Sand and Gravel	45	60	85	110	130	140	160	175	205
Common Earth	40	55	70	95	115	125	145	160	185

Hourly Short Boom Clamshell Handling Capacity, Cubic Meters

Bucket Capacity	0.28	0.38	0.57	0.76	0.96	1.15	1.34	1.53	1.91
Class of Material									
Moist Loam or Sandy Clay	38	50	73	92	107	119	130	145	172
Sand and Gravel	34	45	65	84	99	107	122	134	157
Common Earth	30	42	53	73	88	96	110	122	141

It must be thoroughly understood that the above production is based on the most ideal job and management conditions.

How Degrees of Swing Affect Clamshell Yardage. The output of a clamshell operating at a steady pace is affected by the swing before unloading in the same manner as a dragline. The following table gives the effect of angle of swing on clamshell yardage production.

Angle of Swing in Degrees

30°	45°	60°	75°	90°	120°	150°	180°
1.32	1.19	1.11	1.05	1.00	0.91	0.83	0.77

Job and Management Factors. Ideal conditions seldom, if ever, exist in the field, and it is necessary to rationalize these figures by some other factor to compensate for the fact that actual job conditions differ widely from the perfect conditions assumed so far.

There are two sets of factors on every job that have a great deal to do with the output of equipment on the job. They are the job factor and the management factor.

Job factors are the physical conditions pertaining to a specific job that affect the production rate, other than the class of material to be handled. These may be divided into three general headings:

1.   topography and the dimensions of the work, which include the depth of cut and whether the work will require much moving within the cut or from one cut to another;

2.   surface and weather conditions, which include in some cases the difference between summer and winter work, and also the question of drainage of either surface or underground water;

3.   specification requirements that control the manner in which the work must be handled, or indicate the sequence of various operations, and which also show the amount of bank sloping or cutting close to finished line and grade that is required.

The above conditions are all inherent in the job itself and can be taken into account by the contractor in making up the bid.

Another group of factors which affect output are the management factors. These cover conditions which pertain to the efficiency of the operation and thus affect output. These are grouped under the head of the management factor because, in general, they are made up of items which management can determine and control. These include:

1.   selection, training, and direction of personnel; the quantities given in the tables indicate what can be reasonably expected by an experienced operator who is willing to work; the availability of trained workers and the incentive to produce must be considered;

2.   selection, care, and repair of equipment; the quality of the inspection and preventive maintenance program can have much to do with the lost time on the job;

3.   planning, laying out the job, supervising, and coordinating the operations is a very important factor in increasing efficiency; providing the right number and size of hauling units is one of the first things to show its effect.

Foremanship and supervision are important. When everything is working smoothly, there is added incentive for high production rates.

Because both job and management factors must be considered jointly, that is, the output yardages must be modified by both a job and a management factor, the following table consolidates these factors.

To estimate the yardage, select the proper combined job and management factor from the following table. First, select a job factor classification of excellent, good, fair or poor, whichever applies to your job conditions and follow straight across the management factor classification of excellent, good, fair or poor, whichever applies to the management conditions. The resulting figure is the combination job management factor by which the yardage figures from the basic output tables (as modified by the conversion factors for depth of cut and swing) must be multiplied to give the yardage factor that can reasonably be expected under the job management conditions that will exist on the job.

Usually, pit and quarry and heavy construction jobs will fall into groups 1, 2, or 3, whereas highway grading will usually fall into groups 2, 3, or 4 because of the variations in cuts, required machine travel, close cuts, etc.

Combination Job and Management Factors

| | Management Factors | | | |
Job Factors	1 Excellent	2 Good	3 Fair	4 Poor
1. Excellent	0.84	0.81	0.76	0.70
2. Good	0.78	0.75	0.71	0.65
3. Fair	0.72	0.69	0.65	0.60
4. Poor	0.63	0.61	0.57	0.52

Power Shovel Excavating

Modern shovels, powered by diesel engines, have tremendous capacity when conditions permit working without interruptions.

The cost of power shovel excavating will vary with the size of the job, size of the shovel, and the speed with which the trucks are handled in and out of the hole. Seldom can the shovel work to capacity on building work, due to the congestion of trucks waiting to get in or out of the hole or due to trucks being delayed at the dump.

The size of the job has considerable bearing on the costs, as it costs just as much to get the shovel to the job and remove it at completion for 1,000 cu. yds. (765 cu.m) as for a 10,000 cu. yd. (7645 cu.m) job.

Where it is possible to use a 3/4-cu. yd. (0.60-m) shovel to capacity (figuring a 100% bucket load each cycle), it should excavate 135 cu. yds. (103 cu.m) an hr. but the bucket is not always filled to capacity and there are delays in loading, moving shovel from place to place, cleaning up, bad weather, etc., so the average on building work will probably run 50% to 60% of the rated capacity of the shovel.

Basis for Computing Daily Output of Diesel Powered Shovel. The following method is used in figuring the output of a power shovel in basement excavation, based on a 1 cu. yd. (0.76 cu.m) shovel:

Bucket capacity .... 1 cu. yd. (0.76 cu.m)
Bucket efficiency, average material 90%
Load carried .... 0.9 cu. yds. (0.68 cu.m)
Average operating cycle .... 20 Seconds
Cycles per hour .... 180 Cycles
Cu. Yds. per hour, operating 100% time .... 162 cu. yds. (124 cu.m)
Operating efficiency in basement excavation due to haulage service,
   machine delays and cleanup work, approximately .... 50 to 60%
Average output per 8-hr. day .... 640-760 cu.yds. (490-580 cu.m)
Operating efficiency on heavy construction, roads, dams,
   open cut excavation, etc., including machine delays and
   cleanup work, approximately 66.67%. Average
   output per 8-hr. day .... 864 cu.yds. (660 cu.m)

Dragline Excavating

The cost of dragline excavating will vary in the same manner as shovel excavating, according to the size of the job, size of the machine, and the capacity and speed of the hauling units. But the hauling hazards are reduced substantially by the trucks not having to be loaded in the hole, eliminating the necessity of trucking up a ramp.

As may be determined by comparing the basic output tables for shovels and draglines, the dragline has about 75% to 80% of the basic output capacity of shovels. However, the dragline definitely has a place in excavating for construction work by virtue of its generous digging reach, its ability to dig under extremely wet conditions (it can stand on the top on dry, firm footing), and the easier haul with no ramps to climb, all conditions that under certain circumstances would make shovel digging very expensive if not totally impossible.

In excavating for building construction, the dragline is affected by job and management factors the same as shovels and the same basis for computing daily output may be used as follows.

Basis for Computing Daily Output of Dragline Excavator. The following is based on a dragline having a 1 cu. yd. (0.76 cu.m) bucket.

Bucket capacity .... 1 cu. yd. (0.76 cu.m)
Bucket efficiency, average material 90%
Load carried .... 0.9 cu. yd. (0.68 cu. m)
Average operating cycle .... 24 Seconds
Cycles per hour .... 150 Cycles
Cu. Yds. Per hour, operating 100% time .... 135 cu. yds. (102 cu.m)
Operating efficiency in basement excavation

due to machine delays, traffic delays to hauling
units and clean up work, approximately ....50 to 60%
Average output per 8-hr. day ....540-650 cu. yds. (413-497 cu.m)
Operating efficiency on heavy construction, roads, dams,
open cut excavation, etc., including machine delays and
clean up work, approximately 66.67%
Average output per 8-hr. day .... 720 cu. yds. (550 cu.m)

Backhoe Excavating

The backhoe is a cross between the dragline and the shovel, which is why it is the equipment of choice today. It incorporates some of the characteristics of each and overcomes some of the limitations of each. The backhoe is primarily a unit for digging below machine level, but it will dig harder material than the clamshell or dragline, because the weight of the boom itself may be used to force the dipper into the material. It is, however, limited in digging by the length of the boom and stick. The backhoe dipper can be controlled more accurately than the dragline bucket and is better suited to close-limit work.

In building construction the backhoe is used to dig trenches, footings, and basements. On small residence basements, it offers many advantages. It digs straight, vertical side walls (in soil which will stand); it cuts a level floor; it trims corners neatly and squarely; it can dig sewer and waterline trenches; it always works from the top on dry, safe ground; and it reduces hand trim to a minimum.

In general, the process of digging a basement with a backhoe is to dig a trench around the four sides of the basement and scoop out the center as you go.

On small work it is necessary to have low-cost, simple and easy means of moving the backhoe from job to job. This is best done by using a single-purpose trailer which can be loaded or unloaded in 15 to 20 minutes. The cost of moving the backhoe from job to job will vary with the distance between jobs.

On time studies made on several small basement jobs, where the basements contained from 275 to 350 cu. yds. (210-268 cu.m) of excavation, a backhoe excavator equipped with a 3/4-cu. yd. (0.57-cu.m) bucket averaged from 66 to 88 cu. yds. (50-67 cu.m) per hour.

An example of the cost of excavating a small basement containing approximately 350 cu. yds. (268 cu.m) of excavation, with the soil placed around the excavation, is as follows:

	Hours	Rate	Total	Rate	Total
Move-In Charge		$....	$....	$110.00	$110.00
Excavating					
Hoe operator	5.0	....	....	28.55	142.75

Labor, clean-up 5.0	....	....	21.04	105.20
Hoe charge 5.0	....	....	38.00	190.00
Total Direct Cost...........................		$....		$547.95
Cost per cu. yd			....	1.57
Cost per cu. m			....	2.04

The above costs do not include removal of excavated earth from site or backfilling.

Data on Diesel Power Shovel Costs and Operation

Description	Size of Shovel (in Cu. Yds.)						
	3/4	1	1-1/2	2	3-1/4	4-1/4	5-1/2
Cu. Yds. per Hr. Based on 100% Time	135	180	270	360	585	765	950
Cu. Yds. per Hr. Based on 100% Time and 90% Dipper Efficiency	121	162	243	324	526	688	855
Output per Hr. Based on 66.67% Efficiency, Cu. Yds.	90	120	180	240	390	510	633
Output per Hr. Based on 50% Efficiency, Cu. Yds.	68	90	135	180	293	383	475
Output per 8-Hr. Day, Based on 66.67% Efficiency, Cu. Yds.	720	960	1440	1920	3120	4080	5064
Output per 8-Hr. Day, Based on 50% Efficiency, Cu. Yds.	544	720	1080	1440	2344	3064	3800
Average Cost—Diesel Shovel, including Freight and Unloading	$91,000	96,000	146,000	310,000	547,000	672,000	955,000
Depreciation, Percent per Year	20	16.66	16.66	16.66	12.50	10	10
Depreciation per Hr. Based on 2,000 Hrs. per Year	$9.10	$8.00	$12.16	$25.82	$34.19	$33.60	$47.75
Interest, Taxes, Insurance, 20% per Year of 2,000 Hrs.	$5.50	$5.80	$8.80	$18.60	$32.80	$33.60	$47.80
Hourly Consumption of Fuel, Approximate Gals. per Hour	4.5	4.5	5.2	6.4	12.0	15.5	18.0
Fuel Cost per Hour*	$6.75	$6.75	$7.80	$9.60	$18.00	$23.25	$27.00
Engine Lub. Oil Cost, 15% of Fuel	$1.01	$1.01	$1.17	$1.44	$2.70	$3.48	$4.05
Cost per Hr., Repairs, Maintenance and Supplies, including Rope, Grease	$8.00	$8.00	$8.00	$8.00	$8.00	$8.00	$8.00
Total Hourly Cost, Not Including Labor, Supervision, Compensation, Unemployment, Soc. Sec. etc.	$29.56	$30.36	$37.93	$63.46	$95.69	$101.93	$134.60

*Fuel costs based on diesel fuel at $1.50 per gal.

(metric)
Size of Shovel (in Cubic Meters)

Description	0.60	0.80	1.15	1.50	2.50	3.25	4.20
Cu. M per Hour Based on 100% Time	103	138	207	275	447	585	726
Cu. M per Hour Based on 100% Time and 90% Dipper Efficiency	93	124	186	247	402	527	653
Output per Hour Based on 66.67% Efficiency, Cu. Meters	68	92	138	183	298	390	484
Output per Hour Based on 50% Efficiency, Cu. Meters	52	69	104	137	224	293	363
Output per 8-Hour Day, Based on 66.67% Efficiency, Cu. M	544	736	1104	1464	2384	3120	3872
Output per 8-Hour Day, Based on 50% Efficiency, Cu. M	416	538	832	1096	1792	2344	2904
Average Cost—Diesel Shovel, Including Freight and Unloading	91,000	96,000	146,000	310,000	547,000	672,000	955,000
Depreciation, Percent per Year	20	16.66	16.66	16.66	12.50	10	10
Depreciation per Hour Based on 2,000 Hours per Year	$9.00	$8.00	$12.00	$26.00	$34.00	$34.00	$48.00
Interest, Taxes, Insurance, 20% per Year of 2,000 Hours	$5.50	$5.80	$8.80	$18.60	$32.80	$33.60	$47.80
Hourly Consumption of Fuel, Approximate Liters per Hour	17.00	17.00	20.00	24.00	45.00	60.00	68.00
Fuel Cost per Hour*	$6.80	$6.80	$8.00	$9.60	$18.00	$24.00	$27.20
Engine Lub. Oil Cost, 15% of Fuel	$1.02	$1.02	$1.20	$1.44	$2.70	$3.60	$4.08
Cost per Hour, Repairs, Maintenance and Supplies, Including Rope, Grease,etc	$8.00	$8.00	$8.00	$8.00	$8.00	$8.00	$8.00
Total Hourly Cost, Not Including Labor, Supervision, Compensation, Unemployment, Soc. Sec. etc.	$30.32	$29.62	$38.00	$63.64	$95.50	$103.20	$135.08

*Fuel costs based on diesel fuel at $1.50 per gal.

Below is a digging plan for excavating small basements using a hoe. The order of the letters indicates the order in which the sections are removed. The arrows indicate the direction the hoe travels in taking out each section.

Truck and trailer time to deliver the hoe and removing same at completion of job will vary with length of haul.

In medium to large work and at moderate depths, hoe production can approach that of shovels. Output falls off considerably, however, at greater depths.

For the smaller machines, the size hoe is usually chosen for the width trench it will cut, but for the larger sizes, the capacity of the machine is chosen on a production basis.

Backhoe Excavating Trench
Courtesy Caterpillar, Inc.

Data on typical backhoe bucket dimensions and weight is given in the following table.

Dipper Capacity,		Bucket Outside Width in		Total Width Added for 2 Side Cutters		Weight in	
cu. yd.	cu.m	inches	mm	inches	mm	Lbs.	Kg
3/8	0.28	20-24	500-600	4-6	100-150	850	386
1/2	0.38	24-28	600-700	4-9	100-225	950	430
3/4	0.57	28-39	700-975	4-9	100-225	1400	635
1	0.76	33-45	825-1125	4-9	100-225	1700	771
1-1/4	0.95	39-45	975-1125	4-9	100-225	2000	907
1-1/2	1.15	39-45	975-1125	4-9	100-225	2,300	1043

Production for Power Shovel Excavating
Using a 3/8-Cu. Yd. (0.28 Cu. M) Shovel
Based on an Average Daily Output of 300 Cu. Yds. (230 Cu. M)

Volume of Excavation in Job

Cu. Yds.	1000	1500	2000	2500	3000	4000	5000
Cu. Meters	765	1150	1530	1910	2290	3060	3820

Actual Operating Time of Shovel

Foreman*	27	40	54	67	80	107	134
Shovel Operator	27	40	54	67	80	107	134

*Foreman in charge of job, handling and expediting trucks, etc.

Production for Power Shovel Excavating
Using a 1/2-Cu. Yd. (0.38 Cu. M) Shovel
Based on an Average Daily Output of 400 Cu. Yds. (306 Cu. M)

Volume of Excavation in Job

Cu. Yds.	1000	1500	2000	2500	3000	4000	5000
Cu. M	765	1147	1529	1911	2294	3058	3823

Actual Operating Time of Shovel

Foreman*	20	30	40	50	60	80	100
Shovel Operator	20	30	40	50	60	80	100

*Foreman in charge of job, handling and expediting trucks, etc.

Production for Power Shovel Excavating
Using a 3/4-Cu. Yd. (0.57 Cu. M) Shovel
Based on an Average Daily Output of 600 Cu. Yds. (460 Cu. M)

Volume of Excavation in Job

Cu. Yds.	1000	1500	2000	2500	3000	4000	5000
Cu. M	765	1147	1529	1911	2294	3058	3823

Actual Operating Time of Shovel

Foreman*	14	20	27	34	40	54	68
Shovel Operator	14	20	27	34	40	54	68

*Foreman in charge of job, handling and expediting trucks, etc.

Production for Power Shovel Excavating
Using a 1-Cu. Yd. (0.76 Cu. M) Shovel
Based on an Average Daily Output of 720 Cu. Yds. (550 Cu. M)

Volume of Excavation in Job

Cu. Yds.	1000	1500	2000	2500	3000	4000	5000
Cu. M	765	1147	1529	1911	2294	3058	3823

Actual Operating Time of Shovel

Foreman*	11	17	22	28	34	45	56

Shovel Operator	11	17	22	28	34	45	56

*Foreman in charge of job, handling and expediting trucks, etc.

Combination Backhoe / Loader
Courtesy Caterpillar, Inc.

Production for Dragline Excavating
Using a 1/2-Cu. Yd. (0.38 Cu. M) Dragline
Based on an Average Daily Output of 320 Cu. Yds. (245 Cu. M)

Volume of Excavation in Job

Cu. Yds.	1000	1500	2000	2500	3000	4000	5000
Cu. M	765	1147	1529	1911	2294	3058	3823

Actual Operating Time

Foreman*	25	38	50	63	75	100	125
Dragline Operator	25	38	50	63	75	100	125

*Foreman in charge of job, handling and expediting trucks, etc.

Production for Dragline Excavating
Using a 1-Cu. Yd. (0.76 Cu. M) Dragline
Based on an Average Daily Output of 560 Cu. Yds. (428 Cu. M)

Volume of Excavation in Job

Cu. Yds.	1000	1500	2000	2500	3000	4000	5000
Cu. M	765	1147	1529	1911	2294	3058	3823

Actual Operating Time

Foreman*	18	27	36	45	54	72	90

Dragline
    Operator...............18.............27.........36...........45.........54........72........90
*Foreman in charge of job, handling and expediting trucks, etc.

Sample Excavating Estimate

The following example is based on a basement containing 5,000 cu. yds. (3823 cu. m) excavation, average conditions, using a 1-1/4 c.y. (0.96 cu. m) power shovel, loading into trucks, based on an average of 900 cu. yds. (688 cu. m) per 8-hr. day.

Bringing Shovel to Job	Hours	Rate	Total	Rate	Total
Move-In Charge		$....	$....	$275.00	$275.00
Excavating					
Foreman	48	....	....	29.41	1,411.68
Shovel operator	48	....	....	28.55	1,370.40
Shovel charge	48	....	....	46.00	2,208.00
Total Direct Cost			$....		$5,265.08
Cost per cu. yd				....	1.05
Cost per cu. m				....	1.37

Add cost of trucks for removing excavated soil, overhead expense, and profit. The above method for estimating shovel excavation can also be used for hoe, dragline, and clamshell excavation by merely substituting the proper production values and equipment charges for those pertaining to shovels.

Hauling Excavated Materials in Trucks

Where an excess of excavated materials over and above the amount required for backfill exists, it will have to be removed from the job by trucks.

The usual procedure followed in these cases is to determine beforehand the approximate quantity of material required for backfill, etc., and this amount subtracted from the estimated excavation quantity will give the volume of excavated material to be hauled away. The excess material should be loaded and hauled as the excavation progresses so that no rehandling is necessary.

Types of Trucks. Truck hauling units are mainly of two types: conventional highway hauling units and off-the-highway heavy duty hauling equipment.

The conventional highway hauling unit is the most frequently used. For the average contractor it is the most economical type of truck to use. It is adaptable to practically every job and may be used for other hauling purposes as well.

For the contractor specializing in large earth moving projects, the high production, heavy duty, off-the-highway hauling unit is the most economical equipment to use, because it is specifically designed for this purpose. Not limited by state highway regulations as to weight and size, it has a high ratio of net weight to payload, the additional weight being found in its more rugged construction, such as heavier frame, axles, body, hoist, and engine.

Power Shovel Excavating
Courtesy Caterpillar, Inc.

Hourly Cost of Ownership and Operation of Trucks. To estimate hauling costs, the contractor must know the equipment costs to own and operate. This can be determined by the same method previously described for excavating equipment.

Attention is called to the manner in which tire costs (original, replacement, and repair) are handled. Recognizing the fact that the economic life of tires will not be as long as the life of the vehicle, the original value of the tires is deducted from the total price before computing the hourly charge for depreciation and is then treated as an operating cost.

The hourly cost of tire replacement and tire repairs varies greatly according to operating conditions, road surfaces, and tire loadings. While tire costs in ordinary trucking operations normally are figured on a mileage basis, in construction work, with its rough roads, short hauls, and relatively low speeds (below 30 mph), they are figured on an hourly basis and are computed by dividing the replacement cost by the estimated tire life in hours. The average life expectancy of truck tires in construction work is from 3,000 to 3,500 hours. Tire repairs should run about 15% of the hourly tire cost.

The following tables give data on various sizes of rear dump trucks for both conventional highway hauling units and heavy duty off-highway equipment.

Table Showing How Hourly Cost of Ownership and Operation
May Be Derived for Conventional Highway Rear Dump Trucks
Used in Construction Work Based on an Economic
Life of 3 Yrs. of 2,000 Hours Each
(in Tons)

Capacity in Tons	5	12	15	20	25
Capacity in Cu. Yds., Struck	4	6	10	15	20
Approximate Delivered Price	$30,000	$60,000	$80,000	$90,000	$105,000
Original Tire Value	2,200	3,000	4,000	4,500	5,200
Depreciation, .0167% of Total Investment Less Tires	$5.01	$10.02	$13.36	$15.03	$17.54
Interest, Taxes and Insurance, 0.0067% of Total Investment	2.01	4.02	5.36	6.03	7.04
Hourly Tire Replacement Cost, Based on 3,000-Hr. Life	.70	.80	1.80	2.00	2.40
Hourly Cost of Tire Repairs, 15% of Hourly Tire Replacement Cost	.11	.12	.27	.30	.36
Maintenance and Repairs, .004% of Total Investment	1.20	2.40	3.20	3.60	4.20
Fuel Cost per Hour, .04 Gal. per HP	7.20	7.60	11.00	15.00	16.00
Oil and Grease, 15% of Fuel	1.08	1.14	1.65	2.25	2.40
Total Hourly Cost, Not Including Labor, Supervision, Unemployment, Soc. Sec., Overhead Expense & Profit	$17.31	$26.10	$36.64	$44.21	49.93

Table Showing How Hourly Cost of Ownership and Operation
May Be Derived for Conventional Highway Rear Dump Trucks
Used in Construction Work Based on an Economic
Life of 3 Yrs. of 2,000 Hours Each
(in Kilograms)

Capacity in Kilograms (Kg)	4530	10890	13610	18140	22680
Capacity in Cu. Meters, Struck	3.05	4.60	7.65	11.50	15.30
Approximate Delivered Price	$30,000	$60,000	$80,000	$90,000	$105,000
Original Tire Value	2,420	3,300	4,400	4,950	5,720
Depreciation, 0.0167% of Total Investment Less Tires	$5.01	$10.02	$13.36	$15.03	$17.54
Interest, Taxes and Insurance, 0.0067% of Total Investment	2.01	4.02	5.36	6.03	7.04
Hourly Tire Replacement Cost, Based on 3,000-Hr. Life	.81	1.10	1.47	1.65	1.91
Hourly Cost of Tire Repairs, 15% of Hourly Capacity in Kilograms (Kg)	4530	10890	13610	18140	22680
Tire Replacement Cost	.12	.17	.22	.25	.29
Maintenance and Repairs, .004% of					

Total Investment	1.20	2.40	3.20	3.60	4.20
Fuel Cost per Hour, .15 liters per kilowatt (Kw)	7.50	7.62	10.98	15.00	16.02
Oil and Grease, 15% of Fuel	1.13	1.14	1.65	2.25	2.40
Total Hourly Cost, Not Including Labor, Supervision, Unemployment, Soc. Sec., Overhead & Profit	$17.77	$26.47	$36.23	$43.81	49.39

Table Showing How Hourly Cost of Ownership and Operation
May be Derived for Off-Highway Rear Dump
Trucks Based on an Economic Life of
3 Yrs. of 2,000 Hours Each
(in Tons)

Capacity in Tons	15	20	35	45	50	62
Capacity in Cu. Yds., Struck	10	13	22.5	28	33	40
Horsepower	180	250	430	530	570	700
Approximate Delivered Price	$80,000	$100,000	$165,000	$230,000	$250,000	$310,000
Original Tire Value	4,000	8,200	10,500	17,500	21,000	32,000
Depreciation, .01% of Total Investment Less Tires	8.00	10.00	16.50	23.00	25.00	31.00
Interest, Taxes and Insurance, .0067% of Total Investment	5.36	6.70	11.06	15.41	16.78	20.77
Hourly Tire Replacement Cost, Based on 3,000-Hr. Life	1.33	2.73	3.50	5.83	7.00	10.67
Hourly Cost of Tire Repairs, 15% of Hourly Tire Replacement cost	.20	.41	.53	.88	1.05	1.60
Maintenance and Repairs, .004% of Total Investment	3.20	4.00	6.60	9.20	10.00	12.40
Fuel Cost per Hour, 0.025 Gal. per HP	6.75	9.38	16.12	19.87	21.37	26.25
Oil and Grease, 15% of Fuel	1.01	1.40	2.42	2.98	3.20	3.93
Total Hourly Cost*	$25.85	$34.62	$56.73	$77.17	84.40	106.62

Table Showing How Hourly Cost of Ownership and Operation
May be Derived for Off-Highway Rear Dump
Trucks Based on an Economic Life of
3 Yrs. of 2,000 Hours Each
(in Kilograms)

Capacity in Kilograms (Kg)	13610	18140	31750	40820	45360	56250
Capacity in Cu. M, Struck	7.65	9.95	17.20	21.40	25.25	30.60

Kilowatt (kW)	134	186	321	395	425	522
Approximate Delivered Price	$80,000	$100,000	$165,000	$230,000	$250,000	$310,000
Original Tire Value	4,400	9,200	11,550	19,250	23,100	35,200
Depreciation, 0.01% of Total Investment Less Tires	8.00	10.00	16.50	23.00	25.00	31.00
Capacity in Kilograms (Kg)	13610	18140	31750	40820	45360	56250
Interest, Taxes and Insurance, 0.0067% of Total Investment	5.36	6.70	11.06	15.41	16.75	20.77
Hourly Tire Replacement Cost, Based on 3,000-Hr. Life	1.47	3.07	3.85	6.42	7.70	11.73
Hourly Cost of Tire Repairs, 15% of Hourly Tire Replacement cost	.22	.45	.58	.96	1.16	1.76
Maintenance and Repairs, 0.004% of Total Investment	3.20	4.00	6.60	9.20	10.00	12.40
Fuel Cost per Hour, 0.025 Gal. per HP	6.63	9.21	15.84	19.52	21.00	25.79
Oil and Grease, 15% of Fuel	.99	1.38	2.38	2.93	3.15	3.87
Total Hourly Cost*	$25.87	$65.87	$56.80	$77.44	84.75	107.32

*The foregoing tables are for the purpose of showing the items that must be taken into consideration in figuring hourly costs but do not necessarily reflect current prices of equipment, tires or fuel.

Cost of Hauling Excavated Material in Trucks.

Hauling costs vary with the capacities of the trucks and time of the hauling cycle.

Truck capacity, for proper sizing, should be at least 4 times the dipper or bucket capacity of the excavator. This is important, because the efficiency of the excavator is seriously affected by undersized hauling units due to the increased hazards of truck delays.

The following table gives theoretical spotting time cycles for various sizes of power shovel excavators working at 100% efficiency and loading trucks of 4 times dipper size capacity.

Haul Units Needed to Spot Under Shovel per Hour in Medium Digging

Size of Excavator Dipper	Minimum Haul Unit Capacity at 4 Times Dipper Size	Approximate Shovel Cycle in Seconds 90o Swing No Delays	Loading Time for 4-Dipper Truck in Seconds	Time of Spotting Cycle for Steady Operation in Minutes	Number of Spots Req'd per Hour for Steady Operation
3/8 c.y. (0.28 m³)	1-1/2 (1.15)	19	76	1.26	48
1/2 c.y. (0.38 m³)	2 (2.50)	19	76	1.26	48
3/4 c.y. (0.57 m³)	3 (2.30)	20	80	1.33	45

Size of Excavator Dipper	Minimum Haul Unit Capacity at 4 Times Dipper Size	Approximate Shovel Cycle in Seconds 90o Swing No Delays	Loading Time for 4-Dipper Truck in Seconds	Time of Spotting Cycle for Steady Operation in Minutes	Number of Spots Req'd per Hour for Steady Operation
1 c.y. (0.76 m³)	4 (3.05)	21	84	1.40	43
1-1/4 c.y. (0.96 m³)	5 (3.82)	21	84	1.40	43
1-1/2 c.y. (1.15 m³)	6 (4.58)	23	92	1.53	39
2 c.y. (1.52 m³)	8 (6.11)	25	100	1.66	36
2-1/2 c.y. (1.91 m³)	10 (7.64)	26	104	1.73	35

For dragline operation the above time values may be increased from 20% to 30%.

The hauling time cycle consists of several operations—spotting of truck under excavator; loading truck by excavator; traveling to dumping area; dumping; returning to excavator.

Some of these operations, such as spotting, loading and dumping, can be minimized by good management and supervision. Traveling time, however, will vary with the haul distance, traffic conditions, road conditions, etc., and must be carefully analyzed for each job.

When the hauling time cycle has been estimated, the number of loads each truck is able to haul per day can be determined and multiplying this by the capacity of the truck will give the theoretical daily haul of each truck in cu. yds., assuming 100% job efficiency. This amount should be multiplied by the expected job efficiency factor to obtain the net yardage hauled. Divide the daily truck cost by the net yardage hauled and the result will be the hauling cost per cu. yd.

The number of trucks required to keep the excavator going can also be determined by dividing the hauling time cycle by the loading time cycle using 100% excavator efficiency as a basis. This is done to be sure that a hauling unit is under the excavator at all times. While an excavating job may actually work out to a 50% or 60% efficiency, there usually are times on the job when the excavator is working at full capacity. To take advantage of this, sufficient, properly-sized hauling units must be on the job. In large operations many contractors maintain standby units to step in when one of the hauling fleet trucks breaks down or is otherwise unusually delayed.

In sizing trucks to conform to dipper or bucket capacities, the struck capacity of each is used, assuming that the heaping of loads on both trucks and dippers or buckets will offset each other and produce bank measure quantities.

However, all estimating should use actual job data rather than theoretical figures. Some materials do not follow the usual swell ratios and job efficiency ratings vary widely. Following book figures to the letter in excavating, or any other branch of construction work, can be disastrous. Book values, at best, can only give what can be done under a definite set of conditions. The contractor must exercise his judgment in estimating this work and base his figure on his own particular set of conditions, i.e., equipment, supervision, operators, local materials and local weather.

Tractors and Scrapers

The tractor is probably the most used piece of earth moving equipment on the construction job. It can perform a wide variety of operations, and equipped with the proper accessories, it is a high production and efficient tool for such jobs as site clearing, grading, shallow excavating, etc. In addition the tractor is often called upon to act as the "strong man" in freeing stuck trucks, hauling or dragging equipment, sheds, etc., into place, miscellaneous hoisting using a snatch block and cable, etc. In general, the tractor is almost indispensable on the job.

Equipped with bulldozer apparatus, the tractor, either rubber-tired or crawler type, has demonstrated its ability to do a wider variety of jobs than any other earth moving tool.

The bulldozer usually is employed to open up the job, which may consist of anything from clearing brush and vegetation to the removal of average size trees, up to 30 inches (750 mm) in diameter, stumps, rocks, paving, old floor slabs, old foundations, etc.

Stripping and stockpiling top soil on small quantity jobs with a haul distance from 200 to 400 feet (61 to 122 m) or less is most efficiently done by bulldozer. On larger jobs this is done more economically by scrapers.

On small quantity, shallow excavation jobs with a short haul, tractor dozers are moderately efficient.

Crawler type bulldozers are reasonably economical on pushing distances up to 200 feet (61 meters). Under the same conditions the rubber-tired dozer's economical push distance is about 400 feet (122 meters).

Most backfilling jobs fall to the lot of the bulldozer, which outperforms all other equipment.

Tractors also perform a valuable service when they are called upon to provide the additional power for push loading scrapers. On a production scraper job, dozers helping to load these units more than pay for themselves.

On residence and other small jobs, a loader usually can dig the basement and do all exterior grading cuts and fills with maximum efficiency.

Another application of the bulldozer is spreading fill hauled in and dumped by trucks or cast by other excavating equipment.

Tractors are also the ideal power source for hauling sheepsfoot rollers and pneumatic rollers used in compacting fills.

Scraper Application. Scrapers are high production earth moving units which are capable of digging their own load, hauling the load and then spreading the load in controlled layers. Furthermore, the scraper is a precision tool, as a skilled operator can cut a grade to within 0.1 ft. and can spread fill to the same degree of accuracy.

Scrapers may be powered by three types of prime movers: two wheeled or four wheeled rubber-tired units or crawler type tractors.

Scrapers, as a rule, are not used to any great extent in residence building or on any work in congested city areas but for jobs such as suburban and rural schools, hospitals, factories, housing projects, in fact any construction work with a medium to large site grading problem, scrapers are probably the most efficient pieces of excavating equipment for the job.

The scrapers can be used to strip and stockpile top soil, after the site has been cleared, and reclaim the top soil from the stockpile and spread it where required at the end of the job. Many times basement excavation may be carried on in conjunction with site grading cuts and fills by scrapers.

The basic efficiency of scrapers is relatively unaffected by depth of cut, length of haul or type of soil. Quantities being sufficient to warrant the use of scraper equipment, crawler tractor drawn scrapers are economical at haul distances up to 500 feet (152 meters) and with rubber-tired, high speed prime movers can compete with trucks on longer hauls.

Tandem Power Wheel Tractor-Scraper with Dozer
Courtesy Caterpillar, Inc.

Capacities of scrapers range from about 7 cu. yds. (5.35 cu. m) struck measure to approximately 40 cu. yds. (30 cu. m) heaped measure.

Scrapers powered by crawler tractors are capable of self-loading in most soils but generally it is more economical to use pusher assistance from a tractor to obtain heaped loads in the shortest time. With few exceptions, pusher assistance is essential for self-propelled, rubber-tired scrapers. Pusher loading is generally accomplished in an average of about 1 minute in

100 ft. (30.48 m) of travel. To determine the number of scrapers one pusher can handle, divide the scraper cycle by the pusher cycle.

Estimating the Cost of Owning and Operating Tractors and Scrapers. In order to properly prepare an estimate on work to be done by tractors and/or scrapers, the contractor must know what to charge for the use of the equipment. He should work out an hourly charge for the equipment, to be used in pricing his estimates, so that the cost of ownership and operation is carried proportionately by each job and does not have to come out of profit.

Hourly Cost of Ownership. The cost of ownership has as its basis the total delivered cost of the piece of equipment and is composed of several items—depreciation, interest, insurance and taxes.

In figuring the hourly depreciation cost, most contractors usually take the full delivered price of the machine, less cost of tires, and divide by the expected economic life of the machine in hours. The life expectancy figure commonly accepted is 5 years of 2,000 hours each or 10,000 hours.

Interest, insurance, taxes, and storage are computed on the basis of average investment. The average investment in a piece of equipment which is to be fully depreciated in 5 years is 60% of the full delivered cost, as is shown on previous pages under "Power Shovels and Cranes". The assumed rate for interest, insurance, taxes, and storage is a total of 20%, which when applied to the average investment and then divided by 2,000 hours, amounts to 0.006% of the full delivered cost to give the hourly cost of interest, insurance, taxes, and storage.

Total hourly ownership cost is then obtained by totaling the hourly cost of depreciation and the hourly cost of interest, insurance, taxes, and storage.

Hourly Cost of Operation. This cost includes tire replacement, tire repairs, fuel, lubrication, mechanical repairs, maintenance of blades and cable, etc.

The hourly costs of tire replacements and tire repairs are computed in the same manner as previously described, under "Hourly Cost of Ownership and Operation of Trucks", and are based on a tire life of 3,000 hours of operation with repair costs assumed to be 15% of the hourly tire cost.

Fuel costs may be figured on the following basis: Assuming an average machine in good repair, operating under average conditions, diesel fuel consumption of 0.04 gallons per horsepower per hour is used.

Experience has shown that lubricants and lubricating labor amount to approximately 50% of the fuel costs per hour.

Cost of repairs and maintenance, including major overhauls at the end of the working season, should be determined by the contractor from his own experience record.

Many contractors find that total repairs, maintenance and overhauls for the economic life of the machine runs from 70% of the full delivered price for the larger machines to 90% for the smaller machines. This percentage divided by the economic life of the machine in hours (10,000) will give the hourly cost of mechanical repairs, maintenance, etc., and amounts from 0.007% to 0.009% of the total delivered cost.

The hourly cost of maintenance of blades and cables is figured separately from the cost of general maintenance and repair as this item will vary depending on the soil conditions. This cost can be estimated by dividing the replacement cost of blades and cables by their expected life in hours, which under average conditions can be about 450 hours; under favorable conditions, such as digging soft loam, it may be increased to about 800 hours; under unfavorable conditions, where sharp rocks and shale are being loaded, it might be reduced to about 250 hours.

Approximate Prices of Crawler Type Tractors with Bulldozer Equipment

Horsepower (HP)	Kilowatt (kW)	Lbs.	Kilograms	Approx. Cost
50	37	9,800	4,445	$41,580
65	48	14,600	6,622	62,370
75	56	19,500	8,845	83,160
105	78	30,000	13,608	115,500
200	149	45,900	20,820	231,000
400	298	90,000	40,824	336,000
460	343	112,800	51,166	462,000
520	388	122,000	55,340	565,950
770	574	200,000	90,720	2,079,000

Approximate Prices of Crawler Type Loaders

Capacity in Cu. Yds.	Cu. M	HP	kW	Approx. Cost
1/2	.38	75	56	$109,725
1	.76	150	112	184,800
1-1/2	1.15	180	134	194,040
2-1/4	1.72	260	194	288,750
3	2.30	300	224	577,500
4-1/2	3.44	390	290	635,250

Approximate Prices of Self-Propelled Scrapers

Capacity Struck Cu. Yds.	Cu. M	Heaped Cu. Yd.	Cu. M	HP	kW	Apx. Cost	Original Tire Value
14	10.70	20	15.30	150	112	$319,935	$34,650
21	16.00	31	23.70	265	198	589,050	48,510
28	21.40	38	29.00	330	246	519,750	48,510
32	24.50	44	33.60	450	335	808,500	60,060

Approximate Prices of Tractor-Elevating Scrapers

Capacity Cu. Yds.	Cu. M	HP	kW	Approx. Cost	Orig.Tire Value
11	8.40	150	112	$132,000	$13,200
16	12.20	265	198	275,000	27,500
22	16.80	330	246	319,000	33,000
34	26.00	450	335	583,000	38,500

Ownership and Operation of Bulldozers, Loaders, and Scrapers.
The following data is the hourly cost of ownership and operation for various
sizes of tractor and scraper equipment and is based on preceding price and
horsepower data. It does not include operator's wages, supervision,
compensation, and liability insurance or Social Security taxes.

Hourly Cost of Crawler Type Bulldozers

Horsepower	50	65	75	105	200	400	460
Approx. Delv'd. Cost	$36,000	54,000	72,000	100,000	200,000	291,000	400,000
Depreciation, 0.01%	3.60	5.40	7.20	10.00	20.00	29.10	40.00
Interest, Insurance and Taxes, 0.006%	2.16	3.24	4.32	6.00	12.00	17.46	24.00
Fuel, 0.04 Gal. per HP*	3.00	3.90	4.50	6.30	12.00	16.00	18.40
Lubrication, Including Labor, 50% of Fuel	1.50	1.95	2.25	3.15	6.00	8.00	9.20
Mech. Repairs and Maintenance, 0.007%	2.52	3.78	5.04	7.00	14.00	20.37	28.00
Blade and Cable Replacement	0.45	0.47	0.49	0.51	0.55	0.60	1.00
Total Hourly Cost**	$13.23	$18.74	$23.80	$32.96	$64.55	91.53	120.60

*Based on $1.50 per gallon for diesel fuel.
**Not including Labor, Supervision, Unemployment, Soc. Sec., Overhead Expense

(metric)

Kilowatts	37	48	56	78	149	298	343
Approx. Delv'd. Cost	$36,000	54,000	72,000	100,000	200,000	291,000	400,000
Depreciation, 0.01%	3.60	5.40	7.20	10.00	20.00	29.10	40.00
Interest, Insurance and Taxes. 0.006%	2.16	3.24	4.32	6.00	12.00	17.46	24.00
Fuel Cost per Hour, 0.16 liter per kW	3.08	3.99	4.66	6.49	12.40	24.79	28.54
Lubrication, Including Labor, 50% of Fuel	1.54	2.00	2.33	3.24	6.20	12.40	14.27
Mech. Repairs and Maintenance, 0.007%	2.52	3.78	5.04	7.00	14.00	20.37	28.00
Blade and Cable Replacement	0.49	0.52	0.54	0.56	0.60	0.80	0.95
Total Hourly Cost	$13.39	$18.93	$24.09	$33.30	$65.20	104.92	135.76

Hourly Cost of Crawler Type Loaders

	1	1-1/2	2-1/4	3	4-1/2
Bucket Capacity, cu. yds.	1	1-1/2	2-1/4	3	4-1/2
Horsepower	150	180	260	300	390
Approx. Delv'd Cost	$160,000	$168,000	$250,000	$500,000	$550,000
Depreciation, 0.01%	16.00	16.80	25.00	50.00	55.00
Interest, Insurance and Taxes, 0.006%	9.60	10.08	15.00	30.00	33.00
Fuel, 0.04 Gal. per hp*	9.23	10.80	15.60	18.00	23.40
Lubrication, Including Labor, 50% of Fuel	4.61	5.40	7.80	9.00	11.70
Mech. Repairs and Maintenance, 0.007%	11.20	11.76	17.50	35.00	38.50
Blade Replacement	0.80	0.84	0.88	0.96	1.00
Total Hourly Cost**	$51.44	$55.68	$81.78	$142.96	$162.60

*Based on $1.50 per gallon for diesel fuel.
**Not including Labor, Supervision, Unemployment, Soc. Sec., Overhead Expense

(metric)

Capacity, Struck, Cu. M.	0.76	1.15	1.72	2.30	3.44
kW	112	134	194	224	
Approx. Delv'd Cost	$160,000	$168,000	$250,000	$500,000	$550,000
Depreciation, 0.01%	16.00	16.80	25.00	50.00	55.00
Interest, Insurance and Taxes, 0.006%	9.60	10.08	15.00	30.00	33.00
Fuel Cost per Hr, 0.2 Liters per kW	8.96	10.72	15.52	17.92	23.28
Lubrication, Including Labor, 50% of Fuel	4.48	5.36	7.76	8.96	11.64
Mech. Repairs and Maintenance, 0.007%	11.20	11.76	17.50	35.00	38.50
Blade Replacement	0.80	0.84	0.88	0.96	1.00
Total Hourly Cost	$51.04	$55.56	$81.66	$142.84	$162.42

Hourly Cost of Self-propelled Scrapers

	14	21	28	32
Capacity, Struck in Cu. Yds.	14	21	28	32
Capacity, Heaped in Cu. Yds	20	31	38	44
Horsepower	330	450	500	550
Approx. Delv'd. Cost	$277,000	$510,000	$550,000	$700,000
Original Tire Value	27,000	40,000	40,000	45,000
Depreciation, 0.01% of Total Investment Less Tires	27.70	51.00	55.00	70.00
Interest, Insurance and Taxes, 0.006% of Total Investment	16.62	30.60	33.00	42.00
Tire Replacement, 3,000-Hr. Life	9.00	13.33	13.33	15.00
Tire Repairs, 15%	1.35	2.20	2.20	2.25
Fuel, 0.04 Gal. per hp*	13.20	18.00	22.00	22.00
Lubrication, Incl. Labor, 50% of fuel	6.60	9.00	11.00	11.00

Mech. Repairs and Maintenance,

SITE WORK **2.65**

0.007% of Total Investment	19.39	35.70	36.00	49.00
Blade and Cable Replacement	0.70	0.80	1.00	1.00
Total Hourly Cost**	$94.56	$160.63	$173.53	$215.25

*Based on $1.50 per gallon for diesel fuel.
**Not including Labor, Supervision, Unemployment, Soc. Sec., Overhead Expense

(metric)

Capacity, Struck in Cu. M	10.70	16.00	21.40	24.50
Capacity, Heaped in Cu. M	15.30	23.70	29.00	33.60
kW	246	336	373	410
Approx. Delv'd. Cost	$277,000	$510,000	$550,000	$700,000
Original Tire Value	27,000	40,000	40,000	45,000
Depreciation, .01% of Total	27.70	51.00	55.00	70.00
Interest, Insurance and Taxes,				
0.006% of Total Investment	16.62	30.60	33.00	42.00
Tire Replacement,				
3,000-Hr. Life	9.00	13.33	13.33	15.00
Tire Repairs, 15%	1.35	2.00	2.00	2.25
Fuel Cost per Hr, 0.14 liters per kW	13.78	17.47	19.40	21.32
Lubrication, Incl. Labor, 50% of fuel	6.89	8.74	9.70	10.66
Mech. Repairs and Maintenance,				
0.007% of Total Investment	19.39	35.70	38.50	49.00
Blade and Cable Replacement	0.70	0.80	1.00	1.00
Total Hourly Cost	$95.42	$159.64	$171.93	$211.23

Bulldozer Production

In an average day's work, the bulldozer can be called upon to do several operations. Time will be lost in moving from one task to another. If this time loss is not considered in estimating bulldozer work, the job might not be profitable. Inasmuch as each job has its own peculiar problems, no cost table can reflect the time losses, and the estimator must analyze each job carefully and adjust the production rates to conform to the conditions present. The following table gives the approximate hourly production to be expected from various sized bulldozers and is based on good weather conditions, efficient and well-maintained machinery, and steady operation. It has been established that a well organized and supervised job will result in approximately 50 work-productive minutes each hour, a job efficiency of 83%.

Approximate Hourly Production of Bulldozers
Based on an 83% Job Efficiency

Description of Work	Hp (kW) 50 (37)	Hp (kW) 65 (48)	Hp (kW) 75 (56)	Hp (kW) 105 (78)	Hp (kW) 200 (149)
Simple clearing, vegetation, light brush, tree saplings,					
in sq. ft. per hr.	900-1000	1000-1100	1150-1250	1250-1350	1450-1550
in sq. m per hr.	81-93	93-102	107-116	116-125	135-144

Description of Work	Hp (kW) 50 (37)	Hp (kW) 65 (48)	Hp (kW) 75 (56)	Hp (kW) 105 (78)	Hp (kW) 200 (149)
Moderately difficult clearing thick brush and tree saplings,					
in sq. ft. per hr.	300-350	350-400	400-450	600-650	700-750
in sq. m per hr.	28-33	33-37	37-42	56-60	65-70
Add extra for trees					
Strip top soil or shallow excavation of ordinary soil and placing in piles on premises,					
in cu. yds. per hr.	15-20	17-22	20-25	22-27	27-32
in cu. m per hr.	11-15	13-17	15-19	17-21	21-24
Backfill ordinary soil placed slow enough to permit compacting,					
in cu. yds. per hr.	33-38	37-42	43-48	50-55	60-65
in cu. m per hr.	25-29	28-32	33-37	38-42	46-50
Spread material dumped in piles by trucks or cast by other excavating equipment					
in cu. yds. per hr.	42-47	45-50	55-60	60-65	70-75
in cu. m per hr.	32-36	34-38	42-46	46-50	54-57

Scraper Production

Relative scraper production is mainly governed by the cycle time of operation; that is, the length of time a scraper will take to load, haul, spread-dump, and return to start over again. In properly organized jobs with good, adequately-powered equipment, a portion of this cycle is usually termed the fixed time, because it is relatively unaffected by the length and grades of the haul and consists of the following: loading, approximately 1.0 to 1.5 minutes, spread-dumping, about 0.5 minutes, two turns of 0.25 minutes each, and shifting and acceleration of 0.5 to 1.0 minute or a total fixed time of 2.5 to 3.5 minutes. The other part of the cycle is the actual time of hauling the load to the fill area plus the time of returning empty to the beginning of a new cut. This traveling time is the big variable in the operation. It is affected by length of haul, road conditions, and grades to be negotiated. It must be estimated accurately. Scrapers powered by rubber-tired tractors can reach a speed of 30 mph (48 kph) when empty and on a good level road, but this will be reduced to 10 to 20 mph (32 kph) when loaded. It must also be remembered that adverse grades greatly reduce the speed, and favorable grades increase the speed. The traveling time plus the fixed time is the cycle time of operation.

Yardage production is figured in terms of bank measure. When estimating scraper production, the struck capacity of the machine is normally used as the pay yardage volume, anticipating the scraper will be heap loaded. The difference between struck and heaped capacity is generally assumed to be the amount the material will swell in being disturbed.

The following table gives the daily scraper production for various sized machines operating at several different time cycles and is based on ordinary soil conditions, good weather, good equipment, and an 83% job efficiency.

Dozer, D4 Series
Courtesy Caterpillar, Inc.

Daily Scraper Production for Various Operation Cycles
Based on a 50-Minute Hour and an 8-Hour Day

Operation Cycle in Minutes	Struck Capacity of Scraper					
	7 c.y.	5 m³	14 c.y.	11 m³	24 c.y.	18 m³
3	930	711	1860	1422	3200	2447
4	700	535	1400	1070	2400	1835

Operation Cycle in Minutes	Struck Capacity of Scraper					
	7 c.y.	5 m³	14 c.y.	11 m³	24 c.y.	18 m³
5	560	428	1120	856	1920	1468
6	465	356	930	711	1600	1223
7	400	306	800	612	1370	1047
8	350	268	700	535	1200	917
9	310	237	620	474	1070	818
10	280	214	560	428	960	734

Sample Estimate for Scraper Excavation

Based on a site grading cut containing 8,000 cu. yds. (6116 cu. m), excavated material to be spread-dumped in low areas requiring fill, good conditions permitting a 50-minute hour, an operation cycle of 6 minutes and using two 14 cu. yd. (11 cu. m), struck measure, self-propelled scrapers.

Bringing Scrapers to Job	Hours	Rate	Total	Rate	Total
Move-in charge 2 ea.		$....	$....	$275.00	$550.00
Excavating					
Foreman	35	....	....	29.41	1,029.35
Scraper operators	70	....	....	28.55	1,998.50
Scraper Hours	70	....	....	94.56	6,619.20
Total Direct Cost			$....		$10,197.05
Cost per cu. yd				....	1.27
per cu.m				....	1.67

If a 105-Hp (78 kW) bulldozer is needed to give pusher assistance in loading, to maintain the operation cycle time of 6 minutes, the additional cost per cu. yd. (cu. m) would be as follows:

	Hours	Rate	Total	Rate	Total
Move-in charge		$....	$....	$275.00	$275.00
Excavating					
Bulldozer operator	35	....	....	28.55	999.25
Bulldozer charge	35	....	....	32.96	1,153.60
Additional Direct Cost			$....		$2,427.85
per cu. yd.				....	0.30
per cu.m				....	0.40

Trenching and Ditching Machine Excavation

Designed strictly for the purpose of cutting ditches, the trenching machine is the fastest earthmover for its weight and horsepower. Before a contractor can reap the benefits of this potential output, he must select a trenching machine of a type and size that fits the work to be done, and even more important, must put an operator on the machine who has both skill and experience.

There are two basic types of machine, the wheel type and the ladder type. The wheel type is generally considered to be the fastest and is by far the most common type in use today. Each type has advantages and disadvantages for performing certain classes of work.

Because most trenching or ditching operations in building construction will be foundations, service trenches, etc., the following data covers the cost and operation of a general purpose, crawler-mounted, wheel machine of the type and size generally used for this work.

This machine is equipped with shifting-tilting boom that permits it to dig close to obstructions, such as foundations or curbs, and to dig a vertical trench even when the machine is on a slope.

The cost of excavating with a trenching or ditching machine will vary with the size of the job, width and depth of trenches, kind of soil, accessibility, etc. Most machines are designed to dig certain standard width trenches, so that trenches a trifle narrower or wider will not make any appreciable difference in either output or costs.

While the following table gives the rated capacity of the machine, it will be necessary to allow for factors such as time lost moving machine and bad weather. It is not advisable to figure rated capacity as the average daily output of any machine.

Data on Crawler Mounted Ditching Machine Operation and Cost

Wheel Size	Possible Cutting Widths	Maximum Depth Trench	Weight in Lbs. (Kg)	Approx. Cost FOB Factory
16"x11'-6"	16"-24" (400-600 mm)	11'-0" (3.35 m)	27,000 (12,247 kg)	$162,500
24"x12'-4"	24"-30" (600-750 mm)	15'-6" (4.72 m)	43,000 (19,505 kg)	212,500

Maximum digging speed is 34 fpm (10.36 mpm) Forward and reverse travel speeds are 0.46, 0.95, 1.7, and 2.9 mph (0.74, 1.53, 2.73 and 4.67 kmph). Digging speeds are controlled by a hydraulic transmission, permitting speed to be varied to match soil conditions. Wheel speeds vary from 0 to 9.2 rpm.

Crawler Mounted Ditching Machine Operation and Cost

When figuring operating cost on a ditching machine, include fuel and oil consumption, depreciation, interest on investment, repairs, replacements, and taxes.

Description	Digging Width	
	12" (300 mm)	24" (600 mm)
Approximate Delivered Price	$130,000	$170,000
Depreciation per year, 18%	23,400	30,600
Interest 15% per year	19,500	25,500
Insurance and Taxes per year, 5%	6,500	8,500
Repairs and Replacement per year, 10%	13,000	17,000
Hourly Consumption of Fuel, Gals.	5	6
Daily Cost of Fuel, Oil, Grease, $1.50 gal. (.40 l) + 15%	69.00	82.80
Operating Cost per Mile, Based on 75 mi. (121 km) per Year, 200 Working Days per Year, Avg. Trench Depth, 5'-0" (1.52 m) Depreciation	312.00	408.00
Interest	260.00	340.00
Taxes and Insurance	86.67	113.34
Repairs and Replacements	173.34	226.67
Fuel, Oil and Grease	184.00	220.80
Total machine cost per mile	$1,016.01	$1,308.81
Machine cost per lin. ft. trench	0.193	0.248

Cost of Digging 100 L.F. (30.5 M) of Trench 5'-0" (1.52 m) Deep, Using a Crawler Mounted Ditching Machine, Based on 350 L.F. (106.70 m) per Hour

	Hours	Rate	Total	Rate	Total
Machine operator	0.29	$....	$....	$28.55	$8.28
Machine charge, 12" (300 mm) wide		....	....	0.193	19.30
Machine charge, 24" (600 mm) wide		....	....	0.248	24.80
Cost per 100 lin. ft. (30.5 m), 12" (300 mm) wide					27.58
Cost per 100 lin. ft., 24" (600 mm) wide					33.08
Cost per lin. ft., 12" (300 mm) wide					0.28
Cost per lin. ft., 24" (600 mm) wide					0.33

To the above, add cost of trucking machine to job and removing same at completion, supervision, compensation and liability insurance, social security and unemployment taxes, overhead expense and profit. Add for backfilling trenches, if required.

Volume of Trench Excavation
Cu. Yd. per 100 Lin. Ft.

Trench depth, in.	Trench width, in.						
	12	18	24	30	36	42	48
12	3.7	5.6	7.4	9.3	11.1	13.0	14.8
18	5.6	8.3	11.1	13.9	16.7	19.4	22.3
24	7.4	11.1	14.8	18.5	22.2	26.0	29.6

Trench depth, in.	Trench width, in.						
	12	18	24	30	36	42	48
30	9.3	13.8	18.5	23.2	27.8	32.4	37.0
36	11.1	16.6	22.2	27.8	33.3	38.9	44.5
42	13.0	19.4	30.0	32.4	38.9	45.4	52.0
48	14.8	22.2	29.6	37.0	44.5	52.0	59.2
54	16.7	25.0	33.3	41.6	50.0	58.4	66.7
60	18.6	27.8	37.0	46.3	55.5	64.9	74.1
66	20.4	30.5	40.7	51.0	61.0	71.3	81.6
72	22.2	33.3	44.5	55.5	66.7	77.9	89.0
78	24.1	36.0	48.1	60.2	72.2	84.2	96.5
84	25.9	38.9	51.9	64.8	77.8	90.8	104.0
90	27.8	41.6	55.6	69.5	83.4	97.3	111.4
96	29.6	44.4	59.2	74.0	88.9	102.0	118.6

Volume of Trench Excavation
Cubic Meters per 10 Lineal Meters

Trench depth, mm	Trench width, mm						
	300	450	600	750	900	1050	1200
300	0.90	1.35	1.80	2.25	2.70	3.15	3.60
450	1.35	2.03	2.70	3.38	4.05	4.72	5.40
600	1.80	2.70	3.60	4.50	5.40	6.30	7.20
750	2.25	3.38	4.50	5.63	6.75	7.88	9.00
900	2.70	4.05	5.40	6.75	8.10	9.45	10.80
1050	3.15	4.72	6.30	7.88	9.45	11.03	12.60
1200	3.60	5.40	7.20	9.00	10.80	12.60	14.40
1350	4.05	6.08	8.10	10.13	12.15	14.18	16.20
1500	4.50	6.75	9.00	11.25	13.50	15.75	18.00
1650	4.95	7.43	9.90	12.38	14.85	17.33	19.80
1800	5.40	8.10	10.80	13.50	16.20	18.90	21.60
1950	5.85	8.78	11.70	14.63	17.55	20.48	23.40
2100	6.30	9.45	12.60	15.75	18.90	22.05	25.20
2250	6.75	10.12	13.50	16.88	20.25	23.63	27.00
2400	7.20	10.80	14.40	18.00	21.60	25.20	28.80

Drilling Holes Using an Earth Boring Machine. Where a large number of holes are required for line construction, foundation work, pre-boring, guard rails, piling, fencing, draining, or bridges, an earth boring machine will perform the work much faster and at lower cost than most other methods.

These machines will drill holes through loam, clay, hard pan, and shale and are furnished in various sizes by different manufacturers. Some machines will drill up to 8'-0" (2.43 m) diameter holes to a depth of 120 ft. (36.60 m) or more.

When drilling holes up to 18" (450 mm) in diameter in ordinary soil, a machine will drill at the rate of 1" (25 mm) depth in 1-1/2 to 2 seconds, or drill a 3'-0" (0.91 m) deep hole in 1 to 1-1/4 minutes. Moving the machine and spotting the holes takes some time, so this will govern the number of holes that may be drilled per day. A machine that will drill holes to 24" (600 mm) diameter and 9'-0" (2.74 m) deep costs approximately $100,000.00 f.o.b. factory. This does not include cost of truck, which will have to be added. The larger machines which are used for drilling caisson foundations may cost as much as $300,000.00.

Based on 20 holes, 18" (450 mm) diam. and 3'-0" (0.91 m) deep, per hour or 160 holes per 8-hr. day, the daily cost should average as follows:

	Hours	Rate	Total	Rate	Total
Operator	8.0	$....	$....	$28.55	$228.40
Machine charge	8.0	....	....	20.00	160.00
Truck charge	8.0	....	....	18.00	144.00
Total cost 160 holes			$....		$532.40
Cost per 3'-0" (0.9 m) hole					3.33
Cost per lin. ft.					1.11
Cost per m					3.64

There are also small portable posthole diggers, mounted on two rubber tired wheels and equipped with a 5 to 6 Hp gasoline engine.

This machine can be operated by one person and in ordinary ground it will bore a 9" (225 mm) hole 3'-0" (0.9 m) deep in 1 to 1-1/2 minutes. Such machine costs $2,000.00 to $3,000.00.

Placing And Compacting Fills

Architects and engineers have become more rigid in their requirements as to the manner of placing and compacting fill materials. The usual specification for this work is to place the fill in from 6" to 12" (150 to 300 mm) layers and to compact each layer with some sort of roller equipment.

Placing and Compacting Site Grading Fills. Fill material in site grading work is placed either by scrapers spread-dumping their loads, trucks dumping their loads more or less in piles, or some sort of excavating equipment casting dipper or bucket loads as they are dug. In the case of the scraper spread-dumping fill material, no further spreading is necessary, because it can be controlled with precision by the scraper operator. In the other cases, however, a bulldozer will be required to spread out the fill material into layers of specified depth. Hourly production for this work can be found in the table under "Bulldozer Production".

When the fill has been placed and spread in the specified depth of layer, the compacting operation follows, and this is usually accomplished by rolling with a tractor-drawn sheepsfoot roller. The degree of compaction may be specified by the number of passes to be made by the roller or by a percentage, such as 95%, of maximum density of the material obtained at optimum moisture content in a soil mechanics laboratory. The latter specification is usually called for in government work and for fills under paving. Obtaining 95% of maximum density compaction may take as many as 12 passes of a sheepsfoot roller.

If rolling is not to interfere with production schedules, a rate must be established in cubic yards compacted per hour and applied to the rate of placing and spreading the fill material to determine the size and number of rollers required to keep the job moving. Where frequent turns are made or restricted areas require maneuvering these factors must be considered and the figures adjusted accordingly.

The following table gives rate of compaction, in cu. yds. per hr., for rolling fill with a 5'-0" (1.52 m) wide sheepsfoot roller at 2.5 mph (4.0 kph) with the fill material placed in 12" (300 mm) layers for soil of various compaction factors and for 1 pass of the roller to 12 passes.

Rate of Sheepsfoot Roller Compaction in Cu. Yds. and Cu. M per Hr.
Based on 5'-0" (1.52 m) Wide Roller, 2-1/2 mph (4 kph) Speed,
12" (300 mm) Fill Layers and 100% Job Efficiency

| No. of Passes of Roller | Percentage Factor of Pay Yd. (M) to Loose Yd. (M) | | | | | |
| | Hard Tough Clay 70% | | Medium Clay 80% | | Loam 90% | |
	c.y.	cu.m	c.y.	cu.m	c.y.	cu.m
1	1711	1308	1956	1495	2200	1682
2	856	654	978	748	1100	841
3	570	436	652	498	733	560
4	428	327	489	374	550	421
5	342	261	391	299	440	336
6	285	218	326	249	367	281
7	244	187	279	213	314	240
8	214	164	245	187	275	210
9	190	145	217	166	244	187
10	171	131	196	150	220	168
11	156	119	178	136	200	153
12	143	109	163	125	183	140

For rollers of different widths, adjust the above rates proportionately; for example, for a 10'-0" (3.04 m) wide roller double the rates, and for a 15'-0" (4.57 m) wide roller triple the rates.

For different depths of layers, adjust the above rates proportionately: i.e., for 6" (150 mm) layers reduce the rates to 1/2, and for 9" (225 mm) layers reduce the rates to 3/4.

The above rates must also be adjusted to normal job efficiency and further adjusted to take care of lost time in maneuvering and turning. In small jobs as much as 10% of the time could be lost.

After Fill has been Placed Self-powered Compactor is Used to Tamp Earth
Courtesy Caterpillar, Inc.

Example: Assume a fill to be placed in 9" (225 mm) lifts and rolled 8 passes with a double drum sheepsfoot roller 10'-0" (3.05 m) wide at 2.5 mph (4.02 kph) in soil with a compaction factor of 80%.

From the table, 8 passes in 80% material will compact 245 cu. yds. (187 cu.m) per hr. Adjusting this rate for a 10'-0" (3.05 m) wide roller gives a rate of 490 cu. yds. (374.6 cu. m) per hr. and further adjusting for 9" fill layers results in a compaction rate of 368 cu. yds. (281 cu.m) per hr. Assuming that a 50-minute hour or a job efficiency of 83% is expected, the rate is reduced to 305 cu. yds. (233 cu. m) per hr. and with a time loss of 10% anticipated for turns and maneuvering the final result is a compaction rate of 275 cu. yds. (210 cu.m) per hr. The approximate cost for compacting fill at rate of 275 cu. yds. (210 cu.m) per hr. is as follows:

	Hours	Rate	Total	Rate	Total
Hourly charge for 75-Hp. bulldozer with double drum sheepsfoot roller	1	$....	$....	$23.80	$23.80*
Bulldozer operator	1	....	....	28.55	28.55
Cost per 275 c.y. (210 cu.m)			$....		$52.35
per c.y.				....	0.19
per cu.m				....	0.25

*Does not include hauling equipment to and from job, compensation and liability insurance, social security taxes or overhead and profit.

The sheepsfoot roller method of compaction is applicable to all soils with cohesive qualities, such as loams and clays, and in very dry soils it may be necessary to add moisture to the material to obtain the correct degree of plasticity.

With non-cohesive materials, such as sands and gravels, the sheepsfoot roller usually is ineffective and rubber-tired pneumatic rollers are used instead.

Placing and Compacting Fills for Floors on Grade Inside Buildings. Fill material, necessary to raise the elevation of the subgrade for floors inside buildings, can be placed in a number of ways.

For small jobs this is usually done after the building is enclosed. Fill is brought in through door and window openings, and it is spread and tamped by hand. Three laborers working together should spread and tamp 35 to 40 c.y. (27 to 31 cu.m) of fill per 8-hr. day at the following cost per c.y. (cu.m):

	Hours	Rate	Total	Rate	Total
Labor per c.y.	0.6	$....	$....	$21.04	$12.62
per cu.m			....		16.41

On larger jobs it may be possible to truck the fill material directly into the space and spread-dump it, using a small bulldozer to further spread into even layers. Compaction of material in this case can usually be accomplished by using a single drum sheepsfoot roller or small pneumatic roller for the open areas and by hand tamping at walls and columns. Trucking traffic over the various layers of fill is a big help in compacting the material. The roller can be drawn by the same bulldozer that spreads the material and with 2 laborers spreading and tamping at "hard-to-get" places a daily production of 95 to 100 cu. yds. (73 to 76 cu. m) placed and compacted should be realized at the following cost per cu. yd. (cu. m):

	Hours	Rate	Total	Rate	Total
50-Hp. (37kW) bulldozer and single					
drum sheepsfoot roller	0.08	$....	$....	$13.23	$1.06
Bulldozer operator	0.08	....	....	28.55	2.28
Labor	0.16	....	....	21.04	3.37
Cost per cu. yd.			$....		$6.71
Cost per cu. m			....		8.72

Does not include hauling equipment to and from job, compensation and liability insurance, social security taxes, or overhead and profit.

Grading

Grading consists of dressing up ground surfaces, either by hand or machine, to conform to specified contours or elevations and is usually preparatory to a subsequent operation such as placing a floor slab, walk or drive, or spreading top soil for planting areas.

Grading costs vary according to the degree of accuracy demanded and the method by which it is done. Following are approximate costs on some of the various types of grading usually encountered in construction work.

Rough Grading. After all backfill is in place and the site has been cut and/or filled to the approximate specified contours, rough grading of the site is done, by hand for small jobs and by machine for large areas, usually preparatory to the spreading of top soil. Tolerance for this type of grading is usually plus or minus 0.1 ft. (0.03 m)

On small jobs a laborer should rough grade 800 sq. ft. (74 sq. m) of ground surface per 8-hr. day at the following cost per 100 sq. ft. (9.3 sq. m):

	Hours	Rate	Total	Rate	Total
Labor	1.0	$....	$....	$21.04	$21.04
Cost per sq. ft.			....		0.21
Cost per sq. m			....		2.26

On large jobs a bulldozer is usually employed to do rough grading, and a laborer generally accompanies the machine to check the surface and direct the operation. A 75-Hp (56 kW) bulldozer should rough grade 6,500 to 7,500 sq. ft. (604 to 697 sq. m) per 8-hr. day at the following cost per 1,000 sq. ft. (93 sq. m):

	Hours	Rate	Total	Rate	Total
75-Hp (56 kW) bulldozer charge	1.15	$....	$....	$23.80	$27.37
Bulldozer operator	1.15	....	....	28.55	32.83
Labor	1.15	....	....	21.04	24.20
Cost per 1,000 s.f. (93 sq. m)			$....		$84.40
Cost per sq. ft.			....		0.08
per sq. m			....		0.90

Does not include hauling equipment to and from job, compensation and liability insurance, social security taxes, or overhead and profit.

Grading for Slabs on Ground. Grading for slabs on ground such as floors, walks, and driveways usually is done by hand unless the job is quite large and can be organized so that some of the work is done by machine. This work generally must be done accurately with tolerances of no more than 1/2" (15 mm). A laborer should grade 500 to 600 sq. ft. (46 to 56 sq. m) per 8-hr. day at the following cost per 100 sq. ft. (9.3 sq. m):

	Hours	Rate	Total	Rate	Total
Labor	1.5	$....	$....	$21.04	$31.56
Cost per sq. ft.			....		0.32
Cost per sq. m			....		3.44

For sloping surfaces, add about 50% to the above cost, depending on steepness of pitch.

Finish Grading of Top Soil. After top soil has been spread over areas specified, a finish grading operation, usually including hand-raking, must be performed prior to seeding or sodding. The tolerance on this work usually is plus or minus 1 inch. A laborer should finish grade and hand-rake 600 to 700 sq. ft. (56 to 65 sq. m) per 8-hr. day at the following cost per 100 sq. ft. (9.3 sq. m):

	Hours	Rate	Total	Rate	Total
Labor	1.25	$...	$...	$21.04	$26.30
Cost per sq. ft.				....	0.26
Cost per sq. m				....	2.80

For sloping surfaces, such as berms and ditch sides, add 50% to the above cost.

Grading for Footing Bottoms. When footing pits and trenches are dug by machine, there is always cleanup, squaring, and grading work to be done by hand. Many contractors price this work by the sq. ft. of footing bottom. A laborer should clean up, square, and grade 200 sq. ft. (19 sq. m) of footing bottom per 8-hr. day at the following cost per 100 sq. ft. (9.3 sq. m):

	Hours	Rate	Total	Rate	Total
Labor	4.0	$...	$...	$21.04	$84.16
Cost per sq. ft.				....	0.84
Cost per sq. m				....	9.04

Costs Of Excavating

Digging Fence Post Holes. When digging 250 post holes about 3'-0" (0.91 m) deep in black soil and sand, using an ordinary augur post-hole digger, a worker will dig 4 holes per hour or 32 per 8-hr. day, at the following labor cost per hole:

	Hours	Rate	Total	Rate	Total
Labor	0.25	$...	$...	$21.04	$5.26

Loader Excavating under Favorable Conditions. The following costs are for a theoretical job performed where conditions are favorable, with no time lost because of bad weather or breakdowns.

This is a basement excavation 80'-0"x80'-0" (24 x 24 m) and 13'-0" (4 m) deep. The top 5'-0" (1.5 m) consists of loose clay and loam, while the lower 8'-0" (2.4 m) is blue grey dolomite limestone in beds varying from 6" to 24" (150 to 600 mm).

A 1-1/2 cu. yd. (1.2 cu. m) loader is used, rented at $75.00 per hr. including operator and fuel.

Five 6-cu. yd. (4.60 cu. m) trucks transport the excavated earth and rock to a dump about 1/4-mile (0.4 km) from the job. Trucks are estimated to cost $50.00 per hr. including gasoline and driver.

A 5-1/2"x5" (137.5 x 125 mm) portable air compressor, capacity 900 cu. ft. per minute, and a Type 12 standard I-R jackhammer is used for rock drilling. The rental of the compressor and jackhammer, including gasoline, operator, and repairs to drills, is estimated at $35.00 an hour.

Excavating Loose Clay and Loam. The area of loose clay and loam is 80'-0"x80'-0"x5'-0" (24 x 24 x 1.5 m), a total of 1,185 cu. yds. (864 cu. m) of excavation.

The average on this job is 135 cu. yds. (103 cu. m) an hr. at the following cost per 100 cu. yds. (76 cu. m):

	Hours	Rate	Total	Rate	Total
Loader and Operator	0.75	$....	$....	$82.50	$61.88
Cost per cu. yd.			....		0.62
Cost per cu. m			....		0.81

Five 10-cu. yd. (7.6 cu. m) trucks will haul 1,080 cu. yds. (826 cu. m) of excavated earth 1/4 mile (0.40 km) to a dump per day, or an average of 27 cu. yds. (21 cu. m) an hr. per truck, at the following cost per 100 cu. yds. (76 cu. m):

	Hours	Rate	Total	Rate	Total
Truck and Driver	3.7	$....	$....	$55.00	$203.50
Cost per cu. yd.			....		2.04
Cost per cu. m			....		2.68

Four workers at the dump spread the excavated earth after it has been dumped by the trucks. Very little grading is necessary, because the earth dump is a high-terraced fill around a building. A worker will spread about 10 cu. yds. (7.6 cu. m) per hr. at the following labor cost per cu. yd. (cu. m):

	Hours	Rate	Total	Rate	Total
Labor per cu. yd.	0.10	$....	$....	$21.04	$2.10
per cu. m			...		2.73

The total cost of excavating, hauling and spreading one cu. yd. (cu. m) of loose earth is as follows:

	Rate	Cu. Yd.	Cu. M
Loader excavation	$....	$0.62	$0.81
Trucks hauling excavated earth	....	2.04	2.67
Labor spreading loose earth at dump	....	2.10	2.76
Cost		$4.76	$6.24

Note: Does not include equipment move-in charge.

Rock Excavation

After the top 5'-0" (1.5 m) of soil has been removed, the excavation consists of blue grey dolomite limestone, hard and stratified in beds varying from 6" to 24" (150 to 600 mm). The depth of the cut was 8'-0" (2.4 m), which was taken out in two lifts of 4'-0" (1.2 m) each. The last lift was drilled 5'-0" (1.5 m) deep or 1'-0" (.30 m) below the desired level. Total rock excavation is 80'-0"x80'-0"x8'-0" (24 x 24 x 2.4 m), or 1,900 cu. yds. (1382 cu. m).

Drilling Rock for Blasting. The area drilled is 80'-0"x80'-0" (24 x 24 m), less a 5'-0" (1.5 m) border on 3 sides and a 10'-0" (3.0 m) border on one side. The rows were 5'-0" (1.5 m) apart and the holes were 5'-0" (1.5 m) on centers, staggered, and all of the holes will be 2" (50 mm) in diameter.

There will be 14 rows with 15 holes in a row or a total of 210 holes, as follows: 105 holes 4'-0" (1.2 m) deep, containing 420 lin. ft. (128 m) and 105 holes 5'-0" (1.5 m) deep, containing 525 lin. ft. (160 m), a total of 945 lin. ft. (288 m) of 2" (50 mm) holes.

One worker and a helper will drill 15 lin. ft. (4.6 m), possibly up to 60 lin. ft. (18 m), of 2" (50 mm) hole per hr. or 120 lin. ft. (37 m) per 8-hr. day, at the following cost per 100 lin. ft. (30 m):

	Hours	Rate	Total	Rate	Total
Compressor, jackhammer					
and operator	7	$....	$....	$38.50	$269.50
Labor helping	7	....	....	21.04	147.28
Cost 100 l.f. (30 m)			....		$416.78
Cost per lin. ft.			....		4.17
Cost per cu. yd.(1,900 cy)					0.22
Cost per m			....		13.68
Cost per cu. m			....		0.29

Explosives. The job will require 2.5 cases, 50 lbs. (22.7 kg) each of 40% dynamite, a total of 125 lbs. (57 kg), which is equivalent to 1/8 lb. per lin. ft. (0.4 kg per 0.3 m) of hole or per cu. yd. (0.76 cu. m) of excavation. The cost of explosives is as follows:

	Hours	Rate	Total	Rate	Total
125 lbs. (57 kg) dynamite		$....	$....	0.85	106.25
220 electric detonators					
with 6' (1.8 m) wire lead		....	....	0.45	99.00

75 lin. ft. (23 m) No. 14,................................				
R.C.S.B. lead wire...............................	....	....	0.12	9.00
Total cost.................................	$....			$214.25
Cost per hole		....		1.02
Cost per cu. yd. (1,900 cu. yds.)...................		....		0.11
Cost per cu. m (1452 cu. m)........................		....		0.15

Rock Excavation. There will be 1,900 cu. yds. (1452 cu. m) of rock excavation, taken out in two lifts of 4'-0" (1.2 m) each, and the loader will handle 45 cu. yds. (34 cu. m) per hr. at the following cost per 100 cu. yds. (76 cu. m):

	Hours	Rate	Total	Rate	Total
Loader					
and operator...........................	2.3	$....	$....	$55.00	$126.50
Labor*.....................................	2.3	....	....	21.04	48.39
Powderman............................	2.0	....	....	28.55	57.10
Labor assisting	5.0	....	....	21.04	105.20
Total cost 100 c.y. (76 cu. m)...................		$....			$337.19
Cost per cu. yd.			....		3.37
Cost per cu. m			....		4.38

*One laborer on bull chain, placing chain on rocks too large for shovel to handle.

Two 6-cu. yd. (4.6 cu. m) trucks haul the excavated rock to a dump 1/4 mile (0.4 km) from the job, averaging about 23 cu. yds. (18 cu. m) per hr. per truck, at the following cost per 100 cu. yds. (76 cu. m):

	Hours	Rate	Total	Rate	Total
Truck and driver................................	4.35	$....	$....	$49.50	$215.33
Cost per cu. yd.			....		2.15
Cost per cu. m			....		2.37

Four workers at the dump level 20 cu. yds. (15.3 cu. m) of excavated rock per hr. at the following labor cost per 100 cu. yds. (76 cu. m):

	Hours	Rate	Total	Rate	Total
Labor	20	$....	$....	$21.04	$420.80
Cost per cu. yd.			....		4.21
Cost per cu. m			....		5.47

The total cost of rock excavation, including drilling, blasting, excavating, hauling, and spreading, averages as follows per cu. yd. (cu. m):

	Total	per c.y.	per cu. m
Drilling	$....	$0.22	0.29
Explosives	....	0.12	0.17
Excavating	....	3.38	4.43

Hauling	....	2.16	2.84
Spreading and leveling	....	4.21	5.53
Total	$....	$10.09	13.26

Rock excavation conditions and costs vary widely. Always obtain information from sources at or near the job site before bidding on this work. Check with one or more of the local powdermen--they will know what the conditions are and what production may be expected.

Drilling Rock Using Pneumatic Jackhammer Drills. The hand held jackhammer is a popular tool for rock drilling on all types of construction jobs. They are furnished in sizes from 30 to 80 lbs. (14 to 36 kg), and many rock drilling conditions can be met. Holes up to 20 ft. (6 m) in depth are easily drilled with the heavier machines. Most sizes of jackhammer can be furnished in three styles, depending on the conditions of drilling and the depth of holes being drilled (dry style for shallow holes, blower style for deep holes, and wet for jobs where dust must be kept down to a minimum).

Hollow drill steel of various sizes and shapes are used, the most common being 7/8" and 1" hexagon. Bits are sometimes forged on the steel but the detachable bit is by far the most widely used.

When drilling holes in rock for work such as trenching and blasting, two workers operating drills should drill 125 to 175 lin. ft. (38 to 53 m) of holes per 8-hr. day, at the following cost per 100 lin. ft. (30 m):

	Hours	Rate	Total	Rate	Total
Jackhammerman	11.0	$....	$....	$21.04	$231.44
Compressor operator	5.5	....	....	28.55	157.03
Compressor expense	5.5	....	....	13.00	71.50
Cost per 100 l.f. (30 m)			$....		$459.97
Cost per l.f.				....	4.60
Cost per m				....	15.09

Does not include bit charges.

Rock Drilling Using Wagon Drills. The wagon drill consists of a one-piece tubular frame on pneumatic or steel wheels and a tilting tower, on which is mounted a heavy pneumatic drill. It drills holes downward at any angle from horizontal to vertical, to depths of 25 to 30 feet (7.5-9.0 m). This type of drill is suitable for all kinds of excavation and trenching and usually does the job of at least 2 or 3 heavy handheld jackhammers. It requires only an operator, a helper, and a compressor operator.

Rock Drilling Using Crawler Mounted Drills. An innovation in rock drilling for blast holes, etc. is the completely mechanized, self-propelled drill unit. Many construction equipment manufacturers make these units, which can do the work of two or three wagon drills. The unit can move easily over rugged terrain and can tow its own air supply with it. The set-up time for each hole is greatly reduced because of the self-propelled features. It is not uncommon for a crawler mounted drill to drill more than 500 lin. ft. (152 m) per 8-hour shift in hard sedimentary rock. It can be rented with operator for $800 to $1000 per day.

Excavating Using Pneumatic Diggers. These tools can be used to advantage on all classes of excavation that require picking of dirt or clay, such as trench work, tunneling, caisson sinking, and all kinds of building excavation, in fact, on all kinds of work in stiff clay or hard ground where power shovels, trenching machines, etc., cannot be used.

On work of this kind, two workers using air picks and six workers shoveling will loosen and shovel 35 to 40 cu. yds. (27 to 31 cu. m) per 8-hr. day, at the following labor cost per cu. yd. (cu. m):

	Hours	Rate	Total	Rate	Total
Labor operating picks	0.4	$....	$....	$21.04	$8.42
Labor shoveling	1.2	....	....	21.04	25.25
Compressor operator	0.2	....	....	28.55	5.71
Compressor expense	0.2	....	....	13.00	2.60
Cost per cu. yd.			$....		$41.98
Cost per cu. m				....	54.57

On large excavations providing plenty of room for workers, a worker with a pneumatic pick will loosen 25 to 30 cu. yds. (19.1 to 23 cu. m) per 8-hr. day. After the clay or earth is loosened, a worker will shovel 6 to 7 cu. yds. (4.5 to 5.4 cu. m) per 8-hr. day.

Comparative Cost By Hand

Excavating in stiff clay or tough ground, a worker using a hand pick and shovel will loosen 2-1/2 to 3 cu. yds. (1.9 to 2.3 cu. m) of dirt per 8-hr. day. The same worker, picking only, will loosen 3-1/2 to 4 cu. yds. (2.7 to 3.1 cu. m) of dirt per 8-hr. day, and when shoveling only, will remove 6 to 7 cu. yds. (4.6 to 5.4 cu. m) per 8-hr. day.

Using a hand pick, a worker will loosen and shovel one cu. yd. of dirt at the following cost per cu. yd. (cu. m):

	Hours	Rate	Total	Rate	Total
Labor per c.y.	3	$....	$....	$21.04	$63.12
per cu. m			....		82.06

Backfill Tamping Using a Pneumatic Tamper. A pneumatic tamper enables a worker to do nearly 10 times the amount of work he can do by hand. A backfill tamper strikes over 600 hard, snappy blows a minute, rams the fill hard, and works in and around pipe with far greater thoroughness than hand devices are able to do.

In piers, trenches, etc., where the excavated earth can be shoveled directly into the trench, a worker should backfill 16 to 20 cu. yds. (12.2 to 15.3 cu. m) per 8-hr. day.

Where the trenches have to be tamped, one worker with a pneumatic tamper will tamp as much as 2 workers can backfill, or approximately 35 to 40 cu. yds. (26.8 to 30.6 m) per 8-hr. day.

The cu. yd. (cu. m) cost should average as follows:

	Hours	Rate	Total	Rate	Total
Labor backfilling	0.4	$....	$....	$21.04	$8.42
Labor tamping	0.2	....	....	21.04	4.21
Compressor operator	0.2	....	....	28.55	5.71
Compressor & tamper	0.2	....	....	15.00	3.00
Cost per cu. yd.			$....		$21.34
Cost per cu. m			....		27.74

Comparative Cost By Hand

A worker tamps by hand 8 to 10 cu. yds. (6.1 to 7.6 cu. m) of backfill per 8-hr. day, at the following cost per cu. yd. (cu. m):

	Hours	Rate	Total	Rate	Total
Labor backfilling	0.4	$....	$....	$21.04	$8.42
Labor tamping	0.8	....	....	21.04	16.83
Cost per cu. yd.			$....		$25.25
Cost per cu. m			....		32.83

02250 SOIL TREATMENT

New regulations are being legislated almost on a daily basis. The estimator should become familiar with local EPA regulations and engage the services of a professional who specializes in this area.

02300 PILE FOUNDATIONS

The estimator needs to become familiar with the many types and models of pile hammers available for installing various pile types, such as wood, steel pipe, steel sheets, "H" piles, precast concrete (both round and square), combinations of concrete and steel, corrugated and step taper piles. Pile hammers available include air or steam, diesel, hydraulic, and drop hammers.

Drop Hammer. This is simply a heavy weight that is lifted and dropped on the pile. It operates within a guide that directs the impact. In the past the weight was lifted by human or horse power. Occasionally, this is still the case, but usually, a line on a crane or a winch raises the weight to a predetermined height, where a trip releases it to fall freely on the pile.

The weight of a drop hammer should be about one-third the weight of the pile to be driven. The fall must be regulated so that the pile is not damaged, and a suitable cap, with cushion material, is used on the pile to distribute the driving force. Drop hammers are used much more in other parts of the world than the U.S.

Air and Steam Hammers. *Single-acting pile hammers,* powered by steam or air, are essentially drop hammers with a short stroke and heavy ram. Steam or air is used to raise the weight much more rapidly than for drop hammers. The falling weight opens a port that admits steam or air under the ram to raise it for the next stroke. The rising elements cut off the pressure and open an exhaust port, permitting practically a free fall of the weight on the next stroke.

It is essential that the weight be raised to its full height and fall free to develop the rated capacity. The hammer should be powered at the manufacturer's recommended pressures. It should operate at recommended blows per minute and height of fall. Energy ratings range from 7,000 ft. lbs. up to 60,000 ft. lbs. (967 to 8292 kgm), but they go up to even 1,800,000 ft. lbs. (248,868 kgm) for big offshore hammers.

Double-acting pile hammers use steam or air both to raise the striking hammer and as a force to add energy on the downward stroke. They operate at about twice the speed of single-acting hammers of equal energy but require more steam or air at greater pressure.

A variation is the *differential-acting hammer,* where non-expansive use of the steam develops more effective pressure. The significant difference between double-acting and differential-acting is the manner and sequence of exhausting on the upward and downward strokes of the cycle. In the differential-acting hammer, there is no drop from the entering pressure to the main effective pressure moving the piston on the downward stroke.

Double-acting and differential-acting hammers usually give better results in granular non-cohesive soils or in soft clays. Used in proper soil conditions with the right pile, almost twice the production can be obtained as with a single-acting hammer.

Diesel Hammers. This type of pile hammer gets its energy from the compression blow of a falling weight and the reaction to controlled instantaneous burning and expansion of fuel, which raises the ram for the next stroke. The diesel hammer does not require steam or air under pressure for operation. The only input is a modest amount of diesel fuel, compressed and atomized to ignition readiness. The hammers have a total weight of one-third to one-half of that of comparable energy-rated single- or double-acting steam or air units. Diesel hammers are started by lifting the ram by outside means. In soft overburden there sometimes is a problem where piles move down without developing enough resistance to compress the fuel ignition heat and pressure for the next stroke. Diesels work better in hard driving.

Some diesel hammers have a closed top, designed to give added speed and energy to the down stroke. Contractors like diesel hammers because they are lighter and do not require exterior power.

Vibratory Driver/Extractor. These are like mini-stroke, high blow rate pile hammers, but the vibratory drivers/extractors in use today do not have linearly reciprocating weights or rams. Instead, they use rotating weights set eccentric from their centers of rotation. The result is a machine that is a mechanical sine wave oscillator. It is rigidly connected to the pile, usually by clamps, and it oscillates the pile through the soil. The vibratory driver/extractor incorporates the unique capability of converting from driver to extractor by pulling upon the vibratory hoist line.

Vibratory hammers ordinarily do not operate with leads, so other provisions must be made for holding the piles to plan position for driving. Equipment intended for vibratory or sonic installation or withdrawal of bearing or sheet piles should be capable of adjustment for frequency and amplitude to accommodate to different combinations of soils and piling. For efficient operation the machines must grip the pile firmly, with hydraulic or air actuated rams.

Verification of bearing capacity of piles installed by vibratory machines by load tests or final driving with an impact hammer usually is considered necessary.

Types and Specifications of Pile Hammers

Steam and Air Hammers

Rated Energy ft. lbs.	Rated Energy kg. m.	Manufac	Model	Action	Average Blows per Minute	Striking Weight lbs.	kg
1,800,000	248,868	Vulcan	6300	Single	40	300,000	136,080
1,200,000	165,912	Conmaco	2000	Single	40	200,000	90,720
750,000	103,695	Vulcan	5150	Single	46	150,000	68,040
350,000	48,391	Conmaco	700	Single	43	70,000	31,752
180,000	24,887	Vulcan	360	Single	60	60,000	27,216
150,000	20,739	Conmaco	5300	Single	46	30,000	13,608
120,000	16,591	Vulcan	340	Single	60	40,000	18,144
90,000	12,443	Vulcan	030	Single	55	30,000	13,608
90,000	12,443	Conmaco	300	Single	55	30,000	13,608
60,000	8,296	Vulcan	020	Single	60	20,000	9,072
60,000	8,296	Conmaco	200	Single	60	20,000	9,072
48,750	6,740	Vulcan	016	Single	60	16,250	7,371
48,750	6,740	Conmaco	160	Single	60	16,250	7,371
42,000	5,807	Vulcan	014	Single	60	14,000	6,350
42,000	5,807	Conmaco	140	Single	60	14,000	6,350
37,375	5,167	Conmaco	115	Single	50	11,500	5,216
32,500	4,493	Vulcan	010	Single	50	10,000	4,536

Energy ft. lbs.	Energy kg. m.	Manufac	Model	Action	Blows per Minute	Weight lbs.	kg
32,500	4,493	Conmaco	100	Single	50	10,000	4,536
26,000	3,595	Vulcan	08	Single	50	8,000	3,629
26,000	3,595	Conmaco	80	Single	50	8,000	3,629
19,500	2,696	Vulcan	06	Single	60	6,500	2,948
19,500	2,696	Conmaco	65	Single	60	6,500	2,948
19,150	2,648	Mkt Mfg	11B3	Double	95	5,000	2,268
15,000	2,074	Vulcan	1	Single	60	5,000	2,268
15,000	2,074	Conmaco	50	Single	60	5,000	2,268
13,100	1,811	Mkt Mfg	10B3	Double	105	3,000	1,361
8,750	1,210	Mkt Mfg	9B3	Double	145	1,600	726

Diesel-Powered Hammers

Rated Energy ft. lbs.	Rated Energy kg. m.	Manufac	Model	Action	Average Blows per Minute	Ram Weight lbs.	Est. Fuel Consumption gph	lph
225,000	31,109	Delmag	D80-12	Single	36-45	19,500	6.5	24.6
165,000	22,813	Delmag	D62-23	Single	36-50	14,600	5.5	20.8
127,500	17,628	Mkt Mfg	DE150/110	Single	40-50	11,000	5.5	20.8
117,000	16,176	Delmag	D55	Single	36-50	12,100	5.5	20.8
105,600	14,600	Kobelco	K-60	Single	40-60	13,200	7.3	27.6
105,000	14,517	Delmag	D46-32	Single	37-53	10,140	4.5	17.0
100,000	13,826	Ice	200S	Single	53-70	20,000	5.0	18.9
91,000	12,595	Kobelco	K-45	Single	39-60	9,900	5.0	18.9
84,300	11,655	Mitsubishi	MH-45	Single	42-60	10,210	5.0	18.9
83,100	11,489	Delmag	D36-32	Single	37-53	7,900	3.0	11.4
70,800	9,789	Kobelco	K-35	Single	39-60	7,700	3.5	13.2
70,000	9,678	Ice	70S	Single	38-55	7,000	3.5	13.2
66,100	9,139	Delmag	D30-32	Single	38-52	6,600	2.0	7.6
65,600	9,070	Mitsubishi	MH-35	Single	42-60	7,720	4.5	17.0
59,500	8,226	Mkt Mfg	DE-70B	Single	40-50	7,000	3.3	12.5
50,700	7,010	Kobelco	K-25	Single	39-60	5,510	2.7	10.2
50,000	6,913	Ice	660	Double	84-88	7,564	3.3	12.5
48,500	6,706	Delmag	D25-32	Single	38-52	4,850	2.0	7.6
46,900	6,484	Mitsubishi	MH-25	Single	42-60	5,510	3.0	11.4
42,500	5,876	Mkt Mfg	DE-50B	Double	40-50	5,000	3.3	12.5
40,000	5,530	Ice	40S	Single	38-55	4,000	3.0	11.4
38,200	5,282	Mkt Mfg	DA-55B	Double	78-82	5,000	3.0	11.4
30,000	4,148	Ice	520	Single	80-84	5,070	2.0	7.6
28,100	3,885	Mitsubishi	MH-15	Single	42-60	3,310	2.0	7.6
27,100	3,747	Delmag	D16-32	Single	40-60	3,300	1.7	6.4
25,200	3,484	Kobelco	K-13	Single	40-60	2,860	1.5	5.7
23,800	3,291	Mkt Mfg	DE-30B	Double	40-50	2,800	2.0	7.6
22,500	3,111	Ice	30S	Single	44-67	3,000	1.5	5.7
21,000	2,903	Mkt Mfg	DA-35C	Double	78-82	2,800	2.7	10.2
18,100	2,503	Ice	440	Single	88-92	4,000	1.5	5.7
18,000	2,489	Delmag	D8-22	Single	38-52	1,760	1.0	3.8
17,000	2,350	Mkt Mfg	DE-20B	Double	40-50	2,000	2.0	7.6
9,350	1,293	Mkt Mfg	DA-15C	Double	80-92	1,100	1.5	5.7
9,100	1,258	Delmag	D5	Single	40-60	1,100	1.5	5.7
8,100	1,120	Ice	180	Double	90-95	1,725	0.7	2.6
3,630	502	Delmag	D4	Single	50-60	840	0.2	0.8
1,815	251	Delmag	D3	Single	60-70	485	0.1	0.4

Hydraulic Vibratory Drivers/Extractors

Manufac.	Model	Dynamic Tons	Force Kips	Line Tons	Pull Kips	Suspended Lbs.	Wgt. Kg
Ice	1412	163	326	100	200	26,900	12,202
Ice	815	145	290	50	100	15,600	7,076
Mkt Mfg.	V20	116	232	60	120	10,650	4,831
Ice	612	82	164	40	80	13,700	6,214
Mkt Mfg.	V16	78	156	40	80	9,250	4,196
Ice	416L	65	130	40	80	9,900	4,491
Ice	216	35	70	30	60	4,825	2,189
Mkt Mfg.	V5	30	60	15	30	5,100	2,313

When estimating the cost of piles, one must consider overhead expense before the actual cost of driving the piles can be computed. This includes cost of moving the equipment to the job, setting up the pile rig and being ready to operate, and dismantling and removing at completion.

This expense will always be pretty much the same, regardless of the size of the job. The equipment move charge is as much on a job with 100 piles as on one containing 1,000 piles.

The actual cost of driving the piles will vary with soil conditions, length of piles, and the amount of moving necessary to drive piles at the proper location on the site.

Setting Up and Removing Equipment. The cost of moving the pile driving equipment to the job must be taken into consideration, and then after it arrives at the job, there is the task of setting up the pile driver and getting ready to operate.

After the pile driver has been delivered at the job it should take about one day for the crew to set it up ready to drive and this labor should cost as follows:

Setting up Equipment	Hours	Rate	Total	Rate	Total
Foreman	8	$....	$....	$29.41	$235.28
Pile driver engineer	8	....	....	28.55	228.40
Oiler	8	....	....	21.04	168.32
2 workers, pile driver	16	....	....	26.44	423.04
4 general laborers	32	....	....	21.04	673.28
Total labor cost		$....			$1,738.22

The above item will remain practically the same whether there are 100 or 1,000 piles to be driven.

After the job has been completed, the pile driver must be dismantled and removed from the job and the dismantling should cost as follows:

Dismantling Equipment	Hours	Rate	Total	Rate	Total
Foreman	8	$....	$....	$29.41	$235.28
Engineer on pile driver	8	....	....	28.55	228.40
Oiler	8	....	....	21.04	168.32
2 workers, pile driver	16	....	....	26.44	423.04
4 general laborers	32	....	....	21.04	673.28
Total Labor Cost			$....		$1,738.22

Various Types of Bearing and Friction Piles

Wood	Concrete	Steel	Sheet Pile
Treated	Cast-in-place	HP shape	Wood
Untreated (plain)	Precast/prestressed	Pipe	Steel
	Shell	Shell	Precast Concrete
	Shell-less		

The above types can be used in various combinations, sometimes known as *composite piles.* A good example of composite pile would be heavy-walled pipe piles filled with concrete. Normally, the designer indicates on the drawings or in the written specifications, or on both, the type of pile and the size of hammer required to install.

Driving Wood Piles. After the pile driver is set up and ready to operate, the labor cost of driving the piles will vary with the length of the piles and job conditions.

The following table gives the approximate number of piles of various lengths that should be driven per hour and per day:

Length of Piles	1 Hour	8 Hours
22 feet (6.71 m)	4.7	37
24 feet (7.32 m)	4.5	36
26 feet (7.92 m)	4.2	34
28 feet (8.53 m)	4.0	32
30 feet (9.14 m)	3.7	29
32 feet (9.75 m)	3.5	28
34 feet (10.36 m)	3.3	26
36 feet (10.97 m)	3.1	25
38 feet (11.58 m)	3.0	24
40 feet (12.19 m)	2.8	22
45 feet (13.72 m)	2.5	20
50 feet (15.24 m)	2.3	18
55 feet (16.76 m)	2.1	17
60 feet (18.29 m)	2.0	16

The labor operating the pile driver per 8-hr. day should cost as follows:

	Hours	Rate	Total	Rate	Total
Foreman	8	$....	$....	$29.41	$235.28
Engineer on pile driver	8	....	....	28.55	228.40
Oiler	8	....	....	21.04	168.32
2 workers, pile driver	16	....	....	26.44	423.04
4 general laborers	32	....	....	21.04	673.28
Labor Cost per day			$....		$1,738.22

The above does not include time and costs for pile layout or resurvey, spotting piles, pumping, shoring, excavating, purchasing piles, pile points, or cutting off wood piles after they are driven. It only includes the actual time driving the piles. It also does not include the cost of equipment, fuel, oil, and other supplies for the pile driver. These costs should be added.

Cutting Off Wood Piles. After the wood piles have been driven, the tops projecting above the ground will have to be cut off to receive the foundation that is to be placed on them.

This cost will vary according to the size of the quarters in which workers must operate, but on the average job, it should require 1/6 to 1/4 hrs. to cut off each wood pile 12" or 14" (300 to 350 mm) in diameter with a chain saw, at the following labor cost per pile:

	Hours	Rate	Total	Rate	Total
Labor	0.22	$....	$....	$26.44	$5.82

To the above costs, add the hourly cost of chain saw and the time for removing sawed-off ends, as well as proper removal and disposal.

Computing the Cost of Wood Piles. As described above, the cost of driving wood piles will vary with the number of piles, length, kind of soil, and other job conditions. Pile costs fluctuate so rapidly that current costs must be verified for each project.

The following is a sample estimate, showing a method for arriving at costs for driving 680 treated wood piles 40'-0" (12.2 m) long:

	Local Rate	Total	Total
Trucking pile driver to job	$....	$....	$2,000.00
Setting up pile driver, 1 day for crew		....	1,571.44
680 wood piles, 40'-0" long @ $200.00		....	136,000.00
Driving piles (22 per 8-hr. day) 31 days @ $1,469.28		....	45,547.68
Fuel, oil, misc. supplies, etc. 31 days @ $175.00		....	5,425.00
Dismantling pile driver at completion, 1 day		....	1,571.44
Trucking pile driver, job to yard		....	2,000.00
Equipment rental or depreciation allowance (36 days)		....	18,000.00

Cost 680 piles..	$....	212,115.56
Cost per pile..	....	311.93
Cost per lin. ft.	....	7.80
Cost per m..	....	25.58

Add cost of cutting off piles after driving, and for overhead & profit.

Driving Steel HP Piles. Once the pile driver is set up and ready to operate, labor costs for actually driving the piles will vary with the length of piles and site conditions.

Rolled structural steel shapes are also used as bearing piles. The shape commonly used for this purpose is the HP. This type of pile has proved especially useful for trestle structures in which the pile extends above the ground and serves not only as a pile but also as a column.

Because of their small cross-sectional area, piles of this type can often be driven through dense soils to point bearing where it would be difficult to drive a pile of solid cross-section, such as a wood, cast-in-place, or precast concrete. But the feature that allows it to penetrate dense soils works to its disadvantage in other soil. Where piles are used to support loads by friction, or where they are used primarily for compaction, a considerably longer HP is required to carry the same load that can be supported by a pile of less length but greater cross-sectional area.

The upper ends of HP piles can be encased in concrete to prevent corrosion. Steel piles weighing more than 200 lbs. per l.f. (303 kg per m) have been driven in lengths up to 130 feet (40 m). Lighter sections have been driven to even greater lengths.

Driving HP Section Piles. The following table gives the approximate number of HP piles of various lengths that can be driven per hour and per day, assuming average job site conditions:

Length of Piles	1 Hour	8 Hours
20' (6.1 m)..	4.00	32
30' (9.1 m)..	3.75	30
40' (12.2 m)..	3.00	24
50' (15.2 m)..	2.50	20
60' (18.3 m)..	2.25	18

The labor cost of operating the pile driver per 8-hr. day is as follows:

	Hours	Rate	Total	Rate	Total
Foreman..8		$....	$....	$29.41	$235.28
Engineer on pile driver................8		....	....	28.55	228.40
Oiler..8		....	....	21.04	168.32
2 workers, pile driver 16		....	....	26.44	423.04
4 general laborers.......................32		....	....	21.04	673.28
Labor Cost per 8-hr. day			$....		$1,738.22

The above does not include time and costs for pile layout or resurvey, spotting piles, pumping, shoring, excavation, purchase of piles, pile points and pile splices, or cutting off driven piles. It only includes the actual time driving piles.

It also does not include the cost of equipment, fuel, oil, and other supplies for the pile driver.

Prices of Steel HP Piles. Prices of steel HP piles vary considerably, mainly due to the material cost of the rolled steel sections. Steel sections commonly used for this purpose may range from an 8-inch (200 mm) HP section weighing 36 lbs. per lin. ft. (55 kg per m) to a 14-inch (350 mm) HP section weighing 117 lbs. per lin. ft. (177 kg per m).

Assuming a unit cost of $0.60 per lb. ($1.32 per kg), delivered to job site from warehouse, the HP beam material will cost from $21.60 to $35.10 per l.f. ($70-115 per m). If the job is in an isolated location, additional transportation charges must be figured.

Cutting Off Steel HP Piles. After the steel HP section pile has been driven, the pile tops that project above the ground must be cut off to the required elevation to receive the foundation that will be placed on them.

The cost will vary according to the size of the pile and the quarters in which workers must work. A burning outfit, consisting of gauges, burning torch, hoses, acetylene, and oxygen is normally required to cut HP steel piles at the following cost per pile:

HP	Hours	Rate	Total	Rate	Total
8" (200 mm) section	0.17	$....	$....	$26.44	$4.49
10" (250 mm)	0.21	....	....	26.44	5.55
12" (300 mm)	0.25	....	....	26.44	6.61
14" (350 mm)	0.31	....	....	26.44	8.20

To the above, add the costs for a complete burning outfit, time for removing the cut off pile sections, and the cost for removal from the site and proper disposal.

Splicing Steel HP Piles. When splicing is necessary, steel HP piles can be spliced in the field. When butt welded, splices should be full penetration butt welds across both flanges and the web. Edges on the ends of the upper section, where welding is to be done, should be beveled for penetration welding. The top of the pile length, where driving will be done, should not be beveled.

Steel HP piles are spliced by butt welding at the following cost per pile:

HP	Hours	Rate	Total	Rate	Total
8" (200 mm)	1.33	$....	$....	$26.44	$35.17
10" (250 mm)	1.67	....	....	26.44	44.15
12" 300 mm)	2.00	....	....	26.44	52.88
14" (350 mm)	2.50	....	....	26.44	66.10

The costs vary according to the size of the piles and the quarters in which workers are obliged to operate. Add to the above all costs for additional material and equipment (such as plates, welding machine, and pile equipment), and standby time while welding is being performed, and all required testing.

Prefabricated splicers are available from Dougherty Foundation Products, Inc. of Franklin Lakes, New Jersey. These splicers temporarily hold a length of HP pile until the joint can be completed. The driver can set an extension and then move on to other work while the welding takes place, at the following labor cost per pile:

HP	Hours	Rate	Total	Rate	Total
8" to 14" (200 to 350 mm)........0.50		$....	$....	$26.44	$13.22

Material Cost for DFP* Splicers

HP	
8" (200 mm)...	$24.00 each
10" (250 mm)...	29.00 each
12" (300 mm)...	39.00 each
14" (350 mm)...	48.00 each

*Dougherty Foundation Products, Inc., Franklin Lakes, NJ

The costs vary according to the size of the piles and the quarters in which workers must operate.

CONCRETE PILES

Due to the wide variety of conditions under which concrete piles are installed and used, it is difficult to give dependable cost figures without knowing the details of each installation. The different types of concrete piles vary considerably in cost. When estimating work containing concrete piles, it is always advisable to consult contractors specializing in this class of work. Most of them maintain offices in the principal cities and are usually willing to investigate and quote budget prices on anticipated work. More accurate figures can be obtained than can be given here.

There are two principal types of concrete piles, cast-in-place and precast. The cast-in-place pile is formed in the ground in the position where it is to be used in the foundation. The precast pile is cast above ground, and after it has been properly cured, it is driven or jetted just like a wood pile.

In ordinary building foundation work, the cast-in-place pile is more common, because the required length can be readily adjusted in the field as the job progresses. There is no need to predetermine pile lengths, and the required length is installed at each pile location.

In contrast, it is necessary to predetermine the length of precast piles, and to provide for contingencies, it is generally required that piles be ordered longer than the actual anticipated length. Also, it is not possible to determine the pile length at each location, because subsoil conditions at any construction site will show considerable variation. The cost of waste piling for precast piles can be quite high, as is the cost of cutting them off to the proper grade. Precast piles are prestressed and are manufactured at established plants. Job site casting is relatively rare. It is difficult and costly to handle and transport precast piles especially in long lengths, and a production plant must be relatively close to the job site.

In marine installations, either in salt water or fresh water, the precast pile is used almost exclusively, because of the difficulty of placing cast-in-place piles in open water. For docks and bulkheads, the cast-in-place pile is sometimes used in the anchorage system. On trestle type structures, such as highway viaducts, the precast pile is commonly used. A portion of the pile often extends above the ground and serves as a column for the superstructure. Precast concrete piles are always reinforced internally so that they resist stress from handling and driving. Cast-in-place piles are rarely reinforced, because lateral support of even the poorest of soils is sufficient to overcome any bending moments induced in the pile by column action.

Cast-in-place Concrete Pipe Piles. Pipe can be of any practical diameter. It is almost always filled with concrete, but with adequate wall thickness in non-corrosive ground, it can be used without concrete. Driving usually produces stresses greater than the working load on the pile. Pipe should conform to *ASTM A252, Specifications for Welded and Seamless Steel Pipe Piles.*

Pipe shell that does not contribute to the strength of the pile as a structural member may be of any metal that adequately resists installation stress and maintains an open shaft to receive the concrete, free from water or other foreign matter. Steel shell should meet the applicable requirements of *ASTM Specifications A-245, A-283, A-366, or A-415.*

The end closure for pipe or shell may be cast steel meeting requirements of *ASTM A-27 65/35* or structural steel meeting *ASTM A-36* or better.

Concrete for cast-in-place piles should conform to requirements for cement, aggregate, mixing, placing, and protection specified by ACI or PCA for quality concrete.

Cast-in-place concrete piles can be one of four types: open-end pipe, closed-end pipe, thin-cased and corrugated shell, large-base concrete piles. The closed-end pipe pile is simply a piece of steel pipe, closed at the bottom with a heavy boot, driven into the ground, and filled with concrete. The uses and allowable loads for these piles are about the same as for other types of cast-in-place piles with driven shells. On some jobs the pipe shell is driven all in one piece, but often it is driven in sections that are welded together or fitted together with internal sleeves as the driving progresses. Pipe wall thickness must be adequate to develop the required stiffness and driveability of the pipe.

Open-end steel pipe piles are usually driven to bearing on rock. Since the pipe is open at the bottom during driving, the interior fills with soil which must be removed before concreting. Cleaning is usually done with air and water jets, after which the pipe is re-driven to insure proper seating in the rock. Remaining water is pumped out and the pipe pile is filled with concrete. Open-end pipe installed in this manner usually carry relatively high working loads. They have been installed up to 36 inches (900 mm) in diameter and in lengths exceeding 200 feet (61 meters).

Estimating the Cost of Concrete Pipe Piles. The cost of concrete pipe piles will vary greatly, depending upon the number of piles in the job and the length, because it is the total footage of piles in the job rather than the number of piles that controls the price. For example, the price per foot for 100 piles 35 feet long would be much greater than the price per foot for 1,000 piles 25 feet (7.6 meters) long.

For any type of pile, contractors who specialize in this work are usually willing to submit budget prices, which include all labor, materials, and equipment for the complete job. A detailed estimate would only be required when such budget prices are not available, or for some reason the estimator does not desire to seek out a budget price.

As with all pile driving, regardless of the size of the job, there is a fixed charge for moving equipment on and off the job that will cost from $5,000.00 to $25,000.00, provided the equipment is located in the immediate vicinity of the work.

If it is necessary to ship plant and equipment from one city to another, it will be necessary to add additional freight and trucking amounting to $2,000.00 to $4,000.00, making a total fixed charge of $7,000.00 to $29,000.00, before figuring the cost of the piles.

Labor Driving Cast-in-place Concrete Pipe Piles. After the pile driver has been set up and ready to operate, the number of piles driven per day and the labor cost driving them will vary with the length of the piles and job conditions.

For instance, on a job where the piles were closely spaced and driven from a flat surface, a pile driver averaged 100 piles a day (which is exceptional), while on another job where the piles were more widely scattered and had to be driven from different elevations, it was with difficulty that 20 piles were driven per day.

Labor Placing Concrete For Piles. Practically 90% of all concrete used in concrete piles is ready-mix concrete, which is usually discharged directly into the piles. Ordinarily it requires 1/2 to 1 hour time for one worker per cu. yd. (cu. m) of concrete.

Pipe Sizes Frequently Used for Piling and Concrete Required

Outside Dia. of Pipe, in.	Outside Dia. of Pipe, mm	Thickness Pipe Wall in.	Thickness Pipe Wall mm	Lbs. of Pipe per lin. ft.	Kg of Pipe per meters	Concrete C.Y. per lin. ft.	Concrete Cu. M per meters
8-5/8	210	0.125	3.125	11.35	16.89	0.0142	0.0356
		0.188	4.700	16.90	25.15	0.0137	0.0795
		0.219	5.475	19.64	29.2	0.0135	0.0339
		0.250	6.250	22.36	22.3	0.0133	0.0334
		0.312	7.800	27.74	41.2	0.0129	0.0324
		0.375	9.375	33.00	49.0	0.0125	0.0313
10	250	0.188	4.700	19.65	29.2	0.0187	0.0469
		0.219	5.475	22.85	34.0	0.0185	0.0464
		0.250	6.250	26.03	38.7	0.0182	0.0456
10-3/4	269	0.188	4.700	21.15	31.4	0.0217	0.0544
		0.219	5.475	24.60	36.6	0.0215	0.0539
		0.250	6.250	28.04	41.7	0.0212	0.0532
		0.307	7.675	34.24	50.9	0.0207	0.0519
		0.365	9.125	40.50	60.2	0.0203	0.0509
		0.500	12.500	54.70	81.4	0.0192	0.0481
12	300	0.188	4.700	23.65	35.2	0.0273	0.0685
		0.218	5.450	27.52	40.9	0.0270	0.0677
		0.250	6.250	31.40	46.7	0.0267	0.0670
12-3/4	319	0.188	4.700	25.16	37.4	0.0309	0.0775
		0.219	5.475	29.28	43.5	0.0306	0.0767
		0.250	6.250	33.40	49.7	0.0303	0.0760
		0.312	7.800	41.50	61.7	0.0297	0.0745
		0.375	9.375	49.60	73.8	0.0291	0.0730
		0.500	12.500	65.40	97.3	0.0279	0.0700
14	350	0.188	4.700	27.70	41.2	0.0375	0.0940
		0.219	5.475	32.20	47.9	0.0372	0.0933
		0.250	6.250	36.70	54.6	0.0368	0.0923
		0.312	7.800	45.70	68.0	0.0362	0.0908
		0.375	9.375	54.60	81.2	0.0355	0.0890
		0.500	12.500	72.10	107.2	0.0341	0.0855
16	400	0.188	4.700	31.70	47.1	0.0493	0.1236
		0.219	5.475	36.90	54.9	0.0489	0.1226
		0.250	6.250	42.10	62.5	0.0485	0.1216
		0.312	7.800	52.40	77.9	0.0477	0.1196
		0.375	9.375	62.60	93.1	0.0470	0.1179
		0.500	12.500	82.80	123.2	0.0454	0.1139

Pipe Sizes Frequently Used for Piling
and Concrete Required (Continued)

Outside Dia. of Pipe, in.	Outside Dia. of Pipe, mm	Thickness Pipe Wall in.	Thickness Pipe Wall mm	Lbs. of Pipe per lin. ft.	Kg of Pipe per meters	Concrete C.Y. per lin. ft.	Concrete Cu. M per meters
18	450	0.219	5.475	41.50	61.7	0.0623	0.1562
		0.250	6.250	47.40	70.5	0.0619	0.1552
		0.312	7.800	59.00	87.8	0.0610	0.1530
		0.375	9.375	70.60	105.0	0.0601	0.1507
		0.500	12.500	93.50	139.1	0.0584	0.1465
20	500	0.219	5.475	46.20	68.7	0.0773	0.1938
		0.250	6.250	52.70	78.4	0.0768	0.1926
		0.312	7.800	65.70	97.7	0.0758	0.1901
		0.375	9.375	78.60	116.9	0.0749	0.1878
		0.500	12.500	104.10	154.9	0.0729	0.1828
22	550	0.250	6.250	58.10	86.4	0.0934	0.2342
		0.375	9.375	86.60	128.8	0.0912	0.2287
		0.500	12.500	114.80	170.8	0.0891	0.2234
24	600	0.250	6.250	63.40	94.3	0.1116	0.2799
		0.312	7.800	79.10	117.7	0.1104	0.2769
		0.375	9.375	94.60	140.8	0.1093	0.2741
		0.500	12.500	125.50	186.7	0.1067	0.2676
30	750	0.375	9.375	118.70	176.6	0.1728	0.4333
		0.500	12.500	157.50	234.3	0.1700	0.4263
36	900	0.375	9.375	142.70	212.3	0.2510	0.6294
		0.500	12.500	189.60	282.1	0.2474	0.6204

Quantity of Concrete Required in Various Sizes of Pipe Piles,
in Cu. Ft., Cu. Yds., and Cu. Meters

Length, l.f. (m)	12" OD x 0.250 Wall (300 mm OD x 6.25 mm Wall)			14" OD x 0.250 Wall (350 mm OD x 6.25 mm Wall)		
	Cu. Ft.	Cu. Yd.	Cu. M	Cu. Ft.	Cu. Yd.	Cu. M
1 (0.3)	0.72	0.03	0.02	0.99	0.04	0.03
5 (1.5)	3.60	0.13	0.10	4.96	0.18	0.14
10 (3.0)	7.20	0.27	0.20	9.93	0.37	0.28
15 (4.6)	10.80	0.40	0.31	14.9	0.55	0.42
20 (6.1)	14.40	0.53	0.41	19.8	0.73	0.56
22 (6.7)	15.80	0.59	0.45	21.8	0.81	0.62
24 (7.3)	17.30	0.64	0.49	23.8	0.88	0.67
26 (7.9)	18.70	0.69	0.53	25.8	0.96	0.73
28 (8.5)	20.10	0.74	0.57	27.8	1.03	0.79

Length, l.f. (m)	12" OD x 0.250 Wall (300 mm OD x 6.25 mm Wall)			14" OD x 0.250 Wall (350 mm OD x 6.25 mm Wall)		
	Cu. Ft.	Cu. Yd.	Cu. M	Cu. Ft.	Cu. Yd.	Cu. M
30 (9.1)	21.60	0.80	0.61	29.8	1.10	0.84
32 (9.8)	23.00	0.85	0.65	31.7	1.17	0.90
34 (10.4)	24.50	0.91	0.69	33.7	1.25	0.95
36 (11.0)	10.97	25.90	0.31	19.80	35.70	0.56
38 (11.6)	11.58	27.30	0.33	20.87	37.70	0.59
40 (12.2)	12.19	28.80	0.35	22.02	39.70	0.62
42 (12.8)	12.80	30.20	0.36	23.09	41.70	0.65
44 (13.4)	13.41	31.70	0.38	24.24	43.70	0.69
46 (14.0)	14.02	33.10	0.40	25.31	45.70	0.72
48 (14.6)	14.63	34.60	0.41	26.45	47.60	0.75
50 (15.2)	15.24	36.00	0.43	27.52	49.60	0.78
52 (15.8)	15.85	37.40	0.45	28.59	51.60	0.81
54 (16.5)	16.46	38.90	0.47	29.74	53.60	0.84
56 (17.1)	17.07	40.30	0.48	30.81	55.60	0.87
58 (17.7)	17.68	41.80	0.50	31.96	57.60	0.91
60 (18.3)	18.29	43.20	0.52	33.03	59.60	0.94

Length, l.f. (m)	16" OD x 0.250 Wall (400 mm OD x 6.25 mm Wall)			24" OD x 0.250 Wall (600 mm OD x 6.25 mm Wall)		
	Cu. Ft.	Cu. Yd.	Cu. M	Cu. Ft.	Cu. Yd.	Cu. M
1 (0.3)	1.30	0.05	0.04	3.01	0.11	0.09
5 (1.5)	6.54	0.24	0.19	15.00	0.56	0.42
10 (3.0)	13.00	0.48	0.37	30.10	1.11	0.85
15 (4.6)	19.60	0.73	0.56	45.10	1.67	1.28
20 (6.1)	26.10	0.97	0.74	60.20	2.23	1.70
22 (6.7)	28.80	1.07	0.82	66.20	2.45	1.87
24 (7.3)	31.40	1.16	0.89	72.30	2.68	2.05
26 (7.9)	34.00	1.26	0.96	78.30	2.90	2.22
28 (8.5)	36.60	1.36	1.04	84.30	3.12	2.39
30 (9.1)	39.20	1.45	1.11	90.30	3.34	2.56
32 (9.8)	41.90	1.55	1.19	96.40	3.57	2.73
34 (10.4)	44.50	1.65	1.26	102.00	3.78	2.89
36 (11.0)	47.10	1.74	1.33	108.00	4.00	3.06
38 (11.6)	49.70	1.84	1.41	114.00	4.22	3.23
40 (12.2)	52.30	1.94	1.48	120.00	4.44	3.40
42 (12.8)	54.90	2.03	1.55	126.00	4.67	3.57
44 (13.4)	57.60	2.13	1.63	132.00	4.89	3.74
46 (14.0)	62.20	2.30	1.76	138.00	5.11	3.91
48 (14.6)	62.80	2.33	1.78	144.00	5.33	4.08
50 (15.2)	65.40	2.42	1.85	150.00	5.56	4.25
52 (15.8)	68.00	2.52	1.93	156.00	5.78	4.42
54 (16.5)	70.70	2.62	2.00	162.00	6.00	4.59
56 (17.1)	73.30	2.71	2.08	168.00	6.22	4.76
58 (17.7)	75.90	2.81	2.15	174.00	6.44	4.93
60 (18.3)	78.50	2.91	2.22	180.00	6.67	5.10

Length, l.f. (m)	36" OD x 0.500 Wall (900 mm OD x 12.50 mm Wall)		
	Cu. Ft.	Cu. Yd.	Cu. M
1 (0.3)	6.67	0.25	0.19
5 (1.5)	33.30	1.23	0.94
10 (3.0)	66.70	2.47	1.89
15 (4.6)	100.00	3.70	2.83
20 (6.1)	133.00	4.93	3.77
22 (6.7)	146.00	5.41	4.13
24 (7.3)	160.00	5.93	4.53
26 (7.9)	173.00	6.41	4.90
28 (8.5)	187.00	6.93	5.30
30 (9.1)	200.00	7.41	5.66
32 (9.8)	213.00	7.89	6.03
34 (10.4)	227.00	8.41	6.43
36 (11.0)	240.00	8.89	6.80
38 (11.6)	253.00	9.37	7.16
40 (12.2)	267.00	9.89	7.56
42 (12.8)	280.00	10.37	7.93
44 (13.4)	293.00	10.85	8.30
46 (14.0)	307.00	11.37	8.69
48 (14.6)	320.00	11.85	9.06
50 (15.2)	333.00	12.33	9.43
52 (15.8)	347.00	12.85	9.83
54 (16.5)	360.00	13.33	10.19
56 (17.1)	374.00	13.85	10.59
58 (17.7)	387.00	14.33	10.96
60 (18.3)	400.00	14.81	11.33

Thin Shell Pipe Piles. Monotube (Monotube Pile Company, Canton Ohio) piles are the only uniformly tapered steel shell pipe piles available. Monotubes are strictly a friction pile, where the shell is end driven with conventional pile driving equipment, visually inspected, and filled with concrete.

Monotubes are available in four diameters, four gauges, and three rates of taper, making them suitable for soils with considerable friction resistance, such as sand, silt, and clay. Because of the wedging action of the uniform taper and the factory attached conical nose, they are commonly driven to the same capacity as straight-sided friction piles with less depth. A combination of cold processing and vertical fluting permits the use of lighter gauges that are equal in strength to pipe piles with thicker walls. Tapered sections can be factory-welded to a consent diameter extension or spliced in the field. Field splices are made after the tapered portion has been driven into the soil. The extension is lifted into the leads and inserted 6 to 8 inches into the taper. Factory-crimped extension sections speed up alignment and field splicing.

Specifications for Monotube Piles
Tapered Sections

Type	Size Tip Point Dia. x Butt Dia. x Lngth.	9 Ga. 0.1495"	Weight per Foot 7 Ga. 0.1793"	5 Ga. 0.2092"	3 Ga. 0.2391"	Conc. Vol. Cu. Yds.
	8-1/2" x 12" x 25'	17	20	24	28	0.43
F	8" x 12" x 30'	16	20	23	27	0.55
Taper	8-1/2" x 14" x 40'	19	22	26	31	0.95
0.14 Inch	8" x 16" x 60'	20	24	28	33	1.68
per Foot	8" x 18" x 75'	--	26	31	35	2.59
J	8" x 12" x 17'	17	20	23	27	0.32
Taper	8" x 14" x 25'	18	22	26	30	0.58
0.25 Inch	8" x 16" x 33'	20	24	28	32	0.95
per Foot	8" x 18" x 40'	--	26	30	35	1.37
Y	8" x 12" x 10'	17	20	24	28	0.18
Taper	8" x 14" x 15'	19	22	26	30	0.34
0.40 Inch	8" x 16" x 20'	20	24	28	33	0.56
per Foot	8" x 18" x 25'	--	26	31	35	0.86

Weight per Meter

Type	Size Tip Point Dia. x Butt Dia. x Lngth.	9 Ga. 3.74 mm	7 Ga. 4.48 mm	5 Ga. 5.23 mm	3 Ga. 5.98 mm	Conc. Vol. Cu. M
	212.5 mm x 300 mm x 7.62 m	25.29	29.76	35.71	41.66	0.33
F	200.0 mm x 300 mm x 9.14 m	23.80	29.76	34.22	40.17	0.42
Taper	212.5 mm x 350 mm x 12.19 m	28.27	32.73	38.68	46.12	0.73
3.5 mm	200.0 mm x 400 mm x 18.29 m	29.76	35.71	41.66	49.10	1.28
per meter	200.0 mm x 450 mm x 22.86 m	--	38.68	46.12	52.07	1.98
J	200.0 mm x 300 mm x 5.18 m	25.29	29.76	34.22	40.17	0.24
Taper	200.0 mm x 350 mm x 7.62 m	26.78	32.73	38.68	44.63	0.44
6.25 mm	200 mm x 400 mm x 10.05 m	29.76	35.71	41.66	47.61	0.73
per meter	200 mm x 450 mm x 12.19 m	--	38.68	44.63	52.07	1.05
Y	200.0 mm x 300 mm x 3.05 m	25.29	29.76	35.71	41.66	0.14
Taper	200 mm x 350 mm x 4.57 m	28.27	32.73	38.68	44.63	0.26
10 mm	200 mm x 400 mm x 6.10 m	29.76	35.71	41.66	49.10	0.43
per meter	200 mm x 450 mm x 7.62 m	--	38.68	46.12	52.07	0.66

Extension Sections

Type	Dia. x Length	9 Ga. 0.1495"	7 Ga. 0.1793"	5 Ga. 0.2092"	3 Ga. 0.2391"	Cu. Yds. per Ft.
N12	12" x 12" x 20/40'	20	24	28	33	0.026

Type	Dia. x Length	9 Ga. 0.1495"	7 Ga. 0.1793"	5 Ga. 0.2092"	3 Ga. 0.2391"	Cu. Yds. per Ft.
N14	14" x 14" x 20/40'	24	29	33	38	0.035
N16	16" x 16" x 20/40'	27	32	38	44	0.045
N18	18" x 18" x 20/40'	--	37	43	49	0.058
	(metric)					
N12	300 mm x 300 mm x 6.1/12.2 m	29.76	35.71	41.66	49.10	0.02
N14	350 mm x 350 mm x 6.1/12.2 m	35.71	43.15	49.10	56.54	0.03
N16	400 mm x 400 mm x 6.1/12.2 m	40.17	47.61	56.54	65.46	0.03
N18	450 mm x 450 mm x 6.1/12.2 m	--	55.05	63.97	72.90	0.44

Step-Taper Pile. Shells are made of sheet steel in gauges of 12 to 20 and in basic section lengths of 4, 8, 12, and 16 feet (1.2, 2.4, 3.6, and 4.8 m). Longer lengths are furnished for special conditions. Step-Taper shells are helically corrugated to provide greater strength against collapsing pressures. A driving ring is welded to the bottom of each shell section and joints between sections are screw-connected. The bottom section is closed with a flat steel plate welded to the driving ring. Step-Taper Pile shells are manufactured in standard nominal diameters ranging from 8-5/8 to 18-3/8 inches (207 to 455 mm), but larger diameters can be made to meet special conditions.

When sufficient shell sections are joined together to make up the required pile length, the pile diameter increases from tip to butt at the rate of one inch per section length. Thus the rate of taper or pile shape will vary with the section lengths used. Within practical limits, different section lengths can be combined in a single pile, providing a wide range of available pile shapes. Nominal tip diameters usually range from 8 to 11 inches (200 to 275 mm), but larger tip diameters can be used. With the variety of diameters and section lengths available, the Step-Taper pile provides maximum flexibility in the choice of pile size and configuration to best meet subsoil conditions and loading requirements.

The length of a Step-Taper Pile can be readily adjusted as the work proceeds to meet variable soil conditions, holding waste costs to a minimum. It is unnecessary to install drive-test piles to predetermine pile lengths.

Internal reinforcement is not required for Step-Taper Piles except to resist uplift loads or high lateral loads (where batter piles are not used) or for unsupported pile lengths extending through air, water, or very fluid soil.

Step-Taper Piles are suitable for all types of soils and can function as friction or point bearing piles. In most cases, piles are supported through a combination of friction and direct bearing. In many soils, tapered piles usually develop higher capacity than piles of no taper.

Precast/prestressed Concrete Piles. Precast concrete piles are commonly used in marine structures, such as docks and piers, where they are driven in open water.

Ordinarily they are designed for the particular job and made in a casting yard at or near the job site. The use of precast concrete piles usually delays the starting of actual pile driving because of the time involved in the casting and curing of this type of pile.

Precast concrete piles have the same advantages as other concrete piles over wood and steel piles. They are always reinforced internally to resist stresses produced by handling and driving.

Requirements for precast concrete piles, except for reinforcement requirements, generally apply to prestressed units as well. If prestressed piles are used, the minimum working net prestress in the pile should be 700 psi (5.6 MPa. Prestressing strands are of the ungalvanized, seven-strand type, conforming to the general requirements of ASTM A-416 and may be either regular or high-strength.

Precast piles may be moved when the compressive strength of concrete has reached 4,000 psi (32 MPa), but are not to be driven until 5,000 psi (40 MPa) strength is attained. Piles are to be lifted or supported only at designated points and handled and driven in such a manner as to avoid excessive bending stresses, cracking, or spalling.

Pile heads must be protected from direct impact of the hammer by a cushion head so arranged that strands or bars projecting from the pile head will not be displaced or deformed during driving. Minimum energy per blow of the hammer is established by the engineer. Jetting may be permitted, or required where necessary, to reach the desired depth.

Driving 20'-0" (6.1 m) Concrete Piles. On a 12-story reinforced concrete building, requiring 280 concrete piles 20'-0" (6.1 m) long, the following costs were obtained.

There was considerable broken time on this job, due to bad weather, and while the job required 17-1/2 days, only 112 hrs. actual driving time was required. This included moving the pile driver from one place to another, because some of the piles were driven from the street level and others from the basement floor level.

The pile driving crew averaged two 20'-0" (6.1 m) concrete piles per hr. or 16 piles—320 lin. ft.(97.5 m)—per 8-hr. day, at the following labor costs:

Setting up Pile Driver	Hours	Rate	Total	Rate	Total
Foreman	32	$....	$....	$29.41	$941.12
Engineer	32	....	....	28.55	913.60
Oiler	32	....	....	21.04	673.28
Linesman	32	....	....	26.44	846.08
Four pile driver men	128	....	....	26.44	3,384.32
Driving Piles					
Foreman	140	....	....	29.41	4,117.40
Engineer	140	....	....	28.55	3,997.00
Oiler	140	....	....	21.04	2,945.60

	Hours	Rate	Total	Rate	Total
Linesman	140	....	....	26.44	3,701.60
Four pile driver men	560	....	....	26.44	14,806.40
Dismantling Pile Driver					
Foreman	20	....	....	29.41	588.20
Engineer	20	....	....	28.55	571.00
Oiler	20	....	....	21.04	420.80
Linesman	20	....	....	26.44	528.80
Four pile driver men	80	....	....	26.44	2,115.20
Labor cost 280 piles	1,536		$....		40,550.40

Cost per 20'-0" (6.1 m) pile		....	144.82
Cost per lin. ft		....	7.24
Cost per m		....	23.75

Does not include move charges for equipment, equipment costs, or pile material costs.

Cross-sections of Square Prestressed Piles

Various Types of Precast Concrete Piles
Squared Prestressed Piles

Pile Size[1] in.	Approx. Wgt.[2] lbs./l.f.	Strand per Dia.[3] 7/16"	Pile 1/2"	Peri- meter, in.	Bearing Tons[4] Conc. 5000	Capacity, Strength, psi 6000
10	105	4	4	40	73	90
12	150	6	5	48	105	129
14	205	8	6	56	143	176
16	265	11	8	64	187	229
18	335	13	10	72	237	290
20	415	16	12	80	292	358
22	505	20	15	88	354	433
24	600	23	18	96	421	516
20 HC	320	13	10	80	223	273
22 HC	365	14	11	88	256	314
24 HC	415	16	12	96	291	357

(metric)

Pile Size m	Approx. Wgt. Kg/m	Strand per Dia. 10.94 mm	Pile 12.50 mm	Peri- meter mm	Bearing Kips Conc. 34.5	Capacity Strength, MPa 41.5
250	156	4	4	1000	146	180
300	223	6	5	1200	210	258
350	305	8	6	1400	286	352
400	394	11	8	1600	374	458
450	498	13	10	1800	474	580
500	617	16	12	2000	584	716
550	751	20	15	2200	708	866
600	893	23	18	2400	842	1032
500 HC	476	13	10	2000	446	546
550 HC	543	14	11	2200	512	628
600 HC	617	16	12	2400	582	714

Cross-sections of Octagonal Prestressed Piles

Octagonal Prestressed Piles

Pile Size[1] in.	Approx. Wgt.[2] lbs./l.f.	Strand per Dia.[3] 7/16"	Pile 1/2"	Peri- meter in.	Bearing Tons[4] Conc. 5000	Capacity Strength, psi 6000
10	85	4	4	33	61	74
12	125	5	4	40	87	107
14	170	7	5	46	118	145
16	220	9	7	53	155	190
18	280	11	8	60	196	240
20	345	14	10	66	242	296
22	420	16	12	73	293	359
24	495	19	15	80	348	427
20 HC	245	10	8	66	172	211
22 HC	280	11	8	73	196	240
24 HC	315	12	9	80	219	269

(metric)

Pile Size m	Approx. Wgt. Kg/m	Strand per Dia. 10.94 mm	Pile 12.50 mm	Peri- meter Conc.	Bearing Kips 34.5	Capacity Strength, MPa 41.5
250	126	4	4	825	122	148
300	186	5	4	1000	174	214
350	253	7	5	1150	236	290
400	327	9	7	1350	310	380
450	417	11	8	1500	392	480
500	513	14	10	1650	484	592
550	625	16	12	1850	586	718
600	736	19	15	2000	696	854
500 HC	365	20	8	1650	344	422
550 HC	417	11	8	1825	392	480
600 HC	469	12	9	2000	438	538

Notes (For both square and octagonal piles)

1Voids in 20", 22", and 24" (500, 550, 600 mm) diameter hollow-core (HC) piles are 11", 13", and 15" (275, 325, 375 mm) diameter respectively, providing a minimum 4-1/2" (112.5 mm) wall thickness.

2Weights are based on 150 lbs. (68 kg) per cu. ft. of concrete.

3Based on 7/16" and 1/2" (10.9 and 15 mm) 270 Grade stress-relieved strand with an ultimate strength of 31,000 and 41,300 lbs. (14061 and 18233 kg) respectively.

4Design bearing capacity based on 5,000 and 6,000 psi (34.5 and 41.4 MPa) concrete and an allowable unit stress on the full section of 0.33 fce - 0.27 fce where fce is the concrete stress in the pile due to prestressing, after all losses. These bearing capacity values may be increased if higher strength concrete is used.

Cutting Off Precast/Prestressed Square or Octagonal Piles. After concrete piles have been driven, it might be necessary to cut off the tops of piles to the proper elevation.

Costs for this work vary according to the size of piles and the space in which workers must operate. Before cutting a precast or prestressed pile, make a circumferential cut, with a diamond or carborundum saw, to score the pile and prevent spalling.

Manhours Required to Cut Off Precast/Prestressed Piles

Size	Square	Octagonal
10" (250 mm)	0.70	1.00
12" (300 mm)	1.00	1.30
14" (350 mm)	1.10	1.40
16" (400 mm)	1.30	1.70
18" (450 mm)	1.50	1.90
20" (500 mm)	1.70	2.10
22" (550 mm)	1.80	2.30
24" (600 mm)	2.00	2.50

To costs for the above, add the costs of cutting tools and equipment, as well as costs for removal and disposal of cut off sections of pile.

Prestressed Concrete Cylinder Piles. These are centrifugally cast in 16-foot sections, which are assembled end-to-end and post-tensioned, providing a wide range of available pile shapes. These piles are cast with zero-slump concrete, and strengths from 7,000 to 10,000 psi (48 to 69 MPa) are attained. They are made with the Cen-Vi-Ro process, which is a combination of centrifugal casting, vibrating, and rolling. It produces an extremely dense concrete with an absorption factor about 40% less than bed-cast concrete piles. The use of zero-slump concrete prevents migration of the coarse aggregate during the spinning process so that the aggregates are uniformly distributed across the pile wall.

These piles are available in diameters from 36 to 90 inches (900 to 2250 mm), but the standard sizes are 36, 54, and 66 inch (900, 1350, 1650 mm) O.D. Wall thicknesses are 5 inches (125 mm) for the 36- and 54-inch piles (900 and 1350 mm) and 6 inches (150 mm) for the 66-inch (1650 mm) piles.

Cylinder Piles are driven either open or closed end. They are capable of carrying extremely high compression loads and resisting large lateral forces or bending moments. They are used for elevated highways, bridges, piers, offshore platforms, and breakwaters. For bridges and other trestle-type structures, cylinder piles serve as both foundation pile and pier or column.

Standard Sizes of Cylinder Piles

O.D. in. (mm)	I.D. in. (mm)	W in. (mm)	Lbs./ft. (Kg/m)	Number of Cables
36 (900)	26 (650)	5 (125)	524 (793)	8-16
54 (1350)	44 (1100)	5 (125)	829 (1255)	12-24
66 (1650)	54 (1350)	6 (150)	1217 (1841)	16-32

Prestressed Cylinder Piles

Typical setup for pile load testing in axial compression using a loading platform

Typical setup for pile load testing in axial compression using anchor piles.

Pile Load Testing. When required, load testing might cost anywhere from $5,000 to more than $20,000. Because of the wide variation in piles, and in conditions under which they are tested, it is difficult to give dependable costs without knowing the details of an individual installation. Costs are affected by factors such as the information that is to be obtained and the quarters in which the test is to be undertaken.

There is a number of methods for determining the load-carrying capability of a pile: static load testing with intermittent addition of load; constant rate of penetration; dynamic testing while driving; and others. Before a pile load test is made, the proposed apparatus and structure for reactions and loading should be fully detailed and costs applied. Load tests are normally carried to twice the design load, and pile tests can be in excess of 500 tons (560 m tons).

For a load test, piles have to be driven. Often, reaction piles are also installed to counteract the test force. This requires the same materials and equipment as planned for production driving. Another pile or two, perhaps with a cast-steel point for comparison, can be added with little increase in the cost.

Typical Pile Load Test Setups. Supports for loads or reaction piles should be at least five pile diameters or 5 feet (1.5 m) from the pile under test. In some soils, this should be as much as 10 diameters. The object is to keep the reaction far enough away that it will not affect the pile under test.

Pile load tests vary because of the different types of tests and the various types of piles. The test might vary from $5,000 to $50,000. In some rare cases, the cost has gone as high as $150,000. Not all piles require a load test. Usually, a pile of 20-ton (22 m ton) capacity or less does not require tests.

Some local codes mandate a minimum length that must be driven before a pile is considered a pile (for example, 10 feet (3.1 m) of pile driven tip to cut off). Pile driving contractors commonly quote projects giving several different unit costs:

Item Description	Bid Unit
1) Mobilization	Lump Sum
2) Pile Load Test	Each
3) Pile Costs	Each or Lin. Ft. (Meter)
4) Demobilization	Lump Sum

Some pile driving contractors combine the mobilization and demobilization. On small jobs, where tests are not required, they will quote the project as a lump sum.

02350 CAISSONS

Caisson foundations are normally used to support heavy structures where soil conditions near grade cannot support them. Where such conditions exist, the foundation should be carried down through the unsatisfactory soil to material that can carry the imposed load, without causing any detrimental settlement. The size and depth of caissons will vary with the loads to be carried and the distance to material of the required bearing capacity. For example, caissons bearing on rock in downtown Chicago vary from 90 to 190 feet in depth (27 to 58 m).

Caissons can be dug either with straight shafts to the required depth, or they can be belled at the bottom to provide additional bearing area. Most caissons are installed using various types of drill rigs, but under certain conditions hand dug caissons are used. Drilling machines are capable of not only drilling the shaft but mechanically forming a bell. The cost of excavating varies with the diameter and depth of the caisson, site conditions, and the type of soil.

There are contractors who specialize in installing caissons. On jobs of any size, it is advisable to obtain figures from specialty contractors in this work, because they are equipped to perform it much more economically.

Hand Dug Caissons. The hand method of excavation most commonly used is known as the Chicago method. This consists of excavating with hand tools and lining the hole with wood lagging held in place by steel rings. The wood lagging has beveled edges and is tongue grooved. The lagging is milled from either 2"x6" (50 x 150 mm) or 3"x6" (75 x 150 mm) mixed hardwood in 16' (4.9 m) lengths, which are commonly cut into thirds so that a set of lagging is 5'-4" (1.62 m) long. Three-inch steel channels are formed in a half circle with a plate on each end to permit bolting the half sections together. Two complete rings are required for each set of lagging.

To determine the excavation quantity, multiply the cross-sectional area by the length allowing for the increased diameter due to the lagging.

To determine the quantity of concrete required, multiply the cross-sectional area by lengths making proper corrections if the finished concrete surface is to be below the ground surface. If the caisson has a bell, the volume of it must be computed using the formula for a truncated cone: $V = 1/3(A1 + A2 + A1 + A2)h$, where A1 + A2 are the area of the bases and h is the height of the bell.

Excavation quantity for 6'-0" (1.82 m) caisson 106' (32.3 m) deep
Cross-sectional area for 6'-4" (1.9 m) = 31.5 sq. ft. (2.92 sq. m)
31.5 x 106 x 1/27 = 124 cu. yds.
(9.6 x 32.3 x .3057 = 95 cu. m)

Concrete quantity:
Cross-sectional area for 6'-0" (1.82 m) = 28.3 sq. ft. (2.63 sq. m)
28.3 x 106 x 1/27 = 111 cu. yds.
Allow 5% for waste for a total of 117 cu. yds.
(8.63 x 32.3 x .3057 = 85 cu. m
Allow 5% for waste for a total 89 cu. m)

After determining the excavation quantities and the concrete quantities, it is necessary to determine the amount of lagging and rings. By cutting the bevel and tongue and grooves in the lagging, the width of the lagging is reduced, so that it requires 7 pieces of 2"x6" (50 x 150 mm) per foot of diameter of the caisson. For example, a 4' (1.20 m) diameter caisson requires 28 pieces of lagging around the circumference. Therefore, using 2"x6" (50 x 150 mm) lagging, which is one board foot, it would require 28 board feet per foot of depth for a 4' (1.2 m) caisson or 42 board feet (0.08 cu. m) using 3" thick lagging. To determine the amount of lagging required in a 6' (1.8 m) diameter caisson, 106' (32.3 m) deep using 2" (50 mm) lagging, multiply 6 x 7 or 42 board feet per foot (0.08 cu. m) of depth x the depth, which is 106' (32.3 m), or 4,452 board feet (8.9 cu. m) If the lagging had been 3" (75 mm), it would be 6,678. A caisson 106' (32.3 m) deep would require 20 sections of 5'-4" (1.6 m) lagging. Using 2 rings for each section of lagging, the caisson would require 40 rings.

The rings are usually made from 3" (75 mm) channels, weighing 4.2 lbs. per foot (6.3 kg/m). Each ring will weigh approximately 80 lbs. (36 kg) plus 10 lbs. for the plates at the ends of the half circles, for a total of 90 lbs. (40 kg).

Labor Required for Caissons. Labor costs of caissons excavating varies with the diameter and depth of caissons, whether hand or pneumatic tools are used, the type of soil, and whether or not water or sand is encountered. It is generally assumed that one caisson digger can hand dig approximately 4 to 5 yards (3 to 3.8 cu. m) per 8-hour shift and place the lagging and rings.

Only one worker can work in a 4' (1.2 m) diameter shaft, but two workers can work in a 5' (1.5 m) diameter or larger shaft. In a 6' (1.8 m) caisson, two workers each taking out 4 yards (3 cu. m) should be able to excavate approximately 7' (2.1 m) per shift. At the surface, a pneumatic tugger hoist and tripod would be required with two laborers to operate the hoist and dump the buckets. Generally, several caissons are constructed simultaneously, so the cost involved in furnishing compressed air for the hoist is spread over three or more caissons.

In addition to the above, it will be necessary to allow for time of foreman, timekeepers, material clerks, engineers, and cost of setting up and removing equipment, which will cost from $4.00 to $6.00 per cu. yd. ($5.00 to 8.00 per cu. m) It may be estimated separately under overhead expense and an amount allowed to cover the entire job.

To summarize the costs involved for constructing a caisson 6 ft. (1.8 m) in diameter and 106 ft. (32.3 m) deep, with two others being constructed simultaneously, time to excavate and lag:

124 cu. yds. @ 4 cu. yds./shift/digger = 31 shifts.
Using two diggers, this would be 15-1/2 shifts.
Allow 16 shifts.

Labor	Hours	Rate	Total	Rate	Total
Caisson diggers	256	$....	$....	$21.04	$5,386.24
Hoist operator	128	....	....	28.55	3,654.40
Top laborers	256	....	....	21.04	5,386.24
Compressor operator	43	....	....	28.55	1,227.65
					$15,654.53

Material

Lagging........................... 4.452 mbf	....	....	$760.00	$3,383.52
Rings..............................40	....	....	55.00	2,200.00
				$5,583.52

Equipment

Hoist..................................16 shifts	....	....	$25.00	$400.00
Air tools ....32 shifts	....	....	40.00	1,280.00
Compressor 5-1/3	....	....	50.00	266.65
				$1,946.65

Removal of

Excavated soil ... 124 cy (95 cu. m)	....	....	$10.00	1,240.00
Grand Total.....................................	$....			$ 24,424.70

Cost per cu. yd.	....	$196.97
Cost per cu. m...............................	....	257.10

To the above cost it will be necessary to add for setting up and removing necessary plant and equipment, electric current and wiring, concrete furnishing and placing, overhead expenses, contingencies, miscellaneous expenses, and profit.

Work of this type in the Chicago area generally costs from $200.00 to $275.00 per cu. yd. ($260 to 360 per cu. m) of completed caisson.

Excavating for Caissons with Pneumatic Diggers. When excavating for caissons 5'-0" to 8'-0" (1.5 to 2.4 m) in diameter and up to 100'-0" (30.5 m) deep, a worker with a pneumatic digger will loosen and load into buckets 9 to 11 cu. yds. (6.9 to 8.4 cu. m) per 8-hr. shift, depending upon the toughness of the clay and the working space, at the following cost per cu. yd. (cu. m):

	Hours	Rate	Total	Rate	Total
Labor operating digger					
and loading buckets...............0.90	$....	$....	$21.04	$18.94
Compressor operator 0.45	....	....	28.55	12.85
Compressor expense 0.45	....	....	11.25	5.06
Cost per cu. yd		$....		$36.85
Cost per cu. m		....		47.91

Compressor and operator time at half time (assuming 2 workers digging).

When excavating caissons by hand, two "diggers" will loosen and load into buckets, 6 to 7 cu. yds. (4.6 to 5.4 cu. m) per 8-hr. shift, at the following labor cost per cu. yd. (cu. m):

	Hours	Rate	Total	Rate	Total
Labor per cu. yd. 2.5	$....	$....	$21.04	$52.60
per cu. m......................................		....		68.38

Drilled Caissons. The use of drilling machines for the installation of caisson foundations has eliminated the hand digging method except in situations where adequate clearances to surrounding structures or utilities cannot be maintained. Drilled caissons have proved more economical than hand-dug caissons for several reasons, the most important of which is the fact that smaller shaft diameters may be used. The minimum size of a hand-dug caisson is 4' (1.2 m) in diameter, which often provides a greater cross-sectional area of concrete than is required for the load imposed upon it.

Another important reason is that drilled caissons can be installed in considerably less time. There are many firms specializing in this type of work throughout the country.

A typical job, consisting of 100 caissons with 2'-0" (.6 m) shaft diameters, 5'-0" (1.5 m) bells, and 25' (7.6 m) deep, might be estimated as follows, assuming normal clay digging. First, compute the total volume of excavation, which in this case would be 418 cu. yds. (320 cu. m). To this figure add about 10% overbreak, making a total of 460 cu. yds. (352 cu. m). Assuming each drill rig can complete 5 caissons per day, the job would take 20 days. Allow approximately 10% extra time for weather and total of 23 days to complete the job.

Labor	Hours	Rate	Total	Rate	Total
Drill Rig Operator	184	$....	$....	$28.55	$5,253.20
Oiler	184	....	....	21.04	3,871.36
Caisson Laborer	184	....	....	21.04	3,871.36
Laborer	184		....	21.04	3,871.36
					$ 16,867.28
Equipment					
Drill rig	23 days	$....	$....	$800.00	18,400.00
Material					
Concrete	460 cy (352 cu. m)	....	....	$55.00	25,300.00
Grand Total					$60,567.28
Cost per cu. yd. (418 cu. yds.)					$144.90
Cost per cu. m (320 cu. m)					188.37

These costs are for the complete caisson installation. Allowances must be added for reinforcing steel, additional labor based on union requirements, temporary or permanent casings if necessary, soil and concrete testing, surveying, disposal of excavated materials, and equipment move and set-up charges, if these items are included in the caisson contractor's contract, and overhead and profit.

Churn Drills. In areas where subsurface conditions are such that rotary drilling or augured caissons are difficult or impossible, the use of churn drills is usually effective. This caisson installation is sometimes known as the "drilled in" caisson method.

ESTIMATOR'S REFERENCE BOOK

The cost for a 30" (750 mm) diameter drilled in caisson, complete in place, which includes the concrete, can range from $800 to $1,500 per lineal foot ($2620 to $4920 per m).

If bentonite is used in the installation of the caisson to produce the drillers mud, it is necessary to figure additional cost for the disposal of the bentonite. This can be an additional $200 to $500 per lin. ft. ($600 to $1640 per m). Labor cost for drilling 4 l.f. (1.2 m) is as follows:

	Hours	Rate	Total	Rate	Total
Driller	8	....	....	$28.55	228.40
Helper	8	....	....	21.04	168.32
Cost for 4 lin. ft. (1.2 m)					$ 396.72
Cost per lin. ft.					99.18
Cost per m					325.31
Equipment					
Churn Drill	8	....	....	100.00	800.00
Cost per lin. ft.					200.00
Cost per m					666.67

02400 SHORING

Wood sheet piling and bracing is usually estimated by the square foot (square meter), taking the number of sq. ft. (sq. m) of bank or trench walls to be braced, plus an additional amount for penetration at the bottom of the excavation and an allowance for extension above the top, and then estimating the cost of the work at a certain price per sq. ft. (sq. m) for labor and material required.

Sheet piling or bracing is required where the soil is not self supporting or where the excavation banksides cannot be sloped back. There is usually considerable vibration adjacent to railroad tracks and roadways that can cause the banks to cave in. Unless there is ample space on all sides of the excavation, the banks should be sheet piled to avoid damage to streets, alleys, or adjacent buildings.

Trenches or foundation piers 5'-0" to 8'-0" (1.5 to 2.4 m) deep and 3'-0" to 4'-0" (.9 to 1.2 m) wide may require bracing or sheet piling, but this is much simpler than is ordinarily required for basement excavations.

The cost of this work will vary with the kind of soil, depth of excavation, amount of bracing or sheet piling required, and the method used to place. In ordinary excavations where no water is encountered, 2" or 3" (50 to 75 mm) square edge planks may be used, but if running sand or water is encountered, tongue and grooved planks should be used, as it is easier to keep them in line, and they are more water tight than square edge planks. It is customary to cut the bottom edge of each plank on a slight angle, so that in driving it is wedged against the preceding plank. The upper corners of the plank should also be cut off so that the effect of the driving will be concentrated along the vertical axis of the plank.

It requires 3 to 3-1/2 b.f. (.006 to .007 cu. m) of lumber to sheet pile one sq. ft. of trench bank 5'-0" to 8'-0" (1.5 to 2.4 m) deep, but on large deep excavations it will require 6 to 8 b.f. (0.012 to 0.016 cu. m) of lumber per sq. ft. of wall.

Sheet piling is usually driven with a pneumatic hammer designed especially for this purpose.

On very large jobs an ordinary pile driver is used for driving the sheet piling, but this is only on jobs of sufficient size to warrant the cost of a pile driver and crew.

Bracing and Sheet Piling Trenches and Piers. When excavating trenches and piers 5'-0" to 8'-0" (1.5 to 2.4 m) deep, it is not always necessary to sheathe the banks solid but two or three lines of braces placed along the sides of the trench (as illustrated) will often be sufficient.

Where necessary to sheathe the banks solid, it will require about 3 to 3-1/2 b.f. (.006 to .007 cu. m) of lumber per sq. ft. of trench bank. A method of determining the quantity of lumber required for sheet piling a trench similar to the one illustrated is as follows:

Compute the quantity of lumber necessary to sheet pile a trench 50'-0" (15.2 m) long and 7'-0" (2.1 m) deep, sheathed on both sides. To the depth of 7'-0" (2.1 m) must be added an amount for penetration. If 1'-0" (.3 m) will be sufficient, then figure 8'-0" (2.4 m) lumber. If, however, 2'-0" (.6 m) penetration is required, then 10'-0" (3 m) lumber must be figured leaving 1'-0" (.3 m) to extend above grade. Assuming 2'-0" (.6 m) penetration is required and 10'-0" (3 m) lumber must be used, this is equivalent to 100 lin. ft. (30.5 m) of sheathing 10'-0" (3 m) high, or 1,000 sq. ft. (93 sq. m), and the following lumber will be required:

	B.f. of Lumber
150 pcs. 2"x8"-10'-0" sheathing	2,000
12 pcs. 4"x6"-16'-0" stringers or "wales"	384
26 pcs. 4"x6"-2'-0" braces spaced about 4'-0" apart	104
Total lumber required	2,488
Feet of lumber, b.f. per sq. ft. sheathing, (divide by 1,000)	2.49
Feet of lumber, b.f. per sq. ft. trench bank, (divide by 700)	3.55

	Cu. Meters of Lumber
150 pcs. 50 mm x 200 mm x 3.04 m sheathing	4.56
12 pcs. 100 mm x 150 mm x 4.88 m stringers or "wales"	0.88
26 pcs. 50 mm x 150 mm x .60 m braces spaced about 1200 mm apart	0.23

Total lumber required..5.67
Meter of lumber, cu. m per sq. m, sheathing,
(divide by 93)..0.06
Meter of lumber, cu. m per sq. m, trench bank,
(divide by 20)..0.28

This is based on using 2" (50 mm) plank for sheathing, 2 rows of 4"x6" (100 x 150 mm) stringers or "wales" placed near the top and bottom as illustrated, and 4"x6" (100 x 150 mm) or 6"x6" (150 x 150 mm) braces. Trenches over 10'-0" (3.0 m) deep will require an additional line of stringers for each additional 4'-0" to 5'-0" (1.2 to 1.5 m) depth.

Method of Bracing
Trench Excavation

Method of Sheet Piling
Trench Excavation

Most contractors who do a volume of this work use trench jacks instead of 4"x6" or 6"x6" braces. These jacks cost from $200.00 to $300.00 each and would not be economical for the contractor who does only an occasional job of this sort. The savings in material and labor is so slight that it is practically negligible.

Sheet Piling for Basements and Deep Foundations. On general basement excavation or large piers 8'-0" to 12'-0" (2.4 to 3.6 m) deep, where it is necessary to brace any or all of the outside banks, the following lumber is required for a bank 50'-0" (15 m) long, 8'-0" (2.4 m) deep, and containing 400 sq. ft. (37 sq. m) of bank to be sheathed:

	B.f. of Lumber
80 pcs. 2"x8"-10'-0" sheathing	1,067
6 pcs. 6"x8"-16'-0" (2 lines of stringers or "wales")	384
10 pcs. 8"x8"-12'-0" top braces spaced about 5'-0" apart	640
10 pcs. 8"x8"-6'-0" bottom braces spaced about 5'-0" apart	320
10 pcs. 8"x8"-6'-0" bottom braces spaced about 5'-0" apart	320
Total lumber required	2,731
B.f. per sq. ft. of sheet piling, (divide by 500)	5.46
B.f. per sq. ft. of bank, (divide by 400)	6.83

	Cu. Meter of Lumber
80 pcs. 50 mm x 200 mm x 3.0 m sheathing	2.44
6 pcs. 150 mm x 200 mm x 4.9 m, 3 lines of stringers or wales	0.88
10 pcs. 200 mm x 200 mm x 3.7 m,top braces spaced about 1.5 m apart	0.23
10 pcs. 200 mm x 200 mm x 1.8 m, bottom braces spaced about 1.5 m apart	0.72
10 pcs. 200 mm x 200 mm x 1.8 m, bottom braces spaced about 1.5 m apart	0.72
Total lumber required	4.99
Meters of lumber, cu. m per sq. m, sheet piling, (divide by 46)	0.09
Meters of lumber, cu. m per sq. m, trench bank, (divide by 37)	0.12

Methods of Bracing Basements or Deep Excavations

On deep excavations and piers 14'-0" to 20'-0" (4.3 to 6.1 m) below grade, 3" (75 mm) plank should be used for sheathing and either 6"x8" (150 x 200 mm) or 8"x8" (200 x 200 mm) lumber for stringers or "wales" and 8"x8" (200 x 200 mm) or 10"x10" (250 x 250 mm) timbers for bracing. Lines of stringers or "wales" should be spaced 4'-0" to 5'-0" (1.2 to 1.5 m) apart, depending upon the earth pressure.

An example of lumber required to sheet pile and brace a bank 50'-0" (15 m) long and 16'-0" (4.9 m) deep, containing 800 sq. ft. (74 sq. m) of bank, is as follows:

	B.f. of Lumber
80 pcs. 3"x8"-18'-0" sheathing	2,880
9 pcs. 8"x8"-16'-0" (3 lines of stringers or "wales")	768
10 pcs. 8"x8"-16'-0" top braces spaced 5'-0" apart	853
10 pcs. 8"x8"-12'-0" Intermediate braces spaced 5'-0" apart	640
10 pcs. 8"x8"-8'-0" bottom braces spaced 5'-0" apart	427
10 pcs. 8"x8"-6'-0" stakes spaced 5'-0" apart	20
Total lumber required	5,888
B.f. per sq. ft. of sheet piling, (divide by 900)	6.54
B.f. per sq. ft. of bank (divide by 800)	7.36

	Cu. Meter
80 pcs. 75 mm x 200 mm x 5.5 m sheathing	6.60
9 pcs. 200 mm x 200 mm x 4.9 m, 3 lines of stringers or wales	1.76
10 pcs. 200 mm x 200 mm x 4.9 m, top braces spaced 1.5 m apart	0.60
10 pcs. 200 mm x 200 mm x 3.7 intermediate braces spaced 1.5 m apart	1.08
10 pcs. 200 mm x 200 mm x 2.4 m, bottom braces spaced 1.5 m apart	0.96
10 pcs. 200 mm x 200 mm x 1.8 m, stakes spaced 1.5 m apart	0.72
Total lumber required	11.72
Meters of lumber, cu. m per sq. m, sheet piling, (divide by 84)	0.14
Meters of lumber, cu. m per sq. m, trench bank, (divide by 74)	0.16

Cost of lumber should be computed according to number of times it can be used on the job. Where labor hours are given, either carpenters or laborers may be used, as required.

Lumber Required for Sheet Piling

Size of Timbers	Width Sq. Edge	100 LF Covers SF	BF per 100 SF	Width T & G	100 LF Covers SF	BF per 100 SF
2"x8"	7-1/2"	62-1/2	215	7-1/4	60	220
3"x8"	7-1/2"	62-1/2	320	7-1/4	60	330
2"x10"	9-1/2"	79	210	9-1/4	77	215
3"x10"	9-1/2"	79	315	9-1/4	77	320

(metric)

Size of Timbers	Width Sq. Edge mm	100 M Covers sq. m	Width T & G mm	100 M Covers sq. m
50 mm x 200 mm	187.5	18.75	181.3	18.13
75 mm x 200 mm	187.5	18.75	181.3	18.13
50 mm x 250 mm	237.5	23.75	231.3	23.13
75 mm x 250 mm	237.5	23.75	231.3	23.13

Wood Bracing and Sheet Piling Banks By Hand
Quantities and Production are given per 100 s.f. (sq.m) of bank braced

Class of Work	Depth in Feet	BF Lbr. Req'd 100 SF Bank	No. SF Placed 8-Hr. Day	Hours Placing 100 SF	Hours Removing 100 SF
Bracing Trenches	5-8	100-150	150-200	4-5	1-1 1/2
Trench Sheet Piling	5-8	300-350	80-90	9-10	2 1/2-3
Trench Sheet Piling	10-15	325-375	65-75	11-12	2 1/2-3
Basement Sheet Piling	8-12	675-725	40-50	17-18	5-5 1/2
Basement Sheet Piling	14-20	750-800	35-45	20-21	5 1/2-6

(metric, per 100 sq.m)

Class of Work	Depth in Meters	Cu.M Lbr. Req'd 100 Sq.M Bank	No. Sq.M Placed 8-Hr Day	Hours Placing 10 Sq.M	Hours Removing 10 Sq.M
Bracing Trenches	1.5-2.4	20-30	14-19	12-15	1-1 1/2
Trench Sheet Piling	1.5-2.4	60-70	7-8	27-30	2 1/2-3
Trench Sheet Piling	3.0-4.6	65-75	6-7	33-36	2 1/2-3
Basement Sheet Piling	2.4-3.7	135-145	4-5	52-55	5-5 1/2
Basement Sheet Piling	4.3-6.1	150-160	3-4	61-64	5 1/2-6

Cost of Bracing 100 S.F. (10 Sq.M) of Trench Bank With Plank Bracing

	Hours	Rate	Total	Rate	Total
125 b.f. lumber		$....	$....	$0.76	$95.00
Labor placing	4.50	....	....	21.04	94.68
Labor removing**	1.25	....	....	21.04	26.30
Cost per 100 sq. ft.			$....		$ 215.98
Cost per sq. ft.				....	2.16

(metric)

	Hours	Rate	Total	Rate	Total
0.25 cu. m of lumber		$....	$....	$409.04	$102.26
Labor placing	4.84	....	....	21.04	101.83
Labor removing**	1.35	....	....	21.04	28.40
Cost per 10 sq. m			$....		$ 232.49
Cost per sq. m				....	23.25

Cost of Sheet Piling 100 S.F. (10 Sq.M) of Trench Walls
5'-0" to 8'-0" (1.5 to 2.4 m) Deep

	Hours	Rate	Total	Rate	Total
325 b.f. lumber		$....	$....	$0.76	$247.00*
Labor placing	9.50	....	....	21.04	199.88
Labor removing**	2.75	....	....	21.04	57.86
Cost per 100 sq. ft.			$....		$ 504.74
Cost per sq. ft.				....	5.05

(metric)

	Hours	Rate	Total	Rate	Total
0.65 cu. m, lumber		$....	$....	$409.04	$247.00*
Labor placing	10.22	....	....	21.04	215.03
Labor removing**	2.96	....	....	21.04	62.28
Cost per 10 sq. m			$....		$524.31
Cost per sq. m				....	52.43

Cost of Sheet Piling 100 S.F. (10 Sq.M) of Trench Walls
10'-0" to 15'-0" (3.0 to 4.6 m) Deep

	Hours	Rate	Total	Rate	Total
350 b.f. lumber		$....	$....	$0.76	$266.00*
Labor placing	11.50	....	....	21.04	241.96
Labor removing**	2.75	....	....	21.04	57.86
Cost per 100 sq. ft.			$....		$ 565.82
Cost per sq. ft.				....	5.66

(metric)

	Hours	Rate	Total	Rate	Total
0.7 cu. m, lumber		$....	$....	$409.04	$266.00*
Labor placing	12.38	....	....	21.04	260.48
Labor removing**	2.96	....	....	21.04	62.28
Cost per 10 sq. m			$....		$588.76
Cost per sq. m				....	58.88

Cost of 100 S.F. (10 Sq.M) of Sheet Piling for Basements or Foundations
8'-0" to 12'-0" (2.4 to 3.7 m) Deep

	Hours	Rate	Total	Rate	Total
700 b.f. lumber		$....	$....	$0.76	$532.00*
Labor placing	17.50	....	....	21.04	368.20
Labor removing**	5.25	....	....	21.04	110.46
Cost per 100 sq. ft.			$....		$1,010.66
Cost per sq. ft.				....	10.11

(metric)

	Hours	Rate	Total	Rate	Total
1.4 cu. m, lumber		$....	$....	$409.04	$532.00*
Labor placing	18.84	....	....	21.04	396.39
Labor removing**	5.65	....	....	21.04	118.88
Cost per 10 sq. m			$....		$ 1,047.27
Cost per sq. m				....	104.73

Cost of 100 S.F. (10 Sq.M) of Sheet Piling for Basements or Foundations
14'-0" to 20'-0" (4.3 to 6.1 m) Deep

	Hours	Rate	Total	Rate	Total
775 b.f. lumber		$....	$....	$0.76	$589.00*
Labor placing	20.50	....	....	21.04	431.32
Labor removing**	5.75	....	....	21.04	120.98

	Hours	Rate	Total	Rate	Total
Cost per 100 sq. ft.		$....			$ 1,141.30
Cost per sq. ft.				....	11.41
		(metric)			
1.55 cu. m, lumber		$....	$....	$409.04	$589.00*
Labor placing 22.07		....	....	21.04	464.35
Labor removing** 6.20		....	....	21.04	130.45
Cost per 10 sq. m			$....		$ 1,183.80
Cost per sq. m				....	118.38

*Lumber costs may be reduced if it is removed & used again.
**Omit labor removing bracing or piling if left in place.

Method of Sheet Piling and Bracing Piers and Pits

Driving Sheet Piling by Compressed Air. The pneumatic pile driver is very effective for driving sheet piling. It is essentially a heavy paving breaker equipped with a special fronthead that adjusts for driving 2" to 3" (50 to 75 mm) piling. These machines weigh about 125 lbs. (57 kg) and will drive wood sheet piling into any ground that the piling can penetrate, such as sand, gravel, or shale, and any gradation between soils.

Driving speed varies from 2'-0" (0.61 m) per minute in hard clay or shale to 9'-0" (2.75 m) per minute in sand or gravel.

Under average conditions, two workers together on one machine should drive 100 l.f. (30.5 m)—60 to 79 s.f. (5.6 to 7.3 sq.m)—per hr. at the following labor cost:

	Hours	Rate	Total	Rate	Total
Jackhammerman	1	$....	$....	$21.04	$21.04
Helper	1	....	....	21.04	21.04
Compressor & tools	1	....	....	13.50	13.50
Cost per 100 l.f.			$....		$ 55.58
per l.f.				....	0.56

Add for compressor engineer if required.

(metric)

	Hours	Rate	Total	Rate	Total
Jackhammer operator	3.28	$....	$....	$21.04	$69.01
Helper	3.28	....	....	21.04	69.01
Compressor & tools	3.28	....	....	13.50	44.28
Cost per 30 m			$....		$182.30
per m				....	6.08

Add for compressor engineer if required.

Cost of Sheet Piling 100 S.F. (10 Sq.M) of Trench Banks
5'-0" to 8'-0" (1500 to 2400 mm) Deep, Using Pneumatic Hammers

	Hours	Rate	Total	Rate	Total
325 b.f. lumber		$....	$....	$0.76	$247.00*
Jackhammerman	2.00	....	....	21.04	42.08
Labor	4.00	....	....	21.04	84.16
Labor removing	2.75	....	....	21.04	57.86
Compressor & Tools	2.00	....	....	13.50	27.00
Compressor Engineer	2.00	....	....	28.55	57.10**
Cost per 100 s.f.			$....		$ 515.20
per s.f.				....	5.15

(metric)

	Hours	Rate	Total	Rate	Total
0.65 cu. m, lumber		$....	$....	$409.04	$247.00*
Jackhammer operator	2.00	....	....	21.04	42.08
Labor	4.00	....	....	21.04	84.16
Labor removing	2.75	....	....	21.04	57.86
Compressor & Tools	2.00	....	....	13.50	27.00
Compressor Engineer	2.00	....	....	28.55	57.10**
Cost per 9.3 sq.m				....	515.20
Cost per sq.m				....	55.40

Cost of Sheet Piling 100 S.F. (10 Sq.M) of Trench Banks
10'-0" to 15'-0" (1500 to 2400 mm) Deep, Using Pneumatic Hammers

	Hours	Rate	Total	Rate	Total
350 b.f. lumber		$....	$....	$0.76	$266.00*
Jackhammerman	2.00	....	....	21.04	42.08
Labor	4.50	....	....	21.04	94.68
Labor removing	2.75	....	....	21.04	57.86
Compressor & Tools	2.00	....	....	13.50	27.00
Compressor Engineer	2.00	....	....	28.55	57.10**

			$....		$ 544.72
Cost per 100 s.f.					
per s.f.			....		5.45

	Hours	Rate $....	Total $....	Rate	Total
0.7 cu. m, lumber				$409.04	$266.00*
Jackhammer operator 2.00		....	....	21.04	42.08
Labor		....	....	21.04	94.68
Labor removing...........		....	....	21.04	57.86
Compressor & Tools		....	....	13.50	27.00
Compressor Engineer...........		....	....	28.55	57.10**
Cost per 9.3 sq. m			$....		$ 544.72
per sq. m...........			....		58.57

(metric)

*Lumber cost will vary with number of times it may be used.
**Add only if required.

STEEL SHEET PILING

Steel sheet piling is used for supporting soil in large excavations, cofferdams, caissons, and deep piers or trenches, where wood sheet piling is impractical. It is driven with a pile driver, the same as wood or concrete piles.

There are a number of different types of steel sheet piling on the market, but the following will give an idea of the common sizes and weights.

Sizes and Weights of United States Steel Sheet Piling

Section Number	Width Inches	Web Thick. Inches	Wt. Lbs. per L.F.	Wt. Lbs. per S.F. Wall	Width in mm	Web Thick. mm	Wt Kg per m
MZ38	18	3/8	57.0	38.0	450	9.38	84.80
MZ32	21	3/8	56.0	32.0	525	9.38	83.31
MZ27	18"	3/8	40.5	27.0	450	9.38	60.26
MP102	15"	1/2	40.0	32.0	375	12.50	59.51
MP101	15"	3/8	35.0	28.0	375	9.38	52.07
MP113	16"	1/2	37.3	28.0	400	12.50	55.50
MP112	16"	3/8	30.7	23.0	400	9.38	45.68
MP110	16"	31/64	42.7	32.0	400	12.00	63.53
MP116	16"	3/8	36.0	27.0	400	9.38	53.56
MP115	19-5/8"	3/8	36.0	22.0	510	9.38	53.56

Estimating the Quantity of Steel Sheet Piling. When estimating the quantity of steel sheet piling required for any job, take the entire girth of the basement, cofferdam, or piers to be sheet piled. This gives the number of lineal feet required.

Example: Suppose you have a basement 100' x 125' (30.5 x 38.1 m) to be sheet piled. Adding the four sides of the excavation, 100 + 100 + 125 + 125 = 450 lin. ft. (137.2 m). Using MZ-27 Carnegie Illinois sheet piling 18" (1.5 ft.,

450 mm) wide, 450'-0" (147.2 m) divided by 1.50' (0.45 m) = 300. It requires 300 pcs. to sheet pile the excavation. Using sheet piling 30'-0" (9.1 m) long, 300 x 30 = 9,000 lin. ft. (300 x 9.1 m = 2730 m) weighing 40.5 lbs. per lin. ft. = 364,500 lbs. (165,337 kg).

A shorter method is as follows: 450 lin. ft. (137.2 m) around excavation multiplied by 30'-0" (9.1 m) long, equals 13,500 sq. ft. (1248.52 m) of piling at 27 lbs. per sq. ft. equals 364,500 lbs. (165,337 kg).

Either method is satisfactory. But it is necessary to know the number of pieces to be driven in order to compute the labor costs accurately.

Labor Driving Steel Sheet Piling. The labor cost handling and driving steel sheet piling will vary with the size of the job, length of piling, and kind of soil, together with conditions encountered on the job, as described under "Wood Piles".

Example: Find the cost of driving 300 pcs. of MZ-27 sheet piling 30'-0" (9.1 m) long.

	Total	Total
Trucking pile driver, yard to job......$....		$1,500.00
Setting up pile driver, 3 days for crew.......		3,173.04
300 pcs. MZ-27 piling, 364,500 lbs. (165,337 kg) at $0.40......		145,800.00
Driving piling (30 per day) 10 days @ $1,057.68......		10,576.80
Fuel, oil, misc. supplies, 10 days @ $150.00......		1,500.00
Dismantling pile driver at completion, 2 days......		2,115.36
Trucking pile driver, job to yard.......		1,500.00
Equipment Rental or depreciation allowance		
16 days @ $500.00......		8,000.00
Cost 300 pcs. sheet piling......$....		174,165.20
per pc.......		580.55
per s.f. (13,500 sq. ft.or 1254 sq. m).......		12.90
per sq.m.......		138.88

Sq. ft. costs will vary with weight of piling used, size of job, etc.
Add for overhead and profit.

02500 SITE DRAINAGE

Site drainage includes such items as subdrainage, foundation and underslab drainage, drainage structures, sanitary and storm drainage piping, and dewatering and wellpoints. Some of these will be subcontracted directly with specialists. Most will probably be included in the plumbing subcontract, while others may involve local utilities. Some items the contractor may elect to handle directly.

Drainage trenches must have pitch. To find the cubic yards of earth excavation, first determine the average depth by multiplying the pitch by the required length of run, divide by two and add this to the depth at the start of the run. The total length x required width x the average depth = cubic feet (cubic meters). Divide number of cu. ft. by 27 to find cubic yards.

Subsoil drains are usually vitrified clay pipe, set with lowest point at the same elevation as the bottom of the footing, pitched a minimum of 6" (150 mm) in 100 l.f. (30.5 m), laid 1/4" apart and joint covered with 15# (6.8 kg) asphalt felt or #12 (5.4 kg) copper or aluminum mesh. Four-inch (100 mm) pipe is minimum, but 6" (150 mm) pipe is preferred. One laborer can lay approximately 10 l.f. (3.05 m) of 4" (100 mm) pipe per hour, 8 l.f. (2.44 m) of 6" 150 mm) pipe. The 4" (100 mm) pipe costs about $1.00 per l.f. (3.28 per m); the 6" (150 mm) pipe costs $1.40. Excavation is extra, and it is assumed the trench is properly sloped with a firm bottom before pipe is installed.

All sharp turns are formed with fittings, which cost $2.50 for 4" (100 mm) elbows, $4.75 for 6" (150 mm). Wide angles can be made by bevelling tile ends into easy radius bends.

The clay drain tile is laid on a thin bed of washed gravel. After the joints have been covered with felt or a filter cloth, more gravel is placed around and on top of the drain tile to allow water filtration into the drain tile itself. Approximately 0.75 to 1.0 c.f. per l.f. (0.069 to 0.09 cu.m per m) of gravel of drain is required for proper filtration.

In recent years, vitrified clay drain tile is being displaced by the use of 4" (100 mm) plastic tubing that is already perforated by the manufacturer. This 4" (100 mm) tubing comes in 25', 50', and 100' (7.6, 15.2, and 30.4 m) rolls. It is very flexible, and since you are not handling small individual pieces as with the vitrified clay, this particular material is installed very rapidly. The joints are made by a coupling sleeve very much like that used with plumbing PVC piping. It is wise to lay a strip of filter cloth above the plastic tubing prior to installing the washed gravel to keep all sandy fines from entering the plastic tubing and creating a stoppage. All other grading preparation and gravels are installed around this particular material the same as for vitrified clay tile.

Subsoil drains, whether perimeter or underslab, can be connected to an existing storm sewer (in some localities, to a sanitary sewer), can be trenched to an outfall at grade level away from the structure (depending on the existing terrain), or connected to a sump pump area within the lower level of the building. If subsoil drainage piping terminates at a sump pump, a submergible pump with an automatic float control switch and discharge piping will be needed.

DRAINAGE STRUCTURES

There are a number of types of drainage structures required in storm drain piping systems. These structures can be of masonry, cast-in-place concrete, or precast concrete.

Curb inlets and yard drains are one type of structure that is used to collect surface water run-off and discharge it into the storm water piping system. The size of the inlet is determined by the area of run-off, as well as the size of the discharge pipe leading away from the inlet structure itself. An average price for a curb inlet drainage structure is $250 to $300 per vertical lineal foot ($820-984 per vertical meter), and the average price of a yard drainage structure is $175 to $210 per vertical lineal foot ($574-689 per vertical meter).

Storm drainage manhole structures are required at certain intervals along a storm pipe line, depending on the local codes, and whenever a storm drainage pipe system changes direction. The majority of manholes are precast concrete. It is much quicker to set an entire manhole in the course of the day instead of keeping the open pit shored while constructing a cast-in-place concrete manhole and bottom chamber. Once the manhole structure is in place and the piping connected, the bottom of the manhole structure is channeled using brick and mortar to aid the water flow from the inlet pipe to the outlet pipe.

Headwalls and endwalls are usually cast-in-place reinforced concrete walls at the ends of pipe, where either water enters the pipe (headwall) or where the water discharges from the pipe (endwall). Quite often at a headwall, there is a concrete flume entrance slab, and there might be a trash rack (a grillage screen) at the entrance to collect debris.

At the endwall an outfall slab or rip-rap area is installed that prevents the discharge water from eroding the soil away. The price of headwalls and endwalls varies tremendously, because each is sized individually depending upon the size of the inlet or discharge pipe and the amount of earth and height of the earth bank above the pipe that is being retained.

STORM PIPING

Concrete, asbestos cement, and corrugated metal are often used for piping other than sewer. The relative material costs per lineal foot (meter) are about as follows:

Size of Pipe, in.	Concrete per l.f.	Wgt. of Concrete, lbs.	Corrugated Uncoated per l.f. 20' lengths	PVC per l.f. 10' lengths
6	--	--	$3.75	$1.75
8	--	--	5.25	2.75
10	--	--	7.75	3.75
12	$4.50	92	8.25	4.75
15	5.10	127	9.10	5.75
18	6.50	168	12.50	6.75
24	9.50	265	16.50	--

*30	18.50	384	20.00	--
*36	27.00	524	30.00	--
*42	35.00	685	--	--
*48	44.00	867	39.09	--
*60	66.00	1,296	59.09	--
*72	92.00	1,810	85.45	--
*84	150.00	2,410	--	--
*96	180.00	2,910	--	--

*Reinforced

(metric)

Size of Pipe, mm.	Concrete per m	Wgt. of Concrete, kg	Corrugated Uncoated per m 6 m lengths	PVC per m 3 m lengths
150	--	--	$12.30	$1.75
200	--	--	17.22	2.75
250	--	--	25.43	3.75
300	$14.76	42	27.07	4.75
375	16.73	58	29.86	5.75
450	21.33	76	41.01	6.75
600	31.17	120	54.13	--
*750	60.70	174	65.62	--
*900	88.58	238	98.43	--
*1050	114.83	311	--	--
*1200	144.36	393	128.25	--
*1500	216.54	588	193.87	--
*1800	301.84	821	280.36	--
*2100	492.13	1093	--	--
*2400	590.55	1320	--	--

*Reinforced

One cannot choose the type of pipe by price alone. While the specifications and job conditions usually dictate the selection, wage costs can also bear heavily. Lighter weight pipe will lay much faster and can be handled on the site much more easily.

For 12" (300 mm) pipe, a 4-worker crew will lay about 100 l.f. (30.5 m) of concrete pipe and 200 l.f. (61 m) of corrugated metal per day. For 36" (900 mm) pipe, a crew will lay about 20 l.f. (6.1 m) of concrete pipe and 60 l.f. (18.3 m) of corrugated metal per day.

WELLPOINT SYSTEM OF DE-WATERING

The following is general information only. For specific applications consult a local de-watering contractor. Wellpoint systems are well known for their ability to de-water water bearing soils, such as sand and gravel, so that excavation and construction of foundations can proceed in the dry.

In addition, wellpoint systems are used extensively in such work as soil stabilization, pressure relief for dams and levees, and for water supply for municipalities and industrial plants.

A wellpoint system in construction work is usually a series of properly sized wellpoints, surrounding or paralleling the area to be de-watered and connected to a header pipe by means of risers and swing piping. Header piping, in turn, is connected to one or more centrifugal pumps depending on the volume of water that must be handled.

Wellpoints are usually self-jetted into place to the correct depth and at the proper spacing to meet requirements. Some soil conditions, however, require pre-drilling or "hole-punching" before the wellpoints can be installed.

The discharge from the pumps should be piped to an area where it will not interfere with construction. The entire installation should be located so that it will interfere as little as possible with other divisions of the work.

This system is usually installed in advance of the excavation, but where the water level is low, it is sometimes possible that a portion of the excavation can be at a cheaper rate when done prior to the installation of the system.

Any de-watering problem that is beyond the scope of ordinary pumping, or which the contractor or engineer thinks can be done more economically with a wellpoint system, should be referred to the engineering department of a wellpoint company. In this way the problem will receive expert analysis, at no cost to the contractor or engineer. It is essential for most contractors to obtain expert advice on the layout and installation of wellpoint systems, because frequently, a relatively small excavation with unusual soil presents greater de-watering difficulties than a larger, deeper job in a well-graded medium sand.

After carefully considering the factors involved—such as soil characteristics, site accessibility, hydrology of area, size and depth of area, total dynamic head involved, excavation and construction schedules, labor and working conditions, and power facilities, the wellpoint company can make a layout recommendation for the job.

Layout recommendations are usually accompanied by a quotation giving rental rates and charges for all equipment required, together with an estimate of the labor hours needed to install and remove the system, fuel and lubrication required per operating day, transportation charges for the equipment to and from the job, and daily rates for salary and expenses of company demonstrator. Companies specializing in this type of work will furnish rental quotations on a "sufficient equipment" basis, i.e., a fixed rental rate regardless of the amount of equipment actually required to do the job. Some of these companies will submit lump sum contract proposals for de-watering projects covering furnishing, installation operation, maintenance and removal of all de-watering equipment with a guaranteed result.

It is sound practice to have a demonstrator from the wellpoint company on the job site to supervise the initial installation of the system. The time and expenses is more than compensated for by his working knowledge of wellpoint systems and their installation.

Operation of a wellpoint system is usually performed on a 24-hr. day, 7 day per week basis and must continue until all work, depending on dry conditions, is completed.

The labor cost of operating the system must be determined by the contractor and is based on the length of time operation is estimated to be required, figuring around-the-clock operation.

From the above information the contractor or estimator should be able to arrive at a lump sum figure for the required wellpoint system to be used in his estimate.

As an example, take the case of a job in which an underground tank was to be constructed. The test borings indicated the soil to be sand with a ground water level at 6'-0" (1.82 m) below the surface. The excavation for the circular tank was an average of 84'-0" (25.6 m) in diameter and 18'-6" (5.6 m) deep. To maintain dry conditions, the water level had to be lowered at least 12'-6" (3.8 m). It was also estimated that a dry condition would have to be maintained for a period of 3 months.

The following sample estimate shows how a detailed de-watering estimate might be put together.

Equipment Rental, First Month

	Hours	Rate	Total	Rate	Total
52 wellpoints, 2" (50 mm)		$....	$....	$16.00	$832.00
260 l.f. (79 m) header pipe, 6" (150 mm)		....	....	1.80	468.00
50 l.f. (15 m) discharge pipe, 6" (150 mm)		....	....	1.00	50.00
2 valves, 6" (150 mm)		....	....	40.00	80.00
1 wellpoint pump		....	....	950.00	950.00
1 jetwell pump		....	....	1,025.00	1,025.00

Equipment Rental, Second Month

	Hours	Rate	Total	Rate	Total
52 wellpoints, 2" (50 mm)		....	....	10.00	520.00
260 l.f. (79 m) header pipe, 6" (150 mm)		....	....	0.90	234.00
50 l.f. (15 m) discharge pipe, 6" (150 mm)		....	....	0.50	25.00
2 valves, 6" (150 mm)		....	....	25.00	50.00
1 wellpoint pump		....	....	750.00	750.00
1 jetwell pump		....	....	850.00	850.00

Equipment Rental, Third Month

	Hours	Rate	Total	Rate	Total
52 wellpoints, 2" (50 mm)		....	....	10.00	520.00
260 l.f. (79 m) header pipe, 6" (150 mm)		....	....	.90	234.00
50 l.f. (15 m) discharge pipe, 6" (150 mm)		....	....	.50	25.00
2 valves, 6" (150 mm)		....	....	25.00	50.00
1 wellpoint pump		....	....	750.00	750.00
1 jetwell pump		....	....	850.00	850.00

	Hours	Rate	Total	Rate	Total
2 weeks rental, 200 l.f. (61 m)					
jetting hose		$....	$....	$200.00	$400.00
Labor	160	....	....	19.13	3,060.80
Supervisor, 5 days		....	....	200.00	1,000.00
Supervisor's expenses		....	....	500.00	500.00
Operating Expense					
Diesel fuel 91 days at					
40 gals. (151 liters) per day		....	....	1.30	4,732.00
Lubricating oil, 91 days at					
1 gal. (3.78 liters) per day		....	....	5.00	455.00
Pump operator, 3 shifts,					
regular time	1,560	....	....	25.96	40,497.60
Pump operator, 3 shifts,					
overtime	624	....	....	38.94	24,298.56
Removal of Equipment					
Labor	60	....	....	19.13	1,147.80
Transportation to and					
from job, 16,000 lbs. (7258 kg)					
at $4.00 per cwt (45 kg)		....	....		640.00
Total cost of wellpoint system			$....		$84,994.76

Add for insurance, social security, unemployment compensation, overhead and profit.

The jetwell pump is a double-duty unit and is connected into the system as a stand-by wellpoint pump after installation of wellpoints has been completed.

DEEP WELL SYSTEM OF DE-WATERING

A cheaper alternative to the wellpoint system of de-watering is the use of deep wells, which are 18" to 36" (450 to 900 mm) diameter shafts drilled into the ground and lined with a perforated, corrugated metal pipe in which an electric submergible pump is installed. This pump can be discharged directly onto the ground or into another collector pipe discharge system.

The size (diameter and depth) of the drilled shafts, the distance spaced apart, and the size of the electric submergible pump is determined by the depth of the planned excavation, the elevation of the water table, and the subsurface water flow.

When installing the submergible pump, the use of rigid PVC piping is preferred over flexible canvas discharge hoses. Depending on the depth of the shaft, a pump will be required to have sufficient head pressure to pump the water up to ground elevation. Therefore, any restrictions in collapsed discharge piping would create an additional burden for the pump and losing pump efficiency.

SUBMERSIBLE PUMP INSTALLATION

TWO WIRE SUBMERSIBLE INSTALLATION WITH CONVENTIONAL TANK

Fused Main Disconnect Switch

Pressure Gauge

Air Volume Control

Pressure Tank

Service Line

Check Valve

Pressure Switch

Snifter Valve

Conduit

Well Vent

Well Seal

Bleeder Valve

Tee

THREE WIRE SUBMERSIBLE INSTALLATION

For Two-Wire Installation, conduit control box, by-passes directly into pressure switch.

For Three-Wire Installation, conduit goes into control box, from there to pressure switch.

TWO WIRE SUBMERSIBLE INSTALLATION WITH AIR SEAL TANK (No air volume control required)

Fused Main Disconnect Switch

Pressure Tank

Pressure Gauge

Service Line

Snifter Valve

Check Valve

Pressure Switch

Well Vent

Well Seal

Pump costs vary widely depending upon the size of the pump, the head pressure rating, and the GPM capacity. Pumps can be purchased or rented, and the decision should be based on an evaluation of both methods for the amount of time that the pumps will be in operation.

EJECTOR SYSTEM OF DE-WATERING

The ejector method has a distinct advantage over the wellpoint system. Ejectors are not limited by depth, eliminating the need for multiple stages of wellpoints. When conditions are right, ejectors can cost less than deep wells and can be more effective, but to select the right system of de-watering for a specific project, one must have a good knowledge of the project and a thorough understanding of the various methods.

Installation of ejectors is the same as for wellpoints. The system works by supplying water down a pipe to create a vacuum and then pumping up the ejected water and the ground water. The *two-pipe ejector* system, which is the preferred method, works by pumping water into one pipe and drawing it out the other pipe.

One disadvantage with ejectors is the need for larger pumps with higher operating costs, because one is pumping out more water, both ejected water and ground water.

Ground water quality can create discharge problems. It is always advisable to do a water analysis before deciding on and designing the de-watering system.

Upon completion of the project and termination of the need for the de-watering deep well system, the pump and piping are extracted from the hole and sand or gravel material is used to fill the shaft to prevent potential hazards of future cave-in.

The following two charts give the cost per vertical lineal foot for drilled shafts and perforated corrugated metal pipe liner.

Drilled Shafts

Dia. of shaft, in.	Price per l.f.	Dia. of shaft, mm	Price per m
18	$14.00	450	$45.93
24	17.00	600	55.77
30	22.00	750	72.18
36	30.00	900	98.42
42	38.00	1050	124.67

Installation of Perforated Corrugated Pipe

Dia. of pipe, in.	Price per l.f.	Dia. of pipe, mm	Price per m
18	$16.00	450	$52.49
24	24.00	600	78.74
30	30.00	750	98.42
36	46.00	900	150.92
42	58.00	1050	190.29

02550 SITE UTILITIES

Gas and water are usually brought to the property line or building by the local utility company. It is then picked up by the piping contractor doing the building work. Costs on this work are discussed in Chapter 15, *Mechanical*. The distribution of utilities on new subdivision work is a specialized field subject to many code, health, and sanitation requirements.

Installing Service Pipes. Many public utility companies, instead of breaking up sidewalks and digging trenches, are installing service pipes by means of portable compressors and pneumatic tools. Where no rock is encountered, it is possible to drive pipe directly through the earth into cellars and basements.

This method consists of placing a pointed steel cap over the end of the pipe, which can be removed after the pipe holes through.

DRILLING WATER WELLS

Pollution of streams, rivers, and lakes has led many municipalities to look elsewhere for sources of water. Many have had to install wells as an alternate source of water. Industries sometimes also have to go to wells for their water supply where other sources are not available or are inadequate for their demands.

Large quantities of water are usually found in glacial deposits of sand and gravel. The quality of water from a glacial deposit is usually harder and of lower temperature than water derived from sand rock.

When it is necessary to locate large quantities of water for a town or city, it is customary to explore first by drilling a test hole to locate a water bearing strata, and to determine if there is an ample supply of water at that level. If a strata is found that is determined to be inadequate, blank pipe is then driven to the water bearing strata and a well screen is set in it.

Water travelling through a well screen has a tendency to precipitate the solids from the water on the screening, which eventually will restrict the water flow through it. Voids in the sand and gravel surrounding the pipe and screen may also become filled with the precipitates from the water. Water with a high mineral content may close the screen openings in a relatively short period of time, while water with low mineral content may not restrict the flow of water appreciably for many years. When screens do become clogged up with precipitates they sometimes can be opened by treating them with acid. If this process is not effective, it might be necessary to install a new well.

Rock wells are installed by driving a pipe through the overburden above the rock to the rock surface and continuing by drilling through the rock to the desired depth. It is sometimes possible to increase the quantity of water from a rock well in dense rock by shooting and fracturing the rock in the vicinity of the well. If the water bearing rock is already loose, porous, and creviced, shooting will probably not increase the water quantity.

Wells which extend into glacial sand and gravel deposits, will cost approximately $4.00 to $5.00 per inch (25 mm) in diameter per lineal foot of vertical depth. For example, a 12" (300 mm) diameter well would cost from $55.00 to $70.00 per lin. ft. ($180 to $229 per m) of depth.

Wells drilled into rock formations generally cost approximately $1.70 to $2.25 per inch (25 mm) of diameter per lin. ft. of depth. A 6" (150 mm) diameter rock well would cost about $10.00 to $13.50 per lin. ft. of depth ($32.80 to $44.30 per m). In the rock itself a pipe is not used, but if there is 100 feet (30.5 m) of earth or sand and gravel above the rock surface, it is necessary to include the cost of the pipe for this distance.

For wells with a capacity of 25 to 250 gallons (95 to 950 liters) per minute, submersible, multistage type pumps are generally used. Turbine type pumps are normally used for wells of more than 250 gallons (950 liters) per minute capacity.

Submersible pumps cost from $1,000.00 for a 10-gal. (38-liter) per minute pump to $11,200 for a 250-gal. (950-liter) per minute pump.

Turbine type pumps cost approximately $2,200 to $9,000 depending upon the capacity and the depth of the well.

It is recommended that quotations for drilling a well at a specific site be requested from local well drilling contractors who have knowledge of the area and know the approximate depth to the water bearing strata.

02600 PAVING AND SURFACING

Bituminous Roadways and Parking Lots. This section is confined to on-site roadways and parking lot construction for commercial and residential projects. Street and highway pavements are covered by the chapter on *Road and Highway Construction.*

Roadways and parking lots are constructed in any number of thicknesses, depending on the type of use, amount of traffic, gross vehicle weight anticipated, and the suitability of the subsoil conditions for the intended usage.

After careful attention is given to the designed grade contours and drainage patterns, and the rough grading operation is completed, the paving contractor performs the final grading with a motor grader. This machine is a rubber-tired tractor with a long cutting blade that is variable controlled for depths and angles of cut.

Once the motor grader has completed the final grading, the roadway or parking lot bed is rolled and compacted using a large rubber-tired compactor. While compacting the paving subgrade, the roller operator watches for any soft spots in the subgrade. If any are found, they are excavated and refilled with a select grade of backfill material. If these soft spots are missed and not repaired, the asphalt paving will flex and eventually break up over these soft areas causing costly repairs at a later date.

When the subgrade has been suitably prepared (graded and compacted), the base course of paving materials is laid to the proper thickness and compacted in preparation for the bituminous base course or bituminous wearing course. This gravel base course can vary in thickness from 4" (100 mm) to over 20" (500 mm) depending on the paving composite design and is usually compacted to 98%. After the base course is installed, it may be sprayed with an asphalt tack coat just prior to the first bituminous paving course being laid.

The next step in the paving process is the laying of the bituminous paving materials to the proper thickness and rolling the surface to compact and smooth out the hot asphalt mix. The bituminous course can vary in thickness from 3" to 8" (75 to 200 mm) depending upon the requirements and can be laid in one course or can be split into two courses (binder course and wearing course). When the bituminous is laid in two courses, the top wearing course is usually a finer aggregate mix allowing for a much smoother compacted top surface.

Laying the bituminous paving in two courses will always cost a premium. However, there are some distinct advantages. It allows early paving of a work site without construction abuse of a finished surface. Should soft spots occur, they can be fixed prior to the finish wearing course installation. It allows sufficient time to completely check all areas for low spots (birdbaths) that collect water, and these can then be adjusted in the top wearing course thickness.

When estimating quantities of bituminous paving and base courses for a project, it is customary to figure all quantities in square yards (sq. meters): length in feet x width in feet = area in square feet divided by 9 = square yards; length in meters by width in meters = square meters.

Concrete Curb, Gutter & Paving

This section is not intended to cover road building but only small curb and gutter jobs, and small concrete paving projects. Large jobs of this kind require a vast amount of equipment, and the costs vary widely according to the amount of cut and fill required and the method of receiving and dispersing materials.

The item of plant and equipment should include the cost of transporting equipment to the job and removing it at completion and includes items such as bulldozers, loaders, power rollers, trucks, and wood or metal forms for curb and gutter. These items should be estimated on the basis of total transportation costs both ways and a rental charge for plant and equipment. The total should be divided by the number of sq. yds. (sq.m) of pavement in the job. It is customary to submit bids on paving at a price per sq. yd. (sq.m).

Rolling Streets and Roads. After the excavating and grading has been completed, the subgrade should be rolled to provide a solid foundation to receive the concrete base. A power roller should roll 130 to 150 sq. yds. (109 to 125 sq.m) of subgrade per hr. at the following cost per 100 sq. yds. (84 sq.m):

	Hours	Rate	Total	Rate	Total
Operator	0.7	$....	$....	$28.55	$19.99
Power roller	0.7	....	....	22.00	15.40
Cost per 100 sq. yds.			$....		$35.39
per sq. yd.			....		0.35
per 100 sq.m			....		42.11
per sq. m			...		0.42

Hauling Sand, Gravel, Crushed Stone, Etc. The cost of hauling paving materials will vary with the length of haul from pit or cars to point of delivery. Tables of hauling time are given under "Excavating".

Concrete Curb and Gutter

When estimating the cost of forms for concrete curb or curb and gutter, the size and shape must be considered and the amount of materials estimated.

Combined curb and gutter is usually 12" (300 mm) high at the outside edge (including pavement thickness), 24" to 30" (600 to 750 mm) wide and 6" to 7-1/2" (150 to 187.5 mm) thick.

Separate concrete curb is usually 6" (150 mm) wide at the top, 8" to 9" (200 to 225 mm) wide at the bottom, and 24", 30", or 36" (600, 750, or 900 mm) high.

Forms for Concrete Curb and Gutter. When placing wood forms for separate concrete curb, two mechanics and a helper should set forms for 100 to 110 l.f. (30.5 to 33.5 m) of 24" (600 mm) curb per 8-hr. day, at the following labor cost per 100 l.f. (30 m):

	Hours	Rate	Total	Rate	Total
Mechanics	15.0	$....	$....	$26.44	$396.60
Helper	7.5	....	....	21.04	157.80
Cost per 100 l.f.			$....		$554.40
per l.f.			....		5.54
per m.			....		18.17

Two mechanics and a helper should set forms for 90 to 100 l.f. (27.5 to 30.5 m) of 30" (750 mm) concrete curb per 8-hr. day, at the following labor cost per 100 l.f.:

	Hours	Rate	Total	Rate	Total
Mechanics	17.0	$....	$....	$26.44	$ 449.48
Helper	8.5	....	....	21.04	178.84
Cost per 100 l.f.			$....		$ 628.32
per l.f.			....		6.28
per m.			....		20.60

When setting wood forms for combined concrete curb and gutter, two mechanics and a helper should set forms for 80 to 90 l.f.(24.4 to 27.5 m) of curb and gutter per 8-hr. day at the following labor cost per 100 l.f. (30 m):

	Hours	Rate	Total	Rate	Total
Mechanics	18.8	$....	$....	$26.44	$ 497.07
Helper	9.4	....	....	21.04	197.77
Cost per 100 l.f.			$....		$ 694.84
per l.f.			....		6.95
per m.			....		22.80

Wood forms for concrete curb and gutter are ordinarily built of 2" (50 mm) lumber. Including all bracing, it requires 2-1/4 to 2-1/2 b.f. of lumber per s.f. of forms (0.004 to 0.005 cu.m per sq.m). This cost will run from $0.18 to 0.22 per s.f. ($1.93 to $2.36 per sq.m) and should be added to the above setting costs, taking into consideration the number of uses expected from the forms. These forms can be used many times, if properly taken care of.

Placing Ready-mix Concrete for Curb and Gutter. A two-worker crew should place about 1 c.y. (0.76 cu.m) of concrete per hr. at the following labor cost per c.y.:

	Hours	Rate	Total	Rate	Total
Labor	2	$....	$....	$21.04	$42.08

Combined
Curb and
Gutter

Separate
Concrete
Curb

Concrete Roads and Pavements

Concrete roads and pavements could be thicker in the center than at the sides, so it will be necessary to obtain the average thickness of the pavement when estimating quantities of concrete materials.

Some pavements have uniform thickness. Others have uniform center section and thickened edges. In each case the area of the cross section must be determined and this is multiplied by the length to find the volume of concrete required.

Labor Placing Edge Forms for Concrete Paving. Wood edge forms or concrete paving usually consist of 2" (50 mm) plank of a width equal to the slab thickness, set to line and grade, secured to wood stakes, and braced.

Two mechanics and a helper should set to line and grade from 250 to 300 l.f. (76.2 to 91.4 m) of edge forms per 8-hr day at the following labor cost per 100 l.f. (30.5 m):

	Hours	Rate	Total	Rate	Total
Mechanics	3.0	$....	$....	$26.44	$79.32
Helper	1.5	....	....	21.04	31.56
Cost per 100 l.f.			$....		$ 110.88
per l.f.					1.11
per m.					3.64

The above costs are based on the forms being straight or set to long radius curves. If sharp curves are required, the form must be either made from 1" (25 mm) boards, or the back of a 2" (50 mm) plank must be kerfed. In either case, the stakes must be set much closer together.

Concrete paving edge forms set to short radius curves will cost 50% to 100% more than straight forms.

Approximate Sq. Yds. of Concrete Pavement Obtainable from One Cu. Yd of Concrete

Thick., in in.	No. of Sq. Yds.	Thick. in in.	No. of Sq. Yds.
3	12.0	8.0	4.5
3.5	10.3	8.5	4.2
4	9.0	9.0	4.0
4.5	8.0	9.5	3.8
5	7.2	10.0	3.6
5.5	6.5	10.5	3.4
6	6.0	11.0	3.3
6.5	5.5	11.5	3.1
7	5.1	12.0	3.0
7.5	4.8		

Approximate Sq. Meters of Concrete Pavement Obtainable from One Cu. Meter of Concrete

Thick. in mm	No. of Sq. M	Thick. in mm	No. of Sq. M
75.0	10.0	200.0	3.8
87.5	8.6	212.5	3.5
100.0	7.5	225	3.3
112.5	6.7	237.5	3.2
125.0	6.0	250	3.0
137.5	5.5	262.5	2.9
150.0	5.0	275	2.7
162.5	4.6	287.5	2.6
175.0	4.3	300	2.5
187.5	4.0		

Placing Ready-mix Concrete Pavements. Using ready-mix concrete and wheeling the concrete into place with wheelbarrows, the labor per cu. yd. of concrete should cost as follows:

	Hours	Rate	Total	Rate	Total
Labor	1.2	$....	$....	$21.04	$25.25

If ready-mix concrete can be discharged from truck directly into place without wheeling, deduct 0.7-hr. labor time.

Larger jobs, with 100 c.y. (76.5 cu.m) or more of concrete, can realize better organization. Lower costs are the result. Assuming a pavement containing from 100 to 200 c.y. (76.5 to 152.3 cu.m) of ready-mixed concrete wheeled into place, the labor per c.y. (cu.m) of concrete should cost as follows:

	Hours	Rate	Total	Rate	Total
Labor	0.9	$....	$....	$21.04	$18.94

If ready-mix concrete can be discharged from truck directly into place without wheeling, deduct 0.4-hr. labor time.

Finishing Concrete Pavements. Finishing concrete pavements usually consists of a series of operations performed in the following sequence: strike-off and consolidation, straight-edging, floating, brooming, and edging.

Labor Finishing Concrete Pavements. Assuming an 8" (200 mm) concrete pavement being placed at the rate of 25 c.y. (19 cu.m) per hr. or 1,013 s.f. (94 sq.m) per hr., a crew of 10 cement masons should be able to keep up with the pour and perform all the above described finishing operations at the following labor cost per 100 s.f. (9.3 sq.m):

	Hours	Rate	Total	Rate	Total
Cement Mason	1.0	$....	$....	$27.03	$27.03
Helper	0.4	....	....	21.04	8.42
Cement mason foreman	0.1	....	....	26.80	2.68
Cost per 100 s.f.			$....		$ 38.13
per s.f.				....	0.38
per sq.m				....	4.08

The above costs are based on a job that will run from 7,500 to 8,500 s.f. per 8-hr. day. On smaller jobs the costs will increase from 20% to 25%.

02700 SITE IMPROVEMENTS

02710 FENCES & GATES

Industrial chain link fences usually consist of 3" (75 mm) OD tubular end posts and 2" to 2-1/2" (50 to 62.5 mm) OD intermediate posts depending upon the height of the fencing to be installed. Top and bottom rails are not always used, because having a top rail on a chain link fence gives much more stability for intruders to gain a handhold. Therefore, the spring wire is used for top and bottom stability in preventing chain link fabric sag. This spring wire is at least a 6 gauge material. The wire fabric is normally a 9 gauge wire and can be galvanized, aluminized steel, or vinyl coated in a number of colors.

For chain link fencing over 6' (1.8 m) in vertical height, 6 gauge wire fabric will be used in lieu of the 9 gauge fabric.

Wire arms are available either in single or double sided for the installation of 2 to 4 strands of barbed wire per side.

The maximum line post or intermediate post spacing is 10' (3 m) on center and these posts as well as in the corner posts are usually set in a 9" (225 mm) diameter by 3' (900 mm) deep concrete footing.

Industrial Chain Link Fencing - Cost per Lin. Ft. (Meter)

				Height of Fence				
	4'	1.2 m	5'	1.5 m	6'	1.8 m	8'	2.4 m
Material	9 ga.	9 ga.	9 ga.	9 ga.	9 ga.	9 ga.	9 ga.	9 ga.
Galvanized	$7.80	$25.59	$8.70	$28.54	$9.20	$30.18	$10.60	$34.78
Aluminized Steel	8.10	26.57	9.50	31.17	10.10	33.14	11.80	38.71
Vinyl Coated	9.40	30.84	10.40	34.12	11.10	36.42	12.90	42.32

Add for barbed wire:
Single side - 3 strand...$0.45/lin. ft. - 1.50/m
Double side - 3 strand...0.80/lin. ft. - 2.60/m

	3" (75 mm) Diameter	4" (100 mm) Diameter
Add for corner posts:		
4' (1.2 m)	$50.00	$85.00
5' (1.5 m)	55.00	100.00
6' (1.8 m)	63.00	110.00
8' (2.4 m)	75.00	120.00

Add for gates:
3' x 4' (0.9 x 1.2 m) ....$90.00 each
3' x 5' (0.9 x 1.5 m) ....120.00 each
3' x 6' (0.9 x 1.8 m) ....160.00 each
3' x 8' (0.9 x 2.4 m) ....210.00 each

RESIDENTIAL CHAIN LINK FENCING

The standard corner posts are 2" to 2-1/2" (50 to 62.5 mm) diameter. The standard line posts are 1-5/8" (40 mm) diameter set at 10' (3 m) on center. Normally, with residential fencing, a top rail is added for stability.

Residential Chain Link Fencing (11 ga.)—Cost per Lin. Ft. (Meter)

			Height of Fence			
	3'	(0.9 m)	4'	(1.2 m)	6'	(1.8 m)
Galvanized	$4.40	$14.44	$5.50	18.04	6.50	21.33
Aluminized	4.75	15.58	6.00	19.69	7.10	23.29
Vinyl Coated	5.20	17.06	6.25	20.51	7.90	25.92

Add for gates:
3' x 3' (0.9 x 0.9 m) ....$70.00 Each
3' x 4' (0.9 x 1.2 m) ....84.00 Each

Wood Residential Fences - Cost Per Lin. Ft. (Meter)

Type of Fence	per LF	per Meter	per Each
Cedar Stockade Screen Fence, 4' High	$9.50	--	--
Cedar Stockade Screen Fence, 1200 mm	--	31.20	--
Gate 3' Wide x 4' High	--	--	$61.00
Gate 900 mm high x 1200 mm wide	--	--	61.00
Cedar Stockade Screen Fence 6' High	13.50	--	--
Cedar Stockade Screen Fence 1800 mm	--	44.00	--
Gate 3' Wide x 6' High	--	--	90.00
Gate 900 mm wide x 1800 mm high	--	--	90.00
Cedar Split Rail - 2 Rail 3' High	6.75	--	--
Cedar Split Rail - 2 Rail 900 mm	--	22.00	--
Cedar Split Rail - 2 Rail 3'6" High	7.60	--	--
Cedar Split Rail - 2 Rail 1050 mm	--	24.80	--
Redwood or Cedar Board on Board 4' High	12.60	--	--
Redwood or Cedar Board on Board 1200 mm	--	41.30	--
Redwood or Cedar Board on Board 6' High	15.75	--	--
Redwood or Cedar Board on Board 1800 mm	--	51.40	--
Gates 3' Wide x 4' High	--	--	67.00
Gates 900 mm wide x 1200 mm high	--	--	67.00
Gates 3' Wide x 6' High	--	--	95.00
Gates 900 mm wide x 1800 mm high	--	--	95.00

02720 ROAD AND PARKING APPURTENANCES

Guard rails of formed steel are set on steel, wood, or concrete posts. Rail sections are set 18" (450 mm) to the centerline above grade. Posts are set every 12'-6" (3.8 m) with rails lapped 11-1/2" (450 mm) and attached with 7-5/8" (190 mm) bolts. The 12 ga. sections weigh less than 100 lbs. (45.5 kg), so two workers can handle installation. Special terminal wings are furnished for exposed ends.

A budget price for this type of rail set on steel posts is $14.00 per lin. ft. ($46.00 per m) and about $12.50 if set on wood posts.

The 4'x8' (1.2 x 2.4 m) timber rails set on wood posts will cost about $6.75 per l.f. (22.14 per m). They do not offer the protection of the metal rail but may better fit into the landscape.

Wheel stops for parking lots may be either wood or concrete. The concrete ones are shaped on the bottom to allow drainage; wood ones, unless set on standards, tend to collect water. Costs per lin. ft. installed are $3.50 for 6x6 wood dowelled into paving, $3.50 for concrete dowelled into paving and $2.75 for 4x4 wood set on metal brackets.

The 4" striping in parking lots will cost about $0.40 per l.f. (1.31 per m) or $8.00 per stall.

02760 SITE FURNISHINGS

Playground Equipment

Type of Equipment	Cost Each
Bike Rack, 10' (3.04 m) long	$325.00
Horizontal Monkey Ladders, 12' (3.65 m) long	575.00
Parallel Bars, 10' (3.04 m) long	375.00
See-saw 2 Units	450.00
See-saw 4 Units	750.00
Slides - Stainless Steel, 12' (3.65 m) long, 6' (1.82 m) high	875.00
Swings - 4 seats, 8' (2.43 m) high	800.00
Swings - 8 seats, 8' (2.43 m) high	1,250.00
Whirlers, 10' (3 m) diameter	1,500.00

Benches

Type	Cost Each
Precast concrete with wood backs and seats, 8' (2.43 m) long	$600.00
Fiberglass, 8' (2.43 m) long	310.00
Fiberglass supports with wood backs and seats, 8' (2.43 m) long	240.00
Wood, 8' (2.43 m) long	400.00

Planters

Type of Planter	Cost Each
Precast concrete, 24" dia. x 18" high (600 x 450 mm)	$300.00
Precast concrete, 48" dia. x 30" high (1200 x 750 mm)	375.00
Fiberglass, 36" dia. x 30" high (900 x 750 mm)	520.00
Fiberglass, 60" dia. x 36" high (1500 x 900 mm)	525.00
Wood, 48" dia. x 30" high (1200 x 750 mm)	755.00

02800 LANDSCAPING

Landscaping requirements are influenced to a considerable degree by the aesthetic appreciation of existing vegetation and the economics of retaining and protection costs during the construction period. Some owners and architects are dedicated to the preservation of the maximum number of trees that do not directly interfere with construction in an effort to blend the new construction into the existing environment. This theory produces exceptional immediate beauty to any new construction project but involves additional costs to construction by requiring extensive protective barriers for trees and vegetation and restricts work areas causing a lowering of productivity for automotive and heavy construction equipment.

In instances where existing trees and vegetation are of low quality, or the owner is unwilling to absorb the cost of retention, the site is cleared of all natural growth and provisions are made to provide new landscaping for the site at the conclusion of construction. The beauty of landscaping is subjective and may be limited to a combination of simple sodding, seeding, and minimal plantings. In other instances, it can progress to an elaborate extent, requiring the special talents of a landscape architect.

Most contractors lack the material sources, experienced personnel, and special equipment required for landscaping, and they rely upon landscape contractors to perform these services. Frequently, specifications require landscaping be performed by a certified contractor who is primarily engaged in this type of work and has the resources and capabilities to provide guarantees and certifications in compliance with specification requirements.

Estimators can develop a general knowledge of the costs involved in landscaping in order to evaluate the proposals received from the subcontractor.

Generally, nurseries providing service to homeowners and transit trade are not competitive in prices for commercial and large scale landscape projects. Soliciting proposals from qualified landscaping subcontractors presents a unique problem for the contractor. It has been found that these specialized subcontractors recognize that their performance will not be required until near the completion of a project, which may be many months in the future, and are reluctant to tender firm price proposals prior to the scheduled bid date.

Estimators must take a special effort to obtain these proposals in a timely manner to allow sufficient time to make an analysis of the bid before it is incorporated into the total project estimate.

The establishment of a work performance schedule to coincide with the accepted local planting season will tend to lower the subcontractor's price and will preserve the guarantee period. It is not unusual to provide for landscaping work slightly out of sequence in order to accommodate the planting season.

For the estimator to develop the proper unit of measure for pricing landscape work, or to determine the completeness of a proposal from a subcontractor, the following are required:

1. Area of seeding and fertilizing in square yards or acres on large projects.

2. Topsoil in cubic yards, with allowance for topsoil available from the site and for the amount that must be purchased from external sources.

3. Area of sodding in square yards.

4. Number, size, and species of shrubs.

5. Number, size and, species of trees.

6. Details of any special features, such as rock gardens.

7. Allowance for maintenance for specified periods.

8. Allowance for replacement guarantee, if specified.

After quantities, quality, and features are accurately determined, the estimator is prepared to develop a budget estimate that can be used as a cost comparison with the proposals received from landscaping subcontractors.

For general guidance the table below provides average cost ranges, but due to quantity, cost fluctuation, and geographical conditions, estimators should verify with local landscape subcontractors.

Item	Unit	In-Place Cost	Unit	In-Place Cost
Edging, 1x4 Redwood	L.F.	$1.15	Meter	$3.77
Edging, Brick (3 per foot, no concrete)	L.F.	7.00	Meter	22.96
Erosion Control - Plastic Netting	Sq. Yd.	1.00	Sq. M	1.19
Marble Stone Chips - Hand Placed	100 lbs.	9.20	50 kg	10.14
Move Existing Plants - Replanting, 3'-0" high	Each	50.00	Each	50.00
Seed, Fertilizer – Small Areas	Sq. Yd.	1.50	Sq. M	1.79
Seed, Fertilizer – Large Areas, mechanical feed – 220 lbs/acre	Acre	1,500.00	Sq.M	0.37
Sodding	Sq. Yd.	2.50-4.00	Sq. M	3.00-4.75
Trees,				
Spruce 8' (2.4 m)	Each	175.00	Each	175.00
Birch 8' (2.4 m)	Each	150.00	Each	150.00
Pine-White 8' (2.4 m)	Each	100.00	Each	100.00
Dogwood 8' (2.4 m)	Each	200.00	Each	200.00
Shrubs,				
Honeysuckle 4' (1.2 m)	Each	35.00	Each	35.00
Forsythia 4' (1.2 m)	Each	35.00	Each	35.00
Plant Bed - By Hand	Sq. Ft.	3.75	Sq. M	40.35

Plant maintenance is necessary until rooting begins. The duration of this activity is determined by local weather conditions and the time of planting. It can extend for many months beyond the conclusion of the actual construction period, so allowances have to be made in the landscape estimate to accommodate this requirement. If the maintenance period extends beyond the conclusion of construction, it will necessitate the dispatch of personnel, including transportation of supplies and equipment, necessary to perform the maintenance work. The per trip cost will be considerably higher than if the work had been performed in the construction phase. These excess costs occur from the additional travel time involved for each trip, which would include the driver's wage cost plus a fair and reasonable allowance for truck rental.

Also, the actual work performance would be in an unsupervised, which can mean low productivity.

PLANT GUARANTEE

Should the prime contractor assume the obligation of the guarantee, then the estimate should include an allowance for replanting shrubs and trees that fail to root and grow after transplanting. To arrive at a reasonable allowance for this cost, 10% to 20% of the original planting allowance should be sufficient.

CHAPTER 3

CONCRETE

CSI DIVISION 3
03100 CONCRETE FORMWORK
03200 CONCRETE REINFORCEMENT
03300 CAST-IN-PLACE CONCRETE
03400 PRECAST CONCRETE
03500 CEMENTITIOUS DECKS

When estimating the total cost of placing concrete for footings, foundations, walls, columns, slabs, and other items, in addition to the material and equipment costs, there are five categories of labor operations to consider:

1. *Form Preparation:* Concrete is placed in a fluid consistency and must be contained or formed to the desired shape before the hardening process. The forming operation includes items such as: earth excavation to the desired shape for footings and foundations, together with supplemental wood forms when necessary for proper elevation, steps (changes in elevation), or unstable soil conditions; proper finish grading of earth surfaces for the placement of concrete slabs on grade to the correct thickness and elevation; and cutting, fabricating, placing, aligning, and bracing wood or metal forms for walls, columns, and slabs.

2. *Embedded Items Placement:* Labor will be required for the proper placement of all items to be cast into the concrete mix. Usually steel reinforcing (mesh or bars) will be necessary to insure the structural integrity of the concrete member. Non-structural items that frequently are installed into cast-in-place concrete include: anchor bolts, plates, or inserts for the future connection of other items of work, corner protection angles, and sleeves for the passage of pipes, conduit, and ductwork.

3. *Placement of Concrete:* The ready-mixed concrete that is delivered to the job site by truck must be physically moved from the truck to the location of placement by the use of a chute, wheelbarrow, buggy, or some mechanical means such as a conveyor, pump, or crane. Any means of moving concrete mix involves a labor operation, whether it be a group of laborers, an equipment operator, or a combination. Once the concrete mix is in place, it is then necessary to tamp or vibrate the concrete for proper consolidation and "strike off" or screed the concrete

to its designed level or elevation. In the event that only a small amount of concrete is required, or in remote areas where ready-mixed concrete is unavailable, it will be necessary to estimate the additional labor required to job mix the components (gravel, sand, cement, and water) for the quantity of concrete needed.

4. *Finish, Cure, & Protection:* Once the concrete mix is placed into the forms, vibrated, and screeded to the proper elevation, the next operation is the proper finishing of the exposed surfaces: a steel troweled finish to a slab surface to provide a smooth, hard and consolidated surface, free of voids and unevenness. Concrete requires a chemical action to change it from a fluid consistency to the set or hardened condition, which generates a certain amount of heat. On hot days the heat generated by the setting process and the weather produces a concrete mix that sets too fast and hampers workability for the proper finish. It produces surface cracking, and in some cases, cracking for the full depth of the slab. To help slow the setting process and to seal in moisture, slabs are covered with wet burlap, polyethylene, or a chemical curing compound. Slower curing time produces a stronger product. In cold weather, because of its water content before hardening, concrete must be kept from freezing by either covering with straw or other insulating material to keep in its generated heat, or by a supplemental heating process such as covering the area and installing a temporary heat source. Concrete work, and in particular slabs, must be protected from any form of precipitation prior to the setting process, because the water droplets landing on the slab cause small craters in the finish surface, ruining the quality of finish and diluting the cement ratio of the concrete mix, which weakens at least the top wearing surface of the slab.

5. *Form Removal:* After the concrete has hardened, the forms are removed. This operation entails stripping the nails from the forms, cleaning off the hardened concrete skim particles, preparing the forms for the next forming location by means of applying a form release agent (oiling), and physically moving the forms to the next area of the building for starting the forming process on another scheduled work item. The proper cleaning and oiling (coating the concrete contact surface) of form work is important, because it allows reuses of form materials, reducing the costly expense of new material and fabrication costs.

03100 CONCRETE FORMWORK

Concrete formwork is estimated by computing the actual surface area of the form that comes into contact with the concrete. This is referred to as the "contact surface". The formwork quantities are then categorized into specific groups (wall form, column form, etc.) and in the proper unit of measure—square foot, lineal foot, meter, etc.:

Ribbon footing forms - Lineal Foot (Meter)
Slab edge forms (by height) - Lineal Foot (Meter)
Foundation edge forms - Square Foot (Square Meter)
Wall forms - Square Foot (Square Meter)
Pier forms - Square Foot (Square Meter)
Column forms - Square Foot (Square Meter)
Beam bottom forms - Square Foot (Square Meter)
Beam side forms - Square Foot (Square Meter)
Supported slab forms - Square Foot (Square Meter)

Example: Compute the quantity of formwork required for a free standing wall 50'-0" (15.24 m) long, 4'-0" (1.22 m) high, and 1'-0" (0.30 m) thick. Forms will be required for both sides of the wall. The formwork and quantity is obtained as follows:

Face Form: 50'-0" (15.24 m) x 4'-0" (1.22 m) x 2 sides = 400 s.f. (37.19 m)
End Form: 1'-0" (0.30 m) x 4'-0" (1.22 m) x 2 ends = 8 s.f. (0.73 sq.m)
Total Contact Area = 400 + 8 = 408 s.f. (37.19 + 0.73 = 37.92 sq.m)

WOOD FORMS

Forms for Concrete Footings. There are two main types of footings in most construction work: column footings and wall footings. Column footings are usually square or rectangular masses of concrete designed to spread the column load over an area to accommodate the soil bearing value and are located at the column centers. Wall footings are usually ribbons of concrete, somewhat wider than the walls they carry, and are similarly designed to spread the loading over a larger area than the actual wall bottom thickness.

Forms for footings that are formed completely are usually simple and rough, using 2" (50 mm) planking for sides and 2"x4" (50 x 100 mm) stakes to hold the sides in place and 2"x4" (50 x 100 mm) struts for braces to the banks. This material may be reused numerous times in the course of the work. Plywood is sometimes substituted for the 2" (50 mm) planking for sides.

Ribbon forms for partly formed footings usually consist of 2" (50 mm) thick lumber in lengths as long as are practical, secured at the correct elevation to

2"x4" (50 x 100 mm) stakes, and tied across the tops at intervals of 4'-0" to 6'-0" (1.22 to 1.83 m) with 1"x2" (25 x 50 mm) spreader-ties, except in the case of column footings, which are usually diagonally tied at the corners with 1" boards and tied transversely, if required, with strap iron.

The following are itemized costs of the column and wall footing forms found in the average simple job. Complicated or "cut-up" jobs require up to 25% more material and up to 50% more labor.

Material Cost of 100 S.F. (9 Sq.M) of Column Footing Forms
Footings 5'-0"x5'-0"x1'-6" (1.5 x 1.5 x 0.46 m) Completely Formed

Material	B.F.	Rate	Total	Rate	Total
Sides, 2" (50 mm) planking.............	200	$....	$....	$0.07*	$14.00*
Stakes and braces,					
2"x4" (50x100 mm).........................	100	....	....	0.07*	7.00*
Nails,etc..		....	....	2.10	2.10
Cost per 100 s.f.			$....		$23.10
Material cost per s.f..............................				....	.23
cost per sq.m				....	2.57

Labor Cost of 100 S.F. (9 Sq.M) of Column Footing Forms
Footings 5'-0x"5'-0"x1'-6" (1.5 x 1.5 x 0.46 m) Completely Formed

Placing Forms—18-22					
s.f. (1.6-1.98 sq.m) per hr.	Hours	Rate	Total	Rate	Total
Carpenter..	5.0	$....	$....	$26.44	$ 132.20
Labor helping......................................	2.5	....		21.04	52.60
Stripping Forms—90-110					
s.f. (8.15-9.95 sq.m) per. hr.					
Labor	1.0	....	....	21.04	21.04
Cost per 100 s.f.			$....		$ 205.84
Labor cost per s.f.................................				....	2.06
cost per sq.m				....	22.17

Material Cost of 100 L.F. (30.48 M) of Column Footing Ribbon Forms
Footings 5'-0"x5'-0"x1'-6" (1.5x1.5x0.46 m) Partly Formed

Material	B.F.	Rate	Total	Rate	Total
Ribbons,2"x4" (50x100mm),					
100 l.f. (30.48 m)	67	$....	$....	$0.07*	$4.69*
Stakes, 2"x4" (50x100mm),					
3'-0" (.91m) long	80	....	....	0.07*	5.60*
Corner ties, 1"x6" (25x100mm),					
2'-0" (.60m) long	20	....	....	0.07*	1.40*
Nails, etc.		....	....	1.60	1.60
Cost per 100 l.f.			$....		$13.29
Material cost per l.f.				....	.13
cost per m..				....	0.44

Labor Cost of 100 L.F. (30.48 M) of Column Footing Ribbon Forms
Footings 5'-0"x5'-0"x1'-6" (1.5x1.5x0.46m) Partly Formed

Placing Forms—35 l.f. (10.67m)

per hr.	Hours	Rate	Total	Rate	Total
Carpenter	3.0	$....	$....	$26.44	$79.32
Labor helping	1.5	....	....	21.04	31.56
Stripping Forms—450-500 l.f. (140-150m) per hr.					
Labor	0.21	....	....	21.04	4.42
Labor cost per 100 l.f.			$....		$ 115.30
cost per l.f.				....	1.15
cost per m				....	3.77

Material Cost of 100 S.F. (95 Sq. M) of Wall Footing Forms
Footings 1'-0" (0.30 M) Deep Completely Formed

Material	B.F.	Rate	Total	Rate	Total
Sides, 2"x12" (50x300mm) planks	200	$....	$....	$0.07*	$14.00*
Stakes and braces	75	....	....	0.07*	5.25*
Nails, etc.		....	....	1.30	1.30
Material cost per 100 s.f.			$....		$20.55
cost per s.f.				....	0.21
cost per sq.m				....	2.28

*Based on 6 uses of lumber.

Labor Cost of 100 S.F. (9 Sq.M) of Wall Footing Forms
Footings 1'-0" (0.30 M) Deep Completely Formed

Placing Forms—25-30 s.f. (2.25-2.70 sq.m) per hr.	Hours	Rate	Total	Rate	Total
Carpenter	3.50	$....	$....	$26.44	$ 92.54
Labor	1.75	....	....	21.04	36.82
Stripping Forms—70-75 ft. (21.3-22.8m) per hr.					
Labor	1.50	....	....	21.04	31.56
Labor cost per 100 s.f.			$....		$ 160.92
cost per s.f.				....	1.61
cost per sq.m				....	17.32

Material Cost of 100 L.F. (30.48 M) of Wall Footing Ribbon Forms
Footings 1'-0" (0.30 M) Deep Partly Formed

Material	B.F.	Rate	Total	Rate	Total
Ribbons, 2"x4" (50x100mm), 100 l.f. (30.48m)	67	$....	$....	$0.07*	$4.69*
Stakes, 2"x4" (50x100mm), 2'-0" (.61m) long	33	....	....	0.07*	2.31*

Material	B.F.	Rate	Total	Rate	Total
Ties, 1"x2" (25x50mm) 2'-0"					
(0.60m) long 4		....	....	0.07*	.28*
Nails, etc.		....	....	1.30	1.30
Material cost per 100 l.f.		$....			$8.58
cost per l.f.			....		0.09
cost per m			....		0.28

New Concrete

Female Keyway

Previous Concrete Pour

Female Keyway

2x4/2x6

Stud

Plywood

A Method of Forming Female Keyway

New Concrete

Male Keyway

Existing Concrete Pour

Male Keyway

2x4 (if required)

Stud

Plywood

A Method of Forming Male Keyway

Labor Cost of 100 L.F. (30.48 M) of Wall Footing Ribbon Forms
Footings 1'-0" (0.30 M) Deep Partly Formed

Placing Forms—40 l.f. (12.2m)

per hr.	Hours	Rate	Total	Rate	Total
Carpenter 2.5		$....	$....	$26.44	$ 66.10
Labor helping 1.25		....	....	21.04	26.30
Stripping Forms—450-500					
l.f. (140-150m) per hr.					
Labor 0.21		....	....	21.04	4.42
Labor cost per 100 l.f.		$....			$ 96.82
cost per l.f.			....		0.97
cost per m			....		3.18

*Based on 6 uses of lumber.

On jobs where the wall footings are stepped down to a lower elevation, a riser form is required at each step. The cost of this riser form is about twice the cost of the regular footing side form.

Grooves are usually formed in the tops of footings and are called "footing keys". They are formed by forcing a beveled 2"x4" (50 x 100 mm) into the surface of the wet concrete. After the concrete has hardened, the "key" forms are pried loose, cleaned, and reused. Keyways might be required by the contract drawings, specifications, or by local building codes. The two basic types of keyways are *female* and *male*. Female keyways are usually either a 2x4 (50x100mm) or 2x6 (50x150mm), chamfered and placed in the middle of the new concrete pour. Male keyways are pre-formed with the lower pour and placed in the middle of the next pour location.

The following are itemized labor costs for female keyways, based on using 2"x6" (50 x 150 mm), requiring 1 b.f. per each l.f., plus support of keyway.

Labor Cost of 1,000 L.F. (305 m) of Female Keyway

	Hours	Rate	Total	Rate	Total
Carpenter	8	$....	$....	$26.44	$ 211.52
Cost per l.f.			....		0.21
per m			...		0.69

The following are itemized labor costs for male keyways, based on using 2"x4" (50 x 100 mm), requiring 1.33 b.f. (0.20 cu.m) per each l.f. (0.3 m), plus support of keyway.

Labor Cost of 400 L.F. (122 m) of Male Keyway

	Hours	Rate	Total	Rate	Total
Carpenter	16	$....	$....	$26.44	$423.04
Cost per l.f.					1.06
per m					3.48

Labor Cost of 100 L.F. (30.48 m) 2"x4" (50 x 100 mm) Footing Keys

	Hours	Rate	Total	Rate	Total
Carpenter placing	1.25	$....	$....	$26.44	$ 33.05
Labor stripping	0.50	....	....	21.04	10.52
Labor cost per 100 l.f.			$....		$ 43.57
per l.f.				....	0.44
per m				....	1.44

Forms for Pile Caps. After piles are driven and cut off at the required elevation, they are capped with masses of concrete that are designed to transmit the total loading of the structure to the piles. These pile caps vary in size and shape, from simple square caps covering one pile to complex many-sided caps covering clusters of twenty or more piles.

Forms for pile caps are usually prefabricated in panels at the bench and are then set up and put together in the pile location. Side forms are usually made of 3/4" (20 mm) plyform with 2"x4" (50 x 100 mm) stiffeners nailed to the back face. Five or 6 uses can usually be obtained from these forms.

After panels are set in place, they are secured with 2"x4" braces and tied with strap iron.

The following is an itemized cost of making average pile cap forms, based on 5'-6"x5'-6"x3'-0" (1.68 x 1.68 x 0.9 m), containing 66 s.f. (5.94 sq.m):

Material Costs

Making Pile Cap Side Forms	B.F.	Rate	Total	Rate	Total
3/4" (20mm) Plyform, 66 s.f. (5.94 sq. m)		$....	$....	$0.60 (6.67)	$39.60
20 pcs. 2x4 (50x100mm), 3' (.90 m) long, 41 b.f. (3.54m3)		....	....	$0.40 (4.63)	16.40
8 pcs. 2x4 (50x100mm), 5' (1.52m), 27 b.f. (2.36m3)		....	....	0.40 (4.63)	10.80
Bracing 14 pcs. 2x4 (50x100mm), 8'(2.4m) long, 75 b.f. (6.61m3)		....	....	0.40 (4.63)	30.00
Nails		....	....	1.30	1.30
Material cost per cap, 66 s.f.		$....	$....		$98.10
cost per s.f.		....	....		1.49
cost per m		....	....		16.52

Divide above costs by number of times forms may be used on the job.

Labor Costs

Making Pile Cap Side Forms	Hours	Rate	Total	Rate	Total
Carpenter	4	$....	$....	$26.44	$ 105.76
Labor	2	....	....	21.04	42.08
Labor cost per cap, 66 s.f.			$....		$ 147.84
cost per s.f.				....	2.24
cost per sq.m				....	24.10

Labor Cost Setting and Removing Pile Cap Forms

	Hours	Rate	Total	Rate	Total
Carpenter	3.0	$....	$....	$26.44	$ 79.32
Labor	1.5	....	....	21.04	31.56
Labor stripping, cleaning, moving	1.0	....	....	21.04	21.04
Labor cost per cap, 66 s.f.			$....		$ 131.92
cost per s.f.				....	2.00
cost per sq.m				....	21.52

Forms for Concrete Piers. When column footing tops are lower than the bottom of the floor slab on the ground, piers are used to carry the column loading to the footings.

Forms for this work are usually made up in side panels at the job bench, hauled to the footings, erected, clamped, and braced.

Side panels are usually made of 3/4" (20 mm) plyform with 2"x4" (50 x 100 mm) stiffeners. Clamping may be accomplished by using either 2"x4" (50 x 100 mm) lumber or metal column clamps properly spaced to withstand the pressure developed when placing the concrete.

Bracing is required to hold the form in position and usually consists of 2"x4" (50 x 100 mm) or 2"x6" (50 x 150 mm) lumber, secured to the form at grade level, spanning the pit, and spiked to stakes driven into the ground. Bracing below this point is done by wedging 2"x4" (50 x 100 mm) struts between the form and the earth banks.

The following are itemized costs of building average pier forms, based on a pier 2'-0"x2'-0"x4'-0" (.60 x .60 x 1.22 m) containing 32 s.f. (2.88 sq.m):

Material Costs

Making Pier Forms	B.F.	Rate	Total	Rate	Total
3/4" (20mm) Plyform,					
32 s.f. (2.88 m3)		$....	$....	$0.60 (6.67)	$19.20
12 pcs. 2x4 (50x100mm),					
4'(1.22m), 16 b.f. (2.88 m3)		....	....	0.40 (4.52)	12.80
Clamps 8 pcs. 2x4 (50x100mm),					
3'(0.91m), 16 b.f.(1.41m3)		....	....	0.40 (4.51)	6.40
Nails, etc.		....	....	1.30	1.30
Material Cost per 32 s.f.			$....		$39.70
cost per s.f.			....		1.24
cost per sq.m			....		13.78

Divide above s.f. costs by number of times forms may be used on the job.

Labor Costs

Making Pier Forms	Hours	Rate	Total	Rate	Total
Carpenter	1.0	$....	$....	$26.44	$26.44
Labor	0.5	....	....	21.04	10.52
Labor cost per 32 s.f.			$....		$ 36.96
cost per s.f.			....		1.16
cost per sq.m			....		12.48

Labor Cost Erecting Concrete Pier Forms

	Hours	Rate	Total	Rate	Total
Carpenter	1.0	$....	$....	$26.44	$26.44
Labor	0.5	....	....	21.04	10.52
Stripping, including					
cleaning and moving					
Labor	0.5	....	....	21.04	10.52
Labor cost per pier, 32 s.f.			$....		$ 47.48
cost per s.f.				....	1.48
cost per sq.m				....	15.92

Forms for Concrete Walls

Forms for concrete walls must be designed and constructed to withstand the horizontal pressures exerted by the fluid concrete. The pressure against any given point of the form varies and is influenced by any one or all of the following factors:

1. Rate of vertical rise in filling the forms in feet (meters) per hour.

2. Temperature of the concrete and the weather (ambient temperature).

3. Proportions of the concrete mix and its consistency.

4. Method of concrete placement and degree of vibration.

When wall forms are filled so rapidly that concrete at the bottom does not have enough time to start its initial setting, then regardless of any other factor, the full height of the fluid concrete must be considered as a column of liquid with a weight of approximately 150 lbs. per c.f. (2403 kg per cu.m) with a head pressure equal to its fluid height.

For example, at 10' (3 m) below the top of the concrete, the pressure will amount to 10' x 150 lbs. = 1,500 lbs. per s.f. (3 m x 2403 = 7209 kg per sq.m) exerted against the form sides at the bottom. If the wall form has not been adequately tied and braced to withstand this pressure, it will break open.

To avoid the necessity of overdesigning forms, or form breakage, concrete is placed in layers or "lifts" at the rate of about 4 vertical ft. (1.22 m) per hour. This allows the concrete a chance to start the transformation from the fluid state to a hardened mass that supports itself and relieves pressure on the form surface.

The ambient temperature at which concrete is placed influences the time required for the setting process. Concrete placed at 40°F (4.44°C) takes almost twice the time to set as concrete placed at 80°F (26.67°C). Concrete placed in wall forms during cold weather will require more control over the amount of concrete placed within a given time so that the forms will not be stressed beyond their design capability.

The proportions and consistency of the mix—water/cement ratio per c.y. (cu.m) of concrete—has an effect on the rate of setting time as well as the strength of the concrete mix. The more the cement content and/or the less water needed to provide a workable consistency (slump) per cubic yard, the shorter the time for setting and the stronger the concrete.

When placing concrete into a form, care should be taken to reduce the shock loading of the concrete hitting against the form material. This can be achieved by using a *tremie,* a flexible enclosed chute in the shape of an elephant's trunk, which confines the concrete until it reaches the bottom of the form or the level of the preceding lift. The tremie will also keep the concrete mix from segregating and should be used whenever placing concrete that could "free fall" a distance more than 4 ft. (1.22 m). This segregation is caused by the large aggregate within the concrete mix hitting the reinforcing steel and shaking the cement/sand matrix coating off, the result being an improperly mixed concrete with inferior strength and possible voids (honeycomb).

After each lift is placed, the concrete mix is mechanically vibrated, necessary for the proper consolidation (elimination of voids) and to ensure that the fine matrix of the mix reaches the form surface for a smooth wall finish. Although vibration of the concrete mix is desirable as well as necessary, it can have some adverse effects on formwork if done improperly. If the vibrator is allowed to pass through the just placed concrete lift into the preceding lift, it will disturb the preceding lift's setting process, causing it to liquefy and increasing the head pressure against the formwork. Overvibrating the concrete will also imbalance the large aggregate to fine aggregate proportion within those areas of the mix.

Concrete Wall Forms Made in Panels or Sections. Many contractors, especially those performing smaller work, use foundation wall forms made up into panels 2'x6' (0.6 x 1.8 m), 2'x8' (0.6 x 2.4 m), 4'x6' (1.2 x 1.8 m), 4'x8' (1.2 x 2.4 m) and so on. The frame for these sections is usually 2"x4" (50 x 100 m) lumber, with intermediate braces spaced 12" to 16" (300-400 mm) o.c. The corners are reinforced with galvanized straps or strap iron, and the entire frame is sheathed with 5/8" (15 mm) or 3/4" (20 mm) plyform plywood. Forms of this kind can be used a large number of times, especially where the walls are fairly straight.

The labor cost of making panel forms will vary with the method used and the number made up at one time. A carpenter should complete one panel—28 to 32 s.f. (2.5-2.9 sq.m) in 3/4 to 1 hr. or 8 to 10 panels per 8-hr. day.

The cost of a panel 4'-0"x8'-0" (1.2 x 2.4 sq.m) in size, using 3/4" (20 mm) Plyform plywood should cost as follows:

Material Costs

	B.F.	Rate	Total	Rate	Total
2 pcs. 2"x4"-8'					
(50x100mm-2.4m)	10.67	$....	$....	$0.40	$4.27
7 pcs. 2"x4"-4'					
(50x100mm-1.2m)	18.67	....	....	0.40	7.47
Plyform plywood,					
4'-0"x8'-0"-3/4"					
(1.2m x 2.4m x 20mm)	32 sf	....	....	0.60	19.20
Nails, etc.		....	....	0.85	0.85
Material cost per panel 32 s.f.			$....		$31.79
cost per s.f.				....	0.99
cost per sq.m				....	11.04

Labor Costs

	Hours	Rate	Total	Rate	Total
Carpenter	0.75	$....	$....	$26.44	$ 19.83
Labor	0.25	....	....	21.04	5.26
Labor cost per panel			$....		$ 25.09
cost per s.f.				....	0.78
cost per sq.m				....	8.39

Many contractors specializing in concrete work, especially in foundations for residences, use panel forms exclusively and get up to 25 uses from the forms, with the usual amount of repairs. These conditions should be given consideration when pricing any wall work using panel forms.

Symons Steel-Ply Forms. Symons Steel-Ply Forms (from Symons Corporation, Des Plaines, IL) are lightweight panels with a rolled steel frame completely enclosing a plastic coated plywood face. The plywood has a high strength face, which is 12% stronger than standard plywood. Due to the plastic coating, effects of moisture are reduced 40%. The welded steel frames are painted to prevent rusting and with reasonable care should last for a number of years.

Panels are 2'-0" (0.60 m) wide and come in lengths ranging from 2'-0" to 8'-0" (0.60-2.40 m) high in one foot increments. Fillers are made in widths 4" to 22" (100-550 mm), from 2'-0" to 8'-0" (0.60-2.40 m) high. All steel fillers are available in 1", 1-1/2", and 2" (25, 38, and 50 mm) widths. The average weight of the Steel-Ply Forms is 6 lbs. per s.f. (30 kg per sq.m). Outside corners are steel angles that lock adjoining forms together. Inside corners are 6"x6" (150 x 150 mm).

The side rails of the forms are rolled exclusively for Symons, and have a yield strength in excess of 55,000 psi (379 MPa). "L" shaped cross members on all forms are on 1-ft. centers and have a yield strength in excess of 60,000 psi (414 MPa). The frame is 2-1/2" (64 mm) in width.

*Lumber Walers and Braces used with Steelply® Forming System
In Typical Handset Applications (Courtesy Symons Corp.)*

In the Symons system, only two basic pieces are necessary for erection—the tie and the one-piece connecting hardware called the wedge-bolt. The same wedge-bolt is also used for connecting any Symons accessories to the form. In erecting forms, the panel tie is placed in the dado slot between the forms. A wedge-bolt is inserted through the adjoining forms and tie loop. A second wedge-bolt is inserted through the slot of the first, automatically tying and joining the forms in one operation. The Symons system also uses an easy, efficient method for the erection of corners, pilasters, walers, bracing, and scaffolding. Standard panel ties have 1" (25 mm) breakbacks; 1-1/2" (38 mm) and 2" (50 mm) are also available with wood cones. Safe load capacity of panel ties is 3,000 lbs. (1357 kg) and 4,000 lbs. (1810 kg); flat ties 4,000 lbs. (1810 kg) and 4,500 lbs. (2036 kg) at 1.5 to 1.0 factor of safety. In stripping, panels may be quickly released from the finished wall. Wedge-bolts are easily loosened with a hammer tap, and panel ties then removed by giving tie loops a half twist, or in the case of flat ties, a vertical hammer blow.

Used panels may be rented for about $0.80 per s.f. (8.88 per sq.m) for the first month and $0.70 per s.f. ($7.77 per sq.m) for each subsequent month. New panels rent for about $0.90 per s.f. ($10.00 per sq.m) for the first month and $0.80 per s.f. ($8.88 per sq.m) for each subsequent month. Rental costs should be checked with local suppliers. Minimum rental period is one month. Should the contractor wish to purchase the forms, a portion of the paid rentals are applicable to the purchase price. Rentals are charged from the date shipped from distribution center and stop on date of return. The contractor pays transportation charges both ways.

A typical foundation wall, 8" (200 mm) thick and 8' (2.4 m) high containing 2500 s.f. (225 sq.m) of plywood faced steel wallforms and using a crew of three carpenters to four laborers would average as follows:

Labor Costs

	Hours	Rate	Total	Rate	Total
Erecting, Aligning and Bracing Panels					
Carpenter Foreman	4	$....	$....	27.25	$ 109.00
Carpenters	16	....	....	26.44	$ 423.04
Laborers helping	21	....	....	21.04	441.84
Stripping., Cleaning and Oiling for Reuse					
Carpenters	8	....	....	26.44	211.52
Laborers helping	16	....	....	21.04	336.64
Total Cost, 2,500 s.f.			$....		$ 1,522.04
cost per s.f.					0.61
cost per sq.m					6.56

To the above add form rental, form tie, and form oil costs.

The Symons Company has made on-the-job time studies of placing, aligning, and stripping its forms. A wall similar to the above example, with a crew of three lead and four supporting workers, should work out to a productivity of 1.355 manhours per 100 s.f. (9 sq.m). In five hours this team should complete some 2500 s.f. (225 sq.m) of contact area.

A study of a curved wall forming a sewage treatment structure and using a crew of one carpenter to one laborer showed a productivity of 2.91 manhours per 100 s.f. (9 sq.m) of contact area. Setting of reinforcement and pouring of concrete are not included in the above figures.

Symons *Versiform*. This concrete forming system is a steel framed, plywood faced gang forming system for massive concrete structures and architectural concrete. It adapts to use in walls, columns, piers, culverts, etc., and is available with four different standard facing materials.

Versiform panels are available for rental in 8'x4' (2.4x1.2 m), 8'x2' (2.4x.6 m), 4'x4' (1.2x1.2 m), 4'x2' (1.2x0.6 m), 2'x4' (0.6x1.2 m), and 2'x2' (0.6x0.6 m) panels, with special sizes available for purchase only. A complete line of components and accessory hardware is available for forming virtually any shape or size concrete structure. Versiform panels can be used in conjunction with Symons' other forming systems. A time study of Versiform panels and walers on continuous straight walls showed a productivity of 2.64 manhours per 100 s.f. (9 sq.m) of contact area.

***Uni-Form* Steel Panel Forms.** Uni-Forms (from Universal Form Clamp Co., Chicago, IL) are prefabricated panels consisting of a steel frame 2'x1' (0.6x0.3 m), 2'x2' (0.6x0.6 m), 2'x3' (0.6x0.9 m), 2'x4' (0.6x1.2 m), 2'x5' (0.6x1.5

m), 2'x6' (0.6x1.8 m), 2'x7' (0.6x2.1 m), and 2'x8' (0.6x2.4 m), with a plywood facing. The panels are also furnished in fractional widths. The steel frame is a special rolled tee angle, with an equal legged tee strut to provide additional structural rigidity to the part of frame that acts as a plywood protector.

The plywood face is 5-ply plyform for use in concrete forms. The plywood facing provides a convenient nailing surface. It is advisable to oil the panels after each use. With proper care Uni-Forms have been used as many as 50 times before turning the plywood.

Uni-Form panels require alignment on one side only and employ a simplified method of forming inside corners and fillers.

On a foundation wall job for 47 houses requiring 2,350 s.f. (212 sq.m) of forms per house, using the 7' Uni-Form panels, the following labor costs were obtained:

Erection	Hours	Rate	Total	Rate	Total
Carpenter foreman	6	$....	$....	$27.25	$ 163.50
Carpenters	30	....	....	26.44	793.20
Labor	18	....	....	21.04	378.72
Stripping, Cleaning and Moving					
Carpenters	10	....	....	26.44	264.40
Labor	8	....	....	21.04	168.32
Labor cost 2,350 s.f.			$....		$ 1,768.14
cost per s.f.				....	0.75
cost per sq.m				....	8.07

A carpenter erected and braced 78.3 s.f. (7 sq.m) of forms per hr. and required 0.6 hr. labor time per hr. of carpenter time. A carpenter stripped 235 s.f. (21 sq.m) of forms per hr. A laborer cleaned, oiled, and moved 294 s.f. (26.5 sq.m) of forms an hr.

On another foundation wall job for 50 houses, requiring 1,550 s.f. (140 sq.m) of forms per house (forms 2'-0"x7'-0" and 1'-0"x7'-0"), the following labor costs were obtained per 100 s.f. (9 sq.m) contact area of forms:

Uni-Form Concrete Wall Forms

Erecting and Bracing Panels	Hours	Rate	Total	Rate	Total
Carpenter Foreman	0.65	$....	$....	$27.25	$17.71
Carpenter	2.6	$....	$....	$26.44	$ 68.74
Stripping Forms and Moving					
Carpenter	0.5	....	....	26.44	13.22
Labor	0.8	....	....	21.04	16.83
Oiling Panels					
Labor	0.5	....	....	21.04	10.52
Labor cost per 100 s.f.			$....		$ 127.02
cost per s.f.				....	1.27
cost per sq.m				....	13.67

Gangforming Steel-Ply® Forming System includes 5" Walers, Bracing
And Scaffold Bracket Attachement (Courtesy Symons Corp.)

A carpenter erected and braced 38.75 s.f. (3.5 sq.m) of forms per hr. A carpenter stripped 194 s.f. (17.50 sq.m) of forms per hr. A laborer oiled 194 s.f. (17.50 sq.m) of forms per hr. A laborer carried from one job to the other or loaded into trucks, 129 s.f. (11.60 sq.m) of forms per hr.

Based on the two preceding examples using Uni-Form prefabricated form panels, one finds that even though the panels, type of construction, and classification of labor are similar, the production time on the first example is considerably faster. There are many factors that affect the production time— weather, adaptability of materials used, accessibility to the work area, layout of the work, efficiency and experience of the workmen, and other factors. The estimator must evaluate all of these variables.

Built-In-Place Wood Forms For Concrete Walls

For those situations where the wall forms are to be built-in-place, the estimator should be familiar with form design parameters so that he can ascertain the correct quantity of form materials necessary for the work.

To order plyform (plywood), state the class, number of pieces, width, length, thickness and grade or veneer. Plywood panel veneer for both front and back of the panel should be specified. Example: For a C-D grade panel, one side conforms to the C grade, and the other side conforms to the D grade.

A. Solid Surface. Shims, circular repair plugs and tight knots to 1" across grain permitted.

B. Tight knots to 1" and 1-1/2". Knotholes to 1" across grain and some to 1-1/2", if the total width of knots and knotholes is within specified limits. Synthetic or wood repairs. Discoloration and sanding defects that do not impair strength permitted. Limited splits allowed. Stitching permitted.

C. Knots and knotholes to 2-1/2" width across grain and 1/2" larger within specific limits. Limited splits are permitted. Sitching permitted. Limited to exposure or interior panels.

Grade-Use Guide for Concrete Forms

Use these terms When you specifiy Plywood	Description	Veneer Grade		
		Face	Inner Piles	Back
APA B-B PLYFORM Class I & II	Specifically manufactured for concrete forms. Many reuses. Smooth, solid surfaces. Milled-oiled unless otherwise specified.	B	C	B
APA High Density Overlaid PLYFORM Class I & II	Hard, semi-opague, resin-fiber overlay. Heat-fused to panel faces. Smooth Surface resists abrasion. Up to 200 reuses. Light application of releasing agent recommended between pours.	B	C-Plugged	B
APA STRUCTURAL I PLYFORM	Especially designed for engineered applications. All Group 1 species. Stronger and stiffer than Plyform Class I & II. Recommended for high pressure where face grain is parallel to supports. Also available with high density overlay faces.	B	C or C-Plugged	B
Special Overlays, Proprietary panels And Medium Density Overlaid plyWood specifically Designed for Concrete forming	Produces a smooth, uniform concrete surface. Generally mill treated with form release agent. Check with manufacturer for specification, proper use and surface treatment recommendations for greatest number of reuses.			
APA B-C Exterior (EXT)	Sanded panel often used for concrete forming where only one smooth, solid side is required.	C	C	C
APA – The Engineered Wood Association. Check dealer for availability in your area.				

Concrete Pressure For Column And Wall Forms
Lbs./S.F. and Kg/Sq.M

Pressures of Vibrated Concrete
Lbs. per S.F. (psf), Kg per Sq.M*

Pour Rate, ft./hr.	Pour Rate, mm/hr.	50ºF Columns psf	10ºC Columns kg/sq.m	50º Walls psf	10ºC Walls kg/sq.m	70ºF Columns psf	21ºC Columns kg/sq.m	70ºF Walls psf	21ºC Walls kg/sq.m
1	300	330	1663	330	1663	280	1411	280	1411
2	600	510	2570	510	2570	410	2066	410	2066
3	900	690	3477	690	3477	540	2721	540	2721
4	1200	870	4384	870	4384	660	3326	660	3326
5	1500	1050	5291	1050	5291	790	3981	790	3981
6	1800	1230	6198	1230	6198	920	4636	920	4636
7	2100	1410	7105	1410	7105	1050	5291	1050	5291
8	2400	1590	8012	1470	7407	1180	5946	1090	5493
9	2700	1770	8919	1520	7659	1310	6601	1130	5694
10	3000	1950	9826	1580	7962	1440	7256	1170	5896

*Maximum pressure need not exceed 150h, where h is maximum height of pour. Based on concrete with density of 150 pcf (2262 kg/m³) and 4" (100 mm) slump.

Concrete Pressure For Slab Forms
Lbs./S.F. and Kg/Sq.M

Depth, in.	Depth, mm	Non-Motorized Buggies* lbs./s.f.	kg/sq.m	Motorized Buggies** lbs./s.f.	kg/sq.m
4	100	100	488	125	610
5	125	113	552	138	674
6	150	125	610	150	732
7	175	138	674	163	796
8	200	150	732	175	854
9	225	163	796	188	918
10	250	175	854	200	976

*Includes 50 psf load for workers, equipment, impact, etc.
**Includes 75 psf load for workers, equipment, impact, etc.

Allowable Pressures On Plyform Class I
for Architectural Applications
(deflection limited to 1/360th of the span)
Face Grain Across Supports Allowable Pressures (psf, kg/sq.m)*

Support Spacing, in. (mm)	Plywood Thickness, inches (mm)					
	1/2 (13)	5/8 (16)	3/4 (19)	7/8 (22)	1 (25)	1-1/8 (28)
4 (100)	2935	3675	4495	4690	5075	5645
8 (200)	970	1300	1650	1805	1950	2170
12 (300)	410	575	735	890	1185	1345
16 (400)	175	270	370	475	645	750
20 (500)	100	160	225	295	410	490
24 (600)			120	160	230	280
32 (800)					105	130
36 (900)						115

(metric)
Allowable Pressures, kg/sq. m

Support Spacing, mm	Plywood Thickness (mm)					
	12.5	16	19	22	25	28
100	14330	17943	21946	22899	24778	27561
200	4736	6347	8056	8813	9521	10595
300	2002	2807	3589	4345	5786	6567
400	854	1318	1806	2319	3149	3662
500	488	781	1099	1440	2002	2392
600			586	781	1123	1367
800					513	635
900						561

*Plywood continuous across two or more spans.

Face Grain Parallel to Supports
Allowable Pressures, psf*

Support Spacing, in.	Plywood Thickness (inches)					
	1/2	5/8	3/4	7/8	1	1-1/8
4	1670	2110	2610	3100	4140	4900
8	605	810	1005	1190	1595	1885
12	215	360	620	740	985	1165
16		150	300	480	715	845
20		105	210	290	400	495
24			110	180	225	320

(metric)
Allowable Pressures, kg/sq. m*

Support Spacing, mm	Plywood Thickness (mm)					
	12.5	16	19	22	25	28
100	8154	10302	12743	15135	20213	23924
200	2954	3955	4907	5810	7787	9203
300	1050	1758	3027	3613	4809	5688
400		732	1465	2319	3491	4126
500		513	1025	1416	1953	2417
600			537	879	1099	1562

*Plywood continuous across two or more spans.

Allowable Pressures On Plyform Class I
(deflection limited to 1/270th of the span)
Face Grain Across Supports Allowable Pressures, psf*

Support Spacing, in.	Plywood Thickness (inches)					
	1/2	5/8	3/4	7/8	1	1-1/8
4	2935	3675	4495	4690	5075	5645
8	970	1300	1650	1805	1950	2170
12	430	575	735	890	1185	1345
16	235	325	415	500	670	770
20	135	210	285	350	465	535
24		110	160	215	295	340
32					140	170
36					105	120

(metric)
Allowable Pressures, kg/sq. m*

Support Spacing, mm	Plywood Thickness (mm)					
	12.5	16	19	22	25	28
100	14330	17943	21946	22899	24778	27561
200	4736	6347	8056	8813	9521	10595
300	2099	2807	3589	4345	5786	6567
400	1147	1587	2026	244	3271	3759
500	659	1025	1391	1709	2270	2612
600		537	781	1050	1440	1660
800					684	830
900					513	586

*Plywood continuous across two or more spans.

Face Grain Parallel To Supports Allowable Pressures, psf*

Support Spacing, in.	Plywood Thickness (inches)					
	1/2	5/8	3/4	7/8	1	1-1/8
4	1670	2110	2610	3100	4140	4900
8	605	810	1005	1190	1595	1885
12	270	405	620	740	985	1165
16	115	200	375	525	715	845
20		125	210	290	400	495
24			135	185	255	320

*Plywood continuous across two or more spans.

(metric)
Allowable Pressures, kg/sq.m*

Support Spacing, mm	Plywood Thickness (mm)					
	12.5	15.6	18.8	21.9	25.0	28.1
100	8154	10302	12744	15136	20214	23925
200	2954	3955	4907	5810	7788	9204
300	1318	1977	3027	3613	4809	5688
400	562	977	1831	2563	3491	4126
500		610	1025	1416	1953	2417
600			659	903	1245	1562

*Plywood continuous across two or more spans.

Maximum Spans for Lumber Framing, Inches*
Douglas Fir-Larch No. 2 Or Southern Pine No. 2 (KD)

Equivalent Uniform Load (lb/ft)	Continuous Over 2 or 3 Supports (1 or 2 Spans) Nominal Size								Continuous Over 4 or More Supports (3 or More Spans) Nominal Size							
	2x4	2x6	2x8	2x10	2x12	4x4	4x6	4x8	2x4	2x6	2x8	2x10	2x12	4x4	4x6	4x8
400	36	53	70	90	109	50	79	101	41	60	78	100	122	62	91	118
600	30	43	57	73	89	44	66	88	31	49	64	82	99	51	74	98
800	24	38	50	63	77	39	58	76	25	39	52	66	81	44	64	85
1000	21	33	43	55	67	35	51	68	21	34	44	57	69	39	58	76
1200	19	29	38	49	60	32	47	62	19	30	39	50	61	35	53	69
1400	17	27	35	45	54	30	43	57	17	27	36	46	56	31	49	64
1600	16	25	32	41	50	27	41	54	16	25	33	42	52	28	44	58
1800	15	23	30	39	47	25	38	51	15	24	31	40	48	26	40	53
2000	14	22	29	37	45	23	36	48	14	22	29	38	46	24	37	49
2200	13	21	28	35	43	22	34	45	14	21	28	36	44	22	35	46
2400	13	20	26	34	41	20	32	42	13	20	27	34	42	21	33	44
2600	12	19	26	33	40	19	31	40	13	20	26	33	40	20	31	41
2800	12	19	25	32	38	19	29	38	12	19	25	32	39	19	30	39
3000	12	18	24	31	37	18	28	37	12	19	24	31	38	18	29	38
3200	11	18	23	30	36	17	27	35	12	18	24	30	37	18	28	36
3400	11	17	23	29	36	17	26	34	11	18	23	30	36	17	27	35
3600	11	17	22	29	35	16	25	33	11	17	23	29	35	16	26	34
3800	11	17	22	28	34	16	24	32	11	17	22	29	35	16	25	33
4000	10	16	22	28	34	15	24	31	11	17	22	28	34	15	24	32
4500	10	16	21	27	32	14	22	29	10	16	21	27	33	14	23	30
5000	10	15	20	25	31	13	21	28	10	15	20	26	32	14	22	28

*Spans are based on PS-20 lumber sizes, Single member stresses were multiplied by a 1.25 duration-of-load factor for 7-day loads. Deflection limited to 1/360th of the span with ¼" maximum. Spans are center-to-center of the supports.

Equivalent Uniform Load (Kg./.3048 m)	Continuous Over 2 or 3 Supports (1 or 2 Spans) — Nominal Size								Continuous Over 4 or more Supports (3 or More Spans) — Nominal Size							
	50 mm x 100 mm	50 mm x 150 mm	50 mm x 200 mm	50 mm x 250 mm	50 mm x 300 mm	100 mm x 100 mm	100 mm x 150 mm	100 mm x 200 mm	50 mm x 100 mm	50 mm x 150 mm	50 mm x 200 mm	50 mm x 250 mm	50 mm x 300 mm	100 mm x 100 mm	100 mm x 150 mm	100 mm x 200 mm
181	900	1325	1750	2250	2725	1250	1975	2525	1025	1500	1950	2500	3050	1550	2275	2950
272	750	1075	1425	1825	2225	1100	1650	2200	775	1225	1600	2050	2475	1275	1850	2450
363	600	950	1250	1575	1925	975	1450	1900	625	975	1300	1650	2025	1100	1600	2125
454	525	825	1075	1375	1675	875	1275	1700	525	850	1100	1425	1725	975	1450	1900
544	475	725	950	1225	1500	800	1175	1550	475	750	975	1250	1525	875	1325	1725
635	425	675	875	1125	1350	750	1075	1425	425	675	900	1150	1400	775	1225	1600
726	400	625	800	1025	1250	675	1025	1350	400	625	825	1050	1300	700	1100	1450
816	375	575	750	975	1175	625	950	1275	375	600	775	1000	1200	650	1000	1325
907	350	550	725	925	1125	575	900	1200	350	550	725	950	1150	600	925	1225
998	325	525	700	875	1075	550	850	1125	350	525	700	900	1100	550	875	1150
1089	325	500	650	850	1025	500	800	1050	325	500	675	850	1050	525	825	1100
1179	300	475	650	825	1000	475	775	1000	325	500	650	825	1000	500	775	1025
1270	300	475	625	800	950	475	725	950	300	475	625	800	975	475	750	975
1361	300	450	600	775	925	450	700	925	300	475	600	775	950	450	725	950
1452	275	450	575	750	900	425	675	875	300	450	600	750	925	450	700	900
1542	275	425	575	725	900	425	650	850	275	450	575	750	900	425	675	875
1633	275	425	550	725	875	400	625	825	275	425	575	725	875	400	650	850
1724	275	425	550	700	850	400	600	800	275	425	550	725	875	400	625	825
1814	250	400	550	700	850	375	600	775	275	425	550	700	850	375	600	825
2041	250	400	525	675	800	210	550	725	250	400	525	675	825	350	575	750
2268	250	375	500	625	775	325	525	700	250	375	500	650	800	350	550	700

Maximum Spans for Lumber Framing, Inches*
Hem-Fir No. 2

Equivalent Uniform Load (lb/ft)	Continuous Over 2 or 3 Supports (1 or 2 Spans) Nominal Size								Continuous Over 4 or More Supports (3 or More Spans) Nominal Size							
	2x4	2x6	2x8	2x10	2x12	4x4	4x6	4x8	2x4	2x6	2x8	2x10	2x12	4x4	4x6	4x8
400	33	48	63	80	97	48	73	96	36	53	70	90	109	56	81	107
600	27	39	51	65	80	41	59	78	27	43	57	72	88	45	66	88
800	22	34	44	57	69	35	51	68	22	35	46	59	72	39	58	76
1000	19	29	39	50	60	31	46	61	19	30	40	51	62	35	51	68
1200	17	26	35	44	54	29	42	55	17	27	36	45	55	31	47	62
1400	15	24	32	41	49	27	39	51	16	25	33	42	51	27	43	57
1600	14	23	30	38	46	24	36	48	15	23	30	39	47	25	39	52
1800	14	21	28	36	43	22	34	45	14	22	29	36	44	23	36	47
2000	13	20	27	34	41	21	33	43	13	21	27	35	42	21	33	44
2200	12	19	26	33	40	19	31	40	13	20	26	33	40	20	31	41
2400	12	19	25	31	38	18	29	38	12	19	25	32	39	19	30	39
2600	12	18	24	30	37	18	28	36	12	18	24	31	38	18	28	37
2800	11	18	23	30	36	17	26	35	11	18	24	30	37	17	27	36
3000	11	17	23	29	35	16	25	33	11	17	23	29	36	17	26	34
3200	11	17	22	28	34	16	24	32	11	17	22	29	35	16	25	33
3400	10	16	22	27	33	15	24	31	11	17	22	28	34	15	24	32
3600	10	16	21	27	32	15	23	30	10	16	22	27	33	15	23	31
3800	10	15	20	26	32	14	22	29	10	16	21	27	33	15	23	30
4000	10	15	20	25	31	14	22	29	10	16	21	27	32	14	22	29
4500	10	14	19	24	29	13	21	27	10	15	20	26	31	13	21	28
5000	9	13	18	23	28	12	20	26	9	15	20	25	30	13	20	26

*Spans are based on PS-20 lumber sizes. Single member stresses were multiplied by a 1.25 duration-of-load factor for 7-day loads. Deflection limited to 1/360th of the span with ¼" maximum. Spans are center-to-center of the supports.

Equivalent Uniform Load (Kg./.3048 m)	Contiuous Over 2 or 3 Supports (1 or 2 Spans) Nominal Size								Contiuous Over 4 or more Supports (3 or More Spans) Nominal Size							
	50 mm x 100 mm	50 mm x 150 mm	50 mm x 200 mm	50 mm x 250 mm	50 mm x 300 mm	100 mm x 100 mm	100 mm x 150 mm	100 mm x 200 mm	50 mm x 100 mm	50 mm x 150 mm	50 mm x 200 mm	50 mm x 250 mm	50 mm x 300 mm	100 mm x 100 mm	100 mm x 150 mm	100 mm x 200 mm
181	825	1200	1575	2000	2425	1200	5767	2400	900	3180	1750	2250	2725	1400	2025	2675
272	675	975	1275	1625	2000	1025	1475	1950	675	2107	1425	1875	2200	1125	1650	2200
363	550	850	1100	1425	1725	875	1275	1700	550	875	1150	1475	1800	975	1450	1900
454	475	725	975	1250	1500	775	1150	1525	475	750	1000	1275	1550	875	1275	1700
544	425	650	875	1100	1350	725	1050	1375	425	675	900	1125	1375	775	1175	1550
635	375	600	800	1025	1225	675	975	1275	400	625	8250	1050	1275	675	1075	1425
726	350	575	750	950	1150	600	900	1200	375	575	750	975	1175	625	975	1300
816	350	525	700	900	1075	550	850	1125	350	550	725	900	1100	575	900	1175
907	325	500	675	850	1025	525	825	75	325	525	675	875	1050	525	825	1100
998	300	475	650	825	1000	475	775	1000	325	500	650	825	1000	500	775	1025
1089	300	475	625	775	950	450	725	950	300	475	625	800	975	475	750	975
1179	300	450	600	750	925	450	700	900	300	450	600	775	950	450	700	925
1270	275	450	575	750	900	425	650	875	275	450	600	750	925	425	675	900
1361	275	425	575	725	875	400	625	825	275	425	575	725	900	425	650	850
1452	275	425	550	700	850	400	600	800	275	425	550	725	875	400	625	825
1542	250	400	550	675	825	375	600	775	275	425	550	700	850	375	600	800
1633	250	400	525	675	800	375	575	750	250	400	550	675	825	375	575	775
1724	250	375	500	650	800	350	550	725	250	400	525	675	825	375	575	750
1814	250	375	500	625	775	350	550	725	250	400	525	675	800	350	550	725
2041	250	350	475	600	725	195	525	675	250	375	500	650	775	325	525	700
2268	225	325	450	575	700	300	500	650	225	400	500	625	750	325	500	650

LATERAL CONCRETE PRESSURES FOR VARIOUS TEMPERATURES

2000 PSF MAXIMUM WALL PRESSURE (Rate of placement not more than 10 ft./hr.)

3000 PSF MAXIMUM COLUMN PRESSURE

POUR RATE (FT. PER HR.)

LATERAL PRESSURE (PSF)

—— ALL WALLS. COLUMNS WITH POUR RATE LESS THAN 7 FT. PER HR.

– – COLUMNS WITH POUR RATE GREATER THAN 7 FT. PER HOUR

3.26 ESTIMATOR'S REFERENCE BOOK

Estimating Material Requirements for Wood Forms Built-In-Place for Concrete Walls

The following examples of wall form material estimates assume concrete at 50°F (10°C) with 4" (100 mm) slump, vertical pour rate of 3 to 4 ft. (0.9 to 1.2 m) per hour, and a maximum form pressure of 750 psf (3770 kg/m²).

Lumber Required for Concrete Wall Forms 6'-0" (1.8 m) High

Based on a wall 40'-0" (12 m) long and 6'-0" (1.8 m) high, containing 480 s.f. (43 sq.m) of forms.

	Quan.	Rate	Total	Rate	Total
Studs, 82 pcs. 2"x4"-6' (50x100mm-1.8m), (studs 12" o.c.) .... 328 b.f.(0.77 m3)		$....	$....	$0.40	$131.20
Sheathing 3/4" (18mm) Plyform .... 480 s.f.(43 m2)		....	....	0.60	288.00
Plates, 80 l.f. (24m) 2"x4" (50x100mm) .... 53 b.f. (0.12 m3)		....	....	0.40	21.20
Wales, 3 sets per side 12 pcs. 2"x4"-40' (50x100mm-12m) ..... 320 b.f. (0.75 m3)		....	....	0.40	128.00
Bracing and stakes, 2"x4" (50x100mm) .... 70 b.f. (0.16 m3)		....	....	0.40	28.00
Cost 480 s.f.		$....	$....		$596.40
per s.f.			....		1.24
per sq.m			....		13.87

Material not subject to reuse:

	Quan.	Rate	Total	Rate	Total
Nails .... 20 lbs (9 kg)		$....	$....	$.60 (1.33)	$12.00
Form ties, 3,000 lbs. (1357 kg) safe load 24" o.c. along wales 63		....	....	0.40 (0.87)	25.20
Total Cost 480 s.f. (44.6 sq.m)			$....		$37.20
cost per s.f.			....		0.08
cost per sq.m			....		0.83

Total material cost for one use, 1.24 + 0.08 = $1.32 per s.f. (13.87 + 0.87 = 14.74 per sq.m), assuming 3 uses of form lumber, 1.24 ÷ 3 = 0.41 (14.74 ÷ 3 = 4.91) per use for form lumber, plus 0.08 per s.f. (0.88 per sq.m) for nails and form ties, or a total of 0.49 per s.f. (5.79 per sq.m) for material. Add for bulkheads or end forms if required.

Lumber Required for Concrete Wall Forms 8'-0" (2.4 m) High

Based on a wall 40'-0" (12 m) long and 8'-0" (2.4 m) high, containing 640 s.f. (58 sq.m) of forms.

	Quan.	Rate	Total	Rate	Total
Studs, 82 pcs. 2"x4"-8' (50x100mm-2.4m) 12" (300mm) o.c. ....438 b.f. (1.03 m3)		$....	$....	$0.40	$175.20
Sheathing 3/4" (19mm) Plyform......................640 s.f. (57.6 m2)		....	....	0.60	384.00
Plates, 80 l.f. (24m) 2"x4" (50x100mm)................53 b.f. (0.12m3)		....	....	0.40	21.20
Wales, 4 sets per side, 16 pcs. 2"x4"-40' (50x100mm-12m).......427 b.f. (1.0 m3)		....	....	0.40	170.80
Bracing and stakes, 2"x4" (50x100mm)......80 b.f. (0.19 m3)		....	....	0.40	32.00
Cost per 640 s.f.			$....		$783.20
per s.f.			....		1.22
per sq.m			....		13.50

Material not subject to reuse:

	Quan.	Rate	Total	Rate	Total
Nails ....30 lbs (13.6 kg)		$....	$....	$0.60	$18.00
Form ties, 3,000-lb (1360-kg) safe load 24" (600mm) o.c. wales 84		....	....	0.40	33.60
Total cost 640 s.f.			$....		$51.60
per s.f.			....		0.08
per sq.m			....		0.87

Total material cost for one use, 1.22 + 0.08 = $1.30 per s.f. (13.50 + 0.90 = 14.40 per sq.m). Assuming 3 uses of form lumber, 1.22 ÷ 3 = 0.41 (13.50 ÷ 3 = 4.50) per use of form lumber, plus $0.08 per s.f. for nails and form ties, or a total of 0.49 per s.f. for material. Add for bulkheads or end forms if required.

Lumber Required for Concrete Wall Forms 12'-0" (3.6 m) High

Based on a wall 40'-0" (12 m) long and 12'-0" (3.6 m) high, containing 960 s.f. (89 sq.m) of forms.

	Quan.	Rate	Total	Rate	Total
Plates 4 pcs., 2"x4"-20' (50x100mm-6m)..........53 b.f.(0.12m3)		$....	$....	$0.40 (4.50)	$21.20
Studs, 82 pcs. 2"x4"-12' (50x100mm-3.6m) 12" (300mm) o.c. .... 656 b.f.(1.54m3)		....	....	0.40 (4.50)	262.40

	Quan.	Rate	Total	Rate	Total
Sheathing, 3/4" (19mm) plyform ....960 sf (86.4m2)		....	....	0.60 (6.67)	576.00
Wales, 6 sets per side, 24" (600mm) o.c., 24 pcs. 2"x4"-40' (50x100mm-12m)........640 b.f. (1.5m3)		....	....	0.40 (4.50)	256.00
Bracing & stakes, 2"x4" (50x100mm)....... 98 b.f.(0.23m3)		....	....	0.40 (4.50)	39.20
Cost per 960 s.f.			$....		$1,154.80
cost per s.f.			....		1.20
cost per sq.m			....		12.98

Material not subject to reuse:

	Quan.	Rate	Total	Rate	Total
Nails .... 40 lbs (18kg)		$....	$....	$0.60 (1.33)	$24.00
Form ties, 3,000-lb (1360-kg) safe load 24" (600mm) o.c. along wales.................................... 126		....	....	0.40	50.40
Total cost 960 s.f.			$....		$74.40
cost per s.f.				....	0.08

Total material cost for one use, 1.20 + 0.08 = $1.28 per s.f. (13.37 + .86 = 14.23 per sq.m). Assuming 3 uses of form lumber, 1.20 ÷ 3 = .40 per use for form lumber, plus 0.08 for nail and form ties = 0.48 per s.f. for material. (13.37 ÷ 3 = 4.45 + 0.86 = 5.31 per sq.m).

In comparing the three preceding sample material estimates for wall forming on 6' (1.8 m), 8' (2.4 m), and 12' (3.6 m) high walls, note that all three are in the same cost range. This holds true for even lower or higher walls, as long as the design parameters for temperature, slump, pour rate, and allowable form pressure remain the same.

If the temperature of the concrete is lower, the mix is wetter, or it is advantageous to place the concrete into the forms at a faster pour rate, then the forms will have to be designed to withstand the increased pressure per square foot on the form surface. The design changes would involve any of the following:

1. Stronger form tie (increased tensile strength)
2. Closer horizontal spacing for form ties
3. Increased stud size to counteract bending under increased loads
4. Closer on-center spacing of stud members to counteract deflection of the plyform, or use of thicker plyform material
5. Increased size of wale members
6. Additional bracing and stakes

Basis for Computing Labor Costs. After checking the labor costs on a large number of jobs over a period of years, it has been found that a carpenter can frame and erect about 325 to 400 b.f. (7.58-9.3 cu.m) of lumber per 8-hr. day and requires 3/8 to 1/2 hr. laborer time (carrying lumber, etc.) for each hour of carpenter time.

If there are no laborers employed on the job and all lumber is handled and carried by carpenters, a carpenter should handle, frame, and erect about 225 to 275 b.f. (5.2-6.4 cu.m) of lumber per 8-hr. day.

When removing or "stripping" concrete forms, a carpenter or laborer should remove 750 to 1,000 b.f. (17.5-23.3 cu.m) of lumber per 8-hr. day.

High walls necessitate working from a scaffold and require 10% to 50% more time per sq. ft. of forms because of the additional handling, hoisting of lumber, and additional bracing. The sq. ft. (sq.m) costs on the following pages have been computed on this basis.

Production Times for 100 S.F. (9 Sq.M) of Wood Forms for Concrete Walls
Built-in-Place

Wall Hgt, ft.	Wall Hgt, meters	b.f. lmbr per sf of forms	m³ lmbr per m² of forms	s.f forms per 8-hr Day	sq.m forms per 8-hr Day	Car-penter Hours	Removing Forms Labor Hours	sq. ft. Day	sq. m Day	Labor Hours
4-6	1.2-1.8	2.6	0.06	190-210	17-19	4.0	2.0	360	32	2.25
7-8	2.1-2.4	2.6	0.06	165-190	15-17	4.5	2.3	350	31	2.20
9-10	2.7-3	2.6	0.06	150-160	14-14.5	5.3	2.5	325	29	2.50
11-12	3.3-3.6	2.6	0.06	125-135	11-12	6.0	3.0	300	27	2.70

*If Union rules do not permit laborer working with carpenter, figure labor time as additional carpenter time.

Add foreman time as required. Normal practice on large concrete projects is to allow for one foreman when carpenter crew exceeds 3 carpenters.

When foundation walls rest directly on earth without a concrete footing, mud sills of 2" (50 mm) or 3" (75 mm) plank are usually employed to serve as layout lines and supports for the wall panel forms required. This will increase the cost of 4'-0" (1.2 m) high panel wall forms about 10% for both material and labor. Similarly, 8'-0" (2.4 m) high wall forms should be increased about 5%.

Labor Cost of 100 S.F. (9 Sq.M) Built-in-Place Concrete Wall Forms 4'-0" to 6'-0" (1.2 to 1.8 M) High, Requiring 2.6 BF (0.06 m³) of Lumber per S.F. (Sq.M) of Forms

	Hours	Rate	Total	Rate	Total
Carpenter	4.0	$....	$....	$26.44	$ 105.76
Labor	2.0	....	....	21.04	42.08
Labor stripping & cleaning	2.25	....	....	21.04	47.34
Cost per 100 s.f.			$....		$ 195.18
cost per s.f.				....	1.95
cost per sq.m				....	20.98

Labor Cost of 100 S.F. (9 Sq.M) Built-in-Place Concrete Wall Forms 8'-0" (2.4 M)
High, Requiring 2.6 BF (0.06 m³) of Lumber Per S.F. (Sq.M) of Forms

	Hours	Rate	Total	Rate	Total
Carpenter	4.50	$....	$....	$26.44	$ 118.98
Labor helping	2.30	....	....	21.04	48.39
Labor stripping & cleaning	2.20	....	....	21.04	46.29
Cost per 100 s.f.			$....		$ 213.66
cost per s.f.			....		2.14
cost per sq.m			....		23.03

Labor Cost of 100 S.F. (9 Sq.M) Built-in-Place Concrete Wall Forms 10'-0" (3 m, 3000 mm) High, Requiring 2.6 BF (0.06 m³) of Lumber Per S.F. (Sq.M) of Forms

	Hours	Rate	Total	Rate	Total
Carpenter	5.3	$....	$....	$26.44	$ 140.13
Labor helping	2.5	....	....	21.04	52.60
Labor stripping & cleaning	2.5	....	....	21.04	52.60
Cost per 100 s.f.			$....		$ 245.33
cost per s.f.			....		2.45
cost per sq.m			....		26.36

Labor Cost of 100 S.F. (9 Sq.M) Built-in-Place Concrete Wall Forms 12'-0" (3.6 m, 3600 mm) High, Requiring 2.6 BF (0.06 m³) of Lumber Per S.F. (Sq.M) of Forms

	Hours	Rate	Total	Rate	Total
Carpenter	6.0	$....	$....	$26.44	$ 158.64
Labor helping	3.0	....	....	21.04	63.12
Labor stripping & cleaning	2.7	....	....	21.04	56.81
Cost per 100 s.f.			$....		$ 278.57
cost per s.f.			....		2.79
cost per sq.m			....		30.02

Pilasters. Walls which are thickened for short intervals of 12" to 36" in length and then return to their former thickness are known as pilasters at the thickened area.

The cost of forming the face and sides of pilasters runs from 25% to 50% more than plain wall forms for both material and labor.

Radial Wall Forms. Wall forms that are built to radii or other curves require closer stud spacing, more bracing, and more labor than straight wall forms. Formwork for radial walls will cost from 50% to 100% more than forms for straight walls.

Method of Forming for Radial Concrete Walls

Forming Openings in Concrete Walls. Openings in concrete walls for doors, windows, and louvers are formed by placing bulkheads in the wall forms to form the sills, jambs, and heads of the openings required. These forms usually require from 2-1/4 to 3 b.f. of lumber per s.f. (0.05-0.07 cu.m per sq.m) of contact area, depending upon the size of the opening.

The following is an itemized labor cost of boxed opening forms, based on 4'-0"x4'-0" (1.2 x 1.2 m) openings in a 1'-0" (300 mm) wall requiring 2-1/2 b.f. of lumber per s.f. (0.06 cu.m per sq.m) of contact area.

Labor Cost of 100 S.F. (9 Sq.M) of Boxed Opening Forms

	Hours	Rate	Total	Rate	Total
Setting forms					
Carpenter	7.0	$....	$....	$26.44	$ 185.08
Labor Helping	3.5	....	....	21.04	73.64
Labor stripping forms	3.0	....	....	21.04	63.12
Labor cost 100 s.f.			$....		$ 321.84
cost per s.f.				....	3.22
cost per sq.m				....	34.65

Forming Setbacks in Concrete Walls. When concrete foundation walls extend some height above ground level, it is often required that the portion above grade be faced with masonry. This necessitates the portion above grade being reduced in thickness sufficient to accommodate the facing material. This reduction in thickness is accomplished by placing a "setback" form in the regular wall forms.

Setback forms are usually made up of 2" (50 mm) framing lumber faced with 5/8" (15 mm) or 3/4" (19 mm) plyform, and usually require 1-3/4 to 2-1/4 b.f. per s.f. (0.04-0.05 cu.m per sq.m) of contact area, depending mainly upon the thickness of the setback.

The following is an itemized labor cost of 100 s.f. (9 sq.m) of setback forms requiring 2 b.f. per s.f. (0.047 cu.m per sq.m) per s.f. (sq.m) of contact area:

	Hours	Rate	Total	Rate	Total
Carpenter	6.5	$....	$....	$26.44	$ 171.86
Labor helping	3.3	....	....	21.04	69.43
Labor stripping	2.7	....	....	21.04	56.81
Labor cost per 100 s.f.			$....		$ 298.10
cost per s.f.				....	2.98
cost per sq.m				....	32.06

Forming a Continuous Haunch on Concrete Walls. When an intermediate support on a wall face is required to provide bearing for masonry, slabs, or future construction, and reducing the wall thickness is prohibited by the design, a continuous haunch is commonly used.

A continuous haunch is usually formed of 2" (50 mm) framing lumber faced with 5/8" (15 mm) or 3/4" (19 mm) plyform and requires from 3-1/2 to 4 b.f. (0.08-0.09 m³) of lumber per s.f. (sq.m) of contact area. Following is the itemized labor cost of 100 s.f. (9 sq.m) of continuous haunch forms:

	Hours	Rate	Total	Rate	Total
Carpenter	12.0	$....	$....	$26.44	$ 317.28
Labor helping	6.0	....	....	21.04	126.24
Labor stripping forms	2.5	....	....	21.04	52.60
Labor cost 100 s.f.			$....		$ 496.12
cost per s.f.				....	4.96
cost per sq.m				....	53.37

Form Coating

For clean, easy stripping of forms and to prolong the life of form facings, the contact surface of all forms should receive a coating, which will prevent concrete from bonding to the form face.

The old method of brushing or spraying form faces with paraffin oil is still used by some contractors and gives adequate results for the average job, such as foundation walls, unexposed columns, beams, and slabs. Forms should always be coated before erection, but where this is impractical, such as for built-in-place wall forms, pan, and joist forms for combination slabs, oiling should be done in advance of setting reinforcing steel and care should be taken not to get any oil on concrete, masonry, or steel bearing surfaces.

On large areas, such as flat slab decks, a laborer can coat 400 to 500 s.f. (36-45 sq.m) of surface per hour, but production is considerably less when coating forms for columns, beams, combination slabs, etc. A fair overall average for coating forms is about 300 to 350 s.f. (27-32 sq.m) per hour.

For better class work, when using plywood, plastic-coated plywood, or lined forms, where smooth concrete surfaces are required, there are numerous patented coatings and lacquers available for this purpose. While the initial cost of these coatings is much higher than paraffin oil, better results are obtained and the forms may be used a couple of times without recoating, depending on the care used in applying the coating and in handling the forms when erecting or stripping.

Prices for these coating materials vary from $10.00 to $25.00 per gal. ($2.64-6.60 per liter), but coverage and durability also vary. Some coatings require special thinners before application, while others can be thinned with relatively inexpensive mineral spirits.

Application may be either by brushing, spraying, or dipping. A worker can coat about 400 to 500 s.f. (36-45 sq.m) of form surface per hour.

Formfilm. Formfilm (W.R. Grace & Co.) is a quick-drying plastic type plywood form coating with a high nitro-cellulose content. It is completely water-repellent, resistant to abrasion, and impervious to the alkaline action of concrete. It is recommended for use on untreated plywood forms to produce smooth concrete surfaces free from grain markings. It is not intended for use on metal or plastic forms.

Formfilm must be applied to concrete forms that are free from oil, grease, dirt, and foreign matter. It can be applied by brush, spray, roller, squeegee, or dipping. Formfilm is not ready mixed, because job temperatures vary widely and affect the viscosity. It is usually thinned with *Formfilm Thinner* to the proper working consistency.

Concrete Column Forms

Reinforced concrete columns are square, rectangular, or round. The forms for square or rectangular columns are usually constructed of wood, while metal or fiber molds are used for round columns.

Forms for reinforced concrete columns should be estimated by the square foot (square meter), obtained by multiplying the girth of the column by the height. Example: Obtain the s.f. (sq.m) of forms in a column 18" (0.45 m) square and 12'-0" (3.6 m) high. Proceed as follows: 18 + 18 + 18 + 18 = 72" or 6'-0" (0.45 + 0.45 + 0.45 + 0.45 = 1.80 m), the girth. Multiply the girth by the height, 6'-0" x 12'-0" = 72 s.f. (1.8 m x 3.6 m = 6.48 sq.m) of forms.

The forms for all square or rectangular columns should be estimated in this manner.

To prevent the column forms from spreading while the concrete is placed and until it hardens, wood or metal clamps should be run around the column at intervals, depending upon the height of the column. In recent years wood clamps have been almost entirely eliminated by metal clamps.

The pressure of the fresh concrete has a greater effect on the design of column forms than on wall forms on account of the smaller amount of concrete to be placed. The contractor usually finds it more convenient and economical to fill a group of medium size columns in as short time as possible, perhaps 30 minutes, when the full liquid pressure of the fresh concrete will be acting, 150 lbs. per s.f. for every foot (755 kg/m² for every meter) in depth of the concrete.

Column Form Design

There are two principal methods of erecting column forms; one is to unite all the four sides on two or three horses near the site where the form is to be erected and have 4 to 6 laborers hoist the complete form over the column dowels and stand it up temporarily; or in case of heavier columns, a simple 4"x6" (100 x 150 mm) pole derrick may be used for erection. Sometimes three sides of the column form are united and shoved into place without lifting the heavy box over the dowels, while the other and decidedly more expensive method is to erect each column form side separately.

As an example of the lumber required for a column form, take a concrete column 20" (500 mm) square and 11'-0" (3.3 m) high underneath the lowest beam, and assuming a full liquid pressure of 150 lbs. per c.f. (2400 kg/m³) acting for the entire height of the column. Assume the clear story height as 12'-6" (3.75 m). The area of the column form as usually taken off by the estimator will be 4 sides x 1.667 x 12'-6" high equals 83.33 s.f. (4 x 0.50 m x 3.75 m = 7.5 sq.m).

Using steel column clamps the following lumber would be required for a 20"x20" (500 x 500 mm) concrete column form 12'-6" (3.75 m) high.

	BF	Cu.M
2 pcs. 3/4" (19mm) plywood, 1'-9-1/2"x8'-0"	32	0.06
2 pcs. 3/4" (19mm) plywood, 1'-9-1/2"x4'-6"	18	0.04
2 pcs. 3/4" (19mm) plywood, 1'-8"x8'-0"	32	0.06
2 pcs. 3/4" (19mm) plywood, 1'-8"x4'-6"	18	0.04
Total plywood required for 83.33 s.f. (7.74 sq.m) forms	100	0.20
20 pcs. 1x6x1'-9" cleats	18	0.04
Lumber required for 83.33 s.f. (7.74 sq.m) forms	118	0.24
Lumber required per s.f.	1.42	0.03
Bracing lumber per s.f. forms	0.58	
Total lumber required per s.f. forms	2.00	

In order to brace the column form and hold it plumb, it will require 4 pcs. 2"x4" (50 x 100 mm), 16'-0" to 20'-0" (4.8-6.0 m) long, which will add another 48 b.f. to the lumber required. This bracing lumber may be used at least 3 or

4 times, eventually for purposes other than column forms. Bracing lumber is included in the above table.

The number of steel column clamps required may be somewhat larger than usually recommended by the manufacturers, depending on the speed with which the column forms are filled. The spacing in the lower part of the column should be about 9" (225 mm) and about 1'-0" (300 mm) near the middle. The first three spacings from the top should be 1'-3" to 1'-8" (375-500 mm) apart.

An average of one clamp per foot of column height will be safe in this particular instance.

Because plywood is furnished in panels 4'-0" (1.2 m) wide and 8'-0" (2.4 m) long, there is considerable waste unless columns are of sizes that will cut from the plywood sheets without waste, which is unusual.

Another method of stiffening plywood column form panels is by using 2"x4" (50 x 100 mm) vertical cleats. This method permits greater spacing of clamps as shown in the table and the total lumber requirements for a 20" (500 mm) column using 3/4" (19 mm) plywood sheathing and 2"x4" (50 x 100 mm) vertical cleats would be as follows:

	BF	Cu.M
Plywood sheathing	100	0.20
Cleats	112	0.22
Bracing	48	0.09
Lumber required for 83.33 s.f. forms	260	0.51
Lumber per s.f. (sq.m) of forms	3.1	0.6

When round concrete columns are used, metal or fiber molds are used for supporting the wet concrete until it has set. Some columns consist of a straight shaft extending from floor to ceiling, while others have a cone head at the top of the column. The forms for round concrete columns should be estimated at a certain price per column, stating whether columns have plain shafts or cone heads, because it costs considerably more to erect the latter. There are several firms in the country who make a specialty of renting these column molds and will quote a lump sum price on the job.

Labor Cost of Wood Forms for Square Concrete Columns. The labor cost of wood forms for square or rectangular concrete columns averages about the same regardless of the type of building. Variations are due to the method used in assembling and building up the column section, their size, method of clamping, etc.

On jobs of any size it is customary to have a mill on the job, where the column forms are built to exact size, including cut-outs for beams or girders, and then sent to the floor on which they are to be used, ready for erection.

Here is an important item when figuring column form costs. How many times can they be used on the job without cutting down or extending their length, or must they be reworked for each floor? This is a matter each estimator or contractor must decide from a study of the plans. For this reason separate labor costs are given for making up the column forms and for erection.

Labor Cost of 100 S.F. (9 Sq.M) Concrete Column Forms Using Plywood Sheathing with Adjustable Steel Clamps and 2"x4" (50 x 100 mm) Vertical Cleats

Making Forms	Hours	Rate	Total	Rate	Total
Carpenter	4.0	$....	$....	$26.44	$ 105.76
Labor	2.0	....	....	21.04	42.08
Assembling, Erecting, Plumbing and Bracing					
Carpenter	3.0	....	....	26.44	79.32
Labor helping	3.0	....	....	21.04	63.12
Removing Clamps and Forms					
Labor	2.0	....	....	21.04	42.08
Cost per 100 s.f.			$....		$ 332.36
cost per s.f.				....	3.32
cost per sq.m				....	35.72

Metal Forms for Round Columns

The cost of metal forms for round concrete columns varies according to the diameter and height of the column, whether shaft is plain or with a conical head, and whether it is necessary to cut out for beams and girders framing into the metal column molds.

Where it is necessary to cut the metal molds or "heads" to frame into wood forms for concrete beams and girders, an extra labor allowance should be made.

When plain round metal forms up to 24" (600 mm) in diameter are used without caps, which do not require cutting and fitting at head for concrete beams and girders, the labor erecting, removing, and oiling one column mold should cost as follows:

	Hours	Rate	Total	Rate	Total
Carpenter	2	$....	$....	$26.44	$ 52.88
Laborer	2	....	....	21.04	42.08
Total per column			$....		$ 94.96

Add for scaffolding and/or crane service if required.

Roos Adjustable Metal Column Clamps
Hinged Bar Type—2 Hinged Units=1 Complete Clamp

Clamp Size	Bar Size	Weight Per Clamp	Net Concrete Sizes			
			With 1" Column Lumber		With 2" Column Lumber	
			Minimum	Maximum	Minimum	Maximum
36"	5/16" x 2 1/2" x 36"	40 lbs.	8 1/2" x 8 1/2"	26" x 26"	8 1/2" x 6 1/2"	24" x 24"
48"	3/8" x 2 1/2" x 48"	58 lbs.	12" x 12"	38" x 38"	10" x 10"	36" x 36"
60"	3/8" x 3" x 60"	85 lbs.	22 1/2" x 22 1/2"	48 1/2" x 48 1/2"	20 1/2" x 20 1/2"	46 1/2" x 46 1/2"
36" x 60"	1 Leg, 5/16" x 2 1/2" x 36" / 1 Leg, 3/8" x 3" x 60"	65 lbs.	7 1/2" (36" Leg) / 23 1/2" (Leg)	25" (36" Leg) / 49 1/2" (60" Leg)	5 1/2" (36" Leg) / 21 1/2" (60" Leg)	23" (36" Leg) / 47 1/2" (60" Leg)

Clamp Size	Bar Size	Weight Per Clamp	Net Concrete Sizes			
			With 1" Column Lumber		With 2" Column Lumber	
			Minimum	Maximum	Minimum	Maximum
900 mm	7.8 mm x 62.5 mm x 900 mm	18 Kg.	212 mm x 212 mm	650 mm x 650 mm	162 mm x 162 mm	600 mm x 600 mm
1 200mm	9.4 mm x 62.5 mm x 1 200 mm	26 Kg.	300 mm x 300 mm	950 mm x 950 mm	250 mm x 250 mm	900 mm x 900 mm
1 500mm	9.4 mm x 75 mm x 1 500 mm	39 Kg.	562 mm x 562 mm	1 212 mm x 1 212 mm	512 mm x 512 mm	1 162 mm x 1 162 mm
900 mm x 1 500 MM	1 leg, 7.8 mm x 62.5 mm x 900 mm / 1 leg, 9.4 mm x 75 mm x 1 500 mm	23 Kg.	187 mm (900mm leg) / 587 mm (1 500 mm leg)	625 mm (900 mm leg) / 1 237 mm (1 500 mm)	135 mm (900 mm leg) / 1 237 mm (537 leg)	137 mm (900 mm leg) / 1 187 mm (1 500 mm leg)

If the column has a conical head, necessitating placing a metal cap in addition to the column shaft, the labor erecting, removing, and oiling one column should cost as follows:

	Hours	Rate	Total	Rate	Total
Carpenter	2.5	$....	$....	$26.44	$ 66.10
Laborer	2.5	....	....	21.04	52.60
Total per column		$....			$ 118.70

Add for scaffolding and/or crane service if required.

If necessary to cut and fit at column heads for concrete beams and girders, add 1 to 2 hrs. time per column.

The above labor costs are for columns up to 16'-0" (4.8 m) in length. For columns 16'-0" to 20'-0" (4.8-6.0 m) long, add 1 to 2 hrs. time.

Leasing or Renting Metal Column Molds. Where metal column molds are leased or rented by the manufacturer to the contractor to be used on a specified job and returned in good condition upon completion of the work, the following prices will prove a fair average for columns up to 24" (600 mm) in diameter.

For a plain metal mold for round concrete columns without cone head, the monthly rental should figure about $9.00 per l.f. (29.50 per m) of column form to be rented.

For plain round metal forms for concrete columns having cone head, add $55.00 per column per month rental.

To the above prices add delivery charges to the project site.

C.F. (Cu.M) of Concrete and S.F. (Sq.M) of Forms Required Per L.F. (M) of Height for Round and Square Concrete Columns

Round Columns

Diameter, inches	Area, sq. in.	C.F. of concrete	S.F. of forms	Diameter, mm	Area, sq.m	Cu.M of concrete
12	113	0.79	3.14	300	0.07	0.02
13	133	0.92	3.40	325	0.09	0.03
14	154	1.07	3.67	350	0.10	0.03
15	177	1.23	3.93	375	0.11	0.03
16	201	1.40	4.19	400	0.13	0.04
17	227	1.58	4.45	425	0.15	0.04
18	254	1.77	4.71	450	0.16	0.05
19	284	1.97	4.97	475	0.18	0.06
20	314	2.18	5.24	500	0.20	0.06
21	346	2.41	5.50	525	0.22	0.07
22	380	2.64	5.76	550	0.25	0.07
23	415	2.89	6.02	575	0.27	0.08
24	452	3.14	6.28	600	0.29	0.09
25	491	3.41	6.54	625	0.32	0.10

Diameter, inches	Area, sq. in.	C.F. of concrete	S.F. of forms	Diameter, mm	Area, sq.m	Cu.M of concrete
26	531	3.69	6.81	650	0.34	0.10
27	573	3.98	7.07	675	0.37	0.11
28	616	4.28	7.33	700	0.40	0.12
29	661	4.59	7.59	725	0.43	0.13
30	707	4.91	7.85	750	0.45	0.14
31	755	5.24	8.12	775	0.49	0.15
32	804	5.59	8.38	800	0.52	0.16
33	855	5.94	8.64	825	0.55	0.17
34	908	6.31	8.90	850	0.59	0.18
35	962	6.68	9.16	875	0.62	0.19
36	1018	7.07	9.42	900	0.66	0.20

Square Columns

Inches Square	Area sq.in.	C.F. of Concrete	S.F. of Forms	MM Square	Area Sq.M	Cu.M of Concrete	Sq.M of Forms
12	144	1.00	4.00	300	0.09	0.03	0.37
13	169	1.17	4.33	325	0.11	0.03	0.40
14	196	1.36	4.67	350	0.13	0.04	0.43
15	225	1.56	5.00	375	0.15	0.04	0.46
16	256	1.78	5.33	400	0.17	0.05	0.50
17	289	2.01	5.67	425	0.19	0.06	0.53
18	324	2.25	6.00	450	0.21	0.06	0.56
19	361	2.51	6.33	475	0.23	0.07	0.59
20	400	2.78	6.67	500	0.26	0.08	0.62
21	441	3.06	7.00	525	0.28	0.09	0.65
22	484	3.36	7.33	550	0.31	0.10	0.68
23	529	3.67	7.67	575	0.34	0.10	0.71
24	576	4.00	8.00	600	0.37	0.11	0.74
25	625	4.34	8.33	625	0.40	0.12	0.77
26	676	4.69	8.67	650	0.44	0.13	0.81
27	729	5.06	9.00	675	0.47	0.14	0.84
28	784	5.44	9.33	700	0.51	0.15	0.87
29	841	5.84	9.67	725	0.54	0.17	0.90
30	900	6.25	10.00	750	0.58	0.18	0.93
31	961	6.67	10.33	775	0.62	0.19	0.96
32	1024	7.11	10.67	800	0.66	0.20	0.99
33	1089	7.56	11.00	825	0.70	0.21	1.02
34	1156	8.03	11.33	850	0.75	0.23	1.05
35	1225	8.51	11.67	875	0.79	0.24	1.08
36	1296	9.00	12.00	900	0.84	0.25	1.11

C.F. (Cu.M) of Concrete in Flared Caps For Round Concrete Columns

Diam. of Col. Cap at Top ft (mm)	Diam. of Col. Cap at Bottom in (mm)	CF (Cu.M) Conc. Including Column Diam.	CF (Cu.M) Conc. Excluding Column Diam.
4'-0" (1200)	12" (300)	9.75 (0.28)	8.5 (0.24)
4'-0" (1200)	14" (350)	9.70 (0.27)	8.0 (0.23)
4'-0" (1200)	16" (400)	9.60 (0.27)	7.6 (0.22)
4'-0" (1200)	18" (450)	9.40 (0.27)	7.0 (0.20)
4'-0" (1200)	20" (500)	9.30 (0.26)	6.5 (0.18)
4'-0" (1200)	22" (550)	9.00 (0.25)	5.8 (0.16)
4'-0" (1200)	24" (600)	8.80 (0.25)	5.3 (0.15)
4'-0" (1200)	26" (650)	8.70 (0.25)	4.9 (0.14)
4'-6" (1350)	16" (400)	13.6 (0.39)	11.2 (0.32)
4'-6" (1350)	18" (450)	13.5 (0.38)	10.6 (0.30)
4'-6" (1350)	20" (500)	13.3 (0.38)	10.0 (0.28)
4'-6" (1350)	22" (550)	13.1 (0.37)	9.3 (0.26)
4'-6" (1350)	24" (600)	12.9 (0.37)	8.6 (0.24)
4'-6" (1350)	26" (650)	12.6 (0.36)	7.8 (0.22)
4'-6" (1350)	28" (700)	12.3 (0.35)	7.1 (0.20)
4'-6" (1350)	30" (750)	11.9 (0.34)	6.4 (0.18)
5'-0" (1500)	20" (500)	18.2 (0.52)	14.3 (0.40)
5'-0" (1500)	22" (550)	18.0 (0.51)	13.4 (0.38)
5'-0" (1500)	24" (600)	17.8 (0.50)	12.6 (0.36)
5'-0" (1500)	26" (650)	17.4 (0.49)	11.7 (0.33)
5'-0" (1500)	28" (700)	17.1 (0.48)	10.9 (0.31)
5'-0" (1500)	30" (750)	16.8 (0.48)	10.0 (0.28)
5'-0" (1500)	32" (800)	16.4 (0.46)	9.1 (0.26)
5'-0" (1500)	34" (850)	15.8 (0.45)	8.2 (0.23)
5'-0" (1500)	36" (900)	15.3 (0.43)	7.3 (0.21)

Sonotube Fibre Forms. Sonotube fibre forms (Sonoco Products Co., Hartsville, SC) are spirally wound, laminated fibre tubes that were developed to provide a fast and economical method of forming for round concrete columns, piers, and the like. They are particularly adaptable as forms for encasing wood or steel piling. They provide a fast, economical method of forming all types of round concrete columns. They permit new design possibilities for architects, provide new structural features for engineers, and save time, labor, and money.

Sonotube fibre forms are available in standard sizes from 4" to 48" (100-1200 mm) inside diameter, with wall thickness ranging from 0.200" to 0.500" (5-12 mm). They are available in lengths up to 48 ft. (14.4 m) and can be easily cut to length on the job with ordinary hand or power saw.

There are three types of Sonotube fibre forms to meet virtually all job requirements. The seamless is the premium form, with specially finished inner ply that produces a smoother, continuous concrete surface. Regular A-coated is the standard form for exposed columns. For exposed columns not exceeding 12 ft. (3.6 m) in height, the light wall A-coated is recommended. Sonotube fibre forms are lightweight, easy to handle, and require few workers for erection, manually or with simple block and tackle. Minimum bracing is required, fixed in place at bottom, with light lumber bracing to keep them erect and plumb. No clamping is necessary.

Under normal conditions for average size columns, 12" to 18" (300-450 mm) diameter, 10 to 14 ft. (3-4.2 m) in length, a carpenter and a helper should handle, cut to length, erect, and brace about 12 Sonotube column forms per 8-hr. day at the following labor cost per column:

	Hours	Rate	Total	Rate	Total
Carpenter	0.7	$....	$....	$26.44	$ 18.51
Helper	0.7	....	....	21.04	14.73
Cost per column			$....		$ 33.24

Sonotube fibre forms strip easiest and most economically from 2 to 10 days after concrete is placed. Stripping may be accomplished with the least effort by using an electric hand saw, with blade set to cut slightly less than the wall thickness of Sonotube. Make two or three evenly spaced vertical cuts from bottom to top, after which the form can be pulled free from column. Sonotube forms are for one time use only.

Under normal conditions, a carpenter and a helper should strip 25 to 30 average size Sonotube column forms per 8-hr. day at the following labor cost per column:

	Hours	Rate	Total	Rate	Total
Carpenter	0.3	$....	$....	$26.44	$ 7.93
Helper	0.3	....	....	21.04	6.31
Cost per column			$....		$ 14.24

Approximate Lineal Foot Prices of Sonotube Fibre Forms with "A" Coating

Inside Dia. x Wall Thickness, in.	Inside Dia. x Wall Thickness, mm	Price per L.F.	Price per Meter	Approx. Wgt. lbs./ft.	Approx. Wgt. kg/m
8" x 0.200"	200 x 5 mm	$1.89	$6.18	1.7	2.53
10" x 0.225"	250 x 5.6 mm	2.57	8.41	2.3	3.42
12" x 0.225"	300 x 5.6 mm	3.31	10.82	2.8	4.17
14" x 0.250"	350 x 6.2 mm	4.15	13.56	3.6	5.36
16" x 0.300"	400 x 7.5 mm	5.57	18.20	4.8	7.14
18" x 0.300"	450 x 7.5 mm	6.41	20.95	5.4	8.04
20" x 0.375"	500 x 9.3 mm	8.19	26.78	7.5	11.16
22" x 0.375"	550 x 9.3 mm	9.19	30.04	8.2	12.20
24" x 0.375"	600 x 9.3 mm	10.50	34.34	8.9	13.24

Inside Dia. x Wall Thickness, in.	Inside Dia. x Wall Thickness, mm	Price per L.F.	Price per Meter	Approx. Wgt. lbs./ft.	Approx. Wgt. kg/m
26" x 0.375"	650 x 9.3 mm	12.18	39.83	9.6	14.29
28" x 0.375"	700 x 9.3 mm	13.49	44.12	10.4	15.48
30" x 0.400"	750 x 10 mm	15.12	49.44	11.8	17.56
36" x 0.400"	900 x 10 mm	19.74	64.55	14.5	21.58
42" x 0.500"	1050 x 12.5 mm	36.75	120.18	18.6	27.68
48" x 0.500"	1200 x 12.5 mm	47.25	154.51	23.5	34.97

Sonotube forms are manufactured in plants across the U.S., and distributors maintain stocks in most principal cities. The above prices are f.o.b. the nearest shipping point.

Anchoring Brick, Stone, and Terra Cotta to Concrete Backing

There are a number of methods of anchoring brick, stone, and terra cotta to concrete, some of which consist of metal slots nailed to the column or beam forms, which form a slot in the concrete for placing anchors.

The dovetail anchor slot method of anchoring masonry to concrete consists of a metal slot dovetail in cross section and anchors with dovetail ends for cut stone, brickwork, and terra cotta.

The dovetail anchor slot is nailed to the form with the open side against the wood. The ends are then closed with a wood plug. Or they come with foam filled slots. After the concrete is poured and the forms removed, the slot remains in the concrete, available at any course height to receive the anchors.

The anchors are installed by merely inserting the dovetail end of the anchor edgewise in the slot, then turning the anchor crosswise of the slot with the hand, the other end of the anchor being imbedded in the facing material or mortar joint. The dovetail slot holds the end of anchor securely, making any other means of fastening, drilling or boring unnecessary.

Forms for Concrete Beams, Girders, Spandrel Beams, and Lintels

Forms for reinforced concrete beams, girders, lintels, etc., should be estimated by the square foot (square meter), obtained by adding the dimensions of the three sides of the beam and multiplying by the length. For example, a beam or girder may be marked 12"x24" (300 x 600 mm) on the plans, but maybe 6" (150 mm) of this depth is included in the slab thickness, so the form area should be obtained by taking the dimensions of the beam or girder from the underside of the slab: 18" + 12" + 18" = 48" or 4'-0" (450 + 300 + 450 mm = 1200 mm = 1.2 m) the girth, which multiplied by the length gives the number of s.f. of forms in the beam, 4'-0" x 18'-0" = 72 s.f. (1.2 x 5.4 = 6.48 sq.m).

Spandrel beams, or those projecting above or below the slab, should be measured in the same manner. A good rule to remember when measuring beam forms is to take the area of all forms that come in contact with the concrete.

When preparing an estimate on formwork, beams and girders should always be estimated separately; that is, the columns should be taken off, then the beams, girders and lintels, followed by the floor slabs, stairs.

As an example of the lumber required for beam and girder forms, assume an inside beam 12" (300 mm) wide by 1'-6" (0.45 m) deep by 19'-0" (5.7 m) long.

The contact area of this beam would be 2 sides 1'-6" (0.45 m) deep and a bottom 1'-0" (0.3 m) wide by 19'-0" (5.7 m) long, 1'-6" + 1'-6" + 1'-0" = 4' (0.45 + 0.45 + 0.30 = 1.20 m) girth multiplied by 19'-0" (5.7 m) long, equals 76 s.f. (6.84 sq.m) of forms requiring the following lumber:

Beam soffit, 1 pc. 5/8" plyform 12"x19'-0" s.f. 19
Beam sides, 2 pcs. 5/8" plyform 22"x19'-0" s.f. 70
Beam Bottom Joists 4 pcs. 3"x4"x10'-0" 40
Beam Side Template 4 pcs. 2"x4"x10'-0" 27
Beam Side Ledger 4 pcs. 2"x4"x10'-0" 27
Beam Side Studs 40 pcs. 2"x4"x1'-4" 36
Shores 6 pcs. 4"x4"x10'-0" 80
Shores Cross Head (T) 6 pcs. 4"x4"x3'-0" 24
Braces 12 pcs. 1"x4"x2'-6" 10
Lumber required for 76 s.f. of forms 333
Lumber required per s.f. of "contact area" 4.4

(Metric)

	Cu.M
Beam soffit, 1 pc. 15.6 mm plyform 300 mm x 1.7 sq.m	0.44
Beam sides, 2 pcs. 15.6 mm plyform 550 mm x 1.7 sq.m	1.62
Beam Bottom Joists 4 pcs. 75 mm x 100 mm x 3 m	0.93
Beam Side Template 4 pcs. 75 mm x 100 mm x 3 m	0.63
Beam Side Ledger 4 pcs. 75 mm x 100 mm x 3 m	0.63
Beam Side Studs 40 pcs. 75 mm x 100 mm x 400 mm	0.84
Shores 6 pcs. 100 mm x 100 mm x 3 m	1.86
Shores Cross Head (T) 6 pcs. 100 mm x 100 mm x 900 mm	0.56
Braces 12 pcs. 25 mm x 100 mm x 750 mm	0.23
Lumber required for 7 sq.m of forms	7.73
Lumber required per sq.m of contact area	0.10

When beams or girders are more than 2'-0" (0.6 m) deep, the designs given under "Concrete Wall Forms" should be consulted, which gives spacing and size of cleats or studs required, together with bracing and form ties necessary.

Method of Framing Wood Forms for Concrete Beams

Labor Cost of Wood Forms for Concrete Beams and Girders

The labor cost on beam and girder forms will vary considerably on different jobs, due to the amount of duplication, size of beams and girders, and the methods of framing and placing them.

On typical buildings where the same size beams and girders are used on several floors, the forms can be used several times with only a small amount of labor with each reuse. Larger beams are more economical than very small ones, but one of the most important cost factors is the method of fabricating the units and whether they are built up in a job "shop" using power saws. These are factors that cut costs.

The following labor costs are based on efficient management, economical design, and labor-saving tools and equipment. If the old handsaw methods are to be used, increase the costs given 10 to 20%.

It requires 4 to 4-1/2 b.f. (0.009-0.01 m³) of lumber to complete 1 s.f. (0.09 sq.m) of beam and girder forms, which includes uprights or "shores", beam soffits, bracing, etc.

A carpenter should frame and erect 250 to 275 b.f. (0.58-0.64 m³) of lumber per 8-hr. day, at the following labor cost per 1,000 b.f. (2.3 m³):

	Hours	Rate	Total	Rate	Total
Carpenter	29.0	$....	$....	$26.44	$ 766.76
Helper	14.5	....	....	21.04	305.08
Cost per 1,000 b.f. (2.3 m3)					$ 1,071.84

If laborers are not permitted to carry the lumber and forms for the carpenters, the "helper" time given above should be figured as carpenter time.

A laborer should remove or "strip" 900 to 1,000 b.f. (2.1-2.3 m³) of lumber per 8-hr. day, at the following labor cost per 1,000 b.f. (2.3 m³):

	Hours	Rate	Total	Rate	Total
Labor	9	$....	$....	$21.04	$ 189.36

Labor Cost of 100 S.F. (9 Sq.M) of Inside Beam and Girder Forms Requiring 4-1/2 B.F. per S.F. (0.1 m³ per sq.m) of Forms, 4"x4" (100 x 100 mm) Wood Shores

Making Beam Soffits and Sides	Hours	Rate	Total	Rate	Total
Carpenter	3.50	$....	$....	$26.44	$ 92.54
Labor helping	1.75	....	....	21.04	36.82
Assembling, Erecting, Shoring and Bracing					
Carpenter	5.50	....	....	26.44	$ 145.42
Labor helping	2.75	....	....	21.04	57.86

Removing Forms	Hours	Rate	Total	Rate	Total
Labor	2.75	....	....	21.04	57.86
Cost per 100 s.f.			$....		$ 332.64
per s.f.				....	3.33
per sq.m				....	35.83

Labor Cost of 100 S.F. (9 Sq.M) of Spandrel Beam or Lintel Forms Requiring 4 B.F. per S.F. (0.09 m³ per sq.m) of Forms, 4"x4" (100 x 100 mm) Wood Shores

Making Beam Soffits and Sides	Hours	Rate	Total	Rate	Total
Carpenter	3.50	$....	$....	$26.44	$ 92.54
Labor helping	1.75	....	....	21.04	36.82
Assembling, Erecting, Shoring and Bracing					
Carpenter	8.00	....	....	26.44	211.52
Labor helping	4.00	....	....	21.04	84.16
Removing Forms					
Labor	3.50	....	....	21.04	73.64
Cost per 100 s.f.			$....		$ 498.68
per s.f.				....	4.99
per sq.m				....	53.69

If the same forms can be reused without working over and cutting down, "Labor Making Forms" after first use may be reduced or omitted.

Forms for Upturned Concrete Beams. When the top of a beam is higher than the floor slab level, the beam side forms for that portion above the floor must be supported on temporary legs, which are removed after the concrete is poured and is still wet. It is also necessary to use spreaders to maintain the proper beam width.

While it is true that the additional form material required is negligible, the labor cost on this kind of beam side form will run 50 to 100% more than ordinary beam side forms.

Forms for Reinforced Concrete Floors

Forms for all types of reinforced concrete floors should be estimated by the square foot (square meter or millimeter), taking the actual floor area or the area of the wood or metal forms that come in contact with the concrete.

Estimating the Cost of Form Lumber. When computing the lumber cost per s.f. of forms, the estimator should study the plans carefully to determine just how many times it is possible to use the same form lumber in the construction of the building.

On 8- to 12-story buildings, it may be possible to use the same lumber 4 or 5 times, while on a 3-story building, it might be necessary to buy enough lumber to form almost the entire building. These are items that affect the material costs and can only be determined by the estimator or contractor, who should be in a position to know just how soon the forms can be stripped and how fast they will be required to carry on the work.

On many jobs it will be more economical to use high early strength cement instead of ordinary portland cement, which will enable you to strip forms in 3 days instead of 7 to 10 days, saving on forms needed. Keep in mind that form lumber that has been used 3 or 4 times will have little salvage value.

Types of Reinforced Concrete Floors. There are several types of reinforced concrete floor construction that require wood forms, centering, and shores. There are differences in cost. We will describe some of the more commonly used types in detail.

Flat Slab Construction. Mill, factory, warehouse, and industrial buildings designed for heavy floor loads are usually of flat slab construction, which consists of floors without beams or girders, except spandrel beams between columns on outside walls and around areas such as stair wells and elevator shafts. These buildings usually have square or rectangular columns.

Beam and Girder Type Construction. This is the oldest type of concrete structure, having square or round columns with beams and girders running between each row of columns. The floor slab is carried by the beams and girders and then by the columns.

Pan Construction with Concrete Joists. Apartment buildings, hotels, schools, hospitals, department stores, office buildings, and other structures requiring light floor loads ordinarily use this type of construction, which consists of metal pans in conjunction with concrete joists. The floor is used with round or square columns and beams running at right angles to joists.

Designing Forms for Reinforced Concrete Floors and Estimating Quantity of Lumber Required

The process of designing forms for reinforced concrete floors may be divided into several parts, such as determining the span of the sheathing, the size and spacing of joists and girders or stringers, and the size and spacing of the upright supports or shores.

The spacing of joists for concrete floor forms is governed by the strength of the plywood used. When plywood panels are used for sheathing, the spacing of the joists must coincide with the sizes of the plywood panels, 4'-0"x8'-0" (1.2 x 2.4 m), which means that the joists should be spaced 12", 16", or 19" (300, 400, or 475 mm) on centers, depending on the load to be carried.

Weight of Concrete Floors of Various Thicknesses

In determining the weight per s.f. of floor, wet concrete is figured at 150 lbs. per c.f. (2400 kg/m³). Dead load of form lumber and live load on forms while concrete is being placed is figured at 38 lbs. per s.f. (190 kg/m²), which is sufficient for temporary construction.

Floor Thickness, inches

	2	3	4	5	6	7	8	9	10	11	12
Wgt. per s.f., lbs.	25	37.5	50	62.5	75	87.5	100	112.5	125	137.5	150
Temp. Live Load per s.f., lbs	40	37.5	38	37.5	38	37.5	38	37.5	38	37.5	38
Total Weight per s.f., lbs.	65	75	88	100	113	125	138	150	163	175	188

(metric)

Floor Thickness, mm

	50	75	100	125	150	175	200	225	250	275	300
Wgt. per sq.m, kg	122.1	183.1	244.1	305.2	366.2	427.2	488.2	549.3	610.3	671.3	732.4
Temp. Live Load per sq.m, kg	195.3	183.1	185.5	183.1	185.5	183.1	185.5	183.1	185.5	183.1	185.5
Total Weight per sq.m., lbs.	317.4	366.2	429.7	488.2	551.7	610.3	673.8	732.4	795.8	854.4	917.9

Table Giving Spacing of Stringers and Shores When Supporting Floor Forms for Metal Pan and Concrete Joist Construction

TABLE FOR DESIGNING SLAB FORMS FOR REINFORCED CONCRETE FLOORS BASED ON USING ¾" SHEATHING, WITH JOISTS SPACED 2'-0" ON CENTERS

Thickness of Concrete Floor Slab and Total Dead and Live Load per Sq. Ft.

Thickness Concrete Slab	2"	3"	4"	5"	6"	7"	8"	9"	12"
Total Load per Sq. Ft., Lbs	65	75	88	100	113	125	138	150	188

Distance Between Stringers When Joists are Spaced 2'-0" on Centers

	2"	3"	4"	5"	6"	7"	8"	9"	12"
2"x4" Joists, S4S	6'-0"	5'-6"	5'-6"	5'-0"	5'-0"	4'-6"	4'-6"	4'-0"	4'-0"
2"x6" Joists, S4S	7'-6"	7'-0"	7'-0"	6'-6"	6'-6"	6'-0"	6'-0"	6'-0"	5'-6"
2"x8" Joists, S4S	9'-6"	9'-0"	8'-6"	8'-6"	8'-0"	8'-0"	7'-6"	7'-6"	7'-0"
2"x10" Joists, S4S	11'-0"	10'-6"	10'-0"	10'-0"	9'-6"	9'-6"	9'-0"	9'-0"	8'-6"

Distance Between Shores or Supports Using Stringers of Size and Spacing as Given Below

	2"	3"	4"	5"	6"	7"	8"	9"	12"
4"x4" S4S Stringers, 5-ft. c.c.	5'-0"	4'-6"	4'-0"	4'-0"	4'-0"	3'-6"	3'-6"	....	....
4"x6" S4S Stringers, 6-ft. c.c.	6'-6"	6'-0"	6'-0"	6'-0"	6'-0"	5'-6"	5'-0"	5'-0"	4'-6"
3"x8" S4S Stringers, 6-ft. c.c.	8'-6"	7'-6"	7'-0"	7'-0"	7'-0"	6'-6"	6'-0"	6'-0"	5'-6"
2"x10" S4S Stringers, 7-ft. c.c.	7'-6"	6'-0"	5'-6"	4'-6"	4'-6"	4'-0"	....	....	....

The distance between shores or span of stringers is generally an even fraction of their length. Stringers 14'-0" long will permit a spacing of 4'-8" or 7'-0", while 18'-0" stringers will permit spans of 3'-7", 4'-6", 6'-0" and 9'-0".

Based on using yellow pine, Douglas fir or woods of equal strength.

Using 2" (50 mm) plank to support the concrete joists 2'-1" (0.63 m) or 2'-11" (0.9 m) o.c., with stringers spaced 4'-0" (1.2 m) apart to support the 2" (50 mm) plank.

The table gives the sizes of stringers and the distance between supports or shores when supporting various floor loads.

Size of Stringer	Thickness of Floor in Inches................. Weight of Floor per Sq. Ft.*	6+2½ 86 lbs.	8+2½ 92 lbs.	10+2½ 98 lbs.	12+2½ 105 lbs.	14+2½ 112 lbs.
		Distance Between Shores or Supports				
4"x 4"S4S		5'-0"	4'-9"	4'-8"	4'-8"	4'-6"
3"x 6"S4S		6'-6"	6'-5"	6'-3"	6'-1"	6'-0"
4"x 6"S4S		7'-0"	7'-0"	6'-8"	6'-7"	6'-6"
3"x 8"S4S		8'-0"	8'-0"	7'-6"	7'-6"	7'-6"
2"x10"S4S		8'-0"	8'-0"	7'-6"	7'-6"	7'-6"

The above tables are based on using yellow pine, Douglas fir, or woods of equal strength.

Carrying Capacity of 4"x4" (100 x 100 mm) Uprights or Shores. The carrying capacity of a given size upright or shore is generally limited by two principal factors. The first is compression that the shore exerts on the cross beam or stringer at right angles to the fibers . The stress per sq. in. (sq. mm) on the bearing area of the stringer on the shore should not greatly exceed 500 lbs. per sq. in. (3.45 MPa) for Yellow Pine, Western Fir, or similar lumber. Otherwise, there is a noticeable impression, often as much as 1/8" (3 mm), of the shore into the fibers of the stringer, especially when the lumber is green or water soaked after a long rain. A 4"x4" (100 x 100 mm) S4S, 3-1/2"x3-1/2" (85 x 85 mm) upright should never be loaded to more than 6,000 lbs. (2700 kg), no matter how short it is, and a 4"x4" rough shore should not be loaded to more than 8,000 lbs. (3625 kg).

The other principal limitation of strength is due to the length of the shore and the degree of crookedness in its length. Every experienced contractor knows that shores often show bows of 1" to 2" (25-50 mm), and if not cut off exactly square to their length, the carrying capacity of most shores are greatly limited by the eccentric loading impressed on them by the uneven bearing of the stringer and the bow in the shores.

TABLE FOR DESIGNING SLAB FORMS FOR REINFORCED CONCRETE FLOORS
BASED ON USING 18.75 MM SHEATHING, WITH JOISTS SPACED 600 MM ON CENTERS
Thickness of Concrete Floor Slab and total Dead and Live Load per Square Meter

Thickness Concrete Slab	50 mm	75 mm	100 mm	125 mm	150 mm	175 mm	200 mm	225 mm	300 mm
Total Load per sq.Ft., Kg.	65	75	88	100	113	125	138	150	166
Distance Between Stringers When Joists are Spaced 600 mm on Centers									
50 mm x 100 mm Joists S4S	1.6 M	1.7 M	1.7 M	1.5 M	1.5 M	1.4 M	0.0 M	1.2 M	1.2 M
50 mm x 150 mm Joists S4S	2.3 M	2.1 M	2.1 M	2.0 M	2.0 M	1.8 M	1.8 M	1.8 M	1.7 M
50 mm x 200 mm Joists S4S	2.9 M	2.7 M	2.6 M	2.8 M	2.4 M	2.4 M	2.3 M	2.3 M	2.1 M
50 mm x 250 mm Joists S4S	3.4 M	3.2 M	3.0 M	3.0 M	2.9 M	2.9 M	2.7 M	2.7 M	2.6 M
Distance Between Shores or Supports Using Stringers of Size and Spacing as Given									
100 mm x 100 mm S4S — Stringers 1.5 Meters c.c.	1.5 M	1.4 M	1.2 M	1.2 M	1.2 M	1.1 M	1.1 M		
100 mm x 150 mm S4S — Stringers 1.8 Meters c.c.	2.0 M	1.8 M	1.8 M	1.8 M	1.8 M	1.7 M	1.5 M	1.5 M	1.4 M
75 mm x 200 mm S4S — Stringers 1.8 Meters c.c.	2.6 M	2.3 M	2.1 M	2.1 M	2.1 M	2.0 M	1.8 M	1.8 M	1.7 M
50 mm x 250 mm S4S — Stringers 2.1 Meters c.c.	2.3 M	1.8 M	1.7 M	1.4 M	1.4 M	1.2 M			

The distance between shores or span of stringers is generally an even fraction of their length. Stringers 4.2 meters long will permit a spacing of 1.4 meters or 2.1 meters, while 5.4 meters stringers will permit spans of 1.1 meters, 1.3 meters, 1.8 meters and 2.7 meters.

Based on using yellow pine, Douglas Fir or wood of equal strength.

The following table gives the permissible load on shores for various lengths and eccentricities:

Safe Load on Uprights or Shores of Various Lengths
(in Feet, Inches, and Lbs.)

Length of Upright or Shore Eccentricity	4"x4" S4S Yellow Pine or Fir 8'-0"	10'-0"	4"x4" Rough Yellow Pine or Fir 8'-0"	10'-0"	4"x4" Rough Wisconsin Hemlock 8'-0"	10'-0
(Bow+Top) 0"	6,000	6,000	8,000	8,000	8,000	8,000
(Bow+Top) 1"	5,700	4,800	8,000	7,100	7,300	5,600
(Bow+Top) 1-1/2"	4,700	3,900	7,300	5,800	6,000	5,100
(Bow+Top) 2"	4,200	3,400	6,700	5,500	5,200	4,400
(Bow+Top) 2-1/2"	3,600	3,000	5,300	4,500	4,800	4,100
(Bow+Top) 3"	3,400	2,800	5,100	4,300	4,500	3,800

(in Metric Dimensions and Weights)

Length of Upright or Shore Eccentricity	100x100mm S4S Yellow Pine or Fir 2.4 m	3.0 m	100x100mm Rough Yellow Pine or Fir 2.4 m	3.0 m	100x100mm Rough Wisconsin Hemlock 2.4 m	3.0 m
(Bow+Top) 0 mm	2722	2722	3629	3629	3629	3629
(Bow+Top) 25	2585	2177	3629	3221	3311	2540
(Bow+Top) 37.5	2132	1769	3311	2631	2722	2313
(Bow+Top) 50	1905	1542	3039	2495	2359	1996
(Bow+Top) 62.5	1633	1361	2404	2041	2177	1860
(Bow+Top) 75	1542	1270	2313	1950	2041	1724

Lintels as a rule produce quite large eccentricities on the shores due to the load from the adjoining slab, which comes through on the inside of the lintel forms onto the shore. Never allow a load larger than those given for 3" (75 mm) eccentricity for shores supporting lintels. One frequently sees lintels supported by shores spaced less than 2'-0" (0.6 m) on centers, which is a clear waste of money.

Where the height of the uprights or shores are longer than given in the above table, it is advisable to cross brace the shores, in which case the vertical distance between braces plus 40% may be taken as the post height. Example: You are using a 16'-0" (4.8 m) shore, with 2 rows of cross bracing at 6'-0" (1.8 m) and 12'-0" (3.6 m) above the floor. The distance between braces 6'-0" plus 40%, or 2.40 ft. (0.7 m), makes the capacity the same as a shore 8.40 or practically 8'-6" (2.5 m) long. Thus an 8'-0" (2.4 m) shore good for 8,000 lbs. (3625 kg) would still be good for that weight when 16'-0" long, if cross braced by two rows of bracing 6'-0" (1.8 m) and 12'-0" (3.6 m) above the floor.

*Method of Constructing Forms for Beam and Girder Type
Solid Concrete Floor Slabs*

Estimating the Quantity of Lumber Required for Forms for Beam and Girder Type Solid Concrete Floor Slabs. The accompanying illustrations give detailed designs, from which the lumber required per s.f. (sq.m) of forms may be computed. The illustration shows a 4" (100 mm) concrete floor slab with a 7'-6" (2.25 m) clear span between beams. Using a joist spacing of 2'-0" (600 mm) as given in the table for joist spacing, 2"x8" (50 x 200 mm) joists will carry the required load.

For a floor area 2'-0"x7'-6" (0.6 x 2.25 m) containing 15 s.f. (1.35 sq.m) the following lumber will be required:

	b.f.	metric
Sheathing, 2'-0" (0.6 m) x 7'-6" (2.25 m) =		
15 s.f. (1.35 sq.m) plus 20% waste .... 18 s.f.		1.620 m²
Joist, 1 pc. 2"x8" (50x200mm), 8' (2.4m), plus 10% for joints 12		0.028 m³
Ledger boards, 2 pcs. 1"x4" (25x100mm), 2' (.6m) 1		0.002
Total lumber required for 15 s.f. (1.35 sq.m) forms 31		1.65
B.f. of lumber required per s.f. of forms 2.1		
Cu.m of lumber required per sq.m of forms		1.22

For a 6" (150 mm) solid concrete slab with a clear span of 11'-0" (3.3 m) between girders, usually it is not economical to support joists with this span directly on the girders. It is better to provide a stringer or girt halfway between the concrete beams so that the joists have a span of only 5'-6" (1.65 m). Using a joist spacing of 2'-0" (600 mm), a 2"x6" (50 x 150 mm) joist is strong enough, as shown by the table of joist spacings. A 4"x6" (100 x 150 mm) supported by shores 6'-0" (1.8 m) o.c. will suffice.

A floor area 2'-0"x11'-0" (0.6 x 3.3 m) containing 22 s.f. (2 sq.m) will require the following lumber:

	b.f	metric
Sheathing, 2'-0" (.6m) x 11'-0" (3.3m) =		
22 s.f. (1.98 sq.m) plus 20% waste .... 27 s.f.		2.380 m2
Joists, 1 pc. 2"x6" (50x150mm), 12'(3.6m), plus 10% for joints ... 13		0.030 m3
Stringer, 1 pc. 4"x6" (100x150mm), 2' (.6m) 4		0.009 m3
Shore, 1/3 of a 4"x4" (100x100mm), 10' (3m) S4S,		
plus sill, wedges, 18/3 = ... 6		0.014 m³
Total lumber required for 22 s.f. (1.98 sq.m) forms 50		0.014 m3

B.f. of lumber required per s.f. of forms 2.25
Cu.m of lumber required per sq.m of forms.................................0.027
If shores require bracing, add 0.1 b.f per s.f. of slab.

Assume a solid concrete floor slab 16'-0" (4.8 m) wide, 19'-0" (5.7 m) long, and 12" (300 mm) thick, containing 304 s.f. (27.36 sq.m) of slab forms. Joists will be 2"x6" (50 x 100 mm) spaced 24" (600 mm) apart, supported by 3 lines of 3"x8"-16'-0" (75 x 200 mm - 4.8 m) stringers, which in turn are supported by 4"x4" (100 x 100 mm) shores 5'-4" (1.6 m) apart.

Method of Constructing Forms for Beam and Girder Type
Solid Concrete Floors

The following lumber will be required for 304 s.f. (27.5 sq.m) of forms:

	BF
Sheathing, 16'-0" x 19'-0" = 304 s.f. plus 20% waste	365
Joists, 9 pcs. 2"x6"-14'-0"	126

Joists, 9 pcs. 2"x6"-7'-0" 63
Stringers, 3 pcs. 3"x8"-16'-0" 96
Shores, 9 pcs. 4"x4"-10'-0", including sills and wedges 153
Braces, 6 pcs. 1"x6"-16'-0" 48
Joist support at girders, 2 pcs. 1"x6"-16'-0" 16
  Total lumber required for 304 s.f. forms 867
  B.f. lumber required per s.f. of forms 2.85

(metric)

	m3
Sheathing, 4.8m x 5.7m = 27.36 sq.m plus 20% waste	0.73
Joists, 9 pcs. 50 x 150 mm - 4.2 m	0.25
Joists, 9 pcs. 50 x 150 mm - 2.1 m	0.13
Stringers, 3 pcs. 75 x 200 mm - 4.8 m	0.19
Shores, 9 pcs. 100 x 100 mm - 4.8 m, including sills and wedges	0.31
Braces, 6 pcs. 25 x 150 mm - 4.8 m	0.10
Joist support at girders, 2 pcs. 25 x 150 mm - 4.8 m	0.03
Total lumber required for 28.2 sq.m of forms	0.56
Cu.m of lumber required per sq.m of forms	0.02

The above quantities are based on the assumption that the ends of joists and stringers are supported by the beam and girder forms. If a wall bearing job without beams and girders, the following additional lumber is required:

	BF
Stringers, 2 pcs. 2"x8"-16'-0" at wall	43
Shores, 6 pcs. 4"x4"-10'-0", including sills and wedges	102
Braces, 2 pcs. 1"x6"-16'-0"	16
Additional lumber required for 304 s.f. forms	161
Additional lumber, b.f. required per s.f.	0.53

(metric)

Lumber required for 28.2 Sq.M of Forms	m3
Stringers, 2 pcs. 50 x 200 mm - 4.8m	0.09
Shores, 6 pcs. 100 x 100 mm - 3.0 m, including sills and wedges	0.02
Braces, 2 pcs. 25 x 150 mm - 4.8 m	0.03
Additional lumber required for 28.2 sq.m of forms	0.14
Additional lumber, cu.m required per sq.m	0.005

Estimating the Quantity of Lumber Required for Flat Slab Concrete Floor Forms. Assume a flat slab having a 20'-0" (6 m) span, with 7'-0" (2.1 m) square drop heads, 8" (200 mm) thick as shown in illustration.

The maximum span of the joists is 5'-9" (1.7 m). The table of joist spacing shows that 2"x6" (50 x 150 mm) joists and 3"x8" (75 x 200 mm) stringers are sufficient. The load on the shores of the center stringer will be 5'-9" (1.7 m) x 6'-0" (1.8 m) = 34.5 s.f. (3.06 sq.m) at 138 lbs. per s.f. (695 kg/m²) = 4,761 lbs. Good practice requires a 4"x4" (100 x 100 mm) shore rough, though a 4"x4" (100 x 100 mm) S4S with bracing will do.

Method of Constructing Forms for Flat Slab Concrete Floors

The lumber required for a 20'-0"x20'-0" (6 x 6 m) panel is as follows:

	BF
Sheathing, 400 s.f. plus 20% waste	480
Using 16'-0" long floor sheathing boards requires 9 joists (a double joist being placed at the joint) or for 20' it will require 11-1/4 pcs 2"x6"-12'-0" joists (1-1/4 x 9)	135
Column Strips, 9 joists 2"x6"-10'-0"	90
Stringers, 3 pcs. 3"x8"-20'-0"	120
Stringers, 1 pc. 3"x8"-12'-0"	24
Shores, 13 pcs. 4"x4"-10'-0" plus sills, wedges, etc. @ 17'	221
Horizontal bracing, 7 pcs. 1"x6"-20'-0"	70
X Bracing, not less than 1/3 horizontal bracing	23
Total lumber required for 400 s.f. forms	1,163

For Drop Heads

Joists, 5 pcs. 2"x6"-14'-0"... 70
Stringers, 2 pcs. 2"x8"-8'-0".. 21
Shores, 4 pcs. 4"x4" at 17 b.f. 68
Total lumber required including drop heads 1,322
B.f. of lumber required per s.f. of forms 3.31

(metric)

Lumber required for 37 Sq.M   m3
Sheathing, 37 sq.m plus 20% waste 0.27
Using 4.8 m long floor sheathing boards will require 9
joists (a double joist being placed at the joint) or for
6 m it will require 11.25 pcs
50 x 150 mm - 3.6 m (31 x 225 mm)..................................... 0.27
Column Strips, 9 joists 50 x 150 mm - 3 m.......................... 0.18
Stringers, 3 pcs. 75 x 200 mm - 6 m..................................... 0.24
Stringers, 1 pc. 75 x 200 mm - 3.6 m.................................... 0.05
Shores, 13 pcs. 100 x 100 mm - 3 m plus sills,
wedges, etc. @ 5.1 m 0.44
Horizontal bracing, 7 pcs. 25 x 150 mm - 6 m....................... 0.14
X Bracing, not less than 1/3 horizontal bracing 0.05
Total lumber required for 37 sq.m of forms.......................... 1.64
For Drop Heads
Joists, 5 pcs. 50 x 150 mm - 4.2 m 0.14
Stringers, 2 pcs. 50 x 200 mm - 2.4 m 0.04
Shores, 4 pcs. 100 x 100 mm at .034 cu. m 0.14
Total lumber required including drop heads.......................... 1.95
cu.m of lumber required per sq.m of forms........................... 0.053

A 10'-0" (3 m) shore is sufficient for a 12'-0" (3.6 m) story height. For a 14'-0" (4.2 m) story height, add 2'-0" (0.6 m) to each of 17 shores or 34 x 1.33 ft. (0.4 m) = 45 b.f. (13.6 cu.m), which adds 0.11 b.f. (0.0002 cu.m) per s.f. of floor. A 16'-0" (4.8 m) story height will require 0.23 b.f. (0.0004 cu.m) per s.f. of floor.

For an 18'-0" (5.4 m) story height, it will be necessary to add for another set of cross bracing in addition to the 0.33 b.f. of lumber required for shores. Another set of cross bracing will add 93 b.f. of lumber for 400 s.f. (37 sq.m) of floor or 0.23 b.f. (0.0004 cu.m) additional, which with the 0.33 ft. required for shores, will increase the lumber required per s.f. to 3.31 + 0.33 + 0.23 = 3.87 b.f. per s.f. of forms.

A 20'-0" (6 m) story height will add another 0.11 ft. (0.03 m) for extra length of shores, or a total of 3.98 b.f. (0.008 cu.m) per s.f. of forms.

Where plywood is used for floor sheathing, there is at least a 10% waste in flooring, and while its use saves 480 b.f. (0.96 cu.m) for sheathing, it is necessary to use more joists, because the joists should not be spaced over 19" (475 mm) o.c., and the joists must be doubled every 8'-0" (2.4 m) instead of 16'-0" (4.8 m) where 16'-0" (4.8 m) long sheathing boards are used.

In 400 s.f. (37 sq.m) of forms, the following additional lumber would be required:

	BF	Cu.M
Extra joists, 4 pcs. 2"x6" (50x150mm), 12' (3.6m) 48		0.09
Extra joists, 2 pcs. 2"x6" (50x150mm), 10' (3m)............................20		0.04
Total extra lumber required for 400 s.f. (37 sq.m) forms............68		0.13
Extra lumber required per s.f. (sq.m) of forms..............................0.17		0.0035

If adjustable shores are used instead of 4"x4" (100 x 100 mm) shores, stringers must be supported every 4'-0" (1.2 m) in order to reduce the load to not much more than 3,000 lbs. (1360 kg) per shore although some adjustable shores will carry up to 6,000 lbs. This will require 17 to 23 adjustable shores instead of 17 pcs. of 4"x4" (100 x 100 mm) shores, 247 b.f. (0.49 cu.m), and 9 pcs. of horizontal bracing, 1"x6"-20'-0" (25 x 150 mm - 6m), 90 b.f. (.49 cu.m), instead of 7 pcs. of 1"x6"-20'-0" (25 x 150 mm - 6 m), 70 b.f. (.14 cu.m), and the X bracing will also be increased by 7 b.f. (0.01 cu.m).

Using adjustable shores, 4"x4" (100 x 100 mm) stringers can be used instead of 3"x8" (75 x 200 mm) stringers, which will result in the following saving:

	BF	Cu.M
Shores, 17 pcs. 4"x4"-10' (100x100mm-3m) 247		1.7
Stringers, 4"x4" (100x100mm) instead of		
3"x8" (75x200mm), saving.. 44		0.08
Total saving using adjustable shores...291		0.58
Extra 1"x6" (25x150mm) required for bracing 27		0.05
Extra sills, 6 pcs. 2"x10"-2' (50x250mmx0.6m)" 20		0.04
....47		0.09
Total saving on 400 s.f. (37 sq.m) forms....................................244		0.49
Total saving per s.f. (sq.m) of forms b.f. (cu.m)..........................0.61		0.01

On the other hand, it will be necessary to handle 17 to 23 shores weighing 1,000 to 1,400 lbs. while the 224-ft. savings weighs only 725 lbs.

For a 12" thick flat slab having a 20'-0" span, the lumber required can be approximated as follows:

	BF	Cu.M
Sheathing, per s.f. forms..1.20		0.008
Joists, same spacing but using 2"x8" (50x200mm) joists..............0.75		0.005
Stringers, same as before...0.33		0.002
Shores, spacing will be 4'-6" (1350mm), requiring 21 shores 0.90		0.006
Bracing, 9 pcs. 1"x6" (25x150mm), 20' (6m)................................0.23		0.001
X bracing, 1/3 of horizontal bracing..0.08		0.001
Drop heads, same as before 0.25		0.001

Total lumber, b.f. per s.f. (cu.m/sq.m)
  for 12'-6" (3.8 m) story height..................................3.74   0.008
Each 2' story height, more or less, add or deduct per s.f................0.14

Weight of Combination Metal Pan and Concrete Joist Floors

Concrete joists 4", 5", and 6" (100, 125, and 150 mm) wide, with a 2-1/2" (62 mm) concrete slab on top, using metal pans of various depths and widths.

Conc. Thickness	2-1/2"	2-1/2"	2-1/2"
Conc. Joist Width	4"	5"	6"
Using Metal Pans	Weight per S.F. of Slab, Lbs.		
6" deep, 20" wide	45	48	50
6" deep, 30" wide	41	43	45
8" deep, 20" wide	51	54	57
8" deep, 30" wide	45	48	50
10" deep, 20" wide	56	60	64
10" deep, 30" wide	50	52	55
12" deep, 20" wide	63	67	72
12" deep, 30" wide	54	57	61
14" deep, 20" wide	--	75	80
14" deep, 30" wide	--	62	66

(metric)

Conc. Thickness	62.5 mm	62.5 mm	62.5 mm
Conc. Joist Width	100 mm	125 mm	150 mm
Using Metal Pans	Weight per Sq.M of Slab, Kg		
150 mm deep, 500 mm wide	220	234	244
150 mm deep, 750 mm wide	200	210	220
200 mm deep, 500 mm wide	249	264	278
200 mm deep, 750 mm wide	220	234	244
250 mm deep, 500 mm wide	273	293	312
250 mm deep, 750 mm wide	244	254	269
300 mm deep, 500 mm wide	308	327	352
300 mm deep, 750 mm wide	264	278	298
350 mm deep, 500 mm wide	--	366	391
350 mm deep, 750 mm wide	--	303	322

For each 1/2" (13 mm) variation in slab thickness, add or subtract 6.25 lbs./s.f. (30.5 kg/m²) of floor.

Add 38 lbs./s.f. (186 kg/m²) to the above weights to take care of temporary dead and live load on floors while concrete is being poured.

Lumber Required for Floor Forms Using Metal Pans and Concrete Joist Construction. Assume a slab of 19'-0" (5.7 m) clear span and 16'-0" (4.8 m) long, using 8" (200 mm) metal pans with 5" (125 mm) wide ribs spaced 2'-1" (625 mm) on centers and a 2-1/2" (63 mm) concrete top over the entire slab.

*Method of Constructing Forms for Floor of Combination Metal
Pans and Concrete Joist Construction*

According to the above table this construction weighs 52 lbs./s.f. (254 kg/m²) and including a 38-lb. (186-kg) live load during construction, makes a total load of 90 lbs./s.f. (440 kg/m²).

It is customary to use 2"x8" (50 x 200 mm) S4S planks to support the concrete ribs or joists. If the planks are continuous without patching at the end spans, a span of 4'-0" (1.2 m) will support a load of 265 lbs./l.f. (400 kg/m) with a deflection of .235" (5.8 mm). In this instance the load is 2.09 x 90 lbs. per s.f. = 188 lbs. per l.f., which reduces the deflection to 0.166" or about 1/6", which is permissible.

It will require a stringer or girt every 4'-0" (1.2 m) to support the planks and the load on the stringers will be 360 lbs./l.f. (540 kg/m).

A 2"x10"-16'-0" (50 x 250 mm - 4.8 m) S4S will easily carry this uniform load per l.f. when supported by 4"x4" (100 x 100 mm) shores spaced 8'-0" (2.4 m) apart.

The following lumber will be required for 304 s.f. (27.5 sq.m) of forms:

	BF
Planks, 9 pcs. 2"x8"-20'-0"	240
Stringers, 4 pcs. 2"x10"-16'-0"	107
Shores, 8 pcs. 4"x4"-10'-0", including sills and wedges	136
Joist supports at girders, 2 pcs. 1"x6"-16'-0"	16
Bracing, 4 pcs. 1"x6"-16'-0"	32
Bracing, 2 pcs. 1"x6"-20'-0"	20
Total lumber required for 304 s.f. forms	551
Lumber required per s.f. of forms, b.f.	1.8

(metric)

	Cu. M
Planks, 9 pcs. 50 x 200 mm - 6 m	0.48
Stringers, 4 pcs. 25 x 250 mm - 4.8 m	0.20
Shores, 8 pcs. 100 x 100 mm - 3 m, including sills & wedges	0.27
Joist supports at girders, 2 pcs. 25 x 100 mm - 4.8 m	0.03
Bracing, 4 pcs. 25 x 150 mm - 4.8 m	0.06
Bracing, 2 pcs. 25 x 150 mm - 6 m	0.04
Total lumber required for 27.5 sq.m of forms	1.09
Lumber required per sq.m of forms, cu.m	0.04

The above quantities are based on the assumption that the ends of the joists and stringers are supported by the beam and girder forms. If a wall bearing job without beams or girders, requiring additional stringers and shores at walls, add the following:

	BF
Stringers, 2 pcs. 2"x10"-16'-0"	54
Shores, 4 pcs. 4"x4"-10'-0", including sills and wedges	68
Bracing, 2 pcs. 1"x6"-16'-0"	16
Additional lumber required for 304 s.f. forms	138
Additional lumber required per s.f. of forms	0.45

Lumber Required for Slab Forms. The following tables give the approximate quantity of lumber required per s.f. (sq.m) of forms for floors of different thickness and with varying ceiling heights. These tables are only approximate. Every building is different and the quantities will vary with the span, ceiling height, slab thickness, and weight of the wet concrete to be supported.

Also, the following tables are based on using 4"x4" (100 x 100 mm) shores or uprights. If adjustable shores are used it will be necessary to deduct the lumber required for 4"x4" (100 x 100 mm) and add cost or rental of adjustable shores. Add cost of additional bracing as more shores and bracing are frequently required.

The difference in the quantity of lumber required per s.f. (sq.m) for floors having 10'-0" (3 m, 3000 mm) ceiling heights and those having 20'-0" (6 m, 6000 mm) ceiling heights is principally in the shoring and bracing. If adjustable shores are used, deductions can be made on the following basis: floors 3" to 6" (75-150 mm) thick with 10'-0" (3 m) ceiling heights require 1/3 to 3/8 ft. (112 mm) of lumber per s.f. (sq.m) for shores; 7" to 9" (175-225 mm) slabs require 1/2 to 5/8 ft. (150-188 mm); and 10" to 12" (250-300 mm) slabs require 5/8 to 3/4 ft. (188-228 mm) of shoring lumber per s.f. (sq.m) of floor.

The above quantities can be used for all 10'-0" (3 m) story heights, plus the additional lumber given in the tables for higher ceiling heights.

The forms are designed in yellow pine, Douglas fir, or woods of equal strength, with the loads computed as follows: wet concrete 150 lbs./c.f. (2403 kg/m³), dead load of form lumber, concrete, non-motorized buggies, and workers is an additional 50 lbs./c.f. (800 kg/m²).

Number of Feet of Lumber Required for One Square Foot
of Flat Slab Concrete Floor Forms

	Thickness of Slab in Inches			
Ceiling	6"	8"	10"	12"
Height	Weight Dead and Live Load, psf			
in Ft.	125 lbs.	150 lbs.	175 lbs.	200 lbs.
10	3.0	3.1	3.4	3.6
12	3.1	3.2	3.5	3.7
14	3.3	3.4	3.6	3.9
16	3.4	3.5	3.9	4.0
18	3.4	3.8	4.2	4.4
20	3.5	3.9	4.3	4.6

Note: Dead Load = 150 psf; Live Load = 50 psf

Number of Meters of Lumber Required for One Square Meter
of Flat Slab Concrete Floor Forms

	Thickness of Slab in mm			
Ceiling	150	200	250	300
Height	Weight Dead and Live Load, kg/sq.m			
in meters	610	732	854	977
3.0	9.8	10.2	11.2	11.8
3.7	10.2	10.5	11.5	12.1
4.3	10.8	11.2	11.8	12.8
4.9	11.2	11.5	12.8	13.1
5.5	11.2	12.5	13.8	14.4
6.1	11.5	12.8	14.1	15.1

Note: Dead Load = 150 psf; Live Load = 50 psf

The above quantities include lumber required for forming drop heads. If adjustable shores are used, deduct lumber quantities given on previous pages.

Using Steel Truss Joists Instead of Shoring for Supporting Formwork for Reinforced Concrete Floors. In steel frame buildings where clear floor space is essential, steel truss joists may be used for supporting the wood formwork, as shown in the accompanying illustration.

This is a simple method that is very effective as the salvage value of the truss joists is very high.

Sectional Steel Shoring. Many contractors now use their sectional steel scaffolding units for shoring formwork for reinforced concrete slabs and beams. Prefabricated sectional steel scaffolding sections are as simple to erect for shoring as they are for scaffolding. The same scaffolding components—base plates, adjustable extension legs, welded 5-ft. (1.5 m) wide frames varying in height from 3'-0" to 6'-6" (0.9-1.95 m) and diagonal braces to provide spacing between frames from 2'-6" to 7'-0" (0.75-2.10 m)—are assembled quickly to provide free standing shoring sections suited to the easy placing of stringers, joists, and decking. In most instances, sectional steel shoring can be used wherever conventional shoring methods of adjustable shores or 4"x4" (100 x 100 mm) lumber are used, but is especially adaptable for shoring forms for large expanses of flat slab construction.

Height adjustments are quickly made by the use of 20" (500 mm) adjustable legs, which eliminates the need for cutting, fitting, and wedging. These legs also facilitate stripping operations by the ease with which the shoring units may be lowered.   Sectional steel shoring also provides scaffolding within the shoring for forming and stripping, increasing job safety. This method also minimizes the use of wood, reducing the fire hazard and increasing the number of reuses obtainable from form lumber.

Safe working loads for sectional steel shoring depend on the spacing between frames in the same row, distance between rows of frames, and position of the stringers on the frames. Sectional steel shoring will support the largest loads when the stringers are placed directly over the frame legs by means of inverted base plates or U-heads as shown in the illustration. If this is not practical, stringers should be placed as close as possible to frame legs. If required, additional lateral rigidity may be attained by interlacing frames with standard bracing.

It is available in three types: Standard, 20K Heavy Duty for loads up to 10,000 lbs. (4535 kg) on each frame leg, and Extra Heavy Duty for loads up to 40,000 lbs. (18140 kg) on each leg.

The contractor who does not own this equipment may rent sectional steel shoring, engineered for a specific job, from the manufacturer on the same basis as scaffolding. This service is available nationwide.

Since job conditions and requirements vary considerably, it is practically impossible to give unit labor production values on this method of shoring, but various contractors who have used this method report saving up to 30% on the labor and material handling costs over conventional methods. With shoring labor, including bracing, ranging from 15% to 25% of the total form cost, this results in overall savings of 4-1/2% to 7-1/2% on the entire form cost.

Adjustable Shores for Reinforced Concrete Floors

There are a number of adjustable shores on the market to support reinforced concrete beam and slab forms. These shores are a combination of wood and metal, or all metal, which can be raised or lowered within certain limits.

Adjustable Shores. Adjustable shores in a complete unit, as illustrated are made in 3 sizes: 8 ft. (2.4 m) adjustable to 14 ft. (4.2 m); 7 ft. (2.1 m) adjustable to 13 ft. (3.9 m); 5 ft. (1.5 m) adjustable to 9 ft. (2.7 m).

They may be purchased for about $40.00 for the smaller size and $50.00 for the larger sizes. Extension type shores are also available in two sizes: 8 ft. (2.4 m) adjustable to 14 ft. (4.2 m), at about $45.00 each; and 6 ft. (1.8 m) adjustable to 10 ft. 6 in. (3.15 m), at about $43.00 each. They can be easily extended by inserting a 4"x4" (100 x 100 mm) of any length in the top of the shore.

Rental stocks are maintained in most principal cities and may be rented at about $2.50 per month, per section, f.o.b. warehouse and return.

Each shore will support about 3,000 lbs. (1360 kg) with an adequate factor of safety. In some cases, however, shores are spaced to carry a load of not over 2,200 lbs. (1000 kg), as some forming authorities consider it cheaper to use more shores than to use heavier lumber to prevent excessive deflection between supports. For local prices and availability, the estimator should get a firm quote from the local supplier.

Adjustable Shores Versus 4"x4"s (100 x 100 mm). It is much easier to shore up a floor using adjustable shores instead of 4"x4"s (100 x 100 mm), which require wedging and cutting or adding to for each change in story height. This additional labor all costs money at present wage scales. However, to offset this, the original cost of the adjustable shores is much more than the 4"x4"s (100 x 100 mm), but if they can be used often<%-6> enough they will eventually show a saving. This will have to be considered when contemplating a purchase or rental basis.

Another thing that should be considered is the fact that a 4"x4" (100 x 100 mm) shore will usually carry a larger load than an adjustable shore, so in many cases it will require more adjustable shores than 4"x4"s (100 x 100 mm) to carry the same load. However, on some jobs, especially on light constructed floors, the adjustable shores may carry all the load required. This is an item that should be considered by the estimator before pricing the job.

Shoring Frames from Symons are Used to Support Plywood Decking
For a Multi-story Concrete Structure (Courtesy Symons Corp.)

The form costs on the following pages are computed on a rental basis, based on 10 days for each shore or 3 uses a month.

Labor Cost of Flat Slab Forms. Forms for flat slab floors are continuous—without breaks or offsets—except at column heads, stair wells, elevator shafts, etc., where it is necessary to frame for beams and girders.

The s.f. (sq.m) labor cost will vary with the amount of lumber required per s.f. (sq.m) of forms, as the heavier the slab and the higher the ceiling, the more lumber required per s.f. (sq.m).

Forms for exceedingly light floor slabs cost proportionately more per s.f. (sq.m) than heavy forms, because the preliminary work is the same with both types. However, on work of this class, a carpenter should frame and erect 375 to 425 b.f. (8.75-9.9 cu.m) of lumber per 8-hr. day, at the following labor cost per 1,000 b.f. (2 cu.m):

	Hours	Rate	Total	Rate	Total
Carpenter	20	$....	$....	$26.44	$ 528.80
Labor	10	....	....	21.04	210.40
Cost per 1,000 BF (2 cu. m)		....	$....		$ 739.20

When removing or "stripping" forms, a laborer should remove 1,000 to 1,200 BF (2-2.4 cu.m) of lumber per 8-hr. day, at the following labor cost per 1,000 b.f. (2 cu.m):

	Hours	Rate	Total	Rate	Total
Labor	7	$....	$....	$21.04	$ 147.28

Labor Framing for Slab Depressions at Column Heads. On flat slab jobs there may be a depression at each column head 5'-0" to 7'-0" (1.5-2.1 m) square and 4" to 8" (100-200 mm) deep. The labor cost framing for these depressions will cost about 80% more than for straight slab forms.

A satisfactory method of estimating this additional labor is to figure the entire floor area at the regular price and then compute the area of the depressions and figure them at 80% additional labor.

Example: Floor contains 10,000 s.f. (900 sq.m) of slab forms, including 40 column head depressions 5'-0"x5'-0" (1.5 x 1.5 m), containing 25 s.f. (2.25 sq.m) each.

10,000 s.f. (900 m²) slab forms $2.60/s.f.(28.88/m²) $26,000
Add for 40 depressions (5'-0"x5'-0") 1,000 s.f. (90 m²) at 80%
    more than plain slab forms or $2.08/s.f. (23.11/m²) 2,080

This makes a total of $4.68/s.f. (52.00/m²) for labor framing depressions at column heads.

Labor Cost of 100 S.F. (9 Sq.M) of Flat Slab Concrete Floor Forms Requiring
3 BF of Lumber per S.F. of Forms

4"x4" (100 x 100 mm) Shores

	Hours	Rate	Total	Rate	Total
Carpenter	7.00	$....	$....	$26.44	$ 185.08
Labor	3.50	....	....	21.04	73.64
Labor removing forms	2.25	....	....	21.04	47.34
Cost per 100 s.f. (9 sq.m)			$....		$ 306.06
cost per s.f.			....		3.06
cost per sq.m			....		32.93

Adjustable Shores

	Hours	Rate	Total	Rate	Total
Carpenter	6.00	$....	$....	$26.44	$ 158.64
Labor	3.00	....	....	21.04	63.12
Labor removing forms	2.25	....	....	21.04	47.34
Cost per 100 s.f. (9 sq.m)			....		$ 269.10
cost per s.f.			....		2.69
cost per sq.m			....		28.96

For story heights over 16'-0" (4.8 m), add 25% labor costs. Add labor for framing drop heads.

*Method of Shoring for Reinforced Concrete Construction
Using Section Steel Scaffolding Components*

Labor Cost of 100 S.F. (9 Sq.M) of Flat Slab Forms, 4 B.F. of Lumber Per S.F.

4"x4" (100 x 100 mm) Shores

	Hours	Rate	Total	Rate	Total
Carpenter	9.0	$....	$....	$26.44	$ 237.96
Labor	4.5	....	....	21.04	94.68
Labor removing forms	3.0	....	....	21.04	63.12
Cost per 100 s.f. (9 sq.m)			$....		$ 395.76
cost per s.f.					3.96
cost per sq.m			....		42.61

Adjustable Shores

	Hours	Rate	Total	Rate	Total
Carpenter	8.0	$....	$....	$26.44	$ 211.52
Labor	4.0	....	....	21.04	84.16
Labor removing forms	3.0	....	....	21.04	63.12
Cost per 100 s.f. (9 sq.m)			$....		$ 358.80
cost per s.f.					3.59
cost per sq.m			....		38.63

For story heights over 16'0" (4.8 m), add 25% to labor costs. Add labor for framing drop heads.

Forms for Beam and Girder Type Solid Concrete Floors

The labor cost of wood forms for beam and girder type floors will run somewhat higher than for flat slab floors because of the shorter spans, additional framing around beams, girders, etc.

On forms of this class, a carpenter should frame and erect 325 to 375 BF (0.65-0.75 cu.m) of lumber per 8-hr. day, at the following labor cost per 1,000 b.f. (2.3 cu.m):

	Hours	Rate	Total	Rate	Total
Carpenter	23.0	$....	$....	$26.44	$ 608.12
Labor	11.5	....	....	21.04	241.96
Cost per 1,000 b.f. (2.3 cu.m)			$....		$ 850.08

An experienced laborer should remove or "strip" 1,000 to 1,200 b.f. (2.3-2.4 cu.m) of lumber, per 8-hr. day, at the following labor cost per 1,000 b.f. (2.3 cu.m):

	Hours	Rate	Total	Rate	Total
Labor	7	$....	$....	$21.04	$ 147.28

Number of Feet of Lumber Required for One Square Foot of Beam and Girder Type Solid Concrete Floor Forms

Ceiling Height in Feet	3"	4"	5"	6"	7"	8"	10"	12"
	88 lbs	100 lbs	113 lbs	125 lbs	138 lbs	150 lbs	175 lbs	200 lbs
10	2.0	2.1	2.4	2.5	2.6	2.6	2.7	2.7
12	2.1	2.2	2.4	2.5	2.6	2.6	2.7	2.8
14	2.1	2.2	2.4	2.5	2.6	2.6	2.7	2.8
16	2.2	2.3	2.5	2.6	2.7	2.7	2.8	2.9
18	2.4	2.5	2.6	2.7	2.8	2.8	2.9	3.0
20	2.5	2.6	2.7	2.8	2.9	2.9	3.0	3.1

(Thickness of Slab in Inches; Weight Dead and Live Load, psf)

Note: Dead Load = 150 psf; Live Load = 50 psf

Number of Meters of Lumber Required for One Square Meter of Beam and Girder Type Solid Concrete Floor Forms

Ceiling Height, meters	75	100	125	150	175	200	250	300
	427	488	549	610	671	732	854	977
3.0	6.6	6.9	7.9	8.2	8.5	8.9	8.9	8.9
3.7	6.9	7.2	7.9	8.2	8.5	8.9	8.9	9.2
4.3	6.9	7.2	7.9	8.2	8.5	8.9	8.9	9.2
4.9	7.2	7.5	8.2	8.5	8.9	9.2	9.2	9.5
5.5	7.9	8.2	8.5	8.9	9.2	9.5	9.5	9.8
6.1	8.2	8.5	8.9	9.2	9.5	9.8	9.8	10.2

(Thickness of Slab in mm; Weight Dead and Live Load, kq/sq.m)

Labor Cost of 100 S.F. (9 Sq.M) of Slab Forms Between Beam and Girder Forms Requiring 2 Ft. of Lumber Per S.F. of Floor

4"x4" (100 x 100 mm) Shores

	Hours	Rate	Total	Rate	Total
Carpenter	5.0	$....	$....	$26.44	$ 132.20
Labor	2.5	....	....	21.04	52.60
Labor removing forms	1.5	....	....	21.04	31.56
Cost per 100 s.f. (9 sq.m)			$....		$ 216.36
per s.f.				....	2.16
per sq.m				....	23.24

Adjustable Shores

	Hours	Rate	Total	Rate	Total
Carpenter	4.6	$....	$....	$26.44	$ 121.62
Labor	2.3	....	....	21.04	48.39
Labor removing forms	1.5	....	....	21.04	31.56
Cost per 100 s.f. (9 sq.m)			$....		$ 201.57

per s.f.	....			2.02
per sq.m	....			21.74

Labor Cost of 100 S.F. (9 Sq.M) of Slab Forms Between Beam and Girder Forms Requiring 2-1/2 Ft. of Lumber Per S.F. of Floor

4"x4" (100 x 100 mm) Shores

	Hours	Rate	Total	Rate	Total
Carpenter	6.30	$....	$....	$26.44	$ 166.57
Labor	3.15	....	....	21.04	66.28
Labor removing forms	1.90	....	....	21.04	39.98
Cost per 100 s.f. (9 sq.m)			$....		$ 272.83
per s.f.			....		2.72
per sq.m			....		29.27

Adjustable Shores

	Hours	Rate	Total	Rate	Total
Carpenter	5.70	$....	$....	$26.44	$ 150.71
Labor	2.85	....	....	21.04	59.96
Labor removing forms	1.90	....	....	21.04	39.98
Cost per 100 s.f. (9 sq.m)			$....		$ 250.65
per s.f.			....		2.51
per sq.m			....		27.01

Labor Cost of 100 S.F. (9 Sq.M) of Slab Forms Between Beam and Girder Forms Requiring 3 Ft. of Lumber Per S.F. of Floor

4"x4" (100 x 100 mm) Shores

	Hours	Rate	Total	Rate	Total
Carpenter	7.6	$....	$....	$26.44	$ 200.94
Labor	3.8	....	....	21.04	79.95
Labor removing forms	2.2	....	....	21.04	46.29
Cost per 100 s.f. (9 sq.m)			$....		$ 327.18
per s.f.			....		3.27
per sq.m			....		35.19

Adjustable Shores

	Hours	Rate	Total	Rate	Total
Carpenter	6.8	$....	$....	$26.44	$ 179.79
Labor	3.4	....	....	21.04	71.54
Labor removing forms	2.2	....	....	21.04	46.29
Cost per 100 s.f. (9 sq.m)			$....		$ 297.62
per s.f.			....		2.98
per sq.m			....		32.06

For story heights over 16'-0" (4.8 m), add 25% to labor costs.

Forms for Floors of Metal Pan and Concrete Joist Construction

Wood forms for floors of metal pan and concrete joist construction may be of much lighter construction than solid concrete floors, as the dead load is less. Metal pans are usually furnished 20" to 30" (500-750 mm) wide, with 2"x6" (50 x 150 mm) or 2"x8" (50 x 200 mm) plank spaced 24" to 37" (600-925 mm) on centers, depending upon the width of the joists, which vary from 4" to 7" (100-175 mm). Fiberglass pans, sometimes used in place of steel pans, are usually furnished 41" to 52" (1025-1300 mm) wide. The open deck construction is ordinarily used owing to the wide spacing of the floor boards.

Method of Constructing Wood Slab Forms for Floors of Metal Pans and Concrete Joist Construction

In buildings having one-way beams and long floor spans, a carpenter should frame and erect 350 to 400 b.f. (0.7-0.8 cu.m) of lumber per 8-hr. day, at the following labor cost per 1,000 b.f. (2.3 cu.m):

	Hours	Rate	Total	Rate	Total
Carpenter	21.4	$....	$....	$26.44	$ 565.81
Labor	10.7	....	....	21.04	225.13
Cost per 1,000 b.f. (2.3 cu.m)		$....			$ 790.94

Floors having beams and girders running in both directions with the floor panels averaging 16'-0"x16'-0" (4.8 x 4.8 m) and smaller, a carpenter should frame and erect 325 to 375 BF (0.65-0.75 cu.m) of lumber per 8-hr. day, at the following labor cost per 1,000 b.f. (2.3 cu.m):

	Hours	Rate	Total	Rate	Total
Carpenter	22.8	$....	$....	$26.44	$ 602.83
Labor	11.4	....	....	21.04	239.86
Cost per 1,000 b.f. (2.3 cu.m)			$....		$ 842.69

An experienced laborer should remove or strip 1,000 to 1,200 b.f. (2.3-2.4 cu.m) of lumber per 8-hr. day, at the following labor cost per 1,000 b.f. (2.3 cu.m):

	Hours	Rate	Total	Rate	Total
Labor	7	$....	$....	$21.04	$ 147.28

Number of Feet of Lumber Required for One Sq. Ft. of Floor Forms for Metal Pan and Concrete Joist Construction

Ceiling Height in Feet	Thickness of Floor Slab in Inches				
	8-1/2" 85 lbs.	10-1/2" 90 lbs.	12-1/2" 100 lbs.	14-1/2" 105 lbs.	16-1/2" 115 lbs.
10	1.7	1.7	1.7	1.8	1.8
12	1.7	1.7	1.8	1.9	1.9
14	1.7	1.7	1.8	1.9	2.0
16	1.8	1.8	1.9	2.0	2.1
18	2.1	2.1	2.2	2.3	2.4
20	2.2	2.2	2.3	2.4	2.5

The table assume that ends of joists and stringers are supported by the beam and girder forms.

Number of Meters of Lumber Required for One Sq. Meter of Floor Forms for Metal Pan and Concrete Joist Construction

Ceiling Height, meters	Thickness of Floor Slab in mm				
	212.5 mm 39 kg	262.5 mm 41 kg	312.5 mm 45 kg	362.5 mm 48 kg	412.5 mm 52 kg
3.0	6.9	6.9	6.9	7.2	7.2
3.7	6.9	6.9	6.9	7.2	7.2
4.3	7.2	7.2	7.2	7.5	7.9
4.9	7.5	7.5	7.5	7.9	7.9
5.5	8.2	8.2	8.5	8.9	9.2
6.1	8.5	8.5	8.9	9.2	9.5

The table assume that ends of joists and stringers are supported by the beam and girder forms.

Labor Cost of 100 S.F. (9 Sq.M) of Metal Pan and Concrete Joist Slab
Forms Requiring 1-3/4 Ft. of Lumber Per S.F.

Long spans and one-way beams. For short spans and two-way beams, add 10% to
labor costs.

4"x4" (100 x 100 mm) Shores

	Hours	Rate	Total	Rate	Total
Carpenter	4.2	$....	$....	$26.44	$ 111.05
Labor	2.1	....	....	21.04	44.18
Labor removing forms	1.3	....	....	21.04	27.35
Cost per 100 s.f. (9 sq.m)			$....		$ 182.58
cost per s.f.			....		1.83
cost per sq.m			....		19.69

Adjustable Shores

	Hours	Rate	Total	Rate	Total
Carpenter	3.8	$....	$....	$26.44	$ 100.47
Labor	1.9	....	....	21.04	39.98
Labor removing forms	1.3	....	....	21.04	27.35
Cost per 100 s.f. (9 sq.m)			$....		$ 167.80
cost per s.f.			....		1.68
cost per sq.m			....		18.08

For story heights over 16'-0" (4.8 m), add 25% to labor costs.

Labor Cost of 100 S.F. (9 Sq.M) of Metal Pan and Concrete Joist Slab
Forms Requiring 2 Ft. of Lumber Per S.F.

Long spans and one-way beams. For short spans and two-way beams, add 10% to
labor costs.

4"x4" (100 x 100 mm) Shores

	Hours	Rate	Total	Rate	Total
Carpenter	4.90	$....	$....	$26.44	$ 129.56
Labor	2.45	....	....	21.04	51.55
Labor removing forms	1.50	....	....	21.04	31.56
Cost per 100 s.f. (9 sq.m)			$....		$ 212.67
cost per s.f.			....		2.13
cost per sq.m			....		22.92

Adjustable Shores

	Hours	Rate	Total	Rate	Total
Carpenter	4.30	$....	$....	$26.44	$ 113.69
Labor	2.15	....	....	21.04	45.24
Labor removing forms	1.50	....	....	21.04	31.56
Cost per 100 s.f. (9 sq.m)			$....		$ 190.49
cost per s.f.				....	1.90
cost per sq.m				....	20.44

For story heights over 16'-0" (4.8 m), add 25% to labor costs.

Labor Cost of 100 S.F. (9 Sq.M) of Metal Pan and Concrete Joist Slab
Forms Requiring 2-1/2 Ft. of Lumber Per S.F.

4"x4" (100 x 100 mm) Shores

	Hours	Rate	Total	Rate	Total
Carpenter	6.0	$....	$....	$26.44	$ 158.64
Labor	3.0	....	....	21.04	63.12
Labor removing forms	1.8	....	....	21.04	37.87
Cost per 100 s.f. (9 sq.m)			$....		$ 259.63
cost per s.f.				....	2.60
cost per sq.m				....	27.98

Adjustable Shores

	Hours	Rate	Total	Rate	Total
Carpenter	5.4	$....	$....	$26.44	$ 142.78
Labor	2.7	....	....	21.04	56.81
Labor removing forms	1.8	....	....	21.04	37.87
Cost per 100 s.f. (9 sq.m)			$....		$ 237.46
cost per s.f.				....	2.37
cost per sq.m				....	25.50

For story heights over 16'-0" (4.8 m), add 25% to labor costs.

Labor Cost of 100 S.F. (9 Sq.M) of Metal Pan and Concrete Joist Slab
Forms Requiring 3 Ft. of Lumber Per S.F.

4"x4" (100 x 100 mm) Shores

	Hours	Rate	Total	Rate	Total
Carpenter	7.0	$....	$....	$26.44	$ 185.08
Labor	3.5	....	....	21.04	73.64
Labor removing forms	2.2	....	....	21.04	46.29
Cost per 100 s.f. (9 sq.m)			$....		$ 305.01
cost per s.f.				....	3.05
cost per sq.m				....	32.82

Adjustable Shores

	Hours	Rate	Total	Rate	Total
Carpenter	6.4	$....	$....	$26.44	$ 169.22
Labor	3.2	....	....	21.04	67.33
Labor removing forms	2.2	....	....	21.04	46.29
Cost per 100 s.f. (9 sq.m)			$....		$ 282.84
cost per s.f.				....	2.83
cost per sq.m				....	30.45

For story heights over 16'-0" (4.8 m), add 25% to labor costs.

Forms for Reinforced Concrete Stairs

The labor cost of framing and erecting concrete stair forms varies with the type of stair, open or box string, straight runs from floor to floor or intermediate landing platforms, straight or winding, square or bull nose treads and risers. The story height also affects the labor costs on account of the shoring and bracing necessary to support the wet concrete.

It is difficult to estimate the cost of stair forms on a s.f. (sq.m) basis, because it costs as much to frame a 3'-0" (0.9 m) stair as one 4'-0" (1.2 m) wide, while the latter contains one-third more forms per l.f. (m). The better method is to allow a certain price per flight of stairs of each type. This will result in more accurate estimates than the square foot (square meter) method. However, where it is desirable to estimate the stair forms by the square foot (sq.m), take the area of the soffit of the stairs and platforms. For example, a stair 4'-0" (1.2 m) wide and 18'-0" (5.4 m) long, contains 72 s.f. (6.48 sq.m) of forms.

Some contractors and estimators estimate the cost of stairs at a certain price per l.f. of riser, which includes forms, reinforcing steel, concrete and finishing.

Wood Forms for Straight Concrete Stairs. On straight stairs extending from floor to floor without intermediate landing platforms and having an average story height of 10'-0" to 12'-0" (3-3.6 m), two carpenters working together should lay out the stair, place rough stringers, mark off treads and risers on the rough string boards and set them in place in about 8 to 9 hours.

After the rough strings are in place, it will require another 8 to 10 hrs. for 2 carpenters to sheath the stairs, cut, bevel, and place risers, and place all necessary shoring and bracing ready for concrete.

The forms for an average flight of concrete stairs, 4'-0" (1.2 m) wide and 18'-0" (5.4 m) long, containing 72 s.f. (6.48 sq.m) of forms soffit measurement or 72 l.f. (6.48 m) of risers, should cost as follows:

	Hours	Rate	Total	Rate	Total
Carpenter	36	$....	$....	$26.44	$ 951.84
Labor helping and					
removing forms	9	....	....	21.04	189.36
Cost per flight			$....		$ 1,288.48
per s.f.				....	17.90
per sq.m				....	192.60
per l.f. of riser				....	17.90
per m of riser				....	192.60
per riser (18)				....	71.58

The lumber cost is approximately as follows:

	B.F.	Rate	Total
Stringers, 2 pcs. 2"x12"-20' (50x300mm-6m)	80	$0.40	$32.00
Risers, 18 pcs. 2"x8"-4' (50x200mm-1.2m)	96	0.40	38.40
Soffit sheathing, 4'-0"x18' (1.2x5.4m)	72	0.60	43.20
Joists supporting sheathing,			
4 pcs. 2"x8"-18' (50x200mm-5.4m)	96	0.40	38.40
Shores or uprights,			
6 pcs. 4"x4"-10' (100x100mm-3m)	80	0.40	32.00
Purlins or stringers,			
3 pcs. 4"x6"-4' (100x150mm-1.2m)	24	0.40	9.60
Sills, wedges and bracing	30	0.39	11.70
Cost per flight			$205.30
per s.f. (72)			2.85
per sq.m			30.68

Wood Forms for Concrete Stairs with Intermediate Landing Platforms. For concrete stairs of two short flights with an intermediate landing platform between floors, the labor cost per story will run somewhat higher than for straight run stairs, on account of laying out and framing for two flights instead of one.

Stairs of this type up to 4'-0" (1.2 m) wide and 8'-0" to 10'-0" (2.4-3 m) long from floor to platform, with each short flight containing 36 to 40 s.f. (3.24-3.60 sq.m) of forms (soffit measurement), and 8 to 10 risers, should cost about as follows for labor:

	Hours	Rate	Total	Rate	Total
Carpenter	23	$....	$....	$26.44	$ 608.12
Labor helping and removing forms	7	....	....	21.04	147.28
Cost per flight			$....		$ 755.40
per s.f. (36)				....	20.98
per s.f. riser (36)				....	20.98
per riser (9)				....	83.93
per sq.m				....	225.75

Method of Constructing Wood Forms for Concrete Stairs

Another Method of Constructing Wood Forms for Concrete Stairs

*Method of Constructing Wood Forms for Concrete Stairs
with Intermediate Platforms*

If the stairs are built between masonry walls and have their bearing in slots or chases left in the wall, it will require about 2 hrs. additional carpenter time cutting out stringers, if using a hand saw, or 1/2 to 3/4 hr., if using a power saw.

Winding stairs, stairs having bull nose treads and risers, and other difficult or complicated construction may cost two or three times as much as given above. The estimator will have to use his judgment, depending upon detail of the stair.

When erecting formwork for concrete stairs over 4'-0" (1.2 m) wide, it will be necessary to place additional "cut-out" stringers, to prevent the risers from bulging or giving way until the wet concrete has set. These "cut-out" stringers should be spaced 3'-0" to 4'-0" (0.9-1.2 m) apart.

When figuring labor costs of landing platform, double the labor cost of plain slab forms.

STEELFORMS FOR JOIST CONSTRUCTED FLOORS AND ROOFS

Where the same size steelforms are used on a number of floors, they are removed after the concrete has attained the desired strength and reused again on the same job until the work is complete. The removable steelforms are available from and can be furnished by companies specializing in this type of work. These companies will erect and remove the steelforms with or without the supporting centering at the stipulated lump sum for the entire job.

Removable steelforms in flange and adjustable type are made from 14, 15, and 16 gauge metal that will stand hard use. The flange type is nailed through the flanges while the adjustables are nailed through the sides. The flange type steelforms are furnished in 20" (500 mm) and 30" (750 mm) standard widths in standard depths of 6", 8", 10", 12", 14", 16", and 20" (150, 200, 250, 300, 350, 400, and 500 mm). The intermediate forms are in lengths of 1', 2', and 3' (.3, .6, and .9 m) with end caps as closures. In addition to the straight ends, there are also single tapered end forms in 3' (0.9 m) lengths for the various depths. Also available are adjustable straight-side forms that can be set to a depth of 8", 10", 12", 14", and 16" (200, 250, 300, 350, and 400 mm). These are furnished in standard 20" and 30" (500 and 750 mm) widths and have the same combination of 1", 2", and 3" (25, 50, and 75 mm) intermediates, end forms, and the 3" (75 mm) single tapered ends. For filling out spaces, filler forms are furnished in 10" and 15" (250 and 375 mm) widths matching the steelform depths as described. Tapered lighter gauge metals do not withstand job abuse, particularly with workers working over them before the concrete is placed. They often dent badly so that it requires considerable extra concrete and gives a poor surface finish, and there is added labor expense reconditioning the forms prior to the next use.

For two-way concrete joist construction, one-piece metal domes are also available. The depths furnished in the 30"x30" (750 x 750 mm) void are 8", 10", 12", 14", 16", and 20" (200, 250, 300, 350, 400 and 500 mm), and in the 19"x19" (475 x 475 mm) void are 4", 6", 8",10", 12", 16", and 20" (100, 150, 200, 250, 300, 400, and 500 mm). Flanges for the 30/30 domes are 3" (75 mm) wide to make a 3 ft. (0.9 m) module and for the 19/19 domes are 2-1/2" (62 mm) wide to make a 2 ft. (0.6 m) module. The metal gauge ranges from 14 to 16 gauge.

The steelforms and steeldomes must be nailed to the supporting forms to center and to hold them rigid and in line while the concrete is placed.

It is common practice to oil the forms before installing the reinforcing steel so that they can be removed easier from the ceiling and without concrete adhering to their surfaces.

Estimating Quantities of Steelform. In estimating the area of floor and roof construction requiring removable or permanent forms, the gross floor or roof area is used. No deductions are to made for beams or for tees of beams or for wide joists. Major openings, 50 s.f. (4.5 sq.m) or more, such as atriums, are deducted when applicable. Typical covering systems generally frame through smaller openings, so no area adjustments are made for these openings.

Concrete joist construction can be supported by any one of three types of structural systems. In addition to the aforementioned general estimating rules, the job area is subject to the following special rules. For a reinforced concrete frame, the areas are to be figured out-to-out of concrete frame. For a structural steel frame, the areas are figured center-to-center spandrel beams. For bearing wall construction, the areas are to be figured clear inside brick walls plus a bearing on all walls of 6" (150 mm).

Labor Handling and Placing Steelforms. The labor cost of placing steelforms will vary with the type of form, type of building, quantity of equipment furnished, amount of horizontal and vertical movement of equipment, depressions, openings in the slabs, headers, etc., and the locality in which the work is being executed.

Steelforms are placed usually by carpenters and laborers. In a few localities, iron workers handle the steelforms from the point of removal from the concrete ceiling to the next point of steelform installation on the job. In most localities, however, laborers assist the carpenters and handle the steelforms to the next point of erection on the job ready for the carpenter to install the steelform to the wood centering. Local practice of work assignment greatly affects the labor costs.

The following costs are based on steelforms being placed by experienced workers, such as ones employed by companies specializing in the erection and removal of steelforms.

On straight run work using flange type steelforms such as factory buildings, garages, warehouses, etc. having typical spans without much cutting for headers, openings, etc., an experienced crew consisting of two carpenters and four laborers should handle and place steelforms over approximately 3,500 s.f. (315 sq.m) of floor area in an 8-hr day at the following direct labor cost per 100 s.f. (9 sq.m):

	Hours	Rate	Total	Rate	Total
Mechanic	0.46	$....	$....	$26.44	$ 12.16
Labor	0.91	....	....	21.04	19.15
Cost per 100 s.f. (9 sq.m)	1.37				$ 31.31
cost per s.f.			....		0.31
cost per sq.m			....		3.34

For school buildings, hospitals, apartments, and office buildings, where the concrete joist require fitting and adjusting around pipes, pipe sleeves, chases and conduits, the handling and placing costs of the flange type equipment could increase as much as 40% to 50%, depending on the amount of special work required.

Labor Handling and Removing or Stripping Steelforms

In removing or stripping flange type steelforms after the concrete has set, an experienced crew, usually consisting of two carpenters and three laborers, can remove the steelforms from the ceiling over about 4,000 s.f. (360 sq.m) of floor per 8-hr day at the following direct labor cost per 100 s.f. (9 sq.m):

	Hours	Rate	Total	Rate	Total
Mechanic	0.40	$....	$....	$26.44	$ 10.58
Labor	0.60	....	....	21.04	12.62
Cost per 100 s.f. (9 sq.m)					$ 23.20
cost per s.f.			....		0.23
cost per sq.m			....		2.47

It should be noted that the above costs are based on flange type steelforms. The cost for handling and placing the one-piece dome is slightly higher. The cost for handling and placing adjustable type steelform is about 33.33% higher.

Ceco Steelform Construction. Ceco steelform construction is a combination of concrete joist construction and thin top slabs. The steelforms are formed of 14, 15, or 16 gauge steel, depending on the width and depth of the form. These forms are placed on supporting centering and removed from the concrete after the concrete has reached sufficient strength. The type of forms made available to the industry by the Ceco Corporation are flange forms, the most commonly used type of form, followed by the one-piece metal dome, adjustable forms and the one-piece longforms.

The cost of the steelforms installed in a building will depend on a number of factors such as the type of steelform used, the size of the job, the number of reuses of the equipment, location, availability of the steelforms with respect to warehouse stocks, and so on. Local working conditions also play a very important part in job costs.

The Ceco steelform service includes supplying the necessary steelforms and the labor for their erection and removal. Also offered is a complete service of furnishing and erecting all centering for the support of the steelforms. In steeldome construction, centering is usually included for the concrete solid slab areas that are of the same depth as the dome construction and that are within the floor areas containing the steeldomes. The quotations and contracts are generally on a lump sum basis for specific projects. It is always recommended that the steelform prices be obtained from a steelform supplier before finalizing and submitting a bid on this portion of the job costs.

FORMS OTHER THAN WOOD

Lightweight Steel Forming Material for Concrete. High-strength corrugated steel forming material is often used for forming reinforced concrete floor and roof slabs. It comes either galvanized or uncoated and is manufactured from 100,000 psi (689.5 MPa) tough-temper steel. It has a definite reliable structural strength nearly twice that of ordinary steel of equal weight. Used primarily in floor and roof systems having steel joists, junior beams, or purlins, it is also used over pipe tunnels or similar installations where economies of using permanent forms can be realized. Either structural grade concrete or lightweight insulating concrete may be used with corrugated steel forms, but in either case, reinforcing bars or wire mesh should be added to satisfy flexure and temperature steel requirements.

This material offers many advantages over other methods of flexible centering. The light, rigid sheets are quickly placed. No side pull is exerted on joists, and instead, joist top chords are given lateral support by the stiffness of the material. Sheets in place provide a safe working platform. Mesh is easily and effectively placed. Elimination of sag between joists reduces concrete quantity. Uniform thickness of slab over joists and mid span permits monolithic finish. Little cleanup is required underneath.

For exposed construction, it is available as galvanized. For unexposed joist construction, it can be purchased uncoated.

Sheets are placed with corrugations normal to the supporting joists or beams, with the end lap of sheets occurring over the joists or beams. The sheets are placed with edge lips up and are lapped a minimum of one corrugation with adjacent sheets. The ends of sheets should lap a minimum of 2" (50 mm). The sheets are fastened to the supporting members by clips, arc-welding, or nailing. For steel joist construction, the most satisfactory method of attachment is arc-welding. The manufacturer can provide special curved washers for use in welding the sheets to the steel framework. These provide fast, high-strength welds.

Estimating. Corrugated steel forming material is sold by the square, with the area determination based on sheet width times actual sheet length.

INTERMEDIATE

ENDCAP

TAPERED ENDFORM

3E

Filler widths (10" and 15") are available for filling non-standard spaces only.

CONCRETE QUANTITIES/30" WIDTHS*

Depth of Steelform	Width of Joist	Cubic feet of concrete per square foot for various slab thicknesses			Additional concrete for Tapered Endforms, cu. ft. per lin. foot of bearing wall or beam (One side only)
		2½"	3"	4½"	
8"	5"	.317	.359	.484	.143
	6"	.333	.374	.499	.139
	7"	.347	.389	.514	.135
10"	5"	.348	.390	.515	.179
	6"	.367	.409	.534	.174
	7"	.386	.427	.552	.169
12"	5"	.381	.422	.547	.214
	6"	.404	.445	.570	.208
	7"	.425	.467	.592	.203
14"	5"	.415	.456	.581	.250
	6"	.441	.483	.608	.243
	7"	.467	.508	.633	.236
16"	6"	.481	.522	.647	.278
	7"	.509	.551	.676	.270
	8"	.537	.578	.703	.263
20"	6"	.564	.606	.731	.347
	7"	.599	.641	.766	.338
	8"	.633	.675	.800	.329

*Apply only for areas over FLANGEforms and joists between them. Bridging joists, special headers, beam tees, etc., not included.

CONCRETE QUANTITIES/20" WIDTHS*

Depth of Steelform	Width of Joist	Cubic feet of concrete per square foot for various slab thicknesses			Additional concrete for Tapered Endforms, cu. ft. per lin. foot of bearing wall or beam (One side only)
		2½"	3"	4½"	
8"	4"	.339	.381	.506	.167
	5"	.361	.402	.527	.160
	6"	.380	.422	.547	.154
10"	4"	.377	.419	.544	.208
	5"	.404	.445	.570	.200
	6"	.428	.470	.595	.192
12"	4"	.418	.459	.584	.250
	5"	.449	.491	.616	.240
	6"	.479	.520	.645	.231

*Apply only for areas over FLANGEforms and joists between them. Bridging joists, special headers, beam tees, etc., not included.
**14", 16" and 20" depths are also available. Contact your Ceco Construction Engineer.

VOIDS CREATED BY VARIOUS SIZE FLANGEFORMS

Depth of Steelform	Cubic feet of void created per linear foot (for various widths of FLANGEforms)				**Cubic Feet per Tapered End	
	30" Width	20" Width	15" Width	10" Width	30" Width	20" Width
8"	1.628	1.072	.794	.516	4.465	2.882
10"	2.023	1.329	.982	.634	5.548	3.569
12"	2.414	1.581	1.165	.748	6.617	4.243
14"	2.801	1.829	1.343	.857	7.673	
16"	3.183	2.072	1.516	.961	8.715	
20"	3.933	2.544	1.850	1.155	10.756	

**Total void created by standard 3'-0" length Tapered Endform.

Ceco Flange Type Forms

Concrete Quantities/30" Widths*

Depth of Steelform	Width of Joist	Cubic feet of concrete per square foot of floor for various slab thicknesses		
		2½"	3"	4½"
8"	6"	.333	.375	.500
10"	6"	.367	.409	.534
12"	6"	.403	.445	.570
14"	6"	.440	.482	.607
16"	6"	.479	.521	.646
20"	6"	.560	.602	.727

All sizes shown conform to the Concrete Reinforcing Steel Institute, CRSI, Code of Standard Practice, MSP-2-81.

Apply only for areas over LONGforms and joists between them. Bridging joists, special headers, beam tees, etc., not included.

LONGFORM

Filler widths (10" and 15") are available for filling non-standard spaces only.

Concrete Quantities/20" Widths*

Depth of Steelform	Width of Joist	Cubic feet of concrete per square foot of floor for various slab thicknesses		
		2½"	3"	4½"
8"	5"	.361	.403	.528
10"	5"	.404	.446	.571
12"	5"	.449	.490	.615

Apply only for areas over LONGforms and joists between them. Bridging joists, special headers, beam tees, etc., not included.

**14", 16" and 20" depths are also available. Contact your Ceco Construction Engineer.*

Voids Created by
Various Size Longforms

Depth of Steelform	Cubic feet of void created per linear foot (for various widths of LONGforms)			
	30" Width	20" Width	15" Width	10" Width
8"	1.626	1.070	.793	.515
10"	2.023	1.329	.981	.634
12"	2.416	1.583	1.166	.749
14"	2.804	1.832	1.346	.860
16"	3.188	2.077	1.522	.966
20"	3.944	2.555	1.860	1.166

Shaded areas above indicate standard filler widths.

ENDCAP

Ceco Longform Type

All 3'-0" module sizes (30" x 30" domes) and all 2'-0" module sizes (19" x 19" domes) except 14" depth conform to the Concrete Reinforcing Steel Institute, CRSI, Code of Standard Practice, MSP-2-31.

Voids Created With 2'-0" Design Module

Depth of Dome	OVERALL PLAN SIZE 24" x 24"	
	Plan Size of Void	Cu. Ft. of Void
8"	19" x 19"	1.410
10"	19" x 19"	1.862
12"	19" x 19"	2.200
14"	19" x 19"	2.447

3'-0" MODULE
(30" x 30" Dome System)

Voids Created With 3'-0" Design Module

STANDARD SIZES	OVERALL PLAN SIZE 36" x 36"	
Depth of Dome	Plan Size of Void	Cu. Ft. of Void
8"	30" x 30"	3.877
10"	30" x 30"	4.756
12"	30" x 30"	5.578
14"	30" x 30"	6.662
16"	30" x 30"	7.519
20"	30" x 30"	9.194

Filler Sizes for 3'-0" Module

Depth of Dome	OVERALL PLAN SIZE 36" x 30"		OVERALL PLAN SIZE 26" x 26"	
	Plan Size of Void	Cu. Ft. of Void	Plan Size of Void	Cu. Ft. of Void
8"	20" x 30"	2.535	20" x 20"	1.656
10"	20" x 30"	3.094	20" x 20"	2.011
12"	20" x 30"	3.607	20" x 20"	2.330
14"	20" x 30"	4.334	20" x 20"	2.817
16"	20" x 30"	4.873	20" x 20"	3.152
20"	20" x 30"	5.922	20" x 20"	3.808

U. S. Patent No. 2,830,785 applies to all sizes

Ceco Steel Dome Systems

All sizes shown conform to the Concrete Reinforcing Steel Institute, CRSI, Code of Standard Practice, MSP-2-81.

VOIDS CREATED WITH 4'-0" DESIGN MODULE		
Depth* of Dome	OVERALL PLAN SIZE 48" x 48"	
	Plan Size of Void	Cu. Ft. of Void
14"	41' x 41'	12.45
16"	41' x 41'	14.06
20"	41' x 41'	17.14
24"	41' x 41'	20.05

*18" and 22" depths are also available. Contact your Ceco Construction Engineer.

VOIDS CREATED WITH 5'-0" DESIGN MODULE		
Depth* of Dome	OVERALL PLAN SIZE 60" x 60"	
	Plan Size of Void	Cu. Ft. of Void
14"	52" x 52"	20.42
16"	52" x 52"	23.11
20"	52" x 52"	28.33
24"	52" x 52"	33.33

*18" and 22" depths are also available. Contact your Ceco Construction Engineer.

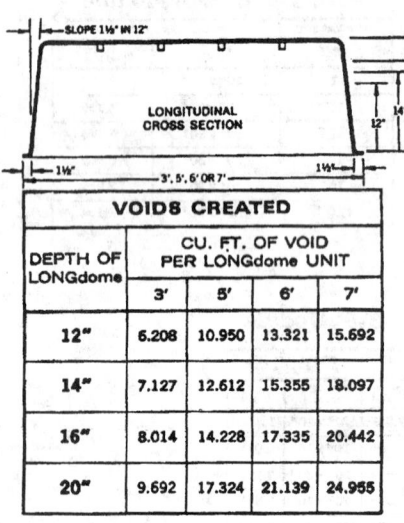

VOIDS CREATED				
DEPTH OF LONGdome	CU. FT. OF VOID PER LONGdome UNIT			
	3'	5'	6'	7'
12"	6.208	10.950	13.321	15.692
14"	7.127	12.612	15.355	18.097
16"	8.014	14.228	17.335	20.442
20"	9.692	17.324	21.139	24.955

FOUR STANDARD LENGTHS may be intermixed in any desired combination. To give you further flexibility, they may be installed in a combination with Ceco's standards 30" x 30" domes, if desired.

Ceco Fiberglass Domes

The reusable forms, constructed of fiberglass or segmented steel, make good use of the top slab for one-way joist construction and carry a 2-hour fire rating as required by fire codes.

6'-0" MODULE

VOIDS CREATED BY 6'-0" MODULE

DEPTH	JOIST WIDTH	CU.FT./L.F.
14"	6"	6.29
16"	6"	7.173
20"	6"	8.92

5'-0" MODULE

VOIDS CREATED BY 5'-0" MODULE

DEPTH	JOIST WIDTH	CU.FT./L.F.
16"	7"	5.733
20"	7"	7.127

Ceco Fiberglass Domes, Wide Module

Erection Costs. Corrugated steel forming material can be rapidly erected in all kinds of weather, providing cover for floors below. The corrugated sheets are large in area but extremely light in weight and can be easily handled by one worker.

A well organized crew, working under favorable conditions, can place and fasten up to 10,000 s.f. (900 sq.m) of corrugated steel forming material per 8-hr day, but most jobs will average 3,000 to 5,000 s.f. (270-450 sq.m) per day.

Using an average figure of 4,000 s.f. (360 sq.m) per 8-hr day for a four-worker crew, the labor costs for placing 1,000 s.f. (90 sq.m) of corrugated steel forming material would be as follows:

	Hours	Rate	Total	Rate	Total
Ironworker	8	$....	$....	$30.28	$ 242.24
Cost per s.f.			...		0.24
cost per sq.m			...		2.58

The above figures are based upon one worker welding or fastening the deck in place and three workers handling and placing the material. Costs do not include equipment, overhead, and profit.

Computing Concrete Quantities. When computing volume of concrete placed over corrugated steel decking, subtract 1/4 in. (6.25 mm) from the slab thickness to allow for corrugations of standard and 3/8 in. (9 mm) for heavy duty corrugated steel forming. Size of corrugations: standard, 1/2" depth x 2-3/16" pitch (12 x 54 mm); heavy duty, 3/4" depth x 3" pitch (19 x 75 mm).

ARCHITECTURAL CONCRETE FORMWORK

Appearance is of paramount importance in architectural concrete, and every operation must be planned and executed with this in mind. The architect may design and specify, but regardless of how detailed the specifications, it is the contractor and the superintendent who must have the know-how to detail and construct the forms and be on constant watch during all operations to see that the correct procedures are followed.

Forms for Architectural Concrete

Form design and construction must be approached differently from ordinary structural concrete. In structural concrete some leakage and irregularities in formwork may be tolerated, so long as the forms are strong enough to carry the weight of concrete and imposed loads and are built fairly true to line and grade. In architectural concrete, there must be no leakage through the forms. Chamfer strips at corners and edges are helpful for this purpose. Wherever possible, the forms should be designed so that pressure of the fresh concrete will tighten joints rather than tend to loosen them.

A common source of trouble is bulging of forms at external corners, resulting in both poor alignment and leakage with subsequent ragged corners, sometimes to the extent of exposing the aggregate.

One method of constructing forms to provide locking of the corner is shown. The wales are extended beyond the intersection far enough to permit two vertical strips to be nailed in the corners. Wedges are driven between these strips and the wales to tighten the corner.

Block slightly loose

Piece of 1"material nailed to wale to support sheathing on side of pilaster

Form Designed so that Pressure of Fresh Concrete Tightens It to Prevent Leakage

Double wales

4"x4" Studs at plywood joints

Plywood joint

Double wales

Double 2"x6' liners at 8 to 10 ft. centers

Double wales

Wedge

Form Construction to Provide Positive Corner Locking

Another method of locking the form corners is to use a tie instead of the wood strips and wedges. Since there are no wedges to work loose, a set of wales can be set and tightened on one level around the building without much chance of subsequent loosening when tightening the wales above or below.

Various materials are used for forming architectural concrete to eliminate imperfections and irregularities in the finished surface. The American Concrete Institute offers a report entitled "Formwork for Concrete" that discusses these forming materials and their application. Plastic coated plywood, hardboard form lining, fiber or laminated pressed paper tubes, plaster waste molds, metal forms, and fiberglass forms are all used as forming material for architectural concrete. Fiberglass has been used on larger building projects with the forms being reused as many as 60 or 70 times.

Form Layout for Use of Structural Plywood

Plywood Form Sheathing. Smooth surfaces required on much present day work are obtained with plywood form sheathing. This should be of structural grade, made especially for form use and preferably should be oiled at the factory. If oiled on the job, sufficient time should be allowed for the oil to penetrate, and any surplus oil should be wiped off before the plywood is used. Plywood 5/8" (15 mm) or 3/4" (19 mm) thick can be used directly against studs without backing when the outer plies are at right angles to the studding. Structural plywood usually has five plies, with the fibers in the two outer and middle plies in one direction and the fibers of the other two plies in the opposite direction, making the plywood stronger and stiffer in one direction. The result may not be serious on the first use of the plywood under average placing speed, but with faster placing the deflection may be pronounced. With the second use of the plywood, there will be noticeable deflection, and on the third and additional uses, a deflection of 1/8" (3 mm) to 1/4" (6 mm) can be expected with 3/4" (19 mm) plywood on studs spaced 16" (400 mm) on centers.

Presdwood is often used as a form liner, which is usually nailed to decking or sheathing lumber as a backing. Board 3/16" (4.75 mm) thick should be used over solid backing while 1/4" (6 mm) board may be used semi-structurally. Ordinary wood working tools are used in cutting and applying Concrete Form Presdwood.

Some manufacturers advertise that plywood may be used 10 times or more in the construction of formwork, but where appearance is important, it is seldom possible to use it more than 3 or 4 times. When using plywood for the facade or facing forms, it must be thoroughly cleaned after each use. Where a special form coating is used to treat the plywood instead of oil, it should be applied with care and allowed to dry thoroughly.

Architectural concrete walls must be straight and plumb. This is essential. It means that the 2"x4" (50 x 100 mm) or preferably 2"x6" (50 x 150 mm) studs and wales must be of uniform depth and perfectly straight. Otherwise, the walls will be crooked, with the possibility of the work being rejected.

The top of the foundation walls on which the architectural concrete wall rests must be perfectly level and straight, using two 2"x6" (50 x 150 mm) wales or similar planks, braced at intervals of not less than 4'-0" (1.2 m) in addition to all other bracing ordinarily required for foundation walls. It requires about 4-1/2 b.f. of lumber per l.f. of wall for these wales and bracing. In order to hold the studs of the form for the architectural concrete wall in line, it is necessary to insert 5/8" greased stud bolts, spaced about 3'-0" (900 mm) on centers, into the top of the foundation wall as illustrated. It requires about 1/8 hr. carpenter and 1/16 hr. labor time per l.f. of wall to brace the top of the foundation wall and to insert the bolts.

As a rule, it is necessary to obtain a perfect horizontal line of concrete at the top of the foundation wall, which will be assured by nailing a 1"x2" (25 x 50 mm) strip on the outside form of the foundation wall and placing the concrete about 1" higher than the bottom of this strip. About two hours after placing the concrete, the surplus concrete, which generally contains a good deal of laitance, is scraped off to the level of the underside of the strip.

After the forms for the foundation walls have been stripped, the wales should be installed and blocked about 3'-0" (900 mm) apart against the foundation wall. The 2"x4" (50 x 100 mm) or 2"x6" (50 x 150 mm) studs should now be installed between wales and the wall. The studs should have 8d nails, protruding about 1" (25 mm), driven into the outside edge exactly 8" (200 mm) from the lower end and 6d nails, protruding less than 3/4" (19 mm) exactly 8" (200 mm) from the lower edge. When inserting the studs between the wales and the foundation wall, the 8d nails form their support on the wales and the 6d nails furnish the support for the 3/4" (19 mm) plywood. The studs are usually spaced 12" (300 mm) or 16" (400 mm) on centers but at every vertical joint in the plywood (usually 8'-0") a double stud or a stud 3-1/2" (88 mm) wide should be used.

The studs are temporarily braced horizontally by nailing a 1"x6" (25 x 150 mm) board higher up and are braced vertically by wedging at the bottom between the studs and foundation wall. The first sheet of plywood is then inserted between the studs and the foundation wall and bolts, holding the wale tight until the lowest part of the form is straight and plumb. The wales are spaced 24" (600 mm) vertically and the oiled tie bolts are spaced 2'-0", 2'-8" and 3'-4" (600, 800, and 1000 mm) apart, so that the holes do not interfere with the 16" (400 mm) stud spacing and are in proper relation to the ends of the 8'-0" (2.4 m) long plywood sheets.

The arrangement of the holes for the tie bolts and the arrangement of the 6d box nails by which the sheets of plywood are held to the studs are also shown in the accompanying illustration.

After 2 or 3 of the 4'-0" (1.2 m, 1200 mm) wide sheets of 3/4" (19 mm) plywood have been fastened to the studs and the proper amount of horizontal walers put in place, the outside form is further strengthened by vertical wales consisting of 2 pcs. of 2"x6" (50 x 150 mm) spaced not more than 8'-0" (2.4 m, 2400 mm) o.c. with about 2'-0" (0.6 m, 600 mm) or more projecting over the top of the outside form. These wales are held in place and plumbed on top by two or more No. 9 or 10 wires, secured to column dowels, special bolts, or other means provided in the foundations or in the floor. These vertical wales are bolted to every pair of horizontal wales they cross using 1/2" (6 mm) bolts. The form is now ready to support the vertical and horizontal reinforcing.

After the reinforcing steel has been placed, the inside form may be erected. The studs are nailed to a 1"x6" (25 x 150 mm) sill or plate, which has been nailed to the floor one day after concreting, and if the inside of the wall is to have a finished appearance, requiring the use of plywood, the plywood should be in sheets 24" (0.6 m, 600 mm) wide and 8'-0" (2.4 m, 2400 mm) long in order to facilitate the nailing of the plywood to the studs and the insertion of the tie bolts and spreaders.

The joints of the plywood should be filled with a mastic consisting of a mixture of 50% tallow and 50% cement for summer use, while for winter work, the tallow should be replaced by pump grease. The surplus filler should then be scraped off using No. 0 sandpaper.

Estimating Quantities for Architectural Concrete Forms. When estimating the quantities for plain architectural concrete walls, do not make any deductions for ordinary door and window openings, because these must be formed with the wall and then boxes built in to allow for the door or window frame to be inserted later.

The only exception is large display windows in the first story of a commercial building or for skeleton type buildings where the piers and lintels are estimated at a higher rate than plain wall surfaces.

All ornamentation should be estimated separately, either by the lin. ft. or sq. ft. (meter or sq.m) depending upon the class of work. Ornamental work will require the use of wood or plaster waste molds for forming, in addition to the structural backing, so that the entire wall area should first be figured as plain wall surfaces and then ornamentation figured extra.

Water tables, belt courses, cornices, door and window trim, copings, and the like should be estimated by the l.f. (m) when less than 12" (0.3 m, 300 mm) wide and by the s.f. when more than 12" (0.3 m, 300 mm) wide. Always mention size and detail for each type of ornamentation. This is necessary for pricing the work.

Columns, pilasters, etc. should be estimated separately, giving width, height, and area in s.f. (sq.m), together with a detail of the forming, such as whether fluted.

Sample Wall Form Calculation

This sample shows how to determine square footage of form contact surface area, size and number of studs and wales needed, and the number of steel ties required. The sample data, table, and charts are provided by Richmond Screw Anchor Company, Ft. Worth, Texas.

Given: Footing poured, 56'-0" (16.8 m) long x 5'-0" (1.5 m) wide x 3'-6" (1.05 m) high

Required: Forms and material required for a concrete wall 50'-0" (15 m) long x 1'0" (0.3 m) wide x 12'-0" (3.6 m) high

Assumptions:

1. Permanent or reusable formwork, Class "A".
2. 3/4" (19 mm) thick Class I plyform with face perpendicular to supports
3. (strong direction).

4. Studs vertical.
5. Wales horizontal.
6. Tie safe working load = 6,000 lbs. (2720 kg)
7. Preferred tie spacing: vertical 2'-0" (0.6 m); horizontal 2'-0" (0.6 m);
8. 2' (0.6) x 2' (0.6) = 4 s.f. (0.36 sq.m)

We want to determine stud and wale sizes using 3/4" (19 mm) plywood.

Determine Concrete Pressure:

Concrete Pressure = $\dfrac{\text{Tie safe working load}}{\text{Area}}$

Concrete Pressure = $\dfrac{\text{6,000 lbs. (2720 kg)}}{\text{4 s.f. (0.36 sq.m)}}$

Concrete Pressure =   1,500 lbs./s.f. (7555 kg/m^2)

Framing Material:

Stud Spacing: Use Table 2, "Spacing of Supports for Class I Plywood, (Strong Direction)—Face Grain Across Supports", or Chart A, "Spacing of Supports for Class I Plywood", which appear at the end of the sample calculation. For Class I plywood with a concrete pressure of 1,500 psf (7560 kg/m^2) and plywood thickness of 3/4" (19 mm), the support (tie) spacing is 7.6" (190 mm), so use 8" (200 mm) center to center (c.c.)

Stud Sizes:

Stud Loading = $\dfrac{\text{Stud Spacing x Design Concrete Pressure}}{\text{12 in./ft. (1000 mm/m)}}$

Stud Loading = $\dfrac{\text{8 in.   (200 mm) x 1,500 lbs/ft.}^2 \text{ (7560 kg/m}^2)}{\text{12 in./ft. (1000 mm/m)}}$

Stud Loading =   1,000 lbs./ft. (1512 kg/m)
Assume studs are continuous or partially continuous.

Using Table 1 "Maximum Wale Spacing Along Stud/Joist Member", and Chart B, "Stud/Joist Loading", stud member sizes can be determined by beginning on Table 1 at 1,000 lbs./ft. (1488 kg/m) and moving horizontal to 24" (0.6 m, 600 mm) vertical tie spacing (wale spacing) and reading lumber sizes above, 2x6 or better.

Wale Size:

Wale Loading = $\dfrac{\text{Wale Spacing}}{12 \text{ in./ft. (1000 mm/m)}}$ x Design Concrete Pressure

Wale Loading = $\dfrac{2 \text{ ft. (0.6 m) x 1,500 lbs./ft.2 (7569 kg/m2)}}{12 \text{ in./ft. (1000 mm/m)}}$

Wale Loading =   3,000 lbs./ft. (4536 kg/m)
Assume wales are continuous or partially continuous.

Using Table 4 and Chart C, begin on Table 4 at 3,000 lbs./ft. (4535 kg/m) wale load and progress horizontally to a span of 2'-0" (0.6 m) horizontal tie spacing and read lumber sizes above, double 2x6 (50 x 150 mm) or better.

Summary:

1. Ties—6,000 lbs. (2720 kg) safe working load
2. Tie Spacing—2'-0" (0.6 m) horizontal; 2'-0" (0.6 m) vertical
3. Concrete Pressure—1,500 lbs./ft.^{2} (7560 kg/m^2) (uniform)
4. Plywood—3/4" (19 mm) Class I perpendicular to supports
5. Stud Spacing—8" (200 mm) c. to c., perpendicular to plywood face grain
6. Stud Member Size—2x6 (50 x 150 mm) or better, continuous or partially continuous
7. Wale Spacing—2'-0" (0.6 m) c. to c. vertically
8. Wale Member Size—Double 2x6 (50 x 150 mm) or better, continuous or partially continuous

Material Calculation List:

3/4" (19 mm) plywood, contact surface area, 50' (15 m) x 12' (3.6 m) = 60 s.f. (5.4 sq.m) one side x 2 sides = 1,200 s.f. (108 sq.m) of contact surface area; add end bulkheads if required

Studs, 50 l.f. (15 m) divided by .667 l.f. (0.2 m) c.c. = 75 each side x 2 sides = 150 pcs. 2x6 (50 x 150 mm), 12'0" (3.6 m) long

Wales, 12' (3.6 m) high divided by 2' (0.6 m) c.c. = 6 each x 2 (doubles) = 12 each side; 12 each x 2 sides = 24 pcs. 2x6 (50 x 150 mm), 50' (15m) —5 @ 10' = 50' (5 @ 3 m = 15m) —each

Ties, 50' x 12' ÷ 4 = 150 each (plus hardware)

Notes:
1. Bottom row of ties, maximum 6" (150 mm) from bottom of form
2. Top row of ties, not to exceed 50% of vertical spacing, and 3" (75 mm) concrete cover required
3. First and last vertical row not to exceed 6" (150 mm) from edge of form
4. Concrete placement shall not exceed recommended placement rate of 7 ft. (2.1 m) per hr. in 4-ft. (1.2-m) layers.
5. Consult tie manufacturer for your special requirements.

Table 1
Maximum Wale Spacing Along Stud/Joist Member

Load on Stud/Joist Lbs/lin. Ft	Lumber Sizes, Nominal				
	2" x 4"	2" x 6"	2" x 8"	2" x 10"	2" x 12"
200	52	82	109	136	158
400	37	58	77	98	119
600	28	44	58	74	90
800	23	36	47	60	73
1000	20	31	41	52	63
1200	17	28	36	46	56
1400	16	25	33	42	51
1600	15	23	31	39	48
1800	14	22	29	37	45
2000	13	21	28	35	43
2200	13	20	26	34	41
2400	12	19	25	32	39
2600	12	19	25	31	38
2800	11	18	24	30	37
3000	11	18	23	30	36

Load on Stud/Joist Lbs/lin. Ft	Lumber Sizes, Nominal					
	3" x 4"	3" x 6"	3" x 8"	4" x 4"	4" x 6"	4" x 8"
200	68	105	129	78	114	140
400	48	75	99	57	89	117
600	39	62	81	46	73	96
800	33	52	69	40	63	83
1000	28	44	58	36	56	74
1200	25	39	51	31	50	65
1400	22	35	46	28	44	58
1600	20	32	42	25	40	53
1800	19	29	39	24	37	48
2000	18	28	36	22	34	45
2200	17	26	34	20	32	42
2400	16	25	33	19	30	40
2600	15	24	31	18	29	38
2800	15	23	30	18	28	36
3000	14	22	29	17	26	35

(Continuous or Partially Continuous Spans)
Table 1.A (metric)
Maximum Wale Spacing Along Stud/Joist Member
(Continuous or Partially Continuous Spans)

Load on Stud/Joist Kg/lin. Meter	Lumber Sizes, Nominal				
	50 x 100 mm	50 x 150 mm	50 x 200 mm	50 x 250 mm	50 x 300 mm
298	1,300	2,050	2,725	3,400	3,950
595	925	1,450	1,925	2,450	2,975
893	700	1,100	1,450	1,850	2,250
1,190	575	900	1,175	1,500	1,825
1,488	500	775	1,025	1,300	1,575
1,786	425	700	900	1,150	1,400
2,083	400	625	825	1,050	1,275
2,381	375	575	775	975	1,200
2,678	350	550	725	925	1,125
2,976	325	525	700	875	1,075
3,274	325	500	650	850	1,025
3,571	300	475	625	800	975
3,869	300	475	625	775	950
4,166	275	450	600	750	925
4,464	275	450	575	750	900

Load on Stud/Joist Kg/lin. Meter	Lumber Sizes, Nominal					
	75 x 100 mm	75 x 150 mm	70 x 200 mm	100 x 100 mm	100 x150 mm	100 x 200 mm
298	3,500	0	0	0	0	0
595	2,925	0	0	0	0	0
893	2,400	0	0	0	0	0
1,190	2,075	0	0	0	0	0
1,488	1,850	0	0	0	0	0
1,786	1,625	0	0	0	0	0
2,083	1,450	0	0	0	0	0
2,381	1,325	0	0	0	0	0
2,678	1,200	0	0	0	0	0
2,976	1,125	0	0	0	0	0
3,274	1,050	0	0	0	0	0
3,571	1,000	0	0	0	0	0
3,869	950	0	0	0	0	0
4,166	900	0	0	0	0	0
4,464	875	0	0	0	0	0

Table 2
Spacing of Supports for Class I Plywood
(Strong Direction)--Face Grain Across Supports*

Pressure (psf)	Plywood Thickness							
	1/4"	3/8"	1/2"	5/8"	3/4"	7/8"	1"	1 1/8"
100	12.2	16.2	19.3	21.9	24.3	26.9	29.4	31.5
200	9.3	13.3	16.1	18.4	20.4	22.9	24.7	26.4
300	7.6	10.7	13.1	15.1	17.1	19.3	21.7	23.4
400	6.6	9.3	11.4	13.1	14.8	16.8	18.1	20.3
500	5.9	8.3	10.2	11.8	13.2	15.0	16.8	18.1
600	5.4	7.6	9.3	10.7	12.1	13.7	15.3	16.5
700	4.6	7.0	8.6	9.9	11.2	12.7	14.2	15.3
800	4.0	6.6	8.1	9.3	10.5	11.8	13.3	14.3
900	3.6	6.2	7.6	8.8	9.9	11.2	12.5	13.5
1000	3.2	5.6	7.2	8.3	9.4	10.6	11.7	12.8
1100	2.9	5.1	6.9	7.9	8.9	10.1	11.3	12.2
1200	2.7	4.6	6.5	7.6	8.5	9.7	10.8	11.7
1300	2.5	4.3	6.0	7.3	8.2	9.3	10.4	11.2
1400	2.3	4.0	5.6	7.0	7.9	9.0	9.8	10.6
1500	2.2	3.7	5.2	6.7	7.6	8.5	9.2	10.2
1600	2.0	3.5	4.9	6.3	7.4	7.9	8.6	9.6
1700	1.9	3.3	4.6	5.9	7.2	7.5	8.1	9.0
1800	1.8	3.1	4.4	5.6	6.9	7.0	7.6	8.5
1900	1.7	2.9	4.1	5.3	6.5	6.7	7.2	8.0
2000	1.6	2.8	3.9	5.0	6.2	6.3	6.9	7.6

* Plywood continuous across two or more spans

Spacing of Supports for Class I Plywood
(Strong Direction)--Face Grain Across Supports*

Pressure (Kg/Sq meter)	Lumber Sizes, Nominal							
	6.25 mm	9.375 mm	12.5 mm	15.625 mm	18.75 mm	21.875 mm	25 mm	28.125 mm
488	305	405	483	548	608	673	735	788
976	233	333	403	460	510	573	618	660
1,465	190	268	328	378	428	483	543	585
1,953	165	233	285	328	370	420	453	508
2,441	148	208	255	295	330	375	420	453
2,929	135	190	233	268	303	343	383	413
3,417	115	175	215	248	280	318	355	383
3,906	100	165	203	233	263	295	333	358
4,394	90	155	190	220	248	280	313	338
4,882	80	140	180	208	235	265	293	320
5,370	73	128	173	198	223	253	283	305
5,858	68	115	163	190	213	243	270	293
6,347	63	108	150	183	205	233	260	280
6,835	58	100	140	175	198	225	245	265
7,323	55	93	130	168	190	213	230	255

* Plywood continuous across two or more spans

Table 3
Spacing of Supports for Class I Plywood
(Weak Direction)--Face Grain Parallel to Supports

Pressure (psi)	Plywood Thickness **					
	1/2"	5/8"	3/4"	7/8"	1"	1 1/8"
100	12.9	16.2	19.7	22.4	25.2	27.6
200	10.5	13.6	16.6	18.8	21.2	23.2
300	8.5	11.7	15.0	17.0	19.1	21.0
400	7.4	10.1	13.2	15.4	17.8	19.5
500	6.6	9.0	11.8	13.3	16.1	18.4
600	6.0	8.2	10.8	12.6	14.7	16.4
700	5.5	7.6	10.0	11.7	13.7	15.2
800	4.8	7.1	8.8	10.9	12.8	14.2
900	4.3	6.3	7.9	10.3	12.1	13.4
1000	3.9	5.7	7.1	9.6	11.4	12.7
1100	3.5	5.2	6.4	8.7	10.3	12.1
1200	3.2	4.7	5.9	8.0	9.5	11.2
1300	3.0	4.4	5.4	7.4	8.8	10.4
1400	2.8	4.1	5.1	6.8	8.1	9.6
1500	2.6	3.8	4.7	6.4	7.6	9.0
1600	2.4	3.5	4.4	6.0	7.1	8.4
1700	2.3	3.3	4.2	5.6	6.7	7.9
1800	2.2	3.2	3.9	5.3	6.3	7.5
1900	2.0	3.0	3.7	5.0	6.0	7.1
2000	1.9	2.8	3.5	4.8	5.7	6.7

* Plywood continuous across two or more spans
** 1/4" and 3/8" plywood not recommended for use in weak direction

Table 3.A (metric)
Spacing of Supports for Class I Plywood (Weak Direction)--Face Grain Parallel to Supports

Pressure Kg/Sq.Meter	Lumber Sizes, Nominal					
	12.5	15.6 mm	18.8 mm	21.9 mm	25.0 mm	28.1 mm
488	323	405	493	560	630	690
976	263	340	415	470	530	580
1,465	213	293	375	425	478	525
1,953	185	253	330	385	445	488
2,441	165	225	295	333	403	460
2,929	150	205	270	315	368	410
3,417	138	190	250	293	343	380
3,906	120	178	220	273	320	355
4,394	108	158	198	258	303	335
4,882	98	143	178	240	285	318
5,370	88	130	160	218	258	303
5,858	80	118	148	200	238	280
6,347	75	110	135	185	220	260
6,835	70	103	126	170	203	240
7,323	65	95	118	160	190	225

* Plywood continuous across two or more spans
** 6.2 mm and 9.4 mm plywood not recommended for use in weak direction

Table 4
Maximum Tie Spacing for Class A (Permanent) Form Design

Single Wale

Load on Wales (Lbs./Ft)	2" x 4"		2" x 6"		2" x 8"		4" x 4"	
	1/8"	1/4"	1/8"	1/4"	1/8"	1/4"	1/8"	1/4"
200	52		76	82	95	109	68	80
400	37		58		77		57	
600	26		44		58		46	
800	23		36		47		40	
1000	20		31		41		36	
1200	17		26		36		31	
1400	16		25		33		28	
1600	15		23		31		25	
1800	14		22		29		24	
2000	13		21		26		22	
2200	13		20		26		20	
2400	12		19		25		19	
2600	12		19		25		18	
2800	11		18		24		18	
3000	11		18		23		17	

Double Wales

Load on Wales (Lbs./Ft)	2" x 4"		2" x 6"		2" x 8"		2" x 10"		2" x 12"	
	1/8"	1/4"	1/8"	1/4"	1/8"	1/4"	1/8"	1/4"	1/8"	1/4"
200	66	74	92	110	113	135	136	162	156	188
400	53		76	82	95	109	115	136	133	156
600	43		67		86	101	104	113	120	138
800	37		58		77	96		112	119	
1000	32		51		61		85		104	
1200	26		44		58		74		90	
1400	25		39		52		66		80	
1600	23		36		47		60		73	
1800	21		33		44		56		68	
2000	20		31		41		52		63	
2200	18		29		38		49		59	
2400	18		28		36		46		56	
2600	17		26		35		44		54	
2800	16		25		33		42		51	
3000	15		24		32		41		50	

Load on Wales (Lbs./Ft)	3" x 10"		3" x 12"		4" x 4"		4" x 10"		4" x 12"	
	1/8"	1/4"	1/8"	1/4"	1/8"	1/4"	1/8"	1/4"	1/8"	1/4"
200	155	184	179	213	81	97	168	200	195	232
400	130	155	151	179	68	78	142	168	164	195
600	118	140	136	162	62	65	126	152	148	176
800	109	127	127	151	57		119	142	138	164
1000	104	113	120	128	51		113	134	130	155
1200	99	103	115	126	46		108	122	125	148
1400	95	96	110	116	43		104	113	120	138
1600	88		107	107	40		100	106	116	129
1800	80		96		38		97	100	113	121
2000	74		90		36		95	95	110	115
2200	69		84		34		89		107	108
2400	65		79		32		83		101	
2600	61		74		30		78		95	
2800	58		71		28		74		90	
3000	56		68		27		70		86	

Table 4.A (metric)
Maximum Tie Spacing for Class A (Permanent) Form Design

Load on Wales (Kg/Meter)	Single Wale 50 mm x 100 mm		50 mm x 150 mm		50 mm x 200 mm		100 mm x 100 mm	
	3.1	6.3	3.1	6.3	3.1	6.3	3.1	6.3
200	1300		1950	2050	2375	2725	1700	2000
400	925		1450		1925		1425	
600	700		1100		1450		1150	
800	675		900		1175		1000	
1000	600		775		1025		900	
1200	425		700		900		775	
1400	400		625		825		700	
1600	375		575		775		625	
1800	350		550		725		600	
2000	325		525		700		550	
2200	325		500		700		500	
2400	300		475		625		475	
2600	300		475		625		450	
2800	275		450		600		450	
3000	275		450		575		425	

Load on Wales (Lbs./Ft.)	Double Wales 50 mm x 100 mm		50 mm x 150 mm		50 mm x 200 mm		50 mm x 250 mm		50 mm x 300 mm	
	3.1	6.3	3.1	6.3	3.1	6.3	3.1	6.3	3.1	6.3
200	1650	1850	2300	2750	2625	3375	3400	4050	3950	4700
400	1325		1950	2050	2375	2725	2675	3400	3325	3950
600	1075		1675		2150	2525	2600	2825	3000	3450
800	825		1450		1925		2400	3400	2800	2975
1000	600		1275		1625		2125		2600	
1200	700		1100		1460		1850		2250	
1400	625		975		1300		1650		2000	
1600	575		900		1175		1600		1825	
1800	525		825		1100		1400		1700	
2000	500		775		1025		1300		1575	
2200	450		725		950		1225		1475	
2400	450		700		900		1150		1400	
2600	425		650		875		1100		1350	
2800	400		625		825		1050		1275	
3000	375		600		800		1025		1250	

Load on Wales (Lbs./Ft.)	75 mm x 250 mm		75 mm x 300 mm		100 mm x 100 mm		100 mm x 250 mm		100 mm x 300 mm	
	3.1	6.3	3.1	6.3	3.1	6.3	3.1	6.3	3.1	6.3
200	3875	4600	4475	5325	2025	2425	4200	5000	4875	5800
400	3250	3875	3775	4475	1700	1950	3650	4200	4100	4875
600	2950	3500	3400	4050	1550	1825	3200	3800	3700	4400
800	2725	3175	3175	3775	1425		2975	3650	3450	4100
1000	2600	2825	3000	3200	1275		2825	3350	3250	3875
1200	2475	2575	2875	3150	1150		2700	3050	3125	3700
1400	2375	2400	2750	2900	1075		2600	2625	3000	3450
1600	2200		2675	2675	1000		2500	2650	2900	3225
1800	2000		2450		950		2425	2500	2825	3025
2000	1850		2250		900		2375	2375	2750	2875
2200	1725		2100		850		2225		2675	2700
2400	1625		1975		800		2075		2525	
2600	1525		1850		750		1950		2375	
2800	1450		1775		700		1850		2250	
3000	1400		1700		675		1750		2150	

SPACING OF SUPPORTS FOR CLASS I PLYWOOD

FACE GRAIN ACROSS SUPPORTS

ALLOWABLE STUD SPACING (CENTER TO CENTER IN INCHES)

FORM PRESSURE (PSF)

1-1/8"
1"
7/8"
3/4"
5/8"
1/2"
3/8"
1/4"

Class A Formwork
(Reuse or Permanent)
Partially Continuous Spans
Δ = Limit Deflection = $l/360$ to 1/16 in.
E = 1.6 x 10^6 psi
F_b = 1600 psi
H or F_v = 75 psi

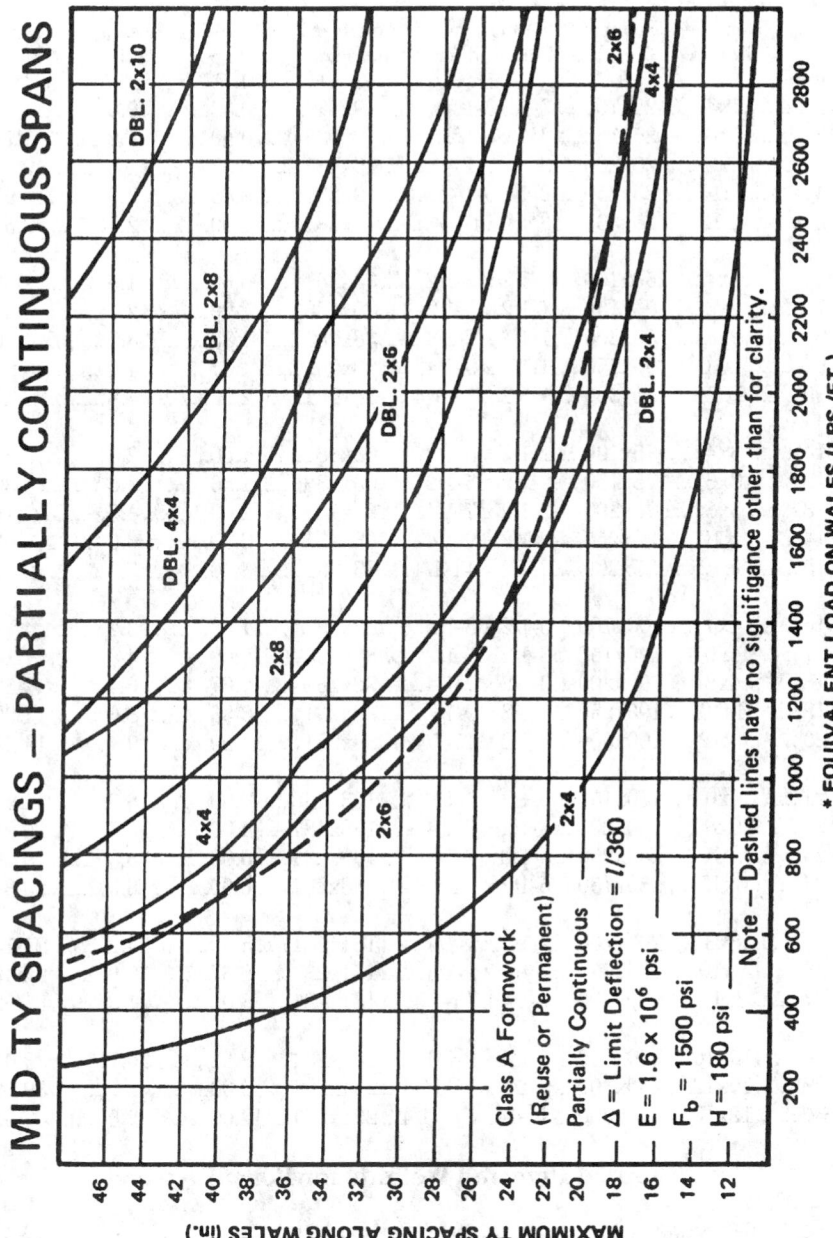

MID TY SPACINGS – PARTIALLY CONTINUOUS SPANS

DBL. 2x10

DBL. 2x8

DBL. 2x6

DBL. 4x4

2x8

4x4

2x6

2x4

DBL. 2x4

2x8

4x4

Class A Formwork
(Reuse or Permanent)
Partially Continuous
Δ = Limit Deflection = $l/360$
E = 1.6 × 10⁶ psi
F_b = 1500 psi
H = 180 psi

Note – Dashed lines have no signifigance other than for clarity.

* EQUIVALENT LOAD ON WALES (LBS./FT.)

MAXIMUM TY SPACING ALONG WALES (in.)

Lumber Size and Weight Table

Item	Nominal Size, in.	Nominal Size, mm	American Standard, in. b x h	American Standard, mm b x h	B.F. per L.F.	Lbs. per L.F. Avg. Wgt.	Kg per Lin.M Avg. Wgt.
1	1x4	25x100	3/4 x 3-1/2	19x88	0.33	0.8	1.19
2	1x6	25x150	3/4 x 5-1/2	19x138	0.50	1.2	1.79
3	1x8	25x200	3/4 x 7-1/4	19x181	0.67	1.6	2.38
4	1x10	25x250	3/4 x 9-1/4	19x231	0.83	2.0	2.98
5	1x12	25x300	3/4 x 11-1/4	19x281	1.00	2.4	3.57
6	2x4	50x100	1-1/2 x 3-1/2	33x88	0.66	1.6	2.38
7	2x6	50x150	1-1/2 x 5-1/2	38x138	1.00	2.4	3.57
8	2x8	50x200	1-1/2 x 7-1/4	38x181	1.33	3.2	4.76
9	2x10	50x250	1-1/2 x 9-1/4	38x231	1.67	4.0	5.95
10	2x12	50x300	1-1/2 x 11-1/4	38x281	2.00	4.8	7.14
11	3x4	75x100	2-1/2 x 3-1/2	63x88	1.00	2.4	3.57
12	3x6	75x150	2-1/2 x 5-1/2	63x138	1.50	3.6	5.36
13	3x8	75x200	2-1/2 x 7-1/4	63x181	2.00	4.8	7.14
14	3x10	75x250	2-1/2 x 9-1/4	63x231	2.50	6.0	8.93
15	3x12	75x300	2-1/2 x 11-1/4	63x281	3.00	7.3	10.86
16	4x4	100x100	3-1/2 x 3-1/2	88x88	1.33	3.2	4.76
17	4x6	100x150	3-1/2 x 5-1/2	88x138	2.00	4.8	7.14
18	4x8	100x200	3-1/2 x 7-1/4	88x181	2.67	6.4	9.52
19	4x10	100x250	3-1/2 x 9-1/4	88x231	3.33	8.0	11.90
20	4x12	100x300	3-1/2 x 11-1/4	88x281	4.00	9.7	14.43
21	6x6	150x150	5-1/2 x 5-1/2	138x138	3.00	8.0	11.90
22	6x8	150x200	5-1/2 x 7-1/4	138x181	4.00	10.7	15.92
23	6x10	150x250	5-1/2 x 9-1/4	138x231	5.00	13.3	19.79
24	6x12	150x150	5-1/2 x 11-1/4	138x281	6.00	16.0	23.81
25	8x8	200x200	7-1/4 x 7-1/4	181x181	5.33	14.2	21.13
26	8x10	200x250	7-1/4 x 9-1/4	181x231	6.67	17.8	26.49
27	8x12	200x300	7-1/4 x 11-1/4	181x281	8.00	21.3	31.69
28	10x10	250x250	9-1/4 x 9-1/4	231x231	8.33	22.2	33.03
29	10x12	250x300	9-1/4 x 11-1/4	231x281	10.00	26.7	39.73
30	12x12	300x300	11-1/4 x 11-1/4	281x281	12.00	32.0	47.62

Nominal Plywood Weights for Form Plywood

Item	Nominal Size, in.	Nominal Size, mm	Weight lbs./s.f.	Weight kg/sq.m	Lbs. per 4'x8' sht. (32 s.f.)
1	1/4	6.30	0.78	3.81	25
2	1/2	12.50	1.53	7.47	49

3	5/8	15.6	1.81	8.84	58
4	3/4	18.8	2.22	10.84	71
5	7/8	21.9	2.60	12.69	83
6	1-1/8	28.1	3.34	16.31	107

Nail Sizes and Weights

Common Nails

Sizes	Gauge	Length in inches	Length in mm	Approx. No. of Pcs. per lb.	Approx. No. of Pcs. per kg
2d	15	1	25.0	840	1852
3d	14	1-1/4	31.3	543	1197
4d	12-1/2	1-1/2	37.5	294	648
6d	11-1/2	2	50.0	167	368
7d	11-1/2	2-1/4	56.3	149	328
8d	10-1/4	2-1/2	62.5	101	223
9d	10-1/4	2-3/4	68.8	91	201
10d	9	3	75.0	66	146
12d	9	3-1/4	81.3	61	134
16d	8	3-1/2	87.5	47	104
20d	6	4	100.0	30	66
30d	5	4-1/2	112.5	20	44
40d	4	5	125.0	17	37
50d	3	5-1/2	137.5	13	29

Double Headers/Duplex Nails

Sizes	Gauge	Length in inches	Length in mm	Approx. No. of Pcs. per lb.	Approx. No. of Pcs. per kg
8d	10-1/4	2-1/4	56.3	8	18
16d	8	3	75.0	44	97

Concrete Nails

Gauge	Length in inches	Length in mm	Approx. No. of Pcs. per lb.	Approx. No. of Pcs. per kg
5	1/2	12.5	185	409
5	5/8	15.6	146	323
5	3/4	18.8	127	280
5	7/8	21.9	112	247
5	1	25.0	100	220
5	1-1/4	31.3	78	171
5	1-1/2	37.5	82	180
5	1-3/4	43.8	61	135
5	2	50.0	59	130
5	2-1/2	62.5	39	87
5	3	75.0	33	74

Concrete Quantities for Wall Forms per Square Foot Contact Area
50'-0" x 8'-0" (15 x 2.4 m), no bulkheads

Wall Thickness in Inches	Sq. Ft. Contact Area, less Bulkheads	Cu. Ft. of Concrete	Cu. Yd. of Concrete	Sq. Ft. per Cu. Yd.	Cu. Yd. per Sq. Ft.
4	800	133.3	4.9	162.0	0.006
6	"	200.0	7.4	108.0	0.009
8	"	266.7	9.9	81.0	0.012
10	"	333.3	12.3	64.8	0.015
12	"	400.0	14.8	54.0	0.019
14	"	466.7	17.3	46.3	0.022
16	"	533.3	19.8	40.5	0.025
18	"	600.0	22.2	36.0	0.028
20	"	666.7	24.7	32.4	0.031
22	"	733.3	27.2	29.5	0.034
24	"	800.0	29.6	27.0	0.037
26	"	866.7	32.1	24.9	0.040
28	"	933.3	34.6	23.1	0.043
30	"	1,000.0	37.0	21.6	0.046
32	"	1,066.7	39.5	20.3	0.049
34	"	1,133.3	42.0	19.1	0.052
36	"	1,200.0	44.4	18.0	0.056

Example: 50'-0" long x 8'-0" high wall, 12" thick, no bulkheads
50' x 8' = 400 s.f. x 2 sides = 800 s.f. contact surface area.
50' x 8' x 1' = 400 c.f. ÷ 27 = 14.81 c.y. concrete.
800 s.f. ÷ 14.81 c.y. = 54 s.f. of forms per c.y.
14.81 c.y. of concrete ÷ 800 = 0.019 c.y. concrete per s.f. of forms.

(Metric)

Wall Thickness in mm	Sq. M Contact Area, less Bulkheads	Cu. M of Concrete	Sq. M per Cu. M	Cu. M per Sq. M
100	72	3.6	20.0	0.050
150	"	5.4	13.3	0.075
200	"	7.2	10.0	0.100
250	"	9.0	8.0	0.125
300	"	10.8	6.7	0.150
350	"	12.6	5.7	0.175
400	"	14.4	5.0	0.200
450	"	16.2	4.4	0.225
500	"	18.0	4.0	0.250
550	"	19.8	3.6	0.275
600	"	21.6	3.3	0.300
650	"	23.4	3.1	0.325

Wall Thickness in mm	Contact Area, less Bulkheads	Cu. M of Concrete	Sq. M per Cu. M	Cu. M per Sq. M
700	"	25.2	2.9	0.350
750	"	27.0	2.7	0.375
800	"	28.8	2.5	0.400
850	"	30.6	2.4	0.425
900	"	32.4	2.2	0.450

Example: 15 m long x 2.4 m high wall, 300 mm thick, no bulkheads
            15 m x 2.4 m = 36 sq.m x 2 sides = 72 sq.m contact surface area.
            15 m x 2.4 m x 0.3 m = 10.8 cu.m of concrete
            72 sq.m ÷ 10.8 cu.m = 6.66 sq.m of forms per cu.m
            10.8 cu.m of concrete ÷ 72 = 0.15 cu.m concrete per sq.m of forms.

Window and Door Openings in Architectural Concrete Walls. It is necessary to form for all door and window openings in concrete walls. The outside wall form is built up solid. Then, boxes or "bucks" are constructed the exact size of the door and window openings. They should be made of 2" (50 mm) lumber and held in place by 1"x4" (25 x 100 mm) strips nailed to both the outside and inside wall sheathing and braced where necessary using 2"x4"s (50 x 100 mm) placed horizontally and diagonally. Where the opening is over 3'-0" (0.9 m, 900 mm) wide, it will be necessary to leave an opening in the bottom of the frame so that concrete can be filled to the proper level. After filling to the correct level, the opening is closed by nailing a piece of plank to the bottom of the frame. A few hours after the concrete has set, the plank is removed and the surplus concrete scraped off. To do this, it is necessary to leave an opening in the inside form at every opening.

Reveals for door and window openings should be designed so that standard size 2" (50 mm) lumber may be used.

Rustication Strips. Rustication strips should be designed as narrow as possible and only about 3/4" (19 mm) deep. When the strips are less than 1" (25 mm) wide, they should have one saw cut. When 2" (50 mm) wide, they should have 2 saw cuts in the back as illustrated. These strips should be nailed to a chalk line, using casing nails long enough to go through the strip and sheathing. It is recommended that the nails be withdrawn by pulling them through the sheathing before the form is removed and allow the wooden strip to remain in place a few days after the form is removed or long enough so that they can be withdrawn without injuring the edges of the concrete.

Ornamentation

Belt courses or other ornamentation where continuous members occur may be formed with wood moldings cut to the proper shape. While some contractors may prefer a waste mold for this particular detail, the wood mold has the advantage of being more easily held in alignment, because the various members are broken at different points, while in a plaster waste mold, the entire form would be cut in two at one point. Soft grained wood that does not warp or split easily should be used. Soft white pine is best, soft-grained Douglas fir is the second choice. In the cornice illustrated, the various members are narrow enough so that common stock sizes can be used throughout. Note the generous use of saw-kerfs in the back side of members to prevent swelling and warping of the lumber. The lumber should be thoroughly oiled on all sides before building the forms as further precaution against swelling and warping.

Large, heavy wood members should be avoided in detailing for ornamental work, because they will swell, and in stripping the corners of the projecting concrete will be fractured. Using several smaller pieces requires less lumber and the members may be arranged so that swelling is away from the concrete, making it easier to strip the form without damage to the concrete corners. Note the illustration where the left-hand detail shows a large member milled from a single wide board in contrast to the smaller pieces shown in the right-hand detail.

In assembling forms, one must constantly keep in mind the steps that can be taken to aid in removing them without injury to the concrete. Boxes, waste molds, wood molds, rustication strips, or anything applied to the face of the forms should be nailed lightly so that when the forms are removed, these members will pull loose from the wall form and remain in the concrete. After these materials have dried out thoroughly and have shrunk, they can be removed without much difficulty and without injury to the concrete corners or edges.

As previously indicated, solid strips of wood, even though well oiled, may swell and result in considerable breakage of concrete corners when the strips are removed. Saw-kerfs on the back of the member will prevent this trouble by relieving any pressure against the concrete. They should be approximately two-thirds the depth of the member, not more than 1-1/2" (37 mm) apart, and in general, there should be one kerf within 3/4" (19 mm) of each edge.

Waste molds are the only solution for some highly ornamental work. They should be made only by experienced ornamental plaster workers who have been given instructions as to how they are to be fitted into the forms. Ordinarily, the waste molds are about 2" (50 mm) thick, reinforced on the back with 2"x4"s (50 x 100 mm) that are attached to the mold with burlap dipped in plaster and then wound around the wood. The 2"x4"s (50 x 100 mm) permit easy handling of the mold and are also used to attach the mold to the form.

Fig.3b- Second Stage
TYPICAL HORIZONTAL CONSTRUCTION JOINT

Fig.4b- Second Stage
CONSTRUCTION JOINT AT FLOOR LEVEL

Fig.3a- First Stage
TYPICAL HORIZONTAL CONSTRUCTION JOINT

Fig.4a- First Stage
CONSTRUCTION JOINT AT FLOOR LEVEL

HORIZONTAL CONSTRUCTION JOINTS

*Method of Forming for Architectural Concrete Waste Molds for Belt
Courses and Cornices*

When waste molds are large and have deep undercuts, it is usually best to
wire them back to the studs or wales with enough wire to be sure that all
points will be pressed firmly against the form. Any openings between the
form and waste mold should be pointed with plaster of Paris or a patching
plaster and nailheads driven through the face and around the edge of the
waste molds should be countersunk and similarly pointed.

Ornamental Concrete Cornices. Where it is necessary to use molds for
belt courses, cornices, door and window reveals, columns, cornices, entrances,
etc., it is more economical to build them of wood, if practicable. However, if
plaster waste molds are required, the following will give some idea of their
cost, although on this work, it is always advisable to obtain definite prices
from a concern specializing in this work.

Plain molds consisting of a few straight lines, $10.00 per s.f. ($111 per
sq.m).

Ordinary molds of stock design will cost from $10.00 to $14.00 per s.f.
($111-155 per sq.m).

Wood Mold Used in Cornice Form

Simple molds will cost from $10.00 to $14.00 per s.f. ($111-155 per sq.m) plus the cost of waste molds.

Cornice molds of stock design about 1'-6" (0.45 m, 450 mm) deep and 1'-3" (0.37 m, 375 mm) wide cost about $18.00 to $25.00 l.f. ($60-83 per m).

Capitals for 1'-6" (0.45 m, 450 mm) pilasters, stock designs, cost $170 to $280 each.

Joints in plaster waste molds should be patched with non-shrinking patching plaster and painted with shellac.

Individual pieces should not weigh more than 150 lbs. (68 kg) to be set without difficulty by 2 workers.

Plaster waste molds are usually given 2 coats of shellac in the shop to make them waterproof and non-absorbent. Before concrete is placed all molds should be greased with a light yellow cup-grease, which may be cut with kerosene if too thick. The grease should be wiped into all angles of the mold and every bit of surplus grease carefully wiped off. Care must be taken not to drop oil, grease or shellac onto hardened concrete or reinforcing.

Construction Joints. Much progress has been made on all types of concrete work in production of satisfactory construction joints, both from the standpoint of good appearance and good bond between successive lifts. On the other hand, some very unsightly joints, for which there is no excuse, occasionally are made on architectural concrete work. It is just as easy to produce a good joint as a poor one. The designer, of course, should indicate on his plans exactly where such joints should be located. Rustications or offsets will mask the joints and are advisable where they can be made to fit into the design.

One requirement that is sometimes neglected is to provide means of holding the form tightly against the hardened concrete. As a result, leakage occurs, discoloring the concrete below, and the upper lift will project slightly beyond the lower lift. The illustration shows the use of 5/8" (15 mm) bolts cast in the concrete for fastening the form for the next lift. Some types of form ties can be used instead of bolts. The bolt or form tie assembly can be used over and over, because only the nut is left in the concrete, if the bolt is used, or the insert, if a form tie is used. In re-erecting the forms for the next lift, the contact surface of the form sheathing should not overlap the hardened concrete more than 1" (25 mm). More overlapping than this only presents more opportunity for leakage due to irregularities in the wall surface against which the sheathing is to be held.

Since sufficient time must be allowed for the concrete between windows or other openings to settle and shrink before the concrete above is placed, the tops of such openings are "must" locations for construction joints. Similarly, it is desirable to locate a joint at the sill line. By so doing, the joints are broken up into short lengths by the opening, making them relatively inconspicuous. Moreover, cracking at corners of openings is thereby minimized.

The construction joint should be only a straight thin line at the surface. If a rustication strip is not used at the joint, a 3/4" (19 mm) wood strip to a point 1/2" (12 mm) above the bottom of this strip and just before the concrete becomes hard, the top should be lightly tamped to be sure that it is tight up against the strip after it has settled or shrunk. All surplus concrete or laitance should then be removed.

In resuming the placing of concrete in the new lift, steps should be taken to get good bond and to avoid honeycomb. The hardened concrete should be clean and thoroughly saturated. A 6" (150 mm) layer of concrete in which the coarse aggregate is reduced about 50%, or a 2" (50 mm) layer of cement-sand grout should then be placed on the hardened concrete followed by the regular concrete.

Control Joints. It is important, both for appearance and performance, that control joints be carefully installed so they will be perfectly straight and vertical. Care is required during placing of concrete to avoid knocking the joint out of alignment or otherwise injuring it. When control joints are to be caulked with a plastic compound, this should be done before the final clean down so that any compound smeared onto the surface is removed. Of course, excessive smearing should be avoided.

Form Ties. Various makes of form ties are available for architectural concrete construction. Important features to look for in selecting form ties are: removal of all metal to a depth of at least 1-1/2" (38 mm) from the face of the wall, minimum strength of 3,000 lbs. (1360 kg) when fully assembled, adjustable length to permit tightening of forms around all openings and inserts and a design that will leave a hole at the surface not more than 7/8" (22 mm) in diameter and not less than 3/4" (19 mm) deep so that it will hold a patch. Ties should not be fitted with cones or washers acting as spreaders, because tapered and shallow holes cannot be patched properly.

Labor Cost of Architectural Forms. The labor fabricating, erecting, and stripping forms is usually estimated on the basis of sq. ft. (sq. m) of contact area. It is customary on a job having a normal amount of ornamentation to take off the total contact area as though the wall were plain. Window and other openings, unless very large, are figured solid. The area thus obtained is priced as though the wall were unornamented. Separate allowance is made for window frames or "bucks", ornamentation, etc.

The average amount of lumber required for plain wall forms is 2-1/4 to 2-1/2 b.f. (0.05-0.06 cu.m) of lumber per s.f. (sq.m) of contact area, plus the sheathing, plywood, or concrete form Presdwood, used for the facing materials. Figure one tie rod for approximately each 10 s.f. (9 sq.m) of form.

On plain walls up to 10 ft. (3 m) story height, it requires about 11 to 13 hrs. carpenter time and 6 to 8 hrs. labor time per 100 s.f. (9 sq.m) of contact area. For walls 12' (3.6 m) to 14' (4.2 m) high, add about 10% to these costs. This includes all time necessary for erecting and stripping the forms and hoisting it up for the next use. In addition, figure about 1 to 1-1/4 hrs. labor time for removing hardened concrete from the plywood and coating plywood between each use.

Door and window frame boxes or "bucks" require 5 to 8 hrs. carpenter time for each opening, depending upon size and detail of same.

Rustication strips should be placed at the rate of 25 to 30 l.f. (7.5-9 m) per hr. for one carpenter.

Ornamental curtain walls, fluted pilasters and piers, reveals, etc., will require about 10 to 12 hrs. carpenter time and 6 hrs. labor time per 100 s.f. (9 sq.m) in addition to the cost of plain wall forms.

Ornamental wood cornices built to special detail will require 24 to 28 hrs. carpenter time and 12 to 14 hrs. labor time per 100 s.f. (9 sq.m) in addition to cost of plain wall forms.

Ornaments projecting from the face of plain walls, such as special sills, balconies, etc., will require about 36 to 40 hrs. carpenter time and 12 to 15 hrs. labor time per 100 s.f. (9 sq.m) in addition to the cost of plain wall forms.

Where plaster waste molds are used, they may be set either by carpenters or plasterers, depending upon local labor union regulations.

When setting simple belt courses of plaster waste mold, allow 12 to 14 hrs. mechanic time and 6 to 7 hrs. helper time per 100 s.f. (9 sq.m).

When erecting and blocking cornices and other elaborate moldings using plaster waste molds, figure 20 to 25 hrs. mechanic time and 10 to 12 hrs. helper time per 100 s.f. (9 sq.m).

When placing letters, a carpenter should set 2 to 3 letters an hr. depending upon size.

*Types of Form Ties Available for Use in
Architectural Concrete Construction*

Labor Cost of 100 S.F. (Sq.M) of Plain Architectural Concrete Walls Using Plywood for Facing Material

	Hours	Rate	Total	Rate	Total
Carpenter					
erecting forms	12.0	$....	$....	$26.44	$ 317.28
Labor helping	6.0	....	....	21.04	126.24
Labor stripping and					
cleaning forms	2.0	....	....	21.04	42.08
Labor coating plywood	0.33	....	....	21.04	6.94
Cost per 100 s.f. (9 sq.m)			$....		$ 492.54
cost per s.f.				....	4.93
cost per sq.m				....	53.05

Details for Architectural Concrete Forms. For the rapid and economical construction of architectural concrete forms, it is necessary to have complete working drawings that show the elevations in large scale, together with every piece of plywood, stud, wale, bolt, lumber for reveals, window boxes or "bucks", and so on, similar to the methods used for detailing cut stone, terra cotta, or structural steel.

03200 CONCRETE REINFORCEMENT

Reinforcing steel is estimated by the pound or the ton, obtained by listing all bars of different sizes and lengths and extending the total to pounds. Reinforcing bars may be purchased from warehouse in stock lengths and all cutting and bending done on the job or they may be purchased cut to length and bent, ready to place in the building.

Example: Assume a retaining wall is reinforced with #5 bars spaced 6" (150 mm) on centers both vertically and horizontally. The wall is 40'-0" long and 15'-0" high and with bars spaced 6" (150 mm) on centers it will require the following:

80 #5 bars 15'-0" (4.5 m) long = 1,200 l.f. (360 m) @ 1.043 lbs. (1.55 kg/m) = 1,252 lbs. (559 kg)

30 #5 bars 40'-0" (12 m) long = 1,200 l.f. (360 m) @ 1.043 lbs. (1.55 kg/m) = 1,252 lbs. (559 kg)

Total weight of steel required = 2,504 lbs. (1118 kg)

If reinforcing steel is worth $60.00 per 100 lbs. (45.35 kg) in place, or $0.60 per lb. ($1.32/kg), the cost of the steel in the wall ready for pouring would be 2,504 lbs. (1118 kg) x 0.60 (1.32) = $1502.40 ($1,476).

When purchased in stock lengths, the labor cost of unloading, sorting, handling, and storing is less than when sent to the job cut to length and bent. The former requires less storage space, because the different lengths are placed in compact piles. Steel cut to length and bent requires considerable storage space, because all bars of each size and length must be placed in separate piles.

However, the shop costs of cutting to length and bending are much less than the job costs, especially where the work is done by iron workers. Practically all reinforcing steel used today is shop fabricated, cut to length and bent prior to delivery to the job.

The total weight of the bars remains the same as previously, but the actual diameter of the bar is slightly less.

For this reason, instead of calling the bars 1/2" (13 mm), 3/4" (19 mm), etc., they now are known as No. 3, 4, 5, 6, etc.

The following table gives the bar classifications:

Standard Sizes and Weights of Concrete Reinforcing Bars

Bar Designation (Numbers)	Unit Weight lbs. per ft.	Nominal Dimensions—Round Sections Diameter in inches	Cross Sectional Area sq. inches	Perimeter Inches
3	0.376	0.375	0.11	1.178
4	0.668	0.500	0.20	1.571
5	1.043	0.625	0.31	1.963
6	1.502	0.750	0.44	2.356
7	2.044	0.875	0.60	2.749
8	2.670	1.000	0.79	3.142
9	3.400	1.128	1.00	3.544
10	4.303	1.270	1.27	3.990
11	5.313	1.410	1.56	4.430
14	7.650	1.693	2.25	5.320
18	13.300	2.257	4.00	7.090

Bar Designation	Unit Weight kg per meter	(metric) Diameter in mm	Cross Sectional Area sq.cm	Perimeter mm
10 M	0.785	11.30	100	35.48
15 M	1.570	16.00	200	50.24
20 M	2.355	19.50	300	61.23
25 M	3.925	25.20	500	79.13
30 M	5.495	29.90	700	93.89
35 M	7.850	35.70	1000	112.10
45 M	11.775	43.70	1500	137.22
55 M	19.625	56.40	2500	177.10

Basis for Estimating Price of Reinforcing Steel. When estimating quantities and weights of reinforcing steel, all bars of each size should be listed separately to obtain the correct weight. No. 6 bars and over take the lowest price per lb. (kg), while bars smaller than No. 6 increase in price on a graduated scale depending on the size of the bars. The smaller the bars, the higher the price per lb. (kg). On large projects the normal practice is to quote the price per lb. (kg), or ton, based on all sites on the project as a whole.

Example: If base price is $60.00 per 100 lbs. ($1.32/kg) what is price of No. 3 bars? The table shows that No. 3 bars take an increase of $5.75 per 100 lbs. (0.13/kg) over base or $65.75 per 100 lbs. ($1.45/kg).

Prices are usually quoted f.o.b. warehouse, with freight allowed to destination.

Size Extras on Reinforcing Bars

Number	Extra per ton	Extra per metric ton
14 and 18	$44.00*	$48.50*
7 to 11	22.00	24.25
6	28.00	30.86
5	33.00	36.37
4	44.00	48.50
3**	127.00	140.00

*Includes cutting extra.
**The #3 bars are not available in some areas.

Prices on rail steel are theoretically the same as for billet steel but will vary with the scrap market on rails.

There are four specifications covering the use of reinforcement steel:

1. ASTM A615: Deformed and plain billet-steel bars for concrete reinforcement.
2. ASTM A616: Rail steel deformed and plain bars for concrete reinforcement.
3. ASTM A617: Axle steel deformed and plain bars for concrete reinforcement.
4. ASTM A706: Low-alloy steel deformed bars for concrete reinforcement.

It is of utmost importance that the quantity take-off be correct. Normal practice is to allow between 2% to 5% of the calculated bar weight for waste and splices. Reinforcement steel up to No. 11 bars may be spliced by overlapping them and wiring them together.

Two standard grades of reinforcement steel available are, ASTM A615, Grade 40 and ASTM A615, Grade 60. Bars are of two minimum yield levels:

40,000 psi (275.8 MPa) and 60,000 psi (413.7 MPa), designated as Grade 40 and Grade 60 respectively.

Both grades are standard in building codes and in the design recommendations of the American Concrete Institute.

Reinforcement Bar Supports. Chairs, sometimes referred to as reinforcement bar support, are required in order to accurately support and locate reinforcement steel in the concrete form. The bar support will hold the bar firmly in its proper location during placement of concrete.

Rust Prevention. Reinforcement steel supports are classified with respect to methods to diminish rust spots or similar imperfections from the concrete surface caused by reinforcement steel supports. There are three classes:

Class 1: Maximum protection; plastic protected reinforcement steel support which are intended for use in situations of moderate to severe exposure and/or situations requiring light grinding or sandblasting of concrete surface.

Class 2: Moderate protection; stainless steel protected reinforcement steel supports, which are intended for use in situations of moderate exposure and/or situations requiring light grinding or sandblasting of the concrete surface. Class 2 protection can be obtained by the use of either Type "A" or Type "B" stainless steel protection reinforcement steel supports. The difference between the two types is in the length of the stainless steel tip attached to the bottom of each leg support.

Type "A" - Stainless steel protected bar support; a tip of stainless steel is attached to the bottom of each leg so that no portion of the non-stainless steel wire lies closer than 1/4" (6 mm) to the form face.

Type "B" - Stainless steel protected bar support, a tip of stainless steel is attached to the bottom of each leg so that no portion of the non-stainless steel wire lies closer than 3/4" (19 mm) to the form face.

Class 3: No protection; bright basic reinforcement steel supports that have no protection against rusting and which are intended for use in situations where surface blemishes can be tolerated and/or where supports do not come in contact with the exposed concrete.

Costs for reinforcement steel supports can vary widely, from less than 1% to more than 10% of the reinforcement steel costs.

Bending Reinforcing Steel. The labor cost of bending reinforcing bars will vary with the size of the bars and the amount of bending necessary; the lighter the bars the higher the bending cost per ton.

Type L Bending. Bending all No. 3 bars; all No. 4 stirrups and column ties; truss bars and all sizes of bars bent at more than 6 points. It also includes all bars bent in more than one plane and all radius bending with more than one radius in any bar, approximately $90.00 per ton ($99/metric ton).

Type H Bending. Bending truss bars for beams and slabs, other than No. 3 and No. 4, radius bending and types not otherwise described, approximately $40.00 per ton ($44/metric ton).

When cut is made by means other than sawing or milling, $0.05 per end. When cut is made by sawing or milling, $0.15 per end.

Spirals for Reinforced Concrete Columns. Spirals should always be coiled and fabricated in the shop, as it is all machine work.

Prices of spirals vary throughout the country in the same manner and for the same reasons as reinforcing steel prices. Quantity extras are figured independently of reinforcing steel tonnages involved.

Cost of Engineering Service. The cost of making working drawings, bending details, setting plans, etc., will vary with the amount of work involved.

Where the engineers prepare the complete design for a reinforced concrete structure, including working drawings, placing plans, bar lists, and bending details, add $45.00 per ton to the detailing charges given below.

For preparing order lists with or without bending details from plans on which all sizes and lengths are shown by others, $0.30 per 100 lbs. (45.35 kg) or $6.00 per ton ($6.60/metric ton).

Hauling Reinforcing Bars From Shops to Job. The cost of hauling reinforcing steel from warehouse to job will vary with the distance of the haul and the size of the city.

For example, in the Chicago area, the trucking charge for deliveries within city limits is about $1.50 per cwt. For jobs located in the Chicago suburbs, the trucking charge is about $1.75 per cwt. Minimum charge for both zones is $100.00.

Unloading Reinforcing Steel From Cars or Trucks. If steel can be unloaded direct from cars or trucks, without sorting or carrying to stock piles, a worker will unload 4-1/2 to 5-1/2 tons (4.1-5 metric tons) per 8-hr. day, at the following labor cost per ton:

	Hours	Rate	Total	Rate	Total
Iron Worker	1.75	$....	$....	$30.28	$ 52.99

If it is necessary to carry the bars up to 60 ft. (18 m), sort, and place in stock piles, figure 2-1/2 to 3 tons (2.27-2.75 metric tons) per 8-hr. day per worker, at the following labor cost per ton:

	Hours	Rate	Total	Rate	Total
Iron Worker	3	$....	$....	$30.28	$ 90.84

PERCENTAGES AND WEIGHTS OF STANDARD STEEL REINFORCING SPIRALS[1]

Weights are in pounds per foot of height exclusive of spacers and ends [2][3]

Spirals above the stepped line are not collapsible for shipment

Labor Placing Reinforcing Bars. The labor cost of placing reinforcing steel will vary with the average weight of the bars, the manner in which it is placed; i.e., whether bars may be laid loose or whether it is necessary to tie them in place, and upon the class of labor employed.

In most of the larger cities there are concerns that make a specialty of setting reinforcing steel. These workers become very adept at this work and under favorable conditions can place steel for 25% less than the average iron worker. For this reason costs are given both ways and the estimator should use his own judgment as to the class of workers obtainable.

Table footnotes:

1 Spirals are sold on the basis of theoretical weights.

2 F = Weight to Add for Finishing. (This includes 1½ turns at top and 1½ turns at bottom of spiral—ACI 318-56.) Total Wt. = (Wt. per Ft. x Height) + (F) + Wt. of Spacers.

3 Two spacers are required for Spirals 20″ or less in diameter, 3 spacers for Spirals 20″ to 30″, and 4 spacers for Spirals over 30″ in diameter and for ALL ⅜″ or larger spiral rods, regardless of spiral diameter. Allow ¾ pound per lin. ft. per spacer.

4 The four spiral rod sizes shown in the table have been approved by the U. S. Department of Commerce Simplified Practice Recommendation R53 and are carried in stock ready for fabrication.

Setting Reinforcing Bars. On jobs using No. 5 bars and smaller, where it is not necessary to tie them in place, a worker should handle and place 700 to 900 lbs. (317-408 kg) per 8-hr. day, at the following labor cost per ton:

	Hours	Rate	Total	Rate	Total
Iron Worker per ton	20.7	$....	$....	$30.28	$ 626.80
per metric ton	22.2				672.22

Experienced reinforcing steel setters should handle and place 900 to 1,100 lbs. (408-500 kg) of steel per 8-hr. day, at the following labor cost per ton:

	Hours	Rate	Total	Rate	Total
Iron worker per ton	16	$....	$....	$30.28	$ 484.48
per metric ton	17.7				535.96

On jobs using No. 6 bars and heavier, where it is not necessary to tie the bars in place, a worker should handle and place 900 to 1,100 lbs. (408-500 kg) of steel per 8-hr. day, at the following labor cost per ton:

	Hours	Rate	Total	Rate	Total
Iron worker per ton	16	$....	$....	$30.28	$ 484.48
per metric ton	17.7				535.96

An experienced reinforcing steel setter should handle and place 1,400 to 1,600 lbs. (635-725 kg) of steel per 8-hr. day, at the following labor cost per ton:

	Hours	Rate	Total	Rate	Total
Iron worker per ton	10.6	$....	$....	$30.28	$ 320.97
per metric ton	11.8				357.30

Setting Reinforcing Bars Tied in Place. On jobs having concrete floors where it is necessary to tie the bars in place with tie wire, and where lightweight bars, (No. 5 and under) are used, a worker should handle, place, and tie 600 to 800 lbs. (270-360 kg) per 8-hr. day, at the following labor cost per ton:

	Hours	Rate	Total	Rate	Total
Iron worker per ton	23	$....	$....	$30.28	$ 696.44
per metric ton	25.4				769.11

An experienced reinforcing steel setter should handle, place, and tie 800 to 1,000 lbs. (360-450 kg) of steel per 8-hr. day, at the following labor cost per ton:

	Hours	Rate	Total	Rate	Total
Iron worker per ton	17.8	$....	$....	$30.28	$ 538.98
per metric ton	19.8				599.54

On jobs using heavy bars, (No. 6 and over) a worker should handle, set, and tie 800 to 1,000 lbs. (360-450 kg) of steel per 8-hr. day, at the following labor cost per ton:

	Hours	Rate	Total	Rate	Total
Iron worker per ton	17.8	$....	$....	$30.28	$ 538.98
per metric ton	19.80				599.54

An experienced reinforcing steel setter should handle, set, and tie 1,300 to 1,500 lbs. (590-680 kg) per 8-hr. day, at the following labor cost per ton:

	Hours	Rate	Total	Rate	Total
Iron worker per ton	11.5	$....	$....	$30.28	$ 348.22
per metric ton	12.6				381.53

Epoxy Coated Reinforcing Bars

Epoxy coated reinforcement steel is used in many concrete structures today. The bars are coated, cut, and bent to the approved configuration of shop drawings at the fabrication and coating shop and then are shipped to the project. Coated reinforcement steel shall meet ASTM A775 designation for the coating.

The estimator should allow as a minimum about 60% additional cost to the base price of reinforcement steel for purchasing epoxy coated reinforcement steel. Additional time in the field is required for the extra care needed to unload and handle the coated steel. Additional care refers to items such as nylon slings for unloading epoxy coated steel bars instead of steel wire slings, not walking on the bars after installation, and possibly covering the steel prior to placement into the form. It is suggested that approximately 5% to 25% additional time be added for the placement of epoxy coated reinforcement steel.

All coated reinforcement bars should be tied with coated tie wire. Coated bar supports should be non-corrosive chairs.

Welded Steel Fabric or Mesh Reinforcing

Welded steel fabric is a popular and economical reinforcing for concrete work of all kinds, especially driveways and floors. It may also be used for temperature reinforcing, beam and column wrapping, road and pavement reinforcing, etc. It is usually furnished in square or rectangular mesh.

It is usually sold at a certain price per s.f. or s.y. (sq.m), depending on the weight. Prices of mesh vary with quantity ordered and freight from mill to destination. Typical prices for some styles usually stocked, with truck job site delivery are as follows. Approximate prices are per 100 s.f. (9 sq.m):

Style	Wt. per 100 SF	Under 5 Tons
6x6-6x6 (W2.9)	42 lbs. (19 kg)	$17.00
6x6-8x8 (W2.1)	30 lbs. (13.6 kg)	12.00
4x4-8x8 (W2.1)	44 lbs. (20 kg)	17.00

Labor Placing Mesh Reinforcing In Walls.

When mesh reinforcing is used, furnished in sheets or rolls, a worker should place 700 to 800 s.f. (63-72 sq.m) in walls per 8-hr. day, at the following labor cost per 100 s.f. (9 sq.m):

	Hours	Rate	Total	Rate	Total
Steel setter	1.0	$....	$....	$30.28	$30.28
Cost per s.f.			....		0.30
per sq.m			....		3.23

Sizes and Weights of Welded Steel Fabric

Style	Wgt. per 100 s.f. based on 60" Width	Wgt. per 9.3 sq.m based on 1500 mm Width	Spacing Wires in Inches Longit. in.	Longit. mm	Trans. in.	Trans. mm	A.S.&W.Co. Steel Wire Gauge No. Trans.	Longit.	Sect. Area Sq. in. per ft. Trans.	Sq. in. per ft. Longit.	Sq. cm per m Trans.	Sq. cm per m Longit.
22-1616	13	5.90	2	50	2	50	16	16	.018	.018	0.15	0.15
22-1414	21	9.53	2	50	2	50	14	14	.030	.030	0.25	0.25
22-1313	28	12.70	2	50	2	50	13	13	.039	.039	0.32	0.32
22-1212	37	16.78	2	50	2	50	12	12	.052	.052	0.43	0.43
22-1111	48	21.77	2	50	2	50	11	11	.058	.058	0.48	0.48
22-1010	60	27.22	2	50	2	50	10	10	.086	.086	0.72	0.72
24-1414	16	7.26	4	100	12	300	14	14	.030	.015	0.25	0.12
24-1314	19	8.62	4	100	12	300	13	14	.039	.016	0.32	0.13
24-1212	28	12.70	4	100	12	300	12	12	.052	.026	0.43	0.22
212-38	105	47.63	2	50	12	300	3	8	.280	.021	2.33	0.17
212-06	166	75.30	2	50	12	300	0	6	.443	.029	3.69	0.24
216-812	46	20.87	2	50	16	400	8	12	.124	.007	1.03	0.06
216-711	55	24.95	2	50	16	400	7	11	.148	.008	1.23	0.07
216-610	65	29.48	2	50	16	400	6	10	.174	.011	1.45	0.09
216-510	75	34.02	2	50	16	400	5	10	.202	.011	1.68	0.09
216-49	89	40.37	2	50	16	400	4	9	.239	.013	1.99	0.11
216-38	104	47.17	2	50	16	400	3	8	.280	.015	2.33	0.12
216-28	119	53.98	2	50	16	400	2	8	.325	.015	2.70	0.12
216-17	139	63.05	2	50	16	400	1	7	.377	.018	3.14	0.15
33-1414	14	6.35	3	75	3	75	14	14	.020	.020	0.17	0.17
33-1212	25	11.34	3	75	3	75	12	12	.035	.035	0.29	0.29
33-1111	32	14.52	3	75	3	75	11	11	.046	.046	0.38	0.38
33-1010	41	18.60	3	75	3	75	10	10	.057	.057	0.47	0.47
33-99	49	22.23	3	75	3	75	9	9	.069	.069	0.57	0.57

Style	Wgt. per 100 s.f. based on 60" Width	Wgt. per 9.3 sq.m based on 1500 mm Width	Spacing Wires in Longit. in.	Longit. mm	Trans. in.	Trans. mm	A.S.&W.Co. Steel Wire Gauge No. Trans.	Longit.	Sect. Area Sq. in. per ft. Trans.	Sq. in. per ft. Longit.	Sq. cm per m Trans.	Sq. cm per m Longit.
33-88	58	26.31	3	75	3	75	8	8	.082	.082	0.68	0.68
316-812	32	14.52	3	75	16	400	8	12	.082	.007	0.68	0.06
316-711	38	17.24	3	75	16	400	7	11	.098	.009	0.82	0.07
316-60	45	20.41	3	75	16	400	6	10	.116	.011	0.97	0.09
316-510	52	23.59	3	75	16	400	5	10	.135	.011	1.12	0.09
316-49	61	27.67	3	75	16	400	4	9	.159	.013	1.32	0.11
316-38	72	32.66	3	75	16	400	3	8	.187	.015	1.56	0.12
316-28	83	37.65	3	75	16	400	2	8	.216	.015	1.80	0.12
316-17	96	43.55	3	75	16	400	1	7	.252	.018	2.10	0.15
316-06	113	51.26	3	75	16	400	0	6	.295	.022	2.45	0.18
44-1414	11	4.99	4	100	4	100	13	13	.015	.015	0.12	0.12
44-1313	14	6.35	4	100	4	100	12	14	.020	.020	0.17	0.17
44-1212	19	8.62	4	100	4	100	12	12	.026	.026	0.22	0.22
44-1010	31	14.06	4	100	4	100	11	12	.043	.043	0.36	0.36
44-88	34	15.42	4	100	4	100	10	12	.062	.062	0.52	0.52
44-77	53	24.04	4	100	4	100	9	12	.074	.074	0.62	0.62
44-66	62	28.12	4	100	4	100	8	12	.087	.087	0.72	0.72
44-44	85	38.56	4	100	4	100	7	11	.120	.120	1.00	1.00
48-1313	11	4.99	4	100	8	200	13	13	.020	.010	0.17	0.08
48-1214	12	5.44	4	100	8	200	12	14	.026	.008	0.22	0.07
48-1212	14	6.35	4	100	8	200	12	12	.026	.013	0.22	0.11
48-1112	17	7.71	4	100	8	200	11	12	.034	.013	0.28	0.11
48-1012	20	9.07	4	100	8	200	10	12	.043	.013	0.36	0.11
48-912	23	10.43	4	100	8	200	9	12	.052	.013	0.43	0.11
48-812	27	12.25	4	100	8	200	8	12	.062	.013	0.52	0.11
48-711	33	14.97	4	100	8	200	7	11	.074	.017	0.62	0.14
412-1212	13	5.90	4	100	12	300	12	12	.026	.009	0.22	0.07
412-1112	16	7.26	4	100	12	300	11	12	.034	.009	0.28	0.07
412-1012	19	8.62	4	100	12	300	10	12	.043	.009	0.36	0.07
412-912	22	9.98	4	100	12	300	9	12	.052	.009	0.43	0.07
412-812	25	9.98	4	100	12	300	8	12	.062	.009	0.52	0.07
412-711	31	11.34	4	100	12	300	7	11	.074	.011	0.62	0.09
412-610	66	14.06	4	100	12	300	6	10	.087	.014	0.72	0.12
412-510	42	29.94	4	100	12	300	5	10	.101	.014	0.84	0.12
412-57	45	19.05	4	100	12	300	5	7	.101	.025	0.84	0.21
412-49	49	20.41	4	100	12	300	4	9	.120	.017	1.00	0.14
416-1012	18	8.16	4	100	16	400	10	12	.043	.007	0.36	0.06
416-912	21	9.53	4	100	16	400	9	12	.052	.007	0.43	0.06
416-812	25	11.34	4	100	16	400	8	12	.062	.007	0.52	0.06
416-711	30	13.61	4	100	16	400	7	11	.074	.009	0.62	0.07
416-610	35	15.88	4	100	16	400	6	10	.087	.011	0.72	0.09
416-510	40	18.14	4	100	16	400	5	10	.101	.011	0.84	0.09
416-49	48	21.77	4	100	16	400	4	9	.120	.013	1.00	0.11
416-38	56	25.40	4	100	16	400	3	8	.140	.015	1.16	0.12
416-28	64	29.03	4	100	16	400	2	8	.162	.015	1.35	0.12

Style	Wgt. per 100 s.f. based on 60" Width	Wgt. per 9.3 sq.m based on 1500 mm Width	Spacing Wires in Inches Longit. in.	Longit. mm	Trans. in.	Trans. mm	A.S.&W.Co. Steel Wire Gauge No. Trans.	Longit.	Sect. Area Sq. in. per ft. Trans.	Sq. in. per ft. Longit.	Sq. cm per m Trans.	Sq. cm per m Longit.
66-1212	13	5.90	6	150	6	150	12	12	.017	.017	0.14	0.14
66-1010	21	9.53	6	150	6	150	10	10	.029	.029	0.24	0.24
66-99	25	11.34	6	150	6	150	9	9	.035	.035	0.29	0.29
66-88	30	13.61	6	150	6	150	8	8	.041	.041	0.34	0.34
66-77	36	16.33	6	150	6	150	7	7	.049	.049	0.41	0.41
66-66	42	19.05	6	150	6	150	6	6	.058	.058	0.48	0.48
66-55	49	22.23	6	150	6	150	5	5	.067	.067	0.56	0.56
66-46	50	22.68	6	150	6	150	4	6	.080	.058	0.67	0.48
66-44	58	26.31	6	150	6	150	4	4	.080	.080	0.67	0.67
66-33	68	30.84	6	150	6	150	3	3	.093	.093	7.74	0.77
66-22	78	35.38	6	150	6	150	2	2	.108	.108	0.90	0.90
66-11	91	41.28	6	150	6	150	1	1	.126	.126	1.05	1.05
66-00	107	48.54	6	150	6	150	0	0	.148	.148	1.23	1.23
612-77	27	12.25	6	150	12	300	7	7	.049	.025	0.41	0.41
612-66	32	14.52	6	150	12	300	6	6	.058	.029	0.48	0.24
612-55	37	16.78	6	150	12	300	5	5	.067	.034	0.56	0.28
612-44	44	19.96	6	150	12	300	4	4	.080	.040	0.67	0.33
612-33	51	23.13	6	150	12	300	3	3	.093	.047	0.77	0.39
612-25	52	23.59	6	150	12	300	2	5	.108	.034	0.90	0.28
612-22	59	26.76	6	150	12	300	2	2	.108	.054	0.90	0.45
612-17	56	25.40	6	150	12	300	1	7	.126	.025	1.05	0.21
612-14	61	27.67	6	150	12	300	1	4	.126	.040	1.05	0.33
612-11	69	31.30	6	150	12	300	1	1	.126	.063	1.05	0.52
612-06	65	29.48	6	150	12	300	0	6	.148	.029	1.23	0.24
612-03	72	32.66	6	150	12	300	0	3	.148	.047	1.23	0.39
612-00	81	36.74	6	150	12	300	0	0	.148	.074	1.23	0.62

Sometimes, a style such as 66-44 is shown as 6x6-4/4.

Labor Placing Steel Fabric on Floors. When used for reinforcing floor slabs or for temperature reinforcing, a worker should place 1,400 to 1,600 s.f. (126-144 sq.m) of fabric per 8-hr. day, at the following labor cost per 100 s.f. (9 sq.m):

	Hours	Rate	Total	Rate	Total
Iron worker	0.53	$....	$....	$30.28	$ 16.05
Cost per s.f.			....		0.16
per sq.m			....		1.72

Reinforcement For Architectural Concrete Walls. It is almost impossible to provide sufficient reinforcement to prevent cracking in a long wall, which is why control joints are used. But it is still necessary to provide adequate reinforcement between joints properly spaced to resist volume change stresses. To be most effective, small bars with a type of deformation to give maximum bond should be placed relatively close together, rather than larger bars at wider spacings. Generally #3 and #4 bars are preferable to larger sizes and the horizontal reinforcement should be at least equal to 0.25% of wall area. In the vertical direction 0.15% reinforcement is sufficient.

Alternate Methods of Placing Reinforcing around Openings

Spreader

Recommended Size and Spacing of Reinforcing Bars in Concrete Walls

Wall Thickness (In.)	Horizontal reinforcement	Vertical reinforcement
(in outside face of wall)		
6	#3 @ 8" (200 mm)	#3 @ 8" (200 mm)
8	#3 @ 6" (150 mm)	#3 @ 8" (200 mm)
(in both faces of wall)		
10	#3 @ 10" (250 mm)	#3 @ 12" (300 mm)
12	#3 @ 8" (200 mm)	#3 @ 12" (300 mm)

The bars should be tied at each intersection with a non-slip tie, and where two curtains of steel are used, it is advisable to include a detail drawing to show how the curtains are to be fastened together with tie spreaders at frequent intervals to make a rigid web of reinforcement.

It is sometimes difficult to place concrete in walls because of a congestion of reinforcing bars. This condition should be avoided insofar as practical.

Labor Placing Reinforcing Bars. The labor cost of placing reinforcing steel in architectural concrete walls is expensive on account of the small size of the rods, the fact they run both horizontally and vertically, must be wired together, and the cramped quarters for the work.

On walls 8 to 12 inches (200-300 mm) thick, a reinforcing steel setter will handle, place, and wire 750 to 850 lbs. (340-385 kg) of reinforcing steel per 8-hr. day, at the following labor cost per ton:

	Hours	Rate	Total	Rate	Total
Iron worker per ton	20	$....	$....	$30.28	$ 605.60
per metric ton					672.22

03300 CAST-IN-PLACE CONCRETE

Reinforced and architectural concrete are among our most common building materials and are used on virtually all types of construction work.

They are used in all types of structures, from residential, apartment, hotel and hospital buildings, requiring comparatively light floor loads, to the heaviest constructed warehouses, factory buildings, wharves, and industrial buildings.

In fact, entire buildings are constructed of concrete, including foundations, foundation walls, columns, beams and girders, floor slabs, ornamental exterior walls, cornices, etc.

Regardless of the fact that it is one of our most common building materials, costs on all types of reinforced concrete construction are subject to wide variation. This is due to a number of causes, such as the function and size of building, whether flat slab or beam and girder type, whether solid concrete construction or used in conjunction with steel, building codes, insurance requirements, etc.

The methods used in erecting and moving the formwork, handling, mixing and placing the concrete, and the ability of the superintendent or foreman, all affect the final cost of the work.

Regardless of how carefully or accurately the job has been estimated, if the superintendent or foreman in charge of the work are not proficient, the actual costs are almost certain to run higher than the estimate--not because the work was estimated too low but because it was poorly handled on the job.

All labor costs given here are based on efficient management; economical layout and design of all formwork; use of modern labor-saving methods, such as bench fabrication of column, beam, and girder forms where possible; and the use of labor saving tools such as power saws, band saws, electric hand saws, etc.

Symons Steel-Ply® Forming System Can Be Moved into Place After Reinforcing Steel is Placed (Courtesy Symons Corp.)

The efficient contractor must use methods and tools that will reduce costs to a minimum, or else be out of the running when it comes to obtaining work and performing it economically. Remember, good management, economical design, and labor-saving tools and methods are absolutely necessary to a profitable contract.

Items to be Included in the Estimate. Reinforced concrete should be estimated in the same manner and in the same units as the work is constructed on the job. This allows comparisons between estimated and actual costs and enables a contractor to tell where variations occur.

The first operation on any reinforced concrete job is the erection of the temporary wood or metal forms to support the weight of the wet concrete until it has set sufficiently to be self-supporting. The members that usually require forms are columns, girders, beams, spandrel beams, lintels, floor slabs, stairs, and platforms. Each class of formwork should be estimated separately and in detail.

All types of formwork, including columns, beams, girders, and the various type of floor slabs, have been designed with the best engineering practices—giving lumber requirements, clamp spacing for columns, and beam and girder lumber requirements, including shoring. There is a complete set of design tables for all types of concrete floor slabs, giving joist spacing, spans of tables for all types of concrete floor slabs, giving joist spacing, spans of stringers, and shores required for any floor load.

After the forms are erected, it is necessary to place the reinforcing steel or wire mesh ready to receive the concrete. This may include unloading steel from trucks at the job, cutting steel to length and bending to special shapes for columns, beams, girders and floor slabs, and placing and tying it in position ready to receive the concrete.

After the reinforcing steel is in place, the job is then ready for concrete placement. This consists of preparing runways, unloading cement and other materials, mixing the cement, sand and gravel into the concrete mass, hoisting, wheeling and placing it in position, removing runways, removing old concrete drippings and spillage, patching, finishing, cleaning and washing wheelbarrows or buggies, and concrete mixer.

Each of these items should be estimated in detail to include all material and labor operations necessary to make up the completed unit.

Weights of Miscellaneous Building Materials
Weight of Cement, Sand, Gravel, Crushed Stone, Concrete, Etc.

The following table gives the approximate weights of cement, concrete, and concrete aggregates, and will assist the estimator in figuring freight or when purchasing materials by weight:

	Wgt., lbs. per c.y.	Wgt., kg per cu. m
Bank sand (dry)	2500	1480
Torpedo sand	2700	1600
Crushed stone	2500	1480
Crushed stone screenings	2500	1480
Gravel	2700	1600
Roofing gravel	2700	1600

Cement weighs 376 lbs. (170 kg) per bbl.
One sack weighs 94 lbs. (1500 kg) and bulks 1 c.f. (1 cu.m)

Cinder concrete	112	1690
Concrete of gravel, limestone, sandstone, trap rock, etc.	150	2265

Quantity of Pit Run Sand and Gravel Required for One Cubic Yard of Concrete. Pit run sand and gravel is not recommended for concrete work where strength is a factor, because it is impossible to grade the aggregate. There is either too much gravel and too little sand or too little gravel and too much sand, usually the latter, to make a dependable concrete.

An aggregate containing approximately 50% sand and 50% gravel (providing it does not contain any humic acid), may be considered acceptable aggregate, and the strength will probably be only 10% less than when using aggregates furnished by washing and grading plants.

It ordinarily requires about 1.25 cu. yds. of aggregate per cu. yd. of concrete (1.25 cu.m aggregate per cu.m of concrete), depending on the coarseness of the sand.

CONCRETE

Concrete should be estimated by the cubic foot, or cubic yard containing 27 c.f. (for metric, by the cu. m). The c.y. is the most convenient unit to use because practically all published tables giving quantities of cement, sand, gravel, or crushed stone are based on the c.y. (cu.m) of concrete. The c.y. (cu.m) is also a convenient unit to use when estimating labor costs.

It was formerly the practice to specify concrete mixtures as so many volumes of fine and coarse aggregate to each volume of cement, as for example 1:1:2, 1:1.5:3, 1:2:4, 1:3:5, etc. The condition of the aggregate when measured was not specified, nor were the consistency of the concrete or the water content.

As used on the job, sand nearly always contains some surface moisture, usually about 4% or 5% by weight of the sand, although it may be considerably more just after rain or if the sand is used soon after washing it. The moisture causes the sand to bulk up in volume, the amount of bulking depending on the amount of moisture and the grading. Bulking, which might be 20% to 30% or more, reduces the true volume of surface dry sand in a given volume of damp sand. For example, if a 1:2:4 mix is specified and 2 volumes of damp sand are measured, the volume of sand on a surface dry basis would be less. Suppose the sand is bulked 25% due to moisture. The volume of dry sand in each cubic foot of damp sand will be 100 divided by 125, or 0.8 c.f. (cu.m) of damp sand would give us 2 x 0.8 or 1.6 c.f. (cu.m) dry sand.

The coarse aggregate carries very little surface moisture, and the moisture has little effect on the volume. The result is that undersanded mixes were often used. In the above example the 1:2:4 mix if based on damp materials would actually be a 1:1.6:4 mixture of surface dry materials. The amount of sand in this case would be less than 29% of the total volume of aggregate. For average mixtures 35% to 45% is usually desirable for good workability when concrete is placed by hand.

To correct for the bulking in the above example the amount of damp sand should be increased to 1.25 x 2 or 2.5 c.f. (cu.m) and the mix would then be 1:2.5:4 based on damp materials as measured in the field.

As a result of the increased knowledge that has been made available, other methods of specifying concrete have been adopted, such as a mix by weight, a minimum cement factor, a maximum water content or the strength at a given age, or a combination of these.

The Importance of Mixing Water. The hardening of concrete mixtures is brought about by chemical reactions between cement and water, the aggregates (sand, gravel, stone, etc.) being inactive ingredients used as fillers. So long as workable mixtures are used, the less water there is in the mix, the stronger, more watertight and more durable will be the concrete. Excess mixing water dilutes the paste made by the cement and water and makes weaker, more porous concrete. For any set of conditions of mixing, placing, and curing and for given materials there is a definite relation between the strength of the concrete and the amount of water used in mixing. This relation can be determined by test and a water-cement ratio-strength curve can be developed representing the actual results on a specific job.

American Concrete Institute (ACI) Standard for Selecting Proportions for Concrete Mixes

Committee 211 of the American Concrete Institute recommends certain practices for selecting proportions for concrete mixes. Following are extracts from this standard.

Basic Relationship

Concrete proportions must be selected to provide necessary placeability, strength, and durability for the particular application. Well established relationships governing these properties are discussed briefly below.

Placeability (including satisfactory finishing properties) encompasses traits loosely accumulated in the terms "workability" and "consistency". For the purpose of this discussion, *workability* is considered to be the property of concrete that determines its capacity to be placed and consolidated properly and to be finished without harmful segregation. It embodies such concepts as moldability, cohesiveness, and compactability. It is affected by the grading, particle shape and proportions of aggregate, the amount of cement, the presence of entrained air, admixtures, and the consistency of the mixture. Procedures in the recommended practice take into account these factors.

Consistency, loosely defined, is the wetness of the concrete mixture. It is measured in terms of slump—the higher the slump the wetter the mixture—and it affects the ease with which the concrete will flow during placement. It is related to, but not synonymous with, workability. In properly proportioned concrete, the unit water content required to produce a given slump will depend on several factors. Water requirement increases as aggregates become more angular and rough textured (but this disadvantage may be offset by improvements in other characteristics such as bond to cement paste). Required mixing water decreases as the maximum size of well graded aggregate is increased. It also decreases with the entrainment of air. Mixing water requirement may often be significantly reduced by certain admixtures.

Strength. Strength is an important characteristic of concrete, but other characteristics such as durability, permeability, and wear resistance are often equally or more important. These may be related to strength in a general way but are also affected by factors not significantly associated with strength. For a given set of materials and conditions, concrete strength is determined by the net quantity of water used per unit quantity of cement. The net water content excludes water absorbed by the aggregates. Differences in strength, for a given water-cement ratio may result from changes in: maximum size of aggregate; grading, surface texture, shape, strength, and stiffness of aggregate particles; differences in cement types and sources; air content; and the use of admixtures which affect the cement hydration process or develop cementitious properties themselves. To the extent that these effects are predictable in the general sense, they are taken into account in this recommended practice. However, in view of their number and complexity, it should be obvious that accurate predictions of strength must be based on trial batches or experience with the materials to be used.

Durability. Concrete must be able to endure those exposures which may deprive it of its serviceability—freezing and thawing, wetting and drying, heating and cooling, chemicals, de-icing agents, and the like. Resistance to some of these may be enhanced by use of special ingredients: low-alkali cement, pozzolans, or selected aggregate to prevent harmful expansion due to the alkali-aggregate reaction which occurs in some areas when concrete is exposed in a moist environment; sulfate resisting cement or pozzolans for concrete exposed to sea water or sulfate-bearing soils; or aggregate free of excessive soft particles where resistance to surface abrasion is required. Use of a low water-cement ratio will prolong the life of concrete by reducing the penetration of aggressive liquids. Resistance to severe weathering, particularly freezing and thawing, and to salts used for ice removal is greatly improved by incorporation of a proper distribution of entrained air. Entrained air should be used in all exposed concrete in climates where freezing occurs.

Background Data

To the extent possible, selection of concrete proportions should be based on test data or experience with the materials to be used. Where such background is limited or not available, estimates given in this recommended practice may be employed. The following information for available materials will be useful:

1. Sieve analyses of fine and coarse aggregates.
2. Unit weight of coarse aggregate.
3. Bulk specific gravities and absorptions of aggregates.
4. Mixing water requirements of concrete developed from experience with available aggregates.
5. Relationships between strength and water-cement ratio for available combinations of cement and aggregate.

Estimates from Tables 3 and 4 may be used when the last two items of information are not available. As will be shown, proportions can be estimated without the knowledge of aggregate specific gravity and absorption (third item above).

Procedure

Estimating the required batch weights for the concrete involves a sequence of logical, straightforward steps, which in effect fit the characteristics of the available materials into a mixture suitable for the work. The question of suitability is frequently not left to the individual selecting the proportions. The job specifications may dictate some or all of the following:

1. Maximum water-cement ratio
2. Minimum cement content
3. Air content
4. Slump
5. Maximum size of aggregate
6. Strength
7. Other requirements relating to such things as strength over-design, admixtures, and special types of cement or aggregate

Regardless of whether the concrete characteristics are prescribed by the specifications or are left to the individual selecting the proportions, establishment of batch weights per cubic yard of concrete can best be accomplished in the following sequence:

Step 1. Choice of slump. If slump is not specified, a value appropriate for the work can be selected from Table 1. The slump ranges shown apply when vibration is used to consolidate the concrete. Mixes of the stiffest consistency that can be placed efficiently should be used.

Table 1—Recommended Slumps
For Various Types of Construction

Type of construction	Slump, in. (mm) Maximum*	Minimum
Reinforced foundation walls and footings	3 (75)	1 (25)
Plain footings, caissons, and substructure walls	3 (75)	1 (25)
Beams and reinforced walls	4 (100)	1 (25)
Building columns	4 (100)	1 (25)
Pavements and slabs	3 (75)	1 (25)
Heavy mass concrete	3 (75)	1 (25)

*May be increased 1 in. (25 mm) for methods of consolidation other than vibration.

Step 2. Choice of maximum size of aggregate. Large maximum sizes of well graded aggregates have less voids than smaller sizes. Hence, concretes with the larger-sized aggregates require less mortar per unit volume of concrete. Generally, the maximum size of aggregate should be the largest that is economically available and consistent with dimensions of the structure. Ordinarily, the ratio of the nominal maximum aggregate size to the minimum dimension within which concrete must be placed should not exceed the value shown in Table 2.

Table 2—Recommended Maximum Ratio
of Nominal Aggregate Size to Minimum
Criterion Dimension

Criterion dimension		Recomm. max. ratio
Narrowest dimension between sides of forms		1/5
Depth of slab		1/3
Clear spacing between individual reinforcing bars, bundles of bars, or pre-tensioning strands	In cols.	2/3
	In all other members	3/4

These limitations are sometimes waived if workability and methods of consolidation are such that the concrete can be placed without honeycomb or void. When high strength concrete is desired, best results may be obtained with reduced maximum sizes of aggregate since these produce higher strengths at a given water-cement ratio.

Step 3. Estimation of mixing water and air content. The quantity of water per unit volume of concrete required to produce a given slump is dependent on the maximum size, particle shape and grading of the aggregates, and on the amount of entrained air. It is not greatly affected by the quantity of cement. Table 3 provides estimates of required mixing water for concretes made with various maximum sizes of aggregate, with and without air entrainment. Depending on aggregate texture and shape, mixing water requirements may be somewhat above or below the tabulated values, but they are sufficiently accurate for the first estimate. Such differences in water demand are not necessarily reflected in strength since other compensating factors may be involved. For example, a rounded and an angular coarse aggregate, both well and similarly graded and of good quality, can be expected to produce concrete of about the same compressive strength for the same cement factor in spite of differences in water-cement ratio resulting from the different mixing water requirements. Particle shape per se is not an indicator that an aggregate will be either above or below average in its strength-producing capacity.

Table 3 indicates the approximate amount of entrapped air to be expected in non-air-entrained concrete, and shows the recommended levels of average air content for concrete in which air is to be purposely entrained for durability. Air-entrained concrete should always be used for structures which will be exposed to freezing and thawing, and generally for structures exposed to sea water or sulfates. When severe exposure is not anticipated, beneficial effects of air entrainment on concrete workability and cohesiveness can be achieved at air content levels approximately half those shown for air-entrained concrete.

When trial batches are used to establish strength relationships or verify strength-producing capability of a mixture, the least favorable combination of mixing water and air content should be used. The air content should be the maximum permitted or likely to occur, and the concrete should be gauged to the highest permissible slump. This will avoid developing an overoptimistic estimate of strength on the assumption that average rather than extreme conditions will prevail in the field. For information on air content recommendations, see ACI 201, 301, and 302.

Table 3—Approximate Mixing Water and
Air Content Requirements for Different
Slumps and Maximum Sizes of Aggregates*

Water, lb. per cu. yd. of concrete for indicated maximum sizes of aggregate

Slump, in.	3/8"	1/2"	3/4"	1"	1-1/2"	2"	3"
			Non-air-entrained concrete				
1 to 2	360	335	315	300	275	260	240
3 to 4	385	365	340	325	300	285	265
6 to 7	410	385	360	340	315	300	285

Water, lb. per cu. yd. of concrete for
indicated maximum sizes of aggregate

Slump, in.	3/8"	1/2"	3/4"	1"	1-1/2"	2"	3"
Air content, percent	3	2.5	2	1.5	1	0.5	0.3
			Air-entrained concrete				
1 to 2	305	295	280	270	250	240	225
3 to 4	340	325	305	295	275	265	250
6 to 7	365	345	325	310	290	280	270
Air content, percent	8	7	6	5	4.5	4	3.5

(metric)
Water, kg per cu.m of concrete for
indicated maximum sizes of aggregate

Slump, mm	9.4	12.5	18.8	25.0	37.5	50.0	75.0
			Non-air-entrained concrete				
1 to 2	214	199	187	178	163	154	142
3 to 4	228	217	202	193	178	169	157
6 to 7	243	228	214	202	187	178	169
Air content, percent	3	2.5	2	1.5	1	0.5	0.3
			Air-entrained concrete				
1 to 2	181	175	166	160	148	142	133
3 to 4	202	193	181	175	163	157	148
6 to 7	217	205	193	184	172	166	160
Air content, percent	8	7	6	5	4.5	4	3.5

*These quantities of mixing water are for use in computing cement factors for trial batches. They are maxima for reasonably well-shaped angular coarse aggregates graded within limits of accepted specifications.

Step 4. Selection of water-cement ratio. The required water-cement ratio is determined not only by strength requirements but also by factors such as durability and finishing properties. Since different aggregates and cements generally produce different strengths at the same water-cement ratio, it is highly desirable to have or develop the relationship between strength and water-cement ratio for the materials actually to be used. In the absence of such data, approximate and relatively conservative values for concrete containing Type I portland cement can be taken from Table 4(a). With typical materials, the tabulated water-cement ratios should produce the strengths shown, based on 28-day tests of specimens cured under standard laboratory conditions. The average strength selected must, of course, exceed the specified strength by a sufficient margin to keep the number of low tests within specified limits.

For severe conditions of exposure, the water-cement ratio should be kept low even though strength requirements may be met with a higher value. Table 4(b) gives limiting values.

Table 4(a)—Relationships Between
Water-Cement Ratio and Compressive Strength of Concrete

	Water-cement ratio, by weight	
Compressive strength at 28 days, psi* (mPa)	Non-air-entrained concrete	Air-entrained concrete
6000 (41)	0.41	--
5000 (34)	0.48	0.40
4000 (27)	0.57	0.48
3000 (21)	0.68	0.59
2000 (14)	0.82	0.74

*Values are estimated average strengths for concrete containing not more than the percentage of air shown in Table 3. For a constant water-cement ratio, the strength of concrete is reduced as the air content is increased.

Strength is based on 6" x 12" (150 x 300 mm) cylinders moist-cured 28 days at 73.4 $\pm$ 3 °F (23 $\pm$ 1.7 °C) in accordance with Section 9(b) of ASTM C 31 for method of Making and Curing Concrete Compression and Flexure Test Specimens in the Field.

Relationship assumes maximum size of aggregate about 3/4" to 1" (18.75-25 mm); for a given source, strength produced for a given water-cement ratio will increase as maximum size of aggregate decreases.

Table 4(b)—Maximum Permissible
Water-Cement Ratios for
Concrete in Severe Exposures*

Type of structure	Structure wet continuously or frequently and exposed to freezing and thawing**	Structure exposed to sea water or sulfates
Thin sections (railings, curbs, sills, ledges, ornamental work) and sections with less than 1 in. cover over steel	0.45	0.40**
All other structures	0.50	0.45***

*Based on report of ACI Committee 201, "Durability of Concrete in Service", previously cited.

**Concrete should also be air-entrained.

***If sulfate resisting cement (Type II or Type V of ASTM C 150) is used, permissible water-cement ratio may be increased by 0.05.

Step 5. Calculation of cement content. The amount of cement per unit volume of concrete is fixed by the determinations made in Steps 3 and 4 above. The required cement is equal to the estimated mixing water content (Step 3) divided by the water-cement ratio (Step 4). If, however, the specification includes a separate minimum limit on cement in addition to requirements for strength and durability, the mixture must be based on whichever criterion leads to the larger amount of cement.

Step 6. Estimation of coarse aggregate content. Aggregates of essentially the same maximum size and grading will produce concrete of satisfactory workability when a given volume of coarse aggregate, on a dry-rodded basis, is used per unit volume of concrete. Appropriate values for this aggregate volume are given in Table 6. It can be seen that, for equal workability, the volume of coarse aggregate in a unit volume of concrete is dependent only on its maximum size and the fineness modulus of the fine aggregate. Differences in the amount of mortar required for workability with different aggregates, due to differences in particle shape and grading, are compensated for automatically by differences in dry-rodded void content.

The volume of aggregate, in cubic feet (cu.m), on a dry-rodded basis, for a cubic yard (cubic meter) of concrete is equal to the value from Table 6 multiplied by 27. This volume is converted to dry weight of coarse aggregate required in a cubic yard of concrete by multiplying it by the dry-rodded weight per cubic foot of the coarse aggregate.

Table 6—Volume of Coarse Aggregate
Per Unit of Volume of Concrete

Maximum Size Aggregate, in.	Volume of dry-rodded coarse aggregate* per unit volume of concrete for different fineness moduli of sand			
	2.40	2.60	2.80	3.00
3/8	0.50	0.48	0.46	0.44
1/2	0.59	0.57	0.55	0.53
3/4	0.66	0.64	0.62	0.60
1	0.71	0.69	0.67	0.65
1-1/2	0.75	0.73	0.71	0.69
2	0.78	0.76	0.74	0.72
3	0.82	0.80	0.78	0.76

*Volumes are based on aggregates in dry-rodded condition as described in ASTM C 29 for Unit Weight and voids in Aggregate.

These volumes are selected from empirical relationships to produce concrete with a degree of workability suitable for usual reinforced construction. For less workable concrete such as required for concrete pavement construction they may be increased about 10%. When placement is to be by pump, they should be reduced about 10%.

Step 7. Estimation of fine aggregate content. At completion of Step 6, all ingredients of the concrete have been estimated except the fine aggregate. Its quantity is determined by difference. Either of two procedures may be employed: the "weight" method or the "absolute volume" method.

If the weight of the concrete per unit volume is assumed or can be estimated from experience, the required weight of fine aggregate is simply the difference between the weight of fresh concrete and the total weight of the other ingredients. Often the unit weight of concrete is known with reasonable accuracy from previous experience with the materials. In the absence of such information, Table 7 can be used to make a first estimate. Even if the estimate of concrete weight per cubic yard (cubic meter) is rough, mixture proportions will be sufficiently accurate to permit easy adjustment on the basis of trial batches as will be shown in the examples.

If a theoretically exact calculation of fresh concrete weight per cubic yard (cubic meter) is desired, the following formula can be used:

$$U = 16.85 \, Ga \, (100 - A)$$
$$+ \, C(1 - Ga/Gc) - W(Gc - 1)$$

where

1. U = weight of fresh concrete per cubic yard, lb. (per cu.m, kg).
2. Ga = weighted average specific gravity of combined fine and coarse aggregate, bulk SSD*
3. Gc = specific gravity of cement (generally 3.15)
4. A = air content, percent
5. W = mixing water requirement, lbs. per c.y. (kg per cu.m)
6. C = cement requirement, lbs. per c.y. (kg per cu.m)

*SSD indicates saturated-surface-dry basis used in considering aggregate displacement. The aggregate specific gravity used in calculations must be consistent with the moisture condition assumed in the basic aggregate batch weights—i.e., bulk dry if aggregate weights are stated on a dry basis, and bulk SSD if weights are stated on a saturated-surface-dry basis.

A more exact procedure for calculating the required amount of fine aggregate involves the use of volumes displaced by the ingredients. In this case, the total volume displaced by the known ingredients—water, air, cement, and coarse aggregate—is subtracted from the unit volume of concrete to obtain the required volume of fine aggregate. The volume occupied in concrete by any ingredient is equal to its weight divided by the density of that material (the latter being the product of the unit weight of water and the specific gravity of the material).

Step 8. Adjustments for aggregate moisture. The aggregate quantities actually to be weighed out for the concrete must allow for moisture in the aggregates. Generally, the aggregates will be moist and their dry weights should be increased by the percentage of water they contain, both absorbed and surface. The mixing water added to the batch must be reduced by an amount equal to the free moisture contributed by the aggregate—i.e., total moisture minus absorption.

Table 7—First Estimate of Weight of Fresh Concrete

Maximum Size of Aggregate		Non-air-entrained Concrete		Air-entrained Concrete	
inches	mm	lbs./c.y.	kg/cu.m	lbs./c.y.	kg/cu.m
3/8	9.4	3840	2278	3690	2189
1/2	12.5	3890	2308	3760	2231
3/4	18.8	3960	2349	3840	2278
1	25.0	4010	2379	3900	2314
1-1/2	37.5	4070	2415	3960	2349
2	50.0	4120	2444	4000	2373
3	75.0	4160	2468	4040	2397

Values calculated by Eq. 7 for concrete of medium richness, 550 lb. of cement per c.y. (325 kg/cu.m), and medium slump with aggregate specific gravity of 2.7. Water requirements based on values for 3" to 4" (75-100 mm) slump in Table 3. If desired, the estimated weight may be refined as follows, if necessary information is available: for each 10 lb. (4.5 kg) difference in mixing water from the Table 3 values for 3" to 4" (75-100 mm) slump, correct the weight 15 lbs. per c.y. (6.8 kg/cu.m) in the opposite direction; for each 100 lb. (45 kg) difference in cement content from 550 lbs., correct the weight 15 lbs. per c.y. (6.8 kg) in the same direction; for each 0.1 by which aggregate specific gravity deviates from 2.7, correct the concrete weight 100 lbs. (45 kg) in the same direction.

Step 9. Trial batch adjustments. The calculated mixture proportions should be checked by means of trial batches prepared and tested in accordance with ASTM C 192, "Making and Curing Concrete Compression and Flexure Test Specimens in the Laboratory", or full-sized field batches. Only sufficient water should be used to produce the required slump, regardless of the amount assumed in selecting the trial proportions. The concrete should be checked for unit weight and yield (ASTM C 138) and for air content (ASTM C 138, C 172, or C 231). It should also be carefully observed for proper workability, freedom from segregation, and finishing properties. Appropriate adjustments should be made in the proportions for subsequent batches in accordance with the following procedure.

Re-estimate the required mixing water per cubic yard of concrete by multiplying the net mixing water content of the trial batch by 27 and dividing the product by the yield of the trial batch in cubic feet (Divide by 0.03 to obtain cubic meters). If the slump of the trial batch was not correct, increase or decrease the re-estimated amount of water by 10 lb. (4.5 kg) for each required increase or decrease of 1" (25 mm) in slump.

If the desired air content (for air-entrained concrete) is not achieved, re-estimate the admixture content required for proper air content and reduce or increase the mixing water content of the above paragraph by 5 lb. (2.25 kg) for each 1% by which the air content is to be increased or decreased from that of the previous trial batch.

If estimated weight per cubic yard of fresh concrete is the basis for proportioning, re-estimate that weight by multiplying the unit weight in lbs. per cu. ft. of the trial batch by 27 and reducing or increasing the result by the anticipated percentage increase or decrease in air content of the adjusted batch from the first trial batch.

Calculate new batch weights starting with Step 4, modifying the volume of coarse aggregate from Table 6 if necessary to provide proper workability.

Sample Computations

Two sample problems illustrate application of the proportioning procedures. The following conditions are assumed:

1. Type I non-air-entraining cement will be used, and its specific gravity is assumed to be 3.15. The specific gravity values are not used if proportions are selected to provide a weight of concrete assumed to occupy 1 cu. yd.
2. Coarse and fine aggregates in each case are of satisfactory quality and are graded within limits of generally accepted specifications, such as the "Specifications for Concrete Aggregates" (ASTM C 33).
3. The coarse aggregate has a bulk specific gravity of 2.68 and an absorption of 0.5%.
4. The fine aggregate has a bulk specific gravity of 2.64, an absorption of 0.7%, and fineness modulus of 2.8.

Example 1. Concrete is required for a portion of a structure which will be below ground level in a location where it will not be exposed to severe weathering or sulfate attack. Structural considerations require it to have an average 28-day compressive strength of 3500 psi (2.44 kg/mm^2).

This is not the specified strength used for structural design but a higher figure expected to be produced on the average. For the method of determining the amount by which average strength should exceed design strength, see "Recommended Practice for Evaluation of Compression Test Results of Field Concrete" (ACI 214-65).

On the basis of information in Table 1, as well as previous experience, it is determined that under the conditions of placement to be employed, a slump of 3" to 4" (75-100 mm) be used and that the available No. 4 to 1-1/2" (38 mm) coarse aggregate will be suitable. The dry-rodded weight of coarse aggregate is found to be 100 lbs. per c.f. (1511 kg/m³). Employing the sequence outlined in the *Procedure* section, the quantities of ingredients per cubic yard (cu.m) of concrete are calculated as follows:

Step 1. As indicated, the desired slump is 3" to 4" (75-100 mm)

Step 2. The locally available aggregate graded from No. 4 to 1-1/2" (38 mm), has been indicated as suitable.

Step 3. Since the structure will not be exposed to severe weathering, non-air-entrained concrete will be used. The approximate amount of mixing water to produce a 3" to 4" slump (75-100 mm) in non-air-entrained concrete with 1-1/2" (38 mm) aggregate is found from Table 3 to be 300 lbs./c.y. (178 kg/m³). Estimated entrapped air is shown as 1%.

Step 4. From Table 4(a), the water-cement ratio needed to produce a strength of 3500 psi (2.44 kg/mm²) in non-air-entrained concrete is found to be about 0.62.

Step 5. From the information derived in Steps 3 and 4, the required cement content is found to be 300/0.62 = 484 lbs./c.y. (178/0.62 = 287 kg/m³)

Step 6. The quantity of coarse aggregate is estimated from Table 6. For a fine aggregate having a fineness modulus of 2.8 and a 1-1/2" (38 mm) maximum size of coarse aggregate, the table indicates that 0.71 c.f. (0.71 cu.m) of coarse aggregate, on a dry-rodded basis, may be used in each c.f. of concrete. For a cubic yard, the coarse aggregate will be 27 x 0.71 = 19.17 c.f. (0.02 cu.m). Since it weighs 100 lb./c.f. (2250 kg/m³) the dry weight of coarse aggregate is 1,917 lb. (2250 kg).

Step 7. With the quantities of water, cement, and coarse aggregate established, the remaining material comprising the cubic yard (cubic meter) of concrete must consist of sand and whatever air will be entrapped. The required sand may be determined on the basis of either weight or absolute volume as shown below:

Weight basis. From Table 7, the weight of a cubic yard (cubic meter) of non-air-entrained concrete made with aggregate having a maximum size of 1-1/2" (38 mm) is estimated to be 4,070 lbs. (2410 kg). For a first trial batch, exact adjustments of this value for usual differences in slump, cement factor, and aggregate specific gravity are not critical. Weights already known are:

Water (net mixing)	300	lbs.	178	kg
Cement	484	lbs.	287	kg
Coarse aggregate	1917	lbs.	1598	kg (dry)*
Total	2701	lbs.	2063	kg

The weight of sand, therefore, is estimated to be 4070 - 2701 = 1369 lbs. (2410 - 2063 = 347 kg) dry*

*Aggregate absorption is disregarded since its magnitude is inconsequential in relation to other approximations.

Absolute volume basis. With the quantities of cement, water, and coarse aggregate established, and the approximate entrapped air content (as opposed to purposely entrained air) taken from Table 3, the sand content can be calculated as follows:

Volume of water $= \dfrac{300}{62.4} = 4.81$ c.f.

Solid volume of cement $= \dfrac{484}{3.15 \times 62.4} = 2.46$ c.f.

Solid volume of coarse aggregate $= \dfrac{1917}{2.68 \times 62.4} = 11.46$ c.f.

Volume of entrapped air $= 0.01 \times 27 = 0.27$ c.f.

Total solid volume of
  ingredients except sand $= 19.00$ c.f.

Solid volume of sand required $= 27 - 19.00 = 8.00$ c.f.

Required weight of
  dry sand $= 8.00 \times 2.64 \times 62.4 = 1318$ lbs.

Batch weights per cubic yard of concrete calculated on the two bases are compared below:

	Based on estimated concrete weight, lbs.	Based on absolute volume of ingredients, lbs.	Based on estimated concrete weight, kg	Based on absolute volume of ingredients, kg
Water (net mixing)	300	300	136	136
Cement	484	484	220	220
Coarse aggregate (dry)	1917	1917	870	870
Sand (dry)	1369	1318	621	598

These batch weights will require adjustment in the field to take into account moisture on aggregates. Also, some adjustment in proportions may be found desirable on the basis of actual field experience.

Example 2. Concrete is required for a heavy bridge pier which will be exposed to fresh water in a severe climate. An average 28-day compressive strength of 3000 psi (20.6 MPa) will be required. Placement conditions permit a slump of 1" to 2" (25-50 mm) and the use of large aggregate, but the only economically available coarse aggregate of satisfactory quality is graded from No. 4 to 1" and this will be used. Its dry-rodded weight is found to be 95 lb./c.f. (1525 kg/m³) Other characteristics are as indicated in the first part of this section.

The calculations will be shown in skeleton form only. Confusion is avoided by following all steps in the *Procedure* section, even when they appear to repeat specified requirements.

Step 1. The desired slump is 1" to 2" (25-50 mm).

Step 2. The locally available aggregate, graded from No. 4 to 1" (25 mm), will be used.

Step 3. Since the structure will be exposed to severe weathering, air-entrained concrete will be used. The approximate amount of mixing water to produce a 1" to 2" (25-50 mm) slump in air-entrained concrete with 1" (25 mm) aggregate is found from Table 3 to be 270 lb./c.y. (160 kg/m³). The recommended air content is 5%.

Step 4. From Table 4(a), the water-cement ratio needed to produce a strength of 3000 psi (20.68 MPa) in air-entrained concrete is estimated to be about 0.59. However, reference to Table 4(b) reveals that, for the severe weathering exposure anticipated, the water-cement ratio should not exceed 0.50. This lower figure must govern and will be used in the calculations.

Step 5. From the information derived in Steps 3 and 4, the required cement content is found to be 270/0.50 = 540 lb./c.y. (160/0.50 = 320 kg/m³).

Step 6. The quantity of coarse aggregate is estimated from Table 6. With a fine aggregate having a fineness modulus of 2.8 and a 1" (25 mm) maximum size of coarse aggregate, the table indicates that 0.67 c.f. (0.67 cu.m) of coarse aggregate, on a dry-rodded basis, may be used in per cu. ft. (cu.m) of concrete. For a cubic yard, therefore, the coarse aggregate will be 27 x 0.67 = 18.09 c.f. (0.67 cu.m). Because it weighs 95 lb./c.f. (1525 kg/m³), the dry weight of coarse aggregate is 18.09 x 95 = 1719 lb. (1022 kg)

Step 7. With the quantities of water, cement, and coarse aggregate established, the remaining material in the cubic yard of concrete is sand and air. The required sand is determined on the basis of either weight or absolute volume:

Weight basis. From Table 7, the weight of a cubic yard (cubic meter) of air-entrained concrete made with aggregate of 1" (25 mm) maximum size is estimated to be 3900 lbs. (2310 kg). For a first trial batch, exact adjustments of this value for differences in slump, cement factor, and aggregate specific gravity are not critical. Weights already known are:

Water (net mixing)	270 lbs.	160 kg
Cement	540 lbs.	320 kg
Coarse aggregate (dry)	1719 lbs.	1022 kg
Total	2529 lbs.	1502 kg

The weight of sand, therefore, is estimated to be 3900 - 2529 = 1371 lbs. dry (2310 - 1502 = 808 kg dry)

Absolute volume basis. With the quantities of cement, water, air, and coarse aggregate established, the sand content is calculated as follows:

Volume of
Water $= \dfrac{270}{62.4} = 4.33$ c.f.

Solid volume
of cement $= \dfrac{540}{(3.15 \times 62.4)} = 2.75$ c.f.

Solid volume
of coarse
aggregate $= \dfrac{1719}{(2.68 \times 62.4)} = 10.28$ c.f.

Volume of
Air $= 0.05 \times 27 = 1.35$ c.f.

Total volume
of ingredients
except sand $= 18.71$ c.f.

Solid volume
of sand
required $= 27 - 18.71 = 8.29$ c.f.

Required
weight of
dry sand $= 8.29 \times 2.64 \times 62.4 = 1366$ lbs.

Batch weight per cubic yard of concrete calculated on the two bases are compared below:

	Based on estimated concrete weight, lbs.	Based on estimated concrete weight, kg	Based on absolute volume of ingredients, lbs.	Based on absolute volume of ingredients kg
Water (net mixing)	270	122.5	270	122.5
Cement	540	244.9	540	244.9
Coarse aggregate (dry)	1719	779.7	1719	779.7
Sand (dry)	1371	621.9	1366	619.6

See Notes at end of Example 1

READY-MIXED CONCRETE

More than 90% of all concrete used in building projects is ready-mixed, which is defined as portland cement concrete manufactured for delivery to a purchaser in a plastic and unhardened state.

Sometimes, it is mixed completely in a stationary mixer and then is transported to the job in trucks. This is known as central-mixed concrete. In other ready-mixed operations, the materials are dry batched and mixed enroute to the job in truck mixers. This is known as transit-mixed concrete. In another procedure, the concrete is mixed in a stationary mixer at the central plant only enough to intermingle the ingredients, usually about 1/2 minute, and the mixing is completed enroute to the job. This is known as shrink-mixed concrete.

Truck mixers consist essentially of a mixer with a separate water tank and water measuring device mounted on a truck chassis. There are other truck conveyances which are similar but are without provisions for water.

ASTM C-94 (ready-mix concrete) requires that when a truck mixer is used either for complete mixing or to finish the partial mixing, each batch of concrete is to be mixed not more than 100 revolutions of the drum or blades at the speed of rotation designated by the manufacturer as the mixing speed. Any additional mixing is to be done at the agitating speed. The specification also requires that the concrete must be delivered and discharged from the truck mixer or agitator truck within 1-1/2 hours after introduction of the water to the cement and aggregate or the cement to the aggregate.

Quality of Ready-Mixed Concrete. ASTM C-94 states that in the absence of applicable general specifications, the purchaser shall select one of the alternate bases for specifying the quality of concrete.

Alternate No. 1: When the purchaser assumes responsibility for the proportioning of the concrete mixture, he shall also specify the following:

1. Cement content in bags or pounds per cubic yard of concrete, or equivalent units.
2. Maximum allowable water content in gallons per cubic yard of concrete, or equivalent units, including surface moisture on the aggregates, but excluding water of absorption.
3. If admixtures are required, the type, name, and dosage to be used. The cement content shall not be reduced when admixtures are used under Alternate No. 1 without the written approval of the purchaser.

Alternate No. 2: When the purchaser requires the manufacturer to assume full responsibility for the selection of the proportions for the concrete mixture, the purchaser shall also specify the requirements for compressive strength as determined on samples taken from the transportation unit at the point of discharge. The purchaser shall specify the requirements in terms of the compressive strength of standard specimens cured under standard laboratory conditions for moist curing. Unless otherwise specified the age at test shall be 28 days.

Alternate No. 3: When the purchaser requires the manufacturer to assume responsibility for the selection of the proportions for the concrete mixture with the minimum allowable cement content specified, the purchaser shall also specify the following:

1. Required compressive strength as determined on samples taken from the transportation unit at the point of discharge. The purchaser shall specify the requirements for strength in terms of tests of standard specimens cured under standard laboratory conditions for moist curing. Unless otherwise specified the age at test shall be 28 days.
2. Minimum cement content in bags or pounds per cubic yard of concrete.
3. If admixtures are required, the type, name, and dosage to be used. The cement content shall not be reduced when admixtures are used.

Alternate No. 3 is useful only if the designated minimum cement content is at about the same level that would ordinarily be required for the strength, aggregate size, and slump specified. At the same time, it must be an amount sufficient to assure durability under expected service conditions, as well as satisfactory surface texture and density, in the event specified strength is attained with it.

The proportions arrived at by Alternates 1, 2, or 3 for each class of concrete and approved for use in a project shall be assigned to a designation to facilitate identification of each concrete mixture delivered to the project. This is the designation required and supplies information on concrete proportions when they are not given separately on each delivery ticket. A certified copy of all proportions as established in Alternates 1, 2, and 3 shall be on file at the batch plant.

Curing. The protection of concrete during the early period to prevent loss of moisture at low temperatures is an important factor in the development of both strength and durability in concrete. Specifications require that the concrete be protected to prevent loss of moisture from the surface and to prevent temperatures at the surface from going below 50°F (10°C) for periods of 5 days where normal portland cement is used, and 3 days where high early strength portland cement is used.

These limits recognize that under average conditions curing of concrete does not cease immediately upon removal of the moisture loss protection. Where conditions are extremely severe, such as for thin sections in hot dry air or very low temperatures, it may be desirable to increase somewhat the protection periods specified.

High Early Strength Portland Cement. Most cement companies manufacturing what is designated as high early strength portland cement, which develops practically the same strength in 72 hours as is obtained with normal portland cement in 7 to 10 days.

High early strength cement is used in the same proportions and in exactly the same manner as ordinary portland cement. This cement is advantageous not only in special rush work but in everyday reinforced concrete work. High early strength cement costs $3.50 per barrel more than standard portland cement, or if we assume that approximately 1-1/2 barrels of cement per c.y. (2 per cu.m) will be used, the cost per c.y. of concrete is increased by $5.25 (6.86 per cu.m). If the average amount of concrete per s.f. (sq.m) of floor construction (including columns, beams, lintels, etc.) is 8" (0.2 m, 200 mm), it takes 40 s.f. of floor for a yard (5 sq.m per cu.m) of concrete. The extra expense of high early strength cement is $0.13 per s.f. ($1.44 per sq.m).

Under favorable conditions using high early strength portland cement, it is possible to strip part or an entire floor in three days so it may be possible to save half of the lumber required for a floor. It takes not less than 3 to 4 b.f. of lumber for each s.f. (sq.m) of floor construction—including columns, girders, and lintels—in addition to the labor costs per s.f. (sq.m) of floor to make up the various sides, panels, shores, wedges, etc. required (see *Concrete Formwork*).

If it is possible to save one-half of the lumber plus half the labor, the resulting savings per s.f. (sq.m) is greater than the additional $0.1313 per s.f. (sq.m) for high-early cement.

During the winter there is the additional saving of two to three days for cold weather protection.

Air Entrained Concrete. Most of the highway departments are specifying air-entrained concrete for pavements and many require it for bridges and other structures. In this concrete, 3% to 5 or 6% of air is incorporated in the concrete in the form of minute, separate air bubbles. Such concrete is more resistant to freezing and thawing and to salt action than normal concrete. It is produced by using air-entraining portland cement or by the addition of an air-entraining agent at the mixer. Originally developed to prevent scaling of pavement where salts are used for ice removal, air-entrained concrete has been adopted for all types of work and in all locations because of its better workability as well as its better resistance to weathering.

In air-entrained concrete the proportion of sand and the amount of water is generally reduced from the normal mix in sufficient quantities to make up for the increased volume of concrete produced by the air. Thus the same cement factor is maintained. In the tables given on the previous pages, the sand can be reduced about 5 lbs. (3 kg) for each 1% of air. Thus, if 4% air is incorporated, the quantities of sand shown in these tables would be reduced by 20 lbs. per c.y. (12 kg/m³) of concrete. For each 1% of air the amount of water would be reduced by about 6 to 7 lbs. per c.y. (3.6-4 kg/m³).

Important Suggestions for Purchasing Materials for Job Mixed Concrete. When purchasing sand, gravel, or crushed stone for concrete, it is important to see that the aggregates and particularly the sand do not contain an abnormal amount of moisture. Otherwise, you will be short on materials, which will reduce profits or may even result in a loss, particularly if there are considerable quantities of aggregates to be purchased. The following suggestions are important when placing orders:

1. Determine whether the sand has a high or low moisture content.
2. Determine whether the sand has a high or low specific gravity.
3. When sand is purchased at a certain price per c.y. (cu.m) make certain that you are not paying for water instead of sand, as it is not uncommon to see settlement of 4" (100 mm) in trucks or railroad cars between the time it is shipped and when it arrives at the job. This is especially true where washed sand is loaded immediately after washing or after hard rains.
4. When sand is purchased by weight, also make certain you are not paying for water instead of sand. Ordinarily sand should not contain over 4% to 5% moisture, while wet sand may contain as high as 18% moisture. This makes a difference in the cost of sand.
5. Watch the grading of sand and percentages from fine to coarse.
6. Determine the grading of the gravel and its specific gravity.
7. When gravel is purchased by the cu. yd. (cu.m), stipulate measurement at job or destination, if possible.
8. When gravel is purchased by weight, weight at destination should govern instead of weight at loading point, when gravel may contain much moisture.
9. When purchasing crushed stone, obtain the grading and specific gravity.
10. When purchasing crushed stone, determine the weight per cu. yd. (cu.m) also the moisture content.

An easy method of determining the moisture content of sand, gravel, or crushed stone, is to measure out 1 c.f. (cu.m) of each and weigh them. Then dry this same material and weigh it. The difference in weight is the moisture content. Example: If one c.f. of loose damp sand weighs 105 lbs. (one cu.m weighs 1690 kg) and after drying it weighs only 90 lbs. (1407 kg/m³) then the moisture content is 15 lbs. (283 kg) or 16.67%.

Quantities of Materials Required for Job-Mixed Concrete. For estimating purposes, material quantities for job mixed concrete given in the following table may be used as a guide. The quantities shown in the table will produce concrete of medium consistency or with a slump of about 3" to 4" (75-100 mm).

Suggested Mixes for Non-Air-Entrained Concrete of Medium Consistency (3" to 4" slump*)

Water-cement ratio, lb. per lb.	Maximum size of aggregate in.	Air Content (entrapped air) per cent	Water, lb. per cu. yd. of concrete	Cement, lb. per cu. yd. of concrete	With fine sand—fineness modulus = 2.50**			With coarse sand—fineness modulus = 2.90**		
					Fine aggregate, per cent of total aggregate	Fine aggregate, lb. per cu. yd. of concrete	Coarse aggregate, lb. per cu. yd. of concrete	Fine aggregate, per cent of total aggregate	Fine aggregate, lb. per cu. yd. of concrete	Coarse aggregate, lb. per cu. yd. of concrete
0.40	⅜	3	385	965	50	1240	1260	54	1350	1150
	½	2.5	365	915	42	1100	1520	47	1220	1400
	¾	2	340	850	35	960	1800	39	1080	1680
	1	1.5	325	815	32	910	1940	36	1020	1830
	1½	1	300	750	29	880	2110	33	1000	1990
0.45	⅜	3	385	855	51	1330	1260	56	1440	1150
	½	2.5	365	810	44	1180	1520	48	1300	1400
	¾	2	340	755	37	1040	1800	41	1160	1680
	1	1.5	325	720	34	990	1940	38	1100	1830
	1½	1	300	665	31	960	2110	35	1080	1990
0.50	⅜	3	385	770	53	1400	1260	57	1510	1150
	½	2.5	365	730	45	1250	1520	49	1370	1400
	¾	2	340	680	38	1100	1800	42	1220	1680
	1	1.5	325	650	35	1050	1940	39	1160	1830
	1½	1	300	600	32	1010	2110	36	1130	1990
0.55	⅜	3	385	700	54	1460	1260	58	1570	1150
	½	2.5	365	665	46	1310	1520	51	1430	1400
	¾	2	340	620	39	1150	1800	43	1270	1680
	1	1.5	325	590	36	1100	1940	40	1210	1830
	1½	1	300	545	33	1060	2110	37	1180	1990
0.60	⅜	3	385	640	55	1510	1260	58	1620	1150
	½	2.5	365	610	47	1350	1520	51	1470	1400
	¾	2	340	565	40	1200	1800	44	1320	1680
	1	1.5	325	540	37	1140	1940	41	1250	1830
	1½	1	300	500	34	1090	2110	38	1210	1990
0.65	⅜	3	385	590	55	1550	1260	59	1660	1150
	½	2.5	365	560	48	1390	1520	52	1510	1400
	¾	2	340	525	41	1230	1800	45	1350	1680
	1	1.5	325	500	38	1180	1940	41	1290	1830
	1½	1	300	460	35	1130	2110	39	1250	1990
0.70	⅜	3	385	550	56	1590	1260	60	1700	1150
	½	2.5	365	520	48	1430	1520	53	1550	1400
	¾	2	340	485	41	1270	1800	45	1390	1680
	1	1.5	325	465	38	1210	1940	42	1320	1830
	1½	1	300	430	35	1150	2110	39	1270	1990

*Increase or decrease water per cu. yd. by 3% for each increase or decrease of 1" in slump, then calculate quantities by absolute volume method. For manufactured fine aggregate, increase the percentage of fine aggregate by 3 and water by 15 lbs. per c.y. of concrete. For less workable concrete, as in pavements, decrease percentage of fine aggregate by 3 and add water by 8 lbs. per c.y. of concrete.

**For definition of fineness modulus, see "Aggregates for Concrete", available from the Portland Cement Association.

Suggested Mixes for Air-Entrained Concrete of Medium Consistency
(3" to 4" slump*)

Water-cement ratio, lb. per lb.	Maximum size of aggregate in.	Air content, per cent	Water, lb. per cu. yd. of concrete	Cement, lb. per cu. yd. of concrete	With fine sand— fineness modulus = 2.50**			With coarse sand— fineness modulus = 2.90**		
					Fine aggregate, per cent of total aggregate	Fine aggregate, lb. per cu. yd. of concrete	Coarse aggregate lb. per cu. yd. of concrete	Fine aggregate, per cent of total aggregate	Fine aggregate, lb. per cu. yd. of concrete	Coarse aggregate, lb. per cu. yd. of concrete
0.55	⅜	7.5	340	620	54	1450	1260	58	1560	1150
	½	7.5	325	590	45	1250	1520	49	1370	1400
	¾	6	300	545	39	1140	1800	43	1260	1680
	1	6	285	520	35	1060	1940	39	1170	1830
	1½	5	265	480	33	1030	2110	37	1150	1990
0.60	⅜	7.5	340	565	54	1490	1260	58	1600	1150
	½	7.5	325	540	46	1290	1520	50	1410	1400
	¾	6	300	500	40	1180	1800	44	1300	1680
	1	6	285	475	36	1100	1940	40	1210	1830
	1½	5	265	440	33	1060	2110	37	1180	1990
0.65	⅜	7.5	340	525	55	1530	1260	59	1640	1150
	½	7.5	325	500	47	1330	1520	51	1450	1400
	¾	6	300	460	40	1210	1800	44	1330	1680
	1	6	285	440	37	1130	1940	40	1240	1830
	1½	5	265	410	34	1090	2110	38	1210	1990
0.70	⅜	7.5	340	485	55	1560	1260	59	1670	1150
	½	7.5	325	465	47	1360	1520	51	1480	1400
	¾	6	300	430	41	1240	1800	45	1360	1680
	1	6	285	405	37	1160	1940	41	1270	1830
	1½	5	265	380	34	1110	2110	38	1230	1990

* and **. See footnotes at bottom of previous table.

Labor Mixing and Placing Concrete

The labor cost of mixing and placing concrete will vary considerably according to the size of the job, the method used in mixing and placing the concrete, the proximity of the stock piles to the mixer, and the distance the concrete has to be transported after mixing.

On large jobs using conveyors and bins for feeding the aggregate to the mixer, the purchase or rental of this equipment, together with the labor cost for placing, should be estimated separately and not included in the cu. yd. price of placing the concrete. This plant and equipment cost may be the same whether the job contains 1,000 or 10,000 c.y. of concrete, and unless it is kept as a separate item, the resulting labor costs will be misleading and valueless as estimating data for future jobs.

For ready-mixed concrete delivered to the job mixed and ready to place in the forms, it is often possible for the truck mixers to deposit the concrete directly into the trenches or wall forms without any handling on the job, other than spading, vibrating, spreading, and leveling the concrete.

However, on many jobs this is not possible. It is necessary to have a receiving hopper where the truck mixers dump the concrete, which is then hauled from receiving hopper to place of deposit either by wheelbarrows, hand, or power buggies. On large operations the concrete might be pumped from receiving hopper to place of deposit.

Another method that is often used on large operations is to use a crane, which picks up a batch of concrete in a bucket and places it directly over the footings, piers, or forms.

When one of the above methods is used, all mixer labor is eliminated, but it usually requires one or two workers at the receiving hopper or at the truck dump to assist the truck mixer, prevent spillage, cleanup, etc.

Most specifications prohibit the use of long spouts for chuting concrete from the mixer to the forms on account of the great danger of segregation. Buggies or buckets are generally desirable. Self-propelled buggies are available and are used on many of the larger jobs.

Weather conditions also influence the labor costs considerably because placing concrete in freezing weather usually necessitates heating the sand, gravel or stone and water, and protecting the concrete from freezing by the use of tarpaulins, straw, oil burning salamanders or other artificial means, all of which add to the overall expense.

With so many small mixers on the market very little concrete is hand mixed today, but occasionally a job presents itself where it is more economical to mix by hand.

Mixes for Small Jobs

For small jobs where time and personnel are not available to determine proportions in accordance with the recommended procedure, mixes in Table 1 will usually provide concrete that is strong and durable enough, as long as too much water is not added at the mixer. These mixes have been predetermined in conformity with the recommended procedure by assuming conditions applicable to the average small job and for aggregate of medium specific gravity. Three mixes are given for each maximum size of coarse aggregate. For the selected size of coarse aggregate, Mix B is intended for initial use. If this mix proves to be oversanded, change to Mix C; if it is undersanded, change to Mix A. It should be noted that the mixes listed in the table are based on dry or surface-dry sand. If the sand is moist or wet, make the corrections in batch weight prescribed in the footnote.

The approximate cement content per cubic foot of concrete listed in the table will be helpful in estimating cement requirements for the job. These requirements are based on concrete that has just enough water in it to permit ready working into forms without objectionable segregation. Concrete should slide, not run, off a shovel.

Concrete Mixes for Small Jobs

Maximum Size of Aggregate Inch	Mix Designation	Approximate Weights of Solid Ingredients Per Cu.Ft. Concrete, Lb.				
			Sand *		Course Aggregate	
		Cement	Air Entrained Concrete **	Concrete Without Air	Gravel or Crushed Stone	Iron Blast Furnace Slag
1/2	A	25	48	51	54	47
	B	25	46	49	56	49
	C	25	44	47	58	51
3/4	A	23	45	49	62	54
	B	23	43	47	64	56
	C	23	41	45	66	58
1	A	22	41	45	70	61
	B	22	39	43	72	63
	C	22	37	41	74	65
1 1/2	A	20	41	45	74	65
	B	20	39	43	77	67
	C	20	37	41	79	69
2	A	19	40	45	79	69
	B	19	38	43	81	71
	C	19	36	41	83	72

Concrete Mixes for Small Jobs (metric)

Maximum Size of Aggregate mm	Mix Designation	Approximate Weights of Solid Ingredients Per Cu.Meter Concrete, Kg.				
			Sand *		Course Aggregate	
		Cement	Air Entrained Concrete **	Concrete Without Air	Gravel or Crushed Stone	Iron Blast Furnace Slag
12.5	A	400	769	817	865	753
	B	400	737	785	897	785
	C	400	705	753	929	817
18.8	A	368	721	785	993	865
	B	368	689	753	1025	897
	C	368	657	721	1057	929
25.0	A	352	657	721	1121	977
	B	352	625	689	1153	1009
	C	352	593	657	1185	1041
37.5	A	320	657	721	1185	1041
	B	320	625	689	1233	1073
	C	320	593	657	1265	1105
50.0	A	304	641	721	1265	1105
	B	304	609	689	1297	1137
	C	304	577	657	1329	1153

*Weights are for dry sand. If damp sand is used, increase tabulated weight of sand 2 lbs.; if very wet sand is used, 4 lbs.

**Air-entrained concrete should be used in all structures that will be exposed to alternate cycles of freezing and thawing. Air-entrainment can be obtained by the use of an air-entraining cement or by adding an air-entraining admixture. If an admixture is used, the amount recommended by the manufacturer will, in most cases, produce the desired air content.

Power Buggy. One method of transporting construction materials, such as wet concrete, earth, sand, gravel, cement, etc., is the power-operated buggy, which requires only steering on the part of the operator.

They are furnished in various sizes by different manufacturers and the cost varies according to size and capacity.

The power buggy is mounted on a power driven three-wheel chassis and is steered by means of a rear caster wheel. The maximum speed over the ground conforms to a fast walking pace either on the level or up to a 20% incline.

This machine has a capacity of 1,500 lbs. (675 kg); a bucket volume of 10 c.f. (0.28 cu.m) and will handle 8 to 9 c.f. (0.22-0.25 cu.m) of wet concrete but a larger volume of light, dry materials. This buggy weighs 560 lbs. (252 kg), is powered by a 7 Hp. gasoline engine with a fuel capacity of 2-3/4 gals. (10.4 liters), which will operate the machine for eight hours. The machine is 5'-5-1/2" (1.6 m) in length overall, 2'-7-1/2" (0.80 m) wide and is furnished with either single or dual wheels and has a turning radius of 34-3/4" (870 mm).

Comparison of Hand and Power Buggy Operation. The smaller power buggy having a capacity of 8 to 9 c.f. (0.22-0.25 cu.m) of wet concrete will probably do the work of three workers pushing hand powered concrete buggies.

A large machine having a capacity of 10 to 13 c.f. (0.28-0.36 cu.m) of wet concrete will probably handle as much concrete as 4 to 5 workers using hand operated buggies, but it may be that heavier runways will be required to handle the larger loads of concrete.

Sizes, Weights and Capacities of Concrete Carts or Buggies

Capacity cu. ft.	Turning Radius	Width Overall	Length Overall	Maximum Height	Approx. Wgt, lbs.
12	72"	34"	87"	54"	950
15	82"	41"	94-1/2"	54"	1,020
21	91"	54"	106-1/2"	56"	1,400

(metric)

Capacity cu. m	Turning Radius	Width Overall	Length Overall	Maximum Height	Approx. Wgt, kg
0.34	1800 mm	850 mm	2175 mm	1350 mm	431
0.42	2050 mm	1025 mm	2362 mm	1350 mm	463
0.59	2275 mm	1350 mm	2663 mm	1400 mm	635

The following is an approximate comparison of cost between the two methods, based on wheeling 100 cu. yds. (76.5 cu.m) of concrete from a 16-S mixer:

Hand Method	Quan.	Rate	Total	Rate	Total
5 Laborers (hours)	40	$....	$....	$21.04	$ 841.60
Cost per c.y.			....		8.42
per cu.m.			....		10.95
Power buggy method					
Power buggy, fixed charge					
(per) Day	1	$....	$....	$75.00	$75.00
Fuel, gals, and oil	5	....	....	1.25	6.25
Operator (hours)	8	....	....	21.04	168.32
Total cost per day			$....		$ 249.57
Cost per c.y.			....		2.50
per cu.m.			....		3.25

On some jobs, it may be that 2 power buggies would be required to keep the mixer operating at capacity, and in such instances would be double that given above. In either case the saving is considerable.

Capacities of Concrete Mixers Based on A.G.C. Standards
(in Sacks of Cement per Batch)

Size of mixer denotes number of cu. ft. (cu.m) of mixed concrete, plus 10% excess, when mixer is level.

Concrete Proportions	Sizes of Standard Mixers				
	3-1/2- S	6 - S	11 - S	16 - S	28 - S
1:1-1/2:3	1/2 (0.014)	1 (0.028)	3 (0.085)	4 (0.113)	8 (0.227)
1:1-1/2:3	1/2 (0.014)	1 (0.028)	2 (0.057)	3 (0.085)	7 (0.198)
1:2:3	1/2 (0.014)	1 (0.028)	2 (0.057)	3 (0.085)	7 (0.198)
1:2:3-1/2	1/2 (0.014)	1 (0.028)	2 (0.057)	3 (0.085)	7 (0.198)
1:2:4	1/2 (0.014)	1 (0.028)	2 (0.057)	3 (0.085)	6 (0.170)
1:2-1/2:4	1/2 (0.014)	1 (0.028)	2 (0.057)	3 (0.085)	6 (0.170)
1:3:5	1/2 (0.014)	1 (0.028)	1 (0.028)	2 (0.057)	5 (0.142)
1:3:6	1/2 (0.014)	1 (0.028)	1 (0.028)	2 (0.057)	4 (0.113)

Mixing Concrete in a One Sack (6-S) Mixer. Foundation concrete is usually mixed in the proportions 1:2:4, 1:3:5, or 1:3:6, which is a one sack mix using a 6-S mixer.

Often a 1:3:5 mix is used for foundation walls, where watertight concrete is not necessary. This requires 1 sack of cement, 3 c.f. (0.8 cu.m) of dry sand and 5 c.f. (0.14 cu.m) of gravel or crushed stone, per batch. Wheelbarrows ordinarily used for dry materials hold 3 to 3-1/2 c.f. (0.08-0.10 cu.m) loaded level full, so each batch will require 1 barrow of sand and 2 barrows of gravel or 9 c.f. (0.25 cu.m) of dry material.

Where possible to start mixing first thing in the morning and continue all day without delays, a mixer should average 30 batches, 6 to 7 c.f. (0.17-0.20 cu.m) of concrete each, per hr. or 240 batches, 54 to 62 c.y. (41-47 cu.m), per

8-hr day. Runs like this on foundation work are exceptional, because in addition to the actual mixing time, it will be necessary to allow for delays, shutdowns, etc., as maximum output cannot be figured for the entire job. Workers must be paid for time getting ready to start work in the morning and cleaning and washing wheelbarrows, buggies, or concrete mixer at noon and at night; cutting and removing old concrete drippings and spillage; building and removing runways; unloading cement; delays caused by non-delivery of materials, etc.

Based on 6 to 7 c.y. (0.17-0.20 cu.m) of concrete per hr., the following crew is required: 2 laborers loading and wheeling gravel or crushed stone; 2 laborers wheeling sand and cement; 1 laborer attending mixer; 3 or 4 laborers wheeling concrete from mixer to place of deposit; 2 or 3 laborers dumping barrows, spreading, leveling, and tamping concrete; 1 mixer engineer if required, and 1 mason foreman in charge of work if required.

Allowing for average conditions, including runways, cleaning and repairing mixer, engine, etc., the labor per c.y. (cu.m) of concrete should cost as follows:

	Hours	Rate	Total	Rate	Total
Labor per c.y.	2.5	$....	$....	$21.04	$ 52.60
per cu.m	3.25		....		68.38

Under extremely favorable conditions, this cost can be reduced 1/2 to 3/4-hr. per c.y. (0.65-1.0 hrs. per cu.m). Add for mason foreman and mixer engineer, if required by union rules.

If a power buggy is used for wheeling concrete from mixer to place of deposit, one power buggy will take away all the concrete a 6-S mixer can turn out. This will eliminate 2 to 3 laborers wheeling concrete and result in a saving of about 1/4-hr. labor time per c.y. (0.33 hrs. per cu.m) of concrete.

Mixing Concrete Using a 11-S Mixer. When using a 11-S mixer of 1 to 2 sack mix capacity, the costs will run about the same as given for 6-S or 16-S mixers, under the same conditions.

Mixing Concrete in a 16-S (3 Sack Batch) Mixer. A mixer of this size will turn out about 16 c.f. of mixed concrete at each batch, depending upon the proportions.

Opinions vary as to amount of mixing required. The time of mixing specified by architects and engineers usually varies between 1-1/2 and 2 minutes per batch. The time specified affects the labor costs to a considerable extent. The longer mixing time restricts the day's output.

If the mixer has a side-loader and is fed by wheelbarrows, with the concrete wheeled direct from the mixer to place of deposit in wheelbarrows or concrete buggies, the mixer should average one batch every 2 minutes or 30 batches, 13 to 15 c.y. (10-11.5 cu.m) per hr.

Using an average of 2 minutes per batch, the mixer should turn out 104 to 120 c.y. (80-92 cu.m) per 8-hr. day, depending on the concrete proportions and the size of each batch. In addition to the actual mixing time, it will be necessary to allow for delays, shutdowns, etc., as maximum output cannot be figured for the entire job. The workers must be paid for time getting ready to start work in the morning, cleaning and washing wheelbarrows and concrete mixer at noon and at night; cutting and removing old concrete drippings and spillage; building and removing runways; delays caused by non-delivery of materials, etc.

Based on an average of 14 c.y. (10.75 cu.m) of concrete per hr., it will require a crew of 25 to 28, distributed as follows: 6 laborers loading and wheeling gravel or crushed stone; 3 laborers wheeling sand; 1 or 2 laborers on cement; 1 laborer feeding mixer; 1 laborer dumping concrete from mixer into wheelbarrows or buggies; 5 to 7 laborers wheeling concrete from mixer to place of deposit, depending upon length of haul; 3 to 4 laborers dumping concrete carts or barrows, spreading, spading, and tamping concrete in place; 3 to 4 laborers for general labor, such as building and removing runways, cleaning concrete mixer, wheelbarrows or buggies, picking up or baling cement sacks and other miscellaneous labor; 1 engineer on mixer and 1 mason or concrete foreman in charge.

On the above basis, the labor per c.y. (0.765 cu.m) of concrete should cost as follows:

	Hours	Rate	Total	Rate	Total
Labor	2.00	$....	$....	$21.04	$ 42.08
Foreman	0.13	....	....	21.77	2.83
Mixer engineer	0.13	....	....	28.55	3.71
Cost per c.y. (0.765 cu.m)			$....		$ 48.62

In many instances concrete will be placed for 65% to 75% of the above costs, but this will merely represent a day's run and not the average for the entire job.

Using a power buggy to transport the concrete from the mixer or hopper to place of deposit, one or two power buggies should handle all the concrete a 16-S mixer can turn out. This will eliminate 3 to 4 laborers wheeling concrete and result in a saving of 1/4 to 1/3 hr. per c.y. (0.33 to 0.50 hr. per cu.m) of concrete.

Mixing Concrete in a 28-S (1 C.Y.) Mixer. If the concrete is mixed in a 28-S mixer fed by wheelbarrows and wheeled to the place of deposit, the labor cost per c.y. (cu.m) will run about the same as for 16-S mixers under similar conditions.

While it is true the daily output will be greatly increased, the labor cost feeding the mixer and distributing the concrete will be increased in proportion.

Labor Mixing and Placing One C.Y. (Cu.M) of Concrete
for Foundations

			Hours per C.Y. (Cu.M)	
Method of Mixing and Placing	Common Labor	Mixer Engr.	Hoist. Engr.	Fore- man
Hand mixing using one mixing board	3.33 (4.36)*	--	--	0.330 (0.432)
Hand mixing using two mixing boards	3.00 (3.92)**	--	--	0.250 (0.327)
Using one sack (6-S) mixer and wheeling to place	2.50 (3.27)***	--	--	0.170 (0.222)
Using two sack (11-S) mixer and wheeling to place	2.00 (2.62)	0.125 (0.163)	--	0.125 (0.163)
Using 16-S mixer and wheeling to place	2.00 (2.62)	0.125 (0.163)	--	0.125 (0.163)
Using 28-S mixer and wheeling to place	2.00 (2.62)	0.083 (0.109)	--	0.080 (0.105)
Using conveyors, batcherplant, hoist and wheeling to place	1.25 (1.63)	0.060 (0.078)	0.060 (0.078)	0.060 (0.078)

*In extreme warm weather or southern states, add 1/2 to 2/3 hr.
**If necessary to load concrete into barrows and wheel from mixing board to place of deposit, add 1/2 to 3/4 hr./c.y. (0.65-1.0 hr./m³).
***Under extremely favorable conditions, this may be reduced 1/4 to 1/2 hr.

Above time includes all lost time in connection with concrete foundations. Add for labor erecting and dismantling hoisting towers, spouts, hoisting engines, and mixers, as this may vary from $550.00 to $5,500.00, depending on the size of job.

If power buggies are used for wheeling concrete from mixer or receiving hopper to place of deposit, it usually results in a saving of 1/4 to 1/3 hrs. per c.y. (0.33-0.42 hrs. per cu.m) of concrete.

Loading Mixer Using a Front-End Loader

Since labor costs are high, many new methods are being used to reduce costs. One method employs a loader to load the sand and gravel from the stock piles and dump it into the loading skip at the mixer. This eliminates the workers loading wheelbarrows with sand and gravel by hand, but there is usually one worker necessary on the stock pile, plus the small loader and operator. When supplying a 6-S mixer, it would eliminate about 5 workers loading and wheeling sand and gravel but would add one worker for the stock piles, plus the loader and operator. When supplying a 16-S mixer with a loader, it would eliminate about 9 workers loading and wheeling sand and gravel but would add 1 worker on the sand stock pile, 1 worker on the gravel stock pile, the loader, and the operator.

When supplying a 11-S mixer in this manner, it would eliminate 6 to 7 workers loading and wheeling sand and gravel but would add 1 worker on the stock piles, the loader, and the operator.

Mixing and Placing Concrete Where Materials Are Delivered to Job Batch Proportioned. Another method of handling concrete work is to have the concrete aggregates delivered to the job in trucks, correctly proportioned for one batch of concrete, including cement, sand, and gravel or crushed stone.

When this method is used, the concrete aggregates for each batch are proportioned in the material yard and delivered to the job ready to mix. The truck backs up to the mixer and discharges the cement, sand, and gravel or crushed stone into the loading hopper. A paving mixer with a large, flat loading hopper is commonly used for this purpose, because it facilitates dumping the aggregates from the trucks into the loading hopper.

This method eliminates the need for workers loading and wheeling cement, sand, and coarse aggregates and the time required per c.y. (cu.m) of concrete should average as given in the following table.

Labor Mixing and Placing One C.Y. (Cu.M) of Concrete for
Foundations When Aggregates Are Yard Proportioned and
Delivered to the Job a Batch to a Load

Method of Mixing and Placing	Common Labor	Mixer Engr.	Hoist. Engr.	Fore-man
16-S mixer and wheeling to place	1.25 (1.63)	0.125 (0.103)	--	0.125 (0.163)
28-S mixer and wheeling to place	1.25 (1.63)	0.083 (0.109)	--	0.083 (0.109)

If power buggies are used for wheeling concrete from mixer to place of deposit, it will usually result in a saving of 1/4 to 1/3 hr. labor time per c.y. (0.33-0.42 hrs. per cu.m) of concrete.

The above includes all lost time in connection with mixing and placing concrete.

Add for labor erecting and dismantling hoisting towers, hoisting engines, mixers, etc. This might vary from $550.00 to $5,500.00, depending on the size of job.

Placing Concrete with a Crane and Bucket. The crane and bucket method eliminates the need for building and removing runways, labor wheeling concrete, etc., especially when used for ready-mixed concrete. Labor is reduced to a minimum.

Labor Placing One C.Y. (Cu.M) of Foundation Concrete
Using Crane and Bucket

Method	Qty. Placed per 8-Hr Day c.y. (m3)	Common Labor	Crane Operator	Mixer Engineer	Fore-man*
16-S mixer, batched material	70-90 (54-69)	0.70 (0.92)	0.10 (0.13)	0.10 (0.13)	0.10 (0.13)
28-S mixer, batched material	150-170 (115-130)	0.50 (0.65)	0.05 (0.07)	0.05 (0.07)	0.05 (0.07)
Ready-mix	150-170 (115-130)	0.40 (0.52)	0.05 (0.07)	--	0.05 (0.07)

Add for equipment cost or rental of crane and bucket.
*Add only if required.

Cold Weather Concreting

During the late fall, winter, and early spring, it is necessary to take special precautions to assure proper hardening of the concrete. Careful contractors will use protection whenever the temperature falls below 50°F (10°C). At such time the most important protection required is to heat the sand and mixing water.

During freezing weather an extra boiler (old hoisting engine boiler, for example) having a rating of 50 to 75 Hp according to the size of the job, 10,000 to 25,000 s.f. (900-2250 sq.m) floor area, with a steam gauge set for a pressure of 2 lbs./sq. in. (0.14/cm²) will be needed with at least 2 perforated pipes for heating the coarse aggregate in addition to that required for heating the sand. On larger jobs or for fast concreting where aggregate bins are used, a grillage of steam pipes 4'-0" (1.2 m) apart at the bottom of the bins will be sufficient. These pipes should have 1/4" (6 mm) perforations about 4" (100 mm) on centers. A 1-1/2" (38 mm) steam line should be run from the boiler to the water tank and another line run up the side of the building with valves and connections for steam hose to be used for cleaning out ice and snow.

Sufficient steam should be available so the concrete has a temperature of 50 to 70°F (10-21.1°C) when leaving the mixer. This will promote the setting of the cement and permit earlier removal of the forms.

On small jobs the flame of an oil heater is allowed to project into the mixer to heat the materials.

Footings are generally protected by covering the fresh concrete with 8" to 12" (200-300 mm) of hay or straw and with tarpaulins or polyethylene thrown over the hay.

Foundation walls should be protected in mild weather by covering the top of the wall with hay and the sides of the forms with tarpaulins. During more severe weather, place oil burning salamanders every 30 to 40 ft. (9-12 m) apart and cover with tarpaulins, care being taken to keep the tarpaulins sufficient distance from the salamanders.

In skeleton construction it will be necessary to curtain the walls with tarpaulins for the story being concreted, and in very severe weather it may be necessary to curtain two stories with tarpaulins and to cover the top floor with hay or straw, held down by planks or timbers, which in turn support a layer of boards and planks with tarpaulins thrown over them. Stair and elevator openings, and openings for steam and plumbing pipes, allow heat to penetrate beneath the protecting layers and to heat the top of the floor.

It will require one salamander for each 200 to 250 s.f. (18-23 sq.m) of floor surface, depending on the severity of the weather. Sufficient fuel should be placed on the floor in the early afternoon to last until next morning. Heat should be supplied for 2 to 5 days or more, depending on whether high-early-strength or normal portland cement is used and on the temperature maintained. The proper fire protection equipment should be available on each floor.

Ready Mixed Concrete

When using ready-mixed concrete for all classes of reinforced construction—floors, beams, girders, columns, etc.—the cost of handling and placing the concrete will run approximately the same as for job-mixed concrete, except for the labor handling materials, supplying and operating mixer, etc., as the trucks dump the ready-mixed concrete into a receiving hopper or bucket, where it is distributed by wheelbarrows, concrete buggies, or hoisting into place.

In addition to saving labor time of about 3/4-hr. per c.y. (1 hr. per cu.m) of concrete for charging mixer, the use of ready-mixed concrete saves the cost of cement sheds and provides more room for storing other materials. It permits higher output per hour, because ready-mixed concrete can usually be furnished in any quantity desired. It also saves the cost of installing and dismantling expensive equipment, which may easily cost $5,500 or more on a job of any magnitude. It is necessary to have a fairly large size hopper on the job to receive the concrete from the trucks. This hopper must be of sufficient capacity to supply the hoisting bucket as fast as it can handle the concrete. Otherwise, the work will be slowed down. With the crew idle part of the time, the labor costs go up.

The labor cost of handling and placing ready-mixed concrete will vary with the amount handled and placed per day, the method of handling, hoisting and placing the concrete, and the class of work. The following tables gives the labor hours necessary to handle and place 1 c.y. (cu.m) of ready-mixed concrete, using various methods:

Labor Placing One C.Y. (Cu.M) of Ready-Mixed Concrete for Concrete Foundations

Method and Quantity Placed per 8-Hr. Day	Labor hours c.y.	cu.m	Hoisting Eng., hrs. c.y.	cu.m	Foreman hours c.y.	cu.m
Placing concrete directly in forms, 1 truck, 50-70 c.y. (38-54 m³)	0.35-0.50	0.46-0.65	--	--	0.13	0.17
Placing concrete directly in forms, 2 trucks, 100-150 c.y. (77-115 m³)	0.27-0.40	0.35-0.52	--	--	0.08	0.105
Wheelbarrows, 50-70 c.y. (38-54 m³)	1.25-1.50	1.63-1.96	0.17*	0.22*	0.17**	0.22**
Wheelbarrows, 100 c.y. (77 m³) & over	1.00-1.25	1.3-1.63	0.13*	0.17*	0.13**	0.17**
Concrete buggies, 50-70 c.y. (38-54 m³)	1.00-1.25	1.3-1.63	0.17*	0.22*	0.17**	0.22**
Concrete buggies, 100 c.y. (77 m³) & over	0.90-1.10	1.18-1.44	0.13*	0.17*	0.13**	0.17**
13 c.f. (.40 m³) power concrete buggies	0.65-0.75	0.85-1.0	0.13*	0.17*	0.13**	0.17**

*Where the concrete is hoisted and it is necessary to use a hoisting engineer, add as given in table.
**If union regulations require a mason or cement finisher foreman in charge of crew, add as given in table.

Labor Placing One C.Y. (Cu.M) of Ready-Mixed Concrete for Reinforced Floors, Beams, Girders, Columns, Etc.

Method of Handling and Placing Concrete per 8-Hr. Day	Labor hours c.y.	cu.m	Hoisting Eng., hrs. c.y.	cu.m	Foreman hours c.y.	cu.m
Hoisting concrete in wheelbarrows on a material hoist & wheeling into place	2.3	3	0.11*	0.14*	0.11*	0.14*
Hoisting concrete in concrete buggies on material hoist & wheeling into place	1.1	1.4	0.08*	0.10*	0.08*	0.10*
Thin floors 2-2-1/2" (50-63mm) thick. Hoisting concrete in wheelbarrows on material hoist & wheeling into place	2.8	3.7	0.11*	0.14*	0.11*	0.14*
Thin floors 2-2-1/2" (50-63mm) thick. Hoisting concrete in concrete buggies on material hoist & wheeling into place	1.6	2.1	0.08*	0.10*	0.08*	0.10*
Hoisting concrete in concrete bucket, discharging into a floor hopper, & wheeling into place using wheelbarrows	1.2	1.6	0.05*	0.06*	0.05*	0.06*
Hoisting concrete in concrete bucket, discharging into a floor hopper, & wheeling using concrete buggies	0.8	1.05	0.04*	0.05	0.04*	0.05*
If power buggies are used for transporting concrete from hopper to place of deposit, deduct per c.y. (cu.m)	.25-.33	.33-.43	--	--	--	--

Method of Handling and Placing Concrete per 8-Hr. Day	Labor hours		Hoisting Eng., hrs.		Foreman hours	
	c.y.	cu.m	c.y.	cu.m	c.y.	cu.m
Mobile crane with bucket on ground, hoisting concrete to floor or forms, & depositing directly into forms	0.5	0.6	--	--	0.04*	0.05*
Add for rental of mobile crane & operator	--	--	0.04*	0.05*	--	--
For thin concrete floors, 2-2-1/2" (50-63mm) thick, add to the above costs	0.5	0.65	--	--	--	--

*Time of hoisting engineer and foreman will vary according to the number of cubic yards of concrete placed per day.

The above costs do not include cost of erecting and dismantling material hoists, concrete hoisting towers, etc., which may run from $550.00 to $5,500.00, depending on size of job and type of equipment used.

PLACING CONCRETE BY PIPELINE

The development and perfection of pumps for transporting freshly mixed concrete through pipelines has revolutionized the placing of substantial yardages of concrete in dams, tunnels, retaining walls, bridge decks, mill, factory, warehouse, and apartment buildings, where concrete is required at rates varying from 15 to 80 c.y. (11.5-61 cu.m) per hr. Pumps force concrete through a pipeline in much the same manner that a piston pump pumps water. It will transport freshly mixed workable concrete to distances of 1,000 ft. (300 m) and to heights in excess of 120 ft. (36 m). Concrete pumps can deliver concrete to forms 1,300 ft. (390 m) away or 200 ft. (60 m) high, when a well designed mix is used.

Concrete pumps can handle efficiently concrete of any slump from 1/2" to 7" (13-175 mm), but the most dependable slump for general conditions is around 3" to 5" (75-125 mm). However, plastic concrete of about 6" (150 mm) slump can generally be pumped the maximum distance and height. The rate of pumping depends much on the consistency of the concrete.

The concrete is delivered to the forms in better condition than when discharged from the mixer, due to the added mixing received in the remixing hopper, which receives the concrete from the mixer and keeps it in a uniform condition as it is fed through the cylinder of the concrete pump to the pipeline. During the pumping operation, the pipeline is full at all times, which keeps the concrete from segregating, thereby eliminating one of the major objections found in some methods used to transport concrete to the forms. The troweling action of the walls of the pipeline in contact with the concrete improves the quality of the concrete.

The smooth, even flow of concrete into the forms reduces the shock loading on the bracing and ties. Danger of honeycomb is reduced, because there is no "slug" of pebbles dumped in, as in the case of a buggy, at the end of a batch that is not be properly spaded. Walls 120 ft. (36 m) long and 15 ft. (4.5 m) high have been made at one pour in 3 ft. (.90 m) lifts without signs of longitudinal jointing or uneven surfaces due to the dumping of wheelbarrows or buggies in such a way as to splash the form above the point of placement.

Runways for wheelbarrows or concrete buggies are not required when placing concrete by pipeline, although it is necessary to provide supports for the pipelines, which usually consists of "horses" or "bridges".

Equipment maintenance varies with the efficiency of the operator, the type of work, abrasiveness of materials, etc., as with all construction machinery.

In addition to replacements for the machine, there will be pipeline maintenance, depending on the abrasiveness of materials. The couplings, which represent approximately three-fourths of the cost of pipe sections, will last indefinitely. The straight pipe will last from 50,000 to 150,000 c.y. (38,250 to 114,750 cu.m) and the elbows will last from 30,000 to 70,000 c.y. (22,950 to 53,550 cu.m), depending upon the abrasiveness of the concrete. The piping can be replaced as it becomes worn out and the old couplings reused.

Cost of Placing Concrete by Pipeline. The cost of placing concrete by pipeline varies widely as do other methods of distribution, and is governed by the size of the job, area covered, the yardage of concrete that can be placed at one setup of the pipelines, distance from the pump, etc.

When placing concrete by ordinary methods, the cost of rental, erecting, and dismantling the hoisting tower and spouts must be considered, along with the cost of raising the concrete hopper and moving spouts for placing the concrete.

The same conditions must be considered with concrete pump distribution, which includes original cost of equipment, depreciation or salvage value, cost of setting up and removing pumping equipment, pipelines, etc.

Always bear in mind, the cost of mixing the concrete will be the same, regardless of the method of distribution, because where the concrete is mixed and dumped into a hoisting bucket using ordinary methods, it is deposited in the hopper. It is only the method of handling, hoisting, and placing the mixed concrete that should be considered. This also applies where ready-mixed concrete is used.

The crew operating the pumping unit will vary with the size, area, and yardage in the job. An operator is required on the concrete pump. A foreman and 6 to 12 laborers should place and remove pipelines and handle, spread, and level the concrete, and the quantity will vary from 15 to 80 c.y. (11.5-61 cu.m) per hr., depending on the size of the pumping equipment and the class of work.

Placing Concrete on a Commercial Building. Assume a 5-story steel frame commercial building, 140' x 250' (42 x 75 m) having floors of metal pan construction, with pans from 6" to 14" (150-350 mm) deep and 2-1/2" (63 mm) of concrete over the pans, the entire job containing approximately 5,200 c.y. (3978 cu.m) of concrete.

Assume the concrete is ready-mixed, using 3/4" (19 mm) gravel aggregate and having a 5" (125 mm) slump.

The concrete is delivered to the job and deposited directly into the remixing hopper on the pumper and pumped to the various floors. The labor per c.y. (cu.m) cost as follows:

	Hour	Rate	Total	Rate	Total
Pump operator	0.07	$....	$....	$28.55	$ 2.00
Hopper attendant	0.07	....	....	21.04	1.47
Labor vibrating, pipeline, etc	0.49	....	....	21.04	10.31
Foreman	0.07	....	....	21.77	1.52
Cost per c.y.			$....		$ 15.30
per cu.m				....	19.89

Average rate of pour 15 c.y. (11.5 cu.m) per hr. Maximum rate of pour 30 c.y. (23 cu.m) per hr. The above costs do not include the cost of the pumper, depreciation, parts and replacements, gas and oil, labor installing and removing equipment, etc.

Placing Concrete by Pipeline on a Sewage Disposal Plant. Assume a sewage disposal plant, size 370' x 380' (111 x 114 m), containing 15,000 c.y. (11,475 cu.m) of concrete consisting of heavily reinforced concrete in slabs and walls, all concrete to be placed by pipeline.

The mixer will discharge into the remixing hopper of a single pumper, which is placed below grade in the excavation made for the plant addition. From this point the concrete is to be pumped through pipelines to the point of placement.

The floors of the aerating basins, 30 ft. (9 m) wide and 120 ft. (36 m) long are poured by radial spouting from the end of the pipeline, which is supported on horses. The piping is taken up as the concreting progresses and is laid in the next basin.

Staging supported on the forms will permit erection of the pipeline, pouring 112 c.y. (85 cu.m) of concrete in the wall and removal of the pipe line in one day. The walls of the aerating and settling basins are 15 ft. (4.5 m) high and are poured in 3 ft. (0.9 m) lifts. Valves at 20 ft. (6 m) intervals are inserted in the 120 ft. (36 m) of distributing line which permits the withdrawal of concrete at these points. Elephant trunks (tremie) direct the concrete into specially constructed chutes leading the concrete down through the maze of steel to the level being poured where it is discharged without spattering the forms above that point. Pouring was started from the valve nearest the pumper. When enough concrete has been withdrawn at a valve, it is closed and the next valve opened for concreting at this point. This procedure is continued until the lift is completed at the end of the pipeline.

With the aid of internal vibration, these high long walls are poured relatively free of blemishes or jointing.

The average labor cost of placing 1 c.y. (cu.m) of concrete is as follows:

Placing Concrete by Pipeline	Hours	Rate	Total	Rate	Total
1 pumper operator	0.071	$....	$....	$28.55	$ 2.03
1 foreman	0.071	....	....	21.77	1.55
8 laborers*	0.86	....	....	21.04	18.09
Cost per c.y.			$....		$ 21.67
per cu.m				....	28.17

*Labor on pumper includes time placing and removing pipeline.

During the pouring of the concrete in the walls, 2 additional workers are required to operate vibrators. Vibrators would be required regardless of the distribution method used.

CAST IN PLACE CONCRETE

Concrete floors, sidewalks, and lightweight concrete floor fill are estimated by the square foot (square meter), taking the number of s.f. (sq.m) of any given thickness—2", 3", 4", etc. (0.05, 0.075, 0.10, 50, 75, 100 mm, etc.)—to obtain the total cubic yards (cubic meters) for placement.

The quantities are easily computed by obtaining the area of the various spaces to have concrete floors or walks and stating the number of s.f. (sq.m) for each thickness.

When preparing an estimate for concrete floors or sidewalks placed on the ground, there are distinct material and labor items to be considered as follows:

1. Labor grading and removing surplus earth under floors or walks or filling where present grade is too low.
2. Cost of sand, gravel, or slag to be placed under concrete floor; labor spreading and tamping same in place.
3. Labor placing wood or metal screeds and forms.
4. Cost of vapor barrier and reinforcing material and placement.
5. Labor mixing and placing concrete.
6. Labor troweling finish surface and blocking off into squares, if required.
7. Labor and materials for curing and hardeners, if required.

Grading for Concrete Floors and Walks. On practically all cement floors and walks placed directly on the ground, it is necessary to do a certain amount of grading before the bed of sand, gravel, or slag can be placed.

If it is not necessary to remove the excavated soil from the premises, a bulldozer or loader is probably the most economical method to use.

A medium size loader and operator should excavate and place in piles, 25 to 30 c.y. (19-23 cu.m) per hour at the following cost per c.y. (cu.m):

	Hours	Rate	Total	Rate	Total
Loader w/Operator per c.y.	0.045	$....	$....	$55.00	$ 2.48
per cu.m.......................................	0.590		....		3.22

This cost will vary with the kind of soil encountered, i.e., sand or loam, ordinary black soil, heavy soil, or clay.

Number of S.F. of Soil of Various Thicknesses Obtainable
From One C.Y. (Sq.M from Cu.M)

Depth, in.	3	4	5	6	7	8	9	10	11	12
Depth, mm	75	100	125	150	175	200	225	250	275	300
Area, s.f.	108	82	64	54	47	40	36	33	29	12
Area, sq.m	13.1	9.9	7.8	6.6	5.7	4.9	4.4	4.0	3.6	1.5

To obtain the cost of grading 1 s.f. of any thickness, divide the c.y. cost by the number of s.f. obtainable from 1 c.y. and the result is the s.f. cost for grading. For example:

1. With excavating at $4.50 per c.y., find the s.f. cost of removing 5" of top soil. One c.y. contains 65 s.f. of 5" fill. $4.50 ÷ 65 = $0.069 per s.f. for grading.

2. With excavating at $5.88 per cu.m, find the sq.m cost of removing 0.13 m of top soil. One cu.m contains 7.7 sq.m of 0.13 m fill. $5.88 ÷ 7.7 = $0.76 per sq.m for grading.

Hand Grading. If the top soil is loaded by hand into wheelbarrows or power buggies, a worker should loosen, shovel and load 5 to 6 c.y. (3.8-4.5 cu.m) per 8-hr. day, at the following labor cost per c.y. (cu.m):

	Hours	Rate	Total	Rate	Total
Labor per c.y.	1.45	$....	$....	$21.04	$ 30.51
per cu.m.......................................	1.89		....		39.77

Grading and Tamping Sand, Gravel, or Slag Fill Under Concrete Floors or Walks. The cost of this work varies with the thickness of the fill and job conditions.

It costs as much to grade a bed of slag or gravel 3" (75 mm) thick as one 9" (0.23 m, 225 mm) thick, because it is only the surface that is graded and the only additional labor on the thicker fill is for handling, tamping, and spreading a larger quantity of material.

Where the fill varies from 3" to 4" (75-100 mm) thick, a worker will spread, grade, and tamp about 6 c.y. (4.6 cu.m) per 8-hr. day, at the following labor cost per c.y. (cu.m):

	Hours	Rate	Total	Rate	Total
Labor per c.y.	1.33	$....	$....	$21.04	$ 27.98
per cu.m	1.74		....		36.61

If the fill varies from 5" to 6" (125-150 mm) thick, a worker will spread, grade and tamp about 8-1/2 c.y. (6.5 cu.m) per 8-hr. day, at the following labor cost per c.y. (cu.m):

	Hours	Rate	Total	Rate	Total
Labor per c.y.	0.94	$....	$....	$21.04	$ 19.78
per cu.m	1.23		....		25.88

If the fill varies from 7" to 9" (175-225 mm) thick, a worker will spread, grade, and tamp about 10-1/2 c.y. (8 cu.m) per 8-hr. day, at the following labor cost per c.y. (cu.m):

	Hours	Rate	Total	Rate	Total
Labor per c.y.	0.76	$....	$....	$21.04	$ 15.99
per cu.m	1.00		....		21.04

A worker will spread, grade, and tamp about 12-1/2 c.y. (9.56 cu.m) of 10" or 12" (250-300 mm) fill per 8-hr. day, at the following labor cost per c.y. (cu.m):

	Hours	Rate	Total	Rate	Total
Labor per c.y.	0.64	$....	$....	$21.04	$ 13.47
per cu.m	0.84		....		17.67

All fill materials will shrink under compaction, some more than others, and the additional material required to make up for shrinkage, together with the additional labor for handling, must be provided for in the estimate.

The best and easiest way to do this is to increase the quantity before pricing by applying the proper shrinkage factor for the material to the net computed volume.

For example: If an area 100'-0" x 50'-0" (30 x 15 m) is to receive a gravel fill 6" (0.15 m) thick, after tamping the gross volume of gravel would be 100'-0" x 50'-0" x 0'-6" x 1.12 = 2,800 c.f. or approximately 104 c.y. (30 x 15 x 0.15 = 75.6 cu.m). In the computation, 1.12 is the shrinkage factor for gravel.

The following table gives approximate values for shrinkage percentages and shrinkage factors for various fill materials:

Material	Shrinkage Percentage	Shrinkage Factor
Cinder	30	1.30
Crushed Limestone	20	1.20
Granulated Slag	15	1.15
Gravel	12	1.12
Sand	8	1.08

To obtain the cost of 1 s.f. (sq.m) of fill of any thickness, divide the cost per c.y. (cu.m) by the number of s.f. (sq.m) obtainable from 1 c.y. (cu.m), multiply by the shrinkage factor and the result is the s.f. (sq.m) cost.

Forms or Screeds for Concrete Sidewalks. It is customary to use 2"x4" (50x100 mm) or 2"x6" (50x150 mm) lumber for forms at each side of a concrete sidewalk, the depth of the forms depending upon the thickness of the concrete. On ordinary sidewalk work, obtain the number of l.f. (m) of walk and multiply by 2, which will be the number of l.f. (m) of forms required. A sidewalk 4" (100 mm) thick will require 2"x4" (50 x 100 mm) lumber; 6" (150 mm) thick, 2"x6" (50 x 150 mm) lumber, etc.

Where concrete floors are placed on the ground it is customary to set wood screeds 6'-0" to 8'-0" (1.8-2.4 m) o.c., so the floor will be level or have a uniform pitch. These screeds are removed as the work progresses and may be used many times. They are usually of 2"x4" (50 x 100 mm) lumber, with stakes placed 3' to 4' (0.9-1.2 m) apart to hold the screeds in place.

Labor Placing Forms and Screeds for Sidewalks and Floors. On all classes of sidewalk work where the forms are either 4", 6", or 8" (100, 150, or 200 mm) wide, a carpenter should place and level 250 to 300 l.f. (75-90 m) per 8-hr. day, at the following labor cost per 100 l.f. (30 m):

	Hours	Rate	Total	Rate	Total
Carpenter	3.0	$....	$....	$26.44	$ 79.32
Helper	1.5	....	....	21.04	31.56
Cost per 100 l.f. (30m)			....		$ 110.88
per l.f. of form			....		1.11
per l.f. of walk			....		2.22
per m of walk			....		7.28

The above includes time removing forms.

When placing screeds for concrete floors on the ground, 2 carpenters working together should drive stakes, place and level 650 to 700 l.f. of screeds per 8-hr. day at the following labor cost per 100 l.f.:

	Hours	Rate	Total	Rate	Total
Carpenter	2.4	$....	$....	$26.44	$ 63.46
Helper	1.2	....	....	21.04	25.25
Cost per 100 l.f.			$....		$ 88.71
per l.f.				....	0.89
per m.				....	2.92

Estimating the Quantity of Concrete Required for Floors and Walks. Compute the number of s.f. (sq.m) of concrete floor or sidewalk that may be obtained from one c.y. (one cu.m) of concrete, laid out into a cube 3'-0" wide, 3'-0" thick and 3'-0" high (1 x 1 x 1 m). The surface of this cube contains 9 s.f. and it is 36" high.

If the rough concrete is 4" (0.10 m) thick, divide 36" (1 m), the height of the cube, by 4" (0.10 m), the thickness of the slab, and the result is 9 (10), showing that it is possible to obtain 9 slabs, each containing 9 s.f. or 81 s.f. of 4" floor or walk per c.y. of concrete (or 10 slabs, each 1 sq.m, 0.1 m thick).

Number of S.F. (Sq.M) of Concrete Floor of Any Thickness
Obtainable From One C.Y. (Cu.M) of Concrete

Thick-ness, in.	No. of s.f.	Thick-ness, mm	No. of sq.m	Thick-ness, in.	No. of s.f.	Thick-ness, mm	No. of sq.m
1	324	25	30.1	7	46	175	4.3
1-1/4	259	31	24.1	7-1/4	44	181	4.1
1-1/2	216	38	20.1	7-1/2	43	188	4.0
1-3/4	185	44	17.2	7-3/4	42	194	3.9
2	162	50	15.1	8	40	200	3.7
2-1/4	144	56	13.4	8-1/4	39	206	3.6
2-1/2	130	63	12.1	8-1/2	38	213	3.5
2-3/4	118	69	11.0	8-3/4	37	219	3.4
3	108	75	10.0	9	36	225	3.3
3-1/4	100	81	9.3	9-1/4	35	231	3.3
3-1/2	93	88	8.6	9-1/2	34	238	3.2
3-3/4	86	94	8.0	9-3/4	33	244	3.1
4	81	100	7.5	10	32	250	3.0
4-1/4	76	106	7.1	10-1/4	31	256	2.9
4-1/2	72	113	6.7	10-1/2	31	263	2.9
4-3/4	68	119	6.3	10-3/4	30	269	2.8
5	65	125	6.0	11	29-1/2	275	2.7
5-1/4	62	131	5.8	11-1/4	29	281	2.7
5-1/2	59	138	5.5	11-1/2	28	288	2.6
5-3/4	56	144	5.2	11-3/4	27-1/2	294	2.6
6	54	150	5.0	12	27	300	2.5
6-1/4	52	156	4.8	12-1/4	26-1/2	306	2.7
6-1/2	50	163	4.6	12-1/2	26	313	2.4
6-3/4	48	169	4.5	12-3/4	25-1/2	319	2.4

Mixing and Placing Concrete for Floors and Walks

Mixing Concrete By Hand. Concrete is seldom, if ever, mixed by hand. It is usually mixed in a small mixer, or ready-mix concrete is used, and discharged direct into the forms or onto the floor. However, for that one-in-a-million job, here is some data on hand mixing.

As with other classes of concrete work, the cost of mixing and placing the concrete depends on the number of times it is turned on the board and the method used in mixing and placing. But on the smaller jobs, it will require 3-1/2 to 4 hrs. labor time to mix and place one c.y. of concrete (4.57-5.23 hrs. per cu.m). This class of work does not proceed as fast as foundation work.

On work of this kind, the crew usually consists of 5 or 6 workers, so it is not possible for one individual to work at the same class of work all day but must alternate between wheeling sand, gravel, cement, and mixing concrete, as the job may require.

Five workers together should handle all materials, mix, place, and tamp 20 to 24 batches—1/2-c.y. (0.38 cu.m)—or 10 to 12 c.y. (7.65-9.18 cu.m) per 8-hr. day, at the following labor cost per c.y. (cu.m):

	Hours	Rate	Total	Rate	Total
Labor per c.y.	3.64	$....	$....	$21.04	$ 76.59
per cu.m			....		99.56

Labor Mixing and Placing Concrete for Floors and Walks Using a Small Mixer. A small concrete mixer of 1/2- or 1-sack (3-1/2-S or 6-S) capacity is ordinarily used for floor and sidewalk work.

Where possible to use a full crew and keep the mixer in operation steadily, a mixer should discharge a batch every 2 minutes, but as in all other kinds of construction work there are always delays, such as engine adjustments, repairing mixer, delays caused by nondelivery of sand, gravel or cement, moving and placing runways, cleaning concrete barrows, etc., so the average for a job will be about 20 to 25 batches per hr. or 160 to 200 batches per 8-hr. day.   On the above basis, either a 3-1/2-S or 6-S concrete mixer, the labor cost per c.y. (cu.m) should run as follows:

	Hours	Rate	Total	Rate	Total
Labor per c.y.	2.5	$....	$....	$21.04	$ 52.60
per cu.m			....		68.38

Using a larger mixer will increase the output but the labor cost will increase in direct proportion. Many contractors have found that a one sack mixer is the most economical unit for floor and sidewalk work.

Materials Required for 1 Cu. Yd. of Concrete

This table is based on use of average wet sand and on 3- to 4-in. slump. Quantities will vary according to grading of aggregate and workability desired. No allowance has been made for waste. 1 sack cement = 1 cu. ft. 4 sacks cement = 1 bbl.

	Mixes					Materials per Cu. Yd. of Concrete			
			Aggregates					Aggregates	
Size of Aggregate	Mixing Water* per sack of Cement U.S. Gal.	Portland Cement Sacks	Sand Cu. Ft.	Gravel or Crushed Stone Cu. Ft.	Yield of concrete per Sack of Cement Cu. Ft.	Water U.S. Gal.	Portland Cement Sacks	Sand Cu. Yd.	Gravel or Crushed Stone Cu. Yd.
Maximum size ¾ in.	5	1	2.50	2.75	4.3	31	6.25	0.58	0.64
Maximum size 1 in.	5	1	2.25	3.00	4.5	30	6.00	0.50	0.67
Maximum size 1½ in.	5	1	2.25	3.50	4.7	29	5.75	0.48	0.75

*Based on 6 gal. of water per sack of cement including water contained in average wet sand.

Materials Required for 1 Cu. Meter of Concrete

This table is based on use of average wet sand and on 75 mm to 100 mm slump. Quantities will vary according to grading of aggregate and workability desired. No allowance hase been made for waste. 1 sack cement = .028 Cu Meters 4 sacks cement = 1 bbl

1 sack cement = .028 Cu Meters 4 sacks cement = 1 bbl

Size of Aggregate	Mixes					Materials per Cu. Meter of Concrete			
	Mixing Water* per Sack of Cement Litres	Portland Cement Sacks	Aggregates		Yield of Concrete per Sack of Cement Cu.Meter	Water Liter	Portland Cement Sacks	Aggregates	
			Sand Cu.Meter	Gravel or Crushed Stone Cu.Meter				Sand Cu.Meter	Gravel or Crushed Stone Cu.Meter
Maximum size 19 mm	18.9	1	0.071	0.078	0.122	117.3	6.25	0.443	0.489
Maximum size 25 mm	18.9	1	0.064	0.085	0.127	113.6	6.00	0.443	0.512
Maximum size 38 mm	18.9	1	0.064	0.099	0.133	109.8	5.75	0.367	0.573

* based on 22.7 liters of water per sack of cement including water contained in average wet sand.

Materials Required for 100 Sq. Ft. of Surface Area for Various Thicknesses of Concrete*

Thickness	Quantity of Concrete	Mixes								
		1:2¼:2¾	Aggregate		1:2¼:3	Aggregate		1:2¼:3½	Aggregate	
In.	Cu. Yd.	Cement Sacks	Sand Cu. Yd.	Gravel or Crushed Stone Cu. Yd.	Cement Sacks	Sand Cu. Yd.	Gravel or Crushed Stone Cu. Yd.	Cement Sack	Sand Cu. Yd.	Gravel or Crushed Stone Cu. Yd.
3	0.93	5.8	0.54	0.60	5.6	0.47	0.62	5.4	0.45	0.70
4	1.23	7.7	0.71	0.79	7.4	0.61	0.82	7.1	0.59	0.92
5	1.54	9.6	0.89	0.99	9.2	0.77	1.03	8.9	0.74	1.16
6	1.85	11.6	1.07	1.18	11.1	0.93	1.24	10.6	0.89	1.39
8	2.46	15.4	1.43	1.57	14.8	1.23	1.65	14.1	1.18	1.85
10	3.08	19.3	1.79	1.97	18.5	1.54	2.06	17.7	1.48	2.31
12	3.70	23.1	2.14	2.37	22.2	1.85	2.48	21.3	1.78	2.78

No allowance has been made for waste.
*100 sq. ft. of portland cement plaster ½ in. thick require 1.6. sacks of cement and 0.15 cu. yd. of sand.

Materials Required for 10 Sq. Meters of Surface Area for Various Thickness of Concrete *

Thickness	Quantity of Concrete	1:2 1/2:2 3/4			1:2 1/4:3			1:2 1/4:3 1/2		
			Aggregate			Aggregate			Aggregate	
		Cement	Sand	Gravel or Crushed Stone	Cement	Sand	Gravel or Crushed Stone	Cement	Sand	Gravel or Crushed Stone
MM	Cu.Meter	Sacks	Cu.Meter	Cu.Meter	Sacks	Cu.Meter	Cu.Meter	Sacks	Cu.Meter	Cu.Meter
75	0.77	6.2	0.44	0.49	6.0	0.39	0.51	5.8	0.37	0.58
100	1.01	8.3	0.58	0.65	8.0	0.50	0.67	7.6	0.49	0.76
125	1.27	10.3	0.73	0.81	9.9	0.63	0.85	9.6	0.61	0.95
150	1.52	12.5	0.88	0.97	11.9	0.77	1.02	11.4	0.73	1.14
200	2.02	16.6	1.18	1.29	15.9	1.01	1.36	15.2	0.97	1.52
250	2.53	20.8	1.47	1.62	19.9	1.27	1.70	19.1	1.22	1.90
300	3.05	24.9	1.76	1.95	23.9	1.52	2.04	22.9	1.46	2.29

Mixes

No allowance has been made for waste

Labor Placing Ready-Mixed Concrete for Floors. To place ready-mixed concrete for floors with the most economy, it is necessary to have a crew large enough to handle the concrete quickly to eliminate waiting time charges by the material company, and the supply of mixed concrete must be steady and of sufficient quantity to keep the crew busy.

A crew of 14 or 15 laborers, using buggies to wheel concrete, should place 22 to 27 c.y. (16.83-20.65 cu.m) of ready-mixed concrete per hr. at the following labor cost per c.y. (cu.m):

	Hours	Rate	Total	Rate	Total
Labor per c.y.	0.6	$....	$....	$21.04	$ 12.62
per cu.m			....		16.41

If the floor is small, such as for a residence, small addition, etc., the work must be organized in a different manner. A job of this sort is too small to keep a number of ready-mix trucks busy. One or two trucks will serve the purpose.

Assuming a floor pour serviced by one 7-c.y. (5.36-cu.m) capacity ready-mix truck, delivering one load of concrete per hr., a crew of 6 or 7 laborers should be kept reasonably busy and still handle the ready-mixed load of concrete fast enough to prevent ready-mix concrete truck waiting time penalties.

Based on the above conditions the labor cost of placing one c.y. (cu.m) of floor concrete will be as follows:

	Hours	Rate	Total	Rate	Total
Labor per c.y.	1.0	$....	$....	$21.04	$21.04
per cu.m	1.3		....		27.35

Estimating the Labor Cost Per S.F. (Sq.M) of Floor or Sidewalk. Labor costs given for mixing and placing concrete are based on cubic yards (cu.m), but floors and sidewalks are usually estimated by the square foot (sq.m).

Once the cost per c.y. (cu.m) has been obtained, divide it by the number of s.f. (sq.m) of floor obtainable from a c.y. (cu.m) of concrete. The result is the labor cost per s.f. (sq.m) of floor. Example: Find the labor cost per s.f. of 4″ (sq.m of 0.1 m) floor. The table on the previous page shows that 81 s.f. of 4″ (7.29 sq.m of 0.1 m) floor can be obtained from one c.y. (cu.m) of concrete. If the labor cost for concrete placement is $21.04 per c.y. (27.35 per cu.m), $21.04 ÷ 81 = $0.26 per s.f. (27.35 ÷ 7.29 = $3.75 per sq.m).

Applying A Finish on Reinforced Concrete Floors. The cost of placing a finish on reinforced concrete floors is subject to wide fluctuations. One reason is the custom of placing the concrete over as large an area as possible and then allowing the concrete to stand until the water disappears from the surface before finishing is started. However, the time required for cement to take its initial set depends on the brand of cement used. Some cements are known for being quick setting or slow setting.

The amount of water used in the concrete mix also has considerable bearing on the finishing cost. If "sloppy" concrete is used, it requires much more time for the floor to set sufficiently to receive the finish than if concrete of a "stiff" consistency is used.

Architects and engineers specify the floor must be finished at the same time the concrete is placed, and in some instances it necessitates a large amount of overtime for finishers. When the crew places concrete right up until quitting time, it is a foregone conclusion that the finishers will have to work far into the night.

The item of overtime should be carefully considered by the estimator, because the cement mason often starts the finishing or floating work when all other trades have quit for the day. Provisions should be made and costs added, if required, for lights and for other trades necessary to support the finishers.

While under favorable conditions, some finishers may finish 175 to 200 s.f. (15.75-18 sq.m) of floor per hr. on certain classes of work, the general average for a reinforced concrete job will vary from 70 to 80 s.f. (6.3-7.2 sq.m) per hr. or 560 to 640 s.f. (50-58 sq.m) per 8-hr. day.

The labor costs given are for straight time, but the estimator should bear in mind that overtime costs time and a half or double time. The labor per 100 s.f. (9 sq.m) of floor should cost as follows:

	Hours	Rate	Total	Rate	Total
Cement mason	1.3	$....	$....	$27.03	$ 35.14
Cost per s.f.			....		0.35
per sq.m			....		3.77

Finishing Concrete Floors by Machine. Finishing machines are used for screeding, floating, and troweling. When placing concrete floors, walks, and pavements on the ground, the wood or metal floor screeds are placed in the usual manner and the concrete is struck off by using a power operated screeding machine, which makes 2-1/2" (63 mm) transverse strokes on the header boards, leveling and compacting the mix. A steady pull forward by the operator advances the machine at the rate of 10 ft. (3 m) a minute or 600 l.f. (180 m) per hour and will strike off a slab 6 to 20 ft. (1.8-6 m) in width.

A four-worker crew is usually required with the machine, two laborers spreading the concrete and two cement masons operating the screeding machine.

After the concrete slab has been screeded ready for troweling, allow sufficient time for the concrete to set hard enough to walk on before starting finishing.

For floating, the machine is equipped with 12-gauge steel trowels, which revolve on a 46" (1.15 m) diameter. It usually requires 3 cement masons—1 finisher to operate the machine and 2 finishers touching up corners, columns, edging, jointing, etc. On a well prepared slab, one finishing crew can float 4,000 s.f. (360 sq.m) of floor per hr., but this rate is not usually maintained.

A smaller finishing machine with a 34" (850 mm) diameter will float up to 3,000 s.f. (270 sq.m) of floor per hr. under the same conditions described above.

After the floor has been floated, the heavy-gauge steel floating trowels are removed and replaced with steel "finishing" trowels. The cement mason then guides the rotating, adjustable pitch trowels over the slab until a smooth, level surface is obtained.

Using the larger machines, and assuming a "four time over" was necessary to complete the job, it would require 4 hrs. of machine operation and assuming a 3/4 to 1 hour waiting period was necessary between each operation, it would require a total of 6-1/4 to 7 hrs. time to finish 4,000 s.f. (360 sq.m) of floor at the following labor cost:

	Hours	Rate	Total	Rate	Total
Cement mason	3	$....	$....	$27.03	$ 81.09
Cost per s.f.			....		0.02
per sq.m			....		0.22

Add machine rental charges.

After the floor has been screeded, the labor floating and troweling (4 times over), 4,000 s.f. (360 sq.m) of cement finish floor should cost as follows:

	Hours	Rate	Total	Rate	Total
Cement mason	32	$....	$....	$27.03	$ 864.96
Cost per s.f.			....		0.22
per sq.m			....		2.37

Add machine rental charges.

Color for Concrete Floors

Color for concrete is used for sidewalks, driveways, porches, tennis courts, floors, etc., the desired color being mixed into the cement topping which should be either 1/2", 3/4", or 1" (13, 19, or 25 mm) thick, depending on service requirements. The amount required varies with the different manufacturers but the following proportions generally will produce good results.

		Approximate quantities required—lbs. per sack of cement		Approximate quantities required—kg per sack of cement	
Color	Commercial names of colors for use with cement	Light Shade	Medium to Dark	Light Shade	Medium to Dark
Pigments for Colored Concrete Floors					
Greys and Blacks	Germantown lampblack* or carbon black* or black oxide of manganese* or mineral black	1/2 1/2 1 1	1 1 12 12	0.23 0.23 0.23 0.23	0.45 0.45 5.44 5.44
Blue Blue-black	Ultramarine blue	5	9	2.27	4.08
Brownish Red to dull brick red	Red oxide of iron	5	9	2.27	4.08
Bright red to Vermillion Red Red sandstone to Purplish red	Mineral turkey red	5	9	2.27	4.08
	Indian red	5	9	2.27	4.08
Brown to reddish-brown	Metallic brown (oxide)	5	9	2.27	4.08
Buff, colonial tint and yellow	Yellow ochre	5	9	2.27	4.08
	Yellow oxide	2	4	0.91	1.81
Green	Chromium oxide or	5	9	2.27	4.08
Green/blue	Greenish blue ultramarine	6	--	2.72	--

*Only first-quality lampblack should be used. Carbon black is light in weight and requires very thorough mixing. Black oxide or mineral black is probably most advantageous for general use. For black use 11 lbs. of oxide for each sack of cement.

Approximate Quantities Required for 100 S.F. (9 Sq.M) of Floor

	1/2" Top 12 mm	3/4" Top 18 mm	1" Top 25 mm
Gray portland cement	1.5 sacks	2.25 sacks	3.00 sacks
Clean sand	150 lbs.(68kg)	225 lbs.(102kg)	300 lbs.(136kg)
Coarse aggregate, 3/8" max. size	263 lbs.(119kg)	394 lbs.(178kg)	525 lbs.(236kg)
Conc. cement color	13.5 lbs.(6.1kg)	20.3 lbs.(9.1kg)	27 lbs.(12.2kg)

For 1" (25 mm) topping, use no more than 5 gals. (19 liters) mixing water, including water in aggregates. Smaller mixes in proportion. Sand and coarse aggregate is based on 100 lbs. per c.f. (1607 kg/cu.m). Colors are usually packed in 9-lb. (4-kg) bags, 100-lb. (45-kg) drums and barrels. When mixed in the proportions of 1:2, use 9 lbs. (4 kg) of cement color to each bag of portland cement used in the topping.

It is advisable to establish the mixture from samples made under job conditions and allowed to dry. Careful measurement of every ingredient, including water, in each batch is essential to insure uniformity of color.

Final troweling should be done after the concrete has stiffened, as this assures a smooth, polished finish. Integral waterproofing or hardener can be added to mix materials, also polishing or a surface treatment can be applied, if desired.

Colorundum Surface Colorant and Hardener for Concrete. Colorundum (from W.R. Grace & Co.) is a ready-to-use powder containing non-crushable hardeners, dispersing agents, and cementitious binders. It provides a durable, integral concrete color that does not wear off as paint and ordinary surface coatings do. When Colorundum is dusted on and trowelled into freshly poured concrete, it produces bright, decorative surfaces with dense coloration and hard, abrasion resistant surfaces with long life.

Colorundum is for interior and exterior concrete floors and sidewalks. The hard wearing, highly decorative finish of Colorundum floors makes them ideal for use in factories, schools, hospitals, office buildings, showrooms, supermarkets, stores or any public building. Colorundum may be used inside or outside on traffic areas such as sidewalks, ramps, patios, terraces, sun decks and playgrounds. For best results, the slump of concrete monolithic slabs should not exceed 2-1/2" (63 mm) and for toppings 3" (75 mm). Excess water causes laitance to rise to the surface, producing discoloration, and also causes separation, which delays the finishing operation.

Do not use air entraining agents or other admixtures in the mix. When excess water disappears and the topping will hold up knee boards, uniformly dust on 2/3 of specified amount of Colorundum per 100 s.f. (9 sq.m) of surface.

After spreading, allow the dry material to wet up. It should then be wood floated and worked into the slab. The first floating should be discontinued as soon as the surface becomes wet. Floating should be resumed when surface moisture has disappeared. Do not steel trowel. Immediately apply the remaining 1/3 of the specified amount per 100 s.f. (9 sq.m) and thoroughly wood float. (For heavy duty finish, the quantity of Colorundum may be increased for the second dust coat.) Apply Colorundum exactly as it comes from the container; do not mix with cement or concrete.

Steel trowel to an even plane free from surface marks and voids. When the floor has obtained its initial set, give it a final steel trowel burnishing. Leave exterior areas, such as sidewalks, under wood float finish after the trowelling.

To cure, use an approved curing and sealing compound according to manufacturer's instructions or cover the surface after finish trowelling with polyethylene. This seals against evaporation of the water so vitally necessary to hydrate the cement, thereby developing optimum strength and density throughout the mass. It also protects the colored surface against spillages.

Use approximately 40 lbs. per 100 s.f. (18 kg per 9 sq.m) for normal traffic areas. For areas subject to extra heavy traffic, the quantity of Colorundum may be increased proportionately.

Gilco Non-Shrink Grouting Compound. Gilco Grout (from Gifford-Hill & Co. Inc.) is a premixed compound containing iron particles, hardening and dispersing agents, binders, and oxidizing agents. When mixed with water, it produces a mass that has controlled expansion to overcome the inherent shrinkage in cementitious mixtures. The metallic aggregate also provides a ductile surface with the "give and take" properties concrete must have to absorb the vibration of machinery, equipment, and columns.

Gilco grout can be used wherever non-shrink concrete is mandatory: setting machinery on concrete foundations; full bearing surface grouting for building columns; and grouting anchor bolts, floor grids, steel sash, and jambs.

Grouting should be continuous once it has started. Do not mix more grout at one time than can be rodded and placed in a period of 20 minutes. Add only enough water to make mix placeable. Avoid an excess of water.

Anchor forms securely to prevent movement during placing and curing. Intersection of form and base slab can be filled with a mix of equal parts of cement and sand before placing grout. Be sure to allow adequate clearance between forms and base plate.

Remove waste material and water from anchor bolt holes. All oil, grease, and paint in contact with grout should be thoroughly removed from base plates before grouting.

It is important to place grout quickly and continuously to avoid all effects of overworking, resulting in segregation, bleeding, and breakdown of initial set. Under no circumstances should the grout be retempered by introducing additional water after it has taken its initial set.

Lightweight Concrete Floor Fill

Concrete made of lightweight aggregate, such as cinders, slag, Haydite, pumice, and vermiculite, is often used for floor fill under marble and tile floors and between wood floor sleepers, screeds, etc.

The labor cost of placing concrete floor fill runs considerably higher than for ordinary concrete, because the fill is usually only 2" or 3" (50 or 75 mm) thick, which necessitates covering a large floor area to place a comparatively small amount of concrete. In many cases it is necessary to pump or hoist the concrete to the upper floors in wheelbarrows or concrete buggies and then wheel it to the different parts of the building where it is to be used.

When placing concrete fill between wood floor sleepers or screeds, it is important that every precaution be taken to prevent the sleepers from being knocked out of level when wheeling over them.

When placing concrete floor fill, it is not advisable to use a large mixer because of the necessity of covering a large floor area to place a small amount of concrete. It usually requires considerable leveling and tamping, making it impractical to handle the concrete in large quantities. A concrete mixer of 1 or 2 sacks capacity is best adapted for this class of work or partial loads of ready-mix can be ordered.

Mixing and Placing Concrete Floor Fill. Because of the additional labor required for spreading, grading, and tamping concrete floor fill, the c.y. labor costs run higher than for foundation or sidewalk work.

There is also more labor required wheeling concrete from the mixer to the hoist and from the hoist to place of deposit, as floor fill is usually placed after the structural portions of the building have been completed, and it is not possible to carry on the work with much speed.

A crew of 13 workers with a 6-S mixer should handle materials, mix, and wheel into place 15 to 20 batches—containing 6 to 7 c.f. (0.17-0.20 cu.m) each or 3-1/2 to 5 c.y. (2.65-3.8 cu.m) of concrete per hour with the crew distributed as follows: 2 workers wheeling coarse aggregate from stock pile to mixer; 2 workers wheeling sand and cement to mixer; 1 worker attending mixer; 3 workers wheeling concrete from mixer to hoist; 3 workers wheeling concrete from hoist to place of deposit; 2 workers dumping barrows and spreading concrete.

The labor per c.y. (cu.m) should cost as follows:

Cost per c.y.	Hours	Rate	Total	Rate	Total
Labor	3.00	$....	$....	$21.04	$ 63.12
Hoisting engineer	0.35	....	....	28.55	9.99
Cost per c.y.			$....		$ 73.11

Cost per cu.m	Hours	Rate	Total	Rate	Total
Labor	3.92	$....	$....	$21.04	$ 82.48
Hoisting engineer	0.46	....	....	28.55	13.13
Cost per cu.m.			$....		$ 95.61

If union rules require a mixer engineer, deduct 0.35 (0.46) hrs. labor time and add 0.35 (0.46) hrs. engineer time per c.y. (cu.m).

If concrete is discharged from mixer directly into a hoist concrete bucket and wheeled from a floor hopper to place of deposit, deduct 0.50-hr. labor time.

If no hoisting is required, deduct 0.70-hr. labor time and omit hoisting engineer.

For concrete roof fill on flat roofs, deduct about 10% from the above costs. On pitch or gable roofs, add 25% to 50%, depending on steepness of slope, accessibility, etc.

After the cubic yard (cubic meter) cost has been computed, to obtain the s.f. (sq.m) cost, divide the c.y. (cu.m) cost by the number of s.f. (sq.m) of fill obtainable from one c.y. (cu.m) and the result will be the labor cost per s.f. (sq.m) of fill. Example: Find the s.f. (sq.m) labor cost of 3" (0.075 m) floor fill with a cost of $43.90 per c.y. (57.39 per cu.m). The tables on previous pages give 108 s.f. of 3" (10 sq.m of 0.075 m) fill obtainable from one c.y. $43.90 ÷ 108 = $0.406 per s.f. (57.39 ÷ 10 = $5.74 per sq.m).

To the above costs must be added the cost of setting screeds and all finishing operations.

Finishing Lightweight Concrete Floor and Roof Fill

All concrete floor and roof fills require some finishing labor, if only setting of screeds and striking off fill at screed height.

Where fill is placed between wood sleepers, screeds are not required, but fill must be struck off and darbied so that no part of it is higher than the top of the sleepers. In addition, the fill should be spaded along both sides of each sleeper so that good anchorage is obtained.

For 2" (50 mm) thick concrete fill between wood floor sleepers, 2 cement masons working with a helper should strike off and darby 5,000 to 6,000 s.f. (450-540 sq.m) per 8-hr. day or about all the fill a small-mixer crew can mix and place in an 8-hr. day. The labor per 100 s.f. (9 sq.m) should cost as follows:

	Hours	Rate	Total	Rate	Total
Cement mason	0.30	$....	$....	$27.03	$ 8.11
Helper	0.15	....	....	21.04	3.16
Cost per 100 s.f. (9 sq.m)			$....		$ 11.27
per s.f.				....	0.11
per sq.m				....	1.18

Finishing Concrete Floor Fill For Resilient Flooring. Where concrete fill is used as a base for cork, asphalt, vinyl, rubber, or vinyl asbestos floor covering, it is necessary to obtain a smooth level finish on the concrete to receive the finish floor material, or all irregularities will show through on the finished surface.

To obtain the required degree of smoothness, screeds must be placed 4'- 0" to 6'-0" (1.2-1.8 m) o.c. and concrete fill must be struck off, darbied, floated, and troweled.

If the fill is not over 2" (50 mm) thick, 2 cement masons working together should strike off, darby, float, and trowel about 1,600 s.f. (144 sq.m) of floor per 8-hr. day at the following labor cost per 100 s.f. (9 sq.m):

	Hours	Rate	Total	Rate	Total
Cement mason	1.0	$....	$....	$27.03	$27.03
Cost per s.f. (9 sq.m)			....		0.27
per sq.m			....		2.91

Finishing Concrete Roof Fill. Concrete fill for flat roofs are usually struck off, darbied, and floated. If fill is not over 2" to 3" (50 to 75 mm) thick, 2 cement masons with a helper should place screeds, strike off, darby, and float 1,500 to 1,700 s.f. (135-150 sq.m) of roof fill per 8-hr. day at the following labor cost per 100 s.f. (9 sq.m):

	Hours	Rate	Total	Rate	Total
Cement mason	1.0	$....	$....	$27.03	$27.03
Helper	0.5	....	....	21.04	10.52
Cost per 100 s.f. (9 sq.m)			$....		$ 37.55
per s.f.			....		0.38
per sq.m			....		4.09

Where concrete fill is placed on pitch or gable roofs, finishing is more difficult. Screeds must be set closer together to help keep the fill from sliding, and an additional helper is usually required. On work of this type, 2 cement masons and 2 helpers should place screeds, strike off, darby, and float 900 to 1,000 s.f. (81-90 sq.m) of fill per 8-hr. day at the following labor cost per 100 s.f. (9 sq.m):

	Hours	Rate	Total	Rate	Total
Cement mason	1.60	$....	$....	$27.03	$ 43.25
Helper	1.60	....	....	21.04	33.66
Cost per 100 s.f. (sq.m)			$....		$ 76.91
per s.f.			....		0.77
per sq.m			....		8.29

CONCRETE SLAB ON GRADE CONSTRUCTION

In the past, this type of floor was built without adequate test data or established construction standards with respect to dampness and floor temperature. Many such floors were cold, especially at the outer edge, and some were damp enough to damage floor coverings and wall construction.

To determine the most satisfactory floor from the standpoint of comfort, the University of Illinois studied the construction of concrete floor slabs that are laid on the ground and designed for use in climates where central heating is necessary. (Small Homes Council, Circular F4.3, "Concrete Floors for Basementless Houses" Printed with permission.)

Nine different types of concrete floors were tested in an effort to determine the proper design for slab on grade houses with respect to heat loss of different types of floors; temperatures at various points throughout the floors for proper placement of floor insulation; and amount of moisture passing from the ground to the top of the concrete slab.

Insulation Requirements for Different Parts of the Country

The floors selected for tests represented standard construction practices and at the same time permitted the use of varying amounts and types of insulation along the edges. The floor slabs were tested simultaneously under similar conditions in a specially constructed laboratory.

When panel heating systems are used with concrete floors, the heat loss to the ground and at the edge is increased due to the higher temperatures maintained in the floors. To prevent excess heat loss in such floors, a minimum of 2" (50 mm) of rigid waterproof insulation should be provided at the edge. In addition, the use of insulation under the entire heated floor area is recommended. Gravel and rock fills are desirable for drainage controls, but they have no insulating value.

The use of concrete floor slabs for basementless homes requires an insulating material around the perimeter of the slab (at the exterior walls) that will not be compressed by imposed loads or soil backfill and that will not be affected by soil acids, moisture, insects, or vermin. Fiberglass perimeter insulation and styrofoam are used successfully for this purpose and come in panels of various widths and thicknesses.

Cleaning Concrete Floors Using a Sanding Machine. To clean droppings of paint, mortar, joint finishing cement, etc., from concrete floors before laying the finish floor surface, a power sanding machine does a satisfactory job at a considerable saving in labor.

This is accomplished by using a floor sander equipped with a sanding disk and open grit silicon carbon grain paper, which cuts through and removes the droppings.

Methods of Constructing Concrete Floors for Basementless Houses

A worker operating a sanding machine will clean about 500 s.f. (45 sq.m) of floor per hour at the following cost per 100 s.f. (9 sq.m):

	Hours	Rate	Total	Rate	Total
Machinery cost	0.2	$....	$....	$5.00	$1.00
Sanding disk		....	....	0.75	0.75
Labor	0.2	....	....	21.04	4.21
Cost per 100 s.f. (9 sq.m)		$....			$ 5.96
per s.f.			....		0.06
per sq.m			....		0.65

When this work is performed by hand using a putty knife or steel scraper, a worker will clean and remove droppings from 150 to 175 s.f. (14-16 sq.m) per hour, at the following cost per 100 s.f. (sq.m):

	Hours	Rate	Total	Rate	Total
Labor	0.6	$....	$....	$21.04	$ 12.62
Cost per s.f.			....		0.13
per sq.m			....		1.40

Lightweight Concrete

There are two types of lightweight concrete. One long-used type is produced by simply combining portland cement and a lightweight aggregate.

The second is known as aerated concrete, which is produced by a process of introducing certain chemicals to generate gases that cause the mass to expand. The weight of aerated concrete can be controlled with a variation from 20 lbs. per c.f. (320 kg/m³) to approximately the weight of ordinary lightweight concrete. For average conditions a weight of 40 to 50 lbs. per c.f. (648-810 kg/m³) with a compressive strength of about 500 lbs. per sq. in. (0.35 kg/mm²) is usually selected.

Aerated concrete is fireproof, has a low moisture absorption rate, and provides good insulation against the transmission of heat and sound. It may be sawed or nailed as readily as other lightweight concrete.

All lightweight concretes are of decided advantage for partition walls and fireproofing, whether in monolithic or precast unit form and for floor fill and roof slabs. The estimator should consider the combined advantages of each type of lightweight concrete before selection.

Haydite Concrete. Haydite concrete uses burnt clay or shale for coarse aggregate instead of gravel or crushed stone. This inert, cellular material weighs less than 50 lbs. per c.f. (810 kg/m³) and when mixed with ordinary torpedo sand and portland cement, produces concrete weighing about 98 lbs. per c.f. (1587 kg/m³) instead of 150 lbs. (2430 kg) for ordinary concrete.

Full structural strengths can be obtained with Haydite concrete and in many instances a redesign based on lighter weight concrete and reduced dead load more than pays for the slight additional cost of the aggregates.

Perlite Lightweight Concrete. For lightweight concrete using Perlite as the aggregate, refer to "Concrete Floors and Walks".

Pottsco Lightweight Aggregate. Pottsco lightweight aggregate is an expanded blast furnace slag. Molten slag is converted into hard, cellular clinker, which is crushed and screened to commercial sizes, ready for use with portland cement and water, without the addition of sand or other materials. It is supposed to be chemically inert and has no corrosive effect on structural or reinforcing steel, plaster, or paint.

Pumice Lightweight Concrete. For lightweight concrete using pumice as the aggregate, refer to "Concrete Floors and Walks".

Vermiculite Lightweight Concrete. Refer to "Vermiculite Concrete" section.

Waylite Lightweight Concrete. For lightweight concrete using Waylite as the aggregate, refer to "Concrete Floors and Walks".

Reinforced Concrete

After the forms and reinforcing steel are in place, the job is ready for concrete. This should include the cost of gravel or crushed stone, sand, and cement, as well as the labor cost of handling materials, mixing, hoisting, placing, and curing the concrete.

For many years concrete used for reinforced concrete was specified to be mixed in the proportions of 1:1:2, 1:1-1/2:3, 1:2:4, etc., depending on the class of work. Nowadays, it is specified that concrete used for specific purposes shall develop a compressive strength of 2,500, 3,000, 3,750 or 4,000 lbs. or more per sq. in. (17.23, 20.69, 25.85, 27.58, or more MPa) at 28 days.

On nearly all types of reinforced concrete building work, such as columns, beams, girders, floor slabs, etc., the coarse aggregate is graded from 1/4" to 1" (6-25 mm) in size. Where conditions permit, however, coarse aggregate graded from 1/4" to 1-1/2" (6-38 mm) is used.

Estimating the Quantity of Concrete Required for Various Types of Concrete Floors

If all concrete floors were merely solid slabs of concrete, it would be an easy matter to compute the quantity of concrete required per s.f. (sq.m) of floor but with so many combination floors consisting of metal pans, etc., having joists 4" to 7" (100-175 mm) wide, 4" to 15" (100-375 mm) deep and from 16" to 35" (400-875 mm) o.c., it requires considerable figuring to obtain accurate concrete quantities.

It is practically impossible to estimate accurately the amount of concrete required per s.f. (sq.m) of floor for the various combination joist floors, which invariably have T beam construction at all beams and girders, solid concrete slabs usually extend in 6" to 1'-6" (150-450 mm) from the outside walls, the joists vary in width and are often doubled under partitions, around stair wells, elevator shafts, etc. The only accurate method is to figure the floors as though they were solid concrete and then deduct for the displacement of the tile or pans.

Quantity of Concrete Required for Steelform and Concrete Joist Floor Construction. Obtain the entire slab volume as though it were of solid concrete, and deduct for the displacement of the steelforms.

The deductions for steelform area is not simple. Steelforms are furnished in a number of different sizes, with tapering sides and ends.

Steelforms furnished by different manufacturers vary slightly in shape but not sufficient to materially affect the concrete quantities. Ceco steelforms (The Ceco Corporation, Oak Brook, IL) are used in the tables, because they are typical.

Sizes and Cu. Ft. Displacement of Flange Type Steelforms

Ceco flange type steelforms slope 1" in each 12" (25 in 300 mm) in height. Example: a 12" (300 mm) pan is 20" (500 mm) wide at the bottom and 18" (450 mm) wide at the top, because of the 1" (25 mm) slope on each side. See illustrations on following pages and in the *Concrete Formwork* section.

The sides of single tapered endforms slope 1" (25 mm) for each 12" (300 mm) in height and one end slopes 1/4" (6 mm) for each 1" (25 mm) in height. For estimating purposes their displacement is the same as for standard steelforms in the following tables.

Sizes and Displacement of Tapered End Forms

Depth of Steelform Inches	Width at Wide End	Width at Narrow End	Average Width	Length in Feet	C.F. Displacement per 3-Ft. Section	Difference Between Tapered end forms and 3'-0" length of straight forms in C.F.
8"	20"	16"	18"	3'-0"	2.882	0.322
10"	20"	16"	18"	3'-0"	3.569	0.399
12"	20"	16"	18"	3'-0"	4.243	0.475
8"	30"	25"	27-1/2"	3'-0"	4.465	0.406
10"	30"	25"	27-1/2"	3'-0"	5.548	0.504
12"	30"	25"	27-1/2"	3'-0"	6.617	0.602
14"	30"	25"	27-1/2"	3'-0"	7.763	0.698
16"	30"	25"	27-1/2"	3'-0"	8.715	1.188
20"	30"	25"	27-1/2"	3'-0"	10.756	

(metric)

Depth of Steelform	Width at Wide End	Width at Narrow End	Average Width	Length in mm	Cu.m Displacement per 1 m Section	Difference Between Tapered end forms and 1 m length of straight forms in cu.m
200mm	500mm	400mm	450mm	900mm	0.089	0.010
250mm	500mm	400mm	450mm	900mm	0.111	0.012
300mm	500mm	400mm	450mm	900mm	0.131	0.015
200mm	750mm	625mm	687.5mm	900mm	0.138	0.013
250mm	750mm	625mm	687.5mm	900mm	0.172	0.016
300mm	750mm	625mm	687.5mm	900mm	0.205	0.019
350mm	750mm	625mm	687.5mm	900mm	0.240	0.022
400mm	750mm	625mm	687.5mm	900mm	0.270	0.037
500mm	750mm	625mm	687.5mm	900mm	0.333	

Sizes and Displacement of Ceco Flange Type Steelforms

Depth of Steelform Inches	Width at Bottom	Width at Top	Average Width	C.F. Displacement per L.F.
8"	10"	8-2/3"	9-1/3"	0.516
10"	10"	8-1/3"	9-1/6"	0.634
12"	10"	8"	9"	0.748
14"	10"	7-2/3"	8-5/6"	0.857
16"	10"	7-1/3"	8-2/3"	0.961
20"	10"			1.155

Depth of Steelform Inches	Width at Bottom	Width at Top	Average Width	C.F. Displacement per L.F.
8"	15"	13-2/3"	14-1/3"	0.794
10"	15"	13-1/3"	14-1/6"	0.982
12"	15"	13"	14"	1.165
14"	15"	12-2/3"	13-5/6"	1.343
16"	15"	12-1/3"	13-2/3"	1.516
20"	15"			1.850
8"	20"	18-2/3"	19-1/3"	1.072
10"	20"	18-1/3"	19-1/6"	1.329
12"	20"	18"	19"	1.581
14"	20"	17-2/3"	18-5/6"	1.829
16"	20"	17-1/3"	18-2/3"	2.072
20"	20"			2.544
8"	30"	28-2/3"	29-1/3"	1.628
10"	30"	28-1/3"	29-1/6"	2.023
12"	30"	28"	29"	2.414
14"	30"	27-2/3"	28-5/6"	2.801
16"	30"	27-1/3"	28-2/3"	3.183

(metric)

Depth of Steelform mm	Width at Bottom, mm	Width at Top, mm	Average Width, mm	Cu.m Displacement per meter
200	250	216.5	233.3	0.048
250	250	212.5	229.2	0.059
300	250	200.0	225.0	0.069
350	250	191.5	220.8	0.080
400	250	183.3	216.5	0.089
500	250			0.107
200	375	341.5	358.3	0.074
250	375	333.3	354.2	0.091
300	375	325.0	350.0	0.108
350	375	316.5	345.8	0.125
400	375	308.3	341.5	0.141
500	375			0.172
200	500	466.5	483.3	0.100
250	500	458.3	479.2	0.123
300	500	450.0	475.0	0.147
350	500	441.5	395.8	0.170
400	500	433.3	466.5	0.192
500	500			0.236
200	750	716.5	733.3	0.151
250	750	708.3	729.2	0.188
300	750	700	725.0	0.224
350	750	691.5	721.7	0.260
400	750	683.3	716.5	0.296

Sizes and Displacement of Adjustable Steelforms

Depth of Steelform Inches	Width at Bottom	Width at Top	C.F. Displacement per L.F.
8"	10"	8"	0.516
10"	10"	8"	0.634
12"	10"	8"	0.748
14"	10"	8"	0.857
15"	10"	8"	0.961
8"	15"	13"	0.794
10"	15"	13"	0.982
12"	15"	13"	1.165
14"	15"	13"	1.343
15"	15"	13"	1.516
8"	20"	18"	1.072
10"	20"	18"	1.329
12"	20"	18"	1.581
14"	20"	18"	1.829
15"	20"	18"	2.072
8"	30"	28"	1.628
10"	30"	28"	2.023
12"	20"	28"	2.414
14"	30"	28"	2.801
15"	30"	28"	3.183

Depth of Steelform mm	Width at Bottom, mm	Width at Top, mm	Cu.m Displacement per meter
200	250	200	0.048
250	250	200	0.059
300	250	200	0.069
350	250	200	0.080
375	250	200	0.089
200	375	325	0.074
250	375	325	0.091
300	375	325	0.108
350	375	325	0.125
375	375	325	0.141
200	500	450	0.100
250	500	450	0.123
300	500	450	0.147
350	500	450	0.170
375	500	450	0.193
200	750	700	0.151
250	750	700	0.188
300	750	700	0.224
350	750	700	0.260
375	750	700	0.296

Cu. Ft. of Concrete Displaced by Adjustable Steelforms and Tapered Endforms

Steelform		Length of Each Row of Steelforms in l.f.									
Depth	Width	10'	11'	12'	13'	14'	15'	16'	17'	18'	19'
8"	30"	15.33	16.96	18.58	20.21	21.83	23.46	25.08	26.71	28.33	29.96
10"	30"	19.33	21.38	23.42	25.46	27.50	29.54	31.59	33.63	35.67	37.71
12"	30"	23.33	25.79	28.25	30.71	33.16	35.62	38.08	40.54	43.00	45.45
14"	30"	27.33	30.21	33.08	35.96	38.83	41.71	44.58	47.46	50.33	53.21
15"	30"	29.33	32.42	35.50	38.58	41.66	44.75	47.83	50.91	54.00	57.08
		20'	21'	22'	23'	24'	25'	26'	27'	28'	29'
8"	30"	31.58	33.21	34.83	36.46	38.08	39.71	41.33	42.96	44.58	46.21
10"	30"	39.75	41.80	43.84	45.88	47.92	49.96	52.01	54.05	56.09	58.13
12"	30"	47.91	50.37	52.83	55.29	57.74	60.20	62.66	65.12	67.58	70.03
14"	30"	56.08	58.96	61.83	64.71	67.58	70.46	73.33	76.21	79.08	81.96
15"	30"	60.16	63.25	66.33	69.41	72.49	75.58	78.66	81.74	84.83	87.91

Cu.M of Concrete Displaced by Adjustable Steelforms and Tapered Endforms

Steelform		Length of Each Row of Steelforms, meters									
Depth	Width										
mm	mm	3 m	3.5 m	4 m	4.5 m	5 m	5.5 m	6 m	6.5 m	7 m	7.5 m
200	750	0.427	0.501	0.575	0.578	0.724	0.799	0.874	0.949	1.024	1.099
250	750	0.539	0.632	0.725	0.728	0.912	1.006	1.101	1.195	1.289	1.383
300	750	0.650	0.762	0.875	0.878	1.100	1.213	1.327	1.440	1.554	1.667
350	750	0.762	0.893	1.024	1.028	1.288	1.421	1.553	1.686	1.818	1.951
375	750	0.817	0.958	1.099	1.103	1.382	1.524	1.666	1.808	1.951	2.093
		8 m	8.5 m	9 m	9.5 m	10 m	10.5 m	11 m	11.5 m	12 m	12.5 m
200	750	1.174	1.249	1.324	1.399	1.474	1.549	1.624	1.700	1.775	1.850
250	750	1.477	1.572	1.666	1.761	1.855	1.949	2.044	2.139	2.233	2.328
300	750	1.780	1.892	2.008	2.122	2.235	2.349	2.463	2.577	2.691	2.804
350	750	2.084	2.217	2.350	2.483	2.616	2.749	2.882	3.016	3.149	3.282
375	750	2.236	2.378	2.521	2.663	2.806	2.949	3.092	3.234	3.378	3.520

Cu. Ft. of Concrete Displaced by Flange Type Steelforms and Straight Endforms

Steelform		Length of Each Row of Steelforms in l.f.									
Depth	Width	1'	2'	3'	4'	5'	6'	7'	8'	9'	10'
8"	10"	0.52	1.04	1.55	2.07	2.59	3.11	3.63	4.14	4.66	5.18
10"	10"	0.64	1.27	1.91	2.54	3.18	3.82	4.45	5.09	5.72	6.36
12"	10"	0.75	1.50	2.25	3.00	3.75	4.50	5.25	6.00	6.75	7.50
14"	10"	0.86	1.72	2.58	3.44	4.30	5.15	6.01	6.87	7.73	8.59
16"	10"	0.96	1.92	2.88	3.84	4.80	5.75	6.71	7.67	8.63	9.59
		11'	12'	13'	14'	15'	16'	17'	18'	19'	20'
8"	10"	5.70	6.22	6.73	7.25	7.77	8.29	8.81	9.32	9.84	10.36
10"	10"	7.00	7.62	8.27	8.90	9.54	10.18	10.81	11.45	12.08	12.72
12"	10"	8.25	9.00	9.75	10.50	11.25	12.00	12.75	13.50	14.25	15.00
14"	10"	9.45	10.31	11.17	12.03	12.89	13.74	14.60	15.46	16.32	17.18
16"	10"	10.55	11.51	12.47	13.43	14.39	15.34	16.30	17.26	18.22	19.18

Steelform		Length of Each Row of Steelforms in Feet									
Depth	Width	1'	2'	3'	4'	5'	6'	7'	8'	9'	10'
8"	15"	0.80	1.59	2.39	3.18	3.98	4.78	5.57	6.37	7.16	7.96
10"	15"	0.98	1.97	2.95	3.94	4.92	5.90	6.89	7.87	8.86	9.84
12"	15"	1.17	2.33	3.50	4.67	5.84	7.00	8.17	9.34	10.50	11.67
14"	15"	1.35	2.69	4.04	5.38	6.73	8.07	9.42	10.76	12.11	13.45
16"	15"	1.48	2.96	4.44	5.92	7.40	8.88	10.36	11.84	13.32	14.80
		11'	12'	13'	14'	15'	16'	17'	18'	19'	20'
8"	15"	8.76	9.55	10.35	11.14	11.94	12.74	13.53	14.33	15.12	15.92
10"	15"	10.82	11.81	12.79	13.78	14.76	15.74	16.73	17.71	18.70	19.68
12"	15"	12.84	14.00	15.17	16.34	17.51	18.67	19.84	21.01	22.17	23.34
14"	15"	14.80	16.14	17.49	18.83	20.18	21.52	22.87	24.21	25.56	26.90
16"	15"	16.28	17.76	19.24	20.72	22.20	23.68	25.16	26.64	28.12	29.60

Steelform		Length of Each Row of Steelforms in Feet									
Depth	Width	1'	2'	3'	4'	5'	6'	7'	8'	9'	10'
8"	20"	1.07	2.15	3.22	4.30	5.37	6.44	7.52	8.59	9.67	10.74
10"	20"	1.33	2.66	3.99	5.32	6.66	7.99	9.32	10.65	11.98	13.31
12"	20"	1.58	3.17	4.75	6.33	7.92	9.50	11.08	12.66	14.25	15.83
14"	20"	1.83	3.66	5.49	7.32	9.16	10.99	12.82	14.65	16.48	18.31
16"	20"	2.07	4.14	6.21	8.28	10.35	12.42	14.49	16.56	18.64	20.70
		11'	12'	13'	14'	15'	16'	17'	18'	19'	20'
8"	20"	11.81	12.89	13.96	15.04	16.11	17.18	18.26	19.33	20.41	21.48
10"	20"	14.64	15.97	17.30	18.63	19.97	21.30	22.63	23.96	25.29	26.62
12"	20"	17.41	19.00	20.58	22.16	23.75	25.33	26.91	28.49	30.08	31.66
14"	20"	20.14	21.97	23.80	25.63	27.47	29.30	31.13	32.96	34.79	36.62
16"	20"	22.77	24.84	26.91	28.98	31.05	33.12	35.19	37.26	39.33	41.40

Steelform		Length of Each Row of Steelforms in Feet									
Depth	Width	1'	2'	3'	4'	5'	6'	7'	8'	9'	10'
8"	30"	1.63	3.26	4.89	6.52	8.15	9.77	11.40	13.03	14.66	16.29
10"	30"	2.03	4.05	6.08	8.10	10.13	12.15	14.18	16.20	18.23	20.25
12"	30"	2.42	4.83	7.25	9.67	12.09	14.50	16.92	19.34	21.75	24.17
14"	30"	2.80	5.61	8.41	11.21	14.02	16.82	19.62	22.42	25.23	28.03
16"	30"	3.30	6.60	9.90	13.20	16.50	19.80	23.10	26.40	29.70	33.00
		11'	12'	13'	14'	15'	16'	17'	18'	19'	20'
8"	30"	17.92	19.55	21.18	22.81	24.44	26.06	27.69	29.32	30.95	32.58
10"	30"	22.28	24.30	26.33	28.35	30.38	32.40	34.43	36.45	38.48	40.50
12"	30"	26.59	29.00	31.42	33.84	36.26	38.67	41.09	43.51	45.92	48.34
14"	30"	30.83	33.64	36.44	39.24	42.05	44.85	47.65	50.45	53.26	56.06
16"	30"	36.30	39.60	42.90	46.20	49.50	52.80	56.10	59.40	62.70	66.00

Cu.M of Concrete Displaced by Flange Type Steelforms and Straight Endforms

Steelform		Length of Each Row of Steelforms in meters									
Depth mm	Width mm	0.5 m	1.0 m	1.5 m	2.0 m	2.5 m	3.0	3.5 m	4.0 m	4.5 m	5.0 m
200	250	0.024	0.048	0.072	0.096	0.120	0.144	0.169	0.192	0.216	0.241
250	250	0.030	0.059	0.089	0.118	0.148	0.177	0.207	0.236	0.266	0.295
300	250	0.035	0.070	0.105	0.139	0.174	0.209	0.244	0.279	0.314	0.348

Steelform		Length of Each Row of Steelforms in meters									
Depth	Width	0.5 m	1.0 m	1.5 m	2.0 m	2.5 m	3.0	3.5 m	4.0 m	4.5 m	5.0 m
350	250	0.040	0.080	0.120	0.160	0.200	0.239	0.279	0.319	0.359	0.399
400	250	0.045	0.089	0.134	0.178	0.223	0.267	0.312	0.356	0.401	0.445

		5.5 m	6.0 m	6.5 m	7.0 m	7.5 m	8.0 m	8.5 m	9.0 m	9.5 m	10.0 m
200	250	0.265	0.289	0.313	0.337	0.361	0.385	0.409	0.433	0.457	0.481
250	250	0.325	0.354	0.384	0.413	0.443	0.473	0.502	0.532	0.561	0.591
300	250	0.383	0.418	0.453	0.488	0.523	0.557	0.592	0.627	0.662	0.697
350	250	0.439	0.479	0.519	0.559	0.599	0.637	0.678	0.718	0.758	0.798
400	250	0.490	0.535	0.579	0.624	0.668	0.713	0.757	0.802	0.846	0.891

Steelform		Length of Each Row of Steelforms in meters									
Depth mm	Width mm	0.5 m	1.0 m	1.5 m	2.0 m	2.5 m	3.0 m	3.5 m	4.0 m	4.5 m	5.0 m
200	375	0.037	0.074	0.111	0.148	0.185	0.222	0.259	0.296	0.333	0.370
250	375	0.046	0.092	0.137	0.183	0.229	0.274	0.320	0.366	0.412	0.457
300	375	0.054	0.108	0.163	0.217	0.271	0.325	0.380	0.434	0.488	0.542
350	375	0.063	0.125	0.188	0.250	0.313	0.375	0.438	0.500	0.563	0.625
400	375	0.069	0.137	0.206	0.275	0.344	0.412	0.481	0.550	0.619	0.687

		5.5 m	6.0 m	6.5 m	7.0 m	7.5 m	8.0 m	8.5 m	9.0 m	9.5 m	10.0 m
200	375	0.407	0.444	0.481	0.517	0.555	0.592	0.628	0.666	0.702	0.740
250	375	0.503	0.549	0.594	0.640	0.686	0.731	0.777	0.823	0.869	0.914
300	375	0.596	0.650	0.705	0.759	0.813	0.867	0.922	0.976	1.030	1.084
350	375	0.687	0.750	0.812	0.875	0.937	1.000	1.062	1.125	1.187	1.250
400	375	0.756	0.825	0.894	0.962	1.031	1.100	1.169	1.237	1.306	1.375

Steelform		Length of Each Row of Steelforms in meters									
Depth	Width	0.5 m	1.0 m	1.5 m	2.0 m	2.5 m	3.0 m	3.5 m	4.0 m	4.5 m	5.0 m
200	500	0.050	0.100	0.150	0.200	0.249	0.299	0.349	0.399	0.449	0.499
250	500	0.062	0.124	0.185	0.247	0.309	0.371	0.433	0.495	0.556	0.618
300	500	0.073	0.147	0.221	0.294	0.368	0.441	0.515	0.588	0.662	0.735
350	500	0.085	0.170	0.255	0.340	0.425	0.511	0.596	0.681	0.766	0.851
400	500	0.096	0.192	0.288	0.385	0.481	0.577	0.673	0.723	0.866	0.962

		5.5 m	6.0 m	6.5 m	7.0 m	7.5 m	8.0 m	8.5 m	9.0 m	9.5 m	10.0 m
200	500	0.549	0.599	0.648	0.699	0.748	0.798	0.848	0.898	0.948	0.998
250	500	0.680	0.742	0.804	0.865	0.928	0.989	1.051	1.113	1.175	1.237
300	500	0.809	0.883	0.956	1.029	1.103	1.177	1.250	1.323	1.397	1.471
350	500	0.936	1.021	1.106	1.191	1.276	1.361	1.446	1.531	1.616	1.701
400	500	1.058	1.154	1.250	1.346	1.442	1.538	1.635	1.731	1.827	1.923

Steelform		Length of Each Row of Steelforms in meters									
Depth mm	Width mm	0.5 m	1.0 m	1.5 m	2.0 m	2.5 m	3.0 m	3.5 m	4.0 m	4.5 m	5.0 m
200	750	0.076	0.151	0.227	0.303	0.379	0.454	0.530	0.605	0.681	0.757
250	750	0.094	0.188	0.282	0.376	0.471	0.564	0.659	0.753	0.847	0.941
300	750	0.112	0.224	0.337	0.449	0.562	0.674	0.786	0.898	1.010	1.123
350	750	0.130	0.261	0.391	0.521	0.651	0.781	0.911	1.041	1.172	1.302
400	750	0.153	0.307	0.460	0.613	0.766	0.920	1.073	1.226	1.380	1.533

		5.5 m	6.0 m	6.5 m	7.0 m	7.5 m	8.0 m	8.5 m	9.0 m	9.5 m	10.0 m
200	750	0.832	0.908	0.984	1.060	1.135	1.211	1.286	1.362	1.438	1.513
250	750	1.035	1.129	1.223	1.317	1.411	0.271	1.599	1.693	1.787	1.881
300	750	1.235	1.347	1.460	1.572	1.684	1.796	1.909	2.021	2.133	2.245
350	750	1.432	1.563	1.693	1.823	1.953	2.083	2.213	2.343	2.474	2.604
400	750	1.686	1.714	1.993	2.146	2.299	2.453	2.606	2.759	2.913	3.066

Cu. Ft. of Concrete Displaced by Adjustable Steelforms and Straight Endforms

Steelform — Length of Each Row of Steelforms in l.f.

Depth	Width	1'	2'	3'	4'	5'	6'	7'	8'	9'	10'
8"	10"	0.51	1.03	1.54	2.06	2.57	3.08	3.60	4.11	4.63	5.14
10"	10"	0.65	1.31	1.96	2.61	3.27	3.92	4.57	5.22	5.88	6.53
12"	10"	0.79	1.58	2.38	3.17	3.96	4.75	5.54	6.34	7.13	7.92
14"	10"	0.93	1.86	2.79	3.72	4.66	5.59	6.52	7.45	8.38	9.31
15"	10"	1.00	2.00	3.00	4.00	5.00	6.00	7.00	8.00	9.00	10.00

Depth	Width	11'	12'	13'	14'	15'	16'	17'	18'	19'	20'
8"	10"	5.65	6.17	6.68	7.20	7.71	8.22	8.74	9.25	9.77	10.28
10"	10"	7.18	7.84	8.49	9.14	9.80	10.45	11.10	11.75	12.41	13.06
12"	10"	8.71	9.50	10.30	11.09	11.88	12.67	13.46	14.26	15.05	15.84
14"	10"	10.24	11.17	12.10	13.03	13.97	14.90	15.83	16.76	17.69	18.62
15"	10"	11.00	12.00	13.00	14.00	15.00	16.00	17.00	18.00	19.00	20.00

Steelform — Length of Each Row of Steelforms in l.f.

Depth	Width	1'	2'	3'	4'	5'	6'	7'	8'	9'	10'
8"	15"	0.79	1.58	2.38	3.17	3.96	4.75	5.54	6.34	7.13	7.92
10"	15"	1.00	2.00	3.00	4.00	5.00	6.00	7.00	8.00	9.00	10.00
12"	15"	1.21	2.42	3.62	4.83	6.04	7.25	8.46	9.66	10.87	12.08
14"	15"	1.42	2.83	4.25	5.67	7.09	8.50	9.92	11.34	12.75	14.17
15"	15"	1.52	3.04	4.56	6.08	7.61	9.13	10.65	12.17	13.69	15.21

Depth	Width	11'	12'	13'	14'	15'	16'	17'	18'	19'	20'
8"	15"	8.71	9.50	10.30	11.09	11.88	12.67	13.46	14.26	15.05	15.84
10"	15"	11.00	12.00	13.00	14.00	15.00	16.00	17.00	18.00	19.00	20.00
12"	15"	13.29	14.50	15.70	16.91	18.12	19.33	20.54	21.74	22.95	24.16
14"	15"	15.59	17.00	18.42	19.84	21.26	22.67	24.09	25.51	26.92	28.34
15"	15"	16.73	18.25	19.77	21.29	22.82	24.34	25.86	27.38	28.90	30.42

Steelform — Length of Each Row of Steelforms in l.f.

Depth	Width	1'	2'	3'	4'	5'	6'	7'	8'	9'	10'
8"	20"	1.07	2.14	3.21	4.28	5.35	6.41	7.48	8.55	9.62	10.69
10"	20"	1.35	2.69	4.04	5.39	6.74	8.08	9.43	10.78	12.12	13.47
12"	20"	1.63	3.25	4.88	6.50	8.13	9.75	11.38	13.00	14.63	16.25
14"	20"	1.90	3.81	5.71	7.61	9.52	11.42	13.32	15.22	17.13	19.03
15"	20"	2.04	4.08	6.13	8.17	10.21	12.25	14.29	16.34	18.38	20.42

Depth	Width	11'	12'	13'	14'	15'	16'	17'	18'	19'	20'
8"	20"	11.76	12.83	13.90	14.97	16.04	17.10	18.17	19.24	20.31	21.38
10"	20"	14.82	16.16	17.51	18.86	20.21	21.55	22.90	24.25	25.59	26.94
12"	20"	17.88	19.50	21.13	22.75	24.38	26.00	27.63	29.25	30.88	32.50
14"	20"	20.93	22.84	24.74	26.64	28.55	30.45	32.35	34.25	36.16	38.06
15"	20"	22.46	24.50	26.55	28.59	30.63	32.67	34.71	36.76	38.80	40.84

		10'	11'	12'	13'	14'	15'	16'	17'	18'	19'
8"	30"	16.25	17.88	19.50	21.13	22.75	24.38	26.00	27.63	29.25	30.88
10"	30"	20.42	22.46	24.50	26.55	28.59	30.63	32.67	34.71	36.76	38.80
12"	30"	24.58	27.04	29.50	31.95	34.41	36.87	39.33	41.79	44.24	46.70
14"	30"	28.75	31.63	34.50	37.38	40.25	43.13	46.00	48.88	51.75	54.63
15"	30"	30.83	33.91	37.00	40.08	43.16	46.25	49.33	52.41	55.49	58.58

		20'	21'	22'	23'	24'	25'	26'	27'	28'	29'
8"	30"	32.50	34.13	35.75	37.38	39.00	40.63	42.25	43.88	45.50	47.13
10"	30"	40.84	42.88	44.92	46.97	49.01	51.05	53.09	55.13	57.18	59.22
12"	30"	49.16	51.62	54.08	56.53	58.99	61.45	63.91	66.37	68.82	71.28
14"	30"	57.50	60.38	63.25	66.13	69.00	71.88	74.75	77.63	80.50	83.38
15"	30"	61.66	64.74	67.83	70.91	73.99	77.08	80.16	83.24	86.32	89.41

Cu.M of Concrete Displaced by Adjustable Steelforms and Straight Endforms

Steelform Depth mm	Width mm	Length of Each Row of Steelforms in meters									
		0.5 m	1.0 m	1.5 m	2.0 m	2.5 m	3.0 m	3.5 m	4.0 m	4.5 m	5.0 m
200	250	0.024	0.048	0.072	0.096	0.119	0.143	0.167	0.191	0.215	0.239
250	250	0.030	0.061	0.091	0.121	0.152	0.182	0.212	0.242	0.273	0.303
300	250	0.037	0.073	0.111	0.147	0.184	0.221	0.251	0.295	0.331	0.368
350	250	0.043	0.086	0.130	0.173	0.216	0.260	0.303	0.346	0.389	0.432
375	250	0.046	0.093	0.139	0.186	0.232	0.279	0.325	0.372	0.418	0.465
		5.5 m	6.0 m	6.5 m	7.0 m	7.5 m	8.0 m	8.5 m	9.0 m	9.5 m	10.0 m
200	250	0.262	0.287	0.310	0.334	0.358	0.382	0.406	0.430	0.454	0.478
250	250	0.334	0.363	0.394	0.425	0.455	0.485	0.516	0.546	0.576	0.607
300	250	0.405	0.441	0.478	0.515	0.552	0.589	0.625	0.662	0.699	0.736
350	250	0.476	0.519	0.562	0.605	0.649	0.692	0.735	0.779	0.822	0.865
375	250	0.511	0.557	0.604	0.650	0.697	0.743	0.790	0.836	0.883	0.929

Steelform Depth mm	Width mm	Length of Each Row of Steelforms in meters									
		0.5 m	1.0 m	1.5 m	2.0 m	2.5 m	3.0 m	3.5 m	4.0 m	4.5 m	5.0 m
200	375	0.037	0.073	0.111	0.147	0.184	0.221	0.257	0.295	0.331	0.368
250	375	0.046	0.093	0.139	0.186	0.232	0.279	0.325	0.372	0.418	0.465
300	375	0.056	0.112	0.168	0.224	0.281	0.337	0.393	0.449	0.505	0.561
350	375	0.066	0.131	0.197	0.263	0.329	0.395	0.461	0.527	0.592	0.658
375	375	0.071	0.141	0.212	0.292	0.353	0.424	0.495	0.565	0.636	0.707
		5.5 m	6.0 m	6.5 m	7.0 m	7.5 m	8.0 m	8.5 m	9.0 m	9.5 m	10.0 m
200	375	0.405	0.441	0.478	0.515	0.552	0.589	0.625	0.662	0.699	0.736
250	375	0.511	0.557	0.604	0.650	0.697	0.743	0.790	0.836	0.883	0.929
300	375	0.617	0.674	0.729	0.785	0.702	0.898	0.954	1.010	1.066	1.122
350	375	0.724	0.790	0.856	0.922	0.988	1.053	1.119	1.185	1.250	1.316
375	375	0.777	0.848	0.918	0.989	1.060	1.131	1.201	1.272	1.342	1.413

Steelform Depth mm	Width mm	Length of Each Row of Steelforms in meters									
		5.5 m	6.0 m	6.5 m	7.0 m	7.5 m	8.0 m	8.5 m	9.0 m	9.5 m	10.0 m
200	375	0.405	0.441	0.478	0.515	0.552	0.589	0.625	0.662	0.699	0.736
250	375	0.511	0.557	0.604	0.650	0.697	0.743	0.790	0.836	0.883	0.929

Steelform Depth mm	Width mm	Length of Each Row of Steelforms in meters

Depth mm	Width mm	5.5 m	6.0 m	6.5 m	7.0 m	7.5 m	8.0 m	8.5 m	9.0 m	9.5 m	10.0 m
300	375	0.617	0.674	0.729	0.785	0.702	0.898	0.954	1.010	1.066	1.122
350	375	0.724	0.790	0.856	0.922	0.988	1.053	1.119	1.185	1.250	1.316
375	375	0.777	0.848	0.918	0.989	1.060	1.131	1.201	1.272	1.342	1.413

Depth	Width	0.5 m	1.0 m	1.5 m	2.0 m	2.5 m	3.0 m	3.5 m	4.0 m	4.5 m	5.0 m
200	500	0.050	0.099	0.149	0.199	0.249	0.298	0.347	0.397	0.447	0.497
250	500	0.063	0.125	0.188	0.250	0.313	0.375	0.438	0.501	0.563	0.626
300	500	0.076	0.151	0.227	0.302	0.378	0.453	0.529	0.604	0.680	0.755
350	500	0.088	0.177	0.265	0.353	0.442	0.530	0.619	0.707	0.796	0.884
375	500	0.095	0.190	0.285	0.380	0.474	0.569	0.664	0.759	0.854	0.949

Depth	Width	5.5 m	6.0 m	6.5 m	7.0 m	7.5 m	8.0 m	8.5 m	9.0 m	9.5 m	10.0 m
200	500	0.546	0.596	0.646	0.695	0.745	0.794	0.844	0.894	0.043	0.994
250	500	0.688	0.751	0.813	0.876	0.939	1.001	1.064	1.126	1.189	1.251
300	500	0.831	0.906	0.982	1.057	1.132	1.208	1.283	1.359	1.434	1.510
350	500	0.972	1.061	1.149	1.237	1.326	1.414	1.503	1.591	1.680	1.768
375	500	1.043	1.138	1.233	1.328	1.423	1.518	1.612	1.708	1.802	1.897

Depth	Width	0.5 m	1.0 m	1.5 m	2.0 m	2.5 m	3.0 m	3.5 m	4.0 m	4.5 m	5.0 m
200	750	0.755	0.831	0.906	0.982	1.057	1.132	1.208	1.283	1.359	1.434
250	750	0.949	1.043	1.138	1.233	1.328	1.423	1.518	1.612	1.708	1.802
300	750	1.142	1.256	1.835	1.484	1.598	1.713	1.827	1.941	2.055	2.169
350	750	1.335	1.469	1.603	1.736	1.870	2.003	2.137	2.271	2.404	2.538
375	750	1.432	1.575	1.719	1.862	2.005	2.148	2.291	2.435	2.578	2.721

Depth	Width	5.5 m	6.0 m	6.5 m	7.0 m	7.5 m	8.0 m	8.5 m	9.0 m	9.5 m	10.0 m
200	750	1.510	1.585	1.661	1.736	1.812	1.887	1.963	2.038	2.114	2.189
250	750	1.897	1.992	2.087	2.182	2.277	2.371	2.466	2.561	2.656	2.751
300	750	2.284	2.398	2.512	2.626	2.740	2.854	2.969	3.083	3.197	3.311
350	750	2.671	2.805	2.938	3.072	3.205	3.339	3.472	3.606	3.739	3.873
375	750	2.864	3.007	3.151	3.294	3.437	3.581	3.724	3.867	4.010	4.153

Sizes and Cu. Ft. (Cu.M) Displacement of Ceco Flange Type Single Tapered Endforms

Endforms are 20" (500 mm) wide at one end tapering to 16" (400 mm) at the other end. The 30" (750 mm) endforms are 30" (750 mm) wide at one end tapering to 25" (625 mm) wide at other end.

Sizes and Cu. Ft. (Cu.M) Displacement of Ceco Adjustable Tapered Steel Endforms

Tapered endforms 3'-0" (0.9 m) long are furnished for 30" (0.75 m, 750 mm) wide steelforms only. Straight endforms 3'-0" (0.9 m, 900 mm) long are furnished for the 30" (0.75 m, 750 mm) standard widths and the 20", 15", and 10" (500, 375, and 250 mm) special widths.

Depth of Steelform, in. (mm)	Width Wide End, Wide End	Width Narrow End, Narrow End	Length in l.f (m)	C.F. (Cu.M) Displacement per 3' (0.91-m) Section
8" (200)	30" (750)	25" (625)	3' (0.91)	4.317 (0.122)
10" (250)	30" (750)	25" (625)	3' (0.91)	5.483 (0.155)
12" (300)	30" (750)	25" (625)	3' (0.91)	6.629 (0.188)
14" (350)	30" (750)	25" (625)	3' (0.91)	7.775 (0.220)
15" (350)	30" (750)	25" (625)	3' (0.91)	8.347 (0.236)

Another Method of Estimating the Concrete Required for Ceco Steelform Construction. The following method of estimating the concrete requirements for Ceco steelform construction (The Ceco Corporation, Oak Brook, IL) is that recommended by the manufacturers.

Figure the concrete as though it were a solid slab and then deduct for displacement of a certain number of lin. ft. of straight metal steelforms and tapered endforms.

Ceco steelform construction is a combination of concrete joists and thin slabs. To compute the amount of concrete in this type of floor construction, multiply the floor area by the equivalent thickness of the floor construction. This is better than separately figuring the concrete in the joists, by multiplying the lin. ft. of joists by their unit of volume, and adding the concrete in the thin slabs.

The cu. ft. of concrete per sq. ft. of floor (or roof) area is a function of the thickness of the thin slab, the depth of the joists, and the width and spacing of the joists. The additional concrete in the joists at their ends as formed by tapered endforms cannot be prorated on a square foot basis. The concrete added to the joists by the tapered endforms must be figured per lin. ft. (m) of beam or wall into which the joists frame, and this unit of volume must be multiplied by the number of lin. ft. (m) of beams and walls along which the tapered endforms are set, to determine the total amount of concrete added by the use of the tapered endforms.

For convenience in estimating, an arrangement in tabular form of the quantities of concrete required for Ceco flange type steelform construction is given. Keep in mind that the metal strength of the Ceco steelforms permits sharp angles in the steelforms and an absolutely flat top, allowing a minimum of concrete. These tables should not be used in computing the concrete for other types of metal form construction.

Typical Cross-section of Flangeforms

All dimensions are outside to outside.

regular widths
FOR FORMING REGULARLY SPACED JOISTS

filler widths
FOR USE ONLY IN FILLING OUT ODD SPACES

Sizes of Ceco Flange Type Steelforms

Typical Cross-sections of Longforms

All widths and depths are outside dimensions.

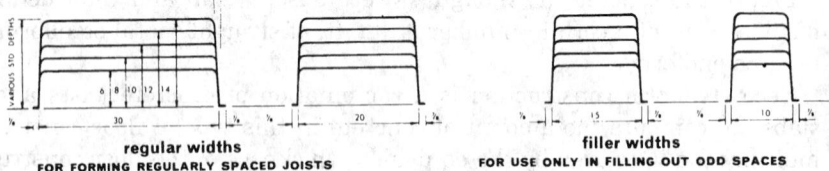

regular widths
FOR FORMING REGULARLY SPACED JOISTS

filler widths
FOR USE ONLY IN FILLING OUT ODD SPACES

Sizes of Ceco Longform Steelforms

Typical Cross-sections of Adjustable Steelforms

All dimensions are outside to outside.

regular widths
FOR FORMING REGULARLY SPACED JOISTS

filler widths
FOR USE ONLY IN FILLING OUT ODD SPACES

Sizes of Ceco Adjustable Type Steelforms

Concrete Required per Sq. Ft. of Floor Using Ceco Flange* Type Steelforms 20 Inches Wide

Depth of Steelform	Width of Joist	c.f. of conc. per s.f. of floor with slab thickness over steelforms of			Add. conc. req'd. (in. c.f.) per l.f. of beam or bearing wall where standard taper endform is used**
		2"	2-1/2"	3"	tapered
	4"	0.298	0.340	0.382	0.17
8"	5"	0.320	0.362	0.404	0.16
	6"	0.339	0.381	0.423	0.16

| Depth of Steelform | Width of Joist | c.f. of conc. per s.f. of floor with slab thickness over steelforms of | | | Add. conc. req'd. (in. c.f.) per l.f. of beam or bearing wall where standard taper endform is used** |
		2"	2-1/2"	3"	tapered
10"	4"	0.336	0.378	0.420	0.21
	5"	0.363	0.405	0.447	0.20
	6"	0.387	0.429	0.471	0.19
12"	4"	0.377	0.419	0.461	0.25
	5"	0.408	0.450	0.492	0.24
	6"	0.437	0.479	0.521	0.23
14"	5"	0.456	0.498	0.540	0.28
	6"	0.490	0.532	0.574	0.27
	7"	0.522	0.564	0.606	0.26
16"	5"	0.507	0.549	0.590	0.32
	6"	0.545	0.587	0.629	0.31
	7"	0.581	0.622	0.664	0.30

Concrete Required per Sq. Meter of Floor Using Ceco Flange* Type Steelforms 500 mm Wide

| Depth of Steelform | Width of Joist | cu.m of conc. per sq.m of floor with slab thickness over steelforms of | | | Add. conc. req'd. (in. cu.m) per meter of beam or bearing wall where standard taper endform is used** |
		50 mm	62.5 mm	75 mm	tapered
200 mm	100 mm	0.091	0.104	0.116	0.016
	125 mm	0.098	0.110	0.123	0.015
	150 mm	0.103	0.116	0.129	0.015
250 mm	100 mm	0.102	0.115	0.128	0.029
	125 mm	0.111	0.123	0.136	0.019
	150 mm	0.118	0.131	0.144	0.018
300 mm	100 mm	0.115	0.128	0.141	0.023
	125 mm	0.124	0.137	0.150	0.022
	150 mm	0.133	0.146	0.159	0.021
350 mm	125 mm	0.139	0.152	0.165	0.026
	150 mm	0.149	0.162	0.175	0.025
	175 mm	0.159	0.172	0.185	0.024
400 mm	125 mm	0.155	0.167	0.180	0.030
	150 mm	0.166	0.179	0.192	0.029
	175 mm	0.177	0.190	0.202	0.028

Concrete required per Sq. Ft. of Floor Using Ceco Adjustable Steelforms 20 Inches Wide

Depth of Steelform	Width of Joist	c.f. of conc. per s.f. of floor with slab thickness over steelforms of			Add. conc. req'd. (in. c.f.) per l.f. of beam or bearing wall where standard taper endform is used** tapered
		2"	2-1/2"	3"	
8"	3-1/2"	0.289	0.329	0.370	0.17
	4-1/2"	0.309	0.350	0.393	0.16
	5-1/2"	0.331	0.372	0.414	0.16
10"	3-1/2"	0.312	0.353	0.395	0.21
	4-1/2"	0.340	0.381	0.424	0.20
	5-1/2"	0.367	0.408	0.450	0.19
12"	3-1/2"	0.336	0.378	0.420	0.25
	4-1/2"	0.371	0.412	0.455	0.24
	5-1/2"	0.403	0.444	0.486	0.23
14"	3-1/2"	0.361	0.403	0.444	0.29
	4-1/2"	0.402	0.443	0.485	0.28
	5-1/2"	0.438	0.480	0.522	0.27

Concrete required per Sq. Meter of Floor Using Ceco Adjustable Steelforms 500 mm Wide

Depth of Steelform	Width of Joist	cu.m of conc. per sq.m of floor with slab thickness over steelforms of			Add. conc. req'd. (in. cu.m) per meter of beam or bearing wall where standard taper endform is used** tapered
		50 mm	62.5 mm	75 mm	
200 mm	87.5 mm	0.088	0.100	0.113	0.016
	112.5 mm	0.094	0.107	0.120	0.015
	137.5 mm	0.101	0.113	0.126	0.015
250 mm	87.5 mm	0.095	0.108	0.120	0.020
	112.5 mm	0.104	0.116	0.129	0.019
	137.5 mm	0.112	0.124	0.137	0.018
300	87.5 mm	0.102	0.115	0.128	0.023
	112.5 mm	0.113	0.126	0.139	0.022
	137.5 mm	0.123	0.135	0.148	0.021
350 mm	87.5 mm	0.110	0.123	0.135	0.027
	87.5 mm	0.110	0.123	0.135	0.027
	112.5 mm	0.123	0.135	0.148	0.026

*The above concrete requirements are also applicable to one piece longforms.
**Amount of concrete given for standard tapered endforms is for one side of beam only.

Concrete Required per Sq. Ft. of Floor Using Ceco
Flange* Type Steelforms 30 Inches Wide

Depth of Steelform	Width of Joist	c.f. of conc. per s.f. of floor with slab thickness over steelforms of			Add. conc. req'd. (in. c.f.) per l.f. of beam or bearing wall where standard taper endform is used**
		2"	2-1/2"	3"	tapered
8"	5"	0.322	0.364	0.405	0.15
	6"	0.338	0.380	0.421	0.15
	7"	0.351	0.392	0.434	0.14
10"	5"	0.353	0.395	0.436	0.19
	6"	0.372	0.414	0.455	0.18
	7"	0.389	0.430	0.472	0.18
12"	5"	0.386	0.427	0.469	0.22
	6"	0.408	0.450	0.491	0.22
	7"	0.430	0.470	0.512	0.21
14"	5"	0.420	0.460	0.503	0.26
	6"	0.447	0.488	0.530	0.25
	7"	0.470	0.511	0.553	0.24
16"	5"	0.451	0.493	0.535	0.29
	6"	0.482	0.523	0.565	0.28
	7"	0.510	0.552	0.594	0.27

Concrete Required per Sq.M of Floor Using Ceco
Flange* Type Steelforms 750 mm Wide

Depth of Steelform	Width of Joist	cu.m of conc. per sq.m of floor with slab thickness over steelforms of			Add. conc. req'd. (in. cu.m) per meter of beam or bearing wall where standard taper endform is used**
		50 mm	62.5 mm	75 mm	tapered
200 mm	200 mm	0.098	0.111	0.123	0.014
	125 mm	0.103	0.116	0.128	0.014
	150 mm	0.107	0.119	0.132	0.013
250 mm	100 mm	0.108	0.120	0.133	0.018
	125 mm	0.113	0.126	0.139	0.017
	150 mm	0.119	0.131	0.144	0.017
300 mm	100 mm	0.118	0.130	0.143	0.020
	125 mm	0.124	0.137	0.150	0.020
	150 mm	0.131	0.143	0.156	0.020
350 mm	125 mm	0.128	0.140	0.153	0.024
	150 mm	0.136	0.149	0.162	0.023
	175 mm	0.143	0.156	0.169	0.022

ESTIMATOR'S REFERENCE BOOK

Depth of Steelform 400 mm	Width of Joist	cu.m of conc. per sq.m of floor with slab thickness over steelforms of			Add. conc. req'd. (in. cu.m) per meter of beam or bearing wall where standard taper endform is used** tapered
		50 mm	62.5 mm	75 mm	
	125 mm	0.137	0.150	0.163	0.027
	150 mm	0.147	0.159	0.172	0.026
	175 mm	0.155	0.168	0.181	0.025

*The above concrete requirements are also applicable to one piece longforms.
**Amount of concrete given for standard tapered endforms is for one side of beam only.

Concrete Required per Sq. Ft. of Floor Using Ceco Adjustable Steelforms 30 Inches Wide

Depth of Steelform	Width of Joist	c.f. of conc. per s.f. of floor with slab thickness over steelforms of			Add. conc. req'd. (in. c.f.) per l.f. of beam or bearing wall where standard taper endform is used* tapered
		2"	2-1/2"	3"	
8"	5-1/2"	0.326	0.368	0.408	0.15
	6-1/2"	0.341	0.383	0.424	0.15
	7-1/2"	0.355	0.397	0.438	0.14
10"	5-1/2"	0.352	0.394	0.434	0.19
	6-1/2"	0.371	0.413	0.454	0.18
	7-1/2"	0.389	0.430	0.472	0.18
12"	5-1/2"	0.378	0.420	0.461	0.22
	6-1/2"	0.401	0.443	0.484	0.22
	7-1/2"	0.422	0.464	0.505	0.21
14"	5-1/2"	0.404	0.446	0.486	0.26
	6-1/2"	0.430	0.472	0.514	0.25
	7-1/2"	0.456	0.497	0.539	0.24

Concrete Required per Sq. Meter of Floor Using Ceco Adjustable Steelforms 750 mm Wide

Depth of Steelform	Width of Joist	cu.m of conc. per sq.m of floor with slab thickness over steelforms of			Add. conc. req'd. (in. cu.m) per meter of beam or bearing wall where standard taper endform is used* tapered
		50 mm	62.5 mm	75 mm	
200 mm	137.5 mm	0.102	0.112	0.124	0.014
	162.5 mm	0.104	0.117	0.129	0.014
	187.5 mm	0.108	0.121	0.134	0.013
250 mm	137.5 mm	0.107	0.120	0.132	0.018
	162.5 mm	0.113	0.126	0.138	0.017
	187.5 mm	0.119	0.131	0.144	0.017

Depth of Steelform	Width of Joist	cu.m of conc. per sq.m of floor with slab thickness over steelforms of			Add. conc. req'd. (in. cu.m) per meter of beam or bearing wall where standard taper endform is used*
		50 mm	62.5 mm	75 mm	tapered
300 mm	137.5 mm	0.115	0.128	0.141	0.020
	162.5 mm	0.122	0.135	0.148	0.020
	187.5 mm	0.129	0.141	0.154	0.020
350 mm	137.5 mm	0.123	0.136	0.148	0.024
	162.5 mm	0.131	0.144	0.157	0.023
	187.5 mm	0.139	0.151	0.164	0.022

*Amount of concrete given for standard tapered endforms is for one side of beam only.

Ceco One Piece Steeldomes for Two-Way Dome Slab Construction.
To estimate the quantity of concrete required for steeldome construction, figure the concrete required for a solid slab of the depth of the dome plus the top slab, and subtract the voids created for each steeldome as given in the table below.

Ceco Steeldomes

Steeldome Sizes

Item	19" x 19" System Standard Size	30" x 30" System Standard Size	Filler Sizes
Plan size of void	19"x19"	30"x30"	20"x30" and 20"x20"
Overall plan size including flanges	24"x24"	36"x36"	26"x36" and 26"x26"
Width of flange	2-1/2"	3"	3"
Depth of domes	4", 6", 8", 10", & 12"	8", 10", 12", 14", 16", & 20"	8", 10", 12", 14", 16", & 20"

(metric)

Item	475 x 475 mm System Standard Size	750 x 750 mm System Standard Size	Filler Sizes
Plan size of void	475x475mm	750x750mm	500x7500mm and 500x500mm
Overall plan size including flanges	600x600mm	900x900mm	650x900mm and 650x650mm
Width of flange	62.5mm	75mm	75mm
Depth of domes	100mm, 150mm, 200mm, 250mm, & 300mm	200mm, 250mm, 300mm, 350mm, 400mm, & 500mm	200mm, 250 mm 300mm, 350mm 400mm, & 500mm

Table of Concrete Voids
Voids Created By Steeldome, Cubic Feet

Depth of Steeldome	19"x19" System Regular 19"x19"	Regular 30"x30"	30"x30" System 20"x30"	20"x20"
8"	1.41	3.85	2.54	1.65
10"	1.90	4.78	3.13	2.06
12"	2.14	5.53	3.63	2.41
14"		6.54	4.27	2.87
16"		7.44	4.85	3.14
20"		9.16	5.90	3.81

Table of Concrete Voids
Voids Created By Steeldome, Cubic Meters

Depth of Steeldome	475 x 475 mm System Regular 475x475mm	Regular 750x750mm	750 x 750 mm System 500x750mm	500x500mm
200 mm	0.040	0.109	0.072	0.047
250 mm	0.054	0.135	0.089	0.058
300 mm	0.061	0.157	0.103	0.068
350 mm		0.185	0.121	0.081
400 mm		0.211	0.137	0.089
500 mm		0.259	0.167	0.108

Lightweight Concrete, Insulating Concrete of Pumice, Perlite, Vermiculite and Other Lightweight Aggregates

Lightweight concrete of crushed slag, foamed slag, haydite, pumice, perlite, and vermiculite are used for floor and roof fill, structural roofs, floors in cold storage rooms, and the like. A comparison of concrete aggregates is as follows:

Type of Aggregate	Lbs. per c.f.	Kg per cu.m	Lbs. per c.f. of conc.	Kg per cu.m of conc.
Gravel	120	3.4	150	4.2
Sand	90-100	2.5-2.8	150	4.2
Crushed stone	100	2.8	145	4.1
Crushed Bank Slag	80	2.3	100-130	2.8-3.7
Haydite	40-60	1.1-1.7	100-130	2.8-3.7
Foamed slag	40-60	1.1-1.7	90-100	2.5-2.8
Cinders	40-50*	1.1-1.4	110-115	3.1-3.3
Pumice	30-60	0.8-1.7	60-90	1.7-2.5
Diatomite	28-40	0.8-1.1	55-70	1.6-2
Perlite	6-16	0.2-0.5	20-50	0.6-1.4
Vermiculite	6-10	0.2-0.3	20-40	0.6-1.1
Waylite	40-60	1.1-1.7	90-100	2.5-2.8
Expanded Shale	44-58	1.1-1.6	92-98	2.6-2.8

*Plus sand.

Vermiculite Insulating Concrete Aggregate. This is a type of mica which is mined, crushed, and screened to size. When subjected to a temperature of about 2,000°F (1,093°C), it expands into a laminated granule containing millions of tiny dead air cells. This produces a lightweight aggregate of 6 to 10 lbs. per c.f. (97-162 kg/m^3), compared with sand or crushed rock aggregate weighing about 100 lbs. per c.f. (1620 kg/m^3). Vermiculite has high insulating value, is fireproof, and will not rot or decay.

Vermiculite insulating concrete is a lightweight building material made like ordinary portland cement concrete, except vermiculite concrete aggregate is used instead of sand, gravel, or crushed stone. Depending on the amount of vermiculite aggregate used, it is possible to make concrete weighing from 20 to 40 lbs. per c.f. (324-648 kg/m^3) Ordinary concrete weighs 145 to 150 lbs. per c.f. (2350-2430 kg/m^3).

Vermiculite concrete is used for poured concrete roof decks over a variety of forming materials; for roof insulation over such surfaces as concrete, steel, wood, etc.; and for poured concrete floor slabs.

Because of its high insulation value, vermiculite concrete is an excellent base for radiant heat floor installations and for grade level floors without radiant heat in homes, farm structures, and all types of commercial buildings. At a density of approximately 25 lbs. per c.f. (405 kg/m^3), 1" (25 mm) of vermiculite concrete has an insulating value equal to approximately 16" (400 mm) of ordinary concrete. The density, insulating value, and strength of vermiculite concrete can be varied to meet a wide range of design requirements.

Vermiculite concrete is incombustible and has earned the highest fire ratings attainable. It carries a 5-hour fire rating for spandrel wall construction and a 1-1/2 hour rating on a steel deck covered with 1-1/2" (38 mm) of vermiculite concrete and completely unprotected on the underside.

Mixing. Vermiculite concrete should be mixed in a mechanical mixer. The required amount of water and cement is placed in the mixer and then the aggregate. Mixing is limited to the minimum time required to obtain a thorough mix and proper fluidity. (Maximum time recommended, 5 minutes).

When transit-mixed vermiculite concrete is used, the operation is as follows:

1. Introduce water and cement into mixer. (Fill auxiliary water tank before leaving plant.)
2. Rotate the mixer slowly until all aggregate has been added.
3. Continue to rotate mixer for approximately one minute after aggregate is in the mixer.
4. Do not rotate drum on way to job site. Mix concrete at job site at fastest speed until it is uniform and flows freely from the mixer.

Placing. Vermiculite concrete is transported and placed immediately after mixing is completed, and the period between completion of the mixing and placing should be short enough so that the mixture does not appreciably change in consistency.

Vermiculite concrete should not be placed when the temperature is less than 40°F (4.4°C). If the temperature expected after placing of the concrete is near or below 40°F (4.4°C), the mixing water should be heated in the temperature range 75° to 100°F (23.8-37.7°C).

Vermiculite concrete can be placed by the use of standard concrete equipment such as bucket hoists, buggies, or specially designed pumping equipment.

Machine Placement. Pumping or spraying the concrete is especially suited to roof decks with curved, sloped, or irregular surfaces and is just as efficient on flat roof decks.

A pump is fast and simple. It can be set up or taken down in about 30 minutes. The long hose lengths that are possible enable the crew to get over obstructions on the job easily.

The Vermiculite Institute annually approves a national roster of roof deck applicators appointed by institute members to assure the highest quality of finished work.

Vermiculite concrete is covered by the United States of America Standards Institute Specification A122.1-1965.

Vermiculite Insulating Concrete Floors-on-Grade. A 1:4 mix of vermiculite concrete is recommended as an insulating base under a sand-gravel concrete topping for floors on grade. It is particularly effective in reducing heat loss where radiant heating coils or ducts are carried in the floor.

d) Do not rotate drum on way to job site. Mix concrete at job site at fastest speed until it is uniform and flows freely from the mixer.

Vermiculite Concrete Roof Insulation Poured Over Structural Concrete Roof Decks

Vermiculite Concrete Roof Deck Poured in Place Over Form Boards

Vermiculite Insulating Concrete Over Vented Metal Roof Decks

Vermiculite Concrete Insulation Over Vented Galvanized Steel Roof Decks,
Precast Concrete Joists (Two-hour Fire Rated)

Vermiculite Concrete Roof Deck
Poured-in-Place Over Paper-Backed Wire Lath

Vermiculite Concrete Insulation Over Precast Prestressed
Concrete Double Tee Roof Slabs

Minimum thickness of the slab should be 3" (75 mm). A reinforcing mesh of 6"x6" (150x150mm) 10/10 ga. welded wire fabric or equal should be placed in the center of the vermiculite concrete, lapping the sides and ends one full mesh.

The sand-gravel topping should be designed to carry the maximum floor load. Minimum thickness should be 1-1/2" (38 mm) except when radiant heating coils are used. In that case minimum thickness over the top of the coils should be 1" (25 mm). The radiant heating coils should be supported on chairs to assure a minimum of 1/2" (13 mm) of sand-gravel concrete below the coils.

Proportion Table for Vermiculite Concrete

Description of Work	Vermiculite Aggregate C.F. (Cu.M)	Portland Cement C.F. (Cu.M)
Poured concrete roofs	4 (0.01)	1 (0.03)
Poured concrete roof insulation	6 or 8 (0.16 or 0.22)	1 (0.03)
Poured concrete slabs		
On grade insulating slab under 1-1/2 ordinary cement topping	4 (0.01)	1 (0.03)
Panel or spandrel walls	4 (0.01)	1 (0.03)

Mixing Table For Vermiculite Concrete

Mix by Vol. Vermiculite Aggregate C.F.	Portland Cement Sacks	Approx. Density in Lbs. per C.F.	Compressive Strength in Lbs. per Sq. In.	Thermal Conductivity (K) in btu/SF/in./ hour/degree F
8	1	20 to 25	100 to 125	0.60
6	1	25 to 30	125 to 225	0.76
4	1	35 to 40	350 to 500	0.97

(metric)

Mix by Vol. Vermiculite Aggregate Cu. M	Portland Cement Sacks	Approx. Density in Kg per Cu. M	Compressive Strength in MPA
0.23	1	321-402	0.69-0.86
0.17	1	402-482	0.86-1.54
0.11	1	562-643	2.40-3.44

Approximate Quantities Required for One C.Y.
of Ready-Mixed Vermiculite Concrete

Mix	Bags (4 C.F.) Vermiculite Concrete Aggregate	Sacks (1 C.F.) Portland Cement	Water Gals.
8-1	7.5	3.75	85 to 95
6-1	7.5	5.0	85 to 95
4-1	7.5	7.5	85 to 95

Approximate Quantities Required for One Cu.M
of Ready-Mixed Vermiculite Concrete

Mix	Bags (.11 Cu.M) Vermiculite Concrete Aggregate	Sacks (.028 Cu.M) Portland Cement	Water Liters
8-1	7.5	3.75	321-359
6-1	7.5	5.0	321-359
4-1	7.5	7.5	321-359

Perlite Concrete Aggregate

Perlite is a siliceous volcanic rock mined in western United States. When crushed and quickly heated to above 1500°F (815.55°C), it expands to form lightweight, non-combustible, glass-like particles of cellular structure. This material, white or light gray in color, weighs 7-1/2 to 15 lbs. per c.f. (121-242 kg/m^3).

Perlite aggregate consists of expanded perlite sized for use in lightweight insulating concrete in place of sand or gravel. It is usually packed in 3 or 4 c.f. (0.08 or 0.11 cu.m) paper bags for easy handling.

The many, tiny glass sealed cells in each particle of expanded perlite make it highly insulating as well as comparatively nonabsorptive. Thus perlite mixes with about 30% less water than comparable lightweight aggregates.

Perlite concrete, in combination with portland cement, forms a very lightweight concrete. Where ordinary concrete weighs 150 lbs. per c.f. (2430 kg/m^3), the dry weight of perlite concrete can be designed from 20 to 40 lbs. per c.f. (324-648 kg/m^3). The extremely light weight and ease of handling make perlite concrete adaptable to practically any shape.

Perlite concrete has received up to 4-hour ratings in Underwriters' Laboratories fire tests. Its light weight makes it suitable for fireproof roof and floor fills, for thin concrete curtain walls (blocks, slabs or monolithic), and for many precast panel and block constructions.

½" COMPRESSIBLE
EXPANSION JOINT IN
TOPPING AT PERIMETER

SAND CONCRETE TOPPING
(1½" Min.)

VERMICULITE MASONRY FILL

VERMICULITE INSULATING CONCRETE
(3" Min.)

Vermiculite Insulating Concrete Floor

— ROOF CONSTRUCTIONS —

3 HOURS:

Roof deck of 2" minimum thickness perlite concrete
on 28 gauge galvanized steel form units supported by
steel joists 4' on center. Ceiling of ⅞" perlite-gypsum
plaster on expanded metal lath attached to ¾" furring
channels wire-tied to lower chord of joists. (For details see Underwriters' Laboratories Class C-3, Design
No. RC1-3.)

2" min. Perlite concrete
28 gauge galvanized steel

7/8 Perlite-gypsum plaster Metal lath

2 HOURS:

Same as 3 hour rated system above except that the
perlite-gypsum plaster ceiling shall be ¾" thick. (For
details see Underwriters' Laboratories Class D-2, Design
No. RC11-2.)

2" min. Perlite concrete
28 gauge galvanized steel

3/4 Perlite-gypsum plaster Metal lath

3 HOURS:

Roof deck of 2½" thickness perlite concrete on paper-backed welded wire mesh supported by steel joists. Ceiling of ¾" perlite-gypsum plaster on high ribbed metal lath attached to lower chord of joists. (For details see Underwriters' Laboratories Class C-3, Design No. RC2-3.)

1 HOUR:

UNPROTECTED STEEL ROOF DECK

Corrugated galvanized steel deck topped with 2⅝" average thickness of perlite concrete reinforced with 48-1214 wire mesh. No ceiling protection. Beams protected with ⅞" perlite-gypsum plaster on metal lath. (For details see Underwriters' Laboratories Class E-I, Design RC2-1.)

— CURTAIN WALL CONSTRUCTIONS —

4 HOURS:

6" non-bearing wall (not including weatherproof facade). Exterior: 4" perlite-portland cement mix sprayed on paper-backed wire mesh. Interior: 1" furred perlite-gypsum plaster on paper-foil backed wire mesh. (For details see Underwriters' Laboratories Class B-4, Design No. 3.)

2 HOURS:

Non load-bearing wall consisting of 3" thickness perlite concrete spray-applied to paper-backed wire lath attached to 4" steel studs 16" on center. (For details see Underwriters' Laboratories Class D-2, Design No. 18.)

For lightweight structural roof construction, where perlite concrete is placed over galvanized steel forms, paper-backed wire mesh, formboards, structural concrete, or other suitable form materials, use a 25 to 29 lb. per c.f. (405-470 kg/m³) density perlite concrete. This offers an ideal balance of low dead weight, adequate compressive and indentation strengths, and good insulating value. For uses where higher strengths are more important than insulating value, i.e., floor fills and certain lightweight structural roof designs, use a 34 to 40 lb. per c.f. (550-648 kg/m³) density perlite concrete.

Perlite concrete should be mixed in a paddle type plaster or a drum type concrete mixer. The required amount of water, air entraining admixture, and portland cement should be placed in the mixer and mixed until a slurry is formed. The proper amount of perlite concrete aggregate should then be added to the slurry and all materials mixed until design wet density is reached. Perlite concrete may also be transit mixed.

Perlite concrete should be carefully deposited and screeded in a continuous operation until a panel or section is completed. Steel troweling should be avoided. Rodding, tamping, and vibrating should not be used unless so specified by the architect.

It is recommended that a 1" (25 mm) air space or expansion joint be provided through the thickness of the perlite concrete at the juncture of all roof projections, such as skylights, penthouses, and parapet walls. Expansion joints are recommended in accordance with good concrete and steel construction practice and are based upon the provision that suitable consideration has been given to the use of adequate through-building expansion joints. Other types of insulating concrete may not require expansion joints because of high initial shrinkage.

Perlite concrete, during the curing period, should be protected for at least the first three days to keep it from drying out too rapidly or freezing. Freshly poured concrete should also be given adequate protection from heavy rain. Allow no traffic until concrete can sustain a man's weight without indentation.

Perlite concrete should not be placed in temperatures under 40°F (4.4°C) nor on frosted surfaces. In near-freezing weather the mixing water should be heated to between 75° and 100°F (23.88-37.77°C).

Perlite concrete is also used as a base for radiant heating coils. For grade level floors a moisture resistant barrier should be used and a suitable wearing surface applied over the perlite insulating concrete.

Considerable savings in dead weight are possible by using perlite concrete as a floor fill over cellular steel or pan-type floors in multiple story buildings. The surface may be covered with concrete topping and other surfacing materials after the perlite concrete has cured several days. Another floor fill material, using perlite blended with other aggregates, is perlite/sand or perlite/expanded shale concrete. These concretes have higher compressive strengths and require only resilient flooring or carpeting as a wearing surface. Following are the results of tests conducted at nationally recognized laboratories. Due to the varying characteristics of naturally occurring aggregates in different geographical areas, trial mixes of aggregate blends are recommended for determination of mix proportions. Perlite concrete can be nailed, sawed, and worked with ordinary carpenter tools.

Density Wet	Density Dry	Compression 28 day, psi	Cement (Sacks)	Perlite (cf)	Heavy Aggregate (cf)	AEA† (oz)	Water U.S. Gal/Sack	Ct. Factor cu. yd.
82	72	1000	1	3	2*	6	8.5	5.82
66	54	1200	1	3	2**	4	8.5	6.12

*Sand. **Rotary Kiln Expanded Shale. +Air Entraining Agent.

CONCRETE

Mix Proportions By Volume for Lightweight Perlite Insulating Concrete

Cement (Sacks)	Perlite C.F.	Water-Gals. per Sack Cement	Air-Entraining Agent, Pints	Density Lbs. per C.F. Oven Dry	Comp. Str. psi-28 Days
1	4	9	1	36	400
1	5	11	1-1/4	30-1/2	250
1	6	12	1-1/2	27	170
1	7	14	1-3/4	24	120
1	8	16	2	22	95

(metric)

Cement (Sacks)	Perlite Cu. M	Water-Liters per Sack Cement	Air-Entraining Agent, Pints	Density Kg per Cu.M Oven Dry	Comp. Str. MPa-28 Days
1	0.11	34	0.237	577	2.75
1	0.14	41	0.296	489	1.72
1	0.17	45	0.355	432	1.17
1	0.20	53	0.414	384	0.83
1	0.23	61	0.473	352	0.65

Material Required for One C.Y. of Placed Perlite Concrete

Proportions Cement Perlite	Cement Sacks	Perlite C.F.	Water Gallons	Air-Entraining Agent, Pints
1:4	6.75	27	61	6-3/4
1:5	5.40	27	59-1/2	6-3/4
1:6	4.50	27	54	6-3/4
1:7	3.85	27	54	6-3/4
1:8	3.38	27	54	6-3/4

Material Required for One Cu. M of Placed Perlite Concrete

Proportions Cement Perlite	Cement Sacks	Perlite Cu.M	Water Liters	Air-Entraining Agent, Pints
1:4	8.83	1	230.9	3.19
1:5	7.06	1	225.2	3.19
1:6	5.89	1	204.4	3.19
1:7	5.04	1	204.4	3.19
1:8	4.42	1	204.4	3.19

Pumice Concrete

Pumice is a natural white or gray glass foam. It is not a volcanic ash. It is mined extensively in the western part of the United States.

Uniform graded pumice is used as aggregate in monolithic concrete, concrete masonry units and precast products. It replaces sand and gravel in any concrete up to strengths of 3,000 psi (20.7 MPa), where thermal insulation, acoustical, and light weight properties are desirable.

All pumice aggregate should be clean, free of all foreign matter, and well graded so as to meet ASTM specifications C-330.

All water used in concrete work shall be free from oil, strong acids, alkali, or organic material.

Mixing. Concrete for columns, walls above ground, floors, window sills, and all self-supporting floor and roof slabs, canopies, and other parts of structure not coming in contact with ground shall be of a mixture of water, portland cement, and pumice aggregate, and shall develop at 28 days a strength of not less than 2,500 psi. (17.2 MPa) and with a slump of not more than 2" (50 mm) by standard test.

Water contained in the aggregate shall be deducted from the amount of water used in the mix except that the water needed to saturate the pumice aggregate shall not be deducted. Pumice aggregate shall be thoroughly saturated either in the stockpile or in the mixer prior to the addition of cement and water for mixing.

SUGGESTED MIX DESIGN DATA FOR PYRAMID PUMICE CONCRETE

Compression lbs/sq. in	Design Weight lbs/cu ft	Sand lbs (dry)	Pyramid Pumice per Cubic Yard Yield						Cement		Water				Recommended slump
			Fine		Coarse		Blend		sax/cu yd yield	lbs/cu yd yield	to saturate		to mix		
			lbs (dry)	cu ft	lbs (dry)	cu ft	lbs (dry)	cu ft			gal	lbs	gal	lbs	
150	40	0	0	0	792	36	0	0	4	376	36.1	300	Determine by sight		0
250	46	0	914	24	264	12	1180	36	3	282	34.8	290	Determine by sight		0
500	55	0	914	24	264	12	1180	36	4	376	34.8	290	Determine by sight		0
1000	60	0	914	24	264	12	1180	36	5	470	34.8	290	50	417	½"
1500	65	0	914	24	264	12	1180	36	6.5	611	34.8	290	52	433	½"
2000	70	0	914	24	264	12	1180	36	7.5	705	34.8	290	53	442	1"
2500	75	0	914	24	264	12	1180	36	8	752	34.8	290	48	400	2"

Pyramid Pumice Aggregate. This should be saturated prior to the addition of cement. It may be done in the stock pile or in the mixer. A 1" (25 mm) slump in pumice concrete will give a consistency equal to a 3" (75 mm) slump in hard rock concrete. Pumice concrete should be mixed with as little water as possible to get the required workability and should be the consistency of thick mud. No free water should be in evidence. Air entraining agents are recommended to cut water-cement ratio and get greater workability. A combination of fine and coarse aggregate or the blended aggregate should be used. Volume measurements are based on loose volume non-compacted pumice.

Placing. All pumice concrete shall be thoroughly vibrated with a mechanical vibrator. Vibration shall be conducted in such a manner as to assure that the vibrator passes through the pumice mass with not more than 6" (150 mm) of space between the successive positions of the vibrator. In all cases where pumice is placed in layers, the vibrator must penetrate into the underlying concrete layer after passing through the newly placed layer. Ample time should be allowed for the settlement of the concrete before final topping or screeding to finished floor level or grade. The pumice aggregate shall be thoroughly saturated prior to mixing with the cement and water used for mixing and ample time shall be allowed for the saturation to take place.

Application. Pumice must be handled carefully if it is to be successful. Because it will float, it must be pre-saturated before use. The stock pile should be sprayed for at least 48 hours or the pumice should be saturated in the mixer before use. Unless this is done, cement particles will be drawn into the cells of the pumice and excessive shrinkage may occur as a result of loss of water from the mix to the pumice particles. One half of the total water required should be added to the aggregate before the cement is introduced.

The concrete should be water cured for at least seven days, after which time the outer fibers will have become strong and dense enough to protect against evaporation and overly rapid drying.

Gradations Available. Pumice aggregate is graded to meet ASTM specifications for properly graded lightweight aggregates for concrete. Designation C-330 in sizes No. 4 to 0 for fine aggregate; 3/8 (9 mm) and 1/2 (13 mm) to 0 for blended aggregate; 1/2" (13 mm) to No. 4 for coarse aggregate.

Shipping Weight. Fine material or 1/4" (6 mm) to 0 aggregate will have an average weight of about 1,200 lbs. per c.y. (711 kg/m^3). Blended aggregate or 3/8" (9 mm) to 0 material will weigh about 1,100 lbs. per c.y. (625 kg/m^3). The coarse aggregate or 1/2" (13 mm) to No. 4 material will have an average shipping weight of about 800 lbs. per c.y. (474 kg/m^3). These weights may increase from 100 to 200 lbs. per c.y. (59-119 kg/m^3) during wet seasons. An open top, side dump, flat bottom (Caswell) car carries approximately 63 c.y. (48 cu.m) of material.

An open top, solid bottom, gondola car will carry approximately 63 c.y. (48 cu.m) of material.

An open top, hopper bottom car will vary in capacity from 60 to 100 c.y. (46-77 cu.m).

Waylite Lightweight Aggregate

A lightweight, cellular aggregate, refined and expanded by a process of agitating and rapidly cooling in an atmosphere of steam molten hot slag as it comes from the blast furnace at temperatures of 2,500 to 3,000°F. In this process the inherent gases expand and fill the material with minute air cells or bubbles completely sealed. These minute air cells are the active insulation and lightweight principle in all Waylite aggregates.

Waylite aggregate crushed and screened is available in three commercial sizes and average dry loose weights per cubic foot.

Aggregate	Grading, in.	Grading, mm	Avg Wgt lbs.	Avg Wgt kg
Coarse	7/16-3/16	10.9-4.6	40	18.14
Acoustical	3/16-3/32	4.6-2.3	45	20.40
Fine	1/8-dust	3-dust	60	27.20

Waylite concrete requires a higher proportion of fine aggregate to coarse than ordinary concrete. This is due to the lower top sizes of the coarse aggregate—7/16" (11 mm)—and the additional angularity or harshness of the crushed fine and coarse aggregates.

The mixing of Waylite concrete requires a somewhat different procedure than for heavy aggregates. To obtain maximum workability with minimum weight, the aggregate and water should be mixed for one-half minute before the addition of the cement, and then mixed for at least an additional two minutes. A mix that appears harsh at first will improve in plasticity with continued mixing without additional water.

Waylite Concrete Mix Data

Mix Loose Volume	Gals. Water per Sack	Cement Sacks	Fine Waylite C.Y.	Coarse Waylite C.Y.	Sand* C.Y.
1 – 2-1/4 – 2-1/2 – 3/4*	9.4	6.4	0.53	0.59	0.18
1 – 2 – 2-1/4 – 3/4*	8.2	7.2	0.53	0.60	0.20
1 – 1-3/4 – 2-1/2*	7.3	8.1	0.54	0.60	0.15
1 – 1-1/2 – 2-1/2*	6.5	8.8	0.49	0.65	0.16
1 – 1-1/2 – 1-3/4 – 1/2*	5.9	9.6	0.53	0.62	0.18

*Sand shall be a fine siliceous bank, lake, or mason sand.

Quantities Required per C.Y., Loose Dry Volume

Mix Loose Volume	Liters Water per Sack	Cement Sacks	Fine Waylite Cu.M	Coarse Waylite Cu.M	Sand* Cu.M
1 – 2-1/4 – 2-1/2 – 3/4*	46.5	8.4	18.83	20.93	6.28
1 – 2 – 2-1/4 – 3/4*	31.0	9.4	18.83	21.19	7.06
1 – 1-3/4 – 2-1/2*	27.6	10.6	18.54	21.19	5.30
1 – 1-1/2 – 2-1/2*	24.6	11.5	17.26	23.02	5.75
1 – 1-1/2 – 1-3/4 – 1/2*	22.3	12.6	18.83	21.97	6.28

*Sand shall be a fine siliceous bank, lake, or mason sand.

Quantities Required per Cu. Meter, Loose Dry Volume

The addition of approved concrete air entraining agents materially improve the workability of Waylite concrete mixes.

PLACING, PATCHING AND
FINISHING ARCHITECTURAL CONCRETE

Architectural concrete demands the use of a plastic, workable mix that can be worked and molded readily in the forms. Surfaces of the hardened concrete should be free of defects such as sand streaking and honeycomb. It is important that sufficient fines be provided in the mixture, and fine aggregate should have at least 15% passing through a 50 mesh sieve and 3% passing the 100-mesh sieve, and the cement factor should not be less than 5-1/2 sacks per c.y. (7.2 sacks per cu.m) of concrete. In other words, the concrete should not be richer than 1:2:3 by loose volumes and preferably 1:2-1/3:3-1/3, containing 6-1/2 or 6 sacks of cement per c.y. (7.85 or 8 per cu.m) of concrete respectively.

Not more than 6-1/2 gals. (24.6 liters) of water per sack of cement should be allowed so that the concrete will have the durability required to withstand weathering.

Mixing and Placing Concrete. Concreting of architectural concrete walls should proceed at a rather slow rate allowing sufficient time for thorough vibrating at the finished surfaces so that the pressure exerted by the concrete on the forms does not exceed 600 lbs. per s.f. (2929 kg/m^2).

A reasonable rate of placing the concrete should be considered when locating joints. Generally a rate of not more than 2 ft. (0.6 m) per hr. is most conducive of good workmanship and lifts of 6 to 8 ft. (1.8-2.4 m) between joints are desirable and 12 ft. (3.6 m) should be the absolute maximum.

The concrete should be spaded or vibrated into all corners and along all form surfaces, but even if vibrated, some spading is necessary in the corners and angles. Vibrators will not fill these places at all times. On many jobs small electric hammers moved along the forms at the point of placement have been found helpful. Vibration should be sufficient only for complete compaction, not to the point where segregation occurs and excess water is forced to the form faces. Such over-vibration may result in sand streaking and discoloration.

When concrete is splashed on the forms above the point of deposit and subsequently hardens, the wall in that area is likely to be rough and discolored. Splash boards are sometimes used but are not completely satisfactory, as they do not prevent segregation of the falling concrete. Tremies of various lengths equipped with a hopper at the top through which the concrete is placed will avoid both splash and segregation. The tremies should be spaced not more than about 9 ft. (2.7 m) apart. Sometimes splashing is the result of mortar being thrown from vibrators as they are lifted out of the concrete and raised to the placing platform. This can be prevented by shutting off the power as the vibrator is withdrawn from the concrete.

Labor Placing Concrete. The labor cost of placing architectural concrete will vary considerably, depending on the size of job, size of sections, amount of ornamentation, quantity of reinforcing steel, etc., but on most work of this class, it will require 2 to 3 hrs. labor time per c.y. (2.6-4 hrs. per cu.m) of concrete, plus the time for the hoisting engineer and foreman.

Based on a 64-c.y. pour per 8-hr. day, the labor cost should average as follows per c.y. (cu.m):

	Hours	Rate	Total	Rate	Total
Foreman	0.13	$....	$....	$21.77	$ 2.83
Hoist engineer	0.13	....	....	28.55	3.71
Labor	2.50	....	....	21.04	52.60
Cost per c.y.			$....		$ 59.14

(metric)

	Hours	Rate	Total	Rate	Total
Foreman	0.17	$....	$....	$21.77	$ 3.70
Hoist engineer	0.17	....	....	28.55	4.85
Labor	3.27	....	....	21.04	68.80
Cost per cu.m			$....		$ 77.35

Curing. On this type of work one of the surest and best ways of curing the concrete is by leaving the forms in place. Whenever it is practical to do so, they should be left in place 4 or 5 days. By that time the concrete is hard enough so that it is not scuffed nor are corners so likely to be broken off during form removal. If the forms are removed earlier, other means must be provided to keep the concrete wet or damp for at least 5 days if a durable concrete is to be obtained.

Patching and Finishing. When the forms are first stripped, the concrete does not always present a pleasing appearance. There may be pock marks, air holes, fins, and some discoloration. Some of the latter will bleach out upon exposure but a simple clean down with cement grout will do wonders for the appearance. The projecting fins and other blemishes extending beyond the face of the wall are removed with a carborundum stone or abrasive hone and then a grout of 1 part portland cement to 1-1/2 parts of very fine sand (sand passing through a fly screen will do) is applied. From 1/3 to 1/2 of the cement should be white cement.

The grout is mixed to the consistency of a heavy paste. After the wall is wet down, the grout is applied uniformly with a stiff fiber brush, completely filling all air voids and holes. Immediately after the grout has been applied, the wall is floated with a wood or cork float. A very fine abrasive hone can also be used. The grout is then allowed to harden partially, the exact time depending upon the weather. When it is hardened sufficiently so it will not be pulled out of the holes, all grout is scraped from the surface of the wall with the edge of a steel trowel. After the wall has dried thoroughly, it is rubbed with a dry piece of burlap to completely remove all of the dried grout or film. There should be no visible film left on the surface. Grout should not be left on the wall overnight. All cleaning of any one panel of concrete should be completed the same day.

Patching mortar used to fill tie holes and imperfections should be of the same mixture of gray and white cements, but should be a stiffer mixture and the cement and sand proportions should be the same as those used in the concrete. Patches should never be steel troweled, but may be finished with wood or cork floats. Allowing the mixed patching mortar to stand for an hour or two before using it reduces the amount of shrinkage but water must not be added in remixing it.

To correct blemishes appearing on the face of the walls, allow about 4 hrs. time for a cement finisher and a helper per 100 s.f. (9 sq.m) of wall. The labor cost per 100 s.f. (9 sq.m) of wall should average as follows:

	Hours	Rate	Total	Rate	Total
Cement mason	4	$....	$....	$27.03	$ 108.12
Helper	4	....	....	21.04	84.16
Cost per 100 s.f (9 sq.m)			$....		$ 192.28
cost per s.f.			....		1.92
cost per sq.m			....		20.70

Cutting Concrete with Portable Concrete Saw

Many states now specify that control joints in highway paving be cut with a concrete saw, rather than produced by forming. The advantages of this method are improved ridability of joint; less sealing material required to seal joint; better aggregate interlock at joint; and no spalling or cracking at joint, which reduces maintenance costs.

This technique is also widely used for cutting floors and paving for repair work, trenches for electric conduit, pipe lines, and sewers, and also for scoring before breaking out concrete for machinery bases and other foundation work.

Clipper concrete saws (from Norton Construction Products Division, Worcester, MA) are a good example. They are available in several electric and gasoline models. The most popular models are the 18 Hp model C-188, the 30 Hp model C-305, the 37 Hp model C-375, and the 65 Hp model C-655, all gasoline powered. Approximate prices range from $2,000.00 for the basic model C-188 manually propelled to $8,000.00 for the model C-655 with self propelling unit and electric starter. The water pump and spotlight are optional features. The self propelling unit and other accessories are optional on the C-188 and C-305 models. The larger C-655 with 65 Hp engine is generally used for sawing contraction joints in airports, highways, and turnpikes. The C-305 and C-188 are widely used in patching, trenching, and plant maintenance work. Electrically powered models are available from 3 Hp and up.

Two types of blades are used in cutting concrete with concrete saws as follows: 1) Green-Con wet-dry abrasive blades for cutting green concrete 2) diamond blades, of various specifications, for cutting cured or old concrete.

The cost of cutting concrete with a concrete saw varies considerably, depending on age of concrete, type of aggregate, blade specification used, accessibility of work, skill of operator, etc.

Assuming proper blade specification is used for the cutting job, green concrete 24 to 48 hours old may be cut 1" deep at the rate of 12 to 14 l.f. (3.6-4.2 m) per minute and 3" (75 mm) deep at the rate of 6 to 8 l.f. (1.8-2.4 m) per minute. These rates cover only actual cutting time. Overall production will be reduced considerably depending on length of each individual cut, number of machine moves, changes in direction of cuts, etc. Blade wear for a Green-Con blade of proper specification should cost from $0.01-0.02/l.f. (0.033-0.066/m) of cut 1" (25 mm) deep. For a 2" (50 mm) cut, blade cost should be $0.015-0.03/l.f.; 3" (75 mm) cut, $0.02-0.04 per l.f. (0.06-0.13 per m).

When cutting old or cured concrete, using a diamond blade of correct specification, cutting time may be double the amount given above for green concrete and cost of blade wear will probably run 5 to 6 times that given above.

CEMENT FLOOR HARDENERS

There are many preparations on the market for waterproofing, dustproofing, and wearproofing cement finish floors. These preparations fall into two classifications, integral treatments in which the entire slab or topping is treated and surface treatments, in which the surface is specially prepared to become waterproof, dustproof and wear resistant.

Integral Treatments. Integral treatments consist of powdered chemicals that are mixed with the cement before sand and water are added and become an integral part of the topping mix throughout its entire thickness. This type of treatment is said to densify and harden the topping and is generally recommended for areas subjected to light pedestrian traffic.

Surface Treatments. There are two types of surface treatments for cement finish floors, dust coat applications which must take place as floors are being finished and liquid chemical applications which may be done at any time after the floor finish is set hard enough for foot traffic.

METALLIC FLOOR TREATMENTS (HARDENERS)

By far the most widespread surface treatment is the metallic dust coat. The term "hardeners" usually applied to metallic floor treatments is somewhat misleading, because iron has a much lower rating on the scale of hardness than even the most common sand or stone aggregate. In this sense iron treatments could be described as "softeners". The basic technique is to provide a floor surface where the hard, brittle sand and stone found in plain concrete is overlaid with a coating of soft, ductile iron. It is reasoned that where hard, brittle aggregates crush under impact and abrasion, iron particles are malleable under these forces and will not fracture or crush out of the floor surface.

Metallic floor treatments consist of a dust coat application of pulverized iron, size graded to pass through mesh screens in the following proportions:

	Percent Passing
4 mesh screen	100%
8 mesh screen	Not less than 90%
14 mesh screen	Not less than 70 nor more than 85%
28 mesh screen	Not less than 35 nor more than 50%
48 mesh screen	Not more than 10%
100 mesh screen	Not more than 5%

Iron particles of the above graduation are mixed with cement and dusted onto freshly floated concrete and then floated and troweled to a smooth even finish. This optimum grading produces a more workable material while plastic and provides a dense surface having great resistance to wear upon hardening.

The iron material should be free from oil and should contain no nonferrous metals. Oil will prevent the cement paste from bonding properly to the iron particles and nonferrous metals might react with the cement to form gases and cause serious surface blistering. Manufacturers, architects, and engineers specify different methods and amounts of iron to be used, which is usually a certain number of pounds of iron mixed with one half that weight of cement. The material is dusted over the floor surface before the finish topping has taken its initial set. These quantities vary from 30 to 120 lbs. (13.6-54 kg) of iron per 100 s.f. (9 sq.m) of floor, depending on the traffic the floor will have to withstand. When the heavier dust coats are used, water-reducing and plasticizing agents are required to produce a dust coat that can be readily worked onto the surface without using an overly wet concrete mix. In wet mixes there is danger that the iron aggregate will sink into the concrete, where it is of no value as surface armoring.

It is not necessary to add anything to the labor cost of applying the regular cement finish top. It merely means that the finisher dusts the floor with the metallic powder while it is still too wet to trowel. By adding this dust coat, it is possible to trowel the floor sooner, because it absorbs a portion of the surface water.

Durocon Metallic Floor Hardener (Castle Chemical). This is a specially prepared metallic hardener for industrial and heavy duty concrete floors. It contains ocon, which provides concrete of increased strength and resistance to acid, alkali, etc.

For average traffic, 30 to 40 lbs. (13.6-18.14 kg) hardener and 30 lbs. (13.6 kg) portland cement; for heavy traffic, 50 lbs. (22.6 kg) hardener and 30 lbs. (13.6 kg) portland cement; and for extra heavy traffic, 60 lbs. (27.2 kg) hardener and 30 lbs. (13.6 kg) portland cement per 100 s.f. (9 sq.m) floor.

Use 40 lbs. (18.14 kg) colored Durocon with 20 lbs. (9 kg) portland cement per 100 s.f. (9 sq.m).

Standard colors, natural, tile red, linoleum brown, French gray, Persian red, battleship gray, Nile green, black, russet and maroon.

Duroplate (Castle Chemical). A specially prepared, size-graded iron particles combined with dispersing agent, it is applied as a dust coat while concrete is still in a plastic condition and machine floated into the surface. It produces a ductile, high impact- and abrasion-resistant, non-absorbent floor that resists oils and grease. It is easily cleaned and maintained. Floor life is extended up to ten times that of normal concrete floors when Duroplate aggregate is used. Use 60 to 120 lbs. (27-54 kg) per 100 s.f. (9 sq.m) of floor depending on degree of duty required.

Durundum Non-Slip Hardener (Castle Chemical). This aggregate floor hardener is composed of powerful coloring mediums and hardening elements plus cementitious binders. It is non-rusting and produces a non-slip surface and is entirely inert, very hard, tough, and resistant to abrasion. Apply as a dust coat, from 30 to 60 lbs. (13.6-27.2 kg) per 100 s.f. (9 sq.m), as conditions require.

Ferro-Fax Metallic Floor Hardener (A. C. Horn, Inc.). Ferro-Fax metallic floor hardener meets government specifications. Free from oil and foreign metals, it produces wearproof, waterproof, dustproof concrete floors. Apply as a dust coat, using 30 lbs. (13.6 kg) per 100 s.f. (9 sq.m) for light traffic; 40 lbs. (18.14 kg) per 100 s.f. (9 sq.m) for moderate traffic; heavy-duty floors, 70 to 100 lbs. (31.7-45 kg) per 100 s.f. (9 sq.m).

It comes in two types: Ferro-Fax Standard, to be mixed with portland cement, and Ferro-Fax Ready-Mixed. Immediately following the leveling, deposit upon the surface 2/3 of a uniform dry mixture consisting of 2 sacks Ferro-Fax Standard and 1 sack portland cement.

Allow the Ferro-Fax to absorb the surface water. Float sufficiently to work the Ferro-Fax into the surface. Discontinue floating as soon as the surface becomes wet. Immediately after floating the first shake, apply the final 1/3 uniformly over the surface. Float just until moisture is brought to the surface. Trowel to the desired finish.

Mastercron Pre-Mixed (Master Builders). A specially graded hard silica aggregate, combined with light-fast and alkali-fast coloring pigments, selected cement and a water-reducing agent which will produce a tough, wear-resistant floor surface.

Ready to use as it comes from the bag, the material is applied as a dust coat to freshly floated concrete floors at the rate of 50 lbs. (23 kg) per 100 s.f. (9 sq.m). It is available plain and in colors.

Masterplate Pre-Mixed (Master Builders). This is a specially prepared metallic aggregate combined with a water reducing and cement dispersing agent. It makes possible the incorporation of 60 to 120 lbs. (27-54 kg) of iron on the surface to provide maximum resistance to abrasion under the heaviest of industrial traffic.

Average duty floors, 60 lbs. (27 kg) Masterplate plus 30 lbs. (13.6 kg) cement per 100 s.f. (9 sq.m); heavy duty floors, 90 lbs. (41 kg) Masterplate plus 45 lbs. (20 kg) cement; extra heavy duty floors, 120 lbs. (54 kg) Masterplate plus 60 lbs. (27 kg) cement.

DPS Masterplate (Master Builders). A complete ready-to-use product containing specially processed iron aggregate, a conductive cement binder and a water-reducing and plasticizing agent, it is applied as a dust coat at the rate of 180 lbs. (82 kg) per 100 s.f. (9 sq.m) of floor surface. It meets federal and industrial requirements for conductive flooring. It retains its conductivity at temperatures from -20º to 150º F (-6.66 to 65.55ºC). and provides an extremely heavy duty floor that is spark-resistant and static-disseminating. it is dark gray in color.

Sparkproof Duroplate Pre-Mixed (Castle Chemical). This is static-disseminating and spark-resistant and meets all requirements of the Navy BuDocks Type Specifications, TS-F-15, Metallic Type (Superseding 48Y) for use in hazardous areas. Will not spark from friction with metal as do ordinary concrete aggregates. A special conductive cement binder is interground in this ready-to-use product. Cure with Conductive Kuraseal, 200 s.f. per gal (4.75 sq.m per liter). It is available in French grey, battleship grey, tile red, maroon, seal brown, black, Nile green, tan, and terra cotta.

Harcol R. M. (Sonneborn). A ready to use, rustproof, dry mix of abrasive resistant aggregates, specially selected for their flint-like hardness, portland cement and limeproof pigments of high and uniform tinctorial strength, it is applied as a dust coat at the rate of 30 to 45 lbs. (13.6-20 kg) per 100 s.f. (9 sq.m), floated into the surface, and troweled. Harcol produces an integrally colored surface highly resistant to abrasion, in bright red, dark red, gray, terra cotta, green, and natural.

Ferrolith H. (Sonneborn). A finely ground metallic water absorbent hardener, free from oils and impurities, it requires 60 to 180 lbs. (27-81 kg) per 100 s.f. (9 sq.m) for ordinary traffic, depending on planned use of floor.

Ferrocon Metallic (Castle Chemical). A finely powdered iron mixed with cement and water, it prevents absorption of moisture and seepage of water. It is for use on old and new structures. Three to 5 coats are recommended for pressure. Use 10 to 15 lbs. (4.5-7 kg) per 100 s.f. (9 sq.m) per coat.

Liquid Chemical Floor Treatments (Hardeners)

As concrete wears down, it is ground into dust, which has an especially detrimental effect in mills and factories where the dust is carried into bearings of machinery, motors, and engines, causing considerable damage and expense.

A number of liquid chemicals that counteract the problem are on the market. They create a chemical action in the cement and cause a complete transformation of the lime so that it solidifies the entire concrete aggregate into a hard, flint-like, homogenous mass that prevents dusting and wearing of the floors.

These liquid chemical compounds are sprayed over the floor until all of the pores in the surface are filled. It is usually necessary to treat the floors two or three times with the liquid to secure satisfactory results. When using these preparations, it is necessary to treat the floors at stated intervals, as the chemical action they create on the cement and lime is not permanent, although some manufacturers claim their product is permanent.

When using these liquid chemicals, care must be exercised to see that the floor is absolutely clean and free from all grease, dust, dirt, oil, or other foreign matter.

Different manufacturers specify varying numbers of applications of their materials but all of them must be applied just as often as the floor shows signs of dusting.

Liquid Chemical Floor Hardeners

Kind of Hardener	No. Coats	S.F. per Gal.	Gals. Reqd. 100 SF	Approx. Price per Gal.	Labor Hours
Granitex, Devoe Paint	2	200	1/2	$7.80	0.8
Hornolith, A. C. Horn	3	150	2/3	5.00	0.8
Lapidolith, Sonneborn	3	150	2/3	4.25	1.0
Flintox, Toch	2	125	3/4	3.75	0.8

(metric)

Kind of Hardener	No. Coats	Sq.M per liter	Liters Reqd. 10 sq.m	Approx. Price per liter	Labor Hours
Granitex, Devoe Paint	2	4.9	2.0	$2.06	0.8
Hornolith, A. C. Horn	3	3.7	2.7	1.32	0.8
Lapidolith, Sonneborn	3	3.7	2.7	1.12	1.0
Flintox, Toch	2	3.1	3.1	0.99	0.8

Labor Cleaning Floors. This cost depends entirely on the judgement of the estimator. Floors that are full of oil and grease will require much more time to clean than a floor free of foreign matter.

The labor applying the liquid is also a variable item, but a worker should spray 700 s.f. (63 sq.m) of floor per hr. on the first application and 800 s.f. (72 sq.m) per hr. on subsequent applications.

The labor applying the first coat should cost as follows per 100 s.f. (9 sq.m):

	Hours	Rate	Total	Rate	Total
Labor	0.15	$....	$....	$19.13	$2.87
Cost per s.f.			....		0.03
per sq.m			....		0.32

On the second and third applications, the labor per 100 s.f. (9 sq.m) should cost as follows:

	Hours	Rate	Total	Rate	Total
Labor	0.13	$....	$....	$21.04	$ 2.74
Cost per s.f.			....		0.03
per sq.m			....		0.32

Concrete Accelerators and Densifiers

There are numerous preparations on the market for controlling the set of concrete, increasing its early strength, densifying and waterproofing the concrete mass, and for preventing freezing during winter weather by lowering the freezing point of the mixing water.

Anti-Hydro Accelerator and Anti-Freeze. (Anti-Hydro Co., Newark, N.J.) This can be mixed with the water or added to the wet mix. As a general rule, the standard proportion of 1.5 gals. of Anti-Hydro per c.y. (7.43 liters per cu.m) of concrete, with reduced water content to compensate for increased slump, gives protection down to 23° F (-5°C) on concrete deposited in forms except in the most exposed locations.

For cold weather work, be sure that sand and aggregate are free of ice.

To protect concrete or mortar against freezing, use the following table for anticipated outside air temperatures as follows:

1. 32 to 25° F (0 to -3.88°C) use 1 part Anti-Hydro to 12 parts water.
2. 25 to 15° F (-3.88 to -9.44°C) use 1 part Anti-Hydro to 10 parts water.
3. 15°F or less (-9.44°C or less) mechanical heat is necessary.

POZZOLITH-HE (Master Builders). This is an accelerator and water-reducing agent recommended for use in concrete where accelerated set, increased early and ultimate strength and improved performance are required or desired. Concrete made with POZZOLITH-HE provides 3-day normal strength in 1 day, 7-day normal strength in 3 days, 28-day normal strength in 7 days and substantial increases in ultimate strength.

Also, non-chloride high early formulations are available for specialized concreting operations where calcium chloride is not permitted.

POZZOLITH-HE admixture is used at the dosage rate of 16-64 fl. oz. per 100 lbs. (1040 ml to 4160 ml per 100 kg) of cement depending on formulation being used and the amount of set and strength acceleration needed or desired.

Plastiment (Sika). Plastiment is a powder added in the proportion of 1/2 to 1 lb. per sack of cement. It will increase workability, density, and surface hardness and decrease shrinkage. It may be used for concrete or mortar and is also available in a concentrated liquid form for use with automatic dispensing equipment. Use 2 to 4 fl. oz. (59-118 ml) per sack of cement.

Dehydratine No. 80 (W. R. Grace & Co.). Liquid integral cement floor hardener, accelerator, and anti-freeze compound, it is used in the proportions of 1 quart per sack of cement.

Sikacrete (Sika). Accelerating and plasticizing liquid for concrete or mortar, used as anti-freeze or rapid hardening compound, it is used in the dilution of 1 part Sikacrete to 3 to 10 parts water. A dilution of 1 to 7 will reduce the hardening rate of mortar or concrete to one-half or protect against freezing to a temperature of 22° F (5.55°C).

Trimix (W. R. Grace & Co.). A chemical solution of uniform strength, which when used according to directions functions as an accelerator, integral hardener, or anti-freeze compound, it is used in mortar or concrete work. As an anti-freeze, use it as follows:

1. 25 to 32° F (-3.88 to 0°C), 1-1/2 qts. (1.42 liters). Trimix to each sack of cement.
2. 20 to 24° F (-6.67 to -4.44°C), 2 qts. (1.89 liters). Trimix to each sack of cement.
3. 15 to 19° F (-9.44 to -7.22°C), 2-1/2 qts. (2.37 liters). Trimix to each sack of cement.
4. Below 15° F (Below -9.44°C), heat is necessary.

BONDING OLD AND NEW CONCRETE

The problem of bonding new concrete to old, so the new concrete will remain permanently in place, is one that has received considerable attention among users of concrete. Unless the utmost care is exercised in preparing the old surface, the new concrete will invariably come loose where it is joined to the old concrete.

Different methods have been used to overcome this fault, but the basic condition of all of them is that the old surface must be thoroughly cleaned and washed, and the old aggregate must be exposed, which means that the thin film of cement that covers the surface of the concrete must be removed. One of the methods of preparing the old surface consists of hacking or picking it, and then washing or turning a steam hose under pressure to remove all dust and dirt from the old floor for a perfectly clean surface.

Another method is to wash the concrete surface with a solution of muriatic acid and water to remove the old film of cement and then thoroughly wash the surface with clear water. Then, the old concrete surface should be given a slush coat of neat portland cement and water, and the new concrete applied.

Hacking and Chipping Old Concrete. The cost of hacking and chipping old concrete surfaces, in preparation for bonding new concrete to old, is a variable item. It depends on the hardness of the old concrete and the condition of the surface. Under average conditions a laborer using a pick should hack and roughen 175 to 200 s.f. (16-18 sq.m) of "green" concrete floor per 8-hr. day, at the following cost per 100 s.f. (9 sq.m):

	Hours	Rate	Total	Rate	Total
Labor	4	$....	$....	$21.04	$ 84.16
Cost per s.f.			....		0.84
per sq.m			....		9.04

Where the concrete floors are of old concrete, a worker using hand tools will do well to hack and roughen 8 to 10 s.f. (0.7-0.9 sq.m) per hour, and even with a compressor and jackhammer or bush-hammering tool, it is often possible to hack and roughen only 20 to 25 s.f. (1.8-2.2 sq.m) per hr. and the cost per 100 s.f. (9 sq.m) should average as follows:

	Hours	Rate	Total	Rate	Total
Compressor expense	2.25	$....	$....	$7.50	$16.88
Labor on jackhammer	4.50		....	21.04	94.68
Cost per 100 s.f. (9 sq.m)			$....		$ 111.56
per s.f.			....		1.12
per sq.m			....		12.05

Anti-Hydro Bonding Coat. Rough and clean the old concrete as previously described. Apply a coat of grout composed of 1/2 to 3/4 sack of portland cement added gradually into a solution of 1 gal. of Anti-Hydro in 3 gals. of water (3.79 liters per 11.37 liters of water) until a thick, creamy consistency is obtained. After applying this grout to the prepared surface, the concrete or mortar should be placed while the grout is still wet. Sufficient Anti-Hydro grout to cover 100 s.f. (9 sq.m) should cost as follows:

	Rate	Total	Rate	Total
0.25 gal. (0.95 l.) Anti-Hydro	$....	$....	$5.00	$1.25
0.25 sack (0.98 l.) portland cement		....	4.25	1.06
Cost per 100 s.f. (9 sq.m)	$....		$....	$2.31
per s.f.		....		0.023
per sq.m		....		0.26

Daraweld-C (W. R. Grace & Co.) Daraweld-C is an emulsion of special internally plasticized high polymer resins uniformly dispersed in water. Mixed with cement mortars or grouts, it forms a durable, highly water-resistant bond. A Daraweld-C bond withstands water immersion without softening or disintegrating, resists the action of heat, cold, oil and gasoline, most acids and other corrosive materials. Its coupling action is so great that grouts containing Daraweld-C will bond not only to concrete and masonry, but to tile, wood, steel, even glass. Ready-to-use, non-settling Daraweld-C will not re-emulsify when subjected to moisture. It is completely compatible with mixes containing calcium chloride.

Daraweld-C is used as a bonding agent admixture in grouts and mortars made from portland or Lumnite cement and assures their lasting adhesion to existing concrete. Formulated for exterior or interior use, it is especially useful for patching concrete floors, repairing cracks in walls, securing an integral bond between successive concrete pours, for waterproofing exteriors below and above grade, in Gunite applications, in bonding concrete to asphalt surfaces. Diluted with water, Daraweld-C is an excellent brush-on sealer for dustproofing concrete floors, and as a primer for painting concrete masonry.

Tremco Floor Bond. A chemically treated metallic powder to use in bonding new concrete toppings to set slabs, where a perfect bond must be secured to prevent cracking.

The old surface must be roughened and cleaned as described previously and the floor bond applied in 2 coats. It requires 25 lbs. (11.33 kg) of Metallic Floor Bond per 100 s.f. (9 sq.m) of surface.

When used for bonding vertical surfaces, 2 applications are required, using 15 lbs. (7 kg) of Floor Bond per 100 s.f. (9 sq.m).

Weld-Crete. (from Larsen Products Corp., Rockville, MD) This product permanently bonds new concrete to old concrete. It also bonds new concrete to many other materials, such as brick, stucco, stone, etc.

The old concrete surface must be structurally sound and free from dust, dirt, loose material, grease, oil, wax, water soluble coatings, etc.

It is not necessary to chip, bush hammer or roughen the old surface in any way.

Weld-Crete may be used on either interior or exterior surfaces and may be applied over "green" concrete and damp or dry surfaces. May be applied with a brush, roller or spray. Spraying is most efficient and should be done with heavy industrial spray equipment.

Coverage, when sprayed, from 200 to 300 s.f. per gal. (4.75-7 sq.m per liter). When brushed or rolled, coverage is somewhat less.

Repairing Leaks Against Hydrostatic Pressure

When repairing leaks in basement walls or floors, tunnels, pits, etc., where water pressure exists, the following methods and materials may be used.

Sika 2. Sika 2, red, fast-setting, sealing liquid is mixed with standard portland cement in small quantities. It is used to plug small infiltrations of water against high pressure. Sika 2 mortar will have an initial set of 15 seconds and a final set of 30 seconds. It is used after surrounding leaking concrete has been sealed with Sika 4A. Quantity required varies with degree of leakage.

Sika 4A. Sika 4A is a clear liquid mixed with neat portland cement in small quantities. The resulting mortar will have an initial set of less than 1 minute and a final set of approximately 5 minutes. Sika 4A may be diluted with water and the cement mixed with sand. This mortar is to be applied against damp surfaces or masonry over which water is running. If pressure is too heavy, bleeder pipes may be installed and after smaller leaks are sealed with Sika 4A, the bleeder holes may be plugged with Sika 2, liquid. Quantity required depends upon the degree of leakage.

CONCRETE ADMIXTURES

Since the introduction of the water-cement ratio law governing the strength and durability of concrete, the necessity for using comparatively low water ratio mixes has been generally recognized. Of greatest importance, however, has been the necessity of producing such mixes with a degree of workability that would insure their easy and economical placing and compacting.

In a cubic yard of concrete, approximately 2-1/2 gallons (9.46 liters) of water per sack of cement are required to hydrate the cement. Any water in excess of that amount must be regarded as "placing" water, merely providing sufficient workability to make the mix placeable. This excess water occupies about 10% of the total space in concrete. As it evaporates, it causes the concrete to shrink. Excess water also reduces the strength of the concrete.

The purpose of admixtures is to improve the plasticity, workability, and finish of the concrete, to prevent segregation of the aggregates and to a certain extent provide integral waterproofing qualities by filling the small voids in the concrete mixture.

Pozzolith (Master Builders). This is a water reducing, set-controlling admixture available in several formulations to facilitate optimum performance benefits with the range of cements, sands, and coarse aggregates used for making concrete.

In the plastic concrete, Pozzolith improves finishing characteristics for flat work and cast surfaces, reduces segregation of the mix, and enhances placement of low slump concrete. In the hardened concrete Pozzolith results in reduced cracking, increased compressive and flexural strengths, improved watertightness, and increased resistance of air entrained concrete to damage from freezing and thawing as well as scaling from deicing salts.

Pozzolith Normal and Retarding formulations are generally used at the rate of 5 ±2 fluid ounce per 100 pound (325 ±130 ml per 100 kg) of cement.

Plastiment (Sika). Plastiment, concrete densifier and retarding agent, is a liquid to be added to concrete or mortar at the rate of 2 to 4 oz. (59-118 ml) per sack of portland cement. For equal workability the water-cement ratio may be reduced approximately 10%. It results in better workability, density, adhesion to old concrete, surface hardness, and flexural and compressive strength. The delayed set results in elimination of cold joints and reduced shrinkage.

Plastocrete (Sika). A water-reducing admixture for ready-mix and performance concrete to produce higher strengths and workability, it improves structural quality of all concrete and can be used throughout the year.

Trimix (Sonneborn). This is a multi-purpose concrete and mortar admixture that reduces water-cement ratio, produces higher early compression strengths, accelerates set, and generally improves quality and workability of portland cement mixtures.

Used in the proportion of 1 qt. (0.95 l) Trimix to each sack of cement in the mix.

Hydrated Lime as a Concrete Admixture. Hydrated lime is often used as an admixture to increase the plasticity and improve the workability of concrete. For this purpose 5 to 8 lbs. (2-3.5 kg) of hydrated lime are usually added to each bag of portland cement used in the concrete mixture.

Air Entraining Admixtures

In air entrained concrete, 3% to 6% of air is incorporated in the concrete in the form of minute separated air bubbles. Such concrete is more resistant to freezing and thawing and to salt action than normal concrete. Originally developed to prevent scaling of pavement where salts are used for ice removal, air entrained concrete is being widely used for all types of work and in all locations because of its better workability as well as its better resistance to weathering even where salts are not used.

MB-VR (Master Builders). MB-VR is a neutralized Vinsol resin type air entraining agent for concrete. It is furnished in water solution form and is ready to use as it comes from the drum. MB-VR is compatible for use with other admixtures commonly used in concrete; however, if more than one admixture is used, each should be dispensed into the mix separately. MB-VR is available via bulk tank delivery or 55-gallon (208-liter) steel drums.

Sika Aer. This air entraining resin solution of neutralized wood resin improves various properties of plastic and hardened concrete.

Add between 1/2 and 2 fl. oz. (15-59 ml) of Sika Aer solution to entrain between 3% and 6% air as air content is in direct proportion to quantity of Sika Aer added.

Darex AEA (W. R. Grace & Co.) Darex AEA is an aqueous solution of highly purified and modified salts of a sulfonated hydrocarbon. It contains a catalyst which promotes more rapid and complete hydration of portland cement. Darex AEA is specifically formulated for use as an air entraining admixture for concrete and is manufactured under rigid control which insures uniform, predictable performance. The addition of 3/4 fluid ounce of Darex AEA per sack of cement will generally entrain 4% to 7% air in a 5-1/2 sack concrete mix.

CONCRETE CURING PROCESS

Efficient methods of curing concrete are of vital importance, because surfaces such as concrete floors, roofs, road, pavements, airport runways, and other surfaces exposed to the sun require protection during the curing period. This used to be accomplished with damp burlap, sand, sawdust, dirt covering, paper, or ponding.

A method that is being used to a large extent involves curing compounds, which provide an air-tight seal over the surface of the concrete and prevent the evaporation of the water from the concrete until it has cured normally. These compounds are usually applied with a brush or spray as soon as the concrete has set sufficiently.

Servicised/Horn Concrete Curing Compounds. (A. C. Horn, Inc.) These are scientifically formulated liquids for spray application to fresh concrete surfaces. The compound quickly forms a vaportight film that seals in 95% or more of the moisture in the concrete for 3 or more days, without changing the concrete color. They can be applied with standard spray equipment. One application should control cure on any size surface area, from mass concrete to a few square feet.

Kure-N-Seal (Sonneborn). Kure-N-Seal cures, seals, and dustproofs in one application, and it increases surface hardness when applied as a curing membrane after final troweling. By locking in essential curing moisture, the hydration process is refined and maximum hardness for the involved mix is assured.

Klearseal (Castle Chemical). This is a clear curing compound for use on surfaces where moderate traffic or corrosive conditions are anticipated, and where a completely clear surface is desired. One gallon covers 300 to 450 s.f. (one liter covers 7-10 sq.m) for smooth trowelled concrete; 250 to 350 s.f. (6-8 sq.m/liter) for broom finished concrete, and 200 to 300 s.f. (4.75-7 sq.m/liter) for rough finished concrete.

03366 POST-TENSIONING CONCRETE

Post-tensioning is the tensioning of concrete after it is placed. It has become increasingly popular in recent years to reduce shrinkage cracks and to enhance the strength, serviceability, and design flexibility of concrete structures. It can provide high quality, watertight floor systems for applications such as parking garage floors, and using corrosion protection systems that are available for post-tensioning tendons protect in corrosive environments.

Many concrete contractors think of post-tensioning as a type of specialty work, but actually, it is just an additional step in producing high-quality reinforced concrete. Of course, the concrete contractor will want the assistance of a qualified post-tensioning engineer and a reliable material supplier to produce a quality job at an economical price.

There are two basic types of post-tensioning: *multi-strand* and *mono-strand.* In the multi-strand type several strands of prestressed wire are placed in a duct, and after the strands are tensioned, the duct is grouted. This method is sometimes referred to as a *grouted* or *bonded* system.

In the mono-strand type, sometimes referred to as *unbonded,* individual strands are made up of seven prestressed wires. Strands are coated with a lubricant grease to prevent rusting and reduce friction during the stressing operation. The greased strand is covered with a plastic coat.

Mono-strand post-tensioning. In the estimate the contractor must allow for certain items that are unique to post-tensioning. Stressing access must be provided, usually a platform or floor for extension at the perimeter of the structure.

Forming requirements must be coordinated with the stressing and anchorage of post-tensioning tendons. Horizontal supporting formwork cannot be removed until the stressing operation is completed in a given pour area.

Post-tensioning anchorages are typically designed to permit the stressing of tendons when the concrete reaches a strength of 3,000 psi (2.1 kg/mm^2). Check with the post-tensioning material supplier for actual concrete strength requirements for the systems.

The contractor must also make provisions for cutting off the stressing tails of the tendons and patching the stressing pockets.

The post-tensioning material supplier will provide the contractor with the placement drawings for the system and will coordinate the pour sequence with post-tensioning anchorage requirements.

Material for a mono-strand system is estimated by the pound. Contract drawings usually specify effective force requirements. The effective force translates into the number of single strands required by dividing the required force by the effective force of each tendon, about 26.5 kips per tendon for 1/2" (13 mm) strand.

TYPICAL DEAD END POST TENSION ANCHORAGE

TYPICAL POST TENSION STRESSING ANCHORAGE

The actual effective force provided by each tendon is calculated for each project and must consider seat loss, friction loss, elastic shortening of concrete, and time-dependent losses. The total length of a tendon is obtained by adding stressing tails to the length of each individual tendon, 1'-6" (450 mm) at each stressing end. To obtain nominal strand weight, multiply the length by 0.525 lbs. per ft. (0.79 kg/m) of 1/2" (13 mm) diameter strand.

Material Costs. Post-tensioning material for mono strand averages about $0.80 per lb. The following are the standard basic steps for placement of post-tensioning tendons:

1. Form concrete slab as usual, but allow about 4'-0" (1.2 m) for post-tensioning set up (jacking) adjacent to the concrete slab side forms.
2. Set side forms, drill holes for post tension wires (tendons), and set grommets.
3. Set reinforcement steel chairs about 3'-0" (0.9 m) on center.
4. Set bottom reinforcement steel.
5. Set post-tension wire in one direction.
6. Set post-tension wire in the other direction.
7. Set top reinforcement steel.
8. Pour concrete.
9. Remove side forms, tension prestress wires (tendons) by jacking.
10. Remove grommet and finish concrete edge.

Labor Costs. The production average for post-tensioning is 40 to 50 s.f. (3.6-4.5 sq.m) per manhour at the following cost per 100 s.f. (9 sq.m):

	Hours	Rate	Total	Rate	Total
Ironworker	2.22	$....	$....	$30.28	$ 67.22
Cost per s.f.				....	0.67
per sq.m				....	7.21

03400 PRECAST CONCRETE

FLEXICORE FLOOR AND ROOF SLABS

Prestressed Flexicore (trademark of The Flexicore Co., Inc.) concrete floor and roof slabs, designed in accordance with the ACI 318 Building Code, are precast in central plants with patented equipment. The cross sections and lengths of these units are indicated below. Filler width slabs are also available.

	Span, feet	Lbs. per s.f. (Based on 150 lbs. Concrete)
6"x24"	15-25	43
8"x24"	20-33	57
10"x20"	28-40	61
10"x24"	28-40	72
12"x24"	35-50	79

(metric)

	Span, meters	Kg per sq.m (Based on 68 kg Concrete)
150 x 600 mm	4.6-7.6	210
200 x 600 mm	6.1-10.1	278
250 x 500 mm	8.5-12.2	298
250 x 600 mm	8.5-12.2	352
300 x 600 mm	10.6-15.2	386

The amount of deflection allowed shall be consistent with the finish to be applied to the underside of the flexicore units, and shall be within the limits specified by ACI 318 Building Code.

These units are manufactured in a rigid steel form made sufficiently strong to resist the pretensioning force applied by the seven wire strand reinforcing steel. Specially constructed rubber tubes are inflated to form circular voids so as to constitute approximately 50% of the cross section. The concrete is thoroughly vibrated to assure maximum density and strength.

The slabs are cast of concrete made of gravel, crushed stone, or lightweight aggregate. When lightweight aggregate is used, the unit weight is approximately 25% less. They are cured in a heated kiln.

Not all Flexicore plants are equipped to make all the sections listed in the above table. Availability of sections and lengths must be checked locally.

Erection. Flexicore units are usually delivered to the job by truck. They are hoisted from the truck to the floor location and often placed directly into their final position by the crane. After the slabs are placed side by side, they are aligned and levelled. The keyways in the sides of the slab are then filled with a grout mixed in a ratio of one to three. The erection crew usually consists of six workers, including a crane operator, an oiler, a foreman, and three laborers. A crew will unload, erect, and grout 2800 s.f. (252 sq.m) per day on smaller jobs to approximately 6000 s.f. (540 sq.m) on some larger jobs.

Cost of Flexicore Floor Slabs. Flexicore slabs are usually quoted on an erected basis by the manufacturer and the price will usually include the cost of grouting and caulking the joints on the ceiling side. The price will vary according to the type of building, size, live load, span, section specified, and location, but usually runs from $4.20 to $5.50 per s.f. ($46.66-61.11 per sq.m) erected, grouted, and caulked.

In localities distant from the manufacturing plant, the slabs are often sold on a "delivered only" basis and erection is performed by the contractor's own personnel.

Caulking the joints underneath is undertaken after the building has been enclosed. A nonstaining type caulking material is used and will average about $0.20 per l.f. (0.67 per m). An underlayment which is not included in the bid is generally used over the Flexicore slabs and is provided by the general contractor.

CONCRETE PLANK

Cantilite (from Concrete Plank Co, Inc., North Arlington, NJ) is a lightweight high-strength concrete plank that is nailable and easily cut in the field. In laying, it is clipped to steel or wood beams. Floor or roof materials are nailed to it. Planks are factory-made in steel forms to give a smooth, even surface on all sides and to assure proper positioning of the reinforcing steel. Side edges are tongued and grooved.

Concrete plank are used for floors and roofs. The plank are 2", 2-3/4" and 3-3/4" (50, 68, 94 mm) thick, 16" (400 mm) wide and made in lengths up to 10'-0" (3.0 m). The 2" (50 mm) plank are used on steel or concrete joists, 4'-0" (1.2 m) o.c. for floors. The 2-3/4" (68 mm) plank are placed on steel or concrete joists up to 5'-0" (1.5 m) o.c. for floors. The 3-3/4" (94 mm) plank are placed on steel or concrete joists up to 6'-0" (1.8 m) o.c. for floors. A 1" (25 mm) or 2" (50 mm) cement finish top is placed over the plank. Roofs having a span up to 7'-0" (2.1 m) will require 2" (50 mm) plank, while 2-3/4" (69 mm) plank will be required for a roof span up to 8'-0" (2.4 m). The 3-3/4" (94 mm) plank will be required for a roof span up to 10'-0" (3.0 m).

Concrete plank costs $2.50 to $3.50 per s.f. (27.70-38.80 per sq.m), f.o.b. factory, depending on quantity in the job. The 2" (50 mm) plank weigh 13 lbs. per s.f. (65 kg/m^2); the 2-3/4" (69 mm) plank, 18 lbs. per s.f. (90 kg/m^2); the 3-3/4" (94 mm) plank, 28 lbs. per s.f. (141 kg/m^2).

Labor Erecting Concrete Plank. A crew consisting of 4 masons and 1 laborer can handle and lay approximately 2,000 s.f. (180 sq.m) of plank per 8-hr. day, on a flat surface.

Labor Cost of 100 S.F. (9 Sq.M) of Concrete Plank Laid on a Flat
Floor or Roof Surface

	Hours	Rate	Total	Rate	Total
Masons	1.6	$....	$....	$28.18	$ 45.09
Labor	0.4	....	....	21.04	8.42
Hoisting engineer	0.4	....	....	28.55	11.42
Cost per 100 s.f. (9 sq.m)			$....		$ 64.93
per s.f.				....	0.65
per sq.m				....	6.99

PRECAST CONCRETE ROOF SLABS

Precast concrete roof slabs are of three general types: rib, flat, and channel. They are adapted to all roof decks, flat or sloping, are fireproof, and with caulked joints, present a smooth surface for the application of built-up roofing. Where slate, ornamental tile, or copper covering is specified, slabs with nailing surfaces can be furnished, permitting direct application.

Precast concrete roof slabs are made with regular concrete aggregates, or where a lightweight slab is desired, they are made of lightweight aggregate.

The rib tile is self-weathering. Its attractive red color provides a desirable architectural feature that is often lacking in industrial buildings. Auxiliary pieces are furnished to suit the particular design required, such as ridges, saw-tooth ridges, gable end finishing tile, monitor flashing tile, and other specials required in connection with hip or valley.

TYPE	RIB TILE	FLAT SLAB	2¼" CHANNEL TILE	3½" CHANNEL TILE	
WIDTH	24"] STOCK SIZE	24"] STOCK SIZE	24"	24"	16"
LENGTH	4'-6"	5'-0"	4'-0" MAX.	6'-0" MAX.	5'-0" MAX
THICKNESS	WEB - 1½"	1⅝"	LEG - 2¼" - WEB-1"	LEG - 3½" - WEB-1"	
WEIGHT PER SQ FT	HAYDITE 16 LBS.	13.5 LBS.	10.5 LBS.	12 LBS.	13.5 LBS.
	SAND 17 LBS.	19 LBS.	16 LBS.	17.5 LBS.	19 LBS.
REINFORCEMENT	WELDED GALV. MESH	WELDED GALV. MESH	MESH & 2 DEF. BARS	MESH & 2 DEF BARS	
ROOF PITCH	SELF WEATHERING	FLAT OR PITCHED	FLAT OR PITCHED	FLAT OR PITCHED	
ROOFING	¼ PITCH OR GREATER	COMP. ROOFING REQ'D	COMP. ROOFING REQ'D	COMP. ROOFING REQ'D	
PURLIN SPACING	4'-0" MAX. - 3'-9"MIN.	5'-0" MAX.	6'-8" MAX.	8'-0" MAX.	9'-0" MAX.

Sizes and Weights of Precast Concrete Roof Tile

Tile of the same standard stock sizes but having a glass insert are also furnished where additional light is required. On this type of roof slab, it is necessary to obtain prices from the manufacturer, because the amount of trim varies with each project.

The other two types of precast concrete roof slabs, flat and channel, are used over all roof decks, whether flat or sloping, and present a smooth surface for the application of any type of built-up roofing. When used with a sloping roof and when slate or other types of ornamental covering are to be used, these slabs can have a concrete nailing surface superimposed and manufactured integrally with the structural slab.

The accompanying illustration gives the sizes and weights of the various types of tile.

Details of Precast Concrete Roof Tile Construction

Labor Placing Precast Concrete Roof or Floor Slabs. Because of their size and weight, it requires 2 workers to handle precast concrete roof or floor slabs. Each piece weighs from 125 to 200 lbs. (57-91 kg). Ordinarily it requires 2 workers on the ground handling the slabs and getting them to the hoist, 2 more on the floor or roof placing them for the mason, and a mason and helper to lay and caulk the joints.

A mason and helper should lay and caulk 960 to 1,050 s.f. (86-95 sq.m) of lightweight concrete tile per 8-hr. day on straight roofs.

PRECAST CONCRETE JOIST FLOORS

Precast concrete joists are widely used to support concrete floors in homes, apartment buildings, office buildings, schools, hospitals, and similar structures. Precast concrete joists are also extensively used for supporting concrete roof slabs in all types of buildings.

The design tables are only for floors and roofs that carry relatively light loads and are not intended for designing floors or roofs carrying heavy or concentrated loads or subject to heavy impact loads or to vibration of mounted machinery. These conditions impose special loading problems, and to meet them floors must be designed accordingly.

Concrete joists are manufactured in a cement products plant and delivered to the job ready to set in place. The joists for each floor of a building should be laid out separately, showing size of joists, spacing, etc., which varies from 20 to 33 inches (500-825 mm), depending on depth of joists, span, floor load, etc.

To develop maximum strength, a reinforced concrete slab 2" to 2-1/2" (50-62 mm) thick is placed over the tops of the joists with the joists embedded into the concrete slab to a depth of 1/2" to 3/4" (13-19 mm). Effective bond shall be obtained between the joists and slab. Concrete for both joists and slab shall have an average compressive strength of not less than 3,750 lbs. per sq. in. (2.63 kg/mm²) at age of 28 days. Floors made of precast joists mortised or embedded into a monolithic floor placed or poured on the job are called *precast joist cast-in-place concrete slab* floors.

Floors made of precast joist over which are laid precast slabs for flooring are called *precast joist and slab* concrete floors.

Concrete joists are manufactured 6, 8, 10, 12, and 14 inches (150, 200, 250, 300, & 350 mm) deep. Various strength requirements of each size may be obtained by increasing the amount of reinforcing steel. For instance, 8-inch (200-mm) concrete joists are manufactured containing from 0.22 to 0.88 sq. in. (142-567 sq.mm) of reinforcing bars at the bottom of the joists and 3/8" (9 mm) round reinforcing bars at the top, with 1/4" (6 mm) round steel stirrups spaced according to the load requirements of the particular job. The strength of 10, 12, and 14-inch (250, 300, and 350 mm) joists is determined in the same manner.

Notes for Table Nos. 1, 2 and 3 on following pages. Superimposed loads shown in tables are the loads that the floor or roof is designed to support in addition to the weights shown for the precast joists and concrete slab.

Load values are limited by maximum moment or shear and will produce deflections less than L/360, except in Table No. 3 where load values shown with an asterisk are maximum loads for a deflection of L/360. In Table No. 3 load values in bold face type are limited only by maximum moment and shear and will produce deflections greater than L/360.

For Table Nos. 1 and 2, one row of bridging should be used at midspan in residential floor construction having spans over 20 ft. and in all other floor construction having spans over 16 ft. (4.8 m). Bridging is not usually required for roof construction.

For Table No. 3, one row of bridging should be used at midspan for all floor construction having spans over 16 ft. (4.8 m). Bridging is usually not required for roof construction.

For floor or roof construction shown in Table Nos. 1 and 3, no shoring is required during construction.

For floor and roof construction shown in Table No. 2, shoring at midspan is required during construction and must remain in place until the concrete has reached its required strength.

Bar sizes are given in numbers based on the number of eighths of an inch included in the nominal diameter of the bars as specified in ASTM A305.

Concrete Joists Designed as Independent Beams. Concrete joists designed to be used with precast concrete slabs, where the slab does not form a part of the beam, use more reinforcing steel in the beams, increasing the cost considerably.

For example, 8" (200 mm) beams with 3/8" (9 mm) top bars and 5/8" (16 mm) bottom bars are used for spans 10 to 15 ft. (3-4.5 m); 1/2" (13 mm) top bars and 3/4" (19 mm) bottom bars for spans 11 to 16 ft. (3.3-4.8 m); and joists using 3/4" (19 mm) top bars and 7/8" (22 mm) bars for spans 12 to 16 ft. (3.6-4.8 m) requiring heavier loads.

Ten-inch joists with 3/8" (9 mm) top bars and 3/4" (19 mm) bottom bars are used for spans 12 to 19 ft. (3.6-5.7 m); 5/8" (16 mm) top bars and 7/8" (22 mm) bottom bars for spans 14 to 20 ft. (4.2-6); and 3/4" (19 mm) top bars and 1" (25 mm) bottom bars for spans 16 to 20 ft. (4.8-6 m) required to carry heavier loads.

Twelve inch joists with 3/8" (9 mm) top bars and 3/4" (19 mm) bottom bars are used for spans 12 to 22 ft. (3.6-6.6 m); 3/8" (9 mm) top bars and 7/8" (22 mm) bottom bars for spans 14 to 24 ft. (4.2-7.2 m); and 1/2" (13 mm) top bars and 1" (25 mm) bottom bars for spans 16 to 24 ft. (4.8-7.2 m) required to carry heavier loads.

These beams require from 1 to 2 lbs. (0.45-0.90 kg) of reinforcing steel per l.f. more than the T-beams and cost $0.30 to 0.40 per l.f. ($1.00-1.33 per m) more than the T-beam joists.

Precast concrete joists are manufactured of Haydite, Waylite, Celocrete, and other lightweight aggregate, also of ordinary aggregate consisting of gravel or crushed stone and sand.

Weight of Precast Concrete Joists, Ordinary Aggregate

Size of Joist In.	10	12	14	16	18	20
8	190	228	266	304		
10	230	276	322	368	414	460
12	350	420	490	560	630	700

Cost of Precast Concrete Joists. The cost of precast concrete joists will vary with the size of joists, amount of reinforcing steel required, size of job, locality, etc. Precast concrete joists are manufactured in a concrete products plant. It is important to obtain definite prices for each particular job.

Approximate Prices per L.F. (M) for Precast Concrete Joists

8" (200 mm)	10" (250 mm)	12" (300 mm)	14" (350 mm)
$4.75 (15.83)	$5.20 (17.33)	$7.00 (23.33)	$8.00 (26.67)

Where the concrete joists are designed as independent beams for use with precast concrete floor or roof slabs, add from $0.25 to 0.30 per l.f. (0.83-1.00 per m) to the above prices to allow for the extra reinforcing steel required.

						Table 1 • Joists and Slab Designed as T-Beams

No shoring used during construction

Superimposed loads = psf/Clear span = ft.

Joist size in.	Top	Bot	Wt. of joist and slab psf	Slab thickness in.	Joist spacing c. to c. in.	10	11	12	13	14	15	16	17	18	19	20	21	22	23	24
8	#3	#5	34	2	20		113	88	69	53	40	30								
			32	2	24	118	90	69	53	40	29									
			32	2	27	101	76	57	42	31										
			37	2½	30	85	61	43	29											
			36	2½	33	74	52	36	23											
	#3	#6	34	2	20				113	91	73	59								
			32	2	24			113	90	72	57	45								
			32	2	27		122	96	72	59	46	36								
			37	2½	30	139	106	81	61	46	33	22								
			36	2½	33	123	93	70	53	38	26									
	#3	#7	34	2	20					137	113	94								
			32	2	24				134	110	90	74								
			32	2	27				115	93	76	61								
			37	2½	30			125	99	78	61	48								
			36	2½	33		128	111	87	68	53	40								
10	#3	#6	37	2	20					124	103	85	70	57	47	38				
			35	2	24				122	99	81	66	53	43	34	27				
			34	2	27				131	105	85	69	55	44	35	27				
			39	2½	30				113	89	70	54	41	31	22					
			38	2½	33			128	100	78	60	46	34	25						
	#3	#7	37	2	20						129	109	92	78	66					
			35	2	24						124	103	87	72	61	50				
			34	2	27					129	107	89	74	61	51	42				
			39	2½	30				137	111	90	73	59	47	37	28				
			38	2½	33				121	98	79	63	50	39	30	22				
	#3	#8	37	2	20							134	115	99						
			35	2	24							125	107	91	78					
			34	2	27							127	108	92	78	66				
			39	2½	30					138	125	110	91	76	63	52				
			38	2½	33					122	111	97	80	66	54	44				
12	#3	#6	44	2	20						126	104	86	71	59	48	39	31		
			41	2	24					123	100	83	68	55	44	35	27			
			39	2	27					131	106	87	70	57	46	36	28			
			43	2½	30					113	90	72	57	44	33	24				
			42	2½	33				126	99	79	62	48	36	27					
	#3	#7	44	2	20								135	115	97	83	70	59	50	42
			41	2	24							128	108	91	77	64	54	45	37	30
			39	2	27							133	111	93	78	65	54	45	37	30
			43	2½	30							115	95	78	63	51	41	32	25	
			42	2½	33						124	101	82	67	54	43	34	26		
	#3	#8	44	2	20									123	107	93	81	69		
			41	2	24								133	114	98	85	73	62	53	
			39	2	27								135	115	99	85	72	62	52	44
			43	2½	30								117	98	83	69	58	48	39	32
			42	2½	33							123	103	86	72	60	49	40	32	25
	#3	#9	44	2	20													131	115	102
			41	2	24											137	120	105	92	80
			39	2	27										137	119	103	90	78	68
			43	2½	30									138	118	101	87	74	63	54
			42	2½	33									122	104	89	75	64	54	45

Table 2 • Joists and Slab Designed as T-Beams

Shoring used at midspan during construction

Superimposed loads = psf/Clear span = ft.

Joist size (in.)	Top	Bot	Wt. of joist and slab (psf)	Slab thickness (in.)	Joist spacing c. to c. (in.)	10	11	12	13	14	15	16	17	18	19	20	21	22	23	24
8	#3	#5	34	2	20		124	99	79	64	51	41								
			32	2	24	128	100	80	63	50	39	30								
			32	2	27	111	86	67	52	41	31									
			37	2½	30	100	76	58	44	33	24									
			36	2½	33	89	67	50	38	28										
	#3	#6	34	2	20				123	102	84	70								
			32	2	24			123	100	82	67	55								
			32	2	27		133	106	86	70	57	46								
			37	2½	30		121	96	76	61	48	38								
			36	2½	33	138	108	85	67	53	42	32								
	#3	#7	34	2	20							124	105							
			32	2	24						120	101	85							
			32	2	27					126	104	87	72							
			37	2½	20					115	94	77	63							
			36	2½	33				126	102	83	68	55							
10	#3	#6	37	2	20					133	112	94	79	66	55	46				
			35	2	24					130	108	89	75	62	51	42	35			
			34	2	27					113	93	77	63	52	43	35	28			
			39	2½	30				125	101	82	66	53	43	34	26				
			38	2½	33				112	90	72	58	46	37	29					
	#3	#7	37	2	20								118	101	87	75				
			35	2	24							132	112	95	81	69	59			
			34	2	27						137	115	97	82	69	59	50			
			39	2½	30							123	102	85	71	59	49	40		
			38	2½	33						134	110	91	75	62	52	42	35		
	#3	#8	37	2	20											124	109			
			35	2	24										133	115	100	87		
			34	2	27									136	116	100	86	75		
			39	2½	30									137	122	104	88	75	64	
			38	2½	33								134	123	109	92	78	66	56	
12	#3	#6	44	2	20						134	112	95	80	67	56	47	39	32	26
			41	2	24						130	109	90	75	63	52	43	35	29	
			39	2	27							114	94	78	65	54	44	36	29	
			43	2½	30					124	101	83	67	55	44	35	28			
			42	2½	33				136	110	89	72	58	47	37	29				
	#3	#7	44	2	20									123	106	92	79	68	59	50
			41	2	24							136	116	99	84	72	62	53	45	38
			39	2	27							119	101	86	73	62	53	44	37	31
			43	2½	30						126	105	89	75	63	52	43	35	29	
			42	2½	33					135	112	93	78	65	54	44	36	29		
	#3	#8	44	2	20											133	116	102	90	79
			41	2	24										123	106	93	81	700	61
			39	2	27									123	106	92	80	69	61	52
			43	2½	30								128	109	94	81	69	59	50	43
			42	2½	33							134	114	97	83	71	60	51	43	36
	#3	#9	44	2	20														125	111
			41	2	24												128	113	100	88
			39	2	27											127	111	98	86	76
			43	2½	30										129	113	98	85	75	65
			42	2½	33									133	115	99	86	75	65	56

Table 3 • Joists Designed as Independent Beams
No shoring used during construction

Superimposed loads = psf/Clear span = ft.

Joist size in.	Top	Bot	Wt. of joist and slab psf.	Joist spac-ing c. to c. in.	10	11	12	13	14	15	16	17	18	19	20	21	22	23	24
8	#3	#5	34	20	112	86	67	52	40	31									
			32	24	89	68	52	40	30	22 / 21*									
			32	27	76	57	43	32	23										
	#4	#6	34	20		106	85	69	56 / 49*	45 / 23*									
			32	24	107	85	67	54	43 / 30*	34 / 8*									
			32	27	92	72	56	44 / 42*	35 / 16*										
	#6	#7	34	20					103	85 / 69*	71 / 39*								
			32	24				100	82 / 79*	67 / 46*	55 / 21*								
			32	27			106	86	69 / 60*	56 / 30*	46 / 8*								
10	#3	#6	37	20					100	82	68	56	46	37 / 26*					
			35	24					98	79	64	52	42	34 / 29*					
			34	27				104	84	68	54	44	35						
	#5	#7	37	20							104	88	74 / 70*	63 / 43*	53 / 21*				
			35	24						99	82	69	58 / 46*	48 / 23*	40 / 5*				
			34	27					102	85	70	59 / 57*	49 / 32*	40 / 12*	33 / —				
	#6	#8	37	20									107 / 85*	93 / 56*	80 / 32*				
			35	24								100 / 89*	85 / 59*	73 / 34*	63 / 14*				
			34	27							102	86 / 70*	73 / 43*	62 / 22*	53 / 4*				
12	#3	#6	44	20						105	87	72	60	49	40	32	26		
			41	24					102	83	68	56	45	36	29	23			
			39	27				109	88	72	58	47	38	30	23				
			43	30					90	71	57	45	35	26					
			42	33			100	79	62	49	38	29							
	#3	#7	44	20									96	81	69	58	49	41 / 39*	34 / 19*
			41	24								106	89	75	63	53	44	37	30 / 20*
			39	27								92	77	65	54	45	37	30 / 28*	24 / 10*
			43	30							91	75	61	50	40	32	25 / 21*		
			42	33						98	80	65	53	43	34	26			
	#4	#8	44	20									103	90	78 / 75*	67 / 50*	58 / 28*		
			41	24										95	81	70	60 / 50*	52 / 28*	44 / 10*
			39	27									96	82	70	60	51 / 37*	44 / 18*	37 / 2*
			43	30								93	78	66	55	46 / 30*	38 / 9*	31 / —	
			42	33							97	82	68	57 / 41*	47 / 19*	39 / —	32		

Labor Setting Precast Concrete Joists. As long as the joist is not too heavy to handle by four workers or in some cases by light derricks, the precast joist is more economical than one which is cast in place. They are obtainable to a depth of 14" (0.35 m) and length of 36'-0" (10.8 m). Such a joist weighs approximately 1,150 to 1,330 lbs. (521-600 kg), if made of lightweight aggregate, falling within the weight where erection by machinery is necessary.

On every job there should be a joist setting plan, prepared by an architect or engineer. This will expedite setting the joists on the job.

Where smaller size joists are used, two workers can carry the joists from the stock pile to the building and set them in place, while with heavier joists, it might require 4 to 6 workers.

Where joists are used on the second floor, two workers will probably be required to carry the joists into the building and 3 workers above to hoist the joists to the second floor. If the joists are not too heavy for two workers to handle, a rope can be slipped around each end of the joist and lifted to the second floor.

After the joists have been placed on the wall, it will require some additional time for exact spacing of the joists, placing headers, etc.

Where the joists can be handled and set by 2 laborers, it will require 6 to 8 minutes to carry one joist from the stock pile, not over 50 ft. (15 m), and set it on the wall, at the following labor cost per joist:

	Hours	Rate	Total	Rate	Total
Labor	0.23	$....	$....	$21.04	$ 4.84

If joists are hoisted to the second floor, double the above costs.

When handling and setting joists requiring 4 workers to carry them to the building and set them on the wall, it will require 6 to 8 minutes for 4 workers, at the following labor cost per joist:

	Hours	Rate	Total	Rate	Total
Labor	0.47	$....	$....	$21.04	$ 9.89

If joists are hoisted to the second floor, double the above costs.

When handling and setting 12" (300 mm) joists up to 24 ft. (7.2 m) long, requiring 6 workers to handle and set, it will require 6 to 8 minutes at the following labor cost per joist:

	Hours	Rate	Total	Rate	Total
Labor	0.7	$....	$....	$21.04	$ 14.73

If joists are hoisted to the second floor, double the above costs.

	8″ Joist	10″ Joist	12″ Joist
Sectional area sq. in.	18.2	22.1	33.4
Average weight (as used in Tables 1-3) lb. per lin. ft.	19	23	35

Sizes and Weights of Precast Concrete Floor Joists

On jobs of sufficient size to warrant the use of a hydraulic crane, and using iron workers to set the joists instead of laborers, the cost of setting one joist should average as follows:

	Hours	Rate	Total	Rate	Total
Crane and operator	0.10	$....	$....	$57.20	$ 5.72
Iron workers	0.20	....	....	30.28	6.06
Cost per joist			$....		11.78

Wood Forms for Precast Concrete Joist Floors. The reinforced concrete floor slab used with precast concrete joists is usually 2″ to 2-1/2″ (50-63 mm) thick, and the joists are spaced from 20″ to 33″ (500-825 mm) apart, so that very light formwork is sufficient.

The formwork usually consists of short pieces of 2 x 4 (50 x 100 mm) wood spreaders as shown in the illustration, cut to bear on the bottom flange of the concrete joists. These spreaders are placed between the concrete joists, and the 5/8" (16 mm) plywood are placed on top of the spreaders to receive the concrete.

Method of Constructing Precast Concrete Joist Floors

Method of Constructing Precast Concrete Joist Floors Using Paper-Backed Floor Lath for Permanent Form Material

On forms of this kind, it will require about 1-1/4 b.f. of lumber per s.f. (0.03 m³ per sq.m) of floor area.

When placing forms for precast concrete joist floors, a carpenter should frame and place forms for 175 to 225 s.f. (16-20 sq.m) of floor per 8-hr. day, at the following cost per 100 s.f. (9 sq.m):

	Hours	Rate	Total	Rate	Total
Carpenter	4	$....	$....	$26.44	$ 105.76
Cost per s.f.			....		1.06
per sq.m			....		11.41

A laborer should remove forms from 600 to 750 s.f. (54-68 sq.m) of floor per 8-hr. day, at the following cost per 100 s.f. (9 sq.m):

	Hours	Rate	Total	Rate	Total
Labor	1.2	$....	$....	$21.04	$ 25.25
Cost per s.f.			....		0.25
per sq.m			....		2.69

Permanent Forms for Precast Concrete Joist Floors. Permanent type forms, not requiring removal, are being widely used in precast concrete joist floor construction, especially for work such as floors over crawl spaces, tunnels, etc., where removal of wood forms would be difficult and expensive. Form materials usually used for this purpose are of two kinds--paper-backed floor lath, such as "Steeltex" (from National Wire Products Corp., Baltimore, MD) or corrugated steel sheets described in the first part of this chapter. With corrugated steel forms, reinforcing steel or wire mesh must be added to satisfy flexure and temperature steel requirements.

Steeltex Forms for Precast Concrete Joist Floors. Steeltex is a welded wire mesh with a waterproofed paper backing attached. The paper backing serves as the form and the mesh as the reinforcing for thin, short span concrete floor slabs. When concrete is placed, the paper backing sags away from the mesh to the limit of the stitch wires thereby placing the reinforcing mesh at the proper position for encasement in the slab.

Steeltex is furnished in rolls, 4'-0" (1.2 m) wide and 125'-0" (37.5 m) long, containing 500 s.f. (45 sq.m). Side laps should be at least 2" (50 mm) and end laps should be a minimum of 12" (300 mm) occurring only over a support with no adjacent splices across the same support. In estimating, compute area of slab and add about 5% for laps and waste. For floor slabs at or near grade level, two ironworkers working together should install about 4000 s.f. (360 sq.m) of Steeltex over steel joists per 8-hr. day. For slabs at higher levels, additional handling labor and hoisting must be added.

Labor Cost of 100 S.F. (9 Sq.M) Steeltex Forms for
Precast Concrete Joist Floors

	Hours	Rate	Total	Rate	Total
Ironworker	0.4	$....	$....	$30.28	$ 12.11
Cost per s.f.			....		0.12
per sq.m			....		1.29

Corrugated Steel Forms for Precast Concrete Joist Floors. There are two methods of using corrugated steel forms for precast concrete joist floors. Corrugated steel forms may be laid continuous over joist tops with end laps occurring only over supports and no adjacent laps over the same joist.

Corrugated steel forms also may be installed between joists, so that tops of joist and protruding stirrups are encased in slab concrete, providing T-beam action. For this method, joists should be precast with ledges on the top edges to furnish support for the corrugated sheets.

Side laps for corrugated steel forms should be one corrugation and end laps a minimum of 2" (50 mm). For estimating purposes, compute area of floor slab placed over corrugated steel sheets and add 5% to 10% for laps and waste, depending upon the method of installation.

Where corrugated steel forms are installed in continuous lengths over precast concrete joists for floors at or near grade level, with an area of 5,000 s.f. (450 sq.m) or more, a 4-worker crew should place and secure 2,500 to 2,700 s.f. (225-243 sq.m) per 8-hr. day. For areas less than 5,000 s.f. (450 sq.m), the crew should install 2,100 to 2,300 s.f. (190-210 sq.m) per 8-hr. day. For slabs at higher levels, additional handling labor and hoisting costs must be added.

Labor Cost of 100 S.F. (9 Sq.M) of Corrugated Steel Slab Forms for
Precast Concrete Joist Floors in Residential Construction

	Hours	Rate	Total	Rate	Total
Ironworker	1.33	$....	$....	$30.28	$ 40.27
Cost per s.f.			....		0.40
per sq.m			....		4.30

Where corrugated steel forms are installed between joists, the above costs should be increased about 10%.

Reinforcing Steel for Precast Concrete Joist Floors. A thin slab of concrete 2" to 2-1/2" (50-64 mm) thick is used with concrete joists and the reinforcing steel consists of 1/4" (6 mm) round rods spaced 10" (250 mm) o.c. at right angles to the joists. It requires about 0.33 lbs. per s.f. (1.66 kg/m²) of steel or 33 lbs. per 100 s.f. of floor. Steel reinforcing mesh of equal effective steel area may be used instead of reinforcing rods, if desired.

An ironworker should place about 400 lbs. (181 kg) of reinforcing steel per 8-hr. day and based on 33 lbs. per 100 s.f. (1.66 kg/m²) the labor per 100 s.f. (9 sq.m) of floor should cost as follows:

	Hours	Rate	Total	Rate	Total
Ironworker	0.67	$....	$....	$30.28	$ 20.29
Cost per s.f.			....		0.20
per sq.m			....		2.15

Where welded wire mesh is used for reinforcing concrete slabs for precast concrete joist floors, an ironworker should place 1,400 to 1,600 s.f. (126-144 sq.m) of mesh per 8-hr. day at the following labor cost per 100 s.f. (9 sq.m):

	Hours	Rate	Total	Rate	Total
Ironworker	0.53	$....	$....	$30.28	$ 16.05
Cost per s.f.			....		0.16
per sq.m			....		1.72

Precast Concrete Joist Floors with Wood Floor Screeds

Wood Sleepers For Use With Precast Concrete Joist Floors. Where wood floors are used with precast concrete joist floors, it will be necessary to place 2"x2" (50 x 50 mm) or 2"x3" (50 x 75 mm) wood strips or "sleepers" in the concrete for nailing the wood floors. These strips should be spaced about 16" (400 mm) o.c. and should rest on top of the steel stirrups that project above the concrete joist. The sleepers are secured to the joists by wiring around the wood strip and through the top of the steel stirrups. Care must be taken when placing the concrete to see that the strips are level to receive the finish wood floors.

The floor strips should be beveled or nails driven into the sides at short intervals to provide anchorage into the concrete, otherwise after the concrete hardens and the wood strips dry out and shrink they become loose and cause squeaking floors.

Where wood strips are placed over the tops of the joists without leveling, a carpenter should place 350 to 450 l.f. (105-135 m) per 8-hr. day, at the following cost per 100 l.f. (30 m):

	Hours	Rate	Total	Rate	Total
Carpenter	2.0	$....	$....	$26.44	$ 52.88
Cost per l.f.			....		0.53
per m			....		1.74

Cost per 100 s.f. (9 sq.m), 16" (400mm) o.c.:

	Hours	Rate	Total	Rate	Total
Carpenter	1.5	....	....	26.44	39.66
Cost per s.f.			....		0.40
per sq.m			....		4.30

Labor Handling and Placing Precast Concrete Slabs. Precast concrete slabs are often used in conjunction with concrete joists instead of cast-in-place concrete slabs.

These slabs are ordinarily used for roofs or for floor slabs in unexcavated areas, where it is either extremely difficult or impossible to remove the forms after the cast-in-place slab has been poured.

When precast slabs are used, it requires joists containing heavier reinforcing steel than where T-beam construction is used. This increases the cost of the concrete joists $0.10-0.15 per l.f. ($0.33-0.50 per m). Precast concrete slabs are furnished in several sizes, including:

2'-1"x 2'-6"x1-1/4" (625 x 750 x 32 mm)
1'-2"x2'-1"x1-1/4" (350 x 625 x 32 mm)
1'-0"x2'-6"x1-1/4" (300 x 750 x 32 mm)
1'-6"x2'-1"x1-1/4" (450 x 625 x 32 mm)

They are reinforced with 4"x4" (100 x 100 mm) No. 10 wire mesh.

When used as roof slabs, they are laid over the concrete joists with about a 1" (25 mm) space between the slabs. After the slabs are placed they are grouted in solid. Insulation board or roofing is applied directly over the slabs.

When used for floors, the precast slabs are placed over the joists and about 1-1/2" (38 mm) of concrete is poured over the slab and either left rough or troweled to a smooth finish to receive the finish flooring.

This method eliminates form cost, but it requires heavier reinforced joists, and it is more difficult for the mechanical trades to place their pipes than where the cast-in-place floor is used.

On smaller jobs, the workers usually work in groups of two carrying and placing the precast slabs over the concrete joists.

On work on the first floor, where the precast slabs are on the floor or convenient to the place where they are to be used, two workers together should handle and set 50 to 60 pcs. an hour, depending upon the size of the slabs used, at the following labor cost per 100 pcs.:

	Hours	Rate	Total	Rate	Total
Labor	3.6	$....	$....	$21.04	$ 75.74
Cost per piece			....		0.76

Add cost of carrying slabs from stock pile to floor. If slabs are to be hoisted to the second floor, double the above costs. On large jobs where it is possible to use a hydraulic crane to lift the slabs from the ground to the second floor or roof, a crane will hoist 7 slabs at one time and place them on the floor. To the above costs, it will be necessary to add cost of material and labor grouting joints between slabs.

PRESTRESSED CONCRETE

The use of prestressed concrete began in Europe some 50 years ago. A French engineer, Eugene Freyssinet, is generally regarded as the developer of the prestressing concept. Many spectacular prestressed concrete structures were designed by Mr. Freyssinet and others in Europe, Latin America, and the United States. The first major application of prestressed concrete in the U.S. was the Walnut Lane Bridge, completed in 1950, in Philadelphia, Pa. Since that time thousands of bridges and buildings have been constructed with structural elements of prestressed concrete.

Prestressed concrete was developed partly conventional building materials were scarce and expensive in Europe during and after World War II. The concept of using available materials to their fullest capacities led to techniques that developed prestressing into a workable and practical method of construction.

The American Concrete Institute defines prestressed concrete as "concrete whose stresses resulting from external loadings are counterbalanced by prestressing reinforcement placed in the structure". This is accomplished by precompressing the concrete by means of internal high tensile strands stressed a predetermined amount. This prestressing introduces compressive stresses into the concrete which counteract the stresses induced when the external loading is applied. The high-strength strands are restrained in the member by either bond between the concrete and the strands, or by end bearing devices. These two basic methods of restraining the strands are known as *pretensioning* and *post-tensioning*.

In pretensioning, the strands are tensioned prior to casting the concrete, and in post-tensioning the strands are tensioned after the concrete is cast. In general, the post-tensioning process is accomplished at the job site when casting large members. The member is cast with the strands properly positioned in the form, and after the concrete has reached sufficient strength, the tensioning members are stressed and secured by means of end bearing devices.

The pretensioning methods are particularly adaptable to the mass production of building elements such as double and single tees, channel sections, I joists and beams, T joists, hollow slabs and flat planks, among others. These members are usually cast in steel casting beds from 100 to 600 ft. (30-180 m) long. With the use of high quality concrete, low slumps and steam curing, a complete manufacturing cycle may be accomplished in a day.

The prestressed and precast industry is based upon the economics of prefabrication and maximum use of the casting facilities. The larger the project and the number of units of identical span and section, the lower the unit cost.

Building Elements of Prestressed Concrete. Precast/Schokbeton Inc. of Kalamazoo, Michigan has a modern prestressing plant for the manufacture of prestressed concrete building elements. The plant is equipped with a variety of casting beds capable of producing a full range of prestressed products, including both structural and architectural components.

Included in the structural items available in the United States are single and double tee slabs, "F" or Monowing slabs, channel slabs, and Flextee slabs for floor and roof construction. Columns, beams and girders are available in a wide range of sizes and shapes.

Architectural products made at the Precast/Schokbeton plant include precast concrete stadium seats, miscellaneous free form concrete made to order, and Schokbeton products produced under a license from N. V. Schokbeton of Zeist, Holland. These include wall panels, window facades and frames, column shells, stair treads, etc.

Typical Prestressed Concrete Products

Cost of Prestressed Concrete Structural Elements. As prestressed concrete building elements are now manufactured generally throughout the country, prices may usually be obtained from local sources. As stated previously, casting beds are in most cases long enough to cast any length member desired: length, width, height, and weight must be acceptable to local trucking regulations.

There are, of course, many variables that can influence the cost of the installed product, such as the size of the job; the number of identical units in a job; the distance from the plant to the job site; the length of the units, as extremely long units will probably require special handling at the plant, in transit and at the job site; accessibility at the site; the size and number of cranes or other equipment required for erection; the condition at the ground, etc.

Erection of Prestressed Concrete Building Units. A typical erection setup for an average job would include a crane, crane operator, oiler, five ironworkers, welding machines, and other miscellaneous equipment and tools. With move-in and move-out time for the cranes, dead time due to the weather or breakdowns, contingencies, etc., the daily cost of an erection setup such as this will average about $2,900 per day. On a large job and under good conditions an experienced crew can probably erect 6,000 to 10,000 s.f. (540-900 sq.m) or more per day. On smaller jobs or where working conditions are below average, a crew might only erect 3,000 to 4,000 s.f. (270-360 sq.m) per day.

The Prestressed Concrete Institute is a source of the latest reference material on prestressed concrete and can provide assistance to the design profession as well as to the purchaser of prestressed concrete products.

03411 TILT-UP WALL PANELS

The tilt-up construction method can have some distinct advantages in reducing costs through utilization of semi-skilled work forces and producing a completed building project in a shorter number of construction days. These advantages provide a recognizable benefit to the contractor, and it would seem that there would result a major effort to adapt all possible structures to this method of construction. This has not been the case, because there are a sufficient number of incidents where contractors have failed to achieve the several advantages and have, in fact, experienced extreme difficulty in controlling costs, maintaining quality control and adhering to the construction schedule. The contractor/estimator should be aware of the disadvantages as well as the advantages and balance one against the other before reaching a final decision on the tilt-up construction method.

The most desirable project is one designed by the architect to adopt this construction method. Under this condition all of the engineering has been performed during the design stage thus requiring the contractor to cast and erect the panels in a sequence and quality established by the contract plans. All imbedded items, provisions for attachment to other structure members, erection details, allowances for flashing, water sealants and the interrelation with other structure members have been provided for in the design, leaving the contractor unobstructed concentration on scheduling, production, erection and finishes. These are the areas where the contractor has experience and expertise. Both office and field staff are accustomed to these situations and the attendant problems. Under these circumstances, the venture into a tilt-up construction project can be approached with a reasonable degree of assurance the project would succeed, but only after there is a complete understanding of advantages and disadvantages.

Converting Projects to Tilt-up Construction. Attempts to convert or adapt a project to incorporate the tilt-up technique are to be approached with extreme caution. Assuming there is sufficient background and experience to warrant the proposing of a design change, the contractor is exposed to a variety of issues, some of which may be obscure during the period of project planning when the compelling objective is developing methods and techniques to increase production, reduce costs, and thereby enjoy more profitability.

Double Row Lift

Double Row Lift

Three Row Lift

Examples of Rigging Configurations for Tilt-up (Courtesy Dayton Superior)

T-38 Twist-Lift Patch Caps

T-37 Twist-Lift Filler Plug

Courtesy of Dayton Superior Corp., Miamisburg, OH
Examples of Hardware Accessories for Lifting Tilt-up Panels

The realities of design change initially demand an indepth exploration of local building codes and restrictions or regulations that may be imposed by the lending institution for construction funding and permanent mortgage money. Some financial institutions have extreme objections to involvement with structures or projects utilizing other than conventional materials and construction methods. This is not because the changes are in any way inferior or detract from the value or stability of the project. More often the objection results from a lack of knowledge or prior experience with the particular construction technique. The same resistance can be encountered in city and county building inspection departments to the point where building permits will not be issued.

Owners usually are more receptive to change, especially in instances where they are the recipients of the ultimate lower cost for the same function level, as this fits with their objective for sponsoring the project. If a reduction in the square foot price is apparent to and understandable by owners, and so long as the project retains the same basic appearance and remains safe for occupancy without addition of excessive maintenance costs, most owners will be receptive to accepting a change in construction methods.

The most serious objection to the conversion process can be expected from the designing architect/engineer who will be reluctant to approve any attempt to modify the design concept which was adopted as the most acceptable solution available. Traditionally, aesthetics and design theory have been entrusted to the professional architect. For outsiders to tamper with these endowed rights is considered an intrusion by persons not meeting the standards, qualifications and professional status enjoyed by the architect. Often such proposed changes are considered a personal affront or even a challenge to ability, regardless of the motivation factor which inspired the contractor to initiate the action.

Should it be found that the designing architect/engineer is inclined to give some degree of consideration to the conversion, the contractor should be prepared to suffer some loss of time in obtaining approval and additional costs in generating engineering data upon which the architect would rely in making the final decision of acceptability. By attempting to modify the original architectural design, the contractor enters an unfamiliar role with many obligations and responsibilities which, if not recognized and accounted for, could transform an otherwise successful venture into a tragedy affecting the financial integrity of the company for many years.

It must be recognized when the contractor assumes even a partial responsibility for the structure design, the services of a registered architect/engineer are necessary to develop and produce the detailed drawings and engineering computations to demonstrate structural integrity and architectural compatibility to the satisfaction of the designing architect. Once design approval is obtained, production and erection drawings have to be provided for use by the field forces.

The loss of construction time resulting from delays in obtaining final approvals and additional engineering costs are strong influences on the feasibility of adapting a structure to the tilt-up method. Some manufacturers of inserts for slab lifting and connection devices offer information on design calculations for their products but this information is not to be relied upon as sufficient or complete in scope to satisfy the engineering responsibility assumed by the contractor. Exposure to design responsibility usually creates the need for a review of insurance coverage to provide adequate protection from the involved risks. Premiums may be exceptionally high or coverage may not exist in instances where this is a first time venture and an experience rating cannot be established.

Once the contractor is aware of the risk element involved, logical solutions can be developed and progress can proceed to accomplish the ultimate objectives of producing a project in a shorter period of time at a lower cost while retaining the functions and quality of the structure.

Evaluation Of The Advantages & Disadvantages

Disadvantages. The disadvantages or special problems provide a base for the contractor/estimator to evaluate the unique situations involved in developing accurate unit costs to support the detailed estimate for tilt-up construction:

1. This type of concrete construction is special and unlike conventional poured-in-place method.

2. Because field personnel will be unfamiliar with the procedure, rejection and opposition undoubtedly will be encountered. To overcome this

reaction and resulting low productivity, additional supervision should be assigned during the training period to avoid costly interruption in the work cycle due to indecision. Strong leadership by knowledgeable supervisors will produce coordinated work crews capable of lowering production and erection costs to an acceptable level.

3. Additional lay-out quality is required to insure precise location of all embedded items and openings within panel area.

4. Installation of reinforcing rods demand more time and precision to accommodate a panel type construction containing numerous embedded components and apertures for windows, doors, grilles, and duct, pipe and conduit penetrations.

5. Concrete pouring of horizontal thin 6" (150 mm) panels consisting of 400 s.f. (36 sq.m), 20' x 20' (6 x 6 m) requires the use of crane or chute facilities to economically distribute the concrete to all panel areas.

6. Vibration of the concrete to produce a homogeneous mass within the panel presents additional problems because the operators do not have free and equal access to all areas of the horizontal panel. Obstructions within the panel form such as opening forms, pipes, conduits, reinforcing and embedded items, all contribute to decreasing productivity in vibration of the concrete.

7. Adequate material storage space convenient to the casting area is mandatory to support the orderly flow of materials to the forming and casting area.

8. A well graded and drained access area is to be provided for movement of materials, delivery of pre-mixed concrete and maneuverability of the crane during distribution and erection of the panels.

9. Provisions and location of casting beds for the panels have a direct impact on the final cost of the operation. If sufficient space is available immediately adjacent to the erection location, cost will be lower as distribution and erection time will be less. However, if excessive time is consumed in distribution of the panels, the erection crew will be non-productive during this period thus escalating erection costs.

10. Development of a well defined casting schedule and erection drawings will eliminate confusion, delays and interruption to the overall pre-casting and erection operations.

Advantages.

1. Pre-casting of concrete panels provides an opportunity to develop work crews having limited prior experience into efficient groups at a reduced cost per hour.

2. This type of construction substantially reduces the investment in large quantities of forming materials and accessories.

3. Using the tilt-up method, most projects can be enclosed more rapidly providing access to interior finishes at an earlier date by avoiding prolonged weather delays.

4. Pre-planning during the casting operation encourages more efficient installation of window and door units in prepared openings.

5. Completion of panel erection supports availability of exterior site areas for storage or site development work.

6. Flexibility of the tilt-up system allows use of precast concrete columns, special wall surface finishes, incorporation of insulation in the wall panels and lightweight aggregate in the concrete.

7. Forming is basically limited to strip side forms at the perimeter of each panel. In instances where openings have to be formed within the panel are for doors and window areas, the cost for forming will be equal to the edge form cost.

Cost Development

Panel size and thickness can vary to accommodate specific job requirements, and engineering design will dictate special features to be found in all non-typical panels.

Cost development for a 20' x 20' x 6" (6 x 6 x 0.15 m) thick solid panel, 400 s.f. (36 sq.m), using 3000# (2.10 kg/mm^2) pre-mixed concrete, #5 reinforcing rods 6" (150 mm) o.c. each way, would be as follows.

Preparing Pouring Beds. The pouring or casting bed for panels is usually provided by the surface of the concrete slab on grade of the structure. When there is insufficient surface area available to accommodate a reasonable form-pour-cure-erect cycle, a stack method can be employed where panels are cast on previously poured panels.

In both methods care is taken to prevent adherence between the two surfaces. Frequently, the cast bed surface is coated with a brush coat of a commercial release agent or a sheet of poly between pours. Cost for either method is minimal, and the allowances included in the table are adequate.

Edge Forms. Edge forms are constructed of 2x6 (50 x 150 mm), with a 1x4 (25 x 100 mm) kick strip secured at the bottom and with the top edge held in alignment by 2x4 (50 x 100 mm) braces 2'-0" (0.6 m) long at 1'-0" (0.3 m) intervals.

Material Cost

80 l.f. (24 m) 2x6 (50x150mm)	=	80 b.f. (0.18 m³)
80 l.f. (24 m) 1x4 (25x100mm)	=	26 b.f. (0.06 m³)
84 pcs. 2"x4"x2' (50x100x600mm)	=	112 b.f. (0.25 m³)
		218 b.f. (0.49 m³)

1/2" (50 mm) strip on edge of 2x6 (50x150mm) not considered.

Cost of Forming Materials: 218 b.f. x 0.40 = $87.20
                               (0.49 m³ x 171.50 = 87.20)

Material cost per s.f. = $0.22 (2.44 per sq.m)

Labor Cost

Edge forms for one 20' x 20' (6 x 6 m) panel = 400 s.f. (36 sq.m)

	Hours	Rate	Total	Rate	Total
Carpenter	4	$....	$....	$26.44	$ 105.76
Helper	4	$....	....	21.04	84.16
			$....		$ 189.92
Cost per s.f. of panels					0.47
per b.f. of forms					0.87
per l.f. of edge forms					2.37
Cost per sq.m of panels					5.06
per m3 of forms					387.20
per m of edge forms					7.77

Labor cost for openings within the panel area to accommodate doors, windows, etc. would equal the labor cost of the edge forms.

Labor cost for stripping and cleaning forms for re-use would be minimal and are included in the above labor allowances.

Reinforcing. Reinforcing Rods - Plain 42 pcs. x 20'-0" (6 m) = 840' (252 m) x 1.043# (1.58 kg) = 876 lbs. (396 kg)

Material Costs

Reinforcing, 876 lbs. (396 kg) @ .60/lb. (1.32/kg)	$525.60
Accessories @ $60./ton (66. per m ton) rebar - Allow	26.28
Tie Wire @ 5 lbs./ton (2.5 kg/ton)	
Allow 2.5 lbs. @ 0.65 (1.4 kg @ 1.43)	2.00

$553.88

Material cost per s.f. 1.38

  per sq.m 15.39

Labor Costs

	Hours	Rate	Total	Rate	Total
2 workers @ 4 hours	8	$....	$....	21.04	168.32
Cost per s.f.					$0.42
per sq.m					4.52

Panel Connectors & Inserts. These costs are predicated on the many available alternates. Design requirements and locality will determine the final cost for these items.

For the purpose of developing the average sq. ft. (sq.m) panel cost, an allowance for material cost of $3.30 per l.f. (11.00 per m) of panel edge is considered adequate. Associated labor costs would be $1.15 per l.f. (3.83 per m) or $92.00.

Material & Labor Costs

80 l.f. @ $3.30 (24 m @ 11.00)...$264.00

  Material cost per s.f. 0.60

  Labor cost per s.f. 0.23

Concrete. The cost of concrete includes 5% waste factor. 20' x 20' x 0.5 x 1.05 = 210; 210 ÷ 27 = 7.8 or 8 c.y. (6 x 6 x 0.15 x 1.05 = 5.67 or 5.75 cu.m)

Concrete, 8 c.y. @ $55.00 per c.y. (5.75 cu.m @ 71.90).................................$440.00

Material Cost Per s.f. 1.10

  per sq.m 12.22

Labor Costs

	Hours	Rate	Total	Rate	Total
5 laborers @ 1 hour	5	$....	$....	21.04	$ 105.20
Cost per s.f.					0.26
per sq.m					2.80

Finish. Finish is either float or broom finish.

Labor Costs

	Hours	Rate	Total	Rate	Total
2 laborers @ 4 hrs	8	$....	$....	21.04	$ 168.32
Cost per s.f.					0.42
per sq.m					4.52

Special surface finishes can increase finish cost an additional $1.10 to $4.50 per s.f. (12.22-50.00 per sq.m). Material unit cost in the table is sufficient to cover the minor material expenditures involved.

Cure. Curing of panel concrete can be accomplished in several ways:

1. Cover with burlap or poly, keeping surface moist and warm;
2. Use of commercial curing compound;
3. Providing a covering of sand or sawdust;
4. Construct canopy to protect from direct sun heat during summer months or to provide enclosure for temporary heat in winter months.

Erection Costs. The labor cost of panel erection is as follows:

	Hours	Rate	Total	Rate	Total
Crane with operator	8	$....	$....	275.00	2,200.00
Erection Supervisor	8	....	....	26.44	211.52
4 man crew	32	....	....	21.04	673.28
Crew Cost Per Day			$....		3,084.80

Panel erection production includes lifting, distribution, positioning, alignment, and bracing:

Production: 1 completed panel per hour equals
400 sq. ft. x 8 = 3200 s.f. (36 x 8 = 288 sq.m) panels per day
Erection cost per s.f., $1.04
Erection cost per sq.m, $10.71

Frequently erection of panels is assigned to subcontractors having the capability to furnish a complete cost package for equipment and personnel.

Patch & Repair. During the stripping of forms, lifting, distribution, and erection of the panels, minor damage might require remedial work to maintain quality control. An allowance has been included in the panel cost development table.

Typical Cost Per S.F. (Sq.M)
For Panels 20'-0"x20'-0"x6" (6 x 6 x 0.15 m)

	Item	Material	Labor	Total
1.	Preparing Pouring Bed (Allowance)	0.04	0.04	0.08
2.	Edge Forms	0.22	0.47	0.69
3.	Reinforcing	1.38	0.42	1.80
4.	Panel Connectors and Inserts	0.60	0.23	0.83
5.	Concrete	1.10	0.26	1.36
6.	Finish - Float or Broom	0.04	0.42	0.46

7. Cure (Allowance)	0.04	0.06	0.10
8. Erection		1.04	1.04
9. Patch & Repair (Allowance)	0.06	0.12	0.18
Total Panel Cost	3.48	3.06	6.54

Tilt panel method is compatible with structural steel, poured-in-place or precast concrete columns. Panels of irregular shape containing an unusual number of openings or requiring substantial modification to accommodate special features will inflict an increase in cost. Proper sequence planning to maximize use of equipment and personnel and work crew efficiency will produce lower cost per square foot.

03500 CEMENTITIOUS DECKS
USG METAL EDGE GYPSUM PLANK

Cementitious deck is lightweight, noncombustible construction for roofs and floors in connection with steel framing. USG (United States Gypsum Co.) Metal Edge Gypsum Plank are made in 2"x15"x10'-0" (50 x 375 mm - 3 m) units and designed for roofs on spans up to 7'-0" (2.1 m). The plank has a water resistant core.

All plank are molded and have steel tongue and groove edging on sides and ends, firmly anchored into the gypsum. They also have an electrically welded galvanized steel fabric as reinforcement for the gypsum. Weight 13.0 lbs. per s.f. (66 kg per sq.m).

On jobs requiring approximately 3,000 s.f. (270 sq.m) or more, the plank cost about $1.35 per s.f. (15.00 per sq.m) in the eastern zone of the United States.

Labor Laying Metal Edge Gypsum Plank. One plank, 12-1/2 s.f. (1.13 sq.m) in area, weighs 160 lbs. (72.5 kg) requiring 2 workers to handle a single plank. Usually, it will require 2 workers on the ground and 2 workers on the roof. This crew should lay approximately 1,200 s.f. (108 sq.m) per 8-hr. day at the following labor costs per 100 s.f.:

	Hours	Rate	Total	Rate	Total
Labor	2.7	$....	$....	$21.04	$ 56.81
Cost per s.f.			....		0.57
per sq.m			....		6.13

The above costs contemplate ordinary flat slab work, on a job containing 3,000 to 4,000 s.f. (270-360 sq.m), with a minimum amount of cutting. Add for extra handling and cutting for pitched roof and double the above labor costs for pitched roofs having hips, valleys, dormers, etc. Add extra for hoisting engineer if required.

Monolithic or Poured-in-Place Gypsum Roof Construction

Monolithic or poured-in-place gypsum roof deck, with its continuous reinforcing and securely welded sub-purlins, becomes an integral part of the main steel construction of a building, adding rigidity to the entire structure and can be designed for any required loading condition.

Permanent formboards, which remain as the undersurface of the completed deck, may be 1/2" (13 mm) gypsum formboard; 1/4" (6 mm) asbestos cement board; and 1" to 2" (25-50 mm) mineral fiber formboard.

Rolled steel sections designed to carry all loads between supports shall be laid on the top of walls, beams or roof supports at sufficient spacing to receive the formboards on the lower flanges; "reinforcing shall be 48-1214 galvanized welded wire fabric having 12 ga. longitudinal wire spaced 4" (100 mm) on centers and 14 ga. transverse wired 8" (200 mm) on centers or "Keydeck", a galvanized woven mesh as manufactured by Keystone Steel and Wire Co., having 16 ga. longitudinal wires and 19 ga. diagonal wires. The effective cross sectional area of reinforcing shall be not less than 0.026 sq. in. per ft. (56 sq.mm per m) of slab width. Standard (mill mixed) gypsum composition containing at least 87-1/2% calcined gypsum is mixed with water to a medium quaking consistency and poured or pumped in place, forming a solid reinforced slab 2" (50 mm) thick, or more, if desired, plus the board undersurface. The top surface of the slab shall be screeded smooth and left ready to receive the specified roofing.

The density of poured gypsum concrete is approximately 50 lbs. per c.f. (810 kg/m³). When cast in the minimum thickness of 2", the total dead weight of the slab is about 10 lbs. per s.f. (50 kg/m²). The weights, excluding sub purlins and insulating values of poured gypsum roof decks are as follows:

Description	Wgt, lbs. per s.f	Wgt, kg per sq.m	"U" Value BTU
2" (50mm) gypsum concrete on 1/2" (12.5mm) gypsum form boards or plastic faced board	11.0	53	0.34
2" (50mm) gypsum concrete on 1" (25mm) mineral fiber formboard	11.2	54	0.16
2" (50mm) gypsum concrete on 2" (50mm) mineral fiber formboard	11.2	54	0.10

The accompanying illustration shows the type of roof construction that is widely used for schools, auditoriums, field houses, gymnasiums, factory, industrial and many commercial buildings, in fact, for most any steel skeleton structure where a permanent, noncombustible highly insulated roof is desired.

Practically all gypsum concrete is now mixed mechanically and pumped from ground to roof level as a means of further increasing the speed of operation and uniformity of quality, with corresponding reduction in cost.

The cost of poured-in-place gypsum roof decks will vary with the size and type of structure, availability of materials, freight charges, labor rates, etc., but the simplicity of design, lightweight and speed of erection, noncombustibility and finished undersurface make the overall cost of this construction very economical.

Poured-in-place gypsum roof decks are applied by contractors approved by the manufacturers of the materials. The following are approximate prices only on the various types of poured-in-place gypsum roof decks. These prices may vary locally due to labor and material costs, freight rates or size of specific job. For definite prices on any project it is always advisable to consult a roof deck contractor specializing in this type of construction.

Details of USG Metal Edge Gypsum Plank

Type of Roof	Approx. price range per s.f.	Approx. price range per sq.m
2" (50mm) gypsum concrete on 1/2" (13mm) gypsum formboards	$1.25-1.40	$13.88-15.55
2" (50mm) gypsum concrete on 1" (25mm) mineral fiber formboard	1.45-1.60	16.11-17.77
For each additional 1/2" (13mm) in thickness of gypsum concrete add	0.12-0.15	1.33-1.67

The accompanying illustration shows a poured gypsum roof on steel joists used in combination with acoustic ceiling boards of mineral fiber to comply with Underwriters Laboratories design No. P207 for a two hour fire rating. This assembly has a "U" value of 0.19, a noise reduction coefficient of 0.55 to 0.65, and a light reflectance of 85%. Other 2 hr. U.L rated systems are also available.

USG Floor Plank Systems

USG (United States Gypsum Co.) Floor Plank Systems are lightweight noncombustible assemblies for use with bar joists and floor loads of the magnitude generally found in residential buildings.

USG floor plank with tongue and grooved metal edges is laid dry without grout and welded to the joists. A topping of Mastical*, Flo-Fill, or 1/4" (6 mm) plywood underlayment and a ceiling of 5/8" (16 mm) Sheetrock Firecode C* gypsum wallboard make up the complete floor system.

1½" Pyrofil Gypsum Concrete

Reinforcing Mesh

Sheetrock Formboard

Sheet metal tees clipped
to bar joists or bulb
tees welded to frame

2"

2½"

Steel bar joist

U/L DESIGN NO. P207
2-HOUR RATING

⅝" Auratone Firecode Ceiling Board

Courtesy of U.S. Gypsum Co.

Two Hour Fire Rated Roof and Ceiling Assembly

USG Floor Plank System

CHAPTER 4

MASONRY

CSI   DIVISION 4
04100 MORTAR
04200 UNIT MASONRY
04210 BRICK
04220 CONCRETE
04240 CLAY TILE
04250 CERAMIC VENEER
04270 GLASS
04280 GYPSUM
04400 STONE
04500 MASONRY RESTORATION

An accurate estimate of the cost of masonry requires a knowledge of various factors, including type of building, thickness of walls, kind of mortar, style of mortar joint, class of workmanship, ability of workers, and the ability of the foreman.

Estimating masonry is more than a matter of referring to a table or chart to obtain the prevailing wage scale and then determining the labor cost of laying a thousand brick. Accuracy is important in the preparation of your estimates, and all of the above items should be considered when determining the unit costs.

The type of building plays an important part in the costs. A mason can lay far more brick on a factory or warehouse building having long straight walls than on a school or office building that is cut up with numerous windows, which require straight walls, plumb jambs, pilasters, etc.

A mason can lay far more brick on a 16" (400 mm) wall than on an 8" (200 mm) wall and can lay more brick in smooth working lime mortar than in coarse portland cement mortar. He can also lay more brick by merely cutting the mortar joints than by striking them with the point of his trowel.

The class of workmanship plays an important part. Many cheaply constructed, speculative buildings have crooked walls and open mortar joints, presenting a slovenly appearance, while the best grade of workmanship requires full mortar joints, straight walls, plumb jambs, and corners—in other words, good workmanship.

Weather conditions also affect labor costs. A mason can lay more brick on a clear, dry day than when it is cold and wet, or when it is necessary to heat materials during freezing weather.

The ability of the foreman in charge to plan, schedule, and lay out the work affects final costs.

Masonry Specifications

In preparing a proposal from plans and specifications, the estimator will usually find masonry units specified according to the American Society for Testing and Materials (ASTM). These specifications set limitations on requirements and properties of the various materials and are generally accepted by manufacturers. ASTM specifications are available for all the following materials:

C 32*	Sewer and Manhole Brick (made from clay or shale)
C 55	Concrete Building Brick
C 62	Building Brick (solid masonry units made from clay or shale)
C 73	Calcium Silicate Face Brick (sand-lime) brick
C 91	Masonry Cement
C 105	Ground Fire Clay or Refractory Mortar for laying up fireclay brickwork
C 126	Ceramic Glazed Structural Clay Facing Tile, Facing Brick and Solid Masonry Units
C 144	Aggregate for masonry mortar
C 150	Portland Cement
C 207	Hydrated Lime for masonry purposes
C 270	Mortar for unit masonry
C 216	Facing Brick (solid masonry units made from clay or shale)
C 279	Chemical-Resistant Masonry Units
C 287	Chemical-Resistant Sulphur Mortar
C 331	Lightweight Aggregates for Concrete Masonry Units
C 476	Grout for Reinforced and Nonreinforced Masonry
C 652	Hollow Brick (hollow masonry units made from clay or shale)
C 902	Paving Brick

*Check for latest edition

The estimator should have a general acquaintance with all the above specifications but will deal primarily with face brick, building (formerly called common) brick, and mortar for unit masonry. Building brick is subdivided into 3 grades: SW, MW, and NW, indicating severe, medium, and negligible weathering. Face brick is subdivided into only SE and MW grades, but each of these grades has 3 types—FBX, FBS, and FBA—which set forth requirements for appearance and size.

Mortar specifications cover both portland cement-lime and masonry cement mortars and are subdivided into types M, S, N, and O, based on 28 day compressive strengths of 2,500, 1,800, 750, and 350 lbs./in.^{2} (17, 12, 5, and 2 MPa). ASTM specifications are also published for all the manufactured ingredients for the mortar.

MASONRY 4.3

Specification based on these ASTM designations allow the estimator to figure within well-defined limits, and the ASTM terminology should be adopted in proposals.

04100 MORTAR

Cements. Type I portland cement is the basic cement used in most mortars. Type II is used where sulphate action is a problem, and Type III is a high early strength cement. All are included under ASTM specifications for mortar C 270.

Air entraining cements should usually not be substituted, because this affects bond strength. Full information on tests and specifications are available in ASTM publications and in information published by the Portland Cement Association.

Masonry cements are included in ASTM specifications, but their chemical makeup is not regulated and varies widely among the manufacturers. It may therefore take some experimenting to find the product most suited to the job.

Masonry cements do, however, contain additives that increase workability and water retentivity, and the single product saves handling both portland cement and lime on the job, and mortar can be used as soon as mixed. However, where a Type S mortar is selected for bonding strength, care should be taken not to select a masonry cement with air entrainment properties. Masonry cements are marketed under such names as Brixment, Atlas, Lehigh, Hy-Test, Medusa-Brisket, etc.

Lime. Lime for construction purposes is available in both the quick (unslaked) and hydrated (slaked) forms, though in recent years hydrated lime has been preferred to quicklime, because it is faster, easier, and safer to use. Building lime is supplied in bulk or packaged in 50-lb. (22.7-kg) multiple-wall, moisture resistant bags. Barrels, once widely used to package lime, are no longer in use as containers for domestic lime shipments. Lime products sold in bulk or bags are generally quoted on a ton—2000 lbs. (907 kg) net—basis.

Quicklime may be purchased in the following forms, the principal difference being in the size of the particles: pebble lime, crushed lime, ground lime, and pulverized lime. All quicklime must be slaked prior to use and manufacturer's directions for slaking should be followed to secure best results. Due to the greater ease with which complete slaking is secured, the finer divided types of quicklime are preferred. Since quicklime is very reactive with water, care should be exercised at all times during the slaking operation to prevent splattering of the lime and potentially serious burns of the eyes and skin. Slaked quicklime putty should be screened and permitted to age until all the lime particles are completely slaked. The aging period may vary from a few hours to several days, depending on the chemical and physical properties of the quicklime used and on the skill of the operator in slaking the lime.

Hydrated lime for structural use is furnished in the dry powdered form usually packaged in 50 lbs. (22.7 kg) net paper bags and sold by the ton (2000 lbs. or 907 kg net). It may also be purchased in bulk, if proper facilities are available for handling and storage. In recent years a new special type of hydrated lime has been marketed known as Type "S" or pressure hydrated lime. This highly hydrated product contains a maximum of 8% combined unhydrated calcium and magnesium oxides. In addition, putties made from this lime develop high plasticities instantly upon mixing. Type "S" hydrated limes meeting all ASTM requirements are commercially available in both the dolomitic and high calcium varieties. Type "N" (normal) hydrated limes should be soaked in a paste or putty form for several hours or overnight prior to use in order to enhance their plasticity and workability. The preferred method for soaking is to sift the hydrated lime evenly into a watertight box or vat that has been previously half filled with clean water. Do not stir, mix, or agitate the mass, but allow the lime to settle naturally; continue to add the hydrate slowly and evenly over the entire surface of the water until the thick paste is formed, after which the mass is permitted to soak until required for use.

All building limes should comply with the requirements of the Standard Specifications for Hydrated Lime for Masonry Purposes (ASTM C-207) or Quicklime for Structural Purposes (ASTM C-5).

Storage of Lime and Lime Putty. Prior to slaking or use, all quicklime and hydrated lime previously delivered to the job site should be stored in a clean, dry place.

Quicklime is shipped in bulk and in waterproofed, multi-wall paper bags. When furnished in bulk, it should be slaked immediately to assure maximum putty yield and to avoid loss. Quicklime furnished in bags may be stored for considerable time, depending on humidity and storage conditions.

Hydrated lime in bulk or packed in bags may be stored for relatively long periods of time provided it is placed in a clean, dry, properly ventilated warehouse.

Lime putty should be aged and stored in clean, tight vats and should be well protected from direct contact with the atmosphere by maintaining a thin film of water over the surface.

Commercial plants for the production of scientifically prepared, well aged putty, are in operation in many markets. The use of commercially prepared lime putty eliminates the need for slaking and aging or soaking of lime on the job. It also has the advantage of providing the estimator with reliable figures for estimating purposes, because purchase order is based on net quantity of lime putty required for the operation. Other advantages of commercially produced lime putty include flexibility during construction, elimination of lost time awaiting preparation of lime putty and admitting of expansion in operations on short notice, reduction in storage space required for materials at the site of the work, and improved quality of putty due to careful supervision during slaking and aging of lime putty.

Lime Putty and Its Preparation. Job specifications generally require mortar to be mixed in definite predetermined proportions by volume. The lime proportion is usually determined in volumes of lime putty, so some knowledge of putty yield of lime is necessary in preparing estimates of quantity of lime required. The putty yield of different limes varies widely, according to chemical and physical properties of the lime, methods of processing, etc. However, the quantities given in the following table may be used as a sound, conservative average for estimating purposes.

Quantity of Lime Putty Obtainable from Various Types of Lime

Type of Lime	Avg C.F. Putty per Ton Lime	Avg Cu.M Putty per M Ton	Lbs. of Lime per CF Putty	Kg of Lime per Cu.M Putty
Hydrated Lime	46	1.45	44	22.18
Pebble or Pulverized Quicklime	80	2.52	25	12.60

It takes 109 lbs. (49 kg) portland cement to produce 1 c.f. (0.028 cu.m) of cement paste of normal working consistency.

It takes 9 c.f. (0.25 cu.m) of putty or paste with 27 c.f. (0.76 cu.m) of ordinary building sand to produce 1 cu. yd. (0.76 cu.m) of 1:3 mortar.

Quantity of Materials Required for Masonry Using Pebble or Pulverized Quicklime at 80 C.F. Lime Putty per Ton

Proportion by Vol			For One Cu.Yd. Mortar			To Lay 1,000 Brick		
Lime Putty	Portland Cement	Sand	Lime Lbs*	Cement Sacks	Sand Cu.Yd.	Lime Lbs*	Cement Sacks	Sand Cu.Yd.
1	0	3	225	0.0	1	150	0	0.67
3	1	12	169	2.25	1	113	1.5	0.67
2	1	9	150	3.0	1	100	2.0	0.67
1 1/2	1	7.5	135	3.6	1	90	2.4	0.67
1	1	6	112.5	4.5	1	75	3.0	0.67
1/2	1	4.5	75	6.0	1	50	4.0	0.67
0	1	3	0	9.0	1	0	6.0	0.67
10%**	1	3	22.5	9.0	1	15	6.0	0.67
15%	1	3	34	9.0	1	23	6.0	0.67

*Dry unslaked lime
**Based on volume of cement required. Add 5 to 15% for waste.

Quantity of Materials Required for Masonry Using
Pebble or Pulverized Quicklime at 2.55 Cu.M Lime Putty per Metric Ton

Proportion by Vol			Quantity of Materials Required					
			For One Cu.Meter Mortar			To Lay 1,000 Brick		
Lime Putty	Portland Cement	Sand	Lime Kg *	Cement Sacks	Sand Cu.Meter	Lime Kg*	Cement Sacks	Sand Cu.Meter
1	0	3	133	0.0	1.31	89	0.0	0.88
3	1	12	100	2.9	1.31	67	2.0	0.88
2	1	9	89	3.9	1.31	59	2.6	0.88
1 1/2	1	7.5	80	4.7	1.31	53	3.1	0.88
1	1	6	67	5.9	1.31	45	3.9	0.88
1/2	1	4.5	44	7.8	1.31	30	5.2	0.88
0	1	3	0	11.8	1.31	0	7.8	0.88
10% **	1	3	13	11.8	1.31	9	7.8	0.88
15%	1	3	20	11.8	1.31	14	7.8	0.88

*Dry unslaked lime
**Based on volume of cement required. Add 5 to 15% for waste.

Where hydrated lime is used in dry powdered form, it is usually assumed that a given volume of dry hydrated lime will produce an equivalent volume of lime putty. A bag of dry hydrated lime—50 lbs. (22.7 kg) net—equals about 1.15 c.f. (0.03 cu.m) of lime putty.

Labor Slaking Lime and Making Mortar. When mixing mortar by hand, a good mortar maker should slake and sand about a ton of quicklime per 8-hr. day, and should make about 4 c.y. (3.0 cu.m) of mortar per 8-hr. day.

Quantity of Materials Required for Masonry Mortar Using
Hydrated Lime at 46 C.F. Lime Putty per Ton

Proportion by Vol			Quantity of Materials Required					
			For One Cu.Yd. Mortar			For 1,000 Brick (18 Cu.Ft.)		
Lime Putty	Portland Cement	Sand	Lime Lbs*	Cement Sacks	Sand Cu.Yd.	Lime Lbs*	Cement Sacks	Sand Cu.Yd.
1	0	3	391	0.0	1	261	0.0	0.67
3	1	12	293.5	2.25	1	196	1.5	0.67
2	1	9	261	3.0	1	174	2.0	0.67
1 1/2	1	7.5	235	3.6	1	157	2.4	0.67
1	1	6	195.5	4.5	1	130	3.0	0.67
1/2	1	4.5	130.5	6.0	1	87	4.0	0.67
0	1	3	0	9.0	1	0	6.0	0.67
10% **	1	3	39	9.0	1	26	6.0	0.67
15%	1	3	59	9.0	1	39	6.0	0.67

*Dry unslaked lime
**Based on volume of cement required. Add 5 to 15% for waste.

The actual labor cost of making 18 c.f. (0.509 cu.m) of mortar, sufficient for 1,000 common brick, should average as follows:

	Hours	Rate	Total	Rate	Total
Mortar maker	1.33	$....	$....	$21.04	$ 27.98

On jobs of any size, the mortar is mixed in a mixer, which produces a more uniform and more easily spread mortar. It not only enables a mason to lay more brick but also results in a saving of 10% to 20% in labor required to mix.

Quantity of Materials Required for Masonry Mortar Using Hydrated Lime at 1.30 Cu.M. Lime Putty per Metric Ton

Proportion by Vol			Quantity of Materials Required					
			For 1.00 Cu. Meters Mortar			For 1,000 Brick(.5 Cu.Meters)		
Lime Putty	Portland Cement	Sand	Lime Kg *	Cement Cu.Meter	Sand Kg	Lime Kg*	Cement Cu.Meter	Sand Kg
1	0	3	232	0.0	1.31	96	0.0	0.88
3	1	12	174	2.9	1.31	116	2.0	0.88
2	1	9	155	3.9	1.31	103	2.6	0.88
1 1/2	1	7.5	139	4.7	1.31	93	3.1	0.88
1	1	6	116	5.9	1.31	77	3.9	0.88
1/2	1	4.5	77	7.8	1.31	52	5.2	0.88
0	1	3	0	11.8	1.31	0	7.8	0.88
10% **	1	3	23	11.8	1.31	15	7.8	0.88
15%	1	3	35	11.8	1.31	23	7.8	0.88

*Dry unslaked lime
**Based on volume of cement required. Add 5 to 15% for waste.

Quik-Slake Masons Lime. This is a lime in pulverized form that slakes the instant it hits water. It comes in 50-lb. (22.7-kg) paper sacks, and the manufacturers say that 4 sacks are sufficient to lay, 1,000 brick under average conditions.

Aggregate. Aggregate may be either natural sand that is clean and sharp, or sand manufactured by crushing stone, gravel, or air-cooled iron blast furnace slag. Sand should be properly graded, as set forth in ASTM Specification C 144, with all sand passing a No. 4 sieve and approximately 10% passing a No. 200 sieve for manufactured sand. Too coarse a sand decreases workability while too fine a sand decreases water retentivity.

Mortar Colors. Mortar coloring agents may be added, either natural, such as white sand, or pigments. While different color effects will require some experimenting, in general, 4 to 8 lbs. (1.8-3.6 kg) of color to one bag of cement, hydrated lime or c.f. (0.02 cu.m) of putty and 3 c.f. (0.08 cu.m) of sand can be figured, or about one 50-lb. (22.7-kg) bag per 1000 bricks with a 3/8" (9.3 mm) joint.

Waterproofing and Shrinkproofing Mortar

Leaky masonry walls have aroused considerable concern, and this has led to extensive investigation and tests by the National Bureau of Standards (Report BMS7), which investigated not only the materials such as brick, tile, concrete blocks, mortar, etc., but also the effects of different grades of workmanship.

After extensive tests, it was determined that workmanship affected the permeability of the walls more than any other factor. Walls with tooled joints were less permeable than similar walls with cut joints. But the quality of the workmanship inside the walls had a greater influence than the kind of surface finish on the joint. In other words, brick as laid on the average contract job, where mortar was spread and furrowed and the joints were not completely filled with mortar, leaked far more than bricks laid in a full bed of mortar and "shoved" into position with full bed and head joints and a complete filling of joints between all brick.

It was also determined that nearly all of the leakage was through the mortar joints regardless of the kind of brick used and that in practically every instance where the "commercial" grade of workmanship was used, leaky walls resulted.

Where brick of high absorptive properties were used, the brick absorbed the water from the mortar, making it difficult to work and resulting in small hair cracks between the mortar joints and the brick. This also caused leakage through the walls.

The use of lime with low plasticity, producing a mortar with low water retentivity, greatly increased the permeability of the walls. This effect was more pronounced when the mortar was used with high absorptive brick.

Water reduction in cement and lime mortars is highly desirable, because with less water there is less evaporation, and mortar shrinkage is reduced below the critical point.

Since these investigations were made, numerous admixtures have been placed on the market. These are to be added to the mortar mixture and are said to give the mortar greater plasticity, to retain the water in the mortar, to prevent shrinkage and leakage in the mortar joints, and to make a smoother mortar that will work better under the trowel. Some of these products are listed below.

Anti-Hydro. Cement mortar for use with brick, block, or stone shall be of 1:3 mix, gauged and tempered with a solution of 1 gal. (3.8 liters) Anti-Hydro to 15 gals. (56 liters) of water, or use 1 qt. (0.9 liters) Anti-Hydro to each sack of cement.

Wet all brick. Push-lay and thoroughly grout all brick at every course. Parge with same mortar between the face brick and back-up masonry.

Castle Aquatite. Mortar to be composed of 1 part portland cement, 3 parts clean and mixed with a solution of 1 part Aquatite to 15 parts water.

Hydratite Plus. A powder in concentrated form, which when mixed with the mortar is dissolved and miscible in the water. It anchors the particle of water intimately to the particle of the cement, sand, and lime with a bond greater than the suction of the brick. With less water it increases the flow or workability of the mortar. Use 2 lbs. (0.9 kg) per sack of cement in 1:1:6 mortar.

Hydrocide Powder. When used for waterproofing masonry mortar, use 1 lb. (0.45 kg) of powder to each sack of cement or per c.f. (0.028 cu.m) of lime putty used in the mixture.

Omicron Mortarproofing. A plasticizing and water reducing agent that increases the workability of mortar, increases water retentivity and minimizes shrinkage. Contains stearate for increased water repellency. Add 1 lb. (0.45 kg) of Omicron for each sack of cement, lime or putty in the mortar mix.

Requires 5 to 6 lbs. (2.2-2.7 kg) per 1,000 brick, depending on mortar proportions and width of joints.

Plastiment (Sika). A powder used in the proportion of up to 1 lb. (0.45 kg) per sack of cement. Increases the workability of mortar, allows a reduction in the water-cement ratio, reduces bleeding and increases the bond between brick and mortar. Of particular advantage in hot weather construction, as it delays the setting time and preserves the workability of the mortar for a longer time.

Toxement IW. Toxement IW powder, paste or concentrated paste may be used for waterproofing and shrinkproofing masonry mortar.

Sikacrete. Accelerating and plasticizing liquid for use in brick mortar. Diluted with the gauging water 1 part Sikacrete to 3 to 7 parts water. Increases the bond and accelerates the hardening rate of mortar so that higher tiers of brick or glass block can be laid and the joints struck sooner.

Types of Mortar

The following mortars are based on proportion set forth in ASTM specifications. These mortars were designated Type A-1, A-2, B, and C prior to 1954.

Type M mortar is a high strength mortar used primarily in foundation masonry, retaining walls, walks, sewers, and manholes. Proportions and costs for a cement-lime mixture are:

1 c.f. portland cement	$5.75
1/4 c.f. lime @ $4.25 per c.f.	1.06
3 c.f. sand @ $0.33 per c.f.	0.99
Total	$7.80
per c.f.	$2.60
(metric)	
1.0 cu.m portland cement	$203.06
0.25 cu.m lime @ $150.09 per cu.m	37.52
3 cu.m sand @ $11.65 per cu.m	34.96
Total	$275.54
per cu.m	$91.85

Proportions and costs for a type M mortar using masonry cement are:

1 bag of portland cement	$5.75
1 bag Type II masonry cement	5.50
6 c.f. sand @ $0.33 per c.f.	1.98

Total $13.23
  per c.f. $2.21
(metric)
1.0 cu.m of portland cement $203.06
1.0 cu.m Type II masonry cement 194.23
6 cu.m sand @ $11.65 per cu.m 69.92
Total $467.21
  per cu.m $77.87

Type S mortar also has a reasonably high compressive strength and develops maximum tensile bond strength between brick and cement-lime mortars. It is recommended for use in reinforced masonry and where flexural strengths are required, such as cavity walls exposed to strong winds, and for maximum bonding power, such as for ceramic veneers. Proportions and costs for cement-lime Type S mixtures are:

1 c.f. portland cement $5.75
1/2 c.f. lime @ $4.25 per c.f. 2.13
4-1/2 c.f. sand @ .33 c.f. 1.49
Total $9.37
  per c.f. 2.08
(metric)
1.0 cu.m portland cement $203.06
0.5 cu.m lime @ $150.09 per cu.m 75.04
4.5 cu.m sand @ 11.65 cu.m 52.44
Total $330.54
  per cu.m 73.45

Costs and proportions for cement masonry Type S Mortar are:

1/2 bag portland cement $2.88
1 bag masonry cement 5.50
4-1/2 c.f. sand @ $0.33 c.f. 1.49
Total $9.87
  per c.f. 2.19
(metric)
0.5 bag portland cement $101.53
1.0 cu.m masonry cement 194.23
4.5 cu.m sand @ $11.65 cu.m 52.44
Total $348.20
  per cu.m 77.38

Type N mortar is a medium strength mortar, generally used in exposed masonry above grade. Proportions and cost of Type N cement-lime mortar are:

1 c.f. portland cement $5.75
1 c.f. lime 4.25

6 c.f. sand @ $0.33 c.f.	1.98
Total	$11.98
per c.f.	$2.00

(metric)

1.0 cu.m portland cement	$203.06
1.0 cu.m lime	150.09
6.0 cu.m sand @ $11.65 cu.m	69.92
Total	$423.07
per cu.m	$70.51

Proportions and cost of Type N masonry cement mortar are:

1 c.f. masonry cement	$5.50
3 c.f. sand @ $0.33 c.f.	0.99
Total	$6.49
per c.f.	$2.16

(metric)

1.0 cu.m masonry cement	$194.23
3.0 cu.m sand @ $11.65 cu.m	34.96
Total	$229.19
per cu.m	$76.40

Type O mortar is a low strength mixture for general interior use where compressive strengths do not exceed 100 psi (0.7 MPa). It may be used elsewhere where exposures are not severe and no freezing will be encountered. Proportions and costs of Type O cement-lime mortar are:

1 c.f. portland cement	$5.75
2 c.f. of lime @ per $4.25 c.f.	8.50
9 c.f. sand @ per $0.33 c.f.	2.97
Total	$17.22
per c.f.	$1.91

(metric)

1.0 cu.m portland cement	$203.06
2.0 cu.m of lime @ $150.09 per cu.m	300.17
9.0 cu.m sand @ $11.65 per cu.m	104.88
Total	$608.12
per cu.m	$67.57

Proportions and cost of Type O masonry cement mortar are:

1 c.f. masonry cement	$5.50
3 c.f. sand @ per $0.33 c.f.	0.99
Total	$6.49
per c.f.	$2.16

(metric)

1.0 cu.m masonry cement	$194.23
3.0 cu.m sand @ per $11.65 cu.m	34.96

Total $229.19
per cu.m...$76.40

In the above costs, lime is figured in an exact proportion. As noted under "Quantity of Materials Required for Masonry Mortar Using Hydrated Lime...", a 50-lb. (22.7-kg) bag of hydrated lime will make 1.15 c.f. (0.03 cu.m) putty. The ASTM specifications for Types N and O mortar allow some leeway in lime proportions and a full bag of lime can be figured. Only with very high strength mortars might you have to figure splitting bags to achieve exact proportions.

C.F. (Cu.M) of Mortar Required to Lay 1,000 Face Brick

Based on standard size brick having 1/4" (6.2 mm) to 3/8" (9.3 mm) end joints and bed joints as follows:

Width of Mortar Joints, inches

1/8"............1/4".............3/8"........1/2".......5/8".............3/4"...........7/8" ....1"
4 ....7.................9.............12.........14..............16 .... 18 20

(metric)

3.1mm....6.3 mm .....9.4 mm.... 12.5 mm... 15.7 mm ...18.9 mm ...21.9 mm... 25.0 mm
0.11 ....0.20 .... 0.25 .... 0.34........0.40..........0.45 .... 0.51 .... 0.57

For Dutch, English, and Flemish Bond, add about 10% to the above quantities on account of additional head joints.

C.F. (Cu.M) of Mortar Required per Thousand Brick
For various thicknesses of walls and joints. No allowance for waste.

Joint Thick., in.	|Wall Thickness, inches					
	4"	8"	12"	16"	20"	24"
1/8"	2.9	5.6	6.5	7.1	7.3	7.5
1/4"	5.7	8.7	9.7	10.2	10.5	10.7
3/8"	8.7	11.8	12.9	13.4	13.7	14.0
1/2"	11.7	15.0	16.2	16.8	17.1	17.3
5/8"	14.8	18.3	19.5	20.1	20.5	20.7
3/4"	17.9	21.7	23.0	23.6	24.0	24.2
7/8"	21.1	25.1	26.5	27.1	27.5	27.8
1"	24.4	28.6	30.1	30.8	31.2	31.5

Joint Thick., mm	Wall Thickness, mm					
	100mm	200mm	300mm	400mm	500mm	600mm
3.1 mm	0.08	0.16	0.18	0.20	0.21	0.21
6.3 mm	0.16	0.25	0.27	0.29	0.30	0.30
9.4 mm	0.25	0.33	0.37	0.38	0.39	0.40
12.5 mm	0.33	0.42	0.46	0.48	0.48	0.49
15.7 mm	0.42	0.52	0.55	0.57	0.58	0.59
18.9 mm	0.51	0.61	0.65	0.67	0.68	0.69

Joint	Wall Thickness, mm					
Thick., mm	100mm	200mm	300mm	400mm	500mm	600mm
21.9 mm	0.60	0.71	0.75	0.77	0.78	0.79
25.0 mm	0.69	0.81	0.85	0.87	0.88	0.89

Labor cost to mix mortars is usually figured as part of labor cost setting brick. However, if figured separately, it will be about $0.92 per c.f. ($32.50 per cu.m), machine mixed, $1.38 per c.f. ($48.75 per cu.m) hand mixed.

Selecting Mortars. Often the estimator has no choice in the makeup of mortar, because the architect's specifications set forth exactly what is required. Because replacing masonry that is found not acceptable by an architect or owner is extremely expensive, it is false economy, not to mention the loss of reputation involved, to alter specifications without approval from the architect.

However, many jobs will leave considerable leeway in the selection of mortars, and an informed estimator may even perform a valuable service in suggesting changes to the mortar proportions that will either save money or make a better job. All too often a mortar may be selected on the basis of high compressive strength when the job may require only a 100 psi (0.69 MPa). This not only wastes cement, but such mortars do not have the adhesive and sealing powers of the weaker mortars, and these two characteristics are the primary functions required on most jobs.

Of further importance to the estimator is the workability of the mortar, which will enable the bricklayer to lay more units. A complete discussion of mortars and recommendations is contained in technical notes on brick construction from the Brick Institute of America (BIA) in Reston, VA.

Mortar for Concrete Masonry. For laying masonry walls subject to average conditions of exposure, use a mortar made in the proportions of 1 volume of masonry cement and between 2 and 3 volumes of damp, loose mortar sand; or 1 volume of portland cement and between 1 and 1-1/4 volumes of hydrated lime or lime putty and between 4 and 6 volumes of damp, loose mortar sand.

Walls that will be subject to extremely heavy loads, severe winds, earthquakes, serious frost, or other conditions requiring extra wall strength should be laid with a mortar made of 1 volume masonry cement plus 1 volume portland cement and between 4 and 6 volumes of damp, loose mortar sand; or 1 volume of portland cement, to which may be added up to 1/4 volume of hydrated lime or lime putty, and between 2 and 3 volumes of damp, loose mortar sand.

Cost of One C.Y. (Cu.M) Cement Mortar for Concrete Masonry
1 volume masonry cement and 3 volumes loose mortar sand

	Rate	Total	Rate	Total
9 sacks masonry cement	$....	$....	$5.50	$49.50
1.00 c.y. sand		....	8.90	8.90
Cost per c.y.		$....		$58.40
per c.f.		....		2.16
(metric)				
11.8 sacks masonry cement	$....	$....	$5.50	$64.75
1.00 cu.m sand		....	11.64	11.64
Cost per cu.m		$....		$76.39

1 volume masonry cement:1 volume hydrated lime:6 volumes loose mortar sand

	Rate	Total	Rate	Total
4.5 sacks portland cement	$....	$....	$5.75	$25.88
196 lbs. hydrated lime		....	0.07	13.72
1.00 c.y. sand		....	8.90	8.90
Cost per c.y.		$....		$48.50
per c.f.		....		1.80
(metric)				
5.9 sacks masonry cement	$....	$....	$5.75	$33.84
116 kg hydrated lime		....	0.15	17.90
1.00 cu.m sand		....	11.64	11.64
Cost per cu.m		$....		$63.39

1 volume portland cement:2 volumes hydrated lime:9 volumes loose mortar sand

	Rate	Total	Rate	Total
3 sacks masonry cement	$....	$....	$5.75	$17.25
261 lbs. hydrated lime		....	0.07	18.27
1.00 c.y. sand		....	8.90	8.90
Cost per c.y.		$....		$44.42
per c.f.		....		1.65
(metric)				
3.9 sacks masonry cement	$....	$....	$5.75	$22.56
155 kg hydrated lime		....	0.15	23.92
1.00 cu.m sand		....	11.64	11.64
Cost per cu.m		$....		$58.12

1 volume masonry cement:1 volume portland cement:6 volume loose mortar sand

	Rate	Total	Rate	Total
4.5 sacks masonry cement	$....	$....	$5.50	$24.75
4.5 sacks portland cement		....	5.75	25.88
1.00 c.y. sand		....	8.90	8.90
Cost per c.y.		$....		$59.53
per c.f.		....		2.20

(metric)

	Rate	Total	Rate	Total
5.9 sacks masonry cement	$....	$....	$5.50	$32.37
5.9 sacks portland cement		....	5.75	33.84
1.00 cu.m sand		....	11.64	11.64
Cost per cu.m		$....		$77.86

1 volume portland cement: 1/4 volume hydrated lime:3 volumes loose mortar sand

	Rate	Total	Rate	Total
7.83 sacks masonry cement	$....	$....	$5.75	$45.02
90 lbs. hydrated lime		....	0.07	6.30
0.86 c.y. sand		....	8.90	7.65
Cost per c.y.		$....		$58.98
per c.f.		....		2.18
(metric)				
10.2 sacks masonry cement	$....	$....	$5.75	$58.89
41 kg hydrated lime		....	0.15	6.33
1.12 cu.m sand		....	11.64	13.09
Cost per cu.m		$....		$78.31

Quantity of Mortar Required to Lay 1,000 Concrete Brick of Various Sizes

		Thickness of Wall		
Width of	1 Brick	2 Bricks	2 Bricks	3 Bricks
Bed Joint,	4" Wall	8" Wall	8" Backup	12" Wall
in. (mm)	100 mm	200 mm	200 mm	300 mm

Modular Size Brick, 2-1/4" x 3-5/8" x 7-5/8" (56 x 90 x 190 mm)

5/12"	9.64 c.f.	12.19 c.f.	14.73 c.f.	13.01 c.f.
10.4 mm	0.272 cu.m	0.345 cu.m	0.417 cu.m	0.37 cu.m

Jumbo Brick, 3-5/8" x 3-5/8" x 7-5/8" (90 x 90 x 190 mm)

3/8"	10.06 c.f.	13.88 c.f.	17.70 c.f.	15.15 c.f.
9.4 mm	0.284 cu.m	0.393 cu.m	0.501 cu.m	0.43 cu.m

Double Brick, 4-7/8" x 3-5/8" x 7-5/8" (122 x 90 x 190 mm)

11/24"	12.68 c.f.	17.77 c.f.	22.86 c.f.	19.47 c.f.
11.5 mm	0.359 cu.m	0.503 cu.m	0.647 cu.m	0.55 cu.m

Roman Brick, 1-5/8" x 3-5/8" x 11-5/8" (41 x 90 x 290 mm)

3/8"	11.79 c.f.	14.65 c.f.	--	--
9.4 mm	0.333 cu.m	0.414 cu.m	--	--

Roman Brick, 1-5/8" x 3-5/8" x 15-5/8" (41 x 90 x 390 mm)

3/8"	15.25 c.f.	19.07 c.f.	--	--
9.4 mm	0.431 cu.m	0.540 cu.m	--	--

All end and back joints figured 3/8" (9.4 mm). Quantities include 10% for waste.

Mortar for glass blocks shall be composed of 1 part waterproof portland cement, 1 part lime, and 4 parts well graded sand, all well mixed to a consistency as stiff and dry as possible.

Cost of Mortar for Laying Glass Blocks

	Rate	Total	Rate	Total
25 lbs. quicklime	$....	$....	$0.07	$1.75
1 bag w'p'f portland cement	....	....	7.50	7.50
4 c.f. sand	....	....	0.33	1.32
Cost 4 c.f. mortar	....	$....	....	$10.57
per c.f.	....	....	....	2.64
(metric)				
14.8 kg quicklime	$....	$....	$0.15	$2.28
1.3 bag w'p'f portland cement	....	....	7.50	9.75
4 cu.m. sand	....	....	11.65	46.73
Cost per cu.m	....	$....	....	$93.23

Mortar Required for Setting Stone. The quantity of mortar required for stone setting will vary with the size of the stone, width of bed, etc., but generally, it will require 4 to 5 c.f. (0.11-0.14 cu.m) of mortar per 100 c.f. (2.83 cu.m) of stone.

If the stone is to be back-plastered, the quantity of mortar will vary with the thickness of the plaster coat, but as most ashlar is 4" to 8" (100-200 mm) thick, it will require the following c.f. of mortar per 100 c.f. (2.83 cu.m) of stone:

C.F. (Cu.M) of Mortar

Plaster Thick., in. (mm)	Thickness of stone, inches (mm)		
	4" (100mm)	6" (150 mm)	8" (200mm)
1/4" (6.3 mm)	6-1/2 (0.18)	4-1/4 (0.12)	3-1/4 (0.09)
1/2" (12.5 mm)	13 (0.37)	8-1/2 (0.24)	6-1/2 (0.18)
3/4" (18.8 mm)	19 (0.54)	13 (0.37)	9-1/2 (0.27)

It is customary to use white non-staining portland cement for setting and backplastering limestone or other porous stone to prevent stains from appearing on the face of the stone. Never use ordinary portland cement for setting limestone.

Cost of One C.Y. (0.76 Cu.M) of 1:3 Non-Staining Cement Mortar With
1/5 Part of Hydrated Lime Based on the Cement Volume

	Rate	Total	Rate	Total
2 bbls. Medusa Stone				
Set cement	$....	$....	$10.25	$20.50
100 lbs. hydrated lime	....	....	0.07	7.00
1 c.y. sand	....	....	8.90	8.90
Cost per c.y.	....	$....	....	$36.40
per c.f.	....	....	....	1.35
(metric)				
2 bbls. Medusa Stone				
Set cement	$....	$....	$10.25	$20.50
45.36 kg hydrated lime	....	....	0.15	7.00
0.76 cu.m sand	....	....	11.64	8.85
Cost per 0.76 cu.m	....	$....	....	$36.35
per cu.m	....	....	....	47.54

If white sand is used for mortar, figure $18.25 per ton (20.27 per metric ton)
and 3,000 lbs. per c.y. (1361 kg per 0.76 cu.m).

If white cement is specified, figure $17.50 per bbl. If white cement
(waterproofed) is specified, figure $18.75 per bbl.

Cost of One C.Y. (0.76 Cu.M) of Cement-Lime Mortar Consisting of 1/2 Part
Lime, 1/2 Part Stone Set Cement, and 3 Parts Sand

	Rate	Total	Rate	Total
1.25 bbls. Stone Set				
cement	$....	$....	$10.25	$12.81
225 lbs. hydrated lime	....	....	0.07	15.75
1 c.y. sand	....	....	8.90	8.90
Cost per c.y.	....	$....	....	$37.46
per c.f.	....	....	....	1.39

If white sand is used for mortar, figure $18.25 per ton and 3,000 lbs. per c.y.
(1361 kg per 0.76 cu.m).

If white cement is specified, figure $17.50 per bbl. If white cement
(waterproofed) is specified, figure $18.75 per bbl.

04200 UNIT MASONRY

04210 BRICK

Brick information is provided by the Brick Institute of America (BIA). The
BIA separates brick walls into 3 general categories: conventional, bonded
walls; drainage type walls; and barrier type walls.

The drainage type walls include brick veneer, the cavity type wall, and the masonry bonded hollow wall. Drainage type walls are recommended for walls subjected to the most severe exposures. The success of this type of wall, of course, depends on the care taken to provide continuous means for water to escape.

The barrier type walls include metal tied walls and reinforced brick masonry. The success of these walls will depend on a solidly filled collar joint. The BIA presently recommends using an 8" (200 mm) metal tied wall in preference to an 8" (200 mm) bonded wall for areas subjected to moderate exposure. Metal tied walls adapt well to brick and block backup.

When estimating the number of brick required for any job, obtain the length, height, and thickness of each wall, stating the totals in cu. ft. (cu.m), or obtain the length and height, stating the totals in sq. ft. (sq.m).

In present day estimating, all openings should be deducted in full, regardless of their size, because the estimate should show as accurately as possible the exact number of brick required for the job. The old method of counting corners twice, doubling the number of brick required for pilaster, chimney breasts, etc., is not used by modern builders. Knowing actual quantities and costs is the only sure method of getting your share of work at a profit.

Brick walls are usually designated on the plans as 4" (100 mm) or 4-1/2" (112.5 mm); 8" (200 mm) or 9" (225 mm); 12" (300 mm) or 13" (325 mm); 16" (400 mm) or 17" (425 mm), etc., increasing by 4" or 4-1/2" (100 or 112.5 mm) in width. This variation in thickness need not be considered by the estimator because a 4" or 4-1/2" (100 or 112.5 mm) wall is 1 brick thick; and 8" or 9" (200 or 225 mm) wall is 2 bricks thick; a 12" or 13" (300 or 325 mm) wall is 3 bricks thick; a 16" or 17" (400 or 425 mm) wall is 4 bricks thick, etc. The easiest method for the estimator is to mark all walls, 4", 8", 12", or 16" (100, 200, 300, or 400 mm) and then reduce the totals to cu. ft. (cu. meters). After the number of cu. ft. (cu. meters) has been obtained, multiply by the number of brick required per cu. ft. (per cu. meter) of wall and the result will be the number of brick required for the job.

When brickwork is estimated on the above basis, it is advisable to add 1-1/2 to 2% to the total to take care of salmon brick, broken brick, bats, etc. The waste should not exceed this amount unless very poor brick or poor job handling methods are used.

Years ago contractors figured their building brickwork on the basis of 7 or 7-1/2 brick per s.f. (0.09 sq.m) of 4" or 4-1/2" (100 or 112.5 mm) wall; 14 or 15 brick per s.f. (0.09 sq.m) of 8" or 9" (200 or 225 mm) wall; 21 or 22-1/2 brick per s.f. (0.09 sq.m) of 12" or 13" (300 or 325 mm) wall, etc.

This method is no longer used, because modern business demands that actual quantities be figured. Work is being estimated too closely today to permit a 10% to 20% overrun on the brickwork, especially when many contractors are figuring on a 5% to 10% margin of profit.

After the number of c.f. (cu.m) of brickwork has been obtained, it will be necessary to determine the number of brick required per c.f. (cu.m) of wall. This will vary with the size of the brick and the width of the mortar joints. The standard non-modular size is 8"x2-1/4"x3-3/4" (200 x 56.2 x 93.7 mm). Modular brick is 8"x2-2/3"x4" (200 x 66.5 x 100 mm). Sizes will vary somewhat depending upon the burning, the brick in the center of the kiln usually being more thoroughly burned.

When estimating the number of face brick required for any job, obtain the length and height of all walls to be faced with brick, and the total will be the number of s.f. (sq.m) of wall.

Always make deductions in full for all openings, regardless of size, because the estimate should show as accurately as possible the exact number of brick required to complete the job.

When making deductions for door and window openings, always note the depth or width of the brick jambs or "reveals". If they are only 4" (100 mm) deep, deduct the full size of the opening as the 4" (100 mm) end of the brick forms the "reveal" or jamb.

If the jamb is more than 4" (100 mm) wide, the full depth of the jamb or "reveal" should be deducted. Example: Deduct for a 5'-0"x7'-0" (1.5 x 2.1 m) opening in a 12" (300 mm) brick wall where the face brick jamb or "reveal" returns the full thickness of the wall. It will be necessary to deduct for the 12" (300 mm) jamb or reveal on each side of the opening, i.e., 2 x 12" = 24" or 2'-0" (2 x 300 mm = 600 mm or 0.6 m). Deduct this from the width of the opening, i.e., 5' - 2' = 3' (1.5 - 0.6 = 0.9 m). The opening deducted should be 3'-0"x7'-0" (0.9 x 2.1 m) instead of 5'-0"x7'-0" (1.5 x 2.1 m), as shown on the plan.

Standard Brick

A standard non-modular brick is 8" (200 mm) long, 2-1/4" (56.2 mm) high, and contains 18 sq. in. (11,613 sq.mm) on the face. A standard modular brick is 8" (200 mm) long, 2-2/3" (66.5 mm) high, and contains 21.28 sq. in. (13,730 sq.mm) on the face. If using a standard non-modular brick, laid dry without a mortar joint, it would require 8 brick per s.f. (0.09 sq.m) of wall: 144 ÷ 18 = 8. However, the thickness of the mortar joint must be added, and this will vary from 3/8" (9.3 mm) to 5/8" (15.6 mm), with 1/2" (12.5 mm) the average width and 3/8" (9.3 mm) the normal thickness. A brick is 8" (200 mm) long, plus 3/8" (9.3 mm) for the vertical or end mortar joint makes a total length of 8-3/8" (209.3 mm). A brick is 2-1/4" (56.2 mm) high, plus 3/8" (9.3 mm) for horizontal or bed mortar joint, making the total height of each brick course 2-5/8" (65.6 mm). 8-3/8" x 2-5/8" = 21.984 or 22 sq. in. (209.3 mm^2 x 65.6 mm^2 = 13,730 mm^2) on the face.

To obtain the number of brick required per s.f. of wall, divide 144 by 22 and the result is 6.54 or 6-1/2 brick per s.f. of 4" (100 mm) or 4-1/2" (112.5 mm) wall. If the wall is 8" (200 mm) or 9" (225 mm)—2 brick—thick, 2 x 6-1/2 = 13 brick per s.f. of 8" (200 mm) or 9" (225 mm) wall. If 12" (300 mm) or 13" (325 mm)—3 brick—thick, 6-1/2 x 3 = 19.5 brick per s.f. of wall, etc.

This same method should be used to obtain the number of brick of any size required for 1 s.f. (0.09 sq.m) of wall of any thickness.

Number of Standard Brick (8" x 2-1/4" x 3-3/4")
Required for One Sq. Ft. of Brick Wall of any Thickness

Vertical or End Mortar Joints Figured as 1/4" Wide

Thick. of Wall	No. of Brick	Width of Horizontal or "Bed" Mortar Joints					
		1/8"	1/4"	3/8"	1/2"	5/8"	3/4"
4 or 4-1/2"	1	7.33	7	6.67	6.33	6.08	5.8
8 or 9"	2	14.67	14	13.33	12.67	12.17	11.6
12 or 13"*	3	22.00	21	20.00	19.00	18.25	17.4
16 or 17"	4	29.33	28	26.67	25.33	24.33	23.2
20 or 21"	5	36.67	35	33.33	31.67	30.42	29.0
24" or 25"	6	44.00	42	40.00	38.00	36.50	34.8

Number of Standard Brick (200 x 56 x 97 mm)
Required for One Sq. Meter of Brick Wall of any Thickness

Vertical or End Mortar Joints Figured as 6 mm Wide

Thick. of Wall	No. of Brick	Width of Horizontal or "Bed" Mortar Joints					
		3.1mm	6.3mm	9.4mm	12.5mm	15.6mm	18.8mm
100 or 112.5mm	1	78.90	75.35	71.80	68.14	65.45	62.43
200 or 225mm	2	157.91	150.70	143.49	136.38	131.00	124.87
300 or 325mm*	3	236.81	226.05	215.29	204.52	196.45	187.30
400 or 425mm	4	315.72	301.40	287.08	272.66	261.89	249.73
500 or 525mm	5	394.73	376.75	358.77	340.90	327.45	312.16
600 or 625mm	6	473.63	452.10	430.57	409.04	392.90	374.60

*Use this column for computing the number of brick required per c.f. (0.028 cu.m) of wall with any width mortar joint.

Variations in Face Brick Quantities. If the brick are laid in running bond without headers, the above quantities will prove sufficient, but if there is a full course of headers every 5th, 6th, or 7th course, it will be necessary to allow for the extra brick required. Also, if the brick are laid in English,

Flemish, English Cross, or Dutch Bond, it will be necessary to make an allowance for extra brick where full header courses are required.

Number of Square Inches Occupied by One 8"x2-1/4" Face Brick With Various Width Mortar Joints

All Vertical Mortar Joints Figured 1/4" Wide
Width of Horizontal or "Bed" Mortar Joints in Inches

1/8"	1/4"	3/8"	1/2"	5/8"	3/4"	7/8"	1"
19.30	20.625	21.656	22.69	23.72	24.75	25.78	26.81

Number of Square Millimeters Occupied by One 200 x 56 mm Face Brick With Various Width Mortar Joints,mm

All Vertical Mortar Joints Figured 6 mm Wide
Width of Horizontal or "Bed" Mortar Joints in mm

3.1	6.3	9.4	12.5	15.6	18.8	21.9	25.0
12,452	13,306	13,972	14,639	15,303	15,968	16,632	17,297

Number of 8"x2-1/4"x3-3/4" Face Brick Required Per S.F. of Wall In Running Bond Without Headers

All Vertical Mortar Joints Figured 1/4" Wide
Width of Horizontal or "Bed" Mortar Joints in Inches

1/8"	1/4"	3/8"	1/2"	5/8"	3/4"	7/8"	1"
7.46	7.0	6.65	6.35	6.08	5.82	5.60	5.38

Number of 200 x 56 x 94 mm Face Brick Required Per Sq. Meter of Wall In Running Bond Without Headers

All Vertical Mortar Joints Figured 6 mm Wide
Width of Horizontal or "Bed" Mortar Joints in mm

3.1	6.3	9.4	12.5	15.6	18.8	21.9	25.0
80	75	72	68	65	63	60	58

The above tables do not include any allowance for waste, breakage, or header courses extending into the building brick backing to form a bond. It is based on using all stretchers or blind headers with metal wall ties. On face brick work, 3% to 5% should be added to the net quantity for waste.

For face brick laid with a row of full "headers" every 5th, 6th, or 7th course, add the following percentages to the above quantities:

Percentage to be Added for Various Brick Bonds

Common (full header course every 5th course)...20% or 1/8
Common (full header course every 6th course)....................................16.66% or 1/8
Common (full header course every 7th course)....................................14.33% or 1/7
English or English Cross* (full headers every other course)....................50% or 1/2
English or English Cross* (full headers every 6th course)..................16.66% or 1/6
Dutch or Dutch Cross* (full headers every other course).........................50% or 1/2
Dutch or Dutch Cross* (full headers every 6th course) ....16.66% or 1/6
Flemish (full headers every course)..33.33% or 1/3
Flemish (full headers every 6th course)..5.6% or 1/18
Double Header (two headers and a stretcher every 6th course) ....8.33% or 1/12
Double Header (two headers and a stretcher every 5th course) ....10% or 1/10
Double Flemish (full headers every other course)..................................10% or 1/10
*Add 10% to 15% extra brick for waste in cutting, unless a masonry saw is used.

Percentages To Be Added for Various Brick Bonds

Double Flemish (full headers every 3rd course)....................................6.33% or 1/15
3 Stretcher Flemish (full headers every other course).........................7.14% or 1/14
3 Stretcher Flemish (full headers every 3rd course).............................4.8% or 1/21
4 Stretcher Flemish (full headers every other course)...........................5.6% or 1/18
4 Stretcher Flemish (full headers every 3rd course)..............................3.7% or 1/27

For garden walls, porch walls, and other places where an 8" (200 mm) wall is used, with face brick on both sides of the wall, no additional brick are required for any type of bond.

For walks and floors with the brick laid on edge, in any pattern except diagonal, calculate as you would for face brick in common bond without headers. For herringbone or other diagonal work, additional brick will be required because of the waste in chipping or cutting the brick for the borders. The additional number of brick will vary with the width of the walk or floor, as the wider the surface, the smaller the average waste per sq. ft. (sq.m). Walks and floors having the brick laid flat require one-third less brick than where they are laid on edge.

Coursing Tables

The following tables contain vertical coursing dimensions for both non-modular and modular brick. For the non-modular brick, vertical coursing dimensions are shown for both 3/8" (9.3 mm) and 1/2" (12.5 mm) mortar joints. In both tables the bricks are assumed to be positioned in the wall as stretchers.

Vertical Coursing Table for Non-modular Brick*

No. of Courses	2-1/4" high		2-5/8" high		2-3/4" high	
			Size of Joint			
	3/8"	1/2"	3/8"	1/2"	3/8"	1/2"
1	0'-2-5/8"	0'-2-3/4"	0'-3"	0'-3-1/8"	0"-3-1/8"	0'-3-1/4"
2	0'-5-1/4"	0'-5-1/2"	0'-6"	0'-6-1/4"	0'-6-1/4"	0'-6-1/2"
3	0'-7-7/8"	0'-8-1/4"	0'-9"	0'-9-3/8"	0'-9-3/8"	0'-9-3/4"
4	0"-10-1/2"	0'-11"	1'-0"	1'-0-1/2"	1'-0-1/2"	1'-1"
5	1'-1-1/8"	1'-1-3/4"	1'-3"	1'-3-5/8"	1'-3-5/8"	1'-4-1/4"
6	1'-3-3/4"	1'-4-1/2"	1'-6"	1'-6-3/4"	1'-6-3/4"	1'-7-1/2"
7	1'-6-3/8"	1'-7-1/4"	1'-9"	1'-9-7/8"	1'-9-7/8"	1' 10-3/4"
8	1'-9"	1'-10"	2'-0"	2'-1"	2'-1"	2'-2"
9	1'-11-5/8"	2'-0-3/4"	2'-3"	2'-4-1/8"	2'- 4-1/8"	2'-5-1/4"
10	2'-2-1/4"	2'-3-1/2"	2'-6"	2'-7-1/4"	2'-7-1/4"	2'-8-1/2"
11	2'-4-7/8"	2'-6-1/4"	2'-9"	2'-10-3/8"	2'-10 -3/8"	2'-11-3/4"
12	2'-7-1/2"	2'-9"	3'-0"	3'-1-1/2"	3'-1-1/2"	3'-3"
13	2'-10-1/8"	2'-11-3/4"	3'-3"	3'-4-5/8"	3'-4-5/8"	3'-6-1/4"
14	3'-0-3/4"	3'-2-1/2"	3'-6"	3'-7-3/4"	3'-7-3/4"	3'-9-1/2"
15	3'-3-3/8"	3'-5-1/4"	3'-9"	3'-10-7/8"	3'-10-7/8"	4'-0-3/4"
16	3'-6"	3'-8"	4'-0"	4'-2"	4'-2"	4'-4"
17	3'-8-5/8"	3'-10-3/4"	4'-3"	4'-5-1/8"	4'-5-1/8"	4'-7-1/4"
18	3'-11-1/4"	4'-1-1/2"	4'-6"	4'-8-1/4"	4'-8-1/4"	4'-10-1/2"
19	4'-1-7/8"	4'-4-1/4"	4'-9"	4'-11-3/8"	4'-11-3/8"	5'-1-3/4"
20	4'-4-1/2"	4'-7"	5'-0"	5'-2-1/2"	5'-2-1/2"	5'-5"
21	4'-7-1/8"	4'-9-3/4"	5'-3"	5'-5-5/8"	5'-5-5/8"	5'-8-1/4"
22	4'-9-3/4"	5'-0-1/2"	5'-6"	5'-8-3/4"	5'-8-3/4"	5'-11-1/2"
23	5'-0-3/8"	5'-3-1/4"	5'-9"	5'-11-7/8"	5'-11-7/8"	6'-2-3/4"
24	5'-3"	5'-6"	6'-0"	6'-3"	6'-3"	6'-6"
25	5'-5-5/8"	5'-8-3/4"	6'-3"	6'-6-1/8"	6'-6-1/8"	6'-9-1/4"
26	5'-8-1/4"	5'-11-1/2"	6'-6"	6'-9-1/4"	6'-9-1/4"	7'-0-1/2"
27	5'-10-7/8"	6'-2-1/4"	6'-9"	7'-0-3/8"	7'-0-3/8"	7'-3-3/4"
28	6'-1-1/2"	6'-5"	7'-0"	7'-3-1/2"	7'3-1/2"	7'-7"
29	6'-4-1/8"	6'-7-3/4"	7'-3"	7'-6-5/8"	7'-6-5/8"	7'-1-1/4"
30	6'-6-3/4"	6'-10-1/2"	7'-6"	7'-9-3/4"	7'-9-3/4"	8'-1-1/2"
31	6'-9-3/8"	7'-1-1/4"	7'-9"	8'-0-7/8"	8'-0-7/8"	8'-4-3/4"
32	7'-0"	7'-4"	8'-0"	8'-4"	8'-4"	8'-8"
33	7'-2-5/8"	7'-6-3/4"	8'-3"	8'-7-1/8"	8"-7-1/8"	8'-11-1/4"
34	7'-5-1/4"	7'-9-1/2"	8'-6"	8'-10-1/4"	8'-10-1/4"	9'-2-1/2"
35	7'-7-7/8"	8'-0-1/4"	8'-9"	9'-1-3/8"	9'-1-3/8"	9'-5-3/4"
36	7'-10-1/2"	8'-3"	9'-0"	9'-4-1/2"	9'-4-1/2"	9'-9"
37	8'-1-1/8"	8'-5-3/4"	9'-3"	9'-7-5/8"	9'-7-5/8"	10'-0-1/4"
38	8'-3-3/4"	8'-8-1/2"	9'-6"	9'-10-3/4"	9'-10-3/4"	10'-3-1/2"
39	8'-6-3/8"	8'-11-1/4"	9'-9"	10'-1-7/8"	10'-1-7/8"	10'-6-3/4"
40	8'-9"	9'-2"	10'-0"	10'-5"	10'-5"	10'-10"
41	8'-11-5/8"	9'-4-3/4"	10'-3"	10'-8-1/8"	10'-8-1/8"	11'-1-1/4"
42	9'-2-1/4"	9'-7-1/2"	10'-6"	10'11-1/4"	10'-11-1/4"	11'-4-1/2"
43	9'-4-7/8"	9'-10-1/4"	10'-9"	11'-2-3/8"	11'-2-3/8"	11'-7-3/4"
44	9'-7-1/2"	10'-1"	11'-0"	11'-5-1/2"	11'-5-1/2"	11'-11"

No. of Courses	2-1/4" high 3/8"	1/2"	2-5/8" high 3/8"	1/2"	2-3/4" high 3/8"	1/2"
45	9'-10-1/8"	10'-3-3/4"	11'-3"	11'-8-5/8"	11'-8-5/8"	12'-2-1/4"
46	10'-0-3/4"	10"-6-1/2"	11'-6"	11'-11-3/4"	11'-11-3/4"	12'5-1/2"
47	10'-3-3/8"	10'-9-1/4"	11'-9"	12'-2-7/8"	12'-2-7/8"	12'-8-3/4"
48	10'-6"	11'-0"	12'-0"	12'-6"	12'-6"	13'-0"
49	10'-8-5/8"	11'-2-3/4"	12'-3"	12'-9-1/8"	12'-9-1/8"	13'-3-1/4"
50	10'-11-1/4"	11'-5-1/2"	12'-6"	13'-0-1/4"	13'-0-1/4"	13'-6-1/2"
100	21'-10-1/2"	22'-11"	25'-0"	26'-0-1/2"	26'-0-1/2"	27'-1"

*Brick positioned in wall as stretchers. Vertical dimensions are from bottom of mortar joint to bottom of mortar joint.

Courtesy of Brick Institute of America

Vertical Coursing Table for Non-modular Brick*
In Millimeters

No. of Courses	56 mm high 9.4 mm	12.5 mm	66 mm high 9.4 mm	12.5 mm	69 mm high 9.4 mm	12.5 mm
1	66.7	69.9	76.2	79.4	79.4	82.6
2	133.4	139.7	152.4	158.8	158.8	165.1
3	200.0	209.6	228.6	238.1	238.1	196.9
4	266.7	279.4	304.8	317.5	317.5	330.2
5	333.4	349.3	381.0	396.9	396.9	412.8
6	400.1	419.1	457.2	476.3	476.3	495.3
7	466.7	489.0	533.4	555.6	555.6	577.9
8	533.4	558.8	609.6	635.0	635.0	660.4
9	600.1	628.7	685.8	714.4	714.4	743.0
10	666.8	698.5	762.0	793.8	793.8	825.5
11	733.4	768.4	838.2	873.1	873.1	908.1
12	800.1	838.2	914.4	952.5	952.5	990.6
13	866.8	908.1	990.6	1031.9	1031.9	1073.2
14	914.4	977.9	1066.8	1111.3	1111.3	1155.7
15	1000.1	1047.8	1143.0	1190.6	1190.6	1219.2
16	1066.8	1117.6	1219.2	1270.0	1270.0	1320.8
17	1133.5	1187.5	1295.4	1349.4	1349.4	1403.4
18	1200.2	1257.3	1371.6	1428.8	1428.8	1485.9
19	1266.8	1327.2	1447.8	1508.1	1508.1	1568.5
20	1333.5	1397.0	1524.0	1587.5	1587.5	1651.0
21	1400.2	1466.9	1600.2	1666.9	1666.9	1733.6
22	1466.9	1536.7	1676.4	1746.3	1746.3	1816.1
23	1533.5	1606.6	1752.6	1825.6	1825.6	1898.7
24	1600.2	1676.4	1828.8	1905.0	1905.0	1981.2
25	1666.9	1746.3	1905.0	1984.4	1984.4	2063.8
26	1733.6	1816.1	1981.2	2063.8	2063.8	2146.3
27	1851.0	1632.0	2057.4	2133.6	2133.6	2228.9
28	1866.9	1955.8	2133.6	2222.5	2222.5	2311.4

No. of Courses	56 mm high		66 mm high		69 mm high	
	Size of Joint					
	9.4 mm	12.5 mm	9.4 mm	12.5 mm	9.4 mm	12.5 mm
29	1933.6	2025.7	2209.8	2301.9	2301.9	2394.0
30	2000.3	2095.5	2286.0	2381.3	2381.3	2476.5
31	2066.9	2165.4	2362.2	2460.6	2460.6	2559.1
32	2133.6	2235.2	2438.4	2540.0	2540.0	2641.6
33	2200.3	2330.5	2514.6	2619.4	2619.4	2724.2
34	2267.0	2374.9	2590.8	2698.8	2698.8	2806.7
35	2333.6	2444.8	2667.0	2778.1	2778.1	2889.3
36	2400.3	2514.6	2743.2	2857.5	2857.5	2971.8
37	2467.0	2584.5	2819.4	2936.9	2936.9	3054.4
38	2533.7	2654.3	2895.6	3016.3	3016.3	3136.9
39	2600.3	2724.2	2971.8	3095.6	3095.6	3219.5
40	2667.0	2794.0	3048.0	3175.0	3175.0	3302.0
41	2733.7	2863.9	3124.2	3254.4	3254.4	3384.6
42	2800.4	2933.7	3200.4	3333.8	3333.8	3467.1
43	2867.0	3003.6	3276.6	3413.1	3413.1	3549.7
44	2933.7	3073.4	3327.4	3492.5	3492.5	3632.2
45	3000.4	3143.3	3429.0	3571.9	3571.9	3714.8
46	3067.1	3213.1	3505.2	3651.3	3651.3	3797.3
47	3133.7	3283.0	3581.4	3730.6	3730.6	3879.9
48	3200.4	3352.8	3657.6	3810.0	3810.0	3962.4
49	3267.1	3422.7	3733.8	3889.4	3889.4	4045.0
50	3333.8	3492.5	3810.0	3968.8	3968.8	4127.5
100	6667.5	6985.0	7620.0	7937.5	7937.5	8255.0

Based on actual conversion from inches to SI, 1" = 25.4 mm. (Most conversions are soft or rounded, i.e., 1" = 25.0 mm.)

Vertical Coursing Table for Modular Brick*

No. of Courses	Nominal Height (h) of Unit				
	2"	2-2/3"	3-1/5"	4"	5-1/3"
1	0'-2"	0'-2-2/3"	0'-3-1/5"	0'-4"	0'-5-1/3"
2	0'-4"	0'-5-1/3"	0'-6-2/5"	0'-8"	0'-10-2/3"
3	0'-6"	0'-8"	0'-9-3/5"	1'-0"	1'-4"
4	0'-8"	0'-10-2/3"	1'-0-4/5"	1'-4"	1'-9-1/3"
5	0'-10"	1'-1-1/3"	1'-4"	1'-8"	2'-2-2/3"
6	1'-0"	1'-4"	1'-7-1/5"	2'-0"	2'-8"
7	1'-2"	1'-6-2/3"	1'-10-2/5"	2'-4"	3'-1-1/3"
8	1'-4"	1'-9-1/3"	2'-1-3/5"	2'-8"	3'-6-2/3"
9	1'-6"	2'-0"	2'-4-4/5"	3'-0"	4'-0"
10	1'-8"	2'-2-2/3"	2'-8"	3'-4"	4'-5-1/3"
11	1'-10"	2'-5-1/3"	2'-11-1/5"	3'-8"	4'-10-2/3"
12	2'-0"	2'-8"	3'-2-2/5"	4'-0"	5'-4"
13	2'-2"	2'-10-2/3"	3'-5-3/5"	4'-4"	5'-9-1/3"
14	2'-4"	3'-1-1/3"	3'-8-4/5"	4'-8"	6'-2-2/3"
15	2'-6"	3'-4"	4'-0"	5'-0"	6'-8"
16	2'-8"	3'-6-2/3"	4'-3-1/5"	5'-4"	7'-1-1/3"

No. of Courses	Nominal Height (h) of Unit				
	2"	2-2/3"	3-1/5"	4"	5-1/3"
17	2'10"	3'-9-1/3"	4'-6-2/5"	5'-8"	7'-6-2/3"
18	3'-0"	4'-0"	4'-9-3/5"	6'-0"	8'-0"
19	3'-2"	4'-2-2/3"	5'-0-4/5"	6'-4"	8'-5-1/3"
20	3'-4"	4'-5-1/3"	5'-4"	6'-8"	8'-10-2/3"
21	3'-6"	4'-8"	5'-7-1/5"	7'-0"	9'-4"
22	3'-8"	4'-10-2/3"	5'-10-2/5"	7'-4"	9'-9-1/3"
23	3'-10"	5'-1-1/3"	6'-1-3/5"	7'-8"	10"-2-2/3"
24	4'-0"	5'-4"	6'-4-4/5"	8'-0"	10'-8"
25	4'-2"	5'-6-2/3"	6'-8"	8'-4"	11'-1-1/3"
26	4'-4"	5'-9-1/3"	6'-11-1/5"	8'-8"	11'-6-2/3"
27	4'-6"	6'-0"	7'-2-2/5"	9'-0"	12'-0"
28	4'-8"	6'-2-2/3"	7'-5-3/5"	9'-4"	12'-5-1/3"
29	4'-10"	6'-5-1/3"	7'-8-4/5"	9'-8"	12'-10-2/3"
30	5'-0"	6'-8"	8'-0"	10'-0"	13'-4"
31	5'-2"	6'-10-2/3"	8'-3-1/5"	10'-4"	13'-9-1/3"
32	5'-4"	7'-1-1/3"	8'-6-2/5"	10'-8"	14'-2-2/3"
33	5'-6"	7'-4"	8'-9-3/5"	11'-0"	14'-8"
34	5'-8"	7'-6-2/3"	9'-0-4/5"	11'-4"	15'-1-1/3"
35	5'-10"	7'-9-1/3"	9'-4"	11'-8"	15'-6-2/3"
36	6'-0"	8'-0"	9'-7-1/5"	12'-0"	16'-0"
37	6'-2"	8'-2-2/3"	9'-10-2/5"	12'-4"	16'-5-1/3"
38	6'-4"	8'-5-1/3"	10'-1-3/5"	12'-8"	16'-10-2/3"
39	6'-6"	8'-8"	10'-4-4/5"	13'-0"	17'-4"
40	6'-8"	8'-10-2/3"	10'-8"	13'-4"	17'-9-1/3"
41	6'-10"	9'-1-1/3"	10'-11-1/5"	13'-8"	18'-2-2/3"
42	7'-0"	9'-4"	11'-2-2/5"	14'-0"	18'-8"
43	7'-2"	9'-6-2/3"	11'-5-3/5"	14'-4"	19'-1-1/3"
44	7'-4"	9'-9-1/3"	11'-8-4/5"	14'-8"	19'-6-2/3"
45	7'-6"	10'-0"	12'-0"	15'-0"	20'-0"
46	7'-8"	10'-2-2/3"	12'-3-1/5"	15'-4"	20'-5-1/3"
47	7'-10"	10'-5-1/3"	12'-6-2/5"	15'-8"	20'-10-2/3"
48	8'-0"	10'-8"	12'-9-3/5"	16'-0"	21'-4"
49	8'-2"	10'10-2/3"	13'-0-4/5"	16'-4"	21'-9-1/3"
50	8'-4"	11'-1-1/3"	13'-4"	16'-8"	22'-2-2/3"
100	16'-8"	22'-2-2/3"	26'-8"	33'-4"	44'-5-1/3"

*Brick positioned in wall as stretchers.
Courtesy of Brick Institute of America

Vertical Coursing Table for Modular Brick*
In Millimeters

No. of Courses	Nominal Height (h) of Unit (mm)				
	50.8	67.73	81.28	101.6	135.46
1	50.8	67.7	81.3	101.6	132.1
2	101.6	135.5	162.6	203.2	264.2

No. of Courses	Nominal Height (h) of Unit (mm)				
	50.8	67.73	81.28	101.6	135.46
3	152.4	203.2	243.8	304.8	406.4
4	203.2	270.9	325.1	406.4	541.9
5	254.0	338.7	406.4	508.0	677.3
6	304.8	406.4	487.7	609.6	812.8
7	355.6	474.1	569.0	711.2	948.3
8	406.4	541.9	650.2	812.8	1083.7
9	457.2	609.6	731.5	914.4	1219.2
10	508.0	677.3	812.8	1016.0	1354.7
11	558.8	745.1	894.1	1117.6	1490.1
12	609.6	812.8	975.4	1219.2	1625.6
13	660.4	880.5	1056.6	1320.8	1761.1
14	711.2	948.3	1137.9	1422.4	1896.5
15	762.0	1016.0	1219.2	1524.0	2032.0
16	812.8	1083.7	1300.5	1625.6	2167.5
17	863.6	1151.5	1376.7	1727.2	2302.9
18	914.4	1219.2	1463.0	1828.8	2438.4
19	965.2	1286.9	1544.3	1930.4	2573.9
20	1016.0	1354.7	1625.6	2032.0	2709.3
21	1066.8	1422.4	1706.9	2133.6	2844.8
22	1117.6	1490.1	1788.2	2235.2	2980.3
23	1168.4	1557.9	1869.4	2336.8	3115.7
24	1219.2	1625.6	1950.7	2438.4	3251.2
25	1270.0	1693.3	2032.0	2540.0	3386.7
26	1320.8	1761.1	2113.3	2641.6	3522.1
27	1371.6	1828.8	2194.6	2743.2	3657.6
28	1422.4	1896.5	2275.8	2844.8	3793.1
29	1473.2	1964.3	2357.1	2946.4	3928.5
30	1524.0	2032.0	2438.4	3048.0	4064.0
31	1574.8	2099.7	2519.7	3149.6	4199.5
32	1625.6	2167.5	2626.4	3251.2	4334.9
33	1676.4	2235.2	2682.2	3352.8	4470.4
34	1727.2	2302.9	2763.5	3454.4	4602.5
35	1778.0	2370.7	2844.8	3556.0	4734.6
36	1828.8	2438.4	2926.1	3657.6	4876.8
37	1828.8	2506.1	3007.4	3759.2	5012.3
38	1930.4	2573.9	3088.6	3860.8	5147.7
39	1981.2	2641.6	3169.9	3962.4	5283.2
40	2032.0	2709.3	3251.2	4064.0	5418.7
41	2082.8	2777.1	3332.5	4165.6	5554.1
42	2133.6	2844.8	3413.8	4267.2	5689.6
43	2184.4	2912.5	3495.0	4368.8	5825.1
44	2235.2	2980.3	3576.3	4470.4	5960.5
45	2286.0	3048.0	3657.6	4572.0	6096.0
46	2336.8	3115.7	3738.9	4673.6	6231.5

47	2387.6	3183.5	3820.2	4876.8	6366.9
48	2438.4	3251.2	3901.4	4978.4	6502.4
49	2489.2	3318.9	3982.7	5080.0	6637.9
50	2540.0	3386.7	4064.0	5181.6	6773.3
100	5080.0	6773.3	8128.0	10160.0	13546.7

Brick Dimensions and Nomenclature

The sizes of brick units shown and discussed below are typical of those currently being produced. However, few manufacturers produce all sizes shown and different manufacturers will vary. It is recommended that the designer consult manufacturers or distributors before proceeding with a design incorporating a specific size of brick that may not be readily available in the particular locality.

The nomenclature indicated, while typical, is not completely standard throughout the industry. Except for the *Standard, Roman,* and *Norman* sizes, individual manufacturers may have their own names for certain sizes listed. It is suggested that in order to avoid confusion, the purchase or specifier first identify the brick by size.

Except for the non-modular Standard, Oversize, and 3-in. (75-mm) units, most bricks are produced in modular sizes. The nominal dimensions of modular brick are equal to the manufactured dimensions plus the thickness of the mortar joint for which the unit is designed. In general, the joint thicknesses used with brick are either 3/8" (9.3 mm) or 1/2" (12.5 mm).

The actual *manufactured* dimensions of the units may vary from the *specified* dimensions by not more than the permissible tolerances for variation in dimensions as prescribed in the applicable ASTM specifications (*Standard Specification for Facing Brick, ASTM Designation C 216; Standard Specifications for Building Brick, ASTM Designation C 62; Standard Specifications for Hollow Brick, ASTM Designation C 652*).

It should be noted that the designated manufacturers heights for the standard brick, the standard modular brick, and all other modular brick designed to be laid 3 courses to 8" (200 mm) are the same—2-1/4" (56.2 mm). There is a very practical reason for this. In 1946-47, with the adoption of modular coordination by the structural clay products industry, brick manufacturers who converted their production completely to modular sizes were faced with a problem in connection with supplying matching brick for additions to existing non-modular buildings. From the standpoint of appearance, most designers required that the vertical coursing in the addition built with modular brick match those in the existing building constructed with non-modular brick. It was agreed that the manufactured face height of the standard modular brick would remain at 2-1/4" (56.2 mm), even though the other dimensions, length and thickness, would become modular. Since that time, the custom has continued. The differences in mortar bed thickness required to maintain the modular coursing of 3 courses in 8" (200 mm) were considered minimal.

Size of Non-Modular Brick

Unit Designation	Manufactured Size, mm thickness	height	length
Three-inch (king size)	3	2-3/4	9-5/8
Three-inch (queen size)	3	2-3/4	8
Three-inch	3	2-3/4	8-5/8
Standard	3-5/8	2-1/4	8
Engineered Standard	3-5/8	2-13/16	8
Closure Standard	3-5/8	3-1/2	8

Metric Sizes of Non-Modular Brick

Unit Designation	Manufactured Size, mm* thickness	height	length
Three-inch (king size)	76	70	245
Three-inch (queen size)	76	70	203
Three-inch	76	70	70
Standard	92	57	203
Engineered Standard	92	71	203
Closure Standard	92	89	203

Sizes of Modular Brick (in inches)*

Unit Designation	Nominal Dimensions t x h x l	Joint size, inches	Manufactured Dimensions t x h x l
Modular	4 x 2-2/3 x 8	3/8	3-5/8 x 2-1/4 x 7-5/8
		1/2	3-1/2 x 2-1/4 x 7-1/2
Engineer, Modular	4 x 3-1/5 x 8	3/8	3-5/8 x 2-13/16 x 7-5/8
		1/2	3-1/2 x 2-11/16 x 7-1/2
Closure Modular	4 x 4 x 8	3/8	3-5/8 x 3-5/8 x 7-5/8
		1/2	3-1/2 x 3-1/2 x 7-1/2
Roman	4 x 2 x 12	3/8	3-5/8 x 1-5/8 x 11-5/8
		1/2	3-1/2 x 1-1/2 x 11-1/2
Norman	4 x 2-2/3 x 12	3/8	3-5/8 x 2-1/4 x 11-5/8
		1/2	3-1/2 x 2-1/4 x 11-1/2
Engineer Norman	4 x 3-1/5 x 12	3/8	3-5/8 x 2-3/4 x 11-5/8
		1/2	3-1/2 x 2-11/16 x 11-1/2

Utility	4 x 4 x 12	3/8	3-5/8 x 3-5/8 x 11-5/8
		1/2	3-1/2 x 3-1/2 x 11-1/2
Triple	4 x 6 x 8	3/8	3-5/8 x 5-5/8 x 7-5/8
		1/2	3-1/2 x 5-1/2 x 7-1/2
Jumbo Utility	4 x 8 x 8	3/8	3-5/8 x 7-5/8 x 7-5/8
		1/2	3-1/2 x 7-1/2 x 7-1/2
6-in. Norwegian	6 x 3-1/5 x 12	3/8	5-5/8 x 2-13/16 x 11-5/8
		1/2	5-1/2 x 2-11/16 x 11-1/2
6-in. Jumbo	6 x 4 x 12	3/8	5-5/8 x 3-5/8 x 11-5/8
		1/2	5-1/2 x 3-1/2 x 11-1/2
8-in. Jumbo	8 x 4 x 12	3/8	7-5/8 x 3-1/2 x 11-1/2
8-in.	8 x 4 x 16	3/8	7-5/8 x 3-1/2 x 15-1/2

*Available as solid units conforming to ASTM C 216 or ASTM C 62, or in a number of cases, as hollow brick conforming to ASTM C652.

Metric Sizes of Modular Brick (in mm)

Unit Designation	Nominal Dimensions t x h x l	Joint size, mm	Manufactured Dimensions t x h x l
Modular	102 x 68 x 203	10	92 x 57 x 194
		13	89 x 57 x 191
Engineer, Modular	102 x 82 x 204	10	92 x 71 x 194
		13	89 x 68 x 191
Closure Modular	102 x 102 x 20	10	90 x 92 x 194
		13	89 x 89 x 191
Roman	102 x 51 x 306	10	92 x 41 x 295
		13	89 x 38 x 292
Norman	102 x 68 x 306	10	92 x 57 x 295
		13	89 x 57 x 292
Engineer Norman	102 x 81 x 306	10	92 x 70 x 295
		13	89 x 68 x 292
Utility	102 x 102 x 306	10	92 x 92 x 295
		13	89 x 89 x 292

Triple	102 x 152 x 204	10	92 x 143 x 194
		13	89 x 140 x 191
Jumbo Utility	102 x 204 x 204	10	92 x 194 x 295
		13	89 x 191 x 191
6-in. Norwegian	152 x 81 x 306	10	143 x 71 x 295
		13	140 x 68 x 292
6-in. Jumbo	152 x 102 x 306	10	143 x 168 x 295
		13	140 x 89 x 292
8-in. Jumbo	204 x 102 x 306	10	194 x 89 x 292
8-in.	204 x 102 x 406	10	194 x 89 x 394

Non-Modular Brick Sizes
(Specified Dimensions)

Position of Brick in the Wall. Most brick are laid as stretchers so that the longer of the face dimensions is horizontal. The drawings below also illustrate terms applied to other brick positions as placed in the wall. The shaded areas indicate the surfaces of the brick that are exposed.

Brick Bonds and Patterns

Bond in brickwork is the overlapping of one brick upon the other, either along the length of the wall or through its thickness, in order to bind them together into a secure structural mass. Units are shifted back and forth so that the vertical joints in two successive layers or "courses" do not come into line; in other words, the brick are laid so as to break the joint, the whole forming a natural bond or a structural unit giving strength to the wall.

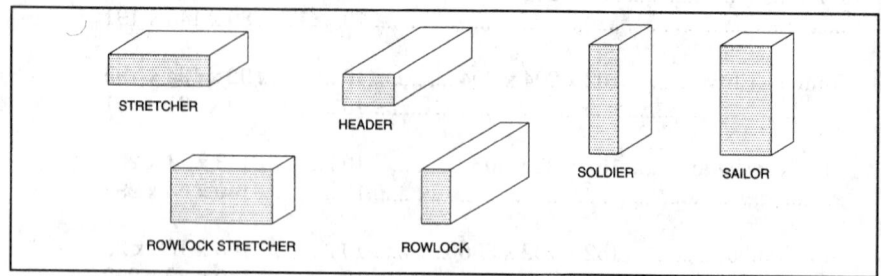

Brick Positions in Wall

Modular Brick (Nominal Dimensions)

Bond. The word bond, when used in reference to masonry, has three meanings:

1. *Structural Bond:* The method by which individual masonry units are interlocked or tied together to cause the entire assembly to act as a single structural unit.

2. *Pattern Bond:* The pattern formed by the masonry units and the mortar joints on the face of a wall. The pattern may result from the type of structural bond used or may be purely a decorative one, unrelated to the structural bonding.

3. *Mortar Bond:* The adhesion of mortar to the masonry units or to reinforcing steel.

Structural Bonds. Structural bonding of masonry walls may be accomplished in three ways: by the overlapping (interlocking) of masonry units, by use of metal ties embedded in connecting joints, and by the adhesion of grout to adjacent wythes of masonry.

The overlapped bond is based on variations of two traditional methods of bonding. The first is known as *English* bond and consists of alternating courses of headers and stretchers (Figure 1). The second is *Flemish* bond and consists of alternating headers and stretchers in every course, so arranged that the headers and stretchers in every other course appear in vertical lines (Figure 1).

The stretchers, laid with the length of the wall, develop longitudinal bonding strength, while the headers, laid across the width of the wall, bond the wall transversely.

Modern building codes require that masonry-bonded brick walls be bonded so that not less than 4% of the wall surface is composed of headers, with the distance between adjacent headers not exceeding 24" (600 mm), vertically or horizontally.

Structural bonding of masonry walls with metal ties is used in both solid wall and cavity wall construction (Figure 2). Most building codes permit the use of rigid steel bonding ties in solid walls.

At least one metal tie should be used for each 4-1/2 s.f. (41.8 sq.m) of wall surface. Ties in alternate courses should be staggered. The distance between adjacent ties should not exceed 24" (600 mm) vertically and 36" (900 mm) horizontally. Additional bonding ties, spaced not more than 3' (0.9 m) apart around the perimeter and within 12" (300 mm) of the opening, should be provided at all openings.

If ties less than 3/16" (4.68 mm) in diameter are used, tie spacing should be reduced so that the tie area per s.f. (sq.m) of wall is not less than specified above.

Structural bonding of solid and reinforced brick masonry walls is sometimes accomplished by grout, which is poured into the cavity or collar joint between wythes of masonry.

The method of bonding will depend on the use requirements, wall type and other factors. However, the metal tie method is generally recommended for exterior walls. Some of the advantages of this method are greater resistance to rain penetration and ease of construction. Metal ties also allow slight differential movements of the facing and backing, which may relieve stresses and prevent cracking.

Pattern Bonds. Frequently, structural bonds, such as English or Flemish, or variations of these, may be used to create patterns in the face of a wall. However, in the strictest sense of the term, pattern refers to the change or varied arrangement of the brick texture or color used in the face. Therefore, it may be possible to secure many patterns using the same structural bond. Patterns also may be produced by the method of handling the mortar joint or by projecting or recessing certain brick from the plane of the wall, thus creating a distinctive wall texture that is not solely dependent on the texture of the individual brick.

ENGLISH CORNER   DUTCH CORNER

ENGLISH BOND

DUTCH CORNER   ENGLISH CORNER

FLEMISH BOND

Figure 1

There are five basic structural bonds, which create typical patterns, commonly used today. These are running bond, common or American bond, Flemish bond, English bond, and block or stack bond. Through the use of these bonds and variations of the color and texture of the brick, and of the joint types and color, an almost unlimited number of patterns can be developed.

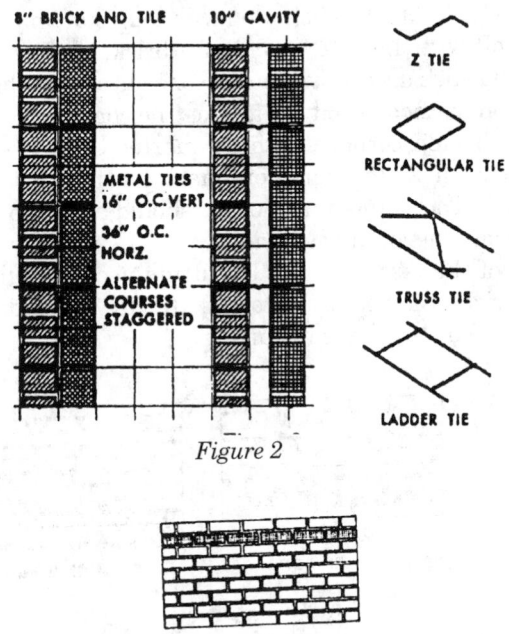

Figure 2

Figure 3

Running Bond. The simplest of the basic pattern bonds, the running bond, consists of all stretchers. Because there are no headers in this bond, metal ties are ordinarily used. Running bond is used largely in cavity wall construction and veneered walls of brick, and often in facing tile walls where the bonding may be accomplished by extra width stretcher tile.

Common or American Bond. Common or American bond (Figure 4) is a variation of running bond with a course of full length headers at regular intervals. This bond is obtained by laying a course of headers every fifth, sixth, or seventh course. To maintain the effect of the running bond, a special double header bond is sometimes used. This method of using headers, as in Common or American Bond, in order to secure transverse strength of wall, can be treated in a way to produce more pleasing effects, as may be seen in Flemish or English Bonds.

English Bond. English Bond (Figure 5) is made up of alternating courses of stretchers and headers. Ordinarily half brick are used for the header courses, except that every sixth course a full header course is used to tie the face and common brick walls together. The allowance for headers was given previously, under *Percentages to be Added for Various Brick Bonds*. Snap headers are used in courses that are not structural bonding courses.

Dutch Bond or English Cross Bond. Dutch or English Cross Bond (Figure 6) is similar to English Bond, except starting courses differ. Each course starts off with 1/4, 1/2, or 3/4 of a brick, alternating as shown in the illustration, and each successive course consists of stretchers and headers.

There are two methods used in starting the corners in Flemish and English bonds. In the "Dutch corner", a three-quarter brick closure is used. In the "English corner", a 2" (50 mm) or quarter brick closure, called a "queen closure", is used. The 2" (50 mm) closure should always be placed 4" (100 mm) in from the corner, never at the corner.

On account of the large amount of cutting necessary to obtain the 1/4 and 3/4 pieces of brick, it is necessary to add 10% to 15% additional brick to allow for waste, caused by breakage in cutting.

Face Brick Laid in Common or American Bond
Figure 4

HEADER COURSES
¾ BRICK
WHOLE BRICK
CLOSURE
BONDING COURSE
BONDING COURSE

Face Brick Laid in English Bond
Figure 5

ENGLISH CROSS OR DUTCH BOND

ENGLISH CORNER   DUTCH CORNER

HEADER COURSES
CLOSURES
BONDING COURSE
BONDING COURSE

Face Brick Laid in English Cross or Dutch Cross Bond
Figure 6

STACK BOND

Figure 7

STRETCHER
HEADER
WHOLE BRICK CLIPPED
CLOSURE
½ BRICK
BAT

BONDING COURSE

BONDING
COURSE

Face Brick Laid in Flemish Bond
Figure 8

Block or Stack Bond. Block or stack bond (Figure 7) is purely a pattern bond. There is no overlapping of units. All vertical joints are aligned. Usually this pattern is bonded to the backing with rigid steel ties, but 8" (200 mm) bonder units are often available. In large wall areas and in load bearing construction, it is advisable to reinforce the wall with steel reinforcement placed in the horizontal mortar joints. In stack bond, it is imperative that prematched or dimensionally accurate masonry units be used if the vertical alignment of the head joints is to be maintained. The architect or contractor might want to use a "gauged" brick, allowing for the manufacturer's acceptable production tolerances for brick used in this type of bond. Check with the supplier for additional costs.

Waste in cutting is greatly reduced with a masonry saw, and a brick can be cut in 10 to 15 seconds.

If brick are laid with full header courses every second or sixth course, add the percentages given on the previous pages.

Face Brick Laid in Double Flemish or Garden Wall Bond
Figure 9

Flemish Bond. Flemish Bond (Figure 8) is effective, because each course is laid in alternate stretchers and headers, the header resting upon the middle of the stretcher in successive courses. When full brick are used for headers, it requires one-half more brick than where half brick or "blind" headers are used.

Double Flemish Bond. This is the arrangement of bricks along the course so that each header is preceded and followed by two stretchers for the entire course (Figure 9). On consecutive courses, the position of the header is directly over the joint between the two stretchers. Each stretcher has a lap of three-fourths of its length over the stretcher beneath.

Three Stretcher Flemish Bond. An arrangement (Figure 10) of bricks along the course so that a header is placed between each sequence of three stretchers. The header of one course is centered over the middle stretcher of the course below.

Four Stretcher Flemish Bond. In this bond (Figure 11), a header occurs between each sequence of four stretchers. The header of one course is centered between the headers of the course below. The allowances for header courses are given on the previous pages.

Garden Wall Bond. Garden Wall Bond (Figure 12) is merely a modification of Flemish Bond, secured by laying courses with two to four stretchers alternating with a header, and requires the same number of brick per square foot as Flemish Bond.

Face Brick Laid in Three Stretcher Flemish Bond
Figure 10

Face Brick Laid in Four Stretcher Flemish Bond
Figure 11

Diamond Bond Patterns. Diamond Bond Patterns (Figure 12) are secured by a modification of the Garden Wall Bond. It is, however, only in case of large wall surfaces that patterns of an elaborate character could be recommended, as any departure from simple bonds adds to the cost of bricklaying.

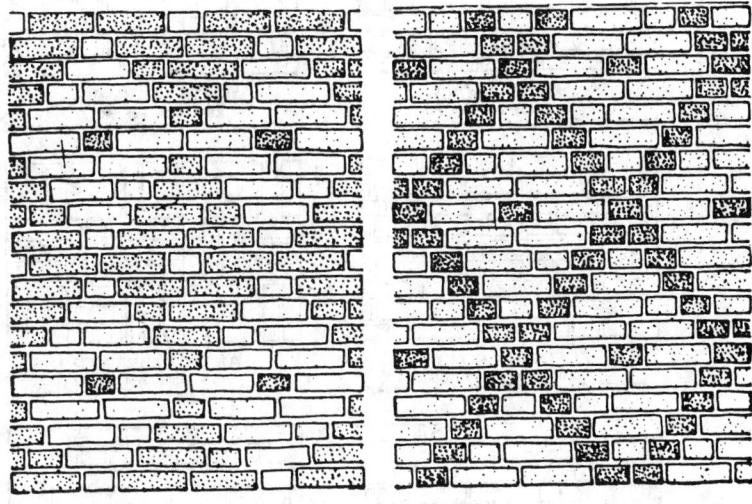

Face Brick Laid in Diamond Bond Patterns
Figure 12

Nominal Modular Sizes of Brick*

Designation	Thick. in. (mm)	Face Hgt. in. (mm)	Dimensions Length in. (mm)	No. of Unit Courses in 16" (400mm)
Standard	4 (100)	2-2/3 (66.7)	8 (200)	6
Engineer	4 (100)	3-1/5 (80.0)	8 200)	5
Economy 8 or Jumbo Closure	4 (100)	4 (100)	8 (200)	4
Double	4 (100)	5-1/3 (133.3)	8 (200)	3
Roman	4 (100)	2 (50)	12 (300)	8
Norman	4 (100)	2-2/3 (66.7)	12 (300)	6
Norwegian	4 (100)	3-1/5 (80)	12 (300)	5
Economy 12 or Jumbo Utility	4 (100)	4 (100)	12 (300)	4
Triple	4 (100)	5-1/3 (133.3)	12 (300)	3
6-in. Norwegian	6 (150)	3-1/5 (80)	12 (300)	5
6-in. Jumbo	6 (150)	4 (100)	12 (300)	4
8-in. Jumbo	8 (200)	4 (100)	12 (300)	4

*Available as solid units conforming to ASTM C 216 or ASTM C 62, or in some cases, as hollow brick conforming to ASTM C 652.

Elevations of Various Brick Bonds

Modular Size Brick

Economy is the prime objective of modular design, which is not a complex and revolutionary plan for coordinating building materials, but rather, a simplified and efficient system of design, layout, and construction resulting in definite economies.

After an exhaustive study, a four-inch (100 mm) module or increment was agreed upon as a unit that would afford the maximum practical standardization and simplification, and at the same time, afford a sufficient flexibility in design.

Conventional Standard Brick

2-2/3" Modular Brick

3" Modular Brick

4" Modular Brick

Modular Design

On the basis of the four-inch (100 mm) module or increment, certain sizes of brick and tile have been recommended. It must be remembered that the four-inch (100 mm) module, governing both the vertical and horizontal dimensions, is used to govern the layout of finished wall areas and to determine heights and widths of doors, windows, and other openings.

Because the module is a unit of wall measurement, it cannot be applied to the individual brick or tile, but must be considered as governing the size of the brick or tile, plus the mortar joint.

In employing the modular system of measurement for masonry walls, the dimension is considered from center to center of the mortar joints. To fit properly into the modular system, the actual sizes of the individual units of either brick or tile are determined by the modular dimension less the mortar joint.

The explanatory details, drawn with 1/2" (12.5 mm) mortar joints, present this matter very clearly. The first illustration indicates a wall section built of standard size brick measuring 2-1/4" x 3-3/4" x 8" (56.25 x 93.75 x 200 mm). It will be noted that a unit of this size does not comply with the modular measurements as indicated by the four-inch (100 mm) modular dimension lines, known as grid lines. Therefore, if standard size brick are used in the construction of a project designed according to the modular system, wasteful cutting and fitting of the brick would be necessary.

Note the difference in the second wall section. In this case a modular brick is employed, having an actual size of 2-1/6" x 3-1/2" x 7-1/2" (54.15 x 87.5 x 187.5 mm). Using a 1/2" (12.5 mm) mortar joint, three courses of brick lay up perfectly to every two horizontal grid lines, and one length of brick fits two vertical grid lines. The same holds true of the other sizes of brick shown.

Exact heights of modular brick will vary, the difference being taken up in the bed joints. It is therefore important to know the exact size of brick. The following table lists standard joint widths.

Modular Face Brick

Three standard modular joint thicknesses are 1/4" (6.25 mm), 3/8" (9.37 mm), and 1/2" (12.5 mm). The number of modular masonry units per square foot (square meter) will remain constant for each nominal size, the difference being taken up in the joint.

Modular Brick Walls Without Headers

Nom. Size of Brick, in. (h x t x l)	Brick per S.F. of Wall	C.F. Mortar per 100 S.F. Wall Joint Thickness			C.F. Mortar per 1,000 Brick Joint Thickness		
		1/4"	3/8"	1/2"	1/4"	3/8"	1/2"
2-2/3 x 4 x 8	6.750	3.81	5.47	6.95	5.65	8.10	10.30

Nom. Size of Brick, in. (h x t x l)	Brick per S.F. of Wall	C.F. Mortar per 100 S.F. Wall Joint Thickness			C.F. Mortar per 1,000 Brick Joint Thickness		
		1/4"	3/8"	1/2"	1/4"	3/8"	1/2"
3-1/5 x 4 x 8	5.625	3.34	4.79	6.10	5.94	8.52	10.84
4 x 4 x 8	4.500	--	4.12	5.24	--	9.15	11.65
5-1/3 x 4 x 8	3.375	--	3.44	4.34	--	10.19	12.87
2 x 4 x 12	6.000	--	6.43	8.20	--	10.72	13.67
2-2/3 x 4 x 12	4.500	3.52	5.06	6.46	7.82	11.24	14.35
3 x 4 x 12	4.000	--	4.60	5.87	--	11.51	14.68
3-1/5 x 4 x 12	3.750	3.04	4.37	5.58	8.11	11.66	14.89
4 x 4 x 12	3.000	2.56	3.69	4.71	8.54	12.29	15.70
5-1/3 x 6 x 12	2.250	--	3.00	3.84	--	13.34	17.05
2-2/3 x 6 x 12	4.500	--	7.85	10.15	--	17.45	22.55
3-1/5 x 6 x 12	3.750	--	6.79	8.77	--	18.10	23.39
4 x 6 x 12	3.000	--	5.72	7.40	--	19.07	24.67

Brick Size, mm	(metric)	Joint Thickness, mm			Joint Thickness, mm		
		6.3	9.4	12.5	6.3	9.4	12.5
67 x 100 x 200	72.66	0.12	0.17	0.21	0.17	0.25	0.31
80 x 100 x 200	60.55	0.10	0.15	0.19	0.18	0.26	0.33
100 x 100 x 200	48.44	--	0.13	0.16	--	0.28	0.36
133 x 100 x 200	36.33	--	0.10	0.13	--	0.31	0.39
50 x 100 x 300	64.59	--	0.20	0.25	--	0.33	0.42
68 x 102 x 305	48.44	0.11	0.15	0.20	0.24	0.34	0.44
76 x 102 x 305	43.06	--	0.14	0.18	--	0.35	0.45
81 x 102 x 305	40.37	0.09	0.13	0.17	0.25	0.36	0.45
102 x 102 x 305	32.29	0.08	0.11	0.14	0.26	0.37	0.48
135 x 102 x 305	24.22	--	0.09	0.12	--	0.41	0.52
68 x 152 x 305	48.44	--	0.24	0.31	--	0.53	0.69
81 x 152 x 305	40.37	--	0.21	0.27	--	0.55	0.71
102 x 152 x 305	32.29	--	0.17	0.23	--	0.58	0.75

Non-Modular Brick and Mortar Required for Single Wythe Walls
in Running Bond (no allowances for breakage or waste)

Brick Size, inches (t x h x l)	3/8" Joints			1/2" Joints		
	No. of Brick per 100 SF	C.F. Mortar per 100 SF	C.F. Mortar per 1000 Brick	No. of Brick per 100 SF	C.F. Mortar per 100 SF	C.F. Mortar per 1000 Brick
2-3/4 x 2-3/4 x 9-3/4	455	3.2	7.1	432	4.5	10.4
2-5/8 x 2-3/4 x 8-3/4	504	3.4	6.8	470	4.1	8.7
3-3/4 x 2-1/4 x 8	655	5.8	8.8	616	7.2	11.7
3-3/4 x 2-3/4 x 8	551	5.0	9.1	522	6.4	12.2

Brick Size, mm (t x h x l)	9.38 mm Joints			12.5 mm Joints		
	No. of Brick per 9 Sq.M	Cu.M Mortar per 9 Sq.M	Cu.M Mortar per 1000 Brick	No. of Brick per 9 Sq.M	Cu.M Mortar per 9 Sq.M	Cu.M Mortar per 1000 Brick
68.75 x 68.75 x 244	455	0.09	0.20	432	0.13	0.29
65.63 x 68.75 x 218.75	504	0.10	0.19	470	0.12	0.25
93.75 x 56.25 x 200	655	0.16	0.25	616	0.20	0.33
93.75 x 68.75 x 200	551	0.14	0.26	522	0.18	0.35

The above quantities include only mortar for bed and vertical joints. The following table gives allowances for backing mortar or collar joints.

C.F. of Mortar Per 100 S.F. of Wall

1/4" joint	3/8" joint	1/2" joint
2.08	3.13	4.17

(metric)

6.4 mm joint	9.5 mm joint	12.7 mm joint
0.06	0.10	0.13

Cubic feet per 1000 units =

$$\frac{10 \times \text{cubic feet per 100 sq. ft. of wall}}{\text{number of units per sq. ft. of wall}}$$

Cubic meters per 1000 units =

$$\frac{10 \times \text{cubic meters per 10 sq.m of wall}}{\text{number of units per sq.m of wall}}$$

Material Quantities per Cubic Foot (Cubic Meter) of Mortar
Quantities by Volume

Material	Mortar Type and Proportions by Volume			
	M 1:1/4:3	S 1:1/2:4-1/2	N 1:1:6	O 1:2:9
Cement	0.333	0.222	0.167	0.111
Lime	0.083	0.111	0.167	0.222
Sand	1.000	1.000	1.000	1.000

Material Quantities per Cubic Foot (Cubic Meter) of Mortar
Quantities by Weight

Material	Mortar Type and Proportions by Volume			
	M	S	N	O
Cement	31.33	20.89	15.67	10.44
Lime	3.33	4.44	6.67	8.89
Sand	80.00	80.00	80.00	80.00

Material Costs of Brick

Brick costs will vary widely among regions and within color ranges:

1. *Common Brick:* Standard modular—4 x 2-1/4 x 7-5/8 (100 x 56.25 x 190.62 mm)—and standard non-modular—4 x 2-1/4 x 8 (100 x 56.25 x 200 mm)—average between $240.00 and $350.00 per thousand.

2. *Jumbo Brick:* 4 x 2-3/4 x 8 (100 x 68.75 x 200 mm) averages between $290.00 and $390.00 per thousand.

3. *Utility Brick:* 4 x 3-5/8 x 11-5/8 (100 x 90.62 x 290.62 mm) averages between $600.00 and $1,100.00 per thousand.

4. *Economy Brick:* 4 x 7-5/8 x 7-5/8 (100 x 190.6 x 190.6 mm) averages between $420.00 and $700.00 per thousand.

5. *Paving Brick:* 4 x 2-1/4 x 8 or 4 x 3-5/8 x 7-5/8 (100 x 56.25 x 200 mm, or 100 x 90.62 x 190.6 mm) averages between $315.00 and $460.00 per thousand.

6. *Glazed Brick:* 4 x 2-3/4 x 8 (100 x 68.75 x 200 mm) averages between $750.00 and $820.00 per thousand.

In general red tones are cheapest, earth tones fall in the middle range, and whites and grays are the most expensive, except for glazed units, which cost $315.00 and up in the more brilliant shades.

Brick Cavity Walls

Brick cavity walls are distinguished from other forms of hollow masonry by a continuous vertical and horizontal air space bridged by masonry ties. The primary purpose of this space is to provide an absolute barrier against the penetration of moisture to the inner side of the wall, including at heads and jambs of openings, joist bearing points, etc., assuring trouble-free masonry and permitting direct plastering without furring. It also increases resistance to heat-flow to about the same extent as the air space created by ordinary furring.

Since this construction affords no supporting material through which moisture penetrating the outer 4" (100 mm) wythe of the wall can soak to the inner wythe, it helps to insure that the inner wythe will remain dry, provided adequate precautions are taken to intercept and deflect moisture running down the inner side of the outer wythe at the heads of all openings, and provided the bottom of the wall is equipped with a suitable flashing to prevent absorption of soil water.

Grid Paper Is Used by the Designer for All Layouts and Details in Modular Design. The Above Plan and Elevation of Brick between Two Window Openings Show How the Brick Fit the Intervening Space Perfectly with No Wasteful Cutting or Fitting.

In place of the usual header courses, the inner and outer wythes of the cavity walls are connected by rust-proof metal ties—usually 3/16" (4.6 mm) diameter, square-end, Z bars—placed in every sixth course and 3'-0" (0.9 m) on center horizontally.

Labor Laying Building Brick

Many factors influence the labor cost of laying building brick. There are no set quantities of brick that a mason will lay per hour or day. With ten masons laying brick on the same wall and on exactly the same class of work, no two of them will lay the same number of brick. Averages must be taken.

Solid brick walls are used much less frequently now than in the past, and where brick is used as a facing, it is most often backed up with cement blocks. Labor costs are much lower for handling and setting cement blocks, because the units are larger than brick. One 8"x8"x16" (200 x 200 x 400 mm) cement block usually displaces 12 building brick in the wall.

The class of work also has a great deal to do with the difference in labor costs. It costs much more to lay brick in an 8" (200 mm) wall with struck joints than in a 16" (400 mm) wall having cut joints on one side of the wall and struck joints on the other. Also, it costs more to lay brick in portland cement mortar than in lime mortar, because the lime mortar spreads far more easily than cement mortar.

Grid paper is used by the designer for all layouts and details in modular design. The above plan and elevation of brick between two window openings show how the brick fit the intervening space perfectly with no wasteful cutting or fitting. It costs more to lay brick in cold, wet weather when the bricks are slippery and the mason's hands cold than in fair weather with all conditions favorable.

It requires more labor to lay brick in the winter, when the brick and mortar must be heated and salamanders placed along the floor or scaffold, requiring additional labor attending salamanders, removing ice, snow, etc., than in the summer when conditions are favorable.

The quantities and costs given are based on a good grade of workmanship applicable to the type of building on which it is to be used.

The quantities given are based on the performances of experienced masons working regularly at their trade under present conditions. They will not hold good in small communities where the mason does the cement work and plastering as well as the bricklaying and stone setting of the community.

The costs are based on efficient management and supervision, the use of modern tools and appliances, machine mortar mixers, etc., on jobs large enough to warrant their use. On jobs of sufficient size to require a mason foreman, it will be necessary to add for this time on the basis of a certain price per 1,000 brick or at a certain price per week, based on the estimated duration of the job.

Labor Production For Laying Building Brick

Class of Work	Mortar Joints Style	Kind of Mortar	Average Number Brick Laid per 8-hr. Day	Aver. Hrs. 1,000 Brick		
				Mason Hours	Labor Hours	Hoist Engr.*
8" walls, 1-story bungalows, garages, two-flat buildings, residences, etc.	Cut	Lime	750–800	10.5	9.0	
	Struck	Lime	700–750	11.0	9.0	
12" walls, ordinary construction, apartment buildings, houses, garages, factories, stores, store and apartment buildings, schools, etc.	Cut	Lime	950–1,050	8.0	8.0	
	Struck	Lime	925–975	8.5	8.0	
	Cut	Cement	825–900	9.3	8.0	
	Struck	Cement	750–850	10.0	8.0	
Hodding brick to second story, add				2.0	2.0	
16" walls, heavy warehouse, factory and industrial work. Straight walls	Cut	Lime	1,100–1,225	7.0	8.0	
	Struck	Lime	1,000–1,125	7.5	8.0	
	Cut	Cement	950–1,050	8.0	8.0	
	Struck	Cement	900–1,000	8.4	8.0	
Backing-up face brick, cut stone, terra cotta on ordinary wall bearing buildings, figure same as given above for 8" and 12" walls						

WINTER WORK. Heating brick, mortar, attending salamanders, removing snow and ice, add extra for this work.

*HOISTING. Add about ¼-hr. hoisting engineer time per 1,000 for reasonable size jobs using 35,000 or more brick per day; ½-hr. per 1,000 for jobs using 15,000 to 20,000 brick per day; and ⅓-hr. per 1,000 for jobs using 25,000 to 30,000 brick per day.

If automatic hoists are used, omit all time for hoisting engineer.

Labor Production For Laying Building Brick—Cont.

Class of Work	Mortar Joints Style	Kind of Mortar	Average Number Brick Laid per 8-hr. Day	Mason Hours	Aver. Hrs. 1,000 Brick	
					Labor Hours	Hoist Engr.*
Backing-up face brick, cut stone, terra cotta on steel or concrete skeleton frame buildings. First Grade workmanship. Walls 8" to 12" thick.	Cut	Lime	825–900	9.3	9.0	
	Struck	Lime	750–850	10.0	9.0	
	Cut	Cement	725–800	10.5	9.0	
	Struck	Cement	700–775	10.8	9.0	
Public buildings, First Grade workmanship, Schools, college and university bldgs. Courthouses, state capitols, public libraries, etc. 12" to 20" walls.	Cut	Lime	850–950	9.0	8.0	
	Struck	Lime	800–900	9.4	8.0	
	Cut	Cement	750–850	10.0	8.0	
	Struck	Cement	700–800	10.6	8.0	
Powerhouses and other structures, having high walls 12" to 20" thick, without intermediate floors.	Cut	Lime	1,100–1,200	7.0	9.0	
	Struck	Lime	1,000–1,100	7.6	9.0	
	Cut	Cement	925–1,050	8.0	9.0	
	Struck	Cement	875–950	8.8	9.0	
Shoved joints. Brick laid in full shoved joints with all vertical joints slushed full of mortar, add to all of the above				1.5		
Basement foundation walls, paving brick.	Cut	Cement	900–1,000	8.5	8.5	
	Struck	Cement	850–950	9.0	9.0	

WINTER WORK. Heating brick, mortar, attending salamanders, removing snow and ice, add extra for this work.
*HOISTING. Add about ¼-hr. hoisting engineer time per 1,000 for reasonable size jobs using 35,000 or more brick per day; ½-hr. per 1,000 for jobs using 15,000 to 20,000 brick per day; and ⅓-hr. per 1,000 for jobs using 25,000 to 30,000 brick per day.
If automatic hoists are used, omit all time for hoisting engineer.

Labor Production For Laying Building Brick—Cont.

Class of Work	Mortar Joints Style	Kind of Mortar	Average Number Brick Laid per 8-hr. Day	Aver. Hrs. 1,000 Brick		
				Mason Hours	Labor Hours	Hoist Engr.*
Building brick foundation walls 8" to 12" thick, ordinary workmanship.	Cut	Lime	1,000–1,100	7.6	8.0	
	Struck	Lime	900–975	8.5	8.0	
	Cut	Cement	900–975	8.5	8.0	
	Struck	Cement	825–900	9.3	8.0	
Chimneys and stacks, building brick, 1'-4" to 2'-0" sq. 15'-0" above roof, 4" to 8" walls. Hodding brick, extra.	Struck	Lime	500–550	16.0	16.0 3.0	
Large chimney and stacks, 3'-0" to 4'-0" Sq. 15' to 30' high above roof, 8" to 12" walls.	Struck	Lime	550–600	14.0	16.0	
Large brick stacks 100' to 150' high, walls from 1'-8" at base to 12" top. Inside put-log scaffold. Outside scaffold extra.	Struck	Lime	650–750	11.5	14.0	
Bricking-in boilers, fire-boxes, etc.	Struck		600–650	12.5	10.0	

WINTER WORK. Heating brick, mortar, attending salamanders, removing snow and ice, add extra for this work.
*HOISTING. Add about ¼-hr. hoisting engineer time per 1,000 for reasonable size jobs using 35,000 or more brick per day; ½-hr. per 1,000 for jobs using 15,000 to 20,000 brick per day; and ⅓-hr. per 1,000 for jobs using 25,000 to 30,000 brick per day.
If automatic hoists are used, omit all time for hoisting engineer.

Labor Laying Face Brick

The labor cost of laying face brick will vary with the grade of workmanship, style of bond in which the brick are laid, width and kind of mortar joint, type of building--whether long, straight walls or walls cut up with pilasters, door and window openings--and last but not least, the ability of the foreman in charge.

Common bond is the most economical method of laying face brick. Other styles, such as English, Flemish Dutch, and Garden Wall bond, require more labor on account of the various patterns and the additional cutting required. This is especially so on First Grade work where all vertical mortar joints must be plumb.

For instance, when laying one of the many variations of Dutch, English, or Flemish brick bonds, it is necessary to do a great deal of cutting, especially where half brick are used for headers. It requires just as much time to cut the brick and spread mortar to lay one-half brick as for a full brick. The labor cost of laying 1,000 face brick in Dutch, English, or Flemish bond will be more if half-brick are used for headers than where full brick are used, because it will take just as much time to cut, spread mortar bed, apply mortar to brick, lay brick, and cut the joints.

Concave   Convex   V-Tooled   Stripped   Rodded

Raked Out   Weathered   Struck   Flush Cut

Styles of Mortar Joints

Flush Cut Joint. The style of mortar joints has considerable bearing on the labor costs. The flush cut joint, sometimes referred to as the *rough cut*, is the most economical, because the bricklayer cuts off the excess mortar with his trowel and the joint is finished. This produces an uncompacted joint with a small hairline crack where the mortar is pulled away from the brick by the cutting action. This joint is not always watertight.

Concave and V-Shaped Joints. These joints are normally kept quite small and are formed by the use of a steel jointing tool. These joints are very effective in resisting rain penetration and are recommended for use in areas subjected to heavy rains and high winds.

Weathered Joint. This joint requires care, as it must be worked from below. However, it is the best of the troweled joints, because it is compacted and sheds water readily.

Struck Joint. This is a common joint in ordinary brickwork. American mechanics often work from the inside of the wall, and this is an easy joint to strike with a trowel. Some compaction occurs, but the small ledge does not shed water readily, resulting in a less watertight joint than those discussed above.

Raked Joint. This joint is made by removing the surface of the mortar while it is still soft. While the joint might be compacted, it is difficult to make weathertight and is not recommended where heavy rain, high wind, or freezing is likely. This joint produces marked shadows and tends to darken the overall appearance of the wall.

Grapevine Joint. This is a concave joint with a drip slot placed horizontal into the mortar, approximately midway between bricks. This joint is very effective in resisting rain penetration.

For concave, V-tooled, weathered, and struck joints, it is necessary first to cut off the excess mortar with the trowel and then to do another operation, using a trowel for the weathered or struck joints or a jointing tool for concave and V-tooled joints. A bricklayer must remove the mortar 1/4" (6.2 mm) to 3/8" (9.3 mm) back from the face of the brick, and for a rodded joint, the bricklayer must remove the excess mortar and then strike the joint with a jointer. These operations all require time and cost money.

The stripped joint is the most expensive of any illustrated, because it is necessary to place narrow wood strips in the mortar joints between each course of brick, and these strips must remain until the mortar has set, so it will not squeeze out toward the face of the brick.

If face brick are laid with 7/8" (21.8 mm) or 1" (25 mm) bed mortar joints, the work proceeds very slowly and requires a stiff mortar, because the weight of the top brick courses squeeze the mortar out of the lower mortar joints, producing an uneven appearance. For this reason the work cannot be carried up many courses at a time. Where the joints are struck, it is necessary for the mortar to take its initial set before the joints can be struck to produce a workmanlike job. Brick work of this kind is very expensive.

The quantities and costs given on the following pages are intended to cover all kinds of face brick work, whether laid in veneer in frame buildings, in solid brick or brick and tile walls.

Labor Laying Face Brick in English, Flemish, Dutch, Garden Wall, and Diamond Bonds

Face brick laid in any of the above bonds with the courses consisting of various arrangements of headers and stretchers requires considerable extra labor cutting brick on account of the large number of "bats" or headers.

Labor Laying 1,000 Standard Size (8"x2¼"x3¾") or Modular Size (7⅝"x2¼"x3⅝") Face Brick in Common Bond

	2:1:9 Lime-Cement Mortar					Cement Mortar		
Style of Mortar Joint	Mason Hours	Labor Hours	Hoist Engr.*	Number Laid per 8-Hr. Day	Number Laid per 8-hr. Day	Mason Hours	Labor Hours	Hoist Engr.*
Ordinary Workmanship								
Flush Cut	16.0	12	0.5	475-525	425-475	17.8	13	0.5
V-Tooled or Concave	17.8	12	0.5	425-475	400-450	19.0	13	0.5
Struck or Weathered	18.0	12	0.5	420-450	390-435	19.5	13	0.5
Raked Out	17.8	12	0.5	425-475	400-450	19.0	13	0.5
Rodded	19.0	12	0.5	400-450	375-425	20.0	13	0.5
Stripped	26.7	15	0.5	275-325	275-300	27.8	16	0.5
First Grade Workmanship								
Flush Cut	19.0	13	0.5	400-450	375-425	20.0	14	0.5
V-Tooled or Concave	20.0	13	0.5	375-425	360-400	21.0	14	0.5
Struck or Weathered	21.0	13	0.5	360-400	350-390	21.5	14	0.5
Raked Out	20.0	13	0.5	375-425	360-400	21.0	14	0.5
Rodded	23.0	13	0.5	325-360	300-340	25.0	14	0.5
Stripped	32.0	16	0.5	225-275	200-240	36.5	18	0.5
⅞" to 1" Flush Cut	32.0	16	0.5	225-275	200-240	36.5	18	0.5
⅞" to 1" Struck	36.0	16	0.5	200-250	190-230	38.0	18	0.5

*Add Hoisting engineer's time only if required.

Labor Laying 1,000 Standard Size (8"x2¼"x3¾") or Modular Size (7⅝"x2¼"x3⅝") Face Brick in English, Flemish**, Dutch and Garden Wall Bond, Laid with Full Headers Every 6th Course**

2:1:9 Lime-Cement Mortar				Ordinary Workmanship		Cement Mortar			
Mason Hours	Labor Hours	Hoist Engr.*	Number Laid per 8-Hr. Day	Style of Mortar Joint	Number Laid per 8-hr. Day	Number Laid per 8-hr. Day	Mason Hours	Labor Hours	Hoist Engr.*
20.0	12	0.5	380-420	Flush Cut	360-400	360-400	21.0	13	0.5
21.3	12	0.5	350-400	V-Tooled or Concave	330-380	330-380	22.5	13	0.5
22.5	12	0.5	335-375	Struck or Weathered	320-360	320-360	23.5	13	0.5
21.3	12	0.5	350-400	Raked Out	330-380	330-380	22.5	13	0.5
22.5	12	0.5	335-375	Rodded	320-360	320-360	23.5	13	0.5
29.0	15	0.5	250-300	Stripped	240-285	240-285	31.0	15	0.5
				First Grade Workmanship					
22.5	13	0.5	335-375	Flush Cut	320-355	320-355	23.7	14	0.5
24.0	13	0.5	315-350	V-Tooled or Concave	300-335	300-335	25.0	14	0.5
25.0	13	0.5	300-340	Struck or Weathered	285-325	285-325	26.5	14	0.5
24.0	13	0.5	315-350	Raked Out	300-335	300-335	25.0	14	0.5
26.7	13	0.5	275-325	Rodded	260-310	260-310	29.0	14	0.5
35.5	16	0.5	200-250	Stripped	190-240	190-240	37.0	18	0.5
35.5	16	0.5	200-250	⅞" to 1" Flush Cut	190-240	190-240	37.0	18	0.5
40.0	16	0.5	175-225	⅞" to 1" Struck	165-215	165-215	42.0	18	0.5

*Add hoisting engineer's time only if required.
**If laid with full headers every course or every 2nd course, increase quantities laid approximately 8 percent.

Labor Laying 1,000 English Size 8⅞"x2⅞"x4" Face Brick

2:1:9 Lime-Cement Mortar				Common Bond		Cement Mortar		
Mason Hours	Labor Hours	Hoist Engr.*	Number Laid per 8-Hr. Day	Style of Mortar Joint	Number Laid per 8-hr. Day	Mason Hours	Labor Hours	Hoist Engr.*
22.5	17	0.6	335-375	Flush Cut	320-360	23.5	18	0.6
23.5	17	0.6	325-360	V-Tooled or Concave	310-340	24.6	18	0.6
26.3	17	0.6	285-325	Struck or Weathered	275-315	27.0	18	0.6
23.5	17	0.6	325-360	Raked Out	310-340	24.6	18	0.6
28.5	17	0.6	260-300	Rodded	250-290	29.5	18	0.6
				English, Flemish or Dutch Bond				
24.7	17	0.6	300-350	Flush Cut	285-325	26.3	18	0.6
26.3	17	0.6	285-325	V-Tooled or Concave	275-310	27.0	18	0.6
28.5	17	0.6	260-300	Struck or Weathered	250-290	29.5	18	0.6
26.3	17	0.6	285-325	Raked Out	275-310	27.0	18	0.6
31.0	17	0.6	240-275	Rodded	230-265	32.2	18	0.6

*Add hoisting engineer's time only if required.

Labor Laying 1,000 Standard Size (8"x2¼"x3¾") Vitrified Face Brick

2:1:9 Lime-Cement Mortar				Common Bond	Cement Mortar			
Mason Hours	Labor Hours	Hoist Engr.*	Number Laid per 8-Hr. Day	Style of Mortar Joint	Number Laid per 8-hr. Day	Mason Hours	Labor Hours	Hoist Engr.*
31.0	17	0.6	235-275	Buttered and Struck	225-265	32.7	18	0.6
22.5	16	0.6	335-375	Flush Cut	320-360	23.5	17	0.6
23.5	16	0.6	320-360	V-Tooled or Concave	300-340	25.0	17	0.6
26.3	16	0.6	285-325	Struck or Weathered	270-310	27.5	17	0.6
23.5	16	0.6	260-300	Rodded	250-285	30.0	17	0.6
28.5	16	0.6	320-360	Racked Out	300-340	25.0	17	0.6
				English, Flemish or Dutch Bond**				
34.8	17	0.6	210-250	Buttered and Struck	200-240	36.4	17	0.6
25.0	16	0.6	300-340	Flush Cut	285-325	26.3	17	0.6
26.3	16	0.6	285-325	V-Tooled or Concave	275-310	27.3	17	0.6
28.5	16	0.6	260-300	Struck or Weathered	250-290	29.7	17	0.6
26.3	16	0.6	285-325	Raked Out	275-310	27.3	17	0.6
31.0	17	0.6	240-275	Rodded	230-265	32.7	17	0.6

*Add hoisting engineer's time only if required.
**Full headers every sixth course. If full headers every other course, increase quantities laid approximately 8 percent.

Roman Face Brick

Roman size face brick, 12"x1½"x4" were extensively used some years ago for facing many types of buildings. Generally, they were laid with thin "buttered" joints and most of the original brick were made by the dry-pressed method and were relatively expensive.

Within the past few years, the modular Roman brick was introduced having a nominal size of 12"x2"x4", including mortar joint, while the actual brick size is 11⅝"x1⅝"x3⅝", requiring a ⅜" mortar joint to lay 2" high and 12" long. These brick are furnished with either standard brick textured faces or a split (rock-like) texture. The split face is generally preferred as it greatly accentuates the horizontal bonding effect unobtainable in any other type of construction. These units are now being produced in practically all sections of the United States and are in great demand, particularly for single story ranch homes, although they are also used for commercial structures, store fronts and for finished masonry interiors.

When laid with a ⅜" mortar joint, it requires 6 brick per sq. ft. of wall. Add for headers, breakage and waste.

When laid with a ⅜" mortar joint, it requires 10¾ cu. ft. of mortar per 1,000 brick, not including any back-up mortar or any allowance for waste.

Labor Laying 1,000 Roman Face Brick 3⅝"x1⅝"x11⅝" Face Brick

Common Bond				Style of Mortar Joint	Flemish Bond			
Mason Hours	Labor Hours	Hoist Engr.*	Number Laid per 8-Hr. Day		Number Laid per 8-hr. Day	Mason Hours	Labor Hours	Hoist Engr.*
29.6	17	0.5	250-290	Buttered and Struck	210-240	35.5	17	0.5
20.5	17	0.5	370-410	Flush Cut	300-350	24.6	17	0.5
23.5	17	0.5	320-360	V-Tooled or Concave	265-300	28.3	17	0.5
25.0	17	0.5	300-340	Struck or Weathered	245-285	30.2	17	0.5
23.5	17	0.5	320-360	Raked Out	265-300	28.3	17	0.5
26.7	17	0.5	280-320	Rodded	230-270	32.0	17	0.5

*Add hoisting engineer's time only if required.

Ordinarily, a full header course extends through the wall every 5th or 6th course, and the other headers are only half brick. On first grade workmanship, it is necessary to plumb all vertical mortar joints in order that the various patterns will work out accurately.

Using the labor laying brick in Common Bond as a base of 100, it will require approximately the following additional time to lay face brick in Dutch, English, and the different variations of Flemish bond, where half brick are used as headers instead of full brick.

Kind of Bond	Percent
Common Bond using Full Headers	100
Dutch Bond using Full Headers every other Course	120
Dutch Bond using Full Headers every Sixth Course	129
English Bond using Full Headers every other Course	120
English Bond using Full Headers every Sixth Course	129
Flemish Bond using Full Headers every Course	120
Flemish Bond using Full Headers every Sixth Course	126
Double Flemish Bond using Full Headers every Course	100
Double Flemish Bond using Full Headers every Sixth Course	116
3-Stretcher Flemish Bond using Full Headers every Other Course	107
3-Stretcher Flemish Bond using Full Headers every Sixth Course	112
4-Stretcher Flemish Bond using Full Headers every Other Course	106
4-Stretcher Flemish Bond using Full Headers every Sixth Course	109

Double Size Building Brick

Double size building brick—3-3/4" x 5" x 8" (93.75 x 125 x 200 mm)—are used in different parts of the country. Their use effects a saving in both mortar and labor over standard size building brick—3-3/4" x 2-1/4" x 8" (93.75 x 56.25 x 200 mm).

Each brick contains 40 sq. in. (25,808 sq.mm or 0.026 sq.m) on the face and when laid with a 1/2" (12.5 mm) bed or horizontal joint and a 1/4" (6.25 mm) vertical or end joint, covers 45.375 or 45-3/8 sq. in. (29,275.95 sq.mm or 0.029 sq.m) and requires 3.2 brick per s.f. (0.09 sq.m) of 4" (100 mm) wall; 6.4 brick per s.f. (0.09 sq.m) of 8" (200 mm) wall and 9.6 brick per s.f. (0.09 sq.m) of 12" (300 mm) or 13" (325 mm) wall, not including any allowance for waste.

Mortar Required for Double Size Building Brick

Based on 1/2" (12.5 mm) bed or horizontal joints and 1/4" (6.25 mm) end or vertical joints.

Mortar Required per 1,000 Double Brick

4" Walls*	8" Walls	12" Walls
12-3/4 c.f.	19-1/4 c.f.	21-1/4 c.f.

Mortar Required per 100 S.F. of Wall

4" Walls*	8" Walls	12" Walls
4 c.f.	12-1/4 c.f.	20-1/2 c.f.

(metric)
Mortar Required per 1,000 Double Brick

100 mm Walls*	200 mm Walls	300 mm Walls
0.36 cu.m	0.55 cu.m	0.60 cu.m

Mortar Required per 9 Sq.M of Wall

100 mm Walls*	200 mm Walls	300 mm Walls
0.12 cu.m	0.37 cu.m	0.62 cu.m

*Freestanding Walls.
No Allowance For Waste Included In Above Quantities.

Labor Laying Double Size Building Brick. A mason will lay about 60% as many double brick—3-3/4" x 5" x 8" (93.75 x 125 x 200 mm)—as standard size building brick—3-3/4" x 2-1/4" x 8" (93.75 x 56.25 x 200 mm). If a mason lays 800 building brick on a certain class of work, he would lay 480 double size building brick on the same class of work. In other words, if it required 10 hrs. mason time per 1,000 building brick, it would require 1-2/3 as many hours per 1,000 double brick, i.e., 10 x 1.667 = 16.67 or 16-2/3 hrs. mason time per 1,000 double brick.

Twin Brick, 8" x 8" x 2-1/4" (200 x 200 x 56.25 mm)

Twin Brick

Twin brick is a double size brick that contains the face brick and a 4" (100 mm) backing brick in one unit, size 8" x 8" x 2-1/4" (200 x 200 x 56.25 mm). It is made with a hollow core so that the twin brick weighs only 7-1/2 lbs. (3.4 kg) and can be picked up and laid in a one-hand operation. It is designed so the wall has no continuous mortar joint from outer to inner wall faces, helping to avoid moisture penetration.

It is intended primarily for residential work having 8" (200 mm) walls that are to be furred and plastered. Extra face brick must be used at openings and corners to insure proper bond.

It requires about 25 c.f. (0.70 cu.m) of mortar per 1,000 units.

On four jobs on which these brick were used, requiring 50,000 units, a mason laid 60 twin brick an hour or 480 per 8 hr. day, at the following labor cost per 1,000 units:

	Hours	Rate	Total	Rate	Total
Mason	17.0	$....	$....	$28.18	$ 479.06
Labor	11.5		....	21.04	241.96
Cost per 1,000 units			$....		$ 721.02
per brick					$ 0.72

Skintled Brickwork

Skintled brickwork is used particularly in storefront work and interiors of the "Olde English Pub" variety.

Skintled brickwork consists of laying common brick irregularly, with random projecting brick at intervals and using clinkers or irregular and other off-colored brick to produce a variegated wall.

These brick are laid in a number of different ways to produce the desired effect. In some of them the mortar projects beyond the face of the brick just as it has been squeezed out of the joint; in others the joints are cut flush or raked out to produce shadows, etc.

The labor cost will vary with the design and care used in laying up the brick. When laying building brick without regard to color of wall and with uncut mortar joints or joints cut flush with the face of the wall, a mason should lay 350 to 400 brick per 8-hr. day, at the following labor cost per 1,000:

	Hours	Rate	Total	Rate	Total
Mason	21	$....	$....	$28.18	$ 591.78
Labor	15		....	21.04	315.60
Cost per 1,000 brick			$....		$ 907.38
per brick					$0.91

If brick joints are raked out, add 2.5 hrs. mason time per 1,000 brick. On skintled brickwork using clinkers and other dark or irregularly shaped brick to produce a textured wall, more care is required in laying the brick, working out the designs and in raking out the joints to produce shadows, etc., and a mason should lay 250 to 300 brick per 8-hr. day, at the following labor cost per 1,000:

	Hours	Rate	Total	Rate	Total
Mason	29	$....	$....	$28.18	$ 817.22
Labor	20	....	....	21.04	420.80
Cost per 1,000 brick			$....		$ 1,238.02
per brick					$ 1.24

English Size Face Brick

English size face brick, 8-7/8" x 2-7/8" x 4" (221.87 x 71.87 x 100 mm) in size, are seldom used in this country any more and while the size is not standard, they are still furnished by a few manufacturers.

They are practically half again as large on the face as the standard brick, and a mason is unable to lay as many as the standard size brick. They also require more labor time handling, hoisting, and wheeling to the masons.

Number of English Size, 8-7/8" x 2-7/8" x 4" (221.87 x 71.87 x 100 mm), Face Brick Required per S.F. of Wall

All Vertical Mortar Joints Figured 1/4" (6.25 mm) Wide
Width of Horizontal or "Bed" Mortar Joints, inches (mm)

Dry	1/8" (3.12)	1/4" (6.25)	3/8" (9.37)	1/2" (12.5)	5/8" (15.62)	3/4" (18.75)
5.65	5.34	5.06	4.88	4.67	4.52	4.37

Add for waste, headers, and the different brick bonds,
as given under "Standard Brick".

C.F. of Mortar Required to Lay 1,000 English Size Face Brick
Based on 1/4" (6.25 mm) End Joints and "Bed" Joints as Given Below

1/8" (3.12)	1/4" (6.25)	3/8" (9.37)	1/2" (12.5)	5/8" (15.62)	3/4" (18.75)
5	9	11	14	17	20

Vitrified Face Brick

There are a number of vitrified face brick on the market. This type is often so hard it is almost impossible to cut. Because of their denseness and brittleness, they do not absorb water from the mortar, and it requires more time for the mortar to take its initial set. It is impossible to lay up many courses of brick at one time, because the weight of the upper brick courses forces the mortar from the joints of the lower courses, causing an unsightly wall. It is also difficult to hold the brick in the wall and keep the wall plumb, because the brick slide out of place very easily.

Brick of this kind are ordinarily laid with "buttered" mortar joints or with 1/4" (6.25 mm) mortar joints struck flush. They are furnished in the standard size, 8" x 2-1/4" x 3-3/4" (200 x 56.25 x 93.75 mm) and require the same number of brick and the same amount of mortar as other standard size face brick.

Brick Floors and Steps

Face Brick Floors Laid in Basket Weave Pattern. Brick floors laid in square or basket weave pattern do not require any cutting and are much easier laid than herringbone designs.

On work of this kind, a mason should lay 180 to 225 brick per 8-hr. day, at the following labor cost per 1,000:

	Hours	Rate	Total	Rate	Total
Mason	40	$....	$....	$28.18	$1,127.20
Labor	20	....	....	21.04	420.80
Cost per 1,000 brick			$....		$ 1,548.00
per brick					$ 1.55

When using brick for stairs, it is recommended that brick pavers be used instead of standard brick.

Face Brick Floors Laid in Herringbone Pattern. Brick floors are usually laid in portland cement mortar, and when laid in herringbone pattern require considerable time cutting and fitting the small pieces of brick at the edge or border of the pattern, as shown in the illustration.

A mason should lay 125 to 150 brick per 8-hr. day, at the following labor cost per 1,000:

	Hours	Rate	Total	Rate	Total
Mason	56	$....	$....	$28.18	$ 1,578.08
Labor	28	....	....	21.04	589.12
Cost per 1,000 brick			$....		$ 2,167.20
per brick					2.17

If there is only one mason on the job, it will require a helper for each mason.

Promenade Tile Floor and Decks. When laying 4" x 8" x 1/2" (100 x 200 x 12.5 mm) promenade tile on porch floors and decks, with the tile laid in portland cement mortar and 1/2" (12.5 mm) mortar joints, struck flush, a mason should lay and point 100 to 125 tile per 8-hr. day, at the following labor cost per 1,000:

	Hours	Rate	Total	Rate	Total
Mason	68	$....	$....	$28.18	$1,916.24
Labor	34	....	....	21.04	715.36
Cost per 1,000 tile			$....		$ 2,631.60
per tile					2.63

BRICK FLOOR PATTERNS

Brick Laid in Basket Weave Pattern.   Brick Laid in Herringbone Pattern.

BRICK STAIR DETAILS

SAND OR
CINDER FILL
SAND CUSHION
CONCRETE SLAB

Brick Steps with End-Set Treads

A tile is 4" x 8" x 1/2" (100 x 200 x 12.5 mm). It requires 4 tile per s.f., not including mortar joints. Figuring a 1/2" (12.5 mm) mortar joint, it requires 3.5 tile per s.f. (0.09 sq.m) of floor.

A mason will lay 30 to 36 s.f. of floor per 8-hr. day, at the following labor cost per 100 s.f.:

	Hours	Rate	Total	Rate	Total
Mason	24	$....	$....	$28.18	$ 676.32
Labor	12	....	....	21.04	252.48
Cost per 100 s.f.			$....		$ 928.80
per s.f.			....		9.29
per sq.m			....		99.94

Brick Steps with Edge-Set Treads

Brick Steps. When laying brick steps in portland cement mortar, similar to those illustrated, a mason should lay 125 to 150 brick per 8-hr. day, at the following labor cost per 1000:

	Hours	Rate	Total	Rate	Total
Mason	56	$....	$....	$28.18	$1,578.08
Labor	28	....	....	21.04	589.12
Cost per 1,000 brick			$....		$ 2,167.20
per brick					2.17

Turning Face Brick Segmental Arches Over Doors and Windows. Where segmental brick arches are laid over wood centers for 3'-0" (0.9 m) wide door and window openings, it will require about 4 hrs. mason time at the following labor cost per arch:

	Hours	Rate	Total	Rate	Total
Mason	4	$....	$....	$28.18	$ 112.72

For a 5'-0" (1.5 m) arch, figure about 6.5 hrs. mason time, as follows:

	Hours	Rate	Total	Rate	Total
Mason	6.5	$....	$....	$28.18	$ 183.17

If the brick are cut or chipped to a radius, add cost of chipping as given below.

Laying Face Brick in Flat or "Jack" Arches. When laying face brick in flat or jack arches, a mason should complete one arch up to 3'-6" (1.0 m) wide in about 5 hrs. at the following labor cost:

	Hours	Rate	Total	Rate	Total
Mason	5	$....	$....	$28.18	$ 140.90

For a 5'-0" (1.5 m) jack arch, figure about 7-1/2 hrs. mason time, as follows:

	Hours	Rate	Total	Rate	Total
Mason	7.5	$....	$....	$28.18	$ 211.35

The above costs do not include chipping or rubbing brick to a radius. If this is necessary, add cost of chipping as given below.

Chipping Brick. Where necessary to chip or cut brick for segmental or jack arches, a mason should chip 400 to 500 brick per 8-hr. day, depending on the texture and density of the brick, at the following labor cost per 1,000:

	Hours	Rate	Total	Rate	Total
Mason	17.5	$....	$....	$28.18	$ 493.15

There is considerable lost motion in cutting brick, and unless there is a great deal of duplication in the cutting, where a mason can work continuously on a few special sizes, a mason will cut only 160 to 200 pieces per 8-hr. day at the following labor cost per 100 pieces:

	Hours	Rate	Total	Rate	Total
Mason	4.5	$....	$....	$28.18	$ 126.81
Labor	1.0	....	....	21.04	21.04
Cost per 100 cuts			$....		$ 147.85
per cut				....	1.48

Fire Resistance

Fire Resistance Ratings. Tables and figures below list fire resistance ratings for various brick walls. Ratings listed are for load bearing walls tested with working loads of 160 psi (1.10 MPa) of gross area, except the 4" (100 mm) wall shown was loaded to 92 psi (0.63 MPa).

The following table lists the fire resistance ratings for steel columns covered with brick:

Fireproofed Column Fire Ratings

Construction*   Ratings
Steel Columns, 6"x6" (150 x 150 mm) or larger
3-3/4" (93.75 mm) brick with brick fill... 4 hr.
2-1/4" (56.25 mm) brick with brick fill... 1 hr.

*Column fire resistance varies with the cross-sectional area of solid material; the larger the area, the greater the fire resistance for a given thickness of protection around the structural steel. Column dimensions are outside dimensions. Smaller columns may require more cover to achieve equal ratings. For columns that are not square, protection should equal that of a square column having equal or lesser cross-sectional area. Thickness does not include plaster.

Cutting Masonry Products With Portable Masonry Saw

On practically every construction job there is a large amount of cutting necessary on masonry materials, and with the high cost of labor, this becomes an expensive item.

This applies particularly to the cross and diamond bonds in brick masonry, arch brick, radial brick, glazed tile, rotary kiln blocks, concrete blocks, natural stone, roofing tile, and conduits.

By using a masonry saw, this work can be done at a fraction of the cost of hand cutting plus the saving of materials.

Masonry saws are obtainable with 12", 14", 18" and 20" (300, 350, 450, and 500 mm) blade capacities, and are designed for both wet, dustless and for dry cutting. The advantage of wet cutting is in the confined areas where dust would be a problem. An exhaust assembly is available for use with dry cutting masonry saws. This assembly consists of a high velocity fan that draws the dust away from the machine and disposes of it through the nearest opening.

A smaller compact model masonry saw has been developed that is light enough to carry, can be set up on the scaffolding, and can be operated from a pickup truck or the back of a station wagon.

There is also a small, highly portable concrete saw that can do the work of many of the heavier models. There are four types of blades that can be used on masonry saws:
1. wet-dry abrasive blades
2. dry "break-resistant" abrasive blades
3. wet "break-resistant" abrasive blades
4. diamond blades

The wet abrasive and diamond blades should be used only with a masonry saw equipped for wet cutting. There are many specifications of each type of blade; it is very important that the correct specification be used for the particular material to be cut.

4-IN. SOLID BRICK WALL
No members framed in

Units at least 75 percent solid laid in mortar containing Sarabond® high bond additive, and backed with a hat shaped metal furring channel ¾ in. thick formed from 0.021-in. sheet metal attached to the brick wall on 24-in. centers using four whistle plugs and diamond pointed 6d common nails; and ½-in. type X gypsum wallboard attached to the metal furring strips with 1½-in. long type S-12 screws spaced 8 in. o.c.

4-IN. SOLID BRICK WALL
Non-combustible or no members framed in

Units at least 75 percent solid
Plaster 2 sides

8-IN. SOLID BRICK WALL
Combustible members framed in
Units at least 75 percent solid
No plaster required

8-IN. HOLLOW BRICK WALL
Combustible members framed in
Units at least 71 percent solid
No plaster required
Combustible members not fully embedded
Plaster 2 sides (⅝-in.)

10-IN. BRICK CAVITY WALL
Combustible members framed in
Units at least 75 percent solid
No plaster required
Walls filled solidly at combustible members

Two-Hour Fire Ratings
Constructions Shown Are Loadbearing

Three-Hour Fire Ratings
Constructions Shown Are Loadbearing

Four-Hour Fire Ratings
Construction Shown Are Loadbearing

Diamond blades for masonry saws vary in price from $240.00 to $470.00 for 14" (350 mm) diameter blades and $425.00 to $905.00 for 18" (450 mm) diameter blades.

The cost of cutting masonry units varies greatly and depends on the type of material, blade specifications, and the operator.

Concrete products vary considerably in hardness, such as sand-gravel blocks compared with Haydite blocks, and blade wear and cutting time will vary accordingly. On average most 8"x8"x16" (200 x 200 x 400 mm) concrete blocks can be sliced in two in 14 to 16 seconds at a cost of less than $0.01 per unit.

Fire brick varies tremendously in hardnesses and types. Ordinary dry press or silica can be cut very rapidly, while Super-duty refractories or Basic, Chrome, and Magnesite require 15 to 18 seconds for each cut completely through, at a cost varying from $0.04 to 0.10 depending on actual density.

All the above costs and cutting speeds are based on cross cuts. Longitudinal cuts will be about the same for equivalent area. The following is an example of the approximate time required making cuts in various masonry products:

Material	Type of Cut	Cut Length, in. (mm)	Cut Depth, in. (mm)	Approx. Time, seconds
Acid Brick	Miter	4-1/2" (112mm)	2-1/4" (56mm)	15
Concrete Block	Straight	8" (200mm)	8" (200mm)	19
Glazed Brick-Tile	Miter	6" (150mm)	4" (100mm)	15
Magnesite Brick	Miter	2-3/4" (68mm)	4" (100mm)	15
Rotary Kiln Block	Straight	8" (200mm)	3" (75mm)	11
Silica Brick	Straight	8" (200mm)	2-1/4" (56 mm)	10
Roofing Tile	Miter	9" (225mm)	2" (50mm)	14
Clay Sewer Pipe	Cross	4" (100mm) dia.	--	18
Concrete Cylinder	Cross	4" (100mm) dia.	--	21

Cleaning Face Brick Work

The cost of cleaning face brick varies with the kind of brick. It is much easier to clean a smooth face vitrified brick than one having a rough texture. Where smooth face brick are used, and it is not necessary to do any pointing but merely wash the wall, the indiscriminate use of muriatic acid or of the wrong proprietary compound can cause unsightly, hard-to-remove stains. Also, chemical cleaning solutions are generally more effective when the outdoor temperature is 50°F or above.

An experienced mechanic should clean 5,000 to 5,500 brick—750 to 825 s.f. (69.6-76.6 sq.m.) of wall—per 8-hr. day at the following labor cost per 1,000 brick, or approximately 150 s.f. (13.9 sq.m) of wall:

	Hours	Rate	Total	Rate	Total
Mechanic	1.6	$....	$....	$28.18	$ 45.09
Helper	0.8	....	....	21.04	16.83
Cost per 1,000 brick			$....		$ 61.92
per s.f. of wall				....	0.41
per sq.m				....	4.41

The labor cost of cleaning and washing rough texture face brick will be higher than for smooth brick because of cement and mortar getting into the surface of the brick.

When cleaning and washing rough textured brick, the same crew should clean and wash 3,500 to 4,000 brick—525 to 600 s.f. (48.7-55.7 sq.m) of wall—per 8-hr. day at the following labor cost per 1,000 brick, or per 150 s.f. (13.9 sq.m) of wall:

	Hour	Rate	Total	Rate	Total
Mechanic	2.2	$....	$....	$28.18	$ 62.00
Helper	1.1	....	....	21.04	21.04
Cost per 1,000 brick			$....		$ 83.04
per s.f. of wall				....	0.55
per sq.m of wall				....	5.92

The labor cost of cleaning face brick with high-pressure water varies considerably, depending on location and whether pointing and patching are required. When cleaning and washing textured brick, a crew can clean from 6,000 to 8,500 brick--900 to 1,275 s.f. (83.6-118.4 sq.m) of wall--per 8-hr. day at the following cost per 1,000 brick or per 150 s.f. (13.9 sq.m) of wall:

	Hour	Rate	Total	Rate	Total
Mechanic	0.55	$....	$....	$28.18	$ 15.50
Helper	0.55	....	....	21.04	11.57
Cost per 1,000 brick			$....		$ 27.07
per s.f. of wall				....	0.18
per sq.m of wall				....	1.94

When estimating cleaning of masonry units, the following precautions apply, and costs must be added when applicable:

1. Saturation of the brick masonry wall surface with water before and after application of chemical or detergent cleaning solutions, add 0% to 10%.

2. Failure to properly use chemical cleaning solutions. This can result in recleaning or permanent staining.

3. Failure to protect windows, doors and trim. Cost will vary in accordance with the amount of material to be protected.

4.   Costs of the chemical or detergents.

5.   Classification and safe disposal of the chemicals.

6.   Failure to protect adjacent property.

The following table is a guide for the estimator to determine the cleaning method to be employed for various new masonry units.

Cleaning Guide for New Masonry

Brick Category	Cleaning Method
Red and red flashed	Bucket and brush hand cleaning
	High pressure water
	Sandblasting

Remarks: Proprietary compounds and emulsifying agents may be used. *Smooth texture:* Mortar stains and smears are generally easier to remove; less surface area exposed; easier to presoak and rinse; unbroken surface, thus more likely to display poor rinsing, acid stain, poor removal of mortar smears. *Rough texture:* Mortar and dirt tend to penetrate deep into textures; additional area for water and acid absorption; essential to use pressurized water during rinsing.

Red, heavy sand finish	Bucket and brush hand cleaning
	High pressure water

Remarks: Clean with plain water and scrub brush, or lightly applied high pressure and plain water. Excessive mortar stains may require use of cleaning solutions. Sandblasting is not recommended.

Light-colored units, white,	Bucket and brush hand cleaning
tan, buff, gray, specks,	High pressure water
pink, brown and black	Sandblasting

Remarks: Do not use muriatic acid! Clean with plain water, detergents, emulsifying agents, or suitable proprietary compounds. Manganese colored brick units tend to react to muriatic acid solutions and stain. Light colored brick are more susceptible to "acid burn" and stains, compared to darker units.

Same as light-colored units,	Bucket and brush hand cleaning
etc., plus sand finish	High pressure water

Remarks: Lightly apply either method. (See notes for light-colored units, etc.) *Sandblasting is not recommended.*

Glazed Brick	Bucket and brush hand cleaning

Wiped glazed surface with soft cloth within a few minutes of laying units. Use soft sponge or brush plus ample water supply for final washing. Use detergents where necessary and acid solutions only for very difficult mortar stain. Do not use acid on salt glazed or metallic glazed brick. Do not use abrasive powders.

Colored mortars..Method generally controlled
....by the brick unit
Many manufacturers of colored mortars do not recommend chemical cleaning solutions. Most acids tend to bleach colored mortars. Mild detergent solutions are generally recommended.

Brick Fireplaces, Mantels, and Hearths

On account of the many sizes, shapes, and styles of brick fireplaces, mantels, and hearths, it is difficult to estimate the cost of the work at a certain price per 1,000 brick. The most satisfactory method is to figure a certain length of time to lay up each fireplace or mantel.

A mason should complete a plain brick fireplace 5'-0" to 6'-0" (1.5-1.8 m) wide and 4'-0" to 5'-0" (1.2-1.5 m) high, requiring 200 to 225 brick, not including hearth, in about 10 hrs. at the following labor cost:

	Hours	Rate	Total	Rate	Total
Mason	10	$....	$....	$28.18	$ 281.80
Labor	5	....	....	21.04	105.20
Cost per fireplace			$....		$ 387.00

On larger and more elaborate brick fireplaces and mantels having raked or rodded mortar joints, a mason should lay about 100 to 125 brick per 8-hr. day, at the following labor cost per 1,000:

	Hours	Rate	Total	Rate	Total
Mason	70	$....	$....	$28.18	$1,972.60
Labor	35	....	....	21.04	736.40
Cost per 1,000 brick			$....		$ 2,709.00

If there is only one mason working on the fireplace, it will require one laborer with each mason.

Brick Fireplace Linings. A mason should complete the lining for a brick fireplace having an opening 3'-0" to 4'-0" (0.9-1.2 m) wide, 2'-6" to 3'-0" (0.76-0.90 m) high and 1'-4" to 1'-9" (0.4-0.5 m) deep, in 6 to 8 hrs. all depending upon the class of work.

If the fireplace linings are laid in common bond and do not require an excessive amount of cutting and fitting, a mason should complete one fireplace lining in about 5 hrs., but where the back-hearths are of irregular shape and require considerable cutting and fitting at angles and at tops for dampers and smoke chambers, it will require about 8 hrs. mason time for each lining.

Fireplaces having an opening 3'-6" to 4'-0" (1050-1200 mm) wide, about 4'-0" (1200 mm) high and 1'-9" (525 mm) deep, lined with firebrick laid in herringbone pattern, requiring cutting and fitting at all edges and corners, will require 20 to 24 hrs. mason time to line one fireplace.

Labor Laying Brick Hearths and Back-Hearths. If brick hearths and back-hearths are laid in square or basket weave pattern, a mason should complete one back-hearth in 3 to 4 hrs. and the front hearth will require the same length of time.

If laid in herringbone pattern with cut brick at all edges, figure 4 to 5 hrs. mason time for both hearth and back-hearth.

Figure about 1/2-hr. laborer time to each hour of mason time, except where only one mason is on the job when it will figure hour for hour.

Fire Brick Work

Fire brick and tile are furnished in innumerable sizes and shapes to meet the requirements of all classes of boilers, furnaces, stacks, etc. It is impossible to list them all, but the most common are listed below. Standard fire brick size is 8"x3-1/2"x2-1/4" (200 x 87.5 x 56.25) and will cost about $480.00 per 1,000 for low duty, $780.00 for high duty.

Fire Clay. When a close thin joint is desired, the clay should be soaked and mixed thin, and the brick should be dipped and rubbed. For work of this kind, 300 lbs. (135 kg) of finely ground fire clay is required per 1,000 brick.

Sizes and Shapes of Standard Fire Brick
Number Required per S.F.

Name & Size	Laid Flat	Laid on Edge
8" Straight, 8"x3-1/2"x2-1/4"	8.0	5.2
9" Straight, 9"x4-1/2"x2-1/2"	6.5	3.5
Small 9" Brick, 9"x3-1/2"x2-1/2"	6.5	4.5
Split Brick, 9"x4-1/2"x1-1/4"	13.0	3.5
2" Brick, 9"x4-1/2"x2"	8.0	3.5
Soap, 9"x2-1/2"x2-1/2"	6.5	7.0
Checker, 9"x2-3/4"x22-3/4"	6.0	6.0

(metric)
Number Required per Sq.M

Name & Size	Laid Flat	Laid on Edge
203 mm Straight, 203x89x64mm	86.1	56.0
228 mm Straight, 228x114x64mm	69.9	37.7
Small 228mm Brick, 228x89x64mm	69.9	48.4
Split Brick, 228x114x32mm	139.9	37.7
51 mm Brick, 228x114x51mm	86.1	37.7
Soap, 228x64x6mm	69.9	75.3
Checker, 203x70x578mm	64.6	64.6

Ordinary fire clay costs about $6.50 per 100 lbs. (45 kg) while high-grade refractory fire clay costs about $7.60 per 100 lbs. (45 kg).

Labor Lining Brick Chimneys and Stacks with Fire Brick. When lining small brick chimneys and stacks from 2'-0" to 3'-0" (600-900 mm) square, laid in fire clay, a mason should lay 450 to 525 brick per 8-hr. day, at the following labor cost per 1,000:

	Hours	Rate	Total	Rate	Total
Mason	16	$....	$....	$28.18	$ 450.88
Labor	12	....	....	21.04	252.48
Cost per 1,000 brick			$....		$ 703.36

On large brick stacks having an inside diameter of 4'-0" to 6'-0" (1.2-1.8 m), a mason should lay 600 to 750 fire brick per 8-hr. day, at the following labor cost per 1,000:

	Hours	Rate	Total	Rate	Total
Mason	11.8	$....	$....	$28.18	$ 332.52
Labor	10.0	....	....	21.04	210.40
Cost per 1,000 brick			$....		$ 542.92

Number of Wedge and Arch Brick Required for Various Circles

No. 1 Wedge Brick 9"x4-1/2"x (2-1/2"-1-7/8")

No. 2 Wedge Brick 9"x4-1/2"x (2-1/2"-1-1/2")

No. 1 Arch Brick 9"x4-1/2"x (2-1/2"-2-1/8")

No. 2 Arch Brick 9"x4-1/2"x (2-1/2"-1-3/4")

(metric)

No. 1 Wedge Brick 228 x 114 mm x (63-48 mm)

No. 2 Wedge Brick 228 x 114 mm x (63-48 mm)

No. 1 Arch Brick 228 x 114 mm x (63-54 mm)

No. 2 Arch Brick 228 x 114 mm x (63-44 mm)

Labor Laying Fire Brick in Fireboxes, Breechings, Etc. When lining fireboxes, breechings, furnaces, etc., with fire brick, requiring arch and radial brick, special shapes, etc., a mason should lay 175 to 225 brick per 8-hr. day, at the following labor cost per 1,000:

	Hours	Rate	Total	Rate	Total
Mason	40	$....	$....	$28.18	$1,127.20
Labor	40	....	....	21.04	841.60
Cost per 1,000 brick			$....		$ 1,968.80

When estimating brick arches, it is advisable to allow additional labor. It is doubtful that a mason thoroughly experienced in boiler work will average more than 25 brick per hr. for the actual arch and skew-backs, due to adverse conditions and close working space. One contractor specializing in this class of work checks the penetrations through the wall, and regardless of size, adds 4 hours mason time for each one.

Labor Bricking in Boilers. When bricking in steel boilers, fireboxes, breechings, etc., with common or fire brick, where the walls are 8" (200 mm) to 12" (300 mm) thick, a mason should lay 600 to 650 brick per 8-hr. day, at the following labor cost per 1,000:

	Hours	Rate	Total	Rate	Total
Mason	13	$....	$....	$28.18	$ 366.34
Labor	13	....	....	21.04	273.52
Cost per 1,000 brick			$....		$ 639.86

Table of Wedge Brick Required for Various Circles / **Number of Arch Bricks Required for Various Circles**

Dia. ft. (mm)	No. 2 Wedge	No. 1 Wedge	Square	Total	No. 2 Arch	No. 1 Arch	9-inch (225mm)	Total
2.0 (600)	...	...	...	...	42	...	...	42
2.5 (750)	63.	...	...	63.	10	40	...	50
3.0 (900)	48.	20.	...	68.	...	57	...	57
3.5 (1050)	36.	39.	...	75.	...	57	7	64
4.0 (1200)	24.	57.	...	81.	...	57	15	72
4.5 (1350)	12.	78.	...	90.	...	57	22	79
5.0 (1500)	...	94.	...	94.	...	57	29	86
5.5 (1650)	...	94.	7.5	102.	...	57	37	94
6.0 (1800)	...	94.	15.	109.	...	57	44	101
6.5 (1950)	...	94.	23.	117.	...	57	52	109
7.0 (2100)	...	94.	30.	124.	...	57	59	116
7.5 (2250)	...	94.	39.	133.	...	57	67	124
8.0 (2400)	...	94.	45.	139.	...	57	74	131
8.5 (2550)	...	94.	53.	147.	...	57	82	139
9.0 (2700)	...	94.	60.	154.	...	57	89	146
9.5 (2850)	...	94.	68.	162.	...	57	97	154
10.0 (3000)	...	94.	75.	169	...	57	104	161

Fire Clay Tile Flue Lining

Flue lining is estimated by the lineal foot (meter). It is furnished in 2'-0" (600 mm) lengths in the following sizes:

Size of flue lining	No. of l.f. per 8-hr. Day	Hrs. to set 100 l.f. Mason	Labor
4"x8"	165-180	4.7	4.7
4"x12"	130-150	5.7	5.7

Size of flue lining	No. of l.f. per 8-hr. Day	Hrs. to set 100 l.f. Mason	Labor
8"x8"	130-150	5.7	5.7
8"x12"	105-125	7.0	7.0
12"x12"	85-100	8.7	8.7
8"x18"	75-90	9.7	9.7
12"x18"	70-85	10.2	10.2
18"x18"	60-70	12.3	12.3
20"x20"	55-65	13.3	13.3
20"x24"	50-60	14.0	14.0
24"x24"	45-55	16.0	16.0
20" round	55-65	13.3	13.3

(metric)

Size of flue lining	No. of meters per 8-hr. Day	Hrs. to set 10 m Mason	Labor
100x200mm	50-55	1.5	1.5
100x300mm	40-46	1.9	1.9
200x200mm	40-46	1.9	1.9
200x300mm	32-38	2.3	2.3
300x300mm	26-30	2.9	2.9
200x450mm	23-27	3.2	3.2
300x450mm	21-26	3.3	3.3
450x450mm	18-21	4.0	4.0
500x625mm	17-20	4.4	4.4
500x600mm	15-18	4.6	4.6
625x600mm	14-17	5.2	5.2
500mm round	17-20	4.4	4.4

Approximate Prices of Tile Flue Lining

Size, inches	Price per l.f.	Size, mm	Price per m
8"x8"	$3.30	200x200mm	$10.83
8"x12"	4.45	200x300mm	14.60
8"x18"	9.10	200x450mm	29.86
12"x12"	5.60	300x300mm	18.37
13"x13"	5.60	325x325mm	18.37
12"x16"	9.45	300x400mm	31.00
12"x18"	9.95	300x450mm	32.64
13"x18"	9.95	325x450mm	32.64
16"x16"	13.25	400x400mm	43.47
16"x20"	19.15	400x500mm	62.83
18"x18"	14.70	450x450mm	48.23
20"x20"	23.00	500x500mm	75.46
20"x24"	25.10	500x600mm	82.35
24"x24"	29.60	600x600mm	97.11
6" Round	$4.00	150mm Round	$13.12
6" Round thimble	3.10	150mm Round thimble	10.17

8" Round	4.45	200mm Round	14.60
8" Round thimble	6.15	200mm Round thimble	20.18
10" Round	6.25	250mm Round	20.51
10" Round thimble	7.00	250mm Round thimble	22.97
12" Round	6.60	300mm Round	21.65
12" Round thimble	8.50	300mm Round thimble	27.89
15" Round	13.85	375mm Round	45.44
18" Round	16.00	450mm Round	52.49
20" Round	23.40	500mm Round	76.77
24" Round	37.30	600mm Round	122.38

Labor Setting Terra Cotta Wall Coping

Size of Coping, in.	L.F. per 8-hr. day	Hrs. to set 100 l.f. Mason hrs.	Labor hrs.
9"	145-160	5.2	5.2
12"	115-130	6.5	6.5
18"	100-120	7.3	7.3

(metric)

Size of Coping, mm		Hrs. to set 10 meters Mason hrs.	Labor hrs.
225	44-49	1.7	1.7
300	35-40	2.1	2.1
450	30-37	2.4	2.4

Approx. Prices of Terra Cotta Wall Coping

	Price per lineal foot			Price per meter	
Size	Double Slant	Single Slant	Size	Double	Single
9"	$2.50	$2.75	225mm	$8.20	$9.02
13"	3.15	5.00	325mm	10.33	16.40
18"	6.30	12.00	450mm	20.66	39.37

Corners, ends, and starters are four times the price of straight coping; angles are six times the price of straight coping.

Brickwork for Incinerators

Incinerators have been used in residences, apartments, apartment hotels, hotels, hospitals, schools, and industrial plants for burning waste. The refuse might be fed into the chimney at each floor, dropped to the incinerator in the basement, or fed direct to the incinerator in the basement.

Practically all incinerators require a separate flue or stack, which should be estimated in the regular way, figuring fire brick, flue lining, and common brick as shown on the plans.

The number of fire and common brick required for the various sizes and types of incinerators will vary with the different manufacturers.

However, the following data will assist the estimator in computing the number of brick and labor hours required to construct the brickwork for the different size incinerators, where they are not detailed on the plans.

If 4" (100 mm) outside walls are used, dimensions below will be reduced in length and height by 4" (100 mm) and in width by 8" (200 mm).

Model	A	B	C	No. of Fire Brick	No. of Common Brick	Minimum Size of Flue	Mason Hours	Labor Hours
		Dimensions						
"A"	6'0"	3'5"	4'0"	300	900	12"x12"	20	14
"B"	5'7"	3'11"	4'7-1/2"	350	1100	14"x14"	24	18
"C"	6'0"	4'3"	4'7-1/2"	400	1300	16"x16"	28	20
"D"	6'10"	4'3"	5'11"	525	1500	20"x20"	33	24
				(metric)				
"A"	1800	1026	1200mm	300	900	300x300mm	20	14
"B"	1674	1176	1386mm	350	1100	350x350mm	24	18
"C"	1800	1275	1386mm	400	1300	400x400mm	28	20
"D"	2049	1275	1773mm	525	1500	500x500mm	33	24

Radial Brick Chimneys

Radial brick chimneys are usually constructed by contractors who specialize in chimney construction. These contractors employ workers with years of experience in this class of work and can build them much more economically than the general contractor. Also, most constructors of radial brick chimneys have their own designs and shapes of radial brick, manufactured especially for them, which are not available to the average contractor.

However, some information on this important subject is of value to all contractors. Radial brick chimneys are constructed of perforated radial brick formed to fit the circular and radial lines of the chimney. The brick are molded with vertical perforations, and should be hard, well burned, acid proof, of necessary refractory powers and crushing strength, and of maximum density.

The perforations serve to form a dead air space in the walls of the chimney that tends to prevent rapid heating and cooling of the walls by conserving the heat inside. Naturally, the higher the temperature of the gases the better the draft.

Four sizes of brick are ordinarily used in the construction of radial brick chimneys. All have the same face dimensions, approximately 6-1/2" (162.5 mm) wide by 4-1/2" (112.5 mm) high. The lengths of the blocks vary in order to make possible the breaking of the joints horizontally and vertically in the walls. The combination of bonds with this type of chimney provides for a lighter chimney than can be produced with ordinary building brick.

Radial chimney brick are generally either red or buff in color. Red radial brick are usually the most economical, because they are in greater demand and there are more sources of supply.

In laying the brick, the mortar is worked into the perforations, locking them together on the principle of a mortise and tenon joint. This produces the strongest bonded wall known to brick construction.

Sizes and Weights of Radial Brick for Radial Brick Chimneys

Type Brick	Outside Face	Inside Face	Depth Inches	Height Inches	Weight Lbs.
No. 4	6-1/2"	5-3/4"	4"	4-1/2"	7
No. 6	6-1/2"	5-1/2"	6"	4-1/2"	9-1/2
No. 7	6-1/2"	5-1/2"	7"	4-1/2"	11
No. 8	6-1/2"	5-1/2"	8"	4-1/2"	12
Pier	9"	9"	4"	4-1/2"	9

(metric)

Type Brick	Outside Face	Inside Face	Depth mm	Height mm	Weight kg
No. 4	165mm	146mm	102mm	115mm	4.3
No. 6	165mm	140mm	152mm	115mm	5.0
No. 7	165mm	140mm	178mm	115mm	5.4
No. 8	165mm	140mm	203mm	115mm	4.1
Pier	229mm	229mm	102mm	115mm	3.2

Each radial brick with an outside face 6-1/2" (162.5 mm) wide and 4-1/2" (112.5 mm) high covers 29-1/4 sq. in. (18,872 mm²) of surface, or it requires 5 (4.923) blocks per s.f. (0.09 sq.m) of wall, not including mortar joints. Allowing 1/2" (12.5 mm) for mortar joints, each brick covers 7"x5" (175 x 125 mm) or 35 sq. in. (21,875 mm²) of surface, or it requires 4.12 blocks per s.f. (0.09 sq.m) of wall. Add about 5% for breakage and waste.

The walls of most radial chimneys taper about 1/4" per 1'-0" (6.25 mm per 300 mm) in height, and the thickness of the walls vary from the bottom to the top, depending upon the height. The wall thickness must be made up of a number of different depth units to provide the proper bond and structural strength. For example, a wall 16-1/4" (406.2 mm) thick might be made up of 1 block 7" (175 mm) deep and 2 blocks 4" (100 mm) deep; a wall 8-5/8" (215.6 mm) thick of 2 blocks 4" (100 mm) deep, etc.

Estimating Quantities of Radial Brick. Radial brick chimney work is frequently estimated by the ton, because wall thicknesses vary, depending on the height. A chimney 125'-0" (38.1 m) high might have walls of the following thicknesses:

Story	Hgt., ft.	Thick., in.	Hgt., m	Thick., mm
First	25'	16-1/4"	7.6m	381.3mm
Second	20'	13-1/4"	6.1m	331.3mm
Third	20'	11-3/4"	6.1m	293.8mm
Fourth	20'	10-1/4"	6.1m	256.3mm
Fifth	20'	8-5/8"	6.1m	215.6mm
Top	20'	7-1/8"	6.1m	178.1mm

These wall thicknesses will be made up of radial brick of different depths to work out the correct wall thickness and proper bond:

Wall Thickness; Size of Units Used	Wgt. 29-1/4 Sq. In. Wall lbs.*	Wgt per s.f., lbs.*
20"-21"; 2 pcs. 8", 1 pc. 4"	31	128
18"-19"; 2 pcs. 7", 1 pc. 4"	29	120
16"-17"; 2 pcs. 6", 1 pc. 4" or 2 pcs. 8"	24-26	99-107
14"-15"; 1 pc. 6", 2 pcs. 4" or 2 pcs. 7"	22-23-1/2	91-97
12"-13"; 1 pc. 7", 1 pc. 4" or 1 pc. 8", 1 pc. 4"	18-19	75-79
10"-11"; 1 pc. 6", 1 pc. 4"	16-1/2	68
8"-9"; 2 pcs. 4" or 1 pc. 8"	12-14	50-58
7"-8"; 1 pc. 7" or 1 pc. 8"	11-12 lbs.	45.5-50

*Net weight of brick. Does not include mortar.

(metric)

Wall Thickness; Size of Units Used	Wgt. 0.1525 Sq.M Wall kg*	Wgt per sq.m, kg*
500-525mm; 2 pcs. 200mm, 1 pc. 100mm	14.1	625.0
450-475mm; 2 pcs. 175mm, 1 pc. 100mm	13.2	585.9
400-425mm; 2 pcs. 150mm, 1 pc. 100mm or 2 pcs. 200mm	10.9-11.8	483.4-522.4
350-375mm; 1 pc. 150mm, 2 pcs. 100mm or 2 pcs. 175mm	10.0-10.7	444.3-473.6
300-325; 1 pc. 175mm, 1 pc. 100mm or 1 pc. 200mm, 1 pc. 100mm	8.2-8.6	366.2-385.7
250-275mm; 1 pc. 150mm, 1 pc. 100mm	7.7	332.0
200-225mm; 2 pcs. 100mm or 1 pc. 200mm	5.4-6.4	244.1-283.2
175-200mm; 1 pc. 175mm or 1 pc. 200mm	5.0-5.4	222.2-244.1

*Net weight of brick. Does not include mortar.

Figure radial brick chimney lining at about 30 lbs. per s.f. (146.5 kg per sq.m) of wall surface.

Mortar Required Laying Radial Brick. It requires about 2/3 ton or 1,334 lbs. (605.5 kg) of mortar to lay one ton of radial brick. Mortar is usually mixed in the proportions of 1 volume portland cement, 1 volume lime putty, and 4 to 6 volumes clean mortar sand.

Unit	A	B	C	D	Inside Radius	Approximate Shipping Weight	Approximate Percent Voids
4 (N)	4-1/2"	6-1/2"	4"	5-20/32"	3'-4"	5.5	32
6 (A)	4-1/2"	6-1/2"	6"	5-21/32"	3'-4"	8.3	32
7 (B)	4-1/2"	6-1/2"	7"	5-1/2"	3'-8"	9.7	32
7-S.R.(B-S.R.)	4-1/2"	6-1/2"	7"	4-15/16"	2'-0"	9.7	32
8 (C)	4-1/2"	6-1/2"	8"	5-1/2"	3'-9"	10.8	32
8-S.R.(C-S.R.)	4-1/2"	6-1/2"	8"	5-1/16"	2'-6"	10.8	32
S (S)	4-1/2"	8"	4"	8"	Straight	7.0	32

(metric)

Unit	in millimeters A	B	C	D	Inside Radius mm	Approximate Shipping Weight	Approximate Percent Voids
4 (N)	112.5	162.5	100	140.6	1000	2.5	32
6 (A)	112.5	162.5	150	141.4	1000	3.8	32
7 (B)	112.5	162.5	175	137.5	1100	4.4	32
7-S.R.(B-S.R.)	112.5	162.5	175	123.4	600	4.4	32
(C)	112.5	162.5	200	137.5	1125	4.9	32
8-S.R.(C-S.R.)	112.5	162.5	200	126.6	750	4.9	32
S (S)	112.5	162.5	100	200.0	straight	3.2	32

Radial chimney brick is priced by the ton and will vary with type of brick and locality. Prices are f.o.b. factory, and the estimator must add freight and trucking from factory to job site and cost of unloading and storing brick at job.

Labor Laying Brick in Radial Brick Chimneys. The labor cost of laying radial brick chimneys will vary with the height of the chimney, thickness of walls, etc., but the following are actual production times on six radial brick chimneys constructed.

Labor Hours per Ton of Brick

Description	Tons of Radial Brick	Mason	Labor	Hoist Engr.
125'-0" high, Walls 16-1/4" to 7-1/8" thick. 12'-9" diam. at bottom, 7'-0" at top. Lined 50'-0"	157	2.7	4.7	0.9
80'-0" high, walls 11-3/4" to 7-1/8" thick. 8'-10" diam. at bottom, 5'-9" at top. Lined 35'-0"	66.5	3.1	5.3	1.0

Description	Tons of Radial Brick	Mason	Labor	Hoist Engr.
60'-0" high, walls 10-1/4" to 7-1/8" thick. 7'-2" diam. at bottom, 4'-4" at top. Lined 21'-0"	31.1	4.7	8.2	1.9
46'-0" high, walls 10-1/4" to 7-1/8" thick. 6'-6" diam. at bottom, 4'-2" at top. Unlined	24.6	3.3	8.0	1.5
46'-0" high walls 10-1/4" to 7-1/8" thick. 6'-6" diam. at bottom, 4'-2" at top. Unlined (2)	49.2	3.0	5.0	1.5
55'-0" high, walls 10-1/4" to 7-1/8" thick. 6'-11" diam. at bottom, 4'-2" at top. Lined 20'-0"	31.6	3.0	5.0	1.5

The above does not include supervision time.

Labor Hours per Metric Ton of Brick

Description	Tons of Radial Brick	Mason	Labor	Hoist Engr.
38 m high, Walls 406 mm to 178 mm thick. 4 m diam. at bottom, 2 m at top. Lined 15 m	142.43	2.98	5.18	0.99
24 m high, walls 294 mm to 178 mm thick. 3 m diam. at bottom, 2 m at top. Lined 11 m	60.327	3.42	5.84	1.10
18 m high, walls 256 mm to 178 mm thick. 2 m diam. at bottom, 1 m at top. Lined 6 m	28.213	5.18	9.04	2.09
14 m high, walls 256 mm to 178 mm thick. 2 m diam. at bottom, 1 m at top. Unlined	22.317	3.64	8.82	1.65
14 m high walls 256 mm to 178 mm thick. 2 m diam. at bottom, 1 m at top. Unlined (2)	44.633	3.31	5.51	1.65
17 m high, walls 256 mm to 178 mm thick. 2 m diam. at bottom, 1 m at top. Lined 6 m	28.667	3.31	5.51	1.65

The above does not include supervision time.

Labor time includes unloading materials from cars into trucks and unloading trucks at site. All mortar joints on the above stacks were struck flush with point of trowel.

Where stacks are lined, labor includes lining with perforated radial fire clay brick laid in high temperature cement. The lining is laid up from inside after stack is topped out.

Miscellaneous Masonry Costs

Labor Costs on High Buildings. On buildings over 10 to 12 stories high, it will be necessary to add an extra allowance for labor handling and hoisting the bricks and mortar. The higher the building the longer it takes the hoist to make a trip from the street level to the floor on which the bricks are being laid and return, extra work raising and handling scaffolding, etc.

Taking a 12-story building, or 120' (36.5 m) in height, as normal or 100%, add about 1% to the labor cost for each story or each 10' (3.0 m) feet in height above 120' (36.5 m).

For instance, a 12-story building or 120' (36.5 m) in height is 100, the 13th story would be 101; the 14th story, 102; the 15th story, 103, etc. To find the average cost on a 15-story building, figure as follows:

The first 12 stories as 100 each or 1,200
The 13th story as 101... 101
The 14th story as 102... 102
The 15th story as 103... 103
Total for 15 stories or 150' (45.7 m) in height....................................... 1,506

Dividing 1,506 by 15 gives 1.004 as the average cost of the masonry labor for a 15-story building. Other story heights would be computed in the same manner. The following table applies to all classes of masonry work on high structures.

Height in Stories	Normal Base	Add Above Normal %	Average for Job
1-12	100.00	None	100.00
13	100.00	1%	100.08
14	100.00	2%	100.22
15	100.00	3%	100.40
16	100.00	4%	100.63
18	100.00	6%	101.17
20	100.00	8%	101.80
22	100.00	10%	102.50
24	100.00	12%	103.25
26	100.00	14%	104.04
28	100.00	16%	104.86
30	100.00	18%	105.70
40	100.00	28%	110.15
50	100.00	38%	114.82

Brick Catch Basins and Manholes

When laying building or sewer brick in catch basins and manholes from 3'-0" to 5'-0" (0.9-1.5 m) in diameter and having 8" to 12" (200-300 mm) walls, a mason should lay 600 to 750 brick per 8-hr. day, at the following labor cost per 1,000:

	Hours	Rate	Total	Rate	Total
Mason	11.5	$....	$....	$28.18	$ 324.07
Labor	11.5	....	....	21.04	241.96
Cost per 1,000 brick			$....		$ 566.03

Pointing Around Steel Sash With Cement Mortar. All of the steel windows in a three story office building were pointed with portland cement mortar between the sash and the brick jambs after they had been set.

There were 39 windows 4'-0"x4'-10" (1.2 x 1.5 m) and 96 windows 4'-0"x7'-0" (1.2 x 2.1 m), making a total of 2,800 l.f. (853 m) that were pointed at the following labor cost per 100 l.f. (30.4 m):

	Hours	Rate	Total	Rate	Total
Bricklayer	1.1	$....	$....	$28.18	$ 40.00
Labor	0.5	....	....	21.04	10.52
Cost per 100 l.f.			$....		$ 50.52
per l.f.					0.51
per m					1.67

Reinforced Brick Masonry

Reinforced brick masonry has been widely used on the west coast because of its resistance to the lateral forces produced by earthquakes. It is now being used throughout the country, because most areas have been classified as seismic. It is also used where high winds or blast conditions may occur and for retaining walls.

These walls are comparable to reinforced concrete construction with the masonry units serving as a form. The exterior wythes are built up in a conventional manner, the reinforcing is set in the interior joints, and the whole is made homogeneous by filling the interior joints with grout.

No special skills or techniques are needed on the part of the mason, but certain care must be taken, including seeing that there is adequate lime in the mortar; a sound base to start on with the aggregate in concrete exposed and wetted to saturation; that brick is wetted so grout will not dry out too quickly; that grout is sufficiently fluid so none will adhere to a trowel; that grout be poured in a layer of about three bricks high and be allowed to set for 15 minutes before another pour is made, and poured from the inside to avoid splashing the exposed surfaces; and that mortar cuts not be spilled into grout or grout allowed to accumulate on bed joints.

The amount of reinforcing will vary widely with the job. In lightly reinforced walls, the reinforcing will fit within the widths of normal brick walls. The more heavily reinforced walls may require wider inner joints or the use of soap courses to accommodate the steel. Headers are not used in reinforced walls, and the contractor should check the drawings to see if header rows, for aesthetic reasons, are shown, because it will require cutting all these brick. This is especially true when figuring reinforced lintels in conventional walls so that coursing can carry through.

Savings, as well as aesthetic advantages, often can be realized through the use of reinforced brick lintels. These are almost always poured in place with temporary shoring for the soffit. The economies result from savings in steel cost and elimination of painting. In better work it may call for the soffit brick joints to be filled temporarily with sand and then repointed later.

A variation on the above has been developed and is known as high lift grouted reinforced brick masonry. It varies in that it is essentially a cavity wall with a 2" (50 mm) or better inner joint. This allows the reinforcing to be preset and grouting delayed until a full 12' (3.65 m) wall height is reached. Mechanical pumping can thus be used.

Precautions listed for standard reinforced walls also apply here with the following additions. Care must be taken to allow cleaning out of grout space by omitting every other unit on one side; brick must be allowed to set for at least 3 days before grouting; vibrating of grout is necessary; once grout pour is started, it must be continuous; and construction dams may be built every 20 or 25 feet (6 or 7.6 meters) to contain the grout laterally.

Handling of Masonry By Palletization

For years, builders considered using palletization for handling masonry materials, and many efforts were made in this direction with mixed results. Now, new systems have overcome most of the problems. Palletized masonry units, whether brick, block, or tile, can be transported from the manufacturing plant, through the job site, up the hoist, and around the scaffold to the mason's station, eliminating all hand handling of units until the mason picks them up for laying. Two companies that manufacture this equipment are the PCM Division of Koehring Corp., Port Washington, Wisc and Prime-Mover Co., Muscatine, Ia.

Time studies on over 300 jobs using the PCM system indicate that with ten or more masons, the conventional ratio of 0.7 to 1.0 labor hour per mason hour can be reduced to 0.3 to 0.5 labor hours per mason hour--a savings of approximately 50% in labor time required for tending masons, or an overall 20% savings on the total masonry labor cost. These ratios include the labor time required to mix mortar, tend mortar, brick, tile, or block, and remove rubbish.

The success of these systems is mainly due to the equipment, which consists of properly sized wood pallets, a tractor with specially design forklift mechanism, special brick buggies, and special mortar buggies.

For efficient use of this system and equipment, certain job conditions must be established from the start and maintained throughout the life of the job. All materials must be purchased on the basis of pallet delivery or palletized as soon as they arrive on the job. In the past, contractors were furnished the wood pallets, but dealers and manufacturers have shown an increasing preference for this type of delivery and in some cases furnish their own pallets. The contractor should specify the quantity of material to be placed on each pallet and how it is arranged. Weight should not exceed 1,000 lbs. (453 kg) per pallet to stay within the capacity of scaffolding and buggies.

The effect on material prices for pallet delivery varies. Some suppliers maintain the same prices, claiming the extra cost of labor and material for palletization is offset by savings in loading and unloading. Other suppliers increase brick prices $2.00 to $3.50 per thousand. Actual cost of a 24"x32" (600 x 800 mm) pallet, for palletization on the job site, is about $4.00 per pallet. One laborer can handle 6 to 7 pallets per hour.

Access roads to the job should be laid out and maintained to facilitate operation of the fork lift tractor. Area surrounding the building should be backfilled, consolidated, and graded before masonry work starts and kept reasonably clear during the work.

Exterior scaffolding must be wide enough to permit maneuvering brick and mortar buggies. Where tubular steel scaffolding is used, this may be accomplished with standard scaffolding components, using a 5' (1.5 m) width tubular scaffold with the addition of a 30" (750 mm) bracket on the inside for the masons and a 20" (500 mm) bracket on the outside for the extra width required for buggies and storage of pallets. No special planking is required, but plank laps should all run in the same direction and a wood easement strip placed at each lap to partially eliminate the irregularity.

Where suspended scaffolding is used, extra width can be obtained by using 8' (2.4 m) wide Wheeling scaffold machines with the addition of 10' (3.0 m) pipe putlogs fastened directly under and through scaffold planking with U-bolts. This method moves the inside drum and cable 2'-0" (0.6 m) out from the wall and creates a clear working space for the mason.

Scaffolding costs will be increased 5% to 15%, depending upon the type of scaffold used.

Standard Pallet 32"x24" (800 x 600 mm). A standard pallet size of 32" x 24" (800 x 600 mm) has been chosen, because it accommodates all common sizes of brick, block and tile, its loaded weight does not exceed safety regulations for scaffolding or light floor construction, and it accommodates just enough material to build 10 l.f. (3.0 m) of wall, 4' (1.2 m) high, 4" (100 mm) thick, allowing proper spacing for stacking on the scaffold according to standard masonry practice. In addition, its loaded weight can be handled by one worker using hand powered equipment, and the pallet load can be split into two 16" x 24" (400 x 600 mm) pallet loads, permitting passage through normal doorways, handling on occasional narrow scaffolds and maneuvering in confined spaces.

Pallets should be made to withstand hard use, preferably from a good grade of 1" (25 mm) lumber--1"x8" (25 x 200 mm) boards for the tops and 2"x4" (50 x 100 mm) for skids. Pallets made on the job, using power saws and a production line set-up, should cost about $2.25 per half pallet—16"x24" (400 x 600 mm)—each, including material and labor.

One standard full size pallet is composed of two 16x24s (400 x 600 mm) strapped together with metal banding. When required, the metal banding can be removed, breaking the package into two halves of 16"x24" (400 x 600 mm) for use where operating space is limited.

Forklift Tractors. Manufacturers provide various capacities and sizes of rubber-tired forklifts, with lifting heights from 18'-6" to 30'-6" (5.6-9.3 m) and with capacities from 2,500 to 4,000 lbs. (1134-1814 kg). The 3,000 lbs. (1361 kg) unit can unload and stockpile as many as 4,000 brick in 10 minutes, and can be used to handle concrete planks and roofing materials as well as masonry.

The forklift tractor unloads pallets of material from delivery trucks and places them in stockpiles. From stockpiles, pallets are then transported to the masons' stations at ground level or on the scaffold up to maximum reach of tractor; to construction hoist for work on scaffolds or floors beyond tractor reach; or to building floors within tractor reach for partition work or exterior masonry performed from inside the scaffold.

The forklift can handle as much palletized masonry material as can 20 laborers using conventional hand methods and wheelbarrows. It can also be equipped with other attachments to perform many other tasks on the job.

The fork lift also delivers mortar to the bricklayers from the mixing point by use of mortar buggies, which are mortar containers with a capacity of 7 c.f. (0.19 cu.m), equipped with wheels and casters so that they can be readily pushed on floors or scaffolds. The tractor handles these mortar containers to and from the mixer the same as packaged materials.

Brick Buggies. Pallet loads are handled on the scaffold and building floors with brick buggies. These come in a variety of sizes, either hand propelled or power driven, with capacities of 1,000 lbs. (453 kg) for half loads—16"x24" (400 x 600 mm) pallets with 120 bricks—and 1,500 lbs. (680 kg) for full loads—24"x32" (600 x 800 mm) pallets with 240 bricks. Lifting forks can be had for 10" (250 mm), 7'-6" (2.2 m), and 10'-0" (3.0 m) lifts, the latter being able to feed double deck scaffolding. Buggies can be operated under average outdoor conditions as well as on up and down ramps.

Lift trucks for 24"x32" (600 x 800 mm) pallets cost about $5,000 and can service 4 to 6 masons on a medium-rise building, as opposed to one laborer keeping each mason supplied with brick. Two laborers, one on the ground and one delivering, should serve 5 masons. Under good conditions, the daily savings in labor cost should amortize the lift and pallet costs in approximately 5 to 6 weeks. Such handling also cuts down the time the hoist must be held for unloading by hand, which can be critical on larger jobs.

04220 CONCRETE UNIT MASONRY

Concrete masonry units, commonly termed concrete block or *concrete brick*, are used extensively in all parts of the country for exterior and interior bearing walls, interior partitions, floor and roof fillers, and the like. Much of the information and data in this section comes from the National Concrete Masonry Association.

Concrete masonry units are widely used for backing up veneers. They are used both for bearing walls on wall bearing structures and for curtain walls in steel or reinforced concrete frame structures. The use of concrete masonry has increased greatly. This is partly due to the lower labor cost of handling and setting the larger concrete units in comparison with building brick, when used as a back-up material, but also because of increased use of architectural facing units.

The term concrete masonry is applied to block, brick, or tile building units molded from concrete and laid by masons into a wall. The concrete is made by mixing portland cement with water and other suitable materials, such as sand, gravel, crushed stone, burned clay or shale, blast furnace slag, and pumice.

Concrete masonry units should always be manufactured in a cement products plant where facilities for manufacturing and curing are uniform so as to obtain high quality products. The quality of all concrete masonry units should conform to the standards set by the American Society for Testing and Materials (ASTM) in its standard specifications for concrete masonry units and concrete brick.

Weight of Concrete Masonry. A 7-5/8"x7-5/8"x15-5/8" (190.62 x 190.62 x 390.62 mm) hollow loadbearing concrete block weighs from 35 to 45 lbs. (15.8-20.0 kg) when made from heavyweight aggregate and from 25 to 35 lbs. (11.3-15.8 kg) when made from lightweight aggregate. Heavyweight aggregates are sand, gravel, crushed stone, air-cooled slag, etc. Lightweight aggregates are coal cinders, expanded shale, clay or slag, pumice, etc.

Estimating Quantities of Concrete Brick. Concrete brick are manufactured, throughout the country generally, in the modular size of 2-1/4"x3-5/8"x7-5/8" (56.2 x 90.6 x 190.62 mm). In some localities, additional sizes are available as follows: Jumbo brick, 3-5/8"x3-5/8"x7-5/8" (90.62 x 90.62 x 190.62 mm); double brick, 4-7/8"x3-5/8"x7-5/8" (115.62 x 90.62 x 190.62 mm); and Roman brick, 1-5/8"x3-5/8"x11-5/8" (40.62 x 90.62 x 290.62 mm) and 1-5/8"x3-5/8"x15-5/8" (40.62 x 90.62 x 181.64).

They should be estimated by the sq. ft. (sq. meter) of wall of any thickness or by the cu. ft. (cu. meter), if all wall thicknesses are combined. For instance, if there are only 8" (200 mm) walls in the job, it is acceptable to take the number of s.f. (sq.m) of walls, but if the job contains 8" (200 mm), 12" (300 mm), and 16" (400 mm) walls, then it will be easier and more satisfactory to reduce the quantities to cu. ft. (cu.m) and multiply by a certain number of brick per cu. ft. (cu.m) of wall.

The number of concrete brick required per sq. ft. or cu. ft. (sq.m or cu.m) of wall will vary with the size of brick and width of mortar joint used. Concrete brick sizes and widths of mortar joints have been standardized so that they will lay up to an even multiple of the 4-inch (100-mm) module.

All concrete brick units are 3-5/8" (92.2 mm) thick, which together with a 3/8" (9.6 mm) mortar back joint equal the modular thickness of 4 inches (101.6 mm). Since there is always one less mortar joint than brick, in walls of multiple brick thickness, the actual wall thickness will always be 3/8" (9.6 mm) less than the modular 4 inches; i.e. 3-5/8" (92.2 mm), 7-5/8" (190.6 mm), 11-5/8" (290.6 mm), 15-5/8" (390.6 mm), etc.

In drawing plans, some architects give actual wall thickness dimensions, while others use nominal dimensions. For estimating purposes, nominal dimensions should be used, i.e. 4" (100 mm), 8" (200 mm), 12" (300 mm), 16" (400 mm), etc., to facilitate computations. When measuring the plans and listing the quantities of brick required, always take exact measurements for length and height and deduct all openings in full. Do not count corners twice but have your estimate show as accurately as possible the actual number of brick required to complete the job. For unexposed backup work, add 1% to 2% for waste. For exposed face work, add 3% to 5% for waste.

Modular Size Concrete Brick

Modular size concrete brick, 2-1/4"x3-5/8"x7-5/8" (56.25 x 90.62 x 190.62 mm) are laid so that 3 courses of brick, with 5/12" (10.4 mm) mortar joints, builds 8" (200 mm) in height, and the 7-5/8" (190.6 mm) length, plus 3/8" (15.6 mm) mortar joint, makes a total length of 8" (200 mm) laid in the wall. In other words, 3 brick lay 8"x8" (200 x 200 mm), or 64 sq. in. (40,000 mm²), so it requires 6-3/4 or 6.75 brick per s.f. (0.09 sq.m) of wall of nominal 4" (100 mm) thickness.

Jumbo brick are 3-5/8"x3-5/8"x7-5/8" (90.62 x 90.62 x 190.62 mm) in size and are laid so that 3 courses of brick, with 3/8" (15.6 mm) mortar joints, builds 12" (300 mm) in height, and the 7-5/8" (190.62 mm) length, plus 3/8" (15.6 mm) mortar joint, makes a total length of 8" (200 mm) laid in the wall. One jumbo brick lays 32 sq. in. in the wall or it requires 4.5 brick per s.f. (0.09 sq.m) of wall of nominal 4" (100 mm) thickness.

Double brick are 4-7/8"x3-5/8"x7-5/8" (121.87 x 90.62 x 190.62 mm) in size, so that 3 courses of brick, with 11/24" (11.4 mm) mortar joint builds 16" (400 mm) high, and the 7-5/8" (190.62 mm) length, plus 3/8" (15.6 mm) mortar joint, makes a total length of 8" (200 mm) in the wall, or one brick lays 42.67 sq. in. in the wall, or it requires 3.4 brick per s.f. (0.09 sq.m) of wall of nominal

4" (100 mm) thickness.   Roman size brick are 1-5/8"x3-5/8"x11-5/8" (40.62 x 90.62 x 290.62 mm) and 1-5/8"x3-5/8"x15-5/8" (40.62 x 90.62 x 390.62) and are laid so that 6 courses of brick, with 3/8" (15.6 mm) mortar joints, builds 12" (300 mm) in height, and the 11-5/8" (290.62 mm) or 15-5/8" (390.62 mm) lengths, plus 3/8" (15.6 mm) mortar joints makes a total length of 12" (300 mm) or 16" (400 mm) respectively laid in the wall. One nominal 12" (300 mm) brick lays 24 sq. in. in the wall or it requires 6 brick per s.f. (0.09 sq.m) of wall of nominal 4" (100 mm) thickness. One nominal 16" (400 mm) brick lays 32 sq. in. in the wall or it requires 4.5 brick per s.f. (0.09 sq.m) of wall of nominal 4" (100 mm) thickness.

The number of modular size brick per s.f. (0.09 sq.m) of wall of various thicknesses is as follows:

per s.f.	1-Brick (4")	Thickness of Wall 2-Bricks (8")	3-Bricks (12")*
2-1/4"x3-5/8"x7-5/8"	6.75	13.50	20.25
3-5/8"x3-5/8"x7-5/8"	4.50	9.00	13.50
4-7/8"x3-5/8"x7-5/8"	3.40	6.80	10.20
1-5/8"x3-5/8"x11-5/8"	6.00	12.00	18.00
1-5/8"x3-5/8"x15-5/8"	4.50	9.00	13.50

(metric)

per sq.m	1-Brick (100mm)	Thickness of Wall 2-Bricks (200mm)	3-Bricks (300mm)
56 x 91 x 191 mm	72.66	145.32	217.98
91 x 91 x 191 mm	48.44	96.88	145.32
122 x 91 x 191 mm	36.60	73.20	109.80
41 x 91 x 291 mm	64.59	129.17	193.76
41 x 91 x 391 mm	48.44	96.88	145.32

*Use this column for number of brick required per c.f. (0.028 cu.m) of wall.
No allowance is included for waste in above quantities.

Labor Handling and Laying Concrete Brick

The labor cost of handling and laying concrete brick varies, depending on factors such as size of brick, kind of mortar, class of work, thickness of walls, and number of openings.

On buildings having long straight walls without many openings, such as basement walls, garages, and factories, a mason should lay 800 to 900 modular size concrete brick, 2-1/4"x3-5/8"x7-5/8" (56.25 x 90.62 x 190.62), per 8-hr. day, while on dwellings and other structures having 8" (200 mm) walls, cut up with numerous openings and pilasters, a mason should lay 650 to 700 brick per 8-hr. day.

Labor Cost of 1,000 Modular Size Concrete Brick
2-1/4"x3-5/8"x7-5/8" (56.25 x 90.62 x 190.62 mm)
Laid in 8" (200 mm) Walls

	Hours	Rate	Total	Rate	Total
Mason	11.8	$....	$....	$28.18	$ 332.52
Labor	8.8	....	....	21.04	185.15
Cost per 1,000 brick			$....		$ 517.67

The above costs are based on flush cut mortar joints. For struck joints, add 1/2-hr. mason time per 1,000 brick.

Estimate the labor cost of laying concrete face brick the same as given for face brick under *Unit Masonry / Brick*.

On similar work, using jumbo brick, 3-5/8"x3-5/8"x7-5/8" (90.62 x 90.62 x 190.62 mm), a mason should lay 675 to 775 brick per 8-hr. day in 12" (300 mm) walls and 550 to 600 brick per day in 8" (200 mm) walls.

Using double brick, 4-7/8"x3-5/8"x7-5/8" (121.87 x 90.62 x 190.62 mm), a mason should lay 450 to 550 brick per 8-hr. day in 12" (300 mm) walls and 375 to 425 brick per day in 8" (200 mm) walls.

Roman size concrete brick, 1-5/8"x3-5/8"x11-5/8" (40.62 x 90.62 x 290.62 mm) or 1-5/8"x3-5/8"x15-5/8" (40.62 x 90.62 x 390.62 mm), are used only for facing or in 8" (200 mm) walls. The labor cost of handling and laying these brick will vary considerably, depending upon the type of joint treatment, flush cut, V-tooled or concave, struck or weathered, or raked out.

On average work with flush cut joints, a mason should lay 300 to 350 nominal 12" (300 mm) Roman brick per 8-hr. day or 225 to 265 nominal 16" (400 mm) Roman brick per 8-hr. day. For tooled, struck, weathered, or raked out joints, reduce by 8% to 12% the above quantities.

Concrete Blocks and Partition Units

Concrete blocks and partition units are manufactured in weights of concrete. Specific information on density of concrete types and on weights of units should be obtained from local manufacturers. When specific information is not available, assume the following values for products:

Type	Lbs. per cu. ft.	Kg per cu.m
Pumice	75	1201
Expanded shale concrete	85	1362
Expanded slag and cinder concrete	95	1522
Air-cooled slag concrete	120	1922
Crushed stone and gravel concrete	135-145	2163-2323

These types of units may be used interchangeably for all purposes, although lightweight units afford a savings of weight to secure economy in design, and they afford increased heat and sound insulation value.

Estimating the Quantity of Concrete Blocks and Partition Units. Concrete blocks and partition units should be estimated by the square foot of wall of any thickness and then multiplied by the number of blocks per 100 s.f. (9.3 sq.m) as given in the following tables.

Always take exact measurements and make deductions in full for all openings, regardless of size. The result will be the actual number of sq. ft. (sq.m) or number of blocks required for the job.

Mortar. The same kind of mortar should be used in laying concrete blocks as given for *Modular Size Concrete Brick.*

Sizes and Shapes of Concrete Building Units

Sizes and Shapes of Concrete Building Units

When laying up exterior walls, face shell bedding shall be used with complete coverage of face shells. Furrowing of the mortar shall not be permitted. Extruded mortar shall be cut off flush with face of wall and the joints firmly compacted, after the mortar has stiffened somewhat.

Tooling is essential in producing tight mortar joints. Mortar has a tendency to shrink slightly and may pull away from the masonry units causing fine, almost invisible cracks at the junction of mortar and masonry units.

Note: The following nomenclature is the same for 2-core and 3-core block. The 2-core has been used here for illustration.

OBE with MG
Open Both Ends with Mortar Grooves

OE with MG-POE
Open End with Mortar Grooves – Plain Other End (Notice thickness of Plain End, unlike Permanent Plain End.)

OE with MG-SSOE
Open End with Mortar Grooves – Steel Sash Other End (Also pictured is a Bullnose Corner: 1" & 2" radius available on any Plain End Corner.)

OE with MG-PPOE
Open End with Mortar Grooves – Permanent Plain Other End (Notice thickness of Permanent Plain End, unlike Plain End.)

PPBE
Permanent Plain Both Ends.

A   Extra Mortar Bed

B   Hand Hold

Hand Hold optional on 2 and 3 core block.

Dimensions of units: In practice, the first dimension of a concrete masonry unit represents the thickness; the second dimension, the height; the third, length.

Example: 8″ x 8″ x 16″, or 200 mm. x 200 mm. x 400 mm.

Sizes and Weights of Concrete Masonry Units

Concrete masonry units are usually made with standard modular face dimensions 7-5/8" (190.6 mm) high and 15-5/8" (390.6 m) long and are available in thicknesses 3/8" (15.6 mm) less than the nominal 3", 4", 6", 8", 10", and 12" (75, 100, 150, 200, and 300 mm) thickness.

When laid up with 3/8" (15.6 mm) mortar joints, the units are 8" (200 mm) high and 16" (400 mm) long, requiring 112.5 units per 100 s.f. (9 sq.m) of wall, not including allowance for waste and breakage.

Hollow and solid partition and furring units have nominal thickness of 2", 3", 4", 6", and 8" (51, 76, 102, 152, and 203 mm).

Hollow and solid loadbearing units have nominal thickness of 4", 6", 8", 10", and 12" (102, 152, 203, 254, and 305 mm) and are also available in half lengths.

Standard specials such as steel and wood sash jambs, bullnose, and closures are also available in full and half lengths. There are also many sizes available on special order.

Sizes, Weights and Quantities of Loadbearing Concrete Blocks and Tile

Actual Size of Units	Thickness	Approx. Weight in lbs. Hvy. Wt. Units	Approx. Weight in lbs. Light Wt. Units	No. of Units 100 s.f. of Wall	Cu.Ft. Mortar 100 s.f. of Wall*
3-5/8" x 4-7/8" x 11-5/8"	4"	11-13	8-10	240	8.0
5-5/8" x 4-7/8" x 11-5/8"	6"	17-19	12-14	240	8.5
7-5/8" x 4-7/8" x 11-5/8"	8"	22-24	14-16	240	9.0
3-5/8" x 7-5/8" x 11-5/8"	4"	17-19	12-14	150	6.0
5-5/8" x 7-5/8" x 11-5/8"	6"	26-28	17-19	150	6.5
7-5/8" x 7-5/8" x 11-5/8"	8"	33-35	21-23	150	7.0
9-5/8" x 7-5/8" x 11-5/8"	10"	42-45	27-29	150	7.5
11-5/8" x 7-5/8" x 11-5/8"	12"	48-51	29-31	150	8.0
3-5/8" x 3-5/8" x 15-5/8"	4"	12-14	9-10	225	9.0
5-5/8" x 3-5/8" x 15-5/8"	6"	17-19	11-13	225	9.5
7-5/8" x 3-5/8" x 15-5/8"	8"	22-24	14-16	225	10.0
3-5/8" x 7-5/8" x 15-5/8"	4"	23-25	16-18	112.5	5.0
5-5/8" x 7-5/8" x 15-5/8"	6"	35-37	24-26	112.5	5.0
7-5/8" x 7-5/8" x 15-5/8"	8"	45-47	29-31	112.5	6.0
9-5/8" x 7-5/8" x 15-1/2"	10"	57-60	36-38	112.5	6.5
11-5/8" x 7-5/8" x 15-5/8"	12"	64-67	40-42	112.5	7.0

(metric)

Actual Size of Units	Thickness	Approx. Weight in kg Hvy. Wt. Units	Approx. Weight in kg Light Wt. Units	No. of Units 10 sq.m of Wall	Cu.M Mortar 10 sq.m of Wall
91 x 122 x 291 mm	100 mm	5-9	4-5	258.3	0.24
141 x 122 x 291 mm	150 mm	8-9	5-6	258.3	0.26
191 x 122 x 291 mm	200 mm	10-11	6-7	258.3	0.27
91 x 191 x 291 mm	100 mm	8-9	5-6	161.5	0.18
141 x 191 x 291 mm	150 mm	12-13	8-9	161.5	0.20
191 x 191 x 291 mm	200 mm	15-16	10-10	161.5	0.21
241 x 191 x 291 mm	250 mm	19-20	12-13	161.5	0.23

Actual Size of Units	Thickness	Approx. Weight in kg Hvy. Wt. Units	Approx. Weight in kg Light Wt. Units	No. of Units 10 sq.m of Wall	Cu.M Mortar 10 sq.m of Wall
291 x 191 x 291 mm	300 mm	22-23	13-14	161.5	0.24
91 x 91 x 391 mm	100 mm	5-6	4-5	242.2	0.27
141 x 91 x 391 mm	150 mm	8-9	5-6	242.2	0.29
191 x 91 x 391 mm	200 mm	10-11	6-7	242.2	0.30
91 x 191 x 391 mm	100 mm	10-11	7-8	121.1	0.15
141 x 191 x 391 mm	150 mm	16-17	11-12	121.1	0.15
191 x 191 x 391 mm	200 mm	20-21	13-14	121.1	0.20
241 x 191 x 391 mm	250 mm	26-27	16-17	121.1	0.21
291 x 191 x 391 mm	300 mm	29-30	18-19	121.1	0.24

Sizes, Weights and Quantities of Concrete Partition Tile

Actual Size of Units	Thickness	Approx. Weight in lbs. Light Wt. Units	No. of Units 100 s.f. of Wall	Cu.Ft. Mortar 100 s.f. of Wall*
3-5/8" x 7-5/8" x 11-5/8"	4"	12-14	150	6.0
5-5/8" x 7-5/8" x 11-5/8"	6"	17-19	150	6.5
7-5/8" x 7-5/8" x 11-5/8"	8"	21-23	150	7.0
1-5/8" x 7-5/8" x 15-5/8"	2"	11-12	112.5	3.0
3-5/8" x 7-5/8" x 15-5/8"	4"	16-18	112.5	5.0
5-5/8" x 7-5/8" x 15-5/8"	6"	24-26	112.5	5.0
7-5/8" x 7-5/8" x 15-5/8"	8"	29-31	112.5	6.0

(metric)

Actual Size of Units	Thickness	Approx. Weight in kg Light Wt. Units	No. of Units 10 sq.m of Wall	Cu.M Mortar 10 sq.m of Wall
91 x 191 x 291 mm	100 mm	5-6	161.5	0.18
141 x 191 x 291 mm	150 mm	8-9	161.5	0.20
191 x 191 x 291 mm	200 mm	10-10	161.5	0.21
41 x 191 x 391 mm	50 mm	5-5	121.1	0.09
91 x 191 x 391 mm	100 mm	7-8	121.1	0.15
141 x 191 x 391 mm	150 mm	11-12	121.1	0.15
191 x 191 x 391 mm	200 mm	13-14	121.1	0.18

*Actual mortar quantities are about half those given in the table, but experience shows that considerably more mortar is required, due to waste, droppings, etc.

Labor Laying Concrete Masonry

The labor cost of laying the various types and sizes of concrete blocks and tile will vary with factors such as the size and weight of the blocks, the class of work, and whether there are long straight walls or walls cut up with numerous openings. Labor costs will also vary depending on whether the blocks are laid above or below grade. Basement walls usually proceed faster than exposed work above grade.

It is usually more economical to use lightweight units, even though they cost a few cents more per piece, because a mason can handle and lay them with less effort. In some localities union regulations require two masons to work together where the blocks weigh more than 35 lbs. (15.8 kg) each.

Where concrete blocks are used for exterior or interior facing walls, with the blocks carefully laid to a line and in various patterns and with neatly tooled mortar joints, the labor costs will run considerably higher than on straight structural walls.

The quantities given in the following tables are based on average conditions with blocks laid in 1:1:6 cement-lime mortar. If portland cement mortar is used above grade, reduce daily output about 5 percent.

All quantities are based on the output of one mason and one laborer or hod carrier per 8-hr. day. If hoisting engineer time is required, add about 1/4-hr. per 100 s.f. (sq.m).

Number of Concrete Partition Tile Laid Per 8-Hr. Day By One Mason

Actual Size of Units	No. of Wall Thick-ness	Light Wt. Units* per 8-Hr Day	Mason Hrs. per 100 Pieces	Labor Hrs. per 100 Pieces
3-5/8" x 7-5/8" x 11-5/8"	4"	195-215	3.9	3.9
5-5/8" x 7-5/8" x 11-5/8"	6"	175-195	4.3	4.3
7-5/8" x 7-5/8" x 11-5/8"	8"	155-175	4.8	4.8
1-5/8" x 7-5/8" x 15-5/8"	2"	180-200	4.2	4.2
3-5/8" x 7-5/8" x 15-5/8"	4"	190-210	4.0	4.0
5-5/8" x 7-5/8" x 15-5/8"	6"	170-190	4.5	4.5
7-5/8" x 7-5/8" x 15-5/8"	8"	150-170	5.0	5.0

(metric)

Actual Size of Units	No. of Wall Thick-ness	Light Wt. Units* per 8-Hr Day	Mason Hrs. per 100 Pieces	Labor Hrs. per 100 Pieces
91 x 191 x 291 mm	100 mm	195-215	3.9	3.9
141 x 191 x 291 mm	150 mm	175-195	4.3	4.3
191 x 191 x 291 mm	200 mm	155-175	4.8	4.8
41 x 191 x 391 mm	50 mm	180-200	4.2	4.2
91 x 191 x 391 mm	100 mm	190-210	4.0	4.0

Actual Size of Units	Wall Thick-ness	No. of Light Wt. Units* per 8-Hr Day	Mason Hrs. per 100 Pieces	Labor Hrs. per 100 Pieces
141 x 191 x 391 mm	150 mm	170-190	4.5	4.5
191 x 191 x 391 mm	200 mm	150-170	5.0	5.0

*For heavyweight concrete units decrease above quantities and increase labor 10 percent.
**For walls below grade.

Number of Concrete Building Units Laid Per 8-Hr. Day by One Mason Light Weight

Actual Size of Units	Wall Thick-ness	No. of Light Wt. Units per 8-Hr Day	Mason Hrs. per 100 Pieces	Labor Hrs. per 100 Pieces
3-5/8" x 4-7/8" x 11-5/8"	4"	215-235	3.5	3.5
5-5/8" x 4-7/8" x 11-5/8"	6"	195-215	3.9	3.9
7-5/8" x 4-7/8" x 11-5/8"	8"	175-195	4.3	4.3
3-5/8" x 7-5/8" x 11-5/8"	4"	195-215	3.9	3.9
5-5/8" x 7-5/8" x 11-5/8"	6"	175-195	4.3	4.3
7-5/8" x 7-5/8" x 11-5/8"	8"	155-175	4.8	4.8
9-5/8" x 7-5/8" x 11-5/8"	10"	135-155	5.5	5.5
11-5/8" x 7-5/8" x 11-5/8"	12"	115-135	6.4	6.4
3-5/8" x 3-5/8" x 15-5/8"	4"	215-235	3.5	3.5
5-5/8" x 3-5/8" x 15-5/8"	6"	195-215	3.9	3.9
7-5/8" x 3-5/8" x 15-5/8"	8"	175-195	4.3	4.3
3-5/8" x 7-5/8" x 15-5/8"	4"	190-210	4.0	4.0
5-5/8" x 7-5/8" x 15-5/8"	6"	170-190	4.5	4.5
7-5/8" x 7-5/8" x 15-5/8"	8"	150-170	5.0	5.0
9-5/8" x 7-5/8" x 15-5/8"	10"	130-150	5.7	5.7
11-5/8" x 7-5/8" x 15-5/8"	12"	110-130	6.7	6.7
7-5/8" x 7-5/8" x 15-5/8"	8"	225-250*	3.4	3.4
11-5/8" x 7-5/8" x 15-5/8"	12"	120-160*	5.7	5.7

(metric)

Actual Size of Units	Wall Thick-ness	No. of Light Wt. Units* per 8-Hr Day	Mason Hrs. per 100 Pieces	Labor Hrs. per 100 Pieces
91 x 122 x 291 mm	100 mm	215-235	3.5	3.5
141 x 122 x 291 mm	150 mm	195-215	3.9	3.9
191 x 122 x 291 mm	200 mm	175-195	4.3	4.3
91 x 191 x 291 mm	100 mm	195-215	3.9	3.9
141 x 191 x 291 mm	150 mm	175-195	4.3	4.3
191 x 191 x 291 mm	200 mm	155-175	4.8	4.8
241 x 191 x 291 mm	250 mm	135-155	5.5	5.5

Actual Size of Units	Wall Thick-ness	No. of Light Wt. Units* per 8-Hr Day	Mason Hrs. per 100 Pieces	Labor Hrs. per 100 Pieces
291 x 191 x 291 mm	300 mm	115-135	6.4	6.4
91 x 91 x 391 mm	100 mm	215-235	3.5	3.5
141 x 91 x 391 mm	150 mm	195-215	3.9	3.9
191 x 91 x 391 mm	200 mm	175-195	4.3	4.3
91 x 191 x 391 mm	100 mm	190-210	4.0	4.0
141 x 191 x 391 mm	150 mm	170-190	4.5	4.5
191 x 191 x 391 mm	200 mm	150-170	5.0	5.0
241 x 191 x 391 mm	250 mm	130-150	5.7	5.7
291 x 191 x 391 mm	300 mm	110-130	6.7	6.7
191 x 191 x 391 mm	200 mm	225-150*	3.4	3.4
291 x 191 x 391 mm	300 mm	120-160*	5.7	5.7

*For walls below grade.

Number of Concrete Building Units
Laid Per 8-Hr. Day by One Mason
Heavy Weight

Actual Size of Units	Wall Thick-ness	No. of Heavy Wt. Units per 8-Hr Day	Mason Hrs. per 100 Pieces	Labor Hrs. per 100 Pieces
3-5/8" x 4-7/8" x 11-5/8"	4"	195-214	3.9	3.9
5-5/8" x 4-7/8" x 11-5/8"	6"	177-195	4.3	4.3
7-5/8" x 4-7/8" x 11-5/8"	8"	159-177	4.7	4.7
3-5/8" x 7-5/8" x 11-5/8"	4"	177-195	4.3	4.3
5-5/8" x 7-5/8" x 11-5/8"	6"	159-177	4.7	4.7
7-5/8" x 7-5/8" x 11-5/8"	8"	141-159	5.3	5.3
9-5/8" x 7-5/8" x 11-5/8"	10"	123-141	6.1	6.1
11-5/8" x 7-5/8" x 11-5/8"	12"	105-123	7.0	7.0
3-5/8" x 3-5/8" x 15-5/8"	4"	195-214	3.9	3.9
5-5/8" x 3-5/8" x 15-5/8"	6"	177-195	4.3	4.3
7-5/8" x 3-5/8" x 15-5/8"	8"	159-177	4.7	4.7
3-5/8" x 7-5/8" x 15-5/8"	4"	173-191	4.4	4.4
5-5/8" x 7-5/8" x 15-5/8"	6"	155-173	5.0	5.0
7-5/8" x 7-5/8" x 15-5/8"	8"	136-155	5.5	5.5
9-5/8" x 7-5/8" x 15-5/8"	10"	118-136	6.3	6.3
11-5/8" x 7-5/8" x 15-5/8"	12"	100-118	7.4	7.4
7-5/8" x 7-5/8" x 15-5/8"	8"	205-227*	3.7	3.7
11-5/8" x 7-5/8" x 15-5/8"	12"	109-145*	6.3	6.3

Actual Size of Units	Wall Thickness	(metric) No. of Light Wt. Units per 8-Hr Day	Mason Hrs. per 100 Pieces	Labor Hrs. per 100 Pieces
91 x 122 x 291 mm	100 mm	195-214	3.9	3.9
141 x 122 x 291 mm	150 mm	177-195	4.3	4.3
191 x 122 x 291 mm	200 mm	159-177	4.7	4.7
91 x 191 x 291 mm	100 mm	177-195	4.3	4.3
141 x 191 x 291 mm	150 mm	159-177	4.7	4.7
191 x 191 x 291 mm	200 mm	141-159	5.3	5.3
241 x 191 x 291 mm	250 mm	123-141	6.1	6.1
291 x 191 x 291 mm	300 mm	105-123	7.0	7.0
91 x 91 x 391 mm	100 mm	195-214	3.9	3.9
141 x 91 x 391 mm	150 mm	177-195	4.3	4.3
191 x 91 x 391 mm	200 mm	159-177	4.7	4.7
91 x 191 x 391 mm	100 mm	173-191	4.4	4.4
141 x 191 x 391 mm	150 mm	155-173	5.0	5.0
191 x 191 x 391 mm	200 mm	136-155	5.5	5.5
241 x 191 x 391 mm	250 mm	118-136	6.3	6.3
291 x 191 x 391 mm	300 mm	110-130	7.4	7.4
191 x 191 x 391 mm	200 mm	205-227*	3.7	3.7
291 x 191 x 391 mm	300 mm	109-145*	6.3	6.3

*For walls below grade.

Mortar Joints. Hollow concrete block or tile should be laid with broken joints to secure the maximum resistance to moisture and heat penetration. Sufficient mortar is spread on the inner and outer shell bed and end joints to cover and to provide a joint of the required thickness when pressed into place. Solid units, brick, etc., are laid in solid beds.

Comparison of Concrete Masonry Sizes With Brick. Allowing for the mortar joints in brickwork, the various sizes of concrete masonry units are equivalent to the following number of modular size brick, i.e., 2-1/4"x3-5/8"x7-5/8" (56.2 x 90.62 x 190.62 mm).

Size of Blocks	Brick Equivalent
4"x8"x16" (102 x 203 x 406 mm)	6
8"x8"x16" (203 x 203 x 406 mm)	12
12"x8"x16" (305 x 203 x 406 mm)	18

Cost of Concrete Masonry Units. The prices quoted below are for index purposes only and are based on 8"x8"x16" (200 x 200 x 400 mm) block delivered to the job. As concrete masonry is widely manufactured, the estimator is always close to a source of supply. Costs on different size blocks are readily obtainable.

Prices for 8"x8"x16" (203x203x406mm) Lightweight Block

City	Price Each
Baltimore	$0.85
Boston	1.05
Chicago	0.80
Cincinnati	1.15
Cleveland	0.70
Dallas	0.80
Denver	0.95
Detroit	1.05
Kansas City	1.10
Los Angeles	1.35
Minneapolis	1.05
New Orleans	1.05
New York	1.45
Philadelphia	0.90
Pittsburgh	0.85
San Francisco	1.85
St. Louis	0.95
Seattle	1.10

Prices for 8"x8"x16" (203x203x406mm) Heavyweight Block

City	Price Each
Baltimore	$0.85
Boston	0.95
Chicago	0.80
Cincinnati	0.80
Cleveland	0.70
Dallas	0.80
Denver	0.80
Detroit	1.05
Kansas City	1.00
Los Angeles	1.50
Minneapolis	0.80
New Orleans	1.05
New York	2.05
Philadelphia	0.80
Pittsburgh	0.80
San Francisco	1.30
St. Louis	0.85
Seattle	1.10

Precast reinforced concrete lintels can usually be obtained from concrete masonry manufacturers.

Control Joints in Concrete Masonry

To control movements in concrete masonry walls caused by various kinds of stresses, control joints are used. Control joints are continuous vertical joints built into walls in such locations and in such a manner as to permit slight wall movement without cracking the masonry.

The spacing and location of control joints depends on a number of factors—length of wall, architectural details, and the experience and records in the particular locality where the structure is to be built. Control joints should be placed at junctions of bearing as well as nonbearing walls; at junctions of walls and columns or pilasters; at construction joints in foundation, roof, and floors; and in walls weakened by chases and openings, columns, and fixtures. In long walls, joints are ordinarily spaced at approximately 20' (6.0 m) intervals, depending on local experience. At return angles in "L", "T", and "U" shaped structures, as a rule, control joint locations are determined by the architect or engineer and are usually indicated on the drawings.

Method of Constructing Control Joints in Concrete Masonry Walls

Control joints can be built with regular full- and half-length stretcher block or full- and half-length offset jamb block. With this type of joint construction, a non-corroding metal Z-tiebar, placed in every other horizontal joint across the control joint will provide lateral support to wall sections on each side of the control joint.

In some localities, special control joint block are available. These block have tongue and groove ends, which provide the required lateral support. These block are also made in full- and half-length units.

Other common methods of constructing control joints are illustrated below. The joints permit free longitudinal movement, but they should have sufficient shear and flexural strength to resist lateral loads. They also must be weather tight when located in exterior walls.

Generally, a control joint is placed at one side of an opening less than 6' (1.8 m) in width and at both jambs of openings over 6' (1.8 m) wide. Control joints can be omitted if adequate tensile reinforcement is placed above and below wall openings.

Typical Control Joint Details

Typical Control Joint Locations

To keep control joints as unnoticeable as possible, care must be taken to build them plumb and of the same thickness as the other mortar joints. If the joint is to be exposed to weather or view, it should be raked out to a depth of at least 3/4" (18.75 mm) and sealed with knife grade caulking compound.

The additional cost of building control joints, over and above the regular wall cost, will be about $0.25 to 0.30 per l.f. ($0.82-0.98 per m) for material and $0.50 to 0.60 per l.f. ($1.64-1.97 per m) for labor.

Reinforced Concrete Masonry

Local conditions frequently demand special construction methods. For example, in earthquake regions it is necessary to provide more than ordinary stability for all types of structures. This is also true in areas where severe wind storms occur or where foundation soils are unstable.

In almost any locality there are structures that are subject to excessive vibrations, very heavy loads, or other stresses. Any of these examples of unusual stress conditions require special attention in design and are often covered by local building codes.

In concrete masonry construction, additional strength is obtained by reinforcing the walls with steel reinforcing rods encased in grout. The walls may be reinforced horizontally, vertically, or both.

Typical Concrete Masonry Bond Beam

Bond Beams. To reinforce concrete masonry walls horizontally, bond beams are frequently used at each story height. Under extreme stress conditions, it may be necessary to use them in every second or third course.

Bond beams may be constructed by forming and pouring a continuous ribbon of reinforced concrete at the course height required, but this method breaks up the continuity of the block pattern and may be objectionable.

In many localities, special bond beam blocks are available that are trough-shaped. When laid open side up, they form a continuous trough in which reinforcing steel and concrete encasement may be placed, eliminating the need for wood side forms, and the block pattern of the wall is preserved.

Bond beams serve both as structural elements and as a means of crack control. They are constructed with special-shaped masonry units that are filled with concrete or grout and reinforced with embedded steel. Their value in crack control is due to the increased strength and stiffness they give a masonry wall. Since they are capable of structural function as well as crack control, bond beams will be found serving the following functions:

1. As lintel beams over doors and windows. Lintels may be pre-constructed on the ground and set in place when they have attained sufficient strength or they may be built in place, using wood centers for support until mortar and concrete is strong enough to permit their removal.

2. Below the sill in walls with openings.

3. At the top of walls and at floor level to distribute vertical loads.

4. As horizontal stiffeners incorporated into masonry to transfer flexural stresses to columns and pilasters when unusually high lateral loading is encountered.

As a means of crack control, the bond beam's area of influence normally is presumed to extend 24 inches (600 mm) above and below its location in the wall. In walls without openings, they are spaced four feet apart and may be any length up to 60 feet (18.2 m) maximum.

Reinforcement for bond beams must satisfy structural requirements but should not be less than two no. 4 steel bars. The beams are always discontinuous at expansion joints, and joints should be designed to transfer lateral force along the wall. Beams may be discontinuous at control joints; practice varies depending on structural requirements. Dummy joints are formed when a bond beam is continuous at a control joint.

Vertical Reinforcement in Concrete Masonry Walls. Where vertical reinforcement is required in concrete masonry walls, it is usually located at building corners, jambs of wall openings, and at regular intervals between wall openings. Size and spacing of reinforcement is usually covered by local building codes. When used in conjunction with bond beams, vertical and horizontal steel should be tied together.

A- Continuous reinforced concrete
bond beam. Lap bars at corners
B- Reinforced concrete studs tied
to footing
C- Reinforced concrete footing
D- Reinforcement in horizontal
mortar joints

Method of Constructing and Reinforcing Concrete Masonry Walls

In placing vertical reinforcement, advantage is taken of the vertical alignment of hollow block cores, which form wells into which the reinforcing bars are placed and grouted solid with poured mortar or concrete. At locations where vertical reinforcement is to occur, the bottom block should be left out for a cleanout hole when wall is laid up. Just prior to setting steel bars in place, the wells should be rodded clean of extruded mortar and debris removed from the cleanout. After cleaning, setting bars, and inspecting, cleanouts are closed with side-forms and the wells are grouted solid.

Concrete Masonry Cavity Walls

A cavity wall consists of two walls separated by a continuous air space and securely tied together with non-corroding metal ties of adequate strength. For each 3 s.f. (0.27 sq.m) of wall surface, a rectangular tie of 3/16" (4.68 mm) or 1/4" (6.25 mm) wire can be used. The ties are embedded in the horizontal joints of both walls. Additional ties are necessary at all openings, with ties spaced about 3' (900 mm) apart around the perimeter and within 12" (300 mm) of the opening. Rectangular cavity wall ties are manufactured 2" (50 mm) and 4" (100 mm) wide and are either mill galvanized or hot dip galvanized. Costs for 2" (50 mm) hot dip galvanized range from $290.00 to 580.00 per 1,000 units. Costs for 4" (100 mm) hot dip galvanized range from $425.00 to 635.00 per 1,000 units.

Typical codes require that 10" (250 mm) cavity walls should not exceed 25' (7.6 m) in height. In residential construction the overall thickness of concrete masonry cavity walls is nominally 10" or 12" (250 or 300 mm).

Neither the inner nor outer walls should be less than 4" (100 mm) thick (nominal dimension), and the space between them should not be less than 2" (50 mm) nor more than 3" (75 mm) wide. Usually the outer wall is nominally 4" (100 mm) thick and the remaining wall thickness made up by the air space and the inner wall. For example, a modular concrete masonry cavity wall nominally 12" (300 mm) thick is composed of a 4" (100 mm) outer wall, a 2" (50 mm) air space and a 6" (150 mm) inner wall.

A simple method of preventing the accumulation of mortar droppings between walls and maintaining a clear cavity is to lay a 1"x 2" (25 x 50 mm) wood strip across a level of ties to catch the droppings. As the masonry reaches the next level for placing ties, the strip is raised, cleaned and laid on the ties placed at this level.

The practice of providing special flashing and weep holes in masonry cavity walls primarily presumes that water will enter the wall from the outside. If concrete masonry walls are properly designed and built with well-compacted mortar joints and are painted or stuccoed, the walls should be weathertight, and there should be no need for special flashing and weep holes.

However, in limited areas subject to severe driving rains, or where experience has shown that sufficient water collects in the wall to make flashing and weep holes necessary, the practice is as follows. The heads of windows, doors, and other wall openings, and the bottom course of masonry immediately above any solid belt course or foundation, are flashed so that any moisture entering the wall cavity will be directed toward the outside walls. Only rust-resisting metal or approved materials treated with asphalt or pitch preparations should be used for flashing.

Weep holes are placed every 2 or 3 units apart in the vertical joints of the bottom course of the outside wall immediately above any solid belt course or foundation. In no case should the weep holes be located below grade. Weep holes can be formed by placing well-oiled rubber tubing in the mortar joint and then extracting it after mortar has become hard. The tubing should extend up into the cavity for several inches to provide a drainage channel through any mortar droppings that might have accumulated.

The labor cost of concrete masonry cavity walls varies the same as ordinary block walls according to the size of unit, pattern of laying, and nature of job.

Concrete Masonry Units Used for Exterior Facing

Concrete masonry walls are extensively used for exterior facing the same as face brick or stone. They may be laid up in any of the attractive designs shown in the accompanying illustrations, ranging from straight ashlar to the many variations of random or broken ashlar.

When used for exterior facing, the walls should be true and plumb and laid with full mortar coverage on vertical and horizontal face shells, no furrowing permitted, with all vertical joints shoved tight. Mortar joints should be 3/8" (9.3 mm) thick.

The mortar joints should be struck off flush with wall surface and when partially set shall be compressed and compacted with a rounded or V-shaped tool. This provides a more waterproof mortar joint. An attractive treatment is obtained by emphasizing the horizontal joints and obscuring the vertical joints in concrete walls. This is done by tooling the horizontal joints and striking the vertical joints flush with the wall surface and then rubbing with carpet or burlap to remove the sheen from the troweled mortar surface.

Varying the bond or joint pattern of a concrete masonry wall can create a wide variety of interesting and attractive appearances, using standard units as well as sculptured-face and other architectural facing units. Due to the increased use of concrete masonry as the finished wall surface, the use of bond patterns other than the typical "running bond" has steadily increased for both loadbearing and nonloadbearing.

After running bond construction, the next most widely used bond pattern with concrete masonry units is stacked bond. Lightweight concrete units should be used where obtainable. They provide better insulation than units made of ordinary concrete aggregates.

The labor costs given on the following pages are based on using lightweight units. If ordinary concrete units are used, add about 10% to labor costs given.

Labor Cost of 100 S.F. (9 Sq.M) of Exterior Facing Concrete Masonry Walls Using 8"x16" (200 x 400 mm) Units, Laid in Regular Ashlar or Running Bond Walls 8" (200 mm) Thick

	Hours	Rate	Total	Rate	Total
Mason	7	$....	$....	$28.18	$ 197.26
Labor	7	....	....	21.04	147.28
Cost per 100 s.f.			$....		$ 344.54
per s.f.				....	3.45
per 10 sq.m				....	371.22
per sq.m				....	37.12

Add 10% to labor costs for 10" (250 mm) thick walls, 20% for 12" (300 mm) thick walls. Deduct 25% from labor costs for 4" (100 mm) veneer facing.

Noncorroding ties
No.6 ga. 16"o.c.vert.
32"o.c. horiz.

Metal ties

Concrete masonry units

Method of Constructing Cavity Walls of Concrete Masonry

Labor Cost of 100 S.F. (9 Sq.M) of Exterior Facing Concrete Masonry Walls
Using 4"x16" (100 x 400 mm) or Half Height Units Laid in Regular Ashlar
or Running Bond Walls 8" (200 mm) Thick

	Hours	Rate	Total	Rate	Total
Mason	10	$....	$....	$28.18	$ 281.80
Labor	7	....	....	21.04	147.28
Cost per 100 s.f.			$....		$ 429.08
per s.f.				....	4.29
per 10 sq.m.				....	461.60
per sq.m.				....	46.16

Add 10% to labor costs for 10" (250 mm) thick walls, 20% for 12" (300 mm) thick walls. Deduct 25% from labor costs for 4" (100 mm) veneer facing.

Labor Cost of 100 S.F. (9 Sq.M) of Exterior Facing Concrete Masonry Walls
Using Coursed Ashlar No. 1, Using Alternate
Courses of 8"x16" (200 x 400 mm) and 4"x16" (100 x 400 mm) Units
Walls 8" (200 mm) Thick

	Hours	Rate	Total	Rate	Total
Mason	8	$....	$....	$28.18	$ 225.44
Labor	6	....	....	21.04	126.24
Cost per 100 s.f.			$....		$ 351.68
per s.f.				....	3.52
per 10 sq.m.				....	378.75
per sq.m.				....	37.88

Add 10% to labor costs for 10" (250 mm) thick walls, 20% for 12" (300 mm) thick walls. Deduct 25% from labor costs for 4" (100 mm) veneer facing.

Labor Cost of 100 S.F. (9 Sq.M) of Exterior Facing Concrete Masonry Walls
Using Coursed Ashlar No. 2, Consisting of Two Courses
of 8"x16" (200 x 400 mm) Units and One Course of 4"x16" (100 x 400 mm) Units
Walls 8" (200 mm) Thick

	Hours	Rate	Total	Rate	Total
Mason	7.5	$....	$....	$28.18	$ 211.35
Labor	5.5	....	....	21.04	115.72
Cost per 100 s.f.			$....		$ 327.07
per s.f.				....	3.27
per 10 sq.m.				....	351.93
per sq.m.				....	35.19

Add 10% to labor costs for 10" (250 mm) thick walls, 20% for 12" (300 mm) thick walls. Deduct 25% from labor costs for 4" (100 mm) veneer facing.

Labor Cost of 100 S.F. (9 Sq.M) of Exterior Facing Concrete Masonry Walls
Using Coursed Ashlar No. 3, Consisting of 8"x16" (200 x 400 mm)
and 4"x16" (100 x 400 mm) Units Walls 8" (200 mm) Thick

	Hours	Rate	Total	Rate	Total
Mason	8.25	$....	$....	$28.18	$ 232.49
Labor	6.50	....	....	21.04	136.76
Cost per 100 s.f.			$....		$ 369.25
per s.f.				....	3.69
per 10 sq.m.				....	397.04
per sq.m.				....	39.70

Add 10% to labor costs for 10" (250 mm) thick walls, 20% for 12" (300 mm) thick walls. Deduct 25% from labor costs for 4" (100 mm) veneer facing.

12-in. wall   8-in. wall

(Bonded every 7ᵗʰ course)

8-in. high block

8-in. high block

5-in. high block

12-in. wall   8-in. wall

(Bonded every 6ᵗʰcourse)   (Bonded every 7ᵗʰcourse)

Examples of Concrete Masonry Used as Backup for Brick

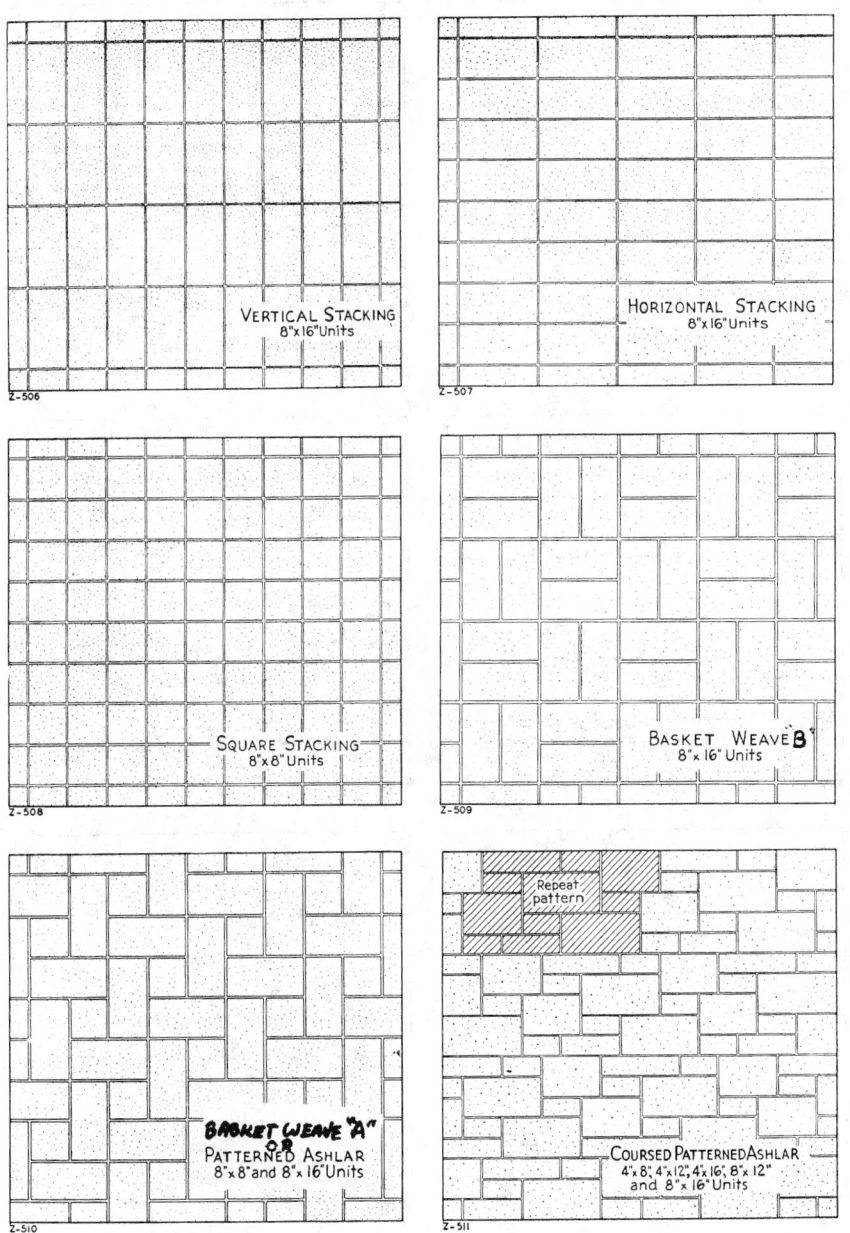

Methods of Laying Concrete Masonry Units for Exterior Facing

Methods of Laying Concrete Masonry Units for Exterior Facing

Labor Cost of 100 S.F. (9 Sq.M) of Exterior Facing Concrete Masonry Walls
Using Coursed Ashlar No. 4, Consisting of
8"x16" (200 x 400 mm) and 4"x16" (100 x 400 mm) Units
Walls 8" (200 mm) Thick

	Hours	Rate	Total	Rate	Total
Mason	7.9	$....	$....	$28.18	$ 222.62
Labor	5.7	....	....	21.04	119.93
Cost per 100 s.f.			$....		$ 342.55
per s.f.				....	3.43
per 10 sq.m.				....	369.10
per sq.m.				....	36.91

Add 10% to labor costs for 10" (250 mm) thick walls and 20% for 12" (300 mm) thick walls. Deduct 25% from labor costs for 4" (100 mm) veneer facing.

Labor Cost of 100 S.F. (9 Sq.M) of Exterior Facing Concrete Masonry Walls
Using Vertical Stacking of 8"x16" (200 x 400 mm) Units
Walls 8" (200 mm) Thick

	Hours	Rate	Total	Rate	Total
Mason	9	$....	$....	$28.18	$ 253.62
Labor	7	....	....	21.04	147.28
Cost per 100 s.f.			$....		$ 400.90
per s.f.				....	4.00
per 10 sq.m.				....	431.37
per sq.m.				....	43.14

Add 10% to labor costs for 10" (250 mm) thick walls and 20% for 12" (300 mm) thick walls. Deduct 25% from labor costs for 4" (100 mm) veneer facing.

Labor Cost of 100 S.F. (9 Sq.M) of Exterior Facing Concrete Masonry Walls
Using Horizontal Stacking of 8"x16" (200 x 400 mm) Units
Walls 8" (200 mm) Thick

	Hours	Rate	Total	Rate	Total
Mason	7.5	$....	$....	$28.18	$ 211.35
Labor	5.5	....	....	21.04	115.72
Cost per 100 s.f.			$....		$ 327.07
per s.f.				....	3.27
per 10 sq.m.				....	351.93
per sq.m.				....	35.19

Add 10% to labor costs for 10" (250 mm) thick walls and 20% for 12" (300 mm) thick walls. Deduct 25% from labor costs for 4" (100 mm) veneer facing.

Labor Cost of 100 S.F. (9 Sq.M) of Exterior Facing Concrete Masonry Walls
Using Square Stacking of 8"x8" (200 x 200 mm) Units
Walls 8" (200 mm) Thick

	Hours	Rate	Total	Rate	Total
Mason	12.50	$....	$....	$28.18	$ 352.25
Labor	8.75	....	....	21.04	184.10
Cost per 100 s.f.			$....		$ 536.35
per s.f.				....	5.36
per 10 sq.m.				....	577.11
per sq.m.				....	57.71

Add 10% to labor costs for 10" (250 mm) thick walls and 20% for 12" (300 mm) thick walls. Deduct 25% from labor costs for 4" (100 mm) veneer facing.

Labor Cost of 100 S.F. (9 Sq.M) of Exterior Facing Concrete Masonry Walls
Using Basket Weave Design Consisting of 8"x16" (200 x 400 mm) Units
Walls 8" (200 mm) Thick

	Hours	Rate	Total	Rate	Total
Mason	7.9	$....	$....	$28.18	$ 222.62
Labor	5.7	....	....	21.04	119.93
Cost per 100 s.f.			$....		$ 342.55
per s.f.				....	3.43
per 10 sq.m.				....	369.07
per sq.m.				....	36.91

Add 10% to labor costs for 10" (250 mm) thick walls and 20% for 12" (300 mm) thick walls. Deduct 25% from labor costs for 4" (100 mm) veneer facing.

Labor Cost of 100 S.F. (9 Sq.M) of Exterior Facing Concrete Masonry Walls
Using Patterned Ashlar Consisting of 8"x16" (200 x 400 mm)
and 8"x8" (200 x 200 mm) Units
Walls 8" (200 mm) Thick

	Hours	Rate	Total	Rate	Total
Mason	8.4	$....	$....	$28.18	$ 236.71
Labor	6.2	....	....	21.04	130.45
Cost per 100 s.f.			$....		$ 367.16
per s.f.				....	3.67
per 10 sq.m.				....	395.06
per sq.m.				....	39.51

Add 10% to labor costs for 10" (250 mm) thick walls and 20% for 12" (300 mm) thick walls. Deduct 25% from labor costs for 4" (100 mm) veneer facing.

Labor Cost of 100 S.F. (9 Sq.M) of Exterior Facing Concrete Masonry Walls Using Coursed Patterned Ashlar Consisting of 4"x8" (100 x 200 mm), 4"x12" (100 x 300 mm), 4"x16" (100 x 400 mm), 8"x12" (200 x 300 mm), and 8"x16" (200 x 400 mm) Units, Walls 8" (200 mm) Thick

	Hours	Rate	Total	Rate	Total
Mason	14	$....	$....	$28.18	$ 394.52
Labor	9	....	....	21.04	189.36
Cost per 100 s.f.			$....		$ 583.88
per s.f.				....	5.84
per 10 sq.m.				....	628.40
per sq.m.				....	62.84

Add 10% to labor costs for 10" (250 mm) thick walls and 20% for 12" (300 mm) thick walls. Deduct 25% from labor costs for 4" (100 mm) veneer facing.

Concrete block is often faced with brick. Three bricks equal the height of one block, and every seventh course of brick can be a header course, or in 12" (300 mm) walls every other block course can be a header block, that is, a block notched to receive a header brick. This allows every sixth course to be a bond.

Labor Cost of 100 S.F. (9 Sq.M) of 8" (200 mm) Brick Faced Block Wall Bonded Every Seventh Course

	Hours	Rate	Total	Rate	Total
Mason	18.5	$....	$....	$28.18	$ 521.33
Labor	13.5	....	....	21.04	284.04
Cost per 100 s.f.			$....		$ 805.37
per s.f.				....	8.05
per 10 sq.m.				....	866.20
per sq.m.				....	86.62

Labor Cost of 100 S.F. (9 Sq.M) of 12" (300 mm) Brick Faced Concrete Block Wall Bonded Every Sixth Course

	Hours	Rate	Total	Rate	Total
Mason	22	$....	$....	$28.18	$ 619.96
Labor	17	....	....	21.04	357.68
Cost per 100 s.f.			$....		$ 977.64
per s.f.				....	9.78
per 10 sq.m.				....	1,052.33
per sq.m.				....	105.23

Insulating Fill for Concrete Masonry Walls. For insulating masonry walls, a water-repellent vermiculite masonry fill is available. This is a free-flowing granular material, processed to provide a high degree of water-repellency. It is designed for the cores of masonry blocks and the voids of cavity walls. It does not bridge and completely fills cores and cavities. Test installations show that the fill reduces up to 50% of heat loss and 25% of air-conditioning costs. Masonry fill is also effective sound-deadening in masonry block partitions.

Vermiculite masonry fill is marketed in 4 c.f. (0.11 cu.m) bags and costs about $5.00 per bag. Approximate coverage is:

	S.F. of Wall Area	Sq.M of Wall Area
2" (50 mm) cavity	24	2.2
2-1/2" (62.5 mm) cavity	20	1.8
4-1/2" (112.5 mm) cavity	11	1.0
6" (150 mm) block, 2 or 3 core	21	1.9
8" (200 mm) block, 2 or 3 core	14-1/2	1.3
12" (300 mm) block, 2 or 3 core	8	0.7

Silicone treated perlite loose fill insulation insulates masonry walls to give greater comfort at lower cost. Silicone treated perlite pours readily into the cores of concrete block or cavity type masonry walls. Tests prove it reduces winter and summer heat transfer 50% or more. Silicone treatment of perlite provides lasting water repellency, prevents moisture penetration, and assures constant insulating efficiency in the most severe weather, including wind driven rain. Performance is rated excellent in accordance with test procedures developed by the National Bureau of Standards.

Silicone treated perlite loose fill insulation is manufactured nationally and marketed in 4 c.f. (0.11 cu.m) bags. Coverage is approximately the same as shown for vermiculite. Installed costs will also be competitive with other loose fill materials.

In addition to perlite and vermiculite, there are rigid insulating materials on the market. *Korfil* and *Korfil II* are made of preformed expanded polystyrene and designed to fit two core masonry units from 6" to 12' (150-300 mm).

Other products available for insulation of masonry units are extruded and expanded polystyrene, as well as batt insulation. The estimator should be aware of the requirements of the owner and price the material that meets the engineers or owners needs and meets the local building code requirements both for insulation and fire. The following is a list of the common materials:

1. Vermiculite
2. 1" (25 mm) extruded polystyrene
3. 2" (50 mm) expanded polystyrene
4. 2" (50 mm) extruded polystyrene

For *Korfil* it is suggested that the local masonry unit manufacturers be contacted for availability and costs.

Table A - R Values
Wall Constructed of 4" (102 mm) Hollow CMU

	60	80	100	120	140
Density, lbs./c.f.	60	80	100	120	140
Density, kg/cu.m	961	1281	1602	1922	2243
No insulation	2.8	2.4	2.2	1.9	1.7
Cores filled with vermiculite	4.9	4.1	3.4	2.7	2.2
Cores filled with perlite	5.4	4.4	3.5	2.8	2.3

Table B - R Values
Walls Construction of 6" (152 mm) Hollow CMU

	60	80	100	120	140
Density, lbs./c.f.	60	80	100	120	140
Density, kg/cu.m	961	1281	1602	1922	2243
No insulation	3.2	2.7	2.4	2.1	1.9
Cores filled with vermiculite	7.1	5.7	4.5	3.6	2.8
Cores filled with perlite	7.8	6.1	4.8	3.7	2.9
Rigid premolded insulation for two core CMU	7.7	6.7	5.9	5.0	4.2

Table C - R Values
Wall Constructed of 8" (203 mm) Hollow CMU

	60	80	100	120	140
Density lbs./c.f.	60	80	100	120	140
Density kg/cu.m	961	1281	1602	1922	2243
No insulation	3.6	3.0	2.6	2.3	2.0
Cores filled with vermiculite	9.6	7.6	6.0	4.7	3.6
Cores filled with perlite	10.4	8.2	6.3	4.8	3.7
Rigid premolded insulation for two core CMU	8.3	7.7	6.3	5.3	4.2

Table D - R Values
Wall Constructed of 10" (254 mm) Hollow CMU

	60	80	100	120	140
Density, lbs./c.f.	60	80	100	120	140
Density, kg/cu.m	961	1281	1602	1922	2243
No insulation	3.8	3.2	2.7	2.4	2.1
Cores filled with vermiculite	11.8	9.3	7.2	5.5	4.2
Cores filled with perlite	12.9	10.0	7.6	5.7	4.3
Rigid premolded insulation for two core CMU	8.3	7.7	6.7	5.6	4.6

Table E - R Values
Wall Constructed of 12" (304 mm) Hollow CMU

	60	80	100	120	140
Density, lbs./c.f.	60	80	100	120	140
Density, kg/cu.m	961	1281	1602	1922	2243
No insulation	3.9	3.3	2.8	2.5	2.2
Cores filled with vermiculite	14.3	11.3	8.8	6.7	5.0

Density, lbs./c.f.	60	80	100	120	140
Density, kg/cu.m	961	1281	1602	1922	2243
Cores filled with perlite	15.7	12.1	9.2	6.9	5.1
Rigid premolded insulation for two core CMU	9.1	8.3	7.1	6.3	5.3

Special Facing Blocks

In many areas, special block are available for the purpose of creating special facing effects. These block range from ordinary cement block, scored across the face to give the appearance of additional joints, to block made or faced with special aggregates which produce various textures and colors.

Scored Block. The simplest departure from ordinary block facing is obtained by scoring regular block one or more times across the face to give the illusion of shallower units and more joints. Most block manufacturers can furnish scored units for about $0.05 per score in addition to the regular block price.

Labor cost for laying up this type of facing is about the same as for a regular running bond facing except when scored grooves are pointed with mortar and tooled to match the actual joints.

Split-Block. Split-block afford another variation in wall finish. Split-block are made by splitting a hardened concrete unit lengthwise. The units are laid in the wall with the fractured face exposed. Many interesting variations can be obtained by introducing mineral colors and by using aggregates of different grades and colors in the concrete mixture.

Split-block can be laid in simple running bond or in any of the other patterns used in concrete masonry construction. The fractured faces of the units produce a wall of rugged appearance.

Prices for 4" (102 mm) thick split-block veneer units run from $1.00 to $1.10 per s.f. ($10.76-11.84 per sq.m) of wall depending on the size of units and pattern used.

Labor costs will be about the same as given on previous pages for the various patterns illustrated.

Slump Block. For a weathered stone effect, slump block is used. When manufacturing slump block, a special consistency mixture is used so that the unit will sag or "slump" when released from the molds before complete setting, producing artistic irregularities in height and texture. Many are integrally colored in varying shades. Standard heights vary from 1-5/8" (41 mm) to 3-5/8" (92 mm), and units are usually laid in a random ashlar pattern, although coursed patterns may be used if desired.

The cost of slump block varies with the locality. Prices should be obtained locally when figuring this work. Prices range from $0.95 to 1.25 per s.f. ($10.22-13.48 per sq.m) of wall facing.

Mortar requirements for slump block facing are high, due to the irregularities of the material and approximate those of random ashlar stone work. On average, it will require about 10 c.f. of mortar per 100 s.f. (0.28 cu.m per 9 sq.m) of 4" (102 mm) veneer facing.

When laying slump block in a random ashlar pattern, a mason and a helper should lay 45 to 55 s.f. (4.1-5.1 sq.m) of 4" (102 mm) veneer facing per 8-hr. day.

Glazed Concrete Block. Concrete block may be ordered with a factory applied glazed surface on one or more sides in 8"x16" (203 x 406 mm), 8"x8" (203 x 203 mm), and 4"x16" (102 x 406 mm) face sizes and in even thicknesses of 2" (51 mm) through 12" (305 mm). Standard cap, base, and finished end units are also furnished. This glaze is a satin finish, factory applied to a 1/8" (3 mm) thickness and is available in a standard range of 48 colors and 11 scored patterns, with special colors and shapes possible. The glazing process is available to local sources through special licensing rather than being centrally manufactured and warehoused, and the local firm should always be consulted for availability and cost.

These units are especially useful where it is desired to have a block finish on one side and glazed on the other, because one unit provides both. This can mean a considerable savings in labor cost and floor space, when compared with building a 4" (102 mm) block partition and adding a 2" (51 mm) glazed furring block. The glazed surface meets both USDA and OSHA requirements for sanitary surfaces. The high performance standards of concrete block for fire resistance and insulation value are retained. Tolerance on face dimensions and face distortions is limited to 1/16" (1.6 mm).

Units are set in standard mortar and struck with a groover but may be pointed or grouted for more finished work. Acid cleaning solutions should not be used, and units must be kept clean as the work progresses. A final cleaning with a masonry cleaning compound should be figured.

Adding the glazed surface to a standard block will add around $1.50 to 1.75 per s.f. ($16.14-18.83 per sq.m) of glazed surface. Some colors will be more expensive than others, and the amount of scoring within the block surface will also affect the price. Two masons and one laborer will set 240 to 250 s.f. (22.2-23.2 sq.m) of 4" (100 mm) partition in running bond per day.

Labor Cost of 100 S.F. (9 Sq.M) of 4" (100 mm) Thick,
8"x16" (200 x 400 mm) Concrete
Block Glazed One Side Laid In Running Bond

	Hours	Rate	Total	Rate	Total
Mason	7	$....	$....	$28.18	$ 197.26
Labor	3.5	....	....	21.04	73.64
Cost per 100 s.f.			$....		$ 270.90
per s.f.					2.71
per 10 sq.m			....		291.60
per sq.m			....		29.16

Add to the above for pointing, grouting, and cleaning as required.

Number of 6"x7-3/4"x6-3/4" (152 x 197 x 171 mm) Double Grooved
Concrete Manhole Blocks and Quantity of Mortar Required to Lay Up
Manholes and Catch Basins of Various Sizes

Depth of Manhole		Internal Diameter of Manhole, in. (mm)			
		30 (750)	36 (900)	42 (1505)	48 (1200)
	No. Block (Cone)	8	16	24	32
4' (1.2m)	No. Block (Bbl.)	35	32	27	20
	Total	43	48	51	52
	Mortar, c.f.	1-1/2	2	2-1/2	3
	Mortar, cu.m	0.042	0.057	0.071	0.085
	No. Block (Cone)	8	16	24	32
6' (1.8 m)	No. Block (Bbl.)	56	56	54	50
	Total	64	72	78	82
	Mortar, c.f.	2	3	4	5
	Mortar, cu.m	0.057	0.085	0.113	0.142
	No. Block (Cone)	8	16	24	32
8' (2.4 m)	No. Block (Bbl.)	77	80	81	80
	Total	85	96	105	112
	Mortar, c.f.	2-1/2	3-1/2	4-1/2	5-1/2
	Mortar, cu.m	0.071	0.099	0.127	0.156
	No. Block (Cone)	8	16	24	32
10' (3.0 m)	No. Block (Bbl.)	99	104	108	110
	Total	107	120	132	142
	Mortar, c.f.	3	4	5	6
	Mortar, cu.m	0.085	0.113	0.142	0.170
	No. Block (Cone)	8	16	24	32
12' (3.7 m)	No. Block (Bbl.)	119	128	135	140
	Total	127	144	159	172
	Mortar, c.f.	3-1/2	5	6-1/2	8
	Mortar, cu.m	0.099	0.142	0.184	0.227
	No. Block (Cone)	8	16	24	32
14' (4.3 m)	No. Block (Bbl.)	140	152	162	179
	Total	148	168	186	202
	Mortar, c.f.	4	5-1/2	7-1/2	9-1/2
	Mortar, cu.m	0.113	0.156	0.212	0.269

Concrete Manhole and Catch Basin Block

Concrete manhole and catch basin block are radial concrete units used for
building circular manhole or catch basins in storm or sanitary sewers. The
best quality units meet or exceed the specifications of the American Concrete
Institute.

MASONRY

4.125

Size and Weight of Units. For manholes or catch basins having 48" (1200 mm) inside diameter, 10 units are usually required for each course; for 42" (1050 mm), 9 units; 36" (900 mm), 8 units; and 30" (750 mm), 7 units. However, one popular type of block requires 12 units per course in 48" (1200 mm) structures and 9 units in 36" (900 mm). A block 7-3/4" (193.7 mm) high and 6" (150 mm) thick weighs about 50 lbs. (22.6 kg) and a block 5-3/4" (143.7 mm) high and 6" (150 mm) thick weighs about 35 lbs. (15.8 kg).

Approximate Prices of Manhole Covers and Frames

Steel 18" (450 mm) Diameter..$200.00
Steel 24" (600 mm) Diameter.. 260.00
Steel 36" (900 mm) Diameter.. 575.00

Method of Construction. The concrete base is placed and then the first course of blocks is embedded at least 3" (75 mm) in the plastic concrete, leveling the units at once, insuring a watertight joint at the base.

Labor Costs for Concrete Block Manholes and Catch Basins. The estimated cost of labor for a concrete block manhole, 48" (1200 mm) inside diameter and 8'-0" (2.4 m) deep, is as follows:

	Hours	Rate	Total	Rate	Total
Mason	7.5	$....	$....	$28.18	$ 211.35
Labor	7.5	....	....	21.04	157.80
Cost per manhole			$....		$ 369.15

Add for excavation, backfill, concrete base, and cover.

Seismic Concrete Masonry

In some regions of the U.S., there are seismic design requirements, and at times, additional material and labor are necessary to meet these local requirements. Additional design requirements should be included in the drawings and outlined in the specifications by the design engineer.

Reinforcement. In seismic zones 2, 3, and 4 on the accompanying map, the minimum area of reinforcement (in either direction) should not be less than 0.07% of the gross cross-sectional area of the wall. The sum of percents of horizontal and vertical reinforcement should be at least 0.2%. In other areas, where reinforcement is used in selected portions of the wall as needed to resist tensile stresses, there is no minimum percentage requirement for reinforcement. The size of bars used in reinforced masonry wall ranges from No. 4 through No. 11.

The minimum and maximum vertical reinforcement in masonry columns is 0.25 and 4% respectively. Minimum size of bar is No. 4 and the minimum number of bars is four.

SEISMIC RISK MAP OF THE UNITED STATES

ZONE 0 - No damage.
ZONE 1 - Minor damage; distant earthquakes may use damage to structures with fundamental periods greater than 1.0 second; corresponds to intensities V and VI of the M.M. Scale.

ZONE 2 - Moderate damage; corresponds to intensity VII of the M.M. Scale.

ZONE 3 - Major damage; corresponds to intensity VIII and higher of the M.M. scale.

ZONE 4 - Those areas within Zone No. 3 determined by the proximity to certain major fault systems.

*Modified Mercalli Intensity Scale of 1931

See also Figures Nos. 2 and 3

Figure 1—SEISMIC ZONE MAP OF THE UNITED STATES

When a building experiences earthquake vibrations, its foundation will move back and forth with the ground. These vibrations can be quite intense, creating stresses and deformations throughout the structure. Flexible structures, such as frame buildings without shear wall stiffening elements, will endure large deflections often resulting in extensive non-structural damage. On the other hand, loadbearing concrete masonry buildings are much stiffer than frame buildings and resist deflection. Non-structural damage is minimized, as is hazard from falling debris. Loadbearing buildings are typically designed to resist twice the lateral force as buildings without stiffening elements. While limited research has been conducted on the damping properties of loadbearing buildings, strength and structural field of performance have been demonstrated.

General. The latest edition of the *Uniform Building Code* requires every structure to be designed and constructed to resist minimum total lateral seismic forces assumed to act nonconcurrently in the direction of each of the main axes of the structure.

LOADBEARING TILE

Loadbearing tile is used to carry part of the structural load of the building and is used for both interior and exterior walls. Some units have finished surfaces, some are scored for plaster or other applied finishes.

The American Society for Testing and Materials (ASTM) has set up three specifications covering structural clay loadbearing wall tile. Specification C-34 covers structural clay loadbearing wall tile in general and sets two grades: LBX for exposed locations; LB for units not exposed to frost or where protected with at least 3" (75 mm) of other masonry facing.

The tile is manufactured as end construction tile, to be placed in walls with axes of cells vertical and as side construction tile, to be placed with axes of cells horizontal. Average compressive strength of end construction tile, Grade LBX, is 1,400 psi (9.64 MPa), and for Grade LB, 1,000 psi (6.9 MPa). Average compressive strength for side construction tile is 700 psi (4.8 MPa), for both grades, based on gross area.

ASTM Specification C-212 covers structural clay facing tile designed for either interior or exterior exposure, including salt glazed and unglazed units. The two grades are FTX for smoothest faces, low absorption, and high degree of mechanical perfection, and FTS, which provides moderate absorption and mechanical perfection.

The compressive strengths given above for LBX grade tile apply here except that a special duty tile is available with 2,500 psi (17.2 MPa) and 1,200 psi (8.3 MPa) average compressive strength for end and side construction types respectively. The Facing Tile Institute's specifications list considerably higher strengths of 3,000 psi (20.7 MPa) and 2,000 psi (13.8 MPa) respectively, for "select" tiles.

ASTM Specifications C-126 covers ceramic glazed structural clay facing tile, facing brick and solid masonry units. Two grades are given: "S" (select) for use with comparatively narrow joints, and "SS" for ground edge tile. Grade S is available in all standard sizes except 7-3/4" x 15-3/4" (197 x 400 mm) and

"G" (or "SS") is available in 7-3/4" x 15-3/4" (197 x 400 mm) units. These units also carry the designation, Types I and II, which indicates single or double finished faces. The Facing Tile Institute also lists a "B" quality ceramic glaze structural tile. These are not produced to meet a specification, but are units that may be available at the yard which fail to meet one or more of the specifications for "S" grade tile. The estimator should be aware, however, that even when "B" quality is acceptable on a job, it may be impossible to get all the special shapes required to complete the job in anything other than select quality.

INTERIOR STRUCTURAL CLAY FACING TILE

Facing tile discussed below includes those units at least one of whose faces—ceramic glazed, salt glazed, or unglazed—are designed to be left exposed, either as a partition, as backup to another type of facing, or as the inner wythe of a cavity wall.

These units are manufactured in accordance with sizes and specifications adopted by the Facing Tile Institute. The four sizes adopted are modular.

Facing tile are available in such a wide variety of colors and finishes as to be adaptable to most any design requirement, and the estimator must be knowledgeable about the product and grade and the desired end result.

While most facing tile is furnished in 4" (102 mm) thicknesses, the estimator should always check the cost of using 2" (51 mm) soaps in combination with a locally produced backing material against the cost of the full width tile, particularly if the wall carries no superimposed loads. Often, structural tile is used where its loadbearing capacity is never utilized.

Mortar design mixes for tile follow those listed in the previous chapter for brick, except interior non loadbearing partitions may utilize a low strength type "O" mortar that calls for 0.111 c.f. (0.003 cu.m) of cement and 0.222 c.f. (0.006 cu.m) of lime for each cubic foot of sand. Note, however, that in much tile work white portland cement and white sand may be called for. Most facing tile is figured for 1/4" (6.2 mm) joints, which limits sand to that passing a No. 16 sieve.

In the figures given below, mortar is figured as two separated 1" wide joints, except for the bed joints of horizontal celled units, which are figured as full. Where walls are built two units thick, metal ties should be figured 16" (400 mm) vertically and 36" (900 mm) horizontally, which will be about 25 per 100 s.f. (9 sq.m). Also, if a solid filled collar joint is desired add 2.6 c.f. of mortar per 100 s.f. (9 sq.m) of wall.

Material Requirements for Facing Tile
8W Series

Size: 8" x 16" x 2" or 4" thick (7-3/4"x15-3/4" actual face size)
  Tile per sq. ft. wall-1.125
  Mortar in cu. ft.:
    For vertical cell tile:
      0.77 per 100 sq. ft.
      6.87 per 1000 units
    For horizontal cell tile:
      1.23 per 100 sq. ft.
      10.92 per 1,000 units
Cost: 2" thick - $2,720 per M, or $3.07 per sq. ft., glazed 1 side
       4" thick - $3,000 per M, or $3.38 per sq. ft., glazed 1 side
       4" thick - $4,580 per M, or $5.16 per sq. ft., glazed 2 sides
       6" thick - $4,255 per M, or $4.80 per sq. ft., glazed 1 side
       8" thick - $5,090 per M, or $5.73 per sq. ft., glazed 1 side

(metric)

Size: 200 x 400 x 50 mm (194 x 394 mm actual face size)
  Tile per sq.m wall-12
  Mortar in cu.m:
    For vertical cell tile:
      0.02 cu.m per 10 sq.m
      0.19 cu.m per 1000 units
    For horizontal cell tile:
      0.03 cu.m per 10 sq.m
      0.31 cu.m per 1,000 units
Cost: 50 mm thick - $2,720 per M, or $33.05 per sq.m, glazed 1 side
       100 mm thick - $3,000 per M, or $36.38 per sq.m, glazed 1 side
       100 mm thick - $4,580 per M, or $55.54 per sq.m, glazed 2 sides
       150 mm thick - $4,255 per M, or $51.67 per sq.m, glazed 1 side
       200 mm thick - $5,090 per M, or $61.68 per sq.m, glazed 1 side

Material Requirements for Facing Tile
6T Series.

Size: 5-1/3 x 12" x 2", 4", 6" and 8" thick (5-1/16" x 11-3/4" actual face size)
  Tile per sq. ft. wall -2.250
  Mortar in cu. ft.:
  For vertical cell tile:
    1.11 per 100 sq. ft.
    4.94 per 1,000 units
  For horizontal cell tile:
    1.80 per 100 sq. ft. for 4" wall
    7.98 per 1,000 units for 4" wall
    2.56 per 100 sq. ft. for 6" wall
    11.45 per 1,000 units for 6" wall

3.36 per 100 sq. ft. for 8" wall
14.93 per 1,000 units for 8" wall
Cost: 2" thick - $1,166 per M, or $3.07 per sq. ft., glazed 1 side
       4" thick - $1,485 per M, or $3.38 per sq. ft., glazed 1 side
       4" thick - $2,065 per M, or $5.16 per sq. ft., glazed 2 sides,
       6" thick - $2,175 per M, or $4.80 per sq. ft., glazed 1 side,
       8" thick - $2,600 per M, or $5.73 per sq. ft., glazed 1 side,

6T Series Structural Facing Tile, 5⅓"x12" Face

8W Series Structural Facing Tile, 8"x16" Face
4S Series 2⅔"x8" Face

(metric)
Size: 133 x 300 x 50 mm, 100 mm, 150 mm, and 200 mm thick
        (127 x 294 mm actual face size)
Tile per sq.m wall - 24.2
Mortar in cu.m:
For vertical cell tile:
  0.03 per 10 sq.m
  0.14 per 1,000 units
For horizontal cell tile:
   0.05 cu.m per 9 sq.m for 100 mm wall
   0.23 cu.m per 1,000 units for 100 mm wall
   0.07 cu.m per 9 sq.m for 150 mm wall
   0.32 cu.m per 1,000 units for 150 mm wall
   0.10 cu.m per 9 sq.m for 200 mm wall
   0.42 cu.m per 1,000 units for 200 mm wall

Cost: 50 mm thick - $1,166 per M, or $33.05 per sq.m, glazed 1 side
100 mm thick - $1,485 per M, or $36.38 per sq.m, glazed 1 side
100 mm thick - $2,065 per M, or $55.54 per sq.m, glazed 2 sides
150 mm thick - $2,175 per M, or $51.67 per sq.m, glazed 1 side
200 mm thick - $2,600 per M, or $61.68 per sq.m, glazed 1 side

Special Shapes. In addition to stretcher, or field units, these series also include many special shapes. These are given as belonging to groups; the higher the number, the more expensive the unit.

In general, the various groups include the following, although not all units will be available in all tile.

Group I:
4" (100 mm) bed return, bullnose or square
2" (50 mm) and 4" (100 mm) soap return, bullnose or square
coved internal corner
octagonal external corner
octagonal interior corner
cost: 8W ....$2,960 per M
6T ....$1,915 per M

Group II:
2" (50 mm) and 4" (100 mm) bed, bullnose cap
2" (50 mm) bed, cove base
cost: 8W ....$4,980 per M
6T ....$2,220 per M

Group III:
6" (150 mm) and 8" (200 mm) bed, bullnose cap
4" (100 mm) bed, cove base
2" (50 mm) bed, round top cove base
4" (100 mm) and 8" (200 mm) bed, full and half ends glazed both sides
4" (100 mm) bed, bullnose cap, glazed both sides
4" (100 mm) bed, sloped sill
Standard radial external or internal field units:
2" (50 mm) and 4" (100 mm) bed, 6-3/8" (159 mm) high, bullnose cap unit
2" (50 mm) and 4" (100 mm) bed, 7-3/4" (194 mm) high, bullnose cap unit
cost: 8W ....$7,060 per M
6T ....$3,500 per M

Group IV:
4" (100 mm) bed, miter unit
4" (100 mm) bed, miter unit with top return
4" (100 mm) bed, miter unit with finished end
2" (50 mm) and 4" (100 mm) bed, bullnose return

4" (100 mm) bed, bullnose return with starter
2" (50 mm) and 4" (100 mm) bed, cove base with return
4" (100 mm) bed, round top cove base
6" (150 mm) bed, bullnose glazed both sides
6" (150 mm) and 8" (200 mm) bed, sloped sill
4" (100 mm) bed, bullnose starter
2" (50 mm) bed, bullnose reveal
2" (50 mm) bed, bullnose coved internal corner
2" (50 mm) bed, bullnose octagonal internal and external corners
4" (100 mm) bed, 6-3/8" (159 mm) high slope sill
4" (100 mm) bed, 7-3/4" (194 mm) high slope sill
cost: 8W .... $11,250 per M
      6T ....$7,440 per M

Group V:

4" (100 mm) bed, bullnose with 4" (100 mm) return
4" (100 mm) bed, miter 7-3/4" (194 mm) high with face return
4" (100 mm) bed, round top cove base with 4" (100 mm) return
2" (50 mm) bed, round top cove base with 4" (100 mm) return
2" (50 mm) bed, bullnose radial
2" (50 mm) bed, bullnose cove base radial
4" (100 mm) bed, bullnose internal corner with 4" (100 mm) reveal
6" (150 mm) bed, 6-3/8" (159 mm) high slope sill
2" (50 mm) bed, round top cove base with square corner
2" (50 mm) bed, round top cove base with coved interior corner
2" (50 mm) bed, round top cove base with octagonal exterior corner
2" (50 mm) bed, round top cove base with octagonal internal corner
4" (100 mm) bed, field radial starter
cost: 8W .... $16,215 per M
      6T ....$7,880 per M

Group VI :

4" (100 mm) bed, bullnose full end, or half end glazed both sides
8" (200 mm) bed, bullnose half end, glazed both sides
cost: 8W .... $20,050 per M
      6T ....$9,725 per M

Group VII :

4" (100 mm) bed, bullnose radial starter
4" (100 mm) bed, cove base radial starter
cost: 8W .... $20,050 per M
      6T $9,725

Estimating Facing Tile Quantities

The most widely used estimating procedure is the wall area method, which is as follows:

Compute gross wall area (height times length). Where coved base or bullnosed caps are specified, do not figure these in overall height.

Determine net area of wall; deduct wall openings. These are computed by using the actual width multiplied by the actual height plus the height of special sill and lintel units. Where openings overlap base or cap units, do not include this in the opening height as it was not included in the gross area. Also deduct for other special shapes by multiplying their height by the following factors:

For exterior bullnose or square corners—1.167 per ft. (3.83 per m) of height.
For Interior coved corners—0.833 per ft. (2.73 per m) of height.
For bullnose or square jambs—0.75 per ft. (2.46 per m) of height.

Determine number of stretchers required by multiplying net area by number of stretchers required per square foot:

	per sq.m
30,58W:1.12	12.06
6T:2.25	24.22
4D:3.37	36.28

Compute number of standard fittings, such as corners and jambs, by the number of courses. Cove bases and caps, which should now be included as well as sills and lintels, are computed by the lineal foot. Miters, cove base corners, etc., should be taken off individually.

Summarize quantities and separate fittings in particular group.

Multiply by the unit cost figures.

Add 2% (or more, depending on job conditions) to cover breakage.

Example:

Step 1
Wall length = 29'-0" (8.83 m)
Wall height, excluding the base course = 8'-0" (2.44 m)
Gross area = 8.00 x 29 = <u>232 s.f.</u> (2.44 x 8.83 = <u>21.54 sq.m</u>)

Step 2

Deductions:

Window area 4 x 4.89 = 19.56 s.f.	1.22 x 1.49 = 1.82 sq.m
Door Area 2 x 6.67 = 13.33 s.f.	0.60 x 2.03 = 1.22 sq.m
Exterior Corner 1.167 x 8.0 = 9.33 s.f.	0.36 x 2.44 = 0.88 sq.m
Interior Coved Corner 0.833 x 8.0 = 6.66 s.f.	0.25 x 2.44 = 0.61 sq.m
Jambs (4+4+6.22+6.22) x 0.75 = 15.33 s.f.	(1.21 + 1.21 + 1.89 + 1.89) x 0.22 = 1.36 sq.m
Total = 64.21 s.f.	Total = 5.95 sq.m
Net wall area = 232 - 64.21 = 167.79 s.f.	21.54 - 5.95 = 15.59 sq.m

Step 3

Number of "6T" stretchers required 167.79 x 2.25 = <u>378</u>   15.59 x 24.22 = <u>378</u>

Step 4

	I	II	Group III	IV	V
Exterior Bullnose corner					
18 courses of 5T4 18					
Interior coved corner					
18 courses of 4T8 18					
Jambs					
24 courses of 6T4 24					
24 courses of 3T4 24					
Lintels					
5 lin. ft. of 6T20 5					
1 miter 6T30R (door)..............1					
1 miter 6T30L (door)..............1					
1 miter 8T31R (window) 1					
1 miter 8T31L (window) 1					
Sill					
3 lin. ft. of 6T20 3					
1 miter 8T31R..............1					
1 miter 8T31L..............1					
Cove Base					
25 lin. ft. of 6T50 25					
1 outside corner 5T54L..............1					
1 inside cove corner 4T58L..............1					
1 bullnose jamb 6T504R..............1					
1 bullnose jamb 6T504L..............1					
Totals..............84		8	25	6	4

Step 5 - Summary

Group	Series	Quantity
Stretcher	6T	378
I	6T	84
II	6T	8
III	6T	25
IV	6T	6
V	6T	4

Add a percentage for breakage between 2% and 5% depending on job conditions and distance of job from the plant. Chippage can frequently be used for special short lengths or other cuts.

A short cut method, devised by Stark Ceramics, is based on the following formula:

(Length x height minus actual openings) x (cost per s.f. or sq.m) + (l.f. x l.f., or m x m, cost factor) = total material cost.

In this method actual total heights are used to figure cross areas, and openings are figured using actual height and width with no allowances for sills, lintels, caps, or bases. Lineal footage includes all special shapes: cove bases, internal coved corners, jambs, sills, lintels, caps, etc. The lineal foot cost factor is an average additional cost to cover these items. It is determined by taking an average of the group costs and deducting the stretcher unit cost already included in the net area figure.

Labor Laying Glazed Structural Facing Tile. To insure the most economical construction, as well as the most attractive appearance in the finished work, all cutting of tile on the job should be done with a power saw using a carborundum blade. The following labor quantities are based on the average job requiring not more than 20% of special shapes, such as bullnose corners, cove base, cap, etc. If the percentage of special pieces vary from the above, increase or decrease labor time proportionately.

Labor Laying Glazed Structural Facing Tile
Labor Hours per 100 Pieces

Nominal Tile Size, in. (mm)*	Sides Glazed	Pcs. per 8-Hr.Day	Mason	Labor
2" x 5-1/3" x 8" (50x133x200mm)	1	155-175	4.8	4.8
4" x 5-1/3" x 8" (100x133x200mm)	1	155-175	4.8	4.8
4" x 5-1/3" x 8" (100x133x200mm)	2	125-155	5.7	5.7
6" x 5-1/3" x 8" (150x133x200mm)	1	155-175	4.8	4.8
6" x 5-1/3" x 8" (150x133x200mm)	2	125-155	5.7	5.7
8" x 5-1/3" x 8" (200x133x200mm)	1	125-155	5.7	5.7
8" x 5-1/3" x 8" (200x133x300mm)	2	105-120	7.0	7.0
2" x 5-1/3" x 12" (50x133x300mm)	1	115-135	6.4	6.4
4" x 5-1/3" x 12" (100x133x300mm)	1	115-135	6.4	6.4
4" x 5-1/3" x 12" (100x133x300mm)	2	90-110	8.0	8.0
6" x 5-1/3" x 12" (150x133x300mm)	1	105-125	7.0	7.0
6" x 5-1/3" x 12" (150x133x300mm)	2	80-100	8.9	8.9
8" x 5-1/3" x 12" (200x133x300mm)	1	85-100	8.7	8.7
8" x 5-1/3" x 12" (200x133x300mm)	2	65-85	10.6	10.6
2" x 8" x 16" (50x200x400mm)	1	65-80	11.0	11.0
4" x 8" x 16" (100x200x400mm)	1	55-65	13.3	13.3

Add for hoisting engineer if required.

*Nominal size includes thickness of standard mortar joint, 1/4" (6.3 mm) for glazed tile, in length, height, and thickness.

Tile 1-3/4" (44.4 mm) thick is usually used for wall furring or partition wainscoting. Tile glazed one side is often used for partitions that are glazed on one side and plastered on the other side. Tile glazed two sides is often used for partition walls glazed on both sides of wall.

Example of Wall Area Method for Determining Facing Tile Quantities

Unglazed Structural Clay Tile

Backup tile is sized for backing up brickwork, but is also used with stone and other facing. These units are available with nominal face sizes of 5"x12" (127 x 305 mm), 8"x12" (203 x 305 mm), and 12"x12" (305 x 305 mm), in thickness of 4" (102 mm), 6" (152 mm), 8" (203 mm), and 12" (305 mm), and are usually furnished scored one side, smooth on the other. These units are often available, especially in larger sizes, in special designs with cast-in handles for greater speed and ease in handling and non-continuous mortar joints to protect against moisture penetration. Generally, backup tile is sized for 1/2" (12.7 mm) joints, the mortar amount for which given below, but certain units will be found to allow for less. Dimensional tolerances are wider in this type than in facing tile. The mortar figures below assume joints consist of two separated 1" (25 mm) wide joints and bed joints are two separated 2" (51 mm) wide joints.

Nominal 5"x12" (127x305mm) tile:
4-7/8" (124 mm) x 11-1/2" (292 mm) x 3-1/2" (89 mm), 5-1/2" (140 mm), 7-1/2" (191 mm), and 11-1/2" (292 mm) actual
Tile per s.f. of wall, 2.25 (per sq.m of wall, 24.22)
Mortar per 100 s.f. of wall, 3.75 c.f. (per 10 sq.m of wall, 0.11 cu.m)
Mortar per 1,000 units, 16.69 c.f. (0.47 cu.m)

Nominal 8"x12" (203x305mm) tile:
7-1/2" (191 mm) x 11-1/2" (292 mm) x 3-1/2" (89 mm), 5-1/2" (140 mm), 7-1/2" (191 mm), and 11-1/2" (292 mm) actual
Tile per s.f. of wall, 1.50 (per sq.m of wall, 16.15)
Mortar per 100 s.f. of wall, 2.73 c.f. (per 10 sq.m of wall, 0.08 cu.m)
Mortar per 1,000 units, 18.23 c.f. (0.51 cu.m)

Labor Laying Backup Tile. The following quantities are based on the average job, having the usual number of openings, pilasters, etc.

Size Tile, inches	Size Tile, mm	No. Pcs. Laid per 8-Hr. Day	Labor Hrs per 100 Pcs.		
			Mason	Labor	Engr.*
4 x 5 x 12	100 x 125 x 300	330-370	2.3	2.3	0.3
8 x 5 x 12	200 x 125 x 300	250-290	3.0	3.0	0.3
4 x 8 x 12	100 x 200 x 300	200-225	3.8	3.8	0.3
8 x 8 x 12	200 x 200 x 300	155-185	4.7	4.7	0.5

*Add for hoisting engineer only if required.

Size Tile, inches	Size Tile, mm	No. S.F. Laid per 8-Hr. Day	No. Sq.M Laid per 8-Hr. Day	Labor Hrs per 100 s.f. (10 sq.m)		
				Mason	Labor	Engr.*
4 x 5 x 12	100x125x300	150-170	13.9-15.8	5.0	5.0	0.6
8 x 5 x 12	200x125x300	115-130	10.7-12.1	6.5	6.5	0.7
4 x 8 x 12	100x200x300	135-150	12.5-13.9	5.7	5.7	0.6
8 x 8 x 12	200x200x300	110-132	14.4-17.2	6.6	6.6	0.7

*Add for hoisting engineer only if required.

Unit Wall Tile

Some tile units are designed to be used for a complete single unit, loadbearing, nominal 8" (203 mm) wall construction, with finished surfaces inside and out. These are usually manufactured under special patents. Some of these units may have one finished surface only and be used in combination with an inside backup.

DENISON BACK UP TILE

8816 SERIES — 8" x 16" face — Scored one side, smooth one side

128-16-BU   8816-BU   6816-BU   4816-BU

88 SERIES — 8" x 12" face — Scored one side, smooth one side

128-BU   88-BU   68-BU   48-BU

HEADER UNITS

128-16-HBU or 128-HBU   8816-HBU or 88-HBU   6816-HBU or 68-HBU   Closure unit 88-CBU Smooth both sides, both ends

85 SERIES — 5" x 12" face — Scored one side, smooth one side

105-BU   85-BU   65-BU   45-BU

85-BU FLATBED   Closure Unit 85-CBU Smooth both sides

ESTIMATING DATA

8816 Series — 1.1 units/sq. ft.
88   Series — 1.5 units/sq. ft.
85   Series — 2.25 units/sq. ft.

Units may be scored two sides or smooth two sides if desired.

DENISON STRUCTURAL CLAY TILE
MASON CITY BRICK AND TILE COMPANY, MASON CITY, IOWA

Faces are available in buff unglazed, either smooth or Ruggtex finish; salt glazed, red textured (to give the appearance of high quality face brick); and ceramic glaze. Certain units are offered with supplemental full and half jamb corners, and lintel and pilaster units along with interior stretcher units to match the interior face of the unit tile.

Units are also available in 4" (100 mm) widths for use as exterior wythes in cavity wall construction. A wall of this type is highly impervious to water, thoroughly fire, termite, and vermin proof, and cannot rot or decay. Textures offer a broad range including brick-like textures and colors suitable for homes and schools; glazed for situations where a sanitary finish is desired, and smooth faced units in natural range shades suitable for larger structures and areas.

The estimator must know exactly the quality of interior finish desired. While it is possible to obtain high grade finishes and workmanship both sides, such high quality is not always required, especially in larger warehouse and manufacturing buildings, and large economies can result where "select one face" units can be used. Often the lower wall may be constructed "select two faces" while "select one face" is used above. The estimator must expect a variation of 1/8" (3.2 mm) in thickness, and if this is not acceptable, should plan to use a built-up wall with a separate interior facing.

Unit wall tile comes in two nominal face sizes: 4"x12" (102 x 305 mm), and 5-1/3"x12" (135 x 305 mm), in 4" (102 mm) and 8" (203 mm) widths.

The 4"x12" (102 x 305 mm) units, which include Uniwall and Tex Dri-wall, have an actual face dimension of 3-5/8" x11-5/8" (92 x 295 mm) with 3-5/8" (92 mm) and 7-5/8" (194 mm) thicknesses. These units will lay up 300 per 100 s.f. (9.3 sq.m) of wall and require 7.5 c.f. (0.21 cu.m) of mortar for 4" (102 mm) walls; 9.75 c.f. (0.28 cu.m) for 8" (203 mm) walls. Labor laying 4"x12" (102 x 305 mm) unit wall tile:

Size of Tile	No. Tiles Laid per 8-Hr. Day	Labor Hrs. per 100 Pcs. Mason	Labor
3-5/8"x3-5/8"x11-5/8" (91x91x291mm)	170-200	4.3	4.3
7-3/4"x3-5/8"x11-5/8" (194x91x291mm)	150-170	5.00	5.00

Size of Tile	S.F. per 8-Hr. Day	Sq.M per Day	Labor Hrs. per 100 S.F. Mason	Labor
3-5/8"x3-5/8"x11-5/8" (91x91x291mm)	57-67	5.3-6.2	13.0	13.0
7-3/4"x3-5/8"x11-5/8" (194x91x291mm)	50-57	4.6-5.3	15.0	15.0

The 5-1/2"x12" (140 x 304 mm) units have actual face dimensions of 5"x11-5/8" (127 x 295 mm) with 3-5/8" (92 mm) and 7-5/8" (194 mm) thicknesses, though they might vary somewhat in various parts of the country from 4-7/8" (124 mm) to 5-1/16" (129 mm) in height and 11-1/2" (292 mm) to 11-3/4" (298 mm) in length. Units can be figured to lay up 225 tile per 100 s.f. (9.3 sq.m) of wall and require 8.5 c.f. (0.24 cu.m) of mortar for 4" (102 mm) walls and 9.25 (0.26 cu.m) for 8" (203 mm) walls, the latter varying as to actual face size. Labor laying 5-1/3"x12" (135 x 305 mm) unit tile:

	No. Tiles Laid per 8-hr. Day	Labor Hrs. per 100 Pcs.	
Size of Tile		Mason	Labor
3-5/8"x5"x11-5/8" (91x125x291mm)	175-200	4.25	1.90
7-5/8"x5"x11-5/8" (191x125x291mm)	130-155	5.50	2.50

	S.F. per Day	Sq.M per Day	Labor Hrs. per 100 S.F.	
Size of Tile			Mason	Labor
3-5/8"x5"x11-5/8" (92x127x295mm)	80-90	30.7-34.4	9.50	4.12
7-5/8"x5"x11-5/8" (194x127x295mm)	60-70	23.2-26.9	12.25	5.50

Clay Masonry Shading Devices

In addition to shades, blinds, and drapes, an effective sun control is the egg crate screen wall design, which baffles the low east and west rays of the sun and the high, direct south rays.

Clay masonry units in glazed and unglazed finishes are available to construct such baffles. They come in a large range of sizes, shapes, and designs. Selection will depend on the degree of shading desired, the wall thickness necessary to provide structural stability, and aesthetic considerations. Glazed solar screen units 4" (102 mm) thick will cost about $7.00 per unit.

Four Patterns of Solar Screen Tile
7-5/8x7-5/8 (194 x 194 mm) for 3/8" (9.5 mm) mortar joints
Thickness: 3-5/8" (92 mm), 5-1/16" (129 mm) and 7-3/4" (197 mm)

UNIT MASONRY/CERAMIC VENEER

Terra cotta is made of a high grade clay (usually a mixture of several different kinds), to which is added calcined clay to uniformly control the shrinkage and prevent undue warping.

Terra cotta is used for the same purposes as materials like sandstone, limestone, granite, and press brick, for interior or exterior trimmings or for entire fronts. It is also used extensively as glazed terra cotta for light courts, for the interior walls of swimming pools, operating rooms in hospitals, clinics and patient rooms, corridors, vestibules and stair halls (full height of wainscot), and in large kitchens where a ceramic material is required for sanitary purposes as well as for fire resistance. It is used for service stations, power houses, and rooms that house machinery, because it can be kept free from dust, dirt, and grease. Its unlimited possibilities in color effects make it an ideal material for store fronts. It is used as a lining for the interior of restaurants, palm courts, hotels, residences, and railroad stations. In stations and public buildings it is also used in the construction of ceilings and domes. In fact, it is a universal building material with the advantage over other materials that it can be produced in almost any color and can be glazed or unglazed with any surface treatment desired. It furnishes unlimited possibilities to the architect not only in color effect but in the beauty of the original modeling in the plastic clay, an effect that cannot be obtained in carved stone except at tremendous cost.

Method of Manufacturing Architectural Terra Cotta. After shop drawings have been approved, full size models are made of plaster to a shrinkage scale, that is, to an increased size to allow for the shrinkage in drying and burning. A model is made for each different shape on the shop drawings. If a piece is ornamental, such ornament as may be required is modeled in clay and attached to the plaster model. From the models, sectional molds of plaster are cast, from which the required number of pieces of terra cotta is produced. The mixture of clays and fusible materials used in forming terra cotta is carefully selected and proportioned to give the desired degree of plasticity when mixed with water. The composition must be such that when fired at high temperatures, it will produce a homogeneous body amply strong to support the required loads.

Forming terra cotta is a manual operation and consists of pressing the soft clay mixture into the molds. The walls of the pieces should be not less than one-inch thick following the contour of the mold, and the partitions should be properly sized and spaced to produce the strength required. The pressed piece remains in the mold until the clay stiffens, after which it is removed, retouched, and placed in the dryer where most of the moisture in the clay is removed.

Following the drying process the exposed surfaces of each piece are sprayed with the ceramic mixture, which when fired develops into the desired glaze color. The pieces are then placed in kilns and fired by gradually increasing the temperature to 2,000°F (1093°C) or more, depending on the temperature of maturity of the clay and glaze. After proper firing the kiln is allowed to cool slowly to normal temperature.

After firing, the terra cotta is fitted for alignment by cutting or grinding and marked to correspond with the piece numbers on the shop drawings.

Method of Manufacturing Ceramic Veneer. Ceramic veneer is the usual term applied to machine made architectural terra cotta and the process of manufacturing is the same as for handmade pieces, except that the material is extruded instead of hand pressed into molds.

A machine mixes the proportioned clay with water and forces the plastic mixture through a steel die, which forms it into the desired shape. In most cases, before entering the die, the plastic mixture passes through a de-airing chamber, where it is subjected to a partial vacuum, and most of the entrapped air is removed. The pieces are then dried and surface finished to a true plane. Subsequent operations are the same as for handmade pieces and include application of ceramic mixture, firing, fitting, and marking.

Cost of Architectural Terra Cotta and Ceramic Veneer. In estimating the cost of architectural terra cotta and ceramic veneer, there is no unit that can be taken as a definite basis for estimating. All estimates are usually made by the manufacturer in a lump sum price for the entire job.

To be able to accurately arrive at the cost of these materials, an estimator must have an intimate knowledge of the manufacturing process. Each job is a custom job. There are no standard sizes, colors, or shapes that are stocked. One manufacturer suggests that the cost of architectural terra cotta will vary from $440.00 to $630.00 and higher per ton ($400-572 per metric ton) and that ceramic veneer will cost from $3.10 to $5.00 and higher per s.f. ($33.30-53.80 per sq.m). The extreme cost range emphasizes the necessity of obtaining firm quotes from manufacturers.

Some of the conditions governing cost of both architectural terra cotta and ceramic veneer are design of work, character and amount of ornamentation, amount of duplication, sizes of pieces, color of glaze, weight, supply and demand, and shipping conditions.

Estimating Quantities of Architectural Terra Cotta and Ceramic Veneer. When estimating quantities of terra cotta from the plans or details, all measurements should be squared; i.e., all molded courses, cornices, round columns, column caps, bases, coursed ashlar, sills, and copings should be figured from the extreme dimensions and reduced to cubic feet (cubic meters).

The weight of terra cotta varies from about 80 lbs. per c.f. (1281 kg per cu.m) for plain 4" (102 mm) thick ashlar to 60 lbs. per c.f. (961 kg per cu.m) for large molded courses. For an average job about 72 lbs. per c.f. (1153 kg per cu.m), equivalent to 28 c.f. per ton of 2,000 lbs. (1153 kg per cu.m) may be used.

After the quantities have been taken off the plans and reduced to cu. ft. (cu.m), they may either be priced in that manner, or multiplying the number of cu. ft. by 72 lbs. (cu.m x 1153 kg) will give the total weight of the terra cotta required for the job.

Ceramic veneer quantities are measured by the square foot area and should include all returns at doors, windows, soffits, etc.

Cost of Handling and Setting Architectural Terra Cotta. The labor cost of handling and setting architectural terra cotta will vary with the sizes of the different pieces, the amount of duplication and the method used in handling and setting same.

Terra cotta weighs about half as much as stone, so it is seldom necessary to use derricks in handling and setting. It is usually furnished in sizes that can be conveniently handled by 2 or 3 workers. Considerable space is required on the job for storing the terra cotta, and as it is sent to the job, each piece is marked to show its exact position in the building and must be carefully handled and placed in convenient piles. When handling the terra cotta on the job, it is customary to load it into wheelbarrows or stone barrows and wheel it to the location where it is to be used. However, if it is necessary to hoist it to the upper floors or roof of the building, the barrows can be placed on the material hoist, raised to the proper floor, and then wheeled to the location.

Because a great deal of terra cotta consists of trimmings or 4" (102 mm) ashlar, it requires considerable handling to set a cubic foot. Most ashlar is 4" (102 mm) thick, and it is necessary to set terra cotta in 3 s.f. of wall to set 1 c.f. (9.8 sq.m to set 1 cu.m). This also applies to sills, lintels, etc.

Based on the average job, consisting of ashlar, sills, lintels, ornamental trimmings, etc., a mason will set about 35 to 40 c.f. (1.0-1.1 cu.m)—105 to 120 s.f. (9.7-11.1 sq.m) of wall—per 8-hr. day, at the following labor cost:

	Hours	Rate	Total	Rate	Total
Terra cotta setter	21	$....	$....	$28.18	$ 591.78
Helper	21	....	....	21.04	441.84
Labor sorting, handling					
wheeling, etc.	44	....	....	21.04	925.76
Cost per 100 c.f.			$....		$ 1,959.38
per c.f.				....	19.59
per s.f. (300)				....	6.53
per ton (28 c.f.)				....	548.85
per cu.m				....	691.55
per sq.m (27.8)				....	70.49
per metric ton				....	604.82

For labor cost of setting plain terra cotta wall ashlar, use same quantities and costs as given on the previous pages under "Labor Laying Glazed Structural Facing Tile".

On more complicated work, consisting of gothic architecture, balusters and balustrade, glazed and enameled terra cotta, and other intricate work that has a great deal of detail and requires the utmost care in setting, a terra cotta setter will set only 20 to 25 c.f. (0.56-0.70 cu.m)—60 to 75 s.f. (5.5-7.0 sq.m) of wall—per 8-hr. day, at the following labor cost:

	Hours	Rate	Total	Rate	Total
Terra cotta setter	36	$....	$....	$28.18	$1,014.48
Helper	36	....	....	21.04	757.44
Labor sorting, handling, wheeling, etc.	66	....	....	21.04	1,388.64
Cost per 100 c.f.			$....		$ 3,160.56
per c.f.				....	31.61
per s.f. (300)				....	10.54
per ton (28 c.f.)				....	884.96
per cu.m				....	1,116.29
per sq.m (28)				....	113.41
per metric ton				....	973.46

Small lintels and similar courses that have to be hung might cost 25% to 50% higher, but these items are exceptional.

In general, the voids in architectural terra cotta should be completely filled with setting mortar or grout to prevent the possible accumulation of moisture in the cavities.

All supporting iron should be firmly embedded in masonry and completely covered with mortar to prevent rusting. Bronze or other non-corrosive metal or cadmium-coated steel anchors should invariably be used for suspended features or for anchoring or bracing freestanding work.

Mortar Required for Setting Architectural Terra Cotta. Under average conditions it will require about 4 c.f. (0.11 cu.m) of mortar to set one ton of terra cotta, or about 1/2 c.y. of mortar per 100 c.f. (0.343 cu.m of mortar per 2.8 cu.m).

The following proportions are recommended as being suitable for making satisfactory mortar for setting terra cotta: 1 part standard portland cement, 1 part lime putty, and 6 parts sand.

Do not under any circumstances use plaster of Paris or salt in the mortar.

All exposed joints in projecting and overhanging features, parapet work, and work standing free should be raked out and pointed with a good asphaltic elastic cement.

Never use a cement that swells in setting for filling the voids in terra cotta. Use only the best quality of cement.

Handling and Setting Ceramic Veneer. Ceramic veneer is available in two types: adhesion type, commonly called "thin" ceramic veneer, and anchored type.

Adhesion Type Ceramic Veneer. Adhesion type ceramic veneer is not over 1-1/8" (28.5 mm) in thickness, and the maximum face areas of individual slabs do not exceed 540 sq. in. Maximum overall face dimensions are 18"x30" (457 x 762 mm) or 20"x27" (508 x 686 mm). Actual face dimensions will vary with the manufacturer.

This type of ceramic veneer requires no metal anchorage. It is held in place by the adhesion of the mortar to the veneer body and the backing wall. The overall thickness from the face of the veneer to the face of the backing wall is 1-3/4" to 2" (44-51 mm).

Adhesion type ceramic veneer may be applied to a variety of backings, such as concrete, masonry, wood, and metal lath. Where the backing is wood, a sheet of waterproof building paper should be installed first, followed by a layer of wire mesh. In the case of metal lath backing or wire mesh on wood backing, 9-1/4" (235 mm) scratch coat of mortar should be applied in advance of the setting of veneer slabs.

Mortar Required for Adhesion Type Ceramic Veneer. Under average conditions, it will require 8 to 10 c.f. (0.22-0.28 cu.m) of mortar to set 100 s.f. (9.3 sq.m) of adhesion type ceramic veneer.

The following proportions are recommended as being suitable for making satisfactory mortar for setting adhesion type ceramic veneer: 2 parts portland cement, 1 part lime putty, and 8 parts sand. A stearate type admixture may be added to the mix in the proportions recommended by the manufacturer.

Setting Adhesion Type Ceramic Veneer. Strict compliance with the manufacturer's recommendations for setting adhesion type ceramic veneer is necessary to assure the development of full shearing strength of veneer to backing. Step-by-step setting directions, recommended by manufacturers are as follows.

All units must be soaked in clean water for at least one hour before setting. The backing wall surface should also be damp at the time of setting. Immediately before setting, the backing wall and the backs of veneer units should both receive a brush coat of portland cement and water. One-half of the setting mortar should be spread on the back of the unit to be set and the other half on the wall surface to be covered. The unit is then set in place, using wood wedges to maintain correct bed joint thickness. Sufficient mortar should be used to create a slight excess, which will be forced out of the joints from the back when the unit is tapped into place, eliminating all air pockets and filling all voids. Face joints may be pointed as soon as mortar has set sufficiently to support the veneer slabs. As a final step, the surface of the veneer should be washed down with clean water.

Cost of Handling and Setting Adhesion Type Ceramic Veneer. On most jobs adhesion type ceramic veneer is delivered to the job in trucks and must be carefully unloaded, stored, and protected in advance of the setting. Two laborers should do this work at the rate of 360 to 400 s.f. (33-37 sq.m) of veneer area per 8-hr. day.

Method of Setting Adhesion Type Ceramic Veneer

Labor Cost Handling and Setting 100 S.F. (9 Sq.M) Adhesion Type
Ceramic Veneer on Concrete or Masonry Backing

Unloading and Storing	Hours	Rate	Total	Rate	Total
Labor	4.2	$....	$....	$21.04	$ 88.37
Setting					
Mason	11.5	....	....	28.18	324.07
Helper	11.5	....	....	21.04	241.96

Cost per 100 s.f.	$....	$ 654.40
per s.f.	....	6.54
per 10 sq.m	....	703.70
per sq.m	....	70.37

On the average job, where most of the veneer consists of typical panels, a mason and a helper should set 65 to 75 s.f. (6.0-7.0 sq.m) of veneer per 8-hr. day.

If veneer is to be applied to a backing of metal lath, add cost of 1/4" (6.3 mm) scratch coat of mortar.

If veneer is to be applied to a wood backing, add cost of one layer of waterproof building paper, one layer of wire mesh, and 1/4" (6.3 mm) scratch coat of mortar.

Above costs do not include any allowance for scaffolding or hoisting. If hoisting is required, add 1 hr. hoisting engineer and 2 hrs. labor per 100 s.f. (9 sq.m) of veneer.

Anchored Type Ceramic Veneer. Anchored type ceramic veneer is recommended where a larger size slab is desired. Anchored type slabs are available in face sizes considerably larger than the adhesion type, again depending on the manufacturer. Ribs or scoring are provided on the backs of such units and the overall thickness of the slabs range from 2" (51 mm) to 2-1/2" (64 mm). Depending on slab thickness, a total of 3" (76 mm) to 4-1/2" (114 mm) is required from rough wall to finished veneer surface to provide adequate grout space between the veneer and the backing. Anchor holes are provided in the bed edges of the slabs for the installation of loose wire anchors, which in turn are fastened to pencil rods anchored to the backing. Once the wire anchoring is in place, the units are bonded to the wall by a poured grout core.

Grout Required for Anchored Type Ceramic Veneer. For ordinary work, it will require 20 to 25 c.f. (0.56-0.70 cu.m) of grout to fill the space back of 100 s.f. (9 sq.m) of anchored type ceramic veneer.

The following proportions are recommended for a suitable grout mix: 1 part portland cement, 1 part sand, and 5 parts graded pea gravel passing a 1/4" (6.3 mm) sieve. Sufficient water should be added to cause the mixture to flow readily.

In addition to the above grout, it may require 1 to 2 c.f. (0.02-0.05 cu.m) of mortar per 100 s.f. (9.3 sq.m) of veneer for buttering the joints in setting to prevent leakage of grout. Any workable mortar mix may be used.

Setting Anchored Type Ceramic Veneer. The following directions for setting anchored type ceramic veneer are recommended by the manufacturers. As in the case of adhesion type veneer, both units and backing wall should be thoroughly soaked with clean water prior to setting.

Assuming the backing wall has had loop tie inserts or eye bolts properly located, the first step in veneer slab erection is the insertion of 1/4" (6.3 mm) diameter pencil rods into the loops projecting from the backing wall. The veneer slabs are then set in place and secured to the pencil rods with wire anchors inserted in the top of each unit. Wood wedges should be used to maintain the proper joint widths and the joints may be buttered with mortar or caulked with rope yarn or other material to prevent grout leakage. Temporary wood wedges should also be used to hold each piece in place at the required distance from the backing wall.

When the setting of each horizontal row of units is completed, grout is poured into the space between the veneer and the backing, puddled thoroughly, and allowed to set. Wedges may be removed and joints pointed when grout has set sufficiently to hold slabs in place. To complete the operation the finished veneer surface should be washed down with clean water.

In structural concrete walls, wire loop dowel anchors are cast in place 1'-0" on vertical centers and not more than 1'-6" on horizontal centers.

1"x6" horizontal groove formed in concrete walls and piers not more than 2'-6" on center. (1'-0" on center in spandrel beams)

1/4" diameter pencil rods placed through eye bolts or loop dowels so as to be held out from wall at least 1"

No. 6 galvanized wire anchors hooked around pencil rods and into holes in edges of CERAMIC VENEER.

Space between wall and CERAMIC VENEER filled with grout

In BRICK, TILE or other types of unit masonry walls, eye bolts are set in place by the mason, not more than 1'-0" on vertical centers or 1'-6" on horizontal centers.

Method of Setting and Securing Anchored Type Ceramic Veneer Terra Cotta to Brick and Concrete Backing

Cost of Handling and Setting Anchored Type Ceramic Veneer. Unloading, storing, and protecting anchored type veneer slabs is carried on in the same manner as for adhesion type slabs except that these units, being larger and heavier, are somewhat more difficult to handle. Two laborers should unload and store 320 to 360 s.f. (29-33 sq.m) anchored type slabs per 8-hr. day.

Setting anchored type veneer slabs depends on the size and weight of slabs and the nature of design, but on an average facing job, a mason and a helper should set, grout, and point 60 to 70 s.f. (5.5-6.5 sq.m) of veneer per 8-hr. day.

Labor Cost Handling and Setting 100 S.F. (9 Sq.M) Anchored Type Ceramic Veneer

Unloading and Storing	Hours	Rate	Total	Rate	Total
Labor	4.7	$....	$....	$21.04	$ 98.89
Setting					
Mason	12.3	....	....	28.18	346.61
Helper	12.3	....	....	21.04	258.79
Cost per 100 s.f.			$....		$ 704.29
per s.f.				....	7.04
per 10 sq.m				....	757.50
per sq.m				....	75.75

The above costs do not include any allowance for scaffolding or hoisting. If hoisting is required, add 1.5 hrs. hoisting engineer and 3 hrs. labor per 100 s.f. (9.3 sq.m) of veneer.

Cleaning Down or Washing Terra Cotta. Broadly speaking, there are three different surface finishes in terra cotta: lustrous glazed, matt glazed, and unglazed. The glazed surfaces are the easiest to clean.

For both lustrous and matt glazed finishes a good abrasive soap or washing powder is best. Where required, there can be added to the latter a slight proportion of sharp sand. A stiff lather should be made and the surface scrubbed hard, allowing the lather to remain long enough to soften the dirt and then rinse off with clean water. Acid is not necessary for cleaning glazed surfaces.

In the past, for cleaning unglazed terra cotta, the addition of a slight proportion of commercial muriatic acid was recommended. One quart (0.94 liter) of acid to 4 gallons (15 liters) of water was sufficient in ordinary cases. In no event did it exceed 1-1/2 pints (0.71 liter) of acid to one gallon of water, and care was taken to rinse off the acid wash shortly after applying.

Acid was used with acid-resistant pails and fiber brushes. Metal pails caused a chemical reaction where the solution became a yellow stain instead of a cleaning fluid. Never use hydrofluoric acid. Many cleaners, in order to save hand labor, employ strong solutions of it, which is highly injurious.

Pointing Terra Cotta. Recommended practice is to strike the joints as work progresses. Pointing mortar applied later is usually of different composition and so of questionable permanence.

Cold Weather Protection. Architectural terra cotta or ceramic veneer should not be set in freezing temperatures or when freezing can be expected within 30 days after setting, without providing enclosures and temporary heat.

GLASS BUILDING BLOCKS

Glass blocks are hollow, partially evacuated units of clear or colored glass. They have been engineered to serve several functions, all providing good thermal and sound insulating values.

Clear and other non-light directing units are available in 5-3/4" (146 mm), 8" (203 mm), and 12" (305 mm) square units, all 3-7/8" (98 mm) thick. These units give high light transmission and are available in clear glass for visibility, wide fluted patterns, and wavy, translucent surfaces. These units are often used in partitions and do not screen out sunlight.

Light directing units are engineered for use on sun exposures of rooms with the type of finished ceiling that efficiently reflects the directed light to produce indirect daylighting, such as in schools and offices. The basic size is 7-3/4" x 7-3/4" x 3-7/8" (197 x 197 x 98 mm). These units also come with white or green fibrous insert to reduce both glare and solar heat gain. They are also available in 11-3/4" (298 mm) square size. The insert reduces the overall light transmission.

Light diffusing units diffuse light in all directions and give maximum light transmission. Size is 7-3/4" (197 mm) square. These units are also available with sun control inserts.

Color glass units have a ceramic coating fused to a single face of each block. They are used for decoration, often as accents with other blocks, and are available in 5-3/4" (146 mm) and 7-3/4" (197 mm) square units and 3-3/4" x 11-3/4" (95 x 298 mm) rectangles.

Sculptured glass modules have various geometric shapes pressed into the glass to a depth of approximately 1-1/2" (38 mm) on both sides of the unit. They are 11-3/4" (298 mm) square and are available in clear glass and in the lighter hues of the color glass units.

Grille wall units are patterned glass with a smooth exterior. They have an outline of grey colored ceramic frit fused at the edges leaving ovals, circles, dots, and hourglass designs exposed in the center. They are also available in solid grey frit. Sizes are 3-3/4"x7-3/4"x3-7/8" (95 x 197 x 98 mm) and 7-3/4" (197 mm) square. The finish is scratch, abrasive, and chemical resistant and will not fade from exposure. They make handsome and practical screen walls.

Mortar for Glass Blocks. Materials used in making mortar for laying glass blocks shall be measured by volume. For this purpose 25 lbs. (11.3 kg) of quicklime or 40 lbs. (18.1 kg) of hydrated lime shall equal 1 cu. ft. (0.028 cu.m).

Lime shall be high-calcium or dolomitic hydrated lime. Dolomitic lime must be pressure hydrated so that it does not contain more than 8% by weight of unhydrated oxides, in accordance with ASTM Specification C207 Type S or Type SA.

A	B	C
36"	4' 4"	14¾"
29¾"	4' 3"	14¾"
27¾"	4' 3"	11¾"
25¾"	4' 3"	11¾"
23¾"	4' 3"	11¾"
21¾"	3' 6"	7¾"
20¾"	3' 6"	7¾"
19¾"	3' 6"	7¾"
18¾"	3' 0"	7¾"
17¾"	3' 0"	7¾"

MAXIMUM SIZES
AVAILABLE IN
ANCHORED TYPE
CERAMIC VENEER

A	B	C
5¾"	2' 0"	3¾"
7¾"	2' 0"	3¾"
11¾"	2' 0"	3¾"
1' 2¾"	2' 2"	3¾"
1' 3¾"	2' 2"	3¾"

MAXIMUM SIZES
AVAILABLE IN
ADHESION TYPE
CERAMIC VENEER

The mortar shall be composed of 1 part waterproof portland cement, 1 part lime, and 4 parts well graded sand. It shall be mixed to a consistency as stiff and dry as possible and still retain good working characteristics. Prepared masonry mortars of high strength and low volume change may be substituted, when approved. Setting accelerators or anti-freeze compounds are not to be used.

Expansion Joints. Expansion joints must be provided at the head and jambs of each glass block panel. This expansion area shall contain expansion strips as designated by the details or specifications. Caulk head and jamb expansion joints and at intermediate reinforcing members with sponge plastic strip or rope and a good grade of non-hardening mastic so that all expansion joints are waterproof.

In residential construction, panels that are not over 5'-0" (1.5 m) in width or 7'-0" (2.1 m) in height, or 25 s.f. (2.3 sq.m) in area, may be mortared in solid without the use of expansion joints at the jambs. The expansion joint at the head must still be used to prevent loads from being carried by the panels. Reinforcing wall ties may be omitted in residential panels that do not exceed the above size limits.

Reinforcement. Reinforcing wall ties shall consist of two No. 9 wires spaced 2" (51 mm) apart to which are welded No. 14 gauge cross wires. Ties are to be 8'-0" (2.4 m) long and not more than 0.20" (5 mm) thick at the weld. They shall be galvanized or treated with some other approved corrosion resisting coating. Ties are to run continuously with ends lapped 6" (152 mm) and are to be installed in horizontal mortar joints approximately 24" (609 mm) on centers.

Ties shall not be laid directly on the block, but shall be completely embedded in the center of the mortar joint.

Glass Building Units--Installation
(Courtesy Pittsburgh Corning)

1. Sill area to be covered by mortar shall first have a heavy coat of PC Asphalt Emulsion and allowed to dry.

2. Adhere PC Expansion Strips to jambs and head with PC Asphalt Emulsion. Expansion strip must extend to sill.

3. When emulsion on sill is dry, place full mortar bed joint. Do not furrow.

4. Set lower course of block. All mortar joints must be full and not furrowed. Steel tools must not be used to tap blocks into position. Mortar shall not bridge expansion joints. Visible width mortar joint shall be 1/4" (6.3 mm) or as specified.

5. Install PC Panel Reinforcing in horizontal joints where required as follows: a) Place lower half of mortar bed joint. Do not furrow. b) Press panel reinforcing into place. c) Cover panel reinforcing with upper half of mortar bed and trowel smooth. Do not furrow. d) Panel reinforcing must run from end to end of panels and where used continuously must lap 6 inches. Reinforcing must not bridge expansion joints.

6. Place full mortar bed for joints not requiring panel reinforcing. Do not furrow.

7. Follow above instructions for succeeding courses. The number of blocks in successive lifts shall be limited to prevent squeezing out of mortar or movement of blocks.

8. Strike joints smoothly while mortar is still plastic and before final set. At this time rake out all spaces requiring caulking to a depth equal to the width of the spaces. Remove surplus mortar from faces of glass blocks and wipe dry. Tool joints smooth and concave, before mortar sets so that exposed edges of blocks have sharp clean lines.

9. After final mortar set pack PC Oakum tightly between glass block panel and jamb and head construction. Leave space for caulking.

10. Caulk panels as indicated on details.

11. Final cleaning of glass block faces shall not be done until after final mortar set. Surplus mortar shall be removed and block faces wiped dry as joints are tooled. Final cleaning shall be done by others after mortar has attained final set.

Labor Laying Glass Blocks. In laying glass blocks, all mortar joints shall be completely filled with mortar. The joints shall not be less than 3/16" (4.7 mm) thick or more than 3/8" (9.5 mm) thick. Blocks shall not be hit with a metal tool but laid by the method known as "shoving", (working into place with the hands) thereby compressing the vertical joints.

When a bricklayer's trowel handle is capped with a common type of rubber ferule, such as used on the bottom of crutches, it prevents damage to the glass.

After the mortar has passed its initial set, the exposed edges of the joints shall be tooled and thoroughly compressed with a round jointer. The finished surface of the joint shall be slightly concave, smooth, and nonporous. The final cleaning shall not be done until after the mortar has reached its final set.

If blocks are in any way disturbed after laying, during the plumbing, tooling, or cleaning process, such blocks shall be removed and relaid.

Inasmuch as glass blocks do not absorb water, the mortar must be stiff as possible and still be workable, otherwise the blocks will be inclined to slide, and it will be difficult to lay up a straight wall and very few courses can be laid at a time as the weight of the glass blocks will squeeze the mortar out of the joints.

Basis for Estimating Labor Costs. The following labor costs are based on one laborer to two masons, which includes time mixing and supplying mortar, glass blocks, etc. It does not include time building and removing scaffolding, placing expansion joint strips, packing joints with sponge plastic rope and caulking, nor does it include time required cleaning glass blocks after erection. These are given as separate items.

Panels containing more than 144 s.f. (13.37 sq.m) and not more than 25'-0" (7.6 m) in length or 20'-0" (6.0 m) in height should be braced by and anchored to structural stiffeners, so a fair average for all classes of work would be panels 10'-0"x10'-0" (3 x 3 m) in size, containing 100 s.f. (9.3 sq.m). Panels of this size have been used as a fair example for determining average costs.

Expansion joint strips 3/8" (9.5 mm) thick, 4-1/8" (105 mm) wide, and 25" (635 mm) long are used at the heads, jambs, and mullions of all openings. These strips are secured in place by an adhesive such as asphalt emulsion, and are butted together end to end to form a continuous cushion around the edges of the panels.

After the panels have been laid and the mortar has set, sponge plastic rope should be packed tightly between the side of the block and the sides of the "chase". Sponge plastic rope should be kept back 3/8" (9.5 mm) from the finished surface. The recess thus formed should then be caulked with a non-hardening waterproof caulking material to a depth of not less than 3/8" (9.5 mm).

In a 10'-0" x10'-0" (3 x 3 m) opening, caulked on both sides of the blocks, there will be 40 l.f. (12.2 m) of caulking on the inside of the wall and 30 l.f. (9.1 m) of caulking on outside of the wall or 70 l.f. (21.3 m) for each opening. A worker should pack rope into joints and caulk 30 to 40 l.f. (9.1-12.2 m), on one side, per hour. To clean both sides of 100 s.f. (9.3 sq.m) of smooth face glass block wall will require 2 to 2-1/2 hrs. labor time.

To clean both sides of 100 s.f. (9.3 sq.m) of ribbed face glass block (light directional type) will require 3 to 3-1/2 hrs. labor time.

Erecting and removing scaffolding on the average job will require 1-1/2 to 2 hrs. labor time per 100 s.f. (9.3 sq.m) of glass area.

Glass blocks must be carefully handled in unloading and handling on the job, as they require the same care as fine enameled brick.

Cost of Glass Blocks. The cost of glass blocks varies with the kind, size, and quantity required. Always obtain delivered prices on blocks, expansion strips, and wall ties for each job figured.

Approximate Prices of Standard Glass Blocks

Size of Blocks	Up to 1,000 s.f. (90 sq.m)	From 1,000 s.f. (90 sq.m)
5-3/4"x5-3/4"x3-7/8" (146x146x98mm)	$2.20	$2.15
7-3/4"x7-3/4"x7-7/8" (197x197x200mm)	3.00	2.90
11-3/4"x11-3/4"x3-7/8" (298x298x98mm)	7.00	6.80

Special block runs considerably more and should be checked locally for both price and availability.

Laying Glass Blocks Using 1/4" Mortar Joints
Cost per 100 S.F. of Wall

Size of Block, inches	5-3/4x5-3/4x3-7/8	7-3/4x7-3/4x7-7/8	11-3/4x11-3/4x3-7/8
No. of blocks per s.f. of wall*	4.00	2.25	1.00
No. of blocks per 100 s.f. wall*	400	225	100
C.F. of mortar required**	5	3.6	2.33
Mason hours	21-23	14-16	10-12
Labor hours	10.5-11.5	7-8	5-6
Labor on scaffolding, hrs.	2	2	2
Mason ramming oakum, caulking, hrs.	1.5	1.5	1.5
Mason cleaning blocks, hrs.	2.5	2.5	2.5
Expansion strips, 3/8"x4-1/8", l.f.	30	30	30
Wall ties, l.f.	60	60	60
Sponge plastic rope joints, l.f.	460	60	60
Caulking joints, l.f.	70	70	70
Asphalt emulsion, 40'-0"x4-1/2", pts.	0.5	0.5	0.5

(metric)
Cost per 10 Sq.M of Wall

Size of Block, inches	143.7x143.7x99.8	193.7x193.7x196.8	293.7x293.7x99.8
No. of blocks per sq.m of wall*	43	24	11
No. of blocks per 10 sq.m wall*	430	240	110
Cu.M of mortar required**	0.15	0.11	0.07
Mason hours	21-23	14-16	10-12
Labor hours	11-12	7-8	5-6
Labor on scaffolding, hrs.	2	2	2
Mason ramming oakum, caulking, hrs.	1.5	1.5	1.5
Mason cleaning blocks, hrs.	2.5	2.5	2.5

Expansion strips, 9 x 103 mm, meters ....9 ....9 9
Wall ties, meters..18 ....18 18
Sponge plastic rope joints, meters.................18 ....18 18
Caulking joints, meters ....21 ....21 ....21.33
Asphalt emulsion,
  12 m x 112 mm, liters...............................0.25 ....0.25 ....0.25

Laying Glass Blocks Using 1/4" Mortar Joints
Cost per 1,000 Glass Blocks

Based on 10'x10' Panels	5-3/4x5-3/4x3-7/8	7-3/4x7-3/4x7-7/8	11-3/4x11-3/4x3-7/8
C.F. of mortar required**	12.5	16	23.3
Mason hours	53-57	62-71	100-120
Labor hours	27-29	31-36	50-60
Labor on scaffolding, hrs.	5	9	20
Mason ramming oakum, caulking, hrs.	3	6	15
Mason cleaning blocks, hrs.	6	11	25
Expansion strips, 3/8"x4-1/8", l.f.	75	135	300
Wall ties, l.f.	110	200	440
Sponge plastic rope joints, l.f.	150	266	600
Caulking joints, l.f.	175	310	700
Asphalt emulsion, 3-1/2" wide, pints	1.25	2.25	2.50
No. pcs. laid per 8-hr. day	140-150	115-125	70-80

Two styles of 8"x8"x4" (203 x 203 x 102 mm) decorative glass block showing three possible patterns formed by each style when arranged into panels of sixteen units.

(metric)
Cost per 1,000 Glass Blocks

Based on 3 x 3 m Panels	146x146x98mm	197x197x200mm	298x298x98mm
Cu.M of mortar required**	0.38	0.49	0.71
Mason hours	53-57	62-71	100-120
Labor hours	27-29	31-36	50-60
Labor on scaffolding, hrs.	5	9	20
Mason ramming oakum, caulking, hrs.	3	6	15
Mason cleaning blocks, hrs.	6	11	25
Expansion strips, 9 x 103 mm, meters	23	41	91
Wall ties, meters	34	61	134
Sponge plastic rope joints, meters	46	81	183
Caulking joints, meters	53	94	213
Asphalt emulsion, 87 mm wide, liters	1.25	2.25	2.50
No. pcs. laid per 8-hr. day	140-150	115-125	70-80

*Does not include allowance for breakage. **Includes 10% for waste.

Two Styles of Decorative Glass Block

GYPSUM TILE PARTITIONS AND FIREPROOFING

Gypsum tile are used for partitions and column fireproofing. They are 12"x30" (305 x 762 mm) in face dimensions. The large units are light in weight and require a minimum of labor for erection.

Estimating Quantities of Gypsum Tile. Gypsum tile should be estimated by the sq. ft. taking the area of all walls or partitions and making deductions in full for all openings. Gypsum tile are easily handled and may be cut with an ordinary hand saw, reducing waste to a minimum. The average mortar joint is 3/8" (9.5 mm) to 1/2" (13 mm) thick.

Sizes and Weights of Gypsum Partition and Furring Tile

Type of Tile	Size of Tile, in. (mm)	Ceiling hgt. to, l.f. (m)	Weight, lbs./s.f. (kg/sq.m)
2" (50mm) Solid	2"x12"x30" (50x300x750)	10' (3.1m)	10 (49)
3" (75mm) Hollow	3"x12"x30" (75x300x750)	13' (4.0m)	10 (49)
4" (100mm) Hollow	4"x12"x30" (100x300x750)	17' (5.2m)	13 (63)
6" (150mm) Hollow	6"x12"x30" (150x300x750)	30 (9.2m)	19 (93)

C.F. of Mortar Required to Lay 100 S.F. of Gypsum Partition Tile

Width of Mortar Joint, inches	2"	3"	4"	6"
1/4	1	1-1/2	2	3
3/8	1-1/2	2	2-1/2	3-1/2
1/2	2*	2 1/2*	3*	4*

*Quantities ordinarily used. Mortar quantities determined through actual experience and include waste due to dropping, filling around pipe, etc.

Cu.M of Mortar Required to Lay 10 Sq.M of Gypsum Partition Tile

Width of Mortar Joint, mm	51	76	102	152
6.4	0.030	--	0.061	0.091
9.5	0.046	--	0.076	0.107
12.7	0.061*	--	0.091*	0.122*

*Quantities ordinarily used. Mortar quantities determined through actual experience and include waste due to dropping, filling around pipe, etc.

Gypsum partition tile cement should always be used for laying up gypsum tile to obtain a secure bond and consequently a rigid wall.

Labor Setting Gypsum Tile. When laying gypsum tile, a worker sets 2-1/2 s.f. (0.23 sq.m) at a time, but on account of the greater footage of tile set, it requires more labor time handling and getting them to the mason.

Gypsum tile are set by bricklayers or tile setters (bricklayers who specialize in tile setting), and the labor costs will vary with the experience of the workers setting them. There are firms that specialize in the erection of gypsum tile partitions and fireproofing, and their workers are skilled in this branch of work. Naturally, they set more tile per day than a bricklayer, who lays brick one day, clay tile the next day, and gypsum tile the day after.

Experienced tile setters should set 15% to 25% more tile than inexperienced workers.

Labor Setting Gypsum Partition Tile

Size of Tile, inches	No. of S.F. per 8-hr. Day	Labor Hrs. per 100 S.F.		
		Bricklayer	Labor	Engr.*
2" Partition	250-300	2.91	2.75	0.25
3" Partition	220-260	3.34	3.00	0.25
4" Partition	195-235	3.72	3.50	0.25
6" Partition	175-215	4.10	3.75	0.50
2" Column	200-230	3.72	3.00	0.25
3" Column	160-190	4.57	4.00	0.25

(metric)
No. of   Labor Hrs. per 10 Sq.M

Size of Tile, in mm	Sq. M per 8-hr. Day	Bricklayer	Labor	Engr.*
50 mm Partition	23-28	3.13	2.96	0.27
75 mm Partition	20-24	3.60	3.23	0.27
100 mm Partition	18-22	4.00	3.77	0.27
150 mm Partition	16-20	4.41	4.04	0.54
50 mm Column	19-21	4.00	3.23	0.27
75 mm Column	15-18	4.92	4.31	0.27

* Add hoisting engineer time only if required.

04400 STONE

Being a natural product, stone varies widely in type and in geographical distribution. Being a very heavy material, except for the lava type stones which have very limited use, the availability of stone locally is of utmost importance to the estimator. Transportation costs can become a high percentage of the total cost. Geologically, stones are classified as follows:

1. *Igneous:* those formed by cooling molten material inside the earth including granite, trap rock, and lava.
2. *Sedimentary:* those consolidated from particles of decayed rocks deposited along streams and including limestone and sandstone.
3. *Metamorphic:* those either igneous or sedimentary which have been altered by deep-seated heat and pressure or intrusion of rock materials and including gneiss, marble, quartzite, and slate.

The Building Stone Institute classifies stones as follows:
1. *Bluestone:* a hard sandstone, blue gray and buff in color, quarried in New York and Pennsylvania, used for paving, treads, copings, sills and stools, hearths, and in random ashlar patterns for walls.
2. *Granite:* a fine to coarse grained igneous rock, grey, red, green, black, or yellow in color, quarried in New England, New York, Virginia, Michigan,

Minnesota, and Colorado, used for paving, curbing and rubble, mosaic, ashlar, and veneered wall construction for almost all building conditions.

3. *Greenstone:* a metamorphosed stone, a distinctive green in color, quarried in Virginia, and used inside and out for flooring, treads, spandrels and curtain walls, sills, copings, hearths.

4. *Limestone:* a sedimentary rock, oolitic, dolomitic, and crystalline, grey, buff, and pinkish in color, quarried in Indiana, Kentucky, Massachusetts, Minnesota, New York, Alabama, Kansas, and Texas, used for wall facings in all patterns, from rubble through the various ashlars to smooth facings in large sheets and for all trim pieces including carved moldings, panels, and sculpture.

5. *Marble:* a metamorphic rock, generally recrystallized limestone, available in a wide range of colors and variegated patterns, quarried in Georgia, Alabama, and Vermont, used generally as interior facings and much of which is imported, but are increasingly used as exterior veneers on precast concrete panels and as brick sized and random ashlar units for exterior walls.

6. *Quartzite:* a compact granular rock composed of quartz crystals in a wide variety of colors, many of which are variegated, quarried in Colorado, Tennessee, and Utah, and used for paving, ashlar, and mosaic wall veneers, copings, sills, hearths, and treads. Quartzite is also highly popular as aggregate for precast panels.

7. *Sandstone:* a sedimentary rock, in warm earth tones from near white to dark browns and reds, quarried in Colorado, Minnesota, Michigan, Indiana, Ohio, New York, and Massachusetts, used for wall facings of all types rubble and mosaics, ashlars, and smooth sheet facings.

8. *Slate:* a fine grained metamorphic rock derived from sedimentary rock shale, predominantly gray and blue-black in color but also available in red, green, mottled green, purple, mottled purple, and weathering green, quarried in Pennsylvania, Virginia, New York, and Vermont, and used in slabs for exterior facings, flooring, paving, treads, bases, sills and stools, copings, fireplace trim and, of course, roofing.

The stone pattern and the type of finish also are of utmost importance to the estimator. Patterns used in laying up stone work in walls include:

1. *Rubble:* including uncoursed field stone, polygonal or mosaic and random coursed.

2. *Ashlar:* including rubble field mixed with coursed ashlar, random range coursed ashlar, random range interrupted coursed, broken range ashlar, and coursed ashlar.

3. *Panelling:* including spandrel and curtain wall units of both natural stone and stone aggregate precast with concrete.

Finishes for stone will vary as to the hardness and evenness of the material, and may be generally classified as natural, rough worked, and smooth.

Natural finishes may be weathered as in field stone, or seam or split faced, when the natural quarry texture is exposed.

Rough finishes are obtained mechanically by cutting with pitching chisels; gang sawing with steel shot or rough chat; plain sawn which still leaves saw marks; planed finish including those where small particles are plucked after planing to roughen the texture; hammered, hand, and machine tooled, where 2 to 10 parallel concave grooved cuts per inch are cut into the exposed surface. A newly devised rough finish for granite is obtained by exposing it to flames.

Smooth finishes include those rubbed wet by hand with carborundum block and water; machine rubbed with carborundum disc; and honed, obtained with a polishing machine and which leaves the stone almost without texture. As this last finish is expensive, it should be limited to interior work or the finest exterior within 30' (9 m) of eye range. Granite and marble may be polished smooth.

Rough faced stone is used in the construction of churches, college and university buildings, chapels, residences, and retaining walls. It is available in granite, quartzite, and sandstone.

Rubble Stone Work

Rubble stone consists of irregularly shaped pieces, only partly trimmed or squared if at all. Finish is usually split faced or natural. Some specifications will restrict the size range. Strip rubble is ledge stone with beds of the natural cleft.

Rubble stone work is estimated by the cu. yd. containing 27 c.f. (0.76 cu.m) or by the perch containing 24-3/4 c.f (0.67 cu.m). A perch of stone is nominally 16-1/2' (4.95 m) long, 1' (0.30 m) high, and 1-1/2' (0.45 m) wide, but is often computed differently in different locations. In many states, especially west of the Mississippi River, rubble work is figured by the perch containing 16-1/2 c.f. (0.46 cu.m). Before submitting prices on rubble work by the perch, be sure to find out the common practice in your locality.

Cost of Rubble Stone. The cost of rubble stone will vary with the locality, kind of stone used, and distance of quarry from the job.

The method of buying and selling rubble stone also varies with the locality. In some places it is sold by the ton containing 2,000 lbs. (907 kg), while in others it is sold by the perch containing 24-3/4 c.f. (0.67 cu.m), or by the cord containing 100 c.f. (2.83 cu.m) and weighing 13,000 lbs. (5897 kg), which is about 130 lbs. per c.f. (2080 kg/cu.m). In some localities it is sold by the cu. yd., containing 27 c.f., or by the cu.m.

Mortar for Rubble Stone Work. The quantity of mortar required for rubble stone will vary with the size and shape of the stone and the thickness of the walls, but a fair average is 7 to 9 c.f. of mortar per c.y. of wall (0.26-0.33 cu.m per cu.m of wall). Mortar costs are given under the mortar section early in this chapter.

Cost of Setting Rubble Stone. When handling and setting rubble stone in walls up to 1'-6" (0.45 m) thick, a mason and laborer should handle and set 70 to 85 c.f. (2.0-2.4 cu.m) per 8-hr. day.

On heavy walls, 2'-0" to 3'-0" (0.6-0.9 m) thick, a mason should set 100 c.f. (2.8 cu.m) per 8-hr. day, and will require about 1-1/2 hrs. labor time to each hr. mason time.

Labor Cost of Rubble Stone for Walls Over 2'-0" (0.6 m) Thick

	Hours	Rate	Total	Rate	Total
Mason	3.0	$....	$....	$28.18	$ 84.54
Labor	4.5	....	....	21.04	94.68
Cost per c.y.			$....		$ 179.22
per c.f.				....	6.64
per cu.m				....	234.49

Ashlar Stone Work

Ashlar Stone has a flat faced surface generally square or rectangular with sawed or dressed beds and joints. Wall patterns are referred to as "coursed" when set to form continuous horizontal joints; "stacked" when set to form continuous vertical joints; and "random" when set with stones of various lengths and heights so that neither horizontal or vertical joints are continuous. The usual restrictions for random ashlar design are that no vertical joint should be higher than the highest course height being used; that no horizontal joint should be more than three stones long; and that no two stones the same height should be placed end to end. Random work will take more planning on the part of the mason than coursed work.

Finish for ashlar work is usually split face, which is the least expensive and brings out the character of the stone. It may also be ordered chat, sand, or shot sawn. The grade of stone for ashlar work is usually a mixture of all the grades available as a full range is part of the desired effect.

Many types of stone are furnished in ashlar stock including granite; limestone in the greys and buffs from Indiana, the creamy pinks from Minnesota and the popular lannon from Wisconsin; both Vermont and Georgia marble; Tennessee quartzite; and the local sandstones, those from Ohio and Pennsylvania being especially popular.

The common heights for ashlar are 2-1/4" (57 mm), 5" (127 mm), 7-3/4" (197mm), and sometimes 10-1/2" (267 mm), 13-1/4" (337 mm), and 16" (406 mm). Lengths are shipped from 1' (0.3 m) to 6' (1.8 m) for cutting on the job.

Most ashlar work today is set as veneer and comes in thicknesses of 3-1/2" to 4" (89-102 mm). Some ashlar comes in varying thicknesses to be set to a uniform back and to create deep shadow lines on the face. Some jobs will still specify that the ashlar be built as part of the back-up wall, not veneered to it. Ashlar must then be ordered with bond blocks. These units will be a nominal 8" (203 mm) deep and usually constitute around 10% of the stone furnished, or one stone in each 10 s.f. (0.93 sq.m) of wall.

Many producers offer stock patterns under their own trade names. These are less expensive than stone cut to specification, but are limited to veneer thickness and a fixed percentage of heights, usually around 15% of 2-1/4" (57 mm), 40% of 5" (127 mm) and 45% of 7-3/4" (197 mm).

Veneer ashlar must be tied to the back-up wall with noncorrosive ties not less than 18" (457 mm) apart horizontally and 24" (610 mm) vertically. The stock heights when used with 1/2" (12.78 mm) mortar joints level well with brick and block back-up, but adjustable anchors and stud anchors are also available.

Costs of ashlar stone will vary widely depending on the stone selected and the distance from the quarry as well as whether stock supplies or special cutting and finishes are required.

The cost of setting this stone also varies widely, depending upon the size of the stone and the amount of cutting required. On some of the more intricate jobs it requires as much mason's time trimming and cutting the stone as is required to set it. The entire success of the use of this random ashlar depends upon the artistic manner in which the work is performed, and the labor costs will be governed accordingly.

Labor time handling stone and helping masons varies widely, depending on method of handling stone at job, access of material piles, size of stone, etc.

To show the variations in the cost of setting, six different classes of work are described.

Estimating Quantities of Random Ashlar. Split faced or other finishes of ashlar stone is usually sold by the ton and is furnished with rough beds, backs, and joints. The stone weighs 160 to 165 lbs. per c.f. (2543-2643 kg per cu.m) and there are 12 to 12-1/2 c.f. (0.34-0.35 cu.m) of stone to the ton.

Stone 4" (102 mm) thick produces 36 to 38 s.f. of wall to the ton (3.7-3.9 sq.m per metric ton) and when laid with a 1/2" (12.7 mm) mortar joint should lay 40 to 42 s.f. (3.7-3.9 sq.m) of wall.

Stone 5" (127 mm) thick produces 29 to 30 s.f. (2.7-2.8 sq.m) of wall to the ton and when laid with a 1/2" (12.7 mm) mortar joint should lay 32 to 33 s.f. (2.9-3.0 sq.m) of wall.

Where the stone is backed with brick or block, the stone is usually furnished 4" (102 mm) to 8" (203 mm) thick to bond with the brick backing. Based on an average 6" thickness, one ton is sufficient for 24 to 25 s.f. (2.2-2.3 sq.m) of wall and when laid with a 1/2" (12.7 mm) mortar joint should lay 26 to 28 s.f. (2.4-2.6 sq.m) of wall.

Cost of 100 S.F. (9.3 Sq.M) of Random Ashlar Used as a Veneer Over
Wood Framing and Sheathing, Stone Veneer 4" (100 mm) Thick

	Hours	Rate	Total	Rate	Total
Mason	24	$....	$....	$28.18	$ 676.32
Labor	12	....	....	21.04	252.48
Cost per 100 s.f.			$....		$ 928.80
per s.f.				....	9.29
per sq.m				....	99.96

Cost Per Ton of Stone

	Hours	Rate	Total	Rate	Total
Mason	10	$....	$....	$28.18	$ 281.80
Labor	5	....	....	21.04	105.20
Cost per ton			$....		$ 387.00
per metric ton				....	425.70

On work above the second story or where large stones are used, it may require an extra laborer bringing stone to the mason.

Setting Random Ashlar in Straight Work Requiring Backing, Cutting, and Fitting. When setting random ashlar in fairly straight walls as described above, but where the stone must be backed, and end joints cut on the job, an experienced mason should set stone in 17 to 25 s.f. (1.5-2.3 sq.m) of wall per 8-hr. day, at the following labor cost per 100 s.f.:

	Hours	Rate	Total	Rate	Total
Mason	38	$....	$....	$28.18	$1,070.84
Labor	38	....	....	21.04	799.52
Cost per 100 s.f.			$....		$1,870.36
per s.f.				....	18.70
per sq.m				....	201.25

The labor time will vary widely according to job conditions, as some jobs may require only one-half as much labor time as mason time, while others may require twice as much.

Setting Random Ashlar in Churches, Chapels, Etc., Where Stone is Cut to Size in Shops. On churches, chapels, and similar structures having numerous corners, offsets, piers, pilasters, etc., where the stone is all backed in the shop and requires very little cutting and fitting by the masons on the job, an experienced mason should set stone in 25 to 30 s.f. (2.3-2.8 sq.m) of wall per 8-hr. day, at the following labor cost per 100 s.f. (9.3 sq.m):

	Hours	Rate	Total	Rate	Total
Mason	29	$....	$....	$28.18	$817.22
Labor	29	....	....	21.04	610.16
Cost per 100 s.f.			$....		$1,427.38
per s.f.			....		14.27
per sq.m			....		153.59

Setting Random Ashlar in Churches, Chapels, Etc., Where Cutting and Fitting is Done on the Job. On churches, chapels, and similar structures having numerous corners, offsets, piers, pilasters, etc., where the rough stone is sent to the job and all backing, jointing, and fitting is done by the masons on the job, an experienced mason should set stone in 14 to 18 s.f. (1.3-1.6 sq.m) of wall per 8-hr. day, at the following labor cost per 100 s.f. (9.3 sq.m):

	Hours	Rate	Total	Rate	Total
Mason	50	$....	$....	$28.18	$1,409.00
Labor	50	....	....	21.04	1,052.00
Cost per 100 s.f.			$....		$ 2,461.00
per s.f.			....		24.61
per sq.m			....		264.80

Random Ashlar in Residence Construction. In a study of a residence faced with 4500 s.f. (418 sq.m) of 5" (127 mm) Lannon stone veneer. The stone varied from 1" (25 mm) to 10" (254 mm) high and consisted of one-third weathered edge, one-third bed face stone, and one-third rock face stone. It was only one to two inches high. Care was taken to produce a natural looking pattern with very little cutting and fitting.

A mason set 34 s.f. (3.2 sq.m) of stone per 8-hr. day, at the following labor cost per 100 s.f. (9.3 sq.m):

	Hours	Rate	Total	Rate	Total
Mason	24	$....	$....	$28.18	$676.32
Labor	14	....	....	21.04	294.56
Cost per 100 s.f.			$....		$ 970.88
per s.f.			....		9.71
per sq.m			....		104.48

Random Ashlar in Memorial Chapel. The memorial chapel illustrated is also built of Lannon (Wisconsin) stone, but on account of the numerous corners, pilasters, piers, etc., and the fact that all backing, jointing, and fitting of stone was done on the job, the labor costs ran rather high, as on work of this kind the time required cutting and fitting the stone is almost as much as the setting time.

On this job, a mason averaged 17 s.f. (1.6 sq.m) of 8" (203 mm) ashlar per 8-hr. day, at the following labor cost per 100 s.f. (9.3 sq.m):

	Hours	Rate	Total	Rate	Total
Mason	47	$....	$....	$28.18	$1,324.46
Labor	40	....	....	21.04	841.60
Cost per 100 s.f.			$....		$2,166.06
per s.f.				....	21.66
per sq.m.				....	233.07

CUT STONE

The term "cut stone" refers here to stone that is cut to an exact dimension from selected quality stone, given a uniform texture, and delivered to the site ready to set in place. In order to provide such quality control, shop drawings must be made and approved in advance. Cut stone is used on the highest quality work and demands the highest quality installation.

Cut stone may be built up with and bonded to a masonry backup, but it is most often applied as a veneer or a facing panel, sometimes preassembled with an insulating backing and frame.

In the past cut stone usually referred to granite or limestone or sometimes sandstone, which was used to face the many "brownstones" that line the older streets of our northern cities. With the shift from slabs to thin veneers, the denser marbles and slate are now often specified.

Shop Drawings. All cut stone work should be cut and set in accordance with approved shop drawings. These should show the bedding, bonding, jointing, and anchoring details. Each stone is dimensioned and given a setting number. The producer will not cut the stone until all details and dimensions are agreed upon. Because jointing in veneer work is generally 1/4" (6.3 mm), extremely accurate measurements are required. Also, the contractor must plan ahead and have cutouts and holes required for lifting large slabs (called Lewis holes), as well as cutouts for posts, door checks, electrical fixtures, etc., as may be required for the work of other trades. Holes for anchorage are also cut by the contractor furnishing the stone.

Mortar for Cut Stone. Three types of mortar are recommended for setting cut stone. The best type for the particular work should be selected.

Non-Staining Cement Mortar. The cement mortar is 1 part non-staining cement to 3 parts sand, with the addition of 1/5 part of hydrated lime based on the cement volume.

Non-Staining Cement-Lime Mortar. Cement-lime mortar shall be composed of 1/2 part non-staining cement, 1/2 part of hydrated lime or quick lime paste, and not to exceed 3 parts of sand.

If quick lime paste is used, first thoroughly mix the quick lime paste and sand and stack to age; the cement shall be added and thoroughly worked into the lime and sand mixture in small batches just prior to use.

Quick lime shall be reduced to a paste by thorough and complete slaking with cold water and screening through mesh screen into a settling box. Lime putty shall stand not less than one week before use.

Sand shall be clean and sharp, free from loam, silt, vegetable matter, salts, and all other injurious substances with grains graded from fine to coarse. Screen through a 6-mesh screen.

Non-Staining Waterproof Cement Mortar. Recommended for use wherever parging is required or when any of the construction materials, such as water, sand, brick, tile, etc., are known to contain salts, which can cause staining or efflorescence; and to walls exposed to excessive amounts of moisture.

Non-staining waterproof cement mortar should be composed of 1 part of non-staining waterproof cement to 3 parts of sand, with the addition of 1/5 part of hydrated lime based on the cement volume.

For quantity and cost of mortar required for setting limestone, refer to the mortar section of this chapter.

Job Conditions. The cut stone contractor must check job conditions carefully. The product is expensive: defective or damaged work is almost always refused and replacements can cause costly job delays.

Carefully unpack and inspect all material to assure it matches samples and is cut to dimensions of shop drawings and tolerances of specifications, and check for shipping damages.

Carefully store all stone on non-staining platforms at least 4" (102 mm) off the ground. Plug all holes during freezing weather.

If hoist and scaffolding are involved, check their condition. Dirty hoists and scaffolding can stain stone.

Check backup construction. Backup must be laid to a true line. Concrete and masonry backup should be dampproofed. If this is not in other contracts, this contractor should provide for it.

If work is to continue alongside or above stone after it is set, add an allowance for protection of exposed surfaces.

Estimating Cut Stone. The cost of cut stone will depend on the kind of stone used (granite, limestone, or marble), the amount of cutting and hand work necessary, and freight from mill to destination.

Cut stone for exterior use is estimated by the cu. ft. (cu.m), taking the cube of the largest dimensions, and any special work such as hand cutting, carving, etc., is estimated separately and added to the cost of the plain stone.

When estimating the cost of plain door and window sills, lintels, steps, platforms, copings, etc., the quantities are sometimes estimated by the lin. ft. (meter) for sills and coping and by the sq. ft. (sq.m) for platforms, but the cu. ft. (cu.m) method is more generally used.

Setting Cut Stone. When estimating the labor cost of handling and setting cut stone, the size of the job should be considered, average size of stone, class of workmanship required, and the method of setting; i.e., whether by hand, hand operated derricks, or by using a mobile crane, stiff leg, guy or other types of derricks.

Setting Cut Stone by Hand. On small cut stone jobs consisting of door and window sills, lintels, wall coping, steps and other light stone that can be handled and set by two workers without the use of derricks, a stone setter should set 50 to 60 c.f. (1.4-1.7 cu.m) of stone per 8-hr. day, at the following labor cost per 100 c.f. (9.3 cu.m):

	Hours	Rate	Total	Rate	Total
Stone setter	14	$....	$....	$28.18	$394.52
Labor	28	....	....	21.04	589.12
Cost per 100 c.f.			$....		$983.64
per c.f.				....	9.84
per cu.m				....	347.37

Setting Cut Stone Using Hand Derricks. Where the stone consists of coursed ashlar, sills, lintels, lightweight cornices, copings, etc., which can be handled on stone barrows or rollers and set with a breast derrick, gin pole, or other hand operated derricks, a stone setter should set 80 to 100 c.f. (2.3-2.8 cu.m) of stone per 8-hr. day, at the following labor cost per 100 c.f. (2.8 cu.m):

	Hours	Rate	Total	Rate	Total
Stone setter	9	$....	$....	$28.18	$ 253.62
Helper	9	....	....	21.04	189.36
2 Derrickmen	18	....	....	28.18	507.24
2 Workers handling stone	18	....	....	21.04	378.72
Cost per 100 c.f.			$....		$ 1,328.94
per c.f.				....	13.29
per cu.m				....	469.31

Setting Heavy Cut Stone Using Hand Derricks. On jobs requiring heavy stone, such as thick ashlar, heavy molded courses, cornices, copings, etc., where the stone is hoisted and set using hand operated derricks, a stone setter should set 115 to 140 c.f. (3.2-3.9 cu.m) of stone per 8-hr. day, at the following labor cost per 100 c.f. (2.8 cu.m):

	Hours	Rate	Total	Rate	Total
Stone setter	7	$....	$....	$28.18	$ 197.26
Helper	7	....	....	21.04	147.28
2 derrickmen	14	....	....	28.18	394.52
3 workers handling stone	21	....	....	21.04	441.84
Cost per 100 c.f.			$....		$ 1,180.90
per c.f.				....	11.81
per cu.m				....	417.03

Setting Cut Stone Using a Power Crane. On many large jobs it will be possible to set the stone using a mobile power crane to lift the stone from the ground to the wall.

This method is usually less expensive than setting up and dismantling stiff-leg or guy derricks, as the crane can cover the entire perimeter of the building without any setup or dismantling cost to speak of.

Where the stone can be set in this manner, it will require one operator for the crane, 2 laborers on the scaffold helping to land the stone and set it, 4 or 5 laborers on the ground sorting and handling stone from stock pile to derricks, plus the stone setter.

On a job of this kind, a crew should handle and set 200 to 240 c.f. (5.7-6.8 cu.m) of stone per 8-hr. day, depending on the size of stone, at the following labor cost per 100 c.f. (2.8 cu.m), excluding crane and operator costs:

	Hours	Rate	Total	Rate	Total
Stone setter	4	$....	$....	$28.18	$ 112.72
2 helpers on scaffold	8	....	....	21.04	168.32
5 workers handling stone	20	....	....	21.04	420.80
Cost per 100 c.f.			$....		$ 701.84
per c.f.				....	7.02
per cu.m				....	247.85

The above does not include any allowance for scaffold. Add for a separate scaffold required for stone setting.

Setting Cut Stone Using Power Derricks. On large jobs where it is necessary to use stiff leg or guy derricks for hoisting and handling the stone, the costs will vary with the size and weight of the stone, quantity, and with the method of handling the stone on the ground.

On work of this kind, it is usually necessary to use stiff leg or guy derricks, or by having the boom of the derrick placed in a boom seat and bolted to the steel frame of the building. On jobs covering a large area, it will be necessary to use a number of derricks. Regardless of the method used in setting the stone, the cost of setting and removing derricks, hoisting engines, etc., should be estimated separately, because this charge will remain the same whether there are 5,000 or 50,000 c.f. (142 or 1416 cu.m) of stone to be set.

After the derricks and equipment are in place, a stone setter should set 200 to 240 c.f. (5.7-6.8 cu.m) of stone per 8-hr. day. It will require 2 laborers helping to land the stone and set it, 4 or 5 laborers on the ground sorting and handling the stone from stock piles to derricks, 3 derrickmen and one hoisting engineer.

In connection with work of this kind, there is always a certain amount of lost time caused by delays in delivery of stone, repairs to derricks and hoisting engines, oiling and attending derricks, moving and raising derricks, inspecting cables, tightening guy lines, and many other items that require attention and must be charged against the cost of stone setting.

Pointing Cut Stone. Cut stone is often pointed at the time the stone is set. Recommended practice is to strike the joints as the work progresses.

Pointing mortar applied afterward is usually of different composition and questionable permanence.

However, on some jobs it may be necessary to point all mortar joints with white or other non-staining cement after all stone has been set. The mortar joints may be struck flush or slightly rounded with a jointer as preferred.

The cost of this work will vary with the size of the individual pieces of stone. The fewer joints the more surface a tuckpointer will be able to point per day.

On an average job, a tuckpointer should point 500 to 600 s.f. (46-56 sq.m) of surface per 8-hr. day, at the following labor cost per 100 s.f. (9.3 sq.m):

	Hours	Rate	Total	Rate	Total
Tuckpointer	1.50	$....	$....	$28.18	$42.27
Helper	0.75	....	....	21.04	15.78
Cost per 100 s.f.			$....		$ 58.05
per s.f.					0.58
per sq.m			....		6.25

Allow one helper on the ground to each 2 tuckpointers or to each scaffold. Add for mortar and scaffold.

Washing and Cleaning Cut Stone. After the stone work has been pointed, it will be necessary to wash and clean it, using a stiff fiber brush and plenty of clean water or soapy water, if necessary, and then rinsed with clean water, which will remove the construction and mortar stains. A wire brush should never be used, as it tends to produce rust marks in the stone.

On new work or work that is fairly clean, an experienced tuckpointer should wash and clean 600 to 800 s.f. (56-74 sq.m) of surface per 8-hr. day, at the following labor cost per 100 s.f. (9.3 sq.m):

	Hours	Rate	Total	Rate	Total
Tuckpointer	1.1	$....	$....	$28.18	$ 31.00
Helper	0.6	....	....	21.04	12.62
Cost per 100 s.f.			$....		$ 43.62
per s.f.					0.44
per sq.m			....		4.73

Allow one helper on the ground to each 2 tuckpointers or to each scaffold. Add for scaffold.

Carving Cut Stone. The cost of carving cut stone is usually figured from full-size details or models furnished by the architect. This cost will vary with the amount of detail and the method used in carving, i.e., whether performed in the shop before setting or after the stone has been set on the job.

Where you are sure of a first class job of setting, it is more economical to have the carving done in the shop. All carved work of a number of pieces is set up with wood strips forming the thickness of mortar joints and accurately fitted before shipping to the job.

Carving Stone by Hand in the Shop. When carving cut stone for medium priced buildings requiring just an average grade of workmanship and no fine detail work, a carver should complete 2-1/2 to 3-1/2 s.f. (0.23-0.32 sq.m) per 8-hr. day, at the following labor cost per s.f. (0.09 sq.m):

	Hours	Rate	Total	Rate	Total
Stone carver	2.7	$....	$....	$31.57	$ 85.24

On jobs having fine detail work and requiring first class workmanship, a carver should complete 1 to 1-1/2 s.f. (0.09-0.14 sq.m) per 8-hr. day, at the following cost per s.f. (0.09 sq.m):

	Hours	Rate	Total	Rate	Total
Stone carver	7	$....	$....	$31.57	$ 220.99

Carving Stone in the Shop Using Pneumatic Tools. The use of air tools will greatly increase the daily output of a stone carver, regardless of the class of workmanship required.

On jobs requiring just an average grade of workmanship and no fine detail work, a carver should complete 5 to 8 s.f. (0.5-0.7 sq.m) per 8-hr. day, at the following labor cost per s.f. (0.09 sq.m):

	Hours	Rate	Total	Rate	Total
Stone carver	1.5	$....	$....	$31.57	$ 47.36
per sq.m			....		509.54

Where first class workmanship and fine detail is required, a carver should complete 2 to 2-1/2 s.f. (0.18-0.23 sq.m) per 8-hr. day, at the following labor cost per s.f. (0.09 sq.m):

	Hours	Rate	Total	Rate	Total
Stone carver	4.0	$....	$....	$31.57	$ 126.28
per sq.m			....		1,358.77

Carving Stone by Hand After It Is Set in the Building. Where the stone is carved on the job after it has been set in place and where first class workmanship is required, a carver should complete 1 to 1-1/4 s.f. (0.09-0.11 sq.m) per 8-hr. day, at the following labor cost per s.f. (0.09 sq.m):

	Hours	Rate	Total	Rate	Total
Stone carver	7	$....	$....	$31.57	$ 220.99
per sq.m			....		2,377.85

If the carving contains considerable fine detail, add 3/4 to 1 hr. per s.f. (0.09 sq.m). If outside scaffold is required, add labor time to above.

Carving Stone on the Job Using Pneumatic Tools. If air tools are used to carve stone on the job, a carver should complete 1-1/2 to 2 s.f. (0.14-0.19 sq.m) per 8-hr. day, at the following labor cost per s.f. (0.09 sq.m):

	Hours	Rate	Total	Rate	Total
Stone carver	4.5	$....	$....	$31.57	$ 142.07
per sq.m			....		1,528.62

If the carving contains considerable fine detail, add 1/2 to 3/4 hr. per s.f. (0.09 sq.m). If outside scaffold is required, add labor time to above.

Cost of Models for Stone Carving. On the better class of work, clay or plaster models for all carving are usually furnished by the architect or a stipulated sum is allowed for models.

Clay models run about $40.00 to $100.00 per s.f. ($430-1076 per sq.m). Plaster models cost about $50.00 to $125.00 per s.f. ($538-1345 per sq.m). A small rosette or head cast in plaster will cost about $100.00. The models quite generally cost more than the carving, and often twice as much where models are made for an individual piece. It will often require a week's work for a skilled artisan to produce a carefully made model for a carved plaque or panel, 3'-6" to 4'-0" (1.0-1.2 m) in diameter.

The sq. ft. (sq.m) price of models should be based on the sq. ft. (sq.m) of model required and not on the sq. ft. (sq.m) surface of carved stone, as one model may often be used for carving 10 or 15 pieces of stone, or one lin. ft. (meter) model of carved molding in plaster may often be used for carving hundreds of lin. ft. (meters) of stone.

Indiana Limestone

Indiana Limestone is probably the best known and most widely used building stone in the United States. This is doubtless due to its appearance, as well as the ease with which it is worked and its freedom from stratification and cleavage plane, and while it is a sedimentary stone, it is to all intents and purposes a free stone, having about equal strength in all directions.

It is used for the exterior facings of buildings, as well as all kinds of trimmings, such as sills, lintels, copings, steps, platforms, etc.

Indiana limestone can be produced in any reasonable quantity and size and is usually cut ready to place in the building by the mills located adjacent to the quarries. The creamy-pink limestone quarried in Minnesota is another high quality material. Limestone is furnished in a variety of different grades and color tones, a brief summary of which is given below:

Select stock is uniform in color and texture, more so than is required for ordinary building construction. It is recommended for entrance work and those portions of a building within ready range of vision, carving, interior, and other exceptional uses.

Standard stock is sound stone with a range of variation in color shades and texture not found in Select but necessarily confined within limits that make it impossible to determine at a distance of a few feet whether it is standard or select stock.

Rustic stock is of more or less coarse grain. Rustic stock is used to a considerable extent for sawed ashlar and various rough textured finishes.

Variegated stock in an irregular mixture of buff and gray produced from the blocks that are quarried where the buff and gray color tones adjoin in the quarry. It shows variation in texture as well as color. Variegated stone, when cut into building units, will exhibit units of each color tone and a small percentage of units with both color tones in one piece. Variegated stone is unusually effective in giving variety to plain surfaces and is a desirable class of material for trim as well as for facing.

There are certain grades of Indiana Oolitic limestone especially adapted for grade courses, steps, buttresses, floor tiling, etc., or any position in the building exposed to abrasion. Some are available in different color tones and may be identified by the following names: special hard gray and special hard buff Indiana limestone.

The following are the trade terms by which the different grades of Indiana limestone are classified: select buff, standard buff, select gray, standard gray, and variegated.

Some of the machine finishes suitable for limestone are sand sawed, chat or shot-sawed finish, smooth machine finish, wet rubbed finish, machine tooled finishes, carborundum finish, and plucked finish. There are also some rough textured finishes. The class of tooling in common practice is understood to cover what is properly termed fluting, that is, concave depressions, four and six parallel, concave grooves to the inch. Convex tooling is sometimes used effectively, but it is never applied unless specifically designated. Tooling and machinery has been developed to produce a variety of rough textured finishes.

It is generally recommended that all molded work be finished smooth. If it is tooled at all, the tooling should run in the direction of the mold and not across it, unless much greater expense is of no consequence. As for tooled ashlar, some architects contend vertical tooling is preferable, others prefer the horizontal.

Indiana Oolitic limestone will take most hand finishes, which under certain conditions are considered preferable to machine finish. Notable examples of this class of work are crandalling and tooth chiselling, but a number of the hand finishes can now be simulated by machine work.

A smooth finish is not commonly furnished other than in the form it comes from the planers. If sand rubbing is desired, it should be specified, and in such event the sand rubbing should be by the wet process or carborundum-and-water rubbed finish.

Indiana limestone weighs 200 lbs. per c.f. (3204 kg per cu.m) when furnished in rough quarry blocks and 180 lbs. per c.f. (2883 kg per cu.m) in scabbled quarry blocks. These figures include the weight of rough excess stone.

Sawed slabs or other squared material weighs 170 lbs. per c.f. (2723 kg per cu.m). The weight of the finished cut stone ready to set in the building is figured at 150 lbs. per c.f. (2403 kg per cu.m). If there are any large round columns in the job, they may be figured at 140 lbs. per c.f. (2243 kg per cu.m) of stock. The actual net weight of a cubic foot of dry seasoned stone is 144 lbs. (2308 kg per cu.m).

Where the cut stone for any job consisting of ashlar, sills, steps, coping, plain machine moldings, etc., is furnished by any of the mills in the Bedford or Indiana limestone district, the price will vary from $12.00 to $20.00 per c.f. ($423-706 per cu.m) f.o.b. cars at the mills.

Considerable variation in price is due to the following reasons: Stone consisting of lightweight pieces, such as 4" (102 mm) or 6" (152 mm) ashlar, small molded courses, small turned columns, bases, sills, coping, etc., require a comparatively small amount of stone but involves considerable cutting and finishing expense, and even higher costs will prevail.

On other jobs consisting of fairly large stone, 8" (203 mm) or more thick, large molded pieces, bases, columns, etc., the cubical contents are a deciding factor, and the cu. ft. (cu.m) price of this stone, even for more elaborate work will usually average less in cost than the lighter stone described above.

Jobs containing many large pieces of stone will take the minimum price. Always remember that the cube of the largest dimensions must always be used when computing the quantity, cost and weights of stone of any kind.

Indiana Limestone Sill Stock

Sill stock is available as a standard item and sold by the lineal foot. The designation S4S means stock will be sent in random lengths of from 5' to 12' (1.5-3.7 m). The S6S designation means stock will be sawed to exact lengths. Sill stock is sold in rectangular shapes. If slopes, lugs, drips, or moldings are required, profile cuts will be made at additional cost. Sand and other special finishes are also extra.

Sill stock will cost about $9.00 per c.f. ($317 per cu.m). If specific lengths are ordered, add $2.00 per c.f. ($70 per cu.m). If custom cut to shop drawing details, then sills, copings, steps, and other simple trim will cost about $25.00 per c.f. ($882 per cu.m). Belt courses, simple cornice molds, and the like cost about $35.00 per c.f. ($123 per cu.m). Elaborately carved detail work will cost $35.00 and up per c.f. ($1236 or more per cu.m).

STANDARD ITEMS

Cutting Beds and Joints. When cutting beds and joints, the cost will vary with the size of the stone. The smaller the stone the higher the price per cu. ft. (cu.m). For example, a piece of ashlar 8" (203 mm) thick contains twice as many cu. ft. (cu.m) of stone as a piece 4" (102 mm) thick, while the cost of cutting the joints will not be much more on the larger piece than on the smaller. Where the joints are cut by hand, the cost will vary from $1.00 to $1.50 per c.f. ($35-53 per cu.m), depending on the conditions described above.

On the second sawing, where sawed slabs are used, and it is necessary to run them through the saw the second time for top and bottom beds, it will cost $1.50 per cu. ft. for this work.

Where the jointing is done on the saw, figure $1.25 per c.f. ($52.97 per cu.m) for cutting joints.

Smooth or Planer Finish. When Indiana limestone is wanted with a smooth finish as it comes from the planer, the cost will vary from $0.75 to $1.00 per s.f. ($8.07-10.76 per sq.m) of surface planed.

The slabs are usually furnished sawed to the correct thickness and then the stone is run through the planer to remove the saw marks, which is not a very expensive process on straight work.

On the cheaper classes of work, and on some of the higher grade work where a sawed finish is desired, the sawed slabs are jointed and the stone is sent to the building without having been run through the planer. Work of this class is a little cheaper than stone having a first class finish.

Tooled Finish. If the stone has a tooled finish, having 4, 6, 8, or 10 bats to the inch (0.16, 0.24, 0.31, or 0.39 bats to the mm), the cost of horizontal tooling on plain surfaces is $0.75 to $1.00 per s.f. ($8.00-10.80 per sq.m) of surface tooled, or about the same as planer finish.

Hand vertical tooling in connection with machine molded work will add $0.75 to $1.00 per s.f. ($8.00-10.80 per sq.m) of surface finished.

Cross or vertical tooling on curved surfaces will add $1.50 per s.f. ($16.15 per sq.m) of surface finished.

Two bat tooling, which is always used vertically, can be done for almost the same cost as given above if it is not made to register. This tooling, however, usually has to be made to register and for that reason is generally more costly than 4, 6, 8, or 10 bat work.

Sand Rubbed Finish. If the stone is to have a sand rubbed finish, making it necessary to rub the surface with sand and water, add $0.75 to $1.00 per s.f. ($8.00-10.80 per sq.m) for all surface to be rubbed.

Rustication. If the stone is to be furnished with rustication or a "rabbet" along the edges of the stone, the cost will vary according to the depth of the rabbet. This is all planer work, so the width of the rabbet does not make as much difference as the depth to which it must be cut. The following will prove a fair average of the cost of rustication:

Depth, inches (mm)	Price per l.f. (m)	Depth, in. (mm)	Price per l.f. (m)
1/2 (12.7)	$0.75 (2.46)	2 (50.8)	$1.50 (4.92)
1 (25.4)	1.00 (3.28)	2-1/2 (63.5)	1.80 (5.90)
1-1/2 (38.1)	1.25 (4.10)	3 (76.2)	1.90 (6.23)

Rounded Edges on Stone. Add about $0.35 per l.f. ($1.15 per m) for stone with edges slightly rounded.

Stone Moldings. The cost of stone moldings will vary greatly according to the size of the stone, the type of mold, breaks, heads, jointing, etc. The two kinds of moldings most commonly used are plain or Classic moldings and Gothic moldings.

When cutting stone moldings on a planer, the smaller the piece of stone the higher the price per c.f. (cu.m), on account of the additional labor involved and the small amount of stone required.

Tools for cutting moldings vary from 3" (76 mm) to 8" (203 mm) wide, so when the moldings are over 8" (203 mm) in width, it is necessary to run them through the planer a second time. The moldings that are the cheapest to run are similar to the illustration of plain moldings, where a large amount of stone is required on which no special work is necessary.

When estimating plain moldings, take the largest dimensions of the stone to obtain the cubical contents and figure at the average price as given on the previous pages, as this will cover most jobs where the molded courses do not form too large a proportion of the total stone in the job.

Gothic Moldings. Gothic moldings are much more expensive than plain or classic moldings, as in almost every instance it is necessary to run the stone through the planer twice or perform a certain amount of hand labor. Gothic moldings cost at least twice as much to run as plain or classic moldings. The cutting of heads for gothic moldings will run much higher than plain moldings, from $3.15 to $15.00 each, depending upon the size and character of the molding.

When an entire job consists of elaborate gothic stone, it frequently will run from $15.00 to $40.00 and more per c.f. ($530 to $1412 and more per cu.m). On work of this kind it is advisable to obtain figures from a reputable concern covering the stone delivered ready to set.

Round Stone Columns. When estimating the cost of plain turned columns, the cubical contents should be obtained by taking the cube of the largest dimension (measured square) and price as follows:

Obtain the cu. ft. cost of the grade of stock to be used, whether select buff, standard buff, select gray, etc., and add seven times the cu. ft. cost of the rough quarry blocks. This will give the cu. ft. price at which the columns should be estimated. For instance, a turned column 3'-0" (0.9 m) in diameter and 36'-0" (11 m) long would contain 3'x3'x36' or 324 c.f. of stone (0.9 x 0.9 x 11 = 9 cu.m of stone). Select buff is worth $15.00 per c.f. ($530 per cu.m) in rough blocks. Multiply $15.00 x 7 gives $105.00 per cu. ft.; 324 cu. ft. @ $105.00 equals $34,020.00, the cost of one column.

A stone column 38'-0" (11.5 m) long is the usual limit of present day machine capacity, so when the column shaft is over 36'-0" (11.0 m) long, it is best to get them out in three sections.

Fluted Stone Columns. Fluted stone columns should be figured on the same basis as plain turned columns described above. After the cost of the plain column has been obtained, the cost of the flutes should be estimated at $0.75 per l.f. ($2.46 per m) of each flute in the column.

Cutting Raised Letters in Indiana Limestone. The cost of cutting raised letters in Indiana limestone will vary according to the size and style of the letters.

On average work, a stone carver should cut one 6" (152 mm) letter per hr. at the following labor cost per letter:

	Hours	Rate	Total	Rate	Total
Stone carver	1	$....	$....	$31.57	$31.57

If the letters are 12" (300 mm) high, it will require about 1-1/2 hrs. time per letter, at the following labor cost per letter:

	Hours	Rate	Total	Rate	Total
Stone carver	1.5	$....	$....	$31.57	$ 47.36

Cutting Sunk Letters in Indiana Limestone. The cost of cutting sunk letters is considerably less than raised ones.

When cutting letters up to 6" (150 mm) high, a cutter should cut one letter in about 20 minutes at the following labor cost per letter:

	Hours	Rate	Total	Rate	Total
Stone cutter	0.33	$....	$....	$31.57	$ 10.42

If the letters are 12" (300 mm) high, a cutter will cut one letter per hr. at the following labor cost per letter:

	Hours	Rate	Total	Rate	Total
Stone cutter	1	$....	$....	$31.57	$31.57

Textured Indiana Limestone Facing Panels

Limestone panels are available as "curtain wall" units, as wall units spanning a full story height, or as spandrels spanning a full bay. Lengths of up to 15' (4.6 m) are possible, but the most practical panels will fall in the range of 40 to 80 s.f. (3.7-7.4 sq.m), 120 s.f. (11.1 sq.m) being the maximum. Thicknesses will depend on panel size and depth of texture but usually fall in the 4" (100 mm) to 5" (125 mm) range. Nine textures are available. The price range for panels in the 40 to 60 s.f. (3.7-5.6 sq.m) size will run from $8.50 to 12.00 per s.f. ($91.50-129.20 per sq.m) plus shipping and erection costs.

Erection costs are estimated to run from $4.40 to $6.25 per s.f. ($43.05-67.28 per sq.m) including cranes or hoists. On a job in Indianapolis where 346 deep textured, full story height panels were used on construction four stories high as continuous piers, a crew of seven using a crane and dual

C-clamps erected 23,000 s.f. (2137 sq.m) in 23 days. However 750 s.f. (70 sq.m) per day is probably a safer average unless the crew is experienced in this type of construction or the job very simple.

Panels are shipped with no backup applied. Conventional masonry units can, of course, be used or rigid sheets of insulation board can be field applied to the back of the panels. Or the backing may be metal studs and plasterboard with either batt or insulating lath attached to the studs. Any "U" factor may be designed for.

Panels can also be preassembled at the plant from several stones with epoxy bond so that deep sills and returns or special pier or spandrel shapes can be delivered ready for erection in one piece at the job site.

Granite

Granite is an igneous stone. It is extremely hard and weighs 160 lbs. per c.f. (2563 kg per cu.m), or 15% to 20% more than cut stone. It is one of the most durable materials available for construction, but because of the high costs of quarrying, fabrication, shipping, handling and setting, considerable time and care must be taken in planning the job and estimating in order to make the best use of the material.

Granite is usually thought of as a facing material for the lower stories of buildings where durability is paramount. However, several tall buildings, such as the 38-story CBS Building in New York City and the Canadian National Bank in Montreal, have been faced with granite full height. It has also been popular as a sidewalk material for the large plazas around skyscrapers, such as the John Hancock Center in Chicago.

Granite can be classified into three basic groups, each having its own design and specification requirements.

Standard building granite is used structurally either self-supporting or as bonded masonry. Granite has a compressive strength of 20,000 psi (138 MPa) and is well suited for bearing walls.

Granite veneer is a facing used for decoration or protection rather than for its structural value and is applied to masonry backup with stainless steel anchors engineered for the job. Veneers will vary from 7/8" (22.2 mm) in thickness up to 8" (200 mm), the majority being in the 2" (50 mm) to 4" (100 mm) range. Also, precast "curtain wall" panels use granite facing. These are supported from the building framing rather than from the backup and present different construction problems altogether.

Masonry granite is granite used as an integral part of a masonry wall in which one of the more rugged finishes are used such as in churches, rambling residences, bridge abutments, and the like.

Granite is available in a wide range of finishes from split faced to mirror finished slabs.

Polished face is a mirror gloss; honed face, a dull gloss; and fine rubbed, a smooth finish with no gloss. Allowable tolerance for these three types is 3/64" (1.2 mm). Next are the rubbed finishes. Sand finish is slightly pebbled with a tolerance of 1/10" (2.5 mm); spot finish has circular markings with an indefinite pattern and 3/32 (2.4 mm) surface tolerance.

Hammered surfaced granite comes in three grades: fine or 8-cut with parallel markings not over 3/32" (2.4 mm); medium or 6-cut with parallel markings not over 1/8" (3.1 mm); and coarse with parallel markings not over 3/32" (2.4 mm). Tolerances are limited to 7/32" (5.6 mm) for the first two, 1/8" (3.1 mm) for the third. These are corrugated finishes which become smoother near arris lines; that is, near the intersections of two dressed faces.

Sawn faces may be had with vertical scorings (VSM), horizontal scorings (HSM), or circular scorings (RSM). The first two are produced by gang saws, the third, by rotary or circular saws. Tolerance is 1/8" (3.1 mm) with scorings 3/32" (2.4 mm) deep. Thermal finishes are produced by flame texturing a sawn surface with torches to flake granite to a dull, uniform, regular rough surface of coarse texture. It is popular for large paving blocks. Tolerance is 1/8" (3.1 mm).

Jet honing is produced by honing previously cut surfaces with a jet of water and compressed air fed with tiny glass spheres under pressure. Tolerance is 1/8" (3.1 mm).

Granite is a natural material and colors are varied, often two to three per material. Color may be selected from samples or specified by quarry. As shipping costs can be a major factor in a material as heavy as granite, local quarries with similar colors can effect major savings.

However, when the architect specifies a color by quarry, any substitution should be made only with his written approval or as an alternate, lest the contractor be held for the difference in shipping costs.

Granites with feldspar produce the red, pink, brown, buff, grey, and off-white varieties; those with horn blende or mica produce the dark greens and blacks.

Graining as well as color is a consideration; a uniformity of texture, whether fine, medium or coarse, being desirable.

The National Building Granite Quarries Assoc., Inc. publishes a list of granite suppliers with colors and graining available from each. Canada and Sweden are also major sources.

Material Cost of Granite. As discussed above, several factors determine the price of granite, including color, graining, and transportation to the site.

The lower priced polished veneers will include the solid blacks and mottled greys. These cost from $10.00 to $13.00 per s.f. ($107.65-140.00 per sq.m) in 1" (25 mm) thickness, $13.00 to $15.00 per s.f. ($139.94-161.46 per sq.m) in 2" (50 mm) thickness, and $15.00 to $18.00 per s.f. ($161.46-193.75 per sq.m) in 4" (100 mm) thickness. The variegated reds, pinks, and bluish and greenish blacks will cost $13.00 to $19.00 per s.f. ($140-205 per sq.m). Sawn finish will cost about $1.75 less per s.f. ($18.83 per sq.m) and thermal finish about $1.00 less.

Random ashlar granites are considerably cheaper, from $5.00 to $10.00 in 4" (100 mm) thicknesses. This is often priced per cu. ft. (cu.m), the range being $15.00 to $31.00 ($530-1095 per cu.m).

Dressed granite, machine cut in structural thicknesses, will be $15.00 to $50.00 per c.f. ($530-1766 per cu.m) depending on the number of openings, washes, and corners. Hand cut granite, like that on a mausoleum, can cost $125.00 or more per s.f. ($1,345 per sq.m).

Hand cutting, except on the most elaborate work, has been replaced by precision machines. Even surface carving and lettering is done today by sandblasting. Carved letters 6" (150 mm) to 10" (250 mm) high will cost about $20.00 per letter if cut into the granite, $31.00 if raised from the granite.

Cutting of granite into shapes other than square blocks adds quickly to the cost. One source suggests that if a simple square step, which might cost $19.00 per c.f. ($671 per cu.m), were to have a wash or 1/8" (3.1 mm) slope on the top surface, the cost would increase 23%; a wash on two surfaces, 30%; a wash on two surfaces plus a drip, 42%; and an ogee molding plus a drip, 55%.

Labor Setting Granite Veneer. One mason and one helper should set from 30 to 40 s.f. (2.8-3.7 sq.m) of veneer per day, polished work with 1/4" (6.3 mm) joints taking more time than straight sawn work. Column enclosures can be figured at around 35 s.f. (3.2 sq.m) per day. Straight run work with no openings and no hoisting involved can be figured at 50 s.f. (4.6 m) per day.

The following figures are based on one mason and one helper. On large jobs this ratio may be cut down with 3 laborers serving 4 masons, provided scaffolding is arranged so workers can move from position to position. In addition, hoisting costs may have to be added.

On a low-rise building, the square foot labor cost for 100 s.f. (10 sq.m) of polished granite veneer would be figured as follows:

	Hours	Rate	Total	Rate	Total
1 mason	20	$....	$....	$28.18	$ 563.60
1 helper	20	....	....	21.04	420.80
Cost per 100 s.f.			$....		$ 984.40
per s.f.				....	9.84
per sq.m				....	105.92

Work involving special trim at openings, copings, and moldings requires more time:

	Hours	Rate	Total	Rate	Total
1 mason	26.5	$....	$....	$28.18	$ 746.77
1 helper	26.5	....	....	21.04	557.56
Cost per 100 s.f.			$....		$ 1,304.33
per s.f.				....	13.04
per sq.m				....	140.35

Artificial Granite. Cast granite is manufactured in slabs 1-1/2" (37.5 mm) to 2-1/2" (62.5 mm) thick as a less expensive substitute for the natural material. It is not as durable and will not take or hold quite the high polish of the true stone. Still, it has been popular on remodeling and store front work. It is set and handled on the job like true granite. The material costs less than half, or in the $5.00 to $8.00 ($53.80-86.11) range. One mason and one helper should set 50 to 60 s.f. (4.6-5.6 sq.m) per day.

Labor Setting Granite Ashlar. The labor cost of setting granite ashlar will vary considerably, depending on type of wall, height of rises, and the skill of the mason. Being thin, sawed bed ashlar is light in weight and easy to handle and when finished or semi-finished, a mason should set 45 to 55 s.f. (4.1-5.1 sq.m) of wall consisting predominantly of low rises per 8-hr. day, and up to 70 s.f. (6.5 sq.m) of wall, where the ashlar consists principally of high rises.

Marble

Marble is classified according to soundness into A, B, C, and D grades. The A grade is in general the cheapest and also the best for exterior use.

Exterior marble veneer varies in thickness from 7/8" (22.2 mm) to 2" (50 mm). Curtain wall material is usually 1-1/4" (38 mm) thick, although if set in a metal frame or precast with a concrete backup 7/8" (22.2 mm) material is used.

The finish for exterior marble is either sawn, sand, or honed. A polished surface is not recommended.

Marble will vary widely in cost with the type selected and the distance it must be shipped. In Chicago 7/8" (22.2 mm) Class A marble costs from $9.00 to $15.00 per s.f.

Jointing for marble is 1/16" (1.6 mm) to 1/8" (3.1 mm) for thin veneer, 3/16" (4.8 mm) for the thicker stock. All bed and face joints must be solidly filled with either a non-staining mastic or mortar. Most work is either pointed or caulked.

Exterior veneer should be anchored to the backup as follows: two anchors for up to 2 s.f. (0.2 sq.m); four for up to 20 s.f. (1.8 sq.m); and two for each additional 10 s.f. (0.9 sq.m) Anchors may be dowels, straps, wire ties or dovetails, but all must be non-corrosive.

In addition to anchors marble veneer must be supported over all openings and at each story height or not more than 20' (6 m) apart vertically.

Marble may be set against a dampproofed, prebuilt wall, in which case it is set out 1" (25 mm) to 1-1/2" (37.5 mm), anchored, and spot mortared 18" (450 mm) on centers; or set along with the backup, in which case it is completely parged on the back with the anchors built into both facing and backup as the work progresses; or it may be solidly grouted to the backup with the grouting carefully paced so as not to dislodge the marble. In all cases the mortar must be non-staining. Soffit marble must have liners.

Marble is cleaned with clear water and fibre brushes. Acids will stain the surface.

Two masons and a laborer will set around 8 s.f. (0.75 sq.m) of 1" (25 mm) veneer per hour in straight runs. Pointing and caulking and hoisting costs must be added.

Labor Cost of 100 S.F. of 1-1/4" Marble Veneer Set Against a Prebuilt, Damproofed Masonry Wall

	Hours	Rate	Total	Rate	Total
Mason	25	$....	$....	$28.18	$ 704.50
Labor	12.5	....	....	21.04	263.00
Cost per 100 s.f.			$....		$ 967.50
per s.f.				....	9.68
per sq.m				....	104.10

Slate

Slate is sometimes used as an exterior veneer. When used in this way, it is usually 1" (25 mm) to 1-1/4" (31.2 mm) thick and furnished with a natural cleft finish. Panel sizes should not exceed 9'-6" (2.85 m) in length, 5' (1.5 m) in width, or about 25 s.f. (2.3 sq.m) in area.

Slate veneer is held in place with anchors and spot mortar. Relief angles should be provided at each story height. All angles and anchors must be non-corrosive. Anchors should be provided four to a slab or every 3 s.f. (0.27 sq.m). Round anchor holes cut in the edge of the slate to receive wire anchors is the recommended procedure. Mortar should be applied in 6" (150 mm) spots, 18" (200 mm) on center. All joints should be full and are usually 3/8" (9.5 mm) wide.

Slate will run around $10.00 per s.f. (107.64 per sq.m) for material. Two masons and one laborer should set around 7 s.f. (0.65 sq.m) per hour.

Labor Cost of 100 S.F. (10 Sq.M) Slate Veneer

	Hours	Rate	Total	Rate	Total
Mason	28	$....	$....	$28.18	$ 789.04
Labor	14	....	....	21.04	294.56
Cost per 100 s.f.			$....		$1,083.60
per s.f.				....	10.84
per sq.m				....	116.60

04500 MASONRY RESTORATION

Cutting out Mortar Joints in Old Brickwork with an Electric Grinder. On old buildings, where it is necessary to cut out the mortar joints in the old brickwork to a depth of 1/2" (12.5 mm) and repoint them, the most efficient method is to use an electric carborundum wheel, about 5" (125 mm) in diameter and 1/4" (6.3 mm) thick, depending upon the width of the mortar joints. The motor is hung above the scaffolding with a flexible shaft connected to the carborundum wheel.

This method is nearly twice as fast as using an ordinary electric hand saw on account of the weight of the saw, approximately 17 lbs. (7.7 kg), which is difficult and awkward to handle for workers on a swinging scaffold.

Either two or four workers work from each scaffold, depending on the size of the job, and it will require one helper on the ground to each scaffold.

An experienced tuckpointer or operator should grind out lime or lime-cement mortar joints in 350 to 450 s.f. (32.5-41.8 sq.m) of brick wall per 8-hr. day, at the following labor cost per 100 s.f. (9.3 sq.m):

	Hours	Rate	Total	Rate	Total
Tuckpointer	2.0	$....	$....	$28.18	$ 56.36
Helper	1.0	....	....	21.04	21.04
Cost per 100 s.f.			$....		$ 77.40
per s.f.			....		0.77
per sq.m			....		8.33

Using ordinary methods, with an electric saw or grinder, a mason will cut out about 25 s.f. (2.3 sq.m) of joints per hr. at the following labor cost per 100 s.f. (9.3 sq.m):

	Hours	Rate	Total	Rate	Total
Mason	4	$....	$....	$28.18	$112.72
Cost per s.f.			....		1.13
per sq.m			....		12.16

Repointing Mortar Joints in Old Brickwork. After the mortar joints have been cut out, two tuckpointers working on the scaffold, with one helper on the ground, should point 400 to 500 s.f. (37.0-46.5 sq.m) of brick wall per 8-hr. day, at the following labor cost per 100 s.f. (9.3 sq.m):

	Hours	Rate	Total	Rate	Total
Tuckpointers	3.5	$....	$....	$28.18	$ 98.63
Helper	1.5	....	....	21.04	31.56
Cost per 100 s.f.			$....		$130.19
per s.f.			....		1.30
per sq.m			....		14.01

The ordinary hanging or swinging scaffold is 25' (7.6 m) long. The tuckpointers working from each scaffold can cover approximately 6' (1.8 m) in height, so that each lift of the scaffold covers 150 s.f. (13.4 sq.m).

Cleaning Paving Brick Using Pneumatic Chipping Hammers. A satisfactory method of cleaning brick with pneumatic chipping hammers is to use a 4"x12" (100 x 300 mm) plank about 10'-0" (3 m) long, placed on 2 carpenter horses and raised at one end 15 to 18 degrees. This is used to hold the brick, which may be placed lengthwise across the plank, so that one side and one end can be cleaned without moving. A plank will hold about 56 brick. Using a pneumatic chipper with a hexagon nozzle and a 2" (50 mm) chisel bit, a worker should clean 150 to 180 brick per hr. or 1,200 to 1,440 per 8-hr. day.

Sandblast Cleaning of Buildings

The portable compressor and the sandblast are widely used for cleaning buildings, bridges, and other structures, either preparatory to repainting or for the purpose of brightening up the surface.

The cost of this work, however, is contingent entirely upon the condition of the old building, as some of them have nothing more than many years' accumulation of dirt while others have been painted numerous times, making it necessary to remove several coats of paint before the surface of the brick or stone can be touched. Also, it is often necessary to repoint all of the old mortar joints. The estimator should consider all of these items when preparing the estimate.

The process of sandblasting is to force sand against the surface to be cleaned through a small hose connected with the air compressor, and the force of the sand striking the object to be cleaned removes the old surface. Where it requires a maximum pressure of sand and air to remove the old dirt, a 3/4" (18.7 mm) hose is used but where volume is desired more than high pressure a 1" (25 mm) hose is used.

It usually requires four workers to operate a sandblasting machine: one worker attending the machine, one at the nozzle, and two on the ground assisting with hose, scaffold, and sand.

When estimating quantities, take the entire area of the surface to be cleaned and make no deductions for door or window openings.

On court houses, state capitols, post offices, public libraries, and other ornamental structures, the quantities are computed as above, and in addition, balustrades, balusters, etc., are measured on both sides, as it is necessary to go around each baluster separately to clean it.

With a portable compressor supplying air, the average operator can cover a strip 25'-0" (7.6 m) wide and 60'-0" (18.3 m) to 75'-0" (23 m) high per 8-hr day. The exact area varies with the quality of the stone encountered but the following table shows the speed at which various stones can be cleaned.

	S.F. per min.	Sq. M per min.
Limestone	7-9	0.65-0.83
Terra Cotta	8-10	0.74-0.92
Marble	3-5	0.27-0.46
Brick or Brownstone	8-10	0.74-0.92
Granite	6-8	0.55-0.74
Sand stone	10-12	0.92-1.11

The quantity of sand required for sandblasting will vary with the method used, but where a canvas screen or covering surrounding the scaffold catches most of the sand, it is then carried back to the tank on the ground. If this is done, about 75% of the sand is saved for use again. This, however, is governed somewhat by the conditions of the work. The following costs are for removing and cleaning only the natural accumulation of dirt and do not include removing paint, oil stains, etc., unless otherwise mentioned.

Maximum quantities are not figured in the tables as average job conditions usually include delays, moving from one job to another, placing and removing scaffold, etc.

Labor Cost of Sandblasting 100 S.F. (9.3 Sq.M)
of Old Brick or Limestone Buildings

	Hours	Rate	Total	Rate	Total
Compressor Operator	0.33	$....	$....	$28.55	$ 9.42
Nozzleman	0.33	....	....	21.77	7.18
Labor	0.67	....	....	21.04	14.10
Cost per 100 s.f.			$....		$ 30.70
per s.f.			....		0.31
per sq.m			....		3.30

Add for equipment rental, sand, and scaffold costs.

Sandblasting Old Brick Work That Has Been Painted. If the old brick work has been painted and the paint can be blasted off without the necessity of repointing the mortar joints, the costs will be about the same as given above for plain work.

Sandblasting Ornamental Buildings, Such as Post Offices, Courthouses, State Capitols, Public Libraries, Etc., Constructed of Granite or Limestone. The cost of sandblasting buildings as enumerated above, having exterior elaborations, such as balustrades, gables, pediments, plain or fluted columns, etc., will be higher than plain buildings on account of the extra labor required for scaffolding, sandblasting balusters, columns, caps, etc., that are not found on the average building.

Repointing Mortar Joints After Sandblasting. On many old buildings that are unusually dirty or that have been painted, it often happens that the mortar is blasted from the joints when removing the paint from the brick. After the job has been cleaned, the tuckpointers must go over the entire surface and repoint all of the brick joints.

CHAPTER 5

METALS

CSI DIVISION 5
STRUCTURAL METAL FRAMING
05200 METAL JOISTS
05300 METAL DECKING
05400 LIGHTGAGE FRAMING
05500 METAL FABRICATIONS

05100 STRUCTURAL METAL FRAMING

The preparation of a structural steel estimate involves a great deal of work because of the large amount of detail required in listing the quantities from the plans, computing the weights of the various sections, making the shop drawings and detailing the steel, and estimating the cost of shop fabrication, freight, trucking, and erecting the steel on the job.

The structural steel framework of a building usually consists of anchor bolts, setting plates, base plates, columns, girders, beams, lintels, roof trusses, etc., which are fabricated from standard shapes, such as angles,"S" beams, channels,"W" beams and columns, plates, rods, etc., in combinations designed to give the required strength.

Estimating Quantities of Structural Steel

When estimating the quantity of structural steel required for any job, each class of work (column bases, columns, girders, beams, lintels, trusses) should be estimated separately, because each involves different labor operations in fabrication and erection.

All standard connections can be taken from the *AISC Manual of Steel Construction.* Be sure to include rivets and bolts for connecting steel to steel. In addition to the structural framing of the building, it is often necessary for the structural steel contractor to furnish numerous miscellaneous items, such as bearing plates, loose lintels, and anchor bolts, even though they are set in place by other contractors.

Items to be Included in a Structural Steel Estimate. When preparing an estimate on structural steel, the following items should be included to obtain the erected price:

1. Cost of structural steel shapes at the mill or warehouse.
2. Freight or trucking of structural steel shapes from rolling mill or warehouse to fabricating shop.

3. Cost of making shop drawings and details for shop fabrication.
4. Cost of assembling and fabricating shapes into columns, girders, beams, trusses, etc. including templates in connection with these items.
5. Cost of applying one or more coats of shop paint.
6. Shop overhead expense and profit.
7. Freight on fabricated steel from shop to destination.
8. Cost of erecting steel, including equipment rental, setting up and removing cranes, hoisting equipment, unloading steel at site, erecting steel, and field riveting, and welded or bolted connections.

Estimating Cost of Structural Shapes at Mill or Warehouse. After the quantities have been listed from the plans, it will be necessary to compute the cost of the steel sections required.

All prices on structural steel are based on mill shipment from rolling mills in Pittsburgh, Birmingham, Chicago, etc., depending on the location of the mill where the steel is purchased, plus freight or trucking from rolling mill to fabricating shop.

Where steel is wanted in a hurry, it is sometimes necessary to purchase it from warehouse stocks, which usually costs from $0.05 to $0.15 per lb. (0.11-0.33 per kg) more than when purchased from mills, depending on quantity.

Another item that must be considered when pricing steel work is the base price on standard sections and the additions to it. For example, assume the base price of structural shapes is $25.00 per 100 lbs. (45 kg) or $500.00 per ton (0.9 metric ton). Find the cost of 4"x 4"x 1/2" (100 x 100 x 12 mm) steel angles. The standard classification of extras on the following pages shows that angles of this size cost $3.10 per 100 lbs. (45 kg) more than the base price or $28.10 per 100 lbs. (45 kg).

Warehouse prices vary from the base price depending on size and quantity ordered. On orders of 30,000 lbs. (13,600 kg) or more, the decrease from base price is $0.70 per 100 lbs. (45 kg), while on orders from 100 to 399 lbs. (45-180 kg), the increase over base price is about $8.00 per 100 lbs. (45 kg).

Structural steel prices change so often that latest quotations from steel suppliers should always be obtained before figuring jobs of any size.

Cost of Making Working Drawings and Detailing Structural Steel. The cost of making shop or working drawings for steel structures is usually figured by the number of sheets estimated to be required for layout, details, erection diagrams, etc. Drawings, 24"x36" (600 x 900 mm) average about $300.00 to $500.00 per sheet, including drafting room overhead.

If drawing costs are reduced to a cost per ton, the unit costs will vary considerably, depending upon the type of building, weight of steel, amount of duplication, etc. The ton cost will run considerably less where fairly heavy shapes are used than on light structures using tube columns, large lightweight trusses, etc., as the latter type of construction requires considerable detailing without involving much tonnage.

Steel skeleton construction, such as office buildings, hotels, etc., without unusual features can usually be detailed for $100.00 to $125.00 per ton (0.9 metric ton).

Ordinary mill and factory buildings consisting of columns, beams, light trusses, and crane runs, requiring considerable detailing without much duplication, will cost from $75.00 to $150.00 per ton (0.9 metric tons).

Theaters, churches, and other structures of this class will cost from $100.00 to $200.00 per ton (0.9 metric ton) for details.

Shop Painting of Structural Steel. After the steel has been fabricated, it is customary to give it a coat of paint before it leaves the shop, except where steel is to be encased in concrete.

The cost of painting is usually estimated at a certain price per ton and the quantity of paint required and the surface area to be covered will vary with the class of steel. Ordinary structural beams contain 225 to 275 s.f. (20.9-25.5 sq.m) of surface per ton; plate girders 125 to 175 s.f. (11.6-16.2 sq.m) of surface per ton, while trusses contain 275 to 325 s.f. (25.5-30.1 sq.m) of surface per ton. The average for a job will vary from 175 to 225 s.f. (16.2-20.9 sq.m) per ton of steel.

A gallon of paint will cover 400 to 500 s.f. (37.1-46.5 sq.m) of surface or about 2 tons (1.8 metric tons) of steel, at a material cost of $15 per gallon, or $7.50 per ton ($4.00 per liter or $8.30 per metric ton).

The labor necessary to paint one ton of structural steel, 200 s.f. (18.5 sq.m), with one shop coat of paint should cost as follows:

	Hours	Rate	Total	Rate	Total
Labor (spray)	0.75	$....	$....	$26.83	$ 20.12

Basis for Estimating All Classes of Structural Steel Work

Every structural shop has constant or fixed costs that must be considered before estimating the detailed fabricating operations. The following items should be computed before adding the various shop operations required for fabricating the different members.

Approximate Constant or Fixed Costs

	Price per 100 lbs.	Price per 50 kg
Base price of steel, f.o.b. Pittsburgh or Chicago	$24-90.00	$26-99.00
Freight or trucking on steel from rolling mill	2.00*	2.20*
Extras	2.90*	3.19*
Cost of structural steel, f.o.b. fabrication shop	29.80	32.85

Shop handling charge which includes unloading steel

from cars or trucks upon arrival and loading cars or
trucks after fabrication... 1.25 .... 1.38
Tonnage Charges   Painting structural steel in shop......... 0.92 .... 1.02
                                   Handling structural steel in shop........ 0.75 .... 0.83
                                   Cutting structural shapes to lengths... 1.10 .... 1.22
Shop Drawings - cost varies $30-60.00 per ton (metric ton)
   see explanation on previous pages, average cost..................... 3.00**..........3.33**
Total Constant or Fixed Costs .... $36.82 .... $40.63
*Varies. **Varies according to type of building.
   (In rare cases this could be as much as $0.10)

The above constant or fixed costs must be determined in every estimate before adding special fabrications that are needed.

Cost of Fabricating Structural Steel

The cost of fabricating structural steel varies widely, depending upon the type of building, amount of duplication, weight of structural members, number of fabricating operations necessary, etc. The cost of fabricating light steel members usually runs considerably higher than heavier members on a per ton basis. The following gives approximate costs, including shop overhead, of the various shop operations and classifications.

Beam and Channel Punching Only. Where beams and channels are punched only, without any other operation, estimate as follows:

Beams and Channels	Price per 100 lbs.	Price per 50 kg
6-10" (150-250 mm)	$5.75	$6.35
12-18" (300-450 mm)	4.50	4.95
Over 18" (over 400 mm)	4.25	4.65

Beams and Channels Framed with Connections. When beam and channel sections are punched and fabricated with the necessary connections, estimate as follows:

Beams and Channels	Price per 100 lbs.	Price per 50 kg
Up to 10" (250 mm)	$9.00	$9.90
12-18" (300-450 mm)	7.50	8.25
Over 18" (over 450 mm)	6.00	6.62

Beam and Angle Framing with Shelf Angles. Where W-beams are fabricated with shelf angles as illustrated, estimate as follows:

Beams with Shelf Angles	Price per 100 lbs.	Price per 50 kg
Up to 10" (250 mm)	$8.50	$9.40
12-18" (300-450 mm)	6.50	7.15
Over 18" (over 450 mm)	6.00	6.60

Beam Framing with Shelf Angles

Beam Framing with Plates or Channels

Built-up Plate Girder

Beam Framing with Plates or Channels. Where beams are fabricated with a plate or channel bolted or welded as illustrated, estimate as follows:

Beams	Price per 100 lbs.	Price per 50 kg
Up to 10" (250 mm)	$12.50	$13.80
12-18" (300-450 mm)	9.50	10.45
Over 18" (over 450 mm)	8.00	8.80

Built-Up Plate Girders. When estimating the cost of built-up plate girders consisting of plates and angles as illustrated, estimate as follows:

	Price per 100 lbs.	Price per 50 kg
	$20.00	$22.00

Built-Up Plate Girders with Cover Plates. Built-up plate girders consisting of plates, angles, and cover plates as illustrated should be estimated per 100 lbs. (per 50 kg):

	Price per 100 lbs.	Price per 50 kg
	$24.00	$26.45

Plate and Angle Columns. Structural steel columns built-up of plates and angles as illustrated should be estimated as follows:

Columns	Price per 100 lbs.	Price per 50 kg
weighing up to 50 lbs. per l.f. (74 kg per m)	$24.00	$26.40

weighing 50 to 100 lbs. per l.f. (742-147 kg per m) .... 21.50 .... 23.70
weighing over 100 lbs. per l.f. (147 kg per m) .... 20.00 .... 22.00

Plate and Angle Columns with Cover Plates. Structural steel columns built-up of plates and angles with cover plates as illustrated, should be estimated as follows:

Not including Weight of Cover Plates Columns	Price per 100 lbs.	Price per 50 kg
weighing up to 50 lbs. per l.f. (74 kg per m)	$28.00	$30.80
weighing 50 to 100 lbs. per l.f. (74-147 kg per m)	25.00	27.55
weighing 100 lbs. per l.f. (147 kg per m)	23.50	25.00

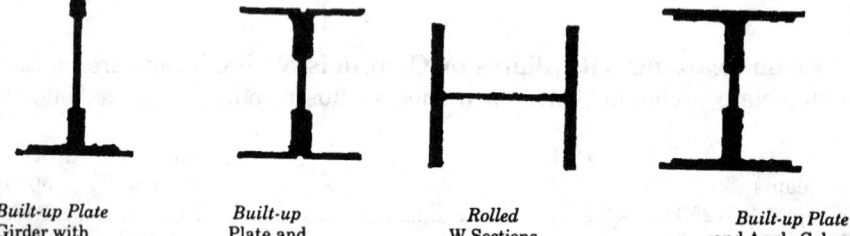

Built-up Plate Girder with Cover Plates	*Built-up Plate and Angle Columns*	*Rolled W Sections*	*Built-up Plate and Angle Column with Cover Plates*

Rolled W Columns. Structural steel columns consisting of rolled wide flange and similar sections as illustrated should be estimated as follows:

WF Columns	Price per 100 lbs.	Price per 50 kg
Up to 6" (150 mm)	$20.00	$22.00
8" (200 mm)	19.00	20.95
Over 8" (over 200 mm)	18.00	19.80

Crane Columns. Structural steel columns for supporting crane runs, etc.

.... $18.00 .... $19.80

Steel Roof Trusses. When estimating the cost of structural steel roof trusses, the fabricating cost will vary with the weight of the trusses as follows:

Steel Roof Trusses	Price per 100 lbs.	Price per 50 kg
weighing up to 1,000 lbs. (453 kg) each	$36.00	$39.60
weighing 1,000 to 2,000 lbs. (453-907 kg) each	33.00	36.30

weighing from 2,000 to 3,000 lbs. (907-1360 kg) each....31.00............34.15
weighing from 3,000 to 4,000 lbs. (1360-1814 kg) each..29.00............31.95
weighing over 4,000 lbs. (1814 kg) each.........................28.00............30.80

Knee Bracing. Fabricating cost.................................$23.00.........$25.30
Angle Struts. Fabricating cost..................................$23.00.........$25.30
"X" Framing and Bracing. Fabricating cost............$14.00.........$15.40
Tie Rods. Tie rods 5/8" (15 mm) rounds, fabricating..$45.00.........$49.60
Bracing Rods. For 3/4" (19 mm) to 1" (25 mm)
rounds, fabricating.. 30.00............33.05

Fabricating Steel Bins, Hoppers and Similar Plate Work. When estimating the cost of fabricating steel plate work, such as steel bins, hoppers, and work of this class, figure the fabricating cost, per 100 lbs. (50 kg)..$29.00.........$31.95

Floor Plates, Plain or Checkered. Estimate the weight of the material in the usual manner. The material usually costs $0.40 to 0.45 per lb. (0.45 kg), depending on market conditions. The cost of fabricating should be figured at $0.23 to 0.30 per lb. (0.45 kg) additional.

Make Up of a Complete Structural Steel Price

The following items should be included to obtain the total price for structural steel delivered f.o.b. cars or job:

	Price per 100 lbs.	Price per 50 kg
Total steel price, f.o.b. fabricating shop, including constant or fixed costs given on previous pages	$36.82*	$40.60
Shop labor and fabrication cost (varies according to operations required, as given on previous pages)	12.00**	13.20**
Total direct shop costs	$48.82	$53.80
Business administration, overhead and sales expense, 20%	9.76	10.76
	$58.58	$64.56
Profit on total cost, including overhead, 10%	5.86	5.46
	$64.44	$70.02
Transportation from shop to job, varies	2.00	2.20
Price of structural steel, delivered at job	$66.44	72.22

*This item varies as described on previous pages.
**Fabricating costs vary according to amount of shop labor required as given under cost of the various fabricating operations.

Mill and Shop Inspection of Structural Steel. Where the structural steel is inspected by one of the testing laboratories, the usual charges for this work are as follows.

Mill inspection is usually figured at $3.00 to $3.50 per ton (0.90 metric ton) and shop inspection from $4.00 to $4.50 per ton (0.90 metric ton). Where both mill and shop inspection are let to one firm, a common price is $6.00 to $7.00 per ton (per metric ton) for both.

Field Inspection of Structural Steel. Where field inspection of the erection of the structural steel is required, figure $1,800.00 per week for salary and expenses of a field inspector, plus transportation to and from job.

Erection of Structural Steel

The field cost of handling and erecting structural steel is difficult to estimate for a number of reasons, such as weather conditions, strikes, delays in receipt of materials, storage facilities, size and type of building, equipment available, labor conditions, number and type of connections.

Any one of the above items may cause a considerable variation in the estimated labor costs and for that reason should be carefully considered when preparing the estimate.

Hauling Structural Steel From Shops to Job. The cost of hauling structural steel from the shop to the job will vary with the facilities for loading trucks, distance of haul, weight of steel members, etc.

The average cost of hauling structural steel is $30 per ton (0.90 metric ton), but varies according to total tonnage.

Plant and Equipment for Steel Erection. The equipment required for steel erection will vary considerably with the size and type of building.

Skeleton frame buildings up to 100'-0" x 100'-0" (30 x 30 m) in size will require one derrick to unload the steel from the trucks and erect it in place, while buildings of irregular shape, such as "L" shape, "U" shape, etc., will probably require additional derricks. One derrick can handle a job 100'-0" x 100'-0" (30 x 30 m), but if the job is 75'-0" x 125'-0" (22 x 38 m), it will probably require two derricks to set the job, while a "U" shape building may require 3 derricks.

Where the buildings have setbacks at certain story heights, as required by many cities, it will often require 2 derricks on the job, one to unload the steel and hoist it from the street and another to set the steel.

It costs $5,000.00 to $10,000.00 to set up and remove a hoisting engine, derrick, compressor outfit, etc., for a one derrick job, with additional derricks at approximately the same cost.

On buildings covering considerable ground area, such as high schools, office buildings, auditoriums, mill and factory buildings up to 8 to 10 stories high, it is often more economical to use a crane, mounted on a truck or crawler, depending upon location of the job. Where a movable crane can be used to advantage, that is the most economical method to use, because the crane can be moved about at slight expense.

Steel Erection Contractors. Most structural steel is erected by specialty contractors, who have an organization and personnel with background and experience. These contractors can erect structural steel much cheaper than the average general contractor, because they have well-organized crews specializing in this type of work. They also have on hand the most modern equipment, such as crawler cranes, truck cranes, lightweight and modern air tools, such as grinders, drills, rivet busters, impact wrenches, welding machines, and other special tools which are furnished to the job as needed and seldom figured in the job cost.

Erecting Structural Steel in Skeleton Frame Buildings. When figuring erection on a job of this type, consider carefully the amount of equipment, size of cranes, etc. required, as described previously.

On a one-crane job, an erection crew should unload, handle, erect, and connect 9 to 10 tons (8.1-9.0 metric tons) of steel per 8-hr. day, at the following labor cost per ton (metric ton):

Erection crew	Hours	Rate	Total	Rate	Total
1 foreman	0.85	$....	$....	$31.20	$ 26.52
2 ir. wkrs."hooking on"	1.70	....	....	30.28	51.48
1 ir. wkr. giving signals	0.85	....	....	30.28	25.74
4 ir. wkrs."connecting"	3.40	....	....	30.28	102.95
Cost per ton			$....		$ 206.69
per metric ton				....	227.36

Add for workmen's compensation and liability insurance. If the structural connections are welded instead of bolted, add $26.00 to $31.00 per ton (metric ton).

Erecting Structural Steel in Wall Bearing Buildings. On wall bearing buildings, the ends of all beams and channels usually rest on the exterior masonry walls while the interior framing is carried on masonry walls or structural steel columns. Field connections are usually bolted.

Eight workers together should handle and erect (including bolted connections) about 8 tons (7.2 metric tons) of steel per 8-hr. day, at the following labor cost per ton (metric ton):

	Hours	Rate	Total	Rate	Total
1 Foreman	1	$....	$....	$31.20	$31.20
7 Iron Workers	7	....	....	30.28	211.96
Cost per ton			$....		$ 243.16
per metric ton				....	267.48

Add for worker's compensation and liability insurance.

Erecting Lightweight Steel for Mill Buildings, Foundries, Auditoriums, Etc. On light-constructed steel buildings, such as machine shops, foundries, mill buildings, garages, auditoriums, etc., where the structural steel consists of steel columns and lightweight roof trusses, requiring considerable handling to set only a small tonnage, the labor costs run higher than on average skeleton frame construction.

A crawler crane or a truck crane should be used for setting the steel. After the crane has been delivered at the job, the labor unloading, handling, erecting, and connecting one ton (0.90 metric ton) of steel should cost:

Erection crew	Hours	Rate	Total	Rate	Total
1 Foreman...1.1		$....	$....	$31.20	$ 34.32
7 iron workers...7.7		....	....	30.28	233.16
Cost per ton			$....		$ 267.48
per metric ton..				....	294.23

Add for worker's compensation and liability insurance.

Erecting Steel Bins, Hoppers, and Similar Plate Work. When erecting steel bins, material hoppers, and similar plate work, an erection crew consisting of 5 iron workers and an engineer should erect in place and bolt 4 to 5 tons of steel a day at the following labor cost per ton (metric ton):

	Hours	Rate	Total	Rate	Total
1 Foreman...2		$....	$....	$31.20	$ 62.40
5 iron workers...10		....	....	30.28	302.80
Cost per ton			$....		$365.20
per metric ton..				....	401.72

Welding the joints or seams in hopper work runs from $60.00 to $100.00 per ton, additional. Add for worker's compensation and liability insurance.

Riveting Steel After Erection. The labor cost of riveting steel after erection will vary with the type of construction and the weight of the structural members. Light-constructed jobs usually require fewer rivets at each connection than heavy-constructed jobs and require more moving of scaffold to drive the same number of rivets. The more rivets that can be driven from one scaffold, the greater the number driven per day.

On light-constructed buildings, a riveting crew should drive 275 to 325 rivets per 8-hr. day, at the following labor cost per 100:

	Hours	Rate	Total	Rate	Total
Ir. wkr. heating rivets.............................2.6		$....	$....	$30.28	$ 78.73
Ir. wkr. catching rivets 2.6		....	....	30.28	78.73
Ir. wkr. driving rivets..............................5.2		....	....	30.28	157.46
Compressor engineer 1.3		....	....	28.55	37.12
Cost per 100 rivets...................................			$....		$ 352.04
Cost per rivet...				....	3.52

The cost of rivets will vary with their size and length. Steel of different thicknesses requires different size rivets. For example, when riveting steel 1/2" (13 mm) or less in thickness, it is customary to use 3/4" (19 mm) rivets, 3/4" (19 mm) steel requires 7/8" (22 mm) rivets, and so on.

Bolting Field Connections. Most structural steel projects have sections bolted together instead of riveted. On work of this kind, an iron worker should place 150 to 200 bolts per 8-hr. day, at the following labor cost per 100:

	Hours	Rate	Total	Rate	Total
Iron worker	.5	$....	$....	$30.28	$151.40
Cost per bolt			....		1.51

Bolted connections normally are completed with a two-worker bolting crew. The cost of bolt units will vary with size, length, and grade of bolt. The most common bolts are ASTM A-325, high strength bolts for structural steel joints, 3/4" diameter x 1-3/4" long (18 x 43 mm), at an approximate cost per bolt unit of $0.65 each. Each bolt unit consists of bolt, washers, and nut. Other grades sometimes required for connections are ASTM 490, heat treated steel structural bolts (add up to 50% to the cost per bolt unit), and for secondary bolted connections, grade A307 carbon steel externally-threaded standard connection (decrease the cost about 15% per bolt unit).

If tension control bolts are used, add additional costs of about 65% per bolt unit.

Instead of counting individual bolts, some estimators use an average count per ton of steel. If one judges that the time involved in making an exact bolt count is not warranted by the cost, then use the following averages.

Average Amount of Field Bolts Required per Ton of Steel

Type of Work	Light	Average	Heavy
	30	25	20

Steel Infrastructure Repairs

Repair of steel infrastructure, which usually refers to bridge and elevated highway structures, presents the estimator with many challenging problems that are not found in new construction. One major difficulty is that repairs are often undertaken in areas where the public has access to the work zone or to adjacent areas. There are many items that might not appear on contract drawings and documents, but which the estimator must consider:

1. Maintenance and protection of traffic - For costing and sample traffic patterns in accordance with federal regulations, see Chapter 22, *Road and Highway Construction.*

2. Maintenance and protection of railroad traffic - Normally, when working within 50 feet of the centerline of a rail track, special precautions must be taken. The estimator should contact local railroad authorities regarding the requirements for flagman, working hours, and the like.

3. Temporary support of existing structures - If temporary supports are not shown on the drawings, the contractor should enlist the services of a professional engineer to design them. The cost of these services will be in direct proportion to the complexity and risk involved. Allow from $65.00 to $100.00 per hour for design services.

4. Activity sequencing - A work sequence must be established. If one has already been set up by the designing engineer, then it is critical that it be followed. In pricing the infrastructure project, the estimator must carefully consider the schedule.

5. Safety - Provisions for netting, rigging, scaffolding, and platforms, as well as environmental conditions, must be considered and priced when required.

6. Work restrictions - Work hours might be restricted, such as during rush hours or holiday periods. In fact, it might be necessary to perform all work outside normal work hours. Area restrictions can result in productivity loss from 10 to 50 %.

7. Hazardous material - Most older steel structures are covered with lead-based paints. Removal of these painted surfaces must comply with regulations set forth by federal, state, and local environmental protection agencies, or other agencies with jurisdiction in this area. For the cost of removing lead-based paints from structural steel, see the "Painting" section in Chapter 9, *Finishes.*

Rivet Removal and Replacement. Rivet heads are snapped off with a hand-held air scaling tool. The use of flame cutting is discouraged and usually not even permitted. The flat end of a chisel point is placed on top of the rivet against the steel member until the rivet head breaks off. The rivet is then hammered out, and a new bolt is placed in the existing hole. Two ironworkers can remove and replace bolts at the rate of about 20 bolts per hr.:

	Hours	Rate	Total	Rate	Total
Ironworkers	2	$....	$....	$30.28	$60.56
Cost per bolt			....		3.03

Drilling Holes into Existing Steel. The best tool for this purpose is a magnetic base drill with a hole bit. Gaining access to the work location and setting up usually take more time than the actual hole drilling.

The set up will take about 15 minutes per hole. Drilling time is between 1 and 5 minutes for a 3/4" to 1-1/4" (18 to 31 mm) hole for steel up to 1-1/2" (33 mm) thick. Two iron workers can drill about 6 holes per hour:

	Hours	Rate	Total	Rate	Total
Ironworkers	2	$....	$....	$30.28	$60.56
Cost per hole			....		10.09

Cutting Existing Steel. When rusted or decayed areas need to be cut out and replaced with new steel, allow for two cuts. The first is cut rough, the second more carefully. After the second cut, additional time is needed for grinding. Production time for this work is difficult to estimate. There are so many types of steel and conditions of existing steel. Assume that 3 l.f. (0.9 m) of burn can be accomplished in about 10 minutes.

Once repairs are complete, it will be necessary to paint the new steel and bolts, and to touch up areas adjacent to the repair work.

ARC WELDING IN BUILDING CONSTRUCTION

Arc welding structural steel in the construction of buildings and bridges has become an established practice. To aid the structural designer in its use, the American Welding Society, the American Institute of Steel Construction, and the American Association of State Highway Officials have established recommended codes and specifications governing design, fabrication, and erection of welded steel structures and applicable codes of most cities, counties and states are in accordance with them. In addition, most fabricating shops now have the equipment, personnel, and experience required to fabricate steel members for welded construction, and in the field an increasing number of qualified operators are available for welding field joints during erection.

Advantages of Arc Welded Steel Design. Advantages of arc welded steel design in building construction are lower overall construction costs, less construction noise, particularly desirable in hospital zones, residential districts, etc., and smooth, clean appearance of erected members which permits leaving structural steel exposed where desirable or necessary.

Overall lower costs are realized through reduced steel material requirements and savings in foundation and exterior wall construction.

When a structure with welded connections is properly designed, savings in steel material are gained from the following:

1. Gross section, rather than effective net section, is used for design, permitting lighter members.
2. Rigid connections provide continuous beam action, permitting a further reduction in member sections.
3. Connection steel, such as splice plates, gussets, etc., are greatly reduced, sometimes eliminated.
4. With beams of less depth, height of building may be reduced, saving column steel and exterior wall construction.
5. In addition, overall costs for detailing, fabrication, freight, trucking, handling and erection are lower due to reduced weight, fewer pieces, and simplified connections.
6. On many structures, where designs and estimates were made for both riveted and welded construction, savings up to 15% on structural steel costs were obtained by using the welded method.

Conditions Affecting Arc Welding Costs

Best results and lowest costs are obtainable only when proper plate preparation and welding procedure is followed. The following conditions have considerable bearing on the cost and quality of welded joints.

Fit-Up. Care in cutting, forming, and handling shapes to be welded to avoid poor fit-up is a major factor in the cost and performance of welded joints. However, a gap of 1/64" to 1/32" (0.39-0.78 mm) is useful in preventing angular distortion and weld cracking. The accompanying graph shows how various sizes of gaps affect welding speeds for square or grooved butt welds. Fillet welds are affected by oversize gaps in a similar manner.

Position of Joints. The position of the joint has considerable effect on the speed and case of welding. Wherever practical, welds should be made in the downhand position with the joint level. Vertical or overhead welds require much more time and skill.

Foreign Matter in Joint. Excessive scale, paint, oil, or rust tend to interfere with welding and should be removed to obtain best speeds and results.

Build-Up or Overwelding. Any amount of weld metal in addition to that actually needed for the specified strength is useless, costly, wasteful, and in some instances, actually harmful. For butt welds, there should be just enough build-up, no more than 1/16" (1.56 mm), to make sure weld is flush with the plate. Excessive build-up not only wastes weld metal but increases welding time.

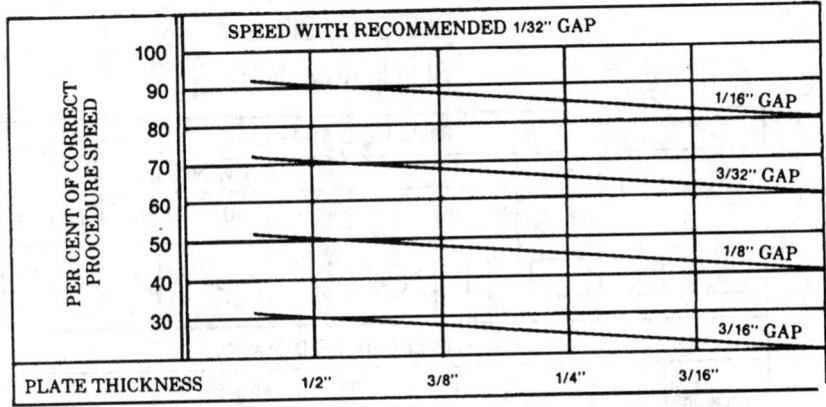

Effect of Gap Size on Welding Speeds for Flat Butt Joints
Welded on Both Sides

Procedures for Making Various Types of Welded Joints

The tables that follow give procedures for making various types of welded joints usually encountered in field erection of structural steel and furnish information for calculating time and material required.

In the tables, electrodes specified are those which will make the joints at the lowest cost.

Arc speed in inches per minute is given for single pass welds and for the first pass in multiple pass welds, as the speed of the first pass is important in obtaining proper penetration.

Feet of joint welded per hour is based on actual welding time. No factor has been included for set-up, electrode changing, cleaning, moving scaffold, etc., which will vary with the job. For overall average quantities, these figures should be multiplied by an operating factor which may range from 25% to 50%, depending upon job conditions.

Pounds of electrode per foot of weld is the amount of electrode required to weld the joint with the recommended plate preparation, fit-up, and a build-up of 1/16" (1.56 mm) or less, including normal spatter loss and 2" (50 mm) stub ends. Any increase in gap or build-up will greatly affect amount of electrode required and time for depositing same.

WELDED JOINTS
Standard symbols

BASIC WELD SYMBOLS

BACK	FILLET	PLUG OR SLOT	GROOVE OR BUTT						
			SQUARE	V	BEVEL	U	J	FLARE V	FLARE BEVEL

SUPPLEMENTARY WELD SYMBOLS

BACKING	SPACER	WELD ALL AROUND	FIELD WELD	CONTOUR		
				FLUSH	CONVEX	

STANDARD LOCATION OF ELEMENTS OF A WELDING SYMBOL

Finish symbol

Contour symbol

Root opening, depth of filling for plug and slot welds

Effective throat

Depth of preparation or size in inches

Reference line

Specification, process or other reference

Tail (ommited when reference is not used)

Basic weld symbol - or detail reference

Groove angle or included angle of countersink for plug welds

Length of weld in inches

Pitch (c. to c. spacing) of welds in inches

Field weld symbol

Weld-all-around symbol

F

A

R

S(E)

T

(Both) (Arrow Side) (Other sides) (Other side)

L @ P

A   B

Arrow connects reference line to arrow side of joint. Use break as at A or B to signify that arrow is pointing to the grooved member in bevel or J-grooved joints.

Note:

Size, weld symbol, length of weld and spacing must read in that order from left to right along the reference line. Neither orientation of reference line nor location of the arrow alter this rule.

The perpendicular leg of weld symbols must be at left.

Arrow and Other Side welds are of the same size unless otherwise shown. Dimensions of fillet welds must be shown on both the Arrow Side and the Other Side Symbol.

The point of the field weld symbol must point toward the tail.

Symbols apply between abrupt changes in direction of welding unless governed by the "all around" symbol or otherwise dimensioned.

These symbols do not explicitly provide for the case that frequently occurs in structural work, where duplicate material (such as stiffeners) occurs on the far side of a web or gusset plate. The fabricating industry has adopted this convention: that when the billing of the detail material discloses the existence of a member on the far side as well as on the near side, the welding shown for the near side shall be duplicated on the far side.

Courtesy of American Institute of Steel Construction, Inc.

Fillet and Lap Welds in Horizontal and Flat Positions

Horizontal Position   Flat Position

Plate Thickness (In.)	Gauge Size of Fillet (In.)	Electrode Size (In.)	Current (Amps.)	Electrode Melt-off Rate (In. per Min.)	Arc. Speed (In. per Min. for First Pass)	Passes or Beads	Ft. of Joint Welded Per Hr. (100% Operating Factor)	Lbs. of Electrode Per Ft. of Weld
3/16	5/32	1/8	170	15.0	15-16	1	75-80	.10
1/4	3/16	5/32	225	14.0	15-16	1	75-80	.15
5/16	1/4	3/16	275	12.0	14-15	1	70-75	.19
3/8	5/16	1/4	350-375	9.6-10.2	12-13	1	60-65	.30
1/2	7/16	1/4	350-375	9.6-10.2	10-11	2	28-31	.57
3/4	5/8	1/4	350-375	9.6-10.2	10-11	3-4	16-14	1.34
1	3/4	1/4	350-375	9.6-10.2	10-11	5	9-10	1.66

Plate thickness in.	Gauge size of fillet in.	Electrode size in.	Currency amps	Electrode melt off rate (in. per pass)	Arc speed (in. per min. for first pass)	passes of beads	Ft. of Joint per Hr. (100% operating factor)	Lbs. of Electrode per ft. of weld
3/16	5/32	1/8	170	15.0	15-16	1	75-80	0.10
1/4	3/16	5/32	225	14.0	15-16	1	75-80	0.15
5/16	1/4	3/16	275	12.0	14-15	1	70-75	0.19
3/8	5/16	1/4	350-375	9.6-10.2	12-13	1	60-65	0.30
1/2	7/16	1/4	350-375	9.6-10.2	10-11	2	28-31	0.57
3/4	5/8	1/4	350-375	9.6-10.2	10-11	3-4	16-14	1.34
1	3/4	1/4	350-375	9.6-10.2	10-11	5	9-10	1.66

(metric)

Plate thickness mm	Gauge size of fillet mm	Electrode size mm	Currency amps	Electrode melt off rate (mm per pass)	Arc speed (mm per min. for first pass)	passes of beads	M of Joint per Hr. (100% operating factor)	Kg of Electrode per m of weld
4.7	3.9	3.1	170	375	375-400	1	22.8-24.3	0.15
6.3	4.7	3.9	225	350	375-400	1	22.8-24.3	0.22
7.8	6.3	4.7	275	300	350-375	1	21.3-22.8	0.28
9.4	7.8	6.3	350-375	240-255	300-325	1	18.2-19.8	0.45
12.5	10.9	6.3	350-375	240-255	250-275	2	8.5-9.4	0.85
18.8	15.6	6.3	350-375	240-255	250-275	3-4	4.8-4.6	1.99
25.0	18.8	6.3	350-375	240-255	250-275	5	2.7-3.0	2.47

Preparation: square edge.

Fit-up: recommended gap, 1/32" (0.78 mm).

Electrode: iron powder E6024 AC. Use E6014 AC electrodes when weldment cannot be positioned close enough to flat for E6024 AC electrodes.

Fillet and Lap Welds in Vertical Position, Welded Up*

Plate Thickness (In.)	Gauge Size of Fillet (In.) †	Electrode Size (In.)	Current (Amp.)	Electrode Melt-off Rate (In. per Min.)	Arc Speed (In. per Min. for First Pass)	Passes or Beads	Ft. of Joint Welded per Hr. (100% Operating Factor)	Lbs. of Electrode Per ft. of Weld
3/16	5/32*	5/32	140	9-1/2	10	1	50	.08
1/4	3/16	3/16	150	8	7	1	35	.14
5/16	1/4	3/16	170	8-1/2	6	1	30	.17
3/8	5/16	3/16	170	8-1/2	3.9	1	19.5	.26
7/16	3/8	3/16	170	8-1/2	2.7	2	13.5	.38
1/2	7/16	3/16	170	8-1/2	6	2	10	.52
5/8	1/2	3/16	170	8-1/2	4	2 or more	7.7	.67
3/4	5/8	3/16	170	8-1/2	4	§	5	1.1
1	7/8	3/16	170	8-1/2	4	§	2.5	2.1

*Weld 5/32 fillet vertically down. T-joint welded both sides with this fillet will have plate strength. Total no. of passes varies with operator.

Preparation: square edge.

Fit-up: recommended gap, 1/32" (0.78 mm); maximum gap 1/16" (1.56 mm). If greater gaps must be welded, use same procedure but add width of gap to fillet size.

Electrode: E6010

Polarity: electrode positive.

Plate thickness in.	Gauge size of fillet in.	Electrode size in.	Currency amps	Electrode melt off rate (in. per pass)	Arc speed (in. per min. for first pass)	passes of beads	Ft. of Joint Welded per Hr. (100% operating factor)	Lbs. of Electrode per ft. of weld
3/16	5/32	5/32	140	9.5	10	1	50	0.08
1/4	3/16	3/16	150	8	7	1	35	0.14
5/16	1/4	3/16	170	8.5	6	1	30	0.17
3/8	5/16	3/16	170	8.5	3.9	1	19.5	0.26
7/16	3/8	3/16	170	8.5	2.7	2	13.5	0.38
1/2	7/16	3/16	170	8.5	6	2	10	0.52
5/8	1/2	3/16	170	8.5	4	2 or more	7.7	0.67

(metric)

Plate thickness mm	Gauge size of fillet mm	Electrode size mm	Currency amps	Electrode melt off rate (mm per pass)	Arc speed (mm per min. for first pass)	passes of beads	Meter Welded per Hr. (100% operating factor)	Kg of Electrode per m of weld
4.7	3.9	3.9	140	238	250	1	16	0.12
6.3	4.7	4.7	150	200	175	1	11	0.21
7.8	6.3	4.7	170	213	150	1	9	0.25
9.4	7.8	4.7	170	213	98	1	6	0.39
10.9	9.4	4.7	170	213	68	2	4	0.57
12.5	10.9	4.7	170	213	150	2	3	0.77
15.6	12.5	4.7	170	213	100	2 or more	2	1.00
18.8	15.6	4.7	170	213	100		2	1.64
25.0	21.9	4.7	170	213	100		1	3.12

The following are actual production quantities of field welding on two different jobs located in different parts of the country and will give the estimator an idea of the amount of welding and labor required.

Description	9 Story Off Bldg.	18 Story Hospital
Size of building on the ground	100'x100'	140'x200'
Total tonnage of steel welded, tons	1,170	2,274
Fillet per ton of steel, lineal inches	63.5	50.68
Electrode used per ton of steel, lbs.	4.0	1.7
Electric current required per ton, kw. hrs.	8.1	6.05
Weld per lb. of electrode bought, lin.inches	16.00	21.34
Welding man hours per ton of steel, hours	1.51	1.80*
Steel welded per month, per machine, tons	128.00	
Lin. ft. fillet per 8-hr. day, per welder	28.0	36.0
Lin ft. fillet per pound of electrode bought	1.33	1.78
Electric current required to melt one lb. of electrode, kw. hrs.	2.90	3.63

(metric)

	9 Story Off Bldg.	18 Story Hospital
Size of building on the ground	30x30 m	42x61 m
Total tonnage of steel welded, metric tons	1,053	2,820
Fillet per ton of steel, mm	1,588	1,267
Electrode used per ton of steel, kg	1.80	0.77
Electric current required per ton, kw. hrs.	8.10	6.05
Weld per kg of electrode bought, mm	1.41	1.88
Welding manhours per ton of steel	1.67	1.62*
Steel welded per month, per machine, tons	115.00	
Meters of fillet per 8-hr. day, per welder	8.50	11.00
Meters of fillet per lb. of electrode bought	0.40	0.54
Electric current required to melt one kg of electrode, kw. hrs.	1.31	1.64

*Includes 0.26 hrs. foreman, 0.91 hrs. welder, 0.26 hrs. helpers, and 0.37 hrs. hoist. engr. time.

Data on Fillet and Lap Welds in Overhead Position

Estimating Data on Fillet and Lap Welds in Overhead Position

TWO PASSES   FOR 3 OR MORE PASSES

Plate Thick-ness (In.)	Gauge Size of Fillet (In.) †	Electrode Size (In.)	Current (Amp.)	Electrode Melt-off Rate (In. per Min.)	Arc Speed (In. per Min. for First Pass)	Passes or Beads	Ft. of Joint Welded per Hr. (100% Operating Factor)	Lbs. of Electrode per ft. of Weld
3/16	5/32	5/32	140	9-1/2	9	1	45	.09
1/4	3/16	3/16	160	8-1/4	7	1	35	.14
5/16	1/4	3/16	160	8-1/4	5.7	1	28.5	.18
3/8	5/16	3/16	160	8-1/4	3.7	2	18.5	.27
7/16	3/8	3/16	160	8-1/4	7-3/4	3	12.5	.40
1/2	7/16	3/16	160	8-1/4	7	§	9.2	.55
5/8	1/2	3/16	160	8-1/4	6	§	7.0	.72
3/4	5/8	3/16	160	8-1/4	6	§	4.5	1.2
1	7/8	3/16	160	8-1/4	6	§	2.2	2.3

*T-joint welded both sides with this fillet will have plate strength.
Total number of passes varies with operators.

Preparation: square edge.

Fit-up: recommended gap, 1/32" (0.78 mm), maximum gap, 1/16" (1.56 mm). If greater gaps must be welded, use same procedure but add width of gap to fillet size.

Plate thick-ness in.	Gauge size of fillet in.	Electrode size in.	Currency amps	Electrode melt off rate (in. per pass)	Arc speed (in. per min. for first pass)	passes of beads	Ft. of Joint Welded per Hr. (100% operating factor)	Lbs. of Electrode per ft. of weld
3/16	5/32	5/32	140	9.5	9	1	45	0.09
1/4	3/16	3/16	160	8.25	7	1	35	0.14
5/16	1/4	3/16	160	8.25	5.7	1	28.5	0.18
3/8	5/16	3/16	160	8.25	3.7	2	18.5	0.27

Plate thick- ness in.	Gauge size of fillet in.	Electrode size in.	Currency amps	Electrode melt off rate (in. per pass)	Arc speed (in. per min. for first pass)	passes of beads	Ft. of Joint Welded per Hr. (100% operating factor)	Lbs. of Electrode per ft. of weld
7/16	3/8	3/16	160	8.25	7.25	3	12.5	0.40
1/2	7/16	3/16	160	8.25	7	*	9.2	0.55
5/8	1/2	3/16	160	8.25	6	*	7.0	0.72
3/4	5/8	3/16	160	8.25	6	*	4.5	1.2
1	7/8	3/16	160	8.25	6	*	2.2	2.3

(metric)

Plate thick ness mm	Gauge size of fillet mm	Electrode size mm	Currency amps	Electrode melt off rate (mm per pass)	Arc speed (mm per min. for first pass)	passes of beads	Meter of Joint Welded per Hr. (100% operating factor)	Kg of Electrode per m of weld
4.7	3.9	3.9	140	238	225	1	14	0.13
6.3	4.7	4.7	160	206	175	1	11	0.21
7.8	6.3	4.7	160	206	143	1	9	0.27
9.4	7.8	4.7	160	206	93	2	6	0.40
10.9	9.4	4.7	160	206	194	3	4	0.60
12.5	10.9	4.7	160	206	175		3	0.82
15.6	12.5	4.7	160	206	150		2	1.07
18.8	15.6	4.7	160	206	150		1	1.79
25.0	21.9	4.7	160	206	150		1	3.42

Approximate Prices of Iron and Steel Products

All prices for structural steel are based on carload shipments from mills in Pittsburgh, Pa., Chicago, Ill., or Birmingham, Ala. Freight must be added from nearest mill point to destination.

Prices on steel should always be obtained from mills or nearest jobber, because steel prices are subject to frequent change.

Structural steel shapes are currently quoted at the basic price of $24.90 per 100 lbs. (45 kg) for ASTM A36 steel.

The estimator should get a price quote from his or her steel supplier for each project.

Extras Over Base Price on Structural Shapes

S Shape Beams

Size, inches	Extra 100 lbs.	Size, mm	Extra 50 kg
6	$2.75	150	$3.03

Size, inches	Extra 100 lbs.	Size, mm	Extra 50 kg
7	2.75	175	3.03
8	2.35	200	2.59
10	2.25	250	2.48
12, 31.8 & 35 lbs.	2.25	300, 14.4 & 15.8 kg.	2.48
12, 40.8 & 50 lbs.	2.25	300, 18.5 & 22.6 kg.	2.48
15	2.25	375	2.48
18	2.25	450	2.48
20	2.25	500	2.48
24	2.25	600	2.48

W Shape Beams

Size, inches	Wgt., lbs.	Extra 100 lbs.	Size, mm	Wgt., kg	Extra 50 kg
4	13	$2.10	100	5.9	$2.31
6	9	2.10	150	4.1	2.31
6	12 & 16	2.10	150	5.4 & 7.2	2.31
6	20 & 25	2.10	150	9 & 11.3	2.31
8	10	2.10	200	4.5	2.31
8	13 & 15	2.10	200	5.8 & 6.8	2.31
8	18 & 21	2.10	200	8.1 & 9.5	2.31
8	24-28	2.10	200	10.8-9.5	2.31
8	31-67	2.10	200	14-30.3	2.31
10	12	2.10	250	5.4	2.31
10	15-19	2.10	250	6.8-7.2	2.31
10	22-30	2.10	250	9.9-16.3	2.31
10	33-45	2.10	250	14.9-20.4	2.31
10	49-112	3.60	250	22.2-50.8	3.97
12	14	2.10	300	6.4	2.31
12	16-22	2.10	300	7.2-9.9	2.31
12	26-35	2.10	300	11.7-15.8	2.31
12	40-50	2.10	300	18.1-22.6	2.31
12	53-58	2.10	300	24.0-26.3	2.31
12	65-190	4.50	300	29.4-86.1	4.96
14	22-26	2.10	350	9.9-11.7	2.31
16	36-57	2.10	400	16.3-25.8	2.31
16	67-100	4.50	400	30.3-45.3	4.96
18	35-46	2.10	450	15.8-20.8	2.31
18	50-71	2.10	450	22.6-32.2	2.31
18	76-119	4.50	450	34.4-53.9	4.96
18	35-46	2.10	450	15.8-20.8	2.31
21	44-57	4.50	525	19.9-25.8	4.96
21	62-93	4.50	525	28.1-42.1	4.96
21	101-147	4.50	525	45.8-66.6	4.96
24	55-162	4.50	600	24.9-73.4	4.96
27	84-178	4.50	675	38.1-80.7	4.96
30	99-132	4.50	750	44.9-59.8	4.96

Size, inches	Wgt., lbs.	Extra 100 lbs.	Size, mm	Wgt., kg	Extra 50 kg
30	173-211	4.50	750	78.4-95.7	4.96
33	118-152	4.50	825	53.5-68.9	4.96
33	201-241	4.50	825	91.1-109.3	4.96
36	135-210	4.50	900	61.2-95.2	
36	230-300	4.50	900	104.3-136.0	4.96

Sample Metric Conversions for Existing W Shapes

Size in. x lbs/ft	Size mm x kg/m	Area mm²	Web Thickness mm	Flange Width mm	Flange Thickness mm
W 24 x 492	610 x 732	93,300	50.00	359.0	89.9
W 24 x 450	610 x 670	85,400	46.00	354.0	83.1
W 24 x 408	610 x 608	77,400	41.90	351.0	75.9
W 24 x 370	610 x 551	70,200	38.60	347.0	69.1
W 24 x 335	610 x 498	63,500	35.10	343.0	63.0
W 24 x 306	610 x 455	57,900	32.00	340.0	57.9
W 24 x 279	610 x 415	52,900	29.50	338.0	53.1
W 24 x 250	610 x 372	47,400	26.40	335.0	48.0
W 24 x 229	610 x 341	43,400	24.40	333.0	43.9
W 24 x 207	610 x 307	39,200	22.10	330.0	39.9
W 24 x 192	610 x 285	36,300	20.60	329.0	37.1
W 24 x 176	610 x 262	33,300	19.00	327.0	34.0
W 24 x 162	610 x 241	30,800	17.90	329.0	31.0
W 24 x 146	610 x 217	27,700	16.50	328.0	27.7
W 24 x 131	610 x 195	24,900	15.40	327.0	24.4
W 24 x 117	610 x 174	22,200	14.00	325.0	21.6
W 24 x 104	610 x 155	19,800	12.70	324.0	19.0
W 24 x 103	610 x 153	19,500	14.00	229.0	24.9
W 24 x 94	610 x 140	17,900	13.10	230.0	22.2
W 24 x 84	610 x 125	15,900	11.90	229.0	19.6
W 24 x 76	610 x 113	14,400	11.20	228.0	17.3
W 24 x 68	610 x 101	12,900	10.50	228.0	14.9
W 24 x 62	610 x 92	11,800	10.90	179.0	15.0
W 24 x 55	610 x 82	10,500	10.00	178.0	12.8

M Shape Beams

Size inches	Wgt., lbs.	Extra 100 lbs.	Size mm	Wgt., kg	Extra 50 kg
5	18.9	$2.10	125	8.6	$2.31
6	20	3.10	150	9.1	3.42
8	6.5	4.10	200	2.9	4.52

W12x45 (300x67)
W=shape
12=depth in inches (300 mm)
45=weight per l.f. (67 kg per m)

Angles

Size inches	Extra 100 lbs.	Size mm	Extra 50 kg
3 x 2 x 3/16	$3.10	75 x 50 x 4.7	$3.42
3 x 2 x 1/4-1/2	3.10	75 x 50 x 6.3-13	3.42
3 x 2-1/2 x 1/4-1/2	3.10	75 x 63 x 6.3-13	3.42
3 x 3 x 3/16	3.10	75 x 75 x 4.7	3.42
3 x 3 x 1/4-1/2	3.10	75 x 75 x 6.3-13	3.42
3-1/2 x 2-1/2 x 1/4-5/16	3.10	88 x 63 x 6.3-7.8	3.42
3-1/2 x 3 x 1/4-5/16	3.10	88 x 75 x 6.3-7.8	3.42
3-1/2 x 3-1/2 x 1/4-5/16	3.10	88 x 88 x 6.3-7.8	3.42
4 x 3 x 1/4	3.10	100 x 75 x 6.3	3.42
4 x 3 x 5/16	3.10	100 x 75 x 7.8	3.42
4 x 3-1/2 x 1/4-5/16	3.10	100 x 88 x 6.3-7.8	3.42
4 x 3-1/2 x 3/8-1/2	3.10	100 x 88 x 9.4-13	3.42
4 x 4 x 1/4-5/16	3.10	100 x 100 x 6.3-7.8	3.42
4 x 4 x 3/8-3/4	3.10	100 x 100 x 9.4-13	3.42
5 x 3 x 1/4	3.10	125 x 75 x 6.3	3.42
5 x 3 x 5/16	3.10	125 x 75 x 7.8	3.42
5 x 3-1/2 x 3/8-3/4	3.10	125 x 88 x 9.4-13	3.42
5 x 5 x 5/16	3.10	125 x 125 x 7.8	3.42
5 x 5 x 3/8-3/4	3.10	125 x 125 x 9.4-19	3.42
6 x 3-1/2 x 5/16-1/2	3.10	150 x 88 x 7.8-13	3.42
6 x 4 x 5/16	3.10	150 x 100 x 7.8	3.42
6 x 4 x 3/8-7/8	3.10	150 x 100 x 9.4-22	3.42
6 x 6 x 3/8-1	3.10	150 x 150 x 9.4-25	3.42
7 x 4 all thick.	3.10	175 x 100 x all thick.	3.42
8 x 4 all thick.	3.10	200 x 100 x all thick.	3.42
8 x 6 all thick.	3.10	200 x 150 x all thick.	3.42
8 x 8 all thick.	3.10	200 x 200 x all thick.	3.42
9 x 4 all thick.	3.10	225 x 100 x all thick.	3.42

Channels—Standard

Size inches	Extra 100 lbs.	Size mm	Extra 50 kg
3	$2.10	75	$2.31
4	2.10	100	2.31
5	2.10	125	2.31
6	2.10	150	2.31
7	2.10	175	2.31
8	2.10	200	2.31
9	2.10	225	2.31
10	2.10	250	2.31
12	2.10	300	2.31
15	2.10	375	2.31

Zees

Size	Extra 100 lbs.	Size	Extra 50 kg
Z 3 x 6.7-12.6	$5.00	Z 75 x 3.0-5.7	$5.50
Z 4 x 8.2-10.3	5.00	Z 100 x 3.7-4.7	5.50

Estimating the Weight of Wrought Iron, Steel, or Cast Iron

When tables of weights are not handy, the following rules will prove of value to the estimator when computing the weights of wrought iron, steel, or cast iron.

Weight of Wrought Iron. One cu. ft. of wrought iron weighs 480 lbs. One sq. ft. of wrought iron 1" thick weighs 40 lbs. One sq. in. of wrought iron one foot long weighs 3-1/3 lbs.

To find the weight of one square foot of flat iron of any thickness, multiply the thickness in inches by 40, and the result will be the weight of the iron in lbs.

To find the weight of one lin. ft. of wrought iron bar of any size, multiply the cross sectional area in sq. in. by 3-1/3, and the result will be the weight per l.f.

Weight of Steel. One cu. ft. of steel weighs 489.6 lbs., or 2% more than wrought iron. One sq. ft. of 1" thick steel weighs 40.8 lbs. A piece of 1" x 1" x 1' steel weighs 3.4 lbs.

To find the weight of one lin. ft. of steel bar of any size, multiply the cross sectional area in square inches by 3.4, and the result will be the weight of the steel in lbs. If the weight per lin. ft. is known, the exact sectional area in square inches may be obtained by dividing the weight by 3.4.

Weight of Cast Iron. One cu. ft. of cast iron weighs 450 lbs (1 cu.m = 5.78 kg). One sq. ft. of cast iron, 1" thick, weighs 37-1/2 lbs. (1 mm = 0.68 kg) A piece of cast iron 1" x 1" x 1' weighs 3-1/8 lbs. (25 x 25 x 300 mm weighs 1.41 kg). One cu. in. of cast iron weighs 0.26 lbs.

Equal Angles
4 x 4 x ½
(100 x 100 x 12.5 mm)
Weight is listed
in AISC manual.

Unequal Angles
4 x 3 x ½
(100 x 75 x 12.5 mm)
Weight is listed
in AISC manual

Standard Channel
6 x 13
(150 x 19)
6" (150 mm)=channel size
13=lbs./l.f. (19 kg/m)

Weight of Square and Round Bars

Thick. or Dia., inches	Lbs. of Square Bar per l.f.	Lbs. of Round Bar per l.f.	Thick or Dia., mm	Kg of Square Bar per m	Kg of Round Bar per m
1/4	0.213	0.167	6.250	0.317	0.248
5/16	0.332	0.261	7.813	0.494	0.388
3/8	0.478	0.376	9.375	0.711	0.559
7/16	0.651	0.511	10.938	0.969	0.760
1/2	0.850	0.668	12.500	1.265	0.994
9/16	1.076	0.845	14.063	1.601	1.257
5/8	1.382	1.040	15.625	2.056	1.547
11/16	1.607	1.260	17.188	2.391	1.875
3/4	1.913	1.500	18.750	2.846	2.232
13/16	2.240	1.760	20.313	3.333	2.619
7/8	2.603	2.040	21.875	3.873	3.035
15/16	--	2.350	23.438	--	3.496
1	3.400	2.670	25.000	5.059	3.972
1-1/16	--	3.010	26.563	--	4.478
1-1/8	4.303	3.380	28.125	6.402	5.029
1-1/4	5.313	4.170	31.250	7.905	6.204
1-5/16	--	4.600	32.813	--	6.844
1-3/8	6.428	5.050	34.375	9.564	7.513
1-7/16	--	5.520	35.938	--	8.213
1-1/2	7.650	6.010	37.500	11.382	8.942
1-5/8	8.928	7.050	40.625	13.283	10.489
1-11/16	--	7.600	42.188	--	11.307
1-3/4	10.410	8.180	43.750	15.488	12.170
1-13/16	--	8.773	45.313	--	13.053
1-7/8	11.950	9.390	46.875	17.779	13.971

Thick. or Dia., inches	Lbs. of Square Bar per l.f.	Lbs. of Round Bar per l.f.	Thick or Dia., mm	Kg of Square Bar per m	Kg of Round Bar per m
2	13.600	10.680	50.000	20.234	15.890
2-1/8	15.350	12.060	53.125	22.838	17.943
2-1/4	17.210	13.520	56.250	25.605	20.115
2-3/8	--	15.060	59.375	--	22.406
2-1/2	21.250	16.690	62.500	31.616	24.832
2-5/8	23.430	18.400	65.625	34.859	27.376
2-3/4	25.710	20.200	68.750	38.252	30.054
2-7/8	--	22.070	71.875	--	32.836
3	30.600	24.030	75.000	45.527	35.752
3-1/8	--	26.080	78.125	--	38.802
3-1/4	35.910	28.210	81.250	53.427	41.971
3-3/8	--	30.42			
3-1/2	--	30.420	87.500	--	45.259
3-5/8	41.650	32.710	90.625	61.967	48.666
3-3/4	47.810	37.550	96.875	71.132	55.867
4	54.400	42.730	100.000	80.937	63.574

Weights of Flat Steel Bars
Lbs. per ft. (Kg per m)

Thick., in.	Thick., mm	1"	25mm	2"	50mm	3"	75mm	4"	100mm	5"	125mm	6"	150mm
3/16	4.69	0.64	0.95	1.28	1.90	1.91	2.84	2.55	3.79	3.19	4.75	3.83	5.70
1/4	6.25	0.85	1.26	1.70	2.53	2.55	3.79	3.40	5.06	4.25	6.32	5.10	7.59
5/16	7.81	1.06	1.58	2.12	3.15	3.19	4.75	4.25	6.32	5.31	7.90	6.38	9.49
3/8	9.38	1.28	1.90	2.55	3.79	3.83	5.70	5.10	7.59	6.38	9.49	7.65	11.38
7/16	10.94	1.49	2.22	2.98	4.43	4.46	6.64	5.95	8.85	7.44	11.07	8.95	13.32
1/2	12.50	1.70	2.53	3.40	5.06	5.10	7.59	6.80	10.12	8.40	12.50	10.20	15.18
9/16	14.06	1.92	2.86	3.83	5.70	5.74	8.54	7.65	11.38	9.57	14.24	11.48	17.08
5/8	15.63	2.12	3.15	4.25	6.32	6.38	9.49	8.50	12.65	10.63	15.82	12.75	18.97
11/16	17.19	2.34	3.48	4.67	6.95	7.02	10.44	9.35	13.91	11.69	17.39	14.03	20.87
3/4	18.75	2.55	3.79	5.10	7.59	7.65	11.38	10.20	15.18	12.75	18.97	15.30	22.76
13/16	20.31	2.76	4.11	5.53	8.23	8.29	12.33	11.05	16.44	13.81	20.55	16.58	24.67
7/8	21.88	2.98	4.43	5.98	8.90	8.93	13.29	11.90	17.70	14.87	22.12	17.85	26.56
15/16	23.44	3.19	4.75	6.38	9.49	9.57	14.24	12.75	18.97	15.94	23.72	19.13	28.46
1	25.00	3.40	5.06	6.80	10.12	10.20	15.18	13.60	20.23	17.00	25.29	20.40	30.35
1-1/16	26.56	3.61	5.37	7.22	10.74	10.84	16.13	14.45	21.50	18.06	26.87	21.68	32.26
1-1/8	28.13	3.83	5.70	7.65	11.38	11.48	17.08	15.30	22.76	19.13	28.46	22.95	34.15
1-3/16	29.69	4.04	6.01	8.08	12.02	12.12	18.03	16.15	24.03	20.19	30.04	24.23	36.05
1-1/4	31.25	4.25	6.32	8.50	12.65	12.75	18.97	17.00	25.29	21.25	31.62	25.50	37.94
1-5/16	32.81	4.46	6.64	8.93	13.29	13.39	19.92	17.85	26.56	22.32	33.21	26.78	39.84
1-3/8	34.38	4.67	6.95	9.35	13.91	14.03	20.87	18.70	27.82	23.38	34.78	28.05	41.74
1-7/16	35.94	4.89	7.28	9.78	14.55	14.66	21.81	19.55	29.09	24.44	36.36	29.33	43.64
1-1/2	37.50	5.10	7.59	10.20	15.18	15.30	22.76	20.40	30.35	25.50	37.94	30.60	45.53
1-9/16	39.06	5.32	7.92	10.63	15.82	15.94	23.72	21.25	31.62	26.57	39.53	31.88	47.43
1-5/8	40.63	5.50	8.18	11.05	16.44	16.58	24.67	22.10	32.88	27.63	41.11	33.15	49.32
1-11/16	42.19	5.74	8.54	11.47	17.07	17.22	25.62	22.95	34.15	28.69	42.69	34.43	51.23
1-3/4	43.75	5.95	8.85	11.90	17.70	17.85	26.56	23.80	35.41	29.75	44.26	35.70	53.11

Thick., in.	Thick., mm	Width of Bars												
		1"	25mm	2"	50 mm	3"	75 mm	4"	100 mm	5"	125 mm	6"	150 mm
1-13/16	45.31	6.16	9.16	12.33	18.34	18.49	27.51	24.65	36.67	30.81	45.84	36.98	55.02
1-7/8	46.88	6.38	9.49	12.75	18.97	19.13	28.46	25.50	37.94	31.87	47.42	38.25	56.91
1-15/16	48.44	6.59	9.80	13.18	19.61	19.77	29.41	26.35	39.20	32.94	49.01	39.53	58.81
2	50.00	6.80	10.12	13.60	20.23	20.40	30.35	27.20	40.47	34.00	50.59	40.80	60.70

Weights and Dimensions of S Shape Beams

Size in inches	Size in mm	Wgt per ft. lbs	Wgt per m kg	Thick. of web, inches	Thick. of web, mm	Width of flange, Inches	Width of flange, mm
3	75	5.70	8.48	0.170	4.25	2.330	58.3
		7.50	11.16	0.349	8.73	2.509	62.7
4	100	7.70	11.46	0.190	4.75	2.660	66.5
		9.50	14.13	0.326	8.15	2.796	69.9
5	125	10.00	14.88	0.210	5.25	3.000	75.0
8	200	18.40	27.38	0.270	6.75	4.000	100.0
		23.00	34.22	0.441	11.03	4.171	104.3
10	250	25.40	37.79	0.310	7.75	4.660	116.5
		35.00	52.07	0.594	14.85	4.944	123.6
12	300	31.80	47.31	0.350	8.75	5.000	125.0
		35.00	52.07	0.428	10.70	5.078	127.0
		40.80	60.70	0.460	11.50	5.250	131.3
		50.00	74.39	0.687	17.18	5.477	136.9
15	375	42.90	63.83	0.410	10.25	5.500	137.5
		50.00	74.39	0.550	13.75	5.640	141.0
18	450	54.70	81.38	0.460	11.50	6.000	150.0
5	125	14.75	21.95	0.494	12.35	3.284	82.1
6	150	12.50	18.60	0.230	5.75	3.330	83.3
		17.25	25.66	0.465	11.63	3.565	89.1
7	175	15.30	22.76	0.250	6.25	3.660	91.5
		20.00	29.76	0.450	11.25	3.860	96.5
18	450	70.00	104.15	0.711	17.78	6.251	156.3
20	500	66.00	98.20	0.505	12.63	6.255	156.4
		75.00	111.59	0.641	16.03	6.391	159.8
		86.00	127.95	0.660	16.50	7.053	176.3
		96.00	142.83	0.800	20.00	7.200	180.0
24	600	90.00	133.90	0.500	12.50	7.000	175.0
		90.00	133.90	0.625	15.63	7.124	178.1
		100.00	148.78	0.745	18.63	7.247	181.2
		106.00	157.71	0.620	15.50	7.875	196.9
		121.00	180.02	0.800	20.00	8.048	201.2

METALS

S24x100 (600x148)
S=shape of beam
24=depth, in. (600 mm)
100=lbs per l.f. (148 kg/m)

C12x20.7 (300x30)
C=shape (channel)
12=depth, in. (300 mm)
20.7=lbs per l.f. (30 kg/m)

Weights of American Standard Channels

Depth of Channels, in.	Depth of Channels, mm	Wgt per l.f., lbs.	Wgt per m, kg	Thick. of web, inches	Thick. of web mm	Width of flange inches	Width of flange mm
3	75	6.00	8.93	0.362	9.05	1.602	40.1
		5.00	0.26	0.264	6.60	1.504	37.6
		4.10	0.17	0.170	4.25	1.410	35.3
4	100	7.25	10.79	0.325	8.13	1.725	43.1
		5.40	8.03	0.180	4.50	1.580	39.5
5	125	9.00	13.39	0.330	8.25	1.890	47.3
		6.70	9.97	0.190	4.75	1.750	43.8
6	150	13.00	19.34	0.440	11.00	2.160	54.0
		10.50	15.62	0.318	7.95	2.038	51.0
		8.20	12.20	0.200	5.00	1.920	48.0
7	175	14.75	21.95	0.423	10.58	2.303	57.6
		12.25	18.23	0.318	7.95	2.198	55.0
		9.80	14.58	0.210	5.25	2.090	52.3
8	200	18.75	27.90	0.490	12.25	2.530	63.3
		13.75	20.46	0.307	7.68	2.347	58.7
		11.50	17.11	0.220	5.50	2.260	56.5
9	225	20.00	29.76	0.452	11.30	2.652	66.3
		15.00	22.32	0.288	7.20	2.488	62.2
		13.40	19.94	0.230	5.75	2.430	60.8
10	250	30.00	44.63	0.676	16.90	3.036	75.9
		25.00	37.20	0.529	13.23	2.889	72.2
		20.00	29.76	0.382	9.55	2.742	68.6
		15.30	22.76	0.240	6.00	2.600	65.0
12	300	30.00	44.63	0.513	12.83	3.173	79.3
		25.00	37.20	0.390	9.75	3.050	76.3
		20.70	30.80	0.280	7.00	2.940	73.5
15	375	50.00	74.39	0.720	18.00	3.720	93.0
		40.00	59.51	0.524	13.10	3.524	88.1
		33.90	50.44	0.400	10.00	3.400	85.0

ESTIMATOR'S REFERENCE BOOK

Weights of Steel Angles

Size in inches	Lbs. per l.f.	Size in mm	Kg per m
1/2x1/2x1/8	0.38	12.5x12.5x3.1mm	0.57
5/8x5/8x1/8	0.48	15.6x15.6x3.1mm	0.71
3/4x3/4x1/8	0.59	18.8x18.8x3.1mm	0.88
3/4x3/4x3/16	0.84	18.8x18.8x4.7mm	1.25
7/8x7/8x1/8	0.70	21.9x21.9x3.1mm	1.04
1x5/8x1/8	0.64	25.0x15.6x3.1mm	0.95
1x1x1/8	0.80	25.0x25.0x3.1mm	1.19
1x1x3/16	1.16	25.0x25.0x4.7mm	1.73
1x1x1/4	1.49	25.0x25.0x6.3mm	2.22
1-1/8x1-1/8x1/8	0.91	28.1x28.1x3.1mm	1.35
1-1/4x1-1/4x1/8	1.01	31.3x31.3x3.1mm	1.50
1-1/4x1-1/4x3/16	1.48	31.3x31.3x4.7mm	2.20
1-1/4x1-1/4x1/2	1.92	31.3x31.3x12.5mm	2.86
1-3/8x7/8x1/8	0.91	34.4x21.9x3.1mm	1.35
1-1/2x1-1/2x-1/8	1.23	37.5x37.5x3.1mm	1.83
1-1/2x1-1/2x3/16	1.80	37.5x37.5x4.7mm	2.68
1-1/2x1-1/2x1/4	2.34	37.5x37.5x6.3mm	3.48
1-1/2x1-1/2x5/16	2.86	37.5x37.5x7.8mm	4.25
1-1/2x1-1/2x3/8	3.35	37.5x37.5x9.4mm	4.98
1-3/4x1-3/4x1/8	1.44	43.8x43.8x3.1mm	2.14
1-3/4x1-3/4x3/16	2.12	43.8x43.8x4.7mm	3.15
1-3/4x1-3/4x1/4	2.77	43.8x43.8x6.3mm	4.12
1-3/4x1-3/4x5/16	3.39	43.8x43.8x7.8mm	5.04
2x1-1/4x3/16	1.96	50.0x31.3x4.7mm	2.92
2x1-1/4x1/4	2.55	50.0x31.3x6.3mm	3.79
2x1-1/2x1/8	1.44	50.0x37.5x3.1mm	2.14
2x1-1/2x3/16	2.12	50.0x37.5x4.7mm	3.15
2x1-1/2x1/8	2.77	50.0x37.5x3.1mm	4.12
2x1-1/2x5/16	3.39	50.0x37.5x7.8mm	5.04
2x2x-1/8	1.65	50.0x50.0x3.1mm	2.45
2x2x3/16	2.44	50.0x50.0x4.7mm	3.63
2x2x1/4	3.19	50.0x50.0x6.3mm	4.75
2x2x5/16	3.92	50.0x50.0x7.8mm	5.83
2x2x3/8	3.70	50.0x50.0x9.4mm	5.50
2x2x1/2	6.00	50.0x50.0x12.5mm	8.92
2-1/4x1-1/2x3/16	2.28	56.3x37.5x4.7mm	3.39
2-1/4x2-1/4x1/8	1.86	56.3x56.3x3.1mm	2.77
2-1/4x2-1/4x3/16	2.75	56.3x56.3x4.7mm	4.09
2-1/4x2-1/4x1/4	3.62	56.3x56.3x6.3mm	5.38
2-1/4x2-1/4x5/16	4.50	56.3x56.3x7.8mm	6.69
2-1/2x1-1/2x3/16	2.44	62.5x37.5x4.7mm	3.63
2-1/2x1-1/2x1/4	3.19	62.5x37.5x6.3mm	4.75
2-1/2x2x3/16	2.75	62.5x50.0x4.7mm	4.09
2-1/2x2x1/4	3.62	62.5x50.0x6.3mm	5.38

Size in inches	Lbs. per l.f.	Size in mm	Kg per m
2-1/2x2x5/16	4.50	62.5x50.0x7.8mm	6.69
2-1/2x2x3/8	5.30	62.5x50.0x9.4mm	7.88
2-1/2x2x1/2	6.74	62.5x50.0x12.5mm	10.03
2-1/2x2-1/2x1/8	2.08	62.5x62.5x3.1mm	3.09
2-1/2x2-1/2x3/16	3.07	62.5x62.5x4.7mm	4.57
2-1/2x2-1/2x1/4	4.10	62.5x62.5x6.3mm	6.10
2-1/2x2-1/2x5/16	5.00	62.5x62.5x7.8mm	7.44
2-1/2x2-1/2x3/8	5.90	62.5x62.5x9.4mm	8.78
2-1/2x2-1/2x1/2	7.70	62.5x62.5x12.5mm	11.45
2x2x3/16	3.07	50.0x50.0x4.7mm	4.57
3x3x1/8	2.50	75.0x75.0x3.1mm	3.72
3x3x<3/16	3.71	75.0x75.0x4.7mm	5.52

Weights of Structural Steel Angles

Size in inches	Lbs. per ft.	Size in mm	Kg per m
3x2x1/4	4.1	75x50.0x6.3mm	6.1
3x2x5/16	5.0	75x50.0x7.8mm	7.4
3x2x3/8	5.9	75x50.0x9.4mm	8.8
3x2-1/2x1/4	4.5	75x62.5x6.3mm	6.7
3x2-1/2x5/16	5.6	75x62.5x7.8mm	8.3
3x2-1/2x3/8	6.6	75x62.5x9.4mm	9.8
3x2-1/2x1/2	8.5	75x62.5x12.5mm	12.6
3x3x1/4	4.9	75x75.0x6.3mm	7.3
3x3x5/16	6.1	75x75.0x7.8mm	9.1
3x3x3/8	7.2	75x75.0x9.4mm	10.7
3x3x7/16	8.3	75x75.0x10.9mm	12.3
3x3x1/2	9.4	75x75.0x12.5mm	14.0
3x3x5/8	11.5	75x75.0x15.6mm	17.1
3-1/2x2-1/2x1/4	4.9	88x62.5x6.3mm	7.3
3-1/2x2-1/2x5/16	6.1	88x62.5x7.8mm	9.1
3-1/2x2-1/2x3/8	7.2	88x62.5x9.4mm	10.7
3-1/2x2-1/2x1/2	9.4	88x62.5x12.5mm	14.0
3-1/2x2-1/2x5/8	11.5	88x62.5x15.6mm	17.1
3-1/2x3x1/4	5.4	88x75.0x6.3mm	8.0
3-1/2x3x5/16	6.6	88x75.0x7.8mm	9.8
4x3x5/8	13.6	100x75.0x15.6mm	20.2
4x3x3/4	16.0	100x75.0x18.8mm	23.8
4x3-1/2x1/4	6.2	100x87.5x6.3mm	9.2
4x3-1/2x5/16	7.7	100x87.5x7.8mm	11.5
4x3-1/2x3/8	9.1	100x87.5x9.4mm	13.5

Size in inches	Lbs. per ft.	Size in mm	Kg per m
4x3-1/2x7/16	10.6	100x87.5x10.9mm	15.8
4x3-1/2x1/2	11.9	100x87.5x12.5mm	17.7
4x4x1/4	6.6	100x100.0x6.3mm	9.8
4x4x5/16	8.2	100x100.0x7.8mm	12.2
4x4x3/8	9.8	100x100.0x9.4mm	14.6
4x4x7/16	11.3	100x100.0x10.9mm	16.8
4x4x1/2	12.8	100x100.0x12.5mm	19.0
4x4x5/8	15.7	100x100.0x15.6mm	23.4
4x4x3/4	18.5	100x100.0x18.8mm	27.5
4-1/2x3x3/8	9.1	113x75.0x9.4mm	13.5
5x3x5/16	8.2	125x75.0x7.8mm	12.2
5x3x3/8	9.8	125x75.0x9.4mm	14.6
5x3x7/16	11.3	125x75.0x10.9mm	16.8
5x3x1/2	12.8	125x75.0x12.5mm	19.0
5x3x3/4	18.5	125x75.0x18.8mm	27.5
6x3-1/2x3/8	11.7	150x87.5x9.4mm	17.4
6x3-1/2x7/16	13.5	150x87.5x10.9mm	20.1
6x3-1/2x1/2	15.3	150x87.5x12.5mm	22.8
6x3-1/2x5/8	18.9	150x87.5x15.6mm	28.1
6x4x5/16	10.3	150x100.0x7.8mm	15.3
6x4x3/8	12.3	150x100.0x9.4mm	18.3
6x4x7/16	14.3	150x100.0x10.9mm	21.3
6x4x1/2	16.2	150x100.0x12.5mm	24.1
6x4x9/16	18.1	150x100.0x14.1mm	26.9
6x4x5/8	20.0	150x100.0x15.6mm	29.8
6x4x3/4	23.6	150x100.0x18.8mm	35.1
6x4x7/8	27.2	150x100.0x21.9mm	40.5
6x6x3/8	14.9	150x150.0x9.4mm	22.2
6x6x7/16	17.2	150x150.0x10.9mm	25.6
6x6x1/2	19.6	150x150.0x12.5mm	29.2
6x6x9/16	21.9	150x150.0x14.1mm	32.6
6x6x5/8	24.2	150x150.0x15.6mm	36.0
6x6x3/4	28.7	150x150.0x18.8mm	42.7
6x6x7/8	33.1	150x150.0x21.9mm	49.2
6x6x1	37.4	150x150.0x25.0mm	55.6
6x6x13/8	7.9	150x150.0x34.4mm	11.8
6x6x11/2	10.2	150x150.0x37.5mm	15.2
3-1/2x3-1/2x1/4	5.8	88x88.0x6.3mm	8.6
3-1/2x3-1/2x5/16	7.2	88x88.0x7.8mm	10.7
3-1/2x3-1/2x3/8	8.5	88x88.0x9.4mm	12.6
3-1/2x3-1/2x7/16	9.8	88x88.0x10.9mm	14.6

Size in inches	Lbs. per ft.	Size in mm	Kg per m
3-1/2x3-1/2x1/2	11.1	88x88.0x12.5mm	16.5
3-1/2x3-1/2x5/8	13.6	88x88.0x15.6mm	20.2
4x3x1/4	5.8	100x75.0x6.3mm	8.6
4x3x5/16	7.2	100x75.0x7.8mm	10.7
4x3x3/8	8.5	100x75.0x9.4mm	12.6
4x3x7/16	9.8	100x75.0x10.9mm	14.6
4x3x1/2	11.1	100x75.0x12.5mm	16.5
5x3-1/2x5/16	8.7	125x88.0x7.8mm	12.9
5x3-1/2x3/8	10.4	125x88.0x9.4mm	15.5
5x3-1/2x7/16	12.0	125x88.0x10.9mm	17.9
5x3-1/2x1/2	13.6	125x88.0x12.5mm	20.2
5x3-1/2x5/8	16.8	125x88.0x15.6mm	25.0
5x4x3/8	11.0	125x100.0x9.4mm	16.4
5x4x1/2	14.5	125x100.0x12.5mm	21.6
5x5x3/8	12.3	125x125.0x9.4mm	18.3
5x5x7/16	14.3	125x125.0x10.9mm	21.3
5x5x1/2	16.2	125x125.0x12.5mm	24.1
5x5x5/8	20.0	125x125.0x15.6mm	29.8
5x5x3/4	23.6	125x125.0x18.8mm	35.1
6x3-1/2x5/16	9.8	150x88.0x7.8mm	14.6
7x3-1/2x3/8	13.0	175x88.0x9.4mm	19.3
7x3-1/2x7/16	15.0	175x88.0x10.9mm	22.3
7x3-1/2x1/2	17.0	175x88.0x12.5mm	25.3
7x3-1/2x5/8	21.0	175x88.0x15.6mm	31.2
8x3-1/2x1/2	18.7	200x88.0x12.5mm	27.8
8x6x1/2	23.0	200x150.0x12.5mm	34.2
8x6x3/4	33.8	200x150.0x18.8mm	50.3
8x8x1/2	26.4	200x150.0x12.5mm	39.3
8x8x5/8	32.7	200x150.0x15.6mm	48.7
8x8x3/4	38.9	200x150.0x18.8mm	57.9
8x8x7/8	45.0	200x150.0x21.9mm	67.0
8x8x1	51.0	200x150.0x25.0mm	75.9
8x8x11/8	56.9	200x150.0x28.1mm	84.7

Weight of Steel Tees

Size flange by stem in inches	1/8	3/16	1/4	5/16	3/8	1/2
		Lbs. per lin. ft.				
		Thickness in inches				
3/4 x 3/4	0.60	--	--	--	--	--
7/8 x 7/8	0.73	--	--	--	--	--

Size flange by stem in inches	Lbs. per lin. ft. Thickness in inches					
	1/8	3/16	1/4	5/16	3/8	1/2
1 x 1	0.90	1.20	--	--	--	--
1-1/4 x 1-1/4	--	1.55	1.98	--	--	--
1-1/2 x 1-1/2	--	1.88	2.43	--	--	--
1-3/4 x 1-3/4	--	2.30	2.90	--	--	--
2 x 1-1/2	--	--	3.10	--	--	--
2 x 2	--	--	3.60	4.40	--	--
2-1/4 x 2-1/4	--	--	4.10	4.90	--	--
2-1/2 x 2-1/2	--	--	4.60	5.50	6.4	--
2-1/2 x 2-3/4	--	--	--	5.9	6.8	--
2-1/2 x 3	--	--	--	6.1	7.2	--
3 x 2-1/2	--	--	--	6.1	7.2	--
3 x 3	--	--	--	6.7	7.8	--
3-1/2 x 3	--	--	--	--	8.5	--
3x4	--	--	--	--	9.3	--
3 x 3-1/2	--	--	--	--	8.5	--
3-1/2 x 3-1/2	--	--	--	--	9.2	11.7
3-1/2 x 4	--	--	--	--	9.8	--
4 x 2-1/2	--	--	--	--	8.5	--
4 x 3	--	--	--	--	9.2	--
4 x 4	--	--	--	--	10.5	13.5
4 x 5	--	--	--	--	11.9	15.3
4-1/2 x 2-1/2	--	--	--	--	9.2	--
4-1/2 x 3	--	--	--	8.6	10.0	--
5 x 3	--	--	--	--	11.0	13.6
6-1/2 x 6-1/2	--	--	--	--	--	19.8

(metric)

Size flange by stem in mm	Kg per meter Thickness in mm					
	3.125	4.69	6.25	7.81	9.38	12.5
18.8 x 18.8	0.89	--	--	--	--	--
21.9 x 21.9	1.09	--	--	--	--	--
25.0 x 25.0	1.34	1.79	--	--	--	--
31.3 x 31.3	--	2.31	2.95	--	--	--
37.5 x 37.5	--	2.80	3.62	--	--	--
43.8 x 43.8	--	3.42	4.31	--	--	--
50.0 x 37.5	--	--	4.61	--	--	--
50.0 x 50.0	--	--	5.36	6.55	--	--
56.3 x 56.3	--	--	6.10	7.29	--	--
62.5 x 62.5	--	--	6.84	8.18	9.52	--
62.5 x 68.8	--	--	--	8.78	10.12	--
62.5 x 75.0	--	--	--	9.08	10.71	--
75.0 x 62.5	--	--	--	9.08	10.71	--
75.0 x 75.0	--	--	--	9.97	11.60	--

Size flange by stem in mm	Kg per meter Thickness in mm					
	3.125	4.69	6.25	7.81	9.38	12.5
87.5 x 75.0	--	--	--	--	12.65	--
75.0 x 100.0	--	--	--	--	13.84	--
75.0 x 87.5	--	--	--	--	12.65	--
87.5 x 87.5	--	--	--	--	13.69	17.41
87.5 x 100.0	--	--	--	--	14.58	--
100.0 x 62.5	--	--	--	--	12.65	--
100.0 x 75.0	--	--	--	--	13.69	--
100.0 x 100.0	--	--	--	--	15.62	20.09
100.0 x 125.0	--	--	--	--	17.70	22.76
112.5 x 62.5	--	--	--	--	13.69	--
112.5 x 75.0	--	--	--	12.80	14.88	--
125.0 x 75.0	--	--	--	--	16.37	20.23
162.5 x 162.5	--	--	--	--	--	29.46

Weights of Standard Diamond Steel Floor Plates

Thickness, inches	Lbs. per sq. ft.	Thickness, mm	Kg per sq. m
1/8	8.00	3.13	39.05
3/16	8.75	4.69	42.71
1/4	11.25	6.25	54.91
5/16	13.75	7.81	67.11
3/8	16.25	9.38	79.31
1/2	21.50	12.50	104.94

Weights of Steel M Shapes

Size	Lbs. per lin. ft.	Size (metric)	Kg per meter
M4 x 13	13.00	M 100 x 5.89	19.34
M5 x 18.9	18.90	M 125 x 8.57	28.12
M6 x 4.4	4.40	M 150 x 2	6.55
M6 x 20	20.00	M 150 x 9.07	29.76
M8 x 6.5	6.50	M 200 x 2.94	9.67
M10 x 9	9.00	M 250 x 4.08	13.39
M12 x 11.8	11.80	M 300 x 5.35	17.56
M14 x 18	18.00	M 350 x 8.16	26.78

Weights of Round Head Structural Rivets

Dia. x Length inches	Lbs. per 100 rivets	Dia. x Length mm	Kg per 100 rivets
1/2 x 1	9.49	12.5 x 25.0 mm	4.30
1/2 x 1-1/2	12.24	12.5 x 37.5 mm	5.55
1/2 x 2	14.99	12.5 x 50.0 mm	6.80
1/2 x 2-1/2	17.74	12.5 x 62.5 mm	8.05
1/2 x 3	20.49	12.5 x 75.0 mm	9.29
1/2 x 3-1/2	23.24	12.5 x 87.5 mm	10.54
1/2 x 4	25.99	12.5 x 100.0 mm	11.79
5/8 x 1	16.38	15.6 x 25.0 mm	7.43
5/8 x 1-1/2	20.68	15.6 x 37.5 mm	9.38
5/8 x 2	24.98	15.6 x 50.0 mm	11.33
5/8 x 2-1/2	29.28	15.6 x 62.5 mm	13.28
5/8 x 3	33.58	15.6 x 75.0 mm	15.23
5/8 x 3-1/2	37.88	15.6 x 87.5 mm	17.18
5/8 x 4	42.18	15.6 x 100.0 mm	19.13
5/8 x 4-1/2	46.76	15.6 x 112.5 mm	21.21
5/8 x 5	51.10	15.6 x 125.0 mm	23.18
5/8 x 5-1/2	55.45	15.6 x 137.5 mm	25.15
5/8 x 6	59.80	15.6 x 150.0 mm	27.13
3/4 x 1	25.81	18.8 x 25.0 mm	11.71
3/4 x 1-1/2	31.99	18.8 x 37.5 mm	14.51
3/4 x 2	38.17	18.8 x 50.0 mm	17.31
3/4 x 2-1/2	44.35	18.8 x 62.5 mm	20.12
3/4 x 3	50.53	18.8 x 75.0 mm	22.92
3/4 x 3-1/2	56.71	18.8 x 87.5 mm	25.72
3/4 x 4	62.89	18.8 x 100.0 mm	28.53
3/4 x 4-1/2	69.07	18.8 x 112.5 mm	31.33
3/4 x 5	75.25	18.8 x 125.0 mm	34.13
3/4 x 5-1/2	82.05	18.8 x 137.5 mm	37.22
3/4 x 6	87.61	18.8 x 150.0 mm	39.74
3/4 x 6-1/2	93.80	18.8 x 162.5 mm	42.55
3/4 x 7	99.97	18.8 x 175.0 mm	45.35
3/4 x 7-1/2	106.15	18.8 x 187.5 mm	48.15
3/4 x 8	112.33	18.8 x 200.0 mm	50.95
7/8 x 1-1/2	46.67	21.9 x 37.5 mm	21.17
7/8 x 2	55.10	21.9 x 50.0 mm	24.99
7/8 x 2-1/2	63.53	21.9 x 62.5 mm	28.82
7/8 x 3	71.96	21.9 x 75.0 mm	32.64
7/8 x 3-1/2	80.39	21.9 x 87.5 mm	36.46
7/8 x 4	88.82	21.9 x 100.0 mm	40.29
7/8 x 4-1/2	97.25	21.9 x 112.5 mm	44.11

Dia. x Length inches	Lbs. per 100 rivets	Dia. x Length mm	Kg per 100 rivets
7/8 x 5	105.68	21.9 x 125.0 mm	47.94
7/8 x 5-1/2	114.11	21.9 x 137.5 mm	51.76
7/8 x 6	122.54	21.9 x 150.0 mm	55.58
7/8 x 6-1/2	130.97	21.9 x 162.5 mm	59.41
7/8 x 7	139.40	21.9 x 175.0 mm	63.23
7/8 x 7-1/2	147.83	21.9 x 187.5 mm	67.06
7/8 x 8	156.26	21.9 x 200.0 mm	70.88
1 x 2-1/2	86.82	25.0 x 62.5 mm	39.38
1 x 3	97.80	25.0 x 75.0 mm	44.36
1 x 3-1/2	108.79	25.0 x 87.5 mm	49.35
1 x 4	119.77	25.0 x 100.0 mm	54.33
1 x 4-1/2	130.76	25.0 x 112.5 mm	59.31
1 x 5	141.74	25.0 x 125.0 mm	64.29
1 x 5-1/2	152.73	25.0 x 137.5 mm	69.28
1 x 6	163.71	25.0 x 150.0 mm	74.26

Weights of Steel Plates in Lbs. per Sq. Ft

Thickness inches	Lbs per sq. ft.	Thickness mm	Kg per sq.m
3/16	7.65	4.69	37.48
1/4	10.20	6.25	49.78
5/16	12.75	7.81	62.23
3/8	15.30	9.38	74.68
7/16	17.85	10.94	87.12
1/2	20.40	12.50	99.57
5/8	25.50	15.63	124.46
3/4	30.60	18.75	149.35
7/8	35.70	21.88	174.24
1	40.80	25.00	199.13
1-1/4	51.00	31.25	248.92
1-1/2	61.20	37.50	298.70
1-3/4	71.40	43.75	348.48
2	81.60	50.00	398.27

Weight of Corrugated Sheets, Lbs per 100 S.F. (Kg per 10 Sq.M)

Gauge Nos.	Black and Painted 2,2-1/2, 50,62.5 & 3" & 75mm		1-1/4" corrug.	31.25mm corrug.	Galvanized 2,2-1/2, 50,62.5 & 3" & 75mm		1-1/4" corrug.	31.25mm corrug.
16	271	133	--	--	286	141	--	--
18	217	107	--	--	232	114	--	--
20	163	80	170	84	178	87	185	91
22	136	67	142	70	151	74	157	77

Gauge Nos.	Black and Painted				Galvanized			
	2,2-1/2, & 3"	50,62.5 & 75mm	1-1/4" corrug.	31.25mm corrug.	2,2-1/2, & 3"	50,62.5 & 75mm	1-1/4" corrug.	31.25mm corrug.
24	110	54	114	56	125	61	129	63
26	83	41	86	42	98	48	101	50
27	76	37	79	39	91	45	94	46
28	68	33	72	35	85	42	87	43

JUNIOR STEEL BEAMS

Junior beams (M shapes) are lightweight, hot-rolled structural beams that are used as secondary floor and roof beams in schools, stores, apartments, hospitals, and other types of light occupancy buildings. They are well adapted for purlins in mill buildings.

They are rolled to ASTM specifications, and are made in 6", 8", 10", 12", and 14" (150, 200, 250, 300, and 350 mm) sizes. Sizes and properties are given below:

Designation	Depth, inches	Lbs. per lin. ft.	Web Thick., inches	Flange Wdth, inches	Area, sq. in.
M6 X 4.4	6	4.4	.114	1.844	1.29
M6 X 20	6	20.0	.250	5.938	5.89
M8 X 6.5	8	6.5	.135	2.281	1.92
M10 X 9	10	9.0	.157	2.690	2.65
M12 X 11.8	12	11.8	.177	3.065	3.47
M14 X 18	14	18.0	.215	4.000	5.10

(metric)

Designation	Depth, mm	Kg per m	Web Thick., mm	Flange Wdth, mm	Area, sq.mm
M150 x 2	150	2.00	2.850	46.10	32.25
M150 x 9	150	9.07	6.250	148.45	147.25
M200 x 3	200	2.94	3.375	57.03	48.00
M250 x 4	250	4.08	3.925	67.25	66.25
M300 x 5.3	300	5.35	4.425	76.63	86.75
M350 x 8	350	8.16	5.375	100.00	127.50

The cost of cutting, punching holes, and coping junior beams will run approximately as given below, but when much fabricating is required, it is advisable to refer plans to a local fabricator or warehouse for a sub bid.

Top cope, per end	$4.00
Bottom cope and seat angles, per end	10.00
Cutting end on diagonal, per end	5.00
Anchor holes in web, per hole	1.00
Conduit holes in web, per hole	1.00

Junior steel beams can be used in conjunction with monolithically finished concrete floors or precast gypsum slabs laid over the top flange of the junior beams. By encasing the top flanges in concrete, the floor is made rigid and does not require bridging in light occupancy buildings.

Where the slab is poured on metal lath or where gypsum plank is used, adjustable steel rigid bridging is furnished.

When a poured-in-place concrete floor is used, the forms are easily constructed by cutting short pieces of lumber to rest on the lower flanges of the beams, which in turn support the floor sheathing.

The reinforcement of the concrete slab usually consists of welded wire fabric, the size and spacing of the wire depending upon the span of the slabs and the loads to be carried.

Where joists rest on walls or bearing partitions, anchors are provided. Where the tops of junior beams are flush with tops of supporting beams, either clip angles or shelf angles are provided. Where the tops of beams are 2-1/2" (62 mm) above the tops of supporting beams, part of the beam is coped out and an angle seat is welded to the junior beam, which rests on the top flange of the supporting beam.

*Method of Constructing Wood Forms for Concrete Floors
Supported by Junior Beams*

050200 METAL JOISTS

Standard and longspan steel joists were developed primarily to take the place of wood construction. The fire-resistance and structural qualities of this type of floor and roof construction has meant that it has not only replaced wood but other types of fire-resistant floor construction as well. Steel joist construction is recognized and accepted by all building codes.

Standard open web and longspan steel joists are manufactured by welding chord members of hot rolled structural or cold formed sections to round bar or angle web members to form a truss. Information on the types of steel joists can be obtained from the Steel Joist Institute and the joist manufacturers.

When steel joists are used in fireproof construction, it is necessary to place a 2" to 2-1/2" (50-62 mm) reinforced concrete slab over the joists, and a fire resistant ceiling of gypsum board or plaster beneath the joists to obtain the required fire rating.

Typical Shortspan Steel Joist

Where a wood floor is required, wooden nailing screeds supported on screed clips are embedded in the concrete slab and the wood flooring is then nailed to the screeds.

Steel joists may also be used in semi-fireproof construction by nailing the wood sheathing directly to nailers attached to the joists without the use of a protective concrete slab. Some manufacturers of steel joists have met the demand for this type of construction by fabricating a nailer joist with the wood nailer attached in the shops by the use of lag screws or bolts.

Open web joists may be used to great advantage in all types of light occupancy buildings. The open web construction greatly facilitates the placing of plumbing, conduits, etc., making unnecessary the need of a furred ceiling or heavy fills on top of the structural slab.

Steel joists are also used in industrial construction to replace structural steel roof purlins with a resulting economy of materials. Metal roof deck is then usually welded to the joists and lightweight insulating material either poured or rigid board applied to the metal deck, and the roofing is installed.

Estimating Steel Joists. Steel joists and related products are generally sold by the manufacturer directly to the contractor. When preparing general contract estimates the most satisfactory method is to send plans or sketches to the manufacturer or his agent thus assuring a sub-bid based on an economical layout and design and one which includes the necessary accessories.

In its quotation the manufacturer will note the tonnage of joists, information which enables the contractor to figure the necessary handling and labor of erection. The cost of joist erection, including placing of accessories, varies from $150.00 to $250.00 per ton (metric ton), excluding material costs, depending on labor conditions, location, and type of building.

Labor Setting Steel Joists. The labor cost of handling and setting steel joists is usually estimated by the ton, based on the total weight of the joists to be placed.

This cost will vary considerably, depending upon the type of building, size of joists, length of spans, amount of handling and hoisting necessary, etc.

It is much cheaper to set joists on long straight spans and large floor areas, such as garage and factory buildings, office buildings, etc., than on small or other irregularly constructed buildings.

All steel joists must be bridged and connected according to the Steel Joist Institute standards.

On an ordinary job of steel floor joists of regular construction and fairly long spans, five workers together should handle and place about 4-1/2 to 5 tons (4.05-4.5 metric tons) of joists, including bridging, per 8-hr. day at the following labor cost per ton:

	Hours	Rate	Total	Rate	Total
Ironworker	40	$....	$....	$30.28	$1,211.20
Cost per ton			....		254.98
per metric ton			....		280.48

On jobs of irregular construction, a five-worker crew together should handle and place about 3-1/2 to 4 tons (3.15-3.60 metric tons) of joists, including bridging, per 8-hr. day, at the following labor cost per ton:

	Hours	Rate	Total	Rate	Total
Ironworker	40	$....	$....	$30.28	$1,211.20
Cost per ton			....		322.99
per metric ton			....		355.29

Prices of Steel Joists. Costs of standard steel joists vary, depending on several factors, such as size of job, the number of joists of one particular size and length, and the job location. For an average job the cost of standard joists will be in the range of $600.00 to $700.00 per ton (0.9 metric ton), which would include nominal bridging members and other accessories.

Costs of longspan steel joists will vary the same as those for standard joists. For an average job the cost of the longspan joists will range from $580.00 to $750.00 per ton (metric ton) including accessories.

05300 METAL DECKING

Steel roof deck is formed from steel sheets in 18, 20, or 22 gauge. Most commercial decks have a standardized cross section with longitudinal ribs spaced 6" (150 mm) on centers. Sections are 30" (750 mm) in width and are manufactured in various lengths to suit job conditions, usually in the 14- to 31-ft. (4-9 m) range, although some can be supplied in greater lengths.

Deck sections are usually 1-1/2" (37 mm) deep, but some manufacturers supply a 1-3/4", 2", or 2-1/2" (44, 50, or 62 mm) deep deck. Sections have interlocking or nesting side laps and telescoping or nesting end laps.

Under ordinary roof live loads, 30 to 40 lbs. per s.f. (146-195 kg/m²), 20-gauge deck is normally used for spans from 6'-0" to 7'-6" (1.8-2.2 m) centers of purlins and 18 gauge deck for spans from 7'-0" to 8'-0" (2.1-2.4 m). Specific information on allowable loads can be obtained from the various manufacturers. By extending a single sheet over two or more purlin spaces, structural continuity, increasing load carrying capacity, can be obtained. Dead load of deck, insulation, and built-up roofing is approximately 7 lbs. per s.f. (34 kg/m²).

Approximate cost of steel roof deck, delivered to job site in the eastern or northern states, on an average size job, 10,000 s.f. (929 sq.m) or more, is as follows:

Gauge of Roof Deck	Price per s.f. of 1-1/2" (37mm) Deck
22 Ga.	$0.75
20 Ga.	0.85
18 Ga.	1.10

Placing Metal Deck on Top of Steel Joists. When metal deck is placed over the top of steel joists to receive the concrete slab or lightweight insulating materials, the metal deck is usually spot welded to the joists.

On work of this type two workers together should place and weld about 200 s.f. (18 sq.m) of deck per hour at the following labor cost per square of 100 s.f. (10 sq.m):

	Hours	Rate	Total	Rate	Total
Iron Worker	1.0	$....	$....	$30.28	$30.28
Cost per s.f.			....		0.30
per sq.m			....		3.26

Steel roof deck can be erected in almost any kind of weather, permitting other trades to work under cover much more rapidly. Deck must be welded to steel support members in accordance with the Steel Deck Institute recommendations. Two tack welds per 18" (450 mm) width of deck section at end laps and one weld on the outside rib on intermediate supports are recommended.

Longspan Deck

Longspan deck is formed from a single 14, 16, 18, or 20 gauge steel sheet into 4-1/2", 6", or 7-1/2" (112.5, 150, or 187.5 mm) deep pans 12" (300 mm) wide. These are rolled sections with integral stiffening ribs formed into the top flange. Interlocking longitudinal side laps are provided on opposite edges for positive side lap attachment and are fastened by welding. Maximum spans allowable vary from 20'-0" to 30'-0" (6.0-9.0 m).

Longspan deck is easily erected--on an average job, a 5-worker crew should erect 2,200 to 2,400 s.f. (204-223 sq.m) per 8-hr. day.

Approximate cost of longspan steel roof deck, delivered to job sites in the eastern or northern states, is $2.00 to $3.50 per s.f. (0.09 sq.m), depending upon gauge and length of sections.

Reinforcing Floor Forms

Steel deck erected in an inverted position (with ribs up) can be used as a reinforcing form for concrete construction. It acts as a form to support the concrete and permits the ribs to act as reinforcing.

Steel deck reinforcing forms can be erected for an entire structure as soon as the steel framework is placed, providing an immediate working platform and protective staging for all trades. Concrete can be placed at any portion of the building without regard for removing, cleaning, and re-setting temporary formwork. Steel deck forms usually provide all necessary reinforcing to satisfy flexure requirements--the only additional reinforcing ordinarily required is temperature mesh to minimize shrinkage cracking. Tight form joints prevent concrete from dripping to lower floors, saving clean-up time.

Steel deck reinforcing forms can also be used in conjunction with composite beam design. Ample area for concrete around stud shear connectors and between deck ribs permits full effectiveness of the connectors. Standard AISC Composite Design Procedure may be followed using the total slab depth in beam property calculations.

Reinforcing floor forms are installed by welding, the same as ordinary steel roof deck, and costs of materials and erection are about the same as given for "Metal Decking".

CELLULAR STEEL DECK AND FLOOR PANELS

Lightgauge cellular steel structural panels are used for longspan structural floors and roofs. Some panels are composed of two identically formed beam sections and a flat plate. Others are formed from two ribbed sections and a flat plate. All panels have interlocking side joints. Components are factory assembled by spot welding. Panels are manufactured in gauges from 20 to 13 and are normally available in 24" (600 mm) widths. Some panels are fabricated in 12" (300 mm) widths. Lengths are available up to 40'-0" (12 m), while depths vary from 1-1/2" to 7-1/2" (37-187 mm).

This type of panel is easily handled and erected, because it is deck and joist combined. Concrete floor forms are eliminated and panels provide a working platform for other trades.

A flat, plate-down installation is often used to provide an attractive ceiling which needs only a finish coat of paint to complete the job. Voids between beam sections may be filled and covered with 2<$E1/2>" (62.5 mm) of concrete for floors or covered with rigid board insulation and built-up roofing for roofs. Installations with beam section forming the ceiling can be merely painted to give a finished fluted ceiling, or a suspended ceiling can be erected to provide a flush type ceiling.

Where panels are covered with a concrete fill, the use of temperature mesh to reduce shrinkage cracks is recommended.

The cost of cellular panels, delivered to job site, ranges from $3.00 per s.f. (0.09 sq.m) for 20 ga. 2" to 3" (50-75 mm) thick panels, $3.25 to $4.00 for 3" to 4" (75-100 mm) 18 ga. units on up to $4.25 for 16 ga. deck.

Cellular Steel Deck and Floor Panel

A crew of four structural iron workers plus a foreman should install around 1600 s.f. (148 sq.m) of light deck per 8-hr. day, 1200 s.f. (111 sq.m) of 18 ga. deck, and around 1000 s.f. (92 sq.m) of the heavy gauge deck under normal job conditions. For example, 100 s.f. (30 sq.m) of 18 ga. labor costs would be as follows:

	Hours	Rate	Total	Rate	Total
Foreman	0.65	$....	$....	$31.20	$ 20.28
Str'l Ironworkers	2.60	....	....	30.28	78.73
Cost per 100 s.f.					$ 99.01
per s.f.					0.99
per sq.m					10.65

05400 LIGHTGAGE FRAMING

The components involved are a complete range of studs, joists, and accessories for the steel framing of buildings. Sections are fabricated from structural-grade, high tensile strip steel by cold forming and are designed specifically for strength, light weight and low cost. Yet, structural framing carries all the benefits of conventional steel framing.

A nailing groove, developed for easy, economical attachment of other materials, is a feature of all double studs and joists. It is obtained by welding two cold-formed steel elements together, so designed that a nail driven into this space is not only held by friction, but is also deformed to provide maximum holding power.

The usual limitations imposed by prefabrication are avoided. Using nailable framing sections architects and engineers have unrestricted freedom in design. All sections are painted at the plant with a coat of oven-dried, rust-resisting red zinc chromate paint, or are galvanized.

Lightgage framing systems can supply complete wall, floor, and roof construction for buildings up to four stories in height, or can be used in combination with other framing systems for interior, load bearing partitions, exterior curtain walls, fire separation walls, parapets, penthouses, trusses, suspended ceilings, and mansard roofs.

Advantages of using these systems include reduced dead loads, uniform fabrication, less on-the-job storage space, no warping, shrinkage or swelling, incombustibility, punched slots for passage of other trades, and secure nailing to eliminate nail popping. When properly engineered and applied, this system may be cheaper than conventional wood or steel construction. But even when comparable in price, the advantages offered to other trades may yield savings on the overall job.

The components can be completely detailed, cut, and assembled in the shop and delivered for erection to the job site. Where on-the-job cutting and assembly is preferred, the material can be cut with a radial saw fitted with a 1/8" highspeed circular blade. Fastening can be by bolts, sheet metal screws, or welding.

Joists come in 6", 8", 10", and 12" (150, 200, 250, and 300 mm) depths and in 12, 14, 16, and 18 gauge material. There are generally three styles: double nailable,"C" joists with 1-5/8" (40.75 mm) flanges and"C" joists with 2-1/2" (62 mm) flanges. Lengths from 6' to 40' (1.8-12 m) are available. Joist webs may be solid, selectively punched, or continuously punched for maximum raceway flexibility. Unnecessary punching should be avoided as the slotting lowers the structural value of the joist.

Joist bridging, which may be by stock "V" units or solid channels, must be supplied in the center of all spans up to 14' (4.2 m); at third points on spans from 14' to 20' (4.2-6.0 m); at quarter points on spans from 26' to 32' (7.2-9.7 m); and at 8' (2.4 m) centers on all spans over 32' (9.7 m).

Studs are available in 2-1/2", 3-5/8", 4", and 6" (62, 90, 100, and 150 mm) depths and in 14, 16, 18, and 20 gauges. Also available are 25 gauge studs, but they are suitable for non-loadbearing partitions only and are discussed under the drywall section of Chapter 14 "Finishes". Lengths from 7' to 40' (2.1-12.1 m) are available with either slotted or solid webs. Studs may be of the double-nailable type, channels, or "C" type with knurled flanges.

Studs mount into standard tracks at floors, ceilings, sills, and fascias. They should be stiffened with lateral bridging at the midpoint of all walls to 10' (3 m) in height; at third points on walls from 10' to 14' (3.0-4.2 m); and every 4' (1.2 m) on walls above 14' (4.2 m). Stock "V" bars, channels or special clipping systems may be ordered for this.

2½" Φ Holes

|← 6" →|← 12" →|← 12" →|

12", 10",   8" and 6" NAILABLE JOISTS

1¼" Φ Holes

|← 3" →|← 6" →|← 6" →|← 6" →|← 6" →|← 6" →|

D = 4"

4" NAILABLE DOUBLE STUDS

1¼" Φ Holes

|← 3" →|← 6" →|← 6" →|← 6" →|← 6" →|← 6" →|

D = 3⅝"

3⅝" NAILABLE DOUBLE STUDS

¼" Φ Holes |← Joist Spacing →| 1"

|←————— 8'2", 9'2" and 10'2" over all —————→|

90°

V-BAR BRIDGING*
(Weight = 0.26 lbs. ft.; gage = 16; t = 0.0568"; net area = 0.075 sq. in.)

UNPUNCHED CHANNEL AND TRACK

⅜" Φ Holes

|← 3" →|← 6" →|← 6" →|← 6" →|← 6" →|

D = 2½"

2½" NAILABLE DOUBLE STUDS

6″ AND 8″ CEE STUDS

4″, 3 5/8″, 3 1/4″ SCREW CEE STUDS

2 1/2″ SCREW CEE STUDS

Dimension A: 0.4″ for all gages.

6″ PUNCHED SCREW STUDS AND CHANNEL STUDS

4″, 3 5/8″, 3 1/4″ PUNCHED SCREW STUDS AND CHANNEL STUDS

2 1/2″ PUNCHED SCREW STUDS AND CHANNEL STUDS

Solid Bridging

"V" Bridging

2 Rows V-Bridging

Cee Stud

Weld Bridging to Stud both sides

"V" Bridging of Studs

Channel Bridging of Studs

Costs will vary because of the various options available. Each job should be priced with the supplier when all the design criteria are known. For general budget purposes joists can be figured at around $0.75 per lb. galvanized. A punched, galvanized "C" joist will vary from 2.24 lbs. per ft. (0.3 m), 6" (150 mm) deep, to 5.85 lbs. (2.65 kg), 12" (300 mm) deep or from around $1.70 to $4.40 per ft (0.3 m). Carload lots will run less.

Studs 3-5/8" (90 mm) deep, punched and galvanized, will run $0.50 per ft. (0.3 m) in 20 gauge material in less than carload lots; $0.75 per ft. (0.3 m) in 18 gauge; and $0.85 per ft. (0.3 m) in 16 gauge. Tracks will approximate stud costs.

Attachment of Other Materials. Steel or aluminum roof decking, metal siding, or corrugated sheets can be quickly and economically attached to sections with self-tapping screws, or by other conventional methods.

When self-tapping screws are used, a hole slightly smaller than the screw
is drilled through the material and the lightgage section. Then, the screw can
be placed and tightened by one person. This is faster and more economical
than most conventional methods of attachment which require two workers.
The 5/8" (15 mm) hex-head self-tapping, cadmium-plated screws are
recommended.

Attachment of materials is generally with standard nails when the
nailable sections are used. When channel studs or other non-nailable sections
are used, the self-drilling screws are recommended for metal siding or
plywood. When drywall is applied to non-nailable sections, self-drilling
screws can be ordered from the drywall manufacturer.

Other types of materials can be readily attached to framing members by
the use of bolts, clips, welds, or other methods consistent with the finish
desired.

05500 METAL FABRICATIONS

Metal Stairs. Metal stairs are usually figured at so much per riser with
stringers, treads, nosings, and railings included. A standard, 3-ft. (0.9-m)
wide simple run metal pan stair can be budgeted at around $85.00 per riser,
material cost. For each additional foot of width up to 5 ft. (1.5 m), add 10%.
Custom stairs will run from $95.00 per riser to twice that amount where any
of the stair components vary from the norm. Simple landings cost about
$8.50 per sq. ft. (0.09 sq.m) without an allowance for railings.

A four-worker crew should erect around 40 to 50 risers or 150 s.f. (14 sq.m)
of landing per 8 hr. day.

The labor erection cost of a three foot wide steel stair consisting of three
flights of sixteen risers and one 3' x 6' (0.9 x 1.8 m) landing each run would
figure as follows:

	Hours	Rate	Total	Rate	Total
Ironworkers					
Stairs - 48 risers	34.0	$....	$....	$30.28	$1,029.52
Landing 54 s.f. (5.0 sq.m)	11.1	....	....	30.28	336.11
Total Cost					$1,365.63
Cost per stair run (3)					455.21
Cost per riser incl.landings					28.45

Pipe Railings. When estimating plain pipe railing, 3'-6" (1.06 m) high,
consisting of two horizontal runs of pipe with uprights 6' to 8' (1.8-2.4 m) on
center, figure as follows for the various sizes, erected in place:

Description	Price per lin. ft.
1-1/4" (31 mm) pipe railing complete as described above per l.f. (0.3 m)	$40.00

1-1/2" (37.5 mm) pipe railing complete as described above per l.f. (0.3 m) . $45.00
2" (50 mm) pipe railing complete as described above per l.f. (0.3 m) $60.00

For each curved section or termination point that is formed to radius, add $35.00 to $45.00.

For following the rake of a stair, add $8.50 per upright.

For each foot (0.3 m) of wall railing, add $8.00.

Aluminum pipe railings can also be ordered from stock and will cost around $65.00 per ft. (0.3 m), anodized. Aluminum wall rails will cost around $9.50 per ft. (0.3 m). A two-worker crew should erect some 100 l.f. (30 m) of straight-run railing per day and 130 ft. (39 m) of wall railing.

Steel Ladders. Straight steel ladders cost from $40.00 to $60.00 per l.f. (0.3 m) erected in place. Add $30.00 to $45.00 for curved handles and platform over coping walls. Add $40.00 per l.f. (0.3 m) if cage is required.

Steel Gratings. Gratings may be welded or of expanded construction and made of steel, aluminum, or stainless steel.

Welded steel gratings weighing 4 lbs run about $4.50 per s.f. (0.09 sq.m); weighing 5 lbs, $4.50; 9 lbs, $6.50; 12 lbs, $8.50; and 16 lbs, $9.25 to $10.25, all depending on the spacing and size of the bars. If the material is galvanized add $0.90 per s.f. (0.09 sq.m).

Aluminum grating will run from $11.00 per s.f. (0.09 sq.m), 3/4" (18 mm) thick, and weighing 1.5 lbs., to $25.00 per s.f. (0.09 sq.m), 2-1/4" (56 mm) thick, and weighing 5.9 lbs in amounts of 75 s.f. (7 sq.m) or more.

Checkered plate is sometimes specified in place of or in combination with gratings and can be figured at about $9.50 per s.f. (0.09 sq.m) for 1/4" (6.2 mm) thickness.

A typical crew laying grating will consist of four workers, who can install about 100 s.f. (9 sq.m) per hour at the following cost per 100 s.f. (9 sq.m):

	Hours	Rate	Total	Rate	Total
Ironworkers	4.0	$....	$....	$30.28	$ 121.12
Cost per s.f.					1.21
per sq.m					13.03

Steel Window Guards. Window guards made with 3/8" x 2" (9 x 50 mm) horizontal bars, top and bottom and round or square upright bars spaced 4" (100 mm) on centers. Delivered to job.

	Approx. Price per s.f.	Approx. Price per sq.m
Window guards with 1/2" (12.5 mm) round upright bars	$7.50	$80.73
Window guards with 1/2" (12.5 mm) square upright bars	8.00	86.11
Window guards with 5/8" (15.6 mm) round upright bars	9.50	102.26
Window guards with 5/8" (15.6 mm) square upright bars	10.00	107.64
Window guards with 3/4" (18.7 mm) round upright bars	12.00	129.17
Window guards with 3/4" (18.7 mm) square upright bars	14.00	150.70

Do not figure any opening as containing less than 15 s.f. (1.4 sq.m)

Window Guards Made of Crimped Diamond Mesh. Window guards, made with 1" channel iron frames and crimped diamond mesh wire, per s.f. (sq.m) erected in place cost $12.00 to $20.00. Do not figure any opening as containing less than 20 s.f. (1.8 sq.m).

Curb Angles, Floor Frames, Trench Frames, and Covers. For miscellaneous items, such as curb angles, floor frames, trench frames, and covers, figure $1.50 per lb. (0.45 kg) delivered to job.

Ornamental Iron Entrance Rails. Fabricated from 1-1/4" x 1/4" or 1-1/4" x 3/8" (31 x 6.2 or 31 x 9.3 mm) bar stock, or 1-1/2" (37.5 mm) bar size channels for rails and 1/2" (12.5 mm) square bar pickets, stock designs for this item, delivered to job, cost about $50.00 per l.f. (0.3 m) of rail, depending on height and amount of ornamentation. For rails following the rake of stairs, add 50 %. Rails made to special design may cost 2 to 3 times the above prices.

Ornamental Porch Columns. Stock design are scrolled, wrought iron columns, with 3/4" x 3/4" (18.7 x 18.7 mm) bar stock, or 15/16" (23.4 mm) square tubing frames and 3/8" x 1/4" (9.3 x 6.2 mm) or 1/2" x 3/16" (12.5 x 4.6 mm) bar or strip stock scrolls. Flat columns, 8-1/2" (212.5 mm) to 12" (300 mm) wide x 8'-0" (2.4 m) high, cost $75.00 to $150.00 each. Corner columns to match flat columns cost $70.00 to $100.00 each.

Stock design columns, with 3/4" x 3/4" (18.7 x 18.7 mm) bar stock frames and cast iron panels, delivered to job, cost as follows: flat columns, 9-1/2" to 11" (237-275 mm) wide x 8'-0" (2.4 m) high are $75.00 to $130.00 each; corner columns to match flat columns are $100.00 to $140.00 each. Brackets to match columns are $35.00 each.

Steel Stacks

Steel stack costs vary widely and are dependent on height, diameter, gauge of material, number of joints, transportation, etc. Steel stacks are usually fabricated by companies specializing in steel boiler and tank manufacturing.

When estimating a job that has a steel stack, always obtain a firm figure from a steel stack fabricator before submitting a bid on the work.

Labor Erecting Steel Stacks. When estimating the labor cost of erecting steel stacks for office buildings or other buildings of this type, where the stack extends from the ground to the top of the building and is either built into the building or placed on the outside of one of the rear walls, there are several items to be taken into consideration—the length and weight of each section of stack, the facilities for hoisting and placing, and the amount of rigging necessary.

Where each section of the stack weighs 500 to 1,000 lbs. (226-453 kg), it is customary to place a small gin pole on the roof of the building and lift each section in place.

Four iron workers should set up all necessary rigging in one day, ready to place the sections. After the derricks and rigging have been placed to start hoisting, the crew should handle, place, and bolt one 15- to 20-ft. (4.5-6.0 m) section of stack in 4 hours, or 2 sections, 30 to 40 l.f. (9.1-12.1 m), per 8-hour day.

The labor cost placing and removing derricks for erecting a steel stack should average as follows:

	Hours	Rate	Total	Rate	Total
Ir. wkrs. placing equip	32	$....	$....	$30.28	$ 968.96
Ir. wkrs. remove equip	32	....	....	30.28	968.96
Labor on equipment			$....		$1,937.92

The labor cost erecting a steel stack up to 30" (750 mm) in diam. and 60 l.f. (18.2 m) long, should average as follows:

	Hours	Rate	Total	Rate	Total
Ir. wkrs.erecting stack	48	$....	$....	$30.28	$1,453.44
Cost per l.f.				....	24.22
per m				....	79.45

To the above costs, add equipment rental and move charges to and from job for equipment.

CHAPTER 6

CARPENTRY

CSI DIVISION 6
06100 ROUGH CARPENTRY
06130 HEAVY TIMBER CONSTRUCTION
06170 PREFABRICATED STRUCTURAL WOOD
06200 FINISH CARPENTRY
06300 WOOD TREATMENT
06400 ARCHITECTURAL WOODWORK

06100 ROUGH CARPENTRY

To estimate carpentry quantities and costs accurately, an estimator should know how the various classes of work are constructed, where joists and studs should be doubled, where to place wood furring strips for walls and floors and openings requiring wood bucks, and where to place wood grounds. These may seem like small items, but they are important.

An inexperienced estimator who is unfamiliar with how work is conducted in the field might figure wood bucks for the exterior window frames. He or she must know how a job goes together and what is required for each item of work.

When taking off lumber quantities from the plans, always estimate as closely as possible the exact quantities of each kind required. Some contractors estimate the cost of wood floors at a certain price per sq. ft. including joists, bridging, subfloor, deadening felt, furring strips, and the finish wood flooring. Entering into the construction of such a floor are materials and labor operations of widely varying costs, yet certain contractors are content to estimate their work in this haphazard manner, and at the completion of the job have no more idea of their labor costs than before they took the job.

When estimating labor costs of rough carpentry, bear in mind the costs will vary with the class of work performed, with the ability of the carpenters employed, and with the experience and ability of the foreman or superintendent.

Modern tools, such as electric hand saws, electric drills, and sanders, play a big part in reducing labor costs. For instance an electric handsaw can cut a 2"x12" (50 x 300 mm) plank in 4 or 5 seconds, while it takes about a minute to do it by hand; it takes 10 to 12 minutes to rip a 12'-0" (3.6 m) plank 2" (50 mm) thick when sawing by hand, while an electric handsaw will do the work in less than a minute. In other words, an electric saw will cut 10 to 20 times as fast.

Another consideration is the cost of filing saws. On average it will require 1/2 hr. to file a saw by hand, while an automatic saw filer will do the same work in 1/4 hr. The same applies to circular saws used on saw rigs or electric hand saws, as these can be filed in 1/2 to 3/4 hr. by machine, while hand filing requires nearly twice as much time.

The same thing applies to the use of electric drills, electric sanders, etc. The contractor who plans to keep up with the procession must use modern, labor-saving tools and equipment wherever possible.

Here is another thing that should be remembered by every contractor. No matter how carefully the plans have been measured and how much consideration they have been given before pricing, if the work is not executed in an efficient manner, careful estimating will not avert losses.

Estimating Lumber Quantities

When estimating the quantity of lumber required for any job, the only safe method is to take off every piece of lumber required to complete that portion of the work. The following tables can simplify the work as much as possible and at the same time provide accurate material quantities.

Lumber is usually sold by the *board foot* (bf), sometimes referred to as *board foot measure*. One board foot is always 144 cubic inches. The formula for figuring the number of board feet in a given length of board or lumber is:

$$\frac{\text{Width (inches) x Thickness (inches) x Length (feet)}}{12}$$

For example, to figure the board feet measure of a 2" x 4" that is 6'-0" long, (2 x 4 x 6) ÷ 12 = 48 ÷ 12 = 4 board feet.

Estimating Wood Joists. When estimating wood joists, always allow 4" to 6" (100-150 mm) on each end of the joist for bearing on the wall.

To obtain the number of joists required for any floor, take the length of the floor in feet, divide by the distance the joists are spaced and add 1 to allow for the extra joist required at end of span.

Example: If the floor is 28 ft. (8.4 m) long and 15 ft. (4.5 m) wide, it will require 16-ft. (4.8-m) joists to allow for wall bearing at each end. Assuming the joists are spaced 16" (400 mm) on centers, one joist will be required every 16" (400 mm) or every 1-1/3 ft. (0.39 m). In other words it will require 3/4 or 0.75 as many joists as the length of the span, plus one. Three-quarters of 28 equal 21, plus 1 extra joist at end, makes 22 joists 16 ft. (4.8 m) long for this space.

The following table gives the number of joists required for any spacing:

Number of Wood Floor Joists Required for any Spacing

Distance Joists are Placed, Inches, o.c.	Distance Joists are Placed, mm, o.c.	Multiply Length of Floor Span by	Add Joists
12"	300 mm	1	1
16"	400 mm	3/4 or 0.75	1
20"	500 mm	3/5 or 0.60	1
24"	600 mm	1/2 or 0.50	1
30"	750 mm	2/5 or 0.40	1
36"	900 mm	1/3 or 0.33	1
42"	1050 mm	2/7 or 0.29	1
48"	1200 mm	1/4 or 0.25	1
54"	1350 mm	2/9 or 0.22	1
60"	1500 mm	1/5 or 0.20	1

Board Feet of Lumber Required per 100 S.F. of Surface When Used for Studs, Joists, Rafters, Wall and Floor Furring Strips, etc

Lumber Size	On Center Spacing			
	12"	16"	20"	24"
1"x2"	16-2/3	12-1/2	10	8-1/3
2"x2"	33-1/3	25	20	16-2/3
2"x4"	66-2/3	50	40	33-1/3
2"x5"	83-1/3	62-1/2	50	41-2/3
2"x6"	100	75	60	50
2"x8"	133-1/3	100	80	66-2/3
2"x10"	166-2/3	125	100	83-1/3
2"x12"	200	150	120	100
2"x14"	233-1/3	175	140	116-2/3
3"x6"	150	112-1/2	90	75
3"x8"	200	150	120	100
3"x10"	250	187-1/2	150	125
3"x12"	300	225	180	150
3"x14"	350	262-1/2	210	175

(metric)
Cubic Meters of Lumber Required per 10 Sq.M of Surface When Used for Studs, Joists, Rafters, Wall and Floor Furring Strips, etc

Lumber Size	On Center Spacing			
	300mm	400mm	500mm	600mm
25x50mm	0.042	0.032	0.025	0.021
50x50mm	0.085	0.064	0.051	0.042
50x100mm	0.169	0.127	0.102	0.085

Lumber Size	On Center Spacing			
	300mm	400mm	500mm	600mm
50x125mm	0.212	0.159	0.127	0.106
50x150mm	0.254	0.191	0.152	0.127
50x200mm	0.339	0.254	0.203	0.169
50x250mm	0.423	0.318	0.254	0.212
50x300mm	0.508	0.381	0.305	0.254
50x350mm	0.593	0.445	0.356	0.296
75x150mm	0.381	0.286	0.229	0.191
75x200mm	0.508	0.381	0.305	0.254
75x250mm	0.635	0.476	0.381	0.318
75x300mm	0.762	0.572	0.457	0.381
75x350mm	0.889	0.667	0.534	0.445

Does not include any allowance for waste in cutting, doubling joists under partitions or around stair wells, extra joists at end of each span, top or bottom plates, etc. These items vary with each job. Add as required.

Number of Wood Joists Required for any Floor and Spacing

Length of Floor		Spacing of Joists, inches (mm)									
l.f.	meters	12 (300)	16 (400)	20 (500)	24 (600)	30 (750)	36 (900)	42 (1050)	48 (1200)	54 (1350)	60 (1500)
6	1.8	7	6	5	4	3	3	3	3	2	2
7	2.1	8	6	5	5	4	4	3	3	3	2
8	2.4	9	7	6	5	4	4	3	3	3	3
9	2.7	10	8	6	6	5	4	4	3	3	3
10	3.0	11	9	7	6	5	4	4	4	3	3
11	3.4	12	9	8	7	5	5	4	4	3	3
12	3.7	13	10	8	7	6	5	4	4	4	3
13	4.0	14	11	9	8	6	5	5	4	4	4
14	4.3	15	12	9	8	7	6	5	5	4	4
15	4.6	16	12	10	9	7	6	5	5	4	4
16	4.9	17	13	11	9	7	6	6	5	5	4
17	5.2	18	14	11	10	8	7	6	5	5	4
18	5.5	19	15	12	10	8	7	6	6	5	4
19	5.8	20	15	12	11	9	7	6	6	5	5
20	6.1	21	16	13	11	9	8	7	6	5	5
21	6.4	22	17	14	12	9	8	7	6	6	5
22	6.7	23	18	14	12	10	8	7	7	6	5
23	7.0	24	18	15	13	10	9	8	7	6	6
24	7.3	25	19	15	13	11	9	8	7	6	6
25	7.6	26	20	16	14	11	9	8	7	7	6
26	7.9	27	21	17	14	11	10	8	8	7	6
27	8.2	28	21	17	15	12	10	9	8	7	6
28	8.5	29	22	18	15	12	10	9	8	7	7
29	8.8	30	23	18	16	13	11	9	8	7	7
30	9.1	31	24	19	16	13	11	10	9	8	7
31	9.4	32	24	20	17	13	11	10	9	8	7
32	9.8	33	25	20	17	14	12	10	9	8	7

Spacing of Joists, inches (mm)

Length of Floor l.f. meters		12 (300)	16 (400)	20 (500)	24 (600)	30 (750)	36 (900)	42 (1050)	48 (1200)	54 (1350)	60 (1500)
33	10.1	34	26	21	18	14	12	10	9	8	8
34	10.4	35	27	21	18	15	12	11	10	9	8
35	10.7	36	27	22	19	15	13	11	10	9	8
36	11.0	37	28	23	19	15	13	11	10	9	8
37	11.3	38	29	23	20	16	13	12	10	9	8
38	11.6	39	30	24	20	16	14	12	11	9	9
39	11.9	40	30	24	21	17	14	12	11	10	9
40	12.2	41	31	25	21	17	14	12	11	10	9

One joist has been added to each of the above quantities to take care of extra joist required at end of span. Add for doubling joists under all partitions.

Estimating Quantity of Bridging. It is customary to place a double row of bridging between joists about 6'-0" to 8'-0" (1.8-2.4 m) on centers. Joists 10'-0" to 12'-0" (3.0-3.6 m) long will require one double row of bridging or 2 pcs. to each joist.

Joists 14'-0" to 20'-0" (4.2-6.0 m) long will require 2 double rows of bridging or 4 pcs. to each joist.

Bridging is usually cut from 1"x3" (25 x 75 mm), 1"x4" (25 x 100 mm), 2"x2" (50 x 50 mm), or 2"x4" (50 x 100 mm) lumber.

The following table gives the approximate number of pcs. and the l.f. (m) of bridging required per 100 s.f. (9 sq.m) of floor.

Joists Up to 12' (3.6m) Long				Joists Up to 20' (6 m) Long			
12" (300mm) o.c.		16" (400mm) o.c.		12" (300mm) o.c.		16" (400mm) o.c.	
Pcs.	l.f. (m)	Pcs.	l.f. (m)	Pcs.	l.f. (m)	Pcs.	l.f. (m)
20	30 (9.1)	16	24 (7.3)	40	60 (18.3)	32	48 (14.6)

Steel Bridging for Wood Joists. Several types of steel bridging are available, fabricated from deformed strip steel varying from 16 to 20 gauge in thickness. Some are galvanized while others are painted with black asphaltum or have a baked enamel finish. Stock sizes fit all joist depths and spacings encountered in average construction work--special sizes are available made to order.

Prices vary according to type, size, finish, and quantity, but a fair average for galvanized bridging is $0.30 to 0.40 per set.

Installation time is about 50% of that required for wood bridging and with many types can be done after subflooring is laid.

Estimating Number of Wood Studs. When estimating the number of wood partition studs, take the length of each partition and then the total length of all partitions.

If a top and bottom plate is required, take the length of the wood partition and multiply by 2. The result will be the number of lin. ft. of plates required.

If a double plate consisting of two top members and a single bottom plate is used, multiply the length of the wood partitions by 3.

Example: Find the quantity of lumber required to build a stud partition 16'-0" (4.8 m) long, 8'-0" (2.4 m) high, with studs spaced 16" (400 mm, 0.4 m) on centers and having single top and bottom plates. 16'-0" = 192" (4.8 m = 4800 mm). Next, 192"/16" = 12 studs (4800/400 = 12), plus 1 extra at the end equals 13 studs 8'-0" long. Two top and bottom plates 16'-0" long equal 32 l.f.

13 pcs. 2x4 @ 8'-0" = 104'-0"	13 pcs. 50x100mm @ 2.4m = 31.2m
2 pcs. 2x4 @ 16'-0" = 32-0"	2 pcs. 50x100mm @ 4.9m = 9.8m
104 + 32 = 136 l.f.	31.2 + 9.8 = 41.0 m
136 l.f. x 2/3 = 90.67 b.f.	41.0 m x .05 x 0.1 = 0.205 cu.m

Convert the total lineal feet of lumber to board feet. For metric, convert the total lineal meters to cubic meters.

B.F. of Lumber Required for Wood Stud Partitions
2"x4" Studs 16" o.c., Single Top and Bottom Plates*

Length of Partition	No. of Studs**	8'-0"	8'-6"	9'-0"	10'-0"	12'-0"
			Height of Partition			
3'-0"	4	25	27	28	31	36
4'-0"	4	27	28	29	32	37
5'-0"	5	33	35	37	40	47
6'-0"	6	40	42	44	48	56
7'-0"	7	47	49	51	56	65
8'-0"	7	48	50	53	57	67
9'-0"	8	55	57	60	65	76
10'-0"	9	61	64	67	73	85
11'-0"	10	68	71	75	81	95
12'-0"	10	69	73	76	83	96
13'-0"	11	76	80	83	91	105
14'-0"	12	83	87	91	99	115
15'-0"	13	89	94	98	107	124
16'-0"	13	91	95	99	108	125
17'-0"	14	97	102	107	116	135
18'-0"	15	104	109	114	124	144
19'-0"	16	111	116	121	132	153
20'-0"	16	112	117	123	133	155
21'-0"	17	119	124	130	141	164
22'-0"	18	125	131	137	149	173
23'-0"	19	132	138	145	157	183
24'-0"	19	133	140	146	159	184
25'-0"	20	140	147	153	167	193
26'-0"	21	147	154	161	175	203

Length of Partition	No. of Studs**	Height of Partition				
		8'-0"	8'-6"	9'-0"	10'-0"	12'-0"
27'-0"	22	153	161	168	183	212
28'-0"	22	155	162	169	184	213
29'-0"	23	161	169	177	192	223
30'-0"	24	168	176	184	200	232
31'-0"	25	175	183	191	208	241
32'-0"	25	176	184	193	209	243
33'-0"	26	183	191	200	217	252
34'-0"	27	189	198	207	225	261
35'-0"	28	196	205	215	233	271
36'-0"	28	197	207	216	235	272
37'-0"	29	204	214	223	243	281
38'-0"	30	211	221	231	251	291
39'-0"	31	217	228	238	259	300
40'-0"	31	219	229	239	260	301

*Add 2/3 b.f. of lumber for each l.f. of double top or bottom plate.
** The formula that precedes the chart is first used to calculate the number of studs, which are then rounded up to the nearest whole stud (i.e., 2.25 studs is rounded to 3 studs.) Board feet figures are determined based on whole studs and then rounded to the nearest board foot (2.25 is rounded to 2, 2.5 to 3, etc.).

Cu.M of Lumber Required for Wood Stud Partitions
50 x 100 mm Studs 400 mm o.c., Single Top and Bottom Plates

Length of Partition	No. of Studs	Height of Partition				
		2.4 m	2.6 m	2.7 m	3.0 m	3.7 m
0.9 m	4	0.059	0.064	0.067	0.073	0.085
1.2 m	4	0.063	0.066	0.069	0.076	0.088
1.5 m	5	0.078	0.083	0.087	0.094	0.111
1.8 m	6	0.094	0.099	0.104	0.113	0.132
2.1 m	7	0.111	0.116	0.120	0.132	0.153
2.4 m	7	0.113	0.118	0.125	0.134	0.158
2.7 m	8	0.130	0.134	0.142	0.153	0.179
3.0 m	9	0.144	0.151	0.158	0.172	0.200
3.4 m	10	0.160	0.167	0.177	0.191	0.224
3.7 m	10	0.163	0.172	0.179	0.196	0.226
4.0 m	11	0.179	0.189	0.196	0.215	0.248
4.3 m	12	0.196	0.205	0.215	0.233	0.271
4.6 m	13	0.210	0.222	0.231	0.252	0.292
4.9 m	13	0.215	0.224	0.233	0.255	0.295
5.2 m	14	0.229	0.241	0.252	0.274	0.318
5.5 m	15	0.245	0.257	0.269	0.292	0.340
5.8 m	16	0.262	0.274	0.285	0.311	0.361
6.1 m	16	0.264	0.276	0.290	0.314	0.366
6.4 m	17	0.281	0.292	0.307	0.333	0.387
6.7 m	18	0.295	0.309	0.323	0.351	0.408

Length of Partition	No. of Studs	Height of Partition				
		2.4 m	2.6 m	2.7 m	3.0 m	3.7 m
7.0 m	19	0.311	0.325	0.342	0.370	0.432
7.3 m	19	0.314	0.330	0.344	0.375	0.434
7.6 m	20	0.330	0.347	0.361	0.394	0.455
7.9 m	21	0.347	0.363	0.380	0.413	0.479
8.2 m	22	0.361	0.380	0.396	0.432	0.500
8.5 m	22	0.366	0.382	0.399	0.433	0.502
8.8 m	23	0.380	0.399	0.417	0.453	0.491
9.1 m	24	0.396	0.415	0.434	0.472	0.547
9.4 m	25	0.413	0.432	0.450	0.491	0.568
9.8 m	25	0.415	0.434	0.455	0.493	0.573
10.1 m	26	0.432	0.450	0.472	0.512	0.594
10.4 m	27	0.446	0.467	0.488	0.531	0.616
10.7 m	28	0.462	0.483	0.507	0.550	0.639
11.0 m	28	0.465	0.488	0.509	0.554	0.642
11.3 m	29	0.481	0.505	0.526	0.573	0.663
11.6 m	30	0.498	0.521	0.545	0.592	0.686
11.9 m	31	0.512	0.538	0.561	0.611	0.707
12.2 m	31	0.517	0.540	0.564	0.613	0.710

Add 0.0005 cu.m of lumber for each lineal meter of double top or bottom plate.

Number of Partition Studs Required for Any Spacing

Distance Apart Studs	Multiply Length of Partition by	Add Wood Studs
12"	1.00	1
16"	0.75	1
20"	0.60	1
24"	0.50	1

Add for top and bottom plates.

(metric)

Distance Apart Studs	Divide Length of Partition by	Add Wood Studs
300 mm	300	1
400 mm	400	1
500 mm	500	1
600 mm	600	1

Multiplying Factors to Determine Number of Feet of Lumber Required Per S.F. of Wood Stud Partition Using 2"x4" Studs Studs Spaced 16" on Centers, with Single Top and Bottom Plates

Length Partition in Feet	No. Studs Req'd	Ceiling Height in Feet			
		8'-0"	9'-0"	10'-0"	12'-0"
2	3	1.16	1.14	1.12	1.10
3	3	0.83	0.81	0.79	0.77

4	4	0.83	0.81	0.79	0.77
5	5	0.83	0.81	0.79	0.77
6	6	0.83	0.81	0.79	0.77
7	6	0.73	0.71	0.70	0.68
8	7	0.74	0.72	0.71	0.69
9	8	0.75	0.73	0.72	0.70
10	9	0.76	0.74	0.73	0.70
11	9	0.71	0.69	0.67	0.65
12	10	0.72	0.70	0.68	0.66
13	11	0.72	0.71	0.69	0.67
14	12	0.73	0.71	0.70	0.68
15	12	0.69	0.67	0.66	0.64
16	13	0.70	0.68	0.67	0.65
17	14	0.71	0.69	0.68	0.65
18	15	0.72	0.70	0.68	0.66
19	15	0.69	0.67	0.65	0.63
20	16	0.69	0.67	0.66	0.64
Dbl. plate, add per sq. ft		0.13	0.11	0.10	0.083

Multiplying Factors to Determine Number of Cubic Meters of Lumber Required Per Sq.M of Wood Stud Partition Using 50 x 100 mm Studs Studs Spaced 400 mm o.c., with Single Top and Bottom Plates

Length Partition in Meters	No. Studs Req'd	Ceiling Height in Meters			
		2.4 m	2.7 m	3.0 m	3.7 m
0.6	3	0.019	0.019	0.019	0.018
0.9	3	0.021	0.020	0.020	0.019
1.2	4	0.022	0.022	0.021	0.020
1.5	5	0.024	0.023	0.022	0.021
1.8	6	0.025	0.024	0.023	0.022
2.1	6	0.026	0.025	0.024	0.023
2.4	7	0.028	0.027	0.026	0.024
2.7	8	0.029	0.028	0.027	0.025
3.0	9	0.031	0.029	0.028	0.026
3.4	9	0.032	0.031	0.029	0.027
3.7	10	0.034	0.032	0.030	0.028
4.0	11	0.035	0.033	0.031	0.029
4.3	12	0.037	0.034	0.033	0.030
4.6	12	0.038	0.036	0.034	0.030
4.9	13	0.039	0.037	0.035	0.031
5.2	14	0.041	0.038	0.036	0.032
5.5	15	0.042	0.039	0.037	0.033
5.8	15	0.044	0.041	0.038	0.034
6.1	16	0.045	0.042	0.039	0.035
Double plate add per sq.m		0.004	0.003	0.003	0.002

For 2"x8" (50 x 200 mm) studs, double above quantities. For 2"x6" (50 x 150 mm) studs, increase above quantities 50%.

Example: Find the number of b.f. of lumber required for a stud partition 18'-0" (5.4 m) long and 9'-0" (2.7 m) high. This partition would contain 18 x 9 = 162 s.f. (5.4 x 2.7 = 14.58 sq.m).

The table gives 0.70 b.f. lumber of partition. Multiply 162 by 0.70 equals 113.4 b.f.

Quantity of Plain End Softwood Flooring Required per 100 S.F. (9 Sq.M) of Floor

Measured Size, inches	Actual Size, inches	Add for Width	B.F. Req. per 100 s.f. Surface	Lbs. per 1000 ft.
1 x 3	3/4 x 2-3/8	27%	132	1800
1 x 4	3/4 x 3-1/4	23%	128	1900

(metric)

Measured Size, mm	Actual Size, mm	Add for Width	Cu.M Req. per 100 sq.m Surface	Kg per 300 m
25 x 75	18.8 x 59	27%	0.335	816
25 x 100	18.8 x 81	23%	0.325	862

The above quantities include 5% for end cutting and waste.

Quantity of End Matched Softwood Flooring Required Per 100 S.F. (9 Sq.M) of Floor

Measured Size, inches	Actual Size, inches	Add for Width	B.F. Req. per 100 s.f. Surface	Lbs. per 1000 ft.
1 x 3	13/16 x 2-3/8	27%	130	1800
1 x 4	13/16 x 3-1/4	23%	126	1900

(metric)

Measured Size, mm	Actual Size, mm	Add for Width	Cu.M Req. per 100 sq.m Surface	Kg per 300 m
25 x 75	18.8 x 59	27%	0.330	816
25 x 100	18.8 x 81	23%	0.320	862

The above quantities include 3% for end cutting and waste.

Quantity of Square Edged (S4S) Boards Required Per 100 S.F. of Surface

Measured Size, inches	Actual Size, inches	Add for Width	B.F. Req. per 100 s.f. Surface	Lbs. per 1000 ft.
1 x 4	3/4 x 3-1/2	14%	119	2300

1 x 6	3/4 x 5-1/2	9%	114	2300
1 x 8	3/4 x 7-1/4	10%	115	2300
1 x 10	3/4 x 9-1/4	8%	113	2300
1 x 12	3/4 x 11-1/4	7%	112	2400

(metric)

Measured Size, mm	Actual Size, mm	Add for Width	Cu.M Req. per 100 sq.m Surface	Kg per 300 m
25 x 100	18.8 x 87.5	14%	0.302	1043
25 x 150	18.8 x 137.5	9%	0.290	1043
25 x 200	18.8 x 181.3	10%	0.292	1043
25 x 250	18.8 x 231.3	27%	0.287	1043
25 x 300	18.8 x 281.3	23%	0.285	1089

The above quantities include 5% for end cutting and waste.

Lineal Foot Table of Board Measure
Number of Board Feet per Lineal Foot of Any Size

2" x 4" = 0.667	4" x 4" = 1.333	8" x 14" = 9.333
2" x 6" = 1.0	4" x 6" = 2.0	8" x 16" = 10.667
2" x 8" = 1.333	4" x 8" = 2.667	10" x 10" = 8.333
2" x 10" = 1.667	4" x 10" = 3.333	10" x 12" = 10.0
2" x 12" = 2.0	4" x 12" = 4.0	10" x 14" = 11.667
2" x 14" = 2.333	4" x 14" = 4.667	10" x 16" = 13.333
2" x 16" = 2.667	4" x 16" = 5.333	10" x 18" = 15.0
2-1/2" x 12" = 2.5	6" x 6" = 3.0	12" x 12" = 12.0
2-1/2" x 14" = 2.917	6" x 8" = 4.0	12" x 14" = 14.0
2-1/2" x 16" = 3.333	6" x 10" = 5.0	12" x 16" = 16.0
3" x 6" = 1.5	6" x 12" = 6.0	12" x 18" = 18.0
3" x 8" = 2.0	6" x 14" = 7.0	14" x 14" = 16.333
3" x 10" = 2.5	6" x 16" = 8.0	14" x 16" = 18.667
3" x 12" = 3.0	8" x 8" = 5.333	14" x 18" = 21.0
3" x 14" = 3.5	8" x 10" = 6.667	16" x 16" = 21.333
3" x 16" = 4.0	8" x 12" = 8.0	16" x 18" = 24.0

Number of Cubic Meters per Lineal Meter of Any Size

500 x 100 mm = 0.005	100 x 100 mm = 0.010	200 x 300 mm = 0.070
50 x 150 mm = 0.008	100 x 150 mm = 0.015	200 x 400 mm = 0.080
50 x 200 mm = 0.010	100 x 200 mm = 0.020	250 x 250 mm = 0.063
50 x 250 mm = 0.013	100 x 250 mm = 0.025	250 x 300 mm = 0.075
50 x 300 mm = 0.015	100 x 300 mm = 0.030	250 x 350 mm = 0.088
50 x 350 mm = 0.018	100 x 350 mm = 0.035	250 x 400 mm = 0.100
50 x 400 mm = 0.020	100 x 400 mm = 0.040	250 x 450 mm = 0.113
62.5 x 300 mm = 0.019	150 x 150 mm = 0.023	300 x 300 mm = 0.090
62.5 x 350 mm = 0.022	150 x 200 mm = 0.030	300 x 350 mm = 0.105
62.5 x 400 mm = 0.025	150 x 250 mm = 0.083	300 x 400 mm = 0.120

Number of Cubic Meters per Lineal Meter of Any Size (Continued)

75 x 150 mm = 0.011	150 x 300 mm = 0.045	300 x 450 mm = 0.135
75 x 200 mm = 0.015	150 x 350 mm = 0.083	350 x 350 mm = 0.123
75 x 250 mm = 0.019	150 x 400 mm = 0.060	350 x 400 mm = 0.140
75 x 300 mm = 0.023	200 x 200 mm = 0.040	350 x 450 mm = 0.158
75 x 350 mm = 0.026	200 x 250 mm = 0.050	400 x 400 mm = 0.160
75 x 400 mm = 0.030	200 x 300 mm = 0.060	400 x 450 mm = 0.180

Lengths of Common, Hip, and Valley Rafters per 12" (300 mm) of Run

1	2	3	4	5*	6**
		Length in In.		% Increase	Length, in.
Roof	Rise and	Common Rafter		Length Com.	Hip or Valley
Pitch	Run or Cut	per 12" Run		Rafter over Run	Rafters
1/12	2 & 12	12.165	0.014	1.014	17.088
1/8	3 & 12	12.369	0.031	1.031	17.233
1/6	4 & 12	12.649	0.054	1.054	17.433
5/24	5 & 12	13.000	0.083	1.083	17.692
1/4	6 & 12	13.417	0.118	1.118	18.000
7/24	7 & 12	13.892	0.158	1.158	18.358
1/3	8 & 12	14.422	0.202	1.202	18.762
3/8	9 & 12	15.000	0.250	1.250	19.209
5/12	10 & 12	15.620	0.302	1.302	19.698
11/24	11 & 12	16.279	0.357	1.357	20.224
1/2	12 & 12	16.971	0.413	1.413	20.785
13/24	13 & 12	17.692	0.474	1.474	21.378
7/12	14 & 12	18.439	0.537	1.537	22.000
5/8	15 & 12	19.210	0.601	1.601	22.649
2/3	16 & 12	20.000	0.667	1.667	23.324
17/24	17 & 12	20.809	0.734	1.734	24.021
3/4	18 & 12	21.633	0.803	1.803	24.739
19/24	19 & 12	22.500	0.875	1.875	25.475
5/6	20 & 12	23.375	0.948	1.948	26.230
7/8	21 & 12	24.125	1.010	2.010	27.000
11/24	22 & 12	25.000	1.083	2.083	27.785
11/12	23 & 12	26.000	1.167	2.167	28.583
Full	24 & 12	26.875	1.240	2.240	29.394

*Use figures in this column to obtain area of roof for any pitch. See explanation below.

**Figures in last column are length of hip and valley rafters in inches for each 12 inches of common rafter run.

To Obtain Area of Roofs For Any Pitch. To obtain the number of sq. ft. (sq.m) of roof area for any pitch, take the entire flat or horizontal area of the roof and multiply by the figure given in the fifth column (*), and the result will be the area of the roof. Always bear in mind that the width of any overhanging cornice must be added to the building area to obtain the total area to be covered.

Example: Find the area of a roof 26'-0" x 42'-0" (7.8 x 12.6 m), with a 1'-0" (300 mm) overhanging cornice and a 1/4 pitch. To obtain roof area,

26'-0" + 1'-0" + 1'-0" = 28'-0" (7.8 + 0.3 + 0.3 = 8.4 m) width
42'-0" + 1'-0" + 1'-0" = 44'-0" (12.6 + 0.3 + 0.3 = 13.2 m) length
28 x 44 = 1,232 s.f.(8.4 x 13.2 = 110.88 sq.m) flat or horizontal area.
Area at 1/4 pitch: 1,232 x 1.12 = 1379.84 or 1,380 s.f. (110.88 x 1.12 = 124.19 or 124 sq.m) roof surface.
Add allowance for overhang on dormer roofs and sides.

Board Feet Content of Joists, Scantlings, and Timbers of Given Lineal Foot Lengths

Size, in.	10	12	14	16	18	20	22	24	26	28	30
1x2	1.67	2.00	2.33	2.67	3.00	3.33	--	--	--	--	--
1x3	2.50	3.00	3.50	4.00	4.50	5.00	--	--	--	--	--
1x4	3.33	4.00	4.67	5.33	6.00	6.67	--	--	--	--	--
1x6	5.00	6.00	7.00	8.00	9.00	10.00	--	--	--	--	--
1x8	6.67	8.00	9.33	10.67	12.00	13.33	--	--	--	--	--
1x10	8.33	10.00	11.67	13.33	15.00	16.67	--	--	--	--	--
1x12	10.00	12.00	14.00	16.00	18.00	20.00	--	--	--	--	--
1-1/4x4	4.17	5.00	5.83	6.67	7.50	8.33	--	--	--	--	--
1-1/4x6	6.25	7.50	8.75	10.00	11.25	12.50	--	--	--	--	--
1-1/4x8	8.33	10.00	11.67	13.33	15.00	16.67	--	--	--	--	--
1-1/4x10	10.42	12.50	14.58	16.67	18.75	20.83	--	--	--	--	--
1-1/4x12	12.50	15.00	17.50	20.00	22.50	25.00	--	--	--	--	--
1-1/2x4	5.00	6.00	7.00	8.00	9.00	10.00	--	--	--	--	--
1-1/2x6	7.50	9.00	10.50	12.00	13.50	15.00	--	--	--	--	--
1-1/2x8	10.00	12.00	14.00	16.00	18.00	20.00	--	--	--	--	--
1-1/2x10	12.50	15.00	17.50	20.00	22.50	25.00	--	--	--	--	--
1-1/2x12	15.00	18.00	21.00	24.00	27.00	30.00	--	--	--	--	--
2x2	3.33	4.00	4.67	5.33	6.00	6.67	7.33	8.00	8.67	9.33	10.00
2x3	5.00	6.00	7.00	8.00	9.00	10.00	11.00	12.00	13.00	14.00	15.00
2x4	6.67	8.00	9.33	10.67	12.00	13.33	14.67	16.00	17.33	18.67	20.00
2x6	10.00	12.00	14.00	16.00	18.00	20.00	22.00	24.00	26.00	28.00	30.00
2x8	13.33	16.00	18.67	21.33	24.00	26.67	29.33	32.00	34.67	37.33	40.00
2x10	16.67	20.00	23.33	26.67	30.00	33.33	36.67	40.00	43.33	46.67	50.00
2x12	20.00	24.00	28.00	32.00	36.00	40.00	44.00	48.00	52.00	56.00	60.00
2x14	23.33	28.00	32.67	37.33	42.00	46.67	51.33	56.00	60.67	65.33	70.00
3x4	10.00	12.00	14.00	16.00	18.00	20.00	22.00	24.00	26.00	28.00	30.00
3x6	15.00	18.00	21.00	24.00	27.00	30.00	33.00	36.00	39.00	42.00	45.00
3x8	20.00	24.00	28.00	32.00	36.00	40.00	44.00	48.00	52.00	56.00	60.00

Size, in.	10	12	14	16	18	20	22	24	26	28	30
3x10	25.00	30.00	35.00	40.00	45.00	50.00	55.00	60.00	65.00	70.00	75.00
3x12	30.00	36.00	42.00	48.00	54.00	60.00	66.00	72.00	78.00	84.00	90.00
3x14	35.00	42.00	49.00	56.00	63.00	70.00	77.00	84.00	91.00	98.00	105.00
4x4	13.33	16.00	18.67	21.33	24.00	26.67	29.33	32.00	34.67	37.33	40.00
4x6	20.00	24.00	28.00	32.00	36.00	40.00	44.00	48.00	52.00	56.00	60.00
4x8	26.67	32.00	37.33	42.67	48.00	53.33	58.67	64.00	69.33	74.67	80.00
4x10	33.33	40.00	46.67	53.33	60.00	66.67	73.33	80.00	86.67	93.33	100.00
4x12	40.00	48.00	56.00	64.00	72.00	80.00	88.00	96.00	104.00	112.00	120.00
4x14	46.67	56.00	65.33	74.67	84.00	93.33	102.67	112.00	121.33	130.67	140.00
6x6	30.00	36.00	42.00	48.00	54.00	60.00	66.00	72.00	78.00	84.00	90.00
6x8	40.00	48.00	56.00	64.00	72.00	80.00	88.00	96.00	104.00	112.00	120.00
6x10	50.00	60.00	70.00	80.00	90.00	100.00	110.00	120.00	130.00	140.00	150.00
6x12	60.00	72.00	84.00	96.00	108.00	120.00	132.00	144.00	156.00	168.00	180.00
6x14	70.00	84.00	98.00	112.00	126.00	140.00	154.00	168.00	182.00	196.00	210.00
6x16	80.00	96.00	112.00	128.00	144.00	160.00	176.00	192.00	208.00	224.00	240.00
8x8	53.33	64.00	74.67	85.33	96.00	106.67	117.33	128.00	138.67	149.33	160.00
8x10	67.67	80.00	93.33	106.67	120.00	133.33	146.67	160.00	173.33	186.67	200.00
8x12	80.00	96.00	112.00	128.00	144.00	160.00	176.00	192.00	208.00	224.00	240.00
8x14	93.33	112.00	130.67	149.33	168.00	186.67	205.33	224.00	242.67	261.33	280.00
8x16	106.67	128.00	149.33	170.67	192.00	213.33	234.67	256.00	277.33	298.67	320.00
10x10	83.33	100.00	116.67	133.33	150.00	167.67	183.33	200.00	216.67	233.33	250.00
10x12	100.00	120.00	140.00	160.00	180.00	200.00	220.00	240.00	260.00	280.00	300.00
10x14	116.67	140.00	163.33	186.67	210.00	233.33	256.67	280.00	303.33	326.67	350.00
10x16	133.33	160.00	186.67	213.33	240.00	266.67	293.33	320.00	346.67	373.33	400.00
12x12	120.00	144.00	168.00	192.00	216.00	240.00	264.00	288.00	312.00	336.00	360.00
12x14	140.00	168.00	196.00	224.00	252.00	280.00	308.00	336.00	364.00	392.00	420.00
12x16	160.00	192.00	224.00	256.00	288.00	320.00	352.00	384.00	416.00	448.00	480.00
14x14	163.33	196.00	228.67	261.33	294.00	326.67	359.33	392.00	424.67	457.33	490.00
14x16	186.67	224.00	261.33	298.67	336.00	373.33	410.67	448.00	485.33	522.67	560.00
14x18	210.00	252.00	294.00	336.00	378.00	420.00	462.00	504.00	546.00	588.00	630.00
14x20	233.33	280.00	326.67	373.33	420.00	466.67	513.33	560.00	606.66	653.33	700.00
16x16	213.33	256.00	298.67	341.33	384.00	426.67	469.33	512.00	554.67	597.33	640.00
16x18	240.00	288.00	336.00	384.00	432.00	480.00	528.00	576.00	624.00	672.00	720.00
16x20	266.67	320.00	373.33	426.67	480.00	533.33	586.67	640.00	693.33	746.67	800.00
18x18	270.00	324.00	378.00	432.00	486.00	540.00	594.00	648.00	702.00	756.00	810.00
18x20	300.00	360.00	420.00	480.00	540.00	600.00	660.00	720.00	780.00	840.00	900.00
20x20	333.33	400.00	466.67	533.33	600.00	666.67	733.33	800.00	866.67	933.33	1000.00

Length, lin. ft.

Cubic Meter Content of Joists, Scantlings, and Timbers of Given Lengths in Meters

Size, mm	3	3.6	4.2	4.8	5.4	6	6.6	7.2	7.8	8.4	9
25x50	0.004	0.005	0.005	0.006	0.007	0.008	--	--	--	--	--
25x75	0.006	0.007	0.008	0.009	0.010	0.011	--	--	--	--	--

Length, meters

Size, mm	Length, meters										
	3	3.6	4.2	4.8	5.4	6	6.6	7.2	7.8	8.4	9
25x100	0.007	0.009	0.010	0.012	0.014	0.015	--	--	--	--	--
25x150	0.011	0.014	0.016	0.018	0.020	0.023	--	--	--	--	--
25x200	0.015	0.018	0.021	0.024	0.027	0.030	--	--	--	--	--
25x250	0.019	0.023	0.026	0.030	0.034	0.037	--	--	--	--	--
25x300	0.023	0.027	0.032	0.036	0.041	0.045	--	--	--	--	--
31x100	0.009	0.011	0.013	0.015	0.017	0.019	--	--	--	--	--
31x150	0.014	0.017	0.020	0.023	0.025	0.028	--	--	--	--	--
31x200	0.019	0.023	0.026	0.030	0.034	0.037	--	--	--	--	--
31x250	0.023	0.028	0.033	0.037	0.042	0.047	--	--	--	--	--
31x300	0.028	0.034	0.039	0.045	0.051	0.056	--	--	--	--	--
37x100	0.011	0.014	0.016	0.018	0.020	0.023	--	--	--	--	--
37x150	0.017	0.020	0.024	0.027	0.030	0.034	--	--	--	--	--
37x200	0.023	0.027	0.032	0.036	0.041	0.045	--	--	--	--	--
37x250	0.028	0.034	0.039	0.045	0.051	0.056	--	--	--	--	--
37x300	0.034	0.041	0.047	0.054	0.061	0.068	--	--	--	--	--
50x50	0.007	0.009	0.010	0.012	0.014	0.015	--	--	--	--	--
50x75	0.011	0.014	0.016	0.018	0.020	0.023	0.025	0.027	0.029	0.032	0.034
50x100	0.015	0.018	0.021	0.024	0.027	0.030	0.033	0.036	0.039	0.042	0.045
50x150	0.023	0.027	0.032	0.036	0.041	0.045	0.050	0.054	0.059	0.063	0.068
50x200	0.030	0.036	0.042	0.048	0.054	0.060	0.066	0.072	0.078	0.084	0.090
50x250	0.037	0.045	0.052	0.060	0.068	0.075	0.082	0.090	0.097	0.105	0.113
50x300	0.003	0.054	0.063	0.072	0.081	0.090	0.099	0.108	0.117	0.126	0.135
50x350	0.052	0.063	0.073	0.084	0.095	0.105	0.115	0.126	0.136	0.147	0.158
75x100	0.023	0.027	0.032	0.036	0.041	0.045	0.050	0.054	0.059	0.054	0.054
75x150	0.034	0.041	0.047	0.054	0.061	0.068	0.074	0.081	0.088	0.081	0.081
75x200	0.045	0.054	0.063	0.072	0.081	0.090	0.099	0.108	0.117	0.108	0.108
75x250	0.056	0.068	0.079	0.090	0.101	0.113	0.124	0.135	0.146	0.135	0.135
75x300	0.068	0.081	0.095	0.108	0.122	0.135	0.149	0.162	0.176	0.162	0.162
75x350	0.079	0.095	0.110	0.126	0.142	0.158	0.173	0.189	0.205	0.189	0.189
100x100	0.030	0.036	0.042	0.048	0.054	0.060	0.066	0.072	0.078	0.072	0.072
100x150	0.045	0.054	0.063	0.072	0.081	0.090	0.099	0.108	0.117	0.108	0.108
100x200	0.060	0.072	0.084	0.096	0.108	0.120	0.132	0.144	0.156	0.144	0.144
100x250	0.075	0.090	0.105	0.120	0.135	0.150	0.165	0.180	0.195	0.180	0.180
100x300	0.090	0.108	0.126	0.144	0.162	0.180	0.198	0.216	0.234	0.216	0.216
100x350	0.105	0.126	0.147	0.168	0.189	0.210	0.231	0.252	0.273	0.252	0.252
150x150	0.068	0.081	0.095	0.108	0.122	0.135	0.149	0.162	0.176	0.162	0.162
150x200	0.090	0.108	0.126	0.144	0.162	0.180	0.198	0.216	0.234	0.216	0.216
150x250	0.113	0.135	0.158	0.180	0.203	0.225	0.248	0.270	0.293	0.270	0.270
150x300	0.135	0.162	0.189	0.216	0.243	0.270	0.297	0.324	0.351	0.324	0.324
150x350	0.158	0.189	0.221	0.252	0.284	0.315	0.347	0.378	0.410	0.378	0.378
150x400	0.180	0.216	0.252	0.288	0.324	0.360	0.396	0.432	0.468	0.432	0.432

Size, mm	3	3.6	4.2	4.8	5.4	6	6.6	7.2	7.8	8.4	9
					Length, meters						
200x200	0.120	0.144	0.168	0.192	0.216	0.240	0.264	0.288	0.312	0.288	0.288
200x250	0.150	0.180	0.210	0.240	0.270	0.300	0.330	0.360	0.390	0.360	0.360
200x300	0.180	0.216	0.252	0.288	0.324	0.360	0.396	0.432	0.468	0.432	0.432
200x350	0.210	0.252	0.294	0.336	0.378	0.420	0.462	0.504	0.546	0.504	0.504
200x400	0.240	0.288	0.336	0.384	0.432	0.480	0.528	0.576	0.624	0.576	0.576
250x250	0.188	0.225	0.263	0.300	0.338	0.375	0.413	0.450	0.488	0.450	0.450
250x300	0.225	0.270	0.315	0.360	0.405	0.450	0.495	0.540	0.585	0.540	0.540
250x350	0.263	0.315	0.368	0.420	0.473	0.525	0.578	0.630	0.683	0.630	0.630
250x400	0.300	0.360	0.420	0.480	0.540	0.600	0.660	0.720	0.780	0.720	0.720
300x300	0.270	0.324	0.378	0.432	0.486	0.540	0.594	0.648	0.702	0.648	0.648
300x350	0.315	0.378	0.441	0.504	0.567	0.630	0.693	0.756	0.819	0.756	0.756
300x400	0.360	0.432	0.504	0.576	0.648	0.720	0.792	0.864	0.936	0.864	0.864
350x350	0.368	0.441	0.515	0.588	0.662	0.735	0.809	0.882	0.956	0.882	0.882
350x400	0.420	0.504	0.588	0.672	0.756	0.840	0.924	1.008	1.092	1.008	1.008
350x450	0.473	0.567	0.662	0.756	0.851	0.945	1.040	1.134	1.229	1.134	1.134
400x400	0.480	0.576	0.672	0.768	0.864	0.960	1.056	1.152	1.248	1.152	1.152
400x450	0.540	0.648	0.756	0.864	0.972	1.080	1.188	1.296	1.404	1.296	1.296
400x500	0.600	0.720	0.840	0.960	1.080	1.200	1.320	1.440	1.560	1.440	1.440
450x450	0.608	0.729	0.851	0.972	1.094	1.215	1.337	1.458	1.580	1.458	1.458
450x500	0.675	0.810	0.945	1.080	1.215	1.350	1.485	1.620	1.755	1.620	1.620
500x500	0.750	0.900	1.050	1.200	1.350	1.500	1.650	1.800	1.950	1.800	1.800

Nails Required for Carpentry Work

The following tables indicates the number of nails in lbs. (kg) for the various kinds of lumber per 1,000 b.f. (2.36 cu.m), or per 1,000 shingles and lath, or per square (100 s.f., 9 sq.m) of asphalt slate surfaced shingle, with the number of nails added for loss of material on account of lap or matching of shiplap, flooring, ceiling, and siding of the various widths. The table gives the sizes generally used for certain purposes with the nailing space 16" (400 mm) on centers, and 1 or 2 nails per board for each nailing space.

Split-less Wood Siding Nails. Supplied either plain or anchor down, split-less nails are designed to reduce splitting of wood siding such as hemlock, cedar, redwood, and other generic sidings. The blunt point helps reduce splitting. The heads are normally checked to help in adhesion of paint and stain.

Split-less Wood Siding Nail Specifications

Size	Length	Gauge	Head	Nails per Lb.
6d	2"	14	3/16"	283
7d	2-1/4"	14	3/16"	248
8d	2-1/2"	13	7/32"	189
9d	2-3/4"	13	7/32"	171
10d	3"	13	7/32"	153

Recommended Sizes and Quantities Commonly Used

Siding	Size	Nails per 1000 B.F.	Lbs. of Nails per 1000 B.F.
1/2" x 4"	6d	2,280	8
1/2" x 6"	7d	1,520	6 1/4
3/4" x 8"	8d	1,140	6
3/4" x 10"	8d	912	5
3/4" x 12"	8d	760	4

Split-less Wood Siding Nail Specifications

Size	Length mm	Gauge	Head mm	Nails per Kg
6d	50	14	4.7	624
7d	56	14	4.7	547
8d	63	13	5.5	417
9d	69	13	5.5	377
10d	225	13	5.5	337

Recommended Sizes and Quantities Commonly Used

Siding mm	Size	Nails per 1000 Bd. Ft.	Kg Nails per 1000 Bd. Ft.
13 x 100	6d	2,280	3.6
13 x150	7d	1,520	2.8
19 x 200	8d	1,140	2.7
19 x 250	8d	912	2.3
19 x 300	8d	760	1.8

Hardboard Siding Nails. Either plain or screw down, hardboard siding nails are made with an extra stiff stock to allow the nail to be driven through two thicknesses of hardboard siding without bending. The head drives flush with the hardboard.

Hardboard Siding Nail Specifications

Size	Length	Gauge	Head	Nails per Lb.
6d	2"	12	3/16"	181
8d	2-1/2"	12	3/16"	146
9d	2-3/4"	12	3/16"	136
10d	3"	11	3/16"	171

Recommended Sizes and Quantities Commonly Used

12" Horizontal Siding

Size	Lbs. Per 1000 Sq. Ft	
6d	6	Allow about 1 nail per Sq. Ft.
8d	7	Allow about 1 nail per Sq. Ft.
10d	10	Allow about 1 nail per Sq. Ft.

Panel Siding

Size	Lbs. Per 1000 Sq. Ft	
6d	17	Allow 3 nails per Sq. Ft.
8d	21	Allow 3 nails per Sq. Ft.
10d	30	Allow 3 nails per Sq. Ft.

Hardboard Siding Nail Specifications

Size	Length mm	Gauge	Head mm	Nails per Kg
6d	50	12	4.69	399.0
8d	63	12	4.69	321.9
9d	69	12	4.69	299.8
10d	75	11	4.69	377.0

Recommended Sizes and Quantities Commonly Used

300 mm Horizontal Siding

Size	Kg. Per 100 Sq. M	
6d	2.9	Allow about 11 nails per Sq. M.
8d	3.4	Allow about 11 nails per Sq. M.
10d	4.9	Allow about 11 nails per Sq. M.

Panel Siding

Size	Kg. Per 100 Sq. M	
6d	8.3	Allow 33 nails per Sq. M.
8d	10.3	Allow 33 nails per Sq. M.
10d	14.6	Allow 33 nails per Sq. M.

Common Nails. Supplied either plain or anchor down, common nails are made for exterior work and for interior work where appropriate. Common nails are used in parts of a building that are resistant to stain or rust. They are also good for fencing and other miscellaneous work.

Common Nail Specifications				
Size	Length	Gauge	Head	Nails per Lb.
6d	2"	11	9/32"	181
8d	2-1/2"	11	9/32"	106
10d	3"	10	9/32"	69
16d	3-1/2"	9	5/16"	49
20d	4"	7	5/16"	31

Comon Nail Specifications				
Size	Length mm	Gauge	Head mm	Nails per Kg
6d	50	11	7	399.0
8d	63	11	7	233.7
10d	75	10	7	152.1
16d	88	9	7.8	108.0
20d	100	7	7.8	68.3

Box Nails – Wood, Siding and Trim. Supplied plain or anchor down, box nails are made from high carbon steel. They drive easily with minimum splitting or bending. The anchor shank increases holding power. Heads of nails are checked to hold paint. Box nails are good for fences, trellises, flower boxes, exterior woodwork, and lawn furniture.

Box Nail Specifications

Size	Length	Gauge	Head	Nails per Lb.
4d	1-1/2"	14	7/32"	392
6d	2"	12-1/2	17/64"	194
7d	2-1/4"	12-1/2	17/64"	172
8d	2-1/2"	11-1/2	19/64"	123
10d	3"	11-1/2	19/64"	103

Recommended Sizes and Quantities Commonly Used

Siding	Size	Nails per 1000 BF	Lbs. per 1000 BF
1/4" x 4"	6d	2280	11-1/2
1/2" x 6"	7d	1520	9
3/4" x 8"	8d	1140	9-1/2
3/4" x 10"	8d	912	7-1/2
3/4" x 12"	8d	760	6-1/4

NOTE: Use one size larger over insulation

Box Nail Specifications

Size	Length mm	Gauge	Head mm	Nails Per Kg
4d	38	14	5.50	864.2
6d	50	12-1/2	6.64	427.7
7d	56	12-1/2	6.64	379.2
8d	63	11-1/2	7.42	271.2
10d	75	11-1/2	7.42	227.1

Recommended Sizes and Quantities Commonly Used

Siding mm	Size	Nails per 1000 BF	Kg per 1000 BF
6 x 11	6d	2280	5.2
13 x 150	7d	1520	4.1
19 x 200	8d	1140	4.3
19 x 250	8d	912	3.4
19 x 300	8d	760	2.8

NOTE: Use one size larger over insulation

Cedar Shake Siding Face Nails. Either plain or anchor, cedar shake nails are designed to drive with minimum splitting or bending. They have a blunt point and checked brad head. The anchor-down type is most popular for use over solid as well as insulation sheathing. Cedar shake nails come in various colors.

Cedar Shake Siding Face Nail Specifications

Size	Length	Gauge	Head	Nails per Lb.
4d	1-1/2"	14	1/8"	402
5d	1-3/4"	14	1/8"	346
6d	2"	14	1/8"	310
6d*	2"	12-1/2	3/16"	209
8d	2-1/2"	13	9/64"	196

*For use over high density fiberboard sheathing.

Recommended Sizes and Quantities Commonly Used

Shingles	Size	Nails per Square	Lbs. per Square
16"	2"	360	1 +
18"	2"	310	1

Cedar Shake Siding Face Nail Specifications

Size	Length mm	Gauge	Head mm	Nails per Kg
4d	38	14	3.1250	886.3
5d	44	14	3.1250	762.8
6d	50	14	3.1250	683.4
6d*	50	12-1/2	4.6900	460.8
8d	63	13	3.5200	432.1

*For use over high density fiberboard sheathing.

Recommended Sizes and Quantities Commonly Used

Shingles	Size	Nails per 10 Sq.M	Kg per Square
400 mm	50 mm	388	0.4535+
400 mm	50 mm	334	0.4535

Cedar Shingle Nails. Supplied plain, cedar shingle nails effectively resist rust and will hold shingles in place as long as the roof lasts.

Cedar Shingle Nail Specifications

Size	Length	Gauge	Head	Nails per Lb.
3d	1-1/4"	14	1/8"	402
4d	1-1/2"	14	1/8"	346
5d	1-3/4"	14	1/8"	310
6d	2"	13	3/16"	209

Recommended Sizes and Quantities Commonly Used

On new roof

16" and 18" - 3d nails average 860 nails per sq. or 1-3/4 lbs.

24" - 4d nails average 570 nails per sq. or 1-1/2 lbs.

Over an old roof

16" and 18" - 5d nails average 860 nails per sq. or 2-3/4 lbs.

24" - 6d nails average 570 nails per sq. or 2-1/2 lbs.

Cedar Shingle Nail Specifications

Size	Length mm	Gauge	Head mm	Nails per Kg
3d	31	14	3.1	886.26
4d	38	14	3.1	762.80
5d	44	14	3.1	683.43
6d	50	13	4.7	460.77

Recommended Sizes and Quantities Commonly Used

On new roof

400 mm and 450 mm - 3d nails average 862 nails per 10 Sq.M or 0.8 Kg

600 mm - 4d nails average 613 nails per 10 Sq.M or 0.7 Kg

Over an old roof

400 mm and 450 mm - 5d nails average 925 nails per 10 Sq.M or 1.2 Kg.

600 mm - 6d nails average 614 nails per 10 Sq.M or 1.3 Kg

Casing Nails. These are supplied plain and used for exterior trim, nailing of window and door frames, cornices, molding, etc.

Casing Nail Specifications

Size	Length	Gauge	Head	Nails per Lb.
6d	2"	12-1/2	9-1/2	236
8d	2-1/2"	11-1/2	8-1/2	145
10d	3"	10	7	94
16d	3-1/2"	10	7	71
20d	4"	9	6	52

Casing Nail Specifications

Size	Length mm	Gauge	Head	Nails per Kg
6d	50	12-1/2	9-1/2	520.29
8d	63	11-1/2	8-1/2	319.67
10d	75	10	7	207.23
16d	88	10	7	156.53
20d	100	9	6	114.64

Siding Nails for Aluminum, Steel and Vinyl. Plain or screw down, siding nails are made with 3/16" (4.7 mm) heads and a stiff 12-gauge shank. Available in three lengths.

Siding Nail Specifications

Size	Length	Gauge	Head	Nails per Lb.
X	1-1/4"	12	5/16"	270
X	1-1/2"	12	5/16"	228
X	2"	12	5/16"	178

Siding Nail Specifications

Size	Length mm	Gauge	Head mm	Nails per Kg
X	31	12	7.8	595.25
X	38	12	7.8	502.65
X	50	12	7.8	392.42

Finishing Nails. These are supplied plain and used on the exterior where required. They can be used with roofing and siding to resist stains and streak over the exterior.

Finish Nail Specifications

Size	Length	Gauge	Head	Nails per Lb.
6d	2"	14	1/8"	309
8d	3-1/2"	13	9/64"	189

Finish Nail Specifications

Size	Length mm	Gauge	Head mm	Nails per Kg
6d	50	14	3.1	681.23
8d	88	13	3.5	417.67

Asphalt Shingle Nails. There are plain, anchor down and screw down. All three types are used for applying asphalt shingle over old roofing. The anchor and screw down type are least likely to loosen. Nails with double-dipped zinc coating have the longest life.

Asphalt Nail Specifications

Size X	Length	Gauge	Head	Nails per Lb.
X	7/8"	11	7/16"	272
X	1"	11	7/16"	250
X	1-1/4"	11	7/16"	202
X	1-1/2"	11	7/16"	180
X	1-3/4"	11	7/16"	156
X	2"	11	7/16"	136
X	2-1/2"	11	7/16"	112

Asphalt Nail Specifications

Size X	Length mm	Gauge	Head mm	Nails per Kg
X	22	11	10.9	599.66
X	25	11	10.9	551.16
X	31	11	10.9	445.33
X	38	11	10.9	396.83
X	44	11	10.9	343.92
X	50	11	10.9	299.83
X	63	11	10.9	246.92

Recommended Sizes and Quantities Commonly Used

On new roof	
1-1/4"	480 nails per square or 2-1/4 Lbs.
Over an old roof	
1-3/4"	480 nails per square or 3-1/8 Lbs.

Recommended Sizes and Quantities Commonly Used

On new roof	
31 mm	517 nails per 10 Sq.M or 1.2 Kg
Over an old roof	
44 mm	517 nails per 10 Sq.M or 1.4 Kg

Insulation Roof Deck Nails. Plain or anchor down, these are used to apply roof deck to rafters. Large flat head holds decking securely without pulling through.

Insulation Roof Deck Nail Specifications

Size	Length mm	Gauge	Head mm	Nails per Kg
10d	75	10	10.9	176.37
16d	88	10	10.9	132.28
30d	113	9	10.9	90.39

Recommended Sizes and Quantities Commonly Used

38 mm Deck - 75 mm nails	188 Kg per 10 Sq.M – 0.91 Kg	
50 mm Deck - 88 mm nails	140 Kg per 10 Sq.M – 0.91 Kg	
75 mm Deck - 113 mm nails	97 Kg per 10 Sq.M – 1.4 Kg	

Insulation Roof Deck Nail Specifications

Size	Length	Gauge	Head	Nails per Lb.
10d	3"	10	7/16"	80
16d	3-1/2"	10	7/16"	60
30d	4-1/2"	9	7/16"	41

Recommended Sizes and Quantities Commonly Used

1-1/2" Deck - 3" nails	175 lbs. per sq. - 2 lbs.	
2" Deck - 3-1/2" nails	130 lbs. per sq. - 2 lbs.	
3" Deck - 4-1/2" nails	90 lbs. per sq. - 3 lbs.	

Gutter Spike Nails. Plain or anchor down (both with flat checkered or oval heads), gutter spike nails are hot dipped galvanized to prevent corrosion to gutters. One spike per every 32" is the recommended minimum.

Gutter Spike Nail Specifications

Size	Length mm	Gauge	Head mm	Nails per Kg
X	150	3/16"	9.4	52.91
X	175	3/16"	9.4	37.48
X	200	3/16"	9.4	33.07
X	200	1/4"	14.1	18.74
X	225	1/4"	14.1	17.64
X	250	1/4"	14.1	16.53

Gutter Spike Nail Specifications

Size	Length	Gauge	Head	Nails per Lb.
X	6"	3/16"	3/8"	24
X	7"	3/16"	3/8"	17
X	8"	3/16"	3/8"	15
X	8"	1/4"	9/16"	8.5
X	9"	1/4"	9/16"	8
X	10"	1/4"	9/16"	7.5

Square Cap Roofing Nails. With a full 1" (25 mm) head and deeply barbed shank, these nails have flat heads to give a full bearing and straight shanks. They are recommended for built-up roofs. Some have a hot dipped zinc coated finish to provide corrosion resistance. They are recommended in high humidity areas for insulation.

Square Cap Roofing Nail Specifications

Size	Length	Gauge	Head	Nails per Lb.
X	5/8"	12	Square	106
X	3/4"	12	Square	100
X	7/8"	12	Square	98
X	1"	12	Square	94
X	1-1/4"	12	Square	89
X	1-1/2"	12	Square	84
X	1-3/4"	12	Square	91
X	2"	12	Square	77
X	2-1/2"	12	Square	74
X	3"	12	Square	64
X	3-1/2"	12	Square	57

Square Cap Roofing Nail Specifications

Size	Length mm	Gauge	Head	Nails per Kg
X	15.6	12	Square	233.69
X	18.8	12	Square	220.46
X	21.9	12	Square	216.05
X	25.0	12	Square	207.23
X	31.3	12	Square	196.21
X	37.5	12	Square	185.19
X	43.8	12	Square	200.62
X	50.0	12	Square	169.76
X	62.5	12	Square	163.14
X	75.0	12	Square	141.10
X	87.5	12	Square	125.66

Drywall Nails. These are anchor-down and have a 1/4" (6.2 mm) head. It is the recommended nail for drywall. All heads are countersunk and have a feathered edge. Points are diamond shaped. Threads are sharp for holding power. Both 1/4" (6.2 mm) and 19/64" (0.3 mm) are available zinc coated to reduce staining.

Drywall Nail Specifications

Size	Length	Gauge	Head	Nails per Lb.
X	1-1/4"	12-1/2	1/4"	362
X	1-3/8"	12-1/2	1/4"	328
X	1-1/2"	12-1/2	1/4"	308
X	1-5/8"	12-1/2	1/4"	279
X	1-1/4"	12-1/2	19/64"	352
X	1-3/8"	12-1/2	19/64"	321
X	1-1/2"	12-1/2	19/64"	302
X	1-5/8"	12-1/2	19/64"	274
X	1-3/4"	10	19/64"	259
X	2"	10	19/64"	228

Recommended Sizes and Quantities Commonly Used

Single layer 3/8" and 1/2" board
1-1/4" Nails — 6 Lbs. Per 1,000 Sq. Ft.

Single layer 5/8" board
1-5/8 Nails — 6 Lbs. Per 1,000 Sq. Ft.

Double Layer 3/8" board
1-1/4" Nails 1st ply — 3 lbs. Per 1,000 Sq.Ft.
1-5/8" Nails 2nd ply — 4 lbs. Per 1,000 Sq.Ft.

Drywall Nail Specifications

Size	Length mm	Gauge	Head mm	Nails per Kg
X	31.3	12-1/2	6.3	798.1
X	34.4	12-1/2	6.3	723.1
X	37.5	12-1/2	6.3	679.0
X	40.6	12-1/2	6.3	615.1
X	31.3	12-1/2	7.4	776.0
X	34.4	12-1/2	7.4	707.7
X	37.5	12-1/2	7.4	665.8
X	40.6	12-1/2	7.4	604.1
X	43.8	10	7.4	571.0
X	50.0	10	7.4	502.7

Recommended Sizes and Quantities Commonly Used

Single layer 9 mm and 13 mm board
31 mm Nails — 2.9 Kg Per 10 Sq.M

Single layer 16 mm board
41 mm Nails — 2.9 Kg Per 10 Sq.M

Double Layer 9 mm board
31 mm Nails 1st ply — 1.5 Kg Per 10 Sq.M
41 mm Nails 2nd ply — 2.0 Kg Per 10 Sq.M

Flooring Spiral Nails with Casing Head. These nails twist in as they are driven to eliminate reduce and squeaks. Nails are manufactured extra stiff to drive straight, and heads are easy to set.

Flooring Spiral Nail Specifications

Size	Length	Gauge	Head	Nails per Lb.
4d	1-1/2"	13	1/8"	369
6d	2"	11-1/2	11/64"	164
7d	2-1/4"	11-1/2	11/64"	152
8d	2-1/2"	11-1/2	11/64"	138

Flooring Spiral Nails with Flat Head. This type is the same as spiral nails above, except that heads are flat for nailing machine use.

Spiral Nails with Flat Heads - Specifications

Size	Length	Gauge	Head	Nails per Lb.
7d	2-1/4"	11-1/2	13/64"	152
8d	2-1/2"	11-1/2	13/64"	138

Recommended Sizes and Quantities Commonly Used

Floor Thickness	Nail Length	Amount per 1,000 Sq.Ft. of flooring — Face width of Flooring			
		1-1/2"	2"	2-1/4"	3-1/4"
3/8"	1-1/2"	20 Lbs.	15 Lbs.	12 Lbs.	
25/32" & 29/32"	2-1/4" to 2-1/2"	60 Lbs.	45 Lbs.	40 Lbs.	36 Lbs.
1-9/32" & 1-21/32"	2-3/4"		50 Lbs.		35 Lbs.

Flooring Spiral Nail Specifications

Size	Length mm	Gauge	Head mm	Nails per Kg
4d	38	13	3.1	813.5
6d	50	11-1/2	4.3	361.6
7d	56	11-1/2	4.3	335.1
8d	63	11-1/2	4.3	304.2

Spiral Nails with Flat Heads - Specifications

Size	Length mm	Gauge	Head mm	Nails per Kg
7d	56	11-1/2	5.1	335.10
8d	63	11-1/2	5.1	304.20

Recommended Sizes and Quantities Commonly Used

Floor Thickness	Nail Length	Amount per 1,000 Sq.Ft. of flooring — Face width of Flooring			
		38 mm	50 mm	56 mm	81 mm
9 mm	38 mm	9.7 Kg	7.3 Kg	5.9 Kg	
20 mm & 23 mm	56 mm to 63 mm	29.3 Kg	22.0 Kg	19.5 Kg	17.6 Kg
32 mm & 41 mm	69 mm		24.4 Kg		17.1 Kg

Underlayment Nails. Supplied anchor down, underlayment nails are good for placement of underlayment materials such as linoleum, rubber, asphalt and vinyl tiles. Nails are stiff stock or hardened nails and recommended for use over hardwood floors.

Underlayment Nail Specifications

Size	Length	Gauge	Head	Nails per Lb.
2d	1-1/4"	12-1/2	1/4"	362
3d	1-3/8"	12-1/2	1/4"	328

Note: When placing nails, always use safety glasses.

Recommended Sizes and Quantities Commonly Used

1/4" Underlayment	
1-1/4" Nails	400 per 100 Sq.Ft. or 3/4 Lb.
3/8" Underlayment	
1-1/2" Nails	400 per 100 Sq.Ft. or 1 Lb.

Underlayment Nail Specifications

Size	Length mm	Gauge	Head mm	Nails per Kg
2d	31	12-1/2	6.3	798.07
3d	34	12-1/2	6.3	723.12

Note: When placing nails, always use safety glasses.

Recommended Sizes and Quantities Commonly Used

6 mm Underlayment	
31 mm Nails	430 per 10 Sq.M or 0.3 Kg
9 mm Underlayment	
38 mm Nails	430 per 10 Sq.M or 0.5 Kg

Plywood Nails. Supplied anchor down, plywood nails are recommended for anchoring down plywood, sheathing or subfloor.

Plywood Nail Specifications

Size	Length mm	Gauge	Head mm	Nails per Kg
4d	38	12-1/2	6.3	628.32
5d	44	12-1/2	6.3	551.16
6d	50	12-1/2	6.3	496.04
8d	63	11-1/2	7.0	297.62
10d	75	9	7.8	154.32

Recommended Sizes and Quantities Commonly Used

6 mm Plywood

31 mm Nails   430 per 10 Sq.M or 0.3 Kg

9 mm Plywood

38 mm Nails   430 per 10 Sq.M or 0.4 Kg

13 mm - 16 mm - 19 mm Plywood subflooring & sheathing

50 mm & 63 mm Nails   161 per 10 Sq.M or 0.3 Kg to 0.6 Kg

25 mm Plywood subflooring

75 mm Nails   161 per 10 Sq.M or 1.0 Kg

Plywood Nail Specifications

Size	Length	Gauge	Head	Nails per Lb.
4d	1-1/2"	12-1/2	1/4"	285
5d	1-3/4"	12-1/2	1/4"	250
6d	2"	12-1/2	1/4"	225
8d	2-1/2"	11-1/2	9/32"	135
10d	3"	9	5/16"	70

Recommended Sizes and Quantities Commonly Used

1/4" Plywood

1-1/4" Nails   400 per 100 Sq.Ft. or 3/4 Lb.

3/8" Plywood

1-1/2" Nails   400 per 100 Sq.Ft. or 1 Lb.

1/2" - 5/8" - 3/4" Plywood subflooring & Sheathing

2" & 2-1/2" Nails   150 per 100 Sq.Ft. or 3/4 to 1-1/4 Lbs.

1" Plywood subflooring

3" Nails   150 per 100 Sq.Ft. or 2-1/4 Lbs.

Prefinished Paneling Nails. These are supplied plain, to be used for prefinished paneling. They are made with stiff wire and come in colors to match prefinished panels.

Prefinished Paneling Nail Specifications

Size	Length	Gauge	Head	Nails per Lb.
X	1"	16-1/2	3/32"	1,260
X	1-1/4"	15-1/2	3/32"	810
X	1-5/8"	15-1/2	3/32"	700

Prefinished Paneling Nail Specifications

Size	Length mm	Gauge	Head mm	Nails per Kg
X	25	16-1/2	2.3	2,777.82
X	31	15-1/2	2.3	1,785.74
X	41	15-1/2	2.3	1,543.24

Duplex Head Nails. These are supplied plain. They are made with two heads, one to secure the nail to the material and the other to allow easy removal of the nail. These are popular for concrete formwork and temporary work that requires withdrawing the nails later.

Duplex Head Nail Specifications

Size	Length Under Lower head	Distance Between Heads	Gauge	Nails per Lb.
8d	2-1/4"	1/4"	10	90
10d	2-3/4"	5/16"	9	59
16d	3"	3/8"	8	45
20d	3-1/2"	3/8"	6	28

Duplex Head Nail Specifications

Size	Length Under Lower head	Distance Between Heads	Gauge	Nails per Kg
8d	56 mm	6.3 mm	10	198.42
10d	69 mm	7.8 mm	9	130.07
16d	75 mm	9.4 mm	8	99.21
20d	88 mm	9.4 mm	6	61.73

Masonry Nails. Masonry nails are a fluted nail that is made of high carbon steel and then heat treated and tempered. They are used to fasten furring strips, cleats, floor plates, etc. directly to masonry.

Masonry Nail Specifications

Size	Length	Gauge	Head	Nails per Lb.
X	1/2"	9	5/16"	324
X	5/8"	9	5/16"	266
X	3/4"	9	5/16"	230
X	1"	9	5/16"	176
X	1-1/4"	9	5/16"	150
X	1-1/2"	9	5/16"	129
X	1-3/4"	9	5/16"	108
X	2"	9	5/16"	95
X	2-1/4"	9	5/16"	85
X	2-1/2"	9	5/16"	77
X	2-3/4"	9	5/16"	70
X	3"	9	5/16"	62
X	3-1/4"	9	5/16"	57
X	3-1/2"	7	3/8"	41
X	4"	7	3/8"	35

Note: Nails should be driven 3" or more from the end Or edge of a masonry unit. Normally only 3/4" penetration is recommended. Use safety glasses when using nails, especially hardened nails.

Masonry Nail Specifications

Size	Length mm	Gauge	Head mm	Nails per Kg
X	13	9	7.8	714.3
X	16	9	7.8	586.4
X	19	9	7.8	507.1
X	25	9	7.8	388.0
X	31	9	7.8	330.7
X	38	9	7.8	284.4
X	44	9	7.8	238.1
X	50	9	7.8	209.4
X	56	9	7.8	187.4
X	63	9	7.8	169.8
X	69	9	7.8	154.3
X	75	9	7.8	136.7
X	81	9	7.8	125.7
X	88	7	9.4	90.4
X	100	7	9.4	77.2

Note: Nails should be driven 75 mm or more from the end Or edge of a masonry unit. Normally only 19 mm penetration is recommended. Use safety glasses when Using nails, especially hardened nails.

Metal Roofing Nails with Neoprene Washers Attached. Supplied anchored down, they have washers that plug the nail hole. All hammering is done on the steel head. Conical washers are used with corrugated metal and flat washers for rib type material. These nails are normally dipped after threading. No raw steel is exposed. They come in colors.

Metal Roofing Nail Specifications
Checkered Head - Conical Neoprene Washer

Size	Length	Gauge	Head	Nails per Lb.
X	1-1/2"	10	7/16"	126
X	1-3/4"	10	7/16"	111
X	2"	10	7/16"	100
X	2-1/2"	10	7/16"	82
X	3"	10	7/16"	70
X	4"	10	7/16"	61
X	4-1/2"	9	7/16"	39

Metal Roofing Nail Specifications
Checkered Head – Flat Royalene Washer

Size	Length	Gauge	Head	Nails per Lb.
X	1-1/2"	10	7/16"	126
X	1-3/4"	10	7/16"	111
X	2"	10	7/16"	100
X	2-1/2"	10	7/16"	82
X	3"	10	7/16"	70

Metal Roofing Nail Specifications
Umbrella Head – Cronical Neoprene Washer

Size	Length	Gauge	Head	Nails per Lb.
X	1-1/2"	10	7/16"	122
X	1-3/4"	10	7/16"	109
X	2"	10	7/16"	96
X	2-1/2"	10	7/16"	70

Recommended Sizes and Quantities Commonly Used

New Roof	
New Roof 1-3/4" nails	100 nails per square - 1 Lb.
Over Existing Roof	
Old Roof 2-1/2" nails	100 nails per square – 1-1/2 Lb.

Metal Roofing Nail Specifications

Umbrella Head – Cronical Neoprene Washer

Size	Length mm	Gauge	Head mm	Nails per Kg
X	38	10	10.9	268.96
X	44	10	10.9	240.30
X	50	10	10.9	211.64
X	63	10	10.9	154.32

Recommended Sizes and Quantities Commonly Used

New Roof	
New Roof 44 mm nails	108 nails per 10 Sq.M – 0.45 Kg
Over Existing Roof	
Old Roof 63 mm nails	108 nails per 10 Sq.M – 0.68 Kg

Metal Roofing Nail Specifications

Checkered Head - Conical Neoprene Washer

Size	Length mm	Gauge	Head mm	Nails per Kg
X	38	10	10.9	277.78
X	44	10	10.9	244.71
X	50	10	10.9	220.46
X	63	10	10.9	180.78
X	75	10	10.9	154.32
X	100	10	10.9	134.48
X	113	9	10.9	85.98

Metal Roofing Nail Specifications

Checkered Head - Flat Royalene Washer

Size	Length mm	Gauge	Head mm	Nails per Kg
X	38	10	10.9	277.78
X	44	10	10.9	244.71
X	50	10	10.9	220.46
X	63	10	10.9	180.78
X	75	10	10.9	154.32

Pole Barn Anchor-Down Nails. This type of nail is good for general applications. Made with washers to plug nail holes, conical washers for corrugated metal and flat washers for rib type material. Nails are usually dipped after threading so that no raw steel is exposed. They come in colors.

Pole Barn Anchor Down Nail Specifications

Size	Length mm	Gauge	Head mm	Nails per Kg
6d	50	11	7.0	317.5
8d	63	11	7.0	253.5
10d	75	10	7.0	172.0
12d	81	10	7.0	160.9
16d	88	9	7.8	125.7
20d	100	7	9.4	77.2
30d	113	7	9.4	66.1
40d	125	5-1/2	10.5	50.7
40d	125	7	9.4	57.3
60d	150	5-1/2	10.5	41.9
60d	150	7	9.4	48.5
70d	175	5-1/2	10.5	35.3
80d	200	5-1/2	10.5	30.9

Pole Barn Anchor Down Nail Specifications

Size	Length	Gauge	Head	Nails per Lb.
6d	2"	11	9/32"	144
8d	2-1/2"	11	9/32"	115
10d	3"	10	9/32"	78
12d	3-1/4"	10	9/32"	73
16d	3-1/2"	9	5/16"	57
20d	4"	7	3/8"	35
30d	4-1/2"	7	3/8"	30
40d	5"	5-1/2	27/64"	23
40d	5"	7	3/8"	26
60d	6"	5-1/2	27/64"	19
60d	6"	7	3/8"	22
70d	7"	5-1/2	27/64"	16
80d	8"	5-1/2	27/64"	14

Truss and Pole Screw-Down Nails. These nails have a spiraled shank that helps prevent racking and twisting of nailed truss rafters.

Truss and Pole Screw-Down Specifications

Size	Length	Gauge	Head	Nails per Lb.
8d	2-1/2"	11	9/32"	115
10d	3"	10	9/32"	78
12d	3-1/4"	10	9/32"	73
16d	3-1/2"	9	5/16"	57
20d	4"	7	3/8"	35
30d	4-1/2"	7	3/8"	30
40d	5"	5-1/2	27/64"	23
60d	6"	5-1/2	27/64"	19
70d	7"	5-1/2	27/64"	16
80d	8"	5-1/2	27/64"	14

Truss and Pole Screw-Down Specifications

Size	Length mm	Gauge	Head mm	Nails per Kg
8d	63	11	7.0	253.53
10d	75	10	7.0	171.96
12d	81	10	7.0	160.94
16d	88	9	7.8	125.66
20d	100	7	9.4	77.16
30d	113	7	9.4	66.14
40d	125	5-1/2	10.5	50.71
60d	150	5-1/2	10.5	41.89
70d	175	5-1/2	10.5	35.27
80d	200	5-1/2	10.5	30.86

NAIL REFERENCE DATA

Specifications

Penny-Length Only		Common Nails No. per Kg.	Guage Specification Only		
Penny Size	Nail Length		Nail Guage	Decimal Equivant	Nearest mm +/-
2d	25 mm	1931.2	5-1/5	0.2000	5.08
3d	31 mm	1252.2	7	0.1770	4.30
4d	38 mm	696.7	9	0.1483	3.91
5d	44 mm	597.5	10	0.1350	3.52
6d	50 mm	399.0	11	0.1205	3.13
7d	56 mm	354.9	12	0.1055	2.73
8d	63 mm	233.7	12-1/2	0.0990	2.34
9d	69 mm	211.6	13	0.0915	2.34
10d	75 mm	152.1	14	0.0800	1.95
12d	81 mm	138.9	14-1/2	0.0760	1.95
16d	88 mm	108.0	16-1/2	0.0580	1.56
20d	100 mm	68.3			
30d	112 mm	52.9			
40d	125 mm	39.7			
50d	206 mm	30.9			
60d	150 mm	24.3			
70d	175 mm				
80d	200 mm				

NAIL REFERENCE DATA

Specifications

Penny-Length Only		Common Nails No. per Lb.	Guage Specification Only		
Penny Size	Nail Length		Nail Guage	Decimal Equivant	Nearest Inch +/-
2d	1"	876	5 1/5	0.2000	13/64"
3d	1-1/4"	568	7	0.1770	11/64"
4d	1-1/2"	316	9	0.1483	5/32"
5d	1-3/4"	271	10	0.1350	9/64"
6d	2	181	11	0.1205	1/8"
7d	2-1/4"	161	12	0.1055	7/64"
8d	2-1/2"	106	12 1/2	0.0990	3/32"
9d	2-3/4"	96	13	0.0915	3/32"
10d	3"	69	14	0.0800	5/64"
12d	3-1/4"	63	14 1/2	0.0760	5/64"
16d	3-1/2"	49	16 1/2	0.0580	1/16"
20d	4"	31			
30d	4-1/2	24			
40d	5"	18			
50d	5-1/2"	14			
60d	6"	11			
70d	7"				
80d	8"				

Recommended Nailing Schedule for Common Applications in Building Construction

Application	Nailed To	Nail Size Inches	Nail Size Penny	Nailing	Spacing C.C.	Nails per Joint
Mudsill, Partition plate, 2"	Concrete	2 1/2 - 2 3/4	8d - 9d	Toe	12" - 24"	---
Ditto in earthquake regions	Concrete	3 1/4 - 3 1/2	12d - 16d	Face	24" - 48"	---
Ditto 3"	Concrete	4 1/2	30d	Face	24" - 48"	---
Furring Strip	Concrete	1 1/2 - 1 3/4	4d - 5d	Toe	12" - 24"	---
Mudsill	Mudsill	2 1/2	8d	Face	---	2
Sleepers	Mudsill	3 1/2	16d	Face	---	2
Joists	Mudsill	3 1/4	12d	Face	---	2 - 3
Subflooring, 1" lumber, Plywood	Mudsill, sleeper, joist	2	6d	Face	6' & 12"	2 - 3
Subflooring, 2" lumber, Plywood	Mudsill, sleeper, joist	3	10d	Face	6' & 12"	3 - 3
Subflooring, 3/8" - 1/2" lumber, Plywood (dph.)	Mudsill, sleeper, joist	1 1/2	4d	Face	6' & 12"	---
Subflooring, 5/8" Plywood (dph.)	Mudsill, sleeper, joist	1 3/4	5d	Face	6' & 12"	---
Subflooring, 3/4" Plywood (dph.)part. bd.	Mudsill, sleeper, joist	2	6d	Face	6' & 12"	---
Subflooring, 1" - 1 1/8" Plywood (dph.) Part. bd.	Mudsill, sleeper, joist	2 1/2	8d	Face	6' & 12"	---
Underlayment, 1/4" - 5/16" plywood	Subfloor	1	2d	Face	6' & 6'-12"	---
Underlayment, 3/8" - 1/2" plywood	Subfloor	1 1/4	3d	Face	6' & 6'-12"	---
Underlayment, 5/8 plywood	Subfloor	1 1/2	4d	Face	6' & 6'-12"	---
Underlayment, 3/4" plywood	Subfloor	1 1/2	4d	Face	6' & 6'-12"	---
Underlayment, 7/8" plywood	Subfloor	1 3/4	5d	Face	6' & 6'-12"	---
Underlayment, 3/16" - 5/8" hardwood	Subfloor	1 - 1 1/2	2d - 4d	Face	6' & 12"	---
Flooring, T & G hardwood	Subfloor, joist, sleeper	2 - 2 1/2	6d - 8d	Face	10" - 18"	---
Flooring, T & G softwood	Subfloor, joist, sleeper	2 - 2 1/2	6d - 8d	Face	10" - 18"	---
Flooring, T & G hardwood, 3/8" & 1/2"	Subfloor, joist, sleeper	1 - 1 1/4	2d - 3d	Toe	10" - 18"	---
Flooring, T & G parquet	Subfloor	1 1/2	4d	Face	---	---
Framing plates	Stud	3 1/4	12d	Face	---	2
Framing studs	Stud, cripple, lintel, sill	2 1/2	8d	Face	16" - 24"	---
Framing studs	Stud, cripple, lintel, sill	2 1/2	8d	Face	---	3
Framing sole plate	Mudsill	3 1/4	12d	Face	16"	---
Framing top plate	Lower top plate	3 1/4	12d	Face	24"	---
Trussed rafter assembly		3 1/4	12d	Face	2 1/2" - 3"	Given
Trussed rafter assembly		2 1/2	8d	Face	2 1/2" - 3"	Given

See Notes, Key to Nail Types and Abbreviations on Later Page

Recommended Nailing Schedule for Common Applications in Building Construction

Application	Nailed To	Nail Size Inches	Nail Size Penny	Nailing	Spacing C.C.	Nails per Joint
Rafter, 4"	Top plate	3 1/4	12d	Toe	—	3
Rafter, 4"	Top plate	6	60d	Face	—	2
Rafter, 4"	Top plate	7	70d	Face	—	2
Rafter, 4",6" & 8"	Top plate	4 - 6	20d - 60d	Toe	—	2 - 3
		7 - 9	70d - 90d	Face	—	
Sheeting, 1" lumber	Framing, rafter	2	6	Face	—	2
Sheeting, 3/8" - 1/2" plywood	Framing, rafter	1 3/4	5d	Face	6" & 12"	—
Sheeting, 5/16" - 1/2" plywood (dph.)	Framing, rafter	1 1/2	4d	Face	6" & 12"	—
Sheeting, 5/8" plywood (dph.)	Framing, rafter	1 3/4	5d	Face	6" & 12"	—
Sheeting, 3/4" plywood (dph.)	Framing, rafter	2	6d	Face	6" & 12"	—
Sheeting, 1 1/8" plywood (dph.)	Framing, rafter	2 1/2	8d	Face	6" & 12"	—
Sheeting, insulated board, gypsum board	Framing, rafter	1 1/2	4d	Face	3" - 4" & 6" - 8"	—
Sheeting, hardboard, 3/8" - 5/8"	Framing, rafter	2	6d	Face	3" - 4" & 6" - 8"	—
Building paper	Sheathing	1/2 - 3/4	x	Face	6"-12"	—
Stripping, 3/4" x 3 5/8"	Framing, joist, rafter	2	6d	Face	—	2
Stripping, 1" x 4"	Framing, joist, rafter	2 1/2	8d	Face	—	2
Stripping, 2" x 3"	Framing, joist, rafter	3 1/2	8d	Face	—	2
Siding, wood 1"	Sheathing and framing	2 1/4	7d	Face	—	1
Siding, wood 1"	Sheathing and framing	2	6d	Face	—	1
Siding, wood 2"	Sheathing and framing	3	10d	Face	—	1
Siding, plywood	Sheathing and framing	2	6d	Face	6" & 12"	—
Siding, T & G wood	Sheathing and framing	1 3/4	5d	Toe	6" & 12"	—
Siding, insulated brick, wood shingle	Sheathing and framing	1 3/4	5d	Face	8" - 12"	—
Siding, wood shingle	Insulated sheathing	1 3/4 - 2	5d - 6d	Face	—	2
Siding, wood shingle	Insulated sheathing	1 3/4 - 2	5d - 6d	Face	—	2
Siding, wood shingle	Plywood	1 1/4	3d	Face	12"	2
Siding, hardboard	Framing	2 - 2 1/2	6d - 8d	Face	12"	2
Siding, hardboard battemboard	Framing	1 1/2	4d	Face	12"	—

See Notes, Key to Nail Types and Abbreviations on Later Page

Recommended Nailing Schedule for Common Applications in Building Construction

Application	Nailed To	Nail Size Inches	Nail Size Penny	Nailing	Spacing C.C.	Nails per Joint
Facia 1"	Framing, rafter	2 1/2	8d	Face	---	2
Facia 2" lumber	Framing, rafter	3 1/4	12d	Face	---	2
Roofing. Built-up	Sheathing	3/4 - 1 1/4	x - 3d	Face	10"	---
Roofing. Built-up	Poured gypsum	1 1/2 - 1 3/4	4d - 5d	Face	10"	---
Roofing. Asphalt shingle	Sheathing	3/4 - 2	x - 6d	Face	---	2 - 3
Roofing. Asphalt shingle	Sheathing	3/4 - 2	x - 6d	Face	---	2 - 3
Roofing. Wood shingle	Sheathing	3/4 - 2	x - 6d	Face	---	2
Roofing. Wood shingle	Sheathing	1 3/4	5d	Face	---	2
Roofing. Aluminum (corrugate or flat)	Rafter, purlin	1 1/2 - 1 3/4	4d - 5d	Face	12"	---
Roofing. Sheet metal (corrugate or flat)	Rafter, purlin	1 - 3	4d - 5d	Face	12"	---
Roofing. Glass fiber (corrugate or flat)	Rafter, purlin	1 1/2 - 3	4d - 10d	Face	12"	---
Roofing. Glass fiber (corrugate or flat)	Rafter, purlin	1 1/2 - 3	4d - 10d	Face	12"	---
Lath, expanded metal, K-Lath	Framing, joist	1 1/2	4d	Face	6" & 12"	---
Lath, gypsum plasterboard	Framing, joist	1 1/4	3d	Face	5"	---
Gypsumboard, 3/8"	Framing, rafter	1 1/4	3d	Face	5" - 8"	---
Gypsumboard, 1/2" - 5/8"	Framing, rafter	1 1/2	4d	Face	5" - 8"	---
Gypsumboard, prefinished	Framing, rafter	1 1/2	4d	Face	5" - 8"	---
Paneling, trim		1 - 1 1/4	2d - 3d	Face	---	---
Paneling, trim, exterior		1 - 1 1/2	2d - 4d	Face	---	---
Paneling, trim, exterior		1	2d	Face	---	---
Paneling, trim, exterior		1 1/2	4d	Face	---	---
Paneling, trim		1	2d	Face	---	---
Paneling, trim		1 1/2 - 1 3/4	4d - 5d	Face	---	---
Paneling, trim		2 1/2	8d	Face	---	---
Acoustic tile		1 - 1 3/4	2d - 5d	Face	---	---
Electric conduit	Wood	1 1/2 - 2	4d - 6d	Face	---	---
Electric conduit	Masonry	1 1/2 - 2	4d - 6d	Face	---	---
Fencing wire	Softwood, treated	1 1/2	4d	Face	---	---
Fencing wire	Hardwood	1 1/2	4d	Face	---	---

See Notes, Key to Nail Types and Abbreviations on Later Page

Abbreviations:
Corr. = Corrugated
Dph. = Diaphragm
Part.bd. = Particle Board

Recommended Nailing Schedule for Common Applications in Building Construction

Application	Nailed To	Nail Size (mm)	Nail Size (Penny)	Nailing	Spacing C.C.	Nails per Joint
Mudsill, Partition plate, 50 mm	Concrete	63 - 69	8d - 9d	Toe	300 - 600 mm	—
Ditto in earthquake regions	Concrete	81 - 88	12d - 16d	Face	600 - 1 200 mm	—
Ditto 75 mm	Concrete	113	30d	Face	601 - 1 200 mm	—
Furring Strip	Concrete	38 - 44	4d - 5d	Toe	300 - 600 mm	—
Mudsill	Mudsill	63	8d	Face	—	2
Sleepers	Mudsill	88	16d	Face	—	2
Joists	Mudsill	81	12d	Face	—	2 - 3
Subflooring, 25 mm lumber, Plywood	Mudsill, sleeper, joist	2	6d	Face	150 & 300 mm	3 - 3
Subflooring, 50 mm lumber, Plywood	Mudsill, sleeper, joist	3	10d	Face	150 & 300 mm	—
Subflooring, 9 mm - 13 mm lumber, Plywood (dph.)	Mudsill, sleeper, joist	38	4d	Face	150 & 300 mm	—
Subflooring, 16 mm Plywood (dph.)	Mudsill, sleeper, joist	44	5d	Face	150 & 300 mm	—
Subflooring, 19 mm Plywood (dph.)part. bd.	Mudsill, sleeper, joist	2	6d	Face	150 & 300 mm	—
Subflooring, 25 mm - 28 mm Plywood (dph.) Part. bd.	Mudsill, sleeper, joist	63	8d	Face	150 & 300 mm	—
Underlayment, 6 mm - 8 mm plywood	Subfloor	25	2d	Face	150 & 150 - 300 mm	—
Underlayment, 9 mm - 13 mm plywood	Subfloor	28	3d	Face	150 & 150 - 300 mm	—
Underlayment, 16 mm plywood	Subfloor	88	4d	Face	150 & 150 - 300 mm	—
Underlayment, 19 mm plywood	Subfloor	88	4d	Face	150 & 150 - 300 mm	—
Underlayment, 22 mm plywood	Subfloor	44	5d	Face	150 & 150 - 300 mm	—
Underlayment, 5 mm - 16 mm hardwood	Subfloor	25 - 88	2d - 4d	Face	150 & 300 mm	—
Flooring, T & G hardwood	Subfloor, joist, sleeper	50 - 63	6d - 8d	Face	250 - 450 mm	—
Flooring, T & G softwood	Subfloor, joist, sleeper	50 - 63	6d - 8d	Face	250 - 450 mm	—
Flooring, T & G hardwood, 9 mm & 13 mm	Subfloor, joist, sleeper	25 - 28	2d - 3d	Toe	250 - 450 mm	—
Flooring, T & G parquet	Subfloor	38	4d	Face	—	—
Framing plates	Stud	81	12d	Face	400 - 600 mm	2
Framing studs	Stud, cripple, lintel, sill	63	8d	Face	—	—
Framing studs	Stud, cripple, lintel, sill	63	8d	Face	—	3
Framing sole plate	Mudsill	81	12d	Face	400 mm	—
Framing top plate	Lower top plate	81	12d	Face	600 mm	—
Trussed rafter assembly		81	12d	Face	63 - 75 mm	Given
Trussed rafter assembly		63	8d	Face	63 - 75 mm	Given

See Notes, Key to Nail Types and Abbreviations on Later Page

6.42   ESTIMATORS REFERENCE BOOK

Recommended Nailing Schedule for Common Applications in Building Construction

Application	Nailed To	Nail Size Inches	Nail Size Penny	Nailing	Spacing C.C.	Nails per Joint
Rafter, 100 mm	Top plate	81	12d	Toe	---	3
Rafter, 100 mm	Top plate	150	60d	Face	---	2
Rafter, 100 mm	Top plate	175	70d	Face	---	2
Rafter, 100, 150 & 200 mm	Top plate	100 - 150 / 175 - 225	20d - 60d / 70d - 90d	Toe / Face	---	2 - 3
Sheeting, 25 mm lumber	Framing, rafter	50	6	Face	---	2
Sheeting, 9 mm - 13 mm plywood	Framing, rafter	44	5d	Face	150 & 300 mm	---
Sheeting, 8 mm - 13 mm plywood (dph.)	Framing, rafter	38	4d	Face	150 & 300 mm	---
Sheeting, 16 mm plywood (dph.)	Framing, rafter	44	5d	Face	150 & 300 mm	---
Sheeting, 19 mm plywood (dph.)	Framing, rafter	50	6d	Face	150 & 300 mm	---
Sheeting, 28 mm plywood (dph.)	Framing, rafter	63	8d	Face	150 & 300 mm	---
Sheeting, insulated board, gypsum board	Framing, rafter	38	4d	Face	75 - 100 mm & 150 - 200	---
Sheeting, hardboard, 9 mm - 16 mm	Framing, rafter	50	6d	Face	75 - 100 mm & 150 - 200	---
Building paper	Sheathing	13 - 19	x	Face	150 - 300 mm	---
Stripping, 19 mm x 91 mm	Framing, joist, rafter	50	6d	Face	---	2
Stripping, 25 mm x 100 mm	Framing, joist, rafter	63	8d	Face	---	2
Stripping, 50 mm x 75 mm	Framing, joist, rafter	88	8d	Face	---	2
Siding, wood 25 mm	Sheathing and framing	56	7d	Face	---	1
Siding, wood 25 mm	Sheathing and framing	50	6d	Face	---	1
Siding, wood 50 mm	Sheathing and framing	75	10d	Face	---	---
Siding, plywood	Sheathing and framing	50	6d	Face	150 & 300 mm	---
Siding, T & G wood	Sheathing and framing	44	5d	Toe	150 & 300 mm	---
Siding, insulated brick, wood shingle	Sheathing and framing	44	5d	Face	200 - 300 mm	---
Siding, wood shingle	Insulated sheathing	44 - 50	5d - 6d	Face	---	2
Siding, wood shingle	Insulated sheathing	44 - 50	5d - 6d	Face	---	2
Siding, wood shingle	Plywood	31	3d	Face	---	2
Siding, hardboard	Framing	50 - 63	6d - 8d	Face	300 mm	---
Siding, hardboard batternboard	Framing	38	4d	Face	300 mm	---

See Notes, Key to Nail Types and Abbreviations on Later Page

Recommended Nailing Schedule for Common Applications in Building Construction

Application	Nailed To	Nail Size Inches	Nail Size Penny	Nailing	Spacing C.C.	Nails per Joint
Facia 25 mm	Framing, rafter	63	8d	Face	—	2
Facia 50 mm lumber	Framing, rafter	78	12d	Face	—	2
Roofing. Built-up	Sheathing	19 - 31	x - 3d	Face	250 mm	—
Roofing. Built-up	Poured gypsum	38 - 44	4d - 5d	Face	250 mm	—
Roofing. Asphat shingle	Sheathing	19 - 50	x - 6d	Face	—	2 - 3
Roofing. Asphat shingle	Sheathing	19 - 50	x - 6d	Face	—	2 - 3
Roofing. Wood shingle	Sheathing	19 - 50	x - 6d	Face	—	2
Roofing. Wood shingle	Sheathing	44	5d	Face	—	2
Roofing. Aluminum (corrugate or flat)	Rafter, purlin	38 - 44	4d - 5d	Face	300 mm	—
Roofing. Sheet metal (corrugate or flat)	Rafter, purlin	25 - 75	4d - 5d	Face	300 mm	—
Roofing. Glass fiber (corrugate or flat)	Rafter, purlin	38 - 75	4d - 10d	Face	300 mm	—
Roofing. Glass fiber (corrugate or flat)	Rafter, purlin	38 - 75	4d - 10d	Face	300 mm	—
Lath, expanded metal, K-Lath	Framing, joist	38	4d	Face	150 & 300 mm	—
Lath, gypsum plasterboard	Framing, joist	28	3d	Face	125 mm	—
Gypsumboard, 9 mm	Framing, rafter	28	3d	Face	125 - 200 mm	—
Gypsumboard, 13 mm - 16 mm	Framing, rafter	38	4d	Face	125 - 200 mm	—
Gypsumboard, prefinished	Framing, rafter	38	4d	Face	125 - 200 mm	—
Paneling, trim		25 - 28	2d - 3d	Face	—	—
Paneling, trim, exterior		25 - 38	2d - 4d	Face	—	—
Paneling, trim, exterior		25	2d	Face	—	—
Paneling, trim, exterior		38	4d	Face	—	—
Paneling, trim		25	2d	Face	—	—
Paneling, trim		38 - 44	4d - 5d	Face	—	—
Paneling, trim		63	8d	Face	—	—
Acoustic tile		25 - 44	2d - 5d	Face	—	—
Electric conduit	Wood	38 - 50	4d - 6d	Face	—	—
Electric conduit	Masonry	38 - 50	4d - 6d	Face	—	—
Fencing wire	Softwood, treated	38	4d	Face	—	—
Fencing wire	Hardwood	38	4d	Face	—	—

See Notes, Key to Nail Types and Abbreviations on Later Page

Abbreviations:
Corr. = Corrugated
Dph. = Diaphragm
Part.bd. = Particle Board

Hardware Accessories Used for Wood Framing

The following items may be used to advantage in all types of wood construction. Developed primarily to provide better joints between wood framing members, in many cases their use has resulted in lower overall costs—the additional material cost being more than offset by increased labor efficiency.

Steel Joist Hangers. Used for framing joists to beams and around openings for stair wells, chimneys, hearths, ducts, etc. Made of galv. steel, varying from 12 ga. to 3/16" (5 mm) in thickness, with square supporting arms, holes punched for nails and with bearing surfaces proportioned to size of lumber. Approximate prices for sizes most commonly used are as follows:

Joist Size, in.(mm)	Gauge	Depth of seat in.(mm)	Opening in Hanger in.(mm)	Price Each
2"x6" (50x150)	12	2" (50)	1-5/8"x5" (41x125)	$0.81
2"x8" (50x200)	12	2" (50)	1-5/8"x5" (41x125)	0.90
2"x10" (50x250)	12	2" (50)	1-5/8"x8-1/2" (41x213)	0.93
2"x12" (50x300)	10	2-1/2" (62.5)	1-5/8"x8-1/2" (41x213)	1.15
4"x8" (100x200)	11	2" (50)	3-5/8"x5-1/4" (91x131)	0.08
4"x10" (100x250)	9	2" (50)	3-5/8"x5-1/4" (91x131)	1.29
4"x12" (100x300)	3/16	2-1/2" (62.5)	3-5/8"x8-1/2" (91x213)	1.99

Framing Accessories. These are used in light wood construction to provide face-nailed connections for framing members. Adaptable to most framing connections, they eliminate the uncertainties and weaknesses of toe-nailing. Manufactured of zinc-coated sheet steel in various gauges and styles. Framing accessories are designed to provide nailing on various surfaces. Special nails, approximately equal to 8d common nails, but only 1-1/4" (31 mm) long, to prevent complete penetration of standard nominal 2" (50 mm) lumber, are furnished with anchors. Approximate prices are as follows:

Type	Price per 100
Trip-L-Grips	$29.00
Du-Al-Clip	23.00
Nail-On Plates	35.00
Post Caps	76.50
H Clips	5.00
Angles	52.00

TECO TRIP-L-GRIPS   TECO DU-AL-CLIPS   TECO TRUSS PLATES

TECO POST CAPS   TECO H-CLIPS   TECO ANGLES

Timber Connectors. Timber connectors are devices for increasing the strength of bolted joints in timber construction. They are placed between adjacent faces of overlapping members, embedded to half-depth in each, to increase the bearing area at the connection. Several types of timber connectors are available, each for a particular purpose, as follows:

Split Ring

Toothed Ring

Spike Grids

Shear Plates

Teco Timber Connectors

Metal Connector Plates (Truss Plates). These are made from structural quality galvanized sheet steel and have integral teeth protruding from one face perpendicular to the plate. They are made in various sizes, thicknesses, and gauges and are designed to laterally transmit load in wood. Metal connector plates are installed by applying uniform pressure over the plate surface, which requires special equipment such as hydraulic or roller presses.

Wedge-Fit Split Rings. Split ring connectors are hot-rolled carbon steel bands, bent to form a ring with a tongue and groove meeting joint. The cross section of the steel band is tapered both ways from center to edges, providing a wedge-fit when connectors are inserted into tapered grooves, conforming to ring section and precut to half-depth in each contact face of overlapping members.

Grooves in wood members must be cut with a special grooving tool, using a power drill press or portable drill. Grooves and bolt holes may be cut in a single operation, using a combination grooving tool and drill bit. If bolt holes are bored first, a smooth shank pilot is used to guide the grooving tool. Split rings are available in two sizes: 2-1/2" (63 mm) diameter rings are used for wood framing involving lumber of 2" (50 mm) nominal dimension and are used in trussed rafter construction for spans up to 50' (15 m), and 4" (100 mm) diameter rings are used for medium to heavy construction, generally with members 3" (75 mm) or more in nominal thickness. Approximate prices of split rings and grooving tools are as follows:

	Approximate Prices	
	2-1/2" Dia.	4" Dia.
Description	63 mm	100 mm
Split rings, per 100 pcs.	$48.00	$78.00
Grooving tool cutter head, complete	138.00	201.00
Replacement cutters,		
Set of 4 for 2-1/2" (63 mm) dia.	80.00	--
Set of 6 for 4" (100 mm) dia.		106.00
Smooth shank pilot,		
9/16" (14 mm) dia.	12.00	--
11/16" (17 mm) dia.	12.00	--
13/16" (20 mm) dia.	--	17.00
15/16" (23 mm) dia.	--	20.00

Shear Plates. Shear plates are used for wood-to-steel connections and for wood structures that are frequently disassembled, where they are used singly for wood-to-steel joints and in pairs for wood-to-wood assemblies. They are also often used in place of split rings for field connected joints. Shear plates are placed in daps precut in the contact faces of wood members with a special grooving tool. Load is transferred from wood member to shear plate, from shear plate to bolt and thence to opposing member, which may be another shear plate in wood, a steel gusset plate, or rolled steel shape. Drilling, grooving, and dapping of wood members are accomplished in the same manner as described for split ring connectors.

Shear plates are available in two sizes—2-5/8" (66 mm) diameter made of pressed steel and 4" (100 mm) diameter made of malleable iron. Approximate prices of shear plates and grooving tools are:

	Approximate Prices	
	2-5/8" Dia.	4" Dia.
	66 mm	100 mm
Shear plates, per 100 pcs.	$72.00	$253.00
Grooving tool cutter head, complete	169.50	270.00

Replacement cutters,
Set of 3 for 2-5/8" (66 mm) dia. ....91.00 ....--
Set of 5 for 4" (100 mm) dia. ....--........................113.00
Smooth shank pilot,
13/16" (20 mm) dia. ....14.00 ....13.00
15/16" (23 mm) dia. ....--.......................17.00

Toothed Rings. Toothed ring connectors are toothed metal bands with each tooth corrugated or curved along its cross section for greater rigidity. Used for wood-to-wood connections between lighter structural members, they are also used for strengthening structures in place and structural repairs.

The function of toothed rings is similar to that of split rings, but the method of installation is not as efficient and requires more labor. Toothed rings are embedded in wood members by applying pressure. Where a large number of joints are to be assembled, a hydraulic jack setup can be used to advantage. It requires a 5 to 7-1/2 ton (4.5- to 6.8-metric ton) jack for single bolt connections and a 10-ton (9072-kg) jack for joints with two bolts. Where fewer joints are involved, the high strength rod assembly is recommended. The assembly consists of a high strength rod with Acme threads on one end, double depth nuts, ball bearing thrust washer, heavy plate washers, lock washer, and nut. Metal sleeve adapters permit the use of 1/2", 5/8", or 3/4" (13, 16, or 19 mm) diameter rods with a standard ball bearing washer. The length of rod required equals the total thickness of wood members plus 1" (25 mm) for each layer of toothed rings plus 5" (125 mm) for nuts and washers. Ratchet and impact wrenches speed assembly. After the joint is drawn closed, the high-strength rod assembly is replaced with ordinary bolt and washers.

Spike Grids. Spike grid connectors are malleable iron castings of a grid-like structure with teeth protruding from both faces. Available in three styles: square flat type, 4-1/8"x4-1/8" (103 x 103 mm); circular type, 3-1/4" (81 mm) dia., used between sawn timbers; and single curve type, 4-1/8"x4-1/8" (103 x 103 mm), used between curved faces of poles or piling and sawn timbers. Spike grids are embedded under pressure, using the high strength rod assembly, as previously described for toothed rings. Approximate prices are as follows:

Type of spike grid	Square Flat	Single Curve	Circular
Price per 100 pcs.	$281.00	$281.00	$218.00

Installation tool for spiked grids, complete with adapter, $213.00.

WOOD FLOOR AND ROOF TRUSSES

Some advantages of wood floor trusses are speed of construction and the ability to install utility lines, such as HVAC ducts, water lines, sprinkler lines, and electrical cables so that they do not interfere with one another. Easy of installation combines with economy.

Floor trusses are used for spans from 12'-0" (3.6 m) to as large as 70'-0" (21 m) and are found in all conventional types of construction, as well as in factories, storage facilities, and any other structure where a large, uninterrupted area is required.

Wood roof trusses, either shop or field fabricated, are used in buildings where clear floor space is a requirement and the width of the building exceeds the economical span of roof joists. Some of the advantages of trussed roof construction are fast erection, early availability of unobstructed weather-protected space, simplified installation of ceiling, floor, mechanical and electrical systems, and use of non-loadbearing movable partitions.

Wood roof trusses are used for spans as short as 25'-0" (7.5 m) and can be used up to 200'-0" (60 m). They are widely used in garages, factories, hangars, gymnasiums, auditoriums, bowling alleys, dance halls, supermarkets, and other structures of this type.

When using wood roof trusses, it is common to span the shorter dimension of the building. One might use any of several truss types, such as mechanically connected or bolted plywood gusset. The type of truss depends on the degree of economy desired, building use, and the general architectural effect desired. Prices of roof trusses are governed by the following conditions:

1.   Cost of material and labor.
2.   Loading conditions and spacing.
3.   Difficulties of installation.
4.   Requirements of local building ordinances.

The last condition causes the price of trusses to vary considerably in different parts of the country. Some cities may require only a 25-lb. (11.25-kg) live load while others require a 50-lb. (22.5-kg) live load, necessitating heavier construction.

The spacing of wood roof trusses that directly support roof sheathing is usually 2'-0" (0.6 m). Where roof loads are light and the installation of a ceiling is not required, spacings of 4'-0" to 4'-10" (1.2-1.4 m) are advantageous. Where snow loads are especially heavy, spacings of 16" (400 mm) and even 12" (300 mm) have been used.

Prices given on following pages are only approximate. The estimator must obtain definite prices—for trusses either delivered at the building site or installed in place—from the manufacturer. Truss suppliers are familiar with conditions in all parts of the country and can quote a definite price.

Bowstring Bolted Roof Truss. For industrial and many commercial structures, the bowstring truss is popular. This type of truss has a curved top chord that starts from a point approximately 10″ (250 mm) high at the end of the truss, rising to a maximum height at centerline of about 1/8 of the total span. This type of truss gives a curved roof shape. Because of its low height at the ends of the truss, a minimum of masonry is required for parapet or fire walls.

The truss may be left exposed, carrying roof load only, as in garages or industrial plant, or it can be designed to carry monorail, floor systems, or other concentrated loads. The bowstring truss is also designed to carry ceiling loads for use in stores, automobile agency salesrooms, and other commercial structures. Bowstring trusses provide wind bracing for walls through diaphragm action and can be used with "knee braces" to develop portal action in mill type buildings. They are efficiently fabricated with bolts, resulting in little reduction in cross-sectional areas of the pieces connected.

Approximate Prices of Wood Bowstring Truss. They are designed with glued laminated or nailed 2″x3″ (50 x 75 mm) or 2″x4″ (50 x 100 mm) top chords, built to conform to a parabolic curve, furnished in lots of 5 or more, spaced 16′-0″ (4.8 m) on centers, with 25-lb. (11.25-kg) live load and 40-lb. (18-kg) total load.

Span, feet	Height, feet	Span, meters	Height meters	Price each
20′-0″	2′-6″	6.10	0.76	$362.00
25′-0″	3′-1″	7.62	0.94	368.00
30′-0″	3′-9″	9.14	1.14	374.00
35′-0″	4′-4″	10.67	1.32	391.00
40′-0″	5′-0″	12.19	1.52	431.00
45′-0″	5′-7″	13.72	1.70	437.00
50′-0″	6′-3″	15.24	1.91	483.00
55′-0″	6′-10″	16.76	2.08	604.00
60′-0″	7′-6″	18.29	2.29	684.00
65′-0″	8′-1″	19.81	2.46	782.00
70′-0″	8′-9″	21.34	2.67	932.00
75′-0″	9′-4″	22.86	2.84	1,029.00
80′-0″	10′-0″	24.38	3.05	1,202.00
85′-0″	10′-7″	25.91	3.22	1,346.00
90′-0″	11′-3″	27.43	3.43	1,489.00
95′-0″	11′-10″	28.96	3.61	1,668.00
100′-0″	12′-6″	30.48	3.81	1,725.00
105′-0″	13′-1″	32.00	3.99	1,875.00
110′-0″	13′-9″	33.53	4.19	1,955.00
115′-0″	14′-4″	35.05	4.37	1,961.00
120′-0″	15′-0″	36.58	4.57	2,162.00
125′-0″	15′-7″	38.10	4.75	2,323.00
130′-0″	16′-3″	39.62	4.95	2,496.00
135′-0″	16′-10″	41.15	5.13	2,645.00

Span, feet	Height, feet	Span, meters	Height meters	Price each
140'-0"	17'-6"	42.67	5.33	2,806.00
145'-0"	18'-2"	44.20	5.54	2,875.00
150'-0"	18'-9"	45.72	5.72	3,025.00

Crescent Type Roof Truss. This is increasingly popular and is identified by its curved lower chord, which affords a higher ceiling at the mid-span of the truss. It produces a curved roof, similar to the bowstring truss. This truss presents a curved ceiling effect and is used in gymnasiums, auditoriums, low-cost churches, and more elaborate stores and restaurants. Recommended span is from 20'-0" to 85'-0" (6.0-25.5 m).

Approximate Prices of Crescent Type Wood Roof Trusses. They are furnished in lots of 5 or more, spaced 16'-0" (4.8 m) on centers, 25-lb. (11.25-kg) live load, and 40 lb. (22.5-kg) total load.

Span, feet	Height, feet	Camber, ft.-in.	Span, meters	Height, meters,	Camber, mm	Price each
20'-0"	5'-0"	1'-0"	6.10	1.52	0.30	$437.00
25'-0"	5'-6"	1'-3"	7.62	1.68	0.38	454.00
30'-0"	6'-0"	1'-6"	9.14	1.83	0.46	483.00
35'-0"	6'-6"	1'-9"	10.67	1.98	0.53	500.00
40'-0"	7'-0"	2'-0"	12.19	2.13	0.61	529.00
45'-0"	7'-6"	2'-3"	13.72	2.29	0.69	575.00
50'-0"	8'-0"	2'-6"	15.24	2.44	0.76	604.00
55'-0"	8'-6"	2'-9"	16.76	2.59	0.84	730.00
60'-0"	9'-0"	3'-0"	18.29	2.74	0.91	805.00
65'-0"	9'-6"	3'-3"	19.81	2.90	0.99	949.00
70'-0"	10'-0"	3'-6"	21.34	3.05	1.07	1,024.00
75'-0"	10'-6"	3'-9"	22.86	3.20	1.14	1,300.00
80'-0"	11'-0"	4'-0"	24.38	3.35	1.22	1,380.00
85'-0"	11'-6"	4'-3"	25.91	3.51	1.30	1,524.00

Belgian Roof Truss. This type is used where it is desirable to give a conventional building a more pleasing appearance through the use of a peaked roof. It is recognized by its sloping top chord and horizontal lower chord. In most instances, a ceiling is applied to the lower chord of these trusses, while the top chord may have shingles or slate as a covering. It is used on some higher class store buildings and low-cost churches and is recommended for spans from 20'-0" to 85'-0" (6.0-25.5 m).

Belgian roof trusses are less efficient than the bowstring type, because the connections generally govern the member sizes. They cost about 50% more than bowstring type trusses.

The Double fink truss is also referred to as a belgian truss and is used for spans from 36'-0" to 60'-0" (10.8-18.0 m).

Flattop Roof Truss. Industrial plants, warehouses, sheds, etc., when built of frame construction, often find use for a truss that absorbs wind loads, in addition to carrying roof and concentrated loads. The flattop truss is ideal for this purpose. It has parallel top and bottom chords, forming a large rectangle. Spans should not exceed 65' where cost is an important factor.

Howe Truss. The Howe truss can be used for spans from 16' to 18' (4.8-5.4 m). It is efficient for loading conditions that balance the top and bottom chords and is often used when the truss is designed as a girder, with loads applied for the bottom chord.

Parallel Chord 4x2 Truss. Normally used in floor truss design, it has parallel chords of stress-related 2"x4" (50 x 100 mm) lumber set flat so that the wider faces are the bearing surface of the truss. The web members are also cut from 2"x4" (50 x 100 mm) and positioned so that they support the chord across their full widths. This type of truss can be manufactured with duct chase openings so that wiring, piping, and ducts can run within the chords.

Parallel Chord 2x4 Truss. Generally used in roof truss design, it is gaining popularity because of the long, clear spans and shallow depth. The slope can be built into the truss, or drainage can be provided by using different opposite side wall heights to slope the truss. Roof slopes should be at least 1/4" (6 mm) per foot of span.

Modified Queen Post Truss. This type of truss is often used for spans the same as those requiring double W but that have load conditions that require fewer bottom chord panels.

Fink Truss. The fink truss is generally suitable for spans from 16' (4.8 m) to as long as 46' (13.8 m) and for all classes of construction. It is an efficient and cost effective truss configuration, 50% to 60% the cost of a comparable steel truss.

Three-Hinged Arch. This type of structure provides good overhead clearance and attractive appearance. However, high forces develop in some joints in the longer spans. It is recommended that the designer consult with the local fabricator before specifying this type.

Cantilever Truss. Trusses with single or double cantilever sections are possible. Cantilevers can approach one-fourth of the distance of the main interior truss span.

Clerestory Truss. This type of truss is used extensively in industrial and agricultural buildings, in spans up to 60' (18 m).

Inverted Truss. This truss is used for variations in interior and exterior roof profile.

Vaulted Ceiling Truss. A partial scissors is used to achieve a vaulted ceiling over a portion of a residence, such as the living room, while maintaining flat ceilings in the other rooms. Designers must consider horizontal movement at the bearing points.

Types of Trusses

King Post Truss

Modified Queen Post Truss

Fink Truss

Howe Truss

Double Fink Truss

Cantilever Truss

Parallel Chord 4x2 Truss

Triple W Truss

Parallel Chord 2x4 Truss

Mono-Pitch Truss. Often used to form a shed roof structure on a residence or garden apartment building where a particular architectural effect is desired. It is also popular and economical for use in back-to-back construction with three or more bearing walls. This gives the same architectural effect as a gabled roof but uses two shorter span mono-pitch trusses. It might vary in configuration but offers a simple solution to some design problems.

Types of Trusses

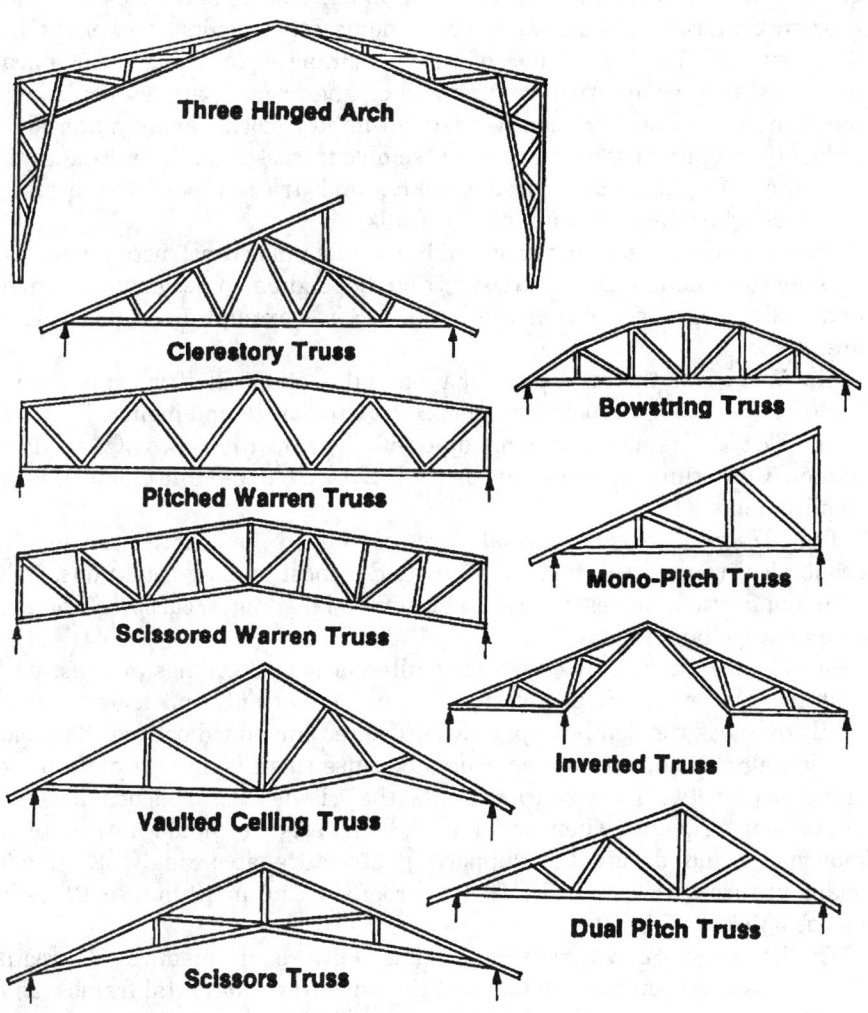

Dual-Pitch Truss. Some design solutions require an asymmetric roof pitch. This type allows different pitch on each side of the roof.

Pitched Warren Truss. This is often specified when a long span and low pitch is required. It is well suited where natural lighting is sought, because vertical windows can be designed into the top section of the loadbearing walls between the truss supports. This truss form is most economical in spans from 30' to 70' (9.0-21.0 m), on center spacings from 2' to 8' (0.6-2.4 m).

Designers and builders of residential construction, multiple unit housing projects, and light commercial construction use trusses to reduce costs and to shorten construction time. Most small home constructors can benefit from this type of framing. Some of the advantages to using mechanically connected trusses are: roof framing and ceiling framing are accomplished at the same time; members can be made of lighter stock, saving material and reducing weight; trusses can be pre-assembled; trusses can be erected rapidly and the job put under cover quickly; and trusses permit the use of non-bearing partitions for all interior walls.

Trussed rafters used in house building and similar construction vary as to truss pattern and method of construction, depending on design requirements and individual preference of the builder. Some examples of truss patterns are:

The W-Type is the most popular type and is adaptable for spans from 18' up to 40' (5.4-12.0 m); roof slope from 2 in 12 to 6 in 12 and higher.

The *Triple-W* is used for spans up to 80' (24 m) with slopes of 3 in 12 and higher. Centerline spacings can be from 2' to 20' (0.6-6.0 m), depending on requirements.

The *Kingpost* truss is usually recommended for shorter spans. The economical range is up to 26' (7.8 m) under most loading conditions. It has wide application, for residential and commercial garages, carports, and short span storage buildings.

The *Scissors* truss is used where vaulted or sloped ceilings are desired for architectural effect. Slope selection is very important, and a designer can easily overlook the depth in span ratio. The recommended bottom chord slope is one-half the slope of the top chord. Because there is no horizontal bottom chord or tie between the bearing points, the scissor truss develops movement at its bearing points when under load. The designer should anticipate this movement and design the supporting structure accordingly. Economical spans are from 18' to 36' (5.4-10.8 m); roof slopes 5 in 12 to 6 in 12; ceiling slopes 2 in 12 to 3 in 12.

The *Scissored Warren* truss has similar depth requirements as the regular chord truss and can achieve many of the advantages of portal frames, as far as internal clearance and appearances are concerned.

Trussed rafter construction varies principally as to the method of assembly, whether members are fastened together at joints by bolting or nailing, with split ring connectors and bolts, or flat, bent, pronged or toothed metal plates, or by gluing.

Using the bolting method, the members at all major joints, i.e., heel joints, apex, splices, etc., are connected by one or more bolts, or by using split rings to transfer the loads. In some cases, one or both ends of diagonal members may be fastened by nailing.

For nailed trussed rafters, all joint connections use metal truss plates, plywood gussets and nails, or ordinary nailed joints.

Other Types of Commonly Used Wood Trusses. The *Double Fink* truss is generally used for spans from 36' to 60' (10.8-18.0 m). It is often used in shorter spans over other types of trusses, because it is easier for workers to handle without special equipment.

The *Piggy Back Sectional* truss is generally used when the slope (or pitch) exceeds the height limits of the fabricating press or by highway restrictions.

The *Long Sectional* is a truss that is fabricated in sections for when the overall length exceeds permissible highway length restrictions, the length of the fabricating jig, or if the overall height is too large to be put through the fabricating press. The height is reduced by producing it in two sections.

An extremely long truss with cantilevered ends can be manufactured in three sections. Trusses of this type, 128' (38.4 m) in overall length, have been fabricated in this manner. The overall height would be limited to the width of the fabrication press, but a fourth piggyback section could be added to the center section.

The *Sectional Scissors* truss is used when overall height exceeds highway regulations or limits of the press. Splices are applied at the job site by general contractor or a subcontractor. Due to its very high center of gravity and unstable configuration, this type requires extreme care in erection.

There are alternate methods of splicing sectional scissors trusses. One such method, when it is feasible, has the advantage of providing manufacturing symmetry.

Roof trusses offer numerous advantages over conventional roof framing: a truss can span greater roof distances; longer clear spans allow complete freedom in the layout of interior partitions; the bottom chord is available for ceiling finish material; and the cost is usually less, when on-site labor is considered.

Special Conditions Associated with Roof Truss. The *parallel chord* roof truss provides a flat roof that can be sloped to one side to drain by using wedge pieces added to the top chord. An alternate method is a slight variation in the bearing plate heights. This technique is acceptable where a slight variation in the interior room height is acceptable or where a hung ceiling will conceal the variation. When there is significant slope and no horizontal bottom chord, one must allow for horizontal movement at the wall.

Truss Openings. The two most widely used floor truss configurations are *Warren* and *Fan*. Each type of truss allows openings for through-truss duct work. The table below shows the allowable openings for each type of truss. Note that the warren and fan trusses both allow a 7" (175 mm) opening when the depth of the truss is 12" (300 mm). However, if the truss is 16" (400 mm) deep, the warren truss allows a 10-1/4" (256 mm) opening, the fan truss an 11" (275 mm) opening.

Center Duct Depth/Opening

Truss Depth, inches	Truss Depth, mm	Maximum Opng Hgt., inches	Maximum Opng Hgt., mm	Width in inches	Width in mm
12	300	9	225	18	450
14	350	11	275	22	550
16	400	13	325	24	600
18	450	15	375	24	600
20	500	17	425	24	600
22	550	19	475	24	600
24	600	21	525	24	600
26	650	23	575	24	600
28	700	25	625	24	600
30	750	27	675	24	600

Warren Truss Depth/Opening

Overall Depth, inches	Overall Depth, mm	Max. Rnd. Opng. in inches	Max. Rnd. Opng. in mm
12	300	7	175
14	350	8-3/4	219
16	400	10-1/4	256
18	450	11-1/2	288
20	500	12-1/2	313
24	600	14-1/2	363
30	750	16-1/2	413

The truss fabricator has dimensions of openings for intermediate depths of trusses.

Overall Depth, inches	Overall Depth, mm	Max. Rnd. Opng. in inches	Max. Rnd. Opng. in mm
12	300	7	175
14	350	9	225
16	400	11	275
18	450	13	325

The designer is cautioned to subtract insulation thickness and working tolerances from these dimensions to determine maximum net duct dimensions.

Installation of Wood Trusses

An estimator must consider the entire process of handling and installing wood trusses. The Truss Plate Institute recommends specific procedures, and the information provided below is excerpted from its publication, *Commentary and Recommendations for Handling Installing & Bracing Metal Plate Connected Wood Trusses (HIB-91).*

Unloading and Lifting. During the unloading process, and throughout all phases of construction, care should be taken to avoid lateral bending, which can cause joint and lumber damage. Use proper equipment to lift trusses. A crane with spreader bar and cables is strongly recommended for trusses with spans greater than 30' (9.0 m). For spans 30' (9.0 m) or less, bridle and chokers may be used.

Do not lift bundles by the strapping, which is not strong enough to safely support the weight of the trusses. It is only meant for shipping purposes. Do not attach cables, chains, or hooks to web members. Thread cables under the top chord of a bundle of trusses near panel points closest to the quarter or third points of the trusses. Lift the bundle slowly in order to determine the load balance. If it is unbalanced, then lower the bundle, readjust the cable, and lift again.

Lifting with a bridle or spreader bar requires the same care in balancing the load. Never lift an unbalanced bundle of trusses. Do not lift single trusses with spans greater than 30' (9.0 m) by the peak.

Storage. Whenever possible, trusses should be unloaded in bundles, on a relatively smooth ground, and picked up by the top chord in a vertical position only. It is common to unload trusses from a trailer one at a time and store them in a stable position to prevent toppling or shifting. When they are stored horizontally, blocking or sleepers should be maximum 10' (3.0 m) centers to prevent lateral bending. Do not store unbraced bundles upright.

If trusses are stored vertically, extreme care should be taken to insure that they are braced and blocked in a stable manner that will prevent toppling. Pitched trusses should be stored with the peak up.

Installation. An open web metal plate connected truss is a manufactured product, comprised of various lumber components that are precisely cut and fitted together with metal connector plates. A metal plate connected wood truss should be installed with greater care than a monolithic item such as sawn wood joist, steel channel, or a beam. Without bracing, or some other type of restraining device, a truss can be unstable.

The installer is responsible for selecting the most suitable method and sequence of installation that is available and which is consistent with the plans and specifications and such information as may be furnished prior to installation. Trusses can be installed by hand or by mechanical means, depending on truss span, installed height above grade, and accessibility by equipment (such as crane or forklift).

The installer must understand truss design drawings, placement plans, and all notes and cautions on them. He should clarify any and all questions with the manufacturer before starting the installation. Use particular care when installing under adverse conditions, such as high winds, uneven or sloping terrain, proximity of high voltage power lines, or constricted job site.

Never walk on trusses that are lying flat. Under no circumstances should construction loads of any description be placed on unbraced trusses.

Hand Installation. Installation by hand should be limited to trusses of a size and configuration that can be carried, raised up to bearing support height, and rotated into position without excessive lateral deflection (bow), which produces strain in the lumber or metal connector plates and weakens the joints. Lateral deflection greater than 3" in 10' (75 mm in 3 m) of span is excessive. Trusses should be handled so as to ensure support at intervals of 25' (7.5 m) or less.

When installed by hand, trusses are positioned over the side walls and rotated into position using a fork-like lifting pole. The longer the span, the more workers will be necessary to avoid lateral strain on the truss. Depending on length, the truss should be supported at the peak for spans less than or equal to 20' and at quarter points for spans less than or equal to 30' (9.0 m).

Mechanical Installation. Trusses that are installed by mechanical means should be handled so as to ensure support at intervals of 25' (7.5 m) or less. The installer should provide adequate rigging (crane, fork lift, slings, tag lines, and spreader bars) for sufficient control during lifting and placement to assure safety to personnel and to prevent damage to trusses and property. Slings, tag lines, spreader bars, etc. should be used in a manner that will not cause any damage to metal connector plates.

Trusses that are lifted in place in banded bundles should be securely supported by temporary means, permitting the safe removal of banding and the sliding of individual trusses. Do not lift bundled trusses by their strapping.

Take care to position truss bundles so that the supporting structure is not overloaded. The bundle straps (banding) can be broken after the bundle has been placed on the supporting structure and prior to the release of the lifting cables.

Under certain conditions, several trusses may be assembled on the ground into structural subcomponents, complete with temporary bracing members and portions of the roof deck or subfloor, and then carefully lifted into place as a self supporting unit. These units are assembled as needed until the roof or floor structure is complete. Usually some decking is omitted so that the stagger-lap is provided between adjacent units for diaphragm continuity. This method requires substantial planning beforehand and special engineering for the trusses, such as location and design of pick-up or liftpoints. This method has been found to be a safe solution to the installation of especially long spans or complex shapes.

Trusses that are mechanically installed one at a time should be held safely in position with the erection equipment until such time as all temporary bracing has been installed. Lines from the end of the spreader bar should "toe-in". Do not permit lines to "toe-out", which will tend to cause trusses to buckle.

For truss spans greater than 60' (18 m), the suggested lift procedure requires the use a strong-back. The strong-back should be attached to the top chord and web members at intervals of approximately 10' (3.0 m) and should be at or above mid height of the truss to prevent overturning. The strong-back should be monolithic with sufficient strength to safely carry the weight of the truss and rigid enough to resist truss bending.

Each truss should be set in position in accordance with the designer's plan and held with the hoisting equipment until the ends of the truss are securely fastened and temporary bracing is installed. Only then should the hoisting equipment be released from the braced truss and used for lifting, placing, and securing the next truss.

Installation tolerances are critical to achieving an acceptable roof or floor line and for effective bracing. A stringline, plumb bob, level, or transit is recommended to attain acceptable installation tolerances.

Trusses should not be installed with an overall bow, or bow in any chord or panel, that exceeds the lesser of L/200 or 2 inches, where L is the span of the truss, chord, or panel length in inches.

L, inches	L, mm	L/200, inches	L/200, mm	L, feet	L, meters
25	625	1/8	3	2.1	0.6
50	1250	1/4	6	4.2	1.3
75	1875	3/8	9	6.3	1.9
100	2500	1/2	13	8.3	2.5
125	3125	5/8	16	10.4	3.1
150	3750	3/4	19	12.5	3.8
175	4375	7/8	22	14.6	4.4
200	5000	1	25	16.7	5.0
225	5625	1-1/8	28	18.8	5.6
250	6250	1-1/4	31	20.8	6.2
275	6875	1-3/8	34	22.9	6.9
300	7500	1-1/2	38	25.0	7.5
350	8750	1-3/4	44	29.2	8.8
400	10,000	2	50	33.3	10

Trusses should not be installed with a variation from plumb (vertical tolerance) at any point along the length of the truss from top to bottom chords that exceeds -1/50> of the depth of the truss at that point (D/50) or 2", whichever is less. This does not apply to trusses specifically designed to be installed out of plumb.

D, inches	D, mm	D/50, inches	D/50, mm	D, feet	D, meters
12	300	1/4	6	1	0.3
24	600	1/2	13	2	0.6
36	900	3/4	19	3	0.9
48	1200	1	25	4	1.2
60	1500	1-1/4	31	5	1.5
72	1800	1-1/2	38	6	1.8
84	2100	1-3/4	44	7	2.1
96	2400	2	50	8	2.4
108	2700	2	50	9	2.7

Location of trusses along the bearing support should be within ±1/4" (6 mm) of plan dimensions. Special hangers or supports should be located to support trusses within ±1/4" (6 mm) of plan dimensions. Trusses are to be located at o.c. spacing specified by the design engineer.

Top chord bearing parallel chord trusses should have a maximum gap 1/2" (13 mm) between the inside of the bearing and the first diagonal or vertical web.

Correction of Errors in Truss Installation. Errors in building lines or dimensions, or by subcontractors and suppliers, must be corrected before installation begins. Alterations, such as cutting overhangs to proper length, might be required and should be made by the builder of record. Any correction that involves the cutting, drilling, or relocation of a truss member or metal connector plate is considered major and should not be made without notifying the truss manufacturer.

A major correction calls for an engineering analysis by qualified registered engineer prior to the field work. Cutting or drilling trusses in the field without the approval of this engineer should be strictly prohibited. Trusses might be designed by the truss design engineer for specific reasons to be field cut by the installer at the designated locations. Such modifications should be indicated on the engineering drawing.

Field Assembly. Trusses that are too long for delivery to the jobsite in one piece can be designed to be delivered in two or more parts and then spliced together at the jobsite. The installer should carefully follow the splicing specifications shown on the truss drawings. Splicing may be performed on the ground before installation, or the truss sections may be supported by temporary shoring and splices installed by workers on a safe working surface.

Trusses that are too high for delivery to the jobsite in one piece may be manufactured in two or more sections and piggy-backed at the jobsite. The installer should observe the permanent bracing and connection details shown on the truss or piggyback design drawings. The supporting trusses should be completely installed, including the temporary and permanent bracing and sheathing when required, before installing top or supported truss sections.

Installation Sequence for Temporary and Permanent Bracing of Wood Truss. Install the first truss with a ground bracing system, which is constructed as follows:

1. Attach verticals to the end wall at the same spacing as the top chord lateral bracing.
2. Attach diagonals to the vertical and ground stake.
3. Attach horizontal ties or backup ground stakes if required.
4. Attach struts to the diagonals if required.
5. Attach lateral braces to the diagonals if required.
6. Attach end braces to the diagonals and lateral braces.

For the first group of three to six trusses, the sequence is as follows:

1. Install three to six trusses with top chord lateral braces affixed to the first (ground-braced) truss. The number of trusses to be initially installed should be specified in the bracing design.
2. Install top chord diagonal bracing between top chord lateral braces at intervals specified in the bracing design.
3. Install bottom chord lateral braces at no greater than 15' (4.5 m) on center.
4. Install web lateral bracing as specified on the truss design.
5. Install bottom chord diagonal braces between bottom chord lateral braces at the same interval as top chord diagonal bracing.
6. Install web cross bracing on web(s) corresponding to the nearest adjacent panel point(s) to the bottom chord lateral brace(s).

Continue installation of trusses with top chord lateral braces as specified on the bracing design or truss layout plan. The number of trusses to be installed between sets of top chord diagonal braces should be specified on the bracing design or truss layout plan.

Repeat 2) and 4) above for each group of trusses and then install sets of web cross bracing at 20-ft. (6.0-m) intervals. Install sets of bottom chord diagonal braces at the same intervals as top chord diagonal bracing and at each end bay. Top chord diagonal braces may remain as permanent bracing per the building design. Install permanent sheathing materials immediately as the temporary bracing is removed. Ground bracing may be removed only after the top chord plane is completely sheathed. Install any other permanent bracing as specified by the building design.

Note that all bracing lumber should be no less than 2'x4'x10' (0.6 x 1.2 x 3 m). A minimum of two 16d double head nails should be used at each connection.

Labor Erecting Wood Roof Trusses. Erection of wood roof trusses in confined areas, or where the supporting structure is not heavy enough to support new loads, has usually been accomplished using a gin pole. However, most truss erection today uses a special hydraulic crane mounted on a truck chassis, with lifting capacity up to 12 tons (10,886 kg).

Before the crane is brought in, area around the building should be clear and all overhead electrical wires removed. Sizes of door openings must be predetermined to permit entrance of the crane.

Where a gin pole is used, the labor cost of erecting will vary with job conditions. Deep excavations, piles of brick, mortar boxes or, lumber piles in the way of derricks prevent erecting crews from making good time.

Mechanical Erection of Trusses. Under average working conditions on jobs requiring 5 or more trusses, a crew consists of 5 workers with an erection machine. Sometimes, the truss fabricator sends trusses with a truck-mounted crane and the cost of unloading and erecting is included in the quoted price. If not, the estimator must add the cost of an erecting machine, such as a rubber-tired cherry picker or a crawler crane. The crew should erect one roof truss of given length at the following rate and costs:

Erecting Roof Trusses Using a Machine

Length of Truss	Hrs. to Erect	Labor Hrs. per Truss	Rate	Total	Rate	Total
30' (9 m)	0.52	2.62	$....	$....	$26.44	$69.27
40' (12 m)	0.75	3.75	....	....	26.44	99.15
50' (15 m)	1	5.00	....	....	26.44	132.20
60' (18 m)	1	5.00	....	....	26.44	132.20
75' (22.5 m)	2	10.00	....	....	26.44	264.40
90' (27 m)	3	15.00	....	....	26.44	396.60
100' (30 m)	4	20.00	....	....	26.44	528.80
120' (36 m)	5	25.00	....	....	26.44	661.00

Erecting Floor Trusses Using a Machine

Length of Truss	Hrs. to Erect	Labor Hrs. per Truss	Rate	Total	Rate	Total
30' (9 m)	0.15	0.75	$....	$....	$26.44	$19.83
40' (12 m)	0.20	1.00	....	....	26.44	26.44
50' (15 m)	0.25	1.25	....	....	26.44	33.05
60' (18 m)	0.35	1.75	....	....	26.44	46.27
75' (22.5 m)	0.45	2.25	....	....	26.44	59.49

Hand Erection of Trusses. Wood roof trusses are erected by hand using a gin pole and normally requires a crew of four workers. The maximum length is 30'-0", and the contractor must be careful to avoid "bowing" or deflection when erecting trusses by hand.

Erecting Roof Trusses by Hand

Length of Truss	Hrs. to Erect	Labor Hrs. per Truss	Rate	Total	Rate	Total
15' (4.5 m)	0.25	1.00	$....	$....	$26.44	$26.44
20' (6 m)	0.30	1.20	....	....	26.44	31.73
25' (7.5 m)	0.30	1.20	....	....	26.44	31.73
30' (9 m)	0.45	1.80	....	....	26.44	47.60

Erecting Floor Trusses by Hand

Length of Truss	Hrs. to Erect	Labor Hrs. per Truss	Rate	Total	Rate	Total
15' (4.5 m)	0.15	0.60	$....	$....	$26.44	15.86
20' (6 m)	0.20	0.80	....	....	26.44	21.15
25' (7.5 m)	0.30	1.20	....	....	26.44	31.73
30' (9 m)	0.45	1.80	....	....	26.44	47.59

Fabrication of Trussed Rafter Members. Fabrication of trussed rafter members is ideally suited for mass production methods, due to the large amount of repetition possible. In most cases, trussed rafters are symmetrical about their center lines so that corresponding members of each half are identical.

Since every member of a truss must do its share of the work, each one must be fabricated to exacting specifications. Design dimensions must be rigidly followed and all joints must have a good mechanical fit.

For efficient fabrication, a full-size layout should be made, either in the shop or on the subfloor of the building, after which one trussed rafter is carefully patterned and constructed with temporary connections. The unit may then be disassembled and the members used as templates for the balance of the trusses. On large projects, templates may be made of plywood or sheet metal.

To insure smooth roof and ceiling lines after erection, layout measurements should be made to upper edges of top chords and lower edges of bottom chords to eliminate variations in stock lumber widths.

A word of caution—the joints and fabrication of various size trusses are similar. It is important to see that the appropriate drawings are consulted and followed when trusses are fabricated.

Split-Ring Connected Trussed Rafters

Members are connected by means of bolts and split rings at all major joints, as the split rings transfer the loads and the bolts hold the members in contact.

In order that each member may function properly, the bolt holes must be located exactly in accordance with design dimensions, and grooves for split rings must be of correct diameter, width, and depth to provide a snug fit. Grooves must be cut with a grooving tool designed especially for this purpose.

Proper procedure for cutting grooves is as follows: after bolt locations have been established, drill bolt holes 1/16" (1.5 mm) larger in diameter than bolt; grooving tool is then fitted with a pilot mandrel of same diameter as drill and cutters are set to cut to a depth equal to half the width of the split ring; using bolt holes as pilot holes, concentric grooves are then cut in the faces of members which will contact each other. The grooving tool may also be fitted with the correct size drill bit, instead of a pilot mandrel, permitting both drilling and grooving operations to be done at one setup. For best results, drilling and grooving should be done with a bench drill or portable drill mounted in a drill stand.

Material Requirements for Split-Ring Connected Trussed Rafters. The material requirements for split-ring connected trussed rafters are based on the following design conditions.

Trussed rafters are designed to support a dead load plus live load on roof of 35 lbs/s.f. (1676 Pa or 171 kgf/sq.m) and a ceiling load of 10 lbs/s.f. (479 Pa or 48.82 kgf/sq.m) with rafters spaced at 2'-0" (0.6 m) centers.

Lumber shall be good grade of sufficient quality to permit the following unit stresses:

c = 900 lbs/sq.in. (6.20 MPa) Compression parallel to grain
f = 900 lbs/sq.in. (6.20 MPa) Extreme fiber in bending
E = 1,600,000 lbs/sq.in. (11,032 MPa) Modulus of elasticity

All split rings to be 2-1/2" (63 mm) diameter. All bolts to be 1/2" (13 mm) diameter machine bolts. Washers may be 2"x2"x1/8" (50 x 50 x 3 mm) plate washers, 2-1/8" (53 mm) diameter cast or malleable iron washers or cut washers.

No allowances have been included for roof overhangs.

Quantities listed for diagonals are most economical lengths to produce a long and short member from a single length.

Material Requirements for One Split-Ring Connected Trussed Rafter of Various Spans for a 4 in 12 Roof Slope

Span	20'-0"	24'-0"	26'-0"	28'-0"	30'-0"	32'-0"
Top Chords	2-2x6	2-2x6	2-2x6	2-2x8	2-2x8	2-2x8
	12'-0"	14'-0"	16'-0"	16'-0"	18'-0"	18'-0"
Bottom Chords	2-2x4	2-2x4	2-2x4	2-2x4	2-2x4	2-2x4
	12'-0"	14'-0"	16'-0"	16'-0"	18'-0"	18'-0"
Diagonals	2-2x4	2-2x4	2-2x4	2-2x4	2-2x4	2-2x4
	8'-0"	10'-0"	10'-0"	12'-0"	12'-0"	12'-0"
Joint Scabs	1-1x4	1-1x4	1-1x4	1-1x4	1-1x4	1-1x4
	2'-0"	2'-0"	2'-0"	3'-0"	3'-0"	3'-0"
Total B.f.	51-1/3	60-2/3	67-1/3	81	89	89
Split Rings	11	11	11	13	13	13
Bolts,						
1/2"x4"	4	4	4	6	6	6
1/2"x6"	2	2	2	2	2	2
1/2"x7-1/2"	1	1	1	1	1	1
Washers	14	14	14	18	18	18
Nails, 8d	16	16	16	16	16	16
(metric)						
Span	6.1m	7.3m	7.9m	8.5m	9.1m	9.8m
Top Chords	2-50x150	2-50x150	2-50x150	2-50x200	2-50x200	2-50x200
	3.6m	4.3m	4.9m	5.5m	6.1m	6.1m
Bottom Chords	2-50x100	2-50x100	2-50x100	2-50x100	2-50x100	2-50x100
	3.6m	4.3m	4.9m	5.5m	6.1m	6.1m
Diagonals	2-50x100	2-50x100	2-50x100	2-50x100	2-50x100	2-50x100
	2.4m	3.0m	3.0m	3.7m	3.7m	3.7m
Joint Scabs	1-25x100	1-25x100	1-25x100	1-25x100	1-25x100	1-25x100
	0.60m	0.60m	0.60m	0.90m	0.90m	0.90m
Total cu.m	0.115	0.136	0.152	0.182	0.200	0.200
Split Rings	11	11	11	11	11	11
Bolts,						
12.5x100mm	4	4	4	6	6	6
12.5x150mm	2	2	2	2	2	2
12.5x188mm	1	1	1	1	1	1
Washers	14	18	14	18	18	18
Nails, 8d	16	16	16	16	16	16

Material Requirements for One Split-Ring Connected Trussed Rafter of Various Spans for a 5 in 12 Roof Slope

Span	20'-0"	24'-0"	26'-0"	28'-0"	30'-0"	32'-0"
Top Chords	2-2x6	2-2x6	2-2x6	2-2x6	2-2x6	2-2x6
	12'-0"	14'-0"	16'-0"	18'-0"	18'-0"	20'-0"
Bottom Chords	2-2x4	2-2x4	2-2x4	2-2x4	2-2x4	2-2x4
	12'-0"	14'-0"	16'-0"	16'-0"	18'-0"	18'-0"
Diagonals	2-2x4	2-2x4	2-2x4	2-2x4	2-2x4	2-2x4
	10'-0"	12'-0"	12'-0"	12'-0"	14'-0"	14'-0"

Joint Scabs	1-1x4	1-1x4	1-1x4	1-1x4	1-1x4	1-1x4
	2'-0"	2'-0"	2'-0"	2-0"	2'-0"	2'-0"
Total B.f.	54	63-1/3	70	74	79-1/3	83-1/3
Split Rings	11	11	11	11	11	11
Bolts,						
1/2"x4"	4	4	4	4	4	4
1/2"x6"	2	2	2	2	2	2
1/2"x7-1/2"	1	1	1	1	1	1
Washers	14	14	14	14	14	14
Nails, 8d	16	16	16	16	16	16

(metric)

Span	6.1m	7.3m	7.9m	8.5m	9.1m	9.8m
Top Chords	2-50x150	2-50x150	2-50x150	2-50x150	2-50x150	2-50x150
	3.6m	4.3m	4.9m	5.5m	6.1m	6.1m
Bottom Chords	2-50x100	2-50x100	2-50x100	2-50x100	2-50x100	2-50x100
	3.6m	4.3m	4.9m	5.5m	6.1m	6.1m
Diagonals	2-50x100	2-50x100	2-50x100	2-50x100	2-50x100	2-50x100
	3.0m	3.6m	3.6m	3.7m	3.7m	3.7m
Joint Scabs	1-25x100	1-25x100	1-25x100	1-25x100	1-25x100	1-25x100
	0.60m	0.60m	0.60m	0.90m	0.90m	0.90m
Total cu.m	0.122	0.142	0.158	0.167	0.178	0.187
Split Rings	11	11	11	11	11	11
Bolts,						
12.5x100	4	4	4	4	4	4
12.5x150	2	2	2	2	2	2
12.5x188	1	1	1	1	1	1
Washers	14	14	14	14	14	14
Nails, 8d	16	16	16	16	16	16

Material Requirements for One Split-Ring Connected
Trussed Rafter of Various Spans for a 6 in 12 Roof Slope

Span	20-0"	24'-0"	26'-0"	28'-0"	30'-0"	32'-0"
Top Chords	2-2x6	2-2x6	2-2x6	2-2x6	2-2x6	2-2x6
	14'-0"	16'-0"	16'-0"	18'-0"	18'-0"	20'-0"
Bottom Chords	2-2x4	2-2x4	2-2x4	2-2x4	2-2x4	2-2x4
	12'-0"	14'-0"	16'-0"	18'-0"	18'-0"	18'-0"
Diagonals	2-2x4	2-2x4	2-2x4	2-2x4	2-2x4	2-2x4
	10'-0"	12'-0"	14'-0"	14'-0"	16'-0"	16'-0"
Joint Scabs	1-1x4	1-1x4	1-1x4	1-1x4	1-1x4	1-1x4
	2'-0"	2'-0"	2'-0"	2'-0"	2'-0"	2'-0"
Total B.f.	58	67-1/3	72-2/3	76-2/3	82	86
Bolts,						
1/2"x4"	4	4	4	4	4	4
1/2"x6"	2	2	2	2	2	2
1/2"x7-1/2"	1	1	1	1	1	1
Washers	14	14	14	14	14	14
Nails, 8d	16	16	16	16	16	16

(metric)					
Span...... 6.1m	7.3m	7.9m	8.5m	9.1m	9.8m
Top Chords......2-50x150	2-50x150	2-50x150	2-50x150	2-50x150	2-50x150
...... 4.3m	4.9m	4.9m	5.5m	5.5m	5.5m
Bottom Chords....2-50x100	2-50x100	2-50x100	2-50x100	2-50x100	2-50x100
...... 3.6m	4.3m	4.9m	4.9m	5.5m	5.5m
Diagonals......2-50x100	2-50x100	2-50x100	2-50x100	2-50x100	2-50x100
...... 3.0m	3.0m	4.3m	4.3m	4.9m	4.9m
Joint Scabs......1-25x100	1-25x100	1-25x100	1-25x100	1-25x100	1-25x100
...... 0.60m	0.60m	0.60m	0.90m	0.90m	0.90m
Total cu.m 0.131	0.151	0.163	0.172	0.185	0.194
Bolts,					
12.5x100...... 4	4	4	4	4	4
12.5x150...... 2	2	2	2	2	2
12.5x188...... 1	1	1	1	1	1
Washers 14	14	14	14	14	14
Nails, 8d...... 16	16	16	16	16	16

Material Requirements for One Split-Ring Connected
Trussed Rafter of Various Spans for a 7 in 12 Roof Slope

Span......20'-0"	24'-0"	26'-0"	28'-0"	30'-0"	32'-0"
Top Chords......2-2x6	2-2x6	2-2x6	2-2x6	2-2x6	2-2x6
......14'-0"	16'-0"	16'-0"	18'-0"	20'-0"	20'-0"
Bottom Chords......2-2x4	2-2x4	2-2x4	2-2x4	2-2x4	2-2x4
......12'-0"	14'-0"	16'-0"	16'-0"	18'-0"	18'-0"
Diagonals......2-2x4	2-2x4	2-2x4	2-2x4	2-2x4	2-2x4
......12'-0"	14'-0"	14'-0"	16'-0"	16'-0"	18'-0"
Joint Scabs......1-1x4	1-1x4	1-1x4	1-1x4	1-1x4	1-1x4
......2'-0"	2'-0"	2'-0"	2'-0"	2'-0"	2'-0"
Total B.f...... 60-2/3	70	72-2/3	79-1/3	86	88-2/3
Split Rings 11	11	11	11	11	11
Bolts,					
1/2"x4" 4	4	4	4	4	4
1/2"x6" 2	2	2	2	2	2
1/2"x7-1/2" 1	1	1	1	1	1
Washers 14	14	14	14	15	14
Nails, 8d...... 16	16	16	16	16	16

(metric)					
Span......6.1m	7.3m	7.9m	8.5m	9.1m	9.8m
Top Chords......2-50x150	2-50x150	2-50x150	2-50x150	2-50x150	2-50x150
......4.3m	4.9m	4.9m	5.5m	6.1m	6.1m
Bottom Chords....2-50x100	2-50x100	2-50x100	2-50x100	2-50x100	2-50x100
......3.6m	4.3m	4.9m	4.9m	5.5m	5.5m
Diagonals......2-50x100	2-50x100	2-50x100	2-50x100	2-50x100	2-50x100
......3.6m	4.3m	4.3m	4.9m	4.9m	4.9m
Joint Scabs......1-25x100	1-25x100	1-25x100	1-25x100	1-25x100	1-25x100

	0.60m	0.60m	0.60m	0.90m	0.90m	0.90m
Total cu.m 0.136	0.158	0.163	0.178	0.194	0.200	
Split Rings 11	11	11	11	11	11	
Bolts,						
12.5x100 4	4	4	4	4	4	
12.5x150 2	2	2	2	2	2	
12.5x188 1	1	1	1	1	1	
Washers 14	14	14	14	14	14	
Nails, 8d 16	16	16	16	16	16	

Plywood Gusset Trusses

Many contractors prefer plywood gusset trusses (PGT) for their simple construction. For small one-house jobs, containing only 20 to 30 rafter units, they can be easily fabricated and assembled on the job using simple hand tools—saw and hammer. They are also adaptable to mass production methods using portable power tools or shop equipment for projects involving thousands of trussed rafters.

Another advantage of the PGT is the fact that only standard dimension lumber and nails are used, in the case of "W" and the scissor-type, and standard plywood for gusset and splice plates.

To function properly, PGT must be fabricated and assembled with the same degree of accuracy as bolted and mechanically-connected trusses. Fabrication and assembly of truss units must adhere closely to design dimensions and joint details. It is important to have good bearing fits for top chord and compression diagonal member joints. Location, number, and size of nails must be in accordance with joint details.

For fast, accurate marking of nail locations, use nailing templates and spray colored paint or shellac through template openings with a paint gun or an aerosol-type container. To permit nailing from one side without nail clinching, use helically threaded nails. For dry, dense lumber, use hardened nails, which are more slender and less likely to split material.

While the use of seasoned lumber is not absolutely essential in the construction of PGT, better results are obtained with dry lumber and its use is recommended. This is particularly true for splice plates over 4" (100 mm) wide and is a design requirement for 8" (200 mm) wide material. The 1" x 8" (25 x 200 mm) splice plates made from green lumber tend to split along nail rows, when undergoing shrinkage, causing the joint to fail.

Material Requirements for Plywood Gusset Trusses. The material requirements for PGT are furnished through the courtesy of Dr. E. George Stern, Research Professor of Wood Construction at the Virginia Polytechnic Institute, Blacksburg, Va. PGT designs for the various spans and roof slopes are based on experimental data obtained in the Wood Research Laboratory of the Virginia Engineering Experimental Station under the direction of Dr. Stern. PGT constructed in accordance with these designs have been subjected to all types of tests and have proved conclusively to be of ample strength for the use intended.

Material requirements are based on the following design conditions: PGT are designed to support a dead load plus live load on roof of 35 lbs/s.f. (1676 Pa or 171 kgf/sq.m) and a ceiling load of 10 lbs/s.f. (479 Pa or 49 kgf/sq.m) with rafters spaced at 2'-0" (0.6 m) centers. Lumber shall be stress-graded No. 1 southern pine or Douglas fir, preferably seasoned 12% to 15% moisture content, permitting the following unit stresses:

c = 1350 lbs/sq.in. (9.30 MPa) Compression parallel to grain
f = 1500 lbs/sq.in. (10.34 MPa) Extreme fiber stress
E = 1,760,000 lbs/sq.in. (12,135 MPa) Modulus of elasticity

Quantities listed for diagonals are most economical lengths to produce a long and short member from a single length.

Material Requirements for One Plywood Gusset Truss of Various Spans for a 2 in 12 Roof Slope

Span	18'-0"	20'-0"	24'-0"	26'-0"	28'-0"	30'-0"
Approx. Overhang	3'-0"	2'-0"	4'-0"	3'-0"	4'-0"	3'-0"
Top Chords	2-2x6	2-2x6	2-2x6	2-2x6	2-2x6	2-2x6
	12'-0"	12'-0"	16'-0"	16'-0"	18'-0"	18'-0"
Bottom Chords	2-2x4	2-2x4	2-2x6	2-2x6	2-2x6	2-2x6
	8'-0"	10'-0"	12'-0"	12'-0"	14'-0"	14'-0"
Diagonals	2-1x6	2-1x6	2-1x6	2-1x6	2-1x6	2-1x6
	8'-0"	10'-0"	10'-0"	10'-0"	12-0"	12'-0"
	1-2x4	1-2x4	1-2x4	1-2x4	1-2x4	1-2x4
	4'-0"	4'-0"	4'-0"	4'-0"	6'-0"	6-0"
Splice Plates	4-1x4	4-1x4	4-1x6	4-1x6	4-1x6	4-1x6
	3'-0"	3'-0"	3'-6"	3'-6"	3'-6"	3'-6"
	2-1x6	2-1x6	2-1x6	2-1x6	2-1x6	2-1x6
	3'-0"	3'-0"	2'-0"	2'-0"	3'-0"	3'-0"
	1-1x8	1-1x8	1-1x8	1-1x8	1-1x8	1-1x8
	4'-0"	4'-0"	4'-0"	4'-0"	4'-0"	4'-0"
Total Board Feet	55	59-2/3	80-1/3	80-1/3	92-2/3	92-2/3
Nails,						
12d	144	144	198	198	242	242
8d	40	40	58	58	72	72

(metric)

Span	5.5m	6.1m	7.3m	7.9m	8.5m	9.1m
Approx. Overhang	0.90m	0.6m	1.2m	0.9m	1.2m	0.9m
Top Chords	2-50x150	2-50x150	2-50x150	2-50x150	2-50x150	2-50x150
	3.6m	3.6m	4.9m	4.9m	5.5m	5.5m
Bottom Chords	2-50x100	2-50x100	2-50x100	2-50x100	2-50x100	2-50x100
	2.4m	0.9m	3.6m	3.6m	4.3m	4.3m

Diagonals	2-50x150	2-50x150	2-50x150	2-50x150	2-50x150	2-50x150
	2.4m	0.9m	0.9m	0.9m	3.6m	3.6m
	1-50x100	1-50x100	1-50x100	1-50x100	1-50x100	1-50x100
	1.2m	1.2m	1.2m	1.2m	1.8m	1.8m
Splice Plates	4-25x100	4-25x100	4-25x100	4-25x100	4-25x100	4-25x100
	0.90m	0.90m	1.1m	1.1m	1.1m	1.1m
	2-25x150	2-25x150	2-25x150	2-25x150	2-25x150	2-25x150
	0.90m	0.90m	0.60m	0.60m	0.90m	0.90m
	1-25x200	1-25x200	1-25x200	1-25x200	1-25x200	1-25x200
	1.20m	1.20m	1.20m	1.20m	1.20m	0.90m
Total cu.m	0.124	0.134	0.181	0.208	0.208	
Nails,						
12d	144	144	198	198	242	242
8d	40	40	58	58	72	72

Material Requirements for One Plywood Gusset Truss of Various Spans for a 3 in 12 Roof Slope

Span	24'-0"	26'-0"	28'-0"	30'-0"	32'-0"	34'-0"
Approx. Overhang	3'-6"	4'-0"	3'-6"	4'-0"	3'-6"	2'-6"
Top Chords	2-2x6	2-2x6	2-2x6	2-2x6	2-2x6	2-2x6
	16'-0"	18'-0"	18'-0"	20'-0"	20'-0"	20'-0"
Bottom Chords	2-2x4	2-2x4	2-2x4	2-2x4	2-2x4	2-2x4
	12'-0"	14'-0"	14'-0"	16'-0"	16'-0"	18'-0"
Diagonals	2-1x6	2-1x6	2-1x6	2-1x6	2-1x6	2-1x6
	10'-0"	10'-0"	12'-0"	12'-0"	12'-0"	14'-0"
	1-2x4	1-2x4	1-2x4	1-2x4	1-2x4	1-2x4
	6'-0"	6'-0"	6'-0"	6'-0"	8'-0"	8'-0"
Splice Plates	4-1x4	4-1x4	4-1x4	4-1x4	4-1x4	4-1x4
	3'-0"	3'-0"	4'-0"	4'-0"	4'-0"	4'-0"
	3-1x6	2-1x6	2-1x6	2-1x6	2-1x6	2-1x6
	3'-0"	3'-0"	3'-0"	3'-0"	3'-0"	3'-0"
		1-1x6	1-1x6	1-1x6	1-1x8	1-1x8
		4'-0"	4'-0"	4'-0"	4'-0"	4'-0"
Total Board Feet	70-1/2	77-2/3	81	87-2/3	89-2/3	94-1/3
Nails,						
12d	138	138	146	146	168	168
8d	38	46	58	60	56	56

(metric)

Span	7.3m	7.9m	8.5m	9.1m	9.8m	10.4m
Approx. Overhang	1.1m	1.2m	1.1m	1.2m	1.1m	0.80m
Top Chords	2-50x150	2-50x150	2-50x150	2-50x150	2-50x150	2-50x150
	4.9m	5.5m	5.5m	6.1m	6.1m	6.1m
Bottom Chords	2-50x100	2-50x100	2-50x100	2-50x100	2-50x100	2-50x100
	3.6m	4.3m	4.3m	4.9m	4.9m	5.5m
Diagonals	2-50x150	2-50x150	2-50x150	2-50x150	2-50x150	2-50x150
	3.0m	3.0m	3.6m	3.6m	3.6m	4.3m
	1-50x100	1-50x100	1-50x100	1-50x100	1-50x100	1-50x100
	1.8m	1.8m	1.8m	1.8m	2.4m	2.4m

Material Requirements for One Plywood Gusset Truss of Various Spans for a 3 in 12 Roof Slope (Continued)

Splice Plates........4-25x100	4-25x100	4-25x100	4-25x100	4-25x100	4-25x100
..............0.90m	0.90m	1.2m	1.2m	1.2m	1.2m
..............3-25x150	2-25x150	2-25x150	2-25x150	2-25x150	2-25x150
..............0.90m	0.90m	0.90m	0.90m	0.90m	0.90m
....	1-25x150	1-25x150	1-25x150	1-25x200	1-25x150
....	1.2m	1.2m	1.2m	1.2m	0.90m
Total cu.m 0.159	0.175	0.182	0.197	0.202	0.212
Nails,					
12d........ 138	138	146	146	168	168
8d........... 38	46	58	50	56	56

Material Requirements for One Plywood Gusset Truss of Various Spans for a 5 in 12 Roof Slope

Span....................24'-0"	26'-0"	28'-0"	30'-0"	32'-0"	34'-0"
Approx. Overhang . 3-1/4	3-1/4	3-1/4	3-1/4	3-1/4	3-1/4
Top Chords............2-2x4	2-2x4	2-2x6	2-2x6	2-2x6	2-2x6
..............14'-0'	16'-0"	16'-0"	18'-0"	18'-0"	20'-0"
Bottom Chords.......2-2x4	2-2x4	2-2x4	2-2x4	2-2x4	2-2x4
..............12'-0"	14'-0"	14'-0"	16'-0"	16'-0"	18'-0"
Diagonals 2-1x6	2-1x6	2-1x6	2-1x6	2-1x6	2-1x6
..............12'-0"	12'-0"	14'-0"	14'-0"	14'-0"	16'-0"
..............1-2x4	1-2x4	1-2x4	1-2x4	1-2x4	1-2x4
..............6'-0"	8'-0"	8'-0"	8'-0"	8-0"	10'-0"
Splice Plates...........6-1x6	6-1x6	4-1x4	4-1x4	4-1x4	4-1x4
..............2'-0"	2'-0"	3'-0"	3'-0"	3'-0"	3'-0"
..............1-1x6	1-1x6	2-1x6	2-1x6	2-1x6	2-1x6
..............3'-0"	3'-0"	2'-0"	2'-0"	2'-0"	2'-0"
....	....	1-1x6	1-1x6	1-1x8	1-1x8
....	....	3'-0"	3'-0"	3'-0"	3'-0"
Total Board Feet .. 57-5/6	64-5/6	77-1/2	84-1/6	84-2/3	94-2/3
Nails,					
12d........ 100	112	110	110	124	124
8d........... 26	18	42	44	34	34

(metric)

Span.................... 7.3m	7.9m	8.5m	9.1m	9.8m	10.4m
Approx. Overhang .56.25	56.25	56.25	56.25	56.25	56.25
Top Chords..........2-50x100	2-50x100	2-50x150	2-50x150	2-50x150	2-50x150
.... 4.3m	4.9m	4.9m	5.5m	5.5m	6.1m
Bottom Chords....2-50x100	2-50x100	2-50x100	2-50x100	2-50x100	2-50x100
.... 3.6m	4.3m	4.3m	4.9m	4.9m	5.5m
Diagonals 2-50x150	2-50x150	2-50x150	2-50x150	2-50x150	2-50x150
.... 3.6m	3.6m	4.3m	4.3m	4.3m	4.9m
.... 1-50x100	1-50x100	1-50x100	1-50x100	1-50x100	1-50x100

	1.8m	2.3m	2.3m	2.3m	2.4m	3.0m
....	1.8m	2.3m	2.3m	2.3m	2.4m	3.0m
Splice Plates	6-25x150	6-25x150	4-25x100	4-25x100	4-25x100	4-25x100
....	0.60m	0.60m	0.90m	0.90m	0.90m	0.90m
....	1-25x150	1-25x150	2-25x150	2-25x150	2-25x150	2-25x150
....	0.90m	0.90m	0.60m	0.60m	0.60m	0.60m
....	....	....	1-25x150	1-25x150	1-25x200	1-25x200
....	....	....	0.90m	0.90m	0.90m	0.90m
Total cu.m	0.130	0.146	0.174	0.189	0.190	0.213
Nails,						
12d	100	112	110	110	124	124
8d	26	18	42	44	34	34

Material Requirements for One Scissors Type Plywood Gusset Truss Rafter of Various Spans for a 5 in 12 Roof Slope and 2 in 12 Ceiling Slope

Span	18'-0"	20'-0"	24'-0"	26'-0"	28'-0"	30'-0"
Overhang	2"	2"	3-3/8"	3-3/8"	3-3/8"	3-3/8"
Top Chords	2-2x4	2-2x4	2-2x4	2-2x6	2-2x6	2-2x6
....	10'-0"	12'-0"	14'-0"	16'-0"	16'-0"	18'-0"
Bottom Chords	2-2x4	2-2x4	2-2x4	2-2x4	2-2x6	2-2x6
& Compr. Diag	12'-0"	14'-0"	18'-0"	18'-0"	20'-0"	22'-0"
Vert. Member	1-1x8	1-1x8	1-1x8	1-1x8	1-1x8	1-1x8
....	3'-0"	3'-0"	4'-0"	4'-0"	4'-0"	5'-0"
Splice Plates	4-1x4	4-1x4	4-1x4	4-1x4	4-1x4	4-1x6
....	2'-6"	2'-6"	3'-0"	3'-0"	3'-0"	3'-0"
....	2-1x4	2-1x4	2-1x4	2-1x4	3-1x6	3-1x6
....	1'-6"	1'-6"	1'-6"	1'-6"	2'-0"	2'-0"
....	1-1x6	1-1x6	1-1x6	1-1x6	1-1x6	1-1x6
....	2'-0"	2'-0"	2'-0"	2'-0"	4'-0"	4'-0"
....	1-1x6	1-1x6	1-1x6	1-1x6	....	....
....	3'-0"	3'-0"	4'-0"	4'-0"	....	....
Total Board Feet	38-1/6	43-1/2	53-1/3	66-2/3	85-2/3	94-1/3
Nails,						
12d	63	63	78	80	104	109
8d	106	106	115	120	14	147

(metric)

Span	5.5m	6.1m	7.3m	7.9m	8.5m	9.1m
Overhang	0.05m	0.05m	0.19m	0.19m	0.19m	0.19m
Top Chords	2-50x100	2-50x100	2-50x100	2-50x150	2-50x150	2-50x150
....	3.0m	3.6m	4.3m	4.9m	4.9m	5.5m
Bottom Chords	2-50x100	2-50x100	2-50x100	2-50x100	2-50x150	2-50x150
& Compr. Diag	3.6m	4.3m	5.5m	5.5m	6.1m	6.1m
Vert. Member	1-25x200	1-25x200	1-25x200	1-25x200	1-25x200	1-25x200
....	0.9m	0.9m	1.2m	1.2m	1.2m	1.5m
Splice Plates	4-25x100	4-25x100	4-25x100	4-25x100	4-25x100	4-25x100
....	0.8m	0.8m	0.9m	0.9m	0.9m	0.9m
....	2-25x100	2-25x100	2-25x100	2-25x100	3-25x150	3-25x150

.... 0.46m	0.46m	0.46m	0.46m	0.61m	0.61m
.... 1-25x150	1-25x150	1-25x150	1-25x150	1-25x150	1-25x150
.... 0.61m	0.61m	0.61m	0.61m	1.2m	1.2m
.... 1-25x150	1-25x150	1-25x150	1-25x150	....	....
.... 0.9m	0.9m	1.2m	1.2m	....	....
Total cu.m 0.086	0.098	0.121	0.150	0.193	0.212
Nails,					
12d 63	63	78	80	104	109
8d 106	106	115	120	14	147

Plywood Gusset Kingpost Truss. PG kingpost trusses are the result of research conducted at Virginia Polytechnic Institute under the sponsorship of the Independent Nail & Packing Company of Bridgewater, Mass. These studies have brought about a new era in nail technology and wood construction.

Test data are available on request from the V.P.I. Wood Research Laboratory at Blacksburg, Va. Samples of the hardened-steel spiral-thread Hi-Load nails used to assemble these plywood gusset trusses can be obtained from the manufacturer.

Material Requirements for Plywood Gusset Kingpost Truss with 1/2" (12 mm) Plywood Gusset and Splice Plates

Span .... 20'-8"	24'-8"	28'-8"	32'-8"	36'-5-1/2"	40'-0"
Roof Slope 2/12	2/12	3/12	2/12	5/12	4/12
Overhang .... 19"	18"	16"	18"	5-1/2"	12"
Top Chord, 2 ea. .... 2x4x12'	2x4x14'	2x4x16'	2x6x18'	2x6x20'	2x6x22'
Bottom Chord, 2 ea.. 2x4x10'	2x4x12'	2x4x14'	2x6x16'	2x6x18'	2x6x20'
Post, 1 or 2 ea. .... 2x4x2'	2x4x2'3-5/8"	2x4x3'10-1/2"	2x4x3'1-7/8	2x6x8'	2-2x4x6'
Heel Gusset, 4 ea.. 2'-9"x1'-3/4"	4'x1'3-1/4"	3'x1'6-1/4"	4'x1'7"	4'x2'7"	4'x2'3-1/2"
Ridge Gusset, 2 ea. ... 2'9"x7"	4'x7-5/8"	3'x10-1/8"	5'x1'0"	4'x1'4"	4'x1'5-1/4"
Center Splice, 2 ea. ..... 3'x4"	4'x4"	4'x4"	4'x6"	4'x5-3/4"	4'x9"
Lumber, b.f. 31-1/3	36	55	71	100	109
Plywood, sq. ft. 16	24	22	36	43	44
(2-1/2"x0.135")					
Hardened Steel Nos. 140	168	218	304	308	342
Spiral Thread 1.39	1.66	2.16	3.01	3.05	3.40
Hi-Load Nails, lbs.					

(metric)

Span .... 6.3m 7.5m	8.7m	10.0m	11.1m	12.1m
Roof Slope2/12 .... 2/12	3/12	2/12	5/12	4/12
Overhang475mm 450	400	450	137	300
Top Chord, 2 ea					
........50x100x3600...50x100x4200	50x100x4800	50x100x5400	50x150x6000	50x150x6600
Bottom Chord, 2 ea					
........50x100x3000...50x100x3600	50x100x4200	50x150x4800	50x150x5400	50x150x6000
Post, 1 or 2 ea					
........50x100x600.....50x100x690	50x100x1262	50x100x928	50x150x2400	2-50x100x1800
Heel Gusset, 4 ea					
.... 600x225x43.75 ....1200x381	900x456	1200x475	1200x475	1200x687

Ridge Gusset, 2 ea	....600x225x175	1200x190	900x456	1500x300	1200x400	1200x431
Center Splice, 2 ea	....900x100	1200x100	1200x100	1200x150	1200x143	1200x225
Lumber, cu.m	....0.070	0.081	0.124	0.160	0.225	0.245
Plywood, sq.m (2-1/2"x0.135")	....1.5	2.2	2.0	3.3	4.0	4.0
Hardened Steel Nos	140	168	218	304	308	342
Spiral Thread Hi-Load Nails, kg	....0.63	0.75	0.98	1.36	1.38	1.54

Labor Required to Fabricate and Assemble Plywood Gusset Trusses. This will vary with the scope of the job, number of truss units, production methods used, and the size of truss units.

On an average one-house job, two carpenters will require about 6 hours to make a layout on the subfloor, place jig stops to guide assembly work, and construct a pilot truss with temporary connections, at the following labor cost:

	Hours	Rate	Total	Rate	Total
Carpenter	12	$....	$....	$26.44	$317.28

For the balance of the truss units, a 2-worker crew, using portable power equipment for cutting, drilling, etc., should fabricate, assemble, and stockpile 10 to 12   24' to 28' (7.2-8.4m) span trussed rafters per 8-hr. day, at the following labor cost per truss:

	Hours	Rate	Total	Rate	Total
Carpenter	1.5	$....	$....	$26.44	$39.66

For larger spans up to 36' (10.8 m), add about 5% to above costs. For smaller spans, reduce costs about 5%.

On large projects, containing several thousand PGT units, where the expense of setting up a shop for fabrication and assembly is warranted, a 2-worker crew should complete 20 to 24 units per 8-hr. day, at the following labor cost per truss:

	Hours	Rate	Total	Rate	Total
Carpenter	0.75	$....	$....	$26.44	$19.83

To the above, add equipment charges and cost of setting up and dismantling fabrication shop. The above costs may be applied to either bolted trussed rafters or nailed trussed rafters.

Labor Erecting Trusses. Erection of trusses for one-story structures can usually be accomplished with a 4-worker crew: 2 laborers carrying units from stockpile to setting location, placing units in an inverted position on the bearing wall plates, and tilting up into an erect position ready for 2 carpenters to anchor and brace in the final position. For efficient operation, the layout of truss locations should be marked on the wall plates and metal framing anchors set in place before erection is started. The 4-worker crew should lay out, set anchors, and erect into place 22 to 26 trussed rafters, up to 28'-0" (8.4 m) span, per 8-hr. day, at the following labor cost per truss:

	Hours	Rate	Total	Rate	Total
Carpenter	0.67	$....	$....	$26.44	$ 17.71
Labor	0.67	....	....	21.04	14.10
Cost per truss			$....		$ 31.81

For spans greater than 28'-0" (8.4 m), it might be necessary to add an extra worker to the crew.

Labor Framing Lumber in Building Construction

The labor quantities and costs that follow are intended to include all classes of frame construction, such as residences, ranch houses, barns, stables, apartment buildings, combined store and apartment buildings, country clubs, schools, and in fact, all buildings that are constructed entirely of wood or have brick, stone, tile or cement block walls and wood joists, rafters, stud walls, partitions, wood subfloors, wall and roof sheathing, or finish wood floors.

Costs are given on two classes of workmanship. "Ordinary workmanship" is encountered in most buildings, where price is the main factor. "First grade workmanship" is where quality is a factor as well, usually found in high-grade residences, apartments, hotel and office buildings, public buildings, high school, and college and university buildings.

The estimator must use his or her own judgment as to the grade of workmanship required, depending upon the type of building, the reputation of the architect, and the requirements of the specifications.

Another thing that is going to govern labor costs is the kind of equipment used. All costs given are based on using an electric saw for cutting all joists, studs, and rafters to length and for cutting off ends of subflooring, roof sheathing, etc. If an ordinary handsaw were used, you would add 1 to 1-1/2 hrs. carpenter time per 1,000 b.f. (2.36 cu.m).

All of the labor quantities and costs are based on laborers or helpers handling and carrying the lumber from the stock piles or benches where the lumber is being cut to length to the building. If the lumber is handled and carried by carpenters, figure labor time at carpenter wages.

Framing and Placing Foundation Wall Plates. Where 2"x4" (50 x 100 mm) or 2"x6" (50 x 150 mm) wood plates are placed on top of foundation walls or concrete slab to receive the floor joists and exterior wall studs, it is customary to place anchor bolts in the walls and floors. The plates are then bored to receive the bolts and placed on the wall (and usually wedged with shingles) ready to receive the joists and exterior wall studs. Special purpose nails may be used for anchoring wall plates to concrete.

Where just an ordinary grade of workmanship is required, two carpenters working together should handle, frame and place 225 to 275 l.f. (67.5-82.5 m) of 2"x4" (50 x 100 mm) or 2"x6" (50 x 150 mm) plates per 8-hr. day, at the following cost per 100 l.f. (30 m):

	Hours	Rate	Total	Rate	Total
Carpenter	6.4	$....	$....	$26.44	$169.22
Cost per l.f.				....	1.69
per m				....	5.55

First Grade Workmanship

On jobs where first grade workmanship is required, with the foundation wall plates drilled for bolts, plates set and bedded absolutely level in a bed of cement mortar, two carpenters working together should handle, frame, and place 175 to 225 l.f. (52.5-67.5 m) of 2"x4" (50 x 100 mm) or 2"x6" (50 x 150 mm) wall plates per 8-hr. day at the following labor cost per 100 l.f. (30 m):

	Hours	Rate	Total	Rate	Total
Carpenter	8	$....	$....	$26.44	$ 211.52
Cost per l.f.				....	2.12
per m				....	6.94

Framing and Placing Box Sills and Plates. On platform framing, where a wood box sill and plate is formed by using a 2" x 4" (50 x 100 mm) or 2"x6" (50 x 150 mm) plate and a 2" x 8" (50 x 200 mm) or 2" x 10" (50 x 250 mm) on the end of the joists to form a box sill, the wall plate should be drilled, leveled, and set the same as described above, and the side or end piece is nailed after the joists are set.

This work should be estimated as given above for "Foundation Wall Plates" and the 2" x 8" (50 x 200 mm) or 2" x 10" (50 x 250 mm) end pieces should be figured in with the floor joists.

Framing and Erecting Exterior Stud Walls for Frame Buildings. The labor cost of framing and erecting stud walls is subject to wide variation, depending upon the type of building, height, regularity of the walls, etc.

The framing on square or rectangular buildings, such as Cape Cod and colonial type houses, costs much less than for English type houses, having walls of irregular shape and height. The average house requires only 1,500 to 2,500 ft. (450-750 m) of lumber for outside stud walls.

On square or rectangular shaped buildings, such as colonial, Georgian houses, etc., a carpenter should frame and erect 350 to 400 b.f. (0.83-0.94 cu.m) of lumber per 8-hr. day, at the following cost per 1,000 b.f. (2.36 cu.m):

	Hours	Rate	Total	Rate	Total
Carpenter	21.4	$....	$....	$26.44	$565.82
Labor	6	....	....	21.04	126.24
Cost 1,000 b.f.			$....		$ 692.06
per cu.m				....	293.25

On English type and other buildings having irregular wall construction, a carpenter should frame and erect 250 to 300 b.f. (0.59-0.70 cu.m) of lumber per 8-hr. day, at the following labor cost per l,000 b.f. (2.36 cu.m):

	Hours	Rate	Total	Rate	Total
Carpenter	29.1	$....	$....	$26.44	$769.40
Labor	6	....	....	21.04	126.24
Cost 1,000 b.f.			$....		$ 895.64
per cu.m				....	379.51

First Grade Workmanship

In high class wood constructed buildings, where every precaution is taken to prevent settling due to shrinkage, where the wood studs rest on masonry walls or steel I beams and are not set on top of the floor joists, and where it is necessary to bridge or truss between all studs and over door and window openings, a carpenter should frame and erect 225 to 275 b.f. (0.53-0.65 cu.m) of lumber per 8-hr. day, on square or rectangular type buildings, at the following labor cost per 1,000 b.f. (2.36 cu.m):

	Hours	Rate	Total	Rate	Total
Carpenter	32	$....	$....	$26.44	$846.08
Labor	10	....	....	21.04	210.40
Cost 1,000 b.f.			$....		$1,056.48
per cu.m				....	447.66

On English type and other buildings having irregular wall construction, a carpenter should frame and erect 150 to 200 b.f. (0.35-0.47 cu.m) of lumber per 8-hr day, at the following labor cost per 1,000 b.f. (2.36 cu.m):

	Hours	Rate	Total	Rate	Total
Carpenter	45.7	$....	$....	$26.44	$1,208.31
Labor	10	....	....	21.04	210.40
Cost 1,000 b.f.			$....		$1,418.71
per cu.m				....	601.15

Framing Interior Stud Partitions. When framing and setting interior stud partitions set on top of rough wood floors that require just the ordinary amount of framing for door openings, a carpenter should frame and erect 375 to 425 b.f. (0.88-1.00 cu.m) per 8-hr. day, at the following labor cost per 1,000 b.f. (2.36 cu.m):

	Hours	Rate	Total	Rate	Total
Carpenter	20	$....	$....	$26.44	$528.80
Labor	6	....	....	21.04	126.24
Cost 1,000 b.f.			$....		$655.04
per cu.m				....	277.56

First Grade Workmanship

Where the wood partition studs are set on masonry walls or steel I beams instead of the wood subfloor, and where it is necessary to brace between studs and truss over all door openings, a carpenter should frame and erect 250 to 300 b.f. (0.59-0.70 cu.m) of lumber per 8-hr. day, at the following labor cost per 1,000 b.f. (2.36 cu.m):

	Hours	Rate	Total	Rate	Total
Carpenter	29.1	$....	$....	$26.44	$ 769.40
Labor	8.0	....	....	21.04	168.32
Cost per 1,000 b.f.			$....		$ 937.72
per cu.m				....	397.34

Framing and Setting Floor Joists. When framing and placing wood floor joists up to 2"x8" (50 x 200 mm) in buildings of regular construction, a carpenter should frame and place 550 to 600 b.f. (1.30-1.42 cu.m) per 8-hr. day, at the following labor cost per 1,000 b.f. (2.36 cu.m):

	Hours	Rate	Total	Rate	Total
Carpenter	13.9	$....	$....	$26.44	$ 367.52
Labor	6	....	....	21.04	126.24
Cost 1,000 b.f. (2.36 cu.m)			$....		$ 493.76
per cu.m				....	209.22

If 2"x10" (50 x 250 mm) or 2"x12" (50 x 300 mm) joists are used, a carpenter should frame and erect 600 to 650 b.f. (1.42-1.53 cu.m) per 8-hr. day, at the following labor cost per 1,000 b.f. (2.36 cu.m):

	Hours	Rate	Total	Rate	Total
Carpenter	12.8	$....	$....	$26.44	$ 338.43
Labor	6	....	....	21.04	126.24
Cost 1,000 b.f.			$....		$ 464.67
per cu.m				....	196.89

First Grade Workmanship

On jobs where the wood joists must be set with the crowning edge up and it is not permissible to block up under the joists with shingles or wood wedges, a carpenter should frame and erect 500 to 550 b.f. (1.18-1.30 cu.m), in sizes up to 2"x8" (50 x 200 mm), per 8-hr. day, at the following labor cost per 1,000 b.f. (2.36 cu.m):

	Hours	Rate	Total	Rate	Total
Carpenter	15.2	$....	$....	$26.44	$ 401.89
Labor	4.5	....	....	21.04	94.68
Cost 1,000 b.f.			$....		$ 496.57
per cu.m				....	210.41

Using 2"x10" (50 x 250 mm) or 2"x12" (50 x 300 mm) joists, a carpenter should frame and erect 550 to 650 b.f. (1.30-1.53 cu.m) of lumber per 8-hr. day, at the following labor cost per 1,000 b.f. (2.36 cu.m):

	Hours	Rate	Total	Rate	Total
Carpenter	13.9	$....	$....	$26.44	$ 367.52
Labor	4.0	....	....	21.04	84.16
Cost 1,000 b.f.			$....		$ 451.68
per cu.m				....	191.39

Framing Panel and Girder Floor Systems. When framing and placing 4"x6" (100 x 150 mm) wood beams and 2"x4" (50 x 100 mm) spacers to form 4-ft. (1.2-m) square grids for panel and girder floor systems, 2 carpenters and a helper should frame and place 1,500 to 1,700 b.f. (3.54-4.0 cu.m) of lumber per 8-hr. day, at the following labor cost per 1,000 b.f. (2.36 cu.m):

	Hours	Rate	Total	Rate	Total
Carpenter	10	$....	$....	$26.44	$ 264.40
Labor	5	....	....	21.04	105.20
Cost per 1,000 b.f.			$....		$ 369.60
per cu.m				....	156.61

Installing Cross Bridging. Under average conditions a carpenter should cut and place 90 to 110 sets (2 pieces) of crossbridging per 8-hr. day at the following labor cost per 100 sets:

	Hours	Rate	Total	Rate	Total
Carpenter	8	$....	$....	$26.44	$ 211.52
Cost per set				....	2.12

Framing and Erecting Rafters for Gable Roofs. When framing and erecting rafters for plain double pitch or gable roofs without dormers or

gables, and where 2"x6" (50 x 150 mm) or 2"x8" (50 x 200 mm) lumber is
used, a carpenter should frame and erect 285 to 335 b.f. (0.67-0.79 cu.m) per
8-hr. day, at the following labor cost per 1,000 b.f. (2.36 cu.m):

	Hours	Rate	Total	Rate	Total
Carpenter	24.6	$....	$....	$26.44	$ 650.42
Labor	7.0	....	....	21.04	147.28
Cost 1,000 b.f.			$....		$ 797.70
per cu.m				....	338.01

On buildings having double pitch or gable roofs cut up with dormers and
gables framing into the main roof, a carpenter should frame and erect 250 to
300 b.f. (0.59-0.70 cu.m) of lumber per 8-hr. day, at the following labor cost
per 1,000 b.f. (2.36 cu.m):

	Hours	Rate	Total	Rate	Total
Carpenter	29.1	$....	$....	$26.44	$ 769.40
Labor	8.0	....	....	21.04	168.32
Cost 1,000 b.f.			$....		$ 937.72
per cu.m				....	397.34

Framing and Erecting Rafters for Hip Roofs. When framing and
erecting plain hip roofs without dormers or gables framing into the main
roof, a carpenter should frame and erect 250 to 300 b.f. (0.59-0.70 cu.m) of
lumber per 8-hr. day, at the following labor cost per 1,000 b.f. (2.36 cu.m):

	Hours	Rate	Total	Rate	Total
Carpenter	29.1	$....	$....	$26.44	$ 769.40
Labor	8.0	....	....	21.04	168.32
Cost 1,000 b.f.			$....		$ 937.72
per cu.m				....	397.34

On difficult hip roofs, where it is necessary to frame for dormers, gables,
and valleys, a carpenter should frame and erect 180 to 220 b.f. (0.42-0.52
cu.m) of lumber per 8-hr. day, at the following labor cost per 1,000 b.f. (2.36
cu.m):

	Hours	Rate	Total	Rate	Total
Carpenter	40	$....	$....	$26.44	$1,057.60
Labor	8	....	....	21.04	168.32
Cost 1,000 b.f.			$....		$1,225.92
per cu.m				....	519.46

Framing Light Timbers for Exposed Roof Beam Construction.
Light timbers, varying from 4"x6" (100 x 150 mm) to 4"x14" (100 x 350 mm)
and larger, depending upon span and spacing, are used for exposed beam

roof construction. This type of construction is widely used for contemporary design—houses, stores, medical groups, small churches—with flat or low rise sloping roofs.

On work of this type, two carpenters and a helper should frame and place 750 to 850 b.f. (1.77-2.0 cu.m) of lumber per 8-hr. day, at the following labor cost per 1,000 b.f. (2.36 cu.m):

	Hours	Rate	Total	Rate	Total
Carpenter	20	$....	$....	$26.44	$528.80
Labor	10	....	....	21.04	210.40
Cost per 1,000 b.f.			$....		$739.20
per cu.m				....	313.22

Framing for Roof Saddles on Flat Roofs. On flat roofs where it is necessary to frame saddles to pitch the water toward the drains, and where the saddles are built up of 2"x4" (50 x 100 mm) framing covered with 1" (25 mm) sheathing, a carpenter should frame and erect 400 to 450 b.f. (0.94-1.06 cu.m) of lumber per 8-hr. day, at the following labor cost per 1,000 b.f. (2.36 cu.m):

	Hours	Rate	Total	Rate	Total
Carpenter	19	$....	$....	$26.44	$502.36
Labor	6	....	....	21.04	126.24
Cost per 1,000 b.f.			$....		$ 628.60
per cu.m				....	266.36

Placing Wood Cant Strips. When cant strips, diagonally cut from 4"x4" (100 x 100 mm) or 6"x6" (150 x 150 mm) lumber, occur in long runs with few breaks, a carpenter should place 475 to 525 l.f. (143-158 m) per 8-hr. day at the following labor cost per 100 l.f. (30 m):

	Hours	Rate	Total	Rate	Total
Carpenter	1.6	$....	$....	$26.44	$ 42.30
Cost per l.f.				....	0.42
per m				....	1.39

If cant strips occur in short runs, such as around roof opening curbs, or if the runs are cut up with pilasters and other breaks, a carpenter should place 300 to 325 l.f. (91-98 m) per 8-hr. day at the following labor cost per 100 l.f. (30 m):

	Hours	Rate	Total	Rate	Total
Carpenter	2.6	$....	$....	$26.44	$ 68.74
Cost per l.f.				....	0.69
per m				....	2.25

Laying Rough Wood Floors. When laying 1"x6" (25 x 150 mm) or 1"x8" (25 x 200 mm) rough wood subflooring, a carpenter should lay 825 to 925 b.f. (1.95-2.18 cu.m) per 8-hr. day, at the following labor cost per 1,000 b.f. (2.36 cu.m):

	Hours	Rate	Total	Rate	Total
Carpenter	9.2	$....	$....	$26.44	$ 243.25
Labor	5.3	....	....	21.04	111.51
Cost 1,000 b.f.			$....		$ 354.76
per cu.m				....	150.32

Laying Rough Wood Floors Diagonally. When laying rough wood subflooring diagonally, making it necessary to cut both ends of the flooring on a bevel, a carpenter should lay 650 to 750 b.f. (1.53-1.77 cu.m) of flooring per 8-hr. day, at the following labor cost per 1,000 b.f. (2.36 cu.m):

	Hours	Rate	Total	Rate	Total
Carpenter	11.5	$....	$....	$26.44	$ 304.06
Labor	5.5	....	....	21.04	115.72
Cost 1,000 b.f.			$....		$ 419.78
per cu.m				....	177.87

Laying Wood Sheathing on Flat Roofs. Figure same as given above for Rough Wood Floors.

Labor Laying Plywood Subflooring. The labor cost of handling and laying plywood subflooring can vary widely, depending on the size of the floor area, the regularity of the joist spacing, and the method of laying.

When plywood is used as subflooring, nailed directly to the joists and with the finish floor applied directly to the plywood, all edges (both longitudinal and crosswise) should be nailed about 6" (150 mm) on centers. Otherwise, the finish floor may develop squeaks caused by deflection of the long edges of the sheets when walked upon.

In other words, with the joists spaced 16" (400 mm) o.c., it is necessary to cut pieces of 2"x4" (50 x 100 mm) approximately 14-3/8" (4.3 m) long and the ends of these 2"x4"s (50 x 100 mm) must be nailed into the sides of the joists to provide nailing for the long edges of the plywood sheets. As the plywood is usually 4'-0" (1.2 m) wide, it means a row of 2"x4" (50 x 100 mm) blocking must be run at right angles to the floor joists every 4'-0" (1.2 m) apart to provide nailing for the long edges of the plywood sheets. This requires a lot of extra work and costs a lot of money.

In buildings with regular spans, where an underlayment is to be provided, and where the long edges of the plywood sheets are not nailed, two carpenters working together should handle, fit, lay, and nail 52 to 60 sheets, 1,664 to 1,920 s.f. (150-173 sq.m) of 4'-0"x8'-0" (1.2 x 2.4 m) plywood 5/8" to

3/4" (16-19 mm) thick per 8-hr. day, at the following labor cost per 100 s.f. (9 sq.m):

	Hours	Rate	Total	Rate	Total
Carpenter	0.9	$....	$....	$26.44	$23.80
Labor	0.3	....	....	21.04	6.31
Cost per 100 s.f. (9 sq.m)			$....		$30.11
per s.f.				....	0.30
per sq.m				....	3.24

If 2"x4" (50 x 100 mm) wood blocking is required between the joists, and the plywood sheets must be nailed both longitudinally and crosswise, two carpenters working together should cut and nail blocks, and handle, fit, lay, and nail 26 to 30 sheets of 4'-0"x8'-0" (1.2 x 2.4 m) plywood, 832 to 960 s.f. (75-86 sq.m) 5/8" to 3/4" (16-19 mm) thick per 8-hr. day, at the following labor cost per 100 s.f. (9 sq.m):

	Hours	Rate	Total	Rate	Total
Carpenter	1.8	$....	$....	$26.44	$47.59
Labor	0.5	....	....	21.04	10.52
Cost per 100 s.f. (9 sq.m)			$....		$ 58.11
per s.f.				....	0.58
per sq.m				....	6.25

Laying Plywood Decking for Panel and Girder Floor Systems. Plywood decking for panel and girder floor construction should be 1" (25 mm) or 1-1/4" (31 mm) thick and laid with staggered joints. To minimize waste, inside dimensions of the building should be laid out on a 4-ft. (1.2 m) module, which also reduces cutting of panels to starters in every other row. In some localities, 4'x4' (1.2 x 1.2 m) plywood panels are available as a stock size.

On large projects, where the work may be highly organized, a carpenter and two helpers should lay 4,500 to 4,700 s.f. (405-423 sq.m) of plywood decking per 8-hr. day, or enough for 3 to 4 average one-story houses, at the following labor cost per 1,000 s.f. (90 sq.m):

	Hours	Rate	Total	Rate	Total
Carpenter	1.75	$....	$....	$26.44	$46.27
Labor	3.50	....	....	21.04	73.64
Cost per 1,000 s.f. (90 sq.m)			$....		$ 119.91
per s.f.				....	0.12
per sq.m				....	1.29

For smaller jobs, where only one or two houses are involved, increase the above costs about 25% to absorb lost motion in laying out work and getting started.

Where plywood decking is to be covered with linoleum or other resilient flooring, requiring a fairly smooth surface, a carpenter should spackle and sand joints and surface cracks and other imperfections at the rate of 200 s.f. (18 sq.m) per hr.

Laying Wood Sheathing on Pitch or Gable Roofs. When laying roof sheathing on plain hip or gable roofs, a carpenter should lay 575 to 615 b.f. (1.36-1.45 cu.m) of lumber per 8-hr. day, at the following labor cost per 1,000 b.f. (2.36 cu.m):

	Hours	Rate	Total	Rate	Total
Carpenter	13.5	$....	$....	$26.44	$356.94
Labor	6.5	....	....	21.04	136.76
Cost 1,000 b.f.			$....		$ 493.70
per cu.m				....	209.19

On roofs that are very steep or cut up with dormers, hips, and valleys, such as English type houses, a carpenter will lay 275 to 325 b.f. (0.65-0.77 cu.m) of sheathing per 8-hr. day at the following labor cost per 1,000 b.f. (2.36 cu.m):

	Hours	Rate	Total	Rate	Total
Carpenter	26.5	$....	$....	$26.44	$700.66
Labor	7.5	....	....	21.04	157.80
Cost 1,000 b.f.			$....		$858.46
per cu.m				....	363.75

Labor Placing Sidewall Sheathing. When sheathing sidewalls of frame buildings with 1"x6" (25 x 150 mm) or 1"x8" (25 x 200 mm) boards laid horizontally, a carpenter should handle and place 625 to 700 b.f. (1.47-1.65 cu.m) per 8-hr. day, at the following labor cost per 1,000 b.f. (2.36 cu.m):

	Hours	Rate	Total	Rate	Total
Carpenter	12	$....	$....	$26.44	$317.28
Labor	5	....	....	21.04	105.20
Cost 1,000 b.f.			$....		$422.48
per cu.m				....	179.02

Labor Placing Diagonal Sidewall Sheathing. When sidewall sheathing is placed diagonally, it usually requires two carpenters working together—one at each end of the board—and usually requires scaffolding the outside walls before the sheathing is placed. The worker at the top must have a scaffold from which to work.

On work of this class, a carpenter should place 450 to 525 b.f. (1.06-1.24 cu.m) of 1"x6" (25 x 150 mm) or 1"x8" (25 x 200 mm) sheathing per 8 hr. day, at the following labor cost per 1,000 b.f. (2.36 cu.m):

	Hours	Rate	Total	Rate	Total
Carpenter	16.5	$....	$....	$26.44	$436.26
Labor	5.5	....	....	21.04	115.72
Cost 1,000 b.f.			$....		$551.98
per cu.m				....	233.89

WOOD BLOCKING, FURRING AND GROUNDS

Placing Wood Furring Strips on Masonry Walls. Where it is necessary to place wood furring strips on brick or tile walls before lathing and plastering, allowing the strips to follow the line of the walls without wedging or blocking out to make them straight or plumb, with the nails driven into dry joints in the brickwork, a carpenter should place 500 to 550 l.f. (150-165 m) of furring per 8-hr. day, at the following labor cost per 100 l.f. (30 m):

	Hours	Rate	Total	Rate	Total
Carpenter	1.5	$....	$....	$26.44	$39.66
Cost per l.f.				....	0.40
per m				....	1.30

Cost per Square (100 s.f. or 9 sq.m)

	Hours	Rate	Total	Rate	Total
Strips 12" (300 mm) centers	1.7	$....	$....	$26.44	$ 44.95
Strips 16" (400 mm) centers	1.3	....	....	26.44	34.37

First Grade Workmanship

Where it is necessary to plug the masonry walls and place all furring strips absolutely straight and plumb to produce a level surface to receive lath and plaster, a carpenter should plug walls and place 200 to 250 l.f. (60-75 m) of strips per 8-hr. day, at the following labor cost per 100 l.f. (30 m):

	Hours	Rate	Total	Rate	Total
Carpenter	3.5	$....	$....	$26.44	$92.54
Cost per l.f.				....	0.93
per m				....	3.05

Cost per Square (100 s.f. or 9 sq.m)

	Hours	Rate	Total	Rate	Total
Strips 12" (300 mm) centers	3.9	$....	$....	$26.44	$103.12
Strips 16" (400 mm) centers	2.9	....	....	26.44	76.68

Placing Wood Grounds. Where wood grounds are nailed directly to wood furring strips, door and window openings, etc., allowing them to follow the rough furring, partitions and wood bucks, without wedging or blocking to make them absolutely straight, a carpenter should place 540 to 590 l.f. (162-177 mm) per 8-hr. day, at the following labor cost per 100 l.f. (30 m):

	Hours	Rate	Total	Rate	Total
Carpenter	1.4	$....	$....	$26.44	$37.02
Cost per l.f.			....		0.37
per m			....		1.21

First Grade Workmanship

Where it is necessary to keep all wood grounds absolutely straight, plumb, and level for the finish plaster and for the interior wood finish, and where it is necessary to plug the masonry walls with wood plugs to hold the nails, a carpenter should place 200 to 250 l.f. (60-75 m) of grounds per 8-hr. day, at the following labor cost per 100 l.f. (30 m):

	Hours	Rate	Total	Rate	Total
Carpenter	3.5	$....	$....	$26.44	$92.54
Cost per l.f.			....		0.93
per m			....		3.05

Wood Blocking and Grounds Around Window Openings. Where the masonry walls are 12" (300 mm) or more in thickness, requiring a wood jamb lining, a carpenter should place all necessary wood blocking and grounds for a window up to 4'-0"x6'-0" (1.2 x 1.8 m) in size in 3/4 to 1-1/4 hr. at the following labor cost:

	Hours	Rate	Total	Rate	Total
Carpenter	1	$....	$....	$26.44	$26.44

Windows 4'-0"x7'-0" (1.2 x 2.1 m) to 5'-0"x8'-0" (1.5 x 2.4 m) in size will require 1-1/8 to 1-3/8 hrs. to place blocking and grounds, at the following labor cost per window:

	Hours	Rate	Total	Rate	Total
Carpenter	1.25	$....	$....	$26.44	$33.05

First Grade Workmanship

In the better class of buildings, where all grounds must be absolutely straight and plumb, and it is necessary to plug the brick and block walls for securing the grounds and blocking, a carpenter should complete one window up to 4'-0"x6'-0" (1.2 x 1.8 m) in size in 1-1/4 to 1-3/4 hrs. at the following labor cost:

	Hours	Rate	Total	Rate	Total
Carpenter..	1.5	$....	$....	$26.44	$39.66

A carpenter should plug walls, place blocking and grounds for one window 4'-0"x7'-0" (1.2 x 2.1 m) to 5'-0"x8'-0" (1.5 x 2.4 m) in 1-1/2 to 2 hrs. at the following labor cost per window:

	Hours	Rate	Total	Rate	Total
Carpenter....................................	1.75	$....	$....	$26.44	$46.27

Wood Blocking for Cabinet and Case Bases. A carpenter should locate, frame and set 115 to 135 b.f. (0.27-0.32 cu.m) of 2"x4" (50 x 100 mm) or 2"x6" (50 x 150 mm) blocking for cabinet or case bases per 8-hr. day at the following labor cost per 1,000 b.f. (2.36 cu.m):

	Hours	Rate	Total	Rate	Total
Carpenter..............................	64	$....	$....	$26.44	$1,692.16
cost per cu.m........................				....	717.02

Wood Furring Strips Over Wood Subfloors. When placing 1"x2" (25 x 50 mm) or 2"x2" (50 x 50 mm) wood furring strips over wood subfloors to receive finish flooring, a carpenter should place 550 to 600 l.f. (165-180 m), without wedging or leveling, per 8-hr. day, at the following labor cost per 100 l.f. (30 m):

	Hours	Rate	Total	Rate	Total
Carpenter..................................	1.4	$....	$....	$26.44	$37.02
Labor...	0.3	....	....	21.04	6.31
Cost per 100 l.f. (30 m)...............			$....		$43.33
per l.f.				....	0.43
per m				....	1.42

Cost per Square (100 s.f. or 9 sq.m)

Strips 12" (300 mm) centers.................. 1.9	$....	$....	$26.44	$ 50.24
Strips 16" (400 mm) centers.................. 1.4	....	....	26.44	37.02

First Grade Workmanship

In the better class of buildings where the wood furring strips must be wedged up and blocked to produce an absolutely level surface to receive finish flooring, a carpenter should place 250 to 300 l.f. (75-90 m) of furring strips per 8-hr. day, at the following cost per 100 l.f. (30 m):

	Hours	Rate	Total	Rate	Total
Carpenter	3.0	$....	$....	$26.44	$79.32
Labor	0.3	....	....	21.04	6.31
Cost per 100 l.f. (30 m)			$....		$85.63
per l.f.				....	0.86
per m				....	2.81

Cost per Square (100 s.f. or 9 sq.m)

	Hours	Rate	Total	Rate	Total
Strips 12" (300 m) centers	3.6	$....	$....	$26.44	$95.18
Strips 16" (400 mm) centers	2.8	....	....	26.44	74.03

Placing Strip Deadening Felt and Wood Furring Strips Over Wood Subfloors. Frequently a narrow strip of deadening felt 1/2"x3" (13 x 75 mm) is placed under the wood furring strips and both are nailed to the wood subfloor. This provides better insulation and deadening than where the furring strips are nailed directly to the subfloor.

On work of this kind, a carpenter should place 425 to 475 l.f. (128-143 m) of deadening felt and furring strips per 8-hr. day, at the following labor cost per 100 l.f. (30 m):

	Hours	Rate	Total	Rate	Total
Carpenter	1.8	$....	$....	$26.44	$47.59
Labor	0.4	....	....	21.04	8.42
Cost 100 l.f. (30 m)			$....		$56.01
per l.f.				....	0.56
per m				....	1.84

Cost per Square (100 s.f. or 9 sq.m)

	Hours	Rate	Total	Rate	Total
Strips 12" (300 mm) centers	2.4	$....	$....	$26.44	$63.46
Strips 16" (400 mm) centers	1.9	....	....	26.44	50.24

First Grade Workmanship

Where it is necessary to wedge and block up the strips to produce an absolutely level surface to receive the finish flooring, a carpenter should place 220 to 250 l.f. (66-75 m) of deadening and furring strips per 8-hr. day, at the following labor cost per 100 l.f. (30 m):

	Hours	Rate	Total	Rate	Total
Carpenter	3.5	$....	$....	$26.44	$92.54
Labor	0.4	....	....	21.04	8.42
Cost 100 l.f. (30 m)			$....		$100.96
per l.f.				....	1.01
per m				....	3.31

Cost per Square (100 s.f. or 9 sq.m)

Strips 12" (300 mm) centers	4.3	$....	$....	$26.44	$ 113.69
Strips 16" (400 mm) centers	3.3	....	....	26.44	87.25

Placing Deadening Quilt Over Rough Wood Floors. Where deadening felt or quilt is laid over wood subfloors, a worker should handle and lay 200 to 250 s.f. (18-22.5 sq.m) per hr. at the following cost per 100 s.f. (9 sq.m):

	Hours	Rate	Total	Rate	Total
Labor	0.5	$....	$....	$21.04	$10.52
Cost per s.f.			....		0.11
per sq.m			....		1.18

The above costs are based on deadening felts or quilts in rolls or sheets that are laid on the rough floor with furring strips placed directly over them.

Cost of Placing Deadening Felt and Wood Furring Strips Over
100 S.F. (9 Sq.M) of Floor

Strips 12" (300 mm) O.C.—Ordinary Workmanship

	Hours	Rate	Total	Rate	Total
Carpenter	2.0	$....	$....	$26.44	$52.88
Labor	0.3	....	....	21.04	6.31
Cost 100 s.f. (9 sq.m)			$....		$59.19
per s.f.			....		0.59
per sq.m			....		6.37

Strips 16" (400 mm) O.C.—Ordinary Workmanship

	Hours	Rate	Total	Rate	Total
Carpenter	1.6	$....	$....	$26.44	$42.30
Labor	0.3	....	....	21.04	6.31
Cost 100 s.f. (9 sq.m)			$....		$ 48.61
per s.f.			....		0.49
per sq.m			....		5.27

Strips 12" (300 mm) O.C.—First Grade Workmanship

	Hours	Rate	Total	Rate	Total
Carpenter	3.8	$....	$....	$26.44	$100.47
Labor	0.3	....	....	21.04	6.31
Cost 100 s.f. (9 sq.m)			$....		$106.78
per s.f.			....		1.07
per sq.m			....		11.51

Strips 16" (400 mm) O.C.—First Grade Workmanship

	Hours	Rate	Total	Rate	Total
Carpenter	3.0	$....	$....	$26.44	$79.32
Labor	0.3	....	....	21.04	6.31
Cost 100 s.f. (9 sq.m)			$....		$ 85.63
per s.f.				....	0.86
per sq.m				....	9.25

Placing Wood Floor Sleepers. When placing 2"x3" (50 x 75 mm) or 2"x4" (50 x 100 mm) wood floor screeds or sleepers over rough tile or concrete floors, to receive finish flooring, a carpenter should place 225 to 275 l.f. (68-83 m) per 8-hr. day, at the following labor cost per 100 l.f. (30 m):

	Hours	Rate	Total	Rate	Total
Carpenter	3.2	$....	$....	$26.44	$ 84.61
Labor	0.8	....	....	21.04	16.83
Cost 100 l.f. (30 m)			$....		$ 101.44
per l.f.				....	1.01
per m				....	3.31

First Grade Workmanship

In the better class of buildings, 2"x3" (50 x 75 mm) or 2"x4" (50 x 100 mm) beveled floor sleepers are placed over the rough concrete floors and wedged or blocked up to provide a perfectly level surface to receive the finish flooring.

The screeds are usually held in place by metal clips placed in the rough concrete, or anchored with special purpose nails.

On work of this class, a carpenter should place, wedge up, and level 130 to 170 l.f. (39-51 m) of sleepers per 8-hr. day, including setting sleeper clips, at the following labor cost per 100 l.f. (30 m):

	Hours	Rate	Total	Rate	Total
Carpenter	5.3	$....	$....	$26.44	$140.13
Labor	0.8	....	....	21.04	16.83
Cost per 100 l.f. (30 m)			$....		$156.96
per l.f.				....	1.57
per m				....	5.15

Lumber Required for Wood Door Bucks
of 2"x4" (50 x 100 mm) Lumber

Size of Opening	Lumber Req'd. l.f./meters	Lumber Req'd. b.f./cu.m	Transom Add l.f./meters	Lumber b.f./cu.m
2'-0" x 4'-0"	12	8	6'-0"	4
600 x 1200 mm	3.66	0.018	1.83	0.009
2'-0" x 4'-6"	12	8	6'-0"	4
600 x 1350 mm	3.66	0.018	1.83	0.009
2'-0" x 5'-0"	13-1/2	9	6'-0"	4
600 x 1500 mm	4.11	0.020	1.83	0.009
2'-6" x 4'-0"	12	8	6'-6"	4-1/3
750 x 1200 mm	3.66	0.018	1.98	0.010
2'-6" x 4'-6"	13	8-2/3	6'-6"	4-1/3
750 x 1350 mm	3.96	0.019	1.98	0.010
2'-6" x 5'-0"	14	9-1/3	6'-6"	4-1/3
750 x 1500 mm	4.27	0.021	1.98	0.010
2'-6" x 5'-6"	15	10	6'-6"	4-1/3
750 x 1650 mm	4.57	0.023	1.98	0.010
2'-6" x 6'-0"	16	10-2/3	6'-6"	4-1/3
750 x 1800 mm	4.88	0.023	1.98	0.010
2'-6" x 6'-6"	17	11-2/3	6'-6"	4-1/3
750 x 1950 mm	5.18	0.025	1.98	0.010
2'-6" x 7'-0"	18	12	6'-6"	4-1/3
750 x 2100 mm	5.49	0.027	1.98	0.010
2'-8" x 6'-8"	17	11-1/3	7'-0"	4-2/3
800 x 2000 mm	5.18	0.025	2.13	0.010
2'-8" x 7'-0"	18	12	7'-0"	4-2/3
800 x 2100 mm	5.49	0.027	2.13	0.010
3'-0" x 7'-0"	19	12-2/3	7'-0"	4-2/3
900 x 2100 mm	5.79	0.028	2.13	0.010
3'-6" x 7'-0"	19	12-2/3	7'-6"	5
1050 x 2100 mm	5.79	0.028	2.29	0.011
4'-0" x 7'-0"	20	13-1/3	8'-0"	5-1/3
1200 x 2100 mm	6.10	0.030	2.44	0.012

Size of Opening	Lumber Req'd. l.f./meters	Lumber Req'd. b.f./cu.m	Transom Add l.f./meters	Lumber b.f./cu.m
2'-0" x 4'-0"	12	8	6'-0"	4
600 x 1200 mm	3.66	0.018	1.83	0.009
5'-0" x 7'-0"	21	14	9'-0"	6
1500 x 2100 mm	6.40	0.032	2.74	0.014
6'-0" x 7'-0"	22	14-2/3	10'-0"	6-2/3
1800 x 2100 mm	6.71	0.033	3.05	0.015
7'-0" x 7'-0"	24	16	11'-0"	7-1/3
2100 x 2100 mm	7.32	0.036	3.35	0.016
3'-6" x 7'-6"	20	13-1/3	7'-6"	5
1050 x 2225 mm	6.10	0.030	2.29	0.011
4'-0" x 7'-6"	20	13-1/3	8'-0"	5-1/3
1200 x 2225 mm	6.10	0.030	2.44	0.012
4'-6" x 7'-6"	21	14	8'-6"	5-2/3
1350 x 2225 mm	6.40	0.032	2.74	0.013
5'-0" x 7'-6"	21	14	9'-0"	6
1500 x 2250 mm	6.40	0.032	2.74	0.014
6'-0" x 7'-6"	22	14-2/3	10'-0"	6-2/3
1800 x 2250 mm	6.71	0.033	3.05	0.015
7'-0" x 7'-6"	24	16	11'-0"	7-1/3
2100 x 2250 mm	7.32	0.036	3.35	0.016
3'-6" x 8'-0"	21	14	7'-6"	5
1050 x 2400 mm	6.40	0.032	2.29	0.011
4'-0" x 8'-0"	22	14-1/3	8'-0"	5-1/3
1200 x 2400 mm	6.71	0.032	2.44	0.012
5'-0" x 8'-0"	22	14-2/3	9'-0"	6
1500 x 2400 mm	6.71	0.033	2.74	0.014
6'-0" x 8'-0"	23	15-2/3	10'-0"	6-1/3
1800 x 2400 mm	7.01	0.035	3.05	0.014
7'-0" x 8'-0"	24	16	11'-0"	7-1/3
2100 x 2400 mm	7.32	0.033	3.35	0.016
9'-0" x 9'-0"	28	18-2/3	9'-0"	6
2700 x 2700 mm	8.53	0.042	2.74	0.014

Lumber quantities in the table above are estimated in lengths that cut with the least waste.

Lumber Required for Wood Door Bucks of 2"x6" (50 x 150 mm) Lumber

Size of Opening	Lumber Req'd. l.f./meters	Lumber Req'd. b.f./cu.m	Transom Add l.f./meters	Lumber b.f./cu.m
2'-0" x 4'-0"	12	12	6'-0"	6
600 x 1200 mm	3.66	--	1.83	0.014
2'-0" x 4'-6"	12	12	6'-0"	6
600 x 1350 mm	3.66	--	1.83	0.014
2'-0" x 5'-0"	13.5	13.5	6'-0"	6
600 x 1500 mm	4.11	--	1.83	0.014
2'-6" x 4'-0"	12	12	6'-6"	6.5
750 x 1200 mm	3.66	--	1.98	0.015
2'-6" x 4'-6"	13	13	6'-6"	6.5
750 x 1350 mm	3.96	--	1.98	0.015
2'-6" x 5'-0"	14	14	6'-6"	6.5
750 x 1500 mm	4.27	--	1.98	0.015
2'-6" x 5'-6"	14	15	6'-6"	6.5
750 x 1650 mm	4.27	--	1.98	0.015
2'-6" x 6'-0"	15	16	6'-6"	6.5
750 x 1800 mm	4.57	--	1.98	0.015
2'-6" x 6'-6"	17	17	6'-6"	6.5
750 x 1950 mm	5.18	--	1.98	0.015
2'-6" x 7'-0"	18	18	6'-6"	6.5
750 x 2100 mm	5.49	--	1.98	0.015
2'-8" x 6'-8"	17	17	7'-0"	7
800 x 2000 mm	5.18	--	2.13	0.016
3'-0" x 7'-0"	19	19	7'-0"	7
900 x 2100 mm	5.79	--	2.13	0.016
3'-6" x 7'-0"	19	16	7'-6"	7.5
1050 x 2100 mm	5.79	--	2.29	0.017

Size of Opening	Lumber Req'd. l.f./meters	Lumber Req'd. b.f./cu.m	Transom Add l.f./meters	Lumber b.f./cu.m
4'-0" x 7'-0"	20	20	8'-0"	8
1200 x 2100 mm	6.10	--	2.44	0.018
5'-0" x 7'-0"	21	21	9'-0"	9
1500 x 2100 mm	6.40	--	2.74	0.020
6'-0" x 7'-0"	22	22	10'-0"	10
1800 x 2100 mm	6.71	--	3.05	0.023
7'-0" x 7'-0"	24	24	11'-0"	11
2100 x 2100 mm	7.32	--	3.35	0.025
3'-6" x 7'-6"	20	20	9'-6"	7.5
1050 x 2250 mm	6.10	--	2.90	0.017
4'-0" x 7'-6"	20	20	8'-0"	8
1200 x 2250 mm	6.10	--	2.44	0.018
2'-0" x 4'-0"	12	12	6'-0"	6
600 x 1200 mm	3.66	--	1.83	0.014
4'-6" x 7'-6"	21	21	8'-6"	8.5
1050 x 2250 mm	6.40	--	2.59	0.019
5'-0" x 7'-6"	21	21	9'-0"	9
1500 x 2250 mm	6.40	--	2.74	0.020
6'-0" x 7'-6"	22	22	10'-0"	10
1800 x 2250 mm	6.71	--	3.05	0.023
7'-0" x 7'-6"	24	24	11'-0"	11
2100 x 2250 mm	7.32	--	3.35	0.025
3'-6" x 8'-0"	21	21	7'-6"	7.5
1050 x 2400 mm	6.40	--	2.29	0.017
4'-0" x 8'-0"	22	22	8'-0"	8
1200 x 2400 mm	6.71	--	2.44	0.018
5'-0" x 8'-0"	22	22	9'-0"	9
1500 x 2400 mm	6.71	--	2.74	0.020
6'-0" x 8'-0"	23	23	10'-0"	10
1800 x 2400 mm	7.01	--	3.05	0.023
7'-0" x 8'-0"	24	24	11'-0"	11
2100 x 2400 mm	7.32	--	3.35	0.025

Size of Opening	Lumber Req'd. l.f./meters	Lumber Req'd. b.f./cu.m	Transom Add l.f./meters	Lumber b.f./cu.m
8'-0" x 8'-0"	25	25	12'-0"	12
2400 x 2400 mm	7.62	--	3.66	0.027
9'-0" x 9'-0"	28	28	9'-0"	9
2700 x 2700 mm	8.53	--	2.74	0.020

Wood Door Bucks

Wood door bucks are made from 2"x4" (50 x 100 mm) or 2"x6" (50 x 150 mm) lumber, depending upon the thickness of the partition in which they are to be used.

Considerable savings are realized by using bucks made from 2"x6" lumber, having the back grooved out the thickness of the partition and 1/2" to 3/4" (13-19 mm) deep. For a 5-1/2" (138 mm) finished partition, this allows for a 4" (100 mm) tile or gypsum partition block and 1-1/2" (37.5 mm) for plaster on both sides of the partition. The partition tile fit into the groove in the wood buck and the 3/4" (19 mm) extension on each side provides a ground for nailing the trim.

The mills furnishing the lumber usually charge $15.00 to $20.00 per 1,000 b.f. (2.36 cu.m) for grooving the backs of 2"x6"s (50 x 150 mm).

The lumber quantities given above are estimated in lengths that cut with the least waste.

Labor Making Rough Door Bucks. A carpenter should make and brace a rough door buck for an opening up to 3'-0"x7'-0" (0.9 x 2.1 m) in about 1 hour. If the door has a transom, add 1/2 hour carpenter time.

When making rough door bucks for door openings 5'-0"x7'-0" (1.5 x 2.1 m) to 6'-0"x9'-0" (1.8 x 2.7 m), a carpenter should make and brace one door buck in 1-1/4 to 1-1/2 hrs. If the door has a transom, add 1/2 to 3/4 hr. carpenter time.

Rough Wood Bucks for Borrowed Lights. When making rough wood bucks for borrowed lights, a carpenter should make and brace one buck for an average size opening in 1 to 1-1/2 hrs.

Setting Rough Door Bucks. After the rough door buck has been made, a carpenter should set, plumb, and brace one door buck for an opening up to 3'-0"x9'-0" (0.9 x 2.7 m) in about 1 to 1-1/4 hr.

For double doors requiring an opening from 5'-0"x7'-0" (1.5 x 2.7 m) to 6'-0"x9'-0" (1.8 x 2.7 m), a carpenter should set, plumb, and brace one large door buck in 1-1/4 to 1-1/2 hrs.

Fireproofing Wood. Where framing lumber and timbers are chemically treated to fireproof the wood, the cost of this treatment almost doubles the cost of wood.

BUILDING AND INSULATING SHEATHING

There are numerous types of insulating sheathing board, which usually consist of felted wood fiber or vegetable fiber products treated to make them moisture resistant.

Some of these sheets are coated with high melting point asphalt or are asphalt impregnated to form a moisture resistant surface which retards moisture penetration.

Insulating sheathing is furnished in sheets 4' (1.2 m) wide and 6' (1.8 m), 7' (2.1 m), 8' (2.4 m), 9' (2.7 m), 10' (3 m) and 12' (3.6 m) long, and 1/2" (13 mm) and 25/32" (20 mm) thick, the same thickness as wood sheathing.

Most manufacturers furnish this type of sheathing 25/32" (20 mm) thick, 2' (0.6 m) wide and 8' (2.4 m) long, with the long edges V-jointed or shiplapped.

Insulating sheathing is usually applied to wood studs under wood siding, shingles, stucco or brick veneer and because of the large size sheets used and the asphalt treatment, building paper is not ordinarily used with it, except under stucco.

Use 2" (50 mm) galvanized nails with 3/8" (9 mm) or 1/2" (13 mm) heads for insulating sheathing.

Place nails 3" (75 mm) apart on all outside edges of board and 6" (150 mm) apart for all intermediate nailing.

Insulating sheathing 25/32" (20 mm) thick costs $170 to $200 per 1,000 s.f. (90 sq.m).

Labor Placing Insulating Sheathing. When placing insulating sheathing on square or rectangular houses of regular construction, a carpenter should place 700 to 900 s.f. (63-81 sq.m) per 8-hr. day, at the following labor cost per 100 s.f. (9 sq.m):

	Hour	Rate	Total	Rate	Total
Carpenter	1.0	$....	$....	$26.44	$26.44
Labor unloading and					
carrying sheets	0.3	....	....	21.04	6.31
Cost per 100 s.f. (9 sq.m)			$....		$ 32.75
per s.f.				....	0.33
per sq.m				....	3.55

On buildings of irregular construction, requiring a great deal of cutting and fitting, it is not possible to take advantage of the large size sheets. Too much cutting and fitting is required. A carpenter will place only 350 to 450 s.f. (32-41 sq.m) of sheathing per 8-hr. day, at the following labor cost per 100 s.f. (9 sq.m):

	Hours	Rate	Total	Rate	Total
Carpenter	2.0	$....	$....	$26.44	$ 52.88
Labor unloading and					
carrying sheets	0.3	....	....	21.04	6.31
Cost per 100 s.f. (9 sq.m)			$....		$ 59.19
per s.f.				....	0.59
per sq.m				....	6.37

Insulating Roof Decking

Several manufacturers produce insulating roof decking made of multiple layers of 1/2" (13 mm) insulating board laminated together with vapor resistant cement and fabricated into tongue and groove planks, 2'-0" (0.6 m) wide, 8'-0" (2.4 m) long, and 1-1/2", 2", or 3" (38, 50, or 75 mm) thick, with a finish painted undersurface. It provides a roof deck, insulation, vapor barrier, and interior ceiling finish using only one material and is especially adaptable to designs with exposed beam ceilings.

Insulating roof decking can be used for flat, pitched, or monosloped roofs. Flat roofs and surfaces with a slope of 3" in 12" (75 in 300 mm) or less are usually covered with built-up roofing. Steeper roofs may be covered with rigid shingles, slate, or tile roofing, providing wood nailing strips are fastened through decking to supporting beams below.

Decking should be laid so that cross joints are staggered and occur only over supports. Decking should be face nailed to all framing members, spacing nails 4" to 6" (100 to 150 mm) apart and keeping back 3/4" to 1" (19 to 25 mm) from edges of plank. Nails should be galvanized common of sufficient length to pass through decking and penetrate supports at least 1-1/2" (38 mm) and should be driven flush but not countersunk. Where underside of decking will be exposed, planks must be handled and laid with care to prevent finished ceiling surface from being marred or damaged. Avoid excessive sliding of plank on roof beams.

Estimating Data on Insulating Roof Decking

Thickness	Max. Distance Between Supports	Size & Type of Nails	Nails Req'd. per 100 s.f. (9 sq.m)	Approx. Price per s.f. (sq.m)
1-1/2" (38 mm)	24" 600 mm)	10d Com. Galv.	6.0 lbs. (2.7 kg)	$0.63 ($6.80)
2" (50 mm)	32" (800 mm)	16d Com. Galv.	3.5 lbs. (1.58 kg)	0.69 (7.40)
3" (75 mm)	48" (1200 mm)	30d Com. Galv.	3.0 lbs. (1.35 kg)	0.81 (8.70)

Labor Placing Insulating Roof Decking. The labor cost of handling and placing insulating roof decking will be about the same for the various thicknesses, because the savings in handling lighter material is offset by the increased nailing due to closer required spacing of supports.

On simple, rectangular, flat or low-rise roofs, two carpenters and a helper should place 750 to 850 s.f. (68-77 sq.m) of decking per 8-hr. day, including stapling paper strips on roof beams for added protection against marring finished surface, at the following labor cost per 100 s.f. (9 sq.m):

	Hours	Rate	Total	Rate	Total
Carpenter	2	$....	$....	$26.44	$52.88
Labor	1	....	....	21.04	21.04
Cost per 100 s.f. (9 sq.m)			$....		$73.92
per s.f.				....	0.74
per sq.m				....	7.95

For steep roofs, with a slope greater than 6" in 12" (150 in 300 mm), add 15% to 25%.

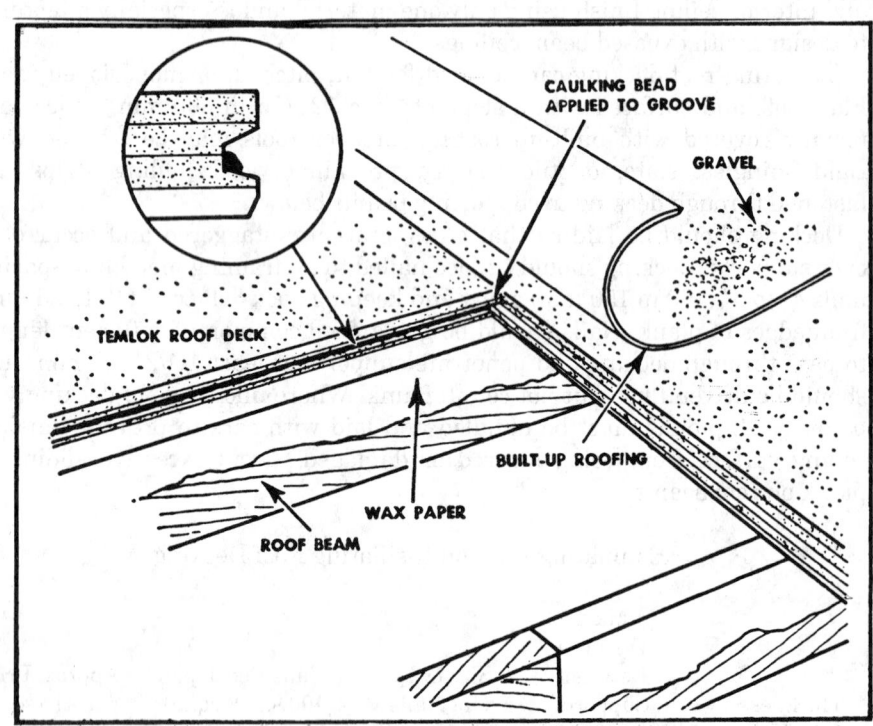

Method of Installing Insulating Roof Decking

PLYWOOD ROOF AND WALL SHEATHING, SUBFLOORING AND UNDERLAYMENT

Roof and Wall Sheathing. Plywood for use in roof and wall sheathing is readily available, easily worked with ordinary tools and skills, and adaptable to almost any light construction application. Because of the large sheet size, installation time is less than for other types of sheathing, and waste is minimal. It may be used under any type of shingle or roofing material, or any type of siding.

Standard size sheets are 4"x8" (100 x 200 mm), but other sizes are available by special order. Most common sheathing thicknesses are 5/16", 3/8", 1/2", 5/8", and 3/4" (8, 9, 13, 16, and 19 mm), which are unsanded and are available with interior, intermediate, or exterior glue.

For most roof and wall sheathing installations, where sheathing is to be covered, interior type plywood is used. Where exposed to the weather, such as for roof overhangs or service building siding, exterior plywood should be used.

Nailing of plywood sheathing should be at 6" (150 mm) o.c. along panel edges and 12" (300 mm) o.c. at intermediate supports. Use 6d common nails for panels of 1/2" (13 mm) or less in thickness, and 8d for greater thickness.

Where plywood is used for wall sheathing, corner bracing is not required because of the rigidity of the plywood. Where other types of sheathing are used, plywood of the same thickness as the sheathing is sometimes used at corners to take the place of corner diagonal bracing.

Subflooring and Underlayment. Plywood subflooring, underlayment, or combined subfloor-underlayment are available in most areas of the country. Some particular panels may not be stocked in all areas, so it is advisable to check with local plywood sources.

Subflooring panels are engineered grades of plywood marked by the American Plywood Association for maximum spans for the various thicknesses. (See table below) Deflection of the panels under concentrated loads at panel edges is the limiting factor in the design of plywood floors. Panel edges should be supported by solid blocking, or tongue and grooved panels should be used, unless a separate layer of underlayment is installed with its joints offset from the joints in the subfloor.

Nailing of subflooring should be at 6" (150 mm) o.c. along the edges and 10" (250 mm) o.c. at intermediate supports. Use 6d common nails for 1/2" (13 mm) plywood, and 8d for 5/8" (16 mm) to 7/8" (22 mm) thick plywood. For 1-1/8" (28 mm) thick panels, use 8d ring shank or 10d common nails spaced 6" (150 mm) o.c. at both panel edges and intermediate supports.

Underlayment grades of plywood are touch sanded panels of 1/4" (6 mm) and 3/8" (9 mm) thickness with a smooth, solid surface for application of nonstructural flooring finished directly to them. This grade or plywood is made with special inner-ply construction which resists punch-through by concentrated point loading.

Joints in the underlayment should be staggered with respect to the joints in the subfloor. Nail with 3d ring shanked underlayment nails at 6" (150 mm) o.c. along the edges and 8" (200 mm) each way in the interior of the panels. Nails should be countersunk, if subfloor and joists are not dry.

Combined Subfloor-Underlayment Plywood. This is a single application installation combining both subfloor and underlayment, eliminating the expense of installing a separate underlayment. Panel edges perpendicular to the joists must be supported by solid blocking if tongue and grooved panels are not used.

Panels should be nailed at 6" (150 mm) o.c. at edges and 10" (250 mm) at intermediate supports with ringshank or spiral-thread nails. Use 6d deformed-shank nails for panel thickness up to 3/4" (19 mm) and 8d for thicker panels. Approximate costs of plywood sheathing, subflooring, and underlayment per 1000 s.f. (90 sq.m) are as follows:

Types of Plywood – Cost per 1,000 Sq. Ft.

Interior Type Standard (C-D INT)	Interior Type Structural 1 C-D	Exterior Type C-C EXT
5/16"............ $276.00	1/2"............$357.00	5/16"............$334.00
3/8"............ 299.00	5/8"............449.00	3/8"............403.00
1/2"............ 356.50	3/4"............541.00	1/2"............449.00
5/8"............ 414.00		5/8"............495.00
3/4"............ 483.00		3/4"............575.00

Plywood Underlayment
Group 1, Interior
1/2"............$345.00
5/8"............ 449.00
3/4"............ 575.00

Types of Plywood – Cost per 100 Sq. Meters

Interior Type Standard (C-D INT)	Interior Type Structural 1 C-D	Exterior Type C-C EXT
7.8 mm............ $297.10	12.5 mm............$384.30	7.8 mm............$359.50
9.4 mm............ 321.90	15.6 mm............483.30	9.4 mm............433.80
12.5 mm............ 383.70	18.8 mm............582.30	12.5 mm............483.30
15.6 mm............ 445.60		15.6 mm............532.80
18.8 mm............ 519.90		18.8 mm............618.90

Plywood Underlayment
Group 1, Interior
12.5 mm............$371.40
15.6 mm............ 483.30
18.8 mm............ 557.60

Prices on all grades and types of plywood should be verified with local suppliers before you submit a proposal for work that includes these materials.

Approximate costs of various materials and labor required for sidewall sheathing are given below.

Labor Cost of 100 S.F. (9 Sq.M) of Structural Insulating Sidewall Sheathing Applied to Square or Rectanglar Buildings of Regular Construction

	Hours	Rate	Total	Rate	Total
Carpenter sheathing	1.0	$....	$....	$26.44	$26.44
Helper unloading and carrying sheets	0.3	....	....	21.04	6.31
Cost per 100 s.f.			$....		$32.75
per s.f.				....	0.33
per sq.m				....	3.55

No paper required as insulating sheathing is treated to resist moisture.
On buildings of irregular construction, add 1 hr. carpenter time.

Labor Cost of 100 S.F. (9 Sq.M) of 1"x6" (25 x 150 mm) Wood Sidewall Sheathing Applied Horizontally to Square or Rectangular Buildings of Regular Construction

	Hours	Rate	Total	Rate	Total
Carpenter sheathing	1.5	$....	$....	$26.44	$39.66
Labor unloading and carrying lumber	0.5	....	....	21.04	10.52
Carpenter applying building paper	0.4	....	....	26.44	10.58
Cost per 100 s.f. (9 sq.m)			$....		$ 60.76
per s.f.				....	0.61
per sq.m				....	6.56

If 1"x8" (25 x 200 mm) shiplap is used instead of 1"x6" (25 x 150 mm) D&M lumber, deduct 5 b.f. (0.01 cu.m) lumber and 0.1 hr. carpenter time per 100 s.f. (9 sq.m) of wall.
On buildings of irregular construction, add 1 hr. carpenter time.

Labor Cost of 100 S.F. (9 Sq.M) of 1"x6" (25 x 150 mm) Wood Sidewall Sheathing Applied Diagonally to Square or Rectangular Buildings of Regular Construction

	Hours	Rate	Total	Rate	Total
Carpenter sheathing	2.0	$....	$....	$26.44	$52.88
Labor unloading and carrying lumber	0.5	....	....	21.04	10.52
Carpenter applying paper	0.4	....	....	26.44	10.58
Cost per 100 s.f.			$....		$73.98
per s.f.				....	0.74
per sq.m				....	7.96

If 1"x8" (25 x 200 mm) shiplap is used instead of 1"x6" (25 x 150 mm) D&M lumber, deduct 5 b.f. (0.01 cu.m) lumber and 0.1 hr. carpenter time per 100 s.f. (9 sq.m) of wall.

On buildings of irregular construction, add 1 hr. carpenter time.

Labor Cost of 100 S.F. (9 Sq.M) of 1/2" (13 mm) Plywood Sidewall Sheathing Applied to Square or Rectangular Buildings of Regular Construction

	Hours	Rate	Total	Rate	Total
Carpenter sheathing	1.0	$....	$....	$26.44	$26.44
Helper unloading and carrying sheets	0.3	....	....	21.04	6.31
Carpenter applying paper	0.4	....	....	26.44	10.58
Cost per 100 s.f.			$....		$43.33
per s.f.			....		0.43
per sq.m			....		4.63

Paper optional; plywood comes in large sheets of 48"x96" (1200 x 2400 mm) treated with asphalt.

On buildings of irregular construction, add 1 hr. carpenter time.

Gypsum Sheathing. Gypsum sheathing is a fireproof solid sheet of gypsum encased in a tough, fibrous, water-resistant covering. The sides and ends are treated to resist moisture. Some manufacturers blend asphalt emulsion into the wet gypsum core mix to provide additional resistance to moisture. It is used for sheathing on frame structures under siding, shingles, stucco, and brick veneer.

Gypsum sheathing has V-joint edges on the long dimension to provide positive assurance of a tight fit at the unsupported joints. It is applied with its length at right angles to the studs.

Gypsum sheathing is 1/2" (13 mm) thick, 2' (0.6 m) wide, and 8' (2.4 m) and 10' (3.0 m) long, and 4' (1.2 m) wide and 8' (2.4 m) and 9' (2.7 m) long to fit supports 16" (400 mm) on centers. Nails should be 1-3/4" (44 mm) long, No. 10-1/2 gauge, galvanized flat head roofing nails, spaced 4" (100 mm) on centers, except under wood siding and stucco, 8" (200 mm). Requires 14 to 21 lbs. (6.3-9.5 kg) per 1,000 s.f. (90 sq.m). Price $85.00 to $125.00 per 1,000 s.f. (90 sq.m). Labor cost same as given for insulating or plywood sheathing.

Labor Cost of 100 S.F. (9 Sq.M) of Gypsum Sidewall Sheathing Applied to
Square or Rectangular Buildings of Regular Construction

	Hours	Rate	Total	Rate	Total
Carpenter sheathing	1.0	$....	$....	$26.44	$26.44
Helper unloading and carrying sheets	0.3	....	....	21.04	6.31
Carpenter applying paper	0.4	....	....	26.44	10.58
Cost per 100 s.f.			$....		$43.33
per s.f.				....	0.43
per sq.m				....	4.63

Paper is optional. Gypsum sheathing is waterproofed sides and edges. Some manufacturers mix asphalt emulsion with wet gypsum to provide a waterproof board.

On buildings of irregular construction, add 1 hr. carpenter time.

Insulating Shingle Backer Strips

A number of manufacturers produce a shingle backer strip, of the same material as insulating sheathing board, that is used as the undercourse in double course shingle work. These strips are 4'-0" (1.2 m) long and either 5/16" (8 mm) or 3/8" (9 mm) thick. They are available in two widths, 13-1/2" (338 mm) for 12" (300 mm) shingle exposure and 15-1/2" (388 mm) for 14" (350 mm) shingle exposure. They provide added insulation, improve the shadow line appearance of the finished wall, eliminate the necessity of building paper over wood sheathing, and are more economical to use than second grade wood shingle undercoursing. Insulating shingle backer strips cost approximately $0.30 per s.f. (0.09 sq.m).

Labor Placing Insulating Shingle Backer Strips. On straight run jobs, with an average number of openings, a carpenter should place 500 to 600 s.f. (45-54 sq.m) of shingle backer strips per 8-hr. day at the following labor cost per 100 s.f. (9 sq.m):

	Hours	Rate	Total	Rate	Total
Carpenter	1.5	$....	$....	$26.44	$39.66
Labor	0.5	....	....	21.04	10.52
Cost 100 s.f.			$....		$50.18
per s.f.				....	0.50
per sq.m				....	5.38

On complicated jobs, requiring a large amount of cutting and fitting, a carpenter should place 400 to 500 s.f. (36-45 sq.m) per 8-hr. day at the following labor cost per 100 s.f. (9 sq.m):

	Hours	Rate	Total	Rate	Total
Carpenter...	1.8	$....	$....	$26.44	$ 47.59
Labor..	0.5	....	....	21.04	10.52
Cost 100 s.f.			$....		$ 58.11
per s.f..				....	0.58
per sq.m				....	6.25

Insulating Lath

Insulating lath is made by several manufacturers of insulating building sheathing. It is made of the same materials as the building boards previously described.

Insulating lath is usually furnished in the 18"x48" (450 x 1200 mm) size, 1/2" (13 mm), 3/4" (19 mm), and 1" (25 mm) thick, although they are obtainable from some sources in 16"x48" (400 x 1200 mm).

Some manufacturers furnish lath both plain and with the back coated with asphalt to form a vapor barrier.

Most insulating lath are manufactured with the long edges shiplapped and all edges beveled to reinforce the plaster against cracking and eliminate unsightly lath marks. Some of these insulating lath are metal reinforced along the edges to reinforce the unsupported horizontal joint.

Insulating gypsum lath consists of regular gypsum board lath with a thin sheet of aluminum attached to one side to provide insulating value. They are furnished in sheets 16"x32" (400 x 800 mm), 16"x48" (400 x 1200 mm), and 3/8" (9 mm) and 1/2" (13 mm) thick.

Estimating Quantities of Insulating Lath

Size of Lath, inches (mm)	No S.F. (Sq.M) per Lath	No. Lath per 100 S.F.	No. Lath per 10 Sq.M
16x48 (400x1200)	5.33 (0.50)	18.76	20.19
18x48 (450x1200)	6.00 (0.56)	16.67	17.94

Approximate prices of insulating lath are as follows per 1,000 s.f. (90 sq.m): 1/2" (13 mm), $200.00; 5/8" (16 mm), $250.00.

Labor Applying Insulating Lath. The labor cost of applying insulating lath will vary with the locality and labor restrictions. Where there are no labor restrictions, an experienced worker should place 125 sq. yds. (105 sq.m) per 8-hr. day, while a less experienced worker will place only 100 sq. yds. (84 sq.m) per 8-hr. day, at the following labor cost per 100 sq. yds. (84 sq.m):

	Hours	Rate	Total	Rate	Total
Lather	8.0	$....	$....	$26.58	$212.64
Cost per sq. yd...............................				....	2.13
per sq.m				....	2.32

WOOD AND WIRE FENCE

Setting Wood Fence Posts. When placing wire fence around the lots of 16 houses, it required 218 cedar posts, 5" (125 mm) in diameter and 7'-0" (2.1 m) long, also 32 clothes posts, 7" (175 mm) in diameter and 10'-0" (3 m) long.

After the holes had been dug, a worker set 7 posts per hr. at the following labor cost per post:

	Hours	Rate	Total	Rate	Total
Carpenter	0.5	$....	$....	$26.44	$13.22

Treated pine and locust posts may also be used for fencing. Locust and cedar posts should be all heartwood.

Placing 2"x4" (50 x 100 mm) Top Bottom Fence Rail Ready to Receive Wire. The posts for these fences are spaced about 6'-0" (1.8 m) apart and a 2"x4" (50 x 100 mm) top rail is nailed flatwise to the top of the post and a 2"x4" (50 x 100 mm) bottom rail is notched into the side of the post so that the wire can be stapled to it.

There is a total 1,400 l.f. (420 m) of fence or 2,800 l.f. (840 m) of top and bottom rail. A carpenter places rail for 110 l.f. (33 m) of fence or 220 l.f. (66 m) of single rail per 8-hr. day, at the following labor cost per 100 l.f. (30 m):

	Hours	Rate	Total	Rate	Total
Carpenter	7.3	$....	$....	$26.44	$193.01
Labor	0.9	....	....	21.04	18.94
Cost per 100 l.f.			$....		$211.95
per l.f. of fence				....	2.12
per m of fence				....	6.95
per l.f. of rail				....	4.24
per m of rail				....	13.90

Stringing Fence Wire. Cyclone wire fencing in 42" (1050 mm) rolls is used for these fences and is stapled to the top and bottom rails and posts about 4" (100 mm) on centers.

There is a total 1,400 l.f. (420 m) of 42" (1050 mm) wire fencing in the job and a carpenter would string and staple about 280 l.f. (84 m) per 8-hr. day, at the following labor cost per 100 l.f. (30 m):

	Hours	Rate	Total	Rate	Total
Carpenter	2.9	$....	$....	$26.44	$76.68
Cost per l.f.				....	0.77
per m				....	2.53

Setting Lally Columns in Basements. In a group of brick houses, 3" (75 mm) iron columns were used to support the 6"x8" (150 x 200 mm) wood girders carrying the floor joists.

These columns were 3" (75 mm) in diameter and 7'-0" (2.1 m) long, and a carpenter set, plumbed, and braced 1 column per hr. at the following labor cost per column:

	Hours	Rate	Total	Rate	Total
Carpenter	1.0	$....	$....	$26.44	$26.44
Labor	1.0	....	....	21.04	21.04
Cost per column			$....		$47.48

MISCELLANEOUS HARDWARE AND ACCESSORIES

Steel Area Walls and Gratings

These are used around basement windows when first floor is at grade level and are made of 16-ga. and 20-ga. galvanized steel, with stiffening ribs and rounded tops. They are attached to masonry walls by screws or bolts.

Size Width x Depth ft./in. (mm)	For Use with Basement Windows ft./in. (mm)	Price each, 16 ga.	Price each, 20 ga.
3'-2"x0'-11-1/2" (950x288mm)	15"x12" (375x300mm)-2-Lt	$10.90	$9.00
3'-2"x1'-5-1/2" (950x438mm)	15"x16" (375x400mm)-2-Lt	14.90	10.90
3'-2"x1'-11-1/2" (950x288mm)	15"x20" (375x500mm)-2-Lt	19.90	14.70

Steel gratings for use with above walls. Frames made of 1-1/4"x1/4" (31 x 6 mm) steel bars with cross bars 1"x13/16" (25 x 20 mm) welded into one piece. Supports on the grating hold it flush with the top of the area wall. Price each painted, $23.50.

Metal Foundation Wall Ventilators

Brick Ventilators. Made of aluminum exactly the size of a brick, 2-1/2"x4"x8" (63 x 100 x 200 mm). The bottom edge is flanged both front and back to insure a positive mortar lock. Louvered face allows 13 sq. in. (8125 sq.mm) of free area. Screened back. Price each, $9.25.

Frame Ventilators. Nailed to sheathing just above the foundation or to ends of joists spaced 16" (400 mm) on centers. Made of sheet aluminum, size 16"x8" (400 x 200 mm), providing 89 sq. in. (55,625 sq.mm) of free area. Screened back. Price each, $23.50.

Concrete Block Ventilators. Made of cast aluminum 16" (400 mm) wide x 8" (200 mm) high, providing 89 sq. in. (55,625 sq.mm) of free area. Screened back. Price each, $25.00.

Metal Roof Ventilators

Metal Roof Ventilators

Metal roof ventilators provide air in the attic space under the roof. They are made of steel or aluminum and usually provide 18 to 75 sq. in. of free area.

Roof Opening, inches (mm)	Free Area, sq. in. (sq.mm)	Price each, Aluminum
12x18 (300x450)	30 (19,355)	$8.60
15x20 (375x500)	50 (32,258)	11.50
16x16 (400x400)	65 (41,935)	18.40

Steel Clothes Chute Doors

Price Each

Furnished in steel, white enameled, opening 9"x12" (225x300mm).........$15.00

Steel Access Doors

For access to piping or wiring installations. Furnished with a flanged frame that may be nailed to studs 16" on centers. Size of door 16" x 24" (400 x 600 mm), overall size 17-1/2"x25-1/2" (438 x 638 mm) ... $24.00

ROUGH HARDWARE

While it is possible to include the cost of rough hardware such as nails and screws in the material unit for each carpentry item figured, most contractors and estimators find it more convenient to make an allowance for this item at the end of the carpentry estimate based on the amount of lumber and other material involved. The usual method of arriving at a rough hardware allowance is to total all the material in the carpentry estimate, taking b.f. (cu.m) of dimension lumber and boards, sq. ft. (sq.m) area of plywood, paneling, etc., and lin. ft. (meters) of grounds, furring strips, trim, etc., and then figure from 25 to 35 lbs. (11 to 16 kg) of nails per 1,000 b.f. (2.36 cu.m). The weight of nails multiplied by the cost per lb. (kg) will give the rough hardware allowance required.

06130 HEAVY TIMBER CONSTRUCTION

The quantities and costs included here cover types such as heavy constructed mill, warehouse, and factory buildings.

Framing and Erecting Heavy Wood Columns or Posts. The labor cost of erecting and framing heavy wood columns or posts, 10"x10" (250 x 250 mm) or larger, will vary according to the number of operations necessary, such as chamfering corners, cutting out and framing for post caps and bases, boring holes through the center, and upon the distance the posts must be hoisted before placing.

On the average mill constructed job, requiring 10"x10" (250 x 250 mm) or 12"x12" (300 x 300 mm) posts 10'-0" to 16'-0" (3.0-4.8 m) long, it will require 3-1/2 to 4 hrs. to chamfer corners, frame for bases and caps, and erect each column, at the following labor cost:

	Hours	Rate	Total	Rate	Total
Carpenter	3.7	$....	$....	$26.44	$97.83
Labor	0.5	....	....	21.04	10.52
Cost per column			$....		$108.35

If hoisted above the second floor, add 1/2 hr. of labor time per column. It is more difficult to estimate the cost of this work by the 1,000 b.f. (2.36 cu.m) of lumber, because it costs just as much to chamfer the 4 corners and frame for post caps and bases for an 8"x8" (200 x 200 mm) as a 12"x12" (300 x 300 mm) column, while the former contains less than half as much lumber per lin. ft. as the latter. The following costs are based on an average 10"x10"-12'-0" (250 x 250 mm - 3.6 m) timber and should be increased or reduced accordingly for larger or smaller timbers.

Based on a 10"x10"-12'-0" (250 x 250 mm - 3.6 m) column, a carpenter should chamfer corners, frame for post caps and bases, handle, and erect 200 to 250 b.f. (0.47-0.59 cu.m) of lumber per 8-hr. day, at the following labor cost per 1,000 b.f. (2.36 cu.m):

	Hours	Rate	Total	Rate	Total
Carpenter	35	$....	$....	$26.44	$925.40
Labor	8	....	....	21.04	168.32
Cost 1,000 b.f. (2.36 cu.m)			$....		$1,093.72
per cu.m				....	463.44

Rounding Corners on Large Timbers. Where the corners of wood columns, beams, or girders must be rounded to a small radius, the yards or mills furnishing the lumber usually charge $16.50 to $21.50 per 1,000 b.f. (2.36 cu.m) for performing this work.

Boring Holes Through Center of Large Timbers. Where the wood columns or posts have a hole bored through the center for their entire length, the yard or mill furnishing the lumber usually charges $32.00 to $37.00 per 1,000 b.f. (2.36 cu.m) for boring these holes.

Framing and Erecting Heavy Wood Beams and Girders for First Floor. The labor cost of framing and erecting heavy wood beams and girders will vary according to the amount of framing necessary and the method used in placing them; i.e., whether they set on top of the wood columns or whether it is necessary to frame for post caps, stirrups, etc.

Placing heavy beams and girders for the first floor of a building does not require as much labor as for the upper floors, because the timbers can be rolled to the place they are to be used without hoisting.

Where heavy wood girders are set on concrete or masonry piers without framing for post caps, joist hangers, etc., a carpenter should frame and erect 1,000 to 1,250 b.f. (2.36-2.95 cu.m) per 8-hr. day, at the following labor cost per 1,000 b.f. (2.36 cu.m):

	Hours	Rate	Total	Rate	Total
Carpenter	7	$....	$....	$26.44	$185.08
Labor	7	....	....	21.04	147.28
Cost 1,000 b.f. (2.36 cu.m)			$....		$332.36
per cu.m				....	140.83

Where necessary to frame for post caps and bases, stirrups, etc., a carpenter should frame and erect 500 to 600 b.f. (1.18-1.42 cu.m) of lumber per 8-hr. day, at the following labor cost per 1,000 b.f. (2.36 cu.m):

	Hours	Rate	Total	Rate	Total
Carpenter	14.5	$....	$....	$26.44	$383.38
Labor	7.0	....	....	21.04	147.28
Cost 1,000 b.f. (2.36 cu.m)			$....		$530.66
per cu.m				....	224.86

Framing and Erecting Heavy Wood Girders Above First Floor. Where heavy wood girders must be hoisted above the first floor, using a timber hoist or derrick, the carpenter time framing and placing the timbers will remain practically the same but the labor time handling, hoisting, and skidding the timbers into place will be increased.

On work of this kind, a carpenter should frame and erect 450 to 550 b.f. (1.06-1.30 cu.m) of lumber per 8-hr. day, at the following labor cost per 1,000 b.f (2.36 cu.m):

	Hours	Rate	Total	Rate	Total
Carpenter	16	$....	$....	$26.44	$423.04
Labor	10	....	....	21.04	210.40
Cost 1,000 b.f. (2.36 cu.m)			$....		$633.44
per cu.m				....	268.41

Framing and Erecting Heavy Wood Joists. When placing heavy wood joists 4"x12" (100 x 300 mm) or 4"x16" (100 x 400 mm) in size, where the joists are set on top of wood girders or in iron stirrups, a carpenter should frame and erect 1,000 to 1,200 b.f. (2.36-2.83 cu.m) per 8-hr. day, at the following labor cost per 1,000 b.f. (2.36 cu.m):

	Hours	Rate	Total	Rate	Total
Carpenter	7.3	$....	$....	$26.44	$193.01
Labor	3.5	....	....	21.04	73.64
Cost 1,000 b.f. (2.36 cu.m)			$....		$266.65
per cu.m				....	112.99

Where smaller joists are used, ranging from 3"x10" (75 x 250 mm) to 4"x10" (100 x 250 mm) in size, a carpenter should frame and erect 800 to 1,000 b.f. (1.88-2.36 cu.m) per 8-hr. day, at the following labor cost per 1,000 b.f. (2.36 cu.m):

	Hours	Rate	Total	Rate	Total
Carpenter	9	$....	$....	$26.44	$237.96
Labor	4	....	....	21.04	84.16
Cost 1,000 b.f. (2.36 cu.m)			$....		$322.12
per cu.m				....	136.49

Framing and Placing Oak Bumpers at Edges of Loading Docks. Oak timbers, varying from 4"x6" (100 x 150 mm) to 12"x12" (300 x 300 mm), are usually called for at loading dock edges. The timbers generally are secured to the edge of the dock slab by bolts set in the concrete or with cinch anchors.

With bolts already set in place, a carpenter should lay out, drill and countersink holes, and secure in place 150 to 175 b.f. (0.35-0.41 cu.m) of oak bumper per 8-hr. day at the following labor cost per 1,000 b.f. (2.36 cu.m):

	Hours	Rate	Total	Rate	Total
Carpenter	48	$....	$....	$26.44	$1,269.12
Labor	24	....	....	21.04	504.96
Cost 1,000 b.f. (2.36 cu.m)			$....		$1774.08
per cu.m				....	751.73

A carpenter should locate, set, and secure in place in the slab edge or wall form about 25 anchor bolts per 8-hr. day at the following labor cost per bolt:

	Hours	Rate	Total	Rate	Total
Carpenter	0.3	$....	$....	$26.44	$7.93

Laying 2"x6" (50 x 150 mm) or 3"x6" (75 x 150 mm) Tongue and Groove Timber Subfloors or Roof Sheathing. When laying 2"x6" (50 x 150 mm) or 3"x6" (75 x 150 mm) tongue and groove flooring or roof decking in large areas without much cutting and fitting, a carpenter should lay 800 to 1,000 b.f. (1.88-2.36 cu.m) per 8-hr. day at the following labor cost per 1,000 b.f. (2.36 cu.m):

	Hours	Rate	Total	Rate	Total
Carpenter	8.9	$....	$....	$26.44	$235.32
Labor	6.0	....	....	21.04	126.24
Cost per 1,000 b.f. (2.36 cu.m)			$....		$361.56
per cu.m				....	153.20

If tongue and groove flooring or roof sheathing is laid in small spaces requiring considerable cutting and fitting around openings, stair wells, elevator shafts, skylights, etc., a carpenter should lay 700 to 800 b.f. (1.65-1.88 cu.m) per 8-hr. day at the following labor cost per 1,000 b.f. (2.36 cu.m):

	Hours	Rate	Total	Rate	Total
Carpenter	10.6	$....	$....	$26.44	$ 280.26
Labor	6.0	$....	....	21.04	126.24
Cost per 1,000 b.f. (2.36 cu.m)			$....		$406.50
per cu.m				....	172.25

LAMINATED WOOD FLOORS

Laminated wood floors consisting of 2" (50 mm) or 3" (75 mm) plank set on edge and spiked together are used for floors capable of carrying heavy loads or supporting heavy machinery with a minimum amount of vibration.

Sometimes the floors are designed to use certain length plank reaching from bearing to bearing, while on others they are spiked together with broken joints, regardless of bearing.

Quantity of Lumber Required for Laminated Wood Floors

Measured Size, inches	Finished Thickness, inches	Add for Waste Percent	To Obtain Qty of Lumber Reqd Mult. Area by	L.F. Lumber Per 100 S.F. of Surface
2 x 4	1-5/8	23	4.90	490 (74 pcs.)
2 x 6	1-5/8	23	7.40	740 (74 pcs.)
2 x 8	1-5/8	23	9.84	984 (74 pcs.)
2 x 10	1-5/8	23	12.30	1230 (74 pcs.)
2 x 12	1-5/8	23	14.76	1476 (74 pcs.)
2 x 14	1-5/8	23	17.22	1722 (74 pcs.)
3 x 6	2-3/4	10	6.60	660 (44 pcs.)

Measured Size, inches	Finished Thickness, inches	Add for Waste Percent	To Obtain Qty of Lumber Reqd Mult. Area by	L.F. Lumber Per 100 S.F. of Surface
3 x 8	2-3/4	10	8.80	880 (44 pcs.)
3 x 10	2-3/4	10	11.10	1110 (44 pcs.)
3 x 12	2-3/4	10	13.20	1320 (44 pcs.)
3 x 14	2-3/4	10	15.40	1540 (44 pcs.)
4 x 6	3-3/4	7	6.42	642 (32 pcs.)
4 x 8	3-3/4	7	8.56	856 (32 pcs.)
4 x 10	3-3/4	7	10.70	1070 (32 pcs.)
4 x 12	3-3/4	7	12.84	1284 (32 pcs.)

(metric)

Measured Size, mm	Finished Thickness, mm	Add for Waste Percent	To Obtain Qty of Lumber Reqd Mult. Area by	No. of Meters Lumber per 10 Sq.M Area
50 x 100	40.6	23	16.1	161 (74 pcs.)
50 x 150	40.6	23	24.3	243 (74 pcs.)
50 x 200	40.6	23	32.3	323 (74 pcs.)
50 x 250	40.6	23	40.4	404 (74 pcs.)
50 x 300	40.6	23	48.4	484 (74 pcs.)
50 x 350	40.6	23	56.5	565 (74 pcs.)
75 x 150	68.8	10	21.7	217 (44 pcs.)
75 x 200	68.8	10	28.9	289 (44 pcs.)
75 x 250	68.8	10	36.4	364 (44 pcs.)
75 x 300	68.8	10	43.3	433 (44 pcs.)
75 x 350	68.8	10	50.5	505 (44 pcs.)
100 x 150	93.8	7	21.1	211 (32 pcs.)
100 x 200	93.8	7	28.1	281 (32 pcs.)
100 x 250	93.8	7	35.1	351 (32 pcs.)
100 x 300	93.8	7	42.1	421 (32 pcs.)

The above quantities do not include any allowance for waste due to end matching, but are based on spans taking lumber of even lengths or laying flooring with broken joints.

Nails Required for 100 S.F. (10 Sq.M) of 2" (50 mm) Laminated Flooring

	Distance Spikes Are Spaced Apart (inches)									
	12		16		20		24		30	
Size of spikes	No. Spikes	No. Lbs.	No. Spikes	No. Lbs.	No. Spikes	No. Lbs.	No. Spikes	No. Lbs.	No. Spikes	No. Lbs.
16d	740	15	600	12-1/4	518	10-1/2	450	9-1/4	370	7-1/2
20d	740	24	600	19-1/4	518	16-3/4	450	14-1/2	370	12
30d	740	31	600	25	518	21-1/2	450	18-3/4	370	15-1/2
40d	740	41	600	33-1/4	518	28-3/4	450	25	370	20-1/2
50d	740	53	600	43	518	37	450	32	370	26-1/2
60d	740	67	600	55	518	47	450	41	370	33-3/4

| | Distance Spikes Are Spaced Apart (mm) | | | | | | | | | |
| | 300 | | 400 | | 500 | | 600 | | 750 | |
Size of spikes	No. Spikes	No. Kg	No. Spikes	No. Kg	No. Spikes	No. Kg	No. Spikes	No. Kg	No. Spikes	No. Kg
16d	797	6.8	646	5.6	558	4.8	484	4.2	398	3.4
20d	797	10.9	646	8.7	558	7.6	484	6.6	398	5.4
30d	797	14.1	646	11.3	558	9.8	484	8.5	398	7.0
40d	797	18.6	646	15.1	558	13.0	484	11.3	398	9.3
50d	797	24.0	646	19.5	558	16.8	484	14.5	398	12.0
60d	797	30.4	646	24.9	558	21.3	484	18.6	398	15.3

Nails Required for 100 S.F. (10 Sq.M) of 3" (75 mm) Laminated Flooring

| | Distance Spikes Are Spaced Apart (inches) | | | | | | | | | |
| | 12 | | 16 | | 20 | | 24 | | 30 | |
Size of Spikes	No. Spikes	No. Lbs.	No. Spikes	No. Lbs.	No. Spikes	No. Lbs.	No. Spikes	No. Lbs.	No. Spikes	No. Lbs.
16d	440	9	352	7-1/4	308	6-1/4	264	5-1/2	220	4-1/2
20d	440	14-1/4	352	11-1/2	308	10	264	8-1/2	220	7
30d	440	18-1/2	352	14-3/4	308	12-3/4	264	11	220	9-1/4
40d	440	24-1/2	352	19-1/2	308	17	264	14-3/4	220	12-1/4
50d	440	31-1/2	352	25-1/4	308	22	264	18-3/4	220	15-3/4
60d	440	40	352	32	308	28	264	24	220	20

(metric)

| | Distance Spikes Are Spaced Apart (mm) | | | | | | | | | |
| | 300 | | 400 | | 500 | | 600 | | 750 | |
Size of Spikes	No. Spikes	No. Kg	No. Spikes	No. Kg	No. Spikes	No. Kg	No. Spikes	No. Kg	No. Spikes	No. Kg
16d	474	4.1	379	3.3	332	0.1	284	2.5	237	2.0
20d	474	6.5	379	5.2	332	4.5	284	3.9	237	3.2
30d	474	8.4	379	6.7	332	5.8	284	5.0	237	4.2
40d	474	11.1	379	8.8	332	7.7	284	6.7	237	5.6
50d	474	14.3	379	11.5	332	10.0	284	8.5	237	7.1
60d	474	18.1	379	14.5	332	12.7	284	10.9	237	9.1

Nails Required for 100 S.F. (10 Sq.M) of 4" (100 mm) Laminated Flooring

| | Distance Spikes are Spaced Apart (inches) | | | | | | | | | |
| | 12 | | 16 | | 20 | | 24 | | 30 | |
Size of Spikes	No. Spikes	No. Lbs.	No. Spikes	No. Lbs.	No. Spikes	No. Lbs.	No. Spikes	No. Lbs.	No. Spikes	No. Lbs.
30d	320	13-1/2	256	10-3/4	224	9-1/2	192	8	160	6-3/4
40d	320	17-3/4	256	14-1/4	224	12-1/2	192	10-3/4	160	9
50d	320	23	256	18-1/4	224	16	192	13-3/4	160	11-1/2
60d	320	29	256	23-1/4	224	20-1/2	192	17-1/2	160	14-1/2

(metric)

| | Distance Spikes are Spaced Apart (mm) | | | | | | | | | |
| | 300 | | 400 | | 500 | | 600 | | 750 | |
Size of Spikes	No. Spikes	No. Kg	No. Spikes	No. Kg	No. Spikes	No. Kg	No. Spikes	No. Kg	No. Spikes	No. Kg
30d	344	6.1	276	4.9	241	4.3	207	3.6	172	3.1
40d	344	8.1	276	6.5	241	5.7	207	4.9	172	4.1
50d	344	10.4	276	8.3	241	7.3	207	6.2	172	5.2
60d	344	13.2	276	10.5	241	9.3	207	7.9	172	6.6

06170 PREFABRICATED STRUCTURAL WOOD

Glued Laminated Beam Construction. Beams of glued laminated construction are popular where price is not the controlling factor.

They are made in uniform or tapered shapes. In the beam shape, similar to illustration, they can be used in spans from 20'-0" to 80'-0" (6-24 m).

They are used in schools, auditoriums, churches, stores, and ranch-style homes and are made of kiln-dried structural woods bonded together by glue, applied under controlled conditions of temperature and pressure. They are planed and sanded to a very smooth surface and later treated with a transparent preservative, making their overall appearance very pleasing. With this type of construction, the ceilings are usually omitted, leaving the beams exposed to the room below. Roof insulation is accomplished through the use of standard insulating boards placed upon the roof deck and covered with roofing material. Because of the purlin construction generally used, 2" or 4" decking is recommended.

Span in ft.	Span in m	Center Hgt., ft.	Center Hgt., m	End Hgt., ft	End Hgt., m	Camber in ft.	Camber in m	Price
20'-0"	6.1	2'-4"	0.71	0'-9-1/2"	0.241	0'-9"	0.229	$334.00
30'-0"	9.1	2'-10"	0.86	0'-11"	0.280	1'-0"	0.305	805.00
40'-0"	12.2	3'-11"	1.19	1'-2-1/2"	0.369	1'-6"	0.457	1,173.00
50'-0"	15.2	5'-1"	1.55	1'-6-1/2"	0.469	2'-0"	0.610	1,863.00
60'-0"	18.3	6'-1"	1.85	1'-9-1/2"	0.546	2'-6"	0.762	2,680.00

Glued Laminated Three Hinged Arch

Another type of glued laminated construction is the three-hinged arch, which gains its support from floor level, incorporating column and beam in one compact design. Fabricated and finished in the manner of the glued beam, they are used where good appearance and function are the major factors and are used quite extensively in churches and schools.

Purlins are generally used to span the resulting bays and are covered with two inch decking and suitable insulating material.

This is the most expensive of the trusses discussed and costs from 5 to 6 times that of the bowstring type truss.

General Notes Regarding Trusses. Where rafters are used between trusses, 1" (25 mm) roof sheathing or 1/2" (13 mm) plywood is generally used. Where purlins are spaced 4'-0" to 6'-0" (1.2-1.8 m) apart, 2" (50 mm) sheathing is common practice. Purlins may be eliminated and wood decking of 4" to 6" may be used to span up to 18'-0" (5.4 m) between trusses.

The use of ceilings supported by the lower chord of the truss should be called to the attention of the truss designer, as it requires a heavier load.

Trusses may bear upon pilasters, steel columns, wood posts and plates, grillages or steel beams. A suitable steel bearing plate should be placed under each end of the truss to distribute the load throughout the masonry. Trusses may afford lateral support through angle irons fastened to truss and column.

Labor Framing Wood Roof Trusses. When framing wood roof trusses, a carpenter experienced in truss construction will handle and frame approximately the following amounts of lumber per hour on the various types of trusses:

Type of Truss	B.f. per hr.	Cu.m per hr.
Bowstring Type Trusses	40	0.09
Scissors and Gothic Type Trusses	30	0.07
Glued Laminated Type Trusses	15	0.04

PREFABRICATED STRUCTURAL TIMBER

Structural timber decking in thicknesses of 2" to 4" (50-100 mm), generally 6" (150 mm) wide and double tongue and grooved is available in hemlock cedar, Douglas fir, and spruce. It can be used for roof decks, structural flooring, and planked walls. Approximate costs are as follows:

	Hemlock per s.f.	Cedar or Fir per s.f	Spruce per s.f.
4"x6" Select	$2.50	$2.80	$3.50
3"x6" Select	2.10	2.40	2.60
	per sq.m	per sq.m	per sq.m
100x150mm Select	$26.91	$30.14	$37.67
75x150mm Select	22.60	25.83	27.99

All prices should be verified with local distributors prior to submitting proposals that include these materials.

06200 FINISH CARPENTRY

For estimating finish carpentry work for a job whose specifications are arranged in accordance with the Construction Specifications Institute (CSI) format, you must check at least five divisions to be certain that all of this type of work is included in the bid. Division 6, the subject of this chapter, covers rough and finish carpentry. But wood and related sidings are in Division 7, Thermal and Moisture Protection. Wood doors and windows are in a section of their own, Division 8. Division 10, *Specialties,* includes such items as tackboards, laminated plastic toilet partitions, signs, folding partitions, telephone booths, and toilet and wardrobe accessories, all of which may be furnished or set by the carpenter. Division 11, *Equipment,* covers kitchen and laundry cabinets and counters for residential work and such unrelated work as wood bank counters, church pews, bars, lab furniture, dock bumpers, and library shelving. Finally Division 12, *Furnishings,* includes wood seating, class room furniture, and similar semi-attached items. In addition, it is prudent to check all the allowances set forth in the specifications. Often the furnishing of some of the items in Divisions 10, 11, and 12 will be covered under a cash allowance, but the installation is to be included in the bid submitted by the carpenter.

EXTERIOR FINISH CARPENTRY

Wood siding and wood doors and windows are discussed under their own CSI divisions. The following items cover other adjacent exterior work.

Placing Corner Boards, Fascia Boards, Etc. When placing wood fascia boards, corner boards, etc., a carpenter should place 175 to 225 l.f. (53-68 m) per 8-hr. day, at the following labor cost per 100 l.f. (30 m):

	Hours	Rate	Total	Rate	Total
Carpenter	4	$....	$....	$26.44	$105.76
Cost per l.f.				....	1.06
per m				....	3.48

Placing Exterior Wood Cornices, Verge Boards, Etc. When placing exterior wood cornices, verge boards, fascia, soffits, etc., consisting of two members, two carpenters working together should place 150 to 175 l.f. (45-53 m) per 8-hr. day, at the following labor cost per 100 l.f. (30 m):

	Hours	Rate	Total	Rate	Total
Carpenter	10	$....	$....	$26.44	$264.40
Cost per l.f.				....	2.64
per m				....	8.67

When the exterior cornices consist of three members (crown mold, bed mold, fascia, etc.), two carpenters working together should place 100 to 125 l.f. (30-38 m) per 8-hr. day, at the following labor cost per 100 l.f. (30 m):

	Hours	Rate	Total	Rate	Total
Carpenter	14.2	$....	$....	$26.44	$375.48
Cost per l.f.				....	3.75
per m				....	12.30

If a four-member wood cornice is used, two carpenters working together should place 60 to 75 l.f. (18-23 m) per 8-hr. day, at the following labor cost per 100 l.f. (30 m):

	Hours	Rate	Total	Rate	Total
Carpenter	23.7	$....	$....	$26.44	$626.63
Cost per l.f.			....		6.30
per m			....		20.66

The above quantities and costs do not include time blocking out for fascia boards, cornices, etc. An extra allowance should be made for all blocking required.

Placing Brick Moldings. A carpenter should fit and set around 32 l.f. (9.6 m) of brick molding per hour at the following cost per 100 l.f. (30 m):

	Hours	Rate	Total	Rate	Total
Carpenter	3.2	$....	$....	$26.44	$ 84.61
Cost per l.f.			....		0.85
per m			....		2.79

Brick moulding costs about $0.50 per l.f. (0.30 m) in random lengths and $0.75 in specified lengths.

Placing Wood Cupolas. One carpenter should set a prefabricated pine cupola in around two hours.

	Hours	Rate	Total	Rate	Total
Carpenter	2	$....	$....	$26.44	$52.88
Labor	2		....	21.04	42.08
Cost per each			$....		$94.96

Porch Work. The labor placing porch work is a variable item owing to the vast difference in the style and construction of porches and the amount of detail involved.

Placing Plain Porch Columns. When placing plain square or turned porch columns, such as commonly used for rear porches and other inexpensive work, a carpenter should place one post in about 1 hr., at the following labor cost:

	Hours	Rate	Total	Rate	Total
Carpenter	1.0	$....	$....	$26.44	$26.44

Placing Porch Top and Bottom Rail and Balusters. When placing wood top and bottom rail and wood balusters, such as used on front porches, a carpenter should complete 15 to 20 l.f. (4.5-6.0 m) of rail per 8-hr. day, at the following labor cost per l.f. (0.30 m):

	Hours	Rate	Total	Rate	Total
Cost per l.f.	0.46	$....	$....	$26.44	$ 12.16
per m			....		39.88

When placing top and bottom rails with open balusters or using matched and beaded ceiling, such as is often used for the cheaper grades of work, a carpenter should complete 35 to 45 l.f. (10.5-13.5 m) of rail per 8-hr. day, at the following labor cost per l.f. (0.30 m):

	Hours	Rate	Total	Rate	Total
Cost per l.f.	0.2	$....	$....	$26.44	$5.29
per m			....		17.35

Framing and Erecting Exterior Wood Stairs for Rear Porches. When framing and erecting outside wood stairs for rear porches on apartment buildings, etc., where the stringers are 2"x10" (50 x 250 mm) or 2"x12" (50 x 300 mm) lumber, with treads and risers nailed on the face of the stringers, it will require 18 to 22 hrs. carpenter time per flight of stairs.

This is for an ordinary stair having 14 to 18 risers extending from story to story, and the labor per flight should cost as follows:

	Hours	Rate	Total	Rate	Total
Carpenter	20	$....	$....	$26.44	$528.80

If the stair consists of two short flights with an intermediate landing platform between stories, it will require 12 to 13 hrs. carpenter time per flight or 24 to 26 hrs. per story, including platform, at the following labor cost:

	Hours	Rate	Total	Rate	Total
Carpenter	25	$....	$....	$26.44	$661.00

Porch Top Rail and Balustrade Cap

Rabbeted Porch Jamb, Plowed Shoe, and Bottom Rail

Framing and Erecting Wood Stairs Having Winders. Where the wood stairs to rear porches have 4 to 6 winders in each story height, figure 24 to 28 hrs. carpenter time, at the following labor cost per story:

	Hours	Rate	Total	Rate	Total
Carpenter	26.0	$....	$....	$26.44	$687.44

Clear Ponderosa Pine Porch Material

Kind of Molding	Size, in.	Size, mm	Price per 100 l.f.	Price per 10 m
Square Baluster Stock	1-1/8x1-1/8	28.1x28.1	$58.00	$19.00
Square Baluster Stock	1-3/8x1-3/8	34.4x34.4	74.00	24.00
Square Baluster Stock	1-5/8x1-5/8	40.6x40.6	86.00	28.00
Balustrade Cap	3-5/8x1-5/8	90.6x40.6	150.00	49.00
Balustrade Shoe	3-5/8x1-5/8	90.6x40.6	150.00	49.00
Rabbeted Porch Jamb	1-1/2x3-1/2	37.5x87.5	150.00	49.00
Plowed Porch Shoe	1-1/2x3-1/2	37.5x87.5	150.00	49.00
Top Rail	3x2-1/4	75.0x56.3	167.00	55.00
Top Rail	3-3/4x1-3/4	93.8x43.8	150.00	49.00
Bottom Rail	3-1/2x1-3/4	87.5x43.8	153.00	50.00

Cornice Boards*
Select Grade

Measured Size, in.	Measured Size, mm	S4S to (in.)	S4S to (mm)	Price per 100 l.f.	Price per 10 m
1x3	25x75	3/4x2-5/8	18.8x65.6	$45.00	$15.00
1x4	25x100	3/4x3-1/2	18.8x87.5	53.00	17.00
1x6	25x150	3/4x5-1/2	18.8x137.5	90.00	30.00
1x8	25x200	3/4x7-1/2	18.8x187.5	107.00	35.00
1x10	25x250	3/4x9-1/2	18.8x237.5	141.00	46.00
1x12	25x300	3/4x11-1/2	18.8x287.5	196.00	64.00

*Prices are per 100 l.f. in random lengths 6'-0" to 16'-0". For metric, prices are per 10 meters in random lengths 1.8 to 4.9 m.

Wood Columns

The finest columns for exterior work are made from clear heart redwood staves glued together under clamp pressure with cold water waterproof glue. Shafts are turned in the lathe with the correct entasis, and to insure uniform stave thickness after turning, the staves are made straight for the lower third of the shaft and swelltapered for the upper two-thirds thus giving the correct entasis in the rough. Any style of fluting or reeding may be had.

The columns receive a prime coat of lead and oil on the exterior surface. Columns over 14" (350 mm) in diameter receive a coat of waterproof asphaltum paint on the hollow core surface.

All columns are furnished with a guarantee as to their workmanship and durability.

No printed price lists are available for these items and each must be priced at the factory. In order of price, the Roman is cheapest followed by the unfluted Ionic, the fluted Attic, the fluted Doric, and finally the fluted Corinthian. A 12" (300 mm) diameter Doric fluted column 10' (3 m) high with a plinth and cap is currently quoted at around $318.00 at the factory.

Stock, tapered column shafts in pine can be figured at around $27.00 per vertical l.f. (0.30 m) in 12" (300 mm) diameter; $50.00 in 14" (350 mm); $56.00 in 18" (450 mm); and $70.00 in 24" (600 mm).

Fiberglass columns cast in a colonial design will run $6.00 per l.f. (0.30 m) in 6" (150 mm) diameter; $7.50 in 8" (200 mm); and $9.00 in 10" (250 mm).

Aluminum columns extruded to a fluted design are stocked in assorted sizes from 6" (150 mm) diameter in 8' (24 m) lengths to 12" (300 mm) diameter in 18' (5.4 m) lengths. They may be used as loadbearing members and come factory finished in a high-gloss, white, baked enamel. An 8" (200 mm) diameter shaft 12' (3.6 m) long will cost about $49.50. If no base or cap is used, it may be attached to the slab and ceiling by cutting 2" (50 mm) lumber to fit the column diameter and bolting it to the adjacent construction. If a base and cap is ordered, they will be furnished in aluminum castings and will be about $11.50 each.

Square porch columns costs about $2.00 per l.f. (0.3 m) in 4x4 (100 x 100 mm) fir. Turned colonial posts in 8' (2.4 m) lengths will cost about $15.00 in 4x4 (100 x 100 mm) stock, $22.00 in 5x5 (125 x 125 mm) stock. It takes two carpenters to set columns, and they should erect a medium sized column in an hour.

Cast Iron Column Bases. Cast iron column bases keep columns off the floor and permit air to circulate on the inside of the column.

Prices of Cast Iron Column Bases

Base Size, inches	Base Size, mm	Height, in. (mm)	Col. Dia. in. (mm)*	Approx. Price
11 x 11	275 x 275	2 (50.0)	8 (200)	$37.00
13-1/2 x 13-1/2	338 x 338	2-1/2 (62.5)	10 (250)	54.00
16 x 16	400 x 400	2-1/2 (62.5)	12 (300)	71.00
18-1/2 x 18-1/2	463 x 463	2-1/2 (62.5)	14 (350)	104.00
21-1/4 x 21-1/4	531 x 531	2-1/2 (62.5)	16 (400)	138.00
24 x 24	600 x 600	3-1/4 (81.3)	18 (450)	161.00
26-1/2 x 26-1/2	663 x 663	3-1/4 (81.3)	20 (500)	184.00

*For Doric columns without base use one size smaller.

Cast iron bases for pilasters are two-thirds the price of bases for corresponding columns.

ROMAN	DORIC	IONIC	CORINTHIAN	ATTIC
Wood Cap and Base	Wood Cap No Base	Compo Cap Wood Base	Compo Cap Wood Base	Wood Cap Wood Base

Placing Plywood Soffits. Two carpenters should set 75 s.f. (6.75 sq.m) of 1/4" (7 mm) and 64 s.f. (5.76 sq.m) of 1/2" (13 mm) plywood on soffits per hour at the following labor cost per 100 s.f. (9 sq.m):

	Hours	Rate	Total	Rate	Total
Carpenter	2.8	$....	$....	$26.44	$74.03
Cost per s.f.			....		0.74
per sq.m			....		2.43

Placing Matched and Beaded Ceiling. Matched and beaded ceiling is usually furnished 4" (100 mm) and 6" (150 mm) wide and 3/8" (9 mm), 1/2" (13 mm), and 3/4" (19 mm) thick. Where the 4" (100 mm) material is used for porch ceilings, overhang soffits, etc., a carpenter should place 285 to 325 b.f. (0.67-0.77 cu.m) per 8-hr. day, at the following labor cost per 1,000 b.f. (2.36 cu.m):

	Hours	Rate	Total	Rate	Total
Carpenter	26	$....	$....	$26.44	$687.44
per cu.m			....		291.29

The labor cost per square—100 s.f. (9 sq.m) surface—requiring 128 b.f. (0.30 cu.m) of ceiling should run as follows:

	Hours	Rate	Total	Rate	Total
Carpenter	3.33	$....	$....	$26.44	$ 88.05

When placing 6" (150 mm) wide matched and beaded ceiling, a carpenter should place 400 to 450 b.f. (0.94-1.06 cu.m) of ceiling per 8-hr. day, at the following labor cost per 1,000 b.f. (2.36 cu.m):

	Hours	Rate	Total	Rate	Total
Carpenter	18.8	$....	$....	$26.44	$497.07
per cu.m				....	210.62

The labor cost per square—100 s.f. (9 sq.m) surface—requiring 120 b.f. (0.28 cu.m) of ceiling should run as follows:

	Hours	Rate	Total	Rate	Total
Carpenter	2.25	$....	$....	$26.44	$59.49

Placing Exterior Door Trim. Much of the architectural type door trim today is of molded material rather than milled wood. It is pre-engineered to fit together with little or no cutting or fitting, providing the door frame itself is set plumb and true. Stock trim can be ordered for widths of 3' (0.9 m), 5' (1.5 m), and 6' (1.8 m) and for either brick or wood wall openings. Side trim is delivered for 7' (2.1 m) door height but designed so that 4" (100 mm) may be cut to fit a 6'-8" (2 m) door without altering the design. Pilasters are nominally 8" (200 mm) in width. Cornices are 6" to 8" (150-200 mm) high and available in several styles. In addition, pediments may be added. A basic cornice piece costs around $53.00 for a 3' (0.9 m) opening, $80.00 for the larger ones. Pediments will run from $53.00 to $80.00 extra, for 3'-0" (0.9 m) openings and up to $125.00 for the double door openings. Pilaster assemblies run $69.00 per pair. Two carpenters should set a cornice in 1/2 hour, the pediment in 1/2 hour, and the pilasters in one hour.

Outside Blinds or Shutters

Shutters are available in wood, aluminum, and molded plastic. Because of the popularity of the latter, it is not always easy to find wood stocked in all sizes. Custom made units will run considerably more than the prices given.

All shutters have stiles, top and middle rails 2-1/4" (56 mm) wide; bottom rail 4-1/2" (113 mm) wide. All slats are stationary.

Shutters with raised panels can be fitted with decorative cutouts in the upper panel at an extra cost of $1.55 per pair. Designs include pine trees, cloverleafs, squirrels, crescents, and roosters.

A carpenter should hang 8 to 10 pair a day depending on size and height off the ground.

Price Per Pair, Hemlock, Preservative Treated

Window Size, ft.in.	Window Size, (mm)	w/Fixed Slats	w/Raised Panels
2'-0"x3'-3"	600 x 975 mm	$16.10	$20.70
2'-0"x3'-11"	600 x 1175 mm	19.00	23.00
2'-0"x4'-7"	600 x 1375 mm	21.90	27.60
2'-0"x5'-3"	600 x 1575 mm	24.20	33.40
2'-4"x3'-3"	700 x 975 mm	16.70	20.70
2'-4"x3'-7"	700 x 1075 mm	17.50	20.70
2'-4"x3'-11"	700 x 1175 mm	19.00	20.70
2'-4"x4'-3"	700 x 1275 mm	20.10	20.70
2'-4"x4'-7"	700 x 1375 mm	23.00	20.70
2'-4"x4'-11"	700 x 1475 mm	24.70	20.70
2'-8"x3'-3"	800 x 975 mm	17.30	20.40
2'-8"x3'-7"	800 x 1075 mm	17.80	21.90
2'-8"x3'-11"	800 x 1175 mm	19.60	23.00
2'-8"x4'-3"	800 x 1275 mm	21.30	25.30
2'-8"x4'-7"	800 x 1375 mm	23.00	26.50
2'-8"x4'-11"	800 x 1475 mm	24.20	28.20
2'-8"x5'-3"	800 x 1575 mm	26.50	29.90
3'-0"x3'-3"	900 x 975 mm	23.00	25.30
3'-0"x3'-7"	900 x 1075 mm	24.20	27.30
3'-0"x3'-11"	900 x 1175 mm	25.30	28.80
3'-0"x4'x3"	900 x 1275 mm	26.50	31.10
3'-0"x4'-7"	900 x 1375 mm	27.60	32.20
3'-0"x4'-11"	900 x 1475 mm	28.80	33.90
3'-0"x5'-3"	900 x 1575 mm	29.90	34.50
3'-4"x3'-3"	1000 x 975 mm	24.20	27.00
3'-4"x3'-11"	1000 x 975 mm	26.50	29.90
3'-4"x4'-7"	1000 x 975 mm	28.80	33.40
3'-4"x4'-11"	1000 x 975 mm	29.90	34.50
3'-4"x5'-3"	1000 x 975 mm	32.20	36.80
3'-8"x4'-7"	1100 x 1375 mm	26.50	35.10
3'-8"x4'-11"	1100 x 1375 mm	28.80	36.80
3'-8"x5'-3"	1100 x 1375 mm	31.10	38.80

Door Blinds
Same as above, except 3 panels of slats in height.

Size, ft.in.	Size meters	Price per pair
2'-8"x6'-9"	0.8 x 2.0 m	$29.00
3'-0"x6'-9"	0.9 x 2.0 m	30.00

Aluminum Shutters

Aluminum shutters are designed to screw permanently on the wall. The louver section is stamped out of one piece of metal. They are furnished in white, black, green, or brown and come packaged with mounting screws.

Size, inches	Size, mm	Price per pair
2'-4"x34"	700 x 850	$17.80
2'-4"x36"	700 x 900	18.40
2'-4"x39"	700 x 975	19.30
2'-4"x43"	700 x 1075	20.10
2'-4"x48"	700 x 1200	20.70
2'-4"x55"	700 x 1375	21.90
2'-4"x64"	700 x 1600	25.30
2'-4"x80"	700 x 2000	29.90
2'-8"x34"	800 x 850	19.00
2'-8"x36"	800 x 900	19.60
2'-8"x39"	800 x 975	20.40
2'-8"x43"	800 x 1075	21.30
2'-8"x48"	800 x 1200	21.90
2'-8"x55"	800 x 1375	23.60
2'-8"x64"	800 x 1600	26.50
2'-8"x80"	800 x 2000	32.20

One carpenter should hang one pair in 0.8 hour.

Molded Shutters

Shutters molded of thermo-formed polymer are prefinished in either black or white but may be painted with an exterior latex paint. They are screwed to siding with 2" (50 mm) aluminum screws, 4 per unit to 55" (1375 mm), 6 per unit over 55" (1375 mm).

Panel, in.	Panel, mm	per pair
14x35	350 x 875	$26.50
14x39	350 x 975	27.30
14x43	350 x 1075	28.80
14x47	350 x 1175	30.20
14x51	350 x 1275	31.10
14x55	350 x 1375	32.80
14x59	350 x 1475	33.90
14x63	350 x 1575	38.00
14x67	350 x 1675	39.10
14x71	350 x 1775	40.30
14x75	350 x 1875	42.00
16x80	400 x 2000	43.10

Labor Costs on Half Timber Work. The following costs on half timber work are taken from a college building. The timber was white oak 3" (75 mm) thick, 10" (250 mm) and 11" (275 mm) wide, and various lengths to meet the requirements of the job.

Nothing was done to the face of the timber and it was left just as received. The backs of all timbers were beveled, the thickness remaining 3" (75 mm). A shop was set up on the job where the timbers were cut, beveled, and fit ready for erection.

There were 14,500 b.f. (34.20 cu.m) of lumber used in this half-timber work and the labor cost per 1,000 b.f. (2.36 cu.m) was as follows:

	Hours	Rate	Total	Rate	Total
Carpenter	66.3	$....	$....	$26.44	$1,752.97
Labor helping	12.0	....	....	21.04	252.48
Cost per 1,000 b.f.			$....		$2,005.45
per cu.m				....	849.77

After the timbers were cut, beveled, and fit in the shop, the labor cost of placing and bracing the timbers in the building was as follows per 1,000 b.f. (2.36 cu.m):

	Hours	Rate	Total	Rate	Total
Carpenter	41.5	$....	$....	$26.44	$1,097.26
Labor helping	2.2	....	....	21.04	46.29
Cost per 1,000 b.f.			$....		$1,143.55
per cu.m				....	484.56

INTERIOR FINISH CARPENTRY

When estimating the labor cost of interior finish, determine the grade of workmanship required by the architect or owner, because the quantity of work a carpenter will perform per hour or per day will vary considerably with the grade of workmanship.

Costs are given for two distinct grades of workmanship. "Ordinary work" is by far the most common and is usually found in medium priced residences, cottages, apartment buildings, factory and warehouse buildings, non-fireproof store, and offices and schools.

"First grade workmanship" is required in fine residences, high-class fireproof apartments, hotels, banks and office buildings, high-class stores, fireproof school and university buildings, courthouses, city halls, state capitols, post offices, and other such buildings. The interior finish in buildings of this class is usually selected birch, selected gumwood, plain or quarter-sawed oak, mahogany, walnut or other first-class hardwoods, and is discussed later in this chapter under "Architectural Woodwork".

Placing Wood Base. This cost will vary with the size of the rooms and whether a single-, two- or three-member base is specified.

Where there are 55 to 60 l.f. (16.5-18 m) of two-member base in each room, without an unusually large amount of cutting and fitting, a carpenter should place 125 to 150 l.f. (37.5-45 m) per 8-hr day, at the following labor cost per 100 l.f. (30 m):

	Hours	Rate	Total	Rate	Total
Carpenter	5.8	$....	$....	$26.44	$153.35
Labor	1.0	....	....	21.04	21.04
Cost per 100 l.f.			$....		$174.39
per l.f.				....	1.74
per m				....	5.72

If there is an unusually large number of miters, such as required in closets and other small rooms, increase the above costs accordingly.

A carpenter should place almost as many l.f. of three-member base (consisting of two base members and a carpet strip), as two-member base (consisting of one member and carpet strip), as it is much easier to fit a small top member against the plastered wall than it is to nail a wide piece of base so that it will fit snug against the wall and follow the irregularities in the plaster.

Where there are 50 to 60 l.f. (15-18 m) of base in a room, a carpenter should place 110 to 130 l.f. (33-39 m) per 8-hr. day, at the following labor cost per 100 l.f. (30 m):

	Hours	Rate	Total	Rate	Total
Carpenter	6.7	$....	$....	$26.44	$177.15
Labor	1.0	....	....	21.04	21.04
Cost per 100 l.f.			....		$ 198.19
per l.f.				....	1.98
per m				....	6.50

First Grade Workmanship

In average size rooms, a carpenter should place 100 to 115 l.f. (30-34.5 m) of two-member hardwood base per 8-hr. day at the following labor cost per 100 l.f. (30 m):

	Hours	Rate	Total	Rate	Total
Carpenter	7.4	$....	$....	$26.44	$195.66
Labor	1.0	....	....	21.04	21.04
Cost per 100 l.f.			$....		$216.70
per l.f.				....	2.17
per m				....	7.11

Where three-member hardwood base is used in average rooms, a carpenter should place 85 to 100 l.f. (25.5-30 m)— two ordinary rooms—per 8-hr. day, at the following labor cost per 100 l.f. (30 m):

	Hours	Rate	Total	Rate	Total
Carpenter	8.7	$....	$....	$26.44	$230.03
Labor	1.0	....	....	21.04	21.04
Cost per 100 l.f.			$....		$ 251.07
per l.f.				....	2.51
per m				....	8.24

On work of this class, the wood grounds should be straight, so that it will not be necessary to "force" the wood base to make it fit tight against the finished wall.

Where a single 1"x4" (25 x 100 mm) pine base is to be fitted to straight runs, a single carpenter should set about 200 l.f. (60 m) per day at the following cost per 100 l.f. (30 m):

	Hours	Rate	Total	Rate	Total
Carpenter	4	$....	$....	$26.44	$ 105.76
Cost per l.f.				....	1.06
per m				....	3.48

Placing Wood Picture Molding. Where just an ordinary grade of workmanship is required, a carpenter should place picture molding in 5 or 6 ordinary rooms per 8-hr. day. This is equivalent to 250 to 275 l.f. (75-82.5 m) at the following labor cost per 100 l.f. (30 m):

	Hours	Rate	Total	Rate	Total
Carpenter	3.0	$....	$....	$26.44	$79.32
Labor	0.5	....	....	21.04	10.52
Cost per 100 l.f.			$....		$89.84
per l.f.				....	0.90
per m				....	2.95

First Grade Workmanship

Where the wood picture molding must fit close to the plastered walls with perfect fitting miters, a carpenter should place molding in 4 to 5 ordinary sized rooms per 8-hr. day. This is equivalent to 175 to 200 l.f. (52.5-60 m), at the following cost per 100 l.f. (30 m):

	Hours	Rate	Total	Rate	Total
Carpenter	4.4	$....	$....	$26.44	$ 116.34
Labor	0.5	....	....	21.04	10.52
Cost per 100 l.f.			$....		$126.86
per l.f.				....	1.27
per m				....	4.17

If the picture molding is placed in fireproof buildings having tile or brick partitions, it will be necessary to place wood grounds for nailing the picture molding, but in non-fireproof buildings the nails may be driven into the plaster so that nails obtain a bearing in the wood studs or wall furring.

Placing Wood Chair or Dado Rail. In large rooms or long, straight corridors, a carpenter should fit and place 275 to 300 l.f. (82.5-90 m) of wood chair rail per 8-hr. day, at the following labor cost per 100 l.f. (30 m):

	Hours	Rate	Total	Rate	Total
Carpenter	2.8	$....	$....	$26.44	$74.03
Labor	0.5	....	....	21.04	10.52
Cost per 100 l.f.			$....		$84.55
per l.f.			....		0.85
per m			....		2.79

In small kitchens, closets, and bathrooms, a carpenter will place only 160 to 180 l.f. (48-54 m) of chair rail per 8-hr. day, at the following labor cost per 100 l.f. (30 m):

	Hours	Rate	Total	Rate	Total
Carpenter	4.7	$....	$....	$26.44	$124.27
Labor	0.5	....	....	21.04	10.52
Cost per 100 l.f.			$....		$134.79
per l.f.			....		1.35
per m			....		4.43

First Grade Workmanship

Where first grade workmanship is required, a carpenter should place 200 to 225 l.f. (60-67.5 m) of chair rail per 8-hr. day, at the following labor cost per 100 l.f. (30 m):

	Hours	Rate	Total	Rate	Total
Carpenter	3.8	$....	$....	$26.44	$100.47
Labor	0.5	....	....	21.04	10.52
Cost per 100 l.f.			$....		$110.99
per l.f.			....		1.11
per m			....		3.64

In small rooms, such as kitchens, pantries, bathrooms, etc., requiring considerable cutting and fitting around medicine cabinets, wardrobes, kitchen cases, etc., a carpenter should place 120 to 135 l.f. (36-40.5 m) of chair rail per 8-hr. day, at the following labor cost per 100 l.f. (30 m):

	Hours	Rate	Total	Rate	Total
Carpenter	6.3	$....	$....	$26.44	$166.57
Labor	0.5	....	....	21.04	10.52
Cost per 100 l.f.			$....		$177.09
per l.f.				....	1.77
per m				....	5.81

Placing Wood Cornices. Where three- or four-member wood cornices are placed in living rooms, reception rooms, dining rooms, etc., a carpenter should place cornice in one average sized room per 8-hr. day, which is equivalent to 50 to 60 l.f. (15-18 m), and the labor cost per 100 l.f. (30 m) would be as follows:

	Hours	Rate	Total	Rate	Total
Carpenter	14.5	$....	$....	$26.44	$383.38
Labor	2.0	....	....	21.04	42.08
Cost per 100 l.f.			$....		$425.46
per l.f.				....	4.25
per m				....	13.94

First Grade Workmanship

Where it is necessary that the wood members fit the plastered walls and ceilings closely, with all miters true and even, two carpenters working together should complete 1 to 1-1/4 rooms per day, at the rate of 35 to 40 l.f. (10.5-12 m) per 8-hr. day for one carpenter and at the following labor cost per 100 l.f. (30 m):

	Hours	Rate	Total	Rate	Total
Carpenter	21	$....	$....	$26.44	$555.24
Labor	2	....	....	21.04	42.08
Cost per 100 l.f.			$....		$597.32
per l.f.				....	5.97
per m				....	19.58

Placing Vertical Wood Panel Strips. When vertical wood panel strips or "battens" are nailed to plastered walls to produce a paneled effect, a carpenter should place 22 to 28 pcs.—175 to 225 l.f.—(52.5-67.5 m) per 8-hr. day, at the following labor cost per 100 l.f. (30 m):

	Hours	Rate	Total	Rate	Total
Carpenter	4.0	$....	$....	$26.44	$105.76
Labor	0.5	....	....	21.04	10.52
Cost per 100 l.f.			$....		$116.28
per l.f.				....	1.16
per m				....	3.80

Placing Wood Strip Paneling. Where panels are formed of wood molding 1-1/2" to 2-1/2" (38-63 mm) wide, making it necessary to cut and miter both ends of each panel strip, the lin. ft. (meter) cost will vary with the size of the panels and the amount of cutting and fitting. Almost as much labor is required on a panel 2'-0"x3'-0" (0.6 x 0.9 m) as for one 3'-0"x6'-0" (0.9 x 1.2 m), even though the former is only half as long.

On small panels up to 2'-0"x4'-0" (0.6 x 1.2 m), requiring 12 l.f. (3.6 m) of molding, a carpenter should complete 9 to 11 panels, containing 110 to 135 l.f. (33-40.5 m) of molding per 8-hr. day, at the following labor cost per 100 l.f. (30 m):

	Hours	Rate	Total	Rate	Total
Carpenter	6.5	$....	$....	$26.44	$171.86
Labor	0.5	....	....	21.04	10.52
Cost per 100 l.f.			$....		$182.38
per l.f.				....	1.82
per m				....	5.97

On larger panels 3'-0"x5'-0" (0.9 x 1.5 m) to 4'-0"x6'-0" (1.2 x 1.8 m), where each panel contains 16 to 20 l.f. (4.8-6 m) of molding, a carpenter should complete 7 to 9 panels, containing 140 to 180 l.f. (42-54 m) of molding per 8-hr. day, at the following labor cost per 100 l.f. (30 m):

	Hours	Rate	Total	Rate	Total
Carpenter	5.0	$....	$....	$26.44	$132.20
Labor	0.5	....	....	21.04	10.52
Cost per 100 l.f.			$....		$142.72
per l.f.				....	1.43
per m				....	4.69

First Grade Workmanship

Where wood panel moldings are used over canvassed or burlap walls, with all strips plumb and level, fitting closely to the plastered walls with perfect fitting miters, a carpenter should complete 7 to 9 small panels, requiring 90 to 115 l.f. (27-34.5 m) of molding per 8-hr. day, at the following labor cost per 100 l.f. (30 m):

	Hours	Rate	Total	Rate	Total
Carpenter	7.8	$....	$....	$26.44	$206.23
Labor	0.5	....	....	21.04	10.52
Cost per 100 l.f.			$....		$216.75
per l.f.				....	2.17
per m				....	7.12

On large panels, from 3'-0"x5'-0" (0.9 x 1.5 m) to 4'-0"x6'-0" (1.2 x 1.8 m), where each panel contains 16 to 20 l.f. (4.8-6.0 m) of panel molding, a carpenter should complete about 6 to 8 panels, containing 120 to 150 l.f. (36-45 m) of molding per 8-hr. day, at the following labor cost per 100 l.f. (30 m):

	Hours	Rate	Total	Rate	Total
Carpenter	6.0	$....	$....	$26.44	$158.64
Labor	0.5	....	....	21.04	10.52
Cost per 100 l.f.			$....		$169.16
per l.f.				....	1.69
per m				....	5.54

Placing Wood Ceiling Beams. In buildings where built-up ceiling beams are used, the labor costs will vary according to the number of intersections of beams in each room and the length of the beams. It is as easy to erect a 12'-0" (3.6 m) built-up beam as an 8'-0" (2.4 m) one.

On average work, a carpenter should place 35 to 45 l.f. (10.5-13.5 m) of built-up wood beams per 8-hr. day, at the following labor cost per 100 l.f. (30 m):

	Hours	Rate	Total	Rate	Total
Carpenter	20	$....	$....	$26.44	$528.80
Labor	3	....	....	21.04	63.12
Cost per 100 l.f.			$....		$591.92
per l.f.				....	5.92
per m				....	19.41

First Grade Workmanship

In the better class of buildings using wood ceiling beams, a carpenter should place 30 to 35 l.f. (9.0-10.5 m) per 8-hr. day, at the following labor cost per 100 l.f. (30 m):

	Hours	Rate	Total	Rate	Total
Carpenter	25	$....	$....	$26.44	$ 661.00
Labor	3	....	....	21.04	63.12
Cost per 100 l.f.			$....		$724.12
per l.f.				....	7.24
per m				....	23.75

Price of Ponderosa Pine Moldings
Per 100 L.F. and 10 M unless otherwise indicated

Description Size in inches	Size in mm	Price per 100 l.f.*	Price per 10 m
Aprons			
11/16"x2-1/4"	17.2x56.3mm	$52..90	$17.36
11/16"x2-1/4" ogee	17.2x56.3mm	52.90	17.36
9/16"x3-1/2"	14.1x87.5mm	97.80	32.09
Astragals			
1-3/8"x7'-0"	34.4x2100mm	3.60 ea.	3.60 ea.
1-1/4"x7'-0"	31x2100mm	4.00 ea.	4.00 ea.
Bases			
9/16"x3"	14.1x75mm	46.00	15.09
1/2"x3-1/2"	12.5x87.5mm	48.30	15.85
1/2"x4-1/4"	12x106mm	55.20	18.11
Base Cap			
11/16"x1-3/8"	17.2x34.4mm	23.00	7.55
Backband			
11/16"x1-1/16"	17.2x26.6mm	18.40	6.04
Balusters			
3/4"x 3/4"	18.8x18.8mm	28.80	9.45
1-1/16"x1-1/16"	26.6x26.6mm	46.00	15.09
1-5/16"x1-5/16"	32.8x32.8mm	74.80	24.54
1-5/8"x1-5/8"	40.6x40.6mm	80.50	26.41
Base Shoe Combination			
3/4"x2-5/8"	18.8x65.6mm	34.50	11.32
Bed Molding			
11/16"x1-3/4"	17.2x43.8mm	40.30	13.22
11/16"x2-1/4"	17.2x56.3mm	57.50	18.86
Base Shoe			
1/2"x 3/4"	12.5x18.8mm	23.00	7.55
Blind Stop			
3/4"x1-3/8"	18.8x34.4mm	46.00	15.09
Brick Molding			
1-5/16"x2"	32.8x50mm	69.00	22.64
Casings			
11/16"x2-1/4"	17.2x56.3mm	42.60	13.98
11/16"x3-1/2"	17.2x87.5mm	78.20	25.66
Chair Rails			
11/16"x2-1/8"	17.2x53.1mm	57.50	18.86
7/16"x2-1/2"	10.9x62.5mm	64.40	21.13
Corner Guards			
3/4"x 3/4"	18.8x18.8mm	24.20	7.94
1-1/8"x1-1/8"	28.1x28.1mm	48.30	15.85
1-5/16"x1-5/16"	32.8x32.8mm	52.90	17.36
Coves			
11/16"x3-1/4"	17.2x81.3mm	80.50	26.41
11/16"x2-1/4"	17.2x56.3mm	55.20	18.11

Description Size in inches	Size in mm	Price per 100 l.f.*	Price per 10 m
11/16"x1-3/4"	17.2x43.8mm	39.10	12.83
3/4"x 7/8"	18.8x21.9mm	29.90	9.81
3/4"x 3/4"	18.8x18.8mm	27.60	9.06
1/2"x 1/2"	12.5x12.5mm	23.00	7.55
11/16"x1-1/8"	17.2x28.1mm	31.10	10.20
Crowns			
11/16"x4-1/4"	17.2x106mm	117.30	38.48
11/16"x3-1/4"	17.2x81.3mm	78.20	25.66
11/16"x2-1/4"	17.2x56.3mm	55.20	18.11
Drip Caps			
1-1/16"x1-5/8"	26.6x40.6mm	55.20	18.11
3/4"x1-5/8"	18.8x40.6mm	48.30	15.85
Full Rounds			
1-5/8"	40.6mm	86.30	28.31
1-5/16"	32.8mm	57.50	18.86
Glazing Beads			
1/2"x 9/16"	12.5x14mm	20.70	6.79
3/8"x 3/8"	9.4x9.4mm	17.30	5.68
Half Rounds			
5/16"x 5/8"	7.8x15.6mm	23.00	7.55
Hand Rails			
1-1/2"x1-3/4"	37.5x43.8mm	66.70	21.88
Hook Strips			
9/16"x2-7/16"	14.1x61mm	48.30	15.85
Jamb Extensions			
3/4"x1-15/16"	18.8x48mm	42.60	13.98
3/4"x 7/8"	18.8x22mm	32.20	10.56
Lattice			
1/4"x1-1/8"	6.3x28mm	15.00	4.92
1/4"x1-3/8"	6.3x34mm	17.30	5.68
1/4"x1-3/4"	6.3x44mm	21.90	7.19
1/4"x2-1/4"	6.3x56mm	29.90	9.81
Mullion Casings			
9/16"x5-1/2"	14.1x138mm	97.80	32.09
3/16x2	4.7x50mm	48.30	15.85
Parting Strip			
1/2"x 3/4"	12.5x18.8mm	18.40	6.04
Picture Molding			
3/4"x1-3/4"	18.8x43.8mm	44.90	14.73
Quarter Rounds			
1/4"x 1/4"	6.3x6.3mm	15.09	2.26
1/2"x 1/2"	12.5x12.5mm	13.80	4.53
3/4"x 3/4"	18.3x18.3mm	23.00	7.55
1-1/16"x1-1/16"	26.6x26.6mm	43.70	14.34
Screen Moldings			
5/16"x 5/8"	7.8x16mm	13.80	4.53
1/4"x 3/4"	6.3x19mm	11.50	3.77

Description Size in inches	Size in mm	Price per 100 l.f.*	Price per 10 m
Screen Stock			
3/4"x1-3/4"	18.8x43.8mm	46.00	15.09
3/4"x2-3/4"	18.8x68.8mm	70.20	23.03
1-1/16"x1-3/4"	26.6x43.8mm	72.50	23.79
1-1/16"x2-3/4"	26.6x68.8mm	86.30	28.31
Shelf Cleat			
11/16"x1-1/2"	17.2x37.5mm	23.00	7.55
Stools			
1-1/16"x3-1/4"	26.6x93.8mm	195.50	64.14
1-1/16"x3-5/8"	26.6x90.6mm	212.80	69.82
1-1/16"x3-3/4"	26.6x93.8mm	109.30	35.86
1-1/16"x2-1/4"	26.6x56.3mm	80.50	26.41
1-1/16"x2-5/8"	26.6x65.6mm	86.30	28.31
1-1/16"x2-3/4"	26.6x68.8mm	103.50	33.96
Stops			
7/16"x2-1/8"	10.9x53mm	55.20	18.11
7/16"x1-5/8"	10.9x40.6mm	48.30	15.85
7/16"x1-3/8"	10.9x34.4mm	42.60	13.98
7/16"x1-1/8"	10.9x28mm	35.70	11.71
7/16"x -7/8"	10.9x21.8mm	34.50	11.32
7/16"x 15/16"	10.9x23.4mm	35.70	11.71
7/16"x 5/8"	10.9x15.6mm	27.60	9.06

PLASTIC LAMINATE COUNTERTOPS

Plastic laminate for countertops will vary in price depending on the number of cutouts, corners, and whether the fabricator must include the cost of taking job measurements. A typical L-shaped kitchen countertop covering 18 l.f. (5.4 m) of inside wall, all edges plastic, laminated to 7/8" (22 mm) particle board, with 4" (100 mm) splashes back and ends, will cost around $40.00 per l.f. (per 0.30 m) in a standard pattern. Rounded edges will cost considerably more; metal banded ones somewhat less. For each cut out add $17.50. For each end splash add $12.00.

Setting Cabinets and Cases. When setting mill assembled cabinets and cases, such as used in the average residence or apartment building, it will require about 1/6 hr. per s.f. of cabinet face area (1.85 hrs. per sq.m) carpenter time at the following labor cost per s.f. (sq.m):

	Hours	Rate	Total	Rate	Total
Carpenter per s.f.	0.16	$....	$....	$26.44	$4.23
per sq.m			....		45.52

A carpenter should fit and hang, complete with pulls and catches, 15 to 17 flush case doors per 8-hr. day, at the following labor cost per door:

	Hours	Rate	Total	Rate	Total
Carpenter..	0.5	$....	$....	$26.44	$13.22

Wood Interior Trim, Casings, Back-band, Stops, Stool, Etc.

Wood Ceiling Cornice and Picture Mold

For lip case doors that do not require fitting, a carpenter should install, complete with pulls and catches, 30 to 34 doors per 8-hr. day, at the following cost per door:

	Hours	Rate	Total	Rate	Total
Carpenter..	0.25	$....	$....	$26.44	$6.61

In the better class buildings, where cases, wardrobes, cabinets, etc., are used, and it is necessary to scribe and fit them against the plastered walls or run a small molding to conceal joints between cases and plastered walls, it requires about 1/3 hr. per s.f. of cabinet face area (3.67 hrs. per sq.m) at the following labor cost per s.f. (sq.m):

	Hours	Rate	Total	Rate	Total
Carpenter per s.f.	0.33	$....	$....	$26.44	$8.73
per sq.m				....	93.93

A carpenter should fit and hang, complete with pulls and catches, 11 to 13 flush case doors per 8-hr. day, at the following labor cost per door:

	Hours	Rate	Total	Rate	Total
Carpenter	0.67	$....	$....	$26.44	$17.71

Often, in restricted areas, such as adding cabinets in narrow lavatory areas, it is necessary to have cabinet work delivered in sections so that it may be turned and properly set within the space. Seldom is a job so organized that cabinet work is delivered before wall finishes are up, so the fit must be preplanned and exact.

Setting Factory Assembled and Finished Kitchen Cabinets. When setting factory assembled and finished kitchen cabinets, such as used in the average apartment or residence, it will require about 0.1 hr. per s.f. (1.11 hrs. per sq.m) of cabinet face area at the following labor cost:

	Hours	Rate	Total	Rate	Total
Carpenter per s.f.	0.1	$....	$....	$26.44	$2.64
per sq.m				....	28.41

First Grade Workmanship

In better class apartments and residences, where a high class job is required, it takes about 1/6 hr. per s.f. (1.85 hrs. per sq.m) of cabinet face area at the following labor cost:

	Hours	Rate	Total	Rate	Total
Carpenter per s.f.	0.16	$....	$....	$26.44	$ 4.23
per sq.m				....	45.52

Setting Wood Fireplace Mantels. If wood fireplace mantels are factory assembled, merely requiring fitting and setting, a carpenter should install 2 to 3 mantels per 8-hr. day at the following labor cost per mantel:

	Hours	Rate	Total	Rate	Total
Carpenter	3.2	$....	$....	$26.44	$84.61
Labor	1.0	....	....	21.04	21.04
Cost per mantel		....	$....		$105.65

Fitting and Placing Closet Shelving. A carpenter should fit and place 115 to 135 s.f. (10.35-12.15 sq.m) of closet shelving per 8-hr. day including setting of shelf cleats at the following labor cost per 100 s.f. (9 sq.m):

	Hours	Rate	Total	Rate	Total
Carpenter	6.4	$....	$....	$26.44	$169.22
Cost per s.f.			....		1.69
per sq.m			....		18.18

Installing Hanging Rods in Closets. A carpenter should install about three hanging rods, including supports, per hr. at the following labor cost per rod:

	Hours	Rate	Total	Rate	Total
Carpenter	0.33	$....	$....	$26.44	$8.73

Setting Metal Medicine Cabinets. A carpenter should set 10 to 12 average size medicine cabinets per 8-hr. day at the following labor cost per cabinet:

	Hours	Rate	Total	Rate	Total
Carpenter	0.75	$....	$....	$26.44	$19.83

Setting Bathroom Accessories. In an 8-hr. day a carpenter should locate and set 30 to 34 bathroom accessories, such as towel bars, soap dishes, paper holders, etc. at the following labor cost per accessory:

	Hours	Rate	Total	Rate	Total
Carpenter	0.25	$....	$....	$26.44	$6.61

Installing Chalkboards, Corkboards, and Accessories. To install chalkboards, before starting, all grounds should be in place and final coat of finish should be thoroughly dry.

Chalkboard should be set so that all edges will be adequately covered by trim moldings. The writing surface should be plumb and in a true plane with 1/4" clearance on all sides. Chalkboard panels should be nailed at the top edges only and the balance secured to the wall with an even coat of spotting cement.

Two carpenters working together should install 125 to 150 s.f. (11.25-13.5 sq.m) of chalkboard per 8-hr. day at the following labor cost per 100 s.f. (9 sq.m):

	Hours	Rate	Total	Rate	Total
Carpenter	11	$....	$....	$26.44	$290.84
Cost per s.f.			....		2.91
per sq.m			....		31.31

Installation of corkboard is similar to that of chalkboard except nailing is eliminated. The labor cost of installing corkboard is approximately the same as for chalkboard.

Installation of aluminum trim moldings in conjunction with chalkboards and corkboards may be accomplished either with exposed wood screws or with concealed snap-on clips. A carpenter should install 90 to 110 l.f. (27-33 m) of molding per 8-hr. day at the following labor cost per 100 l.f. (30 m):

	Hours	Rate	Total	Rate	Total
Carpenter	8	$....	$....	$26.44	$211.52
Cost per l.f.			....		2.12
per m			....		6.95

Labor Placing Finish Hardware

Description of Work	No. Set per 8-Hr. Day	Carp. Hrs. Each
Rim locks or night latches, cheap work	20-24	0.3-0.4
Mortised locks in soft or hardwood doors, ordinary workmanship	12-16	0.5-0.7
Mortised locks in hardwood doors, first grade workmanship	6-8	1.0-1.3*
Cylindrical locks	17-23	0.3-0.5
Front entrance cylinder locks in hardware doors, ordinary workmanship	6-8	1.0-1.3**
Front entrance cylinder locks in hardwood doors, first grade workmanship	4-5	1.6-2.0***
Panic bolts, first grade workmanship	2-3	2.7-4.0
Door closers, exposed	7-9	0.9-1.1
Door closers, concealed	2-3	2.7-4.0
Door holders	20-24	0.3-0.4
Sash lifts and locks, no. of windows	20-24	0.3-0.4
Cremone bolts for casement doors, windows	20-24	0.3-0.4
Kickplates, ordinary workmanship	8-10	0.8-1.0
Kickplates, first grade workmanship	6-8	1.0-1.3

*On this grade of work, the lock must be flush with the face of the door and must operate perfectly.

**On this class of work, the architect or owner can expect some trouble, because time is not sufficient to adjust lock and see that it operates properly.

***This includes front door locks having fancy escutcheon plates, knobs, knockers, etc.

Labor Placing Cabinet Hardware

Description of Work	No. Set per 8-Hr. Day	Carp Hrs. Each
Surface butts	180-200	0.040-0.045
Offset butts	90-100	0.080-0.090
Elbow catches	90-100	0.080-0.090
Friction catches	90-100	0.080-0.090
Drawer and door pulls	140-150	0.053-0.057
Knobs	180-200	0.040-0.045
Rim latches	90-100	0.080-0.090
Mortised locks	20-24	0.300-0.400

Mantels

There are many stock mantel designs from which to choose. A simple molding 6" (150 mm) wide extending out from the wall 2" (50 mm) on 3 sides will cost about $63.00 for an outside dimension of 4'x7' (1.2 x 2.1 m). A full mantel with 7" (175 mm) shelf and of colonial design will cost from $115.00 to $172.00. More elaborate designs in Georgian or French provincial and with end returns to fit over the brick projection will run from $200.00 to $575.00. These costs are for wood members only. Brick or marble trim around the fireplace opening are, of course, extra. It will take a carpenter 3 to 4 hours to set a prefabricated unit in place.

China Cabinets

Corner cabinets can be purchased from stock to add architectural interest to dining, breakfast, or recreation rooms. Mostly of Ponderosa pine with either Colonial or Georgian design, they will vary from $115.00 for a unit 2'-8" (0.8 m) wide with single doors top and bottom to $270.00 for a unit 3'-6" (1.05 mm) wide with double doors. Cabinets are from 7' to 7'-6" (2.1-2.25 m) high. Bottom doors are usually paneled, tops factory glazed. Shipped with front assembled and body knocked down, the unit includes three shelves for the upper portion, one for the lower. Hardware is not included. Allow three hours for one carpenter time to fit and set each unit and apply the hardware.

Ironing Board Cabinet

A built-in feature that appeals to many people is an ironing board cabinet unit complete with adjustable board plus sleeve board, all operating hardware, and flush door which is sized to fit between wall studs 16" (400 mm) on centers. It is shipped in a carton and costs around $92.00. The door is available in either birch or Philippine mahogany. No casing is included, but if the unit is set in a matching paneled wall, it is almost inconspicuous.

Louvered Pine Panels

Shuttered panels are available in stock sizes, complete with hardware, or customized to fit exact openings. They are not only used inside window openings, but for cabinet doors and room separators.

Cost of Stock Single Panel Horizontal Slat Unit

Height	Widths, inches (mm)				
	6 (150)	7 (175)	8 (200)	10 (250)	12 (300)
20 (500)	$3.30	$3.70	$4.10	$5.10	$6.10
23 (575)	4.00	4.30	4.70	5.70	6.70
29 (725)	4.50	4.90	5.60	6.40	7.80
32 (800	5.10	5.50	6.20	7.30	8.70
36 (900)	5.90	6.10	7.00	8.20	10.10

A four-panel hardware kit costs $1.50. If no trimming is required, one carpenter should set 2 panels per hour or 2 window openings consisting of 8 panel units each in a day.

Custom shutters in lengths of 6" to 18" (150-450 mm), and for openings from 12" to 24" (300-600 mm), 2 panels will cost $29.00 unfinished and $40.00 prefinished; for an opening 24" to 30" (600-750 mm), four panels same height, $49.00 and $60.00; for 30" to 42" (750-1050 mm) opening, $45.00 and $74.00; in lengths of 18" to 24" (450-600 mm), openings 12" to 24" (300-600 mm), 2 panels will run $34.00 and $48.00; openings 24" to 30" (600-750 mm), 4 panels, $48.20 and $80.00; openings 30" to 42" (750-1050 mm), 4 panels, $63.50 and $78.00. Custom door panels to 7' (2.1 m) will run $113.00 unfinished, $157.00 finished for 2 panel for an opening to 24" (600 mm) wide. For openings to 30" (750 mm), 2 panels, $116.25 and $161.00; openings 30" to 42" (750-1050 mm), 4 panels, $227.50 and $315.50.

06300 WOOD TREATMENT

Wood and plywood are often given special treatments at the mill to give additional protection against fire, rot, and termites.

Fire protected or fire retardant wood is pressure-impregnated with mineral salts that will react chemically at a temperature below the ignition point of the wood. An Underwriter's Laboratory Flame Spread Rating of 25 can be obtained, and no periodic maintenance is required to keep the rating. However, the lumber must also be kiln dried to 12% moisture content if it is to be painted. Also, wood is not suitable for direct exposure to weather and ground conditions unless further treatment is given. A fire retardant treatment will add $0.14 to 0.17 per b.f. (0.002 cu.m) to lumber costs. If it is necessary to kiln dry the lumber, add $0.07 for soft woods, $0.09 for hard woods.

Rot and termite resistance can be obtained by several methods. Creosote, when permitted by local environmental regulations, is impregnated at the rate of around 8 lbs. per c.f. (127.11 kg/m³) at a cost of $0.23 to 0.60 per b.f. ($84.79 to 211.97 per m³). This finish is not readily paintable and can stain adjacent finishes such as plaster and wallboard. Pentachlorophenol is a popular oil treatment and costs $0.23 to 0.35 per s.f. and leaves the wood with an oil residue. This residue can be eliminated, and the wood left clean and paintable, if the pentachlorophenol is impregnated into the wood by liquid petroleum. This will add $0.12 to the cost.

06400 ARCHITECTURAL WOODWORK

This section covers the assembly and installation of cabinets, wood and wallboard paneling, wood stairs, and railing from components made up in a shop or factory off the job premises.

If the items under this heading are not stock items, it will be necessary to include the cost of making on-the-job measurements and preparing shop drawings. Most cabinet makers are equipped to do this and will include such costs in their proposals. Often they will also install their work, but usually the general contractor handles this work.

There is considerable difference between the grades of workmanship for architectural woodwork and the estimator must be able to determine the grade required.

These same differences apply to the erection and installation of all classes of millwork and interior finish.

The Architectural Woodwork Institute publishes Illustrated Quality Standards, in which the work is divided into three grades: premium, custom, and economy. Premium is the highest grade available in both material and workmanship. It might be used throughout an entire building, but most often, it is for selected spaces or items. Custom, the middle or normal grade in both material and workmanship, is intended for high quality regular work. Economy is the lowest grade and intended for work where price outweighs quality considerations. An estimator may find references to these three grades in an architect's specification.

Methods of Estimating the Labor Cost Millwork and Interior Finish. Different contractors use different methods of estimating the labor cost of erecting and setting millwork and interior finish, but the system generally used is to add a certain percentage of the cost of the millwork to cover the labor cost of erection. While this method is the easiest, it does not produce the best estimates. For example, suppose you are estimating the cost of a building on which you receive four millwork bids, ranging from $4,500 to $5,800. After receiving the bids you add the customary percentage to the lowest acceptable bid to take care of the labor cost of handling and erecting the millwork. This percentage usually runs from 40% to 50% of the cost of the millwork. Taking 40% of $4,500.00 gives $1,850.00 for labor to handle and erect all millwork and interior finish. Suppose a competitor received a bid for $4,025.00 from a mill which did not quote you. If your competitor added 40% to his bid, he would have only $1,610.00 to erect the same amount of millwork on which you figured $1,850.00.

Why this difference of 12-1/2% on such a small item? While the method saves a great deal of time and is almost universally used, it does not produce the most accurate estimates. The only accurate method is to list each item separately; i.e., the number of lin. ft. of wood base, chair rail, picture mold, etc., the number of door jambs to set, the number of sash to fit and hang, together with the number of window openings to be cased or trimmed; the number of wardrobes, linen cases, vanity and kitchen units to set, etc. Estimating labor costs by this method produces more uniform and accurate estimates, reduces risk to a minimum, and enables the contractor to check actual and estimated costs during construction, but it does require a lot more work.

INTERIOR PANELING

Interior paneling used to mean solid wood with stiles and rails and elaborate mouldings for the mansion or bank job; knotty pine for the suburban retreat; and beaver board for the cottage. Today the selection is much wider.

Solid wood paneling is still with us, but it is more apt to be in the form of planks with face-applied mouldings rather than as elaborate paneling. Plywood veneer with hard wood is often favored over solid wood not only for cost savings but because the veneer exposes the beauty of the wood and offers a more easily cleaned surface.

Most paneling today though is sheet board faced to look like wood grain, marble, tile, stucco, leather, or a decorative pattern all its own. The designs come with a selection of moldings that blend into the overall wall.

Insulating wallboard and tiles are still sometimes used as exposed finishes but are selected more for their structural or insulating value rather than their decorative contribution.

Estimating Paneling Quantities. Estimating paneling is more of a task than merely computing the number of sq. ft. (sq.m) of surface to be covered. Each room should be laid out to obtain certain panel effects if the design of the panels will govern to any extent the amount of cutting and fitting necessary. Most paneling will come in stock lengths and widths. The job conditions must be carefully studied to minimize cutting, to make maximum use of studs for attachment, and to eliminate any awkward jointing.

Rough Sketch of Room Showing Location of Doors and Windows

Finished Drawing Showing Location of Doors and Windows

The labor cost of applying wallboard depends upon three things: the cost of placing headers and nailing strips between wood studs to which wallboard and decorative strips are nailed; the cost of installing the wallboard; and the cost of placing decorative panel strips or other means of concealing joints. When installing a ceiling, furring must be applied to provide nailing pieces for all edges of the decorative strips.

Wood decorative or panel strips are usually 1/4" (6 mm) to 1/2" (13 mm) thick and of widths to meet architectural requirements.

There are several ways of covering joints. Today much paneling is grooved so joints are butted and appear as an overall pattern. Most flush panels have cover strips, caps, etc. available to match the basic material or to contrast with it and make a decorative feature of the joint.

A sketch of each room, should be made giving an elevation of each of the four walls and a plan showing the ceiling, similar to that illustrated. From this it is possible to obtain the number of lin. ft. (meters) of "headers" or nailing strips required, the size of each piece of paneling, as well as the number of l.f. (m) of panel strips and moldings necessary to complete the job.

Placing "Headers" or Nailing Strips. When placing "headers" or nailing strips to receive paneling it is necessary to cut them to length and place them flush with the face of the studs, as the paneling must be nailed top and bottom to the "headers" and at the sides to the wood studs.

If the studs are not wide enough to form nailing pieces for the panel strips, a piece of wood, usually 1"x1" (25 x 25 mm), must be nailed to either side of the stud along the entire length and flush with the face of the stud.

When estimating this class of work, care should be taken to obtain the exact number of lin. ft. (meters) of "headers" required as this is an important item in the cost of the work.

With studs spaced 16" (400 mm) o.c., a carpenter should cut and place 225 to 275 l.f. (67.5-82.5 m) per 8-hr. day, at the following labor cost per 100 l.f. (30 m):

	Hours	Rate	Total	Rate	Total
Carpenter	3.2	$....	$....	$26.44	$84.61
Cost per l.f.			....		0.85
per m			....		2.79

Placing Furring Strips for Ceilings. Furring strips should be nailed at right angles to joists. Any convenient width of furring may be used, but it must be wide enough to provide a nailing base to nail all edges of the decorative strips at least every 6" (150 mm). Furring may be any No. 2 material, soft wood being preferable. Make sure that all furring strips are level.

With joists 16" (400 mm) o.c., a carpenter should cut and place 300 to 350 l.f. (90-105 m) of furring strips per 8-hr. day, at the following labor cost per 100 l.f. (30 m):

	Hours	Rate	Total	Rate	Total
Carpenter	2.5	$....	$....	$26.44	$66.10
Cost per l.f.			....		0.66
per m			....		2.17

Placing Sheet Paneling. The actual cost of placing sheet paneling will vary with the size and shape of the room, whether full size sheets may be used or considerable cutting and fitting necessary. It requires practically as much time to place a sheet 16"x96" (400 x 2400 mm) as one 48"x96" (1200 x 2400 mm) on account of the cutting and fitting, while there are three times as many sq. ft. (sq.m) in the latter as in the former.

On straight work, in large rooms, a carpenter should fit and place 450 to 550 s.f. (40.5-49.5 sq.m) per 8-hr. day, at the following labor cost per 100 s.f. (9 sq.m):

	Hours	Rate	Total	Rate	Total
Carpenter	1.6	$....	$....	$26.44	$42.30
Cost per s.f.			....		0.42
per sq.m			....		4.55

When applying sheets in smaller rooms requiring considerable cutting and fitting, a carpenter should fit and place 275 to 300 s.f. (24.75-27.00 sq.m) per 8-hr. day, at the following labor cost per 100 s.f. (9 sq.m):

	Hours	Rate	Total	Rate	Total
Carpenter	2.7	$....	$....	$26.44	$71.39
Cost per s.f.			....		0.71
per sq.m			....		7.68

As previously mentioned, it usually costs more to place a small piece than a large one, as this generally calls for cutting and fitting above and below windows, around door openings, etc., which requires considerably more work than placing a large sheet where no cutting and fitting is necessary. For this reason, estimating the labor cost by the sq. ft. (sq.m) is not the safest method to use, although it is the most common.

Where time in estimating is not a factor, the safest method is to lay out the work as illustrated, dividing it into panels and then allow 20 to 30 minutes to place each piece of wallboard or plywood depending upon the class of work.

06421 HARDWOOD PLYWOOD PANELS

These plywood panels of woods like birch, oak, and mahogany are usually furnished in sheets 4'-0" (1.2 m) wide and 8'-0" to 10'-0" (2.4-3.0 m) high and are used instead of plaster or other wall finishes.

When using 1/4" (6 mm) plywood panels in large sizes for this class of work, it is necessary to have a substantial backing for the thin plywood panels. This may be obtained by applying the plywood direct to a brown or finish plaster coat wall; a 1/2" (13 mm) gypsum board backing, or better still, a 1/2" (13 mm) thick plywood backing with the sheets running in the opposite direction to that of the finish plywood panels.

These plywood panels are usually applied with the panel running from floor to ceiling in one length, i.e., 4'-0"x8'-0" (1.2 x 2.4 m) or 4'-0"x10'-0" (1.2 x 3 m), with the vertical joints butted flush or having a very slight bevel at the edges.

To obtain a first class job, it is usually necessary to apply adhesive to the back of the plywood panels and also use small brads to secure the plywood in place until the adhesive sets.

This is particular work, because the adjoining sheets must be accurately fitted together if a first class job is to result.

If wood panel strips are used to conceal the joints, it is not necessary to use so much care in placing the panels.

Hardwood laminated veneer panels cover a wide variety of products, from the 1/4" (6 mm) thick "grooved" veneers to 3/4" (19 mm) thick flush panels with or without sequence matching of the flitches on up to special burl veneers suitable for the tops of the finest furniture.

Plywood may have a veneer core with a center and at least two intermediate cross banded veneers, a particleboard core, or a lumber core with strips of lumber edge glued into a solid slab with cross banding veneers either side. The back and face of each sheet then receive additional veneers at right angles to the cross banding.

The face veneers may be rotary cut with the log's annular rings exposing a bold variegated grain; flat sliced, which produces a variegated figure similar to that of plain sawn lumber; quarter sliced with growth rings of the log at right angles to the knife, producing a series of stripes in the veneer; and rift cut, which is used with oak logs to produce fairly uniform grains with a combed look caused by the medullary rays.

Then there are various grades of face veneers: premium or architectural panels, which have veneers sequence matched and no patching; custom, which allows no patching on face veneers but is not sequence matched; and economy, which allows "good" face veneers, but no matching of grain or colors.

And, finally, there is the type of tree itself which affects the price. The range from natural yellow birch to Brazilian rosewood is around 800%.

Costs should always be determined locally, since not all veneers are available at all times. The following will serve as a general outline of costs. All types are 4'x8'x 1/4" (1.2 m x 2.4 m x 6 mm) custom.

Natural yellow birch	$34.50
Natural hard maple	34.50
Natural gum	34.50
Select white birch	44.00
Poplar	39.00
Select birch heartwood	40.00
Select white maple	44.00
Plain sawn red oak	52.00
African mahogany	52.00
White ash	55.00
Limba	57.50
Rift sawn white oak	69.00
Pecan	61.00
Cherry	77.00
Butternut	67.00
Quarter sawn white oak	75.00
Walnut	6.00
Primavera	86.50
Avodire	92.50
Teak	138.00
Brazilian rosewood	276.00

If matched sequence panels are ordered, the cost will be about 50% more. Add around $12.00 per sheet for 3/4" (19 mm) thickness.

The stock V-grooved prefinished 4'x8'x 1/4" (1.2 m x 2.4 m x 6 mm) panels are considerably cheaper. Although the range of woods is not so extensive, the finishes vary widely and approximate the colors of more expensive woods still using the grains of the more common species. A random sampling of costs indicates the following range.

		Clear	Colonial
Birch, 4'x8'x1/4" (1.2mx2.4mx6mm)	V-grooved	$18.00	$13.00
Oak, 4'x8'x1/4" (1.2mx2.4mx6mm)	V-grooved	28.00	16.00
Walnut, 4'x8'x1/4" (1.2mx2.4mx6mm)	V-grooved	32.00	19.00
Pecan, 4'x8'x1/4" (1.2mx2.4mx6mm)	V-grooved	27.00	25.00

Inlaid paneling in 4'x8'x1/4" (1.2 m x 2.4 m x 6 mm) sheets consisting of pecan panels striped with 1-1/4" (31 mm) wide walnut stripes costs $34.50 per sheet; with walnut striped in pecan the cost is $40.00 per sheet.

For lower budget jobs, simulated wood grain finishes on 1/4" (6 mm) plywood will be $8.00 per 4'x8' (1.2 m x 2.4 m) sheet.

For the lowest budget jobs, simulated wood finishes printed on hardboard are available in the $7.00 to $8.00 range per 4'x8'x 1/4" (1.2 m x 21.4 m x 6 mm) sheet.

Labor Placing Flush Plywood Wall Panels. When applying plywood panels 4'-0"x8'-0" (1.2 m x 2.4 m) to 4'-0"x10'-0" (1.2 m x 3 m), with the abutting joints fitted tightly together, using adhesive and small brads to secure the plywood to the backing, two carpenters working together should handle, fit, and place one plywood panel in 1-1/2 to 2 hours, or they should place 4 to 5 panels per 8-hr. day.

Many of the panels will have to be cut to half or three-quarter size, and in some instances cut out for doors and windows, so a certain length of time should be allowed for each piece of plywood, rather than figuring on a square foot basis.

06422 SOFTWOOD PLYWOOD PANELING

Softwood plywood consists of an odd number of thin sheets or veneers of Douglas fir of 3, 5, 7, or 9 standard thicknesses. These are laminated with alternating grain direction. Because the grain of each ply composing the material is at right angles to the grain of adjacent plies, the strength of Douglas fir plywood panels is approximately equal in both directions.

All Douglas fir plywood from American Plywood Association mills is made to conform to the moisture-resistance requirements set forth in the revised standards established by the National Bureau of Standards, as U. S. Commercial Standards CS45-55.

Plywall wallboard is the grade of plywood commonly used for interior walls and ceilings instead of lath and plaster or other types of wallboard. It is furnished in sheets 24" to 48" (600-1200 mm) wide and 5' to 12' (1.5-3.6 m) long.

Plywall wallboard is furnished 1/4" (6 mm), 3/8" (9 mm), 1/2" (13 mm), 5/8" (16 mm), 3/4" (19 mm), 1" (25 mm), and 1-1/8" (27 mm) thick. The 1/4" (6 mm) thickness is most widely used for walls and ceilings, although for more substantial construction 3/8" (9 mm) or 1/2" (13 mm) thick is often preferred. Thicker material is used for shelving, wardrobes, cases, etc.

Practically any design may be obtained by the use of plywood as with other types of wallboard. Four penny casing or finishing nails are recommended for paneling up to 1/2" (13 mm) thick with the nails spaced 6" to 10" (150-250 mm) apart. One horizontal or fire stop should be cut in between top and bottom plates as a nailing piece.

Plywood panels may be finished by staining, painting, wall papering, or mechanical surfacing.

Labor Applying Plywood Wallboard. The labor cost of applying plywood wallboard will be about the same as for fiber wallboard covering the same class of work.

Where special jointing of the panels is required to work out wall and ceiling panels, allow extra for grooving, beveling, and any other designs worked into the wallboard. Also add extra for applying panel strips as given on previous pages.

Prices of Douglas Fir Plywood

Douglas fir plywood is made with water resisting glue. All panels are sanded two sides to net thickness shown. Plywood prices can be very volatile. Check local sources.

			Price per 1,000 SF	
	Thick,	No. of	Good(A-D)	Good(A-D)
Size of Panel (inches)	inches	Plies	One side	One Side
24"x60" to 48"x96"	1/4	3	$322.00	$437.00
24"x60" to 48"x96"	3/8	3	402.50	518.50
24"x60" to 48"x96"	1/2	5	448.50	564.50
24"x60" to 48"x96"	5/8	5	494.50	609.50
24"x60" to 48"x96"	3/4	5	603.75	668.00

	(metric)
	Thick,
Size of Panel (mm)	mm
600x1500mm to 1200x2400mm	6
600x1500mm to 1200x2400mm	9
600x1500mm to 1200x2400mm	12
600x1500mm to 1200x2400mm	15
600x1500mm to 1200x2400mm	18

Saw textured plywood siding is popular and relatively inexpensive. It is readily available in Douglas fir and cedar.

Saw Textured Douglas Fir Plywood

Size	per 1000 s.f. (90 sq.m)
3/8" (9 mm) Natural	$600.00
3/8" (9 mm) Grooved	670.00
5/8" (15 mm) Natural	750.00
5/8" (15 mm) Grooved	860.00

Saw Textured Cedar

Size	per 1,000 s.f. (90 sq.m)
3/8" (9 mm) Natural	$830.00
3/8" (9 mm) Grooved	875.00
5/8" (15 mm) Natural	1,265.00
5/8" (15 mm) Grooved	1,320.00

06423 SOLID WOOD PANELING

While most paneling used today is of veneered plywood, nothing matches the beauty of solid wood paneling and its ability to withstand years of use and even improve with age.

Solid paneling is usually quoted in random widths of 4" to 8" (100-200 mm) and random lengths of 6' to 16' (1.8-4.8 m). Specified widths and lengths will run 10% or more additional. Traditional thickness is 3/4" (19 mm), but much paneling today is available at a considerable savings in 1/2" (13 mm) thickness. There is no nationally established grading rules for solid wall paneling, so it is necessary to know the product of the mill quoting to compare prices. Generally, woods are available in two grades: "select", "luxury" or "clear"; and "knotty", "wormy" or "pecky", often designated as "Colonial" grade.

In figuring paneling, add 15% to actual areas for waste and matching.

Paneling is available with V-joint tongue and groove, square joint, shiplap, and Colonial beaded.

The most commonly encountered paneling is knotty pine. Cost is around $1.15 per s.f. ($11.11 per sq.m) Clear pine will run around $1.50.

Clear birch in 3/4" (19 mm) thickness will be about $2.10 per b.f.in lots over 1000 b.f. ($900.00 per cu.m), random width and length. Birch is an excellent cabinet wood with a firm texture, but has a rather strong grain pattern.

White ash is an open grained wood, light in color with contrasting light brown heartwood. It is easy to work and somewhat resembles flat grained white oak. It will cost $2.25 for 3/4" (19 mm) select grade.

Butternut, often called white walnut, is a soft-textured, easy-to-nail wood which takes a beautiful natural finish. It will run $2.50 per b.f. (932.68 per cu.m) select, $1.30 character marked, 3/4" (19 mm) thick.

Cherry, cut from the wild black cherry, is a traditional Colonial paneling. It is a firm textured cabinet wood, but is available only in a rather narrow range of sizes. It runs around $2.45 for 3/4" (19 mm) select down to $1.70 for 3/4" (19 mm) Colonial.

Oak is available from either white or red varieties cut on the quarter grain. Oak is a very hard wood and subject to shrinkage and splitting in adverse conditions, but well worth the extra effort for the right place. Oak runs $1.70 for the red and $2.10 for the white in 3/4" (19 mm) select, $1.30 and $1.55 in 3/4" (19 mm) Colonial.

Walnut has become so popular and the sources for it so restricted that it is now a high priced wood and will run $3.75 or better per s.f. (0.09 sq.m) for 3/4" (19 mm) select. Because it is a hard wood, 1/2" (13 mm) thickness is practical and will run $3.65 for select and $2.65 for colonial.

Other domestic woods available are maple, around $1.40 for 3/4" (19 mm) select; wormy chestnut, at $1.70 for 3/4" (19 mm); and cypress at $1.05 for 3/4" (19 mm) select. Imported woods are also available. The so-called "exotic" woods are more often found in the architectural grade, book-matched plywood where their veneers can be shown off to better advantage.

Some foreign woods stocked in the Chicago area include paldao, a firm textured striped wood for medium brown from the Philippines; African mahogany and the less costly Philippine mahogany which is more difficult to finish; avodire, a light creamy-colored wood which comes from Africa with a roll figure somewhat on the diagonal; limba, or korina, African wood which is cream or tan in color with a darker heartwood, and tigerwood or African lovea, a dullish brown to golden colored wood with a wide or ribbon stripe.

These are all special woods for special jobs. Prices and availability should always be checked, because they vary considerably, but all are expensive woods.

Placing Wood Wainscot or Paneling. The labor cost of placing wood paneling or wainscoting will vary with the height of the wainscot and the manner in which it is delivered, whether assembled or knocked down.

It takes nearly as long to place wood wainscoting 4'-0" (1.2 m) high as 7'-0" (2.1 m) high, as the top and bottom members are identical and the only difference is the height of the panels.

If the wood wainscot is put together in the mill and the carpenters on the job do what little fitting is necessary, set it against the walls and nail it to the grounds, a carpenter should place 12 to 18 l.f. of wainscot per 8-hr. day, at the following labor costs for the different heights:

Height of Wainscot ft.in. (meters)	No. S.F. (Sq.M) 8-Hr. Day	Carp. Hrs. 100 S.F. (9 Sq.M)
4'-0" (1.2 m)	60-70 (5.4-6.3)	11-13
5'-0" (1.5 m)	70-85 (6.3-7.7)	9-11
6'-0" (1.8 m)	80-100 (7.2-9.0)	8-10
7'-0" (2.1 m)	90-110 (8.1-9.9)	7-9
8'-0" (2.4 m)	95-115 (8.6-10.3)	7-8

Placing Wood Wainscot or Paneling (Knocked Down). If wood wainscot or paneling is sent to the job knocked down, making it necessary to assemble the pieces and put them together on the job, a carpenter should place top and bottom rails, cap and intermediate panels for 10 to 15 l.f. (3.0-4.5 m) of wainscoting per 8-hr. day, at the following costs for the different heights:

Height of Wainscot ft.in. (meters)	No. S.F. (Sq.M) 8-Hr. Day	Carp. Hrs. 100 S.F. (9 Sq.M)
4'-0" (1.2 m)	50-60 (4.5-5.4)	13-16
5'-0" (1.5 m)	60-70 (5.4-6.3)	11-13
6'-0" (1.8 m)	70-80 (6.3-7.2)	10-11
7'-0" (2.1 m)	80-90 (7.2-8.1)	9-10
8'-0" (2.4 m)	90-100 (8.1-9.0)	8-9

06424 SHEET BOARD PANELING

An inexpensive prefinished paneling is plastic coated hardboard. This comes in 1/8" (3 mm) and 1/4" (6 mm) thick, 4'x8' (1.2 x 2.4 m) panels, and 16" (400 mm) wide, 1/4" (6 mm) thick grooved planks. The thinner boards should have solid backing for first class work. These panels have a wide range of printed designs from wood grains, solid colors, scenic designs, and marble textures. A full line of aluminum and presdwood moldings is available colored to match or compliment the selected material.

Vinyl surfaced gypsum wallboard comes in 1/2" (13 mm) thickness 8' (2.4 m), 9' (2.7 m), and 10' (3.0 m) lengths. It can be applied with direct nailing using colored nails; glued to furring with a combination adhesive nail-on system, or laminated to a base layer of gypsum backed board. An advantage of this finish is that it can be patched and eventually painted.

Open Work Panels. A popular addition to stock partitioning available today are panels with open work. These vary from 1/8" (3 mm) thick prefinished masonite to solid walnut grilles ranging as high as $23.00 per sq. ft. The amount of openings can vary from 15% to 50%. These materials make good room dividers, help control sun and glare at windows, and cut air conditioning and drapery costs.

The masonite product is known as *Filigree* and is available in a series of sizes up to 4'x8' (1.2 x 2.4 m) sheets. Prefinished frame sections, 3/4" (19 mm) thick and 1-1/2" (38 mm) wide, factory grooved to receive the panel, are available.

Sculptured hardwood panels 3/4" (19 mm) thick are available in many designs and woods including walnut, maple, ash, oak, and poplar, in stock sizes from 2'x4' (0.6 x 1.2 m) to 4'x8' (1.2 x 2.4 m). These may be ordered sanded for finishing on the job or factory finished. Woods available listed in order of ascending cost include sycamore or gumwood; poplar, maple, red oak, ash, and walnut. Some designs, nine are available, are also made in 5/8" (16 mm) clear acrylic.

The least expensive design in the least expensive wood, sycamore, costs $65.00 for a 2'x6' (0.6 x 1.8 m) unfinished, unframed panel. Framed and finished it will cost $210.00, plus shipping charges. The most expensive designs will run $125.00 unfinished and unframed, $270.00 framed and finished.

Wallboard. There are so many kinds and sizes of wall and insulating board that it is impossible to list all of them.

The thinnest boards are made of wood fiber and are usually furnished in sheets 4' (1.2 m) wide, 4' to 12' (1.2-3.6 m) long, and 1/8" (3 mm), 3/16" (5 mm), 1/4" (6 mm), and 3/8" (9 mm) thick. Prices vary from $0.23 per s.f. ($2.55 per sq.m) for the 1/8" (3 mm) to $0.40 per s.f. ($4.30 per sq.m) for the 3/8" (9 mm) board.

Insulating and building board is usually furnished 3/8" (9 mm) and 1/2" (13 mm) thick, 4' (1.2 m) wide and 6', 7', 8', 9', 10', and 12' (1.8, 2.1, 2.4, 2.7, 3.0, and 3.6 m) long, all having square edges. Board of this type will cost $0.22 per s.f. ($2.44 per sq.m) for the 3/8" (9 mm) and $0.32 per s.f. ($3.45 per sq.m) for the 1/2" (13 mm) thickness.

Insulation Board Finish Plank. Used for interior finish, it combines insulation and decoration in one material. Insulating plank are 1/2" (13 mm) thick and furnished in random widths 8", 10", 12", and 16" (200, 250, 300, and 400 mm) wide and 8', 10', and 12' (2.4, 3.0, 3.6 m) long. The long edges may be shiplapped with either a plain bevel or bead and bevel design, or have a blind nailing flange.

Plank 1/2" (13 mm) thick cost about $0.58 per s.f. ($6.24 per sq.m). Wall plank with acoustical properties is also now available in 8' (2.4 m) and 10' (3.0 m) lengths.

Plank Effect Board. The plank effect board is 1/2" (13 mm) thick, 4'-0" (1.2 m) wide, and 8'-0" (2.4 m) high. A series of 5 score lines on each edge and through the center of the board gives the effect of 3 planks 16" (400 mm) wide. It can be applied vertically or horizontally either over old plaster or directly to studs. All edges are flush for butt jointing. The board is painted ivory color on one side and costs $0.70 per s.f. ($7.53 per sq.m).

Concealed Nailing Clips. These provide an invisible attachment of tile or plank to nailing base. Clips fit into tongue and groove. Usually packed 1,000 to a carton. Four pieces are required for each 16"x16" (400 x 400 mm) tile. Prices per 1,000 pcs., $7.50.

Use 2d common coated nails with clips. One pound required per 1,000 clips.

Cement or Mastic. Waterproof cement or mastic is used to install insulating plank. One gallon covers 80 to 100 s.f. (3.79 liters per 7.2-9.0 sq.m) Price per gal. is $3.75 (per liter is $1.13).

Nails Required for Applying Fiberboard, Tile, Plank, Wainscot, and Moldings

For Material 1/2" (12.5 mm) Thick	Gauge of Nails	Length of Nails	Spacing Directions
Board	No. 17	1-1/4" (31.3 mm)	3" (75 mm)
Tile	No. 17	1-1/4" (31.3 mm)	6"-10" (150-250 mm)
Plank	No. 17	1-1/4" (31.3 mm)	6" (150 mm)
Wainscot	No. 17	1-1/4" (31.3 mm)	6" (150 mm)
Moldings	No. 15	1-3/4" (43.8 mm)	6" (150 mm)

Board, plank, and wainscot generally require about 6 brads per s.f. (0.09 sq.m) or 4 to 5 lbs. (1.8-2.25 kg) of 1-1/4" (31 mm) brads per 1,000 s.f. (90 sq.m) Galvanized wire brads are recommended. When they are not available, bright wire brads may be used.

Placing Insulating Plank. The labor cost of placing insulating plank will vary with the surface to which it is applied, length of walls, amount of cutting and fitting necessary, etc. Ceiling work costs more than wall work on account of working on a scaffold and placing the plank overhead.

The most economical method of application is to a solid backing of plywood, wood boards, or other satisfactory nailing surface, because it is necessary to have nailing facilities every 8", 10", 12", or 16" (200, 250, 300, or 400 mm), depending on plank width. If a solid backing is not available, then place wood furring strips, to provide for nailing the boards at edges and ends, and for chair rails, wainscot cap, and other ornamental strips.

For plank without blind nailing flange, nail through face or bead of the plank, never through bevel. Bury nail heads below surface with properly gauged hammer blow or with a nail set.

The cost of wood blocking and furring is given on previous pages of this chapter.

Number of S.F. (Sq.M) of Insulating Plank of Various Widths Placed Per 8-Hr. Day by One Carpenter, When Nailed In Place*

Plank Width	Height of Wainscoting in Feet			
	4'-0"	6'-0"	8'-0"	10'-0"
8"	210-260	275-315	315-365	325-390
10"	240-300	300-360	360-420	375-450
12"	275-340	350-410	410-475	425-500
16"	300-360	375-440	450-525	465-525

(metric)

Plank Width	Height of Wainscoting in Meters			
	1.2 m	1.8 m	2.4 m	3.0 m
200 mm	20-24	26-29	29-34	30-36
250 mm	22-28	28-33	33-39	35-42
300 mm	26-32	33-38	38-44	39-46
400 mm	28-33	35-41	42-49	43-49

*Add for wood furring strips or other backing. When applied to ceilings, reduce above quantities about 25%.

Carpenter Hours Required to Place 100 S.F. of Insulating Plank On Walls When Nailed In Place

Plank Width	Height of Wainscoting in Feet			
	4'	6'	8'	10'
8"	3.5	2.8	2.4	2.3
10"	3.0	2.5	2.0	1.9
12"	2.6	2.1	1.8	1.8
16"	2.5	2.0	1.7	1.7

(metric)

Plank	Carpenter Hours Required to Place 10 Sq.M			
	Height of Wainscoting in mm			
Width	1200	1800	2400	3000
200	3.8	3.0	2.6	2.5
250	4.8	2.7	2.2	2.0
300	5.9	2.3	1.9	1.9
400	7.0	2.2	1.8	1.8

Add for wood furring strips or other backing.

Panel Strips and Moldings For Use With Sheet Paneling. Strips for working out decorative effects on walls or ceilings and for concealing the joints may be of the same material as the sheet or they may be wood or metal, depending upon the effect desired.

The labor cost of placing wood or composition decorative strips over the joints will vary with the design of panel strip used, whether it consists of one, two or three members, the size of the panels, and the design of the walls and ceilings, as the more cutting and fitting required the fewer lin. ft. (meters) of strips a worker will place per day.

Method of Applying Wallboard and Treating Joints

On straight work, using an ordinary panel strip 1/4"x2" (6 x 50 mm) to 1/2"x3" (13 x 75 mm), a carpenter should fit and place 350 to 450 l.f. (105-135 m) per 8-hr. day, at the following labor cost per 100 l.f. (30 m):

	Hours	Rate	Total	Rate	Total
Carpenter	2	$....	$....	$26.44	$52.88
Cost per l.f.			....		0.53
per m			....		1.74

On more complicated work, using a one-member panel or decorative strip, a carpenter should place 275 to 325 l.f. (82.5-97.5 m) per 8-hr. day, at the following labor cost per 100 l.f. (30 m):

	Hours	Rate	Total	Rate	Total
Carpenter	2.7	$....	$....	$26.44	$71.39
Cost per l.f.			....		0.71
per m			....		2.34

Placing Three-Member Decorative or Panel Strips. When the panel strips form a plain wall and ceiling design but consist of three members—a strip 3" (75 mm) or 4" (100 mm) wide with a small molding on each side—a carpenter should fit and place 150 to 175 l.f. (45.0-52.5 m) per 8-hr. day, at the following labor cost per 100 l.f. (30 m):

	Hours	Rate	Total	Rate	Total
Carpenter	5	$....	$....	$26.44	$132.20
Cost per l.f.			....		1.32
per m			....		4.33

Placing Wood Ceiling Beams, Cornices, Etc. In rooms having beam ceiling effects, wood ceiling cornices, etc., a carpenter should fit and place 175 to 200 l.f. (52.5-60.0 m) of each member per 8-hr. day, at the following labor cost per 100 l.f. (30 m):

	Hours	Rate	Total	Rate	Total
Carpenter	4.3	$....	$....	$26.44	$113.69
Cost per l.f.			....		1.14
per m			....		3.74

Three-member cornices would cost three times the above price, four-member cost four times, and so on.

Applying Wallboard Over Existing Plastered Walls and Ceilings. In remodeling work where the wallboard is applied over existing walls and ceilings, it is advisable to remove the existing plaster wherever possible. To obtain a first class job, it is recommended that all interior trim, such as door and window casings, and base, be removed before applying the wallboard. It is also recommended that furring strips be placed over all existing plastered walls, as this will straighten the walls and provide a much better base for securing the wallboard.

After the furring strips have been placed, the cost of applying the wallboard will run about the same as given on the previous pages.

The labor cost of placing wood furring strips is given on the previous pages under "Placing Furring Strips for Ceilings".

Prefinished Panels. New developments in wallboard have been most noticeable in prefinishing and the consequent care needed in concealing nailing.

One of the most popular items is V-grooved hardwood, prefinished with a baked-on finish. Grooves are generally random but with a groove every 16" (400 mm) o.c. to hit a stud. Most boards are 1/4" (6 mm) thick in sheets 4' (1.2 m) wide by 8', 10' or 12' (2.4, 3.0, or 3.6 m) long. These sheets should be applied over a backing board for first class work. Colored nails are available as well as prefinished moldings for cornices, trim, and base. Similar paneling is also available in 7/16" (11 mm) thickness for application directly to studs. A wider range of finishes is generally available in the heavier material.

Method of Placing Wallboard and Horizontal Scoring Marks and Division Strips

Insulating Tile. Tile made of Celotex, Fir-Tex, Insulite, Nu-Wood, Temlok, and other fiber composition are still sometimes used for wall and ceiling finish. They are furnished with beveled edges either square or with a blind nailing or stapling flange. They come in sizes from 12"x12" (300 x 300 mm) to 16"x32" (400 x 800 mm) and may be used for obtaining ashlar effects on walls and tile effects on ceilings.

They can be applied to wood backing or to furring strips with adhesive, nails, staples, or special clips on the walls or ceilings. When furring strips are used, the strips must be spaced to work out with the tile sizes.

In applying square edged tiles, nails should be driven through the tile at an angle to the surface of approximately 60 degrees. Reverse the angle of half the nails and nail through the face of the tile, never through the bevel. Bury nail heads below the surface with properly gauged hammer blow or with nail set.

When applying tiles with blind nailing or stapling flanges, drive the nails or staples in the thick part of the flange perpendicular to the surface.

Insulating fiber tile 1/2" (13 mm) thick costs $10.50 per carton of 64.

Nailing Recommendations for Square Edged Fiber Tile. Where nailing is to be exposed, galvanized wire brads are recommended. When they are not available, bright wire brads may be used. 1-1/4" (31 mm) brads or finishing nails should be used for exposed nailing of 1/2" (13 mm) thick tile.

Nailing Recommendations for Square Edged Fiber Tile

Tile Size, inches (mm)	Location of Brad	No. Brads per Tile	Approx. Lbs. of Brads per 1,000 SF Tile	Approx. Kg of Brads per 100 Sq.M Tile
12x12 (300x300mm)	1 in each corner	4	3-1/2	1.7
12x24 (600x600mm)	4 in each long edge / 3 on 2 opposite edges	8	3-1/2	1.7
16x16 (400x400mm)	3 on center line / 5 on each long edge	9	4	2.0
16x32 (400x800mm)	5 on center line	15	3-1/2	1.7

Nailing Recommendations for Flanged Tile. Where attachment of flanged tile is to be by nails, 3d blued lath nails 1-1/8" (28 mm) long should be used. Where application is by means of staples, rust resistant staples 1/2" (12 mm) or 9/16" (14 mm) in length should be used in conjunction with a gun type stapler.

Tile Size in Inches	No. Nails or Staples per Tile	Approx. No. Staples per 1,000 SF	Approx. Lbs. Nails per 1,000 SF
12x12	4	4,000	6.0
12x24	6	3,000	4.5
16x16	4	2,250	3.4
16x32	6	1,690	2.6

(metric)

Tile Size in mm	No. Nails or Staples per Tile	Approx. No. Staples per 100 Sq.M	Approx. Kg. Nails per 100 Sq.M
300 x 300 mm	4	4,306	2.9
300 x 600 mm	6	3,229	2.2
400 x 400 mm	4	2,422	1.7
400 x 800 mm	6	1,819	1.3

Labor Placing Fiber Tile When Nailed In Place. The labor cost of handling and placing fiber tile will vary considerably with the size of the tile, the skill of the mechanic, the amount of nailing, and the surface to which the tile is applied. A worker will place more tile on a good solid backing than on furring strips. To the quantities given in the following table, add for blocking and furring where necessary.

Quantities and Labor Costs of Applying Fiber Tile to Walls
and Ceilings Where Tile are Nailed In Place

Tile Size, inches	S.F per Tile	No. Tile Req'd per 100 S.F.	No. Placed per 8-Hr. Day	No. of S.F. per 8-Hr. Day	Carp. Hrs. per 100 S.F.
12x12*	1.00	100	180-200	180-200	4.2
12x24*	2.00	50	95-105	190-210	4.0
16x16*	1.77	56*	120-130	215-235	3.6
16x32*	3.54	28*	83-90	295-315	2.7

(metric)

Tile Size, mm	Sq. mm per Tile	No. Tile Req'd per 10 Sq.M.	No. Placed per 8-Hr. Day	No. of Sq.M per 8-Hr. Day	Carp. Hrs. per 10 Sq.M.
300x300	90 000	107.6	180-200	17-19	4.5
300x600	180 000	53.8	95-105	18-20	4.3
400x400	160 000	60.8	120-130	20-22	3.9
410x800	320 000	30.7	83-90	27-29	2.9

*Stock sizes.
Add for wood furring strips or other necessary backing.

Applying Insulating Tile Using Adhesives. Apply the adhesive in spots or dabs about 2" (50 mm) in diameter, 1/4" (6 mm) to 3/8" (9 mm) thick, one on each corner of the square tileboard and two or more additional spots on the rectangular sizes. Keep adhesive about 1" (25 mm) from edges of fiberboard units and use approximately 4 spots per s.f. (0.09 sq.m). Adhesive spots should not be spaced over 10" (250 mm) o.c. in any direction.

After dabs of adhesive have been applied, place the tile in its final position. Slide back and forth for a distance of about 3/4" (19 mm) under uniform pressure of the hands to obtain good contact. When working units larger than 16"x16" (400 x 400 mm), two workers should exert pressure simultaneously for best results. If the surface of the unit is below the level of adjoining pieces, do not pull the unit away from the base to bring surface level, but remove it and add more adhesive. Again work into position as described above to obtain correct surface level.

Brad or shore the first units on either walls or ceilings to prevent sliding out of position during application of surrounding units. Do not bend units over high spots.

Tile and plank may be applied to plasterboard, sound plaster, plywood, or lumber. In each case the use of adhesive or a combination of adhesive and brads is recommended.

Coverage Chart of Adhesive

Tile Size, Inches (mm)	No. Spots per Tile	S.F. per Gal.	Sq.M per Liter	Gals. per 1,000 S.F.	Liters per 100 Sq.M
12x12 (300x300)	4	40	1.0	13	100
12x24 (300x600)	6	40	1.0	13	100
16x16 (400x400)	4	60	1.5	10	66
16x32 (400x800)	6	60	1.5	10	66

Price, $4.00 per gal. ($1.05 per liter)

Labor Placing Fiber Tile Using Adhesive. The labor cost of placing fiber tile using adhesive or adhesive and nails will vary greatly with the skill and experience of the workers, size of tile, etc. For instance, where an experienced mechanic is placing tile and has a helper to build scaffold, unpack tile, and apply adhesive to the backs of tiles, an experienced mechanic and helper should place 40 to 45 s.f. (3.6-4.0 sq.m) of 12"x12" (300 x 300 mm) tile per hour, but if one person is working alone and must stop to apply the adhesive to each tile, set it in place, and then drive several brads through them, 15 to 20 s.f. (1.35-1.8 sq.m) per hour would probably be average. Conditions under which the tile are to be installed must be carefully considered before arriving at the labor cost.

The design in which the tile are laid also affects the labor costs. For instance when tile are laid in herringbone pattern, it requires about 30% to 40% more time than regular square or common bond. Diagonal patterns require about 20% to 25% more time than regular squares or common bond while tile laid in mixed ashlar designs require about 50% more time than straight squares or common bond.

The quantities and costs given in the following table are based on using two workers—an experienced mechanic to place the tile and a worker to build scaffold, unpack the tile, and place them on the scaffold, and place the adhesive on the back of the tile ready for the mechanic.

If the work is performed by inexperienced workers, the labor costs might be as much as double those given here.

Quantities and Labor Costs of Applying Fiber Wall Tile to Walls and Ceilings Using Adhesive and Nails Where Necessary

Add for cleaning plaster or other work necessary for applying tile. The quantities and costs given in the following table are based on using a 4- or 5-ft. scaffold, consisting of horses and planks, and laying the tile in plain squares or common bond.

Tile Size, in.	S.F. per Tile	Tile per 100 S.F.	No. Placed per 8-Hr. Day	S.F. per 8-Hr. Day	Hrs. per 100 S.F.
12x12*	1.00	100	165-185	165-185	4.5
12x24*	2.00	50	88-98	175-195	4.4
16x16*	1.77	56*	112-125	200-220	3.9
16x32*	3.54	28*	65-70	230-250	3.3

(metric)

Tile Size, mm	Sq. mm per Tile	Tile per 10 Sq.M	No. Placed per 8-Hr. Day	Sq.M per 8-Hr. Day	Hrs. per 10 Sq.M
300x300	90 000	107.6	165-185	15-17	4.8
300x600	180 000	53.8	88-98	16-18	4.7
400x400	160 000	60.8	112-125	19-20	4.2
400x800	320 000	30.7	65-70	21-23	3.6

*Standard sizes

If tile are laid in herringbone pattern, add 30% to 40% to the above labor costs; if laid in diagonal patterns, increase the above labor costs 20% to 25%; and if laid in mixed ashlar designs, add about 50% to the above labor costs.

When placing tile in auditoriums and other spaces requiring high scaffolding, add cost of scaffolding. Large jobs are applied proportionately faster than small jobs.

Presdwood, Hardboard, Panelboard, Temwood. There are a number of makes of wood fiber board that are much thinner and yet possess much greater strength than ordinary wallboard.

They are used for both interior and exterior work—applications such as interior finish of walls and ceilings, cupboards, lining of closets, and clothes chutes.

They are usually furnished 1/8" (3 mm), 3/16" (5 mm), 1/4" (6 mm), and 5/16" (8 mm) thick, 4' (1.2 m) wide and 4' to 12' (1.2-3.6 m) long, although they vary somewhat among manufacturers.

Tempered hardboard, Temwood, or Presdwood are the same materials as described above, except they have been subjected to a tempering treatment which further increases the strength, resistance to abrasion, durability, and ease of painting. Tempered hardboards, 1/4" (6 mm) and 5/16" (8 mm) thick are used for the exterior of homes and other small buildings. It is also used quite extensively for concrete forms.

These hardboards or tempered boards are in no way comparable with ordinary wallboard. It is a far denser material, with greater tensile and transverse strength, greater resistance to abrasion, lower moisture absorption, and when properly applied, no warping or buckling.

The labor cost of placing hardboard will run approximately the same as placing other types of board under the same conditions.

Approximate Sq. Ft. (Sq.M) Prices of Hardboard

Type	Thick., in.	Thick., mm	Price per S.F.	Price per Sq.M
Untempered, Plain	1/8"	3.1	$0.15	$1.61
Untempered, Plain	1/4"	6.3	0.21	2.23
Untempered, Pegboard	1/8"	3.1	0.18	1.98
Untempered, Pegboard	1/4"	6.3	0.24	2.60
Tempered Plain	1/8"	3.1	0.17	1.86
Tempered Plain	1/4"	6.3	0.26	2.85
Tempered Pegboard	1/8"	3.1	0.20	2.10
Tempered Pegboard	1/4"	6.3	0.28	2.97
Tempered Plastic Face	1/8"	3.1	0.83	8.91
Tempered Plastic Face	1/4"	6.3	0.92	9.90
Tempered Pl. F. Pegb'd	1/8"	3.1	0.83	8.91
Tempered PL. F. Pegb'd	1/4"	6.3	0.94	10.15

Tileboard

Method of Placing Tileboard in Bathrooms, Kitchens, Etc.

Tileboard has score lines impressed in one surface forming 4"x4" (100 x 100 mm) squares. The board is denser than tempered hardboard and has a lower rate of water absorption.

This material is used for walls in bathrooms, kitchens, lavatories, restaurants, barber shops, and similar places. It is furnished in sheets 4'-0" (1.2 m) wide and 2'-0" to 12'-0" (0.6-3.6 m) long.

Tempered board or tileboard should be cemented to the walls, using Armstrong's Panelboard Cement or a cement of equal quality. The adhesive should be applied to the back of the boards with a saw-tooth trowel, the boards having first been cut to fit. Use supplementary nailing or shoring to hold the boards firmly until the adhesive takes a partial set. Where it is necessary to nail the board to the walls, small brads should be used where scored lines intersect.

Figure about one gallon waterproof cement per 50 s.f. (4.5 sq.m) of wall.

Presdwood Temprtile 1/8" (3 mm) thick costs about $0.25 per s.f. (2.77 per sq.m) unpainted.

Prefinished Tileboard. Tileboard 5/32" (4 mm) thick, made of masonite tempered Presdwood, is furnished scored 4"x4" (100 x 100 mm) to represent tile and finished with multiple coats of enamel or lacquer. Tileboard is also furnished without scoring, if desired.

These boards are furnished with dull or velvet finish, or a high gloss finish in a variety of colors with contrasting joints.

Tileboard is furnished in sheets 4' (1.2 m) wide and 4, 5, 6, 7, 8, 9, 10, and 12 ft. (1.2, 1.5, 1.8, 2.1, 2.4, 2.7, 3.0, and 3.6 m) long.

Tileboard having a semi-gloss or high gloss finish costs about $0.60 to 0.70 per s.f. ($6.66-7.77 per sq.m) depending upon size of board.

Cap molding 1-3/4" (44 mm) wide for use at the top of the wainscoting is furnished in strips 4' (1.2 m) and 8' (2.4 m) long, scored to match tile or unscored, costs $0.35 per l.f. (1.16 per m).

Base molding 3-3/4" (49 mm) high, scored or unscored, in colors to match tileboard costs $0.40 per l.f. ($1.33 per m).

Placing Tempered Tileboard. Where tempered tileboard is used in kitchens, for sink splashboards, bathrooms, shower stalls, and other small places, there is usually considerable cutting and fitting required to place a comparatively small square footage (square meter) of board.

The safest method of estimating is to lay out each wall, showing number and sizes of pieces required, as it will cost almost as much to place a small panel as a large one.

Where the tileboard is secured by nails (without the use of adhesive), figure 3/4 to 1 hr. to fit and place each panel. Taking an average of 12 to 16 s.f. (1.0-1.44 sq.m) per panel, this will equal 125 to 150 s.f. (11.25-13.5 sq.m) per 8-hr. day.

However, if the tileboard is applied to walls, using an adhesive, a worker should fit and place one panel in 1 to 1-1/4 hrs. or 100 to 125 s.f. (9.0-11.25 sq.m) per 8-hr. day. This applies to both plain and prefinished tileboard. One gallon of adhesive will cover about 100 s.f. (9 sq.m) of surface if "spot" glued or about 65 s.f. (5.85 sq.m) of surface if "spread" glued.

Price per Gallon...$10.00
per Liter 2.64

Labor Cost of 100 S.F. (9 Sq.M) Tempered Tileboard, Secured to Walls With Nails

	Hours	Rate	Total	Rate	Total
Carpenter..................	4	$....	$....	$26.44	$105.76
Cost per s.f..................			....		1.06
per sq.m			....		11.41

Labor Cost of 100 S.F. (9 Sq.M) Tempered Tileboard, Secured to Walls Using Adhesive

	Hours	Rate	Total	Rate	Total
Carpenter..................	3.5	$....	$....	$26.44	$92.54
Cost per s.f..................			....		0.93
per sq.m			....		10.01

Wainscot Trim for Wallboard and Wall Tile

Moldings are made of stainless steel, polished finish in plain designs similar to the illustrations, punched for nails, and furnished in 4'-0" (1.2 m), 6'-0" (1.8 m), and 8'-0" (2.4 m) lengths.

Description	for Wallboard	per l.f. (m)	per meter
Cap Mold............................	1/8" (3.1 mm)	$0.25	0.83
Joint Mold............................	1/8" (3.1 mm)........................	0.31	1.02
Outside Corner	1/8" (3.1 mm)........................	0.37	1.21
Inside Corner........................	1/8" (3.1 mm)........................	0.37	1.21
Tub edging or cove base	1/8" (3.1 mm)........................	0.37	1.21

Metal covered wood cap, base or trim, size 5/16" x 1-1/2" (8 x 38 mm), chrome zinc, polished finish. Price per l.f., $0.35 (per m, 1.17).

Labor Placing Metal Moldings. The labor cost of placing metal moldings will vary with the design and size of the panels, cutting and fitting necessary, etc., but on the average job, a carpenter should place 100 to 150 l.f. (30-45 m) per 8-hr. day, at the following labor cost per 100 l.f. (30 m):

	Hours	Rate	Total	Rate	Total
Carpenter.................................6		$....	$....	$26.44	$158.64
Cost per l.f.................................			....		1.59
per m			....		5.22

Cap Mold   *Joint Mold*   *Outside Corner*   *Inside Corner*

Items to Include When Making Up Wallboard Estimates. When making up estimates on wallboard construction, the following items should be included to insure a complete and dependable estimate:

1. Headers or nailing strips to receive wallboard.
2. Layout.
3. Furring strips on walls and ceilings where necessary.
4. Wallboard and labor placing same.
5. Wood or metal panel strips and labor placing same.
6. Labor filling nail holes with wood putty or crevice filler.
7. Beam ceilings, ceiling cornices, metal moldings, etc.
8. Painting and decorating wallboard and panel strips.
9. Nails and adhesive.
10. Overhead expense and profit.

06430 STAIR WORK

Framing and Erecting Wood Stairs of Unassembled Stock Material. When framing and erecting ordinary wood stairs with closed stringers (where the treads and risers are not housed out), it will require 3 to 4 hrs. carpenter time to lay out the work, cut out for treads and risers, and place wood stringers for a flight of ordinary wood stairs up to 4'-0" (1.2 m) wide and 11'-0" (3.3 m) story height, at the following labor cost:

	Hours	Rate	Total	Rate	Total
Carpenter..............................4		$....	$....	$26.44	$105.76
Labor......................................2		....	....	21.04	42.08
			$....		$147.84

After the wood stringers are in place, it will require 4 hrs. carpenter time to place treads and risers for each flight of stairs (providing it is a plain box stair without nosings to form and without newels, balusters, or handrail), at the following labor cost per flight:

	Hours	Rate	Total	Rate	Total
Carpenter	4	$....	$....	$26.44	$105.76

Figure 3 hrs. carpenter time to place 4"x4" newels, 2"x4" rails and 1"x2" balusters for one side of this type stairs, at the following labor cost per flight:

	Hours	Rate	Total	Rate	Total
Carpenter	3	$....	$....	$26.44	$79.32

To lay out the work and erect the stairs complete, including plain wall hand rail secured by brackets (instead of newels, hand rails, and balusters), figure 8 hrs. carpenter time, at the following labor cost per flight:

	Hours	Rate	Total	Rate	Total
Carpenter	8	$....	$....	$26.44	$211.52

For the above stair complete, including newels, handrail, and balusters, one side, figure 10 hrs. carpenter time, at the following labor cost per flight:

	Hours	Rate	Total	Rate	Total
Carpenter	10	$....	$....	$26.44	$264.40

Erecting Wood Stairs Made in Shop. Where wood stairs are made in the shop with stringers housed out to receive treads and risers, it will require 8 hrs. stair builder time to lay out the work and to set stringers, treads, and risers (with plain wall rail attached to brackets), at the following labor cost per flight:

	Hours	Rate	Total	Rate	Total
Stair builder	8	$....	$....	$26.44	$211.52
Labor	2	....	....	21.04	42.08
			$....		$253.60

If the stair consists of two short flights, having an intermediate landing platform between floors, it will require 10 to 12 hrs. stair builder time to lay out work and to place stringers, treads, risers, and plain wall handrail, at the following labor cost:

	Hours	Rate	Total	Rate	Total
Stair builder	11	$....	$....	$26.44	$290.84
Labor	2	....	....	21.04	42.08
			$....		$332.92

Where a plain box stair, 3'-0" to 4'-0" (0.9-1.2 m) wide and 9'-0" to 10'-0" (2.7-3.0 m) story height, with open stringers is used, it will require 10 to 12 hrs. stair builder time to lay out work and place stringers, treads, and risers (not including newels, balusters, or handrail) for each flight of stairs, at the following labor cost:

	Hours	Rate	Total	Rate	Total
Stair builder	11	$....	$....	$26.44	$290.84
Labor	2		....	21.04	42.08
			$....		$332.92

If the stairs have newels, wood handrail, and balusters, add 6 to 8 hrs. stair builder time, at the following labor cost per flight:

	Hours	Rate	Total	Rate	Total
Star builder	7	$....	$....	$26.44	$185.08

Plain box stairs consisting of two short flights with an intermediate landing platform between stories will require 6 to 8 hrs. stair builder time to lay out the work and place stringers, treads, and risers for each short flight of stairs at the following labor cost per flight:

	Hours	Rate	Total	Rate	Total
Stair builder	7	$....	$....	$26.44	$185.08
Labor	2		....	21.04	42.08
			$....		$227.16

To place newels, balusters, and handrail on each short flight of stairs will require about 4 hrs. stair builder time, at the following labor cost per flight:

	Hours	Rate	Total	Rate	Total
Stair builder	4	$....	$....	$26.44	$105.76

The labor cost complete for two short flights of stairs, including newels, handrails, and balusters, should cost as follows per story height:

	Hours	Rate	Total	Rate	Total
Stair builder	22	$....	$....	$26.44	$581.68

Erecting Wood Stairs Having Open Stringers. Wood stairs of the open stringer type, with treads that have a return nosing projecting beyond the face of the stringer, will require about 12 to 14 hrs. stair builder time to lay out the work and place stringers, treads, and risers for one flight of stairs containing 16 to 18 risers, at the following labor cost per flight:

	Hours	Rate	Total	Rate	Total
Stair builder	13	$....	$....	$26.44	$343.72
Labor	2		....	21.04	42.08
			$....		$385.80

If necessary to place newels, handrails, and baluster, add about 8 to 9 hrs. at the following labor cost per flight:

	Hours	Rate	Total	Rate	Total
Stair builder	8.5	$....	$....	$26.44	$224.74

If the stair consists of two short flights with an intermediate landing platform between stories (each short flight contains 8 to 10 risers), it will require about 8 to 9 hrs. per flight or 16 to 18 hrs. per story, at the following labor cost:

	Hours	Rate	Total	Rate	Total
Stair builder	17.0	$....	$....	$26.44	$449.48
Cost per short flight	8.5		....		224.74

To set starting and landing newels, handrails, and wood balusters will require 7 to 9 hrs. per flight, or 14 to 18 hrs. per story, at the following labor cost:

	Hours	Rate	Total	Rate	Total
Stair builder	16	$....	$....	$26.44	$ 423.04
Cost per short flight	8		....		211.52

The above does not include carpenter time framing and erecting intermediate landing platform.

Placing Wood Handrail on Metal Balustrades. A carpenter should place 45 to 55 l.f. (13.5-16.5 m) of wood handrail on metal balustrades per 8-hr. day at the following labor cost per 100 l.f. (30 m):

	Hours	Rate	Total	Rate	Total
Carpenter	16	$....	$....	$26.44	$423.04
Cost per l.f.			....		4.23
per m			....		13.88

Placing Wood Handrail on Wall Brackets. A carpenter should set brackets and install 65 to 70 l.f. (19.5-21.0 m) of wall hung wood handrail per 8-hr. day at the following labor cost per 100 l.f. (30 m):

	Hours	Rate	Total	Rate	Total
Carpenter	12	$....	$....	$26.44	$317.28
Cost per l.f.			....		3.17
per m			....		10.40

Inside Stair Material

Starting Steps		length	Price
Quarter circle	Red Oak	4' (1.2 m)	$38.00
Half circle	Red Oak	4'-6" (1.4 m)	48.00
Bull nose	Red Oak	4'-6" (1.4 m)	44.00
Scroll end	Red Oak	4'-6" (1.4 m)	48.00

Stringers (not housed)

3/4"x11-1/4"x8'-0" (19x281x2400mm)	Red Oak	$32.20
3/4"x11-1/4"x10'-0" (19x281x3000mm)	Red Oak	37.95
3/4"x11-1/4"x12'-0" (19x281x3600mm)	Red Oak	57.50
3/4"x11-1/4"x14'-0" (19x281x4200mm)	Red Oak	64.40
3/4"x11-1/4"x16'-0" (19x281x4800mm)	Red Oak	73.60

Treads (not returned; for returns add $3.00)

1-1/6"x10-1/2"x3'-0" (2x263x900mm)	Red Oak	$12.65
1-1/6"x10-1/2"x3'-6" (2x263x1050mm)	Red Oak	14.38
1-1/6"x10-1/2"x4'-0" (2x263x1200mm)	Red Oak	16.10
1-1/16"x11-1/2"x3'-0" (2x288x900mm)	Red Oak	16.68
1-1/16"x11-1/2"x3'-6" (2x288x1050mm)	Red Oak	16.00
1-1/16"x11-1/2"x4'-0" (2x288x1200mm)	Red Oak	21.85
1-1/16"x11-1/2"x5'-0" (2x288x1500mm)	Red Oak	26.45
1-1/16"x11-1/2"x6'-0" (2x288x1800mm)	Red Oak	33.35

Return Nosing

1-1/8"x1-3/4"x1'-2" (28x44x350mm) with 2" (50 mm) return	Red Oak	$5.75

Tread Nosing

1-1/8"x1-3/4"x1'-2" (28x44x350mm)	Red Oak	2.19
1-1/16"x1-1/8"x1'-2" (27x28x350mm)	Red Oak	2.01

Landing Tread Nosing

1-1/16"x3-1/2"x3'-0" (27x88x900mm)	Red Oak	2.88
1-1/16"x3-1/2"x3'-6" (27x88x1050mm)	Red Oak	3.16
1-1/16"x3-1/2"x4'-0" (27x88x1200mm)	Red Oak	3.57

Tread Brackets

12"x7-3/4"x 1/4" (300x181x6mm) 3 ply	Nat. Birch	2.30

Newel Posts

	Birch	Red Oak
3-1/4"x3-1/4"x3'-2" (27x88x1200mm) tapered starting	$16.10	$16.68
3-1/4"x3-1/4"x3'-10" (27x88x1150mm) tapered, starting	18.40	18.98
3-1/4"x3-1/4"x5'-3" (27x88x1575mm) tapered, landing	25.30	26.45
3-1/4"x3-1/4"x7'-0" (27x88x2100mm) tapered, angle	29.90	31.05

Balusters

		Birch	Red Oak
1-5/16" (33 mm) round or square	30"	$2.53	$2.88
	33"	2.65	3.16
	36"	2.76	3.28
	39 or 42"	3.16	3.45

Railings—Natural birch, random length

1-3/4"x1-5/8" (44x41mm)	$1.96
1-3/4"6x1-11/16" (44x42 mm)	2.30
2-5/8"x1-11/16" (66x42mm)	2.47
2-1/4"x2-3/8" (56x59mm)	4.89

Rail Bolts and Plug	0.50

For 9' finished floors to finished ceiling heights:

Prefabricated Folding Stairways

Premium grade, for 2'-6"x6'-0" (0.75x1.8m) opng	$293.25
Medium grade, for 2'-2"x4'-6" (0.65x1.35m) opng	104.00
Economy grade, for 2'-2"x4'-6" (0.65x1.35m) opng	71.78

Prefabricated Spiral Stairs. Wood stairs are available in 3'6" (1.0 m), 4' (1.2 m), 4'6" (1.35 m), and 5' (1.5 m) diameters with heights made to fit the actual job conditions. The balusters and center column are steel, but the treads, platforms, and railings can be wood to give a finished appearance. Several styles are available and the cost runs from $860.00 for the smallest diameter and the simplest design in an 8' (2.4 m) rise to $1,470.00 for the 5' (1.5 m) diameter. The more elaborate designs will run from $1,350.00 to $1,725.00 for the same sizes. The stairs are factory assembled and finished so that two carpenters should be able to install the unit in one day.

Starting Newels

Volute Newels

Turned Balusters

Inside Stair Materials

Starting Steps and Colonial Newels rail

Railing Fittings

CHAPTER 7

THERMAL AND MOISTURE PROTECTION

CSI DIVISION 7
07100 WATERPROOFING
07150 DAMPPROOFING
07200 INSULATION
07300 SHINGLES AND ROOFING TILES
07400 PREFORMED ROOFING
      AND SIDING
07500 MEMBRANE ROOFING
07600 FLASHING AND SHEET METAL
07800 ROOF ACCESSORIES
07900 SEALANTS

07100 WATERPROOFING

Every building requires water and damp resisting material, plaster bonds, or floor hardeners of one kind or another, and with so many of them on the market, it is impossible even for the best informed contractor to be familiar with all of them.

No attempt has been made here to pass on the relative merits of the various preparations. The data is strictly information that is of interest to the estimator—use of the material, covering capacity, and cost of applying. The prices quoted are approximate and may vary considerably in different locations. Always check with local distributors for current prices.

Water and Damp Resisting Methods. There are several methods used by the different manufacturers for mixing and applying their products to produce a water resisting wall or surface. In some instances, the water repelling compounds are incorporated with the cement, sand, and gravel while the concrete is being mixed, and by filling the voids in the cement and sand, they produce a waterproof mixture. Where the various powders or pastes are mixed in the concrete mass, it is called the *integral method* of waterproofing.

Some water and dampness resisting preparations are furnished in mastic form and are applied with a trowel. Others are mixed with portland cement and sand and applied as a plaster coat to the surfaces to be treated.

There are also numerous water and dampness resisting compounds in the form of heavy paints, which are applied to the concrete or masonry walls, floors, and ceilings with a brush or mop. One or more applications of these materials are supposed to penetrate and fill the pores in the concrete or masonry to such an extent that the treated surfaces will be impervious to dampness or moisture.

Where there is considerable pressure or a large head of water, one of the most satisfactory methods is known as membrane waterproofing. This method consists of applying 2 to 7 hot moppings of pitch or asphalt on 1 to 7 plies of saturated felt or fabric. Waterproofing of this kind is applied in much the same manner as built-up roofing. After the last ply of saturated felt or fabric is applied, the entire surface is mopped with a heavy application of hot pitch or asphalt.

Estimating Quantities of Water and Damp Resisting Preparations. When estimating quantities of the various plaster coats, paints, and membrane waterproofing compounds required for any job, the entire area of the walls or surfaces to be treated should be measured and the quantities stated in sq. ft., sq. yds., sq.m, or by the square (100 s.f. or 9.3 sq.m).

Integral water resisting compounds are usually estimated by the cu. ft., cu. yd., or cu.m of concrete to which the liquid, paste, or powder is added.

Water Required for Mixing Concrete. When mixing concrete, sufficient water should be used to produce a plastic mix—no more and no less. A plastic mix is concrete that can be readily molded, and when the mold is removed, will flow sluggishly without segregation of the water or the fine materials from the coarse. Concrete should be mixed at least 2 minutes after all materials have been placed into the mixer. Remember there is nothing more injurious to good concrete than too much mixing water.

Refer to water required for mixing concrete as given in Step 4. "Selection of water-cement ratio" under *Cast-in Place Concrete.*

Waterproofing by the Integral Method

When using the integral method, the admixtures are usually mixed with the cement and aggregate and form a part of the concrete mass. The quantities are based on the specifications of the various manufacturers.

Material prices given are approximate, as it is impossible to quote prices that will apply to every locality. The contractor or estimator should obtain local prices on the materials specified. Prices are based on material being purchased in drum or barrel quantities. When purchased in one gallon or one pound quantities, the cost increases considerably.

Integral Liquid Admixtures

Integral liquid admixtures consist of calcium chloride solutions, oil water repellent preparations, and other integral mixtures that flow freely from a drum.

The cost of hauling the barrel to the mixer, setting it up on blocks, and drawing off the liquid by spigot into a measuring container will cost from

$0.10 to 0.15 per gal. of waterproofing on small jobs. On large jobs this cost may be halved.

Concrete mixed in the proportions of 1:2:4 or 1:2-1/2:3-1/2 requires 5-1/2 to 6 sacks of cement per c.y. (0.76 cu.m) of concrete.

Trade Name	Qts. per Sack Cement	Concrete Proportion	Gals. per Cu. Yd. Concrete	Price per Gal.
Anti-Hydro*	1	1:2:4	1-1/2	$3.00
Aquatite	2/3	1:2:4	1	1.60
Aquatite**	1	1:2:4	1-1/2	1.60
Hydratite	1	1:2:4	1-1/2	2.35
Toxement IW	1	1:2:4	1-1/2	2.30
Sikacrete	1	1:2:4	1-1/2	3.30
Trimix	1	1:2:4	1-1/2	1.50

(metric)

Trade Name	Liters per Sack Cement	Concrete Proportion	Liters per cu.m Conc.	Price per Gal.
Anti-Hydro*	0.95	1:2:4	5.7	$0.80
Aquatite	0.62	1:2:4	3.8	0.42
Aquatite**	0.95	1:2:4	5.7	0.42
Hydratite	0.95	1:2:4	5.7	0.62
Toxement IW	0.95	1:2:4	5.7	0.60
Sikacrete	0.95	1:2:4	5.7	0.87
Trimix	0.95	1:2:4	5.7	0.40

*A 5-year maintenance guarantee furnished by manufacturer where supervision is supplied at additional cost.
**Use where there is water pressure.

Powdered Integral Admixtures

This material should be added to the portland cement before the addition of aggregates. Then, it is dry mixed thoroughly before adding gauging water. Add about $0.15 to 0.20 per lb. ($0.33-0.44 per kg) to the cost of material for labor handling and mixing.

Trade Name	Lbs. per Sack Cement	Concrete Proportion	Lbs. per Cu. Yd. Concrete	Price per Lb.
Hydratite Powder	1	1:2:4	6	$0.30
Hydrocide Powder	1	1:2:4	6	0.30
Plastiment	1/2-1	1:2:4	3-6	0.55
Toxement IW	1-1/2	1:2:4	7-1/2	0.45

Trade Name	(metric) Liters per Sack Cement	Concrete Proportion	Kg per cu.m Concrete	Price per kg
Hydratite Powder	0.95	1:2:4	96	$0.65
Hydrocide Powder	0.95	1:2:4	96	0.65
Plastiment	0.47-0.95	1:2:4	48-96	1.20
Toxement IW	1.42	1:2:4	120	1.00

Water and Damp-Resisting Plaster Coats

There are numerous preparations on the market to mix with portland cement and sand and apply as a plaster coat for waterproofing concrete and masonry walls. There are other preparations in mastic form that are applied with a trowel, the same as ordinary cement mortar.

Estimating Quantities of Water or Damp-Resisting Plaster Coats. Where water or damp-resisting plaster coats are applied to brick or concrete walls, the water resisting ingredient is usually liquid, paste, or powder.

This compound is usually added to the mortar on the basis of a certain quantity to each sack of cement used and practically all manufacturers specify a mortar mixed in the proportions of 1 part portland cement to 2 parts sand, adding only sufficient water to make a workable mortar.

Quantity of Cement and Sand Required for One C.Y. (0.76 cu.m) of Cement Mortar

Based on damp loose sand containing 5% of moisture and weighing 2,565 lbs. per c.y. (1960 kg per cu.m).

Proportions by Volume Cement Sand	Packed Cement Sacks	Loose Sand, cu. yds.	Loose Sand, cu.m
1:1	18.0	0.70	0.54
1:1-1/2	15.2	0.84	0.64
1:2	12.5	0.95	0.73
1:2-1/2	11.0	1.03	0.79
1:3	9.5	1.10	0.84

Above quantities of sand include 5% for waste.

Cement Plaster Coat—S.F. Obtainable from One C.Y. (Sq.M from One Cu.M) of Cement Mortar

Thickness, in.	Thickness, mm	Square Feet	Square Meters
1/4	6.3	1,296	120
1/2	12.5	648	60
3/4	18.8	432	40
1	25.0	324	30

Cleaning and Washing Old Masonry or Concrete Surfaces. Where old masonry or concrete surfaces are to be given a waterproof plaster coat, it is first necessary to prepare the old surface to receive the cement plaster.

On some jobs it is necessary to rake out the old brick joints to form a key for the plaster, and on others it is necessary to thoroughly wire brush the old surface to remove all surface dirt and then wash the surface with clean water or a mixture of chemical cleaner and water.

Where the surfaces are to be brushed and washed with water, the labor cost of cleaning 100 s.f. of surface should average as follows:

	Hours	Rate	Total	Rate	Total
Labor	1.5	$....	$....	$21.04	$31.56
Cost per s.f.				....	0.32
per sq.m				....	3.44

Roughing and Hacking Old Concrete Walls and Other Vertical Surfaces. Where it is necessary to roughen old, hard concrete surfaces to prepare a suitable bond for the new waterproof plaster coat, this can prove to be very expensive, especially where the old concrete surfaces are very hard and must be hacked and roughened using hand tools, a jackhammer, or bush-hammering tool. Workers often work in cramped quarters, which prevents the best daily output. On work of this kind, a worker will roughen about 20 s.f. (1.8 sq.m) of surface per hour. The following table indicates the labor cost per 100 s.f. (9.3 sq.m) to roughen and clean the surface:

	Hours	Rate	Total	Rate	Total
Labor on Jackhammer	5	$....	$....	$21.04	$105.20
Labor for cleaning	2	....	....	21.04	42.08
Cost per 100 s.f.			$....		$147.28
per s.f.				....	1.47
per sq.m				....	15.82

Cost of tools (purchase or rental), fuels, etc. has to be estimated, in addition to the labor requirements. These costs will vary, depending on the accessibility, depth to be roughened, and quantity of work. For example, light work or small areas can be accomplished with an electric bush-hammer, while heavier work requires an air compressor, hose, and light jackhammer.

Equipment rental and fuel costs for an air compressor and tools to roughen the 100 s.f. (9.3 sq.m) of surface:

	Hours	Rate	Total	Rate	Total
Air Compressor & Tools (Rental)	5	$....	$....	$7.70	$38.50
Fuel Costs 10 gal. (37.8 liters)		....	....	1.35	13.50
Cost per 100 s.f.			$....		$52.00
per s.f.				....	0.52
per sq.m				....	5.60

Applying Slush or Grout Coat of Neat Cement. After the old surfaces are cleaned, it is necessary to give them a slush coat of neat portland cement thoroughly brushed into the pores of the wall, which acts as a bond for the subsequent plaster coats. The labor cost per 100 s.f. of surface should run as follows:

	Hours	Rate	Total	Rate	Total
Labor	2	$....	$....	$21.04	$42.08
Cost per s.f.			....		0.42
per sq.m			....		4.52

The material needed for a slush coat of neat portland cement will require one bag of portland cement, mixed with water, per 100 s.f. (9.3 sq.m) of surface to be coated. A bag of cement costs about $6.00 each, which amounts to the following material costs: $5.50 per 100 s.f. (per 9.3 sq.m), or $0.06 per s.f. ($0.64 per sq.m).

Labor Applying Waterproof Plaster Coats to Vertical Surfaces. When applying waterproof plaster coats to vertical surfaces, the costs run high for a number of reasons.

1. Interference of other trades.
2. Difficulty of mixing and handling.
3. Delay in setting of coats.
4. Difficulty in finishing vertical surfaces of this type with plaster or cement finish.
5. Contingency cost of return to job for breaks and cracks.
6. Waste and spillage of materials.

This work is usually performed by cement masons, plasterers, or bricklayers.

When applying waterproof plaster coats, it is not necessary to give the surface a float or brush finish, but the surface must be reasonably smooth and without uneven projections that would prevent the water from running down the sides of the walls.

Most manufacturers specify the mortar to be mixed in the proportions of 1:1-1/2 or 1:2, and the material is usually applied in one or two coats.

A mason should apply about 350 s.f. (32 sq.m) of first coat per 8-hr. day, at the following labor cost per 100 s.f. (9.3 sq.m):

	Hours	Rate	Total	Rate	Total
Plasterer	2.3	$....	$....	$25.75	$59.23
Helper	2.3	....	....	21.04	48.39
Cost per 100 s.f.			$....		$107.62
per s.f.			....		1.08
per sq.m			....		11.62

On the second or finish coat, a mason should apply about 400 s.f. (37.1 sq.m) per 8-hr. day, at the following labor cost per 100 s.f. (9.3 sq.m):

	Hours	Rate	Total	Rate	Total
Plasterer	2.0	$....	$...	$25.75	$51.50
Helper	2.0	....	....	21.04	42.08
Cost per 100 s.f.		$...			$93.58
per s.f.				....	0.94
per sq.m				....	10.11

If necessary to clean the old masonry or concrete surfaces, or to roughen the old concrete, add for this work as given on previous page.

Water Resisting Plaster Coats Using
Powdered Integral Water Repellents,
per 100 s.f. (9.3 sq.m)

Trade Name	Inches Thick	Mix Prop.	Number Coats	Sack Cement	Cu.Ft. Sand	Quan. W'pf'g	Approx Price/lb.	Mech. Hrs.	Lab. Hrs.
Hydratite	1	1:2	2	4	8	8 lbs.	$0.30	8	8
Hydrocide Powder	1	1:2	2	4	8	4 lbs.	0.30	8	8
Toxement IW	3/4	1:2	2	3	6	4-1/2	0.35	7	7

(metric)

Trade Name	Thick, mm	Mix Prop.	Number Coats	Sack Cement	Cu.M Sand	Quan. W'pf'g	Approx Price/kg	Mech. Hrs.	Lab. Hrs.
Hydratite	25.0	1:2	2	4	0.23	3.6	$0.66	8	8
Hydrocide Powder	25.0	1:2	2	4	0.23	1.8	0.66	8	8
Toxement IW	18.8	1:2	2	3	0.17	2.0	0.77	7	7

Water Resisting Plaster Coats
Using Liquid Integral Admixtures,
per 100 s.f. (9.3 sq.m)

Trade Name	Thick., inches	Mix Prop.	Number Coats	Sack Cement	Cu.Ft. Sand	W'pf'g gals.	Approx Price/lb.	Mech. Hrs.	Lab. Hrs.
Anti-Hydro	3/4	1:2	2	3	6	3/4	$3.00	7	7
Anti-Hydro	1	1:2	2	4	8	1	3.00	8	8
Aquatite	3/4	1:2	2	3	6	3/4	1.60	7	7
Aquatite	1	1:2	2	4	8	1	1.60	8	8
Hydratite	3/4	1:2	2	3	6	3/4	2.35	7	7
Hydratite	1	1:2	2	4	8	1	2.35	8	8
Toxement IW	3/4	1:2	2	3	6	3/4	2.30	7	7
Toxement IW	1	1:2	2	4	8	1	2.30	8	8
Sikacrete	3/4	1:2	2	3	6	3/4	3.30	7	7
Trimix	3/4	1:2	2	3	6	3/4	1.50	7	7

(metric)

Trade Name	Thick., mm	Mix Prop.	Number Coats	Sack Cement	Cu.M Sand	W'pf'g liters	Approx Price/kg	Mech. Hrs.	Lab. Hrs.
Anti-Hydro	18.8	1:2	2	3	0.17	2.8	$6.60	7	7
Anti-Hydro	25.0	1:2	2	4	0.23	3.8	6.60	8	8
Aquatite	18.8	1:2	2	3	0.17	2.8	3.55	7	7
Aquatite	25.0	1:2	2	4	0.23	3.8	3.55	8	8
Hydratite	18.8	1:2	2	3	0.17	2.8	5.20	7	7
Hydratite	25.0	1:2	2	4	0.23	3.8	5.20	8	8
Toxement IW	18.8	1:2	2	3	0.17	2.8	5.10	7	7
Toxement IW	25.0	1:2	2	4	0.23	3.8	5.10	8	8
Sikacrete	18.8	1:2	2	3	0.17	2.8	7.30	7	7
Trimix	18.8	1:2	2	3	0.17	2.8	3.30	7	7

Membrane Waterproofing

Membrane waterproofing is constructed in place by building up a strong, waterproof, and impermeable blanket with overlapping plies of tar, asphalt saturated open mesh cotton fabric, or rag felt. The plies are coated and cemented together with hot coal tar pitch or waterproofing asphalt. There is always one or more applications of pitch or asphalt than plies of felt or fabric.

Estimating the Quantity of Felt or Fabric Required for Membrane Waterproofing. When estimating the quantity of felt or fabric required for waterproofing masonry or concrete walls, pits, floors, etc., it is necessary to add a percentage for lap and waste. If the felt is applied to walls, allow about 6" (150 mm) to lap over the footings at the bottom, with a small allowance at the top of the walls above grade. This will average from 7.5% to 10% of the total area to be waterproofed.

If the walls or floors are covered with a single thickness of felt or membrane, lap each strip of felt 4" to 6" (100-150 mm), depending upon the specifications. As most felts are put up in 36" (900 mm) rolls, it is necessary to add about 12% waste for a 4" (100 mm) lap, and 17% for a 6" (150 mm) lap.

If the lap is 4" (100 mm) wide and only a single thickness of felt is used, add about 20% to the actual area to allow for laps and the additional felt required at the top and bottom of the wall. On floors it is customary to run the felt or membrane up the sides of the walls 4" to 6" (100-150 mm).

If the lap is 6" (150 mm) and only a single thickness of felt is used, add about 25% to the actual wall or floor area for laps, etc.

When 2, 3, 4, or more plies of felt are used, it is not necessary to allow for laps, except at the ends of the wall or floor, as one thickness of felt overlaps the next without waste. However, extra material will be required at the tops and bottoms of the walls and where the paper runs up the sides of the walls 4" to 6" (100-150 mm). On work of this kind, add about 10% to the actual floor or wall area.

Fabric or Felt Required to Cover 100 S.F. (9.3 Sq.M) of Surface

Class of Work	1 Ply 4" lap	1 Ply 6" lap	2-Ply	3-Ply	4-Ply	5-Ply	Ea. Add. Ply
Add percent	20	25	10	10	10	10	10
s.f. of felt	120	125	220	330	440	550	110
sq.m of felt	11	12	20	31	41	51	10

Weight of Tar or Asphalt Felt for Membrane Waterproofing. Tar or asphalt felt for waterproofing is currently furnished in 4 sq. rolls of 432 s.f. (40 sq.m) weighing 60 lbs. (27 kg) per roll.

Double thickness asphalt felt is also furnished in 60 lb. (27 kg) per roll containing 216 s.f. (20 sq.m) and this felt is known as No. 30.

When specifying the grade or weight of felt to be used, it is customary to state that "felt shall weigh not less than 15 lbs. per 108 s.f." (6.8 kg per 10 sq.m). This is known as No. 15 felt. Felt is furnished in 4 sq. rolls of 432 s.f. (40 sq.m), so there are 32 s.f. (3.0 sq.m) per roll or 8 s.f. per 100 s.f. (0.74 sq.m per 9.3 sq.m) allowed for laps.

Tar or asphalt saturated fabric is usually sold by the roll containing 50 sq. yds. (4.6 sq.m) or by the sq. yd. (sq.m).

Estimating the Quantity of Pitch or Asphalt Required for Membrane Waterproofing. The quantity of pitch or asphalt required for mopping 100 s.f. (9.3 sq.m) of surface with one application is approximately 30 to 35 lbs. (14-16 kg).

Many special coatings are sold under trade names. Some are furnished solid, making it necessary to heat and melt them before using, while others are furnished in the form of a heavy black paint that can be either mopped or brushed on cold. The manufacturer usually specifies how the material shall be applied.

Engineers and architects usually specify the quantity of asphalt or compound to be used for each 100 s.f. (9.3 sq.m) of surface.

Weight of Pitch or Asphalt Per 100 S.F. (9.3 Sq.M) of Surface

No. Plies of Saturated Fabric or Tarred Felt	Alternate Moppings of Pitch or Asphalt Required	Lbs. of Pitch or Asphalt Required per 100 s.f. W'p'fing	Kg of Pitch or Asphalt Required per 10 sq.m W'p'fing
2	3	90 - 105	44 - 51
3	4	120 - 140	59 - 68
4	5	150 - 175	73 - 85
5	6	180 - 210	88 - 103
6	7	210 - 245	103 - 120

Applying Membrane Waterproofing. Whether to use saturated fabric or saturated felt in constructing membrane waterproofing depends entirely upon local conditions. Saturated fabric weighing 12 oz. per sq. yd. has

7.10 ESTIMATOR'S REFERENCE BOOK

approximately three times the tensile strength of 15-lb. (6.8-kg) saturated felt, but it costs considerably more per sq. yd. (sq.m) than saturated felt. On flat surfaces, where the waterproofing will be covered with a protection course immediately after it is applied, and where it will not be subjected to any unusual strain, saturated felts can be used. On any surface where greater strength is required, fabric should be used. It is always desirable to provide a protection course over the waterproofing construction. While the labor operations involved in applying membrane waterproofing are similar to those followed in applying built-up roofing, the work is usually more expensive than roofing because of the limited working space and the necessity for using scaffolds on vertical surfaces over 8' (2.4 m) high.

On vertical surfaces the work can be expedited if the plies of felt or fabric are installed up and down the surface like wallpaper rather than across. The pieces of felt or fabric can be cut to the required length and folded so that after the surface has been mopped the top edges can be pushed into the hot bitumen and the remainder of the piece will then fall into place where it can be rubbed into the hot pitch or asphalt. After the required number of plies have been embedded in alternate moppings of hot bitumen all angles at corners, walls, etc., should be reinforced with at least two additional plies of felt or fabric and alternate moppings of bitumen. A protection course of beadboard, fiberboard, or other material should immediately be placed over the waterproofing for best results.

Prior to the application of the first hot bitumen mopping coat, the wall surface is primed, uniformly and completely, with one gallon (3.8 liters) of primer—asphalt primer for asphalt specifications or tarbase primer for pitch specifications—per 100 s.f. (9.3 sq.m) of surface area.

For the application of the primer to the surface, one worker should coat about 100 s.f. (9.3 sq.m) per hour:

	Hours	Rate	Total	Rate	Total
Labor	1.0	$....	$....	$24.99	$24.99
Cost per s.f.			....		0.25
per sq.m			....		2.69

Labor Applying Membrane Waterproofing. When applying membrane waterproofing to walls and floors, it takes three or four workers together. It requires one worker to attend the fire and heat the bitumen, while two or three are mopping the walls and applying the felt or fabric.

If only one application of hot bitumen is applied, two workers together should heat materials and mop 2,000 to 2,200 s.f. (186-204 sq.m) per 8-hr. day, at the following labor cost per 100 s.f. (9.3 sq.m):

	Hours	Rate	Total	Rate	Total
Roofer	0.8	$....	$....	$24.99	$19.99
Cost per s.f.			....		0.20
per sq.m			....		2.15

Water Pressure Table and Waterproofing Required
for Varying Heads of Water

Hydrostatic Head in Feet	Pressure Lbs. per Sq. In.	Lifting Pressure Lbs per SF	Wall Pressure Lbs per SF	Plies of Saturated Felt/Fabric	Asphalt Mastic Thickness
0.5	0.21	39.2	15.6	2-Ply	1/8"
1.0	0.43	62.5	31.2	2-Ply	1/8"
2.0	0.86	125.0	62.5	2-Ply	1/4"
3.0	1.30	187.5	93.7	2-Ply	1/4"
4.0	1.73	250.0	125.0	3-Ply	5/8"
5.0	2.17	312.5	156.2	3-Ply	5/8"
6.0	2.60	375.0	187.5	4-Ply	5/8"
8.0	3.47	500.0	250.0	4-Ply	5/8"
10.0	4.34	625.0	312.5	5-Ply	5/8"
12.0	5.21	750.0	375.0	6-Ply	5/8"
15.0	6.51	937.5	468.7	6-Ply	3/4"
20.0	8.68	1250.0	625.0	7-Ply	3/4"
25.0	10.85	1562.5	781.2	8-Ply	3/4"
30.0	13.02	1875.0	937.5	10-Ply	3/4"
40.0	17.36	2500.0	1250.0	11-Ply	3/4"

(metric)

Hydrostatic Head in meters	Pressure Kg per sq.m	Lifting Pressure Kg per sq.m	Wall Pressure Kg per sq.m	Plies of Saturated Felt/Fabric	Asphalt Mastic Thick., mm
0.15	148	191	76	2-Ply	3.1
0.30	302	305	152	2-Ply	3.1
0.61	605	610	305	2-Ply	6.3
0.91	914	915	475	2-Ply	6.3
1.22	1216	1221	610	3-Ply	15.6
1.52	1526	1526	763	3-Ply	15.6
1.83	1828	1831	915	4-Ply	15.6
2.44	2440	2441	1221	4-Ply	15.6
3.05	3051	3051	1526	5-Ply	15.6
3.66	3663	3662	1831	6-Ply	15.6
4.57	4577	4577	2288	6-Ply	18.8
6.10	6103	6103	3051	7-Ply	18.8
7.62	7629	7629	7628	8-Ply	18.8
9.14	9154	9154	4577	10-Ply	18.8
12.19	12206	12205	6103	11-Ply	18.8

After the surface has been mopped with hot bitumen, two workers must cut the felt to lengths and place it on the wall. Usually it is cut long enough to lap over the top of the wall a few inches, so that it can be held in place by brick or stone until it is self-supporting.

Two workers placing felt or fabric have to work along with the workers mopping, so they will place felt or fabric on 2,000 to 2,200 s.f. (186-204 sq.m) of surface per 8-hr. day, at the following labor cost per 100 s.f. (9.3 sq.m):

	Hours	Rate	Total	Rate	Total
Roofer	0.8	$....	$....	$24.99	$19.99
Cost per s.f.			....		0.20
per sq.m			....		2.15

Labor Applying 100 S.F. (9.3 Sq.M) of Membrane Waterproofing

Desc. of Work	Number of Plies	Mopping Hours	Felt Hours	Total Hours
1-ply fabric, 2 moppings	1	1.6	0.8	2.4
2-ply fabric, 3 moppings	2	2.4	1.6	4.0
3-ply fabric, 4 moppings	3	3.2	2.4	5.6
4-ply fabric, 5 moppings	4	4.0	3.2	7.2
Each additional ply of felt and mopping	1	0.8	0.8	1.6

Prices of Waterproofing Materials

Prices on waterproofing materials must always be checked, because prices vary considerably due to market conditions, location, and manufacturer.

Asphalt primer costs about $2.25 per gal. ($0.59 per liter) in 55-gal. (208-liter) drums; $2.35 per gal. ($0.62 per liter) in 30-gal. (114-liter) drums and $2.50 per gal. ($0.66 per liter) in 5-gal. (19-liter) cans.

Waterproofing pitch or asphalt costs about $270.00 per ton in carload lots; $280.00 per ton in less than carloads.

No. 15 tar or asphalt saturated felt costs $12.00 per roll of 432 s.f. (40 sq.m) in carloads and $12.50 per roll in less than carloads.

Tar or asphalt saturated cotton fabrics costs $1.00 to $1.25 per sq. yd. ($0.13-0.16 per sq.m) in carloads and $1.25 to $1.30 per sq. yd. ($0.16-0.17 per sq.m) in less than carloads.

Material Cost of 100 S.F. (9.3 Sq.M) of 1-Ply Membrane Waterproofing
Consisting of 1-ply 15-lb. (7-kg) felt and 2 moppings of hot bitumen

	Rate	Total	Rate	Total
1 gal. (3.7 liter) asphalt primer	$....	$....	$2.25	$2.25
70 lbs. (31.7 kg) asphalt or pitch*		....	0.13	9.10
0.28 rolls No.15 felt, 120 s.f. (11.1 sq.m)		....	12.00	3.36
Fuel, mops, etc.		....	1.00	1.00
Cost per 100 s.f.		$....		$15.71
per s.f.		....		0.16
per sq.m		....		1.72

Labor Cost of 100 S.F. (9.3 Sq.M) of 1-Ply Membrane Waterproofing
Consisting of 1 ply of felt and 2 moppings of hot bitumen

	Hours	Rate	Total	Rate	Total
Roofer*	2.6	$....	$....	$24.99	$64.97
Roofer placing felt	0.8		....	24.99	19.99
Cost per 100 s.f.			$....		$84.96
per s.f.			....		0.85
per sq.m			....		9.15

*Includes labor cost for primer coat.

Material Cost of 100 S.F. (9 Sq.M) of Each Additional Ply
of Felt and Hot Bitumen

	Rate	Total	Rate	Total
35 lbs. (15.8 kg) asphalt or pitch	$....	$....	$0.13	$4.55
0.24 rolls No. 15 felt, 100 s.f. (9.3 sq.m)		....	12.00	2.88
Fuel, mops, etc.		....	0.50	0.50
Cost per 100 s.f.		$....		$7.93
per s.f.		....		0.08
per sq.m		....		0.86

Labor Cost per 100 S.F. (9.3 Sq.M) for Each Additional Ply
of Felt and Hot Bitumen

	Hours	Rate	Total	Rate	Total
Roofer	1.6	$....	$....	$24.99	$39.98
Cost per s.f.			....		0.40
per sq.m			....		4.30

Material Cost of 100 S.F. (9.3 Sq.M) of 1-Ply Membrane Waterproofing
Using Tar or Asphalt, Saturated Fabric and Hot Pitch or Asphalt
Consisting of 1-ply of Saturated Fabric and 2 Moppings of Hot Bitumen

	Rate	Total	Rate	Total
1 gal. (3.7 liter) asphalt primer	$....	$....	$2.25	$2.25
70 lbs. (32 kg) asphalt or pitch		....	0.13	9.10
13.4 s.y. (11.2 sq.m) satur. fabric		....	1.25	16.75
Fuel, mops, etc.		....	0.85	0.85
Cost per 100 s.f.		$....		$28.95
per s.f.		....		0.29
per sq.m		....		3.12

Material Cost of 100 S.F. (9.3 Sq.M) Additional Ply of Saturated
Fabric and Hot Pitch or Asphalt

	Rate	Total	Rate	Total
35 lbs. (16 kg) asphalt or pitch ....$....		$....	$0.13	$4.55
12 s.y. (10 sq.m) satur. fabric		....	1.25	15.00
Fuel, mops, etc.		....	0.50	0.50
Cost per 100 s.f.		$....		$20.05
per s.f..		....		0.20
per sq.m...		....		

The labor costs for placing fabric and moppings of hot bitumen are the
same as used in placing felts and moppings of hot bitumen.

The Iron Method of Waterproofing

The iron method of waterproofing uses an exceedingly fine metallic
powder, containing no grease, asphalt, oil, or other substances subject to
disintegration.

The waterproofing is applied to the inside or outside surfaces of walls and
to the tops of rough footings and floor slabs, in the form of brush coats, or in a
combination of brush and plaster coats.

Applied either upon the inside or outside of walls, this type of
waterproofing will resist hydrostatic pressure to a considerable degree.

When preparing old concrete vertical surfaces for iron waterproofing, the
vertical surfaces should be given an entirely new bonding exposure by
cutting not less than 1/16" (1.6 mm) with bushhammers or other suitable
tools. See cost of roughing old surfaces on previous pages. The new surface
shall then be thoroughly cleaned by brushing with wire brushes. Surfaces of
brick and stone need only thorough cleaning with the wire brush and
washing.

Holes, cracks, and soft or porous spots in vertical or horizontal surfaces
shall be cut out and pointed with one part iron floor bond, 2 parts portland
cement and 3 parts sand, all by weight. These materials shall be mixed dry,
screened and enough water added to make a stiff mix. Particular care must
be taken at the intersections of all walls and floors, in corners, around pipes
and other projections through the waterproofing and at construction joints in
the concrete.

After the wall surfaces have been cleaned and pointed, they shall be
thoroughly cleaned again by hosing. Excess water shall then be removed.

It usually requires 2 to 6 coats of iron waterproofing, depending on the
density of the surface and the conditions encountered.

Some of these materials are mixed with sand and water and applied as a brush coat while others are furnished already mixed ready to apply by the addition of water.

1-Coat Waterproofing Using the Iron Method, Brushed On, per 100 s.f. (9.3 sq.m)

Trade Name	No. of Coats	Quan. per 100 s.f., lbs.	Quan. per 10 sq.m, kg	Approx. Cost/lb.	Approx. Cost/kg	Labor Hrs.
Ferrolith W	1	13-15	14-16	0.30	0.66	2
Kemox H (Sika)	1	60-100	65-108	0.16	0.35	2
Irontox	1	10-12	11-13	0.25	2.00	2

For each additional application, figure same as given above. Three to five applications are required where there is water pressure to be overcome.

Liquid Membrane Waterproofing

Liquid membrane (elastomeric) waterproofing is a cold, fluid applied, synthetic rubber, seamless coating system. It is available in single or two-component form and can be applied by trowel or spray. The material forms a continuous, flexible, water impervious film that bonds tightly to a wide range of surfaces, including concrete, stone, masonry, wood, and metal. On a troweled surface, an application rate of 5 gals. (19 liters) per 125 s.f. (12 sq.m) of surface will produce a cured coating thickness of approximately 55 mils.

Two workers should coat about 650 s.f. (60 sq.m) per day at the following labor cost:

	Hours	Rate	Total	Rate	Total
Labor	16.0	$....	$....	$24.99	$399.84
Cost per s.f.				....	0.62
per sq.m				....	6.67

Cost of 100 S.F. (9.3 Sq.M) Mastic Waterproof Coatings

The following materials are furnished in the form of a mastic or plastic cement and are applied with a trowel.

Kind of Waterproofing	Thick., inches	No. of Coats	Cover Cap. s.f. per gal.	Quan. per 100 s.f., gals.	Approx. Price per gal.	Labor Hours
Dehydratine No. 10	1/16	1	26	4 Gal.	$1.65	3
Hydrocide Mastic	1/16	1	20-25	5 Gal.	1.13	3
Hydrocide 700	1/16	1	20-25	5 Gal.	1.00	3

STI6MATOR'S REFERENCE BOOK

Tremco 103 Mastic
Asphalt base .... 1/8....1 ....20 ....5 Gal. ....1.00....3
Tar base........ 1/8....1 ....20 ....5 Gal. ....1.70....3

(metric)

Kind of Waterproofing	Thick., mm	No. of Coats	Cover Cap. sq.m per liter	Quan. per 10 sq.m, liters	Approx. Price per liter	Labor Hours
Dehydratine No. 10	1.6	1	6.4	16.29	$0.44	3
Hydrocide Mastic	1.6	1	4.9-6.1	20.37	0.30	3
Hydrocide 700	1.6	1	4.9-6.1	20.37	0.26	3
Tremco 103 Mastic						
Asphalt base	3.1	1	4.9	20.37	0.26	3
Tar base	3.1	1	4.9	20.37	0.45	3

07150 DAMPPROOFING

Dampproof Paints for Exterior Concrete or Masonry Surfaces

Where a heavy head of water is to be overcome, the concrete or masonry should preferably be waterproofed by the membrane method. Where only a dampproofing material is required, there are a number of paints on the market for this purpose. The application of one, two, or more coats of paint is intended to make the walls impervious to dampness.

Number of coats required and covering capacity will vary with porosity and surface of wall.

Heavy Dampproof Paints, Per 100 S.F. (9.3 Sq.M)

Material	Use on	No. Cts.	Sq. Ft. per Gal.	Gals. Reqd. 100 Sq. Ft.	Price per Gal.	Lab. Hrs. Reqd.
Dehydratine 4	Fndtn.	2	33	3	$1.43	2.2
Dehydratine 10	Masonry	1	30	3-1/3	1.65	1.2
Hydrocide Semi-Mastic	Fndtn.	1	30-50	3	1.05	1.2
	Fndtn.	2	15-18	6	1.05	2.2
Hydrocide 600	Fndtn.	1	75-100	1-1/3	1.13	1.2
Hydrocide 648	Fndtn.	2	50	2	1.13	2.2
Marine Liquid	Blw Grd	1	50-75	1-3/4	1.24	1.2
Sikaseal	Fndtn.	2	60-80	1-1/2	2.20	2.2
Tremco 110	Fndtn.	1	100	1	0.95	1.2

(metric)

Material	Use on	No. Cts.	Sq.M per Liter	Liters Reqd. 10 Sq. M	Price per Liter	Lab. Hrs. Reqd.
Dehydratine 4	Fndtn.	2	8.1	12.2	$0.38	2.2
Dehydratine 10	Masonry	1	7.4	13.6	0.44	1.2
Hydrocide Semi-Mastic	Fndtn.	1	7.4-12.3	12.2	0.38	1.2
	Fndtn.	2	3.7-4.4	24.2	0.28	2.2
Hydrocide 600	Fndtn.	1	18.4-24.5	5.4	0.30	1.2
Hydrocide 648	Fndtn.	2	12.3	8.1	0.30	2.2
Marine Liquid	Blw Grd	1	12.3-18.4	7.1	0.33	1.2
Sikaseal	Fndtn.	2	14.7-19.6	6.1	0.58	2.2
Tremco 110	Fndtn.	1	24.5	4.1	0.25	1.2

Labor Applying Water and Dampproofing Paint to Exterior Surfaces Below Grade. Where the water or dampproofing consists of applying one or more coats of heavy paint to exterior concrete or masonry surfaces below grade, it is customary to use an ordinary roofer's brush or a mop. On foundation walls below grade, the workers are usually obliged to work in cramped quarters which reduces output.

To make these paints effective, they must be thoroughly brushed into the surface and all voids and pin holes filled, otherwise water and moisture will seep through.

On work of this kind, a worker should apply 600 to 700 s.f. (56-65 sq.m) per 8-hr. day, at the following labor cost per s.f. (sq.m):

	Hours	Rate	Total	Rate	Total
Roofer	1.3	$....	$....	$24.99	$32.49
Cost per s.f.			....		0.32
per sq.m			....		3.50

On the second and third coats, the labor cost will be slightly less than the first coat on account of the smoother surface to work on.

Dampproof and Plaster Bond Paints

Labor Applying Dampproofing and Plaster Bond Paints. Before applying the first coat of plaster to the interior of exterior brick or masonry walls, the walls are often painted with one or more coats of heavy black paint, which acts as a dampproofing and plaster bond.

The labor cost will vary with the consistency of the paint and the condition of the walls, as the paint must be well brushed in and all pin holes and porous places thoroughly covered with paint, otherwise moisture will seep through the walls.

Where the paint is applied by hand, a worker should cover 700 to 800 s.f. (65-75 sq.m) per 8-hr. day, at the following labor cost per 100 s.f. (9.3 sq.m):

	Hours	Rate	Total	Rate	Total
Roofer	1.1	$....	$....	$24.99	$27.49
Cost per s.f.			....		0.27
per sq.m			....		2.91

When dampproof paints are applied with an air spray, a worker should cover 300 to 350 s.f. (28-32 sq.m) per hr. at the following labor cost per 100 s.f. (9.3 sq.m):

	Hours	Rate	Total	Rate	Total
Roofer	0.33	$....	$....	$24.99	$8.25
Cost per s.f.			....		0.08
per sq.m			....		0.86

Dampproof and Plaster Bond Paints, Per 100 S.F. (9.3 Sq.M)

Heavy black paints are used for dampproofing interior of exterior walls and to form a bond for plaster. Number of coats required will depend upon porosity of wall on which it is used.

Labor costs in the following table are based on hand application. If an air spray is used, deduct as given above.

Name of Material	No. Coats	Sq. Ft. per Gal.	Gals. Reqd. 100 SF	Price per Gal.	Labor Hrs. Applying
Dehydratine 4	1	70-80	1-1/3	$1.43	1.1
Dehydratine 4	2	45-55	2-1/3	1.43	2.0
Plasterbond 232	1	75-80	1-1/3	1.47	1.1
Plasterbond 232	2	35-40	2-1/2	1.47	2.0

(metric)

Name of Material	No. Coats	Sq.M per Liter	Liters per 10 Sq.M	Price per Liter	Labor Hrs.
Dehydratine 4	1	1.7-2.0	5.4	$0.40	1.1
Dehydratine 4	2	1.1-1.3	9.5	0.40	2.0
Plasterbond 232	1	1.8-2.0	5.4	0.45	1.1
Plasterbond 232	2	0.9-1.0	10.2	0.45	2.0

Transparent Dampproofing of Exterior Masonry Walls

There has been a big change in the use of transparent dampproofings and the method of application due to an investigation by the National Bureau of Standards, Publications BMS7. It was determined that "leaky" walls were mainly caused by poor workmanship. In other words, unless the bricks were laid in a full bed of mortar (not furrowed), with all bed, end, and vertical joints completely filled with mortar, the walls were likely to leak within a short period after the completion of the structure. Hairline cracks between brick and mortar joints is also a cause of leakage.

It was determined that painting the surfaces of masonry walls with transparent "paints" or "waxes" alone does not stop the leaks. One method that was efficient was to cut out the old mortar joints to a depth of 1/2" to 3/4" (12.5-18.7 mm) and repoint the joints with new mortar.

This class of work is quite expensive, because the cost depends on the hardness and tenacity of the mortar joints.

The basis for the use of exterior waterproofing, labor and material, is calculated on the ordinary wall constructed with common brick of high absorption with 3/8" to 1/2" (9.3-12.5 mm) lime-cement mortar joints.

On lime-cement mortar joints, two masons can rake out and repoint about 12 s.f. (1.1 sq.m) of brick masonry surface per hour, removing joints to a depth of 3/4" (18.7 mm). Allow about 1/4 hr. helper's time for moving scaffolds, mixing, etc.

On the above basis, the labor cost of cutting out old mortar joints and repointing mortar joints to 100 s.f. (9.3 sq.m) of wall surface will cost:

	Hours	Rate	Total	Rate	Total
Mason	16.0	$....	$....	$28.18	$450.88
Helper	4.0	....	....	21.04	84.16
Cost per 100 s.f.			$....		$535.04
per s.f.				....	5.35
per sq.m				....	57.57

On hard portland cement mortar joints, two masons using a power saw can cut out and repoint joints in about 8 s.f. (0.72 sq.m) of wall per hour and requires 1/4 hr. helper time handling tools and mixing mortar.

On the above basis, the labor cost of cutting out old portland cement mortar joints, and repointing joints to 100 s.f. (9.3 sq.m) of wall will cost:

	Hours	Rate	Total	Rate	Total
Mason	24	$....	$....	$28.18	$676.32
Helper	6	....	....	21.04	126.24
Cost per 100 s.f.			$....		$802.56
per s.f.				....	8.03
per sq.m				....	86.40
Add cost of scaffold.					

Colorless or Transparent Liquid Water Repellent Treatments

There are a number of colorless or transparent liquid preparations on the market used on brick, stone, stucco, cement, and concrete surfaces to render them water repellent and help to prevent rain absorption.

These liquids are colorless and do not affect the original color of the surface to which they are applied.

The covering capacities and quantities given in the following table were furnished by the various manufacturers.

It will be noted that some manufacturers specify one coat and others two coats. Regardless of the material used, sufficient liquid must be applied to completely seal the pores of the surface. Otherwise, the results will be unsatisfactory.

Transparent liquids used for dampproofing brick, stone, stucco, concrete, and cement surfaces are:

Transparent Liquid Water Repellents,
per 100 s.f. (9.3 sq.m)

Name of Material	No. Coats	Sq. Ft. per Gal.	Gals. Reqd. 100 SF	Price per Gal.	Labor Hrs. Applying
Daracone	1	75-100	1-1/4	$4.05	1.6
Supertox	1	125-150	3/4	3.50	0.8
Tremco 141 Invisible	2	75-100	1-1/4	3.25	1.6

(metric)

Name of Material	No. Coats	Sq.M per Liter	Liters per 10 sq.m	Price per liter	Labor Hrs.
Daracone	1	1.8-2.5	4.1	$1.07	1.6
Supertox	1	3.1-3.7	3.1	0.92	0.8
Tremco 141 Invisible	2	1.8-2.5	4.1	0.85	1.6

Colorless or Transparent Silicone Base Liquid Water Repellent Treatments

Silicone base products are truly colorless and permit the masonry to "breathe" while rendering the surface water repellent. They penetrate deeply into the cement or masonry surfaces coating pores, cracks, and fissures with an insoluble, nonoxidizing film of silicone that effectively stops capillary action by which water is absorbed.

Usually one flooding coat is sufficient and can be applied by either brushing or spraying. The covering capacities and quantities given in the following table are furnished by the various manufacturers.

Silicone Base, Transparent Liquid Water Repellents,
per 100 s.f. (9.3 sq.m)

Name of Material	No. Coats	Sq. Ft. Per Gal.	Gals. Req'd. 100 SF	Price Per Gal.	Labor Hours Applying
Dehydratine 22	1	75-100	1-1/4	$3.40	0.8
Hydrocide Colorless SX	1	75-100	1	4.50	0.8
Sika Transparent	1	80-200	2/3	2.70	0.8
Tremco 147-3%	1	100-200	2/3	4.70	0.8
Tremco 147-5%	1	100-200	2/3	6.40	0.8
Devoe Super-Por-Seal	1	150	2/3	5.29	0.8

(metric)

Name of Material	No. Coats	Sq.M per Liter	Liters Req'd. 10 Sq.M	Price per Liter	Labor Hrs.
Dehydratine 22	1	1.8-2.5	5.1	$0.90	0.8
Hydrocide Colorless SX	1	1.8-2.5	4.1	1.20	0.8
Sika Transparent	1	2.0-4.9	2.7	0.70	0.8
Tremco 147-3%	1	2.5-4.9	2.7	1.25	0.8
Tremco 147-5%	1	2.5-4.9	2.7	1.70	0.8
Devoe Super-Por-Seal	1	3.7	2.7	1.40	0.8

Labor Applying Transparent Liquid Waterproofing. When applying transparent liquid dampproof paints by hand, a worker should cover 900 to 1,100 s.f. (84-102 sq.m) of surface per 8-hr. day, at the following labor cost per 100 s.f. (9.3 sq.m):

	Hours	Rate	Total	Rate	Total
Roofer	0.8	$....	$....	$24.99	$19.99
Cost per s.f.				....	0.20
per sq.m				....	2.15

On the second and third coats, a worker should apply 1,000 to 1,200 s.f. (93-111 sq.m) per 8-hr. day, at the following labor cost per 100 s.f. (9.3 sq.m):

	Hours	Rate	Total	Rate	Total
Roofer	0.75	$....	$....	$24.99	$18.74
Cost per s.f.				....	0.19
per sq.m				....	2.04

When applied by spray, a worker should cover 300 to 350 s.f. (27-33 sq.m) per hr. at the following labor cost per 100 s.f. (9.3 sq.m):

	Hours	Rate	Total	Rate	Total
Roofer	0.33	$....	$....	$24.99	$8.25
Cost per s.f.				....	0.08
per sq.m				....	0.89

Add cost of compressor and spray equipment.

Back-Plastering and Back-Painting
Cut Stone and Marble to Prevent Staining

When marble, limestone, brownstone, or sandstone are used in building work, it difficulty is to prevent moisture in the brick or concrete backing from penetrating the surface of the stone or marble and staining or discoloring the face, due to its open grain or porous nature.

To overcome this condition, it is customary to paint the back of the stone, beds, and end joints, up to within 1" (25 mm) of the face, to prevent staining and discoloration. White or non-staining cement is also used for this purpose. The stone is set in non-staining cement mortar, and the back of the stone is plastered or parged with the same mortar.

In many instances the backs, beds, and joints of the stone are painted and then back-plastered with non-staining cement, which acts as a double preventative.

Back-painting of Indiana limestone is not recommended by quarrymen and producers, who say that it tends to create conditions it is supposed to counteract. A full explanation is given under *Cut Stone*.

Where cut stone comes in contact with concrete, it is recommended that the concrete be painted instead of the stone and a one-inch space left between the concrete and cut stone.

Estimating the Cost of Back-Painting Cut Stone and Marble. It is customary to estimate the cost of cut stone by the cu. ft. (cu.m), so the cost of back-painting should be estimated in the same manner.

Plain stone ashlar is usually 4" to 8" (100-200 mm) thick or in alternating courses of 4" (100 mm) and 8" (200 mm) stone so as to provide a bond with the brick or masonry backing. Where just a veneer of cut stone is required, the stone is usually 4" (100 mm) thick and is bonded into the masonry backing by metal anchors.

If cut stone is furnished in alternating courses of 4" (100 mm) and 8" (200 mm), the stone for the entire job will average 6" (150 mm) thick. It will therefore be necessary to back-paint 2 s.f. of surface to each c.f. (0.2 sq.m per cu.m) of stone. Including beds and joints, it will double the above quantities or it will be necessary to back-paint 4-1/3 s.f. per c.f. (0.4 sq.m per cu.m) of stone.

If the stone averages 4" (200 mm) thick, it will be necessary to back-paint about 5 s.f. per c.f. (0.5 sq.m per cu.m) of stone, including the backs, beds, and joints.

If the stone averages 8" (200 mm) thick, it will be necessary to back-paint about 4 s.f. per c.f. (0.4 sq.m per cu.m) of stone.

Back-Painting Cut Stone and Marble. When back-painting limestone, sandstone, marble, etc., the stone is often painted in the yard before delivery, although on many jobs it is painted after delivery. It is customary to paint the stone a few days before setting, although the backs are frequently painted after the stone has been set.

Care must be exercised to thoroughly brush the paint into all the pores and not drop any paint on the face of the stone.

When painted at the building site, a worker should paint 600 to 700 s.f. (55-65 sq.m) of surface per 8-hr. day, at the following labor cost per 100 s.f. (9.3 sq.m):

	Hours	Rate	Total	Rate	Total
Mason	1.3	$....	$....	$28.18	$36.63
Cost per s.f.			....		0.37
per sq.m			....		3.98

Heavy black paints are applied to backs and sides of stone to prevent cement and mortar stains. Quantities are based on a single coat:

Back-Painting 100 S.F. (9.3 Sq.M) Cut Stone or Marble to Prevent Staining

Name of Material	Sq. Ft. per Gal.	Gals. per 100 SF	Price per Gal.	Labor Hours
Dehydratine 4	100	1.00	$1.43	1.3
Hydrocide No. 648	125-175	0.75	1.13	1.3
Sikaseal	120	0.80	2.20	1.3

(metric)

Name of Material	Sq.M per liter	Liters per 10 Sq.M	Price per liter	Labor Hours
Dehydratine 4	2.5	5.1	$1.10	1.6
Hydrocide No. 648	3.1-4.3	3.1	0.90	0.8
Sikaseal	3.0	5.1	0.85	1.6

No. of C.F. (Cu.M) of Stone to be Painted per 100 S.F. (10 Sq.M) of Surface
Includes back, beds and end joints
Average Thickness of Stone

4" (100 mm)	6" (150 mm)	8" (200 mm)	10" (250 mm)	12" (300 mm)
20 (0.61)	23 (0.70)	25 (0.76)	28.5 (0.87)	31 (0.94)

No. of S.F. of Painted Surface per C.F. of Stone
(No. of Sq.M Painted Surface per Cu.M of Stone)
Includes back, beds and end joints
Average Thickness of Stone

4" (100 mm)	6" (150 mm)	8" (200 mm)	10" (250 mm)	12" (300 mm)
5 (0.15)	4.3 (0.13)	4 (0.12)	3.5 (0.10)	3.2 (0.09)

Back-Plastering Cut Stone and Marble. After the stone or marble has been set, it is customary to plaster the back of the stone with non-staining cement, using an ordinary trowel and applying it 3/16" (4.6 mm) to 1/4" (6.2 mm) thick.

The stone must be back-plastered as the setting progresses, as not to delay the bricklayers in backing up the stone, so the time required is rather indefinite. Working steadily, however, a mason should back-plaster 65 to 75 s.f. (5.85-6.75 sq.m) per hr. at the following labor cost per 100 s.f. (9.3 sq.m):

	Hours	Rate	Total	Rate	Total
Mason or stone setter	1.5	$....	$....	28.18	$42.27
Cost per s.f.			....		0.42
per sq.m			....		4.52

07190 VAPOR BARRIERS/RETARDANTS

The combination of high indoor relative humidities and low outside temperatures causes condensation to form within the structure. Serious damage can result—rotting framing members, paint deterioration, and wet walls and ceilings.

Vapor barriers are recommended where these conditions exist. Vapor seal paper should be installed on the warm side of wall, floor, ceiling, or roof. Paper should be installed with joints running parallel to and over framing members. All joints should be lapped about 2" (50 mm). Where paper is to be exposed, nail wood lath strips over paper along framing members to provide a neat and permanent job.

A satisfactory vapor proof paper consists of a 50-lb. (22.7-kg) continuous asphalt film, faced on both sides with 30-lb. (13.6-kg) basis kraft paper. Other vapor proof papers consist of two 30-lb. (13.6-kg) basis sheets of kraft paper cemented together with asphalt and reinforced with strong jute cords, spaced 1/2" (12.5 mm) to 1" (25 mm) on centers and running in both directions. This same reinforced sheet is furnished with two sheets of asphalt coated paper cemented together and reinforced as described above. Another very satisfactory vapor proof paper consists of a heavy kraft paper coated on one side with a thin sheet of aluminum or copper, which prevents the penetration of moisture. Approximate prices on the various types of vapor seal paper are as follows:

	Price per Roll
Description	500 s.f. (46 sq.m)
2 sheets 30-lb (13.6-kg) basis kraft paper, cemented together with asphalt	$5.00
2 sheets asphalt coated 30-lb (13.6-kg) kraft paper, cemented together with asphalt and reinforced with jute cords	$9.00
1 sheet heavy kraft paper coated two sides, with a reflective surface	$12.00

Sisalkraft. A strong, waterproof, windproof building paper consisting of 2 sheets of pretreated kraft paper, cemented together with 2 layers of special asphalt and reinforced with 2 layers of crossed sisal fibers. Costs about $0.20 per s.f.

Copper-Armored Sisalkraft is a combination of Sisalkraft and Anaconda electro-sheet copper bonded under heat and pressure. It is available in 3 weights, 1, 2, or 3 oz. copper per s.f. (0.91, 1.8, or 2.7 oz. per sq.m).

It may be used for waterproofing, flashings, ridge roll flashing, etc. The approximate prices are as follows: 1-oz. costs $0.30 per s.f. ($3.22 per sq.m); 2-oz. costs $0.50 per s.f. ($5.38 per sq.m), and 3-oz. costs $0.70 per s.f. ($7.53 per sq.m) Above prices subject to discount on large orders.

Polyethylene. Film comes in 100' (30 m) rolls of various widths in thicknesses of 0.002" (0.05 mm), 0.004" (0.1 mm), 0.006" (0.15 mm), and 0.008" (0.2 mm). Costs vary from $0.015 to 0.04 per s.f. ($0.16-0.43 per sq.m). Tape for sealing joints costs $3.25 for a roll 2" (50 mm) wide by 100' (30 m) long in the 0.004" (0.10 mm) thickness.

Labor Placing Vapor Seal Paper. Where vapor seal paper is applied to the interior of exterior stud walls, using a stapling machine, to prevent the penetration of moisture and condensation, a carpenter should handle and place 2,000 to 2,500 s.f. (185-232 sq.m) per 8-hr. day, at the following labor cost per 100 s.f. (9.3 sq.m):

	Hours	Rate	Total	Rate	Total
Carpenter	0.4	$....	$....	$26.44	$10.58
Cost per s.f.			....		0.11
per sq.m			....		1.18

07200 INSULATION

To insulate is defined as "to make an island of". This "island" not only excludes the outside elements but retains those created within. Today, insulation has become part of a larger, more inclusive concern—energy conservation. Building design must go beyond the casual approach of letting aesthetic and use-function considerations dictate the design and then turning to technology to provide a satisfactory interior climate regardless of the initial and operating cost. The building envelope itself must be modified to contribute in every possible way to interior conditions of comfort—excluding the extremes of weather but welcoming those that pleasantly modify; retaining our manufactured interior weather but also providing means to allow it naturally to renew and maintain itself without resorting exclusively to mechanical means.

There are three stages where crucial decisions are made about the building and its environment. First, in the preliminary planning; second, in material selection; and finally, in the actual construction.

Preliminary planning considers the basics. First the building form should approximate a cube and have a minimum exposure to the elements or sprawl and take maximum advantage of natural light, solar heat, and breezes. Buildings should be sited to work with nature, not fight it. Fenestration should be concentrated to the south and tempered with overhangs engineered to exclude the sun when it is high in the summer sky, but to allow the rays to penetrate and add their warmth when the sun is low in the winter. Overhangs need not always be eaves but can be louvres, screens, balconies, porches, or a few well-spaced deciduous trees. Roof forms should not be chosen solely to exclude the elements but also to shade and vent the space above the top habitable floor. Wind conditions should be considered, doors protected, and windows placed to provide the optimum natural ventilation supplementing them with clear stories and skylights if need be. The earth itself is both a very cheap and a very good insulator and intelligent use of banking can pay large dividends.

Once the general arrangement has been determined, the second consideration is to select the best wall, ceiling, and floor materials for coping with the conditions they will face, and the best methods of installation to assure that the selected materials will perform to their maximum capabilities.

This discussion has concerned itself with new construction. One of the largest opportunities developing for the contractor is bringing existing buildings up to the energy efficient standards acceptable today. Most of the materials and procedures discussed in the following text apply to both new and remodeling work. However, the latter presents its own problems, and for that type of work, the reader is also referred to the chapter on *Remodeling Work* which deals with work on existing buildings.

The exterior and interior faces of a building will be chosen for appearance and wearability, and will by nature be hard and dense and therefore good conductors of heat. Also by nature insulating materials will be full of air pockets and be soft and easily damaged and need protection. Therefore most exterior envelopes will consist of three plies: an outside wearing surface, and intermediate space designed to interrupt the heat flow, and an interior wearing surface.

Even the exterior and interior plies will contribute some insulating quality. Materials are rated for their thermal resistance or their "R" value, the temperature difference between two exposed faces required to cause one BTU to flow through one square foot of the material per hour. "R" values can be added, one to the other, to arrive at a rating for the total wall, ceiling, or floor construction. The following list gives the "R" ratings for some of the more common building materials.

Material	R Value
4" (100 mm) Face brick	0.44
4" (100 mm) Common brick	0.80
8" (200 mm) Poured concrete	0.64
8" (200 mm) Cinder block	1.11
8" (200 mm) Light weight block	2.00
4" (100 mm) Concrete slab	0.32
1/2" (12.5 mm) Plywood	0.47
1/2" (12.5 mm) Gypsum board	0.45
1/2" (12.5 mm) Gypsum plaster	0.32
1x8 (50 x 200 mm) Drop siding	0.79

Even from the above small sampling, one can see that facing materials can be chosen to contribute to the insulating quality of the envelope. In addition, air spaces of over 3/4" (18.7 mm) should also be added at an average of R = 0.91, depending on the position of the air space and whether the air movement is up or down. Also in choosing exterior facings, colors should be considered for their ability to reflect; the darker the color, the more heat absorbed. And finally the jointing of all materials must be carefully studied to eliminate infiltration. A study of one apartment tower found 50% of the heat loss was due to infiltration around the windows!

Heat transfer through the building enclosure is by three means: convection, conduction, and radiation.

Convection is the thermally produced upward and downward movement of air. Warm air rises, cold air falls. The concern here is to construct a total building envelope that blocks air currents. Hollow walls should be solidly blocked top and bottom at each story, and floor and roof construction should be continuously sealed from wall construction so air conditions in the exterior walls will not be able to spread out over the interior. This can be done either by extending the interior facing or by adding insulation or a combination of both.

Conduction is the transmission of heat through a material. The rate of conductance of a material or combination of materials, known as its "U" factor, is the BTU per hour per inch of thickness per square foot per degree temperature difference. The "R" ratings discussed previously are the reciprocals of conductance or 1 divided by the "U" factor. Most common insulating materials are manufactured to be as poor a conductor of heat as possible. However, their positioning within the wall can also make them block convection and when enclosed in reflective coverings and separated from adjacent construction by at least 3/4" (18.7 mm) they also can cut down on the third means of heat transfer, radiation.

Radiation is the emission of energy from a surface. Bright surfaces, such as aluminum foil, are good reflectors and have low emissive coefficients and are therefore poor absorbers of heat. The interior faces of the two outer plies of building envelope tend to be dark, and the heat will radiate from one surface to the other constantly unless interrupted. By inserting one or more layers of reflective surfaces, the heat will be reflected back to the surface from which i' escaped. Installation is crucial to the success of a radiant barrier, because the reflective surface will be highly conductive. If allowed to come into contac with either side of the enclosing construction, it will speed heat transfer rather than retard it.

Infiltration, as discussed above, will generally be solved by good detailing at the juncture of one material to another and by caulking. Strips of insulation, such as under sill plates, and loose wools stuffed into place, can also plug air leaks.

There is no simple answer as to how much insulation should be added. Both comfort and the additional costs must be considered. Minimum accepted "R" values in the northern regions have been R19 for ceilings, R13 for floors and R11 for walls. These drop to R13, R9, and R8 in milder areas but should never be less than R9 for ceilings and R7 elsewhere. Increases in fuel costs could always make these figures far too low and R24 for ceilings, R19 for floors and R13 for walls might be recommended. But keep in mind that each additional inch (25 mm) of insulation beyond a certain point reduces the heat loss less than the one before it. There is a point of sharply diminishing returns. To determine the best "R" values for a particular project, one must consider the following:

1. Determine all wall, ceiling, and floor construction types and give each its proper "R" value without insulation.
2. Determine the heat loss (or heat gain for cooling) without insulation through each of the above.
3. Choose an "R" rating for walls, ceiling, and floors based on averages for existing climate and consistent with the type of construction contemplated.
4. Refigure heat loss with new "R" values.
5. Figure fuel savings of d) over b). Several alternate fuels might be considered and local utilities can be helpful in determining these costs.
6. Figure the difference in cost of the insulated construction over the uninsulated, remembering to deduct for savings in size of the mechanical equipment, if any.
7. Determine how long it will take savings in fuel to pay for insulation costs. This is the payback period and will have to take many things into consideration such as future fuel costs, and the interest on the additional mortgage to finance the insulation. If the payback period looks attractive, the whole procedure can be repeated with a higher "R" value assigned, and this can be repeated until a point of diminishing returns is evident.

Once the desired "R" value is established, the selection of the best type of insulation can begin. The possibilities include:

1. Blankets and batts, with or without reflective or vapor barriers.
2. Rigid board type, which may also serve as sheathing or lathing.
3. Sprayed on or foamed in place, which may also serve as sound or fire retardants.
4. Poured fill
5. Reflective barriers

Many of these may be used in combination with the other.

07210 BUILDING INSULATION BLANKET AND BATT INSULATION

Insulation formed into batts or rolls is made in a number of materials, such as stone, slag, glass wools, vegetable and cotton fibers, and suspended pulps. They are all light in weight, easily cut and handled, and need support to stay in place. They are usually encased in either kraft paper or foil and can be used as a vapor barrier. The enclosing material is usually made with flapped edges for easy attachment to studs and joists. Units may be ordered unfaced and fitted between studs by friction. Unfaced units are most often used as a second layer of insulation in attic spaces as they do not form a second vapor barrier if the original layer has one. A narrow form of unfaced blanket 1" (25 mm) thick is made for inserting under sill plates. It will compress to 1/32" (0.78 mm).

Most batt and blanket materials have "R" values of around 3.5 to 3.7 per inch (25 mm) of thickness. The 3" (75 mm) units will provide an "R" of 11, 3-1/2" (87.5 mm) of 13 and 6" (150 mm) of 19. The 6" (150 mm) units are commonly used in attics and fitted between the joists; but more and more, they are used in walls with 2x6 (50 x 150 mm) studs placed 24" (600 mm) o.c. in place of the standard 2x4 (50 x 100 mm) 16" (400 mm) o.c. If foil faced units are used and an air space of at least 3/4" (18.7 mm) maintained between the foil and the warm side of the room, the "R" rating can be increased some 15%.

Most glass fiber insulation is distributed on a nationwide basis, but other materials are often produced regionally and prices may vary widely. Widths are 15" (375 mm) and 23" (575 mm) to fit normal 16" (400 mm) and 24" (600 mm) stud and joist spacing. Batts come 48" (1200 mm) and 96" (2400 mm) long, blankets in rolls of from 24' (7.3 m) to 40' (12.2 m), often two rolls per package. The square foot cost for foil faced units 3-1/2" (87.5 mm) thick is $0.16; 6" (150 mm) thick is $0.26. Units bound in kraft paper may run a couple of cents less and unfaced units even less. Sill sealer rolls come in 50' (15 m) rolls 1" (25 mm) thick and in 3-5/8" (90 mm) and 6" (150 mm) widths and cost $0.10 and 0.15 per l.f.

Because these materials are light and precut to fit standard construction, one carpenter can handle the installation. Friction fit type, with no attachments, will install at the rate of 2000 s.f. (186 sq.m) per day. Flanged type bound in kraft paper or foil will install at 1800 s.f. (167 sq.m) per day.

Labor Cost of 100 s.f. (9.3 sq.m) Faced Insulation

	Hours	Rate	Total	Rate	Total
Carpenter	0.45	$....	$....	$26.44	$11.90
Cost per s.f.			....		0.12
per sq.m			....		1.29

07211 LOOSE FILL INSULATION

Loose fill insulation includes mineral wool, which is molten rock extruded by air and steam into fibers, known as blowing wool for machine applications, or nodules for pouring or spreading by hand; and expanded volcanic rocks such as vermiculite or perlite. The latter are most often specified for filling concrete block or cavity walls.

Loose Insulating Wool. Is suitable for any purpose where insulation can be packed by hand between ceiling joists or side wall studding. For maximum results, it is recommended that wool be applied full thickness of side wall studding and approximately 4" (100 mm) to 16" (400 mm) over ceiling areas.

The covering capacity varies considerably with the density to which it is packed. The National Bureau of Standards conductivity figure for glass or rock wool is 0.27 Btu at a density of 10 lbs. per c.f. (162 kg per cu.m).

The covering capacity of bulk wool as given by most manufacturers is based on a density of 6 to 8 lbs. per c.f. (96-128 kg per cu.m).

No. of S.F. (Sq.M) of Surface Covered By One Bag of Loose
Insulating Wool Weighing 40 Lbs. (18 Kg) and Containing 4 C.F. (0.11 Cu.M)
Not Including Area Covered By Studs or Joists

Density, Lbs. per CF	Cu. Ft.	ActualThickness of Loose Insulating Wool in Inches					
		1	2	3	3-1/2	3-5/8	4
6	6.67	80.0	40.0	26.6	22.8	22.0	20.0
7	5.72	68.5	34.4	22.9	19.7	18.8	17.2
8	5.00	60.0	30.0	20.0	17.2	16.6	15.0
9	4.45	53.4	26.7	17.8	15.2	14.7	13.4
10	4.00	48.0	24.0	16.0	13.7	13.3	12.0
Density, kg per cu.m	cu.m	ActualThickness of Loose Insulating Wool, mm					
		25	50	75	87	90	100
96	0.19	7.4	3.7	2.5	2.1	2.0	1.9
112	0.16	6.4	3.2	2.1	1.8	1.7	1.6
128	0.14	5.6	2.8	1.9	1.6	1.5	1.4
144	0.13	5.0	2.5	1.7	1.4	1.4	1.2
160	0.11	4.5	2.2	1.5	1.3	1.2	1.1

No. of S.F. (Sq.M) of Surface Covered By One Bag of Loose Insulating
Wool Weighing 40 Lbs. (18 Kg) and Containing 4 C.F. (0.11 Cu.M)
Including Area Covered By Studs or Joists

Density, Lbs. per CF	Cu. Ft.	ActualThickness of Loose Insulating Wool in Inches					
		1	2	3	3-1/2	3-5/8	4
6	6.67	85.0	42.5	28.4	24.3	23.5	21.2
7	5.72	73.0	36.5	24.3	20.8	20.0	18.3
8	5.00	63.8	31.9	21.3	18.3	17.6	16.0
9	4.45	56.7	28.4	18.9	16.2	15.7	14.2
10	4.00	51.0	25.5	17.0	14.5	14.1	12.7
Density, kg per cu.m	cu.m	ActualThickness of Loose Insulating Wool, mm					
		1	2	3	3-1/2	3-5/8	4
96	0.19	7.9	3.9	2.6	2.3	2.2	2.0
112	0.16	6.8	3.4	2.3	1.9	1.9	1.7
128	0.14	5.9	3.0	2.0	1.7	1.6	1.5
144	0.13	5.3	2.6	1.8	1.5	1.5	1.3
160	0.11	4.7	2.4	1.6	1.3	1.3	1.2

Loose or bulk rock wool is usually furnished in bags weighing 40 lbs. (18
kg) and containing 4 c.f. (0.11 cu.m). Average price $3.00 to $4.50 per bag.

Granule or Pellet Type Insulating Wool. Granule or pellet type
insulating wool is furnished in particles sufficiently large that the material
will not sift or dust through wall or ceiling cracks and is used for insulating
ceiling areas between supporting joists, or void wall spaces where accessible
and which permit pouring the pellets into place, particularly in old house
construction.

Granule or pellet type insulating wool is usually furnished in bags
weighing 40 lbs. (18 kg) and is placed at a density of 7 to 8 lbs. per c.f.
(112-128 kg per cu.m).

The tables for loose or granular insulating wool, based on 7 to 8 lbs. per c.f.
(112-128 kg per cu.m) can be used for estimating quantities obtainable per
bag.

Granule or pellet type insulating wool in bags or cartons weighing 40 lbs.
(18 kg) cost about $4.00 each.

Labor Placing Loose or Bulk Insulating Wool. When placing loose or
bulk insulating wool between wood studs, a worker should place about 350 to
450 s.f. (32.5-41.8 sq.m) per 8-hr. day, at the following labor cost per 100 s.f.
(9.3 sq.m):

	Hours	Rate	Total	Rate	Total
Labor	2	$....	$....	$26.44	$52.88
Cost per s.f.			....		0.53
per sq.m			....		5.70

Labor Placing Granule or Pellet Type Insulating Wool Between Ceiling Joists. Where granule or pellet type insulating wool is poured between ceiling joists and other open spaces to a thickness of 3-1/2" (87.5 mm) or 3-5/8" (90.6 mm), a worker should place 700 to 900 s.f. (65-84 sq.m) per 8-hr. day, at the following labor cost per 100 s.f. (9.3 sq.m):

	Hours	Rate	Total	Rate	Total
Carpenter	1	$....	$....	$26.44	$26.44
Cost per s.f.			....		0.26
per sq.m			....		2.80

Vermiculite Or Perlite Loose-fill Building Insulation. Vermiculite loose-fill insulation is used to insulate attics, lofts, and side walls. It is fireproof, not merely "fire resistant". The fusion point is 2200° to 2400°F (1200° to 1315°C). It is completely mineral and does not decompose, decay, or rot. Because of its granular structure, vermiculite is free-flowing, assuring a complete insulation job without joints or seams. Rodents cannot tunnel into it, and it does not attract termites or other vermin. Vermiculite is a non-conductor and is excellent protection around electrical wiring.

Vermiculite loose-fill is marketed in 4-c.f. bags. Approximate coverage per bag, based on joists spaced 16" (400 mm) o.c., is 14 s.f. (1.3 sq.m) for a 3-5/8" (90.6 mm) thickness; 9 s.f. (0.83 sq.m) for 5-1/2" (137.5 mm) thickness. A bag will cost about $7.00.

Approximate Coverage of Vermiculite
Fill in Cavity and Block Walls

Wall area, s.f.	Wall area, sq.m	No. of Bags Required for Various Wall Types				
		1" (25mm) cavity	2" (50mm) cavity	2.5" (62mm) cavity	8" (200mm) block	12" (300mm) block
100	9	2	4	5	7	13
500	46	10	20	25	34	63
1,000	93	21	42	50	69	125
5,000	464	104	208	250	545	625
10,000	929	208	416	500	1,090	1,250

1 bag = 4 c.f.

One mason can pour around 50 bags, 200 c.f. (5.7 cu.m) per 8-hr. day, at the following labor cost:

	Hours	Rate	Total	Rate	Total
Mason	8	$....	$....	$28.18	$225.44
Cost per c.f.			....		1.13
per cu.m			....		12.16

When adequate ventilation is provided in attics or similar spaces, it is not necessary to use a vapor barrier in the ceiling construction below vermiculite fill insulation. A vapor barrier is recommended on the warm side of insulated exterior walls, except when the insulated space is ventilated to permit movement of water vapor and air from the space.

Reinsulation. Where an existing thickness of blanket or other insulation is inadequate, vermiculite can be poured over it to obtain the thickness desired.

07212 RIGID INSULATION

Rigid insulation boards are made of expanded polystyrene, often called "beadboard"; extruded polystyrene sold as Styrofoam; urethane; glass fibers; glass foams; and wood and vegetable fibers. Many rigid boards serve dual purposes such as sheathing, lath, and even interior finish. Many are also used for roof decks and perimeter insulation, discussed separately below.

Urethane boards have the highest "R" value per inch (mm) of thickness, 7.14, but are flammable and must be covered over, or treated. The cost is around $0.30 per s.f. ($3.22 per sq.m) plain, $0.40 per s.f. ($4.30 per sq.m) treated. The sizes available are many, including the standard 4'x8' (1200 x 2400 mm) sheet and in thicknesses from 1/2" (12.5 mm) to 24" (600 mm). Because of its high insulating value per thickness, it is ideal for use in cavity walls and for applying to the inside face of an exterior masonry wall. In cavity walls it is applied against the outside face of the inside wythe, leaving a clear air space for any penetrating water to drain out.

Where urethane board is used against exterior masonry walls, the surface must be clean and even. Nailers must be applied wherever wood trim will be added, such as at the base and around doors and windows. The board is then applied to the wall with sufficient mastic to cover at least 50% of the back when shoved in place. Some manufacturers supply special channel systems for mechanical attachment as an alternative. The urethane board can then be covered with plasterboard or may be plastered direct. Because of the fire hazard, for whichever method chosen, the interior finish coat must be run up above the finished ceiling to cover all portions of the urethane, as it is the fireproofing for it. Extruded polystyrene, or *Styrofoam*, under which name it is marketed by Dow Chemical Company, also has a high "R" value, 5.4 per inch (25 mm) of thickness. It costs about $0.35 per s.f. ($3.75 per sq.m) in 1" (25 mm) thickness. It is installed in the same locations and by the same methods as urethane board and should also have a continuous interior finish that will protect it from fire.

Polystyrene, which is molded, is known as beadboard. It has a lower "R" rating, 3.85, and is absorbent, but it is much cheaper, about $0.15 per s.f. ($1.61 per sq.m) in 1" (25 mm) thickness. It too must be covered with a fireproofing finish.

Glass fiber boards come in various densities. A 1" (25 mm) board in a medium weight gives an "R" of 4.35 and costs about $0.35 per s.f. ($3.76 per sq.m).

The stronger rigid boards may also be used for sheathing on wood frame construction. But a 1" (25 mm) board in combination with standard 3-1/2" (87.5 mm) batts will give a wall with the highly desirable rating of R19 or better and may be preferable to using 6" (150 mm) batts and 6" (150 mm) studs to gain such a rating.

Rigid board is light, easily cut, and can be installed by one worker, who should erect about 1,500 s.f. (135 sq.m) per day depending on the amount of fitting.

Labor Cost of 100 s.f. (9.3 sq.m) of 1" (25 mm) Rigid Board
Applied to a Masonry Wall

	Hours	Rate	Total	Rate	Total
Mechanic	0.54	$....	$....	$26.44	$14.28
Cost per s.f.				....	0.14
per sq.m				....	1.51

07213 REFLECTIVE INSULATION

Some reflective insulation can be incorporated as the enclosing sheet on batt and blanket insulation and as a backing on rigid insulation. Also available are sheets of insulating material depending entirely on the reflective principle and consisting of alternate layers of trapped air and foil, the outer layers of which are usually laminated to a backing sheet of kraft paper for extra strength and which are extended at the edges to form flaps for stapling. Another type depends on an inner layer or layers of accordion foils. Such insulation must be held away from the inner and outer wall finishes at least 3/4" (19 mm) so that there will be an uninterrupted air space which serves the dual purpose of allowing the foil surface to reflect escaping heat (or entering cold) back to where it came from and keeping the highly conductive foil from transmitting the heat by contact.

A typical lamination would consist of three layers: the two outer sheets of foil laminated to a backing sheet and separated from a center sheet of pure foil, usually aluminum, with sealed air spaces both sides. The unit is held together with flanges extended to allow stapling to studs and joists. The "R" value will depend on the placement of the material, because reflective insulation is more effective with the heat flow down than with heat flow up. Where the heat flows down from a hot roof to a ceiling below, 93% of the heat transfer is by radiation and only 7% by conduction. As heated air always travels up, no convection is directed toward the ceiling. Considering the reverse travel, heated interior air travelling through the ceiling to the cold roof above, 50% will be transmitted by radiation, the same 7% by conduction and the remainder by convection. Heat transfer through the side walls will be 65% to 80% by radiation, 15% to 28% by convection, and 5% to 7% by conduction. The "R" rating then on a three-ply, 2" (50 mm) thick reflective sheet will be 26.18 down, 9.52 up and 14.65 through the wall for a mean "R" of 17.85. When we speak of this as 2" (50 mm) thick, it is the distance from foil to foil. Another 3/4" (19 mm) minimum must be added to each side to isolate the foil from other wall construction so that a full 3-1/2" (87.5 mm) must be reserved for the reflective insulation.

One-inch (25 mm) thick units with just one inner and one outer layer of reinforced foil separated by a 1" (25 mm) air space will give an "R" value of 21.79 down, 7 up, and 10.72 through the wall for a mean "R" of 14.4.

Single sheets with aluminum foil either side of a reinforcing sheet with no integral air space, if properly installed to preserve air spaces to adjacent construction, will give "R" values of 14.28 down, 3.45 up, and 7 through the wall, for a mean "R" of 9.

The cost of two-ply material is $0.12 per s.f. ($1.29 per sq.m); three-ply is $0.14 per s.f. ($1.50 per sq.m); and five-ply is $0.25 per s.f. ($2.69 per sq.m).

The material is preformed to fit average stud and joist spacing and is very light. One worker can install 2200 s.f. (204 sq.m) per day if the wall or ceiling space is clear, at the following labor cost per 100 s.f. (9.3 sq.m):

	Hours	Rate	Total	Rate	Total
Carpenter	0.66	$....	$....	$26.44	$17.45
Cost per s.f.			....		0.17
per sq.m			....		1.88

07214 FOAMED INSULATION

Foamed-in-place insulations are very efficient, and the work is usually subcontracted to an experienced applicator. Foamed insulation will adhere well to most surfaces, will conform to irregular forms, and will set up quickly. The plastic foams flow better into restricted areas than the sprayed-on fibers, which have longer fibers and tend to clog in small spaces.

Application will require special equipment, and the space must be well ventilated. Some foams expand and will set up stresses in restricted areas if care is not taken. Despite these drawbacks, the resulting "R" can be as much as 7.7 per inch (25 mm) and will apply at the rate of 800 s.f. (74 sq.m) per day with a three-worker crew. Material costs for urethane will run about $0.22 per inch per s.f. ($2.36 per 25 mm per sq.m). If subcontracted, a budget figure of $1.00 per s.f. ($10.75 per sq.m) for 1" (25 mm) and $1.65 per s.f. ($17.75 per sq.m) for 2" (50 mm) should be adequate.

07215 SPRAYED INSULATION

Sprayed fibers of rock wool, glass, gypsum, perlite, and cellulose will have "R" values in the 4 to 5 range. Special binders are required as well as special equipment, so this work, too, will be subcontracted. A two-worker crew will apply some 1200 s.f. (111 sq.m) of 1" (25 mm) thickness per day, possibly more on wide open ceilings. Material costs will be about $0.20 to 0.25 per s.f ($2.15-2.70 per sq.m) per 1" (25 mm) of material. If subcontracted, a budget figure of $1.20 per s.f. ($12.90 per sq.m) should be adequate.

07230 HIGH AND LOW TEMPERATURE INSULATION

Rigid foam insulation is composed of expanded polystyrene, glass, or polyurethane, which bound together into rigid board, has largely replaced corkboard for low temperature insulation. It is also used in general building for roofs, because it is quite strong, tough, and resilient; for curtain wall back up because it is lightweight; for core walls as it is self-supporting; and for perimeter and crawl space insulation as it has a low water absorption and a zero rating for capillary action. In addition, it forms a vapor barrier, is dimensionally stable, incombustible, and easily cut. Rigid foam insulation can be placed on concrete and masonry surfaces by adhering units with hot asphalt or with gobs of cement.

Units come in 1" (25 mm), 1-1/2" (37.5 mm), 2" (50 mm), 3" (75 mm), 4" (100 mm), and 6" (150 mm) thicknesses and sizes from 12"x18" (300 x 450 mm) to 36"x36" (900 x 900 mm) in packaged forms and up to 9' (3 m) lengths when shipped in carload lots. There are no standard sizes within the industry, and sizes, costs, and insulating values must be carefully checked with the manufacturers.

Back Plaster. Where insulation is to be applied against masonry walls in hot asphalt, the walls must be made true and even with portland cement plaster. The parge coat or back plaster may be omitted if the wall is of poured concrete.

Back plaster should be made by mixing 1 part portland cement and 3 parts clean sharp sand. Back plaster 1/2" (12.5 mm) thick requires 1.5 bags portland cement and 4.6 c.f. (0.13 cu.m) of sand per 100 s.f. (9.3 sq.m).

Material Cost of One C.Y. (1.76 Cu.M) of 1:3 Portland Cement Back Plaster

	Rate	Total	Rate	Total
9.5 sacks portland cement	$....	$....	$4.00	$38.00
1.10 c.y. (0.84 cu.m) sand		....	6.50	7.15
Cost per c.y.		$....		$45.15
per c.f.		....		1.67
per cu.m		....		58.97

Asphaltic Priming Coat. All concrete or plaster surfaces to which insulation is to be applied in hot asphalt must be primed with asphaltic priming coat made especially for this purpose.

Asphalt. Using the proper grade of asphalt is very important. It must be odorless, as any material with an odor will taint foodstuffs that may be stored in insulated rooms. It should be an oxidized asphalt of 180°F to 200°F (82°C to 93°C) melting point and should preferably be made from a Mexican base oil. The hot asphalt in which corkboard is dipped should be held between 350° and 375°F (176°C and 190°C). Do not heat asphalt to higher temperatures than these. Most manufacturers recommend an asphalt for use with their product.

On dipping floor, wall, and ceiling insulation, allow 0.4 lbs. of asphalt per s.f. (1.95 kg per sq.m) of insulation per layer. Mop coating on floors or top of ceiling requires 0.8 lbs. per s.f. (3.90 kg per sq.m).

Cold Erection Plastic. Where the use of hot asphalt for the erection of foamboard is not practical, use cold erection plastic on floors, walls, and ceilings. A coat 1/16" (1.5 mm) thick requires 1 gal. (3.8 liters) of cold erection plastic per 25 s.f. (2.3 sq.m).

Nails or Impregnated Skewers. Use nails for the attachment of the first layer of insulation to wood and frame construction, for the attachment of the second layer to wood stripped concrete ceilings, or for toenailing the first layer of solid cork walls. All nails should be of galvanized wire and should be 1" (25 mm) longer than the thickness of the insulation. Where two courses of insulation are erected in hot asphalt or cold erection plastic, impregnated hardwood skewers should be used to additionally secure the second course in position. Do not use either nails or skewers when erecting the first course against brick, stone, or concrete walls. Hot asphalt or cold erection plastic is all that is required.

Erection. The requirements (material and labor) for installing one layer of 2" (50 mm) thick insulation, using the hot asphalt dip method of setting, on a masonry wall surface are given in the following tables. Bear in mind that the labor times given are for straight, flat surfaces; and that any irregularities such as columns, pilasters, or beams will require additional labor time due to cutting and setting smaller pieces. If the insulation is thicker than 2" (50 mm), then adjust for the difference in material costs and add about 1/4 hour to both mechanic and helper time per 1" (25 mm) thickness increase over 2" (50 mm).

2" (25 mm) Thick Single Insulation Layer, Hot Dip Method
Requirements per 100 s.f. (9.3 sq.m)

Operation	Material Required	Mechanic Time	Helper Time
Masonry Wall			
Parge Coat- 1/2" (12.5mm)	5.5 c.f. (0.15 cu.m) mortar	3.0 hrs.	3.0 hrs.
Asphalt Prime Coat	1 gal. (3.8 L) primer	1.0	...

Hot Asphalt............................... 40 lbs. (18 kg) ....
Applying Insulation .... 100 s.f. (9.3 sq.m) ....1.5 hrs........ 1.5 hrs.
For applications against concrete walls, the parge coat is not required.

Low Temperature Insulation Using Two Layers Of Pittsburgh Corning Foamglas

Material Cost of 100 S.F. (9.3 Sq.M) of Single Layer Insulation, Set In Hot Asphalt on Masonry Walls

	Rate	Total	Rate	Total
5.5 c.f. (0.15 cu.m) cement plaster	$....	$....	$1.67	$9.19
1 gal. (3.8 L) asphalt primer	....		2.00	2.00
40 lbs. (18 kg) asphalt	....		0.13	5.20
100 s.f. (9.3 sq.m) 2" (50mm) corkboard	....		0.97	97.00
Cost 100 s.f.				$113.39
per s.f.				1.13
per sq.m				12.16

Labor Cost of 100 S.F. (9.3 Sq.M) of Single Layer Insulation, Set in Hot Asphalt on Masonry Walls

	Hours	Rate	Total	Rate	Total
Plaster Coat					
Mechanic	3.0	$....	$....	$25.75	$77.25
Helper	3.0	....	....	21.04	63.12
Prime Coat					
Mechanic	1.0	....	....	26.44	26.44
Insulation					
Mechanic	1.5	....	....	26.44	39.66
Helper	1.5	....	....	21.04	31.56
Cost 100 s.f.		$....			$238.03
per s.f.			....		2.38
per sq.m			....		25.61

Finishes. Corkboard insulation should receive a finish that will give long service under the temperature and moisture conditions common to cold storage rooms. The finish should bond securely to the corkboard and be reasonably resistant to bumps and abrasion. Cost, good appearance, and availability should also be considered. These requirements are best met by asphalt emulsion finishes or portland cement plaster. Asphalt emulsion finishes are generally used on ceilings and portland cement plaster on walls. Rigid foam insulations may be sprayed or brushed with special white vinyl emulsion paints, aluminum paint, tile set with adhesive, or cement plaster over mesh. Direct plastering should not be considered.

Asphalt emulsion finishes shall be applied directly to corkboard in two coats, each 1/8" (3.1 mm) thick. Before applying the first coat, all voids, open joints, or broken corners must be pointed with emulsion finish. The first coat must be hand dry before the second coat is applied. Trowel the second coat smooth. The final thickness of the finish shall be approximately 1/8" (3.1 mm) when dry. Finish all corners true and clean.

The emulsion finishes must be thoroughly dry before any paint is applied, to guard against cracking.

Approximately 5 gals. (19 liters) of asphalt emulsion, at a material cost of about $8.50, will be required for the application of the two coats on 100 s.f. (9.3 sq.m) of surface, at the following labor costs:

	Hours	Rate	Total	Rate	Total
First Coat					
Mechanic	2	$....	$....	26.44	$52.88
Second Coat					
Mechanic	2	....	....	26.44	52.88
Cost 100 s.f.			$....		$105.76
per s.f.				....	1.06
per sq.m				....	11.41

The labor cost of 100 s.f. (9.3 sq.m) of 1/2" (12.5 mm) of 1:3 portland cement plaster should average as follows:

	Hours	Rate	Total	Rate	Total
Mechanic	3	$....	$....	$25.75	$77.25
Helper	3	....	....	21.04	63.12
Cost per 100 s.f.	6		$....		$140.37
per s.f.				....	1.40
per sq.m				....	15.06

Portland cement plaster should be mixed in the proportions of 1 part portland cement, 3 parts clean, screened sand, and 5% hydrated lime.

To reduce cracking to a minimum, finish coat of plaster may be scored each way approximately 4'-0" (1200 mm) on centers.

Some manufacturers state that plaster can be applied directly to insulation; others specify wire mesh must be applied, particularly on ceiling work. As rigid foam insulation does not absorb water, plaster coats will dry very slowly. It is also necessary to add the cost of nailer strips for casings, trim, and bases, if any occur. Light trim such as beads may be stapled on.

07240 ROOF AND DECK INSULATION

Roof and deck insulation is often of the same material as rigid board insulation, but the units are generally furnished in smaller sizes more suitable for handling in exposed conditions. In addition to glass fiber, perlite, urethane and polystyrene boards, wood and mineral fiber boards costing as little as $0.25 per s.f. per 1" of thickness ($2.69 per sq.m per 25 mm), and foam glass blocks are often specified. Poured decks are also common, but these have been covered at the end of the *Concrete* chapter. *Foamglas* (from Pittsburgh Corning) is a unique material that is inorganic, incombustible, dimensionally stable, and water and vapor proof, and weighs only 8.5 lb. per c.f. (136 kg per cu.m) but is strong enough to be used under parking and promenade decks. It can also be ordered with 1/8" per ft. (3.1 mm per 0.3 m) slope. Minimum thickness is 1-1/2" (37.5 mm), and it is available in 1/2" (12.5 mm) increments up to 4" (100 mm). The 1-1/2" (37.5 mm) material has an "R" value of 4.2 and costs about $0.60 per s.f. ($6.45 per sq.m) in 1-1/2" (37.5 mm) thickness. One roofer will install some 800 s.f. (74 sq.m) per day at the following labor cost:

Labor Cost of 100 s.f. (9.3 sq.m) of 1" (25 mm) Deck Insulation

	Hours	Rate	Total	Rate	Total
Roofer	1.0	$....	$....	$24.99	$24.99
Cost per s.f.			....		0.25
per sq.m			....		2.69

07250 PERIMETER INSULATION

Board forms of polystyrene, urethane, fiberglass, and cellular glass are commonly employed as insulation on foundation walls and under slab edges to insulate an on-grade floor construction at the outer wall. Material costs are from $0.15 to 0.25 per s.f. ($1.61-2.69 per sq.m) in 1" (25 mm) thickness.

As perimeter insulation must fit the outline of the interior face of the outside wall and must be run continuously around offsets, perimeter ducts, column foundations and other irregularities, unit costs will vary widely. In general, insulation is run down 24" (600 mm) below outside grade. In addition, it may be run back 24" (600 mm) under the outer edge of the slab.

One carpenter should apply around 800 s.f. (74 sq.m) per day on straight run work, at the following labor cost per 100 s.f. (9.3 sq.m):

	Hours	Rate	Total	Rate	Total
Carpenter	1.14	$....	$....	$26.44	$30.14
Cost per s.f.			....		0.30
per sq.m			....		3.23

07300 ROOF SHINGLES AND ROOFING TILES

Roofing is estimated by the square, containing 100 s.f. (9.3 sq.m). The method used in computing the quantities will vary with the kind of roofing and the shape of the roof.

The labor cost of applying any type of roofing will be governed by the pitch or slope, size, plan (whether cut up with openings, such as skylights, penthouses, gables, dormers, etc.), and on the distance of the roof from the ground.

Rules for Measuring Plain Double Pitch or Gable Roofs. To obtain the area of a plain double pitch or gable roof as shown in Figure 1, multiply the length of the ridge (A to B), by the length of the rafter (A to C). This will give the area of one-half the roof. Multiply this by 2 to obtain the total sq. ft. (sq.m) of roof surface.

Example: Assume the length of the ridge (A to B), is 30'-0" (9.1 m) and the length of the rafter (A to C), 20'-0" (6.1 m):

(A to B) x (A to C)
30'-0" x 20'-0" = 600 s.f.   (9.1 x 6.1 = 55.51 sq.m)
600 s.f. x 2 = 1,200 s.f.   55.51 x 2 = 111.0 sq.m

Rules for Measuring Hip Roofs. To obtain the area of a hip roof as shown in Figure 3, multiply the length of the eaves (C to D) by 1/2 the length of the rafter (A to E). This will give the number of sq. ft. (sq.m) of one end of the roof, which multiplied by 2 gives the area of both ends. To obtain the area of the sides of the roof, add the length of the ridge (A to B) to the length of the eaves (D to H). Divide this sum by 2 and multiply by the length of the rafter (F to G). This gives the area of one side of the roof and when multiplied by 2 gives the number of sq. ft. (sq.m) on both sides of the roof.

To obtain the total number of sq. ft. (sq.m) of roof surface, add the area of the two ends to the area of the two sides. This total divided by 100 (or sq.m divided by 9.3) equals the number of squares in the roof.

Example: Assume the length of the eaves (C to D) is 20'-0" (6.1 m) and the length of the rafter (A to E) is 20'-0" (6.1 m):

Fig 1. Plain Double Pitch
or Gable Roof

Fig 2. Conical Building
and Roof

(C to D) x 1/2 the length of the rafter (A to E) =
20'-0" x 10'-0" = 200 s.f.   (6.1 x 3 = 18.3 sq.m)
200 x 2 = 400 s.f.   (18.3 x 2 = 36.6 sq.m)

To obtain the area of the sides of the roof, the length of the ridge (A to B), is 10'-0" (3.0 m) and the length of the eaves (D to H) is 30'-0" (9.1 m):

10'-0" + 30'-0" = 40'-0"   (9.1 + 3.0 = 12.1 m)

By taking 1/2 the combined length of the ridge and eaves, 1/2 of 40'-0" = 20'-0" (0.5 x 12.1 = 6.2 m), the average length of the roof. Assuming the length of the rafter (F to G) as 20'-0" (6.1 m), 20'-0" x 20'-0" = 400 s.f. (6.1 x 6.1 = 37.2 sq.m), the area of one side of the roof, which multiplied by 2 equals 800 s.f. (74.4 sq.m), the area of both sides of the roof.

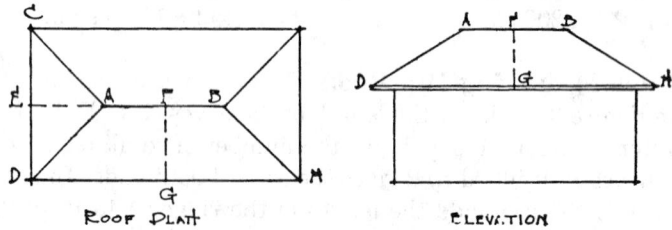

Roof Plan   Elevation

Fig 3. Hip Roof

Adding the area of the two ends to the area of the two sides gives 1,200 s.f. (111 sq.m) of roof area.

The area of a plain hip roof (Fig. 4) running to a point at the top is obtained by multiplying the length of the eaves (B to E) by 1/2 the length of the rafter (A to F). This gives the area of one end of the roof. To obtain the area of all four sides, multiply by 4. Example: Multiply the length of the eaves (B to E), which is 30'-0" (9.1 m) by 1/2 the length of the rafter (A to F), 10'-0" (3.0 m), and the result is 300 s.f. (27.3 sq.m), the area of one end of the roof. 300 x 4 = 1,200 s.f. (27.3 x 42 = 109 sq.m), the area of the 4 sides.

Fig 4. Hip Roof

Rules for Measuring Conical Tower Roofs and Circular Buildings.

To obtain the area of a conical tower roof as shown in Figure 2, multiply 1/2 the length of the rafter (A to B) by the distance around the eaves at B. For example, the length of the rafter (A to B) is 15'-0" (4.6 m) and the diameter of the building at B is 20'-0" (6.1 m).

To obtain the distance around the building, multiply the diameter by 3.1416. If the eaves project beyond the outside walls, and the diameter is given only to the outside walls of the building, add the length of the roof projection on both sides of the building to obtain the correct diameter. Example: If the diameter of the building (C to D) is 20'-0" (6.1 m) and the eaves project 2'-0" (0.6 m) on each side, the diameter of the building at the eaves would be 24'-0" (7.3 m).

Multiplying (B to E) 24'-0" (7.3 m), the diameter of the building at eaves, by 3.1416, gives 75.3984 (22.9336), or approximately 75'-5" (23 m) around the eaves at projection (B to E).

To obtain the area of the roof, multiply 1/2 the length of the rafter (A to B), which is 7-1/2 or 7'-6" (2.3 m), by the distance around the eaves at B and E, which is 75'-5" or 75.4' (23 m). The result is 565.5, or 565-1/2 s.f. (52.9 sq.m), the area of the roof.

To obtain the wall area of a cylindrical or circular building, multiply the height by the circumference, or the distance around the building, and the result will be the number of sq. ft. (sq.m) to be covered. The circumference is obtained by multiplying the diameter (C to D) of the building by 3.1416. To obtain the area of the outside walls of a cylindrical building whose diameter is 20'-0" (6.1 m) and the height 15'-0" (4.6 m), 20 x 3.1416 = 62.832 or 62'-10" (6.1 x 3.1416 = 19.16 or 19.2 m). Multiply 62.832 (19.2 m), the distance around the building, by 15 (4.6 m), the height of the building = 942.48 or 942-1/2 s.f. (88.32 or 88.3 sq.m), the area of the outside walls.

A Short Method of Figuring Roof Areas

To obtain the number of square feet (square meters) of roof area, where the pitch (rise and run) of the roof is known, take the entire flat or horizontal area of the roof and multiply by the factor given below for the roof slope applicable and the result will be the area of the roof.

Always bear in mind that the width of any overhanging cornice must be added to the building area to obtain the total area to be covered.

Example: Find the area of a roof 26'-0" x 42'-0" (7.9 x 12.8 m), with a 12" or 1'-0" (0.3 m) overhanging cornice and a 1/4 pitch or a 6 in 12 rise and run.

To obtain the roof area, 26'-0" + 1'-0" + 1'-0" = 28'-0" width (7.9 + 0.3 + 0.3 = 8.5 m width). 42'-0" + 1'-0" + 1'-0" = 44'-0" length (12.8 + 0.3 + 0.3 = 13.4 m length). 28 x 44 = 1,232 s.f. (8.5 x 13.4 = 113.9 sq.m) flat or horizontal area.

To obtain area at 1/4 pitch or 6 in 12 rise and run, 1,232 x 1.118 = 1377.376 or 1,378 s.f. of roof surface (113.9 x 1.118 = 128 sq.m of roof surface).

Add allowance for overhang on dormer roofs and sides.

Pitch of Roof	Rise and Run	Multiply Flat Area by	LF Hips or Valleys per LF Common Run
1/12	2 in 12	1.014	1.424
1/8	3 in 12	1.031	1.436
1/6	4 in 12	1.054	1.453
5/24	5 in 12	1.083	1.474
1/4	6 in 12	1.118	1.500
7/24	7 in 12	1.158	1.530
1/3	8 in 12	1.202	1.564
3/8	9 in 12	1.250	1.600
5/12	10 in 12	1.302	1.641
11/24	11 in 12	1.357	1.685
1/2	12 in 12	1.413	1.732

Hips and Valleys. The length of hips and valleys, formed by intersecting roof surfaces that run perpendicular to each other and have the same slope, is also a function of the roof rise and run. For full hips or valleys, where both roofs intersect for their full width, the length is determined by taking the square root of the sum of the rise squared plus twice the run squared.

Using the factors given in the last column of the above table, the length of full hips or valleys is obtained by multiplying the total roof run from eave to ridge, not the hip or valley run, by the factor listed for the roof slope.

07311 ASPHALT SHINGLES

Asphalt shingles are furnished in several styles, and the labor cost of laying them varies with the style of shingle used and the type of roof to which they are applied.

Individual asphalt shingles are furnished and laid in exactly the same manner as wood shingles, the only difference being their size.

Strip asphalt shingles are furnished in strips, each containing 3 shingles, so when laying them a roofer handles and lays 3 shingles at a time.

Hexagon asphalt shingles are furnished in strips 11-1/3"x36" (283 x 900 mm).

Estimating Quantities of Asphalt Shingles. Asphalt shingles are sold by the square, containing sufficient shingles to cover 100 s.f. (9.3 sq.m) of roof.

When measuring roofs of any shape, always allow one extra course of shingles for the "starters" at the eaves. The first or *starting* course of shingles must always be doubled.

Obtain the number of lin. ft. (meters) of hips, valleys, and ridges to be covered with asphalt shingles and compute as 1'-0" (300 mm) wide.

Many roofing contractors do not measure hips, valleys, and ridges, preferring to add a percentage of the roof area to cover these items and for waste. The following percentages are commonly used: Gable roofs, 10%; hip roofs, 15%; hip roofs with dormers and valleys, 20%.

Asphalt shingles must be properly nailed—6 nails to a strip and nailed low enough on the shingle (right at the cut-out); otherwise, they will blow off the roof.

Most manufacturers produce a shingle designed for high wind areas that interlocks in such a manner that all of the shingles are integrated into a single unit. Interlocking shingles are available in single coverage for re-roofing and double coverage for new construction.

Self-sealing shingles with adhesive tabs are produced by most manufacturers.

When using asphalt shingles for roofs, roll asphalt roofing of the same materials as the shingles is often used for forming valleys, hips, and roof ridges.

Nails Required for Asphalt Shingles. When laying individual asphalt shingles, use a 12 ga. aluminum nail, 1-1/2" (37.5 mm) long, with a 7/16" (10.9 mm) head. For laying over old roofs, use nails 1-3/4" (43.7 mm) long.

When laying square butt strip shingles, use 11 ga. aluminum nails, 1" (25 mm) long, with a 7/16" (10.9 mm) head. For laying over old roofs, use nails 1-3/4" (43.7 mm) long. It will require about 1 lb. (0.45 kg) of nails per square of shingles laid.

Sizes and Estimating Data on Asphalt Shingles

Kind of Shingle	Size in. (mm)	No. of Shingle per Sq.	Expos. in. (mm)	Nail Length in. (mm)	Nails per Shingle	Lbs. (Kg) Nails per Sq.
3-in.-1 strip ...	12"x36" (300x900)	80	5 (125)	1 (25)	4	1 (0.45)

Approximate Prices on Mineral Surfaced Asphalt Shingles, Seal Tab

Kind of Shingle	Size	Lbs. per Square	Price per square
3-in-1 strip	12"x36"	235	$26.00
3-in-1 strip	12"x36"	340	55.00
3-in-1 strip	12"x36"	350	65.00

(metric)

Kind of Shingle	Size	Kg per Sq. (9.3 sq.m)	Price per square
3-in-1 strip	300 x 900 mm	106	$26.00
3-in-1 strip	300 x 900 mm	154	55.00
3-in-1 strip	300 x 900 mm	158	65.00

Most asphalt shingles carry an Underwriters Class C rating. Incorporating a glass fiber layer results in an "A" rating. Class A shingles weighing 225 lbs. (102 kg) cost about $30.00 per square; 300 lbs. (136 kg), $59.00 per square.

A roofer will lay one square (100 s.f. or 9.3 sq.m) of asphalt shingles in 1 hour on double pitched roofs with no hips, valleys, or dormers. Figure about 35 l.f. (11 m) per hour for fitting hips, valleys, and ridges. Hip and ridge roll material will run about $0.16 per l.f. ($0.52 per m); valley, $0.30 per l.f. ($0.98 per l.f.).

Labor Cost of 1 Square (100 s.f. or 9.3 sq.m) Strip Asphalt
Shingles on Plain Double Pitch or Gable Roofs

	Hours	Rate	Total	Rate	Total
Roofer	1.0	$....	$....	$24.99	$24.99

On roofs with gables, dormers, etc., add 0.2 hr. per square (9.3 sq.m). On difficult constructed hip or English type roofs, add 0.5 hr. per square (9.3 sq.m).

07313 WOOD SHINGLES AND SHAKES

The labor cost of laying wood shingles varies with the type of roof, whether a plain gable roof, a steep roof, or one cut up with gables or dormers, and varies with the manner in which they are laid, whether with regular butts, irregular or staggered butts, or thatched butts.

The costs given on the following pages are based on the actual number of shingles a roofer can lay per day and not upon the number of squares covered, which will vary with the spacing of the shingles. It does not make any difference to the roofer whether the shingles are laid 4" (100 mm), 4-1/2" (112.5 mm), or 5" (125 mm) to the weather. He will lay practically the same number either way. It does make considerable difference in the number of sq. ft. (sq.m) of surface covered, which will vary from 10% to 40%.

The number of shingles laid will vary with the ability of workers and the class of work. Ordinary carpenters cannot lay as many shingles as carpenters who specialize in shingle laying. On the other hand, experienced shinglers usually demand a higher wage rate than ordinary carpenters.

Some carpenters claim to be able to lay 16 bundles (3,200 shingles) per 8-hr. day, but this is unusual and is generally found only on the cheapest grade of work, where only one nail is driven into each shingle instead of two that are necessary for a craftsman-like job.

Estimating the Quantity of Wood Shingles. Ordinary wood shingles are furnished in random widths, but 1,000 shingles are equivalent to 1,000 shingles 4" (100 mm) wide. Dimension shingles are sawed to a uniform width, being either 4" (100 mm), 5" (125 mm), or 6" (150 mm) wide.

Wood shingles are usually sold by the square based on sufficient shingles to lay 100 s.f. (9.3 sq.m) of surface, when laid 5" (125 mm) to the weather, 4 bundles to the square.

When estimating the quantity of ordinary wood shingles required to cover any roof, bear in mind that the distance the shingles are laid to the weather makes considerable difference in the actual quantity required.

There are 144 sq. in. in 1 sq. ft. (0.09 sq.m) and an ordinary shingle is 4" (100 mm) wide. When laid with 4" (100 mm) exposed to the weather, each shingle covers 16 sq. in. or it requires 9 shingles per s.f. of surface. There are 100 s.f. (9.3 sq.m) in a square. 100 x 9 = 900, and allowing 10% to cover the double row of shingles at the eaves, waste in cutting, narrow shingles, etc., it will require 990 shingles (5 bundles) per 100 s.f. of surface.

Number of Shingles and Quantity of Nails Required per 100 s.f. (9.3 sq.m) of surface

Distance Laid to Weather	Area Covered by One Shingle Sq. In.	Percent for Waste	Actual No. per Square without Waste	No. per Square with Waste	No. of 4-Square Bundles Required	Pounds 3d Nails Req'd
4"	16	10	900	990	5.0	3.2
4-1/2"	17	10	850	935	4.7	2.8
4-1/2"	18	10	800	880	4.4	2.5
5"	20	10	720	792	4.0	2.0
5-1/2"	22	10	655	720	3.6	1.6
6"	24	10	600	660	3.3	1.5

Labor Laying One Square (100 Sq. Ft.) Wood Shingles

Class of Work	Mechanic	Number Laid per 8-Hr. Day	Distance Shingles are Laid to Weather					
			4"†	4¼"	4½"	5"	5½"	6"
Plain Gable or Hip Roofs	Carpenter	2,000–2,200	3.8 hrs.	3.6 hrs.	3.4 hrs.	3.0 hrs.	2.8 hrs.	2.5 hrs.
	Shingler	2,750–3,000	2.8	2.6	2.5	2.3	2.0	1.9
Difficult Gable Roofs, cut up with gables, dormers, hips, valleys, etc.	Carpenter	1,700–1,900	4.5	4.2	4.0	3.6	3.3	3.0
	Shingler	2,200–2,500	3.4	3.2	3.0	2.7	2.5	2.3
Difficult Hip Roofs, Steep English Roofs, hips, valleys, etc.	Carpenter	1,300–1,500	5.7	5.4	5.1	4.6	4.2	3.8
	Shingler	2,000–2,200	3.8	3.6	3.4	3.0	2.8	2.5
Shingles Laid Irregularly or with Staggered Butts on Plain Roofs.	Carpenter	1,700–1,900	4.5	4.2	4.0	3.6	3.3	3.0
	Shingler	2,400–2,700	3.1	3.0	2.8	2.5	2.3	2.1
Shingles Laid Irregularly or with Staggered Butts on Difficult Constructed Roofs.	Carpenter	1,100–1,300	6.7	6.3	6.0	5.4	4.9	4.5
	Shingler	1,600–1,800	4.7	4.5	4.2	3.8	3.4	3.1
Shingles with Thatched Butts.* Plain Sidewalls.	Shingler	800–1,000	8.9	8.4	8.0	7.1	6.5	6.0
	Carpenter	1,300–1,500	5.7	5.4	5.1	4.6	4.2	3.8
	Shingler	1,700–1,900	4.5	4.3	4.0	3.6	3.3	3.0
Difficult Sidewalls, having bays, windows, breaks, etc.	Carpenter	1,100–1,250	6.8	6.4	6.0	5.4	5.0	4.5
	Shingler	1,400–1,650	5.2	5.0	4.6	4.2	3.8	3.5

To obtain number of bundles of shingles required, divide number of shingles as given above by 200, and the result will be the number of bundles required, i.e., 2,200 shingles ÷ 200 equals 11 bundles; 1,000 ÷ 200=5 bundles; 800 ÷ 200=4 bundles, etc.

*Shingles with thatched butts require 25% more shingles than when laid regularly.

†Use 4" column for carpenter or shingler time per 1,000 shingles. (5 bundles).

The above table is based on using 2 nails to each shingle and 10 per cent waste.

Distance Laid to Weather	(metric) Area Covered by One Shingle sq.mm	Percent for Waste	Actual No. per 10 sq.m without Waste	Kg 3d Nails Req'd
200 mm	10,323	10	1,066	1.45
212.5 mm	10,968	10	1,006	1.27
212.5 mm	11,614	10	947	1.13
125 mm	12,904	10	853	0.91
135.5 mm	14,194	10	775	0.73
150 mm	15,485	10	710	0.68

How to Apply Shingles for Different Roof Slopes. Roof pitches are computed in fractions, such as 1/8, 1/3, 1/2 pitch. In the following cross section, the steepness of distances AB and BC constitutes *pitch*. Distance AC, extending from one eaveline to the other, is known as the *span*. One-half of this span, distance AD or DC, is called the *run,* and distance BD is called the *rise.* The relationship of the rise to run obviously affects the slope of AB or BD; in fact, roof pitches are computed from the ratio of rise to run. Therefore, the first step is to determine length of the run (AD or DC) and the rise (BD).

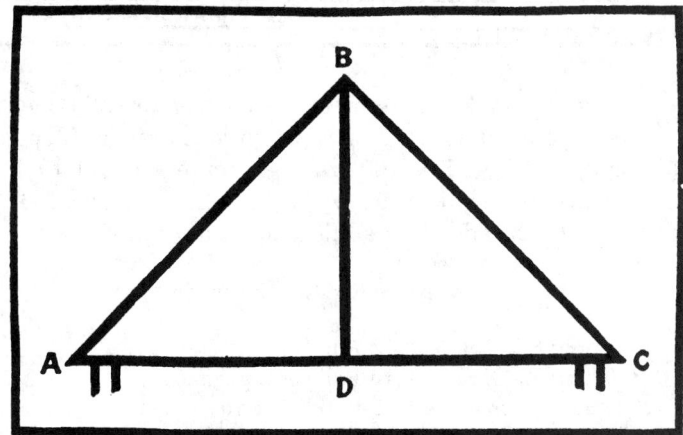

Wood shingles are manufactured in three lengths, 16" (400 mm), 18" (450 mm), and 24" (600 mm). The standard weather exposure, the portion of shingle exposed to weather on roof, for 16" (400 mm) shingles is 5" (125 mm), for 18" (450 mm) shingles it is 5-1/2" (137.5 mm), and for 24" (600 mm) shingles is 7-1/2" (187.5 mm). These standard exposures are recommended on all roofs of 1/4 pitch and steeper, 6" (150 mm) rise in 12" (300 mm) run. On flatter roof slopes, the weather exposure should be reduced to 3-3/4" (93.7 mm) for 16" (400 mm) shingles, 4-1/4" (106 mm) for 18" (450 mm) shingles, and 5-3/4" (143 mm) for 24" (600 mm).

The following diagram shows the weather exposure to be used for various roof pitches. For example, if a roof has a rise of 8" (200 mm) in a run of 12" (300 mm), it is 1/3 pitch and that an exposure of either 5" (125 mm), 5-1/2" (137.5 mm), or 7-1/2" (187.5 mm) should be employed, depending on the length of the shingles used.

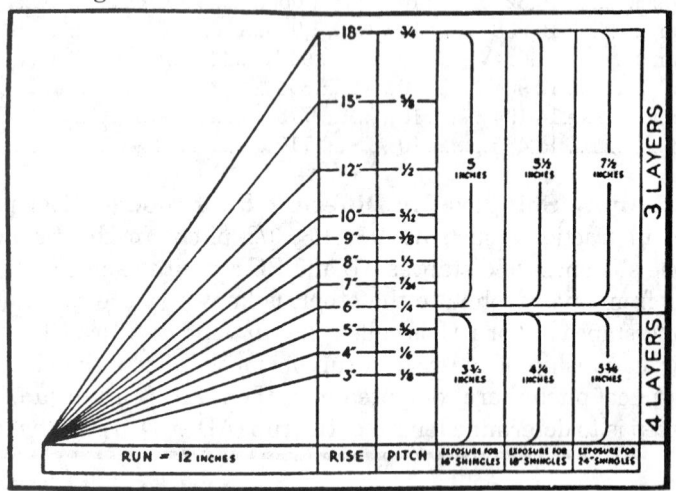

Wood shingles and shakes may be laid over open or solid sheathing, with solid preferred where wind driven snow conditions prevail. If open sheathing is selected, use 1"x4"s (25 x 100 mm) at centers equal to the weather exposure at which the shakes are to be laid, but not over 10" (250 mm). Building paper must be added for additional weather protection.

Western Red Cedar Shingles

144# 16" (400 mm) #1-5X Shingles 4 Bdl. per Sq. $100.00
144# 16" (400 mm) #2-5X Shingles 4 Bdl. per Sq. 90.00
60# 16" (400 mm) Undercourse 14/14 2 Bdl. per Sq. 25.00
158# 18" (450 mm) #1 Perfection 4 Bdl. per Sq. 100.00
350# 24" (600 mm) #1 3/4" (43 mm) to 5/4 Handsplit
Shakes, 5 Bdl. per Sq. 110.00
    For fire retardant shingles add 100%

07314 SLATE ROOFING

Roofing slate is furnished in a number of sizes, thicknesses, and finishes to meet architectural requirements of the buildings on which they are to be used. The four principal classifications are standard, textural, graduated, and flat slate roofs.

Standard Roof. Standard slate roofs are composed of slate approximately 3/16" (4.6 mm) thick (commercial standard slate), of one uniform standard length and width; also one length and random widths or random lengths and widths. Any width less than one-half the length is not recommended. Standard roofs are suitable for any building where a permanent roofing material is desired at a minimum cost. If desired, the butts or corners may be trimmed to give a hexagonal, diamond, or "gothic" pattern for all or part of the roof.

Textural Roof. The term "textural" is used to designate slate of rougher texture than the standard, with uneven tails or butts, and with variations of thickness or size. In general, this term is not applied to slate over 3/8" (9.4 mm) in thickness.

Graduated Roof. The graduated roof combines the artistic features of the textural slate roof with additional variations in thickness, size, and exposure. The slate is so arranged on the roof that the thickest and longest occur at the eaves and gradually diminish in size and thickness until the ridges are reached. Slate for roofs of this type can be obtained in any combination of thicknesses from 3/16" (4.6 mm) to 3/4" (18.7 mm) and heavier when especially desired.

In addition to the usual standard sizes, slate above 1/2" (12.5 mm) thick are produced in lengths up to 24" (600 mm). The graduations in lengths generally range from 24" (600 mm) to 12" (300 mm). The variations in length will at once provide a graduation in exposure by using the standard 3" (75 mm) lap. To illustrate, a suitable range of lengths might be:

	Thickness in. (mm)	Length in. (mm)	Exposure in. (mm)
Course Under Eaves	3/8" (9.4)	14" (350)	No exposure
First Course	3/4" (18.7)	24" (600)	10-1/2" (262.5)
1	3/4" (18.7)	24" (600)	10-1/2" (262.5)
2	1/2" (12.5)	22" (550)	9-1/2" (237.5)
2	1/2" (12.5)	20" (500)	8-1/2" (212.5)
2	3/8" (9.4)	20" (500)	8-1/2" (212.5)
4	3/8" (9.4)	18" (450)	7-1/2" (187.5)
5	1/4" (6.2)	16" (400)	6-1/2" (162.5)
3	1/4" (6.2)	14" (350)	5-1/2" (137.5)
3	3/16" (4.6)	14" (350)	5-1/2" (137.5)
8	3/16" (4.6)	12" (300)	4-1/2" (112.5)

Random widths should be used and so laid that the vertical joints in each course are broken and covered by the slate of the course above.

Flat Roof. Flat roofs offer a wide field for roofing slate and are designated whether or not they are used for "promenade" purposes.

Slate of any thickness may be used instead of slag or gravel as a surfacing material for the usual built-up type of roof. For ordinary roofs the standard 3/16" (4.6 mm) slate are used, but for promenade or extraordinary service, the slate should be 1/4" (6.25 mm) or 3/8" (9.4 mm) thick.

Slate for flat roofs are furnished in the following standard sizes: 6"x6" (150 x 150 mm), 6"x8" (150 x 200 mm), 6"x9" (150 x 225 mm), 10"x6" (250 x 150 mm), 10"x7" (250 x 175 mm), 10"x8" (250 x 200 mm), 12"x6" (300 x 150 mm), 12"x7" (300 x 175 mm), and 12"x8" (300 x 200 mm).

Estimating Quantities of Roofing Slate. Obtain the net area of the roof in sq. ft. as described earlier in this chapter. Add 6" (150 mm) to rafter length to allow for waste of normal roof.

Deduct one-half the area of chimneys and dormers if over 20 s.f. (1.8 sq.m) and less than 80 s.f. (7.4 sq.m). Make no deductions if less than 20 s.f. (1.8 sq.m) and deduct 20 s.f. (1.8 sq.m) less than actual area if more than 80 s.f. (7.4 sq.m).

Include area of dormer sides, sides of dormers if slated, slate saddles, and other places where slate is used in addition to the main roof area. Include overhanging parts of dormers, etc.

Add 1 s.f. for each l.f. (0.30 sq.m per m) of hips and valleys, for loss in cutting and fitting.

Allow 2% to 15% additional slate, depending on the extent to which the roof is intersected by other roofs, dormers, walls, etc.

Divide the total of the above by 100, which will give the number of "squares" of roofing required.

Slate is always sold at the quarry on the basis of the quantity required to cover 100 s.f. (9.3 sq.m) or a "square" of roof when slate is laid with a 3" (75 mm) head lap. If the roof is flat or other than 3" (75 mm) lap is used, the quantity must be corrected to the equivalent amount required as though the 3" (75 mm) lap were used.

Items to be Included in an Estimate for Slate Roofing. The following items should be included to make a complete and accurate estimate on slate roofing:

1. Cost of slate (punched) on cars or trucks at the quarry.
2. Freight or trucking from quarry to destination.
3. Unloading from cars and hauling to job, also unloading and piling at job, unless delivered to the job by truck. Trucking from quarry direct to job within a radius of 500 miles (805 km) is now general practice in the Vermont-New York district.
4. Placing on roof and laying, a) roofing felt, b) elastic cement, c) nails, d) snow guards or snow rails, e) sheet metal, f) labor, g) waste in handling, cutting and fitting.

Size of Slate. The following table gives the different sizes in which roofing slate is furnished, number required per square when laid with 3" (75 mm) lap, and number of nails required per square based on 2 nails to each piece of slate.

Size of slate, inches	Number per square	Number of nails reqd. per square
10x6	686	1,372
10x7	588	1,176
10x8	515	1,030
10x10	414	828
12x6	533	1,066
12x7	457	914
12x8	400	800
12x9	355	710
12x10	320	640
12x12	267	534
14x7	374	748
14x8	327	654
14x9	290	580
14x10	261	522
14x12	218	436
14x14	188	376
16x8	277	554
16x9	246	492
16x10	221	442
16x12	185	370
16x14	159	318
18x9	213	426
18x10	192	384
18x11	175	350
18x12	160	320
18x14	137	274
20x10	169	338
20x11	154	308
20x12	141	282
20x14	121	242
20x16	106	212
22x11	138	276
22x12	126	252
22x14	109	218
22x16	95	190
24x12	115	230
24x14	98	196
24x16	86	172

(metric)

Size of slate, mm	Number per 10 sq.m	Number of nails reqd. per 10 sq.m
250x150	738	1,477
250x175	633	1,266
250x200	554	1,109
250x250	446	891

Size of slate, mm	Number per 10 sq.m	Number of nails reqd. per 10 sq.m
300x150	574	1,147
300x175	492	984
300x200	431	861
300x225	382	764
300x250	344	689
300x300	287	575
350x750	403	805
350x200	352	704
350x225	312	624
350x250	281	562
350x300	235	469
350x350	202	405
400x200	298	596
400x225	265	530
400x250	238	476
400x300	199	398
400x350	171	342
450x225	229	459
450x250	207	413
450x275	188	377
450x300	172	344
450x350	147	295
500x250	182	364
500x275	166	332
500x300	152	304
500x350	130	260
500x400	114	228
550x275	149	297
550x300	136	271
550x350	117	235
550x400	102	205
600x300	124	248
600x350	105	211
600x400	93	185

Slate 3/4" (19 mm) and thicker and 24" (600 mm) or over in length should have 4 nail holes.

Weight of Slate Roofing. A square of slate roofing, i.e., sufficient slate to cover 100 s.f. (9.3 sq.m) of roof surface with a standard 3" (75 mm) lap, will vary from 650 to 8,000 lbs. (295 to 3630 kg) for thicknesses from the commercial standard 3/16" (4.7 mm) to 2" (50 mm).

The weight of slate varies with the size of the slate, color, and quarry. Sometimes, it varies even in the same quarry. The variation may be from 10% above to 15% below the weights given in the following table:

Average Weight of Slate per Square (100 s.f. or 9.3 sq.m)

	Slate Thick., In.	Slate Thick., mm	Sloping Roof, 3" lap, lbs.	Sloping Roof, 25 mm lap, kg	Flat Roof, no lap, lbs.	Flat Roof, no lap, kg
Standard	3/16	4.7	700	318	240	109
Selected full	3/16	4.7	750	340	250	113
	1/4	6.3	900	408	335	152
	3/8	9.4	1,400	635	500	227
	1/2	12.5	1,800	816	675	306
	3/4	18.8	2,700	1,225	1,000	454
	1	25.0	4,000	1,814	1,330	603
	1-1/4	31.3	5,000	2,268	1,670	758
	1-1/2	37.5	6,000	2,722	2,000	907
	1-3/4	43.8	7,000	3,175		
	2	50.0	8,000	3,629		

Nails Required for Slate Roofing. The quantity of nails required for slate roofing will vary with size, kind, etc., but the following table gives quantities of those commonly used:

Approximate Number of Nails to the Pound

Length in Inches	"Copperweld" Slating Nails	Copper Wire Slating Nails	Cut Copper Slating Nails	Cut Brass Nails	Cut Yellow-Metal Slat. Nails
1	386	270	--	--	--
1-1/4	211	144	190	164	154
1-1/2	176	134	135	140	140
1-3/4	133	112	100	108	
2	87	104	--	88	--
2-1/4	--	46	--	80	--
2-1/2	--	--	--	64	--
2-3/4	--	--	--	52	--
3	--	--	--	48	--

			(metric)		
Length in Inches	"Copperweld" Slating Nails	Copper Wire Slating Nails	Cut Copper Slating Nails	Cut Brass Nails	Cut Yellow-Metal Slat. Nails
25.0	175	122	--	--	--
31.3	96	65	86	74	70
37.5	80	61	61	64	64
43.8	60	51	45	49	--
50.0	39	47	--	40	--
56.3	--	21	--	36	--
62.5	--	--	--	29	--
68.8	--	--	--	24	--
75.0	--	--	--	22	--

Labor Punching Slate. It is much cheaper to have slate punched at the quarry. The labor cost of punching slate on the job varies with the size of the slate. The larger the slate, the fewer pieces required per square.

If one slater is working on the job with a helper, the helper should punch the slate and carry them to the slater as fast as required. There should be no additional charge for punching. If, however, there is only one worker on the job who must punch the slate and lay them, he should punch one square of slate in 30 to 40 min. or 12 to 16 sqs. (111-148 sq.m) per 8-hr. day, at the following labor cost per square (100 s.f. or 9.3 sq.m):

	Hours	Rate	Total	Rate	Total
Slater	0.6	$....	$....	$24.99	$14.99

Labor Cost of Laying Roofing Slate. The labor cost of laying roofing slate will vary considerably with the size of the slate, thickness, type of roofs, etc., but when large thick slate are used for graduated roofs, the labor costs will run higher than when the same size slate are used on standard roofs.

The size of slate best adapted for plain roofs are large, wide slates, such as 16"x12" (400 x 300 mm), 18"x12" (450 x 300 mm), 20"x12" (500 x 300 mm), or 24"x12" (600 x 300 mm). Slate from 16"x8" (400 x 200 mm) to 20"x10" (500 x 250 mm) are also popular sizes. For roofs cut up into smaller sections, the 14"x7" (350 x 175 mm) and 16"x8" (400 x 200 mm) are very popular.

A roofing crew consisting of two slaters and one helper can work to advantage and perform the work more economically than one worker alone. In some localities the helper is considered an apprentice and is permitted to lay slate. In others he can only punch and carry the slate to the slaters.

Laying 16"x8" (400 x 200 mm) Roofing Slate. When working on plain roofs that do not require much cutting, 2 slaters and a helper should lay felt, handle, and lay 3-1/2 to 4-1/2 sqs. (350-450 s.f. or 32.5-42 sq.m) of 16"x8" (400 x 200 mm) slate per 8-hr. day. On more complicated hip or gable roofs, requiring considerable cutting and fitting for hips, valleys, dormers, etc., 2 slaters and a helper should handle and lay 1-3/4 to 2-1/4 sqs. (175-225 s.f. or 16-21 sq.m) of 16"x8" (400 x 200 mm) slate per 8-hr. day.

Laying 18"x9" (450 x 225 mm) Roofing Slate. When laying 18"x9" (450 x 225 mm) slate on plain roofs, 2 slaters and a helper should lay felt, handle, and lay 4 to 5 sqs. (400-500 s.f. or 37-46 sq.m) per 8-hr. day.

On more complicated hip or gable roofs, requiring considerable cutting and fitting for hips, valleys, dormers, etc., 2 slaters and a helper should lay 2-1/2 to 3 sqs. (250-300 s.f. or 23-28 sq.m) per 8-hr. day.

Laying 20"x10" (500 x 250 mm) Roofing Slate. When laying 20"x10" (500 x 250 mm) slate on plain roofs, 2 slaters and a helper should lay felt, handle, and lay 5 to 6 sqs. (500-600 s.f. or 46-55 sq.m) of roof per 8-hr. day, but on more complicated roofs, requiring considerable cutting and fitting for hips, valleys, dormers, etc., the same crew would lay only 3 to 3-1/2 sqs. (300-350 s.f. or 28-32 sq.m) per 8-hr. day.

Laying 22"x12" (550 x 300 mm) Roofing Slate. On straight roofs, 2 slaters and a helper should lay felt, handle, and lay 5-1/2 to 6-1/2 sqs. (550-650 s.f. or 51-60 sq.m) of 22"x12" (550 x 300 mm) slate per 8-hr. day, but on the more complicated roofs, requiring cutting and fitting for hips, valleys, dormers, etc., the same crew would lay only 3-1/2 to 4 sqs. (350-400 s.f. or 32-37 sq.m) per 8-hr. day.

Laying Graduated Roofing Slate. When laying graduated roofing slate from 3/16" (4.7 mm) to 3/4" (18.7 mm) thick and 12" (300 mm) to 24" (600 mm) long, considerable care is required in selecting the right slate for each course and laying them to obtain the desired effect. On work of this kind 2 slaters and a helper should lay felt, handle, and lay 1-1/2 to 2 sqs. (150-200 s.f. or 14-19 sq.m) of roof per 8-hr. day.

Labor Cost of One Square (100 s.f. or 9.3 sq.m)
16"x8" (400 x 200 mm) Standard Slate Roofing

	Hours	Rate	Total	Rate	Total
Slater	4	$....	$....	$24.99	$99.96
Helper	2	....	....	21.04	42.08
Cost per 100 s.f.			$....		$142.04
per 10 sq.m				....	154.82
per sq.m				....	15.48

For complicated hip or gable roofs, double the time given above.

Labor Cost of One Square (100 s.f. or 9.3 sq.m)
18"x9" (450 x 225 mm) Standard Roofing Slate

	Hours	Rate	Total	Rate	Total
Slater	3.50	$....	$....	$24.99	$87.47
Helper	1.75	....	....	21.04	36.82
Cost per 100 s.f.			$....		$124.29
per 10 sq.m				....	135.48
per sq.m				....	13.55

For complicated hip or gable roofs, add 2.5 hrs. slater and 1.25 hrs. helper.

Labor Cost of One Square (100 s.f. or 9.3 sq.m)
20"x10" (500 x 250 mm) Standard Roofing Slate

	Hours	Rate	Total	Rate	Total
Slater	3.0	$....	$....	$24.99	$74.97
Helper	1.5	....	....	21.04	31.56
Cost per 100 s.f.			$....		$106.53
per 10 sq.m				....	116.12
per sq.m				....	11.61

For complicated hip or gable roofs, add 2 hrs. slater and 1 hr. helper time.

Labor Cost of One Square (100 s.f. or 9.3 sq.m)
22"x12" (550 x 300 mm) Standard Roofing Slate

	Hours	Rate	Total	Rate	Total
Slater	2.6	$....	$....	$24.99	$64.97
Helper	1.3	....	....	21.04	27.35
Cost per 100 s.f.			$....		$ 92.32
per 10 sq.m				....	100.63
per sq.m				....	10.06

For complicated hip or gable roofs, add 1.6 hrs. slater and 0.8 hr. helper.

Labor Cost of One Square (100 s.f. or 9.3 sq.m) Rough Texture Slate
Roof Random Widths and 3/16" (4.7 mm) to 3/8" (9.4 mm) Thick

	Hours	Rate	Total	Rate	Total
Slater	7.50	$....	$....	$24.99	$187.43
Helper	3.75	....	....	21.04	78.90
Cost per 100 s.f.				....	$266.33
per 10 sq.m				....	290.30
per sq.m				....	29.03

Labor Cost of One Square (100 s.f. or 9.3 sq.m) Graduated Slate Roof, Slate 12" (300 mm) to 24" (600 mm) Long
3/16" (4.7 mm) to 3/4" (18.7 mm) Thick

	Hours	Rate	Total	Rate	Total
Slater	9.0	$....	$....	$24.99	$224.91
Helper	4.5	....	....	21.04	94.68
Cost per 100 s.f.			$....		$319.59
per 10 sq.m				....	348.35
per sq.m				....	34.84

Prices on Slate Roofing Material

Slate shingles come from various parts of the country, and the estimator must obtain a quotation from the quarries in the geographical location, not only for price and shipping costs, but for availability of certain sizes and colors.

Vermont slate shingles in 3/16" (4.7 mm) thicknesses will cost as follows: black or gray, $290.00 per square (9.3 sq.m); green and gray, $300.00; and purples, $375.00. The Pennsylvania black #1 clear and the Buckingham, Virginia black slates will cost about $375.00 per square (9.3 sq.m). Slates thicker than 3/16" (4.7 mm) will cost appreciably more.

Copper nails are usually used for nailing the slates. Use 1-1/4" (31.2 mm) long nails for the 3/16" (4.7 mm) thick slates and 1-3/4" (43.7 mm) long nails for the rough texture and graduated slate roofs where thicker slates are encountered.

07315 PORCELAIN ENAMEL SHINGLES

Porcelain enamel shingles are manufactured to have an exposed surface of 10"x10" (250 x 250 mm) with 144 shingles per square (9.3 sq.m). Weight is 225 lbs. per 100 s.f. (110 kg per 10 sq.m). Finish is fused on at 1500°F (815°C) and provides a long lasting, self-cleaning finish that will not peel or blister.

In ordering these units, it is important to have complete shop drawings based on actual job conditions so that field cutting can be held to a minimum. Shingles are nailed to the roof deck over 2 layers of 30-lb. (13.6-kg) felt. Special nails, ridge and hip sections are furnished by the manufacturer. Valley flashings and drips should be of non-corrosive metal and furnished by the sheet metal contractor.

Material costs will run in the vicinity of $280.00 per square ($30.00 per sq.m), depending on the number of roof irregularities which run up the cost considerably, as all special fittings should be made up in the shop. One roofer can lay a square in about 6-1/2 hours.

07316 METAL SHINGLES

Metal shingles formed with a wood grain texture and available in a wide range of colors can be used for a lightweight, decorative roof, or false mansard. They can be applied in a conventional way on a solid nailable roof deck over 30-lb. (13.6-kg) felt underlayment or supported and clipped to sub-purlins on steel frames. Weights are 36 lbs. per square (1.75 kg per sq.m) for 0.020 aluminum, 54 lbs. (2.64 kg per sq.m) for 0.030 aluminum, and 88 lbs. (4.3 kg per sq.m) for 30 ga. steel.

The shingles are not individual units, but made in strips either 10"x60" (250 x 1500 mm), 12"x36" (300 x 900 mm) or 12"x48" (300 x 1200 mm).

The shingle manufacturer can provide anchor clips, drip caps, starter strips, trim, ridge, and hip laps. Material Costs will run as follows:

	per square	per sq.m
Aluminum, 0.020", mill finish	$72.00	$7.75
Aluminum, 0.030", mill finish	85.00	9.15
Aluminum, 0.020", anodized color	130.00	14.00
Aluminum, 0.030", anodized color	150.00	16.15
Aluminum, 0.020", bonderized	200.00	21.50
Aluminum, 0.030", bonderized	210.00	22.60
Steel, 26 ga., galvanized	33.00	3.55
Steel, 24 ga., galvanized	36.00	3.90
Steel, 26 ga., color galvanized	75.00	8.10
Steel, 24 ga., color galvanized	78.00	8.40

Ridge and cap sections cost $1.80 per l.f. ($5.90 per m) for 0.020" aluminum, $2.10 ($6.90 per m) for .030". Valley sections cost $1.40 and $1.70 ($4.60 and $5.60 per m). One carpenter will take about 4 to 5 hours to lay one square (9.3 sq.m).

07321 CLAY ROOFING TILE

The following information on clay roofing tile has been furnished through the courtesy of the Ludowici-Celadon Company, Chicago, IL.

When estimating the quantities of clay roofing tile, the roof areas are obtained in the same manner as described at the beginning of this chapter.

After the total number of sq. ft. (sq.m) of surface has been obtained, add about 3% to the net measurement to take care of cutting and waste. Material is sold only in full pallets which vary from 3 to 8.7 squares (28-80 sq.m).

Hip and valley fittings should be figured by the piece. Closure and finishing pieces, such as hip starters, closed ridge ends, terminals, etc., should also be figured by the piece. Fittings are sold only in full boxes.

Where interlocking shingle tiles are used, detached gable rake pieces are required for all gable roofs not intersecting with vertical surfaces. These pieces are measured by the total lineal feet of gable rake, keeping separate quantities for lefts and rights as determined by the roof eave to be covered.

For all shingle tile roofs, end bands or half-tile are required on each course, alternating from one end to the other to break joints. End bands are figured by the lineal feet of gable rake, taking half the length at each gable rake.

Laying Interlocking Shingle Tile. On a roof of average difficulty, a crew consisting of 2 roofers and an apprentice or helper should handle and lay 500 to 600 s.f. (46-56 sq.m) of interlocking shingle tile per 8-hr. day, at the following labor cost per square (100 s.f. or 9.3 sq.m):

	Hours	Rate	Total	Rate	Total
Roofer	3.0	$....	$....	$24.99	$74.97
Helper	1.5	....	....	21.04	31.56
Cost per 100 s.f.			$....		$106.53
per 10 sq.m					116.12
per sq.m					11.61

On steep or cut-up roofs, add 15% to 25%. For roofs over 3 stories high to eaves, add 1.5 hrs. helper time.

Laying Tile Shingles (Architectural Patterns). On a roof of average difficulty, 2 roofers and a helper should handle and lay 325 to 375 s.f. (30-35 sq.m) of tile shingles per 8-hr. day, at the following labor cost per square (100 s.f. or 9.3 sq.m):

	Hours	Rate	Total	Rate	Total
Roofer	4.6	$....	$....	$24.99	$114.95
Helper	2.3	....	....	21.04	48.39
Cost per 100 s.f.			$....		$163.34
per 10 sq.m				....	178.04
per sq.m				....	17.80

On steep or cut-up roofs, add 15% to 25%. For roofs over 3 stories high to eaves, add 2.3 hrs. helper time.

Laying Miscellaneous Tile Pattern Roofs. Where Spanish or French pattern tile is used on new work, figure same labor costs as for interlocking shingle tile.

Labor laying Mission tile will vary according to size and exposure of units. On new work of average difficulty, using 15"x8" (375 x 200 mm) tile with 12" (300 mm) exposure, 2 roofers and a helper should handle and lay 350 to 400 s.f. (32-37 sq.m) per 8-hr. day.

Items to Be Included When Estimating Tile Roofs. The following items should be included when estimating the cost of a clay shingle tile roof.

One roll coated felt or roll roofing weighing 40 lbs. per sq. (1.95 kg per sq.m) and 3 lbs. (1.4 kg) roofing nails, 1-3/4" (44 mm) long.

Allow 1 lb. (0.45 kg) colored elastic cement for pointing joints for each 20 l.f. (6 m) of hips and ridges.

Cost of Clay Tile Roofing Materials. The Luduwici-Celadon Company offers a number of different patterns and colors in each of their "Standard" and "Special" series product line. They will also produce custom shapes and colors, if given enough lead time.

Each of the patterns are produced with field tiles and special shapes for the ridge, hip, rake, etc.

Approximate material prices for different patterns and colors are as follows:

	per 100 s.f.	per 10 sq.m
Spanish Red	$180.00	$194.00
Spanish Buff	350.00	377.00
Spanish Blue	550.00	592.00
Mission Granada Red	350.00	377.00
Americana Gray or Green	225.00	242.00
Lanai Black or Brown	225.00	242.00
Classic Red	225.00	242.00
French Red	300.00	323.00
French Blue	650.00	700.00

Details of Clay Interlocking Shingle Tile Roofing

Details of Clay Tile Shingle Roofing

07400 PREFORMED ROOFING & SIDING

07411 PREFORMED METAL SIDING

Corrugated steel is used extensively for roofing and siding steel mills, manufacturing plants, sheds, grain elevators, and other industrial structures.

It is made with various corrugations, varying in width and depth, but the 2-1/2" (62 mm) corrugation width is the most commonly used.

The sheets are usually furnished 26" (650 mm) wide and 6'-0" (1.8 m) to 30'-0" (9.1 m) long. Some manufacturers offer siding protected with vinyl coatings.

When estimating quantities of corrugated siding or roofing, always select lengths that work to best advantage and be sure and allow for both end and side laps.

Corrugated siding or roofing is nailed to the wood framework or siding when used on wood constructed buildings. When used as a wall and roof covering on structural steel framing, the corrugated sheets are fastened to the steel framework, using clip and bolts, which are passed around or under the purlins, which usually consist of channels, angles, or Z bars. When angles are used for purlins, clinch nails are sometimes used for fastening the corrugated sheets.

When used for siding one corrugation lap is usually sufficient, but for roofing two corrugations should be used and if the roof has only a slight pitch, the lap should be three corrugations.

When used for siding, a 1" (25 mm) to 2" (50 mm) end lap is sufficient, but when laid on roofs it should have an end lap of 3" (75 mm) to 6" (150 mm) depending on the pitch of the roof. For a 1/3 pitch, a 3" (75 mm) lap is sufficient; for a 1/4 pitch, a 4" (100 mm) lap should be used; and for a 1/8 pitch, a 5" (125 mm) end lap is recommended.

When applying to wood sheathing or strips, the nails should be spaced about 8" (200 mm) apart at the sides. When applied to steel purlins, the side laps should extend over at least 1-1/2 corrugations, and the sheets should be riveted together every 8" (200 mm) on the sides and at every alternate corrugation on the ends.

Number of S.F. of Corrugated Sheets Required to Cover
100 S.F. of Surface, Using 26"x96" Sheets

2-1/2" Corrugation Width	Length of End Lap in Inches					
	1	2	3	4	5	6
Side lap, 1 corrugation	110	111	112	113	114	115
Side lap, 1-1/2 corrugations	116	117	118	119	120	121
Side lap, 2 corrugations	123	124	125	126	127	128
Side lap, 2-1/2 corrugations	130	131	132	133	134	135
Side lap, 3 corrugations	138	139	140	141	142	143

Number of Sq.M of Corrugated Sheets Required to Cover
10 Sq.M of Surface, Using 650 x 2400 mm Sheets

62 mm Corrugation Width	Length of End Lap in mm					
	25	50	75	100	125	150
Side lap, 1 corrugation	11.0	11.1	11.2	11.3	11.4	11.5
Side lap, 1-1/2 corrugations	11.6	11.7	11.8	11.9	12.0	12.1
Side lap, 2 corrugations	12.3	12.4	12.5	12.6	12.7	12.8
Side lap, 2-1/2 corrugations	13.0	13.1	13.2	13.3	13.4	13.5
Side lap, 3 corrugations	13.8	13.9	14.0	14.1	14.2	14.3

If shorter sheets are used the allowance should be slightly increased.

Material Prices. Corrugated steel siding and roofing panels are manufactured in a number of metal gauges, widths, and depths of corrugations, and in galvanized and prefinished painted surfaces. Average material prices, per 100 s.f. (9.3 sq.m), for sheeting with a 2-1/2" (62.5 mm) corrugation width are as follows:

Metal Gauge	Galvanized	Painted
28	$45.00	$60.00
26	55.00	70.00
24	80.00	100.00
22	90.00	105.00
20	95.00	110.00

Placing Corrugated Steel Roofing or Siding on Wood Framing. If corrugated sheets used for roofing or siding are nailed to wood strips or framing, a carpenter and helper should place one square (100 s.f. or 9.3 sq.m) of 26"x96" (650 x 2400 mm) or larger, in 7/8 to 1-1/8 hr. or 7 to 9 sqs. per 8-hr. day, at the following labor cost per square (100 s.f. or 9.3 sq.m):

Labor Cost of One Square (100 s.f. or 9.3 sq.m) Corrugated Metal Roofing on Wood Framing

	Hours	Rate	Total	Rate	Total
Carpenter	1	$....	$....	$26.44	$26.44
Labor	1	....	....	21.04	21.04
Cost per 100 s.f.			$....		$47.48
per 10 sq.m				....	51.75

Labor Cost of One Square (100 s.f. or 9.3 sq.m) Corrugated Metal Siding on Wood Framing

	Hours	Rate	Total	Rate	Total
Carpenter	1.2	$....	$....	$26.44	$31.73
Labor	1.2	....	....	21.04	25.25
Cost per 100 s.f.			$....		$56.98
per 10 sq.m				....	62.11

Placing Corrugated Steel Roofing on Steel Framing. The labor cost of applying corrugated steel roofing to structural steel framing, where the sheets are fastened with clips or bolts, vary with the size of the roof area, regularity of roof, height above ground, etc., but on a straight job, 2 sheeters and 2 laborers should place 550 to 650 s.f. per 8-hr. day, at the following labor cost per 100 s.f. (9.3 sq.m):

	Hours	Rate	Total	Rate	Total
Labor handling, hoisting	2.67	$....	$....	$21.04	$56.18
Sheeters	2.67	....	....	28.88	77.11
Cost per 100 s.f.			$....		$133.29
per 10 sq.m				....	145.29

On irregular roofs, above costs may be increased 25% to 50%.

Placing Corrugated Steel Siding on Steel Framing. The labor cost of applying corrugated steel siding to walls varies greatly according to length and height of walls, quantity and regularity of openings, etc., but on a straight job, 2 sheeters and 2 laborers should place 450 to 550 s.f. (41-51 sq.m) per 8-hr. day, at the following labor cost per 100 s.f. (9.3 sq.m):

	Hours	Rate	Total	Rate	Total
Labor handling,					
hoisting	3.2	$....	$....	$21.04	$67.33
Sheeters	3.2	....	....	28.88	92.42
Cost per 100 s.f.			$....		$159.75
per 10 sq.m				....	174.13

On coal tipples, grain elevators and other structures having high and irregular walls, the above costs may be doubled.

Protected Metal Roofing and Siding

Protected metal panels consist of a steel core sheet, to which felt is bonded by hot molten zinc. When the zinc has cooled, the felt and the steel core are metallurgically bound together. The felt is then impregnated with asphalt saturant. Finally, a tough, thick, waterproof outer coating is applied. All material is made with protective coating both sides.

There are several colors to choose from. Protected metal panels are manufactured in 4 gauges: 18, 20, 22, and 24, and are available in lengths up to 30 feet (9 meters).

Estimating and Cost Data. Panels are available in several profiles. The H.H. Robertson Company, Pittsburgh, Pa., has the following:

Huski-Rib has a deep fluted profile, designed specifically for high strength long span roofing and siding installations. The Huski-Rib sheet is 31-5/32" (779 mm) wide x 1-1/2" (37.5 mm) deep, and provides a coverage of 28" (700 mm). It is manufactured in 18, 20, 22, and 24 gauge metal, a maximum of 30' (9 m) long, and is suitable for spans up to 15'-9" (4725 mm) for siding, and up to 14'-14" (4550 mm) for roofing depending on local design load requirements.

Box-Rib has a strong, precise vertical texture, best suited for sidings only, either in standard or inverted positions. The standard sheet is 29-1/4" (731 mm) wide x 1-1/2" (37.5 mm) deep and provides a coverage of 28". It is manufactured in 18, 20, 22, and 24 gauge metal and is available in lengths up to 30 ft. (9 m).

Sturdi-Rib sheets come 34" (850 mm) wide x 9/16" (14 mm) deep providing a coverage of 29-7/16" (736 mm), and are made in lengths up to 30 ft (9 m). They are made of 18, 20, 22, and 24 gauge metal and are capable of spanning up to 10'-1" (3025 mm) for siding and 8'-9" (2625 mm) for roofing installations, depending upon local design load requirements.

Profiles of Robertson Galbestos Roofing and Siding

The *Magna-Rib* profile sheets are 26-5/8" (666 mm) wide x 4" (200 mm) deep providing a coverage of 24" (600 mm) installed. They come in 18, 20, and 22 gauge metal in lengths up to 38 ft. (11.5 m). Magna-Rib sheets are capable of spanning up to 22'-8" (6800 mm) for siding in one span.

Corrugated *Galbestos* is the original Robertson protected metal profile and comes in sheets 33" (825 mm) wide x 9/16" (14 mm) deep. Coverage is 29-3/4" (743 mm). Corrugated sheets are available in 18, 20, 22 and 24 gauge metal in lengths up to 30 ft. (9 m). They are capable of spanning up to 9'-3" (2775 mm) for siding and 7'-11" (2375 mm) for roofing installations.

Approximate Price per Sq. Ft. of Corrugated Protected Metal Panels

	Gauge of Sheets			
Description	18	20	22	24
Galbestos C2S (Asphalt) Black	$2.50	$1.90	$1.90	$1.70
Galbestos C2S (Resin) Color	3.00	2.50	2.20	2.00

After figuring the price of sheets from the above, 25% or so should be added to cover cost of flashings, fastenings, closures, corners, etc.

True cost comparisons must include the cost of supporting steel girts or purlins. Longer spans may save up to $0.25 per s.f. ($2.69 per sq.m) or more.

Matching profiled translucent fiberglass reinforced plastic daylight panels are available.

07415 CORRUGATED ALUMINUM ROOFING AND SIDING

When using aluminum roofing and siding, it is very important when applying the sheets that they be insulated against electrogalvanic action. Ordinarily aluminum has a high resistance to corrosion, but when it is contiguous to steel or copper and in the presence of moisture, an electrolytic cell is formed, and the resulting flow of current dissolves the aluminum, causing severe pitting. This condition can be avoided by preventing metal-to-metal contact.

Another significant requirement in using aluminum sheet is to seal the openings made for the fastening devices. Methods include the use of special washers and waterproof roofing compounds. Washers should be used with fasteners to distribute stresses and prevent tearing of the sheet.

To provide adequate drainage, the roof surface should never have a slope less than 2-1/2" per ft. (208.13 mm per m) and preferably not less than 3" per ft. (249.75 mm per m).

For roofing, sheets should have a side lap of 1-1/2 corrugations. For siding, sheets should be lapped 1 corrugation.

A 6" (150 mm) overlap at the ends is recommended for roofing and 4" (100 mm) for siding. When corrugated aluminum sheets are used without sheathing on a steel frame, the method shown in (a) is suggested for fastening to purlins and girts. This method utilizes aluminum straps, 12" (300 mm) apart, with aluminum bolts and nuts, aluminum rivets or cadmium-plated steel bolts.

To prevent galvanic action, direct contact between the top of the purlin and roofing should be prevented, preferably by insertion of an aluminum saddle or by painting the flange of the purlin with aluminum or bitumastic paint. Roofing paper may also be used to separate the metals.

Purlin clips (b) are also widely used as a fastening device. The U bend is slipped over the purlin flange and a leg hammered down against the web to lock the clip in place. The roofing is bolted to the clip. If standard steel purlin clips are used, they should be hot-dip galvanized.

Purlin nails, either single (c) or double (d) are also used for anchoring sheeting. A variation of this method employing aluminum washers, nuts, and base plate under the purlin is shown in (e).

The best method of nailing to sheathing is to use galvanized roofing nails about 1-3/4" (43.7 mm) long with a head about 3/8" (9.4 mm) in diameter, spaced at 8" (200 mm) intervals. These nails should have a hot-dip zinc coating.

Corrugated sheets should always be riveted, bolted, or nailed through the top of a corrugation. The reason is that rain running down the roof tends to collect in the bottom of corrugations and would penetrate imperfectly sealed holes there. At the top, water runs away from the holes.

A washer of non-metallic material, such as zinc chromate impregnated fabric, neoprene, or other rubber should be employed under a nailhead. The purpose of this washer is to prevent metal-to-metal contact between the underside of the head and the roofing sheet, to seal the opening made in the sheet and to permit thermal expansion and contraction of the sheeting.

Hot-dip galvanized nails without washers may be used when washers are not available. Roofing nails with cast-on lead heads can be used in other than industrial or seacoast corrosive atmospheres.

Sizes, Weights, and Coverage Data on Corrugated Aluminum Sheets Used for Roofing

Lengths 5'-0" (1500 mm) to 12'-0" (3600 mm) in 6" (150 mm) increments. Widths, 35" (875 mm) and 48-3/8" (1209 mm). Thicknesses, 0.024" (0.6 mm) and 0.032" (0.8 mm). Corrugation, 2.67" (66.7 mm) pitch and 7/8" (22 mm) deep. Coverage, 32" (800 mm) for 35" (875 mm) width and 45-3/8" (1134 mm) for 48-3/8" (1209 mm).

Data on 0.032" (1 mm) Corrugated Aluminum Roofing Sheets

Sheet Length, ft.	SF per Sheet		Wgt per Sheet, lbs		Approx. No. Sheets per 100 s.f.*	
	35"	48-3/8"	35"	48-3/8"	35"	48-3/8"
5	14.58	20.16	8.04	11.03	6.86	4.96
5-1/2	16.04	22.17	8.85	12.13	6.23	4.51
6	17.50	24.19	9.65	13.24	5.71	4.13
6-1/2	18.96	26.20	10.46	14.34	5.27	3.82
7	20.42	28.22	11.26	15.44	4.90	3.54
7-1/2	21.88	30.23	12.07	16.55	4.57	3.31
8	23.33	32.25	12.87	17.65	4.29	3.10
8-1/2	24.79	34.27	13.68	18.75	4.03	2.92
9	26.25	36.28	14.48	19.86	3.81	2.76
9-1/2	27.71	38.30	15.29	20.96	3.61	2.61
10	29.17	40.31	16.09	22.06	3.43	2.48
10-1/2	30.63	42.33	16.90	23.16	3.26	2.36
11	32.08	44.34	17.70	24.27	3.12	2.26
11-1/2	33.54	46.36	18.51	25.37	2.98	2.16
12	35.00	48.37	19.31	26.47	2.86	2.07

(metric)

Sheet Length, meters	Sq.M per Sheet		Wgt per Sheet, kg		Approx. No. Sheets per 10 sq.m*	
	0.89	1.23	0.89	1.23	0.89	1.23
1.5	1.4	1.9	3.65	5.00	7.38	5.34
1.7	1.5	2.1	4.01	5.50	6.71	4.85
1.8	1.6	2.2	4.38	6.01	6.15	4.45

Sheet Length, meters	Sq.M per Sheet 0.89	1.23	Wgt per Sheet, kg 0.89	1.23	Approx. No. Sheets per 10 sq.m* 0.89	1.23
2.0	1.8	2.4	4.74	6.50	5.67	4.11
2.1	1.9	2.6	5.11	7.00	5.27	3.81
2.3	2.0	2.8	5.47	7.51	4.92	3.56
2.4	2.2	3.0	5.84	8.01	4.62	3.34
2.6	2.3	3.2	6.21	8.51	4.34	3.14
2.7	2.4	3.4	6.57	9.01	4.10	2.97
2.9	2.6	3.6	6.94	9.51	3.89	2.81
3.0	2.7	3.7	7.30	10.01	3.69	2.67
3.2	2.8	3.9	7.67	10.51	3.51	2.54
3.4	3.0	4.1	8.03	11.01	3.36	2.43
3.5	3.1	4.3	8.40	11.51	3.21	2.33
3.7	3.3	4.5	8.76	12.01	3.08	2.23

For 0.024" (0.6 mm) thickness, reduce weights by 25%.
*For side and end lap allowance, add approximately 16% for 35" (875 mm) width and 12% for 48-3/8" (1209 mm) width.

Sizes, Weight, and Coverage Data on Corrugated Aluminum Sheets Used for Siding

Lengths, 5'-0" (1500 mm) to 12'-0" (3600 mm) in 6" (150 mm) increments. Widths, 33-3/4" (844 mm) and 47-1/8" (1178 mm). Thicknesses, 0.024" (0.6 mm) and 0.032" (0.8 mm). Corrugation, 2.67" (66.7 mm) pitch and 7/8" (22 mm) deep. Coverage, 32" (800 mm) for 33-3/4" (844 mm) width and 45-3/8 (1134 mm) for 47-1/8" (1178 mm) width.

Data on 0.032" (0.8 mm) Corrugated Aluminum Siding Sheets

Sheet Length ft.	SF per Sheet 33-3/4"	47-1/8"	Weight per Sheet, lbs 33-3/4"	47-1/8"	Approx. No. Sheets per 100 s.f.* 33-3/4"	47-1/8"
5	14.06	19.64	7.76	10.75	7.11	5.09
5-1/2	15.47	21.60	8.54	11.82	6.46	4.63
6	16.87	23.56	9.31	12.90	5.93	4.24
6-1/2	18.28	25.53	10.09	13.97	5.47	3.92
7	19.69	27.49	10.87	15.05	5.08	3.64
7-1/2	21.09	29.45	11.65	16.12	4.74	3.40
8	22.50	31.42	12.42	17.20	4.44	3.18
8-1/2	23.91	33.38	13.20	18.27	4.18	3.00
9	25.31	35.34	13.97	19.35	3.95	2.83
9-1/2	26.72	37.31	14.75	20.42	3.74	2.68
10	28.12	39.27	15.52	21.50	3.56	2.55
10-1/2	29.53	41.23	16.30	22.57	3.39	2.43
11	30.94	43.20	17.08	23.65	3.23	2.31
11-1/2	32.34	45.16	17.86	24.72	3.09	2.21
12	33.75	47.13	18.63	25.80	2.96	2.12

(metric)

Sheet Length meters	Sq.M per Sheet 0.85	1.2	Weight per Sheet, kg 0.85	1.2	Approx. No. Sheets per 10 sq.m* 0.85	1.2
1.5	1.3	1.8	3.52	4.88	7.65	5.48
1.7	1.4	2.0	3.87	5.36	6.95	4.98
1.8	1.6	2.2	4.22	5.85	6.38	4.56
2.0	1.7	2.4	4.58	6.34	5.89	4.22
2.1	1.8	2.6	4.93	6.83	5.47	3.92
2.3	1.9	2.7	5.28	7.31	5.10	3.66
2.4	2.1	2.9	5.63	7.80	4.78	3.42
2.6	2.2	3.1	5.99	8.29	4.50	3.23
2.7	2.3	3.3	6.34	8.78	4.25	3.05
2.9	2.5	3.5	6.69	9.26	4.03	2.88
3.0	2.6	3.7	7.04	9.75	3.83	2.74
3.2	2.7	3.8	7.39	10.24	3.65	2.62
3.4	2.8	4.0	7.75	10.73	3.48	2.49
3.5	3.0	4.2	8.10	11.21	3.33	2.38
3.7	3.1	4.4	8.45	11.70	3.19	2.28

For 0.024" (0.6 mm) thickness use data given above, except reduce weights 25%.
*For side and end lap allowance, add approximately 9% for 33-3/4" (800 mm) width and 6% for 47-1/8" (1178 mm) width.

Data on Nails Required for Roofing

Kind of Nail	Length, in inches	Length, in mm	No. per lb.	No. per kg	Price per lb.	Price per kg
Galvanized Needle Point	1-3/4	43.8	83	183	$0.85	$1.90
Galvanized Needle Point	2	50.0	77	170	0.85	1.90
Aluminum Nails—Neoprene Washers	1-3/4	43.8	318	701	2.90	6.40
Aluminum Nails—Neoprene Washers	2	50.0	285	628	2.90	6.40

Approximate Prices per S.F. (Sq.M) of Corrugated Aluminum Roofing and Siding Material

Thickness inches (mm)	Finish	Price per s.f.	Price per sq.m
0.0175" (4 mm)	Mill	$0.40	$4.30
0.0175" (4 mm)	Painted	0.45	4.85
0.024" (6 mm)	Mill	0.75	8.10
	Painted	1.00	10.75
0.032" (8 mm)	Mill	1.00	10.75
	Painted	1.25	13.45

After figuring the price of sheets from the above, approximately 20% should be added to cover cost of closures, flashings, fillers, corners, etc.

Labor Placing Corrugated Aluminum Roofing and Siding. Estimate about the same labor cost for erecting aluminum corrugated roofing and siding as given on the previous pages under *Corrugated Steel Roofing and Siding.*
Aluminum is much lighter than steel, but the labor operations are practically the same.

Labor Cost of One Square (100 s.f. or 9.3 sq.m) Corrugated Aluminum Roofing on Wood Framing, Based on using sheets 35" (875 mm) wide, 0.032" (0.8 mm) thick, weighing 0.552 lbs. per s.f. (2.69 kg per sq.m)

	Hours	Rate	Total	Rate	Total
Carpenter	1	$....	$....	$26.44	$26.44
Labor	1	....	....	21.04	21.04
Cost per 100 s.f.			$....		$47.48
per sq.m				....	5.11

Cost of One Square (100 s.f. or 9.3 sq.m) Corrugated Aluminum Siding on Wood Framing, Based on using sheets 33-3/4" (844 mm) wide, 0.032" (0.8 mm) thick, weighing 0.552 lbs. per s.f. (2.69 kg per sq.m)

	Hours	Rate	Total	Rate	Total
Carpenter	1.2	$....	$....	$26.44	$31.73
Labor	1.2	....	....	21.04	25.25
Cost per 100 s.f.			$....		$56.98
per sq.m				....	6.13

Cost of One Square (100 s.f. or 9.3 sq.m) Corrugated Aluminum Siding on Steel Framing, Straight Work, Based on using sheets 33-3/4" (844 mm) wide, 0.032" (0.8 mm) thick, weighing 0.552 lbs. per s.f. (2.69 kg per sq.m)

	Hours	Rate	Total	Rate	Total
Labor handling, hoisting	3.2	$....	$....	$21.04	$67.33
Sheeters	3.2	....	....	28.88	92.42
Cost per 100 s.f.			$....		$159.75
per sq.m				....	17.19

Add extra labor for high and irregular walls such as grain elevators, conveyor housing, etc. Add for closures, flashings, etc.

Cost of One Square (100 s.f. or 9.3 sq.m) Corrugated Aluminum Roofing on Steel Framing, Straight Work, Based on using sheets 35" (875 mm) wide, 0.032" (0.8 mm) thick, weighing 0.552 lbs. per s.f. (2.69 kg per sq.m)

	Hours	Rate	Total	Rate	Total
Labor handling, hoisting	2.67	$....	$....	$21.04	$56.18
Sheeter	2.67	....	....	28.88	77.11

Cost per 100 s.f. .... $.... $133.29
  per 10 sq.m 14.34
  Add extra labor for high or irregular roof areas. Add for closures, flashings, etc.

07460 CLADDING/SIDING

Wood siding is available in plain bevel, plain shiplap, and tongue and groove patterns to be used horizontally in heights from 4" (100 mm) to 12" (300 mm). Bevel siding is either 1/2" (12.5 mm) or 3/4" (19 mm) thick, T&G, and shiplap nominal 1" (25 mm). Cedar is the most common wood for siding but redwood, fir, hemlock, and spruce are also stocked. Wood may also be installed vertically with wood battens at the joints. Bevel siding 1/2" (12.5 mm) thick in clear "A" grade cedar will run about $1.25 per b.f., in redwood about $1.50. T&G red cedar 1x8s (25 x 200 mm) "D" and better will run $1.00 per b.f.; T&G redwood, clear and better, runs $1.70.

Hardboard siding 7/16" (11 mm) thick and factory primed, costs $0.50 per s.f.

Plywood siding in 4x8 (100 x 200 mm) sheets, rough textured and grooved to look like individual boards 5/8" (15.75 mm) thick costs about $0.85 per s.f. in select grade Douglas fir, $1.50 in redwood.

Aluminum siding, factory finished white in 8" (200 mm) widths, will run $0.75 uninsulated, $1.05 insulated per s.f. Flashing strips will cost $0.20 per l.f.; inside corners, $0.55; outside corners $1.10 per l.f. Colored nails cost $2.50 per lb.

Vinyl siding, factory finished in 8" (200 mm) wide strips, costs about $0.75 per s.f. uninsulated, $1.05 with insulating backup. Corner boards run $1.10 for outside positions, $0.55 for inside. Trim moldings run $0.35 per l.f., while soffits run $1.15 per s.f.

Placing Bevel and Drop Siding. The labor cost of placing bevel or drop siding will vary with the class of work and the method of placing same.

On the less expensive type of construction, it is customary to square only one end of the siding, and the ends at the corners are left rough and are later covered with metal corner pieces. This is the cheapest method of placing siding. Metal corners are used extensively in the South, because the long summers and intense heat cause mitered wood corners to open up, making a very unsightly appearance.

On more expensive buildings, the door and window casings and corner boards are placed, and then it is necessary to cut and fit each piece of siding between the corner boards or between the casings and corner boards. On work of this class, two carpenters usually work together, because it is necessary to square one end of the siding and then measure and cut each board separately to insure a snug fit. This increases labor costs considerably but makes for a better appearance than the metal corners.

This also applies where the siding is mitered at the corners. Two carpenters working together must measure and miter each piece of siding so that it will fit the adjoining corner.

The following quantities and production time are based on using ordinary brackets with plank scaffolding. If a more elaborate scaffold is required for placing siding, add extra.

Labor Placing Drop Siding

Measured Size, inches	Actual Size, inches	Class of Work	B.F. Placed per 8-Hr. Day	Carpenter Hours per 1000 b.f.
6	5-1/4	Rough Ends	525-575	14.5*
6	5-1/4	Fitted Ends	350-400	21.4
6	5-1/4	Mitered Corners	285-325	26.3
8	7-1/4	Rough Ends	600-650	12.8*
8	7-1/4	Fitted Ends	415-460	18.2
8	7-1/4	Mitered Corners	325-375	23.0

(metric)

Measured Size, mm	Actual Size, mm	Class of Work	Cu.M Placed per 8-Hr. Day	Carpenter Hours per 2.3 cu.m
150	131.3	Rough Ends	1.2-1.3	14.5*
150	131.3	Fitted Ends	0.8-0.9	21.4
150	131.3	Mitered Corners	0.6-0.7	26.3
200	181.3	Rough Ends	1.4-1.5	12.8*
200	181.3	Fitted Ends	0.9-1.0	18.2
200	181.3	Mitered Corners	0.7-0.8	23.0

*Where an electric saw is used to square both ends of the siding before placing, leaving the exposed corners rough to be covered with metal corner pieces, deduct 1 to 1-1/2 hours time per 1,000 b.f.

Quantity of Bevel Siding Required Per 100 S.F. (10 Sq.M) of Wall

Measured Size, inches	Actual Size, inches	Exposed to Weather	Pattern	Add for Lap	BF Required per SF Surface
1/2x4	1/2x3-1/4	2-3/4	Regular	46%	151
1/2x5	1/2x4-1/4	3-3/4	Regular	33%	138
1/2x6	1/2x5-1/4	4-3/4	Regular	26%	131
1/2x8	1/2x7-1/4	6-3/4	Regular	18%	123
5/8x8	5/8x7-1/4	6-3/4	Regular	18%	123
3/4x8	3/4x7-1/4	6-3/4	Rabbetted	18%	123
5/8x10	5/8x9-1/4	8-3/4	Rabbetted	14%	119
3/4x10	3/4x9-1/4	8-3/4	Rabbetted	14%	119
3/4x12	3/4x11-1/4	10-3/4	Rabbetted	12%	117

(metric)

Measured Size, mm	Actual Size, mm	Exposed to Weather	Pattern	Add for Lap	Cu.M per Sq.M Surface
12.5 x 100	12.5 x 81	68	Regular	46%	3.2
12.5 x 125	12.5 x 106	93	Regular	33%	3.0
12.5 x 150	12.5 x 131	118	Regular	26%	2.8
12.5 x 200	12.5 x 181	168	Regular	18%	2.6
15.6 x 200	15.6 x 181	168	Regular	18%	2.6
18.7 x 200	18.7 x 181	168	Rabbetted	18%	2.6
15.6 x 250	15.6 x 231	218	Rabbetted	14%	2.6
18.7 x 250	18.7 x 231	218	Rabbetted	14%	2.6
18.7 x 300	18.7 x 275	268	Rabbetted	12%	2.5

The above quantities include 5% for end cutting and waste.

Quantity of Drop Siding Required Per 100 S.F. (10 Sq.M) of Wall

Measured Size, inches	Actual Size, inches	Exposed to Weather	Add for Lap	BF Required per 100 SF Surface
1x6	3/4x5-1/4	5-1/4	14%	119
1x8	3/4x7-1/4	7-1/4	10%	115

(metric)

Measured Size, mm	Actual Size, mm	Exposed to Weather	Add for Lap	Cu.M per Sq.M Surface
25 x 150	18.7 x 131.2	131	14%	2.6
25 x 200	18.7 x 181.2	181	10%	2.5

The above quantities include 5% for end cutting and waste.

Quantity of Shiplap Required Per 100 S.F. (10 Sq.M) of Surface

Measured Size, inches	Actual Size, inches	Add for Lap	BF Required per 100 SF Surface
1x8	3/4x7-1/4	10%	115
1x10	3/4x9-1/4	8%	113

(metric)

Measured Size, inches	Actual Size, inches	Add for Lap	BF Required per 100 SF Surface
25 x 150	18.7 x 181.2	10%	2.5
25 x 250	18.7 x 231.2	8%	2.4

The above quantities include 5% for end cutting and waste.

Quantity of Bead Ceiling and Partition Required Per 100 S.F. (10 Sq.M) of Surface

Measured Size, inches	Actual Size, inches	Add for Width	BF Required per 100 SF Surface
1x4	5/8x3-1/4	23%	128
1x4	3/4x3-1/4	23%	128
(metric)			
Measured Size, mm	Actual Size, mm	Add for Width	Cu.M per 10 Sq.M Surface
25 x 100	15.6 x 81.2	23%	2.8
25 x 100	18.7 x 81.2	23%	2.8

The above quantities include 5% for end cutting and waste.

Quantity of Dressed and Matched (D & M) or Tongued and
Grooved (T & G) Boards Required per 100 s.f. (9.3 sq.m) of Surface

Measured Size, in. (mm)	Actual Size, in. (mm)	Add for Width	BF per 100 SF Surface	Cu.M per 10 Sq.M Surface
1x6 (25x150)	3/4x5-1/4 (18x131)	14%	119	0.27
2x6 (50x150)	1-5/8x5-1/4 (40x131)	14%	238	0.54

The above quantities include 5% for end cutting and waste.

Labor Placing Bevel Siding

Measured Size, inches	Actual Size, inches	Exposed to Weather, in.	Class of Workmanship	BF placed per 8-Hr. Day	Carpenter Hrs. per 1000 b.f.
4	3-1/4	2-3/4	Rough Ends	350-400	21.3*
4	3-1/4	2-3/4	Fitted Ends	240-285	30.5
4	3-1/4	2-3/4	Mitred Corners	200-240	36.3
5	4-1/4	3-3/4	Rough Ends	415-460	18.2*
5	4-1/4	3-3/4	Fitted Ends	285-330	25.0
5	4-1/4	3-3/4	Mitred Corners	240-285	30.5
6	5-1/4	4-3/4	Rough Ends	475-525	16.0*
6	5-1/4	4-3/4	Fitted Ends	325-375	23.0
6	5-1/4	4-3/4	Mitred Corners	265-310	28.0
8	7-1/4	6-3/4	Rough Ends	570-620	13.4*
8	7-1/4	6-3/4	Fitted Ends	375-415	20.0
8	7-1/4	6-3/4	Mitred Corners	300-350	24.0
10	9-1/4	8-3/4	Rough Ends	650-700	12.0*
10	9-1/4	8-3/4	Fitted Ends	440-480	17.5
10	9-1/4	8-3/4	Mitred Corners	375-425	20.0
12	11-1/4	10-3/4	Fitted Ends	475-525	16.0
12	11-1/4	10-3/4	Mitred Corners	400-450	19.0

(metric)

Measured Size, mm	Actual Size, mm	Exposed to Weather, mm	Class of Workmanship	BF placed per 8-Hr. Day	Carpenter Hrs. per 0.5 cu.m
100	81	68	Rough Ends	0.8-0.9	42*
100	81	68	Fitted Ends	0.5-0.6	61
100	81	68	Mitred Corners	0.4-0.5	72
125	106	93	Rough Ends	0.9-1.0	36*
125	106	93	Fitted Ends	0.6-0.7	50
125	106	93	Mitred Corners	0.5-0.6	61
150	131	118	Rough Ends	1.0-1.1	32*
150	131	118	Fitted Ends	0.7-0.8	46
150	131	118	Mitred Corners	0.6-0.7	56
200	181	168	Rough Ends	1.3-1.4	26*
200	181	168	Fitted Ends	0.8-0.9	40
200	181	168	Mitred Corners	0.7-0.8	48
250	231	218	Rough Ends	1.4-1.6	24*
250	231	218	Fitted Ends	1.0-1.1	35
250	231	218	Mitred Corners	0.8-1.0	40
300	281	268	Fitted Ends	1.1-1.2	32
300	281	268	Mitred Corners	0.9-1.0	38

*Where an electric handsaw is used to square both ends of the siding before placing, leaving the exposed corners rough to be covered with metal corner pieces, deduct 1 to 1-1/2 hours time per 1,000 b.f.

07500 MEMBRANE ROOFING

Built-up roofing consists of alternate plies of saturated felt and moppings of pitch with tar-saturated felt, or asphalt with asphalt-saturated felt, covered with a top pouring of pitch or asphalt into which slag or gravel is embedded. On flat roofs with slopes of less than 1/4" (6.2 mm) per foot (0.3 m), on which water may collect and stand, coal tar pitch and felt or a low melting point asphalt bitumen and asphalt felt are generally used. Coal tar pitch is not recommended for roofs having an incline in excess of 1" (25 mm) per foot (0.3 m).

On built-up roofs, where slope is over 2" (50 mm) to 4" (100 mm) per foot (0.3 m), steep asphalt, 180°-200°F (82°-93°C) melting point, is more suitable and slag is embedded in preference to gravel, because it remains embedded better than well rounded gravel. Where slag is not available and where a light gray or white surface is desired, hard limestone chips, angular pieces 1/4" (6.2 mm) to 3/4" (18.7 mm) in size, are embedded in the top pouring of bitumen. On slopes over 2" (50 mm) per foot (0.3 m), double coverage mineral surfaced roll roofing with a 19" (475 mm) selvage edge may be used as the top finish.

Smooth surface built-up roofing consists of alternate plies of felt cemented solid to the base sheet and to each other with asphalt, and a top coating of hot or cold asphalt, uniformly distributed.

The specifications on the types of roofing vary widely, depending on the surface to which the built-up roofing is applied and the service required.

The cost of built-up roofing is governed by the incline of the roof, size, plan (whether cut up with openings, skylights, penthouses, irregular roof levels, etc.), and the distance of the roof above the ground. The higher the roof, the higher materials must be hoisted, increasing the labor costs.

The term, "built-up roof" will be commonly heard in the trade, but it is more properly "built-up roofing". Technically, the "roof" is the supporting structure over which the "roofing" is applied. Specification writers, designers, and attorneys will be careful to distinguish between "roof" and "roofing".

Rules for Measuring Flat Roof. When measuring flat roof surfaces that are to be covered with composition, tar and gravel, tin, metal, or prepared roofing, the measurements should be taken from the outside of the walls on all four sides to allow for flashing up the side of each wall. The flashing is usually 8" (200 mm) to 1'-0" (300 mm) high. This applies particularly to brick, stone, or tile buildings having parapet walls above the roof level.

Fig. 1. Flat Roof with Parapet Walls

On flat roofs projecting or overhanging beyond the walls of the building, the measurements should cover the outside dimensions of the roof and not merely to the outside of the walls.

When estimating the area of flat roof surfaces, do not make deductions for openings containing less than 100 s.f. and then deductions should be made for just one-half the size of the opening.

Make deductions in full for all openings having an area of 500 s.f. (46 sq.m) or more.

Example: The length of the building, as shown in Figure 1, is 50'-0" (15 m). The width of the building is 20'-0" (6 m). Multiply the length by the width, 50 x 20 = 1,000 s.f. (15 x 6 = 90 sq.m). Referring to the figure, note that the building is only 48'-0" (14.6 m) long and 18'-8" (5.7 m) wide between walls.

The method of measuring the full length and width of the building is to allow for flashing up the side of the walls, which usually extends 8" (200 mm) to 12" (300 mm) high.

Fig. 2. Flat Roof Overhanging Walls

Figure 2 illustrates a building of the same size with projecting roof. The size of the building proper is 20'-0"x50'-0" (15 x 6 m), while the roof projects beyond the walls 2'-0" (600 mm) on each side, making the size of the roof 24'-0"x54'-0" (7.3 x 16.5 m). The roof area is obtained by multiplying the length of the roof, 54'-0" (16.5 m), by the width of the roof, 24'-0" (7.3 m), making the total roof area, 54'-0" x 24'-0" = 1,296 s.f. (16.5 x 7.3 = 120.5 sq.m).

The skylights on both roofs measure 6'-0"x9'-0" (1.8 x 2.7 m) and contain only 54 s.f. (5.0 sq.m) each. No deduction should be made for these openings, as they contain less than 100 s.f. (9.3 sq.m) each and the extra labor flashing up the sides of the walls more than offsets the cost of the small openings.

Quantity of Pitch or Asphalt Required for Built-Up Roofs. The quantity of roofing pitch or asphalt required for any built-up roof will vary with the number of thicknesses or plies of felt used. All manufacturers of roofing materials indicate in their specifications the amount of materials required for first class roof. On every first class roof, plenty of pitch or asphalt is used in each mopping so that each ply of felt is well cemented to the next and in no instance does felt touch felt.

Approximately 25 lbs. (11.3 kg) of coal tar pitch or approximately 20 lbs. (9 kg) of asphalt should be used for each mopping per 100 s.f. (9.3 sq.m) of surface and the last pouring, in which the gravel or slag is imbedded, should use 60 lbs. (30.6 kg) of asphalt or 75 lbs. (33.75 kg) of pitch per 100 s.f. (9.3 sq.m).

Quantity of Asphalt Required For 100 S.F. (9.3 Sq.M) of Roof

Surface to Which Roof is Applied	No. of Plies	Dry	Mopped	Lbs. per 100 s.f.	Kg per 10 cu.m
Wood, plywood, structural wood fiber	5	2	3	120	129
Poured gypsum, lightweight concrete	4	1	3	120	129
Concrete, precast concrete or gypsum	4		4	140	150

For roofing pitch add about 5 lbs. (2 kg) per mopping each ply per 100 s.f. (9.3 sq.m)—add 15 lbs. (6.8 kg) for top pouring per 100 s.f. (9.3 sq.m).

Estimating the Quantity of Roofing Felt. Asphalt or tarred felt for built-up roofing is furnished in 4 sq. rolls containing 432 s.f. (40 sq.m) and weighing from 56 to 62 lbs. (25-28 kg) per roll for No. 15 felt. No. 30 felt, which is generally used as the base sheet over wood construction when an asphalt specification is used, is furnished in 2 sq. rolls containing 216 s.f. (20 sq.m) and weighing 60 lbs. (27 kg). These weights are those used on a bonded job in the best type of building construction.

When specifying the grade or weight of felt to be used, it is customary to state that "felt shall weigh not less than 15 lbs. per 108 s.f.", and 15 lbs. per 108 s.f. (6.8 kg per 10 sq.m) is standard practice of all roofing manufacturers. Because felt is furnished in 4 sq. rolls of 432 s.f. (40 sq.m), 32 s.f. (3 sq.m) per roll or 8 s.f. (0.7 sq.m) per 100 s.f. (9.3 sq.m) is allowed for laps.

The following table gives the quantity of roofing felt required to cover 100 s.f. (9.3 sq.m) of surface of various thicknesses:

Number of Plies	Add for Waste	S.F. per 100 S.F. of Roof	Lbs. per Sq. of Roof Using No. 15 Felt	Sq.M per 10 Sq.M of Roof	Kg per Sq.M of Roof No. 15 Felt
1	8%	108	15	11	5.4
2	8%	216	30	22	10.7
3	8%	324	45	32	15.6
4	8%	432	60	43	21.0
5	8%	540	75	54	26.4

The general use of No. 30 felt is as a base sheet over wood decks in 1-ply thickness over which subsequent layers of No. 15 felts are mopped in.

Quantity of Roofing Gravel Required for Built-Up Roofs. Roofing gravel should be uniformly embedded into a heavy top pouring of asphalt or pitch so that approximately 400 lbs. (181 kg) of gravel or 300 lbs. (136 kg) of slag is used per 100 s.f. (9.3 sq.m) of roof area.

Labor Applying Built-Up Roofing. The labor cost of handling materials and applying built-up roofing will vary with the specification used and the type of building to which the roofing is applied. On the low type of building, 1, 2, or 3 stories high, where there is no great distance from the ground to the roof deck, the labor for hoisting materials, and supplying hot asphalt or pitch from the ground to the roof deck is considerably less than on a high building, where there is a greater distance from the kettle to the roof.

On low type buildings, a 5-worker crew composed of 1 worker at the kettle, 1 worker carrying "hot stuff", 1 worker rolling in felts, 1 worker nailing in and 1 worker mopping in, can lay the following areas per 8-hr. day:

1. 18 sqs. (167 sq.m) of 5-ply asphalt and gravel or pitch and gravel over wood roof deck, consisting of 1 sheathing paper, 2 dry felts and 3 plies of

felt mopped in with a top pouring of pitch or asphalt and gravel surfacing;

2. 18 sqs. (167 sq.m) of 4-ply asphalt and gravel or pitch and gravel over a concrete deck, all mopped including a top pouring of pitch or asphalt and gravel or slag surfacing;

3. 22 sqs. (204 sq.m) of 4-ply asphalt and gravel or pitch and gravel over wood roof deck, consisting of 1 sheathing paper, 2 dry felts and 2 plies of felt mopped in with a top pouring of pitch or asphalt and gravel or slag surfacing;

4. 22 sqs. (204 sq.m) of 3-ply asphalt and gravel or pitch and gravel over concrete roof deck, all mopped and including a top pouring of pitch or asphalt and gravel or slag surfacing;

5. 24 sqs. (223 sq.m) of 3-ply asphalt and gravel, or pitch and gravel, over wood roof deck, consisting of 1 dry felt, and 2 plies of felt mopped in, with a top pouring of pitch or asphalt, and gravel or slag surfacing.

The above figures apply to buildings up to 3 stories in height.

If the roof surfaces are broken up with skylights, irregular roof level, etc., the above quantities should be reduced about 15% to 20%.

A 5-worker crew should apply the following number of squares (sq.m) of roof per 8-hr. day, based on first class workmanship throughout:

Type of Roofing	Type of Structure "A" sqs. (sq.m)	"B" sqs. (sq.m)
3-ply roof over wood roof deck	24 (223)	18 (167)
3-ply roof over concrete roof deck or insulation	22 (204)	17 (158)
4-ply roof over wood roof deck	22 (204)	17 (158)
4-ply roof over concrete roof deck or insulation	18 (167)	14 (130)
5-ply roof over wood roof deck	18 (167)	14 (130)

Type "A" buildings are the most convenient type of structure for application of built-up roofing. A low type of building from 1 to 3 stories in height, straight, practically flat area not broken up by very many skylights or variations in the deck elevation.

Type "B" buildings are not convenient type of structures for the application of built-up roofing. These would include high buildings which require considerable handling of materials, and buildings with roofs broken up by skylights or penthouses, sawtooth construction, monitors, or considerable variation in deck levels.

An easy and accurate method of computing the labor cost on any type of roof is to take the cost of a 5-worker crew for an 8-hr. day and divide this by the number of squares of roof applied per day:

	Hours	Rate	Total	Rate	Total
Foreman (working)	8	$....	$....	$26.46	$211.68
Roofers (4)	32		....	24.99	799.68
Crew cost per 8-hr. day			$....		$1,011.36
Cost per sq. - 14 sqs. per day				....	72.24
Cost per sq. - 17 sqs. per day				....	59.49
Cost per sq. - 18 sqs. per day				....	56.19
Cost per sq. - 22 sqs. per day				....	45.97
Cost per sq. - 24 sqs. per day				....	42.14

(metric)

	Hours	Rate	Total	Rate	Total
Foreman (working)	8	$....	$....	$26.46	$211.68
Roofers (4)	32		....	24.99	799.68
Crew cost per 8-hr. day			$....		$1011.36
Cost per sq.m - 130 sq.m per day				....	7.78
Cost per sq.m - 158 sq.m per day				....	6.40
Cost per sq.m - 167 sq.m per day				....	6.06
Cost per sq.m - 204 sq.m per day				....	4.96
Cost per sq.m - 223 sq.m per day				....	4.54

On the average roof where the mopping between layers consists of mopping the width of the lap, a roofer should mop 100 s.f. (9.3 sq.m) of surface in 6 to 7 minutes.

On first grade work where the entire roof surface is mopped between each layer of felt, a roofer should mop 100 s.f. (9.3 sq.m) of roof in about 10 minutes.

After the roof surface has been mopped, a roofer should place roofing felt over 100 s.f. (9.3 sq.m) in about 6 minutes.

Where first grade workmanship is required, a roofer should handle pitch and gravel, pour hot pitch or asphalt and spread gravel over one square of roof in 35 to 40 minutes.

Prices on Roofing Materials. Prices on roofing materials such as coal tar pitch, asphalt, felt, etc., vary according to market fluctuations, location of job, etc. Current material prices should always be obtained, before submitting bid.

Material Cost of One Square (100 s.f. or 9.3 sq.m)
3-Ply Tar and Gravel Built-Up Roof
Applied Over Wood Roof Deck

Maximum incline 2" per l.f. (50 mm per 0.3 m) One ply No. 30 felt, 2 plies No. 15 felt (1 dry, nailed and 2 mopped), coal tar pitch surfaced with gravel or slag.

	Rate	Total	Rate	Total
1.1 sqs. (10.2 sq.m) No. 30 felt	$....	$....	$6.00	$6.60
2.2 sqs. (20.4 sq.m) No. 15 felt	....		3.00	6.60
125 lbs. (56 kg) coal tar pitch	....		0.13	16.25

			Rate	Total
400 lbs. (181 kg) roofing gravel	....		0.50	2.00
Nails, fuel, mops, etc.	....		2.50	2.50
Cost per 100 s.f.		$....		$33.95
per 10 sq.m			....	36.54

Material Cost of One Square (100 s.f. or 9.3 sq.m)
3-Ply Tar and Gravel Built-Up Roof Over Poured Concrete, Poured Gypsum Roof Decks, or Insulation Board

Maximum incline 2" per l.f. (50 mm per 0.3 m) Roof consisting of 3 plies of felt, coal tar pitch surfaced with gravel or slag.

	Rate	Total	Rate	Total
3.3 sqs. (30.6 sq.m) No. 15 felt	$....	$....	$3.00	$9.90
150 lbs. (68 kg) coal tar pitch		....	0.13	19.50
400 lbs. (181 kg) roofing gravel		....	0.50	2.00
Fuel, mops, etc.		....	3.00	3.00
Cost per 100 s.f.		$....		$34.40
per 10 sq.m		....		37.02

Material Cost of One Square (100 s.f. or 9.3 sq.m) of Each Additional Ply of No. 15 Felt and Hot Mopping of Pitch

	Rate	Total	Rate	Total
1.1 sqs. (10.2 sq.m) No. 15 Felt	$....	$....	$3.00	$3.30
25 lbs. (11.3 kg) Pitch		....	0.13	3.25
Fuel, Mops, etc.		....	0.50	0.50
Cost per 100 s.f.		$....		$7.05
per 10 sq.m		....		7.59

07520 PREPARED ROLL ROOFING

Ready roofing can be used on small buildings where an inexpensive yet satisfactory roofing is required. It is furnished in various grades and weights and is sold by the roll or in flat sheets packed with the nails and cement necessary for application.

Laying Ready-to-Lay Roll Roofing. When laying ready-to-lay roll roofing, a roofer should handle, lay, and cement or nail in place about 800 s.f. (74 sq.m) per 8-hr. day, at the following cost per 1 square (100 s.f. or 9.3 sq.m):

	Hours	Rate	Total	Rate	Total
Carpenter	1.0	$....	$....	$26.44	$26.44

Asphalt Slate Surfaced Ready-to-Lay Roofing

A ready-to-lay roofing consisting of several plies of asphalt impregnated felt, surfaced with a thick layer of slate granules, either red or green. For roofs having a pitch of 1-1/2" or more per foot (37 mm or more per 0.3 m).

Furnished in rolls 36" (900 mm) wide and 36'-0" (11 m) long, containing 108 s.f. (10 sq.m), sufficient for 100 s.f. (9.3 sq.m) of roof, at $13.00 per 90-lb. (40.5-kg) roll.

Smooth Surfaced Ready-to-Lay Roofing

Mica surface asphalt roofing consists of good quality felt saturated with pure asphalt. Both sides are covered with flake mica. Rolls are 36" (900 mm) wide and 36'-0" (11 m) long.

Description	Lbs per roll	S.F. per roll	Price per roll	Kg per roll	Sq.M per roll	Price per roll
Mica surfaced asphalt roofing	55	108	$11.50	25	10	$11.50
	65	108	13.25	19	10	13.25

7530-40 ELASTIC SHEET AND FLUID APPLIED ROOFING

Elastomeric coatings include neoprene, Hypalon, urethane, butyl, and silicone. Some are applied by spraying, others in sheet form, and all are subcontracted to firms licensed by the manufacturer.

Since these coatings are thin, the roof surface to which they are to be applied must be firm, continuous, smooth, clean, and dry. New concrete should be sealed with a primer.

Thinness and ability to conform to any shape, and the fact that they can be had in most any color including white, make them the choice as the roofing surface for decorative and fluid roof forms. Some materials are also used as traffic decks.

A 1/16" (1.5 mm) thick butyl sheet will cost around $0.65 per s.f. ($7.00 per sq.m) and one roofer can install some 225 s.f. (21 sq.m) per day for a total cost of around $1.25 per s.f. ($13.45 per sq.m).

A 1/16" (1.5 mm) neoprene sheet will have a material cost of some $1.25 per s.f. ($13.55 per sq.m) and will install at the same rate as the butyl for an installed price of around $1.80 per s.f. ($19.40 per sq.m).

Fluid applied Hypalon-neoprene 0.02" thick will be applied at the rate of some 100-110 s.f. (9.3-10.2 sq.m) per day and have a material cost of $0.85 per s.f. ($9.00 per sq.m) for a total figure of around $2.00 per s.f. ($21.50 per sq.m).

EPDM ROOF SYSTEMS

Ethylene Propylene Diene Methylene Rubber is popularly known in the trade as EPDM. EPDM is formulated as an elastomeric polymer synthesized from ethylene, propylene and diene (a small amount). EPDM sheeting is a vulcanized material; that is, the polymer's molecular structure is set as a result of heating during the manufacturing process. EPDM membranes can be produced in various colors, including black, reinforced or non-reinforced, and in thicknesses ranging from 30 to 90 mils.

The material was introduced into the construction industry in the early 1960's and in recent years has become a popular inexpensive flat roof application system. It is estimated that 20% of North America roof installations are EPDM systems. The success of this product during the past several years is due to its ability to withstand a wide range of temperatures, from -50°F to 240°F. (-10°C to 116°C), manufacture and supplier support and because of the stability of the product.

Non-reinforced EPDM is loose-laid over insulation or substrate (roof deck) and attached directly to the substrate using the EPDM manufacturer's approved fasteners and fastening plates, splicing cement and seam sealants. Non-reinforced material normally is available in widths of 7', 10', 10.5', 20' and 30' (2, 3, 5, 6 and 9 m) by 100' (32 m) lengths. Standard or reinforced EPDM, either mechanically fastened, fully-adhered or loose laid, is available in widths up to 10' (3 m) x 200' (61 m). Check with individual manufacturers for sizes and color.

Types of EPDM Installment. *Mechanically fastened* EPDM roofing is secured with manufacturer approved seam fasting plates. The EDPM membrane is rolled into position and overlapping sheets are fastened together with manufacturer approved splicing cement, in-seam sealents.

Fully adhered EPDM roofing is mechanically attached to an approved substrate (roof deck). The substrate is coated with manufacturer approved bonding adhesive, and the EPDM membrane is rolled into place and "brushed" down. Sheets are fastened with approved in-seam sealants or splicing tapes.

Ballasted EPDM roofing is normally installed by first applying approved insulation, which is loose-laid over an approved substrate (roof deck). The EPDM membrane is loose-laid over insulation and secured with a minimum of 10 Lbs. (4.5 Kg) of ballast per square foot.

Warranty. Warranties can vary. The norm is 10 years, and the maximum is 20 years, but 15-year warranties are available. Terms and conditions vary among manufacturers, and the length of a warranty will mandate certain conditions being adhered to during installation. Longer warranty periods will add costs to the installation for items such as certifications, inspections, installation restrictions and applicator certifications.

EPDM - FULLY ADHERED

**EPDM - FULLY ADHERED
MECHANICAL DESIGN**

Materials for EPDM. In addition to the roofing material, the following material supplies are normally required for EDPM roofing installations:

Item	Produced in
Seam Primer	1-gallon containers
Seam Sealer	1/10 gallon cartridges (tubes)
Bonding / Splicing Adhesive	5 gallon pails
Water cut-off mastic	Tubes of 1 gallon cans

Seam tape .... 3" wide
Wood nailers .... to be determined
Preformed pipe boots .... Sized for pipes
Termination bars .... 10' x 1"
Corrosion resistant fasteners .... to be determined

Because the amounts supplied and used will vary with manufacturers, an estimator should consult a specific manufacturer before estimating a project to assure that all materials required and quantities needed are correct for securing a watertight roof and to determine that the manufacturer offers the warranty period that is mandated by the owner.

Typical EPDM Roof Installation. Position the membrane over the substrate (roof deck) without stretching the sheet. When a new membrane is to be sealed to an abutting membrane, overlap the edge with the new membrane a minimum of 5 inches (125 mm) onto the in-place membrane.

Bond the new sheet to the substrate by folding the sheet lengthwise on to itself so that one half of the underside is exposed. Smooth out all the wrinkles and buckles along the edge. Apply the bonding adhesive evenly at the rate of 120 sq.ft. (36.5 m) per gallon per surface, using a 9" (225 mm) paint roller (manufacturer approved). Allow the adhesive to become tacky, then roll the membrane onto the substrate avoiding wrinkling by using a wave air motion at the leading edge to roll the sheet into place. Bond the adhered half of the membrane with an 18" to 24" (450-600 mm) heavy-duty push broom. Pull back the unbonded half of the sheet and repeat the procedure. The adhesive should not be applied to the top surface of any abutting membrane that is to be used as the seam joint.

All lap joints should be a minimum 4' wide (1.2 m) and shingled toward the drains or roof scuppers. Apply splicing adhesives to both bonding surfaces with a 4" (100 mm) paint brush at the rate of 120 lin. ft. (36.5 m) of 4-foot (1.2 m) lap per gallon, allowing the adhesive to dry until tacky. Join the bonding surfaces, avoiding stretching or fishmouths. Roll the lap joint thoroughly with a roller in a direction perpendicular to one length of the joint. Apply a seam sealer to the exposed edge and tool to assure positive contact with both sheets. Apply seam sealer to all lap joints completed during each day.

Mechanically Fastened. The EPDM membrane is fastened using 2" (50 mm) steel disks (or size as determined by manufacturer). The spacing and the type of fastener used depends on the substrate and approval by the EPDM manufacturer.

A roofing crew should be able to place 3,500 or 35 squares per day of partially adhered (mechanically fastened) EPDM roofing at the following labor cost:

	Hours	Rate	Total	Rate	Total
Roofer Foreman	8.0	$.....	$	$25.99	$207.92
Roofers	32.0	.....	.....	24.99	799.68
Laborers	8.0	.....	.....	$21.04	168.32
Total Cost for 35 Squares		.....	$.....		$1,175.92
per Square			.....		33.60
per s.f.			.....		0.34
per sq.m			.....		3.62

A roofing crew should be able to place 2,500 or 25 squares per day of fully adhered EPDM roofing at the following labor cost:

	Hours	Rate	Total	Rate	Total
Roofer Foreman	8.0	$.....	$	$25.99	$207.92
Roofers	32.0	.....	.....	24.99	799.68
Laborers	8.0	.....	.....	$21.04	168.32
Total Cost for 25 Squares		.....	$.....		$1,175.92
per Square			.....		47.04
per s.f.			.....		0.47
per sq.m			.....		5.06

07560 ROOF MAINTENANCE

Ease of periodic preventive maintenance is one of the strong points of a flat roof. Good maintenance programs can eliminate the major cause of roof complaints—leaks. Regardless of who does a roof inspection and how they do it, inspections should be held both in the spring and fall, with special inspections after exceptional storm events. Roof inspection include but are not limited to:

1. Interior walls and ceiling for signs of water staining
2. Cap flashing
3. Edge metal
4. Base flashings
5. Penetrations
6. Field of the roof
7. Ballast
8. Roof adhesives
9. Surface coatings

Normal maintenance includes keeping the drainage system clear and functioning and restricting roof access. If traffic patterns develop, to address the situation the owner should consult the roof manufacturer. A professional roofing contractor should accomplish all roof repairs. Taking these steps will help assure an owner (as best as possible) that a flat roof will be free of leak and will assist in the validation of the warranty.

Maintenance costs depend on many factors. A larger roof does not necessarily mean higher maintenance costs. Typical items that add to the cost of maintenance are access, amount of and type of roof penetrations, drainage system, equipment on roof and foot traffic.

Assuming a roof area of 30,000 sq. ft. (2,787 sq.m), with two annual inspections (one in the spring and one in the fall) inspection cost is as follows:

	Hours	Rate	Total	Rate	Total
Roofer Foreman..................................	32.0	$	$	$25.99	$831.68
Roofer...	32.0	.....	.....	24.99	799.68
Total Cost - two inspections....................					$1,631.36
per s.f..					0.05
per sq. m					0.54

Make allowance for written reports, if required, and for materials for minor and miscellaneous repairs over a 10 to 20 year period (longer if necessary). This allowance can vary greatly (as do opinions concerning roof maintenance) from $0.00 to $1.00 per square foot per year.

07600 FLASHING AND SHEET METAL

07610 SHEET METAL ROOFING

Metal roofing includes galvanized steel, copper, lead, stainless steel, and aluminum plus the many combinations and alloys of these metals such as lead-coated copper, terne (80% lead, 20% tin over copper-bearing carbon steel) microzinc, and terne coated stainless. Terne and aluminum are the least expensive and copper and lead the costliest. Metal prices tend to fluctuate broadly. Many of the ores are imported and at the mercy of the value of the dollar and the political climate of the country where they are mined.

Metal roofs may be applied in many ways. The simplest is the flat seam roof that may be used on slopes as low as 1/4" (6.2 mm) to the foot. Standing seams and battened seams generally need a slope of at least 2-1/2" (62.5 mm). They are more decorative, and in fact batten designs are often selected solely on their decorative value. Some materials may be ordered with prefabricated battens, others are formed in the traditional way over wood strips. Another decorative roof is the "Bermuda" type, where the metal is applied over wood "steps" provided by the carpenter. The step is based on the width of the metal roll to be used and is sloped at least 2-1/2" per foot (62.5 mm per 0.3 m). The roofer interlocks the rolls at each step edge giving a sharp shadowline.

For custom work terne and copper, with or without special coatings, are the usual choices. Many batten roofs today have the batten stamped into the metal and are usually factory finished aluminum or steel.

When estimating quantities of metal roofing, the measurements are taken in the same manner as for other types of roofs.

Where the metal roof is applied over flat roofs having parapet walls, an allowance of 1'-0" (0.3 m) on each side and end of the wall should be made for flashing up the sides of the walls.

If metal roofing is applied to a flat surface or pitch roof with overhanging eaves, the measurement is taken to cover the entire roof surface.

Do not make deductions for openings containing less than 25 s.f. (2.3 sq.m) as the extra flashing costs as much as the roofing omitted.

Deduct in full for all openings containing over 50 s.f. (4.6 sq.m) and add for flashing.

Flat Seam Metal Roofing. The common sizes of tin plates are 10"x14" (250 x 350 mm) or multiples of that size. The sizes generally used for roofing are 14"x20" (350 x 500 mm) and 20"x28" (500 x 700 mm). The larger sizes are more economical to lay, and for flat roofs the 20"x28" (500 x 700 mm) size is preferable.

For a flat seam roof, the edges of the sheets are turned in about 1/2" (12.5 mm), locked together and well soaked with solder. The sheets should be fastened to the roof boards by cleats locked in the seams. These cleats are usually spaced about 8" (200 mm) apart and 2 nails are used in each cleat.

Method of Laying Flat Seam Metal Roofing

For flat seam roofing, a 14"x20" (350 x 500 mm) sheet of tin, with edges turned up 1/2" (12.5 mm) on each side and end, measures 13"x19" (325 x 475 mm) and contains 247 sq. in. (0.16 sq.m); but the covering capacity when locked to other sheets is only 12-1/2"x18-1/2" (312 x 462 mm) or 231-1/4 sq. in. (0.15 sq.m). It requires 62-1/2 sheets of 14"x20" (350 x 500 mm) tin to cover 100 s.f. (9.3 sq.m) of roof.

Sheets measuring 20"x28" (500 x 700 mm) with the edges turned for flat seam roofing measure 19"x27" (475 x 675 mm), and when locked to other sheets, have a covering capacity of 18-1/2"x26-1/2" (462 x 662 mm) or 490-1/4 sq. in. (0.32 sq.m). It requires 20-1/2 sheets to cover 100 s.f. (9.3 sq.m) of roof.

Tin for roofs may also be obtained in rolls 14" (350 mm), 20" (500 mm), 24" (600 mm), and 28" (700 mm) wide. When used for flat seam roofing the loss due to turning edges amounts to 1-1/2" (37.5 mm) on the width of any sheet.

Tin in 14" (350 mm) rolls loses 1-1/2" (37.5 mm) or 11.2% of its width due to turning edges and laps. Another method is to take the length of the roof and divide by the net width of each sheet and the result will be the number of strips of tin required. Multiply the number of strips by the length of each strip to obtain the quantity of tin required for any roof.

The following table gives the covering capacity and allowances for edging and laps to be figured when using tin roofing in rolls for flat seam roofs:

Width of Sheets in. (mm)	Allowance for Edge and Laps	Actual Width in. (mm)	Add for Loss Due to Edges and Laps	No. SF Required Per Sq.
14 (350)	1-1/2 (37)	12-1/2 (312)	12.0%	112
20 (500)	1-1/2 (37)	18-1/2 (462)	8.0%	108
24 (600)	1-1/2 (37)	22-1/2 (562)	7.0%	107
28 (700)	1-1/2 (37)	26-1/2 (662)	6.0%	106

The quantities of galvanized iron or steel sheets required for flat seam roofing will depend upon the size of the sheets, as they may be obtained in sheets from 24"x96" (600 x 2400 mm) to 48"x120" (1200 x 3000 mm) in size.

The actual measurements of the sheets will be 1-1/2" (37.5 mm) less in width and length than the size of the sheet used. For instance, the actual covering capacity of a 24"x96" (600 x 2400 mm) sheet will be only 22-1/2" x 94 = 2,115 sq. in. or 14.76 s.f. (562 x 2350 mm = 1,320,700 sq.mm).

The following table gives the approximate covering capacity of one sheet of the different sizes, allowing 1-1/2" (37.5 mm) in width and length for turning edge and lapping, together with the number of sheets of the different sizes required to cover one sq. (100 s.f. or 9.3 sq.m) of surface:

Size of Sheet, inches	Allowance for Edge of Lap		Actual Size Sheet, in.	Cover Cap. One Sheet	No. Sheets Per Sq.
	Length	Width			
24x96	1-1/2	1-1/2	22-1/2x94-1/2	14.76	6.8
24x120	1-1/2	1-1/2	22-1/2x118-1/2	18.51	5.4
26x96	1-1/2	1-1/2	24-1/2x94-1/2	16.08	6.2
26x120	1-1/2	1-1/2	24-1/2x118-1/2	20.16	5.0
28x72	1-1/2	1-1/2	26-1/2x70-1/2	13.00	7.7
28x84	1-1/2	1-1/2	26-1/2x82-1/2	15.18	6.6
28x96	1-1/2	1-1/2	26-1/2x94-1/2	17.40	5.8
28x108	1-1/2	1-1/2	26-1/2x106-1/2	19.60	5.1
28x120	1-1/2	1-1/2	26-1/2x118-1/2	21.80	4.6
30x96	1-1/2	1-1/2	28-1/2x94-1/2	18.70	5.4
30x120	1-1/2	1-1/2	28-1/2x118-1/2	23.45	4.3

(metric)

Size of Sheet, mm	Allowance for Edge of Lap		Actual Size Sheet, mm	Cover Cap. One Sheet	No. Sheets per Sq.M
	Length	Width			
600x2400	37	37	563x2363	1.4	73
600x3000	37	37	563x2963	1.7	58

Size of Sheet, mm	Allowance for Edge of Lap		Actual Size Sheet, mm	Cover Cap. One Sheet	No. Sheets per Sq.M
	Length	Width			
650x2400	37	37	613x2363	1.5	67
650x3000	37	37	613x2963	1.9	53
700x1800	37	37	663x1763	1.2	83
700x2100	37	37	663x2063	1.4	71
700x2400	37	37	663x2363	1.6	62
700x2700	37	37	663x2663	1.8	55
700x3000	37	37	663x2963	2	49
750x2400	37	37	713x2363	1.7	58
750x3000	37	37	713x2963	2.2	46

Standing Seam Metal Roofing. Standing seam roofing requires a larger allowance for waste than flat seam roofing. The standing seam, edged 1-1/4" (31 mm) and 1-1/2" (37.5 mm) takes 2-3/4" (68.7 mm) off the width and the flat cross seams edged 3/8" (9.3 mm), take 1-1/8" (3 mm) off the length.

If 14"x20" (350 x 500 mm) tin is used, each sheet covers 11-1/4"x18-7/8" (281 x 471 mm) or 212.34 sq. in. (132,351 sq.mm). It requires 68 sheets of 14"x20" (350 x 500 mm) tin per 100 s.f. (9.3 sq.m) of roof.

If 20"x28" (500 x 700 mm) tin is used, each sheet covers 17-1/4x26-7/8" (431 x 671 mm) or 463.59 sq. in. (289,201 sq.mm). It requires 32 sheets of 20"x28" (500 x 700 mm) tin per 100 s.f. (9.3 sq.m).

If roll tin is used, 2-3/4" (68 mm) should be allowed for the two standing seams. The end waste will vary with the length of the sheets.

Tin in 14" (350 mm) rolls loses 2-3/4" (68 mm) off the width due to the standing seams, making it necessary to add 20% for waste. Another method is to take the width of the roof and divide by 11-1/4" (281 mm), the net width of the sheet, and the result is the number of strips of tin required. Multiply the number of strips by the length of the roof to obtain the quantity of tin required.

Method of Laying Standing Seam Metal Roofing

The following table gives the covering capacity and allowances for standing seams, when estimating roll roofing:

Width of Sheets inches	Allowance for Standing Seams	Actual Width inches	Add for Loss Due to Standing Seams	No. Sq. Ft. Required Per Square
14	2-3/4	11-1/4	25%	125
20	2-3/4	17-1/4	16%	116
24	2-3/4	21-1/4	13%	113
28	2-3/4	25-1/4	11%	111

(metric)

Width of Sheets mm	Allowance for Standing Seams	Actual Width mm	Add for Loss Due to Standing Seams	Sq.M Required Per Sq.M
350	68	282	25%	1.25
500	68	432	16%	1.16
600	68	532	13%	1.13
700	68	632	11%	1.11

If galvanized steel or any of the special process metals, such as Armco, Toncan, etc., are used for standing seam roofs, an allowance of 1-1/4" (31.2 mm) and 1-1/2" (37.5 mm) or a total of 2-3/4" (68 mm) should be taken from the width of the sheets to allow for standing seams. The cross seams edged 3/8" (9.3 mm) take 1-1/8" (28 mm) off the length of each sheet. For instance, where 24"x96" (600 x 2400 mm) sheets are used, the actual size of the sheets is 21-1/4"x94-7/8" (531 x 2371 mm), or each sheet will cover 2,016 sq. in. or 14 s.f. (1.3 sq.m). The following table gives the covering capacity of painted or galvanized steel sheets of the different sizes:

Size of Sheet, inches	Standing Seam & End Lap, in. width x length	Actual Size Sheet, inches	Cover Cap. per Sheet, s.f.	No. Sheets per Square (100 s.f.)
24x96	2-3/4x1-1/8	21-1/4x94-7/8	14.00	7.2
24x120	2-3/4x1-1/8	21-1/4x118-7/8	17.54	5.7
26x96	2-3/4x1-1/8	23-1/4x94-7/8	15.33	6.6
26x120	2-3/4x1-1/8	23-1/4x118-7/8	19.20	5.3
30x96	2-3/4x1-1/8	27-1/4x94-7/8	17.92	5.6
30x120	2-3/4x1-1/8	27-1/4x118-7/8	22.50	4.5
36x96	2-3/4x1-1/8	33-1/4x94-7/8	22.00	4.6
36x120	2-3/4x1-1/8	33-1/4x118-7/8	27.45	3.7
42x96	2-3/4x1-1/8	39-1/4x94-7/8	25.85	3.9
42x120	2-3/4x1-1/8	39-1/4x118-7/8	32.40	3.1
48x96	2-3/4x1-1/8	45-1/4x94-7/8	29.80	3.4
48x120	2-3/4x1-1/8	45-1/4x118-7/8	37.40	2.7

(metric)

Size of Sheet, mm	Standing Seam & End Lap, mm width x length	Actual Size Sheet, mm	Cover Cap. per Sheet, sq.m	No. of Sheets per sq.m
600x2400	68x28	532x2372	1.3	7.2
600x3000	68x28	532x2972	1.6	5.7
650x2400	68x28	582x2372	1.4	6.6

Size of Sheet, mm	Standing Seam & End Lap, mm width x length	Actual Size Sheet, mm	Cover Cap. per Sheet, sq.m	No. of Sheets per sq.m
650x3000	68x28	582x2972	1.8	5.3
750x2400	68x28	582x2972	1.7	5.6
750x3000	68x28	682x2372	2.1	4.5
900x2400	68x28	832x2372	2.0	4.6
900x3000	68x28	832x2972	2.6	3.7
1050x2400	68x28	982x2372	2.4	3.9
1050x3000	68x28	982x2972	3.0	3.1
1200x2400	68x28	1132x2372	2.8	3.4
1200x3000	68x28	1132x2972	3.5	2.7

V-Crimped Roofing. This type of roofing is used the same as standing seam roofing. It is usually furnished in sheets covering 24" (600 mm) in width and 6'-0" (1.8 m) to 10'-0" (3.0 m) long.

When estimating quantities of V-crimped roofing, allow for the end lap but there is no waste in the width as only the actual covering capacity is charged for by the manufacturer. For instance, a sheet 26" (650 mm) wide before crimping is 24" (600 mm) wide after crimping, but the sheet is sold as 24" (600 mm) wide.

The following table gives the quantity of V-crimp roofing required to cover 100 s.f. (9.3 sq.m) of roof with end laps 1" (25 mm) to 6" (150 mm):

	End Laps, inches					
Length of Sheet, ft.	1	2	3	4	5	6
	S.F. of V-Crimp Roofing Required					
6	102	103	105	106	108	109
7	102	103	104	105	106	108
8	101	102	103	104	106	107
9	101	102	103	104	105	105
10	101	102	103	104	105	105

(metric)

	End Laps, mm					
Length of Sheet, m	25	50	75	100	125	150
	Sq.M of V-Crimp Roofing Required					
1.8	9.47	9.56	9.75	9.84	10.03	10.12
2.1	9.47	9.56	9.66	9.75	9.84	10.03
2.4	9.38	9.47	9.56	9.66	9.84	9.94
2.7	9.38	9.47	9.56	9.66	9.75	9.75
3.0	9.38	9.47	9.56	9.66	9.75	9.75

Labor Placing 100 S.F. (9.3 Sq.M) Flat and Standing Seam
Tin and Metal Roofing

Description of Roof	SF per 8-Hr. Day	Sq.m per 8-Hr. Day	Hours per 100 s.f. (9.3 sq.m) Tinner	Helper
Flat seam metal, 24"x96" (600x2400mm) sheets or larger	300-350	27-33	2.5	2.5
Flat seam tin, 14"x20" (350x500mm) plates	175-225	16-21	4.0	4.0
Flat seam tin, 20"x28" (500x700mm) plates	225-275	21-26	3.3	3.3
Flat seam tin, 20" (500mm) tin, rolls	325-375	30-35	2.3	2.3
Flat seam tin, 14" (350mm) tin, rolls	275-325	26-30	2.8	2.8
Flat seam tin, 28" (700mm) tin, rolls	425-475	39-44	1.8	1.8
Standing seam metal, 24"x96" (600x2400mm) sheets	250-300	23-27	3.0	3.0
Standing seam tin, 14"x20" (350x500mm) plates	125-175	11-16	5.3	5.3
Standing seam tin, 20"x28" (500x700mm) plates	175-225	16-21	4.0	4.0
Standing seam tin, 14" (350mm) tin, rolls	200-225	18-21	3.8	3.8
Standing seam tin, 20" (500mm) tin, rolls	250-300	23-27	3.0	3.0
Standing seam tin, 24" (600mm) tin, rolls	275-325	26-30	2.8	2.8
Standing seam tin, 28" (700mm) tin, rolls	300-350	27-33	2.4	2.4
V-Crimped metal, 24"x96" (600x2400mm) sheets or larger	425-475	39-44	1.8	1.8

Material Cost of One Square (100 s.f. or 9.3 sq.m)
Flat Seam Tin Roofing
Using 14"x20" (350 x 500 mm) Plates

	Rate	Total	Rate	Total
1.1 sq. (10.2 sq.m) No. 15 felt ....$....		$....	$3.00	$3.30
63 shts. 14"x20" (350x500mm)				
40 lb. (18 kg) coated		....	1.50	94.50
6 lbs. (2.7 kg) solder		....	1.00	6.00
252 cleats		....	.03	7.56
2 lbs. (0.9 kg) nails		....	.80	1.60
Flux and charcoal......................................		....	1.30	1.30
Cost per 100 s.f.		$....		$114.26*
per sq.m...		....		12.30

*Add for ridges, hips, valleys, flashing etc.

Material Cost of One Square (100 s.f. or 9.3 sq.m)
Flat Seam Tin Roofing
Using 20"x28" (500 x 700 mm) Plates

	Rate	Total	Rate	Total
1.1 sq. (10.2 sq.m) No. 15 felt ....$....		$....	$3.00	$3.30
29 shts. 20"x28" (500x700mm)				
40 lb. (18 kg) coated		....	3.00	87.00
5 lbs. (2.2 kg) solder		....	1.00	5.00
145 cleats		....	.03	4.35
1 lb. (0.45 kg) nails...................................		....	.80	.80
Flux and charcoal......................................		....	1.30	1.30
Cost per 100 s.f.		$....		$101.75*
per sq.m...		....		10.95

*Add for ridges, hips, valleys, flashing etc.

Material Cost of One Square (100 s.f. or 9.3 sq.m)
Flat Seam Tin Roofing Using Roll Roofing

	Rate	Total	Rate	Total
1.1 sq. (10.2 sq.m) No. 15 felt ....$....		$....	$3.00	$3.30
108 s.f. (10 sq.m) 20"				
(500mm)				
roll rfg.		....	1.20	129.60
4 lbs. (1.8 kg) solder		....	1.00	4.00
110 cleats		....	0.03	3.30
1 lb. (0.45 kg) nails...................................		....	0.80	0.80
Flux and charcoal......................................		....	1.00	1.00
Cost per 100 s.f.		$....		$142.00*
per sq.m...		....		15.28

*Add for ridges, hips, valleys, flashing, etc.

Material Cost of One Square (100 s.f. or 9.3 sq.m) Flat Seam Roofing Using 24"x96" (600 x 2400 mm) Galvanized Steel Sheets

	Rate	Total	Rate	Total
1.1 sq. (10.2 sq.m) No. 15 felt	$....	$....	$3.00	$3.30
7 shts. 26 ga. steel, 105 lbs. (47.6 kg)		....	0.80	89.60
3 lbs. (1.3 kg) solder		....	1.00	3.00
100 cleats		....	0.03	3.00
1 lb. (0.45 kg) nails		....	0.80	0.80
Flux and charcoal		....	0.90	0.90
Cost per 100 s.f.		$....		$100.60*
per sq.m			....	10.82

*Add for ridges, hips, valleys, flashing, etc.

Material Cost of One Square (100 s.f. or 9.3 sq.m) Standing Seam Roofing Using 14"x20" (350 x 500 mm) Plates

	Rate	Total	Rate	Total
1.1 sq. (10.2 sq.m) No. 15 felt	$....	$....	$3.00	$3.30
68 shts. 14"x20" 40-lb. (18-kg) coated		....	1.50	102.00
100 cleats		....	0.03	3.00
2.5 lbs. (1.1 kg) solder		....	1.00	2.50
1 lb. (0.45 kg) nails		....	0.80	0.80
Flux and charcoal		....	0.90	0.90
Cost per 100 s.f.		$....		$112.50*
per sq.m			....	12.11

*Add for ridges, hips, valleys, flashing, etc.

Material Cost of One Square (100 s.f. or 9.3 sq.m) Standing Seam Roofing Using 20"x28" (500 x 700 mm) Plates

	Rate	Total	Rate	Total
1.1 sq. (10.2 sq.m) No. 15 felt	$....	$....	$3.00	$3.30
32 shts. 20"x28" (500x700mm) 40-lb. (18-kg) coated		....	3.00	96.00
80 cleats		....	0.03	2.40
2 lbs. (0.2 kg) solder		....	1.00	2.00
1 lb. (0.45 kg) nails		....	0.80	0.80
Flux and charcoal		....	0.90	0.90
Cost per 100 s.f.		$....		$105.40*
per sq.m			....	11.35

*Add for ridges, hips, valleys, flashings, etc.

Material Cost of One Square (100 s.f. or 9.3 sq.m)
Standing Seam Roofing Using Roll Roofing

	Rate	Total	Rate	Total
1.1 sq. (10.2 sq.m) No. 15 felt	$....	$....	$3.00	$3.30
116 s.f. (10.7 sq.m) 20" (500mm) roll rfg 40 lb. (18 kg)		....	1.20	139.20
80 cleats		....	0.03	2.40
1 lb. (0.45 kg) solder		....	1.00	1.00
1 lb. (0.45 kg) nails		....	0.80	0.80
Flux and charcoal		....	0.70	0.70
Cost per 100 s.f.		$....		$147.40*
per sq.m		....		15.87

*Add for ridges, hips, valleys, flashing, etc.

Material Cost of One Square (100 s.f. or 9.3 sq.m)
Standing Seam Roofing Using 24"x96" (600 x 2400 mm)
Galvanized Steel Sheets

	Rate	Total	Rate	Total
1.1 sq. (10.2 sq.m) No. 15 felt	$....	$....	$3.00	$3.30
8 shts. 26 ga. steel, 120 lbs. (54 kg)		....	0.80	102.40
60 cleats		....	0.03	1.80
1 lb. (0.45 kg) solder		....	1.00	1.00
1 lb. (0.45 kg) nails		....	0.80	0.80
Flux and charcoal		....	0.70	0.70
Cost per 100 s.f.		$....		$110.00*
per sq.m		....		11.84

*Add for ridges, hips, valleys, flashing, etc.

Material Cost of One Square (100 s.f. or 9.3 sq.m) V-Crimped Metal Roofing

	Rate	Total	Rate	Total
1.1 sq. (10.2 sq.m) No. 15 Felt	$....	$....	$3.00	$3.30
1 sq. (9.3 sq.m) 26 ga. V-crimped rfg	....	....	95.00	95.00
5 lbs. (2.2 kg) lead headed nails	....	....	1.50	7.50
60 l.f. (18 m) wood strips	....	....	0.20	12.00
Cost per 100 s.f.		$....		$117.80*
per sq.m		....		12.68

*Add for ridges, hips, valleys, flashing, etc.

COPPER ROOFING

Copper is used as roofing for any type of roof. There are various methods of applying copper for roofing, but most frequently used are the flat seam and standing seam. The batten seam method is also used, where durability is of more importance than cost, e.g., cathedrals, government buildings, monumental structures, etc.

Flat Seam Roofing. Flat seam roofing is commonly used on dead level or flat surfaces, but is adaptable to any slope. This method is also adaptable for the construction of water cooling roof panels, permitting water to stand up to a depth of 3" (75 mm).

For flat seam roofing, the use of small sheets of 20-oz (0.56-kg), cold-rolled copper, not larger than 16"x18" (400 x 450 mm), with 3/4" (18 mm) seams flat-locked and soldered, is recommended. The sheets are tinned on all edges for soldering and are then formed to lock 3/4" (18 mm) with adjacent sheet. Corners must be clipped to permit folding as shown. Opposite sides of the sheets are folded in opposite directions so they will hook into adjoining sheets.

The sheets are held down by 2" (50 mm) wide copper cleats, two on each of two adjacent sides. The other two sides are held by the edges of adjacent sheets already cleated. For small areas, it is general practice to use only three cleats, two on a long side and one on a short side.

The cross seams are folded in the direction of flow. All seams should be flattened with a mallet and solder sweated to completely fill the seam.

Each 16"x18" (400 x 450 mm) sheet when folded covers 14-1/2"x16-1/2" (362 x 412 mm) or 239-1/4 sq. in. (149,144 sq.mm). It requires 61 sheets of 16"x18" (400 x 450 mm) copper per 100 s.f. (9.3 sq.m) of roof.

Standing Seam Roofing. Standing seam copper roofing may be used for surfaces with a minimum slope of 3" per foot (75 mm per 0.3 m). Standing seams are usually made to finish 1" (25 mm) high. The spacing of the seams is a matter of scale and architectural effect, but for economy, a choice should be made that will use sheets of stock sizes. The use of 20"x96" (500 x 2400 mm) sheets is recommended as maximum for ordinary purposes.

For standing seams to finish 1" (25 mm) high, the vertical bends are made on the long edges of the sheets—1-3/4" (43.7 mm) on one side and 1-1/2" (37.5 mm) on the other; the short bend on the one sheet always adjoins the long bend on the adjacent sheet.   Recommended weights of material are: 16-oz. (0.45 kg) for widths between seams up to 20" (500 mm); 20-oz. (0.56 kg) for widths over 20" (500 mm).

Formed pans are held to the roof deck with 2"x3" (50 x 75 mm) cleats spaced 12" (300 mm) apart and locked into the standing seam as shown in the accompanying illustration. Transverse joints should be staggered and each secured with one cleat locked into the flat seam.

Standing seams should not be riveted or soldered. Cross seams should be left unsoldered whenever conditions permit.

A sheet loses 3-1/4" (81 mm) in width on account of the standing seam, i.e. a sheet 20" (500 mm) wide would cover only 16-3/4" (418 mm). Where 20" (500 mm) sheets are used, add 20% to roof area for standing seams.

The loss at the end of the sheet will depend on the length of the roof and the length of copper sheet used, but allow 1-1/2" (37.5 mm) for each flat seam.

Sizes and Weights of Copper Sheets Commonly Used in Building Work

Lbs. per Sheet for Various Sizes of Copper

Size of Sheet	16-Oz.	20-Oz.	24-Oz.	32-Oz.
20"x96"	13.33	16.67	20.00	26.67
24"x96"	16.00	20.00	24.00	32.00
30"x96"	20.00	25.00	30.00	40.00
36"x96"	24.00	30.00	36.00	48.00
20"x120"	16.67	20.83	25.00	33.33
24"x120"	20.00	25.00	30.00	40.00
30"x120"	25.00	31.25	37.50	50.00
36"x120"	30.00	37.50	45.00	60.00

(metric)
Kg per Sheet for Various Sizes of Copper

Size of Sheet	0.45 kg	0.56 kg	0.68 kg	0.90 kg
500 x 2400 mm	6.0	7.5	9.1	12.1
600 x 2400 mm	7.2	9.0	10.9	14.5
600 x 2400 mm	9.1	11.3	13.6	18.1
900 x 2400 mm	10.8	13.6	16.3	21.7
500 x 3000 mm	7.5	9.4	11.3	15.1
600 x 3000 mm	9.0	11.3	13.6	18.1
750 x 3000 mm	11.3	14.1	17.0	22.7
900 x 3000 mm	13.6	17.0	20.4	27.2

Labor Placing Flat Seam Copper Roofing. When using 16"x18" (400 x 450 mm) copper sheets for flat seam roofing, a tinner and helper should place 175 to 225 s.f. (16-21 sq.m) of roof per 8-hr. day, on average size roofs.

Labor Placing Standing Seam Copper Roofing. When using copper sheets 20" (500 mm) wide and up to 8'-0" (2.4 m) long for standing seam roofing, a tinner and helper should place 200 to 250 s.f. (18-23 sq.m) per 8-hr. day, on average size roofs.

Material Cost of One Square (100 s.f. or 9.3 sq.m)
Flat Seam Copper Roofing Using 16"x18" (400 x 450 mm) Sheets

	Rate	Total	Rate	Total
1.1 sq. (10.2 sq.m) No. 15 felt	$....	$....	$3.00	$3.30
61 shts. 16"x18" (400x450mm) 153 lbs. (69 kg)			3.20	195.20
250 copper cleats		....	0.25	62.50
2 lbs. (0.9 kg) copper nails		....	2.85	5.70
6 lbs. (2.7 kg) solder		....	1.00	6.00
Flux and charcoal		....	1.25	1.25
Cost per 100 s.f.		$....		$273.95*
per sq.m		....		29.49

*Add for ridges, hips, valleys, flashing, etc.

Material Cost of One Square (100 s.f. or 9.3 sq.m)
Standing Seam Copper Roofing Using 20" (500 mm) Copper

	Rate	Total	Rate	Total
1.1 sq. (10.2 sq.m) No. 15 felt ....$....		$....	$3.00	$3.30
120 s.f. (11 sq.m) 120 lbs. (54 kg)		....	1.60	192.00
80 copper cleats..		....	0.25	20.00
1 lb. (0.45 kg) copper nails.......................................		....	2.85	2.85
Cost per 100 s.f.		$....		$218.15*
per sq.m..		....		23.48

*Add for ridges, hips, valleys, flashing, etc.

Stainless steel, type 304, monel and lead are also used for flat, standing, and batten seams.

07620 SHEET METAL FLASHING AND TRIM

Sheet metal work is a highly specialized business, because the greater part of it is done in the shop, and the fabricated materials are sent to the job ready to erect. An estimate on sheet metal work must take into consideration the cost of the finished materials at the shop or delivered to the building site, plus the labor cost of erection.

Estimating Quantities of Sheet Metal Work

The data given below will assist the estimator in measuring quantities from the plans and listing them on the estimate sheet.

When preparing the quantity survey for the different types, shapes, and sizes of metal items, be sure to identify the type and weight, or gauge, of metal—galvanized steel, aluminum, stainless steel, copper, lead, terne, etc. Each one has its own factors for waste, difficulty in shaping, and method of jointing, and all of these affect the production and installation costs.

Metal Gutters and Eaves Trough. Metal gutters and eave trough are furnished in numerous designs, from 2" (50 mm) to 12" (300 mm) in size. When estimating quantities, obtain the number of pieces of each size gutter and the length. The estimate should state the l.f. (m) of each size.

Metal Conductor Pipe or Downspout. Metal conductor pipe is furnished both round and square and from 2" (50 mm) to 6" (150 mm) in size.

When estimating quantities, measure the distance from the roof to the ground, which will be the length of each downspout. After the quantities have been obtained in this manner, list the total number of l.f. (m) of each size.

Metal Conductor Heads. Metal conductor heads are made in various styles, shapes, and sizes.

Conductor heads are generally used on buildings having flat roofs where a cast iron conductor pipe is placed inside the building and installed by the plumbing contractor. When placed on the outside of a wall at the roof level, an opening is usually left in the masonry wall at the low point of the roof to allow the water to pass through the wall opening into the conductor head and down the conductor pipe.

When estimating conductor heads, always note style and size of each head and the material specified.

Metal Flashing and Counterflashing. Flashing and counterflashing are furnished in tin, stainless steel, galvanized steel, aluminum, or copper. They are measured by the l.f. (m) when less than 12" (300 mm) wide and by the s.f. (sq.m) when over 12" (300 mm) wide.

Metal flashing used in connection with a flat roof deck covered with felt or built-up roofing usually extends 6" (150 mm) to 12" (300 mm) up the side of the brick fire wall and projects under the roofing the same distance. A counterflashing is then placed over and above this flashing. The upper edge of the counterflashing is inserted in a reglet (slot) in the brick or stone wall. After the counterflashing has been placed, it is caulked with caulking compound to prevent water from getting in back of the flashing and running under the roofing.

On many jobs using felt or built-up roofing, the felt is run up the wall about 12" (300 mm) and mopped. The metal counterflashing is then placed to extend down over the felt, which prevents water from getting in back of the flashing and running under the roofing.

Where metal counterflashing is used, it is necessary to cut an open joint or reglet in the mortar joint of the masonry wall.

The cost of cutting reglets should be estimated by the lin. ft. (meter).

Metal Valleys and Hips. Metal valleys and hips are frequently used with certain types of shingle and slate roofs. They are either continuous metal strips of a certain size or small shingles of tin, galvanized steel, aluminum, stainless steel, or copper.

They should be estimated by the l.f. (m), stating the width of the valley or hip and the kind of metal used.

Metal Ridge Roll, Hip Roll, and Cap. Metal ridge roll, hip roll and cap should be estimated by the l.f. (m) giving the width, gauge, and kind of metal used.

Metal Ventilators. Metal ventilators for roofs and skylights should be estimated at a certain price for each ventilator, f.o.b. factory or erected. Always state size and type of ventilators required, as there are numerous kinds on the market at varying prices.

Metal Skylights. When estimating metal skylights, consider the following: size, type of skylight, whether single or double pitch or hip skylight; with or without ventilators or side sash; kind of glass, and whether a metal curb flashing is required.

Labor Placing Sheet Metal Work

The costs given below are for the erection of various kinds of sheet metal work on the job. Add cost of the fabricated material, contractor's overhead expense and profit.

Placing Hanging Gutter or Eaves Trough. This cost will vary with the slope of the roof, distance above the ground, and the method used in hanging the gutter.

On wood constructed buildings having hip or gable roofs, metal hangers or supports are placed over the gutters or eave trough and fastened to the eaves under the shingles or other roofing. The eave trough or gutter must always slope in the direction of the downspout, in order to drain satisfactorily.

On hip or gable roofs having eaves 20'-0" to 25'-0" (6.0-7.6 m) above the ground, a tinner and helper should place 180 to 220 l.f. (55-67 m) of gutter per 8-hr. day, at the following labor cost per 100 l.f. (30 m):

	Hours	Rate	Total	Rate	Total
Sht. met. wkr.	4	$....	$....	$28.88	$115.52
Labor	4	....	....	21.04	84.16
Cost 100 l.f.			$....		$199.68
per l.f.				....	2.00
per m				....	6.56

Placing Metal Conductor Pipes or Downspouts. Metal conductor pipes or downspouts are furnished in both round and square and from 1-1/2" (37.5 mm) to 6" (150 mm) diameter. The conductor pipe extends down the side of the building from the eave trough to grade, where it is connected to an elbow that throws the water away from the building or is connected with the drainage system. Metal conductor pipe is usually held in place by hooks fastened to the walls.

On one- or two-story buildings where the conductor pipe is 12'-0" to 25'-0" (3.6-7.6 m) long, a tinner and helper should place 200 to 250 l.f. (60-76 m) per 8-hr. day, at the following labor cost per 100 l.f. (30 m):

	Hours	Rate	Total	Rate	Total
Sht. met. wkr.	3.5	$....	$....	$28.88	$101.08
Labor	3.5	....	....	21.04	73.64
Cost 100 l.f.			$....		$174.72
per l.f.				....	1.75
per m				....	5.74

Where outside conductor pipe is installed on buildings 3 to 4 stories high, a tinner and helper should place 160 to 200 l.f. (48-61 m) per 8-hr. day, at the following labor cost per 100 l.f. (30 m):

	Hours	Rate	Total	Rate	Total
Sht. met. wkr.	4.5	$....	$....	$28.88	$129.96
Labor..	4.5	....	....	21.04	94.68
Cost 100 l.f..................................			$....		$224.64
per l.f.				....	2.25
per m				....	7.38

Placing Metal Flashing. When placing metal roof flashing around parapet walls, etc., it is customary to have the flashing extend under the roofing and up the side of the wall 6" (150 mm) to 12" (300 mm), and then place the counterflashing above to lap over the flashing.

Where it is not necessary to cut reglets in the masonry wall, a tinner should place 140 to 160 l.f. (42-49 m) of flashing per 8-hr. day, at the following cost per 100 l.f. (30 m):

	Hours	Rate	Total	Rate	Total
Sht. met. wkr.	5.4	$....	$....	$28.88	$155.95
Cost per l.f.................................				....	1.56
per m				....	5.12

On frame buildings where the flashing extends under the wood wall siding to protect porch roofs and the like, it is only necessary to place the flashing or metal shingles on the roof and tack them against the sheathing. A tinner should place 165 to 185 l.f. (50-56 m) per 8-hr. day, at the following labor cost per 100 l.f. (30 m):

	Hours	Rate	Total	Rate	Total
Sht. Met Wkr.............................	4.6	$....	$....	$28.88	$132.85
Cost per l.f.................................				....	1.33
per m				....	4.36

Placing Metal Counterflashing. When placing metal counterflashing over felt or metal flashing that extends up on the side of the wall, it will be necessary to place the counterflashing in a reglet or open joint in the masonry that can be caulked or sealed to prevent the water getting in back of the counterflashing and under the roof.

Where reglets have previously been cut, as is usually the case on stone copings, requiring only the sealing of the joint, a tinner should place 140 to 160 l.f. (42-49 m) per 8-hr. day, at the following labor cost per 100 l.f. (30 m):

	Hours	Rate	Total	Rate	Total
Sht. Met. Wkr.	5.4	$....	$....	$28.88	$155.95
Cost per l.f.................................				....	1.56
per m				....	5.12

Cutting Reglets in Masonry Walls. The labor cost of cutting reglets in brick walls will depend entirely upon the kind of mortar used and the condition of the mortar at the time of cutting the reglets.

If the brick are laid in portland cement mortar, it will cost more to cut the reglet than when laid in lime mortar. Also, if the reglets are cut within 24 hrs. after the brick are laid, the labor cost will be much less than cutting them two or three weeks later.

If reglets are cut before the mortar has set, a worker should cut 175 to 225 l.f. (53-69 m) per 8-hr. day, at the following labor cost per 100 l.f. (30 m):

	Hours	Rate	Total	Rate	Total
Labor	4	$....	$....	$21.04	$84.16
Cost per l.f.			....		0.84
per m			....		2.76

If the reglets are cut several days after the brick have been laid, giving the mortar time to harden, it will require a hammer and chisel to cut out the joints, and a worker should cut 65 to 80 l.f. (19-24 m) per 8-hr. day, at the following labor cost per 100 l.f. (30 m):

	Hours	Rate	Total	Rate	Total
Labor	11	$....	$....	$21.04	$231.44
Cost per l.f.			....		2.31
per m			....		7.58

If a portable electric saw with an abrasive blade or cutting wheel is used, a worker cuts 10 to 15 l.f. (3.0-5.0 m) of reglet per hour.

Cutting reglets in stone coping or balustrade is usually performed by stone cutters at about the same cost as given above.

Placing Metal Valleys. When placing metal shingles or valleys in connection with gable or hip roofs, dormers, etc., a tinner should place 115 to 135 l.f. (35-41 m) per 8- hr. day, at the following labor cost per 100 l.f. (30 m):

	Hours	Rate	Total	Rate	Total
Sht. Met. Wkr.	6.4	$....	$....	$28.88	$184.83
Cost per l.f.			....		1.85
per m			....		6.07

Placing Metal Ridge Roll of Cap. If metal ridge roll is placed on the ridge of a hip or gable roof, a tinner should place 90 to 110 l.f. (27-34 m) per 8-hr. day, at the following labor cost per 100 l.f. (30 m):

	Hours	Rate	Total	Rate	Total
Sht. Met. Wkr.	8	$....	$....	$28.88	$231.04
Cost per l.f.			....		2.31
per m			....		7.58

Placing Metal Conductor Heads. On buildings having flat roofs where the conductor pipe or downspout is placed inside of the building, requiring a conductor head at the low point of the roof which is connected to the downspout, a tinner should place one conductor head (made complete in the shop) in 1-1/2 to 2 hrs. at the following labor cost:

	Hours	Rate	Total	Rate	Total
Sht. Met. Wkr.	1.8	$....	$....	$28.88	$51.98

07630 ROOFING SPECIALTIES

Flashings can be formed from a wide variety of materials of varying permanence:

Material	Thickness or weight, inches (mm)	Cost per s.f.	Cost per sq.m	S.F. installed per day	Sq.M installed per day
Aluminum	0.013" (0.33mm)	$0.24	$2.58	150	14
(Mill	0.016" (0.40mm)	0.28	3.01	150	14
Finish)	0.019" (0.48mm)	0.38	4.09	150	14
	0.032" (0.80mm)	0.72	7.75	145	13
	0.040" (1.00mm)	0.86	9.25	145	13
	0.050" (1.25mm)	1.05	11.30	145	13
Copper	16 oz. (0.45 kg)	2.20	23.67	140	13
	20 oz. (0.56 kg)	2.70	29.05	140	13
	24 oz. (0.68 kg)	3.10	33.36	140	13
Stainless Steel	26 ga.	2.00	21.52	150	14
Asph. C't'd Cotton	17 oz. (0.48 kg)	0.15	1.61	350	33
	40 oz. (1.13 kg)	0.22	2.37	350	33
Lead	2.5 lb. (1.13 kg)	1.65	17.75	140	13
Zinc/Copper Alloy		1.35	14.53	140	13
Butyl	1/32" (0.78 mm)	0.52	5.60	275	26
	1/16" (1.56 mm)	0.75	8.07	275	26
Neoprene	1/16" (1.56 mm)	1.25	13.45	275	26
Polyvinyl	0.02" (0.50 mm)	0.14	1.51	275	26
	0.03" (0.75 mm)	0.22	2.37	275	26

Gravel stops may be formed on the job or ordered out in extruded form. If extruded be sure to check if any special finish is specified, which will raise costs considerably. A few costs and rates for installation are:

Material	Cost per l.f. 4" high	Cost per l.f. 6" high	L.F. per Day
Aluminum 0.050" mill finish	$2.50	$2.90	135
Duranodic	3.00	3.40	135

(metric)

Material	Cost per m 100mm high	Cost per m 150mm high	Meters per Day
Aluminum 1.2 mm mill finish	$8.20	$9.51	41
Duranodic	9.85	11.15	41

Preformed gutters can be installed by one sheet metal worker at the rate of 110 to 120 l.f. (33-36 m) per day. Material costs are:

Material	Cost per l.f.		
	4"	5"	6"
Aluminum 0.027" plain	$0.60	$0.65	$0.80
enam.	0.65	0.73	0.90
Copper 16 oz.	3.40	3.80	4.70
Stainless 22 ga.	3.40	3.70	4.00
Vinyl	1.00	1.10	1.25
Galvanized 28 ga.	0.55	0.70	0.90

(metric)

Material	Cost per meter		
	100 mm	125 mm	150 mm
Aluminum 0.67 mm plain	$1.97	$2.13	$2.62
enam.	2.13	2.39	2.95
Copper 0.45 kg	11.15	12.47	15.42
Stainless 22 ga.	11.15	12.14	13.12
Vinyl	3.28	3.61	4.10
Galvanized 28 ga.	1.80	2.30	2.95

Downspouts can be installed by one sheet metal worker at the rate of 175 l.f. (53 m) for small sizes, 150 l.f. (45 m) for medium and 125 l.f. (38 m) for the larger sizes per day. The material costs will run as follows:

Material	Cost per l.f.				
	3" dia.	4" dia.	5" dia.	2"x3"	3"x4"
Aluminum 0.025" plain	$0.50	$0.80	$1.10	$0.75	$0.95
enam.	0.58	0.90	1.15	0.85	1.10
Copper 16 oz.	3.00	3.50	4.30	3.30	4.05
Stainless	7.00	9.00	14.00	7.00	9.00
Vinyl	0.60	--	--	0.65	--
Galv. Steel 28 ga.	0.45	0.60	1.05	0.50	0.65

	Cost per meter				
	75mm	100mm	125mm	50x75mm	75x100mm
Aluminum 0.62 mm plain	$1.64	$2.62	$3.61	$2.46	$3.12
enam.	1.90	2.95	3.77	2.79	3.61
Copper 0.45 kg	9.84	11.48	14.11	10.83	13.29
Stainless	22.97	29.53	45.93	22.97	29.53
Vinyl	1.97	--	--	2.13	--
Galv. Steel 28 ga.	1.48	1.97	3.44	1.64	2.13

For elbows for the above work add the following:

3" (75 mm) diam. or 2"x3" (50 x 75 mm) in plain aluminum $1.00
3" (75 mm) diam. or 2"x3" (50 x 75 mm) in enam. aluminum 1.25
3" (75 mm) diam. or 2"x3" (50 x 75 mm) in 16 oz. (0.45 kg) copper................. 5.00
4" (100 mm) diam. or 3"x4" (75 x 100 mm) in plain aluminum 1.25
4" (100 mm) diam. or 3"x4" (75 x 100 mm) in enam. aluminum 1.50
4" (100 mm) diam. or 3"x4" (75 x 100 mm) 16 oz. (0.45 kg) copper................. 8.00

07800 ROOF ACCESSORIES

Erecting and Glazing Metal Skylights. Metal skylights are manufactured in the shop and sent to the job "knocked down" or in sections ready to erect and glaze.

An average skylight (single pitch, double pitch, or hip) up to 8'-0"x12'-0" (2.4 x 3.6 m) in size, containing 100 s.f. (9.3 sq.m) of area, should be erected and glazed complete by a sheet metal worker and glazier in 8 to 10 hrs. time, at the following labor cost:

	Hours	Rate	Total	Rate	Total
Sheet metal worker	5	$....	$....	$28.88	$144.40
Glazier	4	....	....	25.94	103.76
Cost per skylight			$....		$248.16

When erecting skylights 10'-0"x10'-0" (3 x 3 m) to 10'-0"x15'-0" (3 x 4.5 m), containing 100 to 150 s.f. (9.0-13.5 sq.m), a sheet metal worker and glazier should erect and glaze one skylight in 14 to 16 hrs. at the following labor cost:

	Hours	Rate	Total	Rate	Total
Sheet metal worker	9	$....	$....	$28.88	$259.92
Glazier	6	....	....	25.94	155.64
Cost per skylight			$....		$415.56

When erecting skylights 10'-0"x16'-0" (3 x 5 m) to 10'-0"x20'-0" (3 x 6 m) in size, containing 160 to 200 s.f. (15-18 sq.m), it will require about 24 hrs. for a sheet metal worker and glazier to erect the skylight and glaze it complete, at the following labor cost:

	Hours	Rate	Total	Rate	Total
Sheet metal worker	16	$....	$....	$28.88	$462.08
Glazier	8	....	....	25.94	207.52
Cost per skylight			$....		$669.60

The above costs do not include overhead and profit or the cost of placing a metal curb flashing for skylight. For costs on this work, see below.

Placing Metal Skylight Curb Flashing. Metal skylights are often placed on top of a wood curb projecting 6" (150 mm) or more above the roof level. The space from the top of the curb to the roof level is covered with metal flashing.

When erecting metal skylight curb flashing, a tinner should place 190 to 210 s.f. (17-20 sq.m) of metal per 8-hr day, at the following labor cost per 100 s.f. (9.3 sq.m):

	Hours	Rate	Total	Rate	Total
Sheet metal worker	4	$....	$....	$28.88	$115.52
Cost per s.f.			....		1.16
per sq.m			....		12.48

Erecting Skylights With Side Sash. If the skylights have side sash, the erection cost will vary with the number of sash in the skylight and whether stationary or pivoted. On an average it will require 1 to 1-1/2 hrs. labor time for each sash in the skylight, at the following labor cost per sash:

	Hours	Rate	Total	Rate	Total
Sheet metal worker	1.25	$....	$....	$28.88	$36.10

Erecting Sash Operating Devices for Movable Sash. Skylights furnished with hinged or pivoted sidewall sash require an operating device to open and close them.

An operating device usually consists of a mechanism similar to that illustrated, which is operated from the floor by a worm and gear or a chain attachment, from which all sash on each side of the skylight can be opened and closed.

The labor cost of erecting sash operating devices will vary with the length of each run and the number of runs in each skylight but on an average, a worker should erect 50 to 75 l.f. (15-22.5 m) per 8-hr. day, at the following labor cost per 100 l.f. (30 m):

	Hours	Rate	Total	Rate	Total
Sheet metal worker	13	$....	$....	$28.88	$375.44
Cost per l.f.			....		3.75
per m			....		12.30

Ventilating Skylights. Ventilating skylights are used in factories, foundries, machine shops, garages, dairies, laundries, and other buildings that require economical light and ventilation. They are furnished in any type of metal desired, galvanized copper-bearing steel, aluminum or copper, depending upon the nature of the building on which they are to be used. Normally, for building of ordinary construction, the standard 18-gauge galvanized copper-bearing steel is furnished. Should the building be of reinforced concrete or steel and concrete construction, the non-corrosive aluminum or copper construction should be furnished.

These skylights are of gable type as illustrated and are furnished in standard widths of 4'-0" (1.2 m) to 20'-0" (6 m) and any desired length in multiples of 2'-0" (0.6 m). It is standard practice to interpret the size of a skylight as being outside of curb dimensions.

The length of the ventilating sections depend on the width of the skylights; 4'-0" (1.2 m), 6'-0" (1.8 m) and 8'-0" (2.4 m) widths having maximum ventilating sections of 40'-0" each controlled from one operating station; 10'-0" (3 m) and 12'-0" (3.6 m) wide skylights have maximum ventilation sections of 30'-0" (9 m) in length; 14'-0" (4.2 m) and 16'-0" (4.8 m) widths have 40'-0" (12 m) maximum lengths, and 18'-0" (5.4 m) and 20'-0" (6 m) widths have 30'-0" (9 m) maximum ventilating sections.

A mechanical operating device consisting of either loose rod or chain control is furnished as required, so that the skylight can be readily controlled from the floor. All skylights under 30'-0" (9 m) in length are arranged for full length ventilation. Longer skylights are arranged with ventilating sections as required.

The cost of ventilating skylights vary with their size and number of ventilating sections in each skylight. Prices should be secured from the manufacturer.

In estimating the cost of ventilating skylights, the principal thing to remember is: The smaller the area of the skylight, the higher the price per s.f. (sq.m).

Plastic Dome Skylights. This type skylight consists of a thermo-formed acrylic plastic sheet free blown into square, rectangular and circular domes.

This type skylight is furnished as a factory assembled unit; the plastic is mounted in an extruded aluminum frame, and prefabricated curbs are available. The aluminum frame is generally designed to form counterflashing.

Plastic Domes

Size	Cost Per Roof Dome
16"x16" (400 x 400 mm)	$40.00
24"x24" (600 x 600 mm)	50.00
36"x36" (900 x 900 mm)	65.00
48"x48" (1200 x 1200 mm)	115.00
24" (600 mm) Round	85.00
48" (1200 mm) Round	150.00

Erection cost will vary widely as to the number involved and the preparatory work done by others. As units are vulnerable to breakage and scratching, it takes two workers to handle all but the smallest unit. On a job with several units involved two workers should set 2 units of up to 10 s.f. (0.9 sq.m) in a day if the curbing is already in place.

The dome unit is available as an insulating unit, with a secondary acrylic dome mounted under the primary (exterior) dome. This double dome unit provides a dead air space between the domes that reduces heat loss and condensation formation.

ROOF HATCH

A standard size 2'-6"x3'-0" (750 x 900 mm) steel roof hatch with insulated curb and cover and hardware will cost about $275.00. A mechanic and helper should set this in 2 hrs.

GRAVITY VENTILATORS

The 12" (300 mm) stationary ventilators of the syphon type with a 24" (600 mm) base will cost about $40.00 and take 2 hours to set. Stationary ventilators of the mushroom type with a 24" (600 mm) base will cost $130.00 with the same labor cost.

07900 SEALANTS

No one sealant can solve the requirements across the full range of construction applications. There are sealants for interior or exterior use, gun or pour grade, or ability to expand or contract. There are variations in service temperature range, paintability, and compatability with material to be sealed.

An oil base caulking compound has a life expectancy of about 4 or 5 years. Its restrictions list a joint maximum size of 1/2" (12.5 mm) wide x 3/4" (18.7 mm) deep and is incapable of withstanding any joint movement without rupture. Using this material for exterior applications may not be prudent or cost effective.

Acrylic latex caulk is probably a good starting point for exterior use. It is considered to be the best value among sealants in the middle performance range. It is relatively inexpensive, has a life expectancy of 8 to 12 years, cures quickly, has excellent paintability, is non-staining, and has reasonable elongation and recovery for movement of joints. However, this material has a maximum joint size of 1/2" (12.5 mm) wide and 1/2" (12.5 mm) deep and should not be used for expansion joints.

Polysulfide base sealant, available in one or two part compounds, is among the premiere compounds on the market. It is reasonably expensive and requires a little more labor time to install, but its movement capabilities, bonding strength, wide use range (including expansion joints), and a life expectancy of 15 years make it cost effective while providing excellent joint sealing characteristics.

It is important to have a properly designed joint, both in width and depth. Good sealants will expand and contract with the movement of the joint. As the joint moves, the sealant changes shape while the volume remains constant. It is critical that the width-to-depth ratio be designed to withstand the constant elongation and compression cycles over long periods of time.

Deep beads of sealant wastes material and are more prone to failure than shallow sealant beads. Another important factor to consider is that sealant works best when adhering to the two opposing faces and isolated from the third (back) side of the joint to be sealed. This is accomplished by filling the joint with a material such as oakum or polyethylene foam rod to within 3/8" (9 mm) or 1/2" (13 mm) of the face of the joint. In shallow depth joints, use non-adhering tape on back of the joint to prevent the sealant from bonding.

Sealants are manufactured in colors from white to black and some are clear (silicones). This is necessary to be architecturally compatible with the adjoining surfaces. Also, paint does not adhere at all to some sealants.

The following chart provides a guide for the amount of sealant required to fill various size joints; expressed in l.f. (m) of joint obtained per gallon of material.

Depth	Width, inches							
	1/16"	1/8"	1/4"	3/8"	1/2"	5/8"	3/4"	1"
1/16"	4928	2464	1232	821	616	493	411	307
1/8"	--	1232	616	411	307	246	205	154
3/16"	--	--	411	275	205	164	137	103
1/4"	--	--	307	205	154	123	103	77
3/8"	--	--	--	137	103	82	68	51

Depth	Width, mm							
	1.5	3.1	6.2	9.3	12.5	15.6	18.7	25
1.5 mm	1502	751	376	250	188	150	125	94
3.1 mm	--	376	188	125	94	75	62	47
4.6 mm	--	--	125	84	62	50	42	31
6.2 mm	--	--	94	62	47	37	32	23
9.3 mm	--	--	--	42	31	25	21	16

In using the above chart, if the size of the joint to be sealed is not given on the drawings, use the principle that the depth of the sealant is 1/2 of the width of the joint to be sealed.

When estimating the labor required to caulk a joint, it will depend on the width of the joint, the material used, and the location or accessibility of the joint. One worker should caulk about 600 l.f. (183 m) per 8 hr. day of joints around doors or window frames when they are close to the ground or accessible without the use of scaffolding. This time may be reduced by half when working off ladders or scaffolding. The same person working on sealing wide expansion joints would only finish about 200 l.f. (61 m) per 8-hr. day.

CHAPTER 8

DOORS AND WINDOWS

CSI DIVISION 8
08100 METAL DOORS AND FRAMES
08200 WOOD AND PLASTIC DOORS
08300 SPECIAL DOORS
08400 ENTRANCES AND
STOREFRONTS
08500 METAL WINDOWS
08600 WOOD AND PLASTIC WINDOWS
08700 HARDWARE AND SPECIALTIES
08800 GLAZING
08900 WINDOW WALLS/CURTAIN
WALLS

Most of the items covered in this chapter are furnished by material distributors and set by a carpenter on the job; in the estimate, the cost of labor would be included under the proper trade.

Glazing is an exception, and both material and labor are included under a single item. However, sash and doors usually come to the job factory glazed. The contractor must always check this out so as to eliminate any duplication in the estimate and to make certain the pre-glazing complies with both the specifications and local union rules.

Curtainwall material and labor are usually let to a single subcontractor. It is advantageous to have the metal erection, sealants, and glazing all in one contract so that one source will be responsible for the weather tightness of the wall.

The estimator who will perform the quantity survey for these materials should become very familiar with the various schedules that the project architect includes on the plans.

These schedules are usually laid out in block column form, consisting of a listing of the material involved and all of the particulars for each item, and they might show architectural details for each type of item.

Door schedules will normally list the door openings by "mark" numbers or letters, e.g., #1, #2, #3 or A, B, C. After the mark designation, there will be the door size, door material and the frame, a code designation for the type and size of the frame, fire rating requirements (if any), a code designation for any louver or glass panels required, finish hardware set designations, and a "remarks" column for any special instructions. Bear in mind that the "mark" designation refers to that particular design type and may have nothing to do with the total quantity of doors and frames of that design type required for the project.

Window schedules are laid out in much the same fashion, showing: mark designation, size, material, glass, type, etc.

Finish hardware schedules are normally set up in the hardware section of the specifications, rather than on the drawings. The hardware schedule is a listing of the hardware items required to fit a particular door opening, e.g., hinges, latchset or lockset, closer, kickplates, etc.

Each type of opening, such as for toilets, closets, offices, etc., will have its own "HW Set #" designation; and each will vary from the other in some way. For example, a door that leads to a stairway in an office building could list under its HW Set number: "1-1/2 pair hinges, latchset, and closer". A door that leads to an office in the same building could list under its HW Set number: "1-1/2 pair hinges, lockset, and closer". As you can see from the two examples, they are very similar, but because of the lock requirements for the office door opening, a separate HW Set # will be given.

Here again, keep in mind that the HW Set number has nothing to do with the total quantity of finish hardware required. One set number might be applied to 10 door openings and the next set number applied to only one or two door openings.

08100 METAL DOORS AND FRAMES

Each hollow metal manufacturer produces stock design doors and frames, as well as a custom designs to an architect's requirements. Stock lines vary among manufacturers as to size, width of frame, hardware preparation, etc. A hollow metal door from one manufacturer usually cannot be hung in the hollow metal frame from another manufacturer, due to unmatched hinge or lockset locations.

"Stock" will also refer to the cut-out preparation for hinges, and in particular, for locksets. Most manufacturers prepare the frames for 3-hinge locations each sized 3-1/2" (87.5 mm) or 4-1/2" (112.5 mm) in height and for the standard cylindrical lockset. If the finish hardware specified for the project varies from the manufacturer standards, the hollow metal manufacturer will consider the order "custom" and increase the price accordingly, as well as increase the delivery time to the project.

Material prices for hollow metal items are quoted inclusive of all mortising, reinforcing, drilling, and tapping for mounting the door into the frame and preparation for items of finish hardware. Also included are frame anchors and shop prime coat of paint.

Hollow Metal Frames. Frames are made of 18, 16, and 14 gauge metal and formed to receive a 1-3/8" (34.3 mm) or 1-3/4" (43.7 mm) thick door, depending on which side of the stop the hinge and strike cut-outs are placed.

The following illustration shows a section through a standard hollow metal frame.

TYPICAL FRAME PROFILE

When the manufacturer varies the overall width of the frame to suit the different wall conditions that may be encountered, he simply increases the width of the stop. This allows them to set up any width frame for either thickness of door in lieu of manufacturing frames for 1-3/8" (34.3 mm) doors and 1-3/4" (43.7 mm) doors.

Frames are made in various widths (profile) to suit the thicknesses of the walls to which they are mounted. The standard widths with most manufacturers are 4-3/4" (118.75 mm), 5-3/4" (143.75 mm), 6-3/4" (168.75 mm), and 8-3/4" (218.75 mm).

In addition to the frame width dimensions, frames are ordered according to the door opening size, the standard being 2'-0" (600 mm), 2'-4" (700 mm), 2'-6" (750 mm), 2'-8" (800 mm), 3'-0" (900 mm), 3'-4" (1000 mm), 3'-6" (1050 mm), 3'-8" (1100 mm), and 4'-0" (1200 mm) wide for single swing and 4'-0" (1200 mm), 4'-8" (1400 mm), 5'-0" (1500 mm), 5'-4" (1600 mm), 6'-0" (1800 mm), 6'-8" (2000 mm), 7'-0" (2100 mm), 7'-4" (2200 mm), and 8'-0" (2400 mm) wide for double swing by 6'-8" (2000 mm), 7'-0" (2100 mm), 7'-2" (2150 mm), and 8'-0" (2400 mm) in height.

Frames are available from the factory in welded joint construction, all three sides put together to form a unit, or in knocked down (K.D.) form for use on drywall partitions. The K.D. frame is considered to be a pressure-fit frame that once in place capping the wall exerts a gripping action on the wall. The welded frame is stronger and must be erected prior to the wall erection, whereas the K.D. frame is erected after the wall is built.

Special frames can be ordered that will incorporate a transom panel above the door or sidelites of various heights and widths beside the door.

Average material prices for hollow metal frames of various sizes are as follows:

18 Ga. Welded Hollow Metal Frames

Size	4-3/4" (118.7mm)	5-3/4" (143.7mm)	6-3/4" (168.7mm)	8-3/4" (218.7mm)
2'-6"x7'-0"	60.00	63.00	65.00	67.00
3'-0"x7'-0"	62.00	64.00	66.00	68.00
5'-0"x7'-0"	77.00	80.00	82.00	84.00
6'-0"x7'-0"	80.00	82.00	84.00	86.00

Deduct $10.00 each for knocked down (K.D.) frames.
Add $10.00 for "B" Label, $15.00 for "A" Label.

Labor Erecting Hollow Metal Frames. Two carpenters should erect about 16 welded frames per 8-hr. day at the following labor cost per frame:

	Hours	Rate	Total	Rate	Total
Carpenter	1.0	$....	$....	$26.44	$26.44

One carpenter should erect about 12 K.D. frames per 8-hr. day at the following labor cost per frame:

	Hours	Rate	Total	Rate	Total
Carpenter	0.67	$....	$....	$26.44	$17.71

Hollow Metal Doors. Doors are made with 20, 18, or 16 gauge face sheets and in thicknesses of 1-3/8" (34.3 mm) and 1-3/4" (43.7 mm). Door sizes will correspond to the frame opening sizes given in the *Hollow Metal Frame* section.

Doors may be ordered in a variety of styles including: small vision lites, half glass, full glass, dutch, louvered, and flush. Actually, when a door is ordered with a vision lite, what is received from the manufacturer is a flush door that has been cut out and prepared for future glass installation at the jobsite; the glass is not supplied by the door manufacturer.

The physical construction of the door itself is where manufacturers really differ. Each has its own techniques for assembly—location of seams, inverted channel or flush top and bottom edges, gauge of hinge and closer reinforcements, leading edge reinforcement, core materials, etc. The contractor is obligated to determine the compliance of the product with the requirements contained in the specification. Therefore, be sure of compliance with specification requirements when pricing hollow metal doors, because prices vary widely.

Average material prices for hollow metal doors are as follows:

20 Ga. Hollow Metal Doors

Size	1-3/8" (34.3 mm)	1-3/4" (43.7 mm)
2'-6"x7'-0" (750 x 2100 mm)	132.00	143.00

Size	1-3/8" (34.3 mm)	1-3/4" (43.7 mm)
3'-0"x7'-0" (900 x 2100 mm)	143.00	154.00
3'-6"x7'-0" (1050 x 2100 mm)	159.00	170.00

Add $20.00 for "B" Label, $45.00 for "A" Label
Add $60.00 for louver, $50.00 for lite cut-out.
Add $40.00 for mineral core (for temperature rise rating).

Labor Erecting Hollow Metal Doors. Two carpenters should hang, on hinges, about 12 doors per 8-hr. day at the following labor cost per door:

	Hours	Rate	Total	Rate	Total
Carpenter	1.34	$....	$....	$26.44	$35.43
Laborer (distribution)	0.5	....	....	21.04	10.52
			$....		$45.95

Fire Label Requirements. According to building codes, certain door locations within a building, such as at stairs, trash rooms, furnace rooms, garage entries, storage rooms, etc., will require a fire rating for the door and frame. This rating is expressed as "C Label" for 3/4-hour rating, "B Label" for 1-1/2 to 2-hour rating and "A Label" for 3-hour rating.

The rating is obtained by the manufacturer subjecting its product to actual fire testing by either Underwriters Laboratory or a Factory Mutual testing laboratory. The door and frame is then approved based on the physical construction of the unit. If the manufacturer changes the design after the approval, it must apply for a new test based on the re-designed unit. The approved door and frame then qualifies to have a label affixed, showing that it is a fire rated assembly.

When a door is to be fire rated, it also carries other restrictions, such as that the latch or lockset must be installed with a UL throw bolt, or that glass is either not allowed (as in "A Label" rating) or restricted in size (as in "B or C Label" ratings). If a louver is needed in the door, it must have a fusible link that will melt and allow the louver blades to close when the fire reaches the door, and the door must have a closer device or spring hinges.

Metal Covered Jambs and Trim

Wood jambs and trim covered with No. 24 gauge zinc coated sheet steel, for use with kalamein doors.

Jambs are 1-3/4"x5-1/2" (43.7 x 137.5 mm) rabbeted for 1-3/4" (43.7 mm) doors. Plain casing is 7/8"x3-1/2" (21.8 x 87.5 mm). Round edge and brick mold is 7/8"x7/8" (21.8 x 21.8 mm).

Size Up to	Jamb Width To: 6-3/4 (158.7 mm)	8-3/4 (218.7 mm)	12 (300 mm)
3'-0"x7'-0" (900 x 2100 mm)	$74.00	$80.00	$85.00
4'-0"x7'-0" (1200 x 2100 mm)	80.00	85.00	90.00
5'-0"x7'-0" (1500 x 2100 mm)	85.00	90.00	95.00
6'-0"x7'-0" (1800 x 2100 mm)	90.00	95.00	100.00

For frames over 7'-0" (2100 mm) add $5.00 for each 6" (150 mm) or less.

Exterior Metal Residential Door Assemblies

Prehung metal doors complete with moldings of the styles usually found on traditional wood doors and with a frame, threshold, weatherstripping, and a variety of locks are now available in over fifty stock designs. Prices for a 3' (900 mm) wide unit will vary from $190 to $730.00 depending on the type of lock and the type of glazing.

A plain, flush unit with a double cylinder lock and no glazing runs $210.00; a unit with raised panels runs $240.00; if the upper section is glazed with insulating glass the cost will raise to $270.00. Units similar to the above but in pairs for a 5' (1500 mm) opening will run $400, $465, and $520 respectively. If moldings are ordered for the inside face of the door add $18.00.

08200 WOOD AND PLASTIC DOORS

Flush Veneer Slab Doors for Exterior Use

Furnished with water and weather resistant birch face veneers both sides of the door. All doors 1-3/4" (43.7 mm) thick.

Size	Hol. Core	Sol. Core
2'-6"x6'-8" (750 x 2000 mm)	$40.00	$60.00
2'-8"x6'-8" (800 x 2000 mm)	42.00	62.00
3'-0"x6'-8" (900 x 2000 mm)	44.00	67.00
3'-0"x7'-0" (900 x 2100 mm)	49.00	70.00
3'-0"x8'-0" (900 x 2400 mm)	57.00	102.00

Extras for Altering Solid Core Flush Doors

For V-grooves, add per groove	$ 7.50
For cutting in one square light, add	18.00
For one circle light, approx. 12" (300 mm) diam., add	22.00
For one circle light, approx. 18" (450 mm) diam., add	26.50
For one peek light, add	18.00
For one diamond light, add	18.00

White Pine Entrance Doors

Type	2'-8"x6'-8"x1-3/4" (800 x 2000 x 43.7 mm)	3'-0"x6'-8"x1-3/4" (900 x 2000 x 43.7 mm)
6 Panel Colonial	$95.00	$100.00
4 Panel & Light	97.00	102.00
X Buck & 9 Light Rect.	110.00	115.00
X Buck & 12 Light Diag.	125.00	130.00
12 Panel Spanish	135.00	140.00
21 Panel Spanish	--	150.00

These doors can usually be purchased prehung in rabbeted frames of standard widths with 1-1/2 pair of butts, without sill, lock or weatherstripping, for about $80.00 extra.

Exterior trim sets for front entrances are available. The price range is great, depending on depth of members, elaborateness, and sharpness of carvings and flutings. One manufacturer offers trim resembling entrance "A" shown at $160.00 in the deluxe model, $90.00 for a modified design; trim for doorways "C", "D", and "E" would cost about $90.00. They may be purchased as complete entrance sets with frames, thresholds, and sidelights if desired.

Western Pine Colonial Front Entrances

Exterior wood doors are usually fitted with combination storm and screen units. These are usually of 1-1/8" (28.1 mm) thick Ponderosa pine with aluminum wire. They run as follows for 6'-9" (2025 mm) height:

	2'-6" (750 mm)	2'-8" (800 mm)	3'-0" (900 mm)
4 Light	$42.00	$44.00	$48.00
X Buck & 1 Light	--	54.00	57.00

Combination doors are also available in aluminum, 1" (25 mm) thick for either 6'-8" (2000 mm) or 7'-0" (2100 mm) openings in various qualities.

Standard quality, 2'-6" (750 mm), 2'-8" (800 mm) & 3'-0" (900 mm) wide $42.00
Deluxe quality, 2'-6" (750 mm), 2'-8" (800 mm) & 3'-0" (900 mm) wide 52.00
Deluxe quality with white finish, 3'-0" (900 mm) wide 58.00
X-Buck type with white finish, 3'-0" (900 mm) wide 67.00

Soft Wood French Doors

These doors are made of clear Ponderosa pine. Stiles and top rails are 4-3/4" (118.7 mm) and bottom rails are 9-5/8" (240.6 mm) overall with bead and cove sticking. Muntins are 1/2" (12.5 mm) between glass. Wood stops or glazing beads are mitred and tacked in place.

Type	2'-6"x6'-8" 1-3/8"	1-3/4"	2'-8"x6'-8" 1-3/8"	1-3/4"	3'-0"x6'-8" 1-3/8"	1-3/4"
1 Light French	$70.00	$74.00	$70.00	$74.00	$89.00	$95.00
5 Light French	71.00	75.00	71.00	75.00	90.00	97.00
15 Light French	72.00	76.00	72.00	76.00	91.00	100.00

(metric)

Type	750 x 2000mm 34.3mm	43.7mm	800 x 2000mm 34.3mm	43.7mm	900 x 2000mm 34.3mm	43.7mm
1 Light French	$70.00	$74.00	$70.00	$74.00	$89.00	$95.00
5 Light French	71.00	75.00	71.00	75.00	90.00	97.00
15 Light French	72.00	76.00	72.00	76.00	91.00	100.00

"A"   "B"   "C"

"D"   "E"   "F"

Colonial Front Entrances

Prices of Western Pine Exterior Door Frames

	Drip Cap 4-5/8" (115.6mm)	5-1/4" (131.2mm)	Masonry 4-5/8" (115.6mm)	5-1/4" (131.2mm)
2'-6"x6'-8" (750 x 2000 mm)	$34.00	$35.00	$29.00	$30.50
2'-8"x6'-8" (800 x 2000 mm)	35.00	36.00	30.00	31.00
2'-8"x7'-0" (800 x 2100 mm)	36.00	37.00	30.50	31.50

	Drip Cap		Masonry	
	4-5/8"	5-1/4"	4-5/8"	5-1/4"
	(115.6mm)	(131.2mm)	(115.6mm)	(131.2mm)
3'-0"x6'-8" (900 x 2000 mm)	37.00	38.00	30.50	32.50
3'-0"x7'-0" (900 x 2100 mm)	37.50	38.00	31.50	32.50
4'-0"x6'-8" (1200 x 2000 mm)	40.00	42.00	32.50	33.50
5'-0"x6'-8" (1500 x 2000 mm)	43.00	45.00	33.50	34.50
6'-0"x6'-8" (1800 x 2000 mm)	48.00	49.00	35.50	36.50

Stud Wall Drip Cap Frame.	Brick Wall Frame.	Brick Veneer Wall Frame.

Door Frame Prices with Preglazed Sidelight One Side for 3'-0"x6'-8" (900 x 2000 mm) Doors

	1-1'-0"x6'-8" (1-300 x 2000 mm)		1-1'-4"x6'-8" (1-400 x 2000 mm)		1-1'-8"x6'-8" (1-500 x 2000 mm)	
Glass	Drip Cap	Masonry	Drip Cap	Masonry	Drip Cap	Masonry
Saf. Gl.-1 Lite	$120.00	$114.00	$133.00	$128.00	$147.50	$142.00
3 Light Saf. Gl.	128.00	123.00	136.00	131.00	154.00	149.00
3 Light & Flat Pan.	143.00	138.50	150.50	145.00	158.50	153.00
3 Light & Raised Pan.	154.00	148.00	161.50	156.50	169.50	164.30

Also available in 4 and 5 light with or without flat or raised panel and with fluted, latticed rondel or loxenged, or insulating glass.

Door Frame Prices with Preglazed Sidelights Both Sides for 3'-0"x6'-8" (900 x 2000 mm) Doors

Glass	2-1'-0"x6'-8" (2-300 x 2000 mm)		2-1'-4"x6'-8" (2-400 x 2000 mm)		2-1'-8"x6'-8" (2-500 x 2000 mm)	
	Drip Cap	Masonry	Drip Cap	Masonry	Drip Cap	Masonry
Saf. Gl.-1 Light	$214.00	$211.00	$242.00	$238.50	$270.50	$267.00
3 Light Saf. Gl.	218.50	215.00	234.50	230.50	250.00	246.50
3 Light & Flat Pan.	240.00	237.00	250.00	254.50	272.00	268.00
3 Light & Raised Pan.	263.00	260.00	278.50	275.50	295.00	291.00

Setting Exterior Wood Door Frames

Size of Door Opening	No. Frames Set 8-Hr. Day	Carp. Hrs. per Frame	Add Hrs. for Transom
3'-0"x7'-0" (900 x 2100 mm)	7-9	0.9-1.1	0.3
6'-0"x7'-0" (1800 x 2100 mm)	5-6	1.3-1.6	0.3
8'-0"x8'-0" (2400 x 2400 mm)	5-6	1.3-1.6	0.3
First Grade Workmanship			
3'-0"x7'-0" (900 x 2100 mm)	5-7	1.1-1.6	0.3
6'-0"x7'-0" (1800 x 2100 mm)	4-5	1.6-2.0	0.3
8'-0"x8'-0" (2400 x 2400 mm)	4-5	1.6-2.0	0.3

Interior Flush Doors

Interior hollow core (H.C.) flush doors are manufactured with face veneers of tempered hardboard or plywood of various species, e.g., luan mahogany, birch, or oak, walnut, etc. The core, which is not hollow, will have some material such as honeycomb fiber board or wood stripping to give lateral support to the face veneers.

Interior Hollow Core Flush Doors

Doors 1-3/8" (34.3 mm) Thick	Plain Birch	Luan
2'-0"x6'-8" (600 x 2000 mm)	$28.50	$16.00
2'-4"x6'-8" (700 x 2000 mm)	29.50	17.00
2'-6"x6'-8" (750 x 2000 mm)	31.50	22.00
2'-8"x6'-8" (800 x 2000 mm)	35.00	24.00
3'-0"x6'-8" (900 x 2000 mm)	36.00	25.00

For doors 7'-0" (2100 mm) in height, add $6.00 per door.
For doors 1-3/4" (43.7 mm) thick, add $5.00 per door.

Interior solid core (S.C.) doors have face veneers of plywood in various species, the same as hollow core doors. The cores of the solid door are usually constructed of solid wood blocks (stave) or particleboard.

Interior Solid Core Flush Doors, Particleboard Core

Size	Rotary Birch	Red Oak	Select Birch
2'-8"x6'-8"x1-3/8" (800 x 2000 x 34.3 mm)	$51.00	$54.00	$60.00
3'-0"x6'-8"x1-3/8" (900 x 2000 x 34.3 mm)	53.00	56.00	67.00
2'-8"x6'-8"x1-3/4" (800 x 2000 x 43.7 mm)	54.00	58.00	70.00
3'-0"x6'-8"x1-3/4" (900 x 2000 x 43.7 mm)	56.00	60.00	74.00

Pre-Hung Interior Door Units

Pre-hung interior door units consist of the door, a 1-3/8" (34.3 mm) H.C. frame, stops, face trim, and hinges, all assembled as a unit. The door is bored and mortised for the lockset and bolt, and the frame is mortised for the strike plate.

The frame, called a "split jamb", is in two pieces at the stop for installation in the wall opening.

Prices are for units with adjustable jambs, from 4-3/8" (109.3 mm) to 5-1/2" (137.5 mm) jambs.

Door Size	Luan Doors	Birch Doors
1'-6"x6'-8" (450 x 2000 mm)	$63.50	$76.50
2'-0"x6'-8" (600 x 2000 mm)	64.50	78.50
2'-4"x6'-8" (700 x 2000 mm)	66.50	80.50
2'-6"x6'-8" (750 x 2000 mm)	68.50	83.00
2'-8"x6'-8" (800 x 2000 mm)	71.00	85.00

Interior Paneled Doors

Pine panel doors built up of stiles and rails have largely been supplanted by molded doors consisting of a single piece of resin-impregnated wood fibre. The doors are factory primed for finish painting on the job site and are joint-free and shrink and swell resistant. Colonial style 6 panel doors will cost as follows:

Size	
2'-0"x6'-8"x1-3/8" (600 x 2000 x 34.3 mm)	$32.00
2'-4"x6'-8"x1-3/8" (700 x 2000 x 34.3 mm)	34.00
2'-6"x6'-8"x1-3/8" (750 x 2000 x 34.3 mm)	35.00
2'-8"x6'-8"x1-3/8" (800 x 2000 x 34.3 mm)	37.00
3'-0"x6'-8"x1-3/8" (900 x 2000 x 34.3 mm)	40.00

Sliding and Bi-Folding Closet Doors

With the popularity of sliding doors for closets, many different combinations of doors, materials, and sizes are on the market. With some of them, the doors are the same height as the other doors in the room or building, i.e. 6'-8" (2000 mm) or 7'-0" (2100 mm), while in others the doors extend to the ceiling so that they provide access to the storage shelves at the top.

While there are many different types of doors, hardware, etc., the ones described below are of the type usually found in moderate class housing and apartments.

Frames for sliding doors may be obtained in kits containing two side jambs, head assembly with track attached, and necessary hardware for hanging doors. Available with 5-3/8" (134.3 mm) jambs for lath and plaster construction, or with 4-5/8" (115.6 mm) jambs for dry wall construction. Openings may be arranged for two, three, or four bypassing doors. Any 1-3/8" (34.3 mm) thick doors can be used.

The following table lists bypassing units, doors, and hardware only for a variety of 1-3/8" (34.3 mm) thick doors.

2 Panel Opng.	H. C. Luan	H. C. Birch	W. P. Louver	W. P. Half Louver	6 Pan. W. P. Col.
4'-0"x6'-8" (1200 x 2000 mm)	$47.50	$58.00	$100.50	$127.00	$180.00
5'-0"x6'-8" (1500 x 2000 mm)	52.50	63.00	105.00	132.00	185.00
6'-0"x6'-8" (1800 x 2000 mm)	58.00	68.00	110.50	137.00	190.00

Bi-folding doors offer the advantage of full access to the closet. They are available in kits with doors, hinges, pivots, knobs, and guide assembly and track. Jambs and casing are not included. Openings to 3'-2" (950 mm) panels hinged one side; 4'-0" (1200 mm) and over 4 panels hinged two sides.

Opng.	Panels per Opng.	1-3/8" (34.3mm) Hard-board	1-3/8" (34.3mm) H.C. Birch	1-1/8" (28.1mm) Louver	1-1/8" (28.1mm) Half Louver	1-3/8" (34.3mm) W. P. Col.
2'-0"x6'-8" (600 x 2000 mm)	2	$26.50	$37.00	$47.50	$53.00	$58.00
2'-6"x6'-8" (750 x 2000 mm)	2	29.50	42.00	52.00	58.00	63.00
3'-0"x6'-8" (900 x 2000 mm)	2	33.00	44.00	54.50	62.00	68.00
4'-0"x6'-8" (1200 x 2000 mm)	4	39.00	69.00	78.00	88.00	113.00
5'-0"x6'-8" (1500 x 2000 mm)	4	46.00	75.00	83.50	93.00	118.00
6'-0"x6'-8" (1800 x 2000 mm)	4	52.00	78.00	85.50	98.00	125.00

PLASTIC FACED WOOD DOORS

For some uses, especially institutional, some doors are veneered on all surfaces with plastic laminate. The laminate can be used on both solid and hollow core doors as well as fire, acoustical, x-ray, and other special units. Any standard plastic finish may be applied.

Unless otherwise specified, doors are edged as well as faced with plastic. The doors will be shipped in a carton prefinished and pre-mortised for all hardware and cannot be trimmed on the job. Therefore, special care must be taken in checking all details on the shop drawings to see that dimensions are accurately followed on the job.

Since all mortising and fitting is eliminated, hanging of the doors is simpler. One carpenter and one laborer should hang 2 doors per hour.

These doors are fabricated by manufacturers on special order. The veneers are not applied by contact adhesives but are bonded under pressure. Costs should be obtained from the manufacturer or distributor.

For budget purposes a door 2'-8"x6'-8" (800 x 2000 mm) plastic faced both sides of a 1-3/4" (43.7 mm) solid core will run around $145.00. For one face plastic, one face birch around $110.00.

Interior Door Jambs

The following prices are for plain door jambs 3/4"x4-1/2" (18.7 x 112.5 mm). Side jambs are dadoed at top for head jambs. The prices are for jambs only, not including stops or trim which must be added. Furnished knocked down.

Price Per Set, Dadoed and Sanded, Not Including Stops

Size	Ponderosa Pine	Natural Birch
2'-8"x6'-8" (800 x 2000 mm) or smaller	$16.00	$25.50
3'-0"x7'-0" (900 x 2100 mm) or smaller	17.00	26.50
6'-0"x6'-8" (1800 x 2000 mm) or smaller	20.00	30.00

Add for Door Stops 7/16"x1-3/8" (10.9 x 34.3 mm), Cut to Length

	Pond. Pine	Nat. Birch
3'-0"x7'-0" (900 x 2100 mm) or smaller	$1.20	$3.20
6'-0"x7'-0" (1800 x 2100 mm) or smaller	1.35	4.00

If door jambs are put together in the mill, add $3.00 per set to the above prices.

Setting Interior Wood Door Jambs

Door Size, Desc. of Jamb	No. Set 8-Hr. Day	Carp. Hrs. per Jamb	Add for Transom
3'-0"x7'-0" (900 x 2100 mm), Plain door jambs	8-10	0.8-1.0	0.6
3'-0"x7'-0" (900 x 2100 mm), Paneled door jambs	5-6	1.3-1.6	0.6
6'-0"x7'-0" (1800 x 2100 mm), Plain door jambs	5-6	1.3-1.6	0.6
First Grade Workmanship			
3'-0"x7'-0" (900 x 2100 mm), Plain door jambs	7-9	0.9-1.1	0.6
3'-0"x7'-0" (900 x 2100 mm), Paneled door jambs	4-5	1.6-2.0	0.6
6'-0"x7'-0" (1800 x 2100 mm), Plain door jambs	4-5	1.6-2.0	0.6

Door Jambs

Door Jambs

Stock door casings per 100 l.f. cost $40.00 ($1.31 per m) in Ponderosa pine and $117.00 ($3.84 per m) in birch. Stops will run $13.00 ($0.42 per m) and $26.00 ($0.85 per m).

Oak thresholds, 3-5/8"x5/8" (90.6 x 15.6 mm) will run $111.00 per 100 l.f. ($36.40 per 10 m) and oak sills, 1-5/16"x7-1/4" (32.8 x 181.2 mm) will run $318.00 per 100 l.f. ($104.33 per 10 m). Pre-cut sills will run $3.50 for 2'-8" (800 mm) opening, $4.00 for 3' (900 mm) opening; in the narrow size, $10.00 and $12.50 for the wider.

Interior Door Trim

The labor cost of placing interior door trim varies with the type of trim or casings and the class of workmanship. Production time is given on 4 different types of interior door trim:

Style "A" consists of 1-member casing with either a mitred or square cut head or cap. This is the simplest trim to use.

Style "B" consists of 1-member trim for sides and a built-up cap trim for heads. Caps are either put together in the mill or on the job.

Style "C" is 2-member back-band trim, consisting of a one-piece casing with a back-band of thicker material.

Style "D" is any type of door trim that is assembled and glued in the mill, ready to set in place as a unit.

The class of workmanship has considerable bearing on the labor costs. In ordinary workmanship the casings are nailed to the rough bucks or grounds and the trim might show some hammer marks, but where first grade workmanship is required, the utmost care must be used in cutting and fitting all casings, miters, etc. All nails must be carefully driven and set, the casings must show the same margin on all sides of the jamb and trim must fit close to the walls. In other words, it must be a first class job in every respect.

Fitting and Placing Stationary Transom Sash. Where just an ordinary grade of workmanship is required, a carpenter should fit and place about 12 to 14 stationary transom sash per 8-hr. day, at the following labor cost per sash:

Labor Erecting Interior Door Casings and Trim

Style Trim	Kind of Trim	Size of Opening	No. Sides Trim per 8-Hr. Day	Carp. Hrs. per Side	Carp. Hrs. per Opening	Add for Transom†
"A"	Single Casing	3'-0"x7'-0"	14–16	0.5–0.6	1.0–1.2	0.6
"B"	Cap Trim	3'-0"x7'-0"	14–16	0.5–0.6	1.0–1.2	0.6
"B"	Cap Trim*	3'-0"x7'-0"	7– 8	1.0–1.1	2.0–2.2	0.6
"C"	Back-band Trim	3'-0"x7'-0"	9–10	0.8–0.9	1.6–1.8	0.6
"D"	Mill Assembled**	3'-0"x7'-0"	14–16	0.5–0.6	1.0–1.2	0.6
"A"	Single Casing	6'-0"x7'-0"	9–10	0.8–0.9	1.6–1.8	0.8
"B"	Cap Trim	6'-0"x7'-0"	9–10	0.8–0.9	1.6–1.8	0.8
"B"	Cap Trim*	6'-0"x7'-0"	5– 7	1.1–1.6	2.2–3.2	0.8
"C"	Back-band Trim	6'-0"x7'-0"	7– 8	1.0–1.1	2.0–2.2	0.8
"D"	Mill Assembled**	6'-0"x7'-0"	9–10	0.8–0.9	1.6–1.8	0.8
		First Grade Workmanship				
"A"	Single Casing	3'-0"x7'-0"	9–10	0.8–0.9	1.6–1.8	0.8
"B"	Cap Trim	3'-0"x7'-0"	9–10	0.8–0.9	1.6–1.8	0.8
"B"	Cap Trim*	3'-0"x7'-0"	5– 7	1.1–1.6	2.2–3.2	0.8
"C"	Back-band Trim	3'-0"x7'-0"	8– 9	0.9–1.0	1.8–2.0	0.8
"D"	Mill Assembled**	3'-0"x7'-0"	11–12	0.7–0.8	1.4–1.6	0.8
"A"	Single Casing	6'-0"x7'-0"	7– 8	1.0–1.1	2.0–2.2	1.0
"B"	Cap Trim	6'-0"x7'-0"	7– 8	1.0–1.1	2.0–2.2	1.0
"B"	Cap Trim*	6'-0"x7'-0"	5– 6	1.3–1.6	2.6–3.2	1.0
"C"	Back-band Trim	6'-0"x7'-0"	6– 7	1.1–1.3	2.2–2.6	1.0
"D"	Mill Assembled**	6'-0"x7'-0"	7– 8	1.0–1.1	2.0–2.2	1.0

*Cap trim put together by carpenters on the job.
**Trim assembled and glued together in the mill.
†Includes transom jambs, stops and casings but no transom sash.

	Hours	Rate	Total	Rate	Total
Carpenter	0.6	$....	$....	$26.44	$15.86

First Grade Workmanship

Figure about 9 to 11 transom sash per 8-hr. day, at the following labor cost per sash:

	Hours	Rate	Total	Rate	Total
Carpenter	0.8	$....	$....	$26.44	$21.15

Fitting and Hanging Hinged or Pivoted Transom Sash. If the transom sash are hinged or pivoted, a carpenter should fit and hang 7 to 9 sash per 8-hr. day, at the following labor cost per sash:

	Hours	Rate	Total	Rate	Total
Carpenter	1.0	$....	$....	$26.44	$26.44

First Grade Workmanship

Figure about 5 to 7 hinged or pivoted transom sash per 8-hr. day, at the following labor cost per sash:

	Hours	Rate	Total	Rate	Total
Carpenter	1.3	$....	$....	$26.44	$34.37

Fitting and Hanging Wood Doors. The labor cost of fitting and hanging doors varies with the class of workmanship, weight of door, soft or hard wood, and the tools and equipment used. There are power planes, hinge butt routers, and lock mortisers that enable a carpenter to do about three times as much work per day as where the doors are fitted by hand.

Where just an ordinary grade of workmanship is required, the doors do not always show the same margin at the top and sides, the screws are frequently driven instead of placed with a screw driver, and in many instances, the doors are "hinge-bound" so they will not open and close freely.

Ordinary Workmanship

Where doors are fit and hung by hand, a carpenter should fit and hang 8 doors per 8-hr. day, at the following labor cost per door:

	Hours	Rate	Total	Rate	Total
Carpenter	1.0	$....	$....	$26.44	$26.44

On large production line jobs, where power tools can be used to advantage and where the same class of work is performed using a power plane, an electric hinge butt router for door and jambs, and an electric lock mortiser, a carpenter should fit and hang 12 to 16 doors per 8-hr. day, at the following labor cost per door:

	Hours	Rate	Total	Rate	Total
Carpenter	0.6	$....	$....	$26.44	$15.86

The above costs will vary with the ability of the carpenter and the job setup, as a power planer should plane off a door in 5 to 10 minutes; a hinge butt router should do 60 to 75 openings per 8-hr. day, while a lock router should mortise out for a lock in about 1/2 minute, after it has been set on the door. The labor handling the doors is the same for both hand and machine work.

First Grade Workmanship

In the better class of buildings, the doors must show the same margin between the door and jamb at the top and sides, the hinges must be set even and flush with the jambs, and the edge of the door and the doors must be free from "hinge binding" so they will fit well and work easily at the same time.

On this grade of work a carpenter should fit and hang about 4 doors per 8-hr. day, at the following labor cost per door:

	Hours	Rate	Total	Rate	Total
Carpenter	2.0	$....	$....	$26.44	$52.88

Where power planers, hinge butt routers, and lock mortisers are used, a carpenter should fit and hang 7 doors per 8-hr. day, at the following labor cost per door:

	Hours	Rate	Total	Rate	Total
Carpenter	1.1	$....	$....	$26.44	$29.08

Labor Setting Jambs, Fitting and Hanging Doors, Placing Stops, Hardware and Casing One Door Opening Complete

Size of Door Opening	Style Trim	Kind of Trim	Carp. Hrs. per Opening	Add for Transom†
		Ordinary Workmanship		
3'-0"x7'-0"	"A"	Single Casing	3.1– 3.7	2.2
3'-0"x7'-0"	"B"	Cap Trim	3.1– 3.7	2.2
3'-0"x7'-0"	"B"	Cap Trim*	4.1– 4.7	2.2
3'-0"x7'-0"	"C"	Back-band Trim	3.7– 4.3	2.2
3'-0"x7'-0"	"D"	Mill Assembled**	3.1– 3.7	2.2
6'-0"x7'-0"	"A"	Single Casing	5.6– 6.5	2.8
6'-0"x7'-0"	"B"	Cap Trim	5.6– 6.5	2.8
6'-0"x7'-0"	"B"	Cap Trim*	6.2– 7.9	2.8
6'-0"x7'-0"	"C"	Back-band Trim	6.0– 6.9	2.8
6'-0"x7'-0"	"D"	Mill Assembled**	5.6– 6.5	2.8
		First Grade Workmanship		
3'-0"x7'-0"	"A"	Single Casing	5.5– 6.2	2.7
3'-0"x7'-0"	"B"	Cap Trim	5.5– 6.2	2.7
3'-0"x7'-0"	"B"	Cap Trim*	6.1– 7.6	2.7
3'-0"x7'-0"	"C"	Back-band Trim	5.7– 6.4	2.7
3'-0"x7'-0"	"D"	Mill Assembled**	5.3– 6.0	2.7
6'-0"x7'-0"	"A"	Single Casing	9.2–10.3	3.4
6'-0"x7'-0"	"B"	Cap Trim	9.2–10.3	3.4
6'-0"x7'-0"	"B"	Cap Trim*	9.8–11.3	3.4
6'-0"x7'-0"	"C"	Back-band Trim	9.4–10.7	3.4
6'-0"x7'-0"	"D"	Mill Assembled**	9.2–10.3	3.4

*Cap trim put together by carpenters on the job.
**Trim assembled and glued together in the mill.
†Above is for hinged or pivoted transom. For stationary transom deduct 0.4-hr.

For front door hardware, add 0.5-hr. for ordinary work and 1 to 2 hrs. for first grade work.

Where power planers, hinge butt routers, and lock mortisers are used, deduct 0.4 hr. for single door and 0.8-hr. for double door openings on ordinary workmanship; for first grade workmanship, deduct 0.9-hrs. for single door and 1.8 hrs. for double door openings.

Labor Installing Pre-Hung Door Units. Pre-hung door units are used extensively in the construction of houses and apartments ranging from "economy" grade to middle class.

In the completely assembled package, jambs are assembled, the door is hung in place with hinges applied, and stops are mitred and nailed into place.

Where assembled pre-hung door units are used, a carpenter should install 16 units per 8-hr. day, at the following labor cost per unit:

	Hours	Rate	Total	Rate	Total
Carpenter	0.5	$....	$....	$26.44	$13.22

Fitting and Hanging Hardwood Acoustical Doors. The labor cost of fitting and hanging hardwood acoustical doors is always figured on a first grade workmanship basis, because they must be perfectly installed to insure their efficient operation.

Assuming a carpenter, experienced in hanging acoustical doors, is employed to do the work, the following production should be obtained:

Type of Door	Approx. Wgt. lbs/s.f. (kg/sq.m) per Sq. Ft.	Carpenter Hrs per Opng	Helper Hrs per Opng
Hardwood Class 36	4 (19.5)	6	3
Hardwood Class 41	7 (34.1)	8	4

Fitting and Hanging Toilet Stall Doors. A carpenter should fit and hang 9 to 11, 7/8" (21.8 mm) or 1-1/8" (28.1 mm) toilet doors per 8-hr. day, at the following labor cost per door:

	Hours	Rate	Total	Rate	Total
Carpenter	0.8	$....	$....	$26.44	$ 21.15

First Grade Workmanship

Where hardwood stall doors are used and carefully hung, a carpenter should fit and hang 7 to 9 doors per 8-hr. day, at the following labor cost per door:

	Hours	Rate	Total	Rate	Total
Carpenter	1	$....	$....	$26.44	$26.44

Fitting and Hanging Heavy Wood Swinging Doors Up to 4'-0"x8'-0" (1200 x 2400 mm). When hanging heavy wood swinging doors, such as used in garages, stables, mills and factory buildings, where the doors are 3'-0" (900 mm) to 4'-0" (1200 mm) wide and 7'-0" (2100 mm) to 8'-0" (2400 mm) high, it will require about 4 to 4-1/2 hrs. time to fit and hang one door or 8 to 9 hrs. per pair, at the following labor cost per door:

	Hours	Rate	Total	Rate	Total
Carpenter	4.3	$....	$....	$26.44	$113.69

Fitting and Hanging Heavy Wood Swinging Doors Up to 5'-0"x10'-0" (1500 x 3000 mm). To fit and hang extra heavy wood swinging doors up to 5'-0"x10'-0"x2-1/2" or 3" (1500 x 3000 x 62.5 or 75 mm), will require about 4 to 5 hrs. time for 2 carpenters, at the following labor cost per door:

	Hours	Rate	Total	Rate	Total
Carpenter	9.0	$....	$....	$26.44	$237.96

Fitting and Hanging Wood Sliding Doors Up to 4'-0"x8'-0" (1200 x 2400 mm). When fitting and hanging heavy wood sliding doors, such as in garages and factories, two carpenters working together should place tracks and hang one door complete in 3 to 3-1/2 hrs. at the following labor cost per door:

	Hours	Rate	Total	Rate	Total
Carpenter	6.5	$....	$....	$26.44	$171.86

Placing track and hang a pair of double sliding doors for an opening up to 8'-0"x8'-0" (2400 x 2400 mm) requires 10 to 12 hrs. carpenter time, at the following labor cost per opening:

	Hours	Rate	Total	Rate	Total
Carpenter	11.0	$....	$....	$26.44	$290.84

The above does not include cutting or drilling in masonry walls to place bolts or anchors. It assumes the work is done when the walls are being built.

Fitting and Hanging Heavy Wood Sliding Doors Up to 10'-0"x10'-0" (3000 x 3000 mm). To place track, fit, and hang one sliding door up to 10'-0"x10'-0" (3000 x 3000 mm), 2-1/2" (62.5 mm) to 3" (75 mm) thick, requires about 12 hrs. carpenter time, at the following labor cost per opening:

	Hours	Rate	Total	Rate	Total
Carpenter	12.0	$....	$....	$26.44	$317.28

The above does not include cutting or drilling masonry walls for bolts or anchors.

Fitting and Hanging Heavy Wood Sliding Doors Up to 12'-0"x18'-0"
(3600 x 5400 mm). To place track and hang one large door for an opening up
to 12'-0"x18'-0" (3600 x 5400 mm) requires about 24 to 28 hrs. carpenter time,
at the following labor cost per door:

	Hours	Rate	Total	Rate	Total
Carpenter	26.0	$....	$....	$26.44	$687.44

It requires 4 to 6 workers to handle a door of this size.

Making and Setting Sliding Door Pockets. Where sliding door pockets
are made on the job for door openings up to 3'-0"x7'-0" (900 x 2100 mm), a
carpenter should make the pocket, fit track, and set the pocket in about 4
hrs. at the following labor cost:

	Hours	Rate	Total	Rate	Total
Carpenter	4.0	$....	$....	$26.44	$105.76

Where double doors are used and it is necessary to make and set 2 pockets,
double the above costs.

Setting Sliding Door Pockets. If sliding door pockets are sent to the job
assembled, with track in place or steel pocket door T-frames are used, and
the carpenters on the job set the pocket or steel frame ready to receive the
door, it will require about 1 hr. carpenter time, at the following labor cost per
opening:

	Hours	Rate	Total	Rate	Total
Carpenter	1.0	$....	$....	$26.44	$26.44

If double sliding doors are used and it is necessary to set 2 pockets for each
opening, double the above costs.

Fitting and Hanging Sliding Doors. For an ordinary sliding door up to
3'-0"x7'-0" (900 x 2100 mm), to place hangers and hang the door complete,
not including finish hardware, requires about 1 hr. carpenter time, at the
following labor cost per door:

	Hours	Rate	Total	Rate	Total
Carpenter	1.0	$....	$....	$26.44	$26.44

For double sliding doors, double the above costs.

Labor on Sliding Door Openings Complete with Pockets. Where
necessary to make the sliding door pocket, hang the track, set the pocket,
and fit and hang the door complete, exclusive of finish hardware, figure
about 5 hrs. carpenter time, at the following labor cost per opening:

	Hours	Rate	Total	Rate	Total
Carpenter	5.0	$....	$....	$26.44	$132.20

For double sliding door openings, double the above costs.

If pockets are sent to the job with track fitted in place or steel pocket door T-frames are used, and the job carpenters set the box or steel frame and fit and hang the sliding door, it will require about 2 hrs. carpenter time per opening, exclusive of finish hardware, at the following labor cost per opening:

	Hours	Rate	Total	Rate	Total
Carpenter	2.0	$....	$....	$26.44	$52.88

For double sliding door openings, double the above costs.

Labor Fitting and Hanging Sliding Wood Closet Doors. When hanging and fitting wood closet doors that slide past each other without the necessity for pockets, a carpenter should place track, attach hangers, fit, and hang two doors complete for an opening 3'-0" (900 mm) to 5'-0" (1500 mm) wide and 6'-8" (2000 mm) to 8'-0" (2400 mm) high in about 2 hours. This does not include time setting door jambs and casing openings on two sides. The labor cost per opening should average as follows:

	Hours	Rate	Total	Rate	Total
Carpenter	2.0	$....	$....	$26.44	$52.88

08300 SPECIAL DOORS

TIN CLAD FIRE DOORS

Tin clad fire doors are used in factory, mill, and warehouse buildings. They are either sliding or swinging, with fusible links and automatic closing devices.

Sliding doors and lap type swinging doors in brick walls do not ordinarily require a steel frame, but it might be necessary for tile walls. Flush type swinging doors require a frame.

Standard tin clad fire doors are usually manufactured in two thicknesses depending on the fire rating required. For Class "A" rating, the thickness is 2-5/8" (65.6 mm) consisting of 3-plies of tongue-and-groove lumber clad with tin sheeting on all sides. For Class "B" or "C" rating, the thickness is 1-3/4" (43.7 mm) consisting of 2-plies of tongue-and-groove lumber with cladding.

Method of Calculating the Area of Fire Doors. For flush doors, multiply the width of the opening by the height of the opening.

Square Top Lap Doors. Obtain size of door by adding 8" (200 mm) to width and 4" (100 mm) to height of opening, then multiply width of door by height of door.

Inclined Top Lap Doors. Obtain width of door by adding 8" (200 mm) to width of opening and obtain height by adding 8" (200 mm) to height of door for doors up to 4'-0" (1200 mm) wide and 12" (300 mm) for doors up to 6'-0" (1800 mm) wide. Multiply width of door by height of door.

Prices of Tin Clad Fire Doors and Shutters

The price includes door only, covered both sides with tin of a quality and in a manner specified by the National Board of Fire Underwriters.

	Price per s.f.
Standard 3-ply, 2-5/8" thick, 13 lbs./s.f.	$10.00
Standard 2-ply, 1-3/4" thick, 10 lbs./s.f.	9.50
Standard 2-ply, 1-3/4" thick, shutters, 10 lbs./s.f.	9.00

(metric)

	Price per sq.m
Standard 3-ply, 65.6 mm thick, 63.4 kg/sq.m	$107.64
Standard 2-ply, 43.7 mm thick, 48.8 kg/sq.m	102.26
Standard 2-ply, 48.8 mm thick, shutters, 48.8 kg/sq.m	96.87

Above prices include doors only, no hardware or fixtures of any kind or delivery costs.

Hardware for Swinging, Automatic Closing Tin Clad Doors

Fixture sets include hinges, pintles, weights, chains, catches, latches, complete automatic device with bolts, and screws for attaching. Installation is not included. Door frames are not included.

Door Width & Height Up to and Including	Single Swinging Doors		Double Swinging Doors	
	Flush Type	Lap Type	Flush Type	Lap Type
3'-0"x6'-6" (900 x 1950 mm)	$63.00	$86.00	$146.00	$169.00
3'-0"x8'-6" (900 x 2550 mm)	79.00	108.00	165.00	197.00
3'-0"x10'-6" (900 x 3150 mm)	94.00	128.00	185.00	221.00
4'-0"x6'-6" (1200 x 1950 mm)	67.00	89.00	148.00	171.00
4'-0"x8'-6" (1200 x 2550 mm)	82.00	111.00	168.00	199.00
4'-0"x12'-0" (1200 x 3600 mm)	98.00	133.00	188.00	225.00
5'-0"x6'-6" (1500 x 1950 mm)	70.00	91.00	150.00	174.00
5'-0"x8'-6" (1500 x 2500 mm)	85.00	115.00	171.00	203.00
5'-0"x12'-0" (1500 x 3600 mm)	102.00	137.00	192.00	232.00
6'-0"x6'-6" (1800 x 1950 mm)	74.00	94.00	153.00	176.00
6'-0"x8'-6" (1800 x 2500 mm)	89.00	118.00	174.00	206.00
6'-0"x12'-0" (1800 x 3600 mm)	107.00	141.00	196.00	233.00

If automatic controls are not wanted with swinging tin clad doors, deduct from the above prices, $8.00 for single doors and $12.00 per opening for double swinging doors.

Hardware for Sliding Tin Clad Doors

Prices include non-adjustable hangers, track, weights, and bolts for attaching hardware to doors. Through wall bolts are not included in prices.

Hardware for single sliding door, 3'-0" (900 mm) wide opening..................$191.00
For each 6" (150 mm) additional width of opening, add 16.00
Openings over 6'-0" (1800 mm) wide require 3 hangers on each door,
   add for each additional hanger 11.00
Hardware for double sliding doors, 6'-0" (1800 mm) wide opening..............325.00
For each 6" (150 mm) additional width of opening, add 16.00

Labor Erecting Tin Clad Doors

The following schedules, furnished by the Richmond Fireproof Door Co., are guides for estimating labor to install.

The hours of labor are based on brick walls 13" (325 mm) thick. For thicker walls, add 1/4 hour for one worker for each 4" (100 mm) in thickness per bolt. Concrete and stone walls require slightly more time drilling; terra cotta, slightly less. Where "doors both sides" are noted, this assumes doors back-to-back and tracks bolted together. When tracks cannot be bolted together, use "one side" schedule multiplied by two.

Type	Width, in.	(mm)	Door One Side Mech.	Help.	Door Both Sides Mech.	Help.
Single Slide	2'-0" to 4'-0"	600-1200	7	7	10	10
	4'-0" to 5'-2"	1200-1550	8	8	12	12
	5'-3" to 7'-8"	1575-2300	9	9	14	14
	7'-9" to 8'-8"	2925-2600	11	11	17	17
	8'-9" to 12'-0"	2625-3600	12	12	18	18
Slide in Pairs	4'-0" to 4'-5"	1200-1325	9	9	14	14
	4'-6" to 9'-8"	1350-2900	11	11	16	16
	9'-9" to 10'-6"	2925-3150	13	13	18	18
	10'-7" to 12'-0"	3175-3600	15	15	20	20
	Width					
Vertical	to 4'-10"	1200	13	13	21	21
Sliding	4'-11" to 7'-6"	1475-2250	15	15	24	24
	7'-7" to 10'-2"	2275-3050	17	17	26	26
	10'-3" to 12'-0"	3075-3600	19	19	28	28
	Door Height					
Single Swing	to 5'-9"	1725	5	5	7	7
(Flush or	5'-10" to 8'-9"	1950-2625	5-1/2	5-1/2	8	8
Lap)	8'-10" to 12'	2650-3600	6	6	9	9
	Door Height					
Swing in	to 5'-9"	1725	8	8	12	12
Pairs (Flush	5'-10" to 8'-9"	1950-2625	9	9	14	14
or Lap)	8'-10" to 12'-0"	2650-3600	10	10	16	16

Labor Cost of Installing One Single 3-Ply Sliding Tin Clad Fire Door for 4'-0"x7'-0" (1200 x 2100 mm) Wall Opening

The following estimate includes door, track, hangers, and necessary hardware installation.

	Hours	Rate	Total	Rate	Total
Mechanic	7.0	$....	$....	$26.44	$185.08
Helper	7.0	....	....	21.04	147.28
Cost per opening			$....		$332.36

Labor Cost of Installing One Pair Double Sliding 3-Ply Tin Clad Fire Doors for 6'-0"x7'-0" (1800 x 2100 mm) Wall Opening

The following estimate includes doors, track, hangers, and necessary hardware installation.

	Hours	Rate	Total	Rate	Total
Mechanic	11.0	$....	$....	$26.44	$290.84
Helper	11.0	....	....	21.04	231.44
Cost per opening			$....		$522.28

Labor Cost of Installing One Lap Type Single Swinging 2-Ply Tin Clad Fire Door for 3'-0"x7'-0" (900 x 2100 mm) Wall Opening

The following estimate includes door, hardware, and automatic controls installation.

	Hours	Rate	Total	Rate	Total
Mechanic	5.0	$....	$....	$26.44	$132.20
Helper	5.0	....	....	21.04	105.20
Cost per opening			$....		$237.40

Labor Cost of Installing One Pair Lap Type Double Swinging 2-Ply Tin Clad Fire Doors for 6'-0"x7'-0" (1800 x 2100 mm) Wall Opening

The following estimate includes doors, hardware, and automatic controls installation.

	Hours	Rate	Total	Rate	Total
Mechanic	9.0	$....	$....	$26.44	$237.96
Helper	9.0	....	....	21.04	189.36
Cost per opening			$....		$427.32

Labor Cost of Installing One Flush Type Single Swinging 2-Ply Tin Clad Fire Door for 3'-0"x7'-0" (900 x 2100 mm) Wall Opening

The following estimate includes door, hardware, and automatic controls installation.

	Hours	Rate	Total	Rate	Total
Mechanic	5.0	$....	$....	$26.44	$132.20
Helper	5.0	....	....	21.04	105.20
Cost per opening			$....		$237.40

Labor Cost of Installing One Pair Flush Type Double 3-Ply Swinging Tin Clad Fire Doors for 6'-0"x7'-0" (1800 x 2100 mm) Wall Opening

The following estimate includes doors, hardware, and automatic controls installation.

	Hours	Rate	Total	Rate	Total
Mechanic	9.0	$....	$....	$26.44	$237.96
Helper	9.0	....	....	21.04	189.36
Cost per opening			$....		$427.32

Frames for tin clad doors are made of angles and channels usually furnished under metal fabrications and set by the mason. A 3x3 (75 x 75 mm) angle frame costs about $7.00 per l.f. ($23.00 per m) installed. An 8" (200 mm) channel frame costs about $15.00 per l.f. ($49.00 per m) installed.

08330 COILING DOORS

Coiling doors are usually called rolling shutters and are fully factory assembled units made of steel or aluminum. Often, they are labeled fire doors, and installation is governed by Underwriter's specifications. Steel mounting jambs and heads are usually furnished by the metal fabrications contractor.

A manually operated, 20 gauge door up to 12'x12' (3600 x 3600 mm) costs about $7.00 to $8.00 per s.f. ($75.30-86.10 per sq.m), including hood, guides, and operating hardware.

Where class A fire ratings is required, doors cost $6.50 per s.f. ($70.00 per sq.m) more. If 18-gauge construction, it will add another $1.00 per s.f. ($10.75 per sq.m). Motor operation will cost $800.00 for smaller sizes, $900 for medium. Each pass door will add another $800.00.

Two mechanics should hang and adjust a rolling shutter, up to 12'x12' (3600 x 3600 mm), in one 8-hr. day at the following cost per shutter:

	Hours	Rate	Total	Rate	Total
Mechanics	16.0	$....	$....	$26.44	$423.04

Coiling Grilles

Rolling grilles are used for protection of store fronts, for open fronted stores in enclosed shopping areas, and for closing off counter areas.

Stock commercial grilles will cost about $12.00 per s.f. ($129.00 per sq.m) with all hardware. These are often part of a decorative front, made of anodized aluminum, and cost about $14.00 per s.f. ($150.00 per s.f.). Exact prices can be obtained from the manufacturer, who will also furnish shop drawings upon request to show all the necessary anchoring involved. Steel grilles run about $8.00 per s.f. ($86.00 per sq.m). Motorizing and installation costs are the same as for rolling shutters.

08350 FOLDING DOORS

Steel Bi-fold Closet Doors. These may be installed in openings with drywall, wood, or steel jambs and head frame. Each bi-fold door package comes complete with door panels, track, guides, and pulls for a complete installation.

Doors are available in a variety of styles including flush, panel, half louver, or full louver. The doors are shipped "factory finished" with a baked-on paint finish.

A two-panel unit is used for openings up to 3'-0" (900 mm) wide and costs about $50.00. A four-panel unit is used for openings 4'-0" (1200 mm) to 6'-0" (1800 mm) wide, with the 6'-0" (1800 mm) unit costing about $95.00. Units are sold in 6'-8" (2000 mm) and 8'-0" (2400 mm) heights, with the 8'-0" (2400 mm) heights costing an additional $11 to $22 more per unit.

One carpenter should install a complete bi-fold unit and all hardware in about 45 minutes at the following labor cost per unit:

	Hours	Rate	Total	Rate	Total
Carpenter	0.75	$....	$....	$26.44	$19.83

Folding Doors. These are often folding partitions that can be ordered completely furnished and with all hardware in wood or vinyl. Folding doors for a 3'x6'-8" (900 x 2000 mm) opening will cost about $64.00 in vinyl, $95.00 in wood slats. Doors for 6'x6'-8" (1800 x 2000 mm) openings will run $150 for vinyl, $180 for wood. Deluxe and custom models will run somewhat more. All the above prices include hardware, but the frame must be furnished by others. When the opening is over 150 s.f. (14 sq.m), the units are considered "partitions" rather than doors and cost about $5.00 per s.f. ($54.00 per sq.m) for light models to $7.00 per s.f. ($75.00 per s.f.) for heavy duty units.

08360 OVERHEAD DOORS

Overhead Garage Doors. Overhead garage doors are furnished in several different styles, including the one-piece overhead door, self-balancing overhead door, and the spring-balanced overhead door.

For an average opening up to 8'-0"x7'-0" (2400 x 2100 mm), two carpenters should place tracks, fit, and hang doors and apply the necessary hardware in 4 hours, at the following labor cost per door:

	Hours	Rate	Total	Rate	Total
Carpenter	8.0	$....	$....	26.44	$211.52

For larger openings up to 16'-0" (4800 mm) wide and 7'-0" (2100 mm) high, where one wide overhead folding door is used, two carpenters should place tracks, fit, and hang doors and apply the necessary hardware complete in 5 hours, at the following labor cost per door:

	Hours	Rate	Total	Rate	Total
Carpenter	10.0	$....	$....	$26.44	$264.40

GARAGE, FACTORY AND WAREHOUSE DOORS

Overhead Type Garage Doors

Overhead type garage doors are counterbalanced and controlled by oil tempered torsion springs. Cables are secured at the bottom of the door on each side and are directly connected with the pair of balancing springs located immediately over the door.

Ball bearing rollers carry all sections of door over a continuous channel track. The track has a flange that keeps the rollers in place.

Doors have kiln-dried fir or spruce sections, with hardboard panels. All joints are mortised, tenoned, or doweled with steel pins and glued. All sections are fitted to size, squared for opening, and drilled for hardware.

Doors over 12'-0" (3600 mm) wide have steel struts in intermediate sections to prevent sagging. Doors over 15'-0" (4500 mm) wide have steel struts in all sections.

Standard Size, Sectional Overhead Type, Garage Doors

Furnished complete with hardware, including cylinder lock, springs, and track. Doors are furnished with all wood panels, or with top section open for glass, which is not included. Approximate prices are:

Size Opening	1-3/8" (34.3 mm) Thick	1-3/4" (43.7 mm) Thick
9'-0"x7'-0" (2700 x 2100 mm)	240.00	265.00
14'-0"x7'-0" (4200 x 2100 mm)	500.00	530.00
15'-0"x7'-0" (4500 x 2100 mm)	530.00	580.00
16'-0"x7'-0" (4800 x 2100 mm)	560.00	640.00
10'-0"x10'-0" (3000 x 3000 mm)		425.00
12'-0"x10'-0" (3600 x 3000 mm)		510.00
10'-0"x12'-0" (3000 x 3600 mm)		520.00
12'-0"x12'-0" (3600 x 3600 mm)		610.00

Electric Door Operators

Operators may be used with any type of overhead doors. Motors are suitable for single phase, 60 cycle, 110-120 Volt AC power supply, and are of the reversing type which permit stopping and changing direction at any point of the door travel. Adjustable limit switches stop the door automatically at top and bottom of door openings.

A safety mechanism automatically reverses the door if it meets an obstruction while in operation, or the door will stop and the operator will shut off.

Doors are driven by means of a roller chain fastened to a shoe sliding on 2 steel tracks. The shoe is connected to the door by a detachable arm.

Single door residential operator with 1/4-hp.
  motor for door up to 22' (6600mm) wide, 7' (2100mm) high, complete
  with one toggle type control station...$190.00
For commercial door operator, with 1/4-hp.
  motor for door not over 140 s.f. (13 sq.m) nor over 10' (3000mm) high 265.00
Commercial door operator, with 1/3-hp.
  motor for door not over 168 s.f. (15.6 sq.m) or 12' (3600mm) high...............290.00
Commercial door operator, with 1/2-hp.
  motor for door not over 224 s.f. (20.8 sq.m) or 14' (4200mm) high...............350.00
Extra toggle type control station for commercial operator 11.00
Extra key switch (for residential operator only)...............................22.00

An experienced mechanic should install an electric door operator in 1 to 2 hours after the necessary wiring is in place.

Electronic Door Operators

For use where a transmitter is located in each automobile using the garage. A push of the button operates the door from the car.

For residential doors up to 16'-0"x8'-0" (4800 x 2400 mm)
  approx. price installed...................................$265.00

An experienced mechanic should install an electric door operator in 2 hours after the necessary wiring is in place.

Special Hardware—Add per Door

Low headroom, 6" (150 mm) minimum,
for doors not over 132 s.f. (12.2 sq.m)..$16.00
Low headroom, 6" (150 mm) minimum,
for doors not over 133-175 s.f. (12.3-16.2 sq.m)...........................$28.00
Low headroom, 6" (150 mm) minimum, for doors over 16'-0" (4800 mm) wide
or over 12'-0" (3600 mm) high $40.50
High lift (for lubritoriums) up to 2'-0" (600 mm) above head jamb..............$23.50
High lift (for lubritoriums) up to 5'-0" (1500 mm) above head jamb............$48.00

On commercial jobs steel doors are often specified. A 10'x10' (3000 x 3000 mm) door will cost about $425.00 with manual operator, $750.00 with electric operator.

Fiberglass and aluminum doors are light in weight and let in daylight. A stock door will cost about $240.00 for a single car size, $400.00 for a double car size.

Colonial   *Fan Top*   *Vertical Battens*   *Diagonal Battens*

Doors of Special Designs

Colonial doors, heavily molded outside with raised panels, add $1.50 per s.f. ($16.00 per sq.m).

For fan top on door, add $130.00 extra.

Vertical battens, 2-ply V-jointed or beaded, add $1.50 per s.f. ($16.00 per sq.m).

Diagonal battens, 2-ply V-jointed or beaded, add $1.75 per s.f. ($18.80 per sq.m).

08370 SLIDING GLASS DOORS

Aluminum sliding glass doors are available in several types, as illustrated, for glazing with crystal glass or 5/8" (15.6 mm) insulated glass. "X" indicates sliding panel; "O" indicates a fixed panel.

Frame, glazing bead and door members are heavy aluminum extrusions. Door sections move upon bottom mounted, ball bearing, grooved brass rollers. Frames are designed with provisions for installing bottom roller, horizontally sliding type screens. Weatherstripping is mohair pile, factory mounted in channels to form a continuous double seal. Hardware includes full-grip lucite pulls and touch latch. Cylinder lock sets are available at extra cost.

Screens are made from aluminum extrusions of tubular box type section with bottom rollers and fitted with fiberglass screen.

Prices are for door units set up. Factory assembly includes attachment of all accessory parts, such as weatherstripping, roller mechanisms, latch, corner connecting brackets, and bumpers. Screen units are assembled. Glass and glazing, is included.

Type XO   Type OXO   Type OXXO

Sliding Glass Doors

Material Prices of Aluminum Sliding Doors

Type	Frame Size, ft.in. (mm)	Economy Crystal Gl.	Deluxe Insulating Gl.
XO	6'-0"x7'-0" (1800 x 2100)	$265.00	$530.00
OXO	9'-0"x7'-0" (2700 x 2100)	315.00	630.00
OXXO	12'-0"x7'-0" (3600 x 2100)	415.00	888.00

For cylinder lock set in lieu of latch, add $10.00.

Labor Erecting Aluminum Sliding Glass Doors. Assembly and erection of aluminum sliding glass doors is simple and does not require special tools. All aluminum members are prefit, with holes drilled and connecting brackets attached. The following table gives average labor hours required for installing aluminum sliding glass doors in ordinary residential work.

Type	Number of Doors	Number of Fixed Panels	Mechanic Hours Required for Assembly	for Erection
XO	1	1	2	6
OXO	1	2	3	8
OXXO	2	2	4	10

Performance Standards. There are many suppliers of aluminum windows and sliding doors, and it is often difficult to compare manufacturers. The Architectural Aluminum Manufacturers Association publishes quality specifications, and manufacturers belonging to that association affix the

AAMA "Quality-Certified" label to their products, attesting that they comply with the designated specification. These specifications include the following for sliding glass doors:

SGD-B1 - For residential and limited commercial applications

SGD-A2 - For commercial and residential application requiring increased size or performance characteristics.

HP Series - Either of the above designations followed by "HP" indicates that the door complies with High Performance requirements.

The specifications for aluminum windows are:

DH-B1 ....Double hung residential
DH-A2 ....Double hung commercial
DH-A3 & A4........Double hung monumental
C-B1....................Casement residential
C-A2....................Casement commercial
C-A3....................Casement monumental
P-B1....................Projected residential
P-A2....................Projected commercial
P-A3....................Projected monumental
A-B1....................Awning residential
A-A2....................Awning commercial
HS-B1 & B2 ....Horizontal sliding residential
HS-A2.................Horizontal commercial
HS-A3.................Horizontal monumental
J-B1....................Jalousie residential
JA-B1..................Jalousie-awning residential
VS-B1 ....Vertical sliding residential
TH-A2.................Top hinged commercial
TH-A3.................Top hinged monumental
VP-A2 ....Vertically pivoted commercial
VP-A3 ....Vertically pivoted monumental
HP Series ....Any of the above followed by "HP"
..........................indicates the window complies
..........................with High Performance requirements.

The AAMA also publishes a certified products directory with a current listing of all certified windows and sliding glass doors by company, model and, maximum size approved.

Wood Sliding or Patio Doors

Sliding doors are now available from most of the nationally distributed millwork houses and follow the general designs established for aluminum sliding units. These doors can usually be purchased in two basic widths per panel and in various arrangements including two panels, one fixed, one operating; three panels, two sides fixed and center panel operable; and four panels, two end panels fixed and two center panels operable.

The following price schedule is based on units preglazed with 5/8" safety glass and complete with operating hardware, weatherstripping, sill, and screens for operable sash.

Glass Size	Nominal Door Size	Type	Material Cost
28" (700mm) wide	5'-0"x6'-8" (1500x2000mm)	Double XO	$500.00
34" (850mm) wide	6'-0"x6'-8" (1800x2000mm)	Double XO	538.00
46" (1150mm) wide	8'-0"x6'-8" (2400x2000mm)	Double XO	698.00
34" (850mm) wide	9'-0"x6'-8" (2700x2000mm)	Triple OXO	830.00
46" (1150mm) wide	12'-0"x6'-8" (3600x2000mm)	Triple OXXO	1,050.00

Add for keyed lock, $12.00. Grilles that can be removed for cleaning are available in either diamond or rectangular patterns. Costs are $43.50 each for the 28" (700 mm) width, $44.50 each for the 34" (850 mm) width, $53.00 for the 46" (1150 mm) width in the rectangular pattern; and $45.00, $47.50, and $65.00 each for the diamond patterns.

08380 SOUND RETARDANT DOORS

Hardwood Sound Insulating Doors

Acoustical doors are for use in consultation rooms, conference rooms, hospitals, clinics, music rooms, television and radio studios, and other areas where acoustical levels must be closely maintained.

Depending on the desired decibel reduction or acoustical efficiency, doors are identified and furnished in several models.

Units 1-3/4" (43.7 mm) thick are rated in sound transmission class 36. Units 2-1/2" (62.5 mm) thick are rated in sound transmission class 41. A door 3" (75 mm) thick is also made and has superior acoustical qualities.

Acoustical doors are made in a wide selection of face woods. Custom finishing of doors is available. These are veneered doors and can be furnished with veneers to match those of other doors in the building. Any standard hardware may be used for installation.

The following prices include bottom closers, stops, gaskets, and stop adjusters and are for standard rotary cut veneers of natural birch and red oak.

Material Prices of Hardwood Sound Insulating Doors*

Size	1-3/4" (43.7mm) 35 Decibel	2-1/2" (62.5mm) 40 Decibel	3" (75mm) 50 Decibel
2'-6"x6'-8" (750 x 2000 mm)	$265.00	$345.00	$455.00
2'-8"x6'-8" (800 x 2000 mm)	275.00	355.00	465.00
3'-0"x6'-8" (900 x 2000 mm)	285.00	365.00	475.00
2'-6"x7'-0" (750 x 2100 mm)	275.00	355.00	465.00
2'-8"x7'-0" (800 x 2100 mm)	285.00	365.00	475.00
3'-0"x7'-0" (900 x 2100 mm)	295.00	375.00	485.00
Light openings: rectangular vision light not over 18" (450 mm) square	$35.00	$37.00	$40.00

*f.o.b. factory, uncrated.

Astragals. Astragals for pairs of doors, 2-1/2"x1" (62.5 x 25 mm) in any wood to match face veneer with 1"x1/2" (25 x 12.5 mm) rubber gasket, $30.00 per pair. For rounding meeting edges and grooving one edge with Eveleth bumper strip, add $30.00.

Other woods. For other woods, add to the total price the following percentages:

Sliced natural birch and sliced red oak 8%
Rotary cut red or white birch and rotary cut plain white oak 15%
Sliced red or white birch and sliced plain white oak 20%
Rift sliced white oak and sliced African mahogany 30%
Sliced American black walnut 50%

08400 ENTRANCES AND STORE FRONTS

METAL STORE FRONT CONSTRUCTION

Aluminum storefront and entrance doors are manufactured by a number of firms in a wide variety of styles to suit the design and function of most any type of building.

The glass and glazing subcontractor is the firm to contact for this type of work, and they typically represent more than one manufacturer of these materials. This subcontractor will handle all the necessary functions of bidding, shop drawing preparation, ordering, and erection of the storefront materials, as well as the glass, for a complete installation.

Materials are made from an aluminum alloy and extruded into many different shapes and sizes of various thicknesses. The extrusions are offered in stock lengths of 24' (7200 mm) long, but longer lengths are available upon special order.

To prevent oxidation of the aluminum surfaces, the extrusions are anodized in a number of colors as well as the clear anodized finish. In recent years this material has become available with a paint coating that offers exceptional durability in addition to a controlled color gradation that is difficult to obtain with the color anodized finishes.

Entrance doors are available in numerous sizes and designs—narrow stile, medium stile, and wide or monumental stile—with various styles of applied panels and finish hardware.

The doors can be hung into the frame by using butt hinges, offset pivots, or center pivots. The controlled automatic closing of the doors can be achieved by installing an exposed overhead closer, a closer concealed in the horizontal head frame member, or by a floor closer that is set into a pocket in the floor under the hinge point of the door.

Installation. Once the subcontractor has a contract agreement to provide the storefront construction, he will prepare shop drawings based on the architectural drawings for the project. The shop drawings clearly define all dimensions and jointing details necessary for the installation, and the materials are ordered based on these drawings.

Prior to fabrication of the exact lengths necessary for erection, the subcontractor will verify the shop drawing dimensions with actual field dimensions taken at the jobsite.

After the extrusions are cut to the proper length for field erection, they are delivered to the jobsite, distributed to the proper location, and installation begins.

The extrusions are fastened to each other with concealed, extruded aluminum clips at all joints. The pre-fabricated frames are then set into the opening and fastened, through the glazing pocket, to the substrate material with screws or expansion anchors. Proper assembly allows the vertical extrusions to run top to bottom, with all horizontal extrusions fitted between the verticals.

Method of Estimating Materials. Measure the length of each type and size extrusion and list into proper groupings. Since this material is sold in stock lengths, the estimator will have to combine the measured lengths together to arrive at the total number of stock lengths needed with the least amount of waste. Count the joints to ascertain the required number of concealed clips, one per joint. It will also be necessary to measure the total length of vinyl gasket material needed for glazing, because this material is ordered separately from the aluminum extrusions. After all of the material quantities are measured, grouped, and totaled, the appropriate material prices can be applied for the extension of the material costs.

Doors are listed by size and style, and depending on the architectural requirements, they may be stock units or custom units. The pricing manual, supplied by the aluminum manufacturer, will list the material pricing strategy for the stock units as well as for the component parts for the custom units.

Installed Costs of Storefront Construction. Since there are so many different sizes and shapes of extrusions available, it would be difficult to list the material prices and labor time required. The labor hours are directly related to the size and shapes of the extrusions. For instance, a simple storefront and entrance door installation, 40'-0" (12,000 mm) wide x 10'-0" (3000 mm) high, in one system might take 72 manhours, while a more sophisticated system would require 100 manhours.

For budget purposes only, the following installed costs of aluminum storefront framing, glass not included, are given on a per sq. ft. (per sq.m) basis:

1. Windows without intermediate vertical mullions up to 5'-0"x5'-0" (1500 x 1500 mm) in size, using 1-3/4"x4-1/2" (43.7 x 112.5 mm) tube - $10.00 per s.f. ($107.60 per sq.m).

2. Same as above, but with one vertical mullion - $11.00 per s.f. ($118.40 per sq.m).

3. Windows 40'-0"x10'-0" (12,000 x 3000 mm) in size with vertical mullions spaced 5'-0" (1500 mm) on center, using 1-3/4"x4-1/2" (43.7 x 112.5 mm) tube - $9.00 per s.f. ($96.90 per sq.m).

4. Same as above, but with 3"x6" (75 x 150 mm) tube - $12.50 per s.f. ($134.50 per sq.m).

5. For a narrow stile single door and frame, 3'-0"x7'-0" (900 x 2100 mm) - $580.00 each.

6. For a pair of narrow stile doors and frame, 6'-0"x7'-0" (1800 x 2100 mm) - $900.00 each.

Add 15% for bronze anodized finish, 20% for black anodized finish. Add for transom over single door, $85.00; over pair of doors, $150.00.

Revolving Doors. Revolving doors are made in a variety of styles and sizes, in aluminum, bronze, and stainless steel. The doors are fitted with a mechanism to control the travel speed and are pivoted so that they may collapse together to provide an unobstructed opening. The semi-circular wings at each side of the four-leaf revolving door provide a surface for the door sweep to ride against. At no point during normal operation is there a clear path to the exterior for the escape of the building heat or air conditioning to the outside atmosphere.

Average installed prices for revolving doors are as follows: Aluminum, $15,000; bronze, $26,500; stainless steel, $22,000.

Storefront Members. The detail drawings that follow show some typical cut sections through aluminum storefront extrusions:

08500 METAL WINDOWS

Steel and aluminum windows are available in a great variety of types and sizes to meet almost all building requirements. These products have been standardized by the manufacturers to reduce costs and promote quicker deliveries.

The illustrations included in this chapter are for reference when selecting requirements. For full description and specifications of available products refer to manufacturers' catalogs.

All metal window and door products are either classified as *Standard, Commodity, or Special.*

Commodity are those selected from standards that are subject to the greatest demand and are available from manufacturer warehouse stocks or through dealers engaged in the distribution of these products. Standard and Special units are not available from stock and must be fabricated and shipped from factory. The trend is for distributors to send all orders to the factory for both pricing and fabrication.

The following information deals only with Commodity. For Standard and Special, obtain quotations from manufacturers or dealers.

Labor Installing Steel Windows and Doors. Installation costs vary with the class of labor, type and size of individual units, and job conditions. However, a reasonably accurate estimate of manhours required to install each unit can be made.

Product	Window Opening	Manhours per unit
Pivoted, Commercial	Single Unit*	2.25
Projected and Architectural	Two or more Units*	
Projected Windows	(a) Precast Sills	1.25
	(b) Poured Sills	1.50
Fixed Industrial	Single Unit	1.75
	Two or More Units	
	(a) Precast Sills	0.75
	(b) Poured Sills	1.00
Fixed Architectural	Single Unit	2.00
	Two or More Units	
	(a) Precast Sills	1.00
	(b) Poured Sills	1.25
Intermediate Windows		
In Wood (Casement, Combination and Projected)	(a) Single Unit	2.75
	(b) Two or More Units	2.25
In Stone	(a) Single Unit	3.00
	(b) Two or More Units	2.50
In Masonry	(a) Single Unit	2.25
	(b) Two or More Units	1.25
Mullions Only		0.50

Mullions with Covers .... .75
Residence Casements (Roto) .... Single Unit
                  (a) In Wood 1.75
                  (b) In Masonry or
                      Brick Veneer 1.25
Residence Casements (Roto) .... Two or More Units
                  (a) In Wood 1.50
                  (b) In Masonry or
                      Brick Veneer 1.00
Residence Casements (Fixed) .... Single Unit
                  (a) In Wood 1.50
                  (b) In Masonry or
                      Brick Veneer 1.00
Residence Casements (Fixed) .... Two or More Units
                  (a) In Wood 1.25
                  (b) In Masonry or
                      Brick Veneer 0.75
Casements in Casings .... Single or Multiple Units 1.50
Continuous Top Hung, Fixed .... per l.f. (per m) 0.50 (1.60)
Continuous Top Hung, Swing .... per l.f. (per m) 0.75 (2.40)
Mechanical Operators,
  Rack and Pinion .... per l.f. (per m) 0.75 (2.40)
Mechanical Operator,
  Lever Arm .... per l.f. (per m) 0.50 (1.60)
Mechanical Operators,
  Tension .... per l.f. (per m) 1.50 (4.90)
Industrial Doors and Frames .... up to 35 s.f. (3.5 sq.m)
                  per s.f. (per sq.m) 0.125 (1.40)
                  Over 35 s.f. (3.5 sq.m)
                  per s.f. (3.5 sq.m) 0.10 (1.00)
Double Hung Windows .... Single Units 1.50
                  Two or More Units 1.25
Screen Installation
For Projected Windows .... Net per Screen 1.10
  For Basement and Utility .... Net per Screen 0.75
  For Security Windows .... Net per Screen 1.10
  For Casement Windows .... Net per Screen 0.75
  For Pivoted Windows .... Net per Vent 1.50
*Single ventilator units. For each additional ventilator, add $0.25.

ELEVATIONS VIEWED FROM OUTSIDE

DETAILS SHOWN QUARTER SIZE

NOTE: Details are shown as 2¾" x 5½".
2¾" x 7½" system similar.

ENTRANCE DOORS

CENTER HUNG

OFFSET HUNG

ENTRANCE DOOR

CENTER HUNG

ENTRANCE DOOR

OFFSET HUNG

Prices of Commercial Projected Steel Windows

Window Type	Ventilated Window Price	Screen Price	Fixed Window Type	Window Price
*A12120	$58.00	$11.00	B-22	$31.00
*A13121	59.00	11.00	B-23	40.00
*B22140	70.00	15.00	B-24	45.00
*B23141	77.00	15.00	B-25	57.00
*B24141	85.00	15.00	B-26	69.00
*B25141	94.00	15.00	A-32	40.00
B2522402	99.00	**23.50	A-33	47.00
*B26141	104.00	15.00	A-34	59.00
B2622403	113.00	**23.50	A-35	70.00
*A32160	85.00	18.00	A-36	82.00
*A33161	93.00	18.00	A-42	86.00
*A34161	104.00	18.00	A-43	102.00
*A35161	115.00	18.00	A-44	114.00
*A36161	126.00	18.00	A-45	131.00
A3523602	115.00	**27.50	A-46	145.00
A3623603	126.00	27.50		
A43141	101.00	16.00		
A44141	115.00	16.00		
A45141	132.00	16.00		
A46141	138.00	16.00		

*These types available with either project-out or project-in vents.
**Price includes wicket screen for project-out vent and fixed screen for project-in vent.

Prices include cam handle and pole ring for project out ventilators and cam handle with strike or spring latch with strike for project in ventilators. Hardware is cadmium plated.

Prices also include standard installation fittings, provisions for standard screens, bonderizing, and one factory coat of paint. Glass, putty, glazing, and glazing clips not included.

Prices of Mullions and Covers

Lights High	2	3	4	5	6
Mullions	$7.00	$9.00	$12.00	$15.00	$17.00
Covers	8.00	10.00	13.00	16.00	19.00

Prices of Architectural Projected Steel Windows

Window Type	Window Price	Screen Price	Window Type	Window Price	Screen Price
252A	$69.00	$12.00	215	$137.00	$21.00
252B	69.00	9.00	216	140.00	21.00

Window Type	Window Price	Screen Price	Window Type	Window Price	Screen Price
253	85.00	12.00	218	145.00	21.00
261	$73.00	$9.00	221	$85.00	$10.00
262A	75.00	13.00	222A	94.00	15.00
262B	77.00	10.00	222B	98.00	13.00
263	88.00	13.00	223	100.00	15.00
264	104.00	13.00	224	112.00	15.00
265	123.00	19.00	225	132.00	24.00
266	174.00	19.00	226	142.00	24.00
268	188.00	19.00	228	147.00	24.00
211	81.00	9.00	273	115.00	18.00
212A	92.00	14.00	274	124.00	18.00
212B	93.00	12.00	275	164.00	26.00
213	96.00	14.00	276	181.00	26.00
214	100.00	14.00	278	191.00	26.00

Prices include bronze operating hardware for ventilated units, such as cam handle, strike, and pole ring for project out vents and cam handle with strike or spring latch with strike for project in vents. Also included are standard installation fittings, bonderizing, and one factory coat of paint. Glass, putty, glazing, and glazing clips not included.

Screen prices include screens for all vents in unit. Wicket screens are furnished for project out vents and outside fixed screens for project in vents. Prices also include clips for attachment.

Polished bronze hardware in place of standard bronze hardware,
add per vent..$2.75

Prices of Security Steel Windows

Window Type	Window Price	Screen Price	Mullion Price
32130	$80.00	$9.00	$4.00
33160	98.00	12.00	5.00
62 Fixed	77.00		4.00
62160	109.00	12.00	4.00
631120	112.00	16.00	5.00
641121	140.00	16.00	6.00
931180	177.00	20.00	5.00
941181	198.00	20.00	6.00

Prices include standard operating hardware, installation, fittings, bonderizing, and one factory coat of paint. Glass, putty, glazing, and glazing clips are not included.

Anchors

Anchors for securing frames to wood or masonry walls are included with window units. Four anchors per opening are furnished for heights up to and including 4'-5-1/2" (1337 mm). Six anchors per opening to and including 5'-9-1/2" (1737 mm). Eight anchors per opening to and including 8'-1-1/2" (2437 mm).

Prices of Residential Steel Casements Ventilated Casements

Window Type	Window Price	Head and Jamb Fins Per Set	Inside Casing Price	In/Out Trim Price	Screen Price
1212	$38.00	$4.00	$8.00	$12.00	$6.00
1313	43.00	4.00	10.00	13.00	7.00
1413	50.00	5.00	11.00	14.00	7.00
1414	48.00	5.00	11.00	14.00	8.00
1514	56.00	6.00	12.00	16.00	8.00
2212	43.00	4.00	10.00	14.00	6.00
2313	53.00	5.00	11.00	16.00	7.00
2414	65.00	5.00	12.00	17.00	8.00
2413	62.00	5.00	12.00	17.00	7.00
2514	78.00	6.00	13.00	19.00	7.00
2222	43.00	5.00	10.00	14.00	8.00
2323	50.00	5.00	11.00	16.00	10.00
2424	53.00	6.00	12.00	17.00	11.00
2423	52.00	6.00	12.00	17.00	10.00
2524	56.00	7.00	13.00	19.00	11.00
3222	52.00	5.00	11.00	17.00	8.00
3323	72.00	6.00	13.00	18.00	10.00
3424	96.00	7.00	14.00	20.00	11.00
3423	92.00	7.00	14.00	20.00	10.00
3524	120.00	7.00	15.00	22.00	11.00
4222	56.00	6.00	13.00	19.00	8.00
4323	84.00	7.00	14.00	21.00	10.00
4424	109.00	8.00	15.00	22.00	11.00
4423	108.00	8.00	15.00	22.00	10.00
4524	142.00	9.00	16.00	25.00	11.00
5222	82.00	6.00	14.00	22.00	8.00
5323	114.00	7.00	15.00	23.00	10.00
5424	158.00	9.00	16.00	25.00	11.00
5423	156.00	9.00	16.00	25.00	10.00
5524	192.00	10.00	17.00	27.00	11.00

Fixed Casements

Window Type	Window Price	Window Type	Window Price
12	$15.00	34	$64.00
13	21.00	35	79.00
14	26.00	42	39.00
15	32.00	43	58.00
22	21.00	44	78.00
23	29.00	45	94.00
24	37.00	52	52.00
25	43.00	53	78.00
32	33.00	54	100.00
33	45.00	55	122.00

Prices of ventilated units include locking handles, Roto underscreen operators (lever underscreen operators also available at reduced price) and standard installation fittings, bonderizing and one factory coat of paint for both ventilated and fixed units. Glass, putty, glazing, glazing clips, and mastic not included. Screen prices include screens for all vents in unit and clips for attachment.

Prices of Extras

Bronze hardware for roto casements instead of bronze
lacquer finish hardware, add per side hinged vent.........$6.50
Mastic, use 1 qt. per 40 l.f. (0.94 L per 12 m) of application
  per qt., 4 lbs. (per 0.94 L, 1.8 kg)...........4.75
  per gal., 16.5 lbs. (per 3.78 L, 7.5 kg)...........14.00
  5 gals., 85 lbs. (per 18.90 L, 38.5 kg)...........63.50

Prices of Mullions

Vertical Mullions

2-Lt	$3.50
3-Lt	4.50
4-Lt	5.50
5-Lt	6.50

Horizontal Mullions

1-Lt	$3.50
2-Lt	4.50
3-Lt	5.50
4-Lt	6.50
5-Lt	7.50

Pipe Corners

2-Lt	$6.50
3-Lt	8.50
4-Lt	9.50
5-Lt	11.50

Prices of Residential Steel Casement Picture Windows

For Standard Glazing Window

Type	Price
PW33	$20.00
PW34	22.00
PW35	24.00
PW43	23.00
PW44	26.00
PW45	28.00

For Insulating Glazing Window

Type	Price
DG34	$24.00
DG35	27.00
DG44	28.00
DG45	30.00

Prices are for four sided frames, shipped assembled, with two standard vertical mullions attached. Units for insulating glazing have glazing strips attached. Units are bonderized and painted one prime coat of oven-baked paint. Glass, putty, and glazing are not included.

Prices of Mullions and Fins

	Horizontal Widths			Vertical Heights		
	3'-1-3/4" (943mm)	6'-2-1/2" (1862mm)	9'-2-1/2" (2762mm)	2'-1-3/4" (643mm)	4'-2-1/8" (1253mm)	6'-2-1/2" (1862mm)
Mullions				$2.00	$3.00	$4.00
Fins	$2.00	$3.00	$4.00	2.00	3.00	3.00

Prices of Basement and Utility Steel Windows

Window prices include steel latch with ring and keeper, bonderizing, and one factory coat of paint. Glass, putty, glazing, and glazing clips not included.

Screen and storm window prices include clips for attachment. Storm panels are factory glazed.

Basement Windows

Opening Dimensions	Window Type	Window Price	Mullion Price	Screen Price	Storm Sash
2'-8-1/2"x1'-2-5/8" (812 x 365 mm)	2-Lt.	$27.50	$2.50	$5.00	$10.00
2'-8-1/2"x1'-6-5/8" (812 x 465 mm)	2-Lt.	31.50	2.50	5.00	11.00
2'-8-1/2"x1'-10-5/8" (812 x 565 mm)	2-Lt.	39.50	2.50	6.00	13.00

Utility Windows

Window Type	Opening Dim. Width	Height	Window List Price	Mullion List Price	Screen List Price
22121	2'-8-1/2" (812 mm)	3'-6-5/8" (1065 mm)	$28.00	$3.50	$6.50

ALUMINUM CASEMENT WINDOWS

Prices include aluminum die cast roto operator with aluminum arm. Operator case and crank are brushed to a satin finish and given one coat of clear Duranite phenolic resin baked-on plastic coating. Casements are etched in a hot acid bath and receive a dip coat of methacrylate lacquer. Also included are wood screws and sill anchors when specified. Prices include 1/2" (12.5 mm) insulating glass and screens.

Prices of Aluminum Residential Casements
Ventilated

Type	Price
1212	$38.00
1313	40.00
1413	48.00
1414	48.00
1514	58.00
2212	44.00
2313	52.00
2413	68.00
2414	68.00
2514	82.00
2222	55.00
2323	65.00
2423	72.00
2424	72.00
2524	88.00
3222	90.00
3323	77.00
3423	101.00
3424	101.00
3524	124.00

Type	Price
4222	63.00
4323	110.00
4423	117.00
4424	120.00
4524	149.00
5222	82.00
5323	120.00
5424	159.00
5524	197.00

Aluminum Awning Windows

Prices of Picture Windows for Use with Aluminum Residential Casements
Glazed With 1/2" (12.5 mm) Insulating Glass

Type	Price
33MO	$69.00
34MO	80.00
35MO	90.00
43MO	79.00
44MO	94.00
45MO	106.00

Price of Aluminum Awning Windows

Type	Price
1722	$95.00
1732	149.00
1742	185.00
1753	215.00
3122	118.00
3132	149.00
3142	222.00
3153	258.00
4522	176.00
4532	218.00
4542	248.00
4553	308.00

Prices of Mullions and Fins

Lts. Wide or High	Vertical Mullions	Head Fins Each Head	Jamb Fins Each Jamb
1	$...	$2.50	$...
2	6.00	2.50	2.00
3	9.00	3.50	2.50
4	11.00	4.50	3.50
5	14.00	5.50	3.50

Price includes a mechanical operator with crank handle. Windows are glazed by using snap-in aluminum beads. Price includes integral fin, and glazing with 1/2" (12.5 mm) insulating glass.

Windows are weatherstripped and receive a dip coat of methacrylate lacquer.

Screen prices include screens for all vents in unit. Screens have aluminum frames and 18x14 (450 x 350 mm) mesh aluminum cloth. Storm sash are factory glazed.

Prices of Aluminum Double Hung Windows

Prices of windows include silicone-treated wool-pile weatherstripping, spiral, sash balances, glazing clips, standard anchors, mullion screws, and window hardware consisting of sash lock, keeper, and pull down for upper sash. Prices include 1/2" (12.5 mm) insulating glass and screens.

All windows are finished with a coat of water clear lacquer.

Screens have extruded or rolled aluminum frame sections and are wired with 18x14 (450 x 350 mm) mesh aluminum cloth. Screens consist of full or half panels to allow for screening as desired. Prices include all necessary installation hardware.

Storm sash is furnished glazed with single strength glass set in a plastic seal. All points of contact with window frame are insulated. Prices include all necessary hardware for installation.

Picture windows are shipped knocked down. Picture windows are designed to receive 1/4" (6.2 mm) single glass or 1/2" (12.5 mm), 3/4" (18.7 mm), or 1" (25 mm) insulating glass if specified. Prices include necessary assembly and installation hardware.

Prices of Aluminum Double Hung Windows

Window Type	2-Lt. Top 2-Lt. Bot.	6-Lt. Top 6-Lt. Bot.	Two Light
2030	$80.00	$85.00	$78.00
2830	85.00	90.00	83.00
3030	88.00	93.00	86.00
3430	91.00	96.00	89.00
3830	97.50	103.00	95.00
2038	83.00	88.00	81.00
2838	87.00	93.00	85.00
3038	95.00	100.00	93.00
3438	98.00	103.00	95.00
3838	117.00	127.00	115.00
2044	85.00	90.00	83.00
2844	88.00	93.00	86.00
3044	96.00	101.00	94.00
3444	98.00	103.00	96.00
3844	101.00	106.00	99.00
2050	90.00	95.00	88.00
2850	92.00	95.00	90.00
3050	95.00	100.00	93.00
3450	110.00	115.00	108.00
3850	115.00	120.00	113.00
2060	100.00	105.00	108.00
3060	105.00	110.00	103.00
3860	110.00	115.00	108.00

Prices of Aluminum Fins and Trim or Brick Molds

	Height of Window				
	3'-1-5/8" (940mm)	3'-9-5/8" (1140mm)	4'-5-5/8" (1340mm)	5'-1-5/8" (1540mm)	5'-9-5/8" (1740mm)
Jamb fin and trim or brick mold, per pair	$5.00	$6.00	$6.50	$7.00	$7.50

	Width of Window					
	2'-0-1/2" (612mm)	2'-4-1/2" (712mm)	2'-8-1/2" (812mm)	3'-0-1/2" (921mm)	3'-4-1/2" (1012mm)	3'-8-1/2" (1112mm)
Head Section	$0.60	$0.65	$0.75	$0.80	$0.85	$0.90
Sill Section	0.50	0.55	0.60	0.65	0.75	0.80

Prices of Aluminum Screens, Storm Sash, and Inside Casings for Aluminum Double Hung Windows

Window Type	Half Screen	Full Screen	Storm Sash 2-Piece	Inside Casing
2031	$6.50	$10.00	$39.50	$8.50
2431	7.00	11.00	41.00	8.70
2831	7.50	12.00	43.00	9.10
3031	8.75	14.50	45.00	9.50
3431	9.25	15.50	46.00	9.70
3831	10.00	16.00	50.00	10.10
2039	6.75	10.50	41.00	9.20
2439	8.00	13.00	43.00	9.50
2839	9.00	15.00	45.50	9.80
3039	10.00	17.00	46.50	10.20
3439	11.00	19.00	50.00	10.40
3839	12.00	21.00	54.00	10.80
2045	8.00	13.00	43.00	9.60
2445	9.00	15.00	45.50	9.80
2845	10.00	17.00	46.00	10.20
3045	11.00	19.00	50.00	10.50
3445	12.50	22.00	53.00	10.80
3845	13.50	24.00	55.00	11.20
2051	9.00	15.00	45.50	10.10
2451	10.00	17.00	46.50	10.30
2851	11.00	19.00	50.00	10.70
3051	13.00	23.00	53.00	11.00
3451	14.00	25.00	55.00	11.20
3851	15.00	27.00	64.00	11.60
2059	10.00	17.00	46.50	10.70
2459	11.00	19.00	50.00	10.90
2859	12.00	21.00	53.00	11.30
3059	14.00	25.00	55.00	11.60
3459	15.00	27.00	64.00	11.90
3859	16.00	29.00	65.00	12.30

Prices of Aluminum Picture Windows for Use with Aluminum Double Hung Windows, Glass Not Included

Window Type	Price	Window Type	Price
40P31	$39.00	44P45	$47.00
44P31	40.00	48P45	49.50
40P39	40.50	*410P45	50.00
44P39	42.50	50P45	50.50
48P39	43.00	*56P45	51.50
50P39	44.50	*62P45	52.50
40P45	45.50	*610P45	53.50
42P45*	47.00		

*These sizes are set up (standard) for insulated glazing.

Prices are for units for single glazing. For double glazing, add $3.00 per unit.

Prices of Aluminum Horizontally Sliding Windows

Prices of windows include silicone-treated wool-pile weatherstripping and integral fin. Prices are for frames and sash assembled and glazed with 1/2" (12.5 mm) insulating glass.

All frame and sash members are cleaned, caustic etched, and coated with methacrylate-type lacquer at the factory.

Screens are aluminum, covering half of window opening, and are easily installed and removed from inside.

Storm sash are installed same as screens and are two piece SSB glazed.

Window Type	Window Price	Half Screen	Inside Casing
2020	$34.00	$6.00	$9.60
3020	36.00	7.00	10.50
4020	39.00	6.00	10.10
5020	41.00	7.00	11.20
6020	44.00	7.00	12.20
3030	48.00	7.00	9.80
4030	50.00	7.00	10.90
5030	54.00	8.00	12.00
6030	62.00	8.00	12.90
4040	60.00	8.00	12.00
5040	72.00	9.00	13.00
6040	83.00	9.00	14.50

Prices of Aluminum Picture Windows with Horizontally Sliding Flankers

Picture window prices are for frames and sash assembled and glazed. Screen prices include two per unit. Storm sash prices are for three piece glazed.

Window Type	Window Price	2-Piece Screens	3-Piece Storm Sash	Inside Casing
7020	$70.00	$11.00	$55.00	$16.40
8020	73.00	11.00	57.00	17.00
9020	80.00	11.00	63.00	19.40
7030	86.00	12.00	67.00	17.90
8030	90.00	13.00	71.00	19.20
9030	97.00	13.00	76.00	20.60
7040	101.00	13.00	77.00	19.40
8040	104.00	14.00	81.00	20.50
9040	114.00	14.00	86.00	22.00

08600 WOOD AND PLASTIC WINDOWS

Wood windows generally are sold as complete units with frame, sash, operating hardware, weatherstripping, and glazing assembled at the factory. Frames are usually treated. The exterior of wood windows is generally covered with rigid vinyl (PVC). The interior portion is unfinished, to be finished by painters or others. The following sizes are based on units manufactured by Andersen Window Company, Bayport, Minn. Sizes for other manufacturers vary slightly. Prices, of course, should be checked locally, because they will vary considerably, depending on geographic location, on size of the order, and on whether delivery is to a local dealer or to job site.

The contractor should add or deduct for items such as insect screens, divided light grilles, hardware, window blinds, or window treatments when comparing individual window costs. The prices quoted below are helpful for comparing the various window and glazing types.

Double Hung Window Units

Frame. Wood members are treated with a water repellent preservative and covered with a rigid vinyl (PVC) sheath. Sill ends are prefinished with polyurea or polyester urethane, depending on whether white or color.

Sash. The exterior of the wood sash is protected with a long-lasting patented polyurea finish or a polyester urethane finish coat, depending on color. Interior face of sash is clear for stain or paint finish.

Glazing. Select quality high-performance or high-performance sun insulating glass. Double-pane insulated glass is available for white only. For high altitude and other special glazings, contact the manufacturer.

Glazing System. Sash has a rigid vinyl snap-in-bead.

Weatherstripping. Foam type weatherstripping is applied to top and bottom rails. Rigid vinyl rib on head jamb liner and sills fits into vinyl covered foam weatherstripping on sash. Rigid vinyl leaf weatherstripping is applied to check rail. Polypropylene leaf weatherstrip with foam inserts contacts side jamb liner ribs.

Anchoring Flange and Windbreak. Factory applied rigid vinyl flanges are at head and sides of outer frame members. A flexible vinyl sill windbreak is factory applied to the bottom of sill as flashing.

Sill. This is wood core treated with a water repellent preservative, covered with PVC sheath.

Jamb Liner. This is white PVC for unit in white and complementary color PVC for unit in color.

Sash Lock and Lift. Stone color decorator finish, sash lock, keeper and lift are factory applied.

Double Hung Units High Performance (H.P.) Insulating Glass,
Sun Insulating Glass, and Double Pane Insulating Glass

Window Type	Glass Size		Rough Opening Width x	Height	H.P. Insulating	Insect Screen Add	Grille Add	Combination Screen/Storm Add
1842W	16-1/2"x	22"	1'10-1/8" x	4'5-1/4"	$173.00	$16.40	$13.30	$72.00
1846W		24"		4'9-1/4"	180.00	17.30	13.50	74.00
2042W	20-1/2"x	22"	2'2-1/8" x	4'5-1/4"	192.00	18.00	15.10	77.00
2046W		24"		4'9-1/4"	207.00	18.80	15.60	80.00
2052W		28"		5'5-1/4"	220.00	20.50	16.70	85.00
2062W		68"		6'5-1/4"	233.00	22.00	22.60	104.00
20210W		14"		3'1-1/4"	176.00	14.40	13.30	69.00
2032W		16"		3'5-1/4"	188.00	15.30	13.60	70.00
20310W		20"		4'1-1/4"	201.00	16.60	14.20	75.00
2442W	24-1/2"x	22"	2'6-1/8"	4'5-1/4"	215.00	19.40	15.70	84.00
2446W		24"		4'9-1/4"	224.00	20.40	17.00	88.00
2452W		28"		5'5-1/4"	242.00	22.20	17.80	94.00
24210W		14"		3'1-1/4"	257.00	15.40	14.00	72.00
2432W		16"		3'5-1/4"	205.00	16.60	14.40	72.00
24310W		20"		4'1-1/4"	218.00	17.50	14.80	78.00
2842W	28-1/2"x	22"	2'10-1/8" x	4'5-1/4"	233.00	20.80	17.00	91.00
2846W		24"		4'9-1/4"	248.00	21.80	18.50	93.00
2852W		28		5'5-1/4"	260.00	24.20	19.40	100.00
2856W		60"		5'9-1/4"	247.00	26.00	22.30	104.00
2862W		68"		6'5-1/4"	275.00	30.00	25.20	110.00
28210W		14"		3'1-1/4"	230.00	16.70	15.20	75.00
2832W		16"		3'5-1/4"	260.00	17.50	15.70	76.00
28310W		20"		4'1-1/4"	280.00	19.80	16.00	83.00
3042W	32-1/2"x	22"	3'2-1/8" x	4'5-1/4"	154.00	23.00	18.50	96.00
3046W		24"		4'9-1/4"	160.00	24.10	19.40	96.00

Type	Glass Size		Width x	Height	H.P. Insul.	Screen	Grille	Screen/Storm
3052W		28"		5'5-1/4"	172.00	26.80	21.40	109.00
3056W		60"		5'9-1/4"	182.00	28.00	24.20	114.00
3062W		68"		6'5-1/4"	157.00	30.30	28.30	121.00
30210W		14"		3'1-1/4"	168.00	18.60	16.20	80.00
3032W		16"		3'5-1/4"	177.00	20.60	16.70	82.00
30310W		20"		4'1-1/4"	190.00	22.20	17.80	89.00
3442W	36-1/2"x	22"	3'6-1/8" x	4'5-1/4"	203.00	24.80	20.00	105.00
3446W		24"		4'9-1/4"	175.00	25.90	20.50	107.00
3452W		28"		5'5-1/4"	183.00	28.90	21.90	119.00
3456W		60"		5'9-1/4"	197.00	30.10	25.60	126.00
3432W		16"		3'5-1/4"	208.00	21.00	18.10	88.00
34310W		20"		4'1-1/4"	220.00	23.70	19.50	95.00
3846W	40-1/2"x	24"	3'10-1/8" x	4'9-1/4"	270.00	27.00	11.20	118.00
			(metric)					
1842W	412 x	550 mm	553 x	1131 mm	$173.00	$16.40	$13.30	$72.00
1846W		600 mm		1431 mm	180.00	17.30	13.50	74.00
2042W	512 x	550 mm	653 x	1131 mm	192.00	18.00	15.10	77.00
2046W		600 mm		1431 mm	207.00	18.80	15.60	80.00
2052W		700 mm		1631 mm	220.00	20.50	16.70	85.00
2062W		1700 mm		1931 mm	233.00	22.00	22.60	104.00
20210W		350 mm		931 mm	176.00	14.40	13.30	69.00
2032W		400 mm		1031 mm	188.00	15.30	13.60	70.00
20310W		500 mm		1231 mm	201.00	16.60	14.20	75.00
2442W	612 x	550 mm	753 x	1131 mm	215.00	19.40	15.70	84.00
2446W		600 mm		1431 mm	224.00	20.40	17.00	88.00
2452W		700 mm		1631 mm	242.00	22.20	17.80	94.00
24210W		350 mm		931 mm	257.00	15.40	14.00	72.00
2432W		400 mm		1031 mm	205.00	16.60	14.40	72.00
24310W		500 mm		1231 mm	218.00	17.50	14.80	78.00
2842W	712 x	550 mm	853 x	1131 mm	233.00	20.80	17.00	91.00
2846W		600 mm		1431 mm	248.00	21.80	18.50	93.00
2852W		700 mm		1631 mm	260.00	24.20	19.40	100.00
2856W		1500 mm		1731 mm	247.00	26.00	22.30	104.00
2862W		1700 mm		1931 mm	275.00	30.00	25.20	110.00
28210W		350 mm		931 mm	230.00	16.70	15.20	75.00
2832W		400 mm		1031 mm	260.00	17.50	15.70	76.00
28310W		500 mm		1231 mm	280.00	19.80	16.00	83.00
3042W	812 x	550 mm	953 x	1131 mm	154.00	23.00	18.50	96.00
3046W		600 mm		1431 mm	160.00	24.10	19.40	96.00
3052W		700 mm		1631 mm	172.00	26.80	21.40	109.00
3056W		1500 mm		1731 mm	182.00	28.00	24.20	114.00
3062W		1700 mm		1931 mm	157.00	30.30	28.30	121.00
30210W		350 mm		931 mm	168.00	18.60	16.20	80.00
3032W		400 mm		1031 mm	177.00	20.60	16.70	82.00
30310W		500 mm		1231 mm	190.00	22.20	17.80	89.00
3442W	912 x	550 mm	1053 x	1131 mm	203.00	24.80	20.00	105.00
3446W		600 mm		1431 mm	175.00	25.90	20.50	107.00
3452W		700 mm		1631 mm	183.00	28.90	21.90	119.00
3456W		1500 mm		1731 mm	197.00	30.10	25.60	126.00
3432W		400 mm		1031 mm	208.00	21.00	18.10	88.00
34310W		500 mm		1231 mm	220.00	23.70	19.50	95.00
3846W	1012 x	600 mm	1153 x	1431 mm	270.00	27.00	11.20	118.00

For double hung units in pairs, triple, or quadruple arrangements, multiply cost by two, three, or four respectively.

STATIONARY PICTURE WINDOW UNITS

Picture Window Units Combined with 1-8 Flanking Units
High Performance (H.P.) Insulating Glass

Window Set Up Unit	Glass Size Fixed Unit	Rough Opening Width	Total Unit Height	H.P. Insulating Glass
18 4442 18W	46-1/2" x 43-5/8"	8'1-1/2"	4'5-1/4"	$779.00
18 4446 18W	47-5/8"		4'9-1/4"	862.00
18 5042 18W	54-1/2" x 43-5/8"	8'9-1/2"	4'5-1/4"	907.00
18 5046 18W	47-5/8"		4'9-1/4"	939.00
18 5846 18W	62-1/2" x 47-5/8"	9'5-1/2"	4'9-1/4"	990.00

Window Set Up Unit	Glass Size Fixed Unit	Rough Opening Width	Total Unit Height	H.P. Insulating Glass
		(metric)		
18 4442 20W	1162 x 1090 mm	2437 x	1331 mm	$779.00
18 4446 20W	1190 mm		1431 mm	862.00
18 5042 20W	1362 x 1090 mm	2637 x	1331 mm	907.00
18 5046 20W	1190 mm		1431 mm	939.00
18 5846 20W	1562 x 1190 mm	2837 x	1431	990.00

Picture Window Units Combined with 1-8 Flanking Units
High Performance (H.P.) Sun Insulating Glass

Window Set Up Unit	Glass Size Fixed Unit	Rough Opening Width	Total Unit Height	H.P. Sun Insulating Glass
18 4442 18W	46-1/2" x 43-5/8"	8'1-1/2"	4'5-1/4"	$838.00
18 4446 18W	47-5/8"		4'9-1/4"	862.00
18 5042 18W	54-1/2" x 43-5/8"	8'9-1/2"	4'5-1/4"	907.00
18 5046 18W	47-5/8"		4'9-1/4"	939.00
18 5846 18W	62-1/2" x 47-5/8"	9'5-1/2"	4'9-1/4"	990.00
		(metric)		
18 4442 20W	1162 x 1090 mm	2437 x	1331 mm	$838.00
18 4446 20W	1190 mm		1431 mm	862.00
18 5042 20W	1362 x 1090 mm	2637 x	1331 mm	907.00
18 5046 20W	1190 mm		1431 mm	939.00
18 5846 20W	1562 x 1190 mm	2837 x	1431 mm	990.00

Picture Window Units Combined with 1-8 Flanking Units
Double Pane Insulating Glass

Window Set Up Unit	Glass Size Fixed Unit	Rough Opening Width	Total Unit Height	Double Pane Insulating Glass
18 4442 18W	46-1/2 x 43-5/8"	8'1-1/2"	4'5-1/4"	$705.00
18 4446 18W	47-5/8"		4'9-1/4"	732.00
18 5042 18W	54-1/2 x 43-5/8	8'9-1/2"	4'5-1/4"	745.00
18 5046 18W	47-5/8"		4'9-1/4"	765.00
18 5846 18W	62-1/2 x 47-5/8"	9'5-1/2"	4'9-1/4"	765.00
		(metric)		
18 4442 20W	1162 x 1090 mm	2437 x	1331 mm	$705.00
18 4446 20W	1190 mm		1431 mm	732.00
18 5042 20W	1362 x 1090 mm	2637 x	1331 mm	745.00
18 5046 20W	1190 mm		1431 mm	765.00
18 5846 20W	1562 x 1690 mm	2837 x	1431 mm	803.00

Picture Window Units Combined with 2-0 Flanking Units
High Performance (H.P.) Insulating Glass

Window Set Up Unit	Glass Size Fixed Unit	Rough Opening Width	Total Unit Height	H.P. Insulating Glass
20 4442 20W	46-1/2" x 43-5/8"	8'9-1/2"	4'5-1/4"	$795.00
20 4446 20W	47-5/8"		4'9-1/4"	826.00

Window Set Up Unit	Glass Size Fixed Unit		Rough Opening Width		Total Unit Height	H.P. Insulating Glass
20 4462 20W		67-5/8"			6'5-1/4"	1,068.00
20 5042 20W	54-1/2" x	43-5/8"	9'5-1/2"		4'5-1/4"	858.00
20 5046 20W		47-5/8"			4'9-1/4"	898.00
20 5052 20W		55-5/8"			5'5-1/4"	1,000.00
20 5846 20W	46-1/2" x	67-5/8"	8'9-1/2"		6'5-1/4"	945.00
(metric)						
20 4442 20W	1162 x	1090 mm	2637 x	1331 mm		$795.00
20 4446 20W		1190 mm		1431 mm		826.00
20 4462 20W		1690 mm		1931 mm		1,068.00
20 5042 20W	1362 x	1090 mm	2837 x	1331 mm		858.00
20 5046 20W		1190 mm		1431 mm		898.00
20 5052 20W		1390 mm		1631 mm		1,000.00
20 5846 20W	1162 x	1690 mm	2637 x	1931 mm		945.00

Picture Window Units Combined with 2-0 Flanking Units
High Performance (H.P.) Sun Insulating Glass

Window Set Up Unit	Glass Size Fixed Unit		Rough Opening Width		Total Unit Height	H.P. Sun Insulating Glass
20 4442 20W	46-1/2" x	43-5/8"	8'9-1/2"		4'5-1/4"	$855.00
20 4446 20W		47-5/8"			4'9-1/4"	888.00
20 4462 20W		67-5/8"			6'5-1/4"	1,149.00
20 5042 20W	54-1/2" x	43-5/8"	9'5-1/2"		4'5-1/4"	924.00
20 5046 20W		47-5/8"			4'9-1/4"	966.00
20 5052 20W		55-5/8"			5'5-1/4"	1,077.00
20 5846 20W	46-1/2" x	67-5/8"	8'9-1/2"		6'5-1/4"	1,017.00
(metric)						
20 4442 20W	1162 x	1090 mm	2637 x	1331 mm		$855.00
20 4446 20W		1190 mm		1431 mm		888.00
20 4462 20W		1690 mm		1931 mm		1,149.00
20 5042 20W	1362 x	1090 mm	2837 x	1331 mm		924.00
20 5046 20W		1190 mm		1431 mm		966.00
20 5052 20W		1390 mm		1631 mm		1,077.00

Picture Window Units Combined with 2-0 Flanking Units
Double Pane Insulating Glass

Window Set Up Unit	Glass Size Fixed Unit		Rough Opening Width		Total Unit Height	H.P. Sun Insulating Glass
20 4442 20W	46-1/2" x	43-5/8"	8'9-1/2"		4'5-1/4"	$719.00
20 4446 20W		47-5/8"			4'9-1/4"	755.00
20 4462 20W		67-5/8"			6'5-1/4"	939.00
20 5042 20W	54-1/2" x	43-5/8"	9'5-1/2"		4'5-1/4"	759.00
20 5046 20W		47-5/8"			4'9-1/4"	787.00
20 5052 20W		55-5/8"			5'5-1/4"	874.00
20 5846 20W	46-1/2" x	67-5/8"	8'9-1/2"		6'5-1/4"	825.00

Window Set Up Unit	Glass Size Fixed Unit	Rough Opening Width	Total Unit Height	H.P. Sun Insulating Glass
		(metric)		
20 4442 20W	1162 x 1090 mm	2637 x	1331 mm	$719.00
20 4446 20W	1190 mm		1431 mm	755.00
20 4462 20W	1690 mm		1931 mm	939.00
20 5042 20W	1362 x 1090 mm	2837 x	1331 mm	759.00
20 5046 20W	1190 mm		1431 mm	787.00
20 5052 20W	1390 mm		1631 mm	874.00
20 5846 20W	1162 x 1690 mm	2637 x	1931 mm	825.00

Picture Window Units
High Performance (H.P.) Insulating Glass

Window Set Up Unit	Glass Size Fixed Unit	Rough Opening Width	Total Unit Height	H.P. Insulating Glass
4442W	46-1/2" x 43-5/8"	4'6-1/8"	4'5-1/4"	$356.00
4446W	47-5/8"		4'9-1/4"	420.00
4462W	67-5/8"		6'5-1/4"	369.00
5042W	54-1/2" x 43-5/8"	5'2-1/8"	4'5-1/4"	440.00
5046W	47-5/8"		4'9-1/4"	488.00
5052W	55-5/8"		5'5-1/4"	502.00
5846W	62-1/2" x 47-5/8"	5'10-1/8"	4'9-1/4"	510.00
		(metric equivalent)		
4442W	1162 x 1090 mm	1353 x	1331 mm	$356.00
4446W	1190 mm		1431 mm	420.00
4462W	1690 mm		1931 mm	369.00
5042W	1362 x 1090 mm	1212 x	1331 mm	440.00
5046W	1190 mm		1431 mm	488.00
5052W	1390 mm		1931 mm	502.00
5846W	1562 x 1190 mm	1512 x	1431 mm	510.00

Stationary Picture Window Units
High Performance (H.P.) Sun Insulating Glass

Window Set Up Unit	Glass Size Fixed Unit	Rough Opening Width	Total Unit Height	H.P. Sun Insulating Glass
4442W	46-1/2" x 43-5/8"	4'6-1/8"	4'5-1/4"	$384.00
4446W	47-5/8"		4'9-1/4"	398.00
4462W	67-5/8"		6'5-1/4"	550.00
5042W	54-1/2" x 43-5/8"	5'2-1/8"	4'5-1/4"	453.00
5046W	47-5/8"		4'9-1/4"	475.00
5052W	55-5/8"		5'5-1/4"	543.00
5846W	62-1/2" x 47-5/8"	5'10-1/8"	4'9-1/4"	475.00
		(metric)		
4442W	1162 x 1090 mm	1353 x	1331 mm	$384.00
4446W	1190 mm		1431 mm	398.00
4462W	1690 mm		1931 mm	550.00
5042W	1362 x 1090 mm	1212 x	1331 mm	453.00
5046W	1190 mm		1431 mm	475.00
5052W	1390 mm		1931 mm	543.00
5846W	1562 x 1190 mm	1512 x	1431 mm	475.00

Stationary Picture Window Units
Double Pane Insulating Glass

Window Set Up Unit	Glass Size Fixed Unit	Rough Opening Width	Total Unit Height	H.P. Sun Insulating Glass
4442W	46-1/2" x ....43-5/8"	4'6-1/8"	4'5-1/4"	$313.00
4446W	47-5/8"		4'9-1/4"	332.00
4462W	67-5/8"		6'5-1/4"	423.00
5042W	54-1/2" x ....43-5/8"	5'2-1/8"	4'5-1/4"	353.00
5046W	47-5/8"		4'9-1/4"	365.00
5052W	55-5/8"		5'5-1/4"	413.00
5846W	62-1/2" x ....47-5/8"	5'10-1/8"	4'9-1/4"	403.00

(metric equivalent)

Window Set Up Unit	Glass Size Fixed Unit	Rough Opening Width	Total Unit Height	H.P. Sun Insulating Glass
4442W	1162 x ....1090 mm	1353 x	1331 mm	$313.00
4446W	1190 mm		1431 mm	332.00
4462W	1690 mm		1931 mm	423.00
5042W	1362 x ....1090 mm	1212 x	1331 mm	353.00
5046W	1190 mm		1431 mm	365.00
5052W	1390 mm		1931 mm	413.00
5846W	1562 x ....1190 mm	1512 x	1431 mm	403.00

Gliding Window Units

An advantage of high narrow gliding windows is that furniture can be placed under them, making much better interior arrangements possible.

The gliding windows described below are furnished complete with frames, sash, and weatherstrips. Each sash slides to the opposite side of the opening for ventilation but when fully closed is weatherstripped on all four sides. Prices are for complete gliding window units, assembled and carton packed.

Frame. Wood members are treated with water repellent preservative and covered with white, rigid PVC sheath. Retractable inside head stop allows for operating sash removal. Captive screws release stationary sash.

Sash. Wood core treated with a water repellent preservative and completely covered with white PVC sheath.

Glazing. Select quality double-pane insulating window glass. Installations over 3500' (1067 m) altitude require high altitude glass.

Glazing Bead. Rigid vinyl snap-in glazing bead with flexible vinyl tip against glass.

Weatherstripping. Rigid vinyl spring tension type, factory applied for a tight seal between sash and frame.

Flashing. Continuation of the rigid vinyl sheath on frame forms perimeter flashing and anchoring fin.

Corner Seal. Vinyl frame sheath eliminates corner joints. Sash corner joints are welded to form waterproof joints.

Sill Tank. For added tightness under severe conditions of exposure, a molded vinyl tank has been built in to drain any moisture to the exterior.

Glides. Adjustable chrome-plated steel glides, factory applied.

Locking Latch. Center interlock and two-way spring loaded locking mechanism secures sash at head and sill. Steel locking rods and zinc die-cast handle case with stone color decorator finish. Flexible baffles cover meeting stiles.

Combination Unit. A unit for triple glazing. Features aluminum frame and 1/8" (3.1 mm) tempered glass in storm panels and glass fiber screen cloth. Storm panels and screen with baked-on white decorator finish.

Gliding Windows
Double Pane Insulation Glass

Window Unit Order No	Glass	Size Fixed Unit	Rough Opening Width	Total Unit Height	Double Pane Insulating Glass	Insect Screen
G33	14" x	29-3/8"	3'0-1/2"	3'0-1/2"	$234.00	$14.00
G336		35-3/8"		3'6-1/2"	260.00	15.00
G436	20" x	35-5/8"	4'0-1/2"	3'6-1/2"	291.00	16.00
G44		41-3/8"		4'0-1/2"	315.00	18.00
G536	26" x	35-3/8"	5'0-1/2"	3'6-1/2"	330.00	19.00
G54		41-3/8"		4'0-1/2"	349.00	20.00
G55		53-5/8"		5'0-1/2"	402.00	23.00
G64	32" x	41-3/8"	6'0-1/2"	4'0-1/2"	395.00	23.00
G65		53-5/8"		5'0-1/2"	506.00	25.00
(metric)						
G33	350 x	734 mm	912.5 x	912.5 mm	$234.00	$14.00
G336		884 mm		1062.5 mm	260.00	15.00
G436	500 x	884 mm	1212.5 x	1062.5 mm	291.00	16.00
G44		1034 mm		1212.5 mm	315.00	18.00
G536	650 x	884 mm	1512.5 x	1062.5 mm	330.00	19.00
G54		1034 mm		1212.5 mm	349.00	20.00
G55		1340 mm		1512.5 mm	402.00	23.00
G64	1600 x	1034 mm	1812.5 x	1212.5 mm	395.00	23.00
G65		1340 mm		1512.5 mm	506.00	25.00

Gliding window units may be ordered with a fixed glass in the center and two gliding sash on either end, which when open, pass in front of the fixed panel.

Sash Extras. Glazing with obscure glass, add $3.00 per s.f. ($32.30 per sq.m). Rectangular grilles same as double hung. Estimator should verify that it is available.

Awning Panel Wall Units

Awning units are a system of window panel arrangement ranging from a single individual panel to a multiple 12-panel unit, which is used as a picture frame unit.

Awning units are designed to give a selection of either stationary glass panels, ventilating sash panels, or a combination of both. Both come with high performance (H.P.) sun insulating glass and insulating glass.

Ventilating units have 2" (50 mm) sash, hinged at the top, which can be opened about 35 degrees, closed, or held in intermediate position by rotary operators. Operators have worm gears, twin arms of cadmium plated steel, and removable handles, and they are self-locking.

Insect screens, grilles, and interior blinds are available for all units.

Glass panel wall units come completely set up and glazed. The ventilating units are weatherstripped, and operators are applied.

Awning Windows with High Performance (H.P.) Insulating Glass, Sun Insulating Glass, and Double Pane Insulating Glass

Window Type	Glass Size	Rough Opening Width	Height	H.P. Sun Insulating	Insect Screen Add	Grille Add	Interior Blinds Add
A21W	19-5/16" x 19-3/4"	2'0-5/8"	2'0-5/8"	$154.00	$10.00	$8.00	$58.00
AN31W	31-1/8" x 16-1/8"	3'0-1/2"	1'9"	160.00	11.00	9.00	56.00
A31W	19-3/4"		2'0-5/8"	180.00	12.00	10.00	65.00
AN351W	36" x 16-1/8"	3'5-3/8"	1'9"	177.00	13.00	11.00	59.00
A351W	19-3/4"		2'0-5/8"	194.00	16.00	13.00	63.00
AN41W	43-3/16" x 16-1/8"	4'0-1/2"	1'9"	194.00	22.00	18.00	72.00
A41W	19-3/4"		2'0-5/8"	210.00	11.00	9.00	59.00
AW31W	31-1/8" x 24"	3'0-1/2"	2'4-7/8"	189.00	14.00	10.00	67.00
A330W	31-9/16"		3'0-1/2"	196.00	18.00	17.00	72.00
A335W	36-7/16"		3'5-1/4"	278.00	19.00	18.00	75.00
AP32VW	43-5/8"		4'0-1/2"	306.00	25.00	0.00	80.00
AP321VW	63-3/8"		6'0-3/8"	493.00	37.00	10.00	145.00
A3535W	36" x 36-7/16"	3'5-3/8"	3'5-1/4"	256.00	23.00	21.00	85.00
AP352VW	43-5/8"		4'0-1/2"	250.00	27.00	0.00	91.00
AP3521VW	63-3/8"		6'0-3/8"	448.00	41.00	11.00	164.00
AW41W	43-3/16" x 24"	4'0-1/2"	2'4-7/8"	226.00	16.00	13.00	81.00
AP42VW	43-5/8"		4'0-1/2"	382.00	30.00	0.00	98.00
AP421VW	63-3/8"		6'0-3/8"	598.00	45.00	12.00	175.00
AP530VW	55-1/16" x 31-9/16"	5'0-3/8"	3'0-1/2"	362.00	29.00	0.00	109.00
(metric)							
A21W	482 x 493 mm	615 x	615 mm	$154.00	$10.00	$8.00	$58.00
AN31W	778 x 403 mm	912 x	525 mm	160.00	11.00	9.00	56.00
A31W	493 mm		615 mm	180.00	12.00	10.00	65.00
AN351W	900 x 403 mm	1034 x	525 mm	177.00	13.00	11.00	59.00
A351W	493 mm		615 mm	194.00	16.00	13.00	63.00
AN41W	1079 x 403 mm	1212 x	525 mm	194.00	22.00	18.00	72.00
A41W	493 mm		615 mm	210.00	11.00	9.00	59.00
AW31W	778 x 600 mm	912 x	721 mm	189.00	14.00	10.00	67.00
A330W	789 mm		912 mm	196.00	18.00	17.00	72.00
A335W	910 mm		1031 mm	278.00	19.00	18.00	75.00
AP32VW	1090 mm		1212 mm	306.00	25.00	0.00	80.00
AP321VW	1584 mm		1809 mm	493.00	37.00	10.00	145.00
A3535W	900 x 910 mm	1034 x	1034 mm	250.00	23.00	21.00	85.00
AP352VW	1090 mm		1212 mm	448.00	27.00	0.00	91.00
AP3521VW	1584 mm		1809 mm	544.00	41.00	11.00	164.00

Window Type	Glass Size	Rough Opening Width	Height	H.P. Sun Insulating	Insect Screen Add	Grille Add	Interior Blinds Add
AW41W	1079 x ...600 mm	1212 x	...721 mm	227.00	16.00	13.00	81.00
AP42VW	1090 mm		1212 mm	382.00	30.00	0.00	98.00
AP421VW	1584 mm		1809 mm	598.00	45.00	12.00	175.00
AP530VW	1376 x ...789 mm	1509 x	912 mm	362.00	29.00	0.00	109.00

Andersen *Perma-Shield* awning windows combine slim line sash and frame for maximum exterior view. The vinyl will not chip, flake, peel, blister, rust, pit, or corrode and will not need painting every few years.

Frame. Exterior wood members are treated with a water repellent preservative and covered with a pre-formed rigid vinyl (PVC) sheath in white or color.

Sash. This is wood core treated with a water repellent preservative and completely covered with a PVC sheath in white or color.

Glazing. Select quality high-performance, high-performance sun, or double-pane insulating window glass. For high altitude and other special glazings, contact manufacturer.

Glazing Bead. This is rigid vinyl. It has a snap-in feature with flexible vinyl tip against the glass.

Weatherstripping. This is flexible PVC bulb type, factory applied for tight seal between sash and frame.

Flashing. It is a continuation of the pre-formed rigid vinyl sheath on frame forms perimeter flashing and anchoring fin.

Corner Seal. Pre-formed vinyl frame sheath eliminates corner joints. Sash corner joints are welded to form waterproof joint.

Inside Stops. Western cedar pine natural wood, can be finished to match interior decor.

Awning Roto-Lock Operator. This offers a self-locking underscreen operator. Dual arms pull corners in tightly for maximum weathertightness. It has stone color decorator finish. Sash disengages from operator without tools. Picture window units include same materials as ventilating units except no operating hardware. Recommended additional parts include extension jambs of Western clear pine. Sizes are available for 4-9/16" (114 mm), 5-1/4" (131 mm), 6-9/16" (164 mm), 7-9/16" (189 mm) wall thickness. They are pre-drilled for easy application, except for 7-9/16" (189 mm) size.

Optional Accessories. Two different *Perma-Fit Divided Light Grilles* are available. The white grille is rigid vinyl. Exterior is white and the interior surface has a pine color stainable/paintable polymer finish. The grille in color is polycarbonate. Exterior is color and interior surface has pine color stainable/paintable polymer finish. Perma-Fit grilles without the finish are available in white or color both sides. Grille fits tight against glass and is easily removed for cleaning. Grilles not available for picture window sizes.

Casement Window Units

Casement sash may be ordered in gangs of two, three, four, or five. Prices for each additional unit in a gang arrangement should be quoted at the time of estimate for possible gang arrangement discount. It may be cheaper to use casements in groups than as individual units. Additional savings will also be realized in gang arrangements because of fewer jambs to form and caulk, providing the lintel construction does not become more costly due to the longer span. If some of the sash are ordered as stationary units, a further savings of $30.00 on the basic price per opening can be made. Casements can also be used in combination with larger fixed units or picture windows.

Casement units swing outward for top to bottom ventilation. They can be combined with other styles of windows and patio doors, or can be arranged in bow, angle bay, and box bay windows.

Frame. Wood members are treated with a water repellent preservative and covered with a rigid PVC sheath in white or color.

Sash. This is wood core treated with a water repellent preservative and completely covered with a rigid PVC sheath in white or color.

Glazing. Select quality high-performance, high-performance sun, or double-pane insulating window glass.

Glazing Bead. This is of rigid vinyl and has snap-in features with flexible vinyl tip against glass.

Weatherstripping. This is flexible PVC bulb type, factory applied for tight seal between sash and frame.

Flashing. This is a continuation of the pre-formed rigid vinyl sheath on frame forms perimeter flashing and anchor fin.

Corner Seal. Pre-formed vinyl frame sheath eliminates corner joints. Sash corner joints are welded to form waterproof joint.

Inside Stops. Western clear pine natural wood can be finished to match interior decor.

Sash Lock. The lock pulls sash firmly closed or ejects for opening. Reach-out action eliminates binding of locks and relieves stress on operator. Stone color decorator finish.

Casement Roto Operator. Under-screen operator has arm that attaches to bracket applied to sash. With either left- or right-hand operation and stone color decor finish, sash disengages from operator without tools. Picture window units include same materials as ventilating units except no operating hardware. Recommended additional parts include extension jambs of Western clear pine. Sizes available from Andersen for 4-9/16" (114 mm), 5-1/4" (131 mm), 6-9/16" (164 mm) or 7-9/16" (189 mm) wall thickness. It is pre-drilled for easy application, except for 7-9/16" (189 mm) unit.

Optional Accessories. Two different *Perma-Fit Divided Light Grilles* are available. The white grille is rigid vinyl. Exterior is white and the interior surface has a pine color stainable/paintable polymer finish. The grille in color is polycarbonate. Exterior is color and interior surface has pine color stainable/paintable polymer finish. Perma-Fit grilles without the finish are available in white or color both sides. Grille fits tight against glass and is easily removed for cleaning. Grilles not available for picture window sizes.

Casement Windows with High Performance (H.P.) Insulating Glass, Sun Insulating Glass, and Double Pane Insulating Glass

Window Type	Glass Size	Rough Opening Width x Height		H.P. Insulating	Insect Screen Add	Grille Add	Combination Screen/Storm Add
CR12W	12-5/8" x	19-5/16" ... 1'5-1/2"	2'0-5/8"	$130.00	$10.00	$8.00	$56.00
CR13W		31-1/8"	3'0-1/2"	146.00	11.00	9.00	56.00
CR135W		36"	3'5-3/8"	159.00	12.00	10.00	58.00
CR14W		43-3/16"	4'0-1/2"	171.00	13.00	11.00	59.00
CR15W		55-1/6"	5'0-3/8"	209.00	16.00	13.00	63.00
CR16W		67-1/16"	6'0-3/8"	227.00	22.00	18.00	72.00
CN13W	16-1/8" x	31-1/8" ... 1'9"	3'0-1/2"	147.00	11.00	9.00	59.00
CN135W		36"	3'5-3/8"	160.00	13.00	10.00	61.00
CN14W		43-3/16"	4'0-1/2"	173.00	14.00	11.00	63.00
CN15W		55-1/16"	5'0-3/8"	212.00	17.00	14.00	67.00
CN16W		67-1/16"	6'0-3/8"	230.00	22.00	18.00	74.00
C13W	19-3/4" x	31-1/8" ... 2'0-5/8"	3'0-1/2"	157.00	12.00	10.00	62.00
C135W		36"	3'5-3/8"	170.00	13.00	11.00	64.00
C14W		43-3/16"	4'0-1/2"	187.00	15.00	12.00	66.00
C15W		55-1/16"	5'0-3/8"	215.00	17.00	14.00	71.00
C16W		67-1/16"	6'0-3/8"	254.00	22.00	18.00	76.00
CW12W	24" x	19-5/16" ... 2'4-7/8"	2'0-5/8"	146.00	12.00	9.00	65.00
CW13W		31-1/8"	3'0-1/2"	172.00	14.00	10.00	68.00
CW135W		36"	3'5-3/8"	185.00	15.00	11.00	68.00
CW14W		43-3/16"	4'0-1/2"	210.00	16.00	13.00	71.00
CW15W		55-1/16"	5'0-3/8"	250.00	19.00	15.00	77.00
CW16W		67-1/16"	6'0-3/8"	300.00	24.00	20.00	83.00
CR23W	12-5/8" x	31-1/8" ... 2'10-1/4"	3'0-1/2"	259.00	12.00	9.00	65.00
CR235W		36"	3'5-3/8"	295.00	25.00	20.00	116.00
CR24W		43-3/16"	4'0-1/2"	336.00	26.00	22.00	119.00
CN235W	16-1/8" x	36" ... 3'5-1/4"	3'5-3/8"	304.00	25.00	20.00	122.00
C23W	19-3/4" x	31-1/8" ... 4'0-1/2"	3'0-1/2"	309.00	24.00	19.00	124.00
C235W		36"	3'5-3/8"	334.00	27.00	21.00	128.00
C24W		43-3/16"	4'0-1/2"	367.00	29.00	23.00	133.00
C25W		55-1/16"	5'0-3/8"	424.00	35.00	28.00	142.00
C26W		67-1/16"	6'0-3/8"	484.00	44.00	35.00	152.00
CW24W	24" x	43-3/16" ... 4'9"	4'0-1/2"	410.00	32.00	25.00	142.00
CW25W		55-1/16"	5'0-3/8"	491.00	38.00	29.00	153.00
CW26W		67-1/16"	6'0-3/8"	578.00	48.00	39.00	165.00
(metric)							
CR12W	315 x	428 mm ... 437 x	756 mm	$130.00	$10.00	$8.00	$56.00
CR13W		778 mm	912 mm	146.00	11.00	9.00	56.00
CR135W		900 mm	1034 mm	159.00	12.00	10.00	58.00
CR14W		1079 mm	1212 mm	171.00	13.00	11.00	59.00
CR15W		1376 mm	1509 mm	209.00	16.00	13.00	63.00
CR16W		1676 mm	1809 mm	227.00	22.00	18.00	72.00
CN13W	403 x	778 mm ... 525 x	912 mm	147.00	11.00	9.00	59.00
CN135W		900 mm	1034 mm	160.00	13.00	10.00	61.00

Window Type	Glass Size	Rough Opening Width x Height		H.P. Insulating	Insect Screen Add	Grille Add	Combination Screen/Storm Add
CN14W		1079 mm	1212 mm	173.00	14.00	11.00	63.00
CN15W		1376 mm	1509 mm	212.00	17.00	14.00	67.00
CN16W		1676 mm	1809 mm	230.00	22.00	18.00	74.00
C13W	493 x	778 mm	615 x 912 mm	157.00	12.00	10.00	62.00
C135W		900 mm	1034 mm	170.00	13.00	11.00	64.00
C14W		1079 mm	1212 mm	187.00	15.00	12.00	66.00
C15W		1376 mm	1509 mm	215.00	17.00	14.00	71.00
C16W		1676 mm	1809 mm	254.00	22.00	18.00	76.00
CW12W	600 x	482 mm	721 x 756 mm	146.00	12.00	9.00	65.00
CW13W		778 mm	912 mm	172.00	14.00	11.00	68.00
CW135W		900 mm	1034 mm	185.00	15.00	11.00	68.00
CW14W		1079 mm	1212 mm	210.00	16.00	13.00	71.00
CW15W		1376 mm	1509 mm	250.00	19.00	15.00	77.00
CW16W		1676 mm	1809 mm	300.00	24.00	20.00	83.00
CR23W	315 x	778 mm	856 x 912 mm	259.00	12.00	9.00	65.00
CR235W		900 mm	1034 mm	295.00	25.00	20.00	116.00
CR24W		1079 mm	1212 mm	336.00	26.00	22.00	119.00
CN235W	403 x	900 mm	1031 x 1034 mm	304.00	25.00	20.00	122.00
C23W	493 x	778 mm	1212 x 912 mm	309.00	24.00	19.00	124.00
C235W		900 mm	1034 mm	334.00	27.00	21.00	128.00
C24W		1079 mm	1212 mm	367.00	29.00	23.00	133.00
C25W		1376 mm	1509 mm	424.00	35.00	28.00	142.00
C26W		1676 mm	1809 mm	484.00	44.00	35.00	152.00
CW24W	600 x	1079 mm	1425 x 1212 mm	410.00	32.00	25.00	142.00
CW25W		1376 mm	1509 mm	491.00	38.00	29.00	153.00
CW26W		1676 mm	1809 mm	578.00	48.00	39.00	165.00

Circle Top Window Units

Circle top windows are specially designed with a frame/sash profile. The windows are compatible with casement, awning, and double hung units and can be joined to these or can be installed as single units. When combined, the glazing in the circle top window closely aligns with the window unit below it.

Frame. Wood members are treated with a water repellent preservative and covered with a pre-formed rigid PVC sheath in white or color. The vinyl sheath will not chip, flake, peel, blister, rust, pit, or corrode, and it will not need painting every few years.

Flashing. Continuation of the pre-formed rigid vinyl sheath on frame forms a full perimeter flashing and anchoring fin.

Glazing. Select quality high-performance or high-performance sun insulating glass.

Recommended Optional Accessories. Interior arch casing is unfinished wood for painting or staining. Maple or oak is used for laminated portion, which forms the curved piece. Interior casing is available in ranch or colonial style. Divided light grilles are available in two different types. The white grille is rigid vinyl. Exterior is white and the interior surface has a pine color stainable/paintable polymer finish. The grille in color is glass fiber polycarbonate. Exterior is color and interior surface has pine color stainable/paintable polymer finish. Perma-fit grilles without the finish are available in white or color both sides. Grille fits tight against glass and can be removed for cleaning. Grilles not available for picture window sizes.

Plinth Blocks. Solid oak or maple plinth block with decorative radial sunburst pattern or reversible flush face. Plinth blocks come with arch casing or can be purchased separately.

Extension Jambs. Laminated maple curved extension jambs are available for wall thickness of 4-9/16" (114 mm), 5-1/4" (131 mm), 6-9/16" (164 mm) and 7-9/16" (189 mm) for casement and awning windows and 5-1/4" (131 mm), 6-9/16" (164 mm), and 7-9/16" (189 mm) for double hung windows. They are pre-drilled, except for casement and awning 7-9/16" (189 mm) size. One-piece pine straight extension jamb for sill is available in same sizes as maple parts for casement or awning. Both laminated maple and clear pine parts give uniform appearance when stained.

Circle Top Windows with High Performance (H.P.) Insulating Glass
and Sun Insulating Glass

Window Type	Glass Size		Rough Opening Width x	Height	H.P. Insulating	Grille Add
CTQC1W	19-5/8" x	19-5/8"	2'0-5/8"	2'0-5/8"	301.00	15.00
CTC1W	19-1/2" x	9-3/4"	2'0-5/8"	1'2-7/8"	270.00	14.00
CTCW1W	23-3/4" x	11-7/8"	2'4-7/8"	2'2-3/16	298.00	18.00
CTC2W	43-3/8" x	21-11/16"	4'0-1/2"	2'2-13/16	414.00	27.00
CTCW4W	51-7/8" x	25-15/16"	4'9"	2'7-1/16	489.00	34.00
CTC3W	67-1/8" x	35-9/16"	6'0-3/8"	3'2-3/4"	789.00	89.00
CTN20W	20-3/8" x	10-3/16"	2'2-1/8"	1'3-3/4"	278.00	15.00
CTN24	24-3/8" x	12-3/16"	2'6-1/8"	1'5-3/4"	322.00	18.00
CTN28	28-3/8" x	14-3/16"	2'10-1/8"	1'7-3/4"	348.00	21.00
CTN30W	32-3/8" x	16-3/16"	3'2-1/8"	1'9-3/4"	372.00	23.00
CTN34W	36-3/8" x	18-3/16"	6'6-1/8"	1'11-3/4"	410.00	25.00

(metric)

Window Type	Glass Size		Rough Opening Width x	Height	H.P. Insulating	Grille Add
CTQC1W	491 x	491 mm	616 x	616 mm	301.00	15.00
CTC1W	488 x	244 mm	616 x	372 mm	270.00	14.00
CTCW1W	594 x	297 mm	721 x	654 mm	298.00	18.00
CTC2W	1084 x	542 mm	1212 x	670 mm	414.00	27.00
CTCW4W	1297 x	648 mm	1425 x	776 mm	489.00	34.00
CTC3W	1678 x	889 mm	1809 x	968 mm	789.00	89.00
CTN20W	509 x	255 mm	653 x	393 mm	278.00	15.00
CTN24	609 x	305 mm	75 x	443 mm	322.00	18.00
CTN28	709 x	355 mm	853 x	493 mm	348.00	21.00
CTN30W	809 x	480 mm	953 x	543 mm	372.00	23.00
CTN34W	909 x	455 mm	1953 x	593 mm	410.00	25.00

Bow, Box Bay and Angle Bay Window Units
30 Degree Angle Bay Unit Sizes with High Performance (H.P.)
Insulating Glass, Sun Insulating Glass,
and Double Pane Insulating Glass

Window Type	Glass Size	Rough Opening Width x	Height	H.P. Sun Insulating	Insect Screens Add	Grille Add	Blinds with Hardware
30 C13 20W	5'10" x 3'1-5/16"	5'9-1/8"	3'1-15/16"	$683.00	$24.00	$29.00	$184.00
30 C135 20W	3'6-13/16"		3'6-13/16"	741.00	27.00	32.00	190.00
30 C14 20W	4'2"		4'2"	795.00	29.00	35.00	197.00
30 C15 20W	5'1-7/8"		5'1-7/8"	900.00	35.00	41.00	212.00
30 C16 20W	6'1-7/8"		6'1-7/8"	1,042.00	44.00	55.00	225.00
30 C23 20W	7'9-7/8" x3'1-15/16"	7'9-7/8"	3'1-15/16"	876.00	48.00	39.00	247.00
30 C235 20W	3'6-13/16"		3'6-13/16"	949.00	53.00	42.00	256.00
30 C24 20W	4'2"		4'2"	1,022.00	58.00	46.00	265.00
30 C25 20W	5'1-7/8"		5'1-7/8"	1,158.00	70.00	55.00	284.00
30 C26 20W	6'1-7/8"		6'1-7/8"	1,326.00	88.00	73.00	303.00
30 CP23 20W	7'9-7/8" x3'1-15/16"	7'9-7/8"	3'1-15/16"	754.00	24.00	19.00	207.00
30 CP235 20W	3'6-13/16"		3'6-13/16"	949.00	53.00	42.00	256.00
30 CP24 20W	4'2"		4'2"	1,022.00	58.00	46.00	265.00
30 CP25 20W	5'1-7/8"		5'1-7/8"	1,024.00	35.00	28.00	245.00
30 CP26 20W	6'1-7/8"		6'1-7/8"	1,219.00	44.00	36.00	265.00
30 C33 20W	9'9-3/4" x3'1-15/16"	9'8-7/8"	3'1-15/16"	1,040.00	48.00	48.00	307.00
30 C335 20W	3'6-13/16"		3'6-13/16"	1,130.00	53.00	53.00	318.00
30 C34 20W	4'2"		4'2"	1,212.00	58.00	58.00	330.00
30 C35 20W	5'1-7/8"		5'1-7/8"	1,366.00	70.00	69.00	354.00
30 CP33 20W	9'9-3/4" x3'1-15/16"	9'8-7/8"	3'1-15/16"	874.00	24.00	19.00	236.00
30 CP335 20W	3'6-13/16"		3'6-13/16"	958.00	26.00	21.00	247.00
30 CP34 20W	4'2"		4'2"	1,065.00	29.00	23.00	259.00
30 CP35 20W	5'1-7/8"		5'1-7/8"	1,288.00	34.00	28.00	284.00
			(metric)				
30 C13 20W	1750 x 949 mm	1728 x	949 mm	$683.00	$24.00	$29.00	$184.00
30 C135 20W	1070 mm		1070 mm	741.00	27.00	32.00	190.00
30 C14 20W	1250 mm		1250 mm	795.00	29.00	35.00	197.00
30 C15 20W	1546 mm		1546 mm	900.00	35.00	41.00	212.00
30 C16 20W	1846 mm		1846 mm	1,042.00	44.00	55.00	225.00
30 C23 20W	2346 x 949 mm	2296 x	949 mm	876.00	48.00	39.00	247.00
30 C235 20W	1070 mm		1070 mm	949.00	53.00	42.00	256.00
30 C24 20W	1250 mm		1250 mm	1,022.00	58.00	46.00	265.00
30 C25 20W	1546 mm		1546 mm	1,158.00	70.00	55.00	284.00
30 C26 20W	1846 mm		1846 mm	1,326.00	88.00	73.00	303.00
30 CP23 20W	2346 x 949 mm	2296 x	949 mm	754.00	24.00	19.00	207.00
30 CP235 20W	1070 mm		1070 mm	949.00	53.00	42.00	256.00
30 CP24 20W	1250 mm		1250 mm	1,022.00	58.00	46.00	265.00
30 CP25 20W	1546 mm		1546 mm	1,024.00	35.00	28.00	245.00
30 CP26 20W	1846 mm		1846 mm	1,219.00	44.00	36.00	265.00
30 C33 20W	2793 x 949 mm	2921 x	949 mm	1,040.00	48.00	48.00	307.00
30 C335 20W	1070 mm		1070 mm	1,130.00	53.00	53.00	318.00
30 C34 20W	1250 mm		1250 mm	1,212.00	58.00	58.00	330.00
30 C35 20W	1546 mm		1546 mm	1,366.00	70.00	69.00	354.00
30 CP33 20W	2793 x 949 mm	2921 x	949 mm	874.00	24.00	19.00	236.00
30 CP335 20W	1070 mm		1070 mm	958.00	26.00	21.00	247.00
30 CP34 20W	1250 mm		1250 mm	1,065.00	29.00	23.00	259.00
30 CP35 20W	1546 mm		1546 mm	1,288.00	34.00	28.00	284.00

Note: Sizes of bay windows are the same with different glazing.
Add for casing 3-1/2" (87.5 mm) not installed:

45-C1-20-W	45-C2-20-W	45-C3-20-W
$+18.83	$+30.43	$+30.43

Deduct for no top auxiliary casing:

45-C1-20-W	45-C2-20-W	45-C3-20-W
$-20.64	$-20.64	$-20.64

Bow, Box Bay and Angle Bay Window Units
45 Degree Angle Bay Unit Sizes with High Performance (H.P.) Insulating Glass, Sun Insulating Glass, and High Double Pane Insulating Glass

Window Type	Glass Size	Rough Opening Width x	Height	H.P. Insulating	Insect Screens Add	Grille Add	Blinds with Hardware
45 C13 20W	5'4-1/8" x3'1-15/16"	5'2-5/8"	3'1-15/16"	$702.00	$24.00	$29.00	$184.00
45 C135 20W	3'6-13/16"		3'6-13/16"	760.00	27.00	32.00	190.00
45 C14 20W	4'2"		4'2"	823.00	29.00	41.00	197.00
45 C15 20W	5'1-7/8"		5'1-7/8"	936.00	35.00	55.00	212.00
45 C16 20W	6'1-7/8"		6'1-7/8"	1,079.00	44.00	39.00	225.00
45 C23 20W	7'3-7/8" x3'1-15/16"	7'2-1/2"	3'1-15/16"	906.00	47.00	42.00	247.00
45 C235 20W	3'6-13/16"		3'6-13/16"	979.00	53.00	58.00	256.00
45 C24 20W	4'2"		4'2"	1,060.00	273.00	55.00	265.00
45 C25 20W	5'1-7/8"		5'1-7/8"	1,205.00	70.00	55.00	284.00
45 C26 20W	6'1-7/8"		6'1-7/8"	1,373.00	88.00	19.00	303.00
45 CP23 20W	7'3-7/8" x3'1-15/16"	7'2-1/2"	3'1-15/16"	783.00	23.00T	21.00	207.00
45 CP235 20W	3'6-13/16"		3'6-13/16"	864.00	27.00	23.00	215.00
45 CP24 20W	4'2"		4'2"	950.00	29.00	28.00	226.00
45 CP25 20W	5'1-7/8"		5'1-7/8"	1,071.00	35.00	36.00	245.00
45 CP26 20W	6'1-7/8"		6'1-7/8"	1,267.00	44.00	53.00	265.00
45 C33 20W	9'3-3/4" x3'1-15/16"	9'2-3/8"	3'1-15/16"	1,074.00	47.00	58.00	307.00
45 C335 20W	3'6-13/16"		3'6-13/16"	1,165.00	53.00	69.00	318.00
45 C34 20W	4'2"		4'2"	1,255.00	58.00	19.00	330.00
45 C35 20W	5'1-7/8"		5'1-7/8"	1,417.00	70.00	21.00	354.00
45 CP33 20W	9'3-3/4" x3'1-15/16"	9'2-3/8"	3'1-15/16"	909.00	24.00	23.00	236.00
45 CP335 20W	3'6-13/16"		3'6-13/16"	993.00	27.00	28.00	247.00
45 CP34 20W	4'2"		4'2"	1,106.00	29.00	23.00	259.00
45 CP35 20W	5'1-7/8"		5'1-7/8"	1,338.00	35.00	28.00	284.00
			(metric)				
45 C13 20W	1528 x 949 mm	1565 x	949 mm	$702.00	$24.00	$29.00	$184.00
45 C135 20W	1070 mm		1070 mm	760.00	27.00	32.00	190.00
45 C14 20W	1250 mm		1250 mm	823.00	29.00	41.00	197.00
45 C15 20W	1546 mm		1546 mm	936.00	35.00	55.00	212.00
45 C16 20W	1846 mm		1846 mm	1,079.00	44.00	39.00	225.00
45 C23 20W	2196 x 949 mm	2162 x	949 mm	906.00	47.00	42.00	247.00
45 C235 20W	1070 mm		1070 mm	979.00	53.00	58.00	256.00
45 C24 20W	1250 mm		1250 mm	1,060.00	273.00	55.00	265.00
45 C25 20W	1546 mm		1546 mm	1,205.00	70.00	55.00	284.00
45 C26 20W	1846 mm		1846 mm	1,373.00	88.00	19.00	303.00
45 CP23 20W	2196 x 949 mm	2162 x	949 mm	783.00	23.00	21.00	207.00
45 CP235 20W	1070 mm		1070 mm	864.00	27.00	23.00	215.00
45 CP24 20W	1250 mm		1250 mm	950.00	29.00	28.00	226.00
45 CP25 20W	1546 mm		1546 mm	1,071.00	35.00	36.00	245.00
45 CP26 20W	1846 mm		1846 mm	1,267.00	44.00	53.00	265.00

Window Type	Glass Size	Rough Opening Width x	Height	H.P. Insulating	Insect Screens Add	Grille Add	Blinds with Hardware
45 C33 20W	2787 x 949 mm	2759 x	949 mm	1,074.00	47.00	58.00	307.00
45 C335 20W	1070 mm		1070 mm	1,165.00	53.00	69.00	318.00
45 C34 20W	1250 mm		1250 mm	1,255.00	58.00	19.00	330.00
45 C35 20W	1546 mm		1546 mm	1,417.00	70.00	21.00	354.00
45 CP33 20W	2787 x 949 mm	2759 x	949 mm	909.00	24.00	23.00	236.00
45 CP335 20W	1070 mm		1070 mm	993.00	27.00	28.00	247.00
45 CP34 20W	1250 mm		1250 mm	1,106.00	29.00	23.00	259.00
45 CP35 20W	1546 mm		1546 mm	1,338.00	35.00	28.00	284.00

Add for casing 3-1/2" (87.5 mm) not installed:

45-C1-20-W	45-C2-20-W	45-C3-20-W
$+18.83	$+30.43	$+30.43

Deduct for no top auxiliary casing:

45-C1-20-W	45-C2-20-W	45-C3-20-W
$-20.64	$-20.64	$-20.64

90 Degree Box Bay Window Sizes with High Performance (H.P.) Insulating Glass, Sun Insulating Glass, and High Double Pane Insulating Glass

Window Type	Glass Size	Rough Opening Width x	Height	H.P. Insulating	Insect Screens Add	Grille Add	Blinds with Hardware
90 C23 15W	4'8-3/8" x2'11-15/16"	4'2" x	3'1-15/16"	$883.00	$46.00	$38.00	$236.00
90 C235 15W	3'4-13/16"		3'6-7/8"	897.00	51.00	41.00	244.00
90 C24 15W	4'0"		4'2"	972.00	55.00	45.00	251.00
90 C25 15W	4'11-7/8"		5'1-7/8"	1,135.00	68.00	54.00	267.00
90 C26 15W	5'11-7/8"		6'1-7/8"	1,253.00	88.00	72.00	196.00
90 CP23 15W	4'8-3/8" x2'11-15/16"	4'2"	3'1-15/16"	720.00	22.00	18.00	196.00
90 CP235 15W	3'4-13/16"		3'6-7/8"	795.00	25.00	20.00	203.00
90 CP24 15W	4'0"		4'2"	870.00	26.00	22.00	212.00
90 CP25 15W	4'11-7/8"		5'1-7/8"	1,011.00	33.00	27.00	228.00
90 CP26 15W	5'11-7/8"		6'1-7/8"	1,141.00	43.00	35.00	258.00
90 C33 15W	6'8-1/4" x2'11-15/16"	4'2"	6'1-7/8"	1,005.00	46.00	47.00	296.00
90 C335 15W	3'4-13/16"		3'6-7/8"	1,085.00	51.00	51.00	306.00
90 C34 15W	4'0"		4'2"	1,169.00	55.00	57.00	315.00
90 C35 15W	4'11-7/8"		5'1-7/8"	1,347.00	68.00	68.00	337.00
90 CP33 15W	6'8-1/4" x2'11-15/16"	4'2"	6'1-7/8"	856.00	22.00	18.00	225.00
90 CP3335 15W	3'4-13/16"		3'6-7/8"	935.00	25.00	20.00	235.00
90 CP34 15W	4'0"		4'2"	1,021.00	26.00	22.00	246.00
90 CP35 15W	4'11-7/8"		5'1-7/8"	1,259.00	33.00	26.00	267.00
			(metric)				
90 C23 15W	1409 x 898 mm	1250 x	932 mm	$883.00	$46.00	$38.00	$236.00
90 C235 15W	1020 mm		1071 mm	897.00	51.00	41.00	244.00
90 C24 15W	1200 mm		1250 mm	972.00	55.00	45.00	251.00
90 C25 15W	1496 mm		1546 mm	1,135.00	68.00	54.00	267.00
90 C26 15W	1796 mm		1876 mm	1,253.00	88.00	72.00	196.00
90 CP23 15W	1409 x 898 mm	1250 x	932 mm	720.00	22.00	18.00	196.00
90 CP235 15W	1020 mm		1071 mm	795.00	25.00	20.00	203.00
90 CP24 15W	1200 mm		1250 mm	870.00	26.00	22.00	212.00
90 CP25 15W	1496 mm		1546 mm	1,011.00	33.00	27.00	228.00
90 CP26 15W	1796 mm		1876 mm	1,141.00	43.00	35.00	258.00
90 C33 15W	2006 x 898 mm	1250 x	1876 mm	1,005.00	46.00	47.00	296.00

Window Type	Glass Size	Rough Opening Width x	Height	H.P. Insulating	Insect Screens Add	Grille Add	Blinds with Hardware
90 C335 15W	1020 mm		946 mm	1,085.00	51.00	51.00	306.00
90 C34 15W	1200 mm		1250 mm	1,169.00	55.00	57.00	315.00
90 C35 15W	1496 mm		1546 mm	1,347.00	68.00	68.00	337.00
90 CP33 15W	2006 x 898 mm	1250 x	1876 mm	856.00	22.00	18.00	225.00
90 CP3335 15W	1020 mm		946 mm	935.00	25.00	20.00	235.00
90 CP34 15W	1200 mm		1250 mm	1,021.00	26.00	22.00	246.00
90 CP35 15W	1496 mm		1546 mm	1,259.00	33.00	26.00	267.00

Add for casing 3-1/2" (87.5 mm) not installed:

90-C2-15W	90-C3-15W
$+60.86	$+79.69

Roof Window Units (Skylights)

Roof windows, also called skylights, can enhance the appearance of a living space. However, the estimator must be cautious. Roof window installation is unique and usually requires additional costs. On the exterior, there might be flashing for ice build-up, curbs around the window on a flat roof, or on a sloped roof that is relatively flat, or deflector shield for maximum weather tightness. On the interior, additional finish work might be required.

Roof Windows, Basic and Stationary Unit with High Performance (H.P.) Insulating and Sun Insulating Laminated Tempered Glass

Window Type	Glass Size	Rough opening Width x	Height	H.P. Insulating Lam/Temp	Set Shingle Flashing	Incline Curb Flashing	Set Tile Flashing	Pleated Shade with Hardware
RW2133S	85-1/16" x30-5/16"	1'10"	2'10-1/4"	$299.00	$68.00	$269.00	$79.00	$104.00
RW2144S	40-15/16"		3'8-3/4"	350.00	68.00	274.00	89.00	118.00
RW2172S	68-7/8"		6'0-3/4"	511.00	0.00	123.00	139.00	119.00
RW2944S	26" x 40-15/16"	2'5-7/8"	3'8-3/4"	419.00	82.00	290.00	97.00	131.00
RW2957S	53-15/16"		4'9-3/4"	513.00	82.00	320.00	107.00	155.00
RW2972S	68-7/8"		6'0-3/4"	597.00	199.00	0.00	133.00	177.00
RW4144S	38-3/16" x40-15/16"	3'6"	3'8-3/4"	526.00	91.00	324.00	111.00	166.00
RW4157S	53-15/16"		4'9-3/4"	640.00	91.00	343.00	119.00	189.00
(metric)								
RW2133S	2126.6 x757.8 mm	550 x	856 mm	$299.00	$68.00	$269.00	$79.00	$104.00
RW2144S	1023.4 mm		1118.8 mm	350.00	68.00	274.00	89.00	118.00
RW2172S	1721.9 mm		1818.8 mm	511.00	0.00	123.00	139.00	119.00
RW2944S	650 x 1023.4 mm	746.9 x	1118.8 mm	419.00	82.00	290.00	97.00	131.00
RW2957S	1348.4 mm		1443.8 mm	513.00	82.00	320.00	107.00	155.00
RW2972S	1721.8 mm		1818.8 mm	597.00	199.00	0.00	133.00	177.00
RW4144S	954.7 x1023.4 mm	1050 x	1118.8 mm	526.00	91.00	324.00	111.00	166.00
RW4157S	1348.4 mm		1443.8 mm	640.00	91.00	343.00	119.00	189.00

Roof Windows, Venting Unit with High Performance (H.P.) Insulating and Sun Insulating Tempered Glass

Window Type	Glass Size Width x Height	Rough opening Width x Height	H.P. Sun Insul/Tmp	Set Shingle Flashing	Incline Curb Flashing	Set Tile Flashing	Pleated Shade with Hardware
RW2133V	15-1/16" x27-1/8"	1'10" 2'10-1/4"	$484.00	$68.00	$270.00	$79.00	$94.00
RW2144V	37-3/4"	3'8-3/4"	557.00	68.00	274.00	89.00	107.00
RW2944V	23" x 37-3/4"	2'5-7/8" 3'8-3/4"	622.00	82.00	290.00	97.00	119.00
RW2957V	50-3/4"	4'9-3/4"	756.00	82.00	320.00	107.00	139.00
RW4144V	35-3/16" x37-3/4"	3'6" 3'8-3/4"	794.00	91.00	323.00	112.00	149.00
RW4157V	50-3/4"	4'9-3/4"	965.00	91.00	343.00	119.00	170.00
(metric)							
RW2133V	376.6 x678.1 mm	550 x856.3 mm	$484.00	$68.00	$270.00	$79.00	$94.00
RW2144V	943.8 mm	1118.8 mm	557.00	68.00	274.00	89.00	107.00
RW2944V	575 x 943.8 mm	746.9 x1118.8 mm	622.00	82.00	290.00	97.00	119.00
RW2957V	1268.8 mm	1443.8 mm	756.00	82.00	320.00	107.00	139.00
RW4144V	879.7 x943.8 mm	1050 x1118.8 mm	794.00	91.00	323.00	112.00	149.00
RW4157V	1268.8 mm	1443.8 mm	965.00	91.00	343.00	119.00	170.00

Window Screens

Wood window screens have been largely replaced by aluminum and vinyl units. Most prefabricated windows have insect screens available to fit them and can be ordered as part of the window unit or separately later. The following prices are based on prefabricated window glass sizes.

Glass Size, in.	Glass Size, mm	Price
16-7/16" x 22"	410.9 x 550 mm	$16.40
16-7/16" x 24"	410.9 x 600 mm	17.30
20-7/8" x 22"	521.9 x 550 mm	18.00
20-7/8" x 24"	521.9 x 600 mm	18.80
20-7/8" x 28"	521.9 x 700 mm	20.60
20-7/8" x 68"	521.9 x 1700 mm	22.00
24-7/16" x 22"	610.9 x 550 mm	19.40
24-7/16" x 24"	610.0 x 600 mm	20.50
24-7/16" x 28"	610.9 x 700 mm	22.30
28-7/16" x 22"	710.9 x 550 mm	20.80
28-7/16" x 24"	710.9 x 600 mm	21.80
28-7/16" x 28"	710.9 x 700 mm	24.20
28-7/16" x 60"	710.9 x 1500 mm	26.00
28-7/16" x 68"	710.9 x 1700 mm	30.10
32-7/16" x 22"	810.9 x 550 mm	23.00
32-7/16" x 24"	810.9 x 600 mm	24.20
32-7/16" x 28"	810.9 x 700 mm	26.80
32-7/16" x 60"	810.9 x 1500 mm	28.00
32-7/16" x 68"	810.9 x 1700 mm	30.30
36-7/16" x 22"	910.9 x 550 mm	24.80
36-7/16" x 24"	910.9 x 600 mm	26.00
36-7/16" x 28"	910.9 x 700 mm	28.90

Glass Size, in.	Glass Size, mm	Price
36-7/16" x 60"..	............910.9 x 1500 mm	......30.10
40-7/16" x 24"..	............1010.9 x 600 mm	......27.00

Nailing Wood Window Frames Together. When wood window frames are sent to the job knocked down, it will require 1 to 1-1/4 hr. carpenter time to assemble and nail each frame together, at the following labor cost per frame:

	Hours	Rate	Total	Rate	Total
Carpenter................................	1.2	$....	$....	$26.44	$31.73

It will require 1-3/4 to 2-1/4 hrs. carpenter time to assemble and nail together a double window frame up to 6'-0" (1800 mm) wide and 5'-6" (1650 mm) or 6'-0" (1800 mm) high, at the following labor cost per frame:

	Hours	Rate	Total	Rate	Total
Carpenter................................	2.0	$....	$....	$26.44	$52.88

Setting Single Wood Window Frames. The labor cost of setting wood window frames varies with the size of the frame, amount of bracing necessary, and the distance they have to be carried or hoisted.

Once frames have been delivered to the job, it requires 10 to 15 minutes sorting frames and carrying them to the building and 20 to 30 minutes for 2 carpenters to set, plumb, and brace each frame at the rate of 8 to 12 frames per 8-hr. day. The labor handling and setting each frame should cost as follows:

	Hours	Rate	Total	Rate	Total
Carpenter................................	0.8	$....	$....	$26.44	$21.15
Labor......................................	0.2	....	....	21.04	4.21
Cost per frame...........................			$....		$25.36

Setting Double or Triple Window Frames. It will ordinarily require 10 to 15 minutes for 2 workers to carry a double or triple window frame from the stock pile to place where it is to be used and it will then require 30 to 35 minutes for 2 carpenters to set, plumb, and brace the frame on the wall, at the following labor cost per frame:

	Hours	Rate	Total	Rate	Total
Carpenter................................	1.1	$....	$....	$26.44	$29.08
Labor......................................	0.4	....	....	21.04	8.42
Cost per frame...........................			$....		$37.50

Labor Handling and Setting Complete Window Units. When handling and setting complete window units, consisting of frame, sash, balances, and weatherstrips, all installed complete, figure about 1/4 hr. more than for handling and setting window frames of the same size.

Fitting Wood Sash. When fitting wood sash for double hung windows, a carpenter should fit 24 to 28 single sash per 8-hr. day, at the following labor cost per pair:

	Hours	Rate	Total	Rate	Total
Carpenter	0.6	$....	$....	$26.44	$15.86

First Grade Workmanship

Where the sash must fit perfectly, with neither too much nor too little play, a carpenter should fit 18 to 22 single sash per 8-hr. day, at the following labor cost per pair:

	Hours	Rate	Total	Rate	Total
Carpenter	0.8	$....	$....	$26.44	$21.15

Setting Roof Window (Skylight). When placing frame and window, two workers are usually required for ease of installation at the following cost per frame:

	Hours	Rate	Total	Rate	Total
Carpenter	3.0	$....	$....	$26.44	$79.32

It is important that the estimator include any additional roofing material and labor required.

Hanging Wood Sash. Including the time required placing sash cord, chain, and weights or counterbalances, a carpenter should hang 24 to 28 single sash per 8-hr. day, at the following labor cost per pair:

	Hours	Rate	Total	Rate	Total
Carpenter	0.6	$....	$....	$26.44	$15.86

First Grade Workmanship

Where care must be exercised to see that all sash work fits perfectly, and the cord or chain is the exact length with the correct weights or counterbalances, a carpenter should hang 22 to 26 single sash per 8-hr. day, at the following labor cost per pair:

	Hours	Rate	Total	Rate	Total
Carpenter	0.7	$....	$....	$26.44	$18.51

Fitting and Hanging Wood Sash. A carpenter should fit and hang, complete with sash cord, chain, and weights or counterbalances, 12 to 14 single sash per 8-hr. day, at the following labor cost per pair:

	Hours	Rate	Total	Rate	Total
Carpenter	1.2	$....	$....	$26.44	$31.73

First Grade Workmanship

Where the sash must be fitted to have neither too much nor too little play and hung so they will work with ease and not be too loose, a carpenter should fit and hang, including sash cord, chain, and weights or counterbalances, 10 to 12 single sash per 8-hr. day, at the following labor cost per pair:

	Hours	Rate	Total	Rate	Total
Carpenter	1.5	$....	$....	$26.44	$39.66

Fitting Casement Sash. A carpenter should fit 22 to 26 average size wood casement sash per 8-hr. day, at the following labor cost per pair:

	Hours	Rate	Total	Rate	Total
Carpenter	0.7	$....	$....	$26.44	$18.51

First Grade Workmanship

Where first class workmanship is required with all sash fitted to work easily without binding and hung so that each sash shows the same top, bottom, and side margin, a carpenter should fit 14 to 18 single sash per 8-hr. day, at the following labor cost per pair:

	Hours	Rate	Total	Rate	Total
Carpenter	1.0	$....	$....	$26.44	$26.44

Hanging Casement Sash. The labor cost of hanging casement sash varies, depending on whether they are hung with 2 or 3 butts and style of butts. Where the sash are hung using 2 butts per sash, a carpenter should hang 20 to 24 single sash per 8-hr. day, at the following labor cost per pair:

	Hours	Rate	Total	Rate	Total
Carpenter	0.75	$....	$....	$26.44	$19.83

First Grade Workmanship

In fine residences, apartments, and other buildings where the casement sash must be hung without binding and showing the same margin at the top, bottom, and sides, a carpenter should hang 12 to 14 single sash per 8-hr. day, at the following labor cost per pair:

	Hours	Rate	Total	Rate	Total
Carpenter....................................	1.2	$....	$....	$26.44	$31.73

Fitting and Hanging Casement Sash. Under average conditions, a carpenter should fit and hang 10 to 13 single casement sash per 8-hr. day, at the following labor cost per pair:

	Hours	Rate	Total	Rate	Total
Carpenter....................................	1.4	$....	$....	$26.44	$37.02

First Grade Workmanship

Where casement sash must be fitted and hung to be neither too loose nor so tight as to cause binding when opening and closing and must show the same margin between sash and frame on all sides, a carpenter should fit and hang 7 to 9 single sash per 8-hr. day, at the following labor cost per pair:

	Hours	Rate	Total	Rate	Total
Carpenter....................................	2.2	$....	$....	$26.44	$58.17

Fitting and Hanging Outside Window Shutters. When fitting and hanging outside window shutters on medium priced residences or cottages, a carpenter should fit and hang 7 to 8 pairs per 8-hr. day, at the following labor cost per pair:

	Hours	Rate	Total	Rate	Total
Carpenter....................................	1.10	$....	$....	$26.44	$29.08
Cost per single shutter	0.55		....		14.54

First Grade Workmanship

On the better class of buildings, where more care must be used in fitting and hanging outside window shutters, a carpenter should fit and hang 5 to 6 pairs per 8-hr. day, at the following labor cost per pair:

	Hours	Rate	Total	Rate	Total
Carpenter....................................	1.4	$....	$....	$26.44	$37.02
Cost per single shutter	0.7		....		18.51

Fitting Roof Windows. The labor costs of setting wood, polyvinyl, or steel roof windows varies with the size of the frame, amount of bracing necessary, and the distance that the window must be hoisted or carried.

After frames have been delivered to a project, it will require 10 to 15 minutes for sorting and carrying them to the building, and it will take 20 to 45 minutes for two carpenters to set, plumb, and brace each frame, a rate of 6 to 12 units per day. The most important variable affecting the production rate is the slope of the roof that carpenters are working on.

	Hours	Rate	Total	Rate	Total
Carpenter	1.1	$....	$....	$26.44	$29.08
Labor	0.4			21.04	8.42
					$37.50

If required, add curbs around window unit at approximately 45 minutes per window unit.

Interior Window Trim

The labor cost of placing interior window trim will vary considerably with the type of building, thickness of walls, and style of trim or casing. Frame buildings, stucco buildings, or brick veneer buildings with walls 6" (150 mm) to 8" (200 mm) thick do not require inside wood jamb linings, because the wood window frames usually extend to the inside face of the wall, and the wood casings, window stools, aprons, and stops are nailed directly to the wood frame.

Buildings having walls 10" (250 mm), 12" (300 mm), 16" (400 mm), or more in thickness require a wood jamb lining to extend from the inside of the wood window frame to the face of the finished wall. It will also require wood blocking and grounds to nail the jamb linings, window stools, and aprons.

The style of trim affects labor costs considerably, and for this reason costs are given on four distinct types of interior casings:

Style "A" consists of 1-member trim with either a mitred or square cut head or cap. This is the simplest trim it is possible to use.

Style "B" consists of 1-member trim for sides and a built-up cap trim for window heads. Caps may be put together in the mill or assembled and put together on the job.

Style "C" consists of 2-member back-band trim, consisting of a one-piece casing with a back-band of thicker material.

Style "D" consists of any type of window trim that are assembled and glued together in the mill and set in the building as a unit.

The class of workmanship will also have considerable bearing on the labor costs, as in ordinary structures the casings are nailed to the frames or jamb linings, and the trim may show some hammer marks, but where first grade workmanship is required, the utmost care must be used in cutting and fitting all casings, miters, etc. All nails must be carefully driven and set, casings must show the same margin on all sides and must fit closely to the walls. In other words, it must be a first class job in every respect.

Labor Time Erecting Interior Window Trim

Style & Kind of Trim	No. Windows 8-Hr. Day	Carp. Hrs. per Window	Add for Wood Jamb Lining
"A" Single Casing	78	1.0-1.1	0.6
"B" Cap Trim*	57	1.1-1.6	0.6
"C" Back-band Trim	56	1.3-1.6	0.6
"D" Mill Assembled**	8-10	0.8-1.0	0.6
First Grade Workmanship			
"A" Single Casing	5-7	1.1-1.6	0.8
"B" Cap Trim*	4-6	1.3-2.0*	0.8
"C" Back-band Trim	4-5	1.6-2.0	0.8
"D" Mill Assembled**	7-9	0.9-1.1	0.8

*For cap trim put together on job.
**Trim assembled and glued up in the mill.

On complicated window trim with recessed jamb linings and on trim that has both a casing and a subcasing and a subcasing or panel under the stools, a carpenter will complete only about 1 window per 8-hr. day. This class of work is exceptional and is found only in high grade work.

Labor Setting Window Frames, Fitting and Hanging Wood Sash, And Trimming Interior Window Openings Complete

The following table includes all labor required to set window frame, to fit and hang wood sash, and to place jamb linings, stops, window stool, apron, interior casings, and hardware for one complete single window opening.

Trim Style & Kind	Windows per 8-Hr Day	Hrs. per Window, Carpenter	Labor	Add for Wood Jamb Lining
"A" Single Casing	2-1/2-3	3.0	0.3	0.6
"B" Cap Trim	2-1/2-3	3.0	0.3	0.6
"B" Cap Trim*	2*	3.5	0.3	0.6
"C" Back-band Trim	2*	3.5	0.3	0.6
"D" Mill Assembled**	2-1/2-3	3.0	0.3	0.6
First Grade Workmanship				
"A" Single Casing	2	3.5	0.3	0.8
"B" Cap Trim	2	3.5	0.3	0.8
"B" Cap Trim*	2	4.1	0.3	0.8
"C" Back-band Trim	2	4.4	0.3	0.8
"D" Mill Assembled**	2-1/4-2-1/2	3.3	0.3	0.8

*For cap trim put together on job.
**Trim assembled and glued up in the mill.

08700 HARDWARE AND SPECIALTIES

08710 FINISH HARDWARE

Finish hardware refers to any item that is fitted to a door and frame to perform a specific function—hinges, latch or locksets, closers, stops, and bolts.

Most items of finish hardware can be obtained in various finishes, and each finish has a U.S. Code Symbol Designation as follows:

US P - Prime Paint Coat
US 3 - Polished Brass
US 4 - Satin (dull) Brass
US 9 - Polished Bronze
US 10 - Satin (dull) Bronze
US 10B - Satin Bronze - Oil Rubbed
US 14 - Polished Nickel
US 15 - Satin (dull) Nickel
US 20 - Statuary Bronze
US 26 - Polished Chrome
US 26D - Satin (dull) Chrome
US 28 - Satin Aluminum - Anodized
US 32 - Polished Stainless Steel
US 32A - Satin (dull) Stainless Steel

Keep in mind that the finish will affect two areas of concern for the estimator: cost and delivery time. The US 28 finish on finish hardware is one of the cheapest to buy and readily available from most suppliers, compared with US 32, for example, which will be more expensive and a special order from the manufacturer.

Another item that will affect the cost and the delivery time will be the knob and rose trim-ring specified for latch and locksets. There are numerous styles. However, most suppliers only stock the most popular 3 or 4 patterns.

The specification writer will refer to a specific Federal Specification Series Designation when specifying the type of lockset required for the project, and each of these series relates to the construction and design function of the lockset. For example:

ANSI* Series 1000 - Mortise type latchset or lockset and keyed deadbolt housed in one casing.
ANSI Series 2000 - Mortise type latchset or lockset with integral deadbolt operated by the key in the knob.
ANSI Series 4000 - Cylindrical latch and locksets:
  Grade #1 - Heavy Duty
  Grade #2 - Medium Duty
  Grade #3 - Light Duty
*American National Standards Institute

The following list of finish hardware material prices should be used as a guide only. Always obtain a price quote from a material supplier for the specific type, function, finish, and style of finish hardware specified for the project.

Material Prices For Finish Hardware

	price range
Cylindrical Locksets	
Std. Duty	$20.00-70.00
Heavy Duty	60.00-100.00
Mortise Locksets	
Heavy Duty	60.00-110.00
Push-Pull Bars	50.00-150.00
Anti-Panic Device, Single	175.00-400.00
Hinges (per pair)	
Steel Plain	10.00-20.00
Steel B.B.	25.00-70.00
Bronze B.B.	75.00-150.00
Door Closers	
III	45.00-125.00
IV	55.00-100.00
V	65.00-115.00
Surface Bolt	6.00-25.00
Kick Plate	
Alum	5.00-25.00
Bronze	15.00-50.00
Door Stops:	
Floor Bumper	5.00-10.00
Wall Bumper	10.00-15.00
Bumper Plus Holder	16.00-25.00
Door Plunger	15.00-25.00
Thresholds	
Alum.	10.00-25.00
Bronze	25.00-100.00

The following table gives average production time, in carpenter hours, for the installation of finish hardware items:

Items of Finish Hardware	Hrs. to Install
Cylindrical Locksets (door prepared)	0.5
Mortise Locksets (door prepared)	0.7
Push-Pull Bars	0.4
Anti-Panic Device	1.5
Door Closers	1.0
Surface Bolts	0.3
Flush Bolts	1.0
Door Stops - Wall Type	0.3
Floor Type	0.5
Thresholds	0.5

08730 WEATHERSTRIPPING AND SEALS

The cost of metal weather strips varies with the type of window, whether double hung or casement, size of opening, thickness of sash, and the like. These factors all affect the material and labor costs. It will make considerable difference in the labor costs whether the weather strips are installed in an occupied or unoccupied building. Ordinarily a worker must use more care when working in an occupied building, and it requires 1/4 to 1/2 hour more time per opening than when working in new or unoccupied buildings.

The following quantities and costs are based on experienced workers placing weather strips.

Metal Weather Strips for Double Hung Windows. When weatherstripping double hung windows, standard specifications require a track strip at least 3/4" (18.7 mm) wide at the top and bottom of the window, side strips for the lower sash not less than 1/16" (1.5 mm) wider than the width of the sash runway, and side strips for the upper sash not less than 1/16" (1.5 mm) narrower than the width of the sash runway. Meeting rails to receive two interlocking hook strips or interlocking hook and flat strips.

When estimating the quantities of weather stripping required, take the distance around the window, for the sides, head, and sill strips, plus the width, to take care of the meeting rail between the top and bottom sash. A 4'-0"x6'-0" (1200 x 1800 mm) double hung window will require 20 l.f. (6.0 m) of weather strips for the sides, head, and sill and 4 l.f. (1.2 m) of interlocking material for the meeting rails.

The lin. ft. (meter) cost of metal weather strip for double hung windows averages as follows in zinc:

		Sash Thickness	
Description	1-3/8"	1-3/4"	2-1/4"
Side, head and sill strips	$0.30	$0.35	$0.45
Meeting rail strips	0.70	0.75	1.15
(metric)			
	34.3 mm	43.7 mm	56.2 mm
Side, head and sill strips	$4.98	$1.10	$1.50
Meeting rail strips	2.30	2.40	3.75

The weather strip for a 4'-0"x6'-0"x1-3/8" (1200 x 1800 x 34.3 mm) double hung window would cost 20 x 0.30 plus 4 x 0.70 = $8.80. Bronze will run about twice as much.

Labor Installing Metal Weather Strips for Double Hung Windows. When installing metal weather strips on 1-3/8" (34.3 mm) sash for windows up to 3'-0"x5'-6" (900 x 1650 mm), an experienced carpenter should weatherstrip 1 window in 3/4 to 7/8 hr. or 9 to 10 windows per 8-hr. day, at the following labor cost per window:

	Hours	Rate	Total	Rate	Total
Carpenter	0.9	$....	$....	$26.44	$23.80

When installing metal weather strips on 1-3/4" (43.7 mm) sash for windows up to 4'-0"x7'-0" (1200 x 2100 mm), an experienced carpenter should complete one window in 7/8 to 1 hr. or 8 to 9 windows per 8-hr. day, at the following labor cost per window:

	Hours	Rate	Total	Rate	Total
Carpenter	1.0	$....	$....	$26.44	$26.44

If installed in an occupied building, add 1/4 hr. per window.

Installing Metal Weather Strips for Double Hung Windows Glazed with Plate Glass. When installing metal weather strips on 1-3/4" (43.7 mm) or 2-1/4" (56.2 mm) sash, glazed with plate glass, for openings up to 4'-0"x7'-0" (1200 x 2100 mm), a carpenter should weatherstrip one window in 1-1/8 to 1-1/4 hrs. or 6 to 7 windows per 8-hr. day, at the following labor cost per window:

	Hours	Rate	Total	Rate	Total
Carpenter	1.25	$....	$....	$26.44	$33.05

If installed in occupied buildings, add 1/4 to 1/2 hr. per window.

Weatherstripping In-Swinging Casement Windows. Single or double casement windows that swing in require a 3/4" (18.7 mm) wide track strip on the hinged side, running in a groove in the sash. At the top, on the latch side, and where the sash meet (in the case of double casement windows) interlocking hook and flat strips are generally used. A trough strip and a hook strip are generally used at the bottom of the sash.

Material for jambs, head, and center stiles for in-swinging casement windows costs $0.40 per l.f. ($1.30 per m). Brass trough section and interlocking sillstrip, $1.20 per l.f. ($3.93 per m).

When estimating the number of lin. ft. of strip required for single casement windows, measure the entire distance around the window, but price the sill section separately.

For double casement windows, take the distance around the window, plus the height, which allows for the strip where the two sash meet. A 4'-0"x6'-0" (1200 x 1800 mm) double casement window requires 22 l.f. (6.7 m) of weather strips for jambs, head, and center stiles and 4'-0" (1200 mm) of sill section.

Labor Weatherstripping In-Swinging Single Casement Windows. An experienced carpenter should weatherstrip one single casement window in 1 to 1-1/3 hrs. or 6 to 8 windows per 8-hr. day, depending on size, at the following labor cost per window:

	Hours	Rate	Total	Rate	Total
Carpenter	1.1	$....	$....	$26.44	$29.08

If installed in an occupied building, add 1/4 hr. per window.

Weatherstripping In-Swinging Double Casement Windows. When weatherstripping in-swinging double casement windows, an experienced carpenter should weatherstrip one window in 1-3/4 to 2 hrs. or 4 to 5 windows per 8-hr. day, depending on the size, at the following labor cost per window:

	Hours	Rate	Total	Rate	Total
Carpenter	2.0	$....	$....	$26.44	$52.88

If installed in an occupied building, add 1/4 hr. per window.

Labor Weatherstripping Out-Swinging Casement Windows. Out-swinging single casement windows should have -3/4" (18.7 mm) wide track strip on the hinged side, running in a groove in the sash. At the top, on the latch side, and where the sash meet in the case of double casement sash, interlocking hook and flat strips are required. A heavy sill section with interlocking hook strip is generally used at the bottom of the sash. The material for jambs, heads, and center stiles will cost about $0.40 per l.f. ($1.30 per m), while the sill section will cost about $1.00 per l.f. ($3.30 per m).

Labor Weatherstripping Out-Swinging Single Casement Sash. An experienced carpenter should place weather strips on one single casement sash in 3/4 to 1 hr. or 8 to 10 windows per 8-hr. day, at the following labor cost per window:

	Hours	Rate	Total	Rate	Total
Carpenter	1.0	$....	$....	$26.44	$26.44

If installed in an occupied building, add 1/4 hr. per window.

Labor Weatherstripping Out-Swinging Double Casement Sash. An experienced carpenter should weatherstrip one double casement window in 1-1/4 to 1-1/2 hrs. or 5 to 6 windows per 8-hr. day, at the following labor cost per window:

	Hours	Rate	Total	Rate	Total
Carpenter	1.4	$....	$....	$26.44	$37.02

If installed in an occupied building, add 1/4 hr. per window.

Weather Strips for Single Wood Doors. Consisting of spring bronze strips for jambs and heads and 1-1/2" (37.5 mm) interlocking brass threshold for sill.

DOORS AND WINDOWS 8.81

Door Size Up To	Thick	Price
2'-8"x6'-8"	1-3/8"	$11.20
3'-0"x7'-0"	1-3/8"	11.90
2'-8"x6'-8"	1-3/4"	12.00
3'-0"x7'-0"	1-3/4"	12.80

If aluminum interlocking thresholds are used instead of brass, deduct $1.75 for 2'-8" (800 mm) openings and $2.00 for 3'-0"x7'-0" (900 x 2100 mm) openings.

If interlocking bronze is used instead of spring bronze, add $7.00 per door to the above prices.

Extruded aluminum and felt or neoprene sponge 1-1/8" (28.1 mm) wide and 1/4" (6.2 mm) thick to be set on stops at heads and jambs runs $1.10 per l.f. ($3.60 per m). Extruded aluminum with neoprene gasket 1-1/4" (31.2 mm) wide by 25/32" (19.5 mm) thick to be set on stops runs $2.20 per l.f. ($7.20 per m). A similar gasket for mounting on bottom rail of door costs $1.20 per l.f. ($3.90 per m).

A complete extruded aluminum channel with integral drip and vinyl insert at still costs $1.00 per l.f. ($3.20 per m). A 6" (12.5 mm) wide extruded aluminum thresholds for use with above rail strips costs $2.25 per l.f. ($7.40 per m). A complete extruded aluminum threshold 3-3/4" (93.7 mm) wide by 3/4" (18.7 mm) thick with integral latch track, interlock and hook stop for mounting on outside of door costs $3.00 per l.f. ($9.84 per m) complete. A two-piece extruded aluminum with neoprene gasket astragal for surface mounting on double doors runs $2.60 per l.f. ($8.50 per m).

A one-piece extruded aluminum and neoprene gasket designed for recessing in meeting rail will cost $1.40 per l.f. ($4.60 per m). Adjustable two-piece astragals with pile will run $3.50 per l.f. ($11.50 per m) per each door leaf in extruded aluminum, $4.00 ($13.00) in dull bronze, and $4.80 (15.75) in polished bronze or dull chrome.

Labor Weatherstripping Single Wood Doors. An experienced carpenter should weatherstrip one 3'-0"x7'-0"x1-3/4" (900 x 2100 x 44 mm) door in 1 hour or about 8 doors per 8-hr. day, at the following labor cost per door:

	Hours	Rate	Total	Rate	Total
Carpenter	1.0	$....	$....	$26.44	$26.44

Where interlocking weather strip is used for head and jambs of doors instead of spring bronze, add 1/2 to 3/4 hr. carpenter time to the time given above.

Metal Weatherstripping for French Doors. The material for weatherstripping double doors or French doors consists of spring bronze strips for jambs and head, spring bronze interlocking strip, and threshold for sill. For doors up to 5'-0"x7'-0"x1-3/8" (1500 x 2100 x 34 mm), the material should cost about $13.00; for a 5'-0"x7'-0"-1-3/4 (1500 x 2100 x 44 mm) door, about $14.00.

Labor Weatherstripping French Doors. An experienced carpenter should weatherstrip one pair of 1-3/4" (44 mm) French doors in 2 hrs. or 3 to 4 pair per 8-hr. day, at the following labor cost per pair:

	Hours	Rate	Total	Rate	Total
Carpenter	2.0	$....	$....	$26.44	$52.88

Weatherstripping Double Doors. Where double exterior doors are used for openings up to 5'-0"x7'-0" (1500 x 2100 mm), using spring bronze weather strip for the head, jambs, and between doors, with an interlocking threshold at the bottom of the door, figure 3 to 4 hrs. carpenter time per opening, as follows:

	Hours	Rate	Total	Rate	Total
Carpenter	2.0	$....	$....	$26.44	$52.88

If interlocking weather strip is used for the head, jambs, and meeting rail, add 1 hour per opening.

Metal Thresholds

Price per l.f. (m) or fraction thereof. Drilled and countersunk, with screws, ready for installation.

Metal	Size, inches	Size, mm	Price per l.f	Price per m
Brass	1-1/2"x 1/4"	37.5 x 6.3 mm	$5.00	$16.40
Brass	3-1/2"x 3/4"	87.5 x 18.8 mm	9.00	29.53
Brass	4-1/2"x 3/4"	112.5 x 18.8 mm	12.00	39.37
Brass	5"x 7/8"	125 x 21.9 mm	13.50	44.29
Aluminum	1-1/2"x 1/4"	37.5 x 6.3 mm	$1.50	$4.92
Aluminum	3-1/2"x 3/4"	87.5 x 6.3 mm	2.80	9.19
Aluminum	4-1/2"x 3/4"	112.5 x 6.3 mm	3.90	12.80
Aluminum	5"x 7/8"	112.5 x 6.3 mm	5.50	18.04

All the above thresholds are furnished with interlocking strip.

08800 GLAZING

During the late 1970's, the glass manufacturers changed the process for making certain types of flat glass. Before that time, vertically drawn sheet glass was manufactured in various qualities—AA, A, and B—and thicknesses—S.S., D.S., 3/16" (4.68 mm) and 7/32"(5.4 mm). The 3/16" (4.68 mm) and 7/32" (5.4 mm) glass was known as crystal sheet. Plate glass was made by grinding and polishing both surfaces to a level, parallel plane, thereby removing any distortion lines from the glass.

The new process of making glass consists of floating molten glass over large pools of molten tin. Through automated production stages, the floated glass is fire polished to remove all distortion, allowed to cool, and fed onto the tables in a continuous ribbon where it is cut to sheet size for packing into shipping crates. The glass is classified as "float glass" and is made in glazing quality and mirror quality of various thicknesses.

Clear Float Glass - Glazing Quality

Product	Thickness	Wt. in lb. per Sq. Ft.	Standard Max.Size	Material Cost per Sq. Ft.
Float	3/32"	1.22	40"x100"	$1.25
	1/8"	1.62	80"x120"	1.50
	3/16"	2.43	120"x212"	1.75*
Float/Plate	1/4"	3.24	130"x212"	2.00*
	3/8"	4.92	124"x204"	2.75*
	1/2"	6.56	124"x204"	4.00
	3/4"	9.85	124"x204"	6.25

*Note: Add $0.75 per s.f. for gray or bronze tint.

Float glass is also available in glare and heat reducing tints of gray and bronze, heat absorbing (blue-green), tempered, coated with thin metallic coverings for reflective qualities, laminated, and in insulating glass units consisting of two sheets of glass separated by an internal air space.

Tempered Glass. Tempered glass is a heat treated annealed glass that has high mechanical strength. It is 4 to 5 times as strong as annealed glass of the same thickness. When it is broken, it disrupts into innumerable small fragments of a cubical shape. In recent years, tempered glass has become much more reasonable in cost and should be given more consideration in general construction. However, the nature of its manufacture does not allow cutting or drilling on the job and all fabricating must be done prior to heat treatment. It is available in clear, tinted, and heat absorbing. One-quarter inch (6.25 mm) thick units are available in sizes to 72"x120" (1800 x 3000 mm). Because this glass is all custom ordered, costs should be checked locally, but tempering will add around $2.00 a s.f. ($21.50 per sq.m) premium to base glass prices.

Laminated Glass. Laminated or safety glass is composed of two or more lights of glass with a layer or layers of tough, transparent vinyl plastic sandwiched between glass under heat and pressure to form a single unit. Laminated glass will crack but almost always will hold together. It is available in a wide range of thicknesses and sizes from 7/32" (5.46 mm) thick to 1" (25 mm) thick in sheets up to 80"x120" (2000 x 3000 mm). One-quarter inch (6.25 mm) laminated glass will cost about $4.25 per s.f. ($45.75 per sq.m), 1/2" (12.5 mm) will cost about $10.00 per s.f. ($107.60 per sq.m).

Insulating Glass. Glass units made from two sheets of glass with a sealed air space between are used to decrease heat loss through windows. The seal can be either metal or glass, and air space will vary from 3/16" (4.68 mm) to 1/2" (12.5 mm) in thickness.

Insulating glass units must be made to order in the exact size to be used, although standard sizes are manufactured.

They require a slightly larger allowance for clearance in setting them into the rabbet than single lights of glass.

Each piece weighs about 2-1/2 times what a single lite of the same thickness would weigh and the glazing cost is about double that of single lights.

The price per square foot varies widely, depending upon kind of glass, size, etc., so it is always advisable to obtain definite prices on the sizes required.

For quick estimates two sheets of 1/8" (3.12 mm) for 1/2" (12.5 mm) thick unit will cost $3.50 per s.f. ($37.70 per sq.m).

Two sheets of 1/4" (6.25 mm) for 1" (25 mm) thick unit will cost about $4.50 per s.f. ($48.40 per sq.m).

For tinted glass one side, add 15%.

Spandrel Glass. Heat-strengthened glass is used for spandrels and other opaque panels of curtain walls. The glass is coated on the back with a ceramic frit to give opacity and color.

It is manufactured in several standard colors, plus black and white, but can be ordered in almost any color desired.

Spandrel glass is factory cut to size. Job cutting is not possible due to the heat-strengthening process in manufacturing. It is usually 1/4" (6.25 mm) in thickness but may be ordered up to 3/4" (18.75 mm) thick. Sealing of spandrel glass is usually accomplished by the use of liquid polysulfide sealants and neoprene gaskets. Material prices of 1/4" (6.25 mm) spandrel glass are about $4.00 per s.f. ($43.00 per sq.m) for standard colors, and about $4.50 per s.f. (48.40 per sq.m) for non-standard colors.

Wired Glass. Wire glass is a protective type glass for use in fire doors and windows, skylights and other places where breakage is a problem. It is available in 1/4" (6.25 mm) thicknesses in sheets 60"x144" (1500 x 3600 mm) maximum size in hexagonal, square, diamond, and pin stripe patterns, and comes in clear, hammered, and many other patterned glass. Patterned wire glass will cost around $3.75 per s.f. ($40.30 per sq.m). Clear wire glass will cost $4.50 per s.f. ($48.40 per sq.m) in diamond pattern, $6.75 ($72.63 per sq.m) in pin stripe.

Patterned Glass. Glass can be rolled or figured for decoration and light diffusion. Patterns vary among manufacturers but usually include sandblasted, hammered, ribbed, fluted, pebbled, and rough hammered designs. It is usually available in two thicknesses: 1/8" (3.1 mm) in 48"x132" (1200 x 3300 mm) maximum size and 7/32" (5.5 mm) in 60"x132" (1500 x 3300 mm) maximum size.

Some patterns are illustrated on accompanying pages. Cost of material is about $1.65 per s.f. ($17.25 per sq.m) for 1/8" (3.1 mm) thicknesses, $1.50 ($16.10 per sq.m) for 7/32" (5.4 mm).

Mirrors. Mirrors are manufactured by taking a sheet of "mirror quality" float glass, in 1/8" (3.1 mm) or 1/4" (6.2 mm) thickness, and hermetically sealing a silver coating with a film of electrolytic copper plating onto one side of the glass. The coating is then given a protective paint coat to seal out moisture and reduce the potential of damaging the silver coating.

Average material prices for mirrors will be about $3.50 per s.f. ($37.60 per sq.m).

Any edge work associated with mirrors will cost extra: plain polished edges at $0.25 per inch ($0.01 per mm), bevelled edges at $0.75 per inch ($0.03 per mm).

Estimating The Quantity Of Glass Required. When measuring and listing the quantities of glass for a project, list the glass by:

1. Type of glass and thickness
2. Size, measuring to the next even inch dimension, listing the width first and then the height
3. Type of frame the glass is placed in
4. Note if required to glaze from the inside or outside of the building. If from the outside, ladders or scaffolding may be necessary.
5. Height from the ground, in multi-story buildings, for additional labor requirements of distributing the materials.
6. Be sure not to list any glass for products that will be delivered to the job pre-glazed. It is common practice for sliding door units, aluminum sliding and single or double hung sash, and some wood sash to be glazed at the factory.

Labor Glazing Window Glass. The labor cost of glazing will vary with the size and kind of glass, and whether set in putty or glazing compound, a Thiokol or Silicone sealant, or with wood or metal stops. Putty is now seldom recommended for glazing except on a very small light under 50 united inches (1250 mm), which is length plus width.

Labor Setting Window or Plate Glass in Wood Sash, Using Putty or Glazing Compound

Approximate Size of Glass	Number Lights Set 8-Hr. Day	Glazier Hours per 100 Lights
12"x14" (300 x 350 mm)	65	12.5
20"x28" (500 x 700 mm)	40	20.0
30"x40" (750 x 1000 mm)	30	26.7
40"x48" (1000 x 1200 mm)	20	40.0

Labor Setting Window or Plate Glass Using Wood Stops

Approximate Size of Glass	Number Lights Set 8-Hr. Day	Glazier Hours per 100 Lights
12"x14" (300 x 350 mm)	40	20.0
20"x28" (500 x 700 mm)	20	40.0
30"x40" (750 x 1000 mm)	15	53.3
40"x48" (1000 x 1200 mm)	12	66.7

Putty Required Setting Glass in Wood Sash. Window glass set in 1-3/8" (34.3 mm) wood sash requires 1 lb. (0.45 kg) of putty for each 8 to 8-1/2 l.f. (2.4-2.6 m) of glass or sash rabbet. When set in 1-3/4" (43.7 mm) sash, it requires 1 lb. (0.45 kg) of putty to each 7-1/4 to 7-1/2 l.f. (2.2-2.3 m) of glass or sash rabbet.

The following table gives the quantity of putty in lbs. (kg) required for glazing various size glass:

Sash Thick	Size of Glass Inches				
	12"x14"	14"x20"	20"x28"	30"x36"	40"x48"
1-3/8"	1/2	2/3	1	1-3/8	1-3/4
1-3/4"	3/5	3/4	1-1/8	1-1/2	2

(metric)

Thick	300x350mm	350x500mm	500x700mm	750x900mm	1000x1200mm
34.3mm	0.22	0.30	0.45	0.62	0.79
43.7mm	0.27	0.34	0.51	0.68	0.91

A good grade of wood sash putty will cost about $0.50 per lb. ($1.10 per kg).

Labor Setting Glass in Metal Doors. When glazing metal doors and enclosures, the stops are usually steel and are fastened to the sash with small screws, and a glazier must remove all the screws from the stops, place the glass, and then replace the screws for each small light of glass. On work of this kind a glazier should set about 15 lights per 8-hr. day, at the following labor cost:

	Hours	Rate	Total	Rate	Total
Glazier	8.0	$....	$....	$25.94	$207.52
Cost per light			....		13.83

Labor Glazing Steel Sash. The labor cost of glazing steel sash will vary with the size of the glass, amount of putty required per light, and the class of work; i.e., whether sidewall sash, monitor sash, etc., and also whether it is necessary to hoist the glass and erect special scaffolding to set it.

The time of year in which the work is performed will also influence the labor costs, as a glazier can set more glass in warm weather than during the winter months. The labor costs given below are based on summer glazing. For winter work add up to 25% to costs given.

Putty Required for Glazing Steel Sash. Putty used for glazing steel sash is a special mixture, part of which is red or white lead. One lb. (0.45 kg) should be sufficient to glaze 5 to 5-1/2 l.f. (1.5-1.6 m) of sash rabbet. For instance, a 16"x20" (400 x 500 mm) light would contain 72" (1800 mm) or 6'-0" (1.8 m) to be puttied.

Elastic glazing compound is displacing the regular steel sash putty to a great extent for use in steel sash and must be used in aluminum sash. This compound costs about $0.60 per lb. ($1.34 per kg). Sufficient putty to glaze 100 l.f. (30.5 m) of steel sash should cost as follows:

	Lbs.	Rate	Total	Rate	Total
Putty	20	$....	$....	$0.50	$10.00
Cost per l.f.			....		0.10
per m			....		0.33

	Lbs.	Rate	Total	Rate	Total
Elastic compound	20	$....	$....	$0.60	$12.00
Cost per l.f.			....		0.12
per m			....		0.39

Labor Glazing Steel Sash. On jobs where there are no unusual window heights or other abnormal conditions, a glazier should set the following number of lights of glass in the various types of steel sash per 8-hr. day.

In clerestory, monitor, or outside setting (from scaffold or swing stage), it will require one worker on the ground keeping two workers on a scaffold supplied with glass and putty. This adds approximately 50% to the costs given below. Add up to 25% to the costs for winter work.

Pivoted Steel or Commercial Projected Steel Windows*

Glass Size, in.	Glass Size, mm	No. of Lights per 8-Hr. Day	Glazier Hrs. per 100 Lts.	Lbs. Putty per Light	Kg Putty per Light
22"x16"	550x400mm	50	26.7	1.25	0.57
32"x16"	800x400mm	40	26.7	1.50	0.68
32"x22"	800x550mm	30	32.0	1.67	0.76

Housing and Residence Steel Casements**

9"x12"	225x300mm	60	20.0	0.75	0.34
16"x12"	400x300mm	60	23.0	1.0	0.45
18"x12"	450x300mm	60	23.0	1.0	0.45

Architectural Projected Steel Windows**

18"x16"	450x400mm	30	23.0	1.25	0.57
20"x16"	500x400mm	30	23.0	1.25	0.57
38"x16"	950x400mm	25	40.0	1.75	0.79
40"x16"	1000x400mm	25	50.0	2.00	0.91
46"x16"	1150x400mm	25	50.0	2.00	0.91
48"x16"	1200x400mm	25	50.0	2.25	1.02

*Based on setting glass from inside and working from floor.
**Based on setting glass from outside, using glazing compound or steel sash putty.

Labor Setting Plate Glass in Store Fronts. The labor cost of handling and setting plate glass in store fronts varies with factors such as the size of the glass and the length of the haul.

The average size plate glass used in modern store fronts requires 4 to 6 workers to handle and set each light of glass. No matter how many lights of glass there are in the job, it will be necessary to allow time for all workers going to the job and returning to the warehouse after completion, and time for the truck hauling glass from the warehouse to the job and return.

Unless the job is a considerable distance from the warehouse, an allowance of 1 hour for each worker going to the job and returning to the warehouse should prove sufficient.

It is customary for the truck to remain on the job until the glass is set and return to the warehouse with the workers, so truck time should be figured on this basis.

Number of Workers Required for Various Sizes of Plate Glass in Store Fronts

From 100 to 140 united inches (2500-3500 united mm) 2
From 140 to 160 united inches (3500-4000 united mm) 3
From 160 to 180 united inches (4000-4500 united mm) 4
Not over 210 united inches (5250 united mm) 5
Not over 240 united inches (6000 united mm) 6

On sizes larger than the above, some thought should be given to a mechanical means of hoisting the glass, in addition to the labor requirement.

It is further required that when glass is set 6' (1.82 m) or more over street level, when weather conditions require it, or when insulated glass, such as

Thermopane or Twindow, is installed, additional labor shall be used on the following basis:

First size enumerated in the schedule, add. workers.................................... 1
Next two sizes enumerated in the schedule, add. workers............................ 2
Next two sizes enumerated in the schedule, add. workers............................ 3

LEADED AND ART GLASS

When estimating leaded or art glass, odd or fractional parts of inches are charged as even inches of the next larger size; for example, a 12-1/4"x13-1/4" (306 x 331 mm) lite will be charged as 14"x14" (350 x 350 mm).

Circles, ovals, or irregular patterns will be charged at the rate of a square or rectangle which will contain them.

The following prices are based on leading, using zinc cames not over 3/8" (9.3 mm) wide:

	Price per s.f.	Price per sq.m
Rectangles and squares, Clear sheet glass	$5.75	$61.90
Rectangles and squares, Cathedral glass, white or colored	6.00-25.00	64.60-269.10
Rectangles or squares, Plate glass	8.00	86.10
Diamonds, Clear sheet glass	6.00	64.60
Diamonds, Cathedral glass, white or colored	7.00-35.00	75.30-376.70
Diamonds, Plate glass	10.00	107.60

Colored Art Glass Leaded. Leaded art glass furnished in colors are offered in many designs at $6.00 to $25.00 per s.f. ($64.50-269.10 per sq.m), which includes only straight work.

Landscape and water scenes in colored art glass will cost from $20.00 to $40.00 per s.f. ($215.20-430.60 per sq.m).

Very elaborate designs of art glass in colors with flowers, fruits, etc., can be purchased from $50.00 up per s.f. ($538.21 per sq.m).

The architect usually specifies an allowance of so much per sq. ft. (sq.m) to cover the cost of all art glass in the building.

08900 WINDOW WALLS/CURTAIN WALLS

INSULATED METAL WALL PANELS

Insulated steel, aluminum, and aluminum-steel wall panels were developed to provide the answer to dry wall construction, enabling the builder to enclose large areas in a minimum of time. They are designed to serve as exterior walls and partitions.

One typical wall panel consists of two members pressed together at the side laps to form a structural unit. This type has a strip of felt inserted the full length of the joint to prevent metal-to-metal contact between the two members. These panels are factory filled with borosilicate glass fiber insulation, 2-1/2-lb. (1.13-kg) density. An end closure is attached to both ends of the panel to insure the proper retention of insulation and to seal the panels. Panels are manufactured in standard widths of 12" (300 mm), 16" (400 mm), 24" (600 mm), or 32" (800 mm) and thicknesses vary from 1-9/16" (39 mm) to 3-3/8" (84.3 mm), depending on whether surfaces are flat or fluted and thickness of insulation used.

Lengths vary from 6'-0" (1.8 m) to 30'-0" (9.1 m). Maximum allowable clear span for an all-galvanized steel panel, based on a wind load of 20 lbs. per s.f. (97.64 kg per sq.m), is 14'-0" (4.3 m); for an all aluminum panel, 10'-6" (3.2 m); and for an aluminum-galvanized steel panel, 12'-0" (3.65 m).

The interlocking tongue and groove between adjacent panels offers a joint that unites the series of units into a continuous wall surface. When using the interlocking tongue and groove connection the panels can be erected either vertically or horizontally in walls and partitions. This double-acting tongue and groove joint has three positive bearing surfaces that permit a series of panels to act as a structural unit in resisting loads from either side of a wall. With panels in the horizontal position, the double tongue and groove joint forms a shiplap construction, adding to the weather-resisting properties of the complete wall unit. Where panels are used as semi-permanent constructions enclosing the face of a building to which a future extension is contemplated, details can allow for 100% salvaging of the panels.

Steel Wall Panels. Steel type panels are manufactured of 18 to 26 gauge sheet steel, with either a baked-on coat of shop primer or galvanized finish. The all-steel panel has the advantage of high structural strength and low initial cost. The panels are caulked during erection by buttering a mastic in the female joints prior to putting the panels in place. Welding of panels to supporting steel is the most common method of attachment. The material cost of shop assembled steel wall panels is approximately $2.50 per s.f. ($26.90 per sq.m). For two sheets of 26 ga. metal and 1" (25 mm) insulation to $2.75 per s.f. ($29.60 per sq.m) for two sheets of 22 ga. If the outer sheet is to have a porcelain enamel finish, add $1.25 per s.f. ($13.40 per sq.m).

Aluminum Wall Panels. The extreme light weight of the long length aluminum panels allows easier and faster handling and erecting in the field. A typical aluminum panel, 12'-0" (3.6 m) long, 24" (600 mm) wide, weighs only 60 lbs. (27.2 kg). The erection procedure is the same as for steel panels except that direct contact between the aluminum and steel must be prevented. A coating of bituminous paint is recommended for this purpose. Maintenance costs become negligible with an all-aluminum construction. The original mill finish will eventually darken with weathering. Bolting of panels to supporting steel and use of pre-formed sections are the most common methods of attachments. Aluminum wall panels cost about $3.50 per s.f. ($37.60 per sq.m).

Acoustically Treated Wall Panel

Aluminum-Steel Wall Panels. To supply the need for an economical panel that would have an aluminum exterior surface for high resistance to corrosion and weather but did not require this type surface on the interior, the aluminum-steel panel was developed.

For particular use in conjunction with aluminum-steel panels and all aluminum panels, extrusions have been developed which enable the panel to be placed in position and held through the clamping action of the extrusion at the base of the panel. These extrusions eliminate bolting and accelerate erection. They are also 100% salvageable when placed in buildings that are to be expanded later. At the head of the panels, there is usually no extrusion. Rather, a toggle bolt is used to fasten at this position. The material cost of aluminum-steel wall panels is about $3.50 per s.f. ($37.66 per sq.m), depending on the gauge of metals used.

Labor Erecting Metal Wall Panels. A normal wall panel job with panel lengths ranging between 6' (1.8 m) and 8' (2.4 m) should be constructed at the rate of 12 manhours per 100 s.f. (9.3 sq.m), which includes handling all work from unloading to actual fastening of the panels in place. A difficult job, involving the use of toggle bolts, aluminum panels, short runs, either long or short panels, will take somewhat longer—up to 20 manhours per 100 s.f. (9.3 sq.m). A typical crew would consist of one worker to caulk and hoist, a worker to receive and distribute the panel, a worker to place and weld the bottom, one to plumb and weld the top, and a worker to do the cutting, attach closures, and place flashings and bitumastic. The erection cost for a normal job is about $1.20 per s.f. ($12.90 per sq.m) and $2.25 per s.f. ($24.20 per sq.m) for a difficult job. On large jobs with no openings, an erection cost of

$0.75 per s.f. ($8.10 per sq.m) can be achieved, with 180-190 s.f. (16.7-17.6 sq.m) erected per worker per 8-hour day.

Field Assembled Wall Panels. Steel, aluminum, and combination aluminum-steel wall panels are also available for field assembly—to be assembled and erected simultaneously. Manufactured with a deeply fluted or ribbed exterior surface, they are particularly suited to modern architectural designs.

Panel lengths generally range from 6'-0" (1.8 m) to 40'-0" (12.2 m)—some are available up to 60'-0" (18.3 m). Widths are 12" (300 mm), 18" (450 mm), and 24" (600 mm) and depth varies from 3" (75 mm) to 3-1/2" (87.5 mm).

Steel panels—prime coat, galvanized or stainless—are available in 18, 20, or 22 gauge. Aluminum panels—mill or grained finish—are available in 14, 16 , 18, or 20 B&S gauge. A combination of aluminum exterior face and galvanized steel interior face is also manufactured. Interior faces of panels may be flat or have small V-shaped ribs.

Maximum girt spacings for all-steel panels is 11'-0" (3.3 m) for simple spans and 13'-0" (3.9 m) for double spans; for all-aluminum or aluminum-steel panels, 7'-6" (2.2 m) and 8'-6" (2.6 m).

Material and erection costs for field assembled wall panels total practically the same as for shop assembled panels (given on previous pages) of similar materials.

Movable Fire Partitions. Movable fire partitions, long needed in industrial plants, are available for installation where permanent fire walls are not desired. One such partition has a 2-1/2 hour fire rating.

These lightweight partitions are formed of two 18 gauge galvanized steel, 1-1/2" (37.5 mm) deep fluted sheets of the same type as used for exterior face of field assembled wall panels. our 1/2" (12.5 mm) laminations of gypsum board, containing glass fibers and unexpanded vermiculite, are inserted between the sheets and the build-up is held together by bolts.

Depth of the fire partition is 5" (125 mm) and coverage width per section is 24" (600 mm). Length of section varies with ceiling to floor dimension.

Fire partitions are lightweight and easily erected. Portable attachment to structural steel is by bolting—permanent attachment is by welding.

Cost in place of movable fire partitions is approximately $4.00 per s.f. ($43.00 per sq.m).

Curtain Wall Construction. Buildings of all types are now using the curtain wall method of exterior wall construction, in preference to the conventional masonry, because they are lightweight and permit savings in structural framing and foundations.

Units are thin—varying from 1-1/4" (31.2 mm) to 4-1/2" (112.5 mm)—increasing ratio of usable floor area to total building area.

They provide good insulation with the use of efficient insulating filler materials and smooth reflective exterior and interior surfaces.

DOORS AND WINDOWS 8.93

The large prefabricated and shop assembled panels, containing window units, spandrels, mullions, etc., are erected rapidly, reducing labor costs and permitting earlier completion and occupancy of building.

At the present time, curtain walls must be designed and fabricated to order to fit a particular job. Each job is custom built and costs vary considerably depending on design, materials used, size of units, amount of duplication, type of windows, and secondary framing requirements.

Materials commonly used for curtain wall panels are painted, galvanized or stainless steel, aluminum and porcelain enameled steel, or aluminum. Various combinations of these materials are also used.

Various types of windows are used, including intermediate projected, casement or combination types, awning type, reversible windows, and others.

Panel sizes vary considerably, but most economical designs for multiple story structures have a coverage width of about 4'-0" and a story to story height of 10'-0" (3 m) to 12'-0" (3.6 m). For 2 and 3 story structures, full height panels have been used successfully.

The following average curtain wall costs are given for preliminary estimating purposes only. Always obtain firm figures for curtain walls before submitting bid, unless an allowance has been set up for this work.

Description	Cost per s.f.	Cost per sq.m
1-1/2" (37.5 mm) Steel frames and sash, insul. panels	$10.50-20.00	$113.00-215.30
Same in aluminum	13.50-25.00	145.30-269.10
Same in stainless steel	23.00-35.00	247.50-376.70
Same in bronze	30.00-50.00	322.90-538.20

The curtain wall can be priced out after four elements are agreed on: the framing system, the window system, the spandrel material, and the type of glazing.

CHAPTER 9

FINISHES

CSI DIVISION 9
09000 EXTERIOR INSULATION FINISH
SYSTEM (EIFS)
09100 LATH AND PLASTER
09250 GYPSUM DRYWALL
09300 TILE
09400 TERRAZZO
09500 ACOUSTICAL TREATMENT
09550 WOOD FLOORING
09650 RESILIENT FLOORING
09680 CARPETING
09900 PAINTING
09950 WALL COVERING

09000 EXTERIOR INSULATION AND FINISH SYSTEM (EIFS)

Exterior Insulation and Finish System is a multilayered exterior wall system that is used both for residential and commercial structures of various sizes and shapes. This system is usually referred to by it acronym, EIFS, on drawings and specifications.

Systems vary among the manufacturers but typically consist of the following components:

1. Insulation board (the thickness can vary), normally made of polystyrene or polyisocyanurate foam. Insulation material is secured to the exterior wall surface with an adhesive approved by the EIFS manufacturer.
2. A tough, durable water-resistant base coat is applied on top of the insulation.
3. The water-resistant base coat is reinforced with a fabric wire mesh of fiber glass for added strength.
4. A finish coat is applied. The finish coat can be an acrylic copolymer or other product that is colorfast and crack-resistant.

The greatest advantages of EIFS material is that it is energy-efficient and offers multifarious appilications. Most systems use insulation board with an R-value of R4 to R5 per inch. EIFS can conform to various shapes and can be made to resemble stucco, stone, or other types of expensive finish at a fraction of the cost. EIFS is produced in an almost unlimited selection of colors, finishes and textures.

For warranty purposes the installation of EIFS requires experienced installers who are approved by the manufacturer. Mixing coatings and applying them to the correct thickness is highly specialized work, and certified installers are also necessary to avoid adverse conditions—applying materials while surfaces are wet or just before rain, failing to end a day's work at a joint, not using approved applicators and materials, not caulking properly around openings, flashing and the like.

The following check list of items from a manufacturer's installer will ensure a quality product:

1. Knowledgeable and experienced installers who have been certified by the EIFS manufacturer.
2. Verification from the manufacturer that based on the plans and specifications being used its system will will provide a watertight structure.
3. Validation that all supplied components of the EIFS are approved by the manufacturer.
4. EIFS manufacturer and its approved contractor are aware of any unusual details and conditions.
5. Validation that all materials will come with proper labeling attached for ease of identification.
6. EIFS manufacturer and contractor are aware of and approve the schedule, taking into consideration weather conditions.
7. EIFS manufacturer must examine and approve substrate; if applied by others, substrate must meet the manufacturer's tolerances.
8. EIFS manufacturer must periodically inspect the work while it is being undertaken by the EIFS contractor.
9. At the conclusion of the project, manufacturer must verify that all items have been properly installed and all openings have been caulked correctly, with proper materials around windows, doors, flashing, and penetrations.

The following cost tables show material quantities and labor required for a typical *Exterior Insulation and Finish System* onto an existing substrate (e.g., plywood). The tables assume a 30,000 sq. ft., maximum two-story structure and an R-value of 4.2. For exact material quantities on a job, you would need to check with the manufacture.

Item 1
Insulation Board for 1,000 Sq. Ft

Material Costs		Unit	Rate	Unit	Total
Board (plus 10% waste)	1,100	Sq. Ft.	$0.56	Sq. Ft.	$616.00
Primer	6	Pails	98.00	Pail	588.00
Cement	3	Bags	16.80	Bag	50.40
Total Material Cost					$1,254.40
per s.f.					1.25
per sq.m					13.45

Labor Costs	Hours	Rate	Total	Rate	Total
Carpenter	16.0	$...	$...	$26.44	$423.04
per s.f.					$0.38
per sq.m					$4.09
Total Labor and Material Cost Insulation					$1,639.04
Total per sq. ft.					1.64
Total per sq. m					17.65

Item 2
Fiberglass Reinforcing Mesh

Material Costs		Unit	Rate	Unit	Total
Mesh (plus 10% waste – 150 ± sf per roll)	5	Rolls.	$40.00	Roll.	$200.00
Water resistant base coat	6	Pails	98.00	Pail	588.00
Total Material Cost					$788.00
per s.f.					0.79
per sq.m					3.66

Labor Costs	Hours	Rate	Total	Rate	Total
Carpenter	16.0	$...	$...	$26.44	$423.04
Labor	16.0	...	...	21.04	336.64
Total Labor Cost					759.68
per s.f.					$0.76
per sq.m					$ 7.42
Total Labor and Material Cost Reinforcing Mesh					$1,547.68
Total per sq. ft.					1.55
Total per sq. m					16.65

Item 3
Base and Finish Coat

Material Costs		Unit	Rate	Unit	Total
Primer	6	Pails.	$98.00	Pail.	$588.00
Cement	3	Bags	16.80	Bag	50.40
Topping Material	9	Pails	105.00	Pail	945.00
Total Material Cost					$1,583.40
per s.f.					1.58
per sq.m					17.01

Labor Costs	Hours	Rate	Total	Rate	Total
Carpenter	64.0	$...	$...	$26.44	$1,692.16
Labor	32.0	...	...	21.04	673.28
Total Labor Cost					$2,365.44
per s.f.					$2.37
per sq.m					$25.50
Total Labor and Material Cost Base and Finish Coat					$3,948.84
Total per sq. ft.					3.95
Total per sq. m					42.50

Summary

	Sq. Ft.	Sq.M
Item 1 – Insulation Board	1.64	17.65
Item 2 – Fiberglass Mesh	1.55	16.65
Item 3 – Base and Finish Coat	3.95	42.50
Total Cost per sq. ft.	7.14	
Total Cost per sq.m		76.80

Studs (wood or metal)

Insulation Material

Reinforcement Material with water resistant coat

Base Coat

Finish Coat

No scale

TYPICAL SECTION OF AN EIFS* WALL

* EXTERIOR INSULATION AND FISNISH SYSTEM

09100 LATH AND PLASTER

Plastering costs vary widely depending on the kind of materials used and the class of workmanship performed, and the estimator should consider these items carefully before pricing. There are two distinct grades of plastering, and the estimator should possess a fair knowledge of the class of work required.

The first grade of plastering, the one used on most types of buildings, has the walls rodded and angles and corners fairly straight and true. Work of this class permits variations and "waves" in the finish up to 1/8" (3 mm) to 3/16" (4.5 mm). This is the regular commercial grade of plastering and comes under the heading "Ordinary Work" below.

The highest grade of work is found in high-class residences, hotels, schools, courthouses, federal government buildings, state capitol buildings, public libraries, and the like, where all walls and ceilings, together with all interior and exterior angles, must be absolutely straight and plumb, and not permit variations or "waves" exceeding 1/32" (0.75 mm) to 1/16" (1.5 mm). This is termed "First Grade Workmanship".

Estimating Quantities of Lathing and Plastering. Lathing and plastering are estimated by the square yard (square meter), but the method of deducting openings, etc., varies with the class of work and the individual contractor.

Many plastering contractors make no deductions for door or window openings when estimating new work, while others deduct for one-half of all openings over a certain size.

Both lathing and plastering are estimated by the sq. yd. (sq.m), which is obtained by multiplying the girth of the room by the height, plus the ceiling area. For example: Obtain the number of sq. yds. (sq.m) of plastering in a room 12'-0"x18'-0" (3.6 x 5.4 m), having 9'-0" (2.7 m) ceilings. 18 + 18 + 12 + 12 = 60 l.f. of wall (5.4 + 5.4 + 3.6 + 3.6 = 18 m).

60 x 9 = 540 s.f. wall
12 x 18 = 216 s.f. ceiling
540 + 216 = 756 s.f. walls & ceiling
756 ÷ 9 = 84 sq. yds.

18 x 2.7 = 48.6 sq.m wall
3.6 x 5.4 = 19.4 sq.m ceiling
48.6 + 19.4 = 68 sq.m walls & ceiling

Openings should be measured, and no deductions should be made for openings less than 2'-0" (0.6 m) wide, and then only one-half the area of each opening should be deducted.

The complete contents of large openings should be deducted, and the contractor should figure 1'-6" (0.45 m) wide for each jamb by the height, when the return or "reveal" is less than 1'-0" (0.3 m).

For all beams or girders projecting below ceiling line, allow 1'-0" (0.3 m) in width by the total length as an extra for each internal or external angle, in addition to the actual area.

Corner Beads, Moldings, Etc. All corner beads, quirks, rule joints, moldings, casings, cornerites, screeds, and other lathing trim accessories necessary to complete a plastering job are measured by the lin. ft. (meter), from their longest extension with 1'-0" (0.3 m) extra added for each stop or miter.

Plaster Cornices. Plaster cornices should be measured by the total length of the walls. If the cornices are 1'-0" (0.3 m) or less in girth, measurement should be made by the lin. ft. (meter), stating the girth of each cornice. If the cornices are over 1'-0" (0.3 m) in girth, measurement may be made either by the sq. ft. (sq.m) of exposed face or by the lin. ft. (meter), stating the girth of each cornice.

Allow 1 l.f. (0.3 m) for each internal miter and 3 l.f. (0.9 m) for each external miter. Enriched cornices (cast work) should be measured by the lin. ft. (meter) for each enrichment.

Arches, Corbels, Brackets, Ceiling Frieze Plates, Etc. Arches, corbels, brackets, rings, center pieces, pilasters, columns, capitals, bases, rosettes, pendants, and niches are estimated by the piece. Ceiling frieze plates should be estimated by the sq. ft. (sq.m).

Columns. Plain plastered columns should be measured by the lin. ft. (meter) or by the sq. yd. (sq.m), multiplying the girth by the height.

Cement Wainscot. Cement wainscot should be measured by the sq. ft. or sq. yd. (sq.m).

Circular or Elliptical Work. Circular or elliptical work on a small radius should be estimated at double the rate of straight work.

Gypsum Lath

Gypsum lath is manufactured in three types—plain, perforated, and insulating. Each type is made of a core of gypsum with a tough, fibrous surfacing material. For the usual surfaces, either plain or perforated gypsum lath is used. Where maximum fire protection is required or desired, perforated gypsum lath should be employed. For insulation against heat and cold, and as a vapor barrier, insulating gypsum lath having an aluminum foil back should be used on the inside face of exterior walls and for top floor ceilings.

Gypsum lath is furnished 3/8" (9 mm) thick and face dimension is approximately 16"x48" (400 x 1200 mm), 6 laths per bundle. Each bundle of laths contains 32 s.f. (2.88 sq.m) and weighs about 50 lbs. (22.5 kg). These sizes fit the usual stud or joist spacings without cutting.

Sizes and Weights of Gypsum Lath

Size and Thickness, in. (mm)	Weight per sq.yd. (sq.m)	Approx. Price per 1,000 s.f. (100 sq.m)
16"x48"x3/8" (400x1200x9mm)	14 lbs. (7.5 kg)	$110.00 (118.00)
16"x48"x1/2" (400x1200x13mm)	19 lbs. (10 kg)	115.00 (124.00)

Insulating or aluminum foil backed gypsum lath costs about $0.08 per s.f. ($0.86 per sq.m) more than plain lath.

Gypsum lath is also made 1/2"x16"x48" (13x400x1200mm) for framing over 16" (400 mm) o.c. but not exceeding 24" (600 mm) on center 3/8"x24"x job length (9mm x 600mm x job length) for exterior wall furring with or without foil backing; and 3/8"x16"x96" (9 x 400 x 2400 mm) plain or perforated.

For use in 2" (50 mm) solid studless partitions, gypsum lath is made 1/2" (13 mm) thick, 24" (600 mm) wide, and in lengths up to 12' (3.6 m) These laths are erected with the long dimension vertical.

Nails Required for Gypsum Lath. Nails shall be 1-1/8" (28 mm) long for 3/8" (9 mm) lath and 1-1/4" (31 mm) long for 1/2" (13 mm) lath, 13 gauge with 19/64" (7.5 mm) flathead and blued. For 3/8" (9 mm) lath on 16" (400 mm) spacing, nails shall be spaced approximately 5" (125 mm) apart using 4 nails per stud and for 1/2" (13 mm) lath on 24" (600 mm) spacing, nails shall be spaced approximately 4" (100 mm) apart, using 5 nails per framing member. Lath should be butted lightly together. Gaps over 1/2" (13 mm) wide should be reinforced with metal lath.

With studs spaced 16" (400 mm) o.c., it will require 6 to 7 lbs. (2.7-3.2 kg) of nails per 100 sq. yds. (83.6 sq.m) of gypsum lath and with studs spaced 24" (600 mm) o.c., it will also require 6 to 7 lbs. per 100 sq. yds. (3.3-3.8 kg per 100 sq.m) of 1/2" (13 mm) lath. Lath may also be attached by staples, screws, or clips.

Labor Placing Gypsum Lath. A lather should place 90 to 105 sq. yds. (78-90 sq.m) of gypsum lath per 8-hr. day on average work, at the following labor cost:

Labor Cost of 100 Sq. Yds. (83.6 Sq.M) of 3/8" (9 mm) Gypsum Lath
Applied to Wood Studs or Joists with Nails

	Hours	Rate	Total	Rate	Total
Lather	8.0	$....	$....	$26.58	$212.64
Cost per sq. yd.			....		2.13
per sq.m			....		2.32

Mechanical Stapling of Gypsum Lath. Increased production in placing gypsum lath can be gained where labor conditions permit the use of air-driven staplers for attaching lath to wood studs or joists.

Air-driven stapler heads, using galvanized frozen staples 7/8" (22 mm) long with a 3/8" (9 mm) crown, cost about $250.00 each. Staples cost $0.90 to $1.00 per thousand depending upon quantity purchased and are packed 12,000 to the carton. An air compressor able to handle one stapler head will cost about $370.00 and a 2-stapler head capacity machine will cost about $740.00.

For this type of attachment, it is general practice to tack lath into place, using 2 nails per piece for wall lath and 4 nails per piece for ceiling lath, until enough area is covered to permit continuous stapling. Attachment is completed by stapling 4" (100 mm) o.c. to all wood studs or joists. Approximately 3,500 staples and 1 lb. (0.45 kg) of nails are required per 100 sq. yds. (86 sq.m) of gypsum lath.

Labor Cost of 100 Sq. Yds. (86 Sq.M) of 3/8" (9 mm) Gypsum Lath
Applied to Wood Studs or Joists Using Air-Driven Staples

	Hours	Rate	Total	Rate	Total
Lather	5.3	$....	$....	$26.58	$140.87
Cost per sq. yd.			....		1.41
per sq.m			....		1.54

Two-Inch Solid Gypsum Long Length Lath and Plaster Partition

A studless, non-loadbearing partition consists of 1/2" (13 mm) thick, 24" (600 mm) wide, long length, up to 12'-0" (3.6 m), gypsum lath with V'd tongue and groove edges, held vertically in floor and ceiling runners and plastered to 3/4" (19 mm) grounds each side.

Temporary bracing consists of 3/4" (19 mm) cold rolled channels horizontally, one channel for heights up to 9'-0" (2.7 m) and two channels at the third points for heights up to 12'-0" (3.6 m), is required with adequate vertical braces on partitions over 6'-0" (1.8 m) long, and every 6'-0" (1.8 m) or fraction thereof during initial plastering stages.

A metal ceiling runner is used for attachment at the ceiling and either a 2-1/2" (63 mm) flush metal base or a routed out wood floor runner may be used at the floor to secure the long sheets of gypsum board lath.

On work of this class, it will require 20 to 24 hours lather time to set and place floor and ceiling runners and gypsum lath for 100 sq. yds. (86 sq.m) of partition, after the material has been hoisted and stocked by laborers, which will average about 4 hrs. per 100 sq. yds. (83.6 sq.m).

Material Cost of 100 Sq. Yds. (83.6 Sq.M) 2" (50 mm)
Solid Gypsum Long Length Lath and Plaster Partition
(Partition only, not including plastering)

	Rate	Total	Rate	Rate (metric)	Total
115 l.f. (34.5 m) ceiling runner.................$....	$....		$0.25	$0.83	$28.75
115 l.f. (34.5m) 2-1/2" (63mm) flush metal base*..		....	0.48	1.60	55.20
945 s.f. (85 sq.m) 1/2" (13mm) long length gypsum lath................................		....	0.12	1.33	113.40
50 l.f. (15 m) 3" (75 mm) strip lath and paper backing...............................		....	0.13	0.17	8.00
155 l.f. (46.5 m) temp. bracing channels**......................................		...	0.09		13.95
100 temp. bracing clips..........................		....	0.05		5.00
Cost per 100 sq. yds.	$....				$224.30
per sq. yd...				....	2.24
per s.f...				....	0.25
per sq.m				....	2.68

*Floor runner may be a special wood or steel runner channel at correspondingly lower material and installation cost in carpentry labor, but will require wood base at additional cost.

**115 l.f. (34.5 m) of horizontal temporary bracing channels, plus five 8-l.f. verticals figured on a minimum of 3 re-uses.

Labor Cost of 100 Sq. Yds. (83.6 Sq.M) 2" (50 mm)
Solid Gypsum Long Length Lath and Plaster Partition
(Partition only, not including plastering)

	Hours	Rate	Total	Rate	Total
Lather 22.0		$....	$....	$26.58	$584.76
Labor...4.0		....	....	21.04	84.16
Cost per 100 sq. yds			$....		668.92
Cost per sq. yd.				....	6.69
per s.f.				....	0.74
per sq.m				....	0.81

Add for corner beads, cornerites and other accessories required to complete lathing work. Also available are 1" (25 mm) thick gypsum lath either solid or laminated, which reduce the requirements for bracing and grounds.

Floating Systems of Attachment of Gypsum Lath. In order to eliminate sound transmission through walls and ceilings, the principle of resiliently isolating the interior walls and ceilings from their structural elements (studs and joists) has proven extremely effective. It interposes a shock absorber that reduces the energy transmitted to the structural elements of the building. Interior wall and ceiling finish should be completely and thoroughly isolated from all structural members surrounding the source of sound.

Attachment devices for gypsum lath are designed to isolate the lath from framing members. They are made from sheet metal or wire, and generally include field and corner clips for the attachment to wood, metal furring, and masonry.

In addition to reducing sound transmission, plaster cracking is also often eliminated, often caused by movement of the structural elements, which are stronger than the plastered interior. Floating systems of attachment permit a limited movement of the structural members without imposing undue stresses in the lath and plaster interior.

The spacing of framing members should not exceed 16" (400 mm) o.c. for clip attachments. Plain gypsum lath should be used on ceilings, and either plain or perforated lath may be used on walls.

Suspended Gypsum Lath Ceilings. Suspended gypsum lath ceilings consist of metal hangers, runners, and furring channels, which in turn support gypsum lath as a base for plaster, or the adhesive application of acoustical tile.

The hangers, runners, and furring channels for ceilings are about the same as required for metal lath. Furring channel spacing must not exceed 16" (400 mm) o.c., and some fire resistant ratings require 12" (300 mm) spacing. Starter channels along walls should be within 2" (50 mm) of wall. A 1/2" (13 mm) space should be provided between ends of channels and masonry walls. Each lath is supported transversely under the channels by a long wire clip, with not less than 3 such clips per full length of lath.

Stagger the end joints by starting alternative rows with half lengths of lath, allowing end joints to fall between channels.

Material Cost of 100 Sq. Yds. (83.6 Sq.M) Suspended Gypsum Lath Ceilings Using 16"x48"x3/8" (400 x 1200 x 9 mm) Gypsum Lath with Main Runners 3'-0" (0.9 m) on Centers and Cross Furring Channels 16" (400 mm) o.c.

	Rate	Total	Rate	Rate (metric)	Total
77 pcs. 3/16"x3'-0" (5mm x 0.9m) hangers, 231'-0" (69.3m) ....$....	$....	$.04	$0.13	$9.24
364 l.f. (109.2m) 1-1/2" (38mm) channels		....	0.25	0.83	91.00
700 l.f. (210 m) 3/4" (19mm) channels..........		....	0.17	0.56	119.00
945 s.f. (85 sq.m) 16"x48"x3/8" (400 x 1200 x 9mm) gypsum lath.............		....	0.11	1.22	103.95
500 Brace-tite field clips..........................		....	0.09		45.00
340 Bridjoint clips, B-1		....	0.06		20.40
Cost 100 sq. yds...................................		$....			$388.59
per sq. yd...................................			....		3.89
per sq.m			....		4.64

Labor Cost of 100 Sq. Yds. (83.6 Sq.M) Suspended Gypsum Lath Ceilings
Using 16"x48"x3/8" (400 x 1200 x 9 mm) Gypsum Lath with Main Runners
3'-0" (0.9 m) on Centers and Cross Furring Channels 16" (400 mm) o.c.

	Hours	Rate	Total	Rate	Total
Lather	20.0	$....	$....	$26.58	$531.60
Labor	4.0	....	....	21.04	84.16
Cost 100 sq. yds.			$....		$615.76
per sq. yd.				....	6.16
per sq.m				....	6.71

The above ceiling has a fire rating of one hour. If the job specification calls
for a fire resistance rating of two hours, 14 ga. diagonal wire reinforcement
must be added below the lath; to obtain a 4 hour rating, 1" (25 mm) 20 ga.
galvanized 1" (25 mm) hex wire mesh must be added. All the above ratings
are dependent on the type of floor construction above.

Hollow Steel Stud Partitions Using Gypsum Lath as a Plaster Base.
A non-loadbearing partition of trussed steel studs, track, and shoes to which
gypsum lath is applied to both sides by means of metal clips, nailing, or
stapling.

The construction is similar to hollow steel stud non-loadbearing partitions
except that gypsum lath is applied to the studs by means of special metal
clips and the studs are spaced 16" (400 mm) o. c. to receive the 16"x48" (400 x
1200 mm) gypsum lath.

Material Cost of 100 Sq. Yds. (83.6 Sq.M) of Hollow Steel Stud Non-Bearing
Plastered Partitions Using Gypsum Lath on Both Sides of Partitions,
2-1/2" (63 mm) Studs 16" (400 mm) on Centers, Ready to Receive Plaster

	Rate	Total	Rate	Rate (metric)	Total
105 l.f. (31.5m) top track	$....	$....	$0.25	$0.83	$26.25
105 l.f. (31.5m) bottom track		....	0.25	0.83	26.25
76 9'-0" (2.7m) studs, use 750 l.f. (225 m)*		....	0.34	1.13	255.00
1,900 s.f. (171 sq.m) gypsum lath. 3/8" (9mm)		....	0.11	1.22	209.00
1,000 Trus-Lok field clips		....	0.08		80.00
680 Bridjoint clips		....	0.06		40.80
Cost 100 sq. yds.			$....		$637.30
per sq. yd.			....		6.37
per sq.m			....		7.62

*Allowance made for opening framing.

Labor Cost of 100 Sq. Yds. (83.6 Sq.M) of Hollow Steel Stud Non-Bearing Plastered Partitions Using Gypsum Lath on Both Sides of Partitions, 2-1/2" (63 mm) Studs 16" (400 mm) o.c., Ready to Receive Plaster

	Hours	Rate	Total	Rate	Total
Lather	28.0	$....	$....	$26.58	$744.24
Labor	4.0	....	....	21.04	84.16
Cost 100 sq. yds.			$....		$828.40
per sq. yd.				....	8.28
per sq.m				....	9.03

Add for corner beads, cornerites, and other accessories required to complete lathing work.

Gypsum lath is also used for exterior wall furring and column fireproofing. Special framing support systems are manufactured for this. Foil backed lath may be used for exterior walls.

Wood Lath

Wood lath for plastering is a thing of the past. Gypsum lath and metal lath are commonly used. The following data is included for that one in a thousand job where wood lath may be used for patching in remodeling work.

Wood lath should be estimated by the sq. yd. (sq.m). Because it is customary to deduct only one-half the area of openings, it is not necessary to add any allowance for waste. When placing wood lath, it is customary to allow 3/8" (9 mm) between each lath for a plaster key. A standard wood lath is 1-1/2" (38 mm) wide and 48" (1200 mm) long:

Size of Lath	No. Req'd. 100 sq. yd. (83.6 sq.m)
1-1/2"x48" (38 x 1200 mm)	1,425-1,450

Labor Placing Wood Lath. A lather should place 1,400 to 1,600 1-1/2"x48" (38 x 1200 mm) wood lath, 95 to 110 sq. yds. (80-92 sq.m) per 8-hr. day.

Metal Lath and Furring

Where metal lath is used as a plaster base in partition work, it may be nailed or stapled to wood or steel studs, nailed to masonry, or wire-tied to steel channels and prefabricated steel studs. It may also be used without studs or back-up when attached to floor and ceiling runners and temporarily braced during application and curing of first coat of plaster. Where used as a plaster base in ceiling work, it may be nailed to wood joists, wire-tied to various metal support systems either in contact with or suspended from the building structure, or wire-tied directly to the bottom chords of steel joists. It may also be attached directly to the underside of reinforced concrete floors or joists by means of attachments previously embedded in the concrete.

In addition to its utility as a plaster base, metal lath is also used to reinforce plaster applied to other surfaces. Prefabricated, right-angled strips of metal lath are attached to masonry or other rigid plaster bases, prior to plastering, at the interior angles formed by intersecting walls and along the junction of walls and ceilings. Flat strips of metal lath are attached to solid plaster bases to reinforce the joints in such bases, or to bridge potential crack lines at the corners of door and window openings.

Metal lath is used in the individual fireproofing of beams and columns. It may be adapted to decorative shapes for cornices, coves, curved walls, convoluted ceilings, and various acoustic shapes such as inverted pyramids. In specialty classifications, it is used to form the membrane of hyperbolic paraboloids, or the web section of concrete beams and girders.

Metal Lath Construction. The costs of metal lathing will vary according to the complexity of work, size of the rooms, height above floors, and method of attachment. In the sections below, an attempt is made to cover only the type of lathing that is normal in building construction.

Materials. A number of metal and wire laths and welded wire fabrics is available to suit the various conditions that might be encountered in building construction. In general, there are sheets of slitted and expanded metal, or perforated metal, commonly called metal laths; rolls of netting formed of 19 ga. wire woven into a pattern which provides 2-1/2 strands per inch (25 mm), commonly called wire laths; and sheets of open mesh laths formed of strands of 16 gauge wire welded at right angles to one another and spaced 2" (50 mm) apart, commonly called welded wire fabric. Laths may be of galvanized metal or painted with a rust-inhibitive paint. Some metal laths and all welded wire fabrics are available with a paper backing. Also, some laths may be obtained with a backing of aluminum foil laminated with paper. The backings sometimes perform useful functions insofar as utility and plastering are concerned, but many such laths are often more difficult to install.

Expanded metal lath can be either flat dimpled (self-furring), or ribbed. The flat expanded type is classed as small mesh, having approximately 1000 openings per s.f. (0.09 sq.m). The perforated metal lath, commonly called sheet lath, is fabricated either with furring indentations or with channel-shaped ribs. The openings in sheet lath and in rib lath are classified as small. Although they are somewhat larger than those of the expanded metal laths, they are separated by larger areas of unperforated metal. Stucco mesh has diamond-shaped openings, which are approximately 1-1/2"x3" (38 x 75 mm). Wire laths are flat when installed and the openings are classified as medium size. The welded wire fabrics are either flat, or in the crimped, self-furring variety, all the wires laying in one direction are bent approximately 1/4" (6 mm) out of a flat plane at their intersection with every third wire lying in the other direction.

Installation Procedures. Rather rigid industry standards govern the installation of metal laths, particularly with respect to ceiling work. Complete specifications are prepared by manufacturer associations, by individual manufacturers, or by sectional committees of the American National Standards Institute. The building construction estimator needs a brief outline of the critical factors of installation in order to do a realistic estimate.

Spans of Lath. In all work, except where laths are supported by solid surfaces, and except in the case of studless solid partitions, the weight of metal lath must be suitable to the distance between studs, joists, or other supporting members. The laths are attached to these members by nails, wire-ties, staples, or clips, at intervals not exceeding 6" (150 mm). The following chart gives maximum spans or distance between supports for various weights of metal and wire laths and welded wire fabric.

Type of Lath	Min. Wt. Lath Lbs. per Sq. Yd.	Vertical Supports			Horizontal Support	
		Wood	Metal Solid Partit.	Others	Wood or Concrete	Metal
Flat Expanded	2.5	16	16	12	0	0
Metal Lath	3.4	16	16	16	16	13½
Flat Rib	2.75	16	16	16	16	12
Metal Lath	3.4	19	24	19	19	19
⅜" Rib	3.4	24	..	24	24	24
Metal Lath	4.0	24	..	24	24	24
Sheet Lath*	4.5	24	..	24	24	24
Wire Lath	2.48	16	16	16	13½	13½
V-Stiffened						
Wire Lath	3.3	24	24	24	19	19
Wire Fabric	..	16	0	16	16	16

*Used in studless solid partitions.

Laps of Lath. Industry standards require overlapping at ends and sides of sheets of lath. Most sheets are manufactured slightly larger than nominal size and the estimator need not allow extra quantities on this account. A few extra ties are required in all laps, one between each supporting member, and where laps of the ends of sheets fall between supports, such laps should be laced with tie wire. These extra ties are considered part of normal lathing work and no extra allowance need be made for them.

FINISHES

9.15

Attachment of Lath. Nail-on attachment of metal lath to horizontal or sloping surfaces (ceilings and soffits) requires the use of No. 11 ga. barbed, galvanized roofing nails with a head diameter of 7/16" (11 mm). The length of such nails must permit a minimum penetration into the wood of 1-3/8" (34 mm). In nail-on attachment of metal lath to vertical surfaces (walls and partitions), 4d common nails are permitted. They should penetrate the wood at least 3/4" (19 mm). However, and the remainder should be bent over the strands of lath. Also, in vertical work, 1" (25 mm) long roofing nails with a head diameter of 7/16" (11 mm) are permitted. They are driven home without crushing the strands of metal. In direct attachment to masonry, concrete stub nails with 3/8" (9 mm) flat heads are generally used.

Tie-on attachment of metal lath is generally done with 18 ga. galvanized annealed wire, a strand of wire passing through the mesh, around the support member, and back through the mesh with the two ends twisted together approximately three turns. This operation is more difficult for the lather when paper or foilbacked lath is being used, unless adequate openings in the backing have been provided. When estimating hand tying of lath having heavy backing, a reduction in labor productivity should be allowed. In certain cases 18 ga. tie wire is not permitted. Where metal lath is attached directly to structural elements of the building, such as concrete slabs or joists or the bottom chord of open web steel joists, wire of heavier gauge, or more than one loop of 18 ga. wire, is required (16 ga. or 2-18 ga. for steel joist work; 14 ga. twisted or 10 ga. bent over for concrete work). The extra difficulty of twisting and cutting the heavier gauge wires has been considered in the productivity rates suggested below.

Special devices for attachment of laths are permitted, where they develop a fastening strength equal to the accepted standards. Such devices often result in noticeable labor savings but the estimator should not neglect to adjust his material costs as necessary.

Horizontal Support Systems. Horizontal supports for metal lath fall under the general classifications of contact systems, where the lath is fastened in direct contact with the structural elements of the building, furred systems, or suspended systems. In both of the latter systems, the laths are attached in the usual manner to a framework of pencil rods or channel shapes which have been fastened directly to, or suspended from, structural elements of the building. In a furred system, the channels or pencil rod furring are attached to the structural elements of the building, by means of 16 ga. wire saddle ties—in open web steel joist construction, or by 18 ga. wire saddle ties supported by nails driven in or through wood joists. The forming of saddle ties is considered normal lathing work, and the productivity rates given below consider this type of tying.

In a suspended system, the channels or pencil rods to which the lath is attached are called cross-furring. These members are saddle-tied with 16 ga. wire to the underside of an intersecting framework of heavier channels, which have been hung from the structural elements of the building by means of wire, rod, or flat steel hangers. Wire hangers and rods are saddle-tied around these support channels, which are called main runners, and flats may be welded or bolted to them.

A combination of the furred and suspended systems is usually found where furring members are installed in contact with concrete joists. The furring members, which support the lath, are in contact with the bottoms of the joists, but they are supported by transverse channels, or runners, which are suspended between and run parallel to the joists. The furring members are saddle-tied with 16 ga. wire to the runners, and the runners are suspended by means of 10 ga. wire hangers embedded in the slab above. The wire hangers are saddle-tied to the runners.

In contact, furred or suspended systems, the lath must be supported at spans not exceeding those given in the table of maximum spans, above. In furred and suspended systems, all pencil rods and channels are limited as to distance between supports and center to center spacing. In suspended systems the size of each hanger depends on the area of ceiling that it supports. The following tables list the usual elements and arrangements to be found in furred and suspended ceilings.

Hanger Sizes for Suspended Ceilings

Ceiling area, s.f. max.	Hanger size, min.	Hangers for Supporting up to 25 S.F. Ceiling	
8	12 ga. wire	6 ga. wire	1"x3/16" flat
12	10 ga. wire	5 ga. wire	1-1/4"x1/8" flat
12-1/2	9 ga. wire	3/16" rod	1-1/4"x3/16" flat
16	8 ga. wire	7/32" rod	1-1/2"x1/8" flat
17-1/2	7 ga. wire	1/4" rod	1-1/2"x3/16" flat

(Metric)

Ceiling area, sq.m, max.	Hanger size, min.	Hangers for Supporting up to 2 Sq.M Ceiling	
0.75	12 ga. wire	6 ga. wire	25 x 4.7 mm flat
1.11	10 ga. wire	5 ga. wire	31 x 3.1 mm flat
1.16	9 ga. wire	4.6 mm rod	31 x 4.7 mm flat
1.48	8 ga. wire	5.4 mm rod	37 x 3.1 mm flat
1.62	7 ga. wire	6.2 mm rod	37 x 4.7 mm flat

All wire hangers should be galvanized steel, all flat and rod hangers should be coated with rust-inhibitive paint.

Spans and Spacing for Main Runners in Suspended Ceilings

Min. Size and Type	Max. Span Between Hangers or Supports	Max. CtoC Spacing of Runners
3/4" (19mm) channel, 0.3 lb. (0.14kg)	3'-0" (0.9m)	2'-4" (0.7m)*
3/4" (19mm) channel, 0.3 lb. (0.14kg)	2'-6" (0.75m)	2'-6" (0.75m)
3/4" (19 mm) channel, 0.3 lb. (0.14kg)	2'-0" (0.6m)	3'-0"(0.9m)
1-1/2" (38mm) channel, 0.475 lb. (0.21kg)	5'-0" (1.5m)	2'-0" (0.6m)
1-1/2" (38mm) channel, 0.475 lb. (0.21kg)	4'-0" (1.2m)	3'-0" (0.9m)
1-1/2" (38mm) channel, 0.475 lb. (0.21kg)	3'-6" (1.05m)	3'-6" (1.05m)
1-1/2" (38mm) channel, 0.475 lb. (0.21kg)	3'-0" (0.9m)	4'-0" (0.9m)
2" (50mm) channel, 0.59 lb. (0.27kg)	7'-0" (2.1m)	2'-0" (0.60m)
2" (50mm) channel, 0.59 lb. (0.27kg)	5'-0" (1.5m)	3'-6" (1.05m)
2" (50mm) channel, 0.59 lb. (0.27kg)	4'-6" (1.35m)	4'-0" (1.2m)

All weights of channel are for cold-rolled members.
*This spacing for runner channels supporting furring members against concrete joists, only.

Spans and Spacing for Cross Furring

Min. Size and Type	Max. Span Between Runners or Supports	Max. CtoC Spacing of Cross Furring
3/8" (9mm) pencil rod	2'-6" (0.75m)	12" (300mm)
3/8" (9mm) pencil rod	2'-0" (0.6m)	19" (475mm)
3/4" (19mm) channel, .3 lb. (0.14kg)	4'-0" (1.2m)	16" (400mm)
3/4" (19mm) channel, .3 lb. (0.14kg)	3'-6" (1.05m)	19" (475mm)
3/4" (19mm) channel, .3 lb. (0.14kg)	3'-0" (0.14kg)	24" (600mm)

All weights of channel are for cold-rolled members.

In most horizontal work, the laths, framing members, hangers, and inserts are installed in the usual manner. The estimator should look for the presence of large ducts, which might require additional framing or hangers not detailed on the drawings. He should allow for hung or tied supports at the perimeter of rooms and not plan to set channels into wall pockets. The estimator's checklist of materials should have allowances for lath, furring channels, carrying channels, tie wire, hanger wire, nails, special inserts if any, corner reinforcing, and accessory items, if required, such as expansion bead or casing bead.

Vertical Support Systems. Metal lath may be used in a number of wall and partition framing systems. A generalized classification would include hollow walls and partitions, solid plaster partitions, vertical furring, and wall cladding. Hollow assemblies may have either metal or wood studs. The metal studs are generally used where non-combustible construction is required, where concealed horizontal pipe must be accommodated, or where sound resistive qualities are needed. Although the majority of such studs are rated as non-loadbearing, because they are fabricated from light-gauge steel or rod, tests of completed walls have indicated surprising load carrying capacity and resistance to horizontal impact.

Courtesy Metal Lath Association

Details of Suspended Ceiling

Thin solid plaster partitions may be formed over a core of metal lath alone, or an assembly of metal lath and metal studs. The studless solid partitions are generally 2" (50 mm) thick and may be as much as 10' (3 m) in height. The solid partitions with 3/4" (19 mm) cold rolled channels may range between 1-1/2" (38 mm) thickness and 8-1/2" (213 mm) height to 2-1/2" (63 mm) thickness and 16' (4.8 m) height. Heavier channels may be used for thicker and higher solid partitions.

Vertical, or wall, furring systems may be braced from exterior walls, pilasters, or interior columns, or may be free-standing. They are sometimes decorative, other times functional. Functional aspects include concealment, fireproofing, acoustic treatment, and condensation and temperature control. The lathing installation is essentially that of a partition with studs, lathed on one side only. However, horizontally-placed 3/4" (19 mm) channels are required for stiffening steel studs, and above certain heights, depending on size of the studs, the framing must be braced from the structure behind.

Exterior and interior wall claddings include overcoating (remodeling work), decorative surfacing, acoustic surfacing, and reinforcement.

In most vertical work, the laths are applied in the usual manner and the estimator need not look for hidden items of extra material and labor. However, a few unusual operations worth noting. In erection of studless solid partitions, the laths must be braced temporarily until first, or scratch, coat of plaster has set. In erection of sound resistive partitions, various systems for separating the components of the partition may be encountered. These might include resilient clips and pencil rod furring between the metal studs and metal lath, staggered stud arrangements that require double the normal number of studs, or double rows of small studs with short pieces of cross-tie channels between them. Note also that most soundproofing partitions extend up to the structure itself, and are not terminated at the level of a suspended ceiling or the bottoms of joists. In wall cladding work, some of the more dense back-up materials, such as reinforced concrete or old brick, may be difficult to penetrate with nails, slowing the progress of the lather. In certain areas of the country, the exterior walls of stuccoed dwellings may be constructed without sheathing. In this style of erection, known as line wire construction, the lathing work sometimes includes installation of 18 ga. wires, stretched taut completely around the building, and fastened securely to the studs at 6" intervals vertically. This operation is generally performed only when woven wire or stucco mesh is the plaster base.

(1) Stiffeners placed at mid-height of partitions which are more than 10' long or 9' high, not to exceed 6' on centers vertically.

Courtesy Metal Lath Association

Partition Framing At Door Openings

The estimator's checklist of materials for vertical work should contain allowances for lath, studs if any, four shoes for each prefabricated metal stud (2 top and 2 bottom), ceiling and floor tracks for holding studs and shoes, nails for attachment of either the tracks or bent-over ends (of small studs) to ceiling and floor, nails for attachment of lath to wood or solid back-up, wire for tying lath and shoes to studs, and shoes to tracks, if required, metal base or screeds, if any, and accessory items such as corner, casing or expansion beads, and picture molds, if required. In estimating requirements for studs, allow for double-studding at door frames where necessary, and where splicing is required, allow for an 8" (200 mm) lap and double wire-tying.

Installation—Labor

Nail-On Attachment of Metal Lath. For either horizontal or vertical work, a lather should apply 90 to 100 sq. yds. (75-84 sq.m) of metal lath per 8-hr. day, depending on size of rooms.

Tie-On Attachment of Metal Lath. Usually, the actual tying of metal lath is not an isolated operation, for the lather must also install the studs or channels to which the lath is tied. However, in contact ceiling work, where the lath is sometimes tied directly to open web joists or to hangers previously embedded in concrete, the lather should apply 100 sq. yds. (83.6 sq.m) of metal lath per 8-hr. day.

Furred and Suspended Horizontal Systems. In furred ceiling work, including installation of furring members and lath but not including drilling for and placement of inserts in concrete, the lather should apply 50 sq. yds. (42 sq.m) per 8-hr. day. In suspended ceiling work, including installation of wire and rod hangers (but not concrete inserts for hangers), main runner channels, cross furring channels, and lath, a lather should apply 40 sq. yds. (33 sq.m) per 8-hr. day. Where strap hangers are used (bolted construction), a lather should complete 30 sq. yds. (25 sq.m) per 8-hr day. In drilling concrete and placing hanger inserts, a lather should cover 25 sq. yds. (21 sq.m) of area per 8-hr. day. In closets, toilet rooms, and other small spaces, a lather might complete only 30 sq. yds. (25 sq.m) of suspended ceiling per 8-hr. day.

Beam Fireproofing. In furring and lathing around structural beams, a lather should complete 9 l.f. (2.7 m) of beam per hour.

Vertical Systems Other Than Nail-On Work, Hollow Partitions. In hollow partition work, not including sound-resistive construction, a lather should install top and bottom tracks, tie shoes on studs, tie studs and shoes to tracks, apply lath, and complete 30 sq. yds. (25 sq.m) of partition per 8-hr. day. In 100 sq. yds. (83.6 sq.m) of such partitions, approximately 16 hrs. are required to install runners and studs and the attachment of two surfaces of lath should require approximately 11 hrs. Another version of the steel stud hollow partition, using a framework of 3/4" (19 mm) cold-rolled channels, requires slightly more labor for installation. In this work the lather must erect two walls and tie them together. Productivity is between 24 and 28 sq. yds. (20.08-23.4 sq.m) per 8-hr. day. For estimating sound-resistive hollow partitions with prefabricated steel studs, add 5 hrs. per 100 sq. yds. (83.6 sq.m) to cover installation of resilient clips and pencil rods. No extra allowance need be made in the case of hollow partitions with double rows of 3/4" (19 mm) channels.

Hollow Partitions: Attachments to Suspended Ceilings

Courtesy Metal Lath Association

Hollow Partitions: Attachments to Floor

Vertical Systems, Solid Partitions, and Vertical Furring. In erecting solid partitions with 3/4" (19 mm) cold rolled channels, a lather should complete 22 sq. yds. (18 sq.m) per 8-hr. day, depending on spacing of channels and method of attaching channels at floor and ceiling. Studless solid partitions should be erected at the rate 27 sq. yds. (22.5 sq.m) per 8-hr. day. Labor savings are realized in the studless system due to elimination of wire tying the lath to channel studs. Vertical furring systems should generally be erected at the same rate as solid partitions with channel studs. However, where wall braces are required, add 4 hrs. per 100 sq. yds. (83.6 sq.m).

Vertical Systems, Wall Cladding. In overcoating work, and in other direct-applied lath attachments, a lather should nail metal lath to surfaces of wood and concrete block at the same rate as for nail-on work to wood studs and joists. Lathing should proceed at the rate of 160 sq. yds. (134 sq.m) per 8-hr. day. In attachment to old brick or reinforced concrete, however, a lather should be expected to complete only 125 sq. yds. (105 sq.m) per 8-hr. day.

Accessories. Many metal lathing accessories are available for a variety of functional or trimming purposes. Only a few of the more common items are covered here. The labor rates given below for erection of some of the functional accessories are inclusive of items of work covered as part of the total assembly. The estimator should use the highest productivity rate for erection of an entire assembly, when he lists these functional accessories as separate items. For example, installation of metal base sometimes includes layout of partition and attachment of clips to floor—items of work which are also contemplated in erection of some partition types.

Flush Metal Base. Double for solid partitions—including layout of partition, attachment of clips to floor, cutting and fitting, attachment of side plates: 5 hrs. per 100 l.f. (30 m) of partition.

Single for Masonry Trim. Including drilling for and setting masonry clips, cutting and fitting, attachment of side plates: 3 hrs. per 100 l.f. (30 m).

Single for Wall Furring. Including layout of furring, attachment of clips to floor, cutting and fitting, attachment of side plates: 4 hrs. per 100 l.f. (30 m).

Flush or Applied Metal Base. Single for snap-on to metal studs or for screwing to wood grounds, including cutting and fitting, attaching side plates, and attaching clips to studs, but not including wood grounds: 3 hrs. per 100 l.f. (30 m).

Beads and Molding. Including corner and bullnose beads, picture molds, base screeds, casing bead (except when used for door and window trim), and expansion bead (except when lath and/or framing is severed beneath line of bead):

	Hrs. per 100 l.f. (30m)
wired or stapled to lath bases	2-1/2
nailed or stapled to wood construction	2
nailed to brick, tile or block construction	3
stub-nailed to concrete construction	3-1/2

If on-the-job mitering is required, add up to 1/2 hr. per 100 l.f. (30 m) to above rates.

When windows and doors are being trimmed with casing bead, and the operation is separate from the remainder of the lathing work, the lather should complete 1 door or window per hour. When expansion bead is to be installed to a lath base and the base is to be severed beneath the line of bead, add 3 hrs. per 100 l.f. (30 m) for cutting lath.

Cornerite and Stripite. Placed in interior wall angles and at junctions of walls and ceilings where adjoining constructions are of rigid lath or masonry, and installed along the joints in nonmetallic plaster bases, should be placed at the rate of 500 l.f. (150 m) per 8-hr. day, depending on whether material is applied in isolated or repetitious patterns.

Material Cost of 100 Sq. Yds. (83.6 Sq.M) of 3.4 Lb. (1.5 Kg) Diamond Mesh Metal Lath Applied to Wood Studs or Joists @ 16" (400 mm) O.C., Ready to Receive Plaster

	Rate	Total	Rate	Rate (metric)	Total
105 sq. yds. (87.8 sq.m) diamond mesh.. $....	$....		$1.59	$2.27	$199.95
9 lbs. (4 kg) 1-1/2" (38 mm) roofing nails..		....	1.06	2.39	9.54
Cost 100 sq. yds.			$....		$176.49
per sq. yd.			....		1.76
per sq.m			....		2.11

Labor Cost of 100 Sq. Yds. (83.6 Sq.M) of 3.4 Lb. (1.5 Kg) Diamond Mesh Metal Lath Applied to Wood Studs or Joists @ 16" (400 mm) O.C., Ready to Receive Plaster

	Hours	Rate	Total	Rate	Total
Lather 7.5	7.5	$....	$....	$26.58	$199.35
Cost per sq. yd.			....		1.99
per sq.m			....		2.17

Material Cost of 100 Sq. Yds. (83.6 Sq.M) of 3.4 Lb. (1.5 Kg) 3/8" (9 mm) Rib Metal Lath Applied to Wood Studs or Joists @ 24" (600 mm) O.C., Ready to Receive Plaster

	Rate	Total	Rate	Rate (metric)	Total
105 sq. yds. (87.8 sq.m) 3/8" (9 mm) rib mtl lath .... $....	$....		$2.33	2.78	$244.65
6 lbs. (2.7 kg) 1-1/2" (38 mm) roofing nails...................................		....	1.06	2.38	6.36
Cost 100 sq. yds...............................			$....		$251.01
per sq. yd.................................			....		2.51
per sq.m			....		3.00

Labor Cost of 100 Sq. Yds. (83.6 Sq.M) of 3.4 Lb. (1.5 Kg) 3/8" (9 mm) Rib Metal Lath Applied to Wood Studs or Joists @ 24" (600 mm) O.C., Ready to Receive Plaster

	Hours	Rate	Total	Rate	Total
Lather 7.0	7.0	$....	$....	$26.58	$186.06
Cost per sq. yd.			....		1.86
per sq.m			....		2.03

Material Cost of 100 Sq. Yds. (83.6 Sq.M) of 3.4 Lb. (1.5 Kg)
3/8" (9 mm) Rib Metal Lath Applied Directly to Ties Embedded
in Concrete or to Steel Joists at 24" (600 mm) on Centers

	Rate	Total	Rate	Rate (metric)	Total
105 sq. yds. (87.8 sq.m) 3/8" (9 mm) rib metal lath........................$....		$....	$2.73	$2.78	$244.65
15 lbs. (6.75 kg) 18 ga. tie wire		....	0.95	2.11	14.25
Cost 100 sq. yds...			$....		$258.90
per sq. yd.				....	2.59
per sq.m				....	3.10

Labor Cost of 100 Sq. Yds. (83.6 Sq.M) of 3.4 Lb. (1.5 Kg)
3/8" (9 mm) Rib Metal Lath Applied Directly to Ties Embedded
in Concrete or to Steel Joists at 24" (600 mm) on Centers

	Hours	Rate	Total	Rate	Total
Lather*......................................9.0	9.0	$....	$....	$26.58	$239.22
Cost per sq. yd.				....	2.39
per sq.m				....	2.61

*Lather hours listed are for attachment of lath only. For concrete joist
construction, add for placing wire in forms before pour, or if necessary to drill holes
in concrete.

Material Cost of 100 Sq. Yds. (83.6 Sq.M) of Furred Ceiling Using 3.4 Lb. (1.5 Kg)
Flat Rib Metal Lath Applied to 3/8" (9 mm) Pencil Rods 19" (475 mm) on Centers
Tied to Structure at 2 Ft. (0.6 m) Intervals

	Rate	Total	Rate	Rate (metric)	Total
105 sq. yds. (87.8 sq.m) flat rib metal lath....................$....		$....	$2.33	$2.79	$244.86
660 l.f. (198 m) 3/8" (9 mm) pencil rods		....	0.08	0.26	52.80
6 lbs. (2.7 kg) 18 ga. tie wire (lath to rods)		....	0.95	2.11	5.70
2 lbs. (0.9 kg) 16 ga. Wire for splicing rods		....	0.85	1.89	1.70
4 lbs. (1.8 kg) 16 ga. wire for attaching rods		....	0.85	1.89	3.40
Cost 100 sq. yds.............................			$....		$308.46
per sq. yd...				....	3.08
per sq.m				....	3.69

Labor Cost of 100 Sq. Yds. (83.6 Sq.M) of Furred Ceiling Using 3.4 Lb. (1.5 Kg) Flat Rib Metal Lath Applied to 3/8" (9 mm) Pencil Rods 19" (475 mm) on Centers Tied to Structure at 2 Ft. (0.6 m) Intervals

	Hours	Rate	Total	Rate	Total
Lather*..	16.0	$....	$....	$26.58	$425.28
Cost per sq. yd.				....	4.25
per sq.m				....	4.63

*Lather hours listed are for attachment of lath only. For concrete joist construction, add for placing wire in forms before pour, or if necessary to drill holes in concrete.

Material Cost of 100 Sq. Yds. (83.6 Sq.M) Suspended Metal Lath Ceiling Using 3.4 Lb. (1.5 Kg) 3/8" (9 mm) Rib Metal Lath with Furring Channels 24" (600 mm) on Centers Runner Channels 36" (900 mm) on Centers, Hangers 48" (1200 mm) on Centers

	Rate	Total	Rate	Rate (metric)	Total
100-9 ga. x 3'-0" (0.9 m) hangers 300'-0" (90 m) or 19 lbs. (8.55 kg) .... $....		$....	$0.74	$1.64	$14.06
345 l.f. (103.5 m) 1-1/2" (38 mm) c.r. channels		....	0.25	0.83	86.25
490 l.f. (147 m) 3/4" (19 mm) c.r. channels		....	0.17	0.57	83.80
3 lbs. (1.35 kg) 16 ga. wire		....	0.85	1.88	2.55
105 sq. yds. (87.8 sq.m) 3/8" (9 mm) rib metal lath		....	2.33	2.78	244.65
5 lbs. (2.25 kg) 18 ga. wire...........................		....	0.95	1.90	4.28
Cost per 100 sq. yds.			$....		$435.09
per sq. yd.			....		4.35
per sq.m			....		5.20

Labor Cost of 100 Sq. Yds. (83.6 Sq.M) Suspended Metal Lath Ceiling Using 3.4 Lb. (1.5 Kg) 3/8" (9 mm) Rib Metal Lath With Furring Channels 24" (600 mm) on Centers Runner Channels 36" (900 mm) on Centers, Hangers 48" (1200 mm) on Centers

	Hours	Rate	Total	Rate	Total
Lather*..	50.0	$....	$....	$26.58	$1,329.00
Cost per sq. yd.				....	13.29
per sq.m				....	14.49

*For erection under concrete, add for placing hangers in forms before pour; or, if necessary to drill holes in concrete or tile, and place inserts.

Material Cost of 100 Sq. Yds. (83.6 Sq.M) Suspended Metal Lath Ceiling
Using 3.4 Lb. (1.5 Kg) Diamond Mesh Metal Lath With Furring Channels
13-1/2" (338 mm) on Centers,
Runner Channels 48" (1200 mm) on Centers, Hangers 48" (1200 mm) on Centers

	Rate	Total	Rate	Rate (metric)	Total
85-8 ga. x 3'-0" (0.9 m) hangers 255'-0" (76.5 m) or 19 lbs. (8.55 kg) ....$....	$....	$0.74	$1.64	$14.06
280 l.f. (84 m) 1-1/2" (38 mm) c.r. channels	....	0.25	0.83	70.00
850 l.f. (255 m) 3/4" (19 mm) c.r. channels	....	0.17	0.57	144.50
4 lbs. (1.8 kg) 16 ga. wire	....	0.85	1.88	3.40
105 sq. yds. (87.8 sq.m) diamond mesh	....	1.59	1.90	166.95
8 lbs. (3.6 kg) 18 ga. wire	....	0.95	2.11	7.60
Cost per 100 sq. yds.		$....		$406.51
per sq. yd.		....		4.07
per sq.m		....		4.86

*See above for erection under concrete.

Labor Cost of 100 Sq. Yds. (83.6 Sq.M) Suspended Metal Lath Ceiling
Using 3.4 Lb. (1.5 Kg) Diamond Mesh Metal Lath With Furring Channels
13-1/2" (338 mm) on Centers,
Runner Channels 48" (1200 mm) on Centers, Hangers 48" (1200 mm) on Centers

	Hours	Rate	Total	Rate	Total
Lather*................................54.0	56.0	$....	$....	26.58	$1,488.48
Cost per sq. yd.		....			14.88
per sq.m		....			16.22

*See above for erection under concrete.

Material Cost of 100 L.F. (30 m) Beam or Girder Fireproofing Using 3.4 Lb. (1.5 Kg)
Self-Furring Diamond Mesh Metal Lath Tied to Adjacent Ceiling Lath and
Wrapped Downwards Around 12" (300 mm) WF Beam

	Rate	Total	Rate	Total
33 sq. yds. (27.6 m) 3.4 lb. (1.5 kg) self-furring diamond mesh ....$....	$....	$1.70	$56.10
2.5 lbs. (1.12 kg) 18 ga. wire.............................	....	0.95	2.38
Cost per 100 l.f................................	$....		$58.47
per l.f.	....		.58
per m	....		1.95

Labor Cost of 100 L.F. (30 m) Beam or Girder Fireproofing Using 3.4 Lb. (1.5 Kg)
Self-Furring Diamond Mesh Metal Lath Tied to Adjacent Ceiling
Lath and Wrapped Downwards Around 12" (300 mm) WF Beam

	Hours	Rate	Total	Rate	Total
Lather	12.0	$....	$....	$26.58	$318.96
Cost per l.f.			....		3.19
per m			....		10.46

Material Cost of 100 Sq. Yds. (83.6 Sq.M) of Hollow Steel Stud Partitions;
Lathed Both Sides, Studs 12" (300 mm) on Centers,
Using 2.5 Lb. (1.12 Kg) Diamond Mesh Metal Lath

	Rate	Total	Rate	Rate (metric)	Total
200 l.f. (60 m) snap-in-runner track	$....	$....	$0.25	$0.83	$50.00
Nails or expansion drives-allowance		....	26.50	26.50	26.50
101 9'-0" (2.7 m) 3-5/8" (90 mm) galv. steel studs, 909 l.f. (272.7 m)*		....	0.37	1.23	336.33
210 sq. yds. (175.5 sq.m) diamond mesh		....	1.27		266.70
17 lbs. (7.65 kg) 18 ga. wire		....	0.95		16.15
Cost per 100 sq. yds.			$....		$695.68
per sq. yd.			....		6.96
per sq.m			....		8.32

*Add necessary extras for framing around openings, returns,etc.

Labor Cost of 100 Sq. Yds. (83.6 Sq.M) of Hollow Steel Stud Partitions;
Lathed Both Sides, Studs 12" (300 mm) on Centers,
Using 2.5 Lb. (1.12 Kg) Diamond Mesh Metal Lath

	Hours	Rate	Total	Rate	Total
Lather	27.0	....	....	$26.58	$717.66
Cost per sq. yd.			....		7.18
per sq.m			....		7.83

Material Cost of 100 Sq. Yds. (83.6 Sq.M) of Sound Insulation Partition
Using Prefabricated Steel Studs 24" (600 mm) on Centers, Resilient Clips and
Pencil Rods, Lathed Both Sides With 3.4 Lb. (1.5 Kg) 3/8" (9 mm) Rib Metal Lath

	Rate	Total	Rate	Rate (metric)	Total
200 l.f. (60 m) plain runner track	$....	$....	$0.25	$0.83	$50.00
Nails or expansion drives-allowance		....	26.50	26.50	26.50
51-9'-0" (2.7 m) 3-5/8" (90 mm) galv. steel studs, 459 l.f. (137.7 m)*		....	0.37	1.23	169.83
204 shoes		....	0.08	0.08	16.32

	Rate	Total	Rate	Rate (metric)	Total
210 sq. yds. (175.5 sq.m) 3/8" (9 mm) rib metal lath...............................		....	2.33	2.79	489.30
11 lbs. (4.95 kg) 18 ga. wire		....	0.95	2.11	10.45
Add for sound treatment:					
100 l.f. (30 m) cork, 3-1/2" (88 mm) wide x 1" (25 mm) thick		....	0.53	1.77	53.00
1020 resilient clips........................		....	0.08	0.08	81.60
918 l.f. (275.4 m) 1/4" (6 mm) rod		....	0.07	0.23	64.26
Cost per 100 sq. yds.			$....		$961.26
per sq. yd.........................				....	9.61
per sq.m				....	10.48

*Add necessary extras for framing around openings, etc.

Labor Cost of 100 Sq. Yds. (83.6 Sq.M) of Sound Insulation Partition
Using Prefabricated Steel Studs 24" (600 mm) on Centers, Resilient Clips and
Pencil Rods, Lathed Both Sides With 3.4 Lb. (1.5 Kg) 3/8" (9 mm) Rib Metal Lath

	Hours	Rate	Total	Rate	Total
Lather 27.0		$....	$....	$26.58	$717.66
Add for sound treatment:					
Lather 5.0		....	....	26.58	132.90
Cost per 100 sq. yds					$850.56
per sq. yd.........................				....	8.51
per sq.m				....	9.28

Material Cost of 100 Sq. Yds. (83.6 Sq.M) Metal Lath Solid Partitions, Lathed One
Side Only, Using Channels 16" (400 mm) on Centers, Metal Floor and Ceiling
Runners and 2.5 Lb. (1.12 Kg) Diamond Mesh Metal Lath

	Rate	Total	Rate	Rate, (metric)	Total
690 l.f. (207 m) 3/4" (19 mm) cold rolled channel*.............................$....		$....	$.17	$0.56	$117.30
200 l.f. (60 m) Z-type runner, slotted		....	0.19	0.63	38.00
Nails or screws—allowance.........................		....	26.50	26.50	26.50
105 sq. yds. (87.8 sq.m) diamond mesh		....	1.27	1.52	133.35
8 lbs. (3.6 kg) 18 ga. wire............................		....	0.95	2.11	7.60
Cost per 100 sq. yds.			$....		$322.75
per sq. yd.........................				....	3.23
per sq.m				....	3.86

Labor Cost of 100 Sq. Yds. (83.6 Sq.M) Metal Lath Solid Partitions, Lathed One Side Only, Using Channels 16" (400 mm) on Centers, Metal Floor and Ceiling Runners and 2.5 Lb. (1.12 Kg) Diamond Mesh Metal Lath

	Hours	Rate	Total	Rate	Total
Lather	37.0	$....	$....	$26.58	$983.46
Cost per sq. yd.			....		9.83
per sq.m			....		10.71

Material Cost of 100 Sq. Yds. (83.6 Sq.M) of Hollow Partitions (Concealment) Using Double Rows of 3/4" (19 mm) Channel Studs at 16" (400 mm) on Centers and 3.4 Lb. (1.5 Kg) Diamond Mesh Metal Lath

	Rate	Total	Rate	Rate (metric)	Total
152 pcs. 3/4" (19 mm) studs x 9'-2" (2.75 m) 1390 l.f. (417 m): tops of studs tied to ceiling furring channels*	$....	$....	$0.17	$0.56	$236.30
208 l.f. (62.4 m) 3/4" (19 mm) stiffener		....	0.17	0.56	35.36
152 pcs. 3/4" (19 mm) spacer 512 l.f. (153.6 m): based on 2' (0.6 m) between rows of studs		....	0.17	0.56	87.04
200 l.f. (60 m) snap-on metal base		....	0.48	1.60	96.00
152 single base clips, furring clip		....	0.08	0.08	12.16
Nails or expansion drive-allowance		....	26.50	26.50	26.50
210 sq. yds. (175.5 sq.m) diamond mesh		....	1.59		333.90
17 lbs. (7.65 kg) 18 ga. wire		....	0.95		16.15
Cost per 100 sq. yds.			$....		$843.41
per sq. yd.			....		8.43
per sq.m			....		10.09

*Add necessary extras for framing around opening, returns, etc.

Labor Cost of 100 Sq. Yds. (83.6 Sq.M) of Hollow Partitions (Concealment) Using Double Rows of 3/4" (19 mm) Channel Studs at 16" (400 mm) on Centers and 3.4 Lb. (1.5 Kg) Diamond Mesh Metal Lath

	Hours	Rate	Total	Rate	Total
Lather	32.0	$....	$....	$26.58	$850.56
Cost per sq. yd.			....		8.51
per sq.m			....		9.28

Material Cost of 100 Sq. Yds. (83.6 Sq.M) of Vertical Furring Using 2.75 Lb. (1.24 Kg) Flat Rib Metal Lath and 3/4" (19 mm) Channels 16" (400 mm) on Centers

	Rate	Total	Rate	Rate, (metric)	Total
76 pcs. 3/4" (19 mm) studs x 9'-0" (2.7 m) 684 l.f. (205.2 m)*...........................$....	$....	$0.17	$0.56	$116.28
104 l.f. (31.2 m): 3/4" (19 mm) stiffener		....	0.17	0.56	17.68
51 pcs. 3/4" (19 mm) braces x 2'-0" (0.6 m) 102 l.f. (30.6 m).............................		....	0.17	0.56	17.34
200 l.f. (60 m) double-pronged ceiling-floor runner		....	0.25	0.83	50.00
Nails or expansion drives-allowance		....	26.50	26.50	26.50
105 sq. yds. (87.8 sq.m) flat rib metal lath 166.95		....	....	1.59	1.90
8 lbs. (3.6 kg) 18 ga. wire........................		....	0.95	2.11	7.60
Cost per 100 sq. yds.			$....		$402.35
per sq. yd..................................				....	4.02
per sq.m				....	4.81

*Add necessary extras for framing around opening, for soffits, returns, etc.

Labor Cost of 100 Sq. Yds. (83.6 Sq.M) of Vertical Furring Using 2.75 Lb. (1.24 Kg) Flat Rib Metal Lath and 3/4" (19 mm) Channels 16" (400 mm) on Centers

	Hours	Rate	Total	Rate	Total
Lather*..................................22.0		$....	$....	$26.58	$584.76
Cost per sq. yd.				....	5.85
per sq.m				....	6.38

Material Cost of 100 L.F. (30 M) of Metal Corner Bead Attached to Lath Bases

	Rate	Total	Rate	Total	
100 l.f. (30 m) corner bead.....................$....	$....	$0.16	$0.53	$16.00
Staples or tie wire		....			1.06
Cost per 100 l.f.			$....		$17.06
per l.f..				....	0.17
per m				....	0.57

Labor Cost of 100 L.F. (30 M) of Metal Corner Bead Attached to Lath Bases

	Hours	Rate	Total	Rate	Total
Lather 2.5		$....	$....	$26.58	$66.45
Cost per l.f....................................				....	0.66
per m				....	2.16

In estimating for other beads, molding, or screeds, substitute proper material price, and for other bases, consult discussion of labor on beads and molding for necessary additions to hours required. Also, if beads or molding are to serve as plaster grounds and all plumbing or straightening work is assigned to the installation of the accessory, add 1/2 hrs. per 100 l.f. (30 m)

ESTIMATING PLASTER QUANTITIES

The following covering capacities and quantities of plastering materials are based on manufacturer specifications.

Gypsum plasters are sold by the ton and are usually packed in 100-lb. (45-kg) sacks. When sand is mixed with plaster in the proportion of 1:1, 1:2, or 1:3, it means 100, 200, or 300 lbs. (45, 90, or 135 kg) of sand respectively are to be added to each 100-lb. (45-kg) sack of plaster.

Certain fire-resistant ratings may require an increase in the ratio of plaster to aggregate or an increase in the thickness of the plaster over normal grounds, or both, and the estimator must take this into consideration in using the following data.

Covering Capacity. The covering capacity of plaster necessarily depends upon conditions under which the plaster is applied—the mechanic, thickness of grounds, kind and trueness of the surface to be plastered, type of lath, quality of sand, etc.

The covering capacity given for each type of material and for the various surfaces is based on average conditions and may be accepted as a safe basis for calculating the quantity of each class of material required when applied according to the manufacturer's specifications.

Thickness of grounds has considerable bearing on the covering capacity of plaster. The quantities given are based on the following: 3/4" (19 mm) grounds for metal lath, or about 5/8" (16 mm) of plaster over the face of the lath; 7/8" (22 mm) grounds for 3/8" (9 mm) gypsum lath or 1/2" (13 mm) of plaster; 1" (25 mm) for 1/2" (13 mm) insulating board lath or 1/2" (13 mm) of plaster and 5/8" (16 mm) for unit masonry.

Number of 100 Lb. (45 Kg) Sacks of Gypsum Cement Plaster Required
Per 100 Sq. Yds. (83.6 Sq.M)

Metal Lath	Kind of Plastering Surface Gypsum Lath	Unit Masonry
18 to 20	9 to 11	10 to 12
(Sanded 1:2;1:3)	(Sanded 1:2;1:3)	(Sanded 1:3)

For each 1/8" (3 mm) in Thickness, Proportions 1:3; Add or Deduct Sacks:
2.5 .... 2.5 .... 2.5

For insulation board lath, use same plaster quantities as given for gypsum lath.

Standard specifications specify that the first (scratch) coat on all types of lath shall be mixed in the proportions of 1 part gypsum neat plaster to not more than 2 parts of sand by weight, or 100 lbs. (45 kg) of gypsum neat plaster to 2 c.f. (0.06 cu.m) of perlite or vermiculite aggregate.

First coat on masonry surfaces, except monolithic concrete, and second (brown) coat on all lath bases shall be mixed in the proportions of 1 part gypsum neat plaster to not more than 3 parts of sand by weight, or 100 lbs. (45 kg) of gypsum neat plaster to 2 c.f. (0.06 cu.m) of perlite or vermiculite aggregate.

The sacks of gypsum cement plaster shown in the table above per 100 sq. yds. (83.6 sq.m) for various bases are based on these proportions when used with sand aggregate.

Perlite and Vermiculite Plaster

Vermiculite is a mineral that expands when heated at about 2000°F (1093.33°C). Each ore particle expands to about 12 times its original size. It will not rot, decay, burn, or be attacked by vermin or termites.

Perlite is a siliceous volcanic rock mined in western United States. When crushed and quickly heated to above 1500°F (815.55°C), it expands to form lightweight, noncombustible, inert glass-like particles of cellular structure. This material, white in color, is about 1/10th the weight of sand.

Perlite or vermiculite plaster aggregate is a lightweight material. Compared with 100 lbs. per c.f. (1589 kg per cu.m) for sand aggregate, perlite ranges from 7-1/2 to 15 lbs. per c.f. (119-238 kg per cu.m), and vermiculite 6 to 10 lbs. per c.f. (95-159 kg per cu.m). Plaster using this aggregate will weigh about 30 to 45 lbs. per c.f. (477-715 kg per cu.m), depending on proportions used, against 100 lbs. per c.f. (1589 kg per cu.m), where sand aggregate is used.

When perlite or vermiculite is expanded, each granule has countless small dead air spaces, which in addition to making the material very light, gives it a high insulating value. It has approximately three and one half times as much insulating value as ordinary plaster. It is this insulating quality that provides good fire resistance when combined with gypsum plaster. These aggregates are frequently called for when specific fire ratings are required for the protection of steel framing. As such fire ratings usually require greater plaster thicknesses than normal grounds provide, the estimator is cautioned to determine the proper quantities required per 100 sq. yds. (83.6 sq.m).

Perlite or vermiculite plaster aggregate is usually packed in multi-wall paper bags containing 3 or 4 c.f. (0.08-0.11 cu.m). Since there may be some compression or packing in shipping or storage, to insure correct proportioning do not use the original package. Use a cubic foot measuring box.

Machine-applied vermiculite portland cement plaster is used for panel or spandrel walls and exterior columns. The contractor can apply it to exterior columns simultaneously with the wall, since the paper-backed wire fabric used in wall construction is also wrapped around the columns. The protection is also used for interior columns where more than ordinary wear and tear is expected due to the type of occupancy.

The following table gives the approximate quantities of gypsum plaster and perlite or vermiculite for 100 sq. yds. (83.6 sq.m) of various mixtures.

Recommended Proportions and Quantities

In the following table giving recommended proportions for gypsum plaster and perlite or vermiculite, the proportions 1—2, means 100 lbs. (45 kg) of gypsum plaster to 2 c.f. (0.06 cu.m) of perlite or vermiculite; 1—3 means 100 lbs. (45 kg) of gypsum plaster to 3 c.f. (0.08 cu.m) of perlite or vermiculite.

Type of Construction	Grounds (See Note)	Actual Plaster Thickness	Coat	Recommended Volume Proportions	Required per 100 Sq. Yds.	
					Bags Plaster	Cu. Ft. Aggregate
Metal Lath	3/4"	1/4"	Scratch	1—2	10	20
		7/16"	Brown	1—2	12½	45
Gypsum Lath	7/8"	3/16"	Scratch	1—2	5	10
		1/4"	Brown	1—2	10	20
2" Solid Partition	2"	3/8"	Scratch (One side only)	1—2	15	30
		3/8"	Brown (One side only)	1—2	15	30
		1 1/8"	Brown 2nd Side	1—2	45	90
Unit Masonry	5/8"	3/16"	Scratch	1—3	4	12
		3/8"	Brown	1—3	8	24

Grounds attached to studs, joists, or channels—not the lath.
1/16" allowed for finish coat thickness.

Number of 100-Lb. (45-Kg) Sacks or Wood Fiber Plaster Required
Per 100 Sq. Yds. (83.6 Sq.M)

Kind of Plastering Surface

Metal Lath	Gypsum Lath	Unit Masonry
32 to 34	17 to 18	12 to 14
unsanded	unsanded	Equal parts of sand by weight to be added

For each 1/8" (3 mm) in Thickness, Add or Deduct Sacks:

5	4-1/2	3

For insulation board lath, use same plaster quantities as given for gypsum lath.

For two and three-coat work on all types of lath, gypsum wood fiber plaster may be used without the addition of aggregate.

For two and three-coat work on masonry surfaces and as scratch coat on lath (except monolithic concrete), gypsum wood fiber plaster shall be mixed in the proportions of 1 part of plaster to 1 part of sand, or perlite by weight.

Gypsum Ready-Mix Plasters Made with Lightweight Aggregate

Gypsum ready-mix plasters are generally made with perlite. The regular formula for lath bases is packed in 80-lb. (36-kg) sacks and the masonry formula for application over all masonry bases is packed in 67-lb. (30-kg) sacks. The following table gives the number of sacks required per 100 sq. yds. applied over various surfaces:

Kind of Plastering Surface

Metal Lath	Gypsum Lath	Unit Masonry
32 to 34*	17 to 19*	24 to 26**

For Each 1/8" in Thickness, Add or Deduct Sacks

5	5	5

*80-lb. sacks. **67-lb. sacks.

Quantity of Putty Obtained from Various Plastering Materials

The quantity of lime putty obtainable from one ton of lime varies with the kind of lime, whether pebble quicklime, pulverized quicklime, or hydrated lime, also upon the quality of the lime, as lime produced in different parts of the country will produce varying amounts of lime putty. Under average conditions the following quantities should be obtained.

Plastering Material	C.F. of Putty per Ton of Material	Lbs. Req'd for One C.F. Putty
Pebble quicklime	80	25
Pulverized quicklime	80	25
Hydrated lime	46	43-1/2
Gypsum plaster	37	54
Keene's cement	30	66-2/3
Portland cement	18-1/2	109

Base Coat

Covering Capacity of Lime Plaster. Lime plaster is usually proportioned by volume, using 1 volume of stiff, well-aged lime putty to 2 volumes of dry sand, plus necessary fiber for scratch coat.

For the brown coat, use 1 volume of lime putty to 3 volumes of dry sand, plus fiber as needed.

One c.f. (0.028 cu.m) of sand is considered 100 lbs. (45 kg) or 2,700 lbs. per cu. yd. (1589 kg per cu.m).

Covering Capacity of One Cu. Yd. (Cu.M) of Lime Plaster
Based on same thickness of grounds as given on previous pages

	Plastering Surface Metal Lath	Unit Masonry 5/8" (16 mm) Grounds
Scratch Coat	50-60 sq. yds. (55-66 sq.m)	--
Brown Coat	70-80 sq. yds. (77-87 sq.m)	60-75 sq. yds. (66-68 sq.m)

Bondcrete For Interior Concrete Surfaces. Bondcrete is used as a bond coat on rough concrete surfaces only. If the concrete surfaces are smooth, they must be roughened by bush-hammering or hacking, or a dash coat of portland cement grout, composed of 1 part portland cement and 1-1/2 parts of fine sand mixed to a mushy consistency, shall be applied. Use a stiff fiber brush forcibly dashing the mix onto the surface and damp cure for two days and then allow to dry.

Bondcrete adheres to dry concrete surfaces because of its cementitious qualities and because it maintains a practically unchanging volume during the process of setting and hardening.

On properly prepared rough concrete ceilings, apply a coat of bond plaster scratched in thoroughly, doubled backed, and filled out to a true, even surface and left rough, ready to receive the finish coat. Total thickness of the plaster should not exceed 3/8" (9 mm).

On properly prepared rough concrete columns or walls, scratch in a minimum 1/4" (6 mm) coat of bond plaster and then follow with a brown coat of gypsum plaster—one part gypsum neat plaster to not more than three parts sand, by weight—troweled into the scratch coat before it sets. Brown coat shall be brought out to ground, straightened to a true surface, and left rough, ready to receive the finish coat. Total thickness of plaster should not exceed 5/8" (16 mm).

It usually requires 1,700 to 2,200 lbs. (765-990 kg) of Bondcrete per 100 sq. yds. (83.6 sq.m) depending upon the regularity of the concrete surface and thickness applied.

Plaster-Weld. Plaster-Weld permanently bonds new gypsum plaster, finish or base coat, or acoustical plaster to most surfaces, but cannot be used over calcimine or other water soluble coatings.

No chipping or roughening of surfaces is required. The surface must be structurally sound and free from dust, dirt, grease, oil, or wax.

It may be applied by brush, roller, or spray equipment. After allowing Plaster-Weld to become dry to the touch (one hour on porous surfaces and two hours on non-porous surfaces), plaster can be applied either immediately or later. A time lag in excess of a week or ten days may result in a need for recoating. Plaster-Weld covers up to 500 s.f. per gal. (11.89 sq.m per liter) when sprayed, which gives the most satisfactory results. When applied by brush or roller, coverage is reduced. When sprayed on, the labor cost per 100 sq. yds. (83.6 sq.m) should average as follows:

	Hours	Rate	Total	Rate	Total
Labor	2.25	$....	$....	$25.75	$57.94
Cost per sq. yd.			....		0.58
per s.f.			....		0.06
per sq.m			....		0.63

Plaster-Weld costs approximately $20.00 per gal. ($5.28 per liter) in 1-gal. (3.8-L) cans.

Finishing Lime. Finishing lime is essential to two common types of job mixed finish coats—the white putty coat trowel finish and the sand float finish.

For the trowel finish coat, the proportion is generally 1 part of gypsum gauging plaster to 2 parts of hydrated lime on a dry weight basis, or 100 lbs. (45 kg) of gypsum gauging plaster to 4-1/2 c.f. (0.13 cu.m) of lime putty. One ton of gypsum gauging plaster and 2 tons of hydrated finishing lime, or 90 c.f. (2.55 cu.m) of lime putty will finish approximately 1,200 sq. yds. (1003 sq.m).

While a sand float finish may be mixed with 1 part of lime putty and 2-1/2 to 3 parts of sand, it is generally necessary to add Keene's cement or gypsum gauging plaster to give the finish early hardness and strength. A standard mix for a Keene's cement-lime putty sand float finish is 2 parts of lime putty to 1-1/2 parts of Keene's cement and 4-1/2 parts of sand by volume. This is equivalent to 1 ton of hydrated lime, or 45 c.f. (1.27 cu.m) of putty, to 1-1/4 tons (1.13 metric tons) of Keene's cement and 4-3/4 tons (4.28 metric tons) of sand, and this quantity will cover approximately 1,600 sq. yds. (1338 sq.m).

If type "S" (fully) hydrated limes are used, they must be added to the mix in dry form, as they form immediate plasticity on wetting. If quicklimes are used, they must be slaked and properly aged and added to the mix in the volume indicated above for putty.

Prepared Gypsum Trowel Finishes. There are a number of prepared gypsum trowel finishes, which require only adding water to use. A few of these are Universal White Trowel Finish, Red Top Trowel, and Red Top Imperial (from Larson Products Corp., Bethesda, Md.).

Under average conditions, if mixed and applied in accordance with standard specifications, one ton will cover 350 to 400 sq. yds. (293-334 sq.m) of surface, or 500 to 575 lbs. per 100 sq. yds. (2.69-3.10 kg per sq.m).

Prepared Gypsum Sand Float Finishes. There are a number of gypsum sand float finishes. These finishes are ready for use when water is added, and under average conditions one ton is sufficient for 250 to 275 sq. yds. (209-230 sq.m), or it requires 725 to 800 lbs. per 100 sq. yds. (3.9-4.3 kg per sq.m).

Estimating Labor Costs of Plastering

Before it is possible to prepare an accurate estimate on the cost of any plastering job, the quality of work must be considered. Labor costs may vary up to 25 to 30 percent.

The costs given on the following pages cover two grades of workmanship, which are fully described at the beginning of this chapter.

Applying Scratch Coat of Plaster. A plasterer should apply 140 to 165 sq. yds. (117-138 sq.m) of scratch coat per 8-hr. day, where ordinary grade of workmanship is required, and 135 to 155 sq. yds. (113-130 sq.m) per 8-hr. day where first grade workmanship is required.

Applying Brown Coat. Where an ordinary grade of workmanship is required, a plasterer should apply 90 to 110 sq. yds. (75-92 sq.m) of brown coat per 8-hr. day.

Where first grade workmanship is required, a plasterer should brown out 70 to 85 sq. yds. (59-71 sq.m) per 8-hr. day.

Applying Scratch—Double-Back Basecoat. Over masonry bases and in many markets over gypsum lath base, it is common practice to apply a single plaster mix in a "scratch-in" and "double-back" operation to fill out to grounds.

For ordinary work, a plasterer should apply 90 to 110 sq. yds. (75-92 sq.m) per 8-hr. day, and for first grade workmanship, a plasterer should apply 70 to 85 sq. yds. (59-71 sq.m) per 8-hr. day in this manner.

Applying White Finish or "Putty" Coat. On an ordinary job of white coating, a plasterer should apply 90 to 110 sq. yds. (75-92 sq.m) per 8-hr. day.

Where first grade workmanship is specified, a plasterer will apply 60 to 80 sq. yds. (50-67 sq.m) per 8-hr. day.

General Notes Applying to Finish Plaster Coats

All of the tables on the following pages give costs on a finish white coat. If wanted with sand float finish, deduct quantities given in tables for Finish White Coat and add cost of Sand Float Finish as given on the following pages under "Sand Float Finish".

Applying Sand Float Finish. A plasterer should apply 100 to 120 sq. yds. (84-100 sq.m) of simple sand float finish per 8-hr. day, where an ordinary grade of workmanship is required, or 70 to 90 sq. yds. (58-75 sq.m) when first grade workmanship is required.

Labor Applying 100 Sq. Yds. (83.6 Sq.M) of Plaster to Various Plaster Bases
Ordinary Workmanship

Description of Work	Sq. Yds. 8-Hr Day	Sq.M 8-Hr Day	Plasterer Hours	Tender Hours
Gypsum Mortar on Gypsum lath—1/2" (13 mm)				
Plaster—2-Coat Work 1:2-1/2 Brown Coat	90-110	75-92	8.0	5.0
Gypsum Mortar on Unit Masonry—5/8" (16 mm)				
Plaster—2-Coat Work—1:3 Brown Coat	80-100	67-84	9.0	6.0
Lime Mortar on Unit Masonry—5/8" (16 mm)				
Plaster—2-Coat Work Brown Coat 1:3	90-100	75-84	8.5	5.5
Lime Mortar on Metal Lath—3/4" (19 mm)				
Grounds—5/8" (16 mm) Plaster—3-Coat				
Work Scratch Coat 1:2	150-160	125-134	5.0	5.0
Brown Coat 1:3	90-110	75-92	8.0	5.0
Lime Mortar on Unit Masonry—5/8" (16 mm)				
Plaster—3-Coat Work Scratch Coat 1:2	150-160	125-134	5.0	5.0
Brown Coat 1:3	90-110	75-92	8.0	5.0
Gypsum Mortar on Unit Masonry Tile—5/8" (16 mm)				
Plaster—3-Coat Work				
Scratch Coat 1:3	150-160	125-134	5.0	5.0
Brown Coat 1:3	90-110	75-92	8.0	5.0
Gypsum Mortar on Gypsum Lath—1/2" (13 mm)				
Plaster—3-Coat Work				
Scratch Coat 1:2	150-160	125-134	5.0	5.0
Brown Coat 1:3	90-110	75-92	8.0	5.0
Gypsum Mortar on Metal Lath—3/4" (19 mm)				
Grounds—5/8" (16 mm) Plaster—				
3-Coat Work				
Scratch Coat 1:2	150-160	125-134	5.0	5.0
Brown Coat 1:3	85-100	71-84	9.0	6.0
Two-Inch Solid Plaster				
Partitions—Metal Lath One				
Side—Back Up—Finished				
2 Sides—Gypsum Mortar—3/4" (19 mm)				
Grounds—2" (50 mm) Plaster				
Scratch Coat 1:2	150-160	125-134	5.0	5.0
Back Up Coat 1:2	150-160	125-134	5.0	5.0
2 Brown Coats 1:2	40-50	33-42	18.0	12.0

Labor Applying 100 Sq. Yds. (83.6 Sq.M) of Plaster to Various Plaster Bases
Ordinary Workmanship

Description of Work	Sq. Yds. 8-Hr Day	Sq.M 8-Hr Day	Plasterer Hours	Tender Hours
Hollow Stud Partitions—Lathed				
2 Sides—Plastered to 3/4" (19 mm)				
Grounds				
Scratch Coat 1:2—2 Sides	75-80	63-67	10	10
Brown Coat—2 Sides	40-50	33-42	18	12
Finish White—All Classes of Undercoat	90-110	75-92	8	4
Sand Finish—All Classes of Undercoat	100-120	84-100	7.25	5
First Grade Workmanship—Where First Grade Workmanship is				
Required as described on the previous pages, add to Hours				
Given above				
Brown Coat			2.25	1
Finish White Coat			3.50	1
Sand Finish			2.75	1

Applying Finish Coat of Cement Wainscoting. Where cement plaster finishes are used in kitchens, bathrooms, laundries, corridors, etc., using Keene's cement, or similar finishes applied over a brown coat of gypsum cement plaster or portland cement, a plasterer should apply 50 to 60 sq. yds. (42-50 sq.m) per 8-hr. day.

If the wainscoting is blocked off into squares, 3"x6" (75 x 150 mm), 4"x4" (100 x 100 mm) or 6"x6" (150 x 150 mm) to represent tile, a plasterer should block off 45 to 55 sq. yds. (38-46 sq.m) per 8-hr. day.

To apply the finish coat of plaster and block same off into squares to represent tile, a plasterer should complete 25 to 30 sq. yds. (21-25 sq.m) per 8-hr. day.

Where first grade workmanship is required, a plasterer should apply 35 to 45 sq. yds. (29-38 sq.m) of cement wainscoting per 8-hr. day, and block off 40 to 45 sq. yds. (33-38 sq.m) or will apply cement wainscoting and block off 18 to 22 sq. yds. (15-18 sq.m) per 8-hr. day.

Itemized Plastering Costs

The itemized cost given on the following pages are intended to furnish the estimator and contractor detailed information regarding material quantities and labor costs of all kinds of plastering.

All plaster quantities are based on thickness of grounds as given. If the work is "skimped" and the plaster does not fill out the grounds or is applied thinner than specified, then the quantities in the tables will not be correct. If sand is not added in accordance with manufacturer's specifications, the quantities may be more or less than given. Rates given for quantities in lbs. (kg) are usually for a 100-lb. (100-kg) bag. (Note that metric figures are based on a 100-kg bag, which equals about 220 lbs.)

Labor hours include time required for building and removing ordinary scaffolding, but will not cover scaffolding for high work in churches, auditoriums, etc.

2-Coat Plastering

Material Cost of 100 Sq. Yds. (83.6 Sq.M) 2-Coat Gypsum Cement Plaster, White Finish on Gypsum Lath
Ordinary Workmanship—3/8" (9 mm) Grounds—1/2" (13 mm) Plaster

Brown Coat—1:2.5	Rate	Total	Rate	Rate (metric)	Total
1,000 lbs. (450 kg) gypsum plaster ....$....	$....	$6.40*	$14.22**	$64.00	
25 c.f. (0.70 cu.m) sand	....	0.48	17.14	12.00	
Finish White Coat					
340 lbs. (153 kg) fin. hydrated lime	....	8.50*	18.88**	28.90	
170 lbs. (76.5 kg) gauging plaster	....	9.55*	21.22**	16.24	
Cost 100 sq. yds..	$....			$121.14	
per sq. yd...	....			1.21	
per sq.m	....			1.45	

*Based on 100-lb. bag.
*Metric figures based on 100-kg bag.
Add for gypsum lath.

Labor Cost of 100 Sq. Yds. (83.6 Sq.M) 2-Coat Gypsum Cement Plaster, White Finish on Gypsum Lath
Ordinary Workmanship—7/8" (22 mm) Grounds—1/2" (13 mm) Plaster

Brown Coat—1:2.5	Hours	Rate	Total	Rate	Total
Plasterer 8.0	$....	$....	$25.75	$206.00	
Labor..5.0	....	....	21.04	105.20	
Finish White Coat					
Plasterer 8.0	....	....	25.75	206.00	
Labor..4.0	....	....	21.04	84.16	
Cost 100 sq. yds..................................		$....		$601.36	
per sq. yd..		....		6.01	
per sq.m		....		6.55	

Add for gypsum lath.

Material Cost of 100 Sq. Yds. (83.6 Sq.M) 2-Coat Gypsum Cement Plaster, White Finish on Unit Masonry
Ordinary Workmanship—5/8" (16 mm) Grounds

Brown Coat—1:3	Rate	Total	Rate	Rate (metric)	Total
1,100 lbs. (495 kg) gypsum plaster ....$....	$....	$6.40	$14.22	$70.40	
33 c.f. (0.93 cu.m) sand	....	0.48	17.14	15.84	

Finish White Coat

340 lbs. (153 kg) fin. hydrated lime	....	8.50	18.88	28.90
170 lbs. (76.5 kg) gauging plaster	....	9.55	21.22	16.24
Cost 100 sq. yds.	$....			$131.38
per sq. yd.		....		1.31
per sq.m		....		1.56

Labor Cost of 100 Sq. Yds. (83.6 Sq.M) 2-Coat Gypsum Cement Plaster, White
Finish on Unit Masonry
Ordinary Workmanship—5/8" (16 mm) Grounds

Brown Coat—1:3	Hours	Rate	Total	Rate	Total
Plasterer	9.0	$....	$....	$25.75	$231.75
Labor	6.0	....	....	21.04	126.24
Finish White Coat					
Plasterer	8.0	....	....	25.75	206.00
Labor	4.0	....	....	21.04	84.16
Cost 100 sq. yds.			$....		$648.15
per sq. yd.				....	6.48
per sq.m				....	7.06

Material Cost of 100 Sq. Yds. (83.6 Sq.M) 2-Coat Wood Fiber Plaster, White Finish
on Gypsum Lath
Ordinary Workmanship—7/8" (22 mm) Grounds—1/2" (13 mm) Plaster

Brown Coat	Rate	Total	Rate	Rate (metric)	Total
1,700 lbs. (765 kg) wood fiber .... $....		$....	$6.40	$14.22	$108.80
Finish White Coat					
340 lbs. (153 kg) fin. hydrated lime		....	8.50	18.88	28.90
170 lbs. (76.5 kg) gauging plaster		....	9.55	21.22	16.24
Cost 100 sq. yds.		$....			$153.94
per sq. yd.				....	1.54
per sq.m				....	1.84
Add for gypsum lath.					

Labor Cost of 100 Sq. Yds. (83.6 Sq.M) 2-Coat Wood Fiber Plaster, White Finish
on Gypsum Lath
Ordinary Workmanship—7/8" (22 mm) Grounds—1/2" (13 mm) Plaster

Brown Coat	Hours	Rate	Total	Rate	Total
Plasterer	9.0	$....	$....	$25.75	$231.75
Labor	5.0	....	....	21.04	105.20
Finish White Coat					
Plasterer	8.0	....	....	25.75	206.00
Labor	4.0	....	....	21.04	84.16
Cost 100 sq. yds.			$....		$627.11
per sq. yd.				....	6.27
per sq.m				....	6.84

Bondcrete plaster can only be applied to rough concrete surfaces. If the concrete surfaces are smooth, they must be roughened by bush-hammering or hacking, or a dash coat of portland cement grout, composed of 1 part portland cement and 1-1/2 parts fine sand, mixed to a mushy consistency, must be applied. Use a stiff fiber brush forcibly dashing the mix onto the surface and damp cure for two days and then allow to dry. For costs on this work, refer to section on waterproofing.

Material Cost of 100 Sq. Yds. (83.6 Sq.M) 2-Coat Bondcrete Plaster,
White Finish on Concrete Ceilings
Ordinary Workmanship—Plaster 3/8" (9 mm) Thick

Brown Coat	Rate	Total	Rate	Rate (metric)	Total
2,000 lbs. (900 kg) Bondcrete ....$....	$....	$7.42	$16.48	$148.40
Finish White Coat					
340 lbs. (153 kg) fin. hydrated lime		....	8.50	18.88	28.90
170 lbs. (76.5 kg) gauging plaster		....	9.55	21.22	16.24
Cost 100 sq. yds..			$....		$193.54
per sq. yd..				....	1.94
per sq.m				....	2.32

Labor Cost of 100 Sq. Yds. (83.6 Sq.M) 2-Coat Bondcrete Plaster,
White Finish on Concrete Ceilings
Ordinary Workmanship—Plaster 3/8" (9 mm) Thick

Brown Coat	Hours	Rate	Total	Rate	Total
Plasterer 8.0		$....	$....	$25.75	$206.00
Labor...5.0		....	....	21.04	105.20
Finish White Coat					
Plasterer 8.0		....	....	25.75	206.00
Labor...4.0		....	....	21.04	84.16
Cost 100 sq. yds..			$....		$601.36
per sq. yd..				....	6.01
per sq.m				....	6.55

3-Coat Plastering

Material Cost of 100 Sq. Yds. (83.6 Sq.M) 3-Coat Gypsum Cement Plaster,
White Finish on Gypsum Lath
Ordinary Workmanship—7/8" (22 mm) Grounds—1/2" (13 mm) Plaster

Scratch Coat—1:2	Rate	Total	Rate	Rate (metric)	Total
400 lbs. (180 kg) gypsum plaster ....$....	$....	$6.40	$14.22	$25.60
8 c.f. (0.22 cu.m) sand		....	0.48	17.14	3.84

Brown Coat—1:3	Rate	Total	Rate	Rate (metric)	Total
700 lbs. (315 kg) gypsum plaster		....	6.40	14.22	44.80
21 c.f. (0.59 cu.m) sand		....	0.48	17.14	10.08
Finish White Coat					
340 lbs. (153 kg) fin. hydrated lime		....	8.50	18.88	28.90
170 lbs. (76.5 kg) gauging plaster		....	9.55	21.22	16.24
Cost 100 sq. yds.				$....	$129.46
per sq. yd.				....	1.29
per sq.m				....	1.55

Add for gypsum lath.

Labor Cost of 100 Sq. Yds. (83.6 Sq.M) 3-Coat Gypsum Cement Plaster, White Finish on Gypsum Lath
Ordinary Workmanship—7/8" (22 mm) Grounds—1/2" (13 mm) Plaster

Scratch Coat—1:2	Hours	Rate	Total	Rate	Total
Plasterer	5.0	$....	$....	$25.75	$128.75
Labor	5.0			21.04	105.20
Brown Coat—1:3					
Plasterer	8.0	....	....	25.75	206.00
Labor	5.0	....	....	21.04	105.20
Finish White Coat					
Plasterer	8.0	....	....	25.75	206.00
Labor	4.0	....	....	21.04	84.16
Cost 100 sq. yds.			$....		$835.31
per sq. yd.			....		8.35
per sq.m			....		9.10

If first grade workmanship is required, add 5.75 hrs. plasterer time and 2 hrs. labor time per 100 sq. yds. (83.6 sq.m).

Material Cost of 100 Sq. Yds. (83.6 Sq.M) 3-Coat Gypsum Cement Plaster, White Finish on Unit Masonry
Ordinary Workmanship—5/8" (16 mm) Grounds

Scratch Coat—1:3	Rate	Total	Rate	Rate (metric)	Total
500 lbs. (225 kg) gypsum plaster ....$....		$....	$6.40	$14.22	$32.00
15 c.f. (0.42 cu.m) sand		....	0.48	17.14	7.20
Brown Coat—1 :3					
700 lbs. (315 kg) gypsum plaster		....	6.40	14.22	44.80
21 c.f. (0.59 cu.m) sand		....	0.48	17.14	10.08
Finish White Coat					
340 lbs. (153 kg) fin. hydrated lime		....	8.50	18.88	28.90
170 lbs. (76.5 kg) gauging plaster		....	9.55	21.22	16.24
Cost 100 sq. yds.				$....	$139.22
per sq. yd.				....	1.39
per sq.m				....	1.67

Labor Cost of 100 Sq. Yds. (83.6 Sq.M) 3-Coat Gypsum Cement Plaster, White
Finish on Unit Masonry
Ordinary Workmanship—5/8" (16 mm) Grounds

Scratch Coat—1:3	Hours	Rate	Total	Rate	Total
Plasterer	5.0	$....	$....	$25.75	$128.75
Labor	5.0	....	....	21.04	105.20
Brown Coat—1:3					
Plasterer	8.0	....	....	25.75	206.00
Labor	5.0	....	....	21.04	105.20
Finish White Coat					
Plasterer	8.0	....	....	25.75	206.00
Labor	4.0	....	....	21.04	84.16
Cost 100 sq. yds.			$....		$835.31
per sq. yd.				....	8.35
per sq.m				....	9.10

If first grade workmanship is required, add 5.75 hrs. plasterer time and 2
hrs. labor time per 100 sq. yds. (83.6 sq.m).

Material Cost of 100 Sq. Yds. (83.6 Sq.M) 3-Coat Gypsum Cement Plaster,
White Finish on Metal Lath Walls and Ceilings
Ordinary Workmanship—3/4" (19 mm) Grounds—5/8" (16 mm) Plaster

Scratch Coat—1:2	Rate	Total	Rate	Rate (metric)	Total
800 lbs. (360 kg) gypsum plaster	$....	$....	$6.40	$14.22	$51.20
16 c.f. (0.45 cu.m) sand		....	0.48	17.14	7.68
Brown Coat—1 :3					
1,200 lbs. (540 kg) gypsum plaster		....	6.40	14.22	76.80
36 c.f. (1.01 cu.m) sand		....	0.48	17.14	17.28
Finish White Coat					
340 lbs. (153 kg) fin. hydrated lime		....	8.50	18.88	28.90
170 lbs. (76.5 kg) gauging plaster		....	9.55	21.22	16.24
Cost 100 sq. yds.			$....		$198.10
per sq. yd.				....	1.98
per sq.m				....	2.37

Add for metal lath and furring.

Labor Cost of 100 Sq. Yds. (83.6 Sq.M) 3-Coat Gypsum Cement Plaster,
White Finish on Metal Lath Walls and Ceilings
Ordinary Workmanship—3/4" (19 mm) Grounds—5/8" (16 mm) Plaster

Scratch Coat—1:2	Hours	Rate	Total	Rate	Total
Plasterer	5.0	$....	$....	$25.75	$128.75
Labor	5.0	....	....	21.04	105.20

Brown Coat—1:3	Hours	Rate	Total	Rate	Total
Plasterer	9.0	....	....	25.75	231.75
Labor	6.0	....	....	21.04	126.24
Finish White Coat					
Plasterer	8.0	....	....	25.75	206.00
Labor	4.0	....	....	21.04	84.16
Cost 100 sq. yds.			$....		$882.10
per sq. yd.				....	8.82
per sq.m				....	9.61

Add for metal lath and furring.

If first grade workmanship is required, add 5.75 hrs. plasterer time and 2 hrs. labor time per 100 sq. yds. (83.6 sq.m).

Material Cost of 100 Sq. Yds. (83.6 Sq.M) 2" (50 mm) Solid Plaster Partition on Metal Lath, Ordinary Workmanship

Scratch Coat—1:2	Rate	Total	Rate	Rate (metric)	Total
2,000 lbs. (900 kg) gypsum plaster	$....	$....	$6.40	$14.22	$128.00
40 c.f. (1.13 cu.m) sand		....	0.48	17.14	19.20
Brown Coat—1:3					
2,800 lbs. (1260 kg) gypsum plaster		....	6.40	14.22	179.20
84 c.f. (2.38 cu.m) sand		....	0.48	17.14	40.32
Finish White Coat					
680 lbs. (306 kg) fin. hydrated lime		....	8.50	18.88	57.80
340 lbs. (153 kg) gauging plaster		....	9.55	21.22	32.47
Cost 100 sq. yds.			$....		$456.99
per sq. yd.				....	4.57
per sq.m				....	5.46

Measurement taken on one side only. Add for metal lath and furring.

Labor Cost of 100 Sq. Yds. (83.6 Sq.M) 2" (50 mm) Solid Plaster Partition on Metal Lath, Ordinary Workmanship

Scratch Coat—1:2	Hours	Rate	Total	Rate	Total
Plasterer	10.0	$....	$....	$25.75	$257.50
Labor	10.0	....	....	21.04	210.40
Brown Coat—1:3					
Plasterer	18.0	....	....	25.75	463.50
Labor	12.0	....	....	21.04	252.48
Finish White Coat					
Plasterer	16.0	....	....	25.75	412.00
Labor	8.0	....	....	21.04	168.32
Cost 100 sq. yds.			$....		$1,764.20
per sq. yd.				....	17.64
per sq.m				....	19.23

Measurement taken on one side only. Add for metal lath and furring.

If first grade workmanship is required, add 11.5 hrs. plasterer time and 4 hrs. labor time per 100 sq. yds. (83.6 sq.m).

Material Cost of 100 Sq. Yds. (83.6 Sq.M) 2" (50 mm)
Solid Gypsum Long Length Lath and Plaster Partition
Ordinary Workmanship
Partition 2" (50 mm) Thick; 1/2" (13 mm) Gypsum Lath; 3/4" (19 mm) of Plaster
Each Side of Lath

Scratch Coat—1:2	Rate	Total	Rate	Rate (metric)	Total
800 lbs. (360 kg) gypsum plaster ....$....	$....	$6.40	$14.22	$51.20
16 c.f. (0.45 cu.m) sand	....	0.48	17.14	7.68
Brown Coat—1:3					
2,400 lbs. (1080 kg) gypsum plaster	....	6.40	14.22	153.60
72 c.f. (2.03 cu.m) sand	....	0.48	17.14	34.56
Finish White Coat					
680 lbs. (306 kg) fin. hydrated lime	....	8.50	18.88	57.80
340 lbs. (153 kg) gauging plaster	....	9.55	21.22	32.47
Cost 100 sq. yds.		$....		$337.31
per sq. yd.			....	3.37
per sq.m			....	4.03
Add for gypsum lath.					

Labor Cost of 100 Sq. Yds. (83.6 Sq.M) 2" (50 mm)
Solid Gypsum Long Length Lath and Plaster Partition
Ordinary Workmanship
Partition 2" (50 mm) Thick; 1/2" (13 mm) Gypsum Lath; 3/4" (19 mm) of Plaster
Each Side of Lath

Scratch Coat—1:2	Hours	Rate	Total	Rate	Total
Plasterer 10.0		$....	$....	$25.75	$257.50
Labor 10.0		....	....	21.04	210.40
Brown Coat—1:3					
Plasterer 16.0		....	....	25.75	412.00
Labor 10.0		....	....	21.04	210.40
Finish White Coat					
Plasterer 16.0		....	....	25.75	412.00
Labor 8.0		....	....	21.04	168.32
Cost 100 sq. yds.			$....		$1,670.62
per sq. yd.				....	16.71
per sq.m				....	18.21
Add for gypsum lath.					

Material Cost of 100 Sq. Yds. (83.6 Sq.M) 2" (50 mm)
Solid Studless Metal Lath and Plaster Partition
Ordinary Workmanship
Partition 2" (50 mm) Thick; 3.4-lb. (1.53-kg) Metal Lath; Plaster 2" (50 mm) Thick

Scratch Coat—1:2	Rate	Total	Rate	Rate (metric)	Total
2,000 lbs. (900 kg) gypsum plaster ....$....	$....	$6.40	$14.22	$128.00
40 c.f. (1.13 cu.m) sand		....	0.48		19.20
Brown Coat—1:3					
2,800 lbs. (1260 kg) gypsum plaster		....	6.40	14.22	179.20
84 c.f. (2.38 cu.m) sand		....	0.48	17.14	40.32
Finish White Coat					
680 lbs. (306 kg) fin. hydrated lime		....	8.50	18.88	57.80
340 lbs. (153 kg) gauging plaster		....	9.55	21.22	32.47
Cost 100 sq. yds.			$....		$456.99
per sq. yd.			....		4.57
per sq.m			....		5.46

Add for metal lath and furring.

Labor Cost of 100 Sq. Yds. (83.6 Sq.M) 2" (50 mm)
Solid Studless Metal Lath and Plaster Partition
Ordinary Workmanship
Partition 2" (50 mm) Thick; 3.4-lb. (1.53 kg) Metal Lath; Plaster 2" (50 mm) Thick

Scratch Coat—1:2	Hours	Rate	Total	Rate	Total
Plasterer 10.0		$....	$....	$25.75	$257.50
Labor 10.0		....	....	21.04	210.40
Brown Coat—1:3					
Plasterer 18.0		....	....	25.75	463.50
Labor 12.0		....	....	21.04	252.48
Finish White Coat					
Plasterer 16.0		....	....	25.75	412.00
Labor 8.0		....	....	21.04	168.32
Cost 100 sq. yds.			$....		$1,764.20
per sq. yd.			....		17.64
per sq.m			....		19.23

Add for metal lath and furring.

Material Cost of 100 Sq. Yds. (83.6 Sq.M) 3-Coat Wood Fiber Plaster,
White Finish on Metal Lath Walls and Ceilings
Ordinary Workmanship—3/4" (19 mm) Grounds—5/8" (16 mm) Plaster

Scratch Coat	Rate	Total	Rate	Rate (metric)	Total
1,100 lbs. (495 kg) wood fiber ....$....	$....	$6.40	$14.22	$70.40
Brown Coat					
2,300 lbs. (1035 kg) wood fiber		....	6.40	14.22	147.20

Finish White Coat	Rate	Total	Rate	Rate (metric)	Total
340 lbs. (153 kg) fin. hydrated lime	....		8.50	18.88	28.90
170 lbs. (76.5 kg) gauging plaster	....		9.55	21.22	16.24
Cost 100 sq. yds.			$....		$262.74
per sq. yd.				....	2.63
per sq.m				....	3.14
Add for metal lath and furring.					

Labor Cost of 100 Sq. Yds. (83.6 Sq.M) 3-Coat Wood Fiber Plaster, White Finish on Metal Lath Walls and Ceilings
Ordinary Workmanship—3/4" (19 mm) Grounds—5/8" (16 mm) Plaster

Scratch Coat	Hours	Rate	Total	Rate	Total
Plasterer 5.5		$....	$....	$25.75	$141.63
Labor 5.0		....	....	21.04	105.20
Brown Coat					
Plasterer 10.0		....	....	25.75	257.50
Labor 6.0		....	....	21.04	126.24
Finish White Coat					
Plasterer 8.0		....	....	25.75	206.00
Labor 4.0		....	....	21.04	84.16
Cost 100 sq. yds.			$....		$920.73
per sq. yd.				....	9.21
per sq.m				....	10.04

If first grade workmanship is required, add 5.75 hrs. plasterer time and 2 hrs. labor time per 100 sq. yds. (83.6 sq.m).

On properly prepared rough concrete columns or walls, scratch in a minimum 1/4" (6 mm) coat of bond plaster and then follow with a brown coat of gypsum plaster (1 part gypsum neat plaster to not more than 3 parts sand, by weight) troweled into the scratch coat before it sets. Brown coat shall be brought out to grounds, straightened to a true surface, and left rough ready to receive the finish coat. Total thickness of plaster not to exceed 5/8" (16 mm). Add for preparing concrete surfaces, if required.

Material Cost of 100 Sq. Yds. (83.6 Sq.M) 3-Coat Bondcrete Plaster, White Finish on Concrete Walls and Columns
Ordinary Workmanship—Plaster 5/8" (16 mm) Thick

Scratch Coat—1/4" (6 mm)	Rate	Total	Rate	Rate (metric)	Total
1,500 lbs. (675 kg) Bondcrete .... $....		$....	$7.42	$16.48	$111.30
Brown Coat—1 :3					
800 lbs. (360 kg) gypsum plaster		....	6.40	14.22	51.20
24 c.f. (0.68 cu.m) sand		....	0.48	17.14	11.52

Finish White Coat	Rate	Total	Rate	Rate (metric)	Total
340 lbs. (153 kg) fin. hydrated lime	....		8.50	18.88	28.90
170 lbs. (76.5 kg) gauging plaster	....		9.55	21.22	16.24
Cost 100 sq. yds.			$....		$219.16
per sq. yd.				....	2.19
per sq.m				....	2.62

Labor Cost of 100 Sq. Yds. (83.6 Sq.M) 3-Coat Bondcrete Plaster, White Finish on Concrete Walls and Columns
Ordinary Workmanship—Plaster 5/8" (16 mm) Thick

	Hours	Rate	Total	Rate	Total
Scratch Coat—1/4" (6 mm)					
Plasterer 5.0		$....	$....	$25.75	$128.75
Labor 5.0		....	....	21.04	105.20
Brown Coat—1:3					
Plasterer 8.0		....	....	25.75	206.00
Labor 5.0		....	....	21.04	105.20
Finish White Coat					
Plasterer 8.0		....	....	25.75	206.00
Labor 4.0		....	....	21.04	84.16
Cost 100 sq. yds.			$....		$835.31
per sq. yd.				....	8.35
per sq.m				....	9.10

If first grade workmanship is required, add 5.75 hrs. plasterer time and 2 hrs. labor time per 100 sq. yds. (83.6 sq.m).

Plastering Radiant Panel Heat Ceilings. Where radiant panel heating coils are embedded in ceiling plaster, base coat plastering costs will be higher than normal for both material and labor.

For this type of heating, 3/8" (9 mm) I.D.—1/2" (13 mm) O.D.—copper tubing or electric cable is secured to ceiling framing after lath is installed, in contact with the underside. For proper operation of the heating system, all ceiling coils must be completely embedded in plaster, in perfect contact with plaster (no air pockets permitted) and have a plaster cover of 3/8" (9 mm) or more. This means at least 7/8" (22 mm) of plaster below the face of the lath for copper tubing and 1/2" (13 mm) for electric cable.

Scratch coat application is more difficult, due to the care which must be taken to obtain proper contact with tubing. Brown coat application will require more labor, due to the increased thickness of material required and the difficulty in making it stick to the ceiling. In some cases, a second scratch coat may be necessary to obtain the required thickness.

There should be no change in the method or cost of applying the various plaster finish coats.

Use sand aggregate only for radiant heat ceilings and walls. Lightweight aggregate is an insulator.

Material Cost of 100 Sq. Yds. (83.6 Sq.M)
3-Coat Gypsum Plaster 7/8" (22 mm) Thick,
Applied to Gypsum Lath Ceilings with Radiant Heating Coils
Embedded in the Plaster
Ordinary Workmanship—7/8" (22 mm) Plaster

Scratch Coat—1:2	Rate	Total	Rate	Rate (metric)	Total
400 lbs. (180 kg) gypsum plaster ....$....	$....	$6.40	$14.22	$25.60
8 c.f. (0.22 cu.m) sand		....	0.48	17.14	3.84
Brown Coat—1:3					
1,400 lbs. (630 kg) gypsum plaster		....	6.40	14.22	89.60
42 c.f. (1.19 cu.m) sand		....	0.48	17.14	20.16
Finish White Coat					
340 lbs. (153 kg) fin. hydrated lime		....	8.50	18.88	28.90
170 lbs. (76.5 kg) gauging plaster		....	9.55	21.22	16.24
Cost 100 sq. yds..		$....			$184.34
per sq. yd...			....		1.84
per sq.m			....		2.20

Labor Cost of 100 Sq. Yds. (83.6 Sq.M)
3-Coat Gypsum Plaster 7/8" (22 mm) Thick, Applied
to Gypsum Lath Ceilings with Radiant Heating Coils
Embedded in the Plaster
Ordinary Workmanship—7/8" (22 mm) Plaster

Scratch Coat—1:2	Hours	Rate	Total	Rate	Total
Plasterer 6.0		$....	$....	$25.75	$154.50
Labor...6.0		....	....	21.04	126.24
Brown Coat—1:3					
Plasterer 10.0		....	....	25.75	257.50
Labor...6.0		....	....	21.04	154.50
Finish White Coat					
Plasterer 8.0		....	....	25.75	206.00
Labor...4.0		....	....	21.04	84.16
Cost 100 sq. yds....................................			$....		$982.90
per sq. yd...				....	9.83
per sq.m				....	10.71

Add for gypsum lath.

Material Cost of 100 Sq. Yds. (83.6 Sq.M)
of 3-Coat Gypsum Plaster 5/8" (16 mm) Thick,
Applied to Metal Lath Ceilings With Radiant Electric Heating Cables
Embedded in the Plaster
Ordinary Workmanship—5/8" (16 mm) Plaster

Scratch Coat—1:2	Rate	Total	Rate	Rate (metric)	Total
800 lbs. (360 kg) gypsum plaster .... $....	$....	$6.40	$14.22	$51.20
16 c.f. (0.45 cu.m) sand		....	0.48	17.14	7.68
Brown Coat—1 :3					
1,200 lbs. (540 kg) gypsum plaster		....	6.40	14.22	76.80
36 c.f. (1.01 cu.m) sand		....	0.48	17.14	17.28
Finish White Coat					
340 lbs. (153 kg) fin. hydrated lime		....	8.50	18.88	28.90
170 lbs. (76.5 kg) gauging plaster		....	9.55	21.22	16.24
Cost 100 sq. yds..			$....		$198.10
per sq. yd..			....		1.98
per sq.m			....		2.37

Add for metal lath.

Labor Cost of 100 Sq. Yds. (83.6 Sq.M)
of 3-Coat Gypsum Plaster 5/8" (16 mm) Thick,
Applied to Metal Lath Ceilings With Radiant Electric Heating Cables
Embedded in the Plaster
Ordinary Workmanship—5/8" (16 mm) Plaster

Scratch Coat—1:2	Hours	Rate	Total	Rate	Total
Plasterer 6.0	6.0	$....	$....	$25.75	$154.50
Labor..6.0	6.0	....	....	21.04	126.24
Brown Coat—1:3					
Plasterer 9.0	9.0	....	....	25.75	231.75
Labor..6.0	6.0	....	....	21.04	126.24
Finish White Coat					
Plasterer 8.0	8.0	....	....	25.75	206.00
Labor..4.0	4.0	....	....	21.04	84.16
Cost 100 sq. yds.			$....		$928.89
per sq. yd.			....		9.29
per sq.m			....		10.13

Add for metal lath.

If it is necessary to apply the plaster in four coats instead of three, add labor for a second scratch coat:

	Hours	Rate	Total	Rate	Total
Plasterer 6.0	6.0	$....	$....	$25.75	$154.50
Labor...6.0	6.0	....	....	21.04	126.24
Cost 100 sq. yds..........................			$....		$280.74
per sq. yd....................................			....		2.81
per sq.m			....		3.06

Labor Applying Perlite or Vermiculite Plaster

When using perlite or vermiculite plaster aggregate, a plasterer is applying material weighing 30 to 45 lbs. per c.f. (477-715 kg per cu.m) against 100 lbs. per c.f. (1589 kg per cu.m) for gypsum-sand plaster, so that a plasterer should apply a greater yardage of light aggregate plaster.

Material Cost of 100 Sq. Yds. (83.6 Sq.M) 3-Coat Plaster Applied to Gypsum
Lath, Using Perlite or Vermiculite Aggregate
Ordinary Workmanship—7/8" (22 mm) Grounds—1-1/2" (38 mm) of Plaster

Scratch Coat—1:2	Rate	Total	Rate	Rate (metric)	Total
500 lbs. (225 kg) gypsum plaster ....$....	$....	$6.40	$14.22	$32.00
10 c.f. (0.28 cu.m) aggregate	....	3.18	113.57	31.80
Brown Coat—1:2					
1,000 lbs. (450 kg) gypsum plaster	....	6.40	14.22	64.00
20 c.f. (0.56 cu.m) aggregate	....	3.18		63.60
Finish White Coat...................					
340 lbs. (153 kg) fin. hydrated lime	....	8.50	18.88	28.90
170 lbs. (76.5 kg) gauging plaster	....	9.55	21.22	16.24
Cost 100 sq. yds.............................	$....			$236.54
per sq. yd.............................				2.37
per sq.m	....			2.83
Add for gypsum lath.					

Labor Cost of 100 Sq. Yds. (83.6 Sq.M) 3-Coat Plaster Applied to Gypsum
Lath, Using Perlite or Vermiculite Aggregate
Ordinary Workmanship—7/8" (22 mm) Grounds—1-1/2" (38 mm) of Plaster

Scratch Coat—1:2	Hours	Rate	Total	Rate	Total
Plasterer 4.5	$....	$....	$25.75	$115.88
Labor.....................4.5	....	....	21.04	94.68
Brown Coat—1:2					
Plasterer 7.5	....	....	25.75	193.13
Labor.......................5.0	....	....	21.04	105.20
Finish White Coat					
Plaster.......................8.0	....	....	25.75	206.00
Labor.......................4.0	....	....	21.04	84.16
Cost 100 sq. yds.............................	$....			$799.05
per sq. yd.............................				7.99
per sq.m	....			8.71
Add for gypsum lath.					

Material Cost of 100 Sq. Yds. (83.6 Sq.M) 3-Coat Plaster Applied to Unit Masonry Walls, Using Perlite or Vermiculite Aggregate
Ordinary Workmanship—5/8" (16 mm) Grounds—5/8" (16 mm) of Plaster

Scratch Coat—1:3	Rate	Total	Rate	Rate (metric)	Total
400 lbs. (180 kg) gypsum plaster ....$....	$....	$6.40	$14.22	$25.60
12 c.f. (0.34 cu.m) aggregate	....	3.18	113.57	38.16
Brown Coat—1:3				
800 lbs. (360 kg) gypsum plaster	....	6.40	14.22	51.20
24 c.f. (0.68 cu.m) aggregate	....	3.18	113.57	76.32
Finish White Coat..................................				
340 lbs. (153 kg) fin. gypsum plaster	....	8.50	18.88	28.90
170 lbs. (76.5 kg) gauging plaster	....	9.55	21.22	16.24
Cost 100 sq. yds...		$....		$236.42
per sq. yd..		....		2.36
per sq.m		....		2.83

Labor Cost of 100 Sq. Yds. (83.6 Sq.M) 3-Coat Plaster Applied to Unit Masonry Walls, Using Perlite or Vermiculite Aggregate
Ordinary Workmanship—5/8" (16 mm) Grounds—5/8" (16 mm) of Plaster

Scratch Coat—1:3	Hours	Rate	Total	Rate	Total
Plasterer 4.5		$....	$....	$25.75	$115.88
Labor 4.5		....	....	21.04	94.68
Brown Coat—1:3					
Plasterer 7.5		....	....	25.75	193.13
Labor 5.0		....	....	21.04	105.20
Finish White Coat					
Plasterer 8.0		....	....	25.75	206.00
Labor 4.0		....	....	21.04	84.16
Cost 100 sq. yds..			$....		$799.05
per sq. yd..			....		7.99
per sq.m			....		8.71

Material Cost of 100 Sq. Yds. (83.6 Sq.M) 3-Coat Plaster Applied to Metal Lath, Using Perlite or Vermiculite Aggregate
Ordinary Workmanship—3/4" (19 mm) Grounds—3/4" (19 mm) Plaster

Scratch—1:2	Rate	Total	Rate	Rate (metric)	Total
1,000 lbs. (450 kg) gypsum plaster ....$....	$....	$6.40	$14.22	$64.00
20 c.f. (0.57 cu.m) aggregate	....	3.18	113.57	63.60
Brown Coat—1:2				
1,250 lbs. (562.5 kg) gypsum plaster	....	6.40	14.22	80.00
40 c.f. (1.13 cu.m) aggregate	....	3.18	113.57	127.20

Finish White Coat	Rate	Total	Rate	Rate (metric)	Total
340 lbs. (153 kg) fin. hydrated lime	....		8.50	18.88	28.90
170 lbs. (76.5 kg) gauging plaster	....		9.55	21.22	16.24
Cost per 100 sq. yds.			$....		$379.94
per sq. yd.				....	3.80
per sq.m				....	4.54

Add for metal lath and furring.

Labor Cost of 100 Sq. Yds. (83.6 Sq.M) 3-Coat Plaster Applied to Metal Lath,
Using Perlite or Vermiculite Aggregate
Ordinary Workmanship—3/4" (19 mm) Grounds—3/4" (19 mm) Plaster

Scratch Coat—1:2	Hours	Rate	Total	Rate	Total
Plasterer 4.5		$....	$....	$25.75	$115.88
Labor 4.5		....	....	21.04	94.68
Brown Coat—1:2					
Plasterer 8.0		....	....	25.75	206.00
Labor 6.0		....	....	21.04	126.24
Finish White Coat...............................					
Plasterer 8.0		....	....	25.75	206.00
Labor 4.0		....	....	21.04	84.16
Cost per 100 sq. yds			$....		$832.96
per sq. yd.				....	8.33
per sq.m				....	9.08

Add for metal lath and furring.

Material Cost of 100 Sq. Yds. (83.6 Sq.M) 2" (50 mm) Solid Plaster Partition
Applied to Metal Lath, Using Perlite or Vermiculite Aggregate
(Measurement One Side Only)
Ordinary Workmanship—2" (50 mm) Partition—2" (50 mm) of Plaster

Scratch Coat—1:2	Rate	Total	Rate	Rate (metric)	Total
1,500 lbs. (675 kg) gypsum plaster ....$....		$....	$6.40	$14.22	$96.00
30 c.f. (0.85 cu.m) aggregate		....	3.18	113.57	95.40
Brown Coat—1:2					
6,000 lbs. (2700 kg) gypsum plaster		....	6.40	14.22	384.00
120 c.f. (3.4 cu.m) aggregate		....	3.18	113.57	381.60
Finish White Coat...............................					
680 lbs. (306 kg) fin. hydrated lime		....	8.50	18.88	57.80
340 lbs. (153 kg) gauging plaster		....	9.55	21.22	32.47
Cost 100 sq. yds.			$....		$1,047.27
per sq. yd.				....	10.47
per sq.m				....	12.52

Add for metal lath and furring.

Labor Cost of 100 Sq. Yds. (83.6 Sq.M) 2" (50 mm)
Solid Plaster Partition Applied to Metal Lath,
Using Perlite or Vermiculite Aggregate
(Measurement One Side Only)
Ordinary Workmanship—2" (50 mm) Partition—2" (50 mm) of Plaster

	Hours	Rate	Total	Rate	Total
Scratch Coat—1:2					
Plasterer	5.0	$....	$....	$25.75	$128.75
Labor	5.0	....	....	21.04	105.20
Brown Coat—1:2					
Plasterer	23.0	....	....	25.75	592.25
Labor	17.0	....	....	21.04	357.68
Finish White Coat					
Plasterer	16.0	....	....	25.75	412.00
Labor	8.0	....	....	21.04	168.32
Cost 100 sq. yds.			$....		$1,764.20
per sq. yd.				....	17.64
per sq.m				....	19.23

Add for metal lath and furring.

Material Cost of 100 Sq. Yds. (83.6 Sq.M) 3-Coat Plaster
(Scratch-Double-Back Basecoat) Applied to Gypsum Lath
Using Pre-Mixed Perlite Plaster
Ordinary Workmanship—7/8" (22 mm) Grounds—1/2" (13 mm) of Plaster

	Rate	Total	Rate	Rate (metric)	Total
Scratch Coat					
1180-lb. (531 kg) sacks P. plaster	$....	$....	$6.90	$11.44	$60.72
Double-Back Brown Coat					
1180-lb. (531 kg) sacks P. plaster	....	....	6.90	11.44	81.42
Finish White Coat					
340 lbs. (153 kg) fin. hydrated lime	....	....	8.50	18.88	28.90
170 lbs. (76.5 kg) gauging plaster	....	....	9.55	21.22	16.24
Cost 100 sq. yds.			$....		$187.28
per sq. yd.				....	1.87
per sq.m				....	2.24

Add for gypsum lath.

Labor Cost of 100 Sq. Yds. (83.6 Sq.M) 3-Coat Plaster (Scratch-Double-Back
Basecoat) Applied to Gypsum Lath
Using Pre-Mixed Perlite Plaster
Ordinary Workmanship—7/8" (22 mm) Grounds—1/2" (13 mm) of Plaster

	Hours	Rate	Total	Rate	Total
Scratch Coat					
Plasterer	4.5	$....	$....	$25.75	$115.88
Labor	4.5	....	....	21.04	94.68

Double-Back Brown Coat	Hours	Rate	Total	Rate	Total
Plasterer	7.5	....	....	25.75	193.13
Labor	5.0	....	....	21.04	105.20
Finish White Coat					
Plasterer	8.0	....	....	25.75	206.00
Labor	4.0	....	....	21.04	84.16
Cost 100 sq. yds.			$....		$799.05
per sq. yd.					7.99
per sq.m				....	8.71
Add for gypsum lath.					

Material Cost of 100 Sq. Yds. (83.6 Sq.M) 3-Coat Plaster
(Scratch-Double-Back Basecoat) Applied to Unit Masonry Walls
Using Pre-Mixed Perlite Plaster
Ordinary Workmanship—5/8" (16 mm) Grounds—5/8" (16 mm) of Plaster

Scratch Coat	Rate	Total	Rate	Rate (metric)	Total
1080-lb. (486-kg) sacks P. Plaster	$....	$....	$6.90	$11.44	$74.52
Double-Back Brown Coat					
1680-lb. (756-kg) sacks P. Plaster	....	....	6.90	11.44	115.92
Finish White Coat					
340 lbs. (153 kg) fin. hydrated lime	....	....	8.50	18.88	28.90
170 lbs. (76.5 kg) gauging plaster	....	....	9.55	21.22	16.24
Cost 100 sq. yds.			$....		$235.58
per sq. yd.				....	2.36
per sq.m				....	2.82
Add for gypsum lath.					

Labor Cost of 100 Sq. Yds. (83.6 Sq.M) 3-Coat Plaster
(Scratch-Double-Back Basecoat) Applied to Unit Masonry Walls
Using Pre-Mixed Perlite Plaster
Ordinary Workmanship—5/8" (16 mm) Grounds—5/8" (16 mm) of Plaster

Scratch Coat	Hours	Rate	Total	Rate	Total
Plasterer	4.5	$....	$....	$25.75	$115.88
Labor	4.5	....	....	21.04	94.68
Double-Back Brown Coat					
Plasterer	7.5	....	....	25.75	193.13
Labor	5.0	....	....	21.04	105.20
Finish White Coat					
Plasterer	8.0	....	....	25.75	206.00
Labor	4.0	....	....	21.04	84.16
Cost 100 sq. yds.			$....		$799.05
per sq. yd.				....	7.99
per sq.m				....	8.71

Material Cost of 100 Sq. Yds. (83.6 Sq.M) 3-Coat Plaster Applied to Metal Lath
Using Pre-Mixed Perlite Plaster
Ordinary Workmanship—3/4" (19 mm) Grounds—3/4" (19 mm) Plaster

Scratch Coat	Rate	Total	Rate	Rate (metric)	Total
1580-lb. (711-kg) sacks P. Plaster ....$....	$....		$6.90	$11.44	$109.02
Brown Coat					
1880-lb. (846-kg) sacks P. Plaster	....		6.90	11.44	129.72
Finish White Coat					
340 lbs. (153 kg) fin. hydrated lime	....		8.50	18.88	28.90
170 lbs. (76.5 kg) gauging plaster	....		9.55	21.22	16.24
Cost 100 sq. yds..			$....		$283.88
per sq. yd...				....	2.84
per sq.m				....	3.40

Add for metal lath and furring.

Labor Cost of 100 Sq. Yds. (83.6 Sq.M) 3-Coat Plaster Applied to Metal Lath
Using Pre-Mixed Perlite Plaster
Ordinary Workmanship—3/4" (19 mm) Grounds—3/4" (19 mm) Plaster

Scratch Coat	Hours	Rate	Total	Rate	Total
Plasterer 4.5	$....	$....	$25.75	$115.88	
Labor.......................................4.5	....	....	21.04	94.68	
Brown Coat					
Plasterer 8.0	....	....	25.75	206.00	
Labor.......................................6.0	....	....	21.04	126.24	
Finish White Coat					
Plasterer 8.0	....	....	25.75	206.00	
Labor.......................................4.0	....	....	21.04	84.16	
Cost 100 sq. yds..		$....		$832.96	
per sq. yd...				....	8.33
per sq.m				....	9.08

Add for metal lath and furring.

Material Cost of 100 Sq. Yds. (83.6 Sq.M) 2" (50 mm) Solid Plaster Partition Applied
to Metal Lath Using Pre-Mixed Perlite Plaster (Measurement One Side Only)
Ordinary Workmanship—2" (50 mm) Partition—2" (50 mm) of Plaster

	Rate	Total	Rate	Rate (metric)	Total
1580-lb. (711-kg) sacks P. Plaster ....$....	$....		$6.90	$11.44	$109.02
Brown Coat					
7080-lb. (3186-kg) sacks P. Plaster	....		6.90	11.44	488.52
Finish White Coat					
680 lbs. (306 kg) fin. hydrated lime	....		8.50	18.88	57.80
340 lbs. (153 kg) gauging plaster	....		9.55	21.22	32.47
Cost 100 sq. yds..			$....		$687.81
per sq. yd...				....	6.88

Add for metal lath and furring.

Labor Cost of 100 Sq. Yds. (83.6 Sq.M) 2" (50 mm) Solid Plaster Partition
Applied to Metal Lath Using Pre-Mixed Perlite Plaster
(Measurement One Side Only)
Ordinary Workmanship—2" (50 mm) Partition—2" (50 mm) of Plaster

	Hours	Rate	Total	Rate	Total
Plasterer	5.0	$....	$....	$25.75	$128.75
Labor	5.0	....	....	21.04	105.20
Brown Coat					
Plasterer	23.0	....	....	25.75	592.25
Labor	17.0	....	....	21.04	357.68
Finish White Coat					
Plasterer	16.0	....	....	25.75	412.00
Labor	8.0	....	....	21.04	168.32
Cost 100 sq. yds.			$....		$1,764.20
per sq. yd.				....	17.64
per sq.m				....	19.23

Add for metal lath and furring.

Machine Application of Plaster. Various types of machines are available for the spray application of base coat plaster mortar to walls and ceilings where it is then worked in the normal manner. They vary in size from large machines capable of spraying any plaster mortars, including base coat materials for ordinary plastering and membrane fireproofing, to small machines that handle only the finish coat of acoustical plaster or a sand float finish. Most models have self-contained pump units with gasoline or electric power air compressors. Several small hand gun-hopper units are available for attachment to separate air compressors.

When applying plaster mortars by machine, approximately 20% to 25% additional material will be required to provide equal coverage to hawk and trowel application. This is caused by material densification due to machine action and increased material loss from spattering.

Using the machine method, the labor cost of applying the various plaster coats will be considerably less than for the hawk and trowel method, but an allowance must be made for machine cost and maintenance and additional protection and clean-up labor will be required.

Where scratch coat is machine applied over metal lath, an experienced operator can apply 750 to 850 sq. yds. (627-710 sq.m) per 8-hr. day. For this area, an additional 4 hrs. of plasterer time will be required for touching up and scratching surface.

Material Cost of 100 Sq. Yds. (83.6 Sq.M) of Machine Applied Scratch Coat on Metal
Lath Using Perlite or Vermiculite Aggregate

	Rate	Total	Rate	(metric) Rate	Total
1,200 lbs. (540 kg) gypsum plaster .... $....	$....	$6.40	$14.22	$76.80
24 c.f. (0.68 cu.m) aggregate	....	3.18	113.57	76.32
Machine cost..	....	5.30	5.30	5.30
Cost 100 sq. yds..	$....			$158.42
per sq. yd...	....			1.58
per sq.m	....			1.89

Add for metal lath and furring.

Labor Cost of 100 Sq. Yds. (83.6 Sq.M) of Machine
Applied Scratch Coat on Metal Lath
Using Perlite or Vermiculite Aggregate

	Hours	Rate	Total	Rate	Total
Plasterer.................................1.5	$....	$....	$25.75	$38.63
Labor......................................2.0	....	....	21.04	42.08
Cost 100 sq. yds...........................	$....			$80.71
per sq. yd..................................	....			0.81
per sq.m	....			0.88

Where brown coat is machine applied over scratch coat or gypsum lath, an
experienced operator, with another plasterer rodding and leveling, can apply
375 to 425 sq. yds. (314-355 sq.m) per 8-hr. day.

Material Cost of 100 Sq. Yds. (83.6 Sq.M) of Machine Applied Brown Coat,
7/16" (11 mm) Thick over Scratch Coat or Gypsum Lath,
Using Perlite or Vermiculite Aggregate

	Rate	Total	Rate	(metric) Rate	Total
1,550 lbs. (697.5 kg) gypsum plaster ..... $....	$....	$6.40	$14.22	$99.20
50 c.f. (1.41 cu.m) aggregate	....	3.18	113.57	159.00
Machine cost..	....	10.60	10.60	10.60
Cost 100 sq. yds..	$....			$268.80
per sq. yd...	....			2.69
per sq.m	....			3.21

Labor Cost of 100 Sq. Yds. (83.6 Sq.M) of Machine Applied Brown Coat,
7/16" (11 mm) Thick over Scratch Coat or Gypsum Lath,
Using Perlite or Vermiculite Aggregate

	Hours	Rate	Total	Rate	Total
Plasterer 4.0	$....	$....	$25.75	$103.00
Labor.......................................4.0	....	....	21.04	84.16
Cost 100 sq. yds...........................	$....			$187.16
per sq. yd..................................	....			1.87
per sq.m	....			2.04

Special Prepared Finishes

There are a number of prepared plaster finishes on the market that are ready for use with the addition of water. These may be applied over a brown coat of gypsum cement plaster.

Smooth trowel finishes such as gypsum gauging-lime putty require a much higher degree of rigidity and strength in the lath and basecoat plaster for good visual performance. When the use of high strength basecoat plasters is impractical, consideration should be given to the elimination of restraint at the perimeter angles or the selection of a more rigid lath, or both. They help to compensate for the lower strength of some basecoat plasters. Where a special finish is specified, deduct the cost of the finish white coat and add the cost of the special finish specified.

Material Cost of 100 Sq. Yds. (83.6 Sq.M) Gypsum Trowel Finish
Ordinary Workmanship

	Rate	Total	Rate	Rate (metric)	Total
550 lbs. (247.5 kg) gyp. trowel finish ..... $....	$....		$15.90	$35.33	$87.45
Cost per sq. yd.			....		0.87
per sq.m			....		1.04

Labor Cost of 100 Sq. Yds. (83.6 Sq.M) Gypsum Trowel Finish
Ordinary Workmanship

	Hours	Rate	Total	Rate	Total
Plasterer 8.0		$....	$....	$25.75	$206.00
Labor 4.0		....	....	21.04	84.16
Cost 100 sq. yds.			$....		$290.16
per sq. yd.				....	2.90
per sq.m				....	3.16

For first grade workmanship, add 3.5 hrs. plasterer time and 1-hr. labor time per 100 sq. yds. (83.6 sq.m).

If plaster is wanted with a sand float finish, deduct quantities of plaster and labor given under finish white coat and add the following:

Material Cost of 100 Sq. Yds. (83.6 Sq.M) Sand Float Finish
Ordinary Workmanship

	Rate	Total	Rate	Rate (metric)	Total
200 lbs. (90 kg) unfibered plaster .... $....	$....		$6.90	$15.33	$13.80
4 c.f. (0.11 cu.m) sand		....	0.48	17.45	1.92
Cost per 100 sq. yds.			$....		$15.72
per sq. yd.			....		0.16
per sq.m			....		0.19

FINISHES 9.61

Labor Cost of 100 Sq. Yds. (83.6 Sq.M) Sand Float Finish
Ordinary Workmanship

	Hours	Rate	Total	Rate	Total
Plasterer	7.25	$....	$....	$25.75	$186.69
Labor	5.00	....	....	21.04	105.20
Cost per 100 sq. yds			$....		$291.89
per sq. yd				....	2.92
per sq.m				....	3.18

If first grade workmanship is required, add 3 hours plasterer time and 1 hour labor time per 100 sq. yds. (83.6 sq.m).

Material Cost of 100 Sq. Yds. (83.6 Sq.M) Keene's Cement Smooth Finish, (Medium Hard) Applied Over a Brown Coat of Gypsum Cement Plaster
Ordinary Workmanship

	Rate	Total	Rate	Rate (metric)	Total
530 lbs. (238.5 kg) Keene's cement	$....	$....	$16.96	$37.69	$89.89
270 lbs. (121.5 kg) fin. hydrated lime		....	8.50	18.88	22.95
Cost 100 sq. yds.			$....		$112.84
per sq. yd				....	1.13
per sq.m				....	1.35

Labor Cost of 100 Sq. Yds. (83.6 Sq.M) Keene's Cement Smooth Finish, (Medium Hard) Applied Over a Brown Coat of Gypsum Cement Plaster
Ordinary Workmanship

	Hours	Rate	Total	Rate	Total
Plasterer	14.5	$....	$....	$25.75	$373.38
Labor	5.0	....	....	21.04	105.20
Cost 100 sq. yds.			$....		$478.58
per sq. yd				....	4.79
per sq.m				....	5.22

If blocked off into squares to represent tile, add about 16 hrs. plasterer time per 100 sq. yds. (83.6 sq.m).

If first grade workmanship is required, add 5.5 hrs. plasterer time and 1 hr. labor time per 100 sq. yds. (83.6 sq.m).

Keene's cement sand float finish coat may be applied over a brown coat of gypsum cement plaster, either plain or blocked off into ashlar effects. If blocked, add extra for this work.

Material Cost of 100 Sq. Yds. (83.6 Sq.M) Keene's Cement Sand Float Finish
Ordinary Workmanship

	Rate	Total	Rate	Rate (metric)	Total
160 lbs. (72 kg) Keene's cement ....$....	$....	$16.96	$37.69	$27.14
125 lbs. (56.25 kg) fin. hydrated lime	....	8.50	18.88	10.63
6 c.f. (0.17 cu.m) clean sand	....	0.48	1.69	2.88
Cost 100 sq. yds...............	$....			$40.65
per sq. yd................	....			0.41
per sq.m	....			0.49

Labor Cost of 100 Sq. Yds. (83.6 Sq.M) Keene's Cement Sand Float Finish
Ordinary Workmanship

	Hours	Rate	Total	Rate	Total
Plasterer 13.0	$....	$....	$25.75	$334.75
Labor................6.0	....	....	21.04	126.24
Cost 100 sq. yds............	$....			$460.99
per sq. yd............	....			4.61
per sq.m	....			5.02

If first grade workmanship is required, add 4 hrs. plasterer time and 1 hr. labor time per 100 sq. yds. (83.6 sq.m).

The resistance of gypsum gauging-lime putty trowel finishes to cracking, particularly to check cracking, can be increased by the addition of fine aggregate. The use of not less than 1/2 c.f. (0.01 cu.m) of fine silica sand or perlite to each 100 lbs. (45 kg) of gauging plaster or Keene's cement increases the factor of safety in trowel finishes.

Acoustical Plaster. Acoustical plaster is occasionally specified instead of trowel putty coat or sand float finishes for noise quieting in classrooms and corridors and for acoustical correction of churches and auditoriums. A number of prepared acoustical plasters are available having sound absorption values of 40% to 45% at 512 cycles and a noise reduction coefficient of 55% to 60%. There are two general types of acoustical plaster. Those with a gypsum or lime binder, such as Sabinite, Hi-Lite, Kilnoise, Gold Bond, Perfolite, etc., are for trowel application only with either a troweled, floated, or stippled and perforated finish. Other acoustical plasters, such as Audicote, Sprayolite, Zonolite Acoustical Plaster, etc., are made with an adhesive type binder that shrinks on drying, providing the porosity for sound absorption, and may be applied by machine or trowel.

In general, acoustical plasters are for application over gypsum base coats (scratch and brown) and should be applied in two coats to a total thickness of 1/2" (13 mm)—usually a 3/8" (9 mm) first coat and a 1/8" (3 mm) finish coat.

Because of restrictions due to the asbestos fiber content, acoustical plaster is temporarily out of supply but will probably be back on the market with a new mixture. The prices below are for the old formula material.

Where applied by trowel, a plasterer should complete 65 to 75 sq. yds. (54-63 sq.m) of acoustical plaster, two coats to 1/2" (13 mm) total thickness, per 8-hr. day.

Material Cost of 100 Sq. Yds. (83.6 Sq.M) of Acoustical Plaster
Applied by Trowel

	Rate	Total	Rate	Total
20 bags, 5 sq. yds. (4.18 sq.m), 1/2" (13 mm) thick, acoustical plaster	$....	$....	$7.95	$159.00
Cost per sq. yd		....		1.59
per sq.m		....		1.90
Add for lath and base coats required.				

Labor Cost of 100 Sq. Yds. (83.6 Sq.M) of Acoustical Plaster
Applied by Trowel

	Hours	Rate	Total	Rate	Total
Plasterer	11.5	$....	$....	$25.75	$296.13
Labor	6.0		....	21.04	126.24
Cost 100 sq. yds			$....		$422.37
per sq. yd			....		4.22
per sq.m			....		4.60
Add for lath and base coats required.					

Where application of the two coat acoustical finish is by machine, a plasterer should complete 100 to 110 sq. yds. (84-92 sq.m) per 8-hr. day, but one extra hour of labor time must be figured for the additional cleanup and protection required, and an allowance included for cost and maintenance of equipment.

Material Cost of 100 Sq. Yds. (83.6 Sq.M) of Acoustical Plaster Applied by Machine

	Rate	Total	Rate	Total
14 bags, 7.2 sq. yds. (6 sq.m) 1/2" thick, acoustical plaster	$....	$....	$7.95	$111.30
Machine cost	....		39.75	39.75
Cost per 100 sq. yds		$....		$151.05
per sq. yd		....		1.51
per sq.m		....		1.80
Add for lath and base coats required.				

Labor Cost of 100 Sq. Yds. (83.6 Sq.M) of Acoustical Plaster Applied by Machine

	Hours	Rate	Total	Rate	Total
Plasterer	8.0	$....	$....	$25.75	$206.00
Labor	7.0	....	....	21.04	147.28
Labor (masking)	7.0	....	....	21.04	147.28
Cost per 100 sq. yds			$....		$500.56
per sq. yd				....	5.01
per sq.m				....	5.46

Add for lath and base coats required.

Vermiculite Direct-to-Steel Fireproofing.

Vermiculite Type-MK direct-to-steel fireproofing is a new and economical method of protecting steel columns, beams, girders, and trusses, and steel floor and roof sections. It is applied by machine directly to the metal, without a primer, lath, or reinforcing. It presents a tough, fire-protective, and insulating coating that is free of fissures and can be applied in the early stages of construction, when the building is free of other trades.

This is a mill-mixed material that requires only the addition of water. It is packaged in 25- and 50-lb. (11- and 22.5-kg) bags. Coverage varies with specified thickness.

Colored Interior Finish Plaster

There are a number of colored finish plasters on the market used for textured or float plaster finish.

These special finishes are applied over a brown coat of gypsum cement plaster and may be finished in a number of different ways, such as Spanish Palm Finish, Brush Swirl, Float Finish, English Trowel finish, etc., in addition to a simple float finish.

Labor Applying Special Finish Coats. For colored finish plaster with a simple float finish, a plasterer should apply 100 to 120 sq. yds. (84-100 sq.m) per 8-hr. day, where an ordinary grade of workmanship is permitted.

Oriental Interior Color Finish Plaster. From the U.S. Gypsum Co., this is used for textured or floated plaster finish. It is made in 13 standard shades and white and is ready for use with the addition of water only.

It requires approximately 8 lbs. of Oriental finish per sq. yd. depending on the texture used. Price about $12.00 per 100 lbs. ($26.67 per 100 kg).

Material Cost of 100 Sq. Yds. (83.6 Sq.M) of Oriental Interior Colored Finish
Plaster, with Simple Float Finish, Applied Over a Brown
Coat of Gypsum Cement Plaster
Ordinary Workmanship

	Rate	Total	Rate	Total
Finish Coat				
800 lbs. (363 kg) Oriental finish	$....	$....	$12.72	$101.76
Cost per sq. yd		....		1.01
per sq.m		....		1.21

Labor Cost of 100 Sq. Yds. (83.6 Sq.M) of Oriental Interior Colored Finish
Plaster, with Simple Float Finish, Applied Over a Brown
Coat of Gypsum Cement Plaster
Ordinary Workmanship

Finish Coat	Hours	Rate	Total	Rate	Total
Plasterer	7.25	$....	$....	$25.75	$186.69
Labor	4.50	....	....	21.04	94.68
Cost 100 sq. yds.			$....		$281.37
per sq. yd.					2.81
per sq.m					3.06

If first grade workmanship is required, add 2.75 hrs. plasterer time and 1
hr. labor time per 100 sq. yds. (83.6 sq.m).

Material Costs
Colored Plaster with Smooth or Textured Finish

	Rate	Total	Rate	Rate (metric)	Total
350 lbs. (157.5 kg) Keene's cement	$....	$....	$16.96	$37.69	$59.36
450 lbs. (202.5 kg) fin. hydrated lime		....	8.50	18.88	38.25
16 lbs. (7.2 kg) color		....	1.06	2.35	16.96
Cost 100 sq. yds.		$....			$114.57
per sq. yd.		....			1.15
per sq.m		....			1.37

Material Costs
Colored Plaster with Sand Float Finish

	Rate	Total	Rate	Rate (metric)	Total
100 lbs. (45 kg) Keene's cement	$....	$....	$16.96	$37.69	$16.96
500 lbs. (22.5 kg) fin. hydrated lime		....	8.50	18.88	42.50
1,200 lbs. (540 kg) white silica sand		....	1.06	2.35	12.72
36 lbs. (16.2 kg) color		....	1.06	2.35	40.45
Cost 100 sq. yds.			$....		$112.63
per sq. yd.			....		1.13
per sq.m			....		1.35

The above formulas are standard mixture for permanent colors. Best
results are obtained by soaking lime overnight or for 24 hours.

Ornamental Plastering

Ornamental plastering is not used as extensively as it once was, and
mechanics who can do this type of work are not always available. It is best to

contact a plastering subcontractor for firm prices on intricate ornamental plaster work.

Material Cost of 100 L.F. (30 m) of Plaster Cornice Under 12" (300 mm) in Girth

	Rate	Total	Rate	Rate (metric)	Total
200 lbs. (90 kg) gypsum plaster ....$....	$....	$6.36	$14.13	$12.72
4 c.f. (0.11 cu.m) sand	....	0.48	17.45	1.92
50 lbs. (22.5 kg) molding plaster	....	10.60	23.55	5.30
Cost 100 l.f...	$....			$19.94
per l.f...	....			0.20
per m..	....			0.66

Add for metal lath and furring as given on previous pages.

Labor Cost of 100 L.F. (30 m) of Plaster Cornice Under 12" (300 mm) in Girth

	Hours	Rate	Total	Rate	Total
Plasterer 18.0	$....	$....	$25.75	$463.50
Labor..6.0	....	....	21.04	126.24
Cost 100 l.f...	$....			$589.74
per l.f...	....			5.90
per m..	....			19.35

Add for metal lath and furring as given on previous pages.

Material Cost of 100 S.F. (9 Sq.M) of Plaster Cornice Over 1'-0" (0.30 m) in Girth

	Rate	Total	Rate	Rate (metric)	Total
200 lbs. (90 kg) gypsum plaster ....$....	$....	$6.36	$14.13	$12.72
4 c.f. (0.11 cu.m) sand	....	.48	17.45	1.92
50 lbs. (22.5 kg) molding plaster	....	10.60	23.55	5.30
Cost 100 s.f...	$....			$19.94
per s.f...	....			0.20
per sq.m	....			2.22

Add for metal lath and furring as given on previous pages.

Labor Cost of 100 S.F. (9 Sq.M) of Plaster Cornice Over 1'-0" (0.30 m) Girth

	Hours	Rate	Total	Rate	Total
Plasterer 17.0	$....	$....	$25.75	$437.75
Labor..6.0	....	....	21.04	126.24
Cost 100 s.f..	$....			$563.99
per s.f...	....			5.64
per sq.m	....			60.69

Add for metal lath and furring as given on previous pages.

Material Cost of 100 S.F. (9 Sq.M) of Portland Cement Cornice

	Rate	Total	Rate	Rate (metric)	Total
4 sacks portland cement	$....	$....	$4.24	$4.24	$16.96
40 lbs. (18 kg) hydrated lime		....	8.50	18.88	3.40
10 c.f. (0.28 cu.m) sand		....	0.48	17.14	4.80
Cost 100 s.f.			$....		$25.16
per s.f.				....	0.25
per sq.m				....	2.80

Add for metal lath and furring as given on previous pages.

Labor Cost of 100 S.F. (9 Sq.M) of Portland Cement Cornice

	Hours	Rate	Total	Rate	Total
Plasterer	46.0	$....	$....	$25.75	$1,184.50
Labor	10.0		....	21.04	210.40
Cost 100 s.f.			$....		$1,394.90
per s.f.				....	13.95
per sq.m				....	150.09

Add for metal lath and furring as given on previous pages.

Running Bull Nose Corners, Ceiling Coves, Etc. When running bull nose corners on plaster columns, pilasters, window jambs, etc., and plain ceiling coves and wall angles, a plasterer should run 80 to 100 l.f. (24-30 m) per 8-hr. day, at the following labor cost per 100 l.f. (30 m):

	Hours	Rate	Total	Rate	Total
Plasterer	9.0	$....	$....	$25.75	$231.75
Labor	2.0		....	21.04	42.08
Cost 100 l.f.			$....		$273.83
per l.f.				....	2.74
per m				....	8.99

Portland Cement Plaster

Portland cement plaster is used for both interior and exterior work. It should not be mixed too rich. A too rich mixture has a tendency to check and shrink.

Proper attention to curing is important. After each coat has set, and before the following coat is applied, the mortar should be kept moist for at least three days and then permitted to dry out gradually.

When using portland cement mortar it is customary to add a small percentage of lime (usually about 10%) to make the mortar work easier, because a straight portland cement mortar is very difficult to apply.

For the cost of portland cement plaster applied over metal lath and the cost of metal lath and furring, see previous pages of this chapter.

Labor Applying 100 Sq. Yds. (83.6 Sq.M) Portland Cement Plaster to Various Plaster Bases

Description of Work	S.Y.(Sq.M) per 8-Hr. Day	Hrs. per 100 S.Y. (83.6 Sq.M) Plasterer	Tender
Ordinary Work*			
Interior Work on Brick, Clay Tile or Cement Block—3/4" (19 mm) Thick			
Scratch Coat	120-140 (100-117)	6	4
Brown Coat	75-85 (63-71)	10	5
Float Finish Coat	60-70 (50-59)	12	4
Trowel Finish Coat	45-55 (38-46)	16	4
Interior Portland Cement Plaster on Metal Lath			
Scratch Coat	120-140 (100-117)	6	5
Brown Coat	60-70 (50-59)	12	6
Float Finish Coat	60-70 (50-59)	12	4
Trowel Finish Coat	45-55 (38-46)	16	4
Exterior Portland Cement Stucco on Brick, Clay Tile or Cement Block			
Scratch Coat	120-140 (100-117)	6	5
Brown Coat	75-85 (63-71)	10	6
Scaffold			12
Exterior Portland Cement Stucco on Metal Lath on Frame Construction			
Scaffold			12
Scratch Coat	120-140 (100-117)	6	6
Brown Coat	60-70 (50-59)	12	8
Floated Finish Coat	60-70 (50-59)	12	6
Troweled Finish Coat	45-55 (38-46)	16	6
Textured Finish Coat	50-60 (42-50)	14	6
Blocking Off Portland Cement Plaster into Squares to Represent Tile	45-55 (38-46)	16	4
Portland Cement Straight Base	45-55** (38-46)	16	4
Portland Cement Coved Base	30-35** (25-29)	24	6

*For first grade workmanship as described at the beginning of this chapter, add 30% to 40% to plasterer time for brown and finish coats.
**Lineal feet.

Material Cost of 100 Sq. Yds. (83.6 Sq.M) 2-Coat Portland Cement Plaster
on Brick, Clay Tile or Cement Block
Ordinary Workmanship—5/8" (16 mm) Grounds

Brown Coat	Rate	Total	Rate	Rate (metric)	Total
14 sacks portland cement.....................$....	$....	$4.24	....		$59.36
140 lbs. (63 kg) hydrated lime		....	8.50	18.88	11.90
42 c.f. (1.18 cu.m) sand		....	0.48	17.08	20.16
Finish Coat					
4 sacks portland cement..........................		....	4.24		16.96
40 lbs. (18 kg) hydrated lime		....	8.50	18.88	3.40
12 c.f. (0.34 cu.m) sand		....	0.48	16.94	5.76
Cost 100 sq. yds.................................			$....		$117.54
per sq. yd.				....	1.18
per sq.m				....	1.40

Labor Cost of 100 Sq. Yds. (83.6 Sq.M) 2-Coat Portland Cement Plaster
on Brick, Clay Tile or Cement Block
Ordinary Workmanship—5/8" (16 mm) Grounds

Brown Coat	Hours	Rate	Total	Rate	Total
Plasterer.................................13.0		$....	$....	$25.75	$334.75
Labor.................................6.0		....	....	21.04	126.24
Finish Coat					
Plasterer.................................16.0		....	....	25.75	412.00
Labor.................................4.0		....	....	21.04	84.16
Cost 100 sq. yds...........................			$....		$957.15
per sq. yd...................................				....	9.57
per sq.m				....	10.43

Material Cost of 100 Sq. Yds. (83.6 Sq.M) 3-Coat Portland Cement Plaster
on Brick, Clay Tile or Cement Block
Ordinary Workmanship—5/8" (16 mm) Grounds

Scratch Coat	Rate	Total	Rate	Rate (metric)	Total
6 sacks portland cement ....$....	$....	$4.24	$4.24		$25.44
60 lbs. (27 kg) hydrated lime		....	8.50	18.88	5.10
18 c.f. (0.51 cu.m) sand		....	0.48	16.94	8.64
Brown Coat					
8 sacks portland cement		....	4.24	4.24	33.92
80 lbs. (36 kg) hydrated lime		....	8.50	18.88	6.80
24 c.f. (0.68 cu.m) sand		....	0.48	16.94	11.52
Finish Coat					
4 sacks portland cement		....	4.24	4.24	16.96
40 lbs. (18 kg) hydrated lime		....	8.50	18.88	3.40
12 c.f. (0.34 cu.m) sand		....	0.48	16.94	5.76

Cost 100 sq. yds.		$....		$117.54
per sq. yd.		....		1.18
per sq.m		....		1.40

Labor Cost of 100 Sq. Yds. (83.6 Sq.M) 3-Coat Portland Cement Plaster on Brick, Clay Tile or Cement Block
Ordinary Workmanship—5/8" (16 mm) Grounds

Scratch Coat	Hours	Rate	Total	Rate	Total
Plasterer	6.0	$....	$....	$25.75	$154.50
Labor	4.0	....	....	21.04	84.16
Brown Coat					
Plasterer	10.0	....	....	25.75	257.50
Labor	5.0	....	....	21.04	105.20
Finish Coat					
Plasterer	16.0	....	....	25.75	412.00
Labor	4.0	....	....	21.04	84.16
Cost 100 sq. yds.			$....		$1,097.52
per sq. yd.			....		10.98
per sq.m			....		11.97

If first grade workmanship is required, add 8 hrs. plasterer time and 2 hrs. labor time per 100 sq. yds. (83.6 sq.m).

If blocked off into squares to represent tile, add 16 hrs. plasterer time for ordinary work and 18 hrs. plasterer time for first grade workmanship per 100 sq. yds. (83.6 sq.m).

Material Cost of 100 Sq. Yds. (83.6 Sq.M)
3-Coat Portland Cement Plaster on Metal Lath
Ordinary Workmanship—3/4" (19 mm) Grounds

Scratch Coat	Rate	Total	Rate	Rate (metric)	Total
7 sacks portland cement	$....	$....	$4.24	$4.24	$29.68
70 lbs. (31.5 kg) hydrated lime	....	....	8.50	18.88	5.95
21 c.f. (0.59 cu.m) sand	....	....	0.48	17.08	10.08
Brown Coat					
11 sacks portland cement	....	....	4.24	4.24	46.64
110 lbs. (49.5 kg) hydrated lime	....	....	8.50	18.88	9.35
33 c.f. (0.93 cu.m) sand	....	....	0.48	17.08	15.84
Finish Coat					
4 sacks portland cement	....	....	4.24	4.24	16.96
40 lbs. (18 kg) hydrated lime	....	....	8.50	18.88	3.40
12 c.f. (0.34 cu.m) sand	....	....	0.48	17.08	5.76
Cost 100 sq. yds.			$....		$143.66
per sq. yd.			....		1.44
per sq.m			....		1.72

Add for metal lath and furring.

Labor Cost of 100 Sq. Yds. (83.6 Sq.M)
3-Coat Portland Cement Plaster on Metal Lath
Ordinary Workmanship—3/4" (19 mm) Grounds

	Hours	Rate	Total	Rate	Total
Scratch Coat					
Plasterer	6.0	$....	$....	$25.75	$154.50
Labor	5.0	....	....	21.04	105.20
Brown Coat					
Plasterer	12.0	....	....	25.75	309.00
Labor	6.0	....	....	21.04	126.24
Finish Coat					
Plasterer	16.0	....	....	25.75	412.00
Labor	4.0	....	....	21.04	84.16
Cost 100 sq. yds.			$....		$1,191.10
per sq. yd.				....	11.91
per sq.m				....	12.98

Add for metal lath and furring.

If blocked off into squares to represent tile, add 16 hrs. plasterer time per 100 sq. yds. (83.6 sq.m).

If 5/8" (16 mm) grounds are used instead of 3/4" (19 mm), deduct 2 sacks portland cement, 20 lbs. (9 kg) lime and 6 c.f. (0.17 cu.m) sand.

If first grade workmanship is required, add 8 hrs. plasterer time and 2 hrs. labor time per 100 sq. yds. (83.6 sq.m).

Material Cost of 100 L.F. (30 M) of 6" (150 mm) Portland Cement Straight Base

	Rate	Total	Rate	Rate (metric)	Total
2 sacks portland cement	$....	$....	$4.24	$4.24	$8.48
20 lbs. (9 kg) hydrated lime	....	....	8.50	18.88	1.70
6 c.f. (0.17 cu.m) sand	....	....	0.48	16.94	2.88
Cost 100 l.f.			$....		$13.06
per l.f.				....	0.13
per m				....	0.44

Labor Cost of 100 L.F. (30 M) of 6" (150 mm) Portland Cement Straight Base

	Hours	Rate	Total	Rate	Total
Plasterer	16.0	$....	$....	$25.75	$412.00
Labor	4.0	....	....	21.04	84.16
Cost 100 l.f.			$....		$496.16
per l.f.				....	4.96
per m				....	16.27

For coved base, add 50% to material and labor.

Exterior Stucco

Exterior stucco is generally composed of a portland cement base. The scratch and brown coats should be mixed in the proportions of 1 part of portland cement by weight to 3 parts of clean sand by weight. Do not add more than 8 lbs. (3.6 kg) of lime to each 100 lbs. (45 kg) of portland cement used in the mixture. Portland cement plaster should not be applied when the outside temperature is below 32°F (0°C).

The cost of portland cement stucco will vary according to the materials used and the grade of workmanship required.

Applying Scratch Coat of Portland Cement Stucco. Where just an ordinary grade of workmanship is required, a plasterer should apply 120 to 140 sq. yds. (100-117 sq.m) of scratch coat per 8-hr. day on metal lath.

Applying Brown Coat of Portland Cement Stucco. A plasterer should apply 75 to 85 sq. yds. (63-71 sq.m) of brown coat per 8-hr. day, on brick, tile, or concrete block walls and 60 to 70 sq. yds. (50-59 sq.m) per 8-hr. day on metal lath.

Applying Trowel Finish Coat of Portland Cement. A plasterer should apply 45 to 55 sq. yds. (37-46 sq.m) of portland cement trowel finish per 8-hr. day, where just an ordinary grade of workmanship is required and 35 to 40 sq. yds. (29-33 sq.m) per 8-hr. day, where first grade workmanship is required.

Applying Wet Rough Cast Finish to Portland Cement Stucco. If a wet rough cast finish is used and the mortar and aggregate are thrown against the wall with a paddle or similar tool, a plasterer should complete 30 to 40 sq. yds. (25-33 sq.m) per 8-hr. day.

Applying Pebble Dash or Dry Rough Cast. When applying a pebble dash or dry rough cast finish where the aggregate is thrown against the wet cement or "butter" coat, a plasterer should apply 35 to 40 sq. yds. (29-33 sq.m) per 8-hr. day.

Washing Exterior Stucco With Acid to Expose Crystals. Where the finish coat of stucco contains granite, marble, or crystal screenings, it is washed off with a solution of muriatic acid to expose the crystals. A worker should wash 475 to 525 s.f.—53 to 58 sq. yds. (43-47 sq.m)—per 8-hr. day.

On some jobs it will be necessary to wash the walls two or three times to bring out the crystals satisfactorily, and in such instances the labor cost should be increased accordingly.

Material Cost of 100 Sq. Yds. (83.6 Sq.M) 3-Coat 1:3 Portland Cement Stucco,
Float Finish, Applied to Metal Lath Over Wood Framing
Ordinary Workmanship

Scratch Coat—1:3	Rate	Total	Rate	Rate (metric)	Total
8 sacks portland cement.....................$....		$....	$4.24		$33.92

Scratch Coat—1:3	Rate	Total	Rate	Rate (metric)	Total
60 lbs. (27 kg) hydrated lime	....		8.50	18.88	5.10
24 c.f. (0.68 cu.m) sand	....		0.48	16.94	11.52
Brown Coat—1:3					
16 sacks portland cement	....		4.24	4.24	67.84
120 lbs. (54 kg) hydrated lime	....		8.50	18.88	10.20
48 c.f. (1.36 cu.m) sand	....		0.48	16.94	23.04
Float Finish—1:3					
5 sacks w'p'f. portland cement	....		5.30	5.30	26.50
15 c.f. (0.42 cu.m) sand	....		0.48	16.94	7.20
Cost 100 sq. yds.			$....		$185.32
per sq. yd.					1.85
per sq.m			....		2.22

Add for metal lath.

Labor Cost of 100 Sq. Yds. (83.6 Sq.M) 3-Coat 1:3 Portland Cement Stucco,
Float Finish, Applied to Metal Lath Over Wood Framing
Ordinary Workmanship

Scratch Coat—1:3	Hours	Rate	Total	Rate	Total
Plasterer	6.0	$....	$....	$25.75	$154.50
Labor	6.0	....	....	21.04	126.24
Brown Coat—1:3					
Plasterer	12.0	....	....	25.75	309.00
Labor	8.0	....	....	21.04	168.32
Float Finish—1:3					
Plasterer	12.0	....	....	25.75	309.00
Labor	6.0	....	....	21.04	126.24
Cost 100 sq. yds.			$....		$1,193.30
per sq. yd.			....		11.93
per sq.m			....		13.01

Add for metal lath.

If a smooth troweled finish is desired, add 4 hrs. plasterer time. If a textured finish coat is desired, add 2 hrs. plasterer time. If first grade workmanship is required, add 8 hrs. plasterer time to the above. Add for scaffold.

Material Cost of 100 Sq. Yds. (83.6 Sq.M) 3-Coat 1:3 Portland Cement Stucco,
Float Finish, Applied Over Brick, Clay Tile or Concrete Block Surfaces
Ordinary Workmanship

Scratch Coat—1:3	Rate	Total	Rate	Rate (metric)	Total
6 sacks portland cement	$....	$....	$4.24	$4.24	$25.44
45 lbs. (20.25 kg) hydrated lime	....		8.50	18.88	3.83
18 c.f. (0.51 cu.m) sand	....		0.48	16.94	8.64

Brown Coat—1:3	Rate	Total	Rate	Rate	Total
10 sacks portland cement	....	....	4.24	4.24	42.40
75 lbs. (33.75 kg) hydrated lime		....	8.50	18.88	6.38
30 c.f. (0.85 cu.m) sand		....	0.48	16.94	14.40
Float Finish—1:3					
5 sacks w'p'f. portland cement		....	5.30	5.30	26.50
15 c.f. (0.42 cu.m) sand		....	0.48	16.94	7.20
Cost 100 sq. yds.			$....		$134.79
per sq. yd.				....	1.35
per sq.m				....	1.61

Labor Cost of 100 Sq. Yds. (83.6 Sq.M) 3-Coat 1:3 Portland Cement Stucco,
Float Finish, Applied Over Brick, Clay Tile or Concrete Block Surfaces
Ordinary Workmanship

Scratch Coat—1:3	Hours	Rate	Total	Rate	Total
Plasterer	6.0	$....	$....	$25.75	$154.50
Labor	5.0	....	....	21.04	105.20
Brown Coat—1:3					
Plasterer	10.0	....	....	25.75	257.50
Labor	6.0	....	....	21.04	126.24
Float Finish—1:3					
Plasterer	12.0	....	....	25.75	309.00
Labor	6.0	....	....	21.04	126.24
Cost 100 sq. yds.			$....		$1,078.68
per sq. yd.				....	10.79
per sq.m				....	11.76

For smooth troweled finish, add 4 hrs. plasterer time. If a textured finish coat is desired, add 2 hrs. plasterer time. If first grade workmanship is required, add 8 hrs. plasterer time to the above. Add for scaffold.

Special Finishes for Portland Cement Stucco

If any of the following special finishes are wanted with portland cement stucco, deduct the finish coat given above and add cost of the finish coat.

Material Cost of 100 Sq. Yds. (83.6 Sq.M) of White Cement Float Finish
Ordinary Workmanship

	Rate	Total	Rate	Rate (metric)	Total
5 sacks w'p'f. white cement	$....	$....	$8.50	$8.50	$42.50
1,500 lbs. (675 kg) white silica sand		....	1.06	2.35	15.90
Cost 100 sq. yds.			$....		$58.40
per sq. yd.				....	0.58
per sq.m				....	0.68

Labor Cost of 100 Sq. Yds. (83.6 Sq.M) of White Cement Float Finish
Ordinary Workmanship

	Hours	Rate	Total	Rate	Total
Plasterer	12.0	$....	$....	$25.75	$309.00
Labor	6.0	....	....	21.04	126.24
Cost 100 sq. yds.			$....		$435.24
per sq. yd.				....	4.35
per sq.m				....	4.74

Material Cost of 100 Sq. Yds. (83.6 Sq.M) Finish Coat Colored Stucco, Float Finish
Ordinary Workmanship

	Rate	Total	Rate	Rate (metric)	Total
2 sacks w'p'f white cement	$....	$....	$8.50	$8.50	$17.00
100 lbs. (45 kg) fin. hydrated lime		....	8.50	18.88	8.50
600 lbs. (270 kg) white silica sand		....	1.06	2.35	6.36
18 lbs. (8.1 kg) color		....	1.06	2.35	19.08
Cost 100 sq. yds.			$....		$50.94
per sq. yd.				....	0.51
per sq.m				....	0.61

Labor Cost of 100 Sq. Yds. (83.6 Sq.M) Finish Coat Colored Stucco, Float Finish
Ordinary Workmanship

	Hours	Rate	Total	Rate	Total
Plasterer	12.0	$....	$....	$25.75	$309.00
Labor	6.0	....	....	21.04	126.24
Cost 100 sq. yds.			$....		$435.24
per sq. yd.				....	4.35
per sq.m				....	4.74

If a smooth troweled finish is desired, add 4 hrs. plasterer time. If a textured finish is desired, add 2 hrs. plasterer time. If first grade workmanship is required, add 8 hrs. plasterer time to the above.

Material Cost of 100 Sq. Yds. (83.6 Sq.M) Wet Rough Cast Finish Coat

	Rate	Total	Rate	Rate (metric)	Total
4 sacks w'p'f. portland cement	$....	$....	$5.30	$5.30	$21.20
12 c.f. (0.34 cu.m) sand		....	0.48	16.94	5.76
500 lbs. (225 kg) aggregate		....	6.36	14.13	31.80
Cost 100 sq. yds.			$....		$58.76
per sq. yd.				....	0.59
per sq.m				....	0.70

Labor Cost of 100 Sq. Yds. (83.6 Sq.M) Wet Rough Cast Finish Coat

	Hours	Rate	Total	Rate	Total
Plasterer	20.0	$....	$....	$25.75	$515.00
Labor	10.0	....	....	21.04	210.40
Cost 100 sq. yds.			$....		$725.40
per sq. yd.				....	7.25
per sq.m				....	7.90

Material Cost of 100 Sq. Yds. (83.6 Sq.M) of Dry Rough Cast Finish

	Rate	Total	Rate	Rate (metric)	Total
4 sacks w'p'f. portland cement	$....	$....	$5.30	$5.30	$21.20
12 c.f. (0.34 cu.m) sand		....	0.48	16.94	5.76
750 lbs. (337.5 kg) aggregate		....	6.36	14.13	47.70
Cost 100 sq. yds.			$....		$74.66
per sq. yd.				....	0.75
per sq.m				....	0.89

Labor Cost of 100 Sq. Yds. (83.6 Sq.M) of Dry Rough Cast Finish

	Hours	Rate	Total	Rate	Total
Plasterer	20.0	$....	$....	$25.75	$515.00
Labor	10.0	....	....	21.04	210.40
Cost 100 sq. yds.			$....		$725.40
per sq. yd.				....	7.25
per sq.m				....	7.90

A dry rough cast finish requires a finish or "butter" coat 1/4" (6 mm) thick applied directly over the brown coat. This is brought to a straight, smooth finish and then the aggregate is thrown onto the "butter" coat dry.

If coarse aggregate is used, it requires about 1,000 lbs. per 100 sq. yds. (538 kg per 100 sq.m).

If medium size aggregate is used, it requires 750 lbs. per 100 sq. yds. (897 kg per 100 sq.m). If fine aggregate is used, it requires 500 lbs. per 100 sq. yds. (598 kg per 100 sq.m).

Approximate Prices on Lathing and Plastering Materials
Metal Lath (Expanded Diamond Mesh, flat)

Finish	Wgt., lbs. per sq.yd.	Wgt., kg per sq.m	Price sq.yd.	Price sq.m
Painted	2.5	1.3	$1.30	1.55
Painted	3.4	1.8	1.60	1.90
Galvanized	3.4	1.8	1.85	2.20

Self-furring lath, add $0.10 per sq. yd. to the above prices.

Rib Metal Lath

Description	Finish	Hgt of Rib, in.	Hgt. of Rib, mm	Wgt., lbs. per sq.yd.	Wgt., kg per sq.m	Price sq.yd.	Price sq.m
Flat rib	Painted	1/8	3	2.75	1.5	$1.60	$1.90
Flat rib	Painted	1/8	3	3.4	1.8	1.70	2.00
Rib or	Painted	3/8	9	3.4	1.8	2.30	2.75
Rod ribbed	Painted	3/8	9	4.0	2.15	2.55	3.05

Hot Rolled Channels

Width, inches	Width, mm	Leg, inches	Leg, mm	Lbs. per 1,000 l.f.	Kg per 1,000 m	Price per 1,000 l.f.	Price per 1,000 m
3/4" Standard	19	15/32"	12	300	450	$175.00	$583.00
1" Standard	25	3/8"	9	410	615	210.00	700.00
1-1/2" Standard	38	19/32"	15	650	975	265.00	883.00
2" Standard	50	7/16"	11	1,260	1890	330.00	1100.00

Peforating, add $10.00 per 1,000 l.f. ($33.00 per 1000 m).

Cold Rolled Channels

Width, inches	Width, mm	Leg, inches	Leg, mm	Lbs. per 1,000 l.f.	Kg per 1,000 m	Price per 1,000 l.f.	Price per 1,000 m
3/4"	19 mm	1/2"	13 mm	300	450	$170.00	$567.00
1-1/2"	38 mm	19/32"	15 mm	475	713	255.00	850.00
2"	50 mm	19/32"	15 mm	590	885	320.00	1,067.00

Wire

	Per Cwt.
16 Ga	$85.00
18 Ga	$95.00

Studs & Tracks
20 Ga.

Size, in. (mm)	Price per 1,000 l.f.	Price per 1,000 m
1-5/8" (41 mm)	$320.00	$1,067.00
2-1/2" (63 mm)	340.00	1,133.00
3-1/4" (81 mm)	370.00	1,233.00
4" (100 mm)	390.00	1,300.00
6" (150 mm)	500.00	1,667.00
Shoes (per 1,000 pcs.)	80.00	267.00

Corner Bead

Type	Price per 1,000 l.f.	Price per 1,000 m
Regular	$160.00	$533.00
Expanded	185.00	617.00
Bull Nose	370.00	1,233.00
Cornerite	105.00	350.00

Rods

Type, in. (mm)	Price per 1,000 l.f.	Price per 1,000 m
3/16" (5 mm) plain	$48.00	$160.00
3/16" (5 mm) galvanized	58.00	193.00
1/4" (6 mm) plain	64.00	213.00
1/4" (6 mm) galvanized	74.00	267.00
3/8" (9 mm) plain	80.00	267.00

Gypsum Lath

Type and size, inches (mm)	Price per 1,000 s.f.	Price per 1,000 sq.m
plain or perforated, 16"x48"x3/8" (400x1200x9mm)	$110.00	$1,222.00
foilback, 16"x48"x3/8" (400x1200x9mm)	195.00	2,167.00
plain, 16"x48"x1/2" (400x1200x13mm)	115.00	1,278.00
foilback, 16"x48"x1/2" (400x1200x13mm)	200.00	2,2222.00

Nails and Staples

Type and size, inches (mm)	Price per 100 lbs.	Price per 100 kg
Gypsum lath, 1-1/8" (28 mm) blued	$95.00	$211.00
Gypsum lath, 1-1/4" (31 mm) blued	95.00	211.00
Roofing, barbed, 1-1/2" (38 mm) galv	105.00	233.00
Concrete stub	120.00	267.00
Annular ring nail (blued)	95.00	211.00
Brick—plain 2" (50 mm) and 2-1/2" (63 mm)	105.00	233.00

Lime and Finishing Plaster

Description	Price per 100 lbs.	Price per 100 kg
Pulverized Quicklime	$7.40	$16.44
Plasterer's Hydrated Lime	8.50	18.88
Gypsum Gauging Plaster	9.50	21.11
No. 1 Moulding Plaster	10.60	23.55
Keene's Cement, Regular or Fast	16.96	37.69
Portland Cement	4.24	9.42

FINISHES

Plastering Items

	Price per gal.	Price per liter
Liquid Bonding Agent	$26.50	$7.00

Prepared Finishes

Hair, 4 lbs. (1.8 kg) to Bu.	$8.50 per bu.
Fiber, 2 lbs. (0.9 kg) to Bu.	8.50 per bu.

Gypsum Plaster

	Price per 100 lbs.	Price per 100 kg
Gypsum Cement Plaster	$6.40	$14.22
Wood Fiber (prepared)	6.40	14.22
Bondcrete	7.40	16.44
Perlite Plaster	12.70	28.22
Gypsum Trowel Finish	7.40	14.22
Silica Sand Float	9.50	21.11

09250 GYPSUM WALLBOARD

Gypsum wallboard, or drywall, is manufactured from a gray-white colored rock called gypsum, which is a nonmetallic mineral composed of calcium sulphate chemically combined with crystallized water.

After the gypsum ore is mined or quarried, it is crushed, dried, and ground into a fine powder. It is then heated to remove most of the chemically combined water. The calcined gypsum is then mixed with other ingredients and water to create a slurry that is sandwiched between two sheets of treated paper to form a smooth gypsum wallboard panel. After the gypsum core has set, the wallboard is cut to length, dried, finished, and packaged for shipment.

Advantages. Interior walls and ceilings built with gypsum panels form durable surfaces for most types of decorative treatment. The following are some of the advantages of gypsum wallboard.

1. Fire Protection: Gypsum wallboard is fire resistant and will not support combustion. When one surface is exposed to a flame, the opposite panel surface remains cool until the gypsum core is calcinized. Fire resistant ratings are up to 4 hours for fire partitions, 3 hours for floor-ceiling, and 4 hours for column and shaft fireproofing.

2. Quick Installation: Wallboard panels are easily cut and quickly hung in large sheets, speeding productivity and reducing installation costs.

3. Dry Installation: Because wallboard is a dry panel, less moisture is introduced into the building. Cold weather is less a factor in finishing surfaces. Only the joint finishing must be kept from freezing.

4. Decoration: The strong, smooth face paper of the gypsum wallboard panel is suitable for most decorative treatments, such as paint, wallpaper, and textured coatings.

5. Sound Control: Gypsum wallboard panels are a component in sound-resistive partition and floor-ceiling systems.

6. Expansion: Under average temperature and humidity, gypsum wallboard expands little and does not warp excessively.

7. Availability: Standard gypsum wallboard products are easily obtainable on short notice.

Size and Thickness. Gypsum wallboard panels are manufactured in a number of thicknesses and sizes with various longitudinal edge designs.

Thickness, inches (mm)	Size, l.f.	Size mm	Edge Design
1/4" (6)	4'-0" x 8'-0" to 14'-0"	1200 x 2400 x 4200	S.E.*, T.E.**
3/8" (9)	4'-0" x 8'-0" to 14'-0"	1200 x 2400 x 4200	S.E., T.E.
1/2" (13)	4'-0" x 8'-0" to 14'-0"	1200 x 2400 x 4200	S.E., T.E.
5/8" (16)	4'-0" x 8'-0" to 14'-0"	1200 x 2400 x 4200	S.E., T.E.
*Square Edge **Tapered Edge			

Square edge panels may be used as the first layer of a two layer system or where the joints will be covered by a batten strip. The tapered edge panel should always be used on the finished surface layer, where the joints are to be finished with tape and compound. The tapered edge is a depression along the longitudinal edge that receives the tape and compound build-up to smooth and hide the butting edges of the gypsum panels.

Thicknesses of Gypsum Wallboard Panels and Their Uses. The 1/4" (6 mm) gypsum wallboard is used as a lightweight, low cost utility wallboard. It is also used for curved surfaces, providing increased board flexibility and eliminating the need to score and moisten a panel to achieve a curved surface.

The 3/8" (9 mm) gypsum wallboard is used principally for repair and remodeling work or in double wall construction.

The 1/2" (13 mm) gypsum wallboard is used in single layer new construction, as well as for remodeling installations.

The 5/8" (16 mm) gypsum wallboard is used where a one-hour fire rating is required or where the framing is spaced in excess of 1/2" (13 mm) wallboard limitations. The thickness of these panels not only give increased resistance to fire exposure but also additional resistance to sound transmission. The stiffness decreases the likelihood of sagging.

Tape Joint System for Gypsum Wallboard. A drywall joint treatment, using tape and joint compound, must provide a joint that is as strong as the gypsum wallboard itself. The hand method of concealing joints in gypsum wallboard is backed by more than a quarter century of joint reinforcing experience. It conceals the joints between boards and bonds the gypsum wallboard units together into a single, smooth, even wall and ceiling surface unit.

Vertical Application of Gypsum Wallboard

First the hollow or channel at the edges of the wallboard is filled with joint compound, using a flexible 4", 5", or 6" (100, 125, or 150 mm) joint finishing broadknife.

Apply the joint tape immediately, directly over the compound and press it into place with the broadknife, squeezing excess compound out from under the joint tape but leaving enough compound under tape for proper bond. The depth best not exceed 1/32" (0.75 mm). Apply a thin covering coat of joint compound immediately and let dry.

Next, apply another thin coating of joint compound using a 7" to 10" (175 to 250 mm) flexible broadknife so that the tape will be completely hidden. Feather out edges beyond previous coat as smoothly as possible and let dry thoroughly. Intermediate nail heads should also be carefully filled with joint compound and brought flush with the surface of the wallboard.

Horizontal Application of Gypsum Wallboard

Apply a third thin finish coat of joint compound. It should be of thinner consistency to even up surface and edges of joint. This is usually accomplished using a wide finish coat applicator, wide enough to extend 2" (50 mm) beyond the edges of the second coat.

Dry sand, as needed, between and after coats to insure a smooth, inconspicuous joint and to eliminate scratches, craters, and nicks.

Mechanical Tape Applicator. There are several types of mechanical and semi-mechanical applicators. Joint compound is applied as before, by hand, and then the joint is taped mechanically. Then joint compound is then wiped down with a flexible broadknife and left to dry.

Apply a second fill coat of joint compound over the tape using a hand finisher tool. Spot finish all nails or screws to bring them flush with the surface of the wallboard.

Apply a third and final finish coat, feathering the edges about 2" (50 mm) beyond the last coat, and allow to dry. Dry sand as needed. Remember, drying time between coat applications will vary according to air temperature and humidity.

Fastener Description	Nail Spacing c. to c.	Approx. Lbs. Nails Req'd per MSF Gypsum Panels
1¼″ GWB-54 Annular Ring Nail 12½ ga.; ¼″ dia. head with a slight taper to a small fillet at shank; bright finish; medium diamond point; meets ASTM C380	7″ ceiling 8″ walls	5¼
1¾″ Annular Ring Nail (Same as GWB-54 except for length)	7″ ceiling 8″ walls	5¼

Fastening Application	Fastener Used

GYPSUM PANELS TO STANDARD METAL FRAMING

½″ single-layer panels to standard studs, runners, channels — ⅞″ Type S Bugle Head

⅝″ single-layer panels to standard studs, runners, channels — 1″ Type S Bugle Head

½″ double-layer panels to standard studs, runners, channels — 1⅝″ Type S Bugle Head

⅝″ double-layer panels to standard studs, runners, channels — 1⅝″ Type S Bugle Head

1″ coreboard to metal angle runners in solid partitions — 1¼″ Type S Bugle Head

½″ panels through coreboard to metal angle runners in solid partitions — 1⅞″ Type S Bugle Head

⅝″ panels through coreboard to metal angle runners in solid partitions — 2¼″ Type S Bugle Head

GYPSUM PANELS TO 12-GA. (MAX.) METAL FRAMING

½″ and ⅝″ panels and gypsum sheathing to 20-ga. studs and runners — 1″ Type S-12 Bugle Head

USG Self-Furring Metal Lath through gypsum sheathing to 20-ga. studs and runners — 1¼″ Type S-12 Bugle Head

½″ and ⅝″ double-layer gypsum panels to 20-ga. studs and runners — 1⅝″ Type S-12 Bugle Head

Multi-layer gypsum panels to 20-ga. studs and runners — 1¾″ Type S-12 Bugle Head

Types of Gypsum Panels

Foil-Back Gypsum Panels. These panels are made by laminating a sheet of aluminum foil to the back surface of the gypsum wallboard. The foil reduces outward heat flow in winter and inward heat flow in summer, has a significant thermal insulating value if facing an air space of 3/4" (19 mm) or more, and is effective as a vapor barrier.

In addition, foil-backed panels provide a water vapor retarder to help prevent interior moisture from entering wall and ceiling spaces. Foil-backed wallboard has some limitations. It is not recommended over tile, double layer installations, or where high outside temperatures prevail.

Moisture Resistant Gypsum Panels. MR board has a special asphalt composition gypsum core and is covered with chemically treated face papers to prevent moisture penetration. This board is a little harder to cut, has a brownish color core, and is usually covered with a light green finish face paper. These panels were developed for application in bathrooms, kitchens, utility rooms, and other high moisture areas.

Fire Rated Gypsum Panels. A specially formulated mineral gypsum core is used to make panels in 1/2" (13 mm) and 5/8" (16 mm) thicknesses for application to walls, ceilings, and columns where a fire rated assembly is needed. The formulation must meet or exceed ASTM C36. Based on tests by Underwriters' Laboratories, Inc., certain wall, floor/ceiling, and column assemblies give 45-minute to 4-hour fire resistance ratings.

Cement Boards. These panels are produced with a portland cement core, usually wrapped in a fiber glass mesh and covered with a specially formulated smooth covering. Cement boards are used for interior and exterior construction, where there are areas subject to water, moisture or high humidity, such as bathroom showers and tubs and for kitchen countertops.

These boards require special screws and tape for proper installation. They are not gypsum boards, and they have different stud spacing requirements.

Gypsum Drywall Accessories

Gypsum wallboard can be attached to framing by several methods, depending on the type of framing and the results desired.

Fasteners. Nails or screws can be used to attach gypsum panels to wood framing. However, screws must be used for attachment to metal framing members. There are four basic ways to fasten panels to framing:

1. Single Nailing
2. Double Nailing
3. Screw Attachment
4. Adhesive Attachment

Nail Application: Nails are applied with a hammer, seating the nail so that the head is in a shallow dimple formed by the last blow of the hammer. The dimple should be no deeper than 1/32" (0.75 mm) for gypsum panels. Drive nails at least 3/8" (8 mm) from ends or edges of wallboard.

Screw Application: Screws are applied with an electric power positive-clutch tool called an electric screwgun. Drive screws at least 3/8" (8 mm) from ends or edges of wallboard.

Staple Application: Staples can be used to attach base layer boards to wood framing in a double layer application. Staples should be 16-gauge flattened galvanized wire and provide a minimum penetration of 5/8" (16 mm) into the wood framing.

Double Nailing or Screwing: In this method perimeter screws are spaced 7" (175 mm) on center for ceilings and 8" (200 mm) on center for walls. Nails or screws are not doubled on the panel perimeter. For nails or screws attached to supports, pairs of nails or screws are attached 2" (50 mm) apart, and pairs are spaced every 12" (300 mm).

Adhesive Application: A continuous bead of adhesive is applied to a wood stud. The drywall panel is pressed onto the adhesive and supplementary screws or nails are attached. The main advantage of using adhesive is that it reduces the number of nails or screws required.

Corner Bead. Metal and plastic corner bead is furnished in 8'-0" (2.4 m) and 10'-0" (3.0 m) lengths and is used on all external corners in drywall construction. Corner bead provides a true and straight finish line to corners.

Stop Bead. Metal or vinyl stop beads are used to cap the edge of a gypsum panel at its termination point, whenever that edge is exposed and not covered by either a tape/compound finish or other trim members.

Joint Tape. A strong fiber tape designed with feathered edges and lightly pre-creased for internal corners is used with the joint compound for finishing the butt joints of gypsum panels. Joint tape is packaged and sold in rolls of 75, 250 and 500 l.f. (22.5, 75 and 150 m) lengths and widths of 1-31/32" (50 mm) 2-1/16" (52 mm) wide. Usually, about 0.37 l.f. of tape is required per square foot of drywall panel.

Joint Compound. This material is available in many different forms, e.g., powdered, ready-mixed, all-purpose, fast setting, exterior application, etc. The ready-mixed all-purpose joint compound is used extensively for all interior applications of taping and joint finishing, and is readily available in 5-gal. (18.93-liter) containers weighing about 61.7 lbs. (28 kg) each.

Type of Compound	Approximate Coverage
Ready-mixed joint compound – taping, topping, and all-purpose	0.138 lbs. per square foot 0.674 kg per square meter
Ready-mixed lightweight all-purpose joint compound	0.0094 gal per square foot 0.383 liters per square meter

Powder Joint Compound – taping, topping, and all-purpose	0.083 lbs. per square foot 0.405 kg per square meter
Setting Type Joint Compound, easy sand	0.052 lbs. per square foot 0.253 kg per square meter

Estimating Material Quantities for Gypsum Wallboard and Accessories

The estimator first measures the wall and ceiling surfaces to figure square feet (square meters) of surface to be covered. Do not deduct for openings smaller than 50 s.f. (4.5 sq.m).

When doing the quantity takeoff, an estimator separates the square foot (square meter) quantities into proper categories of types, thicknesses, and sizes of panels. From these figures the estimator can easily determine the number of gypsum wallboards and other materials required, including accessories such as fasteners.

Quantities of Fasteners. For nails or screws, the amount required depends on the framing spacing. If framing is spaced 16" (400 mm) on center, and fasteners are spaced 12" (300 mm) apart on a horizontal 8' x 4' panel, then it would require 1,094 fasteners per 1,000 sq. ft. (92.9 sq.m), or 35 fasteners per panel. "W" is used to denote wood fasteners, and "S" indicates steel fasteners.

Quantities of Joint Tape. Approximately 370 lin. ft. (112.8 m) of joint tape are required per 1,000 sq. ft. (92.9 sq.m) of wallboard.

Quantities of Joint Compound. Approximately 138 lbs. (62.6 kg) of all-purpose joint material is required to complete 1,000 sq. ft. (92.9 sq.m) of wallboard. The amount of compound is reduced if there are very few external corners to be finished.

Drywall Material Prices

Type of Gypsum Panel	Thickness, in. (mm)	Price per 32 s.f.	Price per 1,000 s.f.	Price per 3 sq.m	Price per 100 sq.m
Regular	1/4 (6)	$3.94	$123.00	$3.97	$132.40
Regular	3/8 (9)	4.13	129.00	4.17	138.86
Regular	1/2 (13)	4.29	134.00	4.33	144.24
Regular	5/8 (16)	4.86	152.00	4.91	163.62
Fire Rated	1/2 (13)	4.80	150.00	4.84	161.46
Fire Rated	5/8 (16)	5.22	163.00	5.26	175.46
Moisture Resistant	1/2 (13)	6.21	194.00	6.26	208.83
Moisture Resistant	5/8 (16)	6.66	208.00	6.72	223.90
Add: For Foil Back Panels		2.91	91.00	2.94	97.95

	Price per 1,000 pcs.	Price per 5,000 pcs.
1-1/4" (31 mm) GWB Ring Nails	$2.56	$12.79
1-1/4" (31 mm) Type S Bugle Head Screws	6.90	34.50

	Price per l.f.	Price per m
Corner Bead	$0.12	$0.39
Stop Bead	0.14	0.46

Joint Tape	Per Roll
75' (23 m) roll	$1.15
250' (76 m) roll	2.90
300' (91 m) roll	3.17

Joint Compound – 4.5 gal. (17 liter) container	$9.60

4' x 8' Board – Vertical Attachment
Number of Screws per 1,000 Square Feet

Frame Spacing	Screw Spacing			
	8"	12"	16"	24"
8"	2,844	1,969	1,531	1,094
12"	2,031	1,406	1,094	781
16"	1,625	1,125	875	625
24"	1,219	844	656	469
30"	914	731	569	406

1200 x 2400 mm Board – Vertical Attachment
Number of Screws per 100 Square Meters

Frame Spacing	Screw Spacing			
	200 mm	300 mm	400 mm	600 mm
200 mm	3,160	2,188	1,701	1,215
300 mm	2,257	1,563	1,215	868
400 mm	1,806	1,250	972	694
600 mm	1,354	938	729	521
750 mm	1,174	813	632	451

4' x 10' Board – Vertical Attachment
Number of Screws per 1,000 Square Feet

Frame Spacing	Screw Spacing			
	8"	12"	16"	24"
8"	2,800	1,925	1,488	1,050
12"	2,000	1,375	1,063	750
16"	1,600	1,100	850	600
24"	1,200	825	638	450
30"	1,040	715	553	390

1200 x 3000 mm Board – Vertical Attachment
Number of Screws per 100 Square Meters

Frame	Screw Spacing			
Spacing	200 mm	300 mm	400 mm	600 mm
200 mm	3,111	2,139	1,653	1,167
300 mm	2,222	1,528	1,181	833
400 mm	1,778	1,222	944	667
600 mm	1,333	917	708	500
750 mm	1,156	794	614	433

4' x 12' Board – Vertical Attachment
Number of Screws per 1,000 Square Feet

Frame	Screw Spacing			
Spacing	8"	12"	16"	24"
8"	2,771	1,896	1,458	1,021
12"	1,979	1,354	1,042	729
16"	1,583	1,083	833	583
24"	1,188	813	625	438
30"	1,029	704	542	379

1200 x 3600 mm Board – Vertical Attachment
Number of Screws per 100 Square Meters

Frame	Screw Spacing			
Spacing	200 mm	300 mm	400 mm	600 mm
200 mm	3,079	2,106	1,620	1,134
300 mm	2,199	1,505	1,157	810
400 mm	1,759	1,204	926	648
600 mm	1,319	903	694	486
750 mm	1,144	782	602	421

Typical Weights of Gypsum Board

Panel Size		4' x 8'	4' x 9'	4' x 10'	4' x 12'	4' x 14'
Sq. Ft. per Panel		32	36	40	48	56
	Lbs. per Sq. Ft.	Total Sheet Weight				
Regular Boards						
1/4"	1.2	38.4	--	48.0	--	--
3/8"	1.4	44.8	50.4	56.0	67.2	78.4
1/2"	1.7	54.4	61.2	68.0	81.6	95.2
5/8"	2.3	73.6	82.8	92.0	110.4	128.8
Firecore Boards						
1/2"	1.9	60.8	68.4	76.0	91.2	106.4
5/8"	2.2	70.4	79.2	88.0	105.6	123.2
5/8"	2.5	80.0	90.0	100.0	120.0	140.0

WR Boards

1/2"	1.8	57.6	64.8	72.0	86.4	100.8
5/8"	2.2	70.4	79.2	88.0	105.6	123.2

Typical Weights of Gypsum Board – Metric

Panel Size	1200 x 2400	1200 x 2700	1200 x 3000	1200 x 3600	1200 x 4200
Sq. M per Panel	2.88	3.24	3.60	4.32	5.04

	Kg per Sq. M		Total Sheet Weight			
Regular Boards						
6 mm	5.9	17.0	--	21.2	--	--
9 mm	6.8	19.6	22.0	24.5	29.4	34.3
13 mm	8.3	23.9	26.9	29.9	35.9	41.8
16 mm	11.2	32.3	36.3	40.3	48.4	56.4
Firecore Boards						
13 mm	9.4	27.1	30.5	33.8	40.6	47.4
16 mm	10.6	30.5	34.3	38.2	45.8	53.4
16 mm	12.0	34.6	38.9	43.3	51.9	60.6
WR Boards						
13 mm	8.8	25.3	28.5	31.7	38.0	44.4
16 mm	11.0	31.7	35.6	39.6	47.5	55.4

Example: The following list of material quantities is required for installation and finishing of 300 l.f. of drywall partition, 8'-0" high, covering both sides of the partition: 300 x 8 x 2 sides = 4,800 s.f.

Drywall Material Cost

Item	Rate	Total	Rate	Total
3/4" Drywall Regular – 4,800 s.f.	$....	$....	$0.13	$624.00
Screws 4.8M pcs.	....	....	6.90	33.12
Corner Bead 96 l.f.	....	....	0.08	7.68
Stop Bead 40 l.f.	....	....	0.11	4.40
Joint Tape – 8 rolls	....	....	2.90	23.20
Joint Compound – 6 five-gallon cans	....	....	11.50	74.75
Cost per 4,800 s.f.			$....	$767.15
per s.f.			....	0.16
per sq.m			....	1.72

Labor Placing Gypsum Wallboard–Nailed. When placing 1/2" (13 mm) gypsum wallboard in average size rooms, 2 carpenters experienced in wallboard erection should place about 1,700 s.f. (158 sq.m) of board (one-layer wall and ceiling) per 8-hr. day, at the following labor cost per s.f. (sq.m):

	Hours	Rate	Total	Rate	Total
Carpenter	16	$....	$....	$26.44	$423.04
Cost per s.f.			....		0.25
per sq.m			....		2.68

For walls and ceilings higher than 12'-0", decrease production by 10 to 15%.

When placing 5/8" (16 mm) gypsum wallboard in average size rooms, 2 carpenters experienced in wallboard erection should place about 1,500 s.f. (139 sq.m) of board (one-layer wall and ceiling) per 8-hr. day, at the following labor cost per s.f. (sq.m):

	Hours	Rate	Total	Rate	Total
Carpenter	16	$....	$....	$26.44	$423.04
Cost per s.f.			....		0.28
per sq.m			....		3.04

For walls and ceilings higher than 12'-0", decrease production by 10 to 15%.

Labor Placing Gypsum Wallboard–Screwed. When placing 1/2" (13 mm) gypsum wallboard in average size rooms, 2 carpenters experienced in wallboard erection should place about 2,000 s.f. (186 sq.m) of board (one-layer wall and ceiling) per 8-hr. day, at the following labor cost per s.f. (sq.m):

	Hours	Rate	Total	Rate	Total
Carpenter	16	$....	$....	$26.44	$423.04
Cost per s.f.			....		0.21
per sq.m			....		2.27

When placing 5/8" (16 mm) gypsum wallboard in average size rooms, 2 carpenters experienced in wallboard erection should place about 1,700 s.f. (158 sq.m) of board (one-layer wall and ceiling) per 8-hr. day, at the following labor cost per s.f. (sq.m):

	Hours	Rate	Total	Rate	Total
Carpenter	16	$....	$....	$26.44	$423.04
Cost per s.f.			....		0.25
per sq.m			....		2.68

Labor Placing Gypsum Wallboard in Shafts–Screwed. When placing 1/2" (13 mm) wallboard in shafts, 2 carpenters experienced in wallboard erection in shafts should place about 400 s.f. (37 sq.m) of board per 8-hr. day, at the following labor cost per s.f. (sq.m):

	Hours	Rate	Total	Rate	Total
Carpenter	16	$....	$....	$26.44	$423.04
Cost per s.f.			....		1.06
per sq.m			....		11.43

When placing 5/8" (16 mm) gypsum wallboard in shafts, 2 carpenters experienced in this work should place about 350 s.f. (33 sq.m) of board per 8-hr. day, at the following labor cost per s.f. (sq.m):

	Hours	Rate	Total	Rate	Total
Carpenter	16	$....	$....	$26.44	$423.04
Cost per s.f.				....	1.21
per sq.m				....	12.82

When placing 1-1/2" (37.5 mm) core board in shafts, 2 carpenters experienced in this work should place about 400 s.f. (37 sq.m) of board per 8-hr. day, at the following labor cost per s.f. (sq.m):

	Hours	Rate	Total	Rate	Total
Carpenter	16	$....	$....	$26.44	$423.04
Cost per s.f.				....	1.06
per sq.m				....	11.43

Labor Placing Gypsum Wallboard, Column Enclosures–Screwed. When placing 1/2" (13 mm) gypsum wallboard around columns, 2 carpenters experienced in wallboard erection should place about 400 s.f. (37 sq.m) of board per 8-hr. day, at the following labor cost per s.f. (sq.m):

	Hours	Rate	Total	Rate	Total
Carpenter	16	$....	$....	$26.44	$423.04
Cost per s.f.				....	1.06
per sq.m				....	11.43

When placing 5/8" (16 mm) gypsum wallboard around columns, 2 carpenters experienced in wallboard erection should place about 350 s.f. (33 sq.m) of board per 8-hr. day, at the following labor cost per s.f. (sq.m):

	Hours	Rate	Total	Rate	Total
Carpenter	16	$....	$....	$26.44	$423.04
Cost per s.f.				....	1.21
per sq.m				....	12.82

Labor Placing Gypsum Wallboard, Enclosures Around Beams and Soffits–Screwed. When placing 1/2" (13 mm) gypsum wallboard around beams and soffits, 2 carpenters experienced in wallboard erection should place about 400 s.f. (37 sq.m) of board per 8-hr. day, at the following labor cost per s.f. (sq.m):

	Hours	Rate	Total	Rate	Total
Carpenter	16	$....	$....	$26.44	$423.04
Cost per s.f.				....	1.06
per sq.m				....	11.43

When placing 5/8" (16 mm) gypsum wallboard around beams and soffits, 2 carpenters experienced in wallboard erection should place about 350 s.f. (33 sq.m) of board per 8-hr. day, at the following labor cost per s.f. (sq.m):

	Hours	Rate	Total	Rate	Total
Carpenter	16	$....	$....	$26.44	$423.04
Cost per s.f.				....	1.21
per sq.m				....	12.82

Labor Placing Gypsum Wallboard, Light Wells–Screwed. When placing 1/2" (13 mm) gypsum wallboard in light wells, 2 carpenters experienced in wallboard erection should place about 400 s.f. (37 sq.m) of board per 8-hr. day, at the following labor cost per s.f. (sq.m):

	Hours	Rate	Total	Rate	Total
Carpenter	16	$....	$....	$26.44	$423.04
Cost per s.f.				....	1.06
per sq.m				....	11.43

When placing 5/8" (16 mm) gypsum wallboard in light wells, 2 carpenters experienced in wallboard erection should place about 350 s.f. (33 sq.m) of board per 8-hr. day, at the following labor cost per s.f. (sq.m):

	Hours	Rate	Total	Rate	Total
Carpenter	16	$....	$....	$26.44	$423.04
Cost per s.f.				....	1.21
per sq.m				....	12.82

Labor Finishing Gypsum Wallboard. The finishing of gypsum wallboard is a multi-step process of tape coat, block coat, skim coat, and point-up coat. Sanding between the block and skim coats is not required if the finisher is careful and a first class mechanic. The complete finishing labor for 1,000 s.f. (90 sq.m) of wallboard is as follows:

	Hours	Rate	Total	Rate	Total
Tape Coat	2.0	$....	$....	$26.44	$52.88
Block Coat	1.6	....	....	26.44	42.30
Skim Coat	1.3	....	....	26.44	34.37
Sand	0.4	....	....	26.44	10.58
Point-Up	0.8	....	....	26.44	21.15
Cost per 1,000 s.f.			$....		$161.28
per s.f.				....	0.16
per sq.m				....	1.72

Metal Studs and Furring

Stud framing and wall or ceiling furring on commercial projects is normally accomplished with light gauge metal studs, runner channels, and furring members instead of wood framing.

This material is manufactured from cold rolled galvanized metal and is available in a number of sizes, shapes, and gauges.

Metal Stud Partitions. Runner channels are channel shaped members that are positioned at the top and bottom of the studs, as in top and bottom plates for a wood stud partition. The runner channels are attached to the floor and overhead structure with nails, screws, or powder-actuated drive pins.

The metal studs, which are channel shaped with a backbend to give stiffness, are placed within the web of the runner channels, located for on-center spacing, plumbed, and screwed into place with a 3/8" (9 mm) panhead screw, top and bottom. The web of the metal studs have cutouts for the passage of conduit and piping.

Runner channels and studs are made in 1-5/8" (41 mm), 2-1/2" (63 mm), 3-5/8" (91 mm), 4" (100 mm) and 6" (150 mm) widths in 25 ga., 2-1/2" (63 mm), 3-5/8" (91 mm), 4" (100 mm), 6" (150 mm) widths in 20 ga., and 4" (100 mm), 6" (150 mm) widths in 18 ga. and 16 ga. metals.

The 25 ga. runners and studs are usually specified for non-loadbearing partitions up to 14' (4.2 m) to 16' (4.8 m) in height. Over this height and up to 22' (6.6 m), the 20 ga. material is used. The 18 ga. and 16 ga. materials are used for greater heights, loadbearing walls, and for exterior wall construction where lateral pressure (wind loads) will be imposed, requiring greater stiffness and less deflection.

Metal Furring. This material is made from 25 gauge metal in the following shapes:

1. Z furring channel is used in conjunction with rigid insulation board when furring exterior wall surfaces.

2. The hat shaped furring channel is used for furring walls and ceilings when insulation board is not required.

3. The resilient channel is used primarily on ceilings to provide a separation between the gypsum panels and the framing members for the floor above. It is used extensively in wood framed garden apartments for noise dampening between apartment units.

Estimating Material Quantities. To determine the quantity of runner channel, studs, and fasteners for a partition, the estimator would measure the total length of partition to be erected. Note the thickness and gauge of materials specified, ascertain the height of the partition for length of studs, ascertain the required spacing of studs, and add extras for door or window openings and at corners, and determine what type and spacing of fasteners is required for attachment of the runner channels to the floor and overhead structure. Horizontal stud bridging or fire stops are usually not required as in wood frame partition, but be sure to check, because this item in metal is expensive to install.

Metal Runner

Metal Stud

Metal Z-Furring Channel

Metal Furring Channel

To determine the quantity of furring channel required, ascertain the on-center spacing of the members for quantity of pieces by the length of the members (height of wall, as furring members are usually vertically installed) to give total lineal feet (meters) to be installed. Determine the fastener type and spacing for attachment to the furring channel.

Material Prices

	Price per 1,000 l.f.	Price per 1,000 m
25 ga. - 1-5/8" (41 mm) Runner Channel	$138.00	$460.00
25 ga. - 1-5/8" (41 mm) Stud	148.00	493.00
25 ga. - 2-1/2" (63 mm) Runner Channel	164.00	547.00
25 ga. - 2-1/2" (63 mm) Stud	180.00	600.00
25 ga. - 3-5/8" (91 mm) Runner Channel	201.00	670.00
25 ga. - 3-5/8" (91 mm) Stud	212.00	707.00
25 ga. - 4" (100 mm) Runner Channel	217.00	723.00
25 ga. - 4" (100 mm) Stud	233.00	777.00
25 ga. - 6" (150 mm) Runner Channel	281.00	937.00
25 ga. - 6" (150 mm) Stud	302.00	1,007.00
25 ga. - 7/8" (22 mm) Hat Furring Channel	148.00	493.00

	Price per 1,000 l.f.	Price per 1,000 m
25 ga. - 1" (25 mm) Z Furring Channel	175.00	583.00
25 ga. - Resilient Channel	154.00	513.00
20 ga. - 2-1/2" (63 mm) Runner Channel	323.00	1,077.00
20 ga. - 2-1/2" (63 mm) Stud	339.00	1,077.00
20 ga. - 3-5/8" (91 mm) Runner Channel	360.00	1,200.00
20 ga. - 3-5/8" (91 mm) Stud	387.00	1,290.00
20 ga. - 4" (100 mm) Runner Channel	381.00	1,270.00
20 ga. - 4" (100 mm) Stud	403.00	1,343.00
20 ga. - 6" (150 mm) Runner Channel	488.00	1,627.00
20 ga. - 6" (150 mm) Stud	514.00	1,713.00

3/8" (9 mm) Pan Head Screws 4.75
Powder Actuated Drive Pins (avg. cost inc. shot) .... 0.40 ea.

Example. The following list of material quantities would be required for a metal stud partition 200 l.f. (60 m) long, 10 l.f. (3 m) high, stud at 24" (600 mm) o.c., 25 ga. metal and installed with drive pins at 24" (600 mm) o.c., top and bottom to concrete slabs:

Item	Rate	Total	Rate	Rate, (metric)	Total
2-1/2" (63 mm) 25 ga. Runner, 400 l.f. (120 m)	$....	$....	$0.17	$0.57	$68.00
2-1/2" (63 mm) 25 ga. Stud, 1,010 l.f. (303 m)		....	0.18	0.60	181.80
1" (25 mm) Drive Pins - 202 ea.		....	0.27	0.27	54.54
3/8" (9 mm) Pan Screws - 0.4 M		....	4.77	4.77	1.90
Cost for 200 l.f.			$....		$306.24
per l.f.				....	1.53
per m				....	5.10

Add extra studs for door and window openings and corners.

Labor to Install Metal Stud Partitions. Using the above example given for materials, 2 carpenters should erect the 200 l.f. (60 m) of partition at the following labor cost:

	Hours	Rate	Total	Rate	Total
Layout Wall Line	1.0	$....	$....	$26.44	$26.44
Fasten Bottom Runner	4.0	....	....	26.44	105.76
Fasten Top Runner	4.0	....	....	26.44	105.76
Install Studs	3.0	....	....	26.44	79.32
Plumb & Screw Studs	4.0	....	....	26.44	105.76
Cost for 200 l.f.			$....		$423.04
per l.f.				....	2.11
per m				....	6.92

CERAMIC WALL AND FLOOR TILE

Ceramic tile provides a durable, colorful surface that is virtually maintenance-free. Its applications include interior and exterior finishes for functional and decorative purposes in all types of structures. Special uses include acid resistant and electrically conductive installations. Ceramic tile is available in many sizes, shapes, and finishes.

Glazed Ceramic Wall Tile. This has an impervious facial finish fused onto the body of the tile. The glazed surface comes in a wide variety of colors.

Ceramic Mosaic Tile. This tile, either glazed or unglazed, has a facial area of less than 6 sq. in. and is usually mounted on sheets or mesh, about 2'x 1' (0.6 x 0.3 m), to facilitate setting.

Quarry Tile. This is a rugged ceramic tile used primarily as a finish flooring, interior and exterior, where a long wearing, easily cleanable surface is desired.

Recent developments in the ceramic tile industry make it necessary to stress the importance of relating tile costs to each individual job specification. Glazed ceramic wall tile, for example, can be backmounted or unmounted, and can be installed using conventional portland cement mortar, various types of adhesives, or the more recently developed "dry-set" portland cement mortar. All of these variations can affect costs, both material and labor. Review job requirements carefully before doing the estimate.

However, the estimator's job has been simplified in recent years. Virtually the entire ceramic tile industry has adopted a "simplified practice", prepared under the auspices of the Tile Council of America, which has reduced the number of sizes and shapes and which has established generally recognized standards for the industry.

Estimating Quantities. Ceramic tile is estimated by the square foot, with trim pieces such as base, cap, etc. being estimated by the lineal foot. The estimator should deduct door and window openings, but the trim pieces necessary to finish the openings must be added. The quantities should be related to the type and size of ceramic tile, since these items will affect the cost of the ceramic tile when priced.

Estimate Composition. The finished ceramic tile estimate will include the following items:

1. Cost of ceramic tile delivered to the job site.
2. Cost of accessory materials such as wire mesh, sand, and cement for floor fill under ceramic tile.
3. Cost of mixing and placing floor fill.
4. Direct labor cost of laying and cleaning the ceramic tile.
5. The ceramic tile contractor's overhead and profit.

Variable Factors Influencing Costs. As in the other construction trades, estimating the labor costs of setting ceramic tile requires an intimate knowledge of the labor market in which the work is to be performed. Wage rates vary throughout the country, and it is important the estimator determine the rate in his locality.

Ceramic tile is unique in that it is often necessary to install it in very small quantities (for example, in 1 or 2 bathrooms in a house), and the costs can vary greatly. But such small installations are the exception, and the costs given below are based on the assumption that areas involved are large enough to permit the tile contractor to operate efficiently.

These costs are based on 4-1/4"x4-1/4" (106 x 106 mm) glazed wall tile, 1"x1" (25 x 25 mm) ceramic mosaic tile, and 6"x6"x1/2" (150 x 150 x 13 mm) quarry tile, since these are the sizes most commonly employed. If other sizes are specified, substitute the material price only. For all practical purposes, the labor cost remains constant regardless of size. The developed costs do not include areas where an unusually large amount of trim is required. Make an allowance for extra trim pieces in such special cases.

Setting Methods. Three common methods of adhering ceramic tile to a subsurface are described below.

Conventional Portland Cement Mortar Method. This method is to bond each ceramic tile with a layer of pure portland cement paste to a portland cement setting bed. This is done while the setting bed is still plastic. Wall tile must be soaked in water so that the water needed for curing is not absorbed from the paste. This is the traditionally accepted method, but it is also the costliest method.

Dry-Set Portland Cement Mortar Method. This method uses a dry curing portland cement mortar, accomplished through the use of water retaining additives, and has made ceramic tile installation cheaper and simpler. "Dry-set" is ideally suitable for use with concrete masonry, brick, poured concrete, and portland cement plaster. It should not be used over wood or gypsum plaster. Labor costs are appreciably reduced when this method is used.

Water-Resistant Organic Adhesive. Organic adhesives can be used over smooth base materials, such as wallboards, plywood, and metal. Labor productivity is comparable to that of "dry-set" mortar.

Labor Productivity per S.F. (Sq.M) per Team Day
Team is 1 tile setter and 1 helper
Conventional Mortar

Description	Face-mounted s.f.	Face-mounted sq.m	Back-mounted s.f.	Back mounted sq.m	Un-mounted s.f.	Un-mounted sq.m
Glazed wall tile	--	--	75-90	6.75-8.1	55-65	4.95-5.85

Description	Face-mounted s.f.	Face-mounted sq.m	Back-mounted s.f.	Back-mounted sq.m	Un-mounted s.f.	Un-mounted sq.m
Ceramic mosaic						
tile—walls	45-50	4.05-4.5	55-65	4.95-5.85	--	--
floors	100-125	9-11.25	100-125	9-11.25	--	--
Quarry tile	--	--	--	--	100-125	9-11.25

"Dry-Set" Mortar

Description	Face-mounted s.f.	Face-mounted sq.m	Back-mounted s.f.	Back-mounted sq.m	Un-mounted s.f.	Un-mounted sq.m
Glazed wall tile	--	--	125-150	11.25-13.5	100-125	9-11.25
Ceramic mosaic						
tile—walls	100-125	9-11.25	120-140	10.8-12.6	--	--
floors	125-150	11.25-13.5	125-150	11.25-13.5	--	--
Quarry tile floors	--	--	--	--	125-150	11.25-13.5

Organic Adhesive

Description	Face-mounted s.f.	Face-mounted sq.m	Back-mounted s.f.	Back-mounted sq.m	Un-mounted s.f.	Un-mounted sq.m
Glazed wall tile	--	--	150-175	13.5-15.75	120-140	10.8-12.6
Ceramic mosaic						
tile—walls	120-140	10.8-12.6	125-150	11.25-13.5	--	--
floors	150-175	13.5-15.75	175-200	15.75-18	--	--
Quarry tile floors	--	--	--			

	lin. ft.	meters
Cove Or Base	65-75	5.85-6.75
Cap	80-90	7.2-8.1

Approximate Prices of Ceramic Tile Materials
Ceramic Wall Tile

Description	Size, inches	Size, mm	Price per s.f.	Price per sq.m
Flat tile, all colors, incl. white	4-1/4"x4-1/2"	113x113mm	$1.35	$15.00
	6"x6"	150x150mm	1.45	16.11
	6"x4-1/4"	150x106mm	1.50	16.67
	6"x9"	150x225mm	1.65	18.33
	8-1/2"x4-1/4"	213x106mm	1.55	17.22

For use with 4-1/4"x4-1/4" (106x106mm) tile

			Price per piece*
Cove	6"x3-1/4"	150x81mm	$0.85
Cove	4-1/4"x4-1/4"	106x106mm	0.85
Bullnose	2"x6"	50x150mm	0.85
Bullnose	4-1/4"x4-1/4"	106x106mm	0.90
Double bullnose	4-1/4"x4-1/4"	106x106mm	1.30

For use with 6"x6" (150x150mm) tile

Cove	6"x6"	150x150mm	$1.15
Bullnose	6"x6"	150x150mm	1.20
Double bullnose	6"x6"	150x150mm	1.50

Miscellaneous trim shapes

Bead	6"x3/4"	150x19mm	$0.85
Bullnose	6"x3-3/4"	150x94mm	1.15
Base	6"x4"	150x100mm	1.30
Base	6"x6"	150x150mm	1.35
Sink Trim	6"x2"	150x50mm	1.30
Window Sill	9"x5-1/8"	225x125mm	1.50

*Prices are for stretchers; for angles, multiply prices by two.

Ceramic Mosaic Tile Description
Unglazed, Modular, Solid Color

Size, inches	Size, mm	White & Grays per s.f.	per sq.m	Other Colors per s.f.	per sq.m
1"x1"	25x25mm	$1.30	$14.44	$1.50	$16.67
2"x1"	50x25mm	1.40	15.55	1.60	17.77
2"x2"	50x50mm	1.50	16.67	1.70	18.89
3/4"x3/4"	19x19mm	1.30	14.44	1.50	16.67
1-9/16"x3/4"	39x19mm	1.30	14.44	1.50	16.67
1-9/16"x1-9/16"	39x39mm	1.30	14.44	1.50	16.67
Trim Shapes, price per lin. ft. (meter)					
1"x1" Bullnose	25x25mm			$1.40	15.55
2"x1" Bullnose	50x25mm			1.60	17.77

3/4" Series trim pieces same price as 1" Series

Various manufacturers furnish, as standard items, patterns made from combinations of modular tile sizes and colors. Prices vary depending on number of different sizes, number of colors, and depth of colors used. Patterns such as these are too numerous to itemize, but the general range is from $1.35 to $1.70 per s.f. ($15-18.89 per sq.m). There are exceptions to this range, and when a specific pattern price is needed, check with the specific supplier.

	Price per s.f.	Price per sq.m
Abrasive Ceramic Mosaic Tile	$1.45-1.80	$16.11-20.00
Conductive Floor Tile	2.55-3.20	28.33-35.55
Granitized Ceramic Mosaic Tile	1.35-1.55	15.00-17.22
Abrasive Granitized Ceramic Mosaic Tile	1.45-1.70	16.11-18.89
Glazed Satin Finish Ceramic Mosaic Tile	1.40	15.55
Glazed Textured Finish Ceramic Mosaic Tile	1.60	17.77
Special Decorative Glazed Ceramic Mosaic Tile for Accent Effects	3.20-13.80	35.55-153.33

Venetian glass mosaics, mounted on sheets 12-1/4"x12-1/4" (306 x 306 mm) square, are $2.65 per s.f. ($29.44 per sq.m) for 3/4"x3/4" (19 x 19 mm), $2.90 per s.f. ($32.22 per sq.m) for 1"x1" (25 x 25 mm) and $3.10 per s.f. ($34.44 per sq.m) for 1-1/2"x1-1/2" (38 x 38 mm). Metallic finishes run considerably more. Beads and coves for glass mosaics will run $2.15 per l.f. ($7.17 per m).

Sculptured tiles which can be used for wall surfacing, murals, screens, and sculpture are available in a wide range of ceramic designs and treatments. Design-Technics of New York have tiles in sizes 4-1/4"x8-1/2"x1/4" (106 x 213 x 6 mm) at approximately $4.25 to $5.30 per s.f. ($47.22-58.89 per sq.m), 12"x12"x3/8" (300 x 300 x 9 mm) at $5.30 to $6.65 per s.f. ($58.89-73.89 per sq.m), and 12"x12"x13/8" (300 x 300 x 34 mm) at $6.90 to $8.20 per s.f. ($76.67-91.11 per sq.m).

Quarry Tile

Size, inches	Size, mm	Price per s.f.	Price per sq.m
Deep red, plain surface,			
6"x6"x1/2"	150x150x13mm	$1.30	$14.44
6"x23/4"x1/2"	150x69x13mm	1.35	15.00
6"x6"x3/4"	150x150x19mm	2.00	22.22
9"x9"x3/4"	225x225x19mm	36.11	3.25
Deep red, abrasive,			
6"x6"x3/4"	150x150x19mm	2.25	25.00
6"x6"x3/4"	150x150x19mm	2.75	30.55

Other colors, with the exception of green, are virtually the same price as the red shown above. Green in the 6"x6"x1/2" (150x150x13mm) size costs $2.15 per s.f. ($23.89 per sq.m).

Extras, when specified, are:

	Price per s.f.	Price per sq.m
Tile ground square after burning	$0.10	$1.11
Wax coating	0.22	2.44

	Price per 1,000 pcs.
Floor brick, deep red, plain	
tray packed, 8"x4"x1-3/8 (200x100x34mm)	$175.00
in sealed cartons	206.00
Floor brick, deep red, abrasive	
tray packed, 8"x4"x1-3/8 (200x100x34mm)	190.00
in sealed cartons	222.00

There are many specially cast heavy floor and wall tiles that compliment the popular provincial and country modes. Many of these are quite expensive and are not regularly stocked. Always check locally for availability and prices, and for similar tiles that might be allowed for substitution.

Ceramic Tile Bathroom Accessories

Numerous ceramic tile bathroom accessories are available in a variety of sizes and qualities. They may be recessed or surface mounted, and they come in the full range of tile colors. Some of the more commonly used items are as follows:

Description	Price per piece
Recessed soap holder	$12.75
Recessed glass holder	12.75
Roll paper holder	15.90
Robe hook	4.25
Double robe hook	4.75
Toothbrush holder	6.36
Tile bar brackets (pair)	8.50

Decorative Ceramic Wall Tiles

Glazed ceramic wall tile containing designs as an integral part of each piece of tile is also available. Normally supplied in 4-1/4"x4-1/4" (106 x 106 mm) and 6"x6" (150 x 150 mm) sizes, the price of this material will vary with the design selected and number of colors in the design. If the design is known, the estimator should obtain a firm price from a supplier.

Description, inches	mm	Price per s.f.	Price per sq.m
One color,			
4-1/4"x4-1/4"	106x106mm	$2.65	$29.44
6"x6"	150x150mm	2.50	27.77
Two colors,			
4-1/4"x4-1/4"	106x106mm	3.20	35.55
6"x6"	150x150mm	3.00	33.33
Three colors,			
4-1/4"x4-1/4"	106x106mm	3.70	41.11
6"x6"	150x150mm	3.45	38.33

Ceramic Tile Adhesives and Accessory Materials

	Coverage	Coverage	Price	Price
Organic adhesive	40-50 sf/gal	4.75 sq.m/liter	$8.50/gal	$2.25/liter
Dry-set mortar mix	35 lbs./100 sf	175 kg/100 sq.m	0.50/lb	1.10/kg
Primer (for use in damp areas)	125 sf/gal	2.97 sq.m/liter.	6.50/gal. 1.70/liter	
Plastic underlayment	100 sf/unit*	9 sq.m/unit	8.50/unit*	8.50/unit
Wet tile grout	15-20 sf/lb**	0.33 sq.m/kg	0.30/lb	0.67/kg
Dry tile grout	12-15/sf/lb.**	0.33-0.4 sq.m/kg	0.35/lb	0.77/kg

*Unit consists of 1 gallon (3.785 liters) liquid and 34 lbs. (15.3 kg) aggregate.
**When used with 4-1/4"x4-1/4" (106 x 106 mm) tile.

Material Cost of 100 S.F. (9 Sq.M) of 4-1/4"x4-1/4" (106 x 106 mm)
Glazed Ceramic Wall Tile
Using Conventional Mortar Method of Installation (Unmounted Tile)

	Rate	Total	Rate	Rate (metric)	Total
2 sacks portland cement	$....	$....	$4.24		$8.48
4 c.f. (0.11 cu.m) sand	....		0.48	17.45	1.92
100 s.f. (9 sq.m) glazed wall tile	....		1.33	14.77	133.00
7 lbs. (3.15 kg) wet tile grout mix	....		0.27	0.60	1.89
Cost per 100 s.f.			$....		$145.29
per s.f.				....	1.45
per sq.m				....	16.14

Labor Cost of 100 S.F. (9 Sq.M) of 4-1/4"x4-1/4" (106 x 106 mm)
Glazed Ceramic Wall Tile
Using Conventional Mortar Method of Installation (Unmounted Tile)

	Hours	Rate	Total	Rate	Total
Tile Setter	13.5	$....	$....	$21.04	$284.04
Cost per s.f.				....	2.84
per sq.m				....	30.56

Material Cost of 100 S.F. (9 Sq.M) of 4-1/4"x4-1/4" (106 x 106 mm)
Glazed Ceramic Wall Tile
Using Water Resistant Organic Adhesive (Unmounted Tile)

	Rate	Total	Rate	Rate (metric)	Total
2.5 gals. (9.46 liters) adhesive	$....	$....	$8.50	$2.24	$21.25
100 s.f. (9 sq.m) glazed wall tile	....		1.33	14.77	133.00
9 lbs. (4.05 kg) dry tile grout mix	....		0.37	0.82	3.33
Cost per 100 s.f.			$....		$157.58
per s.f.				....	1.58
per sq.m				....	17.50

Labor Cost of 100 S.F. (9 Sq.M) of 4-1/4"x4-1/4" (106 x 106 mm)
Glazed Ceramic Wall Tile
Using Water Resistant Organic Adhesive (Unmounted Tile)

	Hours	Rate	Total	Rate	Total
Tile Setter	8.0	$....	$....	$21.04	$168.32
Cost per s.f.				....	1.68
per sq.m				....	18.08

Material Cost of 100 S.F. (9 Sq.M) of 4-1/4"x4-1/4" (106 x 106 mm)
Glazed Ceramic Wall Tile
Using "Dry-Set" Portland Cement Mortar (Unmounted Tile)

	Rate	Total	Rate	Rate (metric)	Total
35 lbs. (16.75 kg) "dry-set" mortar mix	$....	$....	$0.48	$1.06	$16.80
100 s.f. (9 sq.m) glazed wall tile		....	1.33	14.77	133.00
9 lbs. (4.05 kg) dry tile grout mix		....	0.37	0.82	3.33
Cost per 100 s.f.				$....	$153.13
per s.f.				....	1.53
per sq.m				....	16.47

Labor Cost of 100 S.F. (9 Sq.M) of 4-1/4"x4-1/4" (106 x 106 mm)
Glazed Ceramic Wall Tile
Using "Dry-Set" Portland Cement Mortar (Unmounted Tile)

	Hours	Rate	Total	Rate	Total
Tile Setter....................	10.0	$....	$....	$21.04	$210.40
Cost per s.f.				....	2.10
per sq.m				....	22.64

Material Cost of 100 S.F. (9 Sq.M) 1"x1" (25 x 25 mm)
Ceramic Mosaic Tile Floors Using
Conventional Mortar Method of Installation (Face-Mounted Tile)

	Rate	Total	Rate	Rate (metric)	Total
2 sacks portland cement	$....	$....	$4.24	$4.24	$8.48
4 c.f. (0.11 cu.m) sand		....	0.48	17.45	1.92
100 s.f. (9 sq.m) ceramic mosaic tile		....	1.27	14.11	127.00
17 lbs. (7.65 kg) wet tile grout mix		....	0.27	0.60	4.59
Cost per 100 s.f.				$....	$141.99
per s.f.				....	1.42
per sq.m				....	15.77

Labor Cost of 100 S.F. 1"x1" (25 x 25 mm) Ceramic Mosaic Tile Floors Using
Conventional Mortar Method of Installation (Face-Mounted Tile)

	Hours	Rate	Total	Rate	Total
Tile setter...................	10.5	$....	$....	$21.04	$220.92
Cost per s.f.				....	2.21
per sq.m				....	23.78

Material Cost of 100 S.F. (9 Sq.M) of 1"x1" (25 x 25 mm)
Ceramic Mosaic Tile Floors Using
Water Resistant Organic Adhesive (Face-Mounted Tile)

	Rate	Total	Rate	Rate (metric)	Total
2.5 gals. (9.46 liters) adhesive ....$....	$....		$8.50	$2.25	$21.25
100 s.f. (9 sq.m) ceramic mosaic tile		....	1.27	14.11	127.00
23 lbs. (10.35 kg) dry tile grout mix		....	0.37	0.82	8.51
Cost per 100 s.f.			$....		$156.76
per s.f.				....	1.57
per sq.m				....	17.62

Labor Cost of 100 S.F. (9 Sq.M) of 1"x1" (25 x 25 mm)
Ceramic Mosaic Tile Floors Using
Water Resistant Organic Adhesive (Face-Mounted Tile)

	Hours	Rate	Total	Rate	Total
Tile setter...8.0		$....	$....	$21.04	$168.32
Cost per s.f.				....	1.68
per sq.m				....	18.11

Material Cost of 100 S.F. (9 Sq.M) of 1"x1" (25 x 25 mm)
Ceramic Mosaic Tile Floor Using
"Dry-Set" Portland Cement Mortar (Face-Mounted Tile)

	Rate	Total	Rate	Rate (metric)	Total
35 lbs. (15.75 kg) "dry-set" mortar mix ....$....	$....		$0.48	$1.07	$16.80
1 c.f. (0.02 cu.m) of sand		....	0.48	15.89	0.48
100 s.f. (9 sq.m) ceramic mosaic tile		....	1.27	14.11	127.00
23 lbs. (10.35 kg) dry tile grout mix		....	0.37	0.82	8.51
Cost per 100 s.f.			$....		$152.79
per s.f.				....	1.53
per sq.m				....	16.98

Labor Cost of 100 S.F. (9 Sq.M) of 1"x1" (25 x 25 mm)
Ceramic Mosaic Tile Floors Using
"Dry-Set" Portland Cement Mortar (Face-Mounted Tile)

	Hours	Rate	Total	Rate	Total
Tile setter..9.0		$....	$....	$21.04	$189.36
Cost per s.f.				....	1.89
per sq.m				....	20.34

Material Cost of 100 S.F. (9 Sq.M) of 1"x1" (25 x 25 mm)
Ceramic Mosaic Tile on Walls
Using Conventional Mortar Method of Installation (Face-Mounted Tile)

	Rate	Total	Rate	Rate (metric)	Total
2 sacks portland cement	$....	$....	$4.24	$4.24	$8.48
4 c.f. (0.11 cu.m) sand		....	0.48	17.45	1.92
100 s.f. (9 sq.m) ceramic mosaic tile		....	1.27	14.11	127.00
17 lbs. (7.65 kg) wet tile grout mix		....	0.27	0.60	4.59
Cost per 100 s.f.			$....		$141.99
per s.f.			....		1.42
per sq.m			....		4.67

Labor Cost of 100 S.F. (9 Sq.M) of 1"x1" (25 x 25 mm)
Ceramic Mosaic Tile on Walls
Using Conventional Mortar Method of Installation (Face-Mounted Tile)

	Hours	Rate	Total	Rate	Total
Tile setter	11.0	$....	$....	$21.04	$231.44
Cost per s.f.				....	2.31
per sq.m				....	24.86

Material Cost of 100 S.F. (9 Sq.M) of 1"x1" (25 x 25 mm)
Ceramic Mosaic Tile on Walls
Using Water Resistant Organic Adhesive (Face-Mounted Tile)

	Rate	Total	Rate	Rate (metric)	Total
2.5 gals. (9.46 liters) adhesive	$....	$....	$8.50	$2.25	$21.25
100 s.f. (9 sq.m) ceramic mosaic tile		....	$1.27	14.11	127.00
23 lbs. (10.35 kg) dry tile grout mix		....	0.37	0.82	8.51
Cost per 100 s.f.			$....		$156.76
per s.f.			....		1.57
per sq.m			....		17.42

Labor Cost of 100 S.F. (9 Sq.M) of 1"x1" (25 x 25 mm) Ceramic Mosaic Tile on Walls
Using Water Resistant Organic Adhesive (Face-Mounted Tile)

	Hours	Rate	Total	Rate	Total
Tile setter	9.0	$....	$....	$21.04	$189.36
Cost per s.f.				....	1.89
per sq.m				....	20.34

Material Cost of 100 S.F. (9 Sq.M) of 1"x1" (25 x 25 mm)
Ceramic Mosaic Tile on Walls
Using "Dry-Set" Portland Cement Mortar (Face-Mounted Tile)

	Rate	Total	Rate	Rate (metric)	Total
35 lbs. (15.75 kg) "dry-set" mortar mix ....$....	$....	$0.48	$1.06	$16.80	
1 c.f. (0.02 cu.m) sand	....	0.48	16.94	0.48	
100 s.f. (9 sq.m) ceramic mosaic tile	....	1.27	14.11	127.00	
23 lbs. (10.35 kg) dry tile grout mix	....	0.37	0.82	8.51	
Cost per 100 s.f.................	$....			$152.79	
per s.f.................	....			1.53	
per sq.m	....			16.97	

Labor Cost of 100 S.F. (9 Sq.M) of 1"x1" (25 x 25 mm) Ceramic Mosaic Tile on Walls
Using "Dry-Set" Portland Cement Mortar (Face-Mounted Tile)

	Hours	Rate	Total	Rate	Total
Tile setter.................	9.0	$....	$....	$21.04	$189.36
Cost per s.f.................			....		1.89
per sq.m			....		20.34

Material Cost of 100 S.F. (9 Sq.M) of 6"x6"x1/2" (150 x 150 x 13 mm)
Quarry Tile Floors
Using Conventional Mortar Method of Installation

	Rate	Total	Rate	Rate (metric)	Total
2 sacks portland cement ....$....	$....	$4.24	$4.24	$8.48	
6 c.f. (0.17 cu.m) sand	....	0.48	16.94	2.88	
100 s.f. (9 sq.m) quarry tile plain surface	....	1.27	14.11	127.00	
35 lbs. (15.75 kg) portland cement grout	....	0.53	1.17	18.55	
Cost per 100 s.f.................	$....			$156.91	
per s.f.................	....			1.57	
per sq.m	....			17.43	

Labor Cost of 100 S.F. (9 Sq.M) of 6"x6"x1/2" (150 x 150 x 13 mm) Quarry Tile Floors
Using Conventional Mortar Method of Installation

	Hours	Rate	Total	Rate	Total
Tile setter.................	16.0	$....	$....	$21.04	$336.64
Cost per s.f.................			....		3.37
per sq.m			....		36.26

Material Cost of 100 S.F. (9 Sq.M) of 6"x6"x1/2" (150 x 150 x 13 mm) Quarry Tile Floors Using "Dry-Set" Portland Cement Mortar

	Rate	Total	Rate	Rate (metric)	Total
35 lbs. (15.75 kg) "dry-set" mortar mix ....$....		$....	$0.48	$1.06	$16.80
2 c.f. (0.06 cu.m) of sand		....	0.48	16.00	0.96
100 s.f. (9 sq.m) quarry tile plain surface		....	1.27	14.11	127.00
35 lbs. (15.75 kg) portland cement grout		....	0.53	1.17	18.55
Cost per 100 s.f..........................			$....		$163.31
per s.f..........................			...		1.63
per sq.m			...		18.15

Labor Cost of 100 S.F. (9 Sq.M) 6"x6"x1/2" (150 x 150 x 13 mm) Quarry Tile Floors Using "Dry-Set" Portland Cement Mortar

	Hours	Rate	Total	Rate	Total
Tile setter..................12.0		$....	$....	$21.04	$252.48
Cost per s.f..........................			....		2.52
per sq.m			....		27.17

Material Cost of 100 L.F. (30 M) of Cove or Base (Ceramic or Quarry Tile)

	Rate	Total	Rate	Rate (metric)	Total
100 l.f. (30 m) cove ....$....		$....	$1.70	$5.67	$170.00
Adhesive or mortar—allow		....	6.36	6.36	6.36
Cost per 100 l.f.			$....		$176.36
per l.f..........................			....		1.76
per m			....		5.88

Labor Cost of 100 L.F. (30 M) of Cove or Base (Ceramic or Quarry Tile)

	Hours	Rate	Total	Rate	Total
Tile setter....................20.0		$....	$....	$21.04	$420.80
Cost per l.f..........................			....		4.21
per m			....		13.81

Material Cost of 100 L.F. (30 M) of Wainscot Cap (Glazed Ceramic Wall Tile)

	Rate	Total	Rate	Rate (metric)	Total
100 l.f. (30 m) bullnose ....$....		$....	$1.70	$5.67	$170.00

	Rate	Total	Rate	Rate (metric)	Total
Adhesive or mortar—allow		....	6.36	6.36	6.36
Cost per 100 l.f.			$....		$176.36
per l.f.				....	1.76
per m				....	5.88

Labor Cost of 100 L.F. (30 M) of wainscot cap (Glazed Ceramic Wall Tile)

	Hours	Rate	Total	Rate	Total
Tile setter	20.0	$....	$....	$21.04	$420.80
Cost per l.f.				....	4.21
per m				....	13.81

Placing Cement Floor Fill. Cement floor fill under ceramic tile floors is usually placed by tile setters and helpers (one tile setter and one or two helpers working together). The fill is placed one or two days in advance of the tile if the overall thickness from rough floor to finished tile surface is over 3" (75 mm). For 3" (75 mm) thickness and under, fill and setting bed may be placed in one operation. When such tile floors are to be placed over wood subfloors, it is necessary to first place a layer of waterproof building paper and a layer of wire mesh reinforcing before placing the fill.

A tile setter and two helpers should place 450 to 500 s.f. (41-45 sq.m) of fill per 8-hour day in areas large enough to permit efficient operations, at the following cost per 100 s.f. (9 sq.m):

Material Costs

	Rate	Total	Rate	Rate (metric)	Total
6 bags portland cement	$....	$....	$4.24	$4.24	$25.44
1 c.y. (0.76 cu.m) sand		....	12.75	16.77	12.75
Cost per 100 s.f.			$....		$38.19
per s.f.				....	0.38
per sq.m				....	4.24

Labor Costs

	Hours	Rate	Total	Rate	Total
Tile setter	1.5	$....	$....	$21.04	$31.56
Helper	3.0		....	21.04	63.12
Cost per 100 s.f.			$....		$94.68
per s.f.				....	0.95
per sq.m				....	10.22

Metal Wall Tile

Wall tile of aluminum, having a baked enamel finish in various colors, or tile of copper or stainless steel are used for the same purposes.

They require the same type of base and are applied with adhesive. A worker will apply about 100 s.f. (9 sq.m) per day.

Metal tile mastic weighs about 16 lbs. per gal. (1.92 kg per liter), and it requires 1/2 lb. per s.f. (2.5 kg per sq.m), or one gal. (3.78 liters) is sufficient for 32 s.f. (2.88 sq.m) of tile. Price about $8.50 per gal. Aluminum metal tile cost about $1.60 per s.f. ($17.77 per sq.m). Copper or stainless steel tile cost about $3.95 per s.f. ($43.89 per sq.m).

MARBLE TILE

To prepare an accurate estimate of cost on marble work, the estimator should understand the correct method of measuring and listing quantities from the plans, the manner in which the marble should be handled and set on the job, as well as the most economical and practical methods of handling the work.

When estimating marble wainscoting, it is incorrect to take the height of the wainscot from the floor to the cap and multiply by the number of lin. ft. (meters), because in all probability the wainscot is made up of three members, (base, die, and cap) and each member should be estimated separately: the base and cap by the lin. ft. (meters) and the die by the square foot (square meter).

The length and size of the pieces of marble has considerable bearing on the labor cost, because it usually costs as much to set a 6" (150 mm) piece as one 2'-0" (0.6 m) long, so that a job made up of short pieces will cost more per foot than a job having a larger percentage of long pieces of marble.

Trade Practice. Trade practice in the marble industry recognizes all marbles and stones commonly used for interior building purposes, as falling within one of four groups or classifications, according to their respective characteristics and working qualities. Therefore, for purposes of standardization and clearer definition, the Marble Institute of America has officially adopted these four classifications:

Group A: Sound marbles and stones that require no sticking, waxing, or filling; characteristically uniform and favorable working qualities.

Group B: Marbles and stones similar in character to the preceding group but somewhat less favorable working qualities; occasional natural faults and requiring a limited amount of waxing and sticking.

Group C: Marbles of uncertain variation in working qualities, geological flaws, voids, veins, and lines of separation common; common shop practice to repair nature's shortcomings by sticking, waxing, and filling; liners and other forms of reinforcement freely employed when necessary.

Group D: Marbles and stones similar to the preceding group and subject to the same methods of finishing and manufacture but embracing those materials that contain a larger proportion of natural faults and a maximum variation in working qualities, etc. This group comprises many of the highly colored marbles prized for their decorative qualities.

Below is a list of some of the more commonly used varieties of marbles and stones and the groups to which they belong:

Group A

Alabama—usual grades
Blanco P
Brocadillo, Vermont
Carthage
Georgia
Italian White
Italian English Vein

Napoleon Gray (Vt.)
Tennessee, Gray and Pink
Vermont, Black
Vermont, White Grades
Vermont Pavonazzo
Westland Green Vein Cream

Group B

Alabama Cream Veined A
Belgian Black
Champville
Cremo Italian

Imperial Black, Tenn.
Mankato Buff
Travertine (Italian)

Group C

Belgian Grand Antique
Bois Jourdan
Botticino
Breche Opal
Escalette
Hauteville
Mankato Pink
Ozark Rouge

Red Levanto
Red Verona
Rosato
Tavernelle, all types
Verdona
Verona, Yellow
Vermont Verde Antique
Westfield Green

Group D

Alps Green
Black and Gold
Bleu Belge
Breche Oriental
Breche Rosora
Forest Green
Grand Antique, French

Grand Antique, Italian
Onyx, Pedrara
Rouge Antique
Rouge Jascolin
Sienna
Verde Antico (Italian)
Vert St. Denis

Estimating Quantities of Marble Work. When measuring and listing quantities of marble work from the plans, there are several general rules that should be followed to insure accuracy.

To estimate accurately the quantity of marble base, obtain the number of lineal feet (meters) of base, listing the various heights separately. All base under 1'-0" (0.3 m) high is extended and priced on a lineal foot (meter) basis. All base 1'-0" (0.3 m) or over is extended and priced on a square foot (square meter) basis. Always bear in mind that no piece of marble should be figured as being less than 1'-0" (0.3 m) long, as it requires just as much time to set a 6" (150 mm) piece as one 2'-0" (0.6 m) long. Short pieces are usually found around pilasters, door and window returns, and the like.

Marble die is estimated by the square foot (square meter), and the labor costs computed in the same manner. When listing quantities, obtain the length of each run and multiply by the height, and the result will be the number of sq. ft. (sq.m) of die. If marble pilasters project beyond the face of the wall, list each projection according to its actual dimensions, and if less than a 6" (150 mm) projection, it should be priced as 6" (150 mm) wide, but if the pilasters are over 6" (150 mm) wide, then use actual dimensions.

Marble wainscot cap under 1'-0" (0.3 m) wide or any marble under 1'-0" (0.3 m) wide, should be measured by the lineal foot (meter).

Marble stair treads should be measured by the square foot (square meter), as they are ordinarily 12" (300 mm) or more in width. Be sure to mention length, width, and thickness of all treads, type of nosing, etc.

Marble stair risers should be estimated by the lin. ft. (meter) giving length, height, and thickness of each riser. It costs practically as much to set a short riser as a long one.

When estimating toilet stall work consisting of toilet backs, partitions, stall fronts, stiles, etc., the fronts should be estimated by the square foot (square meter) if over 12" (300 mm) wide and by the lin. ft. (meter) if less than 12" (300 mm) wide. Partitions and backs should be estimated by the square foot (square meter). Marble cap is usually estimated by the lin. ft. (meter). Some marble contractors figure toilet work at a certain price per stall, including all labor in connection with same, although most contractors recommend pricing on a square or lineal foot (meter) basis.

Marble column bases and caps formed of solid stock over 2" (50 mm) thick, should be estimated at a certain price for each base or cap, mentioning the size and number of cu. ft. (cu.m) in each.

Marble floor tile and border are usually estimated by the square foot (square meter), giving size of tile, such as 6"x6" (150 x 150 mm), 9"x9" (225 x 225 mm), 12"x12" (300 x 300 mm), etc. Small or irregular sized tile, such as dots, etc., are estimated in the same manner, except that a notation should be made of the class of work, as it costs more to lay small and irregular shaped tile.

Marble handrail, moldings, etc., should be estimated by the lin. ft. (meter), giving the size of each, in order that the quantities may also be stated in cu. ft. (cu.m) if over 2" (50 mm) thick.

All classes of circular work, such as base, wainscot, or die, should be estimated in the same units as straight work, except that circular work should be estimated separately, because it usually requires considerably more cutting and fitting than straight work.

All marble work cut and set "on the rake", such as stair wainscoting, wainscoting under stairs, etc., should be estimated by the square foot (sq.m), using the largest dimensions. Work of this kind should be estimated separately as it is more expensive to set than straight work.

Handling and Setting Interior Marble. Estimating the labor cost of handling and setting interior marble will depend much on the conditions under which the work is performed. The amount of labor handling, sorting, and distributing the marble on the job before it is set should be considered, and also whether it is necessary to hoist the marble above the first floor, because this will require considerably more handling. The average size of the pieces should be considered, such as the average length of base, size and thickness of wainscot, length of thresholds, treads, risers, and caps. It costs almost as much to handle and set a short piece of marble as a long one. This is especially true on stock that can be handled by one person.

The labor costs given below are based on the performance of an average crew. If it is necessary to hire temporary workers, daily production may decrease as much as 25%.

On practically all jobs it is necessary to have laborers handle the marble and distribute it ready for the setters, although in some localities, all marble must be handled by marble setters and helpers.

Hoisting Marble. The proposal sheet of practically all marble contractors states that the marble contractor shall have free use of the general contractor's hoisting facilities, runways, etc., so for this reason no allowance has been made for hoisting in the following estimates of cost.

However, if it is necessary for the marble contractor to pay for the use of hoist for hoisting marble to the upper floors of a building, figure about 1/2 hr. hoist time per 100 s.f. (9 sq.m) of marble.

Trucking on Marble From Mill or Cars to Job. If the marble is furnished by an out of town mill and it is necessary to haul it in trucks from cars to the job, the cost will vary with the length of the haul. If the marble is furnished by a mill located in the same town, the price of the marble will probably include delivery to the job.

Marble Contractor's Foreman Expense. It is necessary to have a foreman on all jobs of any size to supervise the arrival, unloading, and distribution of the stock, as well as to lay out the work and supervise the setting. On average allow 1/16 to 1/8 hr. foreman time to each hour setter's time, depending on the size of the job and the number of setters employed. (Most small jobs will not require a foreman.)

Average Labor Cost Setting Interior Marble. While the labor cost of marble setting varies with the kind of material handled, some of the large marble companies estimate the setting cost of the entire job at a certain price per foot. The quantities for the entire job are taken, which include base, cap, wainscot or die, treads, risers, etc., and the labor is priced per s.f. (sq.m); 100 s.f. (9 sq.m) should cost as follows:

	Hours	Rate	Total	Rate	Total
Foreman	1.5	$....	$....	$29.41	$44.12
Marble setter	12.0	....	....	28.18	338.16
Helper	12.0	....	....	21.04	252.48
Labor or helper	8.0	....	....	21.04	168.32
Cost 100 s.f.			$....		$803.08
per s.f.				....	8.03
per sq.m				....	86.41

The above price includes setting on jobs consisting principally of 7/8" (22 mm) and 1-1/4" (31 mm) stock.

If 2" (50 mm) marble is used, the handling and setting cost will increase considerably, and the labor per 100 s.f. (9 sq.m) should cost as follows:

	Hours	Rate	Total	Rate	Total
Foreman	1.5	$....	$....	$29.41	$44.12
Marble setter	16.0	....	....	28.18	450.88
Helper	16.0	....	....	21.04	336.64
Labor or helper	12.0	....	....	21.04	252.48
Cost 100 sq. ft			$....		$1,084.12
per s.f.				....	10.84
per sq.m				....	116.64

On jobs having an average amount of base, wainscot or die, cap, toilet stalls, treads and risers, etc., the above prices will prove close enough, but where there are large quantities of any one class of work, it is advisable to refer to that particular class of work to obtain accurate quantities and costs.

Setting Marble Base. The cost of setting marble base varies with the size of the rooms, whether straight walls or broken up with pilasters, piers, etc., and whether the job consists of long or short pieces of base.

In small corridors and other spaces having numerous pilasters, piers, etc., requiring short pieces of base, a setter and helper should set 60 to 75 l.f. (18.0-22.5 m) per 8-hr. day, at the following labor cost per 100 l.f. (30 m):

	Hours	Rate	Total	Rate	Total
Foreman	1.5	$....	$....	$29.41	$ 44.12
Marble setter	12.0	....	....	28.18	338.16
Helper	12.0	....	....	21.04	252.48
Labor or helper	8.0	....	....	21.04	168.32
Cost 100 l.f.			$....		$803.08
per l.f.				....	8.03
per m				....	26.34

If the base can be set in reasonably long pieces, a marble setter and helper should set 80 to 100 l.f. (24-30 m) per 8-hr. day, at the following labor cost per 100 l.f. (30 m):

	Hours	Rate	Total	Rate	Total
Foreman	1.0	$....	$....	$29.41	$29.41
Marble setter	9.0	....	....	28.18	253.62
Helper	9.0	....	....	21.04	189.36
Labor or helper	8.0	....	....	21.04	168.32
Cost 100 l.f.			$....		$640.71
per l.f.				....	6.41
per m				....	21.02

Setting Circular Base. Setting circular base costs about double the cost of straight work of the same kind set under similar conditions. There is usually considerable cutting and fitting on all circular work.

Setting Marble Wainscot or Die. The labor cost of setting marble wainscot or die will vary with the size of the pieces, the height of the wainscot or die, etc. A marble setter will set almost as many lin. ft. (meters) of 5'-0" (1.5 m) wainscot as 3'-0" (0.9 m) wainscot. The labor cost handling stock will be increased somewhat, as it will require 2 to 4 laborers or helpers to handle each piece of marble and place it on the floor ready for the setter, while lighter stock, such as base, risers, treads, etc., can be handled by one worker.

Setting Marble Wainscot Up to 3'-0" (0.9 m) High. When setting marble wainscot up to 3'-0" (0.9 m) high, a marble setter and helper should set 22 to 26 l.f. (6.6-7.8 m) containing 66 to 78 s.f. (5.9-7.0 sq.m) per 8-hr. day, at the following labor cost per 100 s.f. (9 sq.m):

	Hours	Rate	Total	Rate	Total
Foreman	1.5	$....	$....	$29.41	$44.12
Marble setter	11.0	....	....	28.18	309.98
Helper	11.0	....	....	21.04	231.44
Labor or helper	8.0	....	....	21.04	168.32
Cost 100 s.f.			$....		$753.86
per s.f.				....	7.54
per sq.m				....	81.13

The cost of setting circular wainscot or die will run about double the cost of straight work on account of the extra cutting and fitting.

Setting Marble Wainscot 3'-0" (0.9 m) to 4'-0" (1.2 m) High. A marble setter and helper should set 20 to 25 l.f. (6.0-7.5 m) containing 75 to 90 s.f. (6.75-8.1 sq.m) of wainscot per 8-hr. day, at the following labor cost per 100 s.f. (9 sq.m):

	Hours	Rate	Total	Rate	Total
Foreman	1.2	$....	$....	$29.41	$35.29
Marble setter	10.0	....	....	28.18	281.80
Helper	10.0	....	....	21.04	210.40
Labor or helper	8.0	....	....	21.04	168.32
Cost 100 s.f.			$....		$695.81
per s.f.				....	6.96
per sq.m				....	74.89

The cost of setting circular wainscot or die will run about double the cost of straight work on account of the extra cutting and fitting.

Setting Marble Wainscot 4'-0" (1.2 m) to 5'-0" (1.5 m) High. Where the marble wainscot varies from 4'-0" to 5'-0" (1.2-1.5 m) high, a marble setter and helper should set 20 to 23 l.f. (6.0-6.9 m) containing 85 to 100 s.f. (7.65-9.0 sq.m) per 8-hr. day, at the following labor cost per 100 s.f. (9 sq.m):

	Hours	Rate	Total	Rate	Total
Foreman	1.0	$....	$....	$29.41	$29.41
Marble setter	9.0	....	....	28.18	253.62
Helper	9.0	....	....	21.04	189.36
Labor or helper	8.0	....	....	21.04	168.32
Cost 100 s.f.			$....		$640.71
per s.f.				....	6.41
per sq.m				....	68.97

The cost of setting circular wainscot or die will run about double the cost of straight work on account of the extra cutting and fitting.

Setting Marble Wainscot Over 5'-0" (1.5 m) High. When setting marble wainscot 5'-0" to 7'-0" (1.5-2.1 m) high, a marble setter and helper should set 90 to 110 s.f. (8.1-9.9 sq.m) per 8-hr. day, at the following labor cost per 100 s.f. (9 sq.m):

	Hours	Rate	Total	Rate	Total
Foreman	1.0	$....	$....	$29.41	$29.41
Marble setter	8.0	....	....	28.18	225.44
Helper	8.0	....	....	21.04	168.32
Labor or helper	8.0	....	....	21.04	168.32
Cost 100 s.f.			$....		$591.49
per s.f.				....	5.91
per sq.m				....	63.59

The cost of setting circular wainscot or die will run about double the cost of straight work on account of the extra cutting and fitting.

Wainscot over 7'-0" (2.1 m) high will require working from a scaffold, which decreases the production considerably. In such cases the labor cost of that portion of wainscot over 7'-0" (2.1 m) high should be increased 50% to 100%, and the cost of erecting and removing steel tubular scaffold should be added.

Setting Marble Wainscot Cap. When setting marble wainscot cap, a marble setter and helper should set 60 to 75 l.f. (18-22.5 m) per 8-hr. day, at the following cost per 100 l.f. (30 m):

	Hours	Rate	Total	Rate	Total
Foreman	1.5	$....	$....	$29.41	$44.12
Marble setter	12.0	....	....	28.18	338.16
Helper	12.0	....	....	21.04	252.48
Labor or helper	8.0	....	....	21.04	168.32
Cost 100 l.f.			$....		$803.08
per l.f.				....	8.03
per m				....	26.34

Add for overhead and profit.

Setting Marble Stair Treads. The cost of setting marble stair treads will vary with the thickness and length of the treads. The shorter each piece of marble, the higher the cost per lin. ft. (meter).

If the treads are 3'-0" (0.9 m) to 3'-6" (1.05 m) long and 1-1/4" (31 mm) thick, a marble setter and helper should set 20 to 22 treads containing 60 to 77 l.f. (18.0-23.1 m) per 8-hr. day, at the following labor cost per 100 l.f. (30 m):

	Hours	Rate	Total	Rate	Total
Foreman	1.5	$....	$....	$29.41	$44.12
Marble setter	11.0	....	....	28.18	309.98
Helper	11.0	....	....	21.04	231.44
Labor or helper	8.0	....	....	21.04	168.32
Cost 100 l.f.			$....		$753.86
per l.f.				....	7.54
per m				....	24.73

If the treads vary from 4'-0" to 6'-0" (1.2-1.8 m) long, a marble setter and helper should set 80 to 90 l.f. (24-27 m) per 8-hr. day, at the following labor cost per 100 l.f. (30 m):

	Hours	Rate	Total	Rate	Total
Foreman	1.1	$....	$....	$29.41	$32.35
Marble setter	9.5	....	....	28.18	267.71
Helper	9.5	....	....	21.04	199.88
Labor or helper	8.0	....	....	21.04	168.32
Cost 100 l.f.			$....		$668.26
per l.f.				....	6.68
per m				....	21.91

Setting Marble Stair Risers. The cost of setting marble stair risers will vary with the length of the pieces of marble, width of stairs, etc.

Where the stairs are 3'-0" to 3'-6" (0.9-1.05 m) wide, a marble setter and helper should set 60 to 75 l.f. (18.0-22.5 m) of risers per 8-hr. day, at the following labor cost per 100 l.f. (30 m):

	Hours	Rate	Total	Rate	Total
Foreman	1.5	$....	$....	$29.41	$44.12
Marble setter	11.0	....	....	28.18	309.98
Helper	11.0	....	....	21.04	231.44
Labor or helper	8.0	....	....	21.04	168.32
Cost 100 l.f.			$....		$753.86
per l.f.				....	7.54
per m				....	24.73

If the marble stairs are over 4'-0" (1.2 m) wide, a marble setter and helper should set 80 to 90 l.f. (24-27 m) of risers per 8-hr. day, at the following labor cost per 100 l.f. (30 m):

	Hours	Rate	Total	Rate	Total
Foreman	1.1	$....	$....	$29.41	$32.35
Marble setter	9.5	....	....	28.18	267.71
Helper	9.5	....	....	21.04	199.88
Labor or helper	8.0	....	....	21.04	168.32
Cost 100 l.f.			$....		$668.26
per l.f.				....	6.68
per m				....	21.91

Setting Marble Stair Wainscot on the Rake. Where marble wainscoting is set on the rake of the stairs, the labor cost will run considerably higher than straight work on account of the additional cutting and fitting necessary.

If the wainscoting is 3'-0" to 3'-6" (0.9-1.05 m) high, a marble setter and helper should set 10 to 15 l.f. (3.0-4.5 m) containing 35 to 45 s.f. (3.1-4.1 sq.m) per 8-hr. day, at the following labor cost per 100 s.f. (9 sq.m):

	Hours	Rate	Total	Rate	Total
Foreman	2.0	$....	$....	$29.41	$58.82
Marble setter	20.0	....	....	28.18	563.60
Helper	20.0	....	....	21.04	420.80
Labor or helper	8.0	....	....	21.04	168.32
Cost 100 s.f.			$....		$1,211.54
per s.f.				....	12.12
per sq.m				....	130.41

Where the wainscot is 5'-0" to 7'-0" (1.5-2.1 m) high and follows the rake o¹
the stairs, a marble setter and helper should set 50 to 65 s.f. (4.5-5.9 sq.m
per 8-hr. day, at the following labor cost per 100 s.f. (9 sq.m):

	Hours	Rate	Total	Rate	Total
Foreman	1.5	$....	$....	29.41	$44.12
Marble setter	14.0	....	....	28.18	394.52
Helper	14.0	....	....	21.04	294.56
Labor or Helper	8.0	....	....	21.04	168.32
Cost 100 s.f.			$....		$901.52
per s.f.				....	9.02
per sq.m				....	97.06

Setting Marble Toilet Stalls and Partitions. The cost of handling an¹
setting marble toilet stalls, backs, partitions, stiles, etc., will vary greatl
with the individual job. On work of this kind, the marble partition slabs ar¹
seldom less than 5'-0" (1.5 m) in height and width and each slab contains 2!
to 30 s.f. (2.25-2.70 m) of marble. The backs are usually the same height a¹
the partitions but not so wide and each back contains 16 to 20 s.f. (1.4-1.¹
sq.m) of marble. The stall fronts or stiles are usually 6" (150 mm) to 8" (20¹
mm) wide and 6'-0" (1.8 m) to 7'-0" (2.1 m) high and extend to the floor t¹
support the dividing partitions where metal standards are not used.

As an average on toilet stall work, a good marble setter should complete ¹
stall in 3 to 3-1/2 hours, after the marble backs are in place. Since th¹
average partition contains about 25 s.f. (2.25 sq.m) and each stile and ca¹
about 7 s.f. (0.63 sq.m), a setter and helper should set 75 to 80 s.f. (6.75-7.¹
sq.m) of marble per 8-hr. day, at the following labor cost for one complet¹
stall exclusive of back:

	Hours	Rate	Total	Rate	Total
Foreman	0.4	$....	$....	$29.41	$11.76
Marble setter	3.3	....	....	28.18	92.99
Helper	3.3	....	....	21.04	69.43
Labor or helper	2.6	....	....	21.04	54.70
Setting cost 1 stall			$....		$228.88

Toilet partition work is estimated by the sq. ft. (sq.m) by most marbl¹
contractors.

Setting Marble Toilet Backs. The marble backs for toilet stalls ar¹
usually 6'-0" to 7'-0" (1.8-2.1 m) high. A marble setter and helper should se¹
75 to 85 s.f. (6.75-7.65 sq.m) per 8-hr. day, at the following labor cost per 10¹
s.f. (9 sq.m):

	Hours	Rate	Total	Rate	Total
Foreman	1.2	$....	$....	$29.41	$35.29
Marble setter	10.0	....	....	28.18	281.80
Helper	10.0	....	....	21.04	210.40
Labor or helper	8.0	....	....	21.04	168.32
Cost 100 s.f.			$....		$695.81
per s.f.			....		6.96
per sq.m			....		74.89

Setting Marble Door and Window Trim. A marble setter and helper should set 45 to 55 l.f. (13.5-16.5 m) of marble door and window trim or casings per 8-hr. day, at the following labor cost per 100 l.f. (30 m):

	Hours	Rate	Total	Rate	Total
Foreman	2.0	$....	$....	$29.41	$58.82
Marble setter	16.0	....	....	28.18	450.88
Helper	16.0	....	....	21.04	336.64
Labor or helper	8.0	....	....	21.04	168.32
Cost 100 l.f.			$....		$1,014.66
per l.f.			....		10.15
per m			....		33.29

Add for overhead and profit.

Setting Marble Countertops. When setting marble countertops, a marble setter and helper should set 75 to 85 s.f. (6.75-7.65 sq.m) per 8-hr. day, at the following labor cost per 100 s.f. (9 sq.m):

	Hours	Rate	Total	Rate	Total
Foreman	1.2	$....	$....	$29.41	$35.29
Marble setter	10.0	....	....	28.18	281.80
Helper	10.0	....	....	21.04	210.40
Labor or helper	8.0	....	....	21.04	168.32
Cost 100 s.f.			$....		$695.81
per s.f.			....		6.96
per sq.m			....		74.89

Setting Marble Thresholds. When setting marble thresholds 3'-0" to 3'-6" (0.9-1.05 m) long, a marble setter and helper should set 12 to 14 thresholds per 8-hr. day, at the following labor cost each:

	Hours	Rate	Total	Rate	Total
Marble setter	0.60	$....	$....	$28.18	$16.91
Helper	0.60	....	....	21.04	12.62
Cost per threshold			$....		$29.53

Setting Marble Plinths. When setting marble plinths at door openings, a marble setter and helper should set 20 to 25 plinths per 8-hr. day, at the following labor cost per 100 plinths:

	Hours	Rate	Total	Rate	Total
Foreman	4.5	$....	$....	$29.41	$132.35
Marble setter	36.0	....	....	28.18	1,014.48
Helper	36.0	....	....	21.04	757.44
Labor or helper	8.0	....	....	21.04	168.32
Cost 100 plinths			$....		$2,072.59
per plinth				....	20.73

Setting Marble Stair Railings and Balusters. On jobs having marble balusters up to 6" (150 mm) in diameter and 2'-6" (0.75 m) long, a marble setter and helper should set about 20 balusters per 8-hr. day, at the following labor cost per 100 balusters:

	Hours	Rate	Total	Rate	Total
Foreman	5.0	$....	$....	$29.41	$147.05
Marble setter	40.0	....	....	28.18	1,127.20
Helper	40.0	....	....	21.04	841.60
Labor or helper	16.0	....	....	21.04	336.64
Cost 100 balusters			$....		$2,452.49
per baluster				....	24.52

Setting Marble Ashlar. Where marble ashlar is used for the interior walls of buildings, it is customary to use 7/8" (22 mm) or 1-1/4" (31 mm) marble and the method of handling and setting is similar to that used for setting exterior work, as it is often necessary to use a light breast derrick to set the marble.

On work of this kind, a marble setter and helper should handle and set 60 to 80 s.f. (5.4-7.2 sq.m) of ashlar per 8-hr. day, at the following labor cost per 100 s.f. (9 sq.m):

	Hours	Rate	Total	Rate	Total
Foreman	1.5	$....	$....	$29.41	$44.12
Marble setter	12.0	....	....	28.18	338.16
Helper	12.0	....	....	21.04	252.48
Labor or helper	12.0	....	....	21.04	252.48
Cost 100 s.f.			$....		$887.24
per s.f.				....	8.87
per sq.m				....	95.44

Setting Marble Moldings, Handrail, Etc. On jobs having light marble moldings, stair handrail, well hole rail, etc., where the marble is about 6" (150 mm) square and furnished in reasonably long pieces, a marble setter and helper should set 75 to 85 l.f. (22.5-25.5 m) per 8-hr. day, at the following labor cost per 100 l.f. (30 m):

	Hours	Rate	Total	Rate	Total
Foreman	1.2	$....	$....	$29.41	$35.29
Marble setter	10.0	....	....	28.18	281.80
Helper	10.0	....	....	21.04	210.40
Labor or helper	8.0	....	....	21.04	168.32
Cost 100 l.f.			$....		$695.81
per l.f.				....	6.96
per m				....	22.83

Setting Marble Column Bases. When setting marble column bases 2'-6" to 3'-0" (0.23-0.38 m) square and 0'-9" to 1'-3" (0.75-0.90 m) high, a marble setter and helper should handle and set about one base per hr. at the following labor cost per base:

	Hours	Rate	Total	Rate	Total
Foreman	0.1	$....	$....	$29.41	$ 2.94
Marble setter	1.0	....	....	28.18	28.18
Helper	1.0	....	....	21.04	21.04
Labor or helper	0.5	....	....	21.04	10.52
Cost per base			$....		$62.68

Laying Marble Floor Tile. There are three labor operations to be considered when estimating the labor cost of setting marble floor tile: the cement bed under the marble floor, labor handling and setting floor tile, and the cost of rubbing or smoothing the floors after they have been laid.

As a general rule, the cement bed under the marble floor is 2-1/2" to 3" (63-75 mm) thick, composed of cement and sand mixed rather dry. It requires about 1-1/2 bbls. of portland cement and one cu. yd. (0.76 cu.m) of sand to each 100 s.f. (9 sq.m) of floor.

Material and labor per 100 s.f. (9 sq.m) of floor fill should cost as follows:

Material Costs

	Rate	Total	Rate	Rate (metric)	Total
6 sacks portland cement	$....	$....	$4.24		$25.44
1 c.y. (0.76 cu.m) sand	....	....	12.72	16.74	12.72
Cost 100 s.f.			$....		$38.16
per s.f.				....	0.38
per sq.m				....	4.24

Labor Costs

	Hours	Rate	Total	Rate	Total
Foreman	0.2	$....	$....	$29.41	$ 5.88
Marble setter	1.5	....	....	28.18	42.27
Helper	3.0	....	....	21.04	63.12
Cost 100 s.f.			$....		$111.27
per s.f.					1.11
per sq.m					11.94

If the marble floor tile vary from 6"x6" (150 x 150 mm) to 12"x12" (300 x 300 mm) in size, a marble setter and helper should handle and lay 90 to 110 s.f. (8.1-9.9 sq.m) of floor per 8-hr. day, at the following labor cost per 100 s.f. (9 sq.m):

	Hours	Rate	Total	Rate	Total
Foreman	1.0	$....	$....	$29.41	$29.41
Marble setter	8.0	....	....	28.18	225.44
Helper	8.0	....	....	21.04	168.32
Labor or helper	8.0	....	....	21.04	168.32
Cost 100 s.f.			$....		$591.49
per s.f.					5.91
per sq.m					63.59

When laying marble floors it is customary to place the cement bed just ahead of the tile, and the entire work is performed by the setter and helper laying the floor. If the subfloor is sound and even, marble in residential and light commercial work is only 3/8" (9 mm) thick and set in mastic. Setting costs would average about $0.25 per s.f. less, unless the area was small and irregular, or set with an intricate pattern.

Smoothing Marble Floors. After the marble floors have been laid, it is necessary to rub them with carborundum stone to remove the uneven spots at the joints. It is customary to use a rubbing machine powered by gasoline or electricity, which is operated by one worker and runs back and forth over the floors and grinds them to an even surface. A machine rents for about $7.50 per hour.

The quantity of floor that can be surfaced per day will vary with the kind of marble used, as it is possible to surface considerably more soft marble than hard marble.

When rubbing and surfacing soft marble, a machine should surface 125 to 150 s.f. (11.25-13.5 sq.m) per 8-hr. day, at the following labor cost per 100 s.f. (9 sq.m):

	Hours	Rate	Total	Rate	Total
Machine operator	6.0	$....	$....	$28.18	$169.08
Foreman	0.7	....	....	29.41	20.59
Cost 100 s.f.			$....		$189.67
per s.f.			....		1.90
per sq.m			....		20.44

When rubbing and surfacing hard marble, a rubbing machine should surface 100 to 125 s.f. (9.0-11.25 sq.m) of floor per 8-hr. day, at the following labor cost per 100 s.f. (9 sq.m):

	Hours	Rate	Total	Rate	Total
Machine operator	7.0	$....	$....	$28.18	$197.26
Foreman	0.8	....	....	29.41	23.53
Cost 100 s.f.			$....		$220.79
per s.f.			....		2.21
per sq.m			....		23.78

Setting Marble Floor Border. Where the marble floor has a different colored marble border, the labor setting the border will cost about the same as the balance of the floor.

Setting Structural Slate. The labor cost of setting structural slate work, such as base, treads and risers, toilet stalls and partitions, etc., will run about the same as for the same class of marble work and should be estimated accordingly.

If, however, field cutting should be required, as is often the case with window stools and door thresholds, the labor costs should be increased 20% to 25% as slate is more difficult to work than marble.

Estimating Quantities of Interior Marble

When estimating quantities of interior marble, any fraction of an inch will be treated as a whole inch. No piece of marble will be considered as being less than 6" (150 mm) wide and 1'-0" (0.30 m) long.

Thickness. Thicknesses given are approximate but not exact. For slabs gauged to a given thickness, an extra charge of 10% to the list price is made.

Extra Lengths. On slabs 10'-0" to 12'-0" (3.0-3.6 m) long, add 10% to list price; on slabs over 12'-0" (3.6 m) long, add 20%.

Extra Widths. On slabs over 5'-6" (1.65 m) wide, add 10% to list price.

Polished Both Sides. Prices given are based on marble having one side polished unless otherwise noted. If both sides are to be polished, add $1.50 per s.f. ($16.67 per sq.m) for marbles in Group A and B and $1.25 per s.f. ($13.89 per sq.m) for marbles in Group C and D.

Hone Finish. Hone finish takes the same price as polished work.

Sand Finish. If marble is finished with sand finish on one side only, deduct $0.15 per s.f. If sand finish on both sides, add $0.25 per s.f. ($2.77 per sq.m).

Polished Edges. For each polished edge of 7/8" (22 mm) and 1-1/4" (31 mm) thickness, add $0.50 per l.f. ($1.67 per m) of edge; on 1-1/2" (38 mm) and 2" (50 mm) thickness, add $0.75 per l.f. ($2.50 per m).

Beveled or Rounded Edges. For each beveled or rounded edge on 7/8" (22 mm) and 1-1/4" (31 mm) thickness, add $0.50 per l.f. ($2.50 per m); on 1-1/2" (38 mm) and 2" (50 mm) thickness, add $0.75 per l.f. ($2.50 per m).

Ogee Molding. For single member ogee molding $3.25 per l.f. ($10.83 per m) of edge on 7/8" (22 mm) thick; $3.50 per l.f. ($11.67 per m) on 1-1/4" (31 mm), and $4.25 per l.f. ($14.17 per m) on 1-1/2" (38 mm). For each additional member, add $2.25 per l.f. ($7.00 per m) of edge on 7/8" (22mm) and 1-1/4" (31 mm), and $2.50 ($8.50 per m) on 1-1/2" (38 mm).

Countersinking. For countersinking 1/4" (6 mm) deep or less, $2.00 per s.f. ($22.20 per sq.m) of surface. For each additional 1/4" (6 mm) deeper, or fraction thereof, add extra $2.50 per s.f. ($27.77 per sq.m) of surface.

Grooving. For grooving 1/4" (6 mm) deep or less, $1.50 per l.f. ($5.00 per m) for each groove. For each additional 1/4" (6 mm) deeper, or fraction thereof, add extra $1.50 per l.f. ($5.00 per m) of each groove.

Rabbeting. For rabbeting 1/4" (6 mm) deep or less, $1.50 per l.f. ($5.00 per m) for each rabbet. For each additional 1/4" (6 mm) deeper, or fraction thereof, add extra $1.25 per l.f. ($4.17 per m) of each rabbet.

Special Cutting, Polishing, Drilling, Etc. All special work other than listed above will be charged for extra.

Shipping Weights of Marble. The approximate shipping weights of marble, boxed for shipment, are as follows per sq. ft. (sq.m): 7/8" (22 mm) thick, 15 lbs. (75 kg); 1-1/4" (31 mm) thick 20 lbs. (100 kg); 1-1/2" (38 mm) thick 24 lbs. (120 kg); 2" (50 mm) thick 32 lbs. (160 kg); per c.f., 192 lbs. (per cu.m, 3050 kg).

Approximate Material Prices per S.F. (Sq.M) of 7/8" (22 mm) Marble, Polished One Side

Always check local sources for price and availability. Foreign marbles are always affected by exchange rates. The trend has been to marbles of even color. Many of the quarries for the more colorful marbles have been shut down, and when a block of such marble is found, the price may be out of line because of scarcity.

Type of Marble	Price per s.f.	Price per sq.m
Alabama Clouded "A" or Cream "A"	$8.50	94.44
Brocadillo	10.60	117.77
Carthage Gray	7.95	88.33

Type of Marble	Price per s.f.	Price per sq.m
French Gray	8.50	94.44
Georgia Creole, Cherokee or Mezzotint	8.50	94.44
Gravina	7.95	88.33
Kasota Pink	7.95	88.33
Kasota Yellow	9.55	106.11
Light Cloud Vermont	8.50	94.44
Napoleon Gray	8.50	94.44
Radio Black	9.55	106.11
Tennessee Marbles	8.50	94.44
Verde Antique	10.60	117.77
Westfield Green	11.65	129.44
York Fossil	9.55	106.11
Foreign Marbles		
Alps Green	14.30	158.89
Belgian Black	12.75	141.67
Black and Gold	16.95	188.33
Botticino	12.75	141.67
Bois Jordan	12.75	141.67
Breche Oriental	18.05	200.55
Grand Antique, French	21.20	235.55
Grand Antique, Italian	15.90	176.67

The above prices are for marble in less than carload lots, uncrated, f.o.b. mill. Crating $0.40 to 0.50 per s.f. for marble up to 2" (50 mm) thick.

Approximate Material Prices on Sand Finish Marble Floor
Tile and Border 7/8" (22 mm) Thick
Tile sizes 6"x6" (150 x 150 mm) to 1'-0"x2'-0" (0.3 x 0.6 m)
Prices per Sq. Ft. (Sq.M)

	Sq. Ft.	Sq.M
Alabama Clouded "A"	$5.80	$64.44
Carthage Gray	4.00	44.44
Georgia Creole	5.80	64.44
Georgia Cherokee	3.20	35.55
Italian White	3.70	41.11
Italian English Vein	3.70	41.11
Tennessee Pink	4.80	53.33
Tennessee Gray	4.80	53.33
Tennessee Dark	4.25	47.22
Verde Antique	7.45	82.78
Travernelle	5.85	65.00
Travertine	5.30	58.89

Approximate Material Prices on 3/8" (9 mm)
Thinset Marble Tiles Per Sq. Ft. (Sq.M)

	Polished per s.f.	Polished per sq.m	Honed per s.f.	Honed sq.m
Cremo	$2.70	$30	$2.65	$29.44
Negro-Marquina	3.20	35.55	3.15	$35.00
Rose Aurora	2.55	28.33	2.50	27.77
Statuary Vein	2.25	25.00	2.20	24.44
Travertine Filled	2.15	23.89	2.10	23.33
Travertine Unfilled	2.05	22.77	2.00	22.22
White Italian	2.15	23.89	2.10	23.33

12"x12" (300 x 300 mm) tiles, 3/8" (9 mm) thick, 10 tiles per box.

SLATE TILE

Slate is a natural mica granular crystalline stone. Its main sources are Vermont, Pennsylvania and Virginia. Vermont produces colored slates, including greens; purples; mottled green, purple and red; as well as the standard dark grays.

Slate is usually graded into clear stock and ribbon stock. Ribbon stock has bands of a darker color running through it and is cheaper than clear stock, which has no ribbons but will have some spots and veining in it. Finishes available are as follows:

Natural Cleft: Natural split or cleaved face, moderately rough with some textural variations.

Sand Rubbed Finish: Wet sand on a rubbing bed is used to eliminate any natural cleft, and leave slate in an even plane.

Honed Finish: This is a semi-polished surface.

Slate is used for spandrels, sills, stools, treads and risers, toilet stalls, floors and walks, fireplace facings and hearths.

Spandrels are generally clear slate with either cleft, rubbed or honed finish. Thickness is generally 1" (25 mm), 1-1/4" (31 mm), or 1-1/2" (38 mm). Sizes are somewhat limited, with a length of no more than 6' (1.8 m) and width of 2'-6" (0.75 m) to 4' (1.2 m) being standard recommendations, although larger sizes may be available for certain conditions. Many spandrels today are set in metal grid frames. Otherwise, they must be anchored with bronze or stainless steel anchors in each corner.

Sills and stools may be either ribbon or clear stock. Exterior sills are generally furnished with a sand rubbed finish in lengths up to 4' (1.2 m) and 1" (25 mm) to 2" (50 mm) in thickness. Sills may have drip grooves cut in at additional cost.

Interior stools and fireplace surrounds are usually 1" (25 mm) thick with honed finish, a maximum length of 4' (1.2 m) being most economical.

Treads and risers on the exterior are generally natural cleft finish. Interior work is usually sand rubbed rather than honed, because it is both cheaper and gives a good non-slip surface, although the front edges may be honed. Thickness is generally 1" (25 mm) or 1-1/4" (31 mm) for stairs, 1-1/2" (38 mm) to 2" (50 mm) for landings.

The cost of slate is determined by the grade, the size involved and the finish.

Ribbon grade, often used on stairs, will run around $5.00 per s.f. ($55.55 per sq.m) in 1" (25 mm) thickness and $8.00 per s.f. ($88.89 per sq.m) for 2" (50 mm) thickness, sand finished. Honed finishes will add $0.50 per s.f. ($$5.55 per sq.m) to the cost.

Clear grade, 1" (25 mm) thick and sand finished, will be about $5.50 per s.f. ($61.11 per sq.m) in sizes to 3 s.f. (0.27 sq.m), $7.75 to $8.75 in sizes 3 s.f. (0.27 sq.m) to 6 s.f. (0.54 sq.m), and $11.00 in sizes over 6 s.f. (0.54 sq.m). Lineal foot (meter) costs for sills and stools will vary with both length and width. Up to 6" (150 mm) wide stock will cost about $4.75 per l.f. ($15.83 per m) for 1" (25 mm) thickness; $7.50 per l.f. ($25.00 per m) for 2" (50 mm) thickness. The 10" (250 mm) wide stock will cost $7.50 per l.f. ($15.83 per m) for 1" (25 mm) thickness to $9.25 per l.f. ($30.83 per m) for 2" (50 mm) thickness.

Slate for flooring is available for either mortar bed or thin set application. This slate is 1/4" (6 mm) thick and available in 3" (75 mm) multiples 6"x6" (150 x 150 mm) to 12"x12" (300 x 300 mm). It weighs 3-3/4 lbs. per s.f. (18.75 kg per sq.m). It is set in 1/4" (6 mm) mastic applied with a notched trowel over concrete or plywood. It can have grouted joints or be ordered for butt joints. Cost will run $2.00 per s.f. ($22.22 per sq.m) for material and $2.50 ($27.77 per sq.m) for labor.

Slate for heavier duty floors and for outside applications is generally 1/2" (13 mm), 3/4" (19 mm), or 1" (25 mm) thick, weighing 7-1/2 lbs. (3.4 kg), 11-1/4 lbs. (5 kg) and 15 lbs. (6.75 kg) respectively. It is set in 1" (25 mm) mortar beds on concrete slabs or plywood. Plywood should be covered with felt and metal lath, nailed on. Minimum thickness is 1-1/2" (38 mm). If plywood is set over joists without a subfloor, it should be at least 3/4" (19 mm) thick. Sizes are 6"x6" (150 x 150 mm) to 24"x18" (600 x 450 mm), in multiples of 3" (75 mm). Cost runs from $3.75 per s.f. ($41.67 per sq.m) for 1/2" (13 mm) to $5.00 ($55.55 per sq.m) for 1" (25 mm).

09400 TERRAZZO

The cost of terrazzo work varies with the size of the job, size of rooms or spaces where the floors are to be installed, floor designs, strips of brass or other non-rusting material, and method of laying the floors.

Two methods are used in laying terrazzo floors over concrete construction. One is to bond it to the concrete and the other to separate it from the structural slab.

When the first method is used, the concrete fill is provided for by another contractor and should be left 2" (50 mm) below the finished floor. Before the terrazzo contractor installs his underbed, he must see that this concrete fill is thoroughly cleaned of plaster droppings, wood chips and other debris, and wetted to insure cohesion.

The second method is used in buildings where cracking is anticipated either from settlement, expansion and contraction or vibration. In this case the terrazzo contractor begins his work from the structural floor slab up. This method requires a thickness of at least 3" (75 mm). The concrete slab is covered with a thin bed of dry sand over which a sheet of tar paper is laid. Over the paper the underbed is installed as in the first method except that coarser aggregate can be used in the underbed, such as cinders or fine gravel where its thickness exceeds 2-1/2" (63 mm). When this method is used, the cracks originating in the structural slab are not likely to appear on the surface but terminate at the sand bed.

When terrazzo is laid over wood floors a thickness of not less than 2" (50 mm) is required. The floor should first be covered with tarred paper. Over this paper, nail galvanized wire netting no. 14 gauge, 2" (50 mm) mesh. Then lay the concrete underbed as specified under the second method.

The underbed for terrazzo consisting of 1 part portland cement and 4 parts sharp, screened sand shall be spread and brought to a level not less than 1/2" (13 mm) nor more than 3/4" (19 mm) below the finished floor.

Into this underbed, while still in a semi-plastic state, install metal strips or preassembled decorative units, having proper bonding features.

The terrazzo topping shall be not less than 1/2" (13 mm) nor more than 3/4" (19 mm) in thickness and shall be of granulated marble of colors selected. The topping shall be uniform in composition and the marble granule that appears on the surface shall be used for its entire thickness. The granule to be composed of such proportions of Nos. 1-2-3 sizes as required.

Composition for terrazzo mixture shall be in the proportion of 200 lbs. (90 kg) of marble granules to 100 lbs. (45 kg) of gray or white portland cement, mixed dry, and water added afterward to make the mix plastic, but, not too wet. The mix shall then be placed in the spaces formed by the metal dividing strips and rolled into a compact mass by means of heavy stone or metal rollers until all superfluous cement and water is extracted, after which it must be hand troweled to an even surface, disclosing the lines of the metal strips on a level with the terrazzo filling.

When the floors have set sufficiently hard, they shall be machine rubbed, using coarse carborundum grit stones for the initial rubbing after which a light grouting of pure gray white portland cement shall be applied to the surface, filling all voids, and allowed to remain until the time of final cleaning.

Floors shall have the grouting coat removed by machine, using a fine carborundum grit, after which it must be thoroughly washed. Do not use acids in cleaning terrazzo floors.

Ramped or other surfaces in terrazzo floors so specified, shall be made anti-slip by the addition of an abrasive aggregate. For heavy duty floors, the proportion shall be 2 parts of abrasive to 3 parts of marble granule and the abrasive shall be mixed in the terrazzo topping for its entire thickness. For light traffic floors, the abrasive shall be sprinkled on the surface only, and show a proportion of 5 parts marble granule and 1 part abrasive aggregate.

Terrazzo Base. Terrazzo base is usually coved at the floor and may be any height desired. If the base is to finish flush with the finish plaster, a metal base bead shall be furnished and set by the plasterer. The base shall be divided approximately every 4 l.f. (1.2 m) using metal base dividers. The walls back of the base shall have a scratch coat of cement and sand mortar brought to a line 3/8" (9 mm) back of the finished face of the base, into which the base dividers shall be set. Base shall be finished with a very fine stone, so as to leave the surface at a hone finish.

Estimating the Cost of Terrazzo Work. In addition to the direct cost of labor and materials entering into terrazzo work, an allowance should be made to provide for wood grounds, carborundum stone for rubbing, wiring for electric power, depreciation on machines, use of hoist, cleaning up rubbish, freight and trucking on machines and tools. This will vary with size and location of job.

The costs on the following pages are based on an average job containing about 3,000 s.f. (270 sq.m). The labor on a small job containing 100 to 200 s.f. (9.0-18.0 sq.m) will run twice as much per s.f. (sq.m) as on a job containing 5,000 s.f. (450 sq.m).

Metal Strips in Terrazzo Floors and Base. Brass strips or other non-rusting metals are used in practically all terrazzo floors and add to the cost in proportion to the quantity used.

Standard brass strips, B & S gauge No. 20, cost $0.75 per l.f. ($2.50 per m) including waste.

Metal strips, B & S gauge No. 14, cost 1.45 per l.f. ($4.83 per m) including waste in brass and $0.50 per l.f. ($1.67 per m) in zinc.

The cost of metal strips per sq. ft. of floor will vary with the size of the squares or design and should be estimated accordingly.

Terrazzo base is jointed every 4'-0" (1.2 m) with vertical metal joint dividers.

Estimating Quantities of Terrazzo Work. Terrazzo floors are estimated by the sq. ft. (sq.m) and will vary from 2" (50 mm) to 3" (75 mm) thick, depending upon the subfloors.

Terrazzo floor borders are estimated by the sq. ft. (sq.m) when over 12" (300 mm) wide and by the lin. ft. (meter) when less than 12" (300 mm) wide. Terrazzo base is estimated by the lin. ft. (meter) stating height. An allowance of 1'-0" (0.30 m) should be made for each corner or miter.

Double cove terrazzo base having 2 finished faces, such as ordinarily used under narrow partitions, shower stalls, etc. are estimated by the lin. ft. (meter) and the cost is double that of single cove base.

Labor Placing Terrazzo, Floor and Base. The labor cost of placing terrazzo work will vary with the size of the rooms, design of the floor, size of the squares or pattern requiring brass strips, etc., as the smaller the squares, the more metal strip required and the higher the cost.

The following are approximate quantities of various classes of terrazzo work a crew of 2 mechanics and 3 helpers should install per 8-hr. day:

Description Floor blocked off into	SF per 8-Hr. day	Sq.M per 8-hr. day	Hrs. per 100 SF (9 sq.m) Mechanic	Helpers
5'-0" (1.5 m) squares	400-450	36-41	3.75	5.65
4'-0" (1.2 m) squares	375-425	34-38	4.00	6.00
3'-0" (0.90 m) squares	350-400	32-36	4.25	6.40
2'-0" (0.60 m) squares	325-375	29-37	4.60	6.90
1'-0" (0.30 m) squares	275-325	25-29	5.33	8.00
Border 12" - 24" (300-600mm) wide	275-325	25-29	5.33	8.00

Terrazzo Cove Base			Hrs. per 100 l.f. (30 m)	
Terrazzo cove base, 3" (75mm) high	60-65	5.4-6	13	13
Terrazzo cove base, 6" (150mm) high	50-55	4.5-5	16	16

Double cove base, double time given for single base.

Rubbing and Finishing Terrazzo Work. Terrazzo floors are rubbed by machine and by hand only in corners where the machine cannot reach. On jobs consisting of large and small rooms, a worker rubs and completes about 100 s.f. (9 sq.m) of terrazzo floor per 8-hr. day. A worker should rub and finish about 80 l.f. (24 m) of terrazzo base per 8-hr. day.

Cost of Concrete Subbase Under Terrazzo Floors. There are three methods of applying the concrete underbed for terrazzo floors, as described previously: bonded to the structural concrete slab, separate from the structural slab, or applied over wood floors.

Inasmuch as terrazzo topping is 1/2" (13 mm) thick for all types of floors, cost of finished floor varies with type of underbed.

Material Cost of 100 S.F. (9 Sq.M) of Concrete Underbed 1-1/4" (31 mm) Thick Bonded to Structural Concrete Slab

	Rate	Total	Rate	Rate (metric)	Total (metric)
3 sacks portland cement	$....	$....	$4.24		$12.72
11 c.f. (0.31 cu.m) sand	....		0.48	17.03	5.28
Cost 100 s.f.			$....		$18.00
per s.f.			....		0.18
per sq.m			....		2.00

Material Cost of 100 S.F. (9 Sq.M) of Concrete Underbed, Consisting of 1/4" (6 mm) Dry Sand, 1 Thickness Tarred Felt and 2" (50 mm) of Concrete, Underbed Separated From Structural Slab

	Rate	Total	Rate	Rate (metric)	Total
2.5 c.f. (0.07 cu.m) dry sand	$....	$....	$0.48	$17.14	$1.20
1 sq. No. 15 felt	....		3.18		3.18
4.5 sacks portland cement	....		4.24		19.08
18 c.f. (0.5 cu.m) sand	....		0.48	17.28	8.64
Cost 100 s.f.			$....		$32.10
per s.f.			....		0.32
per sq.m			....		3.57

Material Cost of 100 S.F. (9 Sq.M) of Concrete Underbed 2" (50 mm) Thick, Applied Over Wood Floor

	Rate	Total	Rate	Rate (metric)	Total
1 sq. No. 15 felt	$....	$....	$3.18		$3.18
105 s.f. (9.45 sq.m) 14 ga. galv. wire netting	....		0.13	1.44	13.65
4.5 sacks portland cement	....		4.24		19.08
18 c.f. (0.50 cu.m) sand	....		0.48	17.28	8.64
Cost 100 s.f.			$....		$44.55
per s.f.			....		0.45
per sq.m			....		4.95

Material Cost of 100 S.F. (9 Sq.M) of Terrazzo Floor 1/2" (13 mm) Thick, Applied Over Concrete Underbed

	Rate	Total	Rate	Rate (metric)	Total
3 sacks portable cement	$....	$....	$4.24		$12.72
600 lbs. (270 kg) marble granules	....		0.13*	0.29	78.00
0.5 sack port. cement (grout)	....		4.24		2.12
Cost 100 s.f.			$....		$92.84
per s.f.			....		0.93
per sq.m			....		10.31

*Varies according to kind of granules used. Add for metal strips.

Material Cost of 100 S.F. (9 Sq.M) Terrazzo Floors Blocked Into 5'-0" (1.5 m) Squares, Including 1-1/4" (31 mm) Concrete Underbed

	Rate	Total	Rate	Rate (metric)	Total
Underbed 1-1/4" (31 mm) thick					
3 sacks portland cement ....$....	$....	$4.24	$4.24	$12.72
11 c.f. (0.31 cu.m) sand	....	0.48	17.03	5.28
Terrazzo Floor					
3 sacks portland cement	....	4.24	4.24	12.72
600 lbs. (270 kg) marble granules	....	0.13*	0.29	78.00
0.5 sack cement (grout)................................	....	4.24	4.24	2.12
50 l.f. (15 m) metal strips	....	0.53**	1.76	26.50
Cost 100 s.f.	$....			$137.34
per s.f..	....			1.37
per sq.m	....			15.26

Labor Cost of 100 S.F. (9 Sq.M) Terrazzo Floors Blocked Into 5'-0" (1.5 m) Squares, Including 1-1/4" (31 mm) Concrete Underbed

	Hours	Rate	Total	Rate	Total
Mechanic....................................	3.75	$....	$....	$25.62	$96.08
Helper	5.65	....	....	19.13	108.08
Labor rub and finish.....................	8.00	....	....	25.62	204.96
Cost 100 s.f		$....			$409.12
per s.f.....................................		....			4.09
per sq.m		....			45.46

If terrazzo is placed over wood subfloor or separated from concrete floor, add difference in cost of underbed.

Material Cost of 100 S.F. (9 Sq.M) Terrazzo Floors Blocked Into 4'-0" (1.2 m) Squares, Including 1-1/4" (31 mm) Concrete Underbed

	Rate	Total	Rate	Rate (metric)	Total
Underbed 1-1/4" (31 mm) Thick					
3 sacks portland cement..........................$....	$....	$4.24		$12.72
11 c.f. (0.31 cu.m) sand	....	0.48	17.03	5.28
Terrazzo Floor					
3 sacks portland cement	....	4.24	4.24	12.72
600 lbs. (270 kg) marble granules	....	0.13*	0.29	78.00
0.5 sack cement (grout)................................	....	4.24	4.24	2.12
60 l.f. (18 m) metal strips	....	0.53**	1.76	31.80
Cost 100 s.f.	$....			$142.64
per s.f..	....			1.43
per sq.m	....			15.85

Labor Cost of 100 S.F. (9 Sq.M) Terrazzo Floors
Blocked Into 4'-0" (1.2 m) Squares,
Including 1-1/4" (31 mm) Concrete Underbed

	Hours	Rate $	Total $	Rate	Total
Mechanic	4.0	$....	$....	$28.18	$112.72
Helper	6.0	....	....	21.04	126.24
Labor rub and finish	8.0	....	....	28.18	225.44
Cost 100 s.f.			$....		$464.40
per s.f.					4.64
per sq.m					49.93

If terrazzo is placed over wood subfloor or separated from concrete floor, add difference in cost of underbed.

Material Cost of 100 S.F. (9 Sq.M) Terrazzo Floors
Blocked Into 3'-0" (0.90 m) Squares,
Including 1-1/4" (31 mm) Concrete Underbed

	Rate	Total	Rate	Rate (metric)	Total
Underbed 1-1/4" (31 mm) thick					
3 sacks portland cement	$....	$....	$4.24	$4.24	$12.72
11 c.f. (0.31 cu.m) sand		....	0.48	17.03	5.28
Terrazzo Floor					
3 sacks portland cement		....	4.24	4.24	12.72
600 lbs. (270 kg) marble granules		....	0.13*	0.29	78.00
0.5 sacks cement (grout)		....	4.24	4.24	2.12
80 l.f. (24 m) metal strips		....	0.53**	1.76	42.40
Cost 100 s.f.			$....		$153.24
per s.f.					1.53
per sq.m					17.03

Labor Cost of 100 S.F. (9 Sq.M) Terrazzo Floors
Blocked Into 3'-0" (0.90 m) Squares,
Including 1-1/4" (31 mm) Concrete Underbed

	Hours	Rate $	Total $	Rate	Total
Mechanic	4.25	$....	$....	$28.18	$119.77
Helper	6.40	....	....	21.04	134.66
Labor rub and finish	8.00	....	....	28.18	225.44
Cost 100 s.f.			$....		$479.87
per s.f.					4.80

*Varies according to kind of granules used. **Varies according to kind and gauge of metal.

If terrazzo is placed over wood subfloor or separated from concrete floor, add difference in cost of underbed.

Material Cost of 100 S.F. (9 Sq.M) Terrazzo Floors
Blocked Into 2'-0" (0.60 m) Squares,
Including 1-1/4" (31 mm) Concrete Underbed

	Rate	Total	Rate	Rate (metric)	Total
Underbed 1-1/4" (31 mm) thick					
3 sacks portland cement ....$....	$....	$4.24	$4.24	$12.72
11 c.f. (0.31 m)) sand	....	0.48	17.03	5.28
Terrazzo Floor					
3 sacks portland cement.............................	....	4.24	4.24	12.72
600 lbs. (270 kg) marble granules	....	0.13*	0.29	78.00
0.5 sack cement (grout)...............................	....	4.24	4.24	2.12
120 l.f. (36 m) metal strips	....	0.53**	1.76	63.60
Cost 100 s.f.		$....		$174.44
per s.f..		....		1.74
per sq.m		....		19.38

*Varies according to kind of granules used. **Varies according to kind and gauge of metal.

Labor Cost of 100 S.F. (9 Sq.M) Terrazzo Floors
Blocked Into 2'-0" (0.60 m) Squares,
Including 1-1/4" (31 mm) Concrete Underbed

	Hours	Rate	Total	Rate	Total
Mechanic.................................4.6	$....	$....	$28.18	$129.63
Helper 6.9	....	....	21.04	145.18
Labor rub and finish...............8.0	....	....	28.18	225.44
Cost 100 s.f.		$....		$500.25
per s.f..		....		5.00
per sq.m		....		53.80

If terrazzo is placed over wood subfloor or separated from concrete floor, add difference in cost of underbed.

Material Cost of 100 S.F. (9 Sq.M) Terrazzo Floors
Blocked Into 1'-0" (0.30 m) Squares,
Including 1-1/4" (31 mm) Concrete Underbed

	Rate	Total	Rate	Rate (metric)	Total
Underbed 1-1/4" (31 mm) thick					
3 sacks portland cement ....$....	$....	$4.24	$4.24	$12.72
11 c.f. (0.31 m) sand	....	0.48	17.03	5.28
Terrazzo Floor					
3 sacks portland cement.............................	....	4.24	4.24	12.72
600 lbs. (270 kg) marble granules	....	0.13*	0.29	78.00
0.5 sacks cement (grout).............................	....	4.24	4.24	2.12
210 l.f. (63 m) metal strips	....	0.53	1.76	111.30
Cost 100 s.f.		$....		$222.09

per s.f.	....	2.22
per sq.m	....	24.68

Labor Cost of 100 S.F. (9 Sq.M) Terrazzo Floors
Blocked Into 1'-0" (0.30 m) Squares,
Including 1-1/4" (31 mm) Concrete Underbed

	Hours	Rate $	Total $	Rate	Total
Mechanic	5.35	$....	$....	$28.18	$150.76
Helper	8.00	....	....	21.04	168.32
Labor rub and finish	8.00	....	....	28.18	225.44
Cost 100 s.f.			$....		$544.52
per s.f.				....	5.45
per sq.m				....	58.64

If terrazzo is placed over wood subfloor or separated from concrete floor, add difference in cost of underbed.

Material Cost of 100 L.F. (30 m) of 6" (150 mm) Terrazzo Cove Base

	Rate	Total	Rate	Rate (metric)	Total
Scratch Coat					
1.5 sacks portland cement	$....	$....	$4.24	$4.24	$6.36
3 c.f. (0.08 cu.m) sand		....	0.48	18.00	1.44
Terrazzo Base					
1 sack portland cement		....	4.24	4.24	4.24
200 lbs. (90 kg) marble granules		....	0.13*	0.29	26.00
25 metal dividers		....	0.53**	1.76	13.50
Cost 100 l.f.			$....		$51.54
per l.f.				....	0.52
per m				....	1.72

*Varies according to kind of granules used. **Varies according to kind and gauge of metal.

Labor Cost of 100 L.F. (30 m) of 6" (150 mm) Terrazzo Cove Base

	Hours	Rate $	Total $	Rate	Total
Mechanic	16.0	$....	$....	$28.18	$450.88
Helper	16.0	....	....	21.04	336.64
Labor rub and finish	10.0	....	....	28.18	281.80
Cost 100 l.f.			$....		$1,069.32
per l.f.				....	10.69
per m				....	35.06

Double cove base, double cost of single cove base.

Abrasive Materials for Terrazzo Floors and Stair Treads. Where an anti-slip surface is required on terrazzo floors, platforms, landings, and treads, mix 2 parts abrasive aggregates with 3 parts marble granule. For heavy duty floors, the abrasive aggregate shall be mixed in the terrazzo topping for its entire thickness. For light traffic floors, the abrasive shall be sprinkled on the surface only, to show in the proportion of 3 parts marble granule and 1 part abrasive aggregate.

Abrasive aggregates for 100 s.f. (9 sq.m) of heavy duty floors should cost as follows:

	Rate	Total	Rate	Rate (metric)	Total
240 lbs. (108 kg) abrasive					
aggregates.............................$....	$....	$0.53	$1.18	$127.20
Cost per s.f.			....		1.27
per sq.m			....		14.13

Deduct cost of 240 lbs. (108 kg) of marble granules included in itemized estimates. Abrasive aggregates for 100 s.f. (9 sq.m) of light duty floors should cost as follows:

	Rate	Total	Rate	Rate (metric)	Total
10 lbs. (4.5 kg) abrasive aggregates ....$....	$....	$0.53	$1.18	$5.30
Cost per s.f.			....		0.05
per sq.m			....		0.59

For prices on special terrazzo floors, such as venetian, mosaic, and conductive, it is advisable to contact a local installation contractor.

There are several products on the market that resemble terrazzo floors and are installed by terrazzo mechanics. These are magnisite, latex, epoxy, and polyester resin type terrazzo floors. These floors can be installed over any sound existing floor in 1/2" (13 mm) thickness, with the exception of the epoxy and polyester resin types, which can be installed as thin as the marble granules will allow. The cost of these floors is similar to that of regular terrazzo.

Precast terrazzo is available for base, floor, wainscots, and stair treads. Straight 6" (150 mm) base costs about $4.00 per l.f. ($13.33 per m), and one mechanic sets 125 l.f. (37.5 m) per 8 hr. day. Floor tiles, 1" (25 mm) thick cost $4.30 per s.f. ($47.77 per sq.m) and one mechanic sets 60 to 70 s.f. (5.4-6.3 s.f.) per 8 hr. day. Stair treads cost around $10.50 per l.f. ($35.00 per m), because they are heavy. On a stair of any width, it takes a mechanic and helper to set 100 l.f. (30 m) per 8 hr. day.

09500 ACOUSTICAL TREATMENT

There is a large variety of prefabricated acoustical products available on the market. In general, they fall into three categories: acoustical tiles that

FINISHES 9.137

are mounted on ceilings, on suspended boards, or clipped or splined to suspension systems; acoustical panels for lay-in systems on exposed runners; and metal pan systems for mechanical suspension. Many systems today are also engineered to distribute air and light as well as to control sound; also, some systems are rated as fire resistive assemblies.

Acoustical tiles are applied to ceilings and walls for the purpose of absorbing and deadening sound in offices, banks, and other spaces where the utmost quiet is desired and in auditoriums to obtain optimum reverberation time.

Acoustical tiles are made of many different materials, such as wood fiber, cane fiber, mineral wool, cork, specially processed mineral filaments, perforated metal units, and other insulating materials.

They are made in a variety of sizes, such as 12"x12" (300 x 300 mm), 12"x24" (300 x 600 mm), 24"x24" (600 x 600 mm), 24"x48" (600 x 1200 mm), etc., and vary from 1/2" (13 mm) to 2" (50 mm) thick, depending upon the materials used.

Acoustical tile may be installed over sound, dry plaster or concrete that is thoroughly seasoned, over gypsum board, or nailed or screwed to 1"x3" wood furring strips or by means of mechanical suspension systems.

Acoustical tile work is simple to estimate, but because it is almost always factory cut and prefinished, it is extremely important to see that job conditions are the very best. Once installed, there is little opportunity to "touch up" the work. Call backs can quickly eat up profits as well as ruin reputations.

One of the most important conditions to check is temperature and humidity. These should always approximate the interior conditions that will exist when the building is occupied.

All plastering, poured roof decks, concrete and floor work should be complete and dry; all windows and doors in place and glazed. If weather demands it, the building heating, rather than space heating, should be turned on and kept on as though the building were occupied.

The trend today is to square edge tile, which minimizes joint visibility. However, tile work will always have slight unevenness in it, and if the job conditions include cove lights or high windows where light strikes the tile surface at a small angle, tile with beveled edges will help conceal any unevenness.

Budget Prices on Acoustical Tile. Most acoustical tile jobs are sublet to concerns specializing in this work. They have mechanics experienced in applying their particular brands of tile and can complete the job at less cost than a general contractor who purchases the tile and attempts to install them.

The following are approximate prices on the various classes of tile erected in place on fair size jobs as they might be quoted by a subcontractor including his overhead and profit.

Type of Installation	Price per s.f.	Price per sq.m
1/2" (13 mm) wood fibre tile surface applied	$1.15	$12.77
5/8" (16 mm) mineral tile suspended*	1.50	16.67
3/4" (19 mm) mineral tile suspended*	1.65	18.33
1" (25 mm) pad on steel pan suspended*	2.55	28.33
1" (25 mm) pad on alum. pan suspended*	3.70	41.11
5/8" (16 mm) mineral fibreboard suspended*	1.00	11.11
Air distributing ceilings-board* 2'x2' (0.6x0.6m)	1.45	16.11
Air distributing ceilings tile* 12"x12" (300x300mm)	1.75	19.44

*Including hangers but not additional channel supports.

Acoustical Tile Directly Applied. When acoustical tile is cemented, four spots of cement or adhesive should be applied to each tile on each corner of 12"x12" (300 x 300 mm) tiles. The spots should be placed as uniformly as possible with relation to the corners of the tile. A sufficient area on each corner of the tile should be primed prior to the application of the spot to receive the entire spot when it has been pressed to its final diameter. The spots of cement should be applied to these primed areas using a putty knife or trowel. The spot should be approximately the size and shape of a walnut, and should be nearly uniform in size and shape, except where thicker spots are required to level surfaces. They should stand out from the tile about 1/2" (13 mm) to 3/4" (19 mm) so as to contact the surface at about the same time. The diameter of the spot when pressed in place should be approximately 2-1/2" (62 mm) to 2-3/4" (69 mm) and about -1/8" (3 mm) thick.

In placing, the tile should be held with both hands so that when the pressure is applied it will be between the spots rather than outside them. The tile should be held as nearly parallel to its final position as possible and pressed to position with a firm pressure. A slight lateral sliding motion of the hands while setting is necessary to settle the cement into the ceiling surface. It is important that the tile be pressed to level when setting.

A crew of at least two workers is required to handle the installation efficiently, one on the scaffold applying the adhesive to the tile and the other placing the tiles on the ceiling.

One gallon (3.79 liters) of acoustical cement is sufficient for approximately 60 s.f. (5.4 sq.m) of tile under average conditions, although this varies somewhat with the surface to which it is applied. Adhesive costs about $10.00 per gal. ($2.64 per liter).

The labor cost of placing acoustical tile on ceilings will vary with the size of the rooms or spaces where the tile are to be applied, pattern in which the tile are laid, and height of scaffolding.

The following quantities are based on 2 mechanics working together from a 4' (1.2 m) to 6' (1.8 m) scaffold and laying tile in plain square or ashlar designs:

Size of Acoustical Tile	Pcs. per 8-Hr. Day	SF (Sq.M) per 8-Hr. Day	Mechanic Hrs. per 100 SF (9 sq.m)
12"x12" (300x300mm)	600-700	600-700 (54-63)	2.5
12"x24" (300x600mm)	400-450	800-900 (72-81)	2.0

When acoustical tile are laid in herringbone pattern, it requires 30% to 40% more time than regular squares or common bond. Diagonal patterns require 20% to 25% more time than regular squares or common bond, while mixed ashlar designs require about 50% more time than straight squares or common bond.

When placing tile in auditoriums and other spaces requiring high scaffolding, the additional cost of the scaffolding will have to be added to the cost of installing the tile. Large jobs are applied faster proportionately than small jobs.

Material Cost of 100 S.F. (9 Sq.M) 12"x12" (300 x 300 mm)
Acoustical Tile on Ceilings

	Rate	Total	Rate	Rate (metric)	Total
100 s.f. (9 sq.m) mineral tile					
3/4" (19 mm)	$....	$....	$0.70*	$7.77	$70.00
1.75-gal. (6.62 liters) adhesive		....	7.95	2.10	13.91
Cost 100 s.f.			$....		$83.91
per s.f.			....		0.84
per sq.m			....		9.32

*Cost of tile varies according to kind and thickness.

Labor Cost of 100 S.F. (9 Sq.M) 12"x12" (300 x 300 mm)
Acoustical Tile on Ceilings

	Hours	Rate	Total	Rate	Total
Mechanics	2.5	$....	$....	$26.44	$66.10
Cost per s.f.			....		0.66
per sq.m			....		7.11

Acoustical Tile On Suspended Systems

Lay-in type systems for 2'x2' (0.6 x 0.6 m) and 2'x4' (0.6 x 1.2 m) grids will have a material cost of around $0.55 and $0.50 per s.f. ($6.11 and $5.55 per sq.m) respectively, not including any channel supports if they are needed. One worker should install 700 s.f. (63 sq.m) per 8-hr. day of 2'x4' (0.6 x 1.2 m) grid, 600 s.f. (54 sq.m) for 2'x2' (0.6 x 0.6 m) grid.

Concealed spline systems are more time consuming. Material costs will run about $0.60 per s.f. ($6.67 per sq.m), and one worker will install 500 s.f. (45 sq.m) per day.

If channel carriers are required, the material cost will be about $0.15 per s.f. ($1.67 per sq.m), and one worker installs about 500 s.f. (45 sq.m) per 8-hr. day.

Labor Cost To Install 100 S.F. (9 Sq.M) 2'x2' (0.6 x 0.6 m) Lay-in Grid

	Hours	Rate	Total	Rate	Total
Mechanic	1.3	$....	$....	$26.44	$34.37
Cost per s.f.				....	0.34
per sq.m				....	3.66

Labor Cost To Install 100 S.F. (9 Sq.M) 2'x4' (0.6 x 1.2 m) Lay-in Grid

	Hours	Rate	Total	Rate	Total
Mechanic	1.2	$....	$....	$26.44	$31.73
Cost per s.f.				....	0.32
per sq.m				....	3.44

Labor Cost To Install 100 S.F. (9 Sq.M) 12 x 12 Z Bar System Grid

	Hours	Rate	Total	Rate	Total
Mechanic	1.6	$....	$....	$26.44	$42.30
Cost per s.f.				....	0.42
per sq.m				....	4.52

Labor Cost To Install 100 S.F. (9 Sq.M) Channel Ceiling Supports

	Hours	Rate	Total	Rate	Total
Mechanic	1.6	$....	$....	$26.44	$42.30
Cost per s.f.				....	0.42
per sq.m				....	4.52

Acoustical Tile For Suspended Grid Systems. This material is made "cut to size" ready to install into the grid system. The 24"x24" (600 x 600 mm) and 24"x48" (600 x 1200 mm) boards cost about $0.50 to 0.60 per s.f. ($5.55-6.67 per sq.m) depending on the type of tile, and one worker should install about 100 s.f. (9 sq.m) per hr. The 12"x12" (300 x 300 mm) tiles for the Z bar system costs about $0.75 to 0.85 per s.f. ($8.33-9.44 per sq.m), and one worker should install about 100 s.f. (9 sq.m) in 2 hrs.

Metal Pan Type Acoustical Tile. This type of acoustical treatment consists of metal units 12"x24" (300 x 600 mm) in size, having center grooves which result in the appearance of 12"x12" (300 x 300 mm) units when applied. The edges, which are beveled and returned vertically, are firmly held in place on the 12" (300 mm) sides by special tee runners. Within the metal units and supported on crimped, galvanized wire mesh, or on miniature channels, are mineral wool absorbent pads.

These systems will install at the rate of around 250 s.f. (22.5 sq.m) per day, including sound absorbing pads, but without channel supports if they are required. Materials will cost about $2.00 for steel, $3.00 for aluminum, and $4.00 for stainless.

Labor Cost To Install 100 S.F. (9 Sq.M) Suspended Steel Pan System

	Hours	Rate	Total	Rate	Total
Mechanic	3.2	$....	$....	$26.44	$84.61
Cost per s.f.			....		0.85
per sq.m			....		9.10

09550 WOOD FLOORING

Estimating Quantities of Wood Flooring. When estimating the quantity of board feet (b.f.) of wood flooring required for any job, take the actual number of sq. ft. in any room or space to be floored, and add allowances as given in the following tables:

Measured Size Inches	Finished Size Inches	No. Pcs. in Bundle	Add for Waste Percent	To Obtain Quantity Flooring Repaired Multiply Area by	No. BF Flg. Required for 100 Sq. Ft. Floor
1x1	3/8x7/8	24	16-2/3	1-1/6 or 1.17	117
1x2	3/8x1-1/2	24	33-1/3	1-1/3 or 1.33	133
1x2-1/2	3/8x2	24	25	1-1/4 or 1.25	125
1x2-1/4	25/32x1-1/2	12	501/3	1-1/2 or 1.50	150
1x2-3/4	25/32x2	12	37-1/2	1-3/8 or 1.375	137-1/2
1x3	25/32x2-1/4	12	33-1/3	1-1/3 or 1.33	133
1x4	25/32x3-1/4	8	25	1-1/4 or 1.25	125

(metric)

Measured Size mm	Finished Size mm	No. Pcs. in Bundle	Add for Waste Percent
25 x 25	9.4 x 21.8 mm	24	16.67
25 x 50	9.4 x 37.5 mm	24	33.33
25 x 63	9.4 x 50.0 mm	24	25
25 x 56	19.5 x 37.5 mm	12	50.33
25 x 69	19.5 x 50.0 mm	12	37.5
25 x 75	19.5 x 56.3 mm	12	33.33
25 x 100	19.5 x 81.3 mm	8	25

Amount of Surface 1,000 B.F. of Flooring Will Cover and
Quantity of Nails Required to Lay it

How Measured	Size Flooring	Covers SF Fl.	Nailed Every	Lbs. Nails Req'd. Cut Nails	Helically threaded nails
1x2	3/8x1-1/2	750	8 in.	20 lbs. 4 "d" cut	
1x2-1/2	3/8x2	800	8 in.	17 lbs. 4 "d" cut	
1x2-1/4	25/32x1-1/2	667	12 in.	70 lbs. 8 "d" cut	60 lbs 7 "d"
1x2-3/4	25/32x2	727	12 in.	56 lbs. 8 "d" cut	47 lbs 7 "d"
1x3	25/32x2-1/4	750	10 in.	64 lbs. 8 "d" cut	54 lbs 7 "d"
1x4	25/32x3-1/4	800	10 in.	29 lbs. 8 "d" cut	24 lbs 7 "d"

(metric)

How Measured	Size Flooring	Covers sq.m Fl.	Nailed Every	Kg Nails Req'd. Cut Nails	Helically threaded nails
25 x 50 mm	9.4 m	70	200 mm	9.0 kg 4 "d" cut	
25 x 63 mm	9.4 m	74	200 mm	7.7 kg 4 "d" cut	
25 x 56 mm	20 m	62	300 mm	31.7 kg 8 "d" cut	27 kg 7 "d" cut
25 x 69 mm	20 m	68	300 mm	25.4 kg 8 "d" cut	21 kg 7 "d" cut
25 x 75 mm	20 m	70	250 mm	29.0 kg 8 "d" cut	24 kg 7 "d" cut
25 x 100 mm	20 m	74	250 mm	11 kg 8 "d" cut	11 kg 7 "d" cut

Drill flooring for nails, if possible, for best results. No predrilling is
required for helically threaded nails. The above figures are based on laying
the flooring straight across a rectangular room without producing any design
whatever.

Rules for Grading Hardwood Flooring. In many instances the
specifications of the architects and owners are confusing and not in
accordance with the standard rules for grading as adopted by the
manufacturers. The rules adopted by the flooring manufacturers are given
below so that they may be compared with the architect's specifications and
the proper grade of flooring figured.

First Grade. This grade is 25/32" (20 mm) and thicker, shall have the face
practically free of all defects, but the varying natural color of the wood shall
not be considered a defect. Standard lengths in all widths in this grade shall
be in 2' (0.6 m) bundles and longer as the stock will produce. Not over 30% of
the total footage shall be in bundles under 4' (0.8 m), and not more than 17%
of the total footage shall be in 2' (0.6 m) bundles.

Second Grade. This is 25/32" (20 mm) and thicker, will admit tight, sound
knots and slight imperfections in dressing, but must lay without waste.
Standard lengths in all widths in this grade shall be in 2' (0.6 m) bundles and
longer as the stock will produce. Not over 45% of the total footage shall be in
bundles under 4' (1.2 m), and not more than 25% of the total footage shall be
in 2' (0.6 m) bundles.

Third Grade. This is 25/32" (20 mm) and thicker must be of such character as will lay and give a good serviceable floor. Standard lengths in all widths of this grade shall be in 1-1/4' (0.38 m) bundles and longer as the stock will produce. Not over 65% of the total footage shall be in bundles under 4' (1.2 m), and not more than 27% of the total footage shall be in 1-1/4' (0.38 m) bundles.

Combination Grades

Second and Better Grade. In all thicknesses and widths, it is a combination of first and second grades developing in the strip without crosscutting for each grade. The lowest grade pieces admissible shall not be less than standard second grade. Standard lengths in all widths in this grade shall be in 2' (0.6 m) bundles and longer as the stock will produce. Not over 40% of the total footage shall be in bundles under 4' (1.2 m), and not more than 20% shall be in 2' (0.6 m) bundles.

Third and Better Grade. In all thicknesses and widths, a combination of first, second, and third grades developing in a strip without crosscutting. The lowest grade pieces admissible shall not be less than standard third grade. Standard lengths in all widths of this grade shall be in 1-1/4' (0.38 m) bundles and longer as the stock will produce. Not over 50% of the total footage shall be in bundles under 4' (1.2 m).

Special Color Grades
(The Finest Grades Produced)

Selected First Grade Light Northern Hard Maple. This grade is selected for light color. The color tones in individual strips will vary somewhat, but after laying, it provides a luxurious "light" floor. It costs about $4.00 per s.f. ($44.44 per sq.m) for material.

Selected First Grade Amber Northern Hard Maple. This grade is selected for amber color. The color tones in individual strips will vary somewhat, but after laying, this grade provides a luxurious "amber" appearing floor. The cost is $3.75 per s.f. ($41.67 per sq.m) for material.

Oak Flooring - Quarter Sawed

Clear. The face shall be practically clear, admitting an average 3/8" (9 mm) of bright sap. The question of color shall not be considered. Bundles to be 2' and up. Average length is 3-3/4' (1.13 m).

Select. The face may contain sap, small streaks, pin worm holes, burls, slight imperfections in working, and small tight knots that do not average more than one to every 3' (0.9 m). Bundles to be 2' (0.6 m) and up. Average length is 3-1/4' (0.98 m).

Plain Sawed

Clear. The face shall be practically clear, admitting an average of 3/8" (9 mm) bright sap. The question of color shall not be considered. Bundles to be 2' (0.6 m) and up. Average length 3-3/4' (1.13 m).

Select. The face may contain sap, small streaks, pin worm holes, burls, slight imperfections in working, and small tight knots which do not average more than one to every 3' (0.9 m). Bundles to be 2' (0.6 m) and up. Average length 3-1/4' (0.98 m).

No. 1 Common. Shall be of such nature that will lay a good residential floor and may contain varying wood characteristics, such as flags, heavy streaks and checks, worm holes, knots and minor imperfections in working. Bundles to be 2' (0.6 m) and up. Average length 2-3/4' (0.83 m).

No. 2 Common. May contain sound natural variations of the forest product and manufacturing imperfections. The purpose of this grade is to furnish an economical floor suitable for homes, general utility use, or where character marks and contrasting appearance is desired. Bundles to be 1-1/4' (0.38 m) and up. Average length 2-1/4' (0.68 m).

1-1/4 Shorts. Pieces 9" (225 mm) to 18" (450 mm) long are to be bundled together and designated as 1-1/4' (0.38 m) shorts. Pieces grading No. 1 Common, Select, and Clear to be bundled together and designated No. 1 Common and Better with pieces grading No. 2 Common bundled separately and designated as such. Although pieces 6" (150 mm) under and only 3" (75 mm) over the nominal length of the bundle may be included, the pieces must average 1-1/4" (31 mm), which is achieved through the natural preponderance of longer lengths.

Standard Thicknesses and Widths

25/32" (20 mm) thickness; widths 1-1/2" (38 mm) face, 2" (50 mm) face, 2-1/4" (56 mm) face, and 3-1/4" (81 mm) face.
3/8" (9 mm) thickness; width 1-1/2" (38 mm) face and 2" (50 mm) face.
1/2" (13 mm) thickness; width 1-1/2" (38 mm) face and 2" (50 mm) face.
Above tongue and groove and end matched.

Square Edge Strip Flooring

Grades as shown above but bundling and lineals as follows:
5/16x2" (8 x 50 mm), 5/16x1-1/2" (8 x 38 mm), 5/16x1-1/3" (8 x 33 mm), 5/16x1-1/4" (8 x 31 mm), 5/16x1-1/8" (8 x 28 mm), 5/16x1" (8 x 25 mm), 5/16x7/8" (8 x 22 mm). Also made rough back in 11/32x2" (9 x 50 mm) and 11/32x11/2" (9 x 38 mm).

Clear. Bundled 2' (0.6 m) and up. Average length 5-1/2' (1.65 m).
Select. Bundled 2' (0.6 m) and up. Average length 4-1/2' (1.35 m).

No. 1 Common. Bundled 2' (0.6 m) and up. Average length 3-1/2' (1.05 m).
No. 2 Common. May contain defects of all characters, but will lay a serviceable floor. Bundles to be 1-1/4' (0.38 m) and up. Average length 2-1/2' (0.75 m). All faces shown above in 5/16" (8 mm) square edge are finished 1/64" (0.40 mm) over face.

Approximate Prices of Oak Flooring

Oak flooring prices will vary widely throughout the country. A recent quote gives 25/32"x2-1/4 (20 x 56 mm) oak at $3.25 per s.f. ($36.11 per sq.m), select, $3.00 per s.f. ($33.33 per sq.m) common. Clear will run 30% more than select, if it is at all available.

Grading Rules on Prefinished Hardwood Flooring

Oak. White and Red Oak to be separated in each grade. Grades are established after the flooring has been sanded and finished.
Prime Grade. Face shall be selected for appearance after finishing, but sapwood and the natural variations of color are permitted. Minimum average length 4' (1.2 m). Bundles 2' (0.6 m) and longer.
Standard Grade. Will contain sound wood characteristics which are even and smooth after filling and finishing and will lay a sound floor without cutting. Minimum average length 3' (0.9 m). Bundles 1-1/4' (0.38 m) and longer.
Standard and Better Grade. A combination of Prime and Standard to contain the full product of the board except that no pieces are to be lower than Standard Grade. Minimum average length 3-1/2' (1.05 m). Bundles 1-1/4' (0.38 m) and longer.
Tavern Grade. Shall be of such nature as will make and lay a serviceable floor without cutting, but purposely containing typical wood characteristics, which are to be properly filled and finished. Minimum average length 2-1/2' (0.75 m). Bundles 1-1/4' (0.38 m) and longer.

Beech and Pecan

(Will be furnished only in a combination grade of Tavern and Better)
Tavern and Better. A combination of Prime, Standard, and Tavern to contain the full product of the board, except that no pieces are to be lower than Tavern Grade. Minimum average length 3' (0.9 m). Bundles 1-1/4' (0.38 m) and longer.

General Rules for All Species

Hardwood flooring is bundled by averaging the lengths. A bundle may include pieces from 6" (150 mm) under to 6" (150 mm) over the nominal length of the bundle. No piece shorter than 9" (225 mm) admitted.

The percentages under 4' (1.2 m) referred to apply on total footage in any one shipment of the item.

A 3/4" (19 mm) allowance shall be added to the face length when measuring the length of each piece of flooring.

Flooring shall not be considered of standard grade unless the lumber from which the flooring is manufactured has been properly kiln dried.

Laying and Finishing Wood Floors

The following pages contain detailed itemized costs of laying and finishing all kinds of soft and hardwood floors.

Labor Laying 100 S.F. (9 Sq.M) of Hardwood Floors

Description of Work	SF (Sq.M) per 8-Hr. Day	Carp. Hours	Labor Hours
Ordinary Workmanship			
25/32"x3-1/4" (20x81mm) face softwood floors for porches, kitchens, factories, stores, etc.	400-500 (36-45)	1.8	0.6
25/32"x2-1/4" (20x56mm) face Third Grade Maple, for warehouse, factory and loft building floors	375-425 (34-38)	2.0	0.6
25/32"x2-1/4" (20x56mm) face Oak or Birch in residences, apartments, stores, offices, etc	250-300 (23-27)	2.9	0.6
Same as above laid by experienced floor-layer	300-350 (27-32)	2.5	0.6
25/32"x1-1/2" (20x38mm) face Third Grade Maple for warehouse, factory and loft building floors	300-325 (27-29)	2.3	0.7
25/32"x1-1/2" (20x38mm) face Oak or Birch in residences, apartments, stores, offices, etc	150-200 (14-18)	4.5	0.8
Same as above laid by experienced floor-layer	225-275 (20-25)	3.2	0.8
First Grade Workmanship			
25/32"x2-1/4" (20x56mm) face Oak or Birch in fine residences, apartments, hotels, stores and offices	200-225 (18-20)	3.8	0.6
Same as above laid by experienced floor-layer	225-250 (20-23)	3.3	0.6
25/32"x1-1/2" (20x38mm) face Oak or Birch, same class			

of work as described above .... 120-150 (11-14).......6.0........0.8
Same as above laid by experienced
   floor-layer.. 175-200 ....4.3 ....0.8

In larger cities there are carpenters who make a specialty of laying and finishing floors, and they seldom do any other kind of work. Due to their long experience, they not only do a better job but accomplish more work per day than the average carpenter, and they expect to be paid accordingly. For this reason, production times are given for floors laid and finished by both carpenters and floorlayers.

Another reason some floorlayers will lay more floors than carpenters is because they do not nail the flooring 8" (200 mm) or 16" (400 mm) o.c. but will nail at every other bearing unless watched continually.

The first or Ordinary Workmanship includes the grade of work found in moderate priced homes, apartments, stores, etc., where the floor laying permits some hammer marks. Considerable savings can be obtained by laying the strip flooring all the way through prior to the erection of partitions.

First Grade Workmanship includes all classes of high class buildings where the best in workmanship is required, with all flooring closely driven up, laid free from hammer marks, and with all nails set.

Labor Laying 3/8" (9 mm) or 1/2" (13 mm) Hardwood Flooring. Figure same costs as given for 25/32" (20 mm) flooring.

Setting Sleepers in Mastic on Concrete Floors. A sound, fast, economical way to provide nailing for strip wood flooring installed over concrete subfloors is the sleeper-in-mastic method, used extensively in the construction of slab on grade houses.

Assuming concrete slab is smooth, level, and dampproofed to prevent moisture seepage, 2"x4" (50 x 100 mm) wood sleepers are laid flat in hot or cold mastic and are immediately ready to receive strip flooring.

After priming concrete slab, hot or cold mastic is spread to a 3/32" (2 mm) depth over the entire slab or in 1/4" (6 mm) deep rivers along parallel lines in positions where sleepers are to be placed.

Wood sleepers should be straight, flat, dry 2"x4" (50 x 100 mm) lumber in random lengths from 18" (450 mm) to 48" (1200 mm). For best results, sleepers should be treated with an approved non-creosote wood preservative. Lay sleepers in mastic, flat side down, in staggered rows 12" (300 mm) o.c. at right angles to direction of flooring. Ends should overlap 4" (100 mm) and end joints should be staggered by alternating short and long lengths. Leave 1" (25 mm) clearance between sleepers and all vertical surfaces to allow for normal expansion.

Laying strip wood flooring over sleepers in mastic should follow the same general procedure as for over wood subfloors with the following additional precautions:

1. Use end-matched flooring only.
2. Individual strips of flooring should span at least two sleeper spacings with no adjacent joints occurring in the same spacing.
3. Flooring strips passing over sleeper laps should be nailed to both sleepers.

Material coverages and approximate prices are as follows:

Asphalt primer, 200 to 400 s.f. per gal. (4.90-9.80 liters per sq.m) depending on porosity of concrete surface. Price $4.50 per gal. ($1.19 per liter).

Hot mastic, 25 s.f. per gal. (0.61 liters per sq.m). Price $3.50 per gal. ($0.92 per liter).

Cold mastic, 40 to 50 s.f. per gal. (0.98-1.23 sq.m per liter). Price $3.25 per gal. ($0.86 per liter).

For average size rooms, 100 s.f. (9 sq.m) of floor area will require 120 to 135 l.f. (36.0-40.5 m) or 80 to 90 b.f. (0.19-0.21 cu.m) of 2"x4" (50 x 100 mm) wood sleepers.

Under ordinary conditions, a floorlayer should prime concrete floor, spread mastic, and place wood sleepers at the rate of 100 s.f. (9 sq.m) per hour.

Labor Laying 1,000 b.f (2.25 cu.m) Soft and Hardwood Flooring

Ordinary Workmanship

Description of Work	B.f. (cu.m) Lumber Per 8-Hr. Day	Carp. Hours	Labor Hours
25/32"x3-1/4" (20x81mm) face softwood floors for porches, kitchens, factories, stores, etc	500-600 (1.13-1.35)	14.5	4.0
25/32"x2-1/4" (20x56mm) face Third Grade Maple, used in warehouses, factory and loft buildings, etc	500-575 (1.13-1.29)	15.0	4.0
25/32"x2-1/4" (20x56mm) face Birch or Oak flooring in residences, apartments, stores, etc	350-400 (0.79-0.90)	21.0	4.0
Same as above laid by experienced floorlayer	400-450 (0.90-1.01)	19.0	4.0
25/32"x1-1/2" (20x38mm) face Third Grade Maple, used in warehouses, factory and loft buildings, etc	450-500 (1.01-1.13)	17.0	4.0
25/32"x1-1/2" (20x38mm) face Birch or Oak flooring in residences, apartments, store and office buildings, etc	250-280 (0.56-0.63)	30.0	5.0
Same as above laid by experienced floorlayer	350-400 (0.79-0.90)	21.0	5.0

First Grade Workmanship

Description of Work	B.f. (cu.m) Lumber Per 8-Hr. Day	Carp. Hours	Labor Hours
25/32"x1-1/4" (20x31mm) face Oak or Birch flooring in fine residences, apartments, hotels, stores and offices	275-300 (0.62-0.68)	28	4.0
Same as above laid by experienced floorlayer	325-375	23	4.0
25/32"x1-1/2" (20x38mm) face Oak or Birch, same class of work as described above for 2-1/4" face flooring	175-225 (0.39-0.51)	39	5.0
Same as above laid by experienced floorlayer	275-300 (0.62-0.68)	28	5.0

Sanding Wood Floors by Machine

Practically all wood floors today are sanded and finished by machine. The cost varies, depending on the kind of flooring, whether old or new work, class of workmanship, size of rooms, etc.

The floors are first sanded with a floor sanding machine and then the edges are finished, using a disc type edging machine, which is capable of sanding up to the base shoe or quarter round.

The class of workmanship of number of operations governs production and cost, but where just an ordinary grade of workmanship is required in average size rooms in houses, apartments, offices, etc., a good machine and operator should finish and edge 800 to 900 s.f. (72-81 sq.m) per 8-hr. day, at the following labor cost per 100 s.f. (9 sq.m):

	Hours	Rate	Total	Rate	Total
Operator-finisher	1.0	$....	$....	$26.44	$26.44
Cost per s.f.			....		0.26
per sq.m			....		2.80

When used in store rooms, auditoriums, and other large floor areas, a machine and operator should finish and edge 1,000 to 1,250 s.f. (90-112.5 sq.m) of new floor per 8-hr. day, at the following labor cost per 100 s.f. (9 sq.m):

	Hours	Rate	Total	Rate	Total
Operator-finisher	0.75	$....	$....	$26.44	$19.83
Cost per s.f.			....		0.20
per sq.m			....		2.13

Surfacing Floors by Machine, First Grade Workmanship. In residences, apartments, hotel, store and office buildings, etc., where first grade workmanship is required, eliminating all irregularities and waves in the floor, about four cuts are necessary: first with No. 1-1/2 or 2 grit sandpaper, then with No. 1/2 or 1 grit paper and the last two cuts with No. 0 paper.

On work of this kind an experienced operator should finish and edge 400 to 500 s.f. (36-45 sq.m) of floor per 8-hr. day, at the following labor cost per 100 s.f. (9 sq.m):

	Hours	Rate	Total	Rate	Total
Operator-finisher	1.75	$....	$....	$26.44	$46.27
Cost per s.f.			....		0.46
per sq.m			....		4.95

Resurfacing Old Floors by Machine. When resurfacing old wood floors, including the removal of old varnish, dirt, grease spots, etc., the cost will vary with the condition of the old floors. This variation will run from 300 to 1,000 s.f. (27-90 sq.m) per 8-hr. day and should be determined by the estimator after examining the old floors.

There is little or no difference in resurfacing an old floor or surfacing a new one except that under normal conditions one extra cut with a special open-faced abrasive is usually necessary to remove old varnish or paint.

Sanding Wood Floors

The following specifications list the operations necessary to produce a first class job. Cheaper grades of workmanship will require less sanding and consequently less labor—and lower costs.

Sanding Old Wood Floors Laid in Strips. The first operation is to remove the surface finish by sanding in the direction of the grain of the wood until all surface finish has been removed, using a drum type sanding machine, with extra heavy open coat sandpaper grit No. 3-1/2 or 4.

The second operation is to remove the effects of the first operation and level the floor. Sand as in first operation, using same type machine with sharp sandpaper, grit No. 2 or 2-1/2, to remove the effects of the coarse open coat sandpaper and level the floor surface.

The third operation is to use a spinner along the edges of the floor and hand scrape all corners where the machines could not reach. Use a No. 3-1/2 or 4 grit paper and repeat, using No. 2 or 2-1/2 grit paper.

The fourth operation is to sweep the entire floor to remove the coarse grits and abrasives left by the previous operations.

The fifth operation is to sand as in the first operation using the same type machine with No. 0 or No. 1/2 fine grit.

The sixth operation is the final smoothing of the edges. Sand as in third operation using same type of machine with fine sandpaper, grit 0 or 1/2.

After sweeping the entire floor clean it is ready for any kind of floor finish.

Sanding New Wood Floors Laid in Strips. First, sand to a level surface. Sand in the direction of the grain of the wood until all the high edges or over-wood on the floor has been sanded off to a level surface, using a drum type sanding machine with No. 1-1/2 or 2 grit sandpaper.

The following chart gives the proper abrasives or grits for surfacing all classes of wood floors:

Chart for Floor Surfacing Operations

New Floors			
Kind of Floors	Operation	Floor Conditions	Proper Grit
Oak, Maple and Close- Grained Hardwood Floors	Roughing	Ordinary Floors	2-2½
		Well Laid Floors	1½
		Very Uneven Floors	2½
		Very Hard Floors	2½
	Finishing	Ordinary Finish	½
		Fine Finish	0
		Extra Fine Finish	00
		Rough Finish	1
Parquet Floors	Roughing		1½
	Finishing	Semi-Finish	½
		Finishing	0
Cork Tile Floors	Roughing		1½
	Finshing	Semi-Finish	½
		Finishing	0
Old Floors			
Resurfacing Old Floors— Soft and Hard Woods	Removing Varnish, Paints, etc.	Ordinary Conditions	3½
		Extra Heavy Coat Paint or Varnish	4
	Roughing	Ordinary Floors	2-2½
	Finishing	Ordinary Finish	½
		Fine Finish	0
		For Penetrating Sealer	00

Second, remove the scratch pattern of the first sanding. Sand in the direction of the grain of the wood until the scratch pattern of the first operation is eliminated using the same type sanding machine with No. 1 or 1/2 sandpaper. Frequent changes of sandpaper must be made to insure cutting rather than sliding over the surface.

Third, use a spinner around the edges and scrape by hand where it is impossible to use the machine, without crossing the grain of the wood. Use same grits as on drum sander.

Fourth, sweep the entire floor free from any grit and abrasives and then sand in direction of the grain of the wood, using the drum type machine as in the first operation, with No. 0 sandpaper. Change the sandpaper frequently to insure cutting rather than sliding over the surface.

After sweeping the floor clean it is ready for any kind of floor finish.

Sanding Old Parquet and Herringbone Wood Floors. The first operation is to sand lengthwise of the room or at 45 degrees, using a drum type sanding machine with No. 3-1/2 or 4 open coat grit, using a very light pressure on the drum. Second, remove the coarse marks of the first operation by sanding crosswise of the room using same type machine with No. 1 or 1-1/2 grit. Change paper frequently to assure cutting rather than sliding over the surface. Then sand lengthwise of the room using No. 1/2 or No. 1 grit.

Third, use a portable edging machine or hand scrape all places where the drum machine and edger cannot reach. Sweep the entire floor to remove coarse grits and abrasives left by the previous operations.

Fourth, use a large diameter rotary sanding machine across and then lengthwise of the room with No. 1/2 sandpaper to remove the cross marks of previous operations.

The fifth and final operation is to sweep the entire floor free from all grits, then sand using the large diameter rotary sander with No. 00 or No. 0 sandpaper, going over the floor once across and once lengthwise of the room.

After sweeping the floor clean it is ready for any kind of floor finish.

Finishing Hardwood Floors

The modern and approved method of finishing hardwood floors is to use wood sealers, which are a special compound of ingredients blended so that they penetrate into the fiber and cells of the wood, sealing the pores with tough elastic materials, thereby building up a base resistance to water, dirt, and stains. It wears only as the wood wears down.

Finishing Hardwood Floors in Natural Color. Before finishing hardwood floors, they should be sanded as described above.

After the floor has been completely sanded and swept clean of all dust, a good grade of permanent penetrating seal should be applied. One such seal on the market is Pentra-Seal (from the American-Lincoln Corp., Toledo, OH). It can be applied quickly with a brush, rag, mop, lamb's wool applicator, or squeegee and does not leave brush or lap marks.

For an extra fast job, Fast-Drying Seal is recommended. Apply a thin, even coat and after it has become tack free remove the excess with a No. 2 steel wool pad. Sweep the floor clean and apply a second coat in the same manner. Where a highly glossy finish is desired, apply a third coat and do not use steel wool. This will remain on the surface, since the preceding coats have thoroughly filled every cell and pore, and result in a glossy surface finish.

The floor should now be given a final protective coat of a good grade of paste wax or Lustre Finish. Paste wax should be applied manually and

Lustre Finish can be put on with a mop. Two coats are recommended, followed by buffing with a mixed fiber brush.

Fast-Drying Seal is available in natural as well as seven decorator type colors. Coverage ranges from 400 to 450 s.f. per gal. (9.5-10.7 sq.m per liter) for the first coat and 650 to 750 s.f. per gal. (15.5-17.8 sq.m per liter) for the second coat. Price, Fast-Drying Sealatural, about $15.00 per gal. ($3.96 per liter); colors, slightly higher.

Coverage of Lustre-Finish is about 1,250 s.f. per gal. (29.7 sq.m per liter). Price, about $10.50 per gal. ($2.77 per liter).

Labor Finishing Wood Floors. Assuming the floor has been sanded and is ready for the seal, an experienced worker can finish about 1,000 s.f. (90 sq.m) of natural finish floor per 8-hr. day. Colored seal takes more time as the excess pigment must be removed by hand wiping.

Labor Cost of 100 S.F. (9 Sq.M) of Fast-Drying Seal and Lustre Finish
Floor Finish on Wood Floors, Natural Color

	Hours	Rate	Total	Rate	Total
Mechanic	0.8	$....	$....	$26.44	$21.15
Cost per s.f.			....		0.21
per sq.m			....		2.26

Prefinished Hardwood Plank Flooring

Plank or strip flooring in 3/4" (19 mm) thickness is available in several widths and finishes, most furnished in random lengths and widths. It is generally of red oak and predrilled for nailing. Widths are in the 3" (75 mm) to 8" (200 mm) range. Special distressed finishes and pegging are also available.

Plain flooring in narrower widths will cost about $3.00 per s.f. ($33.33 per sq.m). For pegging, add $0.10 per s.f. Wider planks will run $3.75 per s.f. ($41.67 per sq.m). Premium finishes can raise the cost to as much as $4.00 per s.f. ($44.44 per sq.m).

One carpenter should lay at least 150 s.f. (13.5 sq.m) per day.

Wood Parquet Flooring

Wood parquet is available in stock designs of oak, walnut, teak, cherry, and maple. Standard thickness is 5/16" (8 mm), although heavier parquet is available on custom orders in 11/16" (17 mm) and 13/16" (20 mm) thicknesses. Sizes will vary with patterns running from 9"x9" (225 x 225 mm), 12"x12" (300 x 300 mm), 16"x16" (400 x 400 mm), on up to 36"x36" (900 x 900 mm).

Parquet is set in mastic on any level floor above grade. Most parquet is ordered factory finished, but may be unfinished if desired. Many patterns are available in "prime grade" and "character marked", square or bevelled edges.

Costs of all wood flooring are changing rapidly, but cost approximate $1.70 per s.f. ($18.89 per sq.m) for a standard 5/16" (8 mm) oak, $3.75 per s.f. ($41.67 per sq.m) for walnut, and $3.75 per s.f. ($41.67 per sq.m) for teak. The prime grades in more elaborate patterns will run much higher. If prefinished, add $0.50 per s.f. ($5.55 per sq.m).

Laying Wood Parquet Floors in Mastic Over Concrete. Concrete subfloor shall be primed with one coat of asphalt priming paint and allowed to dry before laying parquet. Average covering capacity of primer 200 to 400 s.f. per gal. (4.75-9.5 sq.m per liter) depending on porosity of cement floor finish. Price $3.25 per gal. in 5-gal. cans ($0.86 per liter in 19-liter can).

Where concrete floors are in direct contact with the ground, below grade, or where there is not sufficient ventilation to prevent condensation, a membrane waterproofing shall be applied to the slabs, extending same up the sides of all walls, 4" (100 mm) to 6" (150 mm), according to manufacturer's specifications. Where wood parquet is to be laid in hot mastic, a hot application, two-ply membrane shall be applied. Where parquet is to be laid in cold mastic, a cold application one-ply polyethylene film shall be applied.

Installation of polyethylene film is as follows: after concrete surface is primed, cold mastic is applied at the rate of 80 to 100 s.f. per gal. (1.9-2.4 sq.m per liter), into which polyethylene film is rolled, lapping joints 4 inches (100 mm). Laps do not have to be sealed.

Parquet shall be laid to a level surface bedded in a suitable mastic without the use of nails or screws. One gallon of mastic will cover 45-50 s.f. (1 liter covers 1.0-1.2 sq.m) and costs about $15.00 per gal. ($4.00 per liter). Expansion joints must be left around all edges of rooms according to manufacturer's specifications.

Labor Laying Wood Parquet Flooring. When laying wood parquet floors in mastic in ordinary size rooms such as residences, apartments, etc., an experienced floorlayer with a helper should lay 300 to 350 s.f. (27-31.5 sq.m) of floor per 8 hr. day, while in large spaces such as schools, auditoriums, gymnasiums, etc., an experienced floorlayer should lay 350 to 400 s.f. (31.5-36 sq.m) of floor per 8-hr. day. This applies to either finished or unfinished parquet, as the same care is required in laying either type of flooring. This does not include any time preparing or leveling old concrete floors, as this must be added extra if required.

Where unfinished parquet is used, add for sanding and finishing.

Laying Wood Parquet Floors in Mastic over Other Surfaces. Wood parquet flooring may also be laid in mastic over plywood or board subflooring, present wood floors, asphalt tile, etc., using the same general procedures as given above for over concrete, except subsurface will not require priming before applying mastic.

Subsurfaces must be sound, smooth, clean, and dry—asphalt tile must be firmly bonded and in good condition—wood subflooring or present wood flooring may require re-nailing. Where installation is over square edged board subflooring, one layer of 15-lb. (6.75-kg) asphalt saturated felt should be laid to keep mastic from seeping through joints.

Prices of Wood Parquet Floors. Owing to the variation in freight rates, dealer's markup, etc., it is practically impossible to give prices that will apply to all parts of the country. Dealer's markup will probably vary from 25% on large jobs to 40% on smaller installations, plus freight from factory to destination.

Material Cost of 100 S.F. (9 Sq.M) of 9"x9"x5/16" (225 x 225 x 8 mm)
Oak Parquet Floors Laid in Mastic

	Rate	Total	Rate	Rate (metric)	Total
0.33 gal. (1.25 L) asphalt primer	$....	$....	$2.65	$0.69	$0.87
100 s.f. (9 sq.m) oak parquet		....	1.80	20.00	180.00
2.5 gal. (9.5 L) hot mastic		....	12.75	3.35	31.88
Cost per 100 s.f.				$....	$212.75
per s.f.				....	2.13
per sq.m				....	23.64

Cost of 100 S.F. (9 Sq.M) of 9"x9"x5/16" (225 x 225 x 8 mm)
Oak Parquet Floors Laid in Mastic

	Hours	Rate	Total	Rate	Total
Floorlayer	3.0	$....	$....	$26.44	$79.32
Helper	3.0	....	....	21.04	63.12
Cost per 100 s.f.			$....		$142.44
per s.f.				....	1.42
per sq.m				....	15.28

If wood parquet is laid over wood or asphalt tile surfaces, omit asphalt primer.

If parquet floors are laid in large spaces, such as auditoriums, etc., deduct 1/4 to 1/2 hr. floorlayer time per 100 s.f. (9 sq.m).

Add for membrane waterproofing, if required.

If unfinished parquet is used, change price of blocks and add for sanding and finish.

Above costs do not include any allowance for preparing subfloor, if necessary.

Laminated Block Flooring. Laminated oak blocks are fabricated from 3-ply all oak plywood. Furnished in 9"x9"x1/2" size, factory finished.

A 27 s.f. (2.4 sq.m) carton—48 pcs.—will cost about $35.00 in a light or dark finish.

Laminated block flooring is for mastic installation only and may be laid over any sound, smooth, clean and dry subfloor, using hot or cold mastic. General installation procedure is the same as for wood parquet floors given on previous pages, except that no allowance for expansion is necessary.

Labor costs for laying laminated block flooring should be about the same as for 9"x9"x5/16" (225 x 225 x 8 mm) wood parquet given above.

Wood Block Industrial Flooring. Creosoted wood blocks 2" (50 mm) to 3" (75 mm) thick will run from $1.70 to $3.80 per s.f. ($18.89-42.22 per sq.m). One carpenter will install around 200 s.f. (18 sq.m) per day.

09650 RESILIENT FLOORING

Resilient flooring includes asphalt tile, rubber, cork, vinyl asbestos, pure vinyl tile, and vinyl sheet flooring. Each type has its own qualities and should be selected on these considerations rather than cost alone. If material is to be installed on or below grade, cork tile and certain sheet vinyls are not suitable.

If grease is a problem, rubber, cork, and asphalt tile perform poorly, while if resilience and quietness are important, cork and rubber tile are excellent, but sheet vinyl is poor.

A quick guide for approximate installed costs per sq. ft. (sq. meter) is as follows:

Type	Thick., inches	Thick., mm	Price per s.f.	Price per sq.m
Asphalt tile	3/32"-1/8"	2-3	$0.70-0.95	$7.77-10.55
Vinyl asbestos	1/16"	2	0.85-0.95	9.44-10.55
	3/32", 1/8"	2, 3	0.90-1.00	10.00-11.11
Cork tile-Prefinished	1/8", 3/16", 5/16"	3, 5, 8	1.30-2.00	14.44-22.22
Vinyl tile	0.050"	1	1.40-2.00	15.55-22.22
Vinyl sheet	0.070-0.075"	1.75-1.88	1.70-2.65	18.89-29.44
Rubber tile	1/8", 3/16"	3-5	1.60-1.95	17.78-21.67

Asphalt Tile Floors

Asphalt tile is one of several types of resilient flooring that can be used satisfactorily on concrete floors that are in direct contact with the ground, at or below grade. It is suitable for basement rooms, because it is not affected by the moisture and alkali always present in concrete in contact with the ground. Asphalt tile is also used in residences, stores, offices, and institutions where a low cost floor is desired.

They are furnished in a variety of colors, and tile sizes are 9"x9"x1/8" (225 x 225 x 3 mm) thick.

When applied over approved wood subfloors, the wood floor shall be finished smooth and level and covered with a layer of lining felt bonded to the wood with mastic and then rolled thoroughly with a linoleum roller. The asphalt tile is then laid over the lining felt, using an approved asphalt emulsion for bonding the tile to the felt.

When applied over concrete floors above, at, or below grade, the concrete floor must be smooth, dry, and even. The tile is laid directly on the concrete, using approved asphalt emulsion. At grade or subgrade, if dampness due to humidity is extreme, the concrete shall be given one coat of asphalt cut-back primer after which the tile is laid in asphalt (solvent type or cutback) cement.

How to Estimate Tile Floors

To Determine and Price the Quantity of Material Required. Compute the actual square footage of surface to be covered, and add a percentage for waste. The percentage will depend on the size of the area and regularity of outline. A fair percentage of waste to add to the actual area is shown in the accompanying table.

After the gross area has been determined, multiply it by the square foot price of tile. The result will be the price of the material.

The square foot (square meter) price of the material will depend on the colors and quantity.

Table of Approximate Percentages of Allowable Waste for Various Areas

Up to 50 s.f. (4.5 sq.m)	14%
50-100 s.f. (4.5-9.0 sq.m)	10%
100-200 s.f. (9-18 sq.m)	7-9%
200-300 s.f. (18-27 sq.m)	6-7%
500-1,000 s.f. (45-90 sq.m)	4-5%
1,000-5,000 s.f. (90-450 sq.m)	4%
5,000-10,000 s.f. (450-900 sq.m)	3%
10,000 (900 sq.m) and up	1-1/2-3%

Quantity of Primer and Asphalt Cement Required. When priming concrete floors, one gallon of primer should cover 200 to 300 s.f. (one liter covers 4.75-7.13 sq.m) of surface, depending on the porosity of the surface. Price $5.00 per gal. ($1.32 per liter).

One gallon of asphalt cement should lay 175 to 225 s.f. (one liter lays 4.16-5.35 sq.m) of asphalt tile over concrete floors. Costs about $12.00 per gal. ($3.17 per liter).

One gallon of asphalt emulsion for installing asphalt tile over lining felt on wood subfloors and direct to above grade concrete floors will cover approximately 140 to 175 s.f. (one liter covers 3.33-4.16 sq.m). Price $5.00 per gal. ($1.32 per liter).

Labor Laying Asphalt Tile. The cost of laying asphalt tile varies with the size and shape of the room or space where the tile are to be laid, size of tile, kind of subfloor, etc., as a tile setter will lay more sq. ft. (square meters) of tile in a large room than one cut up with pilasters, offsets, etc., also a worker lays more sq. ft. (square meters) of large size tile than of the smaller sizes.

When placing lining felt over wood subfloors, an experienced worker should spread paste, lay, and roll 250 s.f. (22.5 sq.m) of felt per hr., depending on the room sizes.

After the lining felt has been laid (over wood floors), an experienced tile setter should lay the following quantities of tile per 8-hr. day:

Size of Tile	SF per 8-Hr. Day	Sq.M per 8-Hr. Day	Hrs. per 100 S.F. (9 sq.m)
9"x9" (225 x 225 mm)	400-425	36-38	2.0

The above quantities are based on fairly large rooms. If laid in small spaces, increase the labor costs accordingly.

Average Sq. Ft. (Sq.M) Prices on Asphalt Floor Tile

Group	1/8	3 mm
A colors	$0.50	$5.55
B colors	0.60	6.66
C colors	0.75	8.33
D colors	0.80	8.88

Diagonal half tile of 9"x9" (225 x 225 mm), add $0.03 per s.f. ($0.33 per sq.m). Feature strips in the standard colors are priced at their group price plus $0.10 per s.f. ($1.11 per sq.m) additional. Standard, die-cut design asphalt tile inserts, size 18"x18" (450 x 450 mm), cost about $3.00 each.

"A" colors—Dark red and black solid patterns.
"B" colors—Dark marbleized patterns.
"C" colors—Medium light marbleized patterns.
"D" colors—Light marbleized patterns.

Greaseproof Asphalt Tile

Greaseproof colors are also acid resistant. Standard size 9"x9" (225 x 225 mm).

	Price per s.f.	Price per sq.m
1/8" (3 mm) Thick Greaseproof Asphalt Tile	$0.80	$8.88

Vinyl Plastic Tile

Vinyl plastic tile is a more expensive resilient flooring material. It is very stable, will not shrink or expand and is highly resistant to oils, fats, greases, acids, alkalis, detergents, soaps, and most solvents.

Furnished in 9"x9" (225 x 225 mm), 12"x12" (300 x 300 mm), and 18"x36" (450 x 900 mm) sizes and in 3/32" (2 mm) and 1/8" (3 mm) thicknesses.

Subfloor requirements are the same as for other resilient flooring. May also be laid over concrete floors, or on grade or below grade, using special adhesives.

Vinyl plastic tile is laid in the same manner as other resilient tile and labor costs are about the same.

Material costs vary according to pattern, ranging from $0.75 per s.f. ($8.33 per sq.m) for marbleized patterns to $3.00 per s.f. ($33.33 per sq.m) for translucent patterns, for 1/8" (3 mm) thick tiles. The 3/32" (2 mm) thick tiles cost about 15% to 20% less.

Rubber Tile Floors and Base

Rubber tile is furnished in 9"x9" (225 x 225 mm), 12"x12" (300 x 300 mm), and 18"x36" (450 x 900 mm) tile, 1/8" (3 mm) thick and in a variety of colors.

Rubber floor tile may be laid over wood or above grade concrete floors. It may also be laid over clean, dry, concrete floors on ground or below grade providing a special adhesive, recommended by the manufacturer of the tile, is used.

When laying rubber tile over an old wood floor, all loose, defective, or badly worn boards shall be replaced and face nailed. Any unevenness in the boards shall be planed or sanded smooth and even.

If the wood floors cannot be placed in proper condition by these methods, install Masonite or 1/4" (6 mm) plywood over the original wood floor.

Over the wood base there shall be installed semi-saturated lining felt which shall be securely affixed to the base with mastic. Depending upon the condition of the subfloor, the spreading capacity should be about 140 s.f. per gal. (3.33 sq.m per liter). After the felt is laid, it should be thoroughly rolled with a three-section 100-lb. (45-kg) linoleum roller.

The tile should be installed over the felt, using mastic.

The finished floor should be thoroughly rolled with a three-section 100-lb. (45 kg) linoleum roller and any high edges smoothed in place with a flat headed hammer.

Laying Rubber Tile. The cost of laying rubber floor tile varies with the size and shape of the room or space where the tile are to be laid, size of tile, kind of subfloor, etc., as it is easier to lay tile in a large square room than one cut up with offsets, pilasters, bays, etc., and a worker will lay more sq. ft. (sq.m) of large size tile than of the smaller sizes.

When laying lining felt over wood subfloors, an experienced worker should spread paste, lay, and roll 250 s.f. (22.5 sq.m) of felt per hr., depending on the room sizes.

After the felt has been laid (over wood floors), an experienced tile setter should lay and finish the following sq. ft. (sq.m) of tile of various sizes per 8-hr. day:

Size of Tile	SF per 8-Hr. Day	Sq.M per 8-Hr. Day	Hrs. per 100 S.F. (9 sq.m)
9"x9" (225x225mm)	400-425	36-38	2.0
12"x12" (300x300mm), 18"x36" (450x900mm)	600-675	54-61	1.3

The above quantities are based on fair sized rooms. If laid in small spaces, increase labor costs accordingly.

Material Prices of Rubber Floor Tile, All Standard Colors

	Price per s.f.	Price per sq.m
1/8" (3 mm) Thick	$1.05-1.80	$11.67-20.00

Rubber tile has lost some of its market to the vinyls but is still used quite often as a stair tread covering.

Stair Treads

Size	Marbleized per s.f.	Plain Marbleized per sq.m	Plain Black per s.f.	Black per sq.m
12-1/2"x36"x1/4" (313x900x6mm)	$10.50	$116.67	$10.00	$111.11
12-1/2"x42"x1/4" (313x1050x6mm)	12.25	136.11	10.50	116.67
12-1/2"x48"x1/4" (313x1200x6mm)	14.25	158.33	13.00	144.44
12-1/2"x54"x1/4" (313x1350x6mm)	15.25	169.44	14.50	161.11
12-1/2"x60"x1/4" (313x1550x6mm)	16.50	183.33	16.00	177.77
12-1/2"x72"x1/4" (313x1800x6mm)	21.00	233.33	20.00	222.22

Landing tile will run about $1.10 per s.f. ($12.22 per sq.m) and 3/32" (2 mm) riser material about $0.65 per l.f. ($2.17 per m).

Prices on Rubber Composition Top-Set Cove Base

Description	Price per l.f.	Price per m
4" (100 mm) base	$0.50	$1.67
6" (150 mm) base	0.55	1.83

Cork Tile

Cork tile are made from high-grade pure cork shavings compressed and baked in molds. It is used for floors and wainscoting in offices, libraries, hospitals, schools, and other places where footsteps must be muffled. They are furnished in 6"x12" (150 x 300 mm), 9"x9" (225 x 225 mm), and 12"x12" (300 x 300 mm) size 1/8" (3 mm), 3/16" (5 mm), and 5/16" (8 mm) thick. Cork tile are supplied beveled or unbeveled. With unbeveled cork tile, sanding the finished floor is usually necessary to offset unevenness of the subfloor. With beveled cork tile, the beveling conceals such irregularities, with the result that sanding is unnecessary and the smooth surface given at the tile factory is retained.

Cork tile should never be placed over a single wood subfloor but should be first covered with Masonite or 3/8" (9 mm) plywood. Never install on grade or subgrade floors.

The wood or above grade concrete subfloor must be smooth and even before cork tile can be placed. Lining felt is used over wood subfloors but this is not necessary when applied over smooth, level concrete floors.

After the floor has been laid and sanded (if unbeveled tile are used), the floor should be given a coat of cork tile filler, followed by two coats of cork tile finish (glossy or flat) and subsequent waxings with an approved water emulsion wax, or the floor may be finished by merely applying 3 applications of water emulsion wax and vigorously buffed by machine.

Approximate Sq. Ft. (Sq.M) Prices of Cork Tile

Kind of Tile	1/8" per s.f.	3mm per sq.m	5/16" per s.f.	8 mm per sq.m
Color according to manufacturer's specs., regular sizes	$1.00	$11.11	$1.50	$16.67

Border feature strips: In any width under 3" (75 mm) and any length between 18" (450 mm) and 36" (900 mm) at $0.10 per s.f. ($1.11 per sq.m) additional.

Factory finished tile: Add $0.15 per s.f. ($1.67 per sq.m) for beveled cork tile finished at factory with sealer, undercoat, and one application of wax. After installation, apply one application of wax and buff.

Material Cost of 100 S.F. (9 Sq.M) 9"x9"x1/8" (225 x 225 x 3 mm)
Cork Tile Floors Over Wood Flooring

	Rate	Total	Rate	Rate (metric)	Total
0.75-gal. (2.84 L) adhesive ....$....		$....	$8.50	$2.24	$6.38
12 sq. yds. (10 sq.m) lining felt		....	0.32	0.38	3.84
105 s.f. (9.45 sq.m) cork tile		....	0.95	10.55	99.75
0.75 gal. (2.84 L) adhesive		....	8.50	2.24	6.38
Finishing Floor					
0.33 gal. (1.25 L) cork tile sealer			3.18	0.84	1.05
0.25 gal. (0.95 L) cork tile undercoat			3.18	0.84	0.80
0.13 gal. (0.49 L) water emul. wax			3.18	0.84	0.41
Cost 100 s.f.		$....			$118.61
per s.f...					1.19
per sq.m					13.18

Labor Cost of 100 S.F. (9 Sq.M) 9"x9"x1/8" (225 x 225 x 3 mm)
Cork Tile Floors Over Wood Flooring

	Hours	Rate	Total	Rate	Total
Mechanic....................................3.5		$....	$....	$26.44	$92.54
Sanding 100 sq. ft 1.0		....	....	26.44	26.44
Finishing Floor					
Clean and wax floor 2.0		....	....	26.44	52.88
Cost 100 s.f.			$....		$171.86
per s.f..					1.72
per sq.m					18.51

Sheet Vinyl Flooring

Sheet vinyl can be installed on concrete floors, below and above grade, and on wooden floors that have a smooth surface. If the wood floor is rough, then install a layer of hardboard or 1/4" (6 mm) plywood to provide a good surface for the sheet vinyl.

This material is available in 6' (1.8 m) and 12' (3.6 m) wide rolls up to 120' (36 m) in length. Thicknesses varies with the product design, type of backing material, and the manufacturer. However, the most recognized gauges will fall in the 0.065" (1.6 mm), .080" (2 mm) and 0.090" (2.25 mm) range for non-cushion backed sheet vinyl.

Most sheet vinyl flooring materials are designed to require a minimum of maintenance, which means that the flooring cleans easily, is resistant to normal abuse, and does not require waxing to maintain the luster appearance.

Installation is by full adhesive coating on the 6' (1.8 m) wide material and adhesive coating on seams only for the 12' (3.6 m) wide rolls.

Material prices will vary from $1.25 to $2.50 per s.f. ($13.89-27.78 per sq.m) depending on the gauge and backing, but the estimator should be aware that this material is sold by the square yard (square meter) and that proper floor layout should be determined to reduce the waste factor.

Labor for installation will vary depending on the size and shape of the area to be covered and the number of cuts to be made. One worker lays about 600 s.f. (54 sq.m) in large areas per 8-hr. day, and about 350 s.f. (31.5 sq.m) in smaller areas per 8-hr. day.

09680 CARPETING

With the development of man-made fibers special padding and backings, carpeting has become competitive with resilient tile in commercial construction. While almost always subcontracted to a firm specializing in the field, the general contractor should be acquainted with the basic choices available today.

Pure wool has set the high standards most people expect from carpet and no man-made fiber has yet equaled it in overall performance and appearance, although in some respects, such as abrasion resistance, texture retention, and mildew resistance, man-made fibers excel. But wool carpeting is in the high cost range, running $16.00 to $27.00 per sq. yd. ($19.14-32.29 per sq.m) in the 36 to 42 oz. (1021-1191 g) commercial grades, without padding.

Nylon is the most often used of the man-made fibers. It has outstanding wear ability and resiliency, is mildew resistant and mothproof. But there are problems with static electricity unless it is specially treated or has grounding mesh woven into the backing. In the lighter weights, it is extremely economical, about $7.00 per sq. yd. ($8.37 per sq.m) for 15 oz. (425 g); $8.50 ($10.17) for 22 oz. (624 g), both without padding.

The olefin (polypropylene) fibers are comparable in price to nylon, but they are found mostly in the indoor-outdoor type of carpet as they offer the lowest moisture absorbency and are exceptionally stain-resistant. Their appearance is dull, resiliency low; they have the lowest melting point, so are subject to cigarette burns. Their range is $5.50 to $9.75 per sq. yd. ($6.58-11.66 per sq.m).

Acrylics surpass the other man-made fibers in looks and feel, most closely resembling wool, but they do not wear as well as nylon, and will pill some. They are $8.00 to $13.00 per sq. yd. ($9.56-15.55 per sq.m) for 36 to 42 oz. (1020-1191 g) weights, without padding.

Polyesters have poor resiliency so are seldom used alone, except for shag carpets, seldom encountered in contract work. They fall in a medium cost range. Padding may be laid separately or come with the carpet. It will add $1.10 to $2.75 per sq. yd. ($1.32-3.29 per sq.m) to the material cost.

Carpets can be taped down or laid in mastic. Installation cost will be about $3.00 per sq. yd. ($3.59 per sq.m) on open work. If not integral with carpet, add another $1.50 per sq. yd. ($1.79 per sq.m) for laying padding.

Estimating Quantities. Carpet is estimated by the square yard (square meter). There are 9 sq. ft. per sq. yd. Since most carpeting is manufactured in 12' (3.6 m) wide material, the estimator must be careful to determine how the carpet will lay on the floor to minimize the waste factor. Bear in mind that carpet has a definite "lay" to the nap and the "lay" must be in the same direction when joining pieces together, or else the carpet pieces will appear to have a different color.

09900 PAINTING

Estimates and bids on painting and interior decorating vary more than for almost any other construction trade. There are several kinds of painters—good, fair, poor, and just plain daubers. Expect wide variation in figures. When preparing painting estimates, consider the quality of materials and workmanship required.

Methods of Measuring and Listing Painting Quantities

When estimating quantities of painting or interior finishing, the actual surface to be painted should be measured as accurately as possible from the plans or from measurements taken at the building. From these measurements it should be possible to estimate the quantity of materials required, but the labor quantities present a much more difficult problem, owing to the different classes of work and the difficulties encountered in applying the paint. For instance, the covering capacities of paint will be the same for a plain surface or a cornice, but the labor cost painting the cornice will be considerably more owing to the height of the cornice above the ground and the amount of "cutting" or "trimming" necessary. Care must be used when pricing any piece of work, as conditions on each job are different.

Clapboard or Drop Siding Walls. Obtain actual area of all walls and gables. Add 10% to actual surface measurement to allow for painting under edge of boards. Do not deduct for openings less than 10'-0"x10'-0" (3 x 3 m).

Shingle Siding. Obtain actual area of all walls and gables and multiply by 1-1/2. Do not deduct for openings less than 10'-0"x10'-0" (3 x 3 m).

Brick, Wood, Stucco, Cement, and Stone Walls. Obtain actual area of all walls and gables. Do not deduct for openings less than 10'-0"x10'-0" (3 x 3 m).

Eaves. Plain eaves painted same colors as side walls, obtain area and multiply by 1-1/2. If eaves are painted a different color than side walls, obtain area and multiply by 2.

Eaves with rafters running through, obtain area and multiply by 3.

Eaves over brick, stucco or stone walls, obtain area and multiply by 3. For eaves over 30'-0" (9 m) above ground, add 1/2 for each additional 10'-0" (3 m) in height.

Cornices, Exterior. Plain cornices, obtain area and multiply by 2. Fancy cornices or cornices containing dentils, etc., obtain area and multiply by 3.

Downspouts and Gutters. Plain downspouts and gutters measure the area and multiply by 2. For fancy downspouts and gutters, obtain area and multiply by 3.

Blinds and Shutters. Plain blinds or shutters, measure one side and multiply by 2. Slatted blinds or shutters, measure one side and multiply by 4.

Columns and Pilasters. Plain columns or pilasters, obtain area in sq. ft. (sq.m); if fluted, obtain area and multiply by 1-1/2; if paneled, obtain area and multiply by 2.

Lattice Work. Measure one side and multiply by 2 if painted one side only; if painted two sides, obtain area and multiply by 4.

Porch Rail, Balustrade, Balusters. If solid balustrade, add one foot to height to cover top and bottom rail. Multiply the length by the height. If painted two sides, multiply by 2.

If balustrade consists of individual balusters and handrail, multiply length by the height, then multiply the result by 4.

Handrail only. If under 1'-0" (0.3 m) girth, figure as 1'-0" (0.3 m) by length and multiply by 2.

Moldings. Cut in on both sides, figure 1 s.f. per l.f. (0.09 sq.m per m) if under 12" (300 mm) girth. If over 12" (300 mm) girth, take actual measurement.

Doors and Frames, Exterior. Inasmuch as it costs almost as much to paint a small door as a large one, do not figure any door less than 3'-0"x7'-0" (0.9 x 2.1 m). Allowing for frame, add 2'-0" (0.6 m) to the width and 1'-0" (0.3 m) to the height. For instance, a 3'-0"x7'-0" (0.6 x 2.1 m) door would be figured as 5'-0"x8'-0" (1.5 x 2.4 m), or 40 s.f. (3.6 sq.m).

If a sash door, containing small lights of glass, add 2 s.f. (0.18 sq.m) for each additional light. A 4-lt. door would contain 8 s.f. (0.72 sq.m) additional; a 12-lt. door, 24 s.f. (2.16 sq.m) additional, etc.

If painted on both sides, obtain s.f. (sq.m) area of one side and multiply by 2.

Door frames only, where no door is hung, allow area of opening to take care of both sides.

Windows, Exterior. Inasmuch as it costs almost as much to paint a small window as a large one, do not figure any window less than 3'-0"x6'0" (0.9 x 1.8 m). Add 2'-0" (0.6 m) to both the width and height of the opening, to take care of the sides and head of the frame and the outside casing or brick mold, and multiply to obtain the area. For instance, a window opening 3'-0"x6'-0" (0.9 x 1.8 m), add 2'-0" (0.6 m) to both the width and height, making 5'-0"x8'-0" (1.5 x 2.4 m), containing 40 s.f. (3.6 sq.m) of surface.

If sash contain more than one light each, such as casement sash, etc., add 2 s.f. (0.18 sq.m) for each additional light. A 6-lt. window would contain 12 s.f. (1.08 sq.m) additional; a 12-lt. window, 24 s.f. (2.16 sq.m) additional, etc.

Roofs. For flat roofs or nearly flat, measure actual area. For roofs having a quarter pitch, measure actual area and add 25%; roofs having a one-third pitch, measure actual area and add 33-1/3%; roofs having a half pitch, measure actual area and add 50 percent.

Fences. Plain fences, measure one side and multiply by 2; picket fences, measure one side and multiply by 4.

Picture Mold and Chair Rail. On picture mold, chair rail, etc., less than 6" (150 mm) wide, obtain the number of lin. ft. (meters) to be painted or varnished and figure 3/4 s.f. per l.f. (0.22 sq.m per m).

Wood and Metal Base. Wood or metal base 6" (150 mm) to 1'-0" (0.30 m) high should be figured as 1'-0" (0.3 m) high; under 6" (150 mm), figure at 0.5 s.f. per l.f. (0.15 sq.m per m).

Wood Panel Strips, Cornices, Etc. When estimating quantities of wood panel strips, wainscot rail, and ceiling cornices, measure the girth of the member and if less than 1'-0" (0.3 m) wide, figure as 1'-0" (0.3 m).

If over 1'-0" (0.3 m) wide, multiply the actual girth by the length: 1'-3" x 200'-0" = 250 s.f. (0.38 x 60 = 22.8 sq.m).

Refer to rule for exterior cornices.

Interior Doors, Jambs, and Casings. When estimating quantities for interior doors, jambs, and casings, add 2'-0" (0.6 m) to the width and 1'-0" (0.3 m) to the height of the opening. This allows for painting or varnishing the edges of the door, the door jambs which are usually 6" (150 mm) wide, and the casings on each side of the door which average from 4" (100 mm) to 6" (150 mm) wide. Example: on a 3'-0"x7'-0" (0.9 x 2.1 m) door opening, add 2'-0" (0.6 m) to the width and 1'-0" (0.3 m) to the height, which gives an opening 5'-0"x8'-0" (1.5 x 2.4 m), containing 40 s.f. (3.6 sq.m) on each side. Some painters figure all single doors at 40 s.f. (3.6 sq.m) per side, or 80 s.f. (7.2 sq.m) for both sides, while others figure them at 50 s.f. (4.5 sq.m) per side or 100 s.f. (9 sq.m) for both sides. Do not deduct for glass in doors.

If a sash door, containing small lights of glass, add 2 s.f. (0.18 sq.m) for each additional light. A 4-lt. door would contain 8 s.f. (0.72 sq.m) additional; a 12-lt. door 24 s.f. (2.16 sq.m) additional, etc.

Interior Windows, Jamb Linings, Sash, and Casings. When estimating painting quantities for windows and window trim, add 2'-0" (0.6 m) to the sides and length to allow for jamb linings, casing at the top and sides, and window stool and apron at the bottom. Example: If the window opening is 3'-0"x6'-0" (0.9 x 1.8 m), adding 2'-0" (0.6 m) to both the width and length gives a window 5'-0"x8'-0" (1.5 x 2.4 m), containing 40 s.f. (3.6 sq.m) of surface.

If sash contain more than one light each, such as casement sash, etc., add 2 s.f. (0.18 sq.m) for each additional light. A 6-lt. window would contain 12 s.f. (1.08 sq.m) additional; a 12-lt. window, 24 s.f. (2.16 sq.m) additional, etc.

Stairs. When estimating quantities of paint or varnish for wood stairs, add 2'-0" (0.6 m) to the length of treads and risers to allow for stair strings on each side of the stairs. A stair tread is 10" (250 mm) to 12" (300 mm) wide, riser is about 7" (175 mm) high, and the average cove and underside of a stair tread 2" (50 mm) in girth, so each tread should be figured 2'-0" (0.6 m) wide. Multiply the width by the length to obtain the number of sq. ft. (sq.m) of paint or varnish required for each step. Multiply the area of each step by the number of treads in the stair, and the result will be the number of sq. ft. (sq.m) of finish. Example: A flight of stairs containing 20 treads, 4'-0" (1.2 m) long. Adding 2'-0" (0.6 m) to the length of the treads to allow for the stringers, gives 6 l.f. (1.8 m) in each tread, which, multiplied by the girth of 2'-0" (0.6 m) gives 12 s.f. (1.08 sq.m) of surface in each step. Since there are 20 treads in the stairs, 12 x 20 = 240 s.f. (1.08 x 20 = 21.6 sq.m).

For painting the soffits of stairs, use same rules of measurement as given above.

Balustrades and Handrails with Balusters. When estimating balustrades around well holes or stair handrail and balusters, measure the distance from the top of the treads to the top of the handrail and add 6" (150 mm) for painting or varnishing the handrail. Multiply the height of the balustrade by the length and the result will be the number of sq. ft. (sq.m) of balustrade. An easy method for computing the length of the stairs is to allow 1 l.f. (0.3 m) in length for each tread in the stairs. Example: If the handrail is 2'-6" (0.75 m) above the treads and the stairs contain 20 treads, add 6" (150 mm) to the height of 2'-6" (0.75 m) to take care of the extra work on the handrail proper, making a total height of 3'-0" (0.9 m). 3 x 20 = 60 s.f. (0.9 x 20 = 18 sq.m) of surface for one side of the balustrade or 120 s.f. (36 sq.m) for two sides. The balustrade around the stair well hole is easily estimated by multiplying the height by the length.

On fancy balustrades having square or turned spindles, which require considerable additional labor, use actual measurements as given above and multiply by 4 to allow for extra labor.

Wood Ceilings. When estimating quantities for wood ceilings, multiply the length by the width. Do not deduct for openings less than 10'-0"x10'-0" (3 x 3 m).

Wainscoting. Plain wainscoting, obtain actual area. Paneled wainscoting, obtain actual area and multiply by 2.

Floors. To compute the quantities of floors to be finished, multiply the length of each room by the width, and the result will be the number of sq. ft. (sq.m) of floor to be finished.

Plastered Walls and Ceilings. To obtain the area of any ceiling, multiply the length by the breadth and the result will be the number of sq. ft. (sq.m) of ceiling. When estimating walls, measure the entire distance around the room and multiply by the room height. The result will be the number of sq. ft. (sq.m) of wall to be decorated. For instance, a 12'-0"x15'-0" (3.6 x 4.5 m) room has two sides 12'-0" (3.6 m) long and two sides 15'-0" (4.5 m) long, giving a total of 54 l.f. (16.2 m) If the ceilings are 9'-0" (2.7 m) high, 54 x 9 = 486 s.f. (16.2 x 2.7 = 43.74 sq.m) area of the walls. Do not deduct for door and window openings.

Cases, Cupboards, Bookcases, Etc. When estimating surfaces of cupboards, wardrobes, bookcases, closets, etc., to be painted or varnished, obtain the area of the front and multiply by 3 if the cases do not have doors.

If the cases have doors, obtain the area of the front and multiply by 5.

This takes care of painting or varnishing doors inside and out, shelves 2 sides, cabinet ends and backs.

Radiators. For each front foot, multiply the face area by 7.

Sanding and Puttying Interior Trim. Sanding and puttying on high grade interior trim should be figured as one coat of paint or varnish; on medium grade trim, figure at 50% of one coat of paint or varnish. For industrial work, figure about 25% of the cost of one coat of paint or varnish.

Emulsified Asphalt

Emulsified asphalt may be used as a floor surfacer, a bonding coat, or a quicksetting, hard-drying and durable patching material. It may be used with satisfactory results over either old or new floors. The following procedure is generally used. First, clean the old surface free from dirt, grease, and foreign matter; second, cut out the worn spots or cracks with perpendicular sides to avoid the formation of feather edges; third, a bonding coat of emulsified asphalt is applied, consisting of 1 part clean water to 4 parts of emulsified asphalt, dry overnight; fourth, the mixture of portland cement, sand, gravel and emulsified asphalt is applied and rolled.

When applied direct to a concrete surface, 1 gal. covers about 75 s.f. one coat or 50 s.f. two coats (1 liter covers 1.78 sq.m one coat, 1.19 sq.m two coats).

When used instead of water as a material for binding cement, sand, and gravel, the quantity will vary with the thickness applied and will require 4 to 5 gals. (15 to 19 liters) per sack of cement.

Painting Concrete and Masonry Surfaces

Hydrocide Colorcoat (Sonneborn). An alkyd-epoxy-silicone-oil base formula, decorative coating especially designed for exterior, highly porous surfaces. Properly applied, it effectively seals and protects such surfaces with a color-fast, decorative coating, which will not run or sag, in one application.

It will fill pores and bridge hairline cracks on concrete or masonry surfaces which are completely cured.
One gallon of material covers 60 to 70 s.f. depending on porosity of the surface. (One liter covers 1.43 to 1.66 sq.m). Price, $9.50 per gal. ($2.50 per liter).

Hydrocide Super Colorcoat (Sonneborn). Super Colorcoat is for use on cured, green, or slightly damp surfaces. It is water dispersed acrylic emulsion formulation. It is also mildew and fungus resistant and non-flammable.
Super Colorcoat will cover approximately 80 to 100 s.f. per gal. (1.9-2.38 sq.m per liter) Price, $10.50 per gal. ($2.77 per liter).

SikaGard Hi-bild (Sika). Tough 100% solids epoxy resin coating for concrete, block, steel, wood, and other structural materials. Can be applied to dry, damp, or wet surfaces. Coverage with short nap roller will be approximately 175 to 200 s.f. per gal. (4.16-4.75 sq.m per liter); when spraying, 250 to 400 s.f. per gal. (5.9-9.5 sq.m per liter). Coverage varies with temperature, substrate, and environment. Available in colors. Price, $22.50 per gal. ($5.94 per liter).

Covering Capacity of Paints

It is difficult to say how many sq. ft. of surface per one gallon (or square meters of surface per one liter) paint will cover, because there are several items that influence the covering capacity. By brushing the paint out very thin, it will cover more surface than when applied thick. Dark paint will hide the surface better than light paint, and for that reason it can be brushed out thinner than the lighter colors. Also, a rough surface will require considerable more paint than a smooth surface.

Soft and porous wood will absorb more oil and require more paint than close grained lumber.

Another thing that must be considered are the ingredients entering into the paint. Different brands of paint vary in covering capacity or hiding power.

Covering Capacity of Oil Base Paint on New Exterior Wood. In figuring the number of square feet a gallon of oil base paint will cover, a great deal depends upon the surface to be painted; that is, the kind of wood, the degree of roughness, etc. Some woods are more porous than others and absorb more paint. Much depends on the way the paint is brushed out as some painters brush the paint out more and thus cover more surface.

The priming coat, properly applied, should cover 450 to 500 s.f. per gal. (10.7-11.8 sq.m per liter). The second coat should cover 500 to 550 s.f. per gal. (11.8-13 sq.m per liter), and the third coat should cover 575 to 625 s.f. per gal. (13.6-14.8 sq.m per liter).

When estimating exterior trim painting, with measurements taken in accordance with standard methods, the priming coat should cover 750 to 850 s.f. per gal. (17.8-20.2 sq.m per liter); the second coat should cover 800 to 900 s.f. per gal. (19-21.4 sq.m per liter), and the third coat should cover 900 to 1,000 s.f. per gal. (21.4-23.8 sq.m per liter).

Bear in mind the above coverages are based on surfaced lumber, as the covering capacity will be greatly reduced when applied to rough boards.

Painting Old Exterior Wood Surfaces. When estimating repaint work on old exterior wood surfaces, if the existing surface is sound and in good condition, two coats should be sufficient to give a good job and paint coverages should be about the same as for the second and third coats on similar new work.

If the old surface shows cracking, blistering, scaling, and peeling, the old paint should be removed by using either a paste or liquid paint remover of the slow drying type, or a blow torch and scraper.

If the old paint is removed completely, three coats of paint should be applied the same as recommended for new work.

Painting Wood Shingle Siding. Wood shingle siding should receive three coats of paint on new work. Paint coverages should be: first coat, 300 to 325 s.f. per gal. (7.1-7.7 sq.m per liter); second coat, 400 to 450 s.f. per gal. (9.5-10.7 sq.m per liter); third coat, 500 to 550 s.f. per gal. (11.9-13 sq.m per liter).

Wood shingles which have been previously painted with an oil paint and are in suitable condition for repainting, should receive two coats of paint. The first coat should cover 400 to 450 s.f. per gal. (9.5-10.7 sq.m per liter), and the second coat 500 to 550 s.f. per gal. (11.9-13 sq.m per liter).

Painting Exterior Wood Floors and Steps. When painting exterior wood floors and steps with oil base paint, material coverages should be: first coat, 325 to 375 s.f. (7.7-8.9 sq.m per liter); second coat, 475 to 525 s.f. (11.3-12.5 sq.m per liter); third coat, 525 to 575 s.f. (12.5-13.7 sq.m per liter).

Painting Interior Wood. New interior wood should receive three coats of paint, a priming coat, a second coat and a final or finish coat.

Paint coverage for interior wood surfaces will vary with the type of work, such as doors and windows, running trim, paneling, etc., in accordance with the standard method of measuring the quantities.

For doors and windows, coverage should be: first coat, 475 to 500 s.f. per gal. (11.3-11.9 sq.m per liter); second coat, 500 to 550 s.f. per gal. (11.9-13 sq.m per liter); third coat, 500 to 550 s.f. per gal. (11.9-13 sq.m per liter).

For running trim up to 6" (150 mm) wide, coverage should be: first coat, 800 to 900 l.f. per gal. (240-270 m per liter); second and third coats, 1,100 to 1,200 l.f. per gal. (330-360 m per liter).

For ordinary flat wood paneling, coverage should be: first coat, 450 to 500 s.f. per gal. (10.7-11.9 sq.m per liter); second and third coats, 500 to 550 s.f. per gal. (11.9-13.0 sq.m per liter).

When painting interior woodwork that has been previously painted and is in good condition, one gallon of paint should cover approximately the same as given for the second and third coats on new wood.

Painting Brick, Stone, Stucco, and Concrete. When painting masonry surfaces that are reasonably smooth, oil base paint should cover 170 to 200 s.f. per gal. (4-4.75 sq.m per liter) for the priming coat; 350 to 400 s.f. per gal. (8.3-9.5 sq.m per liter) for the second coat and 375 to 425 s.f. per gal. (8.9-10.1 sq.m per liter) for the third coat.

These quantities are subject to considerable variation due to the porosity of the surfaces to which the paint is applied, smoothness of walls, etc., as a rough porous surface will require considerably more paint than a smooth hard surface.

Painting Metal Work. The area that any paint may be expected to cover on metal work will vary with the surface to be painted. Badly pitted or rough metal will require more paint than a perfectly smooth surface. The covering capacity will also vary with the temperature, consistency of the paint and the effort behind the brush.

However, for ordinary smooth surfaces, one gallon should cover 550 to 650 s.f. (one liter covers 13-15.5 sq.m), one coat.

Aluminum Paint. Aluminum paint is used for both priming and finishing coats on metal surfaces. The covering capacity varies with the condition of the surface, but on smooth surfaces, one gallon covers 600 to 700 s.f. (one liter covers 14.3-16.6 sq.m) with one coat.

Painting Interior Walls. When painting interior walls, the covering capacity of the paint will depend on the surface to be painted, whether smooth or sand finished plaster, porous or hard finished wallboard, etc.

Three coats are recommended for interior plaster that has never been painted: a priming coat, a second or body coat, and a third or finishing coat. However, if a two-coat job on unpainted plaster is desired, use a good wall primer or sealer for the first coat, followed by the second or finishing coat. To make two coats hide better, tint the first to nearly the same color as the second coat. If the surface has been painted before and the old paint is still in good condition, two coats are sufficient, no priming coat being required.

On smooth plaster or hard finished wallboard, wall primer or sealer should cover 575 to 625 s.f. per gal. (13.7-14.9 sq.m per liter) for the priming coat. For the second coat, figure 500 to 550 s.f. per gal. (11.9-13 sq.m per liter) and for the third coat, 575 to 625 s.f. per gal. (13.7-14.9 sq.m per liter).

On rough, porous sand finished plaster, wall primer or sealer might cover only 275 to 300 s.f. per gal. (6.5-7.1 sq.m per liter), while on very porous wallboard, it may be only 150 to 200 s.f. per gal. (3.5-4.75 sq.m per liter). For the second and third coats on rough surfaced plaster or wallboard, figure 400 to 475 s.f. per gal. (9.5-11.3 sq.m per liter).

One Coat Wall Finishes. There are any number of "one-coat" wall finishes on the market, but when applied to new plastered surfaces or over surfaces that have been previously unpainted, it is advisable to use a primer or wall seal coat, which seals the pores in the plaster or other surface and provides a base suitable to receive the one coat finish. Fully described above.

Latex Base Paints. There are any number of latex base paints on the market that are compatible with most surfaces.

When used over smooth plastered surfaces, it should cover 450 to 500 s.f. per gal. (10.7-11.9 sq.m per liter) first coat and 550 to 650 s.f. per gal. (7.7-8.3 sq.m per liter) second coat.

When applied over rough sand finish plaster, it should cover 325 to 350 s.f. (7.7-8.3 sq.m per liter) per coat.

Covering Capacity of Wall Size. The covering capacity of wall size will vary greatly with the material used.

Wall size consisting of ground or flake glue and water will cover 600 to 700 s.f. per gal. (41.3-16.6 sq. per liter) of surface.

Varnish size consisting of varnish, benzine, or turpentine and a little paint, will cover 450 to 550 s.f. per gal. (10.7-13 sq.m per liter).

Hard oil or gloss oil size consisting of rosin and benzine will cover 450 to 500 s.f. per gal. (10.7-11.9 sq.m per liter).

The covering capacity of prepared wall primers and sealers is given on previous pages under Painting Interior Walls.

Shingle Stain. When staining wood shingles after they have been laid, stain should cover 120 to 150 s.f. per gal. (2.9-3.6 sq.m per liter) of surface, for the first coat.

The second and third coats will go farther, because the wood does not absorb as much stain and should cover 200 to 225 s.f. per gal. (4.75-5.4 sq.m per liter). Stain should cover about 70 s.f. per gal. (1.7 sq.m per liter) of roof with 2 coats, or 50 s.f. per gal. (1.2 sq.m per liter) with 3 coats.

If the shingles are dipped in stain before laying, it will require 3 to 3-1/2 gals. (11.4-13.2 liters) of stain per 1,000 shingles. When dipping shingles in stain, only two-thirds of the shingle need be dipped, as it is unnecessary to dip the end of the shingle that is not exposed.

Covering Capacity of Oil and Spirit Stains. When staining finishing lumber, such as birch, mahogany, oak, gum, etc., oil stain should cover 700 to 725 s.f. per gal. (16.6-17.2 sq.m per liter).

One gallon of spirit stain should cover 500 to 600 s.f. (One liter covers 11.9-14.3 sq.m per liter).

Covering Capacity of Wood Fillers. Liquid filler is normally used with close grained woods, such as pine, birch, etc., Coverage should be 500 to 550 s.f. per gal. (11.9-13 sq.m per liter).

It is necessary to use a paste filler to fill the pores of all open grained wood, such as oak, ash, walnut, mahogany, etc.

If paste filler is used on oak, it should cover about 450 s.f. per gal. (10.7 sq.m per liter).

Covering Capacity of Shellac. The covering capacity of shellac varies with its purity. Shellac should cover 550 to 750 s.f. per gal. (13-17.8 sq.m per liter), depending on whether first, second, or third coat.

Covering Capacity of Varnish. On softwood floors, varnish should cover 400 to 450 s.f. per gal. (9.5-10.7 sq.m per liter) with one coat; 200 to 225 s.f. per gal. (4.75-5.35 sq.m per liter) with 2 coats, and 135 to 150 s.f. per gal. (3.2-3.5 sq.m per liter) with 3 coats.

When applied to hardwood floors, varnish should cover 500 to 550 s.f. per gal. (11.9-13 sq.m per liter) with one coat; 250 to 275 s.f. (5.9-6.5 sq.m per liter) with 2 coats, and 170 to 185 s.f. (4.0-4.4 sq.m per liter) with 3 coats.

When applied to softwood interior finish, spar finishing varnish should cover 400 to 425 s.f. per gal. (9.5-10.1 sq.m per liter) with one coat; 200 to 215 s.f. per gal. (4.75-5.1 sq.m per liter) with 2 coats and 135 to 145 s.f. per gal. (3.2-3.4 sq.m per liter) with 3 coats.

When applied to hardwood interior finish, spar varnish should cover 425 to 450 s.f. per gal. (10.1-10.7 sq.m per liter) with one coat; 210 to 225 s.f. per gal. (5.0-5.35 sq.m per liter) with 2 coats, and 140 to 150 s.f. per gal. (3.3-3.5 sq.m per liter) with 3 coats.

Covering Capacity of Wax. Liquid wax should cover 1,050 to 1,075 s.f. (25.0-25.5 sq.m per liter).

Covering Capacity of Varnish Remover. The amount of varnish remover required to remove old varnish from floors and interior finish depends on the condition of the old work, but varnish remover should cover 150 to 180 s.f. per gal. (3.5-4.3 sq.m per liter).

Enamel Finish. Where interior woodwork is to receive an enamel finish, at least three coats are required: first coat, oil base paint primer and sealer; second coat, prepared enamel undercoat; third coat, enamel finish coat. If a four-coat job is specified, the third coat may be a mixture of 1/2 undercoat and 1/2 enamel, followed by the enamel finish coat.

For flat work, figure material coverages as: paint primer and sealer, 575 to 600 s.f. per gal. (13.7-41.3 sq.m per liter); undercoat, 375 to 400 s.f. (8.9-9.5 sq.m per liter); enamel finish, 475 to 500 s.f. per gal. (11.3-11.9 sq.m per liter).

For running trim, material should cover: paint primer and sealer, 1,100 to 1,200 l.f. per gal. (330-360 m per liter); undercoat, 775 to 800 l.f. per gal. (233-240 m per liter); enamel finish, 775 to 800 l.f. per gal. (233-240 m per liter).

Glazed Finish. Glazing liquid, colored with oil colors, should cover 500 to 550 s.f. per gal. (11.9-13 sq.m per liter).

For glazed finish on interior running trim, not over 6" (150 mm) wide, glazing liquid should cover 1,050 to 1,100 l.f. per gal. (315-330 m per liter).

Estimating Labor Costs for Painting

Perhaps in no other trade do labor costs vary to the extent they do in painting and decorating. On some classes of work the material costs are almost negligible, while the labor costs make it one of the most expensive kinds of work.

Naturally, the cost of the work will depends on the grade of workmanship specified or upon the class of work the painting contractor figures on doing.

In some locales, there are two classes of painter—those who make a specialty of ordinary or commercial work such as factory buildings, flat or apartment buildings, speculative work and work of this class, and those who specialize in high class work. Naturally, a painter who figures the lower grade of workmanship is bound to be lower in cost. The estimator must consider the class of work required.

Another thing to keep in mind is to be sure that all measurements and quantities are listed and computed in strict accordance with the methods of measurement as given at the beginning of this section. This applies particularly to the allowances for labor when estimating such items as sash doors, windows having divided lights, drop siding and shingle siding, fences, trellis and lattice work where the face and edges of all surfaces must be painted, stairs where it is necessary to paint all four sides or the entire circumference of newels, handrails, and balusters.

These are all exceedingly important, because if the quantities are not computed according to the methods given, then the labor costs given below will be far too low in many cases.

Itemized Costs of Painting

The labor and material tables on the following pages give detailed information on material coverages and production times for various types of paint on different kinds of surfaces.

In preparing an estimate for painting work, after obtaining quantities required for the project, the estimator need only to refer to the proper table to determine the required quantity of material and labor hours necessary to perform the work. Once the gallons of paint and man hours are computed, apply the appropriate unit costs applicable for your area.

Do not overlook the items of protection, scaffolding or other equipment costs in preparing the estimate. These are not given in the following tables.

For example, let's assume that the estimator is performing an estimate for staining shingle siding on a residence, and the quantity of work to be performed amounts to 3,300 s.f. (297 sq.m) of surface to be covered with two 2 coats of stain. Refer to the chart for exterior residential work and find the proper work classification, "Stain-Shingle Siding".

(Note that there are two sets of charts for *Labor and Materials Required for Painting Operations*. The first set, which is for the English or Imperial system of measures, is followed by a set of charts giving metric figures for the labor and material for painting.)

The material coverage is stated as 120 to 150 s.f. per gal. (2.8-3.5 sq.m per liter) for the first coat and 200 to 225 s.f. per gal. (4.75-5.4 sq.m per liter) for the second coat. The labor production is 1.20 hrs. per 100 s.f. (9 sq.m) for first coat and 0.82 hours per 100 s.f. (9 sq.m) for the second coat. The sample estimate for this 3,300 s.f. (297 sq.m) of surface would be as follows:

Material Costs

	Rate	Total	Rate (metric)	Total
1st Coat				
Stain 27 gal. (102 L) .... $....		$15.00	$3.97	$405.00
2nd Coat				
Stain 16 gal. (60 L)		15.00	4.00	240.00
Cost of 3,000 s.f. (278 sq.m) .... $....				$645.00
per s.f.				0.22
per sq.m				2.32

Labor Costs

	Hours	Rate	Total	Rate	Total
1st Coat					
Painter 39.6		$....	$....	$26.83	$1,062.47
2nd Coat					
Painter 27.1		....	....	26.83	727.09
Cost of 3,300 s.f. (307 sq.m)			$....		$1,789.56
per s.f.			....		0.54
per sq.m			....		5.81

LABOR AND MATERIALS REQUIRED
FOR PAINTING OPERATIONS
English (Imperial) System

Note: Metric figures are given in a separate set of charts following those for the English system.

Exterior Work - Residential - Brush

Description of Work	Coat	No. of Sq. Ft. per Hr.	Hours per 100 Sq. Ft.	Material Coverage Sq. Ft. per Gal.
Sanding & Puttying, Plain Siding & Trim		200-210	0.50	--
Sanding & Puttying, Outside Trim Only		115-120	0.86	--
Burning Paint Off Plain Surfaces		40-50	2.20	--
Burning Paint Off Wood		30-40	2.85	--
Exterior Painting, Plain Siding & Trim	Priming	100-110	0.95	400-500
	Second	115-125	0.85	500-550
	Third	125-135	0.75	575-625
Exterior House Painting, Rubberized Wood Bond	One Coat	125-135	0.75	450-500
Exterior Trim Only	Priming	65-75	1.40	750-850
	Second	85-95	1.12	800-900
	Third	95-105	1.00	900-1,000
Shingle/Shake Siding* - Oil Paint	First	115-125	0.82	250-300
	Second	150-160	0.65	375-425
Shingle/Shake Roofs* - Oil Paint	First	100-120	0.90	120-150
	Second	155-175	0.60	220-250
Shingle/Shake Siding* - Stain	First	75-85	1.20	120-150
	Second	115-130	0.82	200-225
Shingle/Shake Roofs* - Stain	First	145-155	0.67	100-120
	Second	190-210	0.50	170-200
Shingle/Shake Roofs & Siding* - Sealer	First	200-210	0.49	125-150
	Second	220-250	0.43	150-200

*If surface is extremely dry, increase labor hours and decrease covering capacity.

Description of Work	Coat	No. of Sq. Ft. per Hr.	Hours per 100 Sq. Ft.	Material Coverage Sq. Ft. per Gal.
Brick Walls - Oil Paint	First	100-120	0.93	170-200
	Second	140-160	0.67	350-400
	Third	155-175	0.60	375-425
Brick Walls - Latex Paint	First	95-110	0.98	180-210
	Second	130-150	0.71	350-410
	Third	150-170	0.63	380-430
Porch Floors & Steps - Oil Paint	First	235-245	0.42	325-375
	Second	270-280	0.36	475-525
	Third	285-295	0.35	525-575
Smooth Face Brick - Waterproof Cement Paint	First	175-185	0.55	90-110
	Second	260-270	0.38	140-160
Asphalt Shingle - Clear Waterproof Paint	First	45-55	2.00	120-140
	Second	50-70	1.67	165-185
	Third	60-80	1.43	175-205

Description of Work	Coat	No. of Sq. Ft. per Hr.	Hours per 100 Sq. Ft.	Material Coverage Sq. Ft. per Gal.
Smooth Face Brick - Clear Waterproof Paint	First	200-210	0.50	500-510
	Second	210-230	0.45	590-610
Stucco, Medium Texture - Oil Paint	First	90-100	1.10	140-160
	Second	150-160	0.65	340-360
	Third	150-160	0.65	340-360
Stucco, Medium Texture - Latex Paint	First	110-125	0.85	160-180
	Second	175-200	0.53	180-210
	Third	175-210	0.53	180-220
Stucco, Medium Texture - Waterproof Cement Paint	First	130-140	0.74	90-110
	Second	190-210	0.50	125-145
Exterior Masonry, Stucco, Shingle Siding - Latex Flat Finish	First	100-120	0.91	325-375
	Second	165-175	0.59	375-425
Concrete Walls, Smooth - Waterproof Cement Paint	First	175-180	0.57	110-130
	Second	225-285	0.36	150-170
Concrete Floors & Steps - Floor Enamel	First	250-280	0.38	440-460
	Second	190-210	0.50	575-625
	Third	200-220	0.48	575-625
Cement Floors - Color Stain & Finish	First	325-370	0.28	475-525
	Second	280-300	0.35	450-500
Cement Floors - Epoxy	Primer	100-125	0.89	175-200
	First	115-135	0.80	190-220
	Second	150-180	0.60	200-250
Fences, Plain, Average	First	125-135	0.77	530-550
	Second	190-200	0.51	630-650
Fences, Picket, Average	First	140-150	0.70	630-650
	Second	160-170	0.60	650-675
Fences, Wire Metal, Average	First	100-110	0.95	900-1,000
	Second	140-150	0.70	975-1,125
Shutters, Average, Each Coat	No. Shutter	2-3	0.40	11-13
Downspout & Gutters - Paint	First	170-180	0.57	540-560
	Second	185-195	0.53	575-600
Screens, Wood Only, Average, Each Coat	No. Screens	6-8	0.14	45-55
Storm Sash, 2 Light, Average, Each Coat	No. Sash	3-4	0.29	23-27
Sheet Metal Work	First	230-240	0.43	500-550
	Second	260-270	0.38	550-600
Steel Factory Sash	First	80-90	1.18	950-1,000
For old work, add 10 to 20% to prepare	Second	100-110	0.95	950-1,000
Surface	Third	125-135	0.77	1,250-1,350
Doors & Windows - Paint	First	175-185	0.55	475-500
	Second	130-140	0.75	500-550
	Third	165-175	0.60	500-550
Base, Chair Rail, Picture Mold, & Other Trim less than 6" wide (priced per lin. ft.)	First	275-300	0.35	800-900

Description of Work	Coat	No. of Sq. Ft. per Hr.	Hours per 100 Sq. Ft.	Material Coverage Sq. Ft. per Gal.
....	Second	175-200	0.55	1,100-1,200
....	Third	150-165	0.65	1,100-1,200
Base, Chair Rail, Picture Mold, & Other Trim less than 6" wide - polyurethane (per lin. ft.)	First	260-280	0.37	525-575
....	Second	280-300	0.34	550-600
....	Third	280-300	0.34	550-600
Masonry Walls - Bonding & Penetrating Oil Paint	First	350-360	0.28	225-250
....	Second	350-375	0.28	275-300
Common Brick, Smooth Brick, Concrete or Cinder Blocks, Cement Plaster - Waterproofing	First	100-110	1.00	90-100
....	Second	180-195	0.53	175-200
Wire Cut Tapestry Brick or Very Rough Concrete Block or Stucco - Waterproofing	First	90-100	1.05	50-60
....	Second	150-165	0.65	150-175
Smooth Plaster Walls - Flat Finish Paint	First	450-475	0.22	500-550
....	Second	475-500	0.21	500-550
Plaster Walls - Sand Finish	First	400-425	0.25	450-500
....	Second	425-450	0.24	400-450
Smooth Finish Plaster - Industrial Plaster	First	450-475	0.22	500-550
....	Second	500-550	0.20	400-450
....	Third	500-550	0.20	450-500

Exterior Work - Residential - Roller/Pad

Description of Work	Coat	No. Sq.Ft./Hr	Hrs/100	Coverage
Shingle/Shake Siding* - Oil Paint	First	130-140	0.74	220-260
....	Second	170-185	0.56	330-370
Shingle/Shake Roofs* - Oil Paint	First	115-135	0.80	105-130
....	Second	180-200	0.53	190-215
Shingle/Shake Siding* - Stain	First	85-100	0.74	105-130
....	Second	130-150	0.49	170-195
Shingle/Shake Roofs* - Stain	First	165-180	0.58	85-105
....	Second	220-240	0.43	150-170
Shingle/Shake Roofs & Siding* - Sealer	First	230-240	0.43	110-130
....	Second	250-285	0.37	130-170

*If surface is extremely dry, increase labor hours and decrease covering capacity.

Description of Work	Coat	No. Sq.Ft./Hr	Hrs/100	Coverage
Brick Walls - Oil Paint	First	115-135	0.80	150-170
....	Second	160-180	0.59	305-345
....	Third	180-200	0.53	325-370
Brick Walls - Latex Paint	First	110-125	0.85	155-182
....	Second	150-170	0.63	305-350
....	Third	170-190	0.55	330-370
Porch Floors & Steps - Oil Paint	First	270-280	0.36	285-325
....	Second	310-320	0.32	415-450
....	Third	330-340	0.30	460-500

Description of Work	Coat	No. of Sq. Ft. per Hr.	Hours per 100 Sq. Ft.	Material Coverage Sq. Ft. per Gal.
Smooth Face Brick - Waterproof				
Cement Paint	First	200-210	0.50	80-95
	Second	300-310	0.33	120-140
Asphalt Shingle - Clear Waterproof Paint	First	50-65	1.48	105-120
	Second	60-80	1.42	140-160
	Third	70-90	1.25	150-180
Smooth Face Brick - Clear				
Waterproof Paint	First	230-240	0.43	430-440
	Second	240-265	0.50	515-530
Stucco, Medium Texture - Oil Paint	First	100-115	0.93	120-140
	Second	170-185	0.55	295-310
	Third	170-185	0.55	295-310
Stucco, Medium Texture - Latex Paint	First	130-145	0.72	140-155
	Second	200-230	0.47	155-180
	Third	200-240	0.45	155-190
Stucco, Medium Texture - Waterproof				
Cement Paint	First	150-160	1.00	75-95
	Second	215-240	0.44	110-125
Exterior Masonry, Stucco, Shingle Siding -				
Latex Flat Finish	First	115-135	0.80	280-325
	Second	190-200	0.51	325-370
Concrete Walls, Smooth -				
Waterproof Cement Paint	First	200-210	0.49	95-115
	Second	260-320	0.48	130-150
Concrete Floors & Steps - Floor Enamel	First	290-320	0.37	380-400
	Second	220-240	0.43	500-540
	Third	230-250	0.42	500-540
Cement Floors - Color Stain & Finish	First	370-420	0.25	410-450
	Second	320-345	0.30	390-430
Cement Floors - Epoxy	Primer	115-140	0.78	150-170
	First	130-155	0.70	165-190
	Second	170-210	0.53	170-210
Fences, Plain, Average	First	145-155	0.67	460-480
	Second	220-230	0.45	550-565
Fences, Picket, Average	First	160-170	0.61	550-565
	Second	185-195	0.53	565-585
Fences, Wire Metal, Average	First	115-125	0.83	780-870
	Second	160-170	0.60	850-980
Downspout & Gutters - Paint	First	195-205	0.50	470-485
	Second	210-225	0.46	500-520
Masonry Walls - Bonding &				
Penetrating Oil Paint	First	400-410	0.25	200-220
	Second	400-430	0.24	240-260
Common Brick, Smooth Brick, Concrete or Cinder				
Blocks, Cement Plaster - Waterproofing	First	115-125	0.83	80-85
	Second	205-225	0.53	150-170
Smooth Plaster Walls - Flat Finish Paint	First	520-550	0.19	435-475
	Second	550-575	0.18	435-475
Plaster Walls - Sand Finish	First	460-490	0.21	390-430
	Second	490-515	0.20	350-390

Description of Work	Coat	No. of Sq. Ft. per Hr.	Hours per 100 Sq. Ft.	Material Coverage Sq. Ft. per Gal.
Smooth Finish Plaster -				
Industrial Plaster	First	515-545	0.19	435-475
	Second	575-630	0.17	350-390
	Third	575-630	0.17	390-430

Spray Painting - Residential & Commercial Work

	Coat	No. of Sq. Ft. per Hr.	Hours per 100 Sq. Ft.	Material Coverage
Brick, Tile, & Cement - Latex Paint	First	500-525	0.20	90-110
	Second	525-550	0.19	140-160
Brick, Tile, & Cement - Oil Paint	First	475-500	0.21	250-275
	Second	500-525	0.20	400-450
Brick - Waterproofing	First	550-600	0.17	200-300
	Second	650-700	0.15	225-350

Note: For weathered brick, add 15% labor and 35% to 50% for material.

	Coat			
Rough Brick, Tile, Cement & Stucco - Silicone	One Coat	800-900	0.13	80-90
Stucco - Exterior Latex Paint	First	400-425	0.25	90-110
	Second	450-500	0.22	125-150
Stucco - Oil Paint	First	300-325	0.34	200-225
	Second	325-350	0.30	350-400
Stucco - Waterproofing	First	550-600	0.17	150-180
	Second	650-700	0.15	160-190

Note: For weathered stucco, add 15% for labor and 35% to 50% for material.

	Coat			
Shingle Roofs - Oil Paint	First	400-450	0.25	125-150
	Second	450-500	0.22	200-225
Shingle Roofs - Stain	First	450-475	0.22	125-150
	Second	475-500	0.21	200-225
Shingle Siding - Oil Paint	First	350-375	0.29	125-150
	Second	400-425	0.25	225-250
Shingle Siding - Stain	First	400-450	0.25	125-150
	Second	450-500	0.22	200-225
Smooth Plaster Walls - Flat Finish Paint	First	450-475	0.22	500-550
	Second	475-500	0.21	500-550
Plaster Walls - Sand Finish	First	400-425	0.25	450-500
	Second	425-450	0.24	400-450
Smooth Finish Plaster - Industrial Enamel	First	450-475	0.22	500-550
	Second	500-550	0.20	400-450
	Third	500-550	0.20	400-450
Fence, Picket - Oil or Latex Paint	First	475-500	0.20	500-525
	Second	475-500	0.20	550-600
Fence, Plain - Oil or Latex Paint	First	450-500	0.21	375-400
	Second	570-600	0.17	400-450
Doors, Garage - Oil or Latex Paint	First	540-550	0.22	550-575
	Second	540-550	0.22	560-600

Description of Work	Coat	No. of Sq. Ft. per Hr.	Hours per 100 Sq. Ft.	Material Coverage Sq. Ft. per Gal.
Blinds or Shutters - Paint	First	120-140	0.77	120-140
	Second	130-170	0.67	140-180
	Third	150-190	0.59	150-190

Trim must be figured separately as hand work.

Finishing Interior Trim - Residential

Description of Work	Coat	No. of Sq. Ft. per Hr.	Hours per 100 Sq. Ft.	Material Coverage Sq. Ft. per Gal.
Preparatory Work for Painting	Sanding	290-300	0.35	--
	Sand & Putty	190-200	0.50	--
	Light Sanding	340-350	0.30	--
Back Priming Interior Trim up to 6" Wide	Lin. Ft.	475-500	0.21	1,100-1,200
Doors & Windows, Interior - Paint	First	140-150	0.70	475-500
	Second	115-125	0.85	500-550
	Third	115-125	0.85	500-550
Wood Sash - Oiling & Priming	No. of Sash	9-10	--	600-700
Preparatory Work for Enamel Finish	Sanding	290-300	0.35	--
	Sand & Putty	120-130	0.80	--
	Light Sanding	135-145	0.72	--
Doors & Windows - Enamel Finish	Paint First	140-150	0.70	575-600
	Undercoat Second	90-100	1.10	372-400
	Enamel Third	100-110	1.00	475-500
Four coat work, add 1/2 undercoat, 1/2 enamel	Add'l Coat	90-100	1.10	425-450
Base, Chair Rail, Picture Mold, & Other Trim up to 6" Wide (lin. ft.)	First	250-260	0.40	800-900
	Second	145-155	0.67	1,100-1,200
	Third	125-135	0.77	1,100-1,200
Base, Chair Rail, Picture Mold, & Other Trim up to 6" Wide - Polyurethane (lin. ft.)	First	260-280	0.37	525-575
	Second	280-300	0.34	550-560
	Third	280-300	0.34	550-560
Four coat work, add 1/2 undercoat, 1/2 enamel	Add'l Coat	135-145	0.72	1,100-1,200
Stain Interior Work	One Coat	215-225	0.45	700-725
Stain & Fill Interior Woodwork, Wipe Off	One Coat	65-70	1.50	400-450
Shellac Interior Woodwork	One Coat	215-225	0.45	700-725
Varnish, Gloss, Interior Woodwork	One Coat	165-175	0.60	425-450
Varnish, Flat, Interior Woodwork	One Coat	170-180	0.57	600-625
Standing Trim - Wax & Polish	One Coat	100-110	0.95	550-600
Standing Trim - Penetrating Stainwax	One Coat	170-180	0.57	525-550
	Second	190-200	0.51	600-625
	Polishing Coat	190-200	0.51	--

Description of Work	Coat	No. of Sq. Ft. per Hr.	Hours per 100 Sq. Ft.	Material Coverage Sq. Ft. per Gal.
Sanding for Extra Fine Varnish Finish	Sanding	40-45	2.25	--
Synthetic Resin Finish, Requires Wiping	First	190-200	0.50	550-600
	Second	210-220	0.47	625-675
Spackling or Swedish Putty over Flat Trim	One Coat	60-65	1.60	140-150
Glazing & Wiping over Enamel Trim	One Coat	60-65	1.60	1,050-1,100
Brush Stippling Interior Trim - Painted	--	85-90	1.15	--
Flat Varnishing & Brush Stippling over Glazed Trim	Varnish	170-180	0.57	600-625
	Stipple	240-250	0.42	--

Old Work

Description of Work	Coat	No. of Sq. Ft. per Hr.	Hours per 100 Sq. Ft.	Material Coverage Sq. Ft. per Gal.
Washing Average Enamel Finish	Washing	85-90	1.15	--
Washing Better Grade Enamel Finish	Washing	60-65	1.60	--
Polishing Better Grade Enamel Finish	Polish	170-180	0.57	1,700-1,800
Removing Varnish with Liquid Remover	Flat Surfaces	30-35	3.00	150-180
Wash, Touch Up, One Coat Varnish	Wash-Touch Up	165-175	0.60	--
	Varnish	100-110	0.95	575-600
Wash, Touch Up, One Coat Enamel	Wash-Touch Up	140-150	0.70	--
	Enamel	70-75	1.40	375-400
Wash, Touch Up, One Coat Undercoat	Wash-Touch Up	140-150	0.70	--
and One Coat Enamel	Undercoat	75-85	1.25	475-500
	Enamel	75-85	1.25	475-500
Burning Off Interior Trim	--	20-25	4.00	--
Burning Off Plain Surfaces	--	30-35	3.00	--

Finishing Interior Floors

Description of Work	Coat	No. of Sq. Ft. per Hr.	Hours per 100 Sq. Ft.	Material Coverage Sq. Ft. per Gal.
Wood Floors - Paint	First	290-300	0.35	525-550
	Second	260-270	0.38	425-450
Wood Floors - Filling, Wiping	Fill-Wipe	170-180	0.57	425-450
Hardwood Floors - Penetrating Stainwax	First	390-400	0.25	525-550
	Second	590-600	0.17	1,200-1,250
Floor Seal	First	490-500	0.20	600-625
	Second	590-600	0.17	1,200-1,250
Shellac	First	390-400	0.25	500-550
	Second	475-500	0.21	650-675
Stainfill, 1 Shellac, 1 Varnish	Stainfill	260-270	0.38	475-500
	Shellac	390-400	0.25	650-675
	Varnish	300-310	0.33	500-550

Description of Work	Coat	No. of Sq. Ft. per Hr.	Hours per 100 Sq. Ft.	Material Coverage Sq. Ft. per Gal.
Stainfill, 1 Shellac, Wax & Polish	Stainfill	260-270	0.38	475-500
	Shellac	390-400	0.25	650-675
	Wax/Polish	200-210	0.50	1,050-1,075
Varnish, Each Coat over Shellac	Varnish	300-310	0.33	500-550
Buffing Floors by Machine	Buffing	390-400	0.25	--
Waxing over 2 Coats of Seal & Polish	Wax-Polish	200-210	0.50	1,050-1,075

Old Work

Description of Work	Coat	No. of Sq. Ft. per Hr.	Hours per 100 Sq. Ft.	Material Coverage Sq. Ft. per Gal.
Removing Varnish with Liquid Remover	Flat Surfaces	40-45	2.35	170-180
Clean, Touch Up, & Varnish	Clean-Touch Up	125-135	0.77	--
	Varnish	300-310	0.33	252-550
Clean, Touch Up, Wax, & Polish	Clean-Touch Up	125-135	0.77	--
	Wax-Polish	200-210	0.50	1,050-1,075

Painting Interior Walls

Description of Work	Coat	No. of Sq. Ft. per Hr.	Hours per 100 Sq. Ft.	Material Coverage Sq. Ft. per Gal.
Taping, Beading, Spotting Nail Heads, & Sanding Gypsum Wallboard		90-110	1.00	--
Texture over Gypsum Wallboard	One Coat	250-270	0.39	8-10*
*per lb.				
Casein or Resin Emulsion over Textured Gypsum Wallboard	One Coat	300-320	0.32	325-350
Sizing New Smooth Finish Walls	Sizing	350-400	0.27	600-700
Wall Sealer or Primer on Smooth Walls	Sealer	200-225	0.45	575-625
Wall Sealer or Primer on Sand Finish Walls	Sealer	125-135	0.80	275-300
Smooth Finish Plaster - Flat Finish	Sealer	200-225	0.45	575-625
	Second	165-175	0.60	500-550
	Third	175-185	0.65	575-625
	Stippling	190-200	0.52	--
Sand Finish Plaster - Flat Finish	Sealer	125-135	0.80	275-300
	Second	135-145	0.70	400-450
	Third	125-135	0.80	425-475
Smooth Finish Plaster - Gloss or Semi-Gloss	Sealer	200-225	0.45	575-625
	Second	170-180	0.57	500-550
Third		140-150	0.70	500-550
	Stippling	85-95	1.10	--
Sand Finish Plaster - Gloss or Semi-Gloss	Sealer	125-135	0.80	275-300
	Second	115-125	0.83	275-400
	Third	125-135	0.80	400-450
Texture Plaster, Average - Semi-Gloss	Sealer	100-110	0.95	250-275
	Second	115-125	0.83	320-350
	Third	125-135	0.80	375-400

Description of Work	Coat	No. of Sq. Ft. per Hr.	Hours per 100 Sq. Ft.	Material Coverage Sq. Ft. per Gal.
Smooth Finish Plaster - Latex Rubber Paint	First	140-160	0.67	300-350
	Second	140-160	0.67	300-350
Sand Finish or Average Texture Plaster - Latex Rubber Paint	First	110-130	0.83	250-300
	Second	110-130	0.83	250-300
Glazing & Motting over Smooth Finish Plaster		80-90	1.15	1,050-1,075
Glazing & Motting over Sand Finish Plaster		60-65	1.60	875-900
Glazing & Highlighting Textured Plaster		80-90	1.15	825-850
Starch & Brush Stipple over Painted Glazed Surface		115-125	0.83	--
Flat Varnish & Brush Stipple over Painted Glazed Surface		105-115	0.90	500-550
Textured Oil Paint over Smooth Finish Plaster	Size	215-225	0.45	675-725
	Texture	40-45	2.30	125-150
Water Textured over Smooth Finish Plaster	Size	215-225	0.45	700-725
	Texture	50-55	1.90	85-95
New Smooth Plaster - Latex Base Paint	First	190-200	0.50	500-550
	Second	210-220	0.47	650-700
on Rough Sand Finish	One Coat	150-160	0.65	325-350
on Cement Blocks	One Coat	135-145	0.70	300-325
on Acoustical Surfaces	One Coat	135-145	0.70	200-225
Casein Paint over Smooth Finish Plaster	First	270-280	0.37	500-550
	Second	300-310	0.33	500-550
over Rough Sand Finish Plaster	One Coat	250-260	0.40	325-350
over Cement Blocks	One Coat	165-175	0.60	300-325
over Acoustical Surfaces	One Coat	165-175	0.60	200-225
over Cinder Concrete Blocks	One Coat	125-135	0.70	125-150

When applying paint to walls and ceilings using a roller applicator, increase quantities 10% to 15% and reduce painter time about the same amount.

Old Work

Description of Work	Coat	No. of Sq. Ft. per Hr.	Hours per 100 Sq. Ft.	Material Coverage
Washing Off Calcimine, Average Surfaces		115-125	0.83	--
Washing Smooth Finish Plaster Walls, Average		145-155	0.67	--
Washing Sand Finish Plaster Walls, Average		100-110	1.00	--
Washing Starched Surfaces & Restarching, Smooth Surfaces	Washing	145-155	0.67	--
	Restarching	130-140	0.75	--
Removing Old Wallpaper, 3 layers or less		65-75	1.45	--

Description of Work	Coat	No. of Sq. Ft. per Hr.	Hours per 100 Sq. Ft.	Material Coverage Sq. Ft. per Gal.
Washing Off Glue after Removing Paper (including fixing average cracks & sizing)		125-135	0.77	--
Cutting Hard Oil or Varnish Size Walls (including fixing average cracks)		125-135	0.77	--
Cutting Gloss Painted Walls (incl. fixing cracks)		125-135	0.77	--
Wash, Touch Up, One Coat Gloss Paint to Smooth Plaster Surfaces	Wash-Touch Up	125-135	0.77	--
	One Coat	125-135	0.77	450-500
Synthetic-Resin Emulsion Paint over Old Painted Walls	One Coat	190-200	0.50	400-425

Wallpaper, Canvas, Coated Fabrics, Paper Hanging, Wood Veneer

Description of Work	Coat	No. of Sq. Ft. per Hr.	Hours per 100 Sq. Ft.	Material Coverage Sq. Ft. per Gal.
Canvas Sheeting		50-60	1.82	--
Coated Fabrics	Single Roll	2-2 1/2	--	--
Wallpaper				
One Edge Work	Single Roll	3 1/2-4	--	--
Wire Edge Work	Single Roll	2 1/2-3	--	--
Butt Work	Single Roll	2 1/4-2 3/4	--	--
Scenic Paper 40"x60"	Single Roll	1-1 1/4	--	--
Wood Veneer Flexwood	Sq. Ft.	18	--	--
Lacquer Finish over Wood Veneer	First Coat	100	1.00	275-325
	Second	100	1.00	450-500
Penetrating Wax or Synthetic Resin Application over Wood Veneer	First	100	1.00	550-600
	Second	100	1.00	650-675

LABOR AND MATERIALS REQUIRED
FOR PAINTING OPERATIONS
Metric System

Note: This set of charts presents the metric figures for the same painting operations that appear in the series of charts above.

Exterior Work - Residential - Brush

Description of Work	Coat	No. of Sq.M per Hr.	Hours per 100 Sq.M	Material Coverage Sq.M per Liter
Sanding & Puttying, Plain Siding & Trim		18.6-19.5	5.20	--
Sanding & Puttying, Outside Trim Only		10.7-11.1	9.20	--
Burning Paint Off Plain Surfaces		3.7-4.6	24.0	--

Description of Work	Coat	No. of Sq.M per Hr.	Hours per 100 Sq.M	Material Coverage Sq.M per Liter
Burning Paint Off Wood		2.8-3.7	30.7	--
Exterior Painting, Plain Siding & Trim	Priming	9.3-10.2	10.2	9.8-12.3
	Second	10.7-11.6	8.9	12.3-13.5
	Third	11.6-12.5	8.3	14.1-15.4
Exterior House Painting, Rubberized Wood Bond	One Coat	11.6-12.5	8.3	11.1-12.3
Exterior Trim Only	Priming	6.0-7.0	15.4	18.4-20.9
	Second	7.9-8.8	12.0	19.7-22.1
	Third	8.8-9.8	10.7	22.1-24.6
Shingle/Shake Siding* - Oil Paint	First	10.7-11.6	9.0	6.1-7.4
	Second	13.9-14.9	6.9	9.2-10.4
Shingle/Shake Roofs* - Oil Paint	First	9.3-11.1	9.8	2.9-3.7
	Second	14.4-16.3	6.5	5.4-6.1
Shingle/Shake Siding* - Stain	First	7.0-7.9	13.4	2.9-3.7
	Second	10.7-12.1	8.8	4.9-5.5
Shingle/Shake Roofs* - Stain	First	13.5-14.4	7.1	2.5-2.9
	Second	17.7-19.5	5.3	4.2-4.9
Shingle/Shake Roofs & Siding* - Sealer	First	18.6-19.5	5.2	3.1-3.7
	Second	20.4-23.2	4.6	3.7-4.9

*If surface is extremely dry, increase labor hours and decrease covering capacity.

Description of Work	Coat	No. of Sq.M per Hr.	Hours per 100 Sq.M	Material Coverage Sq.M per Liter
Brick Walls - Oil Paint	First	9.3-11.1	9.8	4.2-4.9
	Second	13.0-14.9	7.1	8.6-9.8
	Third	14.4-16.3	6.5	9.2-10.4
Brick Walls - Latex Paint	First	8.8-10.2	10.5	4.4-5.2
	Second	12.1-13.9	7.7	8.6-10.1
	Third	13.9-15.8	6.7	9.3-10.6
Porch Floors & Steps - Oil Paint	First	21.8-22.8	4.5	8.0-9.2
	Second	25.1-26.0	3.9	11.7-12.9
	Third	26.5-27.4	3.7	12.9-14.1
Smooth Face Brick - Waterproof Cement Paint	First	16.3-17.2	6.0	2.2-2.7
	Second	24.2-25.1	4.0	3.4-3.9
Asphalt Shingle - Clear Waterproof Paint	First	4.2-5.1	21.50	2.9-3.4
	Second	4.6-6.5	18.0	4.1-4.5
	Third	5.6-7.4	15.4	4.3-5.0
Smooth Face Brick - Clear Waterproof Paint	First	18.6-19.5	5.2	12.3-12.5
	Second	19.5-21.4	4.9	14.5-14.7
Stucco, Medium Texture - Oil Paint	First	8.4-9.3	11.2	3.4-3.7
	Second	13.9-14.9	6.9	8.4-8.8
	Third	13.9-14.9	6.9	8.4-8.8
Stucco, Medium Texture - Latex Paint	First	10.2-11.6	9.1	3.9-4.4
	Second	16.3-18.6	5.3	4.4-5.2
	Third	16.3-19.5	5.3	4.4-5.4
Stucco, Medium Texture - Waterproof Cement Paint	First	12.1-13.0	8.0	2.2-2.7
	Second	17.7-19.5	5.4	3.1-3.6

Description of Work	Coat	No. of Sq.M per Hr.	Hours per 100 Sq.M	Material Coverage Sq.M per Liter
Exterior Masonry, Stucco, Shingle Siding - Latex Flat Finish	First	9.3-11.1	9.8	8.0-9.2
	Second	15.3-16.3	6.3	9.2-10.4
Concrete Walls, Smooth - Waterproof Cement Paint	First	16.3-16.7	6.1	2.7-3.2
	Second	20.9-26.5	4.2	3.7-4.2
Concrete Floors & Steps - Floor Enamel	First	23.2-26.0	4.1	10.8-11.3
	Second	17.7-19.5	5.3	14.1-15.4
	Third	18.6-20.4	5.1	14.1-15.4
Cement Floors - Color Stain & Finish	First	30.2-34.4	3.1	11.7-12.9
	Second	26.0-27.9	3.7	11.1-12.3
Cement Floors - Epoxy	Primer	9.3-11.6	9.6	4.3-4.9
	First	10.7-12.5	8.6	4.7-5.4
	Second	13.9-16.7	6.5	4.9-6.1
Fences, Plain, Average	First	11.6-12.5	8.3	13.0-13.5
	Second	17.7-18.6	5.5	15.5-16.0
Fences, Picket, Average	First	13.0-13.9	7.4	15.5-16.0
	Second	14.9-15.8	6.5	16.0-16.5
Fences, Wire Metal, Average	First	9.3-10.2	10.3	22.1-24.6
	Second	13.0-13.9	7.4	24.0-27.6
Shutters, Average, Each Coat	No. Shutter	0.2-0.3	400	0.3-0.3
Downspout & Gutters - Paint	First	15.8-16.7	6.1	13.3-13.8
	Second	17.2-18.1	5.7	14.1-14.7
Screens, Wood Only, Average, Each Coat	No. Screens	0.6-0.7	153.8	1.1-1.4
Storm Sash, 2 Light, Average, Each Coat	No. Sash	0.3-0.4	285.7	0.6-0.7
Sheet Metal Work	First	21.4-22.3	4.6	12.3-13.5
	Second	24.2-25.1	4.1	13.5-14.7
Steel Factory Sash	First	7.4-8.4	12.7	23.3-24.6
For old work, add 10 to 20% to prepare	Second	9.3-10.2	10.2	23.3-24.6
Surface	Third	11.6-12.5	8.3	30.7-33.2
Doors & Windows - Paint	First	16.3-17.2	6.0	11.7-12.3
	Second	12.1-13.0	8.0	12.3-13.5
	Third	15.3-16.3	6.3	12.3-13.5
Base, Chair Rail, Picture Mold, & Other Trim less than 150 mm wide, priced per meter	First	25.5-27.9	3.7	19.7-22.1
	Second	16.3-18.6	5.7	27.0-29.5
	Third	13.9-15.3	6.8	27.0-29.5
Base, Chair Rail, Picture Mold, & Other Trim less than 150 mm wide - polyurethane per meter	First	24.2-26.0	4.0	12.9-14.1
	Second	26.0-27.9	3.7	13.5-14.7
	Third	26.0-27.9	3.7	13.5-14.7
Masonry Walls - Bonding & Penetrating Oil Paint	First	32.5-33.4	3.0	5.5-6.1
	Second	32.5-34.8	2.9	6.8-7.4

Description of Work	Coat	No. of Sq.M per Hr.	Hours per 100 Sq.M	Material Coverage Sq.M per Liter
Common Brick, Smooth Brick, Concrete or Cinder Blocks, Cement Plaster - Waterproofing	First	9.3-10.2	10.3	2.2-2.5
	Second	16.7-18.1	5.7	4.3-4.9
Wire Cut Tapestry Brick or Very Rough Concrete Block or Stucco - Waterproofing	First	8.4-9.3	11.2	1.2-1.5
	Second	13.9-15.3	6.8	3.7-4.3
Smooth Plaster Walls - Flat Finish Paint	First	41.8-44.1	2.3	12.3-13.5
	Second	44.1-46.5	2.2	12.3-13.5
Plaster Walls - Sand Finish	First	37.2-39.5	2.6	11.1-12.3
	Second	39.5-41.8	2.5	9.8-11.1
Smooth Finish Plaster - Industrial Plaster	First	41.8-44.1	2.3	12.3-13.5
	Second	46.5-51.1	2.0	9.8-11.1
	Third	46.5-51.1	2.0	11.1-12.3

Exterior Work - Residential - Roller/Pad

Description of Work	Coat	No. of Sq.M per Hr.	Hours per 100 Sq.M	Material Coverage Sq.M per Liter
Shingle/Shake Siding* - Oil Paint	First	12.1-13.0	8.0	5.4-6.4
	Second	15.8-17.2	6.1	8.1-9.1
Shingle/Shake Roofs* - Oil Paint	First	10.7-12.5	8.6	2.6-3.2
	Second	16.7-18.6	5.7	4.7-5.3
Shingle/Shake Siding* - Stain	First	7.9-9.3	11.6	2.6-3.2
	Second	12.1-13.9	7.7	4.2-4.8
Shingle/Shake Roofs* - Stain	First	15.3-16.7	6.2	2.1-2.6
	Second	20.4-22.3	4.7	3.7-4.2
Shingle/Shake Roofs & Siding* - Sealer	First	21.4-22.3	4.6	2.7-3.2
	Second	23.2-26.5	4.0	3.2-4.2

*If surface is extremely dry, increase labor hours and decrease covering capacity.

Description of Work	Coat	No. of Sq.M per Hr.	Hours per 100 Sq.M	Material Coverage Sq.M per Liter
Brick Walls - Oil Paint	First	10.7-12.5	8.6	3.7-3.7
	Second	14.9-16.7	6.3	7.5-8.5
	Third	16.7-18.6	5.7	8.0-9.1
Brick Walls - Latex Paint	First	10.2-11.6	9.1	3.8-4.5
	Second	13.9-15.8	6.7	7.5-8.6
	Third	15.8-17.7	6.0	8.1-9.1
Porch Floors & Steps - Oil Paint	First	25.1-26.0	3.9	7.0-8.0
	Second	28.8-29.7	3.4	10.2-11.1
	Third	30.7-31.6	3.2	11.3-12.3
Smooth Face Brick - Waterproof Cement Paint	First	18.6-19.5	5.2	2.0-2.3
	Second	27.9-28.8	3.5	2.9-3.4
Asphalt Shingle - Clear Waterproof Paint	First	4.6-6.0	18.9	2.6-2.9
	Second	6.0-7.4	14.9	3.4-3.9
	Third	6.5-8.4	12.4	3.7-4.4
Smooth Face Brick - Clear Waterproof Paint	First	21.4-22.3	4.6	10.6-10.8
	Second	22.3-24.6	4.2	12.7-13.0
Stucco, Medium Texture - Oil Paint	First	9.3-10.7	10.0	2.9-3.4
	Second	15.8-17.2	6.1	7.3-7.6
	Third	15.8-17.2	6.1	7.3-7.6

Description of Work	Coat	No. of Sq.M per Hr.	Hours per 100 Sq.M	Material Coverage Sq.M per Liter
Stucco, Medium Texture - Latex Paint	First	12.1-13.5	7.8	3.4-3.8
	Second	18.6-21.4	5.0	3.8-4.4
	Third	18.6-22.3	4.9	3.8-4.7
Stucco, Medium Texture - Waterproof Cement Paint	First	13.9-14.9	7.0	1.8-2.3
	Second	20.0-22.3	4.7	2.7-3.1
Exterior Masonry, Stucco, Shingle Siding - Latex Flat Finish	First	10.7-12.5	8.6	6.9-8.0
	Second	17.7-18.6	5.5	8.0-9.1
Concrete Walls, Smooth - Waterproof Cement Paint	First	18.6-19.5	5.2	2.3-2.8
	Second	24.2-29.7	3.7	3.2-3.7
Concrete Floors & Steps - Floor Enamel	First	26.9-29.7	3.5	9.3-9.8
	Second	20.4-22.3	4.7	12.3-13.3
	Third	21.4-23.2	4.5	12.3-13.3
Cement Floors - Color Stain & Finish	First	34.4-39.0	2.7	10.1-10.3
	Second	29.7-32.1	3.2	9.6-10.3
Cement Floors - Epoxy	Primer	10.7-13.0	8.4	3.7-4.2
	First	12.1-14.4	7.5	4.1-4.7
	Second	15.8-19.5	5.7	4.2-5.2
Fences, Plain, Average	First	13.5-14.4	7.1	11.3-11.8
	Second	20.4-21.4	4.7	13.5-13.9
Fences, Picket, Average	First	14.9-15.8	6.5	13.5-13.9
	Second	17.2-18.1	5.7	13.9-14.4
Fences, Wire Metal, Average	First	10.7-11.6	9.0	19.2-21.4
	Second	14.9-15.8	6.5	20.9-24.1
Downspout & Gutters - Paint	First	18.1-19.0	5.4	11.6-11.9
	Second	19.5-20.9	6.0	12.3-12.8
Masonry Walls - Bonding & Penetrating Oil Paint	First	37.2-38.1	2.7	4.9-5.4
	Second	37.2-39.9	2.6	5.9-6.4
Common Brick, Smooth Brick, Concrete or Cinder Blocks, Cement Plaster - Waterproofing	First	10.7-11.6	9.0	2.0-2.1
	Second	19.0-20.9	5.0	3.7-4.2
Smooth Plaster Walls - Flat Finish Paint	First	48.3-51.1	2.0	10.7-11.7
	Second	51.1-53.4	1.9	10.7-11.7
Plaster Walls - Sand Finish	First	42.7-45.5	2.3	9.6-10.6
	Second	45.5-47.8	2.1	8.6-9.6
Smooth Finish Plaster - Industrial Plaster	First	47.8-50.6	2.0	10.7-11.4
	Second	53.4-58.5	1.8	8.6-9.6
	Third	53.4-58.5	1.8	9.6-10.6

Spray Painting - Residential & Commercial Work

Description of Work	Coat	No. of Sq.M per Hr.	Hours per 100 Sq.M	Material Coverage Sq.M per Liter
Brick, Tile, & Cement - Latex Paint	First	46.5-48.8	2.1	2.2-2.7
	Second	48.8-51.1	2.0	3.4-3.9
Brick, Tile, & Cement - Oil Paint	First	44.1-46.5	2.2	6.1-6.8
	Second	46.5-48.8	2.1	9.8-11.1

Description of Work	Coat	No. of Sq.M per Hr.	Hours per 100 Sq.M	Material Coverage Sq.M per Liter
Brick - Waterproofing	First	51.1-55.7	1.9	4.9-7.4
	Second	60.4-65.0	1.6	5.5-8.6

Note: For weathered brick, add 15% labor and 35% to 50% for material.

Description of Work	Coat	No. of Sq.M per Hr.	Hours per 100 Sq.M	Material Coverage Sq.M per Liter
Rough Brick, Tile, Cement & Stucco - Silicone	One Coat	74.3-83.6	1.3	2.0-2.2
Stucco - Exterior Latex Paint	First	37.2-39.5	2.6	2.2-2.7
	Second	41.8-46.5	2.3	3.1-3.7
Stucco - Oil Paint	First	27.9-30.2	3.4	4.9-5.5
	Second	30.2-32.5	3.2	8.6-9.8
Stucco - Waterproofing	First	51.1-55.7	1.9	3.7-4.4
	Second	60.4-65.0	1.6	4.4-4.7

Note: For weathered stucco, add 15% for labor and 35% to 50% for material.

Description of Work	Coat	No. of Sq.M per Hr.	Hours per 100 Sq.M	Material Coverage Sq.M per Liter
Shingle Roofs - Oil Paint	First	37.2-41.8	2.5	3.1-3.7
	Second	41.8-46.5	2.3	4.9-5.5
Shingle Roofs - Stain	First	41.8-44.1	2.3	3.1-3.7
	Second	44.1-46.5	2.2	4.9-5.5
Shingle Siding - Oil Paint	First	32.5-34.8	3.0	3.1-3.7
	Second	37.2-39.5	2.6	5.5-6.1
Shingle Siding - Stain	First	37.2-41.8	2.5	3.1-3.7
	Second	41.8-46.5	2.3	4.9-5.5
Smooth Plaster Walls - Flat Finish Paint	First	41.8-44.1	2.3	12.3-13.5
	Second	44.1-46.5	2.2	12.3-13.5
Plaster Walls - Sand Finish	First	37.2-39.5	2.6	11.1-12.3
	Second	39.5-41.8	2.8	9.8-11.1
Smooth Finish Plaster - Industrial Enamel	First	41.8-44.1	2.3	12.3-13.5
	Second	46.5-51.1	2.0	9.8-11.1
	Third	46.5-51.1	2.0	11.1-12.3
Fence, Picket - Oil or Latex Paint	First	44.1-46.5	2.2	12.3-12.9
	Second	44.1-46.5	2.2	13.5-14.7
Fence, Plain - Oil or Latex Paint	First	41.8-46.5	2.3	9.2-9.8
	Second	53.0-55.7	1.8	9.8-11.1
Doors, Garage - Oil or Latex Paint	First	50.2-51.1	2.0	13.5-14.1
	Second	50.2-51.1	2.0	13.8-14.7
Blinds or Shutters - Paint	First	11.1-13.0	8.3	2.9-3.4
	Second	12.1-15.8	7.2	3.4-4.4
	Third	13.9-17.7	6.3	3.7-4.7

Trim must be figured separately as hand work.

Finishing Interior Trim - Residential

Description of Work	Coat	No. of Sq.M per Hr.	Hours per 100 Sq.M	Material Coverage Sq.M per Liter
Preparatory Work for Painting	Sanding	26.9-27.9	3.6	--
	Sand & Putty	17.7-18.6	5.5	--
	Light Sanding	31.6-32.5	3.1	--
Back Priming Interior Trim up to 150 mm wide	meters	44.1-46.5	2.2	27.0-29.5

Description of Work	Coat	No. of Sq.M per Hr.	Hours per 100 Sq.M	Material Coverage Sq.M per Liter
Doors & Windows, Interior - Paint	First	13.0-13.9	7.4	11.7-12.3
	Second	10.7-11.6	9.0	12.3-13.5
	Third	10.7-11.6	9.0	12.3-13.5
Wood Sash - Oiling & Priming	No. of Sash	0.8-0.9	117.6	14.7-17.2
Preparatory Work for Enamel Finish	Sanding	26.9-27.9	3.6	--
	Sand & Putty	11.1-12.1	8.6	--
	Light Sanding	12.5-13.5	7.6	--
Doors & Windows - Enamel Finish	Paint First	13.0-13.9	7.4	14.1-14.7
	Undercoat Second	8.4-9.3	11.3	9.1-9.8
	Enamel Third	9.3-10.2	10.2	11.7-12.3
Four coat work, add 1/2 undercoat, 1/2 enamel	Add'l Coat	8.4-9.3	11.3	10.4-11.1
Base, Chair Rail, Picture Mold, & Other Trim up to 150 mm wide (meters)	First	23.2-24.2	4.2	19.7-22.1
	Second	13.5-14.4	7.1	27.0-29.5
	Third	11.6-12.5	8.3	27.0-29.5
Base, Chair Rail, Picture Mold, & Other Trim up to 150 mm wide - Polyurethane (meters)	First	24.2-26.0	4.0	12.9-14.1
	Second	26.0-27.9	3.7	13.5-13.8
	Third	26.0-27.9	3.7	13.5-13.8
Four coat work, add 1/2 undercoat, 1/2 enamel	Add'l Coat	12.5-13.5	7.7	27.0-29.5
Stain Interior Work	One Coat	20.0-20.9	4.9	17.2-17.8
Stain & Fill Interior Woodwork, Wipe Off	One Coat	6.0-6.5	16.0	9.8-11.1
Shellac Interior Woodwork	One Coat	20.0-20.9	4.9	17.2-17.8
Varnish, Gloss, Interior Woodwork	One Coat	15.3-16.3	6.3	10.4-11.1
Varnish, Flat, Interior Woodwork	One Coat	15.8-16.7	6.2	14.7-15.4
Standing Trim - Wax & Polish	One Coat	9.3-10.2	10.4	13.5-14.7
Standing Trim - Penetrating Stainwax	One Coat	15.8-16.7	10.4	12.9-13.5
	Second	17.7-18.6	5.5	14.7-15.4
	Polishing Coat	17.7-18.6	5.5	--
Sanding for Extra Fine Varnish Finish	Sanding	3.7-4.2	25.3	--
Synthetic Resin Finish, Requires Wiping	First	17.7-18.6	5.5	13.5-16.0
	Second	19.5-20.4	5.0	15.4-16.6
Spackling or Swedish Putty over Flat Trim	One Coat	5.6-6.0	17.2	3.4-3.7
Glazing & Wiping over Enamel Trim	One Coat	5.6-6.0	17.2	25.8-27.0
Brush Stippling Interior Trim - Painted	--	7.9-8.4	12.3	--
Flat Varnishing & Brush Stippling over Glazed Trim	Varnish	15.8-16.7	6.2	14.7-15.4
	Stipple	22.3-23.2	4.4	--

Old Work

Description of Work	Coat	No. of Sq.M per Hr.	Hours per 100 Sq.M	Material Coverage Sq.M per Liter
Washing Average Enamel Finish	Washing	7.9-8.4	12.2	--
Washing Better Grade Enamel Finish	Washing	5.6-6.0	17.2	--
Polishing Better Grade Enamel Finish	Polish	15.8-16.7	6.1	41.8-44.2
Removing Varnish with Liquid Remover	Flat Surfaces	2.8-3.3	32.7	3.7-4.4
Wash, Touch Up, One Coat Varnish	Wash-Touch Up	15.3-16.3	6.3	--
	Varnish	9.3-10.2	10.2	14.1-14.7
Wash, Touch Up, One Coat Enamel	Wash-Touch Up	13.0-13.9	7.4	--
	Enamel	6.5-7.0	14.8	9.2-9.8
Wash, Touch Up, One Coat Undercoat	Wash-Touch Up	13.0-13.9	7.4	--
and One Coat Enamel	Undercoat	7.0-7.9	14.2	11.7-12.3
	Enamel	7.0-7.9	14.2	11.7-12.3
Burning Off Interior Trim	--	1.9-2.3	47.6	--
Burning Off Plain Surfaces	--	2.8-3.3	32.7	--

Finishing Interior Floors

Description of Work	Coat	No. of Sq.M per Hr.	Hours per 100 Sq.M	Material Coverage Sq.M per Liter
Wood Floors - Paint	First	26.9-27.9	3.6	12.9-13.5
	Second	24.2-25.1	4.0	10.4-11.1
Wood Floors - Filling, Wiping	Fill-Wipe	15.8-16.7	6.1	10.4-11.1
Hardwood Floors - Penetrating Stainwax	First	36.2-37.2	2.7	12.9-13.5
	Second	54.8-55.7	1.8	29.5-30.7
Floor Seal	First	45.5-46.5	2.1	14.7-15.4
	Second	54.8-55.7	1.8	29.5-30.7
Shellac	First	36.2-37.2	2.7	12.3-13.5
	Second	44.1-46.5	2.2	16.0-16.6
Stainfill, 1 Shellac, 1 Varnish	Stainfill	24.2-25.1	4.0	11.7-12.3
	Shellac	36.2-37.2	2.7	16.0-16.6
	Varnish	27.9-28.8	3.5	12.3-13.5
Stainfill, 1 Shellac, Wax & Polish	Stainfill	24.2-25.1	4.0	11.7-12.3
	Shellac	36.2-37.2	2.7	16.0-16.6
	Wax/Polish	18.6-19.5	5.2	25.8-26.4
Varnish, Each Coat over Shellac	Varnish	36.2-37.2	2.7	12.3-13.5
Buffing Floors by Machine	Buffing	36.2-37.2	2.7	--
Waxing over 2 Coats of Seal & Polish	Wax-Polish	18.6-19.5	5.2	25.8-26.4

Old Work

Description of Work	Coat	No. of Sq.M per Hr.	Hours per 100 Sq.M	Material Coverage Sq.M per Liter
Removing Varnish with Liquid Remover	Flat Surfaces	3.7-4.2	25.3	4.2-4.4

FINISHES 9.193

Description of Work	Coat	No. of Sq.M per Hr.	Hours per 100 Sq.M	Material Coverage Sq.M per Liter
Clean, Touch Up, & Varnish	Clean-Touch Up	11.6-12.5	8.3	--
	Varnish	27.9-28.8	3.5	6.2-13.5
Clean, Touch Up, Wax, & Polish	Clean-Touch Up	11.6-12.5	8.3	--
	Wax-Polish	18.6-19.5	5.2	25.8-26.4

Painting Interior Walls

Description of Work	Coat	No. of Sq.M per Hr.	Hours per 100 Sq.M	Material Coverage Sq.M per Liter
Taping, Beading, Spotting Nail Heads, & Sanding Gypsum Wallboard		8.4-10.2	10.7	--
Texture over Gypsum Wallboard	One Coat	23.2-25.1	4.1	0.20-0.25
Casein or Resin Emulsion over Textured Gypsum Wallboard	One Coat	27.9-29.7	3.4	8.0-8.6
Sizing New Smooth Finish Walls	Sizing	32.5-37.2	2.8	14.7-17.2
Wall Sealer or Primer on Smooth Walls	Sealer	18.6-20.9	5.1	14.1-15.4
Wall Sealer or Primer on Sand Finish Walls	Sealer	11.6-12.5	8.3	6.8-7.4
Smooth Finish Plaster - Flat Finish	Sealer	18.6-20.9	5.1	14.1-15.4
	Second	15.3-16.3	6.3	12.3-13.5
	Third	16.3-17.2	5.9	14.1-15.4
	Stippling	17.7-18.6	5.5	--
Sand Finish Plaster - Flat Finish	Sealer	11.6-12.5	8.3	6.8-7.4
	Second	12.5-13.5	7.7	9.8-11.1
	Third	11.6-12.5	8.3	10.4-11.7
Smooth Finish Plaster - Gloss or Semi-Gloss	Sealer	18.6-20.9	5.1	14.1-15.4
	Second	15.8-16.7	6.1	12.3-13.5
Third		13.0-13.9	7.4	12.3-13.5
	Stippling	7.9-8.8	12.0	--
Sand Finish Plaster - Gloss or Semi-Gloss	Sealer	11.6-12.5	8.3	6.8-7.4
	Second	10.7-11.6	9.0	6.8-9.8
	Third	11.6-12.5	8.3	9.8-11.1
Texture Plaster, Average - Semi-Gloss	Sealer	9.3-10.2	10.2	6.1-6.8
	Second	10.7-11.6	9.0	7.9-8.6
	Third	11.6-12.5	8.3	9.2-9.8
Smooth Finish Plaster - Latex Rubber Paint	First	13.0-14.9	7.1	7.4-8.6
	Second	13.0-14.9	7.1	7.4-8.6
Sand Finish or Average Texture Plaster - Latex Rubber Paint	First	10.2-12.1	8.9	6.1-7.4
	Second	10.2-12.1	8.9	6.1-7.4
Glazing & Motting over Smooth Finish Plaster		7.4-8.4	12.6	25.8-26.4
Glazing & Motting over Sand Finish Plaster		5.6-6.0	17.2	21.5-22.1
Glazing & Highlighting Textured Plaster		7.4-8.4	12.6	20.3-20.9
Starch & Brush Stipple over Painted Glazed Surface		10.7-11.6	8.9	--

Description of Work	Coat	No. of Sq.M per Hr.	Hours per 100 Sq.M	Material Coverage Sq.M per Liter
Flat Varnish & Brush Stipple over Painted Glazed Surface		9.8-10.7	9.7	12.3-13.5
Textured Oil Paint over Smooth Finish Plaster	Size	20.0-20.9	4.9	16.6-17.8
	Texture	3.7-4.2	25.2	3.1-4.3
Water Textured over Smooth Finish Plaster	Size	20.0-20.9	4.9	17.2-17.8
	Texture	4.6-5.1	20.6	2.1-2.3
New Smooth Plaster - Latex Base Paint	First	17.7-18.6	5.5	12.3-13.5
	Second	19.5-20.4	5.0	16.0-17.2
on Rough Sand Finish	One Coat	13.9-14.9	6.9	8.0-8.6
on Cement Blocks	One Coat	12.5-13.5	7.7	7.4-8.0
on Acoustical Surfaces	One Coat	12.5-13.5	13.0	4.9-5.5
Casein Paint over Smooth Finish Plaster	First	25.1-26.0	3.91	12.3-13.5
	Second	27.9-28.8	3.50	12.3-13.5
over Rough Sand Finish Plaster	One Coat	23.2-24.2	4.2	8.0-8.6
over Cement Blocks	One Coat	15.3-16.3	6.3	7.4-8.0
over Acoustical Surfaces	One Coat	15.3-16.3	6.3	4.9-5.5
over Cinder Concrete Blocks	One Coat	11.6-12.5	8.3	3.1-3.7

When applying paint to walls and ceilings using a roller applicator, increase quantities 10% to 15% and reduce painter time about the same amount.

Old Work

Description of Work	Coat	No. of Sq.M per Hr.	Hours per 100 Sq.M	Material Coverage Sq.M per Liter
Washing Off Calcimine, Average Surfaces		10.7-11.6	9.0	--
Washing Smooth Finish Plaster Walls, Average		13.5-14.4	7.1	--
Washing Sand Finish Plaster Walls, Average		9.3-10.7	10.0	--
Washing Starched Surfaces & Restarching, Smooth Surfaces	Washing	13.5-14.4	7.10	--
	Restarching	12.1-13.0	7.9	--
Removing Old Wallpaper, 3 layers or less		6.0-7.0	15.4	--
Washing Off Glue after Removing Paper (including fixing average cracks & sizing)		11.6-12.5	8.3	--
Cutting Hard Oil or Varnish Size Walls (including fixing average cracks)		11.6-12.5	8.3	--
Cutting Gloss Painted Walls (incl. fixing cracks)		11.6-12.5	8.3	--
Wash, Touch Up, One Coat Gloss Paint to Smooth Plaster Surfaces	Wash-Touch Up	11.6-12.5	8.3	--
	One Coat	11.6-12.5	8.3	11.1-12.3

Description of Work	Coat	No. of Sq.M per Hr.	Hours per 100 Sq.M	Material Coverage Sq.M per Liter
Synthetic-Resin Emulsion Paint over Old Painted Walls	One Coat	17.7-18.6	5.5	9.8-10.6

Wallpaper, Canvas, Coated Fabrics, Paper Hanging, Wood Veneer

Canvas Sheeting		4.65-5.57	19.5	--
Coated Fabrics	Single Roll	0.19-0.23	--	--
Wallpaper				
One Edge Work	Single Roll	0.33-0.37	--	--
Wire Edge Work	Single Roll	0.23-0.28	--	--
Butt Work	Single Roll	0.21-0.26	--	--
Scenic Paper 40"x60"	Single Roll	0.09-0.12	--	--
Wood Veneer Flexwood	sq.m	1.67	--	--
Lacquer Finish over Wood Veneer	First Coat	9.3	1.00	6.8-8.0
	Second	9.3	1.00	11.1-12.3
Penetrating Wax or Synthetic Resin Application over Wood Veneer	First	9.3	10.7	13.5-14.7
	Second	9.3	10.7	16.0-16.6

Prices for Colors in Oil
Colors Ground Fine in Pure Linseed Oil (0.23-Liter Cans)

	per half-pint	per liter
Blacks		
Drop Black	$0.80	$3.38
Lamp Black	1.40	5.92
Blue		
Palco Blue	3.20	13.53
Prussian Blue	2.60	10.99
Cobalt Blue	2.60	10.99
Ultramarine Blue	2.60	10.99
Brown		
Raw Turkey Umber	1.20	5.07
Burnt Turkey Umber	1.20	5.07
Raw Italian Sienna	1.30	5.49
Burnt Italian Sienna	1.30	5.49
Vandyke Brown	1.20	5.07
Green		
Palco Green	3.50	14.79
Chrome Green, L.	2.10	8.88
Chrome Green, M.	2.10	8.88
Chrome Green, D.	2.10	8.88
Red		
Vermilion	2.70	11.41
Rose Lake	2.20	9.30

Venetian Red .... 1.80 .... 7.61
   Yellow
Chrome Yellow, L. .... 3.20 .... 13.53
Chrome Yellow, M. .... 3.20 .... 13.53
Chrome Yellow, O. .... 3.20 .... 13.53
Ochre .... 3.20 .... 13.53

Shellac and Varnish

Kind of Material	per gal.	per liter
Shellac, White, 4-lb. Cut	$18.00-22.00	$4.80-5.
Shellac, Orange, 4-lb. Cut	16.00-20.00	4.20-5.
Cabinet Finish Varnish	20.00-26.00	5.30-6.
Floor Varnish	19.00-25.00	5.00-6.
Flat Varnish	19.00-25.00	5.00-6.
Spar Varnish	19.00-29.00	5.00-7.
Varnish Remover, Liquid	15.00-25.00	4.00-6.
Varnish Remover, Paste	15.00-29.00	4.00-7.
Wallpaper Lacquer	12.00-14.00	3.20-3.
Wallpaper Lacquer Thinner	8.00-10.00	2.10-2.
Lacquer	20.00-25.00	5.30-6.
Lacquer Thinner	10.00-12.00	2.60-3.

Enamels and Enamel Undercoating

Kind of Material	per gal.	per liter
Enamel, Non-Yellowing White	$13.00-15.00	$3.40-4.
Architectural Enamel Paint	13.00-18.00	3.40-4.
Enamel Undercoater	11.00-15.00	2.90-4.
Enamel, White	11.00-15.00	2.90-4.
Enamel, Quick Drying	11.00-16.00	2.90-4.
Enamel, Industrial	13.00-18.00	3.40-4.

Wall Size, Sealer, Paste, Etc.

Kind of Material	Price	Price (metric
Sizing, prepared per gal./liter	$5.00-9.00	$1.32-2.
Glue, Dry for Sizing per lb./kg	0.75-1.25	1.65-2.
Wall Primer and Sealer per gal./liter	8.00-12.00	2.11-3.
Wallpaper Cleaner per lb.	0.75	1.65
Starch, gloss per lb./kg	1.00	2.20
Patching Plaster per lb./kg	1.00	2.20
Spackle per lb./kg	0.75	1.65
Glazing Liquid per gal./liter	8.00-12.00	2.11-3.
Pre-mixed Vinyl Adhesive per lb./kg	1.00	2.20

Wood Fillers and Stains

Kind of Material	per gal.	per liter
Oil Stains	$10.00-15.00	$2.60-4.00
Acid Stains	10.00-15.00	2.60-4.00
Penetrating Stainwax	10.00-15.00	2.60-4.00
Stain, Varnish, all colors	12.00-30.00	3.20-7.90
Stain, Fill, and Seal	10.00-12.00	2.60-3.20
Paste Wood Fillers	8.00-11.00	2.10-2.90
Floor Seal	12.00-15.00	3.20-4.00

Floor Wax

Kind of Material	per gal.	per liter
Liquid Floor Wax	$12.00-20.00	$3.20-5.30
Self-Polishing Wax	7.00-9.00	1.80-240
Paste Wax per lb. (kg)	6.00-8.00	13.20-17.60

Miscellaneous Paints

Kind of Material	per gal.	per liter
Ready Mixed House Paint - Latex	$10.00-15.00	$2.60-4.00
House Paint—One Coat	9.00-12.00	2.40-3.20
Porch and Floor Paint	10.00-15.00	2.60-4.00
Flat Wall Paint - Latex	8.00-18.00	2.10-4.80
One Coat Wall Paint - Oil Base	9.00-15.00	2.40-4.00
Metal Paint for Structural Steel, Bridges, Stacks, etc.	15.00-25.00	4.00-6.60
Cement Floor Paint	8.00-16.00	2.10-4.20
Cement Floor Paint - Rubber Base	8.00-16.00	2.10-4.20
Aluminum Paint	15.00-18.00	4.00-4.80
Vinyl-Chloride Co-Polymer	25.00	6.60
Sand Finish	7.00-10.00	1.80-2.60
Polyurethane	10.00-18.00	2.60-4.80
Drywall Primer	6.00-10.00	1.60-2.60
Waterproof Sealer, Clear	9.00-20.00	2.40-5.30

Lettering On Glass or Wood

The cost of lettering will vary with the style and size of letters and whether painted or gold leaf. Costs given on the following page include only Roman or Gothic style letters, such as are ordinarily used for signs on doors, transoms, and show windows, and are based on all work being performed by experienced sign writers.

Plain Black Letters 2" to 4" (50-100 mm) High. For plain black Gothic letters and numerals, such as appear on windows, transoms, and office doors, an experienced sign writer should average about 25 letters per hr. at the following labor cost per 100 letters:

	Hours	Rate	Total	Rate	Total
Sign painter	4.0	$....	$....	$36.21	$144.84
Cost per letter			....		1.45

Plain Black Letters 6" to 12" (150-300 mm) High. An experienced sign writer should average 12 to 14 plain Gothic letters 6" to 12" (150-300 mm) high, per hr. at the following labor cost per 100 letters:

	Hours	Rate	Total	Rate	Total
Sign painter	7.7	$....	$....	$36.21	$278.82
Cost per letter			....		2.79

Gold Leaf Letters 2" to 4" (50-100 mm) High. An experienced sign writer should complete about 5 plain Roman or Gothic letters in gold leaf per hr. (this class of work proceeds much slower than painted letters) at the following labor cost per 100 letters:

	Hours	Rate	Total	Rate	Total
Sign painter	20.0	$....	$....	$36.21	$724.20
Cost per letter			....		7.24

Gold Leaf Letters 6" to 12" (150-300 mm) High. An experienced sign writer should complete 2 to 3 plain Roman or Gothic letters 6" to 12" (150-300 mm) high, in gold leaf, per hr. at the following labor cost per 100 letters:

	Hours	Rate	Total	Rate	Total
Sign painter	40.0	$....	$....	$36.21	$1,448.40
Cost per letter			....		14.48

Painting Pipe

The cost of painting pipe varies considerably, depending on accessibility of work, layout of piping, scaffolding required, type of paint, whether one color or more, and in the case of old work, the amount of surface preparation required and whether or not pipes are in service.

Quantities are expressed in square feet of surface. When pipe quantities are measured from drawings, determine whether or not pipes are covered and the covering thickness, which greatly affects the diameter and consequently the surface area to be painted, especially in the smaller sizes.

Most estimators convert lin. ft. of pipe to sq. ft. of surface (meters to square meters) according to the following table:

O.D. Pipe or Covering, inches	S.F. area per l.f.	O.D. Pipe or Covering, mm	Sq.M area per meter
3" and under	1	75 and under	0.30
4" - 7"	2	100 - 175	0.61
8" - 11"	3	200 - 275	0.91
12" - 15"	4	300 - 375	1.22
16" - 19"	5	400 - 475	1.52
20" - 22"	6	500 - 550	1.83
23" - 26"	7	575 - 650	2.13
27" - 30"	8	675 - 750	2.44
31" - 34"	9	775 - 850	2.74
35" - 38"	10	875 - 950	3.05
39" - 42"	11	975 - 1050	3.35
42" - 45"	12	1050 - 1125	3.66
46" - 48"	13	1150 - 1200	3.96

Various types of paint materials are specified for pipes, depending on the function, exposure, whether bare or covered, etc. Average material coverages and approximate prices for a few of these materials are as follows:

Description Bare Pipe	S.F. per gal.	Sq.m per liter	Price per gal.	Price per liter
Black asphalt paint, one coat	130-150	39.61-45.71	$9.00-15.00	$2.37-3.96
Aluminum paint, each coat	550-600	167.60-182.83	15.00-18.00	3.96-4.75
Radiator enamel, each coat	550-600	167.60-182.83	18.00-25.00	4.75-6.60
Covered Pipe				
Sizing, one coat	350-400	106.65-121.89	5.00-9.00	1.32-2.37
Paint (oil base), each coat	450-500	137.13-152.36	9.00-19.00	2.37-5.02
Industrial enamel, 1st Coat	450-500	137.13-152.36	15.00-25.00	3.96-6.60
Industrial enamel, 2nd Coat	550-600	167.60-182.83	15.00-25.00	3.96-6.60

Labor Painting Pipe. Labor costs for painting pipes are extremely variable due to the many conditions that might be encountered.

For simple work, consisting of horizontal runs and risers within stepladder reach, all painted one color, a painter should paint 800 s.f. (72 sq.m) of pipe, one coat, per 8-hr. day.

For boiler room or steam power plant piping of average difficulty, where some high ladder work is required, a painter should paint 600 s.f. (54 sq.m) of pipe, one coat, per 8-hr. day.

For difficult industrial process piping or in very congested steam power plants, where most of the work requires high scaffolding, a painter will paint only 350 to 450 s.f. (31-40 sq.m) of pipe, one coat, per 8-hr. day.

If pipes are painted different colors for identification, reduce the above production 10% to 15%.

If pipes are coded by color banding at 10' (3 m) to 15' (4.5 m) intervals, consider this work as another coat for labor, but practically no material will be required.

If painted flow direction indicators are required after the pipes have been painted, then add $7.50 each.

Painting Structural Steel

The cost of painting structural steel after erection will depend upon a number of factors, i.e., the weight of the steel, whether light or heavy sections; access to work and method of painting, whether brush painting by hand or using a spray.

Lightweight members cost more to paint than heavy members, and the less climbing required, the more paint a worker will apply, either by brush or when using a spray.

Structural steel may be estimated, either by the ton of steel or by the number of square feet of surface to be painted.

The ton method is much easier to compute but the square feet of surface method is much more accurate.

The following tables, giving the approximate number of square feet of area on various size beams, angles, and channels, will assist the estimator in computing accurate quantities where the square foot method is used.

S.F. Area per L.F. (Sq.M Area per Meter)
of Structural "S" Beams of Various Sizes

Depth Beam, inches (mm)	Lbs. per l.f.	Kg per meter	S.F. area per l.f.	Sq.M area per meter
3" (75 mm)	5-8	7.4-11.9	1.25	0.381
4" (100 mm)	7-10	10.4-14.9	1.40	0.427
5" (125 mm)	10-15	14.9-22.3	1.90	0.579
6" (150 mm)	12-17	17.9-25.3	2.00	0.609
7" (175 mm)	15-20	22.3-30.0	2.60	0.792
8" (200 mm)	18-25	26.8-37.2	2.70	0.823
9" (225 mm)	21-35	31.2-52.0	2.90	0.884
10" (250 mm)	25-40	37.2-60.0	3.40	1.036
12" (300 mm)	31-55	46.1-81.8	3.70	1.127
15" (375 mm)	42-75	62.5-111.6	4.50	1.371
18" (450 mm)	54-70	80.3-104.2	5.00	1.524
20" (500 mm)	65-100	96.7-148.8	5.40	1.645
24" (600 mm)	80-120	119.0-178.6	6.40	1.950
27" (675 mm)	90-135	134.0-201.0	7.50	2.285

S.F. Area per L.F. (Sq.M Area per Meter)
of Structural Channels of Various Sizes

Depth Channel, inches (mm)	Lbs. per l.f.	Kg per meter	S.F. area per l.f.	Sq.M area per meter
3" (75 mm)	4-6	6.0-9.0	1.00	0.305
4" (100 mm)	5-7	7.4-10.4	1.20	0.366
5" (125 mm)	7-11	10.4-16.3	1.50	0.457
6" (150 mm)	8-16	12.0-23.8	1.70	0.518
7" (175 mm)	10-20	14.9-30.0	2.00	0.609
8" (200 mm)	11-21	16.4-31.2	2.20	0.670
9" (225 mm)	14-25	20.9-37.0	2.35	0.716
10" (250 mm)	15-35	22.3-52.0	2.70	0.823
12" (300 mm)	21-40	31.2-60.0	3.00	0.914
15" (375 mm)	32-50	47.6-74.0	3.70	1.127
18" (450 mm)	34-55	51.1-81.8	4.40	1.341
20" (500 mm)	42-58	62.5-86.3	5.00	1.524

S.F. Area Per L.F. (Sq.M Area per Meter)
of Structural Angles of Various Sizes

Depth Channel, inches (mm)	Lbs. per l.f.	Kg per meter	S.F. area per l.f.	Sq.M area per meter
2"x2" (50 x 50 mm)	2-5	14-36.2	0.70	0.213
3"x3" (75 x 75 mm)	2-11	14-79.5	1.00	0.305
4"x4" (100 x 100 mm)	5-20	36-145	1.35	0.411
5"x5" (125 x 125 mm)	8-31	58-224	1.70	0.518
6"x6" (150 x 150 mm)	13-40	94-289	2.00	0.609
8"x8" (200 x 200 mm)	26-57	188-412	2.70	0.823
4"x3" (100 x 75 mm)	4-17	29-123	1.20	0.366
5"x3" (125 x 75 mm)	6-19	43-137	1.40	0.427
6"x4" (150 x 100 mm)	10-30	72-217	1.70	0.518
7"x4" (175 x 100 mm)	13-34	94-246	1.90	0.579
8"x4" (200 x 100 mm)	17-37	123-286	2.00	0.640
8"x6" (200 x 150 mm)	20-50	145-362	2.40	0.731

Painting Lightweight Structural Steel After Erection. When painting, by hand and using a brush, lightweight structural steel after erection, a painter should paint 600 to 650 s.f. (54-60 sq.m) per 8-hr. day on the first coat; 800 to 900 s.f. (74-83 sq.m) per 8-hr. day on the second and third coats.

On a tonnage basis, this would be at the rate of 2 to 3 tons (1.32-2.73 metric tons) per 8-hr. day on the first coat and 3 to 4 tons (2.73-3.64 metric tons) per 8-hr. day on the second and third coats.

When using a spray gun, a painter should paint 3,600 s.f. (334 sq.m) per 8-hr. day on the first coat and 4,800 s.f. (446 sq.m) on the second and third coats.

Painting Medium to Heavyweight Structural Steel After Erection When painting, by hand using a brush, medium to heavyweight structural steel after erection, a painter should paint 750 to 850 s.f. (70-79 sq.m) on the first coat and 900 to 1,000 s.f. (83-93 sq.m) per 8-hr. day on the second and third coats.

On a tonnage basis, this would be at the rate of 4 to 5 tons (3.64-4.55 metric tons) per 8-hr. day on the first coat and 5 to 6 tons (4.55-5.46 metric tons) per 8-hr. day on the second and third coats.

Using a spray gun, a painter should paint 4,400 s.f. (408 sq.m) per 8-hr day on the first coat and 5,600 s.f. (520 sq.m) per 8-hr. day on the second and third coats.

Labor Cost of 100 S.F. (10 Sq.M) 2-Coat Oil Paint
Applied to Lightweight Structural Steel

	Hours	Rate	Total	Rate	Total
First Coat					
Painter	1.3	$....	$....	$26.83	$34.88
Second Coat					
Painter	1.0	....	....	26.83	26.83
Cost 100 s.f. (1 sq.)			$....		$61.71
per sq.m				....	6.64

For third coat, add same as second coat.

Spray Painting

	Hours	Rate	Total	Rate	Total
First Coat					
Painter	0.23	$....	$....	$26.83	$6.17
Second Coat					
Painter	0.17	....	....	26.83	4.56
Cost 100 s.f. (1 sq.)			$....		$10.73
per sq.m				....	1.15

For third coat, add same as second coat.
Add for air compressor and engineer, depending upon number of spray guns operated.

Labor Cost of 100 S.F. (10 Sq.M) 2-Coat Oil Paint
Applied to Medium to Heavyweight Structural Steel

	Hours	Rate	Total	Rate	Total
First Coat					
Painter	1.0	$....	$....	$26.83	$26.83
Second Coat					
Painter	0.9	....	....	26.83	24.15
Cost 100 s.f. (1 sq.)			$....		$50.98
per sq.m				....	5.49

For third coat, add same as second coat.

Spray Painting

First Coat	Hours	Rate	Total	Rate	Total
Painter	0.18	$....	$....	$26.83	$4.83
Second Coat					
Painter	0.15	....	....	26.83	4.02
Cost 100 s.f. (1 sq.)			$....		$8.85
per sq.m				....	0.95
For third coat, add same as second coat.					

Add for air compressor and engineer, depending on number of spray guns operated. Not included in above costs are equipment such as scaffolding, and costs of complying with hazardous waste regulations.

Preparing Steel Surfaces

Proper protection of steel depends upon proper adhesion of the protective paint. When rust, oils, grease, scale, and other dirt are not completely removed, the paint coat or coats will fail no matter how skillfully they are applied.

The Steel Structures Painting Council (SSPC), Pittsburgh, PA publishes detailed specifications for cleaning methods.

The simplest method is hand cleaning using scrapers, sanders, wire brushes, and chipping tools. These remove only loose rust and scale and effectiveness depends largely on the zeal of the worker. Power tool cleaning, using power operated sanders and brushes, is slightly more expensive and slightly more effective.

For a superior job, one of the blasting methods is suggested: commercial, near white or white. The last specifies the metal be blasted down to a uniform "white" surface and will provide a base good for 7 to 8 years. Near white specifies that at least 95% of the metal will meet the "white" surface test, and its durability almost equals white blast.

Commercial blast is difficult to specify, but the results should leave at least two-thirds of the metal in the "white" stage, with durability of from 6 to 7 years.

With all the above methods, it is essential to use a primer with a good "wetting" power.

In figuring all these methods, cleaning up and protection for adjacent surfaces can add materially to the cost.

Surface Preparation by Hand and Abrasive Method. A properly prepared surface can make a good painting job look even better. A poorly prepared one can make an otherwise good job look bad. Remember that paint is not only to make a surface attractive. It protects a material from exposure and prolongs the useful life of the material. Paint protects steel surfaces from oxidation that results in rust.

Estimators need to understand the types of paints and coatings that an architect or engineer might specify for steel. Knowing the proper procedures for applying a specific painting material will give the estimator a competitive edge. The following checklist indicates major items to consider:

1. Type of surface to be painted.
2. Type of surface preparation - scrape or sandblast.
3. Scaffolding and other equipment requirements (OSHA).
4. Federal, state, and local environmental protection agency requirements.
5. Proper applicators - brushes, rollers, or sprayers.
6. Priming requirements - compatible with final material.
7. Number of paint coats required.
8. Disposal and clean up requirements.

The method of cleaning the surface in preparation for painting is usually dictated by the specifications, which might refer to the paint manufacturer's recommendations. Wood surfaces are usually limited to a few choices of methods, but steel and concrete offer a wide range of possible methods, depending on the condition of the existing surface. Scaled or rusted surfaces can be cleaned by hand, with power tools, or using abrasive methods. A required degree of cleanliness is often stated in the contract documents and is defined by SSPC specifications.

Cleaning Structural Steel with Hand Wire Brush (SSPC-SP 2). When cleaning structural steel surfaces by hand with a wire brush, a worker can clean about 300 s.f. (28 sq.m) of surface per 8-hr. day at the following labor cost per 100 s.f. (9 sq.m):

	Hours	Rate	Total	Rate	Total
Painter	2.67	$....	$....	$26.83	$71.64
Cost per s.f.				....	0.72
per sq.m				....	7.75

Cleaning Piping 14" (350 mm) Diameter or Less with Hand Wire Brush. When cleaning steel pipe surfaces by hand with a wire brush, a worker can clean about 225 s.f. (21 sq.m) of surface per 8-hr. day at the following labor cost per 100 s.f. (9 sq.m):

	Hours	Rate	Total	Rate	Total
Painter	3.20	$....	$....	$26.83	$85.86
Cost per s.f.				....	0.86
per sq.m				....	9.25

Cleaning Structural Steel with Power Tools (SSPC-SP 3). When cleaning rust and scale from structural steel members, bridges, gas holders, oil tanks, and other metal surfaces before repainting, the number of sq. ft. cleaned per 8-hr. day varies with the type of surface and accessibility. A worker using air-powered wire brushes should clean 1,000 to 1,200 s.f. (93-111 sq.m) per day at the following cost per 100 s.f. (9 sq.m):

	Hours	Rate	Total	Rate	Total
Painter	0.72	$....	$....	$26.83	19.32
Compressor Operator	0.36	....	....	28.55	10.28
Compressor Expense & Tools	0.36	....	....	35.00	12.60
Cost per 100 s.f.				$....	$42.20
per s.f.				....	0.42
per sq.m				....	4.52

Cleaning Piping 14" (350 mm) Diameter or Less with Power Tools. When cleaning rust and scale off structural steel pipe, the number of sq. ft. cleaned per day will vary with surface and accessibility. A worker should clean 750 to 900 s.f. (70-84 sq.m) per day using air-powered wire brushes at the following cost per 100 s.f. (9 sq.m):

	Hours	Rate	Total	Rate	Total
Painter	0.97	$....	$....	$26.83	$26.03
Compressor Operator	0.49	....	....	28.55	13.99
Compressor Expense & Tools	0.49	....	....	35.00	17.15
Cost per 100 s.f.				$....	$57.17
per s.f.				....	0.57
per sq.m				....	6.13

Add scaffolding or rigging, clean up, and disposal.

Commercial Blast Cleaning of Structural Steel (SSPC-SP 6). A blast cleaned surface should be free of all visible (viewed without magnification) oil, grease, dirt, mill scale, rust, paint, oxides, corrosion products, and other foreign matter, except staining. This should provide a base good for 7 to 8 years when the surface is properly painted.

A worker should clean 500 to 600 s.f. (46-56 sq.m) of structural steel members, bridges, etc., per day using abrasive blasting methods.

	Hours	Rate	Total	Rate	Total
Painter	1.45	$....	$....	$26.83	$38.90
Compressor Operator	1.45	....	....	28.55	41.40
Compressor Expense & Tools	1.45	....	....	25.00	36.25
Sandblasting Equipment	1.45	....	....	30.00	43.50
Cost per 100 s.f.				$....	$160.05
per s.f.				....	1.60
per sq.m				....	17.22

Add for scaffolding or rigging, abrasive material, clean up, and disposal.

Commercial Blast Cleaning of Piping, 14" (350 mm) Diameter and Less.

A worker should clean 300 to 500 s.f. (28-46 sq.m) of steel pipe per day using abrasive blasting methods at the following cost:

	Hours	Rate	Total	Rate	Total
Painter	2.0	$....	$....	$26.83	$53.66
Compressor Operator	2.0	....	....	28.55	57.10
Compressor Expense & Tools	2.0	....	....	25.00	50.00
Sandblasting Equipment	2.0	....	....	30.00	60.00
Cost per 100 s.f.			$....		$220.76
per s.f.				....	2.21
per sq.m				....	23.78

Add for scaffolding or rigging, abrasive material, clean up, and disposal.

Brush Off Blast Cleaning of Structural Steel (SSPC-SP 7).

A brush off blast cleaned surface should be free of all visible (viewed without magnification) oil, grease, dirt, dust, loose mill scale, loose rust, and loose paint. Tightly adherent mill scale, rust, and paint that cannot be removed by lifting with a dull putty knife may remain on the surface.

On structural steel members, bridges, etc., a worker should clean 900 to 1,100 s.f. (83-102 sq.m) per day using this method:

	Hours	Rate	Total	Rate	Total
Painter	0.80	$....	$....	$26.83	$21.46
Compressor Operator	0.80	....	....	28.55	22.84
Compressor Expense & Tools	0.80	....	....	25.00	20.00
Sandblasting Equipment	0.80	....	....	30.00	24.00
Cost per 100 s.f.			$....		$88.30
per s.f.				....	0.88
per sq.m				....	9.47

Add for scaffolding or rigging, abrasive material, clean up, and disposal.

Brush Off Blast Cleaning Piping 14" (350 mm) Diameter or Less.

On steel pipes, a worker should clean 680 to 820 s.f. (63-76 sq.m) per day using this method at the following cost:

	Hours	Rate	Total	Rate	Total
Painter	1.07	$....	$....	$26.83	$28.71
Compressor Operator	1.07	....	....	28.55	30.55
Compressor Expense & Tools	1.07	....	....	25.00	26.75
Sandblasting Equipment	1.07	....	....	30.00	32.10
Cost per 100 s.f.			$....		$118.11
per s.f.				....	1.18
per sq.m				....	12.70

Add for scaffolding or rigging, abrasive material, clean up, and disposal.

Near White Blast Cleaning of Structural Steel (SSPC-SP 10). A near-white blast cleaned surface, when viewed without magnification, should be free of all visible oil, grease, dirt, dust, mill scale, rust, paint, oxides, corrosion products, and other foreign matter, except for staining.

On structural steel members, bridges, etc., a worker using this method should clean 300 to 400 s.f. (28-37 sq.m) per day at the following cost:

	Hours	Rate	Total	Rate	Total
Painter	2.29	$....	$....	$26.83	$ 61.44
Compressor Operator	2.29	....	....	28.55	65.38
Compressor Expense & Tools	2.29	....	....	25.00	57.25
Sandblasting Equipment	2.29	....	....	30.00	68.70
Cost per 100 s.f.			$....		$252.77
per s.f.				....	2.53
per sq.m				....	27.22

Add for scaffolding or rigging, abrasive material, clean up, and disposal.

Near White Blast Cleaning Piping 14" (350 mm) Diameter and Under. On steel pipes, a worker should clean 225 to 300 s.f. (21-28 sq.m) per day:

	Hours	Rate	Total	Rate	Total
Painter	3.05	$....	$....	$26.83	$ 81.83
Compressor Operator	3.05	....	....	28.55	87.08
Compressor Expense & Tools	3.05	....	....	25.00	76.25
Sandblasting Equipment	3.05	....	....	30.00	91.50
Cost per 100 s.f.			$....		$336.66
per s.f.				....	3.37
per sq.m				....	36.26

Add for scaffolding and rigging, abrasive material, clean up, and disposal.

White Metal Blast Cleaning Structural Steel (SSPC-SP 5). Viewed without magnification, a white metal blast cleaned surface should be free of all visible oil, grease, dirt, dust, mill scale, rust, paint, oxides, corrosion products, and other foreign matter.

On structural steel members, bridges, etc., a worker should clean 230 to 300 s.f. (21-28 sq.m) per day using this method:

	Hours	Rate	Total	Rate	Total
Painter	3.01	$....	$....	$26.83	$ 80.76
Compressor Operator	3.01	....	....	28.55	85.94
Compressor Expense & Tools	3.01	....	....	25.00	75.25
Sandblasting Equipment	3.01	....	....	30.00	90.30
Cost per 100 s.f.			$....		$332.25
per s.f.				....	3.32
per sq.m				....	35.72

Add for scaffolding and rigging, abrasive material, clean up, and disposal.

White Metal Blast Cleaning Piping, 14" (350 mm) Diameter or Less.

On steel pipes, a worker should clean 170 to 230 s.f. (16-21 sq.m) per day using this method:

	Hours	Rate	Total	Rate	Total
Painter	4.00	$....	$....	$26.83	$ 107.32
Compressor Operator	4.00	....	....	28.55	114.20
Compressor Expense & Tools	4.00	....	....	25.00	100.00
Sandblasting Equipment	4.00	....	....	30.00	120.00
Cost per 100 s.f.			$....		$441.52
per s.f.				....	4.42
per sq.m				....	47.56

Add for scaffolding and rigging, abrasive material, clean up, and disposal.

Spray Painting. Almost any painting job can be done quicker with spray equipment. This is particularly true where large surfaces are to be covered or where there are angles and corners that cannot be reached with brushes or rollers.

Practically all surfaces that are painted with a bristle brush can be coated by spray painting. On rough surfaces such as cement, stucco, rough plaster, brick, tile, shingles, and rough boards, spray painting affects a considerable saving in time and labor.

It is now possible to obtain mechanical painting equipment that will apply any paint, varnish, graphite, mineral oxides, or epoxy.

Painting Oil Tanks and other Large Metal Surfaces by Compressed Air. In painting oil tanks and other large metal surfaces, a worker should cover 1,000 s.f. (92 sq.m) per hr. at the following cost per 100 s.f. (9 sq.m):

	Hours	Rate	Total	Rate	Total
Painter	0.10	$....	$....	$26.83	$ 2.68
Compressor operator	0.05	....	....	28.55	1.43
Compressor expense	0.05	....	....	15.00	0.75
Cost per 100 s.f.			$....		$4.86
per s.f.				....	0.05
per sq.m				....	0.54

Does not include paint, scaffold, or preparation.

Comparative Cost by Hand

On the same class of work, a painter will brush about 200 s.f. (19 sq.m) of surface per hr. at the following cost per 100 s.f. (9 sq.m):

	Hours	Rate	Total	Rate	Total
Painter	0.5	$....	$....	$26.83	$13.42
Cost per s.f.				....	0.13
per sq.m				....	1.40

Does not include paint, scaffold, or preparation.

09950 WALLCOVERING

Wallpaper is estimated by the roll containing 36 s.f. (3.3 sq.m). Most rolls are 18" (450 mm) or 20-1/2" (512.5 mm) wide and contain 36 s.f. or 4 sq. yds. (3.24 sq.m) of paper. The double roll contains twice as much as the single roll: 72 s.f. or 8 sq. yds. (6.48 sq.m) of paper.

Estimating Quantity of Wallpaper for Any Room. When estimating the quantity of wallpaper required for any room, measure the entire distance around the room in lin. ft. (meters) and multiply by the height of the walls or the distance from the floor to the ceiling. The result will be the number of sq. ft. (sq. m) of surface in the four walls.

Make deductions in full for the area of all openings such as windows, doors, mantels, and built-in bookcases. The difference between the area of the four walls and the area of the openings will be the actual number of sq. ft. (sq.m) of surface to be papered.

When adding for waste in matching, cutting, fitting, etc., paper hangers use different methods and allow different percentages, contingent upon the height of the ceiling, the design of the paper, and the size of the pattern or figure, but an allowance of 15% to 20% is sufficient in nearly all cases.

When estimating the quantity of paper required for walls, bear in mind it is not necessary to run the side wallpaper more than one or two inches (25-50 mm) above the border, when a border is used. The height of the wood base should also be deducted from the total height of the wall.

In rooms having drop ceilings, the depth of the drop should be deducted from the height of the wall, which will decrease the area of side wallpaper in the same amount as it increases the quantity of ceiling paper.

If a border is required, measure the distance around the room and the result will be the number of lin. ft. (meters) of border. Dividing by 3, gives the number of yards of border required.

When estimating the quantity of wallpaper required for ceilings, multiply the width of the room by the length, and the result will be the number of sq. ft. (sq.m) of ceiling to be papered.

In rooms having a drop ceiling, add the depth of the drop or the distance the ceiling paper extends down the side walls to the length and breadth of the room and multiply as described above.

After the areas of both side walls and ceiling have been computed in sq. ft., divide by 36 (sq.m divide by 3.3), and the result will be the number of single rolls of paper required, or by dividing by 72 (sq.m divided by 6.7) gives the number of double rolls of paper required.

Paste Required Hanging Wallpaper. The quantity of paste required for hanging wallpaper will vary with the weight of the paper and the surface to which it is applied.

Where light or medium weight wallpaper is used, one gallon (3.79 liters) of paste should hang 12 single rolls of paper.

If heavy or rough texture paper is used, it will often be necessary to give it two or three applications of paste to obtain satisfactory results. On work of this kind one gallon of paste should hang 4 to 6 single rolls of paper.

There are prepared pastes on the market that require only the addition of cold water to make them ready for use.

One lb. of prepared cold water dry paste should make 1-1/2 to 2 gals. (5.7-7.6 liters) of ready to use paste.

Labor Hanging Wallpaper. Different methods of hanging wallpaper are used in different sections of the country. In some localities a paper hanger and helper work together, one worker trimming the paper and pasting while the other hangs the paper. However, in many cities the work is performed by one person, who cuts, trims, fits, pastes, and hangs the paper.

The following quantities and costs are based on the performance of one worker but will prove a fair average for use in any locality.

Hanging Wallpaper on Walls and Ceilings, One Edge Work. When hanging light or medium weight paper on ceilings and drops, a paper hanger should hang 28 to 32 single rolls of paper per 8-hr. day, at the following labor cost per roll:

	Hours	Rate	Total	Rate	Total
Paper hanger	0.27	$....	$....	$26.83	$7.24

Hanging Wallpaper on Walls, Butt Work. Where light or medium weight paper is used for bedrooms, halls, etc., and where a good grade of workmanship is required with all paper trimmed on both edges and hung with butt joints, a paper hanger should trim, fit, and hang 20 to 24 single rolls of paper per 8-hr. day, at the following labor cost per roll:

	Hours	Rate	Total	Rate	Total
Paper hanger	0.36	$....	$....	$26.83	$9.66

Another thing that will slow up the work is where new paper is applied over old rough textured wallpaper, such as oatmeal designs, etc.

Hanging Wallpaper, First Grade Workmanship. Where a good grade of medium or heavy weight wallpaper is used, with all paper hung with butt joints, a paper hanger should trim, fit, and hang 16 to 20 single rolls of paper per 8-hr. day, at the following labor cost per roll:

	Hours	Rate	Total	Rate	Total
Paper hanger	0.44	$....	$....	$26.83	$11.81

Labor Hanging Scenic Paper. Where scenic paper or paper having mural designs are used over the wall or up stairways, a paper hanger will hang only 8 to 10 single rolls per 8-hr. day, at the following labor cost per roll:

	Hours	Rate	Total	Rate	Total
Paper hanger	0.9	$....	$....	$26.83	$24.15

Placing Wallpaper Borders. Wallpaper borders are usually estimated by the l.f. or yd. (meter) and as the width varies from 3" to 18" (75-450 mm) there is never over one width required.

A paper hanger should place border at the rate of 100 to 125 l.f. (30-37.5 m) per hr. The labor cost per 100 l.f. (30 m) should run as follows:

	Hours	Rate	Total	Rate	Total
Paper hanger	0.9	$....	$....	$26.83	$ 24.15
Cost per l.f.			....		0.24
per yd.			....		0.72
per m			....		0.78

Hanging Coated Fabrics. Where coated fabrics, such as Walltex, Sanitas, etc., are used, a paper hanger should hang 16 to 20 single rolls per 8-hr. day, at the following labor cost per roll:

	Hours	Rate	Total	Rate	Total
Paper hanger	0.44	$....	$....	$26.83	$ 11.81

Canvassing Plastered Walls. When applying canvas to plastered walls, an average mechanic should apply 50 to 60 s.f. (4.6-5.6 sq.m) per hour.

Labor Cost of 100 S.F. Canvassed Walls

	Hours	Rate	Total	Rate	Total
Paper hanger	1.82	$....	$....	$26.83	$48.83
Cost per s.f.			....		0.49

Flexwood. Flexwood from U.S. Plywood, Louisville, KY, is a genuine wood veneer cut to 1/85" (0.29 mm) thick and glued under heat and hydraulic pressure to cotton sheeting with a waterproof adhesive. A patented flexing operation breaks the cellular unity of the wood to produce a limp, pliable sheet which may be applied to any smooth surface, flat or curved. Waterproof Flexwood 710 Adhesive, which makes a permanent bond, is used to apply Flexwood. Standard sizes of stock material are 18" (450 mm) and 24" (600 mm) wide and 8' (2.4 m) to 12' (3.6 m) long.

The mechanics of installing Flexwood are as follows: the background is sized with Flexwood cement. Another coating of cement is brushed on the Flexwood, which is then hung in the manner of any sheet wall covering. A stiff, broad knife, used with considerable pressure, smoothes out the Flexwood, removes air spaces and furnishes the necessary contact.

Some of the wood available in Flexwood are mahogany, walnut, oak, prima vera, knotty pine, orientalwood, satinwood, zebrawood, rosewood, English oak, maple, lacewood, etc. The cost of the material varies from $1.20 to $3.75 per s.f. ($12.90-40.36 per sq.m) depending upon the kind of wood. Most woods run from $1.50 to $1.75 ($16.10-18.80 per sq.m). Adhesive costs $9.50 per gal. ($2.50 per liter).

It requires a skilled and thoroughly competent paper hanger to apply this material and the labor costs vary considerably with the design to be obtained.

Considerable time is usually required to lay out any room to obtain the desired effect, spacing of strips, etc. This is especially true where certain designs must be obtained on columns, walls, and pilasters and where narrow strips of contrasting woods are used to obtain inlay effects.

On plain walls without inlays, a paper hanger experienced in this class of work should apply 150 s.f. (14 sq.m) of Flexwood per 8-hr. day, at the following labor cost per 100 s.f. (9 sq.m):

	Hours	Rate	Total	Rate	Total
Paper hanger	5.4	$....	$....	$26.83	$144.88
Cost per s.f.			....		1.45
per sq.m			....		15.60

On pilasters, columns, and walls inlaid with narrow strips of contrasting woods, the work is considerably slower, due to the additional cutting and fitting required. On work of this kind, a paper hanger, experienced in this class of work should apply 75 s.f. (9 sq.m) per 8-hr. day, at the following labor cost per 100 s.f. (9 sq.m):

	Hours	Rate	Total	Rate	Total
Paper hanger	10.7	$....	$....	$26.83	$287.08
Cost per s.f.			....		2.87
per sq.m			....		30.88

Additional time for paper hanger or foreman will be required laying out the work into patterns, marking off strips, etc. This will vary with the size of the job.

After the Flexwood is applied, the joints should be sanded lightly to remove any imperfections in the joints.

Finishing Flexwood. Flexwood will take any wood finish, but where the natural color of the wood is desired, the Flexwood is given one coat of lacquer sealer, sanded lightly between coats and then given one coat of lacquer for a finish coat.

One gallon (3.79 liters) of lacquer sealer will cover about 300 s.f. (28 sq.m) of surface, one coat.

After the sealer has been applied, one gallon (3.79 liters) of lacquer should cover 450 to 550 s.f. (42-51 sq.m).

An experienced painter should apply lacquer sealer, sand lightly between coats and apply one coat of lacquer to 90 to 110 s.f. (8-10 sq.m) of surface per hr. at the following labor cost per 100 s.f. (9 sq.m):

	Hours	Rate	Total	Rate	Total
Painter	1.0	$...	$...	$26.83	$26.83
Cost per s.f.			....		0.27
per sq.m			....		2.90

Vinyl Wall Covering. Vinyl wall covering is composed of a woven cotton fabric, to which a compound of vinyl resin, pigment, and plasticizer is electronically fused to one side. It comes 24" (600 mm), 27" (675 mm), and 54" (1350 mm) wide in three weights: heavy, 36 oz. per yd. (1.11 kg per m); medium, 24-33 oz. per yd. (0.74-1.02 kg per m); and light, 22-24 oz. per yd. (0.68-0.74 kg per m).

Some of the patterns available simulate linens, silks, moires, grasses, tweeds, Honduras mahogany, travertine, damasks, and burlaps. Costs range from $0.35 to 0.90 per sq. ft. ($3.76-9.69 per sq.m).

Vinyl fabric is hung using regular wallpaper hanging procedures for hanging fabric baked wallcoverings. A broad knife is used to smooth out any wrinkles, air pockets and insures a good contact. Wash off any excess paste that remains on the surface of the material. No further finishing is necessary.

A competent paper hanger can apply approximately 600 s.f. (56 sq.m) per 8 hr. day at the following costs per 100 s.f. (9 sq.m):

	Hours	Rate	Total	Rate	Total
Paper hanger	1.34	$...	$...	$26.83	$ 35.95
Cost per s.f.			....		0.36
per sq.m			....		3.87

If excessive cutting and fitting is required, the above figures should be increased to allow for the above conditions.

CHAPTER 10

SPECIALTIES

CSI DIVISIONS 10
10100 CHALKBOARDS AND
TACKBOARDS
10150 COMPARTMENTS AND
CUBICLES
10301 PREFABRICATED FIREPLACES
10350 FLAGPOLES
10400 IDENTIFYING DEVICES
10500 LOCKERS
10550 POSTAL SPECIALTIES
10600 PARTITIONS
10650 SCALES
10670 STORAGE SHELVING
10700 SUN CONTROL DEVICES
10750 TELEPHONE ENCLOSURES
10800 TOILET AND BATH
ACCESSORIES

The Construction Specifications Institute (CSI), whose format we use in the general presentation of this text, follow sections covering the usual building trades with four sections covering allied fields, which may or may not be part of the typical general contract.

CSI Division 10 includes specialties that are prefabricated and added to the building, such as chalkboards, toilet partitions, decorative grilles, freestanding fireplaces, flagpoles, directories and signs, lockers, mailboxes and chutes, movable partitions, sun control devices, telephone booths, and the like. Many of these items will be subcontracted and installed under the general contractor's supervision. Some may be covered by allowances.

10100 CHALKBOARDS AND TACKBOARDS

Chalkboards are made of 1/4" (6 mm) or 1/2" (13 mm) hardboard with special finishes applied. These include the following which are available in either black or green:

1/4" (6 mm) Thick - Material Prices

	Price per s.f.	Price per sq.m
Standard, weighing 1.75 lbs. per s.f. (8.75 kg per sq.m)	$1.60	$17.77
Moisture sealed at 1.8 lbs. per s.f. (9 kg per sq.m)	1.85	20.55
Porcelain steel at 2.1 lbs. per s.f. (10.5 kg per sq.m)	4.80	53.33

These are available in 4' (1.2 m) widths in sheets to 16' (4.8 m) long and are applied to the wall with mastic at the rate of one gal. per 70 s.f. (1 liter per 1.67 sq.m). They can be set in custom trim, or the following stock aluminum trim, which comes complete with all necessary attaching clips:

	Price per l.f.	Price per m
Perimeter trim 1-1/2" (38 mm) wide	$2.00	6.56
Maprail headtrim	2.25	7.38
Joint strip	1.18	3.87
Chalktrough 2-1/4" (56 mm) wide	4.70	15.41

Bulletin board materials are 1/4" (6.25 mm) cork and available as follows:

	Price per s.f.	Price per sq.m
Natural cork	$2.10	$23.33
Vinyl cork	3.10	34.44
Cork on 3/8" (9 mm) fiberboard	3.15	35.00
1/8" (3 mm) on 3/8" (9 mm) fiberboard vinyl f'c'd	3.10	34.44

The cork comes in rolls 4' (1.2 m) wide by up to 90' (27 m) long. The panels are 4' (1.2 m) wide up to 16' (4.8 m) long. Stock trim costs $2.00 for 1/4" (6 mm) stock, $2.15 for 1/2" (13 mm) stock.

10150 COMPARTMENTS AND CUBICLES

Toilet Partitions and Screens

Toilet partitions are most often floor-mounted metal units, with baked enamel finish and chrome hardware. One door and side partition for use in a corner or abutting other stall side partitions cost about $185.00. A porcelain enamelled finish adds $300.00 to the cost. Ceiling hung units cost around 10% more, plus of course the cost of ceiling reinforcement if needed. One carpenter and laborer will set a unit in about two hours. Urinal screens hung from the wall cost $115.00 in baked enamel and will take a two-person team half an hour to set. Wall hung toilet stalls will cost $355.00 per unit.

10301 PREFABRICATED FIREPLACES

Metal fireplaces are available in a wide range of styles for burning wood or gas. They are made for placing against walls or in corners, and as freestanding units. However, there are requirements for clearances from walls and for floor protection. They can be vented to an existing flue or have a prefabricated smoke pipe terminating above the roof. A simulated masonry chimney top is available. Finishes are either matte black or porcelain. A 24" (600 mm) wide unit costs $390.00 in matte, $500.00 in porcelain. A 45" (1125 mm) wide unit costs $530.00 in matte and $645.00 in porcelain.

Freestanding units open on all sides will cost $475.00 for matte and $590.00 for porcelain in 36" (900 mm) diameter size; $550.00 and $675.00 in 42" (1050 mm) diameter. If they are to rest on the floor, add $55.00 for the metal base. If hung from the ceiling, add $30.00 for chains.

The 8" (200 mm) metal flues cost about $22.00 per l.f. ($73.33 per m). Chimney top housings for single flues cost $85.00 to $115.00.

10350 FLAGPOLES

Typical prices for several types of flagpoles are as follows:

The 40'-0" (12 m) ground type, swaged sectional steel, including fittings, ground tube, and flashing collar, costs $1,500.00 erected in place and painted.

The 40'-0" (12 m) ground type, cone tapered steel, including fittings, ground tube, and flashing collar, costs $1,700.00 erected in place.

To above prices, add cost of concrete foundation and ornamental base if required.

The 20'-0" (6 m) roof type, swaged sectional steel, including fittings, socket plate, tension braces, and flashing collars, costs $850.00, erected in place and painted. Add cost of setting anchor bolts.

The 12'-0" (3.6 m) outrigger type, cone tapered aluminum, including fittings and base is priced at $900.00 erected in place. Add cost of setting anchor bolts.

10400 IDENTIFYING DEVICES

Directories and Signs

Directory boards are available in glass-enclosed or open-faced models. Glass door units cost $390.00 for 36"x30" (900 x 750 mm), $550.00 for 36"x48" (900 x 1200 mm). Open faced directory boards cost as follows, with 100 letters:

12"X18" (300 x 450 mm) $75.00

24"X18" (600 x 450 mm) 85.00
24"X36" (600 x 900 mm) 115.00

Letters for directory boards are ordered in an assortment, including numbers. An assortment of 200 will cost as follows: 1/2" (13 mm), $33.00; 3/4" (19 mm), $39.00; 1" (25 mm), $50.00; 1-1/2" (38 mm), $60.00.

Plaques

Building committees often will ask the contractor to order a plaque to commemorate a donor, an outstanding citizen, or even themselves. Cast bronze is the ultimate for permanence and costs about $132.00 per s.f. ($1,466.67 per sq.m), aluminum is 20% less, as are built-up plaques.

Signs

Cast letters are often applied to walls and copings to identify the building occupant. Letters are cast from aluminum or bronze and then given a variety of finishes. Prices will cost as follows:

Material & Finish	2"	4"	6"	12"	18"	24"
	50 mm	100 mm	150 mm	300 mm	450 mm	600 mm
Cast Aluminum	$8.00	$13.00	$24.00	$37.00	$66.00	$94.00
Cast Bronze	12.00	24.00	33.00	66.00	100.00	132.00

A worker should install 3 to 4 letters per hr. Plastic letters will cost $12.00 each in 8" (200 mm) height.

10500 LOCKERS

Lockers are made in single, double, and triple high units totaling 72" (1800 mm) high, single tier also comes 60" (1500 mm) high, in widths 9" (225 mm), 12" (300 mm), 15" (375 mm), and 18" (450 mm) and depths of 12" (300 mm), 15" (375 mm), and 18" (450 mm). There are also box lockers 6 units high.

A 12"x15"x72" (300 x 375 x 1800 mm) unit in baked enamel will cost about $40.00 single tier; $55.00 double tier; $75.00 triple tier. For a sloping top add $2.50 per tier. For closed base, add $3.00 per l.f. ($10.00 per m). Open mesh or "athletic" types will cost $90 to $110 per tier 72" (1800 mm) high. Basket racks 8' (2.4 m) high, 5' (1.5 m) wide cost $125.00 plus $6.50 per basket. For a combination lock, add $20.00 per door, $10.00 for a key lock.

Benches for lockers are sold by the lineal foot (meter). Benches made of 1-1/4" (31 mm) thick laminated maple cost $20.00 per foot plus $25.00 for each pedestal.

10550 POSTAL SPECIALTIES

Mail chutes with glass and aluminum enclosures 9"x3-1/2" (225 x 88 mm) cost about $60.00 per l.f., and one worker can install two floor heights per day. A lobby collection box will cost from $1,200.00 on up, depending on the size and design.

Keyed mail delivery boxes in gangs will cost around $55.00 per unit rear loading and $60.00 front loading. One worker can set around 30 units per day.

10600 PARTITIONS

Movable partitioning is available in metal in both flush and panel construction, in laminated gypsum board and made of open wire mesh. Solid units can be specially constructed for sound and fire isolation, and can be prefinished with baked enamel, vinyl, or wood veneers or painted in the shop or on the job.

Metal office type partitioning is usually a nominal 2-1/4" (56 mm) or 3-1/2" (88 mm) thick with a height limit of 12 l.f. (3.6 m). Free standing units have a standard cornice height of 7'-1-1/2" (2.14 m). "Banker" types are usually 1-3/4" (44 mm) thick and 30" (750 mm) high. A standard glass panel can be mounted on the top rail to bring the height to 42" (1050 mm) or 54" (1350 mm).

The base for movable partitioning is designed to form a raceway for both electrical and telephone lines. Switches can usually be built into the door frame panels, which with doors, glazing panels and louver panels can all be ordered as part of the partitioning.

Mesh partitioning is usually made of 10 ga. wire worked into a 1-1/2" (38 mm) diamond mesh set in a channel frame. It will cost about $2.00 per s.f. ($22.22 per sq.m) on straight runs using standard sections. A 3'x7' (0.9 x 2.1 m) door will cost $170.00. A 5'x7' (1.5 x 2.1 m) sliding panel, $300.00. A shop coat of paint is standard. A baked enamel coat will add $0.20 per s.f. ($2.22 per sq.m). Two erectors can install some 100 l.f. (30 m) per day. Add 1.5 hrs. for each door. Ceiling panels will install at the rate of 75 s.f. (6.75 sq.m) per hour with two persons working together.

Office partitions come in so many grades and job conditions are so diverse that it is difficult to give any basic costs, but because the field is highly competitive, it is always a simple matter to get prices together. A standard 2-3/8" 20 ga. flush steel floor to ceiling partition 8' to 9' (2.4-2.7 m) high will cost about $5.00 per s.f. ($55.55 per sq.m), without installation. Two workers should erect about 40 l.f. (12 m) per day. Each door with lock but no closer will cost $230.00. If partitions are glazed add for both labor and material.

Gypsum partitions, 2-1/4" (56 mm) thick, will run about $3.25 per s.f. ($36.11 per sq.m) installed but unfinished. If glazed panels are desired, add $13.00 per s.f. ($144.44 per sq.m).

Freestanding "bank" type partitioning 3' (0.9 m) high cost $30.00 to $40.00 per l.f. ($100.00-133.00 per m) Two workers will set 70 l.f. (21 m) per day.

10650 SCALES

Built-in scales must be very carefully set, adjusted, and maintained and should be set by a contractor licensed by the scale manufacturer. For general budget allowances the following is a guide:

Platform Size, feet (meters)	Capacity, tons (metric tons)	Cost of Mechanical	Cost of Electronic
7'x9' (2.1 x 2.7 m)	5 (4.5)	$9,400	
10'x24' (3.0 x 7.2 m)	10 (9)	10,600	
10'x50' (3.0 x 15 m)	30 (27)	12,500	
10'x70' (3.0 x 21 m)	50 (45)	30,000	35,000
10'x70' (3.0 x 21 m)	60 (54)	31,250	37,500
10'x70' (3.0 x 21 m)	80 (72.5)	35,600	40,000
10'x70' (3.0 x 21 m)	100 (90)		45,000

10670 STORAGE SHELVING

Free standing, baked enamel industrial shelving comes knocked down in 3' (0.9 m) sections, complete with all hardware for erection at the job site. Cost will approximate the following:

Depth, in. (mm)	Price per l.f. (m)			
	4 Shelves per l.f.	4 Shelves per m	7 Shelves per l.f.	7 Shelves per m
12" (300 mm)	$29.00	$96.67	$48.00	$160.00
18" (450 mm)	34.00	113.33	56.00	186.67
24" (600 mm)	40.00	133.33	65.00	216.67

One worker can erect from 9 to 12 l.f. (2.7-3.6 m) per day.

10700 SUN CONTROL DEVICES

Awnings and canopies of canvas, vinyl, and aluminum are prefabricated in the shop and delivered to the job site ready for erection.

Continuous, wall hung, aluminum canopies will cost from $4.75 to $7.00 per s.f. ($52.77-77.77 per sq.m), and two workers can erect about 250 s.f. (22.55 sq.m) per day.

Awnings over individual openings 3'-6" (1.05 m) wide will cost $300.00 in aluminum, $200.00 in cloth. Awnings are often used as decorative features and take forms unrelated to sun control and can easily cost up to $35.00 per s.f. ($388.88 per sq.m). Two workers will take two hours to hang an individual awning.

10750 TELEPHONE ENCLOSURES

The trend in telephone noise control has been to rely on sound absorption rather than physical separation. Where booths are still used, a simple steel and wood unit costs $1,500.00 while an extruded metal unit costs twice as much. Two workers can set three units in two days.

Outdoor post types will cost about $700.00 and take one worker 6 hours to set.

Indoor shelf types will average about $400.00 a station and take one worker 3 hours to install. Directory shelves will cost about $300.00 for a size that holds up to four books, and will take one worker one hour to set.

10800 TOILET AND BATH ACCESSORIES

The following list may serve as a guide in purchasing chrome plated bathroom accessories. Sometimes accessories are of ceramic tile, costs of which are discussed under *Finishes*. Also, some items may be furnished with toilet compartments or will be furnished by service companies who supply the owner. Check to avoid duplication.

Prices of Toilet and Bath Accessories

Ash tray bowl	$35.00
wall-mounted	75.00
recessed	
Clothes hook	3.00
single	3.50
double	
Grab bar	10.00
12" (300 mm)	16.00
24" (600 mm)	225.00
Hand dryer, electric	65.00
Medicine cabinet, 18"x24" (450 x 600 mm), lighted	
Paper towel dispenser	25.00
wall-mounted	240.00
recessed with waste receptacle	
Sanitary napkin dispenser	200.00
coin operated & surface-mounted	18.00
Shelf, stainless steel, 24" (600 mm) long	
Shower rod	9.00
3' (0.9 m)	12.00
5' (1.5 m)	7.00
Soap dish, recessed	

Soap dish with grab bar.. 10.00
Soap dispenser
  powder... 15.00
  liquid 22.00
Toilet tissue dispenser 8.00
Towel bar
  12" (300 mm).. 12.00
  18" (450 mm).. 12.50
  24" (600 mm).. 14.00
Towel ring 9.00
Towel ladder, 16"x30" (400 x 750 mm)... 42.00
Tumbler/toothbrush holder 8.00
Waste receptacle, 10 gal (37.85 liters)... 85.00

10900 WARDROBE SPECIALTIES

Hat and coat checking storage units will cost approximately the following:

| | Price per l.f. (m) | | | |
Material	Single Tier per l.f.	Single Tier per m	Double Tier per l.f.	Double Tier per m
Steel	$20.00	$66.67	$22.00	$73.33
Aluminum	29.00	96.67	35.00	116.67
Stainless	35.00	116.67	40.00	133.33

One worker can set 45 l.f. (13.5 m) of single tier and 32 l.f. (9.6 m) of double tier rack per day.

CHAPTER 11

EQUIPMENT

CSI DIVISION 11
11900 RESIDENTIAL EQUIPMENT

CSI Division 11 includes equipment such as vacuum systems, powered window washers, bank, checkroom, church, school, food service, vending, athletic, lab, laundry, library, medical, parking, waste handling, loading dock, detention, and stage equipment.

These are often purchased by the owner separately and installed by the general contractor. Even if the general contractor is not involved with the installation of such equipment, he should have a list of all equipment going into the building, with cuts and shop drawings, so that he can point out any conflicts between the structure and such equipment. A floor that has to be broken up to conceal a conduit or duct becomes a permanent scar that reflects on the general contractor, even though he might be an innocent partner.

11900 RESIDENTIAL EQUIPMENT

Kitchen Cabinets

The following cabinets are furnished completely assembled and finished. Frames are 3/4" kiln dried hardwood; doors and drawer fronts are 7/8" thick solid lumber; shelves are 1/2" plywood or chipboard; backs of base cabinets are Masonite; backs of wall cabinets and drawer bottoms are plywood; ends of cabinets are flush with hardwood faces.

Cabinets are factory finished with a penetrating sealer coat and clear lacquer. Interiors are finished and drawer bottoms are clear lacquered. Door and drawer hardware come in a variety of styles.

These cabinets are the type found in most residential construction. Economy grade cases for cheaper construction range 25% to 50% less for finished units ready for installation. For top quality kitchen cabinets, the prices will run up to 50% higher for stock sizes and 100% higher or more for custom-made units, depending on the design, materials, hardware, and finish specified. All prices are f.o.b. factory.

Wood Wall Units, 12" (300 mm) Deep

Width	No. Doors	Height 12" (300 mm)	15" (375 mm)	24" (600 mm)	30" (750 mm)
12" (300 mm)	1				$60.00
15" (375 mm)	1				66.00
18" (450 mm)	1				70.00
21" (525 mm)	1				74.00
24" (600 mm)	2			67.00	85.00
27" (675 mm)	2			70.00	95.00
30" (750 mm)	2	53.00	58.00	75.00	104.00
33" (825 mm)	2	58.00	60.00	81.00	109.00
36" (900 mm)	2	60.00	67.00	83.00	114.00
42" (1050 mm)	2			89.00	120.00
48" (1200 mm)	2			100.00	143.00

Corner Shelf and Wall Units, 30" (750 mm) High

Corner wall unit, 23" (575 mm) deep and wide,
  with 2 shelves $100.00
Corner wall unit, 23" (575 mm) deep and wide, with revolving
  3 shelf "Lazy Susan" 143.00

Utility Cabinet and Built-In Oven Cabinets

Utility cabinet, 84" (2100 mm) high, 24" (600 mm) deep,
  18" (450 mm) wide $222.00
Oven cabinet, 84" (2100 mm) high, 24" (600 mm) deep,
  24" (600 mm) wide 265.00
Oven cabinet, 84" (2100 mm) high, 24" (600 mm) deep,
  27" (675 mm) wide 280.00

Wood Cabinet Base Units

Base units are 34-1/2" (863 mm) high and 24" (600 mm) deep without tops.

Description	Width, in. (mm)	Without Tops
4 drawers	15" (375 mm)	$127.00
4 drawers	18" (450 mm)	132.00
4 drawers	21" (525 mm)	140.00
4 drawers	24" (600 mm)	148.00
1 top drawer, 1 door below	12" (300 mm)	91.00
1 top drawer, 1 door below	15" (375 mm)	97.00
1 top drawer, 1 door below	18" (450 mm)	103.00
1 top drawer, 1 door below	21" (525 mm)	113.00
1 top drawer, 1 door below	24" (600 mm)	127.00
1 top drawer, 2 doors below	27" (675 mm)	138.00

EQUIPMENT

Description	Width, in. (mm)	Without Tops
2 top drawers, 2 doors below	36" (900 mm)	170.00
2 top drawers, 2 doors below	42" (1050 mm)	180.00
2 top drawers, 2 doors below	48" (1200 mm)	195.00
Corner base cabinet	42" (1050 mm)	150.00
Corner base cabinet with revolving "Lazy Susan	35" (875 mm)	160.00
3" (75 mm) corner filler		6.00

Wood Sink Fronts

Description	Width	Price
Sink front only, 1 pair doors	27" (675 mm)	$72.00
	30" (750 mm)	74.00
	33" (825 mm)	77.00
	36" (900 mm)	83.00
	42" (1050 mm)	89.00
	48" (1200 mm)	96.00
Sink end	24" (600 mm)	12.00

Wood Sink Base Units

Width, in. (mm)	Price Without Tops or Sinks
27" (675 mm)	$117.00
30" (750 mm)	122.00
33" (825 mm)	127.00
36" (900 mm)	133.00
42" (1050 mm)	143.00
48" (1200 mm)	154.00

Steel Kitchen Units

Prices on steel case units vary considerably with different manufacturers, due to weight of metal used, method of manufacturing, kind of finish, sprayed or baked, and volume of production. Many kinds of steel cases are on a mass production basis, while others are custom made to fit the job. Naturally, the custom cabinets cost more.

The cabinets priced below are completely finished and ready to be installed. They are fabricated from 18, 20, and 22 gauge cold rolled steel and are spot welded, bonderized, and then finished with two coats of enamel baked on at high temperature. They are medium quality.

Base units are mounted on an integral black enameled subbase 4" (100 mm) high and recessed 3" (75 mm) in front for toe space.

Doors are 3/4" (19 mm) thick, double-walled, die formed of one piece of steel, and spot welded. Drawers have 3/4" (75 mm) fronts that make them flush with the doors. Drawers operate on roller bearings.

Hardware consists of semi-concealed hinges. Die-cast chromium plated handles are furnished on doors and drawers.

Steel Wall Cabinets

Wall units are 13" (325 mm) deep, 18" (450 mm) units have 1 removable shelf, and 30" (750 mm) units have 2 shelves.

Width	No. Doors	Height, in. (mm) 18" (300 mm)	30" (750 mm)
12" (300 mm)	1		$39.00
15" (375 mm)	1		41.00
18" (450 mm)	2	39.00	46.00
21" (525 mm)	2	42.00	51.00
24" (600 mm)	2	45.00	53.00
27" (675 mm)	2		57.00
30" (750 mm)	2	48.00	60.00
36" (900 mm)	2	55.00	68.00

Corner Shelf and Wall Units, 30" (750 mm) High

Description	Price Each
Corner shelf unit, 12-1/4" (306 mm) deep, 6" (150 mm) wide, with 2 fixed shelves	$20.00
Corner shelf unit, 12-1/4" (306 mm) deep and wide, with 2 fixed shelves	23.00
Corner wall unit, 21" (525 mm) deep and wide, with revolving 3-shelf "Lazy Susan"	110.00

Steel Base Cabinets
Base units—34-1/2" (863 mm) high and 24" (600 mm) deep

Description	Width, in. (mm)	Without Top
1 drawer above, 1 door below	12" (300 mm)	$72.00
1 drawer above, 1 door below	15" (375 mm)	80.00
1 drawer above, 1 door below	18" (450 mm)	86.00
1 drawer above, 2 doors below	21" (525 mm)	90.00
1 drawer above, 2 doors below	24" (600 mm)	100.00
1 drawer above, 2 doors below	27" (675 mm)	108.00
1 drawer above, 2 doors below	30" (750 mm)	115.00
1 drawer above, 2 doors below	36" (900 mm)	129.00
3 drawers	18" (450 mm)	123.00
Corner base cabinet with revolving "Lazy Susan"	33" (825 mm)	133.00

Steel Sink Fronts

Description	Width	Price
Sink front only—one pair doors	24" (600 mm)	$44.00
Sink front only—one pair doors	30" (750 mm)	45.00
Sink front only—one pair doors	36" (900 mm)	51.00

Fillers and End Panels for Steel Cabinets

Description	Price Each
Corner wall filler, 30" (750 mm) high, 15" (375 mm) deep, 15" (375 mm) wide	$15.00
Straight wall filler, 18" (450 mm) high, 13" (325 mm) deep, 3" (75 mm) wide	7.00
Straight wall filler, 30" (750 mm) high, 13" (325 mm) deep, 3" (75 mm) wide	9.00
Corner base filler, 34-1/2-" (863 mm) high, 27" (675 mm) deep, 27" (675 mm) wide	22.00
Straight base filler, 34-1/2" (863 mm) high, 3" (75 mm) wide	5.00
End panel, 34-1/2" (863 mm) high, 25" (625 mm) deep	12.00

Steel Kitchen Sink Units

Sinks are 36" (900 mm) high, 25" (625 mm) deep. Constructed of 18, 20, or 22 ga. furniture steel with a baked on finish. Sink tops are 14 ga. deep drawn enameling steel to which porcelain enamel is bonded.

Tops have grooved drainboards and 4" (100 mm) backsplash, all insulated to deaden noise. Sink tops are one piece construction, roll rimmed.

Length Unit	Size Bowl	Single or Dbl. Bowl	Price Each
42" (1050 mm)	20"x16-1/2"x7" (500 x 413 x 175 mm)	Single	$220.00
54" (1350 mm)	20"x16-1/2"x7" (500 x 413 x 175 mm)	Single	250.00
66" (1650 mm)	Two 14-1/2"x 17-1/2"x7" (363 x 438 x 175 mm)	Double	300.00

KITCHEN COUNTERTOP

Most all countertops for kitchen cabinets are plastic laminate over a plywood, particleboard, or Masonite subsurface.

The plastic decorative laminate is offered in a wide variety of colors and patterns including simulated wood grain and butcher-block styles.

The tops are usually made in self-edge design (square edges and 90° angle for the backsplash), however they are available in post-form design (rolled edges and angles) at a slight decrease in cost.

To estimate the total lineal feet of countertop, measure the length of the area to be covered along the longest side (usually the wall surface). Plastic laminate tops, made to size, ready to install will cost from $20.00 to $30.00 per l.f. ($66.67-100.00 per m).

Dupont *Corian* is a synthetic countertop material that can be manufactured in one piece. Costs installed range from $110.00 to $200.00 per l.f. ($366.67-666.67 per m).

CHAPTER **12**

FURNISHINGS

CSI DIVISION 12

CSI Division 12 covers furnishings. Separate contracts are usually let for this work, but again, some coordination is necessary, and the general contractor should have a clear understanding regarding damages to the building by other parties working on the premises, especially if painting and decorating are part of the general work. Also, furnishings and draperies can play havoc with a carefully balanced air handling system. If the owner is uncomfortable, the general contractor is the one he calls, even though it is the 10-foot sofa that blocks the baseboard heater or the $5,500.00 chandelier that interrupts the ceiling outlet airflow and causes a draft on the boss's neck. The architect should be the coordinator of all this, but an alert contractor can forestall many problems.

Typical furnishings supplied by the contractor and coordinated with the architect and interior decorator are:

Modular furniture
Planters for plants
Benches
Bulletin boards
Cabinets
Booths
Trash receptacles
Fire extinguishers
Blinds, shades, or other window decorations
Refrigeration
Shredders
Shop equipment
Gym equipment
Recreation equipment
Mailroom equipment
Special lighting
Special sound system

Local labor practices often determine which trades unload and position special furnishings. Laborers might unload benches to be placed loose in an area, but skilled workers might be needed for benches to be positioned and secured.

CHAPTER 13

SPECIAL CONSTRUCTION

CSI DIVISION 13
13700 SPECIAL PURPOSE ROOMS & BUILDINGS
CONCRETE STORAGE BINS AND SILOS

CSI Division 13 covers such items as air supported structures, audiometric rooms, and hyperbaric rooms. We have chosen to include in this chapter a section on concrete storage bins and silos.

13700 SPECIAL PURPOSE ROOMS

Sauna Rooms

The luxury residence or office is often fitted with a prefabricated sauna cabinet. These come in a wide variety of sizes and there are several manufacturers. More than one source should be contacted.

The minimum sauna, for sitting position only, is about 4'x4' (1.2 x 1.2 m), weighs 500 lbs. (225 kg), and needs a separate 120 volt 20 AMP service. The cost complete with door, floor, ceiling, bench, and equipment is about $2,000.00. A 4'x7' (1.2 x 2.1 m) unit, for one lying down or 2 sitting, weighs 750 lbs. (337.5 kg) and requires 220 volt 30 AMP service and costs about $3,300.00. To accommodate several people, units up to 7'x10' (2.1 x 3.0 m) weigh 1200 to 1500 lbs. (540 to 675 kg), require 220 to 250 AMP service, and cost $4,000.00 to $7,500.00.

These units are prefabricated and simple to install but require a carpenter, a helper, and an electrician. Except for the largest units, or when preparatory work is required, the installation should be completed in a day.

CONCRETE STORAGE BINS AND SILOS

Concrete storage bins and silos are used for storing grain, coal, sand, gravel, and the like. In addition there are those used for the processing and preserving of forages for feed on the farm.

Silos are constructed of numerous types of materials such as concrete stave, cast-in-place concrete, concrete block, glazed tile, wood, and steel. The cost of storage silos varies with the storage capacity, the type of structure, and the kind of materials to be stored.

Practically all silos are round or circular, which gives more capacity for the least amount of construction cost, and the stored material packs better.

Farm silos are built for storing corn, sorghum, hay crops, and miscellaneous forages. Research studies reveal that lateral pressures increase with the moisture content of the silage and the size of the silo. All silages exert approximately equal lateral pressures when of the same moisture content. Therefore, the reinforcing requirements are identical. However, hay crop silages are usually stored at excessive moisture content and more reinforcing is recommended.

Excavating and Concrete Foundations for Silos

Footings should be ample to carry the weight of the silo without settlement or tipping. The width and depth are varied according to the height of the silo and the load-carrying capacity of the soil. Recommended footing sizes are given in the accompanying table. These take into consideration the bearing capacities of the different soils as well as the weights of silos of different heights, including the weights of the concrete chutes, roofs, and the weight of the silage carried by the walls.

The depth of the foundation wall below ground level will vary according to the location of the silo. The footings should extend below frost penetration and to firm soil. The bottom of the trench for the footing should be flat and even in width to assure uniform distribution of the load to the soil. Forms will not be required for construction of the footings if care is taken to excavate the trench so that the walls stand vertically. Wall forms are used from the footing up.

Dimensions of Annular Footings for Silos with Walls 6" Thick*

| | Type of Soil** | | | | | |
| Height of Silo, ft. | Sand and Gravel (Type 1) | | Firm clay, wet sand or clay and sand mixture (Type 2) | | Soft Clay (Type 3) | |
	Width Inches	Depth Inches	Width Inches	Depth Inches	Width Inches	Depth Inches
20	12	8	12	8	17	8
25	12	8	12	8	24	8
30	12	8	16	8	32	11
35	12	8	21	10		
40	13	9	26	13		
45	16	11	32	15		
50	19	13				
55	23	16				

Dimensions of Annular Footings for Silos with Walls 150 mm Thick*

Type of Soil**

	Firm clay, wet sand or Sand and Gravel Type 1		clay and sand mixture Type 2		Soft Clay Type 3	
Height of Silo, m	Width mm	Depth mm	Width mm	Depth mm	Width mm	Depth mm
6	300	200	300	200	425	200
7.5	300	200	300	200	600	200
9	300	200	400	200	800	275
10.5	300	200	525	250		
12	325	225	650	325		
13.5	400	275	800	375		
15	475	325				
16.5	575	400				

*ACI-714.
Soil pressures in psf Type 1 - 8,000 (40,000 kg per sq.m); Type 2 - 4,000 (20,000 kg per sq.m); Type 3 - 2,000 (10,000 kg per sq.m).
**Silos higher than 45 ft. (13.5 m) not recommended on Type 2 soil.
Silos higher than 30 ft. (9 m) not recommended on Type 3 soil.

Cu. Ft. of Excavation or Concrete Footing Required for Silo Foundations of Various Sizes per Inch of Depth or Thickness

Footing Width	Diameter in Feet							
	10"	12"	14"	16"	18"	20"	22"	24"
12	2.67	3.17	3.67	4.17	4.75	5.25	5.75	6.33
13	2.89	3.43	3.97	4.51	5.14	5.69	6.23	6.86
16	3.55	4.22	4.88	5.55	6.33	7.00	7.67	8.44
17	3.77	4.48	5.19	5.90	6.72	7.44	8.14	8.97
19	4.21	5.01	5.80	6.59	7.51	8.31	9.10	10.03

Footing Width	Diameter in Feet							
	10"	12"	14"	16"	18"	20"	22"	24"
21	4.66	5.53	6.40	7.28	8.30	9.19	10.06	11.08
23	5.10	6.06	7.01	7.97	9.10	10.06	11.02	12.15
24	5.33	6.33	7.33	8.33	9.50	10.50	11.50	12.67
26	5.55	6.86	7.94	9.03	10.29	11.37	12.45	13.72
27	5.77	7.12	8.25	9.37	10.68	11.81	12.95	14.25
32	7.10	8.44	9.77	11.10	12.65	14.00	15.33	16.90

Diameter and Circumference of Standard Size Silos

Diam. Silo Ft.	Circum. Silo in Ft.	Nominal Circum. in Ft.	Diam. Silo in Ft.	Circum. Silo in Ft.	Nominal Circum. in Ft.
10	31.42	32	18	56.55	57
12	37.70	38	20	62.83	63
14	44.00	44	22	69.10	69
16	50.27	50	24	75.41	76

To obtain the number of cubic feet or cubic yards of excavation or footings required for any job, proceed as follows. Find the cubic feet of concrete required for a silo 16' (4.8 m) in diameter, requiring a footing 19" (475 mm) wide and 13" (325 mm) thick. Refer to above table: a 16' (4.8 m) diameter silo requiring a footing 19" (475 mm) wide, requires 6.59 c.f. of concrete per inch in thickness. Multiply 6.59 by 13" (thickness of footing) and the result is 85.67 c.f. of concrete for the footing.

Silo Wall Forms. The use of commercial steel forms is recommended for the construction of cast-in-place silos. Such forms are quickly set and removed and result in true, smooth walls. Several types of commercial steel forms are available. They differ mainly in the size of the sections or panels and in the manner they are fitted together. In all types, the sections are curved and joined together snugly with clamps to form a complete circle or course. One course of forms is referred to as a lift. Each course or circle is 2' (0.6 m) or more high, according to the forms used. From 6' to 8' (1.8 to 2.4 m) of wall can be built in a day, depending on the size of the crew and the outdoor temperature. In summer weather, forms can usually be removed in 24 hours. In building 6' or 8' (1.8 to 2.4 m) of wall per day, three or four complete courses of forms, using 2' (0.6 m) high sections, are required. The usual practice is to set and fill one lift at a time. Where three or more courses of forms are in use, it is the usual practice to remove the lower course and set it on top. This procedure is continued until the required height of wall is built. Form faces are cleaned and re-oiled for each resetting.

A plumb wall is assured when using commercial forms, provided the first course is carefully leveled. Each manufacturer usually furnishes instructions covering the use of its forms. These companies also make forms for casting chutes and roofs and can furnish erection derricks, hoists, and other construction accessories.

Constructing the Walls. Walls of well-built cast-in-place silos are smooth, plumb, and air and watertight. The walls are 4" to 6" (100 to 150 mm) thick. A 5" to 6" (125 to 150 mm) thickness is recommended for large silos.

Concrete for durable silo walls should be of high quality, using clean sharp sand and gravel. Sand should be from very fine up to particles which pass a No. 4 screen (4 openings per inch). Gravel should be clean, hard, and range in size from 1/4" (6 mm) up to 1-1/2" (38 mm).

A workable concrete mix is one that is mushy but not soupy. It should be somewhat sticky when worked with a shovel or trowel and should be stiff enough to require some spading in the forms. For machine mixing allow about 2 minutes mixing after all materials are in the mixer. Concrete for footing should be mixed in approximately the proportions of 1 bag air entrained portland cement; 2-3/4 c.f. (0.08 cu.m) sand and 4 c.f. (0.11 cu.m) gravel. Use about 5 gals. (18.9 liters) water per bag of cement. Concrete for walls, floor, chute, and roof should be mixed in the proportions of 1 bag of air entrained portland cement; 2-1/4 c.f. (0.06 cu.m) sand and 3 c.f. (0.09 cu.m) gravel. Use about 5 gals. (18.9 liters) water per bag of cement.

Total Weight (Lbs.) of Horizontal and Vertical Bars Required for Cast-in-Place Concrete Silos

Height of silo, ft.	Diameter of Silo, ft.					
	10'	12'	14'	16'	18'	20'
20	547	652				
24	675	796	968	1,263		
28	787	935	1,174	1,492		
32	910	1,047	1,381	1,721	2,014	2,609
36	1,034	1,213	1,587	2,017	2,351	3,112
40		1,400	1,816	2,339	2,836	3,649
44			2,044	2,702	3,291	4,317
48			2,351	3,066	3,780	4,986
52				3,430	4,386	5,654

No allowance made for lapping of bars. Add about 10% when three lengths of bars are used per circle.

Number and Length of Horizontal Reinforcing Bars Required for Cast-in-Place Concrete Silos with Continuous Doorway

	Diameter of Silo, ft.					
	10	12	14	16	18	20
			Length of bars, ft.*			
Height of Silo, ft.	31' 10"	38' 0"	44' 4"	50' 6"	56' 10"	63' 4"

Number of bars required

Diameter	No. 3 Bars		No. 4 Bars		No. 5 Bars	
20	14	17		16		
24	17	21	14	20		
28	20	25	18	24		
32	24	29	22	28	20	26
36	28	33	26	34	24	32
40		39	30	40	30	38
44			34	48	36	46
48			38	56	42	54
52				64	50	62

*Bar lengths do not allow for laps. Add approximately 10% when three lengths of bar are used per circle.

Number and Length of Vertical Reinforcing Bars Required for Cast-in-Place Concrete Silos

Height of Silo, Length of Bars	Diameter of Silo							
	10	12	14	16	18	20	22	24
	Number of Bars Required							
20', 20'-0"	22	26						
24', 13'-0"	44	52	60	68				
28', 15'-0"	44	52	60	68				
32', 17'-0"	44	52	60	68	76	86	92	102
36', 19'-0"	44	52	60	68	76	86	92	102
40', 14'-8"		78	90	102	114	129	138	153
44', 16'-0"		78	90	102	114	129	139	153
48', 17'-4"			90	102	114	129	138	153
52', 18'-8"			90	102	114	129	138	153
60', 16'-6"			120	136	152	168	184	204

All bars are No. 3, spaced 18" apart. Bars are lapped 24" at splices.

Basis for Estimating Labor for Cast-in-Place Concrete Silos. The following data is based on circular type cast-in-place concrete silos.

Labor excavating and placing concrete footings for silos should be estimated separately, because it varies with the depth of excavation and size of silo.

After the footings are in place, an experienced crew of three should build 8 l.f. (2.4 m) of 10'-0" (3 m), 12'-0" (3.6 m), or 14'-0" (4.2 m) diameter silo per 8-hr. day, including the labor on feed room. It will also require about a day to place the roof, and after the concrete is placed, a day will be required to tear down and remove the forms and equipment.

Labor Required to Construct Cast-in-Place Concrete Silos
Labor Required to Construct Each 8'-0" of Wall Height

Diam. of Silo l.f. (m)	Thickness of Wall in. (mm)	Size of Crew	No. of Hours for Crew	Labor Hours for Each 8'-0" Height
10' (3 m)	5"-6" (125-150 mm)	3	8	24
12' (3.6 m)	5"-6" (125-150 mm)	3	8	24
14' (4.2 m)	5"-6" (125-150 mm)	3	8	24
Roof		3	8	24
16' (4.8 m)	5"-6" (125-150 mm)	4	8	32
18' (5.4 m)	5"-6" (125-150 mm)	4	8	32
20' (6.0 m)	5"-6" (125-150 mm)	4	8	32
22'-0"* (6.6 m)	5"-6" (125-150 mm)	4	8	32
24'-0"* (7.2 m)	5"-6" (125-150 mm)	4	8	32
Roof		4	8	32

*Labor required to construct each 6' (1.8 m) of wall height.

A set of metal forms usually includes 4 sets of forms 2'-0" (0.6 m) high. Reinforcing steel may be either rods or mesh but many silo builders prefer mesh reinforcing, because it can be placed at lower cost than rods.

To the above, add labor cost excavating and placing concrete footings.

To the above, add time for 1 day for crew bringing tools and equipment to job and setting up and 1 day tearing down and removing equipment at completion.

Silo Walls Using Slip Forms. For casting monolithic concrete structures of great height, slip forms are often used. This type of form is particularly adaptable to the construction of silos and storage bins. Despite the fact that the sliding forms must be built with more precision and of heavier material than fixed forms, slip forms have proven to be more economical in the long run for high structures. Naturally the higher the construction the less the cost per square foot of wall.

The forms are prefabricated of wood or metal and are usually raised by using screw jacks or a hydraulic jacking system. The yokes, which are usually made of wood and tied together with bolts, are suspended with a screw or jack arrangement on a steel jackrod. The raising is done by a crew, each worker operating from 1 to 5 jacks, depending on the size of the job. This operation must be done on a shift basis, continuing day and night until the structure is completed to prevent the forms from adhering to the concrete.

CHAPTER 14

CONVEYING SYSTEMS

CSI DIVISION 14
14100 DUMBWAITERS
14200 ELEVATORS
14400 LIFTS
14600 MOVING STAIRS & WALKS

14100 DUMBWAITERS

Dumbwaiters must be carefully selected to provide the maximum in convenience and efficiency for the minimum cost. They range from simple hand-operated to computerized models. Standard speeds are 50 fpm, but speeds to 150 fpm are possible on some models. Where speed is not a consideration, lower speeds save some initial cost. Car sizes are directly related to capacities.

Size and Capacity of Dumbwaiters

Capacity, lbs. (kg)	Speed, fpm (mps)	Hp	Stops	Cab Size, s.f. (sq.m)	Price
50 (22.5)	50 (0.25)	1/2	2	3.33 (0.30)	$10,000
100 (45)	50 (0.25)	3/4	2	4.00 (0.36)	12,000
200 (90)	50 (0.25)	1	2	4.70 (0.42)	14,000
300 (135)	50 (0.25)	1-1/2	2	6.25 (0.56)	16,000
500 (225)	50 (0.25)	2	2	9.00 (0.81)	19,000

Dumbwaiters can be mounted on channel guides that are attached to the building construction, but a packaged, self-supporting tower is also available.

14200 ELEVATORS

Elevator construction is such a highly specialized business that it is out of the question to expect anyone except an experienced elevator contractor to prepare an estimate for the installation of such equipment.

There are many items that affect the cost of elevator construction. All elevator manufacturers submit quotations on an installed basis, and the location of the job must be taken into consideration, as well as the intended use, capacity, car speed, size of car, travel, number of stops and openings, design of the car, and kind of current available.

In addition there are many types of controls, such as car switch, continuous pressure push button, single call push button, up down collective (for operation with an operator or automatic operation) signal control, duplex, and others.

All modern passenger elevator installations are partly or completely computerized. In new office buildings computerized systems are available that recognize traffic demands and automatically program elevator service to these demands. With these programming systems, elevator service is made as efficient as possible during peak or down periods.

Elevator Safety Codes

Many states and cities have their own elevator codes, and all elevator installations must comply with the rules and regulations of that state or city, as well as with American National Standards Institute (ANSI) codes. In general, all municipal or state codes follow closely the Code for Elevators, Dumbwaiters, and Escalators (ANSI Standard A17.1 with supplement ANSI A17-1a) and the Code Rules for Moving Walks (ANSI A17.1.13).

It would be a good idea for anyone endeavoring to estimate the cost of an elevator to become familiar with these codes. For example, the capacity of a passenger elevator must be based upon the effective area of the car platform in accordance with the contact load graphs shown in the latest edition of the ANSI codes.

Essential Information for Estimating the Cost of Elevators. The following information is essential to enable the elevator manufacturer to prepare an intelligent estimate on the elevators required for any building.

1. New or renovation project
2. Location of building
3. Number of elevators
4. Use of elevator (passenger, freight, or general purpose)
5. Capacity of elevators
6. Car speed of elevators
7. Controls and operation of elevators
8. Type of motor (single speed, two speed, or variable voltage)
9. Total travel of elevator in feet
10. Number of stops and openings
11. Single or double entrance car
12. Location and size of elevator machinery room or penthouse
13. Kind of current available
14. Size of hatch or platform
15. Design of cab, doors, and frames
16. Car and hatch door power operators
17. Signals

18. Access door location
19. Federal, state, local, and ANSI codes

Costs of Electric and Hydraulic Elevators. The cost of an elevator is a function of its speed, capacity, and type of control, which in turn are determined by type of building to be served, the use of the elevator, and the height of the building or feet of travel and number of stops.

There are so many factors that enter into the cost that an estimator should get definite prices from an elevator contractor whenever possible. The following budget figures can be used for rough estimating only.

Electric Elevators

General Purpose Elevators. There are two basic types of general purpose elevator: gear and gearless. To choose the correct one for a specific need, the contractor should contact a specialist in vertical people moving.

Gearless elevators satisfy most high-rise passenger elevator requirements and are most cost effective for structures 10 floors and greater. The prices given below represent standard platform sizes, with center opening of doors, for a 12-story building.

Capacity, lbs. (kg)	Speed, fpm (mps)	No. of Passengers	Cab Size, s.f. (sq.m)	Price
3,000 (1350)	500 (2.5)	20	60.40 (5.44)	$191,000
3,500 (1575)	700 (3.5)	23	66.00 (5.94)	192,000
3,500 (1575)	1,000 (5.0)	23	68.00 (6.12)	195,000
4,000 (1800)	1,200 (6.0)	27	76.00 (6.84)	200,000

For each additional floor above 12 floors, add $4,200.

Many tall building elevators now operate at speeds up to 1,800 fpm (9 mps), but these require special pricing, depending on express and local runs.

Geared elevators satisfy most low-rise passenger elevator requirements. The prices given below represent standard platform sizes, with center opening of doors, for an 8-story building.

Capacity, lbs. (kg)	Speed, fpm (mps)	No. of Passengers	Cab Size, s.f. (sq.m)	Price
2,000 (900)	200 (1)	13	48.20 (4.34)	$112,000
2,500 (1125)	200 (1)	16	54.80 (4.93)	113,000
3,000 (1350)	200 (1)	20	59.00 (5.31)	115,000
3,500 (1575)	200 (1)	23	66.00 (5.94)	123,000
4,000 (1800)	300 (1.5)	27	74.00 (6.66)	130,000

For each additional floor above 8 floors, add $4,000.

For hospitals, special size elevators are available for increased length requirements and multiple openings necessary for hospitals and modern care facilities.

Capacity, lbs. (kg)	Speed, fpm (mps)	Cab Size, s.f. (sq.m)	Price
4,000 (1800)	125 (0.63)	104 (9.36)	$200,000
4,500 (2025)	200 (1.00)	115 (10.35)	210,000

The above prices are for 5 stops. For each additional stop, add $4,500.

Oil Hydraulic Elevators. There are two types of hydraulic elevators. One type requires the installation of a steel casing into the ground for the piston. Rise is limited to about 60 feet, so they can be used for 6- or 7-story buildings. They are economical to install and require no overhead penthouse. They are available in speeds up to 300 fpm (1.5 mps) and capacity to 5,000 lbs. (2250 kg).

Capacity, lbs. (kg)	Speed, fpm (mps)	No. of Passengers	Cab Size, s.f. (sq.m)	Price
2,000 (900)	120 (0.60)	13	42.10 (3.79)	$85,000
2,100 (945)	120 (0.60)	13	42.10 (3.79)	90,000
2,500 (1125)	120 (0.60)	16	48.00 (4.32)	95,000
3,000 (1350)	120 (0.60)	20	52.00 (4.68)	102,000
3,500 (1575)	120 (0.60)	23	57.60 (5.18)	115,000
4,000 (1800)	200 (1.0)	13	42.00 (3.78)	120,000
4,500 (2025)	250 (1.25)	15	45.00 (4.05)	130,000
5,000 (2250)	300 (1.50)	16	48.70 (4.38)	140,000

The second type of hydraulic elevator, sometimes referred to as a holeless hydraulic, has limited capacity with a maximum travel of 15'-0" (4.5 m) and maximum two stops. It is available in speeds up to 100 fpm (0.5 mps) and capacity to 2,500 lbs. (1125 kg).

Capacity, lbs. (kg)	Speed, fpm (mps)	No. of Passengers	Cab Size, s.f. (sq.m)	Price
2,000 (900)	100 (0.50)	13	42.10 (3.79)	$50,000
2,100 (945)	100 (0.50)	13	42.10 (3.79)	55,000
2,500 (1125)	100 (0.50)	16	47.90 (4.31)	60,000

The cost of an oil hydraulic installation must include the drilling of the hole for the piston. The cost of this hole depends on local conditions such as soil, water, and rock. Elevator prices depend on size, speed, control, and purpose.

Garage Elevator. Freight elevator capacities range from 3,500 to 10,000 lbs. (1575-4500 kg). Speeds vary from 50 to 100 fpm (0.25-0.5 mps), and platforms are from 35 to 112 s.f. (3.15-10.08 sq.m). The following is a typical configuration for this type of equipment:

Freight elevator, single-call push-button control, two-speed motor, automatic leveling, capacity 6,000 lbs. (2700 kg), speed 100 fpm (0.50 mps); total rise 36'-0" (10.8 m) with 4 stops and 4 openings; car platform 10'-0"x20'-0" (3 x 6 m); car built of steel wainscoting on 2 sides 6'-0" (1.8 m) high, a steel car top and two wood or expanded metal lift-up gates with electric contacts. Price installed $75,000. No doors or hatchway gates included.

Residence Type Elevators. During recent years residence type elevators have become quite popular, especially where there are older people or invalids in the house. These elevators are easily installed both in new homes and homes already built. Residential elevator capacities range from 450 to 700 lbs. (203-315 kg), with speeds from 450 to 700 fpm (2.25-3.5 mps).

A typical residential elevator with 450 lbs. (202.5 kg) capacity, 30 fpm (0.15 mps) speed, total travel 11'-0" (3.3 m), 2 stops and 2 openings, 110 or 220 volts, single phase 60 cycles, and platform 36"x42" (900 x 1050 mm), is priced installed approximately $35,000.00. For each additional stop, add $2,500.00.

The same elevator as above with 36"x48" (900 x 1200 mm) platform is priced installed at about $40,000.00. For each additional stop, add $2,500.00.

14400 LIFTS

A wheelchair lift is designed to permit people with physical handicaps to go up and down stairs. The system is usually fastened to a wall but sometimes to the stair treads. Before installing a lift system, the details of walls or stairs must be reviewed to determine whether support members can support a wheelchair lift. A typical wheelchair lift has a travel distance of 27 l.f. (8.1 m), a speed of 25 fpm (0.13 mps), and a 1-Hp motor. The chair is supported on 2" (50 mm) OD steel rails, with platform sizes from 2'-6"x3'-8" (0.75 x 1.1 m) to 3'-2"x4'-2" (0.95 x 1.25 m). For a single story lift adjacent to stairway, the cost is about $30,000.00.

14600 MOVING STAIRS

Escalators. Escalators move large volumes of people from one level to another efficiently. They are available in several widths and speeds.

Width, in. (mm)	Speed, fpm (mps)	People per hr.	Cost
32 (800)	90 (0.45)	5,000	$100,000
32 (800)	120 (0.60)	6,500	110,000
40 (1000)	90 (0.45)	7,000	104,000
40 (1000)	120 (0.60)	9,000	114,000
48 (1200)	90 (0.45)	8,000	116,000
48 (1200)	120 (0.60)	10,000	120,000

To the above price, add about $1,200 per l.f. ($4,000 per m) of rise.

Horizontal People Movers. These are also called power walks, power ramps, or moving walks. Horizontal people movers provide high volume transportation of people within buildings where walking is not advantageous. People movers come in various sizes from 24" to 48" (600-1200 mm) tread width. Larger sizes of 40" (1000 mm) and 48" (1200 mm) accommodate an adult with luggage. For the special requirements for this type of installation, a contractor should consult the manufacturer.

For a 40" (1000 mm) wide, flat horizontal people mover with no curves or offsets, prices range from $120,000 to $150,000 per 50 l.f. (15 m).

CHAPTER 15

MECHANICAL

CSI DIVISION 15
15A PLUMBING AND SEWERAGE
15B HEATING AND AIR
    CONDITIONING
15C SPRINKLER SYSTEMS

15A PLUMBING AND SEWERAGE

To prepare an accurate detailed estimate on plumbing and sewerage requires a working knowledge of the trade. All work must be laid out on paper, and the number of lineal feet of each kind of pipe estimated separately, such as tile or cast iron sewers, soil and vent pipe, water pipe, gas pipe, drains, valves, and pipe fittings. Very few, if any, general building contractors or estimators possess this knowledge, so it makes it doubly difficult for them to prepare more than an approximate estimate on the various kinds of plumbing work.

Budget figures for average plumbing installations can be estimated in one of two ways: as a percentage of the total cost of a project, or as the grand total of a predetermined cost allowance for the installation of each fixture.

Plumbing estimates as a percentage of the total job will vary from a minimum of 3% on up to 12%, depending on the number and type of fixtures required and on their disposition within the building. Percentage estimates would be valid for general plumbing only. Process and other special condition piping would of course have to be entered as a separate item.

Buildings that require a high percentage allowance include motels and nursing homes, which are generally spread out over a large area and require plumbing for each room, and apartment high-rises with many studio units, which require full baths and kitchens for relatively small living areas. On the low side would be assembly, mercantile, warehouse, and industrial buildings where plumbing can be concentrated in a central core. Recent OSHA requirements limiting the travel distance to a rest room have raised the costs on some of these types.

The private residence is difficult to peg. Where kitchens and baths can be back to back, the percentage is on the low side. But in a large, rambling house with a central powder room area but with kitchens, family, and master bedrooms all in separate wings and with hose bibs and lawn sprinkling systems on all four sides and sunken bathtubs, shell-shaped lavatories and gold plated fixtures, even 12% may be too low.

However the average building types will fall in the following ranges:

apartments ... 9% to 12%	motels .... 9% to 12%
assembly .... 4% to 7%	office bldgs. ...... 4% to 7%
banks .... 3% to 6%	shops .... 4% to 7%
dormitories ... 7% to 10%	schools .... 5% to 8%
factories .... 4% to 8%	stores .... 3% to 6%
hospitals .... 8% to 12%	warehouses ...... 3% to 7%

The allowance per fixture estimate is generally more accurate. An experienced plumber can acquire very accurate prices but will have to use some judgment on how much to add for bringing in the main waste and supply lines, depending on how spread out the project.

The per fixture price will be the cost of the fixture itself plus the cost of the immediate piping connections and installation within the room. Some per fixture costs have added to them a prorated cost of the central piping system. It would be more accurate to add it, together with the cost of bringing waste and supply lines from the street to the building and permit costs and overhead and profit, each as separate items, because some buildings require a much more extensive central system to supply fewer fixtures than others. Per fixture allowances are discussed later in this chapter.

15400 PLUMBING SYSTEMS

Estimating Quantities and Costs of Sewer Work. Sewerage is estimated by the lineal foot, obtained by measuring the number of lineal feet of each size sewer pipe required, such as 4" (100 mm), 6" (150 mm), 8" (200 mm), and 12" (300 mm) pipe, and all fittings should also be listed in detail, giving the number of ells, Ys, Ts, etc.

Cast-iron and ductile iron sewer pipe is estimated in the same manner as tile sewer pipe. Most plumbers estimate their sewer work at a certain price per lineal foot, including excavating, sewer pipe, and labor. This is all right on ordinary jobs that do not require more than the usual amount of deep excavation, but if there is an exceptionally large amount of deep excavation, from 5 to 15 feet (1.5 to 4.5 m) deep, then the excavating, shoring, and bracing should be estimated separately.

The best method is to estimate the number of cubic yards of excavating required for sewers, based on a trench 1'-6" (0.45 m) wider than the pipe diameter by the required depth. Refer to chapter on *Site Work* for labor costs excavating for trenches.

Where 4" to 6" (100 to 150 mm) sewer pipe is used, an experienced sewer layer should lay 12 to 15 l.f. (3.6-4.5 m) of pipe per hour or 100 to 125 l.f. (30-37.5 m) per 8 hour day, at the following labor cost per 100 l.f. (30 m):

	Hours	Rate	Total	Rate	Total
Sewer layer	7.0	$....	$....	$29.28	$204.96
per l.f.			....		2.05
per m			....		6.72

Add for excavating and additional time for setting catch basins, gravel basins, triple basins, etc.

Estimating Quantities of Cast Iron or Ductile Iron Soil Pipe, Downspouts, Stacks, and Vents. When estimating soil pipe, stacks, and vents, note the number of stacks and length of each, listing the number of lineal feet of each size pipe, the number of pieces, size, and kind of fittings required, and price them at current market prices.

Estimating Quantity of Pipe and Fittings. When estimating the quantity of black or galvanized pipe, list the number of lin. ft. (meters) of each size pipe required, such as 1/2" (12.5 mm), 3/4" (18.8 mm), 1" (25 mm), 1-1/4" (31.2 mm), 1-1/2" (37.5 mm), 2" (50 mm), 2-1/2" (62.5 mm), 3" (75 mm), and estimate at the current market price.

After the cost of the pipe has been computed, take 60% to 75% of the cost of the pipe to cover the cost of all fittings required.

Brass or copper pipe and fittings should be estimated in the same manner as galvanized pipe.

Estimating the Quantity and Prices of Valves. All valves of the different sizes should be listed on the estimate separately, stating the number of each size required, and pricing them at the current market price.

Estimating the Quantity of Fixtures. Each type of fixture such as sinks, lavatories, laundry trays, water closets, bathtubs, drinking fountains, shower baths, hot water tanks and heaters, house pumps, and bilge pumps, should be listed separately and priced at the current market prices.

Labor Roughing in for Plumbing. The usual practice among plumbers is to allow a certain percentage of the cost of the roughing materials to cover the labor cost of installing them. This will vary with the type of building and grade of work.

The total cost of the roughing-in materials, such as cast-iron soil pipe, downspouts, stacks, vents, and fittings, all black and galvanized pipe and fittings, all valves, increasers, tees, Ys, 1/8 bends, and nipples, in fact, all pipe required to rough in the job, is computed. Then the labor cost is estimated at a certain percentage of the cost of the pipe.

In medium-priced one and two story residences, the labor cost of roughing in the job will vary from 75% to 85% of the cost of the roughing materials, with 80% a fair average.

In apartment buildings of non-fireproof construction, the labor cost of roughing in will average about 90% to 100% of the cost of the roughing in materials, while on high grade fireproof construction, the labor cost roughing in will run from 100% to 120% of the cost of the roughing in materials.

In public garages, the labor cost roughing in the job will average about 90% to 110% of the cost of the roughing in materials. Sewer work is estimated separately from roughing in.

Labor Handling and Placing Plumbing Fixtures. The labor cost handling and placing all kinds of fixtures, such as laundry tubs, kitchen sinks, lavatories, bathtubs, shower baths, water closets, and drinking fountains, is usually estimated at 25% to 30% of the cost of the fixtures.

Preparing Detailed Plumbing Estimates

When preparing detailed estimates on plumbing, it is necessary to list the quantity of each class of work separately, such as the number of lineal feet (meters) of 4" (100 mm), 6" (150 mm), and 8" (200 mm) sewers, and fittings; number of lineal feet of each size soil pipe, vents, and fittings; number of lineal feet of water pipe and fittings of different materials and sizes; number of valves of the various kinds and sizes; number of each type of fixture required, such as hot water heaters, laundry trays, kitchen sinks, slop sinks, bathtubs, lavatories, water closets, shower baths, etc., together with a list of fittings and supplies required for each.

The following is a list of items the plumbing contractor should include in his estimate:

1. Sewerage, including double sewer system
2. House tanks, foundations, etc.
3. Compression tanks, foundations, etc.
4. Sewer ejector and bilge pump and basins
5. Iron, gravel, catch, and other basins
6. Water filters, foundations
7. Water meter
8. Soil and vent pipe and fittings
9. Shower and urinal traps
10. Closet bends
11. Drum traps
12. Lead, solder, sundries
13. Inside downspouts
14. Iron sewer and fittings
15. Floor drains
16. Back water gates
17. Service pipe to building
18. Mason hydrant
19. Stop cock and box
20. Galvanized pipe and fittings
21. Brass or copper pipe and fittings
22. Valves, check valves

23. Pet cocks, sill cocks, hose bibbs
24. Hot water tank and heater
25. Pipe covering, tank covering, heater covering, filter covering, painting
26. Gas fitting, mains, ranges, heaters, etc
27. Manholes and covers
28. Catch basins and covers
29. Surface drain basins and covers
30. Fire system, incl. pumps, hose, Siamese connection, meter, etc.
31. Roughing in material
32. Fixtures, such as water closets, lavatories, bathtubs, shower baths, sinks, laundry tubs, drinking fountains, N. P. fittings
33. Permits, insurance, trucking, freight, telephone, watchman, etc.
34. Overhead expense and profit

15060 PIPE AND PIPE FITTINGS

Approximate Lineal Foot Prices of Cast-Iron Soil Pipe

Prices on all types of pipe and pipe fittings and supplies should be checked with suppliers at the time of the job, because they are subject to continual change.

Size	Single Hub Service	Extra Heavy	Double Hub Service	Extra Heavy
2" (50 mm)	$3.10	$4.30	$4.25	$5.85
3" (75 mm)	4.00	5.05	4.94	6.85
4" (100 mm)	4.75	6.75	6.60	8.98
5" (125 mm)	6.20	8.65	8.40	11.80
6" (150 mm)	7.30	10.10	9.80	13.80
8" (200 mm)	11.25	15.70	15.40	21.35
10" (250 mm)	18.00	25.30	24.70	33.70
12" (300 mm)	25.85	36.30	35.95	48.90

Prices of Cast-Iron Soil Pipe Fittings

Description of Fitting	2" (50)	3" (75)	4" (100)	5" (125)	6" (150)	8" (200)
Quarter bends, service	$3.40	$6.20	$9.00	$11.80	$15.70	$51.70
Extra heavy	3.90	7.30	10.30	14.00	17.60	58.40
Eighth bends, service	2.50	5.05	7.30	10.30	12.35	35.95
Eighth bends						
Extra heavy	2.80	5.70	8.40	11.80	14.00	41.60

Description of Fitting	Size in inches (mm)					
	2" (50)	3" (75)	4" (100)	5" (125)	6" (150)	8" (200)
Sanitary T & Y branches						
Extra heavy	7.00	13.50	16.85	24.70	38.20	78.65
Plain Traps S P or 3/4						
Service	9.60	12.90	18.00	34.80	44.90	102.00
T cleanout						
Extra heavy	30.30	40.45	57.30	103.40	148.30	185.40
Comb. Y & eighth bends						
Extra heavy	10.10	13.50	19.10	37.10	45.00	104.50

Closet bends slip collar type: 4"x4"x12" (100 x 100 x 300 mm), $23.00; 4"x4"x18" (100 x 100 x 450 mm), $27.00.

Lineal Foot Prices of Buttweld Steel Pipe

Size	Standard Weight		Extra Heavy	
	Black	Galvanized	Black	Galvanized
1/4" (6.2 mm)	$0.33	$0.50	$0.50	$0.67
3/8" (9.7 mm)	0.45	0.60	0.65	0.70
1/2" (12.5 mm)	0.47	0.65	0.65	0.80
3/4" (18.7 mm)	0.67	0.80	0.85	1.00
1" (25 mm)	1.00	1.25	1.30	1.45
1-1/4" (31.2 mm)	1.20	1.60	1.65	1.90
1-1/2" (37.5 mm)	1.50	1.85	1.95	2.30
2" (50 mm)	2.00	2.50	2.65	3.10
2-1/2" (62.5 mm)	2.90	3.65	4.10	4.75
3" (75 mm)	3.80	4.75	5.40	6.30
3-1/2" (87.5 mm)	4.70	5.85	6.70	7.85
4" (100 mm)	5.60	6.85	8.00	9.30

Prices on Standard Malleable Iron Fittings

Type of Fitting	Size, in. (mm)								
	1/2" (12.5)	3/4" (18.7)	1" (25)	1-1/4" (31.2)	1-1/2" (37.5)	2" (50)	3" (75)	4" (100)	6" (150)
Tee Black	$0.62	$0.90	$1.60	$2.60	$3.15	$4.65	$11.80	$24.15	$67.40
Galv.	0.90	1.30	2.10	3.70	4.50	6.00	16.30	33.70	95.50
90° El Black	0.50	0.55	1.05	1.70	2.25	3.20	10.70	18.00	53.90
Galv.	0.75	0.80	1.50	2.35	3.15	4.50	14.60	25.30	73.00
45° El Black	0.75	0.95	1.25	2.00	2.50	3.40	11.80	19.10	61.80
Galv.	1.05	1.25	1.75	2.50	3.40	3.60	16.30	26.60	84.30

Lineal Foot Prices of Copper Water Tubing

Type	1/4" (6.2)	3/8" (9.3)	1/2" (12.5)	5/8" (15.6)	3/4" (18.7)	Size, in. (mm) 1" (25)	1-1/4" (31.2)	1-1/2" (37.5)	2" (50)	3" (75)
Light Type M	$0.40	$0.45	$0.67	$0.80	$0.95	$1.30	$1.80	$2.35	$2.60	$6.75
Medium Type L	0.40	0.60	0.85	1.00	1.25	1.70	2.20	2.80	4.20	8.15
Heavy Type K	0.45	0.70	0.95	1.10	1.60	2.00	2.50	3.25	4.90	9.55

Prices of Wrought Copper Soldered Joint Fittings

Type	1/2" (12.5)	3/4" (18.7)	Size, in. (mm) 1" (25)	1-1/4" (31.2)	1-1/2" (37.5)	2" (50)	3" (75)	4" (100)
Tee	$0.22	$0.60	$1.70	$2.65	$3.37	$5.40	$17.65	$35.95
90° Elbow	0.17	0.38	0.80	1.25	1.70	3.00	9.00	20.20
45° Elbow	0.28	0.50	0.85	1.45	1.90	2.80	10.95	22.50

15080 PIPING SPECIALTIES

Cast-Iron Drum Traps

Drum trap with cover and gasket, 4"x8" (100 x 200 mm), 2 inlets $10.50
Drum trap with cover and gasket, 4"x8" (100 x 200 mm), 3 inlets 12.10

Roof Drains

	3" (75mm)	4" (100mm)	5" (125mm)	6" (150mm)	8" (200mm)
Cast Iron	$130.00	$135.00	$195.00	$215.00	$340.00
Galv.	220.00	225.00	280.00	300.00	425.00

Floor Drains

Type	Size of Outlet 3" (75 mm)	4" (100 mm)	5" (125 mm)
Flat round top cast iron	$40.00	$50.00	$85.00
Flat square top cast iron	55.00	55.00	95.00
Funnel type brass	80.00	80.00	
Drain with bucket cast iron	110.00	110.00	115.00
Trench drain 10"x24"	130.00	150.00	

Sizes and Weights of Steel Storage Tanks
Black, Heavy Weight, Welded, 100 Lbs. (45 Kg) Working Pressure

Capacity in gals. (liters)	Dimensions Diam. Length ft.in. (mm, m)	Overall Length in. (mm)	Heavy Weight Welded lbs. (kg)
Tanks with One Convex and One Concave Head			
66 (250)	20"x4' (500x1.2)	52" (1300)	225 (102)
82 (310)	20"x5' (500x1.5)	64" (1600)	265 (120)
120 (454)	24"x5' (600x1.5)	64" (1600)	340 (154)
140 (530)	24"x6' (600x1.8)	76" (1900)	395 (179)
180 (681)	30"x5' (750x1.5)	66" (1650)	480 (218)
220 (833)	30"x6' (750x1.8)	78" (1950)	550 (249)
295 (1117)	30"x8' (750x2.4)	102" (2550)	680 (308)
Tanks with Two Convex Heads			
69 (261)	20"x4' (500x1.2)	55" (1375)	220 (100)
85 (322)	20"x5' (500x1.5)	67" (1675)	260 (118)
125 (473)	24"x5' (600x1.5)	68" (1700)	320 (145)
147 (556)	24"x6' (600x1.8)	80" (2000)	375 (170)
200 (757)	30"x5' (750x1.5)	71" (1775)	450 (204)
235 (889)	30"x6' (750x1.8)	83" (2075)	520 (236)
305 (1154)	30"x8' (750x2.4)	107" (2675)	650 (295)
350 (1325)	36"x6' (900x1.8)	85" (2125)	850 (386)
450 (1703)	36"x8' (900x2.4)	109" (2725)	1050 (476)
560 (2120)	36"x10' (900x3.0)	133" (3325)	1250 (567)
550 (2082)	42"x7' (1050x2.1)	99" (2475)	1200 (544)
770 (2914)	42"x10' (1050x3.0)	135" (3375)	1550 (703)
900 (3407)	42"x12' (1050x3.6)	159" (3975)	1800 (816)
1050 (3974)	42"x14' (1200x4.2)	183" (4575)	2100 (952)
1000 (3785)	48"x10' (1200x3.0)	137" (3425)	2150 (975)
1200 (4542)	48"x12' (1200x3.6)	161" (4025)	2500 (1134)
1500 (5678)	48"x15' (1200x4.5)	197" (4925)	3050 (1383)
1800 (6813)	48"x18' (1200x5.4)	233" (5825)	3550 (1610)

Septic Sewage Disposal Tanks

Septic tanks are used in rural or suburban districts where running water is available but not sewers and are used for buildings such as homes, hotels, summer resorts.

The capacity of a septic tank should be at least equal to the maximum daily flow of sewage. This is normally estimated at 100 gallons (378 liters) per person per day. For part time service in factories, churches, schools, etc., 50 gallons (189 liters) per person per day is generally satisfactory for estimating tank capacities. In some localities, health authorities require at least a 900-gallon (3407-liter) tank for residential installation.

Tanks are furnished in 12 and 14 gauge copper bearing steel, electrically welded and covered with a thick coating of asphalt to protect the tanks against corrosive action.

Size of Tank, in. (mm)	Capacity in gals. (L)	Size Tile Connection in. (mm)	Weight in lbs. (kg)	Home Capacity No. People	Price Each
46"x48" (1150x1200)	300 (1136)	4"-6" (100-150)	275 (126)	56	$67.50
52"x60" (1300x1500)	500 (1893)	4"-6" (100-150)	380 (171)	67	102.00
46"x120" (1150x3000)	750 (2839)	6" (150)	575 (259)	712	140.00
48"x144" (1200x3600)	1000 (3785)	6" (150)	980 (441)	1215	195.00

Precast concrete tanks, not steel tanks, are now the standard in the U.S. The inner surface of the concrete tank is coated with a tar-base material to prevent gases from deteriorating the concrete. Some precast tank suppliers not only manufacture tanks but also set the tank on delivery into an excavated hole using a delivery truck with boom or other lifting device. Size and weight of the tank limits this service. Consult individual suppliers. Sizes range from 900 to 6,000 gallon capacity.

Capacity, gals. (L)	Size ft. (m)	Price	Riser per ft. (0.3m)	Casting	Roof Slab
900 (3407)	4'-6"x8'x6' (1.35x2.4x1.8)	$290.00	$20.00	$150.00	
1,200 (4542)	8' (2.4m) diameter	650.00	20.00	150.00	140.00

For larger sizes, consult local suppliers.

Prices of Brass Valves

Kind of Valve	1/2" (12.5)	3/4" (18.75)	1" (25)	1-1/4" (31.25)	1-1/2" (37.5)	2" (50)	2-1/2" (62.5)	3" (75)
Gate valve, 100 lbs. (45 kg) pressure	$8.00	$10.00	$13.00	$16.50	$19.25	$30.00		
Gate valve, 125 lbs. (56 kg) pressure	11.00	13.75	17.00	22.00	28.00	40.00	80.00	120.00
Globe valve, 100 lbs. (45 kg) pressure	9.50	12.50	17.00	22.00	29.00	43.00		
Globe valve, 125 lbs. (56 kg) pressure	10.00	13.50	18.00	25.00	32.00	48.00	80.00	125.00
Check valve, horizontal, 125 lbs. (56 kg)	15.00	20.00	28.00	36.00	45.00	74.00	100.00	140.00
Check valve, 125 lbs. (56 kg) swing	12.00	14.00	18.00	23.00	27.00	39.00	70.00	100.00

Angle valves, same price as Globe valves.

Gas Heaters

Capacity gals. (L)	Overall Dia. in. (mm)	Overall Ht. in. (mm)	Approx. Wt.,lbs.	Prices 5 Year	Prices 10 Year
20 (75)	14-1/2" (363)	61" (1525)	144 (65)	$100.00	$120.00
30 (114)	16-1/4" (406)	64" (1600)	174 (78)	120.00	170.00
40 (151)	18-1/2" (463)	65" (1625)	195 (88)	150.00	185.00
50 (189)	20-1/4" (506)	65" (1625)	274 (123)	190.00	240.00

Electric Heaters

Capacity gals. (L)	Overall Dia. inches (mm)	Overall Ht. inches (mm)	Approx. Wgt. lbs. (kg)	Prices 10 Yr.
40 (151)	20 (500)	50 (1250)	160 (72)	$170.00
52 (197)	23 (575)	50 (1250)	181 (81)	200.00
66 (250)	23 (575)	62 (1550)	227 (102)	220.00
82 (310)	25 (625)	63 (1575)	270 (122)	250.00
110 (416)	28 (700)	69 (1725)	430 (194)	290.00

15450 PLUMBING FIXTURES AND TRIM

Residential kitchens and bathrooms have tended to go full circle. When plumbing first moved indoors, it was generally installed in a full-sized room and was arranged about the walls as so much furniture. Later sanitary concerns became paramount and kitchens and baths were designed as machines—all chrome and tile and porcelain enamel and reduced to minimum dimensions to facilitate cleaning.

Eventually, the kitchen emerged as an entertainment center, where family and friends could gather. Today it is often difficult to tell where the family room ends and the kitchen begins. All this has had its effect on the materials used with appliances and fixtures made as inconspicuous as possible.

The bathroom now seems to be following suit, especially the bathing area which is often in a separate room together with built in equipment for exercise and relaxation and sometimes even a sauna, greenhouse, and fireplace. Often the contractor will be called in to convert a spare bedroom into a bath and dressing room suite.

The following discussion of fixtures and their prices give figures which include the fittings. Also are included averages for the piping hook-ups to the main risers and stacks that might be added to arrive at per fixture installed budget figures.

Bathtubs

The least expensive, standard, recessed tub is made of enamelled steel. This will cost around $115 for the 4'-6" (1350 mm) length and $145 for the 5'

(1500 mm) length. Units made of cast iron will run $225 and $250 in those sizes in white and $300 and $330 in color. If these units are to be recessed into the floor add another $50. Square recessed tubs 42"x48" (1050 x 1200 mm) will run $490 in cast iron.

Units complete with three enclosing walls can be had in fiberglass for around $400.

A plumber with a helper should set a tub with shower in about two hours. To arrive at a per-fixture installed budget figure add around $250.

Whirlpool bathtubs are not only found in luxury homes these days. They are being installed in place of regular units in average homes. Costs of these units vary depending on the number of jets and other options. Prices range from $1,500.00 to more than $5,000.00. It takes a plumber and helper about one day to install. Whirlpools usually require greater widths than regular tubs for the water jets. An additional cost is the 1-Hp (1.75 kW) electric motor needed to drive the jets.

Shower Stalls

These can range from simple baked enamel on steel units costing around $150 to custom marble enclosures costing six times that. Prefabricated fiberglass units are very popular and can be had with built-in seats and grab bars for the elderly. A 3'x3' (0.9 x 0.9 m) unit will run $450 complete.

Often the plumber will find his work limited to furnishing the receptor and mixing valve and outlet, with other trades providing finished walls. A precast receptor pan in terrazzo together with chrome fittings will cost $110 in 3'x3' (0.9 x 0.9 m) size and $125 in 3'x4' (0.9 x 1.2 m). One worker can install this in half a day.

In industrial, recreational, and educational buildings, group showers with a center outlet column and room for some five persons around it are specified. Such a unit will cost around $750 without subdividing partitions, $1,300 with them.

Two plumbers will set the basic unit in 4 hours but will take a full 2 days if partitions are to be set.

Lavatories

Wall hung lavatories are available in porcelain enamelled cast iron or vitreous china. A 19"x17" (475 x 475 mm) unit will run $95 in white, cast iron, $110 colored; $100 china white, $120 china colored. A 22" (550 mm) long unit will run $100 cast iron white, $115 cast iron colored, and $130 china white and $150 china colored.

Allow two workers one hour to install. Add $230 to arrive at a per fixture installed price.

Drop-in, counter top units are mainly self-rimming today. They come in cast iron, china, steel, stainless, and molded plastic. An 18" (450 mm) round unit in cast iron or vitreous will cost about $65 in white, $75 in color; in stainless, $120 and in enameled steel, $50. A 20"x18" (500 x 450 mm) unit will cost some $10 more. Two workers will take 1-1/2 hours to set a unit.

Washfountains

A 3-ft. (900 mm) round, freestanding unit will serve five to six persons at a cost of $600 per unit in precast terrazzo and $725 in fiberglass or stainless. A 4'-6" (1350 mm) unit will serve up to 10 and costs $700 in terrazzo, $920 in fiberglass and $1,100 in stainless.

Half round units are also available. A 4'-6" (1350 mm) unit in terrazzo will cost $600, in stainless $890. Since installation costs will remain almost the same as for circular units, the cost per person served is much more with these units.

While terrazzo units are cheapest, they are also the heaviest, and one should allow another hour to set these units. Otherwise, allow two workers 2-1/2 hours to set a 36" (900 mm) unit, 3 hours to set a 54" (1350 mm) unit.

Also available for vanity units are precast, one piece, bowl and countertops in a variety of marbleized colors and patterns. These are generally locally manufactured and prices should be checked but a 2' (600 mm) long top and bowl should cost about $100 and take two workers 1-1/2 hrs. to set and a 6' (1800 mm) top with double bowl should cost about $225 and take 3 hours to set.

Water Closets

Water closets can be either wall hung or floor mounted and have tank or flush valve water supply. The tank can be a separate piece, or the unit can be cast in one piece with the tank dropped behind the seat.

A floor mounted one piece will cost $275 in white, $325 in color. A floor mounted two piece unit will cost $125 in white, $165 in color. For areas where water supply is a problem, units designed to flush with as little as two quarts (1.9 liters), aided by compressed air, cost about $600. Some locales now mandate a flush of not greater than 1.6 gals. (6.05 liters).

Floor mounted units will take two workers 1-1/2 hours to set. For a per fixture installed budget figure add $200.

In commercial work wall hung, flush valve units are the norm. These will cost $175 for the fixture, $150 for a standard carrier and take two workers 2 hours to set.

Urinals

Urinals are available in cast iron or vitreous china, floor, pedestal or wall mounted, single or in gangs or as wall mounted troughs.

Vitreous china, floor mounted units 18" (450 mm) wide and 38" (950 mm) high with 4" (100 mm) recessed into the floor will cost about $250. If set in gangs, add $60 for a spacer-cover 6" (150 mm) wide.

Wall hung china units 18" (450 mm) wide and projecting 12-1/2" (312.5 mm) will cost $210.

A women's pedestal type will run $265; men's type, $220.

Two workers should install a urinal in 2-1/2 hours.

For a per fixture installed budget figure, add $250 per unit.

Bidets

For the ultimate touch in the luxury bath, install a bidet. It costs about $325 in white, $450 in color, and takes two workers 2 hours to set.

Sinks

Kitchen sinks are usually self-rimming and set into a cabinet top of plastic or tile. With electric dishwashers, many are only single bowl. Even in luxury kitchens, a single bowl may be set where the dishes are washed, with a second bowl near the refrigerator for washing vegetables and a third near the ice making machine for making drinks.

A single bowl 24"x21" (600 x 525 mm) will cost $130 in enamelled steel, $160 in white cast iron, $250 in colored cast iron, and $300 in stainless. Figure two workers will take 3 hours to set. Add $600 for a per fixture installed price.

A double unit 42"x21" (1050 x 525 mm) will cost $200 in steel, $230 in cast iron white, $320 in cast iron color, and $400 in stainless. Two workers will take 3-1/2 hours to set such a unit.

Cast iron sink tops with drain boards cost $375 for single bowl 42" (1050 mm) unit, $500 for double bowl 60" (1500 mm) unit.

Enamelled steel sink and drain board tops cost $150 for a 42" (1050 mm) length, $380 for a double bowl 60" (1500 mm) length.

Laundry sinks in enamelled cast iron cost $150 for a single compartment, $225 for a double; in cast stone, $100 and $150; in fiberglass, $75 and $110; in stainless steel, $300 and $500. Two workers can set a single unit in 1-1/2 hours; a double unit in 2 hours.

Electric Dishwashers

Dishwashers are installed as separate units, or are formed as part of a complete sink. Typical prices for various units are as follows.

Dishwasher unit only with floor cabinet but without top and backsplasher, where used under a continuous countertop, size 24"x25" (600 x 625 mm) and 34-1/2" (862.5 mm) high will cost about $500.00.

Dishwasher unit with porcelain enamel top and backsplasher, cabinets made of electrically welded rust-resistant steel with baked on enamel finish, size 23"x27" (575 x 675 mm) and 39" (975 mm) high will cost about $475.00.

Add installation charges to all of the above prices.

Garbage Disposal Units. The disposal reduces solid waste but does not do away with the kitchen garbage can. It is a self-contained unit attached to the kitchen sink to form an enlarged drain into which all kitchen food wastes can be placed.

The unit consists of a housing in which is contained a propeller and shredding mechanism. A 1/2-hp (0.37-kW) motor is directly connected to the propeller and shredding mechanism and supplies the power for operating the waste unit.

Waste material can be accumulated in the upper receptacle of the unit until a normal charge is collected, or it may be operated to dispose of the material immediately. The capacity of the waste receptacle is one gallon (3.78 liters).

Waste material passes through a series of shredders where they are reduced to a fine pulp. This pulp then passes into a revolving strainer disc through which it is forced centrifugally into a chamber below and around the flywheel. Fins on the flywheel centrifugally force the pulp into the outlet passage connected to the drain line, which carries the waste to the sewer. Cold water from the faucet flowing through the unit during the grinding operation thoroughly flushes the waste down the drain. The price of this unit averages about $250.00.

Add installation charge to above price.

15B HEATING AND AIR CONDITIONING

Over the years methods of heating and cooling buildings have continually improved, especially in residential work, due to the research programs conducted by heating and air conditioning associations, engineering societies, equipment manufacturers, and research departments of various universities.

Types of Conventional Heating Systems. There are four basic types of conventional heating systems—warm air, hot water, steam, and electric, including both resistance types and heat pumps. These types may be used separately or in combination. Solar heating is also now an option.

For small to medium size residences and for stores, churches, and the like of similar size and construction, warm air systems are most widely used. For medium size to large residences and non-residential structures of comparable size, hot water systems are usually selected. For large scale installations, steam systems are generally required. Because there are no definite rules stating just when to select one system in preference to another, there is much overlapping in their application. An owner's preference for a particular system may be governed by initial cost of installation; operation cost; convenience of operation; adaptability of system to other uses, such as summer cooling using ducts of a warm air system, snow melting using hot water piping connected to a hot water system; fuel availability; and aesthetics.

Estimating the Cost of Heating Systems. There is no fast and easy rule-of-thumb method for estimating the cost of a heating system. This requires expert knowledge, because most jobs must be designed and laid out before they can be estimated. In residential and small commercial work, architects seldom give more information than merely specifying the type of system desired, the performance expected, and the kinds of material to use. It is then up to the heating contractor to figure the heating loads and design a system to fit the conditions before estimating the job. On larger work, the heating system is usually designed and laid out diagrammatically, but the heating contractor must work out details for piping, equipment connections, and controls, all of which influence the cost.

To acquire the necessary technical knowledge for this purpose requires everyday participation in this line of work plus constant study to keep up to date. Heating and air conditioning is too complex a subject to treat in detail in a book of this kind. However, much useful data and information can be gained from manuals published by the HVAC associations (see the List of Associations in Chapter 1). Here, we have provided only the most elementary and approximate data for "roughing" or approximate estimates only.

Boilers are now designed for burning special fuels, such as hard coal, soft coal, oil, and gas, the costs varying with boiler sizes and accessories, such as domestic hot water heaters and thermostatic controls.

The ordinary cast iron radiators standing in the room are pretty much a thing of the past. Heating systems use concealed convection type radiators or baseboard radiation. These are all items that affect costs.

Copper piping is used for both heating and plumbing. The cost of this work depends to a large extent upon the experience of the contractor. Mechanics familiar with sweating joints and installing copper piping can do this work rapidly, while those who are not accustomed will spend considerable time on it.

The estimator should be familiar with the requirements of the trade in order to prepare intelligent estimates.

Determining Heating Loads. The first step in designing and estimating heating work is to determine the maximum heating load. For most residential work, the maximum heating load is the total heat loss of the structure figured at design temperatures. For commercial and industrial structures, special purpose heating loads and pick-up loads may be required which must be added to the total building heat loss to obtain maximum heating load.

Heating loads and heat losses are invariably expressed in British thermal units (Btu) per hour. A Btu is the quantity of heat required to raise the temperature of 1 lb. (0.45 kg) of water 1°F (-17°C).

Building heat losses vary considerably depending upon climatic location of job, exposure, building size, architectural design, purpose of structure, construction materials used, quality of construction workmanship, and other factors.

Heat losses should be figured separately for each room so that the proper amount of radiation may be provided for comfort in all areas. Space limits do not permit giving detailed information and data on figuring building heat losses here. Anything less than a thorough treatment of this subject would be misleading. Complete information and data on this subject are contained in the manuals and guides from the HVAC associations.

Forced Warm Air Heating Systems

Forced warm air heating is the most widely used residential heating method in use at the present time. Generally speaking, the forced warm air heating method differs from the old gravity type by using blower equipment to circulate the air in the system.

It is practically impossible for the average general estimator to compute the cost of a modern forced warm air heating system, unless he has a working knowledge of this subject and is able to calculate building heat losses and design systems which will satisfy heating requirements under design conditions. In addition, the subject has become so complex it cannot be covered in a volume of this kind so as to provide adequate information for the preparation of a cost estimate, other than a rough approximation to be used for preliminary purposes only. Always obtain firm quotations from reputable warm air heating contractors before submitting bids on jobs containing this work.

Winter Air Conditioning or Forced Air Systems. Forced warm air heating systems that are equipped with automatic humidifying devices and air filters are popularly known as winter air conditioning systems. Most forced air installations are of this type.

Each system consists of a direct fired heating unit, a blower, a system of warm air ducts and return air duct system. Air filters are located in the return air duct system just ahead of the blower—humidifying equipment is placed in the warm air plenum chamber over the furnace unit. Air cooling equipment may be added to make this a year-round system.

The unit may be gas, oil, or coal fired. Gas or oil fired units designed for that particular fuel operate at high efficiency, although satisfactory, if less efficient results may be obtained through the use of conversion burners installed in units primarily designed for solid fuel.

Thermostatic control of heating apparatus is highly desirable and invariably used. Automatic controls vary the supply of heat in accordance with demand set up by a thermostat located in the living portion of the home. Actuated by limit controls, the blower commences operation automatically when the heated air in the generating unit has reached a predetermined temperature and cuts out after the source of heat has been shut down and the temperature in the bonnet dropped to a predetermined point.

Design of air conveying systems of this kind presents an engineering problem that can only be solved with a reasonable degree of accuracy by a competent designer. The heat loss of the building, room by room, is established and the volume of heated and conditioned air required for indoor comfort establishes data from which a duct system can be designed with due regard for velocities, static pressures and delivery temperatures.

Items to be Included in the Estimate. Costs vary considerably and methods of estimating also differ, but every estimate for a winter air conditioning system should include the following items:

1. Cartage of material and equipment from shop to job.
2. Winter air conditioning unit, complete with humidifier, air filters, and controls.
3. Labor assembling and setting up unit.
4. Oil storage tank installation, including fill and vent piping, oil booster pump if required, oil gauge, piping to unit, oil filter, etc., for oil fired units, or—piping for gas fired units.
5. Installation of smoke pipe, fittings and accessories.
6. Warm air and return air plenum chambers.
7. Warm air and return air duct systems.
8. Diffusers, registers, intakes, grilles, etc.
9. Special insulation.
10. Electrical work.
11. Labor starting plant in operation and balancing system.
12. Service allowance.
13. Miscellaneous costs.
14. Overhead and profit.

Cost of Winter Air or Forced Air Conditioning Units. Sizes, capacities and prices of units vary with the different manufacturers, but the following listings are representative of models available for residential use.

Approximate Net Prices of Winter Air Conditioning Units
Complete with Automatic Controls, Humidifier, and Air Filters

Bonnet Rating Btu per hr. (Kg-cal per hr.)	Register Rating Btu per Hr. (Kg-cal per hr.) Gas Fired Units	Heating Only	mbh (kg-cal per hr.)	Net Price with Cooling
60,000 (15,120)	48,000 (12,096)	$500.00	24 (6,048)	$1870.00
76,000 (19,152)	60,800 (15,321)	530.00	29 (7,308)	2,295.00
90,000 (22,680)	72,000 (18,144)	560.00	36 (9,072)	2,465.00
108,000 (27,216)	86,400 (21,773)	640.00	47 (11,844)	2,975.00
120,000 (30,240)	96,000 (24,192)	820.00	47 (11,844)	3,145.00
150,000 (37,800)	120,000 (30,240)	1,060.00	58 (14,616)	3,910.00
180,000 (45,360)	144,000 (36,288)	1,190.00		
	Oil Fired Units			
84,000 (21,168)	71,400 (17,993)	890.00	36 (9072)	2,465.00
112,000 (28,224)	95,200 (23,990)	1,020.00	42 (10,584)	2,975.00
140,000 (35,280)	119,000 (29,988)	1,200.00	47 (11,844)	3,400.00
175,000 (44,100)	148,750 (37,485)	1,360.00		
210,000 (52,920)	178,500 (44,982)	1,530.00		

Assembling and Setting up Unit. Most gas fired units can be assembled and set up by a sheet metal worker and a helper in 3 to 4 hours. For an oil fired unit, figure 4 to 5 hours.

Oil Storage Tanks. Common practice for oil storage tanks is the use of one or two inside storage tanks with a capacity of 275 gals. (1040 liters) each. Occasionally, outside buried tanks of 550- or 1,000-gal. (2081- or 3785-liter) capacity are used. Buried tanks must be double wall with fuel leak detectors and must be in accordance with local environmental regulations. On a contract basis, these tanks installed, including all piping, cost as follows:

1-275 gallon (1040 liter) inside tank...$400.00
2-275 gallon (1040 liter) inside tanks...780.00
1-550 gallon (2081 liter) outside tank...600.00
1-1,000 gallon (3875 liter) outside tank...1,000.00

In the use of an outside tank, a basement wall type pump costing from $120.00 to $160.00, must be used with some types of burner equipment.

Gas Piping. An allowance of $200.00 to $300.00 should cover cost of material and labor for connecting gas to unit within the heating room.

Smoke Pipe. Material and labor for smoke pipe connection from unit to flue should cost $80.00 to $120.00 for gas fired and oil fired units.

Plenum Chambers. Plenum chambers must be fabricated to order in the shop and are usually made from light gauge galvanized steel. They are mounted on top of the furnace and serve to connect the warm air and return air duct systems to the unit. Two plenum chambers are required for each installation.

Plenums cost $8.00 per s.f. ($86.00 per sq.m) in 18 ga., $10.00 ($107 per sq.m) in 16 ga., $12.00 ($129 per sq.m) in 14 ga.

Warm Air and Return Air Duct Systems. The cost of duct work will vary with the size of job and type of distribution.

Under ordinary conditions, the warm air heating contractor will measure and list each type and size of duct, fitting, and accessory. These quantities are priced for material, and then an estimate is made of labor required for installation. This is a lengthy operation and requires an experienced heating contractor.

When pressed for time, the duct system cost may be approximated on an outlet basis, using a unit price which covers both material and labor. For ordinary installations, the following costs per outlet are about average where sheet metal workers' wages are from $20.00 to $25.00 per hr.:

Type of System	Cost per Outlet
Conventional system—warm air outlets located on inside walls.	
1st floor outlets	$100.00-125.00
2nd floor outlets	140.00-160.00
Capped stacks	40.00 to 50.00
Radial perimeter system in small 1-sty. buildings with basement or crawl space.	
Low velocity—6" to 8" (150-200 mm) dia. ducts	$100.00-125.00
High velocity—4" and 5" (200-225 mm) dia. ducts	95.00-120.00
Large central return air duct	130.00-140.00
Radial perimeter system in small 1-sty. basementless buildings.	
Galv. steel sheet metal pipe and fittings	$132.00-140.00
Fiber duct with galv. steel fittings	132.00-140.00
Trunk and branch perimeter system in medium-size, 1 sty. buildings w/basement or crawl space.	
Std., galv. steel, rectangular ducts	$150.00-165.00

Add for diffusers, registers, intakes, special insulation, etc. Excavation, backfill and concrete encasement where required, not included.

Diffusers, Registers, and Intakes. Diffusers and registers generally used are of pressed steel construction, with prime coat finish, are fully adjustable and are available in floor, baseboard, or wall types. Return air intakes are of same construction and finish, but are non-adjustable.

Floor type diffusers cost $16.00 to $21.00 for sizes commonly used. Baseboard type diffusers cost $15.00 for 2'-0" (0.6 m) lengths and $21.00 for 4'-0" (1.2 m) lengths. Wall type registers cost $7.00 to $15.00. Wall type intake grilles cost $9.00 to $18.00.

Special Insulation. Ducts in unexcavated spaces or in unheated attic spaces, both supply and return, should be insulated with not less than 1" (25 mm) of adequate insulating material. This material is furnished in flexible form and averages $2.00 to $2.50 per s.f. ($21.50-26.90 per sq.m) applied.

Electrical Work. Custom varies with respect to the supply and installation of electrical circuit from meter to fused safety switch adjacent to unit. Likewise for the connection of blower motor to 110 volt controls and also for the 22 volt wiring to thermostat when used. Most often, the job electrician runs the 110 volt service to the safety switch adjacent to unit. From this point it is handled as a subcontract under the heating contractor and represents a cost from $90.00 to $225.00 for complete wiring installation of all controls and motors.

Labor Starting and Balancing System. It is customary for the heating contractor to start the plant in operation, test all controls and balance the air distribution system. For jobs completed during summer months, this means a comeback call at the beginning of the heating system.

Labor costs for this work will run $100.00 to $150.00 depending upon size of job. For commercial work, an independent certified testing and balancing contractor might be required. The cost can be up to $5,000.00.

Service Allowance. Most jobs carry a one-year free service warranty, and an allowance should be included in the estimate to cover this contingency. Average costs for service during the first year of operation are $200.00 for gas fired and $220.00 for oil fired systems.

Miscellaneous Costs. Under miscellaneous costs are classified such items as federal tax, sales tax, permits for oil burner, or for installation if required.

Approximate Cost of a Complete System Installed. Based on installations in residences of ordinary construction, located in the larger cities, the cost of average winter air conditioning systems ranges as follows for gas fired units. For oil fired units, add 10% to 15%.

Floor Area, s.f. (sq.m)	Heat Loss Btu per Hr. (kg-cal per hr.)	Price for Complete Gas Fired Installation
Conventional Systems, 1-Sty. Buildings with Basements or Crawl Space		
1,000 (92)	65,000 (16,380)	$2,550.00-3,400.00
1,500 (139)	80,000 (20,160)	3,400.00-3,910.00
2,000 (185)	100,000 (25,200)	3,910.00-4,250.00
Conventional Systems, 2-Sty. Buildings with Basements		
1,500 (139)	75,000 (18,900)	$3,400.00-3,910.00
2,000 (185)	90,000 (22,680)	4,250.00-5,100.00
Perimeter Systems, 1-Sty. Buildings with Basements or Crawl Space		
1,000 (92)	65,000 (16,380)	$2,550.00-3,400.00
1,500 (139)	80,000 (20,160)	3,400.00-3,910.00
2,000 (185)	100,000 (25,200)	4,250.00-4,760.00

Perimeter Systems, 1-Sty. Basementless Buildings

1,000 (92)......................................65,000 (16,380)...........................$2,550.00-3,400.00
1,500 (139).....................................80,000 (20,160)..............................3,400.00-3,910.00

The above prices are for individual jobs. For multiple housing projects, where large numbers of similar dwelling units are involved, much better prices may be obtained.

As previously mentioned, equipment, design, and practices vary greatly even within localities. No safe guide may be set down for general use. So much depends on correct design, proven equipment, and experienced installation that it behooves the buyer to carefully investigate all of these items before awarding a contract for this important part of the building.

Summer Air Conditioning

Residential summer air conditioning is available at a price the average home owner can afford. This applies to both existing homes and new homes under construction and is especially true where forced warm air heating is used, because the same air distribution duct system can be used for both heating and cooling. Various warm air heating equipment manufacturers make cooling units for this purpose, and installation work is usually done by heating contractors.

The function of summer air conditioning equipment is the reverse of winter air conditioning, circulating air is cooled and dehumidified, but design problems are similar though more complex. A basic requirement for designing a cooling system is calculating the total cooling load or total heat that must be removed from the structure to achieve a predetermined inside temperature and humidity under design conditions. Heat gain calculations are much more involved than those for heat losses, because many additional factors must be considered, such as sun effect on roof, wall, and glass areas; internal heat gains from human occupancy, cooking, lighting, and appliances; and latent heat energy involved in dehumidification. This is a job for an expert.

Heat gain calculations are always expressed in Btu per hour. Cooling equipment capacities are usually stated per ton of refrigeration. A ton of refrigeration will remove heat at the rate of 12,000 Btu per hr. In a 24-hr. period, this is equivalent to the heat required to melt a ton of ice. With water-cooled compression refrigeration equipment, one horsepower usually equals about one ton of cooling capacity.

Types of Cooling Equipment for Residential Use. There are several types of cooling units available and in popular use for residential summer air conditioning. In general, equipment may be classified as water-cooled or air-cooled, with each type available for several methods of application.

One method employs a remotely located condensing unit, with refrigerant lines connected to a cooling coil installed in the warm air plenum chamber of a forced warm air heating system. This method uses the heating system blower and ducts for air distribution.

Another method uses a self-contained unit, with the condenser, cooling coil, and blower combined in one cabinet. This type may be cut into the heating ducts, or may have its own duct system where hot water or steam heat is used.

Net Prices of Remote System Cooling Units. Approximate prices are for units factory assembled with heating and cooling thermostat included. Pipe, duct materials, or blower assembly is not included.

Capacity Tons (kg)	Rating Btu per hr. (kg-cal per hr.)	Net Price Air-cooled	Labor Hrs. To Install
2 (1814)	24,000 (6048)	$1,000.00	8
3 (2721)	36,000 (9072)	1,450.00	14
5 (4536)	60,000 (15,120)	2,700.00	30

Where a complete air circulation duct system is required, figure costs the same as for winter air conditioning systems.

Approximate Cost of Complete Cooling Unit Installation. Based on conditions of average difficulty, complete residential cooling unit installations should cost approximately as follows:

Capacity Tons (kg)	Rating Btu per Hr. (kg-cal per hr.)	Price for Complete Installation
Remote, Water-cooled System		
2 (1814)	24,000 (6048)	$3,740.00-4,250.00
3 (2721)	36,000 (9072)	4,250.00-4,760.00
Remote, Air-cooled System		
2 (1814)	24,000 (6048)	3,400.00-3,740.00
3 (2721)	36,000 (9072)	4,250.00-4,590.00
5 (4536)	60,000 (15,120)	7,650.00-8,500.00
Self-contained, Water-cooled System		
2 (1814)	24,000 (6048)	3,910.00-4,250.00
3 (2721)	36,000 (9072)	5,950.00-6,290.00
Self-contained, Air-cooled System		
2 (1814)	24,000 (6048)	2,600.00-4,760.00
3 (2721)	36,000 (9072)	7,650.00-8,160.00
5 (4536)	60,000 (15,120)	11,900.00-12,750.00

Room Size Air Conditioning Units. The cost of portable room size air conditioning units varies according to size, ranging from 1/2-Hp. (0.37-kW), which cools room areas up to 400 s.f. (36 sq.m), to 2-Hp. (1.5 kW) for spaces up to 1,200 s.f. (108 sq.m). All models are for window installation and include automatic thermostat controls.

Net prices run from $400.00 for a 4,000 Btu (1008 kg-cal) unit to $1,500.00 for a 20,000 Btu (5040 kg-cal) unit. If electric wiring is necessary, add extra for installation.

Commercial Installations. A number of mechanical equipment manufacturers offer self-contained, one-piece rooftop electric heat pump units for the smaller commercial projects.

These units, which are made in numerous heating and cooling capacities, are pre-wired and precharged at the factory so that when they are delivered to the project site, the installation is relatively simple. The unit is set on a roof curb, power and control wiring connected and duct connections hooked up, and the unit is in operation.

Standard construction of these units includes a manual damper that can be preset to admit up to 25% outdoor ventilation air for year-round fresh air. Some units come equipped with an "economizer cycle" that has a thermostatically-controlled, fully-modulating damper that admits up to 100% outdoor ventilation air to provide low-cost cooling on mild days.

For quick approximation of air cooling costs, cooling units can be estimated from $1,500.00 to $3,000.00 per ton, depending on the ductwork, job conditions, types of grilles, etc. For one ton of air cooling, 12,000 Btus (3024 kg-cal) is required. The following table will give an approximation of final requirements:

Type of Space	Floor area, Btu per s.f. (Kg-cal per sq.m)
Large Stores	40 (112)
Offices	45 (126)
Specialty Shops	50-70 (140-196)
Bars	80 (224)
Restaurants	100-120 (280-336)

Steam and Hot Water Heating

Because of differing conditions, it is impossible to give any set rule that can be accepted without modification for all kinds of buildings. It is necessary to take into consideration all of the conditions in and around a building, and additions or deductions made to suit the requirements.

Methods of Computing Heating. The most advanced method of figuring heating, the one that is generally used by heating engineers, is the Btu Method. This method is based on replacing the heat loss through exposed walls, doors and windows, floors, ceilings, etc., and uses so many factors based on the type of construction that it is impossible to go into detail in a book of this kind. Intricate and elaborate formulas are used for both warm air, steam, and hot water heating. Only approximates are given here.

Flow Control Hot Water Heating Systems. This method of hot water heating has largely superseded the older gravity or "open" type of hot water heating system. It is a complete combined automatic water heating and domestic water supply system for year-round use. Its major advantages are instantaneous heating response, high efficiency, fuel economy, and completely automatic operation.

The flow control valve makes the boiler a reservoir of high temperature water. It prevents circulation through the system when heat is not required. A circulator or pump attached to the main near the boiler forces this hot water through the mains the instant the room thermostat calls for heat. When the thermostat is satisfied, the circulator ceases to operate, and the flow control valve stops all circulation of hot water, thus preventing overheating and needless heat losses.

The flow control hot water heating system requires about 20% more radiation than is figured for steam heat. The mains are smaller, reducing the cost of pipe and fittings and labor to install.

All in all, a flow control hot water heating system costs a trifle less than the older gravity systems requiring large radiators and piping. A gravity hot water heating system can be converted into a flow control system at a nominal expense.

Cost of Special Fittings for Flow Control Hot Water Heating Systems. The cost of fittings varies with the amount of radiation in the job, and whether a complete flow control system is used, or merely a flow control valve and a circulating pump.

Where only the flow control valve and water circulating pump are used, the costs for the special equipment are as follows based on the amount of radiation in the job.

Capacities and Prices of Flow-Control Valves

Valve Size	Radiation Capacity in Square Feet			Weight, lbs.	Price Each
	150 Btu	200 Btu	240 Btu		
1"	354	265	221	5	$20.00
1-1/4"	786	590	492	5	21.00
1-1/2"	1170	875	730	10	24.00
2"	2140	1610	1340	21	29.00
2-1/2"	3680	2750	2300	33	57.00
3"	6660	5000	4160	42	72.00
4"	14000	10000	8750	57	165.00

(metric)

Valve Size	Radiation Capacity in Square Meters			Weight, kg	Price each
	420 kg-cal	560 kg-cal	672 kg-cal		
25	991	742	619	2.2	$20.00
31.2	2200	1652	1378	2.2	21.00

37.5	3276	2450	2044	4.5	24.00
50	5992	4508	3752	9.5	29.00
62.5	10,304	7700	6440	14.9	57.00
75	18,648	14,000	11,648	19.0	72.00
100	39,200	28,000	24,500	25.8	165.00

Sizes and Prices of Water Circulators or Booster Pumps

Recommended for use on modern forced circulation systems or for improving gravity hot water systems.

Direct Radiation Capacity in Sq. Ft.

Size Inches	Motor Hp	Dely Gals. per Min.	150 Btu	200 Btu	240 Btu	Price Each Iron Body	All Bronze
1	1/12	31	750	562	469	$160.00	$155.00
1-1/4	1/12	33	1000	750	675	160.00	180.00
1-1/2	1/12	33	1000	750	675	160.00	180.00
2	1/6	75	3500	2625	2187	280.00	460.00
2-1/2	1/4	120	4500	3375	2820	405.00	735.00
3	1/4	120	4500	3375	2820	405.00	840.00

(metric)
Direct Radiation Capacity in Sq. Meters

Size mm	Motor kW	Dely Liters per Min.	420 kg-cal	560 kg-cal	672 kg-cal	Price Each Iron Body	All Bronze
25	0.06	117	2100	1574	1313	$160.00	$155.00
31.2	0.06	124	2800	2100	1890	160.00	180.00
37.5	0.06	124	2800	2100	1890	160.00	180.00
50	0.12	283	9800	7350	6124	280.00	460.00
62.5	0.18	454	12,600	9450	7896	405.00	735.00
75	0.18	454	12,600	9450	7896	405.00	840.00

Sizes and Prices of Relief Valves

ASME Capacity at 30 lbs. (13.6 kg) in Btu per Hr. (kg-cal per hr.)	Size inches (mm)	Net Price
175,000 (44,100)	3/4 (18.7)	$12.00
250,000 (63,000)	3/4 (18.7)	17.00
350,000 (88,200)	3/4 (18.7)	20.00
480,000 (120,960)	3/4 (18.7)	21.00
750,000 (189,000)	1 (25)	35.00
1,050,000 (264,600)	1-1/4 (31.2)	56.00

Sizes and Prices of Air Control Boiler and Tank Fittings

Air control fittings are used for the purpose of removing air from a hot water heating system and preventing its return.

Air control boiler fittings consist of a tube inserted into a T-shaped fitting. The tube, in effect, extends the supply main down into the boiler water, thus preventing air, which accumulates at the top of the boiler, from rising into the piping and heating units. This air flows around the tube, up into the T-fitting, and then to the pipe leading to the air control tank fitting.

The air control tank fittings provide two separate passages—one through which air can flow to the top of the tank, and the other for the water displaced by the air to flow back into the pipe connections to the system. These will cost from $22.00 to $45.00 in sizes 1" (25 mm) to 3" (75 mm) for boiler fittings, $22.00 for tank controls up to 24" (600 mm).

Coils for Heating Water. Where coils are inserted in the boiler for heating water for domestic purposes, the size of the boiler should be increased, figuring each gallon of water tank capacity as equivalent to 1 s.f. (0.09 sq.m) of steam radiation or 2 s.f. (0.18 sq.m) of hot water radiation. For example, a 30-gal. (113-liter) tank is equivalent to 30 s.f. (2.78 sq.m) of steam or 60 s.f. (5.57 sq.m) of hot water radiation.

Rule for Computing Boiler Sizes. After the total number of square feet of radiation of total heat loss in Btu (kg-cal) per hour has been obtained, the size of the boiler should be computed.

Practically all boiler manufacturers rate their boilers, giving maximum heat output under full drive and also the recommended load representing the attached load, including piping, which may be placed on the boiler in accordance with accepted installation standards for economical operation.

Regarding Steam Boiler Installation. It is recommended on all installations of steam boilers that drain valves be placed on the returns and that the condensation from such returns be discharged into the sewer for a period of 3 days to one week after starting fire, thereby clearing system of grease and dirt.

At the end of this period the boiler should be thoroughly washed and blown out.

Boiler Capacity. Boiler output is defined as the number of sq. ft. of standard direct column radiation or its equivalent in Btu that a boiler will supply at its outlet in one hour, under standard conditions, with 240 Btu (672 kg-cal) representing 1 s.f. (sq.m) of standard direct column steam radiation and 150 Btu (420 kg-cal) representing 1 s.f. (sq.m) of standard direct column water radiation.

Boiler Horsepower. One boiler horsepower is defined as the evaporation of 34-1/2 lbs. (15.6 kg) of water per hour from a feed water temperature of 212°F (100°C) into steam at 212°F (100°C). This is equal to 34.5 x 970.4 = 33478.8 Btu (8436.65 kg-cal) per hour. One boiler horsepower is equal therefore to 33478.8 ÷ 240 = 139 s.f. (8436.5 ÷ by 672 = 12.55 sq.m) of steam radiation. Note that boiler horsepower is not derived from engine horsepower. 1 horsepower hour = 2545 Btu.

To find the Btu output per hour of a steam boiler, multiply the output in s.f. by 240 (multiply sq.m by 672 to find kg-cal).

To find the Btu output per hour of a water boiler, multiply the output in s.f. by 150. (sq.m x 420 = kg-cal).

To find the output in s.f. of steam radiation, divide the output in Btu per hour by 240. (sq.m x 672 = kg-cal).

To find the output in s.f. of water radiation, divide the output in Btu per hour by 150. (sq.m x 420 = kg-cal).

Estimating Pipe and Fittings. To prepare an intelligent estimate on the cost of any heating plant, the entire job should be laid out using the quantity of pipe of each size required; connections, such as elbows, tees, nipples, valves, etc., should all be figured separately. This requires considerable labor and an estimator must thoroughly understand the business to figure with any degree of accuracy.

When estimating the quantities of piping required for any heating plant, the risers, mains, and returns, together with all connections to and from the boilers, must be figured. These vary according to the system used and the type of building.

A one-pipe steam heating system should have the highest point in the main directly over the boiler, and from this point it pitches down around the circuit until it drops to the floor or enters the boiler. The more pitch the better, but there should be at least one inch (25 mm) in 20 feet (6 m). The lowest point in the steam main, where it drops to enter the boiler or to the floor, should not be less than 14 inches above the water line of boiler and the more the better.

Branches for radiators should be taken from top of main, using a nipple and elbow or a nipple and a 45-degree ell, and should pitch up to the radiator.

Connections for steam should never be taken from side of main, as this will cause water hammer and syphoning of the water of condensation in the steam main into the radiator.

Each radiator should be supplied with a good disc radiator valve and a good automatic air valve, and each return main should also have an automatic air valve where it drops to enter the boiler.

If branches are longer than 8 to 10 feet (2.4 to 3.0 m), they should be one size larger than the upright connecting pipes. Where mains drop to floor, a pipe two sizes smaller can be used to return to boiler. Risers for either steam or water should be plumb and straight. If exposed in a room, and it is necessary to use couplings on same, the couplings should be at a uniform height from floor.

Length of Pipe	Size of Pipe											
	¾	1	1¼	1½	2	2½	3	4	5	6	7	8
1	.275	.346	.434	.494	.622	.753	.916	1.175	1.455	1.739	1.996	2.25?
2	.5	.7	.9	1.	1.2	1.5	1.8	2.4	2.9	3.5	4.	4.?
3	.8	1.	1.3	1.5	1.9	2.3	2.7	3.5	4.4	5.2	6.	6.?
4	1.1	1.4	1.7	2.	2.5	3.	3.6	4.7	5.8	7.	8.	9.
5	1.4	1.7	2.2	2.4	3.1	3.8	4.6	5.8	7.3	7.7	10.	11.?
6	1.6	2.1	2.6	2.9	3.7	4.5	5.5	7.	8.7	10.5	12.	13.?
7	1.9	2.4	3.	3.4	4.4	5.3	6.4	8.2	10.2	12.1	14.	15.?
8	2.2	2.8	3.5	3.9	5.	6.	7.3	9.4	11.6	13.9	16.	18.
9	2.5	3.1	3.9	4.4	5.6	6.8	8.2	10.6	13.1	15.7	18.	20.?
10	2.7	3.5	4.3	4.9	6.2	7.5	9.1	11.8	14.6	17.4	20.	22.?
11	3.	3.8	4.8	5.4	6.8	8.3	10.	12.9	16.	19.1	22.	24.?
12	3.3	4.1	5.2	5.9	7.5	9.	11.	14.1	17.4	20.9	24.	27.
13	3.6	4.5	5.6	6.4	8.1	9.8	11.9	15.3	18.9	22.6	26.	29.?
14	3.8	4.8	6.1	6.9	8.7	10.5	12.8	16.5	20.3	24.3	28.	31.?
15	4.1	5.2	6.5	7.4	9.3	11.3	13.7	17.6	21.8	26.1	30.	33.?
16	4.4	5.5	6.9	7.9	10.	12.	14.6	18.8	23.2	27.8	32.	36.
17	4.7	5.9	7.4	8.4	10.6	12.8	15.5	20.	24.7	29.5	34.	38.?
18	5.	6.2	7.8	8.9	11.2	13.5	16.5	21.2	26.2	31.3	36.	40.?
19	5.2	6.6	8.3	9.4	11.8	14.3	17.4	22.3	27.6	33.1	38.	42.?
20	5.5	6.9	8.7	9.9	12.5	15.	18.3	23.5	29.1	34.8	40.	45.?
21	5.8	7.3	9.1	10.4	13.	15.8	19.2	24.7	30.5	36.5	42.	47.?
22	6.	7.6	9.6	10.9	13.7	16.5	20.2	25.9	32.	38.3	44.	49.?
23	6.3	8.	10.	11.3	14.3	17.3	21.1	27.	33.5	40.	46.	52.
24	6.6	8.3	10.4	11.9	14.9	18.	22.	28.2	34.9	41.7	48.	54.?
25	6.9	8.6	10.9	12.3	15.6	18.8	22.9	29.3	36.3	43.5	50.	56.?
26	7.1	9.	11.3	12.8	16.2	19.5	23.8	30.5	37.8	45.2	52.	58.?
27	7.4	9.4	11.7	13.3	16.8	20.3	24.7	31.7	39.3	47.	54.	61.
28	7.7	9.7	12.2	13.8	17.4	21.	25.6	32.9	40.7	48.7	56.	63.?
29	8.	10.	12.6	14.3	18.	21.8	26.6	34.1	42.2	50.4	58.	65.?
30	8.3	10.4	13.	14.8	18.7	22.5	27.5	35.3	43.6	52.1	60.	67.?
31	8.5	10.7	13.5	15.3	19.3	23.3	28.4	36.4	45.1	53.9	62.	70.
32	8.8	11.1	13.9	15.8	19.9	24.1	29.3	37.6	46.5	55.6	64.	72.?
33	9.1	11.4	14.3	16.3	20.5	24.8	30.2	38.8	48.	57.4	66.	74.?
34	9.4	11.7	14.7	16.8	21.2	25.6	31.1	40.	49.5	59.1	68.	76.?
35	9.6	12.1	15.2	17.3	21.8	26.3	32.	41.1	50.9	60.8	70.	79.
36	9.9	12.5	15.6	17.8	22.4	27.	33.	42.3	52.4	62.6	72.	81.?
37	10.2	12.8	16.1	18.3	23.	27.8	33.9	43.5	53.8	64.3	74.	83.?
38	10.5	13.2	16.5	18.8	23.7	28.5	34.8	44.6	55.2	66.	76.	85.?
39	10.7	13.5	16.9	19.3	24.3	29.3	35.7	45.8	56.7	67.8	78.	88.
40	11.	13.8	17.4	19.8	24.9	30.1	36.6	47.	58.2	69.5	80.	90.?
41	11.3	14.2	17.8	20.3	25.5	30.8	37.6	48.2	59.6	71.3	82.	92.?
42	11.5	14.5	18.2	20.8	26.1	31.6	38.5	49.4	61.1	73.	84.	94.?
43	11.8	14.9	18.7	21.3	26.8	32.3	39.4	50.6	62.5	74.8	86.	97.
44	12.1	15.2	19.1	21.8	27.4	33.1	40.3	51.7	64.	76.5	88.	99.?
45	12.4	15.6	19.5	22.2	28.	33.8	41.2	52.9	65.5	78.2	90.	101.6
46	12.7	15.9	20.	22.7	28.6	34.6	42.2	54.	67.	80.	92.	103.?
47	12.9	16.3	20.4	23.2	29.2	35.3	43.	55.2	68.4	81.7	94.	106.
48	13.2	16.6	20.8	23.7	29.9	36.1	43.9	56.4	69.8	83.5	96.	108.?
49	13.5	17.	21.3	24.2	30.5	36.8	44.8	57.6	71.2	85.1	98.	110.?
50	13.8	17.3	21.7	24.7	31.1	37.6	45.8	58.7	72.7	87.	100.	112.?

Table Giving Number of Square Feet of Radiation in Exposed Heating Pipes of Various Sizes and Lengths

Table Giving Number of Square Meters of Radiation in Exposed Heating Pipe of Various Sizes and Lengths

Length of Pipe (M)	Size of Pipe (mm)											
	18.8	25	31.25	37.5	50	62.5	75	100	125	150	175	200
0.30	0.03	0.03	0.04	0.05	0.06	0.07	0.09	0.11	0.14	0.16	0.19	0.21
0.61	0.05	0.07	0.08	0.09	0.11	0.14	0.17	0.22	0.27	0.33	0.37	0.42
0.91	0.07	0.09	0.12	0.14	0.18	0.21	0.25	0.33	0.41	0.48	0.56	0.63
1.22	0.10	0.13	0.16	0.19	0.23	0.28	0.33	0.44	0.54	0.65	0.74	0.84
1.52	0.13	0.16	0.20	0.22	0.29	0.35	0.43	0.54	0.68	0.72	0.93	1.05
1.83	0.15	0.20	0.24	0.27	0.34	0.42	0.51	0.65	0.81	0.98	1.11	1.25
2.13	0.18	0.22	0.28	0.32	0.41	0.49	0.59	0.76	0.95	1.12	1.30	1.47
2.44	0.20	0.26	0.33	0.36	0.46	0.56	0.68	0.87	1.08	1.29	1.49	1.67
2.74	0.23	0.29	0.36	0.41	0.52	0.63	0.76	0.98	1.22	1.46	1.67	1.89
3.05	0.25	0.33	0.40	0.46	0.58	0.70	0.85	1.10	1.36	1.62	1.86	2.10
3.35	0.28	0.35	0.45	0.50	0.63	0.77	0.93	1.20	1.49	1.77	2.04	2.31
3.66	0.31	0.38	0.48	0.55	0.70	0.84	1.02	1.31	1.62	1.94	2.23	2.52
3.96	0.33	0.42	0.52	0.59	0.75	0.91	1.11	1.42	1.76	2.10	2.42	2.73
4.27	0.35	0.45	0.57	0.64	0.81	0.98	1.19	1.53	1.89	2.26	2.60	2.94
4.57	0.38	0.48	0.60	0.69	0.86	1.05	1.27	1.64	2.03	2.42	2.79	3.15
4.88	0.41	0.51	0.64	0.73	0.93	1.11	1.36	1.75	2.16	2.58	2.97	3.35
5.18	0.44	0.55	0.69	0.78	0.98	1.19	1.44	1.86	2.29	2.74	3.16	3.57
5.49	0.46	0.58	0.72	0.83	1.04	1.25	1.53	1.97	2.43	2.91	3.34	3.77
5.79	0.48	0.61	0.77	0.87	1.10	1.33	1.62	2.07	2.56	3.07	3.53	3.99
6.10	0.51	0.64	0.81	0.92	1.16	1.39	1.70	2.18	2.70	3.23	3.72	4.20
6.40	0.54	0.68	0.85	0.97	1.21	1.47	1.78	2.29	2.83	3.39	3.90	4.40
6.71	0.56	0.71	0.89	1.01	1.27	1.53	1.88	2.41	2.97	3.56	4.09	4.62
7.01	0.59	0.74	0.93	1.05	1.33	1.61	1.96	2.51	3.11	3.72	4.27	4.83
7.32	0.61	0.77	0.97	1.11	1.38	1.67	2.04	2.62	3.24	3.87	4.46	5.04
7.62	0.64	0.80	1.01	1.14	1.45	1.75	2.13	2.72	3.37	4.04	4.65	5.24
7.92	0.66	0.84	1.05	1.19	1.50	1.81	2.21	2.83	3.51	4.20	4.83	5.44
8.23	0.69	0.87	1.09	1.24	1.56	1.89	2.29	2.94	3.65	4.37	5.02	5.67
8.53	0.72	0.90	1.13	1.28	1.62	1.95	2.38	3.06	3.78	4.52	5.20	5.87
8.84	0.74	0.93	1.17	1.33	1.67	2.03	2.47	3.17	3.92	4.68	5.39	6.08
9.14	0.77	0.97	1.21	1.37	1.74	2.09	2.55	3.28	4.05	4.84	5.57	6.29
9.45	0.79	0.99	1.25	1.42	1.79	2.16	2.64	3.38	4.19	5.01	5.76	6.50
9.75	0.82	1.03	1.29	1.47	1.85	2.24	2.72	3.49	4.32	5.17	5.95	6.71
10.06	0.85	1.06	1.33	1.51	1.90	2.30	2.81	3.60	4.46	5.33	6.13	6.91
10.36	0.87	1.09	1.37	1.56	1.97	2.38	2.89	3.72	4.60	5.49	6.32	7.13
10.67	0.89	1.12	1.41	1.61	2.03	2.44	2.97	3.82	4.73	5.65	6.50	7.34
10.97	0.92	1.16	1.45	1.65	2.08	2.51	3.07	3.93	4.87	5.82	6.69	7.55
11.28	0.95	1.19	1.50	1.70	2.14	2.58	3.15	4.04	5.00	5.97	6.87	7.76
11.58	0.98	1.23	1.53	1.75	2.20	2.65	3.23	4.14	5.13	6.13	7.06	7.97
11.89	0.99	1.25	1.57	1.79	2.26	2.72	3.32	4.25	5.27	6.30	7.25	8.18
12.19	1.02	1.28	1.62	1.84	2.31	2.80	3.40	4.37	5.41	6.46	7.43	8.38
12.50	1.05	1.32	1.65	1.89	2.37	2.86	3.49	4.48	5.54	6.62	7.62	8.59
12.80	1.07	1.35	1.69	1.93	2.42	2.94	3.58	4.59	5.68	6.78	7.80	8.81
13.11	1.10	1.38	1.74	1.98	2.49	3.00	3.66	4.70	5.81	6.95	7.99	9.01
13.41	1.12	1.41	1.77	2.03	2.55	3.07	3.74	4.80	5.95	7.11	8.18	9.20
13.72	1.15	1.45	1.81	2.06	2.60	3.14	3.83	4.91	6.08	7.26	8.36	9.44
14.02	1.18	1.48	1.86	2.11	2.66	3.21	3.92	5.02	6.22	7.43	8.55	9.64
14.33	1.20	1.51	1.90	2.16	2.71	3.28	3.99	5.13	6.35	7.59	8.73	9.85
14.63	1.23	1.54	1.93	2.20	2.78	3.35	4.08	5.24	6.48	7.76	8.92	10.07
14.94	1.25	1.58	1.98	2.25	2.83	3.42	4.16	5.35	6.61	7.91	9.10	10.27
15.24	1.28	1.61	2.02	2.29	2.89	3.49	4.25	5.45	6.75	8.08	9.29	10.48

The valve on a steam radiator should be either wide open or tightly closed to prevent radiator filling with water.

In a two-pipe gravity hot water heating system the highest return rises from boiler to radiator, the lowest point being at the boiler and the highest point at the end of the main. These pipes may be reduced as the branches are taken off, care being taken not to reduce too rapidly. Each radiator should be fitted with a hot water radiator valve, union elbow and a compression air valve. All pipes should be well supported, so that no pockets or depressions will occur to retard or trap the circulation. All mains, both for steam and hot water, should run about 4 feet from wall to allow for expansion.

Practically all hot water heating systems installed today are flow control systems where the hot water is circulated from the boiler through the mains and pipes using a circulating or booster pump. When this method is used, the size of the mains can be greatly reduced, as will be noted from the following table. With the water being circulated through the system at 30 to 120 gallons (114 to 454 liters) per minute, it is very much faster than the old gravity systems. Also the radiators may be placed above or below the boiler level.

Sizes of Pipes. When estimating steam heating using a rule of thumb method, the size of the main is determined by taking the total amount of direct radiation and adding 25% for piping. From this total extract the square root and divide by 10, which gives the size of the main to use. This is for one-pipe work.

For two-pipe work, one size less is sufficient, and the return can be one or two sizes less than the supply. A steam main should decrease in size but very little according to the area of its branches.

Example: A job contains 400 s.f. of direct steam radiation. By adding 25% for piping, the result is 500 s.f. The square root of 500 is 22.4. Divide 22.4 by 10 equals 2.24; it requires a 2-1/2" steam main.

List of Sizes of Steam Mains. The following table gives the correct sizes of mains for both one- and two-pipe steam heating systems:

Radiation sq. ft. (sq.m)	One-Pipe Work inches (mm)	Two-Pipe Work inches (mm)
125 (11.25)	1-1/2 (37.5)	1-1/4x1 (31.25x25)
250 (22.50)	2 (50)	1-1/2x1-1/4 (37.5x31.25)
400 (36.00)	2-1/2 (62.5)	2x1-1/2 (50x37.5)
650 (58.50)	3 (75)	2-1/2x2 (62.5x50)
900 (81.00)	3-1/2 (87.5)	3x2-1/2 (75x62.5)
1,250 (112.50)	4 (100)	3-1/2x3 (87.5x75)
1,600 (144)	4-1/2 (112.5)	4x3-1/2 (100x87.5)
2,050 (184.5)	5 (125)	4-1/2x4 (112.5x100)
2,500 (225)	6 (150)	5x4-1/2 (125x112.5)
3,600 (324)	7 (175)	6x5 (150x125)
5,000 (450)	8 (200)	7x6 (175x150)
6,500 (585)	9 (225)	8x6 (200x150)
8,100 (729)	10 (250)	9x6 (225x150)

List of Sizes for Hot Water Mains. The following table gives the correct sizes of mains for hot water heating systems:

Radiation s.f. (sq.m)	Gravity Size Pipe inches (mm)	Forced H.W. 2-Pipe System*
75-125 (6.75-11.25)	1-1/4 (31.25)	1 (25)
125-175 (11.25-15.75)	1-1/2 (37.5)	1 (25)
175-300 (15.75-27)	2 (50)	1 (25)
300-475 (27-42.75)	2-1/2 (62.5)	1 (25)
475-700 (42.75-63)	3 (75)	1-1/4 (31.25)
700-950 (63-85.5)	3-1/2 (87.5)	1-1/2 (37.5)
950-1,200 (85.5-108)	4 (100)	1-1/2 (37.5)
1,200-1,575 (108-141.75)	4-1/2 (112.5)	2 (50)
1,575-1,975 (141.75-177.75)	5 (125)	2 (50)
1,975-2,375 (177.75-213.75)	5-1/2 (137.5)	2-1/2 (62.5)
2,375-2,850 (213.75-256.50)	6 (150)	2-1/2 (62.5)

*Based on 200 Btu emission and circulators having 1,725 rpm motors.

Hot water flow mains may be reduced in size in proportion to the branches taken off. They should, however, have areas as large as the sum of all branches beyond that point. Returns should be same as flows.

Tables of Mains and Branches. The following table gives the size of mains and the number of branches of different sizes that each main will supply:

Mains	Branches
1 in. will supply	2-3/4-in.
1-1/4 in. will supply	2 1-in.
1-1/2 in. will supply	2 1-1/4-in.
2 in. will supply	2 1-1/2-in.
2 in. will supply	2 1-1/2-in. and 1 1-1/4-in. or 1 2-in. and 1 1-1/4-in.
3 in. will supply	1 2-1/2-in. and 1 2-in. or 2 2-in. and 1 1-1/2-in.
3-1/2 in. will supply	2 2-1/2-in. or 1 3-in. or 1 2-in. or 3 2-in.
4 in. will supply	1 3-1/2-in. and 1 2-1/2-in. or 2 3-in. and 4 2-in.
4-1/2 in. will supply	1 3-1/2-in. and 1 3-in. or 1 4-in. and 1 2-1/2-in.
5 in. will supply	1 4-in. and 1 3-in. or 1 4-1/2-in. and 1 2-1/2-in.
6 in. will supply	2 4-in. and 1 3-in. or 4 3-in. or 10 2-in.
7 in. will supply	1 6-in. and 1 4-in. or 3 4-in. and 1 2-in.

(metric)

Mains	Branches
25 mm will supply	2 - 68.75 mm
31.25 mm will supply	2 - 25 mm
37.5 mm will supply	2 - 31.25 mm
50 mm will supply	2 - 37.5 mm
50 mm will supply	2 - 37.5 mm and 1 - 31.25 mm or 1 - 50 mm and 1 - 31.25 mm
75 mm will supply	1 - 62.5 mm and 1 - 50 mm or 2 - 50 mm and 1 - 37.5 mm
87.5 mm will supply	2 - 62.5 mm or 1 - 75 mm or 1 - 50 mm or 3 - 50 mm

Mains	Branches
100 mm will supply.......	1 - 87.5 mm and 1 - 62.5 mm or 2 - 75 mm and 4 - 50 mm
112.5 mm will supply....	1 - 87.5 mm and 1 - 75 mm or 1 - 100 mm and 1 - 75 mm
125 mm will supply.......	1 - 100 mm and 1 - 100 mm or 1 - 75 mm and 1 - 112.5 mm
150 mm will supply.......	2 - 100 mm and 1 - 75 mm or 4 - 75 mm or 10 - 50 mm
175 mm will supply	1 - 150 mm and 1 - 100 mm or 3 - 100 mm and 1 - 50 mm

Radiator Tappings. The following is a schedule of tappings or branches required for radiators of different sizes.

Steam

One Pipe

Radiators containing up to 24 s.f. (2.16 sq.m) of radiation.......................1" (25 mm)
Radiators containing 24 to 60 s.f. (2.16-5.4 sq.m) of radiation......1-1/4" (31.25 mm)
Radiators containing 60 to 100 s.f. (5.4-9 sq.m) of radiation...........1-1/2" (37.5 mm)
Radiators containing above 100 s.f. (50 mm) of radiation ....2" (50 mm)

Gravity Hot Water

Two Pipe

Radiators containing up to 40 s.f. (3.6 sq.m) of radiation................. 1x1" (25 x 25 mm)
Radiators containing 40 to 72 s.f. (3.6-6.48 sq.m)
of radiation .... 1-1/4x1-1/4" (31.25 x 31.25 mm)
Radiators containing more than 72 s.f. (6.48 sq.m)
of radiation .... 1-1/2x1-1/2" (37.5 x 37.5 mm)

Forced Flow Hot Water

A 3/4" (18.7 mm) branch will take care of most any ordinary size radiator where forced flow is used.

Size of Expansion Tanks Required for Hot Water Heating Systems

The following table gives the sizes of expansion tanks required for hot water heating systems, complete with trimmings:

Capacity, gals. (liter)	Size, in. (mm)	Radiation, s.f. (sq.m)	Price Each with Trimmings
15 (57)....................	12x30 (200x750)	350 (31.5)	$63.00
18 (68)....................	12x36 (300x900)	500 (45)	70.00
24 (90.8)...................	12x48 (300x1200)	1000 (90)	77.00
30 (114)....................	12x60 (300x1500)	2000 (180).................	88.00
40 (151)....................	14x60 (350x1500)	3000 (270)................	112.00
80 (303)....................	20x60 (500x1500)	4500 (405)................	180.00

For two gauge glass tappings, add $15.00.

Radiation capacities are based on forced circulation and small pipes. On old systems use next larger size tank.

Heights, Sizes and Ratings of Thin Tube Cast Iron Radiators

Cast iron radiations are seldom used these days, but standard measures are 1-3/4" (43.75 mm) in length per section. Add 1/2" (12.5 mm) to the length for each bushing.

S.F. (Sq.M) of Radiation per Section of Various Height Radiators

Number Tubes	Width, in. (mm)	19 (475)	22 (550)	25 (625)	32 (800)
3	3-1/2 (0.32)			1.6 (0.14)	
4	4-3/4 (0.43)	1.6 (0.14)	1.8 (0.16)	2.0 (0.18)	
5	6 (0.54)		2.1 (0.19)	2.4 (0.22)	
6	7-1/4 (0.65)	2.3 (0.20)		3.0 (0.27)	3.7 (0.33)

Approximate Prices of Thin Tube Radiators

Sold only in even number of sections. Price per 100 s.f. (9 sq.m)

No. Tubes	Height			
	19" (475 mm)	22" (550 mm)	25" (625 mm)	32" (800 mm)
3	$380.00	--	--	--
4	380.00	371.00	371.00	--
5	--	364.00	336.00	--
6	385.00	--	364.00	560.00

Radiant Baseboard Panels

Radiant baseboard panels are produced and marketed by a number of concerns. Some are made of cast iron, others with a finned pipe covered with a steel facing or cover.

They are made in various heights from 7" to 9" and usually extend out from the wall 1-3/4" to 2-1/2", depending on the manufacturer.

Rated Heat Emission of Radiant Baseboard Panels

Rating in sq. ft. is based on standard emission of 240 Btu per hour per s.f. (672 kg-cal per sq.m) at average temperature of 215°F (101.67°C).

9-7/8-Inch (246.875-mm) Radiant Panel Baseboard

Rating of Panel s.f.	Water Flow lbs./hr	Heat Output Btu per Hr. per L.F.							
		220°F	210°F	200°F	190°F	180°F	170°F	160°F	150°F
1.95	500	460	420	380	350	310	270	240	200
1.95	2,500	490	450	410	370	330	290	250	210

(metric)

Rating of Panel sq.m.	Water Flow kg/hr	104.4oC	98.88oC	93.3oC	87.7oC	82.2oC	76.67oC	71.1oC	65.5oC
				Heat Output kg-cal per hr per m					
0.18	227	386	353	380	350	310	270	240	200
0.18	1137	412	378	410	370	330	290	250	210

9-7/8-Inch (246.875-mm) Radiant Convector Panel

Rating based on standards given for baseboard panels.

Rating of Panel s.f.	Water Flow lbs/hr	220oF	210oF	200oF	190oF	180oF	170oF	160oF	150oF
				Heat Output Btu per Hour per Lin. Ft.					
3.25	500	780	720	660	600	530	470	420	350
3.25	2,500	830	760	700	640	570	510	440	380

(metric)

Rating of Panel sq.m	Water Flow kg/hr	104.4oC	98.8oC	93.3oC	87.7oC	82.2oC	76.6oC	71.1oC	65.5oC
				Heat Output kg-cal per hr. per m					
0.29	227	655	605	554	504	445	395	353	294
0.29	1137	697	638	588	537	479	428	370	319

Price of Baseboard Radiation and Accessories

Description	Weight lbs. (kg)	Approx. Price
Radiant Baseboard Panels 9-7/8" (247 mm), per l.f. (0.3 m)	12.25 (5.50)	$9.00
Radiant Convector Panels 9-7/8" (247 mm), per l.f. (0.3 m)	14.00 (6.40)	18.00
Plain ends, 9-7/8" (247 mm), each	2.50 (1.10)	6.00
Corner Plate, 9-7/8" (247 mm), each	0.50 (0.23)	6.00
Angle Radiator Valve, 3/4" (19 mm), each	1.00	18.00

Convector-Radiators

Radiant convectors combine convection heating with quick-acting and comfortable radiant heating. They are formed of cast iron with finned sections, or of copper with or without finned sections. There are so many types of convectors. It is impossible to describe all of them in detail.

Convectors described below are 6-1/4" (156 mm) deep overall and may be installed as freestanding or semi-recessed units. Maximum depth of recess should not exceed 4-3/4" (119 mm).

Size, Capacity, and Approximate Price of Convectors

Height Inches	Length Inches	Steam SF EDR	180°	200°	220°	Price Ea.
			Forced Hot Water, 1,000 Btu per Hr.			
20	16	13.0	1.74	2.17	2.60	$73.00

Height	Length	Steam Sq.m EDR	82.2oC	93.3oC	104.4oC	Price Ea.
20	20	17.0	2.28	2.84	3.40	91.00
20	24	21.0	2.81	3.51	4.20	109.00
20	28	26.0	3.48	4.34	5.20	136.00
20	32	29.5	2.95	4.93	5.90	154.00
20	36	33.5	4.48	5.59	6.70	175.00
20	40	38.5	5.15	6.43	7.70	196.00
20	44	43.5	5.82	7.26	8.70	218.00
20	48	46.5	6.22	7.76	9.30	238.00
20	56	55.5	7.43	9.26	11.10	280.00
20	64	63.0	8.44	10.50	12.60	315.00
24	20	18.5	2.48	3.09	3.70	95.00
24	24	23.0	3.08	3.84	4.60	117.00
24	28	28.0	3.75	4.67	5.60	132.00
24	32	32.5	4.35	5.42	6.50	168.00
24	36	36.5	4.88	6.09	7.30	189.00
24	40	42.0	5.62	7.01	8.40	218.00
24	44	47.5	6.36	7.93	9.50	224.00
24	48	51.0	6.83	8.51	10.20	266.00
24	56	60.5	8.18	10.10	12.10	308.00
24	64	69.0	9.24	11.52	13.80	357.00

(metric)

Height mm	Length mm	Steam Sq.m EDR	Forced Hot Water, 252 kg-cal per Hr. 82.2oC	93.3oC	104.4oC	Price Ea.
500	400	1.17	0.16	0.20	0.23	$73.00
500	500	1.53	0.21	0.26	0.31	91.00
500	600	1.89	0.25	0.32	0.38	109.00
500	700	2.34	0.31	0.39	0.47	136.00
500	800	2.65	0.27	0.44	0.53	154.00
500	900	3.01	0.40	0.50	0.60	175.00
500	1000	3.47	0.46	0.58	0.69	196.00
500	1100	3.91	0.52	0.65	0.78	218.00
500	1200	4.18	0.56	0.70	0.84	238.00
500	1400	5.0	0.87	0.83	0.99	280.00
500	1600	5.67	0.76	0.94	1.13	315.00
600	500	1.67	0.22	0.28	0.33	95.00
600	600	2.07	0.27	0.35	0.41	117.00
600	700	2.52	0.34	0.42	0.50	132.00
600	800	2.93	0.39	0.48	0.59	168.00
600	900	3.29	0.44	0.55	0.66	189.00
600	1000	3.78	0.51	0.63	0.76	218.00
600	1100	4.27	0.57	0.71	0.85	224.00
600	1200	4.59	0.61	0.77	0.92	266.00
600	1400	5.44	0.74	0.91	1.09	308.00
600	1600	6.21	0.83	1.03	1.24	357.00

For dampers, add $10.00. For air chambers and accessories less air valve, add $12.00.

Standard Convectors and Cabinets—Non-Ferrous

Convectors listed below are 4" (100 mm), 6" (150 mm), 8" (200 mm), and 10" (250 mm) deep plus 1/4" (6 mm) overall, and may be installed as freestanding or semi-recessed units.

		Height 20"			Height 24"			Height 32"		
Dep.	Len.	EDR	Price	Shpg. Approx. wt. lbs.	EDR	Price	Shpg. Approx. wt. lbs.	EDR	Price	Shpg. Approx. wt. lbs.
4	24	13.1	$67.00	25	15.0	$70.00	25	--	$84.00	--
6	24	19.0	94.00	30	22.4	98.00	30	25.5	115.00	35
	32	27.0	98.00	40	31.6	85.00	35	36.0	126.00	50
6	36	30.8	105.00	40	36.2	112.00	45	41.2	133.00	55
	40	34.9	115.00	45	40.8	98.00	45	46.5	147.00	60
	44	38.8	126.00	50	45.4	140.00	55	51.7	168.00	65
	48	42.7	134.00	55	50.0	147.00	60	56.9	173.00	75
	56	50.6	143.00	60	59.2	154.00	65	67.4	193.00	75
	64	58.3	154.00	65	68.4	168.00	70	77.9	196.00	80
8	32	34.0	115.00	45	37.5	98.00	45	41.6	147.00	55
	36	39.1	123.00	45	43.1	133.00	45	47.8	157.00	60
	40	44.2	143.00	50	48.7	154.00	50	54.0	193.00	65
	48	54.3	168.00	60	59.8	173.00	65	66.4	207.00	75
	56	64.4	182.00	65	70.9	196.00	70	78.8	235.00	115
	64	74.4	203.00	70	82.0	231.00	70	91.0	266.00	120
10	36	45.0	140.00	50	49.0	161.00	50	54.5	185.00	65
	40	50.8	137.00	55	55.3	175.00	55	61.5	207.00	70
	48	62.1	182.00	65	67.7	203.00	75	75.4	238.00	85
	56	73.6	203.00	70	80.2	231.00	80	89.3	280.00	120
	64	85.1	231.00	75	92.7	259.00	85	103.1	301.00	130

(metric)

		Height 20" (500 mm)			Height 24" (600 mm)			Height 32" (800 mm)		
Depth (mm)	Length (mm)	Sq.m EDR	Price	Shpg. Approx. wt. kg	Sq.m EDR	Price	Shpg. Approx. wt. kg	Sq.m EDR	Price	Shpg. Approx. wt. kg
100	600	1.18	$67.00	11.4	1.4	$70.00	11.4	--	$84.00	--
150	600	1.71	94.00	13.5	2	98.00	13.5	2.3	115.00	15.9
	800	2.4	98.00	18	2.8	85.00	16	3.2	126.00	23
150	900	2.7	105.00	18	3.3	112.00	20	3.7	133.00	25
	1000	3.1	115.00	20	3.7	98.00	20	4.2	147.00	27
	1100	3.5	126.00	23	4.1	140.00	25	4.6	168.00	29
	1200	3.8	134.00	25	4.5	147.00	27	5.1	173.00	34
	1400	4.6	143.00	27	5.3	154.00	29	6	193.00	34
	1600	5.2	154.00	29	6.2	168.00	31	7	196.00	36
200	800	3.1	115.00	20	3.4	98.00	20	3.7	147.00	25
	900	3.5	123.00	20	3.9	133.00	20	4.3	157.00	27
	1000	3.9	143.00	23	4.4	154.00	23	4.9	193.00	29
	1200	4.8	168.00	27	5.4	173.00	29	5.9	207.00	34

	1400	5.8	182.00	29	6.4	196.00	31	7.1	235.00	52
	1600	6.7	203.00	31	7.4	231.00	31	8.1	266.00	54
250	900	4	140.00	22.5	4.4	161.00	22.5	4.9	185.00	29
	1000	4.5	137.00	25	5	175.00	25	5.5	207.00	32
	1200	5.6	182.00	29	6.1	203.00	34	6.8	238.00	38
	1400	6.6	203.00	32	7.2	231.00	36	8	280.00	54
	1600	7.7	231.00	34	8.3	259.00	38	9.2	301.00	59

Stock Model Convector Accessories

Description	Price	Shpg. Wt. lbs. (kg)
Damper Package, up to 64" (1600 mm)	$17.00	5 (2.25)
Snap-in Inlet Grille for Stock Models Only	20.00	5 (2.25)
CM Molding Trim, 24-36" (600-900 mm) long	25.00	6 (2.7)
CM Molding Trim, 40-64" (1000-1600 mm) long	32.00	8 (3.6)
Air Vent—Manual	14.00	1 (0.45)
Air Vent—Auto	14.00	1 (0.45)

Hot Water Rating in Btu per Sq. Ft. EDR

Temp. Drop	Average Water Temp. °F		
	180°	190°	200°
10 °F	160	181	201
20 °F	141	159	177
30 °F	131	149	166

(metric)

Hot Water Rating in kg-cal per sq.m EDR

Temp. Drop °C	Average Water Temp. °C		
	82.2°	87.77°	93.33°
-7.77°	448	507	563
-6.67°	395	445	495
-1.11°	367	417	465

Sizes and Approximate Prices of Cast-Iron Sectional Boilers and Gas or Light Oil Burners With Jacket

	Steam		Hot Water	
Installed Radiation Sq. Ft.*	Price Each	Installed Radiation Sq. Ft.* For Oil Firing	Btu Rating	Price Each
845	$1220.00	1500	225,000	$1220.00
1090	1460.00	1910	287,000	1460.00
1340	1675.00	2320	348,000	1675.00
1600	1870.00	2735	410,000	1870.00
1860	1950.00	3145	472,000	1950.00

	Steam		Hot Water	
Installed Radiation Sq. Ft.*	Price Each	Installed Radiation Sq. Ft.* For Oil Firing	Btu Rating	Price Each
2120	2050.00	3560	534,000	2050.00
2380	2350.00	3960	594,000	2350.00
2650	2850.00	4375	656,000	2850.00
2920	2900.00	4785	718,000	2900.00
3200	3000.00	5205	781,000	3000.00
3480	3100.00	5620	843,000	3100.00

(metric)

	Steam		Hot Water	
Installed Radiation sq.m*	Price Each	Installed Radiation sq.m* For Oil Firing	kg-cal Rating	Price Each
76.05	$1220.00	135	56,700	$1220.00
98.1	1460.00	171.9	72,324	1460.00
120.6	1675.00	208.8	87,696	1675.00
144	1870.00	246.1	103,320	1870.00
167	1950.00	283	118,944	1950.00
190.8	2050.00	320.4	134,568	2050.00
214.2	2350.00	356.4	149,688	2350.00
228.5	2850.00	393.7	165,312	2850.00
262.8	2900.00	430.6	180,936	2900.00
288	3000.00	468.4	196,812	3000.00
313.2	3100.00	505.8	212,436	3100.00

*Capacities are based on net installed radiation and on the presence of a quantity of net installed radiation sufficient for the requirements of the building. Nothing need be added for normal piping.

All ratings are based on radiation at 240 Btu per s.f. for steam and 150 Btu per s.f. for water.

Above prices are net to the heating contractor, f.o.b. factory. Prices include boiler trimmings.

Sizes and Approximate Prices of Oil Burning Boiler Units

Price includes boiler complete with oil burner, trombone tank type water heater or tankless type water heater, steel jacket with baked enamel finish, air-cell type insulation furnished with jacket, plain room thermostat, limit control, and combustion control.

Steam boiler prices include low water cut-off, pop safety valve, try cocks, water gauge, pressure and vacuum gauge, and flue brush.

Water boiler prices include combination altitude gauge and thermometer and flue brush.

Installed Radiation s.f.*	Steam Boiler Jacketed price with Trombone Heater	Tankless Heater	Water Boiler Jacketed price with Installed Radiation s.f.*	Net Btu Rating	Trombone Heater	Tankless Heater
460	$1300.00	$1380.00	840	126,000	$1300.00	$1380.00
570	1581.00	1650.00	1030	155,000	1581.00	1650.00
680	1700.00	1785.00	1220	183,000	1700.00	1785.00
790	1870.00	1955.00	1405	211,000	1870.00	1955.00
900	1955.00	2040.00	1590	239,000	1955.00	2040.00

(metric)

Installed Radiation sq.m*	Steam Boiler Jacketed price with Trombone Heater	Tankless Heater	Water Boiler Jacketed price with Installed Radiation sq.m*	Net Kg-cal Rating	Trombone Heater	Tankless Heater
41.4	$1300.00	$1380.00	76	31,752	$1300.00	$1380.00
51.3	1581.00	1650.00	93	39,060	1581.00	1650.00
61.2	1700.00	1785.00	110	46,116	1700.00	1785.00
71.1	1870.00	1955.00	126	53,172	1870.00	1955.00
81.0	1955.00	2040.00	143	60,228	1955.00	2040.00

*Capacities are based on net installed radiation and on the presence of a quantity of net installed radiation sufficient for the requirements of the building. Nothing need be added for normal piping.

All ratings are based on radiation at 240 Btu per s.f. for steam and 150 Btu per s.f. for water.

Above prices are net to the heating contractor, f.o.b. factory. Prices include boiler trimmings. If boilers are furnished without tankless heater, deduct $45.00; less trombone tank type heater, deduct $25.00.

Water Heater Capacities

Tankless type	4 gals. (15 liters) per minute
Trombone tank type	66 gals. (250 liters) in 3 hrs.

Above capacities are based on 40°F (4.44°C) to 140°F (60°C) rise with boiler water at 180°F (82.2°C). Above prices are f.o.b. factory. Prices include boiler trimmings as listed above.

Approximate Prices of Gas Fired Boilers for Hot Water Heating Systems

Furnished with finned cast iron sections to provide staggered heat travel, draft diverter, heavily insulated steel jacket with baked enamel finish.

Boiler price includes gas valve and hydraulic limit control, relay gas valve and recycling manual control switch, transformer, electric safety pilot, gas pressure regulator, manual gas shut-off valve, combination thermometer and altitude gauge, pilot pressure regulator, drain cock, pilot cock, draft diverter, gas manifold, tubing and fittings. Room thermostat not included. Prices are f.o.b. factory.

Radiation s.f	Net I-B-R Rating Output Btu per hr.	A.G.A. Gross Output Btu per hr.	Price Boiler Complete
515	77,500	103,000	$924.00
860	129,000	172,000	1220.00
1205	181,000	241,000	1595.00
1550	233,000	310,000	1820.00
1720	258,000	344,000	2030.00
1960	294,000	392,000	2240.00
2240	336,000	448,000	2660.00
2520	378,000	504,000	2940.00
2800	420,000	560,000	3150.00
3080	462,000	616,000	3640.00
3350	502,500	672,000	4200.00
3920	588,000	784,000	4760.00
4480	672,000	896,000	5040.00
5040	756,000	1,008,000	5670.00
5600	840,000	1,120,000	6020.00
6160	924,000	1,232,000	6440.00
6666	1,000,000	1,344,000	7000.00
7222	1,083,350	1,456,000	7560.00
7777	1,166,650	1,568,000	7840.00
8333	1,250,000	1,680,000	8120.00
8888	1,333,300	1,792,000	8680.00

(metric)

Radiation sq.m	Net I-B-R Rating Output kg-cal per hr.	A.G.A. Gross Output kg-cal per hr.	Price Boiler Complete
46.3	19,530	25,956	$924.00
77.4	32,508	43,344	1220.00
108.5	45,612	60,732	1595.00
139.5	58,716	78,120	1820.00
154.8	65,016	86,688	2030.00
176.4	74,088	98,784	2240.00

201.6	84,672	112,896	2660.00
226.8	95,256	127,008	2940.00
252	105,840	141,120	3150.00
277.2	116,424	155,232	3640.00
301.50	126,630	169,344	4200.00
352.8	148,176	197,568	4760.00
403.2	169,344	225,792	5040.00
453.6	190,512	254,016	5670.00
504	211,680	282,240	6020.00
554.4	232,848	310,464	6440.00
599.9	252,000	338,688	7000.00
650	273,004	366,912	7560.00
700	293,996	395,136	7840.00
750	315,000	423,360	8120.00
800	335,992	451,584	8680.00

Approximate Prices of Gas Fired Boilers for Steam and Vapor Heating Systems

These are furnished with finned cast iron sections to provide staggered heat travel, draft diverter, and heavily insulated steel jacket with baked enamel finish.

Boiler price includes gas valve, relay gas valve and recycling manual control switch, transformer, electric safety pilot, gas pressure regulator, pilot pressure regulator, manual gas shut-off valve, electric hot water cut-off, steam pressure regulator, retard steam gauge, pop safety valve, water gauge, try cocks, drain cock, pilot and burner cocks, draft diverter, gas manifold, tubing and fittings. Room thermostat not included. Prices are f.o.b. factory.

Radiation s.f.	Net I-B-R Rating Output Btu per hr.	A.G.A. Gross Output Btu per hr.	Price Boiler Complete
280	66,750	103,000	$918.00
475	113,500	172,000	1,320.00
670	161,000	241,000	1,955.00
875	210,000	310,000	2,465.00
980	235,000	344,000	2,975.00
1121	269,100	392,000	3,400.00
1293	310,250	448,000	3,570.00
1466	351,700	504,000	3,740.00
1641	393,800	560,000	4,080.00
1996	478,950	672,000	4,420.00
2353	565,250	784,000	5,100.00
2717	652,100	896,000	5,780.00
3081	739,550	1,008,000	6,120.00
3449	827,800	1,120,000	6,460.00
3814	915,300	1,232,000	7,140.00

Radiation s.f.	Net I-B-R Rating Output Btu per hr.	A.G.A. Gross Output Btu per hr.	Price Boiler Complete
4167	1,000,000	1,344,000	7,820.00
4514	1,083,350	1,456,000	8,330.00
4861	1,166,650	1,568,000	8,840.00
5208	1,250,000	1,680,000	9,180.00
5555	1,333,300	1,792,000	11,220.00
5903	1,416,650	1,904,000	13,600.00

(metric)

Radiation sq.m	Net I-B-R Rating Output kg-cal per hr.	A.G.A. Gross Output kg-cal per hr.	Price Boiler Complete
25.2	16,821	25,956	$918.00
42.8	28,602	43,344	1,320.00
60.3	40,572	60,732	1,955.00
78.8	52,920	78,120	2,465.00
88.2	59,220	86,688	2,975.00
100.9	67,813	98,784	3,400.00
116.3	78,183	112,896	3,570.00
131.9	88,628	127,008	3,740.00
147.6	99,238	141,120	4,080.00
179.6	120,695	169,344	4,420.00
211.8	142,443	197,568	5,100.00
244.5	164,329	225,792	5,780.00
277.3	186,366	254,016	6,120.00
310.4	208,605	282,240	6,460.00
343.3	230,655	310,464	7,140.00
375	252,000	338,688	7,820.00
406.3	273,004	366,912	8,330.00
437.5	293,996	395,136	8,840.00
468.7	315,000	423,360	9,180.00
499.9	335,992	451,584	11,220.00
531.3	356,995	479,808	13,600.00

Net boiler is the amount of actual installed radiation that may be attached to the boiler, based on a heat emission rate from the radiators of 240 Btu per s.f. for steam and 150 Btu for gravity hot water. For forced circulation for hot water systems, select the boiler from the net Btu column.

Standard Black and Galvanized Steel Pipe

Inside Diameter inches (mm)	Black Pipe Price per l.f. (0.3 m)	Galv. Pipe Price per l.f. (0.3 m)	Threads Extra Each End
1/4 (6.25)	$0.98	$1.26	$0.98
3/8 (9.38)	1.12	1.40	0.98
1/2 (12.5)	1.12	1.26	0.98

3/4 (18.8)	1.19	1.40	0.98
1 (25)	1.69	1.89	0.98
1-1/4 (31.3)	1.61	2.45	1.12
1-1/2 (37.5)	2.38	2.80	1.00
2 (50)	3.29	3.85	1.68
2-1/2 (62.5)	4.55	5.25	2.80
3 (75)	5.95	7.00	3.00

Above prices are for pipe cut to order. Add for threads as given.

Prices of Steam Radiator Valves

	Sizes in inches (mm)					
	1/2 (12.5)	3/4 (18.8)	1 (25)	1-1/4 (31.3)	1-1/2 (37.5)	2 (50)
Angle steam radiator valve	$29.00	$36.00	$50.00	$69.00	$88.00	$140.00
Gate steam radiator valve	15.40	20.00	24.00	30.00	39.00	56.00

Prices of Water Radiator Valves

	1/2 (12.5)	3/4 (18.8)	1 (25)	1-1/4 (31.3)	1-1/2 (37.5)
Hot water radiator valve	$8.00	$11.00	$15.00	$25.00	$35.00
Hot water radiator union ell	3.50	4.00	6.50	8.50	11.00

Prices on Floor and Ceiling Plates

	1/2 (12.5)	3/4 (18.8)	1 (25)	1-1/4 (31.3)	1-1/2 (37.5)	2 (50)	2-1/2 (62.5)	3 (75)
Pipe size, in. (mm) Chrome plated, ea.	$0.64	$0.64	$0.70	$0.76	$0.81	$0.89	$1.37	$1.71

Floor and ceiling plates made of cold rolled steel, halves securely riveted by a concealed hinge. Can be opened or closed on pipe without effort.

Copper Pipe

The cost of copper pipe on heating work depends to a large extent upon the experience of the steam fitter installing it. The technique of installing copper pipe is entirely different from that of steel pipe. Inexperienced steam fitters figure more for labor than on a steel pipe job, while others figure the same for labor, whether steel or copper. Steam fitters who are thoroughly experienced in installing copper pipe and fittings and some say that they can install copper pipe for 15% to 20% less than steel pipe.

The price of the pipe itself is subject to wide fluctuations, due to the price of copper so that prices on copper pipe and fittings should *always* be obtained separately for each job. Copper pipe is furnished in 3 weights:

1. Heavy Copper Pipe (Government Type "K") for underground service, such as water, gas, steam, oil lines, industrial uses, also interior plumbing.

2. Standard Copper Pipe (Government Type "L") for interior plumbing and heating.

3. Light Copper Pipe (Government Type "M") for low pressure interior plumbing.

Sizes and Weights of Copper Pipe

Nominal Size inches	Type "K" Heavy Wall Thickness inches	Weight per l.f lbs.	Type "L" Standard Wall Thickness inches	Weight per l.f lbs.	Type "M" Light Wall Thickness inches	Weight per l.f. lbs.
3/8	0.049	0.269	0.035	0.198	0.025	0.145
1/2	0.049	0.344	0.040	0.285	0.028	0.204
5/8	0.049	0.418	0.042	0.362		
3/4	0.065	0.641	0.045	0.455	0.032	0.328
1	0.065	0.839	0.050	0.655	0.035	0.465
1-1/4	0.065	1.040	0.055	0.884	0.042	0.681
1-1/2	0.072	1.360	0.060	1.140	0.049	0.940
2	0.083	2.060	0.070	1.750	0.058	1.460
2-1/2	0.095	2.930	0.080	2.480	0.065	2.030
3	0.109	4.000	0.090	3.330	0.072	2.680
3-1/2	0.120	5.120	0.100	4.290	0.083	3.580
4	0.134	6.510	0.110	5.380	0.095	4.660
5	0.160	9.670	0.125	7.610	0.109	6.660
6	0.192	13.870	0.140	10.200	0.122	8.910
8	0.271	25.900	0.200	19.300	0.170	16.460
10	0.338	40.300	0.250	30.100	0.212	25.600
12	0.405	57.800	0.280	40.400	0.254	36.700

(metric)

Nominal Size mm	Type "K" Heavy Wall Thickness mm	Weight per m kg	Type "L" Standard Wall Thickness mm	Weight per m kg	Type "M" Light Wall Thickness mm	Weight per m kg
9.4	1.2	0.40	0.87	0.30	0.63	0.22
12.5	1.2	0.51	1.00	0.43	0.70	0.31
15.6	1.2	0.63	1.05	0.55		
18.8	1.6	0.96	1.13	0.69	0.8	0.49
25	1.6	1.27	1.25	0.99	0.88	0.70
31.2	1.6	1.58	1.38	1.34	1.05	1.03
37.5	1.8	2.06	1.5	1.72	1.23	1.42

50	2.07	3.12	1.8	1.51	1.45	2.21
62.5	2.4	4.43	2.0	3.75	1.63	3.07
75	2.7	6.06	2.3	5.05	1.8	4.06
87.5	3.0	7.75	2.5	6.5	2.08	5.42
100	3.4	9.86	2.8	8.15	2.38	7.06
125	4.0	14.65	3.13	11.53	2.73	10.09
150	4.8	21	3.5	15.45	3.05	13.50
200	6.8	39.2	5.0	29.25	4.25	24.94
250	8.5	61.06	6.3	45.61	5.3	38.79
300	10.1	87.58	7.0	61.22	6.4	55.61

Because of the smooth inside surface of copper pipe, usually one size smaller pipe can be used than where steel pipe is used. Outside diameter of pipe is 3.25 mm more than nominal size of tubing.

Approximate Prices on Copper Tubing

Prices on copper tube and fittings are subject to wide price fluctuations. Always check prices before submitting bids.

Copper Tubing

Nominal Size (in.)

Price per l.f.	1/4"	3/8"	1/2"	5/8"	3/4"	1"	1-1/2"	1-1/2"
Type "K"	$0.56	$0.91	$1.19	$1.40	$2.09	$2.52	$3.15	$4.06
Type "L"	0.52	0.70	1.05	1.26	1.54	2.10	2.03	3.50

(metric)
Nominal Size (mm)

Price per m	6.25	9.38	12.5	15.63	18.8	25	32.5	37.5
Type "K"	$1.84	$2.98	$3.90	$4.59	$6.86	$8.27	$10.33	$13.32
Type "L"	1.71	2.30	3.44	4.13	5.05	6.89	6.66	11.48

Price per m	56	62.5	75	87.5	100	125	150
Type "K"	$19.98	$28.47	$39.03	$55.10	$68.88	$160.72	$229.60
Type "L"	17.22	24.80	33.29	45.00	55.10	128.58	174.50

Prices of Brass Solder-Joint Valves

These valves are made for use with copper tubing and can be used for steam or water service.

Size in inches (mm)

Fitting	1/2" (12.5)	3/4" (18.8)	1" (25)	1-1/4 (31.3)	1-1/2" (37.5)	2" (50)
Gate valve	$15.61	$19.95	$24.15	$32.20	$41.65	$58.50

Approximate Quantity of Solder and Flux Required to Make 100 Joints

Size, in.	3/8	1/2	3/4	1	1-1/4	1-1/2	2	2-1/2	3	3-1/2	4	5	6	8	10
Solder, Lbs	1/2	3/4	1	1-1/2	1-3/4	2	2-1/2	3-1/2	4-1/2	5	6-1/2	9	17	35	45
Flux, Oz.	1	1-1/2	2	3	3-1/2	4	4	7	9	10	13	18	34	70	90

(metric)

Size, mm	9.4	12.5	18.8	25.0	31.3	37.5	50.0	62.5	75.0	87.5	100.0	125.0	150.0	200	250
Solder, kg	0.23	0.34	0.45	0.68	0.79	0.91	1.13	1.59	2.04	2.27	2.95	4.08	7.71	15.88	20.41
Flux, g	28.4	42.5	56.7	85.1	99.2	113.4	113.4	198.5	255.2	283.5	368.6	510.3	963.9	1984.5	2551.5

Estimating the Cost of Pipe and Fittings. To estimate the cost of pipe and fittings required for any heating plant, obtain the total number of sq. ft. of radiation in the job, based on cast iron radiation, and figure at $2.00 to $2.15 per s.f. ($22.22-23.88 per sq.m). This is based on prices of pipe as given on the previous pages.

For one story residences having short pipe runs, figure the cost of the pipe and fittings at $2.00 per s.f. ($22.22 per sq.m) of radiation, based on standard cast iron radiation.

An example of the method used in estimating the cost of pipe and fittings is as follows: For a job containing 1,400 s.f. (126 sq.m) of standard radiation, take 120% of 1,400 (126) which equals 1,680 (151), or the pipe and fittings would cost approximately $3,400.00.

The above allowances for pipe and fittings do not include radiator valves, floor and ceiling plates, boiler fittings, expansion tanks, etc., but only the pipe and fittings required for roughing-in the job.

Pipe Covering

Pipe covering is furnished in several kinds of materials. Fiberglass sectional pipe insulation covering is the most common and cheapest for covering steam and hot water pipes; 85% magnesia pipe covering is for high pressure steam work; and wool felt pipe covering is for protecting pipes against freezing and for general use.

Pipe covering is usually furnished in 3' (0.9 m) lengths, 3/4" (19 mm) to 1" (25 mm) thick and is covered with a canvas or fire safety jacket.

Labor Covering Pipes Up to 3" (75 mm) in Diameter. When placing covering on pipes up to 3" (75 mm) in diameter, a pipe coverer should apply 10 to 15 l.f. (3.0-4.5 m) of covering per hour, or 80 to 120 l.f. (24-36 m) per 8-hr. day, and the labor cost per 100 l.f. (30 m) should average as follows:

	Hours	Rate	Total	Rate	Total
				$29.28	$234.24
Pipe coverer	8.0	$....	$....		
Cost per l.f.				....	2.34
per m				....	7.68

Labor Covering Pipes 4" (100 mm) to 6" (150 mm) in Diameter.

When placing covering on pipes 4" to 6" (100-150 mm) in diameter, a pipe coverer should apply 9 to 11 l.f. (2.7-3.3 m) of covering per hour or 72 to 88 l.f. (21.6-26.4 m) per 8-hr. day, and the labor cost per 100 l.f. (30 m) should average as follows:

	Hours	Rate	Total	Rate	Total
				$29.28	$292.80
Pipe coverer	10.0	$....	$....		
Cost per l.f.				....	2.93
per m				....	9.61

When estimating quantities of pipe covering, add 1 l.f. (0.3 m) of covering for each fitting for pipe up to 2" (50 mm) and 2 l.f. (0.6 m) of covering for each fitting on pipe over 2" thick.

Temperature Regulators for Heating Plants

Temperature regulators and thermostats are furnished in a variety of different models for controlling the temperature.

There are so many different methods of controlling the inside temperature with straight thermostats, day and night thermostats, inside-outside controls, etc., that it is practically impossible to give even a fraction of them in a book of this kind.

Temperature Regulators With Clock. The thermostat with a clock attachment automatically lowers the temperature at night and raises it in the morning. Electric clocks are now usually furnished with all thermostats of this type.

Models of this kind are furnished with thermometer, clock, electric motor, chain, pulleys, transformer and fittings, at the approximate price of $150.00.

Temperature Regulators Without Clock. Many thermostats or regulators are used without the clock attachment and are either manually operated or the same temperature maintained day and night.

There are a number of different types of regulators on the market and they vary in price from $50.00 to $150.00, depending on the method of wiring and number of controls necessary.

Labor Installing Thermostats. An experienced mechanic should install any of the above thermostat outfits in 1 to 2 hours. If they are installed by the manufacturers in cities where they maintain offices, a charge of $150.00 to $200.00 is usually made.

Making Up a Complete Estimate on a Steam Heating Plant. When making up an estimate on a steam heating system, the following items should always be included in order that the estimate may be complete in every detail.

1. Boiler
2. Oil or gas burner
3. Radiators or convectors
4. Radiator Valves
5. Automatic air valves
6. Pipe and fittings
7. Smoke pipe
8. Special foundation, if required
9. Asbestos boiler covering
10. Pipe covering
11. Floor and ceiling plates
12. Painting radiators
13. Thermostat regulator
14. Freight and trucking
15. Water Connection
16. Labor installing system
17. Bond and insurance
18. Overhead and misc. expense
19. Profit

Making Up a Complete Estimate on a Hot Water Heating Plant. When making up an estimate on a hot water heating plant, the following items should always be included in order that the estimate may be complete in every detail.

1. Special boiler foundation, if required
2. Boiler
3. Oil or gas burner
4. Pipe and fittings
5. Radiators or convectors
6. Radiator valves-union ells
7. Circulating pumps
8. Compression air valves
9. Expansion tank and gauge
10. Altitude gauge
11. Thermometer
12. Smoke pipe
13. Painting radiators
14. Boiler covering
15. Pipe covering
16. Floor and ceiling plates
17. Thermostat regulator
18. Labor installing system
19. Freight and trucking
20. Water connection
21. Bond and insurance
22. Overhead and misc. expense
23. Profit

Unit Heaters. Unit heaters are often used to heat commercial and industrial areas, garages, and to supplement other heating units at entrances. They may be direct-fired by gas, oil, or electricity, or have hot water or steam piped to them. Direct-fired are cheaper when only a few units are required and no central plant is available. Hot water and steam are the most common units in use where a central plant is available or its cost justified by the number of units it serves.

Units may have propeller or centrifugal fans blowing either horizontally, up or down. Air may be drawn through or blown through, the latter being used for direct-fired units.

The following costs do not include valves, fittings, or piping, costs for which vary widely with job conditions and are based on units served with 2 lbs. (0.9 kg) steam, 60°F (15.5°C) entering air, single phase 115 volts.

Cost of Unit Heaters

Type	Hp	kW	Btu	kg-cal	1 Speed	2 Speed	Hrs. to Install
Horizontal	1/40	0.02	25,000	6300	$155.00	$--	1.0
Delivery	1/15	0.05	50,000	12600	200.00	245.00	1.5
	1/10	0.07	100,000	25200	300.00	340.00	2.5
	1/5	0.15	200,000	50400	480.00	540.00	4.0
Vertical	1/40	0.02	50,000	12600	190.00	--	1.5
Delivery	1/12	0.06	100,000	25200	260.00	300.00	2.0
	1/3	0.25	200,000	50400	520.00	575.00	5.0
Cabinet	1/20	0.04	25,000	6300	160.00	--	1.5
	1/12	0.06	60,000	15120	230.00	--	2.0
	1/8	0.09	120,000	30240	475.00	--	3.0

For 230 volts, add $16.00. For ceiling suspension, add $25.00.

15C SPRINKLER SYSTEMS

The construction industry and the general public for many years have been aware of the need for and availability of fire protection systems. However, for a long time this protection was restricted to areas where local building codes or extreme business risks dictated a demand for their use. In general, there was great insensitivity toward providing fire protection for the public, even in areas of high concentration such as restaurants, theaters, hotels, and sporting events.

This disregard for public safety changed because of a number of nationally publicized fire disasters, inspiring action by federal and state authorities, insurance companies, building owners, and professional architects.

The fire protection industry responded promptly by providing specialized technical design data and producing trained personnel capable of educating the designers and specification writers in selecting the particular system, or combination of systems, most appropriate to provide an adequate degree of protection.

As the need has developed to provide special protection for costly equipment such as computer rooms and tape storage libraries, for irreplaceable documents, and art objects, for sophisticated research projects, and for the aged and the handicapped, new alarm, control, and protection systems have been developed. Some of the systems are mechanically operated, while others are completely electronic. In some cases, alarm and control systems are designed as a combination of mechanical and electrical devices. The estimator must determine with great precision the cost and installation responsibility that each trade will assume in the overall function of the system.

Wet Pipe Sprinkler System

This type of system meets most fire protection requirements for normal hazards, where piping is not subject to freezing. Water is constantly supplied under pressure to each sprinkler head.

Dry Pipe Sprinkler System

This system can be used in areas subject to freezing, because the piping is filled with air under pressure, which contains the water at the supply source until an activated sprinkler head allows the air pressure to disburse and let the water supply move to the head.

Sprinkler systems are a specialized type of work, requiring the services of a subcontractor who has access to the necessary equipment, which in many instances is a sole source item and not available on the open market. One usually finds that the contract documents require the services of a specialist who has years of experience in this type of work and who has the capabilities to design the sprinkler system.

For the wet and dry pipe systems, unit costs have been developed on a per head basis. It is assumed that an adequate water supply is available and that special alarm and control work is not included. Different heads are used depending on the degree of hazard, and local fire regulations are consulted for a specific project.

For high-rise buildings, booster pumps may be required, which will add from $20,000 to $50,000 to the cost of installation.

Maximum area coverage per head is 130 s.f. (11.7 sq.m). The average area covered per head is from 80 s.f. to 130 s.f. (7.2-11.7 sq.m) under normal hazard. Distance between sprinkler heads cannot exceed 15 l.f. (4.5 m).

Wet System*	Total Cost per Head
Warehouse	90.00
Hospital	150.00
Shopping Center	100.00
Municipal Buildings	125.00

*For dry system, add 15.00 20.00 per head.

From a look at the above costs, it might seem that a fire sprinkler system is costly, but an owner must look at all the benefits. It is an insulation against litigation resulting from fire loss and a savings in fire insurance premiums. Most systems return their initial cost in reduced insurance premiums within 3 to 10 years.

CHAPTER 16

ELECTRICAL

CSI DIVISION 16

Even the largest or tallest building may depend completely on electricity for light, power, building services, tenant equipment, and heat. Heating and lighting must be coordinated so that a maximum use of electric energy may be obtained at a minimum of expense.

Obviously such sophisticated designs are well beyond the scope of this book. But even a modest residence will involve electrical planning and power demands well beyond anything thought of 25 years ago.

An electrical estimate, whether commercial or residential, can be divided into four general categories: service, distribution, branch, and devices/loads. The service comprises the outdoor utility-supplied feeders and connection to the building, as well as the revenue metering and main disconnect switches. Distribution consists of the panelboards and switchgear that immediately follow the main switch and the feeders that serve them and emanate from them. The branch work, also referred to as lighting circuitry, is the raceway and wire between the panelboard breakers and the devices/loads. Finally, there are the devices/loads, such as receptacles and switches, lighting outlets, fixtures, and heaters.

The electrical service is installed in two phases: the contractor's portion, consisting of the setting and wiring of the main disconnect switches and the revenue meter pans, followed by the utility companies portion, consisting of the installation of the feeder, either on poles or underground, its connection to the meter pan, and installation of the meter itself.

For residence construction, the typical service size is 200 amperes, 208 volts, single phase; that is, two phase wires and a neutral, making both 120 and 208 volts available. The main disconnect switch is usually incorporated in the main circuit breaker. The typical cost of such a panelboard, 30 circuits and the usual complement of circuit breakers, is $400.00 plus about 24 hours of labor. The matching meter pan would add $50.00 and an additional 5 hours for installation. Where a 100 ampere service is provided, the panelboard costs would be $200.00 plus about 15 hours of labor for installation, with the meter pan not substantially affected.

In commercial work, a separate disconnect switch connected to a distribution panelboard is common. A 200 ampere fused disconnect switch costs $400.00 plus about 10 hours labor for installation. A matching distribution panel with typical breakers, 42 circuits, no main breaker, costs $800.00 plus about 36 hours of labor for installation. The wiring between the devices—3/0, 2" (50 mm) emt—costs $6.00 plus about 0.30 hours of labor per l.f. for installation. Where necessary, a 400 ampere switch at $650.00 plus about 16 hours of labor for installation and a 400 ampere, 42 circuits no main breaker, panelboard at $500.00 plus about 40 hours of labor for installation may be used. In this case, the wiring between the service switch and the panel would be 500 mom, 3 -1/2" (88 mm) emt, $17.30 plus about 0.43 hours of labor per foot (1.43 per m) of installation.

With the service and distribution completed, the branch work can be analyzed. In residential work, most of the branch work will run as non-metallic sheath cable (romex) or metal sheath cable (bx). Costs are: Romex 2#12 plus ground wire, $0.30 plus about 0.035 hours of labor per l.f. (0.12 hrs. per m) of installation; 2#10 plus ground wire, $0.40 plus about 0.038 hours of labor per l.f. (0.13 hrs. per m) of installation; BX 3#12, $0.30 plus about 0.05 hours of labor per l.f. (0.17 hrs. per m) of installation; BX 3#10, $0.65 plus about 0.06 hours of labor per l.f. (0.20 hrs. per m) of installation. In commercial work, emt and building wire is often used. The equivalent costs are: 3/4" (19 mm) emt 3#12, $0.45 plus about 0.138 labor hours per l.f. (0.46 hrs. per m) of installation; 3/4" (19 mm) emt 3#10, $0.61 plus about 0.146 labor hours per l.f. (0.49 hrs. per m) of installation.

For very preliminary estimates, electrical work can be assumed to run from 6% to 12% of the total job cost for residences, 8% to 11% for office buildings, and 8% to 12% for industrial buildings.

An estimator should make a detailed electrical take off and must have a thorough knowledge of the working drawings and specifications from which the job will be built, as well as of the building code under which it will be inspected and regulations set by the local utility company.

The electrical take off sheet for residential work should have a column for each of the following:

1. Ceiling outlets
2. Bracket outlets
3. Duplex convenience outlets
4. Triplex convenience outlets
5. Waterproof convenience outlets
6. Range outlets
7. Lighting outlet with lampholder
8. Special purpose outlets (dishwashers, clothes dryers, etc.)
9. Single pole switches
10. Three-way switches
11. Length of wiring

The estimator then should check off each item and enter the total on the sheet. This is best done floor by floor, or in large one-story residences, wing by wing, so that checking and changes ordered on the job will be simplified.

Enter next the circuit and feeder control centers, listing them by type and size, and then the branch circuits. List separately two-wire and three-wire runs and empty conduit for telephones and such other special items as intercoms, buzzer systems, television outlets, and the like.

A final list of such items as fixture hangers, studs, locknuts, bushings, cable clamps, and outlet covers complete the take off, and then all these items can be transferred to the pricing sheet item by item.

A pricing sheet should include a line for each type of item; the quantity involved; the material cost by unit and in total; the hours estimated for installation per unit and hourly rate; the total labor cost; and finally, the total labor and material cost.

Service Entrance Equipment

The service entrance equipment consists of the service drop, service entrance conductors, the meter, main control center, and the service ground. The local utility will govern the details of this installation and may or may not furnish the meter. Underground service is highly desirable and may be required by local ordinances. Underground service for residential use costs about $8.00 per l.f. ($26.67 per m). Overhead lines will run around $600.00 per pole. The main control center for residential work consists of a circuit breaker-fuse panel of at least 100 amp. capacity.

Circuit Breakers

Capacity	Cost	Hours to set
100 amp. 3 pole, gen. purpose	$285.00	4
200 amp. 3 pole, gen. purpose	625.00	7
400 amp. 3 pole, gen. purpose	1000.00	11
600 amp. 3 pole, gen. purpose	1700.00	14

16110 RACEWAYS

Rigid galvanized steel conduit is the safest system and in some large cities is required by code. Note that in some areas, when reference is made to rigid pipe, it is understood to mean rigid galvanized pipe. Conduit systems may be galvanized steel rigid, aluminum, or electric metallic tubing, generally referred to as thinwall or EMT.

Material and Labor to Set 100 Lin. Ft. of Conduit
(including supports and fittings, drag wire, raceway, terminations, elbows, and junction boxes)

Size	Galvanized Cost	Hours to Set	Aluminum Cost	Hours to Set	EMT Cost	Hours to Set
1/2" (13 mm)	$60.00	15	$75.00	11	$25.00	6
3/4" (19 mm)	70.00	18	100.00	13	28.00	11
1" (25 mm)	101.00	19	150.00	14	35.00	12
1-1/2" (38 mm)	160.00	26	240.00	19	58.00	16
2" (50 mm)	208.00	30	350.00	22	73.00	17
3" (75 mm)	435.00	38	700.00	28	208.00	19
4" (100 mm)	632.00	45	1100.00	33	338.00	23

Flexible conduit, or Greenfield, is easier to install.

Size	Conduit Cost	Connector Cost	Hours to Set
1/2" (13 mm)	$30.00	$1.35	5
1" (25 mm)	75.00	3.50	8
1-1/2" (38 mm)	120.00	12.30	14
2" (50 mm)	160.00	10.40	18
3" (75 mm)	250.00	32.00	30

Where codes allow, the least expensive method of wiring is pre-wired armored cable (BX) or non-metallic sheathed cable such as Romex.

Material Cost to Install 100 Lin. Ft. of Armored Cable

Gauge of Wire	No. of Wires	Cost	Hours to Set
#10	2	$63.00	5
#10	3	82.00	6
#12	2	36.00	4
#12	3	53.00	5
#14	2	33.00	4
#14	3	42.00	5

Material and Labor to Install 100 Lin. Ft. of Non-Metallic Sheathed Cable with Ground

Gauge of Wire	No. of Wires	Cost	Hours to Set
#12	2	$26.00	3.5
#12	3	43.00	4.0
#14	2	16.00	3.2
#14	3	33.00	3.5

Duct System

Underfloor metal duct systems vary widely in their complexity. Some floor systems have raceways built in, others are added in the floor fill and may be a one or two level system. Standard duct is 1-3/8"x3-1/8" (34 x 78 mm). Super duct is 1-3/8"x7-1/4" (34 x 181 mm). Lengths are 5' (1.5 m), 6' (1.8 m), 10' (3 m), and 12' (3.6 m). Detailed take offs on these systems are beyond the scope of this book. The Square D Company catalog gives a suggested take off method with all the required fittings. Unit do not include receptacle, telephone, or other devices outlets or fittings.

	Material Cost	Standard Hours Per 100 l.f. (30 m)	Material Cost	Super Hours per 100 l.f. (30 m)
Plain Duct per l.f. (per m)	$3.50 (11.67)	10.0	$6.00	14.0
Duct with Inserts per l.f. (per m)	3.75 (12.50)	11.5	6.25	16.0
Single Box	85.00	2.0 ea.	110.00	3.0 ea.
Double Box	100.00	2.5 ea.	200.00	3.5 ea.
Support	5.00	0.25 ea.	6.00	0.5 ea.
Elbow	23.00	0.75 ea.	44.00	1.0 ea.
Connector	4.00	0.30 ea.	8.00	0.3 ea.

Material Cost and Labor to Set 100 L.F. (30 m) of Fiber Duct

Size	Duct Cost	Labor to Set	Coupling Cost	Hours to Set
2" (50 mm)	$75.00	4.00	$12.00	0.10
3" (75 mm)	110.00	5.00	13.00	0.10
4" (100 mm)	115.00	7.00	15.00	0.20
6" (150 mm)	330.00	10.00	20.00	0.30

Material and Labor to Pull 1000 L.F. (300 m) of Wire in Conduit Already in Place (600 V Type THHN/THWN)

Gauge	Material Cost	Hours to Pull
14	$50.00	6.00
12	60.00	7.25
10	100.00	9.00
8	180.00	12.00
6	220.00	13.00
4	330.00	16.00
2	560.00	18.00
1/0	880.00	26.00
2/0	1,020.00	29.00
4/0	1,580.00	38.00

Note: THW has been replaced by THHN/THWN

16130 METALLIC OUTLET BOXES

Type	Cost per Unit	Hrs to Install
Switch Box	$1.50	0.30
3 Gang Box	10.00	0.60
4" (100 mm) Octagon	1.20	0.50
4" (100 mm) Square	1.50	0.50
4 Concrete	2.20	0.40
1 Gang Cast WP	9.00	0.66
3 Gang Cast	20.00	1.00

16140 SWITCHES AND RECEPTACLES

Once the box has been set and connected to the conduit and the wiring pulled, the proper switch or receptacle can be installed which will run as follows:

Type	Cost per Unit	Hours to Install
15 amp single pole toggle	$3.00	0.20
20 " " "	4.00	0.30
15 amp 3-way toggle	5.00	0.35
20 " "	6.00	0.50
15 amp 4-way toggle	14.00	0.60
20 " "	16.00	0.75
600 watt single pole dimmer	7.00	0.50
" " three " "	12.00	0.66
1000 watt single pole dimmer	50.00	0.50
" " three " "	75.00	0.66
15 amp duplex receptacle	2.00	0.20
20 " " "	4.00	0.30
30 amp dryer	8.00	0.50
50 amp range	11.00	0.75

Cost of Toggle Switch and Duplex Receptacle Plates per 100

Description	Cost in Phenolic	Cost in Stainless
1 Gang Switch	$60.00	$142.00
2 Gang Switch	120.00	375.00
3 Gang Switch	230.00	600.00
4 Gang Switch	500.00	950.00
1 Gang Outlet	50.00	142.00
2 Gang Outlet	120.00	375.00
3 Gang Outlet	330.00	600.00
Blank-Single	60.00	142.00
Telephone	60.00	142.00

Weatherproof switches with grey Hypalon presswitch plates cost $600.00 per hundred for the plate not including the switch or box.

16500 LIGHTING

The following is a sampling of some fixtures an electrician might encounter on average work.

Type	Wattage	Size	Shielding	Cost
Industrial Fluorescent	2-40	4' (1.2 m)	Open	$35.00
Pendant Mtg.	4-40	8' (2.4 m)	Open	65.00
	2-40	4' (1.2 m)	Louvers	40.00
	4-40	8' (2.4 m)	Louvers	80.00
Commercial Fluorescent	2-40	4' (1.2 m)	35ox35o	50.00
Surface Mounted	4-40	4' (1.2 m)	35ox35o	80.00
Average Quality	4-40	8' (2.4 m)	35ox35o	150.00
Commercial Fluorescent	8-40	8' (2.4 m)	35ox35o	100.00
Surface Mounted	2-40	4' (1.2 m)	Acrylic	65.00
Premium Quality	4-40	8' (2.4 m)	Acrylic	100.00
Commercial Fluorescent	2-40	4' (1.2 m)	Holophane	80.00
Recessed Troffers	4-40	4' (1.2 m)	Holophane	190.00
Commercial Fluorescent	4-20	2x2	Acrylic	45.00
Square Surface	6-30	3x3	Acrylic	75.00
	6-40	4x4	Acrylic	175.00
Commercial Fluorescent Corridor Unit	1-40	4' (1.2 m)	Acrylic	130.00
Incandescent Industrial	1-100	12" (300 mm) diam.	Open	50.00
Dome Bowl	1-150	14" (350 mm) diam.	Open	60.00
	1-200	16" (400 mm) diam.	Open	65.00
	300-500	18" (450 mm) diam.	Open	75.00
	750-1000	20" (500 mm) diam.	Open	125.00
High Bay Mercury	400	24" (600 mm) diam.	Open	170.00
	1000	24" (600 mm) diam.	Open	225.00
Commercial Incandescent Recessed Fixed Spot	150	10" (250 mm) diam.	Open	55.00
Adj. Spot	150	10" (250 mm) diam.	Open	65.00
Punch Spot	150	8" (200 mm) diam.	Open	65.00
Recessed Square	100	8x8	Opal	80.00
Alum. Wall Bracket	75	5" (125 mm) diam.	Glass	100.00
Pendant Metal Shade	100	11" (275 mm) diam.	Glass	50.00
Pendant Sphere	40	16" (400 mm) diam.	Glass	75.00
Drum	100	18" (450 mm) diam.	Opal	50.00

For Stems add $5.00
6" (150 mm) Drop 10.00
18" (450 mm) Drop 15.00
30" (750 mm) Drop

For plaster rings
  for spots add................................. ....$7.00
Track Lighting
  1 Circuit .... 8' (2.4 m).............................$55.00
  3 Circuit .... 12' (3.6 m)..........................120.00
  150 W Spot...35.00

16720 ALARM AND DETECTION EQUIPMENT

The largest single threat to property and life is fire. While the primary concern is always potential loss of life, damage to or loss of buildings, equipment, materials, files, and records due to fire can put a company out of business. In the case of a museum or other places where historical data is stored, part of history could be lost forever. Consequently, life safety and fire codes are being enacted across the country for all types of buildings, from institutional to office buildings, from hotels to single family residences.

The threat of property damage and loss of life due to fire can be substantially reduced through the proper use of smoke, fire, and gas detectors. These detectors are either wired into the building's electrical system, battery operated, or both. The detector that is wired to the electrical system should be designed to receive backup power from a battery source in the event of a power failure.

Battery operated detectors normally have a small light that flashes when the battery is operative. Some battery systems have a flashing light, short beeps, or both, periodically indicating the battery is not functioning properly. Battery powered systems do not connect to the main electrical system, which makes this device ideal for use in existing buildings (primarily residences) where exposed wiring is objectionable and concealed wiring means costly "cut and patch" work. The one drawback to the battery operated detector is that while it will warn building occupants, it cannot be connected to an exterior alarm system.

Fire detectors and smoke detectors are different. The former can be adjusted to alarm at a particular temperature and the latter is set to alarm at a certain percentage of smoke in the air surrounding the detector. Because most fire fatalities are due to smoke inhalation, in most cases the smoke detector gives the best protection and the quickest alarm.

Placement of detectors will vary depending on the type of building and the interior partitioning. In institutional and commercial buildings, a detector should be located in every room and storage closet, one for every 900 s.f. (81 sq.m) in large open areas, and at approximately 60 l.f. (18 m) on center in corridors. In residential homes a detector should be placed near the kitchen, in hallways leading to bedrooms, and at the top of all stairways at each floor level.

The cost of fire and or smoke detectors varies widely and depends on the amount of sensitivity and sophistication of the alarm system. Residential detectors, either battery or hardwired type, range in price from $10.00 to $55.00 each. Detectors for institutional and commercial buildings will range from $65.00 to $130.00 each.

In addition to the cost of the detectors for non-residential use, there are other associated items of equipment that can be added to the alarm system, such as alarm station pull boxes, test panels, and control console panels, costing several thousand dollars each. Hardwired systems, both commercial and residential, can be provided with an automatic dialers. When the alarm is activated, the system will automatically notify the proper emergency agencies. The automatic dialer installation costs range from $75.00 to $500.00. The wide range of costs is due to the many expenses associated with an automatic dialer. Some emergency units prefer an alarm system wired into a private central station to weed out false alarms. The central control station costs are private and are normally billed directly to the user on an annual basis after installation is completed.

16850 ELECTRICAL HEATING

Heating electrically includes a broad range of methods—radiant units, convector units, baseboard units, electric furnaces with forced air, and heat pumps with forced air. Some of these methods are often included as simple supplemental heaters; others are full plants and involve quite sophisticated design in order to function most efficiently.

Wall heaters are the simplest, most commonly encountered means of heating electrically. Both natural and fan driven units are available and they are usually recessed and have built-in thermostats. A 1500 watt ceiling fan unit will cost around $90.00; a 1500 watt ceiling radiant unit around $150.00; a 1000 watt ceiling radiant unit, $130.00. A 1200 watt fan forced unit will run $100.00 with manual control, $125 with thermostat.

Baseboard heaters are available in standard and high wattage models in surface mounted and semi-recessed designs. The wiring serving these units may be part of the common wiring system or fed from separate branch control centers.

In some areas electricity for heating is separately metered and must be on a separate system. For preliminary figures, baseboard units will run about $30.00 per l.f. installed, not including separate wiring system. Add $32.00 for each thermostat. Baseboards, if used as the sole heating source, are often supplemented by fan units at entrances.

Electric furnaces are similar to conventional warm air furnaces but substitute electric resistance-heating units for flame. They are very compact, require no flue or vent, and can be placed anywhere so that zone heating is simplified. Also, air cooling systems can be added to provide a year-round system. A variation on this are through-the-wall units including both heating and cooling capacities. These are especially suitable for motels, dormitories, and other similar buildings where individual control is important. Electric furnaces will cost about $375.00 for a 35 MBH/hr unit up to $850.00 for a 140 MBH model. Assuming wiring is already in the furnace room, two workers can install the unit in half a day.

Heat pumps are more efficient than resistance type heaters, because they are designed to use existing outside conditions to supplement inside heating and cooling. In the winter, the heat pump extracts heat from outside air, ground, or water and conducts it to the inside. In summer, it removes heat from base to the outside. Heat pumps are generally placed outside to eliminate noise. Units are usually designed to handle air cooling loads. In areas of extreme cold where heating capacity of the pump is inadequate, it can be supplemented by electric resistance heaters. The unit is connected to a duct system within the house.

CHAPTER 17

REMODELING WORK

Remodeling, rehabilitation, and renovation pose many problems not found in new work but if estimated and executed properly can be quite profitable and lead to new clients and future work. This work can also provide the carry-over jobs needed at times to keep crews busy during bad weather or slack periods. Many new structures abut existing structures, requiring a connection between the two and making renovation work necessary.

These estimating suggestions are items encountered in remodeling work. The estimator must investigate and make allowances for all job conditions:

1. Need for protecting adjacent work.

2. Removal of rubbish and salvage.

3. Restrictions during normal working hours, or limited work hours.

4. Extent of daily cleaning up involved. (Will owner occupy work area at the end of the day?)

5. Is there enough work to keep the various trades busy for a full day's work, or will there be callbacks for only fractional periods?

6. Will owner be constantly supervising work?

7. Are plans available of original structure as well as previous renovations that show mechanical and structural details?

8. Is work governed by a strict building code that might require work beyond what is immediately ordered, such as strengthening structural members, adding additional exits to a remodeled third floor, or adding a fire suppression system?

9. Are workers who can form a cooperative team available? There are almost always changes made to fit actual job conditions as they are uncovered?

10. Are there materials present, such as lead paint, that are classified as industrial waste?

11. Are there materials present, such as asbestos floor tiles, exterior siding or boiler covering, that are classified as hazardous waste?

A contractor is often asked whether he thinks it best for the owner to remodel, move, or build anew. On residential construction, a common rule is that if the cost of remodeling plus the original building cost is equal to or less than the cost of a comparable new house, it is best to remodel. Not only are moving costs eliminated, but so are the percentages charged to sell the old house and to buy the new one. Also, investments in landscaping, drapes, carpets, and the like are salvaged.

On commercial work there are other considerations that might be favorable to remodeling rather than rebuilding. First, remodeling may eliminate zoning and building code restrictions that apply to new construction. Second, the expense and time of wrecking is eliminated. Third, if the basic structure is sound, this may be considered as equity for a construction loan. If this is so, financing may be simplified by eliminating preliminary loans before final project is under way.

The savings in time will mean less chance of cost escalation and earlier occupancy. These savings can be reflected in the rental scale, making a remodeled structure more competitive with new. And finally, one might find that preservation of older structures, especially those that are registered with national or local historical societies, is a plus for one's public image.

A visit to the job site is mandatory for remodeling estimating. Existing conditions must be compared with drawings and any differences brought to the owner's attention prior to bid. A site visit is also important to determine the extent of alterations in existing mechanical and electrical work, which is costly, and to match existing materials, patterns, and colors.

Masonry Repairs

Removal of Brick by Hand Chipping. A mason should hand chip around 200 pieces per 8-hr day at the following labor cost per 100 pieces:

	Hours	Rate	Total	Rate	Total
Mason	4.5	$....	$....	$28.18	$126.81
Helper	1.0	....	....	21.04	21.04
Cost per 100 pieces			$....		$147.85
Cost per brick					1.48

Removal of Concrete Block. Block walls can usually be knocked down manually by laborers. For 12" or 8" block walls, figure per 100 s.f. as follows:

	Hours	Rate	Total	Rate	Total
Labor	8.0	$....	$....	$21.04	$168.32
Cost per s.f.					1.68

To remove 4" partition block, figure per 100 s.f. as follows:

	Hours	Rate	Total	Rate	Total
Labor	5.0	$....	$....	$21.04	$105.20
Cost per s.f.					1.05

Where bearing walls are cut through, it will be necessary to provide needling and shoring, and to set in a new lintel. A 3' wide opening in an 8" wall will cost about $100.00 to shore and another $45.00 to build in lintels. A 6' wide opening will cost about $110.00 to shore and $55.00 to set a beam for a lintel.

To Patch Face Brick. To patch 4" face brick, figure one mason and one helper can lay about 65 s.f. per day at the following cost per 100 s.f.:

	Labor	Rate	Total	Rate	Total
Mason	12.0	$....	$....	$28.18	$338.16
Helper	12.0	....	....	21.04	252.48
Cost per 100 s.f.			$....		$590.64
per s.f.					5.91

Material cost per s.f. will add another $1.45 for brick in the $130.00 per 1000 range.

To Patch Building Brick. Where openings are closed up or otherwise patched out in small areas, one mason and one helper should lay about 30 c.f. of building brick per day.

Labor Cost to Patch 100 S.F. (9 sq.m) of 12" Building Brick Wall

	Hours	Rate	Total	Rate	Total
Mason	27.0	$....	$....	$28.18	$760.86
Helper	27.0	....	....	21.04	568.08
Cost per 100 s.f.			$....		$1,328.94
Cost per s.f.				....	13.29
per sq.m				....	143.00

Material cost will add $1.50 per c.f. ($52.96 per cu.m).

To Patch Concrete Block Partitions. One mason and a helper should patch out 150 s.f. (13.5 sq.m) of 4" (100 mm) block partitioning per day at the following cost per 100 s.f. (9 sq.m):

	Hours	Rate	Total	Rate	Total
Mason	5.5	$....	$....	$28.18	$154.99
Helper	5.5	....	....	21.04	115.72
Cost per 100 s.f.			$....		$270.71
per s.f.				....	2.71

per sq.m 29.16

Material costs will add about $0.60 per s.f. ($6.67 per sq.m).

Removing and Replacing (Repointing) Brick Mortar Joint. The mortar joint is usually removed to a depth of 1/2" (13 mm). One mason and a helper should remove and replace about 500 l.f. (150 m) of joint per day at the following cost per 100 l.f. (30 m):

	Hours	Rate	Total	Rate	Total
Mason	1.6	$....	$....	$28.18	$45.09
Helper	1.6	....	....	21.04	33.66
Cost per 100 l.f.			$....		$78.75
per l.f.				....	0.79
per m				....	2.59

To Rebuild Chimneys. On small chimneys one mason with a helper will lay about 500 bricks per day.

To Add Ceramic Brick to Existing Walls. One-inch thick ceramic units formed to the size of standard and Norman bricks may be applied over any nailable surface such as plywood, wood sheathing, furring strips, and various types of concrete and masonry. One sq. ft. (0.09 sq.m) of wall will require 6 of the standard units which cost $95.00 per 1000 in red, $105.00 in golden blend, and $118.50 in gray. On small jobs, one mason (check local jurisdiction on this, as it varies) working alone should install around 700 standard units per 8-hr. day. Mortar requirements are 1-1/2 c.f. per 100 s.f. (0.005 cu.m per sq.m) of wall, and a tuckpointer working alone should point about 450 standard units a day. Special units are available for corners.

Siding

To determine siding costs, it is best to sketch an outline of each elevation. For straight walls, multiply the width by the height and deduct from the total the combined square foot (square meter) area of the doors and windows. Gable area are found by multiplying the width by the height and dividing by two. Deduct window openings. Wall areas for dormers can be figured from the following table, the window areas being already deducted.

Side Wall Area in S.F. (sq.m) for Dormers

Dormer Width	Lift Type	Type of Dormer Gable Type	Hip Type
5'-0" (1.5 m) - One Window	88 (7.9)	60 (5.4)	52 (4.7)
6'-0" (1.8 m)	94 (8.5)	70 (6.3)	58 (5.2)
7'-0" (2.1 m)	100 (9.0)	80 (7.2)	64 (5.8)
8'-0" (2.4 m) - Two Window	88 (7.9)	70 (6.3)	52 (4.7)
9'-0" (2.7 m)	94 (8.5)	75 (6.8)	58 (5.2)

10'-0" (3.0 m)	100 (9.0)	85 (7.7)	64 (5.8)
11'-0" (3.3 m)	106 (9.5)	90 (8.1)	70 (6.3)
12'-0" (3.6 m) - Three Window	93 (8.4)	80 (7.2)	57 (5.1)

Add the net square feet (sq.m) of each elevation, gable, and dormer together, add 8% for waste and figure to the nearest larger half or full square or nearest bundle of shingles for material costs.

Old siding may be stripped off before new is applied, or may be covered over either by applying an underlayment board or by applying wood stripping and felt over present siding. Underlayment board would figure the same quantity as siding. Wood stripping can be figured per square, depending on the exposure of existing siding, as follows:

Exposure, in. (mm)	4" (100)	4-1/2" (113)	5" (125)	5-1/2" (138)	6" (150)	6-1/2" (163)	8" (200)
Stripping l.f. (m)	300 (90)	280 (84)	240 (72)	224 (67)	200 (60)	118 (35)	152 (46)

For corners, use full height for metal corners. For corner boards figure one 3" (75 mm) and one 4" (100 mm) board for each. Add molding, flashing, and caulking costs at doors and windows.

Labor Cost To Add 1/2" (13 mm) Plywood Sheathing Over Existing Boards per 100 S.F. (9 sq.m)

	Hours	Rate	Total	Rate	Total
Carpenter	1.0	$....	$....	$26.44	$26.44
Helper	0.3	....	....	21.04	6.31
Cost Per 100 s.f.			$....		$ 32.75
per s.f.				....	0.33
per sq.m				....	3.55

Corner Boards and Beads

Allow 1/2 hour carpenter time to set one 1x3 (25 x 75 mm) and one 1x4 (25 x 100 mm) corner per story. Beads will take approximately the same time.

Often exterior moldings, corner boards, and base molds must be replaced because of decay. Cedar, cypress, or redwood, preferably the heartwood, will resist future decay.

Costs per l.f. for #2 pine trim and labor would be $0.85 for 1x4 (25 x 100 mm); $1.30 for 1x6 (25 x 150 mm); $1.50 for 1x8 (25 x 200 mm); $2.35 for 1x12 (25 x 300 mm).

Remodeling For Energy Conservation

Most buildings, when appraised for their insulating value, are found wanting. This has opened a whole new field, which is generally referred to as "retrofitting". Under *Thermal & Moisture Protection,* we discussed how deciding it is a simple matter of comparing dollars spent on new materials and equipment against the dollars saved in fuel consumption. Obviously, if a gallon (3.78 liters) of oil increases from $.90 to $1.50 or more in a six-year period, while a 3-1/2" (88 mm) batt of insulation only goes from $0.10 to 0.14 per s.f. ($1.11-1.55 per sq.m), the balance swings in favor of adding more insulation. It will not only pay for itself in a few years but will make a solid return for years to come, will help to preserve natural resources, and will create a more comfortable environment.

As most insulation is hidden within walls and under roofing, one must become a "house detective" before one can become a "house doctor". One of the most reliable aids in determining where the worst leaks is the infrared scanner, on which escaping heat shows up as red. The worse the leak the brighter the color. Such studies can be made into photographs and submitted in reports, and can be retaken after the corrective measures have been taken to determine effectiveness. Even when the original drawings and specifications are available to show what was originally installed, such scans can be useful, because installation may have been poorly done, condensation within the walls may have rendered the insulation useless, or the insulation may have packed down in the case of loose types or shrunk in the case of sprayed or foamed types, leaving substantial gaps for free air passage.

While it is air passage through the walls and roofs that is usually attacked first, infrared scans will often show the greatest heat loss from infiltration around openings, which is probably the first thing to correct.

Caulking

Most caulking materials are available in either knife grade for wide cracks and gun grade for narrow ones. They can be further classified into basic, intermediate, and high performance groups lasting up to five years or so, from five to twenty years, and twenty years and over. On the low end are cord and rope types and those made of oil and resin. These are low in cost and fine for seasonal protection but will need constant replacement. The intermediate caulks are the rubbers, natural and synthetic, starting at the low end with latexes, butyls in the middle, and neoprene at the top. The high performance types are all synthetic including polyurethane, polysulfide, and silicones. A building erected as late as 1980 and caulked with an intermediate grade could have major problems today with infiltration. All openings around windows, doors, trim, sills, cornice, and verge boards and all through the wall piping, air conditioners, vents, and intakes should be carefully checked.

When the space between the door and window frames and a masonry wall are caulked with oakum, an experienced caulker should crank 340 to 400 l.f. (102-120 m) of opening per 8-hr. day at the following labor cost per 100 l.f. (30 m):

	Hours	Rate	Total	Rate	Total
Caulker	2.1	$....	$....	$26.44	$55.52
Cost per l.f.			....		0.56
per m			....		1.84

Where it is not necessary to pack in oakum, one worker using a pressure gun should do about 600 l.f. (180 m) per 8-hr. day at the following labor cost per 100 l.f. (30 m):

	Hours	Rate	Total	Rate	Total
Caulker	1.33	$....	$....	$26.44	$35.17
Cost per l.f.			....		0.35
per m			....		1.15

The cost of caulk sealants, intermediate grade, averages about $1.25 for a 10-oz. (295 milliliters) tube. High performance silicone caulking averages about $4.00 per 10-oz. (295 milliliters) tube.

Refer to the *Thermal & Moisture Protection* division for the quantity of material that will be required for various sizes of joints.

Porch Repairs

To replace an existing porch floor with new 25/32"x3-1/4" (19.5 x 81 mm) face soft wood, allow 2 carpenter hours and 1 laborer hour per 100 s.f. (9 sq.m), painting not included. Cypress, Douglas fir, larch pine, and redwood resist decay and warping and offer medium resistance to wear.

To replace a simple porch column, allow 1/2 hour to remove and shore existing column, 1 hour to replace column.

To replace a simple wood porch railing, figure a carpenter can remove and replace 15 l.f. (4.5 m) per 8-hr. day.

To erect an exterior utility stair of 2x10 (50 x 250 mm) stringers, allow 20 carpenter hours per average story in one run, 25 hours if stair has intermediate landing.

Roof Repairs

To Remove Built-Up Roofing. To remove a 3- to 5-ply roof, figure 1/2 hour each of a roofer's and laborer's time per each square.

To Replace Built-Up Roofing. Figure a crew of 5 per 8-hr. day can install 18 squares of 5 ply roofing, 22 squares of 4 ply, and 24 squares of 3 ply roofing over wood decks on low buildings.

To Replace Shingle Roofing. To remove asphalt roof shingles and underlayment, figure 1 roofer and 1 laborer working together can strip around 20 squares per 8-hr. day. A carpenter can install a square of 12"x36" (300 x 900 mm) strip asphalt shingles on plain double pitch or gable roofs without hips, valleys, or dormers in 2 hours. New shingles are often applied over existing.

To Rehabilitate Existing Roofing. The use of urethane asphalt roofing material is an approved method, but special care must be exercised to insure penetration into all voids, blisters, and cracks. One manufacturer recommends a coating on existing surfaces to be a minimum 50 mil (1.25 mm). Coverage is approximately 1 gal. per 30 s.f. (1.4 L per sq.m) for 50 mil. (1.25 mm) thickness, or 3-1/2 gals. (13.25 L) per square. A roofer should apply 10 to 15 gals. (10-15 L) per hour.

To replace sheet metal gutters, a tinner and a helper should replace 20 l.f. (6 m) of gutter or 25 l.f. (7.5 m) of downspout per hour.

To replace metal counterflashing, figure 1 tinner can replace 15 l.f. (4.5 m) per hour providing a satisfactory reglet exists. If a reglet has to be chiseled into masonry, figure 1 laborer can cut 8 to 10 l.f. (2.4-3.0 m) per hour. If an electric cutting saw can be used, figure 12 to 15 l.f. (3.6-4.5 m) per hour.

Chimney Flashing Table

No. of Bricks Wide	Inches Long	Wide	Long	Lin. Ft. of Flashing Pitch of Roof to 1/3	to 1/2'	to 3/4
2	2	18"	18"	11'	11'	12'
2	2-1/2	18"	22"	12'	12'	13'
2	3	18"	26"	13'	13'	15'
2	3-1/2	18"	30"	13'	14'	16'
2	4	18"	34"	14'	15'	17'
2-1/2	4	22"	34"	15'	16'	18'
2-1/2	4-1/2	22"	38"	16'	17'	20'
2-1/2	5	22"	43"	16'	18'	22'
2-1/2	5-1/2	22"	47"	17'	19'	23'
2-1/2	6	22"	51"	18'	20'	24'

Material Requirements to Add a Gabled Dormer
(Roofing, Siding, Painting, and Interior Finishes Not Included)

Materials & Size	Width of Dormer 5'	6'	8'	10'	12'
Rafter Headers 2"x6"	2-10'	2-12'	2-16'	4-10'	4-12'
Rafters Plates 2"x6"	2-16'	2-16'	2-16'	2-16'	2-16'

Studs and Plates

Side 2"x4"	2-10'	2-10'	2-10'	2-10'	2-10'
Side 2"x4"	2-12'	2-12'	2-12'	2-12'	2-12'
Front 2"x4"	7-10'	8-10'	11-10'	13-10'	11-12'
Gable 2"x4"	1-12'	1-12'	3-10'	2-10'	2-12'
Gable 2"x4"	1-8'	1-8'	1-10'	1-14'	1-14'
Rafters 2"x4"	6-8'	6-10'	6-12'	6-12'	6-14'
Ridge Boards 1"x6"	1-10'	1-10'	1-10'	1-10'	1-10'

Framing

Total Bf	179 Bf	200 Bf	237 Bf	264 Bf	283 Bf
Sheathing 1"x6"	60 Bf	70 Bf	70 Bf	85 Bf	80 Bf
Roof Boards 1"x6"	60 Bf	72 Bf	86 Bf	100 Bf	120 Bf
Cornice Fascia 1"x6"	22 Lf	24 Lf	26 Lf	28 Lf	30 Lf
Crown Mold 1"x3"	22 Lf	24 Lf	26 Lf	28 Lf	30 Lf
Nails	10 lb.	11 lb.	12 lb.	13 lb.	14 lb.
Flashing 12" Wide	33 Lf	35 Lf	39 Lf	42 Lf	46 Lf
Felt 30 lb.	1 RL	1 RL	1 RL	1 RL	2 RL
Window Frame 2/6x4/10	1	1	2	2	3
Window Sash 2/6x4/10	1	1	2	2	3
Labor To Install	26 Hrs.	27 Hrs.	32 Hrs.	36 Hrs.	40 Hrs.

Gabled Dormer Roof and Side Areas for 1/2 Pitch Roofs

	Width				
	5'	6'	8'	10'	12'
Roof	60 SF	72 SF	86 SF	100 SF	120 SF
Wall	60 SF	70 SF	70 SF	85 SF	80 SF
Starter	10 LF	10 LF	10 LF	10 LF	10 LF
Ridge	12 LF	12 LF	12 LF	12 LF	12 LF

Weatherstripping

All doors and windows should be tightly sealed including those leading to unheated attics, garages, or basements. Check windows for sealing gaskets and tight closing latches. Interlocking devices at jambs and heads of doors are preferable to spring or pressure types and select a proper adjustable sill seal and threshold that will fit to floor finish.

Lineal foot cost of material weatherstrip for double hung windows will average as follows in zinc. Bronze will run about 100% more:

	Thickness, in. (mm)		
Description	1-3/8" (34 mm)	1-3/4" (44 mm)	2-1/4" (56 mm)
Side, head & sill strips	$0.40	$0.45	$0.50
Meeting rail strips	0.60	0.65	0.75

Single or double casement windows that swing in require a 3/4" (19 mm) wide track strip on the hinged side, running in a groove in the sash. At the top, on the latch side, and where the sash meet, if in pairs, interlocking hook and flat strips are generally used. A trough strip and a hook strip is generally used at the bottom of the sash.

Weather strips for single wood doors will range from $4.75 for zinc to $7.40 for bronze per opening. If interlocking bronze rather than spring type is used, add $3.20 to the above figures. Extruded aluminum and neoprene spring stop mounted strips run $1.10 per l.f. ($3.67 per m) for 1-1/8"x1/4" (28 x 6 mm) and $2.55 per l.f. ($8.50 per m) for 1-1/4"x25/32" (31 x 20 mm). Bottom rail gasket runs $1.90 per l.f. ($6.33 per m).

Labor Installing Weather Strips

	Hours
1-3/8" (34 mm) DH sash to 3x5'6" (75 x 125 mm)	0.9
1-3/4" (44 mm) DH sash to 4x7 (100 x 175 mm)	1.0
2-1/4" (56 mm) DH sash to 4x7 (100 x 175 mm)	1.25
Single inswinging casement	1.3
Double " "	2.0
Single outswinging casement	1.0
Double outswinging casement	1.4
3x7x1-3/4" (75 x 175 x 44 mm) wood door, spring bronze	1.6
3x7x2-1/4" (75 x 175 x 56 mm) wood door " "	2.0
3x7x1-3/4" (75 x 175 x 44 mm) wood door interlock	2.5
1 pr. 1-3/4" (44 mm) French doors	2.3

It may pay to weatherstrip a door at the head of a stair leading to the attic. Chimneys should also be checked for tight fitting dampers.

Doors and windows in exposed locations should be fitted with storm sash if operable and double glazing if fixed. Openings on the north side and on the side toward prevailing winter winds especially need this extra protection.

A carpenter with helper should set an average preglazed storm window in 1-1/2 hours. Storm windows, up to 12 s.f. (1 sq.m) in area, average about $55.00 per unit installed. Larger units cost about $75.00 installed. Smaller units can be handled by a carpenter working alone.

Doors in exposed locations are best protected by adding a vestibule rather than just a storm door. In commercial work with heavy traffic a revolving door will offer the best protection although the initial cost will be $15,900 or more.

Windows that are not used for ventilation may be permanently sealed, but it should be remembered that if sealing means having to rely completely on running air cooling in the summer months, this could be self-defeating.

Fixed windows can often be fitted with an interior storm window. This is especially adaptable to curtain wall construction where tubular frames can be fitted with a spacer, new metal sash bars, and an additional layer of glass, which can be of the solar type, further cutting down on heat penetration and glare. Such additions can be added entirely from the inside, but because the area between the old and new glass will be hermetically sealed, only experienced installers should be considered. Pittsburgh Plate Glass has a patented system which they will install on a custom basis.

Head

Meeting Rail

Jamb

Insulated Panel Over Double Hung Window
Courtesy Mapes Industries

Several manufacturers offer special replacement windows, which can be pivoted, projected, double hung, or casement types. To install these, one first removes the existing window, but not the frame. A factory assembled subframe is then fitted over the existing frame and secured to it. The new window sash is then installed within the new subframe. These units are usually aluminum, can be had preglazed in either single or double glazing or with part of the window blocked off with insulated panels, and often can be set, window by window, from the inside. A two-worker crew should remove the existing sash, install the new subframe, and add the new sash in approximately three hours per opening.

Unwanted windows are best be eliminated altogether by either bricking them up or covering them with insulated panels. A variation of this, which can retain partial light and ventilation, is to add an insulated panel over part of the window, usually the upper half. These are usually porcelainized metal laminated over a polystyrene core with an aluminum frame factory assembled to fit to the existing window frame. One such manufacturer is

Mapes Industries, Lincoln, Nebraska, which rates its add-on panels 1-1/4" (31 mm) thick at an R value of 6.56.

Solar Heat Controls

Heat gain from sun rays penetrating the building is a blessing in the winter but not in the summer. There are ways to take advantage of it by cooperating with the season. The sun is high in the summer, low in the winter. Exterior screens and shades, such as canopies, awnings, operable shutters, and even a few well placed deciduous trees, can be placed so that the summer rays are excluded while the low winter rays are welcomed in. Excluding the sun from the outside rather than blocking it once it has entered the inside should be the first consideration.

Continuous canopies will cost about $5.30 per s.f. ($58.89 per sq.m) while individual window ones will cost $160 in canvas, $265 in aluminum per unit. With a 3-ft. (0.9-m) overhang, two workers should erect 80 l.f. (24 m) of running canopy or four individual ones per day.

If windows are to be reglazed, the sun effect can be counteracted by using tinted or reflective glass. Glass with a fused reflective coating will cost about $3.00 per s.f. ($33.33 per sq.m) more than plain plate.

Existing glass can be coated with a solar rejecting film for around $1.60 per s.f. ($17.77 per sq.m).

Insulation

The chapter on *Thermal & Moisture Protection* has a discussion on the types of heat loss and the insulation materials available to counteract them. In retrofitting, one's choices are somewhat limited.

Ceilings are prime areas for adding insulation. Few would argue that 6" (150 mm) is probably easily justifiable north of St. Louis, though 3-1/2" (88 mm) was standard just a few years ago. In most residences gabled roofs provide attic space where additional batts and blankets can be added. Where some insulation is already in place, blankets without any cover can be added so that no vapor barrier will be introduced that might cause condensation problems. Three-inch (75 mm) thick rolls cost about $0.20 per s.f. ($2.22 per sq.m) and 6" (150 mm) rolls are $0.26 per s.f. ($2.89 per sq.m). The cost of installing is about the same, with one worker completing 1200 to 1400 s.f. (108-126 sq.m) per 8-hr. day.

Attics can serve as a buffer between outside conditions and interior spaces most effectively if they are ventilated. This is usually designed to allow gravity to carry off unwanted build-ups by providing low inlets, which exhaust through the roof at the high point, in the eaves. Eave vents, 4"x16" (100 x 400 mm), cost about $7.40, and it takes one carpenter some 15 minutes to set. Some metal eaves are perforated to allow air to enter. Ridge vent strips will cost $1.55 per l.f. ($5.17 per m), and it takes 1 worker 1 hour to install 20 l.f. (6 m).

Where no accessible attic space exists, insulation must be added on top of the roofing. Assuming a relatively flat roof in good condition, it is possible to add new non-absorbing rigid insulating board, such as Styrofoam, over the existing roofing, mop it in, and hold it down with a new application of stone. This helps protect the existing roofing, but if there is any doubt as to the condition of what is there, it should be properly repaired. The insulation is not to take the place of roofing membranes. The amount of insulation added can vary from 1-1/2" to 3" (38-75 mm) in single sheets but can also be built up by adding additional plies. Should the old roofing eventually need replacement, the stone can be removed from the insulation and new roofing applied over it. Roof drains and flashings of course would have to be reset to the new roof level.

Another possibility for covering existing roofs, and this is not limited to level ones, is to spray on a polystyrene or urethane insulating coat and then cover this with a butyl or elastomeric roofing coat. Costs of applying such coats and roofing will be similar to new work as discussed in *Thermal & Moisture Protection* with the addition of costs to adapt adjacent drains and flashings to the new levels.

Walls present special problems. If the exterior and interior finishes are to remain, there is little alternative to adding insulation to whatever cavity space exists in the center of the wall. In wood stud construction this is relatively simple. To get into the cavity between the studs it is necessary to cut holes between each stud at least every story height. These are usually cut from the outside by removing a strip of siding. Bevel and shiplap can be pried up easily, but tongue and grooved will have to be broken out. Most walls will have a sheathing course behind the siding. This will have to be drilled with holes large enough to accommodate the nozzle for blowing in the insulation, usually 2" to 3" (50-75 mm). Once the holes are cut the cavity must be plumbed to determine if any obstructions are present, and if any are found additional holes will have to be drilled. Material costs for fiberglass or mineral wool will run from $0.42 to $0.55 per c.f. ($14.83-19.42 per cu.m). One carpenter can install some 400 c.f. (11.33 cu.m) per day. To this must be added the cost of removing and replacing siding.

Where an owner contemplates new interior finish in a building or space, the existing walls can be covered with rigid insulation board and then faced with whatever new finish is contemplated. In older buildings where the plaster is in poor condition this may be the better alternative. It will mean replacing all trim, but in removing the baseboards it will also allow batts to be packed between the floor joists to keep air currents from flowing down the outside walls and out over the floor area. Rigid board of fiberglass or extruded polystyrene will cost about $0.25 per s.f. ($2.77 per sq.m) and will have R values of 4.3 and 5.4 respectively. One carpenter should install 700 to 800 s.f. (63-72 sq.m) per 8-hr. day.

Wood-framed structures can also be re-sided with vinyl or aluminum siding with insulation fitted to the back. Adding the insulation costs about $0.10 per s.f. ($1.11 per sq.m) so that vinyl costs $0.80 per s.f. ($8.88 per sq.m) and aluminum $0.90 per s.f. ($10.00 per sq.m), and one carpenter can install 250 s.f. (22.5 sq.m) per day.

Masonry veneer walls can have the cavities filled much the same as wood sided walls, removing brick to allow access. Patching a masonry wall will be more obvious than a wood one. Where the brick has concrete block as a back-up, it is easier to add rigid insulation to the inside face. If the wall is at all irregular, it will have to be furred, but the air space will contribute to the insulation value.

Block walls can also be foamed by drilling block every other unit on 4-ft. vertical runs. Foaming is tricky and should be subcontracted to someone experienced in its application. Cost will be about $0.50 per inch ($0.02 per mm) of thickness installed per s.f. (0.09 sq.m).

Basement and crawl space walls, especially those with portions above grade, should be insulated. Rigid board or sprayed types are applicable here although the sprayed on type will need protection if the space is subject to much traffic. The same treatments along with batts or blankets can be continued under the floor construction above these spaces. Ducts and piping should also be insulated.

Interior Work

In remodeling work it is especially important that new wood used with old be dry so that shrinkage will be minimized. Kiln-dried wood should be ordered. Edge grained wood, which has a shrinkage just half that of flat grained wood, should be used.

Shrinkage is considerable in birch, gum, maple, and oak; is average in cherry, hemlock, pine, and walnut; and minimal in cedar and redwood.

In extending or closing up doorways and openings in existing walls, often stud dimensions will not match the existing. Special lumber or furring will be required. New studs are 1/8" (3 mm) less than those in used in the past. In much older buildings studs will be found that are not even sanded. In demolishing old work, plan ahead to see if some of the old lumber should not be salvaged. It may be harder to work with, but the size may be irreplaceable.

Often contractors are asked to judge what caused cracks in walls and ceilings and whether to repair, replace, or reface them. Many cracks are not attributable to the workmanship of the cracked surface, but to faults in the materials to which they are applied. Such cracks can be caused by lumber shrinkage, foundation settlement, inadequate bracing, and expansion and contraction.

Plaster can crack, spall, and disintegrate because of poor workmanship. A common condition occurs when the white coat is applied over a base coat that has been allowed to dry out too much. When the white coat is applied, the base coat soaks the moisture from the finish coat and a spidery pattern of cracking occurs. If the surface of the plaster seems soft, the plaster has probably been oversanded. If cracks occur at random on the white coat, it is probably too thin a coat.

Structural cracks usually show up in the first few years of occupancy. Unless major defects are involved, they reach an equilibrium and can be successfully patched if the plaster on either side is not loose. If the crack is less than 1/2" wide, dig out an inverted "V" joint, moisten the adjacent areas to reduce suction and fill with prepared patching compound in two stages. For larger cracks, the crack should be made wider to accommodate a piece of metal lath. The resulting channel is then filled with a three coat application.

When the plaster is soft, or cracking is prevalent over a large area, it is best to cover the plaster. This can be done with plasterboard, canvas, acoustical tile, or any of the decorative hardboards and plywoods. If the plaster is over sound wood lath, the applied board can be nailed through the plaster into the lath or stud. If the backing is unsound, a new furring system should be applied. Furring strips are nailed at right angles to studs or joists at intervals of 12" (300 mm), 18" (450 mm), or 24" (600 mm), depending on module of new finish.

Walls

To strip a wall of plaster and leave ready to replaster if masonry, or relath and plaster, figure 1 laborer will strip from 60 to 70 s.f. (5.4-6.3 sq.m) per hour. Add for cost of clean up and debris removal if this is to be included.

To add 1/2" (13 mm) gypsum board over stripped wood studs, figure 1 carpenter will place around 50 s.f. (4.5 sq.m) per hour in the average room. If large unbroken expanses are involved, double this. If gypsum board is not a pre-finished type but is to be painted, it will require taping. Figure a carpenter will tape 100 l.f. (30 m) of joint, which, in usual applications, would cover the needs of 300 s.f. (27 sq.m) of wall, in about 4 hours.

When pre-finished plywood is specified for first class work, gypsum back-up board should be used as an underlayment.

To add wallboard over an existing wood stud and plaster wall, the wall should first be furred. Furring strips should be nailed at right angles to studs. Any convenient width may be used, but it must be wide enough to provide a nailing base to nail all edges of the decorative strips at least every 6" (150 mm). Furring may be of No. 2 soft wood. The success of the wall will depend largely on the success of leveling the furring strips.

With studs 16" (400 mm) on center, a carpenter should cut and place 450 to 500 l.f. (135-150 m) of furring strips per 8-hr. day at the following cost per 100 l.f. (30 m):

	Hours	Rate	Total	Rate	Total
Carpenter.. 1.8		$....	$....	$26.44	$47.59
Cost per l.f...				....	0.48
per m				....	1.57

When applying wallboard in smaller rooms requiring considerable cutting and fitting around doors and windows, a carpenter should fit and place 400 s.f. (36 sq.m) per 8-hr. day, at the following labor cost per 100 s.f. (9 sq.m):

	Hours	Rate	Total	Rate	Total
Carpenter.. 2.0		$....	$....	$26.44	$52.88
Cost per s.f.				....	0.53
per sq.m				....	5.70

In large rooms with unbroken walls, this cost would be reduced to about $0.25 per s.f. ($2.77 per sq.m).

To add thin board, such as 5/32" (4 mm) prefinished over an existing plaster base with adhesive, figure labor costs as follows per 100 s.f. (9 sq.m):

	Hours	Rate	Total	Rate	Total
Carpenter.. 3.2		$....	$....	$26.44	$84.61
Cost per s.f.				....	0.85
per sq.m				....	9.10

To remove ceramic tile from an existing wall, figure 1 laborer removes about 30 to 40 s.f. (2.7-3.6 sq.m) per hour, not including removal of debris. As old existing tile is undoubtedly set in a mortar base, it will be necessary to set the new in the same way or to replaster the wall to provide a smooth base for applying tile with adhesive. Tile, 4"x4" (100 x 100 mm), applied with mortar costs about $5.00 per s.f. ($55.55 per sq.m). Adhesive set tile runs in the neighborhood of $4.25 per s.f. ($47.22 per sq.m) to which must be added to the cost of a new base.

Ceilings

To remove plaster from ceiling joists, figure 1 laborer clears about 50 s.f. (4.5 sq.m) of area per hour. If the ceiling is suspended and the hangers are also to be removed, figure about 30 s.f. (2.7 sq.m) per hour.

To strip a plastered ceiling of acoustical tile, figure 1 laborer clears 75 s.f. (6.8 sq.m) per hour. These figures do not include removal of debris.

Often a contractor is asked to patch plaster over concrete or old plastered ceilings. This can be accomplished with the use of a product such as *Plaster-Weld* from Larsen Products Corp. It is brushed, rolled, or sprayed on, and in an hour or so, recoated with the desired plaster finish.

To add 12"x12" (300 x 300 mm) wood fiber tile to existing plaster ceiling, figure 2 mechanics working together on a scaffolding can set 800 tiles per day at the following labor cost per 100 s.f. (9 sq.m):

	Hours	Rate	Total	Rate	Total
Mechanics	2.0	$....	$....	$26.44	$52.88
Cost per s.f.			....		0.53
per sq.m			....		5.70

Flooring

To Remove Wood Subflooring. Where existing subfloors are badly worn and must be removed, figure 1 carpenter and 1 laborer can remove 100 s.f. (9 sq.m) per hour at the following cost:

	Hours	Rate	Total	Rate	Total
Carpenter	1.0	$....	$....	$26.44	$26.44
Laborer	1.0	....	....	21.04	21.04
Cost per 100 s.f.			$....		$47.48
per s.f.			....		0.47
per sq.m			....		5.11

To Replace Wood Subfloors. In small cut-up areas, a carpenter should lay between 700 and 800 b.f (1.65-1.89 per cu.m) per 8-hr day of 1"x6" (25 x 150 mm) or 1"x8" (25 x 200 mm) at the following labor cost per 1,000 b.f. (2.36 cu.m):

	Hours	Rate	Total	Rate	Total
Carpenter	10.6	$....	$....	$26.44	$280.26
Laborer	6.0	....	....	21.04	126.24
Cost per 1000 b.f.			$....		$406.50
per b.f.			....		0.41
per cu.m			....		172.25

To Add Plywood Over Existing Subfloors. Labor for 100 s.f. (9 sq.m) of plywood added over existing subfloor would be as follows:

	Hours	Rate	Total	Rate	Total
Carpenter	1.0	$....	$....	$26.44	$26.44
Laborer	0.5	....	....	21.04	10.52
Cost per 100 s.f.			$....		$36.96
per s.f.			....		0.37
per sq.m			....		3.98

To Add New Finished Flooring. Once an existing subfloor is put in good condition, replaced, or given a good underlayment, the following installed costs per sq. ft. (sq.m) may be used for various floor finishes:

1x1 (25 x 25 mm) Ceramic Tile, Adhesive Set	$4.25-5.50	($47.22-61.11)
2 = Asphalt Tile	0.70-0.95	(7.77-10.55)
Vinyl Asbestos 1/16" (1.5 mm)	0.85-1.05	(9.44-11.67)
Cork Tile	1.30-2.00	(14.44-22.22)
Vinyl Tile 0.050" (1.25 mm)	1.05-1.15	(11.67-12.77)
Vinyl Sheet 0.070" (1.75 mm)	1.40-1.85	(15.55-20.55)
Rubber Tile	1.60-1.95	(17.77-21.67)
9x9x5/16" (225 x 225 x 7.8 mm) Oak Parquet	2.85-3.00	(31.67-33.33)

Trim

To Remove Existing Wood Base. A carpenter and helper working together should remove a one-piece wood base at the rate of 100 l.f. (30 m) per hour.

	Hours	Rate	Total	Rate	Total
Carpenter	1.0	$....	$....	$26.44	$26.44
Laborer	1.0			21.04	21.04
Cost per 100 l.f.		....	....		$47.48
per l.f.			$....		0.47
per m			....		1.54

To Replace Wood Base. To place a single member base plus carpet strip, allow $1.30 per l.f. ($4.33 per m) for 3-1/2 (88 mm), $1.60 ($5.33 per m) for 4-1/2 (113 mm). To add new interior trim, the following installed prices may be used.

	per l.f.	per m
Wood picture molding	$1.10	$3.67
Wood chair rail	0.95	3.17
Wood cornices 1"x8"	1.50	5.00
Panel strips	0.60	2.00

To Replace Window Sash and Trim. A carpenter should fit and hang, complete with sash cord, chain, and weights or counterbalances, 10 to 12 single sash per 8-hr. day, at the following labor cost per pair:

	Hours	Rate	Total	Rate	Total
Carpenter	1.5	$....	$....	$26.44	$39.66

A carpenter should fit and hang 10 to 13 casement sash per 8-hr. day at the following labor cost per pair:

	Hours	Rate	Total	Rate	Total
Carpenter	1.4	$....	$....	$26.44	$37.02

A carpenter should set a one member window trim and cap at 1.5 hours per window at the following labor cost:

	Hours	Rate	Total	Rate	Total
Carpenter	1.5	$....	$....	$26.44	$39.66

To Replace Wood Doors and Trim. A carpenter should set a plain wood jamb for a 3'x7' (0.9 x 2.1 m) wood door in an hour and trim both sides in 1.5 hours. To hang the door will take another hour for a total cost as follows:

	Hours	Rate	Total	Rate	Total
Carpenter	3.5	$....	$....	$26.44	$92.54

A frame set for 3'x6'-8" (0.9 x 2 m) opening will run about $16.00 in pine, $21.00 in birch.

If door unit is prefit and knocked down, a carpenter should assemble and install it in an hour.

To Set Factory Assembled and Finished Kitchen Cabinets. When setting factory assembled and finished kitchen cabinets, about 0.15 hours of carpenter time per s.f. (1.67 hrs. per sq.m) of cabinet face area should be allowed.

To Fit and Place Closet Shelving. A carpenter should fit and place approximately 125 s.f. (11.35 sq.m) of closet shelving per 8-hr. day at the following labor cost per 100 s.f. (9 sq.m):

	Hours	Rate	Total	Rate	Total
Carpenter	6.4	$....	$....	$26.44	$169.22
Cost per s.f.			....		1.69
per sq.m			....		18.21

To Install Rod in Closet. A carpenter should install about 3 hanging rods per hour at the following labor cost per rod:

	Hours	Rate	Total	Rate	Total
Carpenter	0.33	$....	$....	$26.44	$8.73

To Set Medicine Cabinets. A carpenter should set 12 face-mounted cabinets per 8-hr. day at the following labor cost per cabinet:

	Hours	Rate	Total	Rate	Total
Carpenter	0.66	$....	$....	$26.44	$17.45

To Set Bathroom Accessories. A carpenter should set 30 to 34 bathroom accessories per 8-hr. day at the following cost per accessory:

	Hours	Rate	Total	Rate	Total
Carpenter	0.25	$....	$....	$26.44	$6.61

To Set Finish Hardware

Type	Carpenter Hrs. per each
Rim Lock	0.5
Mortised Lock	1.0
Cylinder Lock	0.5
Front Entrance Cylinder Lock	2.0
Surface Door Closer	1.0
Concealed Closer	3.0
Sash Lifts and Lock	0.5
Kickplates	1.0

To Set Cabinet Hardware

Type	Carpenter Hrs. per Each
Surface Bolts	0.05
Offset Bolts	0.10
Catches	0.10
Drawer Pulls	0.06
Knobs	0.05
Rim Catch	0.10
Mortise Lock	0.40

Painting

To repaint exterior wood siding and trim:

	SF/Hr.	Hrs./100 SF	SF/Gal.
First Coat	115-125	0.85	500-550
Second Coat	125-135	0.75	575-625

If existing surface is in poor condition, add:

	SF/Hr.	Hrs./100 SF	SF/Gal.
Burning Paint Off	30-40	2.75	
Prime Coat	100-110	0.95	400-450

To add 2 coats to wood shingles:

	SF/Hr.	Hr./100 SF	SF/Gal.
First Coat, Oil	115-125	0.82	250-300
Second Coat, Oil	150-160	0.65	375-425
First Coat, Stain	75-85	1.2	120-150
Second Coat, Stain	115-130	0.82	200-225

To prepare interior work:

Type of Work	SF/Hr.	Hrs./100 SF	SF/Gal.
Sanding & Putty	190-200	0.5	
Washing Enamel	85-90	1.15	
Removing Varnish with Liquid Remover	30-35	3.00	150-180
Wash & Touch Up Varnish	165-175	0.60	
Add 1 Coat Varnish	100-110	0.95	575-600
Burn Off Interior Trim	20-25	4.00	
Burn Off Flat Surfaces	30-35	3.00	
Spackle Over Flat Surface	60-65	1.60	140-150
Filling Wood Floors, Wiping	170-180	0.57	425-450
Taping, Spotting & Sanding Wallboard	90-110	1.00	
Wash Off Calcimine	115-125	0.83	
Wash Smooth Wall	145-155	0.67	
Remove Old Wallpaper	65-75	1.45	
Wash Off Glue	125-135	0.77	

To paint interior work:

Type of work	SF/Hr.	Hrs./100 SF	SF/Gal.
Wood Floors			
1st Coat Paint	290-300	0.35	525-550
2nd Coat Paint	260-270	0.38	425-450
1st Coat Stainwax	390-400	0.25	525-550
2nd Coat Stainwax	480-490	0.21	600-625
Varnish, Each Coat	300-310	0.33	500-550
Buffing	390-400	0.25	
Waxing	200-210	0.50	1,050-1,075
Walls			
Sizing	350-400	0.27	600-700
Sealer	200-225	0.45	575-625
Flat Finish			
1st Coat	165-175	0.60	500-550
2nd Coat	175-185	0.55	575-625
Latex			
1st Coat	140-160	0.67	300-350
2nd Coat	140-160	0.67	300-350

Canvas	50-60	1.82	
Wallpaper			
Per Single Roll	2-1/2-3		
Trim			
Enamel			
Undercoat	90-100	1.10	375-400
Finish Coat	100-110	1.00	475-500
Stain Woodwork	215-225	0.45	700-725
Shellac Woodwork	215-225	0.45	700-725
Varnish Woodwork	165-175	0.60	425-450
Wax and Polish	100-110	0.95	550-600

Plumbing

The following may be used as a guide for replacing existing fixtures. No roughing in or pipe replacement is figured. Fixture cost and roughing in is discussed under *Division 15, Mechanical.*

	Hours
Replace Bathtub	8
Add Shower-Surface Mtd.	2
Replace Single Kitchen Sink	5
Replace Double Kitchen Sink	7
Replace Laundry Tub	5
Replace Lavatory	5
Replace Shower Stall	12
Replace Water Closet	7
Replace Water Heater	7
Replace Central Air Compressor	4
Replace Gas Dryer	4
Replace Automatic, Top-loading Washer	2

Furnish and connect:	Cost
40-gal. (150-L) Natural Gas Water Heater	$330.00
50-gal. (190-L) Natural Gas Water Heater	360.00
40-gal. (150-L) LP Gas Water Heater	290.00

Electrical

To rewire, pulling in old conduit: per outlet, 2.5 hours.
To add new, using BX cable in open wall: per outlet, 1 hour.
Budget figures for remodeling residential work using armored cable and including labor and materials but not fixtures or patching, unless so noted, would run somewhat as follows:

To add a ceiling outlet	$ 82.50

To add a wall switch...66.00
To add a wall outlet..82.50
Wire and connect a dishwasher 99.00
Wire and connect a clothes washer..110.00
Wire and connect a disposal.......................................121.00
Wire and connect wall or window air conditioner 121.00

To Furnish, wire, and connect the following :

40-gal. (150-L) electric water heater............................$ 319.00
52-gal. (197-L) electric water heater............................330.00
82-gal. (310-L) electric water heater............................385.00

CHAPTER 18

EPOXY SYSTEMS

The emphasis on renovation is the result of many factors, one of which is the high cost of replacement. Efforts are underway to restore structures of all types to original, or better, condition to forestall replacement and reduce overall costs of operation and maintenance.

This situation has created a need for new products and methods for the preservation of buildings, bridges, floors, tunnels, airports, runways, in short, all conceivable types of structures. These new products include, among others, polymers in general and epoxies in particular. The epoxy industry has grown substantially over the past 20 years and has matured to the extent that it is providing reliable products, supported by technically proficient manufacturers who provide architects, engineers, owners and contractors with a variety of services. These services include guide specifications, on-site technical assistance and even product design to suit the needs of specialized applications.

Epoxy Compounds

Epoxy compounds are used in a variety of applications in construction. Epoxies are plastics that are thermosetting resins. When resin and a *hardener* (curing agent) are mixed together, a chemical reaction that liberates energy in the form of heat takes place. This reaction is classified as *exothermic*.

Temperature Limitations. Epoxies require some warmth to cure or harden, which explains why low temperature limitations are imposed for application of these items. Conversely, as the temperature rises, they will harden more rapidly, and in some cases, for example, in the sun at 120°F (48.88°C), it might be impossible to install certain types of epoxy compounds, because they will harden too fast.

Pot Life Limitations. The pot life of epoxies, after the resin base and hardener have been mixed, depends on several factors:

1. The type of curing agent used. Some cause reactions to proceed faster than other.
2. The quantity of resin and hardener mixed at one time (total mass).
3. The temperature of epoxy at the time of mixing.
4. Atmospheric temperature.

If the quantity mixed is large, e.g., a 5-gal. (18.9-L) container full, the exothermic chemical reaction proceeds more rapidly, because the heat cannot be readily dissipated to the atmosphere. The large mass "insulates" the compound and causes the reaction to build on itself and proceed more rapidly.

If the mix quantity is small, a half-pint (284 milliliters) container, the heat of the reaction is readily dissipated through the walls of the container to the atmosphere. The mass is small, and the mixture cannot effectively retain the heat of reaction. Also, if the air temperature is low, the reaction will proceed slowly. If the air temperature is high, it will proceed more rapidly.

As an example, let us assume that we mix 4 gals. (15 liters) of *Durabond* epoxy bonding compound and intend to brush it onto preexisting prepared concrete before pouring a new concrete overlay. Assume that only half, about 2 gallons (7.6 liters), of bonding compound is applied to the concrete, while the rest stays in the mixing bucket. What happens?

The epoxy in the mixing bucket (large mass) will react fairly rapidly and harden in about one hour, while that applied to the concrete—a very small mass, about 10 mils (0.25 mm) thick—will take three to five hours, depending on the temperature of the deck and of the atmosphere.

Mixing Ratio Limitations. Unlike other types of polymer, the mixing ratio of epoxy items cannot and must not be altered under any circumstances. Neither the setting nor the curing time can be changed by adding more or less curing agent to the mix. When cold weather approaches, 40 to 50°F (4.44 to 10°C), and an epoxy compound appears to be setting or curing too slowly, do *not* attempt to speed the setting time by adding excess hardener. Also, in the summer when temperatures are high, 90 to 100°F (32.22 to 37.78°C), do *not* try to slow the setting time by using less hardener. Either of these actions will only create a soft and gummy compound.

The chemical ratio of resin to hardener is exact, and the resin requires a fairly precise quantity of curing agent with which to react. If the hardener quantity is insufficient, it is possible that some of the resin will not react and will remain in liquid or semi-liquid state. Also, if an excess of hardener is added, because the hardener can only react with a specific quantity of resin, the excess hardener might not react and possibly remain free in the compound as liquid or semi-liquid. In any event, the physical and chemical properties of the cured epoxy can be severely impaired by these actions.

In summary, one can make an analogy with concrete. There are many similarities, such as in the mass vs. set time and in the exotherm relationship. Consider that cement and water provide the matrix or binder (adhesive) for the aggregate. This is similar to the resin and hardener in epoxies and other thermosetting resin systems.

Both concrete and epoxy are subject to limitations of temperature. Concrete normally cannot be poured when the weather is below freezing, or when it is very hot. Both cannot be placed without consideration of the mass and thickness of the pour. If the pour is excessively large, the heat generated could destroy or weaken the structural value if the exotherm is not controlled.

With concrete, the water/cement ratio is very important, because if it is excessively high, then the concrete will not develop adequate physical properties for a long time, if at all. If it is excessively low, hydration might not be complete, and again, a weak mix will result. The same can be said of the epoxy resin/hardener ratio described above.

Everyone is familiar with the fact that concrete once mixed (sand, stone, cement, and water) begins to cure or harden via chemical reaction and will set normally within a given period of time. Epoxy also has a limited working life after mixing and requires that it be applied within a given period of time.

Cure Time. One final but very important point in our comparison of epoxy and concrete is that concrete must be allowed to cure for a period of time before forms can be stripped and it can be opened to traffic or otherwise placed in service. Curing of concrete is critical.

Curing of epoxy coatings, adhesives, membranes, or patching compounds is also critical. Because temperature affects the cure rate, it is essential during cold weather to provide additional cure time before using the structure, acting on the epoxy, or even just expecting it to perform properly. It is advisable to consult the technical data of the manufacturer for cure times before placing the compound in service.

If one applies the basic theory and ideas inherent in concrete technology, then one will not go astray, providing that he or she does not attempt to add water to epoxy. Adding water to epoxy is a taboo and is like adding water to oil. Epoxy and water do not mix. There are, however, exceptions to every rule. Some epoxy systems are designed to be water dispersible, but these are very specialized products.

Shelf Life. Regardless of the limits imposed by some specifications, epoxies will have almost an unlimited shelf life in their separate containers. To illustrate this point, Dural has retained samples from production batches that are over five years old. Periodically, these samples are re-examined for compliance with specifications and been found to still comply fully with the requirements of the appropriate specification. Pre-dispersion (mixing) of any settled pigments and/or fillers is the only prerequisite for satisfactory performance of out-of-date or expired products.

Surface Preparation

Surfacing cleaning and/or preparation is frequently required to remove laitance or foreign matter so that concrete or masonry surfaces can be restored or prepared to receive some type of finish requiring a positive bond with the existing surface.

Sandblasting is also used extensively in producing an architectural finish for exposed concrete. Sandblasting and/or automatic shotblasting is considered the most effective and economical method of surface cleaning when using epoxy, because the quality of the finished work is predicated upon preparation of the existing surface. While sandblasting is usually performed by a specialty subcontractor, the general contractor and owner should be aware of the need for providing protection for persons, adjacent properties, glass and metal surfaces, ornamental structures, trees, power lines or other items subject to damage by the sandblasting operation.

Clean up at the conclusion of each work day is another issue that should be included. By endorsement on an existing property damage insurance policy, or the issuance of a special policy covering this specific risk, protection can be obtained from the hazards of this work.

Sandblasting

	Price per s.f.	Price per sq.m
Light penetration, including protection and cleaning	$0.55	$6.11
Heavy penetration, including protection and cleaning	1.65-1.75	18.33-19.44
On metal surfaces, light penetration, including protection and cleaning	1.20	13.33
On metal surfaces, heavy penetration, including protection and cleaning	1.90	21.11

Sealers

The following are only some of the many products available to the industry today.

Epoxy Penetrating Sealers. This system is a two-component resin, available in 15%, 30%, and 50% solids. Its use is to waterproof concrete, brick mortar, and cement block and to produce a protective zone that increases the substrate's resistance to damage from fuels, oils, alkali conditions, mild acids, de-icing chemicals, salts, and aggressive waters. When properly applied on concrete bridge decks, skid resistance is not altered and savings can be realized by reducing the amount of de-icing chemicals required per application.

This method is similar to the application of a coat of silicone, but with certain advantages. Epoxy offers a chemical as well as a mechanical adhesion. In addition, chemical resistance and film integrity are superior in epoxy. In general, epoxy provides greater resistance to intrusion of salts and other aggressive chemicals, as well as to water, gasoline, and oils.

Application to bridge decks, where salt applications are generally heavier, should be 2 coats rather than one. It might be desirable to broadcast a silica sand into the final coat to eliminate any skid potential when the surface is wet. This might be necessary where concrete is extremely dense and porosity is minimal.

Penetrating sealer can be applied by rollers, brush, or sprayer at the following cost:

Cost of Applying Penetrating Sealer

	Cost per s.f.	Cost per sq.m
Clean surface by sandblasting methods	$0.61	$ 6.72
Epoxy sealer material	0.17	1.84
Labor applying epoxy sealer	0.50	5.50
Total cost	$1.28	$14.06

Epoxy Injection Systems. This system uses a two-component epoxy resin that is injected under pressure into cracks in concrete decks, columns, piers, and walls. It restores the structural integrity of the concrete, and its principal uses are: bonding cold joints; repairing delaminations; stopping water seepage through foundation walls, floors, beams, and decks. Some injection epoxies are used underwater to repair pilings, water tanks, retaining walls, and tunnel linings.

Cartridge Method: This method is generally used where cracking is extensive. It becomes desirable to use automatic mixing and application equipment.

	Cost per l.f.	Cost per m
Material - 8 oz. tube (5.5 l.f. per tube)	$4.00	$13.33
Labor	24.20	80.66
Total cost	$28.20	$93.99

Injection Machine Method: This method is used when cracking is extensive, and it becomes desirable to use automatic mixing and application. It is strongly suggested that several sample areas first be repaired, and with the cost information developed from these tests, a unit price per gallon (per liter) of epoxy or a lineal foot (meter) price can be established. General costs of "welding" cracks in concrete are as follows:

Material cost is $45.00 per gal. ($11.89 per liter), which under average conditions will "weld" cracks 1/16" (1.5 mm) wide at the rate of 40 l.f. per gal. (347 m per liter).

	Cost per l.f.	Cost per m
Material	$1.13	$ 3.77
Injection machine rental, $500.00 per week, 100 l.f. (30 m)	5.00	16.67
Injection machine operating costs, 40 hrs. @ 4.00 per hr.	1.60	5.33
Injection machine operator, 40 hrs. @ 21.04 per hr. = $841.60 per week (or per 100 l.f.)	8.42	27.62
Labor cost, 40 hrs. @ 21.04 per hr.	8.41	27.58
Total cost	$24.56	80.97

Epoxy Mortar Systems. This system is a two-component resin with an aggregate added to the resin. The aggregate is used for a number of reasons, some of which are:

1. reduce the cost of application;
2. provide a textured surface;
3. increase abrasion resistance;
4. increase mortar compatibility with the physical properties of the concrete.

The epoxy resin, if low viscosity, can be extended with aggregate, up to 4 parts aggregate to 1 part mixed epoxy resin by volume. If the epoxy is a gel or paste for vertical or overhead applications, then up to 1 part aggregate to 1 part mixed epoxy may be added by volume.

Some products can be used as a thin overlayment, as thin as 1/8" (3 mm). Some can be placed 1/4" (6 mm) and thicker and are able to cure to 0ºF (-18ºC) on dry surfaces and 33ºF (1ºC) on moist surfaces. Some products can even cure below freezing without heaters or lamps to initiate the curing. All aggregates, whether silica sand, stone or blends of the two, must be bone dry. Sandblasting sand is often used for this purpose. Larger aggregates must be specially processed to remove latent moisture.

This system is effective as an overlay over existing concrete slabs and requires the following work procedure with the application costs as follows:

Costs of Applying 3/8" (9 mm) Mortar System

	Cost per s.f.	Cost per sq.m
Sandblast or pressure clean surface	$0.61	$ 6.72
Repair spall areas, material 2.20		
and labor 3.30 (if req'd)	5.50	61.11
Crack welding*	27.50–49.50	305.55–550.00
Apply 3/8" overlay	4.95	55.00

*Price per l.f. (per m)

The actual extent of work required can only be determined by a detailed site investigation, and because of the variance that may be encountered in the condition of an existing concrete slab, the tendering of a fixed price proposal for work of this nature is very risky. It should not be attempted by an estimator who is not experienced in this type of work.

Epoxy Coatings. Coatings are generally two-component systems and are used in many applications. Some of the more popular uses are:

1. floor
2. ramp
3. pool
4. tank
5. acid
6. food processing
7. underwater
8. chemical liner
9. abrasion resistant
10. bridge roadway

Many coatings are available in a variety of colors, including green, red, tile red, blue, white, light blue, dark blue, dark gray, and concrete gray. Some epoxy manufacturers will even formulate colors to specification.

Some products can cure on moist surfaces down to 32°F (0°C) and on dry surfaces to 20°F (-7°C).

Epoxy Coatings

Epoxy coatings require basically the same surface preparation as epoxy mortars. If one knows what preparations are required for the latter (sandblasting, steel shotblasting, etc.), then one can use the same cost per s.f. for preparation and cleaning.

Because 100% solids epoxy coatings are free from any solvents, the thickness of coating applied as a wet film will be the same as the dry film thickness. For estimating purposes, 1 gal. (3.79 liters) of mixed epoxy, when applied 1 mil (.001" or 0.025 mm) thick, will cover 1600 s.f. (144 sq.m), or at 10 mils (0.25 mm) thickness, it will cover 160 s.f. (14.4 sq.m).

To figure total gallons (liters) required to cover 1,000 s.f. (90 sq.m), simply determine the mil (mm) thickness specified or required, divide 1,600 (144) by this thickness and then divide 1,000 s.f. (144 sq.m) by that number:

Area to be covered = 1,000 s.f. (90 sq.m)
Thickness specified = 10 mils (0.25 mm)
1,600 divided by 10 = 160 s.f. per gal. (14.4 sq.m per 3.78 liters)
1,000 s.f. divided by 160 = 6.25 gals. (14.4 sq.m per 23.65 liters)

Be aware that the coverage rates are theoretical and do not take into account waste or absorption into concrete.

Cost of Applying 1,000 s.f. (90 sq.m) of 5-mil (0.125 mm) Coating

Epoxy material, $45.00 per gal. ($11.89 per L) x 3.12 gal (11.81 L)* $140.40
Labor, painting (with roller) 4 hrs. @ 21.04 .... 84.16

Total cost per 1,000 s.f. $224.56

per sq.m...24.16

*Coverage for 5 mils (0.125 mm) figured using coverage formula described above.

Epoxy Bonding Compounds. These products are two-compound systems whose principal use is to bond dissimilar materials, such as concrete, steel and wood. Many other uses are possible, but one should always consult the manufacturer prior to using the adhesive for an unusual application.

Another primary construction application for epoxy bonding compounds is adhering new concrete to existing concrete. The aerospace industry uses these to bond aluminum aircraft skins to their structural members.

Product Selection

In order to assist and simplify the product selection process, ASTM C-881 should be consulted by the specifier or user. This specification sets forth a variety of applications and the parameters for the products suited to these applications. Once this is done, the manufacturer should be consulted to determine the product that best meets the specifications for the application intended. Some important factors to consider in product selection are:

Most epoxies do not cure properly below 50°F (10°C). Although the most current ASTM C-881 specifications provide for low temperature curing systems, these are generally limited to 40°F (4.44°C) and rising. Applications at temperatures below 32°F (0°C) are not recommended, because frost and ice crystals may be present on the surface, which would prevent adhesion of the epoxy.

On the other hand, if temperatures are above 85°F (29.44°C), a slower setting product should be selected, because most systems will set too rapidly and will not permit sufficient working time. Most reputable epoxy manufacturers provide products that cover the broadest range of temperatures, from -15°F to 140°F (-26.11 to 60°C) and above.

Moisture-insensitive epoxies (or for that matter, any polymer) should be considered. Moisture-sensitive epoxies will not bond well to damp or wet surfaces. Some manufacturers provide epoxies that will cure underwater as well as on damp and dry surfaces.

Modulus and flexibility. On large-scale applications on concrete and steel, particularly when using coatings, mortars or overlays, flexible products should be used, because the coefficient of expansion for these products is significantly greater than for concrete or steel. This means that they tend to expand and contract at a greater rate than the substrate, and in the case of concrete, could shear off at the concrete interface. Flexible, low modulus products will be able to relieve any stress set up between the epoxy and underlying surface.

Structural applications where the epoxy adhesive or epoxy mortar is to provide structural support calls for high modulus, high strength, and low creep. Again, to assist in simplifying matters, ASTM C-881 should be consulted.

Viscosity. In performing coating or epoxy mortar work on horizontal surfaces, generally a low viscosity epoxy should be used. When applying epoxy mortars to overhead or vertical surfaces, a paste or gel version should be used to prevent sagging or slumping from the surface. It should be noted that low viscosity epoxies, when used to prepare a mortar, may accommodate 3 to 4 volumes of sand, while gel or paste systems may accommodate 1 volume of sand per volume of epoxy.

Equipment

The following is a general list of tools required for epoxy work. It is a guide for the most commonly used tools and should not be considered necessarily a complete list.

Basic Tool and Equipment Guide

Substrate Cleaning Equipment
Sandblaster
High pressure water blaster
Grinders
Scabblers
Broom, mop & pails
Automatic scrubbers

Measuring Tools
Tape measure
Empty 1-, 2-, & 5-gallon (3.79, 7.6 and 18.9 liter) pails

Mixing Equipment
Muller type plastic mixer (stationary drum, blades turn)
Drill and mixing tool (200-600 rpm)
Wooden mixing paddles

Application Equipment and Tools
Airless sprayer
Air sprayer
Power screed
Manual screed
Squeegees
Trowels (flat, various sizes)

Coving trowels - 1/2" (13 mm), 3/4" (19 mm), and 1" (25 mm) radius
Roller pans, rollers
Fiber bristle brushes with wood handles
Paint brushes

Miscellaneous Items
Hand tools, pliers, hammers, screwdrivers, etc.
Kneeling boards or pads
Plaster spiked shoes
Rags and paper towels
Plastic or masking paper and masking tape
Wet film gauge

Safety Equipment
Respirator for organic vapors
Goggles or face shield
Rubber or plastic gloves
Skin cream

Techniques

Injection. There are generally three accepted methods of injecting epoxies into concrete cracks, honeycombing, delaminations, cold joints, and the like. Selection of a method depends in great part on the quantity and quality of injection required.

The basic difference between quantity and quality referred to is that quantity is a primary consideration when filling voids, such as honeycombing, adequately enough to accomplish mass, while quality is important in filling fine hairline cracks, 0.002" (0.05 mm), with a 95% to 100% integrity requirement.

Using Cartridges. Some manufacturers provide an injection resin in plastic cartridges not unlike caulking tubes. They do differ from caulking tubes in that the "A" and "B" components must be mixed immediately within the tube itself prior to use.

Using Pressure Pot. A pressure pot is a paint type pressure vessel used to injection epoxies. The major drawback to this type of procedure is that the epoxies must be mixed in bulk prior to injection and that the injection must be completed prior to the epoxies gelling within the container and lines.

The pressure pot method is practical where it is known that no slowing of the injection will occur, and it is particularly adaptable to the filling of void areas, such as honeycombing, large cracks, and delaminations.

Using Injection Machine. An injection machine is suited to large-scale projects and to when the injection procedure is steady and slow. There are many different injection machines on the market. The use of any type machine with a any given epoxy product should depend on whether it is endorsed by the epoxy manufacturer.

The injection machine must be able to properly proportion the "A" and "B" components, to completely mix them, and to inject the epoxies at the epoxy manufacturer's recommended sustained operating pressure.

Any injection machine considered should be as self contained as possible to avoid having to rent air compressors, power generators, etc.

Epoxy Mortars. The selection of an epoxy for a mortar application should be with the job site conditions in mind. Some job conditions that one should consider are:

1. temperature curing range of the epoxy;
2. moisture insensitivity or sensitivity of the epoxies;
3. whether local aggregates can be used, or will it be necessary to ship aggregates as a part of the epoxy system;
4. whether the aggregate/epoxy system actually performs as an overlay in the environment specified.

When selecting an epoxy for a mortar application, one must be aware that epoxies are available that are capable of curing at a temperature as low as -15°F (-26.11°C) and as high as 140°F (60°C). Some epoxies can be extended with an aggregate using local sands, and for thicker applications, using 3/4" (19 mm) to 1" (25 mm) stone.

By knowing these things, an estimator can introduce economy into an estimate for materials. Some epoxy mortars have very poor workability and require power troweling, while others can be placed with a screed to line and grade. Some epoxy mortars can be self leveling, while others might require power trowels. Knowing this, the estimator can economize by reducing labor and equipment costs.

Estimating Quantities. The common denominator is to reduce all quantities to the cu. ft. (cu.m)—the epoxy manufacturer will provide the contractor with a "mix design", i.e., aggregate size and weights and the weights of epoxy. For example:

	lbs.	kg
Aggregate	120.00	54.00
Epoxy Resin	24.00	10.80
Yield	144.00	64.80
	lbs per c.f.	kg per cu.m

The epoxy weight is 9.2 lbs. per gal. (1.09 kg per liter), and 1 c.f. = 2.6 gal. (1 cu.m = 347.50 liters). So, coverage for a 1/4" (6.25 mm) thick overlay is 12" divided by 0.25 = 48 s.f. per c.f.

If a prime coat is required, add the quantity of epoxy required for it. Usually the coating of epoxy, as a primer, is the mortar resin itself. If a separate resin is recommended as a primer, it should be 100% solids. This means that it should not contain any solvents.

A typical crew size for a mortar overlayment job might be as follows:

1 laborer for mixing
2 laborers for screed
1 finisher for finishing
2 laborers for priming and spreading
6 worker crew = 48 manhours per day

Assuming all work such as substrate cleaning and sandblasting is completed, the crew can apply a 1/8" (3 mm) or 1/4" (6 mm) overlayment as follows:

6'0" (1.8 m) ribbons, 400 l.f. (120 m) = 2,400 s.f. (216 sq.m)
2,400 s.f. (216 sq.m) ÷ 48 manhours = 50 s.f. (4.5 sq.m) per manhour

Important Considerations in Applying Epoxy Mortars

Viscosity. This is a critical factor, normally not considered during applications. However, it partly controls the success of the project. Viscosity must be considered, especially when applications are made in the winter or during any weather that produces cold nights. The following figures indicate typical viscosity variations with temperatures.

Typical Initial Viscosities
"A" & "B" Components Mixed

°F	°C	Centipoises
90	32.0	300
77	25.0	500
68	20.0	1,000
60	15.5	3,000
50	10.0	4,000
33	1.0	7,500

One can see that the warmer the temperature, the lower the viscosity. The chief requirement for adding the filler or silica sand in the proper quantities is that the viscosity of the epoxy be such that all surfaces of the filler are evenly wetted and bonded together. When laboratory mortar formulations are designed, the epoxy is maintained at an even temperature of 72°F (22°C). This is the balancing point for filler addition. When the viscosity rises or becomes thicker, it is not possible to maintain the same filler balance, thus the mortar may exhibit characteristics that are not desirable. During cold weather applications, it is recommended that the resin and curing agent be stored in areas above 40°F (4.4°C) and be heated to a minimum temperature of 90°F (32°C) before use.

Epoxy heating jackets are an easy and fast method of warming the materials. Experience has shown that upon pouring the warm components into the mortar mixer, the components will drop immediately to a temperature of 60 to 70°F (17 to 21°C). This range of temperature at the beginning of the mixing procedure will continue to drop with the addition of the cold sand and gravel. Heating both epoxy components will result in mortar mixtures that provide proper aggregate proportions, giving the desired physical properties when cured. When temperatures are above 50 to 60°F (10 to 15.5°C), heating is generally not required. Of course, heated aggregate will also provide better workability and cure rate.

Reactivity. The reactivity of the mixed system determines such characteristics as pot life, tack free time, cure speed, working schedule, and method of application. The cure reaction starts as soon as components "A" and "B" are mixed. The colder the temperature of the resin system, the slower the reactivity. It shows a longer pot life time and longer time to fully cure.

In cold temperatures applications, slower reactivity can possibly cause application defects or failure, because the substrate can go through a thermal expansion or contraction movement prior to the mortar developing its tack-free stage. If temperatures become sufficiently cold, the reaction could stop entirely and the epoxy not cure at all. Effects could be loss of adhesion or cracks in the mortar, or in the case of very low temperatures, a gooey mess and uncured epoxy. By heating both components to a preselected temperature, desirable lower working viscosities are achieved and the product will reach a "tack-free" state more rapidly. The shorter the time required to reach this desirable stage, the less chance of mortar damage prior to initial curing. Once the tack-free stage develops, common physical properties are obtained in the epoxy mortar that are similar to those of concrete.

Sand. It may be necessary to blend different grades of sand at the job site to insure the proper aggregate blend. The sand should be bagged, clean, and dry. It should be of high quality silica or quartz, with a minimum hardness of 7.0 on the Mohs scale. Sandblasting sand is often used in this application.

Particle shape, as well as gradation, is important in order to have a workable mix at the required sand to binder ratio. All these factors must be checked carefully before specifying a particular sand to assure a mix that will give the greatest ease of application and performance. All sand should be tested by the manufacturer's laboratory prior to use. Remember, different applications require different graded blends to achieve the best results. The manufacturer's representative should assist you in selecting the best and most economical blends.

Preparation of Concrete and Other Substrates for Epoxy Floor Systems

Epoxy flooring systems require a certain amount of preparation of the substrate prior to actual application. The following are general instructions on floor and job preparation. However, each job should be considered individually, and preparation should be estimated for the specific conditions. Very few jobs are exactly alike. You might use only a few of the clean-up steps on many of your jobs.

Preparation

General. Careful attention should be paid to surface preparation of areas to be treated, to mixing and storing of materials area and to adjacent areas not being treated. Epoxy products have excellent adhesion, so it is extremely important to prevent waste and spills. Materials allowed to cure where they are not wanted will be extremely difficult and time-consuming to clean up later. It is a good practice to protect the flooring area where the mixing takes place, if the area is not going to receive a new overlay. The typical method of protecting the floor area would be to spread heavy cardboard or plastic over the floor.

Good working conditions and preparations are essential to a smooth operation and high quality job.

Work Areas. Plan your job well. Provide proper material storage and mixing area. Allow sufficient space and ease of access to the area where the floor is to be treated. Locate the storage and mixing area as close as possible to the area to be treated. On large jobs, mix at many different locations so that the hauling labor is reduced to a minimum. When mortar mixers are used, make sure the mixer is elevated high enough for the mix to be dumped into wheelbarrows. Set out your tools so that they can be cleaned as necessary in the solvent containers. Clean your mixer once every 30 to 40 minutes by adding a gallon of solvent mixture and rotating it until the drum cleans itself. The solvent may be used many times over, but remember to keep a cover on the solvent pail. The mixing area is always a "no smoking" area.

Moisture. Moisture creates no problem with most epoxy flooring systems. However, some require less labor to apply than others when the floor surfaces are dry.

Epoxy flooring systems may be applied to moist or damp surfaces down to 33°F in temperature. However, cure time might be severely retarded at these low temperatures. The practical lower temperature limit should be 40°F. Remember, do not apply epoxies on ice surfaces or frozen concrete.

Temperatures. As a general rule, air temperatures do not govern the concrete temperatures, which are normally related to soil and moisture temperatures. A room with an average air temperature of 70°F (20°C) could still have substrate concrete temperatures of 40 to 50°F (4.5 to 10°C). Cold temperatures will affect the application of some epoxy flooring systems. However, most epoxies require surface temperatures higher than 40°F (4.5°C) to apply their materials.

General Surface Preparation. Epoxy flooring systems may be applied over any sound substrate with the exception of asphalt and petroleum base surfaces. Surfaces must be cleaned and properly prepared, as outlined in this section. The ultimate success or failure of the job depends a great deal on proper bonding to the substrate.

Service life of mortars or coating systems primarily depends upon good surface preparation. The life will be extended considerably by following the suggested recommendations.

Steel Surfaces. These must be dry and clean in accordance with the following requirements:

1. Remove all debris from the surface to be treated.
2. Removal of grease, oil, and contaminants can be accomplished by wiping the surface with a recommended cleaner. The cleaner will dissolve and dilute most contaminants sufficiently to permit them to be wiped or washed off. Before wiping, scrape or wire brush all heavier areas of contaminants to remove thick deposits. Rags or brushes may be required to adequately clean the surface. Wear protective clothing during the cleaning operation.
3. Sandblast the surface to white metal finish, removing all mill scale, rust, rust scale, paints, or other foreign matter. Care must be taken to see that extreme peaks and valleys are not left on the surface, because they require additional coating build-up. These peaks, if too high, also cause thin spots in the coating application and could lead to early coating failure. Vacuum all debris and sand from area to be treated. Immediately apply the epoxy system as recommended.
4. Remove or grind all sharp edges and welds to a rounded surface.
5. Follow manufacturer's recommendations for applying the epoxy system.

New or Non-contaminated Concrete Surfaces. These must be prepared in accordance with the following requirements:

1. Remove all debris from working surface.
2. Surface must be cleaned free of all curing compounds, release oils, and dust. New concrete should be fully cured. Dry surfaces permit easier application, but most epoxy products will adhere to clean, damp surfaces.
3. Determine and record the pH factor of concrete surface.

4. Record the substrate and the ambient temperatures.
5. Etch concrete surface with a 10% solution of commercial hydrochloric acid. Apply 1 gallon of 10% solution per 40 s.f. (3.6 sq.m). The acid solution should be mechanically worked into the surface for 2 or 3 minutes or until bubbling stops. Wash surface with clean water to remove all acid. Rinse second time if any bubbling is noticed. Check pH after rinsing.
6. Sandblast or waterblast surface to remove contaminants, heavy laitance, sharp edges, or protrusions that will interfere with the uniform application.

Old Concrete Surfaces. Old surfaces are usually contaminated or previously coated or sealed. The most economical means of preparing these surfaces is with automatic steel shotblasting equipment. These devices are neither self-propelled, nor are they pushed along the surface. They have self-contained vacuum units that automatically recycle the steel shot, while removing the dust and debris. The degree of shotblasting should be such that aggregate is exposed or at least 1/16" (1.5 mm) to 1/8" (3 mm) of concrete is removed. In cases where heavy oil or grease contamination has occurred, it may be necessary to follow shotblasting with degreasing, thorough rinsing and drying prior to application of the epoxy.

Tile Preparation. Epoxy flooring systems may be applied over old ceramic tile floors. The floor should be prepared by removing all grease, oil, paint, and other contaminants. Scarify the surface of the floor with a sander or grinder. Remove old loose material in the joints. Vacuum all dust and then acid etch as described in the concrete preparation section. Conditions should now be ready for application of the epoxy floor system.

Wood Floor Preparation. Wood floors are the least desirable substrate of all surfaces, but if carefully prepared, an excellent floor system can be produced. Make sure the floor is sound and subject to a minimum of movement. Epoxy floor systems will provide an excellent surface and will hold the wood together. However, if there is a significant amount of floor movement, then cracking could develop.

It is very important to remove all greases, oils, and any other contaminants, including oil paints. A floor sander is often the best method of cleaning the surface. Vacuum the entire area so that no dust is present. The floor must be free of all dust and dirt. Wood that has been subject to weather must be well sanded before epoxy installation. Wood floors that are subject to heavy use must have a 1/2"x1/2" (13 x 13 mm) wire mesh hardware cloth or wire lath nailed to the floor after priming has been completed. The primer need not be dry to install the wire mesh or to start the application of epoxy flooring systems.

Other Coatings or Epoxy Surfaces. All existing epoxy coatings and surfaces must be removed by abrasive blasting or scarifying. Vacuum all dust and debris. If the previous system failed, then it is obviously unacceptable to have it between the substrate and your epoxy floor system. The adhesion will be no better than the surface that previously failed.

Epoxy Floor and Deck Coatings

Epoxy floor coatings are a two-component epoxy system designed to coat concrete, steel floor, and deck areas for extra durability and wear resistance. They provide an attractive, durable finish for industrial space, warehousing areas, and parking deck surfaces, where service is too severe for most paints.

The epoxy floor coating must be resistant to a variety of chemical spillage and exposures normally found in industrial plants and parking structures, such as alkalies, mild non-oxidizing acids, gasoline, and de-icing salts.

The epoxy floor coating must have an excellent resistance to abrasion and be available in smooth or non-skid formulations. Standard colors are clear amber, gray, tile red, and green. One should be able to apply the epoxy floor coating on damp surfaces (no standing water) down to 35°F (2°C), or on dry surfaces down to 20°F (6.6°C).

Outstanding Features. The following is a list of some of the advantageous features of epoxy floor and deck coatings:

1. High film-build.
2. May be applied at low ambient temperatures. Applicable on moist substrates to 35°F (2°C). Applicable on dry substrates down to 20°F (-6.6°C).
3. Rapid drying time.
4. Can be applied in neat form or with a sand filler.
5. Both chemical and mechanical adhesion to concrete.
6. Excellent resistance to chemical and petroleum spillages.
7. Tough, flexible, abrasion resistance to hard rubber wheels and moving vehicular traffic.
8. Excellent colors provide attractive finish, hiding patches and stains.

Typical Uses. Some of the typical uses of these products are:

1. Bridge decks
2. Chemical plants
3. Equipment
4. Paper mills
5. Parking garage decks
6. Pharmaceutical plants
7. Machinery areas
8. Marine uses
9. Water pollution control plants
10. Water treatment plants

Resistance Properties. Epoxy coating products for floors and decks are generally resistant to:

1. Abrasion
2. Fuel spillages
3. Industrial fumes
4. Ultraviolet rays
5. Machine tool cutting oil
6. Natural and synthetic oils
7. Alkalies
8. Gases (hydrogen sulfide, carbon, and sulphur dioxide)
9. Acid fumes
10. Fresh water
11. Salt water
12. Sour crude oil

Limitations. The epoxy floor coatings are usually not recommended where exposure will include direct contact with sulphuric, nitric, chromic, and acetic acids.

Preparation for Special Requirements

In addition to the typical preparation steps that were described earlier, the following are offered for special requirements of the floor coating system.

Concrete. Broom finished surfaces are the easiest to prepare. Lightly abrasive blast or mechanically abrade the surface to remove curing compounds and other contaminants. Remove dust from surface by vacuum or wash with water and air blow or sqeegee off puddles. Puddles indicate low spots and can create unsafe conditions during wet weather.

Aged Concrete. All older concrete must be thoroughly cleaned by an abrasion method, such as sandblasting or some other factory-approved method. All contaminants, such as old sealers, grease, oil, paints, asphalt, tars, and other coatings, must be completely removed before applying the epoxy floor coating. Wash or vacuum all loose debris and dust after abrasive cleaning of the surface (note puddles to locate low areas).

Acid Etching. All concrete should be checked after the cleaning process to determine the correct pH of the substrate. Acid etching is not required unless the concrete indicates a pH of 10 or higher.

Cleaning. Cleaning solutions that have not been approved in writing by the epoxy manufacturer should not be used.

Repair of Structural Cracks. The repair of structurally cracked and delaminated concrete is of paramount importance, because reflective cracking may occur. The crack may transmit contaminants and moisture from one level to another.

Hairline Cracks. Small fissures, often called *hairline cracks,* usually result from concrete shrinkage during curing, or from the lack of proper strength. Routine surface preparation will remove hairline cracks and expose the cracks that are truly a structural problem.

Non-moving Joint Sealing. Saw cuts and construction joints should be filled with a flexible resin (do not cover the exposed surface area of the flexible resin with the epoxy floor coating).

Spalled Areas. Repair all chuckholes and spalled areas with a low modulus, flexible epoxy mortar system.

Typical Application Tools. These include a 9" (225 mm) to 18" (450 mm) wide roller with extension handle, solvent-resistant short cover, squeegee with flat edges, and airless sprayer with extension spray gun.

Typical Rates of Application for Neat Material. The *1st coat* is applied evenly over the surface at an average rate of 1 gal. per 160 to 250 s.f. (1 liter per 3.8 to 5.9 sq.m) with squeegees, rollers, or airless sprayer.

The *2nd coat* is applied within 24 hours of applying the first coat and is applied at an average rate of 1 gal. per 160 to 250 s.f. (1 liter per 3.8 to 5.9 sq.m) with sqeegees, rollers or airless sprayer.

Typical Rates of Application for a Sand Seeded System for Skid Resistance. The *1st coat* is applied at the rate of 100 to 125 s.f. per gal. (2.38 to 2.97 sq.m per liter) with sqeegees, rollers, and or airless sprayer. Immediately seed dry, clean silica sand, sandblasting grade, evenly over the tacky surface. The sand should be seeded until a thin layer of dry sand is on the surface. Use approximately 30 lbs. (13.5 kg) of sand per 100 s.f. (9 sq.m). Allow to become tack-free.

After tack-free stage has developed, remove excess sand and vacuum free of all dust particles. Any areas where the sand is uneven or too high, grind the surface level to adjacent surfaces. Re-vacuum the area.

Immediately apply the *2nd coat* with squeegees, rollers, and airless sprayer. The final anti-skid surface will be determined by the amount of epoxy applied to the surface. Typically, the 2nd coat will cover 80 to 100 s.f. per gal. (1.9 to 2.38 sq.m per liter). For actual quantities of epoxy to be used, a sample should be made, because sand gradations and surface areas vary widely.

CHAPTER **19**

HANDICAPPED FACILITIES

Though clouded by the uncertainty of job conditions that could substantially alter them, the following unit prices have been used as base cost figures in a large number of cases and will serve as a guide.

Site Facilities

Concrete ramps on grade, 1:12 pitch, 4" (100 mm) thick with mesh, broom finish, 4" (100 mm) base	$3.00 s.f.	$33.33 sq.m
Steel pipe handrail, 1-1/2" (38 mm) galv., 2 rails 36" (900 mm) high	31.00 l.f.	103.33 m
Non slip abrasives for walks and ramps	0.40 s.f.	4.44 sq.m
Parking signs for handicapped areas	31.00 ea.	
Pavement painted indicator	5.00 ea.	
Embedded abrasive warning strips	7.50 l.f.	25 m
Exterior hazardous area warning grooves 1" (25 mm) deep	1.90 l.f.	6.33 m
Depressed curb ramps	5.00 s.f.	55.55 sq.m

Interior Stairs and Corridors

Hazardous area warning abrasive strips	$7.50 l.f.	$25.00 m
6" (150 mm) curbs with pipe rail at dangerous areas	32.00 l.f.	106.67 m
Integrated stair nosings, non slip	30.00 ea.	
Corridor wall rails, aluminum pipe	18.00 l.f.	60 m
Balcony railings, aluminum 1-1/2" post with pickets	37.00 l.f.	123.33 m
Stair hand rails, steel, with newel and brackets	31.00 l.f.	103.33 m
Floor grating, aluminum, 1/4" (6 mm) maximum spacing, safety surface	10.20 s.f.	113.33 sq.m
Vinyl tile for sound reflection	0.80-1.10 s.f.	8.88-12.22 sq.m
Ramps, wood not to exceed 12'-0" length and 4'-0" width	6.50 s.f.	72.22 sq.m

For curved railings, add 33%

Information Only

1. Corridors, minimum 36" (900 mm) width single with turn around area every 200 l.f. (60 m)
2. 360-degree wheelchair turnaround is 60" (1500 mm) dia. or 19.62 s.f. (1.76 sq.m).
3. Right angle grid pattern for corridors and public spaces are perceived more easily than curves or serpentines.

4. Diagonal reach from wheelchairs is 48" (1200 mm) above floor level for telephones, etc., with forward reach of 18.5" (463 mm).
5. Overhead obstructions not less than 7'-0" (2.1 m) above finish floor.
6. Parking spaces as specified in 36 CFR 1190.31.

Total Parking in Lot	Required Minimum No. of Accessible Spaces
1 to 25	1
26 to 50	2
51 to 75	3
76 to 100	4
101 to 150	5
151 to 200	6
201 to 300	7
301 to 400	8
401 to 500	9
501 to 1,000	2% of total
over 1,000	20 plus 1 for each add'l 100

7. Passenger loading zones shall provide an access aisle at least 60" (1500 mm) wide by 20' (6 m) long, adjacent, parallel, and level with the vehicle standing space. Vehicle standing spaces and access aisles should have surface slopes not exceeding 1:48 (1/4" per ft., 20.5 mm per m) in all directions.
8. Ramps and curb Ramps, Maximum running slope shall not exceed 1:12 or 8%; maximum rise for any run shall not exceed 2'-6" (0.75 m).
9. Grab bars, like handrails, shall have an outside bar diameter of 1-1/4" (31 mm) to 1-1/2" (38 mm), and the space between the wall and the centerline of the pipe shall be 1-1/2" (38 mm).

Case Work

Counter height should be adjustable, between 28" (625 mm) and 36" (900 mm), with minimum depth of 19" (475 mm) per l.f. (0.3 m) $42.00
General work surfaces 25" to 30" (625 to 750 mm) above finished floor with maximum depth of 25" (625 mm) per l.f. (0.3 m) 50.00-65.00
Built-in case work, with toe clearance of 7" (175 mm) recess and 10" (250 mm) above finished floor per l.f. (0.3 m) 42.00
Closets to be partially accessible for wheelchairs should have adjustable clothes rod (per set) 25.00
All case work to have recessed hardware.

Doors and Openings

Electric operated automatic doors with delay mechanism, complete including hardware (each) $5,500.00
Access door adjacent to revolving door 1,200.00

Vision light in solid doors (each).. 50.00
Lever type hardware (per set)... 30.00
Knurled back handles for dangerous areas (per set)............................. 30.00
Corner protection, stainless steel, 16 ga. per l.f. (0.3 m) 17.00
Corner protection, vinyl, high impact, with stainless
  frame per l.f. (0.3 m).. 21.40

Identifying Devices

International handicapped sign (each)... $31.00
Plaques for rooms with raised letters (per letter)................................. 2.75
  interior sign character 2-1/2" to 4" (63 to 100 mm) 5.15
Exterior sign character 4" to 8" (100 to 400 mm) 10.50

Elevators

Hall lanterns with visual and audible signals indicating
  elevator direction and floor location (each)................................... $250.00
Elevator door jambs equipped with floor designations
  on each side having 2" (50 mm) characters (per pair)................... 15.00
Doors equipped with reopening devices not requiring
  contact to be activated (per car) 500.00
Interior car position indicator with visual and audible
  signals for each floor level (per car).. 250.00
Emergency telephone located 48" (1200 mm) above floor................... N/C
Call buttons with raised characters (each) 40.00

Kitchen

Front loading dishwashers, with rack space accessible
  from the front.. N/C
Front loading washers and dryers... N/C
Cooking surfaces with front controls .... N/C
Wall ovens mounted with top of door 31" (775 mm) above floor......... N/C
Storage and utility areas to have peg boards, shelving, etc.
  at convenient level .... N/C
Refrigerator/freezer with 50% of freezer space below
  54" (1350 mm) and 100% of refrigerator controls below 54 inches 50.00-60.00

Mechanical and Toilet Facilities

Water closets, mounting height 17" to 19" (425 to 475 mm) above floor with
18" (450 mm) clearance each side (additional unit cost 75.00 each)

For complete installation, hook-up to existing
  waste & supply $925.00
Stainless steel drinking fountain, dual level for

handicapped installed complete, waterspouts to be
no higher than 36" (900 mm) from floor to spout outlet 865.00
Drinking fountain, wheelchair handicapped unit installed
complete, waterspout no higher than 36" (900 mm) from floor to
spout outlet 950.00
Shower, fiberglass with cast base 32" x 32" (800 x 800 mm), complete
hook-up to existing waste and supply 610.00
Recessed drinking fountains (add for alcove)... 250.00
Water cooler, non-recessed stainless steel cabinet,
wheelchair type complete with hook-up to existing
waste and supply.. 780.00
Grab bars, 24" (600 mm) stainless steel... 28.00
Handicap shower head controls, shower spray with a
minimum 60" (1500 mm) of hose (each).. 75.00
Shower transfer seat (each).. 70.00
Toilet partitions, handicap units, porcelain enamel (ea.)..........................1,775.00
Toilet partitions, handicap units, stainless steel (each) 1,715.00
Toilet screens, porcelain enamel (each)... 120.00
Toilet screens, stainless steel (each)... 110.00
Type of anchoring system for toilet partitions will influence cost. Unit prices
quoted above are for ceiling hung units.
Tilting mirror, bracket only (each).. 70.00
Medicine cabinet, useable bottom shelf no higher than
44" (1100 mm)..N/C

Electrical

Electric automatic door opening control located near
door jamb (per station).. $400.00
Audible signals at visual electric signals (each) 350.00
Electric door holder 125.00
Public Address System (per speaker)... 225.00
Sound System, AM-FM, 12 outlets... 8,000.00
Fire detection systems, including smoke and heat detectors,
for complete installation of a 10-detector system...................................... 2,500.00
Master clock system, including clocks and bells, complete
for a 20-unit system 14,250.00

Master TV antenna including VHF-UHF antenna and distribution
to a 10-outlet system... 4,800.00
Locating receptacles and switches at lower height
for handicapped ....N/C

Elevator controls, all car buttons to be a minimum 3/4" (19 mm) in the least
dimension, and all buttons shall be between 48" (1200 mm) and 54" (1350
mm) above the floor, depending on the type of door opening.

The following standard design details are shown as examples of minimum guidelines and requirements for accessible design. However, there is no guarantee that these are the latest requirements, and the contractor is strongly advised to check and make sure that these are the most current.

Handrail

Grab Bar

Lavatory

Lavatory

Water Closet

Water Closet

Curb Ramp

Curb Ramp

Crossing

Crossing

Crossing

Unloading Zone

Accessible Parking

CHAPTER 20

ENERGY MANAGEMENT

Energy management systems can no longer be considered a luxury. The impact of energy costs has forced building owners to seek relief through management systems that control energy consumption, optimize comfort, interface with human performance and provide automatic data analysis.

This new market, which has only been lightly penetrated so far, has attracted a great variety of manufacturers, consultants, vendors and suppliers, each attempting to capture a share of this lucrative business, and there has resulted many cases where owners have experienced excessive costs due to over design, or where systems did not provide results because of a lack of knowledge or expertise on the part of the installer.

Although energy management cannot be considered an organized industry as yet, competition and advanced technology are stabilizing factors. It is now possible to make an analysis of the needs of a conservation package for a building and procure responsible bids from several specialists with a feeling of confidence that the system will be cost effective and will produce the desired results.

The first major thrust into energy management was made by the national control and computer manufacturers. Through their established marketing and advertising capabilities, they were able to rapidly develop the large client market of hospitals, commercial and industrial buildings, educational institutions and major office buildings. With the advance of microprocessor technology, which can sell for less than $1,000.00, it is possible to install a system that does not require a $50,000 to $250,000 computer, and the advantages of energy management are available for use in smaller buildings.

In spite of the wide variety of electronic energy management systems available today, there also exists for relatively simple applications the ability to design non-electronic control panels that can provide automatic setback of thermostats and restrictions on electrical or fuel consumption, both of which translate into operating economies of major significance. Other than the economy of initial installation costs, this simple system would be found attractive for use in the more remote areas, where service could be supplied by the ordinary electrical or air conditioning service company. It is estimated that such panels could be installed for approximately $3,000 to $4,000 for an open space store area of about 25,000 sq. ft. (2,250 sq.m) with limited glass area and insulated roof area.

The costs involved in the more complex computer controlled energy management systems cannot be predetermined, because each system is specifically designed for the needs of the owner. It is not uncommon to include additional functions in the system, such as fire and smoke alarms, security controls, or other specialized features.

One of the hidden or ignored costs in building energy management systems is that they share a fundamental need with other complex contemporary technology, the continuous availability of maintenance service. Low cost microprocessors are reliable because of their relatively simple design, but these systems usually provide for service contracts that range in annual cost from 6% to 10% of the system purchase price. In the more complex computer controlled systems, service contracts are offered at approximately 10% of the purchase price with an annual escalation of 10% to 12%.

Consideration must be given to the costs associated with the training of personnel to operate the system, because the maintenance staff will have to take an active part in the installation and commissioning procedure and must actively participate with the operators in the initial training by the manufacturer of the system. While these costs are related to the operational costs that the owner must provide for, they are many times not given consideration by the owner as part of the installation cost of the energy management system. The estimator who is aware of these oftentimes overlooked expenses and their relation to the efficient operation of the system, including minimizing the need for maintenance service by the manufacturer, will be in a position to extend an additional service to the owner by offering certain guidelines for the type and duration of training for the operating personnel. In this way the owner will become conscious of the need for operational training and alerted to the associated costs.

In the absence of a highly developed and organized energy management industry, there is no single source to obtain historical data relating to installation costs, performance reports, reliability studies or cost effectiveness of the various systems available. Much of the information available to the public is obtainable from the large control and computer manufacturers and is essentially sales oriented to their particular equipment and the accomplishments they have achieved. This situation places a burden on the contractor involved in an energy management system for a new building, or in modifying the energy systems in a existing building, because he will be forced to evaluate technical proposals with extreme care before reaching a decision that the offer provides the maximum efficiency for the most reasonable price.

Even though there were some early financial disasters resulting from insufficient knowledge of the capabilities and limitations of energy management systems and from the lack of expertise on the part of some companies, the industry is providing a significant contribution.

It is not unusual to develop systems serving the needs of both small and large clients that will reduce energy consumption in a building by 25% and yield a payback of the initial installation cost in about two years. With such attractive figures as these, it must be contemplated that this industry will continue to mushroom and progressive contractors will recognize the potential of this market by initiating an in-house program to educate their personnel in the arts of this new industry.

From the position of establishing some basic method of determining a cost criteria to serve as an estimating guide, surveys indicate that it is inappropriate to attempt to develop such information at this time due to the absence of sufficient historical data cost bank. It has been found installations costs can be as low as $0.15 per s.f. ($1.67 per sq.m) of floor area for industrial plants containing up to 2 million square feet, while in a 36,000 s.f. 3,240 sq.m) store using a non-electronic system, the cost was found to be $0.07 per s.f. ($0.77 per sq.m). In the instance of the institutional and office complex, installations that are frequently loaded with a series of other functions, in addition to energy management, it is virtually impossible to develop meaningful conceptual costs because of the individuality of the design. Estimating techniques and technical information relating to energy management will eventually be provided to the estimator as it becomes available.

Energy management has rapidly become a major consideration of every business and industry as energy costs continue to rise. To lower energy usage, users are reducing the operation of lighting and HVAC equipment. However, additional help is often needed in the form of *load control*. Load control is defined as automatically controlling electrical equipment loads to reduce the electrical consumption and usage rate.

Most electric utilities base the monthly charges to commercial users on two separate items: a) consumption, the total kilowatt-hours of electricity used during any billing period; and b) demand charges, based on the rate at which electricity is used. Most utility companies measure the electricity being used during a short period of time called a *demand interval*. The monthly demand charge is then based on the maximum peak usage that occurred within the demand interval of each billing period.

Load control systems design requires the input of several important factors before the proper implementation can occur.

Every building has its own set of characteristics. Even for similar buildings, such as chain stores, each has unique situations that change the design and application of a load control system. Variables affecting the control system design include climate, building design and construction, type and capacity of HVAC and electrical equipment, business hours and traffic flow in and out of the building.

The building's electrical usage data (load profile) must be analyzed and all equipment items examined for their operating cycles in order to determine which items are in operation at peak demand times, lengths of continuous operation, etc. This information needs to be obtained for weekday, weekend and seasonal changes.

Local health codes are an important factor should the load control system be installed in food service facilities. It is possible for the load control system to reduce the operation of certain equipment to the point where a violation of the minimum health code requirements may occur. For example, if the load control system turns off the hot water heater too often or for too long a period of time, the temperature of the dishwasher water may fall below the minimum allowable by health codes.

Methods Of Load Control

Duty Cycling. Duty cycling accomplishes load control by turning on or off all loads on a programmed basis. The major function of this method is to reduce the kilowatt-hour consumption. Duty cycling equipment based on temperature (temperature selective duty cycling) increases the equipment's operational efficiency and longevity and lowers energy consumption. An example would be the shut-down of air conditioning equipment once the A/C unit has lowered the temperature to the prescribed level. Needless energy is wasted in allowing the unit to run nearly unloaded.

Each load (equipment item) that is connected to the duty cycler is programmed to turn off for a specific portion of time in a repeating time interval cycle. The amount of time a load is off is directly dependent upon the importance of that item to the overall operation of the facility.

Duty cycling can be accomplished by using two separate cycles Long-Off-Time-Cycle (LOTC) and Short-Off-Time-Cycle (SOTC). When in the LOTC mode, the equipment load is turned off for longer periods of time in the repeating cycle. The duration of the LOTC, SOTC and the repeating time interval cycle is defined by the system user. Temperature set points are also used in conjunction with the LOTC and SOTC to give virtually any combination of temperature regulated on/off mode for the equipment load.

In applications where product temperatures are critical (food service) temperature override devices can be utilized to prevent the degree limit being reached.

The duty cycle function can be overridden by a demand limit control program should the actual demand level exceed the set points of the duty cycle load control center.

Demand Limiting. Demand limiting is a term that describes the process of turning off (shedding) loads temporarily to keep the kilowatt (KW) usage demand below a predetermined level. To accomplish this, the loads connected to the demand limit control are turned off, or shed, when the KW demand level reaches the set level, which is the lowest practical KW usage rate allowable to meet the facilities operating requirements.

The demand limit control is programmed to shed loads in a predetermined order when the demand set level is exceeded. This control will continue to shed loads until the KW demand level is less than the set level minus a *deadband* range. The deadband is intended to prevent unnecessary cycling of loads controlled by the demand limit control when the KW demand level is in an area around the set level. If the demand level is in the deadband range, additional loads are not shed or restored by the demand limit control.

Demand limit controls can monitor KW demand in several ways. One is the instantaneous rate method which compares the instantaneous rate of power usage to the predetermined set level. The other way is the average rate method which continuously samples and averages the KW demand level over a short period of time. The averaged usage range is then compared with the predetermined set level.

Time Of Day Programming. Another method of load control is time of day programming or load scheduling, which is the process of automatically turning on and off specific equipment loads based on the time of day.

Some loads are often left on because of failure to manually disengage the switch. These types of loads are the ones controlled by the load scheduling device.

Most all energy management load control systems incorporate all three methods of load control (duty cycling, demand limiting, and time of day programming) into one integrated package that gives the user a complete system for controlling all equipment loads in his facility.

CHAPTER 21

HAZARDOUS WASTE

This chapter mainly addresses issues of asbestos abatement and control in buildings, but we have also attempted to cover other areas of concern, such as disposing of contaminated soil and removing underground storage tanks. All of these items can directly affect commercial building projects.

ASBESTOS

For many years asbestos materials were used for acoustical insulation, fireproofing spray on structural steel framing members, fireproofing spray on ducts, pipe and duct insulation, and boiler and breeching padding.

Early in the 1970s, scientists determined conclusively the extreme health hazard posed by asbestos fibers. The government, at that point, immediately imposed strict regulations on the manufacture and use of asbestos products. The Occupational Safety and Health Administration (OSHA) has sets of regulations that control such things as permissible asbestos exposure limits and methods of complying with them. Whenever work is performed involving the installation or removal of asbestos materials, it is required that instruments be used to measure and monitor the area's airborne fiber content. Special protective clothing shall be worn by all personnel exposed to the contaminated areas.

In 1978 asbestos was banned from all spray-on materials including those used for decorative purposes. The Environmental Protection Agency (EPA) contracted with Battelle Memorial Institute to conduct a test program to determine whether encapsulation or sealing of asbestos would be effective. As a result, several laws were proposed, and in June, 1980, the Asbestos School Hazard Detection and Control Act was passed. On October 22, 1986, Congress passed the Asbestos Hazard Emergency Response Act (AHERA) for removal of asbestos in private and public schools. Asbestos abatement projects now in progress are being funded at the federal, state, local, and private levels.

Asbestos is considered to be hazardous in a friable state, which means that the material can be easily crushed or turned into powder by using hand pressure. The greatest hazard posed to humans by friable asbestos is airborne fibers. At present, OSHA has set levels of the allowable asbestos fibers in a certain quantity of air. The current Permissible Exposure Limit (PEL) is two fibers, 5 micrometers in length, per cubic centimeters of air. A person inhales about one cubic meter of air per hour which is 1 million cubic centimeters. This amounts to an intake of 2 million fibers per hour and that quantity does not include the fibers that are too small for detection by the current measuring equipment.

OSHA has proposed new standards on exposure to asbestos in the work place. Among other things, the new standard would set the *permissible exposure limit* at either 0.5 fiber or 0.2 fibers per cubic centimeter of air. However, at this writing, the new standard has not yet been accepted as the prescribed standard. Certain states, however, have taken it upon themselves to issue stricter requirements for airborne asbestos fibers. For example, Virginia has recently passed a new guideline that reduces the acceptable asbestos exposure limits from 2 fibers per cubic centimeter of air to 0.5 fibers.

Asbestos is facing ultimate market extinction under a sweeping ban being contemplated by the EPA, and alternative product development is already underway. In late 1983, EPA advised asbestos producers, product manufacturers, and environmental groups that it planned to issue a ruling in July 1984 banning six categories of asbestos containing products and to propose another rule in October 1984 setting production ceilings for remaining products. The limits would decline annually until all products containing asbestos, a known carcinogen, are eventually phased out. The products covered by the proposed ban were saturated asbestos roofing felt, unsaturated roofing felt, asbestos flooring felt, asbestos sheet flooring, vinyl asbestos floor tile, and cement asbestos pipe. A major factor in the schedule of planned phase-out of the remaining asbestos containing products were to be the availability of a suitable substitute material. These proposed rulings were a continuation of the rule making that was begun in 1979 when the EPA announced its intent to ban such products under the Toxic Substance Control Act. Presently, thirty-two states have passed regulations related to asbestos removal.

The estimator must be cautioned that abatement rules, regulations, and restrictions are continually being altered and revised. The estimator must ascertain the latest rules and regulations as of the time of bid for each new project.

Abatement Methods. Once the problem of asbestos content had been discovered and analyzed, then the abatement program was initiated and continues to be expanded. There are presently three methods of controlling asbestos.

The first is *removal*. This is by far the most expensive method, because it takes more time for removal and controlled disposal for final deposit in limited designated disposal sites around the country. Also consider that the material was originally installed for a purpose, whether it be thermal or sound, and therefore, usually needs to be replaced with a new non-asbestos material that performs the function of the removed asbestos.

The second method is the *enclosure system* which is simply installing another material, such as drywall, over the existing asbestos materials.

The third method is known as *encapsulation* which requires the spraying of a sealant material onto the asbestos either to form a bridge over the asbestos insulation, or to penetrate into the asbestos to bind the mass together. The effectiveness of encapsulation depends largely on the condition of the existing asbestos. For example, if the asbestos is very loose and becoming delaminated from the surface, encapsulation will not work, because the material will simply fall off the structure once it is agitated by the application process of encapsulation sealants. These sealants are applied with an airless spray. The EPA has determined that brushes and rollers should not be used, because they create excessive disturbance on the surface of the asbestos materials. Using the airless spray also disturbs the surface of the asbestos, but it is the lesser of evils.

To be effective, the encapsulation sealant should virtually eliminate any asbestos fallout into the atmosphere (airborne materials), provide some resistance to impact and abrasion and be reasonably flexible to handle any normal movement of the building. It needs to have a flamespread resistance for building code compliance, and the material, in the dry state, has to be free of odor.

As stated, the encapsulation sealants basically fall into two categories, *bridging* and *penetrating,* but one important fact to note is that the AHERA Act of 1986 mandates the *removal* of all asbestos material for all school districts, public and private.

A *bridging sealant* is a coating that forms a barrier between the asbestos materials and its surrounding atmosphere. These sealants are applied at the rate of approximately 3 gals. per 100 s.f. for a dry film thickness of about 25 mils. There is very little surface penetration when using a bridging sealant. The bridging sealant is an application of two coats, with half of the material applied in the first coat, and the remaining half on the second coat. Normally, the coats will be color coded for coat identification. Proper coverage is very important; if cracks develop in the finished surface, there is an escape point for the asbestos fibers. The encapsulation is rendered ineffective and the surface area must be retreated.

The *penetrating sealant* is a very watery solution that penetrates into and hardens the asbestos materials. The penetrating sealant method is a little more difficult to control in the initial installation, because the purpose of the penetrating sealant is to absorb into the materials, cure, and bind them together. If insufficient material is used on the first coat, and the coat is allowed to dry, a film is formed that interferes with the penetration of any subsequent coats. The penetrating sealant is normally applied in a three coat application with each coat being applied prior to the drying of the previous coat. It is more difficult to estimate the quantity of materials to be used in the penetrating method because of the porosity and thickness of the asbestos material. Normal coverage can be obtained with 2 to 3 gals. per 100 s.f., but in extreme conditions, 10 to 15 gals. per 100 s.f. may be required. This greatly increases the material cost and the amount of labor time required to spray on the additional material. The easiest method to determine the actual quantity of penetrating sealant to be used is to spray a test area of the asbestos, allowing it to cure, and then coring the portion of asbestos and examine it for proper penetration and bonding characteristics.

In any asbestos abatement program, whether removal, enclosure, or encapsulation, the most expensive and time consuming job is the preparation of the area where the asbestos is to be removed or encapsulated. The walls and floors have to be completely covered with polyethylene. The entrance to the area has to be converted into an air lock, preferably 2 air locks at least 6' apart. Light fixtures have to be removed, other types of equipment and items have to be covered or removed, the return and supply air grills and diffusers have to be removed and the openings sealed in the duct. The work area itself should maintain a negative air pressure, which helps prevent asbestos fibers from escaping into other areas.

Special change rooms are required where the workers can change into their work clothes and exit through an air lock into the work area. Upon return from the work area, the workers will go into a room, strip, dispose of their work suits, enter the shower and wash, exit into the area containing their street clothes, and exit the area. All workers are prohibited from smoking, eating, or drinking in the contaminated work area, or the contaminated portions of the change rooms. Thus the average worker will normally exit the work area, dispose of the work suit, shower, and dress 3 to 4 times each work day.

One of the manufacturers of asbestos encapsulating sealant materials is the Foster Products Division of the H.B. Fuller Co. in Springhouse, Pa. Its bridging encapsulant material is a latex elastomer, white in color, and has the consistency of heavy paint with a weight of 11.4 lbs. per gal. The material is applied by the airless spray method, and has a mild odor while wet but odorless when dry. The recommended coverages for this particular material is 3 gals. per 100 s.f., or if applying in 2 coats, 1-1/2 gals. per 100 s.f. each coat, and this will depend on the texture and porosity of the asbestos.

The *Protektor* sealant material is an acrylic copolymer, comes in colors of yellow, blue, and untinted, and has a consistency of a thin fluid for proper penetration into the asbestos. Its weight per gallon is approximately 8.4 lbs. It is applied with the airless spray method the same as the bridging encapsulant. Its odor is mild while wet, odorless when dry. The recommended coverage of this material is 2 to 10 gals. per 100 s.f. depending on the absorbency and thickness of the asbestos and is best applied in 2 to 3 coats.

During the process of asbestos removal or encapsulation, there are two main pieces of equipment that the contractor will need to have at his disposal during the entire process. The first is an exhaust filtration unit, available from a number of manufacturers, such as the Torit Division of the Donaldson

Company, which has local representatives throughout the country. This unit is an exhaust fan made in varying sizes depending on the amount of air, cubic feet per minute (cfm), that needs to be moved. The air that flows through this exhaust fan is filtered by means of a high efficiency particle air (HEPA) filter and deposits the asbestos fibers in a special drum below the exhaust fan. The typical drum size is a 55-gallon unit, which contains a disposable plastic bag with a special marking, warning the handler of the contaminated hazardous waste material.

The size of the exhaust fan required for the project is determined by the number of air changes per hour that is specified, normally four, and the size of the area that is to be exhausted. An example of this would be if you had a room 50' long x 50' wide x 15' high; the cubic content of this room would be 37,500 c.f. Because the exhaust unit is rated in cubic feet of air moved per minute (cfm), you would divide 37,500 c.f. by 60 minutes, then multiply by 4 air changes, and the results would be 2,500 cfm capacity for the exhaust fan.

In quotations from the Torit Division, a 1,000-cfm exhaust filtration unit, complete with HEPA filter, costs approximately $5,000.00. A 2,000-cfm exhaust filtration unit, complete with HEPA filter, costs approximately $7,000.00. Units can be equipped with automatic alarms and other sensitive detection devices that can double the above prices.

The other item of equipment that will be required is the HEPA vacuum cleaner. One such unit is manufactured by the American Cleaning Equipment Corporation, in Addison, Ill., which is handled by numerous sales representatives throughout the country. This particular vacuum cleaner contains a special five-stage filter unit similar to the exhaust filtration unit. The filter unit alone will cost in excess of $500. The drum that collects the asbestos fibers as they are vacuumed from a surface comes in 15, 30, and 55 gallon sizes. A specially marked plastic bag that the asbestos fibers are deposited into fits in the drum. When the bag is full, it is removed, sealed, and placed ready for transport to an approved disposal site.

To alert the contractor to the specific detailed procedures that all site personnel will be subjected to when involved with asbestos work the following standard "worker protection procedure" program has been included.

It is obvious that extremely low productivity will result from the work restrictions placed on employees. It is estimated that labor production might be reduced by 15% to 60% of normal production rates.

Worker Protection Procedure. *Arrival:* Each worker or authorized visitor arriving at the job site shall remove street clothes in the Clean Change Room and put on a respirator with new filters and protective clothing and shoes before entering the Equipment Room or the Work Area.

Worker Decontamination: Each worker or authorized visitor shall, each time he leaves the Work Area, remove gross contamination from clothing, proceed to the Equipment Room, and remove all clothing except respirators.

Still wearing the respirator, proceed naked to the Shower Room, clean the outside of the respirator with soap and water, remove the respirator, shampoo and wash themselves, remove the respirator filter, wet them and dispose the filters in the container provided for that purpose, and wash and rinse the inside of the respirator.

5 STAGE FILTRATION SYSTEM

3. Final high efficiency filter certified to exceed OSHA requirements

2. impact filter

1. Coarse filter — disposable paper bag

4. Primary cloth filter

5. Paper filter protector

Components of a Typical Asbestos Vacuum

Following showering and drying off, each worker and authorized visitor shall proceed directly to the Clean Change Room and dress in street clothes at the end of each day's work, or in clean coveralls for eating, smoking or drinking. Contaminated work footwear should be stored in the Equipment Room when not in use. Store contaminated work suits in the Equipment Room for re-use or place in container for disposal as asbestos contaminated material.

Upon completion of the asbestos abatement work, dispose of footwear as contaminated waste or clean thoroughly inside and out with soap and water before removing from the Equipment Room. Workers removing waste containers from the Equipment Decontamination Enclosure shall enter the Shower Room from outside wearing a respirator and dressed in clean coveralls.

Workers and visitors shall not eat, drink, smoke, or chew gum or tobacco at the work site except in the established Clean Change Room. Workers shall be fully protected with respirators and protective clothing during preparation of system enclosure prior to commencing the actual asbestos abatement, throughout the entire abatement procedure, and until final clean-up is completed.

Summary Guide to Asbestos Abatement Equipment

Respirators: OSHA asbestos standards for the construction industry list the following types of respirator equipment:

1. half-mask air purifying respirator equipped with high-efficiency filters;
2. full face piece air-purifying respirator equipped with high-efficiency filters;
3. powered air-purifying respirator equipped with high-efficiency filters;
4. supplied air respirators operated in a continuous flow mode;
5. full face piece supplied air respirator operated in a pressure demand mode;
6. full face piece supplied air respirator operated in pressured demand mode equipped with an auxiliary positive pressure self-contained breathing apparatus.

For the least costly types of respirator equipment, prices can range from $25.00 for a) to about $600.00 per unit for c). Sizable systems, consisting of an air compressor with purification and auxiliary back-up tanks, can cost in excess of $20,000.00 per unit.

Disposable Clothing: Disposable coveralls with head covers, disposable gloves, disposable boots, and disposable boot covers. If the boot covers are used, the workers shoes are stored at the site and cleaned at the end of the job, before removal from the project.

High Efficiency Particle Air (HEPA) Vacuum: Regulations require the vacuum to be equipped with a HEPA filter. The filter is to be tested to a capability of retaining a minimum 99.97% of all monodispersed particles of 0.30 micrometers or greater in size.

HEPA Filter Units or Negative Air Units: These are portable air moving systems that draw a larger volume or air through a HEPA filter than a vacuum can. Units moving approximately 1,500 to 2,000 cfm cost about $2,000 to $3,000.00 per unit.

Pressure Monitoring Equipment: HEPA cabinets or negative air units are used to provide a controlled flow of air and to prevent leakage from a work area to an occupied area.

Wetting Agent: These are chemicals added to water, with concentrations stated by individual manufacturers.

Encapsulant: Encapsulation is the application of a liquid sealant onto sprayed or troweled asbestos. Sealants are usually water based material.

Air Monitoring: There are four basic analytical procedures: a) phase contrast microscopy (PCM) is a method for counting fibrous particles of defined size, which are deposited on a filter by drawing air through the filter b) scanning electron microscopy (SEM) is an alternative to the PCM; c) fibrous aerosol monitor (FAM) is a laser based method, developed and calibrated to give the same result as the PCM method. d) transmission electron microscopy (TEM) analyzes and counts every airborne asbestos fiber.

Decontamination Units: Asbestos removal requires the use of decontamination units, both for workers and for waste. These units are constructed on the site or prefabricated and moved to the site. All units are required to be made airtight with tape and plastic.

Glove Bags: These are plastic bags with at least two long-sleeve gloves fastened to them. The glove bag can be fastened onto an asbestos insulated pipe, and the asbestos insulation can be removed from the pipe within the sealed back. There are two types: a) polyethylene (PE) with either *Tyvek*, rubber or other material for the sleeves and gloves, Normally used on horizontal pipes; b) polyvinyl chloride (PVC) type of glove bag normally comes equipped with a zipper. The bags are designed for use on horizontal and vertical pipes, as well as on pipe tees and valves.

Air Monitoring. The contractor shall be responsible for monitoring and testing to determine the fiber count content of the air and on surfaces in and adjoining the work area, during the abatement process and the clean-up phase. All air and surface sampling and testing is normally the expense of the contractor, not the owner. Air and surface samples shall be taken in accordance with OSHA and EPA requirements during each work day. Sampling shall begin by taking samples to establish a "background level" before the start of any construction, including the temporary facilities. Sampling may stop only when final acceptance of the area has been made in writing by the owner.

The laboratory analyzing the samples must be accredited for asbestos analysis by the American Industrial Hygiene Association. Certified copies of the results must be delivered to the owner normally within 24 hours of taking the samples. Continuous air monitoring equipment is used by the contractor to supplement the testing program. However, this air monitoring is not allowed to substitute for the air and surface sampling that is required.

Area Protection. Most contractors use polyethylene to completely cover the walls and floors, assuming the asbestos is on the ceiling, of all areas for decontamination. Normally a 6-mil poly is used for floors, the air locks, and sealing doors and windows. A 4-mil poly is used against wall surfaces. The poly should be as wide as possible to minimize joints, thereby minimizing leakage of airborne asbestos fibers to the outside area.

Worker Decontamination Facilities. The contractor shall construct a worker decontamination facility consisting of 4 chambers as follows:

Equipment Room: Shall have two curtained doorways, one adjacent to the work area and one adjacent to the shower room. The equipment room shall be of sufficient size to accommodate at least one worker, allowing him enough room to remove his disposable uniform, as well as being large enough to house any equipment the contractor may need in the contaminated area during the work process.

Shower Room: Shall have 2 curtained doorways, one adjacent to the equipment room and one adjacent to the clean room. At least one shower with hot and cold water shall be installed in this room. The shower shall be constructed in such a manner that all water will be collected and pumped through a 5-micron filter. From the filter, water may be drained off in any conventional manner to a normal sewer system. Careful attention shall be paid to the construction of the shower to insure against leakage of any kind. The contractor maintains all soap and other drying supplies for the shower area.

Clean Room: Shall have 2 curtained doorways, one adjacent to the shower room and one to the outside of the work area. The clean room shall be of sufficient size to accommodate at least one worker, towels for workers and a supply of clean overalls.

Locker Room: Shall have 2 curtained doorways, one adjacent to the clean room and one to the outside. The locker room shall be of sufficient size to accommodate at least 2 workers and one locker for each employee.

This worker decontamination facility can be constructed of 2"x4" wood framing with plywood for sheathing and completely lined with a 6-mil polyethylene. Doorways and air locks can be constructed using 2 sheets of 6-mil poly hung between the head and the floor and jambs. Careful attention should be paid to the overlapping of poly joints and taping to insure an air tight structure.

On the occasion that you are bidding a project that enables access to the exterior directly from the contaminated area, there are firms such as the Williams Mobile Office Trailer Company that offer a unit specifically designed for asbestos abatement projects. It comes complete with all four rooms including the shower facilities.

Placement of Exhaust Air Filters. A negative air pressure shall be established in the work area and make-up air shall be obtained only through the decontamination chambers. The work area shall be under negative pressure whenever work is being performed, including preparation, removal, clean-up, or re-insulation.

Equipment shall be located so as to maximize air movement throughout the work area. Air movement shall be established in such a way that airborne fibers will be carried away from the immediate worker's area. At the end of each work day, the exhaust filtration unit shall be left on until the fiber count in the work area has reached a safe specified level, at which time the units can be turned off and the employee's decontamination disposal routes sealed. Note that an employee will need to be left at the project to accomplish this particular function.

Removal of Asbestos Materials. The asbestos material shall be sprayed with water containing a wetting agent (amended water) to reduce airborne fibers preceding the removal of the asbestos material. The material shall be sufficiently saturated to prevent emission of airborne fibers in excess of the exposure limits prescribed in the OSHA standards.

The asbestos material shall be removed in small sections by 2-worker teams. Asbestos contaminated materials shall be handled carefully and deliberately. Unnecessary agitation of the material is strictly prohibited. The use of any high rpm power equipment, pressure washers, or hydroblasters is not acceptable. The contractor shall not allow asbestos material to dry out or collect on the floors.

Removed materials shall be immediately placed in 6-mil plastic bags, sealed, and placed into metal drums for transport to the disposal site. Bags and drums shall be marked with the OSHA label prescribed by the OSHA regulation referenced by their specification. The outside of all containers shall be cleaned by wet sponging prior to leaving the work area.

After the wet removal process has been completed, the existing substrate and any surface from which asbestos has been removed shall be thoroughly vacuumed with a HEPA filtered vacuum cleaner. A note of caution: A HEPA filter will fail if used on wet materials.

Clean-up After Decontamination. The following work is performed after decontamination has been accomplished. All plastic coverings shall be vacuumed using a HEPA vacuum cleaner to remove the loose asbestos fibers. The poly is then removed, folded carefully and disposed of in a proper container. The work area is then again vacuumed with a HEPA vacuum cleaner, wet wiped, and in some instances, a second wet wiping is specified. During this particular process, the air monitoring equipment and the exhaust filtration units are in full operation. Once all of the poly is picked up and the room has been vacuumed and wet wiped, the process of re-installing the registers and grilles, the light fixtures, smoke detectors, and any other equipment that was removed prior to the decontamination process can begin.

Estimating Checklist. Because each asbestos removal project has unique job performance parameters, cost data is almost impossible to supply. We offer the following checklist items that the estimator needs to consider in preparing an estimate for asbestos removal work:

1. Work procedures submittal to required agencies.
2. Procedure approvals and permits.
3. Set-up of decontamination rooms.
4. Water, drain, and electrical connections to the decontamination rooms.

5. Maintenance of decontamination rooms during the progress of work.
6. Dismantle and removal of decontamination rooms at completion of work.
7. Air monitoring equipment rental, set-up, and dismantle.
8. Polyethylene covering for floors, walls, and other large objects to remain in the work space; and the subsequent removal of this covering at completion of asbestos removal.
9. Removal or covering for lighting fixtures, smoke detectors, duct grills and registers, etc; and the subsequent reinstallation or uncovering at completion of asbestos removal.
10. Disposable clothing suits.
11. Respirators and filters.
12. Scaffolding, ladders, and work platforms.
13. Surfactant for amended water application.
14. Removal of asbestos materials from substrate.
15. HEPA vacuum equipment rental and use.
16. Exhaust filtration units rental, set-up, and dismantle.
17. Plastic disposal bags and transport drums for protection during haul to disposal site.
18. Transport costs to disposal site and dump fees.
19. Wet cleaning of all surfaces and fixtures.
20. Spray sealants to prevent release of residual asbestos fibers.
21. Replacement thermal or fireproofing materials where asbestos was removed.
22. Record keeping and final submittals.
23. Insurance requirements.
24. Legal liabilities.
25. Training.
26. Development and costs of emergency procedures.
27. Cost for industrial hygiene consultant.
28. Cost to implement an approved safety program.

Sample Costs for Asbestos Removal. Encapsulation of the area where contaminated material is to be removed by negative atmospheric controls; portable showers, portable clean rooms, and portable change rooms. Allow for installation and removal.

	Hours	Rate	Total	Rate	Total
Laborers	32.0	$....	$....	$21.04	$673.28

To the above, add costs for equipment.
Use of glove bags is an ideal method of removing sound asbestos material from around pipes and structural steel. Under normal conditions, a laborer can remove about 40 s.f. per hour:

	Hours	Rate	Total	Rate	Total
Laborer	1.0	$....	$....	$21.04	$21.04
Cost per s.f.			....		0.53

Monitoring devices are portable but require constant calibration and an experienced operator. Cost per project can vary widely, from $1,000.00 to $5,000.00 per day.

Clean up costs will vary with the type of material being removed and its location. It is a standard procedure to check the atmosphere, as well as the surrounding materials, such as the walls, just after completion of the work, and then again 24 hours after the work is completed. The cost can run from $1,500.00 up to $15,000.00 per clean-up operation.

CONTAMINATED SOIL

When removing contaminated soil from any site, precautions are necessary to prevent hazards to human health and to the environment. Specific procedures are required to comply with EPA, DOT, and in some case OSHA regulations regarding hazardous waste site clean up.

Hazardous wastes refer to those listed in the EPA's 40 CFR Part 261 *Identification and Listing of Hazardous Wastes*. Requirements for the transportation of hazardous waste is outlined in DOT 49 CFR Parts 172, 173 and 178 and in some cases in OSHA 29 CFR Part 1910.120 for health and safety.

Estimating Amount to be Removed. When removing contaminated soil from a site, it is important to understand how to determine the magnitude of the work. At most sites, the area of contamination is already defined, and the information is usually available in a site work plan.

Soil samples (borings) are obtained from the site at different locations using split barrel (split spoon) sampling apparatus or a hand auger. The split barrel method is used to obtain deep samples and is usually operated from a drilling rig. Hand augers may be used for shallow soil samples, up to about 5' deep. Samples are then sent to a laboratory for analysis, and the results are used to determine the extent of contaminated soil to be excavated. A plot plan or other diagram may be used to define the length, width, and depth. Calculations should be converted to cubic yards, and prices for disposal should be quoted in cubic yards.

Methods of Excavation, Storage, and Transportation. Common earthmoving equipment is used for excavation, including backhoes, bulldozers, front-end loaders, and cranes. Solid material vacuum trucks and front-end loaders are sometimes necessary to complete the work in a timely manner.

The backhoe is used to excavate and consolidate soil from the ground into a pile. Piles of soil should be consolidated on 6-mil polyethylene sheeting used as ground liner. The front-end loader is then used to load soil from the pile into a roll off container or a dump truck. A typical 30-cubic yard dump truck can only be filled with 15 cubic yards of soil (50% of normal capacity) to remain within weight limits mandated by the Interstate Commerce Commission and the Department of Transportation. But also, overboarding a roll off or dump truck can damage the vehicle.

Using this method, it takes approximately 1 hour to load each roll off or dump truck with about 15 cubic yards of soil.

Once the vehicle is loaded, the material is either stored on site, if roll offs are used, or if dump trucks are used, then it is moved off site as soon as loading is completed. Dump trailers should be lined with 6-mil polyethylene sheeting and covered with standard roll off or dump trailer tarpaulin. The tarpaulin must be securely fastened onto the vehicle and capable of preventing soil from blowing onto the roadway when moving or by the wind when standing or parked. Polyethylene sheeting can be purchased in rolls measuring 100'x20', which is enough to line two standard 30-cubic yard roll offs or trailers.

If the soil is a hazardous waste, an appropriate *hazardous waste manifest* must be used as the shipping paper, and as a minimum, all vehicles must have hazardous waste hauling permits.

When removing small amounts of contaminated soil, it might be appropriate to consolidate excavated soil into 55-gallon drums. It will take about 15 to 30 minutes to fill each drum, depending on soil compaction and on the level of personal protective equipment. Each drum will weigh between 600 and 800 lbs. In order to move and load these drums onto a trailer, a forklift equipped with a drum-moving attachment or a front-end loader is commonly used.

Drums are usually loaded onto a flatbed or 40-foot closed box trailer. Drums containing hazardous waste must be labeled according to EPA, DOT, and possible state regulations. One, two, or more EPA hazardous waste labels and DOT hazardous class labels may be required. It will take between 5 and 10 minutes to label and load each drum, depending on the level of protective equipment.

Personal Protective Equipment. Anyone who enters a hazardous waste site must be protected against potential hazards. The purpose of personal protective clothing and other equipment (PPE) is to shield the individual from chemical, physical, or biological hazards. A proper selection of PPE protects the respiratory tract, skin, eyes, face, hands, feet, head, body, and hearing. It is important to note that when using PPE, workers are at considerable risk of developing heat stress.

There are four levels of PPE. Each worker at a hazardous waste site must have the appropriate level of PPE available for use when working in the contamination zone. The use of PPE can increase the work duration by two or three times. The following gives recommended protective equipment at each level.

Level A. Totally encapsulated chemical protective suit (TECP). Pressure demand full face self contained breathing apparatus (SCBA) or supplied air respirator with escape SCBA. Inner outer chemical resistant gloves. Chemical resistant safety boots/shoes. Two-way radio communications. Hard hat. This level will mean a loss of 60 to 80% in work efficiencies.

Level B. SCBA or supplied air respirator with escape SCBA. Inner outer chemical resistant gloves. Chemical resistant safety boots/shoes. Two-way radio communications. Hard hat. Disposable coveralls may be used. *Tyvek* coveralls, hooded one-piece chemical splash suit. Duct tape should be used to seal ankles, wrists and other openings. Loss of efficiency is 50 to 60%.

Level C. Full face air purifying cartridge filter respirator. Use organic vapor acid gas and high efficiency particulate air filter (HEPA) as necessary for appropriate personal protection. Disposable coveralls, hooded one-piece chemical splash suit. Duct tape should be used to seal ankles, wrists and other openings. Disposable chemical resistant boots to be worn over safety shoes. Hard hat. Loss of efficiency is 15 to 50%.

Level D. Work uniform and possibly dust respirator might be required. Hard hat.

Decontamination. All equipment that comes into contact with hazardous waste must be decontaminated before leaving the contamination zone. There are various methods, including solvent cleaning and high pressure steam cleaning with detergent.

When decontaminating large or heavy equipment, a sump lined with polyethylene sheeting should be constructed. The sump, used to collect cleaning solutions, prevents contaminated runoff from spreading. When all equipment has been decontaminated, the residue must be pumped into drums using a drum pump. In the case where large amounts of cleaning residue are generated, a vacuum trailer might be more appropriate.

Labor Removing Contaminated Soil. When removing large amounts of soil with heavy equipment, it will require one operator for each backhoe, etc. When removing small amounts of soil, a crew of at least three workers might be necessary because of the weight of the drums.

Qualified supervisors must be available. All personnel supervising hazardous waste operations or handling hazardous waste initially must receive training, which must be updated regularly in accordance with federal and state regulations. Federal regulations may also require other personnel to be on site, such as an emergency coordinator or health and safety supervisors, chemists, professional engineers and certified hazardous waste specialists. Specialized direct-reading air monitoring equipment may be needed. When estimating the cost of contaminated soil removal, the following equipment, labor, personal protective equipment, transportation, and disposal costs should be considered.

Excavating, Loading Equipment
Backhoe
Front-end Loader
Solids Vacuum Trailers
Roll off Container, Tractor, and Flatbed Trailer
Dump Trailer with Tractor Equipped
    with Dump Trailer Hydraulic Lines
Forklift

Miscellaneous Equipment
55-Gallon Steel Drums
6-mil Polyethylene Sheeting
Duct Tape to Seal Seams on Trailer Liners
EPA Hazardous Waste Labels
DOT Hazardous Class Labels
Shovels and Picks
Ratchet Wrench to Bolt Drums Closed
Shipping Paper (Hazardous Waste Manifest)
Felt Tip Markers for Drum Labeling
Decontamination Equipment (Pressure Steam Cleaner)

Personal Protective Equipment
(One Ensemble for Each Worker)
Level A
Level B
Level C
Level D
Air Monitoring Equipment

Transportation Costs
Cost per Load for Roll offs and Dump Trailers
Cost per 55-gallon Drums
Cost Vary Depending on Distance to Disposal Site

Disposal Costs
Cost for Disposal of Contaminated Polyethylene
Cost per Cubic Yard of Soil Disposal
Cost per 55-gallon Drum Soil Disposal
Cost per Gallon for Bulk Cleaning Residue from
    Pressure Washing and Decontamination Solids
Cost per Drum for Cleaning Residues from
    Pressure Washing and Decontamination Solids.

Labor Costs

    Heavy Equipment Operators

    Laborers for Liner Placement and Drum Handling

    Supervisors

Average Costs to Remove 100 C.Y. of Contaminated Material Loaded into Dumpsters or Dump Trucks

	Hours	Rate	Total	Rate	Total
Backhoe	6.66	$....	$....	$18.74	$124.81
Loader	6.66	....	....	18.74	124.81
Equipment operators	13.32	....	....	28.55	380.29
Laborers	13.32	....	....	21.04	280.25
Cost per 100 c.y.			$....		$910.16
per c.y.					9.10

The estimator must review the above figures and might possibly want to add the following items: level of protective equipment and clothing required for the specific project and the "lost efficiency" percentage for that level. Trucking and disposal and special liability costs. Cost of permits is not included in the above.

Average Costs to Remove Contaminated Material in Approved 55-Gallon Steel Drums*

	Hours	Rate	Total	Rate	Total
Backhoe	2.0	$....	$....	$18.74	$37.48
Equipment operators	2.0	....	....	28.55	57.10
Laborers	3.0	....	....	21.04	63.12
Cost per c.y.					$157.70
per drum					40.89

*Allow 7 c.f. per drum.

The estimator should add the following items to the estimate: lost efficiency factor for protective equipment and clothing level required. Trucking and disposal and special liability costs. Cost of permits is not included in the above costs.

Removal of contaminated materials in buildings requires special liability insurance coverage. The high liability risks involved, combined with high minimum limits of liability insurance required by owners and agencies, prohibits most contractors from performing this work. The prudent estimator will become knowledgeable about all requirements and ramifications before submitting a bid for contaminated material removal. It is suggested that the estimator leave this operation to a qualified and approved asbestos removal contractor.

Some regulatory agencies are requiring the removal of all asbestos material in public buildings to be completed by the year 1995. Different materials are being continually added, removed, or revised from listings of contaminated material by federal, state, and local OSHA and environmental protection agencies. The estimator is cautioned to check with all agencies that have jurisdiction for a project. Listed below are some of the agencies, both federal and local, that may have jurisdiction over contaminated material removal:

1. Occupational Safety and Health Agency (OSHA)
2. Environmental Protection Agency (EPA)
3. conservation agencies
4. Department of Transportation (DOT)—federal, state, and local
5. health departments
6. United States Coast Guard (USCG)
7. U.S. Army Corps of Engineers (COE)
8. fish and wildlife agencies
9. Consumer Products Safety Commission (CPSC)
10. Mine Safety and Health Administration (MSHA)

UNDERGROUND STORAGE TANKS

Most states have adopted regulations for immediate testing of existing underground storage tanks for leaks and for necessary repairs to be made without delay. In fact, all single-wall steel storage tanks (tanks for fuel, gasoline, oil, heating oil, and industrial storage), even those that do not leak must be replaced by the end of the century.

Tank monitoring regulations for underground storage tanks mandate a secondary containment system, which requires a leak detection monitoring system. The program might employ a continuous system, connected to an audible or visual alarm system, or a manual inspection, performed on a daily basis.

Leak detection systems that have been approved by regulatory agencies include the following: liquid level indicators that detect a change in the height of a fixed volume of liquid in an annular space; liquid sensors that detect the presence of liquid in an annular space; vapor sensors for volatile hazardous substances; and pressure or vacuum loss sensors.

It is strongly recommended that the general contractor contact a specialist in industrial waste to perform the required testing. Testing for leaks varies according to the type, size, and material of the tank. Prices for tank testing range from $1,500 to $10,000.

If a leak is detected, the owner must remove the contents of the tank. It will be necessary to determine the source of the leak; properly remove all substances from the underground storage tank system; repair, replace, or close the system in accordance with the local governing regulatory agency mandates.

The possessor will then be required to determine the extend of the leak. An accepted method is to drive 2-1/2" diameter pipe observation wells about 20 into the ground and take samples of the soil and water if present. The number of observation wells is normally 4 per tank or farm tank. Costs vary due to the many types of soil and subsurface conditions. Installations of wells continues until the extent of soil contamination is determined. The contaminated soil is removed, the tank cleaned and filled, with clean material, and then either left in place or removed.

Removal of the tank is the preferred method, but method of disposal can depend on many extenuating circumstances, such as location of tank with respect to other facilities, or it might be that removal is an insurmountable task. The final determination whether to leave it in place or remove it depends on local regulations.

In-place Disposal of Underground Storage Tanks. The following is a step-by-step procedure for the safe in-place disposal of underground tanks:

1. Drain product from piping into tank, disconnect all piping, and cap and remove all piping.

2. Remove liquids and residues from the tank using explosion-proof or air driven pumps. The last few inches might require a hand pump.

3. Excavate to the top of the storage tank.

4. Remove and cap drop tube, fill pipe, gauge pipe, vapor recovery connections, submersible pumps, and all other tank fixtures and connections, except for the vent line. The vent line remains connected until the tank is purged.

5. Purge the tank of flammable vapors. Vent all vapors a minimum 12' above grade or 3' above adjacent roof lines.

6. Cut tank, if necessary, for the proper introduction of inert, clean fill material to fill the tank.

7. Place suitable solid material, such as sand or other clean fill, through the openings in the top of the tank. It is important to fill the tank as full as possible.

8. Sand or other clean fill is suitable if it is free of rocks or stones, which might limit leveling out of the fill in the tank. Sand may be introduced dry, as long as it flows freely. When the sand cones near the top of the tank, it can be washed into the tank with a nominal amount of water so that it flows to the ends. The use of large amounts of water must be avoided.

9. Fill the tank to about 80% of its calculated capacity. Mix soil and water to make a free-flowing mud and pour the mixture into the tank opening until the tank is full and the mixture overflows the opening.

10. Plug or cap all openings.

11. Disconnect the vent line, and remove and plug the vent line opening.
12. Restore ground surface.

Cost will vary for this type of work. Prices range from $1,500 to $3,000 per ank for tanks up to 2,500 gallon capacity, $3,000 to $5,000 for tanks to 6,000 ;allon capacity.

Removal of Underground Storage Tanks. The following procedures ire followed for the safe removal of underground tanks:

1. Drain product from piping into the tank, disconnect piping from the tank, and cap and remove all piping.
2. Remove liquids and residues from the tank using explosion-proof or air driven pumps. The last few inches might require a hand pump.
3. Excavate to the top of storage tank.
4. Remove and cap the drop tube, fill pipe, gauge pipe, vapor recovery connections, submersible pumps, and all other tank fixtures and connections. One plug shall have a 1/8" vent hole to prevent the tank from being subjected to excessive differential pressure caused by temperature changes.
5. Purge the tank of flammable vapors. Vent all vapors a minimum 12' above grade or 3' above adjacent roof lines.
6. Excavate around tank to uncover it for removal. Remove the tank from the excavation and place it on a level surface. Use wood wedges or blocks to prevent the tank from rolling.
7. After tanks are removed from the ground but before they are removed from the site, they must be labelled. For example, the label might read:

TANK HAS CONTAINED LEADED GASOLINE
NOT VAPOR FREE
NOT SUITABLE FOR STORAGE OF FOOD OR
LIQUIDS INTENDED FOR HUMAN OR ANIMAL
CONSUMPTION

or

TANK HAS CONTAINED LEADED GASOLINE
LEAD VAPORS MIGHT BE RELEASED IF HEAT
IS APPLIED TO THE TANK SHELL

8. Tanks shall be removed from the site as quickly as possible after the tank is vapor free.
9. Tank atmosphere shall be checked with a combustible gas meter to insure that the tank does not exceed 20% of the lowest flammable rate.

10. Tank should be loaded and secured on a truck for transportation to the storage or disposal site with the 1/8" vent hole located at the uppermost point on the tank. The tank shall be transported in accordance with all applicable local, state, and federal regulations.

Costs will vary. Prices range from $2,500 to $6,000 per tank for tanks up to 2,500 gallon capacity, $3,500 to $15,000 for tanks up to 6,000 gallon capacity. Price also varies depending on whether there is one tank to fill or more than one tank at a location.

Removal of a single wall steel tank and replacement with an approved double wall tank, with all the necessary leak and fume detectors, averages about $50,000 for a 1,000 to 1,500 gallon tank; a 2,000 to 6,000 gallon tank costs from $75,000 to $100,000.

The contractor must follow all local, state, and federal regulations. Regulatory agency requirements can be supplemented with information published by the American Petroleum Institute (API).

CHAPTER 22

ROAD AND HIGHWAY CONSTRUCTION

This chapter is for the general contractor who needs to prepare estimates for and engage in paving work associated with building construction projects. It is a reference for solving many of the problems encountered. Much of this information is applicable to major highway work, but the chapter does not specifically address this sophisticated and specialized form of construction.

Supplemental information is available through extensive publications devoted to research data, current practices, and standard specifications that have been produced through the efforts of the American Association of State Highway Officials, the Highway Research Board of the National Research Council, and the Bureau of Public Roads. In addition, several material associations have provided factual publications for the benefit of the highway construction industry. The most notable and active associations are American Concrete Institute, Association of Asphalt Paving Technologists, and The Asphalt Institute.

Road, highway, and paving construction is a high risk business under the most favorable conditions, demanding a thorough knowledge of the technical skills involved, accompanied by managerial and administrative talent to support a totally efficient and coordinated operation. Like other types of contracting, the paving contractor is faced with the obligation of accurately evaluating the capabilities and resources of his organization. This specifically includes the financial capacity of the company and the availability of external sources for interim financing; availability of construction equipment through ownership, lease, or rental with emphasis on condition, suitability, and sufficiency; facilities for maintenance and repair; existence of systems and personnel to efficiently provide cost control, fiscal control, scheduling, purchasing, and expediting; and qualified estimators capable of producing accurate estimates. Once this evaluation is complete, it can be determined the type, size, and complexity of construction projects that would best match the resources and capabilities of the company. This matching process is the basis of selection of projects for bidding.

It should be noted that firms having sufficient financial capacity to qualify for bonding on projects of a certain magnitude may have a deficiency in experienced personnel to efficiently accomplish the work. Contractors are well advised to follow the formula: financial assets + people assets = total capability.

Because of the impact of road and highway construction on the economic, social, and political decisions of a community, there has to be a presumption that many laws and regulations affecting the conduct of a construction program exist. Each geographic location will have its own peculiarities, requiring the contractor's investigation and familiarity with these requirements prior to becoming involved in the bidding process. Prequalification of contractors is often a prerequisite, as is compliance with local contractor license laws or registration as a foreign corporation if the company is not incorporated under that state's law.

The intent of this chapter is not to present extensive engineering and design theory but to provide sufficient examples where contractors, estimators, project managers, and other supervisory personnel have been confronted with problems relating to paving operations, where without a source of some basic knowledge of design, they were placed at a disadvantage in the decision making process. This lack of knowledge can be an extremely important factor when attempting to establish a fair and reasonable cost factor in the initial estimating process, and it can have more impact when attempting to evaluate a remedial solution for defective paving which may be assigned as a contractor's responsibility.

Contractors will generally be working within the requirements of the project specifications. Strict compliance normally will protect the contractor from claims. However, not all specifications are complete and totally descriptive of the precise methods necessary in order to achieve the desired end results. In some instances the designing architect also lacks sufficient understanding of the capabilities and characteristics of the materials involved in the construction. Frequently, the problem is further clouded by including a general reference to the AASHO System (American Association of State Highway Officials) or an attempt to include elements extracted from the Unified Classification System for Soils and fortified with vague references to density, moisture content, compactness, and gradation of materials.

While the road and paving may be a somewhat minor cost factor as it relates to an entire construction project, it would be unacceptable and unprofessional not to establish the most accurate cost possible for this work. Every estimator should be aware that failures are costly and properly developed estimates do not make provisions for re-performance of work. It is obvious that unacceptable work requiring remedial action escalates the cost of the project and lowers the profit margin.

Generally, it is found advantageous to utilize the talent, expertise, and services of a specialty contractor in the paving operation. However, this subcontractor will demand certain supporting performances by the prime or general contractor. This may include the preparation of the subgrade, all line and grade engineering, and any testing that may be required.

More extensive knowledge on the part of the contractor creates a position where methods, schedules, unit costs, and other specific details can be discussed with authority and confidence prior to entering a contract for paving work, with reasonable assurance the ultimate objective will be achieved with minimal problems. This same theory should be applied with the same force and effect on each of the other divisions of work included in the total project.

Construction Documents

Plans. Properly prepared plans provide sufficient information for an estimator to make a quantitative analysis of each discipline of work included in the project. This function should be performed even if the owner has provided the estimated quantities in the *bid form,* because accuracy cannot be determined except by comparison with the contractor's quantity survey.

Specifications. The technical specification for highway work is usually a "guide specification" making many references to the Standard Specification requirements where the details appear for the various types of work to be performed. As an example, the guide specifications may identify curb and gutters as Types A, C, D, and F at certain locations. To determine the specific design of each type, reference is made to the Standard Specification. Copies of Standard Specifications can be purchased from the agency issuing the plans and specifications.

Bid forms. In the case of unit price bidding, which is the usual format in road and highway construction, the form will provide for each classification of work with estimated quantities included. The contractor is required to insert a unit price, including allowance for overhead and profit on each item, and calculate the estimated quantity by the unit price to determine total cost on each line item. A typical example would be as follows:

Bid Form

Code, Item	Quantity	Unit	Unit price	Total
200 Clearing	6	acres	$2,000.00	$12,000.00
303 Excavation, Class 2	100,000	cu. yds.	1.00	100,000.00
310 Seed	500	lbs.	3.00	1,500.00
490 Asphalt Conc. Agg.	2,000	ton	26.00	52,000.00
520 Cement Treated Base	2,000	cu. yds.	31.00	62,000.00
710 Modify Signs	11	each	300.00	3,300.00
730 Pipe Culvert 24"	200	lin. ft.	40.00	8,000.00
732 RC Pipe 48"	200	lin. ft.	65.00	13,000.00
Total Price				$251,800.00

(metric)

Code, Item	Quantity	Unit	Unit price	Total
200 Clearing	24,282	sq.m	$0.49	$12,000.00
303 Excavation, Class 2	76,500	cu.m	1.30	100,000.00

310 Seed	225	kg	6.67	1,500.00
490 Asphalt Conc. Agg.	1,800	m ton	28.89	52,000.00
520 Cement Treated Base	1,530	cu.m	40.52	62,000.00
710 Modify Signs	11	each	300.00	3,300.00
730 Pipe Culvert 600 mm	60	m	133.33	8,000.00
732 RC Pipe 1200 mm	60	m	216.67	13,000.00
Total Price				$251,800.00

Should an error appear in the extension to the total line amount, it has been ruled many times that the owner has the right to use the unit price offered by the contractor and correct the mathematical error. Estimators should use extreme caution in developing the unit price bid item.

Instruction to Bidders. These documents provide the precise information relating to bid guarantee, number of copies of the bid to be furnished, establishment of date and time bids are to be received, documentation to be included in the bid package, and other instructions appropriate to the project. These requirements are to be fulfilled without deviation or the bid may be considered nonresponsive, making the contractor ineligible for award.

Special Instructions. Physically inspect every project site prior to bidding to determine working conditions. Avoid accepting oral interpretation or instructions as they cannot be enforced. If uncertainty or lack of confidence appears during the bidding process, the best decision would be to avoid bidding on the project rather than risk a serious mistake and possible loss.

Soil Stabilization

The need for soil stabilization is widespread in the construction industry to support the designed contour configuration that provides beautification in the landscaping of buildings, parks, and other green areas. In addition, there is an important functional aspect to soil stabilization. This applies to the prevention of soil erosion and providing a sound and stable base to support roads, walks, terraces, parking areas, and other surfaces subjected to concentrated loading and heavy traffic wear.

Without sufficient stability of the subgrade, all finishes placed upon it are, to some degree, influenced by the credibility of the subgrade. It is not uncommon to encounter complete failure of paved areas that can be traced to poor or ineffective preparation of the subgrade prior to the installation of properly designed and installed finish surface.

Soils exist in a broad range of types with wide variation in the value of their properties and they react differently to the several methods available to improve their degree of stabilization. It is generally recognized that the ultimate resistance of a soil to the penetration of concentrated load depends on the following two properties of the soil: 1) *friction* between soil particles and 2) the *cohesion* between the soil particles. Soil resistance due to cohesion is dependent on the external pressure on the soil. However, frictional resistance increases in direct proportion to the external pressure.

Example: Sand has a high frictional resistance without appreciable cohesion and the bearing capacity is increased considerably when a concentrated load is surrounded by a distributed load. Clay, having high cohesive resistance, but without appreciable frictional resistance, reacts quite differently, because the bearing capacity is not increased to any great extent by applying a concentrated load surrounded by a distributed load.

The stability of a soil for subgrade purposes also depends on its *elasticity* or the ability to rebound after a load has been applied and then removed. Excessive elasticity makes a soil less stable because it cannot be permanently compacted.

Water content in the soil can be a significant factor in stabilization, particularly in the expansive soils which are subject to change in volume, because they possess cohesion and capillarity in appreciable amounts.

Often the quality of subgrades can be improved by installing some type of artificial drainage to prevent the accumulation of excessive moisture from the subgrade materials. Such a provision would also provide protection from "pumping" and "heaving" action during freezing and thaw periods.

Excavation

Excavation, moving, and depositing of existing materials to a new location are major cost items for the contractor involved in road and highway construction. All encountered materials are basically rock. However, because of the forces of nature, there are great variances in the size of particles, from materials classified as solid rock continuing down the scale to those identified as sand or clay.

This variation in material is so extensive it would be presumptuous and extremely dangerous for the excavator to attempt calculating the cost involved without engaging in preliminary studies. Each classification retains its own characteristics. In some instances, blasting or ripping is required to separate the materials from the natural position, while other materials can be readily removed by loader or scraper. Dissimilar conditions will be encountered when attempting to compact the various types of materials.

On a given construction project, it would not be unusual to encounter several types or classifications of materials, requiring the contractor to modify methods and equipment to accommodate each condition. An understanding of geology and soils mechanics is necessary to effectively deal with the complexities of excavation. Experience in soils of a geographical location also contributes to the overall knowledge.

In excavating hard materials such as shales, sandstone, and limestone, the strata of rock may be hard and strong, but if lamination thickness and vertical joint spacing are limited in size, production will be improved over instances where these planes of weakness produce greater cubic content of the material.

The most common practices used to determine the blasting and ripping zones for rock excavation is the track drill because of it is widely available, is portable, and can produce data relative to drilling costs and blasting techniques. The seismic timer is an instrumental means to determining the degree of consolidation of the rock which can be related to the cost of excavation.

When excavating the usual non-cementitious materials, a variety of machines and methods are available to perform specific tasks at the most economical cost. Loaders, shovels, cranes, Gradalls, backhoes, draglines, and pans are the most common machines in use, with each one having certain advantages and limitations. It is the duty of the estimator to know the capabilities of these machines and to quantitatively prove their efficiency before making a judgment. The following formula will provide hourly production for a front-end loader with a 3-c.y. (2.30-cu.m) capacity bucket working average soil. Loading time cycle through an angle of 120 degrees is a 0.5 minutes, comprised of the following four elements:

1. Loading - 7 seconds (0.12 minutes)
2. Swing-Position - 12 seconds (0.20 minutes)
3. Dumping - 3 seconds (0.05 minutes)
4. Return-Position - 8 seconds (0.13 minutes) or a total of 30 seconds (0.50 minutes)

Hourly production in cu. yds. (cu.m), bank measurement, per 50-minute working hour is determined by the following formula:

$$P = \frac{50}{CT \times BF \times BC} \text{ where}$$

P = Hourly Production, Cubic Yards or Cubic Meters Bank Measurement
CT = Cycle Time in Minutes
BF = Bucket Factor
BC = Bucket Capacity

Bucket Factor Table

Material	Fill Factor lm* bc	Conversion Factor bm** lm	Bucket Factor bm bc
Sand-Gravel	1.05	0.88	0.92
Weathered Rock-Earth	0.90	0.78	0.72
Blasted Rock	0.80	0.67	0.54

*Loose Measurement **Bank Measurement

The equation becomes:

$$P = \frac{50.0}{0.50 \times BF \times BC}$$

$$P = 100 \times BF \times BC$$

$$P = 100 \times 0.92 \times 3.0$$

$$P = 276 \text{ cybm (210 cu.m) per 50 min. workhour}$$

Assume this 3-c.y. front-end loader could be rented including operator, fuel, maintenance, and transportation charges for $80.00 per hour, then the cost for excavation would be:

$$\text{Cost/c.y. (cu.m) bm} = \frac{\text{Rental cost per hour}}{\text{c.y. (cu.m) bm per hour}}$$

$$\text{Cost/c.y. (cu.m) bm} = \frac{\$80.00}{275\ (210)}$$

$$\text{Cost/c.y. (cu.m) bm} = \$0.29 \text{ per c.y.}$$
$$(\$0.38 \text{ per cu.m})$$

This cost is pure excavation and loading cost. To this must be added hauling, dumping charges, job supervision, establishment and maintenance of haul roads, compaction of fills, and other miscellaneous costs. In making cost production analysis on each type of machine planned for use on a project, other valuable information will be produced. The hourly production rate on the excavating machine will determine the number of trucks required to haul the excavation to other locations within or external to the construction site. Truck costs can be established in similar analysis predicated upon loading time, capacity of the truck, and the haul time to the disposal area.

To determine the number of trucks required to service and support the production rate of the previous front-end loader, it can be established as follows:

P =   275 c.y. (210 cu.m) bm

PM =   c.y. (cu.m) bm per minute of 50 min. work hour

WH =   50 minute/hour

$$PM = \frac{P}{WH}$$

$$PM = \frac{275 \text{ c.y. (210 cu.m)}}{50}$$

PM =   5.5 c.y. (4.2 cu.m) bm

A 10-c.y. (7.65 cu.m) truck could be loaded every 2 minutes of work time, thus 25 trucks would be required to service the loader. Haul time for the trucks would have an influence on the number of trucks required.

Hauling

The hauling of excavated materials to a disposal area is the most important phase of excavation. It is usually the most costly and subject to many variables. In instances of off-site hauling by trucks, the traffic patterns will have a substantial influence on the haul time. More haul time will be consumed during morning and evening peak traffic periods than during the midday normal traffic flow.

On-site hauling by self propelled scrapers can be determined with more precision as the conditions can be controlled by the contractor.

After selection of scrapers of a suitable capacity, there is a need to determine the travel time involved from the point of loading to the disposal destination. This can be accomplished by using the following equation:

$$\text{Travel time (minutes)} = \frac{\text{Length of haul (feet)}}{88 \times \text{Average speed (mph)}}$$

Eighty-eight represents a factor to convert miles per hour to feet per minute. Speed factor tables are available from equipment manufacturers for the conversion of maximum speed to average speed, but these are somewhat complicated to use. On smaller projects, past experience would serve the needs of the contractor. In establishing an average speed factor, use the rated maximum speed of the machine and multiply by realistic percentages reflecting job conditions to determine average speed.

Short haul distances, 200 to 600 feet, would be at a 50% level because of stop frequency and turnaround time loss. For haul distances of 2,000 to 3,000 feet, the percentage increases to 75% to 80%. These percentages would be for average conditions with level grades, no sharp turns or hazards, and without interference of cross traffic.

In using the equation to establish travel time after the loading operation, assume the length of haul to be 3,000 feet with an average speed of 12 mph.

$$\text{Travel time (mins.)} = \frac{3,000}{88 \times 12}$$

$$= \frac{3,000}{1,056}$$

$$= 2.84 \text{ minutes}$$

To establish the number of haul cycles, divide the 50 minute workhour by the travel time:

$$\frac{50}{2.84}$$

$$\text{Number of Haul Cycles/hour} = 17.6$$

Production could be calculated by allowing a 2 minute loading time plus the 2.84 minute haul time for a total 4.84 minute production cycle for a 24-c.y. scraper:

P = (50 ÷ 4.84) x 24
P = 10.33 x 24
P = 247.92 c.y. per hour

Should materials or job conditions be encountered that would require a pusher machine, then it would be an additional cost to the basic scraper.

Fill Compaction

The distribution and compaction of fills to rigid specifications calls for the use of several machines designed for the densifying operation, which is accomplished by spreading-mixing, wetting, or drying, and the use of compactors. Production and costs for spreading materials are complicated by the overlapping and sharing of work with other machines. The following general equations have been found acceptable for hourly production when using the horse power (Hp) rating of the machine as a base:

Dozer   cybm =   3 x Hp
Motor Grader   cybm =   3 x Hp

The use of water wagons is a function of the haul cycle time and can be estimated by the following equation:

Production (gal/h) =   (50 ÷ Cycle time in minutes) x Capacity in gallons

Water wagons range in capacity from 1,500 to 14,000 gallons. The contractor must decide on the most economical based on the anticipated needs at the project. Laboratory test data provides the existing natural moisture content. The moisture-density curve indicates that the desired percent of relative density will be attained with a percentage of moisture by dry weight. With this information the estimator can produce an estimated need for water and project this toward a water wagon cost.

In instances where excessive moisture is encountered, this can be reversed by aeration of existing materials or the importation and mixing of dry materials. In each instance, it becomes a costly operation and delays job progress.

Calculation of Excavation Quantities

In establishing the quantities of excavation, which is subdivided into cuts and fills, the estimator is actually determining the volume or cubic yards involved in each operation through calculations involving areas, distance, and geometric figures.

Volume: 1 c.y.   = 27 c.f.
Distance: 1 l.f.   = 12 in.
          1 station   = 100 l.f.
          1 mile   = 5,280 l.f.

Area: 1 sq. yd.   = 9 s.f.
          1 acre   = 43,560 s.f.
          1 sq. mile   = 640 acres

See Chapter 24, *Mensuration,* for tables that provide the formulas for calculating the quantities of the most frequently found geometric figures.

The usual methods of estimating are the *end area* method and the *cross section* method. These two methods are described in Chapter 2, *Site Work.*

Bituminous Road Paving

This flexible paving method consists of several layers or courses of mineral materials placed upon a properly prepared subgrade of earthwork which performs as a foundation for the paving. Quality, quantity, and methods of performance are governed by the requirements of the contract specifications.

Subbase Course. This is a layer of selected granular material of pre-determined depth applied to the subgrade for the support of the paving materials. It also prevents or reduces frost damage and provides under pavement drainage.

Base Course. A high quality material such as crushed stone which is placed on the subgrade or the subbase. Gradation and depth controlled by specification. Also may be treated with portland cement or bituminous to reduce thickness and improve durability.

Leveling Course. Used to eliminate irregularities in existing paved surfaces prior to resurfacing.

Binder Course. In asphalt paving a binder course, which is often referred to as a leveling course, is installed at a prescribed thickness over the subbase. The strength of the binder course will be substantially equal to the designed surface or wearing course.

Surface Course. The uppermost or exposed surface of the paving designed to withstand wear and polishing by traffic. Smooth surfaces reduce noise level but increase skid action.

Pumping. Commonly referred to as mud pumping. Occurs when the subgrade contains excessive moisture and is a type containing a high percentage of fine particles. The surface becomes flexible and unstable to any load imposed upon it. This condition is unsuitable to support paving and must be removed and replaced with structural support for the paved surface.

Subgrade. Foundation on which the subbase and paving materials are placed. The utilization of locally available materials for use in the subgrade has important economic benefit, unless the material is of a poor quality requiring additives in the form of chemicals or modification in the soil content and gradation to achieve the desired quality of compaction. In such instances, consideration should be given to the economics of removal of a sufficient quantity of the existing soil and replacement by fill material requiring limited compaction effort to produce a suitable surface to receive the subbase material.

For various terms that are used in road and highway construction, including those that apply specifically to asphalt paving, see the Appendix A, *Glossary.*

Standard Design Details

For general guidance and reference, the following standard design details have been approved by the Federal Highway Administration.

Certain states, counties, or municipalities may require some slight modifications, but it is assumed that deviations will not significantly exceed the requirements of these standard details.

Index of Standards

Soils & Soil - Aggregate Mixtures, Characteristics, & Performance
1. Soils & Soil - Aggregate Mixtures, Treatment, & Use
2. Soils & Soil - Aggregate Mixtures, Guide to Classification
3. Standard Timber Barricade
4. Standard Berm Ditches
5. Standard Surface Drain Ditches, Concrete, & Sod
6. Standard Side Ditches Cut Areas
7. Standard Side Ditches, Standard Median Ditches - Flat
8. Standard Median Ditches, V Slope
9. Standard Pipe Outlets, Surface Drainage on Fills
10. Method of Depressing Pavement at Inlets
11. Standard End Support Wall, Round Pipe
12. Standard End Support Wall, Arch Pipe
13. Standard Headwall
14. Standard Headwall
15. Standard Type C Endwall, Round Pipe
16. Standard Type C Endwall, Arch Pipe
17. Standard Type E Endwall, Round Pipe
18. Standard Type E Endwall, Arch Pipe
19. Standard Type F Endwall, Round Pipe
20. Special Type F Endwall, Round Pipe
21. Standard Type G Endwall, Round Pipe
22. Standard Type H Endwall, Round Pipe
23. Standard Type K Inlet, Open-End Grate
24. Concrete Manholes, Catch Basins, and Inlets
25. Standard Underdrains
26. Sediment and Erosion Control, Fiberglass Erosion Stop
27. Standard Concrete Valley Gutter, Flues, Concrete Shoulder, & Rebut
28. Sediment and Erosion Control Placed Rip-Rap Ditch
29. Piled Rip-Rap Detail
30. Concrete Bridge Parapet and Bridge, Median Barrier Details
31. Reinforced Concrete Pavement, Types of Joints
32. Standard Bituminous Concrete Curb
33. Standard Types of Concrete Curb, Combination Concrete Curb & Gutter

GENERAL CLASSIFICATION	GRANULAR MATERIALS					SILT-CLAY MATERIALS							
GROUP CLASSIFICATION	A-2	A-3	A-2-4	A-4-2	A-2-7	A-4	A-7-2	A-4-7	A-7-4	A-7	A-6	A-5	A-8
GENERAL DESCRIPTION	SAND	SAND	SILTY-SAND	SANDY-SILT	CLAYEY-SAND	SILT	SANDY-CLAY	CLAYEY SILT	SILTY CLAY	CLAY	COLLOIDAL CLAY	MICA, DIATOMS & SILT	SWAMP MUCK
STABILITY	WHEN N.P., HIGH; WHEN PLASTIC, GOOD WHEN DRY	IDEAL, WHEN DRY CONFINED	GOOD WHEN DRY	GOOD WHEN DRY	GOOD WHEN DRY & PROPERLY COMPACTED	GOOD WHEN PROPERLY COMPACTED OR UNDISTURBED	GOOD WHEN PROPERLY COMPACTED OR UNDISTURBED	GOOD WHEN PROPERLY COMPACTED OR UNDISTURBED	GOOD WHEN PROPERLY COMPACTED OR UNDISTURBED	GOOD WHEN PROPERLY COMPACTED OR UNDISTURBED	GOOD WHEN PROPERLY COMPACTED OR UNDISTURBED	GOOD TO POOR	NONE
USE AS A BASE	FAIR	EXCELL.	FAIR	FAIR	POOR	POOR	POOR	POOR	POOR	POOR	POOR	POOR	NONE
USE AS A SUB-BASE	EXCELL.	EXCELL.	FAIR	FAIR	FAIR	UNSATISFACTORY	UNSATISFACTORY	UNSATISFACTORY	UNSATISFACTORY	UNSATISFACTORY	UNSATISFACTORY	UNSATISFACTORY	WASTE
USE AS A SUB-GRADE	EXCELL.	EXCELL.	GOOD	GOOD	FAIR	UNSATISFACTORY	UNSATISFACTORY	UNSATISFACTORY	UNSATISFACTORY	UNSATISFACTORY	UNSATISFACTORY	UNSATISFACTORY	WASTE
FILLS UNDER 50'	EXCELL.	GOOD	GOOD	POOR	FAIR	GOOD TO POOR	GOOD TO POOR	VERY POOR	POOR	VERY POOR	VERY POOR	VERY POOR	WASTE
FILLS OVER 50'	GOOD	GOOD TO FAIR	FAIR	POOR	FAIR	POOR	POOR	VERY POOR	VERY POOR	VERY POOR	VERY POOR	VERY POOR	WASTE
FROST ACTION	NONE TO LOW	NONE TO LOW	MEDIUM	HIGH	MEDIUM	HIGH	MEDIUM	MEDIUM TO HIGH	MEDIUM TO HIGH	MEDIUM	MEDIUM	HIGH	MEDIUM
RANGE OF MAX. DRY DENSITY (AASHO T-180) (PCF)	115-135	105-130	110-130	110-135	115-135	110-135	115-130	110-135	105-130	100-120	90-115	100-135	LESS THAN 100
RANGE OF OPTIMUM MOISTURE CONTENTS (AASHO T-180 %)	9-12	8-15	8-15	9-15	6-12	8-15	9-15	10-17	10-15	12-25	14-30	11-18	—
REQUIRED COMPACTION (AASHO T-180) (%)	92-95	92-95	92-95	92-95	92-95	92-95	92-95	≤ 95%	≤ 95%	≤ 95%	≤ 95%	≤ 95%	WASTE
COMPACTION METHODS	ROLLING WITH SMOOTH FACE TAMPING RUBBER-TIRED ROLLER OR VIBRATORY COMPACTOR	TRACTOR DISKING VIBRATION	TAMPING OR RUBBER-TIRED ROLLER	TAMPING OR RUBBER-TIRED ROLLER	TAMPING OR RUBBER-TIRED ROLLER	TAMPING SHEEPFOOT ROLLER OR RUBBER-TIRED ROLLER	TAMPING SHEEPFOOT ROLLER OR RUBBER-TIRED ROLLER	TAMPING SHEEPFOOT ROLLER OR RUBBER-TIRED ROLLER	TAMPING SHEEPFOOT ROLLER OR RUBBER-TIRED ROLLER	TAMPING SHEEPFOOT ROLLER OR RUBBER-TIRED ROLLER	TAMPING SHEEPFOOT ROLLER OR RUBBER-TIRED ROLLER	TAMPING OF RUBBER-TIRED ROLLER	WASTE
COMPACTION ABILITIES	GOOD WITH CLOSE CONTROL	GOOD	GOOD TO POOR	GOOD TO POOR	GOOD TO POOR	GOOD TO POOR	GOOD TO POOR	FAIR TO POOR	FAIR TO POOR	FAIR TO POOR	FAIR TO POOR	VERY POOR	WASTE
PUMPING ACTION	NONE	NONE	SLIGHT TO NONE	SLIGHT TO NONE	SLIGHT TO NONE	GOOD TO FAIR	GOOD TO FAIR	GOOD TO FAIR	GOOD TO FAIR	GOOD TO FAIR	GOOD TO FAIR	GOOD TO FAIR	GOOD TO FAIR
BEARING VALUE	EXCELLENT TO FAIR	EXCELLENT TO FAIR	GOOD TO FAIR	GOOD TO FAIR	GOOD TO FAIR	FAIR TO POOR	FAIR TO POOR	FAIR TO POOR	FAIR TO POOR	FAIR TO POOR	POOR	POOR	POOR
DRAINAGE	GOOD	DRAINS FREELY	FAIR TO PRACTICALLY IMPERVIOUS	FAIR TO PRACTICALLY IMPERVIOUS	FAIR TO PRACTICALLY IMPERVIOUS	FAIR TO PRACTICALLY IMPERVIOUS	FAIR TO PRACTICALLY IMPERVIOUS	POOR	POOR	POOR	IMPERVIOUS	FAIR TO IMPERVIOUS	POOR

NOTES :

A-2 TO A-3 SOILS: WHEN USED AS A BASE, PLASTICITY INDEX AND LIQUID LIMIT SHOULD NOT EXCEED 6 AND 25 RESPECTIVELY. BEST FOR SOIL-CEMENT STABILIZATION, GENERALLY 8 TO 12 % CEMENT BY WEIGHT WILL BE SUFFICIENT.

NON-PLASTIC A-2 TO A-3 SOILS : MAY REQUIRE VIBRATION FOR COMPACTION.

A-4 TO A-7 SOILS: FILLS SHOULD BE PLACED IN DRY SEASON.

A-4 SILTS: SUSCEPTIBLE TO SETTLEMENT AND EROSION.

A-5 SOILS: WHEN MICA IS PRESENT, VERY DIFFICULT TO COMPACT BECAUSE OF EXPANSION AND REBOUND.

A-6 SOILS (CLAY): WILL PUMP IN POROUS BASES FORMING CRACKS. FILLS WILL SETTLE OVER LONG PERIODS OF TIME. HIGH BANKS IN CUTS AND FILLS VERY LIABLE TO SLIDE.

Soils & Soil – Aggregate Mixtures
Characteristics & Performances

TYPE	TREATMENT & USE
A-2 SOILS:	WELL GRADED TO POORLY GRADED SAND AND GRAVELS. GOOD BASE FOR MODERATE FLEXIBLE OR THIN RIGID PAVEMENT. GOOD FILL. FROST HEAVE, BREAK-UP IF PLASTIC. SOFTENS WHEN WET IF PLASTIC. USE BASE COURSE WHEN SUB-GRADE P.I. IS GREATER THAN 6. SUB-DRAINAGE EFFECTIVE. STABILIZE; WITH BITUMEN, CHLORIDES, CEMENT OR ADMIXTURE SOIL.
A-3 SOILS:	CLEAN SANDS AND GRAVELS. IDEAL BASE FOR MODERATE FLEXIBLE OR THIN RIGID PAVEMENT. GOOD FILL. NO FROST HEAVE OR BREAK-UP. SUB-DRAINAGE ONLY THROUGH IMPERVIOUS SHOULDERS. STABILIZE; WITH SOIL BINDER, BITUMINOUS, OR CHEMICAL ADMIXTURES.
A-4 SOILS:	SILTY SOILS. NOT GOOD FOR SURFACE. POOR BASE. ABSORBS WATER. UNSTABLE WHEN WET. BAD FROST HEAVE AND BREAK-UP. USE SUB-DRAINAGE AND/OR BASE AND SUB-BASE WITH FLEXIBLE PAVEMENT. USE BITUMINOUS SUB-GRADE PRIME. USE THICK CONCRETE PAVEMENT (7" TO 10") WITH STEEL REINFORCEMENT AND CRACK CONTROL.
A-5 SOILS:	ELASTIC SILTS. USE SUB-DRAINAGE AND/OR GRANULAR BASE AND SUB-BASE WITH BITUMINOUS SUB-GRADE PRIME. USE THICK CONCRETE PAVEMENT, REINFORCED WITH CRACK CONTROL.
A-6 SOILS:	CLAYS. IMPERMEABLE AND STABLE WHEN DRY AND UNDISTURBED (HARD CLAY). PLASTIC AND ABSORBENT IF DISTURBED. BAD PUMPING INTO POROUS BASE. MACADAM OR PAVEMENT JOINTS. SHRINKS OR CRACKS WHEN DRY. USE GRANULAR BASE AND SUBBASE. USE SUB-DRAINAGE ONLY WHEN MADE PERVIOUS BY CRACKS, ROOT HOLES AND LAMINATIONS. FROST HEAVE SLIGHT WHEN IMPERMEABLE; BAD WHEN PERVIOUS. USE SUB-GRADE PRIME. USE THICK, STRONG, DENSE FLEXIBLE PAVEMENT OR REINFORCED CRACK CONTROLLED CONCRETE.
A-7 SOILS:	EXPANSIVE, PLASTIC CLAYS. EXCESSIVE VOLUME CHANGE. BAD FROST HEAVE AND BREAK-UP. SUB-DRAINAGE NOT EFFECTIVE. USE THICK, DENSE, FLEXIBLE PAVEMENT WITH BASE AND SUB-BASE OVER SUB-GRADE PRIME OR REINFORCED CRACK CONTROLLED CONCRETE PLACED ON IMPERVIOUS PAPER.
A-8 SOILS:	MUCK AND PEAT. UNFIT FOR CONSTRUCTION PURPOSES. EXCAVATE TO SOLID STRATUM AND REPLACE WITH SELECTED FILL. DISPLACEMENT BY SUPERIMPOSED FILL IS DOUBTFUL. DISPLACEMENT BY EXPLOSIVE UNDER SUPERIMPOSED FILL IS SOMETIMES EFFECTIVE.

Soils & Soil - Aggregate Mixtures
Treatment & Use

	SYMBOLS	TYPICAL GRADING	TYPICAL PHYSICALS	REMARKS
A-3 SAND		C.S. =22% F.S. =48% SILT =20% CLAY = 8% COLL. = 2%	L.L. = N.P P.I. = N.P.	SAND-53% MIN. %-#200-20% MAX. P.I.- N.P L.L.-MUST BE N.P
A-2 SAND & FINES		C.S. =20% F.S. =43% SILT =19% CLAY =10% COLL. = 8%	L.L. =22 P.I. = 2 S.L. =18	SAND-53% MIN. %-#200-20% MAX. P.I.-7 MAX. L.L.-34 MAX. (MUST HAVE L.L.)
A-2-4 SILTY SAND		C.S. =25% F.S. =30% SILT =32% CLAY = 7% COLL. = 6%	L.L. =24 P.I. = 2 S.L. =21	SAND-53% MIN. % #200-21% MIN.-30% MAX. P.I.-7 MAX. L.L.-34 MAX.(MAY BE N.P)
A-4-2 SANDY SILT		C.S. =23% F.S. =28% SILT =33% CLAY =10% COLL. = 6%	L.L. =25 P.I. = 3 S.L. =21	SAND-48% MIN %-#200-31% MIN. P.I.-7 MAX. L.L.-40 MAX.(MAY BE N.P)
A-2-7 CLAYEY SAND		C.S =38% F.S. =31% SILT =15% CLAY = 8% COLL. = 8%	L.L. =31 P.I. =10 S.L. =18	SAND-48% MIN. CLAY-29% MAX. P.I.-8-14 L.L.-40 MAX.
A-7-2 SANDY CLAY		C.S. =20% F.S. =29% SILT =17% CLAY =21% COLL. =13%	L.L. =39 P.I. =15 S.L. =16	SAND-48% MIN. CLAY-17%-35% P.I.-15 MIN. L.L.-30 MIN.
A-4 SILT		C.S. =20% F.S. =22% SILT =40% CLAY =10% COLL. = 8%	L.L. =30 P.I. = 6 S.L. =19	SAND-47% MAX. CLAY-29% MAX. P.I.-9 MAX. L.L.-40 MAX.
A-4-7 CLAYEY SILT		C.S. = 8% F.S. = 17% SILT =40% CLAY =23% COLL. =12%	L.L. =33 P.I. =11 S.L. =18	SAND-47% MAX. CLAY-25% MIN. P.I.-14 MAX. L.L.-40 MAX.
A-7-4 SILTY CLAY		C.S. =18% F.S. =20% SILT =35% CLAY =12% COLL. =15%	L.L. =39 P.I. =15 S.L. =16	SAND-47% MAX. CLAY-29% MAX. P.I.-15 MIN. L.L.-30 MIN.
A-7 CLAY		C.S. =18% F.S. =22% SILT =23% CLAY =22% COLL. =15%	L.L. =40 P.I. =17 S.L. =15	SAND-47% MAX. CLAY-30%-59% P.L-15 MIN. L.L.-35 MIN.
A-6 COLLOIDAL CLAY		C.S. = 6% F.S. = 7% SILT =18% CLAY =33% COLL. =36%	L.L. =50 P.I. =33 S.L. =14	CLAY-60% MIN. P.I.-25 MIN. L.L.-45 MIN.
A-5 MICA, DIATOMS, DECOMPOSED ROCK		C.S. =15% F.S. =35% SILT =30% CLAY =15% COLL. = 5%	L.L. =35 P.I. = 4 S.L. =26	GRAD. NOT SIGNIFICANT P.I.- LOW L.L.-HIGH S.L.-26 MIN. VISUAL INSPECTION NECESSARY TO DETERMINE TYPE
A-8 SWAMP MUCK		C.S. =18% F.S. =26% SILT =45% CLAY = 7% COLL. = 4%	L.L. =52 P.I. = 7 S.L. =38	ORGANIC CONTENT-4% MIN. P.I. - LOW L.L.-HIGH, WHEN OBTAINABLE S.L. - 26 MIN.

ROCK REFUSAL

Soils & Soil - Aggregate Mixtures
Guide to Classification

RAILS TO BE UNTREATED TIMBER.

POSTS TO BE TREATED TIMBER. WHERE PAINT IS
CALLED FOR, TWO COATS TO BE APPLIED

ALL LUMBER TO BE SOUTHERN YELLOW PINE OF 1400# STRESS GRADE
OR BETTER.

TO BE PAINTED IN A COLOR SCHEME AND STRIPING DETAILS IN
ACCORDANCE WITH THE LATEST " MANUAL OF UNIFORM TRAFFIC CONTROL DEVICES."

TO BE USED WHERE NOTED ON PLANS

THE COST OF ALL EXCAVATION AND BACKFILL TO BE INCLUDED IN
LUMP SUM PRICE BID FOR TIMBER BARRICADE.

THE LUMP SUM PRICE BID SHALL INCLUDE THE FURNISHING OF
ALL MATERIALS, PRESERVATIVE TREATMENTS, TIMBER CONNECTORS
AND HARDWARE, PAINTING, GALVANIZING AND CONCRETE AS WELL AS ALL LABOR,
TOOLS AND EQUIPMENT AND ALL WORK INCIDENTAL THERETO.

Standard Timber Barricade

Standard Berm Ditches

CONCRETE SURFACE DRAIN DITCH

SOD SURFACE DRAIN DITCH
"A"

CONCRETE SURFACE DRAIN DITCH

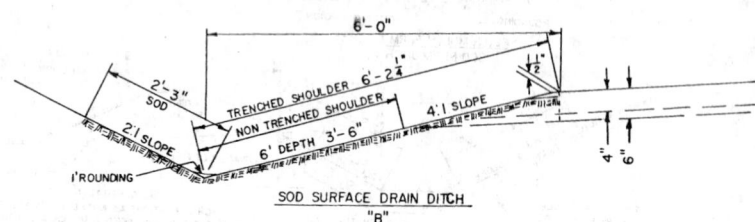

SOD SURFACE DRAIN DITCH
"B"

GENERAL NOTES

1. CONCRETE TO BE EXTENDED AS REQUIRED WHERE DEPTH OF FLOW IN DITCH EXCEEDS LIMIT OF CONCRETE AS INDICATED ABOVE.
2. OMIT SOD STRIP ON SHOULDER SIDE OF DITCH OF NON-TRENCHED SHOULDERS.

Standard Surface Drain Ditches
Concrete & Sod

CONCRETE SIDE DITCH

SOD SIDE DITCH
"A"

CONCRETE SIDE DITCH

SOD SIDE DITCH
"B"

GENERAL NOTES

WHEN DITCHES ARE CONSTRUCTED OTHER THAN WHAT IS
SHOWN, THE ELEVATIONS WILL BE NOTED ON THE PLANS.

Standard Side Ditches Cut Areas

SEED & MULCH MEDIAN DITCH
SOD MEDIAN DITCH

CONCRETE MEDIAN DITCH

"B"	"D"
1'-0"	0.24'
1'-6"	0.36'
2'-0"	0.49'
2'-6"	0.61'
3'-0"	0.73'
3'-6"	0.85'
4'-0"	0.97'

NOTES:
1. DITCH DEPTHS OTHER THAN 6" TO BE NOTED BY ELEVATIONS ON PLANS.
2. DIMENSION "B" MAY BE MODIFIED BY NOTE ON PLANS.
3. 4" TOP SOIL AND SEED & MULCH ALL CONSTRUCTION SLOPES NOT OTHERWISE TREATED.
4. WHERE DITCH IS OTHER THAN 0'-6" DEPTH,THE 4:1 SLOPE RATIO SHALL BE MAINTAINED,EXCEPT AT TRANSITIONS TO INLETS OF DRAINAGE STRUCTURES.
5. WHERE ℄ DITCH IS NOT AT ℄ MEDIAN,OFFSET DISTANCES SHALL BE INDICATED ON PLANS.

SEED & MULCH OR
SOD SIDE DITCH

CONCRETE SIDE DITCH

"B"	"D"
1'-6"	0.67'
1'-9"	0.78'
2'-0"	0.89'
2'-3"	1.01'
2'-6"	1.12'
2'-9"	1.23'
3'-0"	1.34'
3'-3"	1.45'
3'-6"	1.57'
3'-9"	1.68'
4'-0"	1.79'

NOTES:
1. ELEVATIONS AS NOTED ON PLANS.
2. DIMENSIONS "A" AND "B" MAY BE MODIFIED BY NOTE ON PLANS.
3. 2" TOP SOIL AND SEED & MULCH ALL CONSTRUCTION SLOPES NOT OTHERWISE TREATED
4. WHERE DITCH IS NOT ADJACENT TO TOE OF FILL,DISTANCES FROM ℄ DITCH TO A REFERENCE POINT SHALL BE INDICATED ON THE PLANS.

Standard Side Ditches
Standard Median Ditches - Flat

CONCRETE MEDIAN 'V' DITCH

SOD MEDIAN 'V' DITCH

Standard Median Ditches - V Slope

NOTE: IF POSSIBLE, DO NOT CONSTRUCT A CHIMNEY CONNECTION
FROM INLET TO X-DRAIN BUT MAKE A 45° CONNECTION TO
X-DRAIN. WHEN NECESSARY TO USE A CHIMNEY CONNECTION
POUR A 6" CONCRETE COLLAR AROUND THE VERTICAL
PIPE FOR ITS ENTIRE HEIGHT.

Standard Pipe Outlets
Surface Drainage on Fills

PLAN

SECTION A-A

PLAN

SECTION A-A

SECTION B-B

TRANSITION OF CURB FACE AND CROSS
SLOPE AT DEPRESSED INLETS

Method of Depressing Pavement
at Inlets

4-#4 STRAIGHT BARS VERTICAL IN FRONT FACE FOR
12" DIA. TO 18" DIA. PIPE ENDWALLS INCLUSIVE

#4 STRAIGHT BARS HORIZONTAL
@ 1'-7" MAX. C/C BOTH FACES-
BOTTOM BARS TO BE FULL LENGTH-
ALL ENDWALLS

4 BENT BARS @ 1'-0"
C/C - ALL ENDWALLS

#4 STRAIGHT BARS
HORIZONTAL @ 1'-0"
C/C BOTH SIDES
OF OPENING FOR
36" TO 60" DIA.
PIPE ENDWALLS

$\frac{1}{2}$:1, 2:1 OR 4:1 SLOPE

#4 STRAIGHT BARS VERTICAL @ 1'-6" MIN.
TO 2'-0" MAX. FRONT FACE FOR 24" DIA.
TO 60" DIA. PIPE ENDWALLS INCLUSIVE.

1-#4 STRAIGHT BAR
HORIZONTAL- ALL ENDWALLS

2-#4 STRAIGHT BARS HORIZONTAL
FOR 36" TO 60" DIA PIPE
ENDWALLS

SECTION A-A

"S" DISTANCES FROM INSIDE SURFACE OF PIPE
TO VERTICAL BARS IN FRONT AND REAR FACE
4" FOR 12" DIA. TO 18" DIA. PIPES INCLUSIVE
6" FOR 24" DIA. TO 36" DIA. PIPES INCLUSIVE
8" FOR 42" DIA. TO 60" DIA. PIPES INCLUSIVE

ELEVATION

PLAN

DISPOSITION OF BARS DETAIL

GENERAL NOTES:
SPECIFICATIONS : LATEST S.H.A.
CONCRETE   : SEE S.H.A. SPECIFICATIONS
REINFORCING : DEFORMED STEEL BARS #4 BARS ($\frac{1}{2}$"ϕ)
CHAMFER   : ALL EXPOSED EDGES 1"X1" OR AS DIRECTED

OPENINGS		DIMENSIONS						VOLUME	STEEL
D	AREA	A	B	C	E	H	L	CONC.	
INCHES	SQ. FT.							C.Y.	LBS.
12	0.79	9"	6"	6"	1'-9"	0'-10"	4'-0"	0.27	19
15	1.23	9"	6"	6"	1'-9"	1'-0$\frac{1}{2}$"	4'-9"	0.34	23
18	1.78	9"	6"	6"	1'-9"	1'-3"	5'-6"	0.41	29
24	3.14	9"	14"	6"	2'-5"	1'-6"	7'-0"	0.69	38
30	4.91	9"	14"	6"	2'-5"	1'-9"	8'-6"	0.88	58
36	7.07	12"	16"	10"	3'-2"	2'-0"	10'-0"	1.74	90
42	9.62	12"	16"	10"	3'-2"	2'-3"	11'-6"	2.07	99
48	12.57	12"	16"	10"	3'-2"	2'-6"	13'-0"	2.43	119
54	15.90	12"	20"	12"	3'-8"	2'-9"	14'-6"	3.08	132
60	19.64	12"	20"	12"	3'-8"	3'-0"	16'-0"	3.50	158

QUANTITIES IN TABLE TO BE USED FOR ESTIMATING ONLY

**Standard End Support Wall
Round Pipe**

ELEVATION

SECTION A-A

PLAN

DISPOSITION OF BARS DETAIL

"S" DISTANCES
6" FOR 29"X 18" TO 43"X 27" INCLUSIVE
8" FOR 50"X 31" TO 72"X 44" INCLUSIVE

GENERAL NOTES:
SPECIFICATIONS : LATEST S.H.A.
CONCRETE : SEE S.H.A. SPECIFICATIONS
REINFORCING : DEFORMED STEEL BARS #4 BARS ($\frac{1}{2}\phi$)
CHAMFER : ALL EXPOSED EDGES 1"X 1" OR AS DIRECTED

OPENINGS		DIMENSIONS						VOLUME	STEEL
SIZE	AREA	A	B	C	E	H	L	CONC.	LBS.
INCHES	SQ. FT.							C.Y.	
29"X 18"	2.2	9"	14"	6"	2'-5"	1'-0"	6'-0"	.51	34
36"X22"	4.4	9"	14"	6"	2'-5"	1'-3"	7'-6"	.67	41
43"X27"	4.8	12"	16"	10"	3'-2"	1'-6"	8'-11"	1.37	59
50"X31"	6.5	12"	16"	10"	3'-2"	1'-9"	10'-5"	1.66	103
58"X36"	8.7	12"	16"	10"	3'-2"	2'-0"	12'-2"	2.01	133
65"X40"	11.0	12"	20"	12"	3'-8"	2'-3"	13'-8"	2.59	156
72"X44"	13.5	12"	20"	12"	3'-8"	2'-6"	15'-2"	2.97	176

QUANTITIES IN TABLE TO BE USED FOR ESTIMATING ONLY

Standard End Support Wall
Arch Pipe

GENERAL NOTES:
SPECIFICATIONS : LATEST S.H.A.
CONCRETE : SEE LATEST S.H.A. SPECIFICATIONS
REINFORCING: DEFORMED STEEL BARS #4 & 6
CHAMFER: ALL EXPOSED EDGES 1"X1" OR AS DIRECTED

OPENING		DIMENSIONS												VOL. CONC. C.Y.	STEEL LBS.
D INCHES	AREA SQ. FT.	B	C	E	F	G	H	J	K	L	M	N	O		
48	12.57	1'-4"	10"	3'-2"	2'-9"	7'-0¾"	5'-0"	4'-10"	6'-3½"	6'-8½"	5'-9"	2'-10¾"	5'-6"	4.3	262
54	15.9	1'-8"	1'-0"	3'-8"	3'-0"	7'-8½"	5'-6"	5'-4"	6'-10½"	7'-3½"	6'-2¼"	3'-1½"	6'-2"	5.3	301
60	19.64	1'-8"	1'-0"	3'-8"	3'-3"	8'-5"	6'-0"	5'-10"	7'-7¼"	8'-0¼"	6'-11"	3'-7½"	6'-8"	6.0	361

QUANTITIES IN TABLE TO BE USED FOR ESTIMATING ONLY

Standard Headwalls

ELEVATION

SECTION A-A

PLAN

ISOMETRIC VIEW

DISPOSITION OF BARS DETAIL

SECTION B-B

GENERAL NOTES:
SPECIFICATIONS: LATEST S.H.A.
CONCRETE: SEE S.H.A. SPECIFICATIONS
REINFORCING: DEFORMED STEEL BARS #4 & #6
CHAMFER: ALL EXPOSED EDGES 1"X1" OR AS DIRECTED

OPENING		DIMENSIONS													VOL. CONC. C.Y.	STEEL LBS.
D INCHES	AREA SQ. FT.	B	C	E	F	G	H	J	K	L	M	N	O	P		
66	23.80	2'-6"	1'-3"	4'-9"	3'-0"	11'-2½"	6'-8½"	6'-4"	10'-3¼"	10'-8½"	9'-3"	3'-5"	7'-4½"	2'-0"	9.7	585
72	28.27	2'-6"	1'-3"	4'-9"	3'-3"	12'-1"	7'-3"	6'-10"	11'-1¾"	11'-6¾"	10'-1¼"	3'-11"	7'-10½"	2'-0"	10.9	645
78	33.20	3'-0"	1'-6"	5'-6"	3'-6"	13'-0½"	7'-9½"	7'-4"	12'-0"	12'-5"	10'-9"	4'-0"	8'-6¾"	2'-6"	13.3	865
84	38.48	3'-0"	1'-6"	5'-6"	3'-9"	13'-10"	8'-4"	7'-10"	12'-9½"	13'-2½"	11'-6½"	4'-6"	9'-0¾"	2'-6"	14.7	984

QUANTITIES IN TABLE TO BE USED FOR ESTIMATING ONLY

Standard Headwalls

4 #4 STRAIGHT BARS VERTICAL IN FRONT FACE
FOR 12" TO 18" DIA. PIPE ENDWALLS INCLUSIVE

2 #4 STRAIGHT BARS HORIZONTAL 1 EACH FACE
FOR 36" DIAMETER TO 60" DIAMETER PIPE
ENDWALLS INCLUSIVE

#4 STRAIGHT BARS VERTICAL @ 1'-6" MIN. TO 2'-0" MAX.
FRONT FACE FOR 24" DIAMETER TO 60" DIAMETER.
PIPE ENDWALLS INCLUSIVE.

ELEVATION

#4 STRAIGHT BARS HORIZONTAL @ 1'-7" C/C BOTH FACES-
TOP AND BOTTOM BARS TO BE FULL LENGTH -ALL ENDWALLS.

#4 STRAIGHT BARS HORIZONTAL
@ 1'-0" C/C BOTH SIDES
OF OPENING FOR 36" TO
60" DIAMETER PIPE
ENDWALLS.

#4 BENT BARS @ 1'-0"
C/C - ALL ENDWALLS

1-#4 STRAIGHT BAR
HORIZONTAL - ALL
ENDWALLS

2 #4 STRAIGHT BARS HORIZONTAL
FOR 36" TO 60" DIAMETER PIPE ENDWALLS.

SECTION A-A

"S" DISTANCES FROM INSIDE SURFACE OF PIPE
TO VERTICAL BARS IN FRONT AND REAR FACE.
4" FOR 12" DIAMETER TO 18" DIAMETER PIPES INCLUSIVE
6" FOR 24" DIAMETER TO 36" DIAMETER PIPES INCLUSIVE
8" FOR 42" DIAMETER TO 60" DIAMETER PIPES INCLUSIVE

GENERAL NOTES:
SPECIFICATIONS : LATEST S.H.A.
CONCRETE : SEE. S.H.A. SPECIFICATIONS.
REINFORCING: DEFORMED STEEL BARS #4 φ BARS
CHAMFER: ALL EXPOSED EDGES 1"X1" OR
AS DIRECTED.

PLAN

DISPOSITION OF BARS - DETAIL

OPENINGS		DIMENSIONS							VOLUME	STEEL
D	AREA	A	B	C	E	F	H	L	CONC.	LBS.
INCHES	SQ. FT.								C.Y.	
12	0.79	9"	6"	6"	1'-9"	9"	1'-9"	6'-6"	0.61	38
15	1.23	9"	6"	6"	1'-9"	9"	2'-0"	7'-9"	0.78	55
18	1.78	9"	6"	6"	1'-9"	9"	2'-3"	9'-0"	0.95	62
24	3.14	9"	14"	6"	2'-5"	9"	2'-9"	11'-6"	1.56	91
30	4.91	9"	14"	6"	2'-5"	12"	3'-6"	14'-2"	2.19	136
36	7.07	12"	16"	10"	3'-2"	12"	4'-0"	16'-8"	4.18	222
42	9.62	12"	16"	10"	3'-2"	12"	4'-6"	19'-2"	5.13	265
48	12.57	12"	16"	10"	3'-2"	12"	5'-0"	21'-8"	6.12	326
54	15.90	12"	20"	12"	3'-8"	12"	5'-6"	24'-2"	7.68	384
60	19.64	12"	20"	12"	3'-8"	12"	6'-0"	26'-8"	8.86	428

QUANTITIES IN TABLE TO BE USED FOR ESTIMATING ONLY

Standard Type C Endwall
Round Pipe

4 - #4 STRAIGHT BARS VERTICAL IN FRONT FACE FOR
14" X 9" TO 29" X 18" PIPE ENDWALLS INCLUSIVE

2 - #4 STRAIGHT BARS HORIZONTAL, 1 EACH FACE FOR
50" X 31" TO 72" X 44" PIPE ENDWALLS INCLUSIVE

#4 STRAIGHT BARS VERTICAL @ 1'-6" MIN. TO
2'-0" MAX. FROM FACE FOR 36" X 22" TO
72" X 44" PIPE ENDWALLS INCLUSIVE

ELEVATION

#4 STRAIGHT BARS HORIZONTAL @ 1'-7" C/C BOTH FACES –
TOP & BOTTOM BARS TO BE FULL LENGTH - ALL ENDWALLS.

#4 BENT BARS @ 1'-0" C/C
ALL ENDWALLS.

#4 STRAIGHT BARS HORIZONTAL
@ 1'-0" C/C BOTH SIDES OF
OPENING FOR 50" X 31" TO
72" X 44" PIPE ENDWALLS
INCLUSIVE

1- #4 STRAIGHT BAR
HORIZONTAL ALL
ENDWALLS

2 - #4 STRAIGHT BARS HORIZONTAL
FOR 50" X 31" TO 72" X 44" PIPE ENDWALLS INCLUSIVE

SECTION A - A

"S" DISTANCES

4" FOR 14" X 9" TO 22" X 13" INCLUSIVE
6" FOR 29" X 18" TO 43" X 27" INCLUSIVE
8" FOR 50" X 31" TO 72" X 44" INCLUSIVE

GENERAL NOTES:

SPECIFICATION : LATEST S.H.A.
CONCRETE : SEE S.H.A. SPECIFICATIONS
REINFORCING : DEFORMED STEEL BARS. #4 φ BARS
CHAMFER : ALL EXPOSED EDGES 1" X 1" OR
AS DIRECTED.

PLAN

DISPOSITION OF BARS - DETAILS

OPENINGS		DIMENSIONS							VOLUME	STEEL
SIZE	AREA	A	B	C	E	F	H	L	CONC. C.Y.	LBS.
INCHES	SQ. FT.									
14"X 9"	0.8	9"	6"	6"	1'-9"	8"	1'-5"	6'-4"	.54	41
18"X11"	1.1	9"	6"	6"	1'-9"	6"	1'-5"	6'-4"	.53	41
22"X13"	1.6	9"	6"	6"	1'-9"	11"	2'-0"	9'-7"	.95	68
29"X18"	2.8	9"	6"	6"	1'-9"	6"	2'-0"	9'-7"	.92	67
36"X22"	4.4	9"	14"	6"	2'-5"	14"	3'-0"	13'-9"	1.95	120
43"X27"	6.4	9"	14"	6"	2'-5"	9"	3'-0"	13'-9"	1.89	126
50"X31"	8.7	12"	16"	10"	3'-2"	14"	3'-9"	18'-0"	4.29	226
58"X36"	11.4	12"	16"	10"	3'-2"	9"	3'-9"	18'-0"	4.19	226
65"X40"	14.3	12"	20"	12"	3'-8"	13"	4'-5"	21'-10"	6.01	310
72"X44"	17.6	12"	20"	12"	3'-8"	9"	4'-5"	21'-10"	5.89	310

QUANTITIES IN TABLE TO BE USED FOR ESTIMATING ONLY.

Standard Type C Endwall
Arch Pipe

SECTION A-A

SECTION B-B

DISPOSITION OF BARS DETAIL

PLAN

OPENINGS		DIMENSIONS						VOL.	STEEL
D	AREA	A	B	C	E	H	L	CONC.	LBS.
INCHES	SQ.FT.							C.Y.	
12	0.79	9"	6"	6"	1'-9"	1'-9"	3'-6"	0.73	56
15	1.23	9"	6"	6"	1'-9"	2'-0"	4'-3"	0.93	77
18	1.78	9"	6"	6"	1'-9"	2'-3"	5'-0"	1.14	89
24	3.14	9"	14"	6"	2'-5"	2'-9"	6'-6"	1.8	124
30	4.91	9"	14"	6"	2'-5"	3'-6"	8'-0"	2.6	160
36	7.07	12"	16"	10"	3'-2"	4'-0"	9'-6"	5.0	287
42	9.62	12"	16"	10"	3'-2"	4'-6"	11'-0"	6.2	333
48	12.57	12"	16"	10"	3'-2"	5'-0"	12'-6"	7.4	387
54	15.90	12"	20"	12"	3'-8"	5'-6"	14'-0"	9.2	458
60	19.64	12"	20"	12"	3'-8"	6'-0"	15'-6"	10.7	519

QUANTITIES IN TABLE TO BE USED FOR ESTIMATING ONLY

"S" DISTANCES FROM INSIDE SURFACE OF PIPE TO
VERTICAL BARS IN FRONT AND REAR FACE
4" FOR 12"DIA. TO 18" DIA. PIPES INCLUSIVE
6" FOR 24"DIA. TO 36"DIA. PIPES INCLUSIVE
8" FOR 42"DIA. TO 60"DIA. PIPES INCLUSIVE

GENERAL NOTES

SPECIFICATIONS: LATEST S.H.A.
CONCRETE : SEE S.H.A. SPECIFICATIONS.
REINFORCING : DEFORMED STEEL BARS - 4 φ BARS.
CHAMFER : ALL EXPOSED EDGES 1" X 1"

Standard Type E Endwall
Round Pipe

2 #4 BENT BARS, HORIZONTAL, I EACH FACE FOR 50" X 31" TO 72" X 44" ENDWALLS, INCLUSIVE

2 - #4 STRAIGHT BARS VERTICAL IN FRONT FACE FOR 14" X 9" TO 29" X 18" ENDWALLS

#4 STRAIGHT BARS HORIZONTAL @ 1'-7" MAX. C/C BOTH FACES - LAP 1'-3" TOP & BOTTOM BARS @ CORNER - BOTH WINGWALLS - ALL ENDWALLS

#4 BENT BARS @ 1'-0" C/C BOTH WINGWALLS ALL ENDWALLS

2 - 4 STRAIGHT BARS VERTICAL IN FRONT FACE FOR 14" X 9" TO 29" X 18" ENDWALLS, INCLUSIVE

2:1 OR 4:1 SLOPE

#4 STRAIGHT BARS HORIZONTAL @ 1'-0" C/C BOTH SIDES OF OPENING FOR BOTH WINGWALLS FOR 50" X 31" TO 72" X 44" ENDWALL INCLUSIVE

4 #4 STRAIGHT BARS HORIZONTAL, 2 EACH WINGWALL FOR 50" X 31" TO 72" X 44" ENDWALLS INCLUSIVE

#4 STRAIGHT BARS VERTICAL @ 1'-6" MIN. TO 2'-0" MAX. FRONT FACE FOR 36" X 22" TO 72" X 44" ENDWALLS INCLUSIVE

2 - #4 STRAIGHT BARS HORIZONTAL, I EACH WINGWALL - ALL ENDWALLS

#4 STRAIGHT BARS VERTICAL @ 1'-6" MIN. TO 2'-0" MAX. FRONT FACE FOR 36" X 22" TO 72" X 44" ENDWALLS, INCLUSIVE

SPAN + 6"

SECTION A-A

SECTION B-B

DISPOSITION OF BARS DETAIL

PLAN

"S" DISTANCES
4" FOR 14" X 9" TO 22" X 13" INCLUSIVE
6" FOR 29" X 18" TO 43" X 27" INCLUSIVE
8" FOR 50" X 31" TO 72" X 44" INCLUSIVE

GENERAL NOTES
SPECIFICATIONS : LATEST S.H.A
CONCRETE : SEE S.H.A. SPECIFICATIONS
REINFORCING : DEFORMED STEEL BARS - #4 φ BARS
CHAMFER : ALL EXPOSED EDGES 1" X 1" OR AS DIRECTED

| OPENINGS | | DIMENSIONS | | | | | | VOL. | STEEL |
| SIZE | AREA | A | B | C | E | H | L | CONC. C.Y. | LBS. |
INCHES	SQ. FT.								
14"X9"	0.8	9"	6"	6"	1'-9"	1'-5"	3'-10"	.72	56
18"X11"	1.1	9"	6"	6"	1'-9"	1'-5"	3'-10"	.71	56
22"X13"	1.6	9"	6"	6"	1'-9"	2'-0"	5'-11"	1.27	90
29"X18"	2.8	9"	6"	6"	1'-9"	2'-0"	5'-11"	1.23	87
36"X22"	4.4	9"	14"	6"	2'-5"	3'-0"	8'-7"	2.53	173
43"X27"	6.4	9"	14"	6"	2'-5"	3'-0"	8'-7"	2.47	167
50"X31"	8.7	12"	16"	10"	3'-2"	3'-9"	11'-4"	5.68	334
58"X36"	11.4	12"	16"	10"	3'-2"	3'-9"	11'-4"	5.58	331
65"X40"	14.3	12"	20"	12"	3'-8"	4'-5"	13'-10"	7.97	451
72"X44"	17.6	12"	20"	12"	3'-8"	4'-5"	13'-10"	7.85	442

QUANTITIES IN TABLE TO BE USED FOR ESTIMATING ONLY

Standard Type E Endwall
Arch Pipe

ISOMETRIC VIEW

ELEVATION

REINFORCEMENT, DEFORMED STEEL BARS
STEEL IN WINGS:
VERTICAL NO. 6 φ BARS 12" C/C
HORIZONTAL NO. 4 φ BARS 12" C/C
HOOKED ON ONE END
CONCRETE (SEE S.H.A. SPECIFICATIONS)
CHAMFER ALL EXPOSED EDGES 1" X 1" OR AS DIRECTED

TOE OF SLOPE

PLAN

1½ : 1 SLOPE = 9"
2 : 1 SLOPE = 12"
4 : 1 SLOPE = 24"

SLOPE 2:1

| OPENING | | DIMENSIONS FOR CONCRETE | | | | | | | | | | QUANTITIES | |
| DIAM PIPE D | AREA SQ.FT | ENDWALL | | | | WINGS | | | | | | 1 ENDWALL 2 WINGS | |
		L	B	F	H	W	X	Z	K₁	K	P	CONC. CU YDS	STEEL LBS
12"	0.79	1'-9"	9"	9"	1'-7"	2'-3"	1'-6"	2'-0"	9"	7"	1'-3"	0.55	38
15"	1.23	2'-0"	9"	9"	1'-11"	2'-6"	2'-6"	2'-2"	9"	8"	1'-5"	0.73	46
16"	1.39	2'-1"	9"	9"	1'-11"	2'-7"	2'-6"	2'-3"	9"	8"	1'-5"	0.75	46
18"	1.78	2'-3"	9"	9"	2'-2"	3'-0"	3'-0"	2'-7"	9"	8"	1'-7"	0.89	57
24"	3.14	3'-0"	12"	9"	2'-9"	4'-0"	4'-0"	3'-5"	12"	11"	2'-1"	1.64	82
30"	4.91	3'-6"	12"	9"	3'-4"	4'-8"	5'-2"	4'-1"	12"	11"	2'-6"	2.14	113
36"	7.07	4'-0"	12"	9"	3'-10"	5'-6"	6'-2"	4'-7"	12"	11"	2'-10"	2.48	132
42"	9.62	4'-6"	12"	9"	4'-5"	6'-4"	7'-4"	5'-6"	12"	11"	3'-2"	3.30	163
48"	12.57	5'-0"	12	9	4'-11"	7'-0"	8'-4"	6'-1"	12"	11"	3'-6"	3.85	193

SLOPE 1½ : 1

| OPENING | | DIMENSIONS FOR CONCRETE | | | | | | | | | | QUANTITIES | |
| DIAM PIPE D | AREA SQ.FT | ENDWALL | | | | WINGS | | | | | | 1 ENDWALL 2 WINGS | |
		L	B	F	H	W	X	Z	K₁	K	P	CONC. CU YDS	STEEL LBS
12"	0.79	1'-9"	9"	9"	1'-8"	2'-0"	1'-3"	1'-8"	9"	7"	1'-4"	0.51	38
15"	1.23	2'-0"	9"	9"	2'-0"	2'-4"	1'-8"	2'-0"	9"	7"	1'-5"	0.63	42
16"	1.39	2'-1"	9"	9"	2'-0"	2'-4"	1'-9"	2'-0"	9"	7"	1'-5"	0.65	42
18"	1.78	2'-3"	9"	9"	2'-3"	2'-8"	2'-1"	2'-4"	9"	8"	1'-6"	0.77	48
24"	3.14	3'-0"	12"	9"	2'-11"	3'-6"	2'-11"	3'-0"	12"	11"	2'-1"	1.43	73
30"	4.91	3'-6"	12"	9"	3'-6"	4'-4"	3'-9"	3'-5"	12"	11"	2'-4"	1.88	96
36"	7.07	4'-0"	12"	9"	4'-0"	5'-0"	4'-6"	4'-4"	12"	11"	2'-7"	2.32	118
42"	9.62	4'-6"	12"	9"	4'-7"	5'-9"	5'-5"	5'-0"	12"	11"	2'-11"	2.85	144
48"	12.57	5'-0"	12"	9"	5'-1"	6'-4"	6'-2"	5'-6"	12"	11"	3'-3"	3.37	170

SLOPE 4:1

| OPENING | | DIMENSIONS FOR CONCRETE | | | | | | | | | | QUANTITIES | |
| DIAM PIPE D | AREA SQ.FT | ENDWALL | | | | WINGS | | | | | | 1 ENDWALL 2 WINGS | |
		L	B	F	H	W	X	Z	K₁	K	P	CONC. CU YDS	STEEL LBS
12"	0.79	1'-9"	9"	9"	1'-5"	2'-6"	3'-8"	2'-2"	9"	8"	1'-3"	0.81	45
15"	1.23	2'-0"	9"	9"	1'-9"	3'-0"	5'-0"	2'-7"	9"	8"	1'-5"	1.04	68
16"	1.39	2'-1"	9"	9"	1'-9"	3'-0"	5'-0"	2'-7"	9"	8"	1'-5"	1.06	68
18"	1.78	2'-3"	9"	9"	2'-0"	3'-6"	6'-0"	3'-0"	9"	8"	1'-7"	1.26	74
24"	3.14	3'-0"	12"	9"	2'-6"	4'-6"	8'-0"	3'-11"	12"	11"	2'-2"	2.23	104
30"	4.91	3'-6"	12"	9"	3'-1"	5'-6"	10'-4"	4'-9"	12"	11"	2'-7"	3.04	147
36"	7.07	4'-0"	12"	9"	3'-7"	6'-6"	12'-4"	5'-8"	12"	11"	3'-0"	3.75	180
42"	9.62	4'-6"	12"	9"	4'-2"	7'-6"	14'-6"	6'-6"	12"	11"	3'-5"	4.67	218
48"	12.57	5'-0"	12"	9"	4'-8"	8'-3"	16'-8"	7'-4"	12"	11"	3'-0"	5.57	277

Standard Type F Endwall
Round Pipe

ISOMETRIC VIEW

ELEVATION

CONCRETE:(SEE S.H.A. SPECIFICATIONS)
STEEL: DEFORMED BARS
ENDWALLS - NO.6 φ U-BEND
WINGS - VERTICAL NO.6 φ 12" C/C - 18" BEND ONE END
       HORIZONTAL NO.4 φ 12" C/C - 6" HOOK ONE END
FLOOR - NO.4 φ 12" C/C, 2 WAYS
CHAMFER ALL EXPOSED EDGES 1"X1" OR AS DIRECTED

TOE OF SLOPE

PLAN

SLOPE 1½:1

OPENING		DIMENSIONS FOR CONCRETE										QUANTITIES 1-ENDWALL 2-WINGS	
		ENDWALL				WINGS							
DIAM. PIPE D	AREA SQ.FT	L	B	F	H	W	X	Z	K	P		CONC. CU.YDS.	STEEL LBS.
12"	0.79	3-3	9"	9"	1-8	1-11	1-3	1-8	9"	1-0		0.47	33
15"	1.23	3-6	9"	9"	2-0	2-5	1-8	2-1	9"	1-2		0.61	50
16"	1.39	3-7	9"	9"	2-0	2-5	1-9	2-1	9"	1-2		0.63	50
18"	1.78	3-9	9"	9"	2-3	2-8	2-1	2-4	9"	1-4		0.75	53
24"	3.14	5-0	12"	9"	2-11	3-5	2-11	2-11	12"	1-9		1.44	89
30"	4.91	5-6	12"	9"	3-6	4-2	3-8	3-7	12"	2-1		1.93	112
36"	7.07	6-0	12"	9"	4-0	4-0	4-6	4-2	12"	2-4		2.51	154
42"	9.62	6-6	12"	9"	4-7	5-7	5-4	4-0	12"	2-8		3.20	189
48"	12.57	7-0	12"	9"	5-1	6-3	6-2	5-5	12"	2-11		3.92	216

SLOPE 2:1

OPENING		DIMENSIONS FOR CONCRETE										QUANTITIES 1-ENDWALL 2-WINGS	
		ENDWALL				WINGS							
DIAM. PIPE D	AREA SQ.FT	L	B	F	H	W	X	Z	K	P		CONC. CU.YDS.	STEEL LBS.
12"	0.79	3-3	9"	9"	1-7	2-1	1-10	1-0	9"	1-0		0.53	37
15"	1.23	3-6	9"	9"	1-11	2-7	2-5	2-3	9"	1-3		0.70	55
16"	1.39	3-7	9"	9"	1-11	2-7	2-6	2-3	9"	1-3		0.72	55
18"	1.78	3-9	9"	9"	2-2	2-11	2-11	2-6	9"	1-5		0.82	68
24"	3.14	5-0	12"	9"	2-9	3-7	4-0	3-1	12"	1-10		1.63	92
30"	4.91	5-6	12"	9"	3-4	4-5	5-1	3-0	12"	2-2		2.24	125
36"	7.07	6-0	12"	9"	3-0	5-0	6-2	4-6	12"	2-6		2.88	166
42"	9.62	6-6	12"	9"	4-5	6-0	7-3	5-2	12"	2-0		3.74	221
48"	12.57	7-0	12"	9"	4-11	6-9	8-4	5-0	12"	3-2		4.61	262

SLOPE 4:1

OPENING		DIMENSIONS FOR CONCRETE										QUANTITIES 1-ENDWALL 2-WINGS	
		ENDWALL				WINGS							
DIAM. PIPE D	AREA SQ.FT	L	B	F	H	W	X	Z	K	P		CONC. CU.YDS.	STEEL LBS.
12"	0.79	3-3	9"	9"	1-5	2-4	4-2	2-0	9"	1-2		0.76	61
15"	1.23	3-6	9"	9"	1-9	2-11	5-3	2-6	9"	1-5		1.03	81
16"	1.39	3-7	9"	9"	1-9	2-11	5-6	2-6	9"	1-5		1.10	81
18"	1.78	3-9	9"	9"	2-3	3-1	6-4	2-11	9"	1-7		1.33	98
24"	3.14	5-0	12"	9"	2-6	4-0	8-6	3-6	12"	2-0		2.45	161
30"	4.91	5-6	12"	9"	3-1	5-0	10-5	4-4	12"	2-5		3.44	225
36"	7.07	6-0	12"	9"	3-7	5-11	12-0	5-2	12"	2-0		4.59	246
42"	9.62	6-6	12"	9"	4-2	7-0	13-6	6-1	12"	3-4		5.98	367
48"	12.57	7-0	12"	9"	4-8	7-4	17-2	6-0	12"	3-9		7.46	499

Special Type F Endwall
Round Pipe

REINFORCEMENT, DEFORMED STEEL BARS
CONCRETE (SEE S.H.A. SPECIFICATIONS)
CHAMFER ALL EXPOSED EDGES 1"x1"
OR AS DIRECTED.

1½ : 1 SLOPE = 9"
2 : 1 SLOPE = 12"
4 : 1 SLOPE = 24"

ELEVATION   SECTION A-A   PLAN   ISOMETRIC VIEW

QUANTITIES APPROXIMATE

| SLOPE 1½:1 | | | | | | | | | | | SLOPE 2:1 | | | | | | | | | | |
| OPENING | | DIMENSIONS FOR CONCRETE | | | | | | | QUANTITIES 1-ENDWALL 2-WINGS | | OPENING | | DIMENSIONS FOR CONCRETE | | | | | | | QUANTITIES 1-ENDWALL 2-WINGS | |
DIAM OF PIPE D	AREA SQ.FT	L	B	F	H	W	K	K_1	CONC CU.YDS	STEEL LBS	DIAM OF PIPE D	AREA SQ.FT	L	B	F	H	W	K	K_1	CONC CU.YDS	STEEL LBS
12"	0.79	2'-0"	9"	9"	1'-8"	1'-0"	6"	6"	0.33	28	12"	0.79	2'-0"	9"	9"	1'-7	1'-4"	6"	6"	0.36	34
15"	1.23	2'-3"	9"	9"	2'-0"	1'-5"	6"	6"	0.42	31	15"	1.23	2'-3"	9"	9"	1'-11"	1'-11"	6"	6"	0.47	37
16"	1.39	2'-4"	9"	9"	2'-0"	1'-6"	6"	6"	0.44	35	16"	1.39	2'-4"	9"	9"	1'-11"	2'-0"	6"	6"	0.50	38
18"	1.78	2'-6"	9"	9"	2'-3"	1'-10"	6"	6"	0.52	41	18"	1.78	2'-6"	9"	9"	2'-2"	2'-6"	6"	6"	0.61	53
24"	3.14	3'-6"	12"	9"	2'-11"	2'-8"	9"	9"	1.02	53	24"	3.14	3'-6"	12"	9"	2'-9"	3'-6"	9"	9"	1.17	71
30"	4.91	4'-0"	12"	9"	3'-6"	3'-6"	9"	9"	1.39	73	30"	4.91	4'-0"	12"	9"	3'-4"	4'-7"	9"	9"	1.62	91
36"	7.07	4'-6"	12"	9"	4'-0"	4'-3"	9"	9"	1.77	104	36"	7.07	4'-6"	12"	9"	3'-10"	5'-8"	9"	9"	2.12	116
42"	9.62	5'-6"	12"	9"	4'-7"	5'-1"	9"	12"	2.61	126	42"	9.62	5'-6"	12"	9"	4'-5"	6'-9"	9"	12"	3.15	141
48"	12.57	6'-0"	12"	9"	5'-1"	5'-11"	9"	12"	3.21	155	48"	12.57	6'-0"	12"	9"	4'-11"	7'-10"	9"	12"	3.88	174

Standard Type G Endwall
Round Pipe

QUANTITIES IN TABLE TO BE USED FOR ESTIMATING ONLY

D	D	A	B	C	E	H	L	G	VOL. CONC. C.Y.	STEEL LBS.
12"	12"	9"	6"	6"	1'-9"	1'-9"	3'-6"	1'-6"	0.7	62
15"	12"	9"	6"	6"	1'-9"	2'-0"	4'-3"	2'-0"	0.9	76
15"	15"	9"	6"	6"	1'-9"	2'-0"	4'-3"	2'-0"	0.9	76
18"	12"	9"	6"	6"	1'-9"	2'-3"	5'-0"	2'-3"	1.1	88
18"	15"	9"	6"	6"	1'-9"	2'-3"	5'-0"	2'-3"	1.1	85
18"	18"	9"	6"	6"	1'-9"	2'-3"	5'-0"	2'-3"	1.1	85
24"	12"	9"	14"	6"	2'-5"	2'-9"	6'-6"	3'-0"	1.8	123
24"	15"	9"	14"	6"	2'-5"	2'-9"	6'-6"	3'-0"	1.8	122
24"	18"	9"	14"	6"	2'-5"	2'-9"	6'-6"	3'-0"	1.8	117
24"	24"	9"	14"	6"	2'-5"	2'-9"	6'-6"	3'-0"	1.8	116
30"	12"	9"	14"	6"	2'-5"	3'-6"	8'-0"	3'-9"	2.6	172
30"	15"	9"	14"	6"	2'-5"	3'-6"	8'-0"	3'-9"	2.6	171
30"	18"	9"	14"	6"	2'-5"	3'-6"	8'-0"	3'-9"	2.5	171
30"	24"	9"	14"	6"	2'-5"	3'-6"	8'-0"	3'-9"	2.5	166
30"	30"	9"	14"	6"	2'-5"	3'-6"	8'-0"	3'-9"	2.5	164
36"	12"	12"	16"	10"	3'-2"	4'-0"	9'-6"	4'-6"	5.0	279
36"	15"	12"	16"	10"	3'-2"	4'-0"	9'-6"	4'-6"	5.0	279
36"	18"	12"	16"	10"	3'-2"	4'-0"	9'-6"	4'-6"	5.0	274
36"	24"	12"	16"	10"	3'-2"	4'-0"	9'-6"	4'-6"	4.9	272
36"	30"	12"	16"	10"	3'-2"	4'-0"	9'-6"	4'-6"	4.9	262
36"	36"	12"	16"	10"	3'-2"	4'-0"	9'-6"	4'-6"	4.8	260
42"	12"	12"	16"	10"	3'-2"	4'-6"	11'-0"	5'-3"	6.1	328
42"	15"	12"	16"	10"	3'-2"	4'-6"	11'-0"	5'-3"	6.1	328
42"	18"	12"	16"	10"	3'-2"	4'-6"	11'-0"	5'-3"	6.1	327
42"	24"	12"	16"	10"	3'-2"	4'-6"	11'-0"	5'-3"	6.1	317
42"	30"	12"	16"	10"	3'-2"	4'-6"	11'-0"	5'-3"	6.0	315
42"	36"	12"	16"	10"	3'-2"	4'-6"	11'-0"	5'-3"	5.9	308
42"	42"	12"	16"	10"	3'-2"	4'-6"	11'-0"	5'-3"	5.8	307
48"	12"	12"	16"	10"	3'-2"	5'-0"	12'-6"	6'-0"	7.4	404
48"	15"	12"	16"	10"	3'-2"	5'-0"	12'-6"	6'-0"	7.4	404
48"	18"	12"	16"	10"	3'-2"	5'-0"	12'-6"	6'-0"	7.4	393
48"	24"	12"	16"	10"	3'-2"	5'-0"	12'-6"	6'-0"	7.3	391
48"	30"	12"	16"	10"	3'-2"	5'-0"	12'-6"	6'-0"	7.2	383
48"	36"	12"	16"	10"	3'-2"	5'-0"	12'-6"	6'-0"	7.2	381
48"	42"	12"	16"	10"	3'-2"	5'-0"	12'-6"	6'-0"	7.1	373
48"	48"	12"	16"	10"	3'-2"	5'-0"	2'-6"	6'-0"	7.0	371

SECTION A-A

SECTION B-B

'S' DISTANCES FROM INSIDE SURFACE OF PIPE TO VERTICAL BARS IN FRONT & REAR FACE.
4" FOR 12" DIA. TO 18" DIA. PIPES INCLUSIVE
6" FOR 24" DIA. TO 36" DIA. PIPES INCLUSIVE
8" FOR 42" DIA. TO 60" DIA. PIPES INCLUSIVE

ELEVATION OF INVERT AS DIRECTED

PLAN

DISPOSITION OF BARS DETAIL

GENERAL NOTES:
SPECIFICATIONS : LATEST S.H.A.
CONCRETE : SEE S.H.A. SPECIFICATIONS
REINFORCING : DEFORMED STEEL BARS #4 ∅ BARS
CHAMFER : ALL EXPOSED EDGES 1" X 1" OR AS DIRECTED
SOD : 3' OF SOD WILL BE CALLED FOR ON INDIVIDUAL JOBS.

Standard Type H Endwall
Round Pipe

Standard Type K Inlet
Open-End Grate

Concrete Manholes (Double Coursed)
Concrete Catch Basin (lower right)

BATTER UNIT

8

6

6

6

BARREL UNIT

48" INSIDE DIAMETER

48" INSIDE DIAMETER MANHOLE

24" OPENING

BATTER UNITS PER COURSE

1st Course 12
2nd Course 12
3rd Course 10
4th Course 10

30" OPENING

BATTER UNITS PER COURSE

1st Course 12
2nd Course 12
3rd Course 10

The diameter of the opening at the top of the Manhole varies 6" for each course of Batters used. For example, 4 courses of Batters gives a 24" Opening; 3 courses of Batters gives a 30" Opening.

You will note the Barrel Units are 6" high and Batter Units are 8" high.

For Manhole depths of over 12 feet, double courses of Barrel Units are recommended ──

Specifications for Concrete Manholes

UNITS PER COURSE FOR 42"x48" INSIDE DIMENSIONS

Catch Basin Stretchers 9
Catch Basin Corners 4

1st COURSE

2nd COURSE

UNITS PER COURSE FOR 48"x48" INSIDE DIMENSIONS

Catch Basin Stretchers 12
Catch Basin Corners 4

1st COURSE

2nd COURSE

CORNER UNIT

STRETCHER UNIT

Catch Basins and Inlet Units

Standard Underdrains

CUTAWAY VIEW

DITCH LINER MAY BE:
a. SOD
b. JUTE OVER SOD
c. JUTE OVER SEED (AS SHOWN)

SURFACE STAPLES 6" OC IN TWO STAGGERED ROWS ON EITHER SIDE OF EROSION STOP

VARIABLE SLOPES SEEDED & MULCHED

EDGE STAPLES 2' O.C.

CENTER STAPLES 6" O.C.

DITCH SEEDED & FERTILIZED

FIBERGLASS EROSION STOP

FIBERGLASS TO BE CLIPPED, OVERLAPPED, AND STAPLED AT VERTICAL ANGLE CHANGES

TRENCH BACKFILLED & TAMPED WITH ORIGINAL SOIL

PLAN (NOT TO SCALE)

EROSION STOP EXTENDED BEYOND DITCH LINER 2' FOR EACH FOOT OF DITCH LINER WIDTH

EDGE OF DITCH LINER (OR SOD)

SURFACE STAPLES

DETAIL A

JUTE LINER OVERLAPPED 4" IN CENTER (WHEN USED)

FIBERGLASS EROSION STOP - 2" OVERLAP AT ALL JOINTS

SECTION A-A' (NOT TO SCALE)

VARIABLE SLOPE

SURFACE STAPLES

JUTE (OR SOD) DITCH LINER

TYPICAL ROUNDING

DETAIL 'A'

HAND TAMPED BACKFILL

BOTTOM STAPLES - 1 ROW, 24" O.C.

2" CENTER OVERLAP

SECTION B-B' (NOT TO SCALE)

FLOW LEVEL

JUTE (OR SOD)

TRENCH IS PERPENDICULAR TO DITCH GRADIENT

EROSION STOP - TOP EDGE FLUSH WITH SOIL SURFACE

4" TRENCH

Sediment and Erosion Control
Fiberglass Erosion Stop

NOTE: AT INTERSECTIONS, THE THICKNESS
IS TO BE 8", AND THE TYPICAL
SECTION IS SUBJECT TO EXISTING
CONDITIONS

VALLEY GUTTER

CONCRETE SHOULDER & REBUT

CONC. LUG 9" IN DEPTH AND
9" THICK TO BE CONSTRUCTED
ON 6'-0" CENTERS

CONCRETE FLUME

FLUME

PIPE LOCATIONS UNDER DEEP FILL

CONC. FLUME - SEE
DETAIL THIS SHEET

**Standard Concrete Valley Gutter, Flumes
Concrete Shoulder & Rebut**

SECTION A-A

PLACED RIP RAP (SEE PLANS)

PLACED RIP RAP DITCH REQUIRED TO
REDUCE WATER VELOCITY AND CONTROL EROSION.

5" CONCRETE GUTTER (WITH TOE WALL
AS SHOWN ON STANDARD MD-389.02)

NOTE: FOR RIP RAP SEE LATEST
S.H.A. SPECIFICATIONS
OR SPECIAL PROVISIONS.

PLAN

LENGTH AS SHOWN ON PLANS
OR AS DIRECTED BY THE ENGINEER.

GENERAL NOTE:
ALL DIMENSIONS AND LOCATIONS
NOT INDICATED, FOR ITEMS APPEARING
ON THIS SHEET OR ON THE PLANS,
SHALL BE DIRECTED BY THE ENGINEER.

Sediment and Erosion Control
Placed Rip-Rap Ditch

JOINTS OF FACES & TOE WALL TO BE
FILLED WITH POURED MORTAR AS
PER SPECIAL PROVISIONS.

(AS SHOWN ON PLANS
OR AS DIRECTED)

DEPTH

2'-0"

BOTTOM OF PROPOSED STREAM CHANE

2'-0"

1 1/2:1 SLOPE

6"
MIN.

VARIABLE

1'-0"

1:1 SLOPE

2'-0"

TO BE USED WHERE NOTED ON PLANS
OR WHERE DIRECTED BY THE ENGINEER.

PILED RIP-RAP SHALL BE COMPOSED OF ROCK OR BOULDERS OBTAINED
FROM EXCAVATION OR OTHER APPROVED SOURCES. THE SIZES SHALL CONFORM
GENERALLY TO THE SPECIFICATIONS FOR RIP-RAP, EXCEPT THAT 25% OF THE
PIECES CAN BE LESS THAN 9" IN DIAMETER. THERE IS NO TOP LIMIT ON THE SIZE
OF THE STONE OR BOULDERS.

Piled Rip-Rap Detail

Concrete Bridge Parapet and Bridge
Median Barrier Details

Reinforced Concrete Pavement
Types of Joints

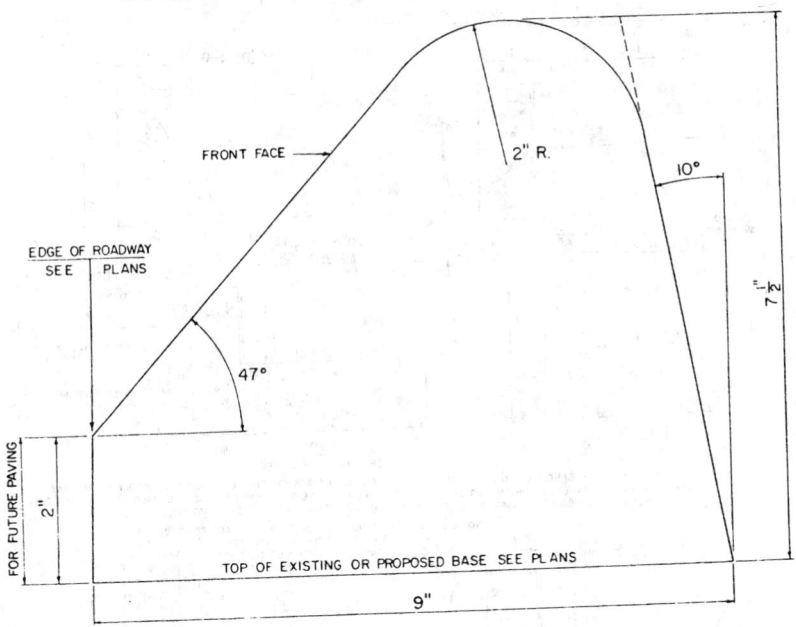

FRONT FACE

2" R.

10°

EDGE OF ROADWAY
SEE PLANS

7½"

47°

FOR FUTURE PAVING

2"

TOP OF EXISTING OR PROPOSED BASE SEE PLANS

9"

Standard Bituminous Concrete Curb

NOTE :-UNLESS OTHERWISE SPECIFIED, LONGITUDINAL TIE BAR DEVICE; TYPE 'A' OR TYPE 'B'; PLACED AT MIDDLE OF KEYWAY (NOT SHOWN) & SPACED ACCORDING TO SPECIFICATIONS FOR CONCRETE PAVEMENT SHALL BE USED AT CONSTRUCTION JOINT BETWEEN COMBINATION CURB & GUTTER AND RIGID PAVEMENT. SEE STANDARD MD-572.61.

NOTE :- JOINT SPACING FOR CONCRETE CURB AND COMB. CURB & GUTTER = 10'. SEE SPECIFICATIONS FOR LOCATION AND DESCRIPTION OF TREATMENT FOR THE TYPES OF JOINTS USED.

Standard Types of Concrete Curb
Combination Concrete Curb and Gutter

Traffic Control

Sometimes traffic must be moved through and around road and street construction, maintenance operations, utility work, curb and sidewalk installation, and any other type of work on or adjacent to roadways. Because there are so many different situations where traffic control is needed, there is no standard sequence of signs or controlling devices for this purpose, but the following provides a guide.

The general principles set forth here are applicable to both urban and rural areas. These guidelines are presented as a minimum, and one cannot emphasize too strongly that the needs of each project must be reviewed carefully in order to provide workers and the public maximum protection.

In general, traffic conditions on city streets are characterized by low speeds, wide ranges in volume, limited maneuvering space, frequent turns and cross movements, and significant pedestrian movement. Rural highways tend to see lower volume of traffic but much higher speeds, less pedestrian traffic, and fewer turns and encroachments.

Motorists must be guided in a clear and positive manner when approaching and traversing construction work areas. Warning signs in construction areas should have a black legend on an orange background. All signs intended for use during hours of darkness should be reflectorized with a material that has a smooth, sealed outer surface, or they must be illuminated to show approximately the same shape and color day or night. Note that street or highway lighting is not considered sufficient to meet the illumination requirements.

As a general rule, signs should be located on the right side of the road or street. In rural areas the edge of the sign shall be a minimum 6'-0" from the roadway edge and a minimum 5'-0" from the ground to the bottom of the sign. In an urban area, the edge of the sign shall be a minimum 2'-0" from the outside face of the curb, and the bottom of the sign shall be at least 7'-0" from the top of the sidewalk.

Temporary Sign Erection. Laborers working in pairs can install 3 single posts with 4'x4' signage in an hour:

	Hours	Rate	Total	Rate	Total
Laborers	2.0	$....	$....	$21.04	$42.08
Cost per Sign			....		14.03

To the above, add for material cost.

Laborers working in pairs should install 2 double posts with 4'x4' signage in one hour:

	Hours	Rate	Total	Rate	Total
Laborers	2.0	$....	$....	21.04	$42.08
Cost per Sign				....	21.04
To the above, add for material cost.					

The following chart shows recommended sizes and typical prices for warning signs. Costs range from $4.00 per s.f. for non-reflectorized to $6.00 per s.f. for reflectorized:

Type of Use	Size	Non-Reflector	Reflector
Person Working	30"x30"	$25.00	$37.50
"	36"x36"	36.00	54.00
"	48"x48"	64.00	96.00
Fresh Oil	48"x48"	25.00	37.50
Loose Gravel	30"x30"	25.00	37.50
Wet Paint	48"x18"	24.00	36.00
Do Not Pass	48"x24"	32.00	48.00
Road Work Ahead	48"x36"	48.00	72.00
Road Work xx Feet	48"x48"	64.00	96.00
Detour xx Feet	48"x48"	64.00	96.00
Road Closed xx Feet	42"x42"	49.00	73.50
"	48"x48"	64.00	96.00
Right Lane Closed xx Feet	48"x48"	64.00	96.00
Single Lane	48"x48"	64.00	96.00
Arrow Boards	48"x24"	32.00	48.00
Flagman Ahead	36"x36"	36.00	54.00
"	48"x48"	64.00	96.00

Wood sign posts may be used in place of steel posts, but the portion of wood to be placed in the ground may require wood preservative.

Type of Post	Length	Price Each
Steel	12'-0"	$12.00
4x4 Wood	12'-0"	6.00

When working adjacent to a roadway, a flagger may be required to stop traffic intermittently as work progresses to protect the work crew and the travelling public. When flagging is required, the estimator needs to add flagger to the work crew.

Advanced Warning Arrow Panel. These are sign panels with a matrix of lights capable of either flashing or sequential display. Necessary signs, barricades, and other traffic control devices shall be used in conjunction with advance warning arrow boards. Arrow panels are generally used for day and night lane closures, roadway diversions, and slow construction on travelled roadways. Placement of arrows is as needed to achieve the desired

ecognition distance. The color of light emitted should be yellow. Prices range
rom $4,000 to $12,000 depending on engine, fuel capacity, and sign
lumination. Other miscellaneous items are:

	Price Each
Cones, 28" High	$8.00
Plastic Barrels	35.00
Blinking Lights	20.00

Roadside Sign Rural District

Roadside Sign Urban District

CHAPTER 23

CITY COST INDEXES

Obviously, wage rates vary from place to place, and from job to job. The hourly labor rates shown throughout the book are U.S. averages for prevailing wages as of mid 1991.

The following table shows labor cost adjustment factors for 97 major U.S. cities and is intended to be used in conjunction with the labor rates used on cost tables throughout this book.

The U.S. average is the base, so its cost factor is theoretically 1.000, and all of the cost adjustment factors below are figured in relation to it. The labor rates for any of these 97 cities are computed by multiplying the labor rates shown in cost tables throughout the book by the appropriate city cost factor from the table below.

For example, the wage rate for general construction labor is shown as $19.13. This is a 1991 U.S. average. Suppose that you want the wage rate for general construction labor in Akron, Ohio. Locate the cost factor below under for Akron and multiply: 19.13 x .995 = 19.03435 or $19.03

City	Index
Akron, OH	0.995
Albany, NY	0.869
Albuquerque, NM	0.756
Anchorage, AK	1.404
Atlanta, GA	0.751
Baltimore, MD	0.863
Billings, MT	0.987
Birmingham, AL	0.633
Boise, ID	0.989
Boston, MA	1.299
Buffalo, NY	1.055
Burlington, VT	0.765
Charleston, SC	0.894
Charleston, WV	0.918
Charlotte, NC	0.854
Cheyenne, WY	0.826
Chicago, IL	1.037
Cincinnati, OH	0.904
Cleveland, OH	1.053
Columbia, SC	0.854
Columbus, OH	0.880
Dallas, TX	0.568
Dayton, OH	0.895

City	Index
Denver, CO	0.683
Des Moines, IA	0.922
Detroit, MI	1.105
Duluth, MN	1.021
El Paso, TX	0.707
Erie, PA	0.866
Fairbanks, AK	1.396
Fargo, ND	0.819
Grand Rapids, MI	0.754
Hartford, CT	1.236
Honolulu, HI	1.181
Houston, TX	0.822
Indianapolis, IN	0.827
Jackson, MS	0.517
Jacksonville, FL	0.841
Kansas City, MO	0.901
Knoxville, TN	0.637
Lansing, MI	0.862
Las Vegas, NV	1.042
Little Rock, AR	0.911
Los Angeles, CA	1.113
Louisville, KY	0.839
Madison, WI	0.831
Manchester, NH	0.765
Memphis, TN	0.729
Miami, FL	0.616
Milwaukee, WI	0.944
Minneapolis, MN	0.893
Montgomery, AL	0.596
Nashville, TN	0.663
Newark, NJ	1.287
New Haven, CT	1.294
New Orleans, LA	0.863
New York, NY	1.491
Norfolk, VA	0.560
Oakland, CA	1.478
Oklahoma City, OK	0.608
Omaha, NE	0.752
Philadelphia, PA	1.169
Phoenix, AZ	0.858
Pittsburgh, PA	0.899
Portland, ME	0.840
Portland, OR	0.887

City	Index
Providence, RI	1.092
Raleigh, NC	0.862
Reading, PA	0.818
Richmond, VA	0.793
Rochester, NY	0.983
Sacramento, CA	1.042
St. Louis, MO	0.993
St. Paul, MN	1.049
St. Petersburg, FL	0.817
Salt Lake City, UT	0.887
San Diego, CA	1.175
San Francisco, CA	1.232
Santa Fe, NM	0.925
Savannah, GA	0.773
Scranton, PA	0.857
Seattle, WA	0.943
Sioux Falls, SD	0.756
South Bend, IN	0.818
Spokane, WA	0.852
Stamford, CT	1.272
Syracuse, NY	0.947
Tampa, FL	0.847
Toledo, OH	0.950
Topeka, KS	0.594
Trenton, NJ	1.155
Tulsa, OK	0.594
Washington, DC	0.816
Wichita, KS	0.820
Wilmington, DE	0.977
York, PA	0.764
Youngstown, OH	0.904

CHAPTER **24**

MENSURATION

The information in this chapter will enable contractors and estimators to estimate quantities and costs accurately and efficiently. A number of tables use decimals in the quantities, because by using decimals, it is possible to state the fractional parts of inches, feet, and yards in a smaller space than when regular fractions are used.

For the estimator who is not thoroughly familiar with the decimal system, complete explanations are given covering the use of all classes of decimal fractions so that they can be used rapidly and accurately.

Estimating is nearly all "figures" of one kind or another, so it is essential that the estimator possess a fair working knowledge of arithmetic for estimates to be accurate and dependable.

Most of the estimator's computations involve measurements of surface and volume and are stated in lineal feet (meters), square feet (square meters), square yards, squares, containing 100 sq. ft. (10 sq.m), cubic feet (cubic meters), and cubic yards. These quantities are often further reduced to thousands of brick, board feet of lumber, etc.

The following abbreviations are used throughout the book to make all tables brief and concise and to insert them in the smallest possible space:

		Abbreviation		Symbols
Inches	=	in.	=	"
Lineal feet	=	lin. ft. or l.f.	=	'
Feet-inches	=	ft. in.	=	2'-5"
Square feet	=	sq. ft. or s.f.		
Square yards	=	sq. yds. or s.y.		
Squares	=	sqs.		
Cubic feet	=	cu. ft. or c.f.		
Cubic yards	=	cu. yds. or c.y.		
Board feet	=	b.f.		

(metric)

millimeters	=	mm		
meters	=	m		
meters & millimeters	=	m, mm		
square meter	=	sq.m	=	m2
cubic meter	=	cu.m	=	m3
Cubic yards	=	cu. yds. or c.y.		

Linear Measures

12 inches	= 12 in.	= 12"	= 1 foot	= 1 ft.	= 1'-0"
3 feet	= 3 ft.	= 3'-0"	= 1 yard	= 1 yd.	
16-1/2 feet	= 16-1/2 ft.	= 16'-6"	= 1 rod	= 1 rd.	
40 rods	= 40 rds.	= 1 furlong	= 1 fur.		
8 furlongs	= 8 fur.	= 1 mile	= 1 mi.		
5,280 feet	= 1,760 yds.	= 1 mile	= 1 mi.		

Square Measure or Measure of Surfaces

144 square inches (sq. in.)	=	1 square foot (sq. ft.)
9 square feet (sq. ft.)	=	1 square yard (sq. yd.)
100 square feet (sq. ft.)	=	1 square (sq.)*
30-1/4 square yards (sq. yds.)	=	1 square rod (sq. rd.)
160 square rods	=	1 Acre = 1 A.
43,560 sq. ft.	=	4,840 sq. yds. = 1 Acre = 1 A.
640 acres	=	640 A. = 1 square mile (sq. mi.)

*Architects' and builders' measure.

Water Quantities

1 cu. ft.	=	7.4805 gal. = 62.4 lbs.
1 gal.	=	8.34 lbs.
1 acre-foot	=	325,850 gal.
1 gal.	=	3.7854 liters
1 cu. meter	=	264.17 gal.
1 cu. ft. salt water	=	64.4 lbs.

Power Units

1 hp	=	550 ft-lbs. per sec.
1 hp	=	0.746 kw

Cubic Measure or Cubical Contents

1,728 cubic inches (cu. in.)	=	1 cubic foot (cu. ft., c.f.)
27 cu. ft.	=	1 cubic yard (cu. yd., c.y.)
128 cu. ft.	=	1 cord (cd.)
24-3/4 cu. ft.	=	1 perch (P.)*

*A perch of stone is nominally 16-1/2' long, 1' high, and 1-1/2' thick, and contains 24-3/4 c.f. However, in some states, especially west of the Mississippi, rubble work is figured by the perch containing 16-1/2 c.f. Before submitting prices on masonry by the perch, find out the common practice in your locality.

Computing Areas and Volumes

To Compute the Area of a Square, Rectangle, or Parallelogram. Multiply the length by the breadth or height. Example : Obtain the area of a wall 22'-0" (6.6 m) long and 9'-0" (2.7 m) high. 22 x 9 = 198 s.f. (6.6 x 2.7 = 17.82 sq.m).

To Compute the Area of a Triangle. Multiply the base by 1/2 the altitude or perpendicular height. Example: Find the area of the end gable of a house 24'-0" (7.2 m) wide and 12'-0" (3.6 m) high from the base to high point of roof. 24 x 6 (half the height) = 144 s.f. (7.2 x 1.8 = 12.96 sq.m).

To Compute the Circumference of a Circle. Multiply the diameter by 3.1416. The diameter multiplied by 3-1/7 is close enough for all practical purposes. Example: Find the circumference or distance around a circle, the diameter of which is 12'-0". 12 x 3.1416 = 37.6992 ft. the distance around the circle or 12 x 3-1/7 = 37.714 ft. (3.6 x 3.1416 = 11.30 m).

To Compute the Area of a Circle. Multiply the square of the diameter by 0.7854 or multiply the square of the radius by 3.1416. Example: Find the area of a round concrete column 24" or 2'-0" in diameter. The square of the diameter is 2 x 2 = 4. 4 x 0.7854 = 3.1416 s.f. the area of the circle (0.6 x 0.6 = 0.36 x 0.7854 = 0.28 sq.m).

The radius is 1/2 the diameter. If the diameter is 2'-0" (0.6 m), then the radius would be 1'-0" (0.3 m). To obtain the square of the radius, 1 x 1 = 1 (0.3 x 0.3 = 0.09). Multiply the square of the radius, 1 x 3.1416 = 3.1416 (0.09 x 3.1416 = 0.28 sq.m), the area of the circle.

To Compute the Cubical Contents of a Circular Column. Multiply the area of the circle by the height. Example: Find the cubical contents of a round concrete column 2'-0" (0.6 m) in diameter and 14'-0" (4.2 m) long.

From the previous example, the area of a circle 2'-0" (0.6 m) in diameter is 3.1416 s.f. (0.28 sq.m); 3.1416 x 14'-0" (the height) = 43.9824 c.f. (3.1416 x 4.2 = 13.19 cu.m) or for all practical purposes 44 c.f. of concrete in each column.

To Compute the Cubical Contents of any Solid. Multiply the length by the breadth or height by the thickness. Computations of this kind are used extensively in estimating all classes of building work, such as excavating, concrete foundations, reinforced concrete, brick masonry, cut stone, granite, etc.

The following charts give illustrations and formulas for all the different geometric shapes of area and volume that the estimator is likely to encounter.

AREA   (A : Area)

NAME FORMULA	SHAPE
Parallelogram A = B x h	
Trapezoid $A = \dfrac{B + C}{2}$ x h	
Triangle $A = \dfrac{B \times h}{2}$	
Trapezium (Divide into 2 triangles) A = Sum of the 2 triangles (See above)	
Regular Polygon $A = \dfrac{\text{Sum of sides (s)}}{2}$ x Inside Radius (R)	
Circle $A : \begin{array}{l} (1)\, \pi R^2 \\ (2)\, .7854 \times D^2 \\ (3)\, .0796 \times C^2 \end{array}$	

AREA (A : Area)

NAME FORMULA	SHAPE
Sector A : (1) $\dfrac{a^{\circ}}{360\,^{\circ}} \times \pi R^2$ (2) Length of area × $\dfrac{R}{2}$	
Segment A : Area of sector minus triangle (See above)	
Ellipse A = M x m x .7854	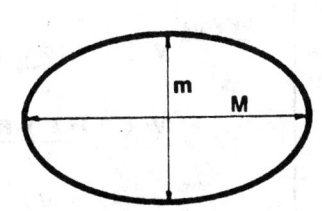
Parabola A = B x $\dfrac{2\ h}{3}$	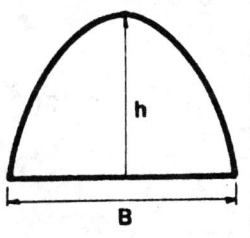

VOLUME (V : volume)

NAME FORMULA	SHAPE
Cube $V = a^3$ (in cubic units)	
Rectangular Solids $V = L \times W \times h$	
Prisms $V(1) = \dfrac{B \times A}{2} \times h$ $V(2) = \dfrac{s \times R}{2} \times 6 \times h$ $V = $ Area of end $\times h$	
Cylinder $V = \pi R^2 \times h$	

VOLUME (V : volume)

NAME FORMULA	SHAPE
Cone $V = \dfrac{\pi R^2 \times h}{3}$	
Pyramide $V\ (1) = L \times W \times \dfrac{h}{3}$ $V\ (2) = \dfrac{B \times A}{2} \times \dfrac{h}{3}$ V = Area of Base $\times \dfrac{h}{3}$	
Sphere $V = \dfrac{1}{6}\ \pi D^3$	
Circular Ring (Torus) $V = 2\pi^2 \times R r^2$ V = Area of section $\times 2\pi R$	

Example: Find the cubical contents of a wall 42'-0" (12.6 m) long, 5'-6" (1.6 m) high, and 1'-4" (0.39 m) thick, 42'-0" x 5'-6" x 1'-4" = 308 c.f. (12.6 x 1.65 x 0.39 = 8.1 cu.m) To reduce cu. ft. to cu. yds., divide 308 by 27. The result is 11-11/27 or 11-1/2 or 11.41 cu. yds.

How to Use the Decimal System in Estimating

A decimal is a fraction whose denominator is not written and is some power of 10. It is often called a decimal fraction but more often simply a decimal. Example: We know 50 cents is 50/100 or 1/2 of a dollar. Writing the same thing in decimals would be $0.50.

Numerator	50
Denominator	100

When written as 0.50 or 0.5, the denominator of the fraction is not written, but it is understood to be 10 from the fact that 5 occupies the first place to the right of the decimal point. Therefore we can have the following:

0.5 means 5/10, for the 5 extends to the 10th place;
0.25 means 25/100, for the 25 extends to the 100th place;
0.125 means 125/1000, for the 125 extends to the 1000th place.

A whole number and a decimal together, form a *mixed decimal*. Example: 2.25 is the same as 2-25/100 or 2-1/4.

The period written at the left of tenth place is called the decimal point. Example: 0.5 = 5/10; 0.25 = 25/100 = 1/4; 0.375 = 375/1000 = 3/8, etc.

It is not necessary to write a zero at the left of the decimal point in the above examples, because 0.5 means the same as .5. The zero is often written to call attention more quickly to the decimal point.

In the construction business, decimals are used chiefly to denote feet and inches, hours and minutes, the fractional working units of the various kinds of materials, and in money, which is dollars and cents or fractional parts of 100.

Table of Feet and Inches Reduced to Decimals

The following table illustrates how feet and inches may be expressed in four different ways, all meaning the same thing.

1 inch	=	1"	=	1/12th foot	=	0.083
1-1/2 inches	=	1-1/2"	=	1/8th foot	=	0.125
2 inches	=	2"	=	1/6th foot	=	0.1667
2-1/2 inches	=	2-1/2"	=	5/24ths ft	=	0.2087

3 inches	=	3"	=	1/4th foot	=	0.25
3-1/2 inches	=	3-1/2"	=	7/24ths ft	=	0.2917
4 inches	=	4"	=	1/3rd foot	=	0.333
4-1/2 inches	=	4-1/2"	=	3/8ths ft	=	0.375
5 inches	=	5"	=	5/12ths foot	=	0.417
5-1/2 inches	=	5-1/2"	=	11/24ths ft	=	0.458
6 inches	=	6"	=	1/2 foot	=	0.5
6-1/2 inches	=	6-1/2"	=	13/24ths ft	=	0.5417
7 inches	=	7"	=	7/12ths foot	=	0.583
7-1/2 inches	=	7-1/2"	=	5/8ths ft	=	0.625
8 inches	=	8"	=	2/3rds foot	=	0.667
8-1/2 inches	=	8-1/2"	=	17/24ths ft	=	0.708
9 inches	=	9"	=	3/4ths foot	=	0.75
9-1/2 inches	=	9-1/2"	=	19/24ths ft	=	0.792
10 inches	=	10"	=	5/6ths foot	=	0.833
10-1/2 inches	=	10-1/2"	=	7/8ths ft	=	0.875
11 inches	=	11"	=	11/12ths foot	=	0.917
11-1/2 inches	=	11-1/2"	=	23/24ths	=	0.958
12 inches	=	12"	=	1 foot	=	1.0

Example : Write 5 feet, 7-1/2 inches, in decimals. It would be written 5.625, which is equivalent to 5-5/8 feet.

Table of Common Fractions Stated in Decimals

The following table gives the decimal equivalents of common fractions frequently used in estimating:

1/16	=	0.0625		9/16	=	0.5625
1/8	=	0.125		5/8	=	0.625
3/16	=	0.1875		11/16	=	0.6875
1/4	=	0.25		3/4	=	0.75
5/16	=	0.3125		15/16	=	0.9375
3/8	=	0.375		13/16	=	0.8125
7/16	=	0.4375		7/8	=	0.875
1/2	=	0.5		8/8	=	1.0

Annexing zeros to a number does not change its value; for example, 0.5 is the same as 0.500.

Table of Hours and Minutes Reduced to Decimals

The following table illustrates how minutes may be reduced to fractional parts of hours and to decimal parts of hours.

No. of Minutes		Fractional Part of an Hour		Decimal Part of an Hour	No. of Minutes		Fraction Part of an Hour		Decimal Part of an Hour
1	=	1/60th	=	0.0167	31	=	31/60ths	=	0.5167
2	=	1/30th	=	0.0333	32	=	8/15ths	=	0.5333
3	=	1/20th	=	0.05	33	=	11/20ths	=	0.55
4	=	1/15th	=	0.0667	34	=	17/30ths	=	0.5667
5	=	1/12th	=	0.0833	35	=	7/12ths	=	0.5833
6	=	1/10th	=	0.10	36	=	3/5ths	=	0.60
7	=	7/60ths	=	0.1167	37	=	37/60ths	=	0.6167
8	=	2/15ths	=	0.1333	38	=	19/30ths	=	0.6333
9	=	3/20ths	=	0.15	39	=	13/20ths	=	0.65
10	=	1/6th	=	0.1667	40	=	2/3rds	=	0.6667
11	=	11/60ths	=	0.1833	41	=	41/60ths	=	0.6833
12	=	1/5th	=	0.20	42	=	7/10ths	=	0.70
13	=	13/60ths	=	0.2167	43	=	43/60ths	=	0.7167
14	=	7/30ths	=	0.2333	44	=	11/15ths	=	0.7333
15	=	1/4th	=	0.25	45	=	3/4ths	=	0.75
16	=	4/15ths	=	0.2667	46	=	23/30ths	=	0.7667
17	=	17/60ths	=	0.2833	47	=	47/60ths	=	0.7833
18	=	3/10ths	=	0.30	48	=	4/5ths	=	0.80
19	=	19/60ths	=	0.3167	49	=	49/60ths	=	0.8167
20	=	1/3rd	=	0.333	50	=	5/6ths	=	0.8333
21	=	7/20ths	=	0.35	51	=	51/60ths	=	0.85
22	=	11/30ths	=	0.3667	52	=	13/15ths	=	0.8667
23	=	23/60ths	=	0.3833	53	=	53/60ths	=	0.8833
24	=	2/5ths	=	0.40	54	=	9/10ths	=	0.90
25	=	5/12ths	=	0.4167	55	=	11/12ths	=	0.9167
26	=	13/30ths	=	0.4333	56	=	14/15ths	=	0.9333
27	=	9/20ths	=	0.45	57	=	19/20ths	=	0.95
28	=	7/15ths	=	0.4667	58	=	29/30ths	=	0.9667
29	=	29/60ths	=	0.4833	59	=	59/60ths	=	0.9833
30	=	1/2	=	0.5	60	=	1	=	1.0

Example: Write 4 hours and 37 minutes in decimals. It would be written 4.6167. Now find the labor cost for 4.6167 hrs. at $7.50 an hr.:

Hours	4.6167
X Hourly Rate	$7.50
	2308350
	323169
	$34.625250
	or $34.63

Similar Decimals. Decimals that have the same number of decimal places are called similar decimals. Thus, 0.75 and 0.25 are similar decimals and so are 0.150 and 0.275; but 0.15 and 0.275 are dissimilar decimals.

To reduce dissimilar decimals to similar decimals, give them the same number of decimal places by annexing or cutting off zeros. Example, 0.125, 0.25, 0.375, and 0.5 may all be reduced to thousandths as follows : 0.125, 0.250, 0.375, and 0.500.

To Reduce a Decimal to a Common Fraction. Omit the decimal point, write the denominator of the decimal, and then reduce the common fraction to its lowest terms. Example: 0.375 equals 375/1000, reduced to its lowest terms, equals 3/8:

375/1000 = 15/40 = 3/8

How to Add Decimals. To add numbers containing decimals, write like orders under one another, and then add as with whole numbers. Example, add 0.125, 0.25, 0.375, and 1.0. The sum is 1.750, which equals 1-750/1000, and which may be further reduced to 1-3/4:

$$
\begin{array}{r}
0.125 \\
0.25 \\
0.375 \\
+1.0 \\
\hline
1.750
\end{array}
$$

How to Subtract Decimals. To subtract one number from another, when either or both contain decimals, reduce to similar decimals, write like orders under one another, and subtract as with whole numbers. Example: Subtract 20 hours and 37 minutes from 27 hours and 13 minutes, both being written in decimals. The difference is 6.6000 or 6.6 hours, which reduced to a common fraction is 6-3/5 hours or 6 hours and 36 minutes.

$$
\begin{array}{r}
27.2167 \\
-20.6167 \\
\hline
6.6000
\end{array}
$$

How to Multiply Decimals. The same method is used as in multiplying other numbers, and the result should contain as many decimal points as there are in both of the numbers multiplied. Example: Multiply 3 feet 6 inches by 4 feet 9 inches.

Multiplicand	4 ft. 9 in. =	4'-9" =	4-3/4 ft. = 4.75	4.75
Multiplier	3 ft. 6 in. =	3'-6" =	3-1/2 ft. = 3.5	3.5
To multiply, proceed as illustrated				2375
				1425
				16,625
Product				or 16.625

In the above example, there are 2 decimals in the multiplicand and 1 decimal in the multiplier. The result or product should contain the same number of decimals as the multiplicand and multiplier combined, which is three. Starting at the right and counting to the left three places, place the decimal point between the two sixes. The result would be 16.625, which equals 16-625/1000 = 16-5/8 sq. ft.

Practical Examples Using Decimals

The quantities of cement, sand and gravel required for one cu. yd. of concrete are ordinarily stated in decimals. For instance, 1 cu. yd. of concrete mixed in the proportions of 1 part cement:2 parts sand:4 parts gravel is usually stated 1:2:4 and requires the following materials:

1.50 bbls. cement = 1-50/100 bbls. = 1-1/2 bbls. = 6 sacks
0.42 cu. yds. sand = 42/100 cu. yds. = 21/50 cu. yds. = 11-1/3 cu. ft.
0.84 cu. yds. gravel = 84/100 cu. yds. = 22-2/3 cu. ft.

There are 4 sacks of cement to the bbl., and each sack weighs 94 lbs. and contains approximately 1 cu. ft. of cement.

There are 27 cu. ft. in a cu. yd.; to obtain the number of cu. ft. of sand required for a yard of concrete, 0.42 cu. yds = 42/100 cu. yds. = 21/50 cu. yds. and 21/50 of 27 cu. ft. =

(21x27) ÷ 50 = 567 ÷ 50 = 11.34 or 11-1/3 cu. ft. sand.

To find the cost of a cu. yd. of concrete based on the above quantities, and assuming it requires 2-1/4 hrs. labor time to mix and place one cu. yd. of concrete, proceed as follows:

1-1/2 bbls. cement=1.50 bbls. @ $9.00 per bbl.	$13.5000
11-1/3 cu. ft. sand=0.42 cu. yd. @ 5.75 per cu. yd.	2.4150
22-2/3 cu. ft. gravel=0.84 cu. yd. @ 5.75 per cu. yd.	4.8300
2-1/4 hrs. labor=2.25 hrs. @ 9.80 per hr.	22.0500
Cost per cu. yd.	$42.7950

You will note the total is carried in four decimal places. This is to illustrate the actual figures obtained by multiplying. The total would be $42.80 per cu. yd. of concrete.

In making the above multiplications you will note there are 2 decimals in both the multiplicand and the multiplier of all the amounts multiplied, so the result should contain as many decimals as the sum of the multiplicand and multiplier, which is 4. You will note all the totals contain 4 decimal places. Always bear this in mind when making your multiplications, because if the lecimal place is wrong it makes a difference of 90 percent in your total. You know the correct result is $42.7950, but suppose by mistake you counted off 5 lecimal places. The result would be $4.27950 or $4.28 per cu. yd. or just about 1/10 enough; or if you made a mistake the other way and counted off just 3 decimal places, the result would be $427.950 or $427.95 per cu. yd., or just about 10 times too much. Watch your decimal places.

The same method is used in estimating lumber. All kinds of framing lumber are ordinarily sold by the 1,000 b.f., so 1,000 is the unit or decimal used when estimating lumber.

Suppose you buy 150 pcs. of 2"x8"-16'-0", which contains 3,200 b.f. This is equivalent to 3-200/1000 = 3-2/10 = 3-1/15 thousandths or stated in decimals it may be either 3.2 or 3.200. The cost of this lumber at $230.00 per 1,000 b.f., would be obtained by multiplying 3.2 ft. at $230.00 per 1,000 b.f. as follows: 3.2 mbf of lumber @ $230.00 = $736.00.

Things to Remember When Using Decimals

0.6 indicates tenths, thus 0.6 = 6/10 = 0.60 = 60/100 = 0.600 = 600/1000
0.06 indicates hundredths, thus 0.06 = 6/100 = 0.060 = 60/1000
0.006 indicates thousandths, thus 0.006 = 6/1000
0.0006 indicates ten thousandths, thus 0.0006 = 6/10000

When multiplying decimals, always remember that the result or product must contain as many decimal places as the sum of the decimal places in both the multiplicand and the multiplier. Find the cost of 3 hrs. 20 min. laborer's time at $9.80 an hr.

3 hrs. 20 min. equal 3-1/3 hrs.	3.3333
9 dollars 80 cents an hr.	9.80
	2666640
	299997
	32666340
	32.666340
	or $32.67

There are 3 decimals in the multiplicand and 2 decimals in the multiplier, so the result or total should contain as many decimals as the sum of the multiplicand and multiplier, which is 5. Be sure your decimal point is in the *right* place and the rest is easy.

Conversion of Feet, Inches to Decimal and Metric Equivalents

Inches	Inches in Decimals	Feet in Decimals	Milli- meters	Centi- meters	Meters
1/16	0.0625	0.005	1.5875	0.15875	0.001588
1/8	0.1250	0.010	3.1750	0.31750	0.003175
3/16	0.1875	0.016	4.7625	0.47625	0.004763
1/4	0.2500	0.021	6.3500	0.63500	0.006350
5/16	0.3125	0.026	7.9375	0.79375	0.007938
3/8	0.3750	0.031	9.5250	0.95250	0.009525
7/16	0.4375	0.036	11.1125	1.11125	0.011113
1/2	0.5000	0.042	12.7001	1.27001	0.012700
9/16	0.5625	0.047	14.2876	1.42876	0.014288
5/8	0.6250	0.052	15.8751	1.58751	0.015875
11/16	0.6875	0.057	17.4626	1.74626	0.017463
3/4	0.7500	0.063	19.0501	1.90501	0.019050
13/16	0.8125	0.068	20.6376	2.06376	0.020638
7/8	0.8750	0.073	22.2251	2.22251	0.022225
15/16	0.9375	0.078	23.8126	2.38126	0.023813
1	1.0000	0.083	25.4001	2.54001	0.025400
1-1/8	1.1250	0.094	28.5751	2.85751	0.028575
1-1/4	1.2500	0.104	31.7501	3.17501	0.031750
1-3/8	1.3750	0.115	34.9251	3.49251	0.034925
1-1/2	1.5000	0.125	38.1002	3.81002	0.038100
1-5/8	1.6250	0.135	41.2752	4.12752	0.041275
1-3/4	1.7500	0.146	44.4502	4.44502	0.044450
1-7/8	1.8750	0.156	47.6252	4.76252	0.047625
2	2.0000	0.167	50.8002	5.08002	0.050800
2-1/8	2.1250	0.177	53.9752	5.39752	0.053975
2-1/4	2.2500	0.188	57.1502	5.71502	0.057150
2-3/8	2.3750	0.198	60.3252	6.03252	0.060325
2-1/2	2.5000	0.208	63.5003	6.35003	0.063500
2-5/8	2.6250	0.219	66.6753	6.66753	0.066675
2-3/4	2.7500	0.229	69.8503	6.98503	0.069850
2-7/8	2.8750	0.240	73.0253	7.30253	0.073025
3	3.0000	0.250	76.2003	7.62003	0.076200
3-1/8	3.1250	0.260	79.3753	7.93753	0.079375
3-1/4	3.2500	0.271	82.5503	8.25503	0.082550
3-3/8	3.3750	0.281	85.7253	8.57253	0.085725
3-1/2	3.5000	0.292	88.9004	8.89004	0.088900
3-5/8	3.6250	0.302	92.0754	9.20754	0.092075
3-3/4	3.7500	0.313	95.2504	9.52504	0.095250
3-7/8	3.8750	0.323	98.4254	9.84254	0.098425
4	4.0000	0.333	101.6004	10.16004	0.101600
4-1/4	4.2500	0.354	107.9504	10.79504	0.107950

Inches	Inches in Decimals	Feet in Decimals	Milli- meters	Centi- meters	Meters
4-1/2	4.5000	0.375	114.3005	11.43005	0.114300
4-3/4	4.7500	0.396	120.6505	12.06505	0.120650
5	5.0000	0.417	127.0005	12.70005	0.127001
5-1/4	4.2500	0.354	107.9504	10.79504	0.107950
5-1/2	5.2500	0.438	133.3505	13.33505	0.133351
5-3/4	5.7500	0.479	146.0506	14.60506	0.146051
6	6.0000	0.500	152.4006	15.24006	0.152401
6-1/4	6.2500	0.521	158.7506	15.87506	0.158751
6-1/2	6.5000	0.542	165.1007	16.51007	0.165101
6-3/4	6.7500	0.563	171.4507	17.14507	0.171451
7	7.0000	0.583	177.8007	17.78007	0.177801
7-1/4	7.2500	0.604	184.1507	18.41507	0.184151
7-1/2	7.5000	0.625	190.5008	19.05008	0.190501
7-3/4	7.7500	0.646	196.8508	19.68508	0.196851
8	8.0000	0.667	203.2008	20.32008	0.203201
8-1/4	8.2500	0.688	209.5508	20.95508	0.209551
8-1/2	8.5000	0.708	215.9009	21.59009	0.215901
8-3/4	8.7500	0.729	222.2509	22.22509	0.222251
9	9.0000	0.750	228.6009	22.86009	0.228601
9-1/2	9.5000	0.792	241.3010	24.13010	0.241301
10	10.0000	0.833	254.0010	25.40010	0.254001
10-1/2	10.5000	0.875	266.7011	26.67011	0.266701
11	11.0000	0.917	279.4011	27.94011	0.279401
11-1/2	11.5000	0.958	292.1012	29.21012	0.292101
12	12.0000	1.000	304.8012	30.48012	0.304801
18	18.0000	1.500	457.2018	45.72018	0.457202
24	24.0000	2.000	609.6024	60.96024	0.609602
30	30.0000	2.50	762.0030	76.20030	0.762003
36	36.0000	3.00	914.4036	91.44036	0.914404
42	42.0000	3.50	1066.8042	106.68042	1.066804
48	48.0000	4.00	1219.2048	121.92048	1.219205
54	54.0000	4.50	1371.6054	137.16054	1.371605
60	60.0000	5.00	1524.0060	152.40060	1.524006
66	66.0000	5.50	1676.4066	167.64066	1.676407
72	72.0000	6.00	1828.8072	182.88072	1.828807
78	78.0000	6.50	1981.2078	198.12078	1.981208
84	84.0000	7.00	2133.6084	213.36084	2.133608

Inches	Inches in Decimals	Feet in Decimals	Milli-meters	Centi-meters	Meters
90	90.0000	7.50	2286.0090	228.60090	2.28600:
96	96.0000	8.00	2438.4096	243.84096	2.43841(
102	102.0000	8.50	2590.8102	259.08102	2.59081(
108	108.0000	9.00	2743.2108	274.32108	2.74321
114	114.0000	9.50	2895.6114	289.56114	2.89561
120	120.0000	10.00	3048.0120	304.80120	3.04801:
126	126.0000	10.50	3200.4126	320.04126	3.20041:
132	132.0000	11.00	3352.8132	335.28132	3.35281:
138	138.0000	11.50	3505.2138	350.52138	3.50521
144	144.0000	12.00	3657.6144	365.76144	3.65761

Conversion Factors
S.I. Metric - English Systems

Multiply	by	to obtain
Acres	0.404687	hectares
	4.04687 x 10⁻³	square kilometers
Ares	1076.39	square feet
Board feet	144 sq.in x 1 in.	cubic inches
	0.0833	cubic feet
Bushels	0.3521	hectoliters
Centimeters	3.28083 x 10⁻²	feet
	0.3937	inches
Cubic centimeters	3.53145 x 10⁻⁵	Cubic feet
	6.102 x 10⁻²	Cubic inches
Cubic feet	2.8317 x 10⁴	
	2.8317 x 10⁻²	Cubic meters
	6.22905	Gallons, imperial
	0.2832	Hectoliters
	28.3170	Liters
	2.38095 x 10⁻²	tons, British shipping
	0.025	tons, U.S. shipping
Cubic inches	16.38716	cubic centimeters
Cubic meters	35.3145	cubic feet
	1.30794	cubic yards
	264.2	gallons, U.S.
Cubic yards	0.764559	cubic meters
	7.6336	hectoliters
Degrees, angular	0.0174533	radians
Degrees, F (less 32 F)	0.5556	degrees, C
Degrees, C	1.8	degrees F (less 32 C)
Foot pounds	0.13826	kilogram meters
Feet	30.4801	Centimeters
	0.304801	Meters
	304.801	Millimeters
	1.64468 x 10⁻⁴	miles, nautical
Gallons, Canadian	0.004546	cubic meters
Gallons, Imperial	0.160538	cubic feet
	1.20091	gallons, U.S.
	4.54596	Liters
Gallons, U.S.	0.832702	gallons, Imperial
	0.13368	cubic feet
	231.0	cubic inches

Multiply	by	to obtain
Gallons, U.S. (con't.)	0.003785	cubic meters
	0.0378	Hectoliters
	3.78543	Liters
Grams, metric	2.20462×10^{-3}	Pounds, avoirdupois
Hectares	2.47104	Acres
	1.076387×10^{5}	Square feet
	3.86101×10^{-3}	Square miles
Hectoliters	3.531	cubic feet
	2.84	Bushels
	0.131	Cubic yards
	26.42	Gallons
Horsepower, metric	0.98632	Horsepower, U.S.
Horsepower, U.S.	1.01387	Horsepower, metric
Inches	2.54001	Centimeters
	2.54001×10^{-2}	Meters
	25.4001	Millimeters
Kilograms	2.20462	Pounds
	9.84206×10^{-4}	Long tons
	9.807	Newtons
Kilogram meters	7.233	foot pounds
Kilograms per m	0.671972	Pounds per ft.
Kilograms per sq cm	14.2234	Pounds per sq. in.
Kilograms per sq.m	0.204817	Pounds per sq. ft.
	9.14362×10^{-5}	Long tons per sq. ft.
Kilograms per sq.mm	1422.34	Pounds per sq. in.
	0.634973	Long tons per sq. in.
Kilograms per cu.m	6.24283×10^{-2}	Pounds per cu. ft.
Kilometers	0.62137	Miles, statute
	0.53959	Miles, nautical
	3280.7	Feet
Kips	4448.0	Newtons
Kips	453.8	Kilograms force
Liters	0.219975	Gallons, Imperial
	0.26417	Gallons, U.S.
	3.53145×10^{-2}	Cubic feet
	61.022	Cubic inches
Meters	3.28083	Feet
	39.37	Inches
	1.09361	Yards
Miles, nautical	6080.204	Feet
	1.85325	Kilometers

Multiply	by	to obtain
Miles, nautical (con't.)	1.1516	Miles, statute
Miles, statute	1.60935	Kilometers
	0.8684	Miles, nautical
Millimeters	3.28083×10^{-3}	Feet
	3.937×10^{-2}	Inches
Ounces, U.S. liquid	29.57	Milliliters
Pounds, avoirdupois	453.592	Grams, metric
	0.453592	Kilograms
	4.464×10^{-4}	Tons, long
	4.53592×10^{-4}	Tons, metric
Pounds force	4.448	Newtons
Pounds per foot	1.48816	Kilograms per m
Pounds per sq. ft.	4.88241	Kilograms per sq.m
Pounds per sq. in.	7.031×10^{-2}	Kilograms per sq.cm
	7.031×10^{-4}	Kilograms per sq. mm
Pounds per cu. ft.	16.0184	Kilograms per cu. m
Radians	57.29578	Degrees, angular
Square centimeters	0.1550	Square inches
Square feet	9.29034×10^{-4}	Ares
	9.29034×10^{-6}	Hectares
	0.0929034	Square meters
Square inches	6.45163	Square centimeters
	645.163	Square millimeters
Square kilometers	247.104	Acres
	0.3861	Square miles
Square meters	10.7639	Square feet
	1.19599	Square yards
Square miles	259.0	Hectares
	2.590	Square kilometers
Square millimeters	1.550×10^{-3}	Square inches
Square yards	0.83613	Square meters
Tons, long	1016.05	Kilograms
	2240.0	Pounds
	1.01605	Tons, metric
	1.120	Tons, short
Tons, long, per sq. ft.	1.09366×10^{-4}	Kilograms per sq.m
Tons, long, per sq. in.	1.57494	Kilograms per sq. mm
Tons, metric	2204.62	Pounds
	0.98421	Tons, long

Multiply	by	to obtain
Tons, metric (con't.)	1.10231	Tons, short
Tons, short	907.185	Kilograms
	0.892857	Tons, long
	0.907185	Tons, metric
Tons, British shipping	42.00	Cubic feet
	0.952381	tons, U.S. shipping
Tons, U.S. shipping	40.00	Cubic feet
	1.050	Tons, British shipping
Yards	0.914402	Meters

Functions of Numbers, 1 to 49

No.	Square	Cube	Square Root	Cubic Root	Logarithm
1	1	1	1.0000	1.0000	0.00000
2	4	8	1.4142	1.2599	0.30103
3	9	27	1.7321	1.4422	0.47712
4	16	64	2.0000	1.5874	0.60206
5	25	125	2.2361	1.7100	0.69897
6	36	216	2.4495	1.8171	0.77815
7	49	343	2.6458	1.9129	0.84510
8	64	512	2.8284	2.0000	0.90309
9	81	729	3.0000	2.0801	0.95424
10	100	1000	3.1623	2.1544	1.00000
11	121	1331	3.3166	2.2240	1.04139
12	144	1728	3.4641	2.2894	1.07918
13	169	2197	3.6056	2.3513	1.11394
14	196	2744	3.7417	2.4101	1.14613
15	225	3375	3.8730	2.4662	1.17609
16	256	4096	4.0000	2.5198	1.20412
17	289	4913	4.1231	2.5713	1.23045
18	324	5832	4.2426	2.6207	1.25527
19	361	6859	4.3589	2.6684	1.27875
20	400	8000	4.4721	2.7144	1.30103
21	441	9261	4.5826	2.7589	1.32222
22	484	10648	4.6904	2.8020	1.34242
23	529	12167	4.7958	2.8439	1.36173
24	576	13824	4.8990	2.8845	1.38021
25	625	15625	5.0000	2.9240	1.39794
26	676	17576	5.0990	2.9625	1.41497
27	729	19683	5.1962	3.0000	1.43136
28	784	21952	5.2915	3.0366	1.44716
29	841	24389	5.3852	3.0723	1.46240
30	900	27000	5.4772	3.1072	1.47712
31	961	29791	5.5678	3.1414	1.49136
32	1024	32768	5.6569	3.1748	1.50515
33	1089	35937	5.7446	3.2075	1.51851
34	1156	39304	5.8310	3.2396	1.53148
35	1225	42875	5.9161	3.2711	1.54407
36	1296	46656	6.0000	3.3019	1.55630
37	1369	50653	6.0828	3.3322	1.56820
38	1444	54872	6.1644	3.3620	1.57978
39	1521	59319	6.2450	3.3912	1.59106
40	1600	64000	6.3246	3.4200	1.60206
41	1681	68921	6.4031	3.4482	1.61278
42	1764	74088	6.4807	3.4760	1.62325
43	1849	79507	6.5574	3.5034	1.63347
44	1936	85184	6.6332	3.5303	1.64345
45	2025	91125	6.7082	3.5569	1.65321
46	2116	97336	6.7823	3.5830	1.66276
47	2209	103823	6.8557	3.6088	1.67210
48	2304	110592	6.9282	3.6342	1.68124
49	2401	117649	7.0000	3.6593	1.69020

Functions of Numbers, 50 to 99

No.	Square	Cube	Square Root	Cubic Root	Logarithm
50	2500	125000	7.0711	3.6840	1.69897
51	2601	132651	7.1414	3.7084	1.70757
52	2704	140608	7.2111	3.7325	1.71600
53	2809	148877	7.2801	3.7563	1.72428
54	2916	157464	7.3485	3.7798	1.73239
55	3025	166375	7.4162	3.8030	1.74036
56	3136	175616	7.4833	3.8259	1.74819
57	3249	185193	7.5498	3.8485	1.75587
58	3364	195112	7.6158	3.8709	1.76343
59	3481	205379	7.6811	3.8930	1.77085
60	3600	216000	7.7460	3.9149	1.77815
61	3721	226981	7.8102	3.9365	1.78533
62	3844	238328	7.8740	3.9579	1.79239
63	3969	250047	7.9373	3.9791	1.79934
64	4096	262144	8.0000	4.0000	1.80618
65	4225	274625	8.0623	4.0207	1.81291
66	4356	287496	8.1240	4.0412	1.81954
67	4489	300763	8.1854	4.0615	1.82607
68	4624	314432	8.2462	4.0817	1.83251
69	4761	328509	8.3066	4.1016	1.83885
70	4900	343000	8.3666	4.1213	1.84510
71	5041	357911	8.4261	4.1408	1.85126
72	5184	373248	8.4853	4.1602	1.85733
73	5329	389017	8.5440	4.1793	1.86332
74	5476	405224	8.6023	4.1983	1.86923
75	5625	421875	8.6603	4.2172	1.87506
76	5776	438976	8.7178	4.2358	1.88081
77	5929	456533	8.7750	4.2543	1.88649
78	6084	474552	8.8318	4.2727	1.89209
79	6241	493039	8.8882	4.2908	1.89763
80	6400	512000	8.9443	4.3089	1.90309
81	6561	531441	9.0000	4.3267	1.90849
82	6724	551368	9.0554	4.3445	1.91381
83	6889	571787	9.1104	4.3621	1.91908
84	7056	592704	9.1652	4.3795	1.92428
85	7225	614125	9.2195	4.3968	1.92942
86	7396	636056	9.2736	4.4140	1.93450
87	7569	658503	9.3274	4.4310	1.93952
88	7744	681472	9.3808	4.4480	1.94448
89	7921	704969	9.4340	4.4647	1.94939
90	8100	729000	9.4868	4.4814	1.95424
91	8281	753571	9.5394	4.4979	1.95904
92	8464	778688	9.5917	4.5144	1.96379
93	8649	804357	9.6437	4.5307	1.96848
94	8836	830584	9.6954	4.5468	1.97313
95	9025	857375	9.7468	4.5629	1.97772
96	9216	884736	9.7980	4.5789	1.98227
97	9409	912673	9.8489	4.5947	1.98677
98	9604	941192	9.8995	4.6104	1.99123
99	9801	970299	9.9499	4.6261	1.99564

Functions of Numbers, 100 to 149

No.	Square	Cube	Square Root	Cubic Root	Logarithm
100	10000	1000000	10.0000	4.6416	2.00000
101	10201	1030301	10.0499	4.6570	2.00432
102	10404	1061208	10.0995	4.6723	2.00860
103	10609	1092727	10.1489	4.6875	2.01284
104	10816	1124864	10.1980	4.7027	2.01703
105	11025	1157625	10.2470	4.7177	2.02119
106	11236	1191016	10.2956	4.7326	2.02531
107	11449	1225043	10.3441	4.7475	2.02938
108	11664	1259712	10.3923	4.7622	2.03342
109	11881	1295029	10.4403	4.7769	2.03743
110	12100	1331000	10.4881	4.7914	2.04139
111	12321	1367631	10.5357	4.8059	2.04532
112	12544	1404928	10.5830	4.8203	2.04922
113	12769	1442897	10.6301	4.8346	2.05308
114	12996	1481544	10.6771	4.8488	2.05690
115	13225	1520875	10.7238	4.8629	2.06070
116	13456	1560896	10.7703	4.8770	2.06446
117	13689	1601613	10.8167	4.8910	2.06819
118	13924	1643032	10.8628	4.9049	2.07188
119	14161	1685159	10.9087	4.9187	2.07555
120	14400	1728000	10.9545	4.9324	2.07918
121	14641	1771561	11.0000	4.9461	2.08279
122	14884	1815848	11.0454	4.9597	2.08636
123	15129	1860867	11.0905	4.9732	2.08991
124	15376	1906624	11.1355	4.9866	2.09342
125	15625	1953125	11.1803	5.0000	2.09691
126	15876	2000376	11.2250	5.0133	2.10037
127	16129	2048383	11.2694	5.0265	2.10380
128	16384	2097152	11.3137	5.0397	2.10721
129	16641	2146689	11.3578	5.0528	2.11059
130	16900	2197000	11.4018	5.0658	2.11394
131	17161	2248091	11.4455	5.0788	2.11727
132	17424	2299968	11.4891	5.0916	2.12057
133	17689	2352637	11.5326	5.1045	2.12385
134	17956	2406104	11.5758	5.1172	2.12710
135	18225	2460375	11.6190	5.1299	2.13033
136	18496	2515456	11.6619	5.1426	2.13354
137	18769	2571353	11.7047	5.1551	2.13672
138	19044	2628072	11.7473	5.1676	2.13988
139	19321	2685619	11.7898	5.1801	2.14301
140	19600	2744000	11.8322	5.1925	2.14613
141	19881	2803221	11.8743	5.2048	2.14922
142	20164	2863288	11.9164	5.2171	2.15229
143	20449	2924207	11.9583	5.2293	2.15534
144	20736	2985984	12.0000	5.2415	2.15836
145	21025	3048625	12.0416	5.2536	2.16137
146	21316	3112136	12.0830	5.2656	2.16435
147	21609	3176523	12.1244	5.2776	2.16732
148	21904	3241792	12.1655	5.2896	2.17026
149	22201	3307949	12.2066	5.3015	2.17319

Functions of Numbers, 150 to 199

No.	Square	Cube	Square Root	Cubic Root	Logarithm
150	22500	3375000	12.2474	5.3133	2.17609
151	22801	3442951	12.2882	5.3251	2.17898
152	23104	3511808	12.3288	5.3368	2.18184
153	23409	3581577	12.3693	5.3485	2.18469
154	23716	3652264	12.4097	5.3601	2.18752
155	24025	3723875	12.4499	5.3717	2.19033
156	24336	3796416	12.4900	5.3832	2.19312
157	24649	3869893	12.5300	5.3947	2.19590
158	24964	3944312	12.5698	5.4061	2.19866
159	25281	4019679	12.6095	5.4175	2.20140
160	25600	4096000	12.6491	5.4288	2.20412
161	25921	4173281	12.6886	5.4401	2.20683
162	26244	4251528	12.7279	5.4514	2.20952
163	26569	4330747	12.7671	5.4626	2.21219
164	26896	4410944	12.8062	5.4737	2.21484
165	27225	4492125	12.8452	5.4848	2.21748
166	27556	4574296	12.8841	5.4959	2.22011
167	27889	4657463	12.9228	5.5069	2.22272
168	28224	4741632	12.9615	5.5178	2.22531
169	28561	4826809	13.0000	5.5288	2.22789
170	28900	4913000	13.0384	5.5397	2.23045
171	29241	5000211	13.0767	5.5505	2.23300
172	29584	5088448	13.1149	5.5613	2.23553
173	29929	5177717	13.1529	5.5721	2.23805
174	30276	5268024	13.1909	5.5828	2.24055
175	30625	5359375	13.2288	5.5934	2.24304
176	30976	5451776	13.2665	5.6041	2.24551
177	31329	5545233	13.3041	5.6147	2.24797
178	31684	5639752	13.3417	5.6252	2.25042
179	32041	5735339	13.3791	5.6357	2.25285
180	32400	5832000	13.4164	5.6462	2.25527
181	32761	5929741	13.4536	5.6567	2.25768
182	33124	6028568	13.4907	5.6671	2.26007
183	33489	6128487	13.5277	5.6774	2.26245
184	33856	6229504	13.5647	5.6877	2.26482
185	34225	6331625	13.6015	5.6980	2.26717
186	34596	6434856	13.6382	5.7083	2.26951
187	34969	6539203	13.6748	5.7185	2.27184
188	35344	6644672	13.7113	5.7287	2.27416
189	35721	6751269	13.7477	5.7388	2.27646
190	36100	6859000	13.7840	5.7489	2.27875
191	36481	6967871	13.8203	5.7590	2.28103
192	36864	7077888	13.8564	5.7690	2.28330
193	37249	7189057	13.8924	5.7790	2.28556
194	37636	7301384	13.9284	5.7890	2.28780
195	38025	7414875	13.9642	5.7989	2.29003
196	38416	7529536	14.0000	5.8088	2.29226
197	38809	7645373	14.0357	5.8186	2.29447
198	39204	7762392	14.0712	5.8285	2.29667
199	39601	7880599	14.1067	5.8383	2.29885

Functions of Numbers, 200 to 249

No.	Square	Cube	Square Root	Cubic Root	Logarithm
200	40000	8000000	14.1421	5.8480	2.30103
201	40401	8120601	14.1774	5.8578	2.30320
202	40804	8242408	14.2127	5.8675	2.30535
203	41209	8365427	14.2478	5.8771	2.30750
204	41616	8489664	14.2829	5.8868	2.30963
205	42025	8615125	14.3178	5.8964	2.31175
206	42436	8741816	14.3527	5.9059	2.31387
207	42849	8869743	14.3875	5.9155	2.31597
208	43264	8998912	14.4222	5.9250	2.31806
209	43681	9129329	14.4568	5.9345	2.32015
210	44100	9261000	14.4914	5.9439	2.32222
211	44521	9393931	14.5258	5.9533	2.32428
212	44944	9528128	14.5602	5.9627	2.32634
213	45369	9663597	14.5945	5.9721	2.32838
214	45796	9800344	14.6287	5.9814	2.33041
215	46225	9938375	14.6629	5.9907	2.33244
216	46656	10077696	14.6969	6.0000	2.33445
217	47089	10218313	14.7309	6.0092	2.33646
218	47524	10360232	14.7648	6.0185	2.33846
219	47961	10503459	14.7986	6.0277	2.34044
220	48400	10648000	14.8324	6.0368	2.34242
221	48841	10793861	14.8661	6.0459	2.34439
222	49284	10941048	14.8997	6.0550	2.34635
223	49729	11089567	14.9332	6.0641	2.34830
224	50176	11239424	14.9666	6.0732	2.35025
225	50625	11390625	15.0000	6.0822	2.35218
226	51076	11543176	15.0333	6.0912	2.35411
227	51529	11697083	15.0665	6.1002	2.35603
228	51984	11852352	15.0997	6.1091	2.35793
229	52441	12008989	15.1327	6.1180	2.35984
230	52900	12167000	15.1658	6.1269	2.36173
231	53361	12326391	15.1987	6.1358	2.36361
232	53824	12487168	15.2315	6.1446	2.36549
233	54289	12649337	15.2643	6.1534	2.36736
234	54756	12812904	15.2971	6.1622	2.36922
235	55225	12977875	15.3297	6.1710	2.37107
236	55696	13144256	15.3623	6.1797	2.37291
237	56169	13312053	15.3948	6.1885	2.37475
238	56644	13481272	15.4272	6.1972	2.37658
239	57121	13651919	15.4596	6.2058	2.37840
240	57600	13824000	15.4919	6.2145	2.38021
241	58081	13997521	15.5242	6.2231	2.38202
242	58564	14172488	15.5563	6.2317	2.38382
243	59049	14348907	15.5885	6.2403	2.38561
244	59536	14526784	15.6205	6.2488	2.38739
245	60025	14706125	15.6525	6.2573	2.38917
246	60516	14886936	15.6844	6.2658	2.39094
247	61009	15069223	15.7162	6.2743	2.39270
248	61504	15252992	15.7480	6.2828	2.39445
249	62001	15438249	15.7797	6.2912	2.39620

Functions of Numbers, 250 to 299

No.	Square	Cube	Square Root	Cubic Root	Logarithm
250	62500	15625000	15.8114	6.2996	2.39794
251	63001	15813251	15.8430	6.3080	2.39967
252	63504	16003008	15.8745	6.3164	2.40140
253	64009	16194277	15.9060	6.3247	2.40312
254	64516	16387064	15.9374	6.3330	2.40483
255	65025	16581375	15.9687	6.3413	2.40654
256	65536	16777216	16.0000	6.3496	2.40824
257	66049	16974593	16.0312	6.3579	2.40993
258	66564	17173512	16.0624	6.3661	2.41162
259	67081	17373979	16.0935	6.3743	2.41330
260	67600	17576000	16.1245	6.3825	2.41497
261	68121	17779581	16.1555	6.3907	2.41664
262	68644	17984728	16.1864	6.3988	2.41830
263	69169	18191447	16.2173	6.4070	2.41996
264	69696	18399744	16.2481	6.4151	2.42160
265	70225	18609625	16.2788	6.4232	2.42325
266	70756	18821096	16.3095	6.4312	2.42488
267	71289	19034163	16.3401	6.4393	2.42651
268	71824	19248832	16.3707	6.4473	2.42813
269	72361	19465109	16.4012	6.4553	2.42975
270	72900	19683000	16.4317	6.4633	2.43136
271	73441	19902511	16.4621	6.4713	2.43297
272	73984	20123648	16.4924	6.4792	2.43457
273	74529	20346417	16.5227	6.4872	2.43616
274	75076	20570824	16.5529	6.4951	2.43775
275	75625	20796875	16.5831	6.5030	2.43933
276	76176	21024576	16.6132	6.5108	2.44091
277	76729	21253933	16.6433	6.5187	2.44248
278	77284	21484952	16.6733	6.5265	2.44404
279	77841	21717639	16.7033	6.5343	2.44560
280	78400	21952000	16.7332	6.5421	2.44716
281	78961	22188041	16.7631	6.5499	2.44871
282	79524	22425768	16.7929	6.5577	2.45025
283	80089	22665187	16.8226	6.5654	2.45179
284	80656	22906304	16.8523	6.5731	2.45332
285	81225	23149125	16.8819	6.5808	2.45484
286	81796	23393656	16.9115	6.5885	2.45637
287	82369	23639903	16.9411	6.5962	2.45788
288	82944	23887872	16.9706	6.6039	2.45939
289	83521	24137569	17.0000	6.6115	2.46090
290	84100	24389000	17.0294	6.6191	2.46240
291	84681	24642171	17.0587	6.6267	2.46389
292	85264	24897088	17.0880	6.6343	2.46538
293	85849	25153757	17.1172	6.6419	2.46687
294	86436	25412184	17.1464	6.6494	2.46835
295	87025	25672375	17.1756	6.6569	2.46982
296	87616	25934336	17.2047	6.6644	2.47129
297	88209	26198073	17.2337	6.6719	2.47276
298	88804	26463592	17.2627	6.6794	2.47422
299	89401	26730899	17.2916	6.6869	2.47567

Appendix **A**

GLOSSARY OF CONSTRUCTION TERMS

This glossary is not intended as a complete and comprehensive dictionary of all the specialized terminology that one will encounter in construction estimating. It is to assist the reader in using this text.

We have drawn on many sources and gratefully acknowledge the Brick Institute of America, the National Concrete Masonry Association, and the Structural Steel Painting Council for permission to reprint entries from their glossaries.

abatement. Procedures to eliminate fiber release from asbestos containing building materials including encapsulation, enclosure or removal methods.

abrasive. A fine graded (sized) granular or spherical material that is used for cleaning, etc.

absorption. The weight of water a masonry unit absorbs when immersed in either cold or boiling water for a stated length of time, expressed as a percentage of the weight of the dry unit.

accelerator. Material such as calcium chloride and compositions predominately of calcium chloride which accelerate hardening and promote early strength development of concrete or mortar.

acceptance testing. Testing of received products to determine that the quality meets required specifications.

additive. Any substance added in small quantities to another substance, usually to improve properties.

admixtures. Materials other than water, aggregates, and hydraulic cement, used as an ingredient of concrete, mortar, or grout, and added to the batch immediately before or during its mixing.

adsorption. Concentration of a substance at a surface or interface of another substance.

aggregate. Granular material such as natural sand, manufactured sand, expanded clay, shale or slate, pumice, volcanic scoria, bituminous or anthracite cinders, gravel, crushed gravel, crushed stone, heavyweight aggregate such as magnetite or ilmenite, and air-cooled or expanded blast-furnace slag, which when bound together into a conglomerate mass by matrix forms concrete, mortar, or grout.

air lock. A system for hazardous substance removal and abatement tha permits ingress and egress from one room to another while permittin minimal air movement between rooms, typically constructed by placing tw overlapping sheets of plastic over an existing or temporarily framed doorwa Two curtained doorways spaced a minimum of 6 ft. apart from an air lock.

air monitoring. The process of measuring asbestos fiber content of specific volume of air in a stated period of time.

alkyd resins. Synthetic resins formed by the condensation of polyhydri alcohols with polybasic acids.

ambient air quality. Average atmospheric purity, as distinguished fror discharge measurements taken at the source of pollution.

amended water. Water to which a surfactant is added.

anchor. A piece or assemblage, usually metal, used to attach building part (i.e., plates, joists, trusses, etc.) to masonry or masonry materials.

anchor pile. A pile connected to a structure by one or more ties to furnis lateral support or resist uplift.

ANSI. American National Standards Institute

arch. A curved compressive structural member, spanning openings o recesses; also built flat.

back arch. A concealed arch carrying the backing of a wall where the exteric facing is carried by a lintel.

jack arch. One having horizontal or nearly horizontal upper and lowe surfaces. Also called flat or straight arch.

major arch. Arch with spans greater than 6 ft. and equivalent to uniforr loads greater than 1000 lb. per ft. Typically known as Tudor, semicircula Gothic, or parabolic arch. Has rise to span ratio greater than 0.15.

minor arch. Arch with maximum span of 6 ft. and loads not exceeding 1,00 lb. per ft. Typically known as jack, segmental, or multicentered arch. Ha rise to span ratio less than or equal to 0.15.

relieving arch. One built over a lintel, flat arch, or smaller arch to divert loads, thus relieving the lower member from excessive loading. Also known as discharge or safety arch.

trimmer arch. An arch, usually a low rise arch or brick, used for supporting a fireplace hearth.

ashlar masonry. Masonry composed of rectangular units of clay or shale, or stone, generally larger in size than brick and properly bonded, having sawed, dressed or squared beds, and joints laid in mortar. Often the unit size varies to provide a random pattern, random ashlar.

asphalt. Solid or semi-solid mixture of bitumens that are found in natural deposits or created as a petroleum by-product and used mainly for paving and roofing.

natural asphalt (or lake asphalt). Derived from petroleum by the natural process of evaporation leaving the asphalt material in open beds or lakes ready for extraction and processing.

petroleum asphalt. Asphalt refined from crude petroleum.

asphalt cement (AC). Refined asphalt for paving purposes.

asphalt joint filler. Liquid asphalt product used to seal cracks and joints in pavement to prevent moisture penetration.

asphalt joint filler, preformed. Asphalt mixed with various fibrous materials and manufactured in solid form to a specific thickness. Widely used as an expansion joint material.

asphalt primer. Liquid asphalt applied to a non-bituminous surface to promote adhesion between existing surface and new surface application.

ASTM. American Society for Testing and Materials

back filling. 1. Replacing of fill materials in an excavated area 2. Rough masonry built behind a facing or between two faces. 3. Filling over the extrados of an arch. 4. Brickwork in spaces between structural timbers, sometimes called brick nogging.

backup. That part of a masonry wall behind the exterior facing.

bat. A piece of brick.

batter. Recessing or sloping masonry back in successive courses; the opposite of corbel.

bed joint. The horizontal layer of mortar on or in which a masonry unit is laid; may cover entire masonry unit or face shell only.

belt course. A narrow horizontal course of masonry, sometimes slightly projected such as window sills, that is made continuous. Sometimes called string or sill course.

bevel. One side of a solid body which is inclined with respect to the other, with the angle between the two sides either greater or less than a right angle.

blast cleaning. Cleaning or roughing of a surface using metallic or non-metallic grit or shot projected with compressed air, centrifugal force, or water.

block, concrete. A hollow or solid unit consisting of portland cement and suitable aggregates combined with water. Other materials such as lime, fly ash, air-entraining agents, or other admixtures may be permitted.

blocking. A method of bonding two adjoining or intersecting walls, not built at the same time, by means of offsets whose vertical dimensions are not less than 8 in.

blown asphalt. Asphalt specially treated at high temperature levels with blown air to obtain specific characteristics necessary in products for roofing, pipe coating, membrane envelopes, and hydraulic application.

bond. 1. Tying various parts of a masonry wall by lapping units one over another or by connecting with metal ties. 2. Patterns formed by exposed faces of units. 3. Adhesion between mortar or grout and masonry units or reinforcement.

bond, grout. The adhesion to and/or interlocking of grout with the masonry units and reinforcement.

bond, mechanical. Tying masonry units together with metal ties or reinforcing steel or keys.

bond, mortar. The adhesion of mortar to masonry units and reinforcement.

bond beam. Course or courses of a masonry wall grouted and usually reinforced in the horizontal direction. Serves as horizontal tie of wall, bearing course for structural members or as a flexural member itself.

bond breaker. A material used to prevent adhesion between two surfaces.

bond course. The course consisting of units which overlap more than one wythe of masonry.

bond strength. Resistance to separation of mortar from concrete masonry units and of mortar and grout from reinforcing steel and other materials with which it is in contact.

bonder. A bonding unit. See Header.

bracing. Any type of horizontal or inclined structural member used to increase stability and to resist lateral forces.

breaking joints. Any arrangement of masonry units that prevents continuous vertical joints from occurring in adjacent courses.

brick. A solid masonry unit of clay or shale, formed into a rectangular prism while plastic, and burned or fired in a kiln.

acid-resistant brick. Brick suitable for use in contact with chemicals, usually in conjunction with acid-resistant mortars.

adobe brick. Large roughly-molded, sun-dried clay brick of varying size.

angle brick. Any brick shaped to an oblique angle to fit a salient corner.

arch brick. 1. Wedge-shaped brick for special use in an arch. 2. Extremely hard-burned brick from an arch of a scove kiln.

building brick. Brick for building purposes not especially treated for texture or color. Formerly called common brick.

clinker brick. A very hard-burned brick whose shape is distorted or bloated due to nearly complete vitrification.

common brick. See Building brick.

dry-press brick. Brick formed in molds under high pressure from relativel dry clay (5% to 7% moisture content).

economy brick. Brick whose nominal dimensions are 4"x4"x8".

engineered brick. Brick whose nominal dimensions are 4"x3.2"x8".

facing brick. Brick made especially for facing purposes, often treated t produce surface texture. They are made of selected clay, or treated, t produce desired color.

fire brick. Brick made of refractory ceramic material that resists high temperatures.

floor brick. Smooth dense brick, highly resistant to abrasion, used as finishe floor surface.

gauged brick. 1. Brick that have been ground or otherwise produced t accurate dimensions. 2. A tapered arch brick.

hollow brick. A masonry unit of clay or shale whose net cross-sectional are in any plane parallel to the bearing surface is not less than 60% of its gros cross-sectional area measured in the same plane.

jumbo brick. A generic term indicating a brick larger in size than th standard. Some producers use this term to describe oversize brick of specifi dimensions manufactured by them.

Norman brick. A brick whose nominal dimensions are 4" x 2-2/3" x 12".

paving brick. Vitrified brick especially suitable for use in pavements wher resistance to abrasion is important.

Roman brick. Brick whose nominal dimensions are 4" x 2" x 12".

salmon brick. Generic term for underburned brick that are more porous slightly larger, and lighter colored than hard-burned brick. Usually pinkish orange color.

SCR brick. See SCR.

sewer brick. Low absorption, abrasive-resistant brick intended for use i drainage structures.

soft-mud brick. Brick produced by molding relatively wet clay (20% to 30% moisture). Often a hand process. When insides of molds are sanded to prevent sticking of clay, the product is sand-struck brick. When molds are wetted to prevent sticking, the product is water-struck brick.

stiff-mud brick. Brick produced by extruding a stiff but plastic clay (12% to 15% moisture) through a die.

brick and brick. A method of laying brick so that units touch each other with only enough mortar to fill surface irregularities.

brick, concrete. A solid unit having a rectangular prismatic shape, usually not larger than 4" x 4" x 12", made from portland cement and suitable aggregates, with or without the inclusion of other materials.

brick grade. Designation for durability of the unit expressed as SW for severe weathering, MW for moderate weathering, or NW for negligible weathering. See ASTM Specifications C 216, C 62, and C 652.

brick type. Designation for facing brick that controls tolerance, chippage, and distortion. Expressed as FBS, FBX, and FBA for solid brick, and as HBS, HBX, HBA, and HBB for hollow brick. See ASTM Specifications C 216 and C 652.

bridging. Wood or metal members that are placed between trusses and joists in an angled position intended to spread the loads.

british thermal unit (btu). The heat energy required to raise the temperature of one pound of water one degree fahrenheit.

built-up roof. Roofing composed of two or more layers of asphalt. Several layers fastened together.

bullnose block. A unit having one or more rounded exterior corners.

buttress. A masonry pilaster decreasing in area from the base to the top, generally used to give greater lateral strength and stability to a wall.

buttering. Placing mortar on a masonry unit with a trowel.

caisson. Also *caisson pile,* a large-diameter shaft, hand or machine excavated to bearing stratum inside a protective casing.

capacity insulation. The ability of masonry to store heat as a result of its mass, density and specific heat.

cathodic protection. A technique to prevent corrosion of a metal surface by making that surface the cathode of an electrochemical cell.

cavity wall. A wall built of two or more wythes of masonry units separated by a continuous air space (with or without insulating materials) and in which the wythes are securely tied together with rigid corrosion resistant metal ties.

c/b ratio. The ratio of the weight of water absorbed by a masonry unit during immersion in cold water to weight absorbed during immersion in boiling water. An indication of the probable resistance of brick to freezing and thawing. Also called saturation coefficient.

cement paint. Supplied in dry powder form, it is based on portland cement to which pigments are sometimes added for decorative purposes.

centering. Temporary formwork for the support of masonry arches or lintels during construction. Also called center(s).

ceramic color glaze. An opaque colored glaze of satin or gloss finish obtained by spraying the clay body with a compound of metallic oxide, chemicals and clays. It is burned at high temperatures, fusing glaze to body and making them inseparable.

chase: A continuous recess built into a wall to receive pipes, ducts, etc.

class of unit. Distinguishes between masonry units of different grade or type in ASTM specifications, being manufactured from different raw materials, or having different specified compressive strengths.

clay. A natural, mineral aggregate consisting essentially of hydrous aluminum silicate; it is plastic when sufficiently wetted, rigid when dried and vitrified when fired to a sufficiently high temperature.

clay mortar-mix. Finely ground clay used as a plasticizer for masonry mortars.

cleanout. An opening in the first course of masonry for cleaning mortar droppings prior to grout placement in grouted masonry. Required in high lift grouting.

clean room. An uncontaminated area or room that is part of the worker decontamination enclosure system in asbestos abatement and removal, with provisions for storage of street clothes and protective equipment.

clear ceramic glaze. Same as Ceramic Color Glaze except that it is translucent or slightly tinted with a gloss finish.

clip. A portion of a brick cut to length.

closer. The last masonry unit laid in a course. It may be whole or a portion of a unit.

closure. Supplementary or short length units used at corners or jambs to maintain bond patterns.

collar beam. Wood member connecting opposite roof rafters.

collar joint. The vertical, longitudinal joint between wythes of masonry.

collar tie. A horizontal member placed between two rafters a specific vertical distance above the very top plate line for the purpose of limiting outward thrust of the rafters.

column. 1. A vertical member whose horizontal dimensions measured at right angles to the thickness does not exceed three times its thickness. 2. (in concrete masonry) a compression member, vertical or nearly vertical, the width of which does not exceed four times its thickness and the height of which exceeds four times its least lateral dimension.

composite lumber. A group of wood materials that contain wood in whole or fiber form bound together with an adhesive.

composite wall. A multiple wythe masonry wall in which at least one of the wythes is dissimilar to the other wythe with respect to type or grade of unit or mortar.

compressive strength. The maximum compressive load in lbs. that a unit will support divided by the gross cross-sectional area of the unit in square inches.

concrete masonry unit, hollow. A unit whose net cross-sectional area in any plane parallel to the bearing surface is less than 75% of its gross cross-sectional area measured in the same plane.

concrete masonry unit, solid: A unit whose net cross-sectional area in every plane parallel to the bearing surface is 75% or more of its gross cross sectional area measured in the same plane.

control joint. A continuous unbonded masonry joint to regulate the location and amount of separation resulting from the dimensional change of different parts of a structure so as to avoid the development of excessively high stresses.

coping: The material or masonry units forming a cap or finish on top of a wall, pier, pilaster, chimney, etc. It protects masonry below from penetration of water from above.

corbel: A shelf of ledge formed by projecting successive courses of masonry out from the face of the wall, pier, or column.

core: The molded open space in a masonry concrete unit.

corrosion. The deterioration of a material by direct or electrochemical reaction with its environment.

course: One of the continuous horizontal layers of units bonded with mortar in masonry.

cross bracing. Any type of bracing that is constructed at right angles to the main or longitudinal axis.

cross-sectional area, gross. In masonry, the total area of a section perpendicular to the direction of the load, including areas within the cell and within reentrant spaces unless these spaces are to be occupied in the masonry by portions of adjacent masonry. (The gross cross-sectional area of scored units is determined to the outside of the scoring.)

cross-sectional area, net. In masonry, the gross cross-sectional area of a section minus the average area of ungrouted cores or cellular spaces. (The cross-sectional area of grooves in scored units is not deducted from the gross cross-sectional area to obtain the net cross-sectional area.)

culls. Masonry units that do not meet the standards or specifications and have been rejected.

curing. To preserve or finish by chemical or physical process.

curing agent. An additive that promotes curing.

curtain wall. A non-loadbearing, usually prefabricated, wall between columns and piers.

curtained doorway. A device used in asbestos removal to allow ingress or egress from one room to another while permitting minimal air movement between the rooms, typically constructed by placing two overlapping sheets of plastic over an existing or temporarily framed doorway. Two curtained doorways spaced a minimum of 6 ft. apart from an air lock.

customized masonry. Architectural masonry units having textured or sculptured surfaces. Methods used to obtain different surface textures including splitting, grinding, forming vertical striations and causing the units to "slump". Sculptured faces are obtained by forming projecting ribs or flutes, either rounded or angular, as well as vertical and horizontal scoring, recesses and curved faces.

cutback asphalt. Asphalt cement blended with petroleum solvents to produce a more liquid or fluid product thus improving handling and workability qualities. Exposure to atmosphere evaporates the solvents and the remaining asphalt performs the intended function. Variation in the types of solvent used produces the following products: a) Rapid Curing Asphalt (RC), b) Medium Curing Asphalt (MC), and c) Slow Curing Asphalt (SC).

dampcheck. An impervious horizontal layer to prevent vertical penetration of water in a wall consisting of either a course of solid masonry, metal or a thin layer of asphaltic or bituminous material. Generally near grade to prevent upward migration of moisture by capillary action. See damp course below.

damp course. A masonry course or layer of impervious material that prevents capillary entrance of moisture from the ground or a lower course. Same as damp check.

dampproofing. Prevention of moisture penetration by capillary action, usually by addition of one or more coats of a compound that is impervious to water.

decontamination enclosure system. In asbestos abatement and removal, a series of areas or connected rooms, with curtained doorways between them, for decontamination of workers or materials and equipment. The system always contains a minimum of one airlock between two curtained doorways.

direct costs. Costs that are charged directly to a project, such as for labor, taxes, insurance, materials, scaffolding, equipment, and inspection.

disposal. Procedures necessary to transport and deposit hazardous waste containing material removed from a building or site in compliance with regulations.

dog's tooth. Brick laid with their corners projecting from the wall face.

double-walled tank. An underground storage tank in which a rigid secondary container is attached to the primary container and which has an annular space.

dovetail anchor. A splayed tenow shaped like a dove's tail, broader at its end than at its base, which fits into the recess of a corresponding mortise.

dowel. Straight metal bars used to connect two sections of masonry.

drilled pile. Also called an *augured pile,* it is a cast-in-place concrete pile in an augured hole, possibly belled at the bottom.

drip. A projecting piece of material, shaped to throw off water and prevent its running down the face of the wall or other surface.

dry stack. Masonry work laid without mortar.

EBM. See Engineered brick masonry.

eccentricity. The normal distance between the centroidal axis of a member and the parallel resultant load.

e1/e2. Ratio of virtual eccentricities occurring at the ends of a column or wall under design. The absolute value is always less than or equal to 1.0.

effective area of reinforcement. The area obtained by multiplying the right cross-sectional area of metal reinforcement by the cosine of the angle between its direction and the direction for which the effectiveness of the reinforcement is to be determined.

effective height. The height of a member to be assumed for calculating the slenderness ratio.

effective thickness. The thickness of a member to be assumed for calculating the slenderness ratio.

efflorescence. A powder or stain sometimes found on the surface of masonry, resulting from deposition of water-soluble salts.

emission standard. The maximum amount of a pollutant that is permitted to be discharged from a single source.

encapsulation. Procedures necessary to coat all spray- or trowel-applied asbestos containing materials with an encapsulant to control the release of asbestos fibers into the ambient air.

empirical design. Design based on applying physical limitations based on experience or observations gained through experience without a structural analysis.

engineered brick masonry. Masonry in which design is based on a rational structural analysis.

engineered design. Design based on a rational analysis considering the interrelationships of the various construction materials, their properties and actual design loads in lieu of empirical design procedures.

EPA. Environmental Protection Agency.

epoxy. Thermosetting resin that is mixed with a curing agent to produce adhesives, coatings, etc.

equipment decontamination system. A decontamination enclosure system for materials and equipment, typically consisting of designated area of the work area, a wash room and an uncontaminated area.

equipment decontamination enclosure system. A decontamination enclosure system in asbestos work for materials and equipment, typically consisting of designated area of the work area, a wash room and an uncontaminated area.

expansion joint. A separation between adjoining parts of a structure that is provided to allow small relative movements, such as those caused by thermal changes, to occur without one part affecting an adjacent part.

face. 1. The exposed surface of a wall or masonry unit. 2. The surface of a unit designed to be exposed in the finished structure.

face shell. The side wall of a hollow concrete masonry unit, generally between 3/4" and 1-1/2" thick.

face shell bedding. Mortar is applied only to the horizontal face of the face shells of hollow masonry units and in the head joints to a depth equal to the thickness of the face shell.

facing. Any material, forming a part of a wall, used as a finished surface.

fascia. The flat surface located at the outer end of a roof overhang or cantilever end.

field. The expanse of wall between openings, corners, etc., principally composed of stretchers.

filter block. A hollow, vitrified clay masonry unit, sometimes salt-glazed, designed for trickling filter floors in sewage disposal plants.

fire clay. A clay that is highly resistant to heat without deforming and used for making brick.

fire-resistive material. See non-combustive material.

fireproofing. Any material or combination protecting structural members to increase their fire resistance.

fire wall. Any wall that subdivides a building so as to resist the spread of fire, by starting at the foundation and extending continuously through all stories to, or above, the roof.

flashing. 1. A thin impervious material placed in mortar joints and through air spaces in masonry to prevent water penetration and/or provide water drainage. 2. Manufacturing methods to produce specific color tones.

flux oil. A thick petroleum product used to soften asphalt to a desired consistency.

frog. A depression in the bed surface of a brick. Sometimes called a panel.

foundation wall. A wall below the floor nearest grade serving as a support for a wall, pier, column, or other structural part of a building.

furring. Strips of wood or metal used to provide a level surface for finishing or to create an air space, or the installation of such material.

furrowing. The practice of striking a v-shaped trough in a bed of mortar (not recommended).

galvanized. Coating of zinc applied to steel in order to protect it from corrosion.

gambrel. Roof having two slopes on each side, the lower slope usually steeper than the upper one. A curb roof of the same section in all parts with a lower steeper slope and an upper flatter roof.

glove bag system. A portable asbestos abatement system designed for isolated and small areas of pipe, fittings, etc. requiring asbestos removal.

grid pavers. Open type masonry units that allow the growing of grass when employed for soil stabilization in parking areas, along the shoulders of highways and airport runways, embankment erosion control, fire engine lines while providing a base to support vehicular traffic.

grounds. Nailing strips placed in masonry walls as a means of attaching trim or furring.

grout. Mixture of cementitious material and aggregate to which sufficient water is added to produce pouring consistency without segregation of the constituents.

high-lift grouting. The technique of grouting masonry in lifts up to 12 ft.

low-lift grouting. The technique of grouting as the wall is constructed.

grout lift. The height to which grout is placed in a cell, collar joint, or cavity without intermission.

grout pour. The total grouted height between masonry lifts. A grout pour may consist of one or more grout lifts.

grouted masonry. Concrete masonry construction composed of hollow units where hollow cells are filled with grout, or multi-wythe construction in which space between wythes is solidly filled with grout.

hacking. 1. The procedure of stacking brick in a kiln or on a kiln car. 2. Laying brick with the bottom edge set in from the plane surface of the wall.

hand cleaning. Any surface preparation using hand tools such as wire brushes, scrapers, and chipping tools.

hard-burned. Nearly vitrified clay products that have been fired at high temperatures. They have relatively low absorptions and high compressive strengths.

head joint. The vertical mortar joint between ends of masonry units. Often called cross joint.

header. 1. Masonry unit that overlaps two or more adjacent wythes of masonry to tie them together. Often called bonder. 2. A floor or roof beam placed between joists or beams to support the ends.

blind header. A concealed brick header in the interior of the wall, not showing on the face.

clipped header. A bat placed to look like a header for purposes of establishing a pattern. Also called false header.

flare header. A header of darker color than the field of the wall.

heading course. A continuous bonding course of header brick. Also called header course.

HEPA filter. A high efficiency particulate absolute (HEPA) filter capable of trapping and retaining 99.9% of asbestos fibers greater than 0.3 microns in length. These filters are used in vacuum cleaners or in air transfer units (exhaust fans).

HEPA vacuum equipment. High efficiency particulate absolute filtered vacuuming equipment with a filter system capable of collecting and retaining asbestos fibers. Filters should be of a 99.9% efficiency for retaining fibers 0.3 microns (3/1000 of a millimeter) or larger.

hot-rolled steel. Steel that is hot reduced; that is, it is formed and shaped when it is hot.

hot specs. New and modified federal specifications that are issued every 15 days on microfilm.

indirect costs. Costs that are not directly attributable to just one job, such as office overhead, cost of capital, equipment depreciation.

initial rate of absorption. The weight of water absorbed expressed in grams per 30 sq. in. of contact surface when a brick is partially immersed for one minute. Also called suction. See ASTM Specification C 67.

interlocking block paver. Solid masonry units capable of transferring loads and stresses laterally by arching or bridging action between units when subjected to vehicular traffic.

IRA. See initial rate of absorption.

joint reinforcement. Steel wires placed in the mortar joint (over the face shell in hollow masonry) and having cross wires welded between them at regular intervals.

kiln. A furnace oven or heated enclosure used for burning or firing brick or other clay material.

kiln run. Brick from one kiln that have not been sorted or graded for size or color variation.

king closer. A brick cut diagonally to have one 2-in. end and one full-width end.

laitance. A milky white deposit on new concrete.

lateral support. Means whereby walls are braced either vertically or horizontally by columns, pilasters, cross walls, beams, floors, roofs, etc.

latex paint. A paint containing a stable aqueous dispersion of synthetic resin, produced by emulsion polymerization, as the principal constituent of the binder.

lead[1]. The section of a masonry wall built up and racked back on successive courses. A line is attached to leads as a guide for constructing a wall between them.

lead[2]. A heavy metal that can be hazardous to health if breathed or swallowed.

lime, hydrated. Quicklime to which sufficient water has been added to convert the oxides to hydroxides.

lime putty. Hydrated lime in plastic form ready for addition to mortar.

lintel. A beam placed over an opening in a wall to carry the superimposed weight of the construction and loads above the opening.

loadbearing. A structural system or element designed to carry loads in addition to its own dead weight.

masonry. Brick, stone, concrete, etc. or masonry combinations thereof bonded with mortar.

masonry cement. 1. A mill-mixed cementitious material to which sand and water must be added. See ASTM C 91. 2. Hydraulic cement produced for use in mortars for masonry construction where greater plasticity and water retention are desired than is obtainable by the use of portland cement alone. Such cements always contain one or more of the following materials: portland cement, portland-pozzolan cement, natural cement, slag cement, hydraulic lime. They usual contain one or more of the following: hydrated lime, pulverized limestone, chalk, talc, pozzolan, clay or gypsum; many masonry cements also include air-entraining and water-repellent additions.

masonry unit. Natural or manufactured building units of burned clay, concrete, stone, glass, gypsum, etc.

hollow masonry unit. One whose net cross-sectional area in any plane parallel to the bearing surface is less than 75% of the gross.

modular masonry unit. One whose nominal dimensions are based on the 4 in. module.

solid masonry unit. One whose net cross-sectional area in every plane parallel to the bearing surface is 75% or more of the gross.

moisture content. The amount of water contained at the time of sampling expressed as a percentage of the total absorption.

mortar. A plastic mixture of cementitious materials, fine aggregate, and water. See ASTM Specifications C 270, C 476 or BIA M1-72.

fat mortar. Mortar containing a high percentage of cementitious components; it is a sticky mortar which adheres to a trowel.

high-bond mortar. Mortar that develops higher bond strengths with masonry units than normally developed with conventional mortar.

lean mortar. Mortar that is deficient in cementitious components; it is usually harsh and difficult to spread.

net section. The minimum cross-section of the member under consideration; in masonry, usually the mortar bedded area plus the grouted area.

nominal dimension. A dimension greater than a specified masonry dimension by the thickness of a mortar joint, but not more than 1/2".

non-combustible material. Any material that will neither ignite nor actively support combustion in air at a temperature of 1200ºF when exposed to fire.

oil paint. A paint that contains drying oil, oil varnish, or oil-modified resin as the basic vehicle ingredient. Not that the common but technically incorrect definition is any paint soluble in organic solvents.

overhand work. Laying brick from inside a wall by workers standing on a floor or on a scaffold.

pargeting. The process of applying a coat of cement mortar to masonry. Also *parging.*

particulates. Fine liquid or solid particles, such as dust, smoke, mist, fumes, or smog, found in the air or in emissions.

panel wall. A non-loadbearing wall constructed between columns or piers and wholly supported at each story.

parapet wall. That part of a wall that extends above the roof level.

partition. An interior wall, one story or less in height.

party wall. A wall on an interior lot line, or any wall used to adapt for joint service between two buildings or adjacent living or work spaces.

phenolic resin. Resin made by the condensation of phenols and aldehydes.

pick and dip. A method of laying brick whereby the bricklayer simultaneously picks up a brick with one hand, and with the other hand, enough mortar on a trowel to lay the brick. Sometimes called the *eastern* or *New England* method.

pier. Any of various vertical supporting structures; an isolated column of masonry.

pilaster. A wall portion projecting from either or both wall faces and serving as a vertical column or beam.

plasticizer. A substance added to paint, varnish, or plaster to give it flexibility.

plumb rule. This is a combination plumb rule and level. It is used in a horizontal position as a level and in a vertical position as a plumb rule. They are made in lengths of 42" and 48" and short lengths from 12" to 24".

pointing. Troweling mortar into a joint after masonry units are laid.

prefabricated brick masonry. Masonry construction fabricated in a location other than its final inservice location in the structure. Also known as preassembled, panelized, and sectional brick masonry.

prism. 1. Any of a number of polyhedron shapes 2. A small masonry assemblage made with masonry units and mortar. Primarily used to predict the strength of full scale masonry members.

queen closer. A cut brick having a nominal 2" horizontal face dimension.

quoin. A projecting right angle masonry corner.

racking. A method entailing stepping back successive courses of masonry.

raggle. A groove in a joint or special unit to receive roofing or flashing.

RBM. Reinforced brick masonry.

reinforced masonry. Masonry units, reinforcing steel, grout, or mortar combined to act together in resisting forces.

return. Any surface turned back from the face of a principal surface.

reveal. That portion of a jamb or recess that is visible from the face of a wall.

rowlock. A brick laid on its face edge so that the normal bedding area is visible in the wall face. Frequently spelled *rolok*.

salt glaze. A gloss finish obtained by thermochemical reaction between silicates of clay and vapors of salt or chemicals.

saturation coefficient: See c/b ratio.

SCR. Structural Clay Research (trademark of the Structural Clay Products Institute, BIA).

SCR acoustile. A side-construction, two-celled facing tile, having a perforated face backed with glass wool for acoustical purposes.

SCR brick. Brick whose nominal dimensions are 6" x 2-2/3" x 12".

SCR building panel. Prefabricated, structural ceramic panels, approximately 2-1/2" thick.

SCR insulated cavity wall. Any cavity wall containing insulation that meets rigid criteria established by the Structural Clay Products Institute (BIA).

SCR masonry process. A construction aid providing greater efficiency, better workmanship, and increased production in masonry construction. It utilizes story poles, marked lines and adjustable scaffolding.

screen block. Open-faced masonry units used for decorative purposes or to partially screen areas from the sun or outside viewers.

sealant. 1. Any type of sealing agent. 2. A material spray applied to the substrate after the wet removal of existing asbestos material. The purpose of this is to prevent the release of residual asbestos fibers left after removal and cleaning operations.

shale. Clay that has been subjected to high pressures until it has hardened.

shear wall. A wall which, in its own plane, carries shear resulting from wind, blast, or seismic forces.

shower room. A room constituting an air lock between the clean room and the equipment room in the worker decontamination enclosure system, with hot and cold or warm running water suitably arranged for complete showering of workers during decontamination. Shower room always includes an air lock.

shoved joints. Vertical joints filled by shoving a brick against the next brick when it is being laid in a bed of mortar.

shrinkage. Volume change due to loss of moisture or decrease in temperature.

single wythe wall. A wall of only one masonry unit in thickness.

slenderness ratio. Ratio of the effective height of a member to its effective thickness.

slushed joints. Vertical joints filled, after units are laid, by "throwing" mortar in with the edge of a trowel (generally not recommended).

soap. A masonry unit of normal face dimensions, having a nominal 2" thickness.

soffit. The underside of a beam, lintel, or arch.

soft-burned. Clay products that have been fired at low temperature ranges, producing relatively high absorptions and low compressive strengths.

solar screen. A perforated wall used as a sunshade.

soldier. A stretcher set on end with face showing on the wall surface.

spall. A small fragment removed from the face of a masonry unit by a blow or by action of the elements.

stack. Any structure or part thereof that contains a flue or flues for the discharge of gases.

stacked bond. A bonding pattern where no unit overlaps either the one above or below, all head joints form a continuous vertical line.

story pole. A marked pole for measuring masonry coursing during construction.

stretcher. A masonry unit laid with its greatest dimension horizontal and its face parallel to the wall face.

stringing mortar. The procedure of spreading enough mortar on a bed to lay several masonry units.

struck joint. Any mortar joint that has been finished with a trowel.

suction. See Initial rate of absorption.

surfactant. In asbestos work, a chemical wetting agent added to water to improve penetration into the asbestos thus reducing the amount of water required for a given operation or area. This surfactant is spray applied to the asbestos to decrease the amount of airborne fibers while removing the asbestos material from the substrate. (Wetting agent-50% polyglycol ether and 50% polyoxyethlene ether or approved equal shall be mixed with water to provide a concentration of 1 oz. surfactant to 5 gallons of water or the manufacturer's recommended concentration.

temper. To moisten and mix clay, plaster, or mortar to a proper consistency.

tie. Any unit of material that connects masonry to masonry or other materials. See *wall tie.*

tooling. Compressing and shaping the face of a mortar joint with a special tool other than a trowel.

toothing. Constructing the temporary end of a masonry wall with the end stretcher of every alternate course projecting. Projecting units are toothers.

traditional masonry. Masonry in which design is based on empirical rules that control minimum thickness, lateral support requirements, and height without a structural analysis.

transformed section. An assumed section of one material having the same elastic properties as the section of two materials.

tuck pointing. The filling in with fresh mortar of cut-out or defective mortar joints in masonry.

veneer: 1. A thin layer of material that is bonded to and serves as the surface for another material. 2. A single wythe of masonry for facing purposes, not structurally bonded.

virtual eccentricity: The eccentricity of a resultant axial load required to produce axial and bending stresses equivalent to those produced by applied axial loads and moments. It is normally found by dividing the moment at a section by the summation of axial loads occurring at that section.

vitrification. The condition resulting when kiln temperatures are sufficient to fuse grains and close pores of a clay product, making the mass impervious.

wall. A vertical member of a structure whose horizontal dimension measured at right angles to the thickness exceeds three times its thickness.

apron wall. That part of a panel wall between window sill and wall support.

area wall. 1. The masonry surrounding or partly surrounding an area. 2. The retaining wall around basement windows below grade.

bearing wall. One that supports a vertical load in addition to its own weight.

cavity wall. A wall built of masonry units so arranged as to provide a continuous air space within the wall (with or without insulating material) and in which the inner and outer wythes of the wall are tied together with metal ties.

composite wall. A multiple-wythe wall in which at least one of the wythes is dissimilar to the other wythe or wythes with respect to type or grade of masonry unit or mortar.

curtain wall. An exterior non-loadbearing wall not wholly supported at each story. Such walls may be anchored to columns, spandrel beams, floors, or bearing walls, but not necessarily built between structural elements.

dwarf wall. A wall or partition that does not extend to the ceiling.

enclosure wall. An exterior non-loadbearing wall in skeleton frame construction. It is anchored to columns, piers, or floors but not necessarily built between columns or piers nor wholly supported at each story.

exterior wall. Any outside wall or vertical enclosure of a building other than a party wall.

faced wall. A composite wall in which the masonry facing and backing are so bonded as to exert a common reaction under load.

fire division wall. Any wall that subdivides a building so as to resist the spread of fire. It is not necessarily continuous through all stories to and above the roof.

fire wall. Any wall that subdivides a building to resist the spread of fire and that extends continuously from the foundation through the roof.

foundation wall. That portion of a loadbearing wall below the level of the adjacent grade, or below first floor beams or joists.

hollow wall. A wall built of masonry units arranged to provide an air space within the wall. The separated facing and backing are bonded together with masonry units.

insulated cavity wall. See SCR insulated cavity wall

loadbearing wall. A wall that supports any vertical load in addition to its own weight.

non-loadbearing wall. A wall that supports no vertical load other than its own weight.

panel wall. An exterior, non-loadbearing wall wholly supported at each story.

parapet wall. That part of any wall entirely above the roof line.

party wall. A wall used for joint service by adjoining buildings.

perforated wall. One that contains a considerable number of relatively small openings. Often called pierced wall or screen wall.

shear wall. A wall that resists horizontal forces applied in the plane of the wall.

single wythe wall. A wall containing only one masonry unit in wall thickness.

solid masonry wall. A wall built of solid masonry units, laid contiguously, with joints between units completely filled with mortar or grout.

spandrel wall. That part of a curtain wall above the top of a window in one story and below the sill of the window in the story above.

veneered wall. A wall having a facing of masonry units or other weather-resisting non-combustible materials securely attached to the backing, but not so bonded as to intentionally exert common action under load.

wall plate. A horizontal member anchored to a masonry wall to which other structural elements may be attached. Also called head plate.

wall tie. A bonder or metal piece that connects wythes of masonry to each other or to other materials.

wall tie, cavity. A rigid, corrosion-resistant metal tie that bonds two wythes of a cavity wall. It is usually steel, 3/16" in a diameter and formed in a "Z" shape or a rectangle.

wall tie, veneer. A strip or piece of metal used to tie a facing veneer to the backing.

water retentivity. That property of a material that prevents the rapid loss of water.

water table. A projection of lower masonry on the outside of the wall slightly above the ground. Often a damp course is placed at the level of the water table to prevent upward penetration of ground water.

waterproofing. Prevention of moisture flow through masonry due to water pressure.

web. The cross wall connecting the face shells of a hollow concrete masonry unit.

weep holes. Openings placed in mortar joints of facing material at the level of flashing to permit the escape of moisture.

wet cleaning. The process of eliminating asbestos contamination from building surfaces and objects by using cloths, mops, or other cleaning tools that have been dampened with water and by afterwards disposing of these cleaning tools as asbestos contaminated waste.

with inspection. Masonry designed with the higher stresses allowed under EBM. Requires the establishment of procedures on the job to control mortar mix, workmanship, and protection of masonry materials.

without inspection. Masonry designed with the reduced stresses allowed under EBM.

worker decontamination enclosure system. A decontamination enclosure system for workers typically consisting of a clean room, a locker room, a shower room, and an equipment room.

wythe. 1. Each continuous vertical section of masonry one unit in thickness. 2. The thickness of masonry separating flues in a chimney. Also called *withe* or *tier*.

Appendix **B**

CONSTRUCTION SAFETY

Construction is the largest industry in the U.S. and one of the most dangerous. The issues of accident prevention and safety in construction are in the limelight, partly due to the number of avoidable disasters in recent years.

The most important reason for developing and carefully monitoring a safety program is that employees' lives depend on it, but there are economic advantages as well. Healthy workers are productive ones, and a good safety record can mean lower workers compensation rates, which for some contractors can be 5% to 10% of total contract amount.

There are many agencies and organizations that provide guidelines and regulations for construction safety. The most visible and probably most important is the federal Occupational Safety and Health Administration (OSHA). OSHA regulations are ever changing, and the contractor must keep abreast of the latest revisions in OSHA standards. This point cannot be stressed too much. Listed below are some of the key OSHA publications that concern construction safety, but it is the contractor's responsibility to determine the latest versions of these publications. The easiest way is by contacting the nearest OSHA field office.

At the same time, the contractor must fulfill state and local regulatory requirements regarding the health and safety of workers and the public as well. It might seem like a daunting task, but it is well worth the time and effort.

The list below outlines the subparts of the Code of Federal Regulations (CRF) Volume 29, Chapter XVII - Occupational Safety and Health Administration, Department of Labor. These are the OSHA Construction Industry Standards. The subparts are self-descriptive and cover all major areas of safety concern for the industry. The revisions date indicated below was the latest when the current edition of our book was published. Once again, the contractor must check for more current revisions of these standards:

Occupational Safety and Health Administration
Construction Industry Standards (29 CFR 1926/1910)
OSHA 2207 Revised 1991

Subpart	Topic
A	General
B	General Interpretations
C	General Safety and Health Provisions
D	Occupational Health and Environmental Controls
E	Personal Protective and Life Saving Equipment

F	Fire Protection and Prevention
G	Signs, Signals, and Barricades
H	Materials Handling, Storage, Use, and Disposal
I	Tools--Hand and Power
J	Welding and Cutting
K	Electrical
L	Scaffolding
M	Floors and Wall Openings
N	Cranes, Derricks, Hoists, Elevators, and Conveyors
O	Motor Vehicles, Mechanized Equipment, and Marine Operations
P	Excavations
Q	Concrete and Masonry Construction
R	Steel Erection
S	Underground Construction, Caissons, Cofferdams, and Compressed Air
T	Demolition
U	Blasting and Use of Explosives
V	Power Transmission and Distribution
W	Rollover Protective Structures; Overhead Protection
X	Effective Dates

In addition to OSHA 2207 - Construction Industry Standards described above, there are a number of other OSHA publications that pertain to the construction industry:

OSHA No.	Title
2098	OSHA Inspections
2201	General Industry Standards
2202	Construction Industry Digest
2209	Handbook for Small Business
2226	Excavation and Trenching Operations
2254	Training Requirements in OSHA Standards and Training Guides
3007	Ground Fault Protection on Construction Sites
3040	Health Hazards of Asbestos
3071	Job Hazard Analysis
3075	Controlling Electrical Hazards
3077	Personal Protective Equipment
3079	Respiratory Protection
3080	Hand and Power Tools
3088	How to Prepare for Workplace Emergencies
3096	Asbestos Standards for Construction Industry
3097	Electrical Standards for Construction
3100	Crane and Derrick Suspended Personnel Platforms
3106	Concrete and Masonry Construction

OSHA 2207 (11-14-90 Edition)

Construction Stds.

Subpt. P, App. B

APPENDIX B TO SUBPART P

Sloping and Benching

(a) *Scope and application.* This appendix contains specifications for sloping and benching when used as methods of protecting employees working in excavations from cave-ins. The requirements of this appendix apply when the design of sloping and benching protective systems is to be performed in accordance with the requirements set forth in § 1926.652(b)(2).

(b) *Definitions.*

Actual slope means the slope to which an excavation face is excavated.

Distress means that the soil is in a condition where a cave-in is imminent or is likely to occur. Distress is evidenced by such phenomena as the development of fissures in the face of or adjacent to an open excavation; the subsidence of the edge of an excavation; the slumping of material from the face or the bulging or heaving of material from the bottom of an excavation; the spalling of material from the face of an excavation; and ravelling, i.e., small amounts of material such as pebbles or little clumps of material suddenly separating from the face of an excavation and trickling or rolling down into the excavation.

Maximum allowable slope means the steepest incline of an excavation face that is acceptable for the most favorable site conditions as protection against cave-ins, and is expressed as the ratio of horizontal distance to vertical rise (H:V).

Short term exposure means a period of time less than or equal to 24 hours that an excavation is open.

(c) *Requirements—(1) Soil classification.* Soil and rock deposits shall be classified in accordance with appendix A to subpart P of part 1926.

(2) *Maximum allowable slope.* The maximum allowable slope for a soil or rock deposit shall be determined from Table B-1 of this appendix.

(3) *Actual slope.* (i) The actual slope shall not be steeper than the maximum allowable slope.

(ii) The actual slope shall be less steep than the maximum allowable slope, when there are signs of distress. If that situation occurs, the slope shall be cut back to an actual slope which is at least ½ horizontal to one vertical (½H:1V) less steep than the maximum allowable slope.

(iii) When surcharge loads from stored material or equipment, operating equipment, or traffic are present, a competent person shall determine the degree to which the actual slope must be reduced below the maximum allowable slope, and shall assure that such reduction is achieved. Surcharge loads from adjacent structures shall be evaluated in accordance with § 1926.651(i).

(4) *Configurations.* Configurations of sloping and benching systems shall be in accordance with Figure B-1.

TABLE B-1

MAXIMUM ALLOWABLE SLOPES

SOIL OR ROCK TYPE	MAXIMUM ALLOWABLE SLOPES (H:V)[1] FOR EXCAVATIONS LESS THAN 20 FEET DEEP [3]
STABLE ROCK TYPE A [2] TYPE B TYPE C	VERTICAL (90°) 3/4 : 1 (53°) 1:1 (45°) 1¼: 1 (34°)

NOTES:

1. Numbers shown in parentheses next to maximum allowable slopes are angles expressed in degrees from the horizontal. Angles have been rounded off.

2. A short-term maximum allowable slope of 1/2H:1V (63°) is allowed in excavations in Type A soil that are 12 feet (3.67 m) or less in depth. Short-term maximum allowable slopes for excavations greater than 12 feet (3.67 m) in depth shall be 3/4H:1V (53°).

3. Sloping or benching for excavations greater than 20 feet deep shall be designed by a registered professional engineer.

Figure B-1

Slope Configurations

(All slopes stated below are in the horizontal to vertical ratio)

B-1.1 *Excavations made in Type A soil.*

1. All simple slope excavations 20 feet or less in depth shall have a maximum allowable slope of ¾:1.

SIMPLE SLOPE—GENERAL

Exception: Simple slope excavations which are open 24 hours or less (short term) and which are 12 feet or less in depth shall have a maximum allowable slope of ½:1.

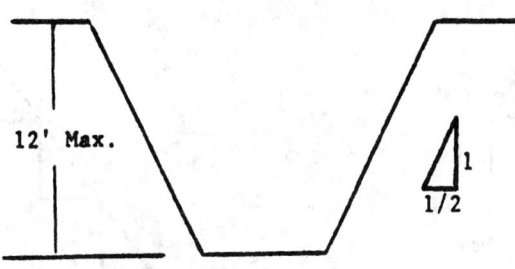

SIMPLE SLOPE—SHORT TERM

2. All benched excavations 20 feet or less in depth shall have a maximum allowable slope ¾ to 1 and maximum bench dimensions as follows:

SIMPLE BENCH

MULTIPLE BENCH

3. All excavations 8 feet or less in depth which have unsupported vertically sided lower portions shall have a maximum vertical side of 3½ feet.

UNSUPPORTED VERTICALLY SIDED LOWER PORTION—MAXIMUM 8 FEET IN DEPTH

All excavations more than 8 feet but not more than 12 feet in depth which unsupported vertically sided lower portions shall have a maximum allowable slope of 1:1 and a maximum vertical side of 3½ feet.

UNSUPPORTED VERTICALLY SIDED LOWER PORTION—MAXIMUM 12 FEET IN DEPTH

All excavations 20 feet or less in depth which have vertically sided lower portions that are supported or shielded shall have a maximum allowable slope of ¾:1. The support or shield system must extend at least 18 inches above the top of the vertical side.

SUPPORTED OR SHIELDED VERTICALLY SIDED LOWER PORTION

4. All other simple slope, compound slope, and vertically sided lower portion excavations shall be in accordance with the other options permitted under § 1926.652(b).

B-1.2 Excavations Made in Type B Soil

1. All simple slope excavations 20 feet or less in depth shall have a maximum allowable slope of 1:1.

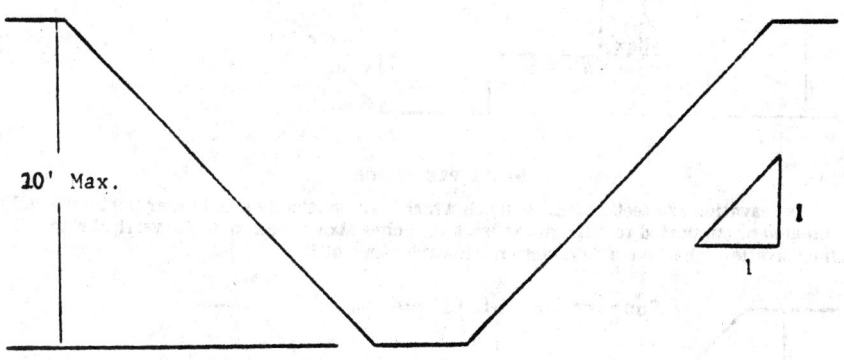

SIMPLE SLOPE

2. All benched excavations 20 feet or less in depth shall have a maximum allowable slope of 1:1 and maximum bench dimensions as follows:

SINGLE BENCH

MULTIPLE BENCH

3. All excavations 20 feet or less in depth which have vertically sided lower portions shall be shielded or supported to a height at least 18 inches above the top of the vertical side. All such excavations shall have a maximum allowable slope of 1:1.

VERTICALLY SIDED LOWER PORTION

4. All other sloped excavations shall be in accordance with the other options permitted in § 1926.652(b).

B-1.3 Excavations Made in Type C Soil

1. All simple slope excavations 20 feet or less in depth shall have a maximum allowable slope of 1½:1.

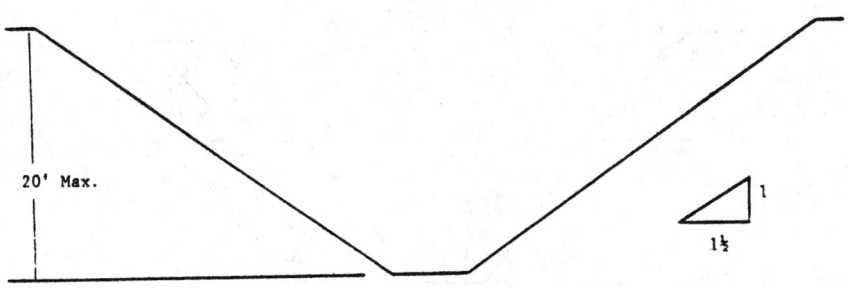

SIMPLE SLOPE

2. All excavations 20 feet or less in depth which have vertically sided lower portions shall be shielded or supported to a height at least 18 inches above the top of the vertical side. All such excavations shall have a maximum allowable slope of 1¼:1.

VERTICAL SIDED LOWER PORTION

3. All other sloped excavations shall be in accordance with the other options permitted in § 1926.652(b).

B-1.4 Excavations Made in Layered Soils

1. All excavations 20 feet or less in depth made in layered soils shall have a maximum allowable slope for each layer as set forth below.

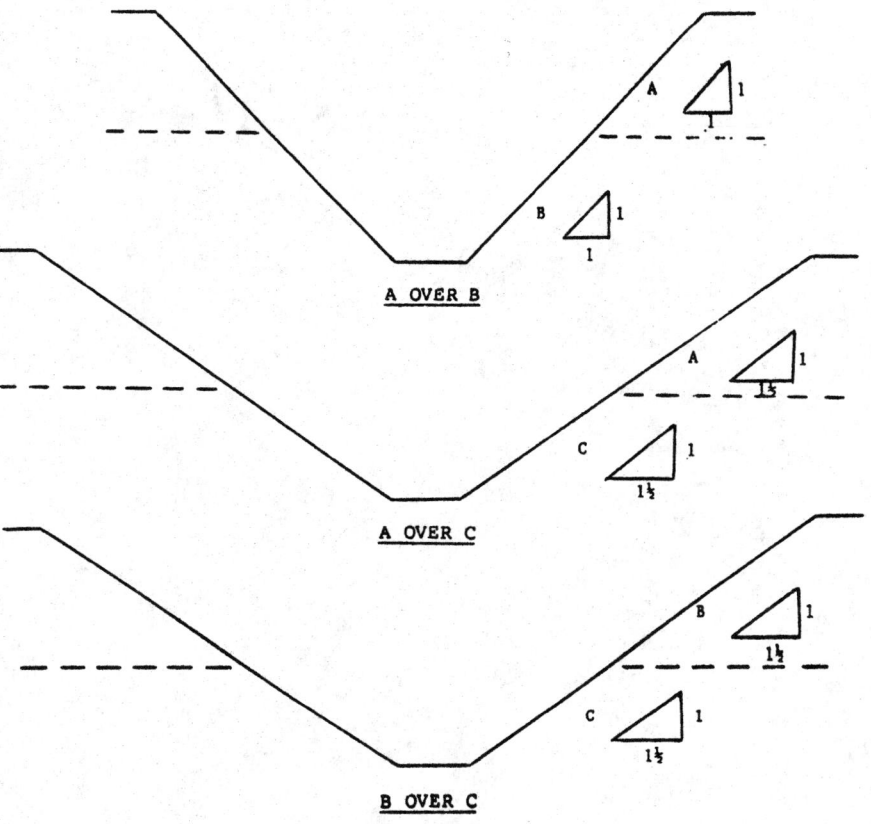

A OVER B

A OVER C

B OVER C

2. All other sloped excavations shall be in accordance with the other options permitted in § 1926.652(b).

INDEX